BECKER'S World of the Cell

Technology Update

EIGHTH EDITION

GLOBAL EDITION

JEFF HARDIN
University of Wisconsin-Madison

GREGORY BERTONI
The Plant Cell

LEWIS J. KLEINSMITH
University of Michigan, Ann Arbor

PEARSON

Senior Acquisitions Editor: *Josh Frost*
Project Manager: *Lori Newman, Margaret Young*
Acquisitions Editor, Global Edition: *Priyanka Ahuja*
Associate Project Editor, Global Edition: *Binita Roy*
Program Manager: *Anna Amato*
Program Management Team Lead: *Michael Early*
Project Management Team Lead: *David Zielonka*
Production Management and Composition: *Progressive Publishing Alternatives, Cenveo Publisher Services*
Design Manager: *Derek Bacchus*
Interior Designer: *Seventeenth Street Studios*
Cover Image: © *Symonenko Viktoriia/Shutterstock*
Cover Designer: *Lumina Datamatics*
Illustrators: *Dartmouth Publishing, Inc.*
Rights & Permissions Project Manager: *Rachel Youdelman*
Photo Researcher: *Sonia Divittorio, Eric Schrader*
Photo Editor: *Donna Kalal*
Manufacturing Buyer: *Stacey Weinberger*
Senior Manufacturing Controller, Production, Global Edition: *Trudy Kimber*
Executive Marketing Manager: *Lauren Harp*
Senior Marketing Manager: *Amee Mosley*

Pearson Education Limited
Edinburgh Gate
Harlow
Essex CM20 2JE
England

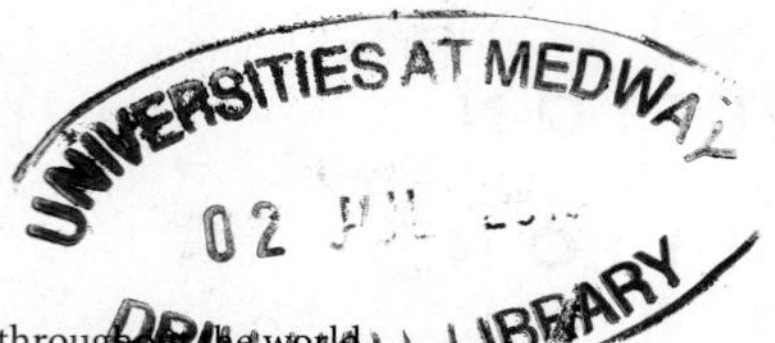

and Associated Companies throughout the world

Visit us on the World Wide Web at:
www.pearsonglobaleditions.com

Authorized adaptation from the United States edition, entitled Becker's World of The Cell Technology Update, 8th edition, ISBN 978-0-133-99939-6, by Jeff Hardin, Gregory Bertoni, and Lewis J. Kleinsmith, published by Pearson Education © 2015.

ISBN 10: 1-292-08166-X
ISBN 13: 978-1-292-08166-3

British Library Cataloguing-in-Publication Data
A catalogue record for this book is available from the British Library

10 9 8 7 6 5 4 3 2 1

Typeset in 10.5 Minion Pro by Cenveo Publisher Services.

Printed and bound by Courier Kendallville in the United States of America.

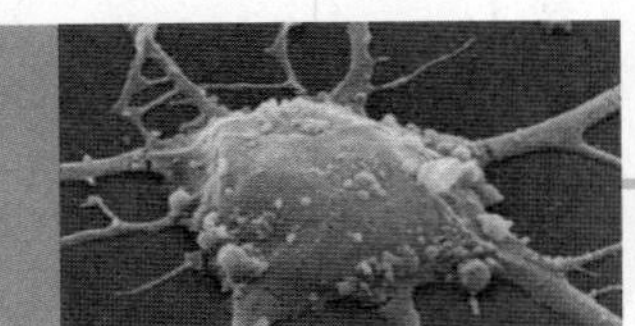

ABOUT THE AUTHORS

JEFF HARDIN is Professor and Chair of the Zoology Department at the University of Wisconsin-Madison. He is also faculty director of the Biology Core Curriculum, a four-semester honors biology sequence for undergraduates. His research interests center on how cells migrate and adhere to one another to change the shape of animal embryos. Dr. Hardin's teaching is enhanced by his extensive use of digital microscopy and his web-based teaching materials, which are used on many campuses in the United States and other countries. As part of his interest in teaching biology, Dr. Hardin was a founding member of the University of Wisconsin Teaching Academy. His teaching awards include a Lily Teaching Fellowship and a National Science Foundation Young Investigator Award. He is also on the editorial board of *CBE: Life Sciences Education*, and is a curator of WormClassroom, a digital initiative that promotes the use of *C. elegans* in college classrooms and laboratories.

GREGORY BERTONI has been active in teaching, research, and scientific writing for over 25 years. He earned a Ph.D. in Cellular and Molecular Biology from the University of Wisconsin-Madison, where he taught students in introductory and graduate-level biochemistry, sophomore cell biology, and plant physiology. He helped to develop a new course entitled "Ways of Knowing" designed to introduce entering freshmen to a variety of academic fields as well as to the learning process itself. His published research includes studies in bacterial pathogenesis, plant-microbe interactions, and plant gene expression. Dr. Bertoni is a science editor for *The Plant Cell,* a leading international research journal in plant cell and molecular biology. He is also responsible for updating the journal's Teaching Tools in Plant Biology, an online resource for biology instructors. He has been teaching biology and medical microbiology at Columbus State Community College in Columbus, Ohio for most of the past 10 years. In addition, Dr. Bertoni is a freelance scientific writer who contributes to text- and web-based projects in biology, physics, and microbiology and assists authors in preparing manuscripts for publication.

LEWIS J. KLEINSMITH is an Arthur F. Thurnau Professor Emeritus of Molecular, Cellular, and Developmental Biology at the University of Michigan, where he has served on the faculty since receiving his Ph.D. from Rockefeller University in 1968. His teaching experiences have involved courses in introductory biology, cell biology, and cancer biology, and his research interests have included studies of growth control in cancer cells, the role of protein phosphorylation in eukaryotic gene regulation, and the control of gene expression during development. Among his numerous publications, he is the author of *Principles of Cancer Biology* as well as several award-winning educational software programs. His honors include a Guggenheim Fellowship, the Henry Russell Award, a Michigan Distinguished Service Award, citations for outstanding teaching from the Michigan Students Association, an NIH Plain Language Award, and a Best Curriculum Innovation Award from the EDUCOM Higher Education Software Awards Competition.

WAYNE M. BECKER taught cell biology at the University of Wisconsin-Madison for 30 years until his retirement. His interest in textbook writing grew out of notes, outlines, and problem sets that he assembled for his students, culminating in *Energy and the Living Cell*, a paperback text on bioenergetics published in 1977, and *The World of the Cell*, the first edition of which appeared in 1986. He earned all his degrees at the University of Wisconsin-Madison. All three degrees are in biochemistry, an orientation that is readily discernible in his textbooks. His research interests were in plant molecular biology, focused specifically on the regulation of the expression of genes that encode enzymes of the photorespiratory pathway. His honors include a Chancellor's Award for Distinguished Teaching, Guggenheim and Fulbright Fellowships, and a Visiting Scholar Award from the Royal Society of London. This text builds on his foundation, and is inspired by his legacy.

ABOUT THE AUTHORS

PREFACE

"Because we enjoy interacting with biology undergrads and think that they should have biology textbooks that are clearly written, make the subject matter relevant to the reader, and help them appreciate not only how much we already know about biology—cell biology, in our case—but also how much more remains to be investigated and discovered." That's how any of the authors of this text would likely respond if asked why we've invested so much time in writing and revising *The World of the Cell*. Each of us has an extensive history of teaching undergraduate courses in cell biology and related areas, and each of us treasures our contact with students as one of the most rewarding aspects of being a faculty member.

As we reflect on the changes we've seen in our courses over the years, we realize that the past several decades have seen an explosive growth in our understanding of the properties and functions of living cells. This enormous profusion of information presents us with a daunting challenge as we confront the task of keeping *Becker's World of the Cell* up to date while simultaneously ensuring that it remains both manageable in length and readily comprehensible to students encountering the field of cell and molecular biology for the first time. This eighth edition represents our most recent attempt to rise to that challenge. As with the previous editions, each of us has brought our own teaching and writing experience to the venture in ways that we have found mutually beneficial—a view that we hope our readers will share.

One major objective for this edition has been to update the content of the text, especially in areas where the pace of research is especially brisk and recent findings are particularly significant. At the same time, we have remained committed to the three central goals that have characterized each preceding edition. As always, our primary goal is to introduce students to the fundamental principles that guide cellular organization and function. Second, we think it is important for students to understand some of the critical scientific evidence that has led to the formulation of these central concepts. And finally, we have sought to accomplish these goals in a book of manageable length that can be easily read and understood by beginning cell biology students—and that still fits in their backpacks! To accomplish this third goal, we have necessarily been selective both in the types of examples chosen to illustrate key concepts and in the quantity of scientific evidence included. We have, in other words, attempted to remain faithful to the overall purpose of each previous edition: to present the essential principles, processes, and methodology of molecular and cell biology as lucidly as possible. We have also given careful attention to accuracy, consistency, vocabulary, and readability to minimize confusion and maximize understanding for our readers.

The Technology Update

The Becker World of the Cell 8e Technology Update brings the power of MasteringBiology to Cell Biology for the first time.

MasteringBiology is an online homework, tutorial and assessment system that delivers self-paced tutorials that provide individualized coaching, hints, and feedback. The Mastering system helps instructors and students with customizable, easy-to-assign, and automatically graded assessments.

Integrated links in every chapter of the textbook will point students to a variety of interactive online materials, including:

- 240 Reading Quiz Questions, encouraging students to read before class
- Over 100 Tutorials and activities that teach complex cell processes
- Over 100 molecular and microscopy videos, providing vivid images of cellular processes
- 17 3D-Structure Tutorials that allow students to manipulate molecular structures, with assessment activities
- All End-of-Chapter questions, now assignable and automatically gradable
- Test Bank Questions for every chapter

The E-text is also available through MasteringBiology, providing access to the complete textbook and providing powerful interactive and customization functions.

- **New Pearson E-text** option is available.
- **New PowerPoint Lecture Tools,** including pre-made lecture outlines containing all of the figures and photos and embedded animations, and 5–10 personal response system (PRS) clicker questions per chapter.

Content Highlights of the Eighth Edition

Updated material and new information has been added throughout the book. Topics that have been altered, updated, or added include the following:

CHAPTER 1: Added a new box essay and figure describing the use of model organisms in cell biology research. Updated discussions of nanotechnology, new types of light and electron microscopes, advances in gene and genome sequencing, new "-omics" fields, and bioinformatic tools at NCBI (PubMed, GenBank, OMIM). Introduced "in silico" research as an extension of traditional in vivo and in vitro research.

CHAPTER 2: Described how viral self-assembly can be used in nanotechnology and biomedicine; described electronegativity in relation to polarity and solubility of biomolecules; and introduced soaps and detergents as amphipathic molecules.

CHAPTER 3: Streamlined discussion of protein function; improved discussion of peptide bond geometry; introduced CASP—the worldwide test of protein structure modeling programs; added miRNA and siRNA to the discussion of RNA; introduced lipid rafts earlier.

CHAPTER 4: Added discussions of the important functions of what had been called "junk" DNA, how mitochondrial DNA analysis is used to trace genetic lineages and the origin of modern humans, construction of an artificial ribosome in vitro, and how prions cause chronic wasting disease in deer and elk. Presented new results of X-ALD gene therapy clinical trial.

CHAPTER 5: Discussion of bioluminescence now includes luciferase, GFP, and YFP as tools for cell biologists; new figure added to show localization of YFP protein fusion by fluorescence microscopy. Strengths in discussions of bioenergetics and thermodynamics have been retained.

CHAPTER 6: Updated discussion on induced fit; added new figures showing formation of lysozyme active site after protein folding and the changes in the active site of carboxypeptidase following substrate binding; added description of aspirin as an irreversible inhibitor of cyclooxygenase; described cryophilic microorganisms such as *Listeria*.

CHAPTER 7: Added a new paragraph and figure describing the common glycolipids MGDG and DGDG; included a more extensive description of lipid raft composition, formation, and proteomics; added descriptions of how antimicrobial peptides disrupt cell membranes, described membrane receptors for nutrients and for the gaseous hormone ethylene; updated roles of caveolae in human physiology and disease.

CHAPTER 8: Added description of aquaporin discovery by Agre and colleagues; added new figure showing structures of bacterial porin and human aquaporin. Updated discussion of all five subclasses of P-type ATPases; added discussion of how the Na^+/glucose symporter can affect the treatment of cholera; introduced additional members of the human glucose transporter family; described use of bacteriorhodopsin in biomolecular electronics.

CHAPTER 9: Added new discussion clarifying resonance stabilization and electron delocalization; expanded discussion of anaerobic respiration; described role of unusual microorganisms in geochemical cycling of nutrients and global biomass production; noted how oxidative stress and free radicals can damage cells.

CHAPTER 10: Included more emphasis on cellular locations of biochemical processes and on similarities between bacterial and eukaryotic respiration; clarified difference between internal and external electron acceptors; introduced pyruvate symporter. In response to reviewer requests, added figure and extensive description of β-oxidation of fatty acids.

CHAPTER 11: Added new paragraph describing newly discovered quantum effects during photosynthetic light harvesting; emphasized similarities between mitochondrial and chloroplast electron transport; introduced pioneering work of van Niel in the 1930s using photosynthetic bacteria. Added an analogy for a potential energy barrier by comparing the thylakoid membrane to a hydroelectric dam, clarified nomenclature of photosystems I and II.

CHAPTER 12: Added new paragraph describing mechanism of action and clinical uses of botulinum toxin. Added new material on N-glycosylation and secretion involving interleukin-31 and p53 and new micrographs showing exocytosis and phagocytosis. Added new paragraph describing reactive oxygen species and their detoxification in the peroxisome; included a more detailed description of the reaction mechanism of P-450 monooxygenases.

CHAPTER 13: Reorganized discussion of squid giant axon and the basic membrane potential measuring technique: how membrane potential is measured in neurons is now discussed prior to what contributes to the membrane potential. Updated depiction of voltage-gated channels to reflect X-ray crystallographic data. Changed terminology from "terminal bulb" to "synaptic bouton," which is more common among neurobiologists. Updated the discussion of neurotransmitters to include endocannabinoids, and created a new table showing chemical structure of selected neurotransmitters. Now mention "kiss-and-run" exocytosis as an additional mechanism of neurotransmitter release.

CHAPTER 14: Added a discussion of co-receptors, a section on how Wnts and Hedgehogs feed into G protein signaling, a new table on hormones, a section on nuclear hormone receptors, and added PTEN to the discussion of insulin signaling. Reorganized content between Chapter 14 and Chapter 19 to streamline discussion: apoptosis and discussion of *C. elegans* has been moved out of this chapter.

CHAPTER 15: Katanins are now discussed in the microtubule section. Added significant new information on formins in the actin section, and much more coverage of Rho GTPases, including RhoGEFs, GAPs, and GDIs.

CHAPTER 16: Added new information and a new part of a figure on hair cells, as well as significant new text on intraflagellar transport (IFT) and ciliopathies to Box 16A. Updated the figure on focal contacts and the leading edge to provide an integrated view of actin polymerization at the leading edge.

CHAPTER 17: Added further description of apical-basal polarity in epithelia, and the epithelial-mesenchymal transition (EMT), and there is now more emphasis on the dynamic nature of cell adhesion. Added new information on the Par3/Par6/aPKC polarity complex and on the dystrophin/dystroglycan complex. Also updated discussion of claudins to include paracellular transport and human diseases associated with claudins. Added text about kindlins to the focal adhesion discussion and updated treatment of integrin-linked kinase (ILK).

CHAPTER 18: Added new information on NTF2 in the Ran/importin/exportin section. Updated section on nuclear bodies to be more explicit about types and functions of nuclear bodies (Cajal, GEMs, speckles, etc.). Added a new figure describing the percentage of various types of DNA in the human genome.

CHAPTER 19: Added new information on spindle midzones, as well as new material and a new figure on myosin and Rho during cytokinesis. Added new material on ATR and checkpoint kinases in the cell cycle control section. Reorganized content between Chapter 14 and Chapter 19 to streamline discussion: much of the Ras discussion and Akt/PI3K has been moved out of this chapter. Apoptosis was removed from Chapter 14, and is now in Chapter 19. In addition, the apoptosis figure was redrawn to show a more accurate depiction of the apoptosome based on cryoEM data.

CHAPTER 20: Added substantially more discussion and a new figure on knockout mice. The section on genetic engineering has been reworked, and there is now better balance between genetic engineering of animals and transgenic plants. A new figure showing pronuclear injection in mice was added. Updated the section on screening of bacterial clones to reflect the "modern" way of doing this by restriction digests and sequencing. The section on gene conversion was shortened to accommodate more modern material.

CHAPTER 21: Added information on regulatory role of the C terminus of RNA pol II. Added a discussion of electrophoretic mobility shift assays (EMSAs) as a technique, while the historical method of R loop detection was streamlined and supplemented with an improved schematic diagram.

CHAPTER 22: Added a new figure showing the results of experiments with microsomes, demonstrating that cotranslational import is required for cleavage of the signal sequence. Added a brief discussion of operons in eukaryotes.

CHAPTER 23: Added further discussion and a new figure regarding the Dolly cloning method. Substantially increased the discussion of embryonic stem (ES) cells and induced pluripotent stem cells (iPS cells), including a new figure. Added two new figures on methylation of DNA and chromatin/histone remodeling, increased coverage of epigenetics, and added discussion of Prader-Willi and Angelman syndromes. Added a box on the yeast two-hybrid system, updated the discussion of the structure of the *lac* repressor/operon to reflect crystallographic data. Revised and expanded the discussion of *Hox* genes (including a new figure). Added discussion of SUMOylation in the post-translational control section.

CHAPTER 24: Added discussion on how cancers evade destruction by the immune system, and how the tumor microenvironment influences tumor growth, invasion, and metastasis. More discussion of the ability of cancer cells to evade apoptosis and of the Rho family in triggering the enhanced cell motility that leads to invasion and metastasis. Added more discussion on how polycyclic aromatic hydrocarbons in tobacco cause unique mutations in the *p53* gene, and added information about the typical mutation patterns in cancer cells obtained from genome sequencing studies. Added information about the role of microRNAs and histone methylation in cancer epigenetics, and more information about cancer vaccines.

APPENDIX: Added mention of photoconvertible and photoswitchable fluorophores. Added a section on various techniques for superresolution microscopy. Streamlined discussion of correlative microscopy. Updated figure on the fluorescence microscope to more accurately reflect modern epifluorescence systems.

Techniques and Methods

Throughout the text, we have tried to explain not only *what* we know about cells but also *how* we know what we know. Toward that end, we have included descriptions of experimental techniques and findings in every chapter, almost always in the context of the questions they address and in anticipation of the answers they provide. For example, polyacrylamide gel electrophoresis is introduced not in a chapter that simply catalogues a variety of methods for studying cells but in Chapter 7, where it becomes important to our understanding of how membrane proteins can be separated from one another. Similarly, equilibrium density centrifugation is described in Chapter 12, where it is essential to our understanding of how lysosomes were originally distinguished from mitochondria and subsequently from peroxisomes as well.

To help readers locate techniques out of context, an alphabetical Guide to Techniques and Methods appears on the inside of the front and back covers, with references to chapters, pages, tables, figures, and boxed essays, as appropriate. To enhance its usefulness, the Guide to Techniques and Methods includes references not just to laboratory techniques but also to the mathematical determination of values such as ΔG (free energy change) and $\Delta E_0'$ (standard reduction potential), bioinformatics techniques such as BLAST searching, and even to clinical procedures such as the treatment of methanol poisoning.

Microscopy is the only exception to our general approach of introducing techniques in context. The techniques of light and electron microscopy are so pervasively relevant to contemporary cell biology that they warrant special consideration as a self-contained unit, which is included as an Appendix entitled *Visualizing Cells and Molecules.* This Appendix gives students ready access to detailed information on a variety of microscopy techniques, including cutting-edge uses of light microscopy for imaging and manipulating molecular processes.

Building on the Strengths of Previous Editions

We have retained and built upon the strengths of prior editions in four key areas:

1. **The chapter organization focuses on main concepts.**
   - Each chapter is divided into sections that begin with a *concept statement heading*, which summarizes the material and helps students focus on the main points to study and review.
   - Chapters are written and organized in ways that allow instructors to assign the chapters and chapter sections in different sequences to make the book adaptable to a wide variety of course plans.
   - Each chapter culminates with a bulleted *Summary of Key Points* that briefly describes the main points covered in each section of the chapter.
2. **The illustrations teach concepts at an appropriate level of detail.**
   - Many of the more complex figures incorporate *minicaptions* to help students grasp concepts more quickly by drawing their focus into the body of an illustration rather than depending solely on a separate figure legend to describe what is taking place.
   - *Overview figures* outline complicated structures or processes in broad strokes and are followed by text and figures that present supporting details.
   - Carefully selected micrographs are usually accompanied by scale bars to indicate magnification.
3. **Important terminology is highlighted and defined in several ways.**
   - **Boldface type** is used to highlight the most important terms in each chapter, all of which are defined in the Glossary.
   - *Italics* are employed to identify additional technical terms that are less important than boldfaced terms but significant in their own right. Occasionally, italics are also used to highlight important phrases or sentences.
   - The Glossary includes definitions and page references for all bold-faced key terms and acronyms in every chapter—more than 1500 terms in all, a veritable "dictionary of cell biology" in its own right.
4. **Each chapter helps students learn the process of science, not just facts.**
   - Text discussions emphasize the experimental evidence that underlies our understanding of cell structure and function, to remind readers that advances in cell biology, as in all branches of science, come not from lecturers in their classrooms or textbook authors at their computers but from researchers in their laboratories.
   - The inclusion of a *Problem Set* at the end of each chapter reflects our conviction that we learn science not just by reading or hearing about it, but by working with it. The problems are designed to emphasize understanding and application, rather than rote recall. Many of the problems are class-tested, having been selected from problem sets and exams we have used in our own courses.
   - Each chapter contains one or more *Boxed Essays* to aid students in their understanding of particularly important or intriguing aspects of cell biology. Some of the essays provide *Deeper Insights* into potentially difficult principles, such as the essay that uses the analogy of monkeys shelling peanuts to explain enzyme kinetics (Box 6A). Other essays

describe *Tools of Discovery*, some of the important experimental techniques used by cell biologists, as exemplified by the description of DNA fingerprinting in Box 18C. And yet another role of the boxed essays is to describe *Human Applications* of research findings in cell biology, as illustrated by the discussion of cystic fibrosis and the prospects for gene therapy in Box 8B.

- A *Suggested Reading* list is included at the end of each chapter, with an emphasis on review articles and carefully selected research publications that motivated students are likely to understand. We have tried to avoid overwhelming readers with lengthy bibliographies of the original literature but have referenced articles that are especially relevant to the topics of the chapter. In most chapters, we have included a few citations of especially important historical publications, which are marked with blue dots to alert the reader to their historical significance.

Supplementary Learning Aids

For Instructors

Instructor Resources (available for download at www.pearsonglobaleditions.com/Hardin).

- PowerPoint Lecture Tools, including pre-made lecture outlines containing all of the figures and photos and embedded animations, and 5–10 Personal Response System (PRS) clicker questions per chapter.
- JPEG images of all textbook figures and photos, including printer-ready transparency acetate masters.
- Videos and animations of key concepts, organized by chapter for ease of use in the classroom.
- The full test bank for *Becker's World of the Cell.*

Computerized Test Bank for *Becker's World of the Cell*

The test bank provides over 1000 multiple-choice, short-answer, and inquiry/activity questions.

MasteringBiology® www.masteringbiology.com

MasteringBiology is an online homework, tutorial and assessment system that delivers self-paced tutorials that provide individualized coaching, focus on your course objectives, and are responsive to each student's progress. The Mastering system helps instructors maximize class time with customizable, easy-to-assign and automatically graded assessments that motivate students to learn outside of class and arrive prepared for lecture. MasteringBiology includes the book's end-of-chapter problems, reading quizzes, animations, and a wide array of tutorials and activities. The E-text is also available through MasteringBiology.

We Welcome Your Comments and Suggestions

The ultimate test of any textbook is how effectively it helps instructors teach and students learn. We welcome feedback and suggestions from readers and will try to acknowledge all correspondence.

Jeff Hardin
Department of Zoology
University of Wisconsin-Madison
Madison, Wisconsin 53706
e-mail: jdhardin@wisc.edu

Gregory Bertoni
The Plant Cell
American Society of Plant Biologists
Rockville, Maryland 20855
e-mail: gbertoni@aspb.org

Lewis J. Kleinsmith
Department of Molecular, Cellular, and Developmental Biology
University of Michigan
Ann Arbor, Michigan 48109
e-mail: lewisk@umich.edu

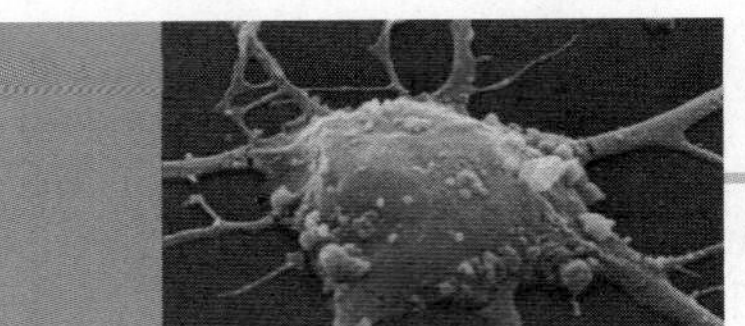

ACKNOWLEDGMENTS

We want to acknowledge the contributions of the numerous people who have made this book possible. We are indebted especially to the many students whose words of encouragement catalyzed the writing of these chapters and whose thoughtful comments and criticisms have contributed much to whatever level of reader-friendliness the text may be judged to have. Each of us owes a special debt of gratitude to our colleagues, from whose insights and suggestions we have benefited greatly and borrowed freely. We also acknowledge those who have contributed to previous editions of our textbooks, including David Deamer, Martin Poenie, Jane Reece, John Raasch, and Valerie Kish, as well as Peter Armstrong, John Carson, Ed Clark, Joel Goodman, David Gunn, Jeanette Natzle, Mary Jane Niles, Timothy Ryan, Beth Schaefer, Lisa Smit, David Spiegel, Akif Uzman, and Karen Valentine. Most importantly, we are grateful to Wayne Becker for his incisive writing and vision, which led to the creation of this book and featured so prominently in previous editions. We have tried to carry on his tradition of excellence. In addition, we want to express our appreciation to the many colleagues who graciously consented to contribute micrographs to this endeavor, as well as the authors and publishers who have kindly granted permission to reproduce copyrighted material.

The many reviewers listed below provided helpful criticisms and suggestions at various stages of manuscript development and revision. Their words of appraisal and counsel were gratefully received and greatly appreciated. Indeed, the extensive review process to which this and the prior editions of the book have been exposed should be considered a significant feature of the book. Nonetheless, the final responsibility for what you read here remains ours, and you may confidently attribute to us any errors of omission or commission encountered in these pages.

We are also deeply indebted to the many publishing professionals whose consistent encouragement, hard work, and careful attention to detail contributed much to the clarity of both the text and the art. Special recognition and sincere appreciation go to Anna Amato in her role as project editor, to Gary Carlson, Josh Frost, Lindsay White, Deborah Gale, Lori Newman, Laura Tommasi, Sonia DiVittorio, and Lee Ann Doctor at Benjamin Cummings, to Stephanie Davidson at Dartmouth Publishing, and to Crystal Clifton and her colleagues at Progressive Publishing Alternatives.

Finally, we are grateful beyond measure to our wives, families, graduate students, and postdoctoral associates, without whose patience, understanding, and forbearance this book could not have been written.

Reviewers for The Eighth Edition

Nihal Altan-Bonnet, *Rutgers University*
Stephen E. Asmus, *Centre College*
Manuel Alejandro Barbieri, *Florida International University*
Kenneth D. Belanger, *Colgate University*
Loren A. Bertocci, *Marian University, Indianapolis*
Annemarie Bettica, *Manhattanville College*
Ann C. Billetz, *Massachusetts College of Liberal Arts*
Robert J. Bloch, *University of Maryland School of Medicine*
Olga Boudker, *Weill Cornell Medical College*
Joshua Brumberg, *Queens College, CUNY*
John G. Burr, *University of Texas, Dallas*
Richard Cardullo, *University of California, Riverside*
Catherine P. Chia, *University of Nebraska-Lincoln*
Francis Choy, *University of Victoria*
Garry Davies, *University of Alaska, Anchorage*
Thomas DiChristina, *Georgia Institute of Technology*
Christy Donmoyer, *Allegheny College*
Scott E. Erdman, *Syracuse University*
Kenneth Field, *Bucknell University*
Theresa M. Filtz, *Oregon State University*
Larry J. Forney, *University of Idaho*
Elliott S. Goldstein, *Arizona State University*
Denise Greathouse, *University of Arkansas*
Richard D. Griner, *Augusta State University*
Mike Harrington, *University of Alberta, Edmonton*
Alan M. Jones, *University of North Carolina Chapel Hill*
Thomas C.S. Keller III, *Florida State University*
Gregory Kelly, *University of Western Ontario*
Karen L. Koster, *University of South Dakota*
Darryl Kropf, *University of Utah*
Charles A. Lessman, *University of Memphis*
Jani Lewis, *State University of New York at Geneseo*
Howard L. Liber, *Colorado State University*
Ryan Littlefield, *University of Washington*
Phoebe Lostroh, *Colorado College*

Michelle Malotky, *Guilford College*
Stephen M. Mount, *University of Maryland*
Leisha Mullins, *Texas A&M University*
Hao Nguyen, *California State University, Sacramento*
Joseph Pomerening, *Indiana University*
Joseph Reese, *Pennsylvania State University*
Nancy Rice, *Western Kentucky University*
Thomas M. Roberts, *Florida State University*
Donald F. Slish, *Plattsburgh State University of New York*
Charlotte Spencer, *University of Alberta, Edmonton*
Lesly A. Temesvari, *Clemson University*
Frances E. Weaver, *Widener University*
Gary M. Wessel, *Brown University*
David Worcester, *University of Missouri*

Reviewers of Previous Editions

Amelia Ahern-Rindell, *University of Portland*
Kirk Anders, *Gonzaga University*
Katsura Asano, *Kansas State University*
Steven Asmus, *Centre College*
Karl Aufderheide, *Texas A&M University*
L. Rao Ayyagari, *Lindenwood College*
William Balch, *Scripps Research Institute*
Tim Beagley, *Salt Lake Community College*
Margaret Beard, *Columbia University*
William Bement, *University of Wisconsin-Madison*
Paul Benko, *Sonoma State University*
Steve Benson, *California State University, Hayward*
Joseph J. Berger, *Springfield College*
Gerald Bergtrom, *University of Wisconsin, Milwaukee*
Karen K. Bernd. *Davidson College*
Maria Bertagnolli, *Gonzaga University*
Frank L. Binder, *Marshall University*
Robert Blystone, *Trinity University*
R. B. Boley, *University of Texas at Arlington*
Mark Bolyard, *Southern Illinois University, Edwardsville*
Edward M. Bonder, *Rutgers, the State University of New Jersey*
David Boone, *Portland State University*
Janet Braam, *Rice University*
James T. Bradley, *Auburn University*
Suzanne Bradshaw, *University of Cincinnati*
J. D. Brammer, *North Dakota State University*
Chris Brinegar, *San Jose State University*
Andrew Brittain, *Hawaii Pacific University*
Grant Brown, *University of Toronto*
David Bruck, *San Jose State University*
Alan H. Brush, *University of Connecticut, Storrs*
Patrick J. Bryan, *Central Washington University*
Brower R. Burchill, *University of Kansas*
Ann B. Burgess, *University of Wisconsin-Madison*
John G. Burr, *University of Texas, Dallas*
Thomas J. Byers, *Ohio State University*
David Byres, *Florida Community College, Jacksonville*
P. Samuel Campbell, *University of Alabama, Huntsville*
George L. Card, *University of Montana*
Anand Chandrasekhar, *University of Missouri*
C. H. Chen, *South Dakota State University*
Mitchell Chernin, *Bucknell University*
Edward A. Clark, *University of Washington*
Philippa Claude, *University of Wisconsin-Madison*
Dennis O. Clegg, *University of California, Santa Barbara*
John M. Coffin, *Tufts University School of Medicine*
J. John Cohen, *University of Colorado Medical School*
Larry Cohen, *Pomona College*
Nathan Collie, *Texas Tech University*
Reid S. Compton, *University of Maryland*
Mark Condon, *Dutchess Community College*
Jonathan Copeland, *Georgia Southern University*
Jeff Corden, *Johns Hopkins University*
Bracey Dangerfield, *Salt Lake Community College*
Garry Davies, *University of Alaska, Anchorage*
Maria Davis, *University of Alabama, Huntsville*
David DeGroote, *St. Cloud State*
Arturo De Lozanne, *University of Texas, Austin*
Douglas Dennis, *James Madison University*
Elizabeth D. Dolci, *Johnson State College*
Aris J. Domnas, *University of North Carolina, Chapel Hill*
Michael P. Donovan, *Southern Utah University*
Robert M. Dores, *University of Denver*
Stephen D'Surney, *University of Mississippi*
Ron Dubreuil, *University of Illinois, Chicago*
Diane D. Eardley, *University of California, Santa Barbara*
Lucinda Elliot, *Shippensburg University*
William Ettinger, *Gonzaga University*
Guy E. Farish, *Adams State College*
Mary A. Farwell, *East Carolina University*
David Featherstone, *University of Illinois, Chicago*
Kenneth Field, *Bucknell University*
Margaret F. Field, *St. Mary's College of California*
James E. Forbes, *Hampton University*
Charlene L. Forest, *Brooklyn College*
Carl S. Frankel, *Pennsylvania State University, Hazleton Campus*
David R. Fromson, *California State University, Fullerton*
David M. Gardner, *Roanoke College*
Craig Gatto, *Illinois State University*
Carol V. Gay, *Pennsylvania State University*
Stephen A. George, *Amherst College*
Nabarun Ghosh, *West Texas A&M University*
Swapan K. Ghosh, *Indiana State University*
Susan P. Gilbert, *University of Pittsburgh*
Reid Gilmore, *University of Massachusetts*
Joseph Gindhart, *University of Massachusetts, Boston*
Michael L. Gleason, *Central Washington University*
T. T. Gleeson, *University of Colorado*
James Godde, *Monmouth College*

Ursula W. Goodenough, *Washington University*
Thomas A. Gorell, *Colorado State University*
James Grainger, *Santa Clara University*
Marion Greaser, *University of Wisconsin-Madison*
Karen F. Greif, *Bryn Mawr College*
Richard Griner, *Augusta State University*
Mark T. Groudine, *Fred Hutchinson Cancer Research Center, University of Washington School of Medicine*
Gary Gussin, *University of Iowa*
Karen Guzman, *Campbell University*
Leah T. Haimo, *University of California, Riverside*
Arnold Hampel, *Northern Illinois University*
Laszlo Hanzely, *Northern Illinois University*
Donna Harman, *Lubbock Christian University*
Bettina Harrison, *University of Massachusetts, Boston*
William Heidcamp, *Gustavus Adolphus College*
John Helmann, *Cornell University*
Lawrence Hightower, *University of Connecticut, Storrs*
Chris Holford, *Purdue University*
James P. Holland, *Indiana University, Bloomington*
Johns Hopkins III, *Washington University*
Nancy Hopkins, *Tulane University*
Betty A. Houck, *University of Portland*
Linda S. Huang, *University of Massachusetts, Boston*
Sharon Isern, *Florida Gulf Coast University*
Kenneth Jacobson, *University of North Carolina, Chapel Hill*
Makkuni Jayaram, *University of Texas, Austin*
William R. Jeffery, *University of Texas, Austin*
Kwang W. Jeon, *University of Tennessee*
Jerry E. Johnson Jr., *University of Houston*
Kenneth C. Jones, *California State University, Northridge*
Patricia P. Jones, *Stanford University*
Cheryl L. Jorcyk, *Boise State University*
David Kafkewitz, *Rutgers University*
Martin A. Kapper, *Central Connecticut State University*
Lon S. Kaufman, *University of Illinois, Chicago*
Steven J. Keller, *University of Cincinnati Main Campus*
Greg Kelly, *University of Western Ontario*
Gwendolyn M. Kinebrew, *John Carroll University*
Kirill Kiselyov, *University of Pittsburgh*
Loren Knapp, *University of South Carolina*
Robert Koch, *California State University, Fullerton*
Bruce Kohorn, *Bowdoin College*
Joseph R. Koke, *Southwest Texas State University*
Irene Kokkala, *Northern Georgia College & State University*
Keith Kozminski, *University of Virginia*
Hal Krider, *University of Connecticut, Storrs*
William B. Kristan, Jr., *University of California, San Diego*
David N. Kristie, *Acadia University (Nova Scotia)*
Frederic Kundig, *Towson State University*
Jeffrey Kushner, *James Madison University*
Dale W. Laird, *University of Western Ontario (London, Ontario, Canada)*
Michael Lawton, *Rutgers University*
Elias Lazarides, *California Institute of Technology*
Wei-Lih Lee, *University of Massachusetts, Amherst*
Esther M. Leise, *University of North Carolina, Greensboro*
Charles Lessman, *University of Memphis*
Daniel Lew, *Duke University*
Carol Lin, *Columbia University*
John T. Lis, *Cornell University*
Kenneth Long, *California Lutheran University*
Robert Macey, *University of California, Berkeley*
Albert MacKrell, *Bradley University*
Roderick MacLeod, *University of Illinois, Urbana-Champaign*
Shyamal K. Majumdar, *Lafayette College*
Gary G. Matthews, *State University of New York, Stony Brook*
Douglas McAbee, *California State University, Long Beach*
Mark McCallum, *Pfeiffer University*
Iain McKillop, *University of North Carolina, Charlotte*
Thomas D. McKnight, *Texas A&M University*
JoAnn Meerschaert, *St. Cloud University*
Trevor Mendelow, *University of Colorado*
John Merrill, *Michigan State University*
Robert L. Metzenberg, *University of Wisconsin-Madison*
Teena Michael, *University of Hawaii, Manoa*
Hugh A. Miller III, *East Tennessee State University*
Jeffrey Miller, *University of Minnesota*
Nicole Minor, *Northern Kentucky University*
James Moroney, *Louisiana State University*
Tony K. Morris, *Fairmont State College*
Deborah B. Mowshowitz, *Columbia University*
Amy Mulnix, *Earlham College*
James Mulrooney, *Central Connecticut State University*
Hao Nguyen, *California State University, Sacramento*
Carl E. Nordahl, *University of Nebraska, Omaha*
Richard Nuccitelli, *University of California, Davis*
Donata Oertel, *University of Wisconsin-Madison*
Joanna Olmsted, *University of Rochester*
Laura Olsen, *University of Michigan*
Alan Orr, *University of Northern Iowa*
Donald W. Ott, *University of Akron*
Curtis L. Parker, *Morehouse School of Medicine*
Lee D. Peachey, *University of Colorado*
Debra K. Pearce, *Northern Kentucky University*
Mark Peifer, *University of North Carolina, Chapel Hill*
Howard Petty, *Wayne State University*
Susan Pierce, *Northwestern University*
Joel B. Piperberg, *Millersville University*
William Plaxton, *Queen's University*
George Plopper, *Rensselaer Polytechnic Institute*
Gilbert C. Pogany, *Northern Arizona University*
Archie Portis, *University of Illinois*
Stephen Previs, *Case Western Reserve University*
Mitch Price, *Pennsylvania State University*
Ralph Quatrano, *Oregon State University*

Ralph E. Reiner, *College of the Redwoods*
Gary Reiness, *Lewis and Clark College*
Douglas Rhoads, *University of Arkansas*
John Rinehart, *Eastern Oregon University*
Tom Roberts, *Florida State University*
Michael Robinson, *North Dakota State University*
Adrian Rodriguez, *San Jose State University*
Donald J. Roufa, *Kansas State University*
Donald H. Roush, *University of North Alabama*
Gary Rudnick, *Yale University*
Donald Salter, *Leuther Laboratories*
Edmund Samuel, *Southern Connecticut State University*
Mary Jane Saunders, *University of South Florida*
John I. Scheide, *Central Michigan University*
David J. Schultz, *University of Louisville*
Mary Schwanke, *University of Maine, Farmington*
David W. Scupham, *Valparaiso University*
Edna Seaman, *University of Massachusetts, Boston*
Diane C. Shakes, *University of Houston*
Joel Sheffield, *Temple University*
Sheldon S. Shen, *Iowa State University*
James R. Shinkle, *Trinity University Texas*
Randall D. Shortridge, *State University of New York, Buffalo*
Maureen Shuh, *Loyola University, New Orleans*
Brad Shuster, *New Mexico State University*
Jill Sible, *Virginia Polytechnic Institute and State University*
Esther Siegfried, *Pennsylvania State University*
Michael Silverman, *California State Polytechnic University, Pomona*
Neil Simister, *Brandeis University*
Dwayne D. Simmons, *University of California, Los Angeles*
Robert D. Simoni, *Stanford University*
William R. Sistrom, *University of Oregon*
Donald F. Slish, *Plattsburgh State University of New York*
Roger Sloboda, *Dartmouth College*
Robert H. Smith, *Skyline College*
Juliet Spencer, *University of San Francisco*
Mark Staves, *Grand Valley State University*
John Sternfeld, *State University of New York, Cortland*
Barbara Y. Stewart, *Swarthmore College*
Margaret E. Stevens, *Ripon College*
Bradley J. Stith, *University of Colorado, Denver*
Richard D. Storey, *Colorado College*
Antony O. Stretton, *University of Wisconsin-Madison*
Philip Stukus, *Denison University*
Stephen Subtelny, *Rice University*
Scott Summers, *Colorado State University*
Millard Susman, *University of Wisconsin-Madison*
Brian Tague, *Wake Forest University*
Elizabeth J. Taparowsky, *Purdue University*
Barbara J. Taylor, *Oregon State University*
Bruce R. Telzer, *Pomona College*
Jeffrey L. Travis, *State University of New York, Albany*
John J. Tyson, *Virginia Polytechnic University*
Akif Uzman, *University of Texas, Austin*
Thomas Vandergon, *Pepperdine University*
Quinn Vega, *Montclair State University*
James Walker, *University of Texas, Pan American*
Paul E. Wanda, *Southern Illinois University, Edwardsville*
Fred D. Warner, *Syracuse University*
James Watrous, *St. Joseph's University*
Andrew N. Webber, *Arizona State University*
Cindy Martinez Wedig, *University of Texas, Pan American*
Fred H. Wilt, *University Of California, Berkeley*
James Wise, *Hampton University*
David Worcester, *University of Missouri, Columbia*
Lauren Yaich, *University of Pittsburgh, Bradford*
Linda Yasui, *Northern Illinois University*
Yang Yen, *South Dakota State University*
James Young, *University of Alberta*
Luwen Zhang, *University of Nebraska, Lincoln*
Qiang Zhou, *University of California, Berkeley*

Pearson wishes to thank and acknowledge the following people for their work on the Global Edition:

Contributor

Katie Smith, *University of York*
Shefali Sabharanjak, *independent science and medical writer*

Reviewers

Pushpa Agarwal, *RV College of Engineering*
P. Lalitha, *GITAM University*
Boitelo Letsolo, *University of the Witwatersrand*

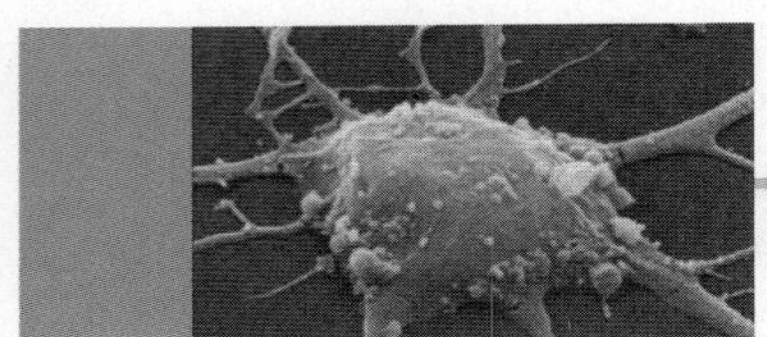

BRIEF CONTENTS

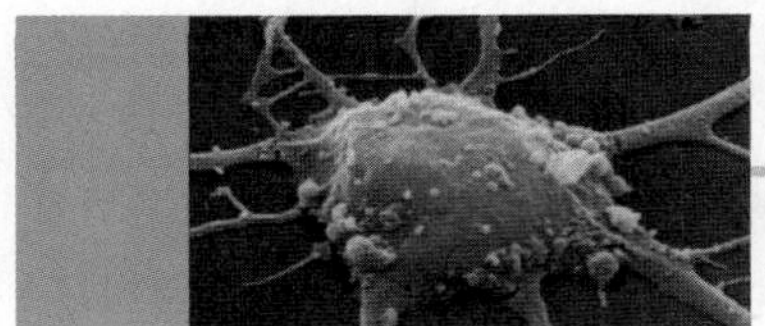

DETAILED CONTENTS

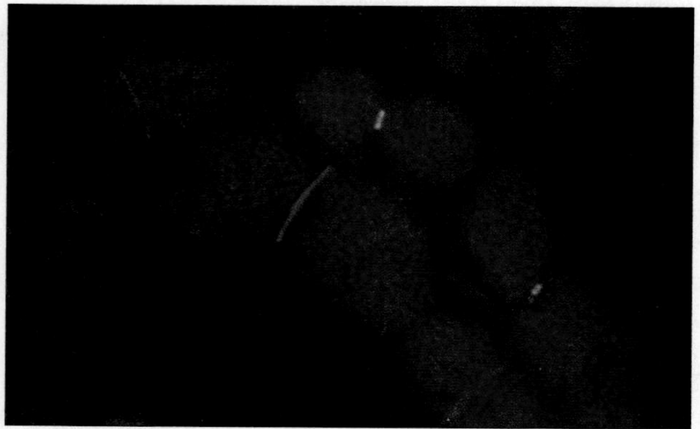

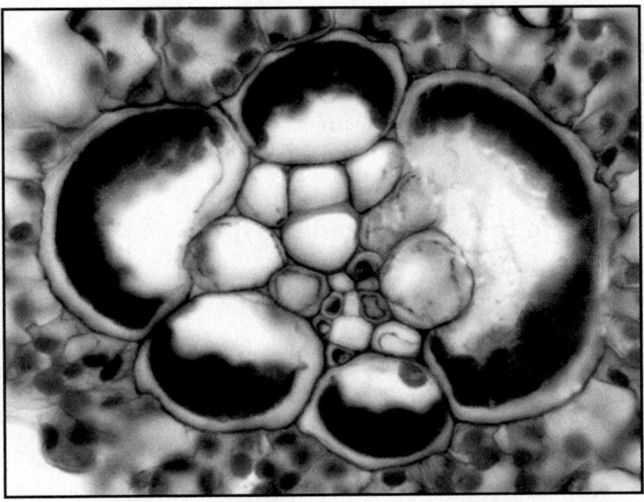

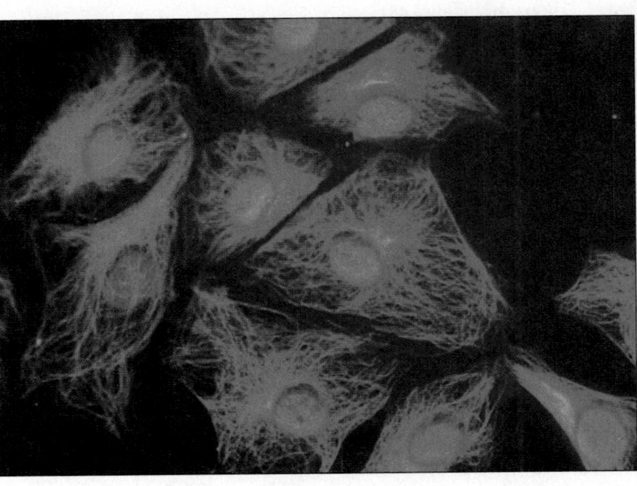

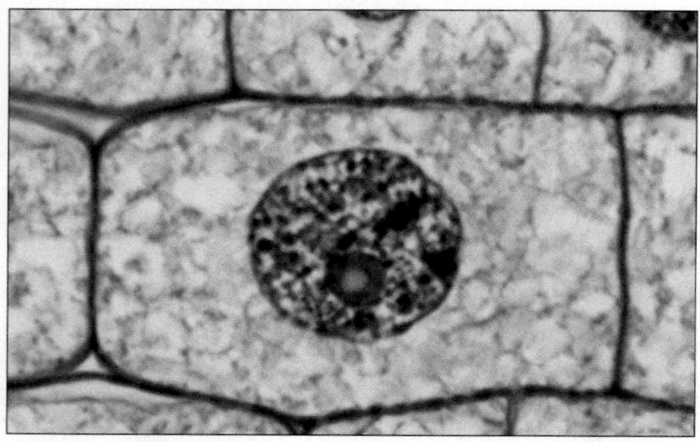

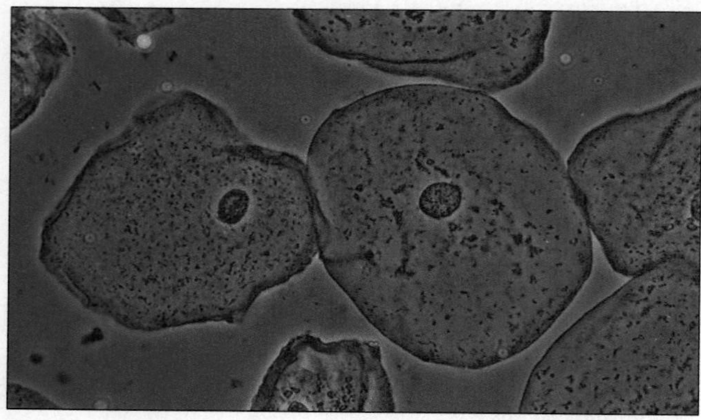

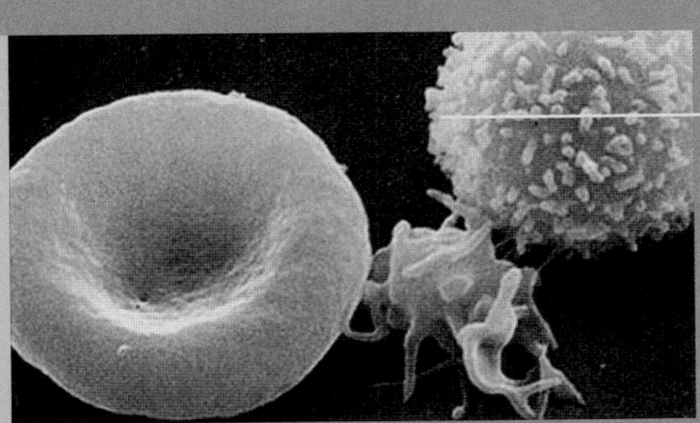

1 A Preview of the Cell

See MasteringBiology® for tutorials, activities, and quizzes.

*The **cell** is the basic unit of biology. Every organism either consists of cells or is itself a single cell. Therefore, it is only by understanding the structure and function of cells that we can appreciate both the capabilities and the limitations of living organisms, whether they are animals, plants, fungi, or microorganisms.*

We are in the midst of an exciting revolution in biology that has brought with it tremendous advances in our understanding of how cells are constructed and how they carry out all the intricate functions necessary for life. Particularly significant is the dynamic nature of the cell, which has the capacity to grow, reproduce, and become specialized and also the ability to respond to stimuli and adapt to changes in its environment. We hope that the static depictions of cells in printed texts do not give you the mistaken impression that cells themselves are static, so we encourage you to view videos and animations of "cells in action" on our website, The Cell Place.

The field of cell biology itself is rapidly changing as scientists from a variety of related disciplines work together to gain a better understanding of how cells work. The convergence of cytology, genetics, and biochemistry has made modern cell biology one of the most exciting and dynamic disciplines in all of biology. If this text helps you to appreciate the marvels and diversity of cellular functions and helps you to experience the excitement of discovery, then one of our main goals in writing this book for you will have been met.

In this introductory chapter, we will look briefly at the origin of cell biology as a discipline. Then we will consider the three main historical strands of cytology, genetics, and biochemistry that have formed our current understanding of what cells are and how they work. We will conclude with a brief discussion of the nature of knowledge itself by considering biological facts, the scientific method, and the use of some common model organisms in cell biology—how do we know what we know?

The Cell Theory: A Brief History

The story of cell biology started to unfold more than 300 years ago, as European scientists began to focus their crude microscopes on a variety of biological material ranging from tree bark to bacteria to human sperm. One such scientist was Robert Hooke, Curator of Instruments for the Royal Society of London. In 1665, Hooke built a microscope and examined thin slices of cork cut with a penknife. He saw a network of tiny boxlike compartments that reminded him of a honeycomb and called these little compartments *cells,* from the Latin word *cellula,* meaning "little room."

We now know that what Hooke observed were not cells at all. Those boxlike compartments were formed by the empty cell walls of dead plant tissue, which is what cork really is. However, Hooke would not have thought of these cells as dead, because he did not understand that they could be alive! Although he noticed that cells in other plant tissues were filled with what he called "juices," he preferred to concentrate on the more prominent cell walls of the dead cork cells that he had first encountered.

Hooke's observations were limited by the *magnification power* of his microscope, which enlarged objects to only 30 times (30×) their normal size. This made it difficult to learn much about the internal organization of cells. A few years later, Antonie van Leeuwenhoek, a Dutch textile merchant, produced small lenses that could magnify objects to almost 300 times (300×) their size. Using these superior lenses, van Leeuwenhoek became the first to observe living cells, including blood cells, sperm cells, bacteria, and single-celled organisms (algae and protozoa) found in pond water. He reported his observations to the Royal Society of London in a series of letters during the late 1600s. His detailed reports attest to both the high quality of his lenses and his keen powers of observation.

Two factors restricted further understanding of the nature of cells. First, the microscopes of the day had limited *resolution (resolving power)*—the ability to see fine

Units of Measurement in Cell Biology

The challenge of understanding cellular structure and organization is complicated by the problem of size. Most cells and their organelles are too small to be seen by the unaided eye. In addition, the units used to measure them are unfamiliar to many students and therefore often difficult to appreciate. The problem can be approached in two ways: by realizing that only two units are really necessary to express the dimensions of most structures of interest to us, and by considering a variety of structures that can be appropriately measured with each of these two units.

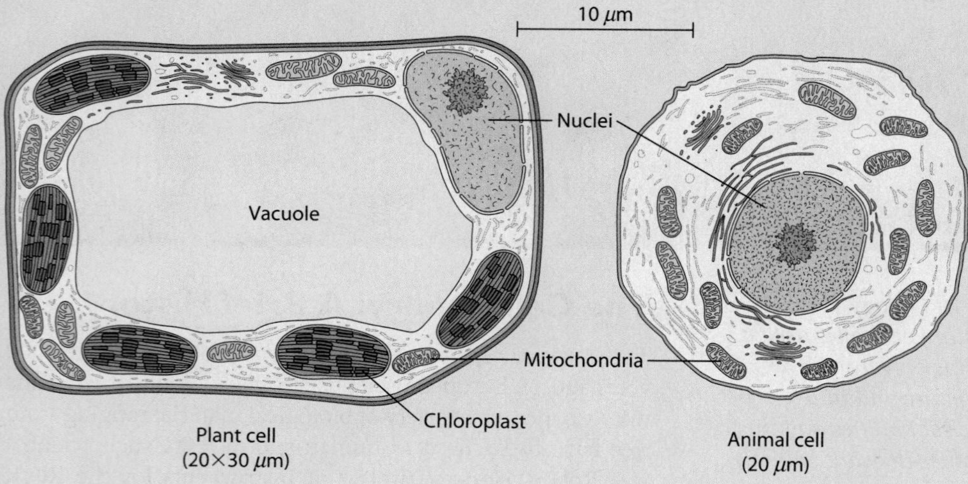

FIGURE 1A-1 The World of the Micrometer. Structures with dimensions that can be measured conveniently in micrometers include almost all cells and some of the larger organelles, such as the nucleus, mitochondria, and chloroplasts. One inch equals approximately 25,000 micrometers.

details of structure. Even van Leeuwenhoek's superior instruments could push this limit only so far. The second factor was the descriptive nature of seventeenth-century biology. It was an age of observation, with little thought given to explaining the intriguing architectural details being discovered in biological materials.

More than a century passed before the combination of improved microscopes and more experimentally minded microscopists resulted in a series of developments that led to an understanding of the importance of cells in biological organization. By the 1830s, important optical improvements were made in lens quality and in the *compound microscope,* in which one lens (the eyepiece) magnifies the image created by a second lens (the objective). This allowed both higher magnification and better resolution. Now, structures only 1 micrometer (μm) in size could be seen clearly. (See **Box 1A** for a discussion of the units of measurement appropriate to cell biology.)

Aided by such improved lenses, the English botanist Robert Brown found that every plant cell he looked at contained a rounded structure, which he called a *nucleus,* a term derived from the Latin word for "kernel." In 1838, his German colleague Matthias Schleiden came to the important conclusion that all plant tissues are composed of cells and that an embryonic plant always arises from a single cell. A year later, German cytologist Theodor Schwann reported similar conclusions concerning animal tissue, thereby discrediting earlier speculations that plants and animals do not resemble each other structurally. These speculations arose because plant cell walls form conspicuous boundaries between cells that are readily visible even with a crude microscope, whereas individual animal cells, which lack cell walls, are much harder to distinguish in a tissue sample. However, when Schwann examined animal cartilage cells, he saw that they were unlike most other animal cells because they have boundaries that are well defined by thick deposits of collagen fibers. Thus, he became convinced of the fundamental similarity between plant and animal tissue. Based on his astute observations, Schwann developed a single unified theory of cellular organization, which has stood the test of time and continues to be the basis for our own understanding of the importance of cells and cell biology.

As originally postulated by Schwann in 1839, the **cell theory** had two basic principles:

1. All organisms consist of one or more cells.
2. The cell is the basic unit of structure for all organisms.

Less than 20 years later, a third principle was added. This grew out of Brown's original description of nuclei, extended by Swiss botanist Karl Nägeli to include observations on the nature of cell division. By 1855 Rudolf Virchow, a German physiologist, concluded that cells arose only by the division of other, preexisting cells. Virchow encapsulated this conclusion in the now-famous Latin phrase *omnis cellula e cellula,* which in translation becomes the third principle of the modern cell theory:

3. All cells arise only from preexisting cells.

The **micrometer (μm)** is the most useful unit for expressing the size of cells and organelles. A micrometer (sometimes also called a *micron*) is one-millionth of a meter (10^{-6} m). In general, bacterial cells are a few micrometers in diameter, and the cells of plants and animals are 10 to 20 times larger in any single dimension. Organelles such as mitochondria and chloroplasts tend to have diameters or lengths of a few micrometers and are thus comparable in size to whole bacterial cells. In general, if you can see it with a light microscope, you can express its dimensions conveniently in micrometers (**Figure 1A-1**).

The **nanometer (nm),** on the other hand, is the unit of choice for molecules and subcellular structures that are too small to be seen with the light microscope. A nanometer is one-billionth of a meter (10^{-9} m, so it takes 1000 nanometers to equal 1 micrometer. A ribosome has a diameter of about 25 nm. Other structures that can be measured conveniently in nanometers are microtubules, microfilaments, membranes, and DNA molecules (**Figure 1A-2**). A slightly smaller unit, the angstrom (Å), is occasionally used in cell biology when measuring dimensions within proteins and DNA molecules. An angstrom equals 0.1 nm, which is about the size of a hydrogen atom.

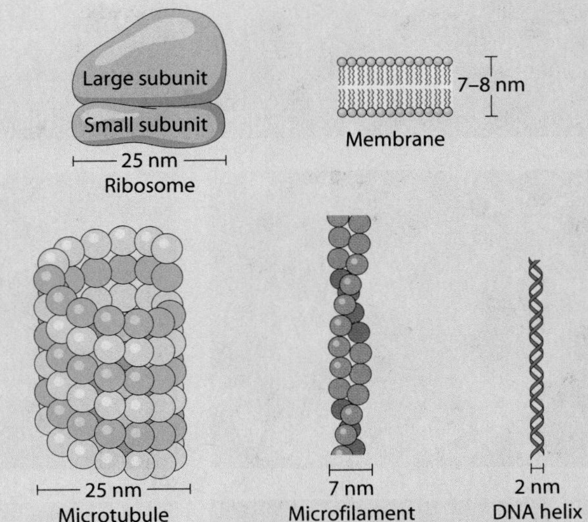

FIGURE 1A-2 The World of the Nanometer. Structures with dimensions that can be measured conveniently in nanometers include ribosomes, membranes, microtubules, microfilaments, and the DNA double helix. One nm equals 10 Å.

Thus, the cell is not only the basic unit of structure for all organisms but also the basic unit of reproduction. No wonder, then, that we must understand cells and their properties to appreciate all other aspects of biology. Because many of you have seen examples of "typical" cells in textbooks that may give the false impression that there are relatively few different types of cells, let's take a look at a few examples of the diversity of cells that exist in our world (**Figure 1-1**).

Cells exist in a wide variety of shapes and sizes, from filamentous fungal cells to spiral-shaped *Treponema* bacteria to the differently-shaped cells of the human blood system (Figures 1-1a–c). Other cells have much more exotic shapes, such as the radiolarian and the protozoan shown in Figures 1-1d and 1-1e. Even the two human single-celled gametes, the egg and the sperm, differ greatly in size and shape (Figure 1-1f). Often, an appreciation of a cell's shape and structure gives clues about its function. For example, the large surface area of the microvilli on our intestinal cells aids in maximizing nutrient absorption (Figure 1-1g), the spiral thickenings in the cell walls of plant xylem tissue give strength to these water-conducting vessels in wood (Figure 1-1h), and the highly branched cells of a human neuron allow it to interact with numerous other neurons (Figure 1-1i). In our studies throughout this textbook, we will see many other interesting examples of diversity in cell structure and function. First, though, let's examine the historical roots leading to the development of contemporary cell biology.

The Emergence of Modern Cell Biology

Modern cell biology results from the weaving together of three different strands of biological inquiry—cytology, biochemistry, and genetics—into a single cord. As the timeline of **Figure 1-2** illustrates, each of the strands has had its own historical origins, and each one makes unique and significant contributions to modern cell biology. Contemporary cell biologists must be adequately informed about all three strands, regardless of their own immediate interests. In addition, they must understand how biological processes follow the laws of chemistry and physics.

Historically, the first of these strands to emerge is **cytology,** which is concerned primarily with cellular structure. In our studies, we will often encounter words containing the Greek prefix *cyto-* or the suffix *-cyte,* both of which mean "hollow vessel" and refer to cells. As we have already seen, cytology had its origins more than three centuries ago and depended heavily on the light microscope for its initial impetus. The advent of electron microscopy and other advanced optical techniques has dramatically increased our understanding of cell structure and function.

The second strand represents the contributions of **biochemistry** to our understanding of cellular structure and function. Most of the developments in this field have occurred over the past 75 years, though the roots go back at least a century earlier. Especially important has been the development of techniques such as ultracentrifugation,

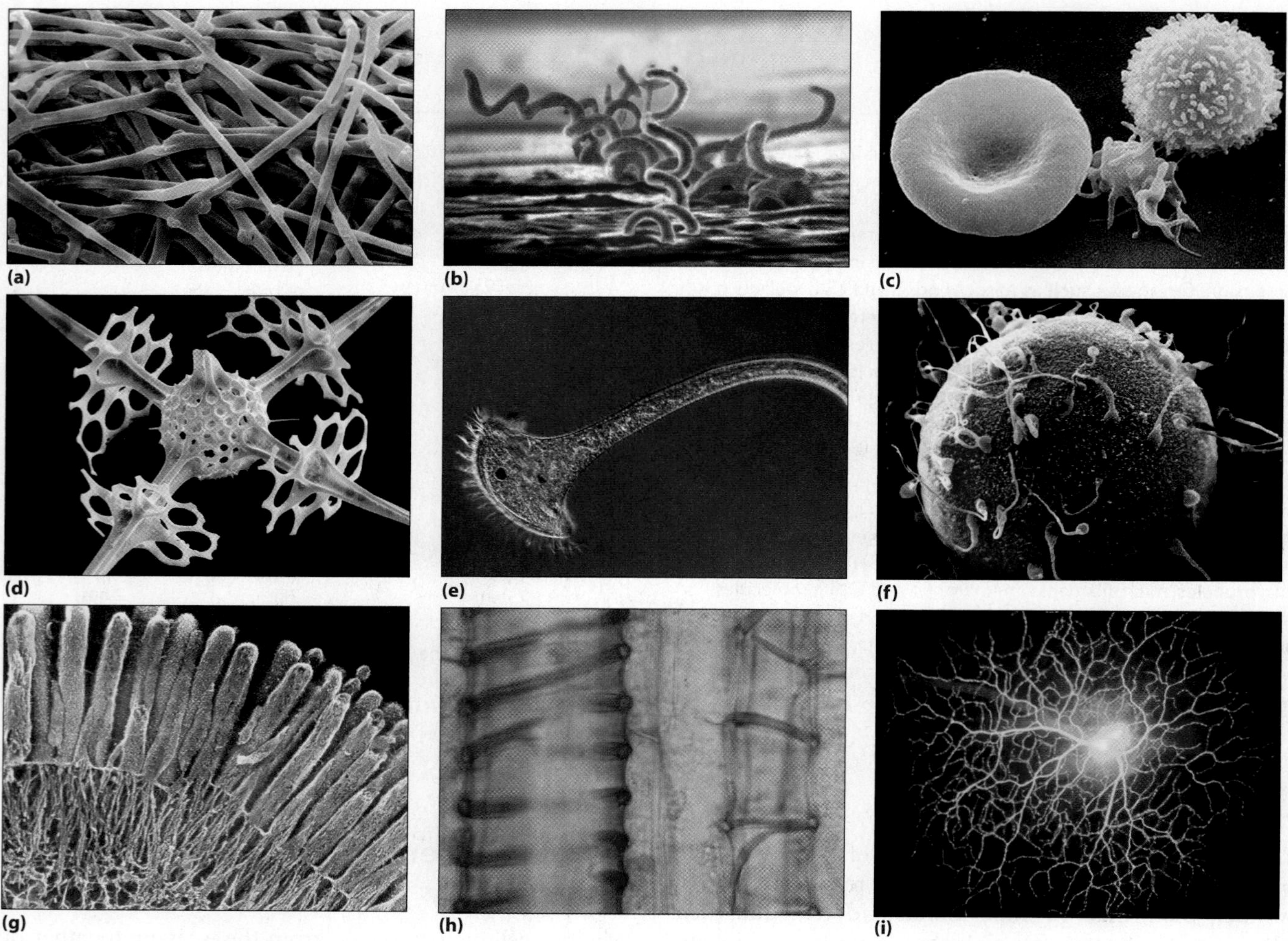

FIGURE 1-1 The Cells of the World. The diversity of cell types existing all around us includes **(a)** filamentous fungal cells; **(b)** *Treponema* bacteria; **(c)** a human red blood cell, a platelet, and a white blood cell (left to right); **(d)** a radiolarian; **(e)** *Stentor* (a protozoan); **(f)** human egg and sperm cells; **(g)** intestinal cells; **(h)** plant xylem cells; and **(i)** a retinal neuron.

chromatography, radioactive labeling, electrophoresis, and mass spectrometry for separating and identifying cellular components. We will encounter these and other techniques later in our studies as we explore how specific details of cellular structure and function were discovered using these techniques. To locate discussions of specific techniques, see the Guide to Techniques and Methods inside the front cover.

The third strand contributing to the development of modern cell biology is **genetics.** Here, the timeline stretches back more than 150 years to Austrian monk and biologist Gregor Mendel. Again, however, we have gained much of our present understanding within the past 75 years. An especially important landmark on the genetic strand is the demonstration that, in all organisms, DNA (deoxyribonucleic acid) is the bearer of genetic information. It encodes the tremendous variety of proteins and RNA (ribonucleic acid) molecules responsible for most of the functional and structural features of cells. Recent accomplishments on the genetic strand include the sequencing of the entire *genomes* (all of the DNA) of humans and other species and the *cloning* (production of genetically identical organisms) of mammals, including sheep, cattle, and cats.

Therefore, to understand present-day cell biology, we must appreciate its diverse roots and the important contributions made by each of its component strands to our current understanding of what a cell is and what it can do. Each of the three historical strands of cell biology is discussed briefly here; a deeper appreciation of each will come in later chapters as we explore cells in detail. Keep in mind also that in addition to developments in cytology, biochemistry, and genetics, the field of cell biology has benefited greatly from advancements in other fields of study such as chemistry, physics, computer science, and engineering.

The Cytological Strand Deals with Cellular Structure

Strictly speaking, cytology is the study of cells. Historically, however, cytology has dealt primarily with cellular structure, mainly through the use of optical techniques.

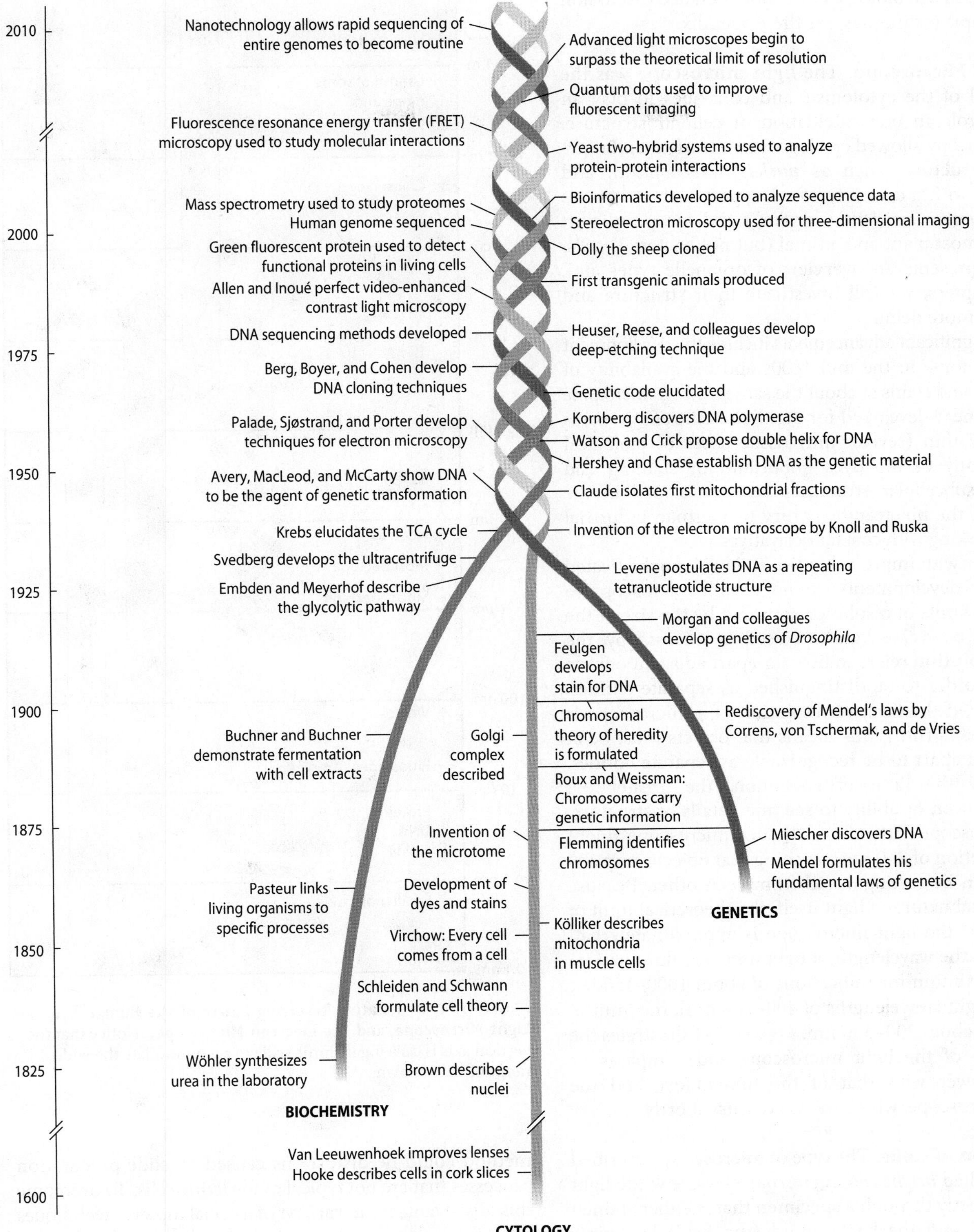
CELL BIOLOGY
2010
Nanotechnology allows rapid sequencing of entire genomes to become routine
Advanced light microscopes begin to surpass the theoretical limit of resolution
Quantum dots used to improve fluorescent imaging
Fluorescence resonance energy transfer (FRET) microscopy used to study molecular interactions
Yeast two-hybrid systems used to analyze protein-protein interactions
Bioinformatics developed to analyze sequence data
Mass spectrometry used to study proteomes
Stereoelectron microscopy used for three-dimensional imaging
Human genome sequenced
2000
Green fluorescent protein used to detect functional proteins in living cells
Dolly the sheep cloned
First transgenic animals produced
Allen and Inoué perfect video-enhanced contrast light microscopy
Heuser, Reese, and colleagues develop deep-etching technique
DNA sequencing methods developed
1975
Berg, Boyer, and Cohen develop DNA cloning techniques
Genetic code elucidated
Kornberg discovers DNA polymerase
Palade, Sjøstrand, and Porter develop techniques for electron microscopy
Watson and Crick propose double helix for DNA
Hershey and Chase establish DNA as the genetic material
1950
Avery, MacLeod, and McCarty show DNA to be the agent of genetic transformation
Claude isolates first mitochondrial fractions
Krebs elucidates the TCA cycle
Invention of the electron microscope by Knoll and Ruska
Svedberg develops the ultracentrifuge
Levene postulates DNA as a repeating tetranucleotide structure
1925
Embden and Meyerhof describe the glycolytic pathway
Morgan and colleagues develop genetics of Drosophila
Feulgen develops stain for DNA
1900
Rediscovery of Mendel's laws by Correns, von Tschermak, and de Vries
Chromosomal theory of heredity is formulated
Buchner and Buchner demonstrate fermentation with cell extracts
Golgi complex described
Roux and Weissman: Chromosomes carry genetic information
1875
Invention of the microtome
Flemming identifies chromosomes
Miescher discovers DNA
Mendel formulates his fundamental laws of genetics
Pasteur links living organisms to specific processes
Development of dyes and stains
GENETICS
Kölliker describes mitochondria in muscle cells
Virchow: Every cell comes from a cell
1850
Schleiden and Schwann formulate cell theory
1825
Wöhler synthesizes urea in the laboratory
Brown describes nuclei
BIOCHEMISTRY
Van Leeuwenhoek improves lenses
Hooke describes cells in cork slices
1600
CYTOLOGY

FIGURE 1-2 The Cell Biology Time Line. Although cytology, biochemistry, and genetics began as separate disciplines, they have increasingly merged since about 1925.

Here we will describe briefly some of the microscopy that is important in cell biology. For a more detailed discussion of microscopic techniques, see the Appendix.

The Light Microscope. The **light microscope** was the earliest tool of the cytologists and continues to play an important role in our elucidation of cellular structure. Light microscopy allowed cytologists to identify membrane-bounded structures such as *nuclei, mitochondria,* and *chloroplasts* within a variety of cell types. Such structures are called *organelles* ("little organs") and are prominent features of most plant and animal (but not bacterial) cells. Chapter 4 presents an overview of organelle types and, in later chapters, we will investigate their structure and function in more detail.

Other significant advancements include the development of the microtome in the mid-1800s and the availability of various dyes and stains at about the same time. A *microtome* is an instrument developed for rapid and efficient preparation of very thin (several μm) tissue slices of biological samples. Many of the dyes important for staining and identifying subcellular structures were developed in the latter half of the nineteenth century by German industrial chemists working with coal tar derivatives.

Together with improved optics and more sophisticated lenses, these developments extended light microscopy to the physical limits of resolution imposed by the size of the wavelengths of visible light. As used in microscopy, the **limit of resolution** refers to how far apart adjacent objects must be in order to be distinguished as separate entities. For example, if the limit of resolution of a microscope is 400 nanometers (nm), this means that objects must be at least 400 nm apart to be recognizable as separate entities. The smaller the limit of resolution, the greater the **resolving power,** or ability to see fine details of structure, of the microscope. Therefore, a better microscope might have a resolution of 200 nm, meaning that objects only 200 nm apart can be distinguished from each other. Because of the physical nature of light itself, the theoretical limit of resolution for the light microscope is approximately half of the size of the wavelength of light used for illumination, allowing maximum magnifications of about 1000–1500×. For visible light (wavelengths of 400–700 nm), the limit of resolution is about 200–350 nm. **Figure 1-3** illustrates the useful range of the light microscope and compares its resolving power with that of the human eye and the electron microscope, which we will discuss shortly.

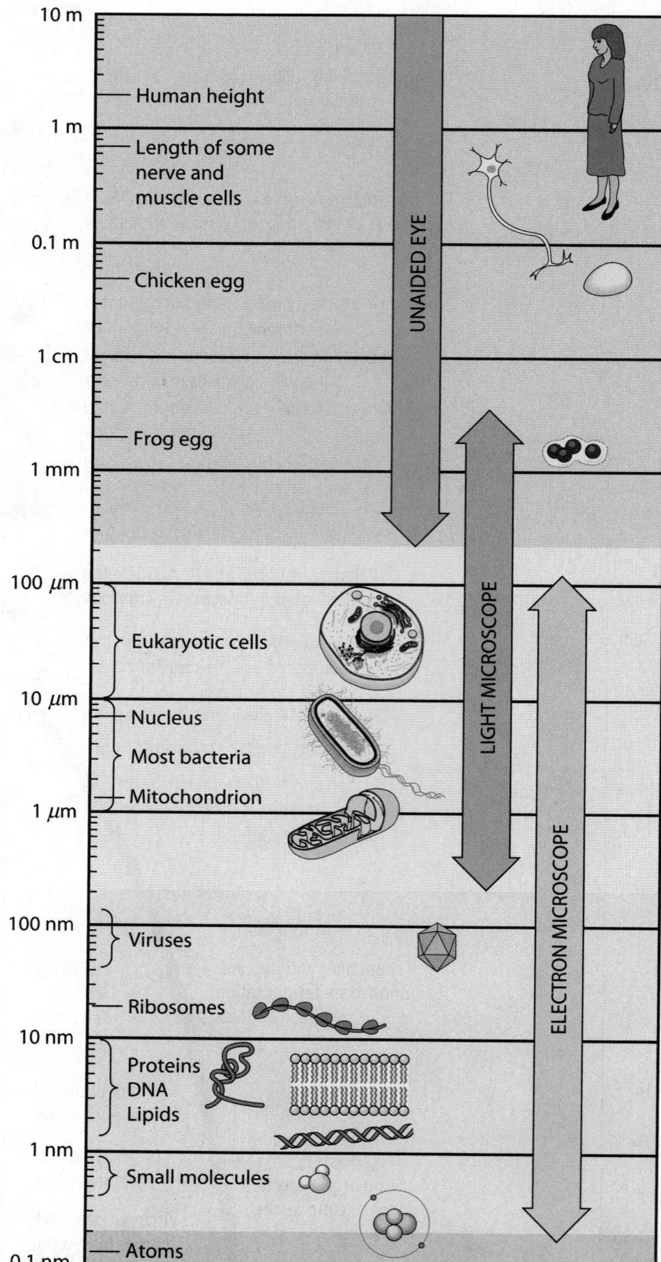

FIGURE 1-3 Relative Resolving Power of the Human Eye, the Light Microscope, and the Electron Microscope. Notice that the vertical axis is on a logarithmic scale to accommodate the wide range of sizes shown.

Visualization of Cells. The type of microscopy described thus far is called *brightfield microscopy* because white light is passed directly through a specimen that is either stained or unstained and the background (the field) is illuminated. A significant limitation of this approach is that specimens often must be chemically fixed (preserved), dehydrated, embedded in paraffin or plastic for slicing into thin sections, and stained to highlight otherwise transparent features. Fixed and stained specimens are no longer alive, and therefore features observed by this method could be distortions caused by slide preparation processes that are not typical of the living cells. To overcome this disadvantage, a variety of special optical techniques have been developed for observing living cells directly. These include phase-contrast microscopy, differential interference contrast microscopy, fluorescence microscopy, and confocal microscopy. In **Table 1-1**, you can see images taken using each of these techniques and compare them with the images seen with brightfield microscopy for both unstained and stained specimens. Because these

Table 1-1 Different Types of Light Microscopy: A Comparison

Type of Microscopy	Light Micrographs of Human Cheek Epithelial Cells		Type of Microscopy
Brightfield (unstained specimen): Passes light directly through specimen; unless cell is naturally pigmented or artificially stained, image has little contrast.	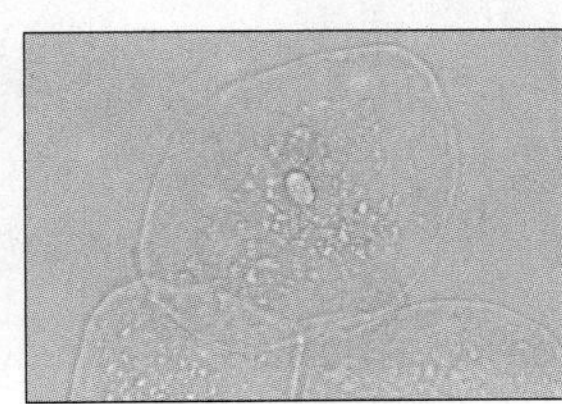	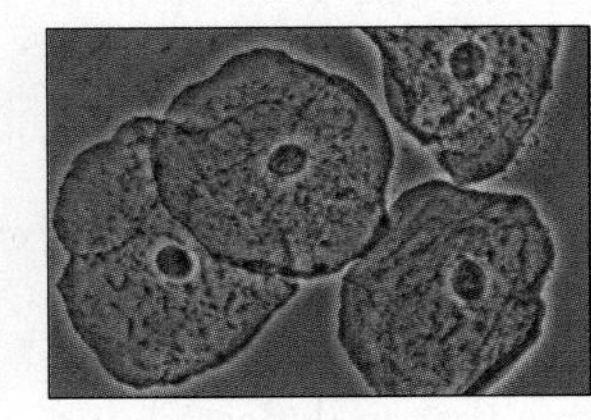	**Phase contrast:** Enhances contrast in unstained cells by amplifying variations in refractive index within specimen; especially useful for examining living, unpigmented cells.
Brightfield (stained specimen): Staining with various dyes enhances contrast, but most staining procedures require that cells be fixed (preserved).		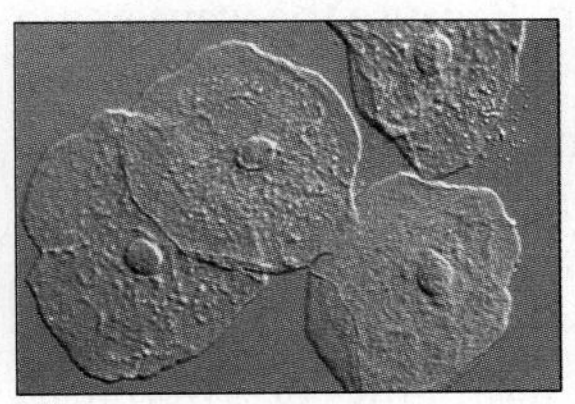	**Differential interference contrast:** Also uses optical modifications to exaggerate differences in refractive index.
Fluorescence: Shows the locations of specific molecules in the cell. Fluorescent substances absorb ultraviolet radiation and emit visible light. The fluorescing molecules may occur naturally in the specimen but more often are made by tagging the molecules of interest with fluorescent dyes or antibodies.	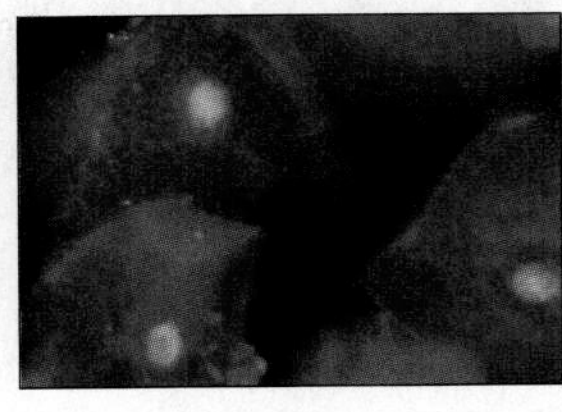	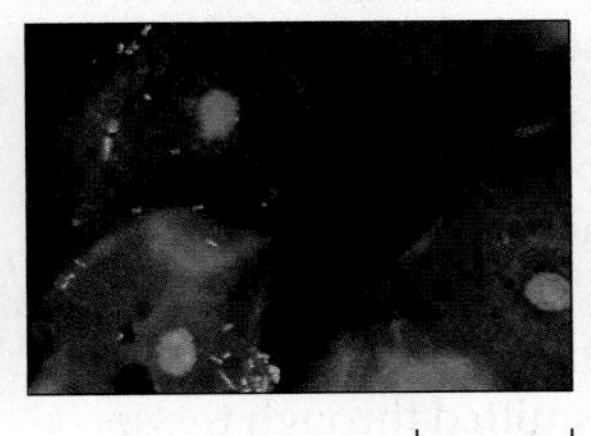 20 μm	**Confocal:** Uses lasers and special optics to focus illuminating beam on a single plane within the specimen. Only those regions within a narrow depth of focus are imaged. Regions above and below the selected plane of view appear black rather than blurry.

Source: Adapted from Campbell and Reece, *Biology,* 6th ed. (San Francisco: Benjamin Cummings, 2002), p. 110.

techniques are discussed in the Appendix in more detail, here we will content ourselves with brief descriptions.

Phase-contrast and *differential interference contrast microscopy* make it possible to see living cells clearly. Both techniques enhance and amplify slight changes in the phase of transmitted light as it passes through a structure having a different refractive index (based in part on density) than the surrounding medium. *Fluorescence microscopy* is a powerful method that enables researchers to detect specific proteins, DNA sequences, or other molecules that are made fluorescent by coupling them to a fluorescent dye or binding them to a fluorescently labeled antibody. By simultaneously using two or more such dyes or antibodies, each emitting light of a different color, researchers can follow the distributions of different kinds of molecules in the same cell. This method can also be used with fluorescently labeled DNA sequences to determine the presence and precise position of specific genes within a cell. In recent years, *green fluorescent protein (GFP)* from the bioluminescent jellyfish *Aequorea victoria* has become an invaluable tool for studying the temporal and spatial distribution of particular proteins in a cell. When a protein of interest is fused with GFP, its synthesis and movement can be followed in living cells. In 2008, the Nobel Prize in Chemistry was awarded to Martin Chalfie, Osamu Shimomura, and Roger Y. Tsien for the discovery and development of GFP.

An inherent limitation of fluorescence microscopy is that the viewer can focus on only a single plane of the specimen at a time, yet fluorescent light is emitted throughout the specimen, blurring the image. This problem is largely overcome by *confocal scanning,* which uses a laser beam to illuminate just one plane of the specimen at a time. When used with thick specimens such as whole cells, this approach gives much better resolution. In the paired fluorescence and confocal images of cheek cells at the bottom of Table 1-1, the confocal image gives much better resolution, and even allows the small bacteria in the cheek cell sample to be seen easily.

Other recent developments in light microscopy are *digital video microscopy,* which uses video cameras to collect digital images for computer storage, and *deconvolution microscopy,* a computational technique that uses complex mathematical algorithms to increase the contrast and resolution of these digital images. By attaching a highly light-sensitive digital video camera to a light microscope, researchers can observe cells for extended periods of time using very low levels of light. This image intensification is particularly useful to visualize fluorescent molecules present at low levels in living cells and even to see and identify individual macromolecules. In fact, extremely powerful light microscopes are currently being developed that use imaging and computational methods so advanced that they can see beyond the theoretical limit of resolution, visualizing structures in the 100-nm range, which, until the past few years, were believed impossible to see with any light microscope.

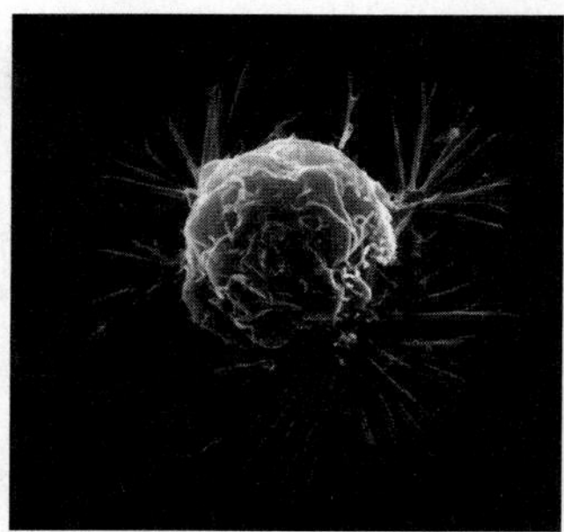

(a) Human cancer cell

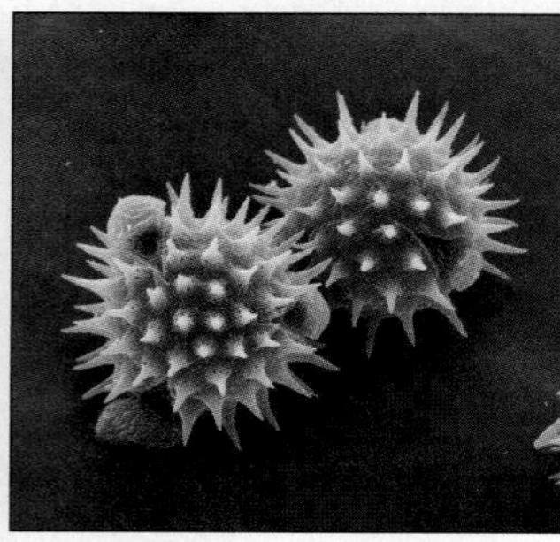

(b) Pollen grains

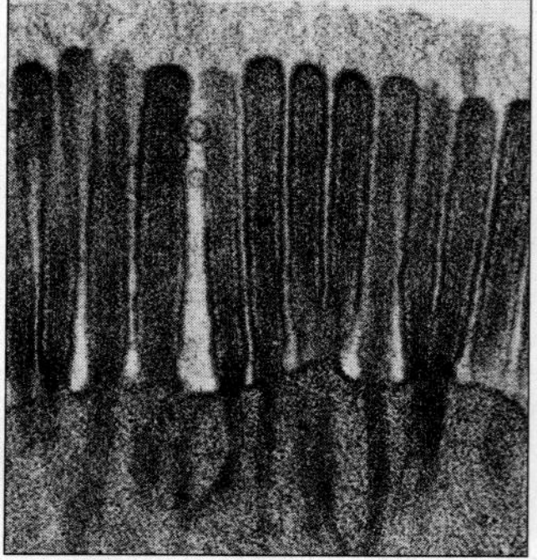

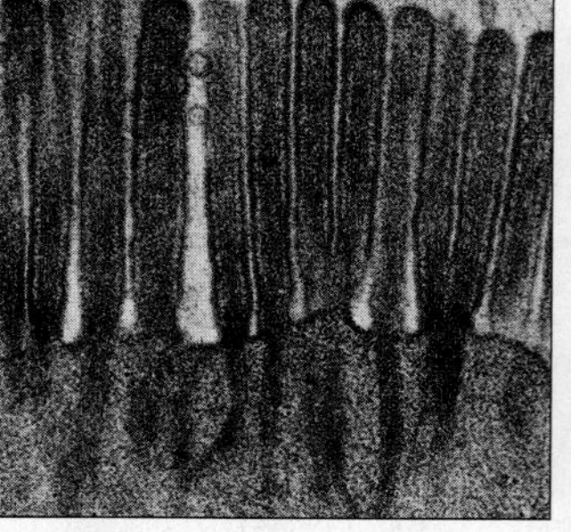

(c) Intestinal cell

(d) Mitochondrion

FIGURE 1-4 Electron Microscopy. A scanning electron microscope was used to visualize **(a)** a breast cancer cell and **(b)** two sunflower pollen grains. A transmission electron microscope was used to obtain images of **(c)** a cat intestinal epithelial cell and **(d)** a mitochondrion from a bat pancreas cell.

The Electron Microscope. However, despite these significant advances, light microscopy is inevitably subject to the limit of resolution imposed by the wavelength of the light used to view the sample. Even the use of ultraviolet radiation, with shorter wavelengths than visible light, increases the resolution by only a factor of two.

A major breakthrough in resolving power came with the development of the **electron microscope,** which was invented in Germany in 1931 by Max Knoll and Ernst Ruska. In place of visible light and optical lenses, the electron microscope uses a beam of electrons that is deflected and focused by an electromagnetic field. Because the wavelength of electrons is so much shorter than the wavelength of photons of visible light, the limit of resolution for the electron microscope is much better—about 0.1–0.2 nm. As a result, the useful magnification of the electron microscope is also much higher—up to 100,000× (see Figure 1-3).

Most electron microscopes have one of two basic designs: the **transmission electron microscope (TEM)** and the **scanning electron microscope (SEM).** Both are described in detail in the Appendix and images from each are shown in **Figure 1-4**. Transmission and scanning electron microscopes are similar because each employs a beam of electrons, but they use quite different mechanisms to form the image. As the name implies, a TEM forms an image from electrons that are transmitted through the specimen. An SEM, on the other hand, scans the surface of the specimen and forms an image by detecting electrons that are deflected from the outer surface of the specimen. Scanning electron microscopy is an especially spectacular technique because of the sense of depth it gives to biological structures.

Several specialized techniques of electron microscopy are in use, including *negative staining, shadowing, freeze fracturing,* and *freeze etching.* Each of these techniques is a useful means of visualizing specimens in three dimensions. Also valuable for this purpose is a technique called *stereo electron microscopy,* in which the same sample is photographed at two slightly different angles to produce a three-dimensional image. *Scanning tunneling microscopy,* developed in the early 1980s, is so sensitive that it allows the visualization of individual atoms. These techniques are described in detail in the Appendix.

Microscopy techniques are constantly evolving. In 2009, a new type of electron microscope, the *scanning transmission electron microscope* was described. Rather than using a vacuum, as in conventional electron microscopy, this microscope allows the visualization of cells in liquid and can be used with whole cells. At about the same time, a *magnetic resonance force microscope* was developed that allows us to image viruses and large macromolecules.

Electron microscopy continues to revolutionize our understanding of cellular architecture by making detailed ultrastructural investigations possible. While organelles such as nuclei or mitochondria are large enough to be seen with a light microscope, they can be studied in much greater detail with an electron microscope. In addition, electron microscopy has revealed cellular structures that are too small to be seen with a light microscope. These include ribosomes, membranes, microtubules, and microfilaments (see Figure 1A-2).

The Biochemical Strand Covers the Chemistry of Biological Structure and Function

At about the same time cytologists started exploring cellular structure with their microscopes, other scientists were making observations that began to explain and clarify cellular function. Much of what is now called biochemistry dates from a discovery reported in 1828 by the German chemist Friedrich Wöhler, a contemporary and fellow countryman of Schleiden and Schwann. He revolutionized our thinking about biology and chemistry by demonstrating that urea, an organic compound of biological origin, could be synthesized in the laboratory from an inorganic starting material, ammonium cyanate.

Until Wöhler reported his results, it had been widely held that living organisms were unique, not governed by the laws of chemistry and physics that apply to the nonliving world. By showing that a compound made by living organisms—a "bio-chemical"—could be synthesized in a laboratory just like any other chemical, Wöhler helped to break down the conceptual distinction between the living and nonliving worlds and dispelled the notion that biochemical processes were somehow exempt from the laws of chemistry and physics.

Another major advance came about 40 years later, when French chemist and biologist Louis Pasteur showed that living yeast cells were responsible for the fermentation of sugar into alcohol. In 1897, German bacteriologists Eduard and Hans Buchner found that fermentation could also take place with isolated extracts from yeast cells—that is, the intact cells themselves were not required. Gradually it became clear that the active agents in the extracts were specific biological catalysts that have since come to be called **enzymes**—from *zyme,* a Greek word meaning "yeast."

Significant progress in our understanding of cellular function came in the 1920s and 1930s as individual steps in the biochemical pathways for fermentation and related cellular processes were elucidated. This was a period dominated by German biochemists such as Gustav Embden, Otto Meyerhof, Otto Warburg, and Hans Krebs. For example, the description of the enzymatic steps in the *Embden-Meyerhof pathway* for glycolysis was a major research triumph of the early 1930s. It was followed shortly by the elucidation of the *Krebs cycle* (also known as the TCA cycle). Both of these pathways are important because of their role in the process by which cells extract energy from glucose and other foodstuffs. We will study these biochemical pathways in detail in Chapters 9 and 10. At about the same time, Fritz Lipmann, an American biochemist, showed that the high-energy compound *adenosine triphosphate (ATP)* is the principal energy storage compound in most cells.

An important advance in the study of biochemical reactions and pathways came as radioactive isotopes such as ^{3}H, ^{14}C, and ^{32}P were first used to trace the metabolic fate of specific atoms and molecules. American chemist Melvin Calvin and his colleagues at the University of California, Berkeley, were pioneers in this field as they traced the fate of ^{14}C-labeled carbon dioxide ($^{14}CO_2$) in illuminated algal cells that were actively photosynthesizing. Their work, carried out in the late 1940s and early 1950s, led to the elucidation of the *Calvin cycle*—the most common pathway for photosynthetic carbon metabolism. The Calvin cycle was the first metabolic pathway to be elucidated using a radioisotope.

Biochemistry took another major step forward with the development of *centrifugation* as a means of separating and isolating subcellular structures and macromolecules based on their size, shape, and/or density—a process called **subcellular fractionation.** Centrifugation techniques used for this purpose include *differential centrifugation, density gradient centrifugation,* and *equilibrium density centrifugation.* These three techniques are described in detail in Box 12A (page 327). Especially useful for resolving small organelles and macromolecules is the **ultracentrifuge,** developed by Swedish chemist Theodor Svedberg in the late 1920s. An ultracentrifuge is capable of very high speeds—over 100,000 revolutions per minute—and can thereby subject samples to forces exceeding 500,000 times the force of gravity. In many ways, the ultracentrifuge is as significant to biochemistry as the electron microscope is to cytology. In fact, both instruments were developed at about the same time, so the ability to see organelles and other subcellular structures coincided with the capability to isolate and purify them.

Other biochemical techniques that have proven useful for isolating and purifying subcellular components include chromatography and electrophoresis. **Chromatography** is a general term describing a variety of techniques by which a mixture of molecules in solution is progressively fractionated as the solution flows over a stationary absorbing phase, usually contained in a column. Chromatographic techniques separate molecules based on their size, charge, or affinity for specific molecules or functional groups. An example of a chromatographic technique is shown in Figure 7-9.

Electrophoresis refers to several related techniques that use an electrical field to separate molecules based on their mobility and is used extensively to determine the sizes of protein, DNA, and RNA molecules. The most common medium for electrophoretic separation of proteins and nucleic acids is a gel of either polyacrylamide or agarose. The use of polyacrylamide gel electrophoresis to resolve proteins is illustrated in Figure 7-22.

After proteins have been separated by electrophoresis, **mass spectrometry** is commonly used to determine the size and composition of individual proteins. This technique, which allows researchers to determine the identity and characteristics of individual proteins in a mix of thousands of different proteins in a cell, has led to significant advances. For example, in the emerging field of *proteomics,* researchers are attempting to understand the functions and interactions of all the proteins present in a particular cell. Mass spectrometry is described in more detail in Chapter 18.

To summarize, with the enhanced ability to see subcellular structures, to fractionate them, and to isolate them, cytologists and biochemists began to realize how well their respective observations on cellular structure and function could complement each other. These scientists were laying the foundations for modern cell biology.

The Genetic Strand Focuses on Information Flow

The third strand in the historical cord of cell biology is genetics. Like the other two, this strand has important roots in the nineteenth century. In this case, the strand begins with Gregor Mendel, whose studies with the pea plants he grew in a monastery garden must surely rank among the most famous experiments in all of biology. His findings were published in 1866, laying out the principles of segregation and independent assortment of the "hereditary factors" that we know today as **genes.** But Mendel was clearly a man ahead of his time. His work went almost unnoticed when it was first published and was not fully appreciated until its rediscovery nearly 35 years later. (It is interesting to note that over 2000 years ago, the Greek philosopher Aristotle referred to a physical entity he called the "germ." He stated that it "springs forth from a definite parent and gives rise to a predictable progeny," and he called it "the ruling influence and fabricator of the offspring.")

In the decade following Mendel's work, the role of the nucleus in the genetic continuity of cells came to be appreciated. In 1880, German biologist Walther Flemming identified **chromosomes,** threadlike bodies seen in dividing cells. Flemming called the division process *mitosis,* from the Greek word for "thread." Chromosome number soon came to be recognized as a distinctive characteristic of a species and was shown to remain constant from generation to generation. That the chromosomes themselves might be the actual bearers of genetic information was suggested by German anatomist Wilhelm Roux as early as 1883 and expressed more formally by his countryman, biologist August Weissman, shortly thereafter.

With the roles of the nucleus and chromosomes established and appreciated, the stage was set for the rediscovery of Mendel's initial observations. This came in 1900, when his studies were cited almost simultaneously by three plant geneticists working independently: Carl Correns in Germany, Ernst von Tschermak in Austria, and Hugo de Vries in Holland. Within three years, the **chromosome theory of heredity** was formulated, following work by American physician Walter Sutton and German biologist Theodor Boveri. The chromosome theory of heredity proposed that the hereditary factors responsible for Mendelian inheritance are located on the chromosomes within the nucleus. This hypothesis received its strongest confirmation from the work of American biologist Thomas Hunt Morgan and his students, Calvin Bridges and Alfred Sturtevant, at Columbia University during the first two decades of the twentieth century. Using *Drosophila melanogaster,* the common fruit fly, as their experimental model organism, they identified a variety of morphological mutants of *Drosophila* and were able to link specific traits to specific chromosomes.

Meanwhile, the foundation for our understanding of the chemical basis of inheritance was slowly being laid. An important milestone was the discovery of DNA by Swiss biologist Johann Friedrich Miescher in 1869. Using such unlikely sources as salmon sperm and human pus from surgical bandages, Miescher isolated and described what he called "nuclein." But, like Mendel, Miescher was ahead of his time. It was about 75 years before the role of his nuclein as the genetic information of the cell came to be fully appreciated.

As early as 1914, DNA was implicated as an important component of chromosomes by German chemist Robert Feulgen's staining technique, a method that is still in use today. But it was considered quite unlikely that DNA could be the bearer of genetic information due to the apparently monotonous structure of DNA. By 1930, DNA was known to be composed of only four different nucleotides—and this did not seem to be enough variety to account for all the diversity seen in living organisms. Proteins, on the other hand, were much more diverse, being composed of 20 different amino acids. In fact, until the middle of the twentieth century, it was widely thought that proteins were the carriers of genetic information from generation to generation, since they seemed to be the only nuclear components with enough variety to account for the obvious diversity of genes.

A landmark experiment clearly pointing to DNA as the genetic material was reported in 1944 by Oswald Avery, Colin MacLeod, and Maclyn McCarty. Their work, discussed in more detail in Chapter 18, showed that DNA could "transform" a nonpathogenic strain of bacteria into a pathogenic strain, causing a heritable genetic change. Eight years later, American biochemists Alfred Hershey and Martha Chase showed that DNA, and not protein, enters a bacterial cell when it is infected and genetically altered by a bacterial virus.

Meanwhile, American biologists George Beadle and Edward Tatum, working in the 1940s with the bread mold *Neurospora crassa,* formulated the "one gene–one enzyme" concept, asserting that each gene controls the production of a single, specific protein. Shortly thereafter, in 1953, the unlikely team of former ornithology student James Watson and physicist Francis Crick, with assistance from X-ray crystallographer Rosalind Franklin, proposed their now-famous *double helix model* for DNA structure, which immediately suggested how replication and genetic mutations could occur. Soon afterward, it was discovered how DNA specifies the order of amino acids in proteins and that several different kinds of RNA molecules serve as intermediates in protein synthesis.

The 1960s brought especially significant developments, including the discovery of the polymerase enzymes that synthesize DNA and RNA and the "cracking" of the genetic code, which specifies the relationship between the order of nucleotides in a DNA or RNA molecule and the order of amino acids in a protein. At about the same time, biochemist Jacques Monod and geneticist François Jacob of France deduced the mechanism responsible for the regulation of bacterial gene expression.

Important techniques along the genetic strand of Figure 1-2 include the separation of DNA molecules and fragments by ultracentrifugation and gel electrophoresis. Of equal, if not greater, importance is *nucleic acid hybridization,* which includes a variety of related techniques that depend on the ability of two single-stranded nucleic acid molecules with complementary base sequences to bind, or *hybridize,* to each other, thereby forming a double-stranded hybrid. These techniques can be applied to DNA-DNA, DNA-RNA, and even RNA-RNA interactions, and they are useful to isolate and identify specific DNA or RNA molecules.

Our current understanding of gene expression has relied heavily on the development of **recombinant DNA technology** since the 1970s. This technology was made possible by the discovery of *restriction enzymes,* which have the ability to cleave DNA molecules at specific sequences so that scientists can create *recombinant DNA molecules* containing DNA sequences from two different sources. This capability led quickly to the development of *gene cloning,* a process for generating many copies of specific DNA sequences for detailed study and further manipulation. These important techniques are explained and explored in detail in Chapters 18 and 20.

At about the same time, **DNA sequencing** methods were devised for rapidly determining the base sequences of DNA molecules. This technology is now routinely applied not just to individual genes but to entire *genomes* (i.e., the total DNA content of a cell). Initially, genome sequencing was applied mainly to bacterial genomes because they are relatively small—a few million bases, typically. But DNA sequencing has long since been successfully applied to much larger genomes, including those from species of yeast, roundworm, plants, and animals that are of special interest to researchers. A major triumph was the sequencing of the entire human genome, which contains about 3.2 billion bases. This feat was accomplished by the *Human Genome Project,* a cooperative international effort that began in 1990, involved hundreds of scientists, and established the complete sequence of the human genome by 2003.

The challenge of analyzing the vast amount of data generated by DNA sequencing has led to a new discipline, called **bioinformatics,** which merges computer science and biology as a means of making sense of sequence data. This approach has led to the recognition that the human genome contains approximately 25,000 protein-coding genes, about half of which were not characterized before genome sequencing. It is also enabling scientists to study the *proteome,* the total protein content of a cell, analyzing thousands of proteins simultaneously. Proteomic studies aim to understand the structure and properties of every protein produced by a genome and to learn how these proteins interact with each other to regulate cellular functions.

Numerous bioinformatic tools are publicly available through the National Center for Biotechnology Information (NCBI), which is operated by the U.S. National Institutes of Health (NIH). In addition to housing PubMed, a searchable archive of over 17 million citations from life science journals, NCBI maintains GenBank, a comprehensive database of all publicly available nucleotide and amino acid sequences (over 100 million as of early 2010). Also available are numerous tools to compare gene and protein sequences from all organisms and to analyze their structure and function. In addition, NCBI provides a wealth of biological information, such as the OMIM (Online Mendelian Inheritance in Man) database, which is an encyclopedic collection of information regarding human genetic disorders and mutations involving over 12,000 genes.

Using yeast as a model organism to isolate and characterize genetic mutants, scientists have learned much about how cells function by studying mutant yeast strains that are deficient in particular genes and their protein products. For example, cell cycle mutants of yeast that display abnormal cell division have been invaluable in helping us understand how normal cells divide. Recently, the use of the yeast *two-hybrid system,* which allows us to determine how specific proteins interact within a living cell, has contributed greatly to our understanding of the complex molecular interactions involved in cellular function.

In the last decade, we have seen major advances both in *nanotechnology,* the development of nanometer-sized machines, tools, and biosensors, and in computer-aided analysis of experimental results. A generation ago, it took a week to sequence a single gene, but now thousands of genes can be sequenced in a few days. The ability to simultaneously analyze thousands of molecules on a global basis throughout the cell has led to a proliferation of "–omics" studies. We now have *transcriptomics,* the study of all transcribed genes in a particular cell, *metabolomics,* the analysis of all metabolic reactions happening at a given time in a cell, and even *lipidomics,* the study of all the lipids in a cell. Keep your eyes open in the next few years, as this explosion of biological information will likely lead to a host of new fields of "–omics" studies!

These and other techniques helped to launch an era of molecular genetics that continues to revolutionize biology. In the process, the historical strand of genetics that dates back to Mendel became intimately entwined with those of cytology and biochemistry, and the discipline of cell biology as we know it today came into being.

"Facts" and the Scientific Method

If asked what you expect to get out of a science textbook, you may reply that you intend to learn the facts about the particular scientific area that the textbook covers—cell biology, in the case of this text. If pressed to explain what a "fact" is, most people would probably reply that a fact is "something that we know to be true." When we say, for example, that "all organisms consist of one or more cells" or that "DNA is the bearer of genetic information," we recognize these statements as facts of cell biology. But we must also recognize that the first of these statements was initially regarded as part of a theory, and the second statement actually replaced an earlier misconception that genes were made of proteins.

Cell biology is rich with examples of facts that were once widely held but have since been altered or even discarded as cell biologists gained a better understanding of the phenomena those facts attempted to explain. As we saw earlier, the early nineteenth century "fact" that living matter consisted of substances quite different from those in nonliving matter, was discredited following work by Wöhler, who synthesized the biological compound urea from an inorganic compound, and Eduard and Hans Buchner, who showed that nonliving extracts from yeast cells could ferment sugar into ethanol. Thus, views held as fact by generations of scientists were eventually discarded and replaced by the new fact that living matter follows the same laws of chemistry and physics as do inorganic materials.

For a more contemporary example, until recently it was regarded as a fact that the sun is the ultimate source of all energy in the biosphere. Then came the discovery of *deep-sea thermal vents* and the thriving communities of organisms that live around them, none of which depends on solar energy. Instead, these organisms depend on energy derived from hydrogen sulfide (H_2S) by bacteria, which use this energy to synthesize organic compounds from carbon dioxide.

BOX 1B TOOLS OF DISCOVERY

Model Organisms in Cell Biology Research

Although we have advanced microscopes and sophisticated methods to isolate and analyze individual cells and cellular components and have volumes of research data describing what we now know about cells, to fully understand how cells work, we need to test our hypotheses in vivo (in living organisms). Cell biologists rely heavily on a variety of *model organisms* representing the major kingdoms of life to help unlock the secrets of the cell.

A **model organism** is a species that is widely studied, well-characterized, easy to manipulate, and has particular advantages making it useful for experimental studies. A few examples are the bacterium *Escherichia coli*, the yeast *Saccharomyces cerevisiae*, and the fruit fly *Drosophila melanogaster* (**Figure 1B-1**). Gregor Mendel, in his classic studies of gene inheritance and segregation, used garden pea plants as his model organism because they were easy to grow, could be used for manual genetic crosses, and displayed easily observable characteristics such as plant height, seed color and shape, etc.

A biochemistry professor once pointed out that his lecture on cellular DNA synthesis was primarily a lecture on *E. coli* DNA synthesis, since the majority of what we knew about this process at the time was based on studies using this single-celled model organism. *E. coli* is easy to grow in the lab, has a short generation time (20 minutes), and is easily mutagenized for studies of gene function. In 1997, it became the first bacterium to have its complete genome sequenced. Because it readily takes up DNA from virtually any organism, it has become the "workhorse" of cellular and molecular biology and is extensively used for the analysis and

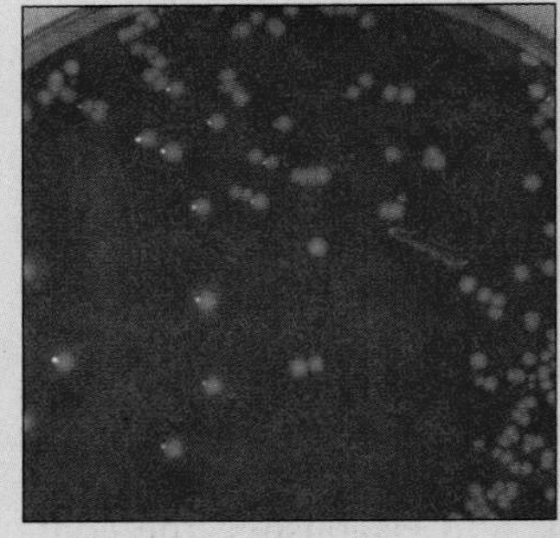

(a) *E. coli*

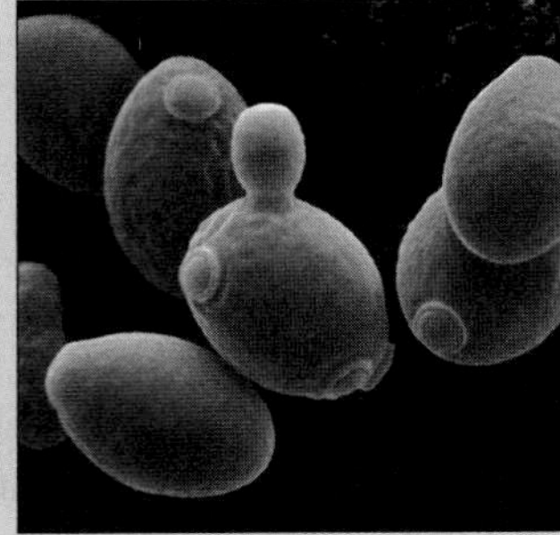

(b) *S. cerevisiae*

(c) *Drosophila*

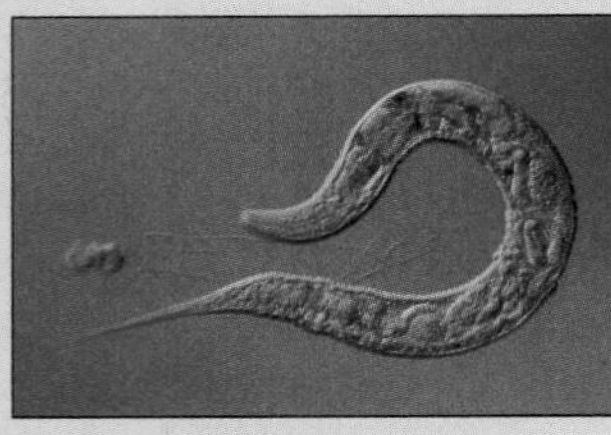

(d) *C. elegans*

(e) *Mus musculus*

(f) *Arabidopsis*

FIGURE 1B-1 Common Model Organisms. Examples of some frequently-used model organisms are shown: **(a)** *Escherichia coli* colonies on blood agar, **(b)** *Saccharomyces cerevisiae* yeast cells (SEM), **(c)** a wild-type *Drosophila melanogaster* fly, **(d)** a *Caenorhabditis elegans* roundworm, **(e)** a laboratory mouse *(Mus musculus)*, and **(f)** an *Arabidopsis thaliana* plant.

Thus, as you can see, a fact is really a much more tenuous piece of information than our everyday sense of the word might imply. Just like cells themselves, facts are dynamic and subject to change, sometimes abruptly. To a scientist, a fact is simply an attempt to state our best current understanding of the natural world around us, based on observations and experiments. A fact is valid only until it is revised or replaced by a better understanding based on more careful observations or more discriminating experiments.

How does new and better information become available? Scientists usually assess new information with a systematic approach called the *scientific method.* After making observations and consulting prior studies, the scientist formulates a *hypothesis,* a tentative explanation or model that can be tested experimentally. Next, the investigator designs a controlled experiment to test the hypothesis by varying specific conditions while keeping other variables constant. The scientist then collects the data, interprets the results, and accepts or rejects the hypothesis, which must be consistent not only with the results of this particular experiment but also with prior knowledge. This often involves the use of *Occam's razor,* the principle stating that, when we try to understand nature, it is often the simplest explanation consistent with the observable facts that is the most likely to be correct.

Research experiments are often conducted in the laboratory using purified chemicals and cellular components. This type of experiment is described as *in vitro,* which literally means "in glass." Experiments using live cells or organisms are referred to as *in vivo* ("in life"), often using one of a variety of popular *model organisms* (**Box 1B**). More recently, experiments using computers to test new hypotheses involv-

production of genes and proteins for research, industrial, and medical uses.

For studies of eukaryotic cell biology, *S. cerevisiae,* a unicellular fungus also known as bakers' or brewers' yeast, has many of the same experimental advantages. Like *E. coli,* yeast grows rapidly on growth media in the lab and is easily manipulated and mutagenized. It is widely used for studies of organelle function, eukaryotic secretion processes, and protein-protein interactions. You may be surprised to learn that much of what we know about human cell division has come from the study of cell cycle mutants in yeast.

However, use of these two unicellular organisms does have its limitations, especially if you want to study processes in multicellular organisms, including cell differentiation, embryonic development, or cell-to-cell communication. For these types of studies, you may have heard biologists talk about experiments using "flies and worms." They are referring to the fruit fly *Drosophila melanogaster,* and the microscopic roundworm *Caenorhabditis elegans,* both of which are extensively used for studies of the genetics and development of multicellular, eukaryotic organisms.

The discovery of genetic recombination by Thomas Hunt Morgan and colleagues in the early 1900s resulted from their use of *Drosophila* and its experimental advantages: it is easy to grow and manipulate in the lab, has a short generation time (about two weeks), produces numerous progeny, and has easily observable physical characteristics (such as eye color and wing shape). Today it is widely used in genetic studies, in part because there are thousands of mutant strains available, each defective in a particular gene, and its genome has been sequenced, Also, it is being increasingly used in studies of embryogenesis, developmental biology, and cell signaling (see Box 14B, page 410).

Similarly, *C. elegans* is a valuable model organism for studies of cell differentiation and development and of apoptosis (programmed cell death) in multicellular organisms. Advantages of this model organism include its ease of manipulation, relatively short life cycle, and small genome, the first of any multicellular organism to be sequenced. It is also one of the simplest animals to possess a nervous system. Its development from a fertilized egg is remarkably predictable, and the origin and fate of each of its approximately 1000 cells has been mapped out, as have the hundreds of connections among the roughly 200 nerve cells. Also, the tiny worms are transparent, and thus it is easy to see individual cells and view fluorescently labeled molecules in the living organism.

For studies of cellular and physiological processes specific to mammals and humans, the common laboratory mouse *(Mus musculus)* has become the primary model organism. It shares many cellular, anatomical, and physiological similarities with humans and is widely used for research in medicine, immunology, and aging. It is subject to, and therefore useful for the study of, a variety of diseases that also affect us, such as cancer, diabetes, and osteoporosis. Numerous mouse strains have been bred or engineered in which particular genes have either been "knocked out" or introduced, making them extremely valuable in biomedical research.

For studies of processes unique to plants, and some processes common to all organisms, *Chlamydomonas reinhardtii,* a unicellular green alga is often used. Like *E. coli* and yeast, "*Chlamy*" is easily grown in the lab on Petri plates and has been used to study photosynthesis, light perception, mating type, cellular motility, and DNA methylation. For flowering plant studies, *Arabidopsis thaliana* is a powerful model organism. It has one of the smallest genomes of any plant and a relatively rapid life cycle (6 weeks), facilitating genetic studies. Its complete genome has been sequenced, and thousands of mutant strains have been created, enabling detailed studies of plant gene function.

Many other model organisms are currently being used in biology to study a wide variety of cellular, genetic, and biochemical processes. As you proceed through your studies in cell biology, keep in mind that much of what we know is based upon research using relatively few of the millions of living organisms on Earth, and remember that is always important to understand how we know what we know.

ing vast amounts of data are described as *in silico,* referring to the silicon used to make computer chips. But all these approaches use the same basic steps in the scientific method.

To a practicing scientist, the scientific method is more a way of thinking than a set of procedures to be followed. Most likely, this is the way our ancestors explained natural phenomena long before scientists were trained at universities—and long before students read about the scientific method! As we consider the scientific method, we need to recognize several important terms that scientists use to indicate the degree of certainty with which a specific explanation or concept is regarded as true. Three terms are especially significant: *hypothesis, theory,* and *law.*

Of the three, the most tentative is a **hypothesis,** a statement or explanation that is consistent with most of the observations and experimental evidence to date. Often, a hypothesis takes the form of a *model* that appears to provide a reasonable explanation of the phenomenon in question. To be useful to scientists, a hypothesis must be *testable* using the scientific method.

If a hypothesis or model has been tested critically under many different conditions—usually by many different investigators using a variety of approaches—and is consistently supported by the evidence, it gradually acquires the status of a **theory.** By the time an explanation or model comes to be regarded as a theory, it is generally and widely accepted by most scientists in the field. The *cell theory* described earlier in this chapter is an excellent example. There is little or no dissent or disagreement among biologists concerning its three principles.

When a theory has been thoroughly tested and confirmed over a long period of time by such a large number of investigators that virtually no doubt of its validity remains, it may come to be regarded as a **law.** The *law of gravity* comes readily to mind, as do the *laws of thermodynamics* that we will encounter in Chapter 5. You may also be familiar with *Fick's law of diffusion,* the *ideal gas laws,* and Mendel's *laws of heredity,* for instance.

However, biologists are generally quite conservative in using the term *law.* Even after more than 150 years of finding no exceptions, the cell theory is still referred to as a "theory." Like the *germ theory,* which proposes that diseases can be caused by microorganisms, and the *theory of evolution,* which describes the origin of biological diversity, there really is no doubt about the validity of the cell theory, and it could easily be considered a law. Perhaps our reluctance to label explanations of biological phenomena as laws is a reflection of the difficulty we have in convincing ourselves that we will never find organisms or cells that are exceptions to even our most well-documented theories.

As you have now seen, the facts presented in biology textbooks such as this one are simply our best current attempts to describe and explain the biological world we live in. As you proceed through this text, watch for applications of the scientific method. You will find that, regardless of the approach, the conclusions from each experiment add to our knowledge of how biological systems work and usually lead to more questions as well, continuing the cycle of scientific inquiry.

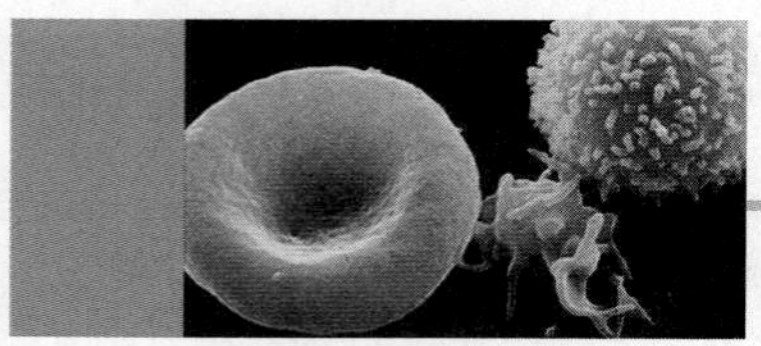

SUMMARY OF KEY POINTS

The Cell Theory: A Brief History

- The biological world is a world of cells. The cell theory states that all organisms are made of cells, the basic units of biological structure, and that cells arise only from preexisting cells.
- The cell theory was developed through the work of many different scientists, including Hooke, van Leeuwenhoek, Brown, Schleiden, Schwann, Nägeli, and Virchow.
- Although the importance of cells in biological organization has been appreciated for about 150 years, the discipline of cell biology as we know it today is of much more recent origin.

The Emergence of Modern Cell Biology

- Modern cell biology has come about by the interweaving of three historically distinct strands—cytology, biochemistry, and genetics—which in their early development probably did not seem at all related.
- The contemporary cell biologist must understand all three strands because they complement one another in the quest to learn what cells are made of and how they function.

The Cytological Strand Deals with Cellular Structure

- The light microscope has allowed us to visualize individual cells, which are approximately 1–50 μm in size. Historically, its limited resolving power did not allow us to see details of structure smaller than about 0.2 μm (200 nm), but modern light microscopes are surpassing that limit.
- Several types of light microscopes allow us to view preserved or living specimens at magnifications of about 1000×. These include brightfield, phase-contrast, differential interference contrast, fluorescence, confocal, and digital video microscopes, each of which offers particular advantages in studying and understanding cells.
- The electron microscope uses a beam of electrons, rather than visible light, for imaging specimens. It can magnify objects up to 100,000× with a resolving power of less than 1 nm, enabling us to view subcellular structures such as membranes, ribosomes, organelles, and even individual DNA and protein molecules.

The Biochemical Strand Covers the Chemistry of Biological Structure and Function

- Discoveries in biochemistry have revealed how many of the chemical processes in cells are carried out, greatly expanding our knowledge of how cells function.
- Major discoveries in biochemistry were the identification of enzymes as biological catalysts, the discovery of adenosine triphosphate (ATP) as the main carrier of energy in living organisms, and the description of the major metabolic pathways cells use to harness energy and synthesize cellular components.
- Several important biochemical techniques that have allowed us to understand cell structure and function are subcellular fractionation, ultracentrifugation, chromatography, electrophoresis, and mass spectrometry.

The Genetic Strand Focuses on Information Flow

- The chromosome theory of heredity states that the characteristics of organisms passed down from generation to generation result from the inheritance of chromosomes carrying discrete physical units known as genes.
- Each gene is a specific sequence of DNA that contains the information to direct the synthesis of one cellular protein.
- DNA itself is a double helix of complementary strands held together by precise base pairing. This structure allows the DNA to be accurately duplicated as it is passed down to successive generations.

- Bioinformatics and nanotechnology allow us to compare and analyze thousands of genes or other molecules simultaneously, causing a revolution in genomic, proteomic, and numerous other fields of "–omics" research.

"Facts" and the Scientific Method

- Science is not a collection of facts but a process of discovering answers to questions about our natural world. Scientists gain knowledge by using the scientific method, which involves creating a hypothesis that can be tested for validity by collecting data through well-designed, controlled experiments.
- Science as a field of study is based on consistency and reproducibility. As a hypothesis becomes more accepted, it is referred to as a theory; and, as a theory becomes indisputable, it becomes a law.
- However, some of the most "proven" theories or laws have been modified, extended, or even refuted. Science is a dynamic, ever-changing field, and it is likely that some of the scientific facts you are now learning will eventually change.
- Scientists use a variety of well-studied model organisms to test new hypotheses, develop new theories, and advance our knowledge of cell biology.

PROBLEM SET

More challenging problems are marked with a •.

1-1 The Historical Strands of Cell Biology. For each of the following events, indicate whether it belongs mainly to the cytological (C), biochemical (B), or genetic (G) strand in the historical development of cell biology.

(a) Schleiden and Schwann describe cells as the building blocks for an organism (1839).

(b) Hoppe-Seyler isolates the protein hemoglobin in crystalline form (1864).

(c) Haeckel postulates that the nucleus is responsible for heredity (1868).

(d) Ostwald proves that enzymes are catalysts (1893).

(e) Morgan and colleagues discover sex-linked mutations in *Drosophila* (1909).

(f) Davson and Danielli postulate a model for the structure of cell membranes (1935).

(g) Krebs defines the metabolic sequence of events in the TCA cycle (1937).

(h) Beadle and Tatum formulate the one gene–one enzyme hypothesis (1940).

(i) Lipmann postulates the central importance of ATP in cellular energy transactions (1940).

(j) Otto Gey cultures the first immortalized human cell line (1951).

(k) Meselson, Stahl, and Vinograd use density gradient centrifugation to separate nucleic acids (1957).

(l) Wilmut, Campbell, and colleagues clone the first mammal (Dolly the sheep) from an adult somatic cell (1997).

1-2 Cell Sizes. To appreciate the differences in cell size illustrated in Figure 1A-1, consider these specific examples. *Escherichia coli,* a typical bacterial cell, is cylindrical in shape, with a diameter of about 1 μm and a length of about 2 μm. As a typical animal cell, consider a human liver cell, which is roughly spherical and has a diameter of about 20 μm. For a typical plant cell, consider the columnar *palisade cells* located just beneath the upper surface of many plant leaves. These cells are cylindrical, with a diameter of about 20 μm and a length of about 35 μm.

(a) Calculate the approximate volume of each of these three cell types in cubic micrometers. (Recall that $V = \pi r^2 h$ for a cylinder and that $V = 4\pi r^3/3$ for a sphere.)

(b) Approximately how many bacterial cells would fit in the internal volume of a human liver cell?

(c) Approximately how many liver cells would fit inside a palisade cell?

1-3 Sizing Things Up. To appreciate the sizes of the subcellular structures shown in Figure 1A-2, consider the following calculations:

(a) All cells and many subcellular structures are surrounded by a membrane. Assuming a typical membrane to be about 8 nm wide, how many such membranes would have to be aligned side by side before the structure could be seen with the light microscope? How many with the electron microscope?

(b) Ribosomes are the cell structures in which the process of protein synthesis takes place. A human ribosome is a roughly spherical structure with a diameter of about 30 nm. How many ribosomes would fit in the internal volume of the human liver cell described in Problem 1-2 if the entire volume of the cell were filled with ribosomes?

(c) The genetic material of the *Escherichia coli* cell described in Problem 1-2 consists of a DNA molecule with a diameter of 2 nm and a total length of 1.36 mm. (The molecule is actually circular, with a circumference of 1.36 mm.) To be accommodated in a cell that is only a few micrometers long, this large DNA molecule is tightly coiled and folded into a *nucleoid* that occupies a small proportion of the cell's internal volume. Calculate the smallest possible volume the DNA molecule could fit into, and express it as a percentage of the internal volume of the bacterial cell that you calculated in Problem 1-2a.

1-4 Limits of Resolution Then and Now. Based on what you learned in this chapter about the limit of resolution of a light microscope, answer each of the following questions. Assume that the unaided human eye has a limit of resolution of about 0.25 mm and that a modern light microscope has a useful magnification of about 1000×.

(a) Define *limit of resolution* in your own words. How has the limit of resolution changed from Hooke's microscope to the modern day light microscope?

(b) What are the approximate dimensions of the smallest structure that Hooke would have been able to observe with his microscope? Would he have been able to see any of the structures shown in Figure 1A-1? If so, which ones? And if not, why not?

(c) What cellular components can typically be observed using a modern light microscope? How does this compare to an electron microscope?

(d) What limits the resolution of a light microscope? Describe why an electron microscope is able to achieve a higher resolution.

(e) Consider the eight structures shown in Figure 1A-1 and 1A-2. Which of these structures would both Hooke and van Leeuwenhoek have been able to see with their respective microscopes? Which, if any, would van Leeuwenhoek have been able to see that Hooke could not? Explain your reasoning. Which, if any, would a contemporary cell biologist be able to see that neither Hooke nor van Leeuwenhoek could see?

1-5 The Contemporary Strands of Cell Biology. For each pair of techniques listed, indicate whether its members belong to the cytological, biochemical, or genetic strand of cell biology (see Figure 1-2). Suggest one advantage that the second technique has over the first technique.

(a) Light microscopy/electron microscopy

(b) Centrifugation/ultracentrifugation

(c) Nucleic acid hybridization/DNA sequencing

(d) Sequencing of a genome/bioinformatics

(e) Transmission electron microscopy/scanning electron microscopy

(f) Chromatography/electrophoresis

1-6 The "Facts" of Life. Each of these statements was once regarded as a biological fact but is now understood to be untrue. In each case, indicate why the statement was once thought to be true and why it is no longer considered a fact.

(a) Animal and plant cells do not share a common nucleus.

(b) Living organisms are not governed by the laws of chemistry and physics, as is nonliving matter, but are subject to a "vital force" that is responsible for the formation of organic compounds.

(c) Genes most likely consist of proteins because the only other likely candidate, DNA, is a relatively uninteresting molecule consisting of only four kinds of monomers (nucleotides) arranged in a relatively invariant, repeating tetranucleotide sequence.

(d) Sunlight is the only source of energy in the biosphere.

• **1-7 More "Facts" of Life.** Each of these statements was regarded as a biological fact until quite recently but is now either rejected or qualified to at least some extent. In each case, speculate on why the statement was once thought to be true, and then try to determine what evidence might have made it necessary to reject or at least to qualify the statement. (Note: This question requires a more intrepid sleuth than Problem 1-6 does, but chapter references are provided to aid in your sleuthing.)

(a) A biological membrane can be thought of as a protein-lipid "sandwich" consisting of an exclusively phospholipid interior coated on both sides with thin layers of protein (Chapter 7).

(b) DNA always exists as a duplex of two strands wound together into a right-handed helix (Chapter 18).

(c) The flow of genetic information in a cell is always from DNA to RNA to protein (Chapter 21).

• **1-8 Pizza, Heartburn, and the Scientific Method.** Although the scientific method may sound rather foreign when described in formal terms, you probably use the scientific method frequently without even realizing it. Suppose, for example, that recently you often have heartburn. You realize that the heartburn is most likely to occur on nights after you have eaten your favorite pizza, with pepperoni, anchovies, and onions. You wonder if the heartburn is caused by eating pizza and, if so, which of the ingredients might be the culprit. Describe how you might go about determining whether the heartburn is due to the pizza and, if so, to which of the ingredients.

1-9 A New Biofuel. As a recent cell biology graduate, you have just been hired by a biotechnology company to develop a biofuel using algal cells that produce an oil very similar to diesel fuel. What model organism(s) might you use for each of the following aspects of this project? What tools and techniques would you use?

(a) Determining which genes and enzymes are required to produce the oil.

(b) Producing large amounts of a certain enzyme for further research.

(c) Studying whether any of the cellular enzymes interact with each other.

(d) Examining the involvement of any cellular organelles in storage or secretion of the oil.

SUGGESTED READING

References of historical importance are marked with a •.

The Emergence of Modern Cell Biology

• Bechtel, W. *Discovering Cell Mechanisms: The Creation of Modern Cell Biology.* New York: Cambridge University Press, 2006.

• Bracegirdle, B. Microscopy and comprehension: The development of understanding of the nature of the cell. *Trends Biochem. Sci.* 14 (1989): 464.

• Bradbury, S. *The Evolution of the Microscope.* New York: Pergamon, 1967.

• Calvin, M. The path of carbon in photosynthesis. *Science* 135 (1962): 879.

• Claude, A. The coming of age of the cell. *Science* 189 (1975): 433.

• de Duve, C., and H. Beaufay. A short history of tissue fractionation. *J. Cell Biol.* 91 (1981): 293s.

• Fruton, J. S. The emergence of biochemistry. *Science* 192 (1976): 327.

• Gall, J. G., K. R. Porter, and P. Siekevitz, eds. Discovery in cell biology. *J. Cell Biol.* 91, part 3 (1981).

• Henry, J. *The Scientific Revolution and the Origins of Modern Science.* New York: Palgrave Press, 2001.

• Judson, H. F. *The Eighth Day of Creation: Makers of the Revolution in Biology.* New York: Simon & Schuster, 1979.

• Kornberg, A. Centenary of the birth of modern biochemistry. *Trends Biochem. Sci.* 22 (1997): 282.

Lederburg, J. *E. coli* K-12. *Microbiol. Today* 31 (2004): 116.

• Minsky, M. Memoir on inventing the confocal scanning microscope. *Scanning* 10 (1988): 128.

• Mirsky, A. E. The discovery of DNA. *Sci. Amer.* 218 (June 1968): 78.

• Palade, G. E. Albert Claude and the beginning of biological electron microscopy. *J. Cell Biol.* 50 (1971): 5D.

• Peters, J. A. *Classic Papers in Genetics.* Englewood Cliffs, NJ: Prentice-Hall, 1959.

• Rasmussen, N. *Picture Control: The Electron Microscope and the Transformation of Biology in America.* Stanford, CA: Stanford University Press, 1997.

• Stent, G. C. That was the molecular biology that was. *Science* 160 (1968): 390.

• Watson, J. *A Passion for DNA.* Cold Spring Harbor, NY: Cold Spring Harbor Laboratory Press, 2000.

• Watson, J. D. *The Double Helix.* New York: Atheneum, 1968.

• Zernike, F. How I discovered phase contrast. *Science* 121 (1955): 345.

Methods in Modern Cell Biology

Celis, J. E. *Cell Biology: A Laboratory Handbook,* 3rd ed., vol. 1. San Diego, CA: Elsevier/Academic Press, 2006.

Cseke, L. J. et al. *Handbook of Molecular and Cellular Methods in Biology and Medicine,* 2nd ed. New York: CRC Press, 2003.

Dashek, W. V., and M. Harrison, eds. *Plant Cell Biology.* Enfield, NH: Science Publishers, 2006.

Goldstein, J. et al. *Scanning Electron Microscopy and X-ray Microanalysis,* 3rd ed. New York: Kluwer Academic/Plenum Publishers, 2003.

Hayat, M. A. *Principles and Techniques of Electron Microscopy: Biological Applications.* New York: Cambridge University Press, 2000.

Hibbs, A. *Confocal Microscopy for Biologists.* New York: Kluwer Academic/Plenum Publishers, 2004.

Jena, B. P., and J. K. Hörber. *Atomic Force Microscopy in Cell Biology.* San Diego, CA: Academic Press, 2002.

Kohen, E., et al. *Atlas of Cell Organelles Fluorescence.* New York: CRC Press, 2004.

Lanza, R., B. Hogan, D. Melton, and R. Pederson. *Essentials of Stem Cell Biology,* 2nd ed. Burlington, MA: Elsevier/Academic Press, 2009.

Lloyd, R. V. *Morphology Methods: Cell and Molecular Biology Techniques.* Totowa, NJ: Humana Press, 2001.

Muzzey, D., and A. van Oudenaarden. Quantitative time-lapse fluorescence microscopy in single cells. *Annu. Rev. Cell Dev. Biol.* 25 (2009):301.

Pruess, M., and R. Apweiler. Bioinformatic resources for *in silico* proteome analysis. In *Cell Biology: A Laboratory Handbook,* 3rd ed., vol. 4 (J. E. Celis, ed.). San Diego, CA: Elsevier/Academic Press, 2006.

Reitdorf, J., and T. W. J. Gadella. *Microscopy Techniques.* New York: Springer, 2005.

Sluder, G., and D. E. Wolf. *Digital Microscopy, Vol. 81 of Methods in Cell Biology.* San Diego, CA: Elsevier/Academic Press, 2007.

Wayne, R. *Light and Video Microscopy.* Boston: Elsevier/Academic Press, 2008.

The Scientific Method

Braben, D. *To Be a Scientist: The Spirit of Adventure in Science and Technology.* New York: Oxford University Press, 1994.

Carey, S. S. *A Beginner's Guide to the Scientific Method.* Belmont, CA: Thomson/Wadsworth, 2004.

Gauch, H. G. *The Scientific Method in Practice.* New York: Cambridge University Press, 2003.

Moore, J. A. *Science as a Way of Knowing.* Cambridge, MA: Harvard University Press, 1993.

2 The Chemistry of the Cell

See MasteringBiology® for tutorials, activities, and quizzes.

Students just beginning in cell biology are sometimes surprised—occasionally even dismayed—to find that almost all courses and textbooks dealing with cell biology involve a substantial amount of chemistry. Yet biology in general and cell biology in particular depend heavily on both chemistry and physics. After all, cells and organisms follow all the laws of the physical universe, so biology is really just the study of chemistry in systems that happen to be alive. In fact, everything cells are and do has a molecular and chemical basis. Therefore, we can truly understand and appreciate cellular structure and function only when we can describe that structure in molecular terms and express that function in terms of chemical reactions and events.

Trying to appreciate cellular biology without a knowledge of chemistry would be like trying to appreciate a translation of Chekhov without a knowledge of Russian. Most of the meaning would probably get through, but much of the beauty and depth of appreciation would be lost in the translation. For this reason, we will consider the chemical background necessary for the cell biologist. Specifically, this chapter will provide an overview of several principles that underlie much of cellular biology, preparing, in turn, for the next chapter, which focuses on the major classes of chemical constituents in cells.

The main points of this chapter can conveniently be structured around five principles:

1. The importance of carbon. *The chemistry of cells is essentially the chemistry of carbon-containing compounds, because the carbon atom has several unique properties that make it especially suitable as the backbone of biologically important molecules.*
2. The importance of water. *The chemistry of cells is also the chemistry of water-soluble compounds, because the water molecule has several unique properties that make it especially suitable as the universal solvent of living systems.*
3. The importance of selectively permeable membranes. *Given that most biologically important molecules are soluble in water, membranes that do not dissolve in water are important in defining cellular spaces and compartments. Because membranes are differentially permeable to specific solutes, they also control the movements of molecules and ions into and out of cells and cellular compartments.*
4. The importance of synthesis by polymerization of small molecules. *Most biologically important molecules are either small, water-soluble organic molecules that can be transported across membranes or large macromolecules that cannot. Biological macromolecules are polymers formed by the linking together of many similar or identical small molecules known as monomers. The synthesis of macromolecules by polymerization of monomers is an important principle of cellular chemistry.*
5. The importance of self-assembly. *Proteins and other biological macromolecules made of repeating monomers are often capable of self-assembly into higher levels of structural organization. Self-assembly is possible because the information needed to specify the spatial configuration of the molecule is inherent in the linear array of monomers present in the polymer.*

A study of these five principles will help us to appreciate the main topics in cellular chemistry that we will need to understand before venturing further into our exploration of what it means to be a cell.

The Importance of Carbon

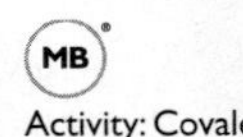
Activity: Covalent Bonds

To study cellular molecules really means to study compounds containing carbon. Almost without exception, molecules of importance to the cell biologist have a backbone, or skeleton, of carbon atoms linked together covalently in chains or rings. Actually, the study of carbon-containing compounds is the domain of **organic chemistry.** In its early days, organic chemistry was synonymous with *biological chemistry* because the carbon-containing compounds that chemists first investigated were obtained from biological sources (hence the word *organic*, acknowledging the organismal origins of the compounds).

The terms *organic chemistry* and *biological chemistry* have long since gone their separate ways, however, because organic chemists have now synthesized an incredible variety of carbon-containing compounds that do not occur naturally in the biological world. Organic chemistry therefore includes all classes of carbon-containing compounds, whereas **biological chemistry** (*biochemistry* for short) deals specifically with the chemistry of living systems and is, as we have already seen, one of the several historical strands that form an integral part of modern cell biology (see Figure 1-2).

The **carbon atom (C)** is the most important atom in biological molecules. Carbon-containing compounds owe their diversity and stability to specific bonding properties of the carbon atom. Especially important are the ways that carbon atoms interact with each other and with other chemical elements of biological importance.

An extremely important property of the carbon atom is its *valence* of four, allowing it to form four chemical bonds with other atoms. This means that the outermost electron orbital of a carbon atom lacks four of the eight electrons needed to fill it completely and make it the most stable (**Figure 2-1a**). Therefore, carbon atoms tend to associate with each other or with other electron-deficient atoms, allowing adjacent atoms to share a pair of electrons, one from each atom, so that each atom has a total of eight electrons. Atoms that share electrons in this way are held together and are said to be joined by a **covalent bond.** *Because four electrons are required to fill the outer orbital of carbon, stable organic compounds have four covalent bonds for every carbon atom.* This gives carbon-containing molecules great diversity in molecular structure and function.

Carbon atoms are most likely to form covalent bonds with one another and with atoms of oxygen (O), hydrogen (H), nitrogen (N), and sulfur (S). The electronic configurations of several of these atoms are shown in Figure 2-1a (sulfur has the same valence as oxygen). Notice that, in each case, one or more electrons are required to complete the outer orbital. The number of "missing" electrons corresponds in each case to the valence of the atom, which indicates, in turn, the number of covalent bonds the atom can form. (Hydrogen is different because its outermost electron orbital can hold only two electrons; and, thus, it has a valence of one and forms only one covalent bond.) Carbon, oxygen, hydrogen, and nitrogen are the lightest elements that form covalent bonds by sharing electron pairs. Their low atomic weight makes the resulting compounds especially stable because the strength of a covalent bond is inversely proportional to the atomic weights of the elements involved in the bond.

The sharing of one pair of electrons between atoms results in a **single bond.** Methane, ethanol, and methylamine are simple examples of carbon-containing compounds containing only single bonds between atoms (Figure 2-1b). Sometimes, two or even three pairs of electrons can be shared by two atoms, giving rise to **double bonds** or **triple bonds.** Ethylene and carbon dioxide are examples of double-bonded compounds (Figure 2-1c). Notice that, in these compounds, each carbon atom still forms a total of four covalent bonds, either one double bond and two single bonds or two double bonds. Triple bonds are rare but can be found in molecular nitrogen, hydrogen cyanide, and acetylene (Figure 2-1d). Thus, both the valence and the low atomic weight of carbon give it unique properties that account for the diversity and stability of carbon-containing compounds, giving carbon a preeminent role in biological molecules.

Carbon-Containing Molecules Are Stable

The stability of organic molecules is a property of the favorable electronic configuration of each carbon atom in the molecule. This stability is expressed as **bond energy**—the amount of energy required to break 1 *mole* (about

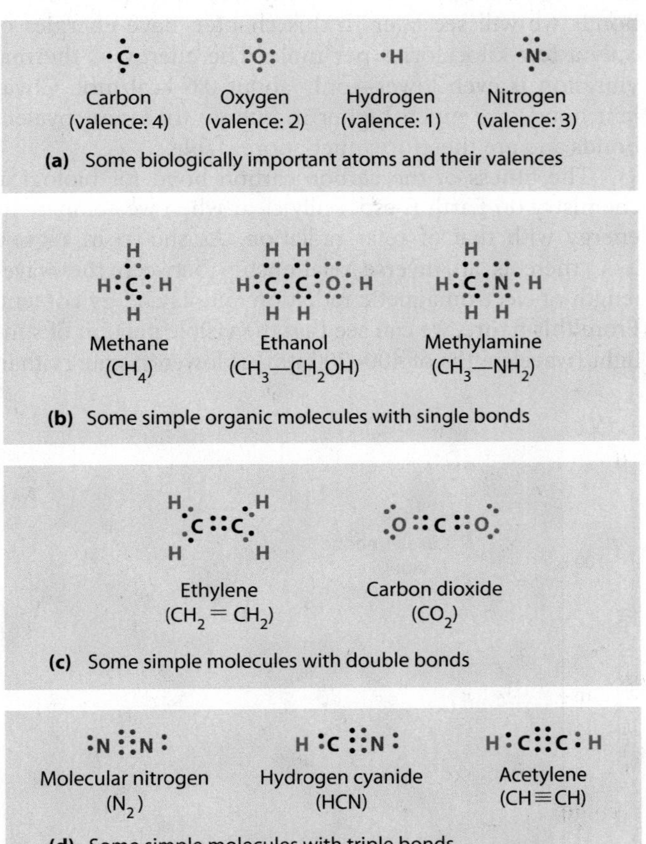

FIGURE 2-1 Electron Configurations of Some Biologically Important Atoms and Molecules. Electronic configurations are shown for **(a)** atoms of carbon, oxygen, hydrogen, and nitrogen and for simple organic molecules with **(b)** single bonds, **(c)** double bonds, and **(d)** triple bonds. Only electrons in the outermost electron orbital are shown. In each case, the two electrons positioned between adjacent atoms represent a shared electron pair, with one electron provided by each of the two atoms.

6 × 10^{23}) of such bonds. (The term *bond energy* is a frequent source of confusion. Be careful not to think of it as energy that is somehow "stored" in the bond but rather as the amount of energy needed to *break* the bond.) Bond energies are usually expressed in *calories per mole (cal/mol),* where a **calorie** is the amount of energy needed to raise the temperature of 1 gram of water by 1°C and a *kilocalorie (kcal)* is equal to 1000 calories.

It takes a large amount of energy to break a covalent bond. For example, the carbon-carbon (C—C) bond has a bond energy of 83 kilocalories per mole (kcal/mol). The bond energies for carbon-nitrogen (C—N), carbon-oxygen (C—O), and carbon-hydrogen (C—H) bonds are all in the same range: 70, 84, and 99 kcal/mol, respectively. Even more energy is required to break a carbon-carbon double bond (C═C; 146 kcal/mol) or a carbon-carbon triple bond (C≡C; 212 kcal/mol), so these compounds are even more stable.

We can appreciate the significance of these bond energies by comparing them with other relevant energy values, as shown in **Figure 2-2**. Most noncovalent bonds in biologically important molecules, such as the hydrogen bonds we will see later in this chapter, have energies of only a few kilocalories per mole. The energy of thermal vibration is even lower—only about 0.6 kcal/mol. Covalent bonds are much higher in energy than noncovalent bonds and are therefore much more stable.

The fitness of the carbon-carbon bond for biological chemistry on Earth is especially clear when we compare its energy with that of solar radiation. As shown in **Figure 2-3**, there is an inverse relationship between the wavelength of electromagnetic radiation and its energy content. From this figure, we can see that the visible portion of sunlight (wavelengths of 400–700 nm) is lower in energy than

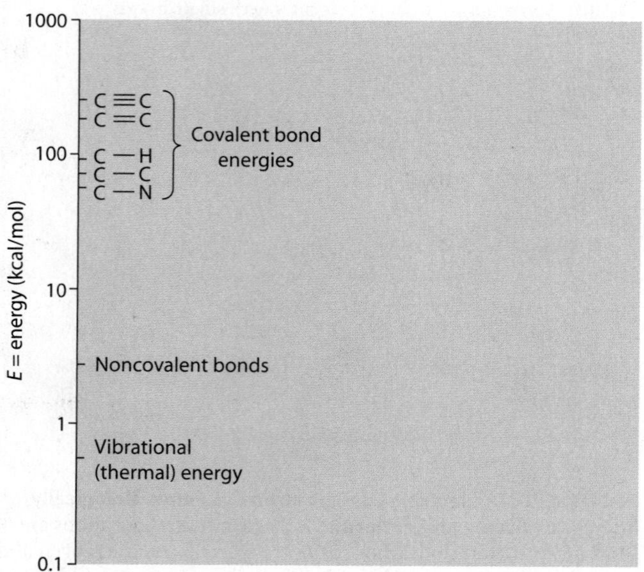

FIGURE 2-2 Energies of Biologically Important Bonds. Notice that energy is plotted on a logarithmic scale to accommodate the wide range of values shown.

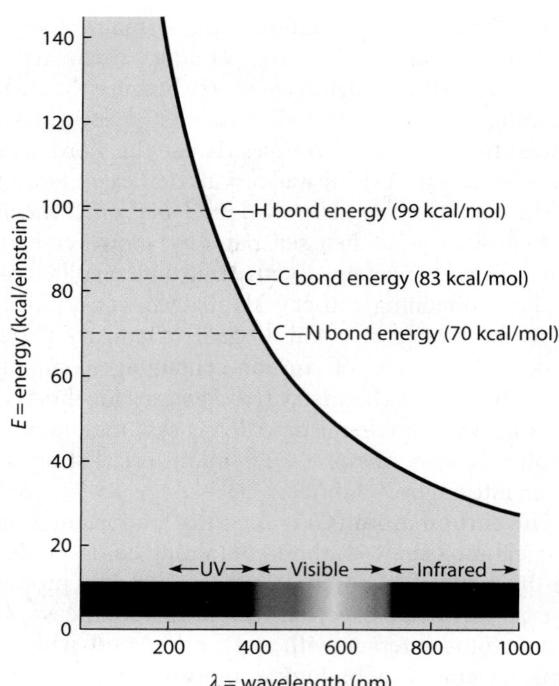

FIGURE 2-3 The Relationship Between Energy (*E*) and Wavelength (λ) for Electromagnetic Radiation. The dashed lines mark the bond energies of the C—H, the C—C, and the C—N single bonds. The bottom of the graph shows the approximate range of wavelengths for ultraviolet (UV), visible, and infrared radiation.

the carbon-carbon bond is. If this were not the case, visible light would break covalent bonds spontaneously, and life as we know it would not exist.

Figure 2-3 illustrates another important point: the hazard that ultraviolet radiation poses to biological molecules due to its high energy. At a wavelength of 300 nm, for example, ultraviolet light has an energy content of about 95 kcal/einstein (one einstein equals one mole of photons). This is enough to break carbon-carbon bonds spontaneously. This threat underlies the current concern about pollutants that destroy the ozone layer in the upper atmosphere because the ozone layer filters out much of the ultraviolet radiation that would otherwise reach Earth's surface and disrupt the covalent bonds that hold biological molecules together.

Carbon-Containing Molecules Are Diverse

In addition to their inherent stability, carbon-containing compounds are characterized by the great diversity of molecules that can be generated from relatively few different kinds of atoms. Again, this diversity is due to the tetravalent nature of the carbon atom and the resulting ability of each carbon atom to form covalent bonds to four other atoms. Because one or more of these bonds can be to other carbon atoms, molecules consisting of long chains of carbon atoms can be built up. Ring compounds are also common. Further variety is possible by the introduction

CH_3-CH_3 Ethane $CH_3-CH_2-CH_3$ Propane

$CH_2=CH_2$ Ethylene $CH\equiv CH$ Acetylene

Benzene

FIGURE 2-4 Some Simple Hydrocarbon Compounds. Compounds in the top row have single bonds only, whereas those in the second row have double or triple bonds. The condensed structure shown on the right for benzene is an example of the simplified structures that chemists frequently use for such compounds.

of branching and the presence of double bonds in the carbon-carbon chains.

When only hydrogen atoms are bonded to carbon atoms in linear or circular molecules, the resulting compounds are called **hydrocarbons** (**Figure 2-4**). Hydrocarbons are important economically because gasoline and other petroleum products—such as hexane (C_6H_{14}), octane (C_8H_{18}), and decane ($C_{10}H_{22}$)—are mixtures of short-chain hydrocarbons with from five to twelve carbon atoms. The natural gas that many of us use for fuel is a mixture of methane, ethane, propane, and butane, which are hydrocarbons with one to four carbon atoms, respectively.

In biology, on the other hand, hydrocarbons play only a limited role because they are essentially insoluble in water, the universal solvent in biological systems. However, they play an important role in the structure of biological membranes. The interior of every biological membrane is a nonaqueous environment consisting of the long hydrocarbon "tails" of phospholipid molecules that project into the interior of the membrane from either surface. This feature of membranes has important implications for their role as permeability barriers, as we will see shortly.

Most biological compounds contain, in addition to carbon and hydrogen, one or more atoms of oxygen and often nitrogen, phosphorus, or sulfur as well. These atoms are usually part of various **functional groups,** which are specific arrangements of atoms that confer characteristic chemical properties on the molecules to which they are attached. Some of the more common functional groups present in biological molecules are shown in **Figure 2-5**. At the near-neutral pH of most cells, several of these groups are ionized because they have gained or lost a proton via protonation or deprotonation, respectively. These groups include the negatively charged *carboxyl* and *phosphate* groups, which have been deprotonated and the positively charged *amino* group, which has become protonated. Other groups, such as the *hydroxyl, sulfhydryl, carbonyl,* and *aldehyde* groups, are uncharged at pH values near neutrality. However, the presence of oxygen or sulfur atoms bound to carbon or hydrogen in these uncharged groups results in a *polar* bond with unequal sharing of electrons. This is due to the higher *electronegativity,* or affinity for electrons, of oxygen and sulfur compared to carbon and hydrogen. The resulting polar bonds have higher water solubility and chemical reactivity than C—C or C—H bonds, in which electrons are equally shared.

Carbon-Containing Molecules Can Form Stereoisomers

Carbon-containing molecules are capable of still greater diversity because the carbon atom is a **tetrahedral** structure (**Figure 2-6**). *When four different atoms or groups of atoms are bonded to the four corners of such a tetrahedral structure, two different spatial configurations are possible.* Although both forms have the same structural formula, they are not superimposable but are, in fact, mirror images of each other. Such mirror-image forms of the same compound are called **stereoisomers.**

A carbon atom that has four different substituents is called an **asymmetric carbon atom.** Because two stereoisomers are possible for each asymmetric carbon atom, a compound with *n* asymmetric carbon atoms will

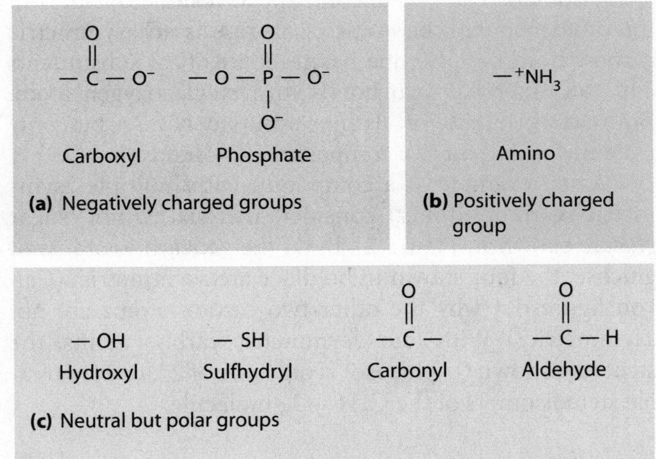

FIGURE 2-5 Some Common Functional Groups Found in Biological Molecules. Each functional group is shown in the form that predominates at the near-neutral pH of most cells. **(a)** The carboxyl and phosphate groups are deprotonated and therefore negatively charged. **(b)** The amino group, on the other hand, is protonated and is therefore positively charged. **(c)** Hydroxyl, sulfhydryl, carbonyl, and aldehyde groups are uncharged at pH values near neutrality but are much more polar than hydrocarbons, thereby conferring greater polarity and water solubility on the organic molecules to which they are attached.

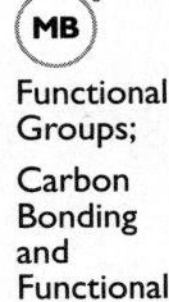

Functional Groups; Carbon Bonding and Functional Groups

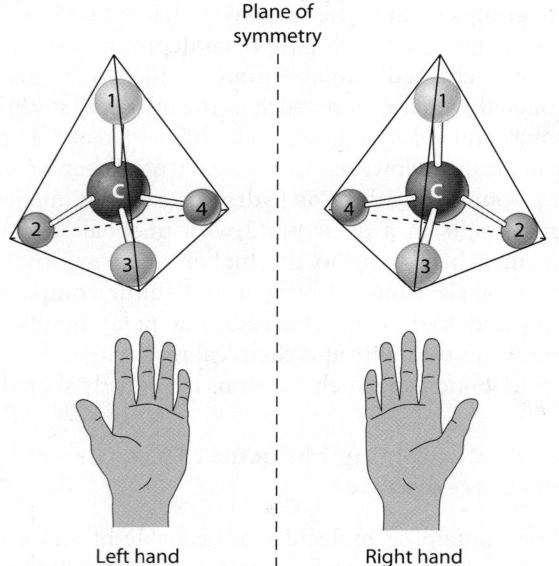

FIGURE 2-6 Stereoisomers. Stereoisomers of organic compounds occur when four different groups are attached to a tetrahedral carbon atom. Stereoisomers, like left and right hands, are mirror images of each other and cannot be superimposed on one another. (The dashed line down the center of the figure is the plane of symmetry, which can be thought of as the plane of the mirror.)

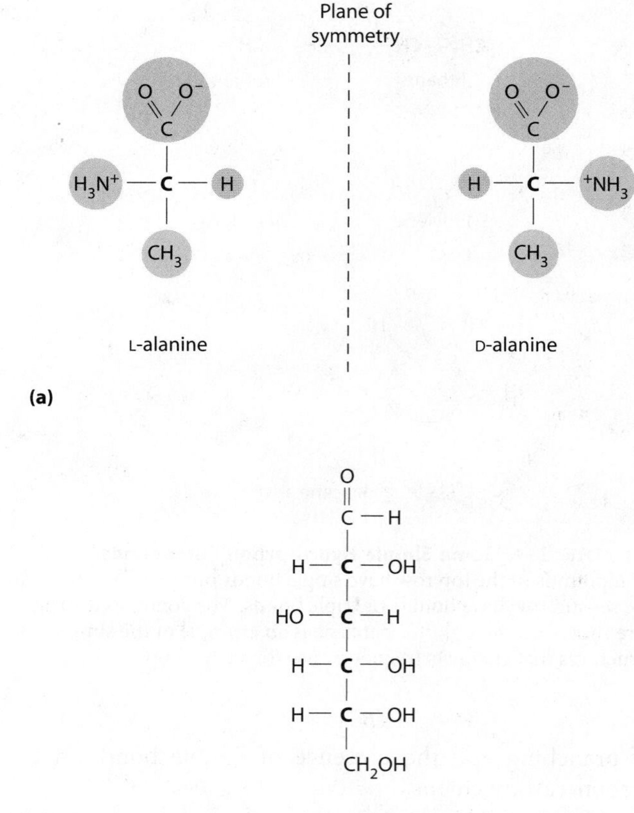

FIGURE 2-7 Stereoisomers of Biological Molecules. (a) The amino acid alanine has a single asymmetric carbon atom (in boldface) and can therefore exist in two spatially different forms, designated as L- and D-alanine. (The dashed line down the center of the figure is the plane of symmetry.) **(b)** The six-carbon sugar glucose has four asymmetric carbon atoms (in boldface), so D-glucose is just one of 16 (2^4) possible stereoisomers of the $C_6H_{12}O_6$ molecule.

have 2^n possible stereoisomers. As shown in **Figure 2-7a**, the three-carbon amino acid *alanine* has a single asymmetric carbon atom (in the center) and thus has two stereoisomers, called L-alanine and D-alanine. Neither of the other two carbon atoms of alanine is an asymmetric carbon atom because one has three identical substituents and the other has two bonds to a single oxygen atom. Both stereoisomers of alanine occur in nature, but only L-alanine is present as a component of proteins.

As an example of a compound with multiple asymmetric carbon atoms, consider the six-carbon sugar *glucose* shown in Figure 2-7b. Of the six carbon atoms of glucose, the four shown in boldface are asymmetric. (Can you figure out why the other two carbon atoms are not asymmetric?) With four asymmetric carbon atoms, the structure shown (D-glucose) is only one of 2^4, or 16, possible stereoisomers of the $C_6H_{12}O_6$ molecule.

The Importance of Water

Just as the carbon atom is uniquely significant because of its role as the universal backbone of biologically important molecules, the water molecule commands special attention because of its indispensable role as the universal solvent in biological systems. Water is, in fact, the single most abundant component of cells and organisms. Typically, about 75–85% of a cell by weight is water, and many cells depend on an extracellular environment that is essentially aqueous as well. In some cases this is a body of water—whether an ocean, lake, or river—where the cell or organism lives, and in other cases, it may be the body fluids with which the cell is bathed.

Water is indispensable for life as we know it. True, there are life forms that can become dormant and survive periods of severe water scarcity. Seeds of plants and spores of bacteria and fungi are clearly in this category. Some plants and animals—notably certain mosses, lichens, nematodes, and rotifers—can also undergo physiological adaptations that allow them to dry out and survive in a highly dehydrated form, sometimes for surprisingly long periods of time. Such adaptations are clearly an advantage in environments that are characterized by periods of drought. Yet all of these are, at best, temporary survival mechanisms. Resumption of normal activity always requires rehydration.

To understand why water is so uniquely suitable for its role, we need to look at its chemical properties. The most critical attribute is its *polarity* because this property accounts for its *cohesiveness*, its *temperature-stabilizing capacity*, and its *solvent properties*, all of which have important consequences for biological chemistry.

Water Molecules Are Polar

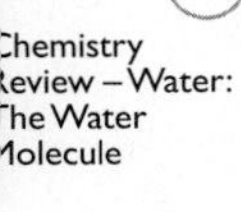

Chemistry Review – Water: The Water Molecule

An unequal distribution of electrons gives the water molecule its **polarity,** which we can define as an uneven distribution of charge within a molecule. To understand the polar nature of water, we need to consider the shape of the molecule. As shown in **Figure 2-8a**, the water molecule is bent rather than linear in shape, with the two hydrogen atoms bonded to the oxygen at an angle of 104.5° rather than 180°. It is no overstatement to say that life as we know it depends critically on this angle because of the distinctive properties that the resulting asymmetry produces in the water molecule. Although the molecule as a whole is uncharged, the electrons are unevenly distributed. The oxygen atom at the head of the molecule is highly **electronegative**—it tends to draw electrons toward it, giving that end of the molecule a partial negative charge and leaving the other end of the molecule with a partial positive charge around the hydrogen atoms.

Water Molecules Are Cohesive

Activity: The Polarity of Water

Because of their polarity, water molecules are attracted to each other and tend to orient themselves spontaneously so that the electronegative oxygen atom of one molecule is associated with the electropositive hydrogen atoms of adjacent molecules. Each such association is called a **hydrogen bond** and is represented by a dotted line, as shown in Figure 2-8b. A hydrogen bond is a type of noncovalent interaction that is about one-tenth as strong as a covalent bond.

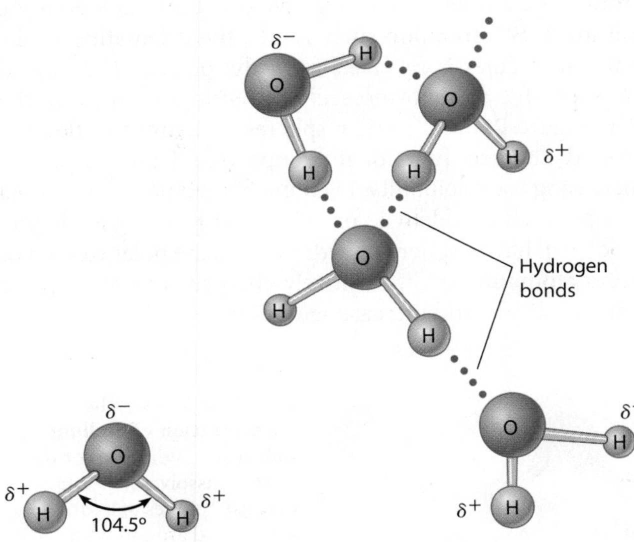

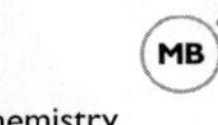

Chemistry Review – Water: Hydrogen Bonding; Activity: Hydrogen Bonds

FIGURE 2-8 Hydrogen Bonding Between Water Molecules. (a) The water molecule is polar because it has an asymmetric charge distribution. The oxygen atom bears a partial negative charge (denoted as δ^-, with the Greek letter delta standing for "partial"), and each of the two hydrogen atoms has a partial positive charge (δ^+). (b) The extensive association of water molecules with one another in either the liquid or the solid state is due to hydrogen bonds (dotted lines) between the highly electronegative oxygen atom of one water molecule and the electropositive hydrogen atoms of adjacent molecules. In ice, the resulting crystal lattice is regular and complete. Every oxygen is hydrogen bonded to hydrogens of two adjacent molecules.

FIGURE 2-9 Walking on Water. The high surface tension of water results from the collective strength of vast numbers of hydrogen bonds, and it enables insects such as this water strider to walk on the surface of a pond without breaking the surface.

Each oxygen atom can bond to two hydrogens, and both of the hydrogen atoms can associate in this way with the oxygen atoms of adjacent molecules. As a result, water is characterized by an extensive three-dimensional network of hydrogen-bonded molecules. Although individual hydrogen bonds are weak, the combined effect of large numbers of them can be quite significant. In liquid water, the hydrogen bonds between adjacent molecules are constantly being broken and reformed, with a typical bond having a half-life of a few microseconds. On the average, however, each molecule of water in the liquid state is hydrogen-bonded to at least three neighbor molecules at any given time. In ice, the hydrogen bonding is still more extensive, giving rise to a rigid, hexagonal crystalline lattice with every oxygen hydrogen-bonded to hydrogens of two adjacent molecules and every water molecule therefore hydrogen-bonded to four neighboring molecules.

It is this tendency to form hydrogen bonds between adjacent molecules that makes water so highly *cohesive.* This cohesiveness accounts for the high *surface tension* of water, as well as for its high *boiling point,* high *specific heat,* and high *heat of vaporization.* The high surface tension of water allows some insects to move across the surface of a pond without breaking the surface (**Figure 2-9**). It is also important in allowing water to move upward through the conducting tissues of plants.

Water Has a High Temperature-Stabilizing Capacity

Hydrogen Bonding and Water

An important property of water that stems directly from the hydrogen bonding between adjacent molecules is the high specific heat that gives water its *temperature-stabilizing capacity.* **Specific heat** is the amount of heat a substance must absorb per gram to increase its temperature 1°C. The specific heat of water is 1.0 calorie per gram.

Due to its extensive hydrogen bonding, the specific heat of water is much higher than that of most other liquids. In other liquids, much of the energy would cause an increase in the motion of solvent molecules and therefore an increase in temperature. In water, the energy is used instead to break hydrogen bonds between neighboring water molecules, buffering aqueous solutions against large changes in temperature. This capability is an important consideration for the cell biologist because cells release large amounts of energy as heat during metabolic reactions. If not for the extensive hydrogen bonding and the resulting high specific heat of water molecules, this release of energy would pose a serious overheating problem for cells, and life as we know it would not be possible.

Water also has a high *heat of vaporization,* which is defined as the amount of energy required to convert 1 gram of a liquid into vapor. This value is high for water because of the hydrogen bonds that must be disrupted in the process. This property makes water an excellent coolant and explains why people perspire, why dogs pant, and why plants lose water through transpiration. In each case, the heat required to evaporate water is drawn from the organism, which is therefore cooled in the process.

MB Water Is an Excellent Solvent

Chemistry Review – Water: Properties of Water; Chemistry Review – Water: Solutions; Hydrophobic versus Hydrophilic

From a biological perspective, one of the most important properties of water is its excellence as a general solvent. A **solvent** is a fluid in which another substance, called the **solute,** can be dissolved. Water is an especially good solvent for biological purposes because of its remarkable capacity to dissolve a great variety of solutes.

It is the polarity of water that makes it so useful as a solvent. Most of the molecules in cells are also polar (or charged) and therefore form hydrogen bonds (or ionic bonds) with water molecules. Solutes that have an affinity for water and therefore dissolve readily in it are called **hydrophilic** ("water-loving"). Most small organic molecules found in cells are hydrophilic. Examples are sugars, organic acids, and some of the amino acids. Molecules that are not very soluble in water are termed **hydrophobic** ("water-fearing"). Among the more important hydrophobic compounds found in cells are the lipids and proteins found in biological membranes. In general, polar molecules and ions are hydrophilic, and nonpolar molecules are hydrophobic. Some biological macromolecules, notably proteins, have both hydrophobic and hydrophilic regions, so some parts of the molecule have an affinity for water while other parts of the molecule do not.

To understand why polar substances dissolve so readily in water, let's consider a salt such as sodium chloride (NaCl) (**Figure 2-10**). Because it is a salt, NaCl exists in crystalline form as a lattice of positively charged sodium *cations* (Na^+) and negatively charged chloride *anions* (Cl^-). For NaCl to dissolve in a liquid, solvent molecules must overcome the attraction of the Na^+ cations and Cl^- anions for each other. When NaCl is placed in water, both the sodium and chloride ions become involved in *electrostatic interactions* with the water molecules instead of with each other, and the Na^+ and Cl^- ions separate and become dissolved. Because of their polarity, water molecules can form *spheres of hydration* around both Na^+ and Cl^-, thus neutralizing their attraction for each other and decreasing their likelihood of reassociation.

As Figure 2-10a shows, the sphere of hydration around a cation such as Na^+ involves water molecules clustered around the ion with their negative (oxygen) ends pointing toward it. For an anion such as Cl^-, the orientation of the water molecules is reversed, with the positive (hydrogen) ends of the solvent molecules pointing in toward the ion (Figure 2-10b). Similar spheres of hydration develop around charged functional groups (see Figure 2-5a, b), increasing their solubility. Even uncharged polar functional groups, such as aldehyde or sulfhydryl groups (see Figure 2-5c), will have a sphere of hydration, as the polar oxygen or sulfur atoms attract the positively-charged ends of the polar water molecule and increase solubility.

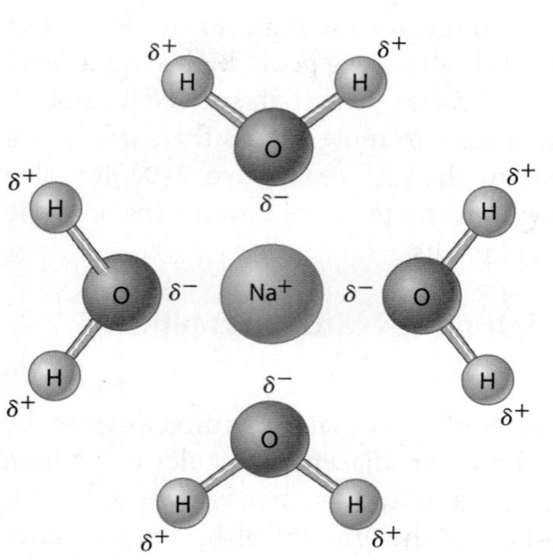

(a) Hydration of sodium ion

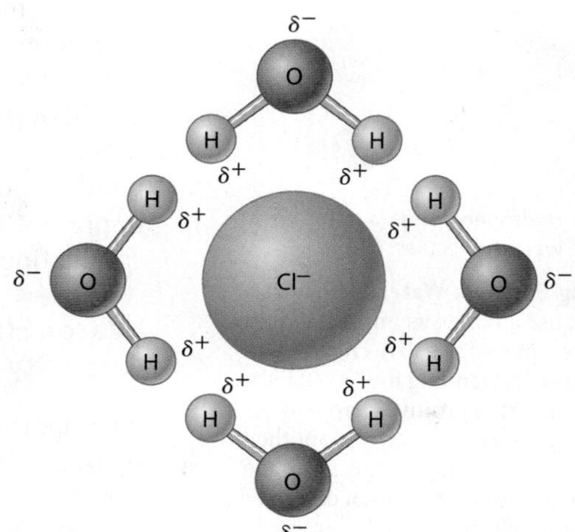

(b) Hydration of chloride ion

FIGURE 2-10 The Solubilization of Sodium Chloride. Sodium chloride (NaCl) dissolves in water because spheres of hydration are formed around both **(a)** the sodium ions and **(b)** the chloride ions. The oxygen atom and the sodium and chloride ions are drawn to scale.

Some biological compounds are soluble in water because they exist as ions at the near neutral pH of the cell and are therefore solubilized and hydrated like the ions of Figure 2-10. Compounds containing carboxyl, phosphate, or amino groups are in this category (see Figure 2-5). Most organic acids, for example, are almost completely ionized by deprotonation at a pH near 7 and therefore exist as anions that are kept in solution by spheres of hydration, as we have just seen with the chloride ion in Figure 2-10b. Amines, on the other hand, are usually protonated at cellular pH and thus exist as hydrated cations, and behave like the sodium ion in Figure 2-10a.

Often, organic molecules have no net charge—that is, they have neither lost nor gained protons and are therefore neutral molecules. However, many organic molecules are nonetheless hydrophilic because they have some regions that are positively charged and other regions that are negatively charged. Water molecules tend to cluster around such regions, and the resulting electrostatic interactions between solute and water molecules keep the solute molecules from associating with one another. Compounds containing the hydroxyl, sulfhydryl, carbonyl, or aldehyde groups shown in Figure 2-5 are usually in this category.

Hydrophobic molecules such as hydrocarbons, on the other hand, have no such polar regions and therefore show no tendency to interact electrostatically with water molecules. In fact, they actually disrupt the hydrogen-bonded structure of water and, for this reason, tend to be excluded by the water molecules. Therefore, hydrophobic molecules tend to coalesce as they associate with one another rather than with the water. This association is driven not so much by any specific affinity of the hydrophobic molecules for one another as by the strong tendency of water molecules to form hydrogen bonds and to exclude molecules that disrupt hydrogen bonding. As we will see later in the chapter, such associations of hydrophobic molecules (or parts of molecules) are a major driving force in the folding of molecules, the assembly of cellular structures, and the organization of membranes.

The Importance of Selectively Permeable Membranes

Every cell and organelle needs some sort of physical barrier to keep its contents in and external materials out. A cell also needs some means of controlling exchange between its internal environment and the extracellular environment. Ideally, such a barrier should be impermeable to most of the molecules and ions found in cells and their surroundings. Otherwise, substances could diffuse freely in and out, and the cell would not really have a defined content at all. On the other hand, the barrier cannot be completely impermeable or else necessary exchanges between the cell and its environment could not take place. The barrier must be insoluble in water so that it will not be dissolved by the aqueous medium of the cell. At the same time, it must be readily permeable to water because water is the basic solvent system of the cell and must be able to flow into and out of the cell as needed.

As you might expect, the membranes that surround cells and organelles satisfy these criteria admirably. A cellular **membrane** is essentially a hydrophobic permeability barrier that consists of *phospholipids, glycolipids,* and *membrane proteins.* In most organisms other than bacteria, the membranes also contain *sterols—cholesterol* in the case of animal cells, *ergosterols* in fungi, and *phytosterols* in the membranes of plant cells. (Don't be concerned if you haven't encountered these kinds of molecules before; we'll meet them all again in Chapter 3.)

Most membrane lipids and proteins are not simply hydrophobic or hydrophilic. They typically have both hydrophilic and hydrophobic regions and are therefore referred to as **amphipathic molecules** (the Greek prefix *amphi–* means "of both kinds"). The amphipathic nature of membrane phospholipids is illustrated in **Figure 2-11**,

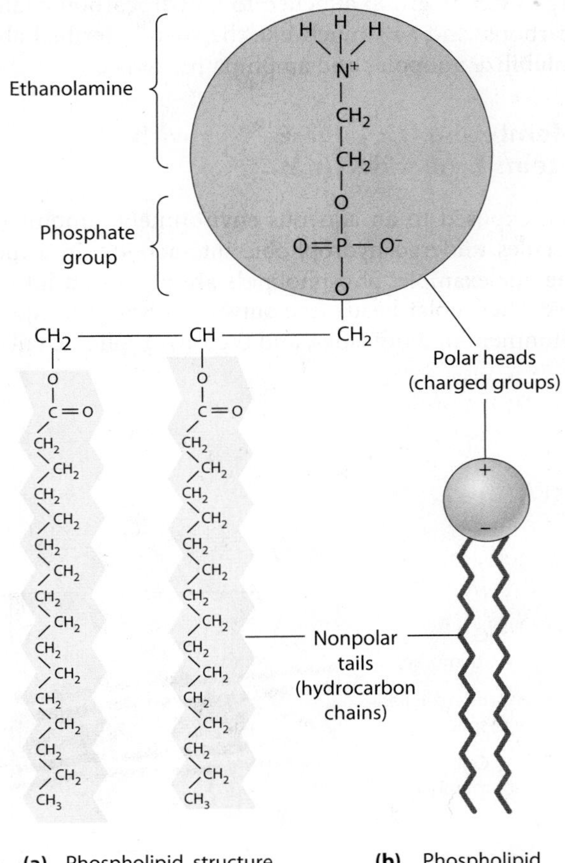

FIGURE 2-11 The Amphipathic Nature of Membrane Phospholipids. **(a)** A phospholipid molecule consists of two long nonpolar tails (yellow) and a polar head (orange). Illustrated here is phosphatidyl ethanolamine, an example of the phosphoglyceride class of membrane phospholipids. The polarity of the head of a phospholipid molecule results from a negatively charged phosphate group linked to a positively charged group—an amino group, in the case of phosphatidyl ethanolamine. **(b)** A phospholipid molecule is often represented schematically by a circle for the polar head (notice the plus and minus charges) and two zigzag lines for the nonpolar hydrocarbon chains.

which shows the structure of *phosphatidyl ethanolamine,* a prominent phospholipid in many kinds of membranes. The distinguishing feature of amphipathic phospholipids is that each molecule consists of a polar *head* and two nonpolar hydrocarbon *tails.* The polarity of the hydrophilic head is due to the presence of a negatively charged phosphate group that is often linked to a positively charged group—an amino group, in the case of phosphatidyl ethanolamine and most other phosphoglycerides. We will learn more about phospholipids in Chapter 3.

Soap is a familiar amphipathic molecule that you all have likely used to dissolve grease, oil, and other nonpolar substances. The nonpolar hydrocarbon tails of the soap molecules interact with and surround the oil or grease, and the polar heads interact with water, enabling the oil or grease to be washed away. In the lab, we often use the amphipathic detergent sodium dodecyl sulfate (SDS) to isolate insoluble proteins and lipids. SDS has a negatively charged sulfate group attached to a hydrocarbon chain of 12 carbons, and acts much like the soap described above to solubilize nonpolar and amphipathic molecules.

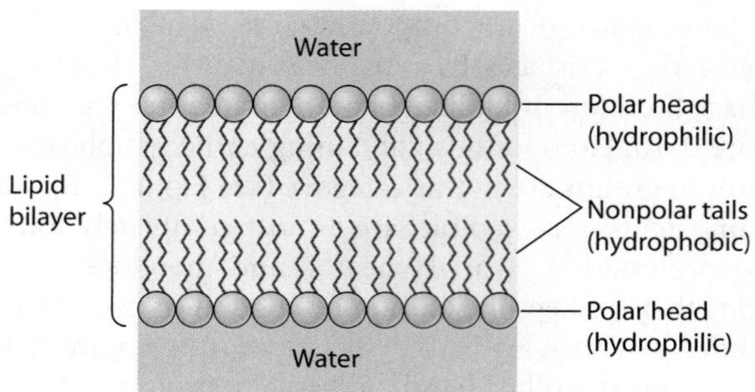

FIGURE 2-12 The Lipid Bilayer as the Basis of Membrane Structure. Because of their amphipathic nature, phospholipids in an aqueous environment orient themselves in a double layer, with the hydrophobic tails (zigzag lines) buried on the inside and the hydrophilic heads (orange) interacting with the aqueous environment on either side of the membrane.

A Membrane Is a Lipid Bilayer with Proteins Embedded in It

When exposed to an aqueous environment, amphipathic molecules undergo hydrophobic interactions. In a membrane, for example, phospholipids are organized into two layers: Their polar heads face outward toward the aqueous environment on both sides, and their hydrophobic tails are hidden from the water by interacting with the tails of other molecules oriented in the opposite direction. The resulting structure is the **lipid bilayer,** shown in **Figure 2-12**. The heads of both layers face outward, and the hydrocarbon tails extend inward, forming the continuous hydrophobic interior of the membrane.

Every known biological membrane has such a lipid bilayer as its basic structure. Each of the lipid layers is typically 3–4 nm thick, so the bilayer has a width of 7–8 nm. It is the lipid bilayer that gives each membrane its characteristic "railroad track" appearance when seen in cross section with the transmission electron microscope (TEM; **Figure 2-13a**). The osmium used to prepare the tissue for electron

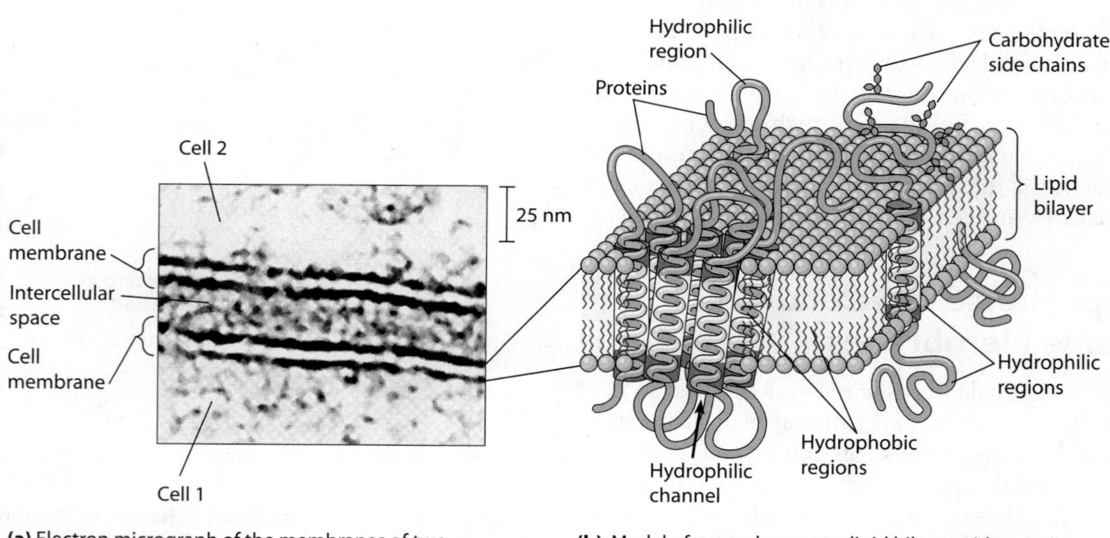

(a) Electron micrograph of the membranes of two adjacent cells

(b) Model of a membrane as a lipid bilayer with proteins embedded in it

FIGURE 2-13 Membranes and Membrane Structure. **(a)** With the transmission electron microscope (TEM), each of the cell membranes surrounding two adjacent cells appears as a pair of dark bands that result when osmium interacts with the hydrophilic heads of the phospholipid molecules but not with their hydrophobic tails. **(b)** Biological membranes consist of amphipathic proteins embedded within a lipid bilayer. Proteins are positioned in the membrane so that their hydrophobic regions are located in the hydrophobic interior of the phospholipid bilayer and their hydrophilic regions are exposed to the aqueous environment on either side of the membrane.

microscopy reacts with the hydrophilic heads of the phospholipid molecules at both surfaces of the membrane but not with the hydrophobic tails in the interior of the membrane, giving each membrane its three-layered appearance.

The structure of biological membranes is illustrated in Figure 2-13b. Embedded within or associated with the membrane lipid bilayer are various membrane proteins. These proteins are almost always amphipathic, and they become oriented in the lipid bilayer accordingly. Hydrophobic regions of the protein associate with the interior of the membrane, whereas hydrophilic regions protrude into the aqueous environment on either or both surfaces of the membrane.

Depending on the particular membrane, the membrane proteins may play any of a variety of roles. Some are *transport proteins,* responsible for moving specific substances across an otherwise impermeable membrane. Others are *enzymes* that catalyze reactions associated with the specific membrane. Still others are the *receptors* on the outer surface of the cell membrane, the *electron transport intermediates* of the mitochondrial membrane, or the *chlorophyll-binding proteins* of the chloroplast. We will encounter each of these kinds of membrane proteins in subsequent chapters, beginning with Chapter 4 (see, for example, Figure 4-9).

Membranes Are Selectively Permeable

Because of its hydrophobic interior, a membrane is readily permeable to nonpolar molecules. However, it is quite impermeable to most polar molecules and is highly impermeable to all ions. Because most cellular constituents are either polar or charged, they have little or no affinity for the membrane interior and are effectively prevented from entering or escaping from the cell. Very small molecules are an exception, however. Compounds with molecular weights below about 100 diffuse across membranes regardless of whether they are nonpolar (O_2 and CO_2) or polar (ethanol and urea). Water is an especially important example of a very small molecule that, although polar, diffuses rapidly across membranes and can readily enter or leave cells.

In contrast, even the smallest ions are effectively excluded from the hydrophobic interior of the membrane. For example, a lipid bilayer is at least 10^8 times less permeable to such small cations as Na^+ or K^+ than it is to water. This striking difference is due to both the charge on an ion and the sphere of hydration surrounding the ion.

Of course, it is essential that cells have ways of transferring not only ions such as Na^+, and K^+, but also a wide variety of polar molecules across membranes that are not otherwise permeable to these substances. As already noted, membranes are equipped with transport proteins to serve this function. A transport protein is a specialized transmembrane protein that serves either as a *hydrophilic channel* through an otherwise hydrophobic membrane or as a *carrier* that binds a specific solute on one side of the membrane and then undergoes a conformational change to move the solute across the membrane.

Whether a channel or a carrier, each transport protein is specific for a particular molecule or ion (or, in some cases, for a class of closely related molecules or ions). Moreover, the activities of these proteins can be carefully regulated to meet cellular needs. As a result, biological membranes are best described as *selectively permeable.* Except for very small molecules and nonpolar molecules, molecules and ions can move across a particular membrane only if the membrane contains an appropriate transport protein.

The Importance of Synthesis by Polymerization

For the most part, cellular structures such as ribosomes, chromosomes, membranes, flagella, and cell walls are made up of ordered arrays of linear *polymers* that are called **macromolecules.** Important macromolecules in cells include *proteins, nucleic acids* (both DNA and RNA), and *polysaccharides* such as starch, glycogen, and cellulose. *Lipids* are sometimes regarded as macromolecules as well; however, they differ somewhat from the other classes of macromolecules in the way they are synthesized and will not be discussed further until Chapter 3. Macromolecules are important in both the structure and the function of cells. To understand the biochemical basis of cell biology, therefore, really means to understand macromolecules—how they are made, how they are assembled, and how they function.

Macromolecules Are Responsible for Most of the Form and Function in Living Systems

The importance of macromolecules in cell biology is emphasized by the cellular hierarchy shown in **Figure 2-14**. We use the term *hierarchy* here to denote how biological molecules and structures can be organized into a series of levels, each level building on the preceding one. Most cellular structures are composed of small, water-soluble *organic molecules* (level 1) that cells either obtain from other cells or synthesize from simple nonbiological molecules such as carbon dioxide, ammonia, or phosphate ions that are available from the environment. Small organic molecules polymerize to form *biological macromolecules* (level 2) such as polysaccharides, proteins, or nucleic acids. These macromolecules may function on their own, or they can then be assembled into a variety of *supramolecular structures* (level 3). These supramolecular structures are themselves components of *organelles* and other *subcellular structures* (level 4) that make up the *cell* itself (level 5).

One of the examples in Figure 2-14 is that of cell wall biogenesis (panels on left). In plants, a major component of the primary cell wall (level 3) is the polysaccharide cellulose (level 2). Cellulose is a repeating polymer of the simple monosaccharide glucose (level 1), a sugar formed by the plant cell from carbon dioxide and water in the process of photosynthesis.

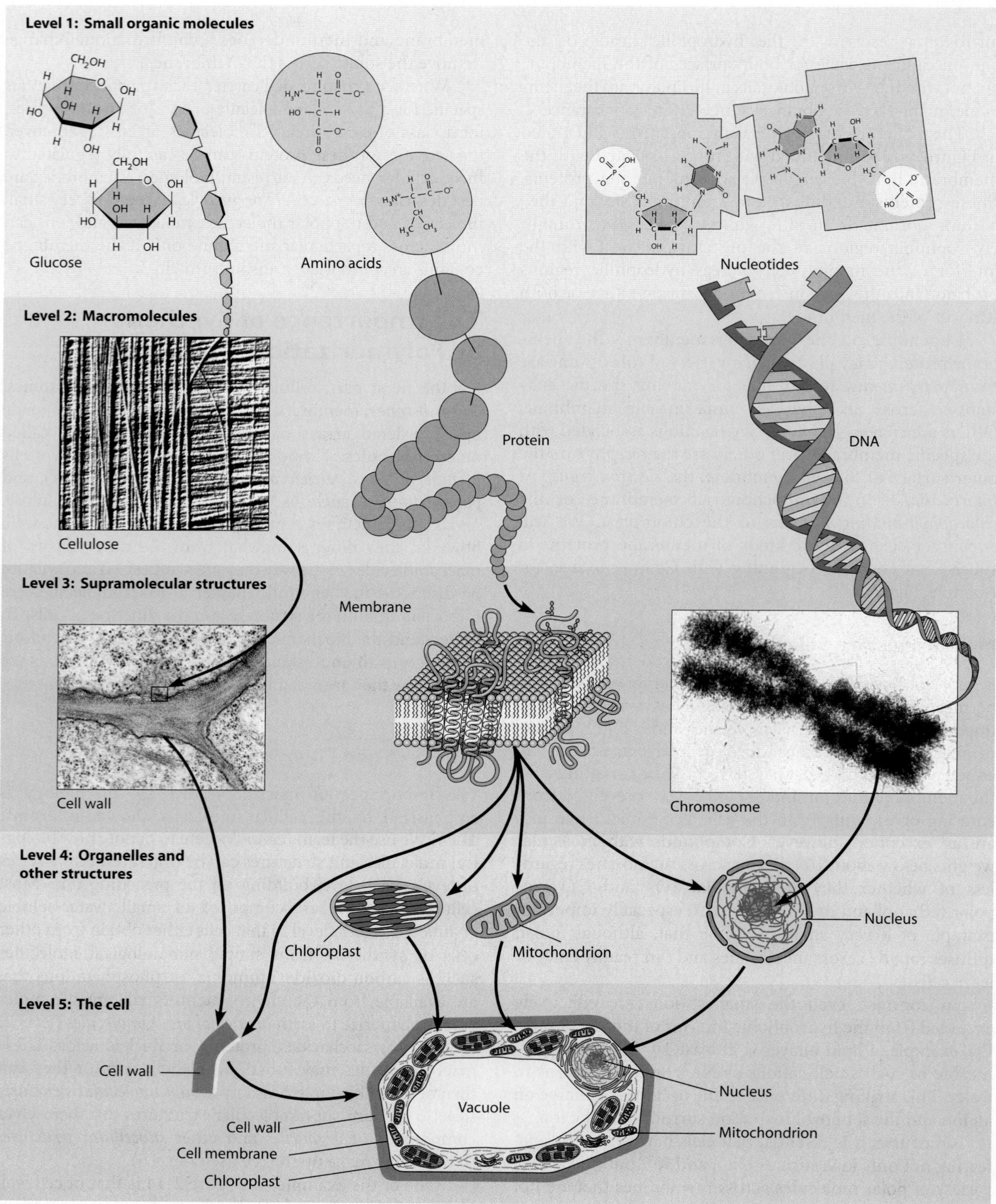

FIGURE 2-14 The Hierarchical Nature of Cellular Structures and Their Assembly. Small organic molecules (level 1) are synthesized from simple inorganic substances and are polymerized to form macromolecules (level 2). The macromolecules then assemble into the supramolecular structures (level 3) that make up organelles and other subcellular structures (level 4) and, ultimately, the cell itself (level 5). The supramolecular structures shown as level 3 are more complex in their chemical composition than the figure suggests. Chromosomes, for example, contain proteins as well as DNA—in about equal amounts, in fact. Similarly, membranes contain not only lipids but also a variety of proteins; and cell walls contain not just cellulose but also other carbohydrates and proteins.

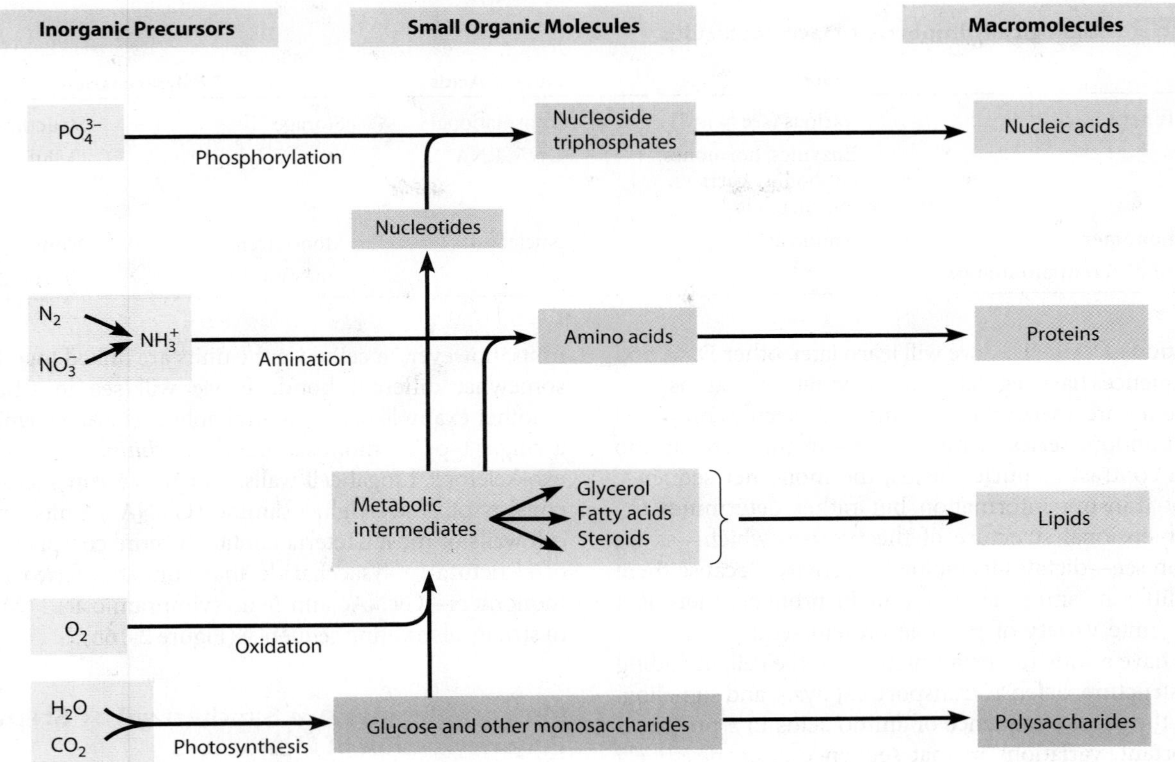

FIGURE 2-15 The Synthesis of Biological Macromolecules. Simple inorganic precursors (left) react to form small organic molecules (center), which are the monomers used in synthesizing the macromolecules (right) that make up most cellular structures.

From such examples, a general principle emerges: *The macromolecules that are responsible for most of the form and order characteristic of living systems are generated by the polymerization of small organic molecules.* This strategy of forming large molecules by joining smaller units in a repetitive manner is illustrated in **Figure 2-15**. The importance of this strategy can hardly be overemphasized because it is a fundamental principle of cellular chemistry. The many enzymes responsible for catalyzing cellular reactions, the nucleic acids involved in storing and expressing genetic information, the glycogen stored by your liver, and the cellulose that gives rigidity to a plant cell wall are all variations on the same design theme. Each is a macromolecule made by the linking together of small repeating units.

Examples of such repeating units, or **monomers,** are the *glucose* present in cellulose or starch, the *amino acids* needed to make proteins, and the *nucleotides* that make up nucleic acids. In general these are small, water-soluble organic molecules with molecular weights less than about 350. They can be transported across most biological membranes, provided the appropriate transport proteins are present in the membrane. By contrast, most macromolecules that are synthesized from these monomers are too large to traverse membranes and therefore must be made in the cell or compartment where they are needed.

Cells Contain Three Different Kinds of Macromolecules

The three major kinds of macromolecular polymers found in the cell are proteins, nucleic acids, and polysaccharides. **Table 2-1** describes the general functions of these macromolecules, gives a few examples of each, and describes the type and variety of monomeric subunits comprising each polymer. For nucleic acids and proteins, the exact order of the different monomers is critical for function. In both nucleic acids and proteins, there are nearly limitless ways in which the different monomers can be arranged in a linear order. Thus, there are millions of different DNA, RNA, and protein sequences. In contrast, polysaccharides are typically composed of a single monomer or of two different monomers. These monomers occur in a repetitious pattern and, thus, there are relatively few different types of polysaccharides.

Nucleic acids (both DNA and RNA) are often called **informational macromolecules** because the order of the four kinds of nucleotide monomers in each is nonrandom and carries important information. A critical role of many DNA and RNA molecules is to serve a *coding* function, meaning that they contain the information necessary to specify the precise amino acid sequences of particular proteins. In this manner, the specific order of the monomeric units in nucleic acids stores and transmits the genetic

Table 2-1 Biologically Important Macromolecules

	Proteins	Nucleic Acids	Polysaccharides	
			Storage	Structural
General function	Various (see below)	Informational	Storage	Structural
Examples	Enzymes, hormones, antibodies, carriers, ion channels	DNA, RNA	Starch, glycogen	Cellulose, chitin
Type of monomer	Amino acids	Nucleotides	Monosaccharides	Monosaccharides
Number of different monomers	20	4	One or a few	One or a few

information of the cell. As we will learn later, other DNA and RNA sequences have regulatory or enzymatic functions.

Proteins are a second type of macromolecule composed of a nonrandom series of monomers—in this case, amino acids. In contrast to nucleic acids, the monomer sequence does not transmit information but rather determines the three-dimensional structure of the protein, which—as we shall soon see—dictates its biological activity. Because there are 20 different amino acids found in proteins, there is a nearly infinite variety of possible protein sequences. Thus, proteins have a wide range of functions in the cell, including roles in structure, defense, transport, catalysis, and signaling. Because the precise sequence of amino acids in a protein is so important, variations in that sequence often negatively affect the ability of the protein to perform its function.

Polysaccharides, on the other hand, typically consist of either a single repeating subunit or two subunits occurring in strict alternation. The order of monomers in a polysaccharide therefore carries no information and is not essential to the function of the polymer. Most polysaccharides are either **storage macromolecules** or **structural macromolecules.** The most familiar storage polysaccharides are the *starch* of plant cells and the *glycogen* found in bacteria and animal cells. As shown in **Figure 2-16a**, both of these storage polysaccharides consist of a single repeating monomer—the simple sugar glucose—and they both function in energy storage.

The best-known example of a structural polysaccharide is the *cellulose* present in plant cell walls and woody tissues. Like starch and glycogen, cellulose consists solely of glucose units; however, in cellulose the units are linked together by a somewhat different bond, as we will see in Chapter 3. Another example of a structural polysaccharide formed from a single type of monosaccharide is *chitin,* found in insect exoskeletons, fungal cell walls, and crustacean shells. Chitin consists of *N*-acetylglucosamine (GlcNAc) units only. The cell walls of most bacteria contain a more complicated kind of structural polysaccharide that consists of two kinds of monomers—GlcNAc and *N*-acetylmuramic acid (MurNAc) in strictly alternating sequence (Figure 2-16b).

Macromolecules Are Synthesized by Stepwise Polymerization of Monomers

MB Activity: Condensation and Hydrolysis Reactions

In Chapter 3, we will look at each of the major kinds of biologically important macromolecules. First, however, it will be useful to consider several important principles that underlie the polymerization processes by which all of these macromolecules arise. Although the chemistry of the monomeric units, and hence of the resulting polymers, differs markedly among macromolecules, the following basic principles apply in each case:

1. Macromolecules are always synthesized by the stepwise polymerization of similar or identical small molecules called *monomers.*
2. The addition of each monomer occurs with the removal of a water molecule and is therefore termed a **condensation reaction.**

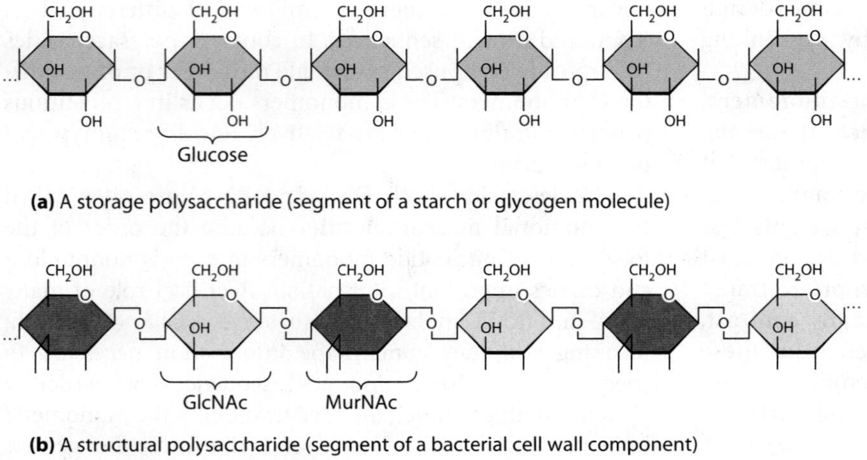

(a) A storage polysaccharide (segment of a starch or glycogen molecule)

(b) A structural polysaccharide (segment of a bacterial cell wall component)

FIGURE 2-16 Storage and Structural Macromolecules. Storage and structural macromolecules contain one or a few kinds of repeating units in a repetitive sequence. **(a)** A portion of the sequence of a linear segment of the storage polysaccharide starch or glycogen, consisting of glucose units linked together by glycosidic bonds. **(b)** A portion of the sequence of a bacterial cell wall polysaccharide, consisting of the sugar derivatives *N*-acetylglucosamine (GlcNAc) and *N*-acetylmuramic acid (MurNAc) in strict alternation.

3. The monomeric units that are to be joined together must be present as **activated monomers** before condensation can occur.
4. Activation usually involves coupling of the monomer to a **carrier molecule,** forming an activated monomer.
5. The energy needed to couple the monomer to the carrier molecule is provided by a molecule called *adenosine triphosphate (ATP)* or a related high-energy compound.
6. Because of the way they are synthesized, macromolecules have an inherent **directionality.** This means that the two ends of the polymer chain are chemically different from each other.

Because the elimination of a water molecule is essential in all biological polymerization reactions, each monomer must have both an available hydrogen (H) and an available hydroxyl group (—OH) elsewhere on the molecule. This structural feature is depicted schematically in **Figure 2-17a**, in which the monomeric units (M) are represented as boxes labeled with the relevant hydrogen and hydroxyl group. For a given kind of polymer, the monomers may differ from one another in other aspects of their structure; but each monomer possesses an available hydrogen and hydroxyl group.

Figure 2-17a also shows monomer activation, an energy-requiring process. Regardless of the polymer type, the addition of each monomer to a growing chain is always energetically unfavorable unless the incoming monomer is in an activated, energized form. Activation usually involves coupling of the monomer to a carrier molecule (C). The energy to drive this activation process is provided by ATP or a closely related *high-energy compound*—one that releases a large amount of energy upon hydrolysis (see pp. 253–254). A different kind of carrier molecule is used for each kind of polymer. For protein synthesis, amino acids are activated by linking them to carriers called *transfer RNA* (or *tRNA*) molecules. Polysaccharides are synthesized from sugar (often glucose) molecules that are activated by linking them to derivatives of nucleotides (*adenosine diphosphate* for starch, *uridine diphosphate* for glycogen). In nucleic acid synthesis, there is no need for specific carrier molecules because the nucleotides themselves (e.g., ATP or GTP) are high-energy molecules.

Once activated, monomers are capable of reacting with each other in a condensation reaction that is followed or accompanied by the release of the carrier molecule from one of the two monomers (see Figure 2-17b). Subsequent elongation of the polymer is a sequential, stepwise process. One activated monomeric unit is added at a time, thus lengthening the elongating polymer by one unit. Figure 2-17c illustrates the *n*th step in such a process, during

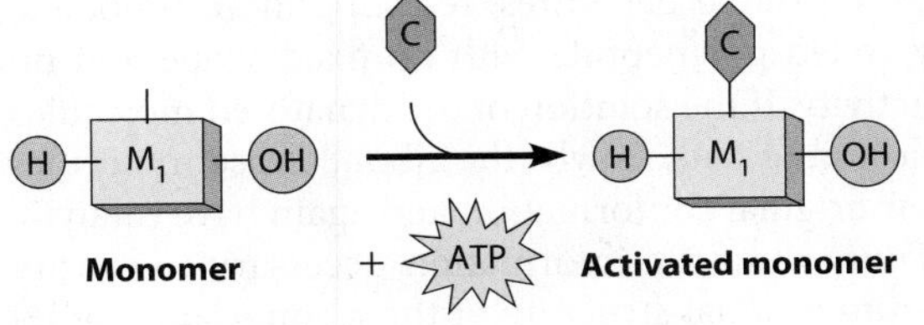

(a) Monomer activation. Monomers (M_1, M_2, etc.) with available H and OH groups (shown in blue) are activated by coupling to the appropriate carrier molecule (C, shown in purple), using energy from ATP or a similar high-energy compound.

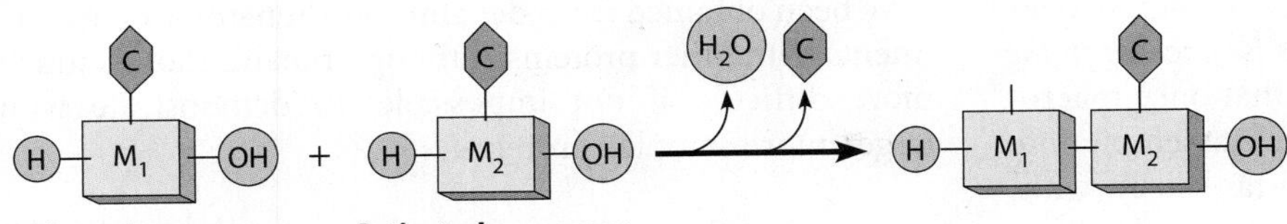

(b) Monomer condensation. The first step in polymer synthesis involves the condensation of two activated monomers, with the release of one of the carrier molecules.

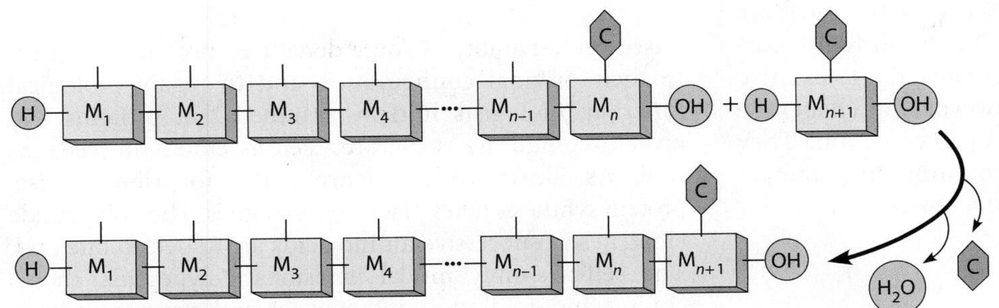

(c) Polymerization. The *n*th step in the polymerization process involves the addition of the next activated monomer (M_{n+1}) to a polymer that already consists of *n* monomeric units.

FIGURE 2-17 Macromolecule Biosynthesis. Biological macromolecules are synthesized in a process involving **(a)** monomer activation, **(b)** condensation, and **(c)** polymerization. Depending on the polymer, the monomers may be identical (some polysaccharides) or may differ from one another (proteins and nucleic acids).

which the next monomer unit is added to an elongating polymer that already contains *n* monomeric units. The chemical nature of the energized monomers, the carrier, and the actual activation process differ for each biological polymer, but the general principle is always the same: *Polymer synthesis always involves activated monomers joined via condensation reactions, and ATP or a similar high-energy compound is required to activate the monomers by linking them to an appropriate carrier molecule.*

The *degradation,* or breakdown, of macromolecules occurs in a manner opposite to the condensation reactions used for synthesis. During degradation, monomers are removed from the polymer by **hydrolysis**—the addition of a water molecule that breaks the bond between adjacent monomers. In the process, one monomer receives a hydroxyl group from the water molecule and the other receives a hydrogen atom, reforming the original monomers.

The Importance of Self-Assembly

So far, we have seen that the macromolecules that characterize biological organization and function are polymers of small, hydrophilic, organic molecules. The only requirements for polymerization are an adequate supply of the monomeric subunits and a source of energy—and, in the case of proteins and nucleic acids, sufficient information to specify the order of the monomers. Still to be considered are the steps that lie beyond—the processes through which these macromolecules are organized into the supramolecular assemblies and organelles that are readily recognizable as cellular structures. Or, in light of Figure 2-14, we need to ask how the macromolecules of level 2 are assembled into the higher-level structures of levels 3, 4, and 5.

Crucial to our understanding of these higher-level structures is the principle of **self-assembly,** which asserts that *the information required to specify the folding of macromolecules and their interactions to form more complex structures with specific biological functions is inherent in the polymers themselves.* This principle says that once macromolecules are synthesized in the cell, their assembly into more complex structures will occur spontaneously, without further input of energy or information. As we will see shortly, this principle has been qualified somewhat by the recognition that proteins called *molecular chaperones* are needed in some cases of protein folding to prevent incorrect molecular interactions that would give rise to inactive structures. Even in such cases, however, the chaperone molecules do not provide additional information. They simply assist the assembly process by inhibiting interactions that would produce incorrect structures.

Many Proteins Self-Assemble

A prototype that is useful for understanding self-assembly processes is the coiling and folding necessary to form a functional three-dimensional protein from one or more linear chains of amino acids. Although the distinction is not always properly made, the immediate product of amino acid polymerization is actually not a protein, but a **polypeptide.** To become a functional protein, one or more such linear polypeptide chains must coil and fold in a precise, predetermined manner to assume the unique three-dimensional structure, or *conformation,* necessary for biological activity.

The evidence for self-assembly of proteins comes largely from studies in which the *native,* or natural, conformation of a protein is disrupted by changing the environmental conditions. Such disruption, or unfolding, can be achieved by raising the temperature, by making the pH of the solution highly acidic or highly alkaline, or by adding certain chemical agents such as urea or any of several alcohols. The unfolding of a polypeptide under such conditions is called **denaturation** because it results in the loss of the natural three-dimensional structure of the protein—and the loss of its function as well. In the case of an enzyme, denaturation results in the loss of catalytic activity.

When the denatured polypeptide is returned to conditions in which the native conformation is stable, the polypeptide may undergo **renaturation,** the return to its correct three-dimensional conformation. In at least some cases, the renatured protein also regains its biological function—catalytic activity, in the case of an enzyme.

Figure 2-18a depicts the denaturation and subsequent renaturation of *ribonuclease,* the protein used by Christian Anfinsen and his colleagues in their classic studies of protein self-assembly. When a solution of ribonuclease is heated, the protein denatures, resulting in an unfolded, randomly coiled polypeptide with no fixed shape and no catalytic activity. If the solution of the denatured molecules is then allowed to cool slowly, the ribonuclease molecules regain their original conformation and again have catalytic activity. Thus, all of the information necessary to specify the three-dimensional structure of the ribonuclease molecule is inherent in its amino acid sequence. Similar results have been obtained from denaturation/renaturation experiments with other proteins, although renaturation is much more difficult, if not impossible, to demonstrate with larger, more complex proteins.

Molecular Chaperones Assist the Assembly of Some Proteins

Based on the ability of some denatured proteins to return to their original configuration and to regain biological function, biologists initially assumed that proteins and protein-containing structures self-assemble in cells as well. As illustrated in Figure 2-18b for ribonuclease, protein synthesis takes place on ribosomes. The polypeptide elongates as successive amino acids are added to one end. The self-assembly model envisions polypeptide chains coiling and folding spontaneously and progressively as polypeptides are synthesized in this way.

By the time the fully elongated polypeptide is released from the ribosome, it is thought to have attained a stable, predictable, three-dimensional structure without any input

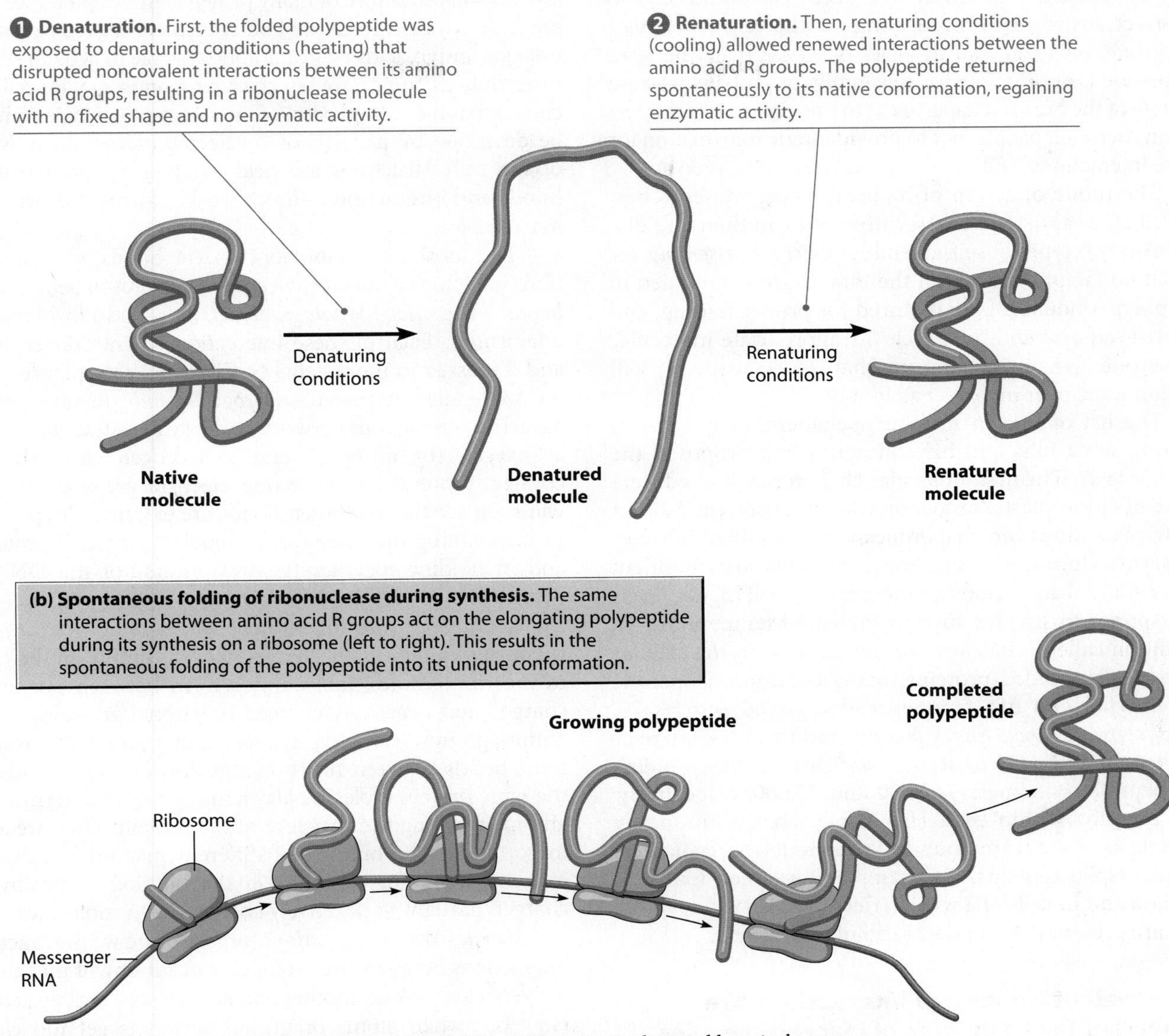

(a) Spontaneous refolding of ribonuclease following denaturation. Anfinsen's experiment showed that all the information needed for the proper folding of a ribonuclease polypeptide into its native three-dimensional conformation is present in its amino acid sequence.
1 Denaturation. First, the folded polypeptide was exposed to denaturing conditions (heating) that disrupted noncovalent interactions between its amino acid R groups, resulting in a ribonuclease molecule with no fixed shape and no enzymatic activity.
2 Renaturation. Then, renaturing conditions (cooling) allowed renewed interactions between the amino acid R groups. The polypeptide returned spontaneously to its native conformation, regaining enzymatic activity.
Denaturing conditions
Renaturing conditions
Native molecule
Denatured molecule
Renatured molecule
(b) Spontaneous folding of ribonuclease during synthesis. The same interactions between amino acid R groups act on the elongating polypeptide during its synthesis on a ribosome (left to right). This results in the spontaneous folding of the polypeptide into its unique conformation.
Completed polypeptide
Growing polypeptide
Ribosome
Messenger RNA

FIGURE 2-18 The Spontaneity of Polypeptide Folding. The protein depicted here is the enzyme ribonuclease, shown **(a)** as used in Anfinsen's classic in vitro protein folding experiments and **(b)** during in vivo synthesis.

of energy or information beyond the basic polymerization process. Furthermore, folding is assumed to be unique, in the sense that each polypeptide with the same amino acid sequence will fold in an identical, reproducible manner under the same conditions. Thus, the self-assembly model assumes that interactions occurring within and between polypeptides are all that is necessary for the biogenesis of proteins in their functional forms.

However, this model for self-assembly in vivo (in the cell) is based entirely on studies with isolated proteins; and, even under laboratory conditions, not all proteins regain their native structure. Based on their work with one such protein (*ribulose bisphosphate carboxylase/oxygenase,* the chloroplast enzyme that catalyzes the conversion of CO_2 to sugar in the process of photosynthesis), John Ellis and his colleagues at Warwick University concluded that the self-assembly model may not be adequate for all proteins. They proposed that the interactions that drive protein folding may need to be assisted and controlled by proteins known as **molecular chaperones** to reduce the probability of the formation of incorrect structures having no biological activity.

Molecular chaperones are proteins that facilitate the correct assembly of proteins and protein-containing structures but are not components of the assembled structures. The molecular chaperones that have been identified to date do not convey information either for polypeptide folding or for the assembly of multiple polypeptides into a single protein. Instead, they function by binding to specific regions that are exposed only in the early stages of assembly, thereby inhibiting unproductive assembly pathways that would lead to incorrect structures. Commenting on the term *molecular chaperone,* Ellis and Van der Vies observed that "the term chaperone is appropriate for this family of proteins because the role of the human chaperone is to prevent incorrect interactions between people, not to provide steric information for those interactions" (Ellis and Van der Vies, 1991, p. 351).

The mode of action of molecular chaperones is best described as **assisted self-assembly.** We can therefore distinguish two types of self-assembly: *strict self-assembly,* for which no factors other than the information contained in the polypeptide itself are required for proper folding; and *assisted self-assembly,* in which the appropriate molecular chaperone is required to ensure that correct assembly will predominate over incorrect assembly.

The list of known molecular chaperones has grown steadily since Ellis and his colleagues first proposed the term in 1987. The first molecular chaperones studied were those of chloroplasts, mitochondria, and bacteria. Within a few years, however, chaperones were identified in other eukaryotic locations. Chaperone proteins are abundant under normal conditions and increase to still higher levels in response to stresses such as increased temperature—a condition called *heat shock*—or an increase in the cellular content of unfolded proteins. Many common chaperone proteins fall into two families, called *Hsp60* and *Hsp70.* "Hsp" stands for *heat-shock protein,* and the numbers refer to the approximate molecular weights of the protein's polypeptide monomers—60,000 and 70,000, respectively. The proteins within each Hsp family are evolutionarily related, and they are found throughout the biological world. Hsp70 proteins, for example, have been found in bacteria and in cells of a wide variety of eukaryotes, where they are present in several intracellular locations.

MB Activity: Ionic Bonds; Nonpolar and Polar Molecules

Noncovalent Bonds and Interactions Are Important in the Folding of Macromolecules

Whether assisted by molecular chaperones or not, polypeptides fold and self-assemble without the input of further information. Also, the three-dimensional structure of a protein or protein complex is remarkably stable once it has been attained. To understand the self-assembly of proteins (and, as it turns out, of other biological molecules and structures as well), we need to consider the covalent as well as noncovalent bonds that hold polypeptides and other macromolecules together.

Covalent bonds are easy to understand. Every protein or other macromolecule in the cell is held together by strong covalent bonds such as those discussed earlier in the chapter. A covalent bond forms whenever two atoms *share* electrons rather than gaining or losing electrons completely. Electron sharing is an especially prominent feature of the carbon atom, which has four electrons in its outer orbital and is therefore at the midpoint between the tendency to gain or lose electrons. Covalent bonds link the monomers of a polypeptide together; they also can stabilize the three-dimensional structure of many proteins. In Chapter 3 we will see how covalent disulfide bonds formed between pairs of cysteine amino acids play an important role in protein structure. But, as important as covalent bonds are in cellular chemistry, the complexity of molecular structure cannot be described by patterns of covalent bonding alone. Most of the cell's structures are held together by **noncovalent bonds and interactions**—much weaker forces that occur in macromolecules.

The most important noncovalent bonds and interactions in biological macromolecules are *hydrogen bonds, ionic bonds, van der Waals interactions,* and *hydrophobic interactions.* Each of these interactions is introduced here and discussed in more detail in Chapter 3 (see Figure 3-5). As we've already noted, *hydrogen bonds* involve weak, attractive interactions between an electronegative atom such as oxygen (or nitrogen) and a hydrogen atom that is covalently bonded to a second electronegative atom. We will soon see that hydrogen bonds are extremely important in maintaining the three-dimensional structure of proteins and in holding together the two strands of the DNA d ouble helix.

Ionic bonds are noncovalent electrostatic interactions between two oppositely charged ions. Typically, in the case of macromolecules, ionic bonds form between positively charged and negatively charged functional groups such as amino groups, carboxyl groups, and phosphate groups. Ionic bonds between functional groups of amino acids in the same protein molecule play a major role in determining and maintaining the structure of the protein. They are also important in the binding of different macromolecules to each other as, for example, in the binding of positively charged proteins to negatively charged DNA molecules.

Van der Waals interactions (or *forces*) are weak attractive interactions between two atoms that occur only if the atoms are very close to one another and are oriented appropriately. However, when atoms or groups of atoms get too close together, they begin to repel each other because of their overlapping outer electron orbitals. The *van der Waals radius* of a specific atom specifies the "private space" around it that limits how close other atoms can come. This radius is typically about 0.12 to 0.19 nm. Van der Waals radii provide the basis for *space-filling models* of biological macromolecules, such as that for the protein insulin shown in **Figure 2-19** and the DNA molecule shown in Figure 3-19b.

The term *hydrophobic interactions* describes the tendency of nonpolar groups within a macromolecule to associate with each other as they minimize their contact with surrounding water molecules and with any hydrophilic groups in the same or another macromolecule. Hydrophobic interactions among nonpolar groups are common in proteins and will cause nonpolar groups to be found in the interior of a protein molecule or imbedded in the nonpolar interior of a membrane (see Figure 2-13b).

FIGURE 2-19 A Space-Filling Model of Insulin. All of the atoms and chemical groups in this model of the protein insulin are represented as spheres with the appropriate van der Waals radii. Notice how tightly the atoms pack against one another. (Color key: C = red, H = white, O = blue, S = yellow.)

Self-Assembly Also Occurs in Other Cellular Structures

The same principle of self-assembly that accounts for the folding and interactions of polypeptides also applies to more complex cellular structures. Many of the characteristic structures of the cell are complexes of two or more different kinds of polymers and involve interactions that are chemically distinct from those of polypeptide folding and association. However, the principle of self-assembly may nonetheless apply. For example, *ribosomes* contain both RNA and proteins, and *membranes* are made up of both phospholipids and proteins. The *primary cell wall* that surrounds a plant cell, although composed mainly of cellulose fibrils, also contains a variety of other carbohydrates as well as a small but crucial protein component. Yet, despite the chemical differences among such polymers as proteins, nucleic acids, and polysaccharides, the noncovalent interactions that drive these supramolecular assembly processes are similar to those that dictate the folding of individual protein molecules.

The Tobacco Mosaic Virus Is a Case Study in Self-Assembly

Some of the most definitive findings concerning the self-assembly of complex biological structures have come from studies of viruses. As we will learn in Chapter 4, a *virus* is a complex of proteins and nucleic acid, either DNA or RNA. A virus is not itself alive; but it can invade and infect a living cell, take over the synthetic machinery of the cell, and use the cell to produce more virus components. When the components—viral nucleic acid and viral proteins—are synthesized, these macromolecules spontaneously assemble into the mature virus particles.

An especially good example is tobacco mosaic virus (TMV), a plant virus that molecular biologists have studied for many years. TMV is a rodlike particle about 18 nm in diameter and 300 nm in length. It consists of a single strand of RNA with 6395 nucleotides and about 2130 copies of a single kind of polypeptide—the *coat protein,* each containing 158 amino acids. The RNA molecule forms a helical core, with a cylinder of protein subunits clustered around it (**Figure 2-20**).

Heinz Fraenkel-Conrat and his colleagues carried out several important experiments that contributed significantly to our understanding of self-assembly. They separated TMV into its RNA and protein components and then allowed them to reassemble in vitro. Functional, infectious viral particles formed spontaneously, providing one of the first and most convincing demonstrations that the components of a complex biological structure can reassemble spontaneously without external information. Especially interesting was the finding that mixing the RNA from one strain of virus strain with the protein from another formed a hybrid virus that was also infectious.

The assembly process has since been studied in detail and is known to be surprisingly complex (**Figure 2-21**). The basic unit of assembly is a two-layered disk of coat protein that undergoes a conformational change to a helical shape as it interacts with a short segment of the RNA molecule. This transition allows another disk to bind, and each successive disk undergoes the same conformational change from a cylinder to a helix as it binds another segment of RNA. The process continues until the end of the RNA molecule is reached, producing the mature virus particle, its RNA completely covered with coat protein.

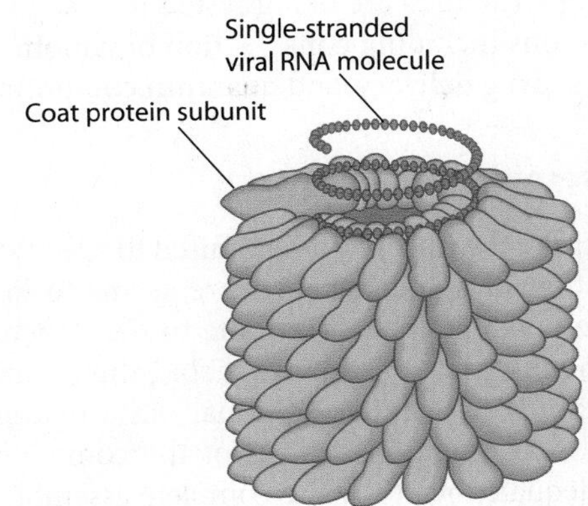

FIGURE 2-20 A Structural Model of Tobacco Mosaic Virus (TMV). A single-stranded RNA molecule is coiled into a helix, surrounded by a coat consisting of 2130 identical protein subunits. Only a portion of the entire TMV virion is shown here, and several layers of protein have been omitted from the upper end of the structure to reveal the helical RNA molecule inside.

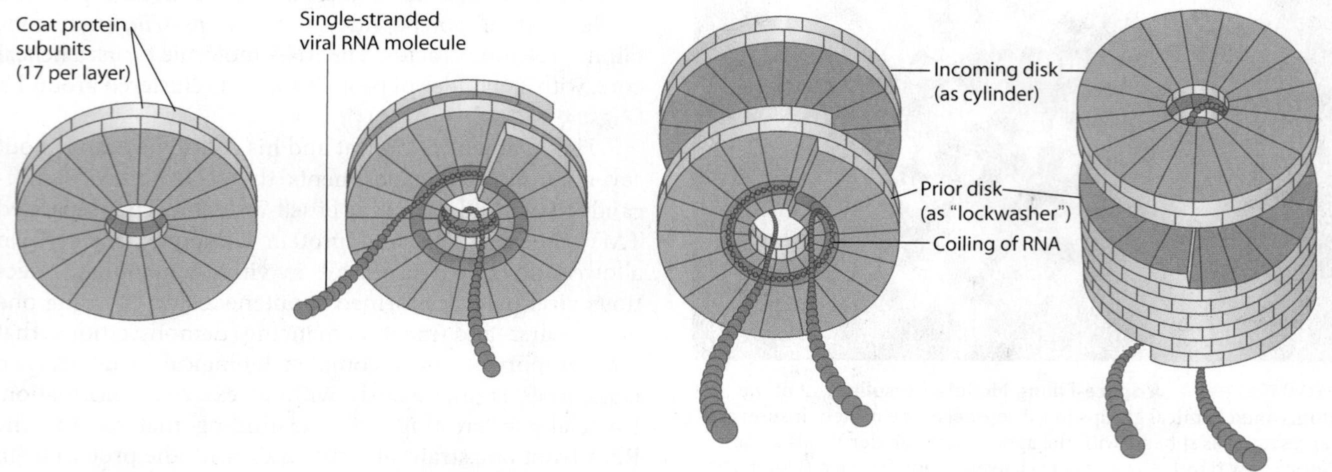

(a) The unit of assembly of the protein coat of TMV consists of 34 identical coat protein subunits that spontaneously assemble to form a double-layered disk with 17 subunits per layer.
(b) The viral RNA molecule associates with a disk, causing the RNA molecule to form a loop of 102 nucleotides as the disk changes from a cylindrical to a helical conformation.
(c) Another disk is then added, and another loop of the RNA associates with it. The prior disk, meanwhile, has changed from a cylindrical to a helical conformation that resembles a lockwasher.
(d) Each incoming disk adds two more layers of protein subunits and associates with another loop of RNA, which forms into a spiral. The process continues until the entire RNA molecule is covered and the virion is complete.
Coat protein subunits (17 per layer)
Single-stranded viral RNA molecule
Incoming disk (as cylinder)
Prior disk (as "lockwasher")
Coiling of RNA

FIGURE 2-21 Spontaneous Self-Assembly of the Tobacco Mosaic Virus (TMV) Virion. Parts **(a)** through **(d)** show how the TMV virion will assemble spontaneously from a mixture of its coat protein and RNA constituents. All the information needed for proper assembly is contained in the proteins and RNA themselves, and no energy input is required.

Recently, scientists have begun exploiting the ability of viral particles to self-assemble in order to produce custom-made synthetic molecular complexes for use in nanotechnology and biomedicine. Individual coat protein molecules can be modified or engineered to have specific properties, such as electrical conductivity, then allowed to self-assemble into predictable structures. These self-assembling structures are being tested for use in a variety of applications including construction of nanometer-sized biosensors, drug delivery, and quantum computing.

Self-Assembly Has Limits

In many cases, the information required to specify the exact conformation of a cellular structure seems to lie entirely within the polymers that contribute to the structure. Such self-assembling systems achieve stable three-dimensional conformations without additional information input because the information content of the component polymers is adequate to specify the complete assembly process. Even in assisted self-assembly, the molecular chaperones that are involved provide no additional information.

However, there appear to be some assembly systems that depend, additionally, on information supplied by a preexisting structure. In such cases, the ultimate structure arises not by assembling the components into a new structure but rather by ordering the components into the matrix of an existing structure. Examples of cellular structures that are routinely built up by adding new material to existing structures include membranes, cell walls, and chromosomes.

On the other hand, such structures are not yet sufficiently characterized to determine whether the presence of a preexisting structure is obligatory or whether, under the right conditions, the components might be capable of self-assembly. Evidence from studies with artificial membranes and with chromatin (isolated chromosomal components), for example, suggests that a preexisting structure, though routinely present in vivo, may not be an indispensable requirement for the assembly process. Additional insight will be necessary before we can say with certainty whether, and to what extent, external information is required or exploited in cellular assembly processes.

Hierarchical Assembly Provides Advantages for the Cell

Each of the assembly processes we have examined exemplifies the basic cellular strategy of **hierarchical assembly** illustrated in Figure 2-14. Biological structures are almost always constructed hierarchically, with subassemblies acting as important intermediates en route from simple starting

BOX 2A DEEPER INSIGHTS

Tempus Fugit and the Fine Art of Watchmaking

"Drat! Another defective watch!" With a look of disgust, Tempus Fugit the watchmaker tossed the faulty timepiece into the wastebasket and grumbled to himself, "That's two out of the last three watches I've had to throw away. What kind of a watchmaker am I, anyway?"

"A good question, Fugit," came a voice from the doorway. Tempus looked up to see Caveat Emptor entering the shop. "Maybe I can help you with it if you'll tell me a bit about how you make your watches."

"Nobody asked for your help, Emptor," growled Tempus testily, wishing fervently he hadn't been caught thinking out loud.

"Ah well, you'll get it just the same," continued Caveat, quite unperturbed. "Now tell me exactly how you make a watch and how long it takes you. I'd be especially interested in comparing your procedure with the way Pluribus Unum does it in his new shop down the street."

At the very mention of his competitor's name, Tempus groaned again. "Unum!" he blustered. "What does he know about the fine art of watchmaking?"

"A good deal, apparently—probably more than you do," replied Caveat. "But tell me exactly how you go about it. How many steps does it take you per watch, and how often do you make a mistake?"

"It takes exactly 100 operations to make a watch. Every step has to be done exactly right, or the watch won't work. It's tricky business, but I've got my error rate down to 1%," said Tempus with at least a trace of pride in his voice. "I can make a watch from start to finish in exactly one hour, but I only make 36 watches each week because I always take Tuesday afternoons off to play darts."

"Well now, let's see," said Caveat, as he pulled out a pocket calculator. "That means, my good Mr. Fugit, that you only make about 13 watches per week that actually work. All the rest you have to throw away just like the one you pitched as I entered your shop."

"How did you know that, you busybody?" Tempus asked defensively, wondering how this know-it-all with his calculator had guessed his carefully kept secret.

"Elementary, my dear Watson," returned Caveat gleefully. "Simple probability is all it takes. You told me that each watch requires 100 operations and that there's a 99% chance that you'll get a given step right. That's 0.99 times itself 100 times, which comes out to 0.366. So only about 37% of your watches will be put together right, and you can't even tell anything about a given watch until it's finished and you test it. Want to know how that compares with Unum's shop?"

Tempus was about to protest, but his tormentor hurried on with scarcely a pause for breath. "He can manage 100 operations per hour, too, and his error rate is exactly the same as yours. He's also off every Tuesday afternoon, but he gets about 27 watches made every week. That's twice your output, Fugit! No wonder he's got so many watches in his window and so many customers at his door. I've heard that he's even thinking of expanding his shop. Want to know how he does it?"

Tempus was too depressed to even attempt a protest. And besides, although he wouldn't like to admit it, he really was dying to know how Unum managed it.

"*Subunit assembly, Fugit, that's the answer! Subunit assembly!* Instead of making each watch from scratch in 100 separate steps, Unum assembles the components into 10 pieces, each requiring 10 steps." Caveat began pushing calculator buttons again. "Let's see, he performs 10 operations with a 99% success rate for each, so I take 0.99 to the tenth power instead of the hundredth. That's 0.904, which means that about 90% of his subunits have no errors in them. So he spends about 33 hours each week making 330 subunits, throws the defective ones away, and still has about 300 to assemble into 30 watches during the last three hours on Friday afternoon. That takes 10 steps per watch, with the same error rate as before, so again he comes up with about 90% success. He has to throw away about 3 of his finished watches and ends up with about 27 watches to show for his efforts. Meanwhile, you've been working just as hard and just as accurately, yet you only make half as many watches. What do you have to say to that, my dear Fugit?"

Tempus managed a weary response. "Just one question, Emptor, and I'll probably hate myself for asking. But do you happen to know what Unum does on Tuesday afternoons?"

"I'm glad you asked," replied Caveat as he slipped his calculator back into his pocket and headed for the door. "But I don't think you'll be able to interest him in darts—he spends every Tuesday afternoon giving watchmaking lessons."

molecules to the end products of organelles, cells, and organisms. Consider how cellular structures are made. First, large numbers of similar, or even identical, monomeric subunits are assembled by condensation into polymers. These polymers then aggregate spontaneously but specifically into characteristic multimeric units. The multimeric units can, in turn, give rise to still more complex structures and eventually to assemblies that are recognizable as distinctive subcellular structures.

This hierarchical process has the double advantage of chemical simplicity and efficiency of assembly. To appreciate the chemical simplicity, we need only recognize that almost all structures found in cells and organisms are synthesized from about 30 small precursor molecules, which George Wald has called the "alphabet of biochemistry." This alphabet includes the 20 amino acids that are found in proteins, the 5 aromatic bases present in nucleic acids, 2 sugars, and 3 lipid molecules. We will encounter each of these in Chapter 3 (see Table 3-1, page 42). Given these building blocks and the polymers that can be derived from them through just a few different kinds of condensation reactions, most of the structural complexity of life can be readily elaborated by hierarchical assembly into successively more complex structures.

The second advantage of hierarchical assembly lies in the "quality control" that can be exerted at each level of assembly, allowing defective components to be discarded at an early stage rather than being built into a more complex structure that would be more costly to reject and replace (see the story in **Box 2A**). Thus, if the wrong subunit becomes inserted into a polymer at some critical point in the chain, that particular molecule may have to be discarded; but the cell will be spared the cost of synthesizing a more complicated supramolecular assembly or even an entire organelle before the defect is discovered. As we proceed through this course, we will often see how cells have exquisite means for ensuring the proper structure and function of cellular components; we will also see how cells can recognize and replace components that are defective.

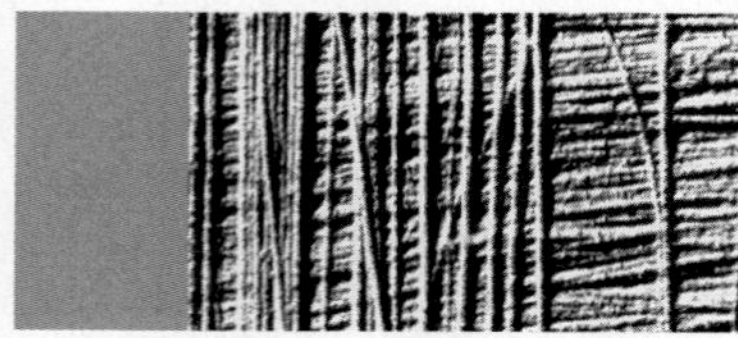

SUMMARY OF KEY POINTS

The Importance of Carbon

- The carbon atom has several properties that make it uniquely suited as the basis for life. Each atom forms four stable covalent bonds and can participate in single, double, or triple bonds with one or more other carbon atoms. This creates a wide variety of linear, branched, and ring-containing compounds.
- Carbon also readily bonds to several other types of atoms commonly found in cellular compounds, including hydrogen, oxygen, nitrogen, phosphorus, and sulfur. These elements are frequently found in functional groups attached to carbon skeletons—groups such as the hydroxyl, sulfhydryl, carboxyl, amino, phosphate, carbonyl, and aldehyde groups.

The Importance of Water

- The unique chemical properties of water help to make it the most abundant compound in cells. These properties include its cohesiveness, its high specific heat, its temperature-stabilizing capacity, and its ability to act as a solvent for most biological molecules.
- These chemical properties are a result of the polarity of the water molecule. Its polarity arises from the unequal sharing of electrons in the oxygen-hydrogen covalent bonds and its bent shape, causing the oxygen end of the atom to be slightly negative and the hydrogen atoms to be slightly positive.
- This polarity leads to extensive hydrogen bonding between water molecules, as well as between water and other polar molecules, rendering them soluble in aqueous solutions.

The Importance of Selectively Permeable Membranes

- All cells are surrounded by a defining cell membrane that is selectively permeable, controlling the flow of materials into and out of the cell.
- Biological membranes have a phospholipid bilayer structure with a hydrophobic interior that blocks the direct passage of large polar molecules, charged molecules, and ions.
- However, large polar and charged molecules and ions can cross membranes via membrane-spanning transport proteins that form hydrophilic channels to allow passage of specific molecules.

The Importance of Synthesis by Polymerization

- The molecular building blocks of the cell are small organic molecules put together by stepwise polymerization to form the macromolecules so important to cellular structure and function. Relatively few kinds of monomeric units comprise most of the polysaccharide, protein, and nucleic acid polymers in cells.
- Polymerization of monomers requires that they first be activated, usually at the expense of ATP. Monomers are then linked together by removal of water in condensation reactions. Polymers are degraded by the reverse reaction—hydrolysis.
- Because they are added to only one specific end of these macromolecules, all biological polymers have directionality.

The Importance of Self-Assembly

- While energy input is required for polymerization of monomers into macromolecules, most macromolecules fold into their final three-dimensional conformations spontaneously. The information needed for this folding is inherent in the chemical nature of the monomeric units and the order in which the monomers are put together.
- For example, the unique three-dimensional conformation of a protein forms by spontaneous folding of the linear polypeptide chain, and this conformation depends only on the specific order of amino acids in the protein. The final structure is the result of several covalent and noncovalent interactions, including disulfide bonding, hydrogen bonding, ionic bonding, van der Waals interactions, and hydrophobic interactions.
- Individual polymers can interact with each other in a unique and predictable manner to generate successively more complex structures. This hierarchical assembly process has the dual advantages of chemical simplicity and efficiency of assembly. It also allows quality control at multiple steps to ensure proper production of cell components.

 **www.masteringbiology.com**

1. MasteringBiology Assignments

Tutorials and Activities

- Molecular Model Structural Tutorials for visualization of 3D structures
- BioFlix Tutorials including 3D animations with hints and detailed feedback for common wrong answers
- Covalent Bonds; Functional Groups; Carbon Bonding and Functional Groups; Water: The Water Molecule; The Polarity of Water; Water: Hydrogen Bonding; Hydrogen Bonds; Hydrogen Bonding and Water; Water: Properties of Water; Water: Solutions; Hydrophobic versus Hydrophilic; Condensation and Hydrolysis Reactions; Ionic Bonds; Nonpolar and Polar Molecules

Questions

- Reading Quiz
- Multiple Choice
- End-of-chapter questions

2. eText Read your book online, search, highlight text, take notes, and more

3. The Study Area

- Chapter Review Quiz, Guide to Techniques and Methods, 3D-Structure Tutorials, Activities, BioFlix, Videos, Biology Review, Word Study Tools, Glossary, Flashcards

PROBLEM SET

More challenging problems are marked with a •.

2-1 The Fitness of Carbon. Each of the properties that follow is characteristic of the carbon atom. In each case, indicate how the property contributes to the role of the carbon atom as the most important atom in biological molecules.

(a) The carbon atom has a valence of four.

(b) The carbon-carbon bond has a bond energy that is above the energy of photons of light in the visible range (400–700 nm).

(c) Carbon is one of the lightest elements to form a covalent bond.

(d) Carbon can form single, double, and triple bonds.

(e) The carbon atom is a tetrahedral structure.

2-2 The Fitness of Water. For each of these statements about water, decide whether the statement is true and describes a property that makes water a desirable component of cells (T); is true but describes a property that has no bearing on water as a cellular constituent (X); or is false (F). For each true statement, indicate a possible benefit to living organisms.

(a) Water is a polar molecule and hence an excellent solvent for polar compounds.

(b) Water can be formed by the reduction of molecular oxygen (O_2).

(c) The density of water is less than the density of ice.

(d) The molecules of liquid water are extensively hydrogen-bonded to one another.

(e) Water does not absorb visible light.

(f) Water is odorless and tasteless.

(g) Water has a high specific heat.

(h) Water has a high heat of vaporization.

2-3 Wrong Again. For each of the following false statements, change the statement to make it true.

(a) Carbon and hydrogen can form a strong double bond.

(b) Water has a higher specific heat than most other liquids because of its low molecular weight.

(c) Oil droplets in water coalesce to form a separate phase because of the strong attraction of hydrophobic molecules for each other.

(d) Most small organic compounds found in biological cells are hydrophobic.

(e) Biological membranes are freely permeable.

2-4 Solubility in Benzene. Consider the chemical structure of benzene that is shown in Figure 2-4. Which cellular molecules would you expect to be soluble in a container of benzene?

2-5 Drug Targeting. Your company has developed a new anti-cancer drug, but it will not cross through the membrane of target cells because it is large and contains many polar functional groups. Can you design a strategy to get this new drug into the cancer cells?

2-6 It's All About Membranes. Answer each of the following questions about biological membranes in 50 words or less:

(a) What is an amphipathic molecule? Why are amphipathic molecules important constituents of membranes?

(b) What does it mean to say that a membrane is selectively permeable?

(c) Why do membranes consist of lipid bilayers rather than lipid monolayers?

(d) If a lipid bilayer is at least 10^8 times less permeable to K^+ ions than it is to water, how do the K^+ ions that are needed for many cellular functions get into the cell?

(e) It is often possible to predict the number of times a membrane protein crosses the membrane by determining how many short sequences of hydrophobic amino acids are present in the amino acid sequence of the protein. Explain.

(f) Why are the several carbohydrate side chains attached to hydrophilic rather than hydrophobic regions of the membrane protein shown in Figure 2-13b?

2-7 The Polarity of Water. Defend the assertion that all of life as we know it depends critically on the fact that the angle between the two hydrogen atoms in the water molecule is 104.5° and not 180°.

2-8 The Principle of Polymers. Polymers clearly play an important role in the molecular economy of the cell. For each statement below, state why it is false and change it to a correct description.

(a) Polymers are assembled from monomers in an extracellular compartment and are transported into the cell when required.

(b) Polysaccharides are one of the three main macromolecular polymers in the cell. A polysaccharide molecule contains a number of different monomers, which give rise to millions of polysaccharide sequences.

• **2-9 TMV Assembly.** Each of these statements is an experimental observation concerning the reassembly of tobacco mosaic virus (TMV) virions from TMV RNA and coat protein subunits. In each case, carefully state a reasonable conclusion that can be drawn from the experimental finding.

(a) When RNA from a specific strain of TMV is mixed with coat protein from the same strain, infectious virions are formed.

(b) When RNA from strain A of TMV is mixed with coat protein from strain B, the reassembled virions are infectious, giving rise to strain A virus particles in the infected tobacco cells.

(c) Isolated coat protein monomers can polymerize into a viruslike helix in the absence of RNA.

(d) In infected plant cells, the TMV virions that form contain only TMV RNA and never any of the various kinds of cellular RNAs present in the host cell.

(e) Regardless of the ratio of RNA to coat protein in the starting mixture, the reassembled virions always contain RNA and coat protein in the ratio of three nucleotides of RNA per coat protein monomer.

• **2-10 Mars Is Alive?** Imagine this futuristic scenario: Life has been discovered on Mars and shown to contain a new type of macromolecule, named marsalive. You have been hired to study this new compound and want to determine whether marsalive is a structural or an informational macromolecule. What features would you look for?

SUGGESTED READING

References of historical importance are marked with a •.

General References and Reviews

Bhagavan, N. Y. *Medical Biochemistry.* San Diego: Harcourt/Academic Press, 2002.

Horton, H. R., et al. *Principles of Biochemistry,* 4th ed. Upper Saddle River, NJ: Pearson Prentice Hall, 2006.

Mathews, C. K., K. E. van Holde, and K. G. Ahern. *Biochemistry,* 3rd ed. Menlo Park, CA: Benjamin/Cummings, 2000.

Nelson, D. L., and M. M. Cox. *Lehninger Principles of Biochemistry,* 5th ed. New York: W. H. Freeman, 2008.

The Importance of Water

Bryant, R. G. The dynamics of water-protein interactions. *Annu. Rev. Biophys. Biomol. Struct.* 25 (1996): 29.

Mathews, R. Wacky water. *New Scientist* 154 (June 21, 1997): 40.

Westof, E., ed. *Water and Biological Macromolecules.* Boca Raton, FL: CRC Press, 1993.

The Importance of Membranes

Baldwin, S. A. *Membrane Transport: A Practical Approach.* Oxford: Oxford University Press, 2000.

Mellman, I., and G. Warren. The road taken: Past and future foundations of membrane traffic. *Cell* 100 (2000): 99.

Rees, D. C. *Membrane Proteins.* Boston: Academic Press, 2003.

Tien, H. T., and A. Ottova-Leitmannova. *Membrane Biophysics: As Viewed from Experimental Bilayer Lipid Membranes.* New York: Elsevier, 2000.

The Importance of Macromolecules

Creighton, T. E. *Proteins: Structure and Molecular Properties,* 2nd ed. New York: W. H. Freeman, 1993.

Ingber, D. E. The architecture of life. *Sci. Amer.* 278 (January 1998): 48.

Jardetzky, O., and M. D. Finucane. *Dynamics, Structure, and Function of Biological Macromolecules.* Washington, DC: IOS Press, 2001.

Jewett, M. C., K. A. Calhoun, A. Voloshin, J. J. Wuu, and J. R. Schwartz. An integrated cell-free metabolic platform for protein production and synthetic biology. *Mol. Syst. Biol.* 4 (2008): 220.

Petsko, G. A., and D. Ringe. *Protein Structure and Function.* Sunderland, MA: Sinauer Assoc., 2004.

Yon, J. M. Protein folding: Concepts and perspectives. *Cell. Mol. Life Sci.* 53 (1997): 557.

The Importance of Self-Assembly

Baker, D. A surprising simplicity to protein folding. *Nature* 405 (2000): 39.

Broadley, S. A., and F. U. Hartl. The role of molecular chaperones in human misfolding diseases. *FEBS Lett.* 583 (2009): 2647.

Creighton, T. E. Protein folding: An unfolding story. *Curr. Biol.* 5 (1995): 353.

• Ellis, R. J. Discovery of molecular chaperones. *Cell Stress Chaperones* 1 (1996): 155.

Ellis, R. J., and S. M. Van der Vies. Molecular chaperones. *Annu. Rev. Biochem.* 60 (1991): 321.

• Fraenkel-Conrat, H., and R. C. Williams. Reconstitution of active tobacco mosaic virus from its inactive protein and nucleic acid components. *Proc. Natl. Acad. Sci. USA* 41 (1955): 690.

Horwich, A. Two families of chaperonin: Physiology and mechanism. *Annu. Rev. Cell Dev. Biol.* 23 (2007): 115.

Ringler, P., and G. E. Schulz. Self-assembly of proteins into designed networks. *Science* 302 (2003): 106.

Steinmetz, N. F., T. Lin, G. P. Lomonossoff, and J. E. Johnson. Structure-based engineering of an icosahedral virus for nanomedicine and nanotechnology. *Curr. Top. Microbiol. Immunol.* 327 (2009): 23.

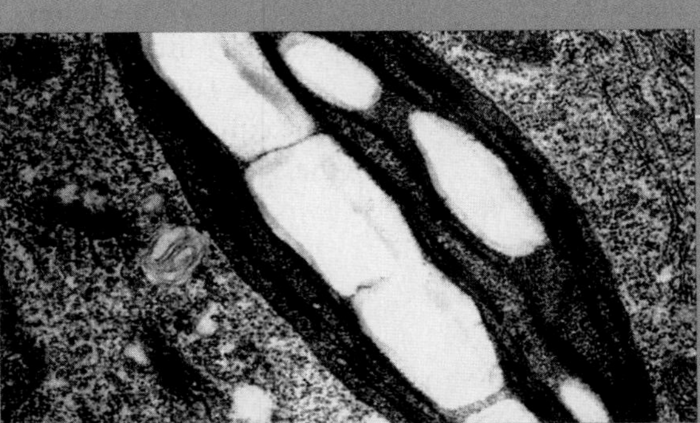

3 The Macromolecules of the Cell

See MasteringBiology® for tutorials, activities, and quizzes.

In Chapter 2, we looked at some of the basic chemical principles of cellular organization. We saw that each of the major kinds of biological macromolecules—proteins, nucleic acids, and polysaccharides—consists of a relatively small number (from 1 to 20) of repeating monomeric units. These polymers are synthesized by condensation reactions in which activated monomers are linked together by the removal of water. Once synthesized, the individual polymer molecules fold and coil spontaneously into stable, three-dimensional shapes. These folded molecules then associate with one another in a hierarchical manner to generate higher levels of structural complexity, usually without further input of energy or information.

We are now ready to examine the major kinds of biological macromolecules. In each case, we will focus first on the chemical nature of the monomeric components and then on the synthesis and properties of the polymer itself. As we will see shortly, most biological macromolecules in cells are synthesized from about 30 common small molecules. We begin our survey with proteins because they play such important and widespread roles in cellular structure and function. We then move on to nucleic acids and polysaccharides. The tour concludes with lipids, which do not quite fit the definition of a polymer but are important cellular components whose synthesis resembles that of true polymers.

Proteins

Chemistry Review-Proteins: Functions of Proteins

Proteins are a class of extremely important and ubiquitous macromolecules in all organisms, occurring nearly everywhere in the cell. In fact, their importance is implied by their name, which comes from the Greek word *proteios*, meaning "first place." Whether we are talking about conversion of carbon dioxide to sugar in photosynthesis, oxygen transport in the blood, the regulation of gene expression by transcription factors, cell-to-cell communication, or the motility of a flagellated bacterium, we are dealing with processes that depend crucially on particular proteins with specific properties and functions.

Based on function, proteins fall into nine major classes. Many proteins are *enzymes*, serving as catalysts that greatly increase the rates of the thousands of chemical reactions on which life depends. *Structural proteins*, on the other hand, provide physical support and shape to cells and organelles, giving them their characteristic appearances. *Motility proteins* play key roles in the contraction and movement of cells and intracellular structures. *Regulatory proteins* are responsible for control and coordination of cellular functions, ensuring that cellular activities are regulated to meet cellular needs. *Transport proteins* are involved in the movement of other substances into, out of, and within the cell. *Hormonal proteins* mediate communication between cells in distant parts of an organism, and *receptor proteins* enable cells to respond to chemical stimuli from their environment. Finally, *defensive proteins* provide protection against disease, and *storage proteins* serve as reservoirs of amino acids.

Because virtually everything that a cell is or does depends on the proteins it contains, it is clear we need to understand what proteins are and why they have the properties they do. We begin our discussion by looking at the amino acids present in the proteins, and then we will consider some properties of proteins themselves.

The Monomers Are Amino Acids

Proteins are linear polymers of **amino acids.** Although more than 60 different kinds of amino acids are typically present in a cell, only 20 kinds are used in protein synthesis, as indicated in **Table 3-1**. Some proteins contain more than 20 different kinds of amino acids, but the additional ones usually are a result of modifications that occur after the protein has been synthesized. Although most proteins contain all or most of the 20 amino acids, the

Table 3-1	Common Small Molecules in Cells			
Kind of Molecules	**Number Present**	**Names of Molecules**	**Role in Cell**	**Figure Number for Structures**
Amino acids	20	See list in Table 3-2.	Monomeric units of all proteins	3-2
Aromatic bases	5	Adenine Cytosine Guanine Thymine Uracil	Components of nucleic acids	3-15
Sugars	varies	Ribose Deoxyribose Glucose	Component of RNA Component of DNA Energy metabolism; component of starch and glycogen	3-15 3-21
Lipids	varies	Fatty acids Cholesterol	Components of phospholipids and membranes	3-27a 3-27e

Source: Adapted from Wald (1994).

proportions vary greatly, and no two different proteins have the same amino acid sequence.

Every amino acid has the basic structure illustrated in **Figure 3-1**, with a carboxyl group, an amino group, a hydrogen atom, and a side chain known as an R group. All are attached to a central carbon atom known as the *α carbon.* Except for glycine, for which the R group is just a hydrogen atom, all amino acids have an asymmetric α carbon atom and therefore exist in two stereoisomeric forms, called D- and L-amino acids (see Figures 2-6 and 2-7). Both kinds exist in nature, but only L-amino acids occur in proteins.

Because the carboxyl and amino groups shown in Figure 3-1 are common features of all amino acids, the specific properties of the various amino acids vary depending on the chemical nature of their R groups, which range from a single hydrogen atom to relatively complex aromatic groups. Shown in **Figure 3-2** are the structures of the 20 L-amino acids found in proteins. The three-letter abbreviations given in parentheses for each amino acid are widely used by biochemists and molecular biologists. **Table 3-2** lists these three-letter abbreviations and the corresponding one-letter abbreviations that are also commonly used.

Nine of these amino acids have nonpolar, *hydrophobic* R groups (Group A). As you look at their structures, you will notice the hydrocarbon nature of the R groups, with few or no oxygen and nitrogen atoms. These hydrophobic amino acids are usually found in the interior of the molecule as a polypeptide folds into its three-dimensional shape. If a protein (or a region of the molecule) has a preponderance of hydrophobic amino acids, the whole protein (or the hydrophobic portion of the molecule) will be excluded from aqueous environments and will instead be found in hydrophobic locations, such as the interior of a membrane.

The remaining 11 amino acids have *hydrophilic* R groups that are either distinctly polar (Group B) or actually charged at the pH values characteristic of cells (Group C). Notice that the two acidic amino acids are negatively charged, and the three basic amino acids are positively charged. Hydrophilic amino acids tend to occur on the

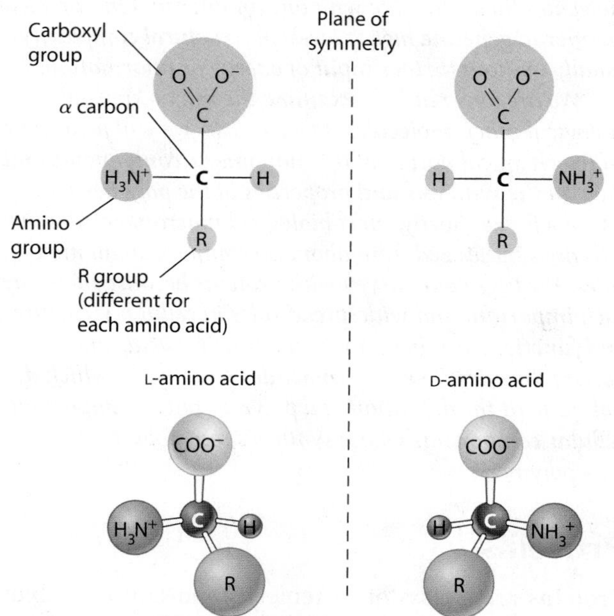

FIGURE 3-1 The Structure and Stereochemistry of an Amino Acid. Because the central carbon atom is asymmetric in all amino acids except glycine, most amino acids can exist in two isomeric forms, designated L and D and shown here as (top) conventional structural formulas and (bottom) ball-and-stick models. The L and D forms are stereoisomers, with the vertical dashed line as the plane of symmetry. Of the two forms, only L-amino acids are present in proteins.

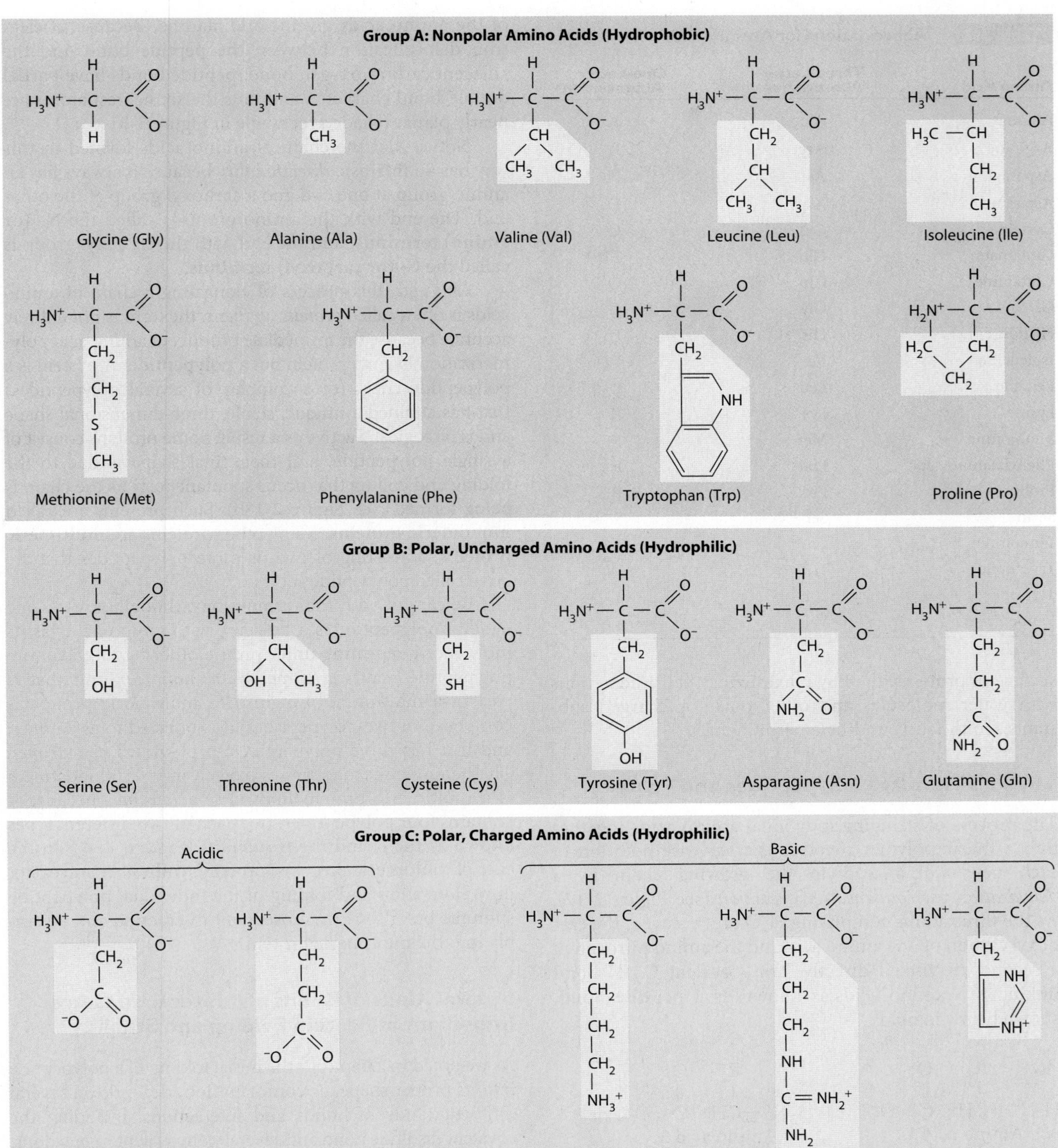

MB Chemistry Review-Organic Molecules: Functional Groups; Amino Acid Functional Groups

FIGURE 3-2 The Structures of the 20 Amino Acids Found in Proteins. All amino acids have a carboxyl group and an amino group attached to the central (α) carbon, but each has its own distinctive R group (light boxes). Those in Group A have nonpolar R groups and are therefore hydrophobic; notice the hydrocarbon nature of their R groups. The others are hydrophilic, either because the R group is polar (Group B) or because the R group is protonated or deprotonated at cellular pH and thus carries a formal electrostatic charge (Group C). Notice the unusual structure of proline—its R group is covalently linked to the amino nitrogen.

Table 3-2 **Abbreviations for Amino Acids**

Amino Acid	Three-Letter Abbreviation	One-Letter Abbreviation
Alanine	Ala	A
Arginine	Arg	R
Asparagine	Asn	N
Aspartate	Asp	D
Cysteine	Cys	C
Glutamate	Glu	E
Glutamine	Gln	Q
Glycine	Gly	G
Histidine	His	H
Isoleucine	Ile	I
Leucine	Leu	L
Lysine	Lys	K
Methionine	Met	M
Phenylalanine	Phe	F
Proline	Pro	P
Serine	Ser	S
Threonine	Thr	T
Tryptophan	Trp	W
Tyrosine	Tyr	Y
Valine	Val	V

surface of proteins, thereby maximizing their interactions with water molecules and other polar or charged substances in the surrounding environment.

The Polymers Are Polypeptides and Proteins

MB Amino Acids and Peptides

The process of stringing individual amino acids together into a linear polymer involves the stepwise addition of each new amino acid to the growing chain by a *dehydration* (or *condensation*) *reaction* (see Figure 2-17). As the three atoms comprising H_2O are removed, the carboxyl carbon of one amino acid and the amino nitrogen of a second are linked directly. This covalent C—N bond linking two amino acids is known as a **peptide bond,** shown below in bold:

$$\underset{\text{Amino acid 1}}{H_3N^+{-}\overset{R_1}{\overset{|}{C}}H{-}\overset{O}{\overset{\|}{C}}{-}O^-} + \underset{\text{Amino acid 2}}{H_3N^+{-}\overset{R_2}{\overset{|}{C}}H{-}\overset{O}{\overset{\|}{C}}{-}O^-} \longrightarrow$$

$$\underset{\text{Peptide}}{H_3N^+{-}\overset{R_1}{\overset{|}{C}}H{-}\overset{O}{\overset{\|}{C}}\boldsymbol{-}\underset{H}{\underset{|}{N}}{-}\overset{R_2}{\overset{|}{C}}H{-}\overset{O}{\overset{\|}{C}}{-}O^-} + H_2O$$

As each new peptide bond is formed by dehydration, the growing chain of amino acids is lengthened by one amino acid. Peptide bond formation is illustrated schematically in **Figure 3-3** using ball-and-stick models of the amino acids glycine and alanine. Because of electron delocalization between the peptide bond and the adjacent carbon-oxygen bond, peptide bonds have partial double-bond character, and thus the six nearest atoms are nearly planar (shaded rectangle in Figure 3-3).

Notice that the chain of amino acids formed in this way has an intrinsic *directionality* because it always has an amino group at one end and a carboxyl group at the other end. The end with the amino group is called the **N- (or amino) terminus,** and the end with the carboxyl group is called the **C- (or carboxyl) terminus.**

Although this process of elongating a chain of amino acids is often called *protein synthesis,* the term is not entirely accurate because the immediate product of amino acid polymerization is not a protein but a **polypeptide.** A protein is a polypeptide chain (or a complex of several polypeptides) that has attained a unique, stable, three-dimensional shape and is biologically active as a result. Some proteins consist of a single polypeptide, and their final shape is due to the folding and coiling that occur spontaneously as the chain is being formed (see Figure 2-18b). Such proteins are called **monomeric proteins.** Many other proteins are **multimeric proteins,** consisting of two or more polypeptides that are often called polypeptide subunits.

Be careful with the terminology, though: On the one hand, a polypeptide is a *polymer,* with amino acids as its monomeric repeating units; on the other hand, the entire polypeptide may sometimes be a *monomer* unit that is part of a multimeric protein. If a multimeric protein is composed of two polypeptides, it is referred to as a *dimer*; and if it has three polypeptides, it is known as a *trimer.* The hemoglobin that carries oxygen in your bloodstream is a multimeric protein known as a *tetramer* because it contains four polypeptides, two each of two different types known as the α and the β *subunits* (**Figure 3-4**). In the case of multimeric proteins, protein synthesis involves not only elongation and folding of the individual polypeptide subunits but also their subsequent interaction and assembly into the multimeric protein.

Several Kinds of Bonds and Interactions Are Important in Protein Folding and Stability

As we noted in Chapter 2, the initial folding of a polypeptide into its proper shape, or **conformation,** depends on several different kinds of bonds and interactions, including the covalent disulfide bond and several noncovalent interactions. In addition, the association of individual polypeptides to form a multimeric protein relies on these same bonds and interactions, which are depicted in **Figure 3-5**. These interactions involve the carboxyl, amino, and R groups of the individual amino acids, which are known as *amino acid residues* once they are incorporated into the polypeptide.

Disulfide Bonds. A special type of covalent bond that contributes to the stabilization of protein conformation is the **disulfide bond,** which forms between the sulfur atoms of two cysteine amino acid residues. These become

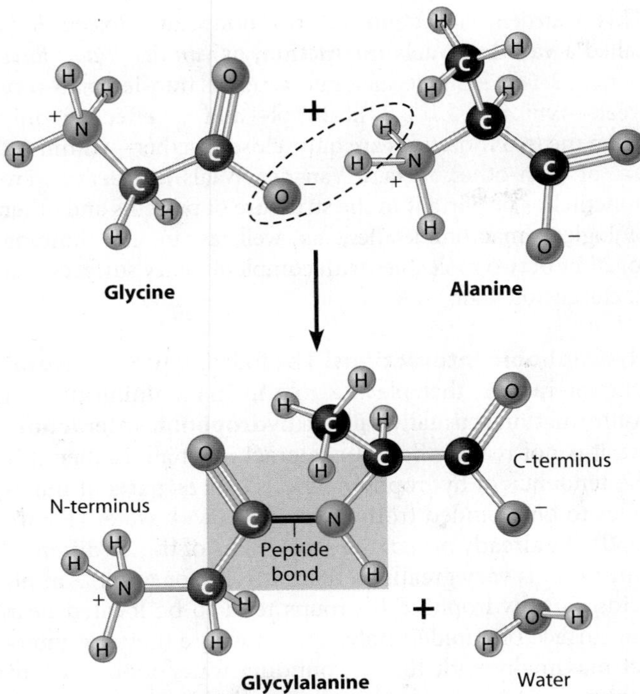

FIGURE 3-3 Peptide Bond Formation. Successive amino acids in a polypeptide are linked to one another by peptide bonds that are formed between the carboxyl group of one amino acid and the amino group of the next as a water molecule is removed (dotted oval). Shown here is the formation of a peptide bond between the amino acids glycine and alanine. The six atoms in the shaded rectangle are nearly planar.

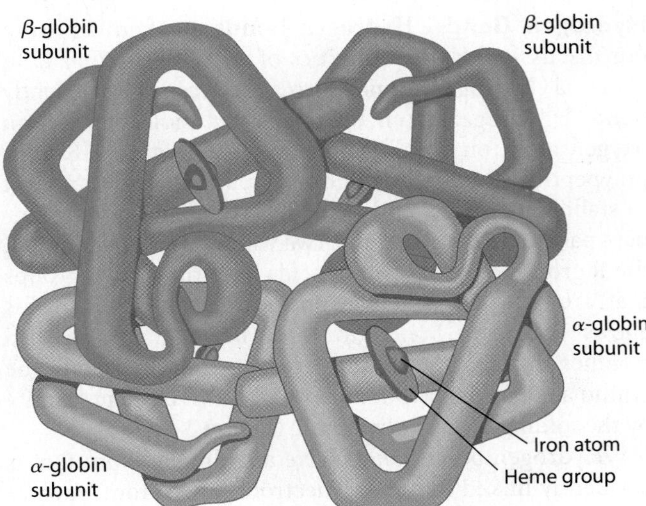

FIGURE 3-4 The Structure of Hemoglobin. Hemoglobin is a multimeric protein composed of four polypeptide subunits (two α chains and two β chains). Each subunit contains a heme group with an iron atom that can bind a single oxygen molecule.

covalently linked following an oxidation reaction that removes the two hydrogen atoms from the sulfhydryl groups of the two cysteines, forming a disulfide bond, shown below in bold and in Figure 3-5a:

$$\text{HC}-\text{CH}_2-\text{SH} \quad + \quad \text{HS}-\text{CH}_2-\text{CH}$$
$$\downarrow \; \rightarrow 2\text{H}$$
$$\text{HC}-\text{CH}_2-\text{S}\boldsymbol{-}\text{S}-\text{CH}_2-\text{CH}$$

Once formed, a disulfide bond confers considerable stability to the structure of the protein because of its covalent nature. It can be broken only by reducing it again—by adding two hydrogen atoms and regenerating the two sulfhydryl groups in the reverse of the reaction above. In many cases, the cysteine residues involved in a particular disulfide bond are a part of the same polypeptide. They may be quite distant from each other along the polypeptide but are brought close together by the folding process. Such *intramolecular disulfide bonds* stabilize the conformation of the polypeptide. In the case of multimeric proteins, a disulfide bond may form between cysteine residues located in two different polypeptides. Such *intermolecular disulfide bonds* link the two polypeptides to one another covalently. The hormone insulin is a dimeric protein that has its two subunits linked in this manner.

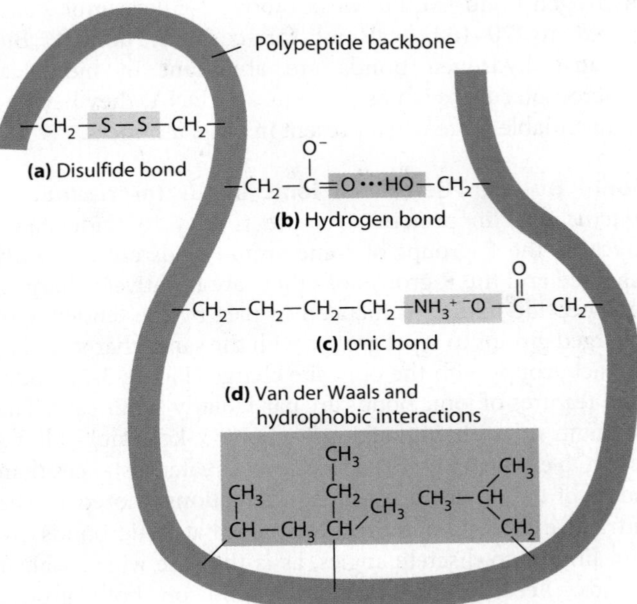

FIGURE 3-5 Bonds and Interactions Involved in Protein Folding and Stability. The initial folding and subsequent stability of a polypeptide depend on **(a)** covalent disulfide bonds as well as on several kinds of noncovalent bonds and interactions, including **(b)** hydrogen bonds, **(c)** ionic bonds, **(d)** van der Waals interactions, and hydrophobic interactions.

In addition to covalent disulfide bonds, **noncovalent bonds and interactions** are also important in maintaining protein structure. Although individually much weaker than covalent bonds, they are diverse and numerous and collectively exert a powerful influence on protein structure and stability. As we noted briefly in Chapter 2, these include *hydrogen bonds, ionic bonds, van der Waals interactions,* and *hydrophobic interactions* (see Figure 3-5).

Hydrogen Bonds. **Hydrogen bonds** are familiar from our discussion of the properties of water in Chapter 2. In water, a hydrogen bond forms between a covalently bonded hydrogen atom on one water molecule and an oxygen atom on another molecule (see Figure 2-8b). In polypeptides, hydrogen bonding is particularly important in stabilizing helical and sheet structures that are prominent parts of many proteins, as we will soon see. In addition, the R groups of many amino acids have functional groups that are able to participate in hydrogen bonding. This allows hydrogen bonds to form between amino acid residues that may be distant from one another along the amino acid sequence but are brought into close proximity by the folding of the polypeptide (Figure 3-5b).

Hydrogen bond *donors* have a hydrogen atom that is covalently linked to a more electronegative atom, such as oxygen or nitrogen, and hydrogen bond *acceptors* have an electronegative atom that attracts this hydrogen atom. Examples of good donors include the hydroxyl groups of several amino acids and the amino groups of others. The carbonyl and sulfhydryl groups of several other amino acids are examples of good acceptors. An individual hydrogen bond is quite weak (about 2–5 kcal/mol, compared to 70–100 kcal/mol for covalent bonds). But because hydrogen bonds are abundant in biological macromolecules such as proteins and DNA, they become a formidable force when present in large numbers.

Ionic Bonds. The role of **ionic bonds** (or *electrostatic interactions*) in protein structure is easy to understand. Because the R groups of some amino acids are positively charged and the R groups of others are negatively charged, polypeptide folding is dictated in part by the tendency of charged groups to repel groups with the same charge and to attract groups with the opposite charge (Figure 3-5c). Several features of ionic bonds are particularly significant. The strength of such interactions—about 3 kcal/mol—allows them to exert an attractive force over greater distances than some of the other noncovalent interactions. Moreover, the attractive force is nondirectional, so that ionic bonds are not limited to discrete angles, as is the case with covalent bonds. Because ionic bonds depend on both groups remaining charged, they will be disrupted if the pH value becomes so high or so low that either of the groups loses its charge. This loss of ionic bonds accounts in part for the denaturation that most proteins undergo at high or low pH.

Van der Waals Interactions. Interactions based on charge are not limited to ions that carry a discrete charge. Even molecules with nonpolar covalent bonds may have transient positively and negatively charged regions. Momentary asymmetries in the distribution of electrons and hence in the separation of charge within a molecule are called *dipoles*. When two molecules that have such transient dipoles are very close to each other and are oriented appropriately, they are attracted to each other, though for only as long as the asymmetric electron distribution persists in both molecules. This transient attraction of two nonpolar molecules is called a **van der Waals interaction,** or *van der Waals force* (Figure 3-5d). A single such interaction is transient and very weak—typically 0.1–0.2 kcal/mol—and is effective only when the two molecules are quite close together—within 0.2 nm of each other, in fact. Van der Waals interactions are nonetheless important in the structure of proteins and other biological macromolecules, as well as in the binding together of two molecules with complementary surfaces that fit closely together.

Hydrophobic Interactions. The fourth type of noncovalent interaction that plays a role in maintaining protein conformation is usually called a **hydrophobic interaction,** but it is not really a bond or interaction at all. Rather, it is the tendency of hydrophobic molecules or parts of molecules to be excluded from interactions with water (Figure 3-5d). As already noted, the side chains of the 20 different amino acids vary greatly in their affinity for water. Amino acids with hydrophilic R groups tend to be located near the surface of a folded polypeptide, where they can interact maximally with the surrounding water molecules. In contrast, amino acids with hydrophobic R groups are essentially nonpolar and are usually located on the inside of the polypeptide, where they interact with one another because they are excluded by water.

Thus, polypeptide folding to form the final protein structure is, in part, a balance between the tendency of hydrophilic groups to seek an aqueous environment near the surface of the molecule and the tendency of hydrophobic groups to minimize contact with water by associating with each other in the interior of the molecule. If most of the amino acids in a protein were hydrophobic, the protein would be virtually insoluble in water and would be found instead in a nonpolar environment. Membrane proteins, which have many hydrophobic residues, are localized in membranes for this very reason. Similarly, if all or most of the amino acids were hydrophilic, the polypeptide would most likely remain in a fairly distended, random shape, allowing maximum access of each amino acid to an aqueous environment. But precisely because most polypeptide chains contain both hydrophobic and hydrophilic amino acids, hydrophilic regions of the molecule are drawn toward the surface, whereas hydrophobic regions are driven toward the interior (see Figure 2-19).

Overall, then, the stability of the folded structure of a polypeptide depends on an interplay of covalent disulfide bonds and four noncovalent factors: hydrogen bonds between R groups that are good donors and good acceptors, ionic bonds between charged amino acid R groups, transient van der Waals interactions between nonpolar molecules in very close proximity, and hydrophobic interactions that drive nonpolar groups to the interior of the molecule.

The final conformation of the fully folded polypeptide is the net result of these forces and tendencies. Individually, each of these noncovalent interactions is quite low in energy. However, the cumulative effect of many of them—

Table 3-3 Levels of Organization of Protein Structure

Level of Structure	Basis of Structure	Kinds of Bonds and Interactions Involved
Primary	Amino acid sequence	Covalent peptide bonds
Secondary	Folding into α helix, β sheet, or random coil	Hydrogen bonds
Tertiary	Three-dimensional folding of a single polypeptide chain	Disulfide bonds, hydrogen bonds, ionic bonds, van der Waals interactions, hydrophobic interactions
Quaternary	Association of multiple polypeptides to form a multimeric protein	Same as for tertiary structure

involving the side groups of the hundreds of amino acids that make up a typical polypeptide—greatly stabilizes the conformation of the folded polypeptide.

Protein Structure Depends on Amino Acid Sequence and Interactions

The overall shape and structure of a protein are usually described in terms of four hierarchical levels of organization, each building on the previous one: the *primary*, *secondary*, *tertiary*, and *quaternary* structures (**Table 3-3**). Primary structure refers to the amino acid sequence, while the higher levels of organization concern the interactions between the amino acid residues. These interactions give the protein its characteristic *conformation*, or three-dimensional arrangement of atoms in space (**Figure 3-6**).

Secondary structure involves local interactions between amino acid residues that are close together along the chain, whereas tertiary structure results from long-distance interactions between stretches of amino acid residues from different parts of the molecule. Quaternary structure concerns the interaction of two or more individual polypeptides to form a single multimeric protein. All three of these higher-level structures are dictated by the primary structure, but each is important in its own right in the overall structure of the protein. Secondary and tertiary structures are involved in determining the conformation of the individual polypeptide, while quaternary structure is relevant for proteins consisting of more than one polypeptide.

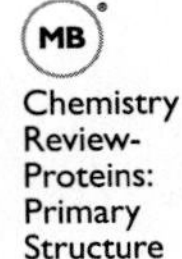

Chemistry Review-Proteins: Primary Structure

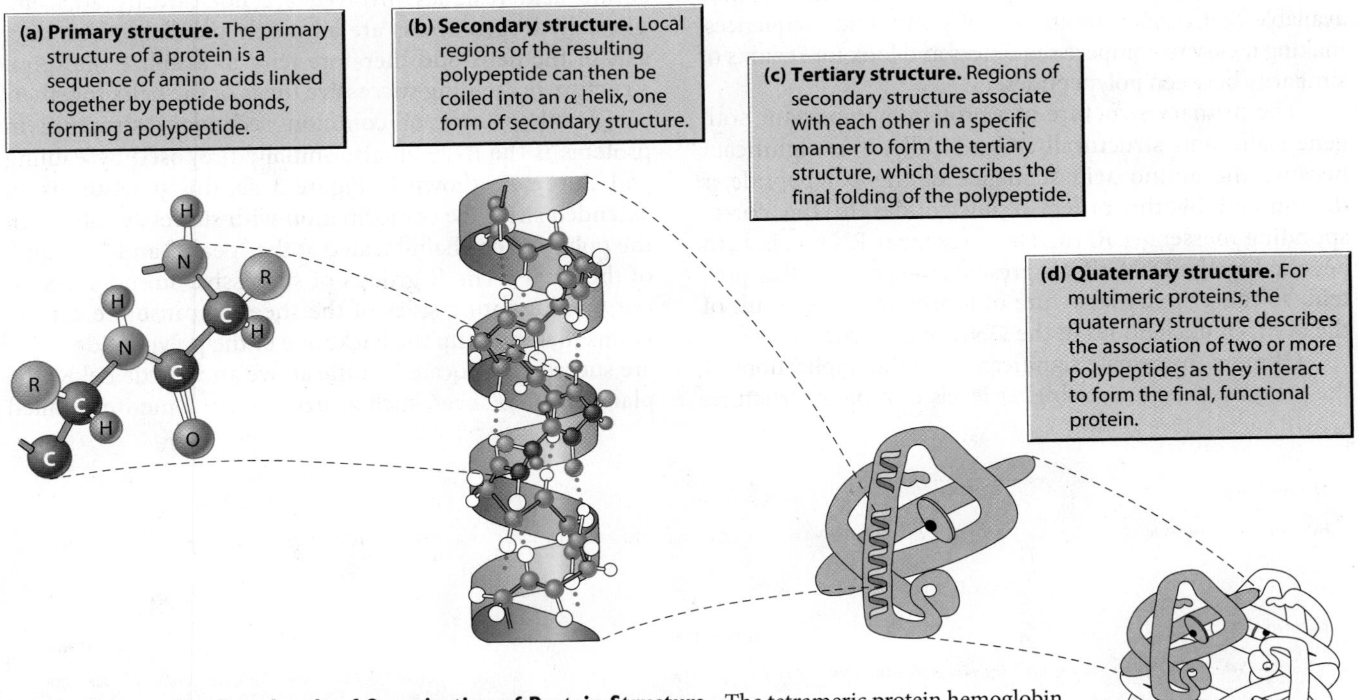

Primary Structure. As already noted, the **primary structure** of a protein is a formal designation for the amino acid sequence (Figure 3-6a). When we describe the primary structure, we are simply specifying the order in which its amino acids appear from one end of the molecule to the other. By convention, amino acid sequences are

(a) Primary structure. The primary structure of a protein is a sequence of amino acids linked together by peptide bonds, forming a polypeptide.

(b) Secondary structure. Local regions of the resulting polypeptide can then be coiled into an α helix, one form of secondary structure.

(c) Tertiary structure. Regions of secondary structure associate with each other in a specific manner to form the tertiary structure, which describes the final folding of the polypeptide.

(d) Quaternary structure. For multimeric proteins, the quaternary structure describes the association of two or more polypeptides as they interact to form the final, functional protein.

MB Chemistry Review-Proteins: Levels of Structure; Activity: Protein Structure; Levels of Structure in Proteins

FIGURE 3-6 The Four Levels of Organization of Protein Structure. The tetrameric protein hemoglobin is used here as an example to illustrate the **(a)** primary, **(b)** secondary, **(c)** tertiary, and **(d)** quaternary levels of protein structure.

always written from the N-terminus to the C-terminus of the polypeptide, which is also the direction in which the polypeptide is synthesized.

The first protein to have its complete amino acid sequence determined was the hormone *insulin.* This important technical advance was reported in 1956 by Frederick Sanger, who eventually received a Nobel Prize for the work. To determine the sequence of the insulin molecule, Sanger cleaved it into smaller fragments and analyzed the amino acid order within individual, overlapping fragments. Insulin consists of two polypeptides, called the *A subunit* and the *B subunit,* with 21 and 30 amino acid residues, respectively. **Figure 3-7** shows the primary structure of insulin, illustrating the primary sequence of each subunit in sequence from its N-terminus (left) to its C-terminus (right). Notice also the covalent disulfide (—S—S—) bond between two cysteine residues within the A chain and the two disulfide bonds linking the A and B chains. As we will see shortly, disulfide bonds play an important role in stabilizing the tertiary structure of many proteins.

Sanger's techniques paved the way for the sequencing of hundreds of other proteins and led ultimately to the design of machines that can determine an amino acid sequence automatically. A more recent approach for determining protein sequences has emerged from the understanding that nucleotide sequences in DNA code for the amino acid sequences of protein molecules. It is now much easier to determine a DNA nucleotide sequence than to purify a protein and analyze its amino acid sequence. Once a DNA nucleotide sequence has been determined, the amino acid sequence of the polypeptide encoded by that DNA segment can be easily determined. Computerized data banks are now available that contain thousands of polypeptide sequences, making it easy to compare sequences and look for regions of similarity between polypeptides.

The primary structure of a protein is important both genetically and structurally. Genetically, it is significant because the amino acid sequence of the polypeptide is determined by the order of nucleotides in the corresponding messenger RNA. The messenger RNA is in turn encoded by the DNA that represents the gene for this protein, so the primary structure of a protein is the result of the order of nucleotides in the DNA of the gene.

Of more immediate significance are the implications of the primary structure for higher levels of protein structure. In essence, all three higher levels of protein organization are direct consequences of the primary structure. Thus, if synthetic polypeptides are made that correspond in sequence to the α and β subunits of hemoglobin, they will assume the native three-dimensional conformations of these subunits and will then interact spontaneously to form the native $\alpha_2\beta_2$ tetramer that we recognize as hemoglobin (see Figure 3-4).

Secondary Structure. The **secondary structure** of a protein describes local regions of structure that result from hydrogen bonding between NH and CO groups along the polypeptide backbone. These local interactions result in two major structural patterns, referred to as the **α helix** and **β sheet** conformations (**Figure 3-8**).

MB Chemistry Review-Proteins: Secondary Structure

The α helix structure was proposed in 1951 by Linus Pauling and Robert Corey. As shown in Figure 3-8a, an α helix is spiral in shape, consisting of a backbone of amino acids linked by peptide bonds with the specific R groups of the individual amino acid residues jutting out from it. A helical shape is common to repeating polymers, as we will see when we get to the nucleic acids and the polysaccharides. In the α helix there are 3.6 amino acids per turn, bringing the peptide bonds of every fourth amino acid in close proximity. The distance between these peptide bonds is, in fact, just right for the formation of a hydrogen bond between the NH group adjacent to one peptide bond and the CO group adjacent to the other, as shown in Figure 3-8a.

As a result, every peptide bond in the helix is hydrogen-bonded through its CO group to the peptide bond immediately "below" it in the spiral and through its NH group to the peptide bond just "above" it, even though the amino acid residues involved are not directly adjacent. These hydrogen bonds are all nearly parallel to the main axis of the helix and therefore tend to stabilize the spiral structure by holding successive turns of the helix together.

Another form of common secondary structure in proteins is the β sheet, also initially proposed by Pauling and Corey. As shown in Figure 3-8b, this structure is an extended sheetlike conformation with successive atoms in the polypeptide chain located at the "peaks" and "troughs" of the pleats. The R groups of successive amino acids jut out on alternating sides of the sheet. Because the carbon atoms that make up the backbone of the polypeptide chain are successively located a little above and a little below the plane of the β sheet, such structures are sometimes called

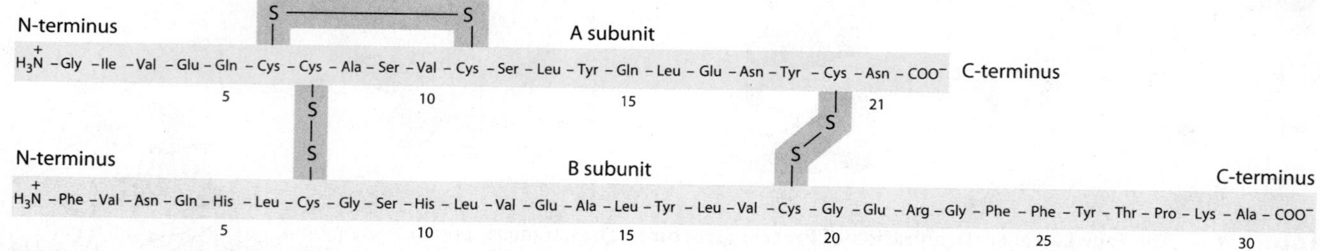

FIGURE 3-7 The Primary Structure of Insulin. Insulin consists of two polypeptides, called the A and B subunits, each shown here from the N-terminus to the C-terminus. The two subunits are covalently linked by two disulfide bonds. (For abbreviations of amino acids, see Table 3-2.)

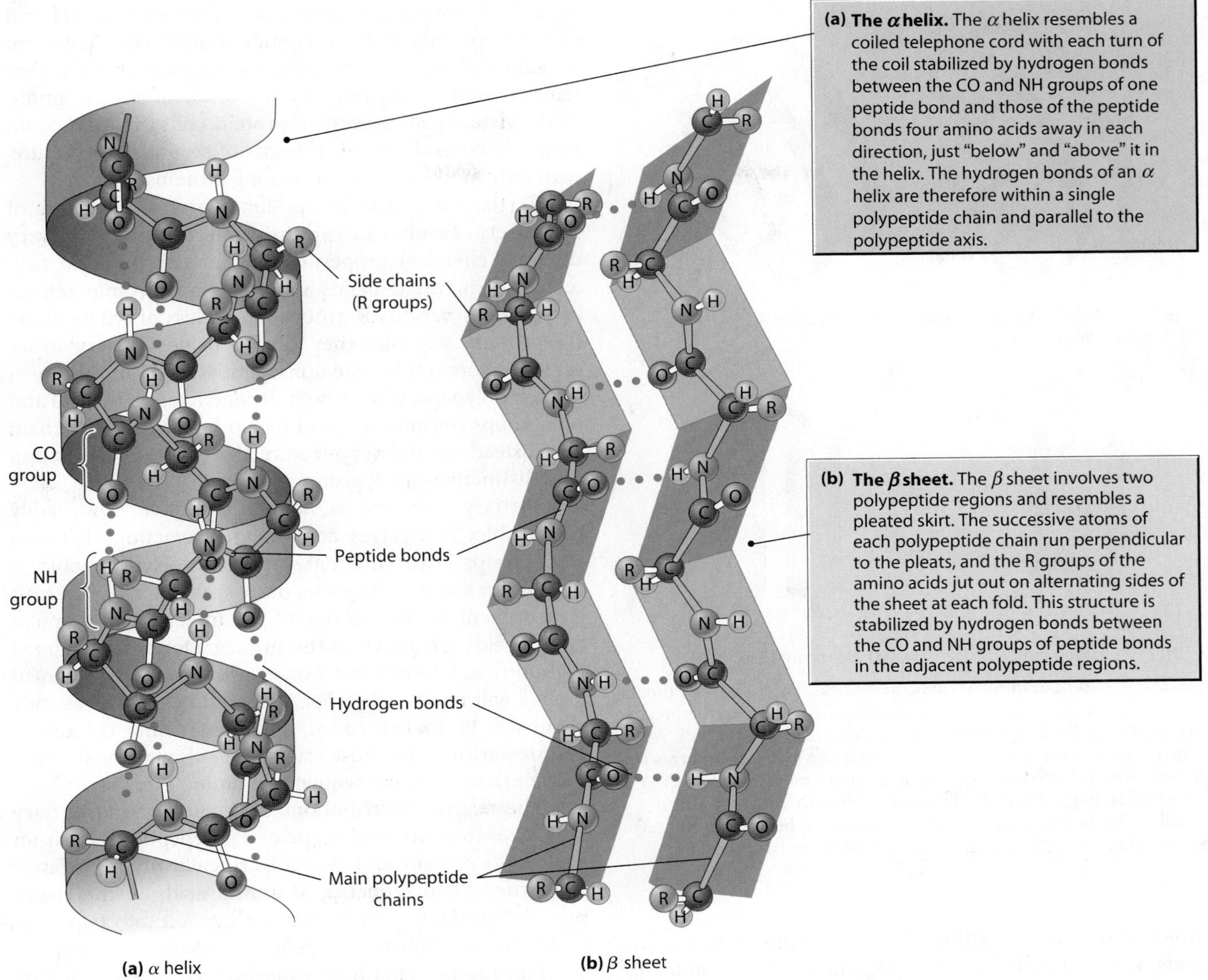

(a) The α helix. The α helix resembles a coiled telephone cord with each turn of the coil stabilized by hydrogen bonds between the CO and NH groups of one peptide bond and those of the peptide bonds four amino acids away in each direction, just "below" and "above" it in the helix. The hydrogen bonds of an α helix are therefore within a single polypeptide chain and parallel to the polypeptide axis.
(b) The β sheet. The β sheet involves two polypeptide regions and resembles a pleated skirt. The successive atoms of each polypeptide chain run perpendicular to the pleats, and the R groups of the amino acids jut out on alternating sides of the sheet at each fold. This structure is stabilized by hydrogen bonds between the CO and NH groups of peptide bonds in the adjacent polypeptide regions.
Side chains (R groups)
CO group
NH group
Peptide bonds
Hydrogen bonds
Main polypeptide chains
(a) α helix
(b) β sheet

FIGURE 3-8 The α Helix and β Sheet. The α helix shown in **(a)** and the β sheet shown in **(b)** are important elements in the secondary structure of proteins. Both are stabilized by hydrogen bonds (blue dots), either within a local region of primary sequence (α helix) or between two separate regions (β sheet).

β-pleated sheets. Whether a local amino acid sequence forms an α helix or a β sheet depends on the particular combination of amino acids present.

Like the α helix, the β sheet is characterized by a maximum of hydrogen bonding. In both cases, all of the CO groups and NH groups adjacent to the peptide bonds are involved. However, hydrogen bonding in an α helix is invariably intramolecular (within the same polypeptide), whereas hydrogen bonding in the β sheet can be either intramolecular (between two segments of the same polypeptide) or intermolecular (linking two different polypeptides). The protein regions that form β sheets can interact with each other in two different ways. If the two interacting regions run in the same N-terminus-to-C-terminus direction, the structure is called a *parallel β sheet*; if the two strands run in opposite N-terminus-to-C-terminus directions, the structure is called an *antiparallel β sheet.*

Whether a specific segment of a polypeptide will form an α helix, a β sheet, or neither depends on the amino acids present in that segment. For example, leucine, methionine, and glutamate are strong "α helix formers," meaning they are commonly found in α-helical regions. Isoleucine, valine, and phenylalanine are strong "β sheet formers," often being found in β-sheet regions. Proline is considered a "helix breaker" because its R group is covalently bonded to its amino nitrogen, which therefore lacks the hydrogen atom needed for hydrogen bonding. Proline is rarely found in an α helix and, when present, introduces a bend in the helix.

To depict localized regions of structure within a protein, biochemists have adopted the conventions shown in **Figure 3-9**. An α-helical region is represented as either a spiral or a cylinder, whereas a β-sheet region is drawn as a flat ribbon or arrow with the arrowhead pointing in the

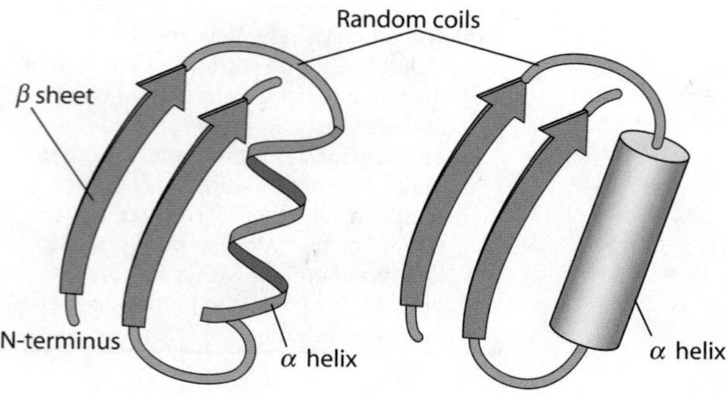

(a) β-α-β motif with α helix represented as a spiral (left) or a cylinder (right)

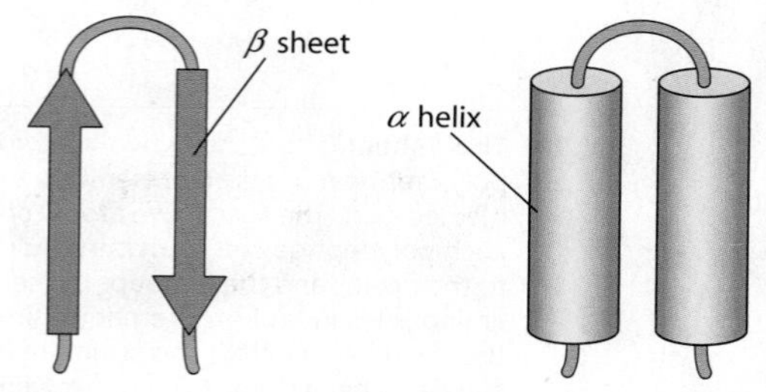

(b) Hairpin loop motif **(c)** Helix-turn-helix motif

FIGURE 3-9 Common Structural Motifs. These short sections of polypeptides show three common units of secondary structure: the **(a)** β-α-β, **(b)** hairpin loop, and **(c)** helix-turn-helix motifs. An α helix can be represented as either a spiral or a cylinder, whereas a β sheet is represented as a flat ribbon or arrow pointing toward the C terminus. In (a), the β sheets are parallel; in (b), they are antiparallel. The short segments (green) connecting α helices and β sheets are called random coils and have no defined secondary structure.

direction of the C-terminus. A looped segment that connects α-helical and/or β-sheet regions is called a *random coil* and is depicted as a narrow cord.

Certain combinations of α helices and β sheets have been identified in many proteins. These units of secondary structure, called **motifs,** consist of small segments of an α helix and/or β sheet connected to each other by looped regions of varying length. Among the most commonly encountered motifs are the β-α-β motif shown in Figure 3-9a and the hairpin loop and helix-turn-helix motifs depicted in Figure 3-9, parts b and c, respectively. When the same motif is present in different proteins, it usually serves the same purpose in each. For example, the helix-turn-helix motif is one of several secondary structure motifs that are characteristic of the DNA-binding proteins we will encounter when we consider the regulation of gene expression in Chapter 23.

Chemistry Review-Proteins: Tertiary Structure

Tertiary Structure. The **tertiary structure** of a protein can probably be best understood by contrasting it with the secondary structure (Figure 3-6b, c). Secondary structure is a predictable, repeating conformational pattern that derives from the repetitive nature of the polypeptide because it involves hydrogen bonding between NH and CO groups adjacent to peptide bonds—the common structural elements along every polypeptide chain. If proteins contained only one or a few kinds of similar amino acids, virtually all aspects of protein conformation could probably be understood in terms of secondary structure, with only modest variations among proteins.

Tertiary structure comes about precisely because of the variety of amino acids present in proteins and the very different chemical properties of their R groups. In fact, tertiary structure depends almost entirely on interactions between the various R groups, regardless of where along the primary sequence they happen to be. Tertiary structure therefore reflects the nonrepetitive and unique aspect of each polypeptide because it depends not on the CO and NH groups common to all of the amino acids in the chain but instead on the very feature that makes each amino acid distinctive—its R group.

Tertiary structure is neither repetitive nor readily predictable; it involves competing interactions between side groups with different properties. Hydrophobic R groups, for example, spontaneously seek out a nonaqueous environment in the interior of the molecule while polar amino acids are drawn to the surface. Oppositely charged R groups can form ionic bonds, while similarly charged groups will repel each other. As a result, the polypeptide chain will be folded, coiled, and twisted into the **native conformation**—the most stable three-dimensional structure for that particular sequence of amino acids.

The relative contributions of secondary and tertiary structures to the overall shape of a polypeptide vary from protein to protein and depend critically on the relative proportions and sequence of amino acids in the chain. Broadly speaking, proteins can be divided into two categories: *fibrous proteins* and *globular proteins.*

Fibrous proteins have extensive secondary structure (either α helix or β sheet) throughout the molecule, giving them a highly ordered, repetitive structure. In general, secondary structure is much more important than tertiary interactions are in determining the shape of fibrous proteins, which often have an extended, filamentous structure. Especially prominent examples of fibrous proteins include the *fibroin* protein of silk and the *keratins* of hair and wool, as well as *collagen* (found in tendons and skin) and *elastin* (present in ligaments and blood vessels).

The amino acid sequence of each of these proteins favors a particular kind of secondary structure, which in turn confers a specific set of desirable mechanical properties on the protein. Fibroin, for example, consists mainly of long stretches of antiparallel β sheets, with the polypeptide chains running parallel to the axis of the silk fiber but in opposite directions. The most prevalent amino acids in fibroin are glycine, alanine, and serine. These amino acids have small R groups that pack together well (see Figure 3-2). The result is a silk fiber that is strong and relatively inextensible because the polypeptide chains in a β-sheet conformation are already stretched to nearly their maximum possible length.

Hair and wool fibers, on the other hand, consist of the protein α-keratin, which is almost entirely α helical. The individual keratin molecules are very long and lie with their helix axes nearly parallel to the fiber axis. As a result, hair is quite extensible because stretching of the fiber is opposed not by the covalent bonds of the polypeptide chain, as in β sheets, but by the hydrogen bonds that stabilize the α-helical structure. The individual α helices in a hair are wound together to form a strong, ropelike structure, as shown in **Figure 3-10**. First, two keratin α helices are coiled around each other, and two of these coiled pairs associate to form a protofilament containing four α helices. Groups of eight protofilaments then interact to form intermediate filaments, which bundle together to form the actual hair fiber. Not surprisingly, the α-keratin polypeptides in hair are rich in hydrophobic residues that interact with each other where the helices touch, allowing the tight packing of the filaments in hair.

As important as fibrous proteins may be, they represent only a small fraction of the kinds of proteins present in most cells. Most of the proteins involved in cellular structure are **globular proteins,** so named because their polypeptide chains are folded into compact structures rather than extended filaments (see Figure 2-19). The polypeptide chain of a globular protein is often folded locally into regions with α-helical or β-sheet structures, and these regions of secondary structure are themselves folded on one another to give the protein its compact, globular shape. This folding is possible because regions of α helix or β sheet are interspersed with random coils, irregularly structured regions that allow the polypeptide chain to loop and fold (see Figure 3-9). Thus, every globular protein has its own unique tertiary structure, made up of secondary structural elements (helices and sheets) folded in a specific way that is especially suited to the particular functional role of that protein.

Figure 3-11 shows the native tertiary structure of ribonuclease, a typical globular protein. We encountered ribonuclease in Figure 2-18, as an example of the denaturation and renaturation of a polypeptide and the spontaneity of its folding. Two different conventions are used in Figure 3-11 to represent the structure of ribonuclease: the *ball-and-stick* model used in Figure 3-8 and the *spiral-and-ribbon* model used in Figure 3-9. For clarity, most of the side chains of ribonuclease have been omitted in both models. The groups shown in gold in Figure 3-11a are the four disulfide bonds that help to stabilize the tertiary structure of ribonuclease.

Globular proteins can be mainly α helical, mainly β sheet, or a mixture of both structures. These categories are illustrated in **Figure 3-12** by the coat protein of tobacco mosaic virus (TMV), a portion of an immunoglobulin (antibody) molecule, and a portion of the enzyme hexokinase, respectively. Helical segments of globular proteins often consist of bundles of helices, as seen for the coat protein of TMV in Figure 3-12a. Segments with mainly β-sheet structure are usually characterized by a barrel-like configuration (Figure 3-12b) or by a twisted sheet (Figure 3-12c).

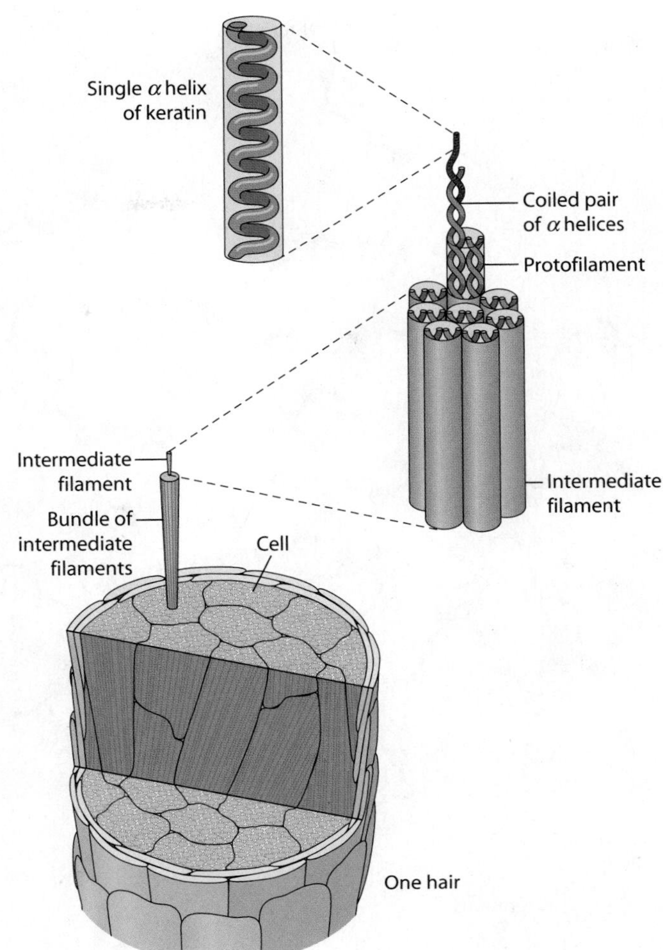

FIGURE 3-10 The Structure of Hair. The main structural protein of hair is α-keratin, a fibrous protein with an α-helical shape. The individual α helices in a hair are wound together to form a strong, ropelike structure. Two keratin α helices are coiled around each other, and two of these coiled pairs associate to form a protofilament containing four α helices. Groups of eight protofilaments then interact to form intermediate filaments, which then bundle together to form the hair fibers (see also Figure 15-23).

Many globular proteins consist of a number of segments called domains. A **domain** is a discrete, locally folded unit of tertiary structure that usually has a specific function. Each domain typically includes 50–350 amino acids, with regions of α helices and β sheets packed together compactly. Small globular proteins are usually folded into a single domain (Figure 3-11b). Large globular proteins usually have multiple domains. The portions of the immunoglobulin and hexokinase molecules shown in Figure 3-12, parts b and c, are, in fact, specific domains of these proteins. **Figure 3-13** shows an example of a protein that consists of a single polypeptide folded into two functional domains.

Proteins that have similar functions (such as binding a specific ion or recognizing a specific molecule) usually have a common domain containing a sequence of identical or very similar amino acid residues. Moreover, proteins

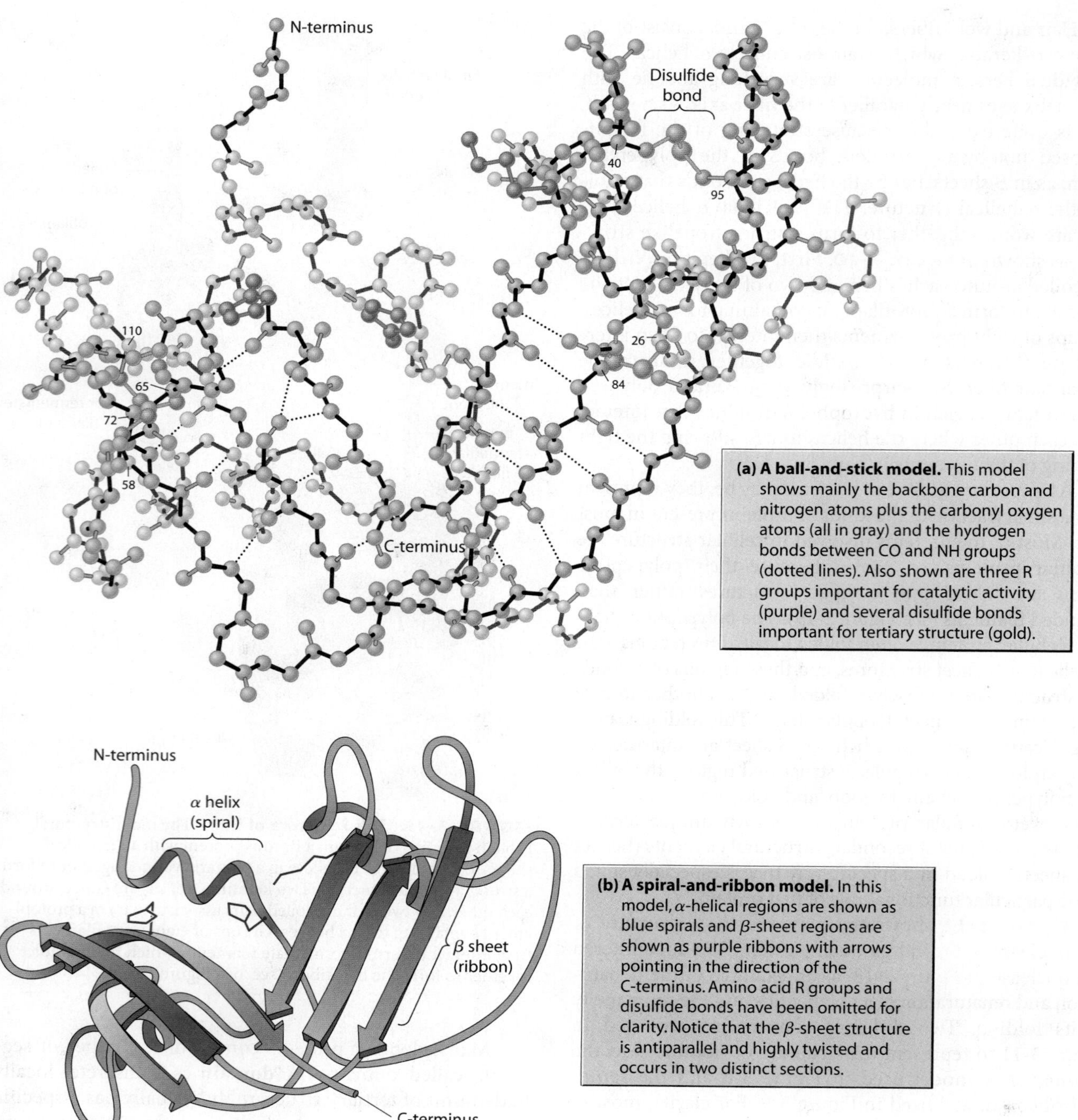

N-terminus
Disulfide bond
40
95
110
26
84
65
72
58
C-terminus
(a) A ball-and-stick model. This model shows mainly the backbone carbon and nitrogen atoms plus the carbonyl oxygen atoms (all in gray) and the hydrogen bonds between CO and NH groups (dotted lines). Also shown are three R groups important for catalytic activity (purple) and several disulfide bonds important for tertiary structure (gold).
N-terminus
α helix (spiral)
β sheet (ribbon)
C-terminus
(b) A spiral-and-ribbon model. In this model, α-helical regions are shown as blue spirals and β-sheet regions are shown as purple ribbons with arrows pointing in the direction of the C-terminus. Amino acid R groups and disulfide bonds have been omitted for clarity. Notice that the β-sheet structure is antiparallel and highly twisted and occurs in two distinct sections.

MB Chemistry Review- Proteins: Models of Proteins

FIGURE 3-11 The Three-Dimensional Structure of Ribonuclease. Ribonuclease is a monomeric globular protein with significant α-helical and β-sheet segments connected by random coils. Its tertiary structure can be represented either by **(a)** a ball-and-stick model or by **(b)** a spiral-and-ribbon model.

with multiple functions usually have a separate domain for each function. Thus, domains can be thought of as the modular units of function from which globular proteins are constructed. Many different types of domains have been described in proteins and given names such as the immunoglobulin domain, the kringle domain, or the death domain. Each type is composed of a particular combination of α-helix and β-sheet secondary structures that give the domain a specific function.

Before leaving the topic of tertiary structure, we should emphasize again the dependence of these higher levels of organization on the primary structure of the polypeptide. The significance of primary structure is exemplified especially well by the inherited condition *sickle-cell anemia.* People with this trait have red blood cells that are distorted from their normal disk shape into a "sickle" shape, which causes the abnormal cells to clog blood vessels and impede blood flow, limiting oxygen availability in the tissues.

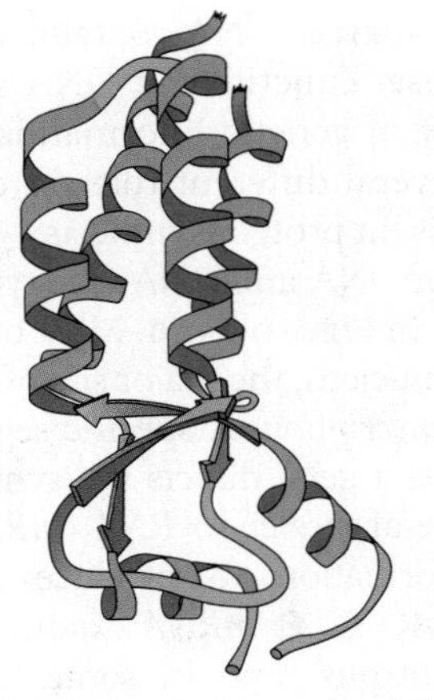

Tobacco mosaic virus coat protein

(a) Predominantly α helix

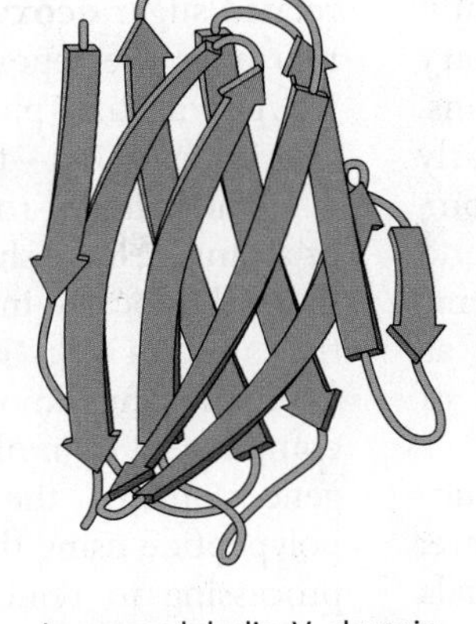

Immunoglobulin, V_2 domain

(b) Predominantly β sheet

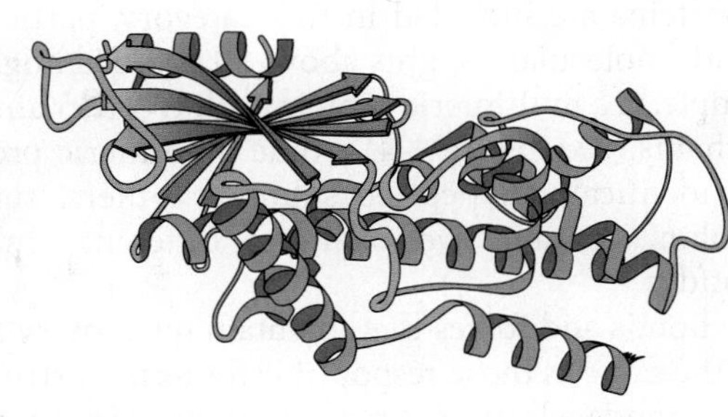

Hexokinase, domain 2

(c) Mixed α helix and β sheet

FIGURE 3-12 Structures of Several Globular Proteins. Shown here are proteins with different tertiary structures: **(a)** a predominantly α-helical structure (blue spirals), the coat protein of tobacco mosaic virus (TMV); **(b)** a mainly β-sheet structure (purple ribbons with arrows), the V_2 domain of immunoglobulin; and **(c)** a structure that mixes α helices and β sheets, domain 2 of hexokinase. The immunoglobulin V_2 domain is an example of an antiparallel β-barrel structure, whereas the hexokinase domain 2 illustrates a twisted β sheet. (Green segments are random coils.)

This condition is caused by a slight change in the hemoglobin molecule within the red blood cells. In people with sickle-cell anemia, the hemoglobin molecules have normal α polypeptide chains, but their β chains have a single amino acid that is different. At one specific position in the chain (the sixth amino acid residue from the N-terminus), the glutamate normally present is replaced by valine. This single substitution (written as E6V) causes enough of a difference in the tertiary structure of the β chain that the hemoglobin molecules tend to crystallize, deforming the cell into a sickle shape. Not all amino acid substitutions cause such dramatic changes in structure and function, but this example underscores the crucial relationship between the amino acid sequence of a polypeptide and the final shape and biological activity of the molecule.

Although we know that the primary sequence of a protein determines its final folded shape, we still are not able to predict exactly how a given protein will fold, especially for large proteins (more than 100 amino acids). In fact, one of the most challenging unsolved problems in structural biochemistry is to predict the final folded tertiary structure of a protein from its known primary structure. Even with all our knowledge of the factors and forces involved in folding, and the availability of supercomputers to do billions of calculations per second, we cannot often predict the most stable conformation for a given protein.

In fact, in every other year since 1994, protein modelers worldwide test their predictive methods in a major modeling experiment known as CASP—the critical assessment of techniques for protein structure prediction. Their predictions are compared to subsequently released three-dimensional protein structures, and the results are published in a special issue of the journal *Proteins: Structure, Function and Bioinformatics.* One of the goals of this modeling research is for drug discovery—the ability to design therapeutic agents able to bind to specific regions of a protein involved in human disease.

FIGURE 3-13 An Example of a Protein Containing Two Functional Domains. This model of the enzyme glyceraldehyde phosphate dehydrogenase shows a single polypeptide chain folded into two domains. One domain binds to the substance being metabolized, whereas the other domain binds to a chemical factor required for the reaction to occur. The two domains are indicated by different shadings.

MB Chemistry Review- Proteins: Quaternary Structure; Proteins

Quaternary Structure. The **quaternary structure** of a protein is the level of organization concerned with subunit interactions and assembly (see Figure 3-6d). Quaternary structure therefore applies only to multimeric proteins. Many proteins are included in this category, particularly those with molecular weights above 50,000. Hemoglobin, for example, is a multimeric protein with two α chains and two β chains (see Figure 3-4). Some multimeric proteins contain identical polypeptide subunits; others, such as hemoglobin, contain two or more different kinds of polypeptides.

The bonds and forces that maintain quaternary structure are the same as those responsible for tertiary structure: hydrogen bonds, electrostatic interactions, van der Waals interactions, hydrophobic interactions, and covalent disulfide bonds. As noted earlier, disulfide bonds may be either within a polypeptide chain or between chains. When they occur within a polypeptide, they stabilize tertiary structure. When they occur between polypeptides, they help maintain quaternary structure, holding the individual polypeptides together (see Figure 3-7). As in the case of polypeptide folding, the process of subunit assembly is often, though not always, spontaneous. Most, if not all, of the requisite information is provided by the amino acid sequence of the individual polypeptides, but often molecular chaperones are required to ensure proper assembly.

In some cases a still higher level of assembly is possible in the sense that two or more proteins (often enzymes) are organized into a **multiprotein complex,** with each protein involved sequentially in a common multistep process. An example of such a complex is an enzyme system called the *pyruvate dehydrogenase complex.* This complex catalyzes the oxidative removal of a carbon atom (as CO_2) from the three-carbon compound pyruvate (or pyruvic acid), a reaction that will be of interest to us when we get to Chapter 10. Three individual enzymes and five kinds of molecules called coenzymes constitute a highly organized *multienzyme complex.* The pyruvate dehydrogenase complex is one of the best understood examples of how cells can achieve economy of function by ordering the enzymes that catalyze sequential reactions into a single multienzyme complex. Other multiprotein complexes we will encounter in our studies include ribosomes, proteosomes, the photosystems, and the DNA replication complex.

Nucleic Acids

Next, we come to the **nucleic acids,** macromolecules of paramount importance to the cell because of their role in storing, transmitting, and expressing genetic information. Nucleic acids are linear polymers of nucleotides strung together in a genetically determined order that is critical to their role as informational macromolecules. The two major types of nucleic acids are **DNA (deoxyribonucleic acid)** and **RNA (ribonucleic acid).** DNA and RNA differ in their chemistry and their role in the cell. As the names suggest, RNA contains the five-carbon sugar **ribose** in each of its nucleotides, whereas DNA contains the closely related sugar **deoxyribose.** Functionally, DNA serves primarily as the repository of genetic information, whereas RNA molecules play several different roles in expressing that information—that is, in protein synthesis.

The primary roles of DNA and RNA in a typical plant or animal cell are shown in **Figure 3-14**. Most of the DNA in a cell is located in the nucleus, the major site of RNA synthesis in the cell. ❶ *Transcription*: A specific segment of a DNA molecule known as a gene directs the synthesis of a complementary molecule of *messenger RNA (mRNA).* Each gene contains the information to produce a specific polypeptide using this mRNA. ❷ *mRNA export*: Following processing to remove introns (and in some cases RNA editing to alter specific bases), the mRNA leaves the nucleus through *nuclear pores*—tiny channels in the nuclear membrane—and enters the cytoplasm. ❸ *Translation (polypeptide synthesis)*: A ribosome, which is a complex of ribosomal proteins and *ribosomal RNA (rRNA)* molecules, attaches to the mRNA to read the coded information. As the ribosome moves down the mRNA, *transfer RNA (tRNA)* molecules bring the correct amino acids to be added to the growing polypeptide chain in the order specified by the information in the mRNA. These roles of DNA and RNA in the storage, transmission, and expression of genetic information will be considered in detail in Chapters 18 –22.

In addition to these three main types of RNA, several others have been discovered in recent years. Many, if not all, eukaryotic cells contain a wide variety of *small RNAs* of 20–30 nucleotides that function as regulatory molecules. *MicroRNAs (miRNAs)* are small endogenous RNAs that down-regulate the expression of specific genes by binding to their mRNAs and either promoting mRNA degradation or inhibiting translation. They are important in regulating genes involved in embryonic development and cell proliferation, and abnormal regulation by miRNAs is associated with certain human diseases. Other small RNAs known as *small interfering RNAs (siRNAs)* are derived from exogenous sources (e.g., infection by an RNA virus) and can inhibit either transcription or translation. Use of siRNAs to target specific mRNAs for destruction is being actively explored as a way to silence genes known to contribute to human disease. These RNAs are discussed in more detail in Chapter 23, but for now we will focus on the chemistry of nucleic acids and nucleotides.

The Monomers Are Nucleotides

Nucleic acids are informational macromolecules and contain nonidentical monomeric units in a specified sequence. The monomeric units of nucleic acids are called **nucleotides.** Nucleotides exhibit less variety than amino acids do; DNA and RNA each contain only four different kinds of nucleotides. (Actually, there is more variety than this suggests, especially in some RNA molecules in which some nucleotides have been chemically modified after insertion into the chain.)

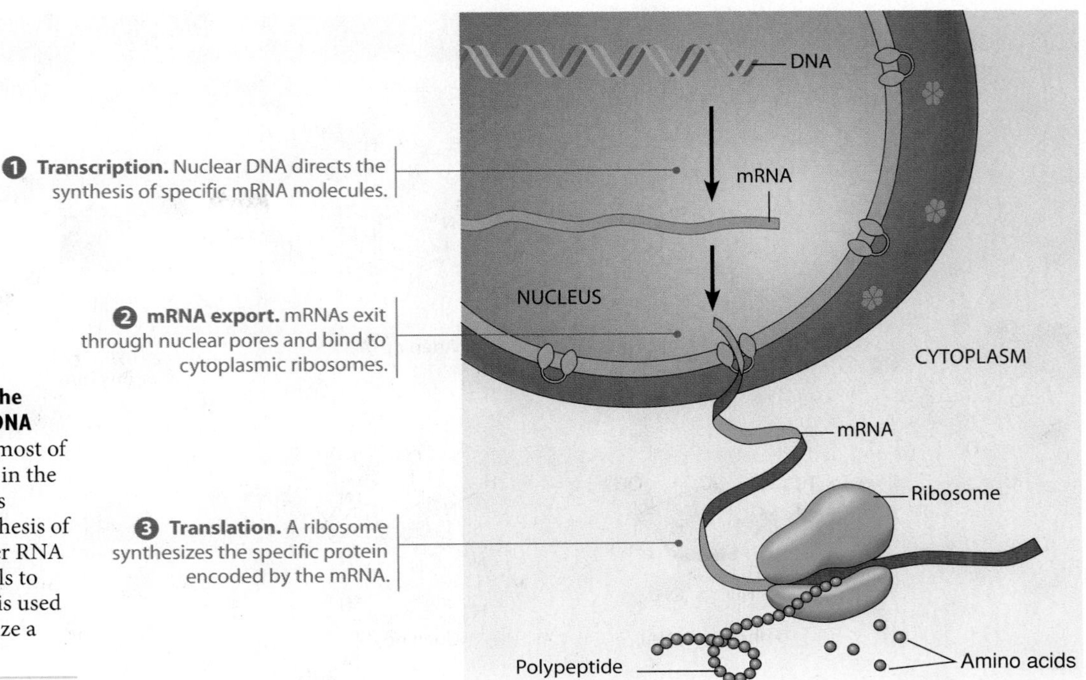
DNA
mRNA
❶ Transcription. Nuclear DNA directs the synthesis of specific mRNA molecules.
❷ mRNA export. mRNAs exit through nuclear pores and bind to cytoplasmic ribosomes.
NUCLEUS
CYTOPLASM
mRNA
Ribosome
❸ Translation. A ribosome synthesizes the specific protein encoded by the mRNA.
Polypeptide
Amino acids

MB Chemistry Review- Nucleic Acids: Functions of Nucleic Acids

FIGURE 3-14 Genetic Information Is Stored in the Nucleotide Sequences of DNA Molecules. In eukaryotes, most of the DNA in a cell is located in the nucleus. This DNA contains instructions for ❶ the synthesis of a complementary messenger RNA (mRNA) that then ❷ travels to the cytoplasm, where ❸ it is used by the ribosome to synthesize a protein.

As shown in **Figure 3-15**, each nucleotide consists of a five-carbon sugar to which is attached a phosphate group and a nitrogen-containing aromatic base. The sugar is either D-ribose (in RNA) or D-deoxyribose (in DNA). The phosphate is joined by a phosphoester bond to the 5′ carbon of the sugar, and the base is attached to the 1′ carbon. The base may be either a **purine** or a **pyrimidine.** DNA contains the purines **adenine (A)** and **guanine (G)** and the pyrimidines **cytosine (C)** and **thymine (T).** RNA also has adenine, guanine, and cytosine, but it contains the pyrimidine **uracil (U)** in place of thymine. Like the 20 amino acids present in proteins, these five aromatic bases are among the most common small molecules in cells (see Table 3-1).

Without the phosphate, the remaining base-sugar unit is called a **nucleoside.** Each pyrimidine and purine may therefore occur as the free base, the nucleoside, or the nucleotide. The appropriate names for these compounds are given in **Table 3-4**. Notice that nucleotides and nucleosides containing deoxyribose are specified by a lowercase d preceding the letters identifying the base.

As the nomenclature indicates, a nucleotide can be thought of as a **nucleoside monophosphate** because it is a nucleoside with a single phosphate group attached to it. This terminology can be readily extended to molecules with two or three phosphate groups attached to the 5′ carbon. For example, the nucleoside adenosine (adenine plus ribose) can have one, two, or three phosphates attached and is designated accordingly as **adenosine monophosphate (AMP), adenosine diphosphate (ADP),** or **adenosine triphosphate (ATP).** The relationships among these compounds are shown in **Figure 3-16**.

You probably recognize ATP as the energy-rich compound used to drive various reactions in the cell, including the activation of monomers for polymer

Table 3-4 The Bases, Nucleosides, and Nucleotides of RNA and DNA

	RNA		DNA	
Bases	**Nucleoside**	**Nucleotide**	**Deoxynucleoside**	**Deoxynucleotide**
Purines				
Adenine (A)	Adenosine	Adenosine monophosphate (AMP)	Deoxyadenosine	Deoxyadenosine monophosphate (dAMP)
Guanine (G)	Guanosine	Guanosine monophosphate (GMP)	Deoxyguanosine	Deoxyguanosine monophosphate (dGMP)
Pyrimidines				
Cytosine (C)	Cytidine	Cytidine monophosphate (CMP)	Deoxycytidine	Deoxycytidine monophosphate (dCMP)
Uracil (U)	Uridine	Uridine monophosphate (UMP)	—	—
Thymine (T)	—	—	Deoxythymidine	Deoxythymidine monophosphate (dTMP)

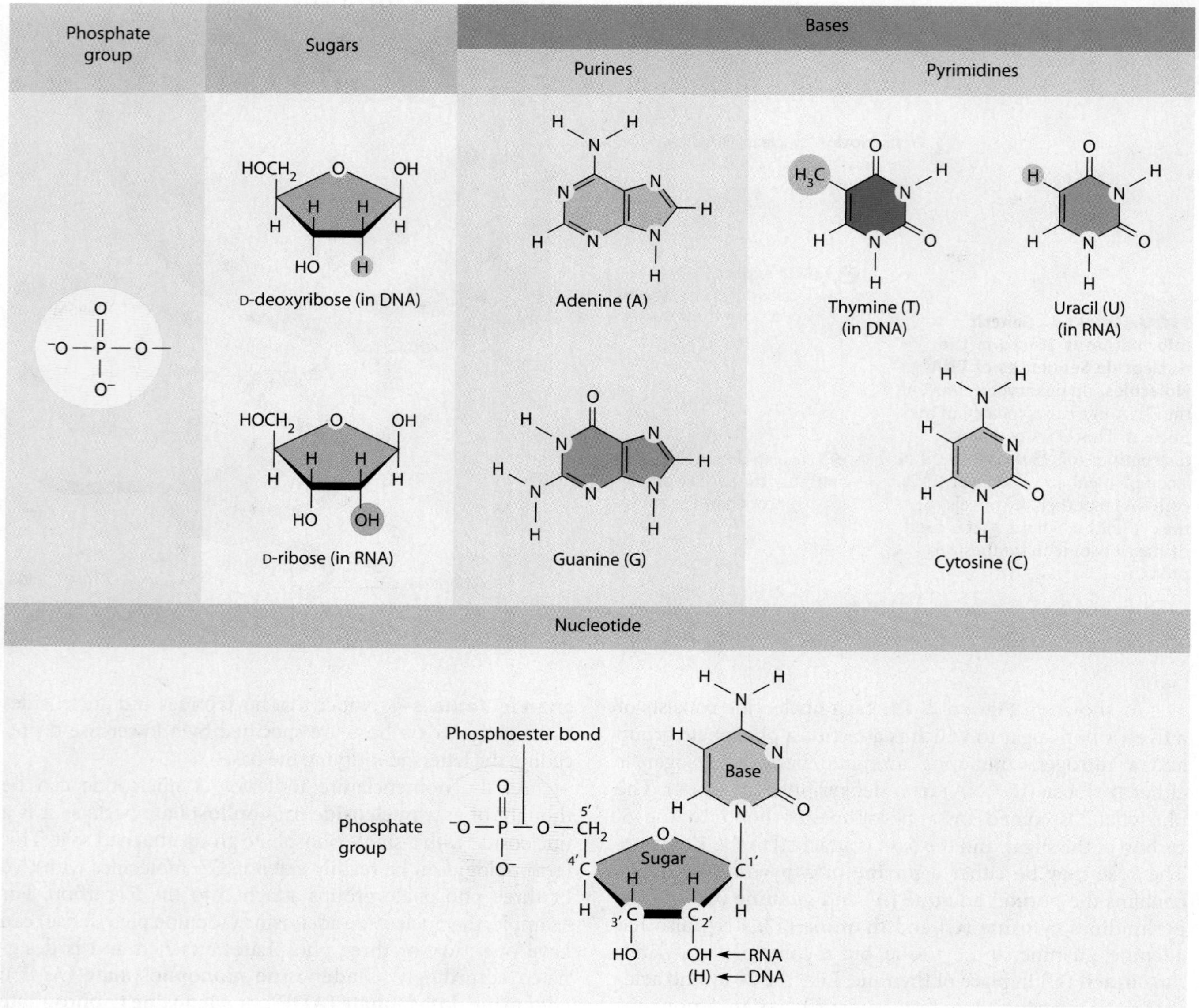

Phosphate group | Sugars | Bases: Purines, Pyrimidines

D-deoxyribose (in DNA)

Adenine (A)

Thymine (T) (in DNA)

Uracil (U) (in RNA)

D-ribose (in RNA)

Guanine (G)

Cytosine (C)

Nucleotide

Phosphoester bond

Phosphate group

Base

Sugar

OH ← RNA

(H) ← DNA

MB

Nucleic Acid Building Blocks

FIGURE 3-15 The Structure of a Nucleotide. In RNA, a nucleotide consists of the five-carbon sugar D-ribose with an aromatic nitrogen-containing base attached to the 1′ carbon and a phosphate group linked to the 5′ carbon by a phosphoester bond. (Carbon atoms in the sugar of a nucleotide are numbered from 1′ to 5′ to distinguish them from those in the base, which are numbered without the prime.) In DNA, the hydroxyl group on the 2′ carbon is replaced by a hydrogen atom, so the sugar is D-deoxyribose. The bases in DNA are the purines adenine (A) and guanine (G) and the pyrimidines thymine (T) and cytosine (C). In RNA, thymine is replaced by the pyrimidine uracil (U).

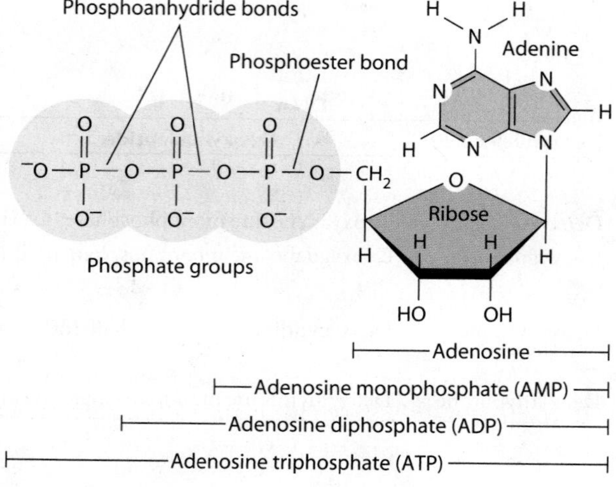

FIGURE 3-16 The Phosphorylated Forms of Adenosine. Adenosine occurs as the free nucleoside and can also form part of the following three nucleotides: adenosine monophosphate (AMP), adenosine diphosphate (ADP), and adenosine triphosphate (ATP). The bond linking the first phosphate to the ribose of adenosine is a phosphoester bond, whereas the bonds linking the second and third phosphate groups to the molecule are phosphoanhydride bonds. As we will see in Chapter 9, the hydrolysis of a phosphoanhydride bond typically liberates two to three times as much free energy as does the hydrolysis of a phosphoester bond.

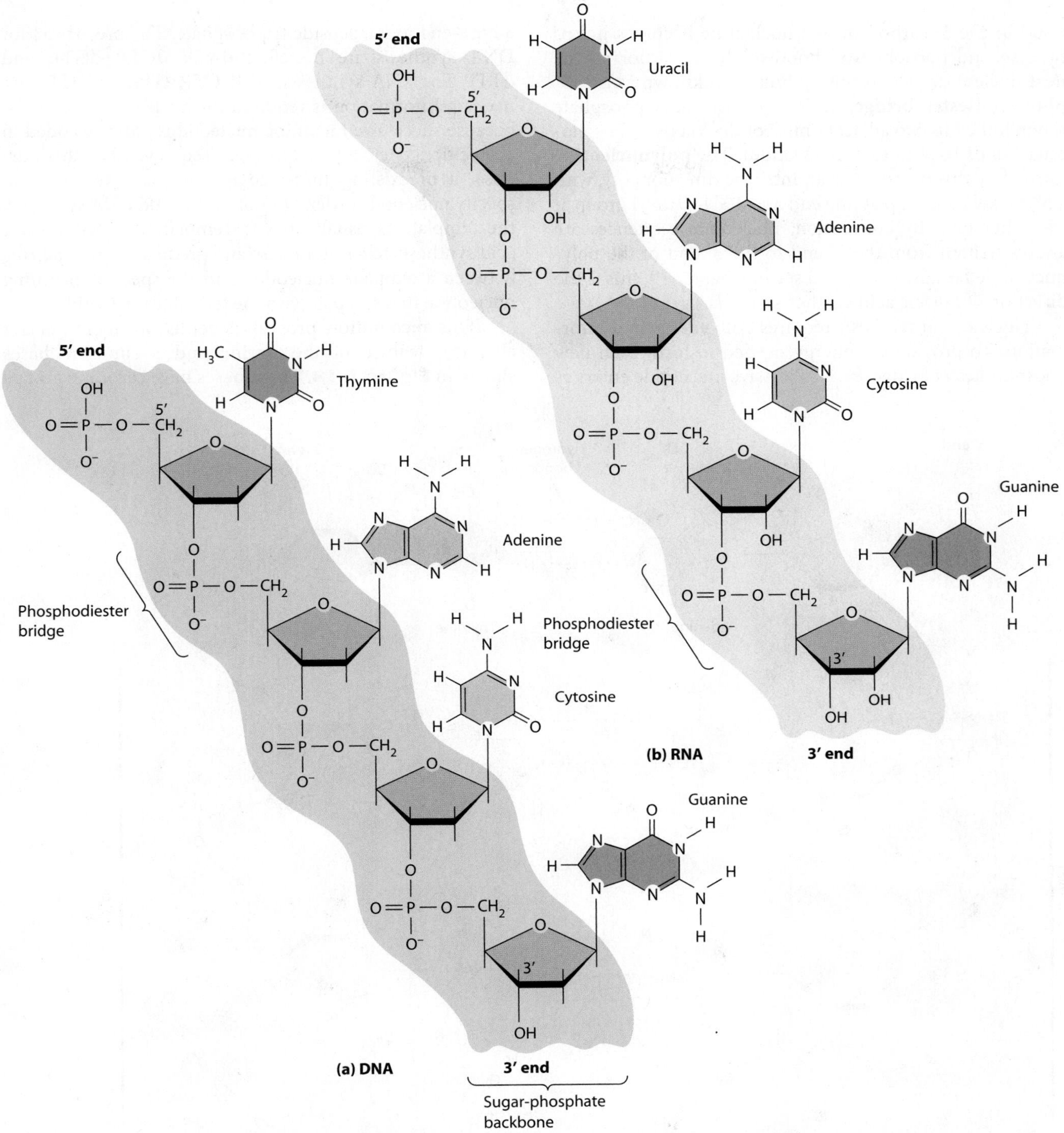

MB Activity: DNA and RNA Structure

FIGURE 3-17 The Structure of Nucleic Acids. Nucleic acids consist of linear chains of nucleotides, each containing a sugar, a phosphate, and a base. The sugar is **(a)** deoxyribose in DNA and **(b)** ribose in RNA. Successive nucleotides in the chain are joined together by 3′, 5′ phosphodiester bridges. The resulting polynucleotide has an intrinsic directionality, with a 5′ end and a 3′ end. For both DNA and RNA, the backbone of the chain is an alternating sugar-phosphate sequence, from which the bases jut out.

formation that we encountered in the previous chapter (see Figure 2-17). As this example suggests, nucleotides play two roles in cells: They are the monomeric units of nucleic acids, and several of them—ATP most notably—serve as intermediates in various energy-transferring reactions.

The Polymers Are DNA and RNA

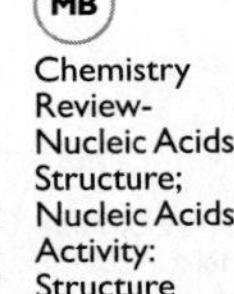

Chemistry Review- Nucleic Acids: Structure; Nucleic Acids; Activity: Structure of RNA and DNA; DNA Structure

Nucleic acids are linear polymers formed by linking each nucleotide to the next through a phosphate group, as shown in **Figure 3-17** (see also Figure 19-7). Specifically, the phosphate group already attached by a phosphoester bond to the 5′ carbon of one nucleotide becomes linked by a second phosphoester bond to the 3′ carbon of the next nucleotide. The resulting linkage is known as a **3′,5′ phosphodiester bridge,** which consists of a phosphate group linked to two adjacent nucleotides via two phosphoester bonds (one to each nucleotide). The **polynucleotide** formed by this process has an intrinsic directionality, with a 5′ phosphate group at one end and a 3′ hydroxyl group at the other end. By convention, nucleotide sequences are always written from the 5′ end to the 3′ end of the polynucleotide because, as we will see in Chapter 19, this is the direction of nucleic acid synthesis in cells.

Nucleic acid synthesis requires both energy and information. To provide the energy needed to form each new phosphodiester bridge, each successive nucleotide enters as a high-energy nucleoside triphosphate. The precursors for DNA synthesis are therefore dATP, dCTP, dGTP, and dTTP. For RNA synthesis, ATP, CTP, GTP, and UTP are needed. Information is required for nucleic acid synthesis because successive incoming nucleotides must be added in a specific, genetically determined sequence. For this purpose, a preexisting molecule is used as a **template** to specify nucleotide order. For both DNA and RNA synthesis, the template is usually DNA. Template-directed nucleic acid synthesis relies on precise and predictable *base pairing* between a template nucleotide and the specific incoming nucleotide that can pair with the template nucleotide.

This recognition process depends on an important chemical feature of the purine and pyrimidine bases shown in **Figure 3-18**. These bases have carbonyl groups

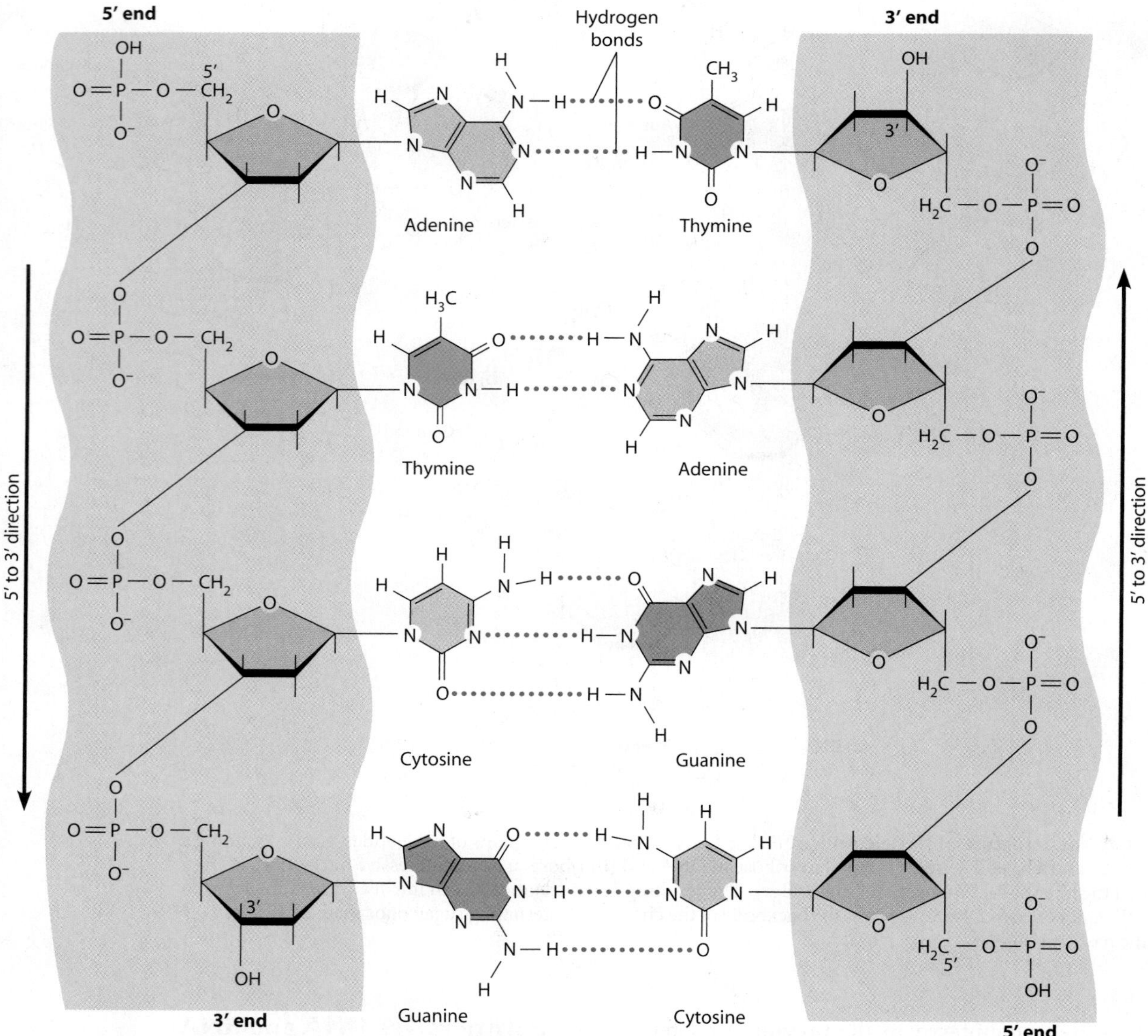

MB The Double Helix; Activity: Nucleic Acid Structure

FIGURE 3-18 Hydrogen Bonding in Nucleic Acid Structure. The hydrogen bonds (blue dots) between adenine and thymine and between cytosine and guanine account for the AT and CG base pairing of DNA. Notice that the AT pair is held together by two hydrogen bonds, whereas the CG pair has three hydrogen bonds. If one or both strands were RNA instead, the pairing partner for adenine would be uracil (U).

and nitrogen atoms capable of hydrogen bond formation under appropriate conditions. Complementary relationships between purines and pyrimidines allow A to form two hydrogen bonds with T (or U) and G to form three hydrogen bonds with C, as shown in Figure 3-18. This pairing of A with T (or U) and G with C is a fundamental property of nucleic acids. Genetically, this **base pairing** provides a mechanism for nucleic acids to recognize one another, as we will see in Chapter 18. For now, however, let's concentrate on the structural implications.

A DNA Molecule Is a Double-Stranded Helix

MB Activity: Heritable Information: DNA; Video Tutor Session Quiz: DNA Structure

One of the most significant biological advances of the twentieth century came in 1953 in a two-page article in the scientific journal *Nature*. In the article, Francis Crick and James Watson postulated a double-stranded helical structure for DNA—the now-famous **double helix**—that not only accounted for the known physical and chemical properties of DNA but also suggested a mechanism for replication of the structure. Some highlights of this exciting chapter in the history of contemporary biology are related in **Box 3A**.

The double helix consists of two complementary chains of DNA twisted together around a common axis to form a right-handed helical structure that resembles a spiral staircase (**Figure 3-19**). The two chains are oriented in opposite directions along the helix, with one running in the $5' \rightarrow 3'$ direction and the other in the $3' \rightarrow 5'$ direction. The backbone of each chain consists of sugar molecules alternating with phosphate groups (see Figure 3-18). The phosphate groups are charged, and the sugar molecules contain polar hydroxyl groups. Therefore, it is not surprising that the sugar-phosphate backbones of the two strands are on the outside of the DNA helix, where their interaction with the surrounding aqueous environment can be maximized. The pyrimidine and purine bases, on the other hand, are aromatic compounds with less affinity for water. Accordingly, they are oriented inward, forming the base pairs that hold the two chains together.

To form a stable double helix, the two component strands must be *antiparallel* (running in opposite directions) as well as *complementary*. By complementary, we mean that each base in one strand can form specific hydrogen bonds with the base in the other strand directly across from it. From the pairing possibilities shown in Figure 3-18, this means that each A must be paired with a T, and each G with a C. In both cases, one member of the pair is a pyrimidine (T or C) and the other is a purine (A or G). The distance between the two sugar-phosphate backbones in the double helix is just sufficient to accommodate one of each kind of base. If we envision the sugar-phosphate backbones of the two strands as the sides of a circular staircase, then each step or rung of the stairway corresponds to a pair of bases held in place by hydrogen bonding (Figure 3-19).

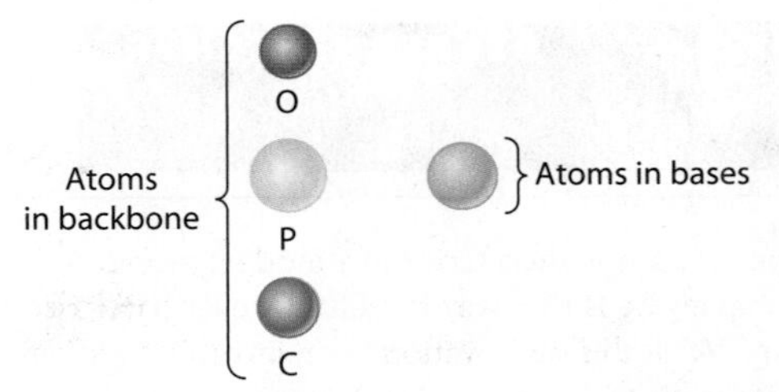

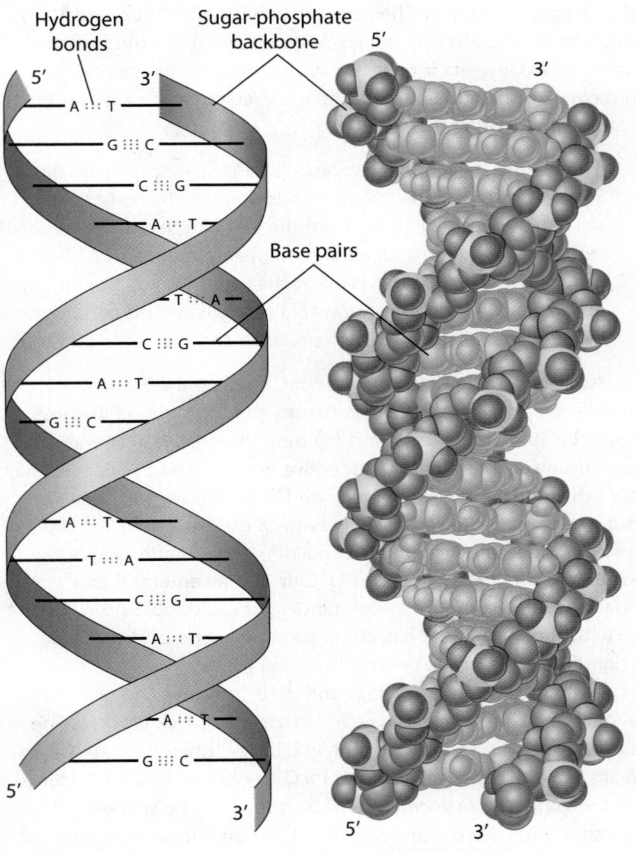

FIGURE 3-19 The Structure of Double-Stranded DNA. (a) A schematic representation of the double-helical structure of DNA. The continuously turning strips represent the sugar-phosphate backbones of the molecule, while the horizontal bars represent paired bases of the two strands. (b) A space-filling model of the DNA double helix, with color-coded atoms as shown at the top of the figure.

MB Chemistry Review- Nucleic Acids: The Double Helix; Activity: DNA Double Helix

The right-handed Watson-Crick helix shown in Figure 3-19 is actually an idealized version of what is called *B-DNA*. B-DNA is the main form of DNA in cells, but two other forms may also exist, perhaps in short segments interspersed within molecules consisting mainly of B-DNA. *A-DNA* has a right-handed, helical configuration that is shorter and thicker than B-DNA. *Z-DNA*, on the other hand, is a left-handed double helix that derives its name from the zigzag pattern of its longer, thinner sugar-phosphate backbone. (For a comparison of the structures of B-DNA and Z-DNA, see Figure 18-5.)

BOX 3A DEEPER INSIGHTS

On the Trail of the Double Helix

"I have never seen Francis Crick in a modest mood. Perhaps in other company he is that way, but I have never had reason so to judge him." With this observation as an introduction, James Watson goes on to describe, in a very personal and highly entertaining way, the events that eventually led to the discovery of the structure of DNA. The account, published in 1968 under the title *The Double Helix,* is still fascinating reading for the personal, unvarnished insights it provides into how an immense scientific discovery came about. In the preface, Watson comments on his reasons for writing the book:

> There remains general ignorance about how science is "done." That is not to say that all science is done in the manner described here. This is far from the case, for styles of scientific research vary almost as much as human personalities. On the other hand, I do not believe that the way DNA came out constitutes an odd exception to a scientific world complicated by the contradictory pulls of ambition and the sense of fair play.

As portrayed in Watson's account, Crick and Watson are about as different from each other in nature and background as they could be. But there was one thing they shared, and that was an unconventional but highly productive way of "doing" science. They did little actual experimentation on DNA, choosing instead to draw heavily on the research findings of others and to use their own considerable ingenuities in building models and exercising astute insights (**Figure 3A-1**). Out of it all emerged, in a relatively short time, an understanding of the double-helical structure of DNA that has come to rank as one of the major scientific events of the twentieth century.

To appreciate their findings and their brilliance, we must first understand the setting in which Watson and Crick worked. The early 1950s was an exciting time in biology. It had been only a few years since Avery, MacLeod, and McCarty had published evidence on the genetic transformation of bacteria, but the work of Hershey and Chase that confirmed DNA as the genetic material had not yet appeared in print. Meanwhile, at Columbia University, Erwin Chargaff's careful chemical analyses had revealed that although the relative proportions of the four bases—A, T, C, and G—varied greatly from one species to the next, it was always the same for all members of a single species. Even more puzzling and portentous was Chargaff's second finding: For a given species, A and T always occurred in the same proportions, and so did G and C (that is, %A = %T and %C = %G).

FIGURE 3A-1 James Watson (left) and Francis Crick (right) at work with their model of DNA.

Important clues came from the work of Maurice Wilkins and Rosalind Franklin at King's College in London. Wilkins and Franklin were using the technique of X-ray diffraction to study DNA structure, and they took a rather dim view of Watson and Crick's strategy of model building. X-ray diffraction is a useful tool for detecting regularly occurring structural elements in a crystalline substance because any structural feature that repeats at some fixed interval in the crystal contributes in a characteristic way to the diffraction pattern that is obtained. From Franklin's painstaking analysis of the diffraction pattern of DNA, it became clear that the molecule was long and thin, with some structural element being repeated every 0.34 nm and another being repeated every 3.4 nm. Even more intriguing, the molecule appeared to be some sort of helix.

This stirred the imaginations of Watson and Crick because they had heard only recently of Pauling and Corey's α-helical structure

RNA structure also depends in part on base pairing, but this pairing is usually between complementary regions within the same strand and is much less extensive than the interstrand pairing of the DNA duplex. Of the various RNA species, secondary and tertiary structures occur mainly in rRNA and tRNA, as we will see in Chapter 21. In addition, some infectious viruses consist of double-stranded RNA held together by hydrogen bonding between complementary base pairs.

Polysaccharides

The next macromolecules we will consider are the **polysaccharides,** which are long-chain polymers of sugars and sugar derivatives. Polysaccharides usually consist of a single kind of repeating unit, or sometimes an alternating pattern of two kinds, and are not informational molecules. We will see in Chapter 7, however, that shorter polymers called *oligosaccharides,* when attached to proteins on the

for proteins. Working with models of the bases cut from stiff cardboard, Watson and Crick came to the momentous insight that DNA was also a helix, but with an all-important difference: It was a *double* helix, with hydrogen-bonded pairing of purines and pyrimidines. The actual discovery is best recounted in Watson's own words:

> When I got to our still empty office the following morning, I quickly cleared away the papers from my desk top so that I would have a large, flat surface on which to form pairs of bases held together by hydrogen bonds. Though I initially went back to my like-with-like prejudices, I saw all too well that they led nowhere. When Jerry [Donohue, an American crystallographer working in the same laboratory] came in I looked up, saw that it was not Francis, and began shifting the bases in and out of various other pairing possibilities. Suddenly I became aware that an adenine-thymine pair held together by two hydrogen bonds was identical in shape to a guanine-cytosine pair held together by at least two hydrogen bonds. All the hydrogen bonds seemed to form naturally; no fudging was required to make the two types of base pairs identical in shape. Quickly I called Jerry over to ask him whether this time he had any objections to my new base pairs.
>
> When he said no, my morale skyrocketed, for I suspected that we now had the answer to the riddle of why the number of purine residues exactly equaled the number of pyrimidine residues. Two irregular sequences of bases could be regularly packed in the center of a helix if a purine always hydrogen-bonded to a pyrimidine. Furthermore, the hydrogen-bonding requirement meant that adenine would always pair with thymine, while guanine could pair only with cytosine. Chargaff's rules then suddenly stood out as a consequence of a double-helical structure for DNA. Even more exciting, this type of double helix suggested a replication scheme much more satisfactory than my briefly considered like-with-like pairing. Always pairing adenine with thymine and guanine with cytosine meant that the base sequences of the two intertwined chains were complementary to each other. Given the base sequence of one chain, that of its partner was automatically determined. Conceptually, it was thus very easy to visualize how a single chain could be the template for the synthesis of a chain with the complementary sequence.
>
> Upon his arrival Francis did not get more than halfway through the door before I let loose that the answer to everything was in our hands. Though as a matter of principle he maintained skepticism for a few moments, the similarly shaped AT and GC pairs had their expected impact. His quickly pushing the bases together in a number of different ways did not reveal any other way to satisfy Chargaff's rules. A few minutes later he spotted the fact that the two glycosidic bonds (joining base and sugar) of each base pair were systematically related by a dyad axis perpendicular to the helical axis. Thus, both pairs could be flip-flopped over and still have their glycosidic bonds facing in the same direction. This had the important consequence that a given chain could contain both purines and pyrimidines. At the same time, it strongly suggested that the backbones of the two chains must run in opposite directions.
>
> The question then became whether the AT and GC base pairs would easily fit the backbone configuration devised during the previous two weeks. At first glance this looked like a good bet, since I had left free in the center a large vacant area for the bases. However, we both knew that we would not be home until a complete model was built in which all the stereochemical contacts were satisfactory. There was also the obvious fact that the implications of its existence were far too important to risk crying wolf. Thus, I felt slightly queasy when at lunch Francis winged into the Eagle to tell everyone within hearing distance that we had found the secret of life.*

The rest is history. Shortly thereafter, the prestigious journal *Nature* carried an unpretentious two-page article entitled simply "Molecular Structure of Nucleic Acids: A Structure for Deoxyribose Nucleic Acid," by James Watson and Francis Crick. Though modest in length, that paper has had far-reaching implications. In fact, near the end, this article contains what some consider to be one of the greatest scientific understatements ever: "It has not escaped our notice that the specific pairing that we have postulated immediately suggests a possible copying mechanism for the genetic material."† The double-stranded model that Watson and Crick worked out in 1953 has proved to be correct in all its essential details, unleashing a revolution in the field of biology.

*Excerpted from *The Double Helix*, pp. 222–225. Copyright © 1968 James D. Watson. Reprinted with the permission of the author and Atheneum Publishers, Inc.
†From *Nature* 171 (1953): 737–738.

cell surface, play important roles in cellular recognition of extracellular signal molecules and of other cells. As noted earlier, polysaccharides include the storage polysaccharides starch and glycogen and the structural polysaccharide cellulose. Each of these polymers contains the six-carbon sugar glucose as its single repeating unit, but they differ in the nature of the bond between successive glucose units as well as in the presence and extent of side branches on the chains.

The Monomers Are Monosaccharides

The repeating units of polysaccharides are simple sugars called **monosaccharides** (from the Greek *mono,* meaning "single," and *sakkharon,* meaning "sugar"). A sugar can be defined as an aldehyde or ketone that has two or more hydroxyl groups. Thus, there are two categories of sugars: the *aldosugars,* with a terminal carbonyl group (**Figure 3-20a**), and the *ketosugars,* with an internal

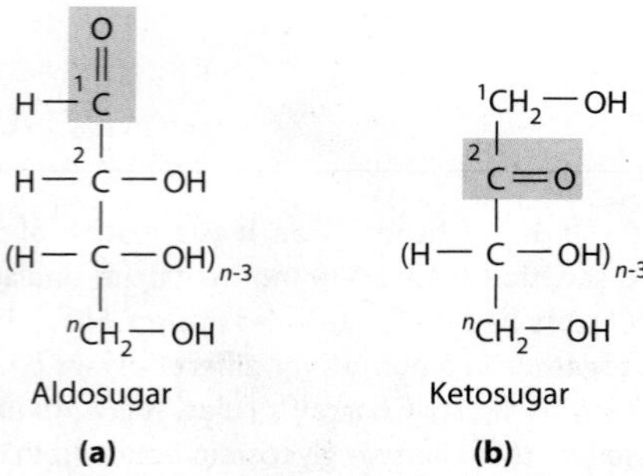

FIGURE 3-20 Structures of Monosaccharides. (a) Aldosugars have a carbonyl group on carbon atom 1. (b) Ketosugars have a carbonyl group on carbon atom 2. The number of carbon atoms in a monosaccharide (n) varies from three to seven.

carbonyl group (Figure 3-20b). Within these categories, sugars are named generically according to the number of carbon atoms they contain. Most sugars have between three and seven carbon atoms and thus are classified as *trioses* (three carbons), *tetroses* (four carbons), *pentoses* (five carbons), *hexoses* (six carbons), or *heptoses* (seven carbons). We have already encountered two pentoses—the ribose of RNA and the deoxyribose of DNA.

The single most common monosaccharide in the biological world is the aldohexose D-glucose, represented by the formula $C_6H_{12}O_6$ and by the structure shown in **Figure 3-21**. The formula $C_nH_{2n}O_n$ is characteristic of sugars and gave rise to the general term **carbohydrate** because compounds of this sort were originally thought of as "hydrates of carbon"—$C_n(H_2O)_n$. Although carbohydrates are not simply hydrated carbons, we will see in Chapter 11 that, for every CO_2 molecule incorporated into sugar, one water molecule is consumed (see Reaction 11-2 on page 295.

(a) Fischer projection (b) Haworth projection

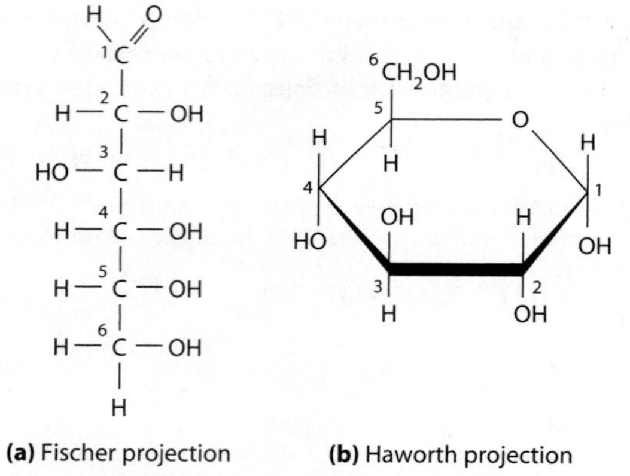

FIGURE 3-21 The Structure of D-Glucose. The D-glucose molecule can be represented by (a) the Fischer projection of the straight-chain form or (b) the Haworth projection of the ring form. In the Fischer projection, the —H and —OH groups are intended to project slightly out of the plane of the paper. In the Haworth projection, carbon atoms 2 and 3 are intended to jut out of the plane of the paper, and carbon atoms 5 and 6 are behind the plane of the paper. The —H and —OH groups then project upward or downward, as indicated. Notice that the carbon atoms are numbered from the more oxidized end of the molecule.

In keeping with the general rule for numbering carbon atoms in organic molecules, the carbons of glucose are numbered beginning with the more oxidized end of the molecule, the carbonyl group. Because glucose has four asymmetric carbon atoms (carbon atoms 2, 3, 4, and 5), there are $2^4 = 16$ different possible stereoisomers of the aldosugar $C_6H_{12}O_6$. Here, we will concern ourselves only with D-glucose, which is the most stable of the 16 isomers.

Figure 3-21a illustrates D-glucose as it appears in what chemists call a **Fischer projection,** with the —H and —OH groups intended to project slightly out of the plane of the paper. This structure depicts glucose as a linear molecule, and it is often a useful representation of glucose for teaching purposes. In reality, however, glucose exists in the cell in a dynamic equilibrium between the linear (or straight-chain) configuration of Figure 3-21a and the ring form shown in Figure 3-21b. This ring forms when the oxygen atom of the hydroxyl group on carbon atom 5 forms a bond with carbon atom 1. Although the bonding of this oxygen atom to carbon atoms 1 and 5 seems unlikely from the Fischer projection, it is actually favored by the tetrahedral nature of each carbon atom in the chain. This ring form is the predominant structure, because it is energetically more stable.

Therefore, the more satisfactory representation of glucose is the **Haworth projection** shown in Figure 3-21b. This shows the spatial relationship of different parts of the molecule and makes the spontaneous formation of a bond between an oxygen atom and carbon atoms 1 and 5 appear more likely. Either of the representations of glucose shown in Figure 3-21 is valid, but the Haworth projection is generally preferred because it indicates both the ring form and the spatial relationship of the carbon atoms.

Notice that formation of the ring structure results in the generation of one of two alternative forms of the molecule, depending on the spatial orientation of the hydroxyl group on carbon atom 1. These alternative forms of glucose are designated α and β. As shown in **Figure 3-22**, α-D-glucose has the hydroxyl group on carbon atom 1

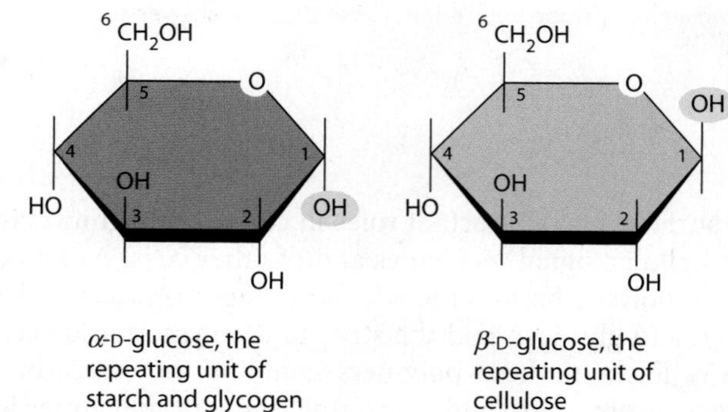

FIGURE 3-22 The Ring Forms of D-Glucose. The hydroxyl group on carbon atom 1 (blue oval) points downward in the α form and upward in the β form.

pointing downward in the Haworth projection, and β-D-glucose has the hydroxyl group on carbon atom 1 pointing upward. Starch and glycogen both have α-D-glucose as their repeating unit, whereas cellulose consists of strings of β-D-glucose.

In addition to the free monosaccharide and the long-chain polysaccharides, glucose also occurs in **disaccharides,** which consist of two monosaccharide units linked covalently. Three common disaccharides are shown in **Figure 3-23**. *Maltose* (malt sugar) consists of two glucose units linked together, whereas *lactose* (milk sugar) contains a glucose linked to a galactose and *sucrose* (common table sugar) has a glucose linked to a fructose. Both galactose and fructose will be discussed in more detail in Chapter 9, where the chemistry and metabolism of several sugars are considered.

Each of these disaccharides is formed by a condensation reaction in which two monosaccharides are linked together by the elimination of water. The resulting **glycosidic bond** is characteristic of linkages between sugars. In the case of maltose, both of the constituent glucose molecules are in the α form, and the glycosidic bond forms between carbon atom 1 of one glucose and carbon atom 4 of the other. This is called an *α glycosidic bond* because it involves a carbon atom 1 with its hydroxyl group in the α configuration. Lactose, on the other hand, is characterized by a *β glycosidic bond* because the hydroxyl group on carbon atom 1 of the galactose is in the β configuration. Some people lack the enzyme needed to hydrolyze this β glycosidic bond and are considered *lactose intolerant* due to their difficulty in metabolizing this disaccharide. The distinction between α and β again becomes critical when we get to the polysaccharides because both the three-dimensional configuration and the biological role of the polymer depend critically on the nature of the bond between the repeating monosaccharide units.

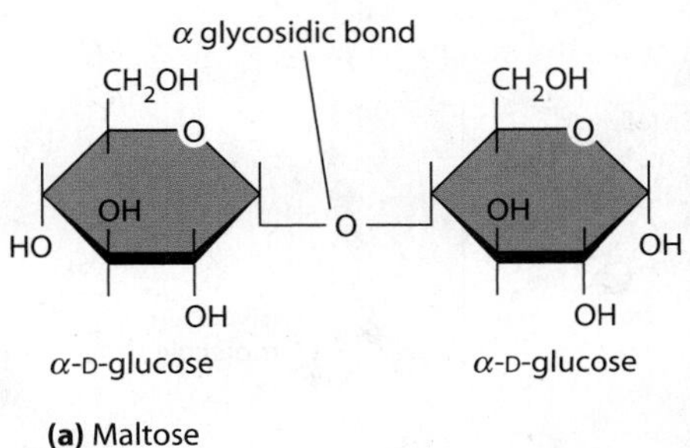

(a) Maltose

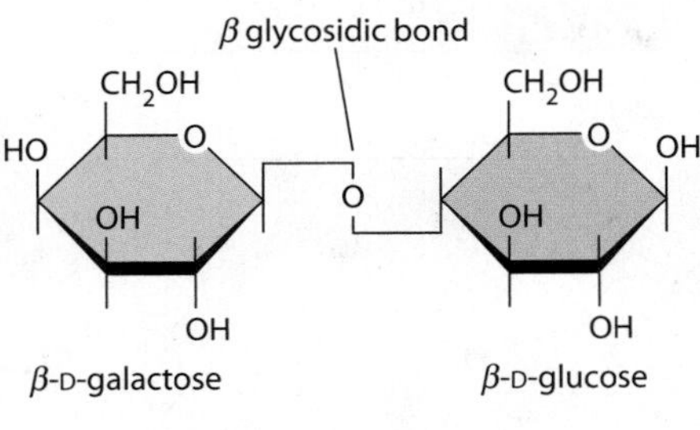

(b) Lactose

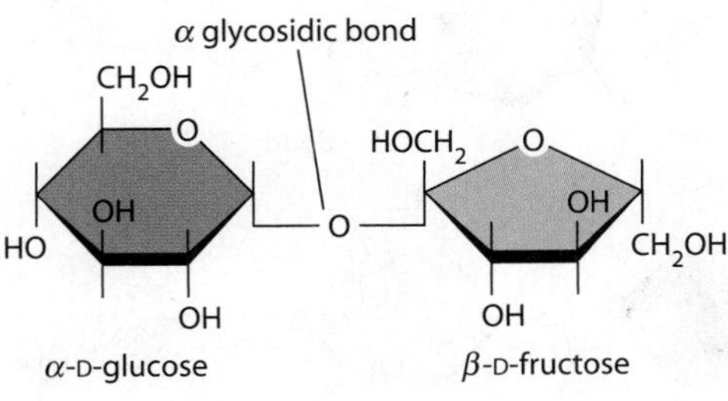

(c) Sucrose

FIGURE 3-23 Some Common Disaccharides. **(a)** Maltose (malt sugar) consists of two molecules of α-D-glucose linked by an α glycosidic bond. **(b)** Lactose (milk sugar) consists of a molecule of β-D-galactose linked to a molecule of β-D-glucose by a β glycosidic bond. **(c)** Sucrose (table sugar) consists of a molecule of α-D-glucose linked to a molecule of β-D-fructose by an α glycosidic bond.

The Polymers Are Storage and Structural Polysaccharides

Polysaccharides typically perform either storage or structural roles in cells. The most familiar *storage polysaccharides* are **starch,** found in plant cells (**Figure 3-24a**), and **glycogen,** found in animal cells and bacteria (Figure 3-24b). Both of these polymers consist of α-D-glucose units linked together by α glycosidic bonds. In addition to $\alpha(1 \rightarrow 4)$ bonds that link carbon atoms 1 and 4 of adjacent glucose units, these polysaccharides may contain occasional $\alpha(1 \rightarrow 6)$ linkages along the backbone, giving rise to side chains (Figure 3-24c). Storage polysaccharides can therefore be branched or unbranched polymers, depending on the presence or absence of $\alpha(1 \rightarrow 6)$ linkages.

Glycogen is highly branched, with $\alpha(1 \rightarrow 6)$ linkages occurring every 8 to 10 glucose units along the backbone and giving rise to short side chains of about 8 to 12 glucose units (Figure 3-24b). In our bodies, glycogen is stored mainly in the liver and in muscle tissue. In the liver, it is used as a source of glucose to maintain blood sugar levels. In muscle, it serves as a fuel source to generate ATP for muscle contraction. Bacteria also commonly store glycogen as a glucose reserve.

Starch, the glucose reserve commonly found in plant tissue, occurs both as unbranched **amylose** and as branched **amylopectin.** Like glycogen, amylopectin has $\alpha(1 \rightarrow 6)$ branches, but these occur less frequently along the backbone (once every 12 to 25 glucose units) and give rise to longer side chains (lengths of 20 to 25 glucose units are common; Figure 3-24a). Starch deposits are usually 10–30% amylose and 70–90% amylopectin. Starch is stored in plant cells as *starch grains* within the plastids—either within the *chloroplasts* that are the sites of carbon fixation and sugar synthesis in photosynthetic tissue or within the *amyloplasts,* which are specialized plastids for starch storage. The potato tuber, for example, is filled with starch-laden amyloplasts.

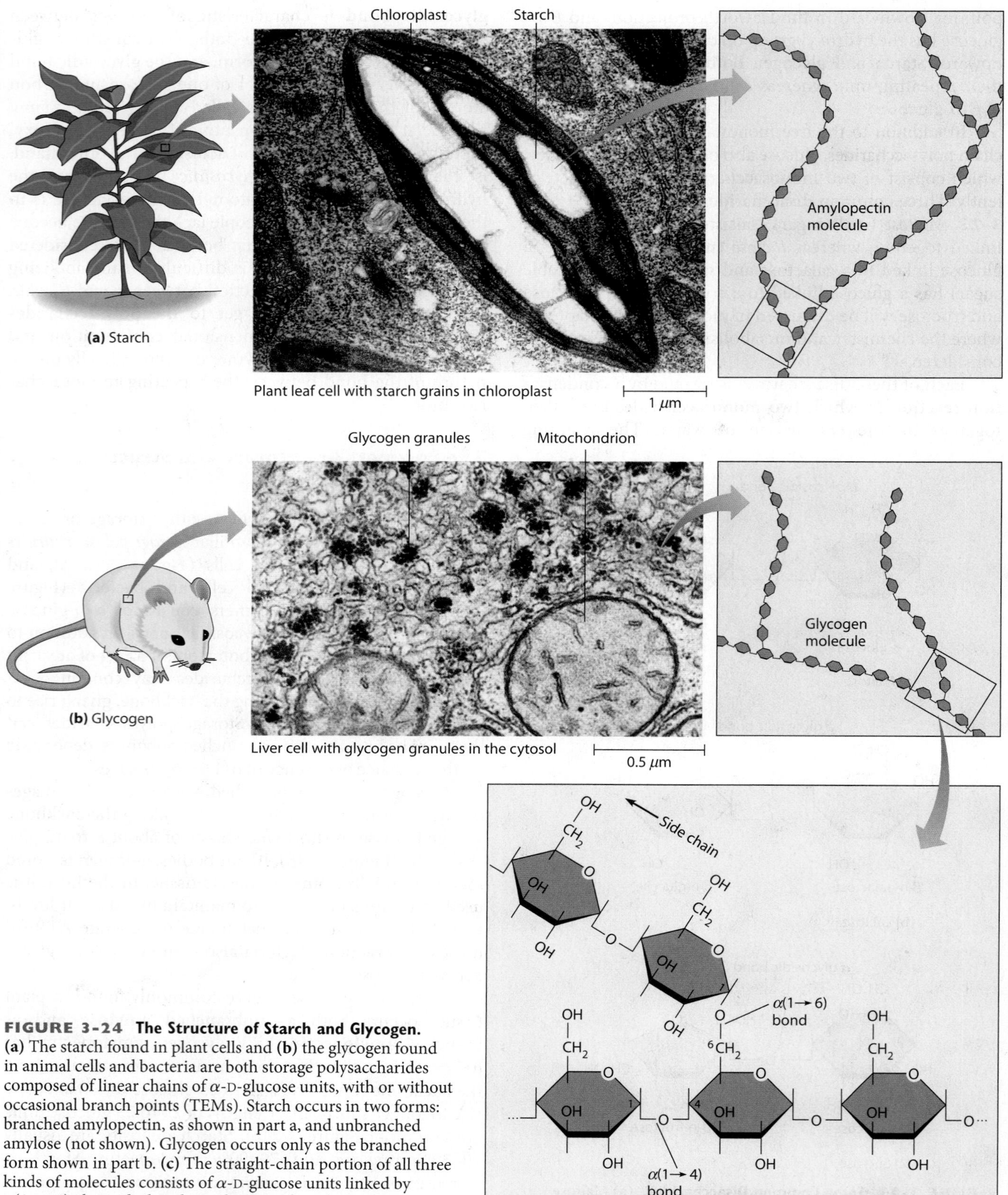

FIGURE 3-24 The Structure of Starch and Glycogen. **(a)** The starch found in plant cells and **(b)** the glycogen found in animal cells and bacteria are both storage polysaccharides composed of linear chains of α-D-glucose units, with or without occasional branch points (TEMs). Starch occurs in two forms: branched amylopectin, as shown in part a, and unbranched amylose (not shown). Glycogen occurs only as the branched form shown in part b. **(c)** The straight-chain portion of all three kinds of molecules consists of α-D-glucose units linked by $\alpha(1 \rightarrow 4)$ glycosidic bonds. In the case of amylopectin and glycogen, branch chains originate at $\alpha(1 \rightarrow 6)$ glycosidic bonds.

The best-known example of a *structural polysaccharide* is the **cellulose** found in plant cell walls (**Figure 3-25**). Cellulose is an important polymer quantitatively—more than half of the carbon in many plants is typically present in cellulose. Like starch and glycogen, cellulose is a polymer of glucose; however, the repeating monomer is β-D-glucose, and the linkage is therefore $\beta(1 \rightarrow 4)$. This bond has structural consequences that we will get to shortly, but it also has nutritional implications. Mammals do not possess an enzyme that can hydrolyze this $\beta(1 \rightarrow 4)$ bond and therefore cannot utilize cellulose as food. As a result, you can digest potatoes (starch) but not grass and wood (cellulose).

Animals such as cows and sheep might seem to be exceptions because they do eat grass and similar plant products. But they cannot cleave β glycosidic bonds either; they rely on microorganisms (bacteria and protozoa) in their digestive systems to do this for them. The microorganisms digest the cellulose, and the host animal then obtains the end-products of microbial digestion, now in a form the animal can use (glucose). Even termites do not actually digest wood. They simply chew it into small pieces that are then hydrolyzed to glucose monomers by microorganisms in their digestive tracts.

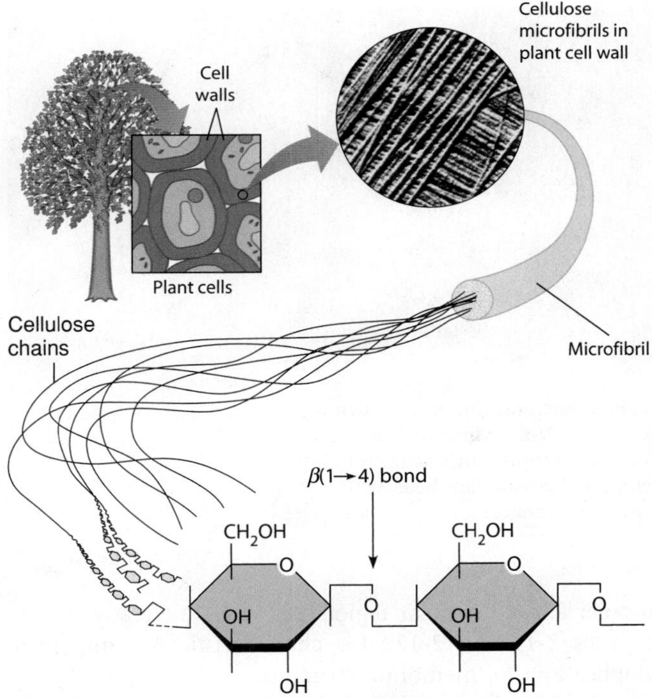

FIGURE 3-25 The Structure of Cellulose. Cellulose consists of long, unbranched chains of β-D-glucose units linked together by $\beta(1 \rightarrow 4)$ glycosidic bonds. Many such chains associate laterally and are held together by hydrogen bonds to form microfibrils. Individual microfibrils can be seen in the micrograph of a primary plant cell wall shown here (TEM). The $\beta(1 \rightarrow 4)$ glycosidic bonds cannot be hydrolyzed by most higher animals.

Although $\beta(1 \rightarrow 4)$-linked cellulose is the most abundant structural polysaccharide, others are also known. The celluloses of fungal cell walls, for example, contain either $\beta(1 \rightarrow 4)$ or $\beta(1 \rightarrow 3)$ linkages, depending on the species. The cell wall of most bacteria is somewhat more complex and contains two kinds of sugars, *N-acetylglucosamine (GlcNAc)* and *N-acetylmuramic acid (MurNAc)*. As shown in **Figure 3-26a**, GlcNAc and MurNAc are derivatives of *β-glucosamine,* a glucose molecule with the hydroxyl group on carbon atom 2 replaced by an amino group. GlcNAc is formed by acetylation of the amino group, and MurNAc requires the further addition of a three-carbon lactyl group to carbon atom 3. The cell wall polysaccharide is then formed by the linking of GlcNAc and MurNAc in a strictly alternating sequence with $\beta(1 \rightarrow 4)$ bonds (Figure 3-26b). Figure 3-26c shows the structure of yet another structural polysaccharide, the **chitin** found in insect exoskeletons, crustacean shells, and fungal cell walls. Chitin consists of GlcNAc units only, joined by $\beta(1 \rightarrow 4)$ bonds.

Polysaccharide Structure Depends on the Kinds of Glycosidic Bonds Involved

The distinction between the α and β glycosidic bonds of storage and structural polysaccharides has more than just nutritional significance. Because of the difference in linkages and therefore in the spatial relationship between successive glucose units, the two classes of polysaccharides differ markedly in secondary structure. The helical shape already established as a characteristic of both proteins and nucleic acids is also found in polysaccharides. Both starch and glycogen coil spontaneously into loose helices, but often the structure is not highly ordered due to the numerous side chains of amylopectin and glycogen.

Cellulose, by contrast, forms rigid, linear rods. These in turn aggregate laterally into *microfibrils* (see Figure 3-25). Microfibrils are about 5–20 nm in diameter and are composed of about 36 cellulose chains. Plant and fungal cell walls consist of these rigid microfibrils of cellulose embedded in a *noncellulosic matrix* containing a rather variable mixture of several other polymers (*hemicellulose* and *pectin,* mainly) and a protein called *extensin* that occurs only in the cell wall. Cell walls have been aptly compared to reinforced concrete, in which steel rods are embedded in the cement before it hardens to add strength. In cell walls, the cellulose microfibrils are the "rods" and the noncellulosic matrix is the "cement."

Lipids

Strictly speaking, **lipids** differ from the macromolecules we have discussed so far in this chapter because they are not formed by the kind of linear polymerization that gives rise to proteins, nucleic acids, and polysaccharides. However, they are commonly regarded as macromolecules because of their high molecular weights and their presence in important cellular structures, particularly membranes. Also, the final steps in the synthesis of triglycerides, phospholipids,

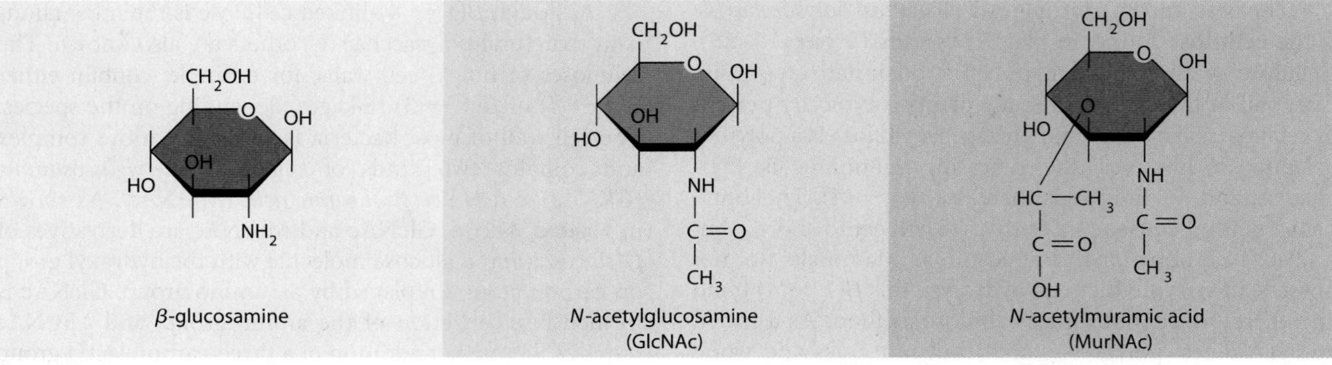

(a) Polysaccharide subunits

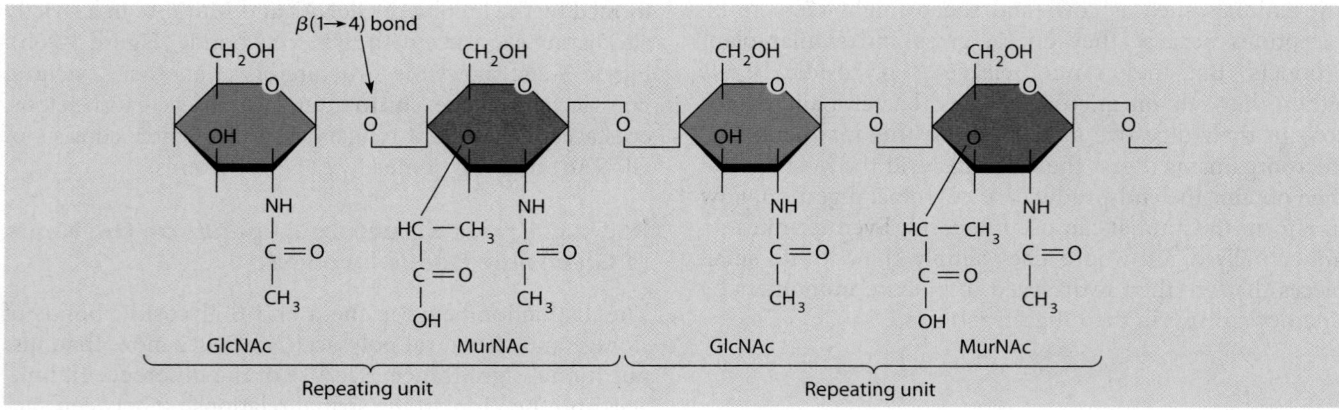

(b) A bacterial cell wall polysaccharide

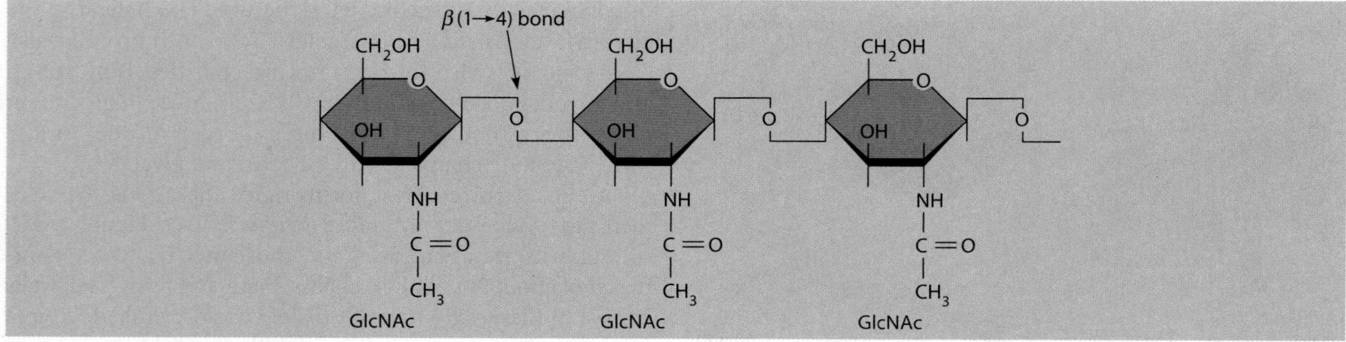

(c) The polysaccharide chitin

FIGURE 3-26 Polysaccharides of Bacterial Cell Walls and Insect Exoskeletons. **(a)** Chemical structures of the monosaccharide subunits glucosamine, *N*-acetylglucosamine (GlcNAc), and *N*-acetylmuramic acid (MurNAc). **(b)** A bacterial cell wall polysaccharide, consisting of alternating GlcNAc and MurNAc units linked by $\beta(1 \rightarrow 4)$ bonds. **(c)** The polysaccharide chitin found in insect exoskeletons and crustacean shells, with GlcNAc as its single repeating unit and successive GlcNAc units linked by $\beta(1 \rightarrow 4)$ bonds.

and other large lipid molecules involve condensation reactions similar to those used in polymer synthesis.

Lipids constitute a rather heterogeneous category of cellular components that resemble one another more in their solubility properties than in their chemical structures. *The distinguishing feature of lipids is their hydrophobic nature.* Although they have little, if any, affinity for water, they are readily soluble in nonpolar solvents such as chloroform or ether. Accordingly, we can expect to find that they are rich in nonpolar hydrocarbon regions and have relatively few polar groups. Some lipids, however, are *amphipathic,* having both a polar and a nonpolar region. As we have already seen in Figures 2-11 and 2-12, this characteristic has important implications for membrane structure.

Because they are defined in terms of solubility characteristics rather than chemical structure, we should not be surprised to find that lipids as a group include molecules that are quite diverse in terms of structure, chemistry, and function. Functionally, lipids play at least three main roles in cells. Some serve as forms of *energy storage,* others are involved in *membrane structure,* and still others have *specific biological functions,* such as the transmission of chemical signals into and within the cell. We will discuss lipids in terms of six main classes, based on their chemical structure: *fatty acids, triacylglycerols, phospholipids, glycolipids, steroids,* and *terpenes.* Note that, because of the wide variety of lipids and the fact that members of different classes sometimes share structural and chemical similarities, this is only one of several different ways to classify lipids. The six main classes of lipids as we will discuss them are illustrated in **Figure 3-27**, which includes representative examples of each class. We will look briefly at each of these six kinds of lipids, pointing out their functional roles in the process.

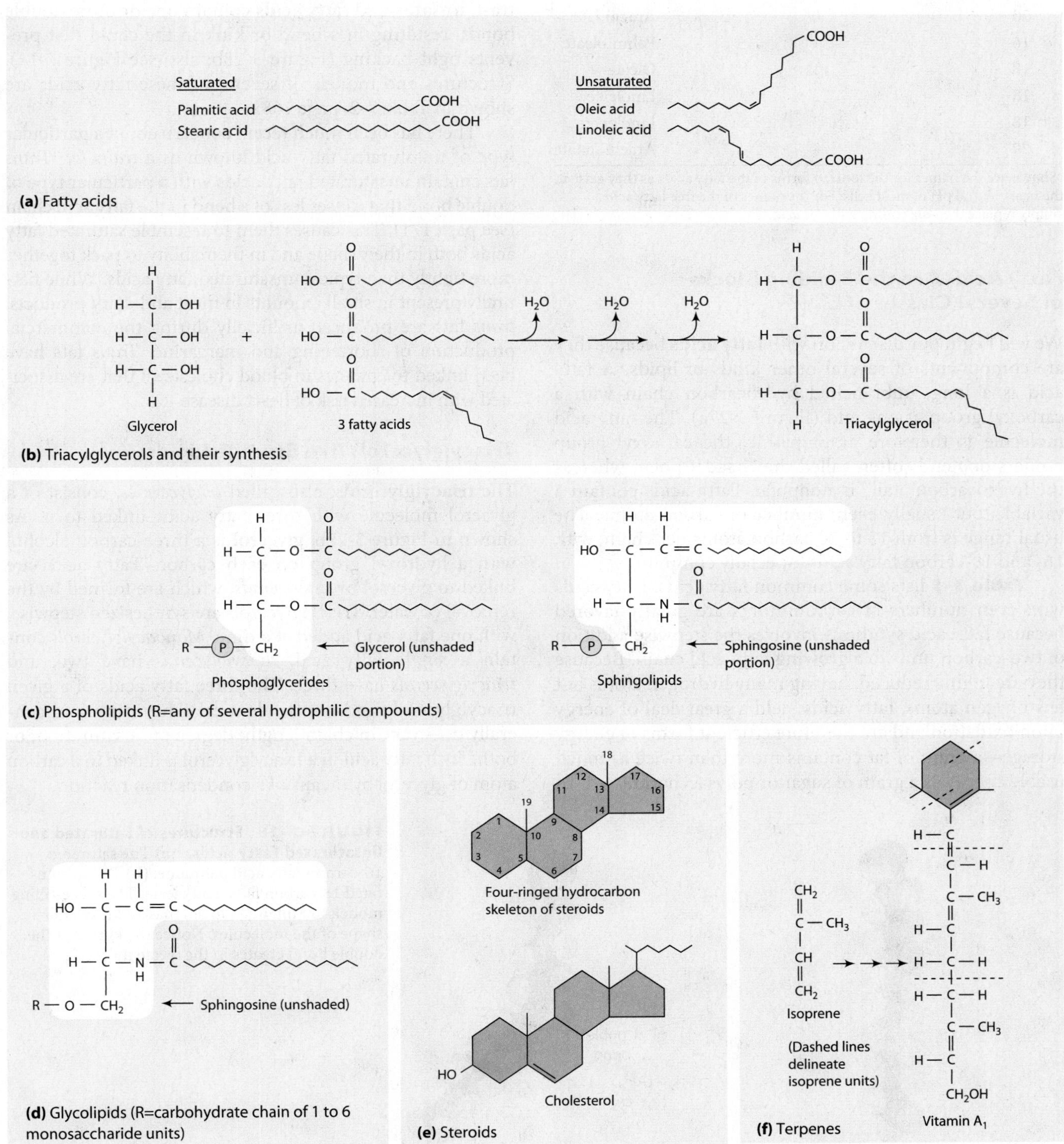

FIGURE 3-27 The Main Classes of Lipids. The zigzag lines in parts a–d represent the long hydrocarbon chains of fatty acids. Each corner of the zigzag lines represents a methylene (—CH_2—) group.

Table 3-5	**Some Common Fatty Acids in Cells**	
Number of Carbons	**Number of Double Bonds**	**Common Name***
12	0	Laurate
14	0	Myristate
16	0	Palmitate
18	0	Stearate
20	0	Arachidate
16	1	Palmitoleate
18	1	Oleate
18	2	Linoleate
18	3	Linolenate
20	4	Arachidonate

*Shown are the names for the ionized forms of the fatty acids as they exist at the near-neutral pH of most cells. For the names of the free fatty acids, simply replace the *-ate* ending with *-ic acid.*

Fatty Acids Are the Building Blocks of Several Classes of Lipids

We will begin our discussion with **fatty acids** because they are components of several other kinds of lipids. A fatty acid is a long, unbranched hydrocarbon chain with a carboxyl group at one end (Figure 3-27a). The fatty acid molecule is therefore amphipathic; the carboxyl group renders one end (often called the "head") polar, whereas the hydrocarbon "tail" is nonpolar. Fatty acids contain a variable, but usually even, number of carbon atoms. The usual range is from 12 to 20 carbon atoms per chain, with 16- and 18-carbon fatty acids especially common.

Table 3-5 lists some common fatty acids. Fatty acids with even numbers of carbon atoms are greatly favored because fatty acid synthesis involves the stepwise addition of two-carbon units to a growing fatty acid chain. Because they are highly reduced, having many hydrogen atoms but few oxygen atoms, fatty acids yield a great deal of energy upon oxidation and are therefore efficient forms of energy storage—a gram of fat contains more than twice as much usable energy as a gram of sugar or polysaccharide.

Table 3-5 also shows the variability in fatty acids due to the presence of double bonds between carbons. Fatty acids without double bonds are referred to as **saturated fatty acids** because every carbon atom in the chain has the maximum number of hydrogen atoms attached to it (**Figure 3-28a**). The general formula for a saturated fatty acid with n carbon atoms is $C_nH_{2n}O_2$. Saturated fatty acids have long, straight chains that pack together well. By contrast, **unsaturated fatty acids** contain one or more double bonds, resulting in a bend or kink in the chain that prevents tight packing (Figure 3-28b; also see Figure 7-14). Structures and models of several of these fatty acids are shown in Table 7-2, page 168.

There has been much recent concern about a particular type of unsaturated fatty acid known as a *trans fat. Trans* fats contain unsaturated fatty acids with a particular type of double bond that causes less of a bend in the fatty acid chain (see page 171). This causes them to resemble saturated fatty acids both in their shape and in their ability to pack together more tightly than typical unsaturated fatty acids. While naturally present in small amounts in meat and dairy products, *trans* fats are produced artificially during the commercial production of shortening and margarine. *Trans* fats have been linked to changes in blood cholesterol that are associated with increased risk of heart disease.

Triacylglycerols Are Storage Lipids

The **triacylglycerols,** also called *triglycerides,* consist of a glycerol molecule with three fatty acids linked to it. As shown in Figure 3-27b, **glycerol** is a three-carbon alcohol with a hydroxyl group on each carbon. Fatty acids are linked to glycerol by *ester bonds,* which are formed by the removal of water. Triacylglycerols are synthesized stepwise, with one fatty acid added at a time. *Monoacylglycerols* contain a single fatty acid, *diacylglycerols* have two, and *triacylglycerols* have three. The three fatty acids of a given triacylglycerol need not be identical. They can—and generally do—vary in chain length, degree of unsaturation, or both. Each fatty acid in a triacylglycerol is linked to a carbon atom of glycerol by means of a condensation reaction.

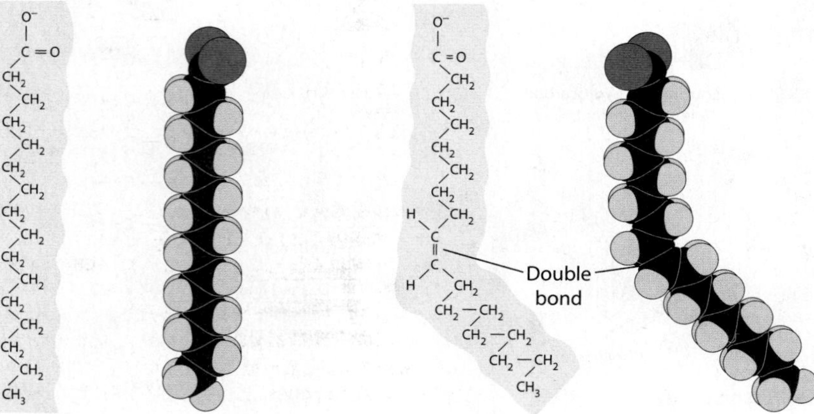

FIGURE 3-28 Structures of Saturated and Unsaturated Fatty Acids. (a) The saturated 16-carbon fatty acid palmitate. **(b)** The unsaturated 18-carbon fatty acid oleate. The space-filling models are intended to emphasize the overall shape of the molecules. Notice the kink that the double bond creates in the oleate molecule.

The main function of triacylglycerols is to store energy, which will be of special interest to our discussion of energy metabolism in Chapter 10. In some animals, triacylglycerols also provide insulation against low temperatures. Animals such as walruses, seals, and penguins that live in very cold climates store triacylglycerols under their skin and depend on the insulating properties of this fat for survival.

Triacylglycerols containing mostly saturated fatty acids are usually solid or semisolid at room temperature and are called *fats*. Fats are prominent in the bodies of animals, as evidenced by the fat that comes with most cuts of meat, by the large quantity of lard that is obtained as a by-product of the meat-packing industry, and by the widespread concern people have that they are "getting fat." In plants, most triacylglycerols are liquid at room temperature, as the term *vegetable oil* suggests. Because the fatty acids of oils are predominantly unsaturated, their hydrocarbon chains have kinks that prevent an orderly packing of the molecules. As a result, vegetable oils have lower melting temperatures than most animal fats do. Soybean oil and corn oil are two familiar vegetable oils. Vegetable oils can be converted into solid products such as margarine and shortening by hydrogenation (saturation) of the double bonds, a process explored further in Problem 3-15 at the end of the chapter.

Phospholipids Are Important in Membrane Structure

Phospholipids make up a third class of lipids (see Figure 3-27c). They are similar to triacylglycerols in some chemical details but differ strikingly in their properties and their role in the cell. First and foremost, phospholipids are important in membrane structure due to their amphipathic nature. In fact, as we saw in Chapter 2, they are critical to the bilayer structure found in all membranes (see Figure 2-12). Based on their chemistry, phospholipids are classified as *phosphoglycerides* or *sphingolipids* (see Figure 3-27c).

Phosphoglycerides are the predominant phospholipids present in most membranes. Like triacylglycerols, a phosphoglyceride consists of fatty acids that are esterified to a glycerol molecule. However, the basic component of a phosphoglyceride is **phosphatidic acid,** which has just two fatty acids and a phosphate group attached to a glycerol backbone (**Figure 3-29a**). Phosphatidic acid is a key intermediate in the synthesis of other phosphoglycerides but is itself not at all prominent in membranes. Instead, membrane phosphoglycerides invariably have, in addition, a small hydrophilic alcohol linked to the phosphate by an ester bond and represented in Figure 3-29a as an R group. The alcohol is usually *serine, ethanolamine, choline,* or *inositol* (see Figure 3-29b), groups that contribute to the polar nature of the phospholipid head group.

The combination of a highly polar head and two long nonpolar chains gives phosphoglycerides the characteristic amphipathic nature that is so critical to their role in membrane structure. As we saw earlier, the fatty acids can vary considerably in both length and the presence and position of sites of unsaturation. In membranes, 16- and 18-carbon fatty acids are most common, and a typical phosphoglyceride molecule is likely to have one saturated and one unsaturated fatty acid. The length and the degree of unsaturation of fatty acid chains in membrane phospholipids profoundly affect membrane fluidity and can, in fact, be regulated by the cells of some organisms.

In addition to the phosphoglycerides, some membranes contain another class of phospholipid called **sphingolipids,** which are important in membrane structure and cell signaling. When sphingolipids were first discovered by Johann Thudicum in the late nineteenth century, their biological role seemed as enigmatic as the Sphinx, after which he named them. As the name suggests, these lipids are based not on glycerol but on the amine alcohol **sphingosine.** As shown in Figure 3-27c, sphingosine has a long hydrocarbon chain with a single site of unsaturation near the polar end. Through its amino group, sphingosine can form an amide bond to a long-chain fatty acid (up to 34 carbons). The resulting molecule is called a *ceramide* and

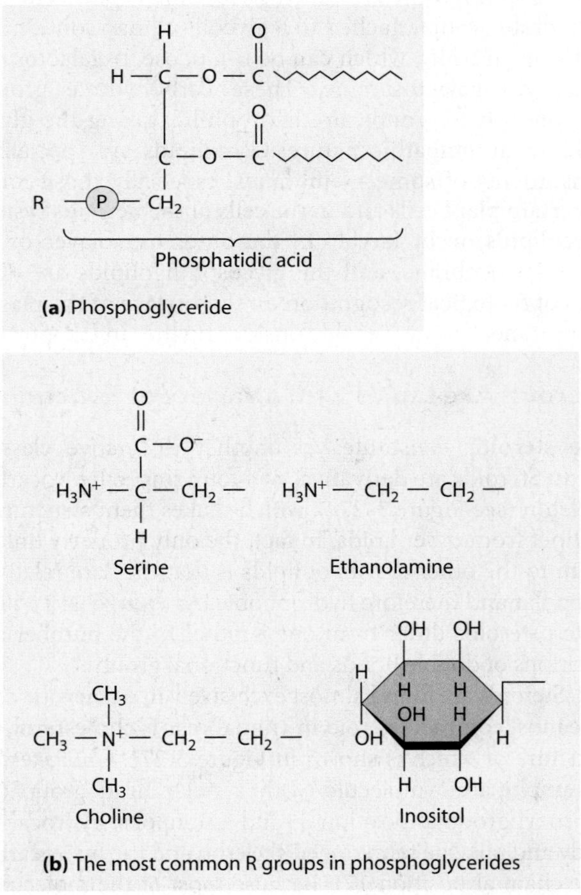

Phosphatidic acid

(a) Phosphoglyceride

Serine

Ethanolamine

Choline

Inositol

(b) The most common R groups in phosphoglycerides

FIGURE 3-29 Structures of Common Phosphoglycerides. **(a)** A phosphoglyceride consists of a molecule of phosphatidic acid (glycerol esterified to two fatty acids and a phosphate group) with a small polar alcohol, represented as R, also esterified to the phosphate group. **(b)** The four most common R groups found in phosphoglycerides are serine, ethanolamine, choline, and inositol, the first three of which contain a positively charged amino group or nitrogen atom.

consists of a polar region flanked by two long, nonpolar tails, giving it a shape approximating that of the phospholipids.

The hydroxyl group on carbon atom 1 of sphingosine juts out from what is effectively the head of this hairpin molecule. A sphingolipid is formed when any of several polar groups becomes linked to this hydroxyl group. A whole family of sphingolipids exists, differing only in the chemical nature of the polar group attached to the hydroxyl group of the ceramide (the R group of Figure 3-27c). Sphingolipids are present predominantly in the outer leaflet of the plasma membrane bilayer, where they often are found in *lipid rafts,* which, as we will see in Chapter 7, are localized microdomains within a membrane that facilitate communication with the external environment of the cell.

Glycolipids Are Specialized Membrane Components

Glycolipids are lipids containing a carbohydrate group instead of a phosphate group and are typically derivatives of sphingosine or glycerol (see Figure 3-27d). Those containing sphingosine are called *glycosphingolipids.* The carbohydrate group attached to a glycolipid may contain one to six sugar units, which can be D-glucose, D-galactose, or *N*-acetyl-D-galactosamine. These carbohydrate groups, like phosphate groups, are hydrophilic, giving the glycolipid an amphipathic nature. Glycolipids are specialized constituents of some membranes, especially those found in certain plant cells and in the cells of the nervous system. Glycolipids occur largely in the outer monolayer of the plasma membrane, and the glycosphingolipids are often sites of biological recognition on the surface of the plasma membrane, a topic we will consider further in Chapter 14.

Steroids Are Lipids with a Variety of Functions

The **steroids** constitute yet another distinctive class of lipids. Steroids are derivatives of a four-ringed hydrocarbon skeleton (see Figure 3-27e), which makes them structurally distinct from other lipids. In fact, the only property linking them to the other classes of lipids is that they are relatively nonpolar and therefore hydrophobic. As **Figure 3-30** illustrates, steroids differ from one another in the number and positions of double bonds and functional groups.

Steroids are found almost exclusively in eukaryotic cells. The most common steroid in animal cells is **cholesterol,** the structure of which is shown in Figure 3-27e. Cholesterol is an amphipathic molecule, with a polar head group (the hydroxyl group at position 3) and a nonpolar hydrocarbon body and tail (the four-ringed skeleton and the hydrocarbon side chain at position 17). Because most of the molecule is hydrophobic, cholesterol is insoluble and is found primarily in membranes. It occurs in the plasma membrane of animal cells and in most of the membranes of organelles, except the inner membranes of mitochondria and chloroplasts. Similar membrane steroids also occur in other cells, including *stigmasterol* and *sitosterol* in plant cells, *ergosterol* in fungal cells, and related sterols in *Mycoplasma* bacteria.

Cholesterol is the starting point for the synthesis of all the **steroid hormones** (Figure 3-30), which include the male and female *sex hormones,* the *glucocorticoids,* and the *mineralocorticoids.* The sex hormones include the *estrogens* produced by the ovaries of females (*estradiol,* for example) and the *androgens* produced by the testes of males (*testosterone,* for example). The glucocorticoids (*cortisol,* for example) are a family of hormones that promote gluconeogenesis (synthesis of glucose) and suppress inflammation reactions. Mineralocorticoids such as *aldosterone* regulate ion balance by promoting the reabsorption of sodium, chloride, and bicarbonate ions in the kidney.

Terpenes Are Formed from Isoprene

The final class of lipids shown in Figure 3-27 consists of the **terpenes.** Terpenes are synthesized from the five-carbon compound *isoprene* and are therefore also called *isoprenoids.* Isoprene and its derivatives are joined together in various combinations to produce such substances as *vitamin A* (see Figure 3-27f), a required nutrient in our bodies, and *carotenoid pigments,* which are involved in light harvesting in plants during photosynthesis. Other isoprene-based compounds are *dolichols,* which are involved in activating sugar derivatives, and electron carriers such as *coenzyme Q* and *plastoquinone,* which we will encounter when we study respiration and photosynthesis in detail in Chapters 10 and 11. Finally, polymers of isoprene units known as polyisoprenoids are found in the cell membranes of Archaea, a unique domain of organisms distinct from eukaryotes and bacteria that we will encounter in Chapter 4.

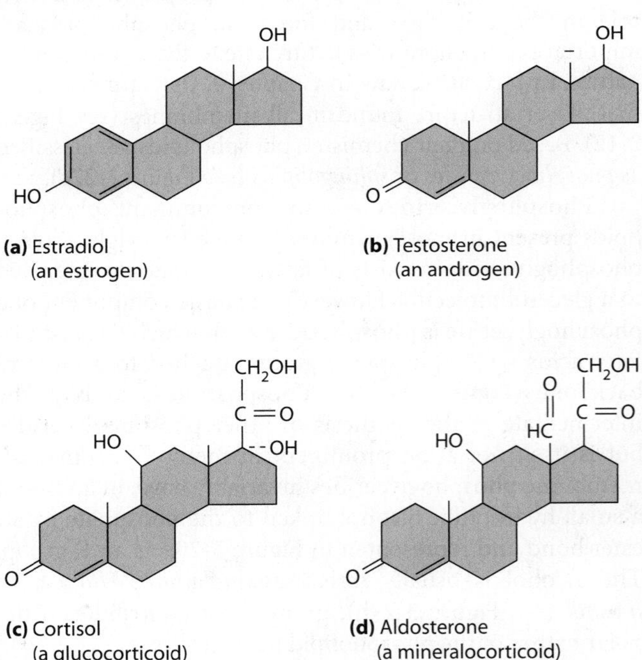

FIGURE 3-30 Structures of Several Common Steroid Hormones. Among the many steroids that are synthesized from cholesterol are the hormones **(a)** estradiol, an estrogen; **(b)** testosterone, an androgen; **(c)** cortisol, a glucocorticoid; and **(d)** aldosterone, a mineralocorticoid.

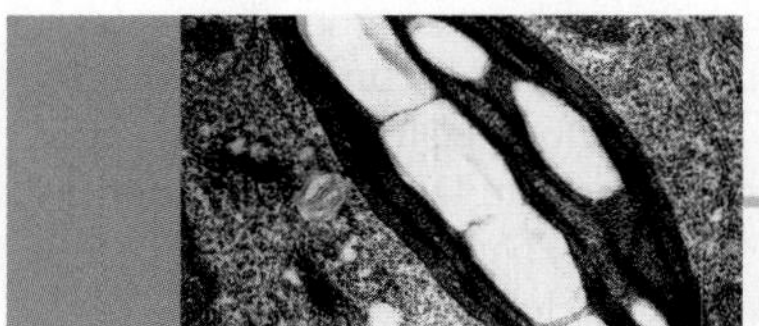

SUMMARY OF KEY POINTS

The Macromolecules of the Cell

- Three classes of macromolecular polymers are prominent in cells: proteins, nucleic acids, and polysaccharides. Lipids are not long, polymeric macromolecules, but they are included in this chapter because of their general importance as constituents of cells (especially membranes) and because their synthesis involves condensation reactions between smaller constituents.
- Proteins and nucleic acids are macromolecules in which the particular ordering of monomers is critical to their roles in the cell. Polysaccharides, on the other hand, usually contain only one or a few kinds of repeating units and play storage or structural roles instead.

Proteins

- All of the thousands of different proteins in a cell are linear chains of amino acid monomers. Each of the 20 different amino acids found in proteins has a different R group, which can be either hydrophobic, hydrophilic uncharged, or hydrophilic charged. These amino acids can be linked together in any sequence via peptide bonds to form the wide variety of polypeptides that make up monomeric and multimeric proteins.
- The amino acid sequence of a polypeptide (its primary structure) usually contains all of the information necessary to specify local folding into α helices and β sheets (secondary structure), overall three-dimensional shape (tertiary structure), and, for multimeric proteins, association with other polypeptides (quaternary structure).
- Major forces influencing polypeptide folding and stability are covalent disulfide bond formation and several types of noncovalent interactions: hydrogen bonds, ionic bonds, van der Waals interactions, and hydrophobic interactions.
- Despite extensive knowledge of the forces involved in protein folding, we are still not able to predict the final folded tertiary structure of a protein from its primary amino acid sequence, except in the case of peptides and relatively small proteins.

Nucleic Acids

- The nucleic acids DNA and RNA are informational macromolecules composed of nucleotide monomers linked together by phosphodiester bridges in a specific order. Each nucleotide is composed of a deoxyribose or ribose sugar, a phosphate, and a purine or pyrimidine base.
- The base sequence of the nucleotides in a particular segment of DNA known as a gene determines the sequence of the amino acids in the protein encoded by that gene. DNA is the carrier of genetic information in a cell, while mRNA, tRNA, and rRNA function mainly in expression of the information in DNA by their involvement in protein synthesis.
- While RNA is mainly single stranded, DNA forms a double-stranded helix based on complementary base pairing (A with T and C with G) that is stabilized by hydrogen bonding. Elucidation of the double-helical structure of DNA was a defining biological breakthrough of the twentieth century.
- In addition to mRNA, tRNA, and rRNA, several other types of RNAs have been discovered in cells, including microRNA (miRNA) and small interfering RNA (siRNA). These RNAs regulate gene expression either by inhibiting translation or by promoting mRNA degradation.

Polysaccharides

- In contrast to nucleic acids and proteins, polysaccharides show little variation in sequence and serve storage or structural roles. They typically consist either of a single type of monosaccharide or of two alternating monosaccharides linked together by either α or β glycosidic bonds. The type of glycosidic bond determines whether the polysaccharide serves as energy storage or as a structural polysaccharide.
- The α linkages are readily digestible by animals and are found in storage polysaccharides such as starch and glycogen, which consist solely of glucose monomers. In contrast, the β glycosidic bonds of cellulose and chitin are not digestible by animals and give these molecules a rigid shape suitable to their functions as structural molecules.

Lipids

- Lipids are not true polymers but are often considered macromolecules due to their high molecular weight and their frequent association with macromolecules, particularly proteins. Lipids vary substantially in chemical structure but are grouped together because they share the common property of being hydrophobic and thus are nearly insoluble in water.
- Fatty acids are lipids consisting of a long hydrocarbon chain of 12–20 carbon atoms with a carboxylic acid group at one end. They are energy-rich molecules found in the triacylglycerols that make up animal fats and vegetable oils, as well as in the phospholipids found in all cellular membranes.
- Phosphoglycerides and sphingolipids are types of phospholipids that make up the lipid bilayer of biological membranes. They are amphipathic molecules with two hydrophobic fatty acid chains and a polar phosphate-containing head group.
- Glycolipids are similar to phospholipids but contain a polar carbohydrate group instead of phosphate. They are often found on the outer surface of membranes, where they play a role in cell recognition.
- Other important cellular lipids are the steroids (including cholesterol and several sex hormones) and the terpenes (including vitamin A and some important coenzymes).

 **www.masteringbiology.com**

1. MasteringBiology Assignments

Tutorials and Activities

- Molecular Model Structural Tutorials for visualization of 3D structures
- BioFlix Tutorials including 3D animations with hints and detailed feedback for common wrong answers
- Proteins: Functions of Proteins; Organic Molecules: Functional Groups; Amino Acid Functional Groups; Amino Acids and Peptides; Proteins: Levels of Structure; Protein Structure; Levels of Structure in Proteins; Proteins: Primary Structure; Proteins: Secondary Structure; Proteins: Tertiary Structure; Proteins: Models of Proteins; Proteins: Quaternary Structure; Proteins; Nucleic Acids: Functions of Nucleic Acids; Nucleic Acid Building Blocks; DNA and RNA Structure; Nucleic Acids: Structure; Nucleic Acids; Structure of RNA and DNA; DNA Structure; The Double Helix; Nucleic Acid Structure; Heritable Information: DNA; Video Tutor Session Quiz: DNA Structure; Nucleic Acids: The Double Helix; DNA Double Helix

Questions

- Reading Quiz
- Multiple Choice
- End-of-chapter questions

2. eText Read your book online, search, highlight text, take notes, and more

3. The Study Area

- Chapter Review Quiz, Guide to Techniques and Methods, 3D-Structure Tutorials, Activities, BioFlix, Videos, Biology Review, Word Study Tools, Glossary, Flashcards

PROBLEM SET

More challenging problems are marked with a •.

3-1 Polymers and Their Properties. For each of the six biological polymers listed, indicate which of the properties apply. Each polymer has multiple properties, and a given property may be used more than once.

Polymers

(a) Cellulose
(b) Messenger RNA
(c) Globular protein
(d) Amylopectin
(e) DNA
(f) Fibrous protein

Properties

1. Branched-chain polymer
2. Extracellular location
3. Glycosidic bonds
4. Informational macromolecule
5. Peptide bond
6. β linkage
7. Phosphodiester bridge
8. Nucleoside triphosphates
9. Helical structure possible
10. Synthesis requires a template

3-2 Stability of Protein Structure. Several different kinds of bonds or interactions are involved in generating and maintaining the structure of proteins. List five such bonds or interactions, give an example of an amino acid that might be involved in each, and indicate which level(s) of protein structure might be generated or stabilized by that particular kind of bond or interaction.

3-3 Amino Acid Localization in Proteins. Amino acids tend to be localized either in the interior or on the exterior of a globular protein molecule, depending on their relative affinities for water.

(a) For each of the following pairs of amino acids, choose the one that is more likely to be found in the interior of a protein molecule, and explain why:

alanine; glycine	glutamate; aspartate
tyrosine; phenylalanine	methionine; cysteine

(b) Explain why cysteine residues with free sulfhydryl groups tend to be localized on the exterior of a protein molecule, whereas those involved in disulfide bonds are more likely to be buried in the interior of the molecule.

3-4 Sickle-Cell Anemia. Sickle-cell anemia (see page 52) is a striking example of the drastic effect a single amino acid substitution can have on the structure and function of a protein.

(a) Given the chemical nature of glutamate and valine, can you suggest why substitution of valine for glutamate at position 6 of the β chain would be especially deleterious?

(b) Suggest several amino acids that would be much less likely than valine to cause impairment of hemoglobin function if substituted for the glutamate at position 6 of the β chain.

(c) Can you see why, in some cases, two proteins could differ at several points in their amino acid sequence and still be very similar in structure and function? Explain.

3-5 Hair Versus Silk. The α-keratin of human hair is a good example of a fibrous protein with extensive α-helical structure. Silk fibroin is also a fibrous protein, but it consists primarily of β-sheet structure. Fibroin is essentially a polymer of alternating glycines and alanines, whereas α-keratin contains most of the common amino acids and has many disulfide bonds.

(a) If you were able to grab onto both ends of an α-keratin polypeptide and pull, you would find it to be both extensible (it can be stretched to about twice its length in moist heat) and elastic (when you let go, it will return to its normal length). In contrast, a fibroin polypeptide has essentially no extensibility, and it has great resistance to breaking. Explain these differences.

(b) Can you suggest why fibroin assumes a pleated sheet structure, whereas α-keratin exists as an α helix and even reverts spontaneously to a helical shape when it has been stretched artificially?

3-6 The "Permanent" Wave That Isn't. The "permanent" wave that your local beauty parlor offers depends critically on rearrangements in the extensive disulfide bonds of keratin that give your hair its characteristic shape. To change the shape of your hair (that is, to give it a wave or curl), the beautician first treats your hair with a sulfhydryl reducing agent, then uses curlers or rollers to impose the desired artificial shape, and follows this by treatment with an oxidizing agent.

(a) What is the chemical basis of a permanent? Be sure to include the use of a reducing agent and an oxidizing agent in your explanation.

(b) Why do you suppose a permanent isn't permanent? (Explain why the wave or curl is gradually lost during the weeks following your visit to the beautician.)

(c) Can you suggest an explanation for naturally curly hair?

3-7 Features of Nucleic Acids. For each of the following features of nucleic acids, indicate whether it is true of DNA only (D), of RNA only (R), of both DNA and RNA (DR), or of neither (N).

(a) Contains the base uracil.

(b) Contains the nucleotide deoxythymidine monophosphate.

(c) Is usually double-stranded.

(d) Is a polymer.

(e) Contains a phosphate group.

(f) Is an inherently directional molecule, with an N-terminus on one end and a C-terminus on the other end.

3-8 Wrong Again. For each of the following false statements, change the statement to make it true, and explain why it was false:

(a) Nucleic acids are polymers consisting of chemically identical repeating nucleotide monomers.

(b) A protein may have an α helical secondary structure. An α helix is spiral in shape and stabilized by covalent bonds between the NH group and the CO group in the polypeptide backbone.

(c) A protein can be denatured by extremes of pH but temperature has little effect on tertiary structure.

(d) Nucleic acids are synthesized from monomers that are activated by linking them to a carrier molecule in an energy-requiring reaction.

(e) The disaccharide sucrose comprises two glucose monomers covalently linked together.

(f) A β-pleated sheet is an extended sheet-like conformation with the R groups of successive amino acids jutting out on the same side of the sheet.

(g) It is easy to predict the final folded structure of a protein from its amino acid sequence using today's powerful supercomputers.

(h) MircoRNAs (miRNAs) are small exogenous RNAs that are derived from exogenous sources (e.g., infection by an RNA virus).

3-9 Storage Polysaccharides. The only common examples of branched-chain polymers in cells are the storage polysaccharides glycogen and amylopectin. Both are degraded exolytically, which means by stepwise removal of terminal glucose units.

(a) Why might it be advantageous for a storage polysaccharide to have a branched-chain structure instead of a linear structure?

(b) Why can't amylopectin be fully degraded by amylose alone?

(c) How are some mammals, like cows, able to eat grass while other animals cannot?

(d) How does the structure of cellulose contribute to the properties of a cell wall?

3-10 Carbohydrate Structure. From the following descriptions of gentiobiose, raffinose, and a dextran, draw Haworth projections of each:

(a) *Gentiobiose* is a disaccharide found in gentians and other plants. It consists of two molecules of β-D-glucose linked to each other by a $\beta(1 \rightarrow 6)$ glycosidic bond.

(b) *Raffinose* is a trisaccharide found in sugar beets. It consists of one molecule each of α-D-galactose, α-D-glucose, and β-D-fructose, with the galactose linked to the glucose by an $\alpha(1 \rightarrow 6)$ glycosidic bond and the glucose linked to the fructose by an $\alpha(1 \rightarrow 2)$ bond.

(c) *Dextrans* are polysaccharides produced by some bacteria. They are polymers of α-D-glucose linked by $\alpha(1 \rightarrow 6)$ glycosidic bonds, with frequent $\alpha(1 \rightarrow 3)$ branching. Draw a portion of a dextran, including one branch point.

3-11 Telling Them Apart. For each of the following pairs of molecules, specify a property that would distinguish between them, and indicate two different chemical tests or assays that could be used to make that distinction:

(a) The protein insulin and the DNA in the gene that encodes insulin

(b) The DNA that encodes insulin and the messenger RNA for insulin

(c) Starch and cellulose

(d) Amylose and amylopectin

(e) The monomeric protein myoglobin and the tetrameric protein hemoglobin

(f) A triacylglycerol and a phospholipid with a very similar fatty acid content

(g) A glycolipid and a sphingolipid

(h) A bacterial cell wall polysaccharide and chitin

• **3-12 Find an Example.** For each of the following classes of proteins, give two examples of specific proteins, briefly state how each one is important in cells, and mention a type of cell in which each protein would be found. You may use any available resources: later chapters in the text, your notes from other classes, or the Internet. Try to find examples that your professor may not be familiar with. (We love when our students can educate us!)

(a) Enzymes

(b) Structural proteins

(c) Motility proteins

(d) Regulatory proteins

(e) Transport proteins

(f) Hormonal proteins

(g) Receptor proteins

(h) Defensive proteins

(i) Storage proteins

3-13 Amylose and Amylopectin. Starch is composed of two polysaccharides, amylose and amylopectin. Both polysaccharides are polymers which consist of repeating units of α-D-glucose.

(a) What is the difference between the structures of amylose and amylopectin?

(b) How do the structures of amylose and amylopectin alter their properties?

• **3-14 Thinking About Lipids.** You should be able to answer each of the following questions based on the properties of lipids discussed in this chapter:

(a) How would you define a lipid? In what sense is the operational definition different from that of proteins, nucleic acids, or carbohydrates?

(b) Arrange the following lipids in order of decreasing polarity: cholesterol, estradiol, fatty acid, phosphatidyl choline, and triglyceride. Explain your reasoning.

(c) Which would you expect to resemble a sphingomyelin molecule more closely: a molecule of phosphatidyl choline containing two molecules of palmitate acid as its fatty acid side chains or a phosphatidyl choline molecule with one molecule of palmitate and one molecule of oleate as its fatty acid side chains? Explain your reasoning.

(d) Assume you and your lab partner Mort determined the melting temperature for each of the following fatty acids: arachidic, linoleic, linolenic, oleaic, palmitic, and stearic acids. Mort recorded the melting points of each but neglected to note the specific fatty acid to which each value belongs. Assign each of the following melting temperatures (in °C) to the appropriate fatty acid, and explain your reasoning: −11, 5, 16, 63, 70, and 76.5.

(e) For each of the following amphipathic molecules, indicate which part of the molecule is hydrophilic: phosphatidyl serine, sphingomyelin, cholesterol, and triacylglycerol.

3-15 Shortening. A popular brand of shortening has a label on the can that identifies the product as "partially hydrogenated soybean oil, palm oil, and cottonseed oil."

(a) What does the process of partial hydrogenation accomplish chemically?

(b) What did the product in the can look like before it was partially hydrogenated?

(c) What is the physical effect of partial hydrogenation?

(d) Why would it be misleading to say that the shortening is "made from 100% polyunsaturated oils"?

SUGGESTED READING

References of historical importance are marked with a •.

General References and Reviews

Jardetzky, O., and M. D. Finucane. *Dynamics, Structure, and Function of Biological Macromolecules.* Washington, DC: IOS Press, 2001.

Murray, R. K. *Harper's Illustrated Biochemistry,* 26th ed. New York: McGraw-Hill, 2003.

Wald, G. The origins of life. *Proc. Natl. Acad. Sci. USA* 52 (1994): 595.

Proteins

• Bernard, S. A., and F. W. Dahlquist. *Classic Papers on Protein Structure and Function.* Sausalito, CA: University Science, 2002.

Ellis, R. J., ed. *The Chaperonins.* New York: Academic Press, 1996.

Ezzell, C. Proteins rule. *Sci. Amer.* 286 (April 2002): 40.

Flannery, M. C. Proteins: The unfolding and folding picture. *Amer. Biol. Teacher* 61 (1999): 150.

Hartl, F. U., and M. Hayer-Hartl. Converging concepts of protein folding in vitro and in vivo. *Nature Struct. Mol. Biol.* 16 (2009): 574.

Kang, T. S., and R. M. Kini. Structural determinants of protein folding. *Cell. Mol. Life Sci.* 66 (2009): 2341.

Kryshtafovych, A., and K. Fidelis. Protein structure prediction and model quality assessment. *Drug Discov. Today* 14 (2009): 386.

Lesk, A. M. *Introduction to Protein Architecture: The Structural Biology of Proteins.* Oxford: Oxford University Press, 2001.

Moult, J., and E. Melamud. From fold to function. *Curr. Opin. Struct. Biol.* 10 (2000): 384.

Petsko, G. A., and D. Ringe. *Protein Structure and Function.* Sunderland, MA: Sinauer Assoc., 2004.

Ringler, P., and G. E. Schulz. Self-assembly of proteins into designed networks. *Science* 302 (2003): 106.

Nucleic Acids

Cao, X., G. Yeo, A. R. Muotri, T. Kuwabara, and F. H. Gage. Noncoding RNAs in the mammalian central nervous system. *Annu. Rev. Neurosci.* 29 (2006): 77.

Chen, X. Small RNAs and their roles in plant development. *Annu. Rev. Cell Dev. Biol.* 25 (2009): 21.

• Crick, F. H. C. The structure of the hereditary material. *Sci. Amer.* (October 1954): 54.

Davies, K. *Cracking the Genome: Inside the Race to Unlock Human DNA.* New York: The Free Press, 2001.

Ezzell, C. The business of the human genome. *Sci. Amer.* (July 2000): 48.

Forsdye, D. R. Chargaff's legacy. *Gene* 261 (2000): 127.

Frouin, I. et al. DNA replication: A complex matter. *EMBO Rep.* 4 (2003): 666.

Liu, Q., and Z. Paroo. Biochemical principles of small RNA pathways. *Annu. Rev. Biochem.* 79 (2010): 295.

• Portugal, F. H., and J. S. Cohen. *The Century of DNA: A History of the Discovery of the Structure and Function of the Genetic Substance.* Cambridge, MA: MIT Press, 1977.

Venter, J. C. et al. The sequence of the human genome. *Science* 291 (2001): 1304.

• Watson, J. *A Passion for DNA.* Cold Spring Harbor, NY: Cold Spring Harbor Laboratory Press, 2000.

Carbohydrates and Lipids

Akoh, C. C., and D. B. Min. *Food Lipids: Chemistry, Nutrition, and Biotechnology.* Boca Raton, FL: CRC Press, 2008.

Garg, H. G., M. K. Cowman, and C. A. Hales. *Carbohydrate Chemistry, Biology and Medical Applications.* New York: Elsevier, 2008.

Lindhorst, T. K. *Essentials of Carbohydrate Chemistry and Biochemistry,* 3rd ed.. Weinheim, Germany: Wiley-VCH, 2007.

Merrill, A. H., Jr., et al. Sphingolipids—the enigmatic lipid class: Biochemistry, chemistry, physiology, and pathophysiology. *Toxicol. Appl. Pharmacol.* 142 (1997): 208.

Shimizu, T. Lipid mediators in health and disease: Enzymes and receptors as therapeutic targets for the regulation of immunity and inflammation. *Annu. Rev. Pharmacol. Toxicol.* 49 (2009): 123.

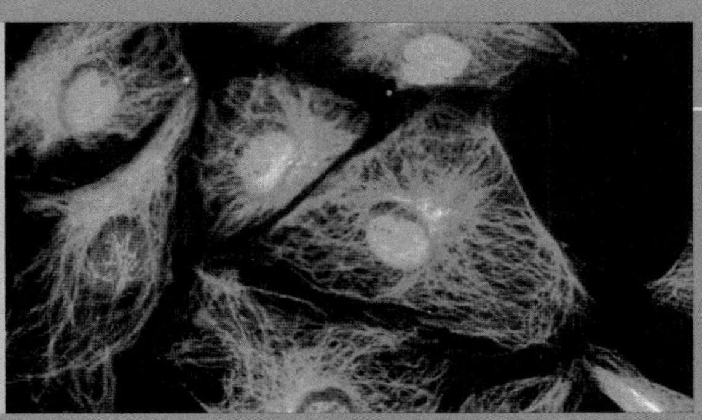

4 Cells and Organelles

See MasteringBiology® for tutorials, activities, and quizzes.

In the previous two chapters, we encountered the major kinds of molecules found in cells, as well as some principles governing the assembly of these molecules into the supramolecular structures that make up cells and their organelles (see Figure 2-14). Now we are ready to focus our attention on cells and organelles directly.

Properties and Strategies of Cells

As we begin to consider what cells are and how they function, several general characteristics of cells quickly emerge. These include the organizational complexity and molecular components of cells, the sizes and shapes of cells, and the specializations that cells undergo.

All Organisms Are Bacteria, Archaea, or Eukaryotes

With improvements in microscopy, biologists came to recognize two fundamentally different types of cellular organization: a simpler one characteristic of bacteria and a more complex one found in all other kinds of cells. Based on the structural differences of their cells, organisms have been traditionally divided into two broad groups, the **prokaryotes** (bacteria) and the **eukaryotes** (plants, animals, fungi, algae, and protozoa). The most fundamental distinction between the two groups is that eukaryotic cells have a true, membrane-bounded nucleus (*eu–* is Greek for "true" or "genuine"; *-karyon* means "nucleus"), whereas prokaryotic cells do not (*pro–* means "before," suggesting an evolutionarily earlier form of life).

Recently, however, the term *prokaryote* is becoming less satisfactory to describe these non-nucleated cells, partly because this is a negative classification based on what cells do not have and partly because it wrongly implies a fundamental similarity among all organisms whose cells lack a nucleus. The fact that two organisms lack a particular gross structural feature does not necessarily imply evolutionary relatedness—cells of humans and *Mycoplasma,* a type of bacterium, both lack cell walls but are not at all closely related. Likewise, sharing a gross structural feature does not necessarily mean a close relationship. Although plants and most bacteria have cell walls, they are only distantly related.

Molecular and biochemical criteria are proving to be more reliable than structural criteria in describing evolutionary relationships among organisms. The more closely related two different organisms are, the more similarities we see in their sequences of particular DNA, RNA, and protein molecules. Especially useful for comparative studies are molecules common to all living organisms that are necessary for universal, basic processes, in which even slight changes in component molecules are not well tolerated. This includes molecules such as the ribosomal RNAs used in protein synthesis and the cytochrome proteins used in energy metabolism.

Based on the pioneering ribosomal RNA sequencing work of Carl Woese, Ralph Wolfe, and coworkers, we now recognize that the group traditionally called the prokaryotes actually includes two widely divergent groups—the **bacteria** and the **archaea,** which are as different from each other as we are from bacteria! Rather than the prokaryote-eukaryote dichotomy, it is more biologically correct to describe all organisms as belonging to one of three *domains*—the bacteria, the archaea, or the **eukarya** (eukaryotes). As shown in **Table 4-1,** cells of each of these domains share some characteristics with cells of the other domains, but all three domains have some unique characteristics.

The bacteria include most of the commonly encountered single-celled, non-nucleated organisms that we have traditionally referred to as bacteria—for example, *Escherichia coli, Pseudomonas,* and *Streptococcus.* The archaea (which were called *archaebacteria* before investigators realized how different they are from bacteria) include many species that live in extreme habitats on Earth and have very diverse metabolic strategies. Members of the archaea include the

Table 4-1 A Comparison of Some Properties of Bacterial, Archaeal, and Eukaryotic Cells*

	Prokaryotes			
Property	**Bacteria**	**Archaea**	**Eukaryotes**	**Refer to:**
Typical size	Small (1–5 μm)	Small (1–5 μm)	Large (10–100 μm)	—
Nucleus and organelles	No	No	Yes	Table 4-2
Microtubules and microfilaments	Actin-like and tubulin-like proteins	Actin-like and tubulin-like proteins	Actin and tubulin proteins	Chapter 15
Exocytosis and endocytosis	No	No	Yes	Chapter 12
Cell wall	Peptidoglycan	Varies from proteinaceous to peptidoglycan-like	Cellulose in plants, fungi; none in animals, protozoa	Chapter 17
Mode of cell division	Binary fission	Binary fission	Mitosis or meiosis plus cytokinesis	Chapter 19
Typical form of chromosomal DNA	Circular, few associated proteins	Circular, associated with histone-like proteins	Linear, associated with histone proteins	Chapter 18
RNA processing	Minimal	Moderate	Extensive	Chapter 21
Transcription initiation	Bacterial type	Eukaryotic type	Eukaryotic type	Chapter 21
RNA polymerase	Bacterial type	Some features of both bacterial, eukaryotic types	Eukaryotic type	Chapter 21
Ribosome size and number of proteins	70S with 55 proteins	70S with 65 proteins	80S with 78 proteins	Chapter 22
Ribosomal RNAs	Bacterial type	Archaeal type	Eukaryotic type	Chapter 21
Translation initiation	Bacterial type	Eukaryotic type	Eukaryotic type	Chapter 22
Membrane phospholipids	Glycerol-3-phosphate + linear fatty acids	Glycerol-1-phosphate + branched polyisoprenoids	Glycerol-3-phosphate + linear fatty acids	Chapter 7

*This table lists many features that we have not yet discussed in detail. Its main purpose is to point out that, despite some sharing of characteristics, each of the three main cell types has a unique set of properties.

methanogens, which obtain energy from hydrogen while converting carbon dioxide into methane; the *halophiles,* which can grow in extremely salty environments; and the *thermacidophiles,* which thrive in acidic hot springs where the pH can be as low as 2 and the temperature can exceed 100°C! Rather than being considered as an evolutionarily ancient form of prokaryote (*archae–* is a Greek prefix meaning "ancient"), archaea are now considered to be descended from a common ancestor that also gave rise to the eukaryotes long after diverging from the bacteria, as shown here:

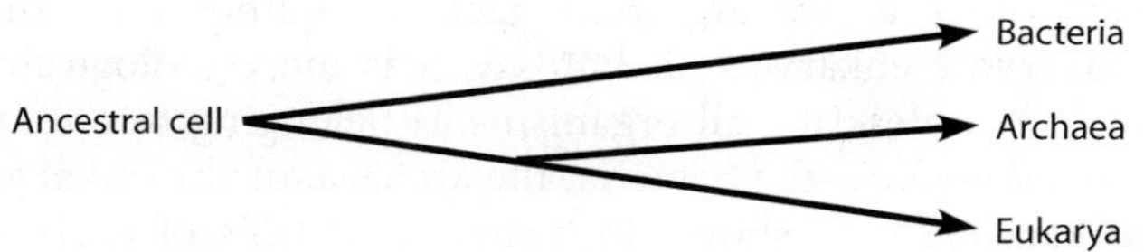

While resembling bacteria in many ways, archaea possess many unique features, as well as features that are found in eukaryotes. They resemble bacteria in cell size and gross structure, in their method of cell division, and in many aspects of basic metabolism and enzyme content. However, they are much more similar to eukaryotes regarding many details of DNA replication, transcription, RNA processing, and initiation of protein synthesis. Unique features of archaea include their ribosomal RNAs and their membrane phospholipids (Table 4-1).

Limitations on Cell Size

Cells come in various sizes and shapes. Some of the smallest bacterial cells, for example, are only about 0.2–0.3 μm in diameter—so small that about 50,000 such cells could fit side by side on your thumbnail. At the other extreme are highly elongated nerve cells, which may extend one or more meters. The nerve cells running the length of a giraffe's neck or legs are especially dramatic examples. So are bird eggs, which are extremely large single cells, although much of their internal volume is occupied by the yolk that nourishes the developing embryo.

Despite these extremes, most cells fall into a rather narrow and predictable range of sizes. Bacterial and archaeal cells, for example, are usually about 1–5 μm in diameter, while most cells of higher plants and animals have dimensions in the range of 10–100 μm. Why are cells so small? Several factors limit cell size, but the three most important are (1) the requirement for an adequate surface area/volume ratio, (2) the rates at which molecules diffuse, and (3) the need to maintain adequate local concentrations of the specific substances and enzymes involved in necessary cellular processes. Let's now look at each of these three factors in turn.

Surface Area/Volume Ratio. In most cases, the main limitation on cell size is set by the need to maintain an adequate **surface area/volume ratio.** Surface area is critical because it is at the cell surface that the needful exchanges between a cell and its environment take place. The cell's internal volume determines the amount of nutrients that will have to be imported and the quantity of waste products that must be excreted. The surface area effectively represents the amount of cell membrane available for such uptake and excretion.

The problem of maintaining adequate surface area arises because the volume of a cell increases with the cube of the cell's length or diameter, whereas its surface area increases only with the square. Thus, large cells have a lower ratio of surface area to volume than small cells do, as illustrated in **Figure 4-1**. This comparison illustrates a major constraint on cell size: As a cell increases in size, its surface area does not keep pace with its volume, and the necessary exchange of substances between the cell and its surroundings becomes more and more problematic. Cell size, therefore, can increase only as long as the membrane surface area is still adequate for the passage of materials into and out of the cell.

Some cells, particularly those that play a role in absorption, have characteristics that maximize their surface area. Effective surface area is most commonly increased by the inward folding or outward protrusion of the cell membrane. The cells lining your small intestine, for example, contain many fingerlike projections called *microvilli* that greatly increase the effective membrane surface area and therefore the nutrient-absorbing capacity of these cells (**Figure 4-2**).

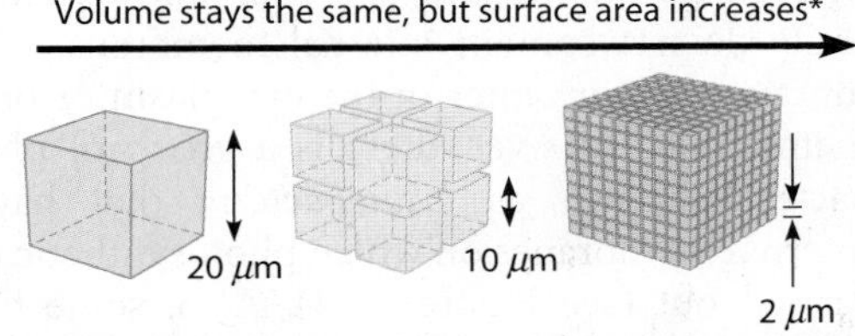

Number of cells	1	8	1000
Length of one side	20 µm	10 µm	2 µm
Total volume	8000 μm^3	8000 μm^3	8000 μm^3
Total surface area	2400 μm^2	4800 μm^2	24,000 μm^2
Surface area to volume ratio	0.3	0.6	3.0

*For a cube having a side with length s, volume = s^3 and surface area = $6s^2$.

FIGURE 4-1 The Effect of Cell Size on the Surface Area/Volume Ratio. The single large cell on the left, the eight smaller cells in the center, and the 1000 tiny cells on the right all have the same total volume (8000 μm^3), but the total surface area increases as the cell size decreases. The surface area/volume ratio therefore increases from left to right as the linear dimension of the cell decreases. Note how the eight small cells in the center have the same total volume as the large cell on the left but twice the surface area. Likewise, the 1000 small bacterial cells on the right have the same total volume but a total surface area ten times that of the single large eukaryotic cell on the left.

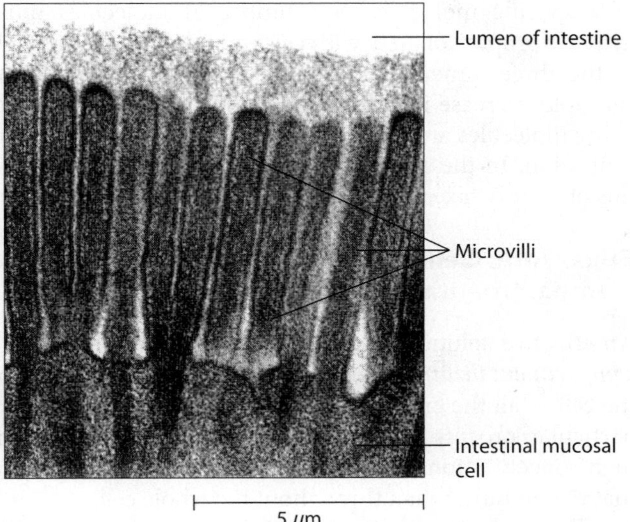

FIGURE 4-2 The Microvilli of Intestinal Mucosal Cells. Microvilli are fingerlike projections of the cell membrane that greatly increase the absorptive surface area of intestinal mucosal cells, such as those lining the inner surface of your small intestine (TEM).

Diffusion Rates of Molecules. Cell size is also limited by how rapidly molecules can move around in the cell to reach sites of specific cellular activities. Many molecules move through the cell by **diffusion,** which is the free, unassisted movement of a substance from a region of high concentration to a region of low concentration. Molecular movement can therefore be limited by the diffusion rates for molecules of various sizes. And because the rate of diffusion decreases as the size of the molecule increases, this limitation is most significant for macromolecules such as proteins and nucleic acids.

Recent work shows that many eukaryotic cells may be able to bypass this limitation by actively transporting ions, macromolecules, and other materials through the cytoplasm using special carrier proteins. Some cells of higher organisms get around this limitation to some extent by *cytoplasmic streaming* (also called *cyclosis* in plant cells), a process that involves active movement and mixing of cytoplasmic contents rather than diffusion. Other cells move specific molecules through the cell using vesicles that are transported along microtubules, as we will see shortly. In the absence of these mechanisms, however, the size of a cell is limited by the diffusion rates of the molecules it contains.

The Need for Adequate Concentrations of Reactants and Catalysts. A third limitation on cell size is the need to maintain adequate concentrations of the essential compounds and enzymes needed for the various processes that cells must carry out. For a chemical reaction to occur in a cell, the appropriate reactants must collide with and bind to a particular enzyme. The frequency of such collisions will be greatly increased by higher concentrations of the reactants and the enzyme. To maintain the concentration of a specific molecule, the number of molecules must increase proportionately with cell volume. Every time each of the three dimensions of the cell doubles, there is an eightfold increase in cell volume, and thus eight times as many molecules are required to maintain the original concentration. In the absence of a concentrating mechanism, this obviously taxes the cell's synthetic capabilities.

Eukaryotic Cells Use Organelles to Compartmentalize Cellular Function

An effective solution to the concentration problem is the *compartmentalization of activities* within specific regions of the cell. If all the enzymes and compounds necessary for a particular process are localized within a specific region, high concentrations of those substances are needed only in that region rather than throughout the whole cell.

To compartmentalize activities, most eukaryotic cells have a variety of **organelles,** which are membrane-bounded compartments that are specialized for specific functions. For example, the cells in a plant leaf have most of the enzymes, compounds, and pigments needed for photosynthesis compartmentalized together into structures called *chloroplasts.* Such a cell can therefore maintain appropriately high concentrations of everything it requires for photosynthesis within its chloroplasts without having to maintain correspondingly high concentrations of these substances elsewhere in the cell. In a similar way, other processes are localized within other compartments.

Bacteria, Archaea, and Eukaryotes Differ from Each Other in Many Ways

Returning to the distinction between bacteria, archaea, and eukaryotes, we recognize many important structural, biochemical, and genetic differences among these groups. Some of these differences are summarized in Table 4-1 and are discussed here briefly—others will be discussed in later chapters. For now, it is important to realize that, despite some sharing of characteristics among cells of each of these three domains, each type of cell has a unique set of distinguishing properties.

Presence of a Membrane-Bounded Nucleus. As already noted, a structural distinction has traditionally been made between eukaryotes and prokaryotes (bacteria and archaea) and is reflected in the nomenclature itself. However, as we learn more about the cellular details of the three domains of living organisms, this distinction is becoming less important than other aspects of structure and function. A eukaryotic cell has a true, membrane-bounded nucleus, whereas a prokaryotic cell does not. Instead of being enveloped by a membrane, the genetic information of a bacterial or archaeal cell is folded into a compact structure known as the *nucleoid,* which is attached to the cell membrane in a particular region of the cytoplasm (**Figure 4-3**). Within a eukaryotic cell, on the other hand, most of the genetic information is localized to the nucleus, which is surrounded not by a single membrane but by a *nuclear envelope* consisting of two membranes (**Figure 4-4**). The nucleus also includes the *nucleolus,* which is the site of ribosomal RNA synthesis and ribosome subunit assembly, and it contains the DNA-bearing *chromosomes,* which are dispersed as chromatin throughout the semifluid *nucleoplasm* that fills the internal volume of the nucleus.

Use of Internal Membranes to Segregate Function. As Figure 4-3 illustrates, bacterial (and archaeal) cells generally do not contain internal membranes; most cellular functions occur either in the cytoplasm or on the plasma membrane. However, there is a group of photosynthetic bacteria known as *cyanobacteria* that have extensive internal membranes on which photosynthetic reactions are carried out (see Figure 11-4). Also, some bacteria have membrane-bound structures that resemble organelles, while others have protein-lined compartments that serve

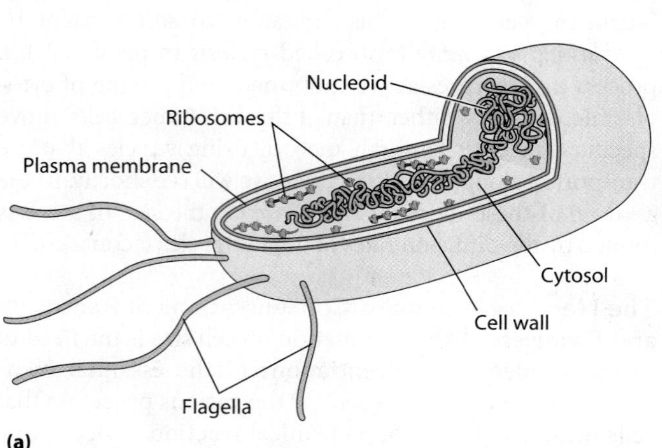

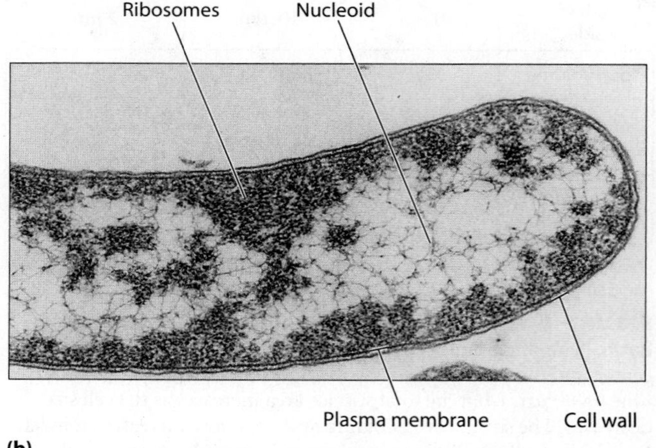

FIGURE 4-3 Structure of a Rod-Shaped Bacterial Cell. **(a)** A three-dimensional model showing the components of a bacterium. **(b)** An electron micrograph of a bacterial cell with several of the same components labeled. Notice that the nucleoid refers to the folded bacterial chromosome, not a membrane-bounded compartment (TEM).

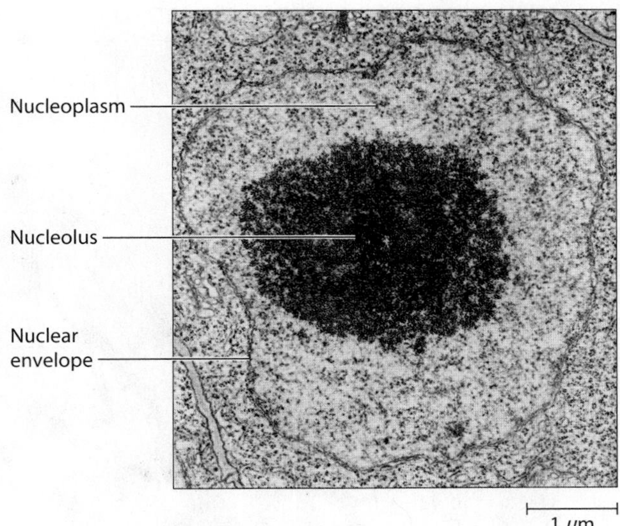

FIGURE 4-4 The Nucleus of a Eukaryotic Cell. The nucleus is enclosed by a pair of membranes called the nuclear envelope. Because this cell is between divisions, the chromosomes are dispersed as chromatin in the nucleoplasm within the nucleus and are not visible. The nucleolus is involved in the synthesis of ribosomal components (TEM).

as organelles by isolating groups of enzymes involved in specific metabolic pathways. In contrast, nearly all eukaryotic cells make extensive use of internal membranes to compartmentalize specific functions (**Figure 4-5** and **Figure 4-6**), and they often have numerous organelles.

Examples of internal membrane systems in eukaryotic cells include the *endoplasmic reticulum,* the *Golgi complex,* and the membranes surrounding and delimiting organelles such as *mitochondria, chloroplasts, lysosomes,* and *peroxisomes,* as well as various kinds of *vacuoles* and *vesicles.* Each of these organelles is surrounded by its own characteristic membrane (or pair of membranes) that may be similar to other membranes in basic structure but can have a distinctive chemical composition. Localized within each such organelle is the molecular machinery needed to carry out the particular cellular functions for which the structure is specialized. We will meet each of these organelles later in this chapter and then return to each one in its appropriate context in succeeding chapters.

The Cytoskeleton. Also found in the cytoplasm of eukaryotic cells are several nonmembranous, proteinaceous structures that are involved in cellular contraction, motility, and the establishment and support of cellular architecture. These include the *microtubules* found in the cilia and flagella of many cell types, the *microfilaments* found in muscle fibrils and other structures involved in motility, and the *intermediate filaments,* which are especially prominent in cells that are subject to stress. Microtubules, microfilaments, and intermediate filaments are key components of the *cytoskeleton,* which imparts structure and elasticity to almost all eukaryotic cells, as we will learn shortly and explore in more detail in Chapter 15. In addition, the cytoskeleton can provide a scaffolding for

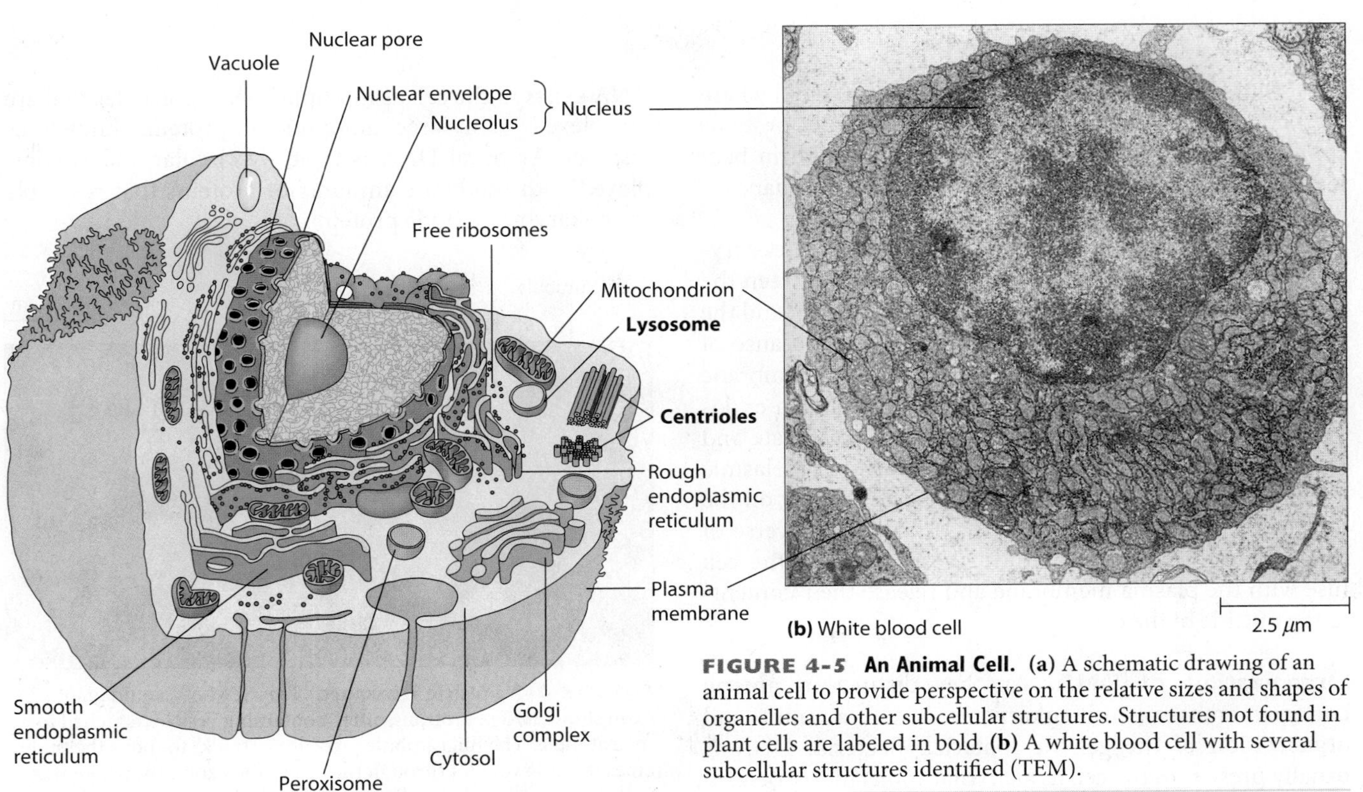

FIGURE 4-5 An Animal Cell. **(a)** A schematic drawing of an animal cell to provide perspective on the relative sizes and shapes of organelles and other subcellular structures. Structures not found in plant cells are labeled in bold. **(b)** A white blood cell with several subcellular structures identified (TEM).

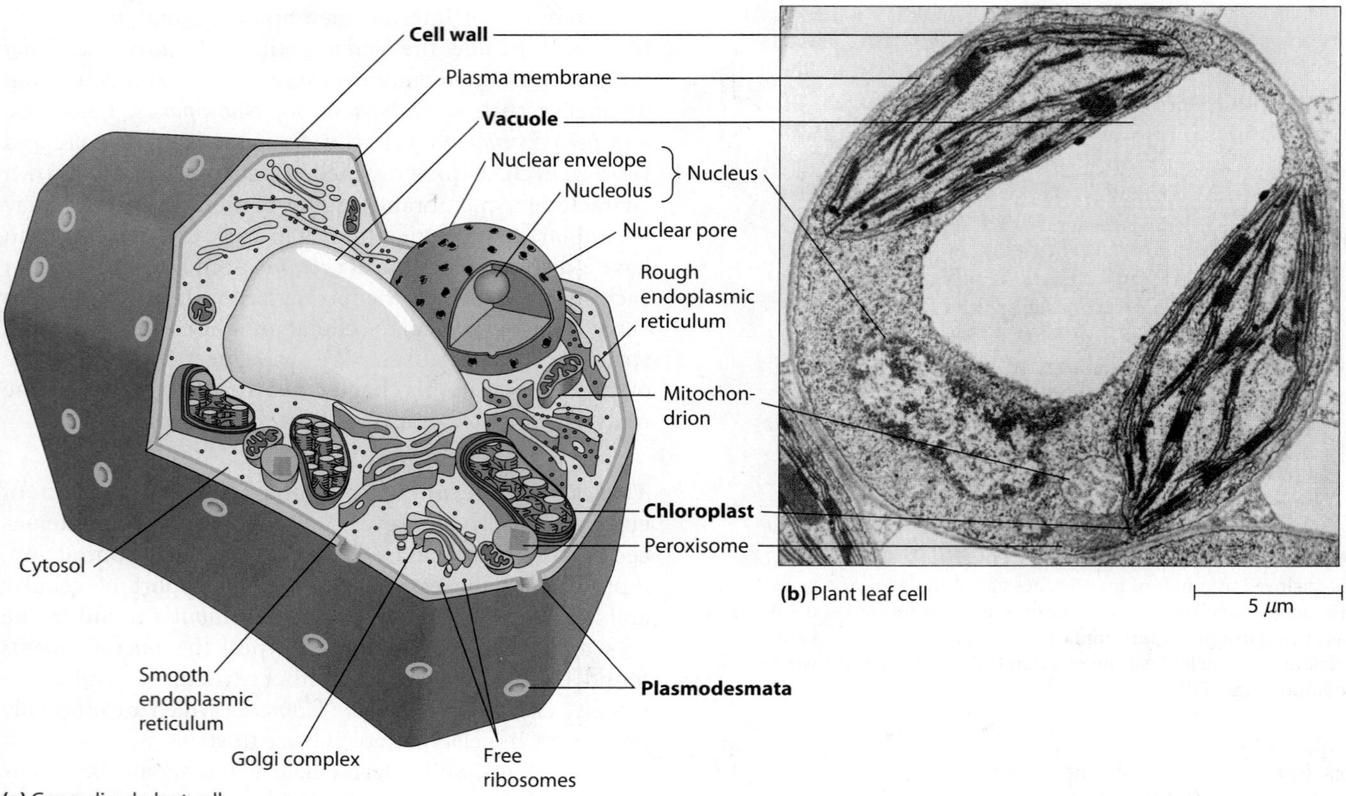

FIGURE 4-6 A Plant Cell. **(a)** A schematic drawing of a plant cell. Compare this drawing with the animal cell in Figure 4-5a, and notice that plant cells are characterized by the absence of lysosomes and the presence of chloroplasts, a cell wall, and a large central vacuole. Structures not found in animal cells are labeled in bold. **(b)** A cell from a *Coleus* leaf, with several subcellular structures identified (TEM).

intracellular transport of vesicles to places in the cell where their contents are needed (**Figure 4-7**). Recently, proteins similar to cytoskeleton proteins have been found in bacteria and appear to have a role in maintaining cell shape.

Exocytosis and Endocytosis. A further feature of eukaryotic cells is their ability to exchange materials between the membrane-bounded compartments within the cell and the exterior of the cell. This exchange is possible because of *exocytosis* and *endocytosis,* processes involving membrane fusion events that are unique to eukaryotic cells. In endocytosis, portions of the plasma membrane invaginate and are pinched off to form membrane-bounded cytoplasmic vesicles containing substances that were previously on the outside of the cell. Exocytosis is essentially the reverse of this process: Membrane-bounded vesicles inside the cell fuse with the plasma membrane and release their contents to the outside of the cell.

Organization of DNA. Another distinction among bacteria, archaea, and eukaryotes is the amount and organization of the genetic material. Bacterial DNA is usually present in the cell as a circular molecule associated with relatively few proteins. On the other hand, eukaryotic DNA exists in the cell as multiple linear molecules that are complexed with large amounts of proteins known as *histones.* Archaeal DNA is typically circular and is complexed with moderate amounts of proteins that resemble the eukaryotic histone proteins.

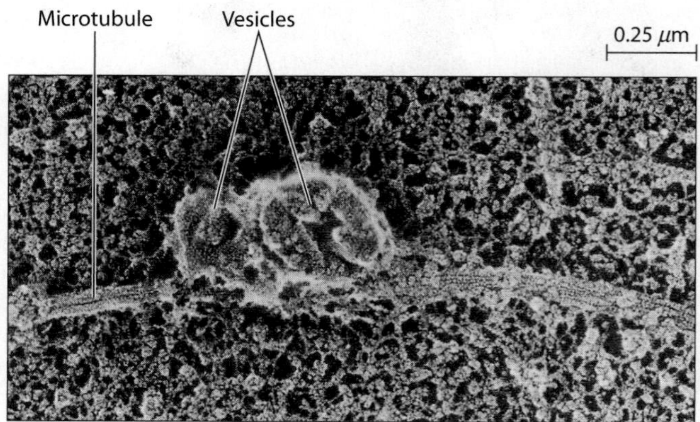

FIGURE 4-7 Vesicle Transport. This SEM of a squid giant axon shows two neurotransmitter-containing vesicles attached to a microtubule. The microtubule provides a "track" to move the molecules in these vesicles through the cell to the axon tips, where they will aid in nerve cell signaling.

The circular DNA molecule of a bacterial or archaeal cell is much longer than the cell itself. It therefore has to be folded and packed together tightly to fit into the cell. For example, the common intestinal bacterium *Escherichia coli* is only about a micrometer or two long, yet it has a circular DNA molecule about 1300 μm in circumference. Clearly, a great deal of folding and packing is necessary to fit that much DNA into such a small cell. By way of analogy, it is roughly equivalent to packing about 60 feet (18 m) of very thin thread into a typical thimble.

But if DNA appears to pose a packaging problem for prokaryotic cells, consider the case of the eukaryotic cell. Most eukaryotic cells have at least 1000 times as much DNA as *E. coli* has but encode only 5–10 times as many proteins. Because scientists did not know the function of the excess, noncoding DNA, it had been often referred to as "junk DNA."

Now, however, it appears that much of this excess DNA may have important functions other than encoding cellular proteins. Some of it is involved in production of regulatory miRNAs, other regions appear to be involved in generation of species diversity during evolution, and some contains repetitive sequences that appear to serve as binding sites for regulatory proteins. A recent study showed that up to 33% of the binding sites for transcription factors with roles in cancer development are found in these regions of repetitive DNA. Thus, we must be careful not to assume that there is no function for certain cellular components just because we currently do not understand their roles in the cell.

Whatever the function of such large amounts of DNA, the problem of packaging all this material is clearly acute. It is solved universally among eukaryotes by the organization of DNA into complex structures called **chromosomes,** which contain at least as much histone protein as DNA (see Figure 18-22). It is as chromosomes that the DNA of eukaryotic cells is packaged, segregated during cell division, and transmitted to daughter cells. **Figure 4-8** shows a chromosome from an animal cell as seen by high-voltage electron microscopy.

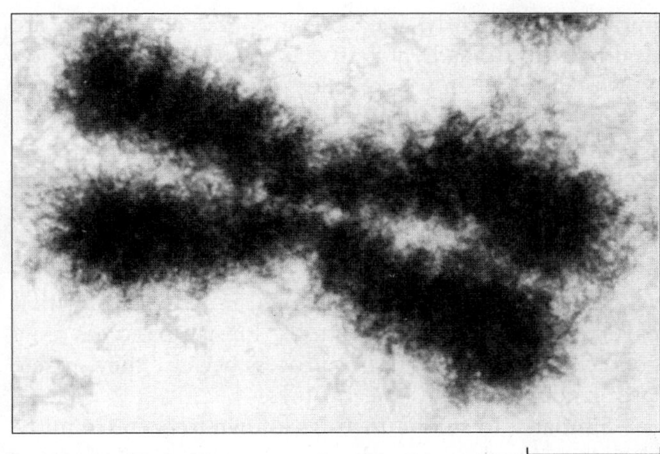

FIGURE 4-8 A Eukaryotic Chromosome. This chromosome was obtained from a cultured Chinese hamster cell and visualized by high-voltage electron microscopy (HVEM). The cell was undergoing mitosis, and the chromosome is therefore highly coiled and condensed.

Segregation of Genetic Information. A further contrast between prokaryotes and eukaryotes is the way they allocate genetic information to daughter cells upon division. Bacterial and archaeal cells merely replicate their DNA and divide by a relatively simple process called *binary fission,* with one molecule of the replicated DNA and half of the cytoplasm going to each daughter cell. Following DNA replication in eukaryotic cells, the chromosomes are distributed equally to the daughter cells by the more complex processes of *mitosis* and *meiosis,* followed by *cytokinesis,* the division of the cytoplasm.

Expression of DNA. The differences among bacterial, archaeal, and eukaryotic cells extend to the expression of genetic information. Eukaryotic cells tend to transcribe genetic information in the nucleus into large RNA molecules and depend on later processing and transport processes to deliver RNA molecules of the proper sizes to the cytoplasm for protein synthesis. Each RNA molecule typically encodes one polypeptide.

By contrast, bacteria transcribe very specific segments of genetic information into RNA messages, and often a single RNA molecule contains the information to produce several polypeptides. In bacteria, little or no processing of RNA occurs; a moderate amount is seen in archaea, though less than in eukaryotes. The absence of a nuclear membrane in bacteria and archaea makes it possible for messenger RNA molecules to become involved in the process of protein synthesis even before they are completely synthesized. Bacteria, archaea, and eukaryotes also differ in the size and composition of the ribosomes and ribosomal RNAs used to synthesize proteins (see Table 4-1). We will explore this distinction in more detail later in the chapter.

Cell Specialization Demonstrates the Unity and Diversity of Biology

In their structure and function, cells are characterized by both unity and diversity, as you can see in the generalized animal and plant cells shown in Figures 4-5 and 4-6. By unity and diversity, we simply mean that all cells resemble one another in fundamental ways, yet they differ from one another in other important ways. In upcoming chapters we will concentrate on those aspects of structure and function common to most cell types. We will find, for example, that virtually all cells oxidize sugar molecules for energy, transport ions across membranes, transcribe DNA into RNA, and undergo division to generate daughter cells.

Much the same is true for structural features. All cells are surrounded by a selectively permeable plasma membrane, all cells have ribosomes for protein synthesis, and all contain double-stranded DNA as their genetic information. Clearly, we can be confident that we are dealing with fundamental aspects of cellular organization and function when we consider processes and structures common to most, if not all, cells.

But sometimes our understanding of cellular biology is enhanced by considering not just the unity but also the diversity of cells—the features that are especially prominent in a particular cell type. For example, to understand how the process of protein secretion works, it is advantageous to consider a cell that is highly specialized for that particular function. Cells from the human pancreas are a good choice for studying this process because they secrete large amounts of digestive enzymes.

Similarly, to study functions known to occur in mitochondria, it is clearly an advantage to select a cell type that is highly specialized in the energy-releasing processes occurring in the mitochondrion. Such a cell would likely have a lot of well-developed, highly active mitochondria. In fact, it was for this very reason—to study a cell that is highly specialized for a particular function—that Hans Krebs chose the flight muscle of the pigeon as the tissue for carrying out the now-classic studies on the cyclic pathway of oxidative reactions that we know as the *tricarboxylic acid (TCA),* or *Krebs, cycle.*

In general, the single cell of unicellular organisms must be capable of carrying out any and all of the functions necessary for survival, growth, and reproduction. It typically does not overemphasize any single function at the expense of others. Multicellular organisms, on the other hand, are characterized by a division of labor among tissues and organs that not only allows for but also depends on specialization of structure and function. Whole groups of cells become highly specialized for a particular task, which then becomes their specific role in the overall functioning of the organism.

The Eukaryotic Cell in Overview: Pictures at an Exhibition

From the preceding discussion, it should be clear that all cells carry out many of the same basic functions and have some of the same basic structural features. However, the cells of eukaryotic organisms are far more complicated structurally than bacterial or archaeal cells, primarily because of the organelles and other intracellular structures that eukaryotes use to compartmentalize various functions. The structural complexity of eukaryotic cells is illustrated by the typical animal and plant cells shown in Figures 4-5 and 4-6.

In reality, of course, there is no such thing as a truly "typical" cell; nearly all eukaryotic cells have features that distinguish them from the generalized cells shown in Figures 4-5 and 4-6. Nonetheless, most eukaryotic cells are sufficiently similar to warrant a general overview of their structural features.

As we have seen, a typical eukaryotic cell has at least four major structural features: an external *plasma membrane* to define its boundary and retain its contents, a *nucleus* to house the DNA that directs cellular activities, *membrane-bounded organelles* in which various cellular functions are localized, and the *cytosol* interlaced by a *cytoskeleton* of microtubules and microfilaments. In addition, plant and fungal cells have a rigid *cell wall* external to the plasma membrane. Animal cells do not have a cell wall; they are usually surrounded by an *extracellular matrix* consisting primarily of proteins that provide structural support.

Our intention here is to look at each of these structural features in overview, as an introduction to cellular architecture. For now, we will simply look at each structure as we might look at pictures at an exhibition, moving through the gallery rather quickly just to get a feel for the overall display. Keep in mind, however, that these introductory "pictures" are only static representations of a dynamic cell that can adapt and respond to its environment. We will study each structure in detail in later chapters when we consider the dynamic cellular processes in which these organelles and other structures are involved (**Table 4-2**).

Table 4-2 **Chapter Cross-References for Cellular Structures and Techniques Used to Study Cells**

For more detailed information about these cellular structures, see the following chapters:	
Cell wall	Chapter 17
Chloroplast	Chapter 11
Cytoskeleton	Chapter 15
Endoplasmic reticulum	Chapter 12
Extracellular matrix	Chapter 17
Golgi complex	Chapter 12
Lysosome	Chapter 12
Mitochondrion	Chapter 10
Nucleus	Chapter 18
Peroxisome	Chapter 12
Plasma membrane	Chapter 7
Ribosome	Chapter 22
For more detailed information about techniques used to study cellular structures, see these chapters and the Appendix:	
Autoradiography	Chapter 18 and Appendix
Centrifugation	Chapter 12
Electron microscopy	Appendix
Light microscopy	Appendix

The Plasma Membrane Defines Cell Boundaries and Retains Contents

Our tour begins with the **plasma membrane** that surrounds every cell (**Figure 4-9**). The plasma membrane defines the boundaries of the cell, ensuring that its contents are retained. The plasma membrane consists of phospholipids, other lipids, and proteins and is organized into two layers (Figure 4-9b). Typically, each phospholipid molecule consists of two hydrophobic "tails" and a hydrophilic "head" and is therefore an *amphipathic molecule* (see

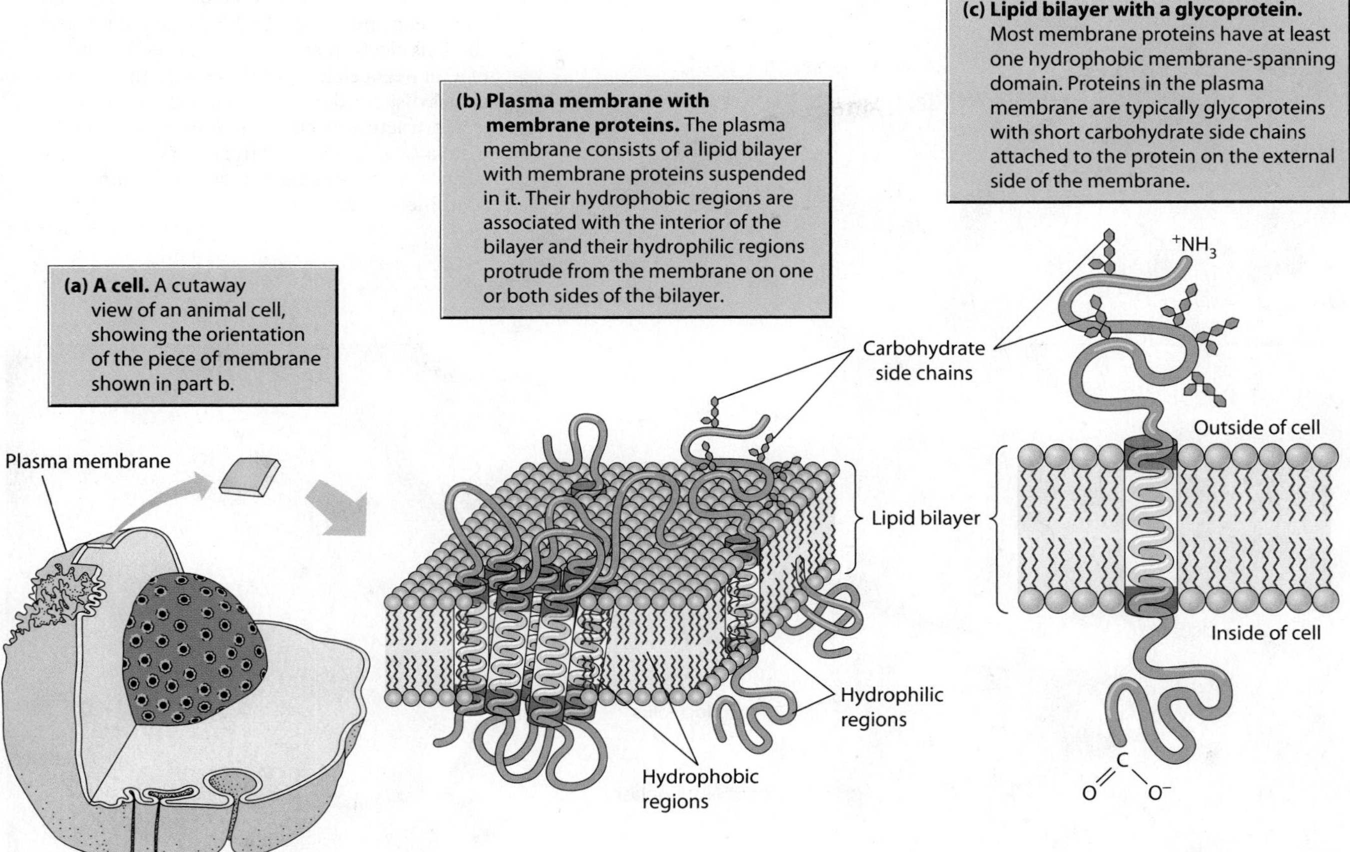

(a) A cell. A cutaway view of an animal cell, showing the orientation of the piece of membrane shown in part b.
(b) Plasma membrane with membrane proteins. The plasma membrane consists of a lipid bilayer with membrane proteins suspended in it. Their hydrophobic regions are associated with the interior of the bilayer and their hydrophilic regions protrude from the membrane on one or both sides of the bilayer.
(c) Lipid bilayer with a glycoprotein. Most membrane proteins have at least one hydrophobic membrane-spanning domain. Proteins in the plasma membrane are typically glycoproteins with short carbohydrate side chains attached to the protein on the external side of the membrane.
Plasma membrane
Carbohydrate side chains
Lipid bilayer
Hydrophilic regions
Hydrophobic regions
$^{+}NH_3$
Outside of cell
Inside of cell
C
O
O⁻

FIGURE 4-9 Organization of the Plasma Membrane. This figure shows successively more detail **(a)–(c)** in a section of the plasma membrane that surrounds all cells.

Figures 2-11 and 2-12). The phospholipid molecules orient themselves in the two layers of the membrane such that the hydrophobic, hydrocarbon tails of each molecule face inward and the hydrophilic, phosphate-containing heads of the molecules face outward (Figure 4-9c). The resulting **lipid bilayer** is the basic structural unit of virtually all membranes and serves as a permeability barrier to most water-soluble substances. Some members of Archaea, however, have an unusual phospholipid membrane that has long hydrophobic tails (twice the normal length) linked to a polar head group on both ends, forming a monolayer.

Membrane proteins are also amphipathic, with both hydrophobic and hydrophilic regions on their surfaces. They orient themselves in the membrane such that hydrophobic regions of the protein are located within the hydrophobic interior of the membrane, whereas hydrophilic regions protrude into the aqueous environment at the surfaces of the membrane. Many of the proteins with hydrophilic regions exposed on the external side of the plasma membrane have carbohydrate side chains known as oligosaccharides attached to them and are therefore called *glycoproteins* (Figure 4-9c).

The proteins present in the plasma membrane play a variety of roles. Some are *enzymes,* which catalyze reactions known to be associated with the membrane—reactions such as cell wall synthesis. Others serve as *anchors* for structural elements of the cytoskeleton that we will encounter later in the chapter. Still others are *transport proteins,* responsible for moving specific substances (ions and hydrophilic solutes, usually) across the membrane. Membrane proteins are also important as *receptors* for external chemical signals that trigger specific processes within the cell. Transport proteins, receptor proteins, and most other membrane proteins are *transmembrane proteins* that have hydrophilic regions protruding from both sides of the membrane. These hydrophilic regions are connected by one or more hydrophobic, membrane-spanning domains.

The Nucleus Is the Information Center of the Eukaryotic Cell

Perhaps the most prominent structure we encounter in a eukaryotic cell is the **nucleus** (**Figure 4-10**). The nucleus serves as the cell's information center. Here, separated from the rest of the cell by a membrane boundary, are the DNA-bearing chromosomes of the cell. Actually, the boundary around the nucleus consists of two membranes, called the *inner* and *outer nuclear membranes.* Taken together, the two membranes make up the **nuclear envelope.** Unique to the membranes of the nuclear envelope are numerous small

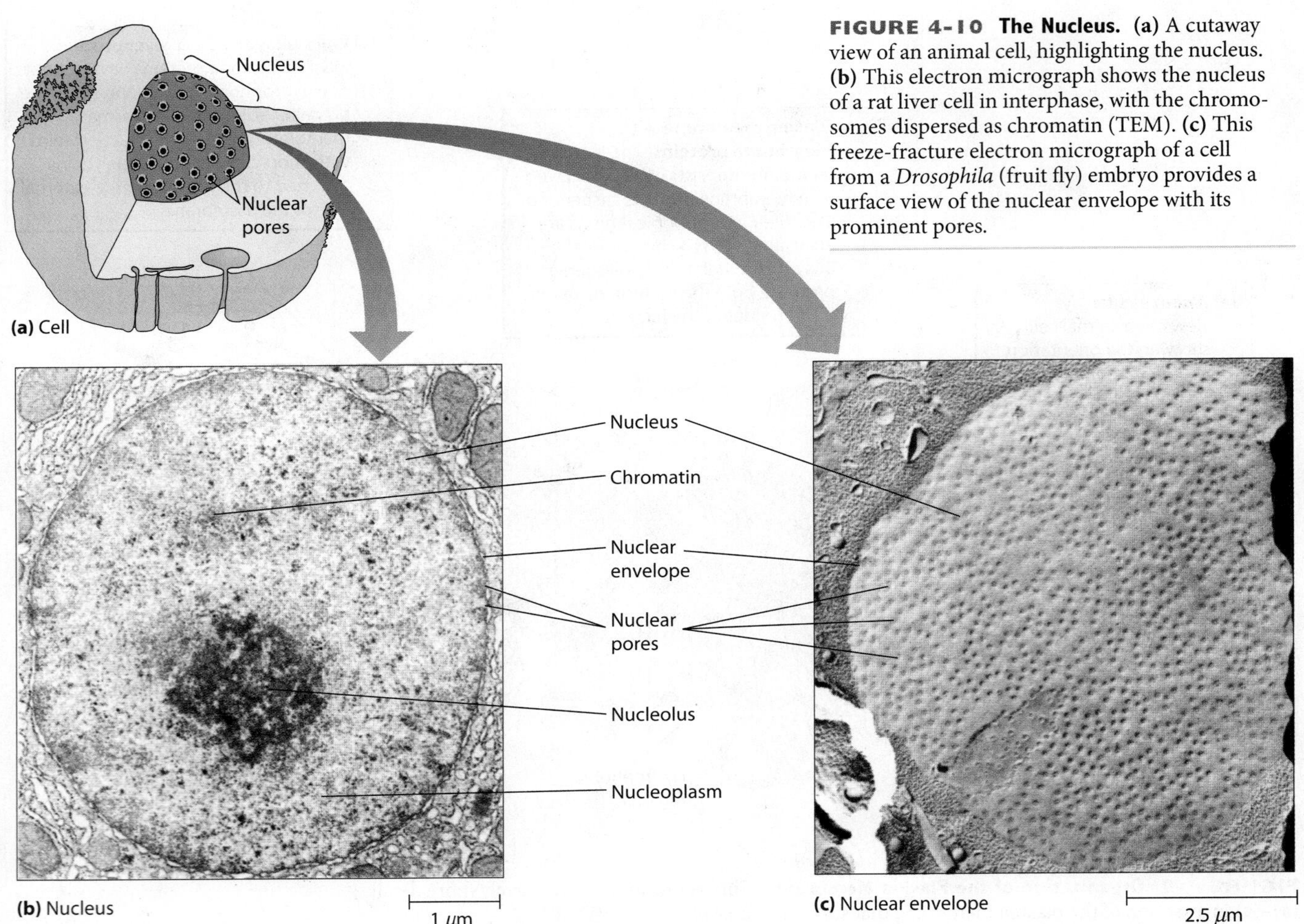

FIGURE 4-10 The Nucleus. **(a)** A cutaway view of an animal cell, highlighting the nucleus. **(b)** This electron micrograph shows the nucleus of a rat liver cell in interphase, with the chromosomes dispersed as chromatin (TEM). **(c)** This freeze-fracture electron micrograph of a cell from a *Drosophila* (fruit fly) embryo provides a surface view of the nuclear envelope with its prominent pores.

openings called *pores* (Figure 4-10). Each pore is a channel through which water-soluble molecules and supramolecular complexes can move between the nucleus and cytoplasm. This channel is lined with transport machinery known as a *pore complex* that regulates the movement of macromolecules through the nuclear envelope and is shown in close-up detail in Figures 18-28 and 18-29. Ribosomal subunits, messenger RNA molecules, chromosomal proteins, and enzymes needed for nuclear activities are transported across the nuclear envelope through these nuclear pores.

The number of chromosomes within the nucleus is characteristic of the species. It can be as low as two (in the sperm and egg cells of some grasshoppers, for example), or it can run into the hundreds. Chromosomes are most readily visualized during mitosis, when they are highly condensed and can easily be stained (see Figure 4-8). During the *interphase* between divisions, on the other hand, chromosomes are dispersed as DNA-protein fibers called **chromatin** and are not easy to visualize (Figure 4-10b).

Also present in the nucleus are **nucleoli** (singular: **nucleolus**), structures responsible for synthesizing and assembling some of the RNA and protein components needed to form the ribosomes. Nucleoli are usually associated with specific regions of particular chromosomes that contain the genes encoding ribosomal RNAs.

Intracellular Membranes and Organelles Define Compartments

The internal volume of the cell exclusive of the nucleus is called the *cytoplasm* and is occupied by *organelles* and by the semifluid *cytosol* in which they are suspended. By "semifluid," we mean that the cytosol is not a thin, watery liquid. Instead, it is believed to be a more viscous material with a consistency closer to that of honey or soft gelatin. In this section, we will look at each of the major eukaryotic organelles. In a typical animal cell, these compartments make up almost half of the cell's total internal volume.

As we continue on our tour of the eukaryotic cell and begin to explore its organelles, you may find it helpful to view these subcellular structures from a human perspective by reading **Box 4A** and acquainting yourself with some of the heritable human diseases that are associated with malfunctions of specific organelles. In most cases, these disorders are caused by genetic defects in specific proteins—enzymes and transport proteins, most commonly—that are localized to particular organelles.

The Mitochondrion. Our tour of the eukaryotic organelles begins with a prominent organelle—the **mitochondrion** (plural: **mitochondria**), shown in **Figure 4-11**. Mitochondria

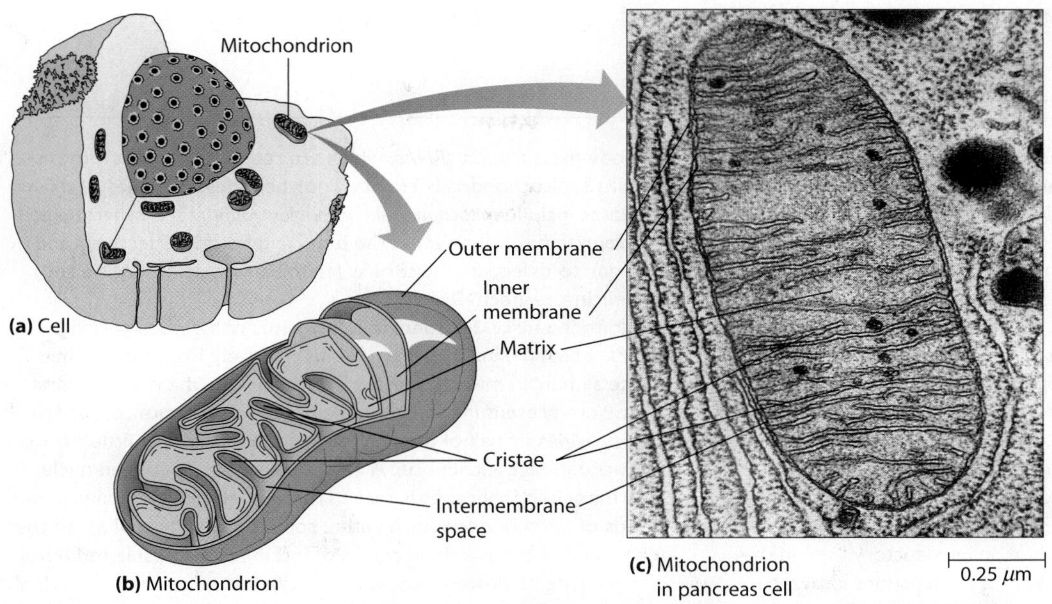

FIGURE 4-11 The Mitochondrion. **(a)** A cutaway view showing the relative numbers and size of the mitochondria within a typical animal cell. Remember that plant cells and all other eukaryotic cells also have mitochondria. **(b)** A schematic illustration of mitochondrial structure. **(c)** A mitochondrion in a rat pancreas cell (TEM).

are found in all eukaryotic cells and are the site of aerobic respiration, the topic of Chapter 10. Mitochondria are large by cellular standards—up to a micrometer across and usually a few micrometers long. A mitochondrion is therefore comparable in size to a whole bacterial cell. Most eukaryotic cells contain hundreds of mitochondria, and each mitochondrion is surrounded by two membranes, designated the *inner* and *outer mitochondrial membranes.* Also found in mitochondria are small, circular molecules of DNA that encode some of the RNAs and proteins needed in mitochondria, along with the ribosomes involved in protein synthesis. In humans and most animals, mitochondria are inherited only through the mother. Therefore, analysis of mitochondrial DNA sequences has been quite useful in tracing genetic lineages in order, for example, to determine the geographic region(s) of origin and subsequent dispersal of modern humans.

Oxidation of sugars and other cellular "fuel" molecules to carbon dioxide in mitochondria extracts energy from food molecules and conserves it as *adenosine triphosphate (ATP).* It is within the mitochondrion that the cell localizes most of the enzymes and intermediates involved in such important cellular processes as the TCA cycle, fat oxidation, and ATP generation. Most of the intermediates involved in transporting electrons from oxidizable food molecules to oxygen are located in or on the **cristae** (singular: **crista**), infoldings of the inner mitochondrial membrane. Other reaction sequences, particularly those of the TCA cycle and those involved in fat oxidation, occur in the semifluid **matrix** that fills the inside of the mitochondrion.

The number and location of mitochondria within a cell can often be related directly to their role in that cell. Tissues with an especially heavy demand for ATP as an energy source can be expected to have cells that are well endowed with mitochondria, and the organelles are usually located within the cell just where the energy need is greatest. This localization is illustrated by the sperm cell in **Figure 4-12**. As the drawing indicates, a sperm cell often

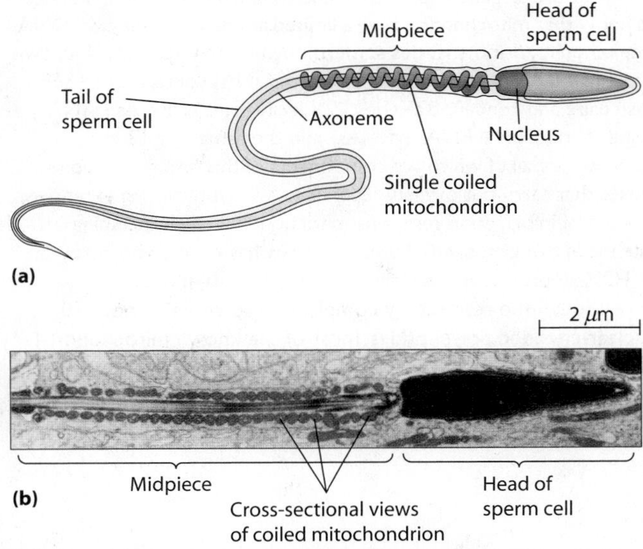

FIGURE 4-12 Localization of the Mitochondrion Within a Sperm Cell. The single mitochondrion present in a sperm cell is coiled tightly around the axoneme of the tail, reflecting the localized need of the sperm tail for energy. **(a)** A schematic drawing of a sperm. **(b)** An electron micrograph of the head and the midpiece of a sperm cell from a marmoset monkey (TEM).

BOX 4A HUMAN APPLICATIONS

Organelles and Human Diseases

Although we may not often acknowledge it—indeed, we may not even be aware of it—many human diseases are actually caused by molecular malfunctions within specific organelles. In fact, several of the organelles that we have encountered on our tour of the "picture exhibition" in this chapter—and that we will encounter in more detail in later chapters—are the sites of a variety of human genetic disorders. Most of them are rare but very serious when they occur. The list of organelle-linked diseases is lengthy, including such diverse mitochondrial disorders as *myopathies* (diseases or disorders of muscle cells), *Leigh syndrome* (a devastating neurodegenerative disorder), and *fatal infantile respiratory defects*. Also included are peroxisomal disorders such as *Zellweger syndrome* and *neonatal adrenoleukodystrophy,* as well as more than 40 *lysosomal storage diseases,* each marked by the harmful accumulation of specific substances. We will consider several of these diseases here, though only at an introductory level. In the process we will anticipate discussions of the functions localized to several organelles, including mitochondria (Chapter 10) as well as peroxisomes and lysosomes (Chapter 12).

Mitochondrial Disorders

Most of the diseases associated with mitochondrial defects are characteristic of either muscle or nerve tissue. This is not surprising, given the high rates of ATP consumption by these tissues and the essential role of the mitochondrion in ATP synthesis. The list includes at least 35 myopathies, as well as a variety of disorders affecting nerve function. Depending on the specific defect, these disorders range greatly in severity. Some lead to infant death; others result in blindness, deafness, seizures, or stroke-like episodes. Milder forms, on the other hand, are characterized by muscular weakness, intolerance of exercise, muscle deterioration, and, in some cases, infertility due to nonmotile sperm.

These are all genetic disorders; and to understand them, we need to know that mitochondria have a limited amount of their own DNA. The mitochondrion encodes some, though by no means all, of its own proteins. Human mitochondrial DNA (mtDNA) consists of 16,568 base pairs and contains 37 genes: 22 specify transfer RNAs (tRNAs), 2 specify ribosomal RNAs (rRNAs), and the remaining 13 encode polypeptides, all of which are components of the respiratory complexes that carry out oxygen-dependent ATP synthesis. An extensive amount of information regarding mitochondrial DNA, including mutations associated with human diseases, has been compiled by the MITOMAP project at the University of California–Irvine.

Although the respiratory complexes also contain about 70 nuclear-encoded polypeptides, most of the known mitochondrial myopathies are due to defects in mitochondrial rather than in nuclear genes, involving either the deletion or mutation of specific mitochondrial genes. Most of these defects occur in the genes that encode *mitochondrial tRNAs,* which are required for the synthesis of all 13 mitochondrially encoded polypeptides. Examples of these diseases include *mitochondrial encephalomyopathy* and *hypertrophic cardiomyopathy,* which affect the brain and heart, respectively, and are due to defects in the tRNAs for the amino acids leucine and isoleucine, respectively.

Mitochondrial disorders follow what is called *maternal inheritance,* which means that they come exclusively from the mother. Since all human mitochondria are derived from the mitochondria that were present in the egg at the time of fertilization, the sperm cell provides its half of the nuclear genome but makes little or no mitochondrial contribution. A further distinction between nuclear and mitochondrial genes is that a typical human cell contains hundreds of mitochondria, each with 2 to 10 copies of mtDNA, so the cell contains thousands of copies of mtDNA. As a result, mtDNAs can be quite heterogeneous within specific tissues, and mitochondrial disorders are likely to arise only when most of the mitochondria within a given tissue contain a particular mutant gene.

Peroxisomal Disorders

Most of the human diseases associated with peroxisomes are due to the absence of a single peroxisomal protein. Considering the variety of cellular functions that are localized to this organelle, it is not surprising that many disorders are known in which specific peroxisomal proteins are either defective or absent. Unlike mitochondria, peroxisomes contain no DNA; thus, all of these defects are due to mutations in nuclear genes.

There are several well-studied peroxisomal disorders: *Zellweger syndrome (ZS), neonatal adrenoleukodystrophy (NALD), X-linked adrenoleukodystrophy (X-ALD),* and *infantile Refsum disease (IRD).* ZS is characterized by a variety of severe neurological, visual, and liver disorders that lead to death during early childhood, often by age one or two. NALD (autosomal recessive) and X-ALD (sex-linked, typically male-only) are less severe than ZS but eventually lead to neurological impairment and death. Patients with NALD or X-ALD usually begin to display symptoms of adrenal failure and neurological debilitation during early childhood. The symptoms of IRD are similar to, but less severe than, those of ZS and NALD.

Although these diseases were discovered independently and not initially considered to be related, we now know that each of these disorders is caused by mutations in any of 11 different human genes. The most severe mutations in these genes cause ZS, moderately severe mutations cause NALD, and the least severe mutations cause IRD.

In some forms of NALD, the defective gene product is a membrane protein involved in the transport of very-long-chain fatty acids into the peroxisomes, where such fatty acids are broken

has a single spiral mitochondrion wrapped around the central shaft, or *axoneme,* of the cell. Notice how tightly the mitochondrion coils around the axoneme, just where the ATP is actually needed to propel the sperm cell. Muscle cells and cells that specialize in the transport of ions also have numerous mitochondria located strategically to meet the special energy needs of such cells (**Figure 4-13**).

The Chloroplast. The next organelle we encounter on our gallery tour is the **chloroplast,** the site of photosynthesis in plants and algae (**Figure 4-14**). Chloroplasts are large organelles, typically a few micrometers in diameter and 5–10 μm long, and can be quite numerous in the leaves of green plants. They are therefore substantially bigger than mitochondria and larger than any other structure in a

down to shorter chain lengths that can be handled by the mitochondrion. When this transport mechanism is impaired or nonfunctional, the very-long-chain fatty acids accumulate in cells and tissues. That accumulation is particularly devastating in the brain, where the very-long-chain fatty acids destroy the myelin sheaths that provide essential insulation for nerve cells, thereby profoundly impairing transmission of neural signals.

Similarly, X-ALD is caused by mutation of a gene encoding a peroxisomal transporter, *ABCD-1*. This transporter is involved in recycling of myelin, a component of the sheath that surrounds neurons. Its deficiency leads to demyelination and neurodegeneration and is typically fatal before adolescence. However, in late 2009, early results of a gene therapy trial showed significant promise: neural degeneration was halted in two X-ALD patients who received an infusion of their own blood cells that had been treated with a modified viral vector containing a functional gene encoding *ABCD-1*.

In ZS, the missing or defective gene product can be any of several proteins that are essential for targeting peroxisomal enzymes for uptake by the organelle. As we will learn in Chapter 12, peroxisomal proteins are encoded by nuclear genes, synthesized on cytoplasmic ribosomes, and then imported into the peroxisome. Individuals with ZS can typically synthesize all of the requisite enzymes, but they have a deficiency in any of several membrane proteins involved in transporting these enzymes into the organelle. As a result, the proteins remain in the cytosol, where they cannot perform their intended functions. Peroxisomes can be detected in the cells of such individuals, but the organelles are empty "ghosts"—membrane-bounded structures without the normal complement of enzymes. Not surprisingly, afflicted individuals develop various neurological, visual, and liver disorders that lead inevitably to death during early childhood.

Lysosomal Disorders

Another organelle subject to a variety of genetic defects is the lysosome, which plays an essential role in the digestion of food molecules and in the recycling of cellular components that are no longer needed. Over 40 heritable *lysosomal storage diseases* are known, each characterized by the harmful accumulation of a specific substance or class of substances—most commonly polysaccharides or lipids—that would normally be catabolized by the hydrolytic enzymes present within the lysosome. In some cases, the defective protein is either a key enzyme in degrading the substance or a protein involved in transporting the degradation products out of the lysosome. In other cases, the requisite enzymes are synthesized in normal amounts but are secreted into the extracellular medium rather than being targeted to the lysosomes.

An example of this latter type of disorder is *I-cell disease,* which is due to a defect in an enzyme called N-acetylglucosamine phosphotransferase. This enzyme is required to correctly process the portion of the protein that targets, or signals, lysosomal enzymes for import into the organelle. Without the necessary signal, the hydrolytic enzymes are not transported into the lysosomes. Thus, the lysosomes become engorged with undegraded polysaccharides, lipids, and other material. This causes irreversible damage to the cells and tissues.

A well-known example is *Tay-Sachs disease,* which is quite rare in the general population but has a higher incidence among Ashkenazi Jews of eastern European ancestry. After about six months, children who are homozygous for this defect show rapid mental and motor deterioration as well as skeletal, cardiac, and respiratory dysfunction. This is followed by dementia, paralysis, blindness, and death, usually within three years. The disease results from the accumulation in nervous tissue of a particular glycolipid called *ganglioside* G_{M2}. (For the structure of gangliosides, see Figure 7-6.) The missing or defective lysosomal enzyme is *β-N-acetylhexosaminidase A,* which cleaves the terminal N-acetylgalactosamine from the "glyco" (carbohydrate) portion of the ganglioside. G_{M2} is a prominent component of the membranes in brain cells. Not surprisingly, lysosomes from children afflicted with Tay-Sachs disease are filled with membrane fragments containing undigested gangliosides.

All of the known lysosomal storage diseases can be diagnosed prenatally. Even more significant are the prospects for *enzyme replacement therapy* and *gene therapy.* Enzyme replacement therapy has been shown to be effective with a particular lysosomal disorder called *Gaucher disease,* characterized by the absence or deficiency of a specific hydrolase called *glucocerebrosidase*. In the absence of this enzyme, lipids called *glucocerebrosides* accumulate in the lysosomes of macrophages, which are the white blood cells that engulf and digest foreign material or invasive microorganisms as well as cellular debris and whole damaged cells. (The structure of cerebrosides is also shown in Figure 7-6.) Glucocerebroside accumulation typically leads to liver and spleen enlargement, anemia, and mental retardation. In the past, treatment depended on the ability to purify glucocerebrosidase from human placental material, treat it so that it could be taken up by macrophages, and infuse it into the bloodstream. Recently, however, a synthetic form has been produced using recombinant DNA technology, simplifying the treatment protocol. Macrophages that take up this enzyme are able to degrade glucocerebrosides as needed, thereby effectively treating what would otherwise be a fatal disease.

Gene therapy is a somewhat more futuristic prospect for the treatment of lysosomal storage diseases as well as other heritable disorders. This approach involves inserting the genes for the missing enzymes into the appropriate cells, thereby effectively curing the disease rather than simply treating it. For a further consideration of gene therapy, see the discussion in Chapter 20.

typical plant cell except the nucleus. (However, in some algae, there is a single large chloroplast that can be larger than the nucleus.) Like mitochondria, chloroplasts are surrounded by both an inner and an outer membrane. Inside the chloroplast there is a third membrane system consisting of flattened sacs called **thylakoids** and the membranes (**stroma thylakoids**), that interconnect them. Thylakoids are stacked together to form the **grana** (singular: **granum**) that characterize most chloroplasts (Figure 4-14c, d).

Chloroplasts are the site of *photosynthesis,* the light-driven process that uses solar energy and carbon dioxide to manufacture the sugars and other organic compounds from which all life is ultimately fabricated. Chloroplasts are found in leaves and other photosynthetic tissues of higher plants,

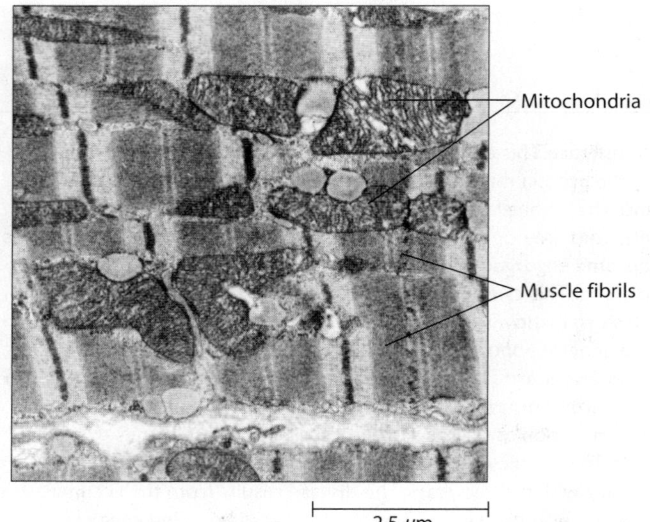

FIGURE 4-13 Localization of Mitochondria Within a Muscle Cell. This electron micrograph of a muscle cell from a cat heart shows the intimate association of mitochondria with the muscle fibrils that are responsible for muscle contraction (TEM).

as well as in the algae. Located within this organelle are most of the enzymes, intermediates, and light-absorbing pigments needed for photosynthesis—the enzymatic reduction of carbon dioxide to sugar, an energy-requiring process. (Note that this reduction process is the reverse of the energy-producing oxidation reactions in mitochondria that convert sugar to carbon dioxide.) Reactions that depend directly on solar energy are localized in or on the thylakoid membrane system. Reactions involved in the reduction of carbon dioxide to sugar molecules occur within the semifluid **stroma** that fills the interior of the chloroplast. Also found in the stroma are chloroplast ribosomes along with small, circular molecules of DNA that encode some of the RNAs and proteins needed in the chloroplast.

Although known primarily for their role in photosynthesis, chloroplasts are also involved in a variety of other chemical processes. They contain enzymes that reduce nitrogen from the oxidation level of soil-derived nitrate ions (NO_3^-) to ammonia (NH_3), the form of nitrogen required for protein synthesis. Enzymes in the chloroplast also catalyze the reduction of sulfate ions (SO_4^{2-}) to hydrogen sulfide (H_2S), which can be incorporated into the amino acid cysteine for use in protein synthesis. Furthermore, the chloroplast is the most prominent example of a class of plant organelles known as the **plastids.** Plastids other than chloroplasts perform a variety of functions in plant cells. *Chromoplasts,* for example, are pigment-containing plastids that are responsible for the characteristic coloration of flowers, fruits, and other plant parts. *Amyloplasts* are plastids that are specialized for the storage of starch (amylose and amylopectin).

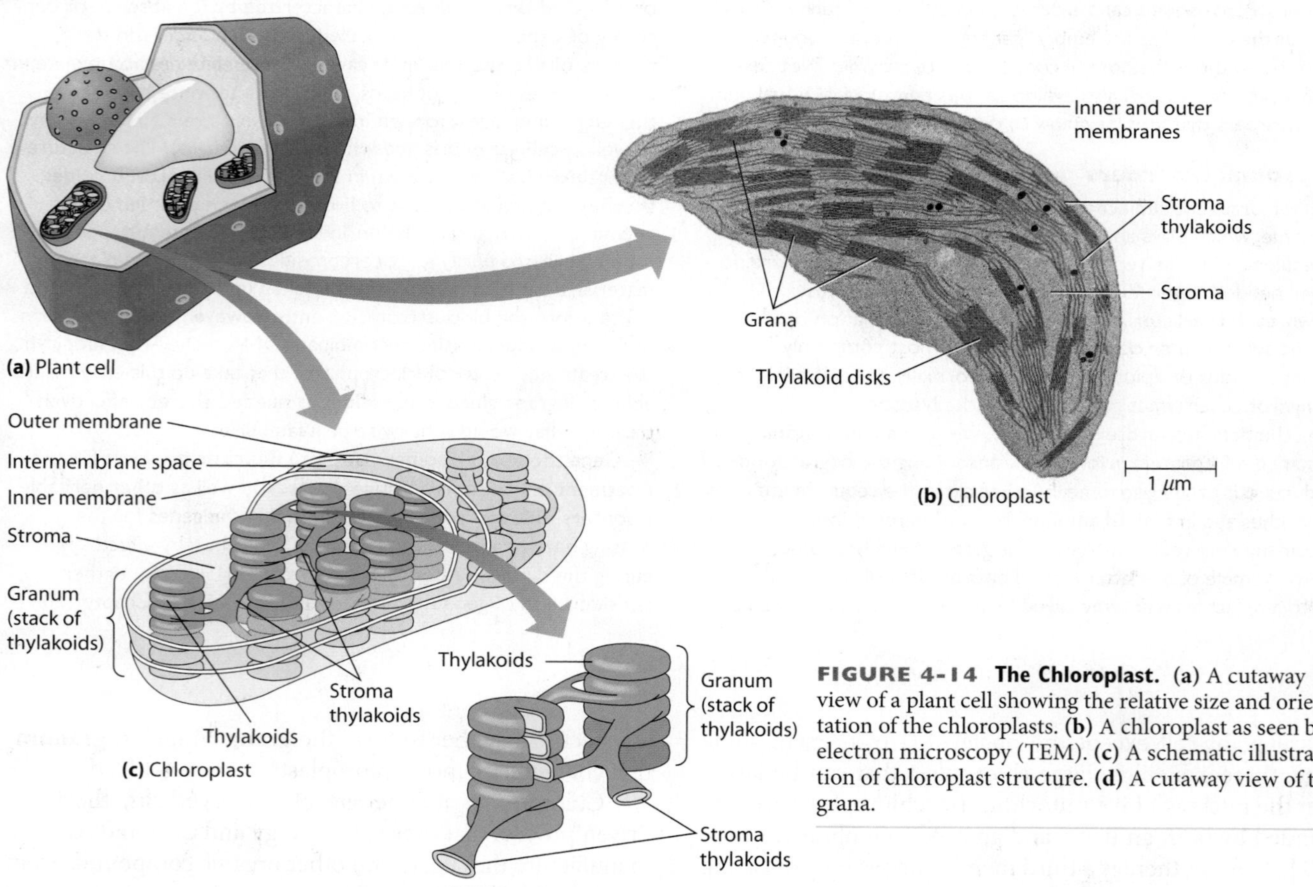

FIGURE 4-14 The Chloroplast. (a) A cutaway view of a plant cell showing the relative size and orientation of the chloroplasts. (b) A chloroplast as seen by electron microscopy (TEM). (c) A schematic illustration of chloroplast structure. (d) A cutaway view of two grana.

The Endosymbiont Theory: Did Mitochondria and Chloroplasts Evolve from Ancient Bacteria? Having just met the mitochondrion and chloroplast as eukaryotic organelles, we pause here for a brief digression concerning the possible evolutionary origins of these organelles. Both mitochondria and chloroplasts contain their own DNA and ribosomes, which enable them to carry out the synthesis of both RNA and proteins (although most of the proteins present in these organelles are in fact encoded by nuclear genes). As molecular biologists studied nucleic acid and protein synthesis in these organelles, they were struck by the many similarities between these processes in mitochondria and chloroplasts and the comparable processes in bacterial cells. Both have circular DNA molecules without associated histones, and both show similarities in rRNA sequences, ribosome size, sensitivity to inhibitors of RNA and protein synthesis, and the type of protein factors used in protein synthesis. In addition to these molecular features, mitochondria and chloroplasts resemble bacterial cells in size and shape, and they have a double membrane in which the inner membrane has bacterial-type lipids.

These similarities led to the **endosymbiont theory** for the evolutionary origins of mitochondria and chloroplasts. This theory, which is described more fully in Box 11A, pages 326–327, proposes that both of these organelles originated from prokaryotes that gained entry to, and established a symbiotic relationship within, the cytoplasm of ancient single-celled organisms called *protoeukaryotes*. The endosymbiont theory proposes that protoeukaryotes ingested bacteria and cyanobacteria by a process known as *phagocytosis* ("cell eating"). Following phagocytosis, these cells were not digested; instead, they took up residence in the cytoplasm and eventually evolved into mitochondria and chloroplasts, respectively.

The Endoplasmic Reticulum. Extending throughout the cytoplasm of almost every eukaryotic cell is a network of membranes called the **endoplasmic reticulum,** or **ER** (**Figure 4-15**). The name sounds complicated, but *endoplasmic* just means "within the plasm" (of the cell), and *reticulum* is simply a fancy word for "network." The endoplasmic reticulum consists of tubular membranes and flattened sacs, or **cisternae** (singular: **cisterna**), that are interconnected. The internal space enclosed by the ER membranes is called the **lumen.** The ER is continuous with the outer membrane of the nuclear envelope (Figure 4-15a).

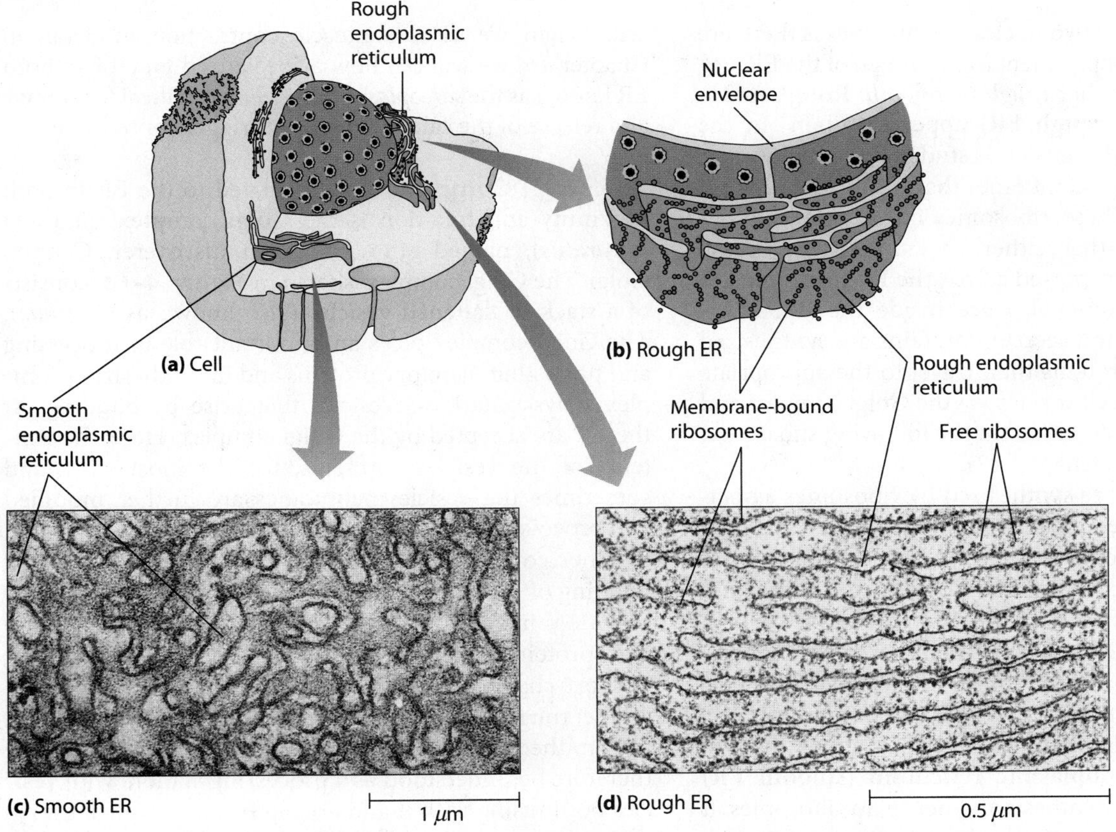

FIGURE 4-15 The Endoplasmic Reticulum. (a) A cutaway view of a typical animal cell showing the location and relative size of the endoplasmic reticulum (ER). **(b)** A schematic illustration depicting the organization of the rough ER as layers of flattened membranes. Rough ER membranes are studded on their outer surface with ribosomes and are continuous with the nuclear envelope. **(c)** An electron micrograph of smooth ER in a cell from guinea pig testis (TEM). **(d)** An electron micrograph of rough ER in a rat pancreas cell (TEM); notice that ribosomes are either attached to the ER or free in the cytosol.

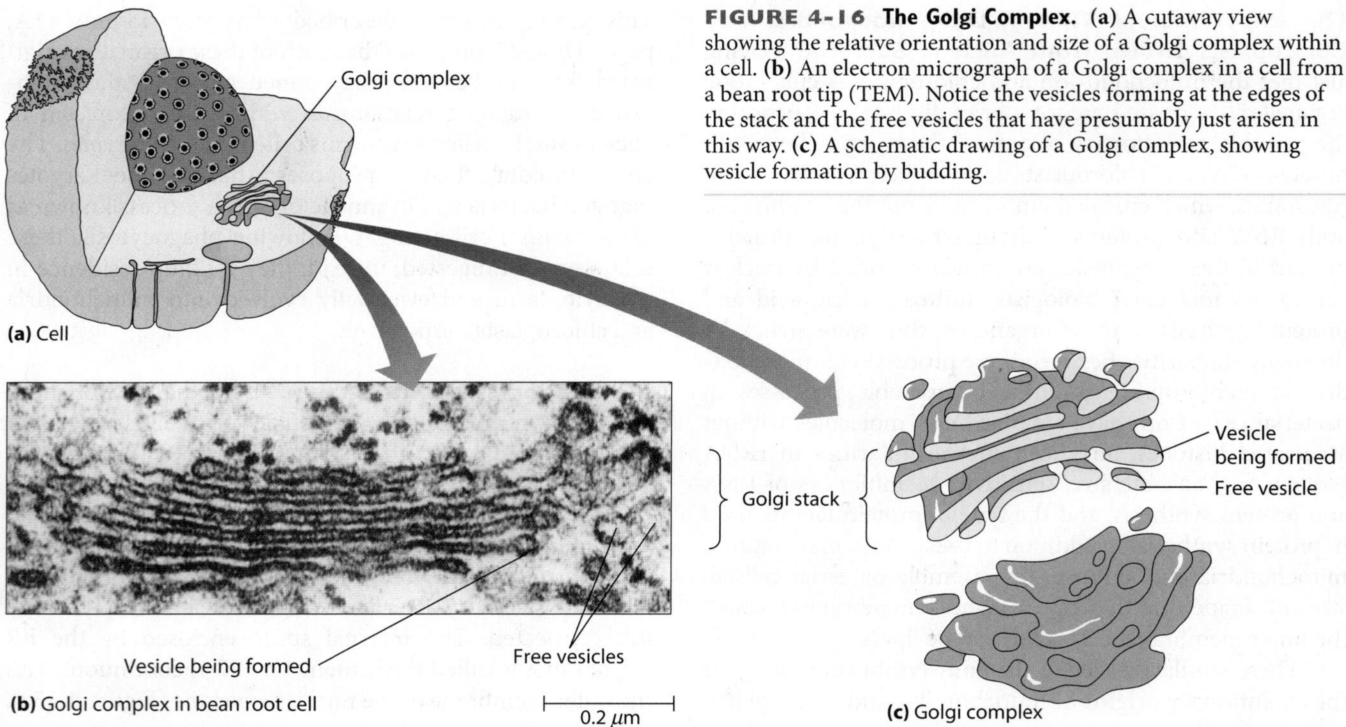

FIGURE 4-16 The Golgi Complex. **(a)** A cutaway view showing the relative orientation and size of a Golgi complex within a cell. **(b)** An electron micrograph of a Golgi complex in a cell from a bean root tip (TEM). Notice the vesicles forming at the edges of the stack and the free vesicles that have presumably just arisen in this way. **(c)** A schematic drawing of a Golgi complex, showing vesicle formation by budding.

The space between the two nuclear membranes is therefore a part of the same compartment as the lumen of the ER.

The ER can be either *rough* or *smooth*. **Rough endoplasmic reticulum (rough ER)** appears "rough" in the electron microscope because it is studded with ribosomes on the side of the membrane that faces the cytosol (Figure 4-15b, d). These ribosomes are actively synthesizing polypeptides that either accumulate within the membrane or are transported across the ER membrane to accumulate in the lumenal space inside the ER. Many membrane proteins and secreted proteins are synthesized in this way. These proteins then travel to the appropriate membrane or to the cell surface via the Golgi complex and secretory vesicles, as described in the following subsection (and in more detail in Chapter 12).

Not all proteins are synthesized by ribosomes associated with membranes of the rough ER, however. Much protein synthesis occurs on free ribosomes that are not attached to the ER but are instead found in the cytosol and are not associated with a membrane (Figure 4-15d). In general, secretory proteins and membrane proteins are made by ribosomes on the rough ER, whereas proteins intended for use within the cytosol or for import into organelles are made on free ribosomes.

The **smooth endoplasmic reticulum (smooth ER)** has no role in protein synthesis and hence no ribosomes. It therefore has a characteristically smooth appearance when viewed by electron microscopy (Figure 4-15c). Smooth ER is involved in the synthesis of lipids and steroids, such as cholesterol and the steroid hormones derived from it. In addition, smooth ER is responsible for inactivating and detoxifying drugs such as barbiturates and other compounds that might otherwise be toxic or harmful to the cell. When we discuss muscle contraction in detail in Chapter 16, we will see how a specialized type of smooth ER known as the *sarcoplasmic reticulum* is critical for storage and release of the calcium ions that trigger contraction.

The Golgi Complex. Closely related to the ER in both proximity and function is the **Golgi complex** (or *Golgi apparatus*), named after its Italian discoverer, Camillo Golgi. The Golgi complex, shown in **Figure 4-16**, consists of a stack of flattened vesicles (also known as *cisternae*). The Golgi complex plays an important role in processing and packaging secretory proteins and in synthesizing complex polysaccharides. Vesicles that arise by budding off the ER are accepted by the Golgi complex. Here, the contents of the vesicles (proteins, for the most part) and sometimes the vesicle membranes are further modified and processed. The processed contents are then passed on to other components of the cell in vesicles that arise by budding off the Golgi complex (Figure 4-16c).

Most membrane proteins and secretory proteins are glycoproteins. The initial steps in *glycosylation* (addition of short-chain carbohydrates) take place within the lumen of the rough ER, but the process is usually completed within the Golgi complex. The Golgi complex should therefore be understood as a processing station, with vesicles both fusing with it and arising from it. Almost everything that goes into it comes back out—but in a modified, packaged form, often ready for export from the cell.

Secretory Vesicles. Once processed by the Golgi complex, secretory proteins and other substances intended for export from the cell are packaged into **secretory vesicles.** The cells of your pancreas, for example, contain many

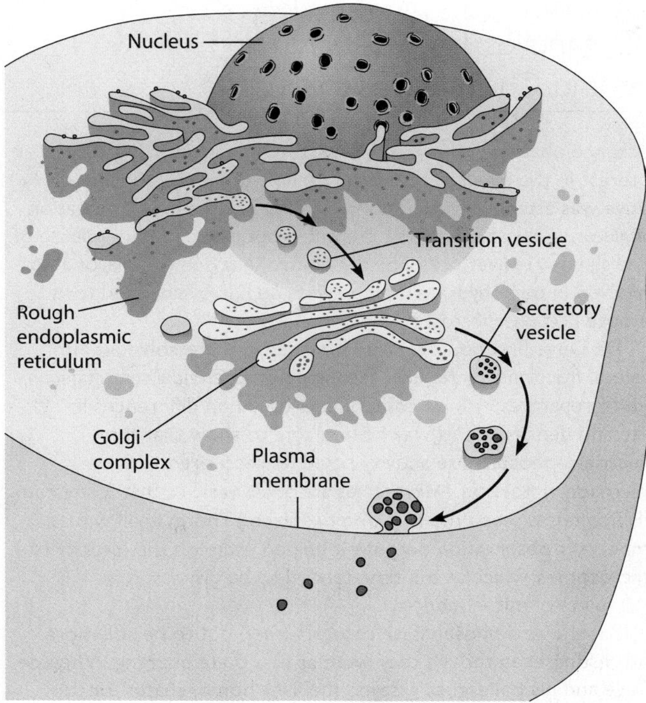

FIGURE 4-17 The Process of Secretion in Eukaryotic Cells. Proteins to be packaged for export are synthesized on the rough ER, passed to the Golgi complex for processing, and eventually compartmentalized into secretory vesicles. These vesicles then make their way to the plasma membrane and fuse with it, releasing their contents to the exterior of the cell.

such vesicles because the pancreas is responsible for the synthesis of several important digestive enzymes. These enzymes are synthesized on the rough ER, packaged by the Golgi complex, and then released from the cell via secretory vesicles, as shown in **Figure 4-17**. These vesicles from the Golgi region move to and fuse with the plasma membrane and discharge their contents to the exterior of the cell by the process of exocytosis. Together, the ER, Golgi, secretory vesicles, and the lysosomes we will discuss shortly constitute the cell's *endomembrane system.* The endomembrane system is responsible for *trafficking* of molecules through the cell, and this process of protein and lipid synthesis, processing, trafficking, and export via the ER, the Golgi complex, and secretory vesicles will be considered in more detail in Chapter 12.

The Lysosome. The next picture at our cellular exhibition is of the **lysosome,** an organelle about 0.5–1.0 μm in diameter and surrounded by a single membrane (**Figure 4-18**). Lysosomes were discovered in the early 1950s by Christian de Duve and his colleagues. The story of that discovery is recounted in **Box 4B** to underscore the significance of chance observations when they are made by astute investigators and to illustrate the importance of new techniques to the progress of science.

Lysosomes are used by the cell as a means of storing *hydrolases,* enzymes capable of digesting specific biological molecules such as proteins, carbohydrates, or fats. While cells need such enzymes, both to digest food molecules and to break down damaged cellular constituents, these enzymes must be carefully sequestered until actually needed, lest they digest the normal components that were not scheduled for destruction in the cell.

Like secretory proteins, lysosomal enzymes are synthesized on the rough ER, transported to the Golgi complex, and then packaged into vesicles that can become lysosomes. A special carbohydrate covering on the inner surface of the lysosomal membrane protects this membrane (and the cell) from the hydrolytic activities of these enzymes until they are needed. Equipped with its repertoire of hydrolases, the lysosome is able to break down virtually any kind of biological molecule. Eventually, the digestion products are small enough to pass through the membrane into the cytosol of the cell, where they can be used to synthesize more macromolecules—recycling at the cellular level.

The Peroxisome. The next organelle at our exhibition is the **peroxisome.** Peroxisomes resemble lysosomes in size and general lack of obvious internal structure. Like lysosomes, they are surrounded by a single rather than a double membrane. Peroxisomes are found in plant and animal

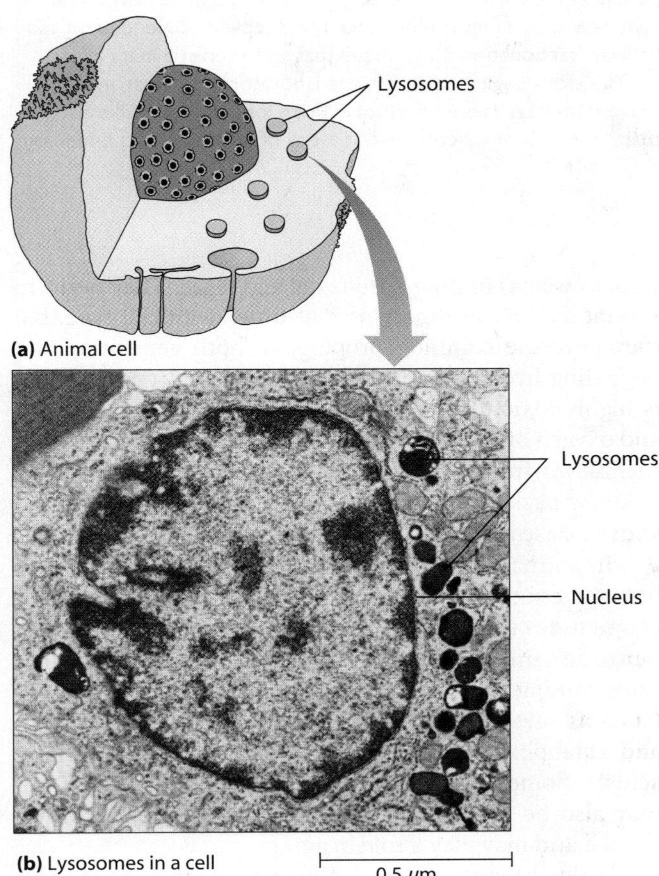

FIGURE 4-18 Lysosomes. (**a**) A cutaway view showing lysosomes within an animal cell. (**b**) Lysosomes in an animal cell stained cytochemically for acid phosphatase, a lysosomal enzyme. The cytochemical staining technique results in dense deposits of lead phosphate at the site of acid phosphatase activity (TEM).

BOX 4B DEEPER INSIGHTS

Discovering Organelles: The Importance of Centrifuges and Chance Observations

Have you ever wondered how the various organelles within eukaryotic cells were discovered? There are almost as many answers to that question as there are kinds of organelles. In general, they were described by microscopists before their role in the cell was understood. As a result, the names of organelles usually reflect structural features rather than physiological roles. Thus, *chloroplast* simply means "green particle" and *endoplasmic reticulum* just means "network within the plasm (of the cell)."

Such is not the case for the *lysosome*, however. This organelle was the first to have its biochemical properties described before it had ever been reported by microscopists. Only after fractionation data had predicted the existence and properties of just such an organelle were lysosomes actually observed in cells. A suggestion of its function is even inherent in the name given to the organelle because the Greek root *lys–* means "to digest." (The literal translation is "to loosen," but that's essentially what digestion does to chemical bonds!)

The lysosome is something of a newcomer on the cellular biology scene; it was not discovered until the early 1950s. The story of that discovery is fascinating because it illustrates how important chance observations can be, especially when made by the right people at the right time. The account also illustrates how significant new techniques can be, since the discovery depended on subcellular fractionation, a technique that was then still in its infancy.

The story begins in 1949 in the laboratory of Christian de Duve, who later received a Nobel Prize for this work. Like so many scientific advances, the discovery of lysosomes depended on a chance observation made by an astute investigator. Because of an interest in the effect of insulin on carbohydrate metabolism, de Duve was attempting at the time to pinpoint the cellular location of *glucose-6-phosphatase*, the enzyme responsible for the release of free glucose in liver cells. For the control enzyme (that is, one not involved in carbohydrate metabolism), de Duve happened to choose *acid phosphatase*.

De Duve first homogenized liver tissue and resolved it into several fractions by the new technique of *differential centrifugation*, which separates cellular components based on differences in size and density. In this way, he was able to show that the glucose-6-phosphatase activity could be recovered with the microsomal fraction. (*Microsomes* are small vesicles that form from ER fragments when tissue is homogenized.) This in itself was an important observation because it helped establish the identity of microsomes, which at the time tended to be dismissed as fragments of mitochondria.

But the acid phosphatase results turned out to be still more interesting, even though they were at first quite puzzling. When de Duve and his colleagues assayed the liver homogenates for this enzyme, they found only a fraction of the expected activity. When assayed again for the same enzyme a few days later, however, the same homogenates had about ten times as much activity.

Speculating that he was dealing with some sort of activation phenomenon, de Duve subjected the homogenates to differential centrifugation to see which subcellular fraction the phenomenon was associated with. He and his colleagues were able to demonstrate

---

cells, as well as in fungi, protozoa, and algae. They perform several distinctive functions that differ with cell type, but they have the common property of both generating and degrading hydrogen peroxide (H_2O_2). Hydrogen peroxide is highly toxic to cells but can be decomposed into water and oxygen by the enzyme *catalase*. Eukaryotic cells protect themselves from the detrimental effects of hydrogen peroxide by packaging peroxide-generating reactions together with catalase in a single compartment, the peroxisome.

In animals, peroxisomes are found in most cell types but are especially prominent in liver and kidney cells (**Figure 4-19**). Beyond their role in detoxifying hydrogen peroxide, animal peroxisomes have several other functions, among them detoxifying other harmful compounds (such as methanol, ethanol, formate, and formaldehyde) and catabolizing unusual substances (such as D-amino acids). Some researchers speculate that peroxisomes may also be involved in regulating oxygen levels within the cell and may play a role in aging.

Animal peroxisomes also play a role in the oxidative breakdown of fatty acids, which are components of triacylglycerols, phospholipids, and glycolipids (see Figure 3-27). As we will see in Chapter 10, fatty acid breakdown occurs primarily in the mitochondrion. However, fatty acids with

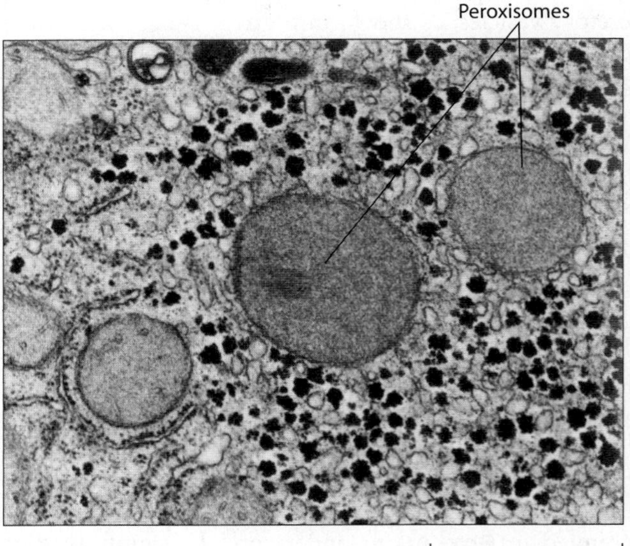

FIGURE 4-19 Animal Peroxisomes. Several peroxisomes can be seen in this cross section of a liver cell (TEM). Peroxisomes are found in most animal cells but are especially prominent features of liver and kidney cells.

that much of the acid phosphatase activity could be recovered in the mitochondrial fraction and that this fraction showed an even greater increase in activity after standing a few days than did the original homogenates.

To their surprise, they then discovered that, upon recentrifugation, this elevated activity no longer sedimented with the mitochondria but stayed in the supernatant. They went on to show that the activity could be increased and the enzyme solubilized by a variety of treatments, including harsh grinding, freezing and thawing, and exposure to detergents or hypotonic conditions. From these results, de Duve concluded that the enzyme must be present in some sort of membrane-bounded particle that could easily be ruptured to release the enzyme. Apparently, the enzyme could not be detected within the particle, probably because the membrane was not permeable to the substrates used in the enzyme assay.

Assuming that particle to be the mitochondrion, they continued to isolate and study this fraction of the liver homogenates. At that point, another chance observation occurred, this time because of a broken centrifuge. The unexpected breakdown forced one of de Duve's students to use an older, slower centrifuge, and the result was a mitochondrial fraction containing little or no acid phosphatase. Based on this unexpected finding, de Duve speculated that the mitochondrial fraction as they usually prepared it might in fact contain two kinds of organelles: the actual mitochondria, which could be sedimented with either centrifuge, and some sort of more slowly sedimenting particle that came down only in the faster centrifuge.

This led them to devise a fractionation scheme that allowed the original mitochondrial fraction to be subdivided into a rapidly sedimenting component and a slowly sedimenting component. As you might guess, the rapidly sedimenting component contained the mitochondria, as evidenced by the presence of enzymes known to be mitochondrial markers. The acid phosphatase, on the other hand, was in the slowly sedimenting component, along with several other hydrolytic enzymes, such as ribonuclease, deoxyribonuclease, β-glucuronidase, and a protease. Each of these enzymes showed the same characteristic of increased activity upon membrane rupture, a property that de Duve termed *latency*.

By 1955, de Duve was convinced that these hydrolytic enzymes were packaged together in a previously undescribed organelle. In keeping with his speculation that this organelle was involved in intracellular lysis (digestion), he called it a *lysosome*.

Thus, the lysosome became the first organelle to be identified entirely on biochemical criteria. At the time, no such particles had been described by microscopy. But when de Duve's lysosome-containing fractions were examined with the electron microscope, they were found to contain membrane-bounded vesicles that were clearly not mitochondria and were in fact absent from the mitochondrial fraction. Knowing what the isolated particles looked like, microscopists were then able to search for them in fixed tissue. As a result, lysosomes were soon identified and reported in a variety of animal tissues. Within six years, then, the organelle that began as a puzzling observation in an insulin experiment became established as a bona fide feature of most animal cells.

more than 12 carbon atoms are oxidized relatively slowly by mitochondria. In the peroxisomes, on the other hand, fatty acids up to 22 carbon atoms long are oxidized rapidly. The long chains are degraded by the removal of two carbon units at a time until they get to a length (10–12 carbon atoms) that the mitochondria can efficiently handle.

The vital role of peroxisomes in breaking down long-chain fatty acids is underscored by the serious human diseases that result when one or more peroxisomal enzymes involved in degrading long-chain fatty acids are defective or absent. One such disease is neonatal adrenoleukodystrophy (NALD), a sex-linked (male-only) disorder that leads to profound neurological debilitation and eventually to death. (See Box 4A, pages 114–115, for a discussion of NALD and other human diseases associated with deficiencies in peroxisome function.)

The best-understood metabolic roles of peroxisomes occur in plant cells. During the germination of fat-storing seeds, specialized peroxisomes called **glyoxysomes** play a key role in converting stored fat into carbohydrates. In photosynthetic tissue, **leaf peroxisomes** are prominent due to their role in *photorespiration,* the light-dependent uptake of oxygen and release of carbon dioxide (to be discussed further in Chapter 11). The photorespiratory pathway is an example of a cellular process that involves several organelles. While some of the enzymes that catalyze the reactions in this sequence occur in the peroxisome, others are located in the chloroplast or the mitochondrion. This mutual involvement in a common cellular process is suggested by the intimate association of peroxisomes with mitochondria and chloroplasts in many leaf cells, as shown in **Figure 4-20**.

Vacuoles. Some cells contain another type of membrane-bounded organelle called a **vacuole.** In animal and yeast cells, vacuoles are used for temporary storage or transport. Some protozoa take up food particles or other materials from their environment by phagocytosis. Phagocytosis is a form of endocytosis that involves an infolding of the plasma membrane around the desired substance. This infolding is followed by a pinching-off process that internalizes the membrane-bounded particle as a type of vacuole, known as a phagosome. Following fusion with a lysosome, the contents of a phagosome are hydrolyzed to provide nutrients for the cell.

Plant cells also contain vacuoles. In fact, a single large vacuole is found in most mature plant cells (**Figure 4-21**). This vacuole, sometimes called the **central vacuole,** may

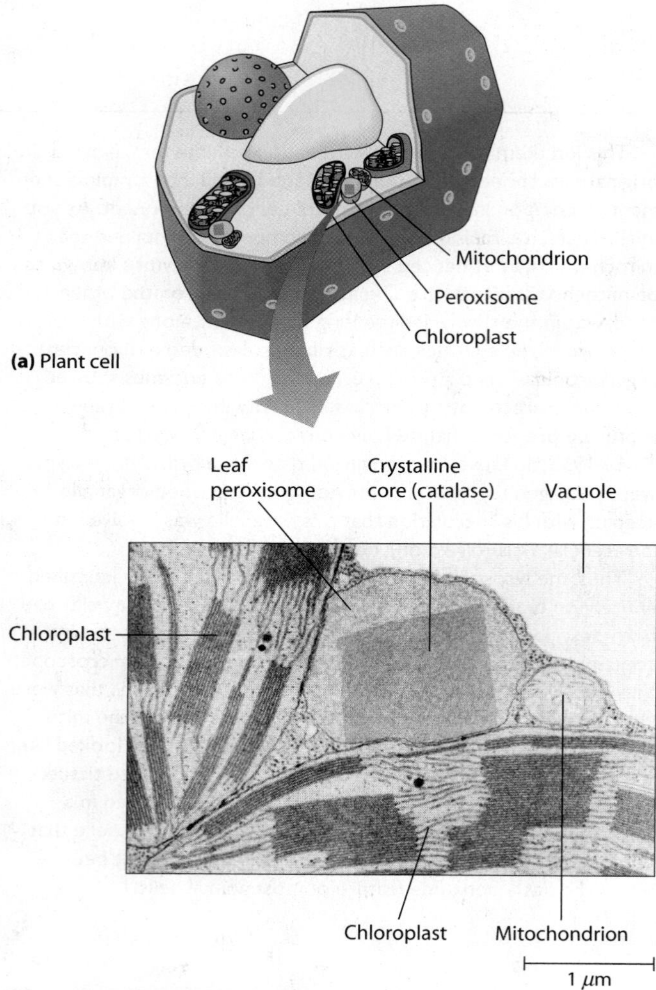

FIGURE 4-20 A Leaf Peroxisome and Its Relationship to Other Organelles in a Plant Cell. **(a)** A cutaway view showing a peroxisome, a mitochondrion, and a chloroplast within a plant cell. **(b)** A peroxisome in close proximity to chloroplasts and to a mitochondrion within a tobacco leaf cell (TEM). This is probably a functional relationship because all three organelles participate in the process of photorespiration. The crystalline core frequently observed in leaf peroxisomes is the enzyme catalase, which catalyzes the decomposition of hydrogen peroxide into water and oxygen.

play a limited role in storage and in intracellular digestion. However, its main importance is its role in maintaining the *turgor pressure* that keeps tissue from wilting. The vacuole has a high concentration of solutes; thus, water tends to move into the vacuole, causing it to swell. As a result, the vacuole presses the rest of the cell constituents against the cell wall, thereby maintaining the turgor pressure. The limp appearance of wilted tissue results when the central vacuole does not provide adequate pressure. We can easily demonstrate this by placing a piece of crisp celery in salt water. The high concentration of salt on the outside of the cells will cause water to move out of the cells. As the turgor pressure decreases, the tissue will quickly become limp and lose its crispness.

Ribosomes. The last portrait in our gallery is the **ribosome.** Strictly speaking, the ribosome is not really an organelle because it is not surrounded by a membrane. We will consider it here, however, because ribosomes, like organelles, are the focal point for a specific cellular activity—in this case, protein synthesis. Ribosomes are found in all cells, but bacteria, archaea, and eukaryotes differ from each other in ribosome size and in the number and kinds of ribosomal protein and RNA molecules (see Table 4-1). Bacteria and archaea contain smaller ribosomes than those found in eukaryotes, and all three have different numbers of ribosomal proteins, although there is some similarity between the ribosomal proteins of archaeal and eukaryotic cells. A defining feature is that each of these three cell types has its own unique type of ribosomal RNA, the very molecule whose characteristics were used to define the archaea as a domain distinct from bacteria.

Compared with even the smallest organelles, ribosomes are tiny structures. The ribosomes of eukaryotic and prokaryotic cells have diameters of about 30 and

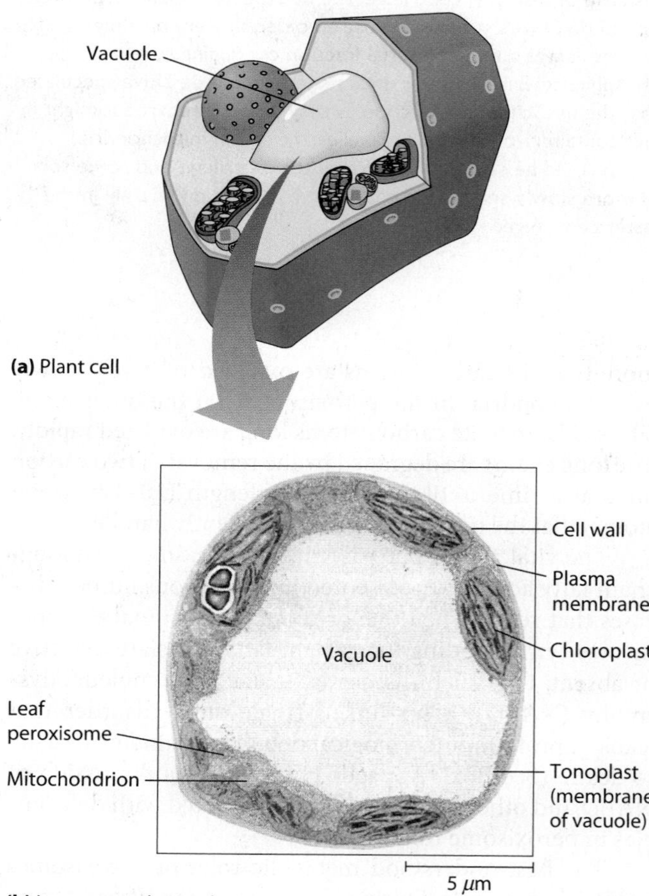

FIGURE 4-21 The Vacuole in a Plant Cell. **(a)** A cutaway view showing the vacuole in a plant cell. **(b)** An electron micrograph of a bean leaf cell with a large central vacuole (TEM). The vacuole occupies much of the internal volume of the cell, with the cytoplasm sandwiched into a thin region between the vacuole and the plasma membrane. The membrane of the vacuole is called the tonoplast.

25 nm, respectively. An electron microscope is therefore required to visualize ribosomes (see Figure 4-15d). To appreciate how small ribosomes are, consider that more than 350,000 ribosomes could fit inside a typical bacterial cell, with room to spare!

Another way to express the size of such a small particle is to refer to its **sedimentation coefficient.** The sedimentation coefficient of a particle or macromolecule is a measure of how rapidly the particle sediments in an ultracentrifuge. The rate of sedimentation is expressed in *Svedberg units (S),* as described in Box 12A, pages 355–357 Sedimentation coefficients are widely used to indicate relative size, especially for large macromolecules such as proteins and nucleic acids and for small particles such as ribosomes. Ribosomes from eukaryotic cells have sedimentation coefficients of about 80S, while those from bacteria and archaea are about 70S (see Table 4-1).

A ribosome consists of two subunits differing in size, shape, and composition (**Figure 4-22**). In eukaryotic cells, the **large** and **small ribosomal subunits** have sedimentation coefficients of about 60S and 40S, respectively. For bacterial and archaeal ribosomes, the corresponding values are about 50S and 30S. (Note that the sedimentation coefficients of the subunits do not add up to that of the intact ribosome. This is because sedimentation coefficients depend on both size and shape and are therefore not linearly related to molecular weight.) In early 2009, scientists successfully extracted all 55 ribosomal proteins (plus the three rRNAs) from an *E. coli* cell and then were able to reconstruct a functional "artificial" ribosome capable of synthesizing protein. This is a significant step toward the creation of artificial cells in vitro that have custom-made ribosomes tailored for industrial uses to produce particular desired proteins.

Ribosomes are far more numerous than most other intracellular structures. Prokaryotic cells usually contain thousands of ribosomes, and eukaryotic cells may have hundreds of thousands or even millions of them. Ribosomes are also found in both mitochondria and chloroplasts, where they function in organelle-specific protein synthesis. The ribosomes of these eukaryotic organelles differ in size and composition from the ribosomes found in the cytoplasm of the same cell, but they are strikingly similar to those found in bacteria and cyanobacteria. This similarity is particularly striking when the nucleotide sequences of ribosomal RNA (rRNA) from mitochondria and chloroplasts are compared with those of bacterial and cyanobacterial rRNAs. These similarities provide further support for the endosymbiotic origins of mitochondria and chloroplasts.

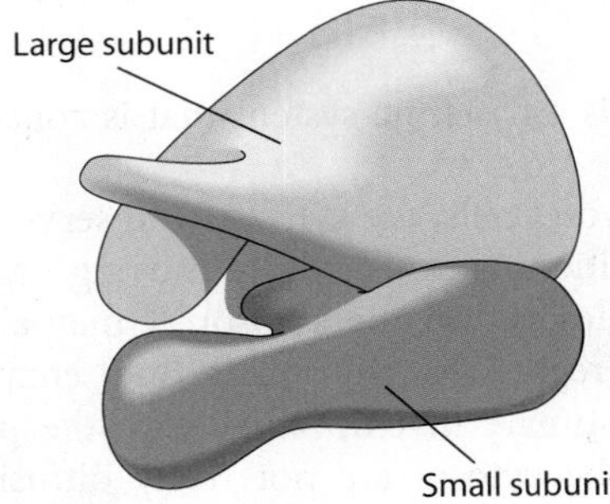

FIGURE 4-22 Structure of a Ribosome. Each ribosome is made up of a large subunit and a small subunit that join together when they attach to a messenger RNA and begin to make a protein. The fully assembled ribosome of a eukaryotic cell has a diameter of about 30 nm. The ribosomes and ribosomal subunits of bacterial and archaeal cells are somewhat smaller than those of eukaryotic cells and consist of their own distinctive protein and RNA molecules.

The Cytoplasm of Eukaryotic Cells Contains the Cytosol and Cytoskeleton

The **cytoplasm** of a eukaryotic cell consists of that portion of the interior of the cell not occupied by the nucleus. Thus, the cytoplasm includes organelles such as the mitochondria; it also includes the **cytosol,** the semifluid substance in which the organelles are suspended. In a typical animal cell, the cytosol occupies more than half of the cell's total internal volume. Many cellular activities take place in the cytosol, including the synthesis of proteins, the synthesis of fats, and the initial steps in releasing energy from sugars.

In the early days of cell biology, the cytosol was regarded as a rather amorphous, gel-like substance. Its proteins were thought to be soluble and freely diffusible. However, several new techniques have greatly changed this view. We now know that the cytosol of eukaryotic cells, far from being a structureless fluid, is permeated by an intricate three-dimensional array of interconnected microfilaments, microtubules, and intermediate filaments called the **cytoskeleton** (**Figures 4-23** and **4-24**). While the cytoskeleton was initially thought to exist only in

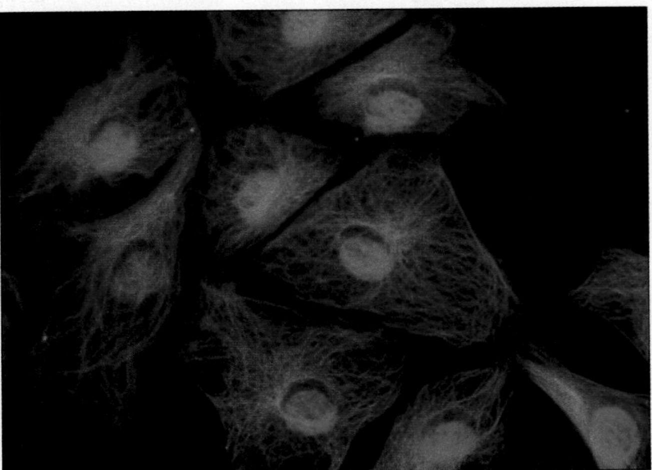

FIGURE 4-23 The Cytoskeleton. This micrograph uses immunofluorescence microscopy to reveal the microtubules in the cytoskeleton of fibroblast cells, which give rise to components of connective tissue and the extracellular matrix (Chapter 17). In this image, one fluorescent antibody is directed against the microtubules (green), and a second highlights the nuclei of the cells (blue).

(a) A microtubule. A diagram of a microtubule, showing 13 protofilaments forming a hollow cylinder. Each protofilament is a polymer of tubulin dimers. All the tubulin dimers are oriented in the same direction, giving polarity to the protofilament and hence to the entire microtubule.

(b) A microfilament. A diagram of a microfilament, showing a strand of F-actin twisted into a helical structure. The F-actin polymer consists of monomers of G-actin, all oriented in the same direction to give the microfilament its inherent polarity.

(c) An intermediate filament. A diagram of an intermediate filament. The structural unit is the tetrameric protofilament, consisting of two pairs of coiled polypeptides. Protofilaments assemble by end-to-end and side-to-side alignment, forming an intermediate filament that is thought to be eight protofilaments thick at any point.

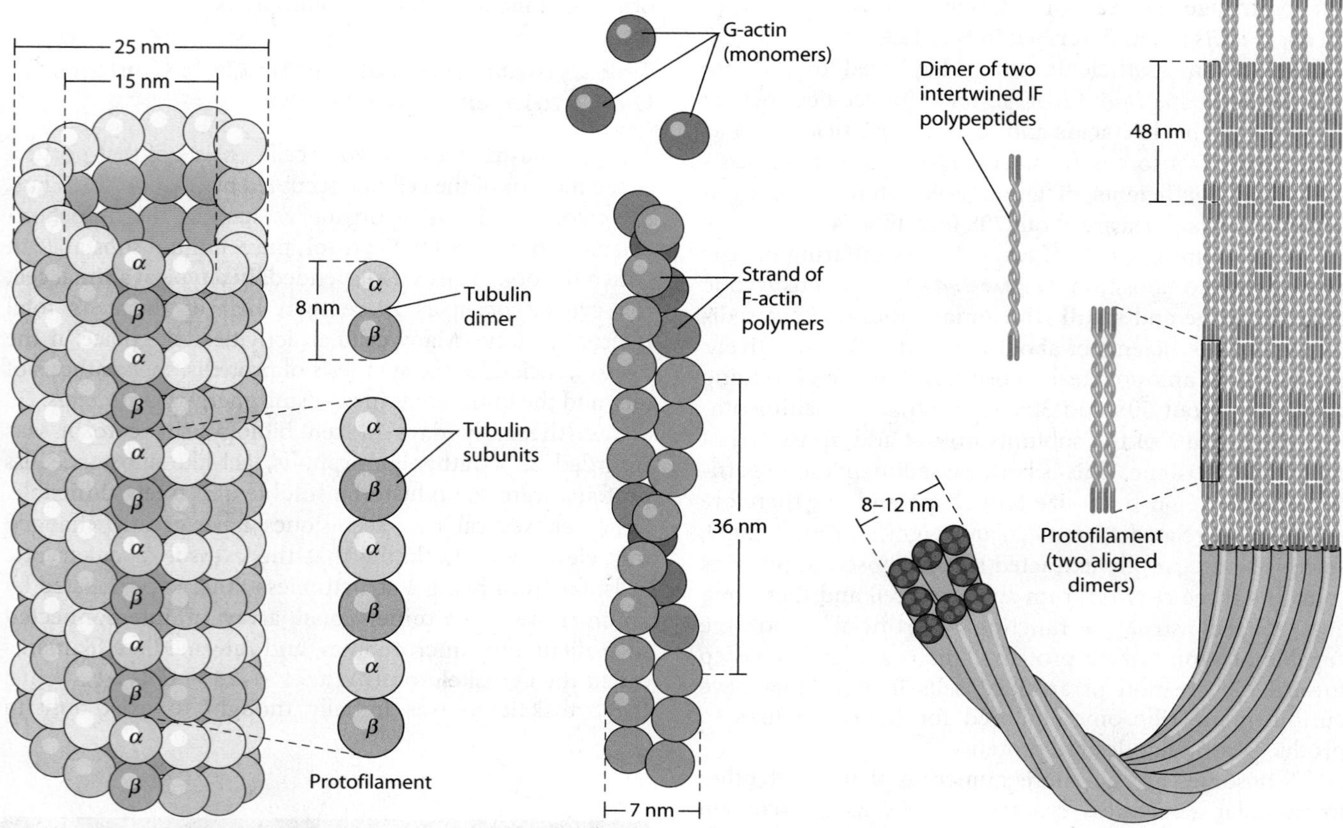

FIGURE 4-24 Structures of Microtubules, Microfilaments, and Intermediate Filaments. This figure illustrates the composition and structural details of the three primary elements making up the cytoskeleton of a eukaryotic cell: **(a)** microtubules, **(b)** microfilaments, and **(c)** intermediate filaments. For more information, see Chapter 15.

eukaryotes, proteins related to the eukaryotic cytoskeletal proteins have recently been discovered in bacteria.

As the name suggests, the cytoskeleton is an internal framework that gives a cell its distinctive shape and high level of internal organization. This elaborate array of microfilaments and microtubules forms a highly structured yet dynamic matrix that helps establish and maintain cell shape. In addition, the cytoskeleton plays an important role in cell movement and cell division. As we will see in later chapters, the microfilaments and microtubules that make up the cytoskeleton function in the contraction of muscle cells, the beating of cilia and flagella, the movement of chromosomes during cell division, and, in some cases, the locomotion of the cell itself. We will also see how, rather than being a static framework inside the cell, the cytoskeleton is a dynamic system that is constantly being remodeled.

In eukaryotic cells, the cytoskeleton serves as a framework for positioning and actively moving organelles and macromolecules within the cytosol. It may also function similarly in regard to ribosomes and enzymes. Some researchers estimate that up to 80% of the proteins and enzymes of the cytosol are not freely diffusible but are instead associated with the cytoskeleton. Even water, which accounts for about 70% of the cell volume, may be influenced by the cytoskeleton. It has been estimated that as much as 20–40% of the water in the cytosol may be bound to the microfilaments and microtubules of the cytoskeleton.

The three major structural elements of the cytoskeleton—*microtubules, microfilaments,* and *intermediate filaments*—are shown in Figure 4-24. They can be visualized by phase-contrast, immunofluorescence, and electron microscopy. Some of the structures they are found in (such as cilia, flagella, or muscle fibrils) can even be seen by ordinary light microscopy. Microfilaments and microtubules are best known for their roles in contraction and motility. In fact, these roles were appreciated well before it became clear that the same structural elements are also integral parts of the pervasive network of microfilaments and microtubules that gives cells their characteristic shape and structure.

Chapter 15 provides a detailed description of the cytoskeleton, followed in Chapter 16 by a discussion of microtubule- and microfilament-mediated contraction and motility. We will also encounter microtubules and microfilaments in Chapter 19 because of their roles in chromosome separation and cell division, respectively. Here, we will focus on the structural features of the three major components of the cytoskeleton: microtubules, microfilaments, and intermediate filaments.

Microtubules. Of the structural elements found in the cytoskeleton, **microtubules** are the largest. A well-known microtubule-based cellular structure is the *axoneme* of cilia and flagella, the appendages responsible for motility of eukaryotic cells. We have already encountered an example of such a structure—the axoneme of the sperm tail shown in Figure 4-12 consists of microtubules. Microtubules also form the *mitotic spindle fibers* that separate chromosomes prior to cell division, as we will see in Chapter 19.

Besides their involvement in motility and chromosome movement, microtubules also play an important role in the organization of the cytoplasm and the intracellular movement of macromolecules and other materials in the cell. They contribute to the overall shape of the cell, the spatial disposition of its organelles, and the distribution of microfilaments and intermediate filaments. Examples of the diverse phenomena governed by microtubules include the asymmetric shapes of animal cells, the plane of cell division in plant cells, the ordering of filaments during muscle development, and the positioning of mitochondria around the axoneme of motile appendages.

As shown in Figure 4-24a, microtubules are hollow cylinders with an outer diameter of about 25 nm and an inner diameter of about 15 nm. Although microtubules are usually drawn as if they are straight and rigid, video microscopy reveals them to be quite flexible in living cells. The wall of the microtubule consists of longitudinal arrays of *protofilaments,* usually 13 of them arranged side by side around the hollow center, called the *lumen.* Each protofilament is a linear polymer of *tubulin.* Tubulin is a dimeric protein consisting of two similar but distinct polypeptide subunits, *α-tubulin* and *β-tubulin.* All of the tubulin dimers in each of the protofilaments are oriented in the same direction, such that all of the subunits face the same end of the microtubule. This uniform orientation gives the microtubule an inherent *polarity.* As we will see in Chapter 15, the polarity of microtubules has important implications for their assembly and for the directional movement of membrane-bounded organelles that microtubules are associated with.

Microfilaments. **Microfilaments** are much thinner than microtubules. They have a diameter of about 7 nm, which makes them the smallest of the major cytoskeletal components. Microfilaments are best known for their role in the contractile fibrils of muscle cells, as we will see in more detail in Chapter 16. However, microfilaments are involved in a variety of other cellular phenomena as well. They can form connections with the plasma membrane and thereby influence locomotion, amoeboid movement, and cytoplasmic streaming. Microfilaments also produce the *cleavage furrow* that divides the cytoplasm of an animal cell after the two sets of chromosomes have been separated by the mitotic spindle fibers. In addition, microfilaments contribute importantly to the development and maintenance of cell shape.

Microfilaments are polymers of the protein *actin* (Figure 4-24b). Actin is synthesized as a monomer called *G-actin* (G for globular). G-actin monomers polymerize into long strands of *F-actin* (F for filamentous), with each strand about 4 nm wide. Each microfilament consists of a chain of actin monomers that are assembled into a filament with a helical appearance and a diameter of about 7 nm. Like microtubules, microfilaments show polarity; all of the subunits are oriented in the same direction. This polarity influences the direction of microfilament elongation; assembly usually proceeds more readily at one end of the growing microfilament, whereas disassembly is favored at the other end.

Intermediate Filaments. **Intermediate filaments** (Figure 4-24c) make up the third structural element of the cytoskeleton. Intermediate filaments have a diameter of about 8–12 nm, larger than the diameter of microfilaments but smaller than that of microtubules. Intermediate filaments are the most stable and the least soluble constituents of the cytoskeleton. Because of this stability, some researchers regard intermediate filaments as a scaffold that supports the entire cytoskeletal framework. Intermediate filaments are also thought to have a tension-bearing role in some cells because they often occur in areas that are subject to mechanical stress.

In contrast to microtubules and microfilaments, intermediate filaments differ in their protein composition from tissue to tissue. Based on biochemical criteria, intermediate filaments from animal cells can be grouped into six classes. A specific cell type usually contains only one or sometimes two classes of intermediate filament proteins. Because of this tissue specificity, animal cells from different tissues can be distinguished on the basis of the intermediate filament proteins present. This *intermediate filament typing* serves as a diagnostic tool in medicine.

Despite their heterogeneity of size and chemical properties, all intermediate filament proteins share common structural features. They all have a central rodlike segment that is remarkably similar from one intermediate filament protein to the other. Flanking the central region of the protein are N-terminal and C-terminal segments that differ greatly in size and sequence, presumably accounting for the functional diversity of these proteins.

Several models have been proposed for intermediate filament structure. One possibility is shown in Figure 4-24c. The basic structural unit is a dimer of two intertwined, intermediate filament polypeptides. Two such dimers align laterally to form a tetrameric *protofilament.* Protofilaments then interact with each other to form an intermediate filament that is thought to be eight protofilaments thick at any point, with protofilaments probably joined end to end in an overlapping manner. We have already seen a similar structure in Figure 3-10 when we looked at the proposed structure of the keratin fibers of hair.

The Extracellular Matrix and the Cell Wall Are "Outside" the Cell

So far, we have considered the plasma membrane that surrounds every cell, the nucleus and cytoplasm within, and the variety of organelles, membrane systems, ribosomes, microtubules, and microfilaments found in the cytoplasm of most eukaryotic cells. While it may seem that our tour of the cell is complete, most cells are also characterized by extracellular structures formed from materials that the cells transport outward across the plasma membrane. These structures often give physical support to the cells making up a particular tissue. For many animal cells, these structures are called the **extracellular matrix (ECM)** and consist primarily of *collagen fibers* and *proteoglycans.* For plant and fungal cells, the extracellular structure is the rigid **cell wall,** which consists mainly of *cellulose microfibrils* embedded in a matrix of other polysaccharides and small amounts of protein.

Most bacteria and archaea are also surrounded by an extracellular structure called a cell wall. However, bacterial cell walls consist not of cellulose but mainly of *peptidoglycans.* Peptidoglycans contain long chains of repeating units of *N*-acetylglucosamine (GlcNAc) and *N*-acetylmuramic acid (MurNAc), amino sugars that we encountered in Chapter 3 (see Figure 3-26). These chains are held together to form a netlike structure by crosslinks composed of about a dozen amino acids linked by peptide bonds—thus the name peptidoglycan. In addition, bacterial cell walls contain a variety of other constituents, some of which are unique to each of the major structural groups of bacteria. Archaeal cell walls vary considerably from species to species—some are mainly proteinaceous, while others have peptidoglycan-like components.

These differences among cells are in keeping with the lifestyles of these groups of organisms. Plants are generally *nonmotile,* a lifestyle that is compatible with the rigidity that cell walls confer on an organism. Animals, on the other hand, are usually *motile,* an essential feature for an organism that needs not only to find food but also to escape becoming food for other organisms! Accordingly, animal cells are not encased in rigid walls; instead, they are surrounded by a strong but elastic network of collagen fibers. Bacteria and archaea may be motile or nonmotile, but they usually live in hypotonic environments (lower concentrations of external solutes than found in the cells), and their cell walls provide rigid support and protection from bursting due to osmotic pressure as water enters the cell.

The ECM of animal cells can vary in structure depending on the cell type. Micrographs of the ECM from bone, cartilage, and connective tissue can be seen in Figure 17-13. The primary function of the ECM is support, but the kinds of extracellular materials and the patterns they are deposited in may regulate such diverse processes as cell motility and migration, cell division, cell recognition and adhesion, and cell differentiation during embryonic development. The main constituents of animal extracellular structures are the collagen fibers and the network of proteoglycans that surround them. In vertebrates, collagen is such a prominent part of tendons, cartilage, and bone that it is the single most abundant protein of the animal body.

Figure 4-25 illustrates the prominence of the cell wall as a structural feature of a typical plant cell. Although the distinction is not made in the figure, plant cell walls are actually of two types. The wall that is laid down during cell division, called the *primary cell wall,* consists mainly of cellulose fibrils embedded in a gel-like polysaccharide matrix. Primary walls are quite flexible and extensible, which allows them to expand somewhat in response to cell enlargement and elongation. As a cell reaches its final size and shape, a much thicker and more rigid *secondary cell wall* may form by deposition of additional cell wall material on the inner surface of the primary wall. The secondary wall usually contains more cellulose than the primary wall and may have a high content of *lignin,* a major component of wood. Deposition of a secondary cell wall renders the cell inextensible and therefore defines the final size and shape of the cell.

Neighboring plant cells, though separated by the wall between them, are actually connected by numerous cytoplasmic bridges, called **plasmodesmata** (singular: **plasmodesma**), which pass through the cell wall. The plasma membranes of adjacent cells are continuous through each plasmodesma, such that the channel is membrane lined. The diameter of a typical plasmodesma is large enough to allow water and small solutes to pass freely from cell to cell. Most of the cells of the plant are interconnected in this way. Primary and secondary cell walls and plasmodesmata will be discussed in more detail in Chapter 17.

Animal cells can also communicate with each other. But instead of plasmodesmata, they have intercellular connections called *gap junctions,* which are specialized for

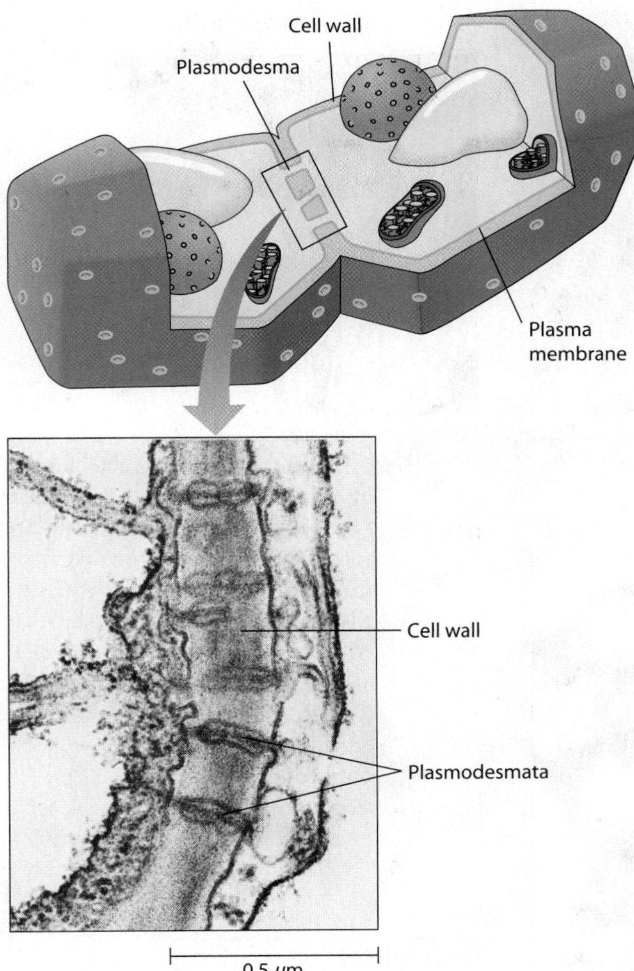

FIGURE 4-25 The Plant Cell Wall. The wall surrounding a plant cell consists of rigid microfibrils of cellulose embedded in a noncellulosic matrix of proteins and sugar polymers. Notice how the neighboring cells are connected by plasmodesmata (TEM).

the transfer of material between the cytoplasms of adjacent cells. Two other types of intercellular junctions are also characteristic of animal cells. *Tight junctions* hold cells together so tightly that the transport of substances through the spaces between the cells is effectively blocked. *Adhesive junctions* also link adjacent cells but for the purpose of connecting them tightly into sturdy yet flexible sheets. Each of these types of junctions is discussed in detail in Chapter 17.

Viruses, Viroids, and Prions: Agents That Invade Cells

Before concluding this preview of cellular biology, we will look at several kinds of agents that invade cells, subverting normal cellular functions and often killing their unwilling hosts. These include the *viruses,* which have been studied for over 100 years, and two other agents we know much less about—the *viroids* and *prions.*

A Virus Consists of a DNA or RNA Core Surrounded by a Protein Coat

Viruses are noncellular, parasitic particles that are incapable of a free-living existence. However, they can invade and infect cells and redirect the synthetic machinery of the infected host cell toward the production of more virus particles. Viruses cannot perform all of the functions required for independent existence and must therefore depend on the cells they invade for most of their needs. They are not considered to be living organisms, nor are they made of cells. A virus particle has no cytoplasm, no ribosomes, no organelles, few or no enzymes, and typically consists of only a few different molecules of nucleic acid and protein.

Viruses are, however, responsible for many diseases in humans, animals, and plants and are thus important in their own right. They are also significant research tools for cell and molecular biologists because they are much less complicated than cells. The tobacco mosaic virus (TMV) that we encountered in Chapter 2 (see Figures 2-20 and 2-21) is a good example of a virus that is important both economically and scientifically—economically because of the threat it poses to tobacco and other crop plants that it can infect and scientifically because it is so amenable to laboratory study. Fraenkel-Conrat's studies on TMV self-assembly described in Chapter 2 are good examples of the usefulness of viruses to cell biologists.

Viruses are typically named for the diseases they cause. Poliovirus, influenza virus, herpes simplex virus, and TMV are several common examples. Other viruses have more cryptic laboratory names (such as T4, Qβ, λ, or Epstein-Barr virus). Viruses that infect bacterial cells are called *bacteriophages,* or often just *phages* for short. Bacteriophages and other viruses will figure prominently in our discussion of molecular genetics in Chapters 18–23.

Viruses are smaller than all but the tiniest cells, with most ranging in size from about 25 to 300 nm. They are small enough to pass through a filter that traps bacteria, and the collected filtrate will still be infectious. Because of this, their existence was postulated by Louis Pasteur 50 years before they were able to be seen in the electron microscope! The smallest viruses are about the size of a ribosome, whereas the largest ones are about one-quarter the diameter of a typical bacterial cell. Each virus has a characteristic shape, as defined by its protein capsid. Some of these varied shapes are shown in **Figure 4-26**.

Despite their morphological diversity, viruses are chemically quite simple. Most viruses consist of little more than a *coat* (or *capsid*) of protein surrounding a *core* that contains one or more molecules of either RNA or DNA, depending on the type of virus. This is another feature distinguishing viruses from cells, which always have both DNA and RNA. The simplest viruses, such as TMV, have a single nucleic acid molecule surrounded by a capsid consisting of proteins of a single type (see Figure 2-20). More complex viruses have cores that contain several nucleic

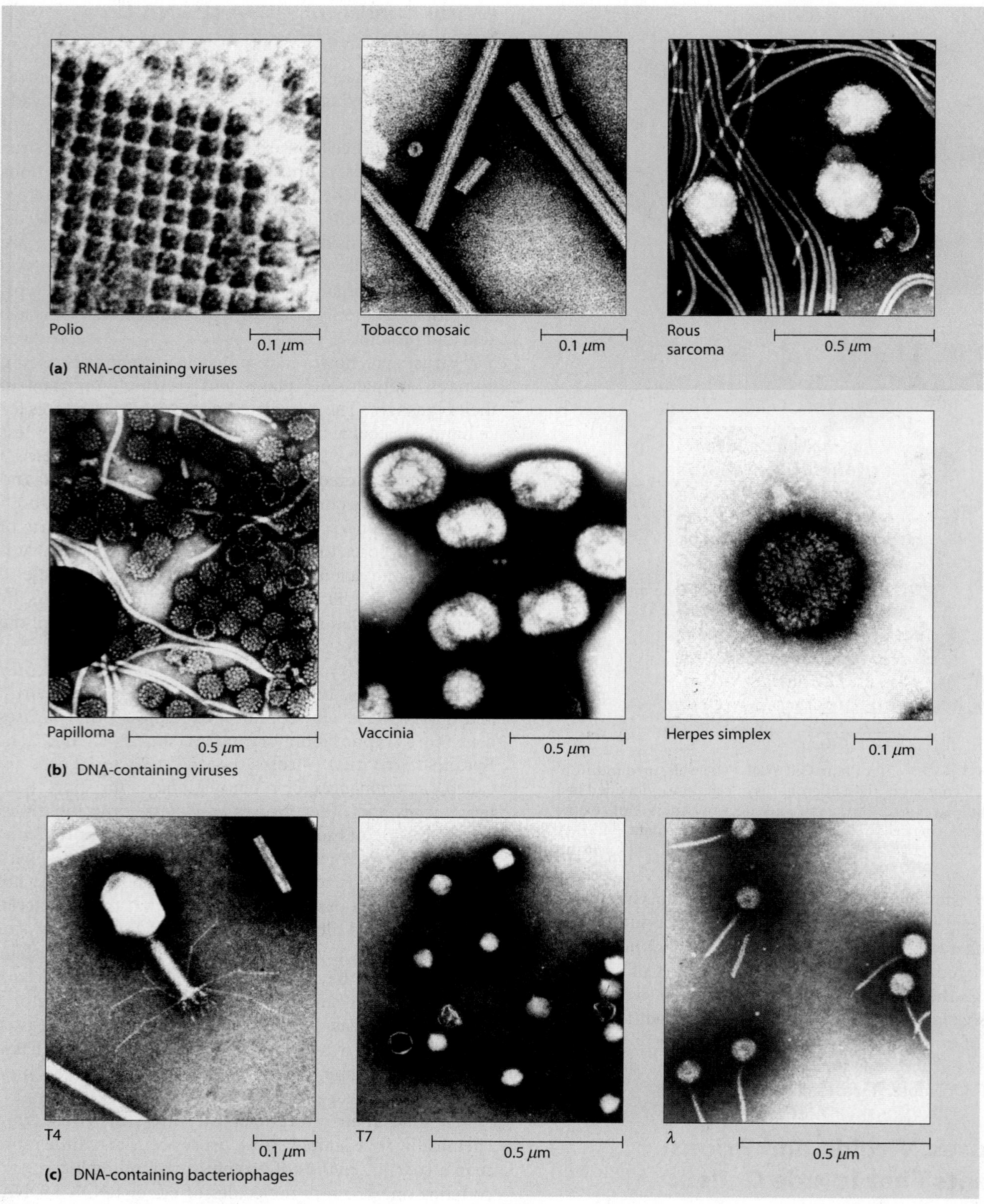

MB Activity: Simplified Viral Reproductive Cycle

FIGURE 4-26 Sizes and Shapes of Viruses. These electron micrographs illustrate the morphological diversity of viruses. **(a)** RNA-containing viruses (left to right): polio, tobacco mosaic, and Rous sarcoma viruses. **(b)** DNA-containing viruses: papilloma, vaccinia, and herpes simplex viruses. **(c)** DNA-containing bacteriophages: T4, T7, and λ (all TEMs).

acid molecules and capsids consisting of several (or even many) different kinds of proteins. Some viruses are surrounded by a membrane that is derived from the plasma membrane of the host cell in which the viral particles were previously made and assembled. Such viruses are called *enveloped viruses.* Human immunodeficiency virus (HIV), the virus that causes AIDS (acquired immune deficiency syndrome), is an enveloped virus; it is covered by a membrane that it received from the previously infected white blood cell.

Students sometimes ask whether viruses are living. The answer depends crucially on what we mean by "living," and it is probably worth pondering only to the extent that it helps us more fully understand what viruses are—and what they are not. The most fundamental properties of living things are *metabolism* (cellular reactions organized into coherent pathways), *irritability* (perception of, and response to, environmental stimuli), and the *ability to reproduce.* Viruses clearly do not satisfy the first two criteria. Outside their host cells, viruses are inert and inactive. They can, in fact, be isolated and crystallized almost like a chemical compound. It is only in an appropriate host cell that a virus becomes functional, undergoing a cycle of synthesis and assembly that gives rise to more viruses.

Even the ability of viruses to reproduce has to be qualified carefully. A basic tenet of the cell theory is that cells arise only from preexisting cells, but this is not true of viruses. No virus can give rise to another virus by any sort of self-duplication process. Rather, the virus must take over the metabolic and genetic machinery of a host cell, reprogramming it for synthesis of the proteins necessary to package the DNA or RNA molecules that arise by copying the genetic information of the parent virus.

Viroids Are Small, Circular RNA Molecules

As simple as viruses are, there are even simpler noncellular agents that can infect eukaryotic cells (though apparently not prokaryotic cells, as far as we currently know). The **viroids** found in some plant cells represent one class of such agents. Viroids are small, circular RNA molecules, and they are the smallest known infectious agents. These RNA molecules are only about 250–400 nucleotides long and are replicated in the host cell even though they do not code for any protein. Some have been shown to have enzymatic properties that aid in their replication.

Viroids are responsible for diseases of several crop plants, including potatoes and tobacco. A viroid disease that has severe economic consequences is *cadang-cadang disease* of the coconut palm. It is not yet clear how viroids cause disease. They may enter the nucleus and interfere with the transcription of DNA into RNA in a process known as *gene silencing.* Alternatively, they may interfere with the subsequent processing required of most eukaryotic mRNAs, disrupting subsequent protein synthesis.

Viroids do not occur in free form but can be transmitted from one plant cell to another when the surfaces of adjacent cells are damaged and there is no membrane barrier for the RNA molecules to cross. There are also reports that they can be transmitted by seed, pollen, or agricultural implements used in cultivation and harvesting. Recent studies have shown that many plant species harbor latent viroids that cause no obvious symptoms.

Prions Are "Proteinaceous Infective Particles"

Prions represent another class of noncellular infectious agents. The term was coined to describe *proteinaceous infective particles* that are responsible for neurological diseases such as scrapie in sheep and goats, kuru in humans, and mad cow disease in cattle. *Scrapie* is so named because infected animals rub incessantly against trees or other objects, scraping off most of their wool in the process. *Kuru* is a degenerative disease of the central nervous system originally reported among native peoples in New Guinea who consumed infected brain tissue. Patients with this or other prion-based diseases suffer initially from mild physical weakness and dementia; but the effects slowly become more severe, and the diseases are eventually fatal. For a discussion of *mad cow disease,* see Box 22A, page 692.

Prion proteins are abnormally folded versions of normal cellular proteins. Both the normal and variant forms of the prion protein are found on the surfaces of neurons, suggesting that the protein may somehow affect the receptors that detect nerve signals. Some biologists speculate that a similar mechanism is responsible for the plaques of misfolded proteins that are found in the brains of deceased Alzheimer syndrome patients.

Prions are not destroyed by cooking or boiling, so special precautions are recommended in areas where prion-caused diseases are known to occur. For example, a prion disease known as *chronic wasting disease* is found in some deer and elk, especially in the Rocky Mountain region of the western United States. Hunters in areas where this prion is known to occur are cautioned to have meat tested for this prion and to avoid eating meat from any animal that tests positive or appears sick.

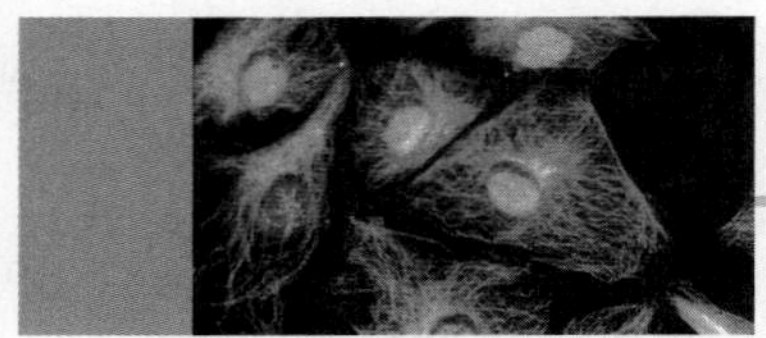

SUMMARY OF KEY POINTS

Properties and Strategies of Cells

- Based on gross morphology, cells have traditionally been described as either eukaryotic (animals, plants, fungi, protozoa, and algae) or prokaryotic (bacteria and archaea) based on the presence or absence of a membrane-bounded nucleus. More recently, analysis of ribosomal RNA sequences and other molecular data suggests a tripartite view of organisms, with eukaryotes, bacteria, and archaea as the three main domains.
- A plasma membrane and ribosomes are the only two structural features common to cells of all three groups. Organelles are found only in eukaryotic cells, where they play indispensable roles in the compartmentalization of function. Bacterial and archaeal cells are relatively small and structurally less complex than eukaryotic cells, lacking the internal membrane systems and organelles of eukaryotes.
- Cell size is limited by the need for an adequate amount of surface area for exchange of materials with the environment and the need for high enough concentrations of the compounds necessary to sustain life. Eukaryotic cells are much larger than prokaryotic cells and compensate for their lower surface area/volume ratio by the compartmentalization of materials within membrane-bounded organelles.

The Eukaryotic Cell in Overview

- Besides the plasma membrane and ribosomes common to all cells, eukaryotic cells have a nucleus that houses most of the cell's DNA, a variety of organelles, and the cytosol with its cytoskeleton of microtubules, microfilaments, and intermediate filaments. Also, plant cells have a rigid cell wall, and animal cells are usually surrounded by a strong but flexible extracellular matrix of collagen and proteoglycans.
- The nucleus contains the cell's DNA complexed with protein in the form of chromatin, which condenses during cell division to form the visible structures we call chromosomes. The nucleus is surrounded by a double membrane called the nuclear envelope, which has pores that allow the regulated exchange of macromolecules with the cytoplasm.
- Mitochondria, which are surrounded by a double membrane, oxidize food molecules to provide the energy used to make ATP. Mitochondria also contain ribosomes and their own circular DNA molecules.
- Chloroplasts trap solar energy and use it to "fix" carbon dioxide into organic form and convert it to sugar. Chloroplasts are surrounded by a double membrane and have an extensive system of internal membranes called the thylakoids, in which most of the components involved in ATP generation are found. Chloroplasts also contain ribosomes and circular DNA molecules.
- The endoplasmic reticulum is an extensive network of membranes that are known as either rough ER or smooth ER. Rough ER is studded with ribosomes and is responsible for the synthesis of secretory and membrane proteins, whereas smooth ER is involved in lipid synthesis and drug detoxification. Proteins synthesized on the rough ER are further processed and packaged in the Golgi complex and are then transported either to membranes or to the surface of the cell via secretory vesicles.
- Lysosomes contain hydrolytic enzymes and are involved in cellular digestion. They were the first organelles to be discovered on the basis of their function rather than their morphology. Because of their close functional relationships, the ER, Golgi complex, secretory vesicles, and lysosomes are collectively called the endomembrane system.
- Peroxisomes are about the same size as lysosomes, and both generate and degrade hydrogen peroxide. Animal peroxisomes play an important role in catabolizing long-chain fatty acids. In plants, specialized peroxisomes known as glyoxysomes are involved in the process of photorespiration and in converting stored fat into carbohydrate during seed germination.
- Ribosomes are sites of protein synthesis in all cells. The striking similarities between mitochondrial and chloroplast ribosomes and those of bacteria and cyanobacteria, respectively, lend strong support to the endosymbiont theory that these organelles are of bacterial origin.
- The cytoskeleton is an extensive network of microtubules, microfilaments, and intermediate filaments that gives eukaryotic cells their distinctive shapes. The cytoskeleton is also important in cellular motility and the intracellular movement of cellular structures and materials.

Agents That Invade Cells

- Viruses satisfy some, though not all, of the basic criteria of living things. Viruses are important both as infectious agents that cause diseases in humans, animals, and plants and as laboratory tools, particularly for geneticists. Viroids and prions are infectious agents that are even smaller (and less well understood) than viruses. Viroids are small self-replicating RNA molecules, whereas prions are misfolded proteins that are thought to be abnormal products of normal cellular genes.

MB **www.masteringbiology.com**

1. MasteringBiology Assignments

Tutorials and Activities

- Molecular Model Structural Tutorials for visualization of 3D structures
- BioFlix Tutorials including 3D animations with hints and detailed feedback for common wrong answers
- Simplified Viral Reproductive Cycle

Questions

- Reading Quiz
- Multiple Choice
- End-of-chapter questions

2. eText Read your book online, search, highlight text, take notes, and more

3. The Study Area

- Chapter Review Quiz, Guide to Techniques and Methods, 3D-Structure Tutorials, Activities, BioFlix, Videos, Biology Review, Word Study Tools, Glossary, Flashcards

PROBLEM SET

More challenging problems are marked with a •.

4-1 Wrong Again. For each of the following false statements, change the statement to make it true.

(a) Bacteria differ from eukaryotes in having no nucleus, mitochondria, chloroplasts, or ribosomes.

(b) Archaea are ancient bacteria that are the ancestors of modern bacteria.

(c) Instead of a cell wall, eukaryotic cells have an extracellular matrix for structural support.

(d) All the ribosomes found in a typical human muscle cell are identical.

(e) Because bacterial cells have no organelles, they cannot carry out either ATP synthesis or photosynthesis.

(f) DNA is found only in the nucleus of a cell.

(g) A large amount of the DNA in eukaryotic cells has no function and is called "junk DNA."

(h) Even the simplest types of infectious agents must have DNA, RNA, and protein.

4-2 It's a matter of size. Arrange these cellular components, or cell invading agents, in the order of size. Start with the smallest and go up to the largest.

(a) The length of a chloroplast	Cell division
(b) The DNA containing the virus Herpes simplex	Absorption
(c) The diameter of a microfilament	Motility
(d) The nucleus	Photosynthesis
(e) The length of a mitochondrion	Secretion
(f) A ribosome of an eukaryotic cell	Transmission of electrical impulses
(g) The DNA containing the bacteriophage T7	

4-3 Toward an Artificial Cell. Scientists have recently constructed an artificial ribosome in vitro from purified ribosomal proteins and rRNAs.

(a) What types of intermolecular forces do you think are holding the individual proteins and rRNAs together in this supramolecular complex?

(b) Describe specifically how high temperature, high salt, or low pH would disrupt its structure, causing the ribosome to fall apart.

(c) If you were asked to determine which organism the ribosomal components were from, how could you do this?

(d) What other molecules would you have to add to the test tube in order for the ribosomes to make polypeptides? (This may require some sleuthing in later chapters.)

4-4 Sentence Completion. Complete each of the following statements about cellular structure in ten words or less.

(a) If you were shown an electron micrograph of a section of a cell and were asked to identify the cell as plant or animal, one thing you might do is . . .

(b) A slice of raw apple placed in a concentrated sugar solution will . . .

(c) A cellular structure that is visible with an electron microscope but not with a light microscope is . . .

(d) Several environments in which you are more likely to find archaea than bacteria are . . .

(e) One reason that it might be difficult to separate lysosomes from peroxisomes by centrifugation techniques is that . . .

(f) The nucleic acid of a virus is composed of . . .

4-5 Telling Them Apart. Suggest a way to distinguish between the two elements in each of the following pairs:

(a) Bacterial cells; archaeal cells

(b) Rough ER; smooth ER

(c) Animal peroxisomes; leaf peroxisomes

(d) Peroxisomes; lysosomes

(e) Viruses; viroids

(f) Microfilaments; intermediate filaments

(g) Polio virus; herpes simplex virus

(h) Eukaryotic ribosomes; bacterial ribosomes

(i) mRNA; miRNA

4-6 Structural Relationships. For each pair of structural elements, indicate with an A if the first element is a constituent part of the second, with a B if the second element is a constituent part of the first, and with an N if they are separate structures with no particular relationship to each other.

(a) Peroxisome; mitochondrion

(b) Golgi complex; nucleus

(c) Microtubes; cytoskeleton

(d) Cell wall; plasmodesmata

(e) Nucleus; nucleoplasm

(f) Smooth ER; ribosome

(g) Plasma membrane; lipid bilayer

(h) Nucleus; cristae

(i) Thylakoid disks; chloroplast

• **4-7 Protein Synthesis and Secretion.** Although we will not encounter protein synthesis and secretion in detail until later chapters, you already have enough information about these processes to order the seven events that are now listed randomly. Order events 1–7 so that they represent the correct sequence corresponding to steps a–g, tracing a typical secretory protein from the initial transcription (readout) of the relevant genetic information in the nucleus to the eventual secretion of the protein from the cell by exocytosis.

Transcription → (a) → (b) → (c) →
(d) → (e) → (f) → (g) → Secretion

1. The protein is partially glycosylated within the lumen of the ER.
2. The secretory vesicle arrives at and fuses with the plasma membrane.
3. The RNA transcript is transported from the nucleus to the cytoplasm.
4. The final sugar groups are added to the protein in the Golgi complex.
5. As the protein is synthesized, it passes across the ER membrane into the lumen of a Golgi cisterna.
6. The protein is packaged into a secretory vesicle and released from the Golgi complex.
7. The RNA message associates with a ribosome and begins synthesis of the desired protein on the surface of the rough ER.

• **4-8 Disorders at the Organelle Level.** Each of the following medical problems involves a disorder in the function of an organelle or other cell structure. In each case, identify the organelle or structure involved, and indicate whether it is likely to be underactive or overactive.

(a) A girl is discovered to have Leigh Syndrome, which causes neurodegeneration due to the disruption of oxidative phosphorylation in nerve cells.

(b) A young boy is diagnosed with Hurler Syndrome, which is characterised by a build-up of glycosaminoglycans (mucopolysaccharides).

(c) A smoker develops lung cancer and is told that the cause of the problem is a population of cells in her lungs that are undergoing mitosis at a much greater rate than is normal for lung cells.

(d) A woman learns that she has suffered an ectopic pregnancy due to defective cilia in her fallopian tubes.

(e) A young child dies of Tay-Sachs disease because her cells lack the hydrolase that normally breaks down a membrane component called ganglioside G_{M2}, which therefore accumulated in the membranes of her brain.

(f) A 60-year-old man is placed on a low-fat diet because he discovers that he has a deficiency in lipoprotein lipase.

• **4-9 Are They Alive?** Biologists sometimes debate whether viruses should be considered as alive. Let's join in the debate.

(a) What are some ways in which viruses resemble cells?

(b) What are some ways in which viruses differ from cells?

(c) Choose either of the two following positions and defend it: (1) Viruses are alive, or (2) Viruses are not alive.

(d) Why do you suppose that viral illnesses are more difficult to treat than bacterial illnesses?

(e) Design a strategy to cure a viral disease without harming the patient.

SUGGESTED READING

References of historical importance are marked with a •.

Eukaryotic, Bacterial, and Archaeal Cells

Bogorad, L. Evolution of early eukaryotic cells: Genomes, proteomes, and compartments. *Photosynth. Res.* 95 (2008): 11.

Brinkmann, H., and H. Philippe. The diversity of eukaryotes and the root of the eukaryotic tree. *Adv. Exp. Med. Biol.* 607 (2007): 20.

Kassen, R., and P. B. Rainey. The ecology and genetics of microbial diversity. *Annu. Rev. Microbiol.* 58 (2004): 207.

Pace, N. R. Time for a change. *Nature* 441 (2006): 289.

Sapp, J. The prokaryote-eukaryote dichotomy: Meanings and mythology. *Microbiol. Mol. Biol. Rev.* 69 (2005): 292.

Tanaka, S., M. R. Sawaya, and T. O. Yeates. Structure and mechanisms of a protein-based organelle in *Escherichia coli*. *Science* 327 (2010): 81.

Trotsenko, Y. A., and V. N. Khmelenina. Biology of extremophilic and extremotolerant methanotrophs. *Arch. Microbiol.* 177 (2002): 123.

• Woese, C. R., and G. E. Fox. Phylogenetic structure of the prokaryotic domain: The primary kingdoms. *Proc. Nat. Acad. Sci. USA* 74 (1977): 5088.

Woese, C. R., O. Kandler, and M. L. Wheelis. Towards a natural system of organisms: Proposal for the domains Archaea, Bacteria, and Eucarya. *Proc. Nat. Acad. Sci. USA* 87 (1990): 4576.

The Nucleus

Burke, B., and C. L. Stewart. The laminopathies: The functional architecture of the nucleus and its contribution to disease. *Annu. Rev. Genomics Hum. Genet.* 7 (2006): 369.

Rippe, K. Dynamic organization of the cell nucleus. *Curr. Opin. Genet. Dev.* 17 (2007): 373.

Tchélidzé P., A. Chatron-Colliet, M. Thiry, N. Lalun, H. Bobichon, and D. Ploton. Tomography of the cell nucleus using confocal microscopy and medium voltage electron microscopy. *Crit. Rev. Oncol. Hematol.* 69 (2009): 127.

Webster, M., K. L. Witkin, and O. Cohen-Fix. Sizing up the nucleus: Nuclear shape, size and nuclear envelope assembly. *J. Cell Sci.* 122 (2009): 1477.

Organelles and Human Diseases

DiMauro, S. Mitochondrial myopathies. *Curr. Opin. Rheumatol.* 18 (2006): 636.

Fan, J., Z. Hu, L. Zeng, W. Lu, X. Tang, J. Zhang, and T. Li. Golgi apparatus and neurodegenerative diseases. *Int. J. Dev. Neurosci.* 26 (2008): 523.

Gould, S. J., and D. Valle. Peroxisome biogenesis disorders: Genetics and cell biology. *Trends Genet.* 16 (2000): 340.

Grabowski, G. A., and R. J. Hopkins. Enzyme therapy for lysosomal storage diseases: Principles, practice, and prospects. *Annu. Rev. Genomics Hum. Genet.* 4 (2003): 403.

Kiselyov, K. et al. Autophagy, mitochondria, and cell death in lysosomal storage diseases. *Autophagy* 3 (2007): 259.

Máximo, V., J. Lima, P. Soares, and M. Sobrinho-Simões. Mitochondria and cancer. *Virchows Arch.* 454 (2009): 481.

Morava, E. et al. Mitochondrial disease criteria: Diagnostic applications in children. *Neurology* 67 (2006): 1823.

Ni, M., and A. S. Lee. ER chaperones in mammalian development and human diseases. *FEBS Lett.* 581 (2007): 3641.

Reeve, A. K., K. J. Krishnan, and D. Turnbull. Mitochondrial DNA mutations in disease, aging, and neurodegeneration. *Ann. NY Acad. Sci.* 1147 (2008):1147.

Wlodkowic, D., J. Skommer, D. McGuinness, C. Hillier, and Z. Darzynkiewicz. ER-Golgi network—a future target for anti-cancer therapy. *Leuk. Res.* 33 (2009): 1440.

Yoshida, H. ER stress and diseases. *FEBS Lett.* 274 (2007): 630.

Intracellular Membranes and Organelles

Allan, V. J., H. M. Thompson, and M. A. McNiven. Motoring around the Golgi. *Nature Cell Biol.* 4 (2002): E236.

Boldogh, I. R., and L. A. Pon. Mitochondria on the move. *Trends Cell Biol.* 17 (2007): 502.

• de Duve, C. The peroxisome in retrospect. *Ann. NY Acad. Sci.* 804 (1996): 1.

Glick, B. S. Organization of the Golgi apparatus. *Curr. Opin. Cell Biol.* 12 (2000): 450.

Jackson, C. L. Mechanisms of transport through the Golgi complex. *J. Cell Sci.* 122 (2009): 443.

Leigh, R. A., and D. Sanders, eds. *The Plant Vacuole.* San Diego: Academic Press, 1997.

Okamoto, K., and J. M. Shaw. Mitochondrial morphology and dynamics in yeast and multicellular eukaryotes. *Annu. Rev. Genet.* 39 (2005): 503.

Reumann, S. The structural properties of plant peroxisomes and their metabolic significance. *Annu. Rev. Genet.* 34 (2000): 623.

• Sabatini, D. D., and G. E. Palade. Charting the secretory pathway. *Trends Cell Biol.* 9 (1999): 413.

Shorter, J., and G. Warren. Golgi architecture and inheritance. *Annu. Rev. Cell Dev. Biol.* 18 (2002): 379.

Voeltz, G. K., M. M. Rolls, and T. A. Rapoport. Structural organization of the endoplasmic reticulum. *EMBO Rep.* 3 (2002): 944.

Waters, M. T., and J. A. Langdale. The making of a chloroplast. *EMBO J.* 28 (2009): 2861.

Yeates, T. O., Crowley, C. S., and S. Tanaka. Bacterial microcompartment organelles: Protein shell structure and evolution. *Annu. Rev. Biophys.* 39 (2010): 185.

The Cytoplasm and the Cytoskeleton

Dillon, C., and Y. Goda. The actin cytoskeleton: Integrating form and function at the synapse. *Annu. Rev. Neurosci.* 28 (2005): 25.

Dinman, J. D. The eukaryotic ribosome: Current status and challenges. *J. Biol. Chem.* 284 (2009): 11761.

Erickson, H. P. Evolution of the cytoskeleton. *Bioessays* 29 (2007): 668.

Hall, A. The cytoskeleton and cancer. *Cancer Metastasis Rev.* 28 (2009): 5.

Howard, J. *Mechanics of Motor Proteins and the Cytoskeleton.* Sunderland, MA: Sinauer Associates, 2001.

Michie, K. A., and J. Lowe. Dynamic filaments of the bacterial cytoskeleton. *Annu. Rev. Biochem.* 75 (2006): 467.

Pogliano, J. The bacterial cytoskeleton. *Curr. Opin. Cell Biol.* 20 (2008): 19.

Schröder M. Engineering eukaryotic protein factories. *Biotechnol. Lett.* 30 (2008): 187.

Strelkov, S. V., H. Herrmann, and U. Aebi. Molecular architecture of intermediate filaments. *BioEssays* 25 (2003): 243.

Strnad, P., C. Stumptner, K. Zatloukal, and H. Denk. Intermediate filament cytoskeleton of the liver in health and disease. *Histochem. Cell Biol.* 129 (2008): 735.

The Extracellular Matrix and the Cell Wall

Bateman, J. F., R. P. Boot-Handford, and S. R. Lamandé. Genetic diseases of connective tissue: Cellular and extracellular effects of ECM mutations. *Nat. Rev. Genet.* 10 (2009): 173.

Bowman, S. M., and S. J. Free. The structure and synthesis of the fungal cell wall. *Bioessays* 28 (2006): 799.

Brett, C. T. Cellulose microfibrils in plants: Biosynthesis, deposition, and integration into the cell wall. *Int. Rev. Cytol.* 199 (2000): 161.

Cosgrove, D. J. Growth of the plant cell wall. *Int. Rev. Mol. Cell Biol.* 6 (2005): 850.

Jamet, E., C. Albenne, G. Boudart, M. Irshad, H. Canut, and R. Pont-Lezica. Recent advances in plant cell wall proteomics. *Proteomics* 8 (2008): 893.

Kreis, T., and R. Vale, eds. *Guidebook to the Extracellular Matrix, Anchor, and Adhesion Proteins,* 2nd ed. Oxford: Oxford University Press, 1999.

Marastoni, S., G. Ligresti, E. Lorenzon, A. Colombatti, and M. Mongiat. Extracellular matrix: A matter of life and death. *Connect. Tissue Res.* 29 (2008): 203.

Viruses, Viroids, and Prions

Aguzzi, A., and M. Polymenidou. Mammalian prion biology: One century of evolving concepts. *Cell* 116 (2004): S109.

Cann, A. J. *Principles of Molecular Virology,* 3rd ed. San Diego: Academic Press, 2001.

Cobb, N. J., and W. K. Surewicz. Prion diseases and their biochemical mechanisms. *Biochemistry* 48 (2009): 2574.

Crozet, C., F. Beranger, and S. Lehmann. Cellular pathogenesis in prion diseases. *Vet. Res.* 39 (2008): 44.

Ding, B., A. Itaya, and X. Zhong. Viroid trafficking: A small RNA makes a big move. *Curr. Opin. Plant Biol.* 8 (2005): 606.

Flores, R. et al. Viroids and viroid-host interactions. *Annu. Rev. Phytopathol.* 43 (2005): 117.

Moore, R. A., L. M. Taubner, and S. A. Priola. Prion protein misfolding and disease. *Curr. Opin. Struct. Biol.* 19 (2009): 14.

Prusiner, S. B., ed. *Prion Biology and Diseases,* 4th ed. Cold Spring Harbor, New York: Cold Spring Harbor Laboratory Press, 2004.

Prusiner, S. B., and M. R. Scott. Genetics of prions. *Annu. Rev. Genet.* 31 (1997): 139.

• Reisner, D., and H. H. Gross. Viroids. *Annu. Rev. Biochem.* 54 (1985): 531.

Stürmer M., H. W. Doerr, and L. Gürtler. Human immunodeficiency virus: 25 years of diagnostic and therapeutic strategies and their impact on hepatitis B and C virus. *Med. Microbiol. Immunol.* 198 (2009): 147.

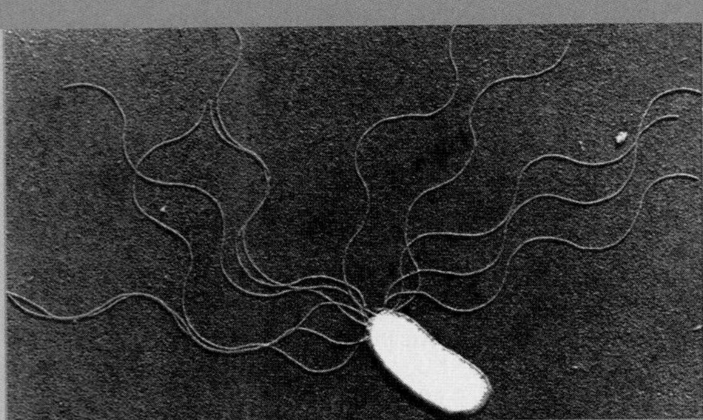

5 Bioenergetics: The Flow of Energy in the Cell

See MasteringBiology® for tutorials, activities, and quizzes.

Broadly speaking, every cell has four essential needs: molecular building blocks, chemical catalysts *called enzymes,* information *to guide all its activities, and* energy *to drive the various reactions and processes that are essential to life and biological function. In Chapters 3 and 4, we saw that cells need amino acids, nucleotides, sugars, and lipids in order to synthesize the macromolecules used to build cellular structures and organelles. Cells also need catalysts known as* enzymes *(Chapter 6) to speed up chemical reactions that would otherwise occur much too slowly to maintain life as we know it. The third general requirement of cells is for* information *to guide and direct their activities. This information is encoded within the nucleotide sequences of DNA and RNA and expressed in the ordered synthesis of specific proteins.*

All cells also require energy *in addition to molecules, enzymes, and information. Energy is needed to drive the chemical reactions involved in the formation of cellular components and to power the many activities that cells carry out. The capacity to obtain, store, and use energy is, in fact, one of the obvious features of most living organisms and helps to define life itself (***Figure 5-1***). Like the flow of information, the flow of energy is a major theme of this text and will be considered in more detail in Chapters 9–11. In this chapter, we will learn the basics of bioenergetics and see how the thermodynamic concept of free energy can allow us to predict whether specific chemical reactions can occur spontaneously in the cell.*

FIGURE 5-1 Energy and Life. The capacity to expend energy is one of the most obvious features of life at both the cellular and organismal levels.

The Importance of Energy

All living systems require an ongoing supply of **energy.** Usually, energy is defined as the capacity to do work. But that turns out to be a somewhat circular definition because work is frequently defined in terms of energy changes. A more useful definition is that *energy is the capacity to cause specific physical or chemical changes.* Since life is characterized first and foremost by change, this definition underscores the total dependence of all forms of life on the continuous availability of energy.

Cells Need Energy to Drive Six Different Kinds of Changes

Now that we have defined energy in this way, you may ask: What kinds of cellular activities give rise to specific physical and chemical changes? Six categories of change come to mind, which in turn define six kinds of work that require an input of energy: synthetic work, mechanical work, concentration work, and electrical work, as well as the generation of heat and light (**Figure 5-2**).

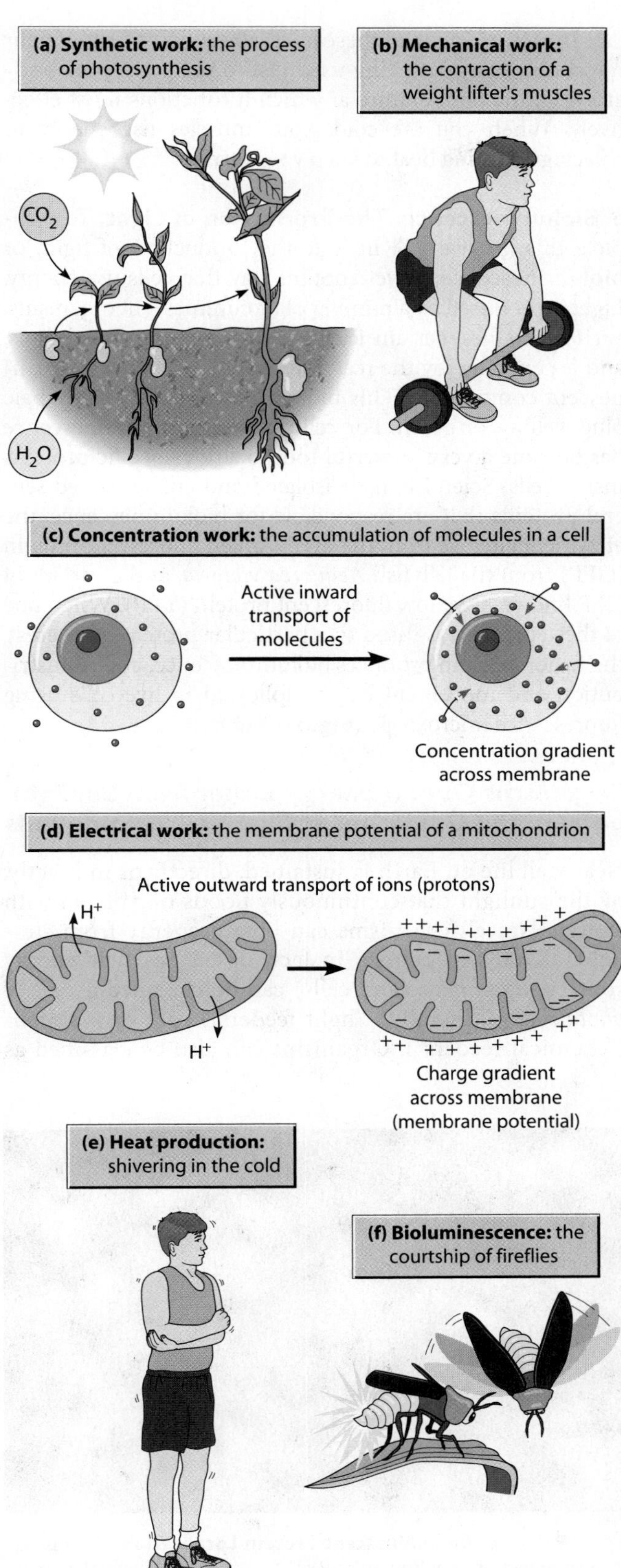

FIGURE 5-2 Several Kinds of Biological Work. The six major categories of biological work are shown here **(a–f)**. All require an input of energy to cause a physical or chemical change.

1. Synthetic Work: Changes in Chemical Bonds. An important activity of virtually every cell at all times is the work of **biosynthesis,** which results in the formation of new chemical bonds and the synthesis of new molecules. This activity is especially obvious in a population of growing cells, where additional molecules must be synthesized for cells to increase in size or number or both. Also, synthetic work is required to maintain existing cellular structures. Because most structural components of the cell are in a state of constant turnover, the molecules that make up these structures are continuously being degraded and replaced. Almost all of the energy that cells require for biosynthetic work is used to make energy-rich organic molecules from simpler starting materials and to incorporate them into macromolecules.

2. Mechanical Work: Changes in the Location or Orientation of a Cell or a Subcellular Structure. Cells often need energy for **mechanical work,** which involves a physical change in the position or orientation of a cell or some part of the cell. An especially good example is the movement of a cell with respect to its environment. This movement often requires one or more appendages such as cilia or flagella (**Figure 5-3**), which require energy in order to propel the cell forward. In other cases, the environment is moved past the cell, as when the ciliated cells that line your trachea beat upward to sweep inhaled particles upward and away from the lungs. This motion also requires energy. Muscle contraction is another good example of mechanical work, involving not just a single cell but a large number of muscle cells working together (see Figure 16-10). Other examples of mechanical work that we will see happening inside cells include the movement of chromosomes along the spindle fibers during mitosis, the streaming of cytoplasm, the movement of organelles and vesicles along microtubules, and the translocation of a ribosome along a strand of messenger RNA.

3. Concentration Work: Moving Molecules Across a Membrane Against a Concentration Gradient. Less conspicuous than either of the previous two categories but every bit as important to the cell is the work of moving molecules or ions against a concentration gradient. The purpose of **concentration work** is either to accumulate substances within a cell or organelle or to remove potentially toxic by-products of cellular activity. As we saw in Chapter 4, diffusion of materials (from areas of high concentration to areas of low concentration) is a spontaneous process and requires no added energy. Therefore, the opposite process, concentration, requires an input of energy. Examples of concentration work include the import of certain sugar and amino acid molecules from low to high concentration across the plasma membrane, the concentration of specific molecules and enzymes within organelles, and the accumulation of digestive enzymes into secretory vesicles to be released as you digest your food.

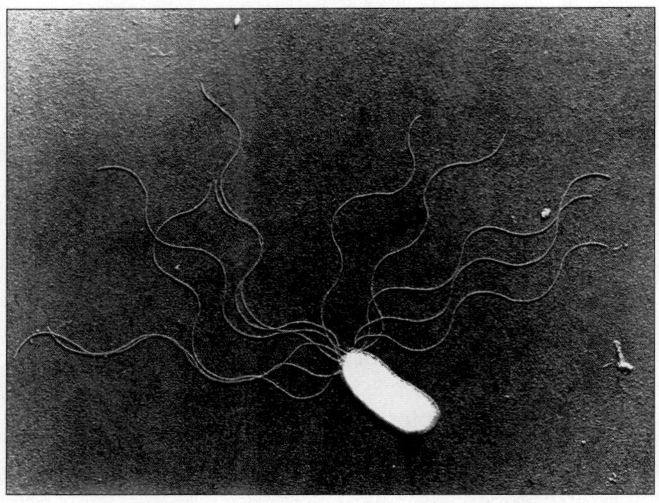

FIGURE 5-3 A Flagellated Bacterium. The spinning motion of bacterial flagella is an example of mechanical work, providing motility for some bacterial species (TEM).

4. Electrical Work: Moving Ions Across a Membrane Against an Electrochemical Gradient. Often, **electrical work** is considered a specialized case of concentration work because it also involves movement across membranes. In this case, however, ions are transported, and the result is not just a change in concentration. A charge difference, known as an *electrical potential* or a *membrane potential,* is also established across the membrane. Every cellular membrane has some characteristic electrical potential that is generated in this way. A difference in the concentration of protons on either side of the mitochondrial or chloroplast membrane forms an electrical potential that is essential for the production of ATP in both cellular respiration (Chapter 10) and photosynthesis (Chapter 11). Electrical work is also important in the transmission of impulses in nerve cells. This involves the establishment of a membrane potential, resulting from pumping Na^+ and K^+ ions into and out of the cell (Chapter 13). An especially dramatic example of electrical work is found in *Electrophorus electricus,* the electric eel. Individual cells in its electric organ use energy to generate a membrane potential of about 150 millivolts (mV) in one cell, and thousands of such cells arranged in series allow the electric eel to develop electric potentials of several hundred volts (ouch!).

5. Heat: An Increase in Temperature That Is Useful to Warm-Blooded Animals. Although living organisms do not use **heat** as a form of energy in the same way that a steam engine does, heat production is a major use of energy in all *homeotherms* (animals that regulate their body temperature independent of the environment). Heat is released as a by-product of many chemical reactions, and we, as homeotherms, take advantage of this "waste product" every day. In fact, as you read these lines, about two-thirds of your metabolic energy is being used just to maintain your body at 37°C, the temperature at which it functions most effectively. When you are cold, your muscles use energy to shiver, generating heat to keep you warm.

6. Bioluminescence: The Production of Light. To complete this list, we will include the production of light, or **bioluminescence,** as yet another way that cells use energy. Light is produced by a number of bioluminescent organisms, such as fireflies, certain jellyfish, and luminous toadstools, and is generated by the reaction of ATP with specific luminescent compounds. This bioluminescence is usually pale blue, yellow, or green. For cell biologists, bioluminescence has become a very powerful tool to study specific proteins inside cells. Scientists have isolated and characterized several proteins that are responsible for bioluminescence: the enzyme luciferase from the firefly, green fluorescent protein (GFP) from the jellyfish *Aequorea victoria,* and a variant of GFP known as yellow fluorescent protein (YFP). When one of these proteins is fused to a particular protein of interest, the fusion protein produces bioluminescence, and its distribution and movement can be followed in live cells using fluorescence microscopy (**Figure 5-4**).

Organisms Obtain Energy Either from Sunlight or from the Oxidation of Chemical Compounds

Nearly all life on Earth is sustained, directly or indirectly, by the sunlight that continuously floods our planet with energy. Not all organisms can obtain energy from sunlight directly, of course. In fact, based on their energy sources, organisms (and cells) can be classified as either *phototrophs* (literally, "light-feeders") or *chemotrophs* ("chemical-feeders"). Organisms can also be classified as

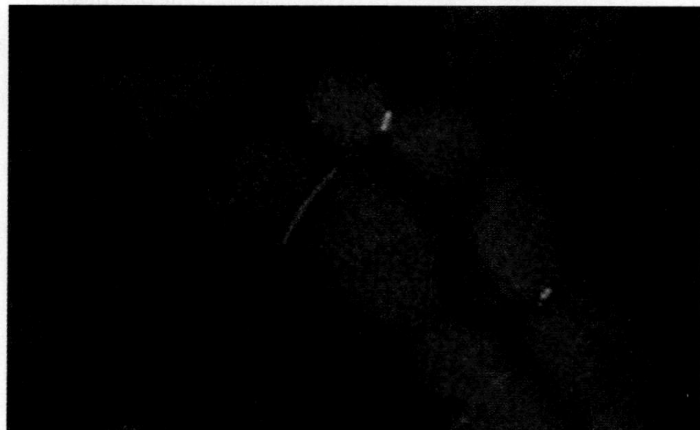

FIGURE 5-4 Bioluminescent Protein Localization. In this experiment, a protein known as PVD2 was fused to the bioluminescent protein YFP in order to study its distribution in *Arabidopsis* leaf cells. This fluorescence microscopy image shows that the PVD2-YFP fusion protein is found in an equatorial ring around dividing chloroplasts (red), demonstrating the involvement of PVD2 in chloroplast division.

autotrophs ("self-feeders") or *heterotrophs* ("other-feeders"), depending on whether their source of carbon is CO_2 or organic molecules, respectively. Most organisms are either photoautotrophs or chemoheterotrophs.

Phototrophs capture light energy from the sun using light-absorbing pigments and then transform this light energy into chemical energy, storing the energy in the form of ATP. *Photoautotrophs* use solar energy to produce all their necessary carbon compounds from CO_2 during photosynthesis. Examples of photoautotrophs include plants, algae, cyanobacteria, and photosynthetic bacteria. *Photoheterotrophs* (some bacteria) harvest solar energy to power cellular activities, but they must rely on the intake of organic molecules for their carbon needs.

In contrast, **chemotrophs** get energy by oxidizing chemical bonds in organic or inorganic molecules. *Chemoautotrophs* (a few bacteria) oxidize inorganic compounds such as H_2S, H_2 gas, or inorganic ions for energy and synthesize all their organic compounds from CO_2 *Chemoheterotrophs,* on the other hand, ingest and use chemical compounds such as carbohydrates, fats, and proteins to provide both energy and carbon for cellular needs. All animals, protozoa, fungi, and many bacteria are chemoheterotrophs.

Keep in mind that, although phototrophs can utilize solar energy when it is available, they must function as chemotrophs whenever they are not illuminated. Most plants are really a mixture of phototrophic and chemotrophic cells. A plant root cell, for example, though part of an obviously phototrophic organism, usually cannot carry out photosynthesis and is every bit as chemotrophic as an animal cell (or a plant leaf cell in the dark).

Energy Flows Through the Biosphere Continuously

Before we continue our discussion of energy flow, first let's review the chemical concepts of oxidation and reduction because they are critical to understanding energy flow in cells and organisms. *Oxidation* is the removal of electrons from a substance and, in biology, it usually involves the removal of hydrogen atoms (a hydrogen ion plus an electron) and the addition of oxygen atoms. Oxidation reactions release energy, as shown below when either glucose or methane is oxidized to carbon dioxide. Note that each carbon atom in glucose or methane has lost hydrogen atoms and has gained oxygen atoms as carbon dioxide is formed and energy is released. *Reduction* is the reverse reaction—the addition of electrons to a substance and usually the addition of hydrogen atoms (and a loss of oxygen atoms). Reduction reactions require an input of energy, as shown below when carbon dioxide is reduced to glucose during photosynthesis. The carbon atoms in six molecules of carbon dioxide have gained hydrogen atoms and lost oxygen atoms during this energy-requiring reduction to glucose.

Glucose oxidation:

$$C_6H_{12}O_6 + 6O_2 \rightarrow 6CO_2 + 6H_2O + \text{energy} \quad \textbf{(5-1)}$$

Methane oxidation:

$$CH_4 + 2O_2 \rightarrow CO_2 + 2H_2O + \text{energy} \quad \textbf{(5-2)}$$

Carbon dioxide reduction:

$$\text{energy} + 6CO_2 + 6H_2O \rightarrow C_6H_{12}O_6 + 6O_2 \quad \textbf{(5-3)}$$

The flow of energy and matter through the biosphere is depicted in **Figure 5-5**. Phototrophs are producers that use solar energy to convert carbon dioxide and water into more reduced cellular compounds such as glucose during photosynthesis. These reduced compounds are converted into other carbohydrates, proteins, lipids, nucleic acids, and all the materials a cell needs to survive. In a sense, we can consider the entire phototrophic organism to be the product of photosynthesis because every carbon atom in every molecule of that organism is derived from carbon dioxide that is fixed into organic form by the photosynthetic process.

Chemotrophs, on the other hand, are consumers that are unable to use solar energy directly and therefore must depend completely on energy that has been packaged into oxidizable food molecules by phototrophs. A world composed solely of chemoheterotrophs would last only so long as food supplies lasted, for even though we live on a planet that is flooded with solar energy each day, this energy is in a form that we humans cannot use to meet our energy needs.

Both phototrophs and chemotrophs use energy to carry out the six kinds of work we have already discussed (see Figure 5-2). An important principle of energy conversion is that no chemical or physical process occurs with 100% efficiency—some energy is released as heat. Although biological processes are remarkably efficient in energy conversion, loss of energy as heat is inevitable in biological processes. Sometimes the heat that is liberated during cellular processes is utilized, as when warm-blooded animals use heat to maintain a constant body temperature. Some plants use metabolically generated heat to attract pollinators or to melt overlying snow (**Figure 5-6**). In general, however, the heat is simply dissipated into the environment, representing a loss of energy from the organism.

Viewed on a cosmic scale, there is a continuous, massive, and unidirectional flow of energy on Earth from its source in the nuclear fusion reactions of the sun to its eventual sink, the environment. We here in the biosphere are the transient custodians of an infinitesimally small portion of that energy. It is precisely that small but critical fraction of energy and its flow through the living systems of the biosphere that is of concern to us. The flow of energy in the biosphere begins with phototrophs, which use light energy to drive electrons energetically "uphill" to create high-energy, reduced compounds. This energy is then released by both phototrophs and chemotrophs in "downhill" oxidation reactions. This flux of energy

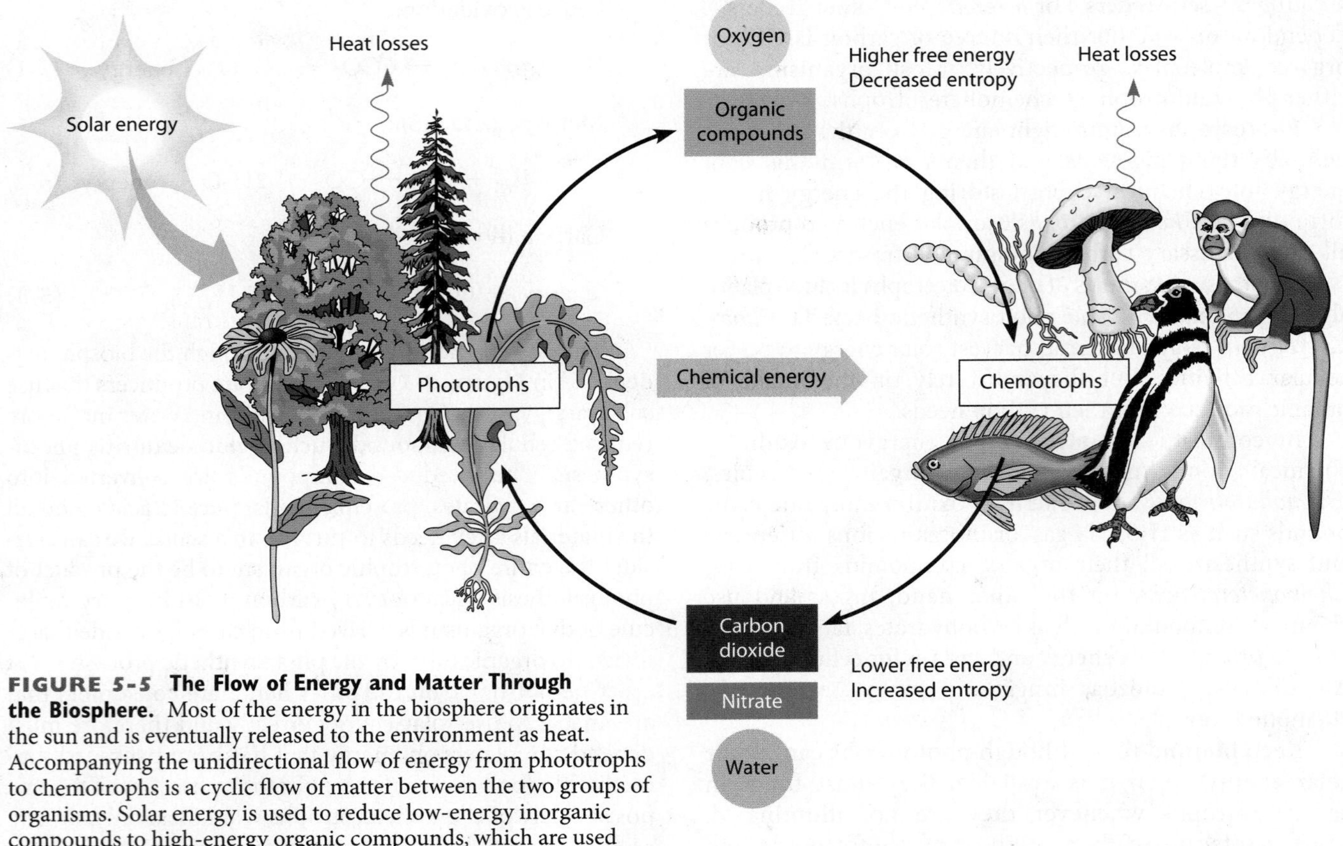

FIGURE 5-5 The Flow of Energy and Matter Through the Biosphere. Most of the energy in the biosphere originates in the sun and is eventually released to the environment as heat. Accompanying the unidirectional flow of energy from phototrophs to chemotrophs is a cyclic flow of matter between the two groups of organisms. Solar energy is used to reduce low-energy inorganic compounds to high-energy organic compounds, which are used by both phototrophs and chemotrophs.

FIGURE 5-6 Skunk Cabbage, a Plant That Depends on Metabolically Generated Heat. The skunk cabbage plant (*Symplocarpus foetidus*) is one of the earliest-flowering plants in the eastern United States. The heat that it generates enables it to melt through overlying snow in late winter and begin growing when most other plants are still dormant.

through living matter—from the sun to phototrophs to chemotrophs to heat—drives all life processes.

The Flow of Energy Through the Biosphere Is Accompanied by a Flow of Matter

Energy enters the biosphere as photons of light unaccompanied by matter and leaves the biosphere as heat similarly unaccompanied by matter. While it is passing through the biosphere, however, energy exists primarily in the form of chemical bond energies of oxidizable organic molecules in cells and organisms. As a result, the flow of energy in the biosphere is coupled to a correspondingly immense flow of matter.

Whereas energy flows unidirectionally from the sun through phototrophs to chemotrophs and to the environment, matter flows in a cyclic fashion between phototrophs and chemotrophs (Figure 5-5). Phototrophs use solar energy to create organic nutrients from inorganic starting materials such as carbon dioxide and water, releasing oxygen in the process. Phototrophs use some of these high-energy, reduced nutrients themselves, and some of them become available to chemotrophs that consume these phototrophs. Chemotrophs typically take in organic nutrients from their surroundings and use oxygen to oxidize them back to carbon dioxide and water, providing energy. Low-energy, oxidized molecules are returned to the environment and then become the raw materials that phototrophic organisms use to make new organic molecules, returning oxygen to the environment in the process and completing the cycle.

In addition, there is an accompanying cycle of nitrogen. Phototrophs obtain nitrogen from the environment in an oxidized, inorganic form (as nitrate from the soil or, in some cases, as N_2 from the atmosphere). They reduce it to ammonia (NH_3), a high-energy form of nitrogen used in the synthesis of amino acids, proteins, nucleotides, and nucleic acids. Eventually, these cellular molecules, like other components of phototrophic cells, are consumed by chemotrophs. The nitrogen in these molecules is then converted back into ammonia and eventually oxidized to nitrate, mostly by soil microorganisms.

Carbon, oxygen, nitrogen, and water thus cycle continuously between the phototrophic and chemotrophic worlds. They enter the chemotrophic sphere as reduced, energy-rich compounds and leave in an oxidized, energy-poor form. The two great groups of organisms can therefore be thought of as living in a symbiotic relationship with each other, with a cyclic flow of matter and a unidirectional flow of energy as components of that symbiosis.

When we deal with the overall macroscopic flux of energy and matter through living organisms, we find cellular biology interfacing with ecology. Ecologists study cycles of energy and nutrients, the roles of various species in these cycles, and environmental factors affecting the flow. In contrast, our ultimate concern in cell biology is how the flux of energy and matter functions on a microscopic and molecular scale. We will be able to understand these energy transactions and the chemical processes that occur within cells as soon as we acquaint ourselves with the physical principles underlying energy transfer in cells. For that we turn to the topic of bioenergetics.

Bioenergetics

The principles governing energy flow are the subject of an area of science known as **thermodynamics.** Although the prefix *thermo*– suggests that the term is limited to heat (and that is indeed its historical origin), thermodynamics also takes into account other forms of energy and processes that convert energy from one form to another. Specifically, thermodynamics concerns the laws governing the energy transactions that inevitably accompany most physical processes and all chemical reactions. **Bioenergetics,** in turn, can be thought of as *applied thermodynamics*—the application of thermodynamic principles to reactions and processes in the biological world.

To Understand Energy Flow, We Need to Understand Systems, Heat, and Work

It is useful to define energy not simply as the capacity to do work, but specifically as the ability to cause change. Without energy, all processes would be at a standstill, including those that we associate with living cells.

Energy exists in various forms, many of them of interest to biologists. Think, for example, of the energy represented by a ray of sunlight, a teaspoon of sugar, a moving flagellum, an excited electron, or the concentration of ions or small molecules within a cell or an organelle. These phenomena are diverse, but they are all governed by certain basic principles of energetics.

Energy is distributed throughout the universe—and for some purposes it is necessary to consider the total energy of the universe, at least in a theoretical way. Whereas the total energy of the universe remains constant, we are usually interested in the energy content of only a small portion of it. We might, for example, be concerned with a reaction or process occurring in a beaker of chemicals, in a cell, or in a particular organism. By convention, the restricted portion of the universe that is being considered at any given moment is called the **system,** and all the rest of the universe is referred to as the **surroundings.** Sometimes the system has a natural boundary, such as a glass beaker or a cell membrane. In other cases, the boundary between the system and its surroundings is just a hypothetical one used only for convenience of discussion, such as the imaginary boundary around one mole of glucose molecules in a solution.

Systems can be either open or closed, depending on whether they can exchange energy with their surroundings (**Figure 5-7**). A *closed system* is sealed from its environment and can neither take in nor release energy in any form. An *open system,* on the other hand, can have energy added to it or removed from it. The levels of organization and complexity that biological systems routinely display are possible only because cells and organisms are open systems, capable of both the uptake and the release of energy. Specifically, biological systems require a constant, large-scale influx of energy from their surroundings both to attain and to maintain the levels of complexity that are characteristic of them. That is essentially why plants need sunlight and you need food.

Whenever we talk about a system, we have to be careful to specify the state of the system. A system is said to be in a specific **state** if each of its variable properties (such as temperature, pressure, and volume) is held constant at a specified value. In such a situation, the total energy content of the system, while not directly measurable, has some unique value. If such a system then changes from one state to another as a result of some interaction between the system and its surroundings, the change in its total energy is determined uniquely by the initial and final states of the system. The magnitude of the energy change is not affected by the mechanism causing the change nor by the intermediate states through which the system may pass. This is a useful property because it allows energy changes to be determined from a knowledge of the initial and final states only, independent of the path taken.

The problem of keeping track of system variables and their effect on energy changes can be simplified if one or more of the variables are held constant. Fortunately, this

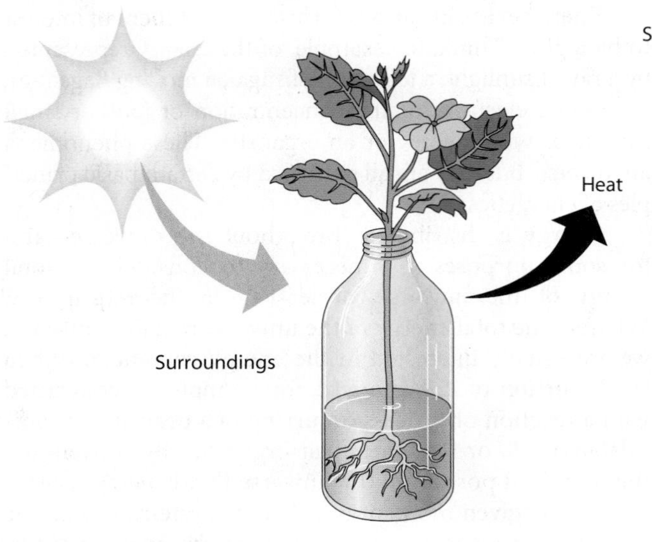

Surroundings

(b) Closed system

FIGURE 5-7 Open and Closed Systems. A system is that portion of the universe under consideration. The rest of the universe is called the surroundings of the system. **(a)** An open system can exchange energy with its surroundings, whereas **(b)** a closed system cannot. All living organisms are open systems, exchanging energy freely with their surroundings.

is the case with most biological reactions because they usually occur in dilute solutions within cells that are at approximately the same temperature, pressure, and volume during the entire course of the reaction. These environmental conditions are generally slow to change compared with the speed of biological reactions. This means that three of the most important system variables that physical chemists need to consider—temperature, pressure, and volume—are essentially constant for most biological reactions.

The exchange of energy between a system and its surroundings occurs either as heat or as work. Heat is energy transfer from one place to another as a result of a temperature difference and occurs spontaneously from the hotter place to the colder place. Heat is an extremely useful form of energy for machines designed either to accomplish mechanical work or to transform heat into other forms of energy. However, heat has only limited biological utility because many biological systems operate under *isothermal* conditions of fixed temperature. Such systems lack the temperature gradients required to convert heat into other forms of energy. As a result, heat is generally not a useful source of energy for cells—although it can be used for such purposes as maintaining body temperature or attracting pollinators, as we noted earlier.

In biological systems, **work** is the use of energy to drive any process other than heat flow. For example, work is performed when the muscles in your arm expend chemical energy to lift this book, when a corn leaf uses light energy to synthesize sugar, or when an electric eel creates the ion concentration gradients to deliver a shock. We will be primarily interested in the amount of useful energy available to do cellular work when we begin calculating energy changes associated with specific reactions in cells.

To quantify energy changes during chemical reactions or physical processes, we need units in which energy can be expressed. In biological chemistry, energy changes are usually expressed in terms of the **calorie (cal)**, which is defined as the amount of energy required to warm 1 gram of water 1 degree centigrade at a pressure of 1 atmosphere. One **kilocalorie (kcal)** equals 1000 calories. An alternative energy unit, the **joule (J)**, is preferred by physicists and is used in some biochemistry texts. Conversion is easy: $1 \text{ cal} = 4.184 \text{ J}$, or $1 \text{ J} = 0.239 \text{ cal}$.

Energy changes are often measured on a per-mole basis, and the most common forms of energy units in biological chemistry are calories or kilocalories per mole (cal/mol or kcal/mol). (Be careful to distinguish between the *calorie* as defined here and the *nutritional Calorie* that is often used to express the energy content of foods. The nutritional Calorie is represented with a capital C and is really a *kilocalorie* as defined here.)

The First Law of Thermodynamics Tells Us That Energy Is Conserved

Much of what we understand about the principles governing energy flow can be summarized by the three laws of thermodynamics. For cell biologists, only the first and second laws are of particular relevance. The **first law of thermodynamics** is called the *law of conservation of energy.* The first law of thermodynamics states that *in every physical or chemical change, the total amount of energy in the universe remains constant, although the form of the energy may change.* Or, in other words, energy can be converted from one form to another but can never be created or destroyed.

Applied to the universe as a whole or to a closed system, the first law means that the total amount of energy present in all forms must be the same before and after any process or reaction occurs. Applied to an open system such as a cell, the first law says that, during the course of any reaction or process, the total amount of energy that leaves the system must be exactly equal to the energy that enters the system minus any energy that remains behind and is therefore stored within the system.

The total energy stored within a system is called the **internal energy** of the system, represented by the symbol ***E***. We are not usually concerned with the actual value of E for a system because that value cannot be measured directly. However, it is possible to measure the *change in internal energy,* ΔE, that occurs during a given process. ΔE is the difference in internal energy of the system before the process (E_1) and after the process (E_2):

$$\Delta E = E_2 - E_1 \tag{5-4}$$

Equation 5-4 is valid for all physical and chemical processes under any conditions. For a chemical reaction, we can write

$$\Delta E = E_{\text{products}} - E_{\text{reactants}} \tag{5-5}$$

In the case of biological reactions and processes, we are usually more interested in the change in **enthalpy,** or *heat content.* Enthalpy is represented by the symbol ***H*** (for heat) and is related to the internal energy E by a term that combines both pressure (P) and volume (V):

$$H = E + PV \tag{5-6}$$

H is dependent on both E and PV because changes in heat content following a process or reaction can affect the total energy as well as the pressure and volume. Unlike many chemical reactions, biological reactions generally proceed with little or no change in either pressure or volume. So, for biological reactions, both ΔP and ΔV are usually zero (or at least negligible), and we can write

$$\Delta H = \Delta E + \Delta(PV) \cong \Delta E \tag{5-7}$$

Thus, biologists routinely determine changes in heat content for reactions of interest, confident that the values are valid estimates of ΔE.

The enthalpy change that accompanies a specific reaction is simply the difference in the heat content between the reactants and the products of the reaction:

$$\Delta H = H_{\text{products}} - H_{\text{reactants}} \tag{5-8}$$

The ΔH value for a specific reaction or process will be either negative or positive. If the heat content of the products is less than that of the reactants, heat is released, ΔH will be negative, and the reaction is said to be **exothermic.** For example, the burning (oxidation) of gasoline in your car is exothermic because the heat content of the products (CO_2 and H_2O) is less than the heat content of the reactants (gasoline and O_2).

If the heat content of the products is greater than that of the reactants, ΔH will be positive, and the reaction is **endothermic.** In an endothermic reaction or process, heat energy is absorbed, as in the melting of an ice cube—the heat content of the resulting liquid water is greater than the heat content of the ice before melting. Thus, the ΔH value for any reaction is simply a measure of the heat that is either liberated from or taken up by that reaction as it occurs under conditions of constant temperature and pressure.

The Second Law of Thermodynamics Tells Us That Reactions Have Directionality

So far, all that thermodynamics has been able to tell us is that energy is conserved whenever a process or reaction occurs—that all of the energy going into a system must either be stored within the system or released again to the surroundings. We have seen the usefulness of ΔH as a measure of how much the total enthalpy of a system would change if a given process were to occur, but we have no way as yet of predicting whether and to what extent the process will in fact occur under the prevailing conditions.

We have, at least in some cases, an intuitive feeling that some reactions or processes are possible, whereas others are not. We are sure that if we set a match to a sheet of paper made of cellulose, it will burn. The oxidation of the glucose monomers of the cellulose to carbon dioxide and water is, in other words, a favorable reaction. Or, to use more precise terminology, it is a *thermodynamically spontaneous* reaction. In the context of thermodynamics, the term *spontaneous* has a specific, restricted meaning that is different from its commonplace usage. **Thermodynamic spontaneity** is a measure of whether a reaction or process *can go,* but it says nothing about whether it *will go.* Our sheet of paper illustrates this point well. The oxidation of cellulose is clearly a *possible* reaction, but we know that it does not "just happen." It needs some impetus—a match, in this particular case.

Not only are we convinced that a sheet of paper will burn if ignited, but we also know—if only intuitively—that there is *directionality* to the process. We are, in other words, equally convinced that the reverse reaction will not occur. If we were to stand around clutching the charred remains, the paper would not spontaneously reassemble in our hands. We have, in other words, a feeling for both the *possibility* and the *directionality* of cellulose oxidation.

You can probably think of other processes that you can make such thermodynamic predictions about with equal confidence. We know, for example, that drops of dye diffuse in water, that ice cubes melt at room temperature, and that sugar dissolves in water. We can therefore label these as thermodynamically spontaneous events. But if we ask why we recognize them as such, the answer has to do with repeated experience. We have seen paper burn, ice cubes melt, and sugar dissolve often enough to know intuitively that these processes really occur—and with such predictability that we can label them as spontaneous, provided only that we know the conditions.

However, when we move from the world of familiar physical processes to the realm of chemical reactions in cells, we quickly find that we cannot depend on experience to

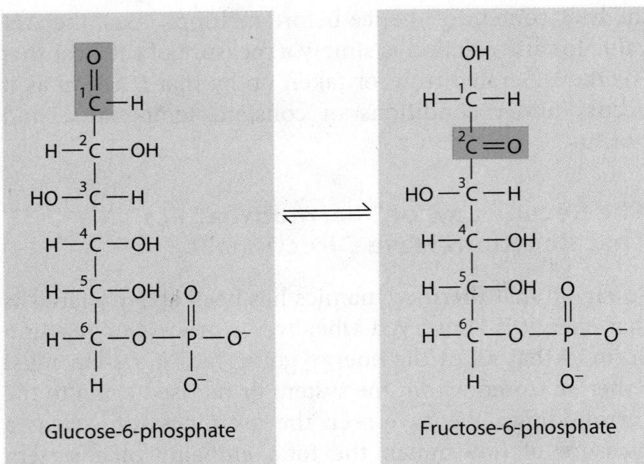

FIGURE 5-8 The Interconversion of Glucose-6-Phosphate and Fructose-6-Phosphate. This reaction involves the interconversion of the phosphorylated forms of an aldosugar (glucose) and a ketosugar (fructose). The reaction is catalyzed by an enzyme called phosphoglucoisomerase and is readily reversible. This reaction is part of the glycolytic pathway, as we will see in Chapter 9.

guide us in our predictions. Consider, for example, the conversion of glucose-6-phosphate into fructose-6-phosphate:

$$\text{glucose-6-phosphate} \rightleftharpoons \text{fructose-6-phosphate} \quad \textbf{(5-9)}$$

This particular interconversion is a significant reaction in all cells, and it is shown in detail in **Figure 5-8**. It is, in fact, the second step in an important and universal energy-yielding pathway called *glycolysis,* which we will return to in Chapter 9. For now, ask yourself what predictions you can make about the likelihood of glucose-6-phosphate being converted into fructose-6-phosphate. You will probably be at a loss to make any predictions at all. We know what will happen with burning paper and melting ice, but we lack the familiarity and experience with phosphorylated sugars even to make an intelligent guess. Clearly, what we need is a reliable means of determining whether a given physical or chemical change can occur under specific conditions without having to rely on experience, intuition, or guesswork.

Thermodynamics provides us with exactly such a measure of spontaneity in the **second law of thermodynamics,** or the *law of thermodynamic spontaneity.* This law tells us that *in every physical or chemical change, the universe always tends toward greater disorder or randomness (entropy).* The second law is useful for our purposes because it allows us to predict in what direction a reaction will proceed under specified conditions, how much energy the reaction will release as it proceeds, and how the energetics of the reaction will be affected by specific changes in the conditions.

An important point to note is that *no process or reaction disobeys the second law of thermodynamics.* Some processes may *seem* to do so because they result in more, rather than less, order. Think, for example, of the increase in order when a house is built, your room is cleaned, or a human being develops from a single egg cell. In each of these cases, however, the increase in order is limited to a specific system (the house, the room, or the embryo). The increase in order is possible only because it is an open system so that energy can be added from the outside, whether by means of an electric saw, your arm muscles, or nutrients supplied by the mother.

Entropy and Free Energy Are Two Alternative Means of Assessing Thermodynamic Spontaneity

Thermodynamic spontaneity—whether a reaction *can* go—can be measured by changes in either of two parameters: *entropy* or *free energy.* These concepts are abstract and can be somewhat difficult to understand. We will therefore limit our discussion here to their use in determining what kinds of changes can occur in biological systems. For further help, see **Box 5A** for an essay that uses jumping beans to introduce the concepts of internal energy, entropy, and free energy.

Entropy. Although we cannot quantify **entropy** directly, we can get some feel for it by considering it to be a measure of *randomness* or *disorder.* Entropy is represented by the symbol *S.* For any system, the *change in entropy,* ΔS, represents a change in the degree of randomness or disorder of the components of the system. For example, the combustion of paper involves an increase in entropy because the carbon, oxygen, and hydrogen atoms of cellulose are much more randomly distributed in space once they are converted to carbon dioxide and water. Entropy also increases as ice melts or as a volatile solvent such as gasoline is allowed to evaporate. Likewise, entropy decreases when a system becomes more ordered, as when the mobile molecules in liquid water become locked in place in ice or when large molecules such as glucose are synthesized from several smaller (randomly moving) carbon dioxide molecules.

Entropy Change as a Measure of Thermodynamic Spontaneity. How can the second law of thermodynamics help predict what changes will occur in a cell? There is an important link between spontaneous events and entropy changes because *all processes or reactions that occur spontaneously result in an increase in the total entropy of the universe.* Or, in other words, the value of $\Delta S_{\text{universe}}$ is positive for every spontaneous process or reaction.

We have to keep in mind, however, that this formulation of the second law pertains to the universe as a whole and may not apply to the specific system under consideration. Every real process, without exception, must be accompanied by an increase in the entropy of the universe. But for a given system, the entropy may increase, decrease, or stay the same as the result of a specific process. For example, the combustion of paper is clearly spontaneous and is accompanied by an *increase* in the system entropy. On the other hand, the freezing of water at $-1°C$ is also a spontaneous event, yet it involves a *decrease* in the system entropy as the water molecules become less randomly arranged in the resulting ice crystals. Thus, while the change in entropy of the universe is a valid measure of the spontaneity of a process, the change in entropy of the system is not.

To express the second law in terms of entropy change is therefore of limited value in predicting the spontaneity of biological processes because it would require keeping track of changes occurring not only within the system but also in its surroundings. Far more convenient would be a parameter that would enable prediction of the spontaneity of reactions by considering the system alone.

Free Energy. As you might guess, a measure of spontaneity for the system alone does in fact exist. It is called **free energy** and is represented by the symbol **G** (after Josiah Willard Gibbs, who first developed the concept). Because of its predictive value and its ease of calculation, *the free energy function is one of the most useful thermodynamic concepts in biology.* In fact, our entire discussion of thermodynamics so far has really been a way of getting us to this concept of free energy because it is here that the usefulness of thermodynamics for cell biologists becomes apparent.

Like most other thermodynamic functions, free energy is defined in terms of mathematical relationships. For biological reactions at constant pressure, volume, and temperature, the **free energy change,** ΔG, is dependent on the free energies of the products and the reactants:

$$\Delta G = G_{products} - G_{reactants} \quad \textbf{(5-10)}$$

This free energy change is related to the changes in enthalpy and entropy by the formula

$$\Delta G = \Delta H - T\Delta S \quad \textbf{(5-11)}$$

where ΔG is the change in free energy, ΔH is the change in enthalpy, T is the temperature of the system in degrees Kelvin ($K = °C + 273$, and ΔS is the change in entropy.

Notice that ΔG is the algebraic sum of two terms, ΔH and $-T\Delta S$, and is therefore influenced by changes in both enthalpy and entropy. As we saw earlier, ΔH will be positive for endothermic reactions and negative for exothermic reactions. Similarly, ΔS for a specific reaction or process can be either positive (increase in entropy) or negative (decrease in entropy). Because of the minus sign, the term $-T\Delta S$ will be negative if entropy increases or positive if entropy decreases. Therefore, you can see that the change in free energy of a reaction, ΔG, will increase when the change in heat content, ΔH, increases or when the change in entropy (randomness), ΔS, decreases.

Free Energy Change as a Measure of Thermodynamic Spontaneity. Free energy is an exceptionally useful concept as a readily measurable indicator of spontaneity, and we will soon see how ΔG can be calculated in real-life situations. ΔG provides exactly what we have been looking for: a measure of the spontaneity of a reaction based solely on the properties of the system in which the reaction is occurring. This will tell us whether a given reaction will proceed spontaneously from left to right as written.

Specifically, every spontaneous reaction is characterized by a *decrease* in the free energy of the system ($\Delta G_{system} < 0$) just as surely as it is characterized by an *increase* in the entropy of the universe ($\Delta S_{universe} > 0$). This is true because with the temperature and pressure held constant, ΔG for the system is related to ΔS for the universe in a simple but inverse way. This gives us a second, equally valid way of expressing the second law of thermodynamics: *All processes or reactions that occur spontaneously result in a decrease in the free energy content of the system.* In other words, *the value of* ΔG_{system} *is negative for every spontaneous process or reaction.* This occurs when the free energy of the products is less than the free energy of the reactants, as shown in Equation 5-10.

Such processes or reactions are called **exergonic,** which means *energy-yielding.* In contrast, any process or reaction that would result in an increase in the free energy of the system is called **endergonic** *(energy-requiring)* and cannot proceed under the conditions for which ΔG was calculated. Note carefully that we are considering only the overall change in free energy and not the specific changes either in the enthalpy or in the entropy of the system. Enthalpy and entropy changes may be negative, positive, or zero for a given reaction and, considered individually, are *not* valid measures of thermodynamic spontaneity.

Because the values for ΔH and $-T\Delta S$ can each be either positive or negative, the value of ΔG for a given reaction will depend on both the signs and numerical values of the ΔH and $-T\Delta S$ terms (**Figure 5-9**, see p. 146). If both the ΔH and $-T\Delta S$ terms for a given reaction are positive (endothermic and a decrease in entropy), ΔG will be positive, and the reaction will be endergonic and not spontaneous (Figure 5-9a). In contrast, a reaction that is exothermic (i.e., ΔH is negative) and results in an increase in entropy (i.e., ΔS is positive and $-T\Delta S$ is negative) has a ΔG value that is the *sum* of these two negative terms and is therefore exergonic and spontaneous (Figure 5-9b).

However, if the ΔH and $-T\Delta S$ terms differ in sign, the ΔG value can be either positive or negative, depending on the magnitudes of the ΔH and $-T\Delta S$ terms. For example, a reaction that is endothermic (i.e., ΔH is positive) and results in an increase in entropy (i.e., ΔS is positive and $-T\Delta S$ is negative) will have a ΔG value that is either positive (Figure 5-9c) or negative (Figure 5-9d), depending on the relative numerical values of ΔH and $-T\Delta S$.

BOX 5A DEEPER INSIGHTS

Jumping Beans and Free Energy

If you are finding the concepts of free energy, entropy, and enthalpy difficult to grasp, perhaps a simple analogy might help.* For this we will need an imaginary supply of jumping beans, which are really seeds of certain Mexican shrubs, with larvae of the moth *Laspeyresia saltitans* inside. Whenever the larvae inside the seed wiggle about, the seeds wiggle, too. The "jumping" action probably serves to get the larvae out of direct sunlight, which could heat them to lethal temperatures.

The Jumping Reaction

For purposes of illustration, imagine that we have some high-powered jumping beans in two chambers separated by a low partition, as shown below. Notice that the chambers have the same floor area and are at the same level, although we will want to vary both of these properties shortly. As soon as we place a handful of jumping beans in chamber 1, they begin jumping about randomly. Although most of the beans jump only to a modest height most of the time, occasionally one of them, in a burst of ambition, gives a more energetic leap, surmounting the barrier and falling into chamber 2. We can write this as the *jumping reaction:*

$$\text{Beans in chamber 1} \rightleftharpoons \text{Beans in chamber 2}$$

We will imagine this to be a completely random event, happening at irregular, infrequent intervals. Occasionally, one of the beans that has reached chamber 2 will happen to jump back into chamber 1, which is the *back reaction.* At first, of course, there will be more beans jumping from chamber 1 to chamber 2 because there are more beans in chamber 1, but things will eventually even out so that, on average, there will be the same number of beans in both compartments. The system will then be at *equilibrium.* Beans will still continue to jump between the two chambers, but the numbers jumping in both directions will be equal.

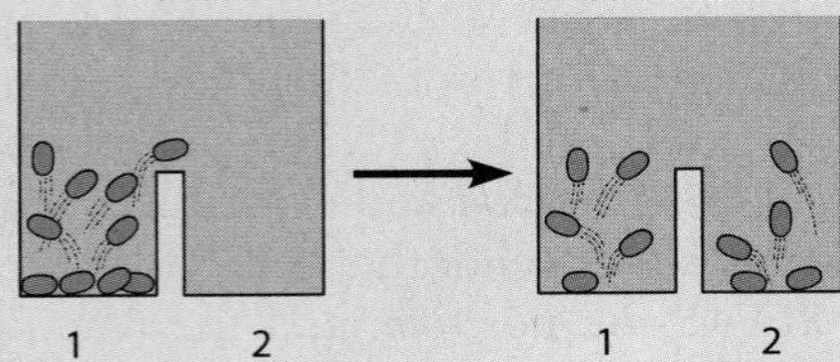

The Equilibrium Constant

Once our system is at equilibrium, we can count up the number of beans in each chamber and express the results as the ratio of the number of beans in chamber 2 to the number in chamber 1. This is simply the *equilibrium constant* (K_{eq}) for the jumping reaction:

$$K_{eq} = \frac{\text{number of beans in chamber 2 at equilibrium}}{\text{number of beans in chamber 1 at equilibrium}}$$

For the specific case shown above, the numbers of beans in the two chambers are equal at equilibrium, so the equilibrium constant for the jumping reaction under these conditions is 1.0.

Enthalpy Change (ΔH)

Now suppose that the level of chamber 1 is somewhat higher than that of chamber 2, as shown in the next diagram. Jumping beans placed in chamber 1 will again tend to distribute themselves between chambers 1 and 2, but this time a higher jump is required to get from 2 to 1 than from 1 to 2, so the latter will occur more frequently. As a result, there will be more beans in chamber 2 than in chamber 1 at equilibrium, and the equilibrium constant will therefore be greater than 1.

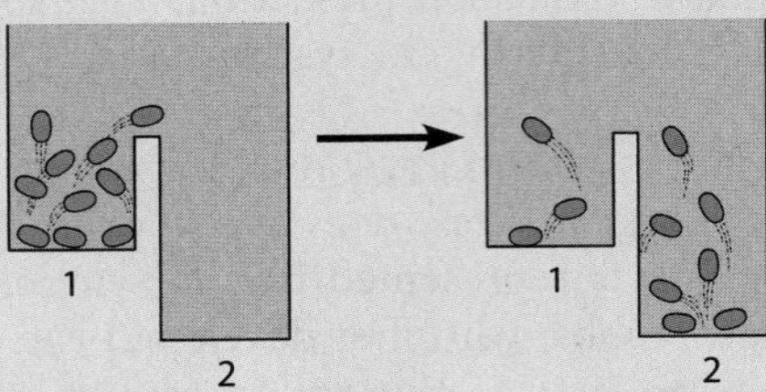

The relative heights of the two chambers can be thought of as measures of the *enthalpy,* or *heat content* (*H*), of the chambers, such that chamber 1 has a higher *H* value than chamber 2, and the difference between them is represented by ΔH. Since it is a "downhill" jump from chamber 1 to chamber 2, it makes sense that ΔH has a negative value for the jumping reaction from chamber 1 to chamber 2. Similarly, it seems reasonable that ΔH for the reverse reaction should have a positive value because that jump is "uphill."

Entropy Change (ΔS)

So far, it might seem as if the only thing that can affect the equilibrium distribution of beans between the two chambers is the difference in enthalpy, ΔH. But that is only because we have kept the floor area of the two chambers constant. Imagine instead the situation shown below, where the two chambers are again at the same height, but chamber 2 now has a greater floor area than chamber 1. The probability of a bean finding itself in chamber 2 is therefore correspondingly greater, so there will be more beans in chamber 2 than in chamber 1 at equilibrium, and the equilibrium constant will be greater than 1 in this case also. This means that the equilibrium position of the jumping reaction has been shifted to the right, even though there is no change in enthalpy.

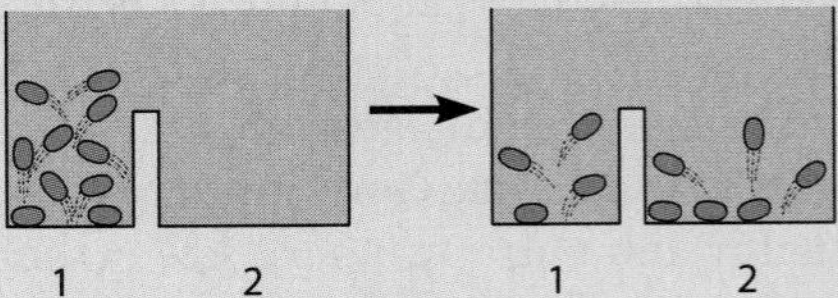

The floor area of the chambers can be thought of as a measure of the *entropy,* or randomness, of the system, *S*, and the *difference* between the two chambers can be represented by ΔS. Since chamber 2 has a greater floor area than chamber 1, the entropy change is positive for the jumping reaction as it proceeds from left to right under these conditions. Note that for ΔH, negative values are associated with favorable reactions, while for ΔS, favorable reactions are indicated by positive values.

Free Energy Change (ΔG)

So far, we have encountered two different factors that affect the distribution of beans: the difference in levels of the two chambers (ΔH) and the difference in floor area (ΔS). Moreover, it should be clear that neither of these factors by itself is an adequate indicator of how the beans will be distributed at equilibrium because a favorable (negative) ΔH could be more than offset by an unfavorable (negative) ΔS, and a favorable (positive) ΔS could be more than offset by an unfavorable (positive) ΔH. You should, in fact, be able to design chamber conditions that illustrate both of these situations, as well as situations in which ΔH and ΔS tend to reinforce rather than counteract each other.

Clearly, what we need is a way of summing these two effects algebraically to see what the net tendency will be. The new measure we come up with is called the *free energy change,* ΔG, which turns out to be the most important thermodynamic parameter for our purposes. ΔG is defined so that *negative* values correspond to favorable (i.e., thermodynamically spontaneous) reactions and *positive* values represent unfavorable reactions. Thus, ΔG should have the *same* sign as ΔH (since a negative ΔH is also favorable) but the *opposite* sign from ΔS (since for ΔS, a positive value is favorable). Based on real-life thermodynamics, the expression for ΔG in terms of ΔH and ΔS is

$$\Delta G = \Delta H - T\Delta S$$

(Notice that the temperature dependence of ΔS is the only feature of this relationship that cannot be readily explained by our model, unless we assume that the effect of changes in room size is somehow greater at higher temperatures.)

ΔG and the Capacity to Do Work

You should be able to appreciate the difficulty of suggesting a physical equivalent for ΔG because it represents an algebraic sum of entropy and energy changes, which may either reinforce or partially offset each other. But as long as ΔG is negative, beans will continue to jump from chamber 1 to chamber 2, whether driven primarily by changes in entropy, enthalpy, or both. This means that if some sort of bean-powered "bean wheel" is placed between the two chambers as shown below, the movement of beans from one chamber to the other can be harnessed to do work until equilibrium is reached, at which point no further work is possible. Furthermore, the greater the difference in free energy between the two chambers (that is, the more highly negative ΔG is), the more work the system can do.

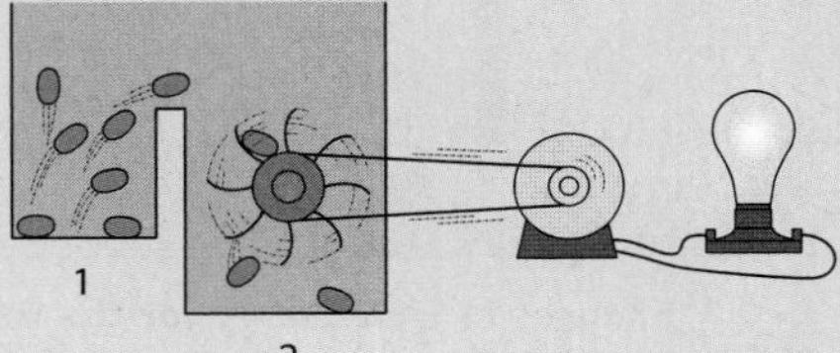

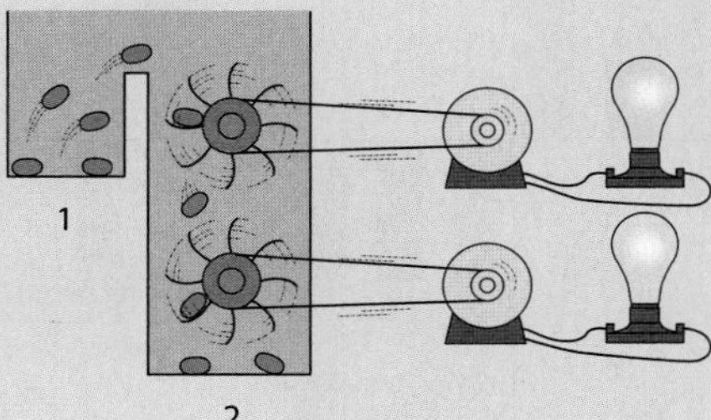

Thus, ΔG is first and foremost a measure of the capacity of a system to do work under specified conditions. You might, in fact, want to think of ΔG as free energy in the sense of *energy that is free or available to do useful work.* Moreover, if we contrive to keep ΔG negative by continuously adding beans to chamber 1 and removing them from chamber 2, we have a dynamic *steady state,* a condition that effectively harnesses the inexorable drive to achieve equilibrium. Work can then be performed continuously by beans that are forever jumping toward equilibrium but that never actually reach it.

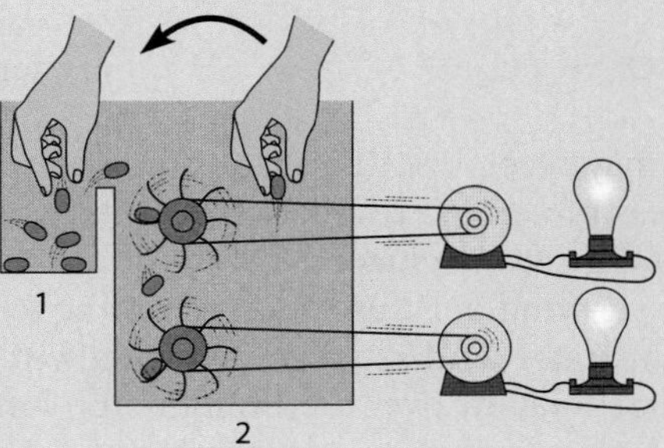

Looking Ahead

To anticipate the transition from the thermodynamics of this chapter to the kinetics of the next, begin thinking about the *rate* at which beans actually proceed from chamber 1 to chamber 2. Clearly, ΔG measures how much energy will be released if beans do jump, but it says nothing at all about the rate. That would appear to depend critically on how high the barrier between the two chambers is. Label this the *activation energy barrier,* and then contemplate how you might get the beans to move over the barrier more rapidly. One approach might be to heat the chambers; this would be effective because the larvae inside the seeds wiggle more vigorously if they are warmed. Cells, on the other hand, have a far more effective and specific means of speeding up reactions: They lower the activation barrier by using catalysts called *enzymes,* which we will meet in the next chapter.

*We are indebted to Princeton University Press for permission to use this analogy, which was first developed by Harold F. Blum in the book Time's Arrow and Evolution (3rd ed., 1968), pp. 45–54.

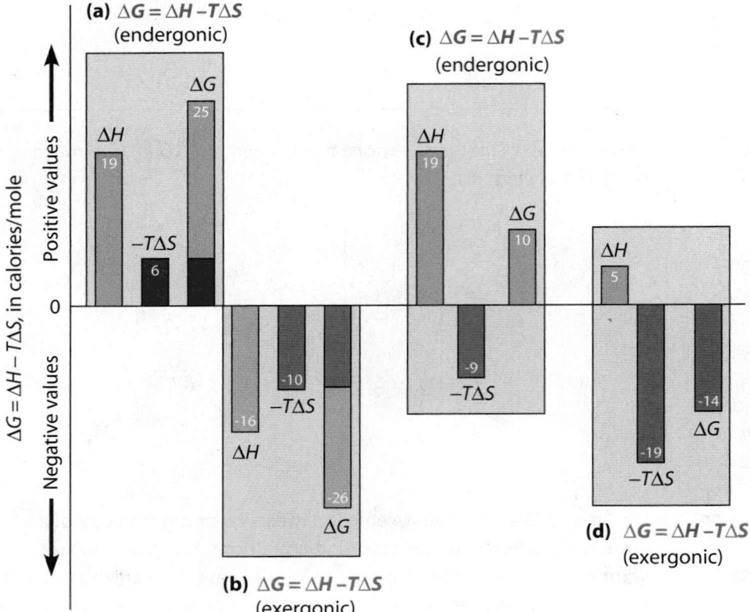

FIGURE 5-9 Dependence of ΔG on the Signs and Numerical Values of ΔH and the Term $-T\Delta S$. The ΔG value for a specific process or reaction is the algebraic sum of the change in enthalpy (ΔH) and the temperature-dependent entropy term ($-T\Delta S$). Green letters and bars identify terms that have negative values (that will increase spontaneity), whereas red letters identify terms that have positive values (that will decrease spontaneity). The numbers shown on the bars in white have been added so you can make the calculations yourself.

Free Energy Change and Thermodynamic Spontaneity: A Biological Example. For a biological example of an exergonic reaction, consider again the oxidation of glucose to carbon dioxide and water:

$$C_6H_{12}O_6 + 6O_2 \rightarrow 6CO_2 + 6H_2O + \text{energy} \quad \textbf{(5-12)}$$

You may recognize this as the summary equation for the process of aerobic respiration, whereby chemotrophs obtain energy from glucose. (Most of the cells in your body are carrying out this process right now.) As **Figure 5-10a** illustrates (and as you already know if you've ever held a marshmallow over the campfire too long), the oxidation of glucose is a highly exergonic process, with a highly negative ΔG value. By combusting glucose under *standard conditions* of temperature, pressure, and concentration (25°C, 1 atmosphere of pressure, and 1 *M* each of reactant and product), we can show that 673 kcal of heat (673,000 cal) are liberated for every mole of glucose that is oxidized, meaning that ΔH for Reaction 5-12 is -673 kcal/mol. The $-T\Delta S$ term can also be determined experimentally and is known to be -13 kcal/mol at 25°C, so this is an example of a reaction in which the ΔH and $-T\Delta S$ terms are additive, with a ΔG of -686 kcal/mol.

Now consider the reverse reaction, by which phototrophs synthesize sugars such as glucose from carbon dioxide and water, with the release of oxygen:

$$6CO_2 + 6H_2O + \text{energy} \rightarrow C_6H_{12}O_6 + 6O_2 \quad \textbf{(5-13)}$$

As you might guess, the values of ΔH, ΔS, and ΔG for this reaction are identical in magnitude, but opposite in sign, when determined under standard conditions and compared with the corresponding values for Reaction 5-12. Specifically, this reaction has a ΔG value of $+686$ kcal/mol,

(a) Energetics of glucose oxidation. The oxidation of glucose to carbon dioxide and water is a highly exergonic reaction, with a ΔG value of -686 kcal/mol under standard conditions. This value is the sum of the ΔH and $-T\Delta S$ terms.

$\Delta H = -673$ kcal/mol
$-T\Delta S = -13$ kcal/mol

(b) Energetics of glucose synthesis. The synthesis of glucose from carbon dioxide and water is exactly as endergonic as its oxidation is exergonic. The ΔG value of $+686$ kcal/mol under standard conditions is the sum of the ΔH and $-T\Delta S$ terms.

$\Delta H = +673$ kcal/mol
$-T\Delta S = +13$ kcal/mol

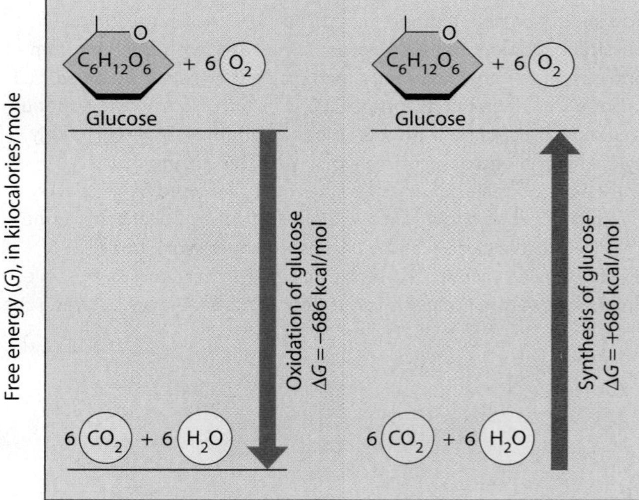

FIGURE 5-10 Changes in Free Energy for the Oxidation and Synthesis of Glucose The exergonic oxidation of glucose shown in **(a)** has a large negative ΔG that is exactly equal in magnitude but opposite in sign to the large positive ΔG for the endergonic synthesis of glucose shown in **(b)**.

which makes it a highly endergonic reaction (Figure 5-10b). Phototrophs must therefore use large amounts of energy to drive this reaction in the direction of glucose synthesis—and that, of course, is where the energy of the sun comes in, as we'll see when we get to photosynthesis in Chapter 11.

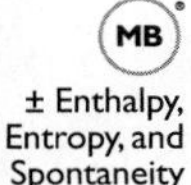

± Enthalpy, Entropy, and Spontaneity

The Meaning of Spontaneity. Before considering how we can actually calculate ΔG and use it as a measure of thermodynamic spontaneity, we need to look more closely at what is—and what is not—meant by the term *spontaneous.* As we noted earlier, spontaneity tells us only that a reaction *can* go; it says nothing at all about whether it *will* go. A reaction can have a negative ΔG value and yet not actually proceed to any measurable extent. The cellulose of paper obviously burns spontaneously once ignited, consistent with a highly negative ΔG value of -686 kcal/mol of glucose units. Yet in the absence of a match, paper is reasonably stable and might require hundreds of years to oxidize.

Thus, ΔG can really tell us only whether a reaction or process is thermodynamically feasible—whether it has the *potential* for occurring. Whether an exergonic reaction will in fact proceed depends not only on its favorable (negative) ΔG but also on the availability of a mechanism or pathway to get from the initial state to the final state. Usually, an initial input of activation energy is required as well, such as the heat energy from the match that was used to ignite the piece of paper.

Thermodynamic spontaneity is therefore a necessary but insufficient criterion for determining whether a reaction will actually occur. In Chapter 6, we will explore the subject of reaction rates in the context of enzyme-catalyzed reactions. For the moment, we need only note that when we designate a reaction as thermodynamically spontaneous, we simply mean it is an energetically feasible event that will liberate free energy if and when it actually takes place.

Understanding ΔG

Our final task in this chapter will be to understand how ΔG is calculated and how it can then be used to assess the thermodynamic feasibility of reactions under specified conditions. For that, we come back to the reaction that converts glucose-6-phosphate into fructose-6-phosphate (Reaction 5-9) and ask what we can learn about the spontaneity of the conversion in the direction written (from left to right). Experience and familiarity provide no clues here, nor is it obvious how the entropy of the universe would be affected if the reaction were to proceed. Clearly, we need to be able to calculate ΔG and to determine whether it is positive or negative under the particular conditions we specify for the reaction. It will be helpful to refer to **Table 5-1** periodically during the following discussion.

The Equilibrium Constant Is a Measure of Directionality

To assess whether a reaction can proceed in a given direction under specified conditions, we must understand the **equilibrium constant K_{eq},** which is the ratio of product concentrations to reactant concentrations at *equilibrium.* When a reversible reaction is at equilibrium, this means there is no net change in the concentrations of either products or reactants with time. For the general reaction in which A is converted reversibly into B, the equilibrium constant is simply the ratio of the equilibrium concentrations of A and B:

$$A \rightleftharpoons B \qquad \textbf{(5-14)}$$

$$K_{eq} = \frac{[B]_{eq}}{[A]_{eq}} \qquad \textbf{(5-15)}$$

where $[A]_{eq}$ and $[B]_{eq}$ are the concentrations of A and B, in moles per liter, when Reaction 5-14 is at equilibrium at

Table 5-1 The Meaning of $\Delta G^{\circ\prime}$ and $\Delta G'$

The Meaning of $\Delta G^{\circ\prime}$		
$\Delta G^{\circ\prime}$ Negative (K'_{eq} > 1.0)	**$\Delta G^{\circ\prime}$ Positive (K'_{eq} < 1.0)**	**$\Delta G^{\circ\prime}$ = 0 (K'_{eq} = 1.0)**
Products predominate over reactants at equilibrium at standard temperature, pressure, and pH.	Reactants predominate over products at equilibrium at standard temperature, pressure, and pH.	Products and reactants are present equally at equilibrium at standard temperature, pressure, and pH.
Reaction goes spontaneously to the right under standard conditions.	Reaction goes spontaneously to the left under standard conditions.	Reaction is at equilibrium under standard conditions.
The Meaning of $\Delta G'$		
$\Delta G^{\circ\prime}$ Negative	**$\Delta G^{\circ\prime}$ Positive**	**$\Delta G^{\circ\prime}$ = 0**
Reaction is thermodynamically feasible as written under the conditions for which $\Delta G'$ was calculated.	Reaction is not feasible as written under the conditions for which $\Delta G'$ was calculated.	Reaction is at equilibrium under the conditions for which $\Delta G'$ was calculated.
Work can be done by the reaction under the conditions for which $\Delta G'$ was calculated.	Energy must be supplied to drive the reaction under the conditions for which $\Delta G'$ was calculated.	No work can be done nor is energy required by the reaction under the conditions for which $\Delta G'$ was calculated.

25°C. Given the equilibrium constant for a reaction, you can easily tell whether a specific mixture of products and reactants is at equilibrium. If it is not, it is easy to tell how far away the reaction is from equilibrium and the direction it must proceed to reach equilibrium.

For example, the equilibrium constant for Reaction 5-9 at 25°C is known to be 0.5. This means that, at equilibrium, there will be one-half as much fructose-6-phosphate as glucose-6-phosphate, regardless of the actual magnitudes of the concentrations:

$$K_{eq} = \frac{[\text{fructose-6-phosphate}]_{eq}}{[\text{glucose-6-phosphate}]_{eq}} = 0.5 \quad \textbf{(5-16)}$$

If the two compounds are present in any other concentration ratio, the reaction will not be at equilibrium and will move toward equilibrium. Thus, a concentration ratio *less* than K_{eq} means that there is too little fructose-6-phosphate present, and the reaction will tend to proceed to the right to generate more fructose-6-phosphate at the expense of glucose-6-phosphate. Conversely, a concentration ratio *greater* than K_{eq} indicates that the relative concentration of fructose-6-phosphate is too high, and the reaction will tend to proceed to the left.

Figure 5-11 illustrates this concept for the interconversion of A and B (Reaction 5-14), showing the relationship between the free energy of the reaction and how far the concentrations of A and B are from equilibrium. (Notice that K_{eq} is assumed to be 1.0 in this illustration; for other values of K_{eq}, the curve would be the same shape but still centered over K_{eq}.) The point of Figure 5-11 is clear: The free energy is lowest at equilibrium and increases as the system is displaced from equilibrium in either direction. Moreover, if we know how the ratio of prevailing concentrations compares with the equilibrium concentration ratio, we can predict in which direction a reaction will tend to proceed *and* how much free energy will be released as it does so. Thus, *the tendency toward equilibrium provides the driving force for every chemical reaction,* and a comparison of prevailing and equilibrium concentration ratios provides one measure of that tendency.

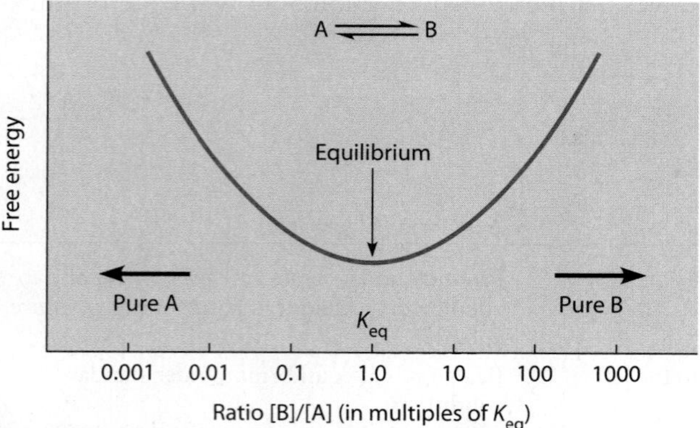

FIGURE 5-11 Free Energy and Chemical Equilibrium. The amount of free energy available from a chemical reaction depends on how far the components are from equilibrium. This principle is illustrated here for a reaction that interconverts A and B and has an equilibrium constant, K_{eq}, of 1.0. The free energy of the system increases as the [B]/[A] ratio changes on either side of the equilibrium point. For a reaction with a K_{eq} value other than 1.0, the graph would have the same shape but would be centered over the K_{eq} value for that reaction.

ΔG Can Be Calculated Readily

It should come as no surprise that ΔG is really just a means of calculating how far from equilibrium a reaction lies under specified conditions and how much energy will be released as the reaction proceeds toward equilibrium. Nor should it be surprising that both the equilibrium constant and the prevailing concentrations of reactants (A) and products (B) are needed to calculate ΔG. For Reaction 5-14, the equation relating these variables is

$$\begin{aligned}\Delta G &= RT\ln\frac{[\text{B}]_{pr}}{[\text{A}]_{pr}} - RT\ln\frac{[\text{B}]_{eq}}{[\text{A}]_{eq}} \\ &= RT\ln\frac{[\text{B}]_{pr}}{[\text{A}]_{pr}} - RT\ln K_{eq} \\ &= -RT\ln K_{eq} + RT\ln\frac{[\text{B}]_{pr}}{[\text{A}]_{pr}} \quad \textbf{(5-17)}\end{aligned}$$

where ΔG is the free energy change, in cal/mol, under the specified conditions; R is the gas constant (1.987 cal/mol-K); T is the temperature in kelvins (use 25°C = 298 K unless otherwise specified); $[\text{A}]_{pr}$ and $[\text{B}]_{pr}$ are the prevailing concentrations of A and B in moles per liter; $[\text{A}]_{eq}$ and $[\text{B}]_{eq}$ are the equilibrium concentrations of A and B in moles per liter; K_{eq} is the equilibrium constant at the standard temperature of 298 K (25°C); and ln stands for the natural logarithm of (i.e., the logarithm of a quantity to the base of the natural logarithm system, *e*, which equals approximately 2.718). Natural logarithms are used because they describe processes in which the rate of change of the process is directly related to the quantity of material undergoing the change (e.g., as in describing radioactive decay).

More generally, for a reaction in which *a* molecules of reactant A combine with *b* molecules of reactant B to form *c* molecules of product C plus *d* molecules of product D,

$$a\text{A} + b\text{B} \rightleftharpoons c\text{C} + d\text{D} \quad \textbf{(5-18)}$$

ΔG is calculated as

$$\Delta G = -RT\ln K_{eq} + RT\ln\frac{[\text{C}]^{c}_{pr}[\text{D}]^{d}_{pr}}{[\text{A}]^{a}_{pr}[\text{B}]^{b}_{pr}} \quad \textbf{(5-19)}$$

where all the constants and variables are as previously defined, and K_{eq} is the equilibrium constant for Reaction 5-18.

Returning to Reaction 5-9, assume that the prevailing concentrations of glucose-6-phosphate and fructose-6-phosphate in a cell are 10 μM (10×10^{-6} M) and 1 μM (1×10^{-6} M), respectively, at 25°C. Since the ratio of prevailing product concentrations to reactant concentrations is 0.1 and the equilibrium constant is 0.5, there is clearly too little fructose-6-phosphate present relative to glucose-6-phosphate for the reaction to be at equilibrium. The reaction should therefore tend toward the right in the direction of fructose-6-phosphate generation. In other words, the reaction is thermodynamically favorable in the direction written. This, in turn, means that ΔG must be negative under these conditions.

The actual value for ΔG is calculated as follows:

$$\begin{aligned}\Delta G &= -(1.987 \text{ cal/mol-K})(298 \text{ K}) \ln(0.5) \\ &\quad + (1.987 \text{ cal/mol-K})(298 \text{ K}) \ln \frac{1 \times 10^{-6}\, M}{10 \times 10^{-6} M} \\ &= -(592 \text{ cal/mol}) \ln(0.5) + (592 \text{ cal/mol}) \ln(0.1) \\ &= -(592 \text{ cal/mol})(-0.693) + (592 \text{ cal/mol})(-2.303) \\ &= +410 \text{ cal/mol} - 1364 \text{ cal/mol} \\ &= -954 \text{ cal/mol} \qquad \textbf{(5-20)}\end{aligned}$$

Notice that our expectation of a negative ΔG is confirmed, and we now know exactly how much free energy is liberated upon the spontaneous conversion of 1 mole of glucose-6-phosphate into 1 mole of fructose-6-phosphate under the specified conditions. The free energy liberated in this or some other exergonic reaction can be either "harnessed" to do work, stored in the chemical bonds of ATP, or released as heat.

It is important to understand exactly what this calculated value for ΔG means and under what conditions it is valid. Because it is a thermodynamic parameter, ΔG can tell us whether a reaction is thermodynamically possible as written, but it says nothing about the rate or the mechanism of the reaction. It simply says that if the reaction does occur, it will proceed to the right and will liberate 954 calories (0.954 kcal) of free energy for every mole of glucose-6-phosphate that is converted to fructose-6-phosphate, *provided* that the concentrations of both the reactant and the product are *maintained at the initial values* (10 and 1 μM, respectively) throughout the course of the reaction.

In upcoming chapters, we will see examples of how some of the energy provided by exergonic reactions can be harnessed by the use of *coupled reactions.* In coupled reactions, an endergonic reaction happens simultaneously with an exergonic reaction, and the free energy of the exergonic reaction is used to permit the endergonic reaction to occur. For example, although the synthesis of ATP is endergonic, often this reaction is coupled to an exergonic oxidative reaction such that the overall sum of the two reactions has a negative ΔG and both reactions can proceed spontaneously. Rather than being released as heat, some of the energy of the oxidation has been conserved in the chemical bonds of the high-energy ATP molecule.

More generally, ΔG *is a measure of thermodynamic spontaneity for a reaction in the direction in which it is written (from left to right), at the specified concentrations of reactants and products.* In a beaker or test tube, this requirement for constant reactant and product concentrations means that reactants must be added continuously and products must be removed continuously. In the cell, each reaction is part of a metabolic pathway, and its reactants and products are maintained at fairly constant, nonequilibrium concentrations by the reactions that precede and follow it in the sequence.

The Standard Free Energy Change Is ΔG Measured Under Standard Conditions

Because it is a thermodynamic parameter, ΔG is independent of the actual mechanism or pathway of a reaction, but it depends crucially on the conditions under which the reaction occurs. A reaction characterized by a large decrease in free energy under one set of conditions may have a much smaller (but still negative) ΔG or may even have a positive ΔG under a different set of conditions. The melting of ice, for example, depends on temperature; it proceeds spontaneously above 0°C but goes in the opposite direction (freezing) below that temperature. It is therefore important to identify the conditions under which a given measurement of ΔG is made.

By convention, biochemists have agreed on certain arbitrary conditions to define the **standard state** of a system for convenience in reporting, comparing, and tabulating free energy changes in chemical reactions. For systems consisting of dilute aqueous solutions, these are usually a standard temperature of 25°C (298 K), a pressure of 1 atmosphere, and all products and reactants present at a concentration of 1 M.

The only common exception to this standard concentration rule is water. The concentration of water in a dilute aqueous solution is approximately 55.5 M and does not change significantly during the course of reactions, even when water is itself a reactant or product. By convention, biochemists do not include the concentration of water in calculations of free energy changes, even though the reaction may indicate a net consumption or production of water.

In addition to standard conditions of temperature, pressure, and concentration, biochemists also frequently specify a standard pH of 7.0 because most biological reactions occur at or near neutrality. The concentration of hydrogen ions (and of hydroxyl ions) is therefore 10^{-7} M, so the standard concentration of 1.0 M does not apply to H^+ or OH^- ions when a pH of 7.0 is specified. Values of K_{eq}, ΔG, or other thermodynamic parameters determined or calculated at pH 7.0 are always written with a prime (as K'_{eq}, $\Delta G'$, and so on) to indicate this exception to standard conditions.

Energy changes for reactions are usually reported in standardized form as the change that would occur if the reactions were run under the standard conditions. More precisely, the standard change in any thermodynamic parameter refers to the conversion of a mole of a specified reactant to products or the formation of a mole of a specified product from the reactants under conditions where the temperature, pressure, pH, and concentrations of all relevant species are maintained at their standard values.

The free energy change calculated under these conditions is called the **standard free energy change,** designated $\Delta G^{\circ\prime}$, where the superscript (°) refers to standard conditions of temperature, pressure, and concentration, and the prime (′) emphasizes that the standard hydrogen ion concentration for biochemists is 10^{-7} *M*, not 1.0 *M*.

It turns out that $\Delta G^{\circ\prime}$ bears a simple linear relationship to the natural logarithm of the equilibrium constant K'_{eq} (**Figure 5-12**). This relationship can readily be seen by rewriting Equation 5-19 with primes and then assuming standard concentrations for all reactants and products. All concentration terms are now 1.0 and the natural logarithm of 1.0 is zero, so the second term in the general expression for $\Delta G^{\circ\prime}$ is eliminated, and what remains is an equation for $\Delta G^{\circ\prime}$, the free energy change under standard conditions:

$$\begin{aligned}\Delta G^{\circ\prime} &= -RT \ln K'_{eq} + RT \ln 1 \\ &= -RT \ln K'_{eq}\end{aligned} \qquad \textbf{(5-21)}$$

In other words, $\Delta G^{\circ\prime}$ can be calculated directly from the equilibrium constant, provided that the latter has also been determined under the same standard conditions of temperature, pressure, and pH. This, in turn, allows Equation 5-19 to be simplified as

$$\Delta G' = \Delta G^{\circ\prime} + RT \ln \frac{[C]^c_{pr}\,[D]^d_{pr}}{[A]^a_{pr}\,[B]^b_{pr}} \qquad \textbf{(5-22)}$$

At the standard temperature of 25°C (298 K), the term *RT* becomes (1.987)(298) = 592 cal/mol, so Equations 5-21 and 5-22 can be rewritten as follows, in what are the most useful formulas for our purposes:

$$\Delta G^{\circ\prime} = -592 \ln K'_{eq} \qquad \textbf{(5-23)}$$

$$\Delta G' = \Delta G^{\circ\prime} + 592 \ln \frac{[C]^c_{pr}\,[D]^d_{pr}}{[A]^a_{pr}\,[B]^b_{pr}} \qquad \textbf{(5-24)}$$

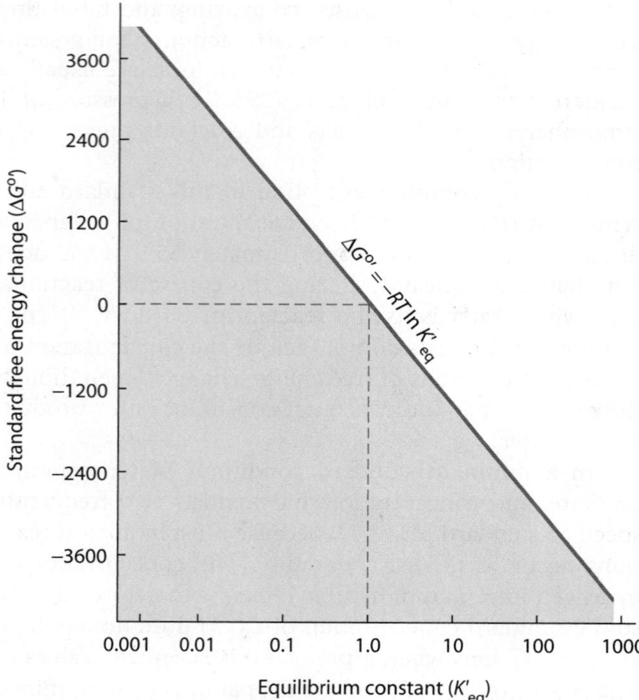

FIGURE 5-12 The Relationship Between $\Delta G^{\circ\prime}$ and K'_{eq}. The standard free energy change and the equilibrium constant are related by the equation $\Delta G^{\circ\prime} = -RT \ln K'_{eq}$. Note that if the equilibrium constant is 1.0, the standard free energy change is zero.

Summing Up: The Meaning of $\Delta G'$ and $\Delta G^{\circ\prime}$

Equations 5-23 and 5-24 represent the most important contribution of thermodynamics to biochemistry and cell biology—a means of assessing the feasibility of a chemical reaction based on the prevailing concentrations of products and reactants and a knowledge of the equilibrium constant. Equation 5-23 expresses the relationship between the standard free energy change $\Delta G^{\circ\prime}$ and the equilibrium constant K'_{eq} and enables us to calculate the free energy change that would be associated with any reaction of interest if all reactants and products were maintained at a standard concentration of 1.0 *M*.

If K'_{eq} is greater than 1.0, then $\ln K'_{eq}$ will be positive and $\Delta G^{\circ\prime}$ will be negative, and the reaction can proceed to the right under standard conditions. This makes sense because if K'_{eq} is greater than 1.0, products will predominate over reactants at equilibrium. A predominance of products can be achieved from the standard state only by the conversion of reactants to products, so the reaction will tend to proceed spontaneously to the right. Conversely, if K'_{eq} is less than 1.0, then $\Delta G^{\circ\prime}$ will be positive and the reaction cannot proceed to the right. Instead, it will tend toward the left because $\Delta G^{\circ\prime}$ for the reverse reaction will have the same absolute value but will be opposite in sign. This is in keeping with the small value for K'_{eq}, which specifies that reactants are favored over products (that is, the equilibrium lies to the left).

The $\Delta G^{\circ\prime}$ values are convenient both because they can easily be determined from the equilibrium constant and because they provide a uniform convention for reporting free energy changes. But bear in mind that a $\Delta G^{\circ\prime}$ value is an arbitrary standard in that it refers to an arbitrary state specifying conditions of concentration that cannot be achieved with most biologically important compounds. $\Delta G^{\circ\prime}$ is therefore useful for standardized reporting, but *it is not a valid measure of the thermodynamic spontaneity of reactions as they occur under real conditions.*

For real-life situations in cell biology, we will use $\Delta G'$, which provides a direct measure of how far from equilibrium a reaction is at the concentrations of reactants and products that actually prevail in the cell (Equation 5-24). Therefore, *$\Delta G'$ is the most useful measure of thermodynamic spontaneity.* If it is negative, the reaction in question is thermodynamically spontaneous and can proceed as written under the conditions for which the calculations were made. Its magnitude serves as a measure of how much free energy will be liberated as the reaction occurs under the specified conditions. This, in turn, determines the maximum amount of work that can be performed on the surroundings, provided a mechanism is available to conserve and use the energy as it is liberated.

A positive $\Delta G'$, on the other hand, indicates that the reaction cannot occur in the direction written under the conditions for which the calculations were made. Such reactions can sometimes be rendered spontaneous, however, by increasing the concentrations of reactants or by decreasing the concentrations of products. Throughout the text, you will often see reactions that are unfavorable under standard conditions but are able to proceed because the products are rapidly removed by a subsequent reaction.

For the special case where $\Delta G' = 0$, the reaction is clearly at equilibrium, and no net energy change accompanies the conversion of reactant molecules into product molecules or vice versa. As we will see shortly, reactions in living cells are rarely at equilibrium. Table 5-1 summarizes the basic features of K'_{eq}, $\Delta G°'$, and $\Delta G'$.

Free Energy Change: Sample Calculations

To illustrate the calculation and utility of $\Delta G'$ and $\Delta G°'$, we return once more to the interconversion of glucose-6-phosphate and fructose-6-phosphate (Reaction 5-9). We already know that the equilibrium constant for this reaction under standard conditions of temperature, pH, and pressure is 0.5 (Equation 5-16). This means that if the enzyme that catalyzes this reaction in cells is added to a solution of glucose-6-phosphate at 25°C, 1 atmosphere, and pH 7.0, and the solution is incubated until no further reaction occurs, fructose-6-phosphate and glucose-6-phosphate will be present in an equilibrium ratio of 0.5. (Notice that this ratio is independent of the actual starting concentration of glucose-6-phosphate and could have been achieved equally well by starting with any concentration of fructose-6-phosphate or any mixture of both, in any starting concentrations.)

The standard free energy change $\Delta G°'$ can be calculated from K'_{eq} as follows:

$$\begin{aligned}\Delta G°' &= -RT \ln K'_{eq} = -592 \ln K'_{eq}\\ &= -592 \ln 0.5 = -592(-0.693)\\ &= +410 \text{ cal/mol}\end{aligned} \quad \textbf{(5-25)}$$

The positive value for $\Delta G°'$ is therefore another way of expressing that the reactant (glucose-6-phosphate) is the predominant species at equilibrium. A positive $\Delta G°'$ value also means that under standard conditions of concentration, the reaction is nonspontaneous (thermodynamically impossible) in the direction written. In other words, if we begin with both glucose-6-phosphate and fructose-6-phosphate present at concentrations of 1.0 *M*, no net conversion of glucose-6-phosphate to fructose-6-phosphate can occur.

As a matter of fact, the reaction will proceed to the *left* under standard conditions. Fructose-6-phosphate will be converted into glucose-6-phosphate until the equilibrium ratio of 0.5 is reached. Alternatively, if both species were added or removed continuously as necessary to maintain the concentrations of both at 1.0 *M*, the reaction would proceed continuously and spontaneously to the left, with the liberation of 410 cal of free energy per mole of fructose-6-phosphate converted to glucose-6-phosphate. In the absence of any provision for conserving this energy, it would be dissipated as heat.

In a real cell, neither of these phosphorylated sugars would ever be present at a concentration even approaching 1.0 *M*. In fact, experimental values for the actual concentrations of these substances in human red blood cells are as follows:

$$[\text{glucose-6-phosphate}]\text{: } 83\ \mu M\ (83 \times 10^{-6}\ M)$$
$$[\text{fructose-6-phosphate}]\text{: } 14\ \mu M\ (14 \times 10^{-6}\ M)$$

Using these values, we can calculate the actual $\Delta G'$ for the interconversion of these sugars in red blood cells as follows:

$$\begin{aligned}\Delta G' &= \Delta G°' + 592 \ln \frac{[\text{fructose-6-phosphate}]_{pr}}{[\text{glucose-6-phosphate}]_{pr}}\\ &= +410 + 592 \ln \frac{14 \times 10^{-6}}{83 \times 10^{-6}}\\ &= +410 + 592 \ln 0.169\\ &= +410 + 592(-1.78) = +410 - 1054\\ &= -644 \text{ cal/mol}\end{aligned} \quad \textbf{(5-26)}$$

The negative value for $\Delta G'$ means that the conversion of glucose-6-phosphate into fructose-6-phosphate is thermodynamically possible under the actual conditions of concentration prevailing in red blood cells and that the reaction will yield 644 cal of free energy per mole of reactant converted to product. Thus, the conversion of reactant to product is thermodynamically impossible under standard conditions, but the red blood cell maintains these two phosphorylated sugars at concentrations adequate to offset the positive $\Delta G°'$, thereby rendering the reaction possible. This adaptation, of course, is essential if the red blood cell is to be successful in carrying out the glucose-degrading process of glycolysis of which this reaction is a part.

Life and the Steady State: Reactions That Move Toward Equilibrium Without Ever Getting There

As this chapter has emphasized, the driving force in all reactions is their tendency to move toward equilibrium. Indeed, $\Delta G^{\circ\prime}$ and $\Delta G'$ are really just convenient means of quantifying how far and in what direction from equilibrium a reaction lies under the specific conditions dictated by standard or prevailing concentrations of products and reactants. But to understand how cells really function, we must appreciate the importance of reactions that move toward equilibrium without ever achieving it. At equilibrium, the forward and backward rates are the same for a reaction, and there is therefore no net flow of matter in either direction. Most importantly, no further energy can be extracted from the reaction because $\Delta G'$ is zero for a reaction at equilibrium. Perhaps you can use Equations 5-21 and 5-22 to prove mathematically that every reaction at equilibrium has a $\Delta G' = 0$.

For all practical purposes, then, a reaction that is at equilibrium is a reaction that has stopped. But a living cell is characterized by reactions that are continuous, not stopped. A cell at equilibrium would be a dead cell. We might, in fact, define life as a continual struggle to maintain many cellular reactions in positions far from equilibrium—because at equilibrium, no net reactions are possible, no energy can be released, no work can be done, and the order of the living state cannot be maintained.

Thus, life is possible only because living cells maintain themselves in a **steady state,** with most of their reactions far from thermodynamic equilibrium. The levels of glucose-6-phosphate and fructose-6-phosphate found in red blood cells illustrate this point. As we have seen, these compounds are maintained in the cell at steady-state concentrations far from the equilibrium condition predicted by the K'_{eq} value of 0.5. In fact, the levels are so far from equilibrium concentrations that the conversion of glucose-6-phosphate to fructose-6-phosphate can occur continuously in the cell, even though the equilibrium state has a positive $\Delta G^{\circ\prime}$ and actually favors glucose-6-phosphate. The same is true of most reactions and pathways in the cell. They can proceed and can be harnessed to perform various kinds of cellular work because reactants, products, and intermediates are maintained at steady-state concentrations far from the thermodynamic equilibrium.

This state, in turn, is possible only because a cell is an open system and receives large amounts of energy from its environment. If the cell were a closed system, all its reactions would gradually run to equilibrium. The cell would come inevitably to a state of minimum free energy, after which no further changes could occur, no work could be accomplished, and life would cease. The steady state so vital to life is possible only because the cell is able to take up energy continuously from its environment, whether in the form of light or preformed organic food molecules. This continuous uptake of energy and the accompanying flow of matter make possible the maintenance of a steady state in which all the reactants and products of cellular chemistry are kept far enough from equilibrium to ensure that the thermodynamic drive toward equilibrium can be harnessed by the cell to perform useful work, thereby maintaining and extending its activities and structural complexity.

We will focus on how this is accomplished in later chapters. In the next chapter, we will look at principles of enzyme catalysis that determine the rates of cellular reactions—that is, we will translate the *can go* of thermodynamics into the *will go* of kinetics. Then we will be ready to move on to subsequent chapters, where we will encounter functional metabolic pathways that result from a series of such reactions acting in concert.

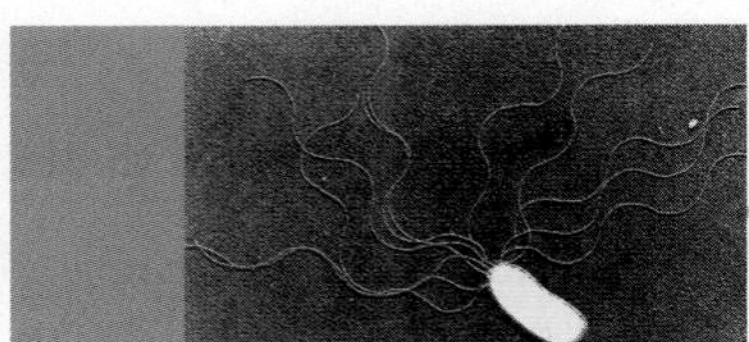

SUMMARY OF KEY POINTS

The Importance of Energy

- The complexity of cells is possible only due to the availability of energy from the environment. All cells require energy, the capacity to cause physical or chemical changes in the cells. These energy-requiring changes include biosynthesis, movement, concentration or electrical work and the generation of heat or bioluminescence.
- Phototrophs obtain energy directly from the sun and use it to reduce low-energy inorganic molecules such as carbon dioxide and nitrate to high-energy molecules such as carbohydrates, proteins, and lipids. These molecules are used both to build cellular structures and to provide energy in the absence of sunlight.
- Chemotrophs cannot harvest solar energy directly but must obtain their energy by oxidizing the high-energy molecules synthesized by phototrophs.
- There is a unidirectional flow of energy in the biosphere as energy moves from the sun to phototrophs to chemotrophs and is ultimately released into the environment as heat. Matter flows in a cyclic manner between phototrophs and chemotrophs as carbon and nitrogen atoms are alternately reduced and oxidized.

Bioenergetics

- All living cells and organisms are open systems that exchange energy with their environment. The flow of energy through these living systems is governed by the laws of thermodynamics.
- The first law specifies that energy can change form but must always be conserved. The second law provides a measure of thermodynamic spontaneity, although this means only that a reaction can occur and says nothing about whether it will actually occur or at what rate.
- Spontaneous processes are always accompanied by an *increase* in the entropy of the universe and a *decrease* in the free energy of the system. Free energy is a far more practical indicator of spontaneity because it can be calculated readily from the equilibrium constant, the prevailing concentrations of reactants and products, and the temperature.

Understanding ΔG

- The equilibrium constant K_{eq} is a measure of the directionality of a particular chemical reaction. It can be used to calculate $\Delta G^{\circ\prime}$, the free energy change under standard conditions.
- $\Delta G'$, which describes the free energy change under specified conditions, is a measure of how far from equilibrium a reaction is. It represents how much energy will be released as the reaction moves toward equilibrium.
- An exergonic reaction has a negative $\Delta G'$ and proceeds spontaneously in the direction written, while an endergonic reaction has a positive $\Delta G'$ and requires the input of energy to proceed as written.
- A negative $\Delta G'$ is a necessary prerequisite for a reaction to proceed spontaneously, but it does not guarantee that the reaction will actually occur at a reasonable rate. Even a reaction with a positive standard free energy change, $\Delta G^{\circ\prime}$, can proceed if the actual concentration of product is kept low enough to provide a negative $\Delta G'$.

Life and the Steady State: Reactions That Move Toward Equilibrium Without Ever Getting There

- Cells obtain the energy they need to carry out their activities by maintaining the many reactants and products of the various reactions at steady-state concentrations far from equilibrium. This allows the reactions to move exergonically toward equilibrium without ever actually reaching it.
- A reaction at equilibrium has a $\Delta G' = 0$, and no useful work can be done by this reaction. Therefore, a cell with all reactions at equilibrium is a dead cell.

MB **www.masteringbiology.com**

1. MasteringBiology Assignments

Tutorials and Activities

- Molecular Model Structural Tutorials for visualization of 3D structures
- BioFlix Tutorials including 3D animations with hints and detailed feedback for common wrong answers
- ± Enthalpy, Entropy, and Spontaneity

Questions

- Reading Quiz
- Multiple Choice
- End-of-chapter questions

2. eText Read your book online, search, highlight text, take notes, and more

3. The Study Area

- Chapter Review Quiz, Guide to Techniques and Methods, 3D-Structure Tutorials, Activities, BioFlix, Videos, Biology Review, Word Study Tools, Glossary, Flashcards

PROBLEM SET

More challenging problems are marked with a •.

5-1 Solar Energy. Although we sometimes hear concerns about a global energy crisis, we actually live on a planet that is flooded continuously with an extravagant amount of energy in the form of solar radiation. Every day, year in and year out, solar energy arrives at the upper surface of the Earth's atmosphere at the rate of 1.94 cal/min/cm^2 of cross-sectional area (the *solar energy constant*).

(a) Assuming the cross-sectional area of the Earth to be about 1.28×10^{18} cm^2, what is the total annual amount of incoming energy?

(b) A substantial portion of that energy, particularly in the wavelength ranges below 300 nm and above 800 nm, never reaches the Earth's surface. Can you suggest what happens to it?

(c) Of the radiation that reaches the Earth's surface, only a small proportion is actually trapped photosynthetically by phototrophs. (You can calculate the actual value in Problem 5-2.) Why do you think the efficiency of utilization is so low?

5-2 Photosynthetic Energy Transduction. The amount of energy trapped and the volume of carbon converted to organic form by photosynthetic energy transducers are mind-boggling: about 5×10^{16} g of carbon per year over the entire Earth's surface.

(a) Assuming that the average organic molecule in a cell has about the same proportion of carbon as glucose does, how many grams of organic matter are produced annually by carbon-fixing phototrophs?

(b) Assuming that all the organic matter in part a is glucose (or any molecule with an energy content equivalent to that of glucose), how much energy is represented by that quantity of organic matter? Assume that glucose has a free energy content (free energy of combustion) of 3.8 kcal/g.

(c) What proportion of the net annual phototrophic production of organic matter calculated in part a do you think is consumed by chemotrophs each year?

(d) Referring to the answer for Problem 5-1a, what is the average efficiency with which the radiant energy incident on the upper atmosphere is trapped photosynthetically on the Earth's surface?

5-3 Energy Conversion. Most cellular activities involve converting energy from one form to another. For each of the following cases, give a biological example and explain the significance of the conversion.

(a) Chemical energy into mechanical energy

(b) Chemical energy into light energy

(c) Chemical energy into electrical energy

(d) Chemical energy into the potential energy of a concentration gradient

(e) Solar (light) energy into chemical energy

5-4 Enthalpy, Entropy, and Free Energy. The oxidation of glucose to carbon dioxide and water is represented by the following reaction, whether the oxidation occurs by combustion in the laboratory or by biological oxidation in living cells:

$$C_6H_{12}O_6 + 6O_2 \rightleftharpoons 6CO_2 + 6H_2O \qquad \textbf{(5-27)}$$

When combustion is carried out under controlled conditions in the laboratory, the reaction is highly exothermic, with an enthalpy change (ΔH) of -673 kcal/mol. As you know from Figure 5-10, ΔG for this reaction at 25°C is -686 kcal/mol, so the reaction is also highly exergonic.

(a) Explain in your own words what the ΔH and ΔG values mean. What do the negative signs mean in each case?

(b) What does it mean to say that the difference between the ΔG and ΔH values is due to entropy?

(c) Without doing any calculations, would you expect ΔS (entropy change) for this reaction to be positive or negative? Explain your answer.

(d) Now calculate ΔS for this reaction at 25°C. Does the calculated value agree in sign with your prediction in part c?

(e) What are the values of ΔG, ΔH, and ΔS for the reverse of the above reaction as carried out by a photosynthetic algal cell that is using CO_2 and H_2O to make $C_6H_{12}O_6$?

5-5 The Equilibrium Constant. The following reaction is one of the steps in the glycolytic pathway, which we will encounter again in Chapter 9. You should recognize it already, however, because we used it as an example earlier (Reaction 5-9):

$$\text{glucose-6-phosphate} \rightleftharpoons \text{fructose-6-phosphate} \qquad \textbf{(5-28)}$$

The equilibrium constant K_{eq} for this reaction at 25°C is 0.5.

(a) Assume that you incubate a solution containing 0.15 *M* glucose-6-phosphate (G6P) overnight at 25°C with the enzyme phosphoglucoisomerase that catalyzes Reaction 5-28. How many millimoles of fructose-6-phosphate (F6P) will you recover from 10 mL of the incubation mixture the next morning, assuming you have an appropriate chromatographic procedure to separate F6P from G6P?

(b) What answer would you get for part a if you had started with a solution containing 0.15 *M* F6P instead?

(c) What answer would you expect for part a if you had started with a solution containing 0.15 *M* G6P but forgot to add phosphoglucoisomerase to the incubation mixture?

(d) Would you be able to answer the question in part a if you had used 15°C as your incubation temperature instead of 25°C? Why or why not?

5-6 Calculating $\Delta G^{\circ\prime}$ and $\Delta G'$. Like Reaction 5-28, the conversion of 3-phosphoglycerate (3PG) to 2-phosphoglycerate (2PG) is an important cellular reaction because it is one of the steps in the glycolytic pathway (see Chapter 9):

$$\text{3-phosphoglycerate} \rightleftharpoons \text{2-phosphoglycerate} \qquad \textbf{(5-29)}$$

If the enzyme that catalyzes this reaction is added to a solution of 3PG at 25°C and pH 7.0, the equilibrium ratio between the two species will be 0.165:

$$K'_{eq} = \frac{[\text{2-phosphoglycerate}]_{eq}}{[\text{3-phosphoglycerate}]_{eq}} = 0.165 \qquad \textbf{(5-30)}$$

Experimental values for the actual steady-state concentrations of these compounds in human red blood cells are 61 μM for 3PG and 4.3 μM for 2PG.

(a) Calculate $\Delta G^{\circ\prime}$. Explain in your own words what this value means.

(b) Calculate $\Delta G'$. Explain in your own words what this value means. Why are $\Delta G'$ and $\Delta G^{\circ\prime}$ different?

(c) If conditions in the cell change such that the concentration of 3PG remains fixed at 61 μM but the concentration of 2PG begins to rise, how high can the 2PG concentration get before Reaction 5-29 will cease because it is no longer thermodynamically feasible?

5-7 Backward or Forward? The interconversion of dihydroxyacetone phosphate (DHAP) and glyceraldehyde-3-phosphate (G3P) is a part of both the glycolytic pathway (see Chapter 9) and the Calvin cycle for photosynthetic carbon fixation (see Chapter 11):

$$\underset{\text{DHAP}}{\text{dihydroxyacetone phosphate}} \rightleftharpoons \underset{\text{G3P}}{\text{glyceraldehyde-3-phosphate}} \qquad \textbf{(5-31)}$$

The value of $\Delta G^{\circ\prime}$ for this reaction is $+1.8$ kcal/mol at 25°C. In the glycolytic pathway, this reaction goes to the *right*, converting

DHAP to G3P. In the Calvin cycle, this reaction proceeds to the *left*, converting G3P to DHAP.

(a) In which direction does the equilibrium lie? What is the equilibrium constant at 25°C?

(b) In which direction does this reaction tend to proceed under standard conditions? What is $\Delta G'$ for the reaction *in that direction*?

(c) In the Calvin cycle, this reaction proceeds to the left. How high must the [G3P]/[DHAP] ratio be to ensure that the reaction is exergonic by at least −3.0 kcal/mol (at 25°C)?

(d) In the glycolytic pathway, this reaction is driven to the right because G3P is consumed by the next reaction in the sequence, thereby maintaining a low G3P concentration. What will $\Delta G'$ be (at 25°C) if the concentration of G3P is maintained at 1% of the DHAP concentration (i.e., if [G3P]/[DHAP] = 0.01)?

5-8 Succinate Oxidation. The oxidation of succinate to fumarate is an important cellular reaction because it is one of the steps in the tricarboxylic acid (TCA) cycle (see Chapter 10). The two hydrogen atoms that are removed from succinate are accepted by a coenzyme molecule called flavin adenine dinucleotide (FAD), which is thereby reduced to $FADH_2$:

```
    O⁻                    O⁻
    |                     |
    C=O                   C=O
    |                     |
  H−C−H                   C−H
    |     + FAD ⇌         ||      + FADH₂
  H−C−H                 H−C
    |                     |
    C=O                   C=O
    |                     |
    O⁻                    O⁻

 Succinate             Fumarate            (5-32)
```

$\Delta G°'$ for Reaction 5-32 is 0 cal/mol.

(a) If you start with a solution containing 0.01 *M* each of succinate and FAD and add an appropriate amount of the enzyme that catalyzes this reaction, will any fumarate be formed? If so, calculate the resulting equilibrium concentrations of all four species. If not, explain why not.

(b) Answer part a assuming that 0.01 *M* $FADH_2$ is also present initially.

(c) Assuming that the steady-state conditions in a cell are such that the $FADH_2$/FAD ratio is 5 and the fumarate concentration is 2.5 μM, what steady-state concentration of succinate would be necessary to maintain $\Delta G'$ for succinate oxidation at −1.5 kcal/mol?

5-9 Protein Folding. In Chapter 2, we learned that a polypeptide in solution usually folds spontaneously into its proper three-dimensional shape. The driving force for this folding is the tendency to achieve the most favored thermodynamic conformation. A folded polypeptide can be induced to unfold (i.e., will undergo denaturation) if the solution is heated or made acidic or alkaline. The denatured polypeptide is a random structure, with many possible conformations.

(a) What is the sign for ΔS for the folding process? What about for the unfolding (denaturation) process? How do you know?

(b) What is the sign of ΔG for the folding process? What about for the unfolding (denaturation) process? How do you know?

(c) Will the contribution of ΔS to the free energy change be positive or negative?

(d) What are the main kinds of bonds and interactions that must be broken or disrupted if a folded polypeptide is to be unfolded? Why do heat and extremes of pH cause unfolding?

5-10 Membrane Transport. As we will learn in Chapter 8, we can treat the transport of molecules across a membrane as a chemical reaction and calculate ΔG as we would for any other reaction. The general "reaction" for the transport of molecule X from the outside of the cell to the inside can be represented as

$$X_{outside} \rightarrow X_{inside} \qquad (5\text{-}33)$$

and the free energy change can be written as

$$\Delta G_{inward} = \Delta G° + RT \ln [X]_{inside}/[X]_{outside} \qquad (5\text{-}34)$$

However, the K_{eq} for the transport of an uncharged solute across a membrane is always 1, so $\Delta G°$ is always 0—and the equation for ΔG is simplified accordingly.

(a) Why is the K_{eq} for the transport of molecules across a membrane always 1? Why is $\Delta G°$ always 0?

(b) What is the simplified equation for ΔG_{inward}? What about for $\Delta G_{outward}$?

(c) Suppose that the concentration of lactose within a bacterial cell is to be maintained at 10 m*M*, whereas the external lactose concentration is only 0.2 m*M*. Do you predict that the ΔG_{inward} for lactose will be positive or negative? Explain your reasoning.

(d) Calculate the ΔG_{inward} for lactose, assuming a temperature of 25°C. Does the sign agree with your prediction in part c?

(e) One of the several sources of energy to drive lactose uptake under these conditions is ATP hydrolysis, which has a $\Delta G°'$ value of −7.3 kcal/mol. Will the hydrolysis of a single ATP molecule be adequate to drive the inward transport of one lactose molecule under these conditions?

• **5-11 Proof of Additivity.** A useful property of thermodynamic parameters such as $\Delta G'$ or $\Delta G°'$ is that they are additive for sequential reactions. Assume that K'_{AB}, K'_{BC}, and K'_{CD} are the respective equilibrium constants for reactions 1, 2, and 3 of the following sequence:

$$\underset{\text{Reaction 1}}{A \rightleftharpoons B} \underset{\text{Reaction 2}}{\rightleftharpoons C} \underset{\text{Reaction 3}}{\rightleftharpoons D} \qquad (5\text{-}35)$$

(a) Prove that the equilibrium constant K'_{AD} for the overall conversion of A to D is the *product* of the three component equilibrium constants:

$$K'_{AD} = K'_{AB} K'_{BC} K'_{CD} \qquad (5\text{-}36)$$

(b) Prove that the $\Delta G°'$ for the overall conversion of A to D is the *sum* of the three component $\Delta G°'$ values:

$$\Delta G°'_{AD} = \Delta G°'_{AB} + \Delta G°'_{BC} + \Delta G°'_{CD} \qquad (5\text{-}37)$$

(c) Prove that the $\Delta G''$ values are similarly additive.

• **5-12 Utilizing Additivity.** The additivity of thermodynamic parameters discussed in Problem 5–11 applies not just to sequential reactions in a pathway, but to *any* reactions or processes. Moreover, it applies to subtraction of reactions. Use this information to answer the following questions:

(a) The phosphorylation of glucose using inorganic phosphate (abbreviated P_i) is endergonic ($\Delta G°' = +3.3$ kcal/mol), whereas the dephosphorylation (hydrolysis) of ATP is exergonic ($\Delta G°' = -7.3$ kcal/mol):

$$\text{glucose} + P_i \rightleftharpoons \text{glucose-6-phosphate} + H_2O \quad \textbf{(5-38)}$$

$$\text{ATP} + H_2O \rightleftharpoons \text{ADP} + P_i \quad \textbf{(5-39)}$$

Write a reaction for the phosphorylation of glucose by the transfer of a phosphate group from ATP, and calculate $\Delta G°'$ for the reaction.

(b) Phosphocreatine is used by your muscle cells to store energy. The dephosphorylation of phosphocreatine (Reaction 5-40), like that of ATP, is a highly exergonic reaction with $\Delta G°' = -10.3$ kcal/mol:

$$\text{phosphocreatine} + H_2O \rightleftharpoons \text{creatine} + P_i \quad \textbf{(5-40)}$$

Write a reaction for the transfer of phosphate from phosphocreatine to ADP to generate creatine and ATP, and calculate $\Delta G°'$ for the reaction.

SUGGESTED READING

References of historical importance are marked with a •.

General References

Horton, H. R., et al. *Principles of Biochemistry,* 4th ed. Upper Saddle River, NJ: Pearson/Prentice Hall, 2006.

Mathews, C. K., K. E. van Holde, and K. G. Ahern. *Biochemistry,* 3rd ed. San Francisco: Benjamin Cummings, 2000.

Murray, R. K. *Harper's Illustrated Biochemistry,* 26th ed. New York: McGraw-Hill, 2003.

Nelson, D. L., and M. M. Cox. *Lehninger Principles of Biochemistry,* 5th ed. New York: W. H. Freeman, 2008.

Historical References

• Becker, W. M. *Energy and the Living Cell: An Introduction to Bioenergetics.* Philadelphia: Lippincott, 1977.

• Blum, H. F. *Time's Arrow and Evolution,* 3rd ed. Princeton, NJ: Princeton University Press, 1968.

• Gates, D. M. The flow of energy in the biosphere. *Sci. Amer.* 224 (September 1971): 88.

• Racker, E. From Pasteur to Mitchell: A hundred years of bioenergetics. *Fed. Proc.* 39 (1980): 210.

Bioenergetics

Anderson, G.M. *Thermodynamics of Natural Systems.* New York: Cambridge University Press, 2005.

Brown, G. C. *The Energy of Life: The Science of What Makes Our Minds and Bodies Work.* New York: Free Press, 2000.

Demetrius, L. Thermodynamics and evolution. *J. Theor. Biol.* 206 (2000): 1.

Demirel, Y., and S. I. Sandler. Thermodynamics and bioenergetics. *Biophys. Chem.* 97 (2002): 87.

Dzeja, P. P., and A. Terzic. Phosphotransfer networks and cellular energetics. *J. Exp. Biol.* 206 (2003): 2039.

Haynie, D. T. *Biological Thermodynamics,* 2nd ed. New York: Cambridge University Press, 2008.

Kurzynski, M. *The Thermodynamic Machinery of Life.* New York: Springer, 2006.

Lewalter, K., and V. Muller. Bioenergetics of archaea: Ancient energy conserving mechanisms developed in the early history of life. *Biochim. Biophys. Acta—Bioenerg.* 1757 (2006): 437.

Moreno-Sánchez R., S. Rodríguez-Enríquez, E. Saavedra, A. Marín-Hernández, and J.C. Gallardo-Pérez. The bioenergetics of cancer: Is glycolysis the main ATP supplier in all tumor cells? *Biofactors* 35 (2009): 209.

Popov, V. Development of cellular bioenergetics: From molecules to physiology and pathology. *Biochim. Biophys. Acta—Bioenerg.* 1757 (2006): 285.

Schäfer, G., and H. S. Penefsky. *Bioenergetics: Energy Conservation and Conversion.* New York: Springer, 2008.

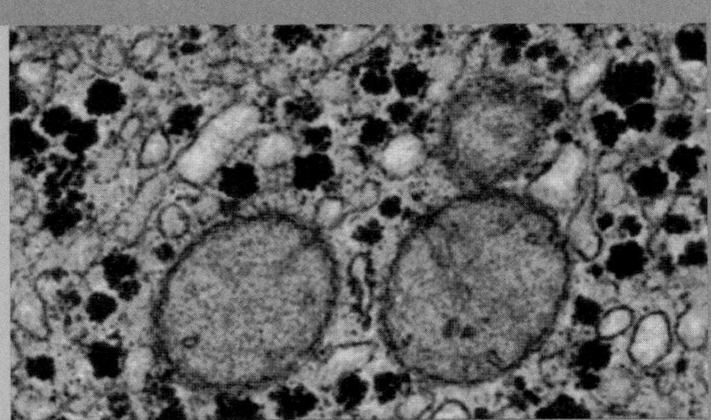

6 Enzymes: The Catalysts of Life

See MasteringBiology® for tutorials, activities, and quizzes.

In Chapter 5, we encountered $\Delta G'$, the change in free energy, and saw its importance as an indicator of thermodynamic spontaneity. Specifically, the sign *of $\Delta G'$ tells us whether a reaction is possible in the indicated direction, and the* magnitude *of $\Delta G'$ indicates how much energy will be released (or must be provided) as the reaction proceeds in that direction. At the same time, we were careful to note that, because it is a thermodynamic parameter, $\Delta G'$ tells us only whether a reaction* can *go but not about whether it actually* will *go. For that distinction, we need to know not just the direction and energetics of the reaction, but something about the reaction mechanism and its rate as well.*

This brings us to the topic of **enzyme catalysis** *because virtually all cellular reactions or processes are mediated by protein (or, in certain cases, RNA) catalysts called* enzymes. *The only reactions that occur at any appreciable rate in a cell are those for which the appropriate enzymes are present and active. Thus, enzymes almost always spell the difference between "can go" and "will go" for cellular reactions. It is only as we explore the nature of enzymes and their catalytic properties that we begin to understand how reactions that are energetically feasible actually take place in cells and how the rates of such reactions are controlled.*

In this chapter, we will first consider why thermodynamically spontaneous reactions do not usually occur at appreciable rates without a catalyst. Then we will look at the role of enzymes as specific biological catalysts. We will also see how the rate of an enzyme-catalyzed reaction is affected by the concentration of available substrate, by the affinity of the enzyme for substrate, and by covalent modification of the enzyme itself. We will also see some of the ways in which reaction rates are regulated to meet the needs of the cell.

Activation Energy and the Metastable State

If you stop to think about it, you are already familiar with many reactions that are thermodynamically feasible yet do not occur to any appreciable extent. An obvious example from Chapter 5 is the oxidation of glucose (see Reaction 5-1). This reaction (or series of reactions, really) is highly exergonic ($\Delta G^{\circ\prime} = -686$ kcal/mol) and yet does not take place on its own. In fact, glucose crystals or a glucose solution can be exposed to the oxygen in the air indefinitely, and little or no oxidation will occur. The cellulose in the paper these words are printed on is another example—and so, for that matter, are *you*, consisting as you do of a complex collection of thermodynamically unstable molecules.

Not nearly as familiar, but equally important to cellular chemistry, are the many thermodynamically feasible reactions in cells that could go but do not proceed at an appreciable rate on their own. As an example, consider the high-energy molecule adenosine triphosphate (ATP), which has a highly favorable $\Delta G^{\circ\prime}$ (−7.3 kcal/mol) for the hydrolysis of its terminal phosphate group to form the corresponding diphosphate (ADP) and inorganic phosphate (P_i):

$$ATP + H_2O \rightleftharpoons ADP + P_i \qquad \textbf{(6-1)}$$

This reaction is very exergonic under standard conditions and is even more so under the conditions that prevail in cells. Yet despite the highly favorable free energy change, this reaction occurs only slowly on its own, so that ATP remains stable for several days when dissolved in pure water. This property turns out to be shared by many biologically important molecules and reactions, and it is important to understand why.

Before a Chemical Reaction Can Occur, the Activation Energy Barrier Must Be Overcome

Molecules that could react with one another often do not because they lack sufficient energy. For every reaction, there is a specific **activation energy (E_A)**, which is the minimum amount of energy that reactants must have before collisions between them will be successful in giving rise to products. More specifically, reactants need to reach an intermediate chemical stage called the **transition state,** which has a free energy higher than that of the initial reactants. **Figure 6-1a** shows the activation energy required for molecules of ATP and H_2O to reach their transition state. $\Delta G^{\circ\prime}$ measures the difference in free energy between reactants and products (−7.3 kcal/mol for this particular reaction), whereas E_A indicates the minimum energy required for the reactants to reach the transition state and hence to be capable of giving rise to products.

The actual rate of a reaction is always proportional to the fraction of molecules that have an energy content equal to or greater than E_A. When in solution at room temperature, molecules of ATP and water move about readily, each possessing a certain amount of energy at any instant. As Figure 6-1b shows, the energy distribution among molecules will be normally distributed around a mean value (a bell-shaped curve). Some molecules will have very little energy, some will have a lot, and most will be somewhere near the average. The important point is that the only molecules that are capable of reacting at a given instant are those with enough energy to exceed the *activation energy barrier,* E_A(Figure 6-1b, dashed line).

The Metastable State Is a Result of the Activation Barrier

For most biologically important reactions at normal cellular temperatures, the activation energy is sufficiently high that the proportion of molecules possessing that much energy at any instant is extremely small. Accordingly, the rates of uncatalyzed reactions in cells are very low, and most molecules appear to be stable even though they are potential reactants in thermodynamically favorable reactions. They are, in other words, thermodynamically unstable, but they do not have enough energy to exceed the activation energy barrier.

Such seemingly stable molecules are said to be in a **metastable state.** For cells and cell biologists, high activation energies and the resulting metastable state of cellular constituents are crucial because life by its very nature is a system maintained in a steady state a long way from equilibrium. Were it not for the metastable state, all reactions would proceed quickly to equilibrium, and life as we know it would be impossible. Life, then, depends critically on the high activation energies that prevent most cellular reactions from occurring at appreciable rates in the absence of a suitable catalyst.

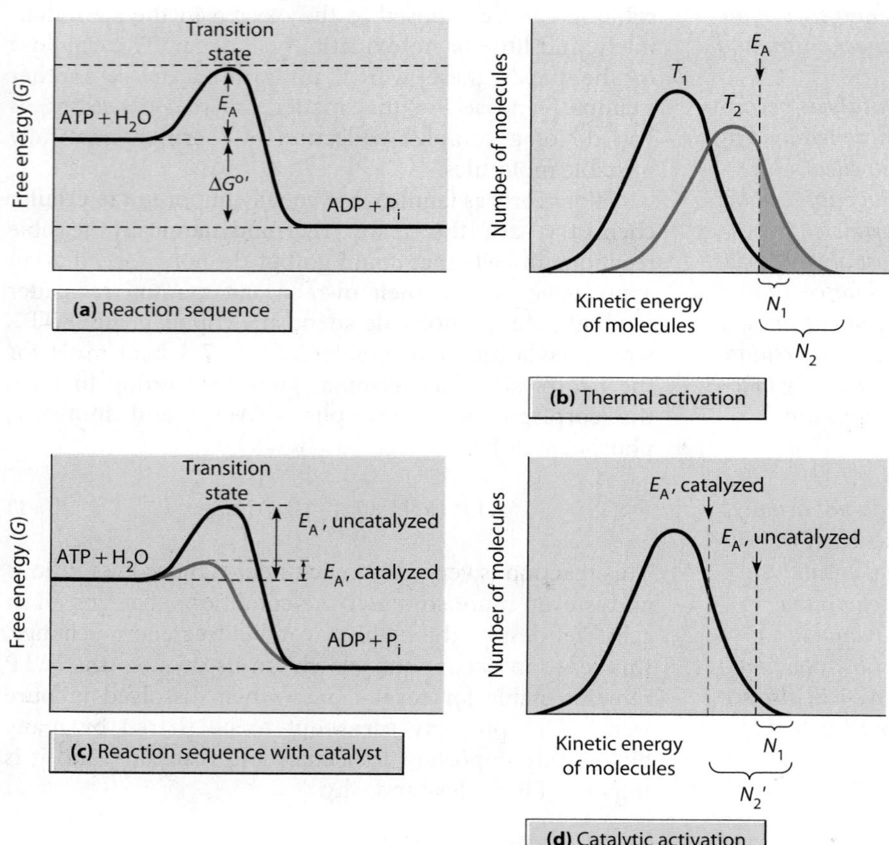

FIGURE 6-1 The Effect of Catalysis on Activation Energy and Number of Molecules Capable of Reaction. (a) The activation energy E_A is the amount of kinetic energy that reactant molecules (here, ATP and H_2O) must possess to reach the transition state leading to product formation. After reactants overcome the activation energy barrier and enter into a reaction, the products have less free energy by the amount $\Delta G^{\circ\prime}$. **(b)** The number of molecules N_1 that have sufficient energy to exceed the activation energy barrier (E_A) can be increased to N_2 by raising the temperature from T_1 to T_2. **(c)** Alternatively, the activation energy can be lowered by a catalyst (blue line), thereby **(d)** increasing the number of molecules from N_1 to N_2' with no change in temperature.

An analogy might help you to understand and appreciate the metastable state. Imagine an egg in a bowl near the edge of a table—its static position represents the metastable state. Although energy would be released if the egg hit the floor, it cannot do so because the edge of the bowl acts as a barrier. A small amount of energy must be applied to lift it up out of the bowl and over the table edge. Then, a much greater amount of energy is released as the egg spontaneously drops to the floor and breaks.

Catalysts Overcome the Activation Energy Barrier

The activation energy requirement is a barrier that must be overcome if desirable reactions are to proceed at reasonable rates. Since the energy content of a given molecule must exceed E_A before that molecule is capable of undergoing reaction, the only way a reaction involving metastable reactants will proceed at an appreciable rate is to increase the proportion of molecules with sufficient energy. This can be achieved either by increasing the average energy content of all molecules or by lowering the activation energy requirement.

One way to increase the energy content of the system is by the input of heat. As Figure 6-1b illustrates, simply increasing the temperature of the system from T_1 to T_2 will increase the kinetic energy of the average molecule, thereby ensuring a greater number of reactive molecules (N_2 instead of N_1). Thus, the hydrolysis of ATP could be facilitated by heating the solution, giving each ATP and water molecule more energy. The problem with using an elevated temperature is that such an approach is not compatible with life because biological systems require a relatively constant temperature. Cells are basically *isothermal* (constant-temperature) systems and require isothermal methods to solve the activation problem.

The alternative to an increase in temperature is to lower the activation energy requirement, thereby ensuring that a greater proportion of molecules will have sufficient energy to collide successfully and undergo reaction. This would be like changing the shape of the bowl holding the egg described earlier into a shallow dish. Now, less energy is needed to lift the egg over the edge of the dish. If the reactants can be bound on some sort of surface in an arrangement that brings potentially reactive portions of adjacent molecules into close juxtaposition, their interaction will be greatly favored and the activation energy effectively reduced.

Providing such a reactive surface is the task of a **catalyst**—an agent that enhances the rate of a reaction by lowering the energy of activation (Figure 6-1c), thereby ensuring that a higher proportion of the molecules possess sufficient energy to undergo reaction without the input of heat (Figure 6-1d). A primary feature of a catalyst is that *it is not permanently changed or consumed as the reaction proceeds.* It simply provides a suitable surface and environment to facilitate the reaction.

Recent work suggests an additional mechanism to overcome the activation energy barrier. This mechanism is known as "quantum tunneling" and sounds like something from a science fiction novel. It is based in part on the realization that matter has both particle-like and wave-like properties. In certain dehydrogenation reactions, the enzyme is believed to allow a hydrogen atom to tunnel *through* the barrier, effectively ending up on the other side without actually going over the top. Unlike most enzyme-catalyzed reactions, these tunneling reactions are temperature independent because an input of thermal energy is not required to ascend the activation energy barrier.

For a specific example of catalysis, let's consider the decomposition of hydrogen peroxide (H_2O_2) into water and oxygen:

$$2H_2O_2 \rightleftharpoons 2H_2O + O_2 \qquad \textbf{(6-2)}$$

This is a thermodynamically favorable reaction, yet hydrogen peroxide exists in a metastable state because of the high activation energy of the reaction. However, if we add a small number of ferric ions (Fe^{3+}) to a hydrogen peroxide solution, the decomposition reaction proceeds about 30,000 times faster than without the ferric ions. Clearly, Fe^{3+} is a catalyst for this reaction, lowering the activation energy (as shown in Figure 6-1c) and thereby ensuring that a significantly greater proportion (30,000-fold more) of the hydrogen peroxide molecules possess adequate energy to decompose at the existing temperature without the input of added energy.

In cells, the solution to hydrogen peroxide breakdown is not the addition of ferric ions but the enzyme *catalase,* an iron-containing protein. In the presence of catalase, the reaction proceeds about 100,000,000 times faster than the uncatalyzed reaction. Catalase contains iron atoms bound to the enzyme, thus taking advantage of inorganic catalysis within the context of a protein molecule. This combination is obviously a much more effective catalyst for hydrogen peroxide decomposition than ferric ions by themselves. The rate enhancement (catalyzed rate ÷ uncatalyzed rate) of about 10^8 for catalase is not at all an atypical value. The rate enhancements of enzyme-catalyzed reactions range from 10^7 to as high as 10^{17} compared with the uncatalyzed reaction. These values underscore the extraordinary importance of enzymes as catalysts and bring us to the main theme of this chapter.

Enzymes as Biological Catalysts

Regardless of their chemical nature, all catalysts share the following three basic properties:

1. A catalyst increases the rate of a reaction by lowering the activation energy requirement, thereby allowing a thermodynamically feasible reaction to occur at a reasonable rate in the absence of thermal activation.

2. A catalyst acts by forming transient, reversible complexes with substrate molecules, binding them in a manner that facilitates their interaction and stabilizes the intermediate transition state.

3. A catalyst changes only the *rate* at which equilibrium is achieved; it has no effect on the *position* of the equilibrium. This means that a catalyst can enhance the rate of exergonic reactions but cannot somehow change the $\Delta G'$ to allow an endergonic reaction to become spontaneous. Catalysts, in other words, are not thermodynamic wizards.

These properties are common to all catalysts, organic and inorganic alike. In terms of our example, they apply equally to ferric ions and to catalase molecules. However, biological systems rarely use inorganic catalysts. Instead, essentially all catalysis in cells is carried out by organic molecules (proteins, in most cases) called **enzymes.** Because enzymes are organic molecules, they are much more specific than inorganic catalysts, and their activities can be regulated much more carefully.

Most Enzymes Are Proteins

The capacity of cellular extracts to catalyze chemical reactions has been known since the fermentation studies of Eduard and Hans Buchner in 1897. In fact, one of the first terms for what we now call enzymes was *ferments.* However, it was not until 1926 that a specific enzyme, *urease,* was crystallized (from jack beans, by James B. Sumner) and shown to be a protein. This established the protein nature of enzymes and put to rest the belief that biochemical reactions in cells occurred via some unknown "vital force." However, since the early 1980s, biologists have recognized that in addition to proteins, certain RNA molecules, known as *ribozymes,* also have catalytic activity. Ribozymes will be discussed in a later section. Here, we will consider enzymes as proteins—which, in fact, most are.

The Active Site. One of the most important concepts to emerge from our understanding of enzymes as proteins is the **active site.** Every enzyme contains a characteristic cluster of amino acids that form the active site where the substrates bind and the catalytic event occurs. Usually, the active site is an actual groove or pocket with chemical and structural properties that accommodate the intended substrate or substrates with high specificity. The active site consists of a small number of amino acids that are not necessarily adjacent to one another along the primary sequence of the protein. Instead, they are brought together in just the right arrangement by the specific three-dimensional folding of the polypeptide chain as it assumes its characteristic tertiary structure.

Figure 6-2 shows the unfolded and folded structures of the enzyme *lysozyme,* which hydrolyzes the peptidoglycan polymer that makes up bacterial cell walls. The active site of lysozyme is a small groove in the enzyme surface into which the peptidoglycan fits. Lysozyme is a single polypeptide with 129 amino acid residues, but relatively few of these are directly involved in substrate binding and catalysis. Four of these are highlighted in Figure 6-2a. Substrate binding depends on amino acid residues from various positions along the polypeptide, including residues from positions 33–36, 46, 60–64, and 102–110. Catalysis involves two specific residues: a glutamate at position 35 (Glu-35) and an aspartic acid at position 52 (Asp-52). The acid side chain of Glu-35 donates an H^+ to the bond about to be hydrolyzed, and Asp-52 stabilizes the transition state, enhancing cleavage of this bond by an OH^- ion from water. Only as the lysozyme molecule folds to attain its stable three-dimensional conformation are these specific amino acids brought together to form the active site (Figure 6-2b).

Of the 20 different amino acids that make up proteins, only a few are actually involved in the active sites of the many proteins that have been studied. Often, these are cysteine, histidine, serine, aspartate, glutamate, and lysine. All of these residues can participate in binding the substrate to the active site during catalysis, and several also serve as donors or acceptors of protons.

Some enzymes contain specific nonprotein cofactors that are located at the active site and are indispensable for catalytic activity. These cofactors, also called **prosthetic groups,** are usually either metal ions or small organic molecules known as *coenzymes* that are derivatives of vitamins. Frequently, prosthetic groups (especially positively charged metal ions) function as electron acceptors because none of the amino acid side chains are good electron acceptors.

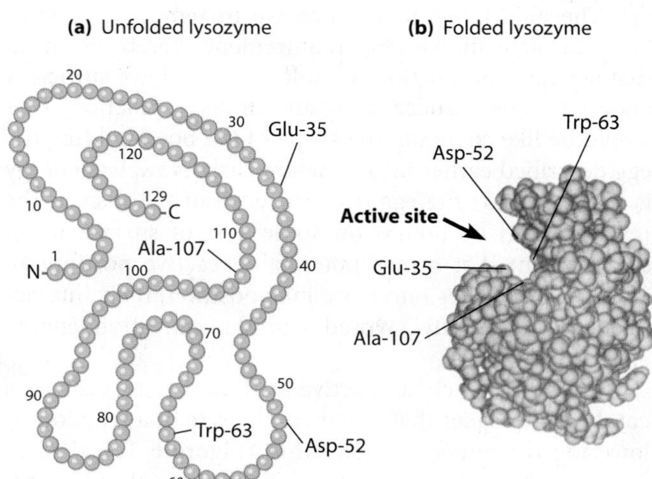

FIGURE 6-2 The Active Site of Lysozyme. **(a)** Four amino acid residues that are important for substrate binding and catalysis are far apart in the primary structure of unfolded lysozyme. **(b)** These residues are brought together to form part of the active site as lysozyme folds into its active tertiary structure.

Where present, prosthetic groups often are located at the active site and are indispensable for the catalytic activity of the enzyme. For example, each catalase enzyme molecule contains a multiring structure known as a *porphyrin* ring, to which an iron atom necessary for catalysis is bound (see Figure 10-13).

The requirement for various prosthetic groups on some enzymes explains our nutritional requirements for trace amounts of vitamins and certain metals. As we will see in Chapters 9 and 10, oxidation of glucose for energy requires two specific coenzymes that are derivatives of the vitamins niacin and riboflavin. Both niacin and riboflavin are essential nutrients in the human diet because our cells cannot synthesize them. These coenzymes, which are bound to the active site of certain enzymes, accept electrons and hydrogen ions from glucose as it is oxidized. Likewise, carboxypeptidase A, a digestive enzyme that degrades proteins, requires a single zinc atom bound to the active site, as we will see later in the chapter. Other enzymes may require atoms of iron, copper, molybdenum, or even lithium. Like enzymes, prosthetic groups are not consumed during chemical reactions, so cells require only minute, catalytic amounts of them.

Enzyme Specificity. Due to the structure of the active site, enzymes display a very high degree of **substrate specificity,** which is the ability to discriminate between very similar molecules. Specificity is one of the most characteristic properties of living systems, and enzymes are excellent examples of biological specificity.

We can illustrate their specificity by comparing enzymes with inorganic catalysts. Most inorganic catalysts are quite nonspecific in that they will act on a variety of compounds that share some general chemical feature. Consider, for example, the *hydrogenation* of (addition of hydrogen to) an unsaturated C=C bond:

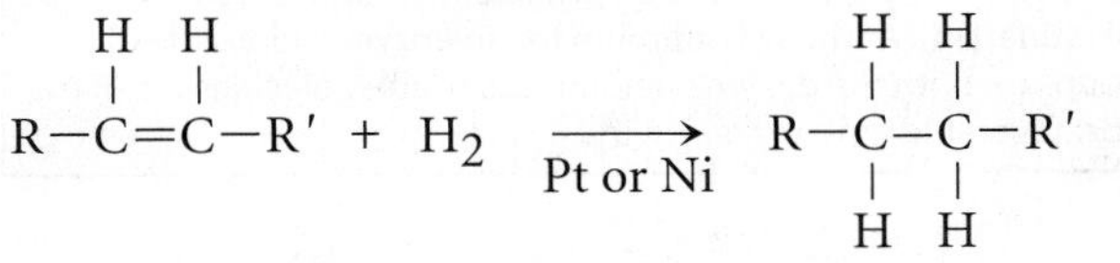

$$\mathrm{R{-}\overset{H}{\overset{|}{C}}{=}\overset{H}{\overset{|}{C}}{-}R' + H_2 \xrightarrow[\text{Pt or Ni}]{} R{-}\underset{H}{\underset{|}{\overset{H}{\overset{|}{C}}}}{-}\underset{H}{\underset{|}{\overset{H}{\overset{|}{C}}}}{-}R'} \qquad \textbf{(6-3)}$$

This reaction can be carried out in the laboratory using a platinum (Pt) or nickel (Ni) catalyst, as indicated. These inorganic catalysts are very nonspecific, however; they can catalyze the hydrogenation of a wide variety of unsaturated compounds. In practice, nickel and platinum are used commercially to hydrogenate polyunsaturated vegetable oils in the manufacture of solid cooking fats or shortenings. Regardless of the exact structure of the unsaturated compound, it can be effectively hydrogenated in the presence of nickel or platinum. This lack of specificity of inorganic catalysts during hydrogenation is responsible for the formation of certain *trans* fats (see Chapter 7) that are rare in nature.

In contrast, consider the biological example of hydrogenation as fumarate is converted to succinate, a reaction we will encounter again in Chapter 10:

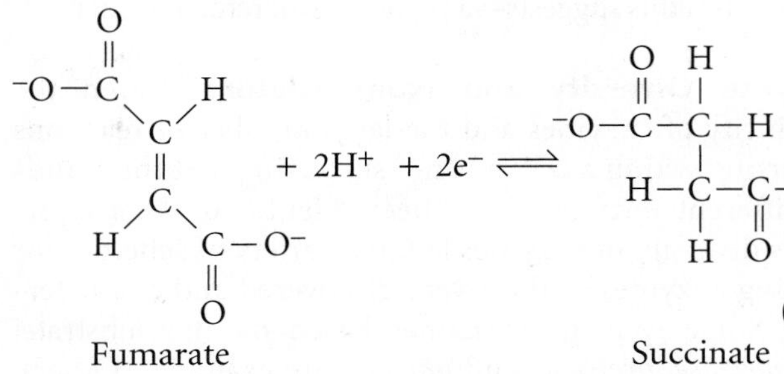

$$\underset{\text{Fumarate}}{\mathrm{{}^-O{-}\overset{O}{\overset{\|}{C}}{-}CH{=}CH{-}\underset{O}{\underset{\|}{C}}{-}O^-}} + 2H^+ + 2e^- \rightleftharpoons \underset{\text{Succinate}}{\mathrm{{}^-O{-}\overset{O}{\overset{\|}{C}}{-}CH_2{-}CH_2{-}\underset{O}{\underset{\|}{C}}{-}O^-}} \qquad \textbf{(6-4)}$$

This particular reaction is catalyzed in cells by the enzyme *succinate dehydrogenase* (so named because it normally functions in the opposite direction during energy metabolism). This dehydrogenase, like most enzymes, is highly specific. It will not add or remove hydrogen atoms from any compounds except those shown in Reaction 6-4. In fact, this particular enzyme is so specific that it will not even recognize maleate, which is an isomer of fumarate (**Figure 6-3**).

Not all enzymes are quite that specific. Some accept a number of closely related substrates, and others accept any of a whole group of substrates as long as they possess some common structural feature. Such **group specificity** is seen most often with enzymes involved in the synthesis or degradation of polymers. Since the purpose of carboxypeptidase A is to degrade dietary polypeptide chains by removing the C-terminal amino acid, it makes sense for the enzyme to accept any of a wide variety of polypeptides as substrates. It would be needlessly extravagant of the cell to require a separate enzyme for every different amino acid residue that has to be removed during polypeptide degradation.

In general, however, enzymes are highly specific with respect to substrate, such that a cell must possess almost as many different kinds of enzymes as it has reactions to catalyze. For a typical cell, this means that thousands of different enzymes are necessary to carry out its full metabolic

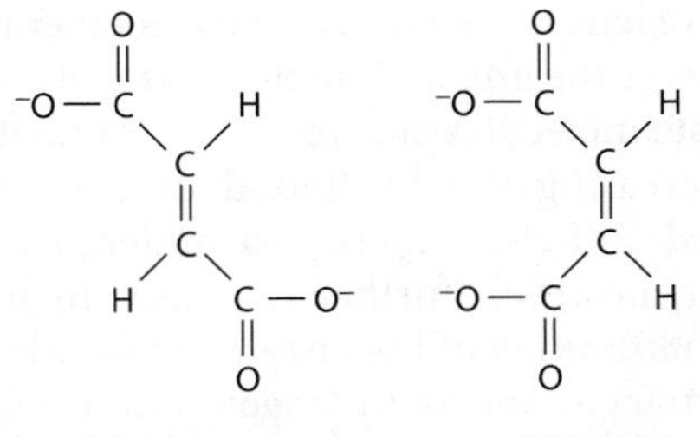

(a) Fumarate: $\mathrm{{}^-O{-}C({=}O){-}CH{=}CH{-}C({=}O){-}O^-}$ (H atoms on opposite sides of the C=C bond)

(b) Maleate: $\mathrm{{}^-O{-}C({=}O){-}CH{=}CH{-}C({=}O){-}O^-}$ (H atoms on the same side of the C=C bond)

FIGURE 6-3 Specificity in Enzyme-Catalyzed Reactions. Unlike most inorganic catalysts, enzymes can distinguish between closely related isomers. For example, the enzyme succinate dehydrogenase uses **(a)** fumarate as a substrate but not **(b)** its isomer, maleate.

program. At first, that may seem wasteful in terms of proteins to be synthesized, genetic information to be stored and read out, and enzyme molecules to have on hand in the cell. But you should also be able to see the tremendous regulatory possibilities this suggests—a point we will return to later.

Enzyme Diversity and Nomenclature. Given the specificity of enzymes and the large number of reactions occurring within a cell, it is not surprising that thousands of different enzymes have been identified. This enormous diversity of enzymes led to a variety of schemes for naming enzymes as they were discovered and characterized. Some were given names based on the substrate; *ribonuclease, protease,* and *amylase* are examples. Others, such as *succinate dehydrogenase,* were named to describe their function. Still other enzymes have names like *trypsin* and *catalase* that tell us little about either their substrates or their functions.

The resulting confusion prompted the International Union of Biochemistry to appoint an Enzyme Commission (EC) to devise a rational system for naming enzymes. Using the EC system, enzymes are divided into the following six major classes based on their general functions: *oxidoreductases, transferases, hydrolases, lyases, isomerases,* and *ligases.* The EC system assigns every known enzyme a unique four-part number based on its function—for example, EC 3.2.1.17 is the number for lysozyme. **Table 6-1** provides one representative example of each class of enzymes and the reaction it catalyzes.

Sensitivity to Temperature. Besides their specificity and diversity, enzymes are characterized by their sensitivity to temperature. This temperature dependence is not usually a practical concern for enzymes in the cells of mammals or birds because these organisms are *homeotherms,* "warm-blooded" organisms that are capable of regulating body temperature independent of the environment. However, many organisms (e.g., insects, reptiles, worms, plants, protozoa, algae, and bacteria) function at the temperature of their environment, which can vary greatly. For these organisms, the dependence of enzyme activity on temperature is significant.

At low temperatures, the rate of an enzyme-catalyzed reaction increases with temperature. This occurs because the greater kinetic energy of both enzyme and substrate molecules ensures more frequent collisions, thereby increasing the likelihood of correct substrate binding and sufficient energy to undergo reaction. At some point, however, further increases in temperature result in *denaturation* of the enzyme molecule. It loses its defined tertiary shape as hydrogen and ionic bonds are broken and the native polypeptide assumes a random, extended conformation. During denaturation, the structural integrity of the active site is destroyed, causing a loss of enzyme activity.

The temperature range over which an enzyme denatures varies from enzyme to enzyme and especially from organism to organism. **Figure 6-4a** contrasts the temperature dependence of a typical enzyme from the human body with that of a typical enzyme from a thermophilic bacterium. Not surprisingly, the reaction rate of the human enzyme is maximum at about 37°C (the *optimal temperature* for the enzyme), which is normal body temperature. The sharp decrease in activity at higher temperatures reflects the denaturation of the enzyme molecules. Most enzymes of homeotherms are inactivated by temperatures above about 50–55°C. However, some enzymes are remarkably sensitive to heat. They are denatured and inactivated at temperatures lower than this—in some cases, even by body temperatures encountered in people with high fevers (40°C). This is thought to be part of the beneficial effect of

(a) Temperature dependence. This panel shows how reaction rate varies with temperature for a typical human enzyme (black) and a typical enzyme from a thermophilic bacterium (green). The reaction rate is highest at the optimal temperature, which is about 37°C (body temperature) for the human enzyme and about 75°C (the temperature of a typical hot spring) for the bacterial enzyme. Above the optimal temperature, the enzyme is rapidly inactivated by denaturation.

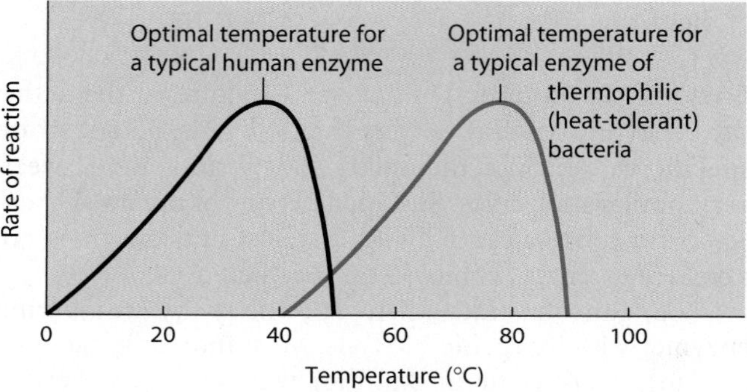

(b) pH dependence. This panel shows how reaction rate varies with pH for the gastric enzyme pepsin (black) and the intestinal enzyme trypsin (red). The reaction rate is highest at the optimal pH, which is about 2.0 for pepsin (stomach pH) and near 8.0 for trypsin (intestinal pH). At the pH optimum for an enzyme, ionizable groups on both the enzyme and the substrate molecules are in the most favorable form for reactivity.

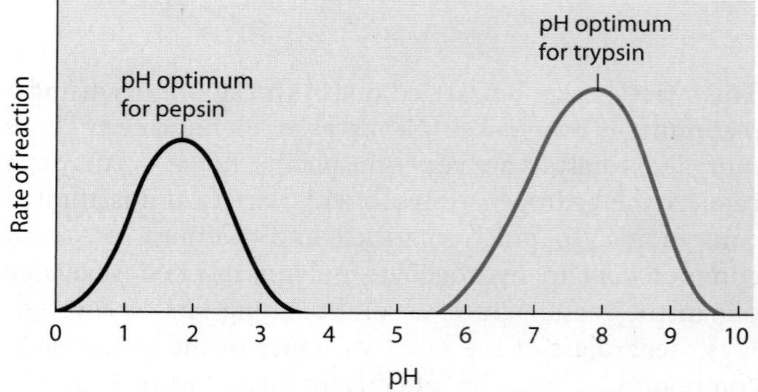

FIGURE 6-4 The Effect of Temperature and pH on the Reaction Rate of Enzyme-Catalyzed Reactions. Every enzyme has an optimum temperature and pH that usually reflect the environment where that enzyme is found in nature.

MB Chemistry Review-Acids, Bases, & pH: pH; Chemistry Review: The pH Scale

Table 6-1 The Major Classes of Enzymes with an Example of Each

Class	Reaction Type	Example: Enzyme Name	Example: Reaction Catalyzed
1. Oxidoreductases	Oxidation-reduction reactions	Alcohol dehydrogenase (oxidation with NAD^+)	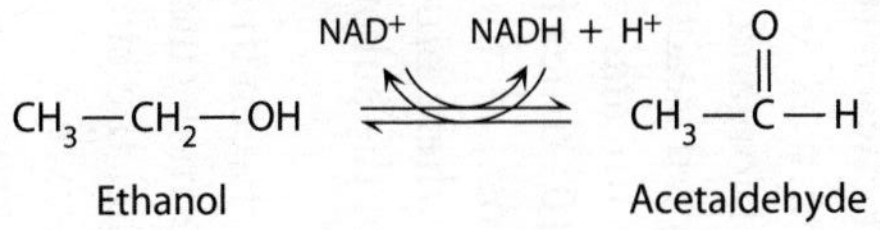 $CH_3{-}CH_2{-}OH$ (Ethanol) $\xrightleftharpoons[]{NAD^+ \; \to \; NADH + H^+}$ $CH_3{-}C(=O){-}H$ (Acetaldehyde)
2. Transferases	Transfer of functional groups from one molecule to another	Glycerokinase (phosphorylation)	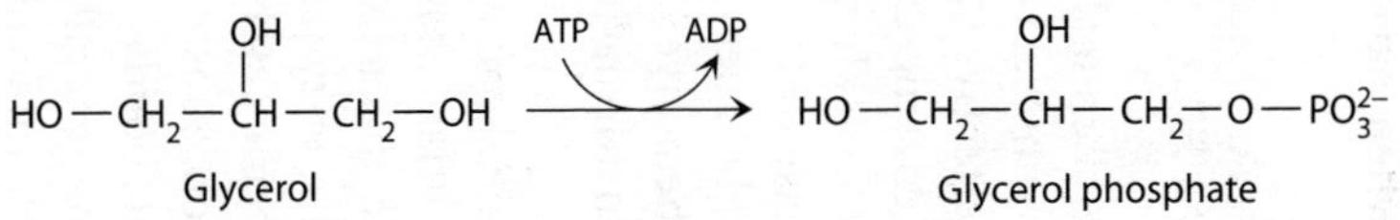 $HO{-}CH_2{-}CH(OH){-}CH_2{-}OH$ (Glycerol) $\xrightarrow{ATP \; \to \; ADP}$ $HO{-}CH_2{-}CH(OH){-}CH_2{-}O{-}PO_3^{2-}$ (Glycerol phosphate)
3. Hydrolases	Hydrolytic cleavage of one molecule into two molecules	Carboxypeptidase A (peptide bond cleavage)	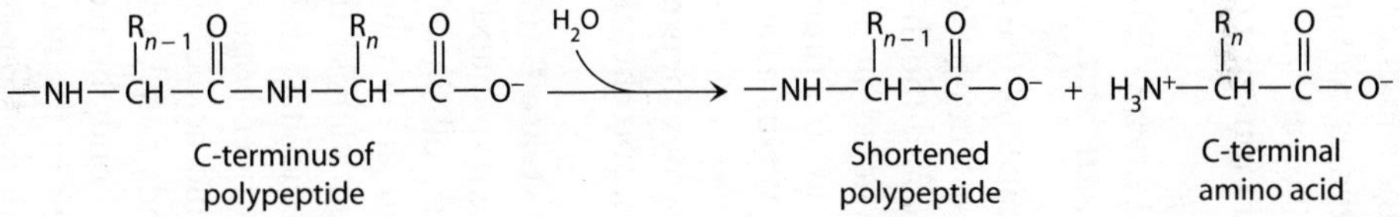 ${-}NH{-}CH(R_{n-1}){-}C(=O){-}NH{-}CH(R_n){-}C(=O){-}O^-$ (C-terminus of polypeptide) $\xrightarrow{H_2O}$ ${-}NH{-}CH(R_{n-1}){-}C(=O){-}O^-$ (Shortened polypeptide) $+\ H_3N^+{-}CH(R_n){-}C(=O){-}O^-$ (C-terminal amino acid)
4. Lyases	Removal of a group from, or addition of a group to, a molecule with rearrangement of electrons	Pyruvate decarboxylase (decarboxylation)	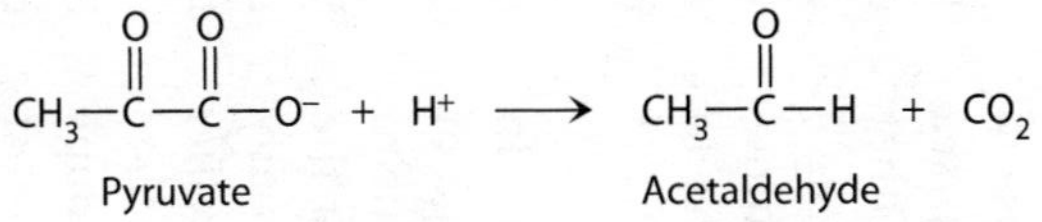 $CH_3{-}C(=O){-}C(=O){-}O^-$ (Pyruvate) $+ H^+ \longrightarrow CH_3{-}C(=O){-}H$ (Acetaldehyde) $+ CO_2$
5. Isomerases	Movement of a functional group within a molecule	Maleate isomerase (*cis-trans* isomerization)	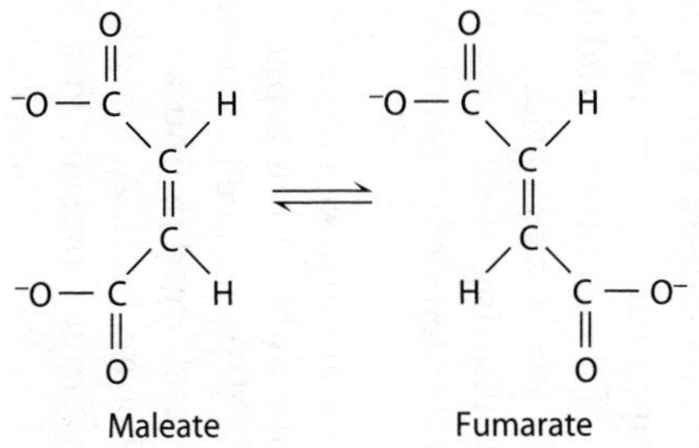 Maleate ($^-O{-}C(=O){-}CH{=}CH{-}C(=O){-}O^-$, *cis*) $\rightleftharpoons$ Fumarate ($^-O{-}C(=O){-}CH{=}CH{-}C(=O){-}O^-$, *trans*)
6. Ligases	Joining of two molecules to form a single molecule	Pyruvate carboxylase (carboxylation)	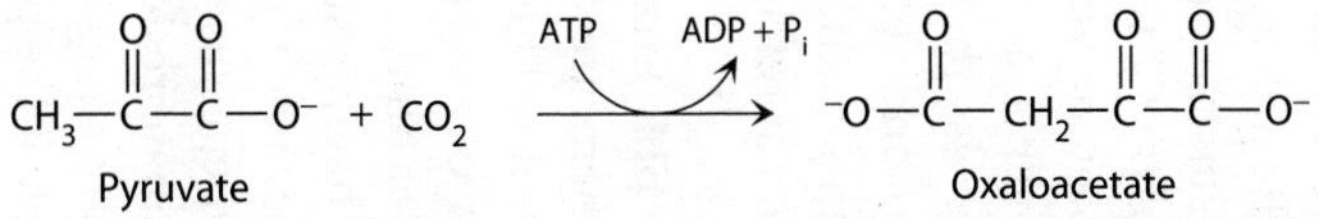 $CH_3{-}C(=O){-}C(=O){-}O^-$ (Pyruvate) $+ CO_2 \xrightarrow{ATP \; \to \; ADP + P_i}$ $^-O{-}C(=O){-}CH_2{-}C(=O){-}C(=O){-}O^-$ (Oxaloacetate)

fever when you are ill—the denaturation of heat-sensitive pathogen enzymes.

Some enzymes, however, retain activity at unusually high temperatures. The green curve in Figure 6-4a depicts the temperature dependence of an enzyme from one of the thermophilic archaea mentioned in Chapter 4. Some of these organisms thrive in acidic hot springs at temperatures as high as 80°C, with optimal temperatures close to the boiling point of water, and others live in deep-sea hydrothermal vents at temperatures over 100°C. Other enzymes, such as those of cryophilic ("cold-loving") *Listeria* bacteria and certain yeasts and molds, can function at low temperatures, allowing these organisms to grow slowly even at refrigerator temperatures (4–6°C).

Sensitivity to pH. Enzymes are also sensitive to pH. In fact, most of them are active only within a pH range of about 3–4 pH units. This pH dependence is usually due to the presence of one or more charged amino acids at the active site and/or on the substrate itself. Activity is usually dependent on having such groups present in a specific, either charged or uncharged form. For example, the active site of carboxypeptidase A includes the carboxyl groups from two glutamate residues. These carboxyl groups must be present in the charged (ionized) form, so the enzyme becomes inactive if the pH is decreased to the point where the glutamate carboxyl groups on the enzyme molecules are protonated and therefore uncharged. Extreme changes in pH also disrupt ionic and hydrogen bonds, altering tertiary structure and function.

As you might expect, the pH dependence of an enzyme usually reflects the environment in which that enzyme is normally active. Figure 6-4b shows the pH dependence of two protein-degrading enzymes found in the human digestive tract. Pepsin (black line) is present in the stomach, where the pH is usually about 2, whereas trypsin (red line) is secreted into the small intestine, which has a pH between 7 and 8. Both enzymes are active over a range of almost 4 pH units but differ greatly in their pH optima, consistent with the conditions in their respective locations within the body.

Sensitivity to Other Factors. In addition to temperature and pH, enzymes are sensitive to other factors, including molecules and ions that act as inhibitors or activators of the enzyme. For example, several enzymes involved in energy production via glucose degradation are inhibited by ATP, which inactivates them when energy is plentiful. Other enzymes in glucose breakdown are activated by adenosine monophosphate (AMP) and ADP, which act as signals that energy supplies are low and more glucose should be degraded.

Most enzymes are also sensitive to the ionic strength (concentration of dissolved ions) of the environment, which affects the hydrogen bonding and ionic interactions that help to maintain the tertiary conformation of the enzyme. Because these same interactions are often involved in the interaction between the substrate and the active site, the ionic environment may also affect binding of the substrate. Several magnesium-requiring chloroplast enzymes required for photosynthetic carbon fixation are active only in the presence of the high levels of magnesium ions that occur when leaves are illuminated.

Substrate Binding, Activation, and Catalysis Occur at the Active Site

Because of the precise chemical fit between the active site of an enzyme and its substrates, enzymes are highly specific and much more effective than inorganic catalysts. As we noted previously, enzyme-catalyzed reactions proceed 10^7 to 10^{17} times more quickly than uncatalyzed reactions do, versus a rate increase of 10^3 to 10^4 times for inorganic catalysts. As you might guess, most of the interest in enzymes focuses on the active site, where binding, activation, and chemical transformation of the substrate occur.

Substrate Binding. Initial contact between the active site of an enzyme and a potential substrate molecule depends on their collision. Once in the active site, the substrate molecules are bound to the enzyme surface in just the right orientation so that specific catalytic groups on the enzyme can facilitate the reaction. Substrate binding usually involves hydrogen bonds or ionic bonds (or both) to charged or polar amino acids. These are generally weak bonds, but several bonds may hold a single molecule in place. The strength of the bonds between an enzyme and a substrate molecule is often in the range of 3–12 kcal/mol. This is less than one-tenth the strength of a single covalent bond (see Figure 2-2). Substrate binding is therefore readily reversible.

For many years, enzymologists regarded an enzyme as a rigid structure, with a specific substrate fitting into the active site like a key fits into a lock. This *lock-and-key model,* first suggested in 1894 by the German biochemist Emil Fischer, explained enzyme specificity but did little to enhance our understanding of the catalytic event. A more refined view of the enzyme-substrate interaction is provided by the **induced-fit model,** first proposed in 1958 by Daniel Koshland. According to this model, substrate binding at the active site distorts both the enzyme and the substrate, thereby stabilizing the substrate molecules in their transition state and rendering certain substrate bonds more susceptible to catalytic attack. In the case of lysozyme, substrate binding induces a conformational change in the enzyme that distorts the peptidoglycan substrate and weakens the bond about to be broken in the reaction.

As shown in **Figure 6-5**, induced fit involves a conformational change in the shape of the enzyme molecule following substrate binding. This alters the configuration of the active site and positions the proper reactive groups of the enzyme optimally for the catalytic reaction. Evidence of such conformational changes upon binding of substrate has come from X-ray diffraction studies of crystallized proteins and nuclear magnetic resonance (NMR) studies of proteins in solution, which can determine the shape of an enzyme molecule with and without bound substrate. Figure 6-5 illustrates the conformational change that

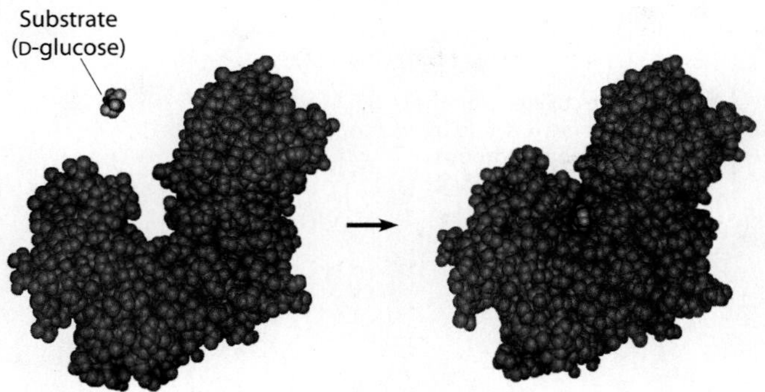

FIGURE 6-5 The Conformational Change in Enzyme Structure Induced by Substrate Binding. This figure shows a space-filling model for the enzyme hexokinase along with its substrate, a molecule of D-glucose. Substrate binding induces a conformational change in hexokinase, known as induced fit, that improves the catalytic activity of the enzyme.

occurs upon substrate binding to hexokinase, which adds a phosphate group to D-glucose. As glucose binds to the active site, the two domains of hexokinase fold toward each other, closing the binding site cleft about the substrate to facilitate catalysis.

Often, the induced conformational change brings critical amino acid side chains into the active site even if they are not nearby in the absence of substrate. In the active site of carboxypeptidase A (**Figure 6-6**), a zinc ion is tightly bound to three residues of the enzyme (Glu-72, His-69, and His-196) and also loosely binds a water molecule (not shown). Substrate binding to the zinc ion replaces the bound water molecule and induces a conformational change in the enzyme that brings other amino acid side chains into the active site, including Arg-145, Tyr-248, and Glu-270. These amino acid residues are then in position to participate in catalysis.

Substrate Activation. The role of the active site is not just to recognize and bind the appropriate substrate but also to *activate* it by subjecting it to the right chemical environment for catalysis. A given enzyme-catalyzed reaction may involve one or more means of **substrate activation.** Three of the most common mechanisms are as follows:

1. *Bond distortion.* The change in enzyme conformation induced by initial substrate binding to the active site not only causes better complementarity and a tighter enzyme-substrate fit but also distorts one or more of its bonds, thereby weakening the bond and making it more susceptible to catalytic attack.
2. *Proton transfer.* The enzyme may also accept or donate protons, thereby increasing the chemical reactivity of the substrate. This accounts for the importance of charged amino acids in active-site chemistry, which in turn explains why enzyme activity is so often pH dependent.
3. *Electron transfer.* As a further means of substrate activation, enzymes may also accept or donate electrons, thereby forming temporary covalent bonds between the enzyme and its substrate.

The Catalytic Event. The sequence of events at the active site is illustrated in **Figure 6-7**, using the enzyme sucrase as an example. Sucrase (also known as invertase or β-fructofuranosidase) hydrolyzes the disaccharide sucrose

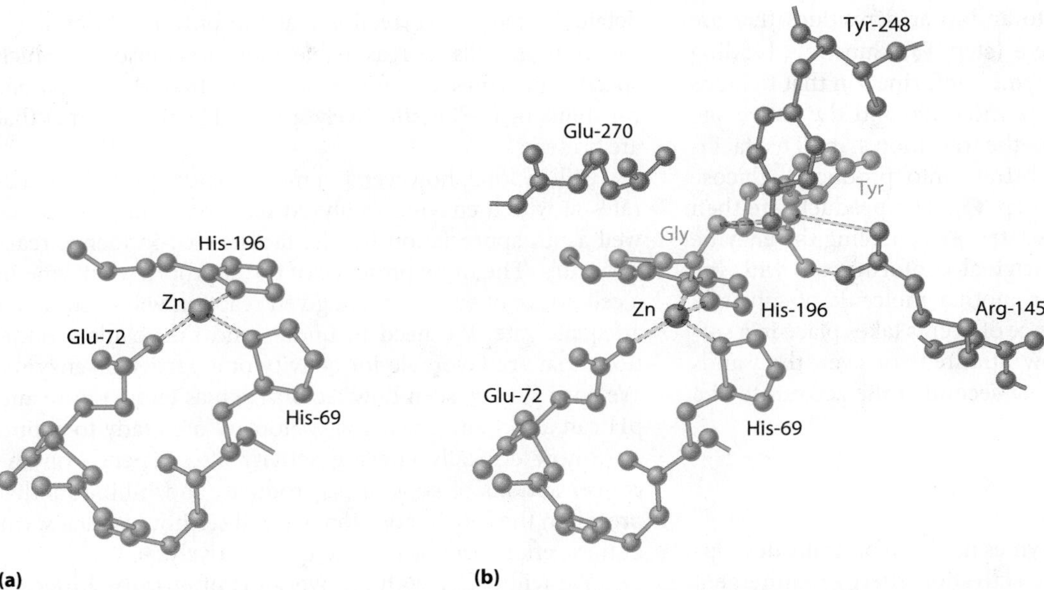

FIGURE 6-6 The Change in Active Site Structure Induced by Substrate Binding. (a) The unoccupied active site of carboxypeptidase A contains a zinc ion tightly bound to side chains of three amino acids (cyan). (b) Binding of the substrate (the dipeptide shown in orange) to this zinc ion induces a conformation change in the enzyme that brings other amino acid side chains (purple) into the active site to participate in catalysis.

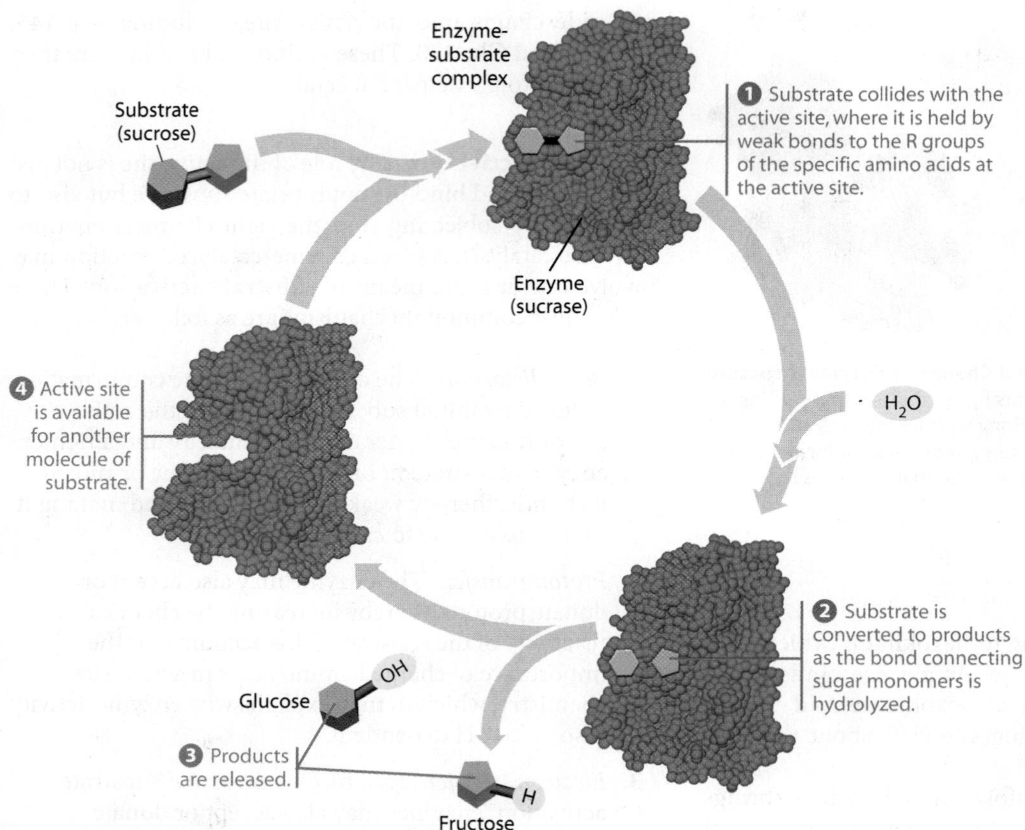
Enzyme-
substrate
complex
Substrate
(sucrose)
1 Substrate collides with the active site, where it is held by weak bonds to the R groups of the specific amino acids at the active site.
Enzyme
(sucrase)
4 Active site is available for another molecule of substrate.
H_2O
2 Substrate is converted to products as the bond connecting sugar monomers is hydrolyzed.
OH
Glucose
3 Products are released.
H
Fructose

MB Chemistry Review-Enzymes & Pathways: Enzyme Action

FIGURE 6-7 The Catalytic Cycle of an Enzyme. In this example, the enzyme sucrase catalyzes the hydrolysis of sucrose to glucose and fructose. The actual structure of this enzyme is shown, but the active site has been modified slightly to emphasize the close fit between enzyme and substrate.

into glucose and fructose. The initial random collision of a substrate molecule—sucrose, in this case—with the active site results in its binding to amino acid residues that are strategically positioned there (step ❶). Substrate binding induces a change in the enzyme conformation that tightens the fit between the substrate molecule and the active site and lowers the free energy of the transition state. This facilitates the conversion of substrate into products—glucose and fructose, in this case (step ❷). The products are then released from the active site (step ❸), enabling the enzyme molecule to return to its original conformation, with the active site now available for another molecule of substrate (step ❹). This entire sequence of events takes place in a sufficiently short time to allow hundreds or even thousands of such reactions to occur per second at the active site of a single enzyme molecule!

Enzyme Kinetics

So far, our discussion of enzymes has been basically descriptive. We have dealt with the activation energy requirement that prevents thermodynamically feasible reactions from occurring and with catalysts as a means of reducing the activation energy and thereby facilitating such reactions. We have also encountered enzymes as biological catalysts and have examined their structure and function in some detail. Moreover, we realize that the only reactions likely to occur in cells at reasonable rates are those for which specific enzymes are on hand, such that the metabolic capability of a cell is effectively specified by the enzymes that are present.

Still lacking, however, is a means of assessing the actual rates at which enzyme-catalyzed reactions will proceed, as well as an appreciation for the factors that influence reaction rates. The mere presence of the appropriate enzyme in a cell does not ensure that a given reaction will occur at an adequate rate. We need to understand the cellular conditions that are favorable for activity of a particular enzyme. We have already seen how factors such as temperature and pH can affect enzyme activity. Now we are ready to appreciate how critically enzyme activity also depends on the concentrations of substrates, products, and inhibitors that prevail in the cell. In addition, we will see how at least some of these effects can be defined quantitatively.

We will begin with an overview of **enzyme kinetics,** which describes quantitative aspects of enzyme catalysis (the word *kinetics* is from the Greek word *kinetikos,* meaning "moving") and the rate of substrate conversion into products. Specifically, enzyme kinetics concerns reaction rates and the manner in which reaction rates are influenced by a variety of factors, including the concentrations of substrates, products, and inhibitors. Most of our attention here will focus on the effects of substrate concentration on the kinetics of enzyme-catalyzed reactions.

We will focus on *initial reaction rates,* the rates of reactions measured over a brief initial period of time during which the substrate concentration has not yet decreased enough to affect the rate of the reaction and the accumulation of product is still too small for the reverse reaction (conversion of product back into substrate) to occur at a significant rate. This resembles the steady-state situation in living cells, where substrates are continually replenished and products are continually removed, maintaining stable concentrations of each. Although this description is somewhat oversimplified compared to the situation in living cells, it nonetheless allows us to understand some important principles of enzyme kinetics.

Enzyme kinetics can seem quite complex at first. To help you understand the basic concepts, **Box 6A** explains how enzymes acting on substrate molecules can be compared to a roomful of monkeys shelling peanuts. You may find it useful to turn to the analogy at this point and then come back to this section.

Most Enzymes Display Michaelis–Menten Kinetics

Here we will consider how the **initial reaction velocity (v)** changes depending on the **substrate concentration ([S]).** The initial reaction velocity is rigorously defined as the rate of change in product concentration per unit time (e.g., m*M*/min). Often, however, reaction velocities are experimentally measured in a constant assay volume of 1 mL and are reported as μmol of product per minute. At low [S], a doubling of [S] will double v. But as [S] increases, each additional increment of substrate results in a smaller increase in reaction rate. As [S] becomes very large, increases in [S] increase only slightly, and the value of v reaches a maximum.

By determining v in a series of experiments at varying substrate concentrations, the dependence of v on [S] can be shown experimentally to be that of a hyperbola (**Figure 6-8**). An important property of this hyperbolic relationship is that as [S] tends toward infinity, v approaches an upper limiting value known as the **maximum velocity (V_{max}).** This value depends on the number of enzyme molecules and can therefore be increased only by adding more enzyme. The inability of increasingly higher substrate concentrations to increase the reaction velocity beyond a finite upper value is called **saturation.** At saturation, all available enzyme molecules are operating at maximum capacity. Saturation is a fundamental, universal property of enzyme-catalyzed reactions. Catalyzed reactions always become saturated at high substrate concentrations, whereas uncatalyzed reactions do not.

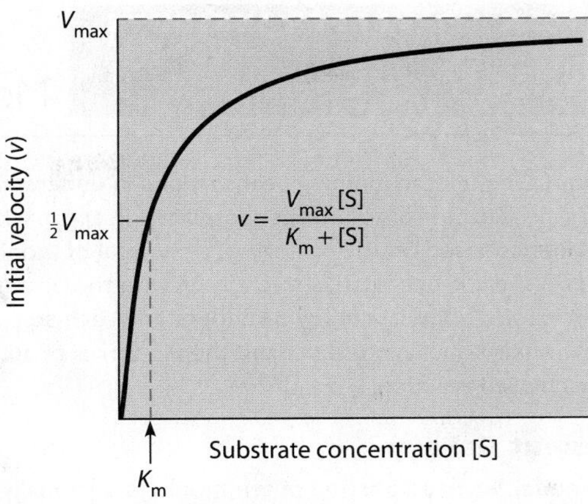

FIGURE 6-8 The Relationship Between Reaction Velocity and Substrate Concentration. For an enzyme-catalyzed reaction that follows Michaelis–Menten kinetics, the initial velocity tends toward an upper limiting velocity V_{max} as the substrate concentration [S] tends toward infinity. The Michaelis constant K_m corresponds to that substrate concentration at which the reaction is proceeding at one-half of the maximum velocity.

MB Chemistry Review- Enzyme and Substrate Concentrations

Much of our understanding of the hyperbolic relationship between [S] and v is due to the pioneering work of two German enzymologists, Leonor Michaelis and Maud Menten. In 1913, they postulated a general theory of enzyme action that has turned out to be basic to the quantitative analysis of almost all aspects of enzyme kinetics. To understand their approach, consider one of the simplest possible enzyme-catalyzed reactions, a reaction in which a single substrate S is converted into a single product P:

$$\text{S} \xrightarrow[\text{Enzyme (E)}]{} \text{P} \tag{6-5}$$

According to the Michaelis–Menten hypothesis, the enzyme E that catalyzes this reaction first reacts with the substrate S, forming the transient enzyme-substrate complex ES, which then undergoes the actual catalytic reaction to form free enzyme and product P, as shown in the sequence

$$\text{E}_f + \text{S} \underset{k_2}{\overset{k_1}{\rightleftharpoons}} \text{ES} \underset{k_4}{\overset{k_3}{\rightleftharpoons}} \text{E}_f + \text{P} \tag{6-6}$$

where E_f is the free form of the enzyme, S is the substrate, ES is the enzyme-substrate complex, P is the product, and k_1, k_2, k_3, and k_4 are the rate constants for the indicated reactions.

Starting with this model and several simplifying assumptions, including the steady-state conditions we described near the end of Chapter 5, Michaelis and Menten arrived at the relationship between the velocity of an enzyme-catalyzed reaction and the substrate concentration, as follows:

$$v = \frac{V_{max}[\text{S}]}{K_m + [\text{S}]} \tag{6-7}$$

BOX 6A DEEPER INSIGHTS

Monkeys and Peanuts

If you found the Mexican jumping beans helpful in understanding free energy in Chapter 5, you might appreciate an approach to enzyme kinetics based on the analogy of a roomful of monkeys ("enzymes") shelling peanuts ("substrates"), with the peanuts present in varying abundance. Try to understand each step first in terms of monkeys shelling peanuts and then in terms of an actual enzyme-catalyzed reaction.

The Peanut Gallery

For our model, we need a troop of ten monkeys, all equally adept at finding and shelling peanuts. We shall assume that the monkeys are too full to eat any of the peanuts they shell but nonetheless have an irresistible compulsion to go on shelling.

Next, we need the Peanut Gallery, a room of fixed floor space with peanuts scattered equally about on the floor. The number of peanuts will be varied as we proceed, but in all cases there will be vastly more peanuts than monkeys in the room. Moreover, because we know the number of peanuts and the total floor space, we can always calculate the "concentration" (more accurately, the density) of peanuts in the room. In each case, the monkeys start out in an adjacent room. To start an assay, we simply open the door and allow the eager monkeys to enter the Peanut Gallery.

The Shelling Begins

Now we are ready for our first assay. We start with an initial peanut concentration of 1 peanut per square meter, and we assume that, at this concentration of peanuts, the average monkey spends 9 seconds looking for a peanut to shell and 1 second shelling it. This means that each monkey requires 10 seconds per peanut and can thus shell peanuts at the rate of 0.1 peanut per second. Then, since there are ten monkeys in the gallery, the rate (let's call it the velocity v) of peanut-shelling for all the monkeys is 1 peanut per second at this particular concentration of peanuts (which we will call [S] to remind ourselves that the peanuts are really the substrate of the shelling action). All of this can be tabulated as follows:

[S] = Concentration of peanuts (peanuts/m^2)	1
Time required per peanut:	
To find (sec/peanut)	9
To shell (sec/peanut)	1
Total (sec/peanut)	10
Rate of shelling:	
Per monkey (peanut/sec)	0.10
Total (v) (peanut/sec)	1.0

The Peanuts Become More Abundant

For our second assay, we herd all the monkeys back into the waiting room, sweep up the debris, and arrange peanuts about the Peanut Gallery at a concentration of 3 peanuts per square meter. Since peanuts are now three times more abundant than previously, the average monkey should find a peanut three times more quickly than before, such that the time spent finding the average peanut is now only 3 seconds. But each peanut, once found, still takes 1 second to shell, so the total time per peanut is now 4 seconds and the velocity of shelling is 0.25 peanut per second for each monkey, or 2.5 peanuts per second for the roomful of monkeys. This generates another column of entries for our data table:

[S] = Concentration of peanuts (peanuts/m^2)	1	3
Time required per peanut:		
To find (sec/peanut)	9	3
To shell (sec/peanut)	1	1
Total (sec/peanut)	10	4
Rate of shelling:		
Per monkey (peanut/sec)	0.10	0.25
Total (v) (peanut/sec)	1.0	2.5

What Happens to v as [S] Continues to Increase?

To find out what eventually happens to the velocity of peanut-shelling as the peanut concentration in the room gets higher and higher, all you need do is extend the data table by assuming ever-increasing values for [S] and calculating the corresponding v. For example, you should be able to convince yourself that a further tripling of the peanut concentration (from 3 to 9 peanuts/m^2) will bring the time required per peanut down to 2 seconds (1 second to find and another second to shell), which will result in a shelling rate of 0.5 peanut per second for each monkey, or 5.0 peanuts per second overall.

Already you should begin to see a trend. The first tripling of peanut concentration increased the rate 2.5-fold, but the next tripling resulted in only a further doubling of the rate. There seems, in other words, to be a diminishing return on additional peanuts. You can see this clearly if you choose a few more peanut concentrations and then plot v on the y-axis (suggested scale: 0–10 peanuts/sec) versus [S] on the x-axis (suggested scale: 0–100 peanuts/m^2).

What you should find is that the data generate a hyperbolic curve that looks strikingly like Figure 6-8. If you look at your data carefully, you should also see the reason your curve continues to "bend over" as [S] gets higher (i.e., why you get less and less additional velocity for each further increment of peanuts): The shelling time is fixed and therefore becomes a more and more prominent component of the total processing time per peanut as the finding time gets smaller and smaller. You should also appreciate that it is this fixed shelling time that ultimately sets the upper limit on the overall rate of peanut processing because, even when [S] is infinite (i.e., in a world flooded with peanuts), there will still be a finite time of 1 second required to process each peanut. This means that the overall maximum velocity, V_{max}, for the ten monkeys would be 10 peanuts per second.

Finally, you should realize that there is something special about the peanut concentration at which the finding time is exactly equal to the shelling time (it turns out to be 9 peanuts/m^2); this is the point along the curve at which the rate of peanut processing is exactly one-half of the maximum rate. In fact, it is such an important benchmark along the concentration scale that you might even be tempted to give it a special name, particularly if your name were Michaelis and you were monkeying around with enzymes instead of peanuts!

Here, v is the initial reaction velocity, [S] is the initial substrate concentration, V_{max} is the maximum velocity, and K_m is the concentration of substrate that gives exactly half the maximum velocity. V_{max} and K_m (also known as the **Michaelis constant**) are important kinetic parameters that we will consider in more detail in the next section. Equation 6-7 is known as the **Michaelis–Menten equation,** a central relationship of enzyme kinetics. (Problem 6–12 at the end of the chapter gives you an opportunity to derive the Michaelis–Menten equation yourself.)

What Is the Meaning of V_{max} and K_m?

To appreciate the implications of the relationship between v and [S] and to examine the meaning of the parameters V_{max} and K_m, we can consider three special cases of substrate concentration: very low substrate concentration, very high substrate concentration, and the special case of $[S] = K_m$.

Case 1: Very Low Substrate Concentration ($[S] \ll K_m$). At very low substrate concentration, [S] becomes negligibly small compared with the constant K_m in the denominator of the Michaelis–Menten equation and can be ignored, so we can write

$$v = \frac{V_{max}[S]}{K_m + [S]} \cong \frac{V_{max}[S]}{K_m} \tag{6-8}$$

Thus, at very low substrate concentration, the initial reaction velocity is roughly proportional to the substrate concentration. This can be seen at the extreme left side of the graph in Figure 6-8. As long as the substrate concentration is much lower than the K_m value, the velocity of an enzyme-catalyzed reaction increases linearly with substrate concentration.

Case 2: Very High Substrate Concentration ($[S] \gg K_m$). At very high substrate concentration, K_m becomes negligibly small compared with [S] in the denominator of the Michaelis–Menten equation, so we can write

$$v = \frac{V_{max}[S]}{K_m + [S]} \cong \frac{V_{max}[S]}{[S]} = V_{max} \tag{6-9}$$

Therefore, at very high substrate concentrations, the velocity of an enzyme-catalyzed reaction is essentially independent of the variation in [S] and is approximately constant at a value close to V_{max} (see the right side of Figure 6-8).

This provides us with a mathematical definition of V_{max}, which is one of the two kinetic parameters in the Michaelis–Menten equation. V_{max} is the upper limit of v as the substrate concentration [S] approaches infinity. In other words, V_{max} is the velocity at saturating substrate concentrations. Under these conditions, every enzyme molecule is occupied in the actual process of catalysis almost all of the time because the substrate concentration is so high that, as soon as a product molecule is released, another substrate molecule arrives at the active site.

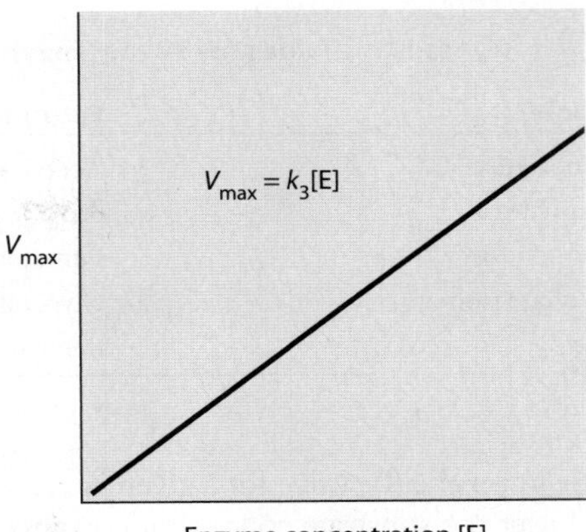

FIGURE 6-9 The Linear Relationship Between V_{max} and Enzyme Concentration. The linear increase in reaction velocity with enzyme concentration provides the basis for determining enzyme concentrations experimentally.

V_{max} is therefore an upper limit determined by (1) the time required for the actual catalytic reaction and (2) how many such enzyme molecules are present. Because the actual reaction rate is fixed, the only way that V_{max} can be increased is to increase enzyme concentration. In fact, V_{max} is linearly proportional to the amount of enzyme present, as shown in **Figure 6-9**, where k_3 represents the reaction rate constant.

Case 3: ($[S] = K_m$). To explore the meaning of K_m more precisely, consider the special case where [S] is exactly equal to K_m. Under these conditions, the Michaelis–Menten equation can be written as

$$v = \frac{V_{max}[S]}{K_m + [S]} \cong \frac{V_{max}[S]}{2[S]} = \frac{V_{max}}{2} \tag{6-10}$$

This equation demonstrates mathematically that K_m is that specific substrate concentration at which the reaction proceeds at one-half of its maximum velocity. The K_m is a constant value for a given enzyme-substrate combination catalyzing a reaction under specified conditions. Figure 6-8 illustrates the meaning of both V_{max} and K_m.

Why Are K_m and V_{max} Important to Cell Biologists?

Now that we understand what K_m and V_{max} mean, it is fair to ask why these kinetic parameters are important to cell biologists. The K_m value is useful because it allows us to estimate where along the Michaelis–Menten plot of Figure 6-8 an enzyme is functioning in a cell (providing, of course, that the normal substrate concentration in the cell is known). We can then estimate at what fraction of the maximum velocity the enzyme-catalyzed reaction is likely to be proceeding in

Table 6-2 K_m and k_{cat} Values for Some Enzymes

Enzyme Name	Substrate	K_m (M)	k_{cat} (s^{-1})
Acetylcholinesterase	Acetylcholine	9×10^{-5}	1.4×10^4
Carbonic anhydrase	CO_2	1×10^{-2}	1×10^6
Fumarase	Fumarate	5×10^{-6}	8×10^2
Triose phosphate isomerase	Glyceraldehyde-3-phosphate	5×10^{-4}	4.3×10^3
β-lactamase	Benzylpenicillin	2×10^{-5}	2×10^3

the cell. The lower the K_m value for a given enzyme and substrate, the lower the substrate concentration range in which the enzyme is effective. As we will soon see, enzyme activity in the cell can be modulated by regulatory molecules that bind to the enzyme and alter the K_m for a particular substrate. K_m values for several enzyme-substrate combinations are given in **Table 6-2** and, as you can see, can vary over several orders of magnitude.

The V_{max} for a particular reaction is important because it provides a measure of the potential maximum rate of the reaction. Few enzymes actually encounter saturating substrate concentrations in cells, so enzymes are not likely to be functioning at their maximum rate under cellular conditions. However, by knowing the V_{max} value, the K_m value, and the substrate concentration in vivo, we can at least estimate the likely rate of the reaction under cellular conditions.

V_{max} can also be used to determine another useful parameter called the **turnover number (k_{cat})**, which expresses the rate at which substrate molecules are converted to product by a single enzyme molecule when the enzyme is operating at its maximum velocity. The constant k_{cat} has the units of reciprocal time (s^{-1}, for example) and is calculated as the quotient of V_{max} over $[E_t]$, the concentration of the enzyme:

$$k_{cat} = \frac{V_{max}}{[E_t]} \tag{6-11}$$

Turnover numbers vary greatly among enzymes, as is clear from the examples given in Table 6-2.

The Double-Reciprocal Plot Is a Useful Means of Linearizing Kinetic Data

The classic Michaelis–Menten plot of v versus [S] shown in Figure 6-8 illustrates the dependence of velocity on substrate concentration. However, it is not an especially useful tool for the quantitative determination of the key kinetic parameters K_m and V_{max}. Its hyperbolic shape makes it difficult to extrapolate accurately to infinite substrate concentration in order to determine the critical parameter V_{max}. Also, if V_{max} is not known accurately, K_m cannot be determined.

To circumvent this problem and provide a more useful graphic approach, Hans Lineweaver and Dean Burk in 1934 converted the hyperbolic relationship of the Michaelis–Menten equation into a linear function by inverting both sides of Equation 6-7 and simplifying the resulting expression into the form of an equation for a straight line:

$$\frac{1}{v} = \frac{K_m + [S]}{V_{max}[S]} = \frac{K_m}{V_{max}[S]} + \frac{[S]}{V_{max}[S]}$$
$$= \frac{K_m}{V_{max}}\left(\frac{1}{[S]}\right) + \frac{1}{V_{max}} \tag{6-12}$$

Equation 6-12 is known as the **Lineweaver–Burk equation.** When it is plotted as $1/v$ versus $1/[S]$, as in **Figure 6-10**, the resulting **double-reciprocal plot** is linear in the general algebraic form $y = mx + b$, where m is the slope and b is the y-intercept. Therefore, it has a slope (m) of K_m/V_{max}, a y-intercept (b) of $1/V_{max}$, and an x-intercept ($y = 0$) of $-1/K_m$. (You should be able to convince yourself of these intercept values by setting first $1/[S]$ and then $1/v$ equal to zero in Equation 6-12 and solving for the other value.) Therefore, once the double-reciprocal plot has been constructed, V_{max} can be determined directly from the reciprocal of the y-intercept and K_m from the negative reciprocal of the x-intercept. Furthermore, the slope can be used to check both values.

Thus, the Lineweaver–Burk plot is useful experimentally because it allows us to determine the parameters V_{max}

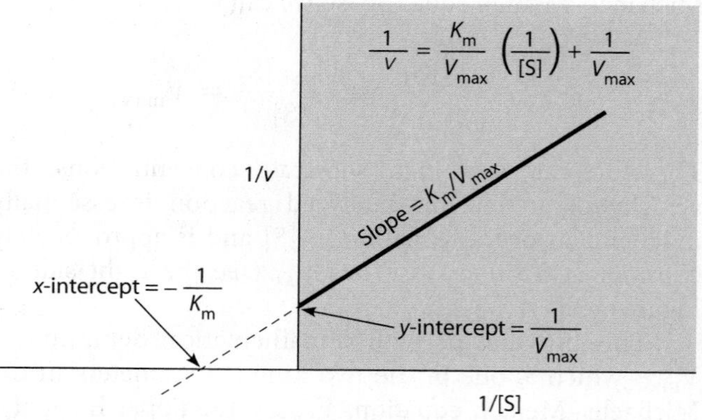

FIGURE 6-10 The Lineweaver–Burk Double-Reciprocal Plot. The reciprocal of the initial velocity, $1/v$, is plotted as a function of the reciprocal of the substrate concentration, $1/[S]$. K_m can be calculated from the x-intercept and V_{max} from the y-intercept.

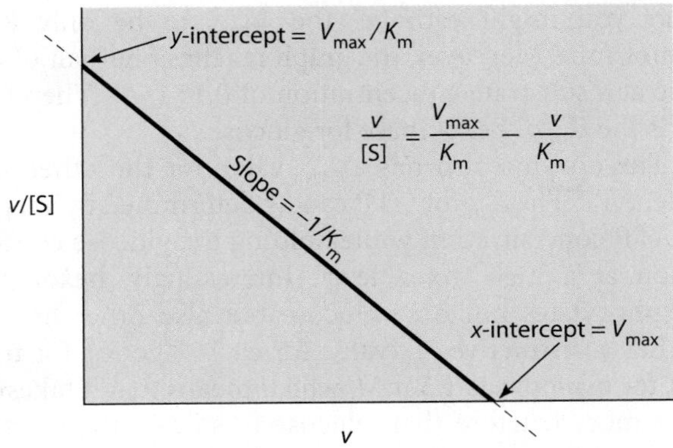

FIGURE 6-11 The Eadie–Hofstee Plot. The ratio $v/[S]$ is plotted as a function of v. K_m can be determined from the slope and V_{max} from the x-intercept.

and K_m without the complication of a hyperbolic shape. It also serves as a useful diagnostic in analyzing enzyme inhibition because the several different kinds of reversible inhibitors affect the shape of the plot in characteristic ways.

The Lineweaver–Burk equation has some limitations, however. The main problem is that a long extrapolation is often necessary to determine K_m, and this may introduce uncertainty in the result. Moreover, the most crucial data points for determining the slope of the curve are the farthest from the y-axis. Because those points represent the samples with the lowest substrate concentrations and lowest levels of enzyme activity, they are the most difficult to measure accurately.

To circumvent these disadvantages, several alternatives to the Lineweaver–Burk equation have come into use to linearize kinetic data. One such alternative is the **Eadie–Hofstee equation,** which is represented graphically as a plot of $v/[S]$ versus v. As **Figure 6-11** illustrates, V_{max} is determined from the x-intercept and K_m from the slope of this plot. (To explore the Eadie–Hofstee plot and another alternative to the Lineweaver–Burk plot further, see Problem 6–13 at the end of this chapter.)

Determining K_m and V_{max}: An Example

To illustrate the value of the double-reciprocal plot in determining V_{max} and K_m, consider a specific example involving the enzyme hexokinase, as illustrated in **Figures 6-12** and **6-13**. Hexokinase is an important enzyme in cellular energy metabolism because it catalyzes the first reaction in the glycolytic pathway, which is discussed in detail in Chapter 9. Using the hydrolysis of ATP as a source of both the phosphate group and the free energy needed for the reaction, hexokinase catalyzes the phosphorylation of glucose on carbon atom 6:

$$\text{glucose} + \text{ATP} \xrightarrow[\text{hexokinase}]{} \text{glucose-6-phosphate} + \text{ADP} \quad \textbf{(6-13)}$$

To analyze this reaction kinetically, we must determine the initial velocity at each of several substrate concentrations.

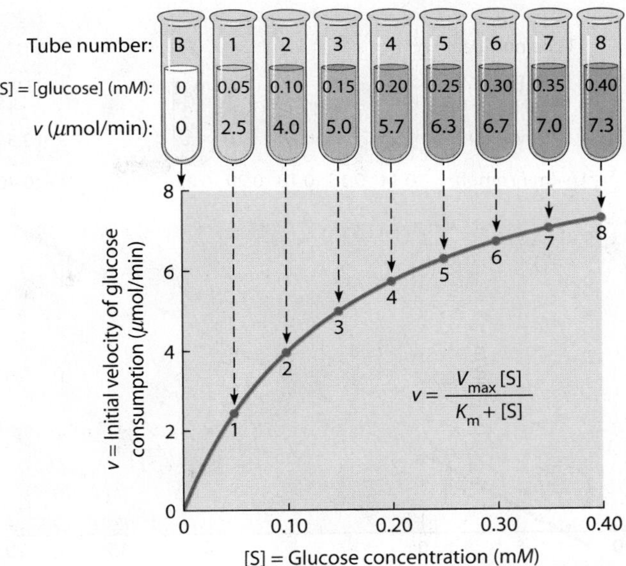

FIGURE 6-12 Experimental Procedure for Studying the Kinetics of the Hexokinase Reaction. Test tubes containing graded concentrations of glucose and a saturating concentration of ATP were incubated with a standard amount of hexokinase. The initial rate of product appearance, v, was then plotted as a function of the substrate concentration [S]. The curve is hyperbolic, approaching V_{max} as the substrate concentration gets higher and higher. For the double-reciprocal plot derived from these data, see Figure 6-13.

When an enzyme has two substrates, the usual approach is to vary the concentration of one substrate at a time while holding that of the other one constant at a level near saturation to ensure that it does not become rate limiting. The velocity determination must be made before either the substrate concentration drops appreciably or the product accumulates to the point that the reverse reaction becomes significant.

In the experimental approach shown in Figure 6-12, glucose is the variable substrate, with ATP present at a saturating concentration in each tube. Of the nine reaction mixtures set up for this experiment, one tube is a negative control designated the reagent blank (B) because it contains no glucose. The other tubes contain concentrations of glucose ranging from 0.05 to 0.40 mM. With all tubes prepared and maintained at some favorable temperature (25°C is often used), the reaction in each is initiated by adding a fixed amount of hexokinase.

The rate of product formation in each of the reaction mixtures can then be determined either by continuous spectrophotometric monitoring of the reaction mixture (provided that one of the reactants or products absorbs light of a specific wavelength) or by allowing each reaction mixture to incubate for some short, fixed period of time, followed by chemical assay for either substrate depletion or product accumulation.

As Figure 6-12 indicates, the initial velocity of the glucose consumption reaction for tubes 1–8 ranged from 2.5 to 7.3 μmol of glucose consumed per minute, with no

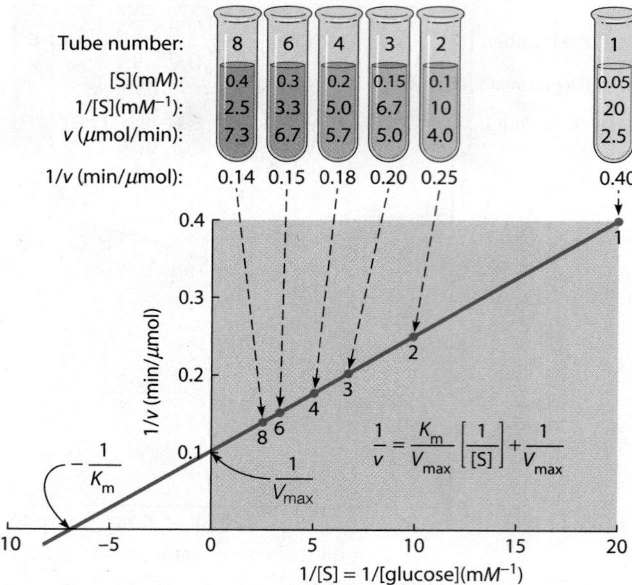

FIGURE 6-13 Double-Reciprocal Plot for the Hexokinase Data of Figure 6-12. For each test tube in Figure 6-12, $1/v$ and $1/[S]$ were calculated, and $1/v$ was then plotted as a function of $1/[S]$. The y-intercept of 0.1 corresponds to $1/V_{max}$, so V_{max} is 10 mM/min. The x-intercept of -6.7 corresponds to $-1/K_m$, so K_m is 0.15 mM. (Some of the tubes depicted in Figure 6-12 are not shown here due to lack of space.)

detectable reaction in the blank. When these reaction velocities are plotted as a function of glucose concentration, the eight data points generate the hyperbolic curve shown in Figure 6-12. Although the data of Figure 6-12 are idealized for illustrative purposes, most kinetic data generated by this approach do, in fact, fit a hyperbolic curve unless the enzyme has some special properties that cause departure from Michaelis–Menten kinetics.

The hyperbolic curve of Figure 6-12 illustrates the need for some means of linearizing the analysis because neither V_{max} nor K_m can be easily determined from the values as plotted. This need is met by the linear double-reciprocal plot shown in Figure 6-13. To obtain the data plotted here, reciprocals were calculated for each value of [S] and v in Figure 6-12. Thus, the [S] values of 0.05–0.40 mM generate reciprocals of 20–2.5 mM^{-1}, and the v values of 2.5–7.3 µmol/min give rise to reciprocals of 0.4–0.14 min/µmol. Because these are reciprocals, the data point representing the lowest concentration (tube 1) is farthest from the origin, and each successive tube is represented by a point closer to the origin.

When these data points are connected by a straight line, the y-intercept is found to be 0.1 min/µmol, and the x-intercept is -6.7 mM^{-1}. From these intercepts, we can calculate that $V_{max} = 1/0.1 = 10$ µmol/min and $K_m = -(1/-6.7) = 0.15$ mM. If we now go back to the Michaelis–Menten plot of Figure 6-12, we can see that both of these values are quite reasonable because we can readily imagine that the plot is rising hyperbolically to a maximum of 10 mM/min. Note that using your eyes alone, you might estimate the V_{max} to be only 8 or 9 µmol/min. Moreover, the graph reaches one-half of this value at a substrate concentration of 0.15 mM. Therefore, this is the K_m of hexokinase for glucose.

The enzyme also has a K_m value for the other substrate, ATP. The K_m for ATP can be determined by varying the ATP concentration while holding the glucose concentration at a high, fixed level. Interestingly, hexokinase phosphorylates not only glucose but also other hexoses and has a distinctive K_m value for each. The K_m for fructose, for example, is 1.5 mM, which means that it takes ten times more fructose than glucose to sustain the reaction at one-half of its maximum velocity.

Enzyme Inhibitors Act Either Irreversibly or Reversibly

Thus far, we have assumed that substrates are the only substances in cells that affect the activities of enzymes in cells. However, enzymes are also influenced by products, alternative substrates, substrate analogues, drugs, toxins, and a very important class of regulators called *allosteric effectors.* Most of these substances have an inhibitory effect on enzyme activity, reducing the reaction rate with the desired substrate or sometimes blocking the reaction completely.

This **inhibition** of enzyme activity is important for several reasons. First and foremost, enzyme inhibition plays a vital role as a control mechanism in cells. Many enzymes are subject to regulation by specific small molecules other than their substrates. Often this is a means of sensing their immediate environment to respond to specific cellular conditions.

Enzyme inhibition is also important in the action of drugs and poisons, which frequently exert their effects by inhibiting specific enzymes. Inhibitors are also useful to enzymologists as tools in their studies of reaction mechanisms and to doctors for treatment of disease. Especially important inhibitors are *substrate analogues* and *transition state analogues.* These are compounds that resemble the real substrate or transition state closely enough to bind to the active site but cannot undergo reaction to create a functional product.

Substrate analogs are important tools in fighting infectious disease, and many have been developed to inhibit specific enzymes in pathogenic bacteria and viruses, usually targeting enzymes that we humans lack. For example, sulfa drugs resemble the folic acid precursor, PABA. They can bind to and block the active site of the bacterial enzyme used to synthesize folic acid, which is required in DNA synthesis. Likewise, azidothymidine (AZT), which is an antiviral medication, resembles the deoxythymidine molecule normally used by the human immunodeficiency virus (HIV) to synthesize DNA using viral reverse transcriptase. However, after binding to the active site, AZT is incorporated into a growing strand of DNA but forms a "dead-end" molecule of DNA that cannot be elongated.

Inhibitors may be either *reversible* or *irreversible.* An **irreversible inhibitor** binds to the enzyme covalently, causing permanent loss of catalytic activity. Not surprisingly, irreversible inhibitors are usually toxic to cells. Ions of heavy metals are often irreversible inhibitors, as are nerve gas poisons and some insecticides. These substances can bind irreversibly to enzymes such as *acetylcholinesterase,* an enzyme that is vital to the transmission of nerve impulses (see Chapter 13). Inhibition of acetylcholinesterase activity leads to rapid paralysis of vital functions and therefore to death. One such inhibitor is *diisopropyl fluorophosphate,* a nerve gas that binds covalently to the hydroxyl group of a critical serine at the active site of the enzyme, thereby rendering the enzyme molecule permanently inactive.

Some irreversible inhibitors of enzymes can be used as therapeutic agents. For example, aspirin binds irreversibly to the enzyme cyclooxygenase-1 (COX-1), which produces prostaglandins and other signaling chemicals that cause inflammation, constriction of blood vessels, and platelet aggregation. Thus, aspirin is effective in relieving minor inflammation and headaches, and has been recommended in low doses as a cardiovascular protectant. The antibiotic *penicillin* is an irreversible inhibitor of the enzyme needed for bacterial cell wall synthesis. Penicillin is therefore effective in treating bacterial infections because it prevents the bacterial cells from forming cell walls, thus blocking their growth and division. And because our cells lack a cell wall (and the enzyme that synthesizes it), penicillin is nontoxic to humans.

In contrast, a **reversible inhibitor** binds to an enzyme in a noncovalent, dissociable manner, such that the free and bound forms of the inhibitor exist in equilibrium with each other. We can represent such binding as

$$E + I \rightleftharpoons EI \qquad \textbf{(6-14)}$$

with E as the active free enzyme, I as the inhibitor, and EI as the inactive enzyme-inhibitor complex. Clearly, the fraction of the enzyme that is available to the cell in active form depends on the concentration of the inhibitor and the strength of the enzyme-inhibitor complex.

The two most common forms of reversible inhibitors are called *competitive inhibitors* and *noncompetitive inhibitors.* A competitive inhibitor binds to the active site of the enzyme and therefore competes directly with substrate molecules for the same site on the enzyme (**Figure 6-14a**). This reduces enzyme activity because many of the active sites of the enzyme molecules are blocked by bound inhibitor molecules and thus cannot bind substrate molecules at the active site. A noncompetitive inhibitor, on the other hand, binds to the enzyme surface at a location *other* than the active site. It does not block substrate binding directly but inhibits enzyme activity indirectly by causing a change in protein conformation that can either inhibit substrate binding to the active site or greatly reduce the catalytic activity at the active site (Figure 6-14b).

Considerable progress has been made in the field of computer-aided drug design. In this approach, the three-dimensional structure of an enzyme active site is analyzed to predict what types of molecules are likely to bind tightly to it and act as inhibitors. Scientists can then design a number of hypothetical inhibitors and test their binding using complex computer models. In this way, we do not have to rely only upon those inhibitors we can discover in

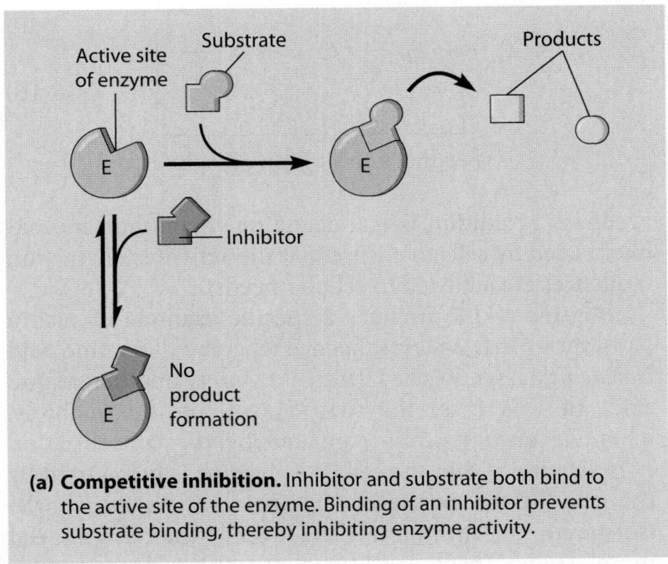

(a) Competitive inhibition. Inhibitor and substrate both bind to the active site of the enzyme. Binding of an inhibitor prevents substrate binding, thereby inhibiting enzyme activity.

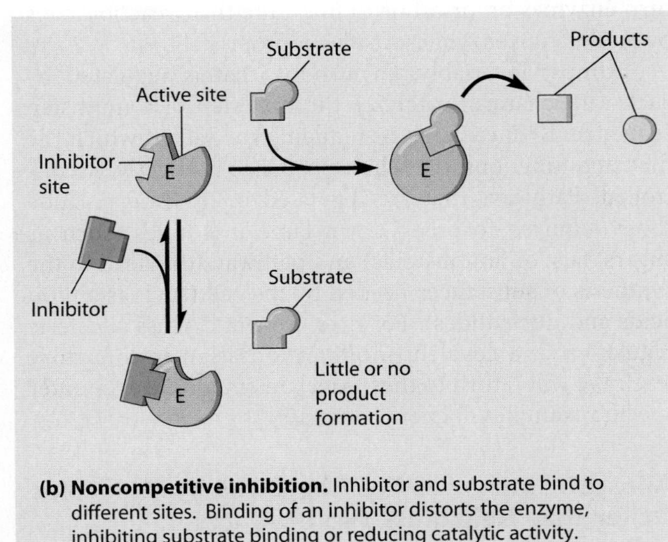

(b) Noncompetitive inhibition. Inhibitor and substrate bind to different sites. Binding of an inhibitor distorts the enzyme, inhibiting substrate binding or reducing catalytic activity.

FIGURE 6-14 Modes of Action of Competitive and Noncompetitive Inhibitors. Both (a) competitive and (b) noncompetitive inhibitors bind reversibly to the enzyme (E), thereby inhibiting its activity. The two kinds of inhibitors differ in which site on the enzyme they bind to.

nature. Hundreds or even thousands of potential inhibitors can be designed and tested, and only the most promising are actually synthesized and evaluated experimentally.

Enzyme Regulation

To understand the role of enzymes in cellular function, we need to recognize that it is rarely in the cell's best interest to allow an enzyme to function at an indiscriminately high rate. Instead, the rates of enzyme-catalyzed reactions must be continuously adjusted to keep them finely tuned to the needs of the cell. An important aspect of that adjustment lies in the cell's ability to control enzyme activities with specificity and precision.

We have already encountered a variety of regulatory mechanisms, including changes in substrate and product concentrations, alterations in temperature and pH, and the presence and concentration of inhibitors. Regulation that depends directly on the interactions of substrates and products with the enzyme is called **substrate-level regulation.** As the Michaelis–Menten equation makes clear, increases in substrate concentration result in higher reaction rates (see Figure 6-8). Conversely, increases in product concentration reduce the rate at which substrate is converted to product. (This inhibitory effect of product concentration is why v needs to be identified as the *initial* reaction velocity in the Michaelis–Menten equation, as given by Equation 6-7.)

Substrate-level regulation is an important control mechanism in cells, but it is not sufficient for the regulation of most reactions or reaction sequences. For most pathways, enzymes are regulated by other mechanisms as well. Two of the most important of these are *allosteric regulation* and *covalent modification.* These mechanisms allow cells to turn enzymes on or off or to fine-tune their reaction rates by modulating enzyme activities appropriately.

Almost invariably, an enzyme that is regulated by such a mechanism catalyzes the first step of a multistep sequence. By increasing or reducing the rate at which the first step functions, the whole sequence is effectively controlled. Pathways that are regulated in this way include those required to break down large molecules (such as sugars, fats, or amino acids) and pathways that lead to the synthesis of substances needed by the cell (such as amino acids and nucleotides). For now, we will discuss allosteric regulation and covalent modification at an introductory level. We will return to these mechanisms as we encounter specific examples in later chapters.

Allosteric Enzymes Are Regulated by Molecules Other than Reactants and Products

The single most important control mechanism whereby the rates of enzyme-catalyzed reactions are adjusted to meet cellular needs is *allosteric regulation.* To understand this mode of regulation, consider the pathway by which a cell converts some precursor A into some final product P via a series of intermediates B, C, and D in a sequence of reactions catalyzed respectively by enzymes E_1, E_2, E_3, and E_4:

$$A \xrightarrow[E_1]{} B \xrightarrow[E_2]{} C \xrightarrow[E_3]{} D \xrightarrow[E_4]{} P \qquad \textbf{(6-15)}$$

Product P could, for example, be an amino acid needed by the cell for protein synthesis, and A could be some common cellular component that serves as the starting point for the specific reaction sequence leading to P.

Feedback Inhibition. If allowed to proceed at a constant, unrestrained rate, the pathway shown in Reaction Sequence 6-15 can convert large amounts of A to P, with possible adverse effects resulting from a depletion of A or an excessive accumulation of P. Clearly, the best interests of the cell are served when the pathway is functioning not at its maximum rate or even some constant rate, but at a rate that is carefully tuned to the cellular need for P.

Somehow, the enzymes of this pathway must be responsive to the cellular level of the product P in somewhat the same way that a furnace needs to be responsive to the temperature of the rooms it is intended to heat. In the latter case, a thermostat provides the necessary regulatory link between the furnace and its "product," heat. If there is too much heat, the thermostat turns the furnace off, inhibiting heat production. If heat is needed, this inhibition is relieved due to the lack of heat. In our enzyme example, the desired regulation is possible because the product P is a specific inhibitor of E_1, the enzyme that catalyzes the first reaction in the sequence.

This phenomenon is called **feedback** (or **end-product**) **inhibition** and is represented by the dashed arrow that connects the product P to enzyme E_1 in the following reaction sequence:

$$A \xrightarrow[E_1]{} B \xrightarrow[E_2]{} C \xrightarrow[E_3]{} D \xrightarrow[E_4]{} P \qquad \textbf{(6-16)}$$

Feedback inhibition of E_1 by P

Feedback inhibition is one of the most common mechanisms used by cells to ensure that the activities of reaction sequences are adjusted to cellular needs.

Figure 6-15 provides a specific example of such a pathway—the five-step sequence whereby the amino acid *isoleucine* is synthesized from *threonine,* another amino acid. In this case, the first enzyme in the pathway, *threonine deaminase,* is regulated by the concentration of isoleucine within the cell. If isoleucine is being used by the cell (in the synthesis of proteins, most likely), the isoleucine concentration will be low and the cell will need more. Under these conditions, threonine deaminase is active, and the pathway functions to produce more isoleucine. If the need for isoleucine decreases, isoleucine will begin to accumulate in the cell. This increase in its

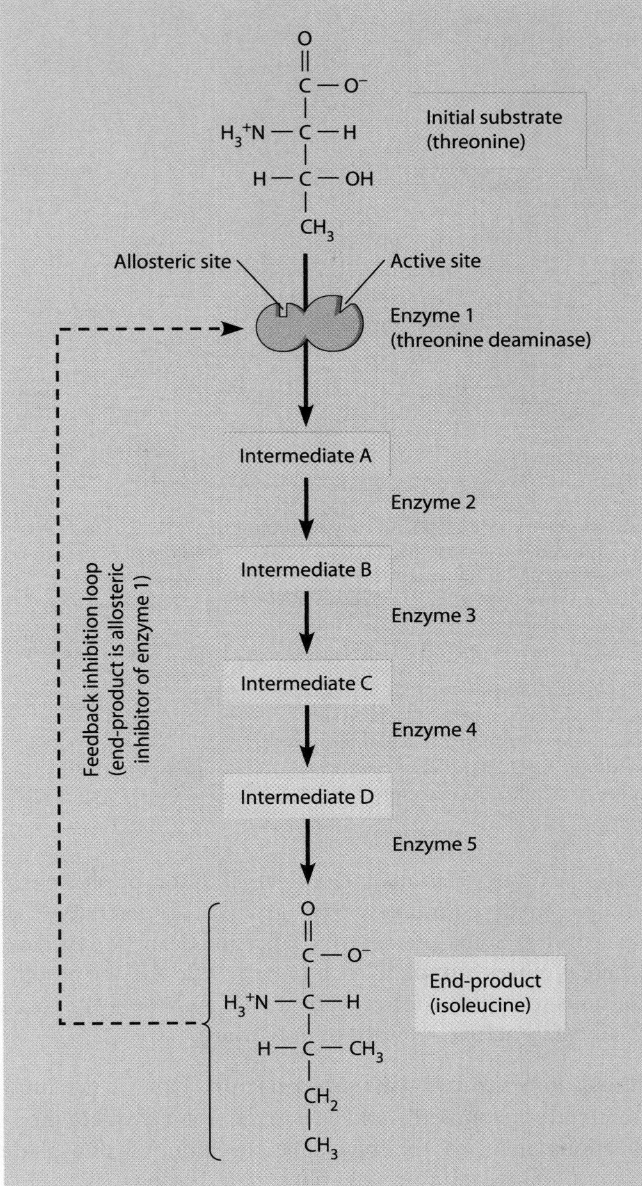

FIGURE 6-15 Allosteric Regulation of Enzyme Activity. A specific example of feedback inhibition is seen in the pathway by which the amino acid isoleucine is synthesized from threonine, another amino acid. The first enzyme in the sequence, threonine deaminase, is allosterically inhibited by isoleucine, which binds to the enzyme at a site *other* than the active site.

concentration will lead to inhibition of threonine deaminase and hence to a reduced rate of isoleucine synthesis.

Allosteric Regulation. How can the first enzyme in a pathway (e.g., enzyme E_1 in Reaction Sequence 6-16) be sensitive to the concentration of a substance P that is neither its substrate nor its immediate product? The answer to this question was first proposed in 1963 by Jacques Monod, Jean-Pierre Changeux, and François Jacob. Their model was quickly substantiated and went on to become the foundation for our understanding of **allosteric regulation.** The term *allosteric* derives from the Greek for "another shape (or form)," thereby indicating that all enzymes capable of allosteric regulation can exist in two different forms.

In one of the two forms, the enzyme has a high affinity for its substrate(s), leading to high activity. In the other form, it has little or no affinity for its substrate, giving little or no catalytic activity. Enzymes with this property are called **allosteric enzymes.** The two different forms of an allosteric enzyme are readily interconvertible and are, in fact, in equilibrium with each other.

Whether the active or inactive form of an allosteric enzyme is favored depends on the cellular concentration of the appropriate regulatory substance, called an **allosteric effector.** In the case of isoleucine synthesis, the allosteric effector is isoleucine and the allosteric enzyme is threonine deaminase. More generally, *an allosteric effector is a small organic molecule that regulates the activity of an enzyme for which it is neither the substrate nor the immediate product.*

An allosteric effector influences enzyme activity by binding to one of the two interconvertible forms of the enzyme, thereby stabilizing it in that state. The effector binds to the enzyme because of the presence on the enzyme surface of an **allosteric** (or **regulatory**) **site** that is distinct from the active site at which the catalytic event occurs. Thus, a distinguishing feature of all allosteric enzymes (and other allosteric proteins, as well) is the presence on the enzyme surface of an *active site* to which the substrate binds and an *allosteric site* to which the effector binds. In fact, some allosteric enzymes have multiple allosteric sites, each capable of recognizing a different effector.

An effector may be either an **allosteric inhibitor** or an **allosteric activator,** depending on the effect it has when bound to the allosteric site on the enzyme—that is, depending on whether the effector is bound to the low-affinity or high-affinity form of the enzyme (**Figure 6-16**). The binding of an allosteric inhibitor shifts the equilibrium between the two forms of the enzyme to favor the low-affinity state (Figure 6-16a). The binding of an allosteric activator, on the other hand, shifts the equilibrium in favor of the high-affinity state (Figure 6-16b). In either case, binding of the effector to the allosteric site stabilizes the enzyme in one of its two interconvertible forms, thereby either decreasing or increasing the likelihood of substrate binding.

Most allosteric enzymes are large, multisubunit proteins with an active site or an allosteric site on each subunit. Thus, quaternary protein structure is important for these enzymes. Typically, the active sites and allosteric sites are on different subunits of the protein, which are referred to as **catalytic subunits** and **regulatory subunits,** respectively (notice the C and R subunits of the enzyme molecules shown in Figure 6-16). This means, in turn, that the binding of effector molecules to the allosteric sites affects not just the shape of the regulatory subunits but that of the catalytic subunits as well.

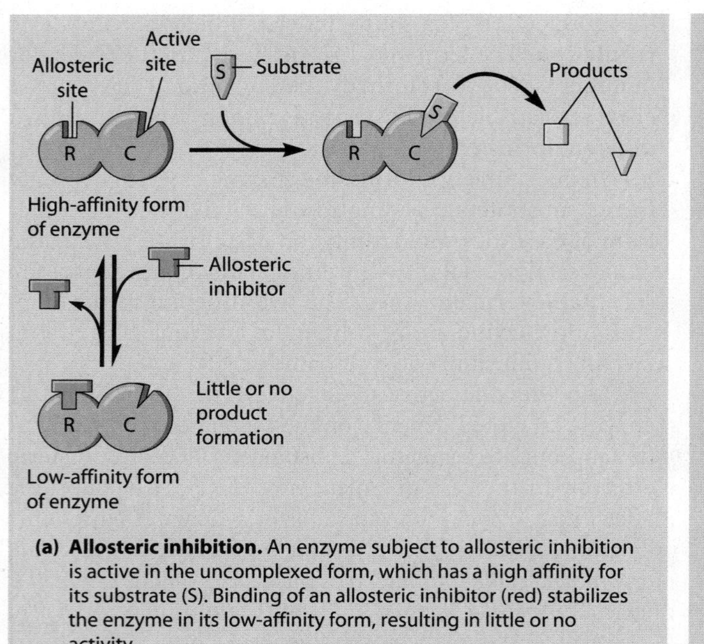

(a) Allosteric inhibition. An enzyme subject to allosteric inhibition is active in the uncomplexed form, which has a high affinity for its substrate (S). Binding of an allosteric inhibitor (red) stabilizes the enzyme in its low-affinity form, resulting in little or no activity.

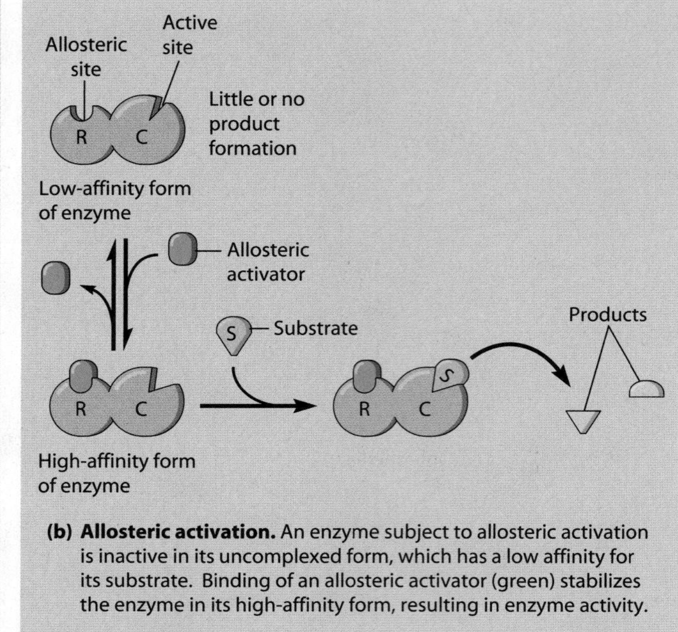

(b) Allosteric activation. An enzyme subject to allosteric activation is inactive in its uncomplexed form, which has a low affinity for its substrate. Binding of an allosteric activator (green) stabilizes the enzyme in its high-affinity form, resulting in enzyme activity.

MB Chemistry Review-Enzymes & Pathways: Controlling Enzymes

FIGURE 6-16 Mechanisms of Allosteric Inhibition and Activation. An allosteric enzyme consists of one or more catalytic subunits (C) and one or more regulatory subunits (R), each with an active site or an allosteric site, respectively. The enzyme exists in two forms, one with a high affinity for its substrate (and therefore a high likelihood of product formation) and the other with a low affinity (and a correspondingly low likelihood of product formation). The predominant form of the enzyme depends on the concentration of its allosteric effector(s).

Allosteric Enzymes Exhibit Cooperative Interactions Between Subunits

Many allosteric enzymes exhibit a property known as **cooperativity.** This means that, as the multiple catalytic sites on the enzyme bind substrate molecules, the enzyme undergoes conformational changes that affect the affinity of the remaining sites for substrate. Some enzymes show *positive cooperativity,* in which the binding of a substrate molecule to one catalytic subunit increases the affinity of other catalytic subunits for substrate. Other enzymes show *negative cooperativity,* in which the substrate binding to one catalytic subunit reduces the affinity of the other catalytic sites for substrate.

The cooperativity effect enables cells to produce enzymes that are more sensitive or less sensitive to changes in substrate concentration than would otherwise be predicted by Michaelis–Menten kinetics. Positive cooperativity causes an enzyme's catalytic activity to increase faster than normal as the substrate concentration is increased, whereas negative cooperativity means that enzyme activity increases more slowly than expected.

Enzymes Can Also Be Regulated by the Addition or Removal of Chemical Groups

In addition to allosteric regulation, many enzymes are subject to control by **covalent modification.** In this form of regulation, an enzyme's activity is affected by the addition or removal of specific chemical groups via covalent bonding. Common modifications include the addition of phosphate groups, methyl groups, acetyl groups, or derivatives of nucleotides. Some of these modifications can be reversed, whereas others cannot. In each case, the effect of the modification is to activate or to inactivate the enzyme—or at least to adjust its activity upward or downward.

Phosphorylation/Dephosphorylation. One of the most frequently encountered and best understood covalent modifications involves the reversible addition of phosphate groups. The addition of phosphate groups is called **phosphorylation** and occurs most commonly by transfer of the phosphate group from ATP to the hydroxyl group of a serine, threonine, or tyrosine residue in the protein. Enzymes that catalyze the phosphorylation of other enzymes (or of other proteins) are called **protein kinases.** The reversal of this process, **dephosphorylation,** involves the removal of a phosphate group from a phosphorylated protein, catalyzed by enzymes called **protein phosphatases.** Depending on the particular enzyme, phosphorylation may activate or inhibit the enzyme.

Enzyme regulation by reversible phosphorylation/dephosphorylation was discovered by Edmond Fischer and Edwin Krebs (not Hans Krebs, namesake of the Krebs cycle) at the University of Washington in the 1950s. They were awarded the 1992 Nobel Prize in Physiology or Medicine for their groundbreaking work with *glycogen phosphorylase,* a glycogen-degrading enzyme found in liver and skeletal muscle cells (**Figure 6-17**). In muscle cells, it provides glucose as an energy source for muscle

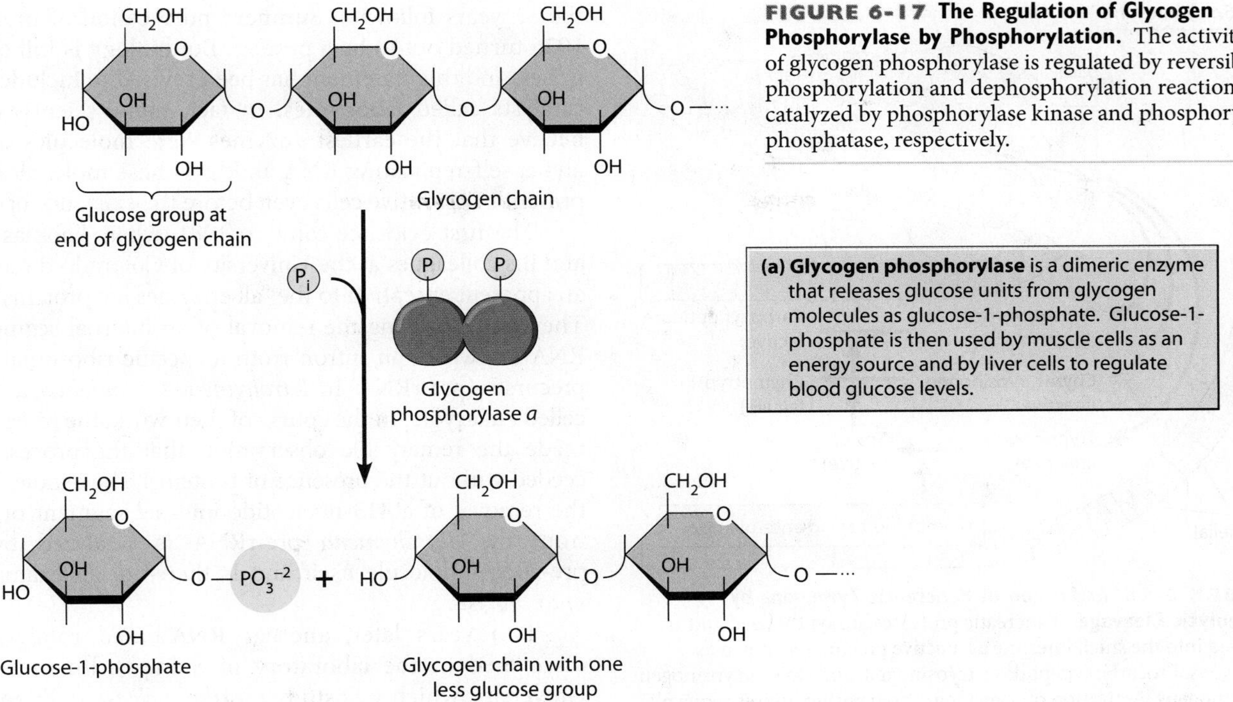

FIGURE 6-17 The Regulation of Glycogen Phosphorylase by Phosphorylation. The activity of glycogen phosphorylase is regulated by reversible phosphorylation and dephosphorylation reactions catalyzed by phosphorylase kinase and phosphorylase phosphatase, respectively.

(a) Glycogen phosphorylase is a dimeric enzyme that releases glucose units from glycogen molecules as glucose-1-phosphate. Glucose-1-phosphate is then used by muscle cells as an energy source and by liver cells to regulate blood glucose levels.

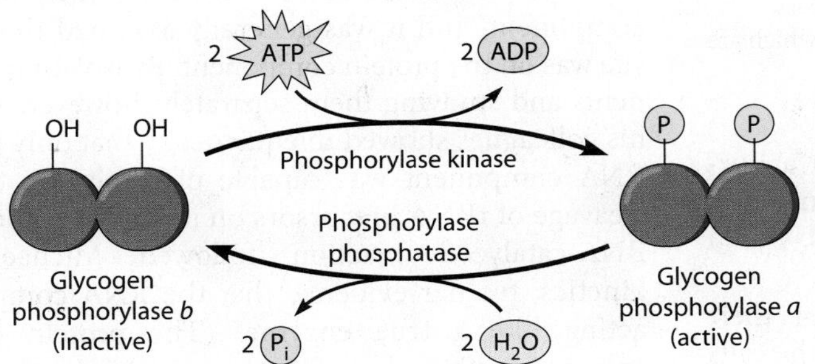

(b) Glycogen phosphorylase is regulated in part by a phosphorylation/dephosphorylation mechanism. The inactive form of the enzyme, phosphorylase *b*, can be converted to the active form, phosphorylase *a*, by a dual phosphorylation reaction catalyzed by the enzyme phosphorylase kinase. Removal of the phosphate groups by phosphorylase phosphatase returns the phosphorylase molecule to the inactive *b* form.

contraction and in liver cells it provides glucose for secretion to help maintain a constant blood glucose level. Glycogen phosphorylase breaks down glycogen by successive removal of glucose units as glucose-1-phosphate (Figure 6-17a). Regulation of this dimeric enzyme is achieved in part by the presence of two interconvertible forms of the enzyme—an active form called *phosphorylase a* and an inactive form called *phosphorylase b* (Figure 6-17b).

When glycogen breakdown is required in the cell, the inactive *b* form of the enzyme is converted into the active *a* form by the addition of a phosphate group to a particular serine on each of the two subunits of the phosphorylase molecule. The reaction is catalyzed by *phosphorylase kinase* and results in a conformational change of phosphorylase to the active form. When glycogen breakdown is no longer needed, the phosphate groups are removed from phosphorylase *a* by the enzyme *phosphorylase phosphatase*.

The muscle and liver forms of glycogen phosphorylase, known as isozymes, have subtle differences in their manner of regulation. Besides being regulated by the phosphorylation/dephosphorylation mechanism shown in Figure 6-17, liver glycogen phosphorylase, an allosteric enzyme, is inhibited by glucose and ATP and activated by AMP. Glucose binding to active liver glycogen phosphorylase *a* will inactivate it, blocking glycogen breakdown when glucose accumulates faster than needed. The existence of two levels of regulation for glycogen phosphorylase illustrates an important aspect of enzyme regulation. Many enzymes are controlled by two or more regulatory mechanisms, thereby enabling the cell to make appropriate responses to a variety of situations.

Proteolytic Cleavage. A different kind of covalent activation of enzymes involves the one-time, irreversible removal of a portion of the polypeptide chain by an appropriate proteolytic (protein-degrading) enzyme. This kind of modification, called **proteolytic cleavage,** is exemplified especially well by the proteolytic enzymes of the pancreas, which include trypsin, chymotrypsin, and carboxypeptidase. After being synthesized in the pancreas, these enzymes are secreted in an inactive form into the

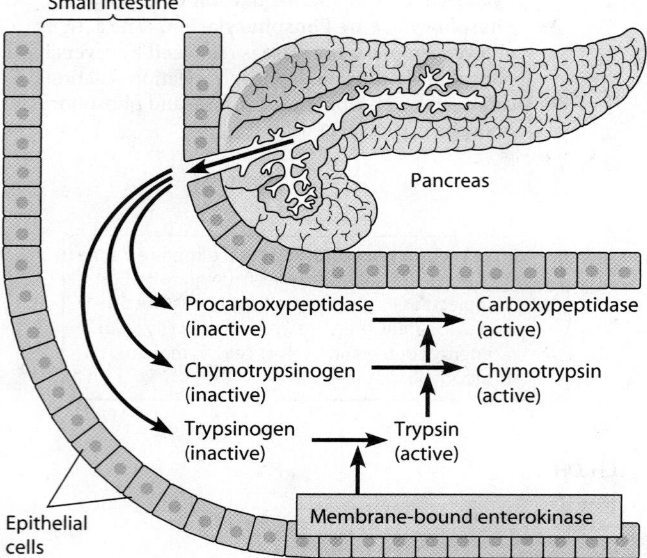

FIGURE 6-18 Activation of Pancreatic Zymogens by Proteolytic Cleavage. Pancreatic proteases are synthesized and secreted into the small intestine as inactive precursors known as zymogens. Procarboxypeptidase, trypsinogen, and chymotrypsinogen are zymogens. Activation of trypsinogen to trypsin requires removal of a hexapeptide segment by enterokinase, a membrane-bound duodenal enzyme. Trypsin then activates other zymogens by proteolytic cleavage. Procarboxypeptidase is activated by a single cleavage event, whereas the activation of chymotrypsinogen is a somewhat more complicated two-step process, the details of which are not shown here.

duodenum of the small intestine in response to a hormonal signal (**Figure 6-18**). These proteases can digest almost all ingested proteins into free amino acids, which are then absorbed by the intestinal epithelial cells.

Pancreatic proteases are not synthesized in their active form. That would likely cause problems for the cells of the pancreas, which must protect themselves against their own proteolytic enzymes. Instead, each of these enzymes is synthesized as a slightly larger, catalytically inactive molecule called a *zymogen*. Zymogens must themselves be cleaved proteolytically to yield active enzymes. For example, trypsin is synthesized initially as a zymogen called *trypsinogen*. When trypsinogen reaches the duodenum, it is activated by the removal of a hexapeptide (a string of six amino acids) from its N-terminus by the action of *enterokinase*, a membrane-bound protease produced by the duodenal cells. The active trypsin then activates other zymogens by specific proteolytic cleavages.

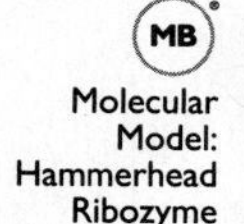

Molecular Model: Hammerhead Ribozyme Structure

RNA Molecules as Enzymes: Ribozymes

Until the early 1980s, it was thought that all enzymes were proteins. Indeed, that statement was regarded as a fundamental truth of cellular biology and was found in every textbook. Cell biologists became convinced that all enzymes were proteins because every enzyme isolated in the 55 years following Sumner's purification of urease in 1926 turned out to be a protein. But biology is full of surprises, and this statement has been revised to include RNA catalysts called **ribozymes.** In fact, many scientists now believe that the earliest enzymes were molecules of catalytic, self-replicating RNA and that these molecules were present in primitive cells even before the existence of DNA.

The first evidence came in 1981, when Thomas Cech and his colleagues at the University of Colorado discovered an apparent exception to the "all enzymes are proteins" rule. They were studying the removal of an internal segment of RNA known as an intron from a specific ribosomal RNA precursor (pre-rRNA) in *Tetrahymena thermophila*, a single-celled eukaryote. In the course of their work, the researchers made the remarkable observation that the process proceeded without the presence of proteins! They showed that the removal of a 413-nucleotide internal segment of RNA from the *Tetrahymena* pre-rRNA is catalyzed by the pre-rRNA molecule itself and is therefore an example of *autocatalysis*.

Two years later, another RNA-based catalyst was discovered in the laboratory of Sidney Altman at Yale University, which was studying *ribonuclease P*, an enzyme that cleaves transfer RNA precursors (pre-tRNAs) to yield functional RNA molecules. It had been known that ribonuclease P consisted of a protein component and an RNA component, and it was generally assumed that the active site was on the protein component. By isolating the components and studying them separately, however, Altman and his colleagues showed unequivocally that only the isolated RNA component was capable of catalyzing the specific cleavage of tRNA precursors on its own. Furthermore, the RNA-catalyzed reaction followed Michaelis–Menten kinetics, further evidence that the RNA component was acting like a true enzyme. (The protein component enhances activity but is not required for either substrate binding or cleavage.)

The significance of these findings was recognized by the Nobel Prize that Cech and Altman shared in 1989 for their discovery of ribozymes. Since these initial discoveries, additional examples of ribozymes have been reported. Of special significance is the active site for a crucial step in protein synthesis by ribosomes. The large ribosomal subunit (see Figure 4-22) is the site of the peptidyl transferase activity that catalyzes peptide bond formation.

The active site for this peptidyl transferase activity was for a long time assumed to be located on one of the protein molecules of the large subunit, with the rRNA providing a scaffold for structural support. However, in 1992, Harry Noller and his colleagues at the University of California, Santa Cruz, demonstrated that, despite the removal of at least 95% of the protein from the large ribosomal subunit, it retained 80% of the peptidyl transferase of the intact subunit. This strongly suggested that one of the rRNA molecules was the catalyst.

Furthermore, the activity was destroyed by treatment with ribonuclease, an enzyme that degrades RNA, but was not affected by proteinase K, an enzyme that degrades protein. Thus, the peptidyl transferase activity responsible for peptide bond formation in ribosomal protein synthesis is due to the rRNA, now known to be a ribozyme. It appears the function of the ribosomal proteins is to provide support and stabilization for the catalytic RNA, not the other way around! This supports the idea that RNA-based catalysis preceded protein-based catalysis.

The discovery of ribozymes has markedly changed the way we think about the origin of life on Earth. For many years, scientists had speculated that the first catalytic macromolecules must have been amino acid polymers resembling proteins. But this concept immediately ran into difficulty because there was no obvious way for a primitive protein to carry information or to replicate itself, which are two primary attributes of life. However, if the first catalysts were RNA rather than protein molecules, it becomes conceptually easier to imagine an "RNA world" in which RNA molecules acted both as catalysts and as replicating systems capable of transferring information from generation to generation.

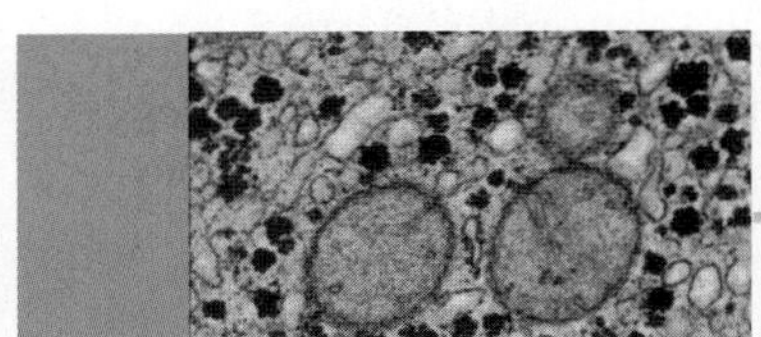

SUMMARY OF KEY POINTS

Activation Energy and the Metastable State

- While thermodynamics allows us to assess the feasibility of a reaction, it says nothing about the likelihood that the reaction will actually occur at a reasonable rate in the cell.
- For a given chemical reaction to occur in the cell, substrates must reach the transition state, which has a higher free energy than either the substrates or products. Reaching the transition state requires the input of activation energy.
- Because of this activation energy barrier, most biological compounds exist in an unreactive, metastable state. To ensure that the activation energy requirement is met and the transition state is achieved, a catalyst is required, which is always an enzyme in biological systems.

Enzymes as Biological Catalysts

MB Activity: How Enzymes Work; How Enzymes Function

- Catalysts, whether inorganic or organic, act by forming transient complexes with substrate molecules that lower the activation energy barrier and rapidly increase the rate of the particular reaction.
- Chemical reactions in cells are catalyzed by enzymes, which in some cases require organic or inorganic cofactors for activity. The vast majority of enzymes are proteins, but a few are composed of RNA and are known as ribozymes.
- Enzymes are exquisitely specific, either for a single specific substrate or for a class of closely related compounds. This is because the actual catalytic process takes place at the active site—a pocket or groove on the enzyme surface that only the correct substrates will fit into.
- The active site is composed of specific, noncontiguous amino acids that become positioned near each other as the protein folds into its tertiary structure. These amino acids are responsible for substrate binding, substrate activation, and catalysis.
- Binding of the appropriate substrate at the active site causes a change in the shape of the enzyme and substrate known as induced fit. This facilitates substrate activation, often by distorting one or more bonds in the substrate, by bringing necessary amino acid side chains into the active site, or by transferring protons and/or electrons between the enzyme and substrate.

Enzyme Kinetics

- An enzyme-catalyzed reaction proceeds via an enzyme-substrate intermediate. Most reactions follow Michaelis–Menten kinetics, characterized by a hyperbolic relationship between the initial reaction velocity v and the substrate concentration [S].
- The upper limit on velocity is called V_{max}, and the substrate concentration needed to reach one-half of this maximum velocity is termed the Michaelis constant, K_m. The hyperbolic relationship between v and [S] can be linearized by a double-reciprocal equation and plot, from which V_{max} and K_m can be determined graphically.
- Enzyme activity is sensitive to temperature, pH, and the ionic environment. Enzyme activity is also influenced by substrate availability, products, alternative substrates, substrate analogues, drugs, and toxins, most of which have an inhibitory effect.
- Irreversible inhibition involves covalent bonding of the inhibitor to the enzyme surface, permanently disabling the enzyme. A reversible inhibitor, on the other hand, binds noncovalently to an enzyme in a reversible manner, either at the active site (competitive inhibition) or elsewhere on the enzyme surface (noncompetitive inhibition).

Enzyme Regulation

- Enzymes must be regulated to adjust their activity levels to cellular needs. Substrate-level regulation involves the effects of substrate and product concentrations on the reaction rate. Additional control mechanisms include allosteric regulation and covalent modification.

- Most allosterically regulated enzymes catalyze the first step in a reaction sequence and are multisubunit proteins with both catalytic subunits and regulatory subunits. Each of the catalytic subunits has an active site that recognizes substrates, whereas each regulatory subunit has one or more allosteric sites that recognize specific effector molecules.
- Allosteric enzymes exist in an equilibrium between a low-activity and a high-activity form. A given effector binds to and stabilizes one of these two forms and therefore will either inhibit or activate the enzyme, depending on which form of the enzyme is bound by the effector.
- Enzyme inhibitors can act either irreversibly or reversibly. Irreversible inhibitors bind covalently to the enzyme, causing a permanent loss of activity. Reversible inhibitors bind noncovalently either to the active site (competitive inhibitors such as substrate analogs) or to a separate site on the enzyme (noncompetitive inhibition).
- Enzymes can also be regulated by covalent modification. The most common covalent modifications include phosphorylation, as seen with glycogen phosphorylase, and proteolytic cleavage, as occurs in the activation of proteolytic pancreatic zymogens.

RNA Molecules as Enzymes: Ribozymes

- Although it was long thought that all enzymes were proteins, we now recognize the catalytic properties of certain RNA molecules called ribozymes.
- These include some rRNA molecules that are able to catalyze the removal of their own introns, RNA components of enzymes that also contain protein components, and the rRNA component of the large ribosomal subunit.
- The discovery of ribozymes has changed the way we think about the origin of life on Earth because RNA molecules, unlike proteins, can both carry information and replicate themselves.

MB **www.masteringbiology.com**

1. MasteringBiology Assignments

Tutorials and Activities

- Molecular Model Structural Tutorials for visualization of 3D structures
- BioFlix Tutorials including 3D animations with hints and detailed feedback for common wrong answers
- Acids, Bases, & pH: pH; The pH Scale; Enzymes & Pathways: Enzyme Action; Enzyme and Substrate Concentrations; Enzymes & Pathways: Controlling Enzymes; Hammerhead Ribozyme Structure; How Enzymes Work; How Enzymes Function

Questions

- Reading Quiz
- Multiple Choice
- End-of-chapter questions

2. eText Read your book online, search, highlight text, take notes, and more

3. The Study Area

- Chapter Review Quiz, Guide to Techniques and Methods, 3D-Structure Tutorials, Activities, BioFlix, Videos, Biology Review, Word Study Tools, Glossary, Flashcards

PROBLEM SET

More challenging problems are marked with a •.

6-1 The Need for Enzymes. You should now be in a position to appreciate the difference between the thermodynamic feasibility of a reaction and the likelihood that it will actually proceed.

(a) Define the terms activation energy and transition state.

(b) Describe the effect of heat on enzyme activity and explain why using heat to alter enzyme activity is problematic in cells.

(c) An alternative solution is to lower the activation energy. What does it mean in molecular terms to say that a catalyst lowers the activation energy of a reaction?

(d) Organic chemists often use inorganic catalysts such as nickel, platinum, or cations in their reactions, whereas cells use proteins called enzymes. What advantages can you see to the use of enzymes? Can you think of any disadvantages?

(e) Describe the term 'quantum tunnelling.' Why is this reaction independent of temperature?

6-2 Activation Energy. As shown in Reaction 6-2, hydrogen peroxide, H_2O_2, decomposes to H_2O and O_2. The activation energy, E_A, for the uncatalyzed reaction at 20°C is 18 kcal/mol. The reaction can be catalyzed either by ferric ions (E_A = 13 kcal/mol) or by the enzyme *catalase* (E_A = 7 kcal/mol).

(a) Draw an activation energy diagram for this reaction under catalyzed and uncatalyzed conditions, and explain what it means for the activation energy to be lowered from 18 to 13 kcal/mol by ferric ions but from 18 to 7 kcal/mol by catalase.

(b) Suggest two properties of catalase that make it a more suitable intracellular catalyst than ferric ions.

(c) Suggest yet another way that the rate of hydrogen peroxide decomposition can be accelerated. Is this a suitable means of increasing reaction rates within cells? Why or why not?

6-3 Rate Enhancement by Catalysts. The decomposition of H_2O_2 to H_2O and O_2 shown in Reaction 6-2 can be catalyzed either by an inorganic catalyst (ferric ions) or by the enzyme catalase. Compared with the uncatalyzed rate, this reaction proceeds about 30,000 times faster in the presence of ferric ions but about 100,000,000 times faster in the presence of catalase, an iron-containing enzyme. Assume that 1 μg of catalase can decompose a given quantity of H_2O_2 in 1 minute at 25°C and that all reactions are carried out under sterile conditions.

(a) How long would it take for the same quantity of H_2O_2 to be decomposed in the presence of an amount of ferric ions equivalent to the iron content of 1 μg of catalase?

(b) How long would it take for the same quantity of H_2O_2 to decompose in the absence of a catalyst?

(c) Explain how these calculations illustrate the indispensability of catalysts and the superiority of enzymes over inorganic catalysts.

6-4 Temperature and pH Effects. Figure 6-4 illustrates enzyme activities as functions of temperature and pH. In general, the activity of a specific enzyme is highest at the temperature and pH that are characteristic of the environment in which the enzyme normally functions.

(a) Explain the shapes of the curves in Figure 6-4 in terms of the major chemical or physical factors that affect enzyme activity.

(b) For each enzyme in Figure 6-4, suggest the adaptive advantage of having the enzyme activity profile shown in the figure.

(c) Some enzymes have a very flat pH profile—that is, they have essentially the same activity over a broad pH range. How might you explain this observation?

6-5 Michaelis–Menten Kinetics. Figure 6-19 represents a Michaelis–Menten plot for a typical enzyme, with initial reaction velocity plotted as a function of substrate concentration. Three regions of the curve are identified by the letters A, B, and C.

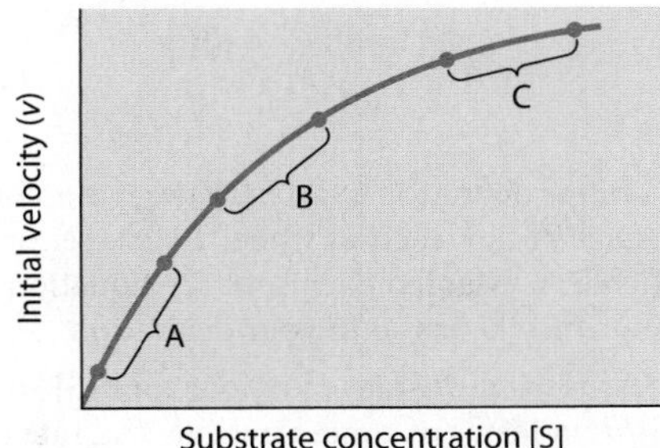

FIGURE 6-19 Analysis of the Michaelis–Menten Plot. See Problem 6-5.

For each of the statements that follow, indicate with a single letter which one of the three regions of the curve fits the statement best. A given letter can be used more than once.

(a) The active site of an enzyme molecule is occupied by substrate most of the time.

(b) The active site of an enzyme molecule is free most of the time.

(c) This is the range of substrate concentration in which most enzymes usually function in normal cells.

(d) This includes the point (K_m, $V_{max}/2$).

(e) Reaction velocity is limited mainly by the number of substrate molecules present.

(f) Reaction velocity is limited mainly by the number of enzyme molecules present.

6-6 Enzyme Kinetics. The enzyme β-galactosidase catalyzes the hydrolysis of the disaccharide lactose into its component monosaccharides:

$$\text{lactose} + H_2O \xrightarrow[\beta\text{-galactosidase}]{} \text{glucose} + \text{galactose} \quad \textbf{(6-17)}$$

To determine V_{max} and K_m of β-galactosidase for lactose, the same amount of enzyme (1 μg per tube) was incubated with a series of lactose concentrations under conditions where product concentrations remained negligible. At each lactose concentration, the initial reaction velocity was determined by assaying for the amount of lactose remaining at the end of the assay. The following data were obtained:

Lactose concentration (mM)	Rate of lactose consumption (μmol/min)
1	10.0
2	16.7
4	25.0
8	33.3
16	40.0
32	44.4

(a) Why is it necessary to specify that product concentrations remained negligible during the course of the reaction?

(b) Plot *v* (rate of lactose consumption) versus [S] (lactose concentration). Why is it that when the lactose concentration is doubled, the increase in velocity is always less than twofold?

(c) Calculate $1/v$ and $1/[S]$ for each entry on the data table, and plot $1/v$ versus $1/[S]$.

(d) Determine K_m and V_{max} from your double-reciprocal plot.

(e) On the same graph as part b, plot the results you would expect if each tube contained only 0.5 μg of enzyme. Explain your graph.

6-7 More Enzyme Kinetics. The galactose formed in Reaction 6-17 can be phosphorylated by the transfer of a phosphate group from ATP, a reaction catalyzed by the enzyme galactokinase:

$$\text{galactose} + \text{ATP} \xrightarrow[\text{galactokinase}]{} \text{galactose-1-phosphate} + \text{ADP} \quad \textbf{(6-18)}$$

Assume that you have isolated the galactokinase enzyme and have determined its kinetic parameters by varying the concentration of galactose in the presence of a constant, high (i.e., saturating) concentration of ATP. The double-reciprocal (Lineweaver–Burk) plot of the data is shown as **Figure 6-20**.

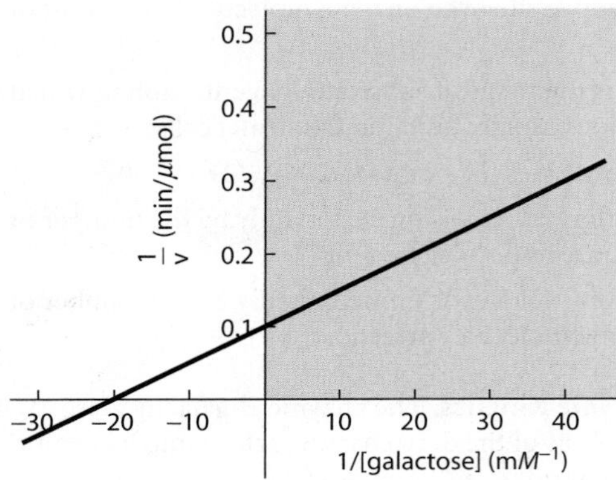

FIGURE 6-20 Double-Reciprocal Plot for the Enzyme Galactokinase. See Problem 6-7.

(a) What is the K_m of galactokinase for galactose under these assay conditions? What does K_m tell us about the enzyme?

(b) What is the V_{max} of the enzyme under these assay conditions? What does V_{max} tell us about the enzyme?

(c) Assume that you now repeat the experiment, but with the ATP concentration varied and galactose present at a constant, high concentration. Assuming that all other conditions are maintained as before, would you expect to get the same V_{max} value as in part b? Why or why not?

(d) In the experiment described in part c, the K_m value turned out to be very different from the value determined in part b. Can you explain why?

6-8 Turnover Number. Carbonic anhydrase catalyzes the reversible hydration of carbon dioxide to form bicarbonate ion:

$$CO_2 + H_2O \rightleftharpoons HCO_3^- + H^+ \quad \textbf{(6-19)}$$

As we will discover in Chapter 8, this reaction is important in the transport of carbon dioxide from body tissues to the lungs by red blood cells. Carbonic anhydrase has a molecular weight of 30,000 and a turnover number (k_{cat} value) of $1 \times 10^6 \text{sec}^{-1}$. Assume that you are given 1 mL of a solution containing 2.0 μg of pure carbonic anhydrase.

(a) At what rate (in millimoles of CO_2 consumed per second) will you expect this reaction to proceed under optimal conditions?

(b) Assuming standard temperature and pressure, how much CO_2 is that in mL per second?

6-9 Enzyme Kinetics: Wrong Again. For each of the following false statements, change the statement to make it true and explain your reasoning.

(a) Glycogen phosphorylase is responsible for the breakdown of glycogen. The enzyme is converted from the inactive *b* form, to the active *a* form by the removal of a phosphate group.

(b) The enzyme hexokinase can phosphorylate a number of hexoses displaying identical K_m values for each.

(c) Carboxypeptidase A is an enzyme which degrades dietary peptides. This digestive enzyme requires an iron ion to induce conformational change to facilitate catalysis.

(d) An enzyme that is subject to allosteric activation is most likely to catalyze the final reaction in a biosynthetic pathway.

(e) If researchers claim that an enzyme is allosterically activated by compound A and allosterically inhibited by compound B, one of these claims must be wrong.

6-10 What Type of Inhibition? A new mucinase enzyme was recently discovered that breaks down a glycoprotein in mucous membranes and contributes to bacterial vaginosis (*J. Clin. Microbiol.* 43:5504). You are a research pathologist testing a new inhibitor of this enzyme that you have discovered and want to design experiments to understand the nature of this inhibition. You have a supply of the normal glycoprotein substrate, the inhibitor, and an assay to measure product formation.

(a) How might you determine whether inhibition is reversible or irreversible?

(b) If you find that the inhibition is reversible, how would you determine whether the inhibition is competitive or noncompetitive?

6-11 Biological Relevance. Explain the biological relevance of each of the following observations concerning enzyme regulation.

(a) When you need a burst of energy, the hormones epinephrine and glucagon are secreted into your bloodstream and circulated to your muscle cells, where they initiate a cascade of reactions that leads to the phosphorylation of the inactive *b* form of glycogen phosphorylase, thereby converting the enzyme into the active *a* form.

(b) Even in the *a* form, glycogen phosphorylase is allosterically inhibited by a high concentration of glucose or ATP within a specific liver cell.

(c) Your pancreas synthesizes and secretes the proteolytic enzyme carboxypeptidase in the form of an inactive precursor called procarboxypeptidase, which is activated as a result of proteolytic cleavage by the enzyme trypsin in the duodenum of your small intestine.

• **6-12 Derivation of the Michaelis–Menten Equation.** For the enzyme-catalyzed reaction in which a substrate S is converted into a product P (see Reaction 6-5), velocity can be defined as the disappearance of substrate or the appearance of product per unit time:

$$v = -\frac{d[S]}{dt} = +\frac{d[P]}{dt} \quad \textbf{(6-20)}$$

Beginning with this definition and restricting your consideration to the initial stage of the reaction when [P] is essentially zero, derive the Michaelis–Menten equation (see Equation 6-7). The following points may help you in your derivation:

(a) Begin by expressing the rate equations for $d[S]/dt$, $d[P]/dt$, and $d[ES]/dt$ in terms of concentrations and rate constants.

(b) Assume a steady state at which the enzyme-substrate complex of Reaction 6-6 is being broken down at the same rate as it is being formed such that the net rate of change, $d[ES]/dt$, is zero.

(c) Note that the total amount of enzyme present, E_t, is the sum of the free form, E_f, plus the amount of complexed enzyme ES: $E_t = E_f + ES$.

(d) When you get that far, note that V_{max} and K_m can be defined as follows:

$$V_{max} = k_3[E_t] \quad K_m = \frac{k_2 + k_3}{k_1} \tag{6-21}$$

• **6-13 Linearizing Michaelis and Menten.** In addition to the Lineweaver–Burk plot (see Figure 6-10), two other straight-line forms of the Michaelis–Menten equation are sometimes used. The Eadie–Hofstee plot is a graph of $v/[S]$ versus v (see Figure 6-11), and the Hanes–Woolf plot graphs $[S]/v$ versus $[S]$.

(a) In both cases, show that the equation being graphed can be derived from the Michaelis–Menten equation by simple arithmetic manipulation.

(b) In both cases, indicate how K_m and V_{max} can be determined from the resulting graph.

(c) Make a Hanes–Woolf plot, with the intercepts and slope labeled as for the Lineweaver–Burk and Eadie–Hofstee plots (see Figures 6-10 and 6-11). Can you suggest why the Hanes–Woolf plot is the most statistically satisfactory of the three?

SUGGESTED READING

References of historical importance are marked with a •.

General References

Colowick, S. P., and N. O. Kaplan, eds. *Methods Enzymol.* New York: Academic Press, 1955–present (ongoing series).

Copeland, R. A. *Enzymes: A Practical Introduction to Structures, Mechanisms, and Data Analysis.* New York: Wiley, 2000.

Mathews, C. K., K. E. van Holde, and K. G. Ahern. *Biochemistry,* 3rd ed. San Francisco: Benjamin Cummings, 2000.

Nelson, D. L., and M. M. Cox. *Lehninger Principles of Biochemistry,* 5th ed. New York: W. H. Freeman, 2008.

Purich, D. L., and R. D. Allison. *The Enzyme Reference: A Comprehensive Guidebook to Enzyme Nomenclature, Reactions, and Methods.* Boston: Academic Press, 2002.

Historical References

• Changeux, J. P. The control of biochemical reactions. *Sci. Amer.* 212 (April 1965): 36.

• Cori, C. F. James B. Sumner and the chemical nature of enzymes. *Trends Biochem. Sci.* 6 (1981): 194.

• Friedmann, H., ed. *Benchmark Papers in Biochemistry. Vol. 1: Enzymes.* Stroudsburg, PA: Hutchinson Ross, 1981.

• Lineweaver, H., and D. Burk. The determination of enzyme dissociation constants. *J. Amer. Chem. Soc.* 56 (1934): 658.

• Monod, J., J. P. Changeux, and F. Jacob. Allosteric proteins and cellular control systems. *J. Mol. Biol.* 6 (1963): 306.

• Phillips, D. C. The three-dimensional structure of an enzyme molecule. *Sci. Amer.* 215 (November 1966): 78.

• Sumner, J. B. Enzymes. *Annu. Rev. Biochem.* 4 (1935): 37.

Structure and Function of Enzymes

Benkovic, S. J., and S. Hammes-Schiffler. A perspective on enzyme catalysis. *Science* 301 (2003): 1196.

Erlandsen, H., E. E. Abola, and R. C. Stevens. Combining structural genomics and enzymology: Completing the picture in metabolic pathways and enzyme active sites. *Curr. Opin. Struct. Biol.* 10 (2000): 719.

Fersht, A. *Structure and Mechanism in Protein Science: A Guide to Enzyme Catalysis and Protein Folding.* New York: W.H. Freeman, 1999.

Hatzimanikatis, V., C. Li, J. A. Ionita, and L. J. Broadbelt. Metabolic networks: Enzyme function and metabolite structure. *Curr. Opin. Struct. Biol.* 14 (2004): 300.

Minshull, J. et al. Predicting enzyme function from protein sequence. *Curr. Opin. Chem. Biol.* 9 (2005): 202.

Page, M. J., and E. DiCera. Role of Na^+ and K^+ in enzyme function. *Physiol. Rev.* 86 (2006): 1049.

Syed, U., and G. Yona. Enzyme function prediction with interpretable models. *Methods Mol. Biol.* 541 (2009): 373.

Walsh, C. Enabling the chemistry of life. *Nature* 409 (2001): 226.

Mechanisms of Enzyme Catalysis

Fitzpatrick, P. Special issue on enzyme catalysis. *Arch. Biochem. Biophys.* 433 (2005): 1.

Frey, P. A., and D. B. Northrop, eds. *Enzymatic Mechanisms.* Washington, DC: IOS Press, 1999.

Maragoni, A. G. *Enzyme Kinetics: A Modern Approach.* Hoboken, NJ: Wiley Interscience, 2003.

Matthews, J. N., and G. C. Allcock. Optimal designs for Michaelis–Menten kinetic studies. *Stat. Med.* 23 (2004): 477.

Mulholland, A. J. Modeling enzyme reaction mechanisms, specificity and catalysis. *Drug Disc. Today* 10 (2005): 1393.

Murphy, E. F., M. J. C. Crabbe, and S. G. Gilmour. Effective experimental design: Enzyme kinetics in the bioinformatics era. *Drug Disc. Today* 7 (2002): S187.

Sigel, R. K., and A. M. Pyle. Alternative roles for metal ions in enzyme catalysis and the implications for ribozyme chemistry. *Chem. Rev.* 107 (2007): 97.

Ribozymes: RNA as an Enzyme

Baghen, S., and M. Kashani-Sabet. Ribozymes in the age of molecular therapeutics. *Curr. Mol. Med.* 4 (2004): 489.

• Cech, T. R. The chemistry of self-splicing RNA and RNA enzymes. *Science* 236 (1987): 1532.

Khan, A. U. Ribozyme: A clinical tool. *Clin. Chim. Acta* 367 (2006): 20.

Krupp, G., and R. K. Gaur. *Ribozymes: Biochemistry and Biotechnology.* Natick, MA: Eaton Publishing, 2000.

Li, Q. X., P. Tan, N. Ke, and F. Wong-Staal. Ribozyme technology for cancer gene target identification and validation. *Adv. Cancer Res.* 96 (2007): 103.

Lilley, D. M. Structure, folding and mechanisms of ribozymes. *Curr. Opin. Struct. Biol.* 15 (2005): 313.

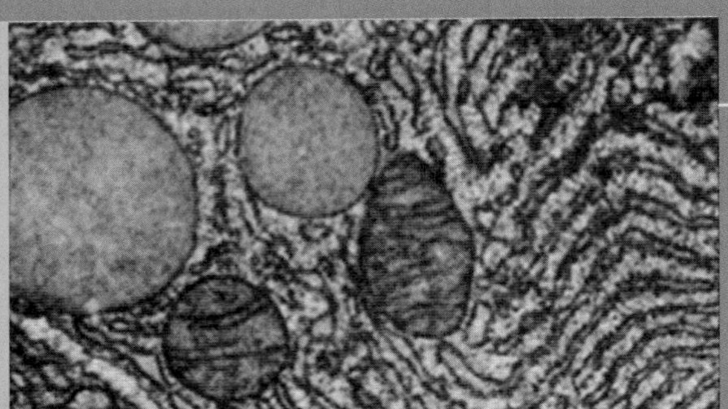

7 Membranes: Their Structure, Function, and Chemistry

See MasteringBiology® for tutorials, activities, and quizzes.

An essential feature of every cell is the presence of **membranes** *that define the boundaries of the cell and its various internal compartments. Even the casual observer of electron micrographs is likely to be struck by the prominence of membranes around and within cells, especially those of eukaryotic organisms (***Figure 7-1***). In Chapter 3, we encountered the structural molecules that allow membranes to be formed and began to study membranes and membrane-bounded organelles in Chapter 4. Now we are ready to look at membrane structure and function in greater detail. In this chapter, we will examine the molecular structure of membranes and explore the multiple roles that membranes play in the life of the cell. In Chapter 8, we will discuss the transport of solutes across membranes, with an emphasis on the mechanisms involved.*

The Functions of Membranes

We begin our discussion by noting that biological membranes play five related yet distinct roles, as illustrated in **Figure 7-2**. ❶ They define the boundaries of the cell and its organelles and act as permeability barriers. ❷ They serve as sites for specific biochemical functions, such as electron transport during mitochondrial respiration or protein processing in the ER. ❸ Membranes also possess transport proteins that regulate the movement of substances into and out of the cell and its organelles. ❹ In addition, membranes contain the protein molecules that act as receptors to detect external signals. ❺ Finally, they provide mechanisms for cell-to-cell contact, adhesion, and communication. Each of these functions is described briefly in the following five sections.

Membranes Define Boundaries and Serve as Permeability Barriers

One of the most obvious functions of membranes is to define the boundaries of the cell and its compartments and to serve as permeability barriers. The interior of the cell must be physically separated from the surrounding environment, not only to keep desirable substances in the cell but also to keep undesirable substances out. Membranes serve this purpose well because the hydrophobic interior of the membrane is an effective permeability barrier for hydrophilic molecules and ions. The permeability barrier for the cell as a whole is the **plasma** (or **cell**) **membrane,** a membrane that surrounds the cell and regulates the passage of materials both into and out of cells. In addition to the plasma membrane, various **intracellular membranes** serve to compartmentalize functions within eukaryotic cells.

Recently, there has been much interest in a class of *antimicrobial peptides (AMPs),* which are small molecules of 10–50 amino acids that can affect this permeability barrier. Over 1200 different AMPs are known, with more than 20 being produced by human skin alone. One class of AMPs consists of cationic, amphipathic molecules that disrupt bacterial membrane structure by interacting with the negatively charged phospholipids in their cell membranes. These AMPs thus act like detergents and disrupt the membrane structure, causing holes to form that destroy the cell's permeability barrier and kill the cells. Some AMPs have shown promise as antiviral agents in disrupting the outer covering of membrane-enclosed viruses such as the human immunodeficiency virus (HIV).

Membranes Are Sites of Specific Proteins and Therefore of Specific Functions

Membranes have specific functions associated with them because the molecules and structures responsible for those functions—proteins, in most cases—are either embedded in or localized on membranes. One of the most useful ways to characterize a specific membrane, in fact, is to describe the particular enzymes, transport proteins, receptors, and other molecules associated with it.

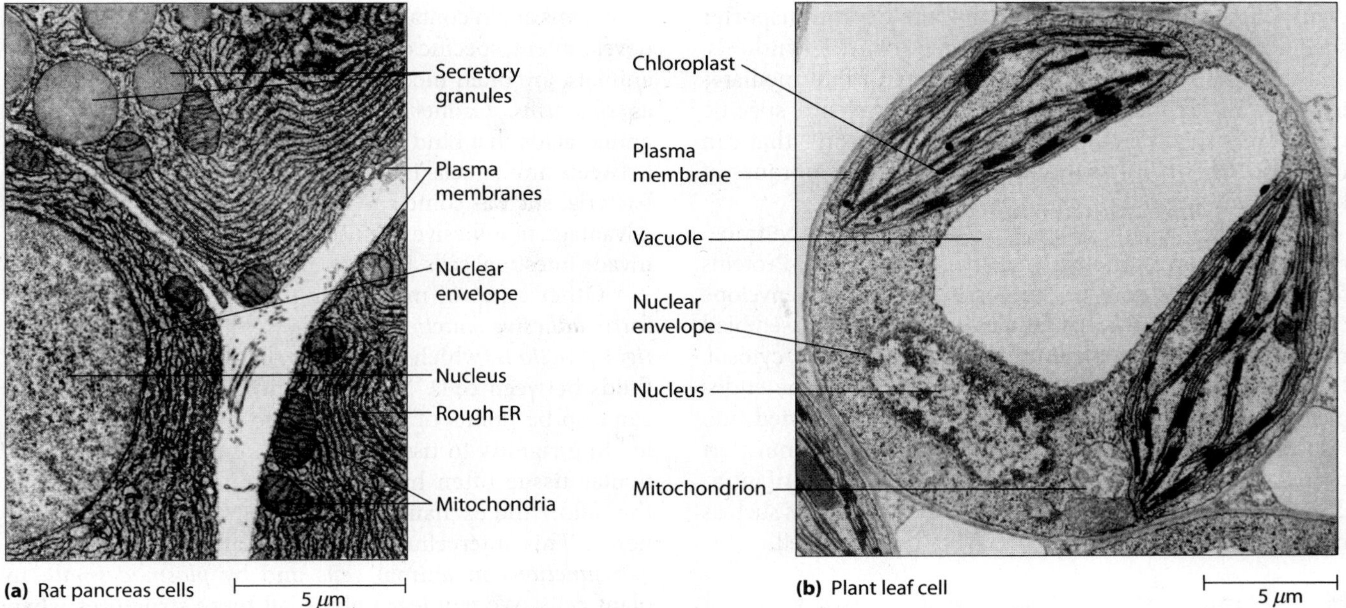

FIGURE 7-1 The Prominence of Membranes Around and Within Eukaryotic Cells. Among the structures of eukaryotic cells that involve membranes are the plasma membrane, nucleus, chloroplasts, mitochondria, endoplasmic reticulum (ER), secretory granules, and vacuoles. These structures are shown here in **(a)** portions of three cells from a rat pancreas and **(b)** a plant leaf cell (TEMs).

For example, many distinctive enzymes are present in or on the plasma membrane or the membranes of particular organelles. Such enzymes are often useful as *markers* to identify particular membranes during the isolation of organelles and organelle membranes from suspensions of disrupted cells (see Box 12A, pages 355–357). For example, *glucose-6-phosphatase* is a membrane-bound enzyme found in the endoplasmic reticulum, and its presence in, say, a preparation of mitochondria would demonstrate contamination with ER membranes.

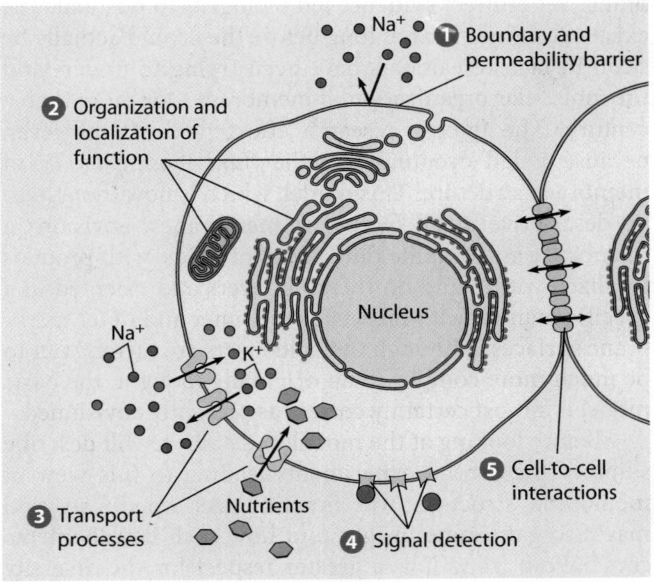

FIGURE 7-2 Functions of Membranes. Membranes not only define the cell and its organelles but also have a number of important functions, including transport, signaling, and adhesion.

Other functions associated with specific membranes are a direct result of the particular proteins present in these membranes. For example, the plasma membrane contains the enzymes that synthesize the cell wall of plants, fungi, and bacteria. In vertebrate cells, the plasma membrane contains enzymes that secrete the materials that make up the extracellular matrix. Other membrane proteins, such as those in chloroplast and mitochondrial membranes or in the bacterial plasma membrane, are critical for energy-generating processes such as photosynthesis and respiration.

Membrane Proteins Regulate the Transport of Solutes

Another function of membrane proteins is to carry out and regulate the *transport* of substances into and out of cells and their organelles. Nutrients, ions, gases, water, and other substances are taken up into various compartments, and various products and wastes must be removed. While lipophilic molecules, very small molecules, and gases can typically diffuse directly across cellular membranes, most substances needed by the cell require transport proteins that recognize and transport a specific molecule or a group of similar molecules.

For example, cells may have specific transporters for glucose, amino acids, or other nutrients. Your nerve cells transmit electrical signals as Na^+ and K^+ ions are transported across the plasma membrane of neurons by specific ion channel proteins. Transport proteins in muscle cells move calcium ions across membranes to assist in muscle contraction. The chloroplast membrane has a transporter specific for the phosphate ions needed for ATP synthesis, and the mitochondrion has transporters for intermediates involved in aerobic respiration. There is even a specific transporter for water—known as an aquaporin—that can rapidly transport water molecules through membranes of kidney cells to facilitate urine production.

Molecules as large as proteins and RNA can be transferred across membranes by transport proteins. Proteins form the nuclear pore complexes in the nuclear envelope through which mRNA molecules and partially assembled ribosomes can move from the nucleus to the cytosol. In some cases, proteins that are synthesized on the endoplasmic reticulum or in the cytosol can be imported into lysosomes, peroxisomes, or mitochondria via transport proteins. In other cases, proteins in the membranes of intracellular vesicles facilitate the movement of molecules such as neurotransmitters either into the cell or out of the cell.

Membrane Proteins Detect and Transmit Electrical and Chemical Signals

Cells receive information from their environment, usually in the form of electrical or chemical signals that impinge on the outer surface of the cell. The nerve impulses being sent from your eyes to your brain as you read these words are examples of such signals, as are the various hormones present in your circulatory system. *Signal transduction* is the term used to describe the specific mechanisms used to transmit such signals from the outer surface of cells to the cell interior.

Many chemical signal molecules bind to specific membrane proteins known as *receptors* on the outer surface of the plasma membrane. Binding of these signal molecules to their receptors triggers specific chemical events on the inner surface of the membrane that lead to changes in cell function. For example, muscle and liver cell membranes contain insulin receptors and can therefore respond to this hormone, which helps cells take in glucose. White blood cells have specific receptors that recognize chemical signals from bacteria and initiate a cellular defense response.

Many plant cells have a transmembrane receptor protein that detects the gaseous hormone ethylene and transmits a signal to the cell that can affect a variety of processes including seed germination, fruit ripening, and defense against pathogens. Bacteria often have plasma membrane receptors that sense nutrients in the environment and can signal the cell to move toward these nutrients. Thus, membrane receptors allow cells to recognize, transmit, and respond to a variety of specific signals in nearly all types of cells.

Membrane Proteins Mediate Cell Adhesion and Cell-to-Cell Communication

Membrane proteins also mediate adhesion and communication between adjacent cells. Although textbooks often depict cells as separate, isolated entities, most cells in multicellular organisms are in contact with other cells. During embryonic development, specific cell-to-cell contacts are critical and, in animals, are often mediated by membrane proteins known as *cadherins.* Cadherins have extracellular sequences of amino acids that bind calcium ions and stimulate adhesion between similar cells in a tissue. However, some pathogenic bacteria, such as some species of *Listeria* and *Shigella,* take advantage of adhesive membrane proteins to attach to and invade intestinal cells and cause disease.

Other types of membrane proteins in animal tissues form *adhesive junctions,* which hold cells together, and *tight junctions,* which form seals that block the passage of fluids between cells. Membrane proteins such as *ankyrin* can also be points of attachment to the cell cytoskeleton, lending rigidity to tissues. In addition, cells within a particular tissue often have direct cytoplasmic connections that allow the exchange of at least some cellular components. This intercellular communication is provided by *gap junctions* in animal cells and by *plasmodesmata* in plant cells. We will learn about all these structures when we move "beyond the cell" in Chapter 17.

All the functions we have just considered—compartmentalization, localization of function, transport, signal detection, and intercellular communication—depend on the chemical composition and structural features of membranes. It is to these topics that we now turn as we consider how our present understanding of membrane structure developed.

Models of Membrane Structure: An Experimental Perspective

Until electron microscopy was applied to the study of cell structure in the early 1950s, no one had ever seen a membrane. Yet indirect evidence led biologists to postulate the existence of membranes long before they could actually be seen. In fact, researchers have been trying to understand the molecular organization of membranes for more than a century. The intense research effort paid off, however, because it led eventually to the *fluid mosaic model* of membrane structure. This model, which is now thought to be descriptive of all biological membranes, envisions a membrane as two quite fluid layers of lipids, with proteins localized within and on the lipid layers and oriented in a specific manner with respect to the inner and outer membrane surfaces. Although the lipid layers are turning out to be much more complex than originally thought, the basic model is almost certainly correct as presently envisioned.

Before looking at the model in detail, we will describe some of the central experiments leading to this view of membrane structure and function. As we do so, you may also gain some insight into how such developments come about, as well as a greater respect for the diversity of approaches and techniques that are often important in advancing our understanding of biological phenomena. **Figure 7-3** presents a chronology of membrane studies

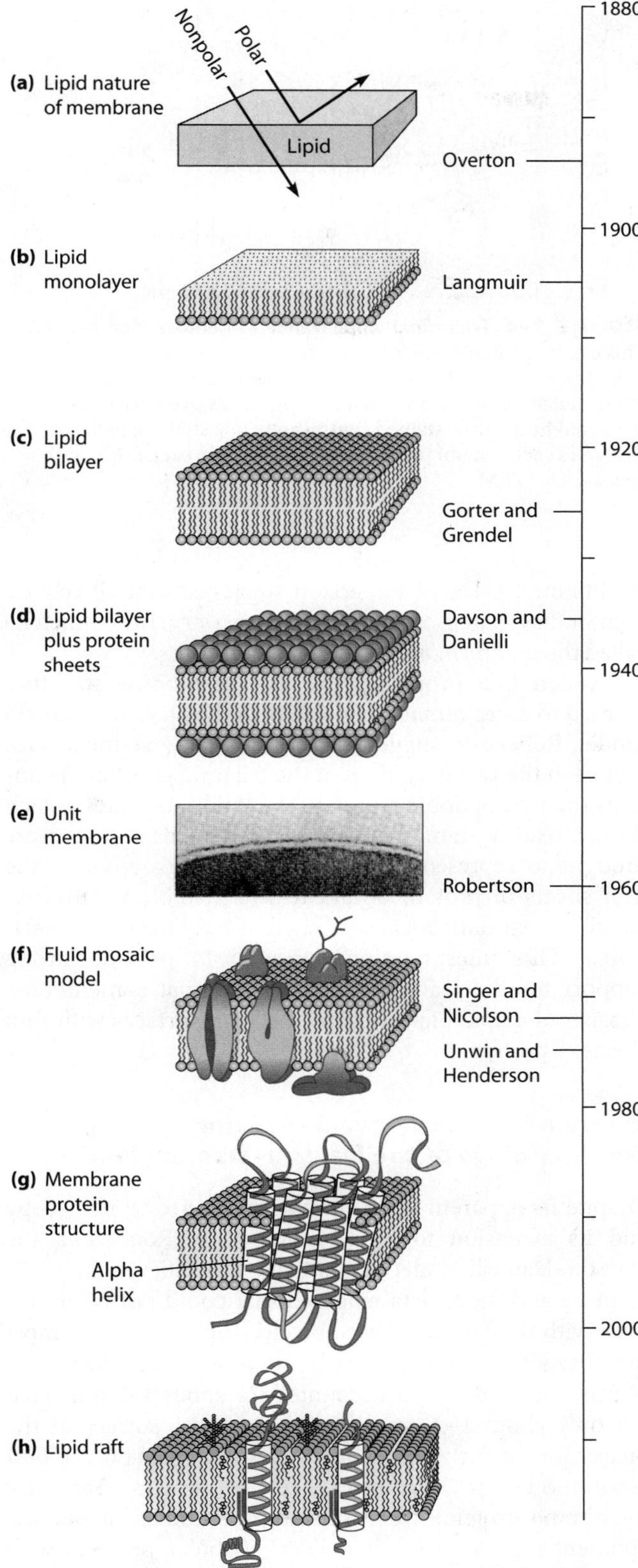

FIGURE 7-3 Timeline for Development of the Fluid Mosaic Model. The fluid mosaic model of membrane structure that Singer and Nicolson proposed in 1972 was the culmination of studies dating back to the 1890s **(a)**–**(e)**. This model **(f)** has been significantly refined by subsequent studies **(g** and **h)**.

that began over a century ago and led eventually to our current understanding of membranes as fluid mosaics.

Overton and Langmuir: Lipids Are Important Components of Membranes

A good starting point for our experimental overview is the pioneering work of German scientist Charles Ernest Overton in the 1890s. Working with cells of plant root hairs, he observed that lipid-soluble substances penetrate readily into cells, whereas water-soluble substances do not. From his studies, Overton concluded that lipids are present on the cell surface as some sort of "coat" (Figure 7-3a). He even suggested that cell coats are probably mixtures of cholesterol and lecithin, an insight that proved to be remarkably farsighted in light of what we now know about the prominence of sterols and phospholipids as membrane components.

A second important advance came about a decade later through the work of Irving Langmuir, who studied the behavior of purified phospholipids by dissolving them in benzene and layering samples of the benzene-lipid solution onto a water surface. As the benzene evaporated, the molecules were left as a lipid film one molecule thick—that is, a "monolayer." Because phospholipids are *amphipathic* molecules (see Figure 2-11), Langmuir reasoned that the phospholipids orient themselves on water such that their hydrophilic heads face the water and their hydrophobic tails protrude away from the water (Figure 7-3b). Langmuir's lipid monolayer became the basis for further thought about membrane structure in the early years of the twentieth century.

Gorter and Grendel: The Basis of Membrane Structure Is a Lipid Bilayer

The next major advance came in 1925 when two Dutch physiologists, Evert Gorter and F. Grendel, extracted the lipids from a known number of erythrocytes (red blood cells) and used Langmuir's method to spread the lipids as a monolayer on a water surface. They found that the area of the lipid film on the water was about twice the estimated total surface area of the erythrocyte. Therefore, they concluded that the erythrocyte plasma membrane consists of not one but *two* layers of lipids.

Hypothesizing a bilayer structure, Gorter and Grendel reasoned that it would be thermodynamically favorable for the nonpolar hydrocarbon chains of each layer to face inward, away from the aqueous milieu on either side of the membrane. The polar hydrophilic groups of each layer would then face outward, toward the aqueous environment on either side of the membrane (Figure 7-3c). Gorter and Grendel's experiment and their conclusions were momentous because this work represented the first attempt to understand membranes at the molecular level. Moreover, the **lipid bilayer** they envisioned became the basic underlying assumption for each successive refinement in our understanding of membrane structure.

Davson and Danielli: Membranes Also Contain Proteins

Shortly after Gorter and Grendel proposed their bilayer model in 1925, it became clear that a simple lipid bilayer, could not explain all the properties of membranes—particularly those related to *surface tension, solute permeability,* and *electrical resistance.* For example, the surface tension of a lipid film was significantly higher than that of cellular membranes but could be lowered by adding protein to the lipid film. Moreover, sugars, ions, and other hydrophilic solutes readily moved into and out of cells even though pure lipid bilayers are nearly impermeable to water-soluble substances.

To explain such differences, Hugh Davson and James Danielli suggested that proteins are present in membranes. They proposed in 1935 that biological membranes consist of lipid bilayers that are coated on both sides with thin sheets of protein (Figure 7-3d). Their model, a protein-lipid-protein "sandwich," was the first detailed representation of membrane organization and dominated the thinking of cell biologists for the next several decades.

The original model was later modified to accommodate additional findings. Particularly notable was the suggestion, made in 1954, that hydrophilic proteins might penetrate into the membrane in places to provide polar pores through an otherwise hydrophobic bilayer. These proteins could then allow water-soluble substances to cross the cell membrane. Specifically, the lipid interior accounted for the hydrophobic properties of membranes, and the protein components explained their hydrophilic properties.

The real significance of the Davson–Danielli model, however, was that it recognized the importance of proteins in membrane structure. This feature, more than any other, made the Davson–Danielli sandwich the basis for much subsequent research on membrane structure.

Robertson: All Membranes Share a Common Underlying Structure

With the advent of electron microscopy in the 1950s, cell biologists could finally verify the presence of a plasma membrane around each cell. They could also observe that most subcellular organelles are bounded by similar membranes. Furthermore, when membranes were stained with osmium, a heavy metal, and then examined closely at high magnification, they were found to have extensive regions of "railroad track" structure that appeared as two dark lines separated by a lightly stained central zone, with an overall thickness of 6–8 nm. This pattern is seen in **Figure 7-4** for the plasma membranes of two adjacent cells that are separated from each other by a thin intercellular space. Because this same staining pattern was observed with many different kinds of membranes, J. David Robertson suggested that all cellular membranes share a common underlying structure, which he called the *unit membrane* (see Figure 7-3e).

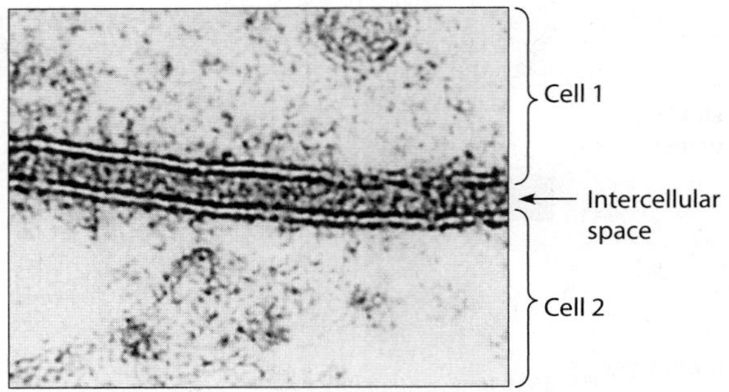

FIGURE 7-4 Trilaminar Appearance of Cellular Membranes. This electron micrograph of a thin section through two adjacent cells shows their plasma membranes separated by a small intercellular space. Each membrane appears as two dark lines separated by a lightly stained central zone in a staining pattern that gives each membrane a trilaminar, or "railroad track," appearance (TEM).

When first proposed, the unit membrane structure seemed to agree remarkably well with the Davson–Danielli model. Robertson suggested that the lightly stained space (between the two dark lines of the trilaminar pattern) contains the hydrophobic region of the lipid molecules, which do not readily stain. Conversely, the two dark lines were thought to represent phospholipid head groups and the thin sheets of protein bound to the membrane surfaces, which appear dark because of their affinity for heavy metal stains. This interpretation appeared to provide strong support for the Davson–Danielli view that a membrane consists of a lipid bilayer coated on both surfaces with thin sheets of protein.

Further Research Revealed Major Shortcomings of the Davson–Danielli Model

Despite its apparent confirmation by electron microscopy and its extension to all membranes by Robertson, the Davson–Danielli model encountered difficulties in the 1960s as more and more data emerged that could not be reconciled with their model. Based on electron microscopy, most membranes were reported to be about 6–8 nm thick—and, of this, the lipid bilayer accounted for about 4–5 nm. That left only about 1–2 nm of space on either surface of the bilayer for the membrane protein, a space that could at best accommodate a thin monolayer of protein. Yet after membrane proteins were isolated and studied, it became apparent that most of them were globular proteins with sizes and shapes that are inconsistent with the concept of thin sheets of protein on the two surfaces of the membrane.

As a further complication, the Davson–Danielli model did not readily account for the distinctiveness of different kinds of membranes. Depending on their source, membranes vary considerably in chemical composition and

Table 7-1 Protein, Lipid, and Carbohydrate Content of Biological Membranes

Membrane	Approximate Percentage by Weight			
	Protein	Lipid	Carbohydrate	Protein/Lipid Ratio
Plasma membrane				
Human erythrocyte	49	43	8	1.14
Mammalian liver cell	54	36	10	1.50
Amoeba	54	42	4	1.29
Myelin sheath of nerve axon	18	79	3	0.23
Nuclear envelope	66	32	2	2.06
Endoplasmic reticulum	63	27	10	2.33
Golgi complex	64	26	10	2.46
Chloroplast thylakoids	70	30	0	2.33
Mitochondrial outer membrane	55	45	0	1.22
Mitochondrial inner membrane	78	22	0	3.54
Gram-positive bacterium	75	25	0	3.00

especially in the ratio of protein to lipid (**Table 7-1**), which can vary from 3 or more in some bacterial cells to only 0.23 for the myelin sheath surrounding nerve axons. Even the two membranes of the mitochondrion differ significantly: The protein/lipid ratio is about 1.2 for the outer membrane and about 3.5 for the inner membrane, which contains all the enzymes and proteins related to electron transport and ATP synthesis. Yet all of these membranes look essentially the same when visualized using electron microscopy.

The Davson–Danielli model was also called into question by studies in which membranes were exposed to *phospholipases,* enzymes that degrade phospholipids by removing their head groups. According to the model, the hydrophilic head groups of membrane lipids should be covered by a layer of protein and therefore protected from phospholipase digestion. However, up to 75% of the membrane phospholipid can be degraded when the membrane is exposed to phospholipases, suggesting that many of the phospholipid head groups are exposed at the membrane surface and not covered by a layer of protein.

Moreover, the surface localization of membrane proteins specified by the Davson–Danielli model was not supported by the experience of scientists who tried to isolate such proteins. Most membrane proteins turned out to be quite insoluble in water and could be extracted only by using organic solvents or detergents. These observations indicated that many membrane proteins are hydrophobic (or at least amphipathic) and suggested that they are located, at least partially, within the hydrophobic interior of the membrane rather than on either of its surfaces.

Singer and Nicolson: A Membrane Consists of a Mosaic of Proteins in a Fluid Lipid Bilayer

The preceding problems with the Davson–Danielli model stimulated considerable interest in the development of new ideas about membrane organization, culminating in 1972 with the **fluid mosaic model** proposed by S. Jonathan Singer and Garth Nicolson. This model, which now dominates our view of membrane organization, has two key features, both implied by its name. Simply put, the model envisions a membrane as a *mosaic* of proteins embedded in, or at least attached to, a *fluid* lipid bilayer (Figure 7-3f). This model retained the basic lipid bilayer structure of earlier models but viewed membrane proteins in an entirely different way—not as thin sheets on the membrane surface but as discrete globular entities within the lipid bilayer (**Figure 7-5a**).

This way of thinking about membrane proteins was revolutionary when Singer and Nicolson first proposed it, but it turned out to fit the data quite well. Three classes of membrane proteins are now recognized based on differences in how the proteins are linked to the bilayer. *Integral membrane proteins* are embedded within the lipid bilayer, where they are held in place by the affinity of hydrophobic segments of the protein for the hydrophobic interior of the lipid bilayer. *Peripheral proteins* are much more hydrophilic and are therefore located on the surface of the membrane, where they are linked noncovalently to the polar head groups of phospholipids and/or to the hydrophilic parts of other membrane proteins. *Lipid-anchored proteins* are essentially hydrophilic proteins and therefore reside on membrane surfaces, but they are covalently attached to lipid molecules that are embedded within the bilayer.

The fluid nature of the membrane is the second critical feature of the Singer–Nicolson model. Rather than being rigidly locked in place, most of the lipid components of a membrane are in constant motion, capable of lateral mobility (i.e., movement parallel to the membrane surface). Many membrane proteins are also able to move laterally within the membrane, although some proteins are anchored to structural elements such as the cytoskeleton on one side of the membrane or the other and are therefore restricted in their mobility.

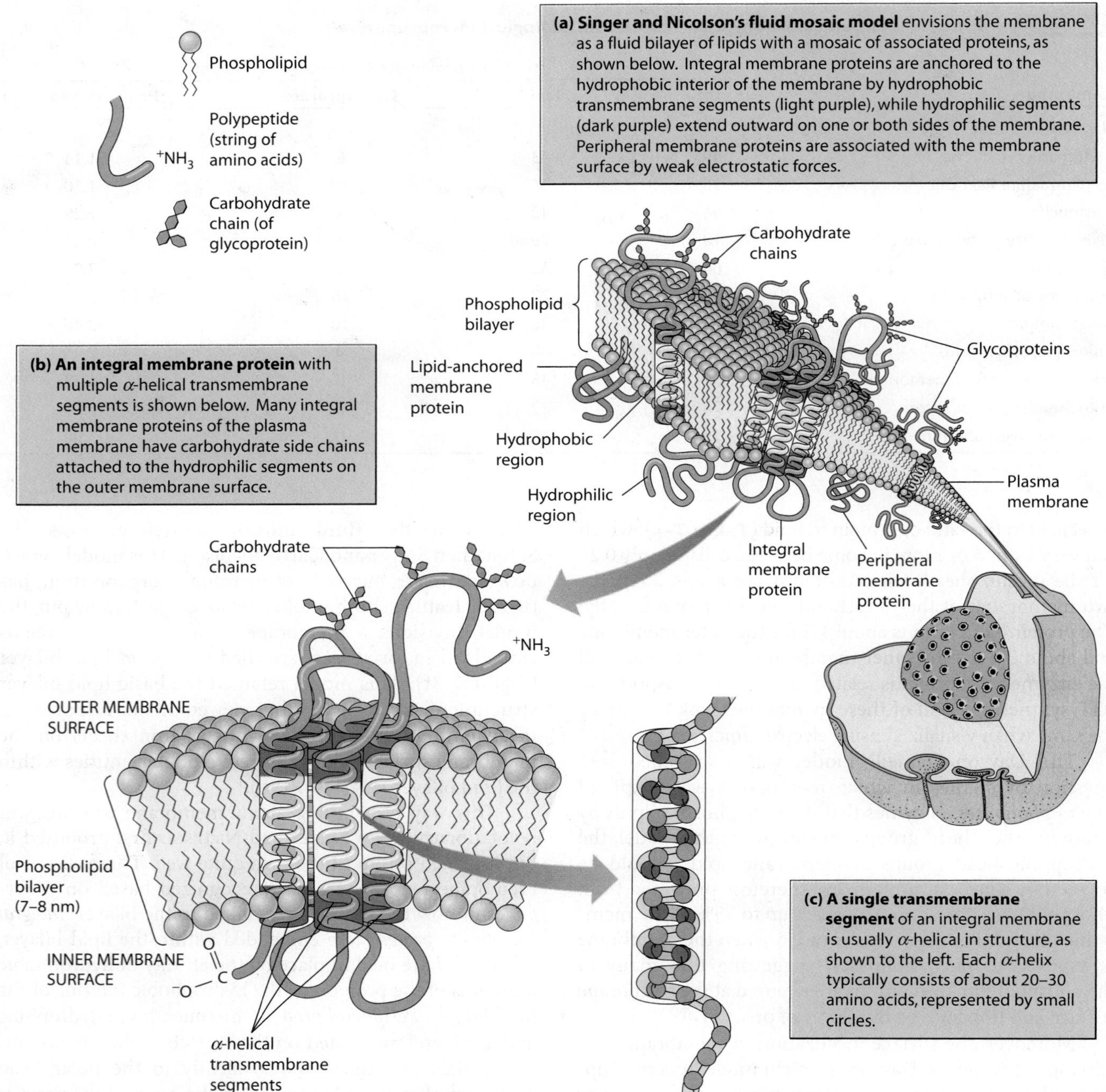

Phospholipid
Polypeptide (string of amino acids)
$^{+}NH_3$
Carbohydrate chain (of glycoprotein)
(a) Singer and Nicolson's fluid mosaic model envisions the membrane as a fluid bilayer of lipids with a mosaic of associated proteins, as shown below. Integral membrane proteins are anchored to the hydrophobic interior of the membrane by hydrophobic transmembrane segments (light purple), while hydrophilic segments (dark purple) extend outward on one or both sides of the membrane. Peripheral membrane proteins are associated with the membrane surface by weak electrostatic forces.
Carbohydrate chains
Phospholipid bilayer
Glycoproteins
Lipid-anchored membrane protein
Hydrophobic region
Hydrophilic region
Plasma membrane
Integral membrane protein
Peripheral membrane protein
(b) An integral membrane protein with multiple α-helical transmembrane segments is shown below. Many integral membrane proteins of the plasma membrane have carbohydrate side chains attached to the hydrophilic segments on the outer membrane surface.
Carbohydrate chains
$^{+}NH_3$
OUTER MEMBRANE SURFACE
Phospholipid bilayer (7–8 nm)
INNER MEMBRANE SURFACE
$^{-}O-C=O$
α-helical transmembrane segments
(c) A single transmembrane segment of an integral membrane is usually α-helical in structure, as shown to the left. Each α-helix typically consists of about 20–30 amino acids, represented by small circles.

FIGURE 7-5 The Fluid Mosaic Model of Membrane Structure. These drawings show **(a)** representative phospholipids and proteins in a typical plasma membrane, with closeups of **(b)** an integral membrane protein and **(c)** one of its transmembrane segments.

The major strength of the fluid mosaic model is that it readily explains most of the criticisms of the Davson–Danielli model. For example, the concept of proteins partially embedded within the lipid bilayer accords well with the hydrophobic nature and globular structure of most membrane proteins and eliminates the need to accommodate membrane proteins in thin surface layers of unvarying thickness. Moreover, the variability in the protein/lipid ratios of different membranes simply means that different membranes vary in the amount of protein they contain. Also, the exposure of lipid head groups at the membrane surface is obviously compatible with their susceptibility to phospholipase digestion, while the fluidity of the lipid layers and the intermingling of lipids and proteins within the membrane make it easy to envision the mobility of both lipids and proteins.

Unwin and Henderson: Most Membrane Proteins Contain Transmembrane Segments

The next illustration in the timeline (see Figure 7-3g) depicts an important property of integral membrane proteins that cell biologists began to understand in the 1970s: Most such proteins have in their primary structure one or more hydrophobic sequences that span the lipid bilayer (Figure 7-5b, c). These *transmembrane segments* anchor the protein to the membrane and hold it in proper alignment within the lipid bilayer.

The example in Figure 7-3g is *bacteriorhodopsin,* the first membrane protein shown to possess this structural feature. Bacteriorhodopsin is a plasma membrane protein found in archaea of the genus *Halobacterium,* where its presence allows cells to obtain energy directly from sunlight, as we will see in Chapter 8. Nigel Unwin and Richard Henderson used electron microscopy to determine the three-dimensional structure of bacteriorhodopsin and to reveal its orientation in the membrane. Their remarkable finding, reported in 1975, was that bacteriorhodopsin consists of a single peptide chain folded back and forth across the lipid bilayer a total of seven times. Each of the seven transmembrane segments of the protein is a closely packed α helix composed mainly of hydrophobic amino acids. Successive transmembrane segments are linked to each other by short loops of hydrophilic amino acids that extend into or protrude from the polar surfaces of the membrane. (For the detailed three-dimensional structure of bacteriorhodopsin, see Figure 8-14.) Based on subsequent work in many laboratories, membrane biologists currently believe that all transmembrane proteins are anchored in the lipid bilayer by one or more transmembrane segments.

Recent Findings Further Refine Our Understanding of Membrane Structure

Almost from the moment Singer and Nicolson proposed it, the fluid mosaic model revolutionized the way scientists think about membrane structure. The model launched a new era in membrane research that not only confirmed the basic model but also refined and extended it. Moreover, our understanding of membrane structure continues to expand as new research findings further refine and modify the basic model.

Recent developments emphasize the concept that membranes are not homogenous, freely mixing structures. Both lipids and proteins are ordered within membranes, and this ordering often occurs in dynamic microdomains known as *lipid rafts* (Figure 7-3h), which we will discuss later in this chapter. In fact, most cellular processes that involve membranes depend critically on specific structural complexes of lipids and proteins within the membrane. This interaction between a membrane protein and a particular lipid can be highly specific and is often critical for proper membrane protein structure and function.

So, to understand membrane-associated processes, we need more than the original fluid mosaic model with lipids and proteins simply floating around randomly. But the fluid mosaic model is still basic to our understanding of membrane structure, so it is important for us to closely examine its essential features. These features include the chemistry, the asymmetric distribution, and the fluidity of membrane lipids; the relationship of membrane proteins to the bilayer; and the mobility of proteins within the bilayer. We will discuss each of these features in turn, focusing on both the supporting evidence and the implications of each feature for membrane function.

Membrane Lipids: The "Fluid" Part of the Model

We will begin our detailed look at membranes by considering membrane lipids, which are important components of the "fluid" part of the fluid mosaic model.

Membranes Contain Several Major Classes of Lipids

One feature of Singer and Nicolson's fluid mosaic model is that it retains the lipid bilayer initially proposed by Gorter and Grendel, though with a greater diversity and fluidity of lipid components than early investigators recognized. The main classes of membrane lipids are *phospholipids, glycolipids,* and *sterols.* **Figure 7-6** lists the main lipids in each of these categories and depicts the structures of several.

Phospholipids. As we already know from Chapter 3, the most abundant lipids found in membranes are the **phospholipids** (Figure 7-6a). Membranes contain many different kinds of phospholipids, including both the glycerol-based **phosphoglycerides** and the sphingosine-based **sphingolipids.** The most common phosphoglycerides are *phosphatidylcholine, phosphatidylethanolamine, phosphatidylserine,* and *phosphatidylinositol.* A common sphingolipid is *sphingomyelin* (Figure 7-6a), which is one of the main phospholipids of animal plasma membranes but is absent from the plasma membranes of plants and most bacteria. The kinds and relative proportions of phospholipids present vary significantly among membranes from different sources (**Figure 7-7**).

Glycolipids. As their name indicates, **glycolipids** are formed by adding carbohydrate groups to lipids. Some glycolipids are glycerol based, and others are derivatives of sphingosine and are therefore called *glycosphingolipids.* The most common examples are **cerebrosides** and **gangliosides.** Cerebrosides are called *neutral glycolipids* because each molecule has a single uncharged sugar as its head group—galactose, in the case of the galactocerebroside shown in Figure 7-6b. A ganglioside, on the other hand, always has an oligosaccharide head group that

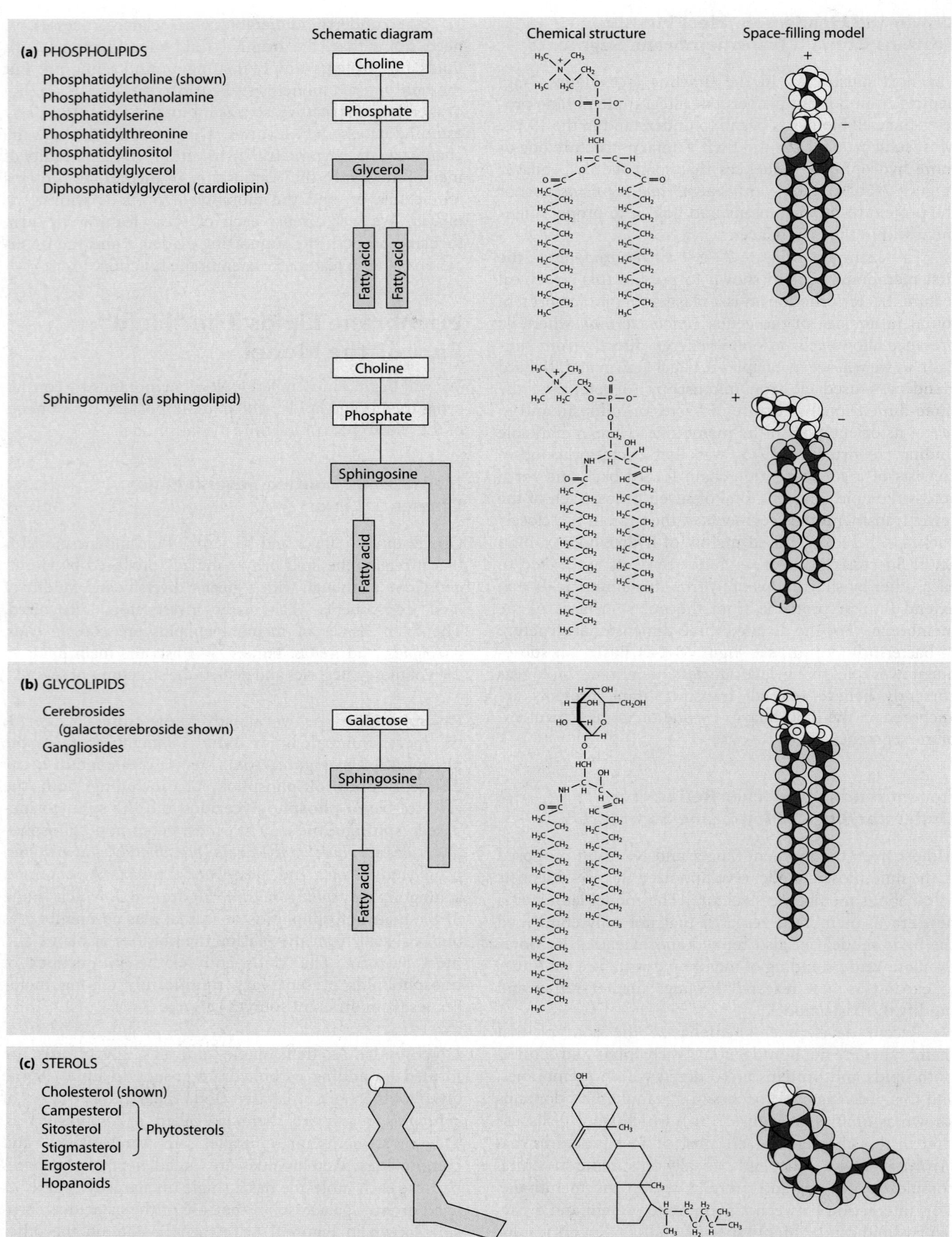

Schematic diagram
Chemical structure
Space-filling model
(a) PHOSPHOLIPIDS
Phosphatidylcholine (shown)
Phosphatidylethanolamine
Phosphatidylserine
Phosphatidylthreonine
Phosphatidylinositol
Phosphatidylglycerol
Diphosphatidylglycerol (cardiolipin)
Choline
Phosphate
Glycerol
Fatty acid
Fatty acid
+
−
Sphingomyelin (a sphingolipid)
Choline
Phosphate
Sphingosine
Fatty acid
+
−
(b) GLYCOLIPIDS
Cerebrosides
(galactocerebroside shown)
Gangliosides
Galactose
Sphingosine
Fatty acid
(c) STEROLS
Cholesterol (shown)
Campesterol
Sitosterol
Stigmasterol
Phytosterols
Ergosterol
Hopanoids

FIGURE 7-6 The Three Major Classes of Membrane Lipids. Each class of lipids is illustrated by a schematic diagram (on the left), a chemical structure (in the middle), and a space-filling model (on the right). **(a)** Phospholipids consist of a small polar head group (such as choline, ethanolamine, serine, or inositol) attached to a lipid backbone, forming either phosphoglycerides (glycerol-based) or sphingolipids (sphingosine-based). (For the structures of choline, ethanolamine, serine, and inositol, see Figure 3-29.) **(b)** Glycolipids are also either glycerol- or sphingosine-based, with the latter more common. A cerebroside has a neutral sugar as its head group, whereas a ganglioside has an oligosaccharide chain containing one or more sialic acid residues and therefore carries a negative charge. **(c)** The most common membrane sterols are cholesterol in animals and several related phytosterols in plants.

contains one or more negatively charged sialic acid residues and gives the molecule a net negative charge.

Cerebrosides and gangliosides are especially prominent in the membranes of brain and nerve cells. Gangliosides exposed on the surface of the plasma membrane also function as antigens recognized by antibodies in immune reactions, including those responsible for blood group interactions. The human ABO blood groups, for example, involve glycosphingolipids known as A antigen and B antigen that serve as specific cell surface markers of the different groups of red blood cells. Cells of blood type A have the A antigen, and cells of blood type B have the B antigen. Type AB blood cells have both antigen types, and type O blood cells have neither.

Several serious human diseases are known to result from impaired metabolism of glycosphingolipids. The best-known example is *Tay-Sachs disease,* which is caused by the absence of a lysosomal enzyme, β-*N*-acetylhexosaminidase A, that is responsible for one of the steps in ganglioside degradation. As a result of the genetic defect, gangliosides accumulate in the brain and other nervous tissue, leading to impaired nerve and brain function and eventually to paralysis, severe mental deterioration, and death.

Two common glycolipids that do not contain sphingosine are derivatives of glycerol that are abundant in plant and algal chloroplasts. Monogalactosyldiacylglycerol (MGDG) and digalactosyldiacylglycerol (DGDG) have either one or two galactose molecules, respectively, attached to a glycerol backbone that contains two polyunsaturated fatty acid groups (**Figure 7-8**). These two glycolipids constitute up to 75% of the total membrane lipids in leaves, where they are believed to play a role in stabilizing membrane proteins of the photosynthetic apparatus. Together, these two lipids are sometimes considered to be the most abundant glycolipids on Earth. Recently, several studies have shown that MGDG has an inhibitory effect on inflammation and cell proliferation in some human cells and tissues.

Sterols. Besides phospholipids and glycolipids, the membranes of most eukaryotic cells contain significant amounts of **sterols** (Figure 7-6c). The main sterol in animal cell membranes is **cholesterol,** which is necessary for maintaining and stabilizing membranes in our bodies by acting as a fluidity buffer. The membranes of plant cells contain small amounts of cholesterol and larger amounts of **phytosterols,** including campesterol, sitosterol, and stigmasterol. Fungal membranes contain a sterol known as ergosterol that is similar in structure to cholesterol but is not found in humans. Ergosterol is the target of antifungal

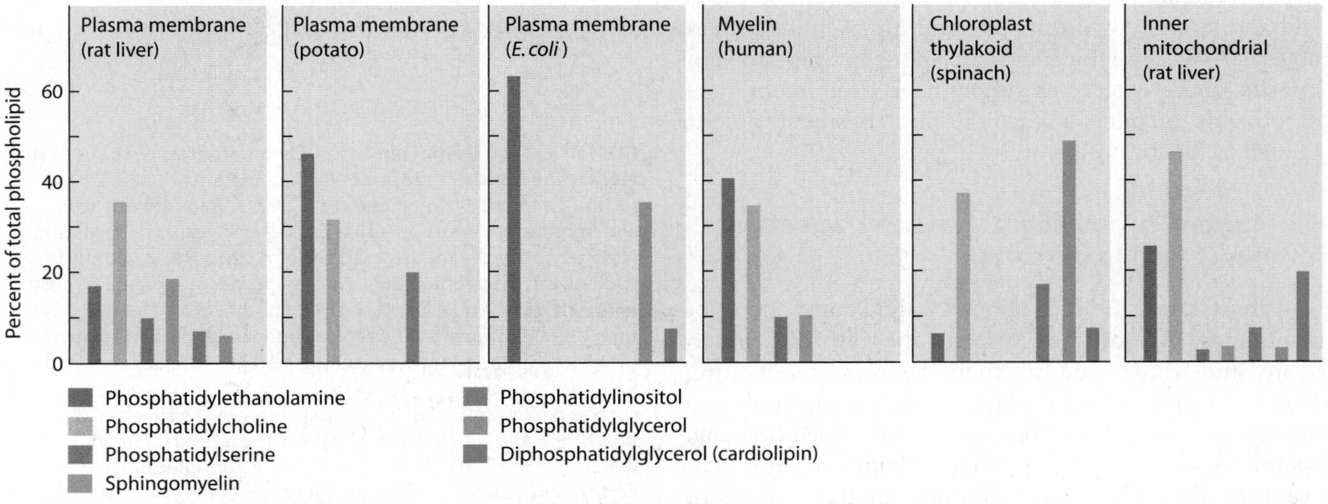

FIGURE 7-7 Phospholipid Composition of Several Kinds of Membranes. The relative abundance of different kinds of phospholipids in biological membranes varies greatly with the source of the membrane.

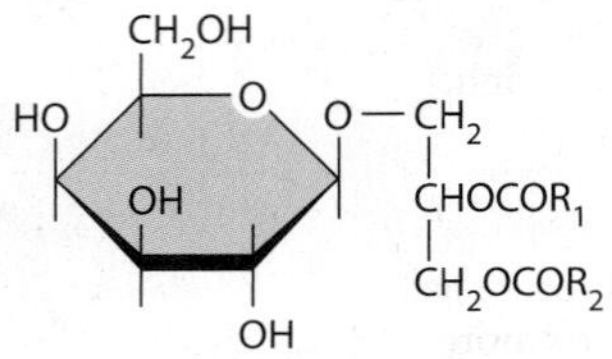

(a) Monogalactosyldiacylglycerol (MGDG)

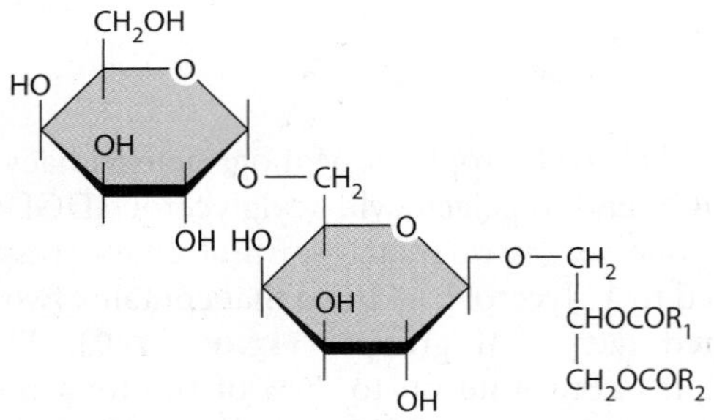

(b) Digalactosyldiacylglycerol (DGDG)

FIGURE 7-8 The Structure of Two Common Glycolipids. MGDG and DGDG are abundant in plant cell membranes, and are similar in structure to the animal fat triacylglycerol (see Figure 3-27b). They have one or two galactose molecules attached to glycerol in addition to the two fatty acid groups (R_1 and R_2)

medications such as nystatin, which selectively kill fungi but do not harm human cells because they lack ergosterol.

Sterols are not found in the membranes of most bacterial cells. They are, however, found in the plasma membrane of *Mycoplasma* species, which, unlike most bacteria, lack a cell wall. Presumably these *Mycoplasma* bacteria have sterols to add stability and strength to the membrane. Sterols are also absent from the inner membranes of both mitochondria and chloroplasts, which are believed to be derived evolutionarily from the plasma membranes of bacterial cells. However, the plasma membranes of at least some bacteria contain sterol-like molecules called *hopanoids* that appear to substitute for sterols in membrane structure. The hopanoid molecule is rigid and strongly hydrophobic, and it closely resembles cholesterol. Because hopanoids are abundant in petroleum deposits, these molecules might have been membrane components of ancient bacteria that presumably contributed to the formation of fossil fuels.

Thin-Layer Chromatography Is an Important Technique for Lipid Analysis

How do we know so much about the lipid components of membranes? As you may suspect, they are difficult to isolate and study due to their hydrophobic nature. However, using nonpolar solvents such as acetone and chloroform, biologists and biochemists have been isolating, separating, and studying membrane lipids for more than a century. One important technique for the analysis of lipids is **thin-layer chromatography (TLC)**, depicted schematically in **Figure 7-9**. This technique is used to separate different kinds of lipids based on their relative polarities.

In this procedure, the lipids are solubilized from a membrane preparation using a mixture of nonpolar organic solvents and separated using a glass plate coated with silicic acid, a polar compound that dries to form a thin film on the glass plate. A sample of the extract is applied to one end of the TLC plate by spotting the extract onto a small area called the *origin* (Figure 7-9a). After the solvent in the sample has evaporated, the edge of the plate is dipped into a solvent system that typically consists of chloroform, methanol, and water. As the solvent moves past the origin and up the plate by capillary action, the lipids are separated based on their polarity—that is, by their relative affinities for the polar silicic acid plate and the less polar solvent.

Nonpolar lipids such as cholesterol have little affinity for the polar silicic acid (the stationary phase) and therefore move up the plate with the solvent system (the mobile phase). Lipids that are more polar, such as phospholipids, interact more strongly with the silicic acid, which slows their movement. In this way the various lipids are separated progressively as the leading edge of the mobile phase continues to move up the plate. When the leading edge, or *solvent front,* approaches the top, the plate is removed from the solvent system and dried. The separated lipids are then recovered from the plate by dissolving each spot or band in a nonpolar solvent such as chloroform for identification and further study.

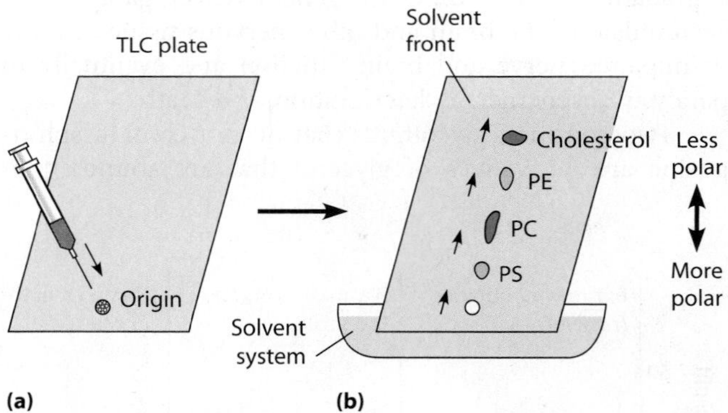

FIGURE 7-9 Using Thin-Layer Chromatography to Analyze Membrane Lipids. Thin-layer chromatography (TLC) is a useful technique for analyzing membrane lipids. Lipids are extracted from a membrane preparation with a mixture of organic solvents and separated according to their degree of polarity. **(a)** A sample is spotted and dried onto a small area of a glass or metal plate coated with a thin layer of silicic acid. **(b)** Components of the sample are then carried upward by the solvent into which the TLC plate is placed. As the solvent moves up the plate by capillary action, the lipids are separated according to their polarity: Less polar lipids such as cholesterol do not adhere strongly to the silicic acid and move further up the plate, while more polar lipids remain closer to the origin. The pattern shown is for lipids of the erythrocyte plasma membrane. The main components are cholesterol, phosphatidylethanolamine (PE), phosphatidylcholine (PC), and phosphatidylserine (PS).

Figure 7-9b shows the TLC pattern seen for the lipids of the erythrocyte plasma membrane. The main components of this membrane are cholesterol (25%) and phospholipids (55%), with phosphatidylethanolamine (PE), phosphatidylcholine (PC), and phosphatidylserine (PS) as the most prominent phospholipids. Other minor components, such as phosphatidylinositol and sphingolipids, are not shown. Control plates are run simultaneously using the same methods and small amounts of known lipids for comparison and identification.

Fatty Acids Are Essential to Membrane Structure and Function

Fatty acids are components of all membrane lipids except the sterols. They are essential to membrane structure because their long hydrocarbon tails form an effective hydrophobic barrier to the diffusion of polar solutes. Most fatty acids in membranes are between 12 and 20 carbon atoms in length, with 16- and 18-carbon fatty acids especially common. This size range appears to be optimal for bilayer formation because chains with fewer than 12 or more than 20 carbons are less able to form a stable bilayer. Thus, the thickness of membranes (about 6–8 nm, depending on the source) is dictated primarily by the chain length of the fatty acids required for bilayer stability.

In addition to differences in length, the fatty acids found in membrane lipids vary considerably in the presence and number of double bonds. **Table 7-2** shows the structures of several fatty acids that are especially common in membrane lipids. *Palmitate* and *stearate* are saturated fatty acids with 16 and 18 carbon atoms, respectively. *Oleate* and *linoleate* are 18-carbon unsaturated fatty acids with one and two double bonds, respectively. Other common unsaturated fatty acids in membranes are *linolenate,* with 18 carbons and three double bonds, and *arachidonate,* with 20 carbons and four double bonds (see Table 3-5, page 68). All unsaturated fatty acids in membranes are in the *cis* configuration, resulting in a sharp bend, or kink, in the hydrocarbon chain at every double bond. Due to the bent nature of their side chains, fatty acids with double bonds do not pack tightly in the membrane.

Membrane Asymmetry: Most Lipids Are Distributed Unequally Between the Two Monolayers

Are the various lipids in a membrane randomly distributed between the two monolayers of lipid that constitute the lipid bilayer? Chemical studies involving membranes derived from a variety of cell types have revealed that most lipids are unequally distributed between the two monolayers. This **membrane asymmetry** includes differences in both the kinds of lipids present and the degree of unsaturation of the fatty acids in the phospholipid molecules.

For example, most of the glycolipids present in the plasma membrane of an animal cell are restricted to the outer monolayer. As a result, their carbohydrate groups protrude from the outer membrane surface, where they are involved in various signaling and recognition events. Phosphatidylethanolamine, phosphatidylinositol, and phosphatidylserine, on the other hand, are more prominent in the inner monolayer, where they are involved in transmitting various kinds of signals from the plasma membrane to the interior of the cell. Further details of signal detection and transduction await us in Chapter 14.

Membrane asymmetry is established during membrane biogenesis by the insertion of different lipids, or different proportions of the various lipids, into each of the two monolayers. Once established, asymmetry tends to be maintained because the movement of lipids from one monolayer to the other requires their hydrophilic head groups to pass through the hydrophobic interior of the membrane—an event that is thermodynamically unfavorable. While such "flip-flop," or **transverse diffusion**, of membrane lipids does occur occasionally, it is relatively rare. For instance, a typical phospholipid molecule flip-flops less than once a week in a pure phospholipid bilayer. This movement contrasts strikingly with the **rotation** of phospholipid molecules about their long axis and with the **lateral diffusion** of phospholipids in the plane of the membrane, both of which occur freely, rapidly, and randomly. **Figure 7-10** illustrates these three types of lipid movements.

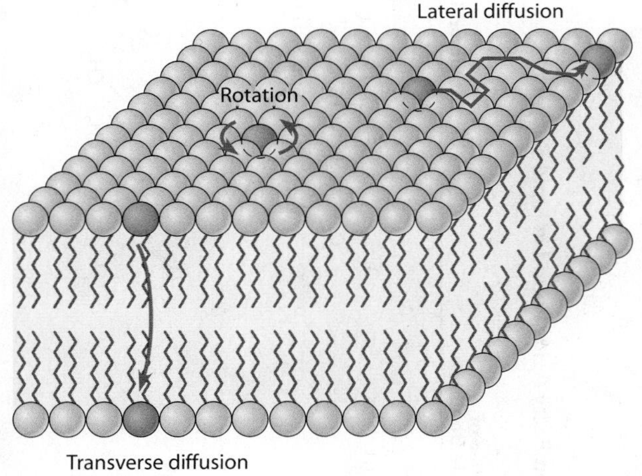

FIGURE 7-10 Movements of Phospholipid Molecules Within Membranes. A phospholipid molecule is capable of three kinds of movement in a membrane: rotation about its long axis; lateral diffusion by exchanging places with neighboring molecules in the same monolayer; and transverse diffusion, or "flip-flop," from one monolayer to the other. In a pure phospholipid bilayer at 37°C, a typical lipid molecule exchanges places with neighboring molecules about 10 million times per second and can move laterally at a rate of about several micrometers per second. By contrast, an individual phospholipid molecule flip-flops from one layer to the other at a rate ranging from less than once a week in a pure phospholipid bilayer to once every few hours in some natural membranes. The more rapid movement in natural membranes is due to the presence of enzymes called phospholipid translocators, or flippases, that catalyze the transverse diffusion of phospholipid molecules from one monolayer to the other.

Table 7-2 **Structures of Some Common Fatty Acids Found in Lipid Bilayers**

Name of Fatty Acid	Number of Carbon Atoms	Number of Double Bonds	Structural Formula	Space-Filling Model
Saturated				
Palmitate	16	0	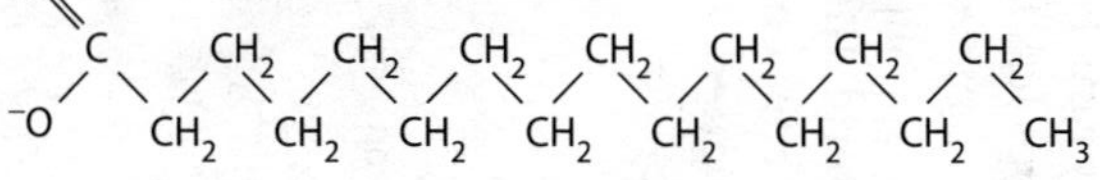 $^-O-C(=O)-CH_2-CH_2-CH_2-CH_2-CH_2-CH_2-CH_2-CH_2-CH_2-CH_2-CH_2-CH_2-CH_2-CH_2-CH_3$	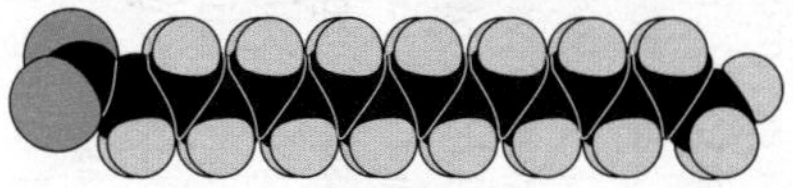
Stearate	18	0	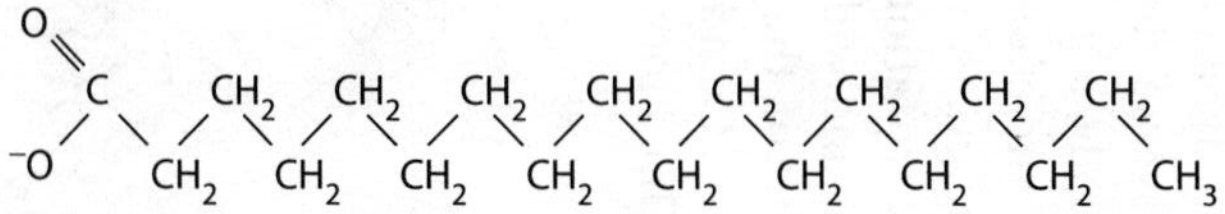 $^-O-C(=O)-CH_2-CH_2-CH_2-CH_2-CH_2-CH_2-CH_2-CH_2-CH_2-CH_2-CH_2-CH_2-CH_2-CH_2-CH_2-CH_2-CH_3$	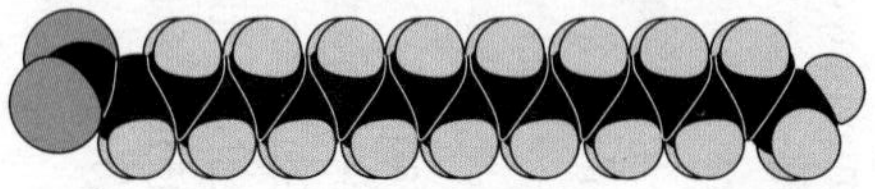
Unsaturated				
Oleate	18	1	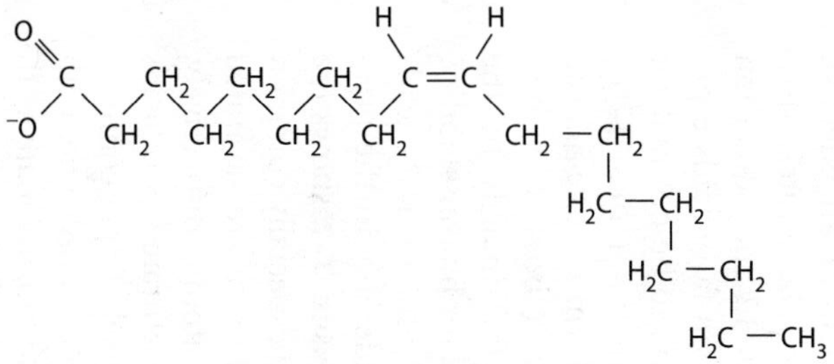 $^-O-C(=O)-CH_2-CH_2-CH_2-CH_2-CH_2-CH_2-CH_2-CH{=}CH-CH_2-CH_2-H_2C-CH_2-H_2C-CH_2-H_2C-CH_3$	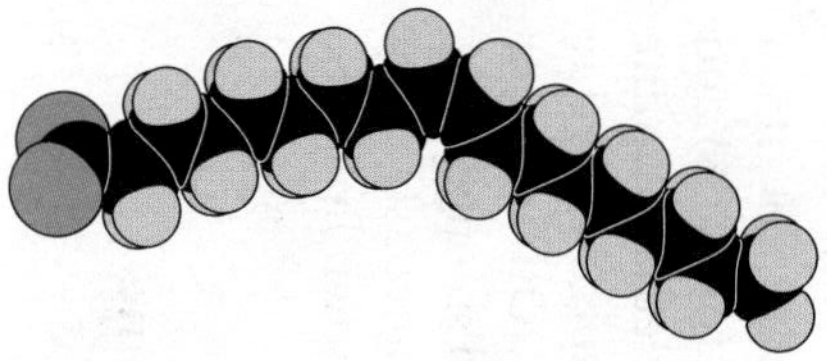
Linoleate	18	2	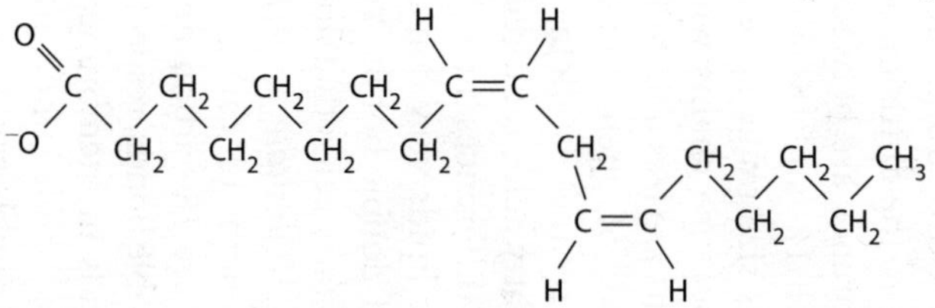 $^-O-C(=O)-CH_2-CH_2-CH_2-CH_2-CH_2-CH_2-CH_2-CH{=}CH-CH_2-CH{=}CH-CH_2-CH_2-CH_2-CH_2-CH_3$	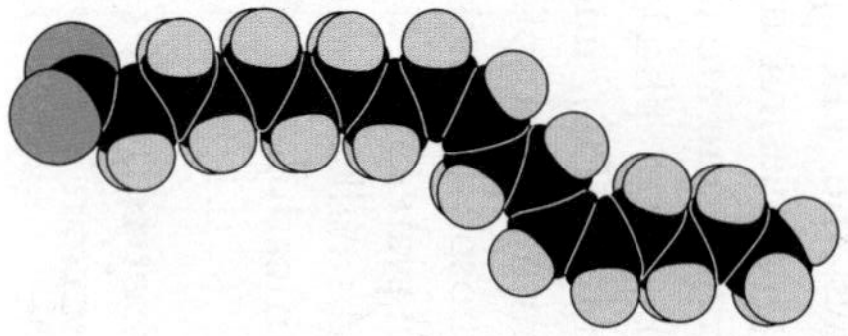

While phospholipid flip-flop is relatively rare, it occurs more frequently in natural membranes than in artificial lipid bilayers. This is because some membranes—the smooth endoplasmic reticulum (ER), in particular—have proteins called **phospholipid translocators,** or **flippases,** that catalyze the flip-flop of membrane lipids from one monolayer to the other. Such proteins act only on specific kinds of lipids. For example, one of these proteins in the smooth ER membrane catalyzes the translocation of phosphatidylcholine from one side of the membrane to the other but does not recognize other phospholipids. This ability to move lipid molecules selectively from one side of the bilayer to the other contributes further to the asymmetric distribution of phospholipids across the membrane. The role of smooth ER in the synthesis and selective flip-flop of membrane phospholipids is a topic we will return to in Chapter 12.

The Lipid Bilayer Is Fluid

One of the most striking properties of membrane lipids is that rather than being fixed in place within the membrane, they form a fluid bilayer that permits lateral diffusion of membrane lipids as well as proteins. Lipid molecules move especially fast because they are much smaller than proteins. A typical phospholipid molecule, for example, has a molecular weight of about 800 and can travel the length of a bacterial cell (a few micrometers, in most cases) in one second or less! Proteins move much more slowly than lipids, partly because they are much larger molecules and partly due to their interactions with cytoskeletal proteins on the inside of the cell.

The lateral diffusion of membrane lipids can be demonstrated experimentally by a technique called **fluorescence recovery after photobleaching** (**Figure 7-11**). The investigator *tags,* or labels, lipid molecules in the membrane of a living cell by covalently linking them to a fluorescent dye. A high-intensity laser beam is then used to bleach the dye in a tiny spot (a few square micrometers) on the cell surface. If the cell surface is examined immediately thereafter with a fluorescence microscope, a dark, nonfluorescent spot is seen on the membrane. Within seconds, however, the edges of the spot become fluorescent as bleached lipid molecules diffuse out of the laser-treated area and fluorescent lipid molecules from adjoining regions of the membrane diffuse in. Eventually, the spot is indistinguishable from the rest of the cell surface. This technique demonstrates that membrane lipids are in a fluid rather than a static state, and it provides a direct means of measuring the lateral movement of specific molecules.

Membranes Function Properly Only in the Fluid State

As you might guess, membrane fluidity changes with temperature, decreasing as the temperature drops and increasing as it rises. In fact, we know from studies with artificial lipid bilayers that every lipid bilayer has a characteristic **transition temperature (T_m)** at which it becomes fluid ("melts") when warmed from a solid gel-like state. This change in the state of the membrane is called a **phase transition,** and you have probably seen this yourself if you have ever accidentally left a stick of butter on the stove! To function properly, a membrane must be maintained in the fluid state—that is, at a temperature above its T_m value. At a temperature below the T_m value, all functions that depend on the mobility or conformational changes of membrane proteins will be impaired or disrupted. This includes such vital processes as transport of solutes across the membrane, detection and transmission of signals, and cell-to-cell communication (see Figure 7-2).

The technique of **differential scanning calorimetry** is one means of determining the transition temperature of a given membrane. This procedure monitors the uptake of

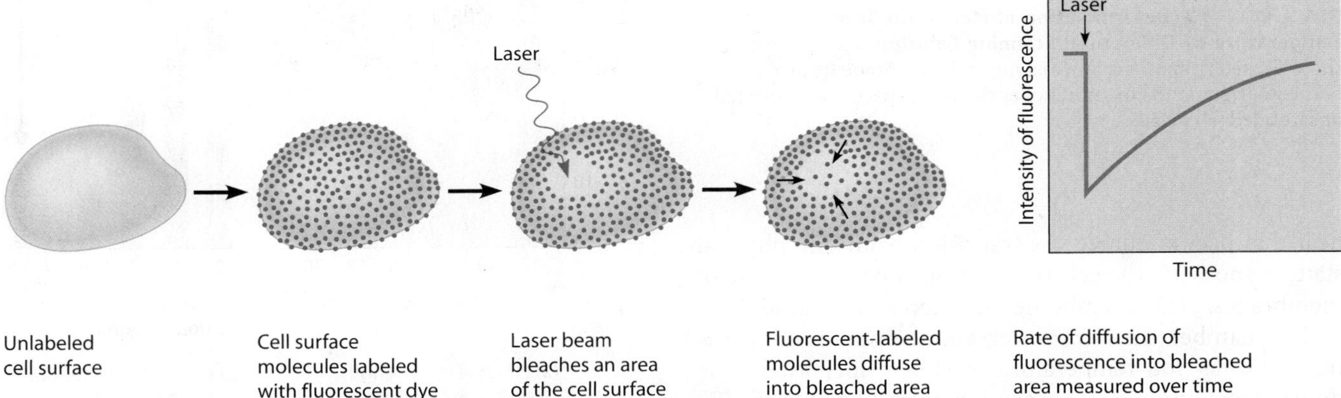

FIGURE 7-11 Measuring Lipid Mobility in Membranes by Fluorescence Recovery After Photobleaching. Membrane lipids are labeled with a fluorescent compound, and the fluorescence in a local area is then bleached by irradiating the cell with a laser beam. As fluorescent molecules from surrounding regions diffuse into the bleached area, fluorescence will reappear in the laser-bleached spot. Membrane fluidity is measured by determining the rate of this reappearance of fluorescence. Similar experiments can be carried out to measure the mobility of membrane proteins, as shown in Figure 7-28.

(a) Normal membrane. When the temperature of a typical membrane preparation is increased slowly in a calorimeter chamber, a peak of heat absorption marks the gel-to-fluid transition temperature, T_m.

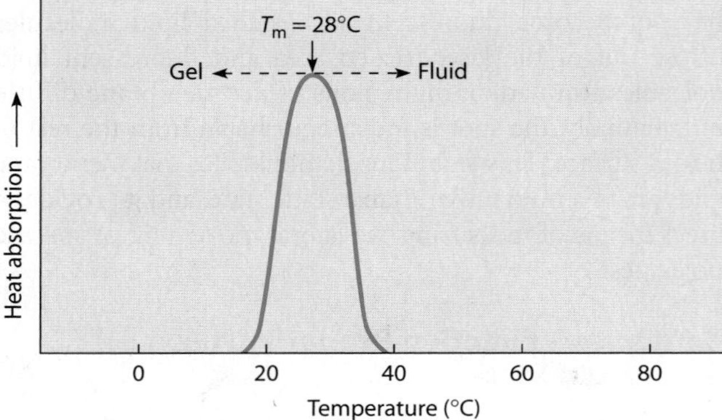

(b) Membranes enriched in unsaturated or saturated fatty acids. Membranes from cells grown in media enriched in the unsaturated fatty acid oleate (left) are more fluid than normal membranes (lower T_m). Membranes from cells grown in media enriched in the saturated fatty acid stearate (right) are less fluid than normal membranes (higher T_m).

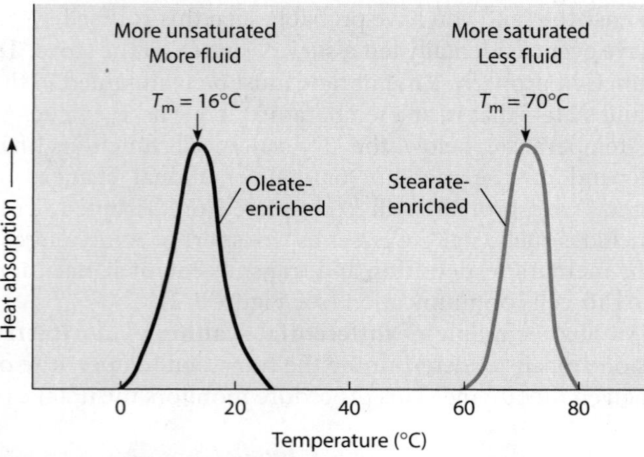

FIGURE 7-12 Determination of Membrane Transition Temperature by Differential Scanning Calorimetry. These graphs show T_m determinations for **(a)** a normal membrane from a homeotherm and **(b)** membranes enriched in specific unsaturated or saturated fatty acids.

heat that occurs during the transition from one physical state to another—the gel-to-fluid transition, in the case of membranes. The membrane of interest is placed in a sealed chamber, the *calorimeter,* and the uptake of heat is measured as the temperature is slowly increased. The point of maximum heat absorption corresponds to the T_m (**Figure 7-12**).

Effects of Fatty Acid Composition on Membrane Fluidity. A membrane's fluidity depends primarily on the kinds of lipids it contains. Two properties of a membrane's lipid makeup are especially important in determining fluidity: the length of the fatty acid side chains and their degree of unsaturation. Long-chain fatty acids have higher transition temperatures than do short-chain fatty acids, which means that membranes enriched in long-chain fatty acids tend to be less fluid.

For example, as the chain length of saturated fatty acids increases from 10 to 20 carbon atoms, the T_m rises from 32°C to 76°C, and the membrane thus becomes progressively less fluid (**Figure 7-13a**). The presence of unsaturation affects the T_m even more markedly. For fatty acids with 18 carbon atoms, the transition temperatures are 70, 16, 5, and −11°C for zero, one, two, and three double bonds, respectively (Figure 7-13b). As a result, membranes containing many unsaturated fatty acids tend to have lower transition temperatures and thus are more fluid than membranes with many saturated fatty acids. Figure 7-12b illustrates this increased fluidity for membranes enriched in oleate (18 carbons, one double bond) versus membranes enriched in stearate (18 carbons, saturated).

The effect of unsaturation on membrane fluidity is so dramatic because the kinks caused by double bonds in fatty acids prevent the hydrocarbon chains from fitting together snugly. Membrane lipids with saturated fatty

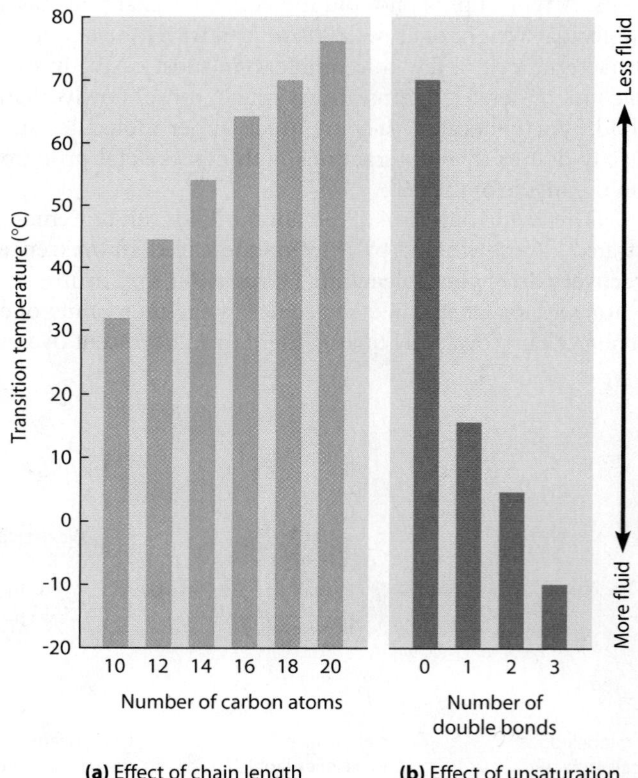

FIGURE 7-13 The Effect of Chain Length and the Number of Double Bonds on the Melting Point of Fatty Acids. The transition temperature of fatty acids **(a)** increases with chain length for saturated fatty acids, becoming less fluid with longer chains. **(b)** The transition temperature decreases dramatically with the number of double bonds for fatty acids with a fixed chain length, becoming more fluid as more double bonds are present.

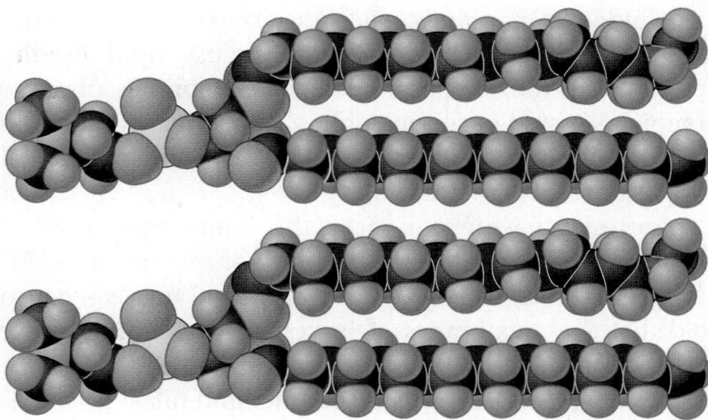

(a) Lipids with saturated fatty acids pack together well in the membrane

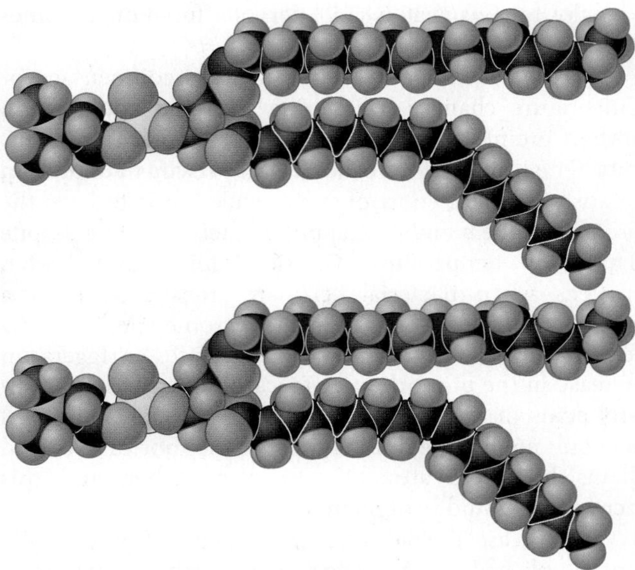

(b) Lipids with a mixture of saturated and unsaturated fatty acids do not pack together well in the membrane

FIGURE 7-14 The Effect of Unsaturated Fatty Acids on the Packing of Membrane Lipids. **(a)** Membrane phospholipids with no unsaturated fatty acids fit together tightly because the fatty acid chains are parallel to each other. **(b)** Membrane lipids with one or more unsaturated fatty acids do not fit together as tightly because the *cis* double bonds cause bends in the chains that interfere with packing. Each structure shown is a phosphatidylcholine molecule, with either two 18-carbon saturated fatty acids (stearate; part a) or two 18-carbon fatty acids, one saturated (stearate) and the other with one *cis* double bond (oleate; part b).

acids pack together tightly (**Figure 7-14a**), whereas lipids with unsaturated fatty acids do not (Figure 7-14b). The lipids of most plasma membranes contain fatty acids that vary in both chain length and degree of unsaturation. In fact, the variability is often intramolecular because membrane lipids commonly contain one saturated and one unsaturated fatty acid. This property helps to ensure that membranes are in the fluid state at physiological temperatures.

Most unsaturated fatty acids found in nature contain *cis* double bonds. In contrast, many commercially processed fats and hydrogenated oils contain significant numbers of *trans* double bonds. The *trans* double bond does not introduce as much of a bend in the fatty acid chain, as shown below (H atoms not shown):

R/C=C\R — *cis* configuration

R/C=C/R — *trans* configuration

Thus, in their overall shape and ability to pack together closely, *trans* fats resemble saturated fats more than *cis* unsaturated fats do. The presence of *trans* fats in membranes increases the transition temperature and decreases the membrane fluidity. As with saturated fats, consuming *trans* fats has been correlated with high blood cholesterol levels and increased risk of heart disease.

Effects of Sterols on Membrane Fluidity. For eukaryotic cells, membrane fluidity is also affected by the presence of sterols—mainly cholesterol in animal cell membranes and phytosterols in plant cell membranes. Sterols are prominent components in the membranes of many cell types. A typical animal cell, for example, contains large amounts of cholesterol—up to 50% of the total membrane lipid on a molar basis. Cholesterol molecules are usually found in both layers of the plasma membrane, but a given molecule is localized to one of the two layers (**Figure 7-15a**). The molecule orients itself in the layer with its single hydroxyl group—the only polar part of an otherwise hydrophobic molecule—close to the polar head group of a neighboring phospholipid molecule, where it can form a hydrogen bond (Figure 7-15b). The rigid hydrophobic steroid rings and the hydrocarbon side chain of the cholesterol molecule interact with the portions of adjacent hydrocarbon chains that are closest to the phospholipid head groups.

This intercalation of rigid cholesterol molecules into the membrane of an animal cell makes the membrane less fluid at higher temperatures than it would otherwise be. However, cholesterol also effectively prevents the hydrocarbon chains of phospholipids from fitting snugly together as the temperature is decreased, thereby reducing the tendency of membranes to gel upon cooling. Thus, cholesterol acts as a fluidity buffer: It has the moderating effect of *decreasing* membrane fluidity at temperatures above the T_m and *increasing* it at temperatures below the T_m. The sterols in the membranes of other eukaryotes and hopanoids in prokaryotes presumably function in the same way.

Besides their effects on membrane fluidity, sterols decrease the permeability of a lipid bilayer to ions and small polar molecules. They probably do so by filling in spaces between hydrocarbon chains of membrane phospholipids, thereby plugging small channels that ions and small molecules might otherwise pass through. In general, a lipid bilayer containing sterols is less permeable to ions and small molecules than is a bilayer lacking sterols.

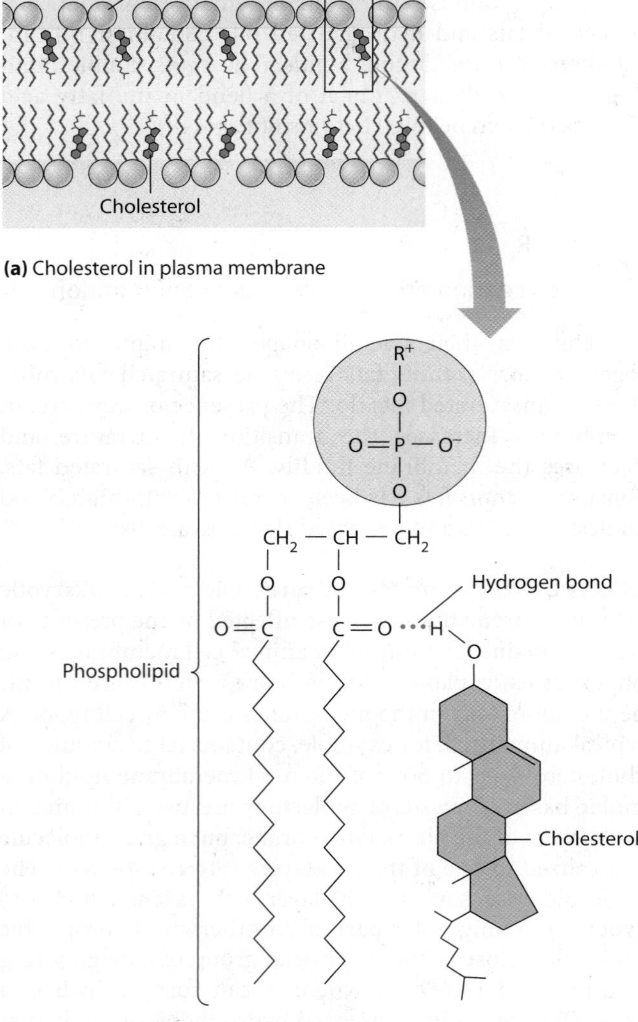

FIGURE 7-15 Orientation of Cholesterol Molecules in a Lipid Bilayer. (a) Cholesterol molecules are present in both lipid monolayers in the plasma membranes of most animal cells, but a given molecule is localized to one of the two layers. **(b)** Each molecule orients itself in the lipid layer so that its single hydroxyl group is close to the polar head group of a neighboring phospholipid molecule. The hydroxyl group of cholesterol forms a hydrogen bond with the oxygen of the ester bond between the glycerol backbone and a fatty acid. The nonpolar steroid rings and hydrocarbon side group of the cholesterol molecule interact with adjacent hydrocarbon chains of the membrane phospholipids.

Most Organisms Can Regulate Membrane Fluidity

Most organisms, whether prokaryotic or eukaryotic, are able to regulate membrane fluidity, primarily by changing the lipid composition of the membranes. This ability is especially important for *poikilotherms*—organisms such as bacteria, fungi, protozoa, algae, plants, invertebrates, and "cold-blooded" animals such as snakes that cannot regulate their own temperature. Because lipid fluidity decreases as the temperature falls, membranes of these organisms would gel upon cooling if the organism had no way to compensate for decreases in environmental temperature.

You may have experienced this compensating effect even though you are a *homeotherm,* or "warm-blooded" organism: On chilly days, your fingers and toes can get so cold that the membranes of sensory nerve endings cease to function, resulting in temporary numbness. At high temperatures, on the other hand, the lipid bilayers of poikilotherms become so fluid that they no longer serve as an effective permeability barrier. For example, most cold-blooded animals are paralyzed by temperatures that are much above 45°C because nerve cell membranes become so leaky to ions that overall nervous function becomes disabled.

Fortunately, most poikilotherms can compensate for temperature changes by altering the lipid composition of their membranes, thereby regulating membrane fluidity. This capability is called **homeoviscous adaptation** because the main effect of such regulation is to keep the viscosity of the membrane approximately the same despite changes in temperature. Consider, for example, what happens when bacterial cells are transferred from a warmer to a cooler environment. In some species of the genus *Micrococcus,* a drop in temperature triggers an increase in the proportion of 16-carbon versus 18-carbon fatty acids in the plasma membrane. This helps the cell maintain membrane fluidity because the shorter fatty acid chains have less attraction for each other, and this increases the fluidity of membranes.

In this case, the desired increase in membrane fluidity is accomplished by activating an enzyme that removes two terminal carbons from 18-carbon hydrocarbon tails. In other bacterial species, adaptation to environmental temperature involves an alteration in the extent of unsaturation of membrane fatty acids rather than in their length. In the common intestinal bacterium *Escherichia coli,* for example, a decrease in environmental temperature triggers the synthesis of a *desaturase* enzyme that introduces double bonds into the hydrocarbon chains of fatty acids. As these unsaturated fatty acids are incorporated into membrane phospholipids, they decrease the transition temperature of the membrane, thereby ensuring that the membrane remains fluid at the lower temperature.

Homeoviscous adaptation also occurs in yeasts and plants. In these organisms, temperature-related changes in membrane fluidity appear to depend on the increased solubility of oxygen in the cytoplasm at lower temperatures. Oxygen is a substrate for the desaturase enzyme system involved in the generation of unsaturated fatty acids. With more oxygen available at lower temperatures, unsaturated fatty acids are synthesized at a greater rate and membrane fluidity increases, thereby offsetting the temperature effect.

This capability has great agricultural significance because plants that can adapt in this way are cold hardy (resistant to chilling) and can thus be grown in colder environments. Animals such as amphibians and reptiles also adapt to lower temperatures by increasing the proportion of unsaturated fatty acids in their membranes. Furthermore, these animals can increase the proportion of cholesterol in the membrane, thereby decreasing the interaction between hydrocarbon chains and reducing the tendency of the membrane to gel.

Although homeoviscous adaptation is generally most relevant to poikilothermic organisms, it is also important to mammals that hibernate. As an animal enters hibernation, its body temperature often drops substantially—a decrease of more than 30°C for some rodents. The animal adapts to this change by incorporating a greater proportion of unsaturated fatty acids into membrane phospholipids as its body temperature falls.

Lipid Rafts Are Localized Regions of Membrane Lipids That Are Involved in Cell Signaling

Until recently, the lipid component of a membrane was regarded as uniformly fluid and relatively homogeneous within a given monolayer. In recent years, however, the discovery of localized regions of membrane lipids that sequester proteins involved in cell signaling has generated much excitement. These regions are called either *lipid microdomains* or, more popularly, **lipid rafts**—and represent areas of lateral heterogeneity within a membrane monolayer. Lipid rafts are dynamic structures that change in composition as individual lipids and proteins move in and out of them. They were first identified in the outer monolayer of the plasma membrane of eukaryotic cells but have since been detected in the inner monolayer also.

Lipid rafts in the outer membrane monolayer of animal cells are characterized by elevated levels of cholesterol and glycosphingolipids. The glycosphingolipids have longer and more saturated fatty acid tails than those seen in most other membrane lipids. Moreover, the phospholipids present in lipid rafts are more highly saturated than those in the surrounding membrane. These properties, plus the rigidity and hydrophobic nature of cholesterol, allow tight packing of the cholesterol and the hydrocarbon tails of the glycosphingolipids and the phospholipids. As a result, lipid rafts are thicker and less fluid than the rest of the membrane, thereby distinguishing them as discrete lipid microdomains. In addition, these regions are less able to be solubilized by nonionic detergents, a characteristic that facilitates their separation from the rest of the membrane components as intact structures for subsequent analysis.

Initial models of lipid raft formation proposed that localized regions of tightly-associated cholesterol and glycosphingolipid molecules attracted particular raft-associated proteins. Some of these raft-associated proteins are lipoproteins containing a fatty acid such as palmitate attached to a specific cysteine residue, a feature that may facilitate their targeting to lipid rafts. Recent proteomic studies of lipid raft regions have identified over 200 proteins enriched in lipid rafts. Some of these raft-associated proteins can capture and organize particular raft lipids, suggesting an active role of these proteins in lipid raft formation. In addition, lipid rafts contain actin-binding proteins, and some studies have suggested a role of the cytoskeleton in forming and organizing lipid rafts. Studies using inhibitors have shown that both depletion of cholesterol in cells and disruption of the actin cytoskeleton interfere with targeting of proteins to lipid rafts.

Much of the excitement surrounding lipid rafts relates to their role in the detection of, and responses to, extracellular chemical signals. For example, lipid rafts are involved in the transport of nutrients and ions across cell membranes, the binding of activated immune system cells to their microbial targets, and the transport of cholera toxin into intestinal cells. Many receptor proteins involved in the detection of external chemical signals are localized to the outer lipid monolayer of the plasma membrane. When a receptor binds its specific ligands, it can move into particular lipid rafts that are also located in the outer monolayer. Receptor-containing lipid rafts in the outer monolayer are thought to be coupled functionally to specific lipid rafts in the inner monolayer. Some lipid rafts contain specific *kinases,* enzymes that generate second messengers within the cell by catalyzing the phosphorylation of specific substances. In this way, signals that are detected by the receptor proteins in the outer monolayer can be transmitted to the interior of the cell by the functional links between the lipid rafts in the two membrane monolayers.

Closely related to lipid rafts in structure and perhaps in function are *caveolae* (Latin for "little caves"), which are small, flask-shaped invaginations of the plasma membrane of mammalian cells that were first observed more than 50 years ago. They contain the cholesterol-binding protein *caveolin,* which contributes to their curved morphology, and are enriched in cholesterol, sphingolipids, and lipid-anchored proteins. Proposed cellular roles for caveolae include participation in endocytosis and exocytosis, redox sensing, and regulation of airway function in the lungs. In addition, caveolae have been shown to contain proteins important in calcium signaling in heart muscle cells and have been discussed as potential targets for treatment of cardiovascular disease.

Membrane Proteins: The "Mosaic" Part of the Model

Having looked in some detail at the "fluid" aspect of the fluid mosaic model, we come now to the "mosaic" part. That may include lipid rafts and other lipid domains, but the main components of the membrane mosaic are the many membrane proteins as initially envisioned by Singer and Nicolson. We will look first at the confirming evidence that microscopists provided for the membrane as a mosaic of proteins and then consider the major classes of membrane proteins.

The Membrane Consists of a Mosaic of Proteins: Evidence from Freeze-Fracture Microscopy

Strong support for the fluid mosaic model came from studies in which artificial bilayers and natural membranes were prepared for electron microscopy by **freeze fracturing.** In this technique, a lipid bilayer or a membrane (or a cell containing membranes) is frozen quickly and then subjected to a sharp blow from a diamond knife. Because the nonpolar interior of the bilayer is the path of least resistance through the frozen specimen, the resulting fracture often follows the plane between the two layers of membrane lipid. As a result, the bilayer is split into its inner and outer monolayers, revealing the inner surface of each (**Figure 7-16a**).

Electron micrographs of membranes prepared in this way provide striking evidence that proteins are actually suspended within membranes. Whenever a fracture plane splits the membrane into its two layers, particles having the size and shape of globular proteins can be seen adhering to one or the other of the inner membrane surfaces, called the *E* (for *exoplasmic*) and *P* (for *protoplasmic*) *faces* (Figure 7-16b). Moreover, the abundance of such particles correlates well with the known protein content of the particular membrane under investigation. The electron micrographs in **Figure 7-17** illustrate this well: The erythrocyte plasma membrane has a rather low protein/lipid ratio (1.14; see Table 7-1) and a rather low density of particles when subjected to freeze fracture (Figure 7-17a), whereas a chloroplast membrane has a higher protein/lipid ratio (2.33) and a correspondingly higher density of intramembranous particles, especially on the inner lipid layer (Figure 7-17b).

Confirmation that the particles seen in this way really *are* proteins came from work by David Deamer and Daniel Branton, who used the freeze-fracture technique to examine artificial bilayers with and without added protein. Bilayers formed from pure phospholipids showed no

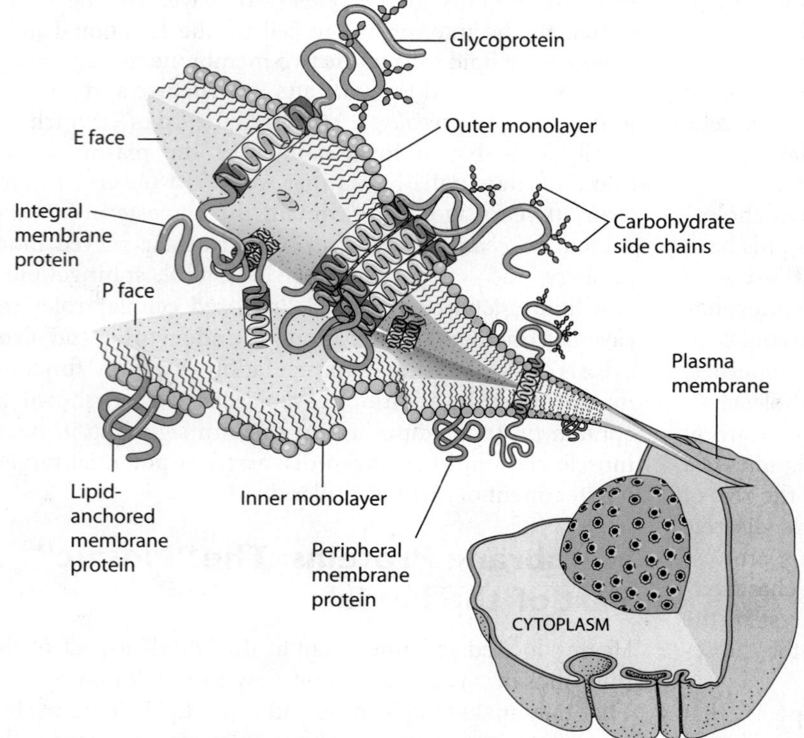

(a) Separation of membrane monolayers. Notice how the fracture plane has passed through the hydrophobic interior of the membrane, revealing the inner surfaces of the two monolayers. Integral membrane proteins that remain with the outer monolayer are seen on the E (exoplasmic) face, whereas those that remain with the inner monolayer are seen on the P (protoplasmic) face.
Glycoprotein
E face
Outer monolayer
Integral membrane protein
Carbohydrate side chains
P face
Plasma membrane
Lipid-anchored membrane protein
Inner monolayer
Peripheral membrane protein
CYTOPLASM

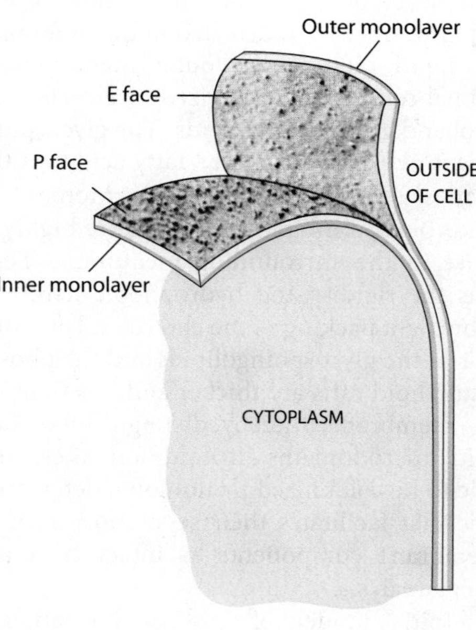

(b) Surface view of monolayers. This sketch of a freeze-fractured membrane shows electron micrographs of the E and P faces from the plasma membrane of a mouse kidney tubule cell. Individual proteins imbedded in either face show up as small particles (TEMs).
Outer monolayer
E face
P face
OUTSIDE OF CELL
Inner monolayer
CYTOPLASM

FIGURE 7-16 Freeze-Fracture Analysis of a Membrane. Sketches of a freeze-fractured membrane showing separation of the two lipid monolayers: **(a)** drawings of representative membrane components and **(b)** an electron micrograph of the surface of each monolayer.

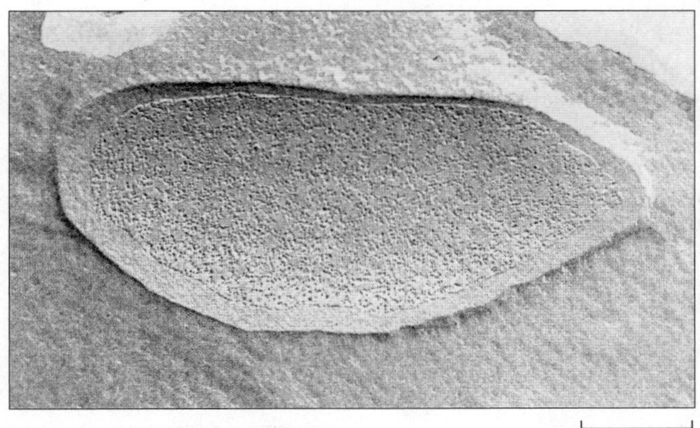

(a) Erythrocyte plasma membrane

(b) Chloroplast membrane

FIGURE 7-17 Membrane Proteins Visualized by Freeze-Fracture Electron Microscopy. Membrane proteins appear as discrete particles embedded within the lipid bilayer. The lower density of particles in **(a)** the erythrocyte membrane compared with **(b)** the chloroplast membrane agrees well with the protein/lipid ratios of the two membranes—1.14 and 2.33, respectively (TEMs).

evidence of particles on their interior surfaces (**Figure 7-18a**). When proteins were added to the artificial bilayers, however, particles similar to those seen in natural membranes were readily visible (Figure 7-18b).

Membranes Contain Integral, Peripheral, and Lipid-Anchored Proteins

Membrane proteins differ in their affinity for the hydrophobic interior of the membrane and therefore in the extent to which they interact with the lipid bilayer. That difference in affinity, in turn, determines how easy or difficult it is to extract a given protein from the membrane. Based on the conditions required to extract them—and thus, by extension, on the nature of their association with the lipid bilayer—membrane proteins fall into one of three categories: integral, peripheral, or lipid-anchored. We will consider each of these in turn, referring in each case to the diagrams shown in **Figure 7-19**.

The plasma membrane of the human erythrocyte, shown in **Figure 7-20**, provides a cellular context for our discussion. This membrane has been one of the most widely studied since the time of Gorter and Grendel, due to the ready availability of red blood cells and how easily pure plasma membrane preparations can be made from them. We will refer to the erythrocyte plasma membrane at several points in the following discussion to note examples of different types of proteins and their roles within the membrane.

Integral Membrane Proteins. Most membrane proteins are amphipathic molecules possessing one or more hydrophobic regions that exhibit an affinity for the hydrophobic interior of the lipid bilayer. These proteins

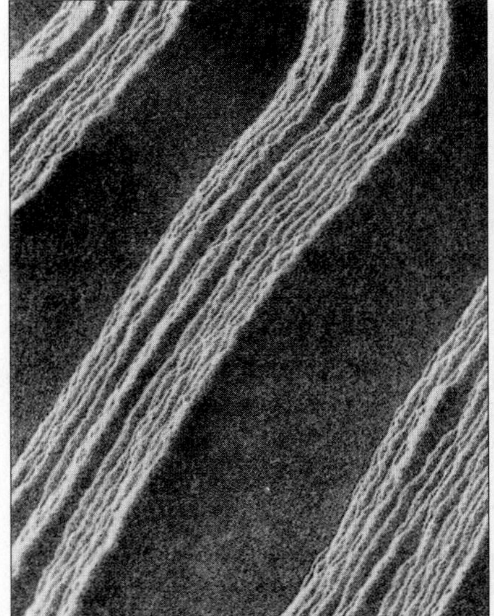

(a) Artificial bilayers without proteins

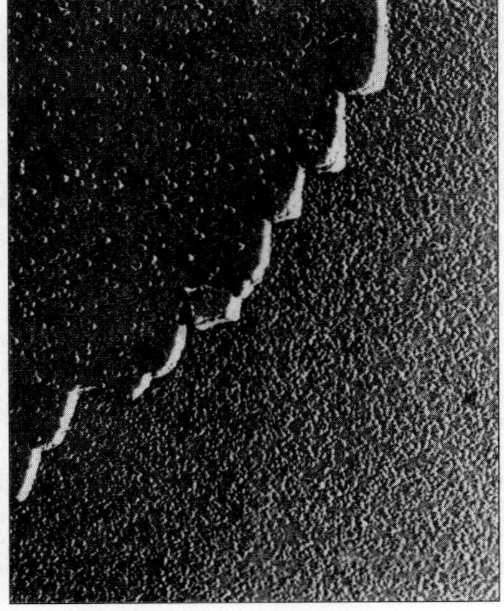

(b) Artificial bilayers with proteins

FIGURE 7-18 Freeze-Fracture Comparison of Lipid Bilayers With and Without Added Proteins. This figure compares the appearance by freeze-fracture electron microscopy of **(a)** artificial lipid bilayers without proteins and **(b)** artificial bilayers with proteins added. The white lines in the artificial membranes of part a represent individual lipid bilayers in a multilayered specimen, and the gray regions show where single bilayers have split to reveal smooth surfaces. In contrast, the artificial membrane of part b shows large numbers of globular particles in the fracture surface. These are the proteins that were added to the membrane preparation (TEMs).

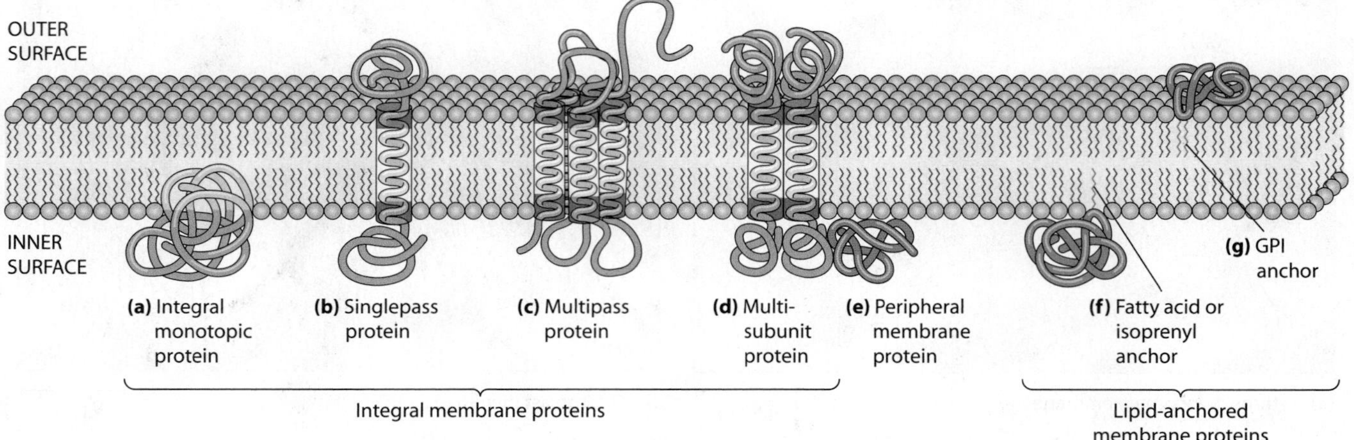

FIGURE 7-19 The Main Classes of Membrane Proteins. Membrane proteins are classified according to their mode of attachment to the membrane. Integral membrane proteins **(a–d)** contain one or more hydrophobic regions that are embedded within the lipid bilayer. **(a)** A few integral proteins appear to be embedded in the membrane on only one side of the bilayer (integral monotopic proteins). However, most integral proteins are transmembrane proteins that span the lipid bilayer either **(b)** once (singlepass proteins) or **(c)** multiple times (multipass proteins). Multipass proteins may consist of either a single polypeptide, as in part c, or **(d)** several associated polypeptides (multisubunit proteins). **(e)** Peripheral membrane proteins are too hydrophilic to penetrate into the membrane but are attached to the membrane by electrostatic and hydrogen bonds that link them to adjacent membrane proteins or to phospholipid head groups. Lipid-anchored proteins **(f–g)** are hydrophilic and do not penetrate into the membrane; they are covalently bound to lipid molecules that are embedded in the lipid bilayer. **(f)** Proteins on the inner surface of the membrane are usually anchored by either a fatty acid or an isoprenyl group. **(g)** On the outer membrane surface, the most common lipid anchor is glycosylphosphatidylinositol (GPI).

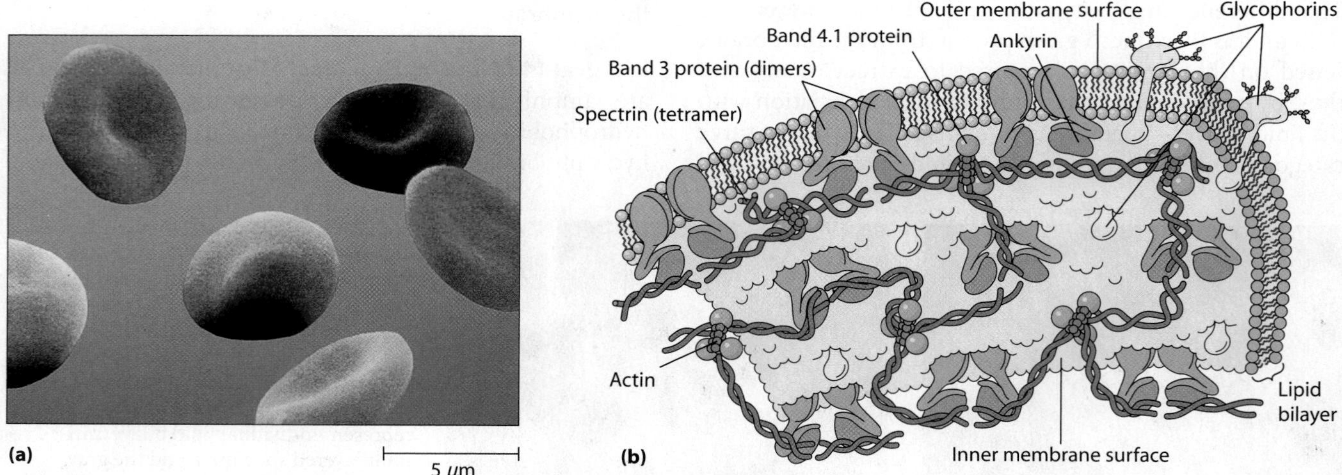

FIGURE 7-20 Structural Features of the Erythrocyte Plasma Membrane. **(a)** An erythrocyte is a small, disk-shaped cell with a diameter of about 7 μm. A mammalian erythrocyte contains no nucleus or other organelles, which makes it easy to obtain very pure plasma membrane preparations without contamination by organelle membranes, as often occurs with plasma membrane preparations from other cell types. **(b)** The erythrocyte plasma membrane as seen from inside the cell. The membrane has a relatively simple protein composition. The two major integral membrane proteins are glycophorin and an anion exchange protein known from its electrophoretic mobility as band 3. The membrane is anchored to the underlying cytoskeleton by long, slender strands of tetrameric spectrin, $(\alpha\beta)_2$, that are linked to glycophorin molecules by band 4.1 protein and to band 3 protein by ankyrin, another peripheral membrane protein. The free ends of adjacent spectrin tetramers are held together by short chains of actin and band 4.1. Not shown is another protein, band 4.2, that assists ankyrin in linking spectrin to band 3 protein. For the electrophoretic fractionation of these proteins, see Figure 7-22.

are called **integral membrane proteins** because their hydrophobic regions are embedded within the membrane interior in a way that makes these molecules difficult to remove from membranes. However, such proteins also have one or more hydrophilic regions that extend outward from the membrane into the aqueous phase on one or both sides of the membrane. Because of their affinity for the lipid bilayer, integral membrane proteins are difficult to isolate and study by standard protein purification techniques, most of which are designed for water-soluble proteins. Treatment with a detergent that disrupts the lipid bilayer is usually necessary to solubilize and extract integral membrane proteins.

A few integral membrane proteins are known to be embedded in, and therefore to protrude from, only one side of the bilayer. These are called **integral monotopic proteins** (Figure 7-19a). However, most integral membrane proteins are **transmembrane proteins,** which means that they span the membrane and have hydrophilic regions protruding from the membrane on both sides. Such proteins cross the membrane either once (*singlepass proteins;* Figure 7-19b) or several times (*multipass proteins;* Figure 7-19c). Some multipass proteins consist of a single polypeptide (Figure 7-19c), whereas others have two or more polypeptides (*multisubunit proteins;* Figure 7-19d).

Most transmembrane proteins are anchored to the lipid bilayer by one or more hydrophobic **transmembrane segments,** one for each time the protein crosses the bilayer. In most cases, the polypeptide chain appears to span the membrane in an α-helical conformation consisting of about 20–30 amino acid residues, most—sometimes even all—of which have hydrophobic R groups. In some multipass proteins, however, several transmembrane segments are arranged as a β sheet in the form of a closed β sheet—the so-called *β barrel.* This structure is especially prominent in a group of pore-forming transmembrane proteins called *porins* that are found in the outer membrane of many bacteria as well as chloroplasts and mitochondria. Regardless of their conformation, transmembrane segments are usually separated along the primary structure of the protein by hydrophilic sequences that protrude or loop out on the two sides of the membrane. Loop regions containing positively charged amino acid residues are more likely to be found on the cytoplasmic side of the membrane. This "positive-inside rule" helps to ensure that all molecules of a particular transmembrane protein are oriented the same way.

Singlepass membrane proteins have just one transmembrane segment, with a hydrophilic carboxyl (C-) terminus extending out of the membrane on one side and a hydrophilic amino (N-) terminus protruding on the other side. Depending on the particular protein, the C-terminus may protrude on either side of the membrane. An example of a singlepass protein is *glycophorin,* a prominent protein in the erythrocyte plasma membrane (Figure 7-20b). Glycophorin is oriented in the membrane so that its C-terminus is on the inner surface of the membrane and its N-terminus is on the outer surface (**Figure 7-21a**).

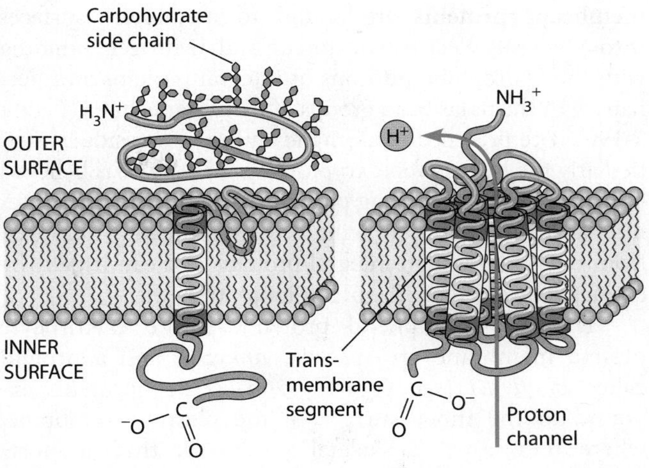

FIGURE 7-21 The Structures of Two Integral Membrane Proteins. **(a)** Glycophorin is a singlepass integral membrane protein in the erythrocyte plasma membrane. Its α-helical transmembrane segment consists entirely of hydrophobic amino acids. The N-terminus protrudes on the outer surface, the C-terminus on the cytoplasmic surface. Glycophorin is a glycoprotein, with 16 carbohydrate chains attached to its outer surface. **(b)** Bacteriorhodopsin is a multipass integral membrane protein in the plasma membrane of *Halobacterium.* Its seven transmembrane segments, which account for about 70% of its 248 amino acids, are organized into a proton channel. The C- and N-termini of the protein have short hydrophilic segments that protrude on the inner and outer surfaces of the plasma membrane, respectively. Short hydrophilic segments also link each of the transmembrane segments.

Multipass membrane proteins have several transmembrane segments, ranging from 2 or 3 to 20 or more such segments. An example of a multipass protein in the erythrocyte plasma membrane is a dimeric transport protein called *band 3 protein* (also known as the *anion exchange protein*). Each of its two polypeptides spans the lipid bilayer at least six times, with both the C-terminus and the N-terminus on the same side of the membrane. Current models of the dimeric protein assume a total of 12 transmembrane segments.

One of the best-studied examples of a multipass protein is bacteriorhodopsin, the plasma membrane protein that serves halobacteria as a proton pump (see Figure 7-3g). Its three-dimensional structure was reported by Unwin and Henderson in 1975, based on electron microscopy. Bacteriorhodopsin turned out to have seven α-helical, membrane-spanning segments, each corresponding to a sequence of about 20 hydrophobic amino acids in the primary structure of the protein (Figure 7-21b). The seven transmembrane segments are positioned in the membrane to form a channel and thereby to facilitate the light-activated pumping of protons across the membrane, a topic we will return to in Chapter 8.

Peripheral Membrane Proteins. In contrast to integral membrane proteins, some membrane-associated proteins lack discrete hydrophobic sequences and therefore do not penetrate into the lipid bilayer. Instead, these **peripheral**

membrane proteins are bound to membrane surfaces through weak electrostatic forces and hydrogen bonding with the hydrophilic portions of integral proteins and perhaps with the polar head groups of membrane lipids (Figure 7-19e). The presence of aromatic amino acid residues, particularly the hydrophobic tryptophan side chain, is believed to play a role in anchoring proteins at the membrane-water interface. Peripheral proteins are more readily removed from membranes than integral proteins and can usually be extracted by changing the pH or ionic strength.

The main peripheral proteins of the erythrocyte plasma membrane are *spectrin, ankyrin,* and a protein called *band 4.1* (see Figure 7-20b). These proteins are bound to the inner surface of the plasma membrane, where they form a skeletal meshwork that supports the plasma membrane and helps maintain the shape of the erythrocyte (see Figure 7-20a).

Lipid-Anchored Membrane Proteins. When Singer and Nicolson initially proposed the fluid mosaic model, they regarded all membrane proteins as either peripheral or integral membrane proteins. However, we now recognize a third class of proteins that are neither specifically peripheral nor integral but have some of the characteristics of both. The polypeptide chains of these **lipid-anchored membrane proteins** are located on one of the surfaces of the lipid bilayer but are covalently bound to lipid molecules embedded within the bilayer (Figure 7-19, parts f and g).

Several mechanisms are employed for attaching lipid-anchored proteins to membranes. Proteins bound to the inner surface of the plasma membrane are attached by covalent linkage either to a fatty acid or to an isoprene derivative called an *isoprenyl* group (Figure 7-19f). In the case of **fatty acid-anchored membrane proteins,** the protein is synthesized in the cytosol and then covalently attached to a saturated fatty acid embedded within the membrane bilayer, usually *myristic acid* (14 carbons) or *palmitic acid* (16 carbons). **Isoprenylated membrane proteins,** on the other hand, are synthesized as soluble cytosol proteins before being modified by addition of multiple 5-carbon isoprenyl groups (see Figure 3-27f), usually in the form of a 15-carbon *farnesyl* group or 20-carbon *geranylgeranyl* group. After attachment, the farnesyl or geranylgeranyl group is inserted into the lipid bilayer of the membrane.

Many lipid-anchored proteins attached to the external surface of the plasma membrane are covalently linked to *glycosylphosphatidylinositol (GPI),* a glycolipid found in the outer monolayer of the plasma membrane (Figure 7-19g). Lipid rafts are enriched in these **GPI-anchored membrane proteins,** which are made in the endoplasmic reticulum as singlepass transmembrane proteins that subsequently have their transmembrane segments cleaved off and replaced by GPI anchors. The proteins are then transported from the ER to the exterior of the plasma membrane. Once at the cell surface, GPI-anchored proteins can be released from the membrane by the enzyme *phospholipase C,* which is specific for phosphatidylinositol linkages.

Proteins Can Be Separated by SDS–Polyacrylamide Gel Electrophoresis

Before continuing our discussion of membrane proteins, it is useful to consider briefly how membrane proteins are isolated and studied. We will look first at the general problem of solubilizing and extracting proteins from membranes, and we will then learn about an electrophoretic technique that is very useful in the fractionation and characterization of proteins.

Isolation of Membrane Proteins. A major challenge to protein chemists has been the difficulty of isolating and studying membrane proteins, many of which are hydrophobic. Peripheral membrane proteins are in general fairly straightforward to isolate because they are loosely bound to the membrane by weak electrostatic interactions and hydrogen bonding with either the hydrophilic portions of integral membrane proteins or the polar head groups of membrane lipids. Peripheral proteins can be extracted from the membrane by changes in pH or ionic strength. Peripheral membrane proteins can also be solubilized by the use of a chelating (cation-binding) agent to remove calcium or by addition of urea, which breaks hydrogen bonds. Lipid-anchored proteins are similarly amenable to isolation, though with the requirement that the covalent bond to the lipid must first be cleaved. Once extracted from the membrane, most peripheral and lipid-anchored proteins are sufficiently hydrophilic to be purified and studied with techniques commonly used by protein chemists.

Integral membrane proteins, on the other hand, are difficult to isolate from membranes, especially in a manner that preserves their biological activity. In most cases, these proteins can be solubilized only by using detergents that disrupt hydrophobic interactions and dissolve the lipid bilayer. As we will now see, the use of strong ionic detergents such as *sodium dodecyl sulfate (SDS)* allows integral membrane proteins not just to be isolated, but to be fractionated and analyzed by the technique of electrophoresis. However, the use of such strong detergents can affect the function of the protein, and may be suitable only for analytical purposes.

SDS–Polyacrylamide Gel Electrophoresis. Cells contain thousands of different macromolecules that must be separated from one another before the properties of individual components can be investigated. One of the most common approaches for separating molecules from each other is **electrophoresis,** a group of related techniques that utilize an electrical field to separate charged molecules. How quickly any given molecule moves during electrophoresis depends upon its charge as well as its size. Electrophoresis can be carried out using various support media, such as paper, cellulose acetate, starch, polyacrylamide, or agarose (a polysaccharide obtained from seaweed). Of these media, gels made of polyacrylamide or agarose provide the best separation and are most commonly employed for the electrophoresis of nucleic acids and proteins.

When electrophoresis is used to study membrane proteins, membrane fragments are first solubilized with the anionic detergent SDS, which disrupts most protein-protein and protein-lipid associations. The proteins denature, unfolding into stiff polypeptide rods that cannot refold because their surfaces are coated with negatively charged detergent molecules. The solubilized, SDS-coated polypeptides are then layered on the top of a polyacrylamide gel. An electrical potential is then applied across the gel, such that the bottom of the gel is the positively charged anode (**Figure 7-22**). Because the polypeptides are coated with negatively charged SDS molecules, they migrate down the gel toward the anode. The polyacrylamide gel can be thought of as a fine meshwork that

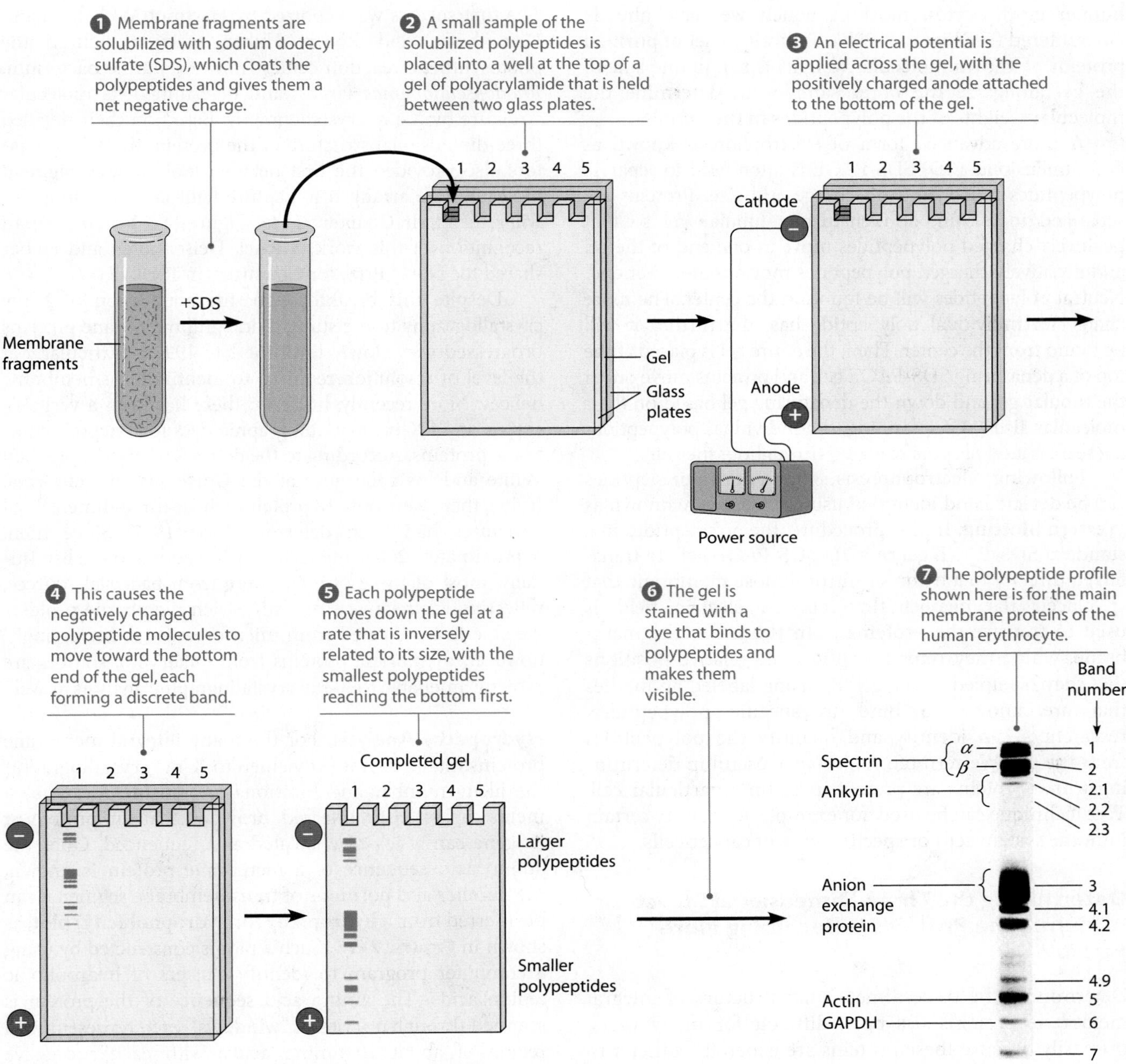

1 Membrane fragments are solubilized with sodium dodecyl sulfate (SDS), which coats the polypeptides and gives them a net negative charge.
2 A small sample of the solubilized polypeptides is placed into a well at the top of a gel of polyacrylamide that is held between two glass plates.
3 An electrical potential is applied across the gel, with the positively charged anode attached to the bottom of the gel.
+SDS
Membrane fragments
Gel
Glass plates
Cathode
Anode
Power source
4 This causes the negatively charged polypeptide molecules to move toward the bottom end of the gel, each forming a discrete band.
5 Each polypeptide moves down the gel at a rate that is inversely related to its size, with the smallest polypeptides reaching the bottom first.
Completed gel
Larger polypeptides
Smaller polypeptides
6 The gel is stained with a dye that binds to polypeptides and makes them visible.
7 The polypeptide profile shown here is for the main membrane proteins of the human erythrocyte.
Band number
Spectrin α β
Ankyrin
Anion exchange protein
Actin
GAPDH
1
2
2.1
2.2
2.3
3
4.1
4.2
4.9
5
6
7

FIGURE 7-22 SDS–Polyacrylamide Gel Electrophoresis (SDS-PAGE) of Membrane Proteins. Steps 1 through 6 show the general procedure used to separate membrane proteins using SDS-PAGE. Following step 1, the procedure is the same as is used for soluble, nonmembrane proteins. Step 7 shows the individual polypeptides separated from an erythrocyte membrane preparation. The separated bands of protein were initially identified using sequential numbers, as shown on the right. We now know the identity of some of these separated proteins, such as glyceraldehyde-3-phosphate dehydrogenase (GAPDH), shown on the left.

impedes the movement of large molecules more than that of small molecules. As a result, polypeptides move down the gel at a rate that is inversely related to the logarithm of their size.

When the smallest polypeptides approach the bottom of the gel, the process is terminated. The gel is then stained with a dye that binds to polypeptides and makes them visible. (*Coomassie brilliant blue* is commonly used for this purpose.) The particular polypeptide profile shown in Figure 7-22 is for the membrane proteins of human erythrocytes, most of which we have already encountered (see Figure 7-20b). Typically, a set of purified proteins of known molecular weights is run in one lane of the gel alongside the other samples to determine the molecular weights of the polypeptides in the samples.

A more advanced form of electrophoresis known as two-dimensional (2D) SDS-PAGE is often used to separate polypeptides based on both charge and size. Proteins are separated in a thin, nondenaturing, tubular gel so that positively charged polypeptides move to one end of the gel and negatively charged polypeptides move to the other end. Neutral polypeptides will be found in the center. The more charge an individual polypeptide has, the further it will be found from the center. Then, the entire gel is placed at the top of a denaturing SDS-PAGE gel, and proteins move out of the tubular gel and down the denaturing gel based on their molecular sizes. After staining, the individual polypeptides are seen as a set of spots scattered throughout the gel.

Following electrophoresis, individual polypeptides can be detected and identified using a procedure known as **Western blotting.** In this procedure, the polypeptides in a standard SDS-PAGE gel or a 2D SDS-PAGE gel are transferred directly to a nylon or nitrocellulose membrane that is placed flat against the gel. An electric field is used to transfer the proteins from the gel to the membrane, where they remain in the same relative positions that they occupied in the gel. By using labeled antibodies that are known to bind to specific polypeptides, researchers can identify and quantify the polypeptides from the gel. Western blotting is very useful in determining which proteins are present in or on a particular cell. This technique can be used, for example, to identify certain immune system cells or specific types of cancer cells.

Determining the Three-Dimensional Structure of Membrane Proteins Is Becoming More Feasible

Determining the three-dimensional structure of integral membrane proteins has been difficult for many years, primarily because these proteins are generally difficult to isolate and purify due to their hydrophobicity. However, they are proving increasingly amenable to study by *X-ray crystallography,* which determines the structure of proteins that can be isolated in crystalline form. For the many membrane proteins for which no crystal structure is available, an alternative approach called *hydropathic analysis* can be used, provided that the protein or its gene can at least be isolated and sequenced. We will look briefly at each of these techniques.

X-Ray Crystallography. X-ray crystallography is widely used to determine the three-dimensional structure of proteins. A description of this technique is included in the Appendix (see page A-27). The difficulty of isolating integral membrane proteins in crystalline form virtually excluded these proteins from crystallographic analysis for many years. The first success was reported by Hartmut Michel, Johann Deisenhofer, and Robert Huber, who crystallized the photosynthetic reaction center from the purple bacterium, *Rhodopseudomonas viridis,* and determined its molecular structure by X-ray crystallography. Based on their detailed three-dimensional structure of the protein, these investigators also provided the first detailed look at how pigment molecules are arranged to capture light energy, a topic we will return to in Chapter 11 (see Figure 11B-1 on p. 330). In recognition of this work, Michel, Deisenhofer, and Huber shared the Nobel Prize for Chemistry in 1988.

Despite this breakthrough, the application of X-ray crystallography to the study of integral membrane proteins progressed very slowly until the late 1990s, particularly at the level of resolution required to identify transmembrane helices. More recently, however, there has been a veritable explosion of X-ray crystallographic data for integral membrane proteins. According to the data assembled by Stephen White and his colleagues at the University of California, Irvine, there were only 18 proteins whose three-dimensional structures had been determined by 1997. Since then, approximately 200 proteins have been added to the list. Initially, most of these proteins were from bacterial sources, reflecting the relative ease with which membrane proteins are able to be isolated from microorganisms. Increasingly, however, membrane proteins from eukaryotic sources are proving amenable to X-ray crystallographic analysis as well.

Hydropathy Analysis. For the many integral membrane proteins that have not yet yielded to X-ray crystallography, the likely number and locations of transmembrane segments can often be inferred, provided that the protein or its gene can at least be isolated and sequenced. Once the amino acid sequence of a membrane protein is known, the number and positions of transmembrane segments can be inferred from a **hydropathy** (or **hydrophobicity**) **plot**, as shown in **Figure 7-23**. Such a plot is constructed by using a computer program to identify clusters of hydrophobic amino acids. The amino acid sequence of the protein is scanned through a series of "windows," each representing a region of about 10 amino acids, with each successive window one amino acid further along in the sequence.

Based on the known hydrophobicity values for the various amino acids, a **hydropathy index** is calculated for each successive window by averaging the hydrophobicity values of the amino acids in the window. (By convention, hydrophobic amino acids have positive hydropathy values

(a) Hydropathy plot of connexin. The hydropathy index on the vertical axis is a numerical measure of the relative hydrophobicity of successive segments of the polypeptide chain based on its amino acid sequence.

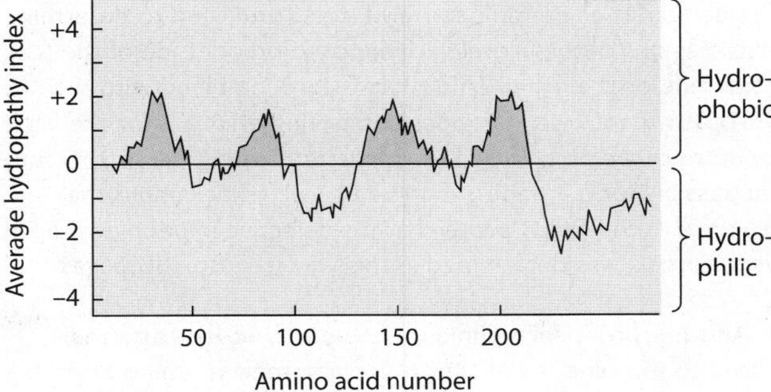

(b) Transmembrane structure of connexin. Connexin has four distinct hydrophobic regions, which correspond to the four α-helical segments that span the plasma membrane.

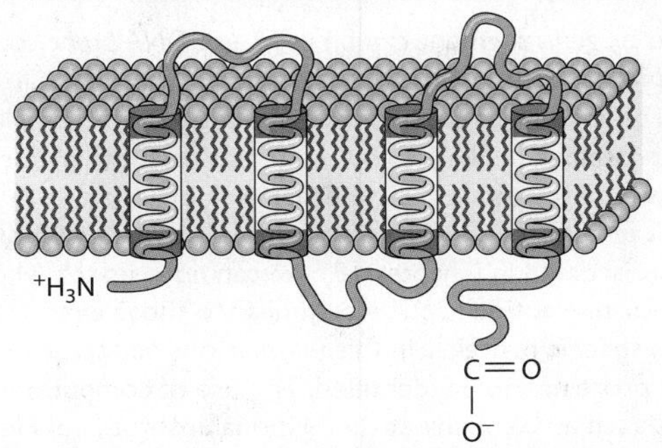

FIGURE 7-23 Hydropathy Analysis of an Integral Membrane Protein. A hydropathy plot is a means of representing hydrophobic regions (positive values) and hydrophilic regions (negative values) along the length of a protein. This example uses hydropathy data to analyze the plasma membrane protein *connexin*.

and hydrophilic residues have negative values.) The hydropathy index is then plotted against the positions of the windows along the sequence of the protein. The resulting hydropathy plot predicts how many membrane-spanning regions are present in the protein, based on the number of positive peaks. The hydropathy plot shown in Figure 7-23a is for a plasma membrane protein called *connexin*. The plot shows four positive peaks and therefore predicts that connexin has four stretches of hydrophobic amino acids and hence four transmembrane segments, as shown in Figure 7-23b.

Molecular Biology Has Contributed Greatly to Our Understanding of Membrane Proteins

Membrane proteins have not yielded as well as other proteins to biochemical techniques, mainly because of the problems involved in isolating and purifying hydrophobic proteins in physiologically active form. Procedures such as SDS-PAGE and hydropathy analysis have certainly been useful, as have labeling techniques involving radioisotopes or fluorescent antibodies, which we will discuss later in the chapter. Within the past three decades, however, the study of membrane proteins has been revolutionized by the techniques of molecular biology, especially DNA sequencing and recombinant DNA technology. DNA sequencing makes it possible to deduce the amino acid sequence of a protein without the need to isolate the protein in pure form for amino acid sequencing.

In addition, sequence comparisons between proteins often reveal evolutionary and functional relationships that might not otherwise have been appreciated. DNA pieces can also be used as probes to identify and isolate sequences that encode related proteins. Moreover, the DNA sequence for a particular protein can be altered at specific nucleotide positions to determine the effects of changing specific amino acids on the activity of the mutant protein it codes for. **Box 7A** describes these exciting developments in more detail.

Membrane Proteins Have a Variety of Functions

What functions do membrane proteins perform? Most of what we summarized about membrane function at the beginning of the chapter is relevant here because the functions of a membrane are really just those of its chemical components, especially its proteins.

Some of the proteins in membranes are *enzymes*, which accounts for the localization of specific functions to specific membranes. As we will see in coming chapters, each of the organelles in a eukaryotic cell is in fact characterized by its own distinctive set of membrane-bound enzymes. We encountered an example earlier when we noted the association of glucose-6-phosphatase with the ER. Another example is glyceraldehyde-3-phosphate dehydrogenase (GAPDH), an enzyme involved in the catabolism of blood glucose. GAPDH is a peripheral plasma membrane protein in erythrocytes and other cell types (see Figure 7-22, step ❼). Closely related to enzymes in their function are *electron transport proteins* such as the cytochromes and iron-sulfur proteins that are involved in oxidative processes in mitochondria, chloroplasts, and the plasma membranes of prokaryotic cells.

Other membrane proteins function in solute transport across membranes. These include *transport proteins*, which facilitate the movement of nutrients such as sugars and amino acids across membranes, and *channel proteins*, which provide hydrophilic passageways through otherwise hydrophobic membranes. Also in this category are *transport ATPases*, which use the energy of ATP to pump ions across membranes.

Numerous membrane proteins are *receptors* involved in recognizing and mediating the effects of specific chemical signals that impinge on the surface of the cell. Hormones, neurotransmitters, and growth-promoting substances are examples of chemical signals that interact with specific

Revolutionizing the Study of Membrane Proteins: The Impact of Molecular Biology

Because membrane proteins mediate a remarkable variety of cellular functions, cell biologists are very interested in these proteins. The study of membrane proteins has begun to yield definitive insights and answers as biochemical techniques commonly used to isolate and analyze cellular proteins have been applied to membrane proteins. This chapter describes several such applications, including SDS–polyacrylamide gel electrophoresis, hydropathy analysis, and procedures for labeling membrane proteins with radioactivity or fluorescent antibodies. Two other biochemical approaches that can be used to study membrane proteins are affinity labeling and membrane reconstitution.

Affinity labeling utilizes radioactive molecules that bind to specific proteins because of known functions of the proteins. For example, *cytochalasin B* is known to be a potent inhibitor of glucose transport. Membranes that have been exposed to radioactive cytochalasin B are therefore likely to contain radioactivity bound specifically to protein molecules involved in glucose transport.

Membrane reconstitution involves the formation of artificial membranes from specific purified components. In this approach, proteins are extracted from membranes with detergent solutions and separated individually. The purified proteins are then mixed together with phospholipids to form liquid-filled membrane vesicles called *liposomes* that can be "loaded" with particular molecules. These reconstituted vesicles can then be tested for their ability to carry out specific membrane protein functions, such as nutrient transport or cell-to-cell communication.

Despite some success with these approaches, membrane biologists are often stymied in their attempts to isolate, purify, and study membrane proteins. Biochemical techniques that work well with soluble proteins are not often useful with hydrophobic proteins. However, the study of membrane proteins has been revolutionized by the techniques of molecular biology, especially *DNA sequencing* and *recombinant DNA technology*. We will consider these techniques in detail in Chapters 18 and 20, but we need not wait until then to appreciate the enormous impact of molecular biology on the study of membranes and membrane proteins. **Figure 7A-1** summarizes several approaches that have proven especially powerful for both membrane and nonmembrane proteins.

Vital to these approaches is the isolation (cloning) of a gene or gene fragment that encodes all or part of a specific protein (Figure 7A-1, top). A top priority is to determine the nucleotide sequence of the cloned gene ❶. *DNA sequencing* is one of the triumphs of molecular biology. Determining the nucleotide sequence of a DNA molecule is now far easier than determining the amino acid sequence of the protein it encodes. Moreover, most of the sequencing procedure is carried out quickly and automatically by DNA sequencing machines. Once the DNA for a particular protein has been sequenced, the predicted *amino acid sequence* of the protein can be deduced by using the genetic code (see Figure 21-6) ❷. The predicted amino acid sequence can then be studied using *hydropathy analysis* (see Figure 7-23) to identify likely transmembrane segments of the protein ❸.

Knowing the amino acid sequence of the protein also allows the investigator to prepare synthetic peptides that correspond to specific segments of the protein ❹. Antibodies made against these peptides can then be radioactively labeled and used to determine which regions of the protein are exposed on one side of the membrane or the other. This information, combined with the hydropathy data, often provides compelling evidence for the likely structure of the protein and its orientation within the membrane—and possibly for its mode of action as well. For example, the structure of the CFTR protein that is defective in people with cystic fibrosis was determined in this way (see Box 8B, pages 234–235).

Another powerful technique, **site-specific mutagenesis,** is used to examine the effect of changing specific amino acids in a protein ❺. The DNA sequence encoding the protein is altered by changing the nucleotides corresponding to a particular amino acid. The mutant DNA is then introduced into living cells, which then synthesize a mutant protein having the altered amino acid. Then, the functional properties of this mutant protein can be studied to determine whether the amino acid is required for proper protein function.

A gene or gene segment can be used as a *DNA probe* to isolate similar DNA sequences ❻. DNA identified in this way is likely to encode proteins similar to the protein that the probe DNA codes for. Such proteins are likely to be related to each other both in evolutionary origin and in their mechanisms of action.

Thanks to the advent of techniques for sequencing whole genomes (discussed in Chapter 18), we can now search whole genomes for nucleotide sequences similar to those already known to encode specific proteins. In this way, various *families,* or groups, of related proteins can be identified. The use of computerized databases, such as GenBank at the National Institutes of Health, has been extremely valuable in suggesting roles for proteins based entirely on their gene sequences.

From studies based on these and other techniques, we now know that cells in the human body need more than 30 families of membrane proteins to facilitate the transport of the great variety of solutes that must be moved across membranes. Each member of such a family may be present in a variety of *isoforms* that differ in such properties as time of expression during development, tissue distribution, or location within the cell. Perhaps it is not so surprising, then, to learn that the genes known to encode transport proteins represent about 10% of the human genome!

Most of these molecular approaches are indirect in the sense that they allow scientists to deduce properties and functions of proteins rather than proving them directly. Still, these techniques are powerful tools that have already significantly expanded our understanding of membrane proteins. And certainly in the future, these and newer techniques of molecular biology, such as proteomics, will continue to revolutionize the study of membranes and their proteins.

FIGURE 7A-1 Application of Molecular Biology Techniques to the Study of Membrane Proteins. These techniques are invaluable for studying membrane proteins, which are often very difficult to isolate.

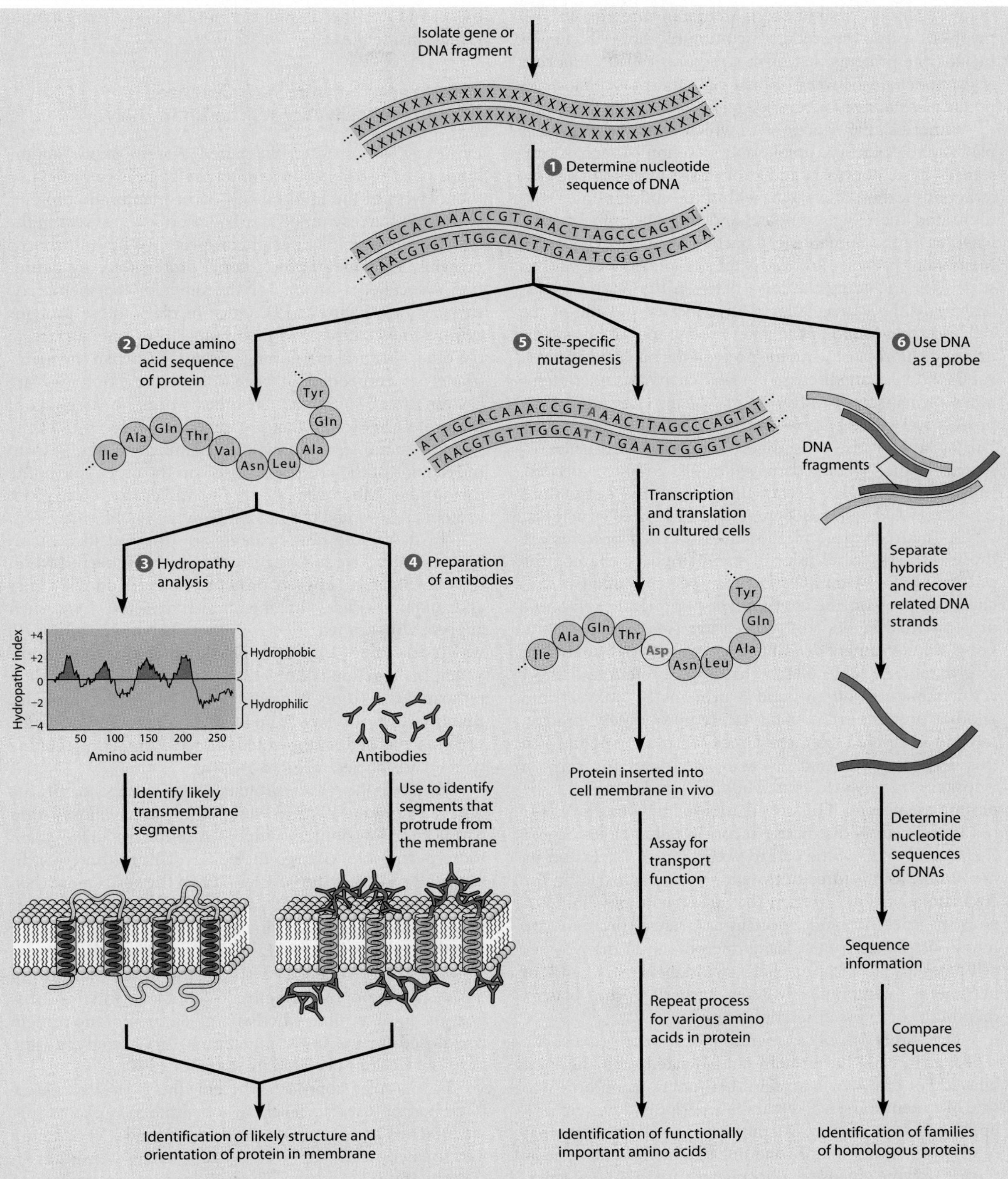
Isolate gene or DNA fragment
1 Determine nucleotide sequence of DNA
2 Deduce amino acid sequence of protein
5 Site-specific mutagenesis
6 Use DNA as a probe
Ile
Ala
Gln
Thr
Val
Asn
Leu
Ala
Gln
Tyr
DNA fragments
3 Hydropathy analysis
4 Preparation of antibodies
Transcription and translation in cultured cells
Separate hybrids and recover related DNA strands
Hydropathy index
Hydrophobic
Hydrophilic
Amino acid number
Antibodies
Asp
Identify likely transmembrane segments
Use to identify segments that protrude from the membrane
Protein inserted into cell membrane in vivo
Assay for transport function
Determine nucleotide sequences of DNAs
Sequence information
Repeat process for various amino acids in protein
Compare sequences
Identification of likely structure and orientation of protein in membrane
Identification of functionally important amino acids
Identification of families of homologous proteins

protein receptors on the plasma membrane of target cells. In most cases, the binding of a hormone or other signal molecule to the appropriate receptor on the membrane surface triggers some sort of intracellular response, which in turn elicits the desired effect. Membrane proteins are also involved with intercellular communication. Examples include the proteins that form structures called *connexons* at *gap junctions* between animal cells and those that make up the *plasmodesmata* between plant cells.

Other cellular functions in which membrane proteins play key roles include uptake and secretion of various substances by endocytosis and exocytosis; targeting, sorting, and modification of proteins within the endoplasmic reticulum and the Golgi complex; and the detection of light, whether by the human eye, a bacterial cell, or a plant leaf. Membrane proteins are also vital components of various structures, including the links between the plasma membrane and the extracellular matrix located outside of the cell, the pores found in the outer membranes of mitochondria and chloroplasts, and the pores of the nuclear envelope. All these topics are discussed in later chapters. Other membrane proteins are involved in *autophagy* ("self eating"), a process we will learn about in more detail in Chapter 12. During autophagy, cells digest their own organelles or structures that become damaged or are no longer needed. In this way, the molecular components of these structures can be recycled and reused in newly synthesized structures.

A final group of membrane-associated proteins are those with structural roles in stabilizing and shaping the cell membrane. Examples include spectrin, ankyrin, and band 4.1 protein, the erythrocyte peripheral membrane proteins that we encountered earlier (see Figure 7-20b). Long, thin tetramers of α and β spectrin, $(\alpha\beta)_2$, are linked to glycophorin molecules by band 4.1 protein and short actin filaments and to band 3 proteins by ankyrin and another protein, called band 4.2. (Appropriately enough, *ankyrin* is derived from the Greek word for "anchor.") In this way, spectrin and its associated proteins form a cytoskeletal network that underlies and supports the plasma membrane. This spectrin-based network gives the red blood cell its distinctive biconcave shape (see Figure 7-20a) and enables the cell to withstand the stress on its membrane as it is forced through narrow capillaries in the circulatory system. Proteins that are structurally homologous to spectrin and spectrin-associated proteins are found just beneath the plasma membrane in many other cell types also, indicating that a cytoskeletal meshwork of peripheral membrane proteins underlies the plasma membrane of many different kinds of cells.

The function of a membrane protein is usually reflected in how the protein is associated with the lipid bilayer. For example, a protein that functions on only one side of a membrane is likely to be a peripheral protein or a lipid-anchored protein. Membrane-bound enzymes that catalyze reactions on only one side of a membrane, such as the ER enzyme glucose-6-phosphatase are in this category. In contrast, the tasks of transporting solutes or transmitting signals across a membrane clearly require transmembrane proteins. In Chapter 8 we will examine the role of transmembrane proteins in transport of materials across membranes. In Chapter 14 we will see how transmembrane receptors bind signaling molecules (such as hormones) on the outside of the plasma membrane and then generate signals inside the cell.

Membrane Proteins Are Oriented Asymmetrically Across the Lipid Bilayer

Earlier in the chapter, we noted that most membrane lipids are distributed asymmetrically between the two monolayers of the lipid bilayer. Most membrane proteins also exhibit an asymmetric orientation with respect to the bilayer. For example, peripheral proteins, lipid-anchored proteins, and integral monotopic proteins are by definition associated with one or the other of the membrane surfaces (see Figure 7-19). Once in place, these proteins cannot move across the membrane from one surface to the other. Integral membrane proteins that span the membrane are embedded in both monolayers, but they are asymmetrically oriented. In other words, the regions of the protein molecule that are exposed on one side of the membrane are structurally and chemically different from the regions of the protein exposed on the other side of the membrane. Moreover, all of the molecules of a given protein are oriented the same way in the membrane.

To determine how proteins are oriented in a membrane, radioactive labeling procedures have been devised that distinguish between proteins exposed on the inner and outer surfaces of membrane vesicles. One such approach makes use of the enzyme *lactoperoxidase (LP)*, which catalyzes the covalent binding of iodine to proteins. When the reaction is carried out in the presence of ^{125}I, a radioactive isotope of iodine, LP labels the proteins. Because LP is too large to pass through membranes, only proteins exposed on the outer surface of intact membrane vesicles are labeled (**Figure 7-24a**).

To label only those proteins that are exposed on the inner membrane surface, vesicles are first exposed to a hypotonic (low ionic strength) solution to make them more permeable to large molecules. Under these conditions, LP can enter the vesicles. When the vesicles are then transferred to an isotonic solution that contains ^{125}I but no external LP, the ^{125}I, being a small molecule, can diffuse into the vesicle where the LP is trapped. Thus, the enzyme labels the proteins exposed on the inner surface of the vesicle membrane (Figure 7-24b, c). In this way, it is possible to determine whether a given membrane protein is exposed on the inner membrane surface only, on the outer surface only, or on both surfaces.

In a similar approach, the enzyme *galactose oxidase (GO)* can be used to label carbohydrate side chains that are attached to membrane proteins or lipids. Vesicles are first treated with GO to oxidize galactose residues in carbohydrate side chains. The vesicles are then exposed to tritiated borohydride ($^3H-BH_4$), which reduces the galactose groups, introducing labeled hydrogen atoms in the

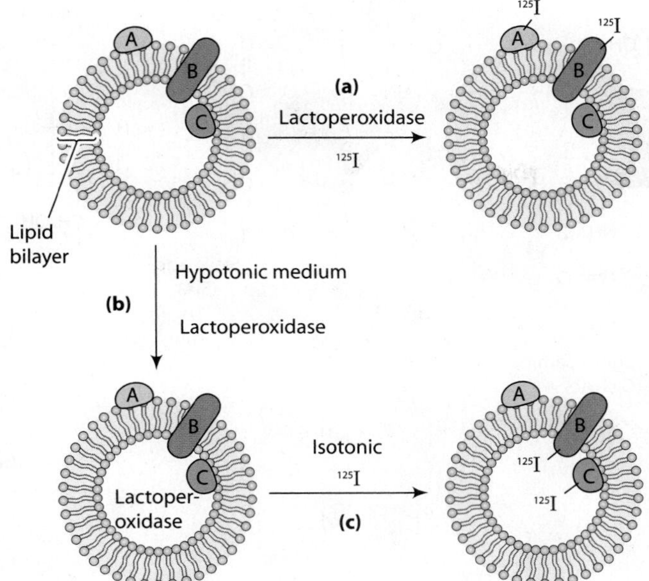

FIGURE 7-24 A Method for Labeling Proteins Exposed on One or Both Surfaces of a Membrane Vesicle. **(a)** When lactoperoxidase (LP) and ^{125}I are present in the solution outside a membrane vesicle, LP catalyzes the labeling of membrane proteins exposed on the outer membrane surface (i.e., proteins A and B). If membrane vesicles are **(b)** first incubated in a hypotonic medium to make them permeable to LP and **(c)** then transferred to an isotonic solution containing ^{125}I but no external LP, proteins exposed on the inner membrane surface (i.e., proteins B and C) become labeled.

process. In both of these approaches, the labeled proteins are typically separated on an SDS-PAGE gel that is then dried and exposed to X-ray film to reveal the locations of the labeled polypeptides.

The orientation of a protein within a membrane can also be determined using antibodies that are designed to recognize specific parts of the protein. Intact cells (or organelles) are exposed to these antibodies and are then tested to determine whether antibodies have bound to the membrane. If so, it can be concluded that the antibody-binding site, or *epitope*, must be on the outer membrane surface.

Many Membrane Proteins Are Glycosylated

In addition to lipids and proteins, most membranes contain small but significant amounts of carbohydrates, except for chloroplast, mitochondrial, and bacterial membranes. The plasma membrane of the human erythrocyte, for example, contains about 49% protein, 43% lipid, and 8% carbohydrate by weight. The glycolipids that we encountered earlier account for a small portion of membrane carbohydrate, but most of the carbohydrate in membranes is found as part of **glycoproteins**—membrane proteins with carbohydrate chains covalently linked to amino acid side chains.

The addition of a carbohydrate side chain to a protein is called **glycosylation**. This process occurs in the ER and Golgi compartments of the cell soon after synthesis. As **Figure 7-25** shows, glycosylation involves linkage of the carbohydrate either to the nitrogen atom of an amino group

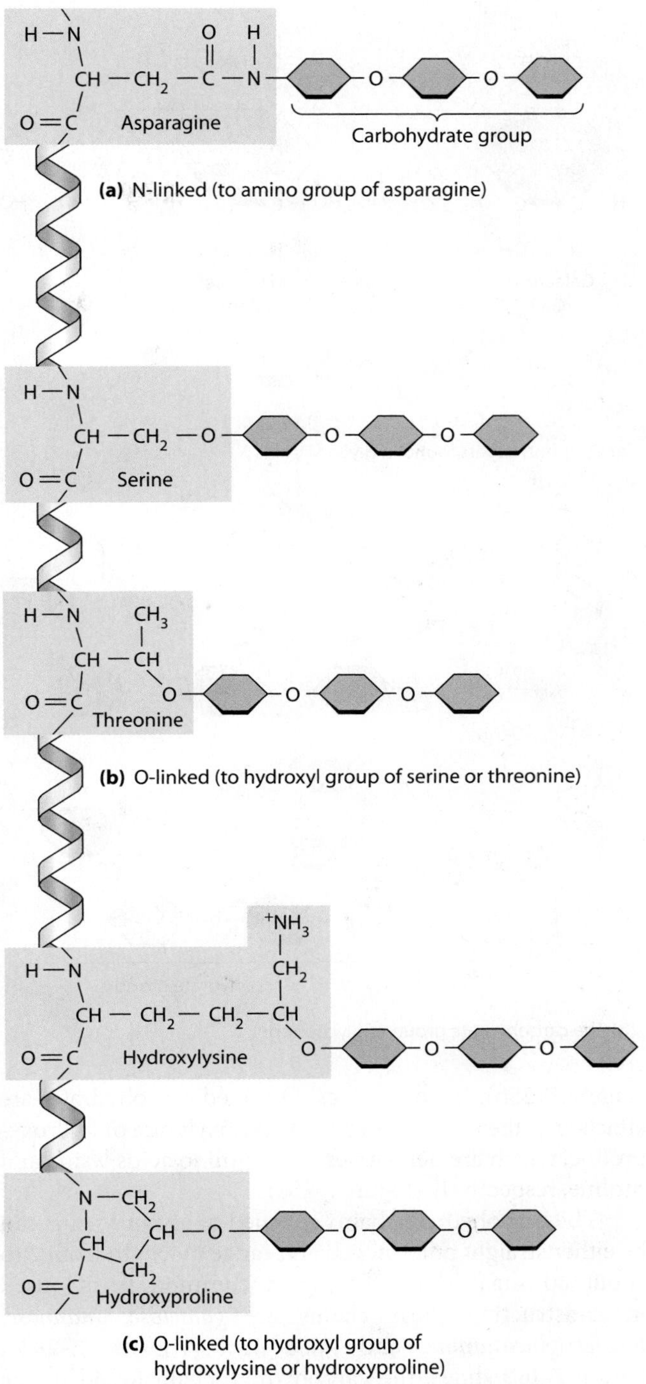

Asparagine — Carbohydrate group

(a) N-linked (to amino group of asparagine)

Serine

Threonine

(b) O-linked (to hydroxyl group of serine or threonine)

Hydroxylysine

Hydroxyproline

(c) O-linked (to hydroxyl group of hydroxylysine or hydroxyproline)

FIGURE 7-25 N-Linked and O-Linked Glycosylation of Membrane Proteins. Carbohydrate groups are linked to specific amino acid residues in membrane proteins in two different ways. **(a)** N-linked carbohydrates are attached to the amino group of asparagine side chains. **(b)** O-linked carbohydrates are attached to the hydroxyl group of serine or threonine side chains. **(c)** In some cases, O-linked carbohydrates are attached to the hydroxyl group of the modified amino acids hydroxylysine and hydroxyproline.

(N-linked glycosylation) or to the oxygen atom of a hydroxyl group *(O-linked glycosylation)*. N-linked carbohydrates are attached to the amino group on the side chain of asparagine (Figure 7-25a), whereas O-linked carbohydrates are usually bound to the hydroxyl groups of either serine or threonine

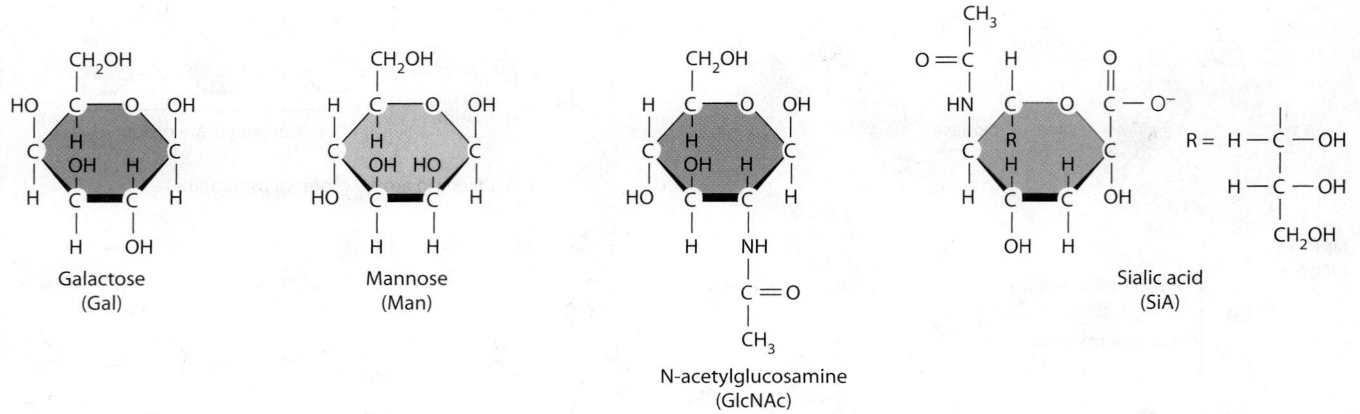

Galactose (Gal)

Mannose (Man)

N-acetylglucosamine (GlcNAc)

Sialic acid (SiA)

(a) Common sugars found in glycoproteins

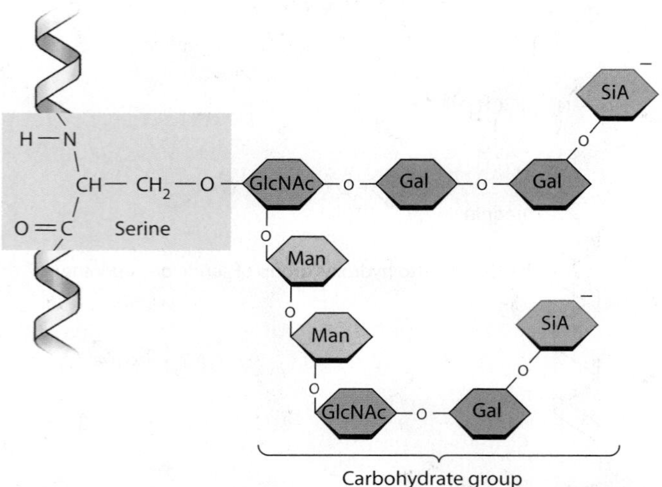

(b) The carbohydrate group of glycophorin

FIGURE 7-26 Examples of Carbohydrates Found in Glycoproteins. **(a)** The four most common carbohydrates in glycoproteins are galactose (Gal), mannose (Man), *N*-acetylglucosamine (GlcNAc), and sialic acid (SiA). **(b)** For example, 15 of the 16 carbohydrate chains of glycophorin are linked to the amino acid serine in the hydrophilic portion of the protein on the outer surface of the erythrocyte plasma membrane. Each of these carbohydrate chains consists of three units of Gal and two units each of GlcNAc, Man, and SiA. The two terminal sialic acid groups are negatively charged.

(Figure 7-25b). In some cases, O-linked carbohydrates are attached to the hydroxyl group of hydroxylysine or hydroxyproline, which are derivatives of the amino acids lysine and proline, respectively (Figure 7-25c).

The carbohydrate chains attached to glycoproteins can be either straight or branched and range in length from 2 to about 60 sugar units. The most common sugars used in constructing these chains are *galactose, mannose, N-acetylglucosamine,* and *sialic acid* (**Figure 7-26a**). Figure 7-26b shows the carbohydrate chain found in the erythrocyte plasma membrane protein, glycophorin. This integral membrane protein has 16 such carbohydrate chains attached to the portion of the molecule (the N-terminus) that extends outward from the erythrocyte membrane. Of these 16 chains, 1 is N-linked and 15 are O-linked. Notice that both branches of the carbohydrate chain terminate in a negatively charged sialic acid. Because of these anionic groups on their surfaces, erythrocytes repel each other, thereby reducing blood viscosity.

Glycoproteins are most prominent in plasma membranes, where they play an important role in cell-cell recognition. Consistent with this role, glycoproteins are always positioned so that the carbohydrate groups protrude on the external surface of the cell membrane. This arrangement, which contributes to membrane asymmetry, has been shown experimentally using *lectins,* plant proteins that bind specific sugar groups very tightly. For example, *wheat germ agglutinin,* a lectin found in wheat embryos, binds very specifically to oligosaccharides that terminate in *N*-acetylglucosamine, whereas *concanavalin A,* a lectin from jack beans, recognizes mannose groups in internal positions. Investigators visualize these lectins by linking them to *ferritin,* an iron-containing protein that shows up as an electron-dense spot when viewed with an electron microscope. When such ferritin-linked lectins are used as probes to localize the oligosaccharide chains of membrane glycoproteins, binding is always specifically to the outer surface of the plasma membrane.

In many animal cells, the carbohydrate groups of plasma membrane glycoproteins and glycolipids protrude from the cell surface and form a surface coat called the **glycocalyx** (meaning "sugar coat"). **Figure 7-27** shows the prominent glycocalyx of an intestinal epithelial cell. The carbohydrate groups on the cell surface are important components of the recognition sites of membrane receptors such as those involved in binding extracellular signal

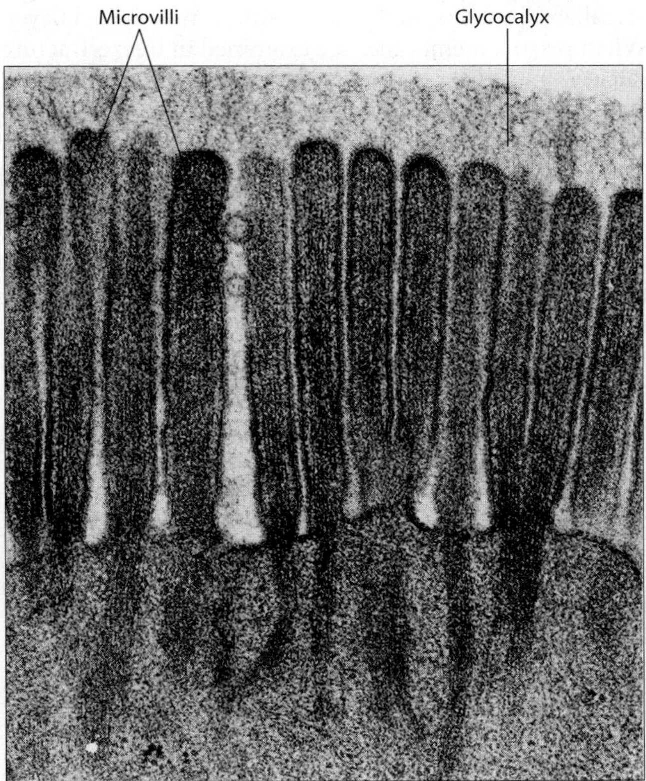

Microvilli
Glycocalyx

FIGURE 7-27 The Glycocalyx of an Intestinal Epithelial Cell. This electron micrograph of a cat intestinal epithelial cell shows the microvilli (fingerlike projections that are involved in absorption) and the glycocalyx on the cell surface. The glycocalyx on this cell is about 150 nm thick and consists primarily of oligosaccharide chains about 1.2–1.5 nm in diameter (TEM).

molecules, in antibody-antigen reactions, and in intercellular adhesion to form tissues. In some pathogenic bacteria, such as *Streptococcus pneumoniae,* the presence of a glycocalyx can hide the antigenic surface proteins that would normally stimulate the immune response. This allows these bacteria to escape detection and subsequent destruction by immune system cells and thus enables them to adhere to target cells and cause disease in the host organism.

Membrane Proteins Vary in Their Mobility

Earlier in the chapter, we noted that lipid molecules can diffuse laterally within the plane of a membrane (see Figure 7-11). Now we can ask the same question about membrane proteins: Are they also free to move within the membrane? In fact, membrane proteins are much more variable than lipids in their mobility. Some proteins appear to move freely within the lipid bilayer. Others are constrained, often because they are anchored to protein complexes located adjacent to one side of the membrane or the other.

Experimental Evidence for Protein Mobility. Particularly convincing evidence for the mobility of at least some membrane proteins has come from *cell fusion experiments* such as those summarized in **Figure 7-28**. In these studies, David Frye and Michael Edidin took advantage of two powerful techniques, one that enabled them to fuse cells from two different species and another that made it possible for them to label specific proteins on the surfaces of cells with antibodies containing fluorescent dye molecules. *Antibodies* are immune system proteins that recognize and bind to specific molecular *antigens* such as cell surface proteins.

Frye and Edidin prepared two **fluorescent antibodies,** each one having a differently-colored dye linked to it, so that the human and mouse proteins could be distinguished. The anti-mouse antibodies were linked to a green fluorescent dye called *fluorescein,* whereas the anti-human antibodies were linked to a red fluorescent dye, *rhodamine.*

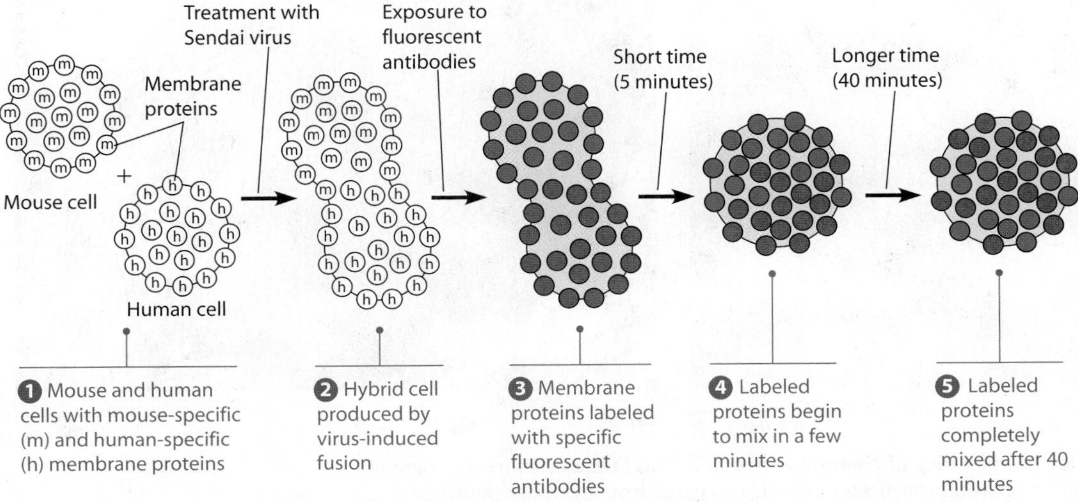

Treatment with Sendai virus
Exposure to fluorescent antibodies
Short time (5 minutes)
Longer time (40 minutes)
Membrane proteins
Mouse cell
Human cell
1 Mouse and human cells with mouse-specific (m) and human-specific (h) membrane proteins
2 Hybrid cell produced by virus-induced fusion
3 Membrane proteins labeled with specific fluorescent antibodies
4 Labeled proteins begin to mix in a few minutes
5 Labeled proteins completely mixed after 40 minutes

FIGURE 7-28 Demonstration of the Mobility of Membrane Proteins by Cell Fusion. The mobility of membrane proteins can be shown experimentally by the mixing of membrane proteins that occurs when cells from two different species (mouse and human) are fused and the membrane proteins are labeled with specific fluorescent antibodies.

Thus, under a fluorescence microscope, the mouse cells appeared green and the human cells appeared red due to each antibody recognizing and binding to its specific protein antigens on the surface of the cells (Figure 7-28).

Frye and Edidin fused mouse and human cells using Sendai virus, exposed them to the red and green fluorescent antibodies, and observed the fused cells by fluorescence microscopy. At first, the green fluorescent membrane proteins from the mouse cell were localized on one-half of the hybrid cell surface, and the red fluorescent membrane proteins derived from the human cell were restricted to the other half (see Figure 7-28). In a few minutes, however, the proteins from the two parent cells began to intermix. After 40 minutes, the separate regions of green and red fluorescence were completely intermingled.

If the fluidity of the membrane was depressed by lowering the temperature below the transition temperature of the lipid bilayer, this intermixing could be prevented. Frye and Edidin therefore concluded that the intermingling of the fluorescent proteins had been caused by lateral diffusion of the human and mouse proteins through the fluid lipid bilayer of the plasma membrane. Compared to most membrane lipids, however, membrane proteins diffuse through the lipid bilayer much more slowly due to their larger size.

If proteins are completely free to diffuse within the plane of the membrane, then they should eventually become randomly distributed. Support for the idea that at least some membrane proteins behave in this way has emerged from freeze-fracture microscopy, which directly visualizes proteins embedded within the lipid bilayer. When plasma membranes are examined in freeze-fracture micrographs, their embedded protein particles often tend to be randomly distributed. Such evidence for protein mobility is not restricted to the plasma membrane. It has also been found, for example, that the protein particles of the inner mitochondrial membranes are randomly arranged (**Figure 7-29a**). If isolated mitochondrial membrane vesicles are exposed to an electrical potential, the protein particles, which bear a net negative charge, all move to one end of the vesicle (Figure 7-29b). Removing the electrical potential causes the particles to become randomly distributed again, indicating that these proteins are free to move within the lipid bilayer.

Experimental Evidence for Restricted Mobility. Although many types of membrane proteins have been shown to diffuse through the lipid bilayer, their rates of movement vary. A widely used approach for quantifying the rates at which membrane proteins diffuse is fluorescence photobleaching recovery, a technique discussed earlier in the context of lipid mobility (see Figure 7-11). The rate at which unbleached molecules from adjacent parts of the membrane move back into the bleached area can be used to calculate the diffusion rates of various kinds of fluorescent lipid or protein molecules.

Membrane proteins are much more variable in their diffusion rates than lipids are. A few membrane proteins diffuse almost as rapidly as lipids, but most diffuse more

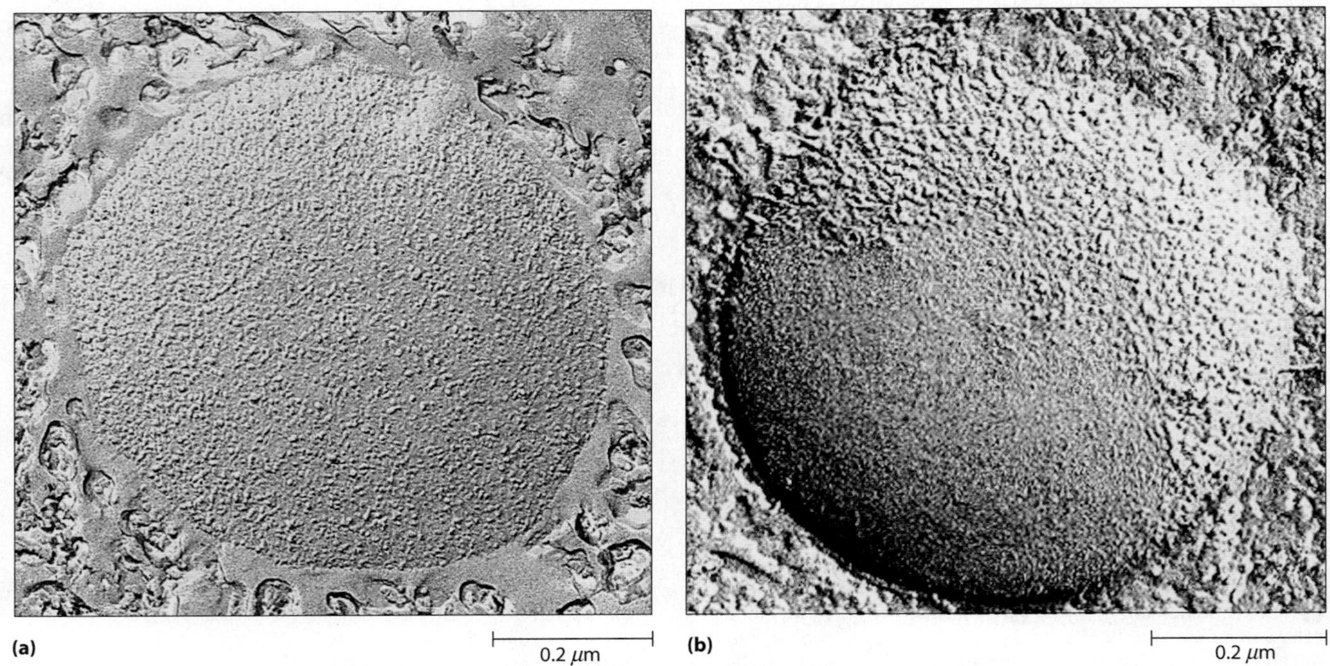

FIGURE 7-29 Evidence for the Mobility of Membrane Proteins. (a) Freeze-fracture micrograph showing the random distribution of protein particles in vesicles prepared from the inner mitochondrial membrane. (b) Exposure of the vesicles to an electrical field causes the membrane particles to migrate to one end of the vesicle (upper right of micrograph). If membrane proteins were not mobile, this movement of proteins to one side of the vesicle would not happen.

slowly than would be expected if they were completely free to move within the lipid bilayer. Moreover, the diffusion of many membrane proteins is restricted to a limited area of the membrane, indicating that at least some membranes consist of a series of separate *membrane domains* that differ in their protein compositions and hence in their functions. For example, the cells lining your small intestine have membrane proteins that transport solutes such as sugars and amino acids out of the intestine and into the body. These transport proteins are restricted to the side of the cell where the corresponding type of transport is required.

Mechanisms for Restricting Protein Mobility. Several different mechanisms account for restricted protein mobility and hence for cell polarization. In some cases, membrane proteins aggregate within the membrane, forming large complexes that move only sluggishly, if at all. In other cases, membrane proteins form structures that become barriers to the diffusion of other membrane proteins, thereby effectively creating specific membrane domains. The *tight junctions* to be discussed in Chapter 17 are one example. The most common restraint on the mobility of membrane proteins, however, is imposed by the binding, or *anchoring*, of such proteins to structures located adjacent to one side of the membrane or the other. For example, many proteins of the plasma membrane are anchored either to elements of the cytoskeleton on the inner surface of the membrane or to extracellular structures such as the extracellular matrix of animal cells. We will encounter both of these anchoring mechanisms in Chapter 17.

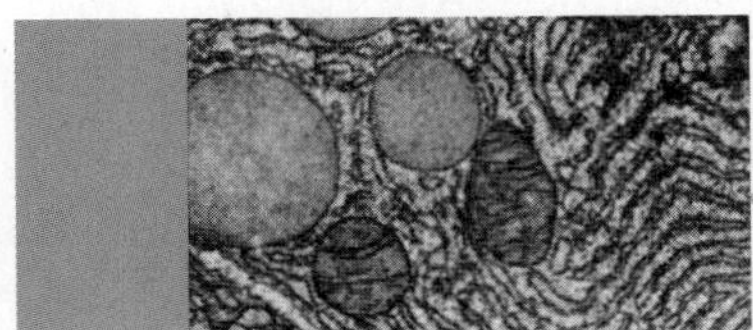

SUMMARY OF KEY POINTS

The Functions of Membranes

- Cells have a variety of membranes that define the boundaries of the cell and its internal compartments. All biological membranes have the same general structure: a fluid phospholipid bilayer containing a mosaic of embedded proteins.
- While the lipid component of membranes provides a permeability barrier, specific proteins in the membrane regulate transport of materials into and out of cells and organelles.
- Membranes serve as sites for specific proteins and thus can have specific functions. They can detect and transduce external signals, mediate contact and adhesion between neighboring cells, or participate in cell-to-cell communication. They also help to produce external structures such as the cell wall or extracellular matrix.

Models of Membrane Structure: An Experimental Perspective

- Our current understanding of membrane structure represents the culmination of more than a century of studies, beginning with the recognition that lipids are an important membrane component.
- Once proteins were recognized as important components, Davson and Danielli proposed their "sandwich" model—a lipid bilayer surrounded on both sides by layers of proteins. As membranes and membrane proteins were examined in more detail, however, this model was eventually discredited.
- In place of the sandwich model, Singer and Nicolson's fluid mosaic model emerged and is now the universally accepted description of membrane structure. According to this model, proteins with varying affinities for the hydrophobic membrane interior float in and on a fluid lipid bilayer.
- We now know that lipids and proteins are not distributed randomly in the membrane but are often found in microdomains known as lipid rafts that are involved in cell signaling and other interactions.

Membrane Lipids: The "Fluid" Part of the Model

- Prominent lipids in most membranes include numerous types of phospholipids and glycolipids. The proportion of each lipid type can vary considerably depending on the particular membrane or monolayer.
- In eukaryotic cells, sterols are also important membrane components, including cholesterol in animal cells and phytosterols in plant cells. Sterols are not found in the membranes of most prokaryotes, but some bacterial species contain similar compounds called hopanoids.
- Proper fluidity of a membrane is critical to its function. Cells often can vary the fluidity of membranes by changing the length and degree of saturation of the fatty acid chains of the membrane lipids or by the addition of cholesterol or other sterols.
- Long-chain fatty acids pack together well and decrease fluidity. Unsaturated fatty acids contain *cis* double bonds that interfere with packing and increase fluidity.
- Most membrane phospholipids and proteins are free to move within the plane of the membrane unless they are specifically anchored to structures on the inner or outer membrane surface. Transverse diffusion, or "flip-flop," between monolayers is not generally possible, except for phospholipids when catalyzed by enzymes called phospholipid translocators, or flippases.
- As a result, most membranes are characterized by an asymmetric distribution of lipids between the two monolayers and

an asymmetric orientation of proteins within the membranes so that the two sides of the membrane are structurally and functionally dissimilar.

Membrane Proteins: The "Mosaic" Part of the Model

- Proteins are major components of all cellular membranes. Membrane proteins are classified as integral, peripheral, or lipid anchored, based on how they are associated with the lipid bilayer.
- Integral membrane proteins have one or more short segments of predominantly hydrophobic amino acids that anchor the protein to the membrane. Most of these transmembrane segments are α-helical sequences of about 20–30 predominantly hydrophobic amino acids.
- Peripheral membrane proteins are hydrophilic and remain on the membrane surface. They are typically attached to the polar head groups of phospholipids by ionic and hydrogen bonding.
- Lipid-anchored proteins are also hydrophilic in nature but are covalently linked to the membrane by any of several lipid anchors that are embedded in the lipid bilayer.
- Membrane proteins function as enzymes, electron carriers, transport molecules, and receptor sites for chemical signals such as neurotransmitters and hormones. Membrane proteins also stabilize and shape the membrane and mediate intercellular communication and cell-cell adhesion.
- Many proteins in the plasma membrane are glycoproteins, with carbohydrate side chains that protrude from the membrane on the external side, where they play important roles as recognition markers on the cell surface.
- Thanks to current advances in SDS-PAGE, molecular biology X-ray crystallography, affinity labeling, and the use of specific antibodies, we are learning much about the structure and function of membrane proteins that were difficult to study in the recent past.

PROBLEM SET

More challenging problems are marked with a •.

7-1 Functions of Membranes. For each of the following statements, specify which one of the five general membrane functions (permeability barrier, localization of function, regulation of transport, detection of signals, or intercellular communication) the statement illustrates.

(a) The membrane of a plant root cell has an ion pump that exchanges phosphate inward for bicarbonate outward.

(b) All of the acid phosphatase in a mammalian cell is found within the lysosomes.

(c) On their outer surface tissue, cells of multicellular organisms carry specific glycoproteins that are responsible for cell-cell adhesion.

(d) Photosystems I and II are embedded in the thylakoid membrane of the chloroplast.

(e) When cells are disrupted and fractionated into subcellular components, the enzyme cytochrome P-450 is recovered with the endoplasmic reticulum fraction.

(f) Adjacent plant cells frequently exchange cytoplasmic components through membrane-lined channels called plasmodesmata.

(g) Insulin does not enter a target cell but instead binds to a specific membrane receptor on the external surface of the membrane, thereby activating the enzyme adenylyl cyclase on the inner membrane surface.

(h) The inner mitochondrial membrane contains an ATP-ADP carrier protein that couples outward ATP movement to inward ADP movement.

(i) The interior of a membrane consists primarily of the hydrophobic portions of phospholipids and amphipathic proteins.

7-2 Elucidation of Membrane Structure. Each of the following observations played an important role in enhancing our understanding of membrane structure. Explain the significance of each, and indicate in what decade of the timeline shown in Figure 7-3 the observation was most likely made.

(a) Some membrane proteins can be readily extracted with 1 *M* NaCl, whereas others require the use of an organic solvent or a detergent.

(b) When artificial lipid bilayers are subjected to freeze-fracture analysis, no particles are seen on either face.

(c) When halobacteria are grown in the absence of oxygen, they produce a purple pigment that is embedded in their plasma membranes and has the ability to pump protons outward when illuminated. If the purple membranes are isolated and viewed by freeze-fracture electron microscopy, they are found to contain patches of crystalline particles.

(d) Ethylurea penetrates much more readily into a membrane than does urea, and diethylurea penetrates still more readily.

(e) When a membrane is observed in the electron microscope, both of the thin, electron-dense lines are about 2 nm thick, but the two lines are often distinctly different from each other in appearance.

(f) The electrical resistivity of artificial lipid bilayers is several orders of magnitude greater than that of real membranes.

(g) The addition of phospholipase to living cells causes rapid digestion of the lipid bilayers of the membranes, which suggests that the enzyme has access to the membrane phospholipids.

7-3 Wrong Again. For each of the following false statements, change the statement to make it true and explain your reasoning.

(a) Because membranes have a hydrophobic interior, polar and charged molecules cannot pass through membranes.

(b) Different cellular organelles have membranes with an identical chemical composition.

(c) Glycoproteins are proteins containing oligosaccharide chains which protrude from the inner membrane.

(d) Membrane fluidity is affected by temperature. When temperature decreases, membrane fluidity increases, and the temperature at which this occurs is known as the transition temperature (T_m).

(e) You would expect membrane lipids from tropical plants such as palm and coconut to have short-chain fatty acids with multiple C=C double bonds.

7-4 Gorter and Grendel Revisited. Gorter and Grendel's classic conclusion that the plasma membrane of the human erythrocyte consists of a lipid bilayer was based on the following observations: (i) the lipids that they extracted with acetone from 4.74×10^9 erythrocytes formed a monolayer 0.89 m^2 in area when spread out on a water surface; and (ii) the surface area of one erythrocyte was about 100 μm^2, according to their measurements.

(a) Show from these data how they came to the conclusion that the erythrocyte membrane is a bilayer.

(b) We now know that the surface area of a human erythrocyte is about 145 μm^2. Explain how Gorter and Grendel could have come to the right conclusion when one of their measurements was only about two-thirds of the correct value.

7-5 Martian Membranes. Imagine a new type of cell was discovered on Mars in an organism growing in benzene, a nonpolar liquid. The cell had a lipid bilayer made of phospholipids, but its structure was very different from that of our cell membranes.

(a) Draw what might be a possible structure for this new type of membrane. What might be characteristic features of the phospholipid head groups?

(b) What properties would you expect to find in membrane proteins embedded in this membrane?

(c) How might you isolate and visualize these unusual membranes?

7-6 That's About the Size of It. From chemistry, we know that each methylene (—CH_2—) group in a straight-chain hydrocarbon advances the chain length by about 0.13 nm. And from studies of protein structure, we know that one turn of an α helix includes 3.6 amino acid residues and extends the long axis of the helix by about 0.56 nm. Use this information to answer the following:

(a) How long is a single molecule of palmitate (16 carbon atoms) in its fully extended form? What about molecules of laurate (12C) and arachidate (20C)?

(b) How does the thickness of the hydrophobic interior of a typical membrane compare with the length of two palmitate molecules laid end to end? What about two molecules of laurate or arachidate?

(c) Approximately how many amino acids must a helical transmembrane segment of an integral membrane protein have if the segment is to span the lipid bilayer defined by two palmitate molecules laid end to end?

(d) The protein bacteriorhodopsin has 248 amino acids and seven transmembrane segments. Approximately what portion of the amino acids are part of the transmembrane segments? Assuming that most of the remaining amino acids are present in the hydrophilic loops linking the transmembrane segments together, approximately how many amino acids are present in each of these loops, on the average?

7-7 Temperature and Membrane Composition. Which of the following responses are *not* likely to be seen when a bacterial culture growing at 37°C is transferred to a culture room maintained at 25°C? Explain your reasoning.

(a) Initial decrease in membrane fluidity

(b) Gradual replacement of shorter-chain fatty acids by longer-chain fatty acids in the membrane phospholipids

(c) Gradual replacement of stearate by oleate in the membrane phospholipids

(d) Enhanced rate of synthesis of unsaturated fatty acids

(e) Incorporation of more cholesterol into the membrane

7-8 Membrane Fluidity and Temperature. The effects of temperature and lipid composition on membrane fluidity are often studied by using artificial membranes containing only one or a few kinds of lipids and no proteins. Assume that you and your lab partner have made the following artificial membranes:

Membrane 1: Made entirely from phosphatidylcholine with saturated 16-carbon fatty acids.

Membrane 2: Same as membrane 1, except that each of the 16-carbon fatty acids has a single *cis* double bond.

Membrane 3: Same as membrane 1, except that each of the saturated fatty acids has only 14 carbon atoms.

After determining the transition temperatures of samples representing each of the membranes, you discover that your lab partner failed to record which membranes the samples correspond to. The three values you determined are –36°C, 23°C, and 41°C. Assign each of these transition temperatures to the correct artificial membrane, and explain your reasoning.

7-9 The Little Bacterium That Can't. *Acholeplasma laidlawii* is a small bacterium that cannot synthesize its own fatty acids and must therefore construct its plasma membrane from whatever fatty acids are available in the environment. As a result, the *Acholeplasma* membrane takes on the physical characteristics of the fatty acids available at the time.

(a) If you give *Acholeplasma* cells access to a mixture of saturated and unsaturated fatty acids, they will thrive at room temperature. Can you explain why?

(b) If you transfer the bacteria of part a to a medium containing only saturated fatty acids but make no other changes in culture conditions, they will stop growing shortly after the change in medium. Explain why.

(c) What is one way you could get the bacteria of part b growing again without changing the medium? Explain your reasoning.

(d) If you were to maintain the *Acholeplasma* culture of part b under the conditions described there for an extended period of time, what do you predict will happen to the bacterial cells? Explain your reasoning.

(e) What result would you predict if you were to transfer the bacteria of part a to a medium containing only unsaturated fatty acids without making any other changes in the culture conditions? Explain your reasoning.

• **7-10 Hydropathy: The Plot Thickens.** A hydropathy plot can be used to predict the structure of a membrane protein based on its amino acid sequence and the hydrophobicity values of the amino acids. Hydrophobicity is measured as the standard free energy change, $\Delta G^{\circ\prime}$, for the transfer of a given amino acid residue from a hydrophobic solvent into water, in kilojoules/mole (kJ/mol). The hydropathy index is calculated by averaging the hydrophobicity values for a series of short segments of the polypeptide, with each segment displaced one amino acid further from the N-terminus. The hydropathy index of each successive segment is then plotted as a function of the location of that segment in the amino acid sequence, and the plot is examined for regions of high hydropathy index.

(a) Why do scientists try to predict the structure of a membrane protein by this indirect means when the technique of X-ray crystallography would reveal the structure directly?

(b) Given the way it is defined, would you expect the hydrophobicity index of a hydrophobic residue such as valine or isoleucine to be positive or negative? What about a hydrophilic residue such as aspartic acid or arginine?

(c) Listed below are four amino acids and four hydrophobicity values. Match the hydrophobicity values with the correct amino acids, and explain your reasoning.

Amino acids: alanine; arginine; isoleucine; serine
Hydrophobicity (in kJ/mol): +3.1; +1.0; −1.1; −7.5

(d) Shown in **Figure 7-30** is a hydropathy plot for a specific integral membrane protein. Draw a horizontal bar over each transmembrane segment as identified by the plot. How long is the average transmembrane segment? How well does that value compare with the number you calculated in Problem 7–6c? How many transmembrane segments do you think the protein has? Can you guess which protein this might be?

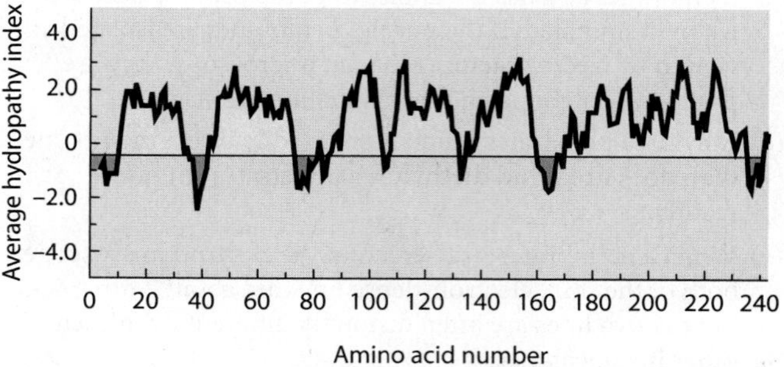

FIGURE 7-30 Hydropathy Plot for an Integral Membrane Protein. See Problem 7-10d.

• **7-11 Inside or Outside?** From Figure 7-24, we know that exposed regions of membrane proteins can be labeled with ^{125}I by the lactoperoxidase (LP) reaction. Similarly, carbohydrate side chains of membrane glycoproteins can be labeled with ^{3}H by oxidation of galactose groups with galactose oxidase (GO) followed by reduction with tritiated borohydride ($^{3}H{-}BH_4$). Noting that both LP and GO are too large to penetrate into the interior of an intact cell, explain each of the following observations made with intact erythrocytes.

(a) When intact cells are incubated with LP in the presence of ^{125}I and the membrane proteins are then extracted and analyzed on SDS–polyacrylamide gels, several of the bands on the gel are found to be radioactive.

(b) When intact cells are incubated with GO and then reduced with $^{3}H{-}BH_4$, several of the bands on the gel are found to be radioactive.

(c) All of the proteins of the plasma membrane that are known to contain carbohydrates are labeled by the GO/$^{3}H{-}BH_4$ method.

(d) None of the proteins of the erythrocyte plasma membrane that are known to be devoid of carbohydrate is labeled by the LP/^{125}I method.

(e) If the erythrocytes are ruptured before the labeling procedure, the LP procedure labels virtually all of the major membrane proteins.

• **7-12 Inside-Out Membranes.** It is technically possible to prepare sealed vesicles from erythrocyte membranes in which the original orientation of the membrane is inverted. Such vesicles have what was originally the cytoplasmic side of the membrane facing outward.

(a) What results would you expect if such inside-out vesicles were subjected to the GO/$^{3}H{-}BH_4$ procedure described in Problem 7-11?

(b) What results would you expect if such inside-out vesicles were subjected to the LP/^{125}I procedure of Problem 7-11?

(c) What conclusion would you draw if some of the proteins that become labeled by the LP/^{125}I method of part b were among those that had been labeled when intact cells were treated in the same way in Problem 7-11a?

(d) Knowing that it is possible to prepare inside-out vesicles from erythrocyte plasma membranes, can you think of a way to label a transmembrane protein with ^{3}H on one side of the membrane and with ^{125}I on the other side?

SUGGESTED READING

References of historical importance are marked with a •.

General References

Peirce, M. J., and R. Wait, eds. *Membrane Proteomics: Methods and Protocols*. New York: Humana Press/Springer Science, 2009.

Quinn., P. J., ed. *Membrane Dynamics and Domains*. New York: Kluwer Academic/Plenum Publishers, 2004.

Vance, D. E., and J. E. Vance, eds. *Biochemistry of Lipids, Lipoproteins, and Membranes*, 5th ed. New York: Elsevier Science, 2008.

Yeagle, P. L., ed. *The Structure of Biological Membranes,* 2nd ed. Boca Raton, Florida: CRC Press LLC, 2005.

Historical References

Davson, H., and J. F. Danielli, eds. *The Permeability of Natural Membranes.* Cambridge, England: Cambridge University Press, 1943.

• Deamer, D. W., A. Kleinzeller, and D. M. Fambrough, eds. *Membrane Permeability: 100 Years Since Ernest Overton*. San Diego: Academic Press, 1999.

• Deisenhofer, J. et al. Structure of the protein subunits in the photosynthetic reaction centre of *Rhodopseudomonas viridis* at 3 Å resolution. *Nature* 318 (1985): 618.

• Gorter, E., and F. Grendel. On bimolecular layers of lipids on the chromocyte of the blood. *J. Exp. Med.* 41 (1925): 439.

• Robertson, J. D. The ultrastructure of cell membranes and their derivatives. *Biochem. Soc. Symp.* 16 (1959): 3.

• Singer, S. J., and G. L. Nicolson. The fluid mosaic model of the structure of cell membranes. *Science* 175 (1972): 720.

• Unwin, N., and R. Henderson. The structure of proteins in biological membranes. *Sci. Amer.* 250 (February 1984): 78.

Membrane Structure and Function

Cossins, A. R. *Temperature Adaptations of Biological Membranes.* London: Portland Press, 1994.

Hanzal-Bayer, M. F., and J. F. Hancock. Lipid rafts and membrane traffic. *FEBS Lett.* 581 (2007): 2098.

Hunte, C., and S. Richers. Lipids and membrane protein structures. *Curr. Opin. Struct. Biol.* 18 (2008): 406.

Ishitsuka, R., S. B. Sato, and T. Kobayashi. Imaging lipid rafts. *J. Biochem (Tokyo)* 137 (2005): 249.

Lai, Y., and R. L. Gallo. AMPed up immunity: How antimicrobial peptides have multiple roles in immune defense. *Trends Immunol.* 30 (2009): 131.

Lindner, R., and H. Y. Naim. Domains in biological membranes. *Exp. Cell Res.* 315 (2009): 2871.

Michel, V., and M. Bakovic. Lipid rafts in health and disease. *Biol. Cell* 99 (2007): 129.

Pani, B., and B. B. Singh. Lipid rafts/caveolae as microdomains of calcium signaling. *Cell Calcium* 45 (2009): 625.

Patra, S. K. Dissecting lipid raft facilitated cell signaling pathways in cancer. *Biochim. Biophys. Acta* 1785 (2008): 182.

Veenhoff, L. M., E. H. M. L. Heuberger, and B. Poolman. Quaternary structure and function of transport proteins. *Trends Biochem. Sci.* 27 (2002): 242.

Yi, F., S. Jin, and P. L. Li. Lipid raft-redox signaling platforms in plasma membrane. *Meth. Mol. Bio.* 580 (2009): 93.

Membrane Lipids

Dowhan, W. Molecular basis for membrane phospholipid diversity: Why are there so many lipids? *Annu. Rev. Biochem.* 66 (1997): 199.

Hawkes, D. J., and J. Mak. Lipid membrane: A novel target for viral and bacterial pathogens. *Int. Rev. Cytol.* 255 (2006): 1.

Katsaros, J., and T. Gutberlet, eds. *Lipid Bilayers: Structure and Interactions*. New York: Springer-Verlag, 2000.

Li, X. A., W. V. Everson, and E. J. Smart. Caveolae, lipid rafts, and vascular disease. *Trends Cardiovasc. Med.* 15 (2005): 92.

Menon, A. Flippases. *Trends Cell Biol.* 5 (1995): 355.

Ohvo-Rekila, H. B. et al. Cholesterol interactions with phospholipids in membranes. *Prog. Lipid Res.* 41 (2002): 66.

Pororsky, T., and A. K. Menon. Lipid flippases and their biological functions. *Cell. Mol. Life Sci.* 63 (2006): 2908.

Stroud, R. M. The state of lipid rafts: From model membranes to cells. *Annu. Rev. Biophys. Biomol. Struct.* 32 (2003): 257.

Zepic, H. H., P. Walde, E. L. Kostoryz, J. Code, and D. M. Yourtee. Lipid vesicles for toxicological assessment of xenobiotics. *Crit. Rev. Toxicol.* 38 (2008): 1.

Zhang, S. C., and C. O. Rock. Membrane lipid homeostasis in bacteria. *Nat. Rev. Microbiol.* 6 (2008): 222.

Membrane Proteins

Bates, I. R., P. W. Wiseman, and J. W. Hanrahan. Investigating membrane protein dynamics in living cells. *Biochem. Cell Biol.* 84 (2006): 825.

Carpenter, E. P., K. Beis, A. D. Cameron, and S. Iwata. Overcoming the challenges of membrane protein crystallography. *Curr. Opin. Struct. Biol.* 18 (2008): 581.

Elofsson, A., and G. von Heijne. Membrane protein structure: Prediction versus reality. *Annu. Rev. Biochem.* 76 (2007): 125.

Engel, A., and H. E. Gaub. Structure and mechanics of membrane proteins. *Annu. Rev. Biochem.* 77 (2008): 127.

Nyholm, T. K., S. Ozdirekcan, and J. A. Killian. How protein transmembrane segments sense the lipid environment. *Biochemistry* 46 (2007): 1457.

Peirce, M. J., J. Saklatvala, A. P. Cope, and R. Wait. Mapping lymphocyte plasma membrane proteins: A proteomic approach. *Methods Molec. Med.* 136 (2007): 361.

Sussman, M. R. Molecular analysis of proteins in the plant plasma membrane. *Annu. Rev. Plant Physiol. Plant Mol. Biol.* 45 (1994): 211.

Werten, P. J. L. et al. Progress in the analysis of membrane protein structure and function. *FEBS Lett.* 529 (2002): 65.

White, S. H. *Membrane Proteins of Known Structure.* http://blanco.biomol.uci.edu/Membrane_Proteins_xtal.html (2010; updated regularly).

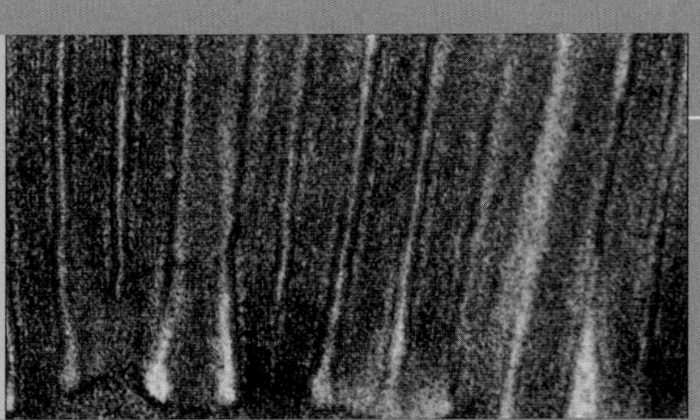

8 Transport Across Membranes: Overcoming the Permeability Barrier

See MasteringBiology® for tutorials, activities, and quizzes.

In Chapter 7, we focused on the structure and chemistry of membranes. We noted that its hydrophobic interior makes a membrane an effective barrier to the passage of most molecules and ions, thereby keeping some substances inside the cell and others out. Within eukaryotic cells, membranes also delineate organelles by retaining the appropriate molecules and ions needed for specific functions.

However, it is not enough to think of membranes simply as permeability barriers. Crucial to the proper functioning of a cell or organelle is the ability to overcome the permeability barriers for specific molecules and ions so that they can be moved into and out of the cell or organelle selectively. In other words, membranes are not simply barriers to the indiscriminate movement of substances into and out of cells and organelles. They are also selectively permeable, *allowing the controlled passage of specific molecules and ions from one side of the membrane to the other. In this chapter we will look at the ways substances are moved selectively across membranes, and we will consider the significance of such transport processes to the life of the cell.*

Cells and Transport Processes

An essential feature of every cell and subcellular compartment is its ability to accumulate a variety of substances at concentrations that are often strikingly different from those in the surrounding environment. Some of these substances are macromolecules—such as DNA, RNA, and proteins—that are moved into and out of cells and organelles by mechanisms we will consider in later chapters. Specifically, Chapter 12 includes a discussion of *endocytosis* and *exocytosis,* bulk transfer processes whereby substances are moved into and out of cells enclosed within membrane-bounded vesicles. Mechanisms for the secretion of proteins from cells and for the import of proteins into organelles are discussed in Chapter 22.

As important as these topics are, most of the substances that move across membranes are not macromolecules but dissolved gases, ions, and small organic molecules—*solutes,* in other words. Some of the more common ions transported across membranes are sodium (Na^+), potassium (K^+), calcium (Ca^{2+}), chloride (Cl^-), and hydrogen (H^+) ions. Most of the small organic molecules are *metabolites*—substrates, intermediates, and products in the various metabolic pathways that take place within cells or specific organelles. Sugars, amino acids, and nucleotides are some common examples. Such solutes are almost always present at higher concentrations on the inside of the cell or organelle than on the outside. Very few cellular reactions or processes could occur at reasonable rates if they had to depend on the low concentrations at which essential substrates are present in the cell's surroundings. In some cases, such as electrical signaling in nerve and muscle tissue, the controlled movement of ions across the membrane is central to the function of the cell. Also, many prescribed medications have intracellular targets and therefore must be able to cross membranes to enter the cell.

A central aspect of cell function, then, is **transport**—the ability to move ions and organic molecules across membranes selectively. The importance of membrane transport is evidenced by the fact that about 20% of the genes that have been identified in the bacterium *Escherichia coli* are involved in some aspect of transport. **Figure 8-1** summarizes a few of the many transport processes that occur within eukaryotic cells.

Solutes Cross Membranes by Simple Diffusion, Facilitated Diffusion, and Active Transport

Three fundamentally different mechanisms are involved in the movement of solutes across membranes, as shown in **Table 8-1**. A few types of molecules move across membranes by *simple diffusion*—direct, unaided movement of

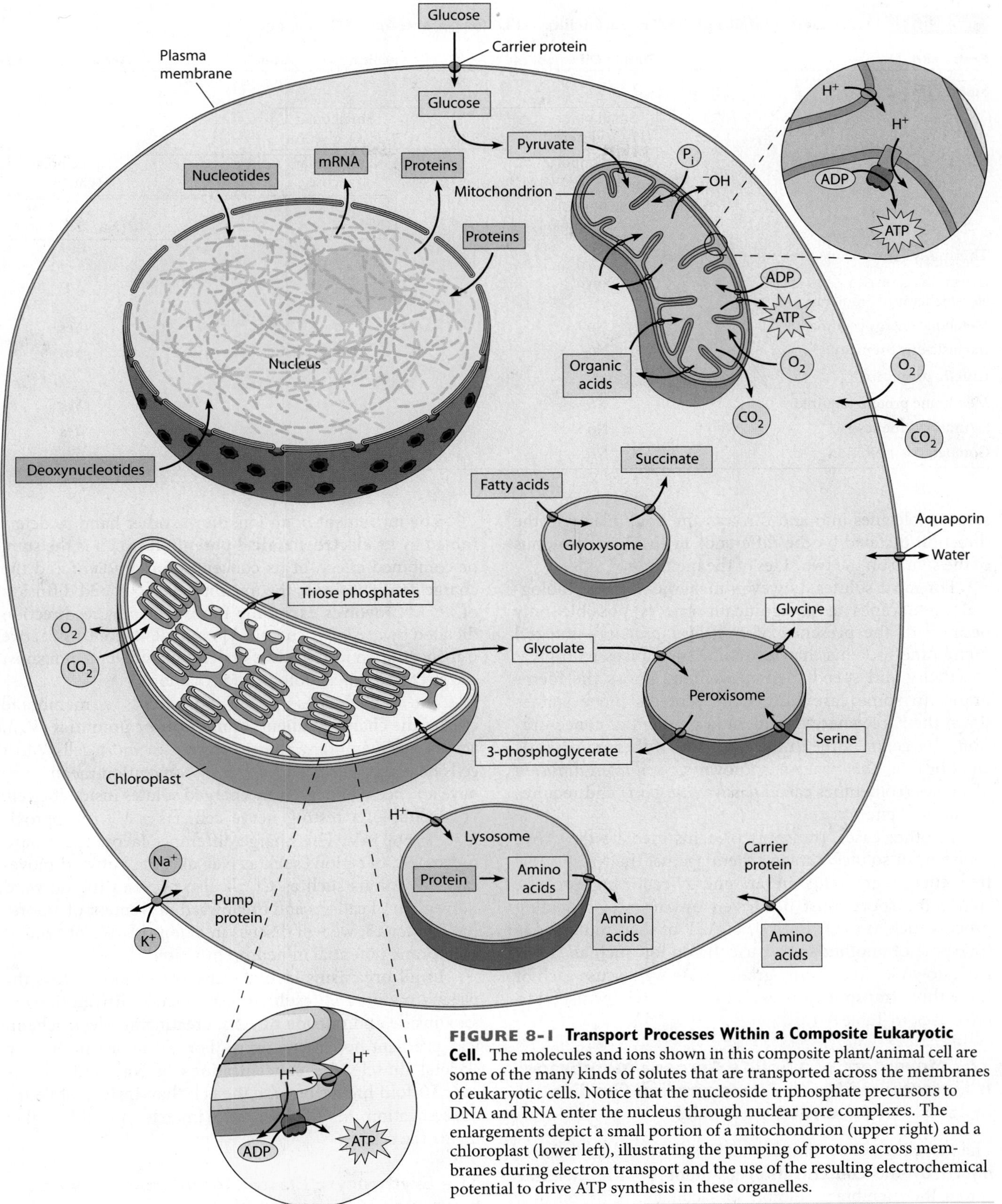

FIGURE 8-1 Transport Processes Within a Composite Eukaryotic Cell. The molecules and ions shown in this composite plant/animal cell are some of the many kinds of solutes that are transported across the membranes of eukaryotic cells. Notice that the nucleoside triphosphate precursors to DNA and RNA enter the nucleus through nuclear pore complexes. The enlargements depict a small portion of a mitochondrion (upper right) and a chloroplast (lower left), illustrating the pumping of protons across membranes during electron transport and the use of the resulting electrochemical potential to drive ATP synthesis in these organelles.

Table 8-1 Comparison of Simple Diffusion, Facilitated Diffusion, and Active Transport

Properties	Simple Diffusion	Facilitated Diffusion	Active Transport
Solutes transported			
	Small polar (H_2O, glycerol)	Small polar (H_2O, glycerol)	
	Small nonpolar (O_2, CO_2)	Large polar (glucose)	Large polar (glucose)
	Large nonpolar (oils, steroids)	Ions (Na^+, K^+, Ca^{2+})	Ions (Na^+, K^+, Ca^{2+})
Thermodynamic properties			
Direction relative to electrochemical gradient	Down	Down	Up
Metabolic energy required	No	No	Yes
Intrinsic directionality	No	No	Yes
Kinetic properties			
Membrane protein required	No	Yes	Yes
Saturation kinetics	No	Yes	Yes
Competitive inhibition	No	Yes	Yes

solute molecules into and through the lipid bilayer in the direction dictated by the difference in the concentrations of the solute on the two sides of the membrane.

For most solutes, however, movement across biological membranes at a significant rate is possible only because of the presence of *transport proteins*—integral membrane proteins that recognize substances with great specificity and speed their movement across the membrane. In some cases, transport proteins move solutes down their free energy gradient (a gradient of concentration, charge, or both) in the direction of thermodynamic equilibrium. This process is known as *facilitated diffusion* of solutes (sometimes called *passive transport*) and requires no input of energy.

In other cases, transport proteins mediate the *active transport* of solutes, moving them against their respective free energy gradients in an energy-requiring process. Active transport must be driven by an energy-yielding process such as the hydrolysis of ATP or the simultaneous transport of another solute, usually an ion such as H^+ or Na^+, down its free energy gradient. As we discuss each of these three transport processes in turn, it will be useful to refer back to Table 8-1.

The Movement of a Solute Across a Membrane Is Determined by Its Concentration Gradient or Its Electrochemical Potential

The movement of a molecule that has no net charge is determined by the **concentration gradient** of that molecule across the membrane. Simple or facilitated diffusion of a molecule involves exergonic movement "down" the concentration gradient (negative ΔG), whereas active transport involves endergonic movement "up" the concentration gradient (positive ΔG) and requires some driving force.

The movement of an ion, on the other hand, is determined by its **electrochemical potential,** which is the sum, or combined effect, of its concentration gradient and the charge gradient across the membrane. Facilitated diffusion of an ion involves exergonic movement in the direction dictated by its electrochemical potential. In contrast, active transport of an ion involves endergonic movement against the electrochemical potential for that ion.

The active transport of ions across a membrane creates the charge gradient, or **membrane potential (V_m),** across the membrane that is present in most cells. Most cells have a negative V_m, which by convention means they have an excess of negatively charged solutes inside the cell. For example, a resting nerve cell has a V_m of approximately -60 mV. This charge difference favors the inward movement of cations such as Na^+ and the outward movement of anions such as Cl^-. It also opposes the outward movement of cations and the inward movement of anions. In Chapter 13, we will discuss in detail the role of cellular membrane potential in neuron function.

In all organisms, active transport of ions across the plasma membrane results in asymmetric distributions of ions inside and outside of cells, creating an electrochemical gradient for many ions. For example, in human skeletal muscle, the concentrations of Na^+ and Cl^- are over 10-fold higher outside the cell than inside, while the concentration of K^+ is approximately 40-fold higher inside the cell.

The Erythrocyte Plasma Membrane Provides Examples of Transport Mechanisms

In our discussion of membrane transport processes, we will use as examples the transport proteins of the erythrocyte. These are some of the most extensively studied and best understood of all cellular transport proteins. Vital to the erythrocyte's role in providing oxygen to body tissues is the movement across its plasma membrane of O_2, CO_2, and bicarbonate ion (HCO_3^-), as well as glucose, which serves as the cell's main energy source. Also important is the membrane potential maintained across the plasma membrane by the active transport of potassium ions inward and sodium ions outward. In addition, special pores, or *channels,* allow water and ions to enter and leave the cell rapidly in response to cellular needs. These transport activities are summarized in **Figure 8-2** and will be used as examples in the following discussion.

(a) Simple diffusion. Oxygen, carbon dioxide, and water diffuse directly across the plasma membrane in response to their relative concentrations inside and outside the cell.

(b) Facilitated diffusion mediated by carrier proteins. The movement of glucose across the plasma membrane is facilitated by a specific glucose transporter called GLUT1. An anion exchange protein facilitates the reciprocal transport of chloride (Cl^-) and bicarbonate (HCO_3^-).

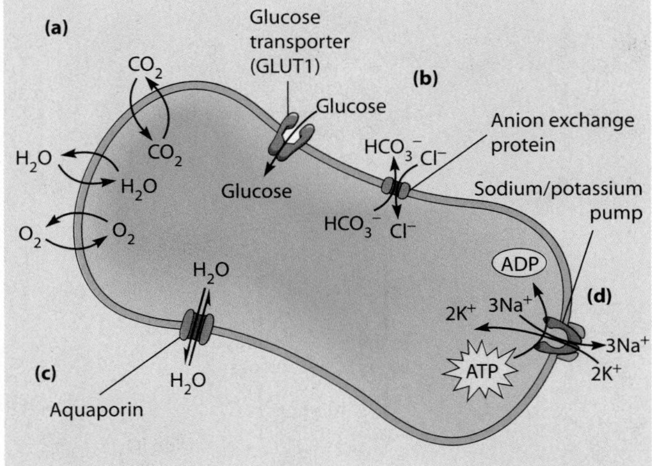

(c) Facilitated diffusion mediated by channel proteins. Aquaporin channel proteins can facilitate the rapid inward or outward movement of water molecules.

(d) Active transport. Driven by the hydrolysis of ATP, the Na^+/K^+ pump moves three sodium ions outward for every two potassium ions moved inward, establishing an electrochemical potential across the plasma membrane for both ions.

FIGURE 8-2 Important Transport Processes of the Erythrocyte. Depicted here are several types of transport processes that are vital to erythrocyte function.

Simple Diffusion: Unassisted Movement Down the Gradient

The most straightforward way for a solute to get from one side of a membrane to the other is **simple diffusion,** which is the unassisted net movement of a solute from a region where its concentration is higher to a region where its concentration is lower (Figure 8-2a). Because membranes have a hydrophobic interior, simple diffusion is typically a means of transport only for gases, nonpolar molecules, or small polar molecules such as water, glycerol, or ethanol.

Oxygen is a gas that traverses the hydrophobic lipid bilayer readily and therefore moves across membranes by simple diffusion. This behavior enables erythrocytes in the circulatory system to take up oxygen in the lungs and release it in body tissues. In the capillaries of body tissues, where oxygen concentration is low, oxygen is released from hemoglobin and diffuses passively from the cytoplasm of the erythrocyte into the blood plasma and from there into the cells lining the capillaries (**Figure 8-3a**).

In the capillaries of the lungs, the opposite occurs: Oxygen diffuses from the inhaled air in the lungs, where its concentration is higher, into the cytoplasm of the erythrocytes, where its concentration is lower (Figure 8-3b). Carbon dioxide is also able to cross membranes by simple diffusion. However, most CO_2 is actually transported in the form of bicarbonate ion (HCO_3^-), as we will see later in the chapter. Not surprisingly, carbon dioxide and oxygen move across the erythrocyte membrane in opposite directions. Carbon dioxide diffuses inward in body tissues and outward in the lungs.

Diffusion Always Moves Solutes Toward Equilibrium

No matter how a population of molecules is distributed initially, diffusion always tends to create a random solution in which the concentration is the same everywhere. To illustrate this point, consider the apparatus shown in **Figure 8-4a**, consisting of two chambers separated by a membrane that is freely permeable to molecules of S, an uncharged solute represented by the black dots. Initially, the concentration of S is higher in chamber A than in chamber B. Other conditions being equal, random movements of solute molecules back and forth through the membrane will lead to a net movement of solute S from chamber A to chamber B. When the concentration of S is equal on both sides of the membrane, the system is at equilibrium. Random back-and-forth movement of individual molecules continues, but no further net change in concentration occurs. Thus, *diffusion is always movement toward equilibrium* and is therefore a spontaneous process.

Another way to express this is to say that diffusion always tends toward minimum free energy. As we learned in Chapter 5, chemical reactions and physical processes always proceed in the direction of decreasing free energy, in accordance with the second law of thermodynamics. Diffusion through membranes is no exception: Free energy is minimized as molecules move down their

(a) In the capillaries of body tissues (low $[O_2]$ and high $[CO_2]$ relative to the erythrocytes), O_2 is released by hemoglobin within the erythrocytes and diffuses outward to meet tissue needs. CO_2 diffuses inward and is converted to bicarbonate by carbonic anhydrase in the cytosol. Bicarbonate ions are transported outward by the anion exchange protein, accompanied by the inward movement of chloride ions to maintain charge balance. Carbon dioxide therefore returns to the lungs as bicarbonate ions.

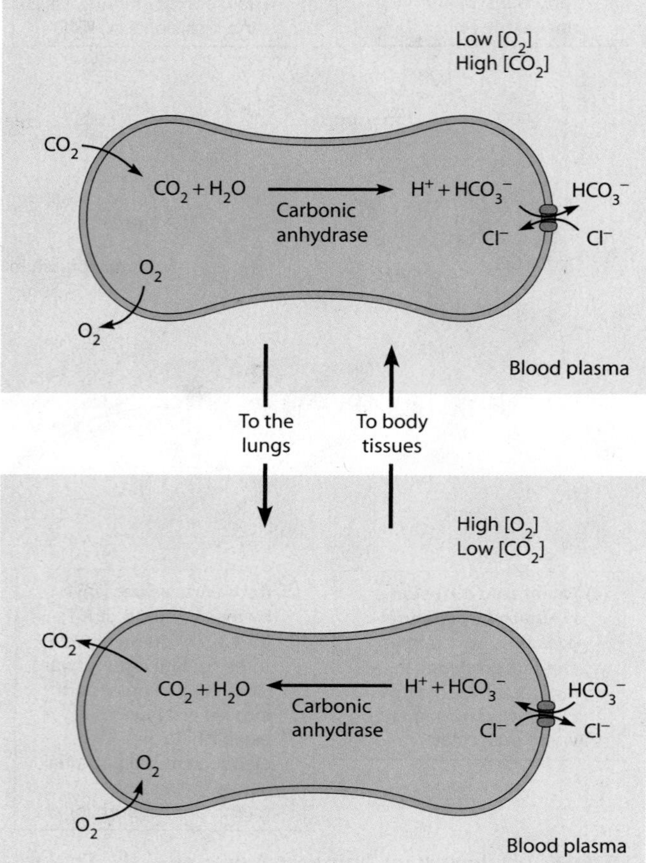

(b) In the capillaries of the lungs (high $[O_2]$ and low $[CO_2]$ relative to the erythrocytes), O_2 diffuses inward and binds to hemoglobin. Bicarbonate moves inward from the blood plasma, accompanied by an outward movement of chloride ions. Incoming bicarbonate is converted into CO_2, which diffuses out of the erythrocytes and into the cells lining the capillaries of the lungs. The CO_2 is now ready to be expelled from the body.

FIGURE 8-3 Directions of Oxygen, Carbon Dioxide, and Bicarbonate Transport in Erythrocytes. The directions in which O_2, CO_2, and HCO_3^- move across the plasma membrane of the erythrocyte depend on the location of the erythrocyte in the body.

concentration gradient. The driving force for diffusion is entropy, the randomization of molecules as concentrations equalize on both sides of the membrane (like the jumping beans in Box 5A, page 116). We will apply this principle later in the chapter when we calculate ΔG, the free energy change that accompanies the transport of molecules and ions across membranes. Strictly speaking, therefore, *diffusion always proceeds from regions of higher to lower free energy*. At thermodynamic equilibrium, no further net movement occurs because the free energy of the system is at a minimum.

(a) Simple diffusion takes place when the membrane separating chambers A and B is permeable to molecules of dissolved solute, represented by the black dots. Net movement of solute molecules across the membrane is from chamber A to B (high to low solute concentration). Equilibrium is reached when the solute concentration is the same in both chambers.

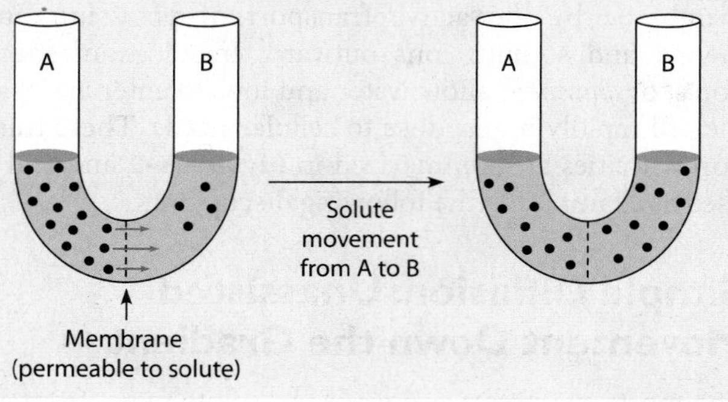

(b) Osmosis occurs when the membrane between the two chambers is not permeable to the dissoved solute, represented by the black triangles. Because solute cannot cross the membrane, water diffuses from chamber B where the solute concentration is lower (more water) to chamber A where the solute concentration is higher (less water). At equilibrium, the solute concentration will be equal on both sides of the membrane.

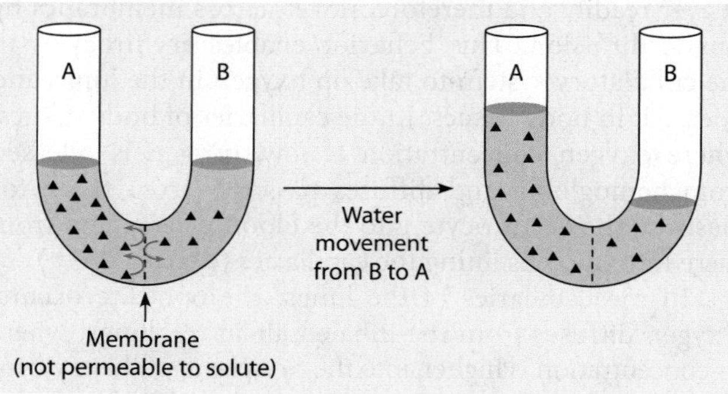

FIGURE 8-4 Comparison of Simple Diffusion and Osmosis. In both examples shown, there is initially more solute in chamber A than in chamber B. The membrane in example **(a)** is permeable to the solute and the membrane in example **(b)**, like a typical cell membrane, is not.

Osmosis Is the Diffusion of Water Across a Selectively Permeable Membrane

Several properties of water cause it to behave in a special way. First of all, water molecules are not charged, so they are not affected by the membrane potential. Moreover, the concentration of water is not appreciably different on opposite sides of a membrane. What, then, determines the direction in which water molecules diffuse? When a solute is dissolved in water, the solute molecules disrupt the ordered three-dimensional interactions that normally occur between individual molecules of water, thereby increasing the entropy and decreasing the free energy of the solution. Water, like other substances, tends to diffuse from areas where its free energy is higher to areas where its free energy is lower. Thus, water tends to move from regions of lower solute concentration (higher free energy) to regions of higher solute concentration (lower free energy).

This principle is illustrated by Figure 8-4b. Solutions of differing solute concentrations are placed in chambers A and B, as in Figure 8-4a, but with the two chambers now separated by a *selectively permeable membrane* that is permeable to water but *not* to the dissolved solute. Under these conditions, water moves, or diffuses, across the membrane from chamber B to chamber A. Such movement of water in response to differences in solute concentration is called **osmosis.** Osmotic movement of water across a membrane is always from the side with the higher free energy (that is, with the lower solute concentration) to the side with the lower free energy (that is, with the higher solute concentration). For most cells, this means that water will tend to move inward because the concentration of solutes is almost always higher inside a cell than outside. If not controlled, the inward movement of water would cause cells to swell and perhaps to burst. For further discussion of osmotic water movement and how organisms from different kingdoms control water content and movement, see **Box 8A**.

Simple Diffusion Is Limited to Small, Nonpolar Molecules

To investigate the factors influencing the diffusion of solutes through membranes, scientists frequently use membrane models. An important advance in developing such models was provided in 1961 by Alec Bangham and his colleagues, who found that when lipids are extracted from cell membranes and dispersed in water, they form *liposomes*. Liposomes are small vesicles about 0.1 μm in diameter, each consisting of a closed, spherical, lipid bilayer that is devoid of membrane proteins. Bangham showed that it is possible to trap solutes such as potassium ions in the liposomes as they form and then measure the rate at which the solute escapes by diffusion across the liposome bilayer.

The results were remarkable. Ions such as potassium and sodium were trapped in the vesicles for days, whereas small uncharged molecules such as oxygen exchanged so rapidly that the rates could hardly be measured. The inescapable conclusion was that the lipid bilayer represents the primary permeability barrier of a membrane. Small uncharged molecules can pass through the barrier by simple diffusion, whereas sodium and potassium ions can barely pass at all. Based on subsequent experiments by many investigators using a variety of lipid bilayer systems and thousands of different solutes, we can predict with considerable confidence how readily a solute will diffuse across a lipid bilayer. The three main factors affecting diffusion of solutes are *size, polarity,* and *charge.* We will consider each of these factors in turn.

Solute Size. Generally speaking, lipid bilayers are more permeable to smaller molecules than to larger molecules. The smallest molecules relevant to cell function are water, oxygen, and carbon dioxide. Membranes are quite permeable to these molecules, which require no specialized transport processes for moving into and out of cells, although we will see later that there is a specialized transporter in some cells allowing rapid transport of water molecules across certain membranes. But, in the absence of a transporter, even such small molecules do not move across membranes freely. Water molecules, for example, diffuse across a bilayer 10,000 times more slowly than they move when allowed to diffuse freely in the absence of a membrane.

Still, water diffuses across membranes at rates much higher than would be expected for such a polar molecule. The reason for this behavior is not well understood. One proposal is that membranes contain tiny pores that allow the passage of water molecules but are too small for any other polar substance. An alternative suggestion is that in the continual movements of membrane lipids, transient "holes" are created in the lipid monolayers that allow water molecules to move first through one monolayer and then through the other. There is little experimental evidence to support either of these hypotheses, however, and simple diffusion of water across membranes remains an enigma.

In addition to water, small polar molecules with a molecular weight (MW) of up to about 100—such as ethanol (CH_3CH_2OH; MW = 46) and glycerol ($C_3H_8O_3$; MW = 92)—are able to diffuse across membranes. However, larger polar molecules such as glucose ($C_6H_{12}O_6$; MW = 180) cannot. Cells therefore need specialized proteins in their plasma membranes to facilitate the entry of glucose and most other polar solutes.

Solute Polarity. There is a correlation between the lipid solubility of a molecule and its membrane permeability. In general, lipid bilayers are more permeable to nonpolar molecules and less permeable to polar molecules. This is because nonpolar molecules dissolve more readily in the hydrophobic phase of the lipid bilayer and can therefore cross the membrane much more rapidly than can polar molecules of similar size. For example, the steroid hormones estrogen and testosterone are largely nonpolar and therefore can diffuse across membranes despite having molecular weights of 370 and 288, respectively.

A simple measure of the polarity (or nonpolarity) of a solute is its *partition coefficient,* which is the ratio of its solubility in an organic solvent (such as vegetable oil or octanol)

BOX 8A DEEPER INSIGHTS

Osmosis: The Diffusion of Water Across a Selectively Permeable Membrane

Most of the discussion in this chapter focuses on the transport of *solutes*—ions and small molecules that are dissolved in the aqueous environment of cells, their organelles, and their surroundings. That emphasis is quite appropriate because most of the traffic across membranes involves ions such as K^+, Na^+, and H^+, and hydrophilic molecules such as sugars, amino acids, and a variety of metabolic intermediates. But to understand solute transport fully, we also need to understand the forces that act on water, thereby determining its movement within and between cells.

Because most solutes cannot cross cell membranes, water tends to move across membranes in response to differences in solute concentrations on the two sides of the membrane. Specifically, water tends to diffuse from the side of the membrane with the lower solute concentration to the side with the higher solute concentration. This diffusion of water, called *osmosis,* is readily observed when a selectively permeable membrane separates two compartments, one of them containing a solute that cannot cross the membrane (see Figure 8-4b). Water will move across the membrane to equalize the concentration of solutes on both sides of the membrane.

Osmotic movement of water into and out of a cell is related to the relative **osmolarity,** or solute concentration, of the extracellular solution. A solution with a higher solute concentration than that inside a cell is called a *hypertonic solution,* whereas a solution with a solute concentration lower than that inside a cell is referred to as a *hypotonic solution.* Hypertonic solutions cause water molecules to diffuse out of the cell, dehydrating it. In contrast, hypotonic solutions cause water to diffuse into the cell, increasing the internal pressure. A solution with the same solute concentration as the cell is called an *isotonic solution,* and there will be no net movement of water in either direction.

Osmosis accounts for a well-known observation: Cells tend to shrink or swell as the solute concentration of the extracellular medium changes. Consider, for example, the scenario of **Figure 8A-1**. An animal cell that starts out in an isotonic solution will shrink and shrivel if it is transferred to a hypertonic solution. On the other hand, the cell will swell if placed in a hypotonic solution. And it will actually lyse—or burst—if placed in a very hypotonic solution, such as pure water containing no solutes.

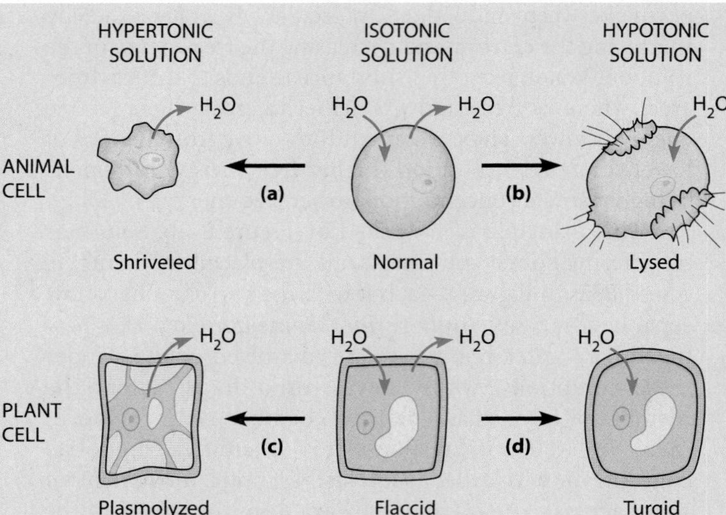

FIGURE 8A-1 Responses of Animal and Plant Cells to Changes in Osmolarity. **(a)** If an animal cell (or other cell not surrounded by a cell wall) is transferred from an isotonic solution to a hypertonic solution, water leaves the cell and the cell shrivels. **(b)** If the cell is transferred to a hypotonic solution, water enters the cell and the cell swells, sometimes to the point that it bursts. **(c)** Plant cells (or other cells with rigid cell walls) also shrivel (plasmolyze) in a hypertonic solution, but **(d)** they will only become turgid—and will not burst—in a hypotonic solution.

Osmolarity: A Common Problem with Different Solutions

The osmotic movements of water shown in Figure 8A-1 occur because of differences in the osmolarity of the cytoplasm and the extracellular solution. Usually, the solute concentration is greater inside a cell than outside. This is due both to the high concentrations of ions and small organic molecules required for normal cellular functions and to the large numbers of macromolecules dissolved in the cytosol. In addition, most of these molecules are charged, and the *counterions* that balance these charges contribute significantly to the intracellular osmolarity. Thus, most cells are hypertonic compared with their surroundings, so water will move inward across the plasma membrane, causing the cells to swell.

How do cells cope with the problem of high osmolarity and the resulting influx of water due to osmosis? Cells of plants, algae, fungi, and many bacteria have rigid cell walls that keep the cells from swelling and bursting in a hypotonic solution (Figure 8A-1d). Instead, the cells become very firm from the **turgor pressure** that builds up due to the inward movement of water. The resulting turgidity accounts for the firmness, or *turgor,* of fully hydrated plant tissue. Without this turgidity, the tissue will wilt.

In a hypertonic solution, on the other hand, the outward movement of water causes the plasma membrane to pull away from the cell wall by a process called **plasmolysis** (Figure 8A-1c). The wilting of a plant or a plant part during water deprivation is due to the plasmolysis of its cells. You can demonstrate plasmolysis readily by dropping a piece of celery into a solution with a high concentration of salt or sugar. Plasmolysis can be a practical problem when plants are grown under conditions of high salinity, as is sometimes the case in locations near an ocean.

Cells without cell walls solve the osmolarity problem by continuously and actively pumping out inorganic ions, thereby reducing the intracellular osmolarity and thus minimizing the difference in solute concentration between the cell and its surroundings. Animal cells continuously remove sodium ions. This is, in fact, one important purpose of the *Na^+/K^+ pump*. Animal cells swell and sometimes even lyse when treated with *ouabain,* an inhibitor of the Na^+/K^+ pump. When medications are given intravenously in a hospital, they are typically dissolved in *phosphate-buffered saline,* which has the same osmolarity as the blood, avoiding potential problems of cell lysis or dehydration.

to its solubility in water. In general, the more nonpolar—or hydrophobic—a substance is, the higher its partition coefficient and the more readily and rapidly it can move across (or dissolve in) a membrane. For example, the partition coefficients of the various amino acids were used in Chapter 7 to calculate the hydropathy index of a protein. Amino acids with nonpolar side chains (see Figure 3-2), such as tryptophan, leucine, and valine, have high partition coefficients and are likely to be found in transmembrane regions of a membrane protein, in contrast to those with those with polar side chains and low partition coefficients (serine and threonine).

Solute Charge. The relative impermeability of polar substances in general and of ions in particular is due to their strong association with water molecules, forming a *shell of hydration*. For such solutes to move into the membrane, the water molecules must be stripped off, and an input of energy is required to eliminate the bonding between the ions and the water molecules. Therefore, the association of ions with water molecules to form shells of hydration dramatically restricts ion transport across membranes.

The impermeability of membranes to ions is important to cell activity because in order to function, every cell must maintain an electrochemical potential across its plasma membrane. In most cases this potential is a gradient either of sodium ions (animal cells) or protons (most other cells). On the other hand, membranes must also allow ions to cross the barrier in a controlled manner. As we will see later in the chapter, the proteins that facilitate ion transport serve as hydrophilic channels that provide a low-energy pathway for movement of the ions across the membrane.

The Rate of Simple Diffusion Is Directly Proportional to the Concentration Gradient

So far, we have focused on qualitative aspects of simple diffusion. We can be more quantitative by considering the thermodynamic and kinetic properties of the process (see Table 8-1). Thermodynamically, simple diffusion is always an exergonic process, requiring no input of energy. Individual molecules simply diffuse randomly in both directions, but net flux will always be in the direction of minimum free energy—which in the case of uncharged molecules means down the concentration gradient.

Kinetically, a key feature of simple diffusion is that the net rate of transport for a specific substance is directly proportional to the concentration difference for that substance across the membrane over a broad concentration range. For the diffusion of solute S from the outside to the inside of a cell, the expression for the rate, or velocity, of inward diffusion through the membrane, v_{inward}, is

$$v_{inward} = P\Delta[S] \qquad \textbf{(8-1)}$$

where v_{inward} is the rate of inward diffusion (in moles/second·cm^2 of membrane surface), and $\Delta[S]$ is the concentration gradient of the solute across the membrane

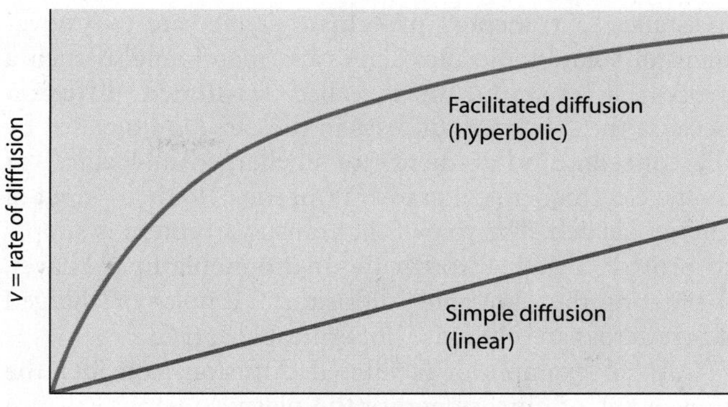

FIGURE 8-5 Comparison of the Kinetics of Simple Diffusion and Facilitated Diffusion. For simple diffusion across a membrane, the relationship between v, the rate of diffusion, and $\Delta[S]$, the solute concentration gradient, is linear over a broad concentration range (red line). For facilitated diffusion, the relationship is linear when the concentration gradient is small, but exhibits saturation kinetics and is therefore hyperbolic (green line), eventually reaching a maximum value at very high $\Delta[S]$.

($\Delta[S] = [S]_{outside} - [S]_{inside}$). P is the permeability coefficient, an experimentally determined parameter that depends on the thickness and viscosity of the membrane; the size, shape, and polarity of S; and the equilibrium distribution of S in the membrane and aqueous phases. As either the permeability or the concentration gradient increases, the rate of inward transport increases.

As Equation 8-1 indicates, simple diffusion is characterized by a linear relationship between the inward flux of the solute across the membrane and the concentration gradient of the solute, with no evidence of saturation at high concentrations. This relationship is seen as the red line in **Figure 8-5**. Simple diffusion differs in this respect from facilitated diffusion, which is subject to saturation and generally follows Michaelis–Menten kinetics, as we will see shortly. Simple diffusion can therefore be distinguished from facilitated diffusion by its kinetic properties, as indicated in Table 8-1.

We can summarize simple diffusion by noting that it is relevant only to molecules such as ethanol and O_2 that are small enough and/or nonpolar enough to cross membranes without the aid of transport proteins. Simple diffusion proceeds exergonically in the direction dictated by the concentration gradient, with a linear, nonsaturating relationship between the diffusion rate and the concentration gradient.

Facilitated Diffusion: Protein-Mediated Movement Down the Gradient

Most substances in cells are too large or too polar to cross membranes at reasonable rates by simple diffusion, even if the process is exergonic. Such solutes can move into and out of cells and organelles at appreciable rates only with the assistance of **transport proteins** that mediate the movement of solute molecules across the membrane. If such a process is exergonic, it is called **facilitated diffusion** because the solute still diffuses in the direction dictated by the concentration gradient (for uncharged molecules) or by the electrochemical gradient (for ions), with no input of energy needed. The role of the transport protein is simply to provide a path through the hydrophobic lipid bilayer, facilitating the "downhill" diffusion of a polar or charged solute across an otherwise impermeable barrier.

As an example of facilitated diffusion, consider the movement of glucose across the plasma membrane of a cell in your body. The concentration of glucose is typically higher in the blood than in the cell, so the inward transport of glucose is exergonic—that is, it does not require the input of energy. However, glucose is too large and too polar to diffuse across the membrane unaided. A transport protein is required to facilitate its inward movement.

Carrier Proteins and Channel Proteins Facilitate Diffusion by Different Mechanisms

Transport proteins involved in facilitated diffusion of small molecules and ions are integral membrane proteins that contain several, or even many, transmembrane segments and therefore traverse the membrane multiple times. Functionally, these proteins fall into two main classes that transport solutes in quite different ways. **Carrier proteins** (also called *transporters* or *permeases*) bind one or more solute molecules on one side of the membrane and then undergo a conformational change that transfers the solute to the other side of the membrane. In so doing, a carrier protein binds the solute molecules in such a way as to shield the polar or charged groups of the solute from the nonpolar interior of the membrane.

Channel proteins, on the other hand, form hydrophilic *channels* through the membrane that allow the passage of solutes without a major change in the conformation of the protein. Some of these channels are relatively large and nonspecific, such as the *pores* found in the outer membranes of bacteria, mitochondria, and chloroplasts. Pores are formed by transmembrane proteins called *porins* and allow selected hydrophilic solutes with molecular weights up to about 600 to diffuse across the membrane. However, most channels are small and highly selective. Most of these smaller channels are involved in the transport of ions rather than molecules and are therefore referred to as *ion channels*. The movement of solutes through ion channels is much more rapid than transport by carrier proteins, presumably because complex conformational changes are not necessary.

Carrier Proteins Alternate Between Two Conformational States

An important topic of contemporary membrane research concerns the mechanisms whereby transport proteins facilitate the movement of solutes across membranes. For carrier proteins, the most likely explanation is the **alternating conformation model,** in which the carrier protein is an allosteric protein that alternates between two conformational states. In one state the solute-binding site of the protein is open or accessible on one side of the membrane. Following solute binding, the protein changes to an alternate conformation in which the solute-binding site is on the other side of the membrane, triggering its release. We will encounter an example of this mechanism shortly when we discuss the facilitated diffusion of glucose into erythrocytes.

Carrier Proteins Are Analogous to Enzymes in Their Specificity and Kinetics

As we noted earlier, carrier proteins are sometimes called *permeases*. This term is apt because the suffix *-ase* suggests a similarity between carrier proteins and enzymes. Like an enzyme-catalyzed reaction, facilitated diffusion involves an initial binding of the "substrate" (the solute to be transported) to a specific site on a protein surface (the solute's binding site on the carrier protein) and the eventual release of the "product" (the transported solute), with an "enzyme-substrate" complex (solute bound to carrier protein) as an intermediate. Like enzymes, carrier proteins can be regulated by external factors that bind and modulate their activity.

Specificity of Carrier Proteins. Another property that carrier proteins share with enzymes is *specificity*. Like enzymes, transport proteins are highly specific, often for a single compound or a small group of closely related compounds and sometimes even for a specific stereoisomer. A good example is the carrier protein that facilitates the diffusion of glucose into erythrocytes (see Figure 8-2b). This protein recognizes only glucose and a few closely related monosaccharides, such as galactose and mannose. Moreover, the protein is *stereospecific:* it accepts the D- but not the L-isomer of these sugars. This specificity is presumably a result of the precise stereochemical fit between the solute and its binding site on the carrier protein.

Thus, the properties of carrier proteins explain the characteristic features of facilitated diffusion: movement of polar molecules and ions down a concentration gradient, specificity for the particular substrate transported, the ability to be saturated at high levels of substrate, and sensitivity to specific inhibitors of transport.

Kinetics of Carrier Protein Function. As you might expect from the analogy with enzymes, carrier proteins become saturated as the concentration of the transportable solute is raised. This is because the number of transport proteins is limited, and each transport protein functions at some finite maximum velocity. As a result, carrier-facilitated transport, like enzyme catalysis, exhibits *saturation kinetics*. This type of transport has an upper limiting velocity V_{max} and a constant K_m corresponding to the concentration of transportable solute needed to achieve one-half of the maximum rate of transport. This means that the initial rate of solute transport, v, can be described mathematically by the same equation we used to describe enzyme kinetics (Equation 6-7, page 139):

$$v = \frac{V_{max}[S]}{K_m + [S]} \tag{8-2}$$

where [S] is the initial concentration of solute on one side of the membrane (e.g., on the outside of the membrane if the initial rate of inward transport is to be determined). A plot of transport rate versus initial solute concentration is therefore hyperbolic for facilitated diffusion instead of linear as for simple diffusion (see Figure 8-5, green line). This difference in saturation kinetics is an important means of distinguishing between simple and facilitated diffusion (see Table 8-1).

A further similarity between enzymes and carrier proteins is that proteins are often subject to *competitive inhibition* by molecules or ions that are structurally related to the intended "substrate." For example, the transport of glucose by a glucose carrier protein is competitively inhibited by the other monosaccharides that the protein also accepts—that is, the rate of glucose transport is reduced in the presence of other transportable sugars.

Carrier Proteins Transport Either One or Two Solutes

Although carrier proteins are similar in their kinetics and their presumed mechanism of action involving alternate conformations, they may differ in significant ways. The most important differences concern the number of solutes transported and the direction they move. When a carrier protein transports a single solute across the membrane, the process is called **uniport** (**Figure 8-6a**). The glucose carrier protein we will discuss shortly is a *uniporter*. When two solutes are transported simultaneously and their transport is coupled such that transport of either stops if the other is absent, the process is called **coupled transport** (Figure 8-6b). Coupled transport is referred to as **symport** (or *cotransport*) if the two solutes are moved in the same direction or as **antiport** (or *countertransport*) if the two solutes are moved in opposite directions across the membrane. The transport proteins that mediate these processes are called *symporters* and *antiporters*, respectively. As we will see later, these same terms apply whether the mode of transport is facilitated diffusion or active transport.

The Erythrocyte Glucose Transporter and Anion Exchange Protein Are Examples of Carrier Proteins

Now that we have described the general properties of carrier proteins, let us briefly consider two specific examples: the uniport carrier for glucose and the antiport anion carrier for Cl^- and HCO_3^-. Both of these transporters are present in the cell membrane of erythrocytes.

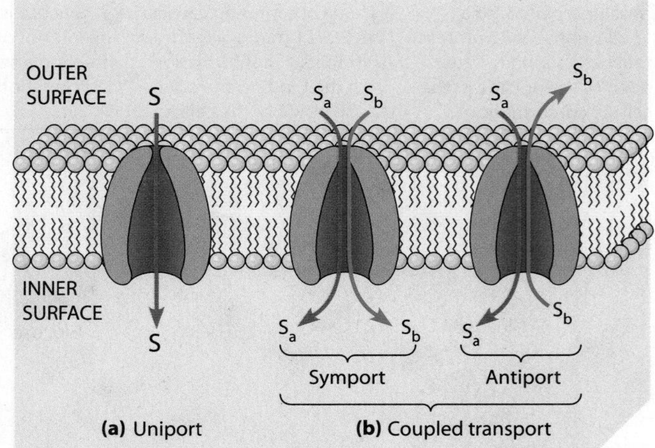

FIGURE 8-6 Comparison of Uniport, Symport, and Antiport Transport by Carrier Proteins. **(a)** In uniport, a membrane transport protein moves a single solute across a membrane. **(b)** Coupled transport involves the simultaneous transport of two solutes, S_a and S_b. Coupled transport may be either symport (both solutes moved in the same direction) or antiport (the two solutes moved in opposite directions). Note that transporters are not simply open channels, as depicted in this simplified illustration, but alternate between two conformations as solutes are transported.

The Glucose Transporter: A Uniport Carrier. As we noted earlier, the movement of glucose into an erythrocyte is an example of facilitated diffusion mediated by a uniport carrier protein (see Figure 8-2b). The concentration of glucose in blood plasma is usually in the range of 65–90 mg/100 mL, or about 3.6–5.0 m*M*. The erythrocyte (or almost any other cell in contact with the blood, for that matter) is capable of glucose uptake by facilitated diffusion because of its low intracellular glucose concentration and the presence in its plasma membrane of a glucose carrier protein, or **glucose transporter (GLUT).** The GLUT of erythrocytes is called GLUT1 to distinguish it from related GLUTs in other mammalian tissues. GLUT1 allows glucose to enter the cell about 50,000 times faster than it would enter by free diffusion through a lipid bilayer.

GLUT1-mediated uptake of glucose displays all the classic features of facilitated diffusion: It is specific for glucose (and a few related sugars, such as galactose and mannose), it proceeds down a concentration gradient without energy input, it exhibits saturation kinetics, and it is susceptible to competitive inhibition by related monosaccharides. GLUT1 is an integral membrane protein with 12 hydrophobic transmembrane segments. These are presumably folded and assembled in the membrane to form a cavity lined with hydrophilic side chains that form hydrogen bonds with glucose molecules as they move through the membrane.

GLUT1 is thought to transport glucose by an alternating conformation mechanism, as illustrated in **Figure 8-7**. The two conformational states are called T_1, which has the binding site for glucose open to the outside of the

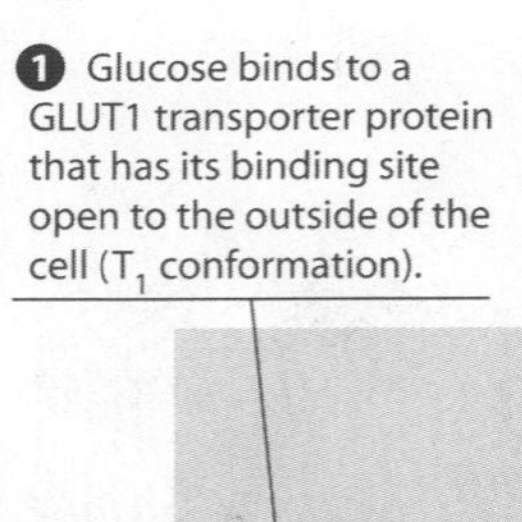

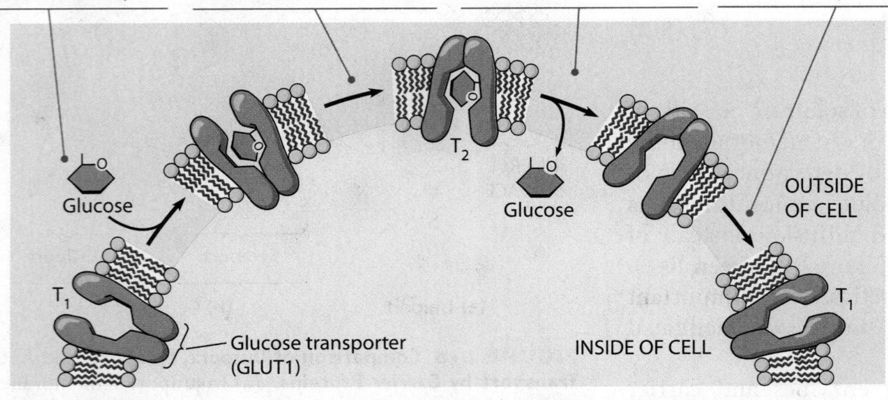

1 Glucose binds to a GLUT1 transporter protein that has its binding site open to the outside of the cell (T_1 conformation).
2 Glucose binding causes the GLUT1 transporter to shift to its T_2 conformation with the binding site open to the inside of the cell.
3 Glucose is released to the interior of the cell, initiating a second conformational change in GLUT1.
4 Loss of bound glucose causes GLUT1 to return to its original (T_1) conformation, ready for a further transport cycle.
Glucose
T_2
Glucose
OUTSIDE OF CELL
T_1
Glucose transporter (GLUT1)
INSIDE OF CELL
T_1

FIGURE 8-7 The Alternating Conformation Model for Facilitated Diffusion of Glucose by the Glucose Transporter GLUT1 in the Erythrocyte Membrane. The inward transport of glucose by GLUT1 is shown here in four steps, arranged at the periphery of a cell.

cell, and T_2, with the binding site open to the interior of the cell. The process begins when a molecule of D-glucose collides with and binds to a GLUT1 molecule that is in its T_1 conformation ❶. With glucose bound, GLUT1 now shifts to its T_2 conformation ❷. The conformational change allows the release of the glucose molecule to the interior of the cell ❸, after which the GLUT1 molecule returns to its original conformation, with the binding site again facing outward ❹.

The example shown in Figure 8-7 is for inward transport, but the process is readily reversible because carrier proteins function equally well in either direction. A carrier protein is really just a gate in an otherwise impenetrable wall, and, like most gates, it facilitates traffic in either direction. Individual solute molecules may be transported either inward or outward, depending on the relative concentrations of the solute on the two sides of the membrane. If the concentration is higher outside, net flow will be inward. If the higher concentration occurs inside, net flow will be outward.

The low intracellular glucose concentration that makes facilitated diffusion possible for most animal cells exists because incoming glucose is quickly phosphorylated to glucose-6-phosphate by the enzyme hexokinase, with ATP as the phosphate donor and energy source:

$$\text{glucose} \xrightarrow[\text{hexokinase}]{\text{ATP} \;\; \text{ADP}} \text{glucose-6-phosphate} \qquad \textbf{(8-3)}$$

This hexokinase reaction is the first step in glucose metabolism, which we will discuss in Chapter 9. The low K_m of hexokinase for glucose (1.5 mM) and the highly exergonic nature of the reaction ($\Delta G°' = -4.0$ kcal/mol) ensure that the concentration of glucose within the cell is kept low, maintaining the concentration gradient across the cell membrane. For many mammalian cells, the intracellular glucose concentration ranges from 0.5 to 1.0 mM, which is about 15–20% of the glucose level in the blood plasma outside the cell.

The phosphorylation of glucose also has the effect of locking glucose in the cell, because the plasma membrane of the erythrocyte does not have a transport protein for glucose-6-phosphate. (GLUT1, like most sugar transporters, does not recognize the phosphorylated form of the sugar.) Phosphorylation is in fact a general strategy for retaining molecules within the cell because most cells do not have plasma membrane proteins capable of transporting phosphorylated compounds.

The GLUT1 of erythrocytes is just one of several glucose transporters in mammals. In humans there are 14 different GLUT proteins, each encoded by a separate gene. Each transporter has distinct physical and kinetic characteristics appropriate for its function in the particular cell type where it is found. For example, GLUT3 and GLUT4 are found in cerebral neurons and skeletal muscle, respectively, where they import glucose for energy. On the other hand, GLUT2 is the glucose transporter in liver cells, which break down glycogen to produce glucose for the blood. GLUT2 has properties that facilitate glucose transport *out* of liver cells to maintain blood glucose at a constant level.

The Erythrocyte Anion Exchange Protein: An Antiport Carrier. Another well-studied example of facilitated diffusion is the **anion exchange protein** of the erythrocyte plasma membrane (see Figure 8-2b). This antiport protein, also called the *chloride-bicarbonate exchanger,* facilitates the reciprocal exchange of chloride (Cl^-) and bicarbonate

(HCO_3^-) ions in opposite directions across the plasma membrane. The coupling of chloride and bicarbonate transport is obligatory, and transport will stop if either anion is absent. Moreover, the anion exchange protein is very selective. It exchanges bicarbonate for chloride in a strict 1:1 ratio, and it accepts no other anions.

The anion exchange protein is thought to function by alternating between two conformational states in what is termed a "ping-pong" mechanism. In one conformation, the anion exchange protein binds chloride on one side of the membrane. Chloride binding causes a change in conformation such that chloride is released on the other side of the membrane, where the protein binds bicarbonate. This causes a second conformational change, releasing bicarbonate on the other side of the membrane, where the protein again binds chloride. Repeated cycles of binding and release transport the two ions in opposite directions.

Either anion can bind to the protein on either side of the membrane, so the direction of transport depends upon the relative concentrations of the ions on opposite sides of the membrane. In cells with a high bicarbonate concentration, bicarbonate leaves the cell and chloride enters. In cells with a low bicarbonate concentration, the reciprocal process occurs—bicarbonate enters the cell and chloride leaves.

The anion exchange protein plays a central role in the process by which waste CO_2 produced in metabolically active tissues is delivered to the lungs to be expelled. In these tissues, CO_2 diffuses into erythrocytes, where the cytosolic enzyme *carbonic anhydrase* converts it to bicarbonate ions (see Figure 8-3). As the concentration of bicarbonate in the erythrocyte rises, it moves out of the cell. To prevent a net charge imbalance, the outward movement of each negatively charged bicarbonate ion is accompanied by the uptake of one negatively charged chloride ion.

In the lungs, this entire process is reversed: Chloride ions are transported out of erythrocytes accompanied by the uptake of bicarbonate ions, which are then converted back to CO_2 by carbonic anhydrase. The net result is the movement of CO_2 (in the form of bicarbonate ions) from tissues to the lungs, where the CO_2 is exhaled from the body. In addition, the import of bicarbonate into erythrocytes in the lungs increases the cellular pH, which enhances oxygen binding to hemoglobin in the lungs. When the erythrocytes reach the tissues and bicarbonate is exported, the cellular pH drops and oxygen binding decreases, allowing more rapid release to the tissues.

Channel Proteins Facilitate Diffusion by Forming Hydrophilic Transmembrane Channels

While some transport proteins facilitate diffusion by functioning as carrier proteins that alternate between different conformational states, others do so by forming hydrophilic *transmembrane channels* that allow specific solutes to move across the membrane directly. We will consider three kinds of transmembrane protein channels: *ion channels, porins,* and *aquaporins.*

Ion Channels: Transmembrane Proteins That Allow Rapid Passage of Specific Ions. Despite their apparently simple design—a tiny pore lined with hydrophilic atoms—**ion channels** are remarkably selective. Most allow passage of only one kind of ion, so separate channels are needed for transporting such ions as Na^+, K^+, Ca^{2+}, and Cl^-. This selectivity is remarkable given the small differences in size and charge among some of these ions. Selectivity results both from ion-specific binding sites involving specific amino acid side chains and polypeptide backbone atoms inside the channel and from a constricted center of the channel that serves as a size filter. The rate of transport is equally remarkable: In some cases, a single channel can conduct almost a million ions of a specific type per second!

Most ion channels are *gated,* which means that the pore opens and closes in response to some stimulus. In animal cells, three kinds of stimuli control the opening and closing of gated channels: *Voltage-gated channels* open and close in response to changes in the membrane potential; *ligand-gated channels* are triggered by the binding of specific substances to the channel protein; and *mechanosensitive channels* respond to mechanical forces that act on the membrane. We will discuss gated channels in detail in Chapter 13.

Regulation of ion movement across membranes plays an important role in many types of cellular communication. Muscle contraction and many cellular responses require regulation of Ca^{2+} levels via calcium-specific ion channels. Also, the transmission of electrical signals by nerve cells depends critically on rapid, controlled changes in the movement of Na^+ and K^+ ions through their respective channels. These changes are so rapid that they are measured in milliseconds. In addition to such short-term regulation, most ion channels are subject to longer-term regulation, usually in response to external stimuli such as hormones.

Ion channels are also necessary for maintaining the proper salt balance in the cells and airways lining our lungs. In lung epithelial cells, a specific chloride ion channel known as the *cystic fibrosis transmembrane conductance regulator (CFTR)* protein helps to maintain the proper Cl^- concentration in these airways. Defects in CFTR lead to excessive buildup of mucus in the lungs, causing a life-threatening condition known as cystic fibrosis, a topic we will discuss in detail in **Box 8B**.

Porins: Transmembrane Proteins That Allow Rapid Passage of Various Solutes. Compared with ion channels, the pores found in the outer membranes of mitochondria, chloroplasts, and many bacteria are somewhat larger and much less specific. These pores are formed by multipass transmembrane proteins called **porins.** However, the

BOX 8B HUMAN APPLICATIONS

Membrane Transport, Cystic Fibrosis, and the Prospects for Gene Therapy

Transport proteins located in the plasma membrane play critical roles in speeding up and controlling the movement of molecules and ions into and out of cells. To remain healthy, our bodies depend on the proper functioning of many such membrane proteins. If any of these proteins is defective, the movement of a particular ion or molecule across cell membranes is likely to be impaired, and disease may result.

An example that has attracted the attention of researchers and doctors alike is **cystic fibrosis (CF),** a fatal disease caused by genetic defects in a transport protein in the plasma membrane. The parts of the body that are most noticeably affected are the lungs, pancreas, and sweat glands. Complications in the lungs are the most severe medical problems because they are difficult to treat and can become life-threatening. The airways of a CF patient are often obstructed with abnormally thick mucus and are vulnerable to chronic bacterial infections, especially by *Pseudomonas aeruginosa.*

Using the tools of molecular and cellular biology, researchers have achieved a detailed understanding of this disease. During the 1980s, cells from CF patients were shown to be defective in the secretion of chloride ions (Cl^-). The cells that line unaffected lungs secrete chloride ions in response to a substance called cyclic AMP, whereas cells from CF patients do not. (Cyclic AMP is a form of AMP that is involved in a variety of regulatory roles in cells; for its structure, see Figure 14-6.) Experiments with tissue from CF patients suggested that this difference might be due to a defect in a membrane protein that normally serves as a channel for the movement of chloride ions across the membrane.

Many symptoms of CF can be explained by faulty Cl^- secretion. In the lungs of a healthy, unaffected person (**Figure 8B-1a**), chloride ions are secreted from the cells lining the airways and enter the lumen, the space inside the airway passage. The movement of Cl^- out of the cell and into the lumen provides the driving force for the concurrent movement of sodium ions into the lumen in order to maintain charge balance. Osmotic pressure causes water to follow the sodium and chloride ions, resulting in the secretion of a dilute salt solution. The water that moves into the lumen in this way provides vital hydration to the mucous lining of the air passages. In the cells of a person with CF, Cl^- ions cannot exit into the lumen, so sodium ions and water do not move outward either (Figure 8B-1b). As a result, the mucus is insufficiently hydrated and becomes very thick, a condition that favors bacterial growth.

An exciting breakthrough in CF research came in 1989 when investigators in the laboratories of Francis Collins at the University of Michigan and of Lap-Chee Tsui and John Riordan at the University of Toronto isolated the gene that is defective in CF patients. The gene encodes a protein called the **cystic fibrosis transmembrane conductance regulator (CFTR).** The sequence of nucleotide bases in the gene was determined by using methods that are described in Chapter 18 (see Figure 18-14). Knowing the base sequence of the gene, scientists were able to predict the amino acid sequence and the structure of the CFTR protein. As shown in Figure 8B-1c, the protein has two sets of *transmembrane domains* that anchor the protein in the plasma membrane and provide an anion-selective pore. It also has two *nucleotide-binding domains* that serve as binding sites for ATP, which provides the energy to drive transport of chloride ions across the membrane. In addition, the protein has a large cytoplasmic *regulatory domain,* which has several serine hydroxyl groups that can be reversibly phosphorylated. The CFTR protein has since been shown to function as a chloride channel in cells, and channel function is known to be affected when the phosphorylation sites in the regulatory domain are changed due to a mutation in the *CFTR* gene.

By sequencing the *CFTR* genes from CF patients, investigators have identified more than 600 mutations in the gene. The most common of these mutations causes the deletion of a single amino acid in the first nucleotide-binding domain (ΔPhe-509). The question of how this mutation causes CF remained unanswered until researchers examined the location of the CFTR protein in cells with and without the mutation. Normal CFTR was found in the cell membrane, as predicted. In contrast, mutant CFTR was not detected in the plasma membrane.

Currently, the most likely explanation is that normal CFTR is synthesized on the rough endoplasmic reticulum (ER), moves through the Golgi complex, and is eventually inserted in the plasma membrane by a route that is explained in Chapter 12. Mutant CFTR, on the other hand, is apparently trapped in the ER, perhaps because it is folded improperly. It is therefore recognized as a defective protein and degraded. Consequently, CFTR is not present in the plasma membrane of CF cells, chloride ion secretion cannot take place, and disease results.

Armed with information about the gene and the protein associated with CF, researchers are now trying to develop new treatments or perhaps even a cure for the disease. One such approach is *gene therapy,* in which a normal copy of a gene is introduced into affected cells of the body. Investigators would like to direct normal copies of the *CFTR* gene into the cells that line the airways of CF patients. These cells should then be able to synthesize a correct CFTR protein that, unlike mutant CFTR, would be located in the plasma membrane. The CFTR proteins would then allow proper Cl^- secretion and correct the disease.

Two kinds of practical problems must be overcome if gene therapy is to work: The *CFTR* gene must be delivered efficiently to the affected tissue, and its expression must be regulated to achieve and maintain normal production of the CFTR protein. In most clinical studies to date, the normal *CFTR* gene has been introduced

transmembrane segments of porin molecules cross the membrane not as α helices but as a closed cylindrical β *sheet* called a β *barrel* (**Figure 8-8**). The β barrel has a water-filled pore at its center. Polar side chains (not shown) line the inside of the pore, whereas the outside of the barrel consists mainly of nonpolar side chains that interact with the hydrophobic interior of the membrane. The pore allows passage of various hydrophilic solutes. The upper size limit for the solute molecules is determined by the pore size of the particular porin—only solutes smaller than

into CF patients by one of two means: Either the *CFTR* is incorporated into the DNA of a virus called *adenovirus* or it is mixed with fat droplets called *liposomes*. The viral or liposomal preparation is sprayed as an aerosol into the nose or lungs of CF patients, who are then monitored for the correction of the chloride transport abnormality. Another strategy involves treating patients with a highly compacted DNA molecule containing the normal *CFTR* gene. This DNA is readily taken up into cells, and this procedure eliminates concerns regarding variability and possible side effects due to the use of a modified viral vector.

In initial experiments with CF mutant mice, several groups of British investigators showed that the chloride ion channel defect could be corrected by spraying the respiratory tract with liposomes containing the normal human *CFTR* gene. The same approach is now being used in clinical trials with human patients. In one such study, a "gene spray" of the *CFTR* gene mixed with liposomes was administered nasally. Gene delivery and expression was demonstrated in most of the patients, but expression was short-lived and the ion channel defect was only partially corrected.

More recently, Eric Alton and his colleagues in London reported a well-controlled experiment in which they administered liposomes with or without the *CFTR* gene in the form of an aerosol to the lungs and the nose. A short-term improvement in chloride transport was seen in the patients who received the *CFTR* gene but not in the patients who received the placebo (liposome-only) treatment. The researchers also reported reduced bacterial adherence to respiratory epithelial cells in the *CFTR*-treated patients, an observation that may have significant clinical relevance. Overall, however, progress has been slow in coming, though researchers and clinicians alike remain hopeful that gene therapy will eventually become a realistic therapeutic option for treating cystic fibrosis.

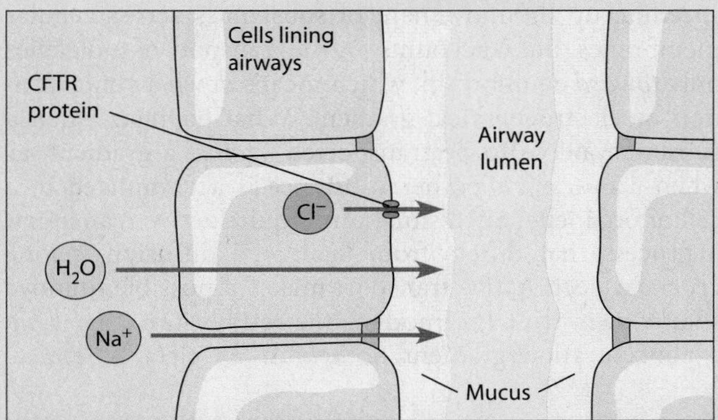

(a) Normal cells lining airways; hydrated mucus

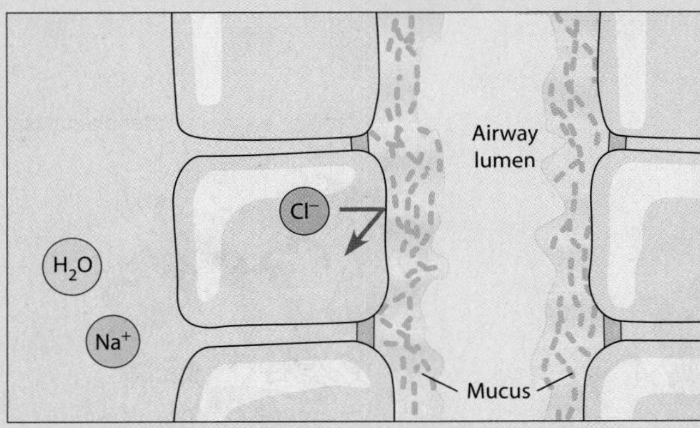

(b) Cells of a person with cystic fibrosis; dehydrated mucus infected with bacteria

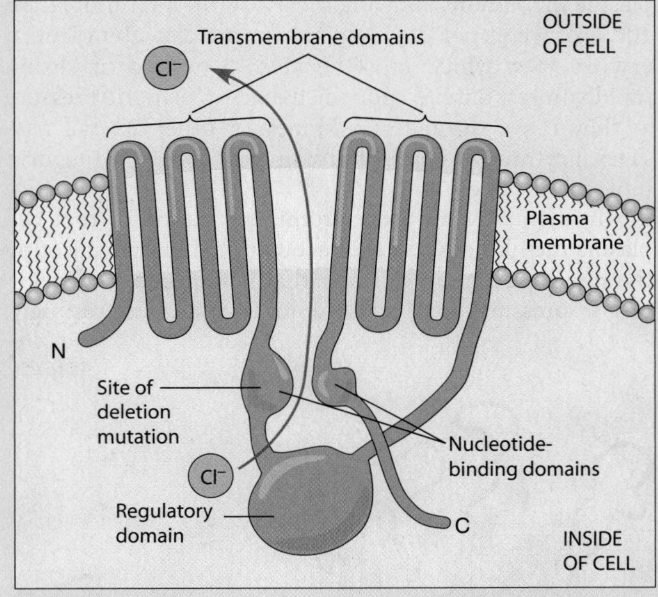

(c) CFTR protein

FIGURE 8B-1 Cystic Fibrosis and Chloride Ion Secretion. **(a)** Normal lungs have hydrated mucus. **(b)** Cystic fibrosis is caused by a defect in the secretion of chloride ions in cells lining the lungs, leading to insufficient hydration and the promotion of bacterial growth. **(c)** The cystic fibrosis transmembrane conductance regulator (CFTR) is an integral membrane protein that functions as a chloride ion channel. Deletion of a single amino acid in the first nucleotide-binding domain of the CFTR protein blocks its transport function, resulting in cystic fibrosis.

about 600 d can pass through the *E. coli* porin shown in the figure. Mutations in certain bacterial porins can lead to antibiotic resistance in these bacteria by effectively blocking entry of the antibiotics that would normally be used to fight an infection.

Aquaporins: Transmembrane Channels That Allow Rapid Passage of Water. Whereas water can diffuse slowly across cell membranes in the absence of a protein transporter, movement of water across membranes in some tissues is much more rapid than can be accounted for by diffusion alone. In fact, the existence of water channels in cell membranes was postulated as early as the mid- to late 1800s. Despite a century of experiments suggesting their existence, water channels remained elusive, and their very existence was sometimes in doubt.

It was not until 1992 that Peter Agre and colleagues at Johns Hopkins University finally isolated the long-sought-after water channel protein, which they named **aquaporin (AQP).** They found that this particular erythrocyte protein, when introduced into membranes of relatively water-impermeable frog egg cells, would cause the cells to explode when placed in pure water due to the rapid influx of water. Control eggs without the protein placed in the pure water were unchanged. In 2003, Agre shared the Nobel Prize in chemistry with Roderick MacKinnon, who determined the first three-dimensional structure of an ion channel.

Aquaporins facilitate the rapid movement of water molecules into or out of cells in specific tissues that require this capability. For example, the proximal tubules of your kidneys reabsorb water as part of urine formation, and cells in this tissue have a high density of AQPs in their plasma membrane, allowing the kidney to filter more than 100 L of water per day. Aquaporins are also abundant in erythrocytes, which must be able to expand or shrink rapidly in response to sudden changes in osmotic pressure as they move through the kidney or other arterial passages. Erythrocytes have approximately 200,000 aquaporin molecules per cell.

In plants, AQPs are a prominent feature of root cell plasma membranes and the vacuolar membrane, reflecting the rapid movement of water that is required to develop turgor pressure, as discussed in Box 8A. Interestingly, prokaryotes appear not to contain aquaporins, possibly because their small size—and hence their large surface area/volume ratio—makes facilitated transport of water unnecessary.

All aquaporins described to date are tetrameric integral membrane proteins that have four identical monomers, with each monomer containing six helical transmembrane segments. The four monomers associate side by side in the membrane with their 24 transmembrane segments oriented to form four central channels lined with hydrophilic side chains (Figure 8-8c). The space in the center of the tetramer is blocked by a lipid molecule. The diameter of each of the four water channels is about 0.3 nm, just large enough for water molecules to pass through one at a time in single file. Even with this constraint, water molecules flow through an aquaporin channel at the rate of several billion per second.

Active Transport: Protein-Mediated Movement Up the Gradient

Facilitated diffusion is an important mechanism for speeding up the movement of substances across cellular membranes. But it accounts for the transport of molecules only *toward* equilibrium, which means *down* a concentration or electrochemical gradient. What happens when a substance needs to be transported *against* a gradient, as when a nutrient or other substance is accumulated in a cell or organelle? Such situations require **active transport,** a process that differs from facilitated diffusion in one crucial aspect: Active transport makes it possible to move solutes *away* from thermodynamic equilibrium (that is, *up* a concentration gradient or *against* an electrochemical

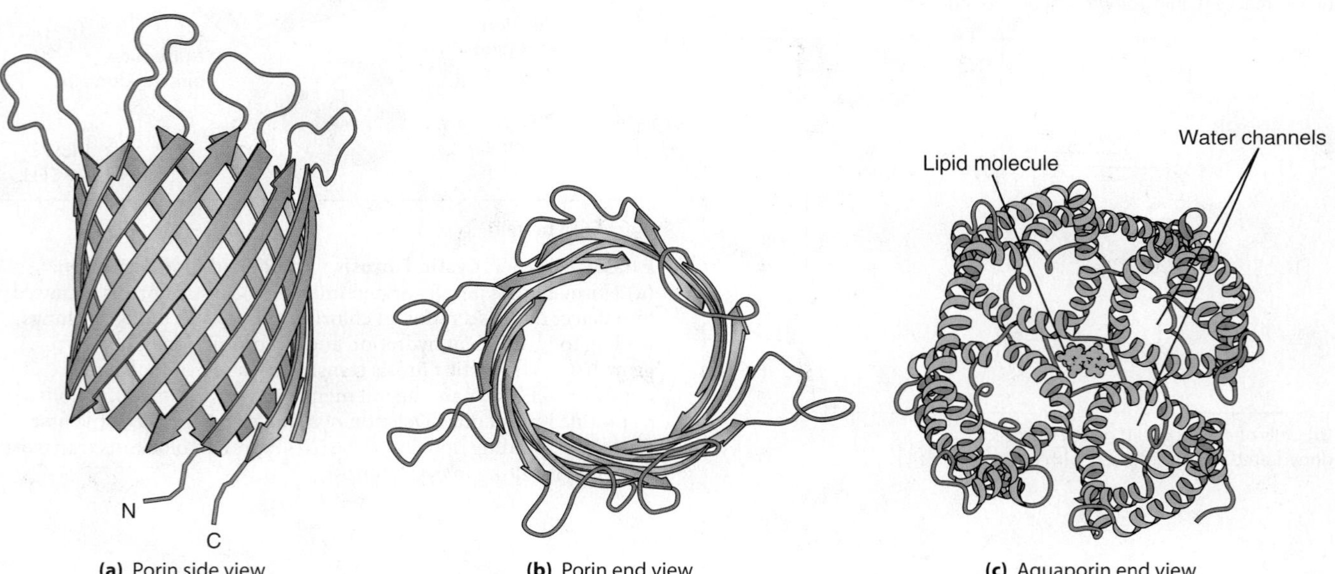

MB Molecular Model: Aquaporin-I

FIGURE 8-8 Porin and Aquaporin Channel Proteins. (a) Side view and (b) end view of an *E. coli* porin protein showing the 14-stranded transmembrane β-barrel. (c) End view through a human aquaporin, showing the six α helices in each of the four identical monomers. Two of the four water channels in the tetramer are labeled.

potential). Therefore, it always requires an input of energy. In other words, active transport couples a thermodynamically unfavorable process (movement up a concentration gradient) to an exergonic process (usually ATP hydrolysis). As a result, membrane proteins involved in active transport must provide mechanisms not only for moving desired solute molecules across the membrane but also for coupling such movements to energy-yielding reactions.

Active transport performs three major functions in cells and organelles. First, it makes possible the uptake of essential nutrients from the environment or surrounding fluid, even when their concentrations in the cell are already higher. Second, it allows various substances, such as secretory products and waste materials, to be removed from the cell or organelle, even when the concentration outside is greater than the concentration inside. Third, it enables the cell to maintain constant, nonequilibrium intracellular concentrations of specific inorganic ions, notably K^+, Na^+, Ca^{2+}, and H^+.

This ability to create an internal cellular environment whose solute concentrations are far removed from equilibrium is a crucial feature of active transport. In contrast to simple or facilitated diffusion, which create conditions that are the same on opposite sides of a membrane, active transport is a means of establishing differences in solute concentration and/or electrical potential across membranes. The end result is a nonequilibrium steady state, without which life as we know it would be impossible. Many membrane proteins involved in active transport are called *pumps* to emphasize that an input of energy is required to move materials against their concentration or electrochemical gradients.

An important distinction between active transport and simple or facilitated diffusion concerns the direction of transport. Simple and facilitated diffusion are both *nondirectional* with respect to the membrane. For both types of diffusion, solutes can move in either direction, depending entirely on the prevailing concentration or electrochemical gradient. Active transport, on the other hand, usually has intrinsic **directionality.** An active transport system that moves a solute across a membrane in one direction will not usually move that solute in the other direction. Active transport is therefore said to be a *unidirectional* process.

(a) Direct active transport involves a transport system coupled to an exergonic chemical reaction, most commonly the hydrolysis of ATP. As shown here, ATP hydrolysis drives the outward transport of protons, thereby establishing an electrochemical potential for protons across the membrane.

(b) Indirect active transport involves the coupled transport of a solute S and ions—protons, in this case. The exergonic inward movement of protons provides the energy to move the transported solute, S, against its concentration gradient or electrochemical potential.

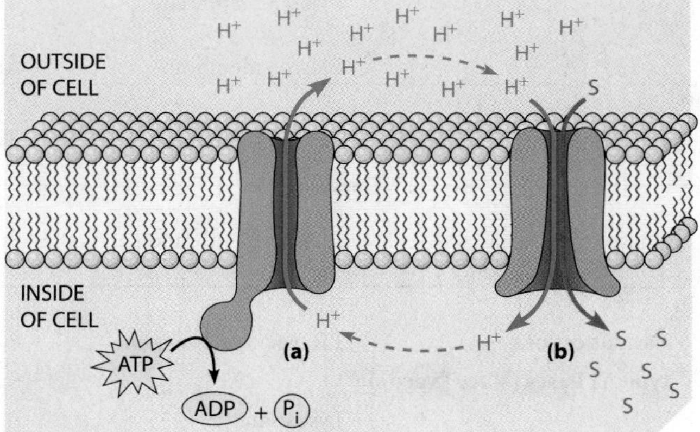

FIGURE 8-9 Comparison of Direct and Indirect Active Transport. Note the circulation of protons across the membrane that results from the coupling of direct and indirect mechanisms of active transport. Note also that transporters are not simply open channels, as depicted in the simplified illustration.

The Coupling of Active Transport to an Energy Source May Be Direct or Indirect

Active transport mechanisms can be divided into two related categories that differ primarily in the source of energy and whether or not two solutes are transported simultaneously. Depending on the energy source, active transport is regarded as being either *direct* or *indirect* (**Figure 8-9**). In **direct active transport** (also called *primary active transport*), the accumulation of solute molecules or ions on one side of the membrane is coupled *directly* to an exergonic chemical reaction, most commonly the hydrolysis of ATP (Figure 8-9a). Transport proteins driven directly by ATP hydrolysis are called *transport ATPases* or *ATPase pumps*.

Indirect active transport also requires energy but depends on the simultaneous transport of two solutes, with the favorable movement of one solute *down* its gradient driving the unfavorable movement of the other solute *up* its gradient. The coupling of a favorable process or reaction with an unfavorable one allows both to proceed with an overall decrease in free energy. This dual transport process can be described as either a symport or an antiport, depending on whether the two solutes move in the same or opposite directions. In most cases, one of the two solutes is an ion that moves exergonically down its electrochemical gradient—Na^+ in animals and H^+ in most other organisms. As it moves, it drives the simultaneous endergonic transport of the second solute (often a monosaccharide or an amino acid) against its concentration gradient or, in the case of ions, its electrochemical potential (Figure 8-9b). Thus indirect active transport is often called *secondary active transport*.

Direct Active Transport Depends on Four Types of Transport ATPases

The most common mechanism employed for direct active transport involves transport ATPases that link active transport to the hydrolysis of ATP. Four main types of transport

Table 8-2 Main Types of Transport ATPases (Pumps)

Solutes Transported	Kind of Membrane	Kind of Organisms	Example of ATPase function
P-type ATPases (P for "phosphorylation")			
P_1			
K^+, Cu^+, Zn^{2+}, Cd^{2+}, Pb^{2+}	Plasma membrane	Bacteria, archaea, plants, fungi, animals	Transport of potassium or heavy metal ions
P_2			
Ca^{2+}/H^+	SR* or plasma membrane	Eukaryotes	Keeps [Ca^{2+}] low in cytosol
Na^+/K^+	Plasma membrane	Animals	Maintains membrane potential (−60 mV)
H^+/K^+	Plasma membrane	Animals	Pumps H^+ to acidify stomach
P_3			
H^+	Plasma membrane	Plants, fungi	Pumps protons out of cell to generate membrane potential (−180 mV)
P_4			
Phospholipids	Plasma membrane	Eukaryotes	Flippases that maintain asymmetry in the lipid bilayer
P_5			
Various cations	ER, vacuole, lysosome	Eukaryotes	Not well characterized
V-type ATPases (V for "vacuole")			
H^+	Lysosomes, secretory vesicles	Animals	Keeps pH of compartment low, which activates hydrolytic enzymes
	Vacuolar membrane	Plants, fungi	
F-type ATPases (F for "factor"); also called ATP synthases			
H^+	Inner mitochondrial membrane	Eukaryotes	Uses H^+ gradient to drive ATP synthesis
	Plasma membrane	Bacteria	
	Thylakoid membrane	Plants	
ABC-type ATPases (ABC for "ATP-binding cassette")			
Importers A variety of solutes**	Plasma membrane, organellar membranes	Bacteria	Nutrients such as vitamin B_{12}
Exporters Antitumor drugs, toxins, antibiotics, lipids	Plasma membrane	Bacteria, archaea, eukaryotes	Multidrug resistance transporter removes drugs and antibiotics from cell

*Sarcoplasmic reticulum, a specialized type of ER found in animal muscle cells

**Solutes include ions, sugars, amino acids, carbohydrates, vitamins, peptides, and proteins.

ATPases have been identified: *P-type, V-type, F-type,* and *ABC-type ATPases* (**Table 8-2**). These four types of transport proteins differ in structure, mechanism, localization, and physiological roles, but all of them use the energy of ATP hydrolysis to transport solutes against a concentration gradient or an electrochemical potential.

P-type ATPases. P-type ATPases (P for "phosphorylation") are members of a large family of proteins that are reversibly phosphorylated by ATP as part of the transport mechanism, with a specific aspartic acid residue becoming phosphorylated in each case. P-type ATPases have 8–10 transmembrane segments in a single polypeptide that crosses the membrane multiple times (see Figure 7-5b). Also, they are sensitive to inhibition by the vanadate ion, VO_4^{3-}, which closely resembles the phosphate ion (PO_4^{3-}) and thus interferes with phosphorylation. Vanadate sensitivity can therefore be used by researchers as a means of identifying P-type ATPases.

Most P-type ATPases are located in the plasma membrane and, based on sequence and structural similarities, fall into one of five subfamilies. P_1-ATPases have been found in all organisms and are mainly responsible for transporting ions of heavy metals. Several kinds of P_2-ATPases are responsible for maintaining gradients of ions such as Na^+, K^+, H^+, and Ca^{2+} across the plasma membrane of many eukaryotic cells. The best-known example is the Na^+/K^+ pump found in almost all animal cells. We will consider this ATPase in more detail shortly. Another example is the Ca^{2+}/H^+ ATPase involved in muscle contraction (see Problem 8-10 and Chapter 16). A third P_2-ATPase is the H^+/K^+ ATPase responsible for acidification of the gastric juice in your stomach. This protein is the target of medications known as *proton pump inhibitors* that are used to combat excess stomach acid. Similar to these are the P_3-H^+ ATPases of plants and fungi that pump protons outward across the plasma membrane of these cells, acidifying the external medium.

The P_4-ATPases differ from the preceding types in that they do not pump ions but instead pump relatively hydrophobic molecules such as cholesterol and fatty acids. Also, they do not transport materials completely across membranes but transport lipids from one leaflet of the membrane bilayer to the other, acting as flippases that help to maintain membrane asymmetry. P_5-ATPases are less well characterized but several are known to transport cations. They share sequence homology but not solute specificity. Some are found in the ER, where they appear to function in protein processing, and others are associated with the vacuole (yeast) or lysosome (animals) and have been implicated in hereditary neuronal diseases in humans.

V-type ATPases. **V-type ATPases** (V for "vacuole") pump protons into such organelles as vacuoles, vesicles, lysosomes, endosomes, and the Golgi complex. Typically, the proton gradient across the membranes of these organelles ranges from tenfold to over 10,000-fold. V-type pumps are not inhibited by vanadate because they do not undergo phosphorylation as part of the transport process. They have two multisubunit components: an integral component embedded within the membrane and a peripheral component that juts out from the membrane surface. The peripheral component contains the ATP-binding site and hence the ATPase activity.

F-type ATPases. **F-type ATPases** (F for "factor") are found in bacteria, mitochondria, and chloroplasts. F-type ATPases are involved in proton transport and have two components, both of them multisubunit complexes. The integral membrane component, called F_o, is a transmembrane pore for protons. The peripheral membrane component, called F_1, includes the ATP-binding site. F-type ATPases can use the energy of ATP hydrolysis to pump protons against their electrochemical potential.

These transport proteins can also facilitate the reverse process to synthesize ATP, as we will see when we study respiration and photosynthesis in Chapters 10 and 11, respectively. In the reverse direction, the exergonic flow of protons down their gradient is used to drive ATP synthesis. When they function in this latter mode, these F-type ATPases are more appropriately called **ATP synthases.** As ATP synthases, these proteins function to use either the energy from sugar oxidation (respiration) or the energy of solar radiation (photosynthesis) to produce a transmembrane proton gradient that drives ATP synthesis.

F-type ATPases illustrate an important principle: *Not only can ATP be used as an energy source to generate and maintain ion gradients, but such gradients can be used as an energy source to synthesize ATP.* This principle, which was discovered in studies of F-type pumps, is the basis of ATP-synthesizing mechanisms in all eukaryotic organisms and in most prokaryotes as well.

ABC-type ATPases. The fourth major class of ATP-driven pumps is the **ABC-type ATPases,** also called **ABC transporters.** The ABC designation is for "ATP-*b*inding *c*assette," where the term *cassette* describes the catalytic domain of the protein that binds ATP as an integral part of the transport process. The more than 150 known ABC-type ATPases comprise a very large family of transport proteins that are found in all organisms. They are related to each other in sequence and probably also in molecular mechanism. Most of the ABC-type ATPases discovered initially were from bacterial species and were *importers* that are involved in uptake of nutrients. But increasing numbers of ABC-type ATPases known as *exporters,* some of great clinical importance, are being reported in eukaryotes as well. As we will soon see, some have been found in human tumor cells, where, unfortunately, they can export antitumor medications. At last count, 48 different genes for ABC-type ATPases have been identified in the human genome.

The typical ABC-type ATPase has four protein domains. Two of these domains are highly hydrophobic and are embedded in the membrane, while the other two are peripheral and are associated with the cytoplasmic side of the membrane. Each of the two embedded domains consists of six membrane-spanning segments that form the channel through which solute molecules pass. The two peripheral domains are the cassettes that bind ATP and couple its hydrolysis to the transport process. These four domains are separate polypeptides in most cases, especially in bacterial cells. However, examples are also known in which the four domains are part of a large, multifunctional polypeptide.

Although the other three classes of ATPases transport only cations, the different ABC-type ATPases handle a remarkable variety of solutes. Most ABC transporters are specific for a particular solute or class of closely related solutes. But the variety of solutes transported by the many members of this superfamily is great, including ions, sugars, amino acids, and even peptides and polysaccharides.

ABC transporters are of considerable medical interest because some of them pump antibiotics or other drugs out of the cell, thereby making the cell resistant to the drug. For example, some human tumors are remarkably resistant to a variety of drugs that are normally quite effective at arresting tumor growth. Cells of such tumors have unusually high concentrations of a large protein called the **multidrug resistance (MDR) transport protein,** which was in fact the first ABC-type ATPase to be identified in humans. The MDR transport protein uses the energy of ATP hydrolysis to pump hydrophobic drugs out of cells, thereby reducing the cytoplasmic concentration of the drugs and hence their effectiveness as therapeutic agents. Unlike most ABC transporters, the MDR protein has a remarkably broad specificity: It can export a wide range of chemically dissimilar drugs commonly used in cancer chemotherapy, so that a cell with the MDR protein in its plasma membrane becomes resistant to a wide variety of therapeutic agents. Similarly, MDR proteins of some bacteria can transport a variety of antibiotics out of the cell, giving these cells resistance to multiple antibiotics.

Medical interest in this class of transport proteins was heightened when cystic fibrosis was shown to be caused by a genetic defect in a plasma membrane protein that is structurally related to the ABC transporters. We have long known that people with cystic fibrosis accumulate unusually thick mucus in their lungs—a condition that often leads to pneumonia and other lung disorders. Now we understand that the underlying problem is an inability to secrete chloride ions and that the genetic defect is in a protein that functions as a chloride ion channel.

The protein, called the *cystic fibrosis transmembrane conductance regulator (CFTR),* is similar in sequence and likely in topology to the core domains of ABC transporters. However, CFTR is an ion channel; and, unlike most of the ABC-type ATPases, it does not use ATP to drive transport. Instead, ATP hydrolysis appears to be involved in opening the channel. Recent developments in our understanding of cystic fibrosis at the molecular level and attempts at gene therapy to treat this disease are reported in Box 8B.

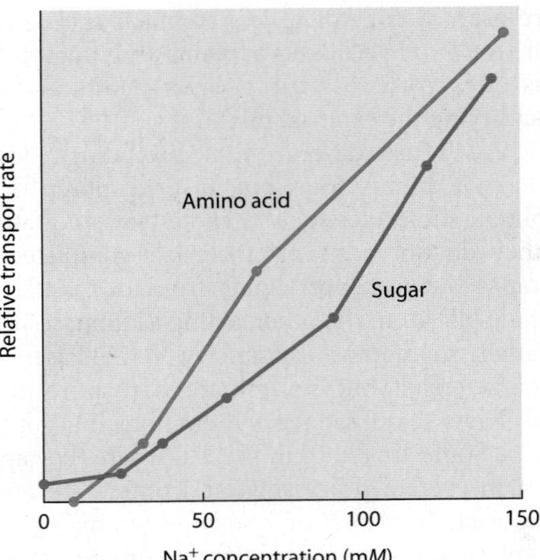

FIGURE 8-10 Effect of External Sodium Ion Concentration on Amino Acid and Sugar Transport. In this experiment, investigators varied the extracellular concentration of sodium ions and measured the transport rate of the amino acid glycine into erythrocytes or the sugar 7-deoxy-D-glucoheptose into intestinal lining cells. Studies of this type provided the first evidence that transport of amino acids and sugars into cells is stimulated by sodium ions present in the extracellular medium.

Indirect Active Transport Is Driven by Ion Gradients

In contrast to direct active transport, which is powered by energy released from a chemical reaction such as ATP hydrolysis, indirect active transport (also called *secondary active transport*) is driven by the movement of an ion down its electrochemical gradient. This principle has emerged from studies of the active uptake of sugars, amino acids, and other organic molecules into cells: The inward transport of molecules *up* their electrochemical gradients is often coupled to, and driven by, the simultaneous inward movement of either sodium ions (for animal cells) or protons (for most plant, fungal, and bacterial cells) *down* their respective electrochemical gradients.

The widespread existence of such symport mechanisms explains why most cells continuously pump either sodium ions or protons out of the cell. In animals, for example, the relatively high extracellular concentration of sodium ions maintained by the Na^+/K^+ pump serves as the driving force for the uptake of a variety of sugars and amino acids (**Figure 8-10**). The uptake of such compounds is regarded as indirect active transport because it is not directly driven by the hydrolysis of ATP or a related "high-energy" compound. Ultimately, however, uptake still depends on ATP because the Na^+/K^+ pump that maintains the sodium ion gradient is itself driven by ATP hydrolysis. The continuous outward pumping of Na^+ by the ATP-driven Na^+/K^+ pump and inward movement of Na^+ by symport (coupled to the uptake of another solute) establishes a circulation of sodium ions across the plasma membrane of every animal cell, similar to the cycling of H^+ ions shown in Figure 8-9.

While animal cells use sodium ions to drive indirect active transport, most other organisms rely on a proton gradient instead. For example, fungi and plants utilize proton symport for the uptake of organic solutes, with an ATP-driven proton pump responsible for the generation and maintenance of the proton electrochemical potential. Many kinds of bacterial cells also make extensive use of proton cotransport to drive the uptake of solute molecules, as do mitochondria. In all such cases, however, the proton gradient is maintained by an electron transfer process that accompanies cellular respiration, as we will see in Chapter 10.

In addition to the symport *uptake* of organic molecules such as sugars and amino acids, sodium ions or proton gradients can be used to drive the *export* of other ions, including Ca^{2+} and K^+. This type of indirect active transport is usually antiport and may involve the exchange of potassium ions for protons or the exchange of calcium ions for sodium ions, for example.

Examples of Active Transport

Having considered some general features of active transport, we are now ready to look at three specific examples: one example each of direct and indirect active transport from animal cells, and an unusual type of light-driven transport in a bacterium. In each case we will note what kinds of solutes are transported, what the driving force is, and how the energy source is coupled to the transport mechanism. We will focus first on the *Na^+/K^+ ATPase* (or *pump*) present in all animal cells because it is a well-understood example of direct active transport by a P-type ATPase. Then we will consider a second example from animal cells: the indirect active transport of glucose by a Na^+/*glucose symporter,* using the energy of the sodium ion gradient established by the Na^+/K^+ ATPase. Finally, we will look briefly at light-driven *proton transport* in certain bacteria.

Direct Active Transport: The Na^+/K^+ Pump Maintains Electrochemical Ion Gradients

A characteristic feature of most animal cells is a high intracellular level of potassium ions and a low intracellular level of sodium ions, such that in a typical animal cell the $[K^+]_{inside}/[K^+]_{outside}$ ratio is about 35:1 and the $[Na^+]_{inside}/[Na^+]_{outside}$ ratio is about 0.08:1. The resulting electrochemical potentials for potassium and sodium ions are essential as the driving force for coupled transport as well as for the transmission of nerve impulses. Both the inward pumping of potassium ions and the outward pumping of sodium ions are therefore energy-requiring processes, as both ions are being moved up their electrochemical gradients.

The protein responsible for this process was discovered in 1957 by Jens Skou, a Danish physiologist, who was awarded the Nobel Prize in chemistry in 1997. The discovery of this ion-transporting enzyme was in fact the first documented case of active transport. This **Na^+/K^+ ATPase,** or **Na^+/K^+ pump,** as this P-type ATPase is usually called, uses the exergonic hydrolysis of ATP to drive the endergonic inward transport of potassium ions and the outward transport of sodium ions against their concentration gradients. The Na^+/K^+ pump is primarily responsible for the asymmetric distribution of ions across the plasma membrane in animal cells. Like most other active transport systems, this pump has inherent directionality: Potassium ions are always pumped inward and sodium ions are always pumped outward. In fact, sodium and potassium ions activate the ATPase only on the side of the membrane they are transported from—sodium ions from the inside, potassium ions from the outside. Three sodium ions are moved out and two potassium ions are moved in per molecule of ATP hydrolyzed.

Figure 8-11 is a schematic illustration of the Na^+/K^+ pump. The pump is a tetrameric transmembrane protein with two α and two β subunits. The α subunits contain binding sites for sodium ions and ATP on the cytoplasmic side and for potassium ions on the external side of the membrane. We know that the β subunits are located outside the α subunits and are glycosylated, but their function is not yet clear.

The Na^+/K^+ pump is an allosteric protein exhibiting two alternative conformational states, referred to as E_1 and E_2. The E_1 conformation is open to the inside of the cell and has a high affinity for sodium ions, whereas E_2 is open to the outside, with a high affinity for potassium ions. Phosphorylation of the enzyme by ATP, a sodium-triggered event, stabilizes it in the E_2 form. Dephosphorylation, on

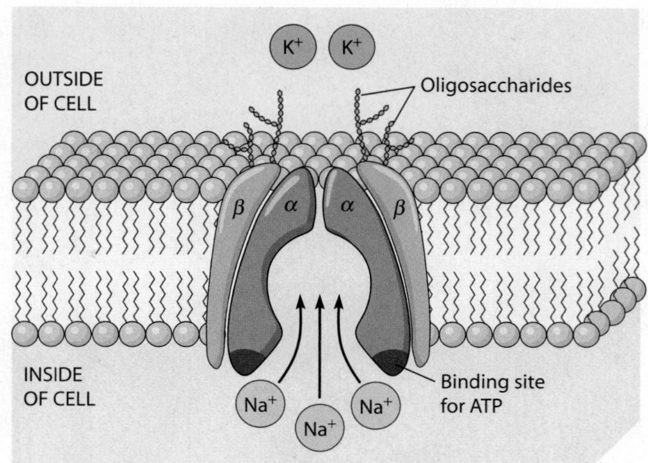

FIGURE 8-11 The Na^+/K^+ Pump. The Na^+/K^+ pump found in most animal cells consists of two α and two β subunits. The α subunits are transmembrane proteins, with binding sites for ATP on the cytoplasmic side. The β subunits are located on the outside of the α subunits and are glycosylated. The pump is shown in the E_1 conformation, which is open to the inside of the cell. Binding of sodium ions causes a conformational change to the E_2 form, which opens to the outside. The transport mechanism in this pump is shown in Figure 8-12.

the other hand, is triggered by K^+ and stabilizes the enzyme in the E_1 form.

As illustrated schematically in **Figure 8-12**, the actual transport mechanism involves an initial binding of three sodium ions to E_1 on the inner side of the membrane ❶, upper right. The binding of sodium ions triggers phosphorylation of the α subunit of the enzyme by ATP ❷, resulting in a conformational change from E_1 to E_2. As a result, the bound sodium ions are moved through the membrane to the external surface, where they are released to the outside ❸. Then, potassium ions from the outside bind to the α subunits ❹, triggering dephosphorylation and a return to the original conformation ❺. During this process, the potassium ions are moved to the inner surface, where they dissociate, leaving the carrier ready to accept more sodium ions ❻.

The Na^+/K^+ pump is not only one of the best-understood transport systems but also one of the most important for animal cells. Besides maintaining the appropriate intracellular concentrations of both potassium and sodium ions, it is responsible for maintaining the membrane potential that exists across the plasma membrane. The Na^+/K^+ pump assumes still more significance when we take into account the vital role that sodium ions play in the inward transport of organic substrates, a topic we now come to as we consider sodium symport.

Indirect Active Transport: Sodium Symport Drives the Uptake of Glucose

As an example of indirect active transport, let us consider the uptake of glucose by the **Na^+/glucose symporter.**

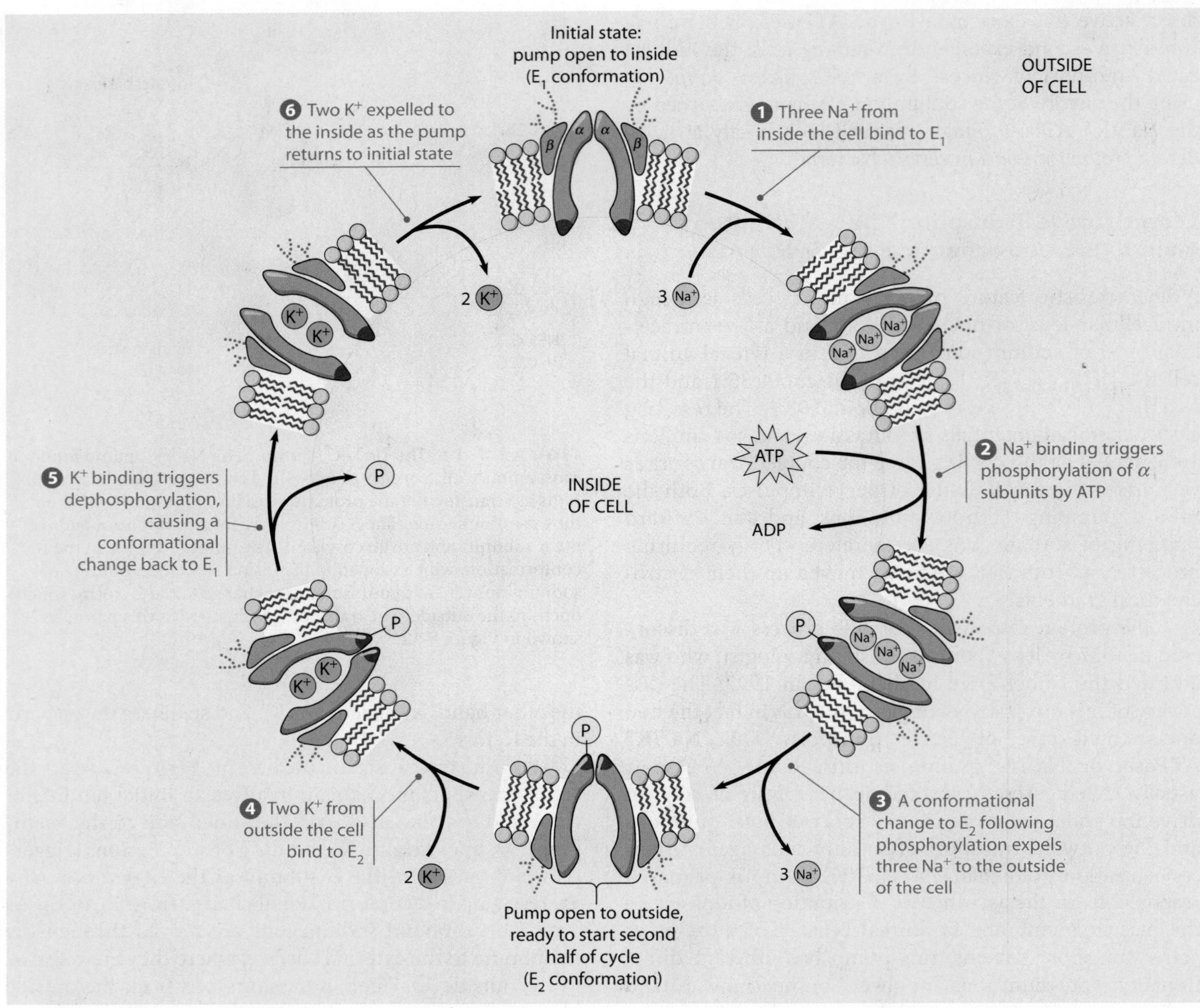
Initial state:
pump open to inside
(E_1 conformation)
OUTSIDE OF CELL
6 Two K+ expelled to the inside as the pump returns to initial state
1 Three Na+ from inside the cell bind to E_1
2 K+
3 Na+
INSIDE OF CELL
ATP
ADP
P
2 Na+ binding triggers phosphorylation of α subunits by ATP
5 K+ binding triggers dephosphorylation, causing a conformational change back to E_1
4 Two K+ from outside the cell bind to E_2
3 A conformational change to E_2 following phosphorylation expels three Na+ to the outside of the cell
Pump open to outside, ready to start second half of cycle
(E_2 conformation)

FIGURE 8-12 Model Mechanism for the Na^+/K^+ Pump. The transport process is shown here in six steps arranged at the periphery of a cell. The outward transport of sodium ions is coupled to the inward transport of potassium ions, both against their respective electrochemical potentials. The driving force is provided by ATP hydrolysis, which is required for phosphorylation of the α subunit of the pump at step ❷. E_1 and E_2 are the conformation states of the protein with the channel open to the inside (top of figure) and to the outside (bottom of figure) of the cell, respectively.

Although most glucose transport into or out of cells in your body occurs by facilitated diffusion, as shown in Figure 8-7, the epithelial cells lining your intestines have transport proteins that are able to take up glucose and certain amino acids from the intestines even when their concentrations there are much lower than in the epithelial cells. This energy-requiring process is driven by the simultaneous uptake of sodium ions, which is exergonic because of the steep electrochemical gradient for sodium that is maintained across the plasma membrane by the Na^+/K^+ pump (higher $[Na^+]$ outside the cell). These *sodium-dependent glucose transporters* are often referred to as *SGLT* proteins.

Figure 8-13 depicts the transport mechanism for the Na^+/glucose symporter, which requires the inward movement of two sodium ions to drive the simultaneous uptake of one glucose molecule. Transport is initiated by the binding of two external sodium ions to their binding sites on the symporter, which is open to the outer surface of the membrane ❶. This allows a molecule of glucose to bind ❷, followed by a conformational change in the protein that exposes the sodium ions and the glucose molecule to the inner surface of the membrane ❸. There the two sodium ions dissociate in response to the low intracellular sodium ion concentration ❹. This

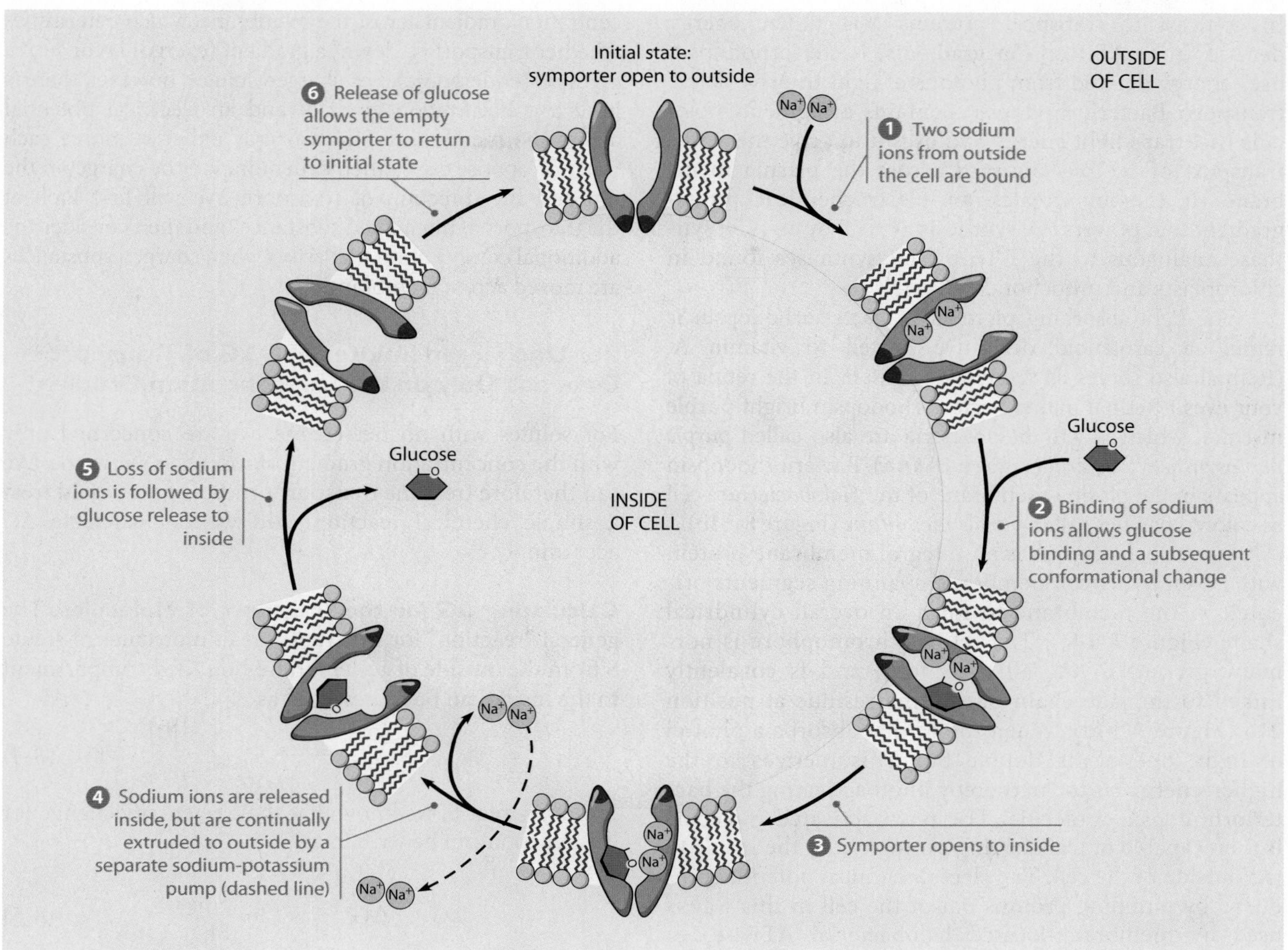

Initial state:
symporter open to outside
OUTSIDE
OF CELL
Na+ Na+
1 Two sodium
ions from outside
the cell are bound
6 Release of glucose
allows the empty
symporter to return
to initial state
Na+
Na+
Glucose
2 Binding of sodium
ions allows glucose
binding and a subsequent
conformational change
INSIDE
OF CELL
Glucose
5 Loss of sodium
ions is followed by
glucose release to
inside
Na+
Na+
Na+ Na+
4 Sodium ions are released
inside, but are continually
extruded to outside by a
separate sodium-potassium
pump (dashed line)
Na+ Na+
Na+
Na+
3 Symporter opens to inside

FIGURE 8-13 Model Mechanism for the Na^+/Glucose Symporter. The transport process is shown here in six steps arranged at the periphery of a cell. The inward transport of glucose against its concentration gradient is driven by the simultaneous inward transport of sodium ions down their electrochemical gradient. The sodium ion gradient is in turn maintained by the continuous outward transport of sodium ions (dashed arrow) by the Na^+/K^+ pump of Figure 8-12.

locks the transporter in its inward-facing conformation until the glucose molecule dissociates ❺. Then, the empty transporter is free to return to its outward-facing conformation ❻.

The sodium ion gradient is in turn maintained by the continuous outward extrusion of sodium ions (dashed arrow) by the Na^+/K^+ pump of Figure 8-12. As a result, sodium ions circulate across the plasma membrane, pumped outward by the Na^+/K^+ pump and flowing back into the cell as the driving force for sodium symport of molecules such as glucose. Similar mechanisms are involved in the uptake of amino acids and other organic substrates by sodium symport in animal cells and by proton symport in plant, fungal, and bacterial cells.

Knowledge of the Na^+/glucose cotransport mechanism has helped in the treatment of the disease cholera, caused by the bacterium *Vibrio cholera,* which produces a toxin that paralyzes intestinal cells and can lead to death via dehydration. Oral rehydration is the standard treatment for cholera, and it has been shown that rehydration with a salt and sugar solution is the most effective remedy. Administration of NaCl helps the body to retain water as it strives to maintain salt balance in the tissues, and providing glucose along with the salt therefore allows for more efficient uptake of salt via the Na^+/glucose symporter.

The Bacteriorhodopsin Proton Pump Uses Light Energy to Transport Protons

The final active transport system we will consider is the simplest. Nothing more is involved than a small integral membrane protein called **bacteriorhodopsin.** This protein, briefly introduced in Chapter 7, is a proton pump found in the plasma membrane of halophilic ("salt-loving") archaea belonging to the genus *Halobacterium.*

In contrast to transport proteins that utilize energy derived from ATP or ion gradients, bacteriorhodopsin uses energy derived from photons of light to drive active transport. Bacteriorhodopsin contains a pigment molecule that traps light energy and uses it to drive the active transport of protons outward across the plasma membrane. It thereby creates an electrochemical proton gradient that powers the synthesis of ATP by an ATP synthase analogous to the F-type ATP synthases found in chloroplasts and mitochondria.

The light-absorbing pigment of bacteriorhodopsin is *retinal,* a carotenoid derivative related to vitamin A. (Retinal also serves as the visual pigment in the retina of your eyes.) Retinal makes bacteriorhodopsin bright purple in color, which is why halobacteria are also called *purple photosynthetic bacteria* (**Figure 8-14a**). Bacteriorhodopsin appears in the plasma membrane of the *Halobacterium* cell as colored patches called *purple membrane* (Figure 8-14b).

Bacteriorhodopsin is an integral membrane protein with seven α-helical membrane-spanning segments oriented in the membrane to form an overall cylindrical shape (Figure 8-14c). The retinal chromophore is normally present in the all-*trans* form and is covalently linked to the side chain of a lysine residue at position 216 (Figure 8-14d). When the retinal absorbs a photon of light, one of its double bonds isomerizes to the higher-energy *cis* form, thereby photoactivating the bacteriorhodopsin molecule. The photoactivated molecule is then capable of transferring protons from the inside to the outside of the cell. The electrochemical potential produced by pumping protons out of the cell in this way is used by membrane-located halobacterial ATPases to synthesize ATP as the protons flow down their concentration gradient back into the cell. Recently, because it is a light-driven proton pump, the bacteriorhodopsin molecule is being researched as a conducting medium for use in *biomolecular electronics*. This is a relatively new field in which biological molecules with electron- and ion-conducting properties are being studied and used as structural elements of biooptical, biocomputing, and biosensor devices.

Energy-dependent proton pumping is one of the most basic concepts in cellular energetics. Proton pumping, which occurs in all bacteria, mitochondria, and chloroplasts, represents the driving force for all of life on Earth because it is an absolute requirement for the efficient synthesis of ATP. We will discuss the mechanisms underlying the generation of proton gradients and the use of the energy of such gradients in more detail when we discuss respiration and photosynthesis in Chapters 10 and 11.

The Energetics of Transport

Every transport event in the cell is an energy transaction. Energy either is released as transport occurs or is required to drive transport. To understand the energetics of transport, we must recognize that two different factors may be involved. For uncharged solutes, the only variable is the concentration gradient across the membrane, which determines whether transport is "down" a gradient (exergonic) or "up" a gradient (endergonic). For charged solutes, however, there is both a concentration gradient and an electrical potential across the membrane. The two may either reinforce each other or oppose each other, depending on the charge on the ion and the direction of transport. We will first look at the transport of uncharged substances and then consider the additional complication that arises when charged substances are moved across membranes.

For Uncharged Solutes, the ΔG of Transport Depends Only on the Concentration Gradient

For solutes with no net charge, we are concerned only with the concentration gradient across the membrane. We can therefore treat the transport process as we would treat a simple chemical reaction, and we can calculate ΔG accordingly.

Calculating ΔG for the Transport of Molecules. The general "reaction" for the transport of molecules of solute S from the outside of a membrane-bounded compartment to the inside can be represented as

$$\mathrm{S_{outside}} \longrightarrow \mathrm{S_{inside}} \tag{8-4}$$

From Chapter 5, we know that the free energy change for this reaction can be written as

$$\Delta G = \Delta G^\circ + RT\ln\frac{[\mathrm{S}]_{\mathrm{inside}}}{[\mathrm{S}]_{\mathrm{outside}}} \tag{8-5}$$

where ΔG is the free energy change, ΔG° is the standard free energy change, R is the gas constant (1.987 cal/mol·K), T is the absolute temperature, and $[\mathrm{S}]_{\mathrm{inside}}$ and $[\mathrm{S}]_{\mathrm{outside}}$ are the prevailing concentrations of S on the inside and outside of the membrane, respectively. However, the equilibrium constant K_{eq} for the transport of an uncharged solute is always 1 because at equilibrium the solute concentrations on the two sides of the membrane will be the same:

$$K_{eq} = \frac{[\mathrm{S}]_{\mathrm{inside}}}{[\mathrm{S}]_{\mathrm{outside}}} = 1.0 \tag{8-6}$$

This means that ΔG° is always zero:

$$\Delta G^\circ = -RT\ln K_{eq} = -RT\ln 1 = 0 \tag{8-7}$$

So the expression for ΔG of inward transport of an uncharged solute simplifies to

$$\Delta G_{\mathrm{inward}} = +RT\ln\frac{[\mathrm{S}]_{\mathrm{inside}}}{[\mathrm{S}]_{\mathrm{outside}}} \tag{8-8}$$

Notice that if $[\mathrm{S}]_{\mathrm{inside}}$ is less than $[\mathrm{S}]_{\mathrm{outside}}$, ΔG is negative, indicating that the inward transport of substance S is exergonic. It can therefore occur spontaneously, as would be

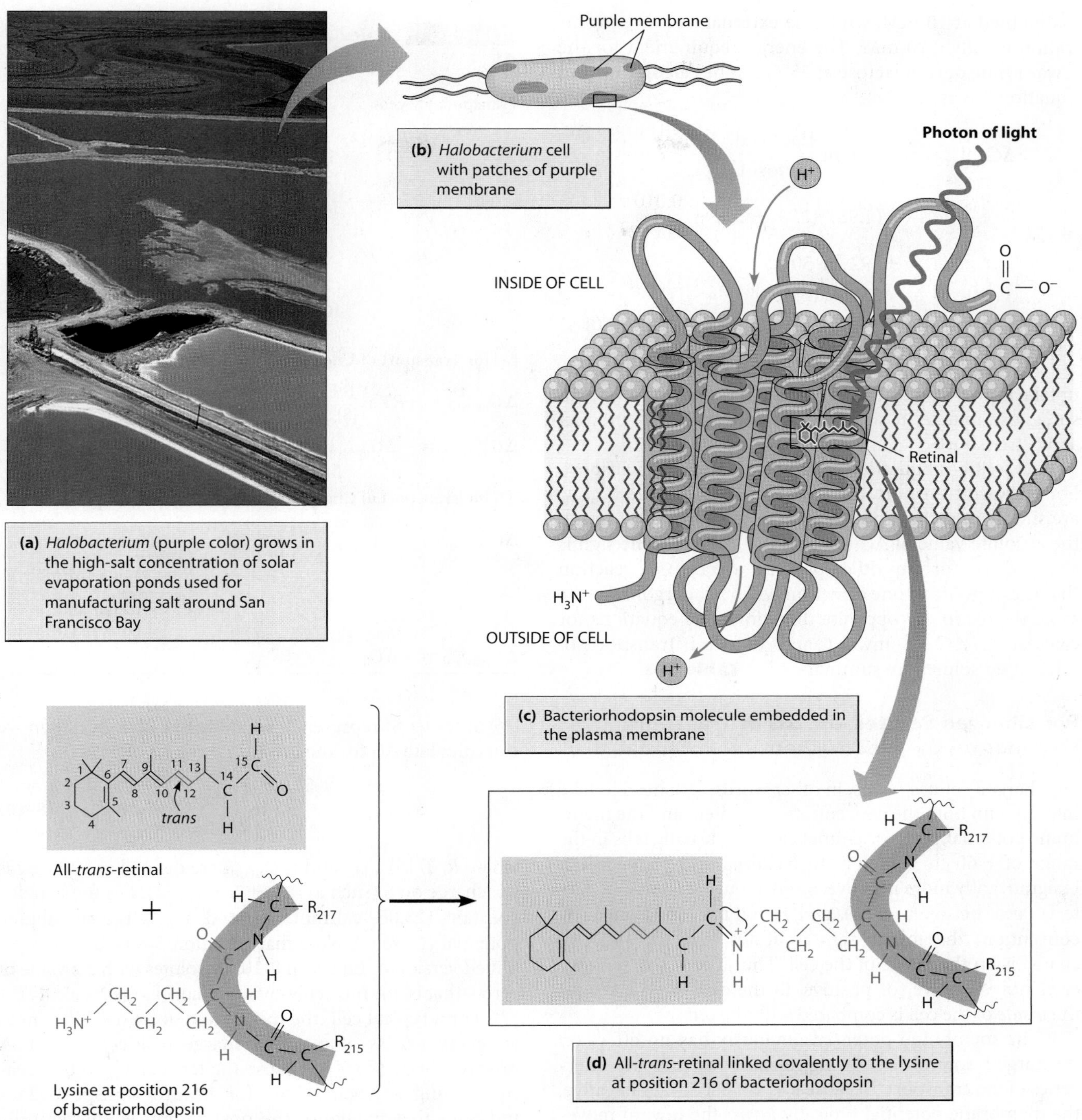

FIGURE 8-14 The Bacteriorhodopsin Proton Pump of Halobacteria. (a) Archaea belonging to the genus *Halobacterium* are characterized by a purple color that is due to the protein bacteriorhodopsin. (b) Bacteriorhodopsin is a light-activated proton pump that is present in the plasma membrane of *Halobacterium* cells as bright purple patches known as purple membrane. (c) The seven α-helical transmembrane segments of bacteriorhodopsin are separated by short, nonhelical segments and are oriented in the membrane to form an overall cylindrical shape. (d) The chromophore, all-*trans*-retinal, is linked to the lysine at position 216 in the seventh transmembrane segment of the protein.

expected for facilitated diffusion down a concentration gradient. But if $[S]_{inside}$ is greater than $[S]_{outside}$, inward transport of S is against the concentration gradient, and ΔG will be positive. In this case, inward transport of S is endergonic, and the amount of energy required to drive the active transport of S into the cell is indicated by the magnitude of the positive value of ΔG.

An Example: The Uptake of Lactose. Suppose that the concentration of lactose within a bacterial cell is to be maintained at 10 m*M*, while the external lactose concentration is only 0.20 m*M*. The energy requirement for the inward transport of lactose at 25°C can be calculated from Equation 8-8 as

$$\begin{aligned}\Delta G_{\text{inward}} &= +RT \ln \frac{[\text{lactose}]_{\text{inside}}}{[\text{lactose}]_{\text{outside}}} \\ &= +(1.987)(273 + 25) \ln \frac{0.010}{0.0002} \\ &= +592 \ln 50 \\ &= +2316 \text{ cal/mol} \\ &= +2.32 \text{ kcal/mol}\end{aligned} \qquad \textbf{(8-9)}$$

In many bacterial cells, the energy to drive lactose uptake is provided by the electrochemical proton gradient, so lactose uptake in such cells is an example of indirect active transport.

As written, Equation 8-8 applies to inward transport. For outward transport, the positions of S_{inside} and $S_{outside}$ are simply interchanged within the logarithm. As a result, the absolute value of ΔG remains the same, but the sign is changed. As for any other process, a transport reaction that is exergonic in one direction will be endergonic to the same degree in the opposite direction. The equations for calculating ΔG of inward and outward transport of uncharged solutes are summarized in **Table 8-3**.

For Charged Solutes, the ΔG of Transport Depends on the Electrochemical Potential

For charged solutes—ions, in other words—we need to take into account both the concentration gradient and the membrane potential, V_m. For animal cells, V_m usually falls in the range of −60 to −90 mV. In bacterial and plant cells it is significantly more negative, often about −150 mV in bacteria and between −200 and −300 mV in plants. By convention, the minus sign indicates that the negative charge is on the inside of the cell. Thus, the V_m value indicates how negative (or positive, in the case of a plus sign) the *inside* of the cell is compared with the *outside*.

The membrane potential obviously has no effect on uncharged solutes, but it significantly affects the energetics of ion transport. Because it is almost always negative, the membrane potential typically *favors* the inward movement of cations and *opposes* their outward movement. As we mentioned earlier, the net effect of both the concentration gradient and the potential gradient for an ion is called the *electrochemical potential* for that ion.

Calculating ΔG for the Transport of Ions. Both components of the electrochemical potential must be considered when determining the energetics of ion transport. To calculate ΔG for the transport of ions therefore requires an equation with two terms, one to express the effect of the concentration gradient across the membrane and the other to take into account the membrane potential.

Table 8-3 Calculation of ΔG for the Transport of Charged and Uncharged Solutes

Transport Process

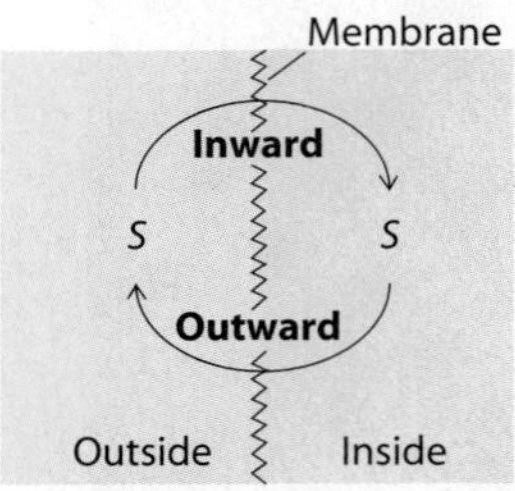

ΔG for Transport of Uncharged Solutes:

$$\Delta G_{\text{inward}} = +RT \ln \frac{[\text{S}]_{\text{inside}}}{[\text{S}]_{\text{outside}}}$$

$$\Delta G_{\text{outward}} = -\Delta G_{\text{inward}}$$

$R = 1.987$ cal/mol · K

$T = K = °C + 273$

ΔG for Transport of Charged Solutes:

$$\Delta G_{\text{inward}} = +RT \ln \frac{[\text{S}]_{\text{inside}}}{[\text{S}]_{\text{outside}}} + zFV_m$$

$$\Delta G_{\text{outward}} = -\Delta G_{\text{inward}}$$

z = charge on ion

$F = 23{,}062$ cal/mol · V

V_m = membrane potential (in volts)

If we let S^z represent a solute with a charge z, then we can calculate ΔG for the inward transport of S^z as

$$\Delta G_{\text{inward}} = +RT \ln \frac{[\text{S}]_{\text{inside}}}{[\text{S}]_{\text{outside}}} + zFV_m \qquad \textbf{(8-10)}$$

where *R*, *T*, $[\text{S}]_{\text{inside}}$, and $[\text{S}]_{\text{outside}}$ are defined as before; z is the charge on S (such as +1, +2, −1, −2); F is the Faraday constant (23,062 cal/mol · V); and V_m is the membrane potential (in volts). Note that Equation 8-8 is just the simplified version of Equation 8-10 for solutes with a z value of zero—that is, for molecules with no net charge.

For a typical cell, the membrane potential V_m is negative (an excess of negative charge inside the cell). A positive ion ($z > 0$) will cause the term zFV_m to be negative, giving a negative ΔG for inward transport. This indicates that uptake of the positive ion is energetically favorable, as you would expect if the interior of the cell has a net negative charge. The uptake of positive ions becomes more exergonic and more favorable as the membrane potential of the cell decreases. The opposite is true for a negative ion ($z < 0$), which gives a positive value for zFV_m and a positive change in free energy, indicating an unfavorable process that will not occur spontaneously.

For the outward transport of S, ΔG has the same value as for inward transport but is opposite in sign, so we can write

$$\Delta G_{\text{outward}} = -\Delta G_{\text{inward}} \qquad \textbf{(8-11)}$$

An Example: The Uptake of Chloride Ions. To illustrate the use of Equation 8-10 and to point out that intuition does not always serve us well in predicting the direction of ion transport—consider what happens when nerve cells with an intracellular chloride ion concentration of 50 m*M* are placed in a solution containing 100 m*M* Cl^-. Since the Cl^- concentration is twice as high outside the cell as inside, you might expect chloride ions to diffuse passively into the cell without the need for active transport. However, this expectation ignores the membrane potential of about −60 mV (−0.06 V) that exists across the plasma membrane of nerve cells. The minus sign reminds us that the inside of the cell is negative with respect to the outside, which means that the inward movement of negatively charged anions such as Cl^- will be *against* the membrane potential. Thus, the inward movement of chloride ions is *down* the concentration gradient but *up* the charge gradient.

To quantify the relative magnitudes of these two opposing forces at 25°C, we can use Equation 8-10 with the relevant values inserted:

$$\begin{aligned} \Delta G_{\text{inward}} &= +RT \ln \frac{[\text{S}]_{\text{inside}}}{[\text{S}]_{\text{outside}}} + zFV_{\text{m}} \\ &= +(1.987)(273 + 25) \ln \left(\frac{0.05}{0.10}\right) \\ &\quad +(-1)(23{,}062)(-0.06) \\ &= +592 \ln (0.5) + (23{,}062)(0.06) \\ &= -410 + 1384 \\ &= 974 \text{ cal/mol} \\ &= +0.97 \text{ kcal/mol} \end{aligned} \tag{8-12}$$

The sign of ΔG is positive, meaning that even though the chloride ion concentration is twice as high outside the cell as inside, energy will still be required to drive the movement of chloride ions into the cell. This is due to the excess of negative charge inside the cell, represented by the negative value of V_m. The movement of a mole of chloride ions up the charge gradient represented by the membrane potential requires more energy (+1384 calories) than is released by the movement of the mole of chloride ions down their concentration gradient (−410 calories). To guide you in such calculations, the equations for inward and outward transport of charged solutes are included along with those for the transport of uncharged solutes in Table 8-3, which summarizes the thermodynamic properties of each of these processes.

In this chapter, we have focused on the movement of ions and small molecules into and out of cells and organelles and have seen that these solutes pass directly through the membrane, though often with the aid of specific transport proteins. In addition to such traffic, however, many cells are able to take up and release substances that are too large to pass through a membrane regardless of its permeability properties. *Exocytosis* is the process whereby cells release proteins that are synthesized within the cell and sequestered within membrane-bounded vesicles, while *endocytosis* involves the uptake of macromolecules and other substances by trapping and engulfing them within an invagination of the plasma membrane. Both of these processes are unique to eukaryotic cells. We will consider exocytosis and endocytosis in detail when we get to Chapter 12.

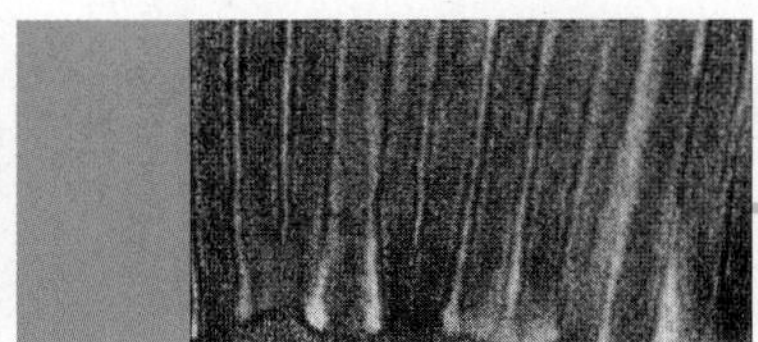

SUMMARY OF KEY POINTS

Cells and Transport Processes

- The selective transport of molecules and ions across membrane barriers ensures that necessary substances are moved into and out of cells and cell compartments at the appropriate time and at useful rates.
- Nonpolar molecules and small, polar molecules cross the membrane by simple diffusion. Transport of all other solutes, including ions and many molecules of biological relevance, is mediated by specific transport proteins that provide passage through an otherwise impermeable membrane.
- Each such transport protein has at least one, and frequently several, hydrophobic membrane-spanning sequences that embed the protein within the membrane and often act as the channel itself. Typically, a separate regulatory domain controls channel opening and closing.

Simple Diffusion: Unassisted Movement Down the Gradient

- Simple diffusion through biological membranes is limited to small or nonpolar molecules such as O_2, CO_2, and lipids. Water molecules, although polar, are small enough to diffuse across membranes in a manner that is not entirely understood.
- Membranes are permeable to lipids, which can pass through the nonpolar interior of the lipid bilayer. Membrane permeability of most compounds is directly proportional to their partition coefficient—their relative solubility in oil versus water.
- The direction of diffusion of a solute across a membrane is determined by its concentration gradient and always moves toward equilibrium. The solute will diffuse down the gradient from a region of high concentration to a region of low concentration.

- If the membrane is impermeable to the solute, water will move by osmosis from the area of low solute concentration (higher [H_2O]) to the area of high solute concentration (lower [H_2O]).

Facilitated Diffusion: Protein-Mediated Movement Down the Gradient

- Transport can either be downhill or uphill in relation to an uncharged solute's concentration gradient. For an ion, we must consider its electrochemical potential—the combined effect of the ion's concentration gradient and the charge gradient across the membrane.
- Downhill transport of large, polar molecules and ions, called facilitated diffusion, must be mediated by carrier proteins and channel proteins because these molecules and ions cannot diffuse through the membrane directly.
- Carrier proteins function by alternating between two conformational states. Examples include the glucose transporter and the anion exchange protein found in the plasma membrane of the erythrocyte.
- Transport of a single kind of molecule or ion is called uniport. The coupled transport of two or more molecules or ions at a time may involve movement of both solutes in the same direction (symport) or in opposite directions (antiport).
- Channel proteins facilitate diffusion by forming transmembrane channels lined with hydrophilic amino acids. Three important categories of channel proteins are ion channels (used mainly for transport of H^+, Na^+, K^+, Ca^{2+}, Cl^-, and HCO_3^-), porins (for various high-molecular-weight solutes), and aquaporins (for water).

Active Transport: Protein-Mediated Movement Up the Gradient

- Active transport—the uphill transport of large, polar molecules and ions—requires a protein transporter and an input of energy. It may be powered by ATP hydrolysis, the electrochemical potential of an ion gradient, or light energy.
- Active transport powered by ATP hydrolysis utilizes four major classes of transport proteins: P-type, V-type, F-type, and ABC-type ATPases. One widely encountered example is the ATP-powered Na^+/K^+ pump (a P-type ATPase), which maintains electrochemical potentials for sodium and potassium ions across the plasma membrane of animal cells.
- Active transport driven by an electrochemical potential usually depends on a gradient of either sodium ions (animal cells) or protons (plant, fungal, and many bacterial cells). For example, the inward transport of nutrients across the plasma membrane is often driven by the symport of sodium ions that were pumped outward by the Na^+/K^+ pump. As they flow back into the cell, they drive inward transport of sugars, amino acids, and other organic molecules.
- In *Halobacterium,* active transport is powered by light energy. As photons of light are absorbed by bacteriorhodopsin, protons are pumped across the cell membrane and out of the cell. As the protons flow back into the cell, ATP is synthesized.

The Energetics of Transport

- The ΔG for transport can be readily calculated. If $\Delta G < 0$, transport will be spontaneous. If $\Delta G > 0$, an input of energy will be required to drive transport. If $\Delta G = 0$, there will be no net movement of the solute.
- For uncharged solutes, ΔG depends only on the concentration gradient. For charged solutes, both the concentration gradient and the membrane potential must be taken into account.

PROBLEM SET

More challenging problems are marked with a •.

8-1 True or False? Indicate whether each of the following statements about membrane transport is true (T) or false (F). If false, reword the statement to make it true.

(a) Facilitated diffusion of glucose occurs rapidly because the concentration gradient is maintained by packaging intracellular glucose into vesicles.

(b) The exergonic movement of an ion coupled with the movement of a solute down a concentration gradient is an example of primary active transport.

(c) The K_{eq} value for the diffusion of polar molecules out of the cell is less than one because membranes are essentially impermeable to such molecules.

(d) Aquaporins facilitate the rapid movement of water molecules into or out of cells.

(e) Oxygen can move freely across the plasma membrane by simple diffusion.

(f) In simple diffusion, the net rate of transport for a specific substance is indirectly proportional to the concentration difference for that substance across the membrane.

(g) ABC transporters are of medical interest because they are known to be involved in drug resistance.

(h) Transport channel proteins have a high level of specificity for a solute.

8-2 Telling Them Apart. From the following list of properties, indicate which one(s) can be used to distinguish between each of the following pairs of transport mechanisms.

Transport Mechanisms

(a) Simple diffusion; facilitated diffusion

(b) Facilitated diffusion; active transport

(c) Simple diffusion; active transport

(d) Direct active transport; indirect active transport

(e) Symport; antiport

(f) Uniport; coupled transport

(g) P-type ATPase; V-type ATPase

Properties

1. Directions in which two transported solutes move
2. Direction the solute moves relative to its concentration gradient or its electrochemical potential
3. Kinetics of solute transport
4. Requirement for metabolic energy
5. Requirement for simultaneous transport of two solutes
6. Intrinsic directionality
7. Competitive inhibition
8. Sensitivity to the inhibitor vanadate

8-3 Mechanisms of Transport. For each of the following statements, answer with a D if the statement is true of simple diffusion, with an F if it is true of facilitated diffusion, and with an A if it is true of active transport. Any, all, or none (N) of the choices may be appropriate for a given statement.

(a) Requires the presence of transport proteins.

(b) Solutes move down their free energy gradient in the direction of thermodynamic equilibrium.

(c) Is not subject to saturation.

(d) Requires the hydrolysis of ATP.

(e) Is a way of establishing a difference in the concentration gradient of solutes across a membrane.

(f) Is limited to small, nonpolar solutes.

(g) Applies only to ions.

(h) Transport can occur in either direction across the membrane, depending on the prevailing concentration gradient.

(i) Has a positive ΔG.

(j) Usually has intrinsic directionality.

8-4 Discounting the Transverse Carrier Model. At one time, membrane biologists thought that transport proteins might act by binding a solute molecule or ion on one side of the membrane and then diffusing across the membrane to release the solute molecule on the other side. We now know that this transverse carrier model is almost certainly wrong. Suggest two reasons that argue against such a model. One of your reasons should be based on our current understanding of membrane structure and the other on thermodynamic considerations.

8-5 Potassium Ion Transport. Most of the cells of your body pump potassium ions inward to maintain an internal K^+ concentration that is 35 times the external concentration.

(a) What is ΔG (at 37°C) for the transport of potassium ions into a cell that maintains no membrane potential across its plasma membrane?

(b) For a nerve cell with a membrane potential of −60 mV, what is ΔG for the inward transport of potassium ions at 37°C?

(c) For the nerve cell in part b, what is the maximum number of potassium ions that can be pumped inward by the hydrolysis of one ATP molecule if the ATP/ADP ratio in the cell is 5:1 and the inorganic phosphate concentration is 10 m*M*? (Assume $\Delta G^{\circ\prime} = -7.3$ kcal/mol for ATP hydrolysis.)

8-6 Ion Gradients and ATP Synthesis. The ion gradients maintained across the plasma membranes of most cells play a significant role in cellular energetics. Ion gradients are often either generated by the hydrolysis of ATP or used to make ATP by the phosphorylation of ADP.

(a) Cite an example in which ATP is used to generate and maintain an ion gradient. What is another way that an ion gradient can be generated and maintained?

(b) Cite an example in which an ion gradient is used to make ATP. What is another use for ion gradients?

(c) Assume that the sodium ion concentration is 12 m*M* inside a cell and 145 m*M* outside the cell and that the membrane potential is −90 mV. Can a cell use ATP hydrolysis to drive the outward transport of sodium ions on a 2:1 basis (two sodium ions transported per ATP hydrolyzed) if the

ATP/ADP ratio is 5:1, the inorganic phosphate concentration is 50 m*M*, and the temperature is 37°C? What about on a 3:1 basis? Explain your answers.

(d) Assume that a bacterial cell maintains a proton gradient across its plasma membrane such that the pH inside the cell is 8.0 when the outside pH is 7.0. Can the cell use the proton gradient to drive ATP synthesis on a 1:1 basis (one ATP synthesized per proton transported) if the membrane potential is +180 mV, the temperature is 25°C, and the ATP, ADP, and inorganic phosphate concentrations are as in part c? What about on a 1:2 basis? Explain your answers.

8-7 Sodium Ion Transport. A marine protozoan is known to pump sodium ions outward by a simple ATP-driven Na^+ pump that operates independently of potassium ions. The intracellular concentrations of ATP, ADP, and P_i are 20, 2, and 1 m*M*, respectively, and the membrane potential is −75 mV.

(a) Assuming that the pump transports three sodium ions outward per molecule of ATP hydrolyzed, what is the lowest internal sodium ion concentration that can be maintained at 25°C when the external sodium ion concentration is 150 m*M*?

(b) If you were dealing with an uncharged molecule rather than an ion, would your answer for part a be higher or lower, assuming all other conditions remained the same? Explain.

8-8 The Case of the Acid Stomach. The gastric juice in your stomach has a pH of 2.0. This acidity is due to the secretion of protons into the stomach by the epithelial cells of the gastric mucosa. Epithelial cells have an internal pH of 7.0 and a membrane potential of −70 mV (inside is negative) and function at body temperature (37°C).

(a) What is the concentration gradient of protons across the epithelial membrane?

(b) Do you think that proton transport can be driven by ATP hydrolysis at the ratio of one molecule of ATP per proton transported?

(c) If protons were free to move back into the cell, calculate the membrane potential that would be required to prevent them from doing so.

(d) Calculate the free energy change associated with the secretion of 1 mole of protons into gastric juice at 37°C.

8-9 Charged or Not: Does It Make a Difference? Many solutes that must move into and out of cells exist in either a protonated or ionized form, or they have functional groups that can be either protonated or ionized. Simple molecules such as CO_2, H_3PO_4 (phosphoric acid), and NH_3 (ammonia) are in this category, as are organic molecules with carboxylic acid groups, phosphate groups, and/or amino groups.

(a) Consider ammonia as a simple example of such compounds. What is the charged form of ammonia called? What is its chemical formula?

(b) Which of these two forms will predominate in a solution with a highly acidic pH? Explain your answer.

(c) For which of these two forms will the uptake across the plasma membrane of a cell be affected by the concentration gradient of that form on the inside versus the outside of the cell? For which, if either, of these two forms will uptake be affected by the membrane potential of the plasma membrane? Explain your answers.

(d) For a cell that must take up ammonia from its environment, will uptake of the charged form require more or less energy than uptake of the uncharged form, assuming that the cell has a negative membrane potential? Explain your answer.

(e) Instead of ammonia, consider the uptake of acetic acid, CH_3COOH, an important intermediate in several biological pathways. What is the charged form in this case? For which of the two forms will the uptake across the plasma membrane of a cell be affected by the concentration gradient of that form on the inside versus the outside of the cell? For which, if either, of these two forms will uptake be affected by the membrane potential of the plasma membrane? Explain your answers.

(f) For a cell that must take up acetic acid from its environment, will uptake of the charged form require more or less energy than uptake of the uncharged form, assuming that the cell has a negative membrane potential? Explain your answer.

• **8-10 The Calcium Pump of the Sarcoplasmic Reticulum.** Muscle cells use calcium ions to regulate the contractile process. Calcium is both released and taken up by the sarcoplasmic reticulum (SR). Release of calcium from the SR activates muscle contraction, and ATP-driven calcium uptake causes the muscle cell to relax afterward. When muscle tissue is disrupted by homogenization, the SR forms small vesicles called *microsomes* that maintain their ability to take up calcium. In the experiment shown in **Figure 8-15**, a reaction medium was prepared to contain 5 m*M* ATP and 0.1 *M* KCl at pH 7.5. An aliquot of SR microsomes containing 1.0 mg protein was added to 1 mL of the reaction mixture, followed by 0.4 mmol of calcium. Two minutes later, a calcium ionophore was added. (An ionophore is a substance that facilitates the movement of an ion across a membrane.) ATPase activity was monitored during the additions, with the results shown in the figure.

(a) What is the ATPase activity, calculated as micromoles of ATP hydrolyzed per milligram of protein per minute?

(b) The ATPase is calcium-activated, as shown by the increase in ATP hydrolysis when the calcium was added and the decrease in hydrolysis when all the added calcium was taken up into the vesicles 1 minute after it was added. How many calcium ions are taken up for each ATP hydrolyzed?

(c) The final addition is an ionophore that carries calcium ions across membranes. Why does ATP hydrolysis begin again?

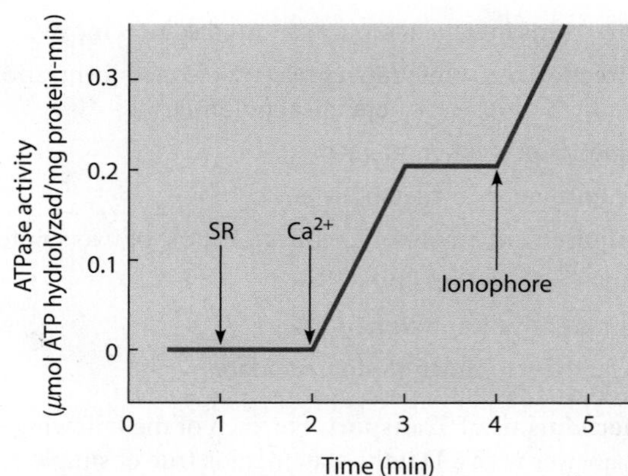

FIGURE 8-15 Calcium Uptake by the Sarcoplasmic Reticulum. See Problem 8-10.

• **8-11 Inverted Vesicles.** An important advance in transport research was the development of methods for making closed membrane vesicles that retain the activity of certain transport systems. One such system uses resealed vesicles from red blood cell membranes, in which the orientation of membrane proteins may be either the same as in the intact cell (right side out) or inverted (inside out). Such vesicles have demonstrable ATP-driven Na^+/K^+ pump activity. By resealing the vesicles in one medium and then placing them in another, it is possible to have ATP, sodium ions, and potassium ions present inside the vesicle, present outside the vesicle, or not present at all.

(a) Suggest one or two advantages that such vesicles might have compared to intact red blood cells for studying the Na^+/K^+ pump. Can you think of any possible disadvantages?

(b) For the inverted vesicles, indicate whether each of the following should be present inside the vesicle (I), outside the vesicle (O), or not present at all (N) in order to demonstrate ATP hydrolysis: Na^+, K^+, and ATP.

(c) If you were to plot the rate of ATP hydrolysis as a function of time after initiating transport in such inverted vesicles, what sort of a curve would you expect to obtain?

• **8-12 Ouabain Inhibition.** Ouabain is a specific inhibitor of the active transport of sodium ions out of the cell and is therefore a valuable tool in studies of membrane transport mechanisms. Which of the following processes in your own body would you expect to be sensitive to inhibition by ouabain? Explain your answer in each case.

(a) Active uptake of lactose by the bacteria in your intestine

(b) Facilitated diffusion of glucose into a muscle cell

(c) Active transport of dietary phenylalanine across the intestinal mucosa

(d) Uptake of potassium ions by red blood cells

SUGGESTED READING

References of historical importance are marked with a •.

General References

Baldwin, S. A., ed. *Membrane Transport: A Practical Approach.* New York: Oxford University Press, 2000.

Bernhardt, I. C., and J. C. Ellory, eds. *Membrane Transport in Red Blood Cells in Health and Disease.* Berlin: Springer-Verlag, 2003.

DeFelice, L. J., and T. Goswami. Transporters as channels. *Annu. Rev. Physiol.* 69 (2007): 87.

Hedrich, R., and I. Marten. 30-year progress of membrane transport in plants. *Planta* 224 (2006): 725.

Facilitated Diffusion

Jiang, Y. et al. The open pore conformation of potassium channels. *Nature* 417 (2002): 523.

Hoffert, J. D., C. L. Chou, and M. A. Knepper. Aquaporin-2 in the "-omics" era. *J. Biol. Chem.* 284 (2009): 14683.

King, L. S., D. Kozono, and P. Agre. From structure to disease: The evolving tale of aquaporin biology. *Nature Rev. Mol. Cell Biol.* 5 (2004): 687.

Ludewig, U., and M. Dynowski. Plant aquaporin selectivity: Where transport assays, computer simulations and physiology meet. *Cell Mol. Life Sci.* 66 (2009): 3161.

Maurel, C. et al. The cellular dynamics of plant aquaporin expression and functions. *Curr. Opin. Plant Biol.* 12 (2009): 690.

Verkman, A. S., and A. K. Mitra. Structure and function of aquaporin water channels. *Amer. J. Physiol.* 278 (2000): F13.

Verrey, F. et al. Novel renal amino acid transporters. *Annu. Rev. Physiol.* 67 (2005): 557.

Zeuthen, T., and W. D. Stein. *Molecular Mechanisms of Water Transport Across Biological Membranes.* San Diego: Academic Press, 2002.

Active Transport

Chang, G. Multidrug resistance ABC transporters. *FEBS Lett.* (2003): 102.

Efferth, T. Adenosine triphosphate-binding cassette transporter gene in ageing and age-related diseases. *Ageing Res. Rev.* 2 (2003): 11.

Folmer, D. E., R. P. Elferink, and C. C. Paulusma. P4 ATPases—Lipid flippases and their role in disease. *Biochim Biophys. Acta* 1791 (2009): 628.

Higgins, C. F., and K. J. Linton. The xyz of ABC transporters. *Science* 293 (2001): 1782.

Hirai, T., and S. Subramaniam. Structural insights into the mechanism of proton pumping by bacteriorhodopsin. *FEBS Lett.* 545 (2003): 2.

Huang, Y. Pharmacogenetics/genomics of membrane transporters in cancer chemotherapy. *Cancer Metastasis Rev.* 26 (2007): 183.

• Jardetzky, O. Simple allosteric model for membrane pumps. *Nature* 211 (1966): 969.

Jin, Y. et al. Bacteriorhodopsin as an electronic conduction medium for biomolecular electronics. *Chem. Soc. Rev.* 37 (2008): 2422.

Kawasaki-Nishi, S., T. Nishi, and M. Forgac. Proton translocation driven by ATP hydrolysis in V-ATPases. *FEBS Lett.* 545 (2003): 76.

Lanyi, J. K. Bacteriorhodopsin. *Annu. Rev. Physiol.* 66 (2004): 665.

Oldham, M. L., A. L. Davidson, and J. Chen. Structural insights into ABC transporter mechanism. *Curr. Opin. Struct. Biol.* 18 (2008): 726.

Pedersen, P. L. Transport ATPases into the year 2008: a brief overview related to types, structures, functions and roles in health and disease. *J. Bioenerg. Biomembr.* 39 (2007): 349.

Pusch, M. Cl^- channels: A journey from Ca^{2+} sensors to ATPases and secondary active ion transporters. *Annu. Rev. Physiol.* 67 (2005): 697.

Rea, P. Plant ATP-binding cassette transporters. *Annu. Rev. Plant Biol.* 58 (2007): 347.

Stevens, T., and M. Forgac. Structure, function, and regulation of the vacuolar (H^+)-ATPase. *Annu. Rev. Cell Biol.* 13 (1997): 779.

Wright, E. M., B. A. Hirayama, and D. F. Loo. Active sugar transport in health and disease. *J. Intern. Med.* 261 (2007): 32.

Yatime, L. et al. P-type ATPases as drug targets: tools for medicine and science. *Biochim Biophys. Acta* 1787 (2009): 207.

Box 8B: *Membrane Transport, Cystic Fibrosis, and the Prospects for Gene Therapy*

Alton, E. W. et al. Cationic lipid-mediated *CFTR* gene transfer to the lungs and nose of patients with cystic fibrosis: A double-blind placebo-controlled trial. *Lancet* 353 (1999): 947.

Hwang, T-C., and D. N. Sheppard. Gating of the CFTR Cl^- channel by ATP-driven nucleotide-binding domain dimerisation. *J Physiol* 587 (2009): 2151.

Rossnecker, J., S. Huth, and C. Rudolph. Gene therapy for cystic fibrosis lung disease: Current status and future perspectives. *Curr. Opin. Mol. Ther.* 8 (2006): 439.

Welsh, M. J., and A. E. Smith. Cystic fibrosis. *Sci. Amer.* 273 (December 1995): 52.

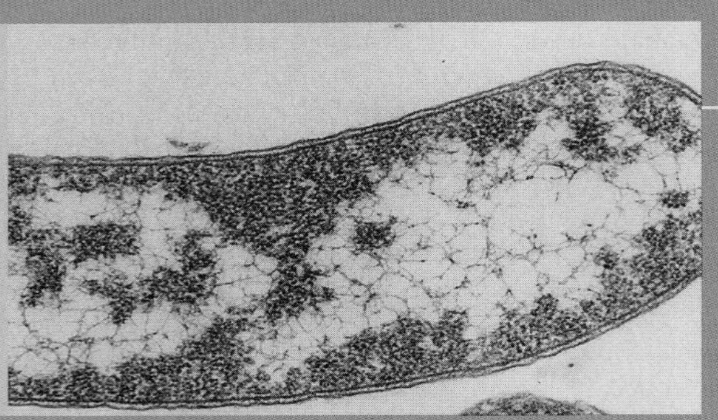

9 Chemotrophic Energy Metabolism: Glycolysis and Fermentation

See MasteringBiology® for tutorials, activities, and quizzes.

As we learned in earlier chapters, cells cannot survive without a source of energy and a source of chemical "building blocks"—the small molecules from which macromolecules such as proteins, nucleic acids, and polysaccharides are synthesized. In many organisms, including you and me, these two requirements are related. The desired energy and small molecules are both present in the food molecules that these organisms produce or ingest.

In this chapter and the next, we will consider how chemotrophs, *such as animals and most microorganisms, obtain energy from the food they engulf or ingest, focusing especially on the oxidative breakdown of sugar molecules. Remember that oxidation reactions involve the loss of electrons and hydrogens and release energy. Then, in Chapter 11, we will discuss the process by which* phototrophs, *such as green plants, algae, and some bacteria, tap the solar radiation that is the ultimate energy source for almost all living organisms. They will use this energy to reduce carbon dioxide (add electrons and hydrogens) in order to produce sugar molecules. Keep in mind that the reactions whereby cells obtain energy also can provide the various small molecules that cells need for synthesis of macromolecules and other cellular constituents.*

Metabolic Pathways

Chemistry Review-Enzymes & Pathways: Metabolic Pathways

When we encountered enzymes in Chapter 6, we considered individual chemical reactions catalyzed by individual enzymes functioning in isolation—but that is not the way cells really operate. To accomplish any major task, a cell requires a series of reactions occurring in an ordered sequence. This, in turn, requires many different enzymes because most enzymes catalyze only a single reaction, and many such reactions are usually needed to accomplish a major biochemical operation.

When we consider all the chemical reactions that occur within a cell, we are talking about **metabolism** (from the Greek word *metaballein,* meaning "to change"). The overall metabolism of a cell consists, in turn, of many specific **metabolic pathways,** each of which accomplishes a particular task. From a biochemist's perspective, *life at the cellular level can be defined as a network of integrated and carefully regulated metabolic pathways, each contributing to the sum of activities that a cell must carry out.*

Metabolic pathways are of two general types. Pathways that synthesize cellular components are called **anabolic pathways** (using the Greek prefix *ana–*, meaning "up"), whereas those involved in the breakdown of cellular constituents are called **catabolic pathways** (using the Greek prefix *kata–*, meaning "down"). Anabolic pathways usually involve a substantial increase in molecular order (and therefore a local decrease in entropy) and are *endergonic* (energy-requiring). Polymer synthesis and the biological reduction of carbon dioxide to sugar are examples of anabolic pathways. Often, anabolic pathways synthesize polymers such as starch and glycogen from glucose units in order to store energy for future use. Certain steroid hormones, for example, are called anabolic steroids because they stimulate the synthesis of muscle proteins from amino acids.

Catabolic pathways, by contrast, are degradative pathways that typically involve a decrease in molecular order (increase in entropy) and are *exergonic* (energy-liberating). These reactions often involve hydrolysis of macromolecules or biological oxidations. Catabolic pathways play two roles in cells: They release the free energy needed to drive cellular functions, and they give rise to the small organic molecules, or *metabolites,* that are the building blocks for biosynthesis. However, a catabolic pathway is not simply the reverse of the corresponding anabolic pathway. For example, the catabolic pathway for glucose degradation and the anabolic pathway for glucose synthesis use slightly different enzymes and intermediates.

As we will see shortly, catabolism can be carried out either in the presence or absence of oxygen (i.e., under either *aerobic* or *anaerobic* conditions). The energy yield per glucose molecule is much greater in the presence of oxygen. However, anaerobic catabolism is also important, not only for organisms in environments that are always devoid of oxygen but also for organisms and cells that are temporarily deprived of oxygen.

ATP: The Universal Energy Coupler

The anabolic reactions of cells are responsible for growth and repair processes, whereas catabolic reactions release the energy needed to drive the anabolic reactions and to carry out other kinds of cellular work. The efficient linking, or *coupling,* of energy-yielding processes to energy-requiring processes is therefore crucial to cell function. This coupling is made possible by specific kinds of molecules that conserve the energy derived from exergonic reactions and release it again when and where energy is needed.

In virtually all cells, the molecule most commonly used as an energy intermediate is the phosphorylated compound **adenosine triphosphate (ATP).** ATP is, in other words, the primary energy "currency" of the biological world. Keep in mind, however, that ATP synthesis is not the only way that cells store chemical energy. Other high-energy molecules, such as GTP and creatine phosphate, store chemical energy that can be converted to ATP. In addition, chemical energy is stored as *reduced coenzymes* such as NADH that are a source of reducing power in cells. These molecules are very important in shuttling energy between different metabolic pathways and processes in cells. Because ATP is involved in most cellular energy transactions, it is essential that we first understand its structure and function and appreciate the properties that make this molecule so suitable for its role as the universal energy coupler.

ATP Contains Two Energy-Rich Phosphoanhydride Bonds

MB Activity: Chemical Reactions and ATP

As we learned in Chapter 3, ATP is a complex molecule containing the aromatic base adenine, the five-carbon sugar ribose, and a chain of three phosphate groups. The phosphate groups are linked to each other by **phosphoanhydride bonds** and to the ribose by a **phosphoester bond,** as shown for the ATP molecule in **Figure 9-1**. The compound formed by linking adenine and ribose is called *adenosine.* Adenosine may occur in the cell in the unphosphorylated form or with one, two, or three phosphates attached to carbon atom 5 of the ribose, forming *adenosine monophosphate (AMP), diphosphate (ADP),* and *triphosphate (ATP),* respectively.

The ATP molecule serves well as an intermediate in cellular energy metabolism because energy is released when ATP undergoes *hydrolysis*—water is used to break the phosphoanhydride bond that links the third (outermost)

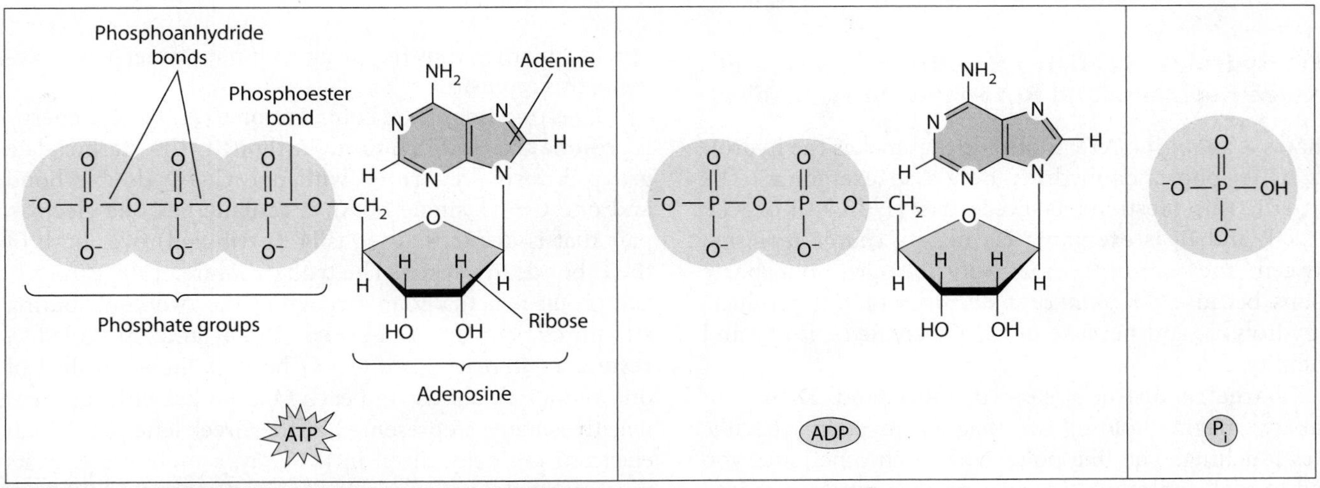

(a) Structures of ATP, ADP, and inorganic phosphate (at pH 7)

Reaction 1: Hydrolysis ($\Delta G^{\circ\prime} = -7.3$ kcal/mol)

$$ATP^{4-} + H_2O \rightleftharpoons ADP^{3-} + P_i^{2-} + H^+$$

Reaction 2: ATP synthesis ($\Delta G^{\circ\prime} = +7.3$ kcal/mol)

(b) Balanced chemical equation for ATP hydrolysis and synthesis

FIGURE 9-1 ATP Hydrolysis and Synthesis. **(a)** ATP consists of adenosine (adenine + ribose) plus three phosphate groups attached to carbon atom 5 of the ribose. **(b)** *Reaction 1:* ATP hydrolysis to ADP and inorganic phosphate (P_i) is highly exergonic, with a standard free energy change of −7.3 kcal/mol. *Reaction 2:* ATP synthesis by phosphorylation of ADP is highly endergonic, with a standard free energy change of +7.3 kcal/mol.

ATP and Energy

phosphate to the second. Two products are formed: The terminal phosphate receives an —OH from the water molecule and is released as inorganic phosphate (HPO_4^{2-}, often written as P_i), and the resulting ADP molecule gets an —H and immediately loses a proton by ionization. Thus, the hydrolysis of ATP to form ADP and P_i is highly exergonic, with a standard free energy change ($\Delta G^{\circ\prime}$) of −7.3 kcal/mol (Figure 9-1, Reaction 1). The reverse reaction, whereby ATP is synthesized from ADP and P_i with the loss of a water molecule by condensation, is correspondingly endergonic, with a $\Delta G^{\circ\prime}$ of +7.3 kcal/mol (Figure 9-1, Reaction 2). As you can see, energy is required to drive ATP synthesis from ADP and P_i, and energy is released upon ATP hydrolysis.

Biochemists sometimes refer to bonds such as the phosphoanhydride bonds of ATP as "high-energy" or "energy-rich" bonds, a very useful convention introduced in 1941 by Fritz Lipmann, a leading bioenergetics researcher of the time. However, these terms need to be understood correctly to avoid the erroneous impression that the bond somehow contains energy that can be released. All chemical bonds *require* energy to be broken and *release* energy when they form. What we really mean by "*energy-rich* bond" is that free energy is released when the bond is hydrolyzed. The energy is therefore a characteristic of the reaction the molecule is involved in and not of a particular bond within that molecule. Thus, to call ATP or any other molecule a "high-energy or energy-rich compound" should always be understood as a shorthand way of saying that the hydrolysis of one or more of its bonds is highly exergonic.

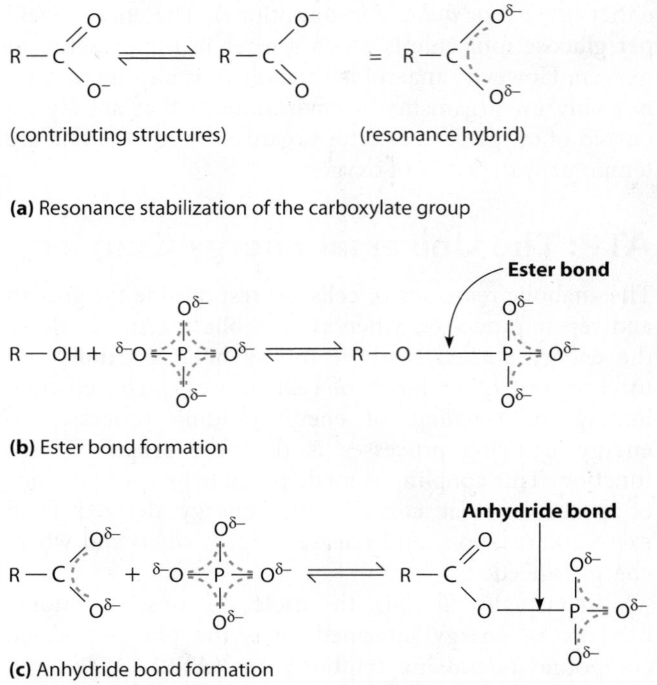

(a) Resonance stabilization of the carboxylate group

(b) Ester bond formation

(c) Anhydride bond formation

FIGURE 9-2 Decreased Resonance Stabilization of the Carboxyl and Phosphate Groups Following Bond Formation. Resonance stabilization due to electron delocalization (blue dashed lines) is an important feature of both the carboxylate group in **(a)** and the phosphate groups in **(b)** and **(c)**. Creation of either **(b)** an ester bond or **(c)** an anhydride bond (by removal of water) decreases the opportunity for electron delocalization. As a result, the ester or anhydride product is a higher-energy compound than the reactants, and energy will be released when the bond is broken by hydrolysis.

ATP Hydrolysis Is Highly Exergonic Because of Charge Repulsion and Resonance Stabilization

What is it about the ATP molecule that makes the hydrolysis of its phosphoanhydride bonds so exergonic? The answer to this question has three parts: Hydrolysis of ATP to ADP and P_i is exergonic because of *charge repulsion* between the adjacent negatively charged phosphate groups, because of *resonance stabilization* of both products of hydrolysis, and because of their *increased entropy* and solubility.

Charge repulsion is easy to understand. By way of analogy, imagine holding two magnets together with like poles touching. The like poles repel each other, and you need to make an effort (i.e., you need to put energy into the system) to force them together. If you let go, the magnets spring apart, releasing the energy. Now consider the three phosphate groups of ATP. As Figure 9-1 shows, each group bears at least one negative charge due to its ionization at the near neutral pH of the cell. These negative charges tend to repel one another, thereby straining the covalent bond linking the phosphate groups together. Similarly, ATP synthesis requires the joining of two negatively charged molecules (ADP^{3-} and P_i^{2-}) that naturally repel each other, thus requiring an input of energy to overcome this repulsion.

A second important contribution to ATP bond energy is **resonance stabilization.** Although the carboxylate group is formally written with one C═O double bond and one C—O single bond, it actually has one electron pair that is *delocalized* (equally distributed) over both of the C bonds to oxygen. The true structure of the carboxylate group is actually an average of the two contributing structures shown in **Figure 9-2a** and is called a **resonance hybrid.** Each C—O bond is the equivalent of one and a half bonds, and each O atom has only a partial negative charge (represented by the Greek letter δ). When electrons are delocalized in this way, a molecule is in its most stable (lowest-energy) configuration and is said to be *resonance-stabilized.*

Similarly, the phosphate ion is resonance-stabilized because the extra electron pair formally shown as part of a P═O double bond is delocalized over all four O atoms adjacent to the central P. When a phosphoester bond is formed between a phosphate ion and an alcohol group, the extra electrons are only delocalized over three O atoms (Figure 9-2b). The resulting molecule is less resonance stabilized and thus has higher energy. A similar decrease in electron delocalization during anhydride bond formation (Figure 9-2c) also results in a higher energy product.

A third important factor contributing to the exergonic nature of ATP hydrolysis is the overall increase in entropy as a phosphate group is removed from ATP and is no longer fixed in position. The spatial randomization of the ADP and phosphate decreases their free energy and makes the reaction more exergonic. Although a water molecule is added during hydrolysis (a decrease of entropy), it loses a proton, which also has increased entropy as it is randomized in solution. In addition, the ADP and phosphate become more soluble because they are more highly hydrated, and the increased interactions with water molecules lead to a decrease in free energy, adding to the exergonic nature of ATP hydrolysis.

For esters, only a moderate amount of energy is liberated upon hydrolysis, while for anhydrides, the hydrolysis reaction is highly exergonic. In addition to increased resonance stabilization, both products of anhydride hydrolysis are charged and therefore repel each other, which is not the case with ester hydrolysis. Hydrolysis of anhydride and phosphoanhydride bonds releases roughly twice the amount of free energy as does hydrolysis of ester and phosphoester bonds. ATP illustrates this difference well: Hydrolysis of either of the phosphoanhydride bonds that link the second and third phosphate groups to the rest of the molecule has a standard free energy change of about −7.3 kcal/mol, whereas hydrolysis of the phosphoester bond that links the first (innermost) phosphate group to the ribose has a $\Delta G^{\circ\prime}$ of only about −3.6 kcal/mol:

$$\text{ATP} + H_2O \longrightarrow \text{ADP} + P_i + H^+ \qquad \Delta G^{\circ\prime} = -7.3 \text{ kcal/mol} \tag{9-1}$$

$$\text{ADP} + H_2O \longrightarrow \text{AMP} + P_i + H^+ \qquad \Delta G^{\circ\prime} = -7.3 \text{ kcal/mol} \tag{9-2}$$

$$\text{AMP} + H_2O \longrightarrow \text{adenosine} + P_i \qquad \Delta G^{\circ\prime} = -3.6 \text{ kcal/mol} \tag{9-3}$$

Thus, ATP and ADP are both "higher-energy compounds" than AMP is, to use the shorthand of biochemists.

In fact, because the standard $\Delta G^{\circ\prime}$ value of −7.3 kcal/mol is based on equal concentrations of ATP and ADP (1 *M*), it typically underestimates the actual free energy change associated with the hydrolysis of ATP to ADP under most biological conditions, in which the concentration of ATP is greater. As we learned in Chapter 5, the actual free energy change, $\Delta G'$, depends on the prevailing concentrations of reactants and products (see Equation 5-22). For the hydrolysis of ATP (Reaction 9-1), $\Delta G'$ is calculated as

$$\Delta G' = \Delta G^{\circ\prime} + RT \ln \frac{[\text{ADP}][P_i]}{[\text{ATP}]} \tag{9-4}$$

Table 9-1 Standard Free Energies of Hydrolysis for Phosphorylated Compounds Involved in Energy Metabolism

Phosphorylated Compound and Its Hydrolysis Reaction	$\Delta G^{\circ\prime}$ (kcal/mol)
Phosphoenolpyruvate (PEP)	
$+ H_2O \longrightarrow$ pyruvate + P_i	−14.8
1,3-bisphosphoglycerate	
$+ H_2O \longrightarrow$ 3-phosphoglycerate + P_i	−11.8[1]
Phosphocreatine	
$+ H_2O \longrightarrow$ creatine + P_i	−10.3
Adenosine triphosphate (ATP)	
$+ H_2O \longrightarrow$ adenosine diphosphate + P_i	−7.3
Glucose-1-phosphate	
$+ H_2O \longrightarrow$ glucose + P_i	−5.0
Glucose-6-phosphate	
$+ H_2O \longrightarrow$ glucose + P_i	−3.3
Glycerol phosphate	
$+ H_2O \longrightarrow$ glycerol + P_i	−2.2

[1]The $\Delta G^{\circ\prime}$ value for 1,3-bisphosphoglycerate is for the hydrolysis of the phosphoanyhydride bond on carbon atom 1.

In most cells, the ATP/ADP ratio is significantly greater than 1:1, often in the range of about 5:1. As a result, the term $\ln([\text{ADP}][P_i]/[\text{ATP}])$ is negative, and $\Delta G'$ is therefore more negative than −7.3 kcal/mol, usually in the range of −10 to −14 kcal/mol.

ATP Is an Important Intermediate in Cellular Energy Metabolism

ATP occupies an intermediate position in the overall spectrum of energy-rich phosphorylated compounds in the cell. Some of the more common phosphorylated intermediates in cellular energy metabolism are ranked according to their $\Delta G^{\circ\prime}$ values in **Table 9-1**. The values are negative, so the compounds closest to the top of the table release the most energy upon hydrolysis of the phosphate group.

This means that, under standard conditions, a compound is capable of exergonically phosphorylating any compound below it but none of the compounds above it. Thus, ATP can be formed from ADP by the transfer of a phosphate group from phosphoenolpyruvate (PEP) but not from glucose-6-phosphate, as **Figure 9-3** illustrates. Similarly, ATP can be used to phosphorylate glucose but not pyruvate. Reactions 9-5 and 9-6 have negative $\Delta G^{\circ\prime}$ values, which we will designate as $\Delta G^{\circ\prime}_{transfer}$ values to emphasize that they represent the standard free energy change that accompanies the transfer of a phosphate group from a donor to an acceptor molecule. $\Delta G^{\circ\prime}_{transfer}$ values

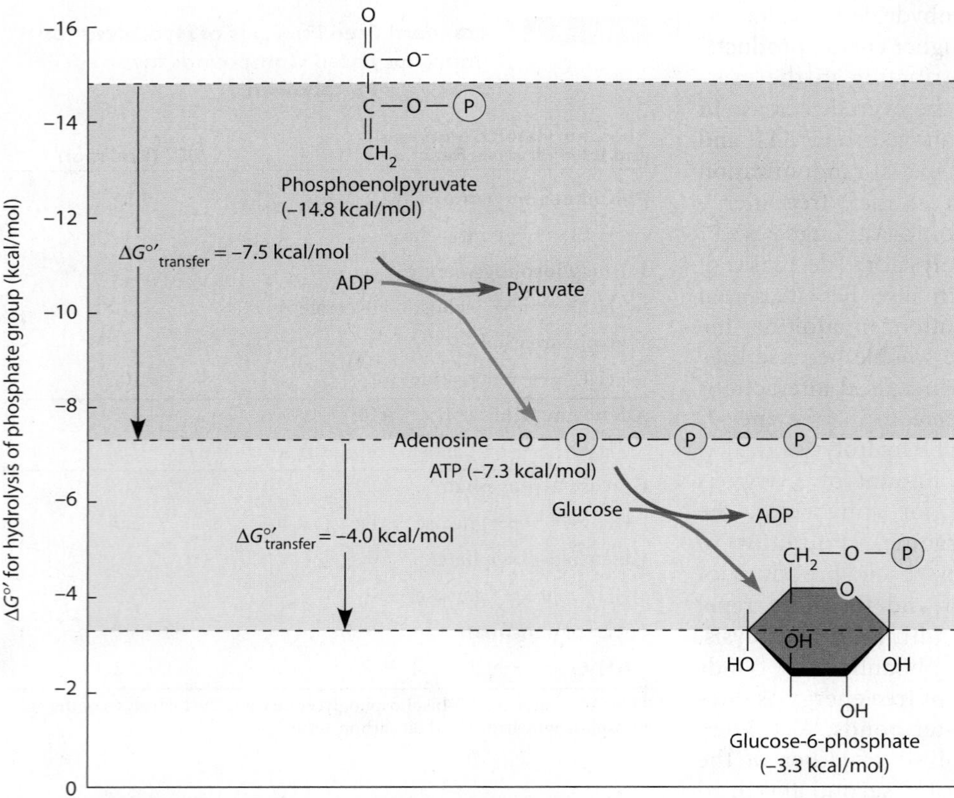

FIGURE 9-3 Examples of Exergonic Transfer of Phosphate Groups. These examples are based on the $\Delta G^{\circ\prime}$ values for the hydrolysis of the compounds shown in Table 9-1. For the hydrolysis of phosphate groups, phosphoenolpyruvate (PEP; −14.8 kcal/mol) is considered a high-energy compound, ATP (−7.3 kcal/mol) an intermediate-energy compound, and glucose-6-phosphate (−3.3 kcal/mol) a low-energy compound. PEP can therefore transfer its phosphate group exergonically onto ADP to form ATP [$\Delta G^{\circ\prime}_{\text{transfer}} = \Delta G^{\circ\prime}_{\text{donor}} - \Delta G^{\circ\prime}_{\text{acceptor}} = -14.8 - (-7.3) = -7.5$ kcal/mol], and ATP can be used to phosphorylate glucose exergonically [$\Delta G^{\circ\prime}_{\text{transfer}} = -7.3 - (-3.3) = -4.0$ kcal/mol], but the reverse reactions are not possible under standard conditions. In fact, the $\Delta G^{\circ\prime}$ values for these transfers are so highly negative that both reactions are irreversible under typical cellular conditions.

can be predicted from Table 9-1 and calculated as shown:

$$\text{glucose} + \text{ATP} \longrightarrow \text{glucose-6-phosphate} + \text{ADP} + \text{H}^+$$
$$\Delta G^{\circ\prime}_{\text{transfer}} = \Delta G^{\circ\prime}_{\text{donor}} - \Delta G^{\circ\prime}_{\text{acceptor}}$$
$$= -7.3 - (-3.3) = -4.0 \text{ kcal/mol} \qquad \textbf{(9-5)}$$

$$\text{PEP} + \text{ADP} + \text{H}^+ \longrightarrow \text{pyruvate} + \text{ATP}$$
$$\Delta G^{\circ\prime}_{\text{transfer}} = -14.8 - (-7.3) = -7.5 \text{ kcal/mol} \qquad \textbf{(9-6)}$$

Reactions such as 9-5 and 9-6 that involve the movement of a chemical group from one molecule to another are called **group transfer reactions.** Group transfer reactions represent one of the most common processes in cellular metabolism, and the phosphate group is one of the most frequently transferred groups, especially in energy metabolism.

The most important point to understand from Table 9-1 and Figure 9-3 is that the ATP/ADP pair occupies a crucial *intermediate* position in terms of bond energies. This means that ATP can serve as a phosphate *donor* in some biological reactions and that its dephosphorylated form, ADP, can serve as a phosphate *acceptor* in other reactions because there are compounds both above and below the ATP/ADP pair in energy.

In summary, the ATP/ADP pair represents a reversible means of conserving, transferring, and releasing energy within the cell (**Figure 9-4**). As catabolic processes occur in the cell, whether anaerobically (Figure 9-4a) or aerobically (Figure 9-4b), the energy-liberating reactions in the sequence are coupled to the ATP/ADP system such that the available free energy drives the formation of ATP from ADP. The free energy released upon hydrolysis of ATP then provides the driving force for the many processes (such as biosynthesis, active transport, charge separation, and muscle contraction) that are essential to life and require the input of energy.

(a) Anaerobic conditions. Under anaerobic (no oxygen) or hypoxic (oxygen-deficient) conditions, a modest amount of ATP is generated by fermentation. Lactate is the most common end-product in some organisms, and ethanol plus carbon dioxide are the most common end-products in other organisms.

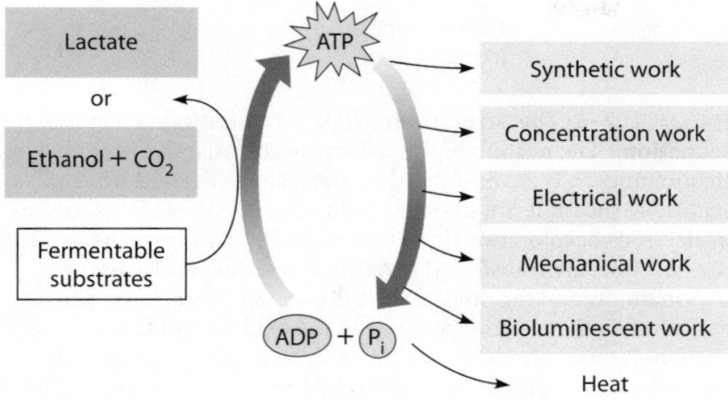

(b) Aerobic conditions. In the presence of oxygen, ATP is generated by aerobic respiration as oxidizable nutrients are catabolized completely to carbon dioxide and water. Aerobic respiration yields approximately 20 times more ATP per glucose molecule than does anaerobic fermentation.

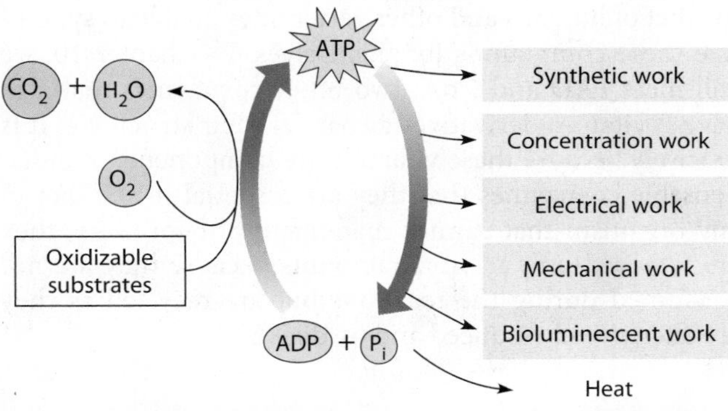

MB Chemistry Review- Enzymes & Pathways: Biosynthesis

FIGURE 9-4 The ATP/ADP System as a Means of Conserving and Releasing Energy Within the Cell. ATP is generated during **(a)** anaerobic conditions or **(b)** aerobic conditions by the oxidative catabolism of nutrients (left side) and is used to do cellular work (right side).

Chemotrophic Energy Metabolism

Now we have the essential concepts in hand to take up the main theme of this chapter and the next. We are ready, in other words, to discuss **chemotrophic energy metabolism**—the reactions and pathways by which cells catabolize nutrients and conserve as ATP some of the free energy that is released in this breakdown. Or, to put it more personally, we will be looking at the specific metabolic processes by which the cells of your own body make use of the food you eat to meet your energy needs. We begin our discussion by considering *oxidation* because much of chemotrophic energy metabolism involves energy-yielding oxidative reactions.

Biological Oxidations Usually Involve the Removal of Both Electrons and Protons and Are Highly Exergonic

To say that nutrients such as carbohydrates, fats, or proteins are sources of energy for cells means that these are oxidizable organic compounds and that their oxidation is highly exergonic. Recall from chemistry that **oxidation** is the removal of electrons. Thus, for example, a ferrous ion (Fe^{2+}) is oxidizable because it readily gives up an electron as it is converted to a ferric ion (Fe^{3+}):

$$Fe^{2+} \longrightarrow Fe^{3+} + e^- \qquad \textbf{(9-7)}$$

The only difference in biological chemistry is that the oxidation of organic molecules frequently involves the removal not just of electrons but of hydrogen ions (protons) as well, so that the process is often also one of **dehydrogenation.** Consider, for example, the oxidation of ethanol to the corresponding aldehyde:

$$\underset{\text{Ethanol}}{CH_3{-}CH_2{-}OH} \xrightarrow{\text{oxidation}} \underset{\text{Acetaldehyde}}{CH_3{-}\overset{\overset{\displaystyle H}{|}}{C}{=}O} + 2e^- + 2H^+ \qquad \textbf{(9-8)}$$

Electrons are removed, so this is clearly an oxidation. But protons are liberated as well, and an electron plus a proton is the equivalent of a hydrogen atom. Therefore what happens, in effect, is the removal of the equivalent of two hydrogen atoms:

$$\underset{\text{Ethanol}}{CH_3{-}CH_2{-}OH} \xrightarrow[\text{(dehydrogenation)}]{\text{oxidation}} \underset{\text{Acetaldehyde}}{CH_3{-}\overset{\overset{\displaystyle H}{|}}{C}{=}O} + [2H] \qquad \textbf{(9-9)}$$

Thus, for cellular reactions involving organic molecules, oxidation is almost always manifested as a dehydrogenation reaction. Many of the enzymes that catalyze oxidative reactions in cells are in fact called *dehydrogenases.*

None of the preceding oxidation reactions can take place in isolation, of course; the electrons must be transferred to another molecule, which is *reduced* in the process. **Reduction,** the opposite of oxidation, is defined as the addition of electrons and is an endergonic process. We will soon see that reduction of coenzymes is an important way that cells store chemical energy as reducing power. In biological reductions, as with oxidations, the electrons transferred are frequently accompanied by protons. The overall reaction is therefore a **hydrogenation:**

$$\underset{\text{Acetaldehyde}}{CH_3{-}\overset{\overset{\displaystyle H}{|}}{C}{=}O} + [2H] \xrightarrow[\text{(hydrogenation)}]{\text{reduction}} \underset{\text{Ethanol}}{CH_3{-}CH_2{-}OH} \qquad \textbf{(9-10)}$$

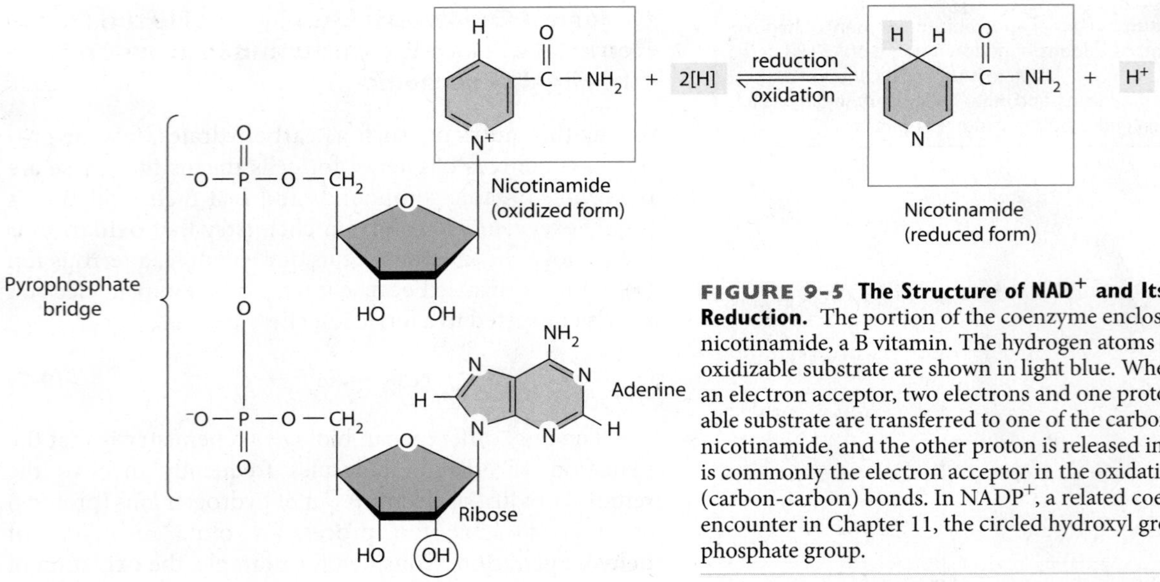

FIGURE 9-5 The Structure of NAD$^+$ and Its Oxidation and Reduction. The portion of the coenzyme enclosed in the red box is nicotinamide, a B vitamin. The hydrogen atoms derived from an oxidizable substrate are shown in light blue. When NAD$^+$ is used as an electron acceptor, two electrons and one proton from the oxidizable substrate are transferred to one of the carbon atoms of nicotinamide, and the other proton is released into solution. NAD$^+$ is commonly the electron acceptor in the oxidation of C—C (carbon-carbon) bonds. In NADP$^+$, a related coenzyme we will encounter in Chapter 11, the circled hydroxyl group is replaced by a phosphate group.

Reactions 9-9 and 9-10 also illustrate the general feature that biological oxidation-reduction reactions almost always involve *two-electron* (and therefore two-proton) transfers.

As written, Reactions 9-9 and 9-10 are only *half reactions,* representing an oxidation and a reduction event, respectively. In real reactions, however, *oxidation and reduction always take place simultaneously.* Any time an oxidation occurs, a reduction must occur as well because the electrons (and protons) removed from one molecule must be added to another molecule. The brackets around the 2H in Reactions 9-9 and 9-10 are meant to show that hydrogen atoms are never actually released into solution but are instead transferred to another molecule.

Coenzymes Such as NAD$^+$ Serve as Electron Acceptors in Biological Oxidations

In most biological oxidations, electrons and hydrogens removed from the substrate being oxidized are transferred to one of several **coenzymes.** In general, coenzymes are small molecules that function along with enzymes (hence the name), usually by serving as carriers of electrons or small functional groups. As we will see shortly, coenzymes are not consumed but are recycled within the cell, so the relatively low intracellular concentration of a given coenzyme is adequate to meet the needs of the cell.

The most common coenzyme involved in energy metabolism is **nicotinamide adenine dinucleotide NAD$^+$.** Its structure is shown in **Figure 9-5**. Despite its formidable name and structure, its function is straightforward: NAD$^+$ serves as an electron acceptor by adding two electrons and one proton to its aromatic ring, thereby generating the reduced form, NADH, plus a proton:

$$\underset{\text{(oxidized)}}{\text{NAD}^+} + 2[\text{H}] \longrightarrow \underset{\text{(reduced)}}{\text{NADH}} + \text{H}^+ \qquad \textbf{(9-11)}$$

As a nutritional note, the nicotinamide of NAD$^+$ is a derivative of *niacin,* which we recognize as a *B vitamin*—one of a family of water-soluble compounds essential in the diet of humans and other vertebrates unable to synthesize these compounds for themselves. In Chapter 10, we will meet FAD and CoA, two other coenzymes that also have B vitamin derivatives as part of their structures. It is precisely because these vitamins are components of indispensable coenzymes that they are essential in the diet of any organism that cannot manufacture them. Also, they are required only in small amounts because they are not consumed during the reactions but are recycled as they are alternatively reduced and oxidized.

Most Chemotrophs Meet Their Energy Needs by Oxidizing Organic Food Molecules

We are interested in oxidation because it is the means by which chemotrophs such as humans meet their energy needs. Many different kinds of substances serve as substrates for biological oxidation. For example, a wide variety of microorganisms can use inorganic compounds such as hydrogen gas or reduced forms of iron, sulfur, or nitrogen as their energy sources. These organisms, which utilize rather specialized oxidative pathways, play important roles in the geochemical cycling of nutrients in the biosphere. They also produce a significant amount of biomass—by some estimates, 50% as much as global plant production—and they are important food sources for numerous other organisms in the food chain. However, we and most other chemotrophs depend on organic food molecules as oxidizable substrates, namely, carbohydrates, fats, and proteins. Keep in mind that oxidation of these organic food molecules produces energy for the cell as both ATP and reduced coenzymes.

Glucose Is One of the Most Important Oxidizable Substrates in Energy Metabolism

To simplify our discussion initially and to provide a unifying metabolic theme, we will concentrate on the biological oxidation of the six-carbon sugar **glucose** ($C_6H_{12}O_6$). Glucose is a good choice for several reasons. In many vertebrates, including humans, glucose is the main sugar in the blood and hence the main energy source for most of the cells in the body. Blood glucose comes primarily from dietary carbohydrates such as sucrose or starch and from the breakdown of stored glycogen (see Figures 3-21 to 3-24). Current guidelines recommend a diet of approximately 50% carbohydrate, 30% lipid, and 20% protein. Glucose is therefore an especially important molecule for you personally.

Glucose is also important to plants because it is the monosaccharide released upon starch breakdown. In addition, glucose makes up one-half of the disaccharide sucrose (glucose + fructose), the major sugar in the vascular system of most plants. Moreover, the catabolism of most other energy-rich substances in plants, animals, and microorganisms alike begins with their conversion into one of the intermediates in the pathway for glucose catabolism. Rather than looking at the fate of a single compound, then, we are considering a metabolic pathway that is at the very heart of chemotrophic energy metabolism.

The Oxidation of Glucose Is Highly Exergonic

Glucose is a good potential source of energy because its oxidation is a highly exergonic process, with a $\Delta G^{\circ\prime}$ of -686 kcal/mol for the complete conversion of glucose to carbon dioxide and water using oxygen as the *final electron acceptor:*

$$C_6H_{12}O_6 + 6O_2 \longrightarrow 6CO_2 + 6H_2O \qquad \textbf{(9-12)}$$

As a thermodynamic parameter, $\Delta G^{\circ\prime}$ is unaffected by the route from substrates to products. Therefore, it will have the same value whether the oxidation is by direct combustion, with all of the energy released as heat, or by biological oxidation, with some of the energy conserved as ATP. Thus, oxidation of the sugar molecules in a marshmallow will release the same amount of free energy whether you burn the marshmallow over a campfire or eat it and catabolize the sugar molecules in your body. Biologically, however, the distinction is critical: *Uncontrolled combustion occurs at temperatures that are incompatible with life, and most of the free energy is lost as heat. Biological oxidations involve enzyme-catalyzed reactions that occur without significant temperature changes, and much of the free energy is conserved in chemical form as ATP.*

Glucose Catabolism Yields Much More Energy in the Presence of Oxygen than in Its Absence

Access to the full 686 kcal/mol of free energy in glucose is possible only if glucose is completely oxidized to carbon dioxide and water. Even then, because no energy conversion process is 100% efficient, only part of the energy can be recovered. The complete oxidation of glucose (or other organic nutrients such as proteins and lipids) to carbon dioxide and water in the presence of oxygen is called *aerobic respiration,* a complex, multistep process we will discuss in detail in Chapter 10. Many organisms, typically bacteria, can carry out *anaerobic respiration,* using inorganic electron acceptors other than oxygen. Examples of alternative acceptors include elemental sulfur (S), protons (H^+), and ferric ions (Fe^{3+}).

Even in the absence or scarcity of oxygen or other inorganic electron acceptors, most organisms can still extract limited amounts of energy from the partial oxidation of glucose but with lower energy yields per glucose molecule. They do so by means of *glycolysis,* a pathway that does not require oxygen. Instead, electrons that are removed during glucose oxidation are returned to an organic molecule later in the same pathway. Such an anaerobic process is called **fermentation** and is identified in terms of the principal end-product. In some animal cells and many bacteria, the end-product is lactate, so the process of anaerobic glucose catabolism is called *lactate fermentation.* In most plant cells and in microorganisms such as yeast, the process is termed *alcoholic fermentation* because the end-product is ethanol, an alcohol plus carbon dioxide.

Based on Their Need for Oxygen, Organisms Are Aerobic, Anaerobic, or Facultative

Organisms can be classified in terms of their need for and use of oxygen as an electron acceptor in energy metabolism. Most organisms we see on a daily basis have an absolute requirement for oxygen and are called **obligate aerobes.** You look at such an organism every time you look in the mirror. On the other hand, some organisms, including many bacteria, cannot use oxygen as an electron acceptor and are called **obligate anaerobes.** In fact, oxygen is toxic to these organisms. Not surprisingly, such organisms occupy environments from which oxygen is excluded, such as deep puncture wounds or the sludge at the bottoms of ponds. Most strict anaerobes are bacteria, including organisms responsible for gangrene, food poisoning, and methane production.

Facultative organisms can function under either aerobic or anaerobic conditions. Given the availability of oxygen, most facultative organisms carry out the full aerobic respiratory process. However, they can switch to anaerobic respiration or fermentation if oxygen is limiting or absent. Many bacteria and fungi are facultative organisms, as are most molluscs and annelids (worms). Some cells or tissues of otherwise aerobic organisms can function in the temporary absence or scarcity of oxygen if required to do so. Your skeletal muscle cells are an example; they normally function aerobically but switch to lactate fermentation whenever the oxygen supply becomes limiting—during periods of prolonged or strenuous exercise, for example.

The rest of this chapter is devoted mainly to exploring the anaerobic generation of ATP by the fermentation of glucose. Aerobic energy metabolism then becomes the focus of the next chapter. We will consider fermentation processes first because the glycolytic pathway is common to both fermentation and aerobic respiration. By beginning with fermentation, we will be considering the ways that energy can be extracted from glucose without net oxidation. We will also be laying the foundation for the aerobic processes of the next chapter.

Glycolysis and Fermentation: ATP Generation Without the Involvement of Oxygen

Whether it is an obligate or facultative anaerobe, any organism or cell that obtains its energy using fermentation carries out energy-yielding oxidative reactions without using oxygen as an electron acceptor. The six-carbon glucose molecule is split into two three-carbon molecules, each of which is then partially oxidized by a reaction sequence that is sufficiently exergonic to generate two ATP molecules per molecule of glucose fermented. For most cells, this is the maximum possible energy yield that can be achieved without access to oxygen or to an alternative electron acceptor. However, there are exceptions to this limit. Certain microorganisms can get up to five ATP molecules per glucose by using specialized enzyme systems. Also, some plants and animals that are adapted to anoxia (oxygen deprivation) can produce more than two ATP molecules per glucose.

Glycolysis Generates ATP by Catabolizing Glucose to Pyruvate

The process of **glycolysis,** also called the **glycolytic pathway,** is a ten-step reaction sequence that converts one molecule of glucose into two molecules of pyruvate, a three-carbon compound. During the partial oxidation of glucose to pyruvate in glycolysis, energy and reducing power are conserved in the form of ATP and NADH, respectively. An overview of the three major phases of glycolysis is shown in **Figure 9-6**, and the detailed pathway is shown in **Figure 9-7**. Glycolysis is common to both aerobic and anaerobic glucose metabolism and is present in virtually all organisms. In most cells, these enzymes occur in the cytosol. In some parasitic protozoans called trypanosomes, however, the first seven enzymes are compartmentalized in membrane-bounded organelles called *glycosomes.*

Historically, the glycolytic pathway was the first major metabolic sequence to be elucidated. Most of the definitive work was done in the 1930s by the German biochemists Gustav Embden, Otto Meyerhof, and Otto Warburg. In fact, an alternative name for the glycolytic pathway is the *Embden–Meyerhof pathway.*

Glycolysis in Overview. In the absence of oxygen, glycolysis leads to fermentation. In the presence of oxygen, glycolysis leads to aerobic respiration, the subject of

(a) Phase 1: Preparation and cleavage. The six-carbon glucose molecule is phosphorylated twice by ATP and split to form two molecules of glyceraldehyde-3-phosphate. This requires an input of two ATP per glucose.

(b) Phase 2: Oxidation and ATP generation. The two molecules of glyceraldehyde-3-phosphate are oxidized to 3-phosphoglycerate. Some of the energy from this oxidation is conserved as two ATP and two NADH molecules are produced.

(c) Phase 3: Pyruvate formation and ATP generation. The two 3-phosphoglycerate molecules are converted to pyruvate, with accompanying synthesis of two more ATP molecules.

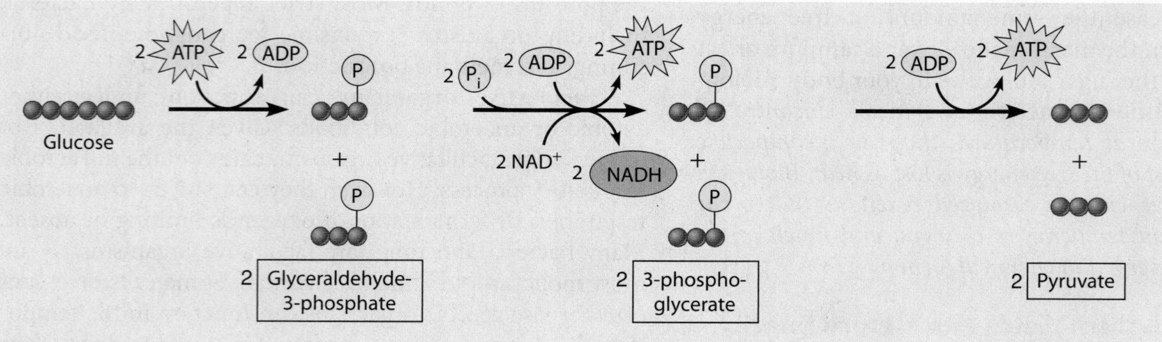

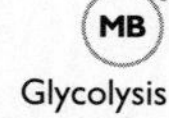

Glycolysis

FIGURE 9-6 An Overview of the Glycolytic Pathway. During glycolysis, one molecule of glucose is split and partly oxidized, generating two molecules of pyruvate. In the process, energy is conserved as a net gain of two molecules of ATP and two molecules of NADH. This ten-step process occurs in three main phases **(a–c)**, as shown above. Simplified structures show only carbon atoms (gray) and phosphate groups (yellow).

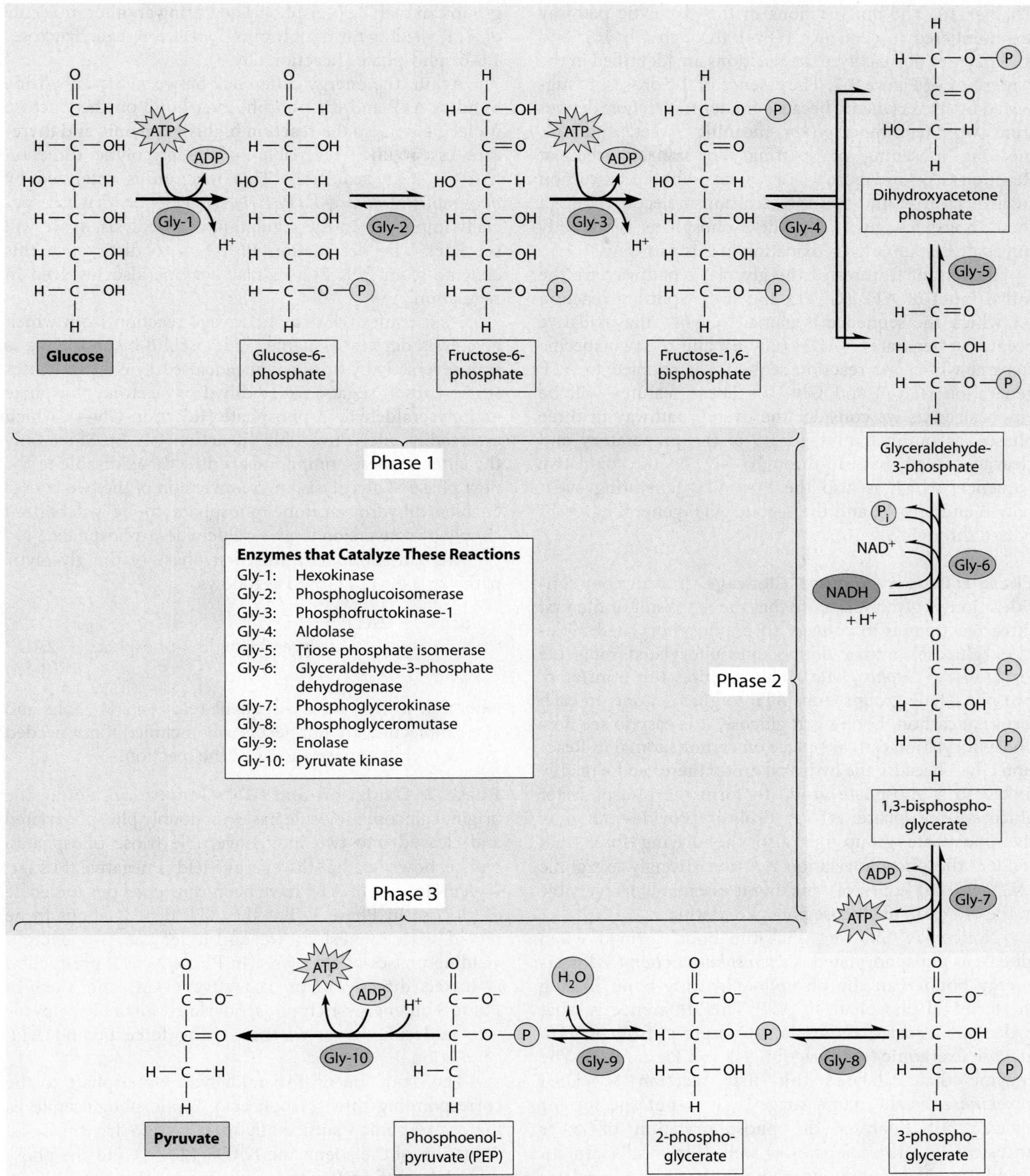

FIGURE 9-7 The Glycolytic Pathway from Glucose to Pyruvate. Glycolysis is a sequence of ten reactions in which glucose is catabolized to pyruvate, with a single oxidative reaction (Gly-6) and two ATP-generating steps (Gly-7 and Gly-10). The enzymes that catalyze these reactions are identified in the center box.

Chapter 10. The ten reactions in the glycolytic pathway are numbered in sequence (Gly-1 through Gly-10), and the enzymes that catalyze the reactions are identified in the center box of Figure 9-7. The essence of the process is suggested by its very name because the term *glycolysis* derives from two Greek roots: *glykos,* meaning "sweet," and *lysis,* meaning "loosening" or "splitting." The splitting occurs at Reaction Gly-4 in Figure 9-7, at which point the six-carbon sugar is cleaved into two three-carbon molecules. One of these molecules, glyceraldehyde-3-phosphate, is the only molecule that undergoes oxidation in this pathway.

Important features of the glycolytic pathway are the initial input of ATP (Gly-1), the sugar-splitting reaction for which the sequence is named (Gly-4), the oxidative event that generates NADH (Gly-6), and the two specific steps at which the reaction sequence is coupled to ATP generation (Gly-7 and Gly-10). These features will be emphasized as we consider the overall pathway in three phases, as outlined in Figure 9-6: the preparatory and cleavage steps (Gly-1 through Gly-5); the oxidative sequence, which is also the first ATP-generating event (Gly-6 and Gly-7); and the second ATP-generating event (Gly-8 through Gly-10).

Phase 1: Preparation and Cleavage. To begin our consideration of glycolysis, note that the net result of the first three reactions is to convert an unphosphorylated molecule (glucose) into a doubly phosphorylated molecule *(fructose-1,6-bisphosphate).* This requires the transfer of two phosphate groups from ATP to glucose, one on each terminal carbon. Looking at glucose, it is easy to see how phosphorylation can take place on carbon atom 6 in Reaction Gly-1 because the hydroxyl group there can be readily linked to a phosphate group to form the phosphoester glucose-6-phosphate. ATP hydrolysis provides not only the phosphate group but also the driving force that renders the phosphorylation reaction strongly exergonic ($\Delta G^{\circ\prime} = -4.0$ kcal/mol), making it essentially irreversible in the direction of glucose phosphorylation.

Notice, by the way, that the bond formed when glucose is phosphorylated is a *phosphoester bond,* a lower-energy bond than the phosphoanhydride bond linking the terminal phosphate to ATP. This difference is what makes the transfer of the phosphate group from ATP to glucose exergonic (see Reaction 9-5 and Figure 9-3). The enzyme that catalyzes this first reaction is called *hexokinase;* as the name suggests, it is not specific for glucose but catalyzes the phosphorylation of other hexoses (six-carbon sugars) as well. (Liver cells contain an additional enzyme, *glucokinase,* that phosphorylates only glucose.)

The carbonyl group on carbon atom 1 of the glucose molecule is not as readily phosphorylated as the hydroxyl group on carbon atom 6. But in the next reaction (Gly-2), the aldosugar glucose-6-phosphate is converted to the corresponding ketosugar, fructose-6-phosphate, which has a hydroxyl group on carbon atom 1. That hydroxyl group can then be phosphorylated using another molecule of ATP, yielding the doubly phosphorylated sugar, fructose-1,6-bisphosphate (Reaction Gly-3).

Again, the energy difference between the anhydride bond of ATP and the phosphoester bond on the fructose molecule renders the reaction highly exergonic and therefore essentially irreversible in the glycolytic direction ($\Delta G^{\circ\prime} = -3.4$ kcal/mol). This reaction is catalyzed by *phosphofructokinase-1 (PFK-1),* an enzyme that is especially important in the regulation of glycolysis, as we will see later. The designation PFK-1 is to distinguish this enzyme from PFK-2, a similar enzyme also involved in regulation.

Next comes the actual cleavage reaction from which glycolysis derives its name. Fructose-1,6-bisphosphate is split reversibly by the enzyme *aldolase* to yield two trioses (three-carbon sugars) called dihydroxyacetone phosphate and glyceraldehyde-3-phosphate (Reaction Gly-4), which are readily interconvertible (Reaction Gly-5). Since only the latter of these compounds is directly oxidizable in the next phase of glycolysis, interconversion of the two trioses enables dihydroxyacetone phosphate to be catabolized simply by conversion to glyceraldehyde-3-phosphate.

We can summarize this first phase of the glycolytic pathway (Gly-1 to Gly-5) as follows:

$$\text{glucose} + 2\text{ATP} \longrightarrow 2\ \text{glyceraldehyde-3-phosphate} + 2\text{ADP} \qquad \textbf{(9-13)}$$

Note that in this and subsequent reactions, H^+ ions and H_2O molecules are not necessarily included if not needed for the overall understanding of the reaction.

Phase 2: Oxidation and ATP Generation. So far, the original glucose molecule has been doubly phosphorylated and cleaved into two interconvertible triose phosphates. Notice, however, that the energy yield is negative thus far: Two molecules of ATP have been *consumed* per molecule of glucose in Phase 1. But the ATP debt is about to be repaid with interest as we encounter the two energy-yielding phases of glycolysis. In Phase 2, ATP production is linked directly to an oxidative event, and then in Phase 3 an energy-rich phosphorylated form of the pyruvate molecule serves as the driving force behind ATP generation.

The oxidation of glyceraldehyde-3-phosphate to the corresponding three-carbon acid, 3-phosphoglycerate, is highly exergonic—sufficiently so, in fact, to drive both the reduction of the coenzyme NAD^+ (Gly-6) and the phosphorylation of ADP with inorganic phosphate, P_i (Gly-7). Historically, this was the first example of a reaction sequence in which the coupling of ATP generation to an oxidative event was understood.

The important features of this highly exergonic sequence are the involvement of NAD^+ as the electron acceptor and the coupling of the oxidation to the formation of a high-energy, doubly phosphorylated intermediate,

1,3-bisphosphoglycerate. The phosphoanhydride bond on carbon atom 1 of this intermediate has such a highly negative $\Delta G^{\circ\prime}$ of hydrolysis (−11.8 kcal/mol; see Table 9-1) that the transfer of the phosphate to ADP, catalyzed by the enzyme *phosphoglycerate kinase,* is a highly exergonic reaction. ATP generation by the direct transfer of a high-energy phosphate group to ADP from a phosphorylated substrate such as 1,3-bisphosphoglycerate is called **substrate-level phosphorylation.**

To summarize the substrate-level phosphorylation of Reactions Gly-6 and Gly-7, we can write an overall reaction that accounts for one of the two glyceraldehyde-3-phosphate molecules generated from each glucose molecule in the first phase of glycolysis:

$$\text{glyceraldehyde-3-phosphate} + \text{NAD}^+ + \text{ADP} + \text{P}_i \longrightarrow \text{3-phosphoglycerate} + \text{NADH} + \text{H}^+ + \text{ATP} \quad \textbf{(9-14)}$$

Keep in mind that each reaction in the glycolytic pathway beyond glyceraldehyde-3-phosphate occurs twice per starting molecule of glucose so that, on a per-glucose basis, two molecules of NADH need to be reoxidized in order to regenerate the NAD^+ that is needed for continual oxidation of glyceraldehyde-3-phosphate. It also means that the initial investment of two ATP molecules in Phase 1 is recovered here in Phase 2, so the net ATP yield to this point is now zero.

Next, in the final phase of glycolysis, we will see the generation of two more ATP molecules. Thus, for glycolysis overall, there is a net gain of two ATP molecules per glucose metabolized to pyruvate.

Phase 3: Pyruvate Formation and ATP Generation. Generating another molecule of ATP from 3-phosphoglycerate depends on the phosphate group on carbon atom 3. At this stage, the phosphate group is linked to the carbon atom by a phosphoester bond with a low free energy of hydrolysis ($\Delta G^{\circ\prime}$ = −3.3 kcal/mol). In the final phase of the glycolytic pathway, this phosphoester bond is converted to a *phosphoenol bond,* the hydrolysis of which is highly exergonic ($\Delta G^{\circ\prime}$ = −14.8 kcal/mol; see Table 9-1). This increase in the amount of free energy released upon hydrolysis involves a rearrangement of internal energy within the molecule. To accomplish this, the phosphate group of 3-phosphoglycerate is moved to the adjacent carbon atom, forming 2-phosphoglycerate (Reaction Gly-8). Water is then removed from 2-phosphoglycerate by the enzyme enolase (Reaction Gly-9), thereby generating the high-energy compound phosphoenolpyruvate (PEP).

If you look carefully at the structure of PEP (see Figure 9-7, Reaction Gly-9), you will notice that, unlike the phosphoester bonds of either 3- or 2-phosphoglycerate, the phosphoenol bond of PEP has a phosphate group on a carbon atom that is linked by a double bond to another carbon atom. This characteristic makes the hydrolysis of the phosphoenol bond of PEP one of the most exergonic hydrolytic reactions known in biological systems.

PEP hydrolysis is exergonic enough to drive ATP synthesis in Reaction Gly-10, which involves the transfer of a phosphate group from PEP to ADP, generating another molecule of ATP in another substrate-level phosphorylation. This transfer, catalyzed by the enzyme *pyruvate kinase,* is highly exergonic ($\Delta G^{\circ\prime}$ = −7.5 kcal/mol; see Figure 9-3 and Reaction 9-6) and is therefore essentially irreversible in the direction of pyruvate and ATP formation.

To summarize the third phase of glycolysis, Gly-8 to Gly-10, we can write an overall reaction for pyruvate formation:

$$\text{3-phosphoglycerate} + \text{ADP} \longrightarrow \text{pyruvate} + \text{ATP} \quad \textbf{(9-15)}$$

Summary of Glycolysis. Two molecules of ATP were initially invested in Reactions Gly-1 and Gly-3, and two were returned in the first phosphorylation event (Gly-7), so the two molecules of ATP formed per molecule of glucose by the second phosphorylation event (Gly-10) represent the net ATP yield of the glycolytic pathway. This becomes clear when we add up the three reactions that summarize the three phases of the pathway (Reactions 9-13, 9-14, and 9-15). The latter two reactions are multiplied by 2 to account for both triose molecules generated in Reaction 9-13. The result is an overall expression for the pathway from glucose to pyruvate:

$$\text{glucose} + 2\text{NAD}^+ + 2\text{ADP} + 2\text{P}_i \xrightarrow{\text{reactions Gly-1 through Gly-10}} 2\ \text{pyruvate} + 2\text{NADH} + 2\text{H}^+ + 2\text{ATP} \quad \textbf{(9-16)}$$

This pathway is highly exergonic in the direction of pyruvate formation. Under typical intracellular conditions in your body, for example, $\Delta G'$ for the overall pathway from glucose to pyruvate with the generation of two molecules each of ATP and NADH is about −20 kcal/mol.

The glycolytic pathway is one of the most common and highly conserved metabolic pathways known. Virtually all cells possess the ability to extract energy from glucose by oxidizing it to pyruvate. Some of this energy is conserved in the form of two molecules of ATP per molecule of glucose. What happens next, however, usually depends on the availability of oxygen because catabolism beyond pyruvate is quite different under aerobic conditions than it is under anaerobic conditions.

The Fate of Pyruvate Depends on Whether Oxygen Is Available

Pyruvate occupies a key position as a branching point in chemotrophic energy metabolism (**Figure 9-8**). Its fate depends on the kind of organism involved, the specific cell type, and whether oxygen is available. In the presence of oxygen, pyruvate undergoes further oxidation to a molecule called acetyl coenzyme A (abbreviated as acetyl

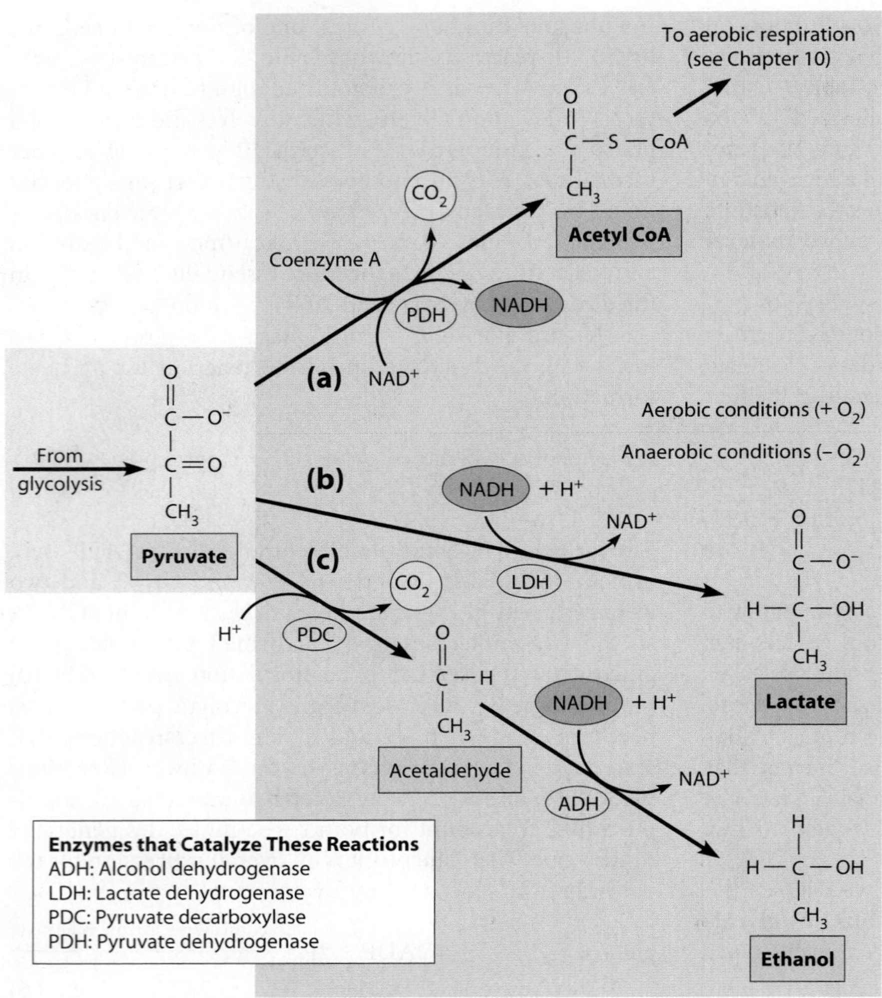
To aerobic respiration (see Chapter 10)
(a) Aerobic conditions. In the presence of oxygen, many organisms convert pyruvate to an activated form of acetate known as acetyl CoA. In this reaction, pyruvate is both oxidized (with NAD being reduced to NADH) and decarboxylated (liberation of a carbon atom as CO2). Acetyl CoA then becomes the substrate for aerobic respiration, where NADH is oxidized back to NAD+ by molecular oxygen (see Chapter 10).
CO2
Acetyl CoA
Coenzyme A
PDH
NADH
NAD+
(a)
From glycolysis
Pyruvate
Aerobic conditions (+ O2)
Anaerobic conditions (– O2)
(b)
NADH + H+
NAD+
LDH
(c)
CO2
H+
PDC
Lactate
Acetaldehyde
NADH + H+
NAD+
ADH
Ethanol
(b) and (c) Anaerobic conditions. When oxygen is absent, pyruvate is reduced so that NADH can be oxidized to NAD, the form of this coenzyme required in Reaction Gly-6 of glycolysis. Common products of pyruvate reduction are (b) lactate (in most animal cells and many bacteria) or (c) ethanol and CO2 (in many plant cells and in yeasts and other microorganisms).
Enzymes that Catalyze These Reactions
ADH: Alcohol dehydrogenase
LDH: Lactate dehydrogenase
PDC: Pyruvate decarboxylase
PDH: Pyruvate dehydrogenase

FIGURE 9-8 The Fate of Pyruvate Under Aerobic and Anaerobic Conditions. The fate of pyruvate depends on the organism involved and on whether oxygen is available. The enzymes that catalyze these reactions are identified in the box.

CoA), which in turn can be completely oxidized to CO_2, generating more than 30 ATP per glucose.

An important feature of glycolysis, however, is that it can also take place in the absence of oxygen. Under anaerobic conditions, no further oxidation of pyruvate occurs, no acetyl CoA is formed, and no additional ATP can be generated. Instead, the energy needs of the cell are met by the modest ATP yield of two ATP per glucose in the glycolytic pathway, and cells must consume glucose much more rapidly in order to maintain steady-state cellular ATP levels. In anaerobic conditions, rather than being oxidized, pyruvate is reduced by accepting the electrons (and protons) that must be removed from NADH. As Figure 9-8 illustrates, the most common products of pyruvate reduction are lactate (part b) or ethanol and carbon dioxide (part c).

In the Absence of Oxygen, Pyruvate Undergoes Fermentation to Regenerate NAD^+

As usually defined, the glycolytic pathway ends with pyruvate. Fermentative processes cannot end there, however, *because of the need to regenerate NAD^+*, the oxidized form of the coenzyme. As Reaction 9-16 indicates, the conversion of glucose to pyruvate requires that one molecule of NAD^+ is reduced to NADH per molecule of pyruvate generated. Coenzymes are present in cells at only modest concentrations, however, so the conversion of NAD^+ to NADH during glycolysis would cause cells to run out of NAD^+ very quickly if there were not some mechanism for regenerating NAD^+. Cells also have mechanisms to continually monitor and stabilize the NAD^+/NADH ratio, which is an indicator of the cell's *redox* state, the general level of oxidation of cellular components. Excessive oxidation of cellular components can be damaging to the cell, as free radicals and other harmful compounds are produced. Large changes in this ratio are a signal that the cell is under oxidative stress, and cellular mechanisms will try to keep it relatively constant.

In the presence of oxygen, NADH is reoxidized by the transfer of its electrons to oxygen, as we will see in Chapter 10. Under anaerobic conditions, however, the electrons are transferred to pyruvate, which has a carbonyl group that can be readily reduced to a hydroxyl group (see Figure 9-8 b, c). The two most common pathways for fermentation use pyruvate as the electron acceptor, converting it either to lactate or to CO_2 and ethanol. We will consider both of these alternatives briefly.

Lactate Fermentation. The anaerobic process that terminates in lactate is called **lactate fermentation.** As Figure 9-8b indicates, lactate is generated by the direct transfer of electrons from NADH to the carbonyl group of pyruvate, reducing it to the hydroxyl group of lactate. On a per-glucose basis, this reaction can be represented as

$$2\text{ pyruvate} + 2\text{NADH} + 2\text{H}^+ \rightleftharpoons 2\text{ lactate} + 2\text{NAD}^+ \qquad \textbf{(9-17)}$$

This reaction is readily reversible; in fact, the enzyme that catalyzes it is called *lactate dehydrogenase* because of its ability to catalyze the oxidation, or dehydrogenation, of lactate to pyruvate.

By adding Reactions 9-16 and 9-17, we can write an overall reaction for the metabolism of glucose to lactate under anaerobic conditions:

$$\text{glucose} + 2\text{ADP} + 2\text{P}_i \longrightarrow 2\text{ lactate} + 2\text{ATP} \qquad \textbf{(9-18)}$$

Lactate fermentation is the major energy-yielding pathway in many anaerobic bacteria, as well as in animal cells operating under anaerobic or hypoxic conditions. Lactate fermentation is important to us commercially because the production of cheese, yogurt, and other dairy products depends on microbial fermentation of lactose, the main sugar found in milk.

A more personal example of lactate fermentation involves your own muscles during periods of strenuous exertion. Whenever muscle cells use oxygen faster than it can be supplied by the circulatory system, the cells become temporarily hypoxic. Pyruvate is then reduced to lactate instead of being further oxidized, as it is under aerobic conditions. The lactate produced in this way is transported by the circulatory system from the muscle to the liver. There it is converted to glucose again by the process of *gluconeogenesis.* As we will see later in this chapter, gluconeogenesis is essentially the reverse of lactate fermentation but with several critical differences that enable it to proceed exergonically in the direction of glucose formation.

Alcoholic Fermentation. Under anaerobic conditions, plant cells can carry out **alcoholic fermentation** (in waterlogged roots, for example), as do yeasts and other microorganisms. In this process, pyruvate loses a carbon atom (as CO_2) to form the two-carbon compound acetaldehyde. Acetaldehyde reduction by NADH gives rise to ethanol, the alcohol for which the process is named. This reductive sequence is catalyzed by two enzymes, *pyruvate decarboxylase* and *alcohol dehydrogenase* (Figure 9-8c). The overall reaction can be summarized as follows:

$$2\text{ pyruvate} + 2\text{NADH} + 4\text{ H}^+ \longrightarrow 2\text{ ethanol} + 2\text{CO}_2 + 2\text{NAD}^+ \qquad \textbf{(9-19)}$$

By adding this reductive step to the overall equation for glycolysis (Reaction 9-16), we arrive at the following summary equation for alcoholic fermentation:

$$\text{glucose} + 2\text{ADP} + 2\text{P}_i + 2\text{ H}^+ \longrightarrow 2\text{ ethanol} + 2\text{CO}_2 + 2\text{ATP} \qquad \textbf{(9-20)}$$

Alcoholic fermentation by yeast cells is a key process in the baking, brewing, and winemaking industries. The yeast cells in bread dough break down glucose anaerobically, generating both CO_2 and ethanol. Carbon dioxide is trapped in the dough, causing it to rise, and the alcohol is driven off during baking and becomes part of the pleasant aroma of baking bread. For the brewer, both CO_2 and ethanol are essential; ethanol makes the product an alcoholic beverage, and CO_2 accounts for the carbonation.

Other Fermentation Pathways. Although lactate and ethanol are the fermentation products of greatest physiological or economic significance, they by no means exhaust the microbial fermentation repertoire. In *propionate fermentation,* for example, bacteria reduce pyruvate to propionate ($CH_3—CH_2—COO^-$), an important reaction in the production of Swiss cheese. Many bacteria that cause food spoilage do so by *butylene glycol fermentation.* Other fermentation processes yield acetone, isopropyl alcohol, or butyrate, the latter of which is responsible for the rotten smell of rancid food or vomit. However, all these reactions are just metabolic variations on the common theme of reoxidizing NADH by the transfer of electrons to some organic acceptor.

Fermentation Taps Only a Fraction of the Substrate's Free Energy but Conserves That Energy Efficiently as ATP

An essential feature of every fermentative process is that *no external electron acceptor is involved and no net oxidation occurs.* In both lactate and alcoholic fermentation, for example, the NADH generated by the single oxidative step of glycolysis (Reaction Gly-6) is reoxidized in the final reaction of the sequence (Reactions 9-17 and 9-19). Because no net oxidation occurs, fermentation gives a modest ATP yield—two molecules of ATP per molecule of glucose, in the case of either lactate or alcoholic fermentation.

Most of the free energy of the glucose molecule is still present in the two lactate or ethanol molecules. In the case of lactate fermentation, for example, the two lactate molecules produced from every glucose molecule contain most of the 686 kcal of free energy present per mole of glucose because the complete aerobic oxidation of lactate has a $\Delta G^{\circ\prime}$ of -319.5 kcal/mol. In other words, about 93% (639 kcal) of the original free energy of glucose is still present in the two lactate molecules, and only about 7% (47 kcal/mol) of the free energy potentially available from glucose was obtained during fermentation.

Although the energy yield from lactate fermentation is low, the available free energy is conserved efficiently as ATP. Using standard free energy changes, these two molecules of ATP represent $2 \times 7.3 = 14.6$ kcal/mol. This corresponds to an efficiency of energy conservation of about 30% ($14.6/47 \times 100\%$). Based on actual $\Delta G'$ values for ATP hydrolysis under cellular conditions (often in the range of -10 to -14 kcal/mol), two molecules of ATP represent at least 20 kcal/mol, which means that the efficiency of energy conservation probably exceeds 40%.

Alternative Substrates for Glycolysis

So far, we have assumed glucose to be the starting point for glycolysis and thus, by implication, for all of cellular energy metabolism. Glucose is certainly a major substrate for both fermentation and respiration in a variety of organisms and tissues. It is not the only such substrate, however. For many organisms and some tissues within organisms, glucose is not significant at all. So it is important to ask two questions: What are some of the major alternatives to glucose, and how are they handled by cells?

One principle quickly emerges: *Regardless of the chemical nature of the alternative substrate, it is often converted into an intermediate in the main pathway for glucose catabolism.* Most carbohydrates, for example, are converted to intermediates in the glycolytic pathway. To emphasize this point, we will briefly consider two classes of alternative carbohydrate substrates—other sugars and storage carbohydrates. In Chapter 10, we will see how proteins and lipids can be converted into intermediates in the tricarboxylic acid (TCA) cycle, the next stage of aerobic respiration.

Other Sugars and Glycerol Are Also Catabolized by the Glycolytic Pathway

Many sugars other than glucose are available to cells, depending on the food sources of the organism in question. Most of them are either monosaccharides (usually hexoses or pentoses) or disaccharides that can be readily hydrolyzed into their component monosaccharides. Ordinary table sugar (sucrose), for example, is a disaccharide consisting of the hexoses glucose and fructose, and milk sugar (lactose) contains glucose and galactose (see Figure 3-23). Besides glucose, fructose, and galactose, mannose is another relatively common dietary hexose.

Figure 9-9 illustrates the reactions that bring various carbohydrates into the glycolytic pathway. In general, disaccharides are hydrolyzed into their component monosaccharides, and each monosaccharide is converted to a glycolytic intermediate in one or a few steps. Glucose and fructose enter most directly after phosphorylation on carbon atom 6. Mannose is converted to mannose-6-phosphate and then to fructose-6-phosphate, a glycolytic intermediate. The entry of galactose requires a somewhat more complex reaction sequence involving five steps to convert it to glucose-6-phosphate (Figure 9-9).

Phosphorylated pentoses can also be channeled into the glycolytic pathway but only after being converted to hexose phosphates. That conversion is accomplished by a metabolic sequence called the *phosphogluconate pathway,* also known as the *pentose phosphate pathway.* Glycerol, a three-carbon molecule resulting from lipid breakdown, enters glycolysis after conversion to dihydroxyacetone phosphate (Figure 9-9). Thus, the typical cell has metabolic capabilities to convert most naturally occurring sugars (and a variety of other compounds as well) to one of the glycolytic intermediates for further catabolism under either anaerobic or aerobic conditions.

Polysaccharides Are Cleaved to Form Sugar Phosphates That Also Enter the Glycolytic Pathway

Although glucose is the immediate substrate for both fermentation and respiration in many cells and tissues, the concentration of the free monosaccharide in cells is low. Instead, it occurs primarily in the form of storage polysaccharides, most commonly starch in plants and glycogen in animals. One advantage of storing glucose as starch and glycogen is that these two glucose polymers are insoluble in water and thus do not overload the limited solute capacity of the cell. As indicated in **Figure 9-10**, these storage polysaccharides can be mobilized by a process called *phosphorolysis.* Phosphorolysis resembles hydrolysis but uses inorganic phosphate rather than water to break a chemical bond. Inorganic phosphate is used to break the $\alpha(1 \rightarrow 4)$ bond between successive glucose units, liberating the glucose monomers as glucose-1-phosphate. The glucose-1-phosphate that is formed in this way can be converted to glucose-6-phosphate, which is then catabolized by the glycolytic pathway.

Notice that glucose stored in polymerized form enters the glycolytic pathway as glucose-6-phosphate, without the input of the ATP that would be required for the initial phosphorylation of the free sugar. Consequently, the overall energy yield for glucose is greater by one molecule of ATP when it is catabolized from the polysaccharide level than when it is catabolized with the free sugar as the starting substrate. This is not a case of getting something for nothing, however, because energy was required to activate the glucose units that were added to the polysaccharide chain during starch or glycogen synthesis.

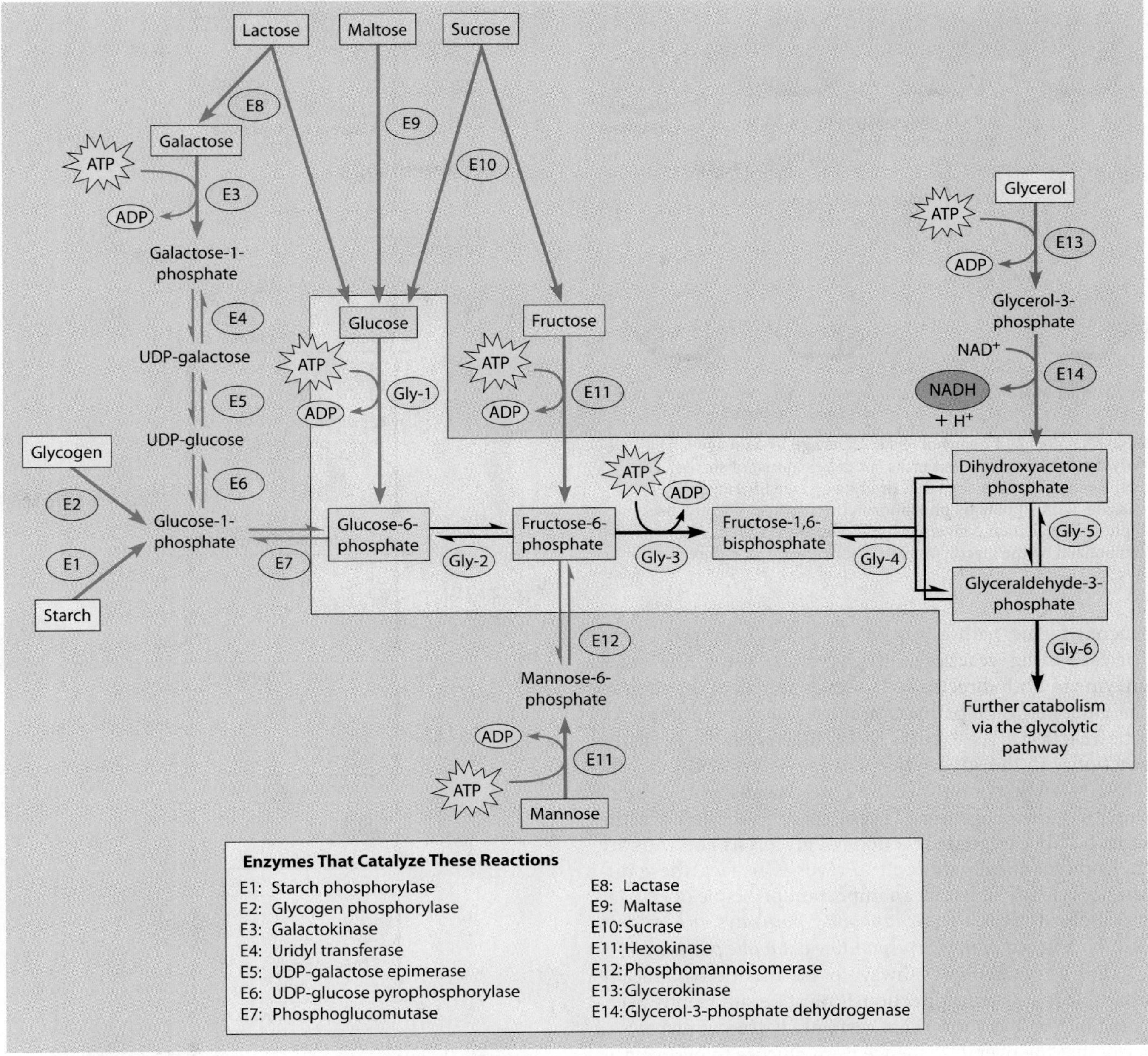

FIGURE 9-9 Carbohydrate Catabolism by the Glycolytic Pathway. Carbohydrate substrates that can be metabolized by conversion to an intermediate in the glycolytic pathway are enclosed in colored boxes. These include the hexoses galactose, glucose, fructose, and mannose; the disaccharides lactose, maltose, and sucrose; the polysaccharides glycogen and starch; and the three-carbon compound glycerol. The conversion reactions are shown by blue arrows. The enzymes that catalyze these reactions are identified in the box at the bottom. The first six reactions of the glycolytic pathway are highlighted in tan; for the names of the enzymes that catalyze these reactions, see Figure 9-7. In some cases, other enzymes or reaction sequences may be involved, depending on the organism and tissue.

Gluconeogenesis

Cells are able to catabolize glucose and other carbohydrates to meet their energy needs, and they can synthesize sugars and polysaccharides needed for other purposes. The process of glucose synthesis is called **gluconeogenesis,** which literally means "the genesis (formation) of new glucose." More specifically, gluconeogenesis is defined as the process by which cells synthesize glucose from three-carbon and four-carbon precursors that are usually noncarbohydrate in nature. The most common starting materials are pyruvate and its fermentation product lactate. Gluconeogenesis occurs in all organisms and in animals occurs mainly in the liver and kidneys.

The gluconeogenesis and glycolysis pathways share much in common; in fact, seven of the ten reactions in the

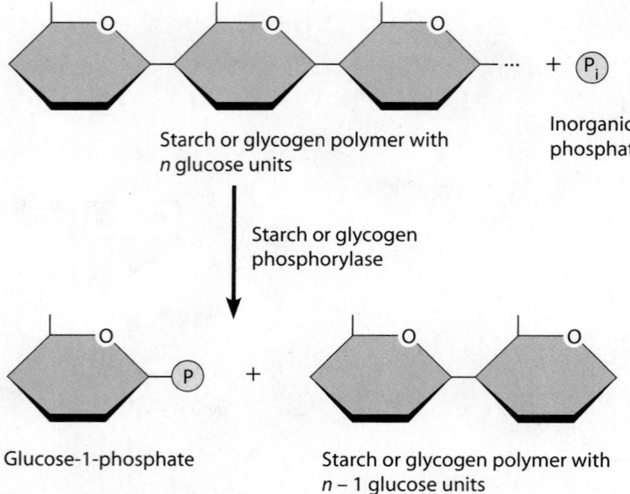

FIGURE 9-10 Phosphorolytic Cleavage of Storage Polysaccharides. Glucose units (pink hexagons) of storage polysaccharides such as starch or glycogen are liberated as glucose-1-phosphate by phosphorolytic cleavage. The glucose-1-phosphate is then converted to glucose-6-phosphate and catabolized by the glycolytic pathway, as shown in Figure 9-7.

gluconeogenic pathway occur by simple reversal of the corresponding reactions in glycolysis, using the same enzyme in both directions. However, not all of the steps of the gluconeogenic pathway are just the reversal of glycolytic reactions. As **Figure 9-11** illustrates, three of the reactions of the glycolytic pathway—Gly-1, Gly-3, and Gly-10—are accomplished by other means in the direction of gluconeogenesis. These three reactions are the most highly exergonic reactions of glycolysis and thus are thermodynamically difficult to reverse. In fact, these differences clearly illustrate an important principle of cellular metabolism: *Biosynthetic anabolic pathways are seldom just the reversal of the corresponding catabolic pathways.*

For a metabolic pathway to be thermodynamically favorable in a specific direction, it must be sufficiently exergonic in that direction. That certainly is true of glycolysis; recall that the overall sequence from glucose to pyruvate as summarized by Reaction 9-16 has a $\Delta G'$ value of about -20 kcal/mol under typical intracellular conditions in the

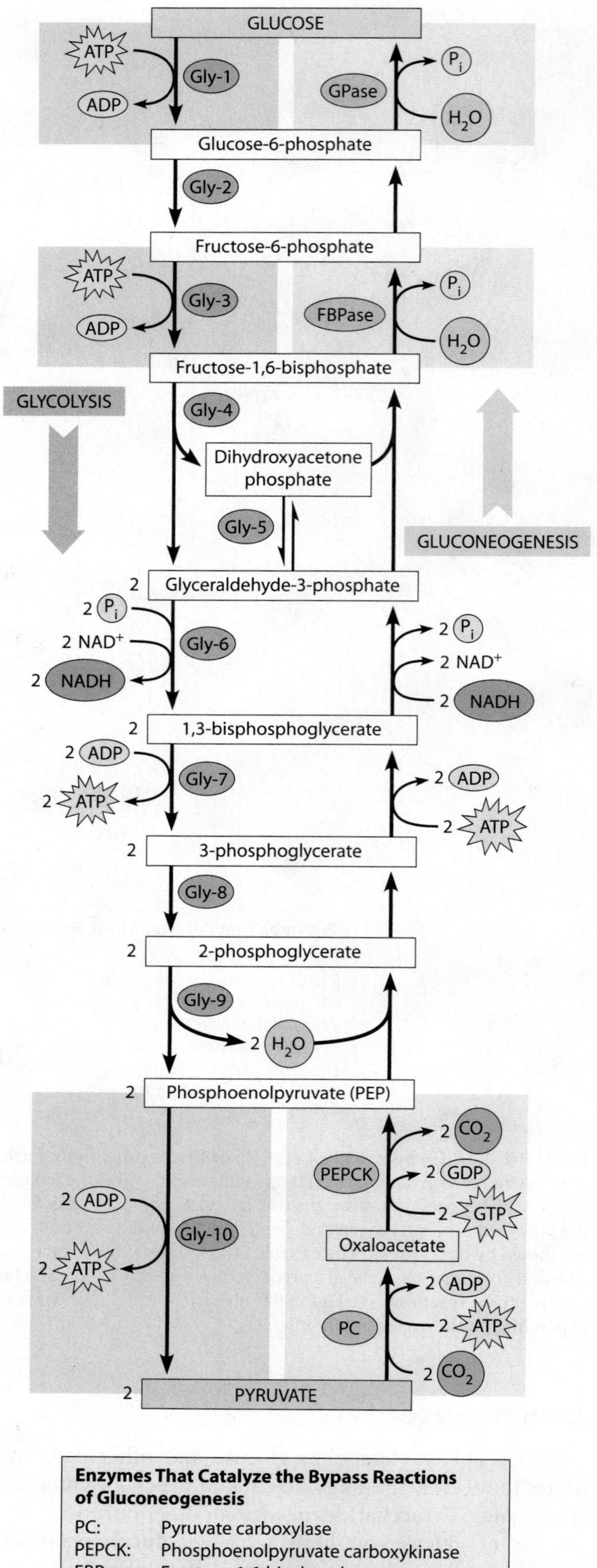

Enzymes That Catalyze the Bypass Reactions of Gluconeogenesis	
PC:	Pyruvate carboxylase
PEPCK:	Phosphoenolpyruvate carboxykinase
FBPase:	Fructose-1,6-bisphosphatase
GPase:	Glucose-6-phosphatase

FIGURE 9-11 Pathways for Glycolysis and Gluconeogenesis Compared. The pathways for glycolysis (left) and gluconeogenesis (right) have nine intermediates and seven enzyme-catalyzed reactions in common. The three essentially irreversible reactions of the glycolytic pathway (in green shading) are circumvented in gluconeogenesis by four bypass reactions (in yellow shading). Gluconeogenesis, on the other hand, is an anabolic pathway, requiring the coupled hydrolysis of six phosphoanhydride bonds (four from ATP, two from GTP) to drive it in the direction of glucose formation. The enzymes that catalyze the bypass reactions are shown in gold and are identified in the box. (For the names of the glycolytic enzymes, see Figure 9-7.) In animals, glycolysis occurs in muscle and various other tissues, whereas gluconeogenesis occurs mainly in the liver and to a lesser degree in the kidneys.

human body. Clearly, then, $\Delta G'$ for the reverse process would be about $+20$ kcal/mol, making glucose synthesis by the direct reversal of glycolysis highly endergonic and therefore thermodynamically impossible.

Gluconeogenesis is possible because the three most exergonic reactions in the glycolytic pathway (Gly-1, Gly-3, and Gly-10) do not simply "run in reverse" in the gluconeogenic direction. Instead, the gluconeogenic pathway has *bypass reactions* at each of those three sites—alternative reactions that circumvent the three glycolytic reactions that would be the most difficult to drive in the reverse direction. In fact, the three reactions of the glycolytic pathway that are bypassed in gluconeogenesis in Figure 9-11 are shown as unidirectional in Figure 9-7.

In the case of Gly-1 and Gly-3, using the exact reverse reaction would require synthesis of ATP, so instead it is bypassed by a simple hydrolytic reaction that liberates inorganic phosphate (Figure 9-11). Notice how effectively this simple metabolic ploy overcomes the thermodynamic hurdle. In the case of the conversion of glucose to glucose-6-phosphate in glycolysis, the reaction is exergonic because of the input of an ATP molecule. And in the gluconeogenic direction, it is exergonic due to the hydrolysis of the phosphoester bond, which has a $\Delta G^{\circ\prime}$ of -3.3 kcal/mol.

The third site of irreversibility in the glycolytic pathway, Reaction Gly-10, is bypassed in gluconeogenesis by a two-reaction sequence (Figure 9-11). Both of these reactions are driven by the hydrolysis of a phosphoanhydride bond, from ATP in one case and from the related compound GTP in the other. (GTP is the abbreviation for guanosine triphosphate; see Figure 3-15 for the structure of guanine.) First, CO_2 is added to pyruvate in a *carboxylation* reaction, forming a four-carbon compound called oxaloacetate. Next, the carboxyl group is removed in a *decarboxylation* reaction to form phosphoenolpyruvate (PEP). In this case, both the phosphate group and the energy are provided by GTP, which is energetically the equivalent of ATP.

What these bypass reactions accomplish becomes clear when the glycolytic and gluconeogenic pathways are compared directly (Figure 9-11). As a catabolic pathway, glycolysis is inherently exergonic, producing two ATPs per glucose. Gluconeogenesis, on the other hand, is an anabolic pathway, requiring the equivalent of six ATP molecules consumed per molecule of glucose synthesized. The difference of four ATP molecules per glucose represents enough energy to ensure that gluconeogenesis proceeds exergonically in the direction of glucose synthesis.

While it is important to acquire an academic understanding of processes like glycolysis and gluconeogenesis, it is also important to appreciate what this knowledge means for you as a person. **Box 9A** describes what is happening right now to the sugars you ate at breakfast this morning if you had a bowl of cereal with milk and will help you understand what happens in the cells of your body following your next meal.

The Regulation of Glycolysis and Gluconeogenesis

Because cells have enzymes to catalyze the reactions of both the glycolytic and the gluconeogenic pathways, it is crucial to keep both pathways from proceeding simultaneously in the same cell in an obviously futile cycle. How can the synthesis and breakdown of glucose be controlled to keep this from happening? One solution in our bodies is spatial regulation—having these pathways operate in separate cells, as shown for our muscle cells (glycolysis) and liver cells (gluconeogenesis) in Figure 9A-2. We will later examine another solution—temporal regulation, in which glycolysis and gluconeogenesis operate at different times within a single cell.

Like all metabolic pathways, glycolysis and gluconeogenesis are regulated to function at rates that are responsive to cellular and organismal needs for their products, which are ATP and glucose, respectively. Not surprisingly, glycolysis and gluconeogenesis are regulated in a reciprocal, or inverse, manner: Intracellular conditions known to stimulate one pathway usually have an inhibitory effect on the other. In addition, glycolysis is closely coordinated with other major pathways of energy generation and utilization in the cell, especially the pathways involved in aerobic respiration that we will be considering in Chapter 10.

Key Enzymes in the Glycolytic and Gluconeogenic Pathways Are Subject to Allosteric Regulation

Recall from Chapter 6 that **allosteric regulation** of enzyme activity involves the interconversion of an enzyme between two forms, one of them catalytically active (or more active) and the other inactive (or less active). An enzyme molecule will be active or inactive, depending on whether a specific allosteric effector is bound to the allosteric site as well as on whether that effector is an allosteric activator or an allosteric inhibitor (see Figure 6-16).

Figure 9-12 shows the key regulatory enzymes of the glycolytic and gluconeogenic pathways and the allosteric effectors that regulate each enzyme. For glycolysis, the key enzymes are hexokinase, phosphofructokinase-1 (PFK-1), and pyruvate kinase. For gluconeogenesis, fructose-1, 6-bisphosphatase and pyruvate carboxylase are the key regulatory enzymes. Based primarily on studies with liver cells, each allosteric effector shown in Figure 9-12 is identified as either an activator (+) or an inhibitor (−) of the enzyme(s) it binds to. Several points become apparent from the figure. Notice, for instance, that *each of the regulatory enzymes is unique to its pathway,* so each pathway can be regulated independently of the other. Notice also *the reciprocal nature of the regulation of the two pathways:* AMP and acetyl CoA, the two effectors to which both pathways are sensitive, have opposite effects in the two directions. AMP, for example, activates glycolysis but inhibits gluconeogenesis. Acetyl CoA activates gluconeogenesis but inhibits glycolysis.

BOX 9A DEEPER INSIGHTS

"What Happens to the Sugar?"

Now that you have studied the pathways of glycolysis and gluconeogenesis in detail, we hope you can use this information to understand how your body meets its energy needs and what it does with the nutrients that you eat. To put it more specifically, can you relate what you are learning in these chapters to what the cells in your body are doing with the food you had for breakfast this morning? As you add sugar to your coffee or cereal, can you answer the question, "What happens to the sugar?" This essay will address these questions and might help you appreciate the relevance of these chapters by letting you see how the information they contain applies to your daily life.

To keep the topic manageable, let's focus on a bowl of cereal (**Figure 9A-1**), considering the disaccharide sucrose (from the sugar bowl), the disaccharide lactose (in the milk), and the polysaccharide starch (in the cereal). First we'll follow the sugars and starch through your digestive tract; then we'll consider the glucose in your blood and the several ways the cells in different parts of your body use it.

Let's start with a spoonful of cereal you've just eaten. The sucrose and lactose remain intact until they reach your small intestine, but the digestion of starch begins in your mouth because saliva contains salivary amylase, an enzyme that splits starch into smaller polysaccharides. Further digestion occurs in your small intestine, where pancreatic amylase completes the breakdown of starch to the disaccharide maltose.

The maltose generated from starch is hydrolyzed to glucose in your intestine by the enzyme maltase. Maltase is one of a family of intestinal disaccharidases, each specific for a different disaccharide. The lactose from the milk and the sucrose that you sprinkled on the cereal are hydrolyzed by other members of this family—lactase and sucrase, respectively. Lactose yields one molecule each of glucose and galactose, whereas sucrose is hydrolyzed to one molecule each of glucose and fructose (see Figure 9-9). In some people, intestinal lactase disappears gradually after age 4 or so, when milk drinking usually decreases. If such people ingest milk or other dairy products, they are likely to experience cramps and diarrhea, a condition called *lactose intolerance*.

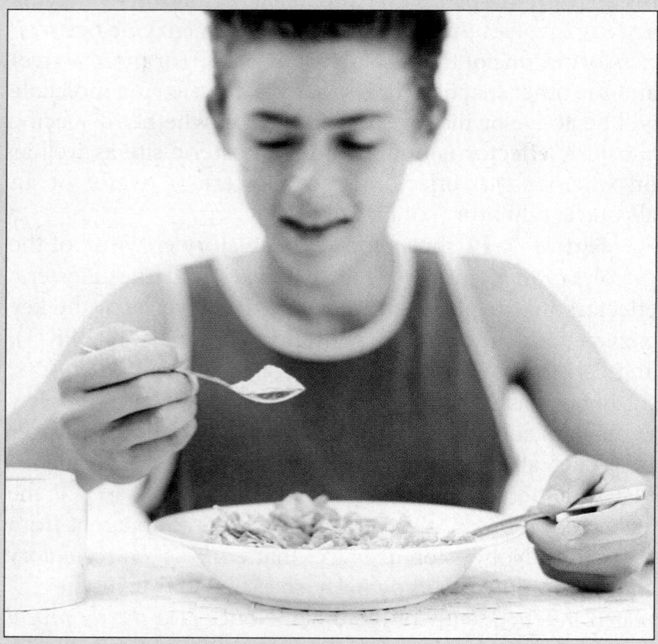

FIGURE 9A-1 What Happens to the Sugar? In this essay we will see what happens in different parts of your body to the sugars and starch in a breakfast of milk and cereal.

Glucose, galactose, and fructose molecules are absorbed by intestinal epithelial cells. These cells have numerous microvilli that project into the lumen of the intestine, thereby greatly increasing the absorptive surface of the cell (see Figure 1-1g and Figure 4-2). Moreover, only two layers of epithelial cells separate nutrients in the lumen of your intestine from the blood in your capillaries. Some sugars, such as fructose, move across the plasma membrane of an epithelial cell by facilitated diffusion because the concentrations of these sugars are lower in capillary blood than in the intestinal lumen. Glucose, however, is moved by active transport because of its high concentration in blood.

Fructose and galactose are transported by your bloodstream to the various tissues of your body. These sugars are eventually absorbed by body cells and converted to intermediates in the glycolytic pathway, as shown in Figure 9-9. The pathway for galactose utilization is more complex than that of most other simple sugars, with five reactions required to convert a molecule of galactose into glucose-6-phosphate. A genetic defect in this pathway may result in an inability to metabolize galactose, resulting in high levels of galactose in the blood and high levels of galactose-1-phosphate in tissues. This disorder, called *galactosemia,* has serious consequences, including mental retardation. Not surprisingly, it occurs most commonly in infants because the major dietary source of galactose is milk. Provided the condition is detected early, the symptoms can be prevented or alleviated by removing milk and dairy products from the diet.

The main sugar in the blood, of course, is glucose. Its concentration in your blood a few hours after a meal is probably about 80 mg/dL (80 mg per deciliter [100 mL] of blood, or about 4.4 m*M*). The level may rise to 120 mg/dL (6.6 m*M*) shortly after you've eaten. In general, however, the blood glucose level is maintained within rather narrow limits. Maintenance of the blood glucose level is one of the most important regulatory functions in your body, particularly for proper functioning of your brain and nervous system. Your blood glucose level is under the control of several hormones, including insulin, glucagon, epinephrine, and norepinephrine (see Chapter 14).

Once in your bloodstream, glucose is transported to cells in all parts of your body, where it has four main fates: It can be oxidized completely by aerobic respiration to CO_2, it can be fermented anaerobically to lactate, it can be used to synthesize the polysaccharide glycogen, or it can be converted to body fat.

Aerobic respiration is the most common fate of blood glucose because most of the tissues in your body function aerobically most of the time. Your brain is particularly noteworthy as an aerobic organ. It needs large amounts of energy to maintain the membrane potentials essential for the transmission of nerve impulses, and it normally depends solely on glucose to meet this need. In fact, your brain needs about 120 g of glucose per day, which is about 15% of your total energy consumption. When you are at rest, your brain accounts for about 60% of your glucose usage. The brain also accounts for about 20% of your total oxygen consumption. As the brain has no significant stores of glycogen, the supply of both oxygen and glucose must be continuous. Even a short interruption of either has dire consequences. Your heart has similar requirements because it is also a completely aerobic organ and has little or no energy reserves. The supply of oxygen and fuel molecules must therefore be constant, though the heart—unlike the brain—can use a variety of fuel molecules, including glucose, lactate, and fatty acids.

In addition to aerobic respiration in a wide variety of tissues, glucose can be *catabolized anaerobically* (fermented to lactate), especially in red blood cells and skeletal muscle cells. Red blood cells have no mitochondria and depend exclusively on glycolysis to meet their energy needs. Skeletal muscle can function in either the presence or the absence of oxygen. When you exert yourself strenuously, oxygen becomes limiting, so the rate of glycolysis exceeds that of aerobic respiration and excess pyruvate is converted to lactate. Lactate is released into the blood and taken up not only by your heart for use as fuel but also by gluconeogenic tissues, especially your liver. When lactate molecules enter a liver cell they are reoxidized to pyruvate, which is then used to make glucose by gluconeogenesis (see Figure 9-11). The glucose is returned to the bloodstream, where it can be taken up by muscle (or any other) cells again.

Skeletal muscle is the main source of blood lactate, and the liver is the primary site of gluconeogenesis, so a cycle is set up, as shown in **Figure 9A-2**. Lactate produced by glycolysis in hypoxic (oxygen-deficient) muscle cells is transported via the blood to the liver. There, gluconeogenesis converts the lactate to glucose, which is released into the blood. This process is called the *Cori cycle,* for Carl and Gerti Cori, whose studies in the 1930s and 1940s described it. The next time you are resting after strenuous exercise, think of what is happening: The lactate your muscle cells have just released into the bloodstream is being taken up by liver cells and converted back to glucose. The reason you are breathing heavily is to provide the oxygen your body needs to return your muscle cells to aerobic conditions and to generate all the ATP and GTP needed for gluconeogenesis in your liver and for rebuilding body glycogen stores.

Glycogen storage is the third significant fate of blood glucose. Glycogen is stored primarily in the cells of your liver and skeletal muscle. Muscle glycogen is used to supply glucose during times of strenuous exertion. Liver glycogen, on the other hand, is used as a source of glucose when the liver is stimulated hormonally to release glucose into the bloodstream to maintain the blood glucose level.

The fourth possible fate of blood glucose is its use for the *synthesis of body fat*. The route to fat is via pyruvate to acetyl CoA, just as in the initial phase of aerobic respiration. Whenever you eat more food than your body needs for energy and for the biosynthesis of other molecules, excess glucose is oxidized to acetyl CoA, used for the synthesis of triacylglycerols, and then stored as body fat, especially in adipose tissue that is specialized for this purpose. Thus, your body has three sources of energy at all times: the glucose in your blood, the glycogen in your liver and skeletal muscle cells, and the triacylglycerols stored in adipose tissue.

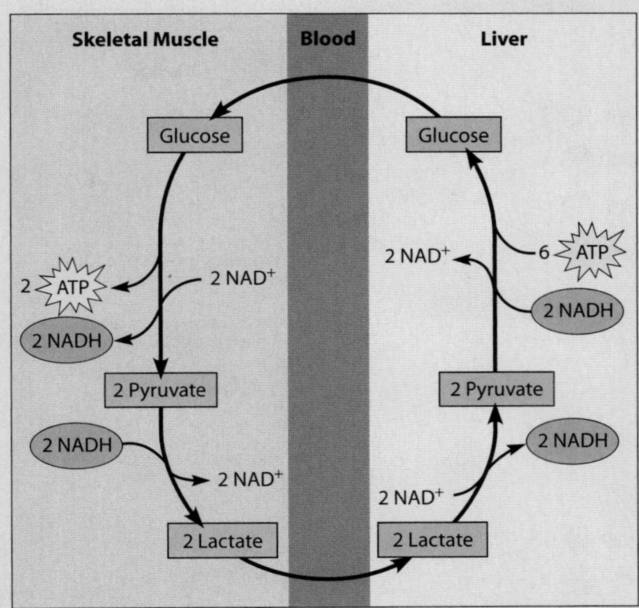

FIGURE 9A-2 The Cori Cycle: The Link Between Glycolysis in Muscle Cells and Gluconeogenesis in the Liver. Skeletal muscle cells derive much of their energy from glycolysis, especially during anaerobic periods of strenuous exercise. The lactate produced in this way is transported by the bloodstream to the liver, where it is reoxidized to pyruvate. Pyruvate is used as the substrate for gluconeogenesis within the liver, generating glucose that is then returned to the blood.

To conclude, let's come back to our original question: "What happens to the sugar?" All the glucose and other sugars in your body come originally from the food you eat—either as monosaccharides directly or from the breakdown of disaccharides and polysaccharides in your intestinal tract. The ultimate fate of that glucose is oxidation to CO_2 and water, which you then exhale and excrete. But in the meantime, glucose molecules can circulate in your bloodstream or be stored as glycogen in liver or muscle cells. In its circulating form, glucose can be oxidized immediately by aerobic tissues such as the brain, it can be converted to lactate and become a part of the Cori cycle, or it can be used to synthesize glycogen or fat for storage.

So although it may look like just a modest spoonful of sugar as you sprinkle it on your cereal or add it to your coffee, it plays an important role in the energy metabolism of all of the cells in your body.

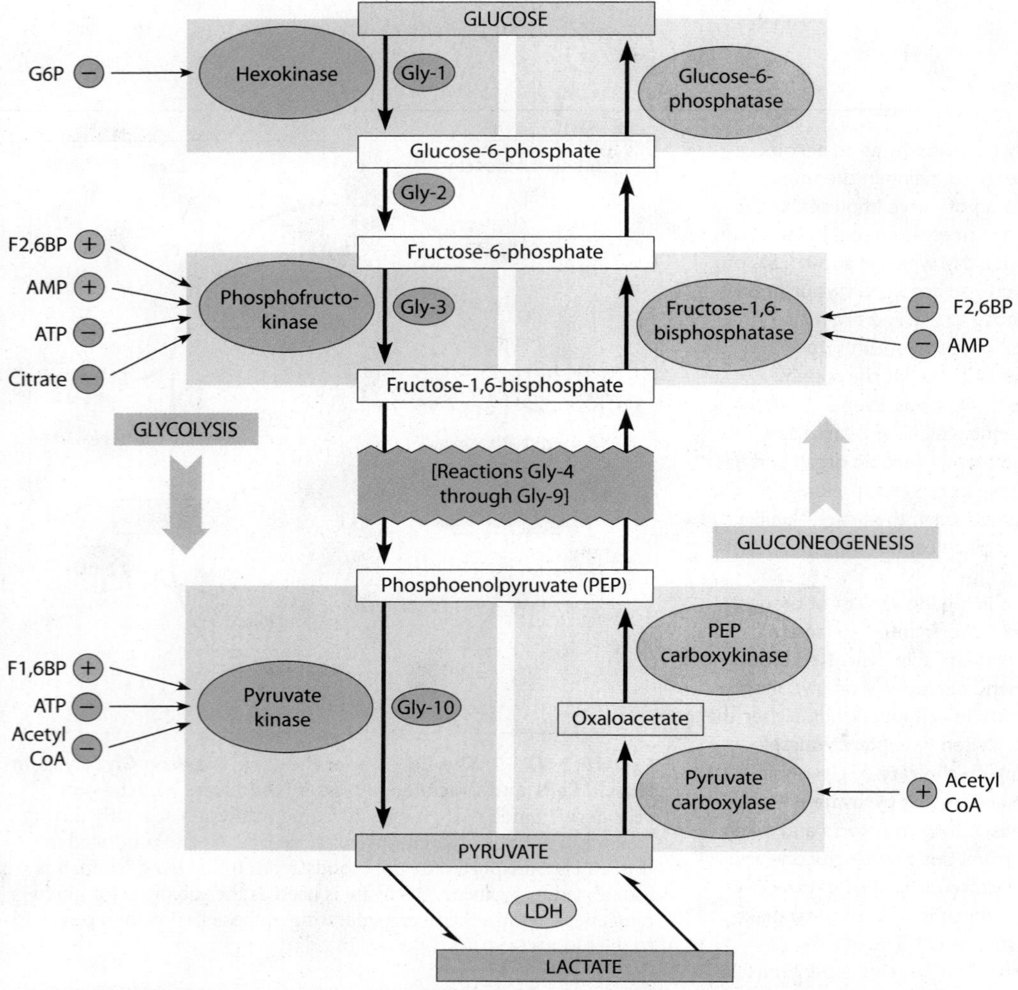

FIGURE 9-12 The Regulation of Glycolysis and Gluconeogenesis. Glycolysis and gluconeogenesis are regulated in a reciprocal manner. In both cases, regulation involves allosteric activation (+) or inhibition (−) of enzymes that catalyze reactions unique to the pathway. For glycolysis, the key regulatory enzymes are those that catalyze the three irreversible reactions unique to this pathway (green). For gluconeogenesis, two of the four bypass enzymes (gold) that are unique to this pathway are the main sites of allosteric regulation. Allosteric regulators include acetyl CoA, AMP, ATP, citrate, fructose-1,6-bisphosphate (F1,6BP), fructose-2,6-bisphosphate (F2,6BP), and glucose-6-phosphate (G6P). Acetyl CoA and citrate are intermediates in aerobic respiration. F2,6BP is synthesized by phosphofructokinase-2 (PFK-2), as shown in Figure 9-13b.

Moreover, the effects of the regulatory agents make sense—that is, they are invariably in the direction one would predict, based on an understanding of the role each pathway plays in the cell. Consider, for example, the effects of ATP and AMP. When the concentration of ATP is low and the AMP concentration is high, the cell is clearly low on energy, so it is reasonable for AMP to activate glycolysis. Conversely, as the ATP concentration increases and the AMP concentration decreases, the stimulatory effects of AMP on glycolysis lessen. The inhibitory effect of ATP on both PFK-1 and pyruvate kinase comes into play, reducing the rate of glycolysis.

You may be surprised to learn that ATP is an allosteric inhibitor of PFK-1 because this enzyme uses ATP as a substrate. That seems contradictory because increases in substrate concentration should *increase* the rate of an enzyme-catalyzed reaction. This apparent contradiction is readily explained, however. As an allosteric enzyme, PFK-1 has both an active site and an allosteric site. The active site of PFK-1 has a high affinity for ATP, whereas the allosteric site has a low affinity for ATP. Thus, at low ATP concentrations, binding occurs at the catalytic site but not at the allosteric site, so most of the PFK-1 molecules remain in the active form and glycolysis proceeds. As the ATP concentration increases, however, binding is enhanced at the allosteric site, stabilizing the inactive form of PFK-1 and thereby serving to slow down the whole glycolytic sequence.

Both the glycolytic and the gluconeogenic pathways (Figure 9-12) are also subject to allosteric regulation by compounds involved in respiration. As we will learn in the next chapter, acetyl CoA and citrate are key intermediates in an aerobic pathway called the *tricarboxylic acid cycle*.

High levels of acetyl CoA and citrate indicate that the cell is well supplied with substrate for the next phase of respiratory metabolism beyond pyruvate. Thus, it is not surprising to find that acetyl CoA and citrate both have inhibitory effects on glycolysis, thereby decreasing the rate at which pyruvate is formed. Similarly, the stimulatory effect of acetyl CoA on gluconeogenesis is consistent with the availability of pyruvate for conversion to glucose.

Fructose-2,6-Bisphosphate Is an Important Regulator of Glycolysis and Gluconeogenesis

Although each of the preceding mechanisms plays a significant role in the regulation of glycolysis and gluconeogenesis, the most important regulator of both pathways is **fructose-2,6-bisphosphate (F2,6BP).** F2,6BP is synthesized by the ATP-dependent phosphorylation of fructose-6-phosphate on carbon number 2. This is the same type of reaction that produces fructose-1,6-bisphosphate in Reaction Gly-3 in the glycolytic pathway by phosphorylation of fructose-6-phosphate on carbon number 1. However, synthesis of F2,6BP is catalyzed by a separate form of phosphofructokinase, which is called **phosphofructokinase-2 (PFK-2)** to distinguish it from PFK-1, the glycolytic enzyme. As we saw in Figure 9-12, F2,6BP activates the glycolytic enzyme (PFK-1) that phosphorylates fructose-6-phosphate, and it inhibits the gluconeogenic enzyme (FBPase) that catalyzes the reverse reaction.

Figure 9-13 depicts these regulatory roles of PFK-2 and F2,6BP in the human body in more detail. The activity of PFK-2 depends on the phosphorylation status of one of its subunits. The kinase activity of the enzyme is high when that subunit is in the unphosphorylated form and low when it is phosphorylated. The phosphorylation of PFK-2 by ATP is catalyzed by a protein kinase (Figure 9-13b). The activity of this enzyme depends, in turn, on cyclic AMP (cAMP), which is a key intermediate in many cellular signal transduction pathways (see Figure 14-6).

In addition to its *phosphofructokinase* activity responsible for synthesis of F2,6BP, PFK-2 has a *fructose-2,6-bisphosphatase* activity that removes the phosphate group from F2,6BP, converting the compound back to fructose-6-phosphate (Figure 9-13). This activity is also regulated by cAMP-stimulated phosphorylation, but in this case phosphorylation increases the activity of the enzyme. Because it has two separate catalytic activities, PFK-2 is called a *bifunctional enzyme.*

As noted earlier, F2,6BP activates the glycolytic enzyme PFK-1 (Figure 9-13d) and inhibits the gluconeogenic enzyme FBPase (Figure 9-13e). cAMP, in turn, affects the F2,6BP concentration in two ways: It inactivates the PFK-2 kinase activity and stimulates the F2,6BPase phosphatase activity. Both of these effects tend to decrease the concentration of F2,6BP in the cell. This change leads, in turn, to less stimulation of PFK-1 and less inhibition of fructose-1,6-bisphosphatase, thereby decreasing the glycolytic activity and increasing the gluconeogenic activity.

The effects of cAMP shown in Figure 9-13 are important in hormonal regulation because the cAMP level in liver cells is controlled primarily by the hormones glucagon and epinephrine (adrenalin). These hormones cause an increase in cAMP concentration, thereby stimulating gluconeogenesis when more glucose is needed. Moreover, the increase in cAMP concentration stimulates a regulatory cascade that increases the rate of glycogen breakdown (see Figure 14-22). Not surprisingly, the effect of cAMP on glycogen synthesis is just the opposite: Whether triggered by glucagon or epinephrine, an increase in liver cAMP concentration leads to a decrease in the rate of glycogen formation. For further details on hormonal regulation and the role of cAMP in mediating hormonal effects, see the discussion on hormonal signal transduction in Chapter 14.

Novel Roles for Glycolytic Enzymes

It is easy to get the impression that, for a fundamental and ubiquitous pathway like glycolysis, we have fully described its individual steps and learned "all there is to know" about the enzymes involved. But, as we often discover in biology, there may be surprises in store. One such surprise is the discovery that several of the glycolytic enzymes have regulatory functions that, at first glance, seem unrelated to their roles as catalysts in specific reactions in glycolysis. These functions include transcriptional regulation, stimulation of cell motility, and regulation of *apoptosis*—the process of programmed cell death.

For example, the enzyme hexokinase, which catalyzes the ATP-dependent phosphorylation of glucose to initiate the glycolytic sequence (Gly-1 in Figure 9-7), has been shown to be a transcriptional regulator in yeast cells. An isoform known as hexokinase 2 becomes localized in the nucleus in response to high levels of glucose. In the nucleus, it acts as a transcriptional repressor to down-regulate the expression of genes required for catabolism of sugars other than glucose. In mammals, four isoforms of hexokinase have been discovered. One isoform displays increased expression in tumor cells, which show increased glycolysis even in the presence of oxygen. Another isoform binds to mitochondria and is believed to coordinate glycolysis with mitochondrial respiration. It also appears to act as an inhibitor of apoptosis in cells with a plentiful supply of glucose. As you can see, we have much more to learn about a gene once thought to serve merely a "housekeeping" function in supplying cellular energy.

Other glycolytic enzymes, including glyceraldehyde-3-phosphate dehydrogenase (GAPDH; Gly-6) and enolase (Gly-9), also have DNA-binding abilities and can act as transcriptional regulators. GAPDH—along with lactate dehydrogenase, the enzyme that converts pyruvate to lactate in fermentation—can act as a transcriptional activator during cell division. GAPDH activity in this role is stimulated by NAD^+ and inhibited by NADH. Thus, GAPDH is thought to serve as a link between energy metabolism and regulation of cell division, a process requiring a plentiful

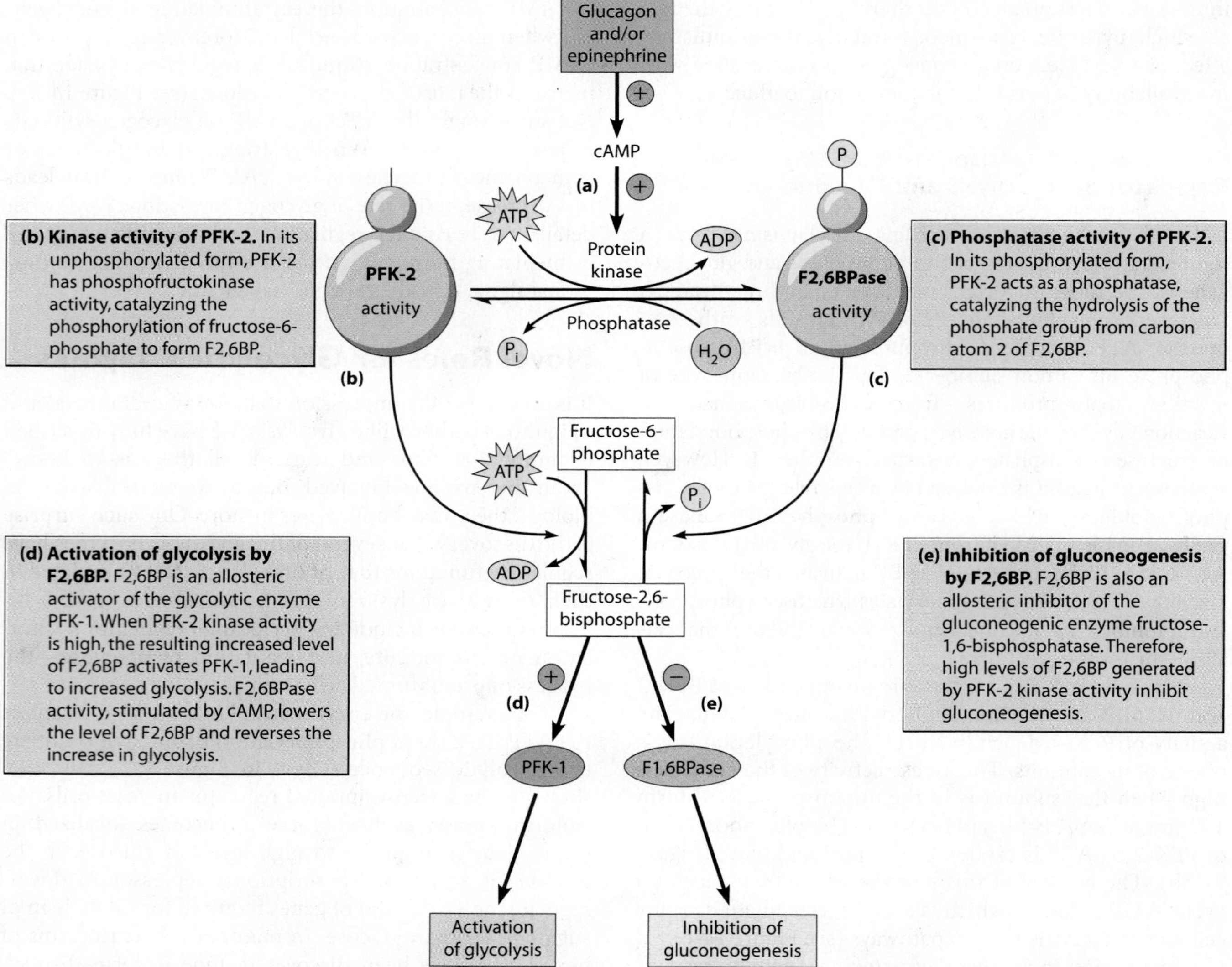

(a) Regulation of PFK-2 activities by reversible phosphorylation. In response to glucagon and epinephrine, cAMP levels rise and PFK-2 is phosphorylated by a cAMP-activated protein kinase. In the absence of cAMP, it is dephosphorylated by a phosphatase. The phosphorylation state of PFK-2 determines which of its two catalytic activities it will have.
Glucagon and/or epinephrine
cAMP
(a)
ATP
ADP
Protein kinase
Phosphatase
P_i
H_2O
P
PFK-2 activity
F2,6BPase activity
(b) Kinase activity of PFK-2. In its unphosphorylated form, PFK-2 has phosphofructokinase activity, catalyzing the phosphorylation of fructose-6-phosphate to form F2,6BP.
(c) Phosphatase activity of PFK-2. In its phosphorylated form, PFK-2 acts as a phosphatase, catalyzing the hydrolysis of the phosphate group from carbon atom 2 of F2,6BP.
(b)
(c)
Fructose-6-phosphate
ATP
ADP
P_i
Fructose-2,6-bisphosphate
(d) Activation of glycolysis by F2,6BP. F2,6BP is an allosteric activator of the glycolytic enzyme PFK-1. When PFK-2 kinase activity is high, the resulting increased level of F2,6BP activates PFK-1, leading to increased glycolysis. F2,6BPase activity, stimulated by cAMP, lowers the level of F2,6BP and reverses the increase in glycolysis.
(e) Inhibition of gluconeogenesis by F2,6BP. F2,6BP is also an allosteric inhibitor of the gluconeogenic enzyme fructose-1,6-bisphosphatase. Therefore, high levels of F2,6BP generated by PFK-2 kinase activity inhibit gluconeogenesis.
(d)
(e)
PFK-1
F1,6BPase
Activation of glycolysis
Inhibition of gluconeogenesis

FIGURE 9-13 The Regulatory Roles of PFK-2 and F2,6BP. Phosphofructokinase-2 (PFK-2) is a bifunctional enzyme with two opposing catalytic activities that depend upon its own phosphorylation state **(a–c)**. Which of the two activities it has depends on the presence of hormonal signals and determines the level of fructose-2,6-bisphosphate (F2,6BP), an activator of glycolysis **(d)** and inhibitor of gluconeogenesis **(e)**.

supply of ATP. GAPDH is overexpressed in cells undergoing programmed cell death in response to cellular injury and accumulates in the nuclei of these cells. It is proposed to be an intracellular sensor of oxidative stress, with a possible link to neurodegenerative diseases.

Other recent work suggests that enolase (Gly-9) can also act as a transcriptional repressor. It inhibits expression of the cancer-inducing *MYC* oncogene, which encodes a protein overexpressed in a wide variety of tumor cells. Enolase expression was shown to be down-regulated in lung cancer cells, and low expression of enolase correlates with poor survival of lung cancer patients.

Another surprising role for a glycolytic enzyme was revealed in work showing the involvement of phosphoglucoisomerase (PGI; Gly-2) in cell motility and migration during cancer cell *metastasis* (metastasis is the release of cells from malignant tumors into the bloodstream, where they can be transported throughout the body to establish secondary tumors). As an enzyme in the cell cytoplasm, PGI catalyzes the conversion of glucose-6-phosphate to fructose-6-phosphate. However, PGI has also been shown to be secreted from tumor cells, and the secreted form of PGI stimulates high levels of cell migration and proliferation. Moreover, PGI promotes the growth of fibroblasts (connective tissue cells), and the overexpression of PGI induces transformation of fibroblasts to cancer cells.

Clearly, there is much more to learn about the glycolytic enzymes, which for many years were thought to play relatively mundane roles as catalysts in energy-producing reactions. We now know that several of these enzymes have regulatory roles as well. They may help to integrate metabolic pathways with energy-dependent cell functions, such as cell division, apoptosis, and cell motility. So, as you learn about aerobic respiration and photosynthesis in the next two chapters, we hope you will avoid the urge to think that you must simply memorize "all there is to know" about these well-characterized biochemical pathways. Instead, keep in mind that some of the enzymes involved are revealing surprising new secrets.

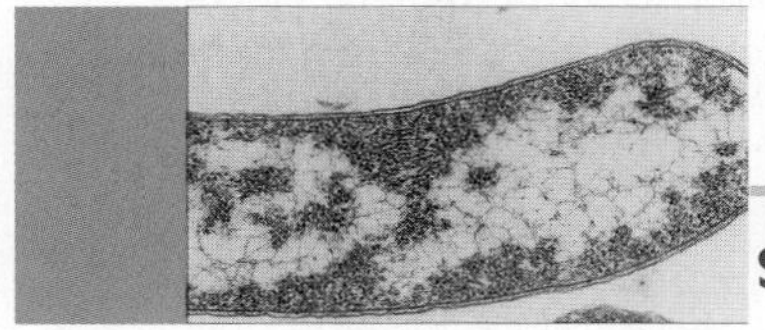

SUMMARY OF KEY POINTS

Metabolic Pathways

- Metabolic pathways in cells are usually either anabolic (synthetic) or catabolic (degradative). Catabolic reactions provide the energy necessary to drive the anabolic reactions.

ATP: The Universal Energy Coupler

- ATP is useful for storing chemical energy in cells because its terminal anhydride bond has an intermediate free energy of hydrolysis. This allows ATP to serve as a donor of phosphate groups to a number of biologically important molecules such as glucose. It also allows ADP to serve as an acceptor of phosphate groups from molecules such as PEP.

Chemotrophic Energy Metabolism

- Most chemotrophs derive the energy needed to generate ATP from the catabolism of organic nutrients such as carbohydrates, fats, and proteins. They do so either by fermentative processes in the absence of oxygen or by aerobic respiratory metabolism in the presence of oxygen.
- Although glycolysis may seem overly complex, it represents a mechanism by which glucose can be degraded in dilute solution at temperatures compatible with life, with a large portion of the free energy yield conserved as ATP.

Glycolysis and Fermentation: ATP Generation Without the Involvement of Oxygen

- Using glucose as a prototype substrate, catabolism under both anaerobic and aerobic conditions begins with glycolysis, a ten-step pathway that converts glucose into pyruvate. In most cases, this leads to the production of two molecules of ATP per molecule of glucose.
- In the absence of oxygen, the reduced coenzyme NADH generated during glycolysis must be reoxidized at the expense of pyruvate, leading to fermentation end-products such as lactate or ethanol plus carbon dioxide.

Alternative Substrates for Glycolysis

- Although usually written with glucose as the starting substrate, the glycolytic sequence is also the mainstream pathway for catabolizing a variety of related sugars, such as fructose, galactose, and mannose.
- Glycolysis is also used to metabolize the glucose-1-phosphate derived by phosphorolytic cleavage of storage polysaccharides such as starch or glycogen.

Gluconeogenesis

- Gluconeogenesis is, in a sense, the opposite of glycolysis because it is the pathway used by some cells to synthesize glucose from three- and four-carbon starting materials such as pyruvate. However, the gluconeogenic pathway is not just glycolysis in reverse.
- The two pathways have seven enzyme-catalyzed reactions in common, but the three most exergonic reactions of glycolysis are bypassed in gluconeogenesis by reactions that render the pathway exergonic in the gluconeogenic direction by the input of energy from ATP and GTP.

The Regulation of Glycolysis and Gluconeogenesis

- Glycolysis and gluconeogenesis are regulated by modifying the activity of enzymes that are unique to each pathway. These enzymes are regulated by one or more key intermediates in aerobic respiration, including ATP, ADP, AMP, acetyl CoA, and citrate.
- An important allosteric regulator of both glycolysis and gluconeogenesis is fructose-2,6-bisphosphate. Its concentration depends on the relative kinase and phosphatase activities of the bifunctional enzyme PFK-2.
- PFK-2 in turn is regulated by the hormones glucagon and epinephrine via their effects on the cyclic AMP concentration in cells.

Novel Roles for Glycolytic Enzymes

- In addition to their roles as catalysts, several well-known glycolytic enzymes have recently been shown to play regulatory roles in cells, affecting processes such as cell division, programmed cell death, and cancer cell migration.

MB www.masteringbiology.com

1. MasteringBiology Assignments

Tutorials and Activities

- Molecular Model Structural Tutorials for visualization of 3D structures
- BioFlix Tutorials including 3D animations with hints and detailed feedback for common wrong answers
- Enzymes & Pathways: Metabolic Pathways; Chemical Reactions and ATP; ATP and Energy; Enzymes & Pathways: Biosynthesis; Glycolysis

Questions

- Reading Quiz
- Multiple Choice
- End-of-chapter questions

2. eText Read your book online, search, highlight text, take notes, and more

3. The Study Area

- Chapter Review Quiz, Guide to Techniques and Methods, 3D-Structure Tutorials, Activities, BioFlix, Videos, Biology Review, Word Study Tools, Glossary, Flashcards

PROBLEM SET

More challenging problems are marked with a •.

9-1 High-Energy Bonds. When first introduced by Fritz Lipmann in 1941, the term *high-energy bond* was considered a useful concept for describing the energetics of biochemical molecules and reactions. However, the term can lead to confusion when relating ideas about cellular energy metabolism to those of physical chemistry. To check out your own understanding, indicate whether each of the following statements is true (T) or false (F). If false, reword the statement to make it true.

(a) The synthesis of ATP from ADP (with the loss of a water molecule by condensation) is an exergonic reaction.

(b) Each phosphate group in a molecule of ATP bears at least one negative charge due to its ionization at the near neutral pH of the cell.

(c) Free energy is released when an "energy-rich bond" is hydrolyzed.

(d) The terminal phosphate of ATP is a high-energy phosphate that takes its high energy with it when it is hydrolyzed.

(e) Phosphoester bonds are low-energy bonds because they require less energy to break than the high-energy bonds of phosphoanhydrides.

(f) When a phosphate group is removed from ATP, the result is an overall increase in entropy.

9-2 The History of Glycolysis. Following are several historical observations that led to the elucidation of the glycolytic pathway. In each case, suggest a metabolic basis for the observed effect, and explain the significance of the observation for the elucidation of the pathway.

(a) Alcoholic fermentation in yeast extracts requires a heat-labile fraction originally called *zymase* and a heat-stable fraction *(cozymase)* that is necessary for the activity of zymase.

(b) Alcoholic fermentation does not take place in the absence of inorganic phosphate.

(c) In the presence of iodoacetate, a known inhibitor of glycolysis, fermenting yeast extracts accumulate a doubly phosphorylated hexose.

(d) In the presence of fluoride ion, another known glycolytic inhibitor, fermenting yeast extracts accumulate two phosphorylated three-carbon acids.

9-3 Glycolysis in 25 Words or Less. Complete each of the following statements about the glycolytic pathway in 25 words or less.

(a) Although the brain is an obligately aerobic organ, it still depends on glycolysis because . . .

(b) During glycolysis, when glucose is partially oxidized to pyruvate, energy and reducing power are conserved in . . .

(c) Skeletal muscle cells normally function aerobically, but during exercise . . .

(d) In yeast, the anaerobic fermentation process produces the end products . . .

(e) Two organs in your body that can use lactate are . . .

(f) The synthesis of glucose from lactate in a liver cell requires more molecules of nucleoside triphosphates (ATP and GTP) than are formed during the catabolism of glucose to lactate in a muscle cell because . . .

9-4 Energetics of Carbohydrate Utilization. The anaerobic fermentation of free glucose has an ATP yield of two ATP molecules per molecule of glucose. For glucose units in a glycogen molecule, the yield is three ATP molecules per molecule of glucose. The corresponding value for the disaccharide sucrose is two ATP molecules per molecule of monosaccharide if the sucrose is eaten by an animal, but two and a half ATP molecules per molecule of monosaccharide if the sucrose is metabolized by a bacterium.

(a) Explain why the glucose units present in glycogen have a higher ATP yield than free glucose molecules.

(b) Based on what you know about the process of glycogen breakdown, suggest a mechanism for bacterial sucrose metabolism that is consistent with an energy yield of two and a half ATP molecules per molecule of monosaccharide.

(c) What is the likely mechanism for sucrose breakdown in the gut of an animal to explain the energy yield of two ATP molecules per molecule of monosaccharide?

(d) What energy yield (in molecules of ATP per molecule of monosaccharide) would you predict for the bacterial catabolism of raffinose, a trisaccharide?

9-5 Glucose Phosphorylation. The direct phosphorylation of glucose by inorganic phosphate is a thermodynamically unfavorable reaction:

$$\text{glucose} + P_i \longrightarrow \text{glucose-6-phosphate} + H_2O \qquad \Delta G^{\circ\prime} = +3.3 \text{ kcal/mol} \tag{9-21}$$

In the cell, glucose phosphorylation is accomplished by coupling the reaction to the hydrolysis of ATP, a highly exergonic reaction:

$$\text{ATP} + H_2O \longrightarrow \text{ADP} + P_i \qquad \Delta G^{\circ\prime} = -7.3 \text{ kcal/mol} \tag{9-22}$$

Typical concentrations of these intermediates in yeast cells are as follows:

[glucose-6-phosphate] = 0.08 m*M*	[ATP] = 1.8 m*M*
[P_i] = 1.0 m*M*	[ADP] = 0.15 m*M*

Assume a temperature of 25°C for all calculations.

(a) What minimum concentration of glucose would have to be maintained in a yeast cell for direct phosphorylation (Reaction 9-21) to be thermodynamically spontaneous? Is this physiologically reasonable? Explain your reasoning.

(b) What is the overall equation for the coupled (ATP-driven) phosphorylation of glucose? What is its $\Delta G^{\circ\prime}$ value?

(c) What minimum concentration of glucose would have to be maintained in a yeast cell for the coupled reaction to be thermodynamically spontaneous? Is this physiologically reasonable?

(d) By about how many orders of magnitude is the minimum required glucose concentration reduced when the phosphorylation of glucose is coupled to the hydrolysis of ATP?

(e) Assuming a yeast cell to have a glucose concentration of 5.0 m*M*, what is $\Delta G'$ for the coupled phosphorylation reaction?

9-6 Ethanol Intoxication and Methanol Toxicity. The enzyme alcohol dehydrogenase was mentioned in this chapter because of its role in the final step of alcoholic fermentation. However, the enzyme also occurs commonly in aerobic organisms, including humans. The ability of the human body to catabolize the ethanol in alcoholic beverages depends on the presence of alcohol dehydrogenase in the liver. One effect of ethanol intoxication is a dramatic decrease in the NAD^+ concentration in liver cells, which decreases the aerobic utilization of glucose. Methanol, on the other hand, is not just an intoxicant; it is a deadly poison due to the toxic effect of the formaldehyde to which it is converted in the liver.

(a) Why does ethanol consumption lead to a reduction in NAD^+ concentration and to a decrease in aerobic respiration?

(b) The medical treatment for methanol poisoning usually involves administration of large doses of ethanol. Why is this treatment effective?

(c) Most of the unpleasant effects of hangovers result from an accumulation of acetaldehyde and its metabolites. Where does the acetaldehyde come from?

9-7 Propionate Fermentation. Although lactate and ethanol are the best-known products of fermentation, other pathways are also known, some with important commercial applications. Swiss cheese production, for example, depends on the bacterium *Propionibacterium freudenreichii,* which converts pyruvate to propionate (CH_3—CH_2—COO^-). Fermentation of glucose to propionate always generates at least one other product as well.

(a) Why is it not possible to devise a scheme for the fermentation of glucose with propionate as the sole end-product?

(b) Suggest an overall scheme for propionate production that generates only one additional product, and indicate what that product might be.

(c) If you know that Swiss cheese production actually requires both propionate and carbon dioxide and that both are produced by *Propionibacterium* fermentation, what else can you now say about the fermentation process that this bacterium carries out?

9-8 Glycolysis and Gluconeogenesis. As Figure 9-11 indicates, gluconeogenesis is accomplished by what is essentially the reverse of the glycolytic pathway but with bypass reactions in place of the first, third, and tenth reactions in glycolysis.

(a) Explain why it is not possible to accomplish gluconeogenesis by a simple reversal of all the reactions in glycolysis.

(b) Write an overall reaction for gluconeogenesis that is comparable to Reaction 9-16 for glycolysis.

(c) Explain why gluconeogenesis requires the input of six molecules of nucleoside triphosphates (four ATPs and two GTPs) per molecule of glucose synthesized, whereas glycolysis yields only two molecules of ATP per molecule of glucose.

(d) Assuming concentrations of ATP, ADP, and P_i are such that $\Delta G'$ for the hydrolysis of ATP is about −10 kcal/mol, what is the approximate $\Delta G'$ value for the overall reaction for gluconeogenesis that you wrote in part b?

(e) With all of the enzymes for glycolysis and gluconeogenesis present in a liver cell, how does the cell "know" whether it should be synthesizing or catabolizing glucose at any given time?

• **9-9 Trypanosomes, Glycosomes, and the Compartmentalization of Glycolysis.** In most organisms, all of the glycolytic enzymes occur in the cytosol. However, in certain parasitic protozoa known as *trypanosomes,* the enzymes that catalyze the first seven steps of glycolysis are compartmentalized in membrane-bounded organelles called *glycosomes.*

(a) What experimental evidence most likely led to the discovery that seven of the ten glycolytic enzymes are localized in an organelle rather than in the cytosol?

(b) What benefit do you think trypanosomes derive from this compartmentalization of glycolysis?

(c) What specific transport proteins do you predict are present in the glycosomal membrane? Explain.

(d) Glycosomes are usually regarded as a specialized kind of peroxisome. What other enzymes would you therefore expect to find in this organelle? Explain.

• **9-10 You've Got Some Explaining to Do.** Explain each of the following observations:

(a) In his classic studies of glucose fermentation by yeast cells, Louis Pasteur observed that the rate of glucose consumption by yeast cells was much higher under anaerobic conditions than under aerobic conditions.

(b) Under experimental conditions, a suspension of red blood cells does not require oxygen in the solution.

(c) A runner is taking part in a marathon. His pace starts off steady, but midway through, the runner undergoes a short sprint. Shortly after, the runner returns to a steady jog but is unable to continue and must exit the race.

(d) Fermentation of glucose to lactate is an energy-yielding process, although it involves no net oxidation (i.e., even though the oxidation of glyceraldehyde-3-phosphate to glycerate is accompanied by the reduction of pyruvate to lactate and no net accumulation of NADH occurs).

• **9-11 Arsenate Poisoning.** Arsenate ($HAsO_4^{2-}$) is a potent poison to almost all living systems. Among other effects, arsenate is known to uncouple the phosphorylation event from the oxidation of glyceraldehyde-3-phosphate. This uncoupling occurs because the enzyme involved, glyceraldehyde-3-phosphate dehydrogenase, can utilize arsenate instead of inorganic phosphate, forming glycerate-1-arseno-3-phosphate. This product is a highly unstable compound that immediately undergoes nonenzymatic hydrolysis into glycerate-3-phosphate and free arsenate.

(a) In what sense might arsenate be called an *uncoupler* of substrate-level phosphorylation?

(b) Why is arsenate such a toxic substance for an organism that depends critically on glycolysis to meet its energy needs?

(c) Can you think of other reactions that are likely to be uncoupled by arsenate in the same way as the glyceraldehyde-3-phosphate dehydrogenase reaction?

9-12 Life without phosphofructokinase. Many bacteria do not have phosphofructokinase-1 (Gly-3) and thus cannot convert glucose to fructose-1,6-bisphosphate. Instead they use a pathway known as the Entner–Doudoroff pathway to partially oxidize glucose and convert it to two three-carbon molecules. See if you can draw the products of the first three steps of the pathway based on the following description:

(a) In the first step, the ring form of glucose-6-phosphate is oxidized at carbon 1 to form 6-phosphogluconolactone, as the coenzyme $NADP^+$ is reduced to NADPH + H^+.

(b) Next, the ring is broken by hydrolysis to form the carboxylic acid 6-phosphogluconate, which resembles glucose-6-phosphate but is more oxidized at carbon 1.

(c) After a molecule of water is removed, a molecule of 2-keto-3-deoxy-6-phosphogluconate is formed.

(d) Because an aldolase splits this six-carbon molecule into pyruvate plus glyceraldehyde-3-phosphate (which is then converted to another molecule of pyruvate), how will the final ATP yield per glucose compare with typical glycolysis?

(e) Given an alternate source of cellular ATP, how would you predict lactic acid production might be affected by the addition of arsenate (see Problem 9-11) in these bacteria compared to bacteria performing standard glycolysis?

• **9-13 Regulation of Phosphofructokinase-1.** Shown in **Figure 9-14** are plots of initial reaction velocity (expressed as % of V_{max}) vs. fructose-6-phosphate concentration for liver phosphofructokinase (PFK-1) in the presence and absence of fructose-2,6-bisphosphate (F2,6BP) (Figure 9-14a) and in the presence of a low or high concentration of ATP (Figure 9-14b).

(a) Explain the effect of F2,6BP on enzyme activity as shown in Figure 9-14a.

(b) Explain the effect of the ATP concentration on the data shown in Figure 9-14b.

(c) What assumptions do you have to make about the concentration of ATP in Figure 9-14a and about the concentration of F2,6BP in Figure 9-14b? Explain.

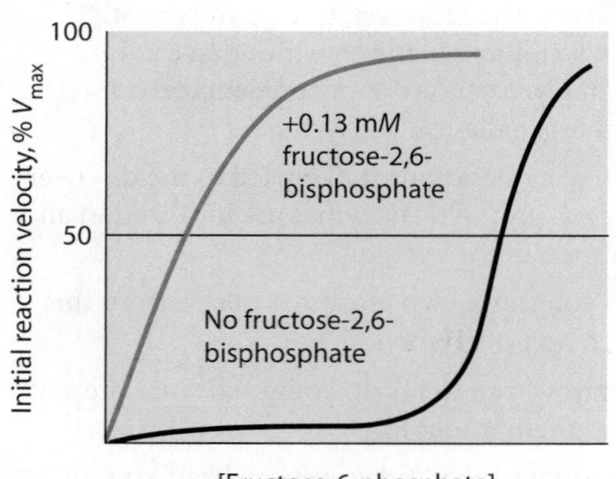

(a) Kinetics of PFK–1 in the presence or absence of F2,6BP

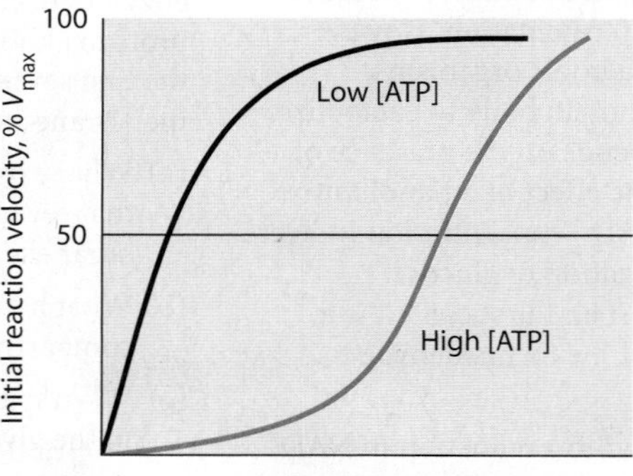

(b) Kinetics of PFK–1 in the presence of a high or low concentration of ATP

FIGURE 9-14 Allosteric Regulation of Phosphofructokinase-1. Shown here are Michaelis–Menten plots of liver phosphofructokinase (PFK-1) activity, depicting **(a)** the dependence of initial reaction velocity on concentration of the substrate fructose-6-phosphate in the presence (red line) or absence (black line) of fructose-2,6-bisphosphate and **(b)** the dependence of initial reaction velocity on fructose-6-phosphate concentration at high (red line) or low (black line) ATP concentrations. In both cases, initial reaction velocity is expressed as a percentage of V_{max}, the maximum velocity. See Problem 9-13.

SUGGESTED READING

References of historical importance are marked with a •.

General References

Berg, J. M., J. L. Tymoczko, and L. Stryer. *Biochemistry,* 5th ed. New York: W. H. Freeman, 2002.

Canback, B., S. G. E. Andersson, and C. G. Kurland. The global phylogeny of glycolytic enzymes. *Proc. Natl. Acad. Sci. USA* 99 (2002): 6097.

Horton, H. R. et al. *Principles of Biochemistry,* 4th ed. Upper Saddle River, NJ: Pearson/Prentice Hall, 2006.

Nelson, D. L., and M. M. Cox. *Lehninger Principles of Biochemistry,* 5th ed. New York: W. H. Freeman, 2008.

Mathews, C. K., K. E. van Holde, and K. G. Ahern. *Biochemistry,* 3rd ed. Menlo Park, CA: Benjamin Cummings, 2000.

Metzler, D.E. *Biochemistry: The Chemical Reactions of Living Cells,* 2nd ed. San Diego: Academic Press, 2001.

ATP and ATP Generation

• Hinkle, P. C., and R. E. McCarty. How cells make ATP. *Sci. Amer.* 238 (March 1978): 104.

Ingwall, J. S. *ATP and the Heart.* New York: Kluwer Academic Publishers, 2002.

• Lipmann, F. *Wanderings of a Biochemist.* New York: Wiley, 1971.

Wang, P., Z. Guan, and S. Saraswati. A *Drosophila* temperature-sensitive seizure mutant in phosphoglycerate kinase disrupts ATP generation and alters synaptic function. *J. Neurosci.* 24 (2004): 4518.

• Westheimer, F. Why nature chose phosphates. *Science* 235 (1987): 1173.

Glycolysis and Fermentation

Boels, I. C., M. Kleerebezem, and W. M. de Vos. Engineering of carbon distribution between glycolysis and sugar nucleotide biosynthesis in *Lactococcus lactis. Appl. Environ. Microbiol.* 69 (2003): 1129.

Chuang, D.-M., C. Hough, and V. V. Senatorov. Glyceraldehyde-3-phosphate dehydrogenase, apoptosis, and neurodegenerative diseases. *Annu. Rev. Pharmacol. Toxicol.* 45 (2005): 269.

Downs, D. Understanding microbial metabolism. *Annu. Rev. Microbiol.* 60 (2006): 533.

Giege, P., J. L. Heazlewood, and U. Roessner-Tunali. Enzymes of glycolysis are functionally associated with the mitochondrion in *Arabidopsis* cells. *Plant Cell* 15 (2003): 2140.

Jeffries, T. W., and Y. S. Jin. Metabolic engineering for improved fermentation of pentoses by yeast. *Appl. Microbiol. Biotechnol.* 63 (2004): 495.

Kim, J., and C. Dang. Multifaceted roles of glycolytic enzymes. *Trends Biochem. Sci.* 30 (2005): 142.

Kondoh, H. et al. Protection from oxidative stress by enhanced glycolysis: A possible mechanism of cellular immortalization. *Histol. Histopathol.* 22 (2007): 85.

Peters, D. Carbohydrates for fermentation. *Biotechnol. J.* 1 (2006): 806.

Rolland, F., E. Baena-Gonzales, and J. Sheen. Sugar sensing and signaling in plants: Conserved and novel mechanisms. *Annu. Rev. Plant Biol.* 57 (2006): 675.

Siebers, B., and P. Schonheit. Unusual pathways and enzymes of central carbohydrate metabolism in Archaea. *Curr. Opin. Microbiol.* 8 (2005): 695.

Teusink, B., J. Passarge, and C. A. Reijenga. Can yeast glycolysis be understood in terms of in vitro kinetics of the constituent enzymes? Testing biochemistry. *Eur. J. Biochem* 267 (2000): 5313.

Gluconeogenesis

Battezzati, A., A. Caumo, and F. Martino. Nonhepatic glucose production in humans. *Amer. J. Physiol.* 286 (2004): E129.

Boden, G. Gluconeogenesis and glyconeogenesis in health and diabetes. *J. Investig. Med.* 52 (2004): 375.

Dobson, G. P., S. Hitchins, and W. E. Teague, Jr. Thermodynamics of the pyruvate kinase reaction and the reversal of glycolysis in heart and skeletal muscle. *J. Biol. Chem.* 277 (2002): 27176.

Kalhan, S. C., P. Parima, and R. van Beek. Estimation of gluconeogenesis in newborn infants. *Amer. J. Physiol.* 281 (2001): E991.

Previs, S. F., and H. Brunengraber. Methods for measuring gluconeogenesis in vivo. *Curr. Opin. Clin. Nutr. Metab. Care* 1 (1998): 461.

Previs, S. F., D. Z. Brunengraber, and H. Brunengraber. Is there glucose production outside of the liver and kidney? *Annu. Rev. Nutr.* 29 (2009): 43.

Regulation of Glycolysis and Gluconeogenesis

Boden, G. Effects of free fatty acids on gluconeogenesis and glycogenolysis. *Life Sciences* 72 (2003): 977.

Chesney, J. 6-phosphofructo-2-kinase/fructose-2,6-bisphosphatase and tumor cell glycolysis. *Curr. Opin. Clin. Nutr. Metab. Care* 9 (2006): 535.

Even, S., M. Cocaign-Bousquet, and N. D. Lindley. Transcriptional, translational and metabolic regulation of glycolysis in *Lactococcus lactis* subsp. *cremoris* MG 1363 grown in continuous acidic cultures. *Microbiology* 149 (2003): 1935.

Gatenby, R. A., and R. J. Gillies. Why do cancers have high aerobic glycolysis? *Nature Rev. Cancer* 4 (2004): 891.

Koebmann, B. J., J. L. Snoep, and H. Westerhoff. The glycolytic flux in *Escherichia coli* is controlled by the demand for ATP. *J. Bacteriol.* 184 (2002): 3909.

Muoio, D. M., and C. B. Newgard. Obesity-related derangements in metabolic regulation. *Annu. Rev. Biochem.* 75 (2006): 367.

Nordlie, R. C., J. D. Foster, and A. J. Lange. Regulation of glucose production by the liver. *Annu. Rev. Nutr.* 19 (1999): 379.

Pelicano, H. et al. Glycolysis inhibition for anticancer treatment. *Oncogene* 25 (2006): 4633.

Plaxton, W. C. The organization and regulation of plant glycolysis. *Annu. Rev. Plant Physiol. Plant Mol. Biol.* 47 (1996): 185.

Ramirez, J-M., L. P. Folkow, and A. S. Blix. Hypoxia tolerance in mammals and birds: From the wilderness to the clinic. *Annu. Rev. Physiol.* 69 (2007): 113.

Shaw, R. J. Glucose metabolism and cancer. *Curr. Opin. Cell Biol.* 18 (2006): 598.

Webster, K. A. Evolution of the coordinate regulation of glycolytic enzyme genes by hypoxia. *J. Exp. Biol.* 206 (2003): 2911.

Zaman, S., S. I. Lippman, X. Zhao, and J. R. Broach. How *Saccharomyces* responds to nutrients. *Annu. Rev. Genet.* 42 (2008): 27.

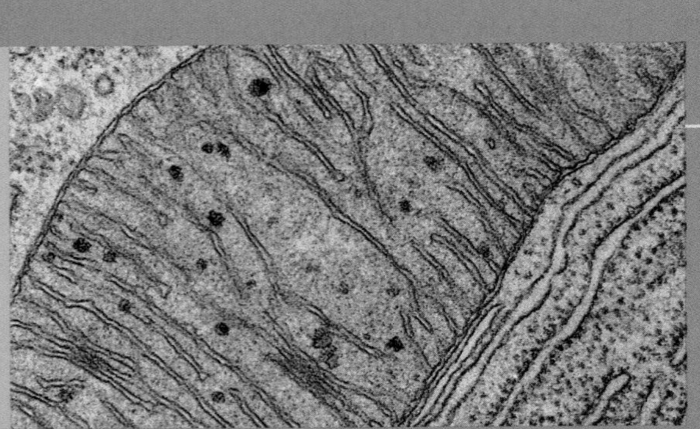

10 Chemotrophic Energy Metabolism: Aerobic Respiration

See MasteringBiology® for tutorials, activities, and quizzes.

In the previous chapter, we learned that some cells meet their energy needs by anaerobic fermkentation, either because they are strict anaerobes or because they are facultative cells functioning temporarily in the absence or scarcity of oxygen. However, we also noted that fermentation yields only modest amounts of energy due to the absence of an external electron acceptor. The electrons that are removed from glucose as it is partially oxidized during fermentation are transferred to pyruvate, and only two molecules of ATP can be generated per molecule of glucose.

In short, fermentation can meet energy needs, but the ATP yield is low because the cell has access to only a limited portion of the total free energy potentially available from the oxidizable molecules it uses as substrates. In addition, fermentation often results in the accumulation of waste products such as ethanol or lactate, which can be toxic to the cells if they accumulate.

Cellular Respiration: Maximizing ATP Yields

All of this changes dramatically when we come to **cellular respiration,** or **respiration** for short. With an *external* electron acceptor available, complete oxidation of substrates to CO_2 becomes possible, and ATP yields are much higher. By "external" we mean an electron acceptor that is not a by-product of glucose catabolism. When an "internal" electron acceptor such as pyruvate or acetaldehyde is used (as during fermentation) to accept electrons from NADH (see Figure 9-8), the by-products are not completely oxidized to CO_2.

As a formal definition, *cellular respiration is the flow of electrons, through or within a membrane, from reduced coenzymes to an external electron acceptor, usually accompanied by the generation of ATP.* We have already encountered NADH as the reduced coenzyme generated by the glycolytic catabolism of sugars or related compounds. As we will see shortly, two other coenzymes, *FAD* (for *flavin adenine dinucleotide*) and *coenzyme Q* (or *ubiquinone*), also collect the electrons that are removed from oxidizable organic substrates and pass them to the terminal electron acceptor via a series of electron carriers, generating ATP in the process.

For many organisms, including us, the terminal electron acceptor is *oxygen,* the reduced form of this terminal electron acceptor is *water,* and the overall process is called **aerobic respiration.** Note that, in medical terminology at an organismal level, respiration refers to breathing, the uptake of oxygen. However, while many chemotrophs on Earth carry out aerobic respiration, a variety of terminal electron acceptors other than molecular oxygen can be used by other organisms, especially bacteria and archaea. Examples of alternative acceptors and their reduced forms include elemental sulfur (S/H_2S), protons (H^+/H_2), and ferric ions (Fe^{3+}/Fe^{2+}). Respiratory processes that involve electron acceptors such as these require no molecular oxygen and are therefore examples of **anaerobic respiration.** Anaerobic respiratory processes play important roles in nutrient cycling and the overall energy economy of the biosphere. Here, however, we will concentrate on aerobic respiration because it is the basis of energy metabolism in the aerobic world of which we and most other familiar organisms are a part.

We will focus much of our attention on the *mitochondrion* (**Figure 10-1**) because most aerobic ATP production in eukaryotic cells takes place within this organelle. In bacterial cells, the roles of the plasma membrane and cytoplasm in energy production are analogous to the roles of the mitochondrial inner membrane and matrix, respectively.

Aerobic Respiration Yields Much More Energy than Fermentation Does

With oxygen available as the terminal electron acceptor, pyruvate can be oxidized completely to CO_2 instead of being used to accept electrons from NADH. Therefore,

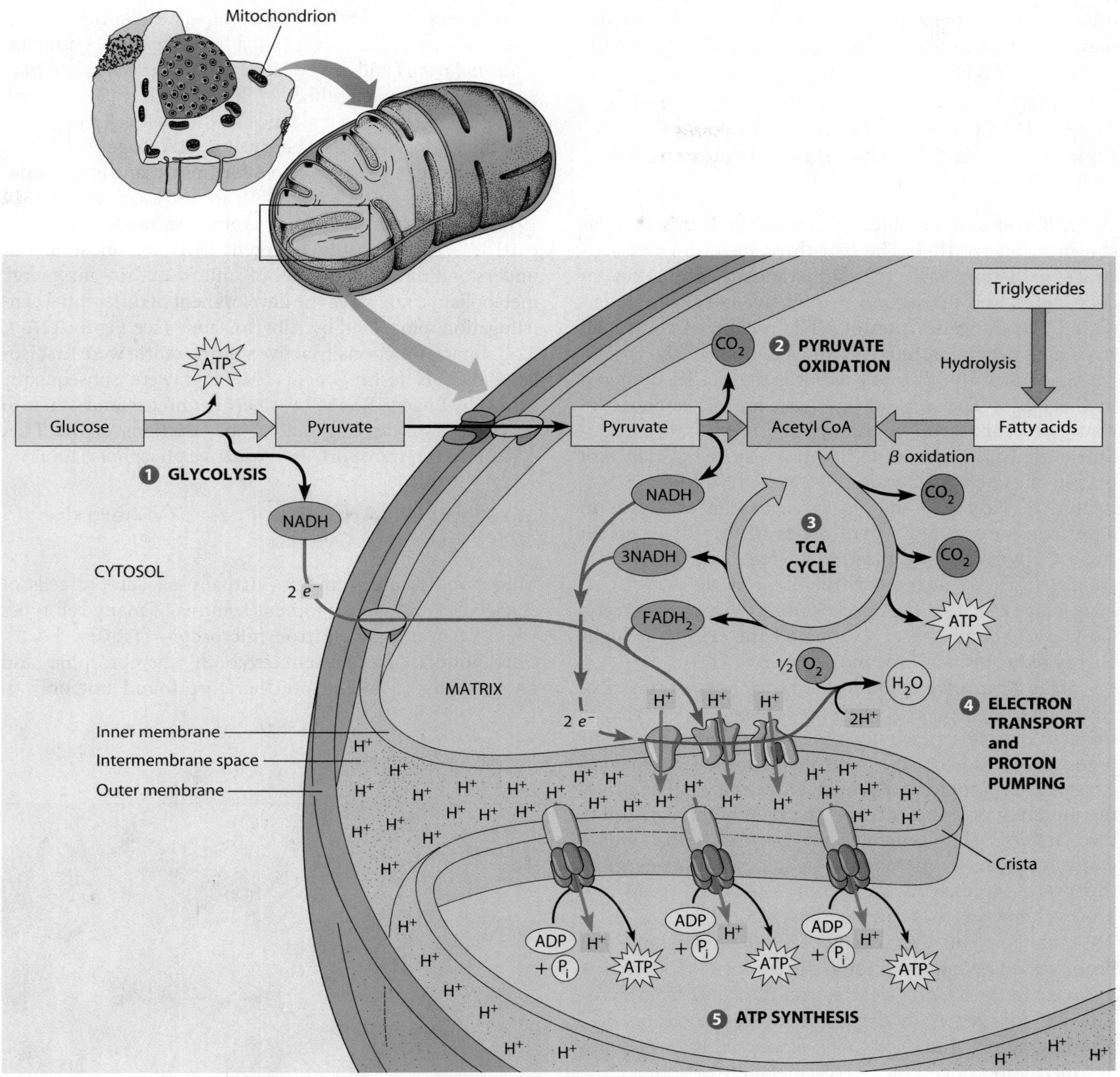

FIGURE 10-1 The Role of the Mitochondrion in Aerobic Respiration. The mitochondrion plays a central role in aerobic respiration. Most respiratory ATP production in eukaryotic cells occurs in this organelle. Oxidation of glucose and other sugars begins in the cytosol with glycolysis (stage ❶), producing pyruvate. Pyruvate is transported across the inner mitochondrial membrane and is oxidized within the matrix to acetyl CoA (stage ❷), the primary substrate of the tricarboxylic acid (TCA) cycle (stage ❸). Acetyl CoA can also be formed by *β* oxidation of fatty acids. Electron transport is coupled to proton pumping (stage ❹), with the energy of electron transport conserved as an electrochemical proton gradient across the inner membrane of the mitochondrion (or across the plasma membrane, in the case of prokaryotes). The energy of the proton gradient is used in part to drive the synthesis of ATP from ADP and inorganic phosphate (stage ❺).

aerobic respiration has the potential of generating up to 38 ATP molecules per glucose.

Oxygen makes all of this possible by serving as the terminal electron acceptor, thereby providing a means for the continuous reoxidation of NADH and other reduced coenzymes. After these coenzyme molecules accept electrons from pyruvate and other oxidizable substrates, they can then transfer these electrons to oxygen. Aerobic respiration therefore involves oxidative pathways in which electrons are removed from organic substrates and transferred to coenzyme carriers, which then transfer these electrons to oxygen, accompanied indirectly by the generation of ATP.

Respiration Includes Glycolysis, Pyruvate Oxidation, the TCA Cycle, Electron Transport, and ATP Synthesis

We will consider aerobic respiration in five stages, as shown in Figure 10-1. The first three stages involve substrate oxidation and the simultaneous reduction of coenzymes, and the second two involve coenzyme reoxidation and the generation of ATP. In cells, of course, all five stages occur continuously and simultaneously.

Stage ❶ is the *glycolytic pathway* that we encountered in Chapter 9. The steps of glycolysis are the same under aerobic and anaerobic conditions and result in the oxidation of glucose to pyruvate. However, in the presence of oxygen, the fate of the pyruvate is different (see Figure 9-8). Instead of serving as an electron acceptor, as occurs in anaerobic fermentation, pyruvate is further oxidized. In stage ❷, pyruvate is oxidized to generate *acetyl coenzyme A (acetyl CoA),* which enters the *tricarboxylic acid (TCA) cycle* (stage ❸). The TCA cycle completely oxidizes acetyl CoA to CO_2 and conserves most of the energy as high-energy reduced coenzyme molecules.

Stage ❹ involves *electron transport,* or the transfer of electrons from reduced coenzymes to oxygen, coupled to the *active transport (pumping)* of protons across a membrane. The transfer of electrons from coenzymes to oxygen is exergonic and provides the energy that drives the pumping of protons across the membrane containing the carriers. This generates an *electrochemical proton gradient* across the membrane. In stage ❺, the energy of this proton gradient is used to drive ATP synthesis in a process known as *oxidative phosphorylation.*

Our goal in this chapter is to understand the processes following glycolysis—the complete oxidation of pyruvate in the TCA cycle, electron transport from pyruvate via coenzymes to oxygen, and the formation of a proton gradient that drives ATP synthesis. We will begin our discussion of aerobic energy metabolism with a close look at the mitochondrion because of its prominent role in eukaryotic energy metabolism.

The Mitochondrion: Where the Action Takes Place

Our discussion of aerobic respiration starts with a description of the **mitochondrion** because most of aerobic energy metabolism in eukaryotic cells takes place within this organelle. Because of this, the mitochondrion is often called the "energy powerhouse" of the eukaryotic cell. As early as 1850, the German biologist Rudolph Kölliker described the presence of what he called "ordered arrays of particles" in muscle cells. Isolated particles were found to swell in water, leading Kölliker to conclude that each particle was surrounded by a semipermeable membrane. These particles are now called *mitochondria* (singular: *mitochondrion*) and are believed to have arisen when bacterial cells were engulfed by larger cells but survived, taking up permanent residence in the host cell cytoplasm (see the Endosymbiotic Theory, Box 11A, page 298–299).

Evidence suggesting a role for this organelle in oxidative events began to accumulate almost a century ago. In 1913, for example, Otto Warburg showed that these particles could consume oxygen. However, most of our understanding of the role of mitochondria in energy metabolism came since the development of differential centrifugation, pioneered by Albert Claude (see Figure 12A-1, 327). Intact, functionally active mitochondria were first isolated by this technique in 1948 and were subsequently shown by Eugene Kennedy, Albert Lehninger, and others to be capable of carrying out all the reactions of the TCA cycle, electron transport, and oxidative phosphorylation.

Mitochondria Are Often Present Where the ATP Needs Are Greatest

Mitochondria are found in virtually all aerobic cells of eukaryotes and are prominent features of many cell types when examined by electron microscopy (**Figure 10-2**). Mitochondria are present in both chemotrophic and phototrophic cells and are therefore found not only in

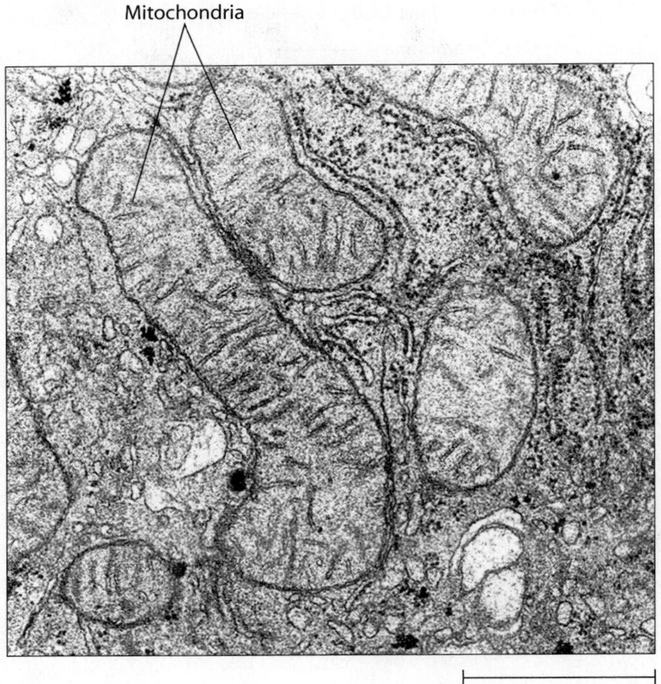

FIGURE 10-2 Mitochondria: One or Many? Mitochondria are prominent features in this cross section of a rat liver cell and were traditionally thought to be discrete oval structures. However, current research suggests that these may be separate slices through a single large multilobed mitochondrion (TEM).

animals but also in plants. Their occurrence in phototrophic cells reminds us that even photosynthetic organisms rely on respiration to meet energy and carbon needs during periods of darkness—and at all times in nonphotosynthetic tissue, such as the roots of plants.

The crucial role of the mitochondrion in meeting cellular ATP needs is often reflected in the localization of mitochondria within the cell. Frequently, mitochondria are clustered in regions of cells with the most intense metabolic activity and the greatest need for ATP. An especially good example is in muscle cells (see Figure 4-13). The mitochondria in these muscle cells are organized in rows along the fibrils responsible for contraction. A similar strategic localization of mitochondria occurs in flagella and cilia, as well as in sperm tails (see Figure 4-12).

Are Mitochondria Interconnected Networks Rather than Discrete Organelles?

When seen in an electron micrograph such as that in Figure 10-2, mitochondria typically appear as oval, sausage-shaped structures measuring one or more μm in length and 0.5–1.0 μm across. However, they can be seen in various shapes and sizes, depending on the cell type. This appearance has fostered the widely accepted view that mitochondria are discrete entities and, as such, are large and often very numerous organelles. In fact, a mitochondrion with these dimensions is similar in size to an entire bacterial cell and is the largest organelle in most animal cells other than the nucleus (see Figure 1A-1). (In plant cells, the chloroplasts and some of the vacuoles are also typically larger than the mitochondria.)

Calculations based on cross-sectional views of mitochondria in electron micrographs indicate that the number of mitochondria per cell is highly variable, ranging from one or a few per cell in many protists, fungi, and algae (and in some mammalian cells as well) up to a few thousand per cell in some tissues of higher plants and animals. Mammalian liver cells, for example, are thought to contain about 500–1000 mitochondria each, though only a small fraction of these are seen in a typical thin section prepared for electron microscopy, such as that shown in Figure 10-2.

The notion that the mitochondrial profiles seen in electron micrographs represent discrete organelles of known size and abundance has been challenged by the work of Hans-Peter Hoffman and Charlotte Avers. After examining a complete series of thin sections through an entire yeast cell, these investigators concluded that the oval profiles observed in individual micrographs all represent slices through a single large, extensively branched mitochondrion. Such results suggest that the number of mitochondria present in a cell may be considerably smaller than generally believed—and that the size of a mitochondrion may be much larger than can be calculated from individual cross-sectional profiles seen in thin-section electron micrographs.

Further support for the concept of mitochondria as interconnected networks (rather than many separate organelles) comes from studies in which intact living cells were examined by phase-contrast microscopy or by fluorescence microscopy using mitochondria-specific fluorescent dyes. Such investigations revealed that living cells contain large branched mitochondria in a dynamic state of flux, with segments of one mitochondrion frequently pinching off and fusing with another mitochondrion.

Most of the discussion and illustrations in this chapter will presume the conventional view of mitochondria as discrete entities, but keep in mind that what we regard as individual mitochondria may well be parts of a large, dynamic network instead, at least in some types of cells.

The Outer and Inner Membranes Define Two Separate Compartments and Three Regions

Figure 10-3 shows both an electron micrograph and an illustration of a typical mitochondrion—or perhaps a small segment of a much larger network. In either case, a distinctive feature is the presence of two membranes, called the outer and inner membranes. The **outer membrane** is not a significant permeability barrier for ions and small molecules because it contains transmembrane channel proteins called **porins** that permit the passage of solutes with molecular weights up to about 5000. Similar proteins are found in the outer membrane of gram-negative bacteria. Because porins allow the free movement of small molecules and ions across the outer membrane, the **intermembrane space** between the inner and outer membranes of the mitochondrion is continuous with the cytosol with respect to small solutes. However, enzymes targeted to the intermembrane space are effectively confined there because enzymes and other soluble proteins are too large to pass through the porin channels.

In contrast to the outer membrane, the **inner membrane** of the mitochondrion presents a permeability barrier to most solutes, thereby partitioning the mitochondrion into two separate compartments—the *intermembrane space* and the interior of the organelle (the mitochondrial *matrix*). The inner and outer membranes, shown in **Figure 10-4**, are each about 7 nm in thickness and are typically separated by an intermembrane space of about 7 nm. However, in certain spots the membranes are in contact, and it is in these regions that proteins destined for the mitochondrial matrix pass through the two membranes (see Figure 22-19).

At the perimeter of the mitochondrion, the portion of the inner membrane adjacent to the intermembrane space is known as the *inner boundary membrane.* In addition, the inner membrane of most mitochondria has many distinctive infoldings called **cristae** (singular: **crista**) that greatly increase its surface area. In a typical liver mitochondrion, for example, the area of the inner membrane (including the cristae) is about five times greater than that of the outer membrane. Because of its large surface area, the inner membrane can accommodate large numbers of

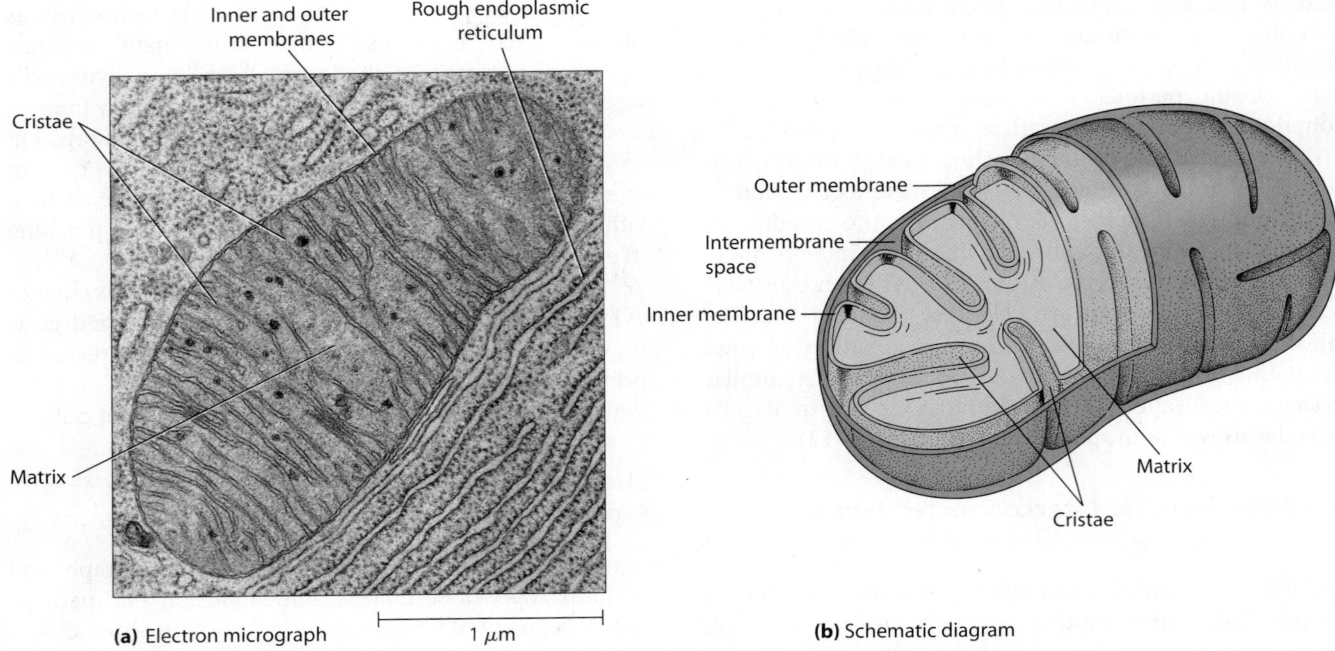

FIGURE 10-3 Mitochondrial Structure. (a) A mitochondrion of a bat pancreas cell as seen by electron microscopy (TEM). The cristae are formed by infoldings of the inner membrane. (b) A mitochondrion is illustrated schematically in this cutaway view that shows the traditional "baffle" model of cristae structure. As noted in the text, however, this model is currently being reconsidered. Recent analysis using EM tomography suggests that the connections between the intracristal spaces and the intermembrane space are more limited tubular openings known as crista junctions.

the protein complexes needed for electron transport and ATP synthesis, thereby enhancing the mitochondrion's capacity for ATP generation. The inner membrane is about 75% protein by weight, which is a higher proportion of protein than in any other cellular membrane. These inner membrane proteins include the transmembrane portions of proteins involved in solute transport, electron transport, and ATP synthesis. The cristae also provide numerous localized regions, the *intracristal spaces,* where protons can accumulate between the folded inner membranes during the electron transport process that we will discuss later in the chapter.

It was long thought that cristae were wide, flattened structures with broad connections to the inner boundary membrane, giving rise to the "baffle" model of cristae structure shown in Figure 10-3b. This view has been challenged recently due to microscopic evidence from EM (electron microscope) tomography, a microscopic technique that, like a CAT (computer-aided tomography) scan, provides detailed three-dimensional representations of cellular structures by combining successive serial images from thick (1 μm) sections. It is now believed that

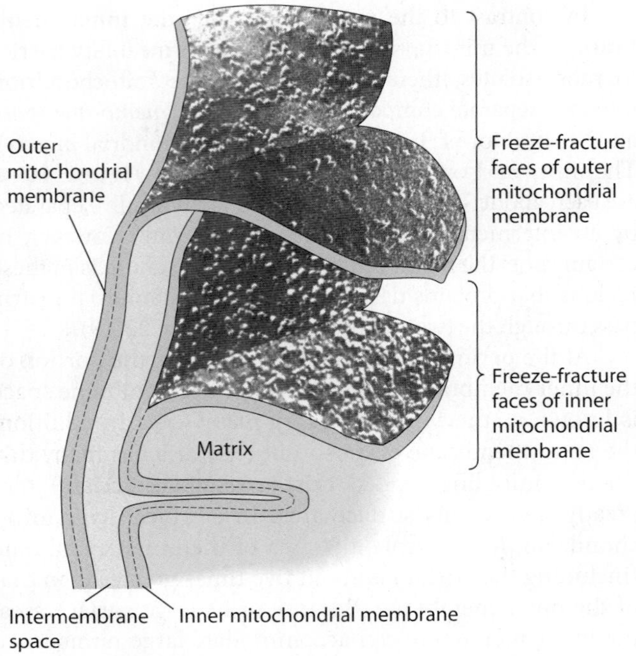

Outer mitochondrial membrane
Freeze-fracture faces of outer mitochondrial membrane
Freeze-fracture faces of inner mitochondrial membrane
Matrix
Intermembrane space
Inner mitochondrial membrane

FIGURE 10-4 Structure of the Inner and Outer Mitochondrial Membranes. When the inner and outer mitochondrial membranes are subjected to freeze fracturing, each of the membranes splits along its hydrophobic interior, separating each membrane into two fracture faces. Segments of electron micrographs of the two fracture faces for both inner and outer membranes are superimposed here on a schematic diagram of the freeze-fractured membranes to illustrate the density of protein particles in each membrane.

cristae in many tissues may be tubular structures that associate in layers to form lamellar cristae of irregular size and shape. In some respects their morphology is similar to the grana stacks of thylakoids that define a third compartment in chloroplasts. Cristae appear to have only limited connections to the inner boundary membrane through small tubular openings known as *crista junctions.* The small size of these openings is thought to limit diffusion of materials between the intracristal space and the intermembrane space, effectively creating a third, nearly enclosed region in the mitochondrion.

The relative prominence of cristae within the mitochondrion frequently reflects the relative metabolic activity of the cell or tissue in which the organelle is located. Heart, kidney, and muscle cells have high respiratory activities, and thus their mitochondria have correspondingly large numbers of prominent cristae. The flight muscles of birds are especially high in respiratory activity and have mitochondria that are exceptionally well endowed with cristae. Plant cells, by contrast, have lower rates of respiratory activity than most animal cells do and have correspondingly fewer cristae within their mitochondria.

The interior of the mitochondrion is filled with a semifluid **matrix.** Within the matrix are many of the enzymes involved in mitochondrial function as well as DNA molecules and ribosomes. In most mammals, the mitochondrial genome consists of a circular DNA molecule of about 15,000–20,000 base pairs that code for ribosomal RNAs, transfer RNAs, and about a dozen polypeptide subunits of inner-membrane proteins. Besides the proteins encoded by the mitochondrial genome, mitochondria contain numerous nuclear-encoded proteins that are synthesized on cytoplasmic ribosomes and then imported into mitochondria (see Chapter 22).

Mitochondrial Functions Occur in or on Specific Membranes and Compartments

Specific functions and pathways have been localized within the mitochondrion by disruption of the organelle and fractionation of the various components. **Table 10-1** lists some of the main functions localized to each compartment of the mitochondrion.

Most of the mitochondrial enzymes involved in pyruvate oxidation, in the TCA cycle, and in the catabolism of fatty acids and amino acids are matrix enzymes. In fact, when mitochondria are disrupted very gently, six of the eight enzymes of the TCA cycle are released as a single large multiprotein complex, suggesting that the product of one enzyme can pass directly to the next enzyme without having to diffuse through the matrix.

On the other hand, most of the intermediates in the electron transport system are integral components of the inner membrane, where they are organized into large complexes. Protruding from the inner membrane into the matrix are knoblike spheres called **F_1 complexes** that are involved in ATP synthesis (**Figure 10-5**). Each complex is an assembly of several different polypeptides. Individual F_1 complexes can be seen in Figure 10-5a, an electron micrograph taken at high magnification using a technique called *negative staining,* which results in a light image against a dark background. F_1 complexes are about 9 nm in diameter and are especially abundant along the cristae (Figure 10-5b).

Table 10-1 Localization of Metabolic Functions Within the Mitochondrion

Membrane or Compartment	Metabolic Functions
Outer membrane	Phospholipid synthesis Fatty acid desaturation Fatty acid elongation
Inner membrane	Electron transport Oxidative phosphorylation Pyruvate import Fatty acyl CoA import Metabolite transport
Matrix	Pyruvate oxidation TCA cycle β oxidation of fats DNA replication RNA synthesis (transcription) Protein synthesis (translation)

Each F_1 complex is attached by a short protein stalk to an **F_o complex,** an assembly of hydrophobic polypeptides that are embedded within the mitochondrial inner membrane (Figure 10-5c) or in the plasma membrane of bacteria. (Note that the subscript in F_o is the letter "o" rather than a zero. Historically, this is because the F_o complex was the site of sensitivity to the antibiotic oligomycin.) The combination of an F_1 complex linked to an F_o complex is called an **F_oF_1 complex** and is regarded as an *ATP synthase* because that is its normal role in energy metabolism. The F_oF_1 complex is in fact responsible for most of the ATP generation that occurs in the mitochondrion and in bacterial cells—and, as we will see in Chapter 11, in chloroplasts as well. In each case, ATP generation by F_oF_1 complexes is driven by an electrochemical gradient of protons across the membrane in which the F_oF_1 complexes are anchored, a topic we will explore in detail later in this chapter.

In Bacteria, Respiratory Functions Are Localized to the Plasma Membrane and the Cytoplasm

Bacteria do not have mitochondria, yet most bacterial cells are capable of aerobic respiration. Where in the bacterial cell are the various components of respiratory metabolism localized? Essentially, the cytoplasm and plasma membrane of a bacterial cell perform the same functions as the mitochondrial matrix and inner membrane, respectively. Therefore, in bacteria, most of the enzymes of the TCA cycle are found in the cytoplasm, whereas the electron

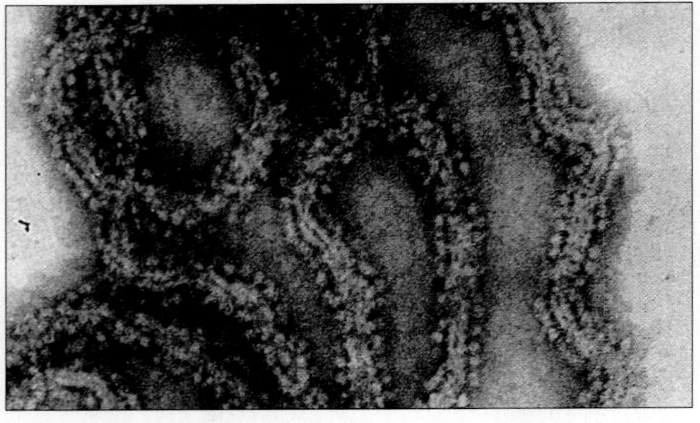

(a) Mitochondrial inner membrane

FIGURE 10-5 The F_1 and F_o Complexes of the Inner Mitochondrial Membrane. **(a)** This electron micrograph was prepared by negative staining to show the spherical F_1 complexes that line the matrix side of the inner membrane of a bovine heart mitochondrion (TEM). **(b)** Cross section of a mitochondrion, showing major structural features. **(c)** An enlargement of a small portion of a crista, showing the F_1 complexes that project from the inner membrane on the matrix side and the F_o complexes embedded in the inner membrane. Each F_1 complex is attached to an F_o complex by a short protein stalk. Together, an F_oF_1 pair constitutes a functional ATP synthase.

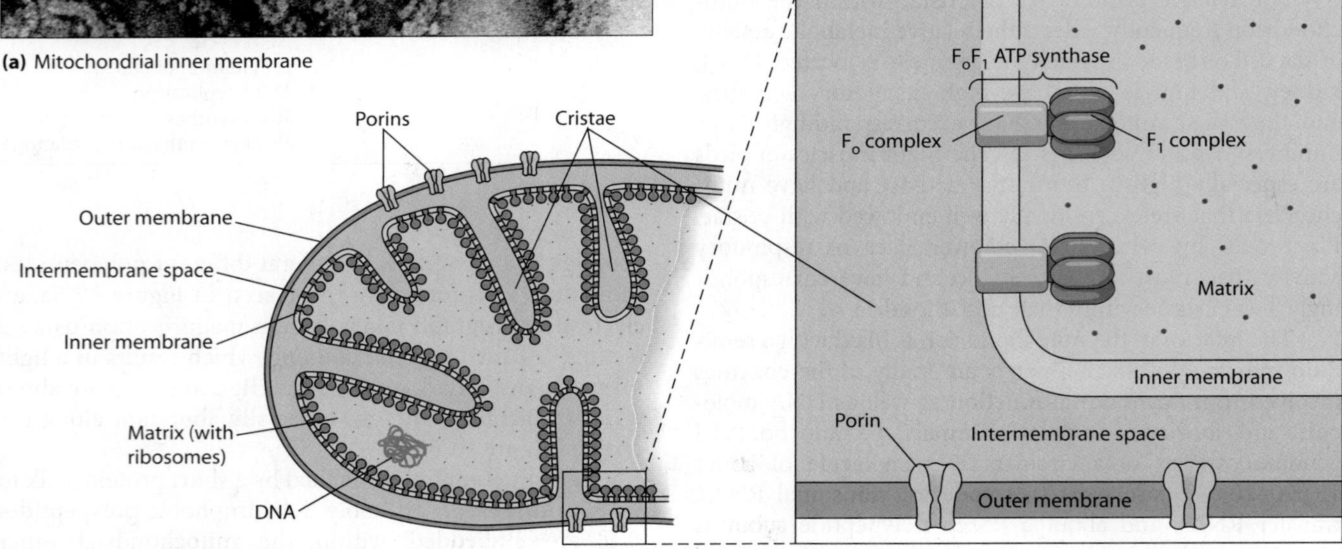

(b) Cross-sectional diagram of a mitochondrion

(c) Cross-sectional diagram of a portion of a crista showing F_oF_1 complexes

transport proteins are located in the plasma membrane. The F_oF_1 complex is also localized to the plasma membrane in bacteria, with the F_o component embedded in the membrane and the F_1 component protruding from the membrane into the cytoplasm.

The Tricarboxylic Acid Cycle: Oxidation in the Round

Having considered localization of respiratory functions in mitochondria and in bacterial cells, we will now return to the eukaryotic context and follow a molecule of pyruvate across the inner membrane of the mitochondrion to see what fate awaits it inside.

In the presence of oxygen, pyruvate is oxidized fully to carbon dioxide, and the energy released in the process is used to drive ATP synthesis. Pyruvate oxidation involves a cyclic pathway that is a central feature of energy metabolism in almost all aerobic chemotrophs. An important intermediate in this cyclic series of reactions is citrate, which has three carboxylic acid groups and is therefore a tricarboxylic acid. For this reason, this pathway is usually called the **tricarboxylic acid (TCA) cycle.** It is also commonly referred to as the *Krebs cycle* in honor of Hans Krebs, whose laboratory played a key role in elucidating this metabolic sequence in the 1930s.

The TCA cycle metabolizes acetyl coenzyme A (usually abbreviated as acetyl CoA), a compound produced from pyruvate decarboxylation. Acetyl CoA consists of a two-carbon acetate group from pyruvate linked to a carrier called *coenzyme A*. (Coenzyme A was discovered by Fritz Lipmann, who shared a Nobel Prize with Krebs in 1953 for their work on aerobic respiration.) Acetyl CoA arises either by oxidative decarboxylation of pyruvate or by the stepwise oxidative breakdown of fatty acids (see Figure 10-1). Regardless of its origin, acetyl CoA transfers its acetate group to a four-carbon acceptor called oxaloacetate, thereby generating citrate. Citrate is then subjected to two successive decarboxylations and several oxidations, regenerating the oxaloacetate that will accept two more carbons from another molecule of acetyl CoA in the next round of the cycle.

Figure 10-6 shows an overview of the TCA cycle. Each round of TCA cycle activity involves the entry of two carbons, the release of two carbons as carbon dioxide, and the regeneration of oxaloacetate. Oxidation occurs at five steps: four in the cycle itself and one in the reaction that converts pyruvate to acetyl CoA. In each case, electrons

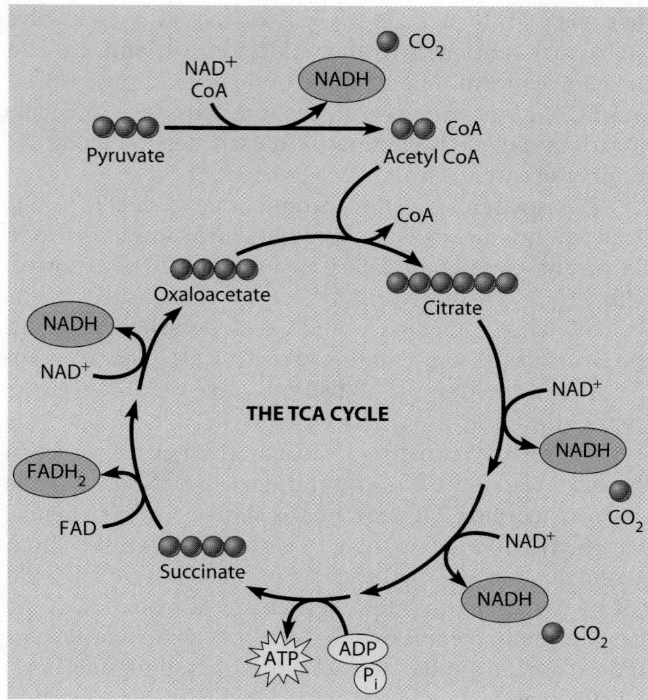

FIGURE 10-6 Overview of the TCA Cycle. Pyruvate from glycolysis is oxidatively decarboxylated to acetyl CoA, generating NADH and CO_2. Acetyl CoA enters the TCA cycle by combining with oxaloacetate to form citrate. Two molecules of CO_2 are released and 2 NADH are formed as citrate is converted to succinate by two oxidative decarboxylation steps plus an ATP-generating step. Succinate is then oxidized and converted to oxaloacetate, generating $FADH_2$ and NADH. Overall, for each pyruvate metabolized to 3 CO_2, there are 4 NADH, 1 $FADH_2$, and 1 ATP generated.

are accepted by coenzyme molecules. Substrates for the TCA cycle are therefore acetyl CoA, oxidized coenzymes, ADP, and P_i, and the products are carbon dioxide, reduced coenzymes, and a molecule of ATP.

With this brief overview in mind, let's look at the TCA cycle in more detail, focusing on what happens to the carbon molecules that enter as acetyl CoA and how the energy released by each of the oxidations is conserved as reduced coenzymes.

Pyruvate Is Converted to Acetyl Coenzyme A by Oxidative Decarboxylation

As we just noted, carbon enters the TCA cycle in the form of acetyl CoA, but from Chapter 9 we know that the glycolytic pathway ends with pyruvate, which is formed in the cytoplasm. The pyruvate molecule, being relatively small, passes through porins in the mitochondrial outer membrane into the intermembrane space. At the inner mitochondrial membrane, a specific pyruvate symporter transports pyruvate across the membrane into the matrix along with a proton.

Once inside the mitochondrial matrix, pyruvate is converted to acetyl CoA by the *pyruvate dehydrogenase complex (PDH),* which consists of three different enzymes, five coenzymes, and two regulatory proteins. These components work together to catalyze the *oxidative decarboxylation* of pyruvate:

$$\underset{\text{Coenzyme A}}{\text{CoA—SH}} + \underset{\text{Pyruvate}}{{}^{-}\text{O}-\overset{\text{O}}{\overset{\|}{{}^{1}\text{C}}}-\overset{\text{O}}{\overset{\|}{{}^{2}\text{C}}}-{}^{3}\text{CH}_3} \xrightarrow{\text{NAD}^+ \;\; \text{NADH}} \underset{\text{Acetyl CoA}}{\text{CoA—S}-\overset{\text{O}}{\overset{\|}{{}^{1}\text{C}}}-{}^{2}\text{CH}_3} + \text{CO}_2 \qquad \textbf{(10-1)}$$

This reaction is a *decarboxylation* because one of the carbon atoms of pyruvate (carbon atom 1) is liberated as carbon dioxide (grey shading). As a result, carbon atoms 2 and 3 of pyruvate become, respectively, carbon atoms 1 and 2 of acetate. In addition, this reaction is an *oxidation* because two electrons (and a proton as well) are removed from the substrates and transferred to the coenzyme NAD^+, reducing it to NADH. The electrons carried by NADH represent potential energy that is tapped when the NADH is later reoxidized by the electron transport system.

The oxidation of pyruvate occurs on carbon atom 2, which is oxidized from an α-keto to a carboxylic acid group. This oxidation is possible due to the simultaneous elimination of carbon atom 1 as carbon dioxide. The oxidation is highly exergonic ($\Delta G^{\circ\prime} = -7.5$ kcal/mol), with the free energy used to energize, or activate, the acetate molecule by linking it to the sulfhydryl group of **coenzyme A (CoA),** forming **acetyl CoA.**

As shown in **Figure 10-7**, CoA is a complicated molecule containing the B vitamin *pantothenic acid.* Like the nicotinamide of NAD^+, pantothenic acid is classified as a vitamin because humans and other vertebrates need it as a part of an essential coenzyme but cannot synthesize it themselves. The free sulfhydryl, or *thiol,* group at the end of the CoA molecule can form a thioester bond with organic acids such as acetate. The thiol group is important enough that we will abbreviate coenzyme A not just as CoA, but as CoA—SH. Compared with an ester bond, a thioester bond is a higher-energy bond because significantly more free energy is released upon its hydrolysis. Just as NAD^+ is adapted for transfer of electrons, so coenzyme A is well suited as a carrier of acetate (hence the "A" in CoA). Thus, the acetyl group that is transferred to coenzyme A from pyruvate is in a higher-energy, or activated, form.

The TCA Cycle Begins with the Entry of Acetate as Acetyl CoA

The TCA cycle (**Figure 10-8**) begins with the entry of acetate in the form of acetyl CoA. With each round of TCA cycle activity, two carbon atoms enter in organic form (as acetate), and two carbon atoms leave in inorganic form (as carbon dioxide). In the first reaction (TCA-1), the two-carbon acetate group of acetyl CoA is added onto the four-carbon compound oxaloacetate to form citrate, a

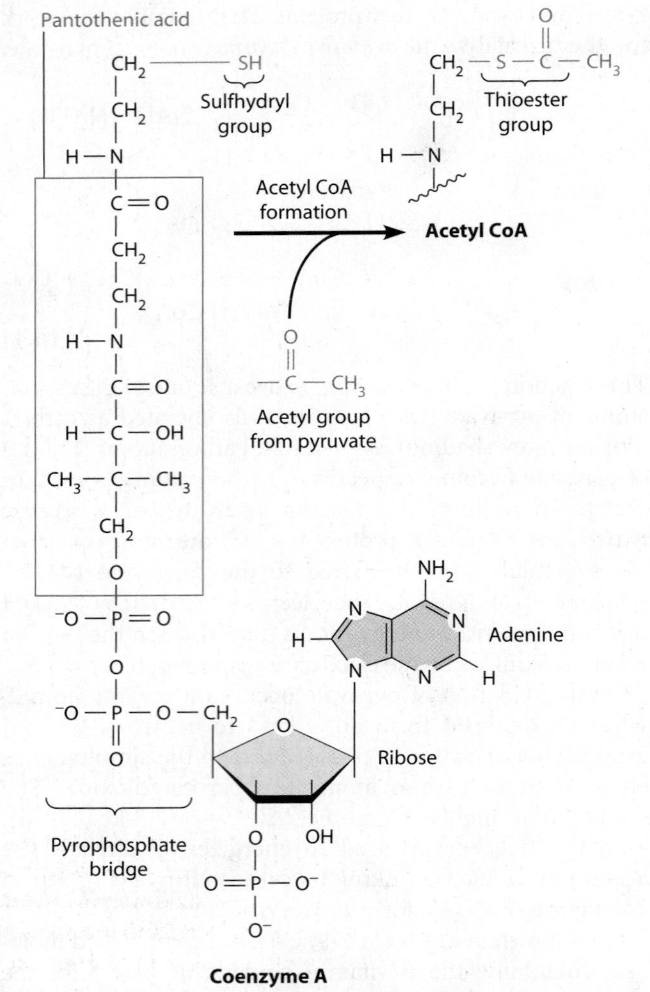

FIGURE 10-7 Structure of Coenzyme A and Acetyl CoA Formation. The portion of the coenzyme enclosed in the red box is pantothenic acid, a B vitamin. Formation of a thioester bond between CoA and an acetyl group generates acetyl coenzyme A. The acetyl group is formed by the oxidative decarboxylation of pyruvate (Reaction PDH in Figure 10-8). It is transferred to CoA from one of the enzymes of the pyruvate dehydrogenase complex.

six-carbon molecule. This condensation is driven by the free energy of hydrolysis of the thioester bond and is catalyzed by the enzyme *citrate synthase.* Note that citrate is a tricarboxylic acid—the class of compounds giving the TCA cycle its name. (The two incoming carbon atoms are shown highlighted in pink in Figure 10-8 to help you keep track of them in subsequent reactions.) In the next reaction (TCA-2), citrate is converted to the related compound isocitrate, using the enzyme *aconitase.* Isocitrate has a hydroxyl group that can be quite easily oxidized. This hydroxyl group of isocitrate is now the target of the first oxidation, or dehydrogenation, of the cycle (TCA-3).

Two Oxidative Decarboxylations Then Form NADH and Release CO_2

Next, we will see that four of the eight steps in the TCA cycle are oxidations. This is evident in Figure 10-8 because four steps (TCA-3, TCA-4, TCA-6, and TCA-8) involve coenzymes that enter in the oxidized form and leave in the reduced form. The first two of these reactions, TCA-3 and TCA-4, are also decarboxylation steps. One molecule of carbon dioxide is eliminated in each step, reducing the number of carbons from six to five to four.

The isocitrate generated from citrate in step TCA-2 is oxidized by the enzyme *isocitrate dehydrogenase* to a six-carbon compound called *oxalosuccinate* (not shown), with NAD^+ as the electron acceptor. Oxalosuccinate is unstable and immediately undergoes decarboxylation to the five-carbon compound α-ketoglutarate. This reaction (TCA-3) is the first of the two oxidative decarboxylation steps of the cycle.

The second oxidative decarboxylation event occurs in the next reaction (TCA-4), and also uses NAD^+ as the electron acceptor. This reaction is similar to the oxidative decarboxylation of pyruvate shown previously. Both α-ketoglutarate and pyruvate are α-keto acids, so it should not be surprising that the mechanism of oxidation is the same for both, complete with decarboxylation and linkage of the oxidized product to coenzyme A as a thioester. Thus, α-ketoglutarate is oxidized to succinyl CoA, in a reaction catalyzed by the enzyme *α-ketoglutarate dehydrogenase.*

Direct Generation of GTP (or ATP) Occurs at One Step in the TCA Cycle

At this point, the carbon balance of the cycle is already satisfied: two carbon atoms entered as acetyl CoA, and two carbon atoms have now been released as carbon dioxide. (But notice carefully from Figure 10-8 that the two carbon atoms released in a given cycle *are not the same two* that entered in Reaction TCA-1 of that cycle). We have also encountered two of the four oxidation reactions of the TCA cycle and have generated two molecules of NADH. In addition, we recognize succinyl CoA as an activated compound that, like acetyl CoA, has a high-energy thioester bond.

The energy of this thioester bond is used to generate a molecule of either ATP (in bacterial cells and plant mitochondria) or GTP (in animal mitochondria). GTP and ATP are energetically equivalent because their terminal phosphoanhydride bonds have identical free energies of hydrolysis. Thus, the net result of succinyl CoA hydrolysis is the generation of one molecule of ATP, whether directly or via GTP, as depicted in Figure 10-8.

The Final Oxidative Reactions of the TCA Cycle Generate $FADH_2$ and NADH

Of the remaining three steps in the TCA cycle, two are oxidations. In Reaction TCA-6, the succinate formed in the previous step is oxidized to fumarate. This reaction is unique in that both electrons come from adjacent carbon atoms, generating a C=C double bond. Other oxidations we have encountered thus far involve the removal of electrons either from adjacent carbon and oxygen atoms

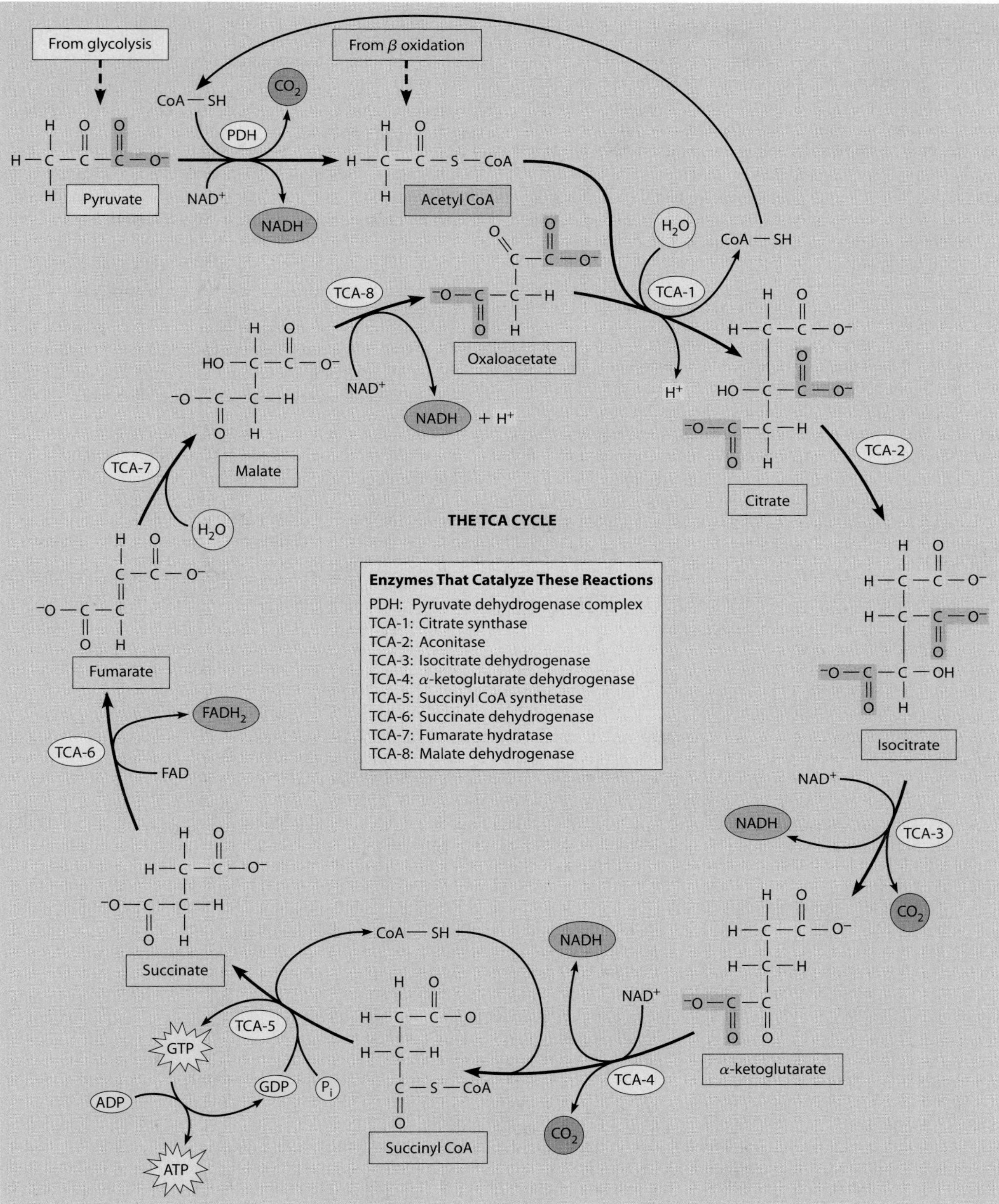

FIGURE 10-8 The Tricarboxylic Acid (TCA) Cycle. The two carbon atoms of pyruvate that enter the cycle via acetyl CoA are shown in pink in citrate and subsequent molecules until they are randomized by the symmetry of the fumarate molecule. The carbon atom of pyruvate that is lost as CO_2 is shown in gray, as are the two carboxyl groups of oxaloacetate that give rise to CO_2 in Reactions TCA-3 and TCA-4. Five of the reactions are oxidations, with NAD^+ as the electron acceptor in four reactions (PDH, TCA-3, TCA-4, and TCA-8) and FAD as the electron acceptor in one case (TCA-6). The reduced form of the coenzyme is shown in purple in each case. Note that when CO_2 is released, no H^+ is given off during NAD^+ reduction, thereby maintaining the charge balance of these reactions. The generation of GTP shown in Reaction TCA-5 is characteristic of animal mitochondria. In bacterial cells and plant mitochondria, ATP is formed directly.

(TCA-3) or from a carbon atom and a sulfur atom of separate molecules (PDH, TCA-4), generating either a C═O double bond or a C—S bond, respectively. The oxidation of a carbon-carbon bond releases less energy than does the oxidation of a carbon-oxygen bond—not enough energy to transfer electrons exergonically to NAD^+. Accordingly, the electron acceptor for this dehydrogenation is not NAD^+ but a lower-energy coenzyme, **flavin adenine dinucleotide (FAD).** Like NAD^+ and coenzyme A, FAD contains a B vitamin as part of its structure—riboflavin, in this case (**Figure 10-9**). FAD accepts two protons and two electrons, so the reduced form is written as $FADH_2$. As we will see later, the maximum ATP yield upon coenzyme oxidation is about three for NADH but only about two $FADH_2$.

In the next step of the cycle, the double bond of fumarate is hydrated, producing malate (Reaction TCA-7, catalyzed by the enzyme *fumarate hydratase*). Because fumarate is a symmetric molecule, the hydroxyl group of water has an equal chance of adding to either of the internal carbon atoms. As a result, the carbon atoms of Figure 10-8 that were color-coded pink (to keep track of the most recent acetate group to enter the cycle) are randomized at this step and are therefore not color-coded from this point on. In Reaction TCA-8, the hydroxyl group of malate becomes the target of the final oxidation in the cycle. As electrons are removed from adjacent carbon and oxygen atoms to form a C═O double bond, NAD^+ serves as the electron acceptor, producing NADH as malate is converted to oxaloacetate.

Summing Up: The Products of the TCA Cycle Are CO_2, ATP, NADH, and $FADH_2$

With the regeneration of oxaloacetate, one turn of the cycle is complete. We can summarize what has been accomplished by noting the following properties of the TCA cycle:

1. Two carbons enter the cycle as acetyl CoA, which is joined to the four-carbon acceptor molecule oxaloacetate to form citrate, a six-carbon compound.
2. Decarboxylation occurs at two steps in the cycle so that the input of two carbons as acetyl CoA is balanced by the loss of two carbons as carbon dioxide.
3. Oxidation occurs at four steps, with NAD^+ as the electron acceptor in three cases and FAD as the electron acceptor in one case.
4. ATP is generated at one point, with GTP as an intermediate in animal cells.
5. One turn of the cycle is completed upon regeneration of oxaloacetate, the original four-carbon acceptor.

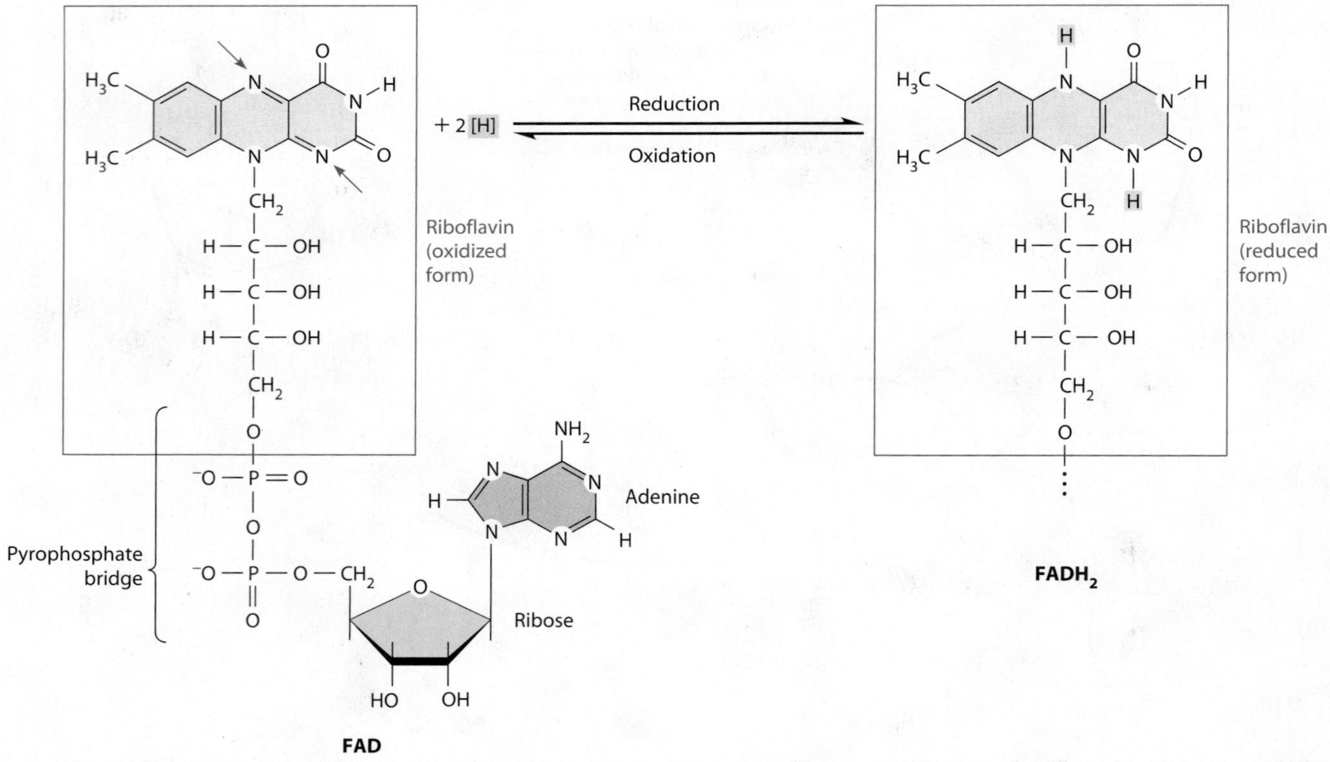

FIGURE 10-9 Structure of FAD and Its Oxidation and Reduction. The portion of the coenzyme enclosed in the red box is riboflavin, a B vitamin. The arrows point to the two nitrogen atoms of riboflavin that acquire one proton and one electron each when FAD is reduced to $FADH_2$. The half of the molecule that includes riboflavin and one phosphate group represents the structure of flavin mononucleotide (FMN), a closely related coenzyme.

By summing the eight component reactions of the TCA cycle as shown in Figure 10-8, we arrive at an overall reaction. (In this and subsequent reactions, protons and water molecules are not explicitly shown if they would be present only for charge or chemical balancing.) This reaction is written as

$$\text{acetyl CoA} + 3\text{NAD}^+ + \text{FAD} + \text{ADP} + \text{P}_i \longrightarrow 2\text{CO}_2 + 3\text{NADH} + \text{FADH}_2 + \text{CoA—SH} + \text{ATP} \quad \textbf{(10-2)}$$

Because the cycle must, in effect, occur twice to metabolize both of the acetyl CoA molecules derived from a single molecule of glucose, the summary reaction on a per-glucose basis can be obtained by doubling Reaction 10-2. If we then add to this reaction the summary reactions for glycolysis through pyruvate (Reaction 9-16) and for the oxidative decarboxylation of pyruvate to acetyl CoA (Reaction 10-1, also multiplied by 2), we arrive at the following overall reaction for the entire sequence from glucose through the TCA cycle:

$$\text{glucose} + 10\text{NAD}^+ + 2\text{FAD} + 4\text{ADP} + 4\text{P}_i \longrightarrow 6\text{CO}_2 + 10\text{NADH} + 2\text{FADH}_2 + 4\text{ATP} \quad \textbf{(10-3)}$$

As you consider this summary reaction, two points may strike you: how modest the ATP yield is thus far and how many coenzyme molecules are reduced during the oxidation of glucose. However, despite the low ATP yield, we must also recognize the reduced coenzymes NADH and $FADH_2$ as high-energy compounds. As we will see later in this chapter, the transfer of electrons from these coenzymes to oxygen is highly exergonic. Thus, the reoxidation of the 12 reduced coenzyme molecules shown on the right side of Reaction 10-3 provides the energy needed to drive the synthesis of most of the 38 ATP molecules produced during the complete oxidation of glucose.

For the release of that energy, we must look to the remaining stages of respiratory metabolism—electron transport and oxidative phosphorylation. Before doing so, however, we will consider several additional features of the TCA cycle: its regulation, its central position in energy metabolism, and its role in other metabolic pathways.

Several TCA Cycle Enzymes Are Subject to Allosteric Regulation

Like all metabolic pathways, the TCA cycle must be carefully regulated to ensure that its level of activity reflects cellular needs for its products. The main regulatory sites are shown in **Figure 10-10**. Most of the control involves **allosteric regulation** of four key enzymes by specific *effector molecules* that bind reversibly to them. As you may recall from Chapter 6, effector molecules can be either inhibitors or activators, which are indicated in Figure 10-10 as red minus signs and green plus signs, respectively. In addition to allosteric regulation, the pyruvate dehydrogenase complex (PDH) is reversibly inactivated by phosphorylation and activated by dephosphorylation of one of its protein components.

To appreciate the logic of TCA cycle regulation, remember that it uses acetyl CoA, NAD^+, FAD, and ADP as substrates and generates as its products NADH, $FADH_2$, CO_2, and ATP (see Reactions 10-3 and 10-4). Three of these products—NADH, ATP, and acetyl CoA—are important allosteric effectors of one or more enzymes, as shown in Figure 10-10. In addition, NAD^+, ADP, and AMP each activate at least one of the regulatory enzymes. In this way, the cycle is highly sensitive to the redox and energy status of the cell, as assessed by both the $NADH/NAD^+$ ratio and the relative concentrations of ATP, ADP, and AMP.

All four of the NADH-generating dehydrogenases shown in Figure 10-10 are inhibited by NADH. An increase in the NADH concentration of the mitochondrion therefore decreases the activities of these dehydrogenases, leading to a reduction in TCA cycle activity. In addition, PDH is inhibited by ATP, which is more abundant when energy is plentiful, and both PDH and isocitrate dehydrogenase are activated by AMP and ADP, which are more abundant when energy is needed.

The overall availability of acetyl CoA is determined primarily by the activity of the PDH complex (see Reaction 10-1), which is allosterically inhibited by NADH, ATP, and acetyl CoA and is activated by NAD^+, AMP, and free CoA (Figure 10-10). In addition, this enzyme complex is inactivated by phosphorylation of one of its protein components when the [ATP]/[ADP] ratio in the mitochondrion is high and is activated by removal of the phosphate group when the [ATP]/[ADP] ratio is low. These phosphorylation and dephosphorylation reactions are catalyzed by *PDH kinase* and *PDH phosphatase*, respectively. Not surprisingly, ATP is an activator of the kinase and an inhibitor of the phosphatase.

Due to these multiple control mechanisms, the generation of acetyl CoA is sensitive to the [acetyl CoA]/[CoA] and $[NADH]/[NAD^+]$ ratios within the mitochondrion and to the mitochondrial ATP status as well. In addition to these regulatory effects on reactions of the TCA cycle, feedback control from the cycle to the glycolytic pathway is provided by the inhibitory effects of citrate and acetyl CoA on phosphofructokinase and pyruvate kinase, respectively (see Figures 9-12 and 10-10).

The TCA Cycle Also Plays a Central Role in the Catabolism of Fats and Proteins

It is essential to understand the central role of the TCA cycle in all of aerobic energy metabolism. Thus far, we have regarded glucose as the main substrate for cellular respiration. In addition to glucose (and other carbohydrates), we must also note the roles of alternative fuel molecules in cellular energy metabolism and the TCA cycle, especially fats and proteins, which constitute roughly half of a diet based on the most recent dietary guidelines

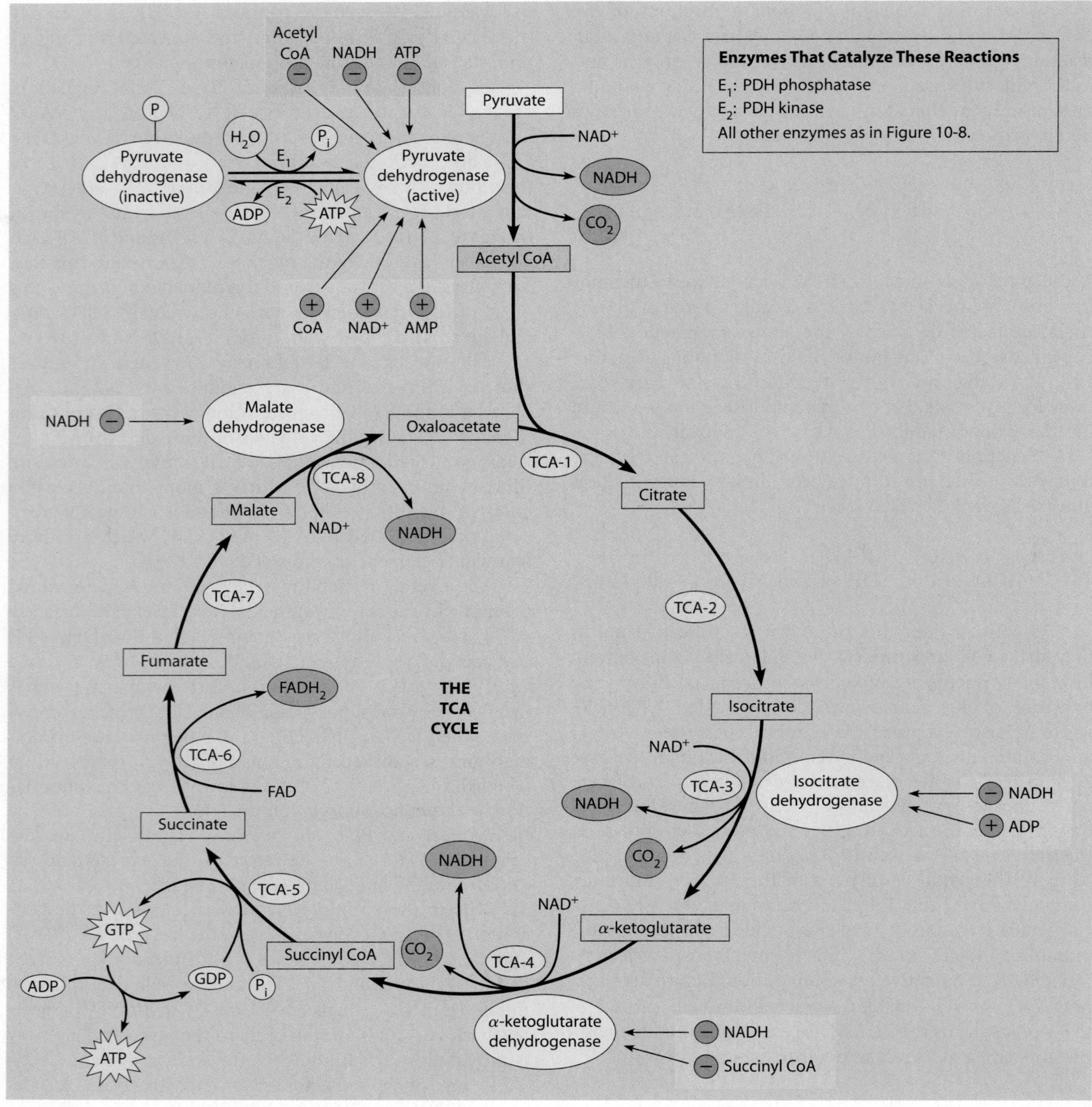

FIGURE 10-10 Regulation of the TCA Cycle. The pyruvate dehydrogenase reaction and the TCA cycle are shown here in outline form, with full names given for regulatory enzymes. Major regulatory effects are indicated as either activation (+) or inhibition (−). Allosteric regulators include CoA, NAD^+, AMP, and ADP as activators and acetyl CoA, NADH, ATP, and succinyl CoA as inhibitors. In addition to its allosteric effect on pyruvate dehydrogenase activity, ATP activates PDH kinase (E_2), the enzyme that phosphorylates one component of the PDH complex, thereby converting it to an inactive form. The enzyme PDH phosphatase (E_1) removes the phosphate group, returning the enzyme to its active form.

(carbohydrates = 50%, fats = 30%, proteins = 20%). Far from being a minor pathway for the catabolism of a single sugar, the TCA cycle represents the main conduit of aerobic energy metabolism in a broad spectrum of organisms ranging from microbes to higher plants and animals.

Fat As a Source of Energy. When we first encountered fats in Chapter 3, we noted their role in energy storage and observed that they are highly reduced compounds that liberate more energy per gram upon oxidation than do carbohydrates. For this reason, fats are an important long-term energy storage form for many organisms. Fat reserves are especially important in hibernating animals and migrating birds. In plant seeds, fats are a common form of energy and carbon storage. Fats are well suited for this storage function because they allow a maximum number of calories to be stored compactly.

Most fat is stored as deposits of **triacylglycerols,** which are neutral triesters of *glycerol* and long-chain *fatty acids* that we encountered earlier (see Figure 3-27b). Catabolism of triacylglycerols begins with their hydrolysis to glycerol and free fatty acids. The glycerol is channeled into the glycolytic pathway by oxidative conversion to dihydroxyacetone phosphate, as shown in Figure 9-9. The fatty acids are linked to coenzyme A to form fatty acyl CoAs, which are then degraded by **β oxidation,** a catabolic process that generates acetyl CoA and the reduced coenzymes NADH and $FADH_2$.

In bacteria, β oxidation occurs in the cytoplasm, and in eukaryotes it occurs both in the mitochondria and in peroxisomes. In plants and other eukaryotes that do not depend upon fatty acids as an energy source, β oxidation occurs in the peroxisome and can function as a way to recycle membrane fatty acids. Here we will focus on the process of β oxidation as it occurs in the mitochondrion of animals using saturated fatty acids with an even number of carbons as an energy source.

In animals, most fatty acids derived from dietary fats, like the pyruvate derived from carbohydrates, are oxidatively converted into acetyl CoA in the mitochondrion, which then can be further catabolized by the TCA cycle. The fatty acids are degraded in a series of repetitive cycles, each of which removes two carbons from the fatty acid until it is completely degraded. This process of fatty acid catabolism to acetyl CoA is called β oxidation because the oxidative events in each cycle occur on the carbon atom in the β position of the fatty acid (i.e., the second carbon in from the carboxylic acid group). Each cycle involves the same four steps—oxidation, hydration, reoxidation, and thiolysis (**Figure 10-11**)—and results in the production of one molecule each of $FADH_2$, NADH, and acetyl CoA as the fatty acid is shortened by two carbons in each cycle.

β oxidation of a fatty acid begins with an energy-requiring activation step in the cytosol. The energy of hydrolysis of ATP drives the attachment of a CoA molecule to the fatty acid (Reaction FA-1 in Figure 10-11). This forms a *fatty acyl CoA*, which is transported into the mitochondrion by a specific translocase located in the inner membrane. The next four enzymatic steps are repeated in a series of cycles until the fatty acid is completely degraded, losing two carbons per cycle as acetyl CoA. The first three of these steps closely resemble the oxidation, hydration, and reòxidation steps that convert succinate to oxaloacetate in the TCA cycle (TCA-6, 7, and 8).

First, an integral membrane protein acting as a *dehydrogenase* oxidizes the fatty acyl CoA, forming a double bond between the α and β carbons (FA-2). The

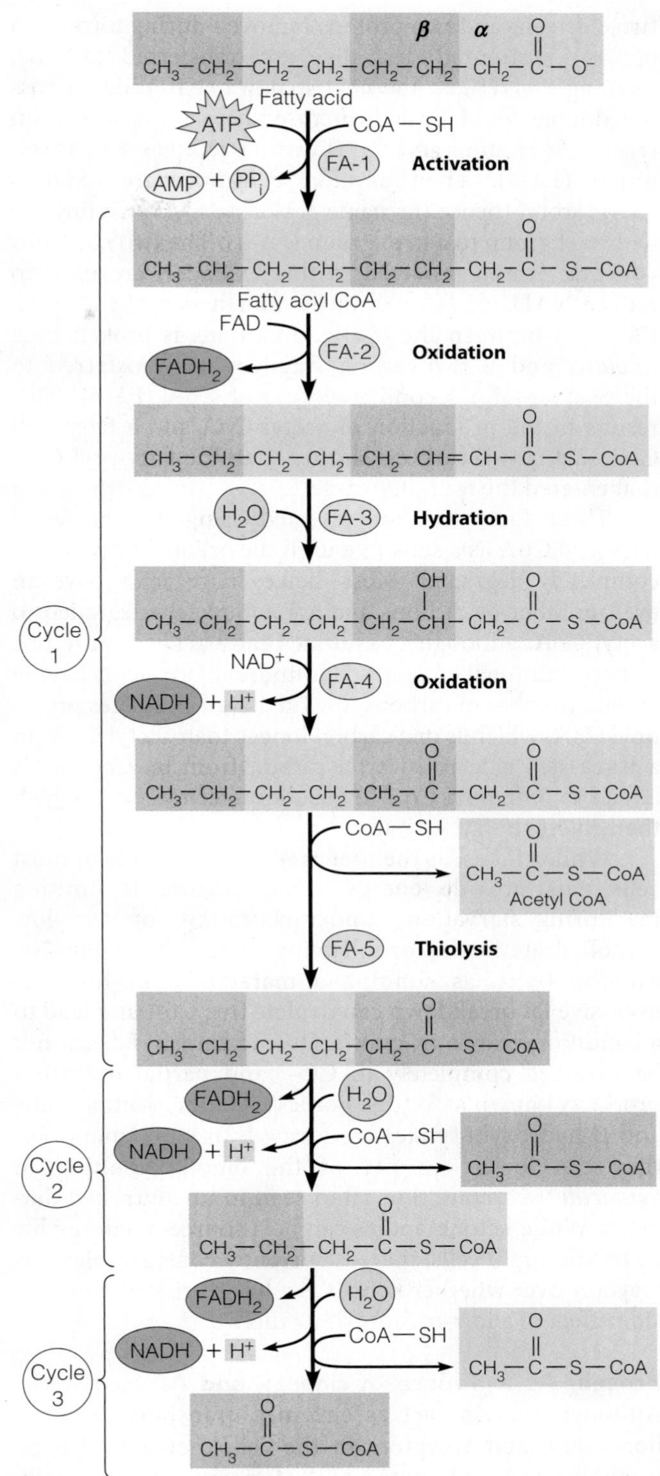

FIGURE 10-11 The B-Oxidation Pathway. Following an ATP-dependent activation step that links a fatty acid to CoA, the fatty acyl CoA derivative is transported into the mitochondrion (or peroxisome) and degraded in a series of repetitive cycles of oxidation, hydration, reoxidation, and thiolysis. Each cycle generates one $FADH_2$ and one NADH in the oxidation steps and releases acetyl CoA in the thiolysis step, shortening the fatty acid by two carbons. The example above is for the eight-carbon fatty acid octanoate, which requires three cycles of β oxidation. (Reaction numbers and chemical details are shown for the first cycle only.)

two electrons and two protons removed during formation of the unsaturated derivative are transferred to FAD, forming $FADH_2$. In the next step, water is added across the double bond by a *hydratase* so that the α carbon receives a H atom and the β carbon receives a hydroxyl group (FA-3). Then another *dehydrogenase* oxidizes the β carbon (hence the name β oxidation), converting the hydroxyl group to a keto group (FA-4). The two electrons and one proton removed in this oxidation are used to reduce NAD^+ to NADH. In the fourth step of the cycle, the bond between the α and β carbons is broken by a *thiolase*, and a two-carbon fragment is transferred to the S atom of a second molecule of CoA (FA-5). This results in the production of acetyl CoA and a fatty acyl CoA that is two carbons shorter than the fatty acyl CoA that entered the four-step cycle.

These four steps are repeated using the shortened fatty acyl CoA as a substrate until the original fatty acid is completely degraded. Most dietary fatty acids have an even number of carbons and are completely degraded to acetyl CoA, although unsaturated fatty acids require one or two additional enxymes. For unusual fatty acids having an odd number of carbons, the final cycle produces propionyl CoA, which is one carbon longer than acetyl CoA. In a three-step side pathway, a carbon from bicarbonate is added to propionyl CoA to generate succinyl CoA, which then enters the TCA cycle.

While glucose is the preferred energy source of most cells, fats provide energy when glucose is limiting (as during starvation), under conditions of very low carbohydrate intake, or following extremely demanding exercise (such as running a marathon). In humans, excessive fat breakdown can deplete free CoA and lead to a condition known as *ketosis*. During ketosis, fats cannot be oxidized completely to CO_2, and partial oxidation products known as ketone bodies (acetone, acetoacetate, and β-hydroxybutyrate) are formed. In large quantities, they can lower the pH of the blood, resulting in *ketoacidosis*, a condition often seen in uncontrolled diabetes. While ketone bodies can be a source of energy for heart and brain cells, there is currently considerable controversy over whether ketosis is a health risk or simply a side effect of a low-carbohydrate diet.

Protein As a Source of Energy and Amino Acids. Although proteins act as enzymes, transport proteins, hormones, and receptors in the cell, they can also be catabolized to generate ATP if necessary, especially during fasting or starvation conditions when carbohydrates and lipid stores are depleted. In plants, catabolism of proteins to free amino acids provides building blocks for protein synthesis during the germination of protein-storing seeds. In addition, all cells eventually undergo turnover of proteins and protein-containing structures, and the resulting amino acids can either be used to synthesize new proteins or be degraded oxidatively to yield energy.

Protein catabolism begins with hydrolysis of the peptide bonds that link amino acids together in the polypeptide chain. The process is called **proteolysis,** and the enzymes responsible for it are called *proteases.* The products of proteolytic digestion are small peptides and free amino acids. Further digestion of peptides is catalyzed by peptidases, which either hydrolyze internal peptide bonds *(endopeptidases)* or remove successive amino acids from the end of the peptide *(exopeptidases).*

Free amino acids, whether ingested as such or obtained by the digestion of proteins, can be catabolized for energy. Generally, these alternative substrates are converted to intermediates of mainstream catabolism in as few steps as possible. Despite their number and chemical diversity, all of the pathways for amino acid catabolism eventually lead to pyruvate, acetyl CoA, or a few key intermediates in the TCA cycle, notably α-ketoglutarate, oxaloacetate, fumarate, and succinyl CoA. The mitochondrion also participates in the *urea cycle,* a pathway prominent in the liver in which excess amino groups from degraded amino acids are converted to *urea* for excretion.

Of the 20 amino acids found in proteins, three of them give rise to pyruvate or TCA cycle intermediates directly: alanine, aspartate, and glutamate can be directly converted to pyruvate, oxaloacetate, and α-ketoglutarate, respectively (**Figure 10-12**). All the other amino acids require more complicated pathways, but ultimately all of them have end products that are TCA cycle intermediates.

The TCA Cycle Serves as a Source of Precursors for Anabolic Pathways

In addition to the catabolic function of the TCA cycle, there is a considerable flow of four-, five-, and six-carbon intermediates into and out of the cycle in most cells. These side reactions can replenish the supply of intermediates in the cycle if needed or use intermediates in the cycle for the synthesis of other compounds in anabolic pathways. Because the TCA cycle is a central link between catabolic and anabolic pathways, it is often called an **amphibolic pathway** (from the Greek prefix *amphi–*, meaning "both").

Besides its central role in catabolism, the TCA cycle is involved in various anabolic processes. For example, the three reactions shown in Figure 10-12 convert α-keto intermediates of the TCA cycle into the amino acids alanine, aspartate, and glutamate. These amino acids are constituents of proteins, so the TCA cycle is indirectly involved in protein synthesis by providing several of the amino acids required for the process. Other metabolic precursors provided by the TCA cycle include succinyl CoA and citrate. Succinyl CoA is the starting point for the biosynthesis of heme, whereas citrate can be transported out of the mitochondrion and used as a source of acetyl CoA for the stepwise synthesis of fatty acids in the cytosol.

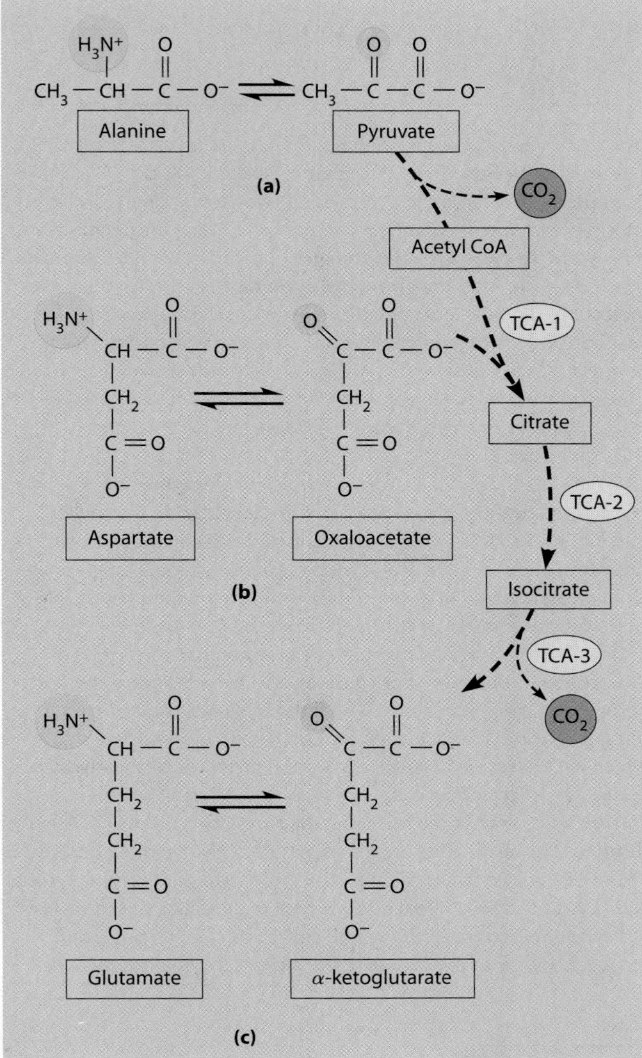

FIGURE 10-12 Interconversion of Several Amino Acids and Their Corresponding Keto Acids in the TCA Cycle. The amino acids **(a)** alanine, **(b)** aspartate, and **(c)** glutamate can be converted into the corresponding α-keto acids: pyruvate, oxaloacetate, and α-ketoglutarate, respectively. Each of these keto acids is an intermediate in the TCA cycle, a portion of which is shown to provide the metabolic context for these reactions. In each case, the amino group is shown in blue and the keto group in yellow. These reactions are readily reversible and can occur in either catabolism (oxidizing amino acids to CO_2 and H_2O) or anabolism (producing amino acids for protein synthesis).

The Glyoxylate Cycle Converts Acetyl CoA to Carbohydrates

The *glyoxylate cycle* has several intermediates in common with the TCA cycle but performs a specialized anabolic function in some germinating seeds and fungal spores, as described in **Box 10A**. This cycle occurs in a special type of peroxisome called the *glyoxysome*, which is common in fat-storing plant seeds that must convert this fat to sucrose. The glyoxylate cycle has several of the same reactions (and enzymes) as the TCA cycle but lacks the two oxidative decarboxylation reactions where CO_2 is released in the TCA cycle. Instead, two acetate molecules (which enter the pathway as acetyl CoA) are used to generate succinate, a four-carbon compound (see Figure 10A-2). The succinate is then converted to phosphoenolpyruvate, from which sugars can be synthesized by gluconeogenesis.

Organisms possessing this pathway can synthesize sugars from two-carbon compounds such as acetate. The glyoxylate cycle also makes possible the conversion of stored fat to carbohydrate, with acetyl CoA as the intermediate. This capability is vital to seed germination in those plant species that store significant amounts of carbon reserves in their seeds as fats. Mammals, on the other hand, lack a functional glyoxylate cycle and therefore cannot convert fats into carbohydrates.

Electron Transport: Electron Flow from Coenzymes to Oxygen

Having considered the first three stages of aerobic respiration—glycolysis, pyruvate oxidation, and the TCA cycle—let's pause briefly to ask what has been achieved thus far. As Reaction 10-3 indicates, chemotrophic energy metabolism through the TCA cycle accounts for the synthesis of four ATP molecules per glucose, two from glycolysis, and two from the TCA cycle. Complete oxidation of glucose to CO_2 could yield 686 kcal/mol, but we have recovered less than 10% of that amount (only 4 ATP $\times$ approximately 10 kcal/mol each, based on the $\Delta G'$ value for ATP synthesis in a typical cell). Where is the rest of the free energy? And when will we get to the substantially greater ATP yield that is characteristic of respiration?

The answer is straightforward: The free energy is right there in Reaction 10-3, represented by the reduced coenzyme molecules NADH and $FADH_2$. As we will see shortly, large amounts of free energy are released when these reduced coenzymes are reoxidized by transfer of their electrons to molecular oxygen. In fact, about 90% of the potential free energy present in a glucose molecule is conserved in the 12 molecules of NADH and $FADH_2$ that are formed when a molecule of glucose is oxidized to CO_2.

The Electron Transport System Conveys Electrons from Reduced Coenzymes to Oxygen

The process of coenzyme reoxidation by the transfer of electrons to oxygen is called **electron transport.** Electron transport is the fourth stage of respiratory metabolism (see Figure 10-1). The accompanying process of ATP synthesis (stage ❺) will be discussed later in this chapter. Keep in mind, however, that electron transport and ATP synthesis are not isolated processes. They are both integral parts of cellular respiration, functionally linked to each other by the electrochemical proton gradient that is the result of electron transport as well as the source of the energy that drives ATP synthesis.

BOX 10A DEEPER INSIGHTS

The Glyoxylate Cycle, Glyoxysomes, and Seed Germination

Plant species that store carbon and energy reserves in their seeds as fats face a special metabolic challenge when their seeds germinate: They must convert the stored fat to sucrose, the immediate source of carbon and energy for most cells in the seedling. Many plant species are in this category, including such well-known oil-bearing species as soybeans, peanuts, sunflowers, castor beans, and maize. The fat consists mainly of triacylglycerols and is stored in the cell as *lipid bodies*. The electron micrograph in **Figure 10A-1** shows the prominence of lipid bodies in the cotyledon (embryonic leaf) of a cucumber seedling.

The advantage of storing fat rather than carbohydrate is that one gram of triacylglycerol contains more than twice as much energy as one gram of carbohydrate. This difference enables fat-storing species to pack the greatest amount of carbon and calories into the least amount of space. However, it also means that such species must be able to convert the stored fat into sugar when the seeds germinate.

While many organisms readily convert sugars and other carbohydrate to stored fat (as people do), most eukaryotic organisms, including us, cannot carry out the conversion of fat to sugar. For the seedlings of fat-storing plant species, however, the conversion of storage triacylglycerols to sucrose is essential because sucrose is the form in which carbon and energy are transported to the growing shoot and root tips of the developing seedling.

The metabolic pathways that make this conversion possible are β oxidation and the **glyoxylate cycle.** The function of β oxidation is to degrade the stored fat to acetyl CoA. The acetyl CoA then enters the glyoxylate cycle (**Figure 10A-2**), a five-step cyclic pathway that is named for one of its intermediates, the two-carbon keto acid called *glyoxylate*. The glyoxylate cycle is related to the TCA cycle and uses three of the same reactions. A critical difference, however, is the presence of two glyoxysome-specific enzymes, *isocitrate lyase* and *malate synthase*. Using these enzymes, the glyoxylate cycle bypasses the two decarboxylation reactions of the TCA cycle. Also, the glyoxylate cycle takes in not one but two molecules of acetyl CoA per turn of the cycle, generating succinate, a four-carbon compound. Thus, the glyoxylate cycle is anabolic (carbon enters as two 2-carbon molecules and leaves as a four-carbon molecule), whereas the TCA cycle is catabolic (carbon enters as a two-carbon molecule and leaves as two CO_2 molecules).

In the seedlings of fat-storing species, the enzymes of β oxidation and the glyoxylate cycle are localized in organelles called **glyoxysomes,** which are found in the seedlings of fat-storing species of plants (Figure 10A-1) and in spores of some fungi. The intimate association of glyoxysomes with lipid bodies presumably facilitates the transfer of fatty acids from the lipid bodies.

Figure 10A-2 shows the relevant metabolism in an intracellular context. Stored triacylglycerols are hydrolyzed in the lipid bodies, releasing fatty acids. The fatty acids are transported into the glyoxysome and are degraded by β oxidation to acetyl CoA, which is converted to succinate by the enzymes of the glyoxylate cycle. The succinate moves to the mitochondrion, where it is converted to malate by reactions that are a part of the TCA cycle. (Notice that mitochondria are adjacent to glyoxysomes in the cucumber cotyledon in Figure 10A-1.) Malate goes to the cytosol and is oxidized to oxaloacetate, which is decarboxylated to form phosphoenolpyruvate (PEP). PEP serves as the starting point for gluconeogenesis in the cytosol, ultimately yielding sucrose,

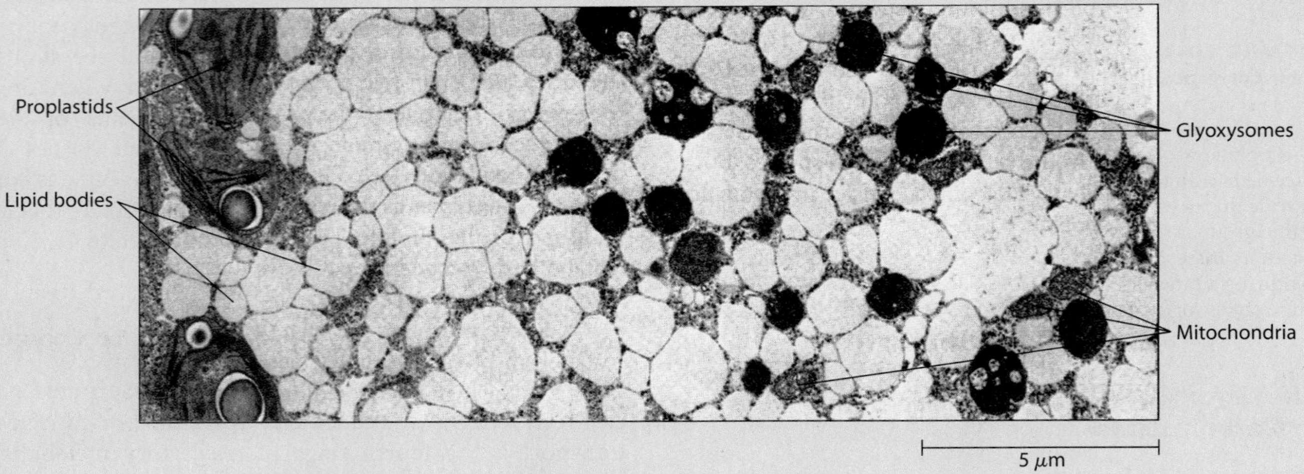

FIGURE 10A-1 The Association of Glyoxysomes and Lipid Bodies in Fat-Storing Seedlings. Shown here is a cell from the cotyledon of a cucumber seedling during early postgerminative development. The glyoxysomes and mitochondria involved in fat mobilization and gluconeogenesis are intimately associated with the lipid bodies in which the fat is stored. The fat is present primarily as triacylglycerols (also called triglycerides). Evidence that the cotyledon is not yet photosynthetically active and is therefore still heterotrophic in its nutritional mode can be seen in the presence of proplastids instead of mature chloroplasts (TEM).

the major carbohydrate transported to growing tissues in plants.

The route from stored triacylglycerols to sucrose is obviously quite complex, involving enzymes located in lipid bodies, glyoxysomes, mitochondria, and the cytosol, but it is the metabolic lifeline on which the seedlings of all fat-storing plant species depend. It complements much of the metabolism we've been learning about in Chapters 9 and 10, including gluconeogenesis, the TCA cycle, β oxidation—and now the glyoxylate cycle as well.

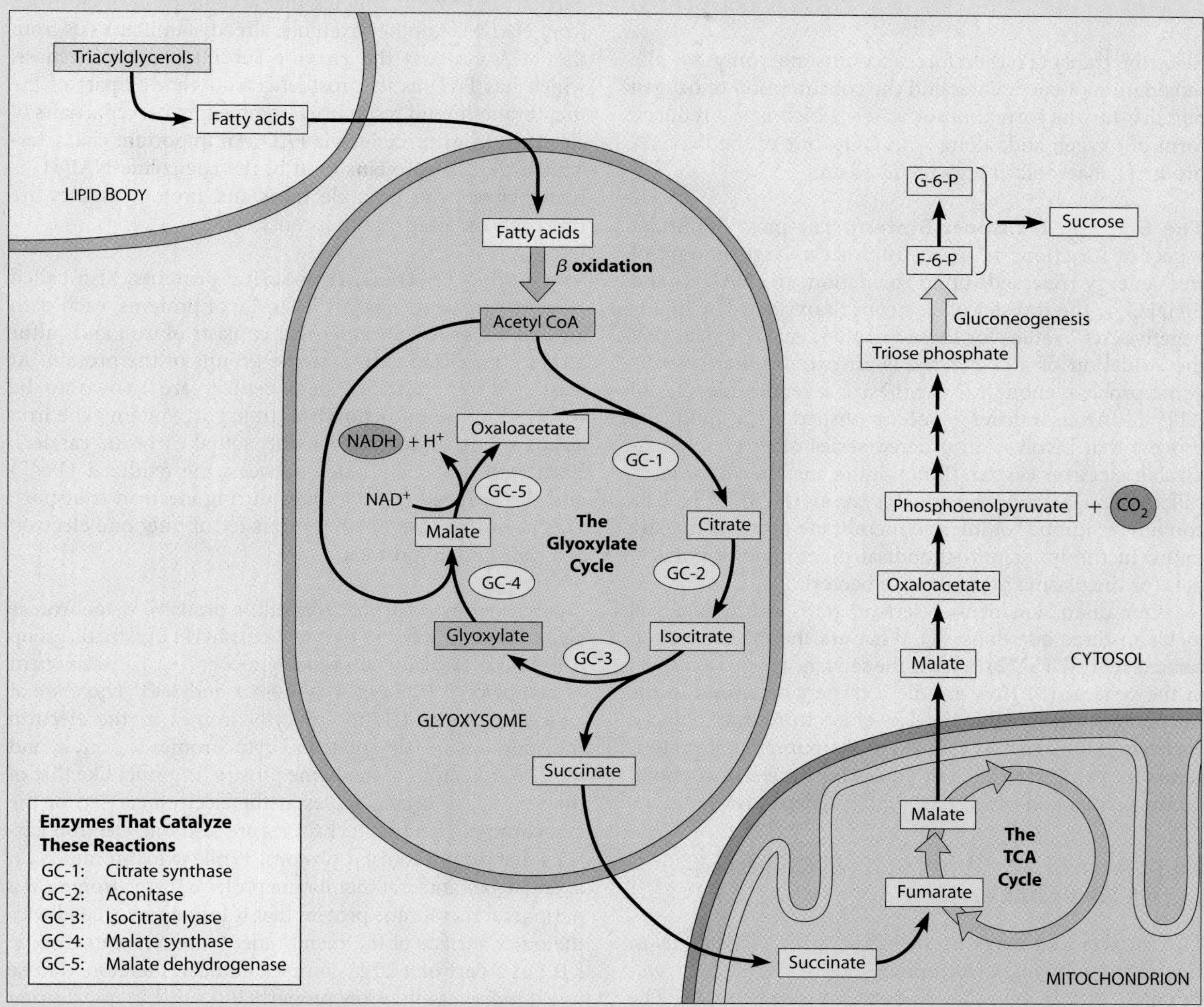

FIGURE 10A-2 The Glyoxylate Cycle and Gluconeogenesis in Fat-Storing Seedlings. Seedlings of fat-storing plant species can convert stored fat into sugar. Fatty acids derived from the hydrolysis of storage triacylglycerols are oxidized to acetyl CoA by the process of β oxidation. Acetyl CoA is then converted into succinate by the glyoxylate cycle, a five-reaction anabolic sequence named for glyoxylate, the molecule that is generated and consumed by Reactions GC-3 and GC-4, respectively. All the enzymes of β oxidation and the glyoxylate cycle are located in the glyoxysome. Conversion of succinate to malate occurs within the mitochondrion, whereas the further metabolism of malate via phosphoenolpyruvate to hexoses and hence to sucrose takes place in the cytosol. In seedlings of fat-storing plant species such as soybean, peanut, maize, and castor bean, acetyl CoA is derived from β oxidation of fatty acids. In bacteria and eukaryotic microorganisms capable of growing on two-carbon substrates such as ethanol or acetate, acetyl CoA is generated from acetyl phosphate, which is formed by ATP-dependent phosphorylation of free acetate.

Electron Transport and Coenzyme Oxidation. Electron transport involves the highly exergonic oxidation of NADH and $FADH_2$ with O_2 as the *terminal electron acceptor,* so we can write summary reactions as follows:

$$\text{NADH} + \text{H}^+ + \tfrac{1}{2}\text{O}_2 \longrightarrow \text{NAD}^+ + \text{H}_2\text{O} \qquad \Delta G^{\circ\prime} = -52.4 \text{ kcal/mol} \quad \textbf{(10-4)}$$

$$\text{FADH}_2 + \tfrac{1}{2}\text{O}_2 \longrightarrow \text{FAD} + \text{H}_2\text{O} \qquad \Delta G^{\circ\prime} = -45.9 \text{ kcal/mol} \quad \textbf{(10-5)}$$

Electron transport therefore accounts not only for the reoxidation of coenzymes and the consumption of oxygen but also for the formation of water, which is the reduced form of oxygen and, along with CO_2, one of the two end products of aerobic energy metabolism.

The Electron Transport System. The most important aspect of Reactions 10-4 and 10-5 is the large amount of free energy released upon oxidation of NADH and $FADH_2$ by the transfer of electrons to oxygen. The highly negative $\Delta G^{\circ\prime}$ values for these reactions make it clear that the oxidation of a coenzyme is an extraordinarily exergonic process, enough to synthesize several molecules of ATP. Electron transfer is accomplished as a multistep process that involves an ordered series of reversibly oxidizable electron carriers functioning together in what is called the **electron transport system (ETS).** The ETS contains a number of integral membrane proteins that are found in the inner mitochondrial membrane of eukaryotes (or the plasma membrane of bacteria).

Our discussion of the electron transport system will focus on three questions: (1) What are the major electron carriers in the ETS? (2) What is the sequence of these carriers in the system? (3) How are these carriers organized in the membrane to ensure that the flow of electrons from reduced coenzymes to oxygen is coupled to the pumping of protons across the membrane, thereby producing the electrochemical proton gradient on which ATP synthesis depends?

The Electron Transport System Consists of Five Kinds of Carriers

The carriers that make up the ETS include *flavoproteins, iron-sulfur proteins, cytochromes, copper-containing cytochromes,* and a quinone known as *coenzyme Q.* The flavoproteins and coenzyme Q transport protons along with electrons. Except for coenzyme Q, all the carriers are proteins with specific prosthetic groups capable of being reversibly oxidized and reduced. Almost all the events of electron transport occur within membranes, so it is not surprising that most of these carriers are hydrophobic molecules. In fact, most of these intermediates occur in the membrane as parts of large assemblies of proteins called *respiratory complexes.* We will first look briefly at the chemistry of these electron carriers and then see how they are organized into respiratory complexes and ordered into a sequence of carriers that transfer electrons from reduced coenzymes to oxygen.

Flavoproteins. Several membrane-bound **flavoproteins** participate in electron transport, using either *flavin adenine dinucleotide (FAD)* or *flavin mononucleotide (FMN)* as the prosthetic group. FMN is essentially the flavin-containing half of the FAD molecule shown in Figure 10-9. An example of a flavoprotein is *NADH dehydrogenase,* which is part of the protein complex that accepts pairs of electrons from NADH. Another example, already familiar to us from the TCA cycle, is the enzyme succinate dehydrogenase, which has FAD as its prosthetic group and is part of the membrane-bound respiratory complex that accepts pairs of electrons from succinate via FAD. An important characteristic of the flavoproteins (and of the coenzyme NADH) is that they transfer both electrons and protons as they are reversibly oxidized and reduced.

Iron-Sulfur Proteins. Iron-sulfur proteins, also called *nonheme iron proteins,* are a family of proteins, each with an *iron-sulfur (Fe-S) center* that consists of iron and sulfur atoms complexed with cysteine groups of the protein. At least a dozen different Fe-S centers are known to be involved in the mitochondrial transport system. The iron atoms of these centers are the actual electron carriers. Each iron atom alternates between the oxidized (Fe^{3+}) and the reduced (Fe^{2+}) states during electron transport, which, in this case, involves transfer of only one electron at a time and no protons.

Cytochromes. Like the iron-sulfur proteins, **cytochromes** also contain iron but as part of a porphyrin prosthetic group called *heme,* which you probably recognize as a component of hemoglobin (see **Figures 10-13** and 3-4). There are at least five different kinds of cytochromes in the electron transport system, designated as cytochromes *b, c,* c_1, *a,* and a_3. The iron atom of the heme prosthetic group, like that of the iron-sulfur center, serves as the electron carrier for the cytochromes. Thus, cytochromes are also one-electron carriers that do not transfer protons. While cytochromes *b,* c_1, *a,* and a_3 are integral membrane proteins, cytochrome *c* is a peripheral membrane protein that is loosely associated with the outer surface of the membrane. Moreover, cytochrome *c* is not a part of a large complex and can therefore diffuse much more rapidly, a key property in its role in transferring electrons between protein complexes.

Copper-Containing Cytochromes. In addition to their iron atoms, cytochromes *a* and a_3 also contain a single copper atom bound to the heme group of the cytochrome, where it associates with an iron atom to form a **bimetallic iron-copper (Fe-Cu) center.** Like iron atoms, copper ions can be reversibly converted from the oxidized (Cu^{2+}) to the reduced (Cu^+) form by accepting or donating single electrons. The iron-copper center plays a critical role in keeping an O_2 molecule bound to the cytochrome oxidase

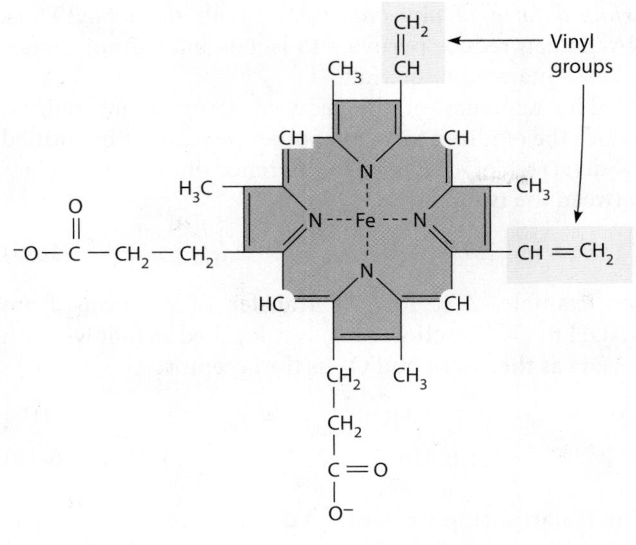

HEME

FIGURE 10-13 The Structure of Heme. Heme, also called iron-protoporphyrin IX, is the prosthetic group in cytochromes *b*, *c*, and c_1. A similar molecule, called heme A, is present in cytochromes a_1 and a_3. The heme of cytochromes *c* and c_1 is covalently attached to the protein by thioether bonds between the sulfhydryl groups of two cysteines in the protein and the vinyl ($-CH{=}CH_2$) groups of the heme (highlighted in yellow). In other cytochromes, the heme prosthetic group is linked noncovalently to the protein.

complex until the O_2 molecule has picked up the requisite four electrons and four protons, at which point the oxygen atoms are released as two molecules of water. (A nutritional note: If you have ever wondered why you require the elements iron and copper in your diet, their roles in electron transport are a major part of the reason.)

Coenzyme Q. The only nonprotein component of the ETS is **coenzyme Q (CoQ),** a quinone with the structure shown in **Figure 10-14**. Because of its ubiquitous occurrence in nature, coenzyme Q is also known as *ubiquinone.* Figure 10-14 also illustrates the reversible reduction, in two successive one-electron (plus one-proton) steps, from the quinone form (CoQ) via the *semiquinone* form (CoQH) to the *dihydroquinone* form ($CoQH_2$).

Unlike the proteins of the ETS, most of the coenzyme Q is freely mobile in the nonpolar interior of the inner mitochondrial membrane (or of the plasma membrane, in the case of bacteria). CoQ molecules are the most abundant electron carriers in the membrane and occupy a central position in the ETS, serving as a collection point for electrons from the reduced prosthetic groups of FMN- and FAD-linked dehydrogenases in the membrane. Although most of the CoQ is mobile, recent work suggests that a portion of it is tightly bound to specific respiratory complexes and may participate in the mechanism of proton pumping.

Note that coenzyme Q accepts electrons as well as protons when it is reduced and that it releases both electrons and protons when it is oxidized. This property is vital to the role of coenzyme Q in the active transport, or pumping, of protons across the inner mitochondrial membrane. When CoQ is reduced to $CoQH_2$, it always accepts protons from one side of the membrane then diffuses across the membrane to the outer surface, where it is oxidized to CoQ, with the protons ejected to the other side of the membrane. This motion provides a proton pump coupled to electron transport, which is thought to be one of the mechanisms whereby mitochondria, chloroplasts, and bacteria establish and maintain the electrochemical proton gradients that are used to store the energy of electron transport.

The Electron Carriers Function in a Sequence Determined by Their Reduction Potentials

Now that we are acquainted with the kinds of electron carriers that make up the ETS, our next question concerns the sequence in which these carriers operate. To answer that question, we need to understand the **standard reduction potential,** E_0', which is a measure, in volts (V), of the affinity a compound has for electrons. It describes how easily a compound will gain electrons and become reduced.

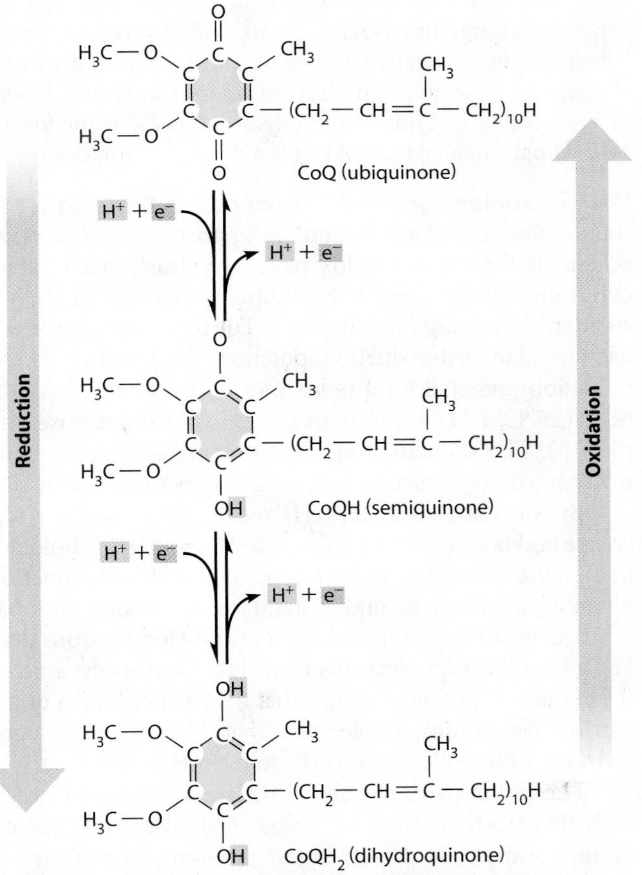

CoQ (ubiquinone)

CoQH (semiquinone)

$CoQH_2$ (dihydroquinone)

FIGURE 10-14 Oxidized and Reduced Forms of Coenzyme Q. Coenzyme Q (also called ubiquinone) accepts both electrons and protons as it is reversibly reduced in two successive one-electron steps to form first CoQH (the semiquinone form) and then $CoQH_2$ (the dihydroquinone form).

Reduction potentials are determined experimentally for a **redox (reduction-oxidation) pair,** which consists of two molecules or ions that are interconvertible by the loss or gain of electrons. For example, NAD^+ and NADH constitute a redox pair, as do Fe^{3+} and Fe^{2+} or $\frac{1}{2}O_2$ and H_2O, as shown in Reactions 10-6 to 10-8:

$$NAD^+ + H^+ + 2e^- \longrightarrow NADH \qquad \textbf{(10-6)}$$

$$Fe^{3+} + e^- \longrightarrow Fe^{2+} \qquad \textbf{(10-7)}$$

$$\tfrac{1}{2}O_2 + 2H^+ + 2e^- \longrightarrow H_2O \qquad \textbf{(10-8)}$$

Relative E_0' values allow us to compare redox pairs and to predict the direction electrons will tend to flow when several redox pairs are present in the same system, as is clearly the case for the electron transport system.

As described above, the reduction potential is a measure of the affinity that the oxidized form of a redox pair has for electrons. For a redox pair to have a positive E_0' means that the oxidized form has a high affinity for electrons and is therefore a good electron *acceptor.* For example, the E_0' value for the O_2/H_2O couple is highly positive, meaning that O_2 is a good electron acceptor. On the other hand, the NAD^+/NADH couple has a highly negative E_0' value, meaning that NAD^+ is a poor electron acceptor. Alternatively, a negative E_0' value can be thought of as a measure of how good an electron *donor* the reduced form of a redox pair is. Thus, the highly negative E_0' value for the NADH pair means that NADH is a good electron donor.

Understanding Standard Reduction Potentials. To standardize calculations and comparisons of reduction potentials for various redox pairs, we clearly need values determined under specified conditions, much as we did for standard free energy in Chapter 5. For this purpose, we will use the standard reduction potential (E_0'), which is the reduction potential for a redox pair under standard conditions (25°C, 1 *M* concentration, 1 atmosphere pressure, and pH 7.0). The standard reduction potentials of redox pairs relevant to energy metabolism are given in **Table 10-2**.

By convention, the $2H^+/H_2$ redox pair is used as a reference and is assigned the value 0.00 V (Table 10-2, boldface line). For a redox pair to have a positive standard reduction potential means that, under standard conditions, the oxidized form of the pair has a higher affinity for electrons than H^+ and will accept electrons from H_2. Conversely, a negative reduction potential means that the oxidized form of the pair has less affinity for electrons than H^+, and its reduced form will donate electrons to H^+ to form H_2.

The redox pairs of Table 10-2 are arranged in order, with the most negative E_0' values (i.e., the best electron donors and hence the strongest reducing agents) at the top. Any redox pair shown in Table 10-2 can undergo a redox reaction with any other pair. The direction of such a reaction under standard conditions can be predicted by inspection because, under standard conditions, *the reduced form of any redox pair will spontaneously reduce the oxidized form of any pair below it on the table.* Thus, NADH can reduce pyruvate to lactate but cannot reduce α-ketoglutarate to isocitrate.

The tendency of the reduced form of one pair to reduce the oxidized form of another pair can be quantified by determining $\Delta E_0'$, the difference in the E_0' values between the two pairs:

$$\Delta E_0' = E_{0,\,\text{acceptor}}' - E_{0,\,\text{donor}}' \qquad \textbf{(10-9)}$$

For example, $\Delta E_0'$ for the transfer of electrons from NADH to O_2 (Reaction 10-4) is calculated as follows, with NADH as the donor and O_2 as the acceptor:

$$\Delta E_0' = E_{0,\,\text{acceptor}}' - E_{0,\,\text{donor}}'$$
$$= +0.816 - (-0.32) = +1.136\text{V} \qquad \textbf{(10-10)}$$

The Relationship Between $\Delta G°'$ and $\Delta E_0'$. As you may already have guessed, $\Delta E_0'$ is a measure of thermodynamic spontaneity for the redox reaction between any two redox pairs under standard conditions. The spontaneity of a redox reaction under standard conditions can therefore be

Table 10-2 Standard Reduction Potentials for Redox Pairs of Biological Relevance*

Redox Pair (oxidized form → reduced form)	No. of Electrons	E_0' (V)
Acetate → pyruvate	2	−0.70
Succinate → α-ketoglutarate	2	−0.67
Acetate → acetaldehyde	2	−0.60
3-phosphoglycerate → glyceraldehyde-3-P	2	−0.55
a-ketoglutarate → isocitrate	2	−0.38
NAD^+ → NADH	2	−0.32
FMN → $FMNH_2$	2	−0.30
1,3-bisphosphoglycerate → glyceraldehyde-3-P	2	−0.29
Acetaldehyde → ethanol	2	−0.20
Pyruvate → lactate	2	−0.19
FAD → $FADH_2$	2	−0.18
Oxaloacetate → malate	2	−0.17
Fumarate → succinate	2	−0.03
$2H^+$ → H_2	**2**	**0.00****
CoQ → $CoQH_2$	2	+0.04
Cytochrome *b* (Fe^{3+} → Fe^{2+})	1	+0.07
Cytochrome *c* (Fe^{3+} → Fe^{2+})	1	+0.25
Cytochrome *a* (Fe^{3+} → Fe^{2+})	1	+0.29
Cytochrome a_3 (Fe^{3+} → Fe^{2+})	1	+0.55
Fe^{3+} → Fe^{2+} (inorganic iron)	1	+0.77
$\frac{1}{2}O_2$ → H_2O	2	+0.816

*Each $\Delta E_0'$ value is for the following half-reaction, where *n* is the number of electrons transferred:

$$\text{oxidized} + nH^+ + ne^- \rightarrow \text{reduced form.}$$

**By definition, this redox pair is the reference point for determining values of all other redox pairs. It requires that $[H^+] = 1.0\ M$ and therefore specifies pH 0.0. At pH 7.0, the value for the $2H^+/H_2$ pair is −0.42 V.

expressed as either $\Delta G^{\circ\prime}$ or $\Delta E_0^{\prime}$. The sign convention for $\Delta E_0^{\prime}$ is the opposite of that for $\Delta G^{\circ\prime}$, so an exergonic reaction is one with a negative $\Delta G^{\circ\prime}$ and a positive $\Delta E_0^{\prime}$. For any oxidation-reduction reaction, $\Delta G^{\circ\prime}$ is related to $\Delta E_0^{\prime}$ by the equation

$$\Delta G^{\circ\prime} = -nF\,\Delta E_0^{\prime} \tag{10-11}$$

where n is the number of electrons transferred, and F is the Faraday constant (23,062 cal/mol · V). For example, the reaction of NADH with oxygen (Reaction 10-4) involves the transfer of two electrons, so $\Delta G^{\circ\prime}$ for the reaction can be calculated as

$$\begin{aligned}\Delta G^{\circ\prime} &= -2F\,\Delta E_0^{\prime} = -2(23{,}062)(+1.136)\\ &= -52{,}400 \text{ cal/mol} = -52.4 \text{ kcal mol}\end{aligned} \tag{10-12}$$

The $\Delta G_0^{\prime}$ for this reaction is highly negative, so the transfer of electrons from NADH to O_2 is thermodynamically spontaneous under standard conditions. This difference in reduction potentials between the NAD^+/NADH and O_2/H_2O redox pairs drives the ETS and, as we will soon see, creates a proton gradient whose electrochemical potential will drive ATP synthesis.

Ordering of the Electron Carriers. We now have the information needed to put the pieces of the ETS together. As we already know, the respiratory ETS consists of several FMN- and FAD-linked dehydrogenases, iron-sulfur proteins with a total of 12 or more Fe-S centers, five cytochromes (including two with Fe-Cu centers), and a pool of coenzyme Q molecules. Shown in **Figure 10-15** are the major ETS components from free NADH

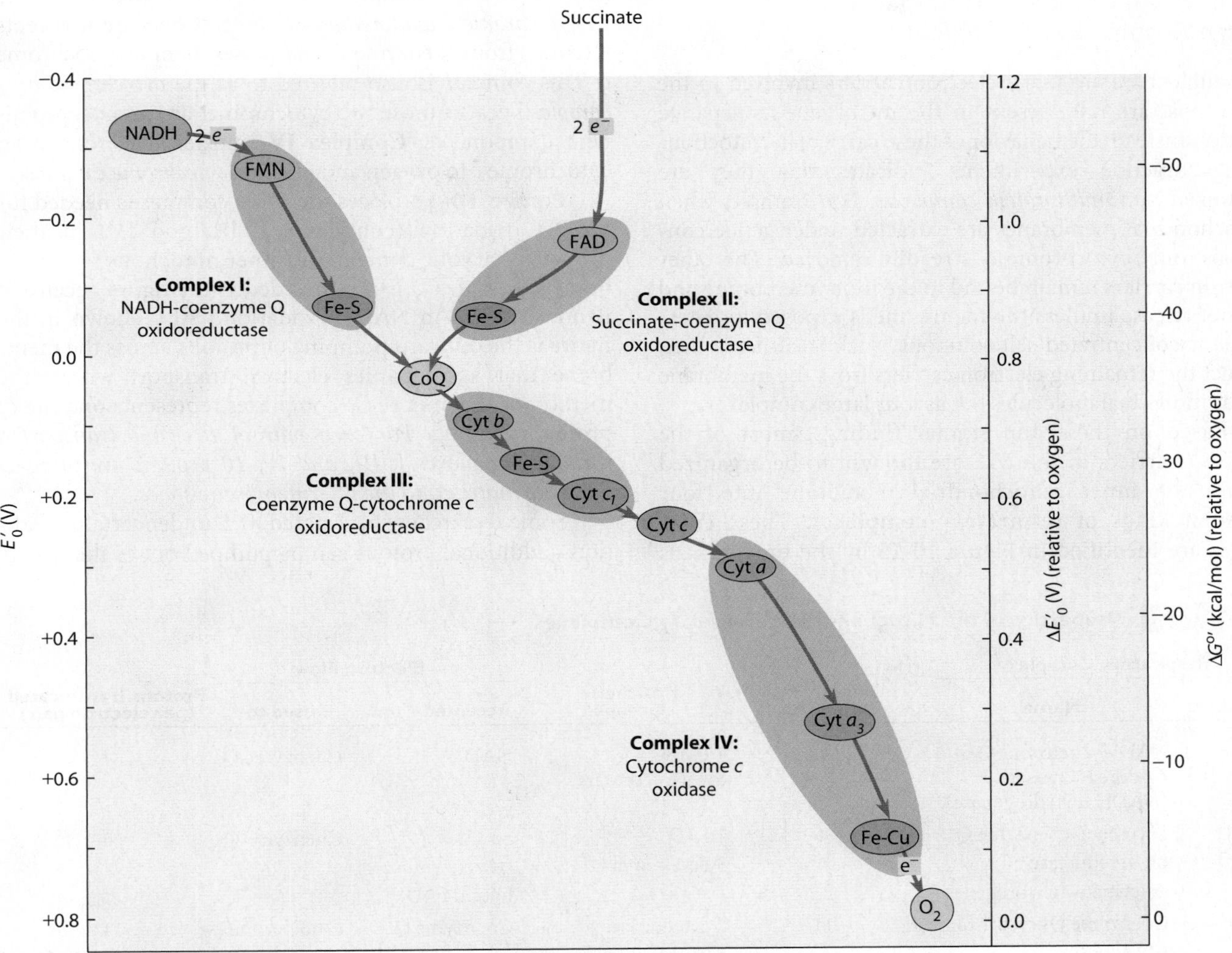

FIGURE 10-15 Major Components of the Respiratory Complexes and Their Energetics. Major intermediates in the transport of electrons from NADH (−0.32 V) and $FADH_2$ (−0.18 V) to oxygen (+0.816 V) are positioned vertically according to their energy levels, as measured by their standard reduction potentials ($E_0^{\prime}$, left axis). The four respiratory complexes are shown as large brown ovals, with the major electron carriers in each complex enclosed within inset ovals. Coenzyme Q and cytochrome *c* are small, mobile intermediates that transfer electrons between the several complexes. The red lines trace the exergonic flow of electrons through the system. On the right axes are the $\Delta E_0^{\prime}$ and $\Delta G^{\circ\prime}$ values relative to oxygen (i.e., the changes in the standard reduction potential and the standard free energy for the transfer of two electrons to O_2).

($E_0' = -0.32$ V) and the $FADH_2$ of succinate dehydrogenase ($E_0' = -0.18$ V) to oxygen ($E_0' = +0.816$ V), arranged according to their standard reduction potentials (E_0' values as obtained from Table 10-2).

In terms of energy, the key point of Figure 10-15 is that *the position of each carrier is determined by its standard reduction potential.* The electron transport system, in other words, consists of a series of chemically diverse electron carriers, with their order of participation in electron transfer determined by their relative reduction potentials. This means, in turn, that electron transfer from NADH or $FADH_2$ at the top of the figure to O_2 at the bottom is spontaneous and exergonic, with some of the transfers between successive carriers characterized by quite large differences in E_0' values and hence by large changes in free energy.

Most of the Carriers Are Organized into Four Large Respiratory Complexes

Although there are many electron carriers involved in the ETS, most are not present in the membrane as separate entities. Instead, the behavior of these carriers in mitochondrial extraction experiments indicates that they are organized into *multiprotein complexes.* For example, when mitochondrial membranes are extracted under gentle conditions, only cytochrome *c* is readily removed. The other electron carriers remain bound to the inner membrane and are not released unless the membrane is exposed to detergents or concentrated salt solutions. Such treatments then extract the remaining electron carriers from the membrane not as individual molecules but as four large complexes.

Based on these and similar findings, most of the electron carriers in the ETS are thought to be organized within the inner mitochondrial membrane into four different kinds of **respiratory complexes.** These complexes are identified in Figure 10-15 by the brown ovals and Roman numerals, which indicate the ordering of the various electron carriers based on their E_0' values as well as their organization within the membrane.

Properties of the Respiratory Complexes. Each respiratory complex consists of a distinctive assembly of polypeptides and prosthetic groups, and each complex has a unique role to play in the electron transport process. **Table 10-3** summarizes some properties of these complexes.

Complex I transfers electrons from NADH to coenzyme Q and is called the *NADH-coenzyme Q oxidoreductase complex* (or *NADH dehydrogenase complex*). **Complex II** transfers to CoQ the electrons derived from succinate in Reaction TCA-6. This complex is called the *succinate–coenzyme Q oxidoreductase complex,* although it is often also referred to by its more common name, *succinate dehydrogenase.* **Complex III** is called the *coenzyme Q–cytochrome c oxidoreductase complex* because it accepts electrons from coenzyme Q and passes them to cytochrome *c*. This complex is also referred to as the *cytochrome b/c_1 complex* because those two cytochromes are its most prominent components. **Complex IV** transfers electrons from cytochrome *c* to oxygen and is called *cytochrome c oxidase.*

Figure 10-16 places the three complexes needed for NADH oxidation (complexes I, III, and IV) in their proper eukaryotic context, the inner mitochondrial membrane. (Complex II is not included in the figure because it is not involved in NADH oxidation.) Also shown in the figure is the outward pumping of protons across the membrane that accompanies electron transport within the membrane. Each of these complexes represents one site of proton pumping. *For each pair of electrons transported through complexes I, III, and IV, 10 protons are pumped from the matrix into the intermembrane space.*

Some researchers have noted that, under certain conditions, additional protons can be pumped out of the matrix

Table 10-3 Properties of the Mitochondrial Respiratory Complexes

Respiratory Complex				Electron Flow		
Number	**Name**	**Number of Polypeptides***	**Prosthetic Groups**	**Accepted from**	**Passed to**	**Protons Translocated (per electron pair)**
I	NADH–coenzyme Q oxidoreductase (NADH dehydrogenase)	43 (7)	1 FMN 6–9 Fe-S centers	NADH	Coenzyme Q	4
II	Succinate–coenzyme Q oxidoreductase (succinate dehydrogenase)	4 (0)	1 FAD 3 Fe-S centers	Succinate (via enzyme-bound FAD)	Coenzyme Q	0
III	Coenzyme Q–cytochrome *c* oxidoreductase (cytochrome b/c_1 complex)	11 (1)	2 cytochrome *b* 1 cytochrome c_1 1 Fe-S center	Coenzyme Q	Cytochrome *c*	4**
IV	Cytochrome *c* oxidase	13 (3)	1 cytochrome *a* 1 cytochrome a_3 2 Cu centers (as Fe-Cu centers with cytochrome a_3)	Cytochrome *c*	Oxygen (O_2)	2

*The number of polypeptides encoded by the mitochondrial genome is indicated in parentheses for each complex.

**The value for complex III includes two protons translocated by coenzyme Q.

(a) Complex I receives 2 electrons from NADH and passes them to CoQ via FMN and an Fe-S protein. During this process, 4 H^+ are pumped out of the matrix by complex I.

(b) Complex III passes electrons from $CoQH_2$ to cytochrome *c* via cytochromes *b* and c_1 and an Fe-S protein. $CoQH_2$ carries 2 H^+ across the inner membrane and 2 more H^+ are pumped out of the matrix.

(c) Complex IV receives electrons from cytochrome *c* and, via cytochrome *a* and a_3, passes them to molecular oxygen, which is reduced to water as 2 more H^+ are pumped from the matrix by complex IV.

(d) ATP synthase uses the energy from the proton gradient generated during electron transport to synthesize ATP from ADP and P_i.

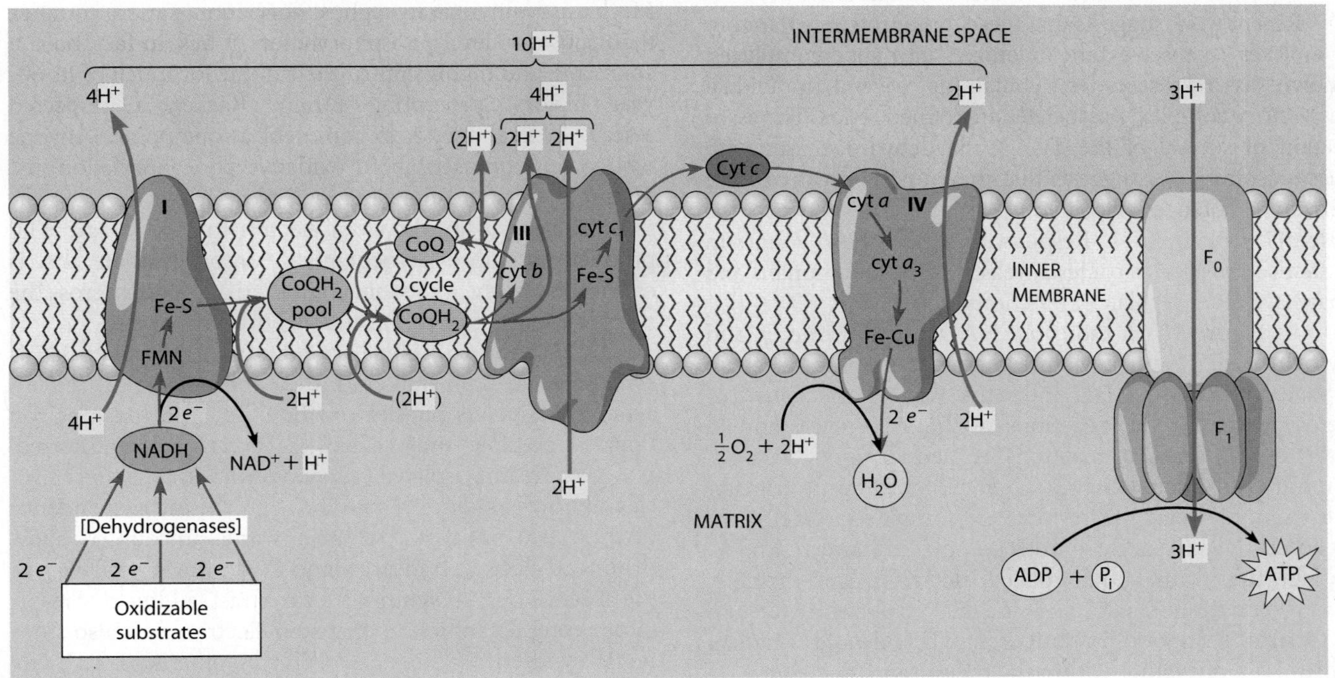

FIGURE 10-16 The Flow of Electrons Through Respiratory Complexes I, III, and IV Causes Directional Proton Pumping. **(a–c)** Electrons derived from oxidizable substrates in the mitochondrial matrix flow exergonically from NADH to oxygen via respiratory complexes I, III, and IV. **(d)** During the transport of two electrons, 10 H^+ are pumped across the inner membrane, and 3 ATP are synthesized by the F_oF_1 ATP synthase. Two extra H^+ (shown in parentheses) are pumped when the Q cycle operates.

without requiring additional electrons to be transported. This can be explained by the operation of a mitochondrial **Q cycle,** first proposed by Peter Mitchell in 1975. In this model, for every electron transferred from $CoQH_2$ to the Fe-S protein of complex III, one electron from $CoQH_2$ is used in the Q cycle to reduce a second molecule of CoQ via cytochrome *b*. As this CoQ is reduced to $CoQH_2$, two additional protons (shown in parentheses in Figure 10-16) are taken up from the matrix and then released into the intermembrane space as $CoQH_2$ is reoxidized.

The Role of Cytochrome c Oxidase. Of the several respiratory complexes involved in aerobic respiration in humans, cytochrome *c* oxidase (complex IV) is the **terminal oxidase,** transferring electrons directly to oxygen. This complex is therefore the critical link between aerobic respiration and the oxygen that makes it all possible. Cyanide and azide ions are highly toxic to nearly all aerobic cells because they bind tightly to the Fe-Cu center of cytochrome *c* oxidase, thereby blocking electron transport. Transport of four electrons from cytochrome *c* to O_2 plus the addition of four protons results in the production of two molecules of H_2O. However, studies have shown that complexes I and III also can transfer electrons directly to O_2, resulting in its incomplete reduction. This can generate toxic *superoxide anion* (O_2^-) or *hydrogen peroxide* (H_2O_2), compounds that can contribute to cellular aging under both normal and pathological conditions.

The Respiratory Complexes Move Freely Within the Inner Membrane

Unlike some integral proteins of the plasma membrane, the protein complexes of the mitochondrial inner membrane are mobile, free to diffuse within the plane of the membrane. This diffusion can be demonstrated experimentally. In one study, for example, inner mitochondrial membranes were placed in an electric field and the distribution of protein complexes was assessed subsequently by freeze-fracture analysis. Particles corresponding to protein complexes were found to accumulate at one end of the membrane. When the electrical field was turned off, the particles returned to a random distribution within seconds, clearly demonstrating their freedom to diffuse in a fluid lipid bilayer.

The results of this and similar studies make it clear that the respiratory complexes are not lined up in the membrane in the orderly fashion often seen in textbook diagrams but exist in the membrane as mobile complexes. In fact, the inner mitochondrial membrane has a high ratio of unsaturated to saturated phospholipids and virtually no cholesterol, so its fluidity is very high, and the mobility of the respiratory complexes is correspondingly high.

Recent work suggests that these multiprotein respiratory complexes themselves are organized into supercomplexes, known as **respirasomes,** containing several individual respiratory complexes associated in defined ratios. The association of several of the TCA cycle dehydrogenases with these respirasomes suggests that they function to minimize diffusion distances, facilitating electron flow between the respiratory complexes. Respirasomes have been found in systems as diverse as bacteria, yeast, plants, and humans and may have other regulatory functions as well.

As Figure 10-15 indicates, NADH, coenzyme Q, and cytochrome *c* are key intermediates in the electron transfer process. NADH links the ETS to the dehydrogenation (oxidation) reactions of the TCA cycle and to most other oxidation reactions in the matrix of the mitochondrion. Coenzyme Q and cytochrome *c* transfer electrons between the respiratory complexes. Coenzyme Q accepts electrons from both complexes I and II, and is, in fact, the "funnel" that collects electrons from virtually every oxidation reaction in the cell. Coenzyme Q and cytochrome *c* are both relatively small molecules that can diffuse rapidly, either within the membrane (coenzyme Q) or on the membrane surface (cytochrome *c*). They are also quite numerous—about 10 molecules of cytochrome *c* and 50 molecules of coenzyme Q for every complex I. Because of their abundance and mobility, these carriers are able to transfer electrons between the major complexes frequently enough to account for the observed rates of electron transfer in actively respiring mitochondria.

The Electrochemical Proton Gradient: Key to Energy Coupling

So far, we have learned that coenzymes are reduced during the oxidative events of glycolysis, pyruvate oxidation, and the TCA cycle, the first three stages of aerobic respiration (see Figure 10-1). We have also learned that reduced coenzymes are reoxidized by the exergonic transfer of electrons to oxygen via a system of reversibly oxidizable intermediates located within the inner mitochondrial membrane (stage ❹). Now we will consider how the free energy released during electron transport is used to generate an electrochemical proton gradient and how the energy of the gradient is then used to drive ATP synthesis in stage ❺.

Because this means of ATP synthesis involves phosphorylation events that are linked to oxygen-dependent electron transport, the process is called **oxidative phosphorylation,** thereby distinguishing it from *substrate-level phosphorylation,* which occurs as an integral part of a specific reaction in a metabolic pathway. (Reactions Gly-7 and Gly-10 of the glycolytic pathway and Reaction TCA-5 of the TCA cycle are examples of substrate-level phosphorylation; see Figures 9-7 and 10-8.)

Electron Transport and ATP Synthesis Are Coupled Events

Mechanistically, oxidative phosphorylation is more complex than substrate-level phosphorylation. It has, in fact, been a confusing and highly controversial topic for much of its 60-year history, prompting Efraim Racker, a respected researcher in the field, to comment at one point, "Anyone who is not confused about oxidative phosphorylation just doesn't understand the situation." Now, however, we know that the crucial link between electron transport and ATP production is an **electrochemical proton gradient** that is established by the directional pumping of protons across the membrane in which electron transport is occurring.

Under normal cellular conditions, ATP synthesis is *coupled* to electron transport, meaning not only that ATP synthesis depends on electron flow but also that electron flow is possible only when ATP can be synthesized. However, treating isolated mitochondria with certain chemicals—known as *uncouplers*—abolishes the interdependence of these two processes. In other words, uncouplers allow continued electron transport (and O_2 consumption) even in the absence of ATP synthesis. In contrast, addition of respiratory complex inhibitors that stop electron flow also stops ATP synthesis. Therefore ATP synthesis is strictly dependent on electron transport, but electron transport is not necessarily dependent on ATP synthesis. In brown adipose (fat) tissue in the bodies of newborn mammals, a similar uncoupling mechanism operates, using exergonic electron transport to generate heat rather than ATP.

Respiratory Control of Electron Transport. Because electron transport is coupled to ATP synthesis, the availability of ADP regulates the rate of oxidative phosphorylation and therefore of electron transport. This is called **respiratory control,** and its physiological significance is easy to appreciate: Electron transport and ATP generation will be favored when the ADP concentration is high (i.e., when the ATP concentration is low) and inhibited when the ADP concentration is low (when the ATP concentration is high). Oxidative phosphorylation is therefore regulated by cellular ATP needs, such that electron flow from organic fuel molecules to oxygen is adjusted to the energy needs of the cell. This regulatory mechanism becomes apparent during exercise, when the accumulation of ADP in muscle tissue causes an increase in electron transport rates, followed by a dramatic rise in the need for oxygen.

The Chemiosmotic Model: The "Missing Link" Is a Proton Gradient. How can ATP synthesis, a dehydration reaction, be tightly coupled to electron transport, which involves the sequential oxidation and reduction of various protein complexes in a lipid bilayer? Scientists puzzled over this question, but the answer eluded their best efforts for several decades. Most were convinced that some as yet undiscovered high-energy intermediate must be involved in oxidative phosphorylation just as in substrate-level phosphorylation. (Recall the ATP-generating steps of the glycolytic pathway and the TCA cycle in Reactions Gly-7 and Gly-10 in Figure 9-7 and Reaction TCA-5 in Figure 10-8.)

But while others continued their feverish search for high-energy intermediates, the British biochemist Peter Mitchell made the revolutionary suggestion that such intermediates might not exist at all. In 1961, Mitchell proposed an alternative explanation, which he called the **chemiosmotic coupling model.** According to this model, the exergonic transfer of electrons through the respiratory complexes is accompanied by the unidirectional pumping of protons across the membrane in which the transport system is localized. The electrochemical proton gradient that is generated in this way represents potential energy that then provides the driving force for ATP synthesis. In other words, the "missing link" between electron transport and ATP synthesis was not a high-energy chemical intermediate but an electrochemical proton gradient. This mechanism is illustrated in Figure 10-16, which depicts both the outward pumping of protons that accompanies electron transport through complexes I, III, and IV and the proton-driven synthesis of ATP by the F_oF_1 complex.

As you might expect, Mitchell's theory met with considerable initial skepticism and resistance. Not only was it a radical departure from conventional wisdom on coupling, but Mitchell proposed it in the virtual absence of experimental data. Over the years, however, a large body of evidence has been amassed in its support, and in 1978 Mitchell was awarded a Nobel Prize for his pioneering work. Today the chemiosmotic coupling model is a well-verified concept that provides a unifying framework for understanding energy transformations not just in mitochondrial membranes but in chloroplast and bacterial membranes as well.

The essential feature of the chemiosmotic model is that *the link between electron transport and ATP formation is an electrochemical potential across a membrane.* It is, in fact, this feature that gives the model its name: The "chemi" part of the term refers to the chemical processes of oxidation and electron transfer and the "osmotic" part comes from the Greek word *osmos,* meaning "to push"—to push protons across the membrane, in this case. The chemiosmotic model has turned out to be exceptionally useful not only because of the very plausible explanation it provides for coupled ATP generation but also because of its pervasive influence on the way we now think about energy conservation in biological systems.

Coenzyme Oxidation Pumps Enough Protons to Form 3 ATP per NADH and 2 ATP per $FADH_2$

Before considering the experimental evidence that has confirmed the chemiosmotic model, notice that Figure 10-16 provides us with estimates of the numbers of protons pumped outward by the three respiratory complexes involved in NADH oxidation, as well as the number of protons required to drive the synthesis of ATP by the F_oF_1 complex. The word *estimate* is important here because investigators still disagree on some of these numbers. This disagreement arises partly due to a structural feature that varies in different ATP synthase complexes. As we will see shortly, the number of protons required to produce 3 ATP is directly related to the number of *c* subunits in the F_o portion of the F_oF_1 ATP synthase, and this number can differ among organisms.

As we have already seen, most of the dehydrogenases in the mitochondrial matrix transfer electrons from oxidizable substrates to NAD^+, generating NADH. NADH, in turn, transfers electrons to the FMN component of complex I, thereby initiating the electron transport system. As Figure 10-16 shows, transfer of two electrons from NADH down the respiratory chain to oxygen is accompanied by the transmembrane pumping of 4 protons by complex I, 4 protons by CoQ plus complex III, and 2 protons by complex IV. This gives a total of 10 protons per NADH (12 protons per NADH if the Q cycle is operating).

The number of protons required to drive the synthesis of one molecule of ATP by the F_oF_1 ATP synthase is thought to be 3 or 4, with 3 generally regarded as more likely. If we assume that 10 protons are pumped per NADH oxidized and that 3 protons are required per ATP molecule, then we can conclude that oxidative phosphorylation yields about 3 molecules of ATP synthesized per molecule of NADH oxidized. This agrees well with evidence dating back to the 1940s that a fixed relationship usually exists between the number of molecules of ATP generated and the number of oxygen atoms consumed in respiration as a pair of electrons passes along the sequence of carriers from reduced coenzyme to oxygen. Racker and his colleagues, for example, showed that synthetic phospholipid vesicles containing complex I, complex III, or complex IV could generate one molecule of ATP per pair of electrons that passed through the complex. These results are consistent with our current understanding that each of these three complexes serves as a proton pump, thereby contributing to the proton gradient that drives ATP synthesis.

For purposes of our discussion, we will assume the ATP yields to be 3 for NADH and 2 for $FADH_2$, recognizing that the values on which these yields are based are still imprecisely known. These are reasonable numbers, considering that the $\Delta G^{\circ\prime}$ for the oxidation of NADH by molecular oxygen is −52.4 kcal/mol. This is enough of a free energy change to produce 3 ATP (requiring about 10 kcal/mole each under cellular conditions), even assuming an efficiency of only 50%.

The Chemiosmotic Model Is Affirmed by an Impressive Array of Evidence

Since its initial formulation in 1961, the chemiosmotic model has come to be universally accepted as the link between electron transport and ATP synthesis. To understand why, we will first consider several of the most important lines of experimental evidence that support this model. In the process, we will also learn more about the mechanism of chemiosmotic coupling.

1. Electron Transport Causes Protons to Be Pumped Out of the Mitochondrial Matrix. Shortly after proposing the chemiosmotic model, Mitchell and his colleague Jennifer Moyle demonstrated experimentally that the flow of electrons through the ETS is accompanied by the movement of protons across the inner mitochondrial membrane. They first suspended mitochondria in a medium in which electron transfer could not occur because oxygen was lacking. The proton concentration (i.e., the pH) of the medium was then monitored as electron transfer was stimulated by the addition of oxygen. Under these conditions, the pH of the medium declined rapidly with either NADH or succinate as the electron source (**Figure 10-17**). Because a decline in pH reflects an increase in proton concentration, Mitchell and Moyle concluded that electron transfer within the mitochondrial inner membrane is accompanied by **unidirectional pumping of protons** from the mitochondrial matrix into the external medium.

The mechanism whereby the transfer of electrons from one carrier to another is coupled to the directional transport of protons is still not well understood. There may, in fact, be more than one such mechanism. A possible mechanism is the one discussed earlier for coenzyme Q, which moves two protons across the mitochondrial membrane as it carries two electrons from respiratory complex I to complex III (Figure 10-16).

In the case of proton-pumping complexes I and IV, current evidence suggests that a second pumping mechanism may involve allosteric changes in protein conformation. In one conformational state, a polypeptide in the complex could bind a proton at the inner surface of the membrane. Transfer of one or more electrons through the complex might then cause the polypeptide to assume an alternative conformation and release the proton at the outer surface of the membrane. Thus, the pumping of protons may be achieved either by the physical movement of a carrier molecule such as coenzyme Q or by a conformational change in a polypeptide within a respiratory complex.

2. Components of the Electron Transport System Are Asymmetrically Oriented Within the Inner Mitochondrial Membrane. The unidirectional pumping demonstrated by Mitchell and Moyle requires that the electron carriers making up the ETS be asymmetrically oriented within the membrane. If not, protons would presumably be pumped randomly in both directions. Studies with a variety of antibodies, enzymes, and labeling agents designed to interact with various membrane components have shown clearly that some constituents of the respiratory complexes face the matrix side of the inner membrane, others are exposed on the opposite side, and

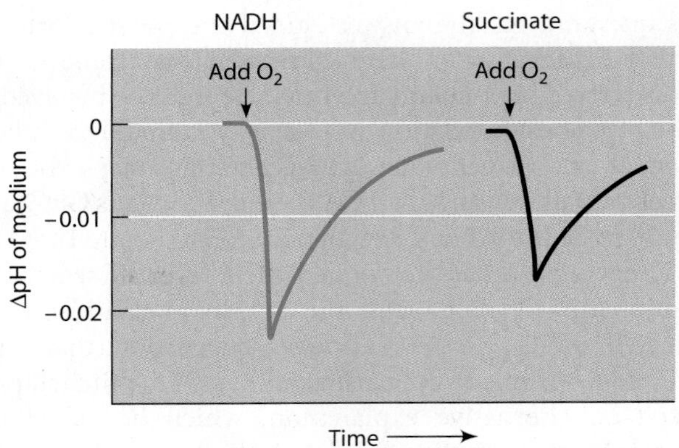

FIGURE 10-17 Experimental Evidence That Electron Transport Generates a Proton Gradient. Isolated mitochondria were incubated with either NADH or succinate as an electron source. The mitochondria were deprived of oxygen, and the pH of the medium was then monitored as electron transport was stimulated by the addition of a known quantity of oxygen. The rapid decline in pH reflects an increase in proton concentration, indicating that protons were pumped out of the mitochondrial matrix when oxygen was available. The pH change was reproducibly greater with NADH than with succinate, which is consistent with the greater ATP yield that accompanies the oxidation of NADH.

some are transmembrane proteins. These findings confirm the prediction that ETS components are asymmetrically distributed across the inner mitochondrial membrane, as shown schematically in Figure 10-16.

3. Membrane Vesicles Containing Complexes I, III, or IV Establish Proton Gradients. Further support for the chemiosmotic model has come from experiments involving the reconstitution of membrane vesicles from mixtures of isolated components. Since each of respiratory complexes I, III, and IV has a coupling site for ATP synthesis (see Figure 10-16), the chemiosmotic model predicts that each of these complexes should be capable of pumping protons across the inner mitochondrial membrane. This prediction has been confirmed experimentally by reconstituting artificial phospholipid vesicles containing complex I, III, or IV. When provided with appropriate oxidizable substrates, each of these three complexes is able to pump protons across the membrane of the vesicle. As noted before, these vesicles are also capable of ATP synthesis.

4. Oxidative Phosphorylation Requires a Membrane-Enclosed Compartment. An obvious prediction of the chemiosmotic model is that oxidative phosphorylation occurs only within a compartment enclosed by an intact mitochondrial membrane. Otherwise, the proton gradient that drives ATP synthesis could not be maintained. This prediction has been verified by experimental demonstration that electron transfer carried out by isolated complexes cannot be coupled to ATP synthesis unless the complexes are incorporated into membranes that form enclosed, intact vesicles.

5. Uncoupling Agents Abolish Both the Proton Gradient and ATP Synthesis. Additional evidence for the role played by proton gradients in ATP formation has come from studies using agents such as dinitrophenol (DNP) that are known to uncouple ATP synthesis from electron transport. For example, Mitchell showed in 1963 that membranes become freely permeable to protons in the presence of DNP. Dinitrophenol, in other words, abolished both ATP synthesis and the capability of a membrane to maintain a proton gradient. This finding is clearly consistent with the concept of ATP synthesis driven by an electrochemical proton gradient.

6. The Proton Gradient Has Enough Energy to Drive ATP Synthesis. For the chemiosmotic model to be viable, the proton gradient generated by electron transport must store enough energy to drive ATP synthesis. We can address this point with a few thermodynamic calculations. The electrochemical proton gradient across the inner membrane of a metabolically active mitochondrion involves both a membrane potential (the "electro" component of the gradient) and a concentration gradient (the "chemical" component). As you may recall from Chapter 8, the equation used to quantify this electrochemical gradient has two terms, one for the membrane potential, V_m, and the other for the concentration gradient, which in the case of protons is a pH gradient (see Table 8-3, page 218).

A mitochondrion actively involved in aerobic respiration typically has a membrane potential of about 0.16 V (positive on the side that faces the intermembrane space) and a pH gradient of about 1.0 pH unit (higher on the matrix side). This electrochemical gradient exerts a **proton motive force (pmf)** that tends to drive protons back down their concentration gradient—back into the matrix of the mitochondrion, that is. The pmf can be calculated by summing the contributions of the membrane potential and the pH gradient using the following equation:

$$\text{pmf} = V_m + 2.303\, RT\, \Delta\text{pH}/F \qquad \textbf{(10-13)}$$

where pmf is the proton motive force in volts, V_m is the membrane potential in volts, ΔpH is the difference in pH across the membrane ($\Delta\text{pH} = \text{pH}_{\text{matrix}} - \text{pH}_{\text{cytosol}}$), R is the gas constant (1.987 cal/mol · K), T is the temperature in kelvins, and F is the Faraday constant (23,062 cal/mol · V).

For a mitochondrion at 37°C with a V_m of 0.16 V and a pH gradient of 1.0 unit, the pmf can be calculated as follows:

$$\text{pmf} = 0.16 + \left(\frac{2.303(1.987)(37 + 273)(1.0)}{23{,}062}\right)$$
$$= 0.16 + 0.06 = 0.22 \text{ V} \qquad \textbf{(10-14)}$$

Notice that the membrane potential accounts for more than 70% of the mitochondrial pmf.

Like the redox potential, pmf is an electrical force in volts and can be used to calculate $\Delta G^{\circ\prime}$, the standard free energy change for the movement of protons across the membrane, using the following equation:

$$\Delta G^{\circ\prime} = -nF(\text{pmf}) = -(23.062)(0.22) = -5.1 \text{ kcal/mol} \qquad \textbf{(10-15)}$$

Thus, a proton motive force of 0.22 V across the inner mitochondrial membrane corresponds to a free energy change of about −5.1 kcal/mole of protons. This is the amount of energy that will be released as protons return to the matrix.

Is this enough to drive ATP synthesis? Not surprisingly, the answer is yes, though it depends on how many protons are required to drive the synthesis of one ATP molecule by the F_oF_1 complex. Mitchell's original model assumed two protons per ATP, which would provide about 10.2 kcal/mol, barely enough to drive ATP formation, assuming the $\Delta G'$ for phosphorylation of ADP to be about 10 to 14 kcal/mol under mitochondrial conditions. Differences of opinion still exist concerning the number of protons per ATP, but the real number is probably closer to three or four. That number would provide about 15 to 20 kcal of energy per ATP, enough to ensure that the reaction is driven strongly in the direction of ATP formation.

7. Artificial Proton Gradients Are Able to Drive ATP Synthesis in the Absence of Electron Transport. Direct evidence that a proton gradient can in fact drive ATP synthesis has been obtained by exposing mitochondria or inner membrane vesicles to artificial pH gradients. When mitochondria are suspended in a solution in which the external proton concentration is suddenly increased by the addition of acid, ATP is generated in response to the artificially created proton gradient. Because such artificial pH gradients induce ATP formation even in the absence of oxidizable substrates that would otherwise pass electrons to the ETS, it is evident that ATP synthesis can be induced by a proton gradient even in the absence of electron transport.

ATP Synthesis: Putting It All Together

We are now ready to take up the fifth, and final, stage of aerobic respiration: ATP synthesis. We have thus far seen that: (1) some of the energy of glucose is transferred to reduced coenzymes during the oxidation reactions of glycolysis and the TCA cycle; and (2) how this energy is used to generate an electrochemical proton gradient across the inner membrane of the mitochondrion. Now we can ask how the pmf of that gradient is harnessed to drive ATP synthesis. For that, we return to the F_1 complexes that can be seen along the inner surfaces of cristae (see Figure 10-5a) and ask about the evidence that these particles are capable of synthesizing ATP.

F_1 Particles Have ATP Synthase Activity

Key evidence concerning the role of F_1 particles came from studies conducted by Racker and his colleagues to test the prediction of the chemiosmotic hypothesis: that a reversible, proton-translocating ATPase is present in membranes that are capable of coupled ATP synthesis. Beginning with intact mitochondria (**Figure 10-18a**), the investigators were able to disrupt the mitochondria in such a way that fragments of the inner membrane formed small vesicles, which they called *submitochondrial particles* (Figure 10-18b). These submitochondrial particles were capable of carrying out electron transport and ATP synthesis. By subjecting these particles to mechanical agitation or protease treatment, they were able to dislodge the F_1 structures from the membranous vesicles (Figure 10-18c).

When the F_1 particles and the membranous vesicles were separated from each other by centrifugation, the membranous fraction could still carry out electron transport but could no longer synthesize ATP; the two functions had become uncoupled (Figure 10-18d). The isolated F_1 particles, on the other hand, were not capable of either electron transport or ATP synthesis but had ATPase activity (Figure 10-18e), a property consistent with the F-type ATPase that we now know the mitochondrial F_oF_1 complex to be (see Table 8-2, page 210). The ATP-generating capability of the membranous fraction was restored by adding the F_1 particles back to the membranes, suggesting that the spherical projections seen on the inner surface of the inner mitochondrial membrane are an important part of the ATP-generating complex of the membrane. These F_1 particles were therefore referred to as *coupling factors* (the "F" in F_1) and are now known to be the structures responsible for the ATP-synthesizing activity of the inner mitochondrial membrane or the bacterial plasma membrane.

The F_oF_1 Complex: Proton Translocation Through F_o Drives ATP Synthesis by F_1

Although the F_1 portion of the F_oF_1 ATP synthase complex is not directly membrane bound, it is attached to the F_o complex that is embedded in the inner mitochondrial membrane (see Figure 10-5c). We now know that the F_o complex serves as the **proton translocator,** the channel through which protons flow across the mitochondrial inner membrane (or the bacterial plasma membrane as in the case of the *E. coli* complex discussed below). Thus, the F_oF_1 complex is the complete, functional **ATP synthase.** The F_o component provides a channel for the

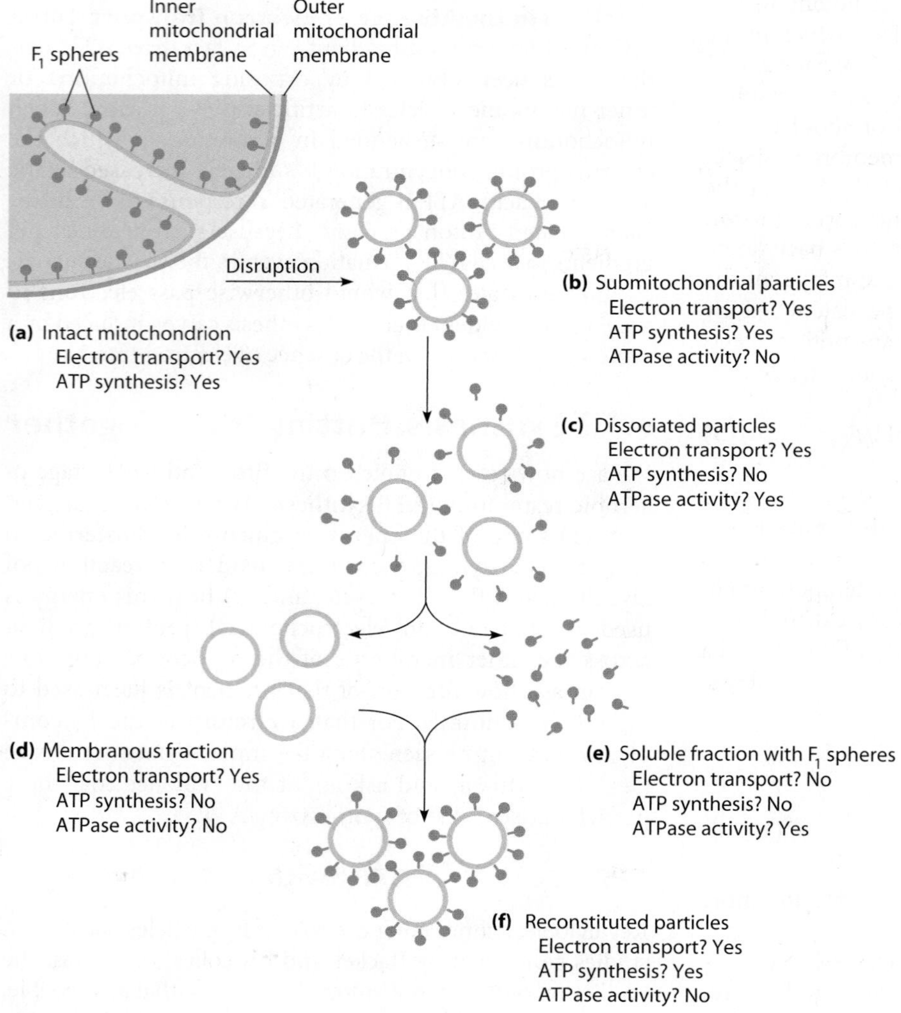

FIGURE 10-18 Dissociation and Reconstitution of the Mitochondrial ATP-Synthesizing System. (a) Intact mitochondria were disrupted so that fragments of the inner membrane formed (b) submitochondrial particles, capable of both electron transport and ATP synthesis. (c) When these particles were dissociated by mechanical agitation or enzyme treatment, the components could be separated into (d) a membranous fraction devoid of ATP-synthesizing capacity and (e) a soluble fraction with F_1 spheres having ATPase activity. (f) Mixing the two fractions reconstituted the structure and restored ATP synthase activity.

Table 10-4 Polypeptide Composition of the *E. coli* F_oF_1 ATP Synthase (ATPase)*

Structure	Polypeptide	Molecular Weight	Number Present	Function
F_o	*a*	30,000	1	Proton channel
	b	17,000	2	Peripheral stator stalk connecting F_o and F_1
	c	8,000	10	Rotating ring that turns γ subunit of F_1
F_1	α	52,000	3	Promotes activity of β subunit
	β	55,000	3	Catalytic site for ATP synthesis
	δ	19,000	1	Anchors $\alpha_3\beta_3$ ring to stator stalk of F_o
	γ	31,000	1	Rotates to transmit energy from F_o to F_1
	ε	15,000	1	Anchors γ subunit to c_{10} ring of F_o

*Mitochondrial F_oF_1 complexes are similar to the bacterial complex but with somewhat different polypeptide compositions for F_o and F_1.

exergonic flow of protons through the membrane, and the F_1 component carries out the actual synthesis of ATP, driven by the energy of the proton gradient.

Table 10-4 presents the polypeptide composition of the bacterial F_oF_1 complex from *Escherichia coli,* and **Figure 10-19** illustrates the four major structural components in this functional complex. Each of the F_o and F_1 complexes contains both a static component that remains stationary and a mobile component that moves during proton translocation. Together, these components of the F_oF_1 complex form one of the wonders of the molecular world—a miniature motor in which proton flow turns a microscopic gear that drives ATP synthesis.

The F_o complex is embedded in the bacterial membrane (or the inner mitochondrial membrane in eukaryotic cells) and consists of 1 *a* subunit, 2 *b* subunits, and 10 *c* subunits. The *a* and *b* subunits comprise the static component that is immobilized in the membrane. The ten *c* subunits form a ring that acts as a miniature gear that can rotate in the membrane relative to the *a* and *b*

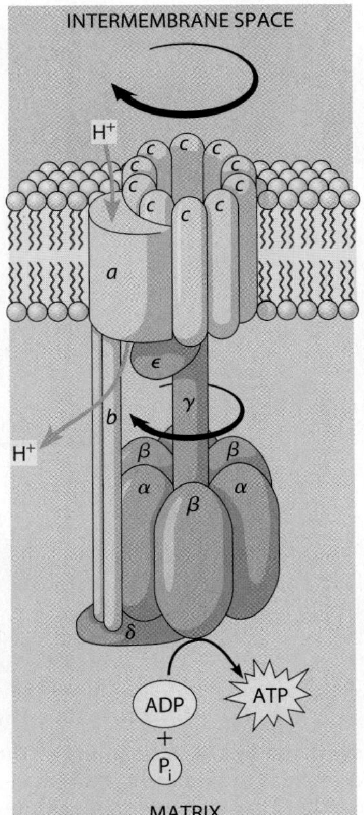

(a) The F_o static component consists of one *a* and two *b* subunits. The *a* subunit forms the proton channel and is immobilized in the membrane. The *b* subunits form the peripheral stalk and are attached both to the *a* subunit and to F_1.

(b) The F_o mobile component consists of a ring of 10 *c* subunits. Only one *c* subunit can form an ionic bond with the *a* subunit at a time. For each proton translocated, the ring rotates one-tenth of a turn as the adjacent *c* subunit in the ring bonds with the *a* subunit.

(c) The F_1 static component consists of the δ subunit plus a catalytic ring formed by a hexagon of alternating α and β subunits. The $\alpha_3\beta_3$ ring is the site of ATP synthesis and is immobilized by the δ subunit, which connects it to the b_2 peripheral stalk of F_o.

(d) The F_1 mobile component consists of the ϵ and γ subunits, which form the central stalk that is firmly attached to the c_{10} ring of F_o. As proton translocation turns the c_{10} ring, the γ subunit rotates inside the $\alpha_3\beta_3$ catalytic ring of F_1.

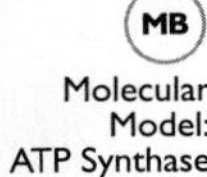

Molecular Model: ATP Synthase

FIGURE 10-19 F_o and F_1 Components of the *E. coli* F_oF_1 ATP Synthase. This illustration shows the subunit composition of the static and mobile components of the F_o **(a)** and **(b)** and F_1 **(c)** and **(d)** complexes that comprise the functional F_oF_1 ATP synthase in *E. coli.* As 10 H^+ move through the F_o proton translocator, the ring of 10 *c* subunits in F_o rotates once, resulting in the synthesis of 3 ATP by the $\alpha_3\beta_3$ catalytic ring of F_1.

subunits. The *a* subunit functions as the proton channel, and the *b* subunits form the *stator stalk* that connects the surfaces of the F_o and F_1 complexes.

The F_1 complex protrudes into the bacterial cytoplasm (or into the mitochondrial matrix in eukaryotic cells) and consists of 3 α subunits, 3 β subunits, plus 1 delta (δ), 1 gamma (γ), and 1 epsilon (ε) subunit. ATP is synthesized by a catalytic ring of three $\alpha\beta$ complexes that form a hexagon of alternating subunits. The δ subunit of F_1 anchors the $\alpha_3\beta_3$ catalytic ring to the b_2 stator stalk of F_o, immobilizing the catalytic ring. The mobile component of F_1 consists of the γ and ε subunits, which are attached to (and move with) the ring of *c* subunits in F_o. As protons move through the F_o proton channel, the c_{10} ring rotates, spinning the γ subunit within the $\alpha_3\beta_3$ catalytic ring. This results in ATP synthesis by the catalytic ring, using an intriguing mechanism that is described further in the next section.

ATP Synthesis by F_oF_1 Involves Physical Rotation of the Gamma Subunit

Once the link between electron transport within the membrane and proton pumping across the membrane had been established, the next piece of the puzzle was almost as daunting: How does the exergonic flux of protons through F_o drive the otherwise endergonic synthesis of ATP by the three β subunits of F_1? A novel answer to this question was suggested in 1979 by Paul Boyer, who proposed the **binding change model** shown in **Figure 10-20**.

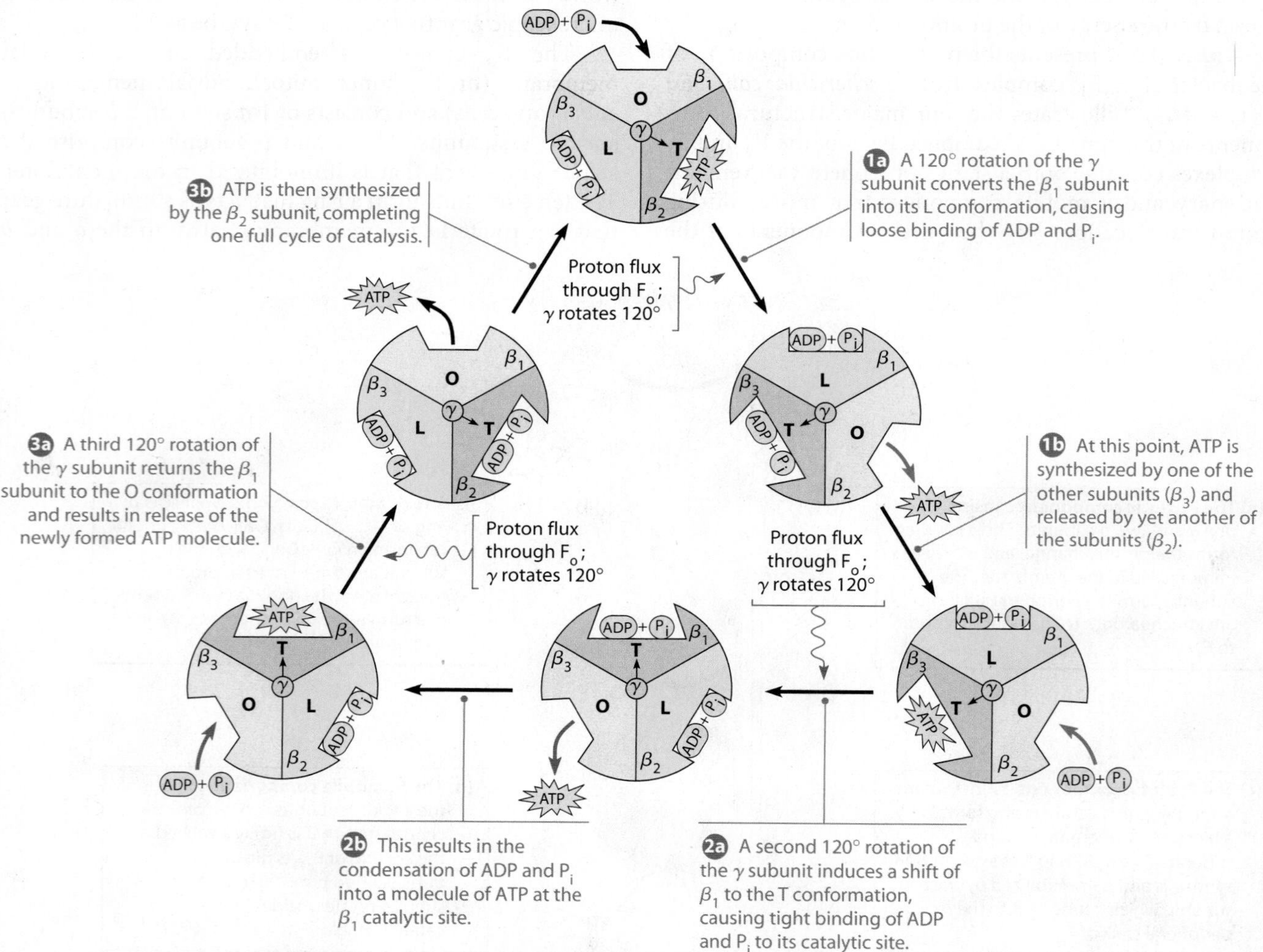

FIGURE 10-20 The Binding Change Model for ATP synthesis by the B Subunits of the F_oF_1 Complex. According to this model, each of the β subunits of the F_1 complex is in a different conformation at any instant. Each undergoes a sequence of conformational changes from the O (open) conformation through the L (loose) conformation to the T (tight) conformation. These conformational changes are driven by the rotation of the γ subunit. The process of ATP synthesis begins with one of the β subunits (arbitrarily identified as β_1) in its O conformation and involves the six steps shown, involving three 120° rotations and synthesis of 3 ATP per full 360° rotation. Notice that the same sequence of events occurs at each of the other sites, but it is offset temporally.

Boyer's model envisioned the catalytic site on each of the three β subunits of the F_1 complex as progressing through three distinctly different conformations with quite different affinities for the substrates (ADP and P_i) and the product (ATP). Boyer identified these as the L (for loose) conformation, which binds ADP and P_i loosely; the T (for tight) conformation, which binds ADP and P_i tightly and catalyzes their condensation into ATP; and the O (for open) conformation, which has a very low affinity for substrates or product and is unoccupied most of the time. Boyer proposed that, at any point in time, each of the three active sites is in a different conformation and that the hexagonal ring of α and β subunits actually *rotates* with respect to the central stalk containing the γ subunit. This rotation was proposed to be driven by the flux of protons through the F_o complex. (As mentioned earlier, we now know that it is the γ subunit that rotates within the stationary $\alpha_3\beta_3$ catalytic ring, but the concept of relative motion was quite accurate.)

Though not widely accepted at first, Boyer's model was substantially confirmed in 1994 by John Walker and co-workers, who used X-ray crystallography to provide the first detailed atomic model of the F_1 complex. Their model identified structures in the active sites of the β subunits that corresponded to Boyer's O, L, and T conformations, and they confirmed that each of the three active sites is in a different conformation at any given time. Even more remarkably, the γ subunit within the stalk that connects the F_1 and F_o complexes (see Figure 10-19) was shown to extend into the center of the F_1 structure in an asymmetric manner, such that its contact with and effect upon each of the β subunits is different at any given instant.

The Binding Change Model in Action. The sequence of events shown in Figure 10-20 is a current model of how the exergonic flux of protons through the F_o complex drives the synthesis of ATP by the catalytic sites on the three β subunits of the F_1 complex. As protons flow through a channel in the *a* subunit of F_o, they drive the rotation of the ring of *c* subunits in the F_o structure—and therefore also the rotation of the γ subunit of the stalk, which is attached to the *c* ring. The asymmetry of the γ subunit not only results in characteristically different interactions with the three β subunits at any point in time but also causes each β subunit to pass successively through the O, L, and T conformations as the γ subunit rotates 360 degrees.

The way that proton flow through F_o causes the c_{10} ring to rotate is truly a marvel of protein structure and function. Each of the 10 *c* subunits has an aspartate residue capable of forming an ionic bond with an arginine residue on the immobile *a* subunit. Only one aspartate from one *c* subunit can bind to the arginine at a time—and only if the aspartate is unprotonated (charged). As a proton is taken up from the external side of the membrane, it protonates and neutralizes the aspartate residue, disrupting the ionic bond. This causes a conformational change that rotates the ring one-tenth of a turn, causing the aspartate residue in the adjacent residue to lose a proton and form an ionic bond with the arginine in the *a* subunit. Thus, as 10 protons pass through the membrane via the *a* subunit, the c_{10} ring (and the attached F_1 γ subunit) makes one complete rotation.

Meanwhile, in the $\alpha_3\beta_3$ catalytic ring, the β_1 subunit starts out in the O conformation, with the substrates ADP and P_i free to enter the catalytic site (see Figure 10-20). The catalytic site has little affinity for them in this conformation. As protons flow through the F_o subunit in the membrane, the *c* ring and the attached γ subunit rotate 120° degrees (step 1a). This induces a shift in the β_1 site from the O to the L conformation and results in the loose binding of ADP and P_i to the catalytic site.

Following the synthesis of an ATP molecule by the β_3 subunit (step 1b), the γ subunit rotates another 120° (step 2a), inducing a shift in the β_1 subunit to the T conformation. This increases the affinity for ADP and P_i, which become tightly bound to the catalytic site in an orientation that allows them to condense spontaneously to form ATP (step 2b). The γ subunit now rotates a further 120°, returning the β_1 catalytic site to the O conformation, thereby releasing the ATP from the catalytic site (step 3a). Following the generation of an ATP molecule by one of the other subunits (the β_2 subunit; step 3b), one full cycle is completed, and the β_1 subunit is available for another round of ATP synthesis. Note that the same sequence of events has occurred at the other two subunits as well. Thus, one complete revolution of the γ subunit results in the synthesis of 3 ATP molecules, one by each of the three β subunits.

Spontaneous Synthesis of ATP? Close scrutiny of Figure 10-20 reveals that each of the three steps at which ATP is formed occurs without a direct input of energy. Yet we know that the synthesis of ATP is a highly endergonic reaction. How, we might then ask, does ATP synthesis proceed spontaneously at the catalytic site of a β subunit in its T conformation? Actually, what we really know is that ATP synthesis is a highly endergonic reaction *in dilute aqueous solution*. However, the reaction *at the catalytic site of a β subunit* involves enzyme-bound intermediates in a drastically different environment—so different, in fact, that the reaction under these conditions has a $\Delta G^{\circ\prime}$ close to zero (**Figure 10-21**) and can therefore occur spontaneously, without any immediate energy input. It is thought that the immediate environment at the catalytic site may serve to minimize charge repulsion between ADP and P_i, favoring their condensation into ATP.

This doesn't mean that ATP synthesis proceeds without thermodynamic cost, though. It simply means that the input of energy is required at other points in the cycle. Prior energy input is needed to drive the transition of the catalytic site from the L to the T conformation, thereby packing the ADP and P_i together tightly and facilitating their interaction (step 2a of Figure 10-20). Subsequent energy input is required for release of the tightly bound ATP product from the catalytic site (step 3a). This energy

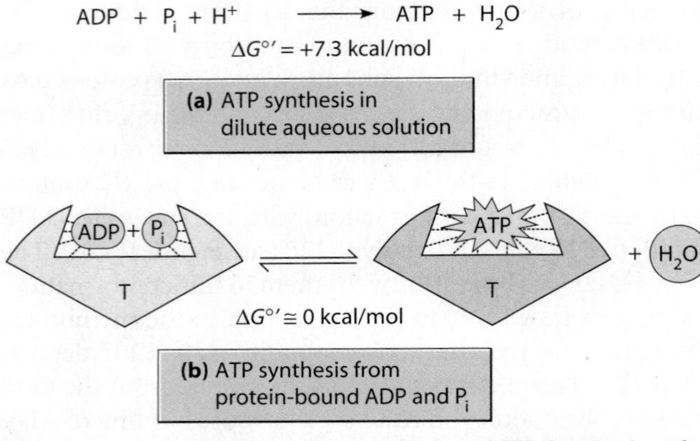

FIGURE 10-21 Comparative Energetics of ATP Synthesis. The energy requirement for ATP synthesis differs greatly, depending on the environment. **(a)** In dilute aqueous solution, ATP synthesis from soluble ADP and P_i is highly endergonic, with a $\Delta G^{\circ\prime}$ of +7.3 kcal/mol. **(b)** At the catalytic site of a β subunit in its T conformation, however, the environment is drastically different and the reaction has a $\Delta G^{\circ\prime}$ close to 0 (i.e., the K_{eq} is close to 1) and can therefore proceed spontaneously with no immediate energy requirement.

comes from the proton gradient generated by electron transport and transmitted through the rotation of the *c* ring and γ subunit as mechanical energy. Thus, the F_oF_1 ATP synthase provides us with a marvelous example of how the cell can convert electrical work (electron transport) to concentration work (proton gradient) to mechanical work (subunit rotation) to biosynthetic work (ATP synthesis).

The Chemiosmotic Model Involves Dynamic Transmembrane Proton Traffic

The dynamics of the chemiosmotic model are summarized in **Figure 10-22**. Complexes I, III, and IV of the ETS pump protons outward across the inner membrane of the mitochondrion, where they accumulate in the intracristal space. The resulting electrochemical gradient then drives ATP generation by means of the F_oF_1 complexes associated with the inner membrane. There is, in other words, continuous, dynamic two-way proton traffic across the inner membrane. First let's count the number of protons extruded by each of the respiratory assemblies shown in Figure 10-22: 10 for complexes I + III + IV (using NADH) and 6 for complexes II + III = IV (using $FADH_2$). Assuming that, and that 3 protons are required to drive the synthesis of 1 ATP molecule, then the production of 3 ATP per NADH and 2 ATP per $FADH_2$ seems quite reasonable.

Aerobic Respiration: Summing It All Up

To summarize aerobic respiration, let's now return to Figure 10-1 and review the role of each of the components. As substrates such as carbohydrates and fats are oxidized to generate energy, coenzymes are reduced. These reduced coenzymes represent a storage form of much of the free energy that was released by the original substrate molecules during oxidation. This energy can be used to drive ATP synthesis as the coenzymes are themselves reoxidized by the ETS. As electrons are transported from NADH or $FADH_2$ to oxygen, they pass through several respiratory complexes, where electron transport is coupled to the directional pumping of protons across the membrane. *The resulting electrochemical gradient exerts a pmf that serves as the driving force for ATP synthesis.* Under most conditions, a steady-state pmf will be maintained across the membrane. The transfer of electrons from coenzymes to oxygen is carefully and continuously adjusted so that the outward pumping of protons balances the inward proton flux necessary to synthesize ATP at the desired rate.

The Maximum ATP Yield of Aerobic Respiration Is 38 ATPs per Glucose

Now we can return to the question of the **maximum ATP yield** per molecule of glucose under aerobic conditions. Recall from Reaction 10-3 that the complete oxidation of glucose to carbon dioxide by glycolysis and the TCA cycle results in the generation of 4 molecules of ATP by substrate-level phosphorylation, with most of the remaining free energy of glucose oxidation stored in the 12 coenzyme molecules—10 of NADH and 2 of $FADH_2$. In prokaryotes and some eukaryotic cells, electrons from all the NADH molecules pass through all three ATP-generating complexes of the ETS, yielding 3 ATP molecules per molecule of coenzyme. Electrons from $FADH_2$, on the other hand, traverse only two of the three complexes, yielding only 2 ATP molecules per molecule of coenzyme. The maximum theoretical ATP yield obtainable upon reoxidation of the 12 coenzyme molecules formed per glucose can therefore be represented as follows:

$$10NADH + 10H^+ + 5O_2 + 30ADP + 30P_i \longrightarrow 10NAD^+ + 10H_2O + 30ATP \quad \textbf{(10-16)}$$

$$2FADH_2 + O_2 + 4ADP + 4P_i \longrightarrow 2FAD + 2H_2O + 4ATP \quad \textbf{(10-17)}$$

Summing these reactions gives us an overall reaction for electron transport and ATP synthesis:

$$10NADH + 10H^+ + 2FADH_2 + 6O_2 + 34ADP + 34P_i \longrightarrow 10NAD^+ + 2FAD + 12H_2O + 34ATP \quad \textbf{(10-18)}$$

Addition of Reaction 10-18 to the summary reaction for glycolysis and the TCA cycle (Reaction 10-3) leads to the following overall expression for the maximum theoretical ATP yield obtainable by the complete aerobic respiration of glucose or other hexoses:

$$C_6H_{12}O_6 + 6O_2 \xrightarrow{38ADP + 38P_i \;\; 38ATP} 6CO_2 + 6H_2O \quad \textbf{(10-19)}$$

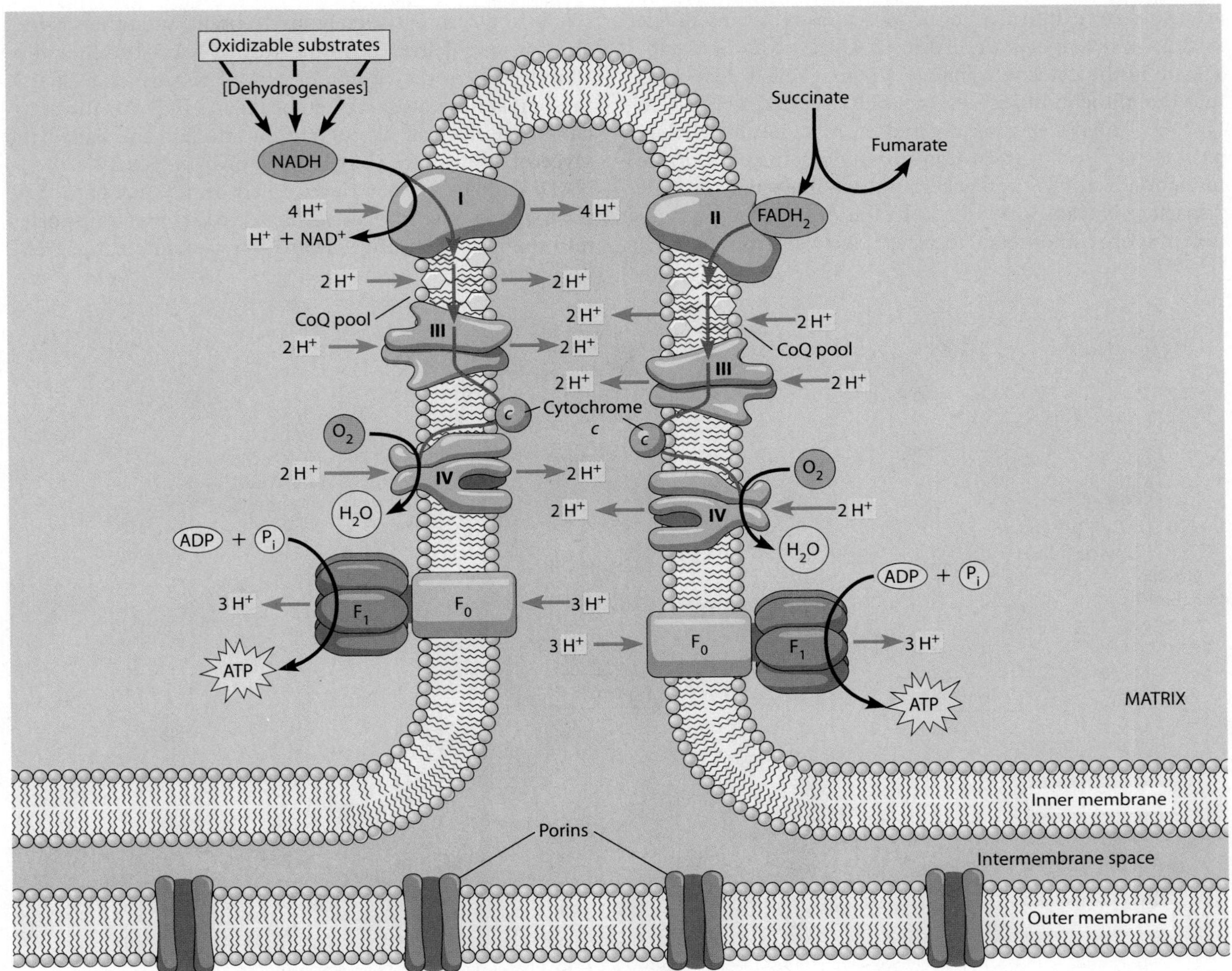

FIGURE 10-22 Dynamics of the Electrochemical Proton Gradient. Respiratory complexes I through IV are integral components of the inner mitochondrial membrane. Complexes I, III, and IV (but not complex II) couple the exergonic flow of electrons (red lines) through the complexes with the outward pumping of protons (blue) across the membrane. The proton motive force of the resulting electrochemical proton gradient drives ATP synthesis by F_1 as protons are translocated back across the membrane by the F_o complex, which is also embedded in the inner membrane. ATP synthesis also consumes protons in the mitochondrial matrix (see Figure 9–1b), a depletion that contributes to the transmembrane proton gradient driving ATP synthesis.

This summary reaction is valid for most prokaryotic cells and for some types of eukaryotic cells. Depending on the cell type, however, the maximum ATP yield for a eukaryotic cell may be only 36 instead of 38 because of the lesser ATP yield from NADH molecules generated in the cytosol rather than in the mitochondrion (see question 1 below).

Before leaving Reaction 10-19 and the aerobic energy metabolism that it summarizes, we will consider two questions that are often asked, with the hope that the answers will further enhance your understanding of aerobic respiration.

1. Why Does the Maximum ATP Yield in Eukaryotic Cells Vary Between 36 and 38 ATPs Per Glucose?

Recall that when glucose is catabolized aerobically in a eukaryotic cell, glycolysis gives rise to two molecules of NADH per glucose in the cytosol, while the catabolism of pyruvate generates another eight molecules of NADH in the matrix of the mitochondrion. This spatial distinction is important because the inner membrane of the mitochondrion does not have a carrier protein for NADH or NAD^+, so NADH generated in the cytosol cannot enter the mitochondrion to deliver its electrons to complex I of the ETS. Instead, the electrons and H^+ ions are passed inward by one of several *electron shuttle systems* that differ in the number of ATP molecules formed per NADH molecule oxidized.

An **electron shuttle system** consists of one or more electron carriers that can be reversibly reduced, with transport proteins present in the membrane for both the oxidized and the reduced forms of the carrier. In liver, kidney, and heart cells, electrons from cytosolic NADH are transferred into the mitochondrion by means of the *malate-aspartate shuttle.* After NADH reduces oxaloacetate in the cytosol to malate, the malate is transported into the mitochondrial matrix, where it reduces NAD^+ to NADH. Thus, electrons derived from cytosolic NADH pass through all three proton-pumping complexes of the mitochondrial ETS and generate three molecules of ATP. Aspartate is transported back to the cytosol for production of more oxaloacetate to accept electrons from NADH.

In contrast, in skeletal muscle, brain, and other tissues, electrons are delivered from cytosolic NADH to the mitochondrial respiratory complexes by means of a shuttle mechanism that uses FAD rather than NAD^+ as the mitochondrial electron acceptor. This mechanism, called the **glycerol phosphate shuttle,** is shown in **Figure 10-23**. NADH in the cytosol reduces dihydroxyacetone phosphate (DHAP) to glycerol-3-phosphate, which is transported into the mitochondrion. There, it is reoxidized to DHAP

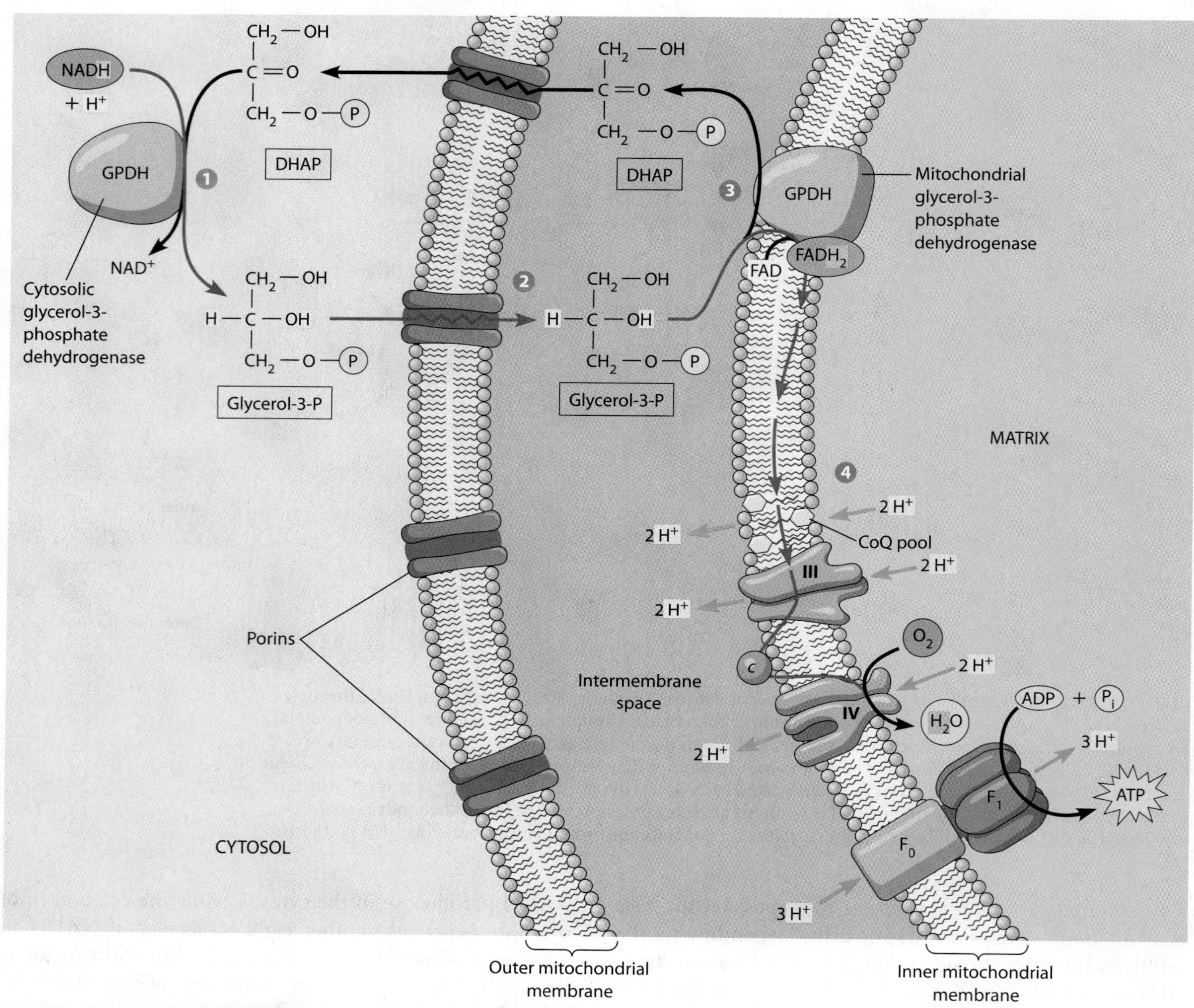

FIGURE 10-23 The Glycerol Phosphate Shuttle. The inner mitochondrial membrane is impermeable to NADH, so electrons from cytosolic NADH are moved into the mitochondrion by one of two shuttle mechanisms. Cells of skeletal muscle, brain, and other tissues use the glycerol phosphate shuttle for this purpose. The red line traces the path of electrons from cytosolic NADH to oxygen. ❶ The cytosolic enzyme glycerol-3-phosphate dehydrogenase (GPDH) uses electrons (and protons) from NADH to reduce dihydroxyacetone phosphate (DHAP) to glycerol-3-phosphate (glycerol-3-P), which ❷ can cross the outer mitochondrial membrane by diffusing through a porin channel. ❸ Glycerol-3-P is then reoxidized to DHAP by an FAD-linked GPDH in the inner membrane, with concomitant reduction of FAD to $FADH_2$. Because NADH is a more energy-rich coenzyme than $FADH_2$, the inward transport of electrons is exergonic, driven by the difference in the reduction potentials of the two coenzymes. The cost of this inward transport is a decreased ATP yield because ❹ electrons from the $FADH_2$ generated by the mitochondrial GPDH bypass complex I, thereby reducing the number of protons pumped across the inner membrane.

by an enzyme that uses FAD instead of NAD^+ as its electron acceptor. As a result, these electrons bypass complex I of the ETS, generating only two molecules of ATP instead of three. This reduces the maximum theoretical yield by one ATP per cytosolic NADH and therefore by two ATP molecules per molecule of glucose (from 38 to 36).

2. Why Is the ATP Yield of Aerobic Respiration Referred to as the "Maximum Theoretical ATP Yield"? This wording is a reminder that yields of 36 or 38 ATP molecules per molecule of glucose are possible only if we assume that the energy of the electrochemical proton gradient is used solely to drive ATP synthesis. Such an assumption may be useful in calculating maximum possible yields, but it is not otherwise realistic because the pmf of the proton gradient provides the driving force not only for ATP synthesis, but for other energy-requiring reactions and processes. For example, some of the energy of the proton gradient is used to drive the transport of various metabolites and ions across the membrane and to drive the import of proteins into mitochondria. Several of these transport processes are illustrated in **Figure 10-24**.

Depending on the relative concentrations of pyruvate, fatty acids, amino acids, and TCA-cycle intermediates in the cytosol and the mitochondrial matrix, variable amounts of energy may be needed to ensure that the mitochondrion has adequate supplies of oxidizable substrates and TCA-cycle intermediates. Moreover, the inward transport of phosphate ions needed for ATP synthesis is accompanied by the concomitant outward movement of hydroxyl ions, which are neutralized by protons in the intermembrane space, thereby also diminishing the proton gradient. In some cells, the phosphate transporter can act as a symporter, transporting a phosphate ion and a proton inward simultaneously.

Aerobic Respiration Is a Highly Efficient Process

To determine the overall efficiency of ATP production by aerobic respiration, we need to ask what proportion of the energy of glucose oxidation is preserved in the 36 or 38

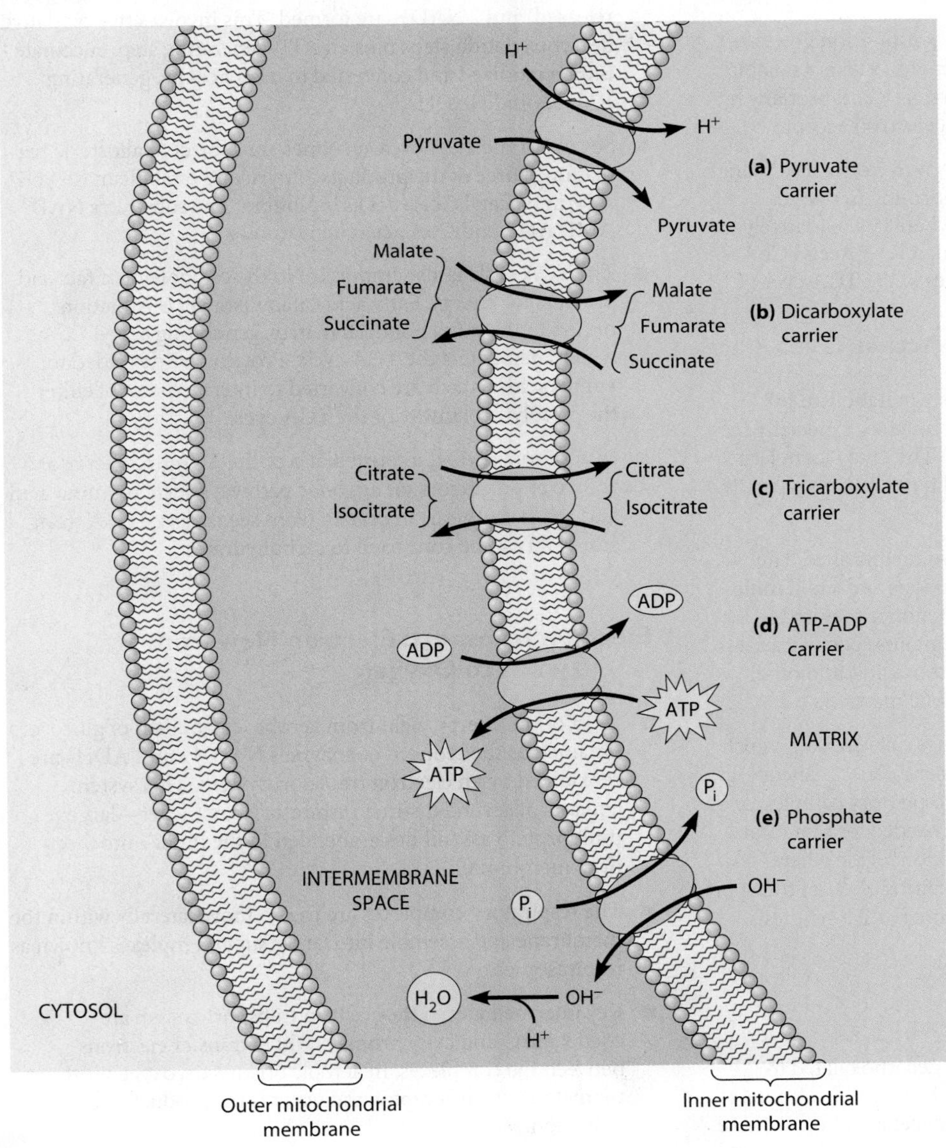

FIGURE 10-24 Major Transport Systems of the Inner Mitochondrial Membrane. The major transport proteins localized in the inner mitochondrial membrane are shown here. **(a)** The pyruvate carrier cotransports pyruvate and protons inward, driven by the pmf of the electrochemical proton gradient. The **(b)** dicarboxylate and **(c)** tricarboxylate carriers exchange organic acids across the membrane, with the direction of transport depending on the relative concentrations of dicarboxylic and tricarboxylic acids on the inside and outside of the inner membrane, respectively. **(d)** The ATP-ADP carrier exchanges ATP outward for ADP inward, and **(e)** the phosphate carrier couples the inward movement of phosphate with the outward movement of hydroxyl ions, which are neutralized by protons in the intermembrane space.

molecules of ATP generated per molecule of glucose. The $\Delta G^{\circ\prime}$ value for the complete oxidation of glucose to CO_2 and H_2O is −686 kcal/mol. ATP hydrolysis has a standard $\Delta G^{\circ\prime}$ of about −7.3 kcal/mol, but the actual $\Delta G'$ under cellular conditions is typically in the range of −10 to −14 kcal/mol. Assuming a value of 10 kcal/mol as we did earlier in the chapter, the 36–38 moles of ATP generated by aerobic respiration of 1 mole of glucose in an aerobic cell correspond to about 360–380 kcal of energy conserved per mole of glucose oxidized. This is an efficiency of about 52–55%, well above that obtainable with the most efficient machines we are capable of creating.

This, then, is aerobic energy metabolism. No transistors, no mechanical parts, no noise, no pollution—and all done in units of organization that require an electron microscope to visualize. Yet the process goes on continuously in living cells with a degree of efficiency, fidelity, and control that we can scarcely understand well enough to appreciate fully, let alone aspire to reproduce in our test tubes.

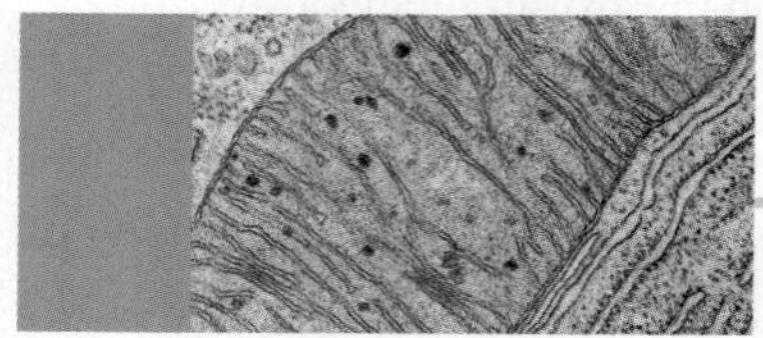

SUMMARY OF KEY POINTS

Cellular Respiration: Maximizing ATP Yields

- Compared with fermentation, aerobic respiration gives the cell access to much more of the free energy that is available from organic substrates such as sugars, fats, and proteins by using molecular oxygen as a terminal electron acceptor.
- The complete catabolism of glucose derived from carbohydrates begins with glycolysis in the cytosol, forming pyruvate. Pyruvate enters the mitochondrion, where it is oxidatively decarboxylated to acetyl CoA, releasing CO_2. Acetyl CoA is then oxidized fully to CO_2 by enzymes of the TCA cycle.

The Mitochondrion: Where the Action Takes Place

- Mitochondria are the site of respiratory metabolism in eukaryotic cells and are prominent organelles, especially in cells requiring a large amount of ATP. They may form large, interconnected networks in some cell types but are typically regarded as discrete organelles.
- A mitochondrion is surrounded by two membranes. The outer membrane is freely permeable to ions and small molecules due to the presence of porins. The inner membrane is a significant permeability barrier that contains specific carriers for the inward transport of pyruvate, fatty acids, and other organic molecules into the mitochondrial matrix.
- The inner membrane has many infoldings called cristae, which greatly increase the surface area of the membrane and hence its ability to accommodate the numerous respiratory complexes, F_oF_1 complexes, and transport proteins needed for respiratory function. Spaces within the cristae provide a region where protons pumped across the membrane during electron transport are concentrated in order to drive subsequent ATP synthesis.

The Tricarboxylic Acid Cycle: Oxidation in the Round

- Pyruvate from glycolysis is oxidatively decarboxylated to acetyl CoA, generating NADH and CO_2. Acetyl CoA enters the TCA cycle by combining with oxaloacetate to form citrate.
- As citrate is converted to succinate, two molecules of CO_2 are released, and 2 NADH are formed. This involves two oxidative decarboxylation steps plus an ATP-generating step. Succinate is then oxidized and converted to oxaloacetate, generating $FADH_2$ and NADH.
- Several of the TCA cycle enzymes are subject to allosteric regulation. Three of the products of pyruvate catabolism (NADH, ATP, and acetyl CoA) act as inhibitors. The substrates NAD^+, AMP, ADP, and CoA act as activators.
- The TCA cycle is also important in the catabolism of fats and proteins for energy. Fatty acid catabolism via β oxidation occurs in the mitochondrial matrix, generating acetyl CoA, which then enters the TCA cycle. Proteins are degraded to amino acids, which are converted to intermediates of either the glycolytic pathway or the TCA cycle.
- In addition, several intermediates of the TCA cycle serve as a source of precursors for anabolic pathways such as amino acid and heme synthesis. In certain plant seeds, acetyl CoA from stored fat can be converted to carbohydrates.

Electron Transport: Electron Flow from Coenzymes to Oxygen

- Most of the energy yield from aerobic catabolism of glucose is obtained as the reduced coenzymes NADH and $FADH_2$ are reoxidized by an electron transport system. This system consists of several distinct respiratory complexes—large multiprotein assemblies embedded in the inner mitochondrial membrane.
- The respiratory complexes are free to move laterally within the membrane and assemble into larger supercomplexes known as respirasomes.
- Key intermediates in the electron transport system are coenzyme Q and cytochrome *c*, which transfer electrons between the complexes. In aerobic organisms, oxygen is the ultimate electron acceptor and water is the product of oxygen reduction.

The Electrochemical Proton Gradient and ATP Synthesis

- Of the four main respiratory complexes, three (complexes I, III, and IV) couple the transfer of electrons to the outward pumping of protons. This establishes an electrochemical proton gradient that is the driving force for ATP generation.
- The ATP-synthesizing system consists of a proton translocator, F_o, embedded in the membrane and an ATP synthase, F_1, that projects from the inner membrane on the matrix side.
- ATP is synthesized by F_1 as the proton gradient powers the movement of protons through F_o. Thus, the electrochemical proton gradient and ATP are, in effect, interconvertible forms of stored energy.

Aerobic Respiration: Summing It All Up

- Complete oxidation of glucose to six molecules of CO_2 yields a total of 10 NADH, 2 $FADH_2$, and 4 ATP. Each NADH oxidized in the electron transport system results in synthesis of 3 ATP, and each $FADH_2$ oxidized gives 2 ATP.
- Therefore, the maximum ATP yield of aerobic respiration is 38 ATP per molecule of glucose. However, in some eukaryotic cells, the ATP yield is reduced to 36 as NADH generated by glycolysis in the cytosol passes its electrons to mitochondrial $FADH_2$ via the glycerol phosphate shuttle.

MB **www.masteringbiology.com**

1. MasteringBiology Assignments

Tutorials and Activities
- Molecular Model Structural Tutorials for visualization of 3D structures
- BioFlix Tutorials including 3D animations with hints and detailed feedback for common wrong answers
- ATP Synthase

Questions
- Reading Quiz
- Multiple Choice
- End-of-chapter questions

2. eText Read your book online, search, highlight text, take notes, and more

3. The Study Area
- Chapter Review Quiz, Guide to Techniques and Methods, 3D-Structure Tutorials, Activities, BioFlix, Videos, Biology Review, Word Study Tools, Glossary, Flashcards

PROBLEM SET

More challenging problems are marked with a •.

10-1 Localization of Molecules and Functions Within the Mitochondrion. Indicate whether you would expect to find each of the following molecules or functions in the matrix (MA), the inner membrane (IM), the outer membrane (OM), the intermembrane space (IS), or not in the mitochondrion at all (NO).

(a) Dicarboxylate carrier
(b) Crista junctions
(c) Succinate dehydrogenase
(d) Coenzyme A
(e) Accumulation of a high proton concentration
(f) Malate dehydrogenase
(g) Fatty acyl CoA translocase
(h) Fatty acid elongation
(i) Respirasomes
(j) Coenzyme Q
(k) Nucleotide phosphorylation
(l) Conversion of lactate into pyruvate
(m) ATP synthase

10-2 Localization of Molecules and Functions Within the Prokaryotic Cell. Repeat Problem 10-1, but indicate where in a prokaryotic cell you would expect to find each of the listed molecules or functions. Choices: cytoplasm (CY), plasma membrane (PM), exterior of cell (EX), or not present at all (NO).

10-3 True or False. Indicate whether each of the following statements is true (T) or false (F). If false, reword the statement to make it true.

(a) The orderly flow of carbon through the TCA cycle is possible because each of the enzymes of the cycle is embedded in the inner mitochondrial membrane in such a manner that their order in the membrane is the same as their sequence in the cycle.
(b) Most of the control of the TCA cycle involves allosteric regulation of key enzymes by specific effector molecules that bind reversibly to them.
(c) Respiration can proceed without the use of oxygen as the terminal electron acceptor. Other terminal electron acceptors include sulphur (S/H_2S), protons (H^+/H_2) and ferric ions (Fe^{3+}/Fe^{2+}).
(d) We can predict that the flow of electrons through the electron transport system is exergonic because the $NAD^+/NADH$ redox pair has a highly negative $\Delta E_0'$ and the O_2/H_2O redox pair has a highly positive $\Delta E_0'$.

(e) Compared with NAD+, FAD is a higher-energy coenzyme.

(f) For fatty acids with an odd number of carbons, the final cycle of β oxidation produces propionyl CoA, which is one carbon longer than acetyl CoA.

10-4 Mitochondrial Transport. For aerobic respiration, a variety of substances must be in a state of flux across the inner mitochondrial membrane. Assuming a brain cell in which glucose is the sole energy source, indicate for each of the following substances whether you would expect a net flux across the membrane and, if so, how many molecules will move in which direction per molecule of glucose catabolized.

(a) $FADH_2$

(b) Electrons

(c) Oxaloacetate

(d) Acetyl CoA

(e) Oxygen

(f) Protons

(g) Pyruvate

(h) ATP

(i) Glycerol-3-phosphate

(j) ADP

(k) NADH

(l) Water

10-5 Completing the Pathway. In each of the following cases, complete the pathway by indicating the structures and order of the intermediates.

(a) The conversion of citrate to α-ketoglutarate by Reactions TCA-2 and TCA-3 involves as intermediates not only isocitrate but also molecules identified (but not shown in Figure 10-8) as aconitate and oxalosuccinate. Illustrate the pathway from citrate to α-ketoglutarate by showing the structures and order of all three intermediates.

(b) Synthesis of the amino acid glutamate can be effected from pyruvate and alanine by a metabolic sequence that illustrates the amphibolic role of the TCA cycle. Devise such a pathway, assuming the availability of whatever additional enzymes may be needed.

(c) If pyruvate-2-^{14}C (pyruvate with the middle carbon atom radioactively labeled) is provided to actively respiring mitochondria, most of the radioactivity will be incorporated into citrate. Trace the route whereby radioactively labeled carbon atoms are incorporated into citrate, and indicate where in the citrate molecule the label will first appear.

10-6 The Calculating Cell Biologist. Use Table 10-2, page 272 as the basis for the calculations needed to answer the following questions.

(a) Without doing any calculations initially, predict whether isocitrate can pass electrons exergonically to NAD^+ under standard conditions. Why do you predict that?

(b) What is the $\Delta E_0'$ for the oxidation of isocitrate by NAD^+ under standard conditions? Does this calculation support the prediction you made in part a? Explain your answer.

(c) Calculate the $\Delta G^{\circ\prime}$ for the oxidation of isocitrate by NAD^+ under standard conditions. In what way is this calculation relevant to aerobic energy metabolism?

(d) Repeat parts a–c for the oxidation of lactate to pyruvate by NAD^+. In what way is this calculation relevant to aerobic energy metabolism?

(e) Now repeat parts a–c for the oxidation of succinate to fumarate with NAD^+ as the electron acceptor. You should find that the $\Delta G^{\circ\prime}$ is highly positive. What does this tell us about the likelihood that NAD^+ could serve as the electron acceptor for the succinate dehydrogenase reaction of the TCA cycle?

(f) Finally, repeat parts a–c for the oxidation of succinate to fumarate with coenzyme Q as the electron acceptor. Why does it make sense to regard coenzyme Q as the electron acceptor when succinate dehydrogenase is shown in Figure 10-8 with FAD as the immediate electron acceptor?

10-7 Calculating Maximum ATP Yields. **Table 10-5** is intended as a means of summarizing the ATP yield during the aerobic oxidation of glucose.

(a) Complete Table 10-5 for an aerobic bacterium. What is the maximum ATP yield?

(b) Indicate on Table 10-5 the changes that are necessary to calculate the maximum ATP yield for a eukaryotic cell that uses the glycerol phosphate shuttle to move electrons from the cytosol into the matrix of the mitochondrion.

10-8 Regulation of Catabolism. Explain the advantage to the cell of each of the following regulatory mechanisms.

(a) Isocitrate dehydrogenase (Figure 10-8, Reaction TCA-3) is allosterically activated by ADP.

Table 10-5 Calculation of the Maximum ATP Yield from Aerobic Oxidation of Glucose

Stage of Respiration	Glycolysis (glucose → 2 pyruvate)	Pyruvate Oxidation (2 pyruvate → 2 acetyl CoA)	TCA Cycle (2 turns)
Yield of CO_2			
Yield of NADH			
ATP per NADH			
Yield of $FADH_2$			
ATP per $FADH_2$			
ATP from substrate-level phosphorylation			
ATP from oxidative phosphorylation			
Maximum ATP yield			

(b) The dehydrogenases that oxidize isocitrate, α-ketoglutarate, and malate (Reactions TCA-3, TCA-4, and TCA-8) are all allosterically inhibited by NADH.

(c) Pyruvate dehydrogenase (Figure 10-8, Reaction 10–1, page 261) is allosterically inhibited by ATP.

(d) Phosphofructokinase (Figure 9-7, Reaction Gly-3) is allosterically inhibited by citrate.

(e) Pyruvate dehydrogenase kinase (Figure 10-10, enzyme E_1) is allosterically activated by NADH.

(f) α-ketoglutarate dehydrogenase (Figure 10-8, Reaction TCA-4) is allosterically inhibited by succinyl CoA.

10-9 Lethal Synthesis. The leaves of *Dichapetalum cymosum,* a South African plant, are very poisonous. Animals that eat the leaves have convulsions and usually die shortly thereafter. One of the most pronounced effects of poisoning is a marked elevation in citrate concentration and a blockage of the TCA cycle in many organs of the affected animal. The toxic agent in the leaves of the plant is fluoroacetate, but the actual poison in the tissues of the animal is fluorocitrate. If fluoroacetate is incubated with purified enzymes of the TCA cycle, it has no inhibitory effect on enzyme activity.

(a) Why might you expect fluorocitrate to have an inhibitory effect on one or more of the TCA cycle enzymes if incubated with the purified enzymes in vitro, even though fluoroacetate has no effect?

(b) Which enzyme in the TCA cycle do you suspect is affected by fluorocitrate? Give two reasons for your answer.

(c) How could fluoroacetate be converted to fluorocitrate?

(d) Why is this phenomenon referred to as *lethal synthesis?*

10-10 Oxidation of Saturated Fatty Acids.

(a) Calculate the number of $FADH_2$, NADH, and acetyl CoA produced when a molecule of palmitate (16 carbons) is degraded by β oxidation.

(b) Compare that to the number produced from β oxidation of stearate (18 carbons).

(c) Write an equation for the number of acetyl CoA, $FADH_2$, and NADH molecules produced from a fatty acid of n carbons (n = an even number).

(d) Following complete oxidation of the acetyl CoA molecules in the TCA cycle and electron transport from $FADH_2$ and NADH to oxygen, how many ATP will be produced from palmitate and stearate?

(e) Write an equation for the number of ATP produced from complete oxidation of a fatty acid of n carbons (n = an even number).

10-11 Oxidation of Cytosolic NADH. In some eukaryotic cells, the NADH generated by glycolysis in the cytosol is reoxidized by the glycerol phosphate shuttle shown in Figure 10-23.

(a) Write balanced reactions for the reduction of dihydroxyacetone phosphate (DHAP) to glycerol-3-phosphate (glycerol-3-P) by cytosolic NADH and for the oxidation of glycerol-3-P to DHAP by the FAD-linked glycerol-3-P dehydrogenase in the inner membrane of the mitochondrion.

(b) Add the two reactions in part a to obtain a summary reaction for the transfer of electrons from cytosolic NADH to mitochondrial FAD. Calculate $\Delta E_0'$ and $\Delta G^{\circ\prime}$ for this reaction. Is the inward movement of electrons thermodynamically feasible under standard conditions?

(c) Write a balanced reaction for the reoxidation of $FADH_2$ by coenzyme Q within the inner membrane, assuming that CoQ is reduced to $CoQH_2$. Calculate $\Delta E_0'$ and $\Delta G^{\circ\prime}$ for this reaction. Is this transfer thermodynamically feasible under standard conditions?

(d) Write a balanced reaction for the transfer of electrons from cytosolic NADH to mitochondrial CoQ, and calculate $\Delta E_0'$ and $\Delta G^{\circ\prime}$ for this reaction. Is this transfer thermodynamically feasible under standard conditions?

(e) Assume that the $[NADH]/[NAD^+]$ ratio in the cytosol is 5.0 and that the $[CoQH_2]/[CoQ]$ ratio in the inner membrane is 2.0. What is $\Delta G'$ for the reaction in part d at 25°C and pH 7.0?

(f) Is $\Delta G'$ for the inward transfer of electrons from NADH to CoQ affected by the ratio of the reduced to the oxidized forms of the enzyme-bound FAD in the inner membrane? Why or why not?

• **10-12 Brown Fat and Thermogenin.** Most newborn mammals, including human infants, have a special type of adipose tissue called brown fat, in which a naturally occurring uncoupling protein called *thermogenin* is present in the inner mitochondrial membrane. Thermogenin uncouples ATP synthesis from electron transport so that the energy released as electrons flow through the electron transport chain is lost as heat instead.

(a) What happens to the energy that is released as electron transport continues but ATP synthesis ceases? Why might it be advantageous for a baby to have thermogenin present in the inner membrane of the mitochondria that are present in brown fat tissue?

(b) Given its location in the cell, suggest a mode of action for thermogenin. What kind of an experiment can you suggest to test your hypothesis?

(c) What would happen to a mammal if all of its mitochondria were equipped with uncoupling protein, rather than just those in brown fat tissue?

(d) Some adult mammals also have brown fat. Would you expect to find more brown fat tissue and more thermogenin in a hibernating bear or in a physically active bear? Explain your reasoning.

• **10-13 Dissecting the Electron Transport System.** To determine which segments of the electron transport system (ETS) are responsible for proton pumping and hence for ATP synthesis, investigators usually incubate isolated mitochondria under conditions such that only a portion of the ETS is functional. One approach is to supply the mitochondria with an electron donor and an electron acceptor that are known to tap into the ETS at specific points. In addition, inhibitors of known specificity are often added. In one such experiment, mitochondria were incubated with β-hydroxybutyrate, oxidized cytochrome c, ADP, P_i, and cyanide. (Mitochondria have an NAD^+-dependent dehydrogenase that is capable of oxidizing β-hydroxybutyrate to β-ketobutyrate.)

(a) What is the electron donor in this system? What is the electron acceptor? What is the most likely pathway of electron transport in this system?

(b) Based on what you already know about the ETS, how many moles of ATP would you expect to be formed per mole of β-hydroxybutyrate oxidized? Write a balanced equation for the reaction that occurs in this system.

(c) Cyanide is the only reagent added to the system that is not a part of the balanced equation for the reaction. What was the purpose of adding cyanide to the system? What result would you expect if the cyanide had not been added?

(d) Would you expect the enzymes of the TCA cycle to be active in this assay system? Explain your reasoning.

(e) Why is it important that β-ketobutyrate cannot be further metabolized in this system? Lactate is quite similar in structure to β-ketobutyrate; what effect would it have had if the investigators had used lactate as the oxidizable substrate instead of β-hydroxybutyrate?

SUGGESTED READING

References of historical importance are marked with a •.

General References

Metzler, D. E. *Biochemistry: The Chemical Reactions of Living Cells.* 2nd ed. San Diego: Academic Press, 2001.

Morowitz, H., J. Kostelnik, and J. D. Yang. The origin of intermediary metabolism. *Proc. Natl. Acad. Sci. USA* 97 (2000): 7704.

Orgel, L. E. Self-organizing biochemical cycles. *Proc. Natl. Acad. Sci. USA* 97 (2000): 12503.

Papa, S., F. Guerrieri, and J. M. Tager, eds. *Frontiers of Cellular Bioenergetics: Molecular Biology, Biochemistry, and Physiopathology.* New York: Kluwer Academic/Plenum Publishers, 1999.

• Racker, E. From Pasteur to Mitchell: A hundred years of bioenergetics. *Fed. Proc.* 39 (1980): 210.

Mitochondrial Structure and Function

• Ernster, L., and G. Schatz. Mitochondria: An historical overview. *J. Cell Biol.* 91 (1981): 227s.

Frey, T. G., and C. A. Mannella. The internal structure of the mitochondria. *Trends. Biochem. Sci.* 25 (2000): 319.

Gena, P., E. Fanelli, C. Brenner, M. Svelto, and G. Calamita. News and views on mitochondrial water transport. *Front. Biosci.* 14 (2009): 4189.

Legros, F., P. Frachon, and F. Malka. Organization and dynamics of human mitochondrial DNA. *J. Cell Sci.* 117 (2004): 2653.

Lenaz, G. et al. New insights into structure and function of mitochondria and their role in aging and disease. *Antioxid. Redox Signal.* 8 (2006): 417.

Swayne, T. C., A. C. Gay, and L. A. Pon. Visualization of mitochondria in budding yeast. *Methods Cell Biol.* 80 (2007): 591.

Tanji, K., and E. Bonilla. Optical imaging techniques to visualize mitochondria. *Methods Cell Biol.* 80 (2007): 135.

Wallace, D. C., W. Fan, and V. Procaccio. Mitochondrial energetics and therapeutics. *Ann. Rev. Pathol.* 5 (2010): 297.

The Tricarboxylic Acid Cycle and Electron Transport

Beinert, H., R. Holm, and E. Münck. Iron-sulfur clusters: Nature's modular, multipurpose structures. *Science* 277 (1997): 653.

Cocchini, G. Function and structure of complex II of the respiratory chain. *Annu. Rev. Biochem.* 72 (2003): 77.

Hederstedt, L. Complex II is complex too. *Science* 299 (2003): 671.

Huynen, M. A., T. Dandekar, and P. Bork. Variation and evolution of the citric-acid cycle: A genomic perspective. *Trends Microbiol.* 7 (1999): 281.

• Kornberg, H. L. Tricarboxylic acid cycles. *BioEssays* 7 (1987): 236.

• Krebs, H. A. The history of the tricarboxylic acid cycle. *Perspect. Biol. Med.* 14 (1970): 154.

• Mitchell, P. Keilin's respiratory chain concept and its chemiosmotic consequences. *Science* 206 (1979): 1148.

Schnarrenberger, C., and W. Martin. Evolution of the enzymes of the citric acid cycle and the glyoxylate cycle of higher plants. A case study of endosymbiotic gene transfer. *Eur. J. Biochem.* 269 (2002): 868.

Schultz, B. E., and S. I. Chan. Structures and proton pumping strategies of mitochondrial respiratory enzymes. *Annu. Rev. Biophys. Biomol. Struct.* 30 (2001): 23.

Oxidative Phosphorylation and ATP Synthesis

Aksimentiev, A. et al. Insights into the molecular mechanism of rotation in the F_o sector of ATP synthase. *Biophys. J.* 86 (2004): 1332.

Itoh, H., K. Adachi, and A. Takahashi. Mechanically driven ATP synthesis by F_1-ATPase. *Nature* 427 (2004): 465.

Junge, W., H. Sielaff, and S. Engelbrecht. Torque generation and elastic power transmission in the rotary F_oF_1-ATPase. *Nature* 459 (2009): 364.

Leyva, J. A., M. A. Bianchet, and L. M. Amzel. Understanding ATP synthesis: Structure and mechanism of the F_1-ATPase. *Mol. Membr. Biol.* 20 (2003): 27.

• Mitchell, P. Coupling of phosphorylation to electron and hydrogen transfer by a chemiosmotic type of mechanism. *Nature* 191 (1961): 144.

Rubinstein, J. L., R. Henderson, and J. E. Walker. Structure of the mitochondrial ATP synthase by electron cryomicroscopy. *EMBO J.* 22 (2003): 6182.

von Ballmoos, C., A. Wiedenmann, and P. Dimroth. Essentials for ATP synthesis by F_1F_0 ATP synthases. *Annu. Rev. Biochem.* 78 (2009):649.

Regulation of Respiratory Metabolism

Blank, L. M., and U. Sauer. TCA cycle activity in *Saccharomyces cereviseae* is a function of the environmentally determined specific growth and glucose uptake rates. *Microbiology* 150 (2004): 1085.

Das, J. The role of mitochondrial respiration in physiological and evolutionary adaptation. *BioEssays* 9 (2006): 890.

de Meirleir, L. Defects of pyruvate metabolism and the Krebs cycle. *J. Child Neurol.* 17, Supp 3 (2002): 3S26.

Lascaris, R., J. Piwowarski, and H. van der Spek. Overexpression of HAP4 in glucose-derepressed yeast cells reveals respiratory control of glucose-regulated genes. *Microbiology* 150 (2004): 929.

Sheehan, T. E., D. A. Hood, and P. A. Kumar. Tissue-specific regulation of cytochrome *c* oxidase subunit expression by thyroid hormone. *Amer. J. Physiol.* 286 (2004): E968.

Weinberg, F., and N. S. Chandel. Mitochondrial metabolism and cancer. *Ann. NY Acad. Sci.* 1177 (2009): 66.

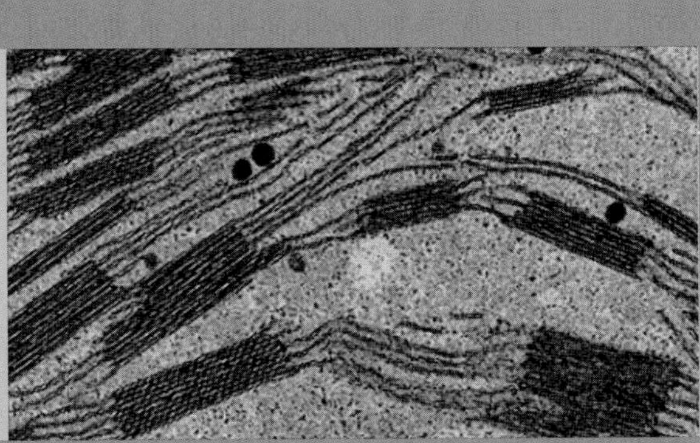

11 Phototrophic Energy Metabolism: Photosynthesis

See MasteringBiology® for tutorials, activities, and quizzes.

In the two preceding chapters, we studied fermentation and respiration, two chemotrophic solutions to the universal challenge of meeting the energy and carbon needs of a living cell. Most chemotrophs (us included) depend on an external source of organic substrates for survival. As you may recall, chemotrophs obtain energy by oxidizing high-energy, reduced compounds such as carbohydrates, fats, and proteins in catabolic pathways. They can also use intermediates of these oxidative pathways as sources of carbon skeletons for biosynthesis via anabolic pathways. However, left alone, chemotrophs would soon perish without an ongoing supply of these reduced organic compounds.

In this chapter, we will learn how photosynthetic organisms produce the chemical energy and organic carbon that are drained from the biosphere by chemotrophs. These photosynthetic organisms use solar energy to drive the reduction of CO_2 *to produce carbohydrates, fats, and proteins—the reduced forms of carbon that all chemotrophs depend upon.*

The use of solar energy to drive the anabolic pathways that produce these building blocks of life is aptly named **photosynthesis**—*the conversion of light energy to chemical energy and its subsequent use in synthesizing organic molecules. Nearly all life on Earth is sustained by the energy that arrives at the planet as sunlight.* **Phototrophs** *are organisms that convert solar energy to chemical energy in the form of ATP and the reduced coenzyme NADPH. NADPH is the electron and hydrogen carrier for a large number of anabolic pathways and is closely related to NADH, which we have previously encountered in catabolic pathways.*

Some phototrophs, such as the halobacteria described in Chapter 7, are **photoheterotrophs,** *organisms that acquire energy from sunlight but depend on organic sources of reduced carbon. Most other phototrophs—including plants, algae, and most photosynthetic bacteria—are known as* **photoautotrophs,** *organisms that use solar energy to synthesize energy-rich organic molecules from simple inorganic starting materials such as carbon dioxide and water. Many photoautotrophs release molecular oxygen as a by-product of photosynthesis. Thus, phototrophs not only replenish reduced carbon in the biosphere but also provide the oxygen in the atmosphere used by aerobic organisms to oxidize these reduced compounds for energy.*

In this chapter, we will study two general aspects of photosynthesis: how photoautotrophs capture solar energy and convert it to chemical energy and how this energy is used to transform energy-poor carbon dioxide and water into energy-rich organic molecules such as carbohydrates, fats, and proteins.

An Overview of Photosynthesis

Photosynthesis involves two major biochemical processes: energy transduction and carbon assimilation (**Figure 11-1**). During the **energy transduction reactions,** light energy is captured by chlorophyll molecules and converted to chemical energy in the form of ATP and NADPH. The ATP and NADPH generated by the energy transduction reactions subsequently provide energy and reducing power, respectively, for the **carbon assimilation reactions,** also known as *carbon fixation reactions.* During the carbon assimilation reactions, commonly known as the *Calvin cycle,* fully oxidized carbon atoms from carbon dioxide are fixed (reduced and covalently joined), forming carbohydrates. These processes take place in the chloroplast of eukaryotic phototrophs or in specialized membrane systems of photosynthetic bacteria.

The Energy Transduction Reactions Convert Solar Energy to Chemical Energy

Light energy from the sun is captured by a variety of green pigment molecules called *chlorophylls,* which are present in the green leaves of plants and in the cells of algae and photosynthetic bacteria. Light absorption by a chlorophyll molecule excites one of its electrons, which is then ejected

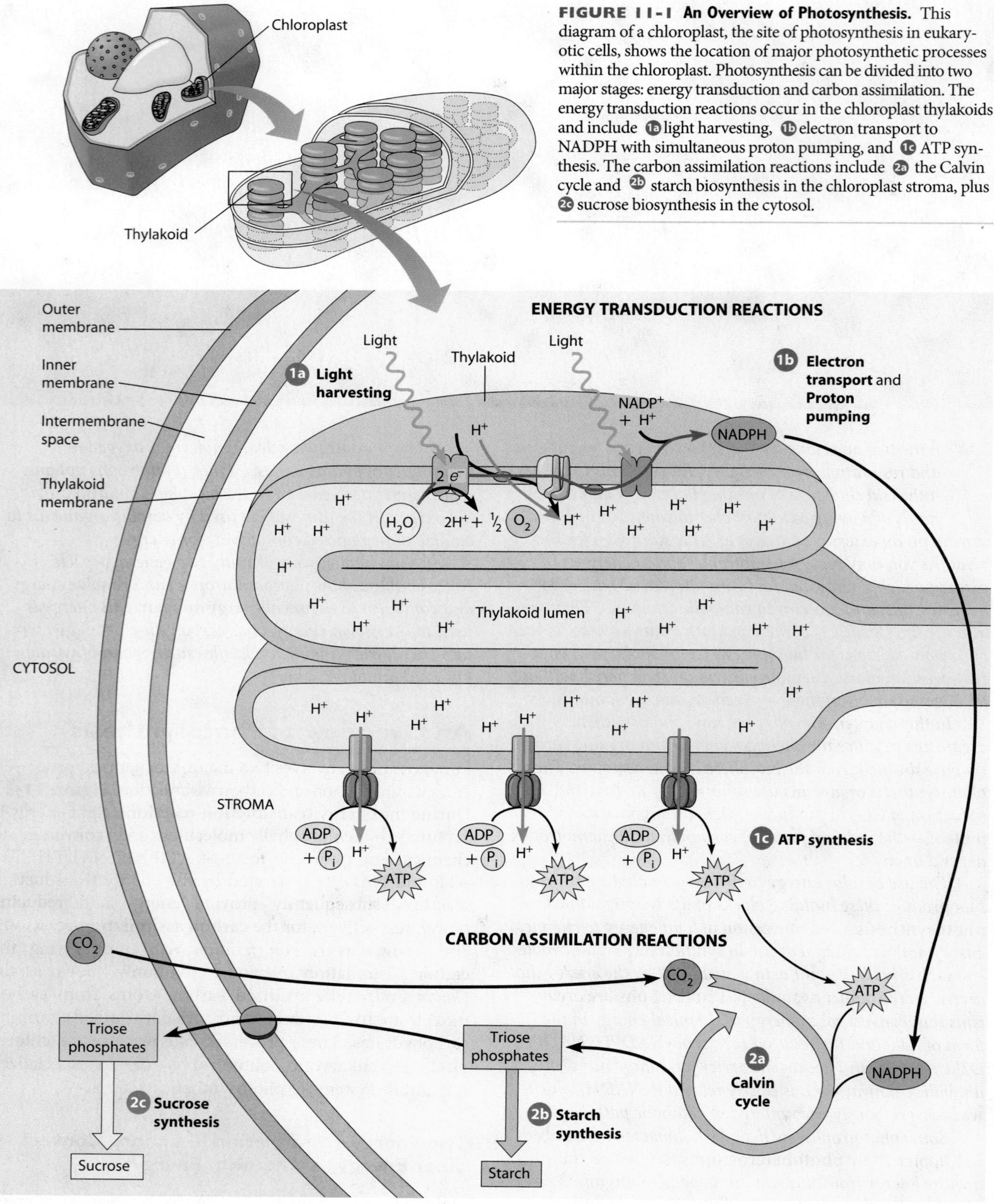

FIGURE 11-1 An Overview of Photosynthesis. This diagram of a chloroplast, the site of photosynthesis in eukaryotic cells, shows the location of major photosynthetic processes within the chloroplast. Photosynthesis can be divided into two major stages: energy transduction and carbon assimilation. The energy transduction reactions occur in the chloroplast thylakoids and include (1a) light harvesting, (1b) electron transport to NADPH with simultaneous proton pumping, and (1c) ATP synthesis. The carbon assimilation reactions include (2a) the Calvin cycle and (2b) starch biosynthesis in the chloroplast stroma, plus (2c) sucrose biosynthesis in the cytosol.

from the molecule and flows energetically downhill through an electron transport system (ETS) much like the ETS we saw previously in mitochondria. As in mitochondria, this flow of electrons is coupled to unidirectional proton pumping, which first stores energy in an electrochemical proton gradient that subsequently drives an ATP synthase. As you may recall, in mitochondria this process is known as *oxidative phosphorylation*—ATP synthesis driven by energy derived from the oxidation of organic compounds. In photosynthetic organisms, ATP synthesis driven by energy derived from the sun is called *photophosphorylation.*

To incorporate fully oxidized carbon atoms from carbon dioxide into organic molecules, these carbon atoms must be reduced. Therefore, photoautotrophs need not only energy in the form of ATP but also reducing power, which is provided by NADPH, a coenzyme related to NADH (see Figure 9-5). In **oxygenic phototrophs**—plants, algae, and cyanobacteria—water is the electron donor, and light energy absorbed by chlorophyll powers the movement of two electrons from water to $NADP^+$, which is reduced to NADPH. Molecular oxygen is released as water is oxidized. In **anoxygenic phototrophs**—green and purple photosynthetic bacteria—compounds such as sulfide (S^{2-}), thiosulfate ($S_2O_3^{2-}$), or succinate serve as electron donors, and oxidized forms of these compounds are released. In both oxygenic and anoxygenic phototrophs, the light-dependent generation of NADPH is called *photoreduction.*

Usually it is the oxygenic phototrophs such as trees, grasses, and other common plants that come to mind when we think of photosynthesis. However, it was work with the anoxygenic green and purple photosynthetic bacteria conducted by C. B. van Niel in the 1930s and 1940s that led to our understanding that photosynthesis is a light-driven oxidation-reduction process.

The Carbon Assimilation Reactions Fix Carbon by Reducing Carbon Dioxide

Most of the energy accumulated within photosynthetic cells by the light-dependent generation of ATP and NADPH is used for carbon dioxide fixation and reduction. A general reaction for the complete process of photosynthesis may be written as

$$\text{light} + CO_2 + 2H_2A \rightarrow [CH_2O] + 2A + H_2O \quad \textbf{(11-1)}$$

where H_2A is a suitable *electron donor,* $[CH_2O]$ represents a carbohydrate, and A is the oxidized form of the electron donor. By expressing photosynthesis in this way, we avoid perpetuating the incorrect notion that all phototrophs use water as an electron donor.

When we focus on oxygenic phototrophs, which do use water as an electron donor, we can rewrite Reaction 11-1 by substituting H_2O for H_2A, multiplying all reactants and products by six, and subtracting water molecules that appear on both sides of the reaction. This gives the following more specific and familiar form of the general reaction for photosynthesis:

$$\text{light} + 6CO_2 + 6H_2O \rightarrow C_6H_{12}O_6 + 6O_2 \quad \textbf{(11-2)}$$

We will soon see, however, that the *immediate* product of photosynthetic carbon fixation is a three-carbon sugar (a triose), not a hexose as shown above. These trioses are then used for a variety of biosynthetic pathways, including the biosynthesis of glucose (as shown in Reaction 11-2), sucrose, and starch. Sucrose is the major transport carbohydrate in most plant species. It conveys energy and reduced carbon through the plant from photosynthetic cells to nonphotosynthetic cells. Starch (or glycogen in photosynthetic bacteria) is the major storage carbohydrate in phototrophic cells.

The Chloroplast Is the Photosynthetic Organelle in Eukaryotic Cells

Our model organisms for studying photosynthesis in this chapter are the most familiar oxygenic phototrophs—green plants. Therefore, we will look at the structure and function of the **chloroplast,** the organelle responsible for most of the events of photosynthesis in eukaryotic phototrophs. In plants and algae, the primary events of photosynthetic energy transduction and carbon assimilation are confined to these specialized organelles.

Because chloroplasts are usually large (1–5 μm wide and 1–10 μm long) and opaque, they were described and studied early in the history of cell biology by Antonie van Leeuwenhoek and Nehemiah Grew in the seventeenth century. The prominence of chloroplasts within a *Coleus* leaf cell is shown in **Figure 11-2a**. A mature leaf cell usually contains 20–100 chloroplasts, whereas an algal cell typically contains only one or a few chloroplasts. The shapes of these organelles vary from the simple flattened spheres common in plants to the more elaborate forms found in green algae. A cell of the filamentous green alga *Spirogyra,* for example, contains one or more ribbon-shaped chloroplasts, as shown in Figure 11-2b.

Not all plant cells contain chloroplasts. Newly differentiated plant cells have smaller organelles called **proplastids,** which can develop into any of several kinds of **plastids** equipped to serve different functions. Chloroplasts are only one example of a plastid. Some proplastids differentiate into *amyloplasts,* which are sites for storing starch. Other proplastids acquire red, orange, or yellow pigments, forming the *chromoplasts* that give flowers and fruits their distinctive colors. Proplastids can also develop into organelles for storing protein (*proteinoplasts*) and lipids (*elaioplasts*).

Chloroplasts Are Composed of Three Membrane Systems

A closer view of a plant leaf cell chloroplast is shown in the electron micrographs of **Figure 11-3a and b**. A chloroplast, like a mitochondrion, has both an **outer membrane**

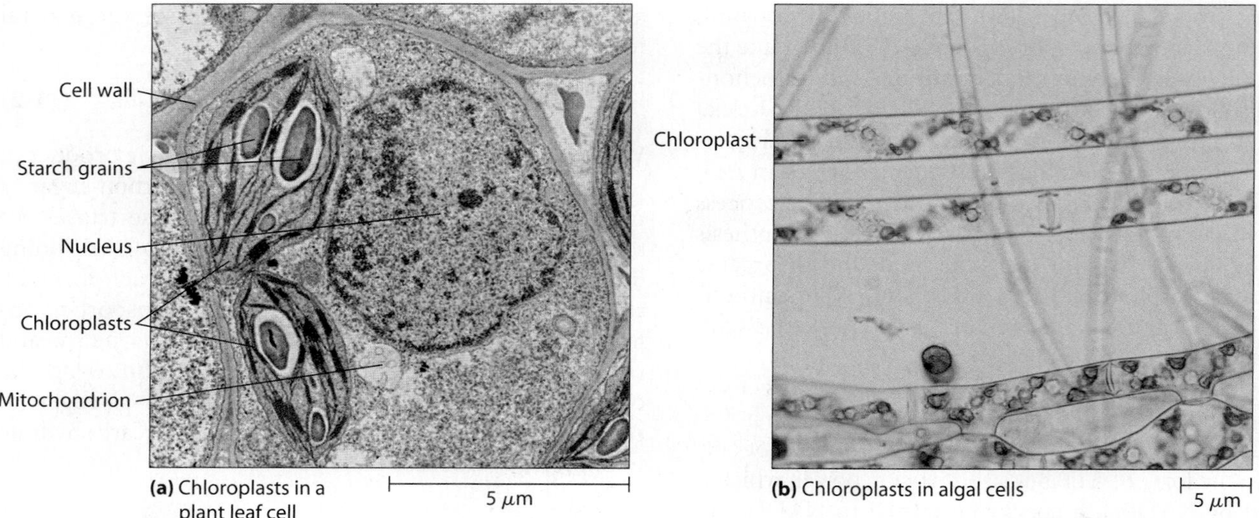

FIGURE 11-2 Chloroplasts. (a) The prominence of chloroplasts in a leaf cell of a plant is demonstrated by this electron micrograph of a parenchyma cell from a *Coleus* leaf. The cell contains many chloroplasts, three of which are seen in this particular cross section. The presence of large starch granules in the chloroplasts indicates that the cell was photosynthetically active just prior to fixation for electron microscopy (TEM). (b) This light micrograph reveals the unusual ribbon-shaped chloroplasts in cells of the filamentous green alga *Spirogyra*.

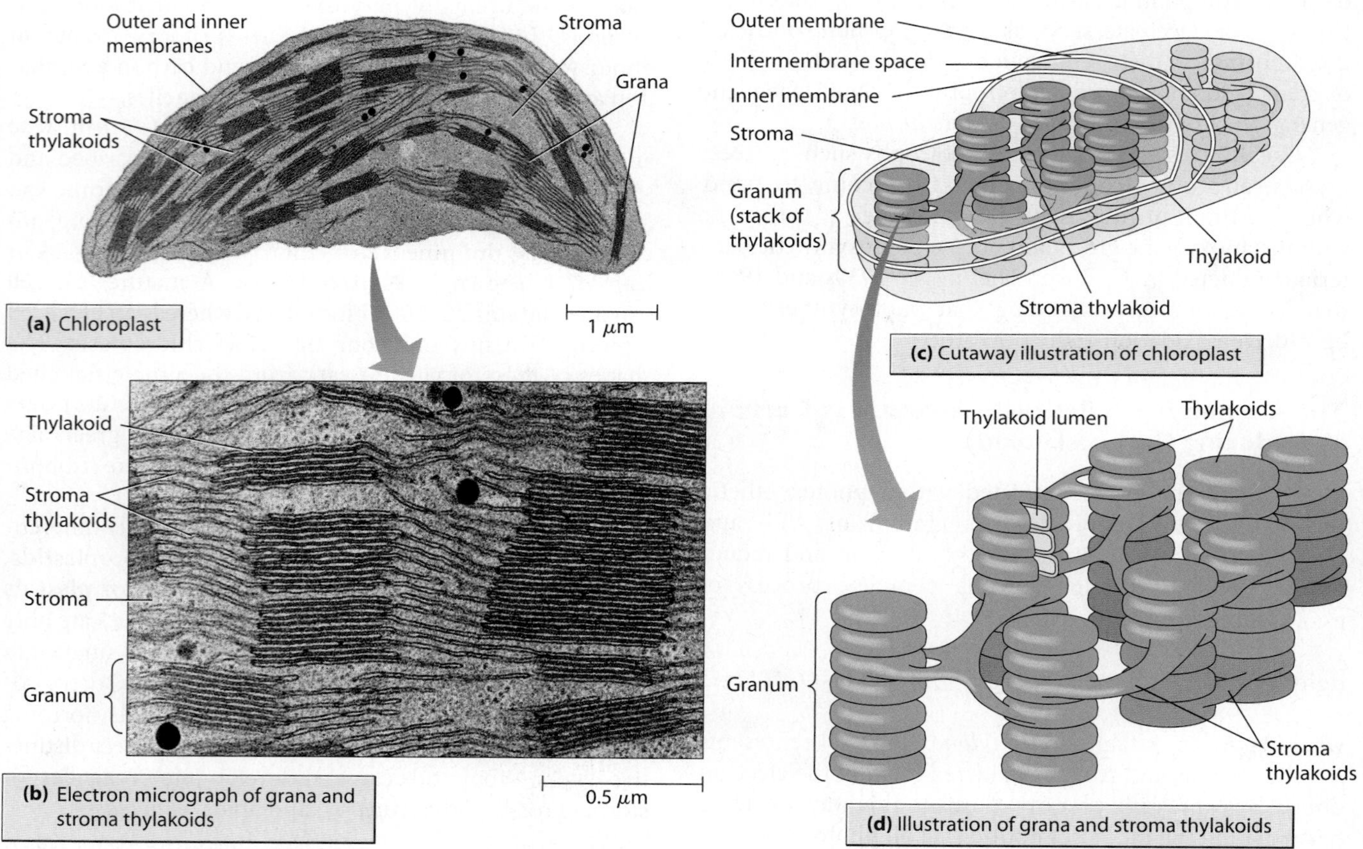

FIGURE 11-3 Structural Features of a Chloroplast. (a) An electron micrograph of a chloroplast from a leaf of timothy grass *(Phleum pratense)* (TEM). (b) A more magnified electron micrograph of the chloroplast in part a shows the arrangement of grana and stroma thylakoids (TEM). (c) An illustration showing the three-dimensional structure of a typical chloroplast. (d) An illustration depicting the continuity of the thylakoid membranes, the arrangement of thylakoids into stacks called grana, and the stroma thylakoids that interconnect the grana. The thylakoid membranes enclose a separate compartment called the thylakoid lumen.

and an **inner membrane,** often separated by a narrow **intermembrane space** (Figure 11-3c). The inner membrane encloses the **stroma,** a gel-like matrix teeming with enzymes for carbon, nitrogen, and sulfur reduction and assimilation. The outer membrane contains transmembrane proteins called **porins** that are similar to those found in the outer membrane of mitochondria. Because porins permit the passage of solutes with molecular weights up to about 5000, the outer membrane is freely permeable to most small organic molecules and ions. However, the inner membrane forms a significant permeability barrier. Transport proteins in the inner membrane control the flow of most metabolites between the intermembrane space and the stroma. Three important metabolites that are able to diffuse freely across both the outer and the inner membranes, however, are water, carbon dioxide, and oxygen.

In addition, chloroplasts have **thylakoids,** a third membrane system inside the chloroplast that creates an internal compartment. Thylakoids, illustrated in Figure 11-3c and d, are flat, saclike structures suspended in the stroma. They are usually arranged in stacks called **grana** (singular: **granum**). Grana are interconnected by a network of longer thylakoids called **stroma thylakoids.** Essentially all of the photosynthetic components, such as pigments, enzymes, and electron carriers, are localized on or in the thylakoid membranes.

Electron micrographs of serial sections show that the grana and stroma thylakoids enclose a single continuous compartment, the **thylakoid lumen.** The semipermeable barrier formed by the thylakoid membrane separates the lumen from the stroma and plays an important role in the generation of an electrochemical proton gradient and the synthesis of ATP. The high concentration of protons pumped across this membrane barrier into the lumen during light-driven electron transport stores potential energy that later drives ATP synthesis when the protons return to the stroma. This is analogous to the situation when water in a reservoir is held back by a dam, which acts as a physical barrier to the water. The elevated water level stores potential energy (gravitational in this case) that can be converted to mechanical energy when water is released through controlled openings in the dam, thereby turning the turbines that generate electrical energy.

Photosynthetic bacteria do not have chloroplasts. In some of them, however, such as the cyanobacteria, the plasma membrane is folded inward and forms *photosynthetic membranes.* Such structures, shown in **Figure 11-4**, are analogous to the thylakoids. Indeed, to some extent cyanobacteria appear to be free-living chloroplasts. Similarities among mitochondria, chloroplasts, and bacterial cells have led biologists to formulate the *endosymbiont theory,* which suggests mitochondria and chloroplasts evolved from bacteria that were engulfed by

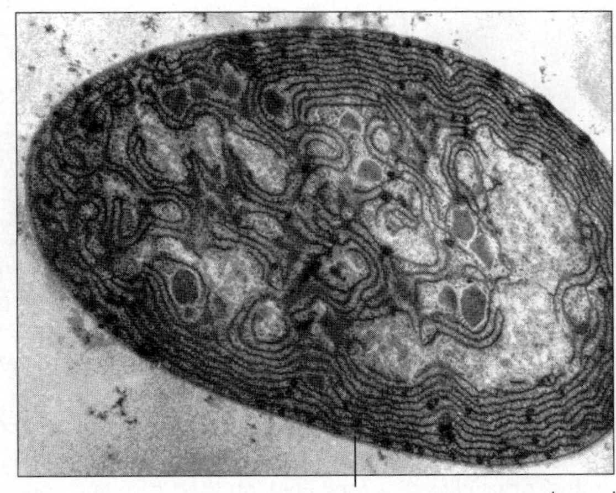

FIGURE 11-4 The Photosynthetic Membranes of a Cyanobacterium. Cyanobacteria, like other photoautotrophs, are capable of photosynthetic carbon fixation. This electron micrograph of a thin section of *Anabaena azollae* reveals the extensive folded membranes of cyanobacterial cells that resemble the thylakoids of chloroplasts (TEM).

primitive cells 1 to 2 billion years ago. A summary of this theory is presented in **Box 11A**.

Photosynthetic Energy Transduction I: Light Harvesting

The first stage of photosynthetic energy transduction is the capture of light energy from the sun. As with all electromagnetic radiation, light has both wave-like and particle-like properties. The portion of the electromagnetic spectrum visible to us, for example, consists of light having wavelengths ranging from about 380 to 750 nm. However, light also behaves as a stream of discrete particles called **photons,** each photon carrying a **quantum** (indivisible packet) of energy. The wavelength of a photon and the precise amount of energy it carries are inversely related (see Figure 2-3). A photon of ultraviolet or blue light, for example, has a shorter wavelength and carries a larger quantum of energy than a photon of red or infrared light does.

When a photon is absorbed by a **pigment** (light-absorbing molecule), such as chlorophyll, the energy of the photon is transferred to an electron, which is energized from its *ground state* in a low-energy orbital to an *excited state* in a high-energy orbital. This event, called **photoexcitation,** is the first step in photosynthesis. Because each pigment has a different configuration of atoms and electrons, pigments display characteristic **absorption spectra** that describe the particular wavelengths of light absorbed by a pigment. The absorption spectra of several common pigments found in photosynthetic organisms, along with the spectrum of solar

BOX 11A DEEPER INSIGHTS

The Endosymbiont Theory and the Evolution of Mitochondria and Chloroplasts from Ancient Bacteria

The debate on the evolutionary origins of mitochondria and chloroplasts has a long history. As early as 1883, Andreas F. W. Schimper suggested that chloroplasts arose from a symbiotic relationship between photosynthetic bacteria and nonphotosynthetic cells. By the mid-1920s, other investigators had extended Schimper's idea by proposing a symbiotic origin for mitochondria. Such ideas encountered ridicule and neglect for decades, however, until the 1960s, when it was discovered that mitochondria and chloroplasts contain their own DNA. Further research revealed that mitochondria and chloroplasts are **semiautonomous organelles,** organelles that can divide on their own and contain not only DNA but also mRNA, tRNAs, and ribosomes.

We now realize that the processes of DNA, RNA, and protein synthesis in mitochondria and chloroplasts are more similar to the analogous processes in bacterial cells than they are to those in eukaryotic cells. This led biologists to formulate the **endosymbiont theory.** This theory, developed most fully by Lynn Margulis, proposes that mitochondria and chloroplasts evolved from ancient bacteria that established a **symbiotic relationship** (a mutually beneficial association) with primitive nucleated cells 1 to 2 billion years ago. The proposed sequence of events leading to mitochondria and chloroplasts is outlined in **Figure 11A-1**.

A preliminary assumption of the endosymbiont theory is that the absence of molecular oxygen in Earth's primitive atmosphere limited early cells to *anaerobic* mechanisms for acquiring energy. A few anaerobic cells subsequently developed pigments capable of converting light energy to chemical energy, allowing them to use sunlight as a source of energy. Some of these photosynthetic organisms developed mechanisms for using water as an electron donor. As a result, oxygen was released as a by-product and the composition of Earth's atmosphere was dramatically altered.

As oxygen accumulated in Earth's atmosphere, some anaerobic bacteria evolved into *aerobic* organisms by developing oxygen-dependent electron transport and oxidative phosphorylation

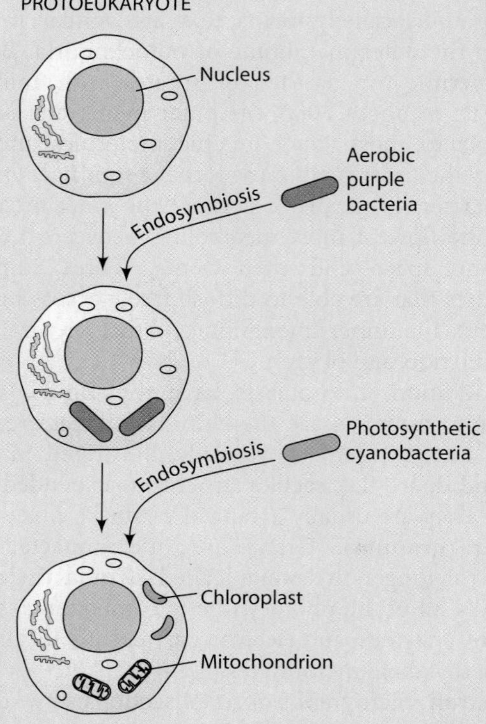

FIGURE 11A-1 Major Events That Might Have Occurred During the Evolution of Eukaryotic Cells. Considerable evidence exists for an endosymbiotic origin for mitochondria and chloroplasts. Most biologists agree that the primitive cells that were ingested by protoeukaryotes and then evolved into mitochondria and chloroplasts were, respectively, purple bacteria and cyanobacteria. This is based on similarities in size and membrane lipid composition, comparisons of rRNA base sequences, the presence of circular DNA molecules, and the ability to reproduce autonomously.

radiation reaching the Earth's surface, are shown in **Figure 11-5**. Note how the overlapping absorption spectra of various chlorophylls and accessory pigments effectively utilize almost the entire spectrum of sunlight that reaches the Earth.

A photoexcited electron in a pigment molecule is unstable and must either return to its ground state in a low-energy orbital or undergo transfer to a relatively stable high-energy orbital, usually in a different molecule. When the electron returns to a low-energy orbital in the pigment molecule, the absorbed energy is lost either as heat released or as light emitted (fluorescence). Alternatively, the absorbed energy can be transferred from the photoexcited electron to an electron in an adjacent pigment molecule in a process known as **resonance energy transfer.** This process is very important for moving captured energy from light-absorbing molecules to molecules such as chlorophyll that are capable of passing an excited electron to an organic acceptor molecule. The transfer of the photoexcited electron itself to another molecule is called **photochemical reduction.** This transfer is essential for converting light energy to chemical energy.

Chlorophyll Is Life's Primary Link to Sunlight

Chlorophyll, which is found in nearly all photosynthetic organisms, is the primary energy-transduction pigment that channels solar energy into the biosphere. The structures of two types of chlorophyll—chlorophylls *a* and *b*—are shown in **Figure 11-6**. The skeleton of each molecule consists of a central *porphyrin ring* and a strongly

pathways. The endosymbiont theory suggests that the ancestor of eukaryotic cells (called a **protoeukaryote**) developed at least one important feature distinguishing it from other primitive cells: the ability to ingest nutrients and particles from the environment by *phagocytosis*. This characteristic enabled protoeukaryotes to establish endosymbiotic relationships with primitive bacteria. How these primitive cells acquired a membrane-bound nucleus is still unclear.

Mitochondria Apparently Evolved from Ancient Purple Bacteria

The first step toward the evolution of mitochondria may have occurred when an anaerobic protoeukaryote that depended upon glycolysis for energy ingested smaller aerobic bacteria by phagocytosis (Figure 11A-1). The ingested aerobic bacteria were capable of aerobic respiration and would have provided larger amounts of useful energy to the anaerobic host cell. In turn, the host cell provided protection and nutrients to the bacteria residing in its cytoplasm. Thus, the ingested bacteria and the protoeukaryote established a mutually beneficial symbiotic relationship. Gradually, over hundreds of millions of years, the bacteria lost functions that were not essential in their new cytoplasmic environment and developed into mitochondria. When the base sequences of mitochondrial ribosomal RNAs (rRNAs) are compared with those of various bacterial rRNAs, the closest matches occur among *purple bacteria,* suggesting that the ingested ancestor of mitochondria was an ancient member of this group.

Chloroplasts Apparently Evolved from Ancient Cyanobacteria

According to the endosymbiont theory, the first step toward the evolution of chloroplasts occurred when members of a subgroup of early eukaryotes, already equipped with aerobic bacteria or primitive mitochondria, ingested primitive photosynthetic cells. As just described for the evolution of mitochondria, the ingested organisms probably provided energy for the host cell in exchange for shelter and nutrients. The photosynthetic cells gradually lost functions that were not essential in their new environment and evolved into an integral component of the eukaryotic host. When the base sequences of chloroplast rRNAs are compared with those of various bacterial rRNAs, the closest matches occur among *cyanobacteria,* suggesting that the ancestor of chloroplasts was an ancient member of this group.

The endosymbiont theory is based primarily on biochemical similarities observed among mitochondria, chloroplasts, and bacteria, but support is also provided by contemporary symbiotic relationships that resemble what might have occurred in the distant past. Algae, dinoflagellates, diatoms, and photosynthetic bacteria live as endosymbionts in the cytoplasm of cells occurring in more than 150 kinds of existing protists and invertebrates. The cell wall of the ingested organism is often no longer present, and in a few instances the cell structure is even further reduced, with only the chloroplasts of the endosymbiont remaining.

A striking example of endosymbiosis occurs in certain marine slugs and related mollusks, where the cells lining the animal's digestive tract contain clearly identifiable chloroplasts. These chloroplasts, which originate from the green algae the mollusks feed on, continue to carry out photosynthesis long after being incorporated into the animal's cells. The carbohydrates produced by the photosynthetic process are even distributed as a source of nutrients to the rest of the organism. Although the chloroplasts do not grow and divide, the organelles continue to function in the cytoplasm of the animal cell for several months, and these host organisms have been referred to as "green animals."

In the final analysis, our ideas about how eukaryotic cells acquired their complex array of organelles over billions of years of evolution must remain speculative because the events under consideration are inaccessible to direct laboratory experimentation. However, one strength of the proposed role of endosymbiosis in the evolutionary origins of eukaryotic organelles is that it involves interactions and events that are observed in contemporary cells.

hydrophobic *phytol* side chain. The alternating double bonds in the porphyrin ring are responsible for absorbing visible light, while the phytol side chain interacts with lipids of the thylakoid or cyanobacterial membranes, anchoring the light-absorbing molecules in these membranes.

The magnesium ion (Mg^{2+}) found in chlorophylls *a* and *b* affects the electron distribution in the porphyrin ring and ensures that a variety of high-energy orbitals are available. As a result, several specific wavelengths of light can be absorbed. Chlorophyll *a*, for example, has a broad absorption spectrum, with maxima at about 420 and 660 nm (Figure 11-5). Chlorophyll *b* is distinguished from chlorophyll *a* by the presence of a *formyl* (—CHO) group in place of one of the *methyl* ($-CH_3$) groups on the porphyrin ring. This minor structural alteration shifts the absorption maxima toward the center of the visible spectrum (Figure 11-5).

All plants and green algae contain both chlorophyll *a* and *b*. The combined absorption spectra of the two forms of chlorophyll provide access to a broader range of wavelengths of sunlight and enable such organisms to collect more photons. Because chlorophylls absorb mainly blue and red wavelengths of light, they appear green. Other oxygenic photosynthetic organisms supplement chlorophyll *a* with either chlorophyll *c* (brown algae, diatoms, and dinoflagellates), chlorophyll *d* (red algae), or phycobilin (red algae and cyanobacteria).

Bacteriochlorophyll is a subfamily of chlorophyll molecules restricted to anoxygenic phototrophs (photosynthetic bacteria) and is characterized by a saturated site not found in other chlorophyll molecules (indicated by an

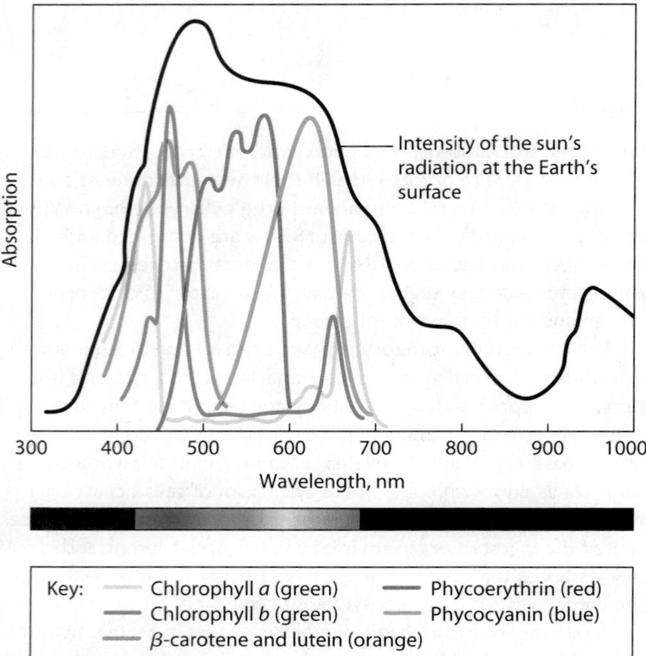

FIGURE 11-5 The Absorption Spectra of Common Photosynthetic Pigments. The graph shows absorption spectra of various chlorophylls and accessory pigments (colored lines) and compares them with the spectral distribution of solar energy reaching the Earth's surface (black line). The small reference strip below the graph shows the color that each wavelength appears to the human eye.

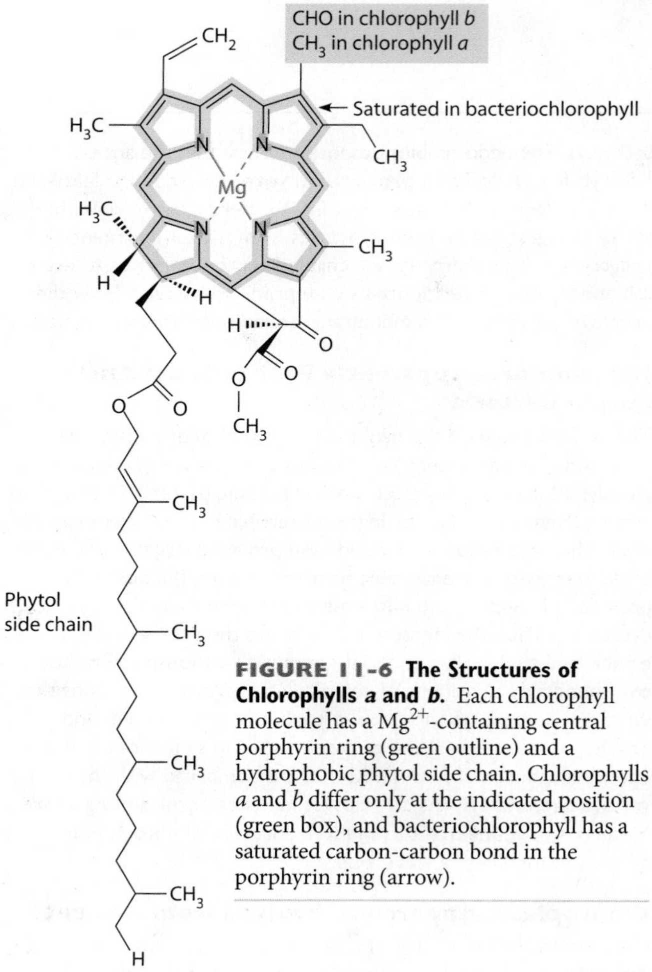

FIGURE 11-6 The Structures of Chlorophylls *a* and *b*. Each chlorophyll molecule has a Mg^{2+}-containing central porphyrin ring (green outline) and a hydrophobic phytol side chain. Chlorophylls *a* and *b* differ only at the indicated position (green box), and bacteriochlorophyll has a saturated carbon-carbon bond in the porphyrin ring (arrow).

arrow in Figure 11-6). In this case, structural alterations shift the absorption maxima of bacteriochlorophylls toward the near-ultraviolet and far-red regions of the spectrum.

Accessory Pigments Further Expand Access to Solar Energy

Most photosynthetic organisms also contain **accessory pigments,** which absorb photons that cannot be captured by chlorophyll. They then transfer the energy of these photons to a chlorophyll molecule by resonance energy transfer. This feature enables organisms to collect energy from a much larger portion of the sunlight reaching the Earth's surface, as shown in Figure 11-5, which shows the emission spectrum of solar energy in black and the absorption spectra of various accessory pigments in color. Two types of accessory pigments are **carotenoids** and **phycobilins.** Two carotenoids that are abundant in the thylakoid membranes of most plants and green algae are *β-carotene* and *lutein,* which have very similar absorption spectra. When sufficiently abundant and not masked by chlorophyll, these pigments confer an orange or yellow tint to leaves (or to carrots, from which they got their name). With absorption maxima between 420 and 480 nm, carotenoids absorb photons from a broad range of the blue region of the spectrum and thus appear yellow.

Phycobilins are found only in red algae and cyanobacteria. Two common examples are *phycoerythrin* and *phycocyanin.* Phycoerythrin absorbs photons from the blue, green, and yellow regions of the spectrum (thus appearing red) and enables red algae to utilize the dim light that penetrates an ocean's surface water. Phycocyanin, on the other hand, absorbs photons from the orange region of the spectrum, thus appearing blue, and is characteristic of cyanobacteria living close to the surface of a lake or on land. Thus, variations in the amounts and properties of accessory pigments often reflect a phototroph's adaptation to a specific environment.

Light-Gathering Molecules Are Organized into Photosystems and Light-Harvesting Complexes

Chlorophyll molecules, accessory pigments, and associated proteins are organized into functional units called **photosystems,** which are localized to thylakoid or photosynthetic bacterial membranes. The chlorophyll molecules are anchored to the membranes by the long hydrophobic phytol side chains. **Chlorophyll-binding**

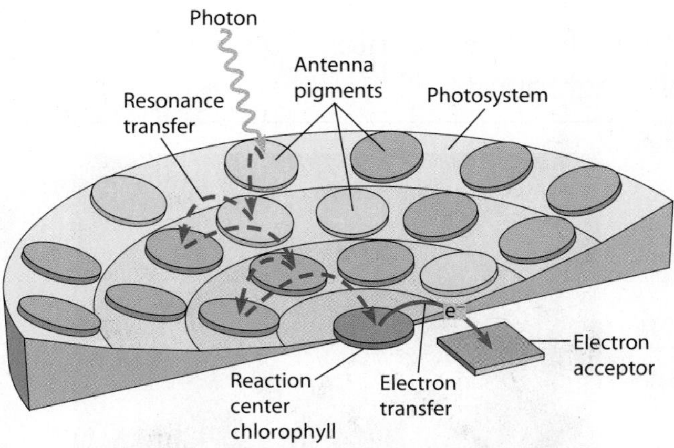

FIGURE 11-7 The Transfer of Energy to the Reaction Center of a Photosystem. Light energy absorbed by antenna pigments is passed along by resonance energy transfer until it reaches a specific chlorophyll *a* molecule at the reaction center of a photosystem. The energy is conserved as an excited electron that is ejected from the chlorophyll and passed along to an organic acceptor molecule.

proteins stabilize the chlorophyll within a photosystem and modify the absorption spectra of specific chlorophyll molecules. Other proteins in the photosystems bind components of electron transport systems or catalyze oxidation-reduction reactions.

Most pigments of a photosystem serve as light-gathering **antenna pigments,** which collect light energy much like a radio antenna collects radio waves. These antenna pigments absorb photons and pass the energy to a neighboring chlorophyll molecule or accessory pigment by resonance energy transfer, as shown in **Figure 11-7**. However, recent work with photosynthetic algae suggests that this energy transfer involves quantum mechanical effects not previously seen in biology. Rather than simply a stepwise linear process from molecule to molecule, energy transfer can proceed in a wave-like manner, following several paths simultaneously! This suggests that some antenna pigments are "wired" together by quantum-mechanical probability effects that can increase the efficiency of light harvesting.

The photochemical events that drive electron flow and proton pumping do not begin until the energy reaches the **reaction center** of a photosystem, where two distinct chlorophyll *a* molecules known as the *special pair* reside. Surrounded by other components of the reaction center, this pair of chlorophyll molecules catalyzes the conversion of solar energy into chemical energy. The structure of a bacterial reaction center is shown in **Box 11B**.

Each photosystem is generally associated with a **light-harvesting complex (LHC),** which, like a photosystem, collects light energy. The LHC does not, however, contain a reaction center. Instead, it passes the collected energy to a nearby photosystem by resonance energy transfer. These LHCs are mobile in the thylakoid membrane and can move to take advantage of changing light conditions. Plants and green algae have LHCs composed of about 80–250 chlorophyll *a* and *b* molecules, along with carotenoids and pigment-binding proteins. Red algae and cyanobacteria have a different type of LHC, called a **phycobilisome,** which contains phycobilins rather than chlorophyll and carotenoids. Together, a photosystem and the associated LHCs are referred to as a **photosystem complex.**

Oxygenic Phototrophs Have Two Types of Photosystems

In the 1940s, Robert Emerson and his colleagues at the University of Illinois discovered that two separate photoreactions are involved in oxygenic photosynthesis. Using the green alga *Chlorella,* they measured the rate of photosynthesis versus wavelength by monitoring the rate of oxygen release as they changed the wavelength of light supplied to the cells. Initially, they observed a dramatic drop in photosynthesis rate above a wavelength of about 690 nm. This seemed odd to them because *Chlorella* contains chlorophyll molecules that strongly absorb light at wavelengths above 690 nm. When they supplemented the longer wavelengths of light with a shorter wavelength (about 650 nm), the decrease in photosynthesis was nearly eliminated. Indeed, they discovered that photosynthesis driven by a combination of long and short wavelengths of red light exceeded the sum of activities obtained with either wavelength alone. This synergistic phenomenon became known as the **Emerson enhancement effect.**

We now know that the Emerson enhancement effect is the result of two distinct photosystems working together in series. **Photosystem I (PSI)** has an absorption maximum of 700 nm, whereas **photosystem II (PSII)** has an absorption maximum of 680 nm. Each electron that passes from water to $NADP^+$ must be photoexcited twice, once by each photosystem. When illumination is restricted to wavelengths above 690 nm, PSII is not active, and photosynthesis is severely impaired.

As we will soon see, each electron is first excited by PSII and then by PSI. But don't be confused by the nomenclature. The photosystems were named based on the order of their discovery, so although PSI was discovered first, it was later shown that the initial photoexcitation event occurs in PSII.

In the reaction center of each photosystem is a distinctive **special pair** of chlorophyll *a* molecules that have the electrons that will be photoexcited following photon absorption. These chlorophyll molecules are designated **P680** for PSII and **P700** for PSI to reflect their specific absorption maxima. The granal and stromal thylakoid membranes have differing amounts of the two photosystems. As we saw earlier with the associated light-harvesting complexes, these photosystems are somewhat mobile in the membrane and thus able to move to allow the cell to adapt to changing conditions of light quantity (brightness) and quality (wavelength).

BOX 11B DEEPER INSIGHTS

A Photosynthetic Reaction Center from a Purple Bacterium

Much of our knowledge of photosynthetic reaction centers and light harvesting has come from studies of reaction center complexes from photosynthetic bacteria. In the late 1960s, Roderick Clayton and co-workers purified the reaction center complex of a purple bacterium, *Rhodopseudomonas sphaeroides* (now *Rhodobacter sphaeroides*). More recently, Hartmut Michel, Johann Deisenhofer, and Robert Huber crystallized a reaction center complex from a different purple bacterium, *Rhodopseudomonas viridis* (now *Blastochloris viridis*) and determined its molecular structure by X-ray crystallography. They not only provided the first detailed look at how pigment molecules are arranged to capture light energy but also were the first group to crystallize any membrane protein complex. For their exciting and ground-breaking contributions, Michel, Deisenhofer, and Huber shared a Nobel Prize in 1988.

As shown in **Figure 11B-1**, the reaction center of *R. viridis* includes four protein subunits. The first subunit is a cytochrome *c* molecule bound to the outer surface of the bacterial membrane. The L and M subunits span the membrane and stabilize a total of four bacteriochlorophyll *b* molecules, two bacteriopheophytin molecules, and two quinones. Subunits L and M are homologous to proteins D1 and D2, respectively, of PSII in oxygenic phototrophs (see Figure 11-9). The H subunit is bound to the cytoplasmic surface of the membrane and is homologous to a PSII protein called CP43. Electron flow through this bacterial photosystem resembles the flow of electrons through PSII oxygenic phototrophs, with one major difference: The bacterial photosystem does not include an oxygen-evolving complex.

Two of the bacteriochlorophyll *b* molecules, designated *P960* to indicate their absorption maxima at 960 nm, have a direct role in catalyzing the light-dependent transfer of an electron to an ETS. Absorption of a photon, either directly or by accessory pigments, lowers the reduction potential of P960 from about +0.5 to −0.7 V. The photoexcited electron is immediately transferred to bacteriopheophytin, thereby stabilizing the charge separation. From bacteriopheophytin, the electron flows exergonically through two quinones and a cytochrome b/c_1 complex to cytochrome *c*. This electron flow is coupled to unidirectional proton pumping across the bacterial membrane. Cytochrome *c* then returns the electron to oxidized P960. As in chloroplasts, the electrochemical proton gradient across the membrane drives ATP synthesis by a CF_0CF_1 complex.

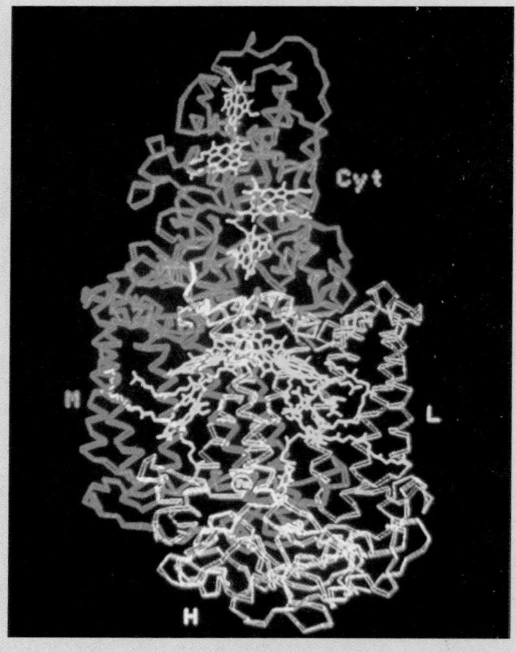

FIGURE 11B-1 A Bacterial Photosynthetic Reaction Center. The structure for the reaction center from *Rhodopseudomonas viridis* shows the cytochrome at the bacterial surface (cyan) with four bound heme groups (yellow), the transmembrane L and M subunits containing bacteriochlorophyll (yellow), and the cytoplasmic H subunit.

Notice that the flow of electrons just described is cyclic, with no net gain of reducing power. How, then, does the cell generate reductant? In *R. viridis,* the cytochrome b/c_1 complex and cytochrome *c* accept electrons from donors such as hydrogen sulfide, thiosulfate, or succinate. The ATP generated by cyclic photophosphorylation is then used to push electrons energetically uphill from the cytochrome b/c_1 complex or from cytochrome *c* to NAD^+. The cyclic flow of electrons described earlier is very different from the cyclic flow through PSI in oxygenic phototrophs. Thus, the bacterial photosystem is most accurately described as a somewhat simplified version of PSII in plants, green algae, and cyanobacteria.

Photosynthetic Energy Transduction II: NADPH Synthesis

The second stage of photosynthetic energy transduction uses a series of electron carriers to transport excited electrons from chlorophyll to the coenzyme **nicotinamide adenine dinucleotide phosphate ($NADP^+$)**, forming NADPH, the reduced form of $NADP^+$. This process is known as **photoreduction** and involves a chloroplast **electron transport system (ETS)** that resembles the mitochondrial ETS we studied in Chapter 10. The complete photoreduction pathway (**Figure 11-8**) includes several components, and many of the molecules of the chloroplast ETS are similar to those found in the mitochondrial ETS—cytochromes, iron-sulfur proteins, and quinones.

Recall from Equation 10-11 that $\Delta G^{\circ\prime}$ (standard free energy) and E_0' (standard reduction potential) are opposite in sign, meaning that electrons will spontaneously flow toward a compound with a higher reduction potential. Absorption of light energy by each photosystem boosts electrons to the "top" of an ETS (lower E_0'). As the electrons flow exergonically from PSII to PSI, a portion of their energy is conserved in a proton gradient across the thylakoid membrane. From PSI, electrons are passed to *ferredoxin* and then to $NADP^+$, generating reducing power in the form of NADPH. $NADP^+$ differs from NAD^+ only by having an additional phosphate group attached to its adenosine (see Figure 9-5). $NADP^+$ is the coenzyme of choice for a large number of anabolic pathways, whereas NAD^+ is usually involved in catabolic pathways.

Photosystem II Transfers Electrons from Water to a Plastoquinone

Our detailed tour of the photoreduction pathway begins at photosystem II, which uses electrons from water to reduce a plastoquinone (Q_B) to a plastoquinol (Q_BH_2). The PSII reaction center, part of which is shown in **Figure 11-9**, contains approximately 25 polypeptides, including D1 and

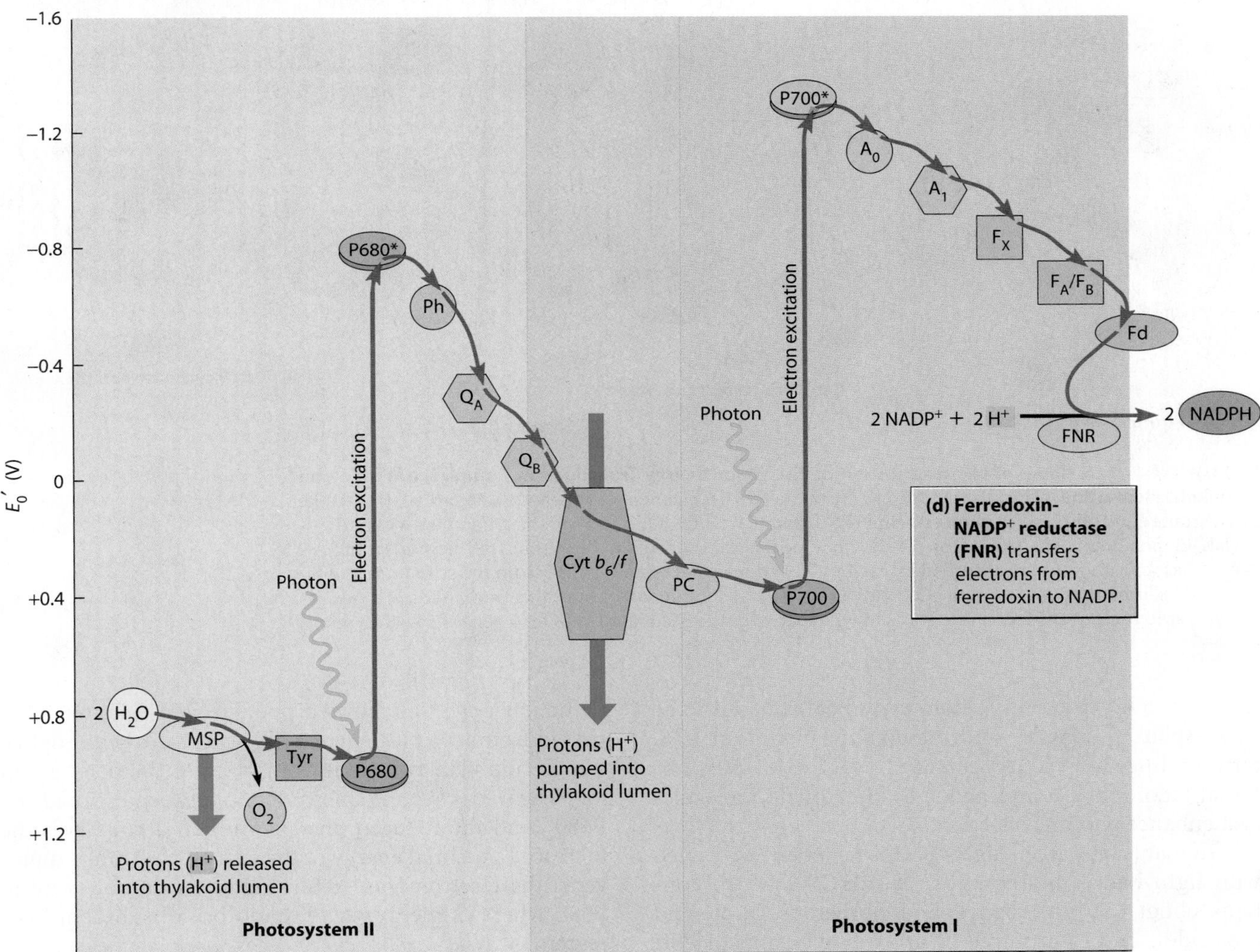

(a) Photosystem II removes electrons from water and passes them sequentially to a manganese-stabilizing protein (MSP), a tyrosine residue (Tyr) on protein D1, a special pair of chlorophyll a molecules (P680), pheophytin (Ph), and two plastoquinones (QA and QB). Oxygen is released as water is oxidized.
(b) The cytochrome b6/f complex receives electrons from PSII and passes them to PS-I via plastocyanin (PC). Simultaneously, protons are pumped from the chloroplast stroma into the thylakoid lumen.
(c) Photosystem I accepts electrons from PC and passes them to ferredoxin (Fd) via another special pair of chlorophyll a molecules (P700), a modified chlorophyll a molecule (A0), phylloquinone (A1), and three iron-sulfur centers (FX, FA, and FB).
(d) Ferredoxin-NADP+ reductase (FNR) transfers electrons from ferredoxin to NADP.
−1.6
−1.2
−0.8
−0.4
0
+0.4
+0.8
+1.2
E0′ (V)
P700*
A0
A1
FX
FA/FB
Fd
2 NADP+ + 2 H+
FNR
2 NADPH
P680*
Ph
QA
QB
Electron excitation
Photon
Cyt b6/f
PC
P700
2 H2O
MSP
Tyr
P680
O2
Protons (H+) released into thylakoid lumen
Protons (H+) pumped into thylakoid lumen
Photosystem II
Photosystem I

FIGURE 11-8 Noncyclic Electron Flow in Oxygenic Phototrophs. The flow of electrons from water to NADPH (red arrows) involves a linear pathway through **(a)** photosystem II (PSII), **(b)** the cytochrome b_6/f complex, **(c)** photosystem I (PSI), and **(d)** ferredoxin-$NADP^+$ reductase (FNR). Note how absorption of photons by chlorophylls P680 and P700 causes a large decrease in their reduction potentials (vertical axis), enabling them to donate excited electrons to an acceptor with a highly negative reduction potential. Figure 11-9 shows the orientation of these components plus the ATP synthase complex within a thylakoid membrane.

(a) PSII includes 25 polypeptides, two of which (D1 and D2) bind chlorophyll P680 and components of the electron transport system. The cluster of four manganese atoms aids in the oxidation of water.

(b) The cytochrome b_6/f complex contains seven polypeptides and includes two cytochromes (b_6 and f) and an iron-sulfur protein (Fe-S).

(c) PSI contains more than ten polypeptides and binds chlorophyll P700 and several additional electron carriers. Electrons are transferred to NADPH via soluble ferredoxin (Fd) using ferredoxin-NADP$^+$ reductase (FNR).

(d) The ATP synthase complex uses the proton gradient generated by electron transport to synthesize ATP.

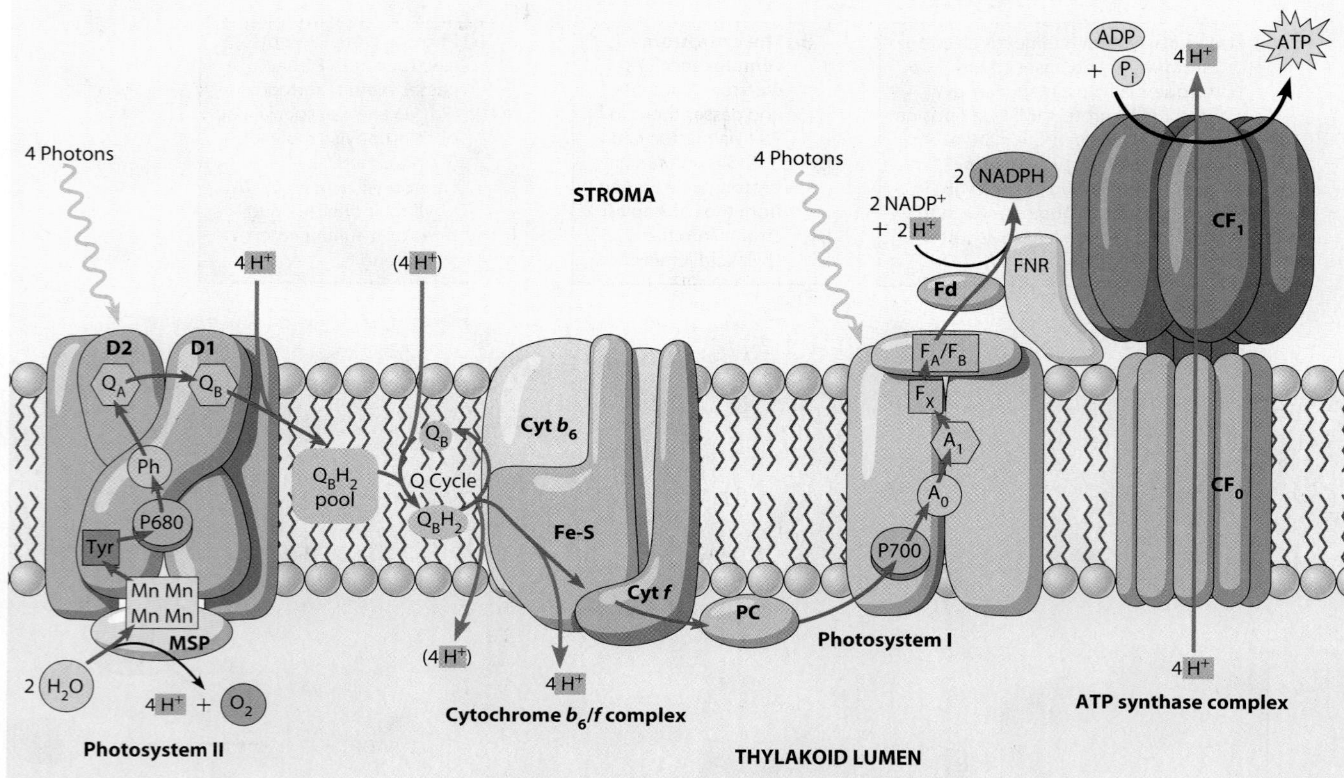

FIGURE 11-9 A Model of the Orientation of the Major Energy Transduction Complexes Within the Thylakoid Membrane. This diagram shows the predicted arrangement of **(a)** photosystem II (PSII), **(b)** the cytochrome b_6/f complex, **(c)** photosystem I (PSI), and **(d)** the CF_oCF_1-ATP synthase complex within the thylakoid membrane. The stoichiometry shown is for the transport of four electrons (red arrows) from H_2O to NADP$^+$ as four photons are absorbed by each photosystem. This removes 6 H^+ from the stroma and adds 8 H^+ to the lumen. Operation of the Q cycle can transfer an additional four protons from the stroma to the lumen per four electrons transported, as shown in parentheses. See Figure 11-8 for a general key to symbols.

D2, that bind not only chlorophyll P680 but also the water-splitting complex and the components of an ETS. Surrounding the reaction center of PSII are 40–50 additional chlorophyll *a* and about 10 β-carotene molecules that enhance photon collection.

In plants and green algae, PSII is generally associated with **light-harvesting complex II (LHCII),** which contains about 250 chlorophyll and numerous carotenoid molecules. Energy captured by antenna pigments of PSII or LHCII is funneled to the reaction center by resonance energy transfer. When the energy reaches the reaction center, it lowers the reduction potential of a P680 chlorophyll *a* molecule to about −0.80 V, making it a better electron donor (see Figure 11-8). The lowering of P680's reduction potential allows a photoexcited electron to be passed from it to *pheophytin (Ph),* a modified chlorophyll *a* molecule with two protons in place of the magnesium ion. The *charge separation* produced between the oxidized P680$^+$ and the reduced pheophytin (Ph$^-$) conserves the increased potential energy of the excited electron and prevents the electron from returning to its ground state in P680, where its energy would simply be lost as heat or fluorescence. *Thus, solar energy has been harvested and converted into electrochemical potential energy in the form of this charge separation.*

Next the electron is passed to Q_A, a **plastoquinone** that is tightly bound to protein D2 (Figure 11-9). Plastoquinone is similar to *coenzyme Q,* a component of the mitochondrial electron transport system (see Figure 10-15). Another plastoquinone, Q_B, receives two electrons from Q_A and picks up two protons from the stroma, thereby becoming reduced to **plastoquinol,** Q_BH_2. It then enters a mobile pool of Q_BH_2 in the interior lipid phase of the photosynthetic membrane. Q_BH_2 can then pass two electrons and two protons to the *cytochrome b_6/f complex* as it is oxidized back to Q_B. Because a chlorophyll molecule transfers one electron per photon absorbed, the formation of one mobile plastoquinol molecule depends on two sequential photoreactions at the same reaction center:

$$2 \text{ photons} + Q_B + 2H^+_{\text{stroma}} + 2e^- \longrightarrow Q_BH_2 \quad \textbf{(11-3)}$$

To replace the electron lost to plastoquinone, oxidized $P680^+$ is reduced by an electron obtained from water. To do this, PSII includes an **oxygen-evolving complex (OEC),** an assembly of proteins and manganese ions that catalyzes the splitting and oxidation of water, producing molecular oxygen (O_2), electrons, and protons. Two water molecules donate four electrons, one at a time by way of a tyrosine residue on protein D1, to four molecules of oxidized $P680^+$ (see Figures 11-8 and 11-9). A manganese-stabilizing protein (MSP) holds a cluster of four manganese ions that serve to accumulate the four electrons released as the two water molecules are oxidized. This prevents the formation of partly oxidized intermediates, such as hydrogen peroxide (H_2O_2) or superoxide anion (O_2^-), that can be toxic to the cell if released. In the process, four protons and one oxygen molecule are released within the thylakoid lumen as shown here:

$$2H_2O \longrightarrow O_2 + 4e^- + 4H^+_{\text{lumen}} \quad \textbf{(11-4)}$$

The protons accumulating in the lumen contribute to an electrochemical proton gradient across the thylakoid membrane, and the oxygen molecule diffuses out of the chloroplast. Because the complete oxidation of two water molecules to molecular oxygen depends on four photoreactions, the net reaction catalyzed by four photoexcitations at PSII can be summarized by doubling Reaction 11-3 and adding it to Reaction 11-4:

$$4 \text{ photons} + 2H_2O + 2Q_B + 4H^+_{\text{stroma}} \longrightarrow O_2 + 2Q_BH_2 + 4H^+_{\text{lumen}} \quad \textbf{(11-5)}$$

Note that the protons removed from the stroma are actually still in transit to the lumen as part of Q_BH_2, while the protons added to the lumen at this point are derived from oxidation of water. The light-dependent oxidation of water to protons and molecular oxygen, called *water photolysis,* is believed to have appeared in cyanobacteria between 2 and 3 billion years ago, thereby permitting exploitation of water as an abundant electron donor. The oxygen released by the process dramatically changed the Earth's early atmosphere, which did not originally contain free oxygen, and allowed the development of aerobic respiration.

The Cytochrome b_6/f Complex Transfers Electrons from a Plastoquinol to Plastocyanin

Electrons carried by Q_BH_2 flow through an ETS coupled to unidirectional proton pumping across the thylakoid membrane into the lumen. This happens by way of the **cytochrome b_6/f complex,** which is analogous to the mitochondrial *respiratory complex III* described in Chapter 10 (see Figures 10-15 and 10-16). The cytochrome b_6/f complex is composed of seven distinct integral membrane proteins, including two cytochromes and an iron-sulfur protein.

Mobile Q_BH_2 donates the two electrons received from PSI, one at a time via *cytochrome b_6* and the iron-sulfur protein, to *cytochrome f.* Oxidation of each Q_BH_2 releases two protons into the thylakoid lumen and enables Q_B to return to the pool of plastoquinone, where it is available for accepting more electrons from PSII and more protons from the stroma.

Recent work supports a model explaining how additional protons can be pumped from the stroma into the lumen by operation of a proposed chloroplast **Q cycle,** first suggested for mitochondria by Peter Mitchell in 1975. In this model, for every electron transferred from Q_BH_2 to plastocyanin (PC), one electron from Q_BH_2 is used in the Q cycle to reduce a second molecule of Q_B that is associated with cytochrome b_6. As this Q_B is reduced to Q_BH_2, additional protons from the stroma are taken up and eventually released into the lumen once Q_BH_2 is reoxidized, as shown here:

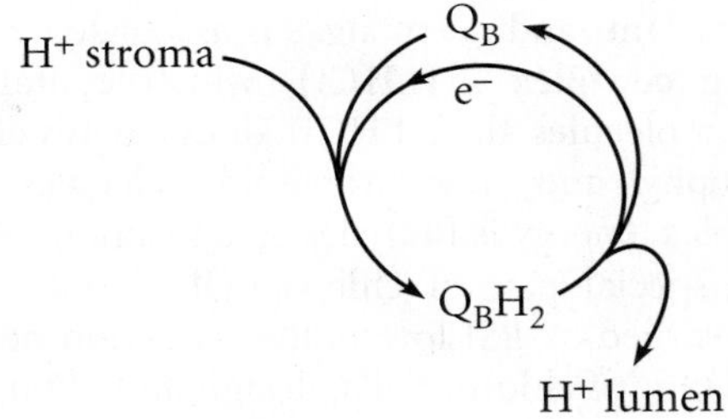

Because half of the electrons from Q_BH_2 are recycled back to Q_B, the operation of this proposed Q cycle would double the number of protons translocated from the stroma to the lumen per electron passed to PC, as shown by the additional $4H^+$ enclosed in parentheses in Figure 11-9. This result agrees with recent work showing that, under certain circumstances, such as low light intensity or induction of photosynthesis in dark-adapted plants, for every electron transported through the photosystems, three H^+ are accumulated in the lumen (one from water splitting, one during electron transfer to PC via cytochrome b_6/f, and one from the Q cycle).

Next, reduced cytochrome f donates electrons to a copper-containing protein called **plastocyanin (PC).** Plastocyanin, like plastoquinol, is a mobile electron carrier, and it will carry electrons to PSI. Unlike plastoquinol, however, it is a peripheral membrane protein found on the lumenal side of the thylakoid membrane, and it carries only one electron at a time. Starting with the two Q_BH_2 molecules generated at PSII (see Reaction 11-5), the net reaction catalyzed by the cytochrome b_6/f complex for every electron directly transferred to PC can be summarized as

$$2Q_BH_2 + 4PC(Cu^{2+}) \rightarrow 2Q_B + 4H^+_{lumen} + 4PC(Cu^+) \quad \textbf{(11-6)}$$

In addition, operation of the proposed Q cycle would double the number of protons moved into the lumen to eight because half of the electrons move through the Q cycle on their way to PC. Thus, four photoreactions at PSII add a total of eight (to 12) protons to the thylakoid lumen: four from the oxidation of water to oxygen and four (to eight) carried to the lumen by plastoquinol (Figure 11-9).

Photosystem I Transfers Electrons from Plastocyanin to Ferredoxin

The task of photosystem I is to transfer photoexcited electrons from reduced plastocyanin to a protein known as *ferredoxin,* the immediate electron donor to $NADP^+$. As indicated in Figure 11-9, the reaction center of PSI includes a third chlorophyll *a* molecule, called A_0, instead of a pheophytin molecule. Other components of the reaction center include a *phylloquinone* (A_1) and three *iron-sulfur centers* (F_X, F_A, and F_B) that form an ETS linking A_0 to ferredoxin.

PSI in plants and green algae is associated with **light-harvesting complex I (LHCI),** which contains fewer antennae molecules than LHCII does—between 80 and 120 chlorophyll and a few carotenoid molecules. As in the PSII complex, energy is funneled to a reaction center containing a special pair of chlorophyll *a* molecules. The energy absorbed by PSI lowers the reduction potential of its special pair of chlorophylls, designated P700, to about −1.30 V (see Figure 11-8). A photoexcited electron is then rapidly passed from P700 to A_0, and charge separation between oxidized $P700^+$ and the reduced A_0 prevents the electron from returning to its original ground state. As in PSII, this charge separation preserves part of the light energy absorbed as electrochemical potential energy. The electron lost by P700 is replaced by an incoming electron from reduced plastocyanin.

From A_0, electrons flow exergonically through the ETS to ferredoxin, the final electron acceptor for PSI. **Ferredoxin (Fd)** is a mobile iron-sulfur protein found in the chloroplast stroma. Ferredoxin is an important reductant in several other metabolic pathways in the chloroplast, including those for nitrogen and sulfur assimilation. Starting with the four reduced plastocyanin molecules generated at the cytochrome b_6/f complex (Reaction 11-6), the net reaction catalyzed by PSI can be summarized as

$$4\text{ photons} + 4PC(Cu^+) + 4Fd(Fe^{3+}) \longrightarrow 4PC(Cu^{2+}) + 4Fd(Fe^{2+}) \quad \textbf{(11-7)}$$

Ferredoxin-NADP⁺ Reductase Catalyzes the Reduction of NADP⁺

The final step in the photoreduction pathway is the transfer of electrons from ferredoxin to $NADP^+$, thereby providing the NADPH essential for photosynthetic carbon reduction and assimilation. This transfer is catalyzed by the enzyme **ferredoxin-NADP⁺ reductase (FNR),** a peripheral membrane protein found on the stromal side of the thylakoid membrane. Starting with the four reduced ferredoxin molecules generated at PSI (see Reaction 11-7), we obtain the following:

$$4Fd(Fe^{2+}) + 2NADP^+ + 2H^+_{stroma} \longrightarrow 4Fd(Fe^{3+}) + 2NADPH \quad \textbf{(11-8)}$$

Notice that the reduction of one $NADP^+$ molecule consumes one H^+ from the stroma plus two electrons, each from a single reduced ferredoxin. While not actually moving protons from the stroma to the lumen, this reaction lowers the proton concentration of the stroma and contributes to the electrochemical proton gradient across the thylakoid membrane.

Functioning together within the chloroplast, the various components of the ETS provide a continuous, unidirectional flow of electrons from water to $NADP^+$, as indicated in Figures 11-8 and 11-9. This is referred to as **noncyclic electron flow,** primarily to distinguish it from the *cyclic electron flow* from ferredoxin back to cytochrome b_6/f that we will encounter shortly. The net result of the complete light-dependent oxidation of two water molecules to molecular oxygen can be obtained by summing Reactions 11-5 through 11-8, assuming ferredoxin does not donate electrons to other pathways:

$$8\text{ photons} + 2H_2O + 6H^+_{stroma} + 2NADP^+ \longrightarrow 8H^+_{lumen} + O_2 + 2NADPH \quad \textbf{(11-9)}$$

For every eight photons absorbed (four by PSII and four by PSI), two NADPH molecules are generated. Furthermore, the ETS for photoreduction is coupled to a mechanism for unidirectional proton pumping across the thylakoid membrane from the stroma to the lumen. Operation of the proposed Q cycle would increase the number of protons accumulating in the lumen to 12. Thus, solar energy has been captured and stored as potential energy in two forms: the reductant NADPH and an electrochemical proton gradient.

Next we will see how the potential energy of this proton gradient is used to synthesize ATP in a manner resembling mitochondrial ATP synthesis. Despite the differences between chloroplasts and mitochondria, both have an outer membrane, an inner membrane, and an internal space in which protons are accumulated—the intracristal space in mitochondria and the thylakoid in chloroplast. Both involve exergonic electron transport through an ordered array of membrane-bound carriers that are alternatively reduced and oxidized as they create the proton gradient used for ATP synthesis. However, in mitochondria, electrons travel from high-energy reduced coenzymes to oxygen, forming water, as chemical energy is converted from one form (NADH) to another (ATP). In chloroplasts, solar energy is used to remove electrons from water, forming oxygen, as solar energy is converted into chemical energy in the form of NADPH and ATP.

Photosynthetic Energy Transduction III: ATP Synthesis

Now we will see how, in the final stage of photosynthetic energy transduction, the potential energy now stored in the proton gradient is used to synthesize ATP from ADP and P_i. Because the energy used to phosphorylate ADP originated in the sun, this process is known as **photophosphorylation.** Since the thylakoid membrane is virtually impermeable to protons, a substantial electrochemical proton gradient can develop across the membrane in illuminated chloroplasts. As described in Chapter 10 for mitochondria, we can calculate the total proton motive force (pmf) across the thylakoid membrane by summing a proton concentration (pH) term and a membrane potential (V_m) term (see Equation 10-13). This can be used to determine the standard free energy change ($\Delta G^{\circ\prime}$) for the movement of protons from the lumen to the stroma.

In mitochondria, the pH difference (ΔpH) across the inner membrane is only about 1 unit, contributing about 0.06 V (30%) of the total pmf of 0.22 V. The other 70% of the pmf comes from the membrane potential difference of about 0.16 V (see Equation 10-14). In chloroplasts, however, ΔpH is more important than ΔV_m and contributes about 80% of the total pmf. Light-induced proton pumping into the thylakoid lumen causes its pH to drop to about 6. Coupled with a rise in stromal pH to about 8 due to proton depletion, this creates a ΔpH of about 2 units, which is enough to generate a pmf of about 0.12 V at 25°C. Assuming current estimates of a membrane potential difference of approximately 0.03 V, this gives a total pmf of 0.15 V, representing a $\Delta G^{\circ\prime}$ of about 3.5 kcal/mol of protons moving across the thylakoid membrane.

The ATP Synthase Complex Couples Transport of Protons Across the Thylakoid Membrane to ATP Synthesis

In chloroplasts as in mitochondria and bacteria, the movement of protons across a membrane from high to low concentration drives the synthesis of ATP by an **ATP synthase** complex. The ATP synthase complex found in chloroplasts, designated the **CF_oCF_1 complex,** is remarkably similar to the F_oF_1 complexes of mitochondria and bacteria described in Chapter 10 (see Figure 10-19). The **CF_1** component is a hydrophilic assembly of polypeptides that protrudes from the stromal side of the thylakoid membrane and contains three catalytic sites for ATP synthesis. Like the bacterial F_1, CF_1 is composed of five distinct polypeptides with a stoichiometry of $\alpha_3\beta_3\gamma\delta\epsilon$. These CF_1 polypeptides have similar structures and functions as the analogous F_1 polypeptides.

The **CF_o** component is a hydrophobic assembly of polypeptides anchored to the thylakoid membrane, much like the F_o component of the F_oF_1-ATP synthase. CF_o subunits I and II form a peripheral stalk that connects CF_o with CF_1, thus serving the same function as the two *b* subunits of F_1 (see Figure 10-19). CF_o subunit IV (similar to the *a* subunit of F_o) is the **proton translocator** through which protons flow from the lumen back to the stroma under the pressure of the pmf. Proton flow through subunit IV causes rotation of an adjacent ring composed of subunit III polypeptides (similar to the *c* subunits in the *c* ring of F_o). This rotation is coupled to ATP synthesis by the CF_1 component as described in Chapter 10 for the F_oF_1-ATP synthase.

In the past, results of biochemical experiments suggested that one ATP was synthesized for every three protons passing through the CF_oCF_1 complex. Other experimental models, however, indicate that four protons are translocated for every ATP synthesized. Recent structural evidence that there are 14 copies of subunit III in the rotating ring of spinach chloroplast CF_oCF_1 suggests that translocation of 14 protons through this CF_oCF_1 causes one rotation and leads to the synthesis of 3 ATP (greater than 4 H^+/ATP). Perhaps differences in the number of subunit III polypeptides in different organisms explains the different H^+/ATP ratios obtained by researchers using different experimental systems.

Assuming a H^+/ATP ratio of four, the reaction catalyzed by the ATP synthase complex can be summarized as

$$4H^+_{lumen} + ADP + P_i \longrightarrow 4H^+_{stroma} + ATP \quad \textbf{(11-10)}$$

This result is quite reasonable in terms of thermodynamics. The $\Delta G'$ value for the synthesis of ATP within the chloroplast stroma is usually 10–14 kcal/mol, whereas the energy available is about 14 kilocalories per 4 moles of protons passing through the ATP synthase complex (3.5 kcal/mol $\times$ 4 mol = 14 kcal). Thus, the flow of four electrons through the noncyclic pathway shown in Figures 11-8 and 11-9 (without the Q cycle) not only generates two NADPH molecules but also leads to the synthesis of two ATP molecules (4 electrons $\times$ 2 protons/electron $\times$ 1 ATP/4 protons = 2 ATP molecules).

Cyclic Photophosphorylation Allows a Photosynthetic Cell to Balance NADPH and ATP Synthesis

Before we conclude our discussion of the photosynthetic energy transduction reactions, we must consider how phototrophs might balance NADPH and ATP synthesis to meet the precise energy needs of a living cell. Notice that noncyclic electron flow leads to the generation of two ATP for every two NADPH molecules. Because both ATP and NADPH are consumed by a variety of metabolic pathways, it is very unlikely that a photosynthetic cell will always require them in the precise ratio generated by noncyclic electron flow. Typically, cells will require more ATP than NADPH because many cellular activities, such as active transport across membranes, require ATP but not NADPH.

When NADPH consumption is low and/or when additional ATP is needed, an optional process known as **cyclic electron flow** can divert the reducing power generated at PSI into ATP synthesis rather than $NADP^+$ reduction. Cyclic electron flow is illustrated in **Figure 11-10** and is distinct from the Q cycle discussed previously. In cyclic electron flow, the reduced ferredoxin generated by PSI transfers electrons back to a cytochrome b_6/f complex instead of donating them to $NADP^+$. The exergonic flow of electrons from ferredoxin through the cytochrome b_6/f complex to plastocyanin is coupled to proton pumping, thereby contributing to the pmf across the thylakoid membrane. From plastocyanin, electrons return to an oxidized $P700^+$ molecule in PSI, completing a closed circuit and allowing P700 to absorb another photon. The excess ATP synthesis resulting from this cyclic electron flow is called *cyclic photophosphorylation*. No water is oxidized and no oxygen is released since the flow of electrons from PSII is not involved.

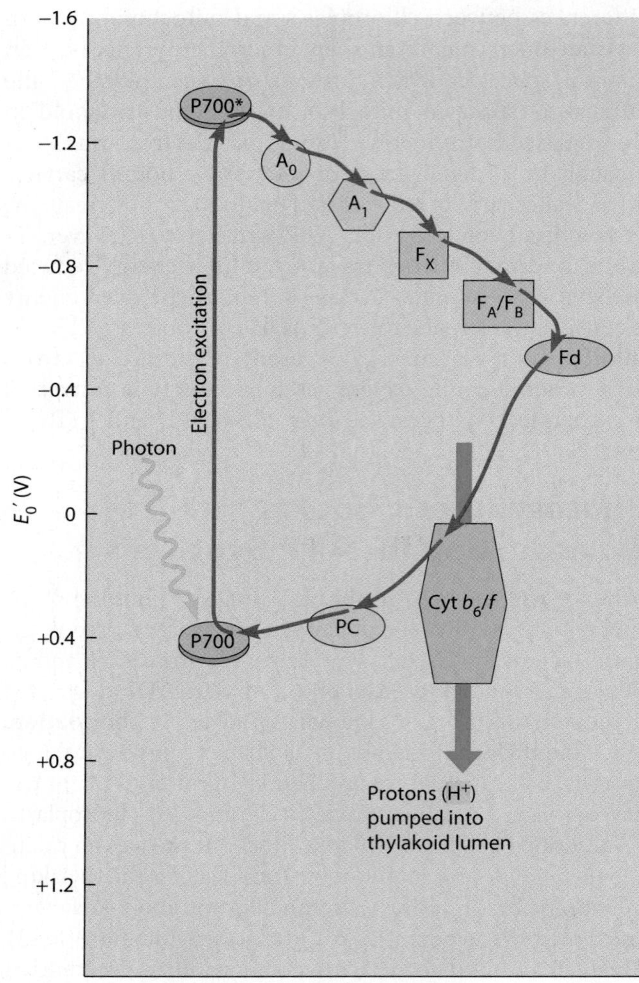

FIGURE 11-10 Cyclic Electron Flow. Cyclic electron flow through PSI enables oxygenic phototrophs to increase the ratio of ATP/NADPH production within photosynthetic cells. When the concentration of $NADP^+$ is low (i.e., when the concentration of NADPH is high), ferredoxin (Fd) donates electrons to the cytochrome b_6/f complex. Electrons then return to P700 via plastocyanin (PC). Because this cyclic electron flow is coupled to unidirectional proton pumping across the thylakoid membrane, excess reducing power is channeled into ATP synthesis. See Figure 11-8 for a general key to symbols.

A Summary of the Complete Energy Transduction System

The model shown in Figure 11-9 represents the entire photosynthetic energy transduction system within a thylakoid membrane. The essential features of the complete system for the transfer of electrons from water to $NADP^+$ and the resulting ATP synthesis can be summarized in terms of the following component parts:

1. *Photosystem II complex:* An assembly of chlorophyll molecules, accessory pigments, and proteins that contains the P680 reaction center chlorophyll. Water is oxidized and split by the oxygen-evolving complex, and electrons flow from water to P680. Following photon absorption, photoexcited P680 molecules donate electrons to plastoquinone, reducing it to plastoquinol, a mobile electron carrier.
2. *Cytochrome b_6/f complex:* An electron transport system that transfers electrons from plastoquinol to plastocyanin, thereby linking PSII with PSI. Electron flow through this complex is coupled to unidirectional proton pumping across the thylakoid membrane, establishing an electrochemical proton gradient that drives ATP synthesis. Optional cyclic electron flow allows additional ATP synthesis.
3. *Photosystem I complex:* An assembly of chlorophyll molecules, accessory pigments, and proteins that contains the P700 reaction center chlorophyll, which accepts electrons from plastocyanin. Following photoexcitation, P700 donates electrons to ferredoxin, a stromal protein.
4. *Ferredoxin-$NADP^+$ reductase:* An enzyme on the stromal side of the thylakoid membrane that catalyzes

the transfer of electrons from two reduced ferredoxin proteins along with a proton to a single $NADP^+$ molecule. The NADPH generated in the stroma is an essential reducing agent in many anabolic pathways.

5. *ATP synthase complex (CF_oCF_1):* A proton channel and ATP synthase that couples the exergonic flow of protons from the thylakoid lumen to the stroma with the synthesis of ATP. Like NADPH, the ATP accumulates in the stroma, where it provides energy for carbon assimilation.

Within a chloroplast, both noncyclic and cyclic pathways of electron flow operate (with or without the Q cycle), thereby providing flexibility in the relative amounts of ATP and NADPH generated. ATP can be produced on a close to equimolar basis with respect to NADPH if the noncyclic pathway is operating alone, or ATP can be generated in excess by using either the Q cycle or the cyclic pathway of PSI.

Photosynthetic Carbon Assimilation I: The Calvin Cycle

With information about chloroplast structure and photosynthetic energy transduction in mind, we are now prepared to look closely at photosynthetic carbon assimilation. More specifically, we will look at the primary events of carbon assimilation: the initial fixation and reduction of carbon dioxide to form simple three-carbon carbohydrates. The fundamental pathway for the movement of inorganic carbon into the biosphere is the **Calvin cycle,** which is found in all oxygenic and most anoxygenic phototrophs.

This pathway is named after Melvin Calvin, who received a Nobel Prize in 1961 for the work he and his colleagues Andrew Benson and James Bassham did to elucidate the process. Taking advantage of the availability of radioactive isotopes following World War II, they were able to use $^{14}CO_2$ to show that the primary products of photosynthetic carbon fixation are triose phosphates. As mentioned earlier, these triose phosphates enter a variety of metabolic pathways, the most important of these being sucrose and starch biosynthesis. An overview of the Calvin cycle is shown in **Figure 11-11**, with more biochemical and structural detail shown in **Figure 11-12**.

In plants and algae, the Calvin cycle is confined to the chloroplast stroma, where the ATP and NADPH generated by photosynthetic energy transduction reactions accumulate. In plants, carbon dioxide generally enters a leaf through special pores called **stomata** (singular: **stoma**). Once inside the leaf, carbon dioxide molecules diffuse into **mesophyll cells** and, in most plant species, travel unhindered into the chloroplast stroma, the site of carbon fixation.

For convenience, we will divide the Calvin cycle into three stages:

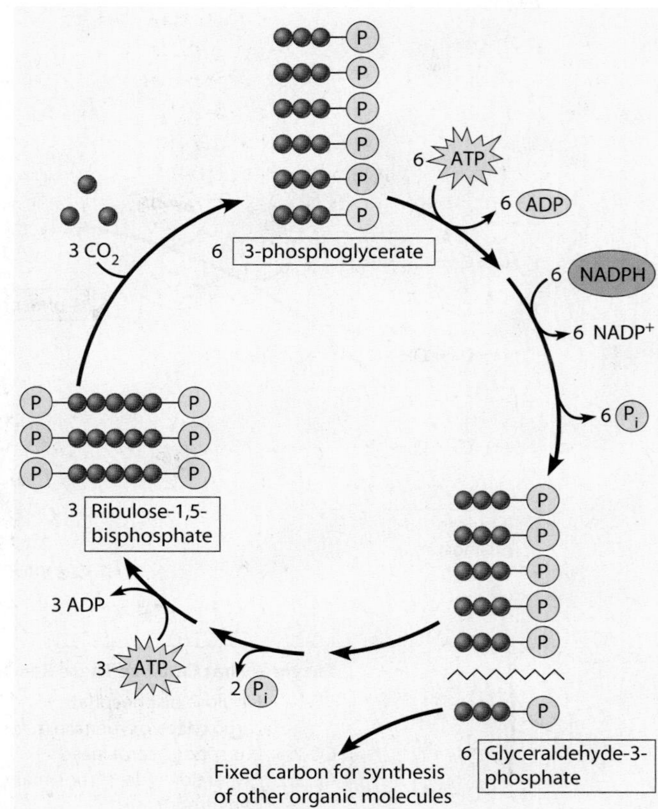

FIGURE 11-11 Overview of the Calvin Cycle. Starting with three molecules of ribulose-1,5-bisphosphate, three molecules of CO_2 are fixed and the products are rearranged to form six molecules of 3-phosphoglycerate. Using ATP and NADPH, these six molecules of 3-phosphoglycerate are reduced to six molecules of glyceraldehyde-3-phosphate. More ATP is used as five of these glyceraldehyde-3-phosphate molecules are rearranged to regenerate three molecules of Ru-P_2, which will accept three more CO_2. The sixth glyceraldehyde-3-phosphate, representing the three CO_2 that were fixed, leaves the cycle for synthesis of other organic compounds in the cell.

1. The carboxylation of the organic acceptor molecule *ribulose-1,5-bisphosphate* and immediate hydrolysis to generate two molecules of *3-phosphoglycerate.*
2. The reduction of 3-phosphoglycerate to form the triose phosphate *glyceraldehyde-3-phosphate.*
3. The regeneration of the initial acceptor molecule to allow continued carbon assimilation.

Carbon Dioxide Enters the Calvin Cycle by Carboxylation of Ribulose-1,5-Bisphosphate

The first stage of the Calvin cycle begins with the covalent attachment of carbon dioxide to the carbonyl carbon of ribulose-1,5-bisphosphate (Reaction CC-1 in Figure 11-12). Carboxylation of this five-carbon acceptor molecule would seem to generate a six-carbon product, but no one has

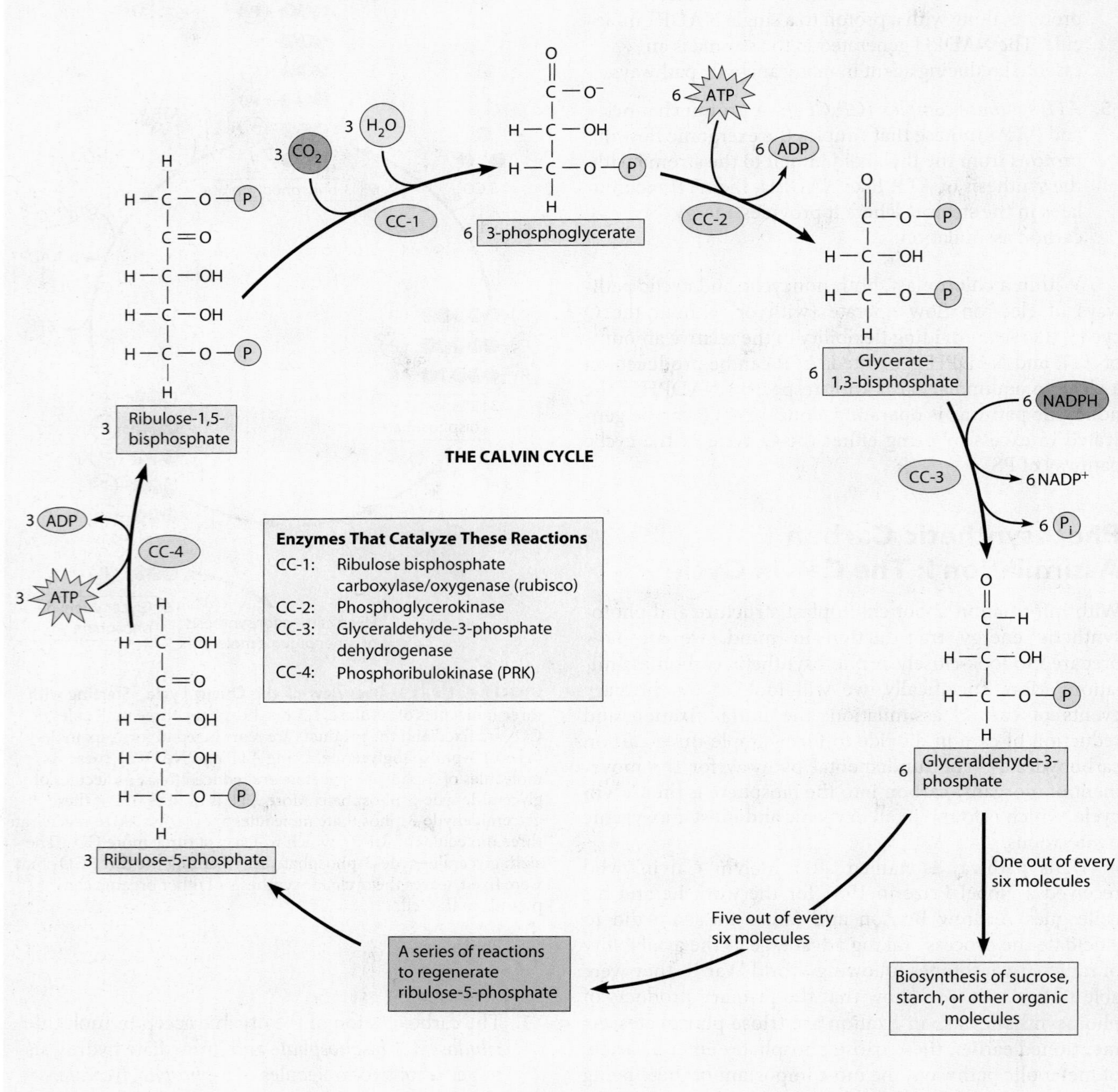

FIGURE 11-12 The Calvin Cycle for Photosynthetic Carbon Assimilation. During one turn of the Calvin cycle (CC), 3 CO_2 molecules are fixed, forming one molecule of triose phosphate. CO_2 is fixed in Reaction CC-1, forming 3-phosphoglycerate. Reactions CC-2 and CC-3 use ATP and NADPH to reduce 3-phosphoglycerate to glyceraldehyde-3-phosphate. One out of every six glyceraldehyde-3-phosphate molecules is used for the biosynthesis of sucrose, starch, or other organic molecules. The other five glyceraldehyde-3-phosphate molecules are used to form three molecules of ribulose-5-phosphate. The ribulose-5-phosphate is then phosphorylated using ATP in Reaction CC-4 to regenerate ribulose-1, 5-bisphosphate, the acceptor molecule for Reaction CC-1, thus completing one turn of the cycle.

isolated such a molecule. As Calvin and his colleagues showed by exposing the alga *Chlorella* to radiolabeled $^{14}CO_2$, the first detectable product of carbon dioxide fixation by this pathway is a three-carbon molecule, 3-phosphoglycerate. Presumably, the six-carbon compound exists only as a transient enzyme-bound intermediate, which is immediately hydrolyzed to generate two molecules of 3-phosphoglycerate as shown below:

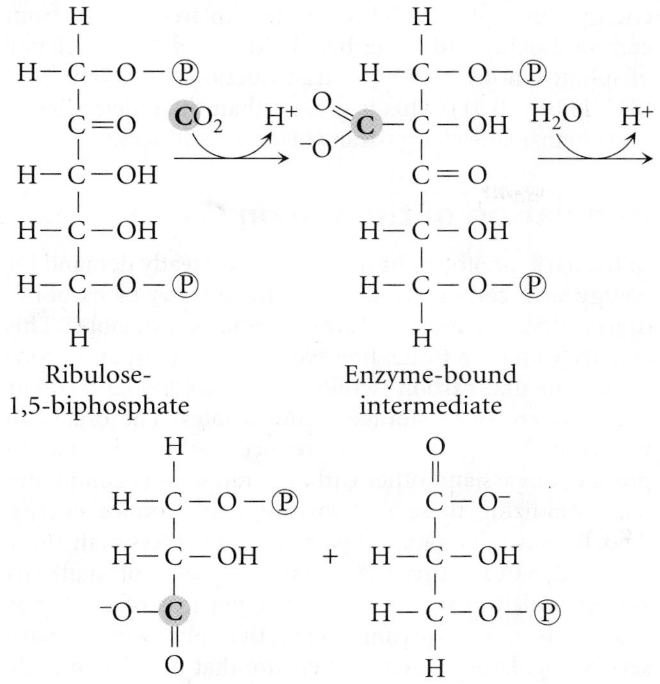

CO_2 H^+ H_2O H^+

Ribulose-1,5-biphosphate

Enzyme-bound intermediate

Two molecules of 3-phosphoglycerate **(11-11)**

The newly fixed carbon dioxide (grey shading) appears as a carboxyl group on one of the two 3-phosphoglycerate molecules.

The enzyme that catalyzes the capture of the carbon dioxide and the formation of 3-phosphoglycerate is called **ribulose-1,5-bisphosphate carboxylase/oxygenase (rubisco).** This relatively large enzyme (a molecular weight of about 560,000) is unique to phototrophs and is found in all photosynthetic organisms except for a few photosynthetic bacteria. Considering its essential role in carbon dioxide fixation for virtually the entire biosphere, it is hardly surprising that rubisco is thought to be the most abundant protein on the planet. About 10–25% of soluble leaf protein is rubisco, and one estimate puts the total amount of rubisco on the Earth at 40 million tons, or almost 15 lb (about 7 kg) for each living person.

3-Phosphoglycerate Is Reduced to Form Glyceraldehyde-3-Phosphate

In the second stage of the Calvin cycle, the 3-phosphoglycerate molecules formed during carbon dioxide fixation are reduced to glyceraldehyde-3-phosphate. This involves a sequence of reactions that is essentially the reverse of the oxidative sequence of glycolysis (Reactions Gly-6 and Gly-7 in Figure 9-7) except that the coenzyme involved is NADPH, not NADH. The steps for NADPH-mediated reduction are shown in Figure 11-12 as Reactions CC-2 and CC-3. In the first reaction, *phosphoglycerokinase* catalyzes the transfer of a phosphate group from ATP to 3-phosphoglycerate. This reaction generates an activated intermediate, glycerate-1,3-bisphosphate. In the second reaction, *glyceraldehyde-3-phosphate dehydrogenase* catalyzes the transfer of two electrons and one proton from NADPH to glycerate-1,3-bisphosphate, reducing it to glyceraldehyde-3-phosphate (G3P).

Some accounting is in order at this point. For every carbon dioxide molecule that is fixed by rubisco (Reaction CC-1), two 3-phosphoglycerate molecules are generated. The reduction of both of these molecules to glyceraldehyde-3-phosphate requires the hydrolysis of two ATP molecules and the oxidation of two NADPH molecules—all for a net gain of only one carbon atom. The net synthesis of one triose phosphate molecule requires the fixation and reduction of *three* carbon dioxide molecules to maintain carbon balance and will therefore consume six ATP and six NADPH:

$$\text{3 ribulose-1,5-bisphosphate} + 3CO_2 + 3H_2O + 6ATP + 6NADPH \longrightarrow 6G3P + 6ADP + 6P_i + 6NADP^+ \quad \textbf{(11-12)}$$

Regeneration of Ribulose-1,5-Bisphosphate Allows Continuous Carbon Assimilation

One out of every six triose phosphate molecules—the net gain following three carboxylations in the Calvin cycle—is used for the biosynthesis of sucrose, starch, or other organic molecules. The five remaining triose phosphates are required in the third and final stage of the Calvin cycle for regeneration of three molecules of the acceptor pentose ribulose-1,5-bisphosphate (Reaction CC-4). Four basic reactions—catalyzed by *aldolases, transketolases, phosphatases,* and *isomerases*—transform five molecules of glyceraldehyde-3-phosphate to three molecules of *ribulose-5-phosphate.* After this, *phosphoribulokinase (PRK)* phosphorylates each ribulose-5-phosphate to form the ribulose-1,5-bisphosphate that accepts carbon dioxide in Reaction CC-1. Regeneration of ribulose-1,5-bisphosphate from glyceraldehyde-3-phosphate consumes three more ATP molecules. The series of events may be summarized by the following net reaction:

$$5G3P + 3ATP + 2H_2O \longrightarrow \text{3 ribulose-1,5-bisphosphate} + 3ADP + 2P_i \quad \textbf{(11-13)}$$

The Complete Calvin Cycle and Its Relation to Photosynthetic Energy Transduction

Primary carbon assimilation by the Calvin cycle may be summarized by combining Reactions 11-12 and 11-13, which encompass all the chemistry of the cycle from carbon dioxide fixation and reduction through triose phosphate synthesis and regeneration of the organic acceptor molecule. The resulting net reaction is

$$3CO_2 + 9ATP + 6NADPH + 5H_2O \longrightarrow G3P + 9ADP + 6NADP^+ + 8P_i \quad \textbf{(11-14)}$$

Note the requirements for ATP and NADPH: The Calvin cycle consumes nine ATP molecules and six NADPH molecules for every three-carbon carbohydrate synthesized, which is three ATP and two NADPH for each carbon dioxide molecule fixed. Thus, activity of the cyclic pathway of PSI and/or the Q cycle is required to produce this excess ATP.

We are now ready to write an overall reaction for photosynthesis that takes both energy transduction and carbon assimilation into account. Considering the needs of the Calvin cycle and assuming no diversion of ATP or NADPH to other pathways, the requirements are clear. According to Reaction 11-14, the synthesis of one molecule of glyceraldehyde-3-phosphate requires nine ATP and six NADPH molecules. This demand can be met by the flow of 12 electrons through the noncyclic pathway, which requires 24 photons to be absorbed by the two photosystems in series. This provides all six of the NADPH molecules (3 × Reaction 11-9) and eight of the nine ATP molecules (8 × Reaction 11-10). The flow of two electrons through the cyclic pathway of PSI will require two more photons and will provide one more ATP molecule (1 × Reaction 11-10). Thus, the net reaction for energy transduction is

$$\text{26 photons} + 9\text{ADP} + 9\text{P}_i + 6\text{NADP}^+ \longrightarrow \text{O}_2 + 9\text{ATP} + 6\text{NADPH} + 3\text{H}_2\text{O} \quad \textbf{(11-15)}$$

The absorption of 26 photons is accounted for by 12 photoexcitation events at PSI and 12 at PSII during noncyclic electron flow and 2 photoexcitation events at PSI during cyclic electron flow. (If the Q cycle is operating, somewhat fewer photons would be needed.)

By adding the summary reaction for the Calvin cycle (Reaction 11-14) to the net reaction for energy transduction (Reaction 11-15), we obtain the following overall expression:

$$\text{26 photons} + 3\text{CO}_2 + 5\text{H}_2\text{O} + \text{P}_i \longrightarrow \text{G3P} + 3\text{O}_2 + 3\text{H}_2\text{O} \quad \textbf{(11-16)}$$

Because the phosphate group is generally removed by hydrolysis when glyceraldehyde-3-phosphate is incorporated into more complex organic compounds, we may rewrite Reaction 11-16 as

$$\text{26 photons} + 3\text{CO}_2 + 6\text{H}_2\text{O} \longrightarrow \text{glyceraldehyde} + 3\text{O}_2 + 3\text{H}_2\text{O} \quad \textbf{(11-17)}$$

This reaction is almost identical to the net photosynthetic reaction we introduced near the beginning of this chapter (Reaction 11-2). We are now, however, in a much better position to understand some of the photochemical and metabolic complexity behind what might otherwise appear to be a simple reaction. Moreover, we have replaced the vague terms *light* and $C_3H_6O_3$ with a specific number of photons and a primary product of carbon dioxide fixation.

This information enables us to calculate the maximum efficiency of photosynthetic energy transduction and carbon assimilation. For red light, assuming a wavelength of 670 nm, 26 moles of photons represent 26 × 43 kcal/mol of photons, or a total of 1118 kcal of energy. Since glyceraldehyde differs in free energy from carbon dioxide and water by 343 kcal/mol, the efficiency of photosynthetic energy transduction is about 31% (343/1118 = 0.31). This is greater than the efficiencies of most man-made energy-transducing machinery.

Regulation of the Calvin Cycle

In the dark, phototrophs must meet the steady demand for energy and carbon by tapping the surplus of carbohydrates that accumulates when light is available. This activity would be futile, however, if the Calvin cycle continued to fix carbon while the glycolytic and other pathways consume storage carbohydrates. The organism would be using energy to reduce carbon dioxide to produce sugars and other carbohydrates while simultaneously oxidizing these carbohydrates to produce energy. And, because all biological pathways run at less than 100% thermodynamic efficiency, this futile cycle of synthesis and degradation would result in a net loss of energy as heat. It is not surprising, then, that phototrophs have several regulatory systems to ensure that the Calvin cycle does not operate unless light is available.

The Calvin Cycle Is Highly Regulated to Ensure Maximum Efficiency

The first level of control is the regulation of synthesis of key enzymes of the Calvin cycle. Enzymes that are important only for light-dependent carbon dioxide fixation and reduction are generally not present in plant tissues that are not exposed to light. Plastids in root cells, for example, contain less than 1% of the amount of rubisco activity found in the chloroplasts of leaf cells.

Enzymes of the Calvin cycle are also regulated by the levels of key metabolites. Consider the changes that occur in the chloroplast stroma as light drives the movement of electrons from water to ferredoxin. As protons are pumped from the stroma to the lumen, the pH of the stroma rises from about 7.2 (a typical value in the dark) to about 8.0. Simultaneously, magnesium ions diffuse in the opposite direction from the lumen to the stroma, and the concentration of magnesium ions in the stroma rises about fivefold. Eventually, high levels of reduced ferredoxin, NADPH, and ATP also accumulate. Each of these factors serves as a signal that can activate enzymes of the Calvin cycle as well as enzymes of other metabolic pathways.

Three enzymes that are unique to the Calvin cycle are logical points for metabolic control. Rubisco is an obvious candidate because it catalyzes the carboxylation reaction of the Calvin cycle. The others are *sedoheptulose bisphosphatase* and *PRK*, which have roles in regenerating the organic acceptor molecule ribulose-1,5-bisphosphate. These three enzymes are all stimulated by a high pH and a high concentration of magnesium ions. In the dark, pH and magnesium ion levels in the stroma decline, rendering these enzymes less active. Each of these enzymes not only is unique to the Calvin cycle but also catalyzes an

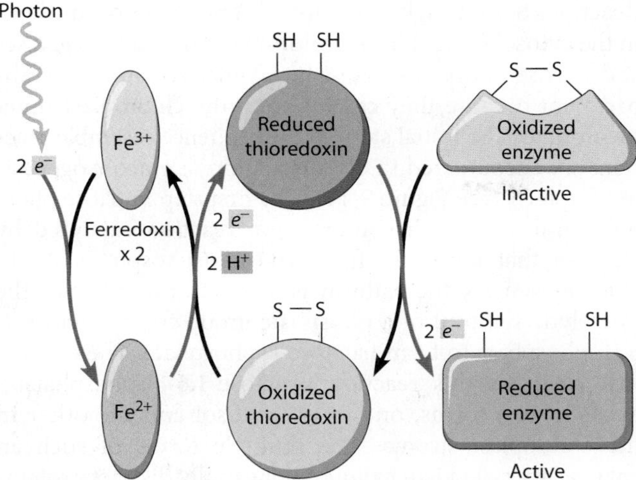

FIGURE 11-13 Thioredoxin-Mediated Activation of a Calvin Cycle Enzyme. Two reduced ferredoxin molecules generated by the photoreduction pathway donate electrons to thioredoxin, an intermediate electron carrier. Thioredoxin subsequently activates a target enzyme by reducing a disulfide bond to two sulfhydryl groups. In other cases, thioredoxin may inactivate enzymes.

essentially irreversible reaction. Recall from Chapter 9 that the glycolytic pathway is also regulated at sites of unique and irreversible reactions. This method of regulation of irreversible, pathway-specific enzymes is a common theme in metabolism.

Another system for coordinating photosynthetic energy transduction and carbon assimilation, shown in **Figure 11-13**, depends on ferredoxin. During illumination, electrons donated by water are used to reduce ferredoxin. An enzyme known as *ferredoxin-thioredoxin reductase* then catalyzes the transfer of electrons from ferredoxin to *thioredoxin,* another mobile electron carrier. Thioredoxin affects enzyme activity by reducing disulfide (S—S) bonds to sulfhydryl (SH) groups, which causes a protein to undergo a conformational change. Glyceraldehyde-3-phosphate dehydrogenase (Reaction CC-3 in Figure 11-12), sedoheptulose bisphosphatase, and PRK are all activated by the conformational change resulting from reduction of disulfide bonds by thioredoxin. In the dark, no reduced ferredoxin is available, and the sulfhydryl groups spontaneously reoxidize to disulfide bonds, thereby inactivating the enzymes. The CF_oCF_1-ATP synthase described earlier is also affected by the thioredoxin system.

Because the Calvin cycle produces three-carbon molecules that can directly enter the glycolytic pathway, the breakdown of complex carbohydrates is not necessary in the light. Not surprisingly, the same mechanisms that activate enzymes of the Calvin cycle inactivate enzymes of degradative pathways. An example is *phosphofructokinase,* the most important control point in glycolysis. While the Calvin cycle is operating in the light, inhibition of phosphofructokinase by high levels of ATP ensures that the early steps of the glycolytic pathway are bypassed, thereby preventing the development of another potentially futile cycle.

Rubisco Activase Regulates Carbon Fixation by Rubisco

Calvin cycle activity is also regulated by **rubisco activase,** a protein that can modulate the activity of rubisco. Rubisco activase functions by removing inhibitory sugar-phosphate compounds from the rubisco active site, thus stimulating rubisco's ability to fix carbon. This allows rubisco to have maximum catalytic activity and promotes carbon fixation at otherwise suboptimal concentrations of CO_2. Rubisco activase also has an inherent ATPase activity, which is essential for its ability to activate rubisco. This ATPase activity is sensitive to the ADP/ATP ratio and is highest when this ratio is low (in the light). In the dark, accumulation of ADP inhibits the ATPase activity of rubisco activase, down-regulating rubisco and conserving supplies of ATP when rubisco is not needed.

In addition to sensing the energy status of the cell, rubisco activase is responsive to the redox state of the cell via interaction with thioredoxin. Recent work by Archie Portis and co-workers shows that the ADP-induced inhibition of activity is greatly diminished in the light by thioredoxin-mediated reduction of a disulfide bridge in rubisco activase, allowing enhanced activity. Thus, rubisco activase helps to regulate the level of carbon fixation by rubisco by responding to the level of light, the ADP/ATP ratio, and the redox state of the cell.

Photosynthetic Carbon Assimilation II: Carbohydrate Synthesis

Because triose phosphates generated by the Calvin cycle are consumed by metabolic pathways in the cytosol as well as the chloroplast stroma, there must be a mechanism for transporting them across the chloroplast inner membrane. The most abundant protein in the chloroplast inner membrane is a *phosphate translocator* that catalyzes the exchange of dihydroxyacetone phosphate, glyceraldehyde-3-phosphate, or 3-phosphoglycerate in the stroma for P_i in the cytosol. This *antiport* system (see Figure 8-6) ensures that triose phosphates will not be exported unless P_i—which is required for synthesizing new triose phosphates—returns to the stroma. Moreover, the phosphate translocator is specific enough to prevent other intermediates of the Calvin cycle from leaving the stroma. Within the cytosol, triose phosphates may be used for sucrose synthesis, as we will describe shortly, or they may enter the glycolytic pathway to provide ATP and NADH in the cytosol. The triose phosphates remaining in the stroma are generally used for starch synthesis, which we will also describe shortly.

Glucose-1-Phosphate Is Synthesized from Triose Phosphates

The key hexose phosphate required for both starch and sucrose synthesis is glucose-1-phosphate, which is formed from two triose phosphates as shown in **Figure 11-14**,

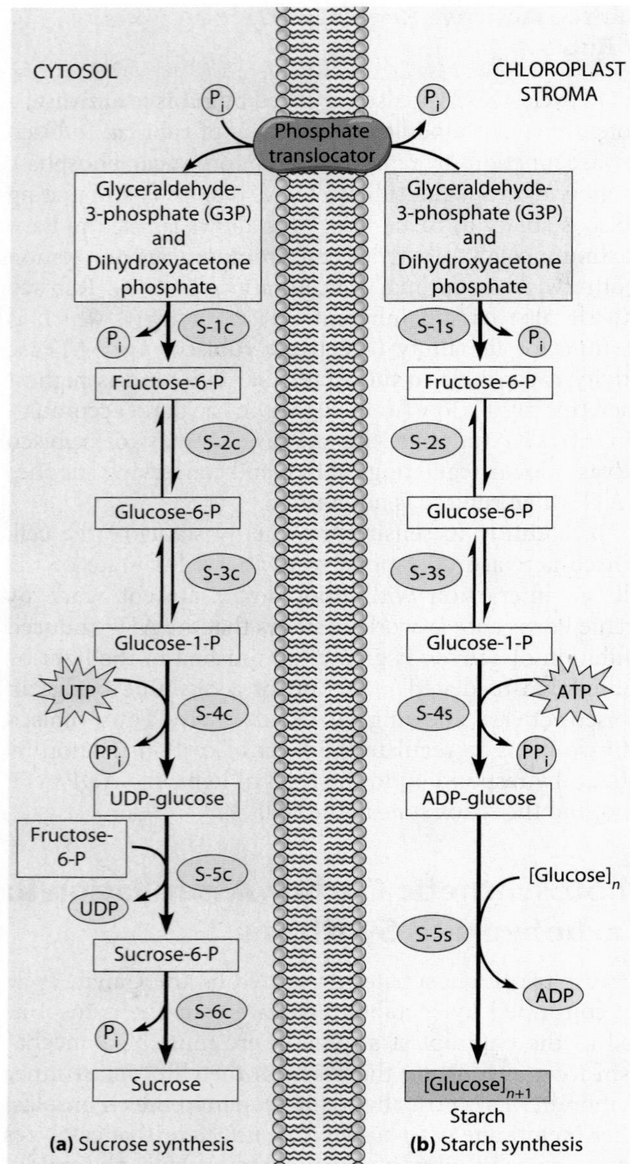

Enzymes That Catalyze These Reactions

S-1:	Aldolase and fructose-1,6-bisphosphatase
S-2:	Phosphoglucoisomerase
S-3:	Phosphoglucomutase
S-4c:	UDP-glucose pyrophosphorylase
S-5c:	Sucrose phosphate synthase
S-6c:	Sucrose phosphatase
S-4s:	ADP-glucose pyrophosphorylase
S-5s:	Starch synthase

FIGURE 11-14 The Biosynthesis of Sucrose and Starch from Products of the Calvin Cycle. The triose phosphates glyceraldehyde-3-phosphate and dihydroxyacetone phosphate from the chloroplast stroma are exchanged for inorganic phosphate from the cytosol via a phosphate translocator. **(a)** Sucrose synthesis is confined to the cytosol, whereas **(b)** starch synthesis occurs only in the chloroplast stroma. The enzymes and isoenzymes that catalyze these reactions are restricted to either the cytosol (c) or the stroma (s).

Reactions S-1 through S-3. This set of reactions occurs both in the cytosol (c) and in the chloroplast stroma (s). Hexoses and hexose phosphates must be synthesized in each compartment because they cannot cross the chloroplast inner membrane. The initial steps in the sequence resemble reactions we encountered in our discussion of gluconeogenesis in Chapter 9 (see Figure 9-11). The first step in triose phosphate utilization is a condensation reaction catalyzed by aldolase that generates fructose-1,6-bisphosphate. As in gluconeogenesis, the pathway is rendered exergonic by the hydrolytic removal of a phosphate group from fructose-1, 6-bisphosphate to form fructose-6-phosphate. The enzyme that catalyzes this reaction, fructose-1,6-bisphosphatase, exists in two forms, one in the cytosol and the other in the chloroplast stroma. The multiple forms of such an enzyme are called **isoenzymes**—physically distinct proteins that catalyze the same reaction. The fructose-6-phosphate can then be converted, via glucose-6-phosphate, to glucose-1-phosphate (Reactions S-2 and S-3). Again, separate isoenzymes catalyze each of these reactions in the cytosol and in the chloroplast stroma.

Textbooks often depict glucose as the end-product of photosynthetic carbon assimilation. But this portrayal is more a definition of convenience for writing summary reactions than a metabolic fact, as very little free glucose actually accumulates in photosynthetic cells. Most glucose is converted either into transport carbohydrates (such as sucrose) or into storage carbohydrates (such as starch in plants or glycogen in photosynthetic bacteria). In reality, the designation of an end-product for photosynthetic carbon assimilation becomes rather arbitrary once you get past the pool of triose phosphates produced by the Calvin cycle. Indeed, we could regard the whole phototrophic organism—plant, algal cell, or bacterium—as the "end-product" of photosynthesis.

The Biosynthesis of Sucrose Occurs in the Cytosol

Recall from Chapter 3 that sucrose is a disaccharide consisting of one molecule each of glucose and fructose linked by a glycosidic bond (see Figure 3-23). Sucrose is of interest because it is the major carbohydrate used for transporting stored energy and reduced carbon in most plant species. Moreover, in some species, such as sugar beets and sugar cane, sucrose also serves as a storage carbohydrate. As shown in Figure 11-14a, sucrose synthesis is localized in the cytosol of a photosynthetic cell. Triose phosphates exported from the chloroplast stroma that are not consumed by other metabolic pathways are converted to fructose-6-phosphate and glucose-1-phosphate. Glucose from glucose-1-phosphate is then activated by the reaction of glucose-1-phosphate with *uridine triphosphate (UTP)*, generating *UDP-glucose* (Reaction S-4c). Finally, the glucose is transferred to fructose-6-phosphate to form the phosphorylated disaccharide *sucrose-6-phosphate,* and hydrolytic removal of the phosphate group generates free sucrose (Reactions S-5c and S-6c). (In some plant species, glucose may be transferred directly from UDP-glucose to free fructose.) The sucrose is exported from leaves and conveys energy and reduced carbon to nonphotosynthetic tissues of the organism.

Like the Calvin cycle, sucrose synthesis is precisely controlled to prevent conflict with degradation pathways. *Cytosolic* fructose-1,6-bisphosphatase, for example, is inhibited by *fructose-2,6-bisphosphate,* a metabolite introduced in Chapter 9 as an important regulator of glycolysis and gluconeogenesis in liver cells. In plant cells, fructose-2,6-bisphosphate accumulates in response to high levels of fructose-6-phosphate and P_i (signals that indicate low sucrose demand) or low levels of 3-phosphoglycerate and dihydroxyacetone phosphate (signals that indicate high triose phosphate demand). Another control point for sucrose synthesis is *sucrose phosphate synthase,* the enzyme that catalyzes the transfer of glucose from UDP-glucose to fructose-6-phosphate (Reaction S-5c). This enzyme is stimulated by glucose-6-phosphate and inhibited by sucrose-6-phosphate, UDP, and P_i.

The Biosynthesis of Starch Occurs in the Chloroplast Stroma

Starch synthesis in plant cells is confined to plastids. In photosynthetic plant cells, starch synthesis is generally restricted to chloroplasts, which are essentially photosynthetic plastids. When sufficient energy and carbon are available to meet the metabolic needs of a plant, the triose phosphates within the chloroplast stroma are converted to glucose-1-phosphate, which is then used for starch synthesis. As shown in Figure 11-14b, glucose is activated by the reaction of glucose-1-phosphate with ATP, generating *ADP-glucose* (Reaction S-4s). The activated glucose is then transferred directly to a growing starch chain by *starch synthase,* leading to elongation of the polysaccharide (Reaction S-5s). As shown in Figure 11-2a, starch may accumulate in large storage granules within the chloroplast stroma. When photosynthesis is limited by darkness or other factors, starch is degraded to triose phosphates, which can then enter glycolysis or be converted to sucrose in the cytosol and exported from the cell.

Like the Calvin cycle and sucrose synthesis, starch synthesis is precisely controlled. Chloroplast fructose bisphosphatase, which channels fructose-1,6-bisphosphate into glucose and starch biosynthesis in the stroma, is activated by the same thioredoxin system that affects enzymes of the Calvin cycle. This regulation ensures that starch synthesis occurs only when there is sufficient illumination for photoreduction. The key enzyme for regulation, though, is *ADP-glucose pyrophosphorylase,* which catalyzes the activation of glucose and commits it to starch synthesis by forming ADP-glucose (Reaction S-4s). ADP-glucose pyrophosphorylase is stimulated by glyceraldehyde-3-phosphate and inhibited by P_i. Thus, when triose phosphates are diverted to the cytosol and ATP is hydrolyzed to ADP and P_i (signals that indicate high energy demand), starch synthesis is blocked.

Photosynthesis Also Produces Reduced Nitrogen and Sulfur Compounds

Photosynthesis encompasses more than carbon dioxide fixation and carbohydrate synthesis. In plants and algae, the ATP and NADPH generated by photosynthetic energy transduction reactions are consumed by a variety of other anabolic pathways found in chloroplasts. Carbohydrate synthesis is only one example of carbon metabolism; the synthesis of fatty acids, chlorophyll, and carotenoids also occurs in chloroplasts. Moving beyond carbon metabolism, several key steps of nitrogen and sulfur assimilation are localized in chloroplasts. The reduction of nitrite (NO_2^-) to ammonia (NH_3), for example, is catalyzed by a reductase enzyme in the chloroplast stroma, with reduced ferredoxin serving as an electron donor. The ammonia is then channeled into amino acid and nucleotide synthesis, portions of which also occur in chloroplasts. Furthermore, much of the reduction of sulfate (SO_4^{2-}) to sulfide (S^{2-}) is catalyzed by enzymes in the chloroplast stroma. In this case, ATP and reduced ferredoxin provide energy and reducing power. The sulfide, like ammonia, may then be used for amino acid synthesis.

Rubisco's Oxygenase Activity Decreases Photosynthetic Efficiency

The primary reaction catalyzed by rubisco—acting as a *carboxylase*—is the addition of carbon dioxide and water to ribulose-1,5-bisphosphate, forming two molecules of 3-phosphoglycerate. However, rubisco can also function as an *oxygenase.* Through this activity, rubisco catalyzes the addition of molecular oxygen, rather than carbon dioxide, to ribulose-1,5-bisphosphate:

H
H—C—O—Ⓟ
C=O
H—C—OH
H—C—OH
H—C—O—Ⓟ
H

Ribulose-1,5-biphosphate

$\xrightarrow{O_2 \quad 2H^+}$

H
H—C—O—Ⓟ
$^-$O—C
O (double bond)

Phosphoglycolate

O (double bond)
C—O$^-$
H—C—OH
H—C—O—Ⓟ
H

3-Phosphoglycerate

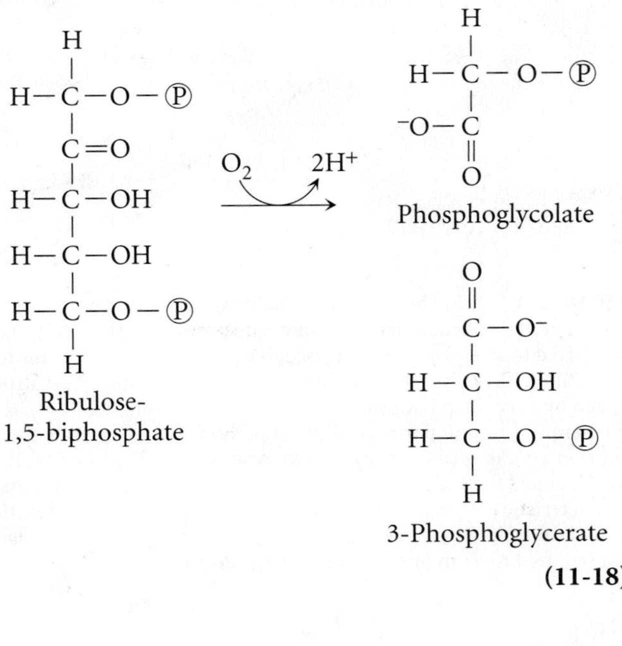

(11-18)

The result of rubisco's oxygenase activity is a single three-carbon product—3-phosphoglycerate—and one two-carbon product, **phosphoglycolate.** Because phosphoglycolate cannot be used in the next step of the Calvin cycle, it appears to be a wasteful diversion of material from carbon assimilation. No alternative functions for rubisco's oxygenase activity have been clearly demonstrated.

Efforts to reduce rubisco's oxygenase activity through alteration of the amino acid sequence of the enzyme have been largely unsuccessful, as a decrease in oxygenase activity is typically accompanied by a compensatory decline in carboxylase activity. Why then does rubisco have this detrimental oxygenase activity? According to one theory, the oxygenase activity is an evolutionary relic from a time when oxygen did not make up a large part of the Earth's atmosphere, and it cannot be eliminated without seriously compromising the carboxylase function. Not even natural selection appears to be up to the task of altering this enzyme. Instead, phototrophs that depend on rubisco have developed three alternative strategies for coping with the enzyme's apparently wasteful oxygenase activity. We will briefly consider each strategy in the following discussions of the *photorespiratory glycolate pathway*, *C_4 photosynthesis*, and *crassulacean acid metabolism*.

The Glycolate Pathway Returns Reduced Carbon from Phosphoglycolate to the Calvin Cycle

In all photosynthetic plant cells, phosphoglycolate generated by rubisco's oxygenase activity is channeled into the **glycolate pathway** (**Figure 11-15**). This pathway disposes of phosphoglycolate and returns about 75% of the reduced carbon present in phosphoglycolate to the Calvin cycle as 3-phosphoglycerate, with the other 25% released as CO_2. Because the glycolate pathway is characterized by light-dependent uptake of oxygen and evolution of carbon dioxide, it is also referred to as **photorespiration.**

Several steps of the glycolate pathway are localized in a specific type of peroxisome called a **leaf peroxisome.** As we learned in Chapter 4, peroxisomes are organelles containing oxidase enzymes that generate hydrogen peroxide.

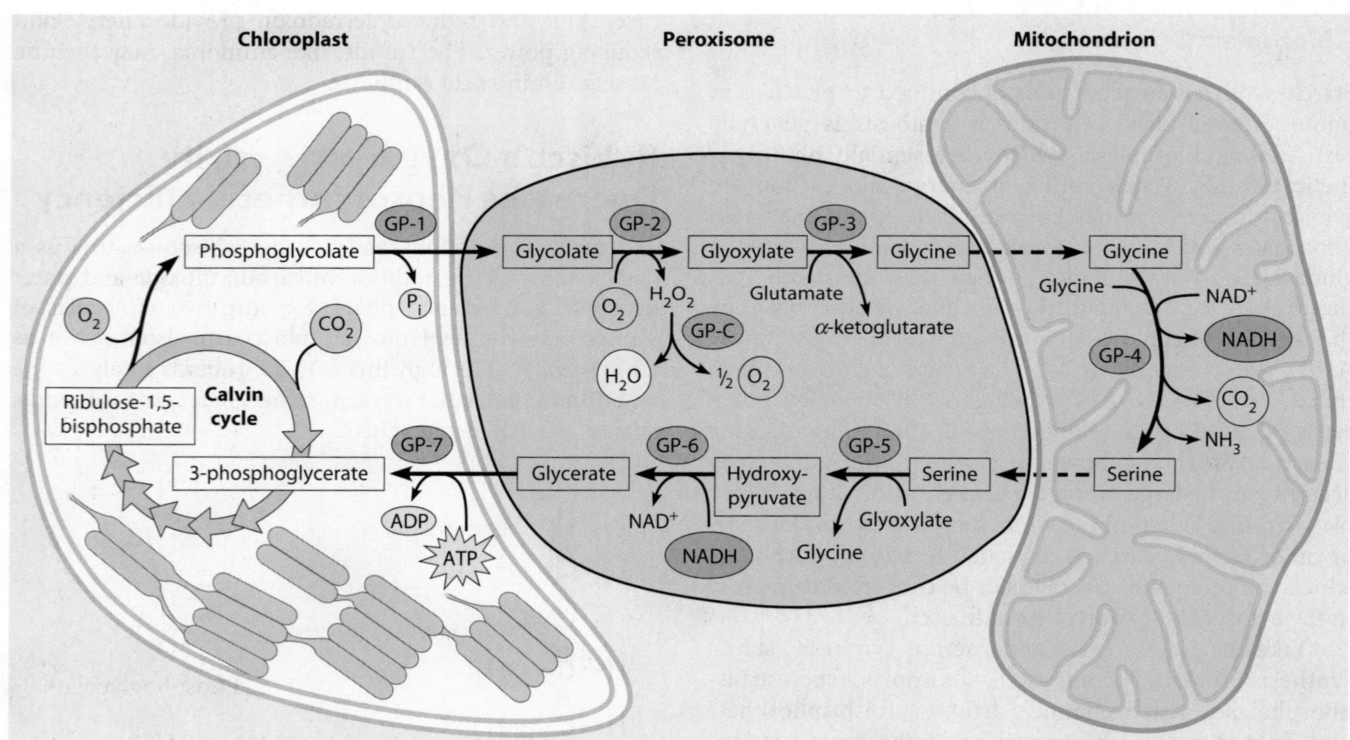

FIGURE 11-15 The Glycolate Pathway. Glycolate arises as a result of the oxygenase activity of rubisco. The immediate product is phosphoglycolate, which is converted to free glycolate by a phosphatase localized in the chloroplast membrane (Reaction GP-1). Free glycolate diffuses out of the chloroplast stroma and is metabolized by a five-step pathway (Reactions GP-2 through GP-6) that occurs partially in the peroxisome and partially in the mitochondrion. Glycerate then diffuses into the chloroplast and is phosphorylated to form 3-phosphoglycerate (Reaction GP-7), which enters the Calvin cycle. The oxygen uptake and carbon dioxide evolution characteristic of photorespiration occur in the peroxisome (Reaction GP-2) and mitochondrion (Reaction GP-4), respectively. Note that two molecules of glycolate are required to form one molecule of 3-phosphoglycerate.

Enzymes That Catalyze These Reactions

GP-1: Phosphoglycolate phosphatase
GP-2: Glycolate oxidase
GP-3: Glutamate: glyoxylate aminotransferase
GP-4: Glycine decarboxylase and serine hydroxymethyl transferase
GP-5: Serine: glyoxylate aminotransferase
GP-6: Hydroxypyruvate reductase
GP-7: Glycerate kinase
GP-C: Catalase

The potentially destructive hydrogen peroxide is then eliminated by another peroxisomal enzyme, catalase, which degrades it to water and oxygen. Because of their essential role in the glycolate pathway, peroxisomes are found in all photosynthetic plant tissues.

A typical leaf peroxisome from a mesophyll cell is shown in the electron micrograph of Figure 4-20b. The crystalline core within the matrix of the organelle is composed of catalase, a marker enzyme for peroxisomes. Notice the close proximity of the leaf peroxisome to a chloroplast and a mitochondrion. This association is frequently found in photosynthetic plant cells and most likely reflects a way to facilitate metabolite transfer among the three organelles involved in the glycolate pathway.

In the chloroplast, the phosphoglycolate generated by rubisco is rapidly dephosphorylated by a phosphatase in the stroma (Reaction GP-1). The product is *glycolate,* which diffuses to a nearby leaf peroxisome, where an *oxidase* converts it to glyoxylate (Reaction GP-2). The oxidation of glycolate is accompanied by the uptake of oxygen and the generation of hydrogen peroxide, which is immediately degraded to oxygen and water by catalase (Reaction GP-C). During the next reaction in the peroxisome, an *aminotransferase* catalyzes the transfer of an amino group from glutamate to glyoxylate, forming glycine (Reaction GP-3).

Glycine diffuses from the leaf peroxisome to a mitochondrion, where two enzyme activities working in series—a *decarboxylase* and a *hydroxymethyl transferase*—convert two glycine molecules to a single *serine,* concomitant with the generation of NADH and the release of carbon dioxide and ammonia (Reaction GP-4). Rubisco's oxygenase activity therefore leads not only to a loss of carbon but also to a potential loss of nitrogen. To prevent depletion of nitrogen reserves, the ammonia must be reassimilated at the expense of ATP and reductant.

Serine diffuses back to the peroxisome, where another aminotransferase removes the amino group, generating *hydroxypyruvate* (Reaction GP-5). A *reductase,* using NADH as an electron donor, then reduces hydroxypyruvate to *glycerate* (Reaction GP-6). Finally, glycerate diffuses to the chloroplast, where it is phosphorylated by *glycerate kinase* to generate 3-phosphoglycerate (Reaction GP-7), a key intermediate of the Calvin cycle.

What is the benefit of this long salvage pathway, winding through several organelles? Three out of every four carbon atoms that exit the Calvin cycle as part of phosphoglycolate are recovered as 3-phosphoglycerate. Without this pathway, phosphoglycolate would accumulate to toxic levels, and the triose phosphates essential for the regeneration of ribulose-1,5-bisphosphate and the continuation of the Calvin cycle would be depleted. In terms of energy and reduced carbon, however, phosphoglycolate metabolism is expensive. For every three carbon atoms salvaged, an ammonia molecule must be reassimilated at the expense of one ATP and two reduced ferredoxin molecules, and the glycerate generated by Reaction GP-6 must be phosphorylated at the expense of one ATP molecule. With rubisco's apparently unavoidable oxygenase activity, however, photorespiration is a net gain for the plant. Just consider the value of the three carbon atoms salvaged: Nine ATP and six NADPH were consumed when each carbon dioxide was originally fixed and reduced.

C_4 Plants Minimize Photorespiration by Confining Rubisco to Cells Containing High Concentrations of CO_2

Plants in hot, arid environments under intense illumination are particularly affected by rubisco's oxygenase activity. As the temperature increases, the solubility of carbon dioxide declines more rapidly than the solubility of oxygen, thereby lowering the CO_2:O_2 ratio in solution. Another problem occurs when plants respond to drought by closing their stomata during the day to reduce water loss. When the stomata are closed, carbon dioxide cannot enter the leaf, and the concentration of carbon dioxide in leaf cells may decline. Moreover, water photolysis continues to generate oxygen, which accumulates because it cannot diffuse out of the leaf when the stomata are closed.

In some cases, the potential for energy and carbon drain through photorespiration is so overwhelming that plants must depend on adaptive strategies for solving the problem. One general approach is to confine rubisco to cells that contain a high concentration of carbon dioxide, thereby minimizing the enzyme's inherent oxygenase activity.

In many tropical grasses, including economically important plants such as maize, sorghum, and sugarcane, the isolation of rubisco is accomplished by a short carboxylation/decarboxylation pathway referred to as the **Hatch–Slack cycle,** after Marshall D. Hatch and C. Roger Slack, two plant physiologists who played key roles in the elucidation of the pathway. Plants containing this pathway are referred to as **C_4 plants** because the immediate product of carbon dioxide fixation by the Hatch-Slack cycle is the four-carbon organic acid oxaloacetate. This term distinguishes such plants from **C_3 plants,** in which the first detectable product of carbon dioxide fixation is the three-carbon compound 3-phosphoglycerate.

To appreciate the advantage of the Hatch–Slack cycle, we must first consider the arrangement of the Hatch–Slack and Calvin cycles within the leaf of a C_4 plant. As shown in **Figure 11-16**, C_4 plants, unlike C_3 plants, have in their leaves two distinct types of photosynthetic cells—*mesophyll cells* and *bundle sheath cells*—that differ in their enzyme composition and hence their metabolic activities. The carbon dioxide fixation step within a C_4 plant is accomplished by an enzyme other than rubisco in mesophyll cells, which are exposed to the carbon dioxide and oxygen that enter a leaf through its stomata. The carbon dioxide that is fixed in mesophyll cells is subsequently released in **bundle sheath cells,** which are relatively isolated from the atmosphere. The entire Calvin cycle, including rubisco, is confined to chloroplasts in the bundle sheath cells. Due to the activity of the Hatch–Slack

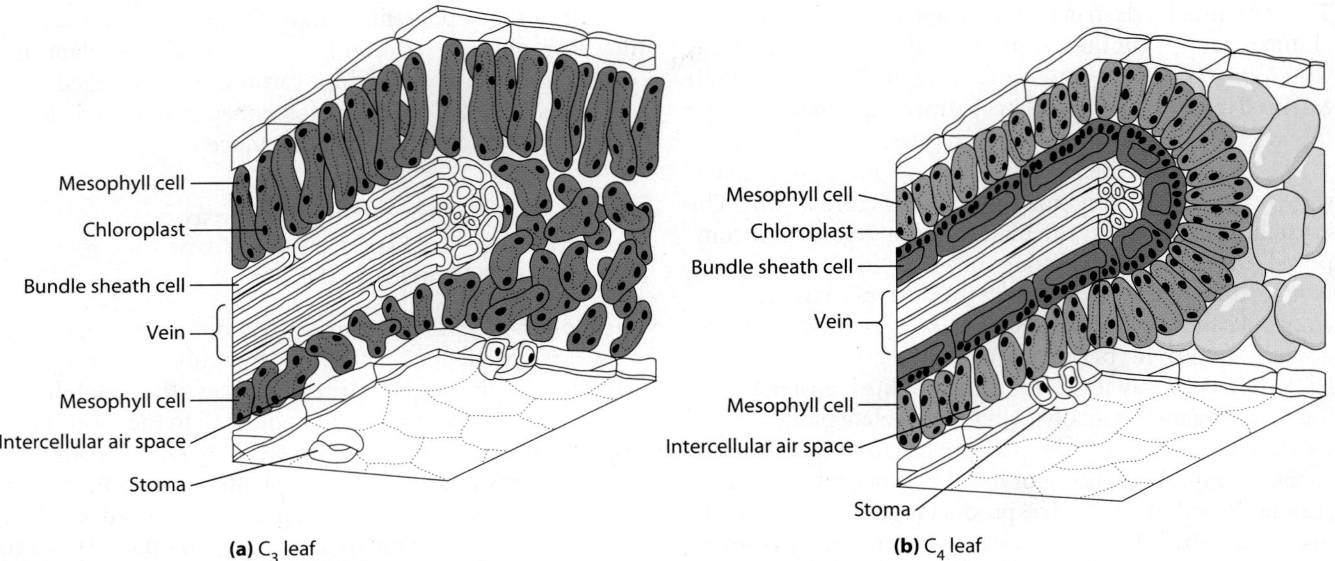

FIGURE 11-16 Structural Differences Between Leaves of C_3 and C_4 Plants. **(a)** In C_3 plants, the Calvin cycle occurs in mesophyll cells. **(b)** In C_4 plants, the Calvin cycle is confined to bundle sheath cells, which are relatively isolated from atmospheric carbon dioxide and oxygen. C_4 plants utilize the Hatch–Slack cycle for collecting carbon dioxide in mesophyll cells and concentrating it in bundle sheath cells. The bundle sheath cells surround the vascular bundles (veins) of the leaf, which carry carbohydrates to other parts of the plant. This concentric arrangement is called Kranz (German for "halo" or "wreath") anatomy and is essential to the photosynthetic efficiency of C_4 plants.

cycle, the carbon dioxide concentration in bundle sheath cells may be as much as ten times the level in the atmosphere, strongly favoring rubisco's carboxylase activity and minimizing its oxygenase activity.

As detailed in **Figure 11-17**, the Hatch–Slack cycle begins with the carboxylation of *phosphoenolpyruvate (PEP)* to form oxaloacetate (Reaction HS-1). Carboxylation is catalyzed by a specific cytosolic form of *PEP carboxylase,* which is particularly abundant in mesophyll cells of C_4 plants. Not only does this carboxylase lack rubisco's oxygenase activity, it is an excellent scavenger for carbon dioxide in the form of the bicarbonate that is produced when carbon dioxide dissolves in water.

In one version of the Hatch–Slack pathway, the oxaloacetate generated by PEP carboxylase is rapidly converted to malate by an *NADPH-dependent malate dehydrogenase* (Reaction HS-2 in Figure 11-17). Malate is a stable four-carbon acid that carries carbon from mesophyll cells to chloroplasts of bundle sheath cells, where decarboxylation by *NADP$^+$ malic enzyme* releases CO_2 (Reaction HS-3). The liberated carbon dioxide is then refixed and reduced by the Calvin cycle. Because decarboxylation of malate is accompanied by the generation of NADPH, the Hatch–Slack cycle also conveys reducing power from mesophyll to bundle sheath cells. This decreases the demand for noncyclic electron flow from water to NADP$^+$ in the bundle sheath cells, thereby minimizing the formation of oxygen by PSII complexes and further favoring rubisco's carboxylase activity.

The pyruvate generated by decarboxylation of malate diffuses into a mesophyll cell, where it is phosphorylated at the expense of ATP to regenerate PEP (Reaction HS-4), the original carbon dioxide acceptor of the Hatch–Slack cycle. Thus, the overall process is cyclic, and the net result is a feeder system that captures carbon dioxide in mesophyll cells and passes it to the Calvin cycle in bundle sheath cells. The Hatch–Slack cycle is not a substitute for the Calvin cycle: It is simply a preliminary carboxylation/decarboxylation sequence that concentrates CO_2 in the bundle sheath cells where there is less oxygen to compete with rubisco's carboxylation activity.

Because ATP is hydrolyzed to AMP in Reaction HS-4, the actual cost of moving carbon from mesophyll to bundle sheath cells is equivalent to two ATP molecules per carbon dioxide molecule. Carbon assimilation within a C_4 plant therefore consumes a total of five ATP molecules per carbon atom, rather than the three required in C_3 plants. In an environment that enhances rubisco's oxygenase activity, however, the energy required to prevent the formation of phosphoglycolate may be far less than the energy that would otherwise be lost through photorespiration.

When temperatures exceed about 30°C, the photosynthetic efficiency of a C_4 plant exposed to intense sunlight may be twice that of a C_3 plant. This is one reason that crabgrass, a C_4 plant, often outgrows other C_3 lawn grasses. While the higher efficiency of a C_4 plant is largely due to reduced photorespiration, other factors are also important. For example, because of their ability to concentrate CO_2, C_4 plants are less affected by the lowered CO_2 concentrations that can exist in dense growth where CO_2 is being rapidly consumed.

Enrichment of carbon dioxide in the vicinity of rubisco by the Hatch–Slack cycle confers an additional advantage

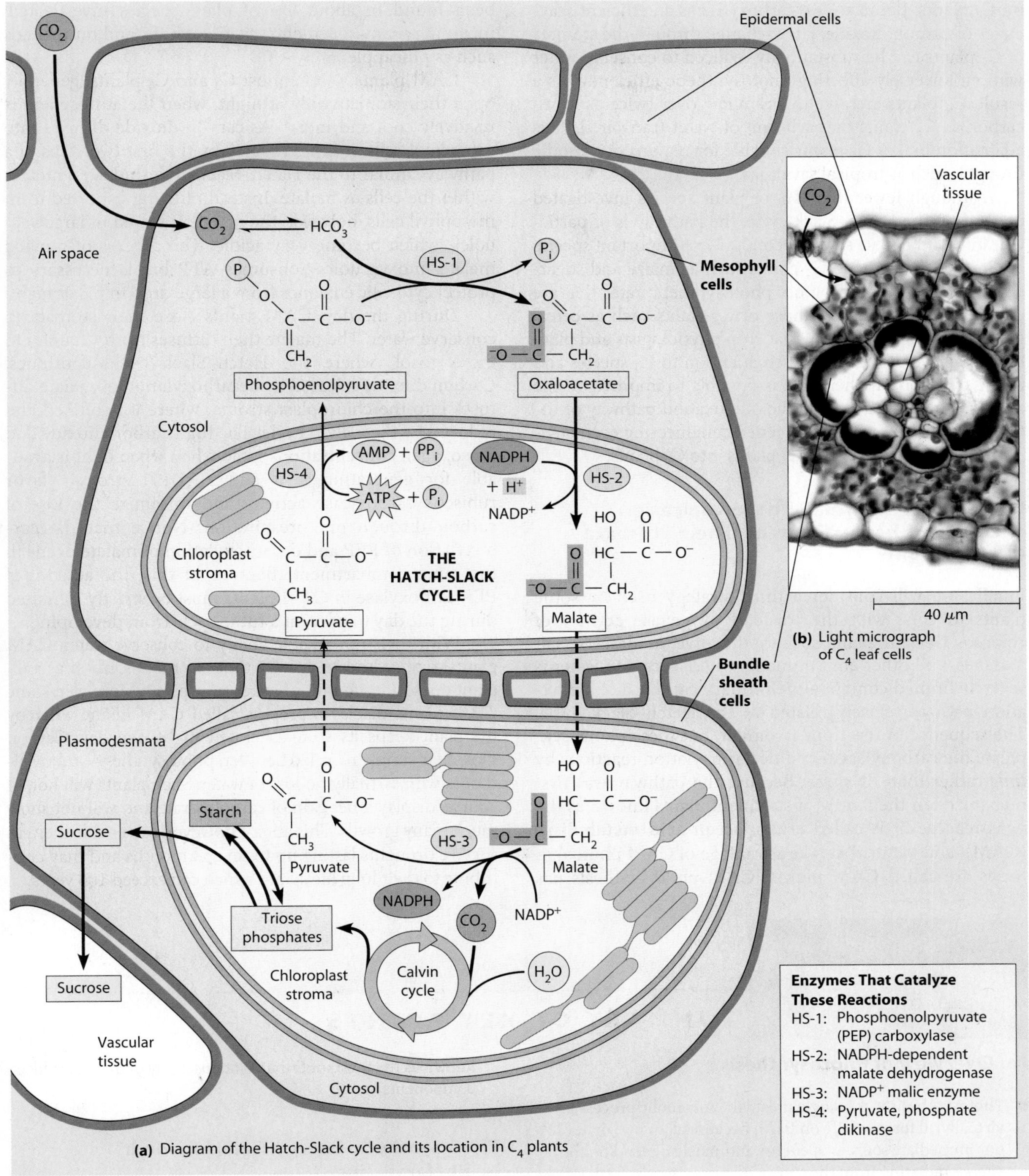

FIGURE 11-17 Localization of the Hatch–Slack Cycle Within Different Cells of a C_4 Leaf. (a) Carbon dioxide fixation in C_4 plants occurs by the Hatch–Slack cycle within mesophyll cells, initially forming oxaloacetate. (The path of incoming carbon is indicated by heavy black arrows.) The malate formed by reduction of oxaloacetate is then passed inward to the bundle sheath cells, where it is decarboxylated. The carbon dioxide is refixed by the Calvin cycle, eventually yielding sucrose, which passes into the adjacent vascular tissue for transport to other parts of the plant. The enzymes that catalyze the Hatch–Slack reactions are listed in the box at lower right. The enzyme pyruvate, phosphate dikinase, which catalyzes phosphorylation of pyruvate, is unique to the Hatch–Slack cycle. (b) The C_4 leaf cells shown in cross section in this light micrograph are of maize, *Zea mays.*

on C_4 plants. Because PEP carboxylase is an efficient scavenger of carbon dioxide, gas exchange through the stomata of C_4 plants can be substantially reduced to conserve water without adversely affecting photosynthetic efficiency. As a result, C_4 plants are able to assimilate over twice as much carbon as C_3 plants for each unit of water transpired. This adaptation makes C_4 plants suitable for regions of periodic drought, such as tropical savannas.

Although fewer than 1% of plant species investigated depend on the Hatch–Slack cycle, the pathway is of particular interest because several economically important species are in this group. Also, C_4 plants such as maize and sugarcane are characterized by net photosynthetic rates that are often two or three times those of C_3 plants such as cereal grains. Little wonder, then, that crop physiologists and plant breeders have devoted so much attention to C_4 species and to the question of whether it is possible to improve on the relatively inefficient carbon dioxide fixation pathway of the C_3 plants. Some scenarios of genetic engineering even envision genetically converting C_3 plants into C_4 plants.

CAM Plants Minimize Photorespiration and Water Loss by Opening Their Stomata Only at Night

Finally, we will consider a third strategy used by some plants to cope with the wasteful oxygenase activity of rubisco. Certain plant species that live in deserts, salt marshes, and other environments where access to water is severely limited contain a preliminary carbon dioxide fixation pathway closely related to the Hatch–Slack cycle. The sequence of reactions is similar, but these plants segregate the carboxylation and decarboxylation reactions by *time* rather than by *space.* Because the pathway was first recognized in the family of succulent plants known as the Crassulaceae, it is called **crassulacean acid metabolism (CAM),** and plants that take advantage of CAM photosynthesis are called **CAM plants.** CAM photosynthesis has been found in about 4% of plant species investigated, including many succulents, cacti, orchids, and bromeliads such as pineapple.

CAM plants, unlike most C_3 and C_4 plants, generally open their stomata only at night, when the atmosphere is relatively cool and moist. As carbon dioxide diffuses into mesophyll cells, it is assimilated by the first two steps of a pathway similar to the Hatch–Slack cycle and accumulates within the cells as malate. Instead of being exported from mesophyll cells, however, the malate is stored in large vacuoles, which become very acidic. The process of moving malate into vacuoles consumes ATP but is necessary to protect cytosolic enzymes from a large drop in pH at night.

During the day, CAM plants close their stomata to conserve water. The malate then diffuses from vacuoles to the cytosol, where the Hatch–Slack cycle continues. Carbon dioxide released by decarboxylation of malate diffuses into the chloroplast stroma, where it is refixed and reduced by the Calvin cycle. The high carbon dioxide and low oxygen concentrations established when light is available for generating ATP and NADPH strongly favor rubisco's carboxylase activity and minimize the loss of carbon through photorespiration. Notice that the carboxylation of PEP and decarboxylation of malate occur in the same compartment. Because of this, the activity of PEP carboxylase in CAM plants must be strictly inhibited during the day to prevent a futile cycle from developing.

With their remarkable ability to conserve water, CAM plants may assimilate over 25 times as much carbon as a C_3 plant does for each unit of water transpired. Moreover, some CAM plants display a process called *CAM idling,* whereby the plant keeps its stomata closed night *and* day. Carbon dioxide is simply recycled between photosynthesis and respiration, with virtually no loss of water. Such plants will not, of course, display a net gain of carbohydrate and will not show much if any growth. This ability, however, can enable them to survive droughts lasting up to several months and may contribute to their long life spans, which can exceed 100 years.

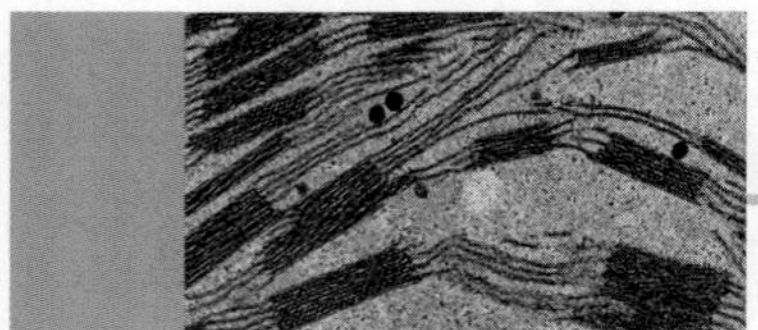

SUMMARY OF KEY POINTS

An Overview of Photosynthesis

- Photosynthesis is the single most vital metabolic process for virtually all forms of life on Earth because all of us, whatever our immediate sources of energy, ultimately depend on the energy radiating from the sun.
- The energy transduction reactions of photosynthesis convert solar energy into chemical energy in the form of NADPH and ATP. The carbon assimilation reactions use this chemical energy to fix and reduce carbon dioxide to carbohydrates.
- In eukaryotic phototrophs, photosynthesis occurs in the chloroplasts, which contain an internal membrane system known as the thylakoids that contains many of the required components.

Photosynthetic Energy Transduction I: Light Harvesting

- Photons of light are absorbed by chlorophyll or accessory pigment molecules within the thylakoid or photosynthetic bacterial membranes. Their energy is rapidly passed to a special pair of chlorophyll molecules at the reaction center of a photosystem.

- At the reaction center, the energy is used to excite and eject an electron from chlorophyll and induce charge separation. In oxygenic phototrophs, this electron is replaced by an electron obtained from a water molecule, generating oxygen.

Photosynthetic Energy Transduction II: NADPH Synthesis

- Electron transfer from water to $NADP^+$ relies on two photosystems acting in series, with photosystem II responsible for the oxidation of water and photosystem I responsible for the reduction of $NADP^+$ to NADPH in the stroma.
- Electron flow between the two photosystems passes through a cytochrome b_6/f complex, which pumps protons into the thylakoid lumen. The resulting proton gradient represents the stored energy of sunlight.

Photosynthetic Energy Transduction III: ATP Synthesis

- The proton motive force across the thylakoid membrane is used to drive ATP synthesis by the CF_oCF_1 complex embedded in the membrane.
- As protons flow back from the lumen to the stroma through the CF_o proton channel in the membrane, ATP is synthesized by the CF_1 portion of the complex that extends into the stroma.

Photosynthetic Carbon Assimilation I: The Calvin Cycle

- In the stroma, ATP and NADPH are used for the fixation and reduction of carbon dioxide into organic form by enzymes of the Calvin cycle.
- The Calvin cycle involves three main stages: fixation of carbon dioxide by rubisco to form 3-phosphoglycerate, reduction of 3-phosphoglycerate to glyceraldehyde-3-phosphate, and regeneration of ribulose-1,5-bisphosphate, the initial carbon dioxide acceptor.
- The net synthesis of one triose phosphate molecule requires the fixation of three CO_2 molecules and uses nine ATP and six NADPH molecules. For each three CO_2 fixed, one molecule of glyceraldehyde-3-phosphate leaves the cycle to be used for further carbohydrate synthesis.

Regulation of the Calvin Cycle

- Key enzymes of the Calvin cycle are regulated to ensure maximum efficiency. Several of them are regulated at the level of synthesis, being made only in photosynthetic tissues that are exposed to the light. They are also activated by the high stromal pH and magnesium concentrations that occur in the light.
- Additional means of regulation involve thioredoxin, which senses the redox state of the cell, and rubisco activase, which removes inhibitors from the rubisco active site in the light.

Photosynthetic Carbon Assimilation II: Carbohydrate Synthesis

- The initial product of carbon dioxide fixation is glyceraldehyde-3-phosphate, which can be converted to a second triose phosphate called dihydroxyacetone phosphate. Some of these triose phosphate molecules are used for the biosynthesis of more complex carbohydrates, such as glucose, sucrose, starch, or glycogen. Others are used as sources of energy or carbon skeletons for other metabolic pathways.
- In addition, ATP produced in the chloroplast is used for fatty acid and chlorophyll synthesis and for nitrogen and sulfur reduction and assimilation.

Rubisco's Oxygenase Activity Decreases Photosynthetic Efficiency

- Rubisco can use oxygen as well as CO_2, resulting in the production of phosphoglycolate, which cannot be used in the Calvin cycle. The glycolate pathway converts two molecules of phosphoglycolate into one molecule of 3-phosphoglycerate, which can enter the Calvin cycle.
- The glycolate pathway involves the chloroplast, the peroxisome, and the mitochondrion. Because CO_2 is released and oxygen is consumed in a light-dependent manner, this process is also called photorespiration.
- In C_4 and CAM plants, carbon dioxide is fixed by a preliminary carboxylation that does not involve rubisco. Then it is decarboxylated under conditions of low oxygen and concentrated CO_2—either in a different cell type or at a different time of day—conditions that favor rubisco's carboxylase activity.

MB **www.masteringbiology.com**

1. MasteringBiology Assignments

Tutorials and Activities
- Molecular Model Structural Tutorials for visualization of 3D structures
- BioFlix Tutorials including 3D animations with hints and detailed feedback for common wrong answers

Questions
- Reading Quiz
- Multiple Choice
- End-of-chapter questions

2. eText Read your book online, search, highlight text, take notes, and more

3. The Study Area
- Chapter Review Quiz, Guide to Techniques and Methods, 3D-Structure Tutorials, Activities, BioFlix, Videos, Biology Review, Word Study Tools, Glossary, Flashcards

PROBLEM SET

More challenging problems are marked with a •.

11-1 True, False, or Insufficient Information. Indicate whether each of the following statements is true (T), false (F), or does not provide enough information for you to make a decision (I).

(a) The inner membrane of a chloroplast contains porins, which allows the free passage of most small organic molecules and ions.

(b) Chlorophyll *a* and *b* have combined absorption spectra which allow access to a broader range of light wavelengths for organisms to collect.

(c) The energy requirement expressed as ATP consumed per molecule of carbon dioxide fixed is higher for a C_4 plant than for a C_3 plant.

(d) The ultimate electron donor for the photosynthetic generation of NADPH is always water.

(e) The enzyme rubisco is unusual in that, depending on conditions, it exhibits two different enzymatic activities.

11-2 The *hcef* mutant. Recently, Livingston et al. (*The Plant Cell* 22: 221, 2010) isolated an *Arabidopsis thaliana* mutant with unusually high cyclic electron flow (*hcef*). See if you can predict the results they observed when comparing the *hcef* mutant plants with normal *Arabidopsis* plants.

(a) How was noncyclic electron flow affected in *hcef* mutant plants?

(b) What happened to the light-driven proton flux across the thylakoid membrane?

(c) How was the activity of PSII affected?

(d) The researchers observed a large increase in the level of fructose-1,6-bisphosphate in the stroma of the *hcef* mutant. What enzyme do you predict is defective?

(e) What would you predict is the effect on starch synthesis in the stroma?

11-3 The Role of Sucrose. A plant uses solar energy to make ATP and NADPH, which then drive the synthesis of carbohydrates in the leaves. At least one carbohydrate, sucrose, is translocated to nonphotosynthetic parts of the plant (stems, roots, flowers, and fruits) for use as a source of energy. Thus, ATP is used to make sucrose, and the sucrose is then used to make ATP. It would seem simpler for the plant just to make ATP and translocate the ATP itself directly to other parts of the plant, thereby completely eliminating the need for a Calvin cycle, a glycolytic pathway, and a TCA cycle (and making life a lot easier for cell biology students). Suggest at least two major reasons that plants do not manage their energy economies in this way.

11-4 Effects on Photosynthesis. Assume that you have an illuminated suspension of *Chlorella* cells carrying out photosynthesis in the presence of 0.1% carbon dioxide and 20% oxygen. What will be the short-term effects of the following changes in conditions on the levels of 3-phosphoglycerate and ribulose-1,5-bisphosphate? Explain your answer in each case.

(a) Light is restricted to green wavelengths (510–550 nm).

(b) Oxygen concentration is reduced from 20% to 1%.

(c) Carbon dioxide concentration is suddenly reduced 1000-fold.

(d) An inhibitor of photosystem II is added.

11-5 Energy Flow in Photosynthesis. A portion of the solar energy that arrives at the surface of a leaf is eventually converted to chemical energy and appears in the chemical bonds of carbohydrates that are generally regarded as the end-products of photosynthesis. In between the photons and the carbohydrate molecules, however, the energy exists in a variety of forms. Trace the flow of energy from photon through ATP to starch molecule, assuming the wavelength of light to be in the absorption range of one of the accessory pigments rather than that of chlorophyll.

11-6 The Mint and the Mouse. Joseph Priestley, a British clergyman, was a prominent figure in the early history of research in photosynthesis. In 1771, Priestley wrote these words:

> *One might have imagined that since common air is necessary to vegetable as well as to animal life, both plants and animals had affected it in the same manner; and I own that I had that expectation when I first put a sprig of mint into a glass jar standing inverted in a vessel of water; but when it had continued growing there for some months, I found that the air would neither extinguish a candle, nor was it at all inconvenient to a mouse which I put into it.*

Explain the basis of Priestley's observations, and indicate their relevance to the early understanding of the nature of photosynthesis.

11-7 Photosynthetic Efficiency. On page 312 we estimated the maximum photosynthetic efficiency for the conversion of red light, carbon dioxide, and water to glyceraldehyde. Under laboratory conditions, a photosynthetic organism *might* convert 31% of the light energy striking it to chemical bond energy of organic molecules. In reality, however, photosynthetic efficiency is far lower, closer to 5% or less. Considering a plant growing in a natural environment, suggest four reasons for this discrepancy.

11-8 Chloroplast Structure. Where in a chloroplast are the following substances or processes localized? Be as specific as possible.

(a) Plastoquinol

(b) Starch synthase

(c) P700

(d) Carotenoid molecules

(e) Ferredoxin-$NADP^+$ reductase

(f) Reduction of 3-phosphoglycerate

(g) Cyclic electron flow

(h) Proton pumping

(i) Oxygen-evolving complex

(j) Light-harvesting complex I

11-9 Metabolite Transport Across Membranes. For each of the following metabolites, indicate whether you would expect it to be in steady-state flux across one or more membranes in a photosynthetically active chloroplast and, if so, indicate which membrane(s) the metabolite must cross.

(a) ATP

(b) NADPH

(c) P_i

(d) O_2

(e) Pyruvate

(f) Protons

(g) Starch

(h) Electrons

(i) CO_2

(j) Glyceraldehyde-3-phosphate

11-10 Crassulacean Acid Metabolism. A CAM plant uses a pathway very similar to the Hatch–Slack cycle for preliminary carbon dioxide fixation. Trace the flow of carbon from the atmosphere to glyceraldehyde-3-phosphate within a CAM plant. How does this minimize water loss by such plants?

11-11 The Endosymbiont Theory. The endosymbiont theory suggests that mitochondria and chloroplasts evolved from ancient bacteria that were ingested by primitive nucleated cells. Biologists have proposed that endosymbiosis led to the evolution of other cellular structures, such as flagella and peroxisomes, as well. Over hundreds of millions of years, the ingested bacteria lost features not essential for survival inside the host cell.

(a) What features of mitochondria and chloroplasts made scientists believe that they might be derived from ingested bacteria?

(b) Describe one structure or metabolic process that purple bacteria might have dispensed with once they became endosymbionts in a eukaryotic cell. How would loss of this feature prevent the bacteria from living outside the host?

(c) Describe one structure or metabolic process that cyanobacteria might have dispensed with once they became endosymbionts in a eukaryotic cell. How would loss of this feature prevent the bacteria from living outside the host?

(d) It has also been suggested that flagella evolved from ancient bacteria. Provide one observation which supports this, and one reason that goes against it.

SUGGESTED READING

References of historical importance are marked with a •.

General References

Aro, E.-M., and B. Andersson, eds. *Regulation of Photosynthesis, Advances in Photosynthesis and Respiration,* Vol. 11. Boston: Kluwer Academic Publishers, 2001.

• Emerson, R. Photosynthesis. *Annu. Rev. Biochem.* 6 (1937): 535.

Laisk, A., L. Nedbal, and Govindjee, eds. *Photosynthesis in Silico: Understanding Complexity from Molecules to Ecosystems.* Dordrecht, The Netherlands: Springer, 2009. Leegood, R. C., T. D. Sharkey, and S. von Caemmerer, eds. *Photosynthesis: Physiology and Metabolism. Advances in Photosynthesis,* Vol. 9. Boston: Kluwer Academic Publishers, 2000.

The Chloroplast

Argyroudi-Akoyunoglou, J. H., and H. Senger. *The Chloroplast: From Molecular Biology to Biotechnology.* Boston: Kluwer Academic Publishers, 1999.

Leister, D. Chloroplast research in the genomic age. *Trends Genet.* 19 (2003): 47.

Li, H.-M., and C.-C. Chiu. Protein transport into chloroplasts. *Annu. Rev. Plant Biol.* 61 (2010): 157.

Photosynthetic Energy Transduction

Eberhard, S., G. Finazzi, and F. A. Wollman. The dynamics of photosynthesis. *Annu. Rev. Genet.* 42 (2008): 463.

McCarty, R. E., Y. Evron, and E. A. Johnson. The chloroplast ATP synthase: A rotary enzyme? *Annu. Rev. Plant Physiol. Plant Mol. Biol.* 51 (2000): 83.

Nelson, N., and C. F. Yocum. Structure and function of photosystems I and II. *Annu. Rev. Plant Biol.* 57 (2006): 521.

Richter, M. L. et al. Coupling proton movement to ATP synthesis in the chloroplast ATP synthase. *J. Bioenerg. Biomembr.* 37 (2005): 467.

Scholes, G. D. Quantum-coherent electronic energy transfer: Did nature think of it first? *J. Phys. Chem. Lett.* 1 (2010): 2.

Seelert, H., N. A. Dencher, and D. J. Müller. Fourteen protomers compose the oligomer III of the proton rotor in spinach chloroplast ATP synthase. *J. Mol. Biol.* 333 (2003): 337.

Shikanai, T. Cyclic electron transport around photosystem I: Genetic approaches. *Annu. Rev. Plant Biol.* 58 (2007): 199.

The Calvin Cycle and Carbohydrate Synthesis

• Calvin, M. The path of carbon in photosynthesis. *Science* 135 (1962): 879.

Hügler, M., and S. M. Sievert. Beyond the Calvin cycle: Autotrophic carbon fixation in the ocean. *Annu. Rev. Marine Sci.* 3 (2011).

Portis, A. R., Jr. Rubisco activase—Rubisco's catalytic chaperone. *Photosynth. Res.* 75 (2003): 11.

Spreitzer, R. J., and M. E. Salvucci. Rubisco: Structure, regulatory interactions, and possibilities for a better enzyme. *Annu. Rev. Plant Biol.* 53 (2002): 449.

Photorespiration, C_4 Plants, and CAM Plants

Cushman, J. C., and H. J. Bohnert. Crassulacean acid metabolism: Molecular genetics. *Annu. Rev. Plant Physiol. Plant Mol. Biol.* 50 (1999):305.

Foyer, C. H., A. J. Bloom, G. Queval, and G. Noctor. Photorespiratory metabolism: Mutants, energetics, and redox signaling. *Annu. Rev. Plant Biol.* 60 (2009): 455.

Matsuoka, M. et al. Molecular engineering of C_4 photosynthesis. *Annu. Rev. Plant Physiol. Plant Mol. Biol.* 52 (2001): 297.

Tipple, B. J., and M. Pagani. The early origins of terrestrial C_4 photosynthesis. *Annu. Rev. Earth Planetary Sci.* 35 (2007): 435.

Wingler, A. et al. Photorespiration: Metabolic pathways and their role in stress protection. *Philos. Trans. R. Soc. Lond. B Biol. Sci.* 355 (2000): 1517.

Box 11A: The Endosymbiont Theory

Margulis, L., and D. Sagan. *Acquiring Genomes: A Theory of the Origins of Species.* New York: Basic Books, 2003.

Poole, A. M., and D. Penny. Evaluating hypotheses for the origin of eukaryotes. *BioEssays* 29 (2007): 500.

12 The Endomembrane System and Peroxisomes

See MasteringBiology® for tutorials, activities, and quizzes.

A full appreciation of eukaryotic cells depends on an understanding of the prominent role of intracellular membranes and the compartmentalization of function within organelles*—intracellular membrane-bounded compartments that house various cellular activities. Whether we consider the storage and transcription of genetic information, the biosynthesis of secretory proteins, the breakdown of long-chain fatty acids, or any of the numerous other metabolic processes occurring within eukaryotic cells, many of the reactions of a particular pathway occur within a distinct type of organelle. Also, the movement of molecules between organelles, known as* trafficking, *must be tightly regulated to ensure that each organelle has the correct components for its proper structure and function.*

We briefly encountered the major organelles found in eukaryotic cells in Chapter 4, and we then learned more about the mitochondrion and chloroplast in Chapters 10 and 11, respectively. We are now ready to consider several other individual organelles in more detail. We will begin with the rough endoplasmic reticulum, *the* smooth endoplasmic reticulum, *and the* Golgi complex, *which are sites for protein synthesis, processing, and sorting. Next, we will look at* endosomes, *organelles that are important for carrying and sorting material brought into the cell. Endosomes help to form* lysosomes, *which are organelles responsible for digestion of both ingested material and unneeded intracellular components. We will conclude with a look at* peroxisomes, *which house hydrogen peroxide–generating reactions and perform diverse metabolic functions.*

As you study the role of each organelle, keep in mind that the endoplasmic reticulum, the Golgi complex, endosomes, and lysosomes (but not peroxisomes) comprise the **endomembrane system** *of the eukaryotic cell, as shown in* **Figure 12-1**. *(The* nuclear envelope, *which we will study in Chapter 18, is closely associated with the endomembrane system.) Material flows from the endoplasmic reticulum to and from the Golgi complex, endosomes, and lysosomes by means of* transport vesicles *that shuttle between the various organelles. These transport vesicles carry membrane lipids and membrane-bound proteins to their proper destinations in the cell, and they also carry soluble materials destined for secretion. Thus, these organelles and the vesicles connecting them make up a single dynamic system of membranes and internal spaces. Currently, one of the most exciting questions in modern cell biology concerns* endomembrane trafficking: *How does each of the multitude of proteins and lipids in a cell manage to reach its proper destination at the proper time?*

The Endoplasmic Reticulum

The **endoplasmic reticulum (ER)** is a continuous network of flattened sacs, tubules, and associated vesicles that stretches throughout the cytoplasm of the eukaryotic cell. Although the name sounds formidable, it is actually quite descriptive. *Endoplasmic* simply means "within the (cyto)plasm," and *reticulum* is a Latin word meaning "network." The membrane-bounded sacs are called **ER cisternae** (singular: **ER cisterna**), and the space enclosed by them is called the **ER lumen** (Figure 12-1). Of the total membrane in a mammalian cell, up to 50–90% surrounds the ER lumen. Unlike more prominent organelles, such as the mitochondrion or chloroplast, however, the ER is not visible by light microscopy unless one or more of its components are stained with a dye or labeled with a fluorescent molecule.

The ER was first observed in the late nineteenth century, when it was noted that some eukaryotic cells, particularly those involved in secretion, contained regions that stained intensely with basic dyes. The significance of these regions remained in doubt until the 1950s, when the resolving power of the electron microscope was improved dramatically. This allowed cell biologists to visualize for the first time the ER's elaborate network of intracellular membranes and to investigate the role of the ER in cellular processes. This is a common

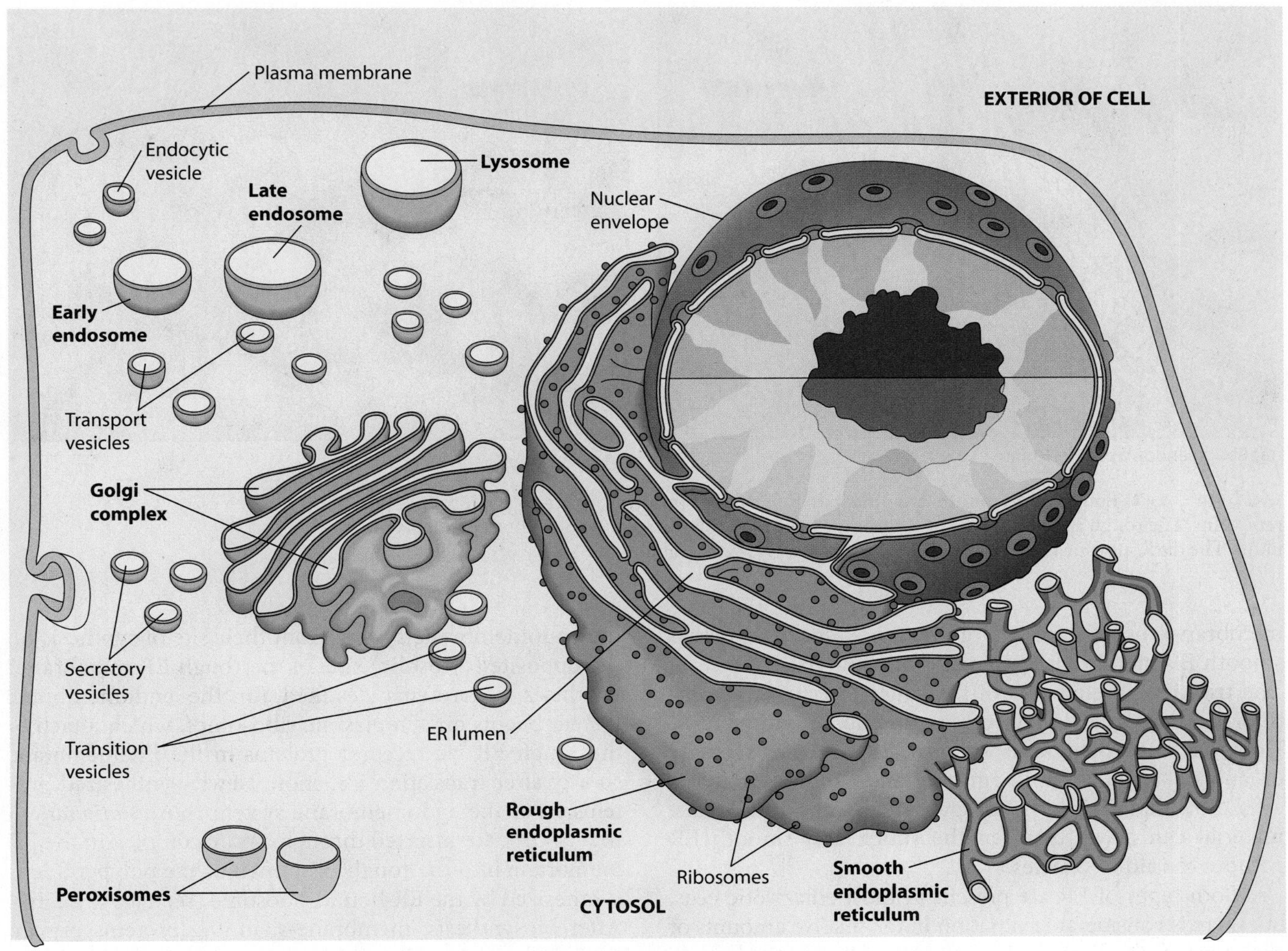

FIGURE 12-1 The Endomembrane System. The endomembrane system of the eukaryotic cell consists of the endoplasmic reticulum (ER), the Golgi complex, endosomes, and lysosomes (but not peroxisomes). It is associated with both the nuclear envelope and the plasma membrane. The ER lumen is linked to the interiors of the Golgi complex, endosomes, and lysosomes by transport vesicles that shuttle material between organelles, as well as to and from the plasma membrane.

theme in scientific discovery—conceptual advances in one field often follow technological advances in a related (or even an unrelated) field.

We now know that enzymes associated with the ER are responsible for the biosynthesis of proteins destined for incorporation into the plasma membrane or into organelles of the endomembrane system and for synthesis of proteins destined for export from the cell. The ER also plays a central role in the biosynthesis of lipids, including triacylglycerols, cholesterol, and related compounds. The ER is the source of most of the lipids that are assembled to form intracellular membranes and the plasma membrane.

The Two Basic Kinds of Endoplasmic Reticulum Differ in Structure and Function

The two basic kinds of endoplasmic reticulum typically found in eukaryotic cells are distinguished from one another by the presence or absence of ribosomes attached to the ER membrane (**Figure 12-2**). **Rough endoplasmic reticulum (rough ER)** is characterized by ribosomes attached to the cytosolic side of the membrane (the side that faces away from the ER lumen; Figure 12-2a). Translation by these ribosomes occurs in the cytosol, but the newly-synthesized proteins will enter the ER lumen shortly. Because ribosomes contain RNA, it was this RNA that reacted strongly with the basic dyes originally used to identify the rough ER. A subdomain of rough ER, the **transitional elements (TEs),** plays an important role in the formation of **transition vesicles** that shuttle lipids and proteins from the ER to the Golgi complex. In contrast, **smooth endoplasmic reticulum (smooth ER)** appears smooth due to the absence of ribosomes attached to the membrane (Figure 12-2b) and has other roles in the cell.

Rough and smooth ER are easily distinguished morphologically. As illustrated in Figure 12-1, rough ER

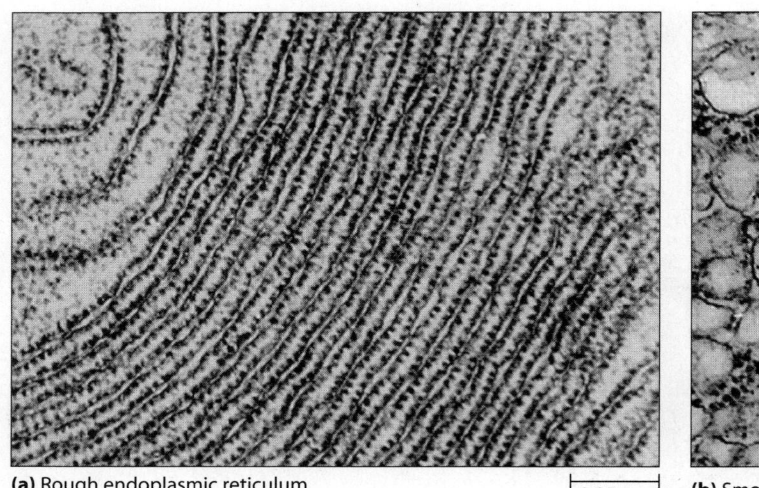

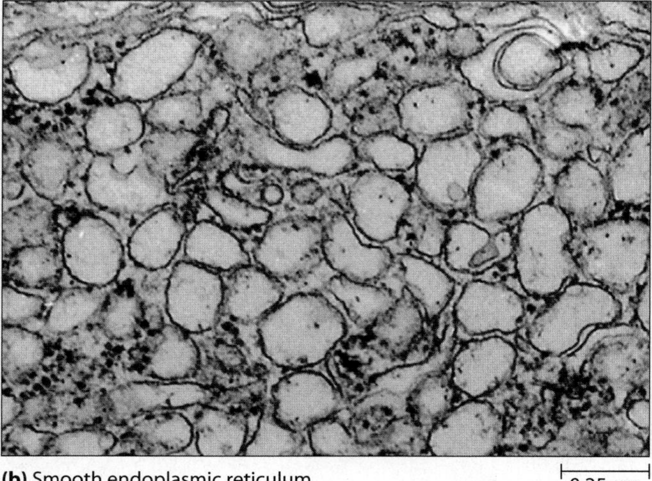

FIGURE 12-2 Rough and Smooth Endoplasmic Reticulum. **(a)** Electron micrograph of endoplasmic reticulum. The rough ER is studded with ribosomes. **(b)** Electron micrograph of smooth endoplasmic reticulum. The dark spots near the smooth ER appear to be glycogen granules (TEMs).

membranes usually form large flattened sheets, whereas smooth ER membranes generally form tubular structures. The transitional elements of the rough ER are an exception to this rule; they often resemble smooth ER. However, the rough and smooth ER are not separate organelles—electron micrographs and studies in living cells show that their lumenal spaces are continuous. Thus, material can travel between the rough and smooth ER without the aid of vesicles.

Both types of ER are present in most eukaryotic cells, but there is considerable variation in the relative amounts of each type, depending on the activities of the particular cell. Cells involved in the biosynthesis of secretory proteins, such as liver cells and cells producing digestive enzymes, tend to have very prominent rough ER networks. On the other hand, cells producing steroid hormones, such as in the testis or ovary, contain extensive networks of smooth ER.

When tissue is homogenized for subcellular fractionation, the ER membranes often break into smaller fragments that spontaneously close to form sealed vesicles known as **microsomes.** Fractions can be isolated with and without attached ribosomes, depending on whether the membrane originated from rough or smooth ER, respectively. Such preparations are tremendously useful for exploring both types of ER. Keep in mind, however, that microsomes do not exist in the cell; they are simply an artifact of the fractionation process. **Box 12A** presents more detailed information about subcellular fractionation by differential centrifugation.

Rough ER Is Involved in the Biosynthesis and Processing of Proteins

The ribosomes attached to the cytosolic side of the rough ER membrane are responsible for synthesizing both membrane-bound and soluble proteins for the endomembrane system. So how do proteins enter the ER lumen and the endomembrane system from their site of synthesis on the opposite (cytosolic) side of the rough ER membrane? Synthesis of proteins destined for the endomembrane system begins on cytoplasmic ribosomes, which attach to the rough ER via receptor proteins in the ER membrane shortly after translation initiation. Newly-synthesized proteins enter the endomembrane system *cotranslationally*—that is, they are inserted through a pore complex in the ER membrane into the rough ER lumen as the polypeptide is synthesized by the ER-bound ribosome (see Figure 22-16). After biosynthesis, membrane-spanning proteins remain anchored to the ER membrane either by hydrophobic regions of the polypeptide or by covalent attachment to membrane lipids. Soluble proteins, including secretory proteins, are released into the ER lumen.

In addition to its role in the biosynthesis of polypeptide chains, the rough ER is the site for several other processes, including the initial steps of addition and processing of carbohydrate groups to glycoproteins, the folding of polypeptides, the recognition and removal of misfolded polypeptides, and the assembly of multimeric proteins. Thus, ER-specific proteins include a host of enzymes that catalyze cotranslational and posttranslational modifications. These modifications include glycosylation, which is important for sorting of proteins to their proper destinations, and disulfide bond formation, which is essential for proper protein folding. We will discuss the topics of protein biosynthesis, targeting, and folding in more detail in Chapter 22.

The ER is also a site for quality control. In *ER-associated degradation (ERAD),* proteins improperly modified, folded, or assembled are exported from the ER for degradation by cytosolic *proteasomes* before they can move on to the Golgi complex. Several human diseases, including cystic fibrosis and familial hypercholesterolemia, are associated with defects in these processes.

BOX 12A TOOLS OF DISCOVERY

Centrifugation: An Indispensable Technique of Cell Biology

Centrifugation is a procedure used for the isolation and purification of organelles and macromolecules. This method is based on the fact that when a particle is subjected to centrifugal force by spinning a cellular extract at extremely rapid rates in a laboratory centrifuge, the rate of movement of the particle through a specific solution depends on its *size* and *density,* as well as the solution's density and viscosity. The larger or denser a particle is, the higher its **sedimentation rate,** or rate of movement through the solution. Centrifugation is used routinely in labs throughout the world to separate dissolved molecules from cell debris in suspensions of broken cells, to collect precipitated macromolecules such as DNA and proteins from cell suspensions, and to separate different cellular components based on their sedimentation rate.

Because most organelles and macromolecules differ significantly from one another in size and/or density, centrifuging a mixture of cellular components will separate the faster-moving components from the slower-moving ones. This procedure, called **subcellular fractionation,** enables researchers to isolate and purify specific organelles and macromolecules for further manipulation and study in vitro.

Albert Claude, George Palade, and Christian de Duve shared a Nobel Prize in 1974 for their pioneering work in centrifugation and subcellular fractionation. Claude played a key role in developing differential centrifugation as a method of isolating organelles. Palade was quick to use this technique in studies of the endoplasmic reticulum and the Golgi complex, establishing the roles of these organelles in the biosynthesis, processing, and secretion of proteins (see Figure 12-10). De Duve, in turn, used centrifugation in his discovery of two entirely new organelles—lysosomes and peroxisomes.

Centrifugation is also routinely used for isolating and purifying DNA and proteins. Following cell lysis, insoluble cell debris and organelles are removed by centrifugation, and the soluble cytoplasmic portion containing the DNA or protein is recovered. Treatment with the appropriate precipitating agent makes the DNA or protein insoluble so that it can be isolated in a second round of centrifugation.

Centrifuges and Sample Preparation

In essence, a **centrifuge** consists of a rotor—often housed in a refrigerated chamber—spun extremely rapidly by an electric motor. The rotor holds tubes containing solutions or suspensions of particles for fractionation. Centrifugation at very high speeds—above 20,000 revolutions per minute (rpm)—requires an **ultracentrifuge** equipped with a vacuum system to reduce friction between the rotor and air. Some ultracentrifuges reach speeds over 100,000 rpm, subjecting samples to forces exceeding 500,000 times the force of gravity (*g*).

Tissues must first undergo **homogenization,** or disruption, before the cellular components can be separated by centrifugation. To preserve the integrity of organelles, homogenization is usually done in a cold isotonic solution such as 0.25 *M* sucrose. Disruption can be achieved by grinding tissue in a mortar and pestle, by forcing cells through a narrow orifice (French press) or by subjecting tissue to ultrasonic vibration, osmotic shock, or enzymatic digestion. The resulting **homogenate** is a suspension of organelles, smaller cellular components, membranes, and molecules. If tissue is homogenized gently enough, most organelles and other structures remain intact and retain their original biochemical properties.

Differential Centrifugation

Differential centrifugation separates organelles based on their size and/or density differences. As illustrated in **Figure 12A-1**, particles that are large or dense (purple spheres) sediment rapidly, those that are intermediate in size or density (blue spheres) sediment less rapidly, and the smallest or least dense particles (black spheres) sediment very slowly.

We can express the relative size and/or density of an organelle or macromolecule in terms of its **sedimentation coefficient,** a measure of how rapidly the particle sediments when subjected to centrifugation. Sedimentation coefficients are expressed in **Svedberg units (S),** in honor of Theodor Svedberg, the Swedish chemist who developed the ultracentrifuge between 1920 and

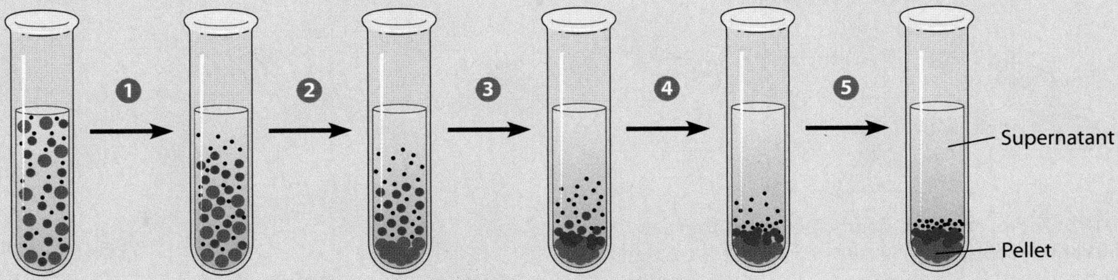

● Particles with large sedimentation coefficients
• Particles with intermediate sedimentation coefficients
· Particles with small sedimentation coefficients

FIGURE 12A-1 Differential Centrifugation. Differential centrifugation separates particles based on differences in sedimentation rate, which reflect differences in size and/or density. The technique is illustrated here for three particles that differ significantly in size. The particles, which are initially part of a homogeneous mixture, are subjected to a fixed centrifugal force for five successive time intervals (circled numbers). Particles that are large or dense (purple spheres) sediment rapidly, those that are intermediate in size or density (blue spheres) sediment less rapidly, and the smallest or least dense particles (black spheres) sediment very slowly. Eventually, all of the particles reach the bottom of the tube, unless the process is interrupted after a specific length of time.

1940. The sedimentation coefficients of some organelles, macromolecules, and viruses are shown in **Figure 12A-2**.

An example of differential centrifugation is illustrated in **Figure 12A-3**. The tissue of interest is first homogenized (step ❶). Subcellular fractions are then isolated by subjecting the homogenate and subsequent supernatant fractions to successively higher centrifugal forces and longer centrifugation times (steps ❷–❺). The **supernatant** is the clarified suspension of homogenate that remains after particles of a given size and density are removed as a **pellet** following each step of the centrifugation process. In each case, the supernatant from one step can be poured off into a new centrifuge tube and then returned to the centrifuge and subjected to greater centrifugal force to obtain the next pellet. In successive steps, the pellets are enriched in nuclei, unbroken cells, and debris (step ❷); mitochondria, lysosomes, and peroxisomes (step ❸); ER and other membrane fragments (step ❹); and free ribosomes and large macromolecules (step ❺). The material in each pellet can be resuspended and used for electron microscopy or biochemical studies. The final supernatant consists mainly of soluble cellular components such as DNA and protein.

Each fraction obtained in this way is enriched for the respective organelles but is also likely contaminated with other organelles and cellular components. Often, most of the contaminants in a pellet can be removed by resuspending the pellet in an isotonic solution and repeating the centrifugation procedure.

Density Gradient Centrifugation

In the previous example of differential centrifugation, the particles about to be separated were uniformly distributed throughout the

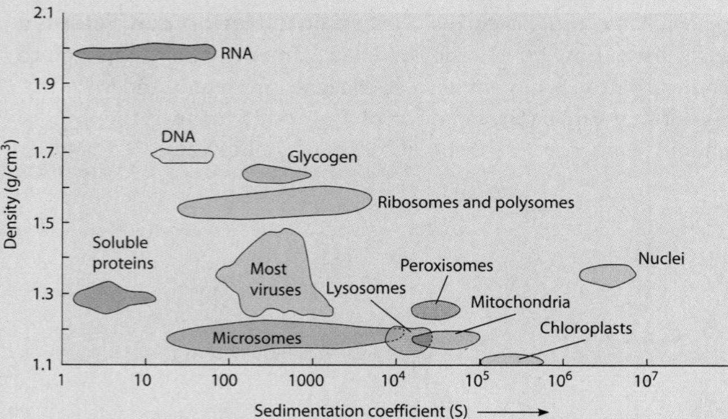

FIGURE 12A-2 Sedimentation Coefficients and Densities of Organelles, Macromolecules, and Viruses. A particle's sedimentation coefficient, expressed in Svedberg units (S), indicates how rapidly it sediments when subjected to a centrifugal force. A higher S value represents more rapid sedimentation. A particle's density (expressed in g/cm^3) determines how far it will move during equilibrium density centrifugation.

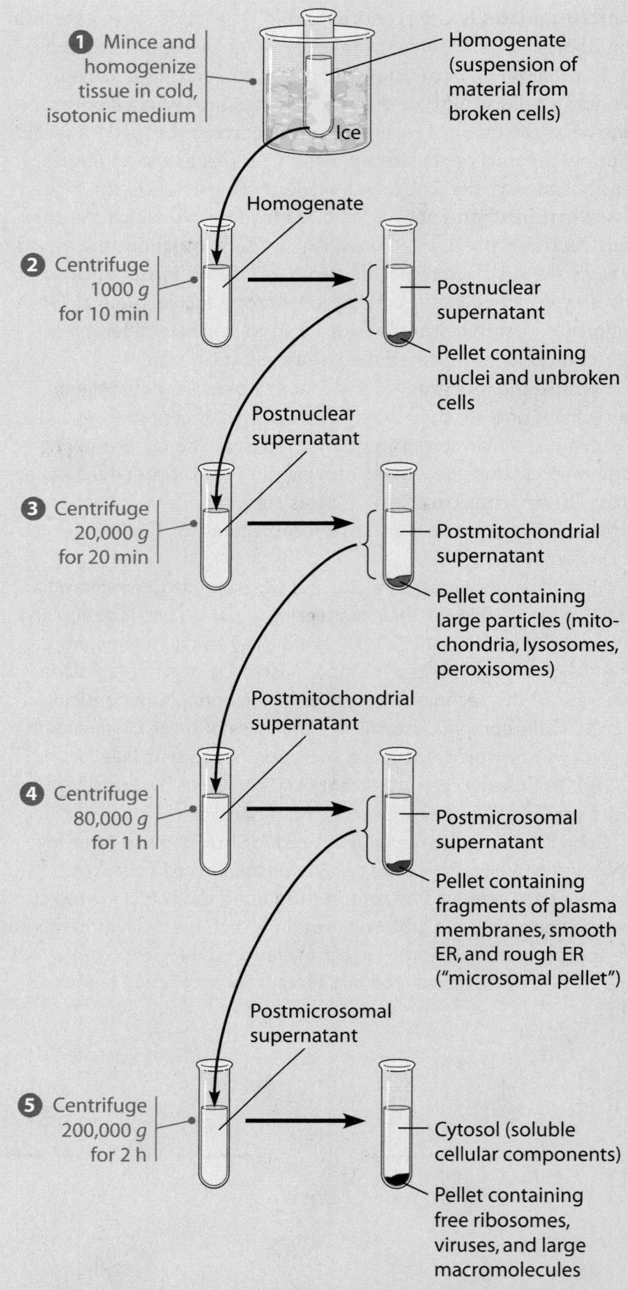

FIGURE 12A-3 Differential Centrifugation and the Isolation of Organelles. To isolate specific cellular components, homogenized tissue is subjected to a series of centrifugation steps. Each step uses the supernatant from the previous step, and subjects it to a higher *g* force for a longer time.

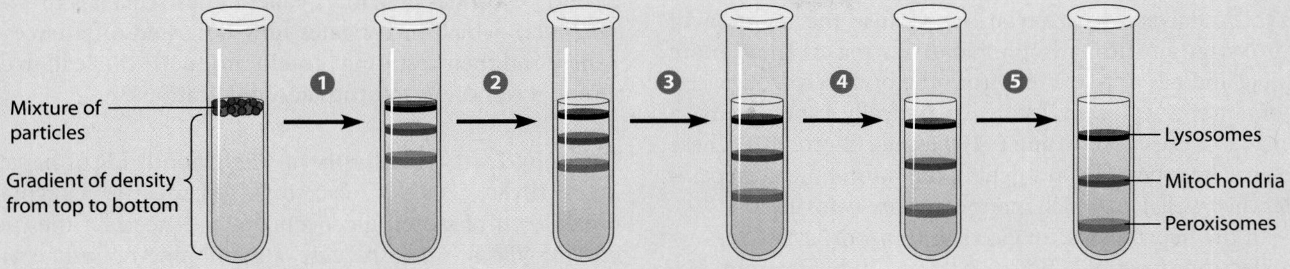

FIGURE 12A-4 Density Gradient Centrifugation. Density gradient centrifugation, like differential centrifugation (see Figure 12A-1), is a technique for separating particles such as organelles based on differences in sedimentation rate. For this centrifugation method, however, the sample for fractionation is placed as a thin layer on top of a gradient of solute that increases in density from the top of the tube to the bottom. The effect is illustrated here for three organelles that differ significantly in size and/or density. Subjected to a fixed centrifugal force for five successive time intervals (circled numbers), the organelles migrate through the gradient as distinct bands.

solution prior to centrifugation. **Density gradient** (or **rate-zonal**) **centrifugation** is a variation of differential centrifugation in which the sample for fractionation is placed as a thin layer on top of a *gradient of solute*. The gradient consists of an increasing concentration of solute—and therefore density—from the top of the tube to the bottom. When subjected to a centrifugal force, particles differing in size and/or density move downward as discrete *zones*, or bands, that migrate at different rates. Due to the gradient of solute in the tube, the particles at the leading edge of each zone continually encounter a slightly denser solution and are thus slightly impeded. As a result, each zone remains very compact, maximizing the ability to separate different particles.

This process is illustrated in **Figure 12A-4**. Particles that are large or dense (purple) move into the gradient as a rapidly sedimenting band, particles that are intermediate in size or density (blue) sediment less rapidly, and the smallest particles (black) move quite slowly. The centrifuge is stopped after the bands of interest have moved far enough into the gradient to be resolved from each other but *before* any of the bands reach the bottom of the tube. Stopping at this point is essential because all of the particles are denser than the solution in the tube, even at the very bottom of the tube. If centrifugation continues too long, the bands will reach the bottom of the tube one after another, piling up and negating the very purpose of the process.

Density gradient centrifugation can be used to separate lysosomes, mitochondria, and peroxisomes, each of which has a slightly different density. The tissue of interest is first homogenized, and a pellet enriched in these organelles is layered over a gradient of solute that increases in concentration and density from the top of a tube to the bottom. As shown in Figure 12A-4, the dense peroxisomes move into the gradient as a rapidly sedimenting band. The mitochondria, having intermediate density, form a band above the peroxisomes. The lysosomes are the smallest and least dense and form a band nearer the top of the tube. By assaying each fraction for *marker enzymes* that are unique to each organelle, the fractions containing these organelles can be identified and the extent of cross-contamination can be determined.

Equilibrium Density Centrifugation

Equilibrium density (or **buoyant density**) **centrifugation** is a related method for resolving organelles and macromolecules based on density differences (see Figure 12A-2). Like density gradient centrifugation, this procedure includes a gradient of solute that increases in concentration and density, but in this case the solute is concentrated so that the density gradient spans the range of densities of the organelles or macromolecules about to be separated. Therefore, the material to be purified does not move to the bottom of the tube during centrifugation but instead forms a tight band in a stable equilibrium position where its density is exactly equal to the density of the gradient at that position.

For organelle separation using equilibrium density centrifugation, a gradient of sucrose is often used, and the density range is 1.10–1.30 g/cm^3 (0.75–2.3 *M* sucrose). This method can also be used to separate different forms of DNA and RNA based on their differing densities. It was used to determine the mechanism of replication of DNA in the classic experiments of Meselson and Stahl (see Figure 19-3).

Smooth ER Is Involved in Drug Detoxification, Carbohydrate Metabolism, Calcium Storage, and Steroid Biosynthesis

Drug Detoxification. Drug detoxification often involves enzyme-catalyzed **hydroxylation** because the addition of hydroxyl groups to hydrophobic drugs makes them more soluble and easier to excrete from the body. Hydroxylation of organic acceptor molecules is typically catalyzed by a member of the **cytochrome P-450** family of proteins. These proteins are especially prevalent in the smooth ER of hepatocytes (liver cells), in which many drugs are detoxified.

In the hepatocytes, an electron transport system transfers electrons from NADPH or NADH to a heme group in a cytochrome P-450 protein, which then donates an electron to molecular oxygen. One atom of molecular oxygen gains two electrons and two H^+, forming H_2O. The other oxygen atom is added to the organic substrate molecule as part of a hydroxyl group. Because one of the two oxygen atoms of O_2 is incorporated into the reaction product, these cytochrome P-450 enzymes are often called *monooxygenases.* The net reaction is shown below, where R represents the organic hydroxyl acceptor:

$$\text{RH} + \text{NAD(P)H} + \text{H}^+ + \text{O}_2 \longrightarrow \text{ROH} + \text{NAD(P)}^+ + \text{H}_2\text{O} \qquad \textbf{(12-1)}$$

The elimination of hydrophobic barbiturate drugs, for example, is enhanced by hydroxylation enzymes in the smooth ER. Injection of the sedative phenobarbital into a rat causes a rapid increase in the level of barbiturate-detoxifying enzymes in the liver, accompanied by a dramatic proliferation of smooth ER. However, this means that increasingly higher doses of the drug are necessary to achieve the same sedative effect, an effect known as *tolerance* that is seen in habitual users of phenobarbital. Furthermore, the enzyme induced by phenobarbital can hydroxylate and therefore solubilize a variety of other drugs, including such useful agents as antibiotics, anticoagulants, and steroids. As a result, the chronic use of barbiturates decreases the effectiveness of many other clinically useful drugs.

Another cytochrome P-450 protein found in the smooth ER is part of an enzyme complex called *aryl hydrocarbon hydroxylase.* This complex is involved in metabolizing *polycyclic hydrocarbons,* organic molecules composed of two or more linked benzene rings that are often toxic. Hydroxylation of such molecules is important for increasing their solubility in water, but the oxidized products are often more toxic than the original compounds. Aryl hydrocarbon hydroxylase converts some potential carcinogens into their chemically active forms. Mice synthesizing high levels of this hydroxylase have a higher incidence of spontaneous cancer than normal mice do, whereas mice treated with an inhibitor of aryl hydrocarbon hydrolase develop few tumors. Significantly, cigarette smoke is a potent inducer of aryl hydrocarbon hydroxylase.

Recent work shows that differences in activities and side effects of certain medications can result from differences in the presence or activity of particular cytochrome P-450 genes in different patients. This has led to a new field of study known as **pharmacogenetics** (also called *pharmacogenomics*), which investigates how inherited differences in genes (and their resulting protein products) can lead to differential responses to drugs and medications.

Carbohydrate Metabolism. The smooth ER of hepatocytes (liver cells) is also involved in the enzymatic breakdown of stored glycogen, as evidenced by the presence of *glucose-6-phosphatase,* a membrane-bound enzyme that is unique to the ER. Thus, its presence is used as a marker to identify the ER during subcellular fractionation or to visualize the ER using fluorescent antibodies. Glucose-6-phosphatase hydrolyzes the phosphate group from glucose-6-phosphate to form free glucose and inorganic phosphate (P_i):

$$\text{glucose-6-phosphate} + \text{H}_2\text{O} \longrightarrow \text{glucose} + \text{P}_\text{i} \qquad \textbf{(12-2)}$$

This enzyme is abundant in the liver because a major role of the liver is to keep the level of glucose in the blood relatively constant. The liver stores glucose as glycogen in granules associated with smooth ER (**Figure 12-3a**). When glucose is needed by the body, especially between meals and in response to increased muscular activity, liver glycogen is broken down by phosphorolysis (see Figure 9-10), producing glucose-6-phosphate (Figure 12-3b). Because membranes are generally impermeable to phosphorylated sugars, the glucose-6-phosphate must be converted to free glucose by glucose-6-phosphatase in order to leave the cell and enter the bloodstream. Free glucose then leaves the liver cell via a glucose transporter (GLUT2) and moves into the blood for transport to other cells that need energy. Significantly, glucose-6-phosphatase activity is present in liver, kidney, and intestinal cells but not in muscle or brain cells. Muscle and brain cells retain glucose-6-phosphate and use it to meet their own substantial energy needs.

Calcium Storage. The *sarcoplasmic reticulum* found in muscle cells is an example of smooth ER that specializes in the storage of calcium. In these cells, the ER lumen contains high concentrations of calcium-binding proteins. Calcium ions are pumped into the ER by *ATP-dependent calcium ATPases* and are released in response to extracellular signals to aid in muscle contraction (see Figure 14-12). Binding of neurotransmitter molecules to receptors on the surface of the muscle cell triggers a signal cascade that leads to the release of calcium from the sarcoplasmic reticulum and causes the contraction of muscle fibers. We will discuss nerve impulse transmission and muscle contraction in more detail in Chapters 13 and 16.

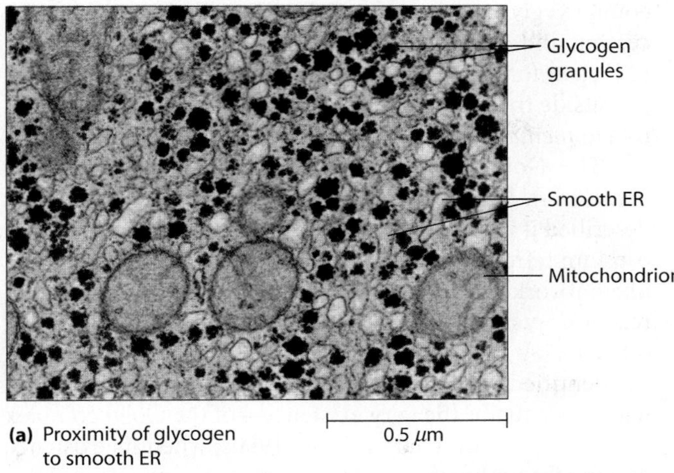

(a) Proximity of glycogen to smooth ER

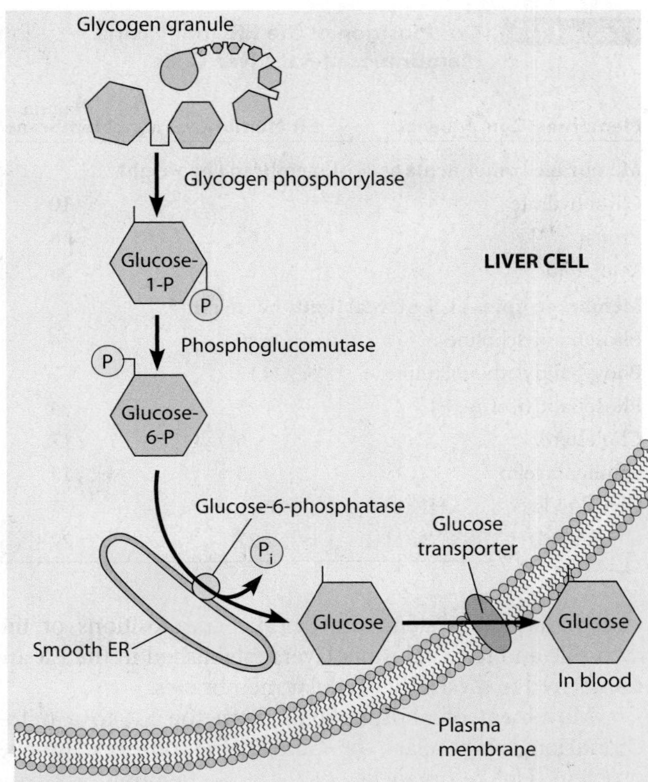

(b) Process of glycogen breakdown in liver

FIGURE 12-3 The Role of the Smooth ER in the Catabolism of Liver Glycogen. **(a)** This electron micrograph of a monkey liver cell shows numerous granules of glycogen closely associated with smooth ER (TEM). **(b)** The breakdown of liver glycogen involves the stepwise removal of glucose units as glucose-1-phosphate, followed by the conversion of glucose-1-phosphate to glucose-6-phosphate by enzymes in the cytosol. Removal of the phosphate group depends on glucose-6-phosphatase, an enzyme associated with the smooth ER membrane. Free glucose is then transported out of the liver cell into the blood by a glucose transporter in the plasma membrane.

Steroid Biosynthesis. The smooth ER in certain cells is the site of biosynthesis of cholesterol and steroid hormones such as cortisol, testosterone, and estrogen. Large amounts of smooth ER are found in the cortisol-producing cells of the adrenal gland; the Leydig cells of the testes, which produce testosterone; the cholesterol-producing cells of the liver; and the follicular cells of the ovary, which produce estrogen. Smooth ER has also been found in close association with plastids in some plants, where it may be involved in phytohormone synthesis.

Cholesterol, cortisol, and the male and female steroid hormones just described share a common four-ring structure but differ in the number and arrangement of carbon side chains and hydroxyl groups (see Figure 3-27e and Figure 3-30). *Hydroxymethylglutaryl-CoA reductase* (*HMG-CoA reductase*), the committed step in cholesterol biosynthesis, is present in large amounts in the smooth ER of liver cells. This enzyme is targeted for inhibition by a class of cholesterol-lowering drugs known as *statins.* In addition, the smooth ER contains a number of P-450 monooxygenases that are important not only in the synthesis of cholesterol but also in its conversion into steroid hormones by hydroxlyation.

The ER Plays a Central Role in the Biosynthesis of Membranes

In eukaryotic cells, the ER is the primary source of membrane lipids, including phospholipids and cholesterol. Indeed, most of the enzymes required for the biosynthesis of membrane phospholipids are found nowhere else in the cell. There are, however, important exceptions. Mitochondria synthesize phosphatidylethanolamine by decarboxylating imported phosphatidylserine. Peroxisomes have enzymes to synthesize cholesterol, and chloroplasts contain enzymes for the synthesis of chloroplast-specific lipids.

Biosynthesis of fatty acids for membrane phospholipid molecules occurs in the cytoplasm and incorporation is restricted to the monolayer of the ER membrane facing the cytosol. Cellular membranes, of course, are phospholipid *bilayers,* with phospholipids distributed to both sides. Thus, there must be a mechanism for transferring phospholipids from one layer of the membrane to the other. Because it is thermodynamically unfavorable for phospholipids to flip spontaneously at a significant rate from one side of a bilayer to the other, transfer depends on **phospholipid translocators,** also called **flippases,** which catalyze the translocation of phospholipids through ER membranes (see Figure 7-10).

Phospholipid translocators, like other enzymes, are quite specific and affect only the rate of a process. Therefore, the type of phospholipid molecules transferred across a membrane depends on the particular translocators present, contributing to the *membrane asymmetry* described in Chapter 7. For example, the ER membrane contains a translocator for phosphatidylcholine, and thus it is found in both monolayers of the ER membrane. In contrast, there is no translocator for phosphatidylethanolamine, phosphatidylinositol, or phosphatidylserine, which are therefore confined to the cytosolic monolayer. When vesicles from the ER membrane fuse with other organelles of the

Table 12-1 Composition of the ER and Plasma Membranes of Rat Liver Cells

Membrane Components	ER Membrane	Plasma Membrane
Membrane components as % of membrane by weight		
Carbohydrate	10	10
Protein	62	54
Total lipid	27	36
Membrane lipids as % of total lipids by weight		
Phosphatidylcholine	40	24
Phosphatidylethanolamine	17	7
Phosphatidylserine	5	4
Cholesterol	6	17
Sphingomyelin	5	19
Glycolipids	trace	7
Other lipids	27	22

endomembrane system, the distinct compositions of the cytosolic and lumenal monolayers established in the ER are transferred to these other cellular membranes.

Movement of phospholipids from the ER to a mitochondrion, chloroplast, or peroxisome poses a unique problem. Unlike organelles of the endomembrane system, these organelles do not grow by fusion with ER-derived vesicles. Instead, cytosolic **phospholipid exchange proteins** (also called *phospholipid transfer proteins*) convey phospholipid molecules from the ER membrane to the outer mitochondrial and chloroplast membranes. Each exchange protein recognizes a specific phospholipid, removes it from one membrane, and carries it through the cytosol to another membrane. Such transfer proteins also contribute to the movement of phospholipids from the ER to other cellular membranes, including the plasma membrane.

Although the ER is the source of most membrane lipids, the compositions of other cellular membranes vary significantly from the composition of the ER membrane (**Table 12-1**). A striking feature of the plasma membrane of hepatocytes is the relatively low amount of phosphoglycerides and high amounts of cholesterol, sphingomyelin, and glycolipids. Researchers have observed an increasing gradient of cholesterol content from the ER through the compartments of the endomembrane system to the plasma membrane. This correlates with an increasing gradient of membrane thickness. ER membranes are about 5 nm thick, whereas plasma membranes are about 8 nm thick. The observed change in membrane thickness has implications for sorting and targeting integral membrane proteins, which we will discuss after we look at the Golgi complex and its role in protein processing.

The Golgi Complex

We now turn our attention to the Golgi complex, a component of the endomembrane system that is closely linked, both physically and functionally, to the ER. In the Golgi complex, glycoproteins from the ER undergo further processing and, along with membrane lipids, are sorted and packaged for transport to their proper destinations inside or outside the cell. Thus, the Golgi complex plays a central role in *membrane* and *protein trafficking* in eukaryotic cells.

The **Golgi complex** (or *Golgi apparatus*) derives its name from Camillo Golgi, the Italian biologist who first described it in 1898. He reported that nerve cells soaked in osmium tetroxide showed deposits of osmium in a thread-like network surrounding the nucleus. The same staining reaction was demonstrated with a variety of cell types and other heavy metals. However, no cellular structure could be identified that explained the staining. As a result, the nature—actually, the very existence—of the Golgi complex remained controversial until the 1950s, when its existence was finally confirmed by electron microscopy.

The Golgi Complex Consists of a Series of Membrane-Bounded Cisternae

The Golgi complex is a series of flattened membrane-bounded *cisternae,* disk-shaped sacs that are stacked together as illustrated in **Figure 12-4a**. A series of such cisternae is called a *Golgi stack* and can be visualized by electron microscopy (Figure 12-4b). Usually, there are 3–8 cisternae per stack, though the number and size of Golgi stacks vary with the cell type and with the metabolic activity of the cell. Some cells have one large stack, whereas others—especially active secretory cells—have hundreds or even thousands of Golgi stacks.

The static view of the ER and the Golgi complex presented by electron micrographs such as Figure 12-4b can be misleading. These organelles are actually dynamic structures. Both the ER and Golgi complex are typically surrounded by numerous *transport vesicles* that carry lipids and proteins from the ER to the Golgi complex, between the cisternae of a Golgi stack, and from the Golgi complex to various destinations in the cell, including endosomes, lysosomes, and secretory granules. Thus, the Golgi complex lumen, or *intracisternal space,* is part of the endomembrane system's network of internal spaces (see Figure 12-1).

The Two Faces of the Golgi Stack. Each Golgi stack has two distinct sides, or *faces* (Figure 12-4). The *cis face* is oriented toward the ER. The Golgi compartment closest to the ER is a network of flattened, membrane-bounded tubules referred to as the ***cis*-Golgi network (CGN).** Vesicles containing newly synthesized lipids and proteins from the ER continuously arrive at the CGN, where they fuse with CGN membranes. The opposite side of the Golgi complex is called the *trans face.* The compartment on this side of the Golgi complex has similar morphology and is referred to as the ***trans*-Golgi network (TGN).** Here, proteins and lipids leave the Golgi in **transport vesicles** that continuously bud from the tips of TGN cisternae. These transport vesicles carry lipids and proteins from the Golgi complex to secretory granules, endosomes, lysosomes, and the plasma membrane. The central sacs between the CGN

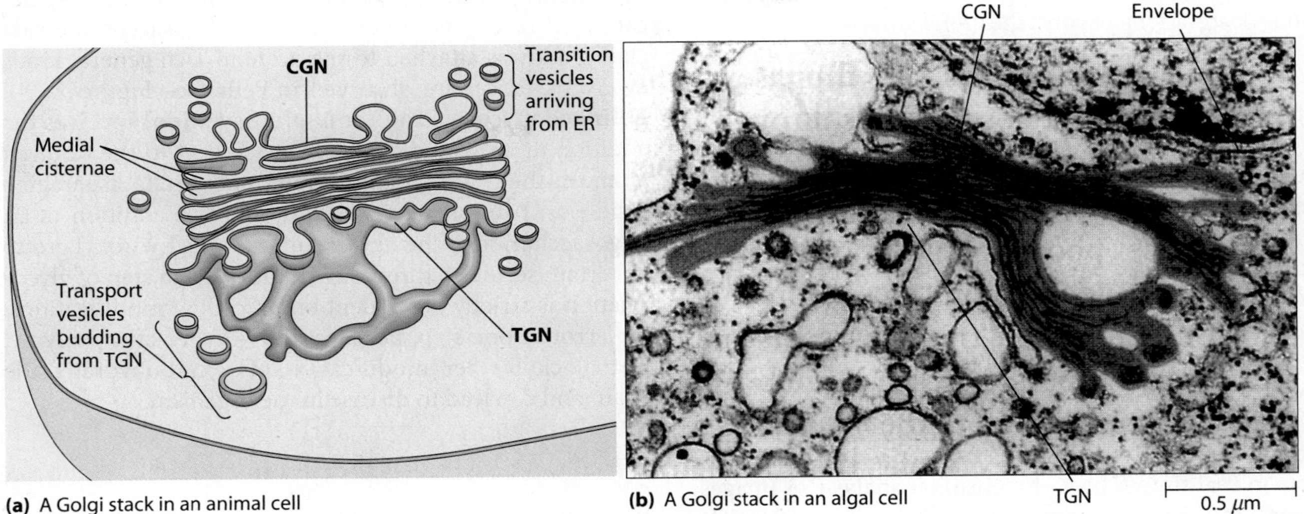

FIGURE 12-4 Golgi Structure. A Golgi stack consists of a small number of flattened cisternae. **(a)** At the *cis* face, transition vesicles arriving from the ER fuse with membranes of the *cis*-Golgi network (CGN). At the *trans* face, transport vesicles arise by budding from the *trans*-Golgi network (TGN). The transport vesicles carry lipids and proteins to other components of the endomembrane system or form secretory vesicles. **(b)** This electron micrograph shows a Golgi stack lying next to the nuclear envelope of an algal cell (TEM).

and TGN comprise the **medial cisternae** of the Golgi stack, in which much of the processing of proteins occurs.

The CGN, TGN, and medial cisternae of the Golgi complex are biochemically and functionally distinct. Each compartment contains specific receptor proteins and enzymes necessary for specific steps in protein and membrane processing, as shown by immunological and cytochemical staining techniques. This biochemical *polarity* is illustrated in **Figure 12-5**, which shows a Golgi stack in a rabbit kidney cell. Staining to detect *N-acetylglucosamine transferase I,* an enzyme that modifies carbohydrate side chains of glycoproteins, shows that the enzyme is concentrated in medial cisternae of the Golgi complex.

Two Models Depict the Flow of Lipids and Proteins Through the Golgi Complex

Two models have been proposed to explain the movement of lipids and proteins from the CGN to the TGN via the medial cisternae of the Golgi complex. According to the **stationary cisternae model,** each compartment of the Golgi stack is a stable structure. Trafficking between successive cisternae is mediated by *shuttle vesicles* that bud from one cisterna and fuse with the next cisterna in the *cis*-to-*trans* sequence. Proteins destined for the TGN are simply carried forward by shuttle vesicles, while molecules that belong in the ER and successive Golgi compartments are actively retained or retrieved.

According to the second model, known as the **cisternal maturation model,** the Golgi cisternae are transient compartments that gradually change from CGN cisternae through medial cisternae to TGN cisternae. In this model, transition vesicles from the ER converge to form the CGN, which accumulates specific enzymes for the early steps of protein processing. Step by step, each *cis* cisterna is transformed first into an intermediate medial cisterna and then into a *trans* cisterna as it acquires additional enzymes. Enzymes that are no longer needed in late compartments return in vesicles to early compartments. In both models, the TGN forms transport vesicles or

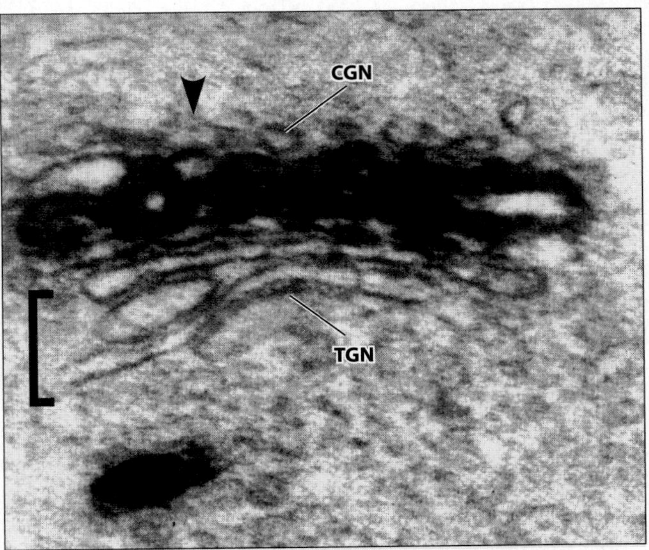

FIGURE 12-5 Immunochemical Staining of a Golgi Complex. This electron micrograph shows a Golgi stack in a rabbit kidney cell. The cell section has been stained to detect the enzyme *N-acetylglucosamine transferase I,* which plays a role in terminal glycosylation of proteins. The arrow and bracket indicate unlabeled cisternae, showing that the enzyme is concentrated in a few medial cisternae close to the *cis* face of the Golgi stack. This indicates that *N*-acetylglucosamine is added to existing oligosaccharides on glycoproteins shortly after the proteins enter the Golgi stack (TEM).

secretory granules containing sorted cargo targeted for various destinations beyond the Golgi complex.

Experimental results suggest that these two models are not necessarily mutually exclusive. It is likely that both apply to some degree, depending on the organism and the role of the cell. While the stationary cisternae model is supported by substantial evidence, some cellular components observed in medial compartments of the Golgi complex are clearly too large to travel by the small shuttle vesicles found in cells. For example, polysaccharide scales produced by some algae first appear in early Golgi compartments. Too large to fit inside transport vesicles, the scales nevertheless reach late Golgi compartments on their way to the plasma membrane for incorporation into the cell wall.

Recently, time-lapse fluorescence microscopy has been used in live yeast cells to study individual Golgi cisternae in real time. Three-dimensional analysis of images supports the cisternal maturation model and suggests that the cisternae mature at a constant rate. In addition, the rate of movement of labeled secretory proteins through the Golgi complex was measured and was shown to match the rate of cisternal maturation.

Anterograde and Retrograde Transport. The movement of material from the ER through the Golgi complex toward the plasma membrane is called **anterograde transport** (*antero* is derived from a Latin word meaning "front," and *grade* is related to a word meaning "step"). Every time a secretory granule fuses with the plasma membrane and discharges its contents by exocytosis, a bit of membrane that originated in the ER becomes a part of the plasma membrane. To balance the flow of lipids toward the plasma membrane and to ensure a supply of components for forming new vesicles, the cell recycles lipids and proteins no longer needed during the late stages of anterograde transport. This is accomplished by **retrograde transport** (*retro* is a Latin word meaning "back"), the flow of vesicles from Golgi cisternae back toward the ER.

In the stationary cisternae model, retrograde flow facilitates both the recovery of ER-specific lipids and proteins that are passed from the ER to the CGN and the transport of compartment-specific proteins back to distinct medial cisternae of the Golgi stack. Material destined for the TGN continues forward. Whether such retrograde traffic occurs directly from all medial cisternae of the Golgi stack back to the ER or by reverse flow through successive cisternae is not yet clear. In the cisternal maturation model, retrograde flow carries material back toward newly forming compartments after receptors and enzymes are no longer needed in the more mature compartments.

Roles of the ER and Golgi Complex in Protein Glycosylation

Much of the protein processing carried out within the ER and Golgi complex involves **glycosylation**—the addition of carbohydrate side chains to specific amino acid residues of proteins, forming **glycoproteins.** Subsequent enzyme-catalyzed reactions then modify the oligosaccharide side chain that was attached to the protein. Two general kinds of glycosylation are observed in cells (see Figure 7-25). **N-linked glycosylation** (or **N-glycosylation**) involves the addition of a specific oligosaccharide unit to the *nitrogen* atom on the terminal amino group of certain asparagine residues. **O-linked glycosylation** involves addition of an oligosaccharide to the *oxygen* atom on the hydroxyl group of certain serine or threonine residues. Each step of glycosylation is strictly dependent on preceding modifications. An error at one step, perhaps due to a defective enzyme, can block further modification of a carbohydrate side chain and can lead to disease in the organism.

Initial Glycosylation Occurs in the ER

We will focus here on N-glycosylation. **Figure 12-6** describes the steps of glycosylation that may occur as a glycoprotein travels from the ER to the CGN and through the Golgi complex to the TGN. Note that specific enzymes that catalyze various steps of glycosylation and subsequent modifications are present in specific compartments of the ER and Golgi complex. The initial steps of N-glycosylation take place on the cytosolic surface of the ER membrane and later steps occur in the ER lumen (**Figure 12-7**). Despite the variety of oligosaccharides found in mature glycoproteins, all the carbohydrate side chains added to proteins in the ER initially have a common **core oligosaccharide** consisting of two units of N-acetylglucosamine (GlcNAc, see Figure 3-26a), nine mannose units, and three glucose units.

Glycosylation begins as *dolichol phosphate,* an oligosaccharide carrier, is inserted into the ER membrane (Figure 12-7, step ❶). GlcNAc and mannose groups are then added to the phosphate group of dolichol phosphate (step ❷). The growing core oligosaccharide is then translocated from the cytosol to the ER lumen by a *flippase* (step ❸). Once inside the ER lumen, more mannose and glucose units are added (step ❹). The completed core oligosaccharide is then transferred as a single unit from dolichol to an asparagine residue of the recipient protein (step ❺). Finally, the core oligosaccharide attached to the protein is trimmed and modified (step ❻).

Usually, the core oligosaccharide is added to the protein as the polypeptide is being synthesized by a ribosome bound to the ER membrane. We know that this *cotranslational glycosylation* helps to promote proper protein folding because experimental inhibition of glycosylation leads to the appearance of misfolded, aggregated proteins. Addition of a single glucose unit allows other ER proteins to interact with the newly synthesized glycoprotein to ensure its proper folding. One of two ER proteins known as **calnexin** (CNX, membrane-bound) and **calreticulin** (CRT, soluble) can bind to the monoglucosylated glycoprotein and promote disulfide bond formation by forming a complex with the glycoprotein and a thiol oxidoreductase known as *ERp57,* which catalyzes disulfide bond formation. The

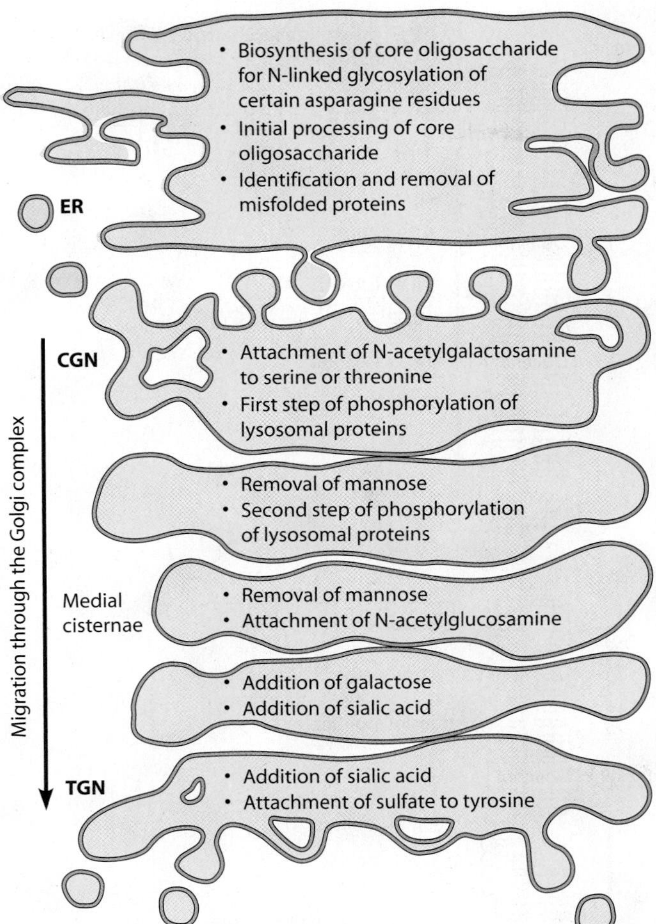

FIGURE 12-6 Compartmentalization of the Steps of Glycosylation and Subsequent Modification of Proteins. Enzymes that catalyze specific steps of glycosylation and further modification of proteins reside in different compartments of the ER and Golgi complex. Processing occurs sequentially as proteins travel from compartment to compartment. The steps listed in the figure are examples of potential modifications and do not necessarily occur with all glycoproteins.

protein complex then dissociates, and the final glucose unit is removed by an enzyme named *glucosidase II.*

At this point, a specific glucosyl transferase in the ER known as *UGGT (UDP-glucose:glycoprotein glucotransferase)* acts as a sensor for proper folding of the newly synthesized glycoprotein. UGGT binds to improperly folded proteins and adds back a single glucose unit, making the protein a substrate for another round of CNX/CRT binding and disulfide bond formation. Once the proper conformation is achieved, UGGT no longer binds the new glycoprotein, which is then free to exit the ER and move to the Golgi.

Further Glycosylation Occurs in the Golgi Complex

Further processing of N-glycosylated proteins happens in the Golgi complex as the glycoproteins move from the *cis* face through the medial cisternae to the *trans* face of the Golgi stack. These **terminal glycosylations** in the Golgi show remarkable variability among proteins and account for much of the great diversity in structure and function of protein oligosaccharide side chains.

Terminal glycosylation always includes the removal of a few of the carbohydrate units of the core oligosaccharide. In some cases, no further processing occurs in the Golgi complex. In other cases, more complex oligosaccharides are generated by the further addition of GlcNAc and other monosaccharides, including galactose, sialic acid, and fucose (see Figure 7-26a). Some glycoproteins contain galactose units that are added by *galactosyl transferase,* a marker enzyme unique to the Golgi.

Given the role of the Golgi complex in glycosylation, it is not surprising that two of the most important categories of enzymes present in Golgi stacks are *glucan synthetases,* which produce oligosaccharides from monosaccharides, and *glycosyl transferases,* which attach carbohydrate groups to proteins. The ER and Golgi complex contain hundreds of different glycosyl transferases, which indicates the potential complexity of oligosaccharide side chains. Within the Golgi stack, each cisterna contains a distinctive set of processing enzymes.

Notice in the preceding discussion that the mature oligosaccharides in glycoproteins are found only on the lumenal side of the ER and Golgi complex membranes and thus contribute to membrane asymmetry. Because the lumenal side of the ER membrane is topologically equivalent to the exterior surface of the cell, it is easy to see why all plasma membrane glycoprotein oligosaccharides are found on the extracellular side of the membrane.

Roles of the ER and Golgi Complex in Protein Trafficking

Membrane-bound and soluble proteins synthesized in the rough ER must be directed to a variety of intracellular locations, including the ER itself, the Golgi complex, endosomes, and lysosomes. Moreover, once a protein reaches an organelle where it is to remain, there must be a mechanism for preventing it from leaving. Other groups of proteins synthesized in the rough ER are destined for incorporation into the plasma membrane or for release to the outside of the cell. Therefore, each protein contains a specific "tag" targeting the protein to a transport vesicle that will carry material from one specific cellular location to another. Depending on the protein and its destination, the tag may be a short amino acid sequence, an oligosaccharide side chain, a hydrophobic domain, or some other structural feature. Tags may also be involved in excluding material from certain vesicles.

Membrane lipids may also be tagged to help vesicles reach their proper destinations. This tag can be one or more phosphate groups attached to positions 3, 4, and/or 5 of a membrane phosphatidylinositol (PI) molecule by a specific kinase. For example, a functional PI 3-kinase is required for proper sorting of vesicles to the vacuole in yeast. In mammalian cells, inhibition of inositol kinases

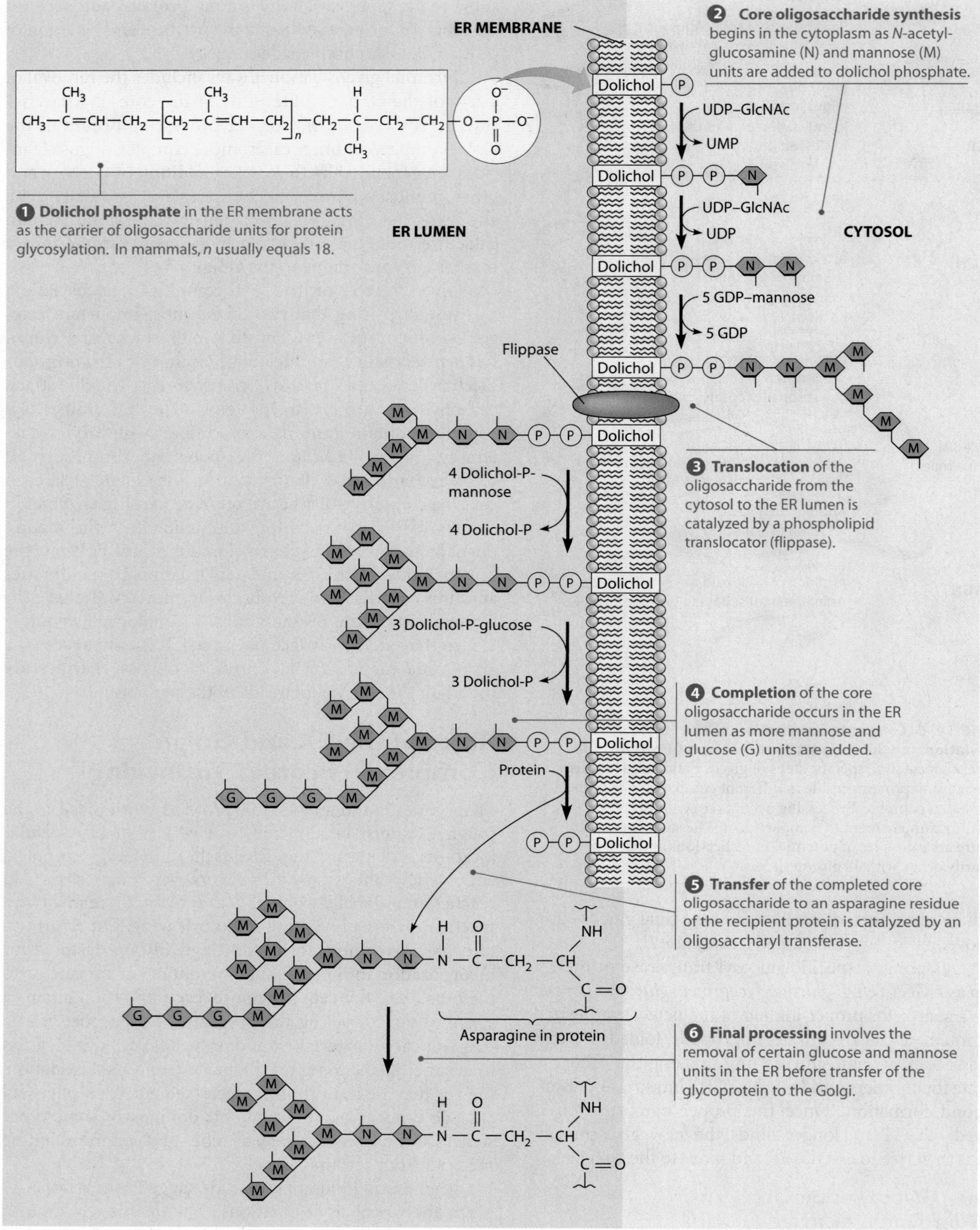

FIGURE 12-7 N-linked Glycosylation of Proteins in the ER. Synthesis of core oligosaccharides begins in the cytoplasm, using a dolichol phosphate molecule as a carrier. The partially synthesized oligosaccharide is translocated to the ER lumen, where additional monosaccharides are added. The completed oligosaccharide is then transferred to the target protein, and several monosaccharides are removed in final processing.

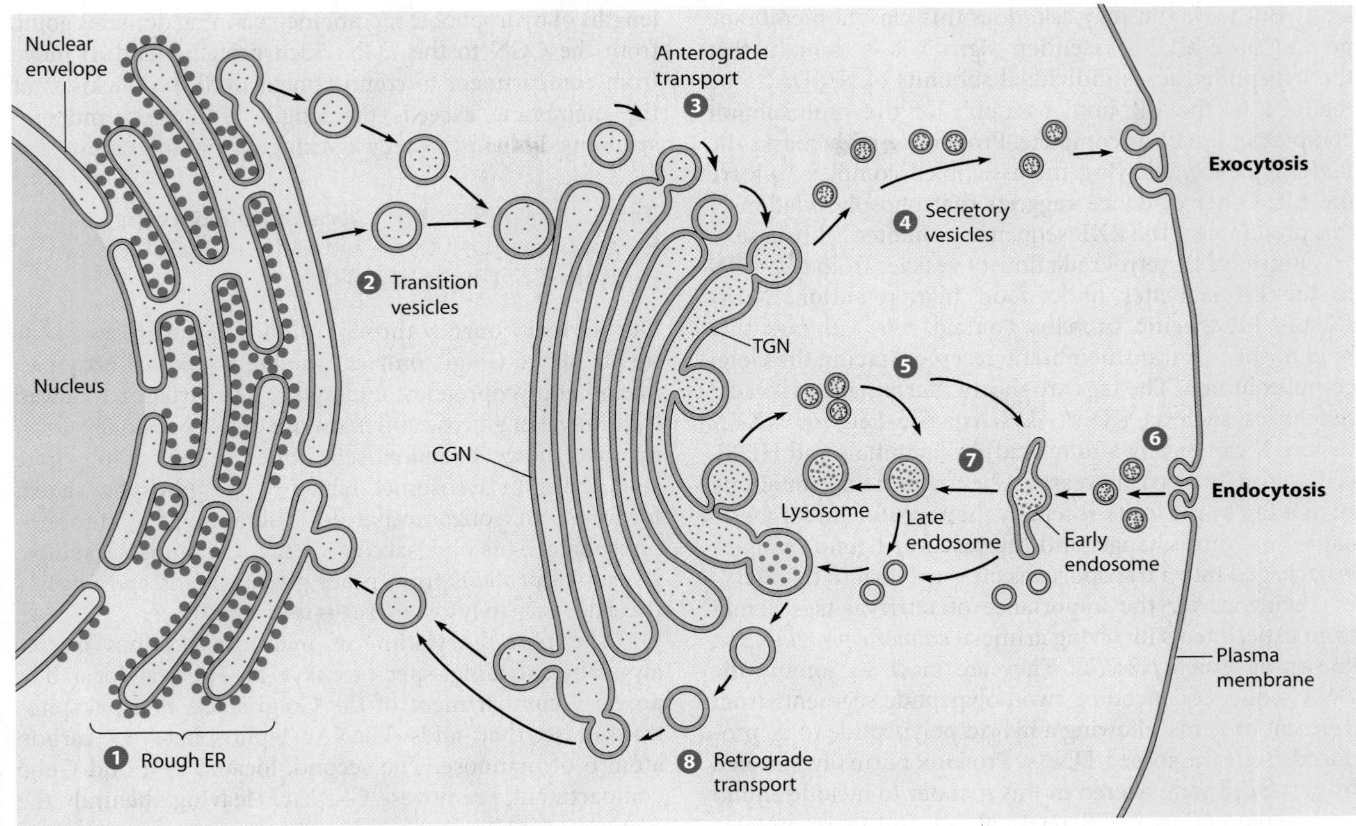

FIGURE 12-8 Trafficking Through the Endomembrane System. Vesicles carry lipids and proteins along several routes from the ER through the Golgi complex to various destinations, including secretory vesicles, endosomes, and lysosomes. ❶ As ribosomes of the rough ER synthesize proteins, the proteins enter the ER lumen, where initial glycosylation steps occur. ❷ Transition vesicles carry glycosylated proteins (and newly synthesized lipids) to the CGN. ❸ Lipids and proteins move through the cisternae of the Golgi stack via shuttle vesicles and cisternae mature as new enzymes are added via vesicles. ❹ At the TGN, some vesicles bud off to form secretory vesicles, which move to the plasma membrane and release their contents by exocytosis. ❺ Other vesicles bud from the TGN to form endosomes that, in turn, help to make lysosomes. ❻ Proteins and other materials are taken into the cell by endocytosis, forming endocytic vesicles that fuse with early endosomes. ❼ Early endosomes containing material for digestion mature to form late endosomes and then lysosomes. ❽ Retrograde traffic returns compartment-specific proteins to earlier compartments.

disrupts vesicle trafficking to the lysosome. Aside from specific tags on some lipids, the length and degree of saturation of certain membrane lipids have also been shown to be important in vesicle trafficking.

An overview of the trafficking involving the ER and Golgi complex is presented in **Figure 12-8**. Sorting of proteins begins in the ER and early compartments of the Golgi stack, which contain mechanisms for retrieving or retaining compartment-specific proteins. This important step preserves the compartment-specific functions needed to maintain the integrity of glycosylation and processing pathways. The final sorting of material that will leave the Golgi complex occurs in the TGN, where lipids and proteins are selectively packaged into distinct populations of transport vesicles that are destined for different locations in the cell. In some cells, the Golgi complex is also involved in the processing of proteins that enter the cell by endocytosis.

ER-Specific Proteins Contain Retention and Retrieval Tags

The protein composition required in the ER is maintained both by preventing some proteins from escaping when vesicles bud from the ER membrane and by retrieving other proteins that have left the ER and reached the CGN. It is not entirely clear how proteins that never leave the ER are retained, but one theory proposes that they form extensive complexes that are physically excluded from vesicles budding from the ER.

Several proteins localized to the ER contain the tripeptide sequence RXR (Arg-X-Arg, where X is any amino acid), which appears to promote retention in the ER. This *retention tag* is also found in some multisubunit proteins that are destined for the plasma membrane. The *N*-methyl-D-aspartate (NMDA) receptor, which is important in neurotransmission in the mammalian brain, contains such a tag. But why, you may ask, does this plasma membrane protein have an ER retention signal? It is thought that the tripeptide causes individual subunits of NMDA to be retained in the ER until assembly of the multisubunit complex in the ER is complete. Proper assembly masks the RXR sequence, allowing the assembled complex to leave the ER. Other evidence suggests that phosphorylation of this protein near the RXR sequence promotes ER release.

Retrieval by retrograde flow of vesicles from the CGN to the ER is better understood than retention. Many soluble ER-specific proteins contain *retrieval tags* that bind to specific transmembrane receptors facing the Golgi complex lumen. The tags are short C-terminal amino acid sequences such as KDEL (Lys-Asp-Glu-Leu) or KKXX (where X can be any amino acid) in mammals and HDEL (His-Asp-Glu-Leu) in yeast. When a protein containing such a tag binds to its receptor, the receptor undergoes a conformational change, and the receptor-ligand complex is packaged into a transport vesicle for return to the ER.

Evidence for the importance of retrieval tags comes from experiments involving artificial *chimeric proteins,* also known as *fusion proteins.* They are made by joining the DNA sequences encoding two polypeptide segments from different proteins, allowing a hybrid polypeptide to be produced from the joined DNAs. Proteins normally secreted from the cell were altered in this manner to include amino acids representing an ER retrieval tag, and these proteins were subsequently found in the ER rather than being secreted. Interestingly, some of these chimeric proteins found in the ER had undergone partial processing by enzymes found only in the Golgi complex. This finding indicates that the suspected tag did not simply prevent escape of the protein from the ER but actively promoted the retrieval of ER-specific proteins from the Golgi complex.

Golgi Complex Proteins May Be Sorted According to the Lengths of Their Membrane-Spanning Domains

Like resident proteins required in the ER, some resident proteins of the Golgi complex contain retention or retrieval tags. Moreover, in the Golgi complex as in the ER, the formation of large complexes that are excluded from transport vesicles may play a role in maintaining the protein composition of the Golgi complex. In addition, we will next see that a third, distinctly different mechanism is also likely at work in this organelle—a mechanism involving hydrophobic regions of Golgi proteins.

All known Golgi-specific proteins are integral membrane proteins that have one or more hydrophobic membrane-spanning domains anchoring them to Golgi membranes. The length of the hydrophobic domains may determine into which cisterna of the Golgi complex each membrane-bound protein is incorporated as it moves through the organelle. Recall that the thickness of cellular membranes increases progressively from the ER (about 5 nm) to the plasma membrane (about 8 nm). Among Golgi-specific proteins, there is a corresponding increase in the lengths of hydrophobic membrane-spanning domains going from the CGN to the TGN. Such proteins tend to move from compartment to compartment until the thickness of the membrane exceeds the length of their membrane-spanning domains, thereby blocking further migration.

Targeting of Soluble Lysosomal Proteins to Endosomes and Lysosomes Is a Model for Protein Sorting in the TGN

During their journey through the ER and early compartments of the Golgi complex, soluble lysosomal enzymes, like other glycoproteins, undergo N-glycosylation followed by removal of glucose and mannose units. Within the Golgi complex, however, mannose residues on the carbohydrate side chain of lysosomal enzymes are phosphorylated, forming an oligosaccharide containing mannose-6-phosphate. This oligosaccharide tag distinguishes soluble lysosomal proteins from other glycoproteins and ensures their delivery to lysosomes (**Figure 12-9**).

The phosphorylation of mannose residues is catalyzed by two Golgi-specific enzymes. The first, located in an early compartment of the Golgi stack, is a phosphotransferase that adds GlcNAc-1-phosphate to carbon atom 6 of mannose. The second, located in a mid-Golgi compartment, removes GlcNAc, leaving behind the mannose-6-phosphate residue.

The interior surface of the TGN membrane has mannose-6-phosphate receptors (MPRs) that bind to the mannose-6-phosphate residues of lysosomal proteins. The pH of the TGN is around 6.4, which favors binding of soluble lysosomal enzymes to these receptors. Following binding of tagged lysosomal proteins to the MPRs, the receptor-ligand complexes are packaged into transport vesicles and conveyed to an endosome. In animal cells, lysosomal enzymes needed for degradation of material brought into the cell by endocytosis are transported from the TGN to organelles known as **late endosomes.** These develop from **early endosomes,** which are formed by the coalescence of vesicles from the TGN and plasma membrane.

As an early endosome matures to form a late endosome, the pH of the lumen decreases to about 5.5, causing the bound lysosomal enzymes to dissociate from the MPRs. This prevents the retrograde movement of these enzymes back to the Golgi along with the receptors that are recycled in vesicles that return to the TGN. Finally, the late endosome either matures to form a new lysosome or delivers its contents to an active lysosome.

Strong support for this model of lysosomal enzyme targeting came from studies of a human genetic disorder called *I-cell disease.* Cultured fibroblast cells from patients with I-cell disease synthesize all the expected lysosomal enzymes but then release most of the soluble proteins to the extracellular medium instead of incorporating them into lysosomes. The distinguishing feature of the misdirected proteins is the absence of mannose-6-phosphate residues on their oligosaccharide side chains. The

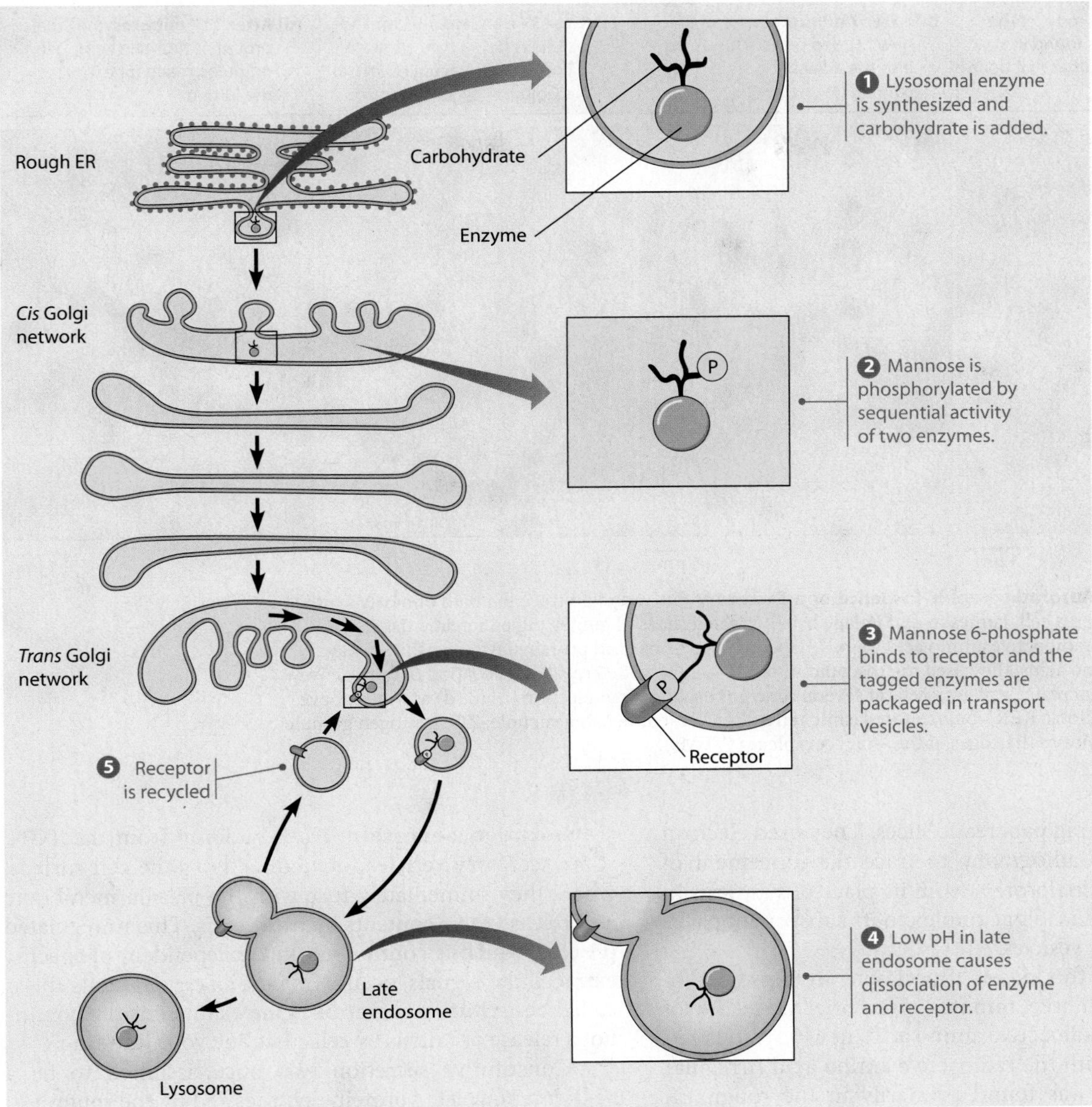
Rough ER
Carbohydrate
Enzyme
1 Lysosomal enzyme is synthesized and carbohydrate is added.
Cis Golgi network
P
2 Mannose is phosphorylated by sequential activity of two enzymes.
Trans Golgi network
3 Mannose 6-phosphate binds to receptor and the tagged enzymes are packaged in transport vesicles.
Receptor
5 Receptor is recycled
4 Low pH in late endosome causes dissociation of enzyme and receptor.
Late endosome
Lysosome

FIGURE 12-9 Targeting of Soluble Lysosomal Enzymes to Endosomes and Lysosomes by a Mannose-6-Phosphate Tag. ❶ In the ER, soluble lysosomal enzymes undergo *N*-glycosylation followed by removal of glucose and mannose units. ❷ Within the Golgi complex, mannose residues on the lysosomal enzymes are phosphorylated by two enzymes. The first one adds *N*-acetylglucosamine-1-phosphate to carbon atom 6 of mannose. The second one removes *N*-acetylglucosamine, leaving behind the phosphorylated mannose residue. ❸ The tagged lysosomal enzymes bind to mannose-6-phosphate receptors in the TGN and are packaged into coated transport vesicles that convey the enzymes to a late endosome. ❹ The acidity of the late endosomal lumen causes the enzymes to dissociate from their receptors. ❺ The receptors are recycled in vesicles that return to the TGN. The late endosome either matures to form a lysosome or transfers its contents to an active lysosome.

mannose-6-phosphate tag is clearly essential for targeting soluble lysosomal glycoproteins to the lysosome. As you may have guessed, I-cell disease results from a defective phosphotransferase that is needed to add the mannose-6-phosphate to oligosaccharide chains on lysosomal enzymes. However, while mannose-6-phosphate residues are important for directing traffic from the ER to the lysosome, other pathways appear to be involved as well. For example, in patients with I-cell disease, lysosomal acid hydrolases somehow still reach the lysosomes in liver cells.

Secretory Pathways Transport Molecules to the Exterior of the Cell

Integral to the vesicular traffic shown in Figure 12-8 are **secretory pathways** by which proteins move from the ER through the Golgi complex to **secretory vesicles** and **secretory granules,** which then discharge their contents to the exterior of the cell. The concerted roles of the ER and the Golgi complex in secretion were demonstrated in 1967 by James Jamieson and George Palade in secretory

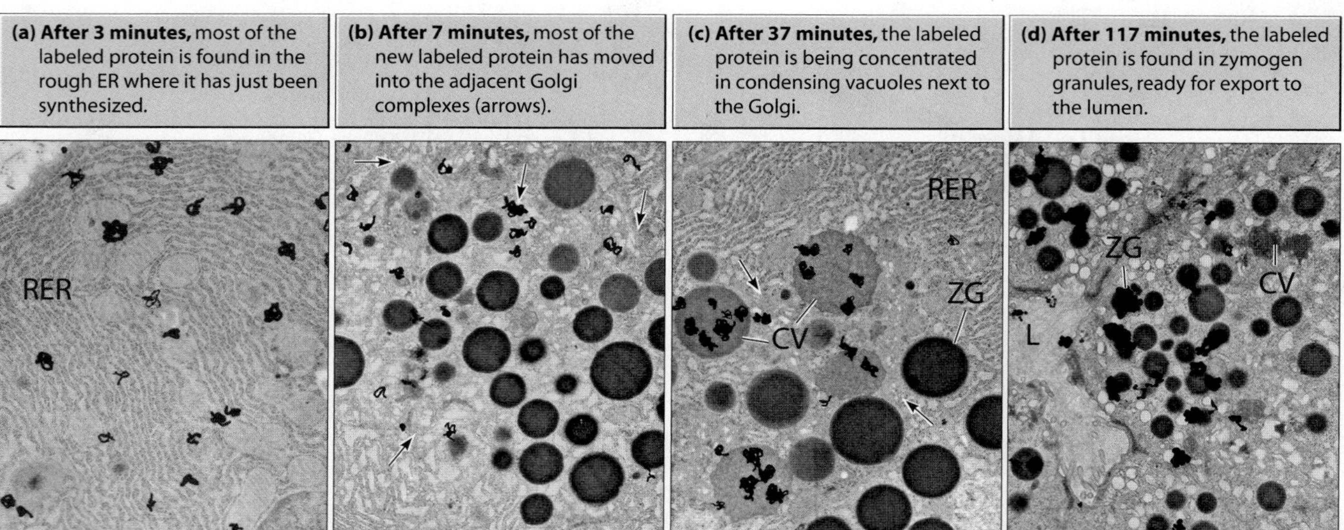

(a) After 3 minutes, most of the labeled protein is found in the rough ER where it has just been synthesized.
(b) After 7 minutes, most of the new labeled protein has moved into the adjacent Golgi complexes (arrows).
(c) After 37 minutes, the labeled protein is being concentrated in condensing vacuoles next to the Golgi.
(d) After 117 minutes, the labeled protein is found in zymogen granules, ready for export to the lumen.
RER
RER
RER
CV
ZG
ZG
CV
L
RER
1 μm
1 μm
1 μm
1 μm

FIGURE 12-10 Autoradiographic Evidence of a Secretory Pathway. To trace the path of newly synthesized proteins through a cell, Jamieson and Palade briefly treated slices of guinea pig pancreatic tissue with a small amount of a radioactive amino acid to pulse label newly synthesized proteins. After washing away unincorporated amino acids, they used microscopic autoradiography to determine the location of radioactively labeled protein in pancreatic secretory cells at several time points following injection (**a**) to (**d**) as noted above the panels. Abbreviations: RER, rough endoplasmic reticulum; CV, condensing vacuole; ZG, zymogen granule; L, lumen. Arrows point to the edges of the Golgi complexes (TEMs).

cells from guinea pig pancreatic slices. They used electron microscopic autoradiography to trace the movement of radioactively labeled protein from its place of synthesis in the ER through the Golgi complex to its subsequent packaging in secretory vesicles.

Results from this classic experiment are presented in **Figure 12-10**. Three minutes after brief exposure of tissue slices to a radioactive amino acid, newly synthesized protein labeled with the radioactive amino acid (irregular dark coloration) was found primarily in the rough ER (panel a). A few minutes later, labeled protein began to appear in the Golgi complex (panel b). By 37 minutes, labeled protein was detected in vesicles budding from the Golgi that Jamieson and Palade called *condensing vacuoles* (panel c). After 117 minutes, labeled protein began to accumulate in dense *zymogen granules,* vesicles that discharge secretory proteins to the exterior of the cell (panel d). Some of the radioactively labeled protein was observed in the exterior lumen adjacent to the cell, demonstrating that some of the secretory granules had released their contents from the cell.

Based on this classic experiment and numerous similar studies since then, the secretory pathways shown in Figure 12-8 are now understood in considerable detail. Moreover, we now distinguish two different modes of secretion by eukaryotic cells. *Constitutive secretion* involves the continuous discharge of vesicles at the plasma membrane surface, whereas *regulated secretion* involves controlled, rapid releases that happen in response to an extracellular signal.

Constitutive Secretion. After budding from the TGN, some secretory vesicles move directly to the cell surface, where they immediately fuse with the plasma membrane and release their contents by exocytosis. This unregulated process, which is continuous and independent of specific extracellular signals, occurs in most eukaryotic cells and is called **constitutive secretion.** One example is the continuous release of mucus by cells that line your intestine.

Constitutive secretion was once assumed to be a *default pathway* for proteins synthesized by the rough ER. According to this model, all proteins destined to remain in the endomembrane system must have a tag that diverts them from constitutive secretion. Otherwise, they will move through the endomembrane system and be released outside the cell by default. Support for this model came from studies in which removal of the KDEL retrieval tags on resident ER proteins led to secretion of the modified protein. However, more recent evidence suggests that a variety of short amino acid tags may identify specific proteins for constitutive secretion.

For some proteins, constitutive secretion may require N-glycosylation of the protein, as was recently shown for mouse interleukin-31, a secreted protein involved in extracellular cell-to-cell signaling. Other recent work showed that addition of an N-glycosylation site to the human p53 tumor suppressor protein, a protein that is not normally secreted, caused this altered p53 to be secreted. As you might expect, the secreted protein was N-glycosylated.

Regulated Secretion. While vesicles containing constitutively secreted proteins move continuously and directly from the TGN to the plasma membrane, secretory vesicles involved in **regulated secretion** accumulate in the cell and then fuse with the plasma membrane only in response to specific extracellular signals. An important example is the release of neurotransmitters, which will be described in Chapter 13 (see Figure 13-18). Two additional examples of regulated secretion are the release of insulin from pancreatic β cells in response to glucose and the release of *zymogens*—inactive precursors of hydrolytic enzymes—from pancreatic acinar cells in response to calcium or endocrine hormones.

Regulated secretory vesicles form by budding from the TGN as immature secretory vesicles, which undergo a subsequent maturation process. Maturation of secretory proteins involves concentration of the proteins—referred to as *condensation*—and frequently also some proteolytic processing. The mature secretory vesicles then move close to the site of secretion and remain near the plasma membrane until receiving a hormonal or other chemical signal that triggers release of their contents by fusion with the plasma membrane.

Zymogen granules are a type of mature regulated secretory vesicle that are usually quite large and contain highly concentrated protein, as shown in Figure 12-10 and **Figure 12-11**. Note that the zymogen granules are concentrated in the region of the cell between the Golgi stacks from which they arise and the portion of the plasma membrane bordering the lumen into which the contents of the granules are eventually discharged.

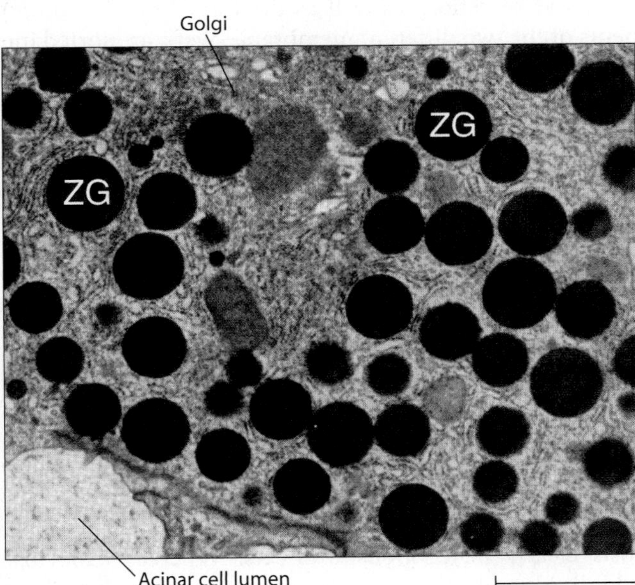

FIGURE 12-11 Zymogen Granules. This electron micrograph of an acinar (secretory) cell from the exocrine pancreas of a rat illustrates the prominence of zymogen granules (ZG), which contain enzymes destined for secretion. They are usually concentrated in the region of the cell between the Golgi complex from which they arise and the portion of the plasma membrane bordering the acinar lumen into which the zymogens will be discharged (TEM).

The information needed to direct a protein to a regulated secretory vesicle is presumably inherent in the amino acid sequence of the protein, though the precise signals and mechanisms are not yet known. Current evidence suggests that high concentrations of secretory proteins in secretory granules promote the formation of large *protein aggregates* that exclude nonsecretory proteins. This could occur in the TGN, where only aggregates would be packaged in vesicles destined for secretory granules, or it could occur in the secretory granule itself. The pH of the TGN and the secretory granule lumens may serve as a trigger favoring aggregation as material leaves the TGN. The soluble proteins that do not become part of an aggregate in the TGN or a secretory granule would be carried by transport vesicles to other locations.

Exocytosis and Endocytosis: Transporting Material Across the Plasma Membrane

Two methods of transporting materials across the plasma membrane are *exocytosis,* the process by which secretory granules release their contents to the exterior of the cell, and *endocytosis,* the process by which cells internalize external materials. Both processes are unique to eukaryotic cells and are also involved in the delivery, recycling, and turnover of membrane proteins. We will first consider exocytosis because it is the final step in a secretory pathway that began with the ER and the Golgi complex.

Exocytosis Releases Intracellular Molecules Outside the Cell

In **exocytosis,** proteins in a vesicle are released to the exterior of the cell as the membrane of the vesicle fuses with the plasma membrane. A variety of proteins are exported from both animal and plant cells by exocytosis. Animal cells secrete peptide and protein hormones, mucus, milk proteins, and digestive enzymes in this manner. Plant and fungal cells secrete enzymes and structural proteins associated with the cell wall, and carnivorous plants secrete hydrolytic enzymes that are used to digest trapped insects.

The process of exocytosis is illustrated schematically in **Figure 12-12a**. Vesicles containing cellular products destined for secretion move to the cell surface (step ❶), where the membrane of the vesicle fuses with the plasma membrane (step ❷). Fusion with the plasma membrane discharges the vesicle contents to the exterior of the cell (step ❸). In the process, the membrane of the vesicle becomes integrated into the plasma membrane, with the *inner* (lumenal) surface of the vesicle becoming the *outer* (extracellular) surface of the plasma membrane (step ❹). Thus, glycoproteins and glycolipids that were originally formed in the ER and Golgi lumens will face the extracellular space.

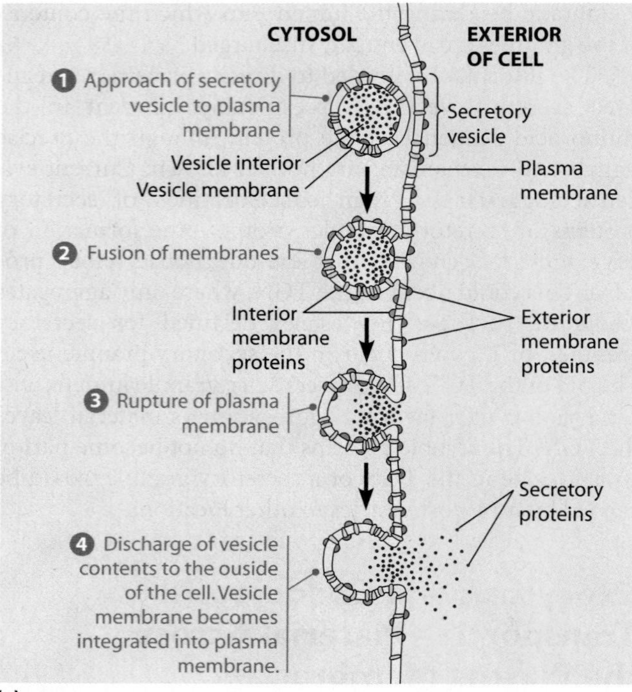

(a)

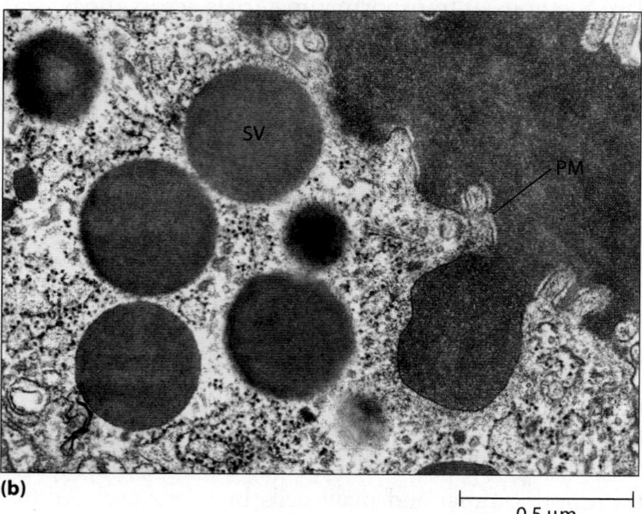

(b)

FIGURE 12-12 Exocytosis. **(a)** This illustration shows how a secretory vesicle approaches the plasma membrane, fuses with it, and releases its contents to the exterior of the cell during exocytosis. **(b)** In this monkey pancreatic cell, secretory vesicles (SV) approach the plasma membrane (PM), where one vesicle has fused with the membrane and secreted its contents.

The mechanism underlying the movement of exocytic vesicles to the cell surface is not yet clear. Current evidence points to the involvement of microtubules in vesicle movement. For example, in some cells, vesicles appear to move from the Golgi complex to the plasma membrane along "tracks" of microtubules that are oriented parallel to the direction of vesicle movement. Moreover, vesicle movement stops when the cells are treated with *colchicine,* a plant alkaloid that prevents microtubule assembly. We will discuss the intracellular movement of vesicles along microtubules in Chapter 16.

The Role of Calcium in Triggering Exocytosis. Fusion of regulated secretory vesicles with the plasma membrane is generally triggered by a specific extracellular signal. In most cases, the signal is a hormone or a neurotransmitter that binds to specific receptors on the cell surface and triggers the synthesis or release of a *second messenger* within the cell, as we will see in Chapter 14. During regulated secretion, a transient elevation of the intracellular concentration of calcium ions often appears to be an essential step in the signal cascade leading from the receptor on the cell surface to exocytosis. For example, microinjection of calcium into pancreatic cells induces mature secretory granules to discharge their contents to the extracellular medium. The specific role of calcium is not yet clear, but it appears that an elevation in the intracellular calcium concentration leads to the activation of protein kinases whose target proteins are components of either the vesicle membrane or the plasma membrane.

Polarized Secretion. In many cases, exocytosis of specific proteins is limited to a specific surface of the cell. For example, the secretory cells that line your intestine release digestive enzymes only on the side of the cell that faces the interior of the intestine. This phenomenon, called **polarized secretion,** is also seen in nerve cells, which secrete neurotransmitter molecules only at junctions with other nerve cells (see Figure 13-16). Proteins destined for polarized secretion, as well as lipid and protein components of the two different membrane layers, are sorted into vesicles that bind to localized recognition sites on subdomains of the plasma membrane.

Endocytosis Imports Extracellular Molecules by Forming Vesicles from the Plasma Membrane

Most eukaryotic cells carry out one or more forms of **endocytosis** for uptake of extracellular material. A small segment of the plasma membrane progressively folds inward (**Figure 12-13**, step ❶), and then it pinches off to form an **endocytic vesicle** containing ingested substances or particles (steps ❷–❹). Endocytosis is important for several cellular processes, including ingestion of essential nutrients by some unicellular organisms and defense against microorganisms by white blood cells.

In terms of membrane flow, exocytosis and endocytosis clearly have opposite effects. Whereas exocytosis adds lipids and proteins to the plasma membrane, endocytosis removes them. Thus, the steady-state composition of the plasma membrane results from a balance between exocytosis and endocytosis. Through endocytosis and retrograde transport, the cell can recycle and reuse molecules deposited in the plasma membrane by secretory vesicles during exocytosis.

The magnitude of the resulting membrane exchange is impressive. For example, the secretory cells in your

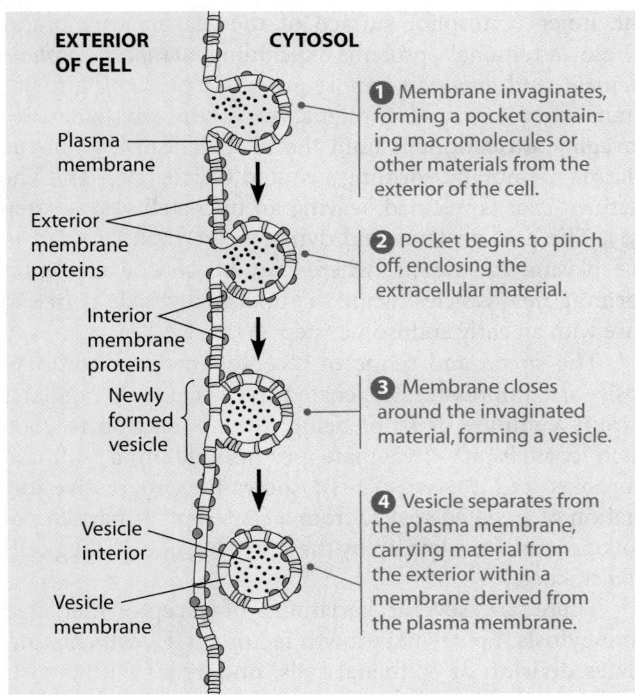

EXTERIOR OF CELL
CYTOSOL
Plasma membrane
Exterior membrane proteins
Interior membrane proteins
Newly formed vesicle
Vesicle interior
Vesicle membrane
1 Membrane invaginates, forming a pocket containing macromolecules or other materials from the exterior of the cell.
2 Pocket begins to pinch off, enclosing the extracellular material.
3 Membrane closes around the invaginated material, forming a vesicle.
4 Vesicle separates from the plasma membrane, carrying material from the exterior within a membrane derived from the plasma membrane.

FIGURE 12-13 Endocytosis. This illustration shows the uptake of materials from the exterior of the cell during endocytosis. For clarity, the coat proteins at the site of invagination and around the endocytic vesicle have been omitted from this diagram.

pancreas recycle an amount of membrane equal to the whole surface area of the cell within about 90 minutes. Cultured macrophages (large white blood cells) are even faster, replacing an amount of membrane equivalent to the entire plasma membrane within about 30 minutes!

During endocytosis, the membrane of an endocytic vesicle isolates the internalized substances from the cytosol. Most endocytic vesicles develop into early endosomes, which fuse with vesicles from the TGN, acquiring digestive enzymes and maturing to form new lysosomes. A distinction is usually made between *phagocytosis* (Greek for "cellular eating"), in which large solid particles are ingested, and *pinocytosis* ("cellular drinking"), in which liquids containing soluble or suspended molecules are taken up.

Phagocytosis. The ingestion of large particles (>0.5 μm diameter), including aggregates of macromolecules, parts of other cells, and even whole microorganisms or other cells, is known as **phagocytosis.** For many unicellular eukaryotes, such as amoebas and ciliated protozoa, phagocytosis is a routine means for acquiring food. Phagocytosis is also used by some primitive animals, notably flatworms, coelenterates, and sponges, as a means of obtaining nutrients.

In more complex organisms, however, phagocytosis is usually restricted to specialized cells called **phagocytes.** For example, your body contains two classes of white blood cells—*neutrophils* and *macrophages*—which use phagocytosis for defense rather than nutrition. These cells engulf and digest foreign material or invasive microorganisms found in the bloodstream or in injured tissues. Macrophages have an additional role as scavengers, ingesting cellular debris and whole damaged cells from injured tissues. Under certain conditions, other mammalian cells engage in phagocytosis. For example, fibroblasts found in connective tissue can take up collagen to allow remodeling of the tissue, and dendritic cells in the mammalian spleen can ingest bacteria as part of an immune response.

Phagocytosis has been studied most extensively in the amoeba, which uses it for nutrition. Contact with food particles or smaller organisms triggers the onset of phagocytosis, as shown in **Figure 12-14**. Folds of membrane called *pseudopods* gradually surround the object and then meet and engulf the particle, forming an intracellular **phagocytic vacuole.** This endocytic vesicle, also called a *phagosome,* then fuses with a late endosome or matures directly into a lysosome, forming a large vesicle in which the ingested material is digested. As part of their role in the immune system, human phagocytes generate toxic concentrations of hydrogen peroxide, hypochlorous acid, and other oxidants in the phagocytic vacuole to kill microorganisms.

Receptor-Mediated Endocytosis. Cells can acquire certain soluble and suspended materials by a process known as **receptor-mediated endocytosis** (also called **clathrin-dependent endocytosis**). For this process, cells use specific receptors that are found on the outer surface of the plasma membrane. Receptor-mediated endocytosis is the primary mechanism for the specific internalization of most macromolecules by eukaryotic cells. Depending on the cell type, mammalian cells can ingest hormones, growth factors, enzymes, serum proteins, cholesterol, antibodies, iron, and even some viruses and bacterial toxins by this mechanism.

The discovery of receptor-mediated endocytosis and its role in the internalization of *low-density lipoproteins (LDL)* is highlighted in **Box 12B** (see p. 374). Receptor-mediated endocytosis of LDL carries cholesterol into mammalian cells. An interest in familial hypercholesterolemia, a hereditary predisposition to high blood cholesterol levels and hence to atherosclerosis and heart disease, led Michael Brown and Joseph Goldstein to the discovery of receptor-mediated endocytosis, for which they shared a Nobel Prize in 1986.

Receptor-mediated endocytosis is illustrated in **Figure 12-15**. The process begins with the binding of specific molecules (referred to as ligands) to their *receptors*—specific ligand-binding proteins found on the outer surface of the plasma membrane (step ❶). As the receptor-ligand complexes diffuse laterally in the membrane, they encounter specialized membrane regions called *coated pits* that serve as sites for the collection and internalization of these complexes (step ❷). In a typical mammalian cell, coated pits occupy about 20% of the total surface area of the plasma membrane.

Accumulation of receptor-ligand complexes within the coated pits triggers accumulation of additional proteins on

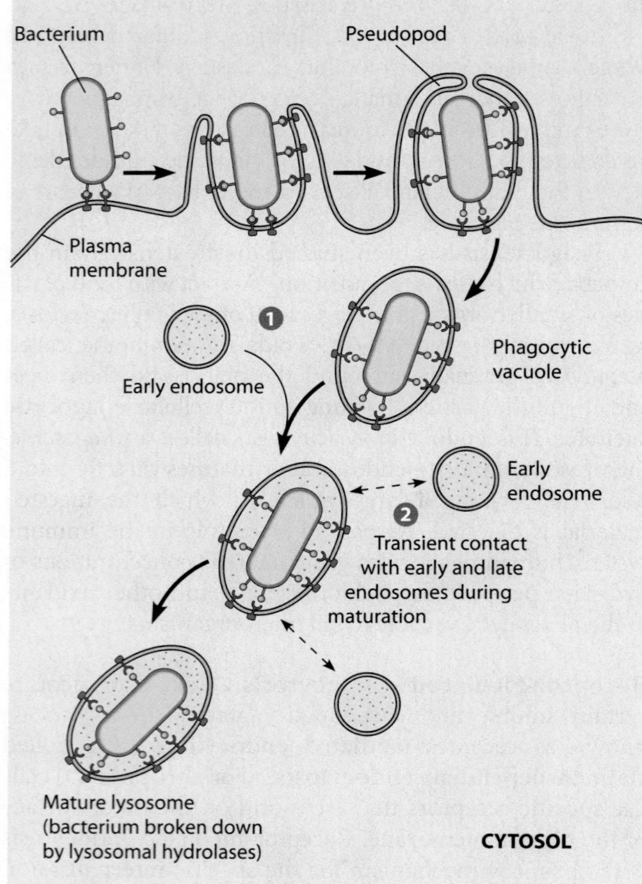

(a)

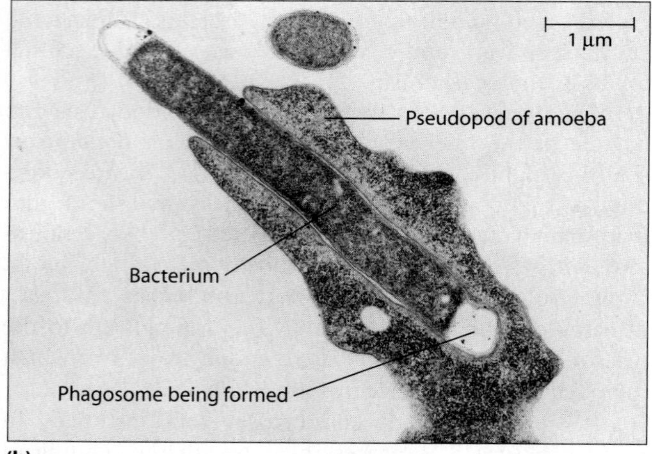

(b)

FIGURE 12-14 Phagocytosis. **(a)** Particles or microorganisms bind to receptors on the cell surface, triggering the onset of phagocytosis. Folds of membrane called *pseudopods* gradually surround the particle. Eventually, the pseudopods meet and engulf the particle, forming a *phagocytic vacuole.* The vacuole then ❶ fuses with an early endosome or ❷ forms transient connections (indicated by dashed lines) with early and late endosomes and matures into a lysosome, in which digestion of the internalized material occurs. **(b)** This micrograph shows a phagosome being formed as an amoeba engulfs a rod-shaped bacterial cell (TEM).

the inner (cytosolic) surface of the plasma membrane. These additional proteins—including *adaptor protein, clathrin,* and *dynamin*—are required for promoting membrane curvature and invagination of the pit (step ❸). Invagination continues until the pit pinches off from the plasma membrane, forming a **coated vesicle** (step ❹). The clathrin coat is released, leaving an uncoated vesicle (step ❺). The coat proteins and dynamin are then recycled to the plasma membrane, where they become available for forming new vesicles, while the uncoated vesicle is free to fuse with an early endosome (step ❻).

The speed and scope of receptor-mediated endocytosis are impressive. A coated pit usually invaginates within a minute or so of being formed, and up to 2500 such coated pits invaginate per minute in a cultured fibroblast cell. **Figure 12-16** shows the progressive formation of a coated vesicle from a coated pit as particles of yolk protein are taken up by the maturing oocyte (egg cell) of a chicken.

There are several variations of receptor-mediated endocytosis. Epidermal growth factor (EGF), which stimulates division of epithelial cells, undergoes endocytosis by the mechanism shown in Figure 12-15. Here, endocytosis plays an important role in cell signaling. As EGF receptors are internalized, the cell becomes less responsive to EGF, a process known as *desensitization.* Deficiencies in desensitization caused by defective endocytosis can lead to overstimulation by EGF, resulting in excessive cell growth, cell division, and possible tumor formation.

In another variation of receptor-mediated endocytosis, receptors are concentrated in coated pits independent of formation of receptor-ligand complexes. Binding of ligands to receptors simply triggers internalization. In yet another variation, the receptors are not only constitutively concentrated, but they are also constitutively internalized regardless of whether ligands have bound to the receptors. For example, the LDL receptors described in Box 12B are constitutively internalized.

Following receptor-mediated endocytosis, the uncoated vesicles fuse with vesicles budding from the TGN to form early endosomes in peripheral regions of the cell. *Early endosomes* are sites for the sorting and recycling of extracellular material brought into the cell by endocytosis. Protein molecules essential for new rounds of endocytosis are often—but not always—recycled after separation from the material fated for digestion. The early endosome continues to acquire lysosomal proteins from the TGN and matures to form a late endosome, which then develops into a lysosome. The roles of endosomes in digestion will be discussed in more detail when we examine lysosomes later in the chapter.

Recycling of plasma membrane receptor molecules is facilitated by acidification of the early endosome. The interior of an endocytic vesicle has a pH of about 7.0, whereas the interior of an early endosome has a pH of 5.9–6.5. The lower pH is maintained by an *ATP-dependent proton pump* in the endosomal membrane. The

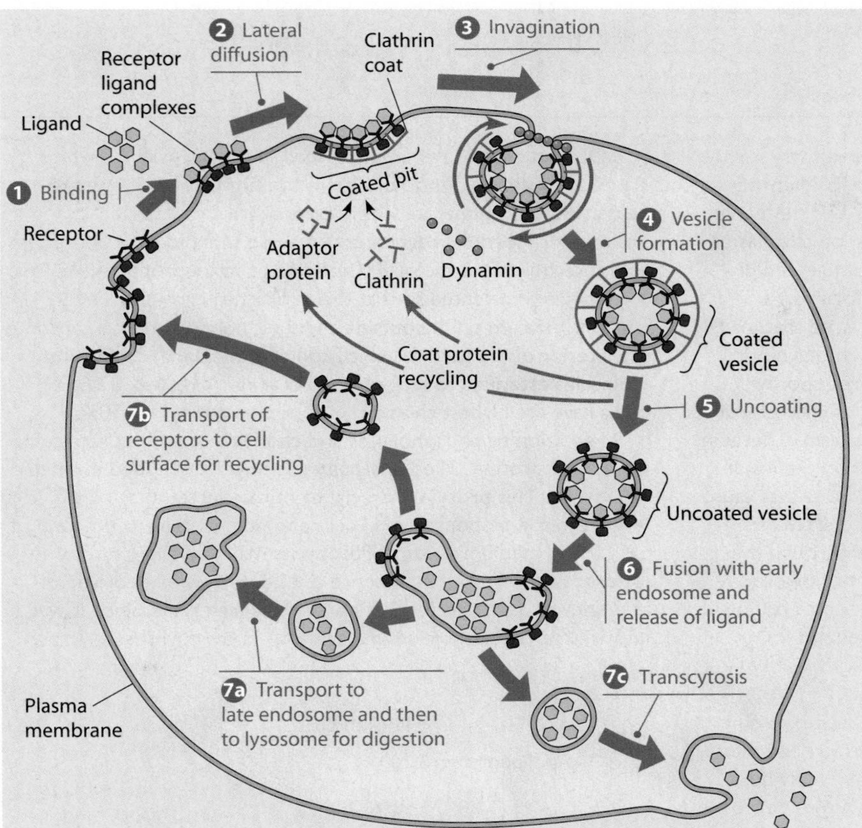

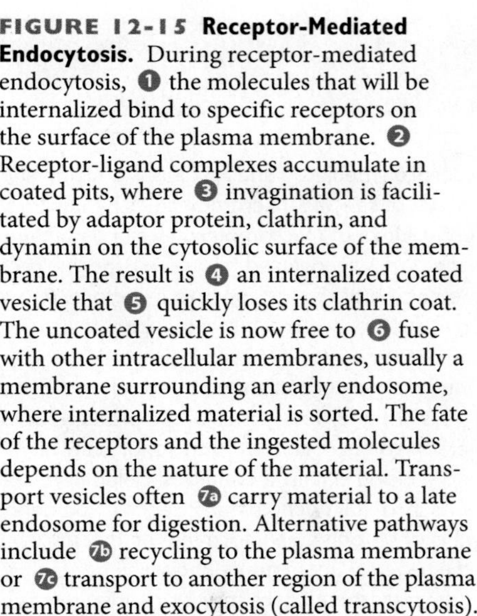

FIGURE 12-15 Receptor-Mediated Endocytosis. During receptor-mediated endocytosis, ❶ the molecules that will be internalized bind to specific receptors on the surface of the plasma membrane. ❷ Receptor-ligand complexes accumulate in coated pits, where ❸ invagination is facilitated by adaptor protein, clathrin, and dynamin on the cytosolic surface of the membrane. The result is ❹ an internalized coated vesicle that ❺ quickly loses its clathrin coat. The uncoated vesicle is now free to ❻ fuse with other intracellular membranes, usually a membrane surrounding an early endosome, where internalized material is sorted. The fate of the receptors and the ingested molecules depends on the nature of the material. Transport vesicles often 7a carry material to a late endosome for digestion. Alternative pathways include 7b recycling to the plasma membrane or 7c transport to another region of the plasma membrane and exocytosis (called transcytosis). For clarity, the nucleus is not shown.

MB Receptor-mediated Endocytosis

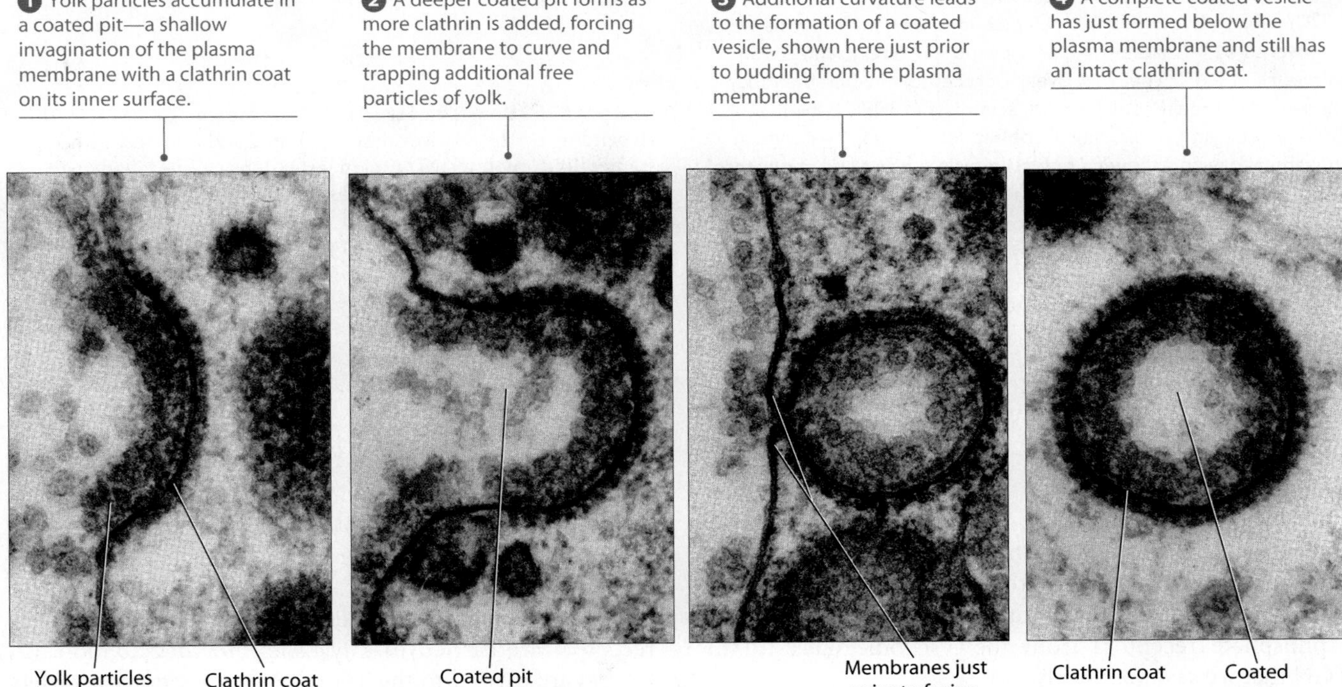

1 Yolk particles accumulate in a coated pit—a shallow invagination of the plasma membrane with a clathrin coat on its inner surface.
2 A deeper coated pit forms as more clathrin is added, forcing the membrane to curve and trapping additional free particles of yolk.
3 Additional curvature leads to the formation of a coated vesicle, shown here just prior to budding from the plasma membrane.
4 A complete coated vesicle has just formed below the plasma membrane and still has an intact clathrin coat.
Yolk particles in coated pit
Clathrin coat
Coated pit
Membranes just prior to fusion
Clathrin coat
Coated vesicle

FIGURE 12-16 Receptor-Mediated Endocytosis of Yolk Protein by a Chicken Oocyte. This series of electron micrographs illustrates the formation of a coated vesicle from a coated pit during receptor-mediated endocytosis (TEMs).

Cholesterol, the LDL Receptor, and Receptor-Mediated Endocytosis

Receptor-mediated endocytosis is a highly efficient pathway for the uptake of specific macromolecules by eukaryotic cells. Many different kinds of macromolecules can be taken up by this means, and each one is recognized by its own specific receptor on the plasma membrane of the appropriate cell types. As we consider the discovery of receptor-mediated endocytosis, we will focus on a specific receptor—and, as it turns out, on a health issue that many of us are concerned about: the level of cholesterol in our blood.

As you may know, one of the primary factors predisposing a person to heart attacks is an abnormally high level of cholesterol in the blood serum, a condition called *hypercholesterolemia.* Because of its insolubility, cholesterol tends to be deposited on the inside walls of blood vessels, forming the *atherosclerotic plaques* that cause *atherosclerosis,* commonly known as hardening of the arteries. Ultimately, the plaques may block the flow of blood through the vessels, causing strokes and heart attacks. Remember, however, that cholesterol is a normal component of healthy animal cell membranes—a moderate level of cholesterol is required for maintenance of these membranes. Therefore, the body will synthesize cholesterol if dietary amounts are inadequate.

Although a high blood cholesterol level is often linked to diet, some people are genetically predisposed to high blood cholesterol levels and hence to atherosclerosis and heart disease. Individuals with this hereditary predisposition, called **familial hypercholesterolemia (FH),** can have grossly elevated levels of serum cholesterol (about 650–1000 mg/100 mL of blood serum, compared with the normal range of about 130–200 mg/100 mL). They typically develop atherosclerosis early in life and often die from heart disease before the age of 20.

The link between FH and receptor-mediated endocytosis was discovered by Michael Brown and Joseph Goldstein, who began this work in 1972. Their discovery of receptor-mediated endocytosis led to Nobel Prizes for both scientists in 1986. Brown and Goldstein began by culturing fibroblast cells from FH patients in the laboratory and showing that such cells synthesized cholesterol at abnormally high rates compared with normal cells. Their next key observation was that normal cells also synthesized cholesterol at abnormally high rates when they were deprived of the **low-density lipoproteins (LDLs)** that were usually present in the culture medium. LDL is one form in which cholesterol is transported in the blood and taken up into cells.

LDL is one of several types of *blood lipoprotein particles,* which are classified by density. Another class is the **high-density lipoproteins (HDLs),** moderately high levels of which are considered healthy. A lipoprotein particle consists of a monolayer of phospholipid and cholesterol molecules and one or more protein molecules, with the lipids oriented so that their polar head groups face the aqueous medium on the outside and their nonpolar tails extend into the interior of the particle, which contains additional cholesterol molecules esterified to long-chain fatty acids (**Figure 12B-1**). LDLs have the highest cholesterol content—more than 50%.

In addition to phospholipids and cholesterol, each LDL particle has a large protein called *apolipoprotein B-100* embedded in its lipid monolayer. This protein is crucial to our understanding of the difference in the response of FH cells and normal cells to the level of LDL in the medium. Normal fibroblasts maintain a low rate of cholesterol synthesis in the presence of LDL, but when deprived of LDL they make cholesterol at abnormally high rates. Since it was known that cholesterol down-regulates its own synthesis, Brown

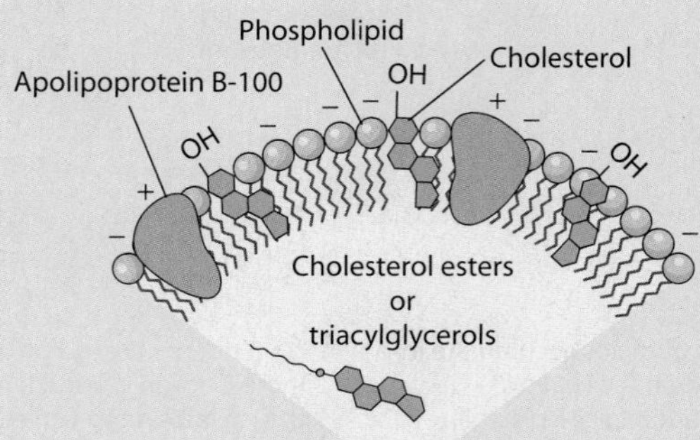

FIGURE 12B-1 LDL Structure. Lipoproteins differ in density depending on the relative amounts of lipid and proteins, with higher lipid amounts resulting in lower density. The low-density lipoprotein (LDL) shown here has a density of 1.02–1.06 g/mL. It contains about 800 phospholipid molecules and 500 free cholesterol molecules in the lipid monolayer plus about 1500 esterified cholesterol molecules in the interior. Apolipoprotein B-100 is embedded in the lipid monolayer and mediates the binding of the LDL to the LDL receptors on the surfaces of cells.

slightly acidic environment of the early endosome decreases the affinity of most receptor-ligand complexes (for example, LDL and its receptor), thereby freeing receptors to be recycled to the plasma membrane while newly ingested material is diverted to other locations. This is similar to the acid-induced recycling of mannose-6-phosphate receptors from the lysosome back to the TGN that we saw previously.

Sorting receptors and ligands is not always as simple as sending the receptors to the plasma membrane and retaining the ligands in the endosome. Depending on the ligand, some receptor-ligand complexes do not dissociate in the early endosome. While dissociated ligands are swept along to their fate in a lysosome, intact receptor-ligand complexes are still subject to sorting and packaging into transport vesicles. There are at least three alternative fates for these complexes: (1) Some receptor-ligand complexes (for example, epidermal growth factor and its receptor) are carried to a lysosome for degradation. (2) Others are carried to the TGN, where they enter a variety of pathways transporting material throughout the endomembrane system. (3) Receptor-ligand complexes can also travel by transport vesicles to a different region of the plasma membrane, where they are secreted as part of a

and Goldstein hypothesized that LDL was involved in the transport of cholesterol into the cell. Because FH fibroblasts, on the other hand, synthesized cholesterol at a high rate regardless of whether LDL was present in the medium, they thought that these cells might be defective in LDL-dependent cholesterol uptake.

Based on these observations, Brown and Goldstein postulated that the uptake of cholesterol into cells requires a specific receptor for LDL particles on the cell surface. They then tested whether this receptor is absent or defective in FH patients. In a brilliant series of experiments, these investigators and their colleagues demonstrated the existence of an LDL-specific membrane protein, called the **LDL receptor,** and they showed that it recognizes the apolipoprotein B-100 molecule that is present in every LDL particle. They also showed that the cells from FH patients either lacked the LDL receptor entirely or had defective LDL receptors.

To visualize the LDL particles, these scientists conjugated, or linked, them to molecules of ferritin, a protein that binds iron atoms. Because iron atoms are electron dense, they appear as dark dots in the electron microscope (**Figure 12B-2**). Using this technique, Brown and Goldstein showed that the ferritin-conjugated LDL particles bound to the surface of the cell and clustered at specific locations (Figure 12B-2a). We now recognize these sites as *coated pits,* which are localized regions of the plasma membrane characterized by the presence of *clathrin* on the cytoplasmic side of the membrane and by the accumulation of membrane-bound receptor-ligand complexes on the exterior of the membrane.

Dark dots were also seen on the inside of vesicles that formed by invagination and pinching off of coated pits (Figure 12B-2b). The receptors, in other words, not only bound the LDL on the cell surface but also were apparently involved in the internalization of LDL within vesicles. In short, these workers had discovered a new mechanism by which cells can take up macromolecules from their environment. And since it was an endocytic process involving specific receptors, Brown and Goldstein gave it the name we know it by today—receptor-mediated endocytosis.

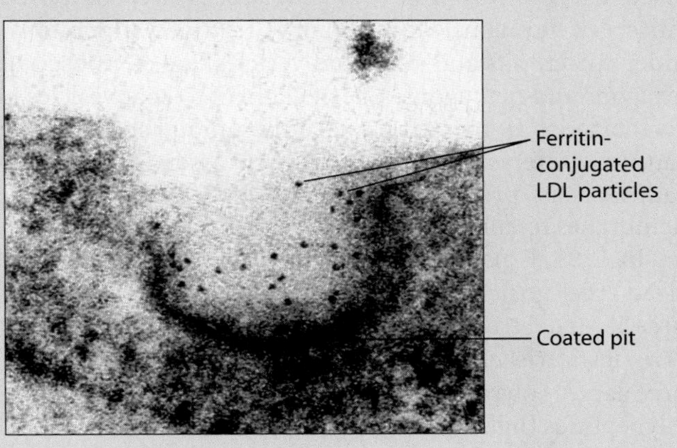

(a) Particles bind to receptors in coated pit

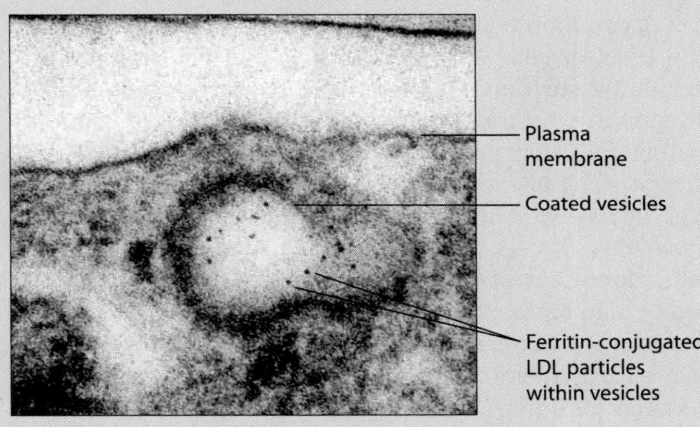

(b) Coated vesicle forms in pit

FIGURE 12B-2 Visualization of LDL Binding. Conjugation of LDL particles with ferritin, an iron-binding protein, allows visualization of the LDL-ferritin complexes by electron microscopy because of the density of the iron atoms bound to the ferritin. **(a)** LDL-ferritin conjugates, visible as dark dots, bind to receptors concentrated in a coated pit on the surface of a cultured human fibroblast cell. **(b)** The LDL-ferritin conjugates are internalized when a coated vesicle forms from the coated pit region following invagination of the plasma membrane (TEMs).

process called **transcytosis.** This pathway accommodates the transfer of extracellular material from one side of the cell, where endocytosis occurs, through the cytoplasm to the opposite side, where exocytosis occurs. For example, immunoglobulins are transported across epithelial cells from maternal blood to fetal blood by transcytosis.

Clathrin-Independent Endocytosis. An example of a clathrin-independent endocytic pathway is **fluid-phase endocytosis,** a type of pinocytosis for nonspecific internalization of extracellular fluid. Fluid-phase endocytosis, unlike receptor-mediated endocytosis, does not concentrate ingested material. Because the cell engulfs fluid without a mechanism for collecting or excluding particular molecules, the concentration of material trapped in vesicles reflects its concentration in the extracellular environment. In contrast to other forms of endocytosis, fluid-phase endocytosis proceeds at a relatively constant rate in most eukaryotic cells. Because it compensates for the membrane segments that are continuously added to the plasma membrane by exocytosis, it is a means for controlling a cell's volume and surface area. Once inside the cell, fluid-phase endocytic vesicles, like clathrin-dependent endocytic vesicles, are routed to early endosomes.

Coated Vesicles in Cellular Transport Processes

Most of the vesicles involved in lipid and protein transfer are referred to as *coated vesicles* because of the characteristic coats, or layers, of proteins covering their cytosolic surfaces as they form. Coated vesicles are a common feature of most cellular processes that involve the transfer or exchange of substances between specific membrane-bounded compartments of eukaryotic cells or between the inside and the outside of a cell.

Coated vesicles were first reported in 1964 by Thomas Roth and Keith Porter, who described their involvement in the selective uptake of yolk protein by developing mosquito oocytes. Since then, coated vesicles have been shown to play vital roles in diverse cellular processes. Coated vesicles are involved in vesicular traffic throughout the endomembrane system, as well as transport during exocytosis and endocytosis. It is probable that such vesicles participate in most, if not all, vesicular traffic connecting the various membrane-bounded compartments and the plasma membrane of the eukaryotic cell.

A common feature of coated vesicles is the presence of a layer, or coat, of protein on the cytosolic side of the membrane surrounding the vesicle. The most studied coat proteins are *clathrin, COPI,* and *COPII* (COP is an abbreviation for "coat protein."). Coat proteins participate in several steps of the formation of transport vesicles. The type of coat protein on a vesicle helps in the sorting of molecules that are fated for different destinations in the cell. More general roles for COPs may include forcing nearly flat membranes to form spherical vesicles, preventing premature, nonspecific fusion of a budding vesicle with nearby membranes, and regulating the interactions between budding vesicles and microtubules that are important for moving vesicles through the cell.

The specific set of proteins covering the exterior of a vesicle is an indicator of the origin and destination of the vesicle within the cell (**Table 12-2**). *Clathrin-coated vesicles* are involved in the selective transport of proteins from the TGN to endosomes and in the endocytosis of receptor-ligand complexes from the plasma membrane. *COPI-coated vesicles,* on the other hand, facilitate retrograde transport of proteins from the Golgi back to the ER, as well as between cisternae of the Golgi complex. *COPII-coated vesicles* are involved in the transport of material from the ER to the Golgi.

A fourth, more recently discovered coat protein is *caveolin.* The precise role of *caveolin-coated vesicles,* called **caveolae,** is still controversial. Caveolae are small invaginations of the plasma membrane characterized by the presence of the protein *caveolin.* They are a type of lipid raft that is rich in cholesterol and sphingolipids, and they may be involved in cholesterol uptake by cells. Mice deficient in caveolin show dramatic abnormalities in the cardiovascular system, which then becomes enriched in cholesterol. Other studies suggest that these cholesterol-carrying caveolae play a role in signal transduction.

Clathrin-Coated Vesicles Are Surrounded by Lattices Composed of Clathrin and Adaptor Protein

Clathrin-coated vesicles are surrounded by coats composed of two multimeric proteins, *clathrin* and *adaptor protein (AP).* The term **clathrin** comes from *clathratus,* the Latin word for "lattice," and is well chosen because clathrin and AP assemble to form protein lattices composed of polygons. Flat clathrin lattices are composed entirely of hexagons, whereas curved lattices, which form under coated pits and surround vesicles, are composed of hexagons and pentagons. Use of clathrin to coat vesicles is advantageous to the cell because the unique shape of the clathrin proteins and the way they assemble to form clathrin coats provides the driving force that causes flat membranes to curve and form spherical vesicles.

In 1981, Ernst Ungewickell and Daniel Branton visualized the basic structural units of clathrin lattices, three-legged structures called **triskelions** (**Figure 12-17a**). Each triskelion is a multimeric protein composed of three large polypeptides (heavy chains) and three small polypeptides (light chains) radiating from a central vertex, as illustrated in Figure 12-17b. Antibodies that recognize clathrin light chains bind to the legs of the triskelion near the central vertex, suggesting that the light chains are associated with the inner half of each leg.

By combining information gathered from electron microscopy and X-ray crystallography, researchers have assembled the following model for the organization of triskelions into the characteristic hexagons and pentagons of clathrin-coated pits and vesicles (Figure 12-17c). One clathrin triskelion is located at each vertex of the polygonal lattice. Each polypeptide leg extends along two edges of the

Table 12-2 Coated Vesicles Found Within Eukaryotic Cells

Coated Vesicle	Coat Proteins*	Origin	Destination
Clathrin	Clathrin, AP1, ARF	TGN	Endosomes
Clathrin	Clathrin, AP2	Plasma membrane	Endosomes
COPI	COPI, ARF	Golgi complex	ER or Golgi complex
COPII	COPII (Sec13/31 and Sec23/24), Sar1	ER	Golgi complex
Caveolin	Caveolin	Plasma membrane	ER?

*ARF designates ADP ribosylation factor 1; AP1 and AP2 designate different adaptor protein complexes (also called assembly protein complexes).

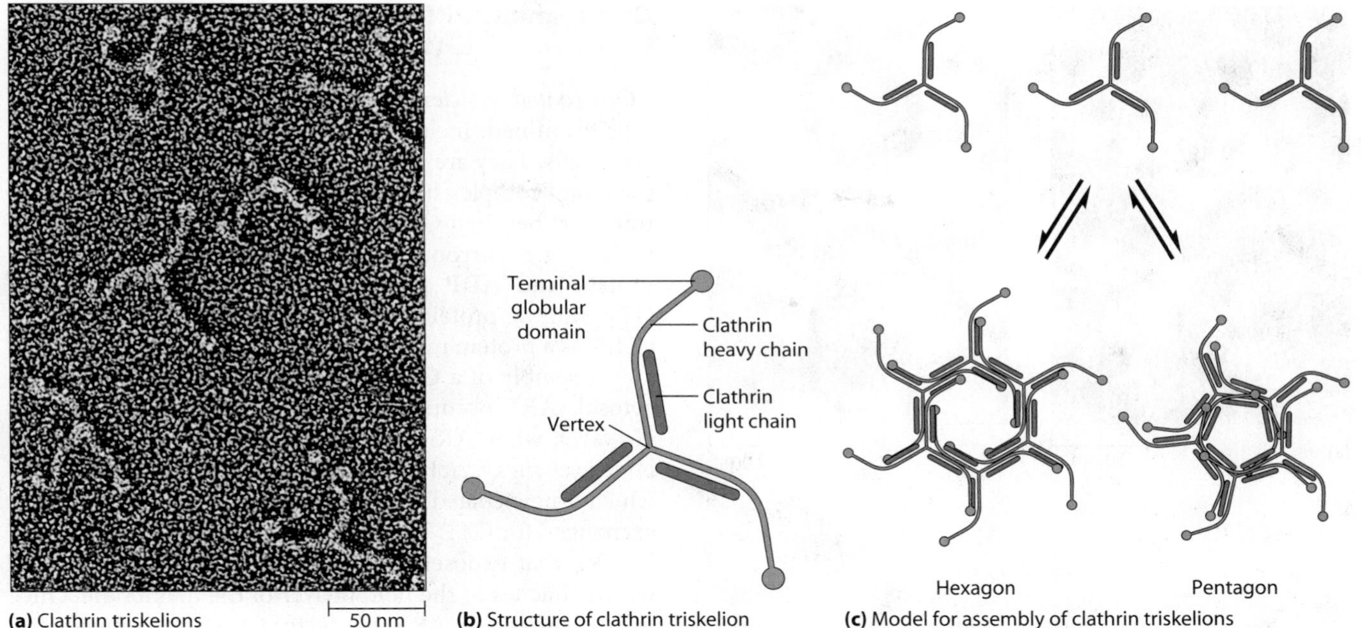

FIGURE 12-17 Clathrin Triskelions. Shown above are **(a)** a micrograph of clathrin triskelions (SEM), **(b)** an illustration showing how each triskelion is composed of three clathrin heavy chains and three clathrin light chains, and **(c)** a model for the assembly of triskelions into the characteristic clathrin pentagons and hexagons found in coated pits and vesicles.

lattice, with the knee of the clathrin heavy chain located at an adjacent vertex. This arrangement of triskelions into overlapping networks ensures extensive longitudinal contact between clathrin polypeptides and may confer the mechanical strength needed when a coated vesicle forms.

The second major component of clathrin coats—**adaptor protein (AP)**—was originally identified simply by its ability to promote the assembly of clathrin coats around vesicles and is sometimes called *assembly protein.* We now know that eukaryotic cells contain at least four types of AP complexes, each composed of four polypeptides—two adaptin subunits, one medium chain, and one small chain. The four polypeptides, which are slightly different in each type of AP complex, bind to different transmembrane receptor proteins and confer specificity during vesicle budding and targeting.

In addition to ensuring that appropriate macromolecules will be concentrated in coated pits, AP complexes mediate the attachment of clathrin to proteins embedded in the plasma membrane. Considering the central role of APs, it is not surprising that AP complexes are sites for regulation of clathrin assembly and disassembly. For example, the ability of AP complexes to bind to clathrin is affected by pH, phosphorylation, and dephosphorylation.

The Assembly of Clathrin Coats Drives the Formation of Vesicles from the Plasma Membrane and TGN

The binding of AP complexes to the plasma membrane and the concentration of receptors or receptor-ligand complexes in coated pits require ATP and GTP—though perhaps only for regulation of the process. The assembly of a clathrin coat on the cytosolic side of a membrane appears to provide part of the driving force for formation of a vesicle at the site. Initially, all clathrin units are hexagonal and form a planar, two-dimensional structure (**Figure 12-18a, c**). As more clathrin triskelions are incorporated into the growing lattice, a combination of hexagonal and pentagonal units allows the new clathrin coat to curve around the budding vesicle (Figure 12-18b, d).

As clathrin accumulates around the budding vesicle, at least one more protein—**dynamin**—participates in the process. Dynamin is a cytosolic GTPase required for coated pit constriction and closing of the budding vesicle. This essential protein was first identified in *Drosophila,* a fruit fly used extensively as a genetic model organism. Flies expressing a temperature-sensitive form of dynamin were instantly paralyzed after a temperature shift disrupted the dynamin function. Further investigation revealed an accumulation of coated pits in the membranes of neuromuscular junctions in the affected flies. Formation of the closed vesicle occurs as dynamin forms helical rings around the neck of the coated pit. As GTP is hydrolyzed, the dynamin rings tighten and separate the fully sealed endocytic vesicle from the plasma membrane.

Some mechanism is also required to *uncoat* clathrin-coated membranes. Moreover, uncoating must be done in a regulated manner because, in most cases, the clathrin coat remains intact as long as the membrane is part of a coated pit or budding vesicle but dissociates rapidly once the vesicle is fully formed. Like assembly, dissociation of the clathrin coat

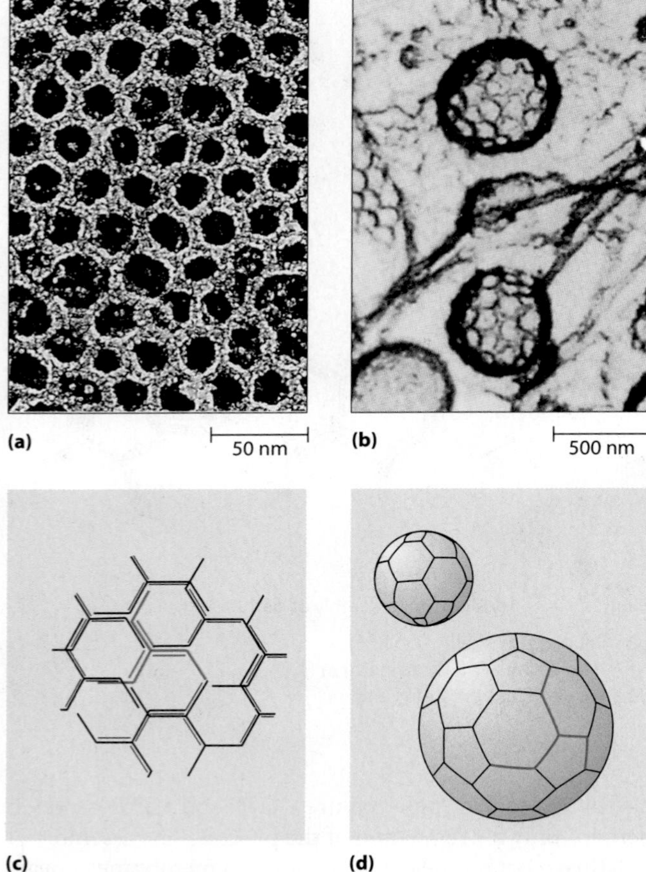

FIGURE 12-18 Clathrin Lattices. Each vesicle is surrounded by a cage of overlapping clathrin complexes. **(a)** Freeze-etch electron micrograph of a clathrin lattice in a human carcinoma cell. This flat lattice is composed of hexagonal units. **(b)** Electron micrograph of clathrin cages isolated from a calf brain. Cages include both pentagonal and hexagonal units (TEMs). **(c)** and **(d)** Interpretive drawings of the lattice and clathrin cages shown in (a) and (b).

is an energy-consuming process, accompanied by the hydrolysis of about three ATP molecules per triskelion. At least one protein, an *uncoating ATPase,* is essential for this process, though the uncoating ATPase releases only the clathrin triskelions from the APs. The factors responsible for releasing APs from the membrane have not yet been identified.

Clathrin-coated vesicles readily dissociate into soluble clathrin complexes, adaptor protein complexes, and uncoated vesicles that can spontaneously reassemble under appropriate conditions. In a slightly acidic solution containing calcium ions, clathrin complexes will reassemble independently of adaptor protein and membrane-bounded vesicles, resulting in empty shells called *clathrin cages.* Assembly occurs remarkably fast—within seconds, under favorable conditions. Ease of assembly and disassembly is an important feature of the clathrin coat because fusion of the underlying membrane with the membrane of another structure appears to require partial or complete uncoating of the vesicle.

COPI- and COPII-Coated Vesicles Travel Between the ER and Golgi Complex Cisternae

COPI-coated vesicles have been found in all eukaryotic cells examined, including mammalian, insect, plant, and yeast cells. They are involved in retrograde transport from the Golgi complex back to the ER as well as bidirectional transport between Golgi complex cisternae. COPI-coated vesicles are surrounded by coats composed of **COPI** protein and **ADP ribosylation factor (ARF),** a small GTP-binding protein. The major component of the coat, COPI, is a protein multimer composed of seven subunits.

Assembly of a COPI coat is mediated by ARF. In the cytosol, ARF occurs as part of an ARF-GDP complex. However, when ARF encounters a specific *guanine nucleotide exchange factor* associated with the membrane (from which a new coated vesicle is about to form), the GDP is exchanged for GTP. This induces a conformational change in ARF that exposes its hydrophobic N-terminal region, which attaches to the lipid bilayer of the membrane. Once firmly anchored, ARF binds to COPI multimers, and assembly of the coat drives the formation and budding of a new vesicle. After the formation of a free vesicle, a protein in the donor membrane triggers hydrolysis of GTP, and the resulting ARF-GDP releases the coat proteins for another cycle of vesicle budding. Our knowledge of COP-mediated vesicle transport has been aided considerably by using the fungal toxin *brefeldin A,* which inhibits this process by interfering with the ability of the guanine nucleotide exchange factor to produce ARF-GTP from ARF-GDP.

COPII-coated vesicles were first discovered in yeast, where they have a role in transport from the ER to the Golgi complex. Mammalian and plant homologues of some of the components of **COPII** coats have been identified, and the COPII-mediated mechanism of ER export appears to be highly conserved between organisms as different as yeast and humans. The COPII coat found in yeast is assembled from two protein complexes—called *Sec13/31* and *Sec23/24*—and a small GTP-binding protein called *SarI,* which is similar to ARF. By a mechanism resembling formation of a COPI coat, a SarI molecule with GDP bound to it approaches the membrane from which a vesicle is about to form. A peripheral membrane protein then triggers exchange of GTP for GDP, enabling SarI to bind to Sec13/31 and Sec23/24. After the formation of a free vesicle, a component of the COPII coat triggers GTP hydrolysis, and SarI releases Sec13/31 and Sec23/24.

SNARE Proteins Mediate Fusion Between Vesicles and Target Membranes

Much of the intracellular traffic mediated by coated vesicles is highly specific. As we have seen, the final sorting of proteins synthesized in the ER occurs in the TGN when lipids and proteins are packaged into vesicles for transport to various destinations. Recall that when clathrin-coated vesicles form from the TGN, the adaptor complexes include two adaptin subunits. The two adaptin subunits are partly

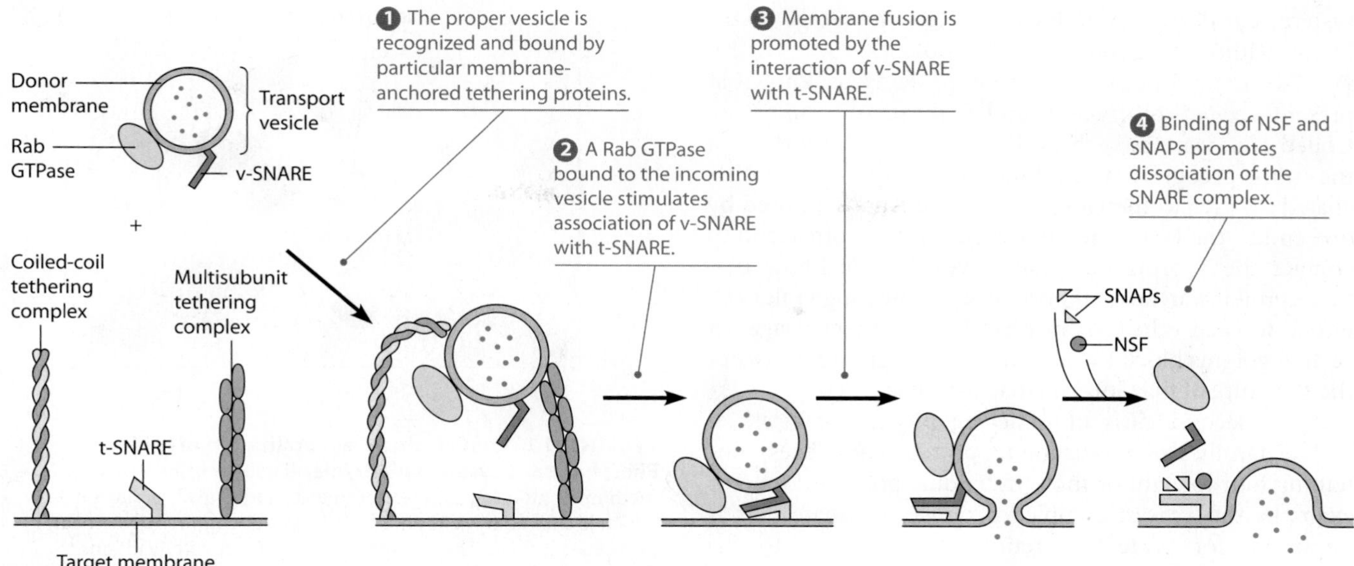

MB Transport Vesicle Targeting and Fusion

FIGURE 12-19 The SNARE Hypothesis for Transport Vesicle Targeting and Fusion. The basic molecular components that mediate sorting and targeting of vesicles in eukaryotic cells include tethering proteins, v-SNAREs on transport vesicles, t-SNAREs on target membranes, Rab GTPase, NSF, and several SNAPs. The exact timing of GTP or ATP hydrolysis is still unclear, but it most likely occurs after vesicle fusion.

responsible for the specificity displayed when receptors are concentrated for inclusion in a budding vesicle.

Once a vesicle forms, however, additional proteins are needed to ensure delivery of the vesicle to the appropriate destination. Therefore, there must be a mechanism to keep the various vesicles in the cell from accidentally fusing with the wrong membrane. The **SNARE hypothesis** provides a working model for this important sorting and targeting step in intracellular transport (**Figure 12-19**).

According to the SNARE hypothesis, the proper sorting and targeting of vesicles in eukaryotic cells involves two families of **SNARE (SNAP receptor) proteins**: the **v-SNAREs (vesicle-SNAP receptors)** found on transport vesicles and the **t-SNAREs (target-SNAP receptors)** found on target membranes. The v-SNAREs and t-SNAREs are complementary molecules that, along with additional tethering proteins, allow a vesicle to recognize and fuse with a target membrane. Both v-SNAREs and t-SNAREs were originally investigated because of their role in neuronal exocytosis. Since their discovery in brain tissue, both families of proteins have also been implicated in transport from the ER to the Golgi complex in yeast and other organisms.

When a vesicle reaches its destination, a third family of proteins, the **Rab GTPases,** comes into play. Rab GTPases are also specific: Vesicles fated for different destinations have distinct members of the Rab family associated with them. As illustrated in Figure 12-19, the affinity of complementary v-SNAREs and t-SNAREs for one another enables them to form a stable complex. This ensures that, when these proteins collide, they will remain in contact long enough for a Rab protein associated with the vesicle to lock the complementary t-SNARE and v-SNARE together, facilitating membrane fusion.

Following vesicle fusion, **N-ethylmaleimide-sensitive factor (NSF)** and a group of **soluble NSF attachment proteins (SNAPs)** mediate release of the v- and t-SNAREs of the donor and target membranes. ATP hydrolysis may be involved at this step, but its precise role is unclear. NSF and SNAPs are involved in fusion between a variety of cellular membranes, indicating they are not responsible for specificity during targeting.

SNARE proteins are required for the fusion of neurotransmitter-containing vesicles with the plasma membrane of nerve cells. This fusion event releases the neurotransmitter by exocytosis, leading to an electrical impulse that will initiate muscle contraction. Botulinum toxin (Botox), produced by the bacterium *Clostridium botulinum,* is a protease that cleaves a SNARE protein that is required for this fusion. Thus, the toxin interferes with muscle contraction and can cause paralysis. Although it is one of the most potent biological toxins known, in very small doses Botox can be used therapeutically to control muscle spasms or correct crossed eyes. It is also used cosmetically to remove wrinkles caused by muscle contractions in the skin. Currently, it is being investigated for possible use as a treatment for migraine headaches.

Recent research suggests that SNARE proteins alone cannot account for the specificity observed in vesicle targeting. A different class of proteins known as **tethering proteins** acts over longer distances and provides specificity by connecting vesicles to their target membranes prior to v-SNARE/t-SNARE interaction (Figure 12-19). We know this because experimental toxin-induced cleavage of SNARE proteins in vivo can block SNARE complex formation without blocking vesicle association with the target membrane. Also, in an in vitro reconstituted system, ER-derived vesicles can attach to Golgi membranes without addition of SNARE proteins.

Two main groups of tethering proteins are known at present—*coiled-coil proteins* and *multisubunit complexes.* Coiled-coil proteins such as the **golgins** are important in the initial recognition and binding of COPI- or COPII-coated vesicles to the Golgi. The golgins are anchored by one end to the Golgi membrane and use the other end to contact the appropriate passing vesicle. We know that these proteins are also important in connecting Golgi cisternae to each other because antibodies directed against certain golgins block the action of the golgins and disrupt the structure of the Golgi medial cisternae.

The second class of tethering proteins consists of several families of multisubunit protein complexes containing four to eight or more individual polypeptides. For example, the *exocyst* complex of yeast and mammals is important for protein secretion, binding both to the plasma membrane and to vesicles from the TGN whose contents are destined for export. Other types of multisubunit tethering complexes such as the *COG (conserved oligomeric Golgi)* complex, the *GARP (Golgi-associated retrograde protein)* complex, and the *TRAPP (transport protein particle)* complex are implicated in the initial recognition and specificity of vesicle–target membrane interaction. Most of the proteins in these complexes are highly conserved among organisms as different as yeast and humans. Identifying the functions of these complexes and the roles of their individual subunits is currently one of the most intriguing frontiers of modern cell biology.

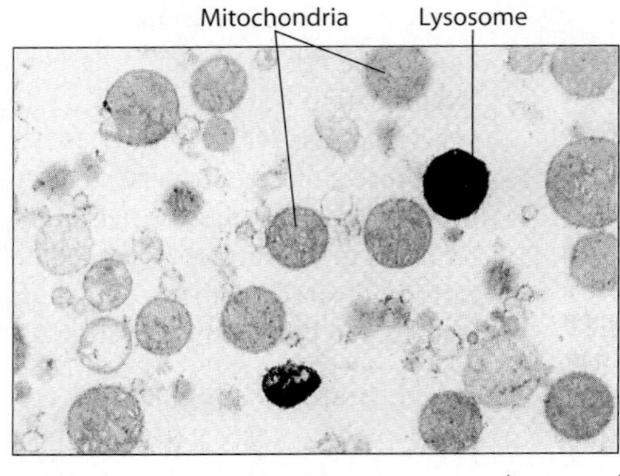

FIGURE 12-20 Cytochemical Localization of Acid Phosphatase, a Lysosomal Enzyme. Tissue was incubated in a medium containing soluble lead nitrate and β-glycerophosphate, which is cleaved by acid phosphatase, producing free glycerol and phosphate anions. The phosphate anions react with lead ions to form insoluble lead phosphate, which precipitates at the site of enzyme activity and reveals the location of acid phosphatase within the cell. The darkly stained organelles shown here are lysosomes highlighted by deposits of the electron-dense lead phosphate. They are surrounded by mitochondria, which lack acid phosphatase and do not become electron dense (TEM).

Lysosomes and Cellular Digestion

The **lysosome** is an organelle of the endomembrane system that contains digestive enzymes capable of degrading all the major classes of biological macromolecules—lipids, carbohydrates, nucleic acids, and proteins. These hydrolytic enzymes degrade extracellular materials brought into the cell by endocytosis and digest intracellular structures and macromolecules that are damaged or no longer needed. We will first look at the organelle itself and then consider lysosomal digestive processes, as well as some of the diseases that result from lysosomal malfunction.

Lysosomes Isolate Digestive Enzymes from the Rest of the Cell

As we learned in Chapter 4, lysosomes were discovered in the early 1950s by Christian de Duve and his colleagues (see Box 4B, page 92). Differential centrifugation led the researchers to realize that an acid phosphatase initially thought to be located in the mitochondrion was in fact associated with a class of particles that had never been reported before. Along with the acid phosphatase, the new organelle contained several other hydrolytic enzymes, including β-glucuronidase, a deoxyribonuclease, a ribonuclease, and a protease. Because of its apparent role in cellular lysis, de Duve called this newly discovered organelle a *lysosome.*

Only after the lysosome's existence had been predicted, its properties described, and its enzyme content specified was the organelle actually observed by electron microscopy and recognized as a normal constituent of most animal cells. Final confirmation came from cytochemical staining reactions capable of localizing the acid phosphatase and other lysosomal enzymes to specific structures that can be seen by electron microscopy (**Figure 12-20**).

Lysosomes vary considerably in size and shape but are generally about 0.5 μm in diameter. Like the ER and Golgi complex, the lysosome is bounded by a single membrane. This membrane protects the rest of the cell from the hydrolytic enzymes in the lysosomal lumen. The lumenal side of lysosomal membrane proteins is highly glycosylated, forming a nearly continuous carbohydrate coating that appears to protect membrane proteins from lysosomal proteases. ATP-dependent proton pumps in the membrane maintain an acidic environment (pH 4.0–5.0) within the lysosome. This favors enzymatic digestion of macromolecules both by activating acid hydrolases and by partially denaturing the macromolecules targeted for degradation. The products of digestion are then transported across the membrane to the cytosol, where they enter various synthetic pathways or are exported from the cell.

The list of lysosomal enzymes has expanded considerably since de Duve's original work, but all have the common property of being *acid hydrolases*—hydrolytic enzymes with a pH optimum around 5.0. The list includes at least 5 phosphatases, 14 proteases and peptidases, 2 nucleases, 6 lipases, 13 glycosidases, and 7 sulfatases. Taken together, these lysosomal enzymes can digest all the major classes of biological molecules. No wonder, then, that they are sequestered from the rest of the cell, where they cannot quickly destroy the cell itself.

Lysosomes Develop from Endosomes

Lysosomal enzymes are synthesized by ribosomes attached to the rough ER and are translocated through a pore in the ER membrane into the ER lumen before transport to the Golgi complex. After modification and processing in the ER and Golgi complex compartments, the lysosomal enzymes are sorted from other proteins in the TGN. Earlier in the chapter, we described the addition of a unique mannose-6-phosphate tag to soluble lysosomal enzymes. Distinctive sorting signals are also present on membrane-bound lysosomal proteins. The lysosomal enzymes are packaged in clathrin-coated vesicles that bud from the TGN, lose their protein coats, and travel to one of the endosomal compartments (see Figure 12-9).

Lysosomal enzymes are delivered from the TGN to endosomes in transport vesicles, as shown in **Figure 12-21**. Recall that early endosomes are formed by the coalescence of vesicles from the TGN and vesicles from the plasma membrane. Over time, the early endosome matures to form a late endosome, an organelle having a full complement of acid hydrolases but not engaged in digestive activity. As the pH of the early endosomal lumen drops from about 6.0 to 5.5, the organelle loses its capacity to fuse with endocytic vesicles. The late endosome is essentially a collection of newly synthesized digestive enzymes as well as extracellular and intracellular material fated for digestion, packaged in a way that protects the cell from hydrolytic enzymes.

The final step in lysosome development is the activation of the acid hydrolases, which occurs as the enzymes and their substrates encounter a more acidic environment. There are two ways eukaryotic cells accomplish this step. ATP-dependent proton pumps may lower the pH of the late endosomal lumen to 4.0–5.0, transforming the late endosome into a lysosome, thereby generating a new organelle. Alternatively, the late endosome may transfer material to the acidic lumen of an existing lysosome.

Lysosomal Enzymes Are Important for Several Different Digestive Processes

Lysosomes are important for cellular activities as diverse as nutrition, defense, recycling of cellular components, and differentiation. We can distinguish the digestive processes that depend on lysosomal enzymes by the site of their activity and by the origin of the material that is digested, as shown in Figure 12-21. Usually, the site of activity is intracellular. In some cases, though, lysosomes may release their enzymes to the outside of the cell by exocytosis. The materials to be digested are often of extracellular origin, although there are also important processes known to involve lysosomal digestion of internal cellular components.

To distinguish between mature lysosomes of different origins, we refer to those containing substances of extracellular origin as **heterophagic lysosomes,** whereas those with

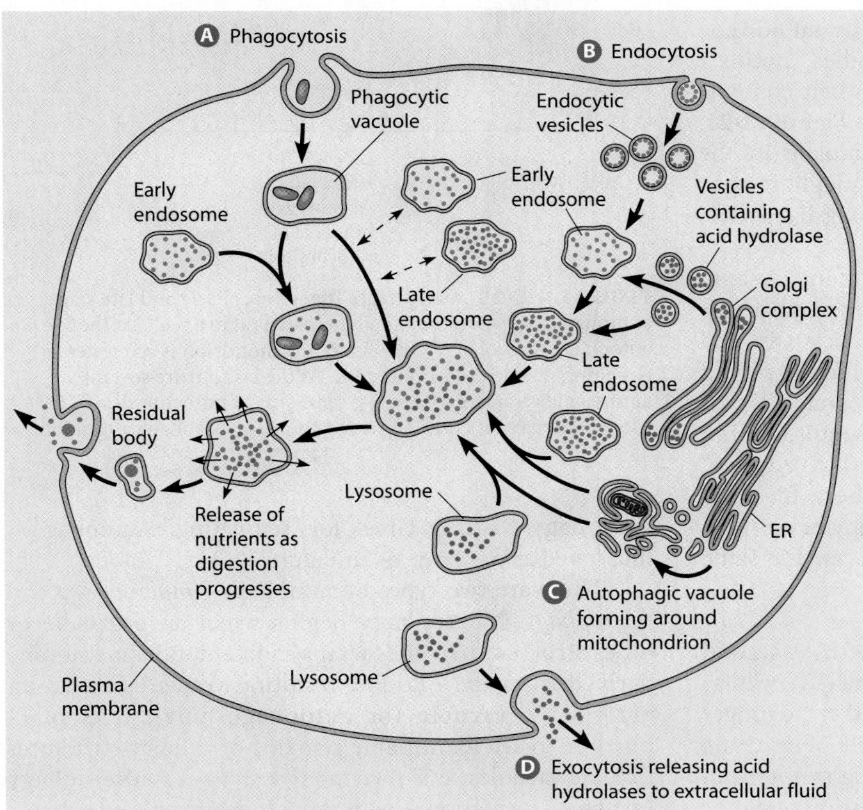

FIGURE 12-21 The Formation of Lysosomes and Their Roles in Cellular Digestive Processes. Illustrated in this composite cell are the major processes in which lysosomes are involved. The pathways depicted are (A) phagocytosis, (B) receptor-mediated endocytosis, (C) autophagy, and (D) extracellular digestion. For clarity, the nucleus is not shown.

materials of intracellular origin are called **autophagic lysosomes.** The specific processes in which lysosomal enzymes are involved are *phagocytosis, receptor-mediated endocytosis, autophagy,* and *extracellular digestion.* These are illustrated in Figure 12-21 as pathways Ⓐ, Ⓑ, Ⓒ, and Ⓓ, respectively.

Phagocytosis and Receptor-Mediated Endocytosis: Lysosomes in Defense and Nutrition. One of the most important functions of lysosomal enzymes is the degradation of foreign material brought into eukaryotic cells by *phagocytosis* and *receptor-mediated endocytosis* (see Figures 12-14 and 12-15). Phagocytic vacuoles are transformed into lysosomes by fusing with early endosomes (Figure 12-21, pathway Ⓐ). Depending on the material ingested, these lysosomes can vary considerably in size, appearance, content, and stage of digestion. Vesicles containing material brought into a cell by receptor-mediated endocytosis also form early endosomes (pathway Ⓑ of Figure 12-21). As early endosomes fuse with vesicles from the TGN containing acid hydrolases, they mature to form late endosomes and lysosomes, in which the ingested material is digested.

Soluble products of digestion, such as sugars, amino acids, and nucleotides, are then transported across the lysosomal membrane into the cytosol and are used as a source of nutrients by the cell. Some may cross by facilitated diffusion, whereas others undergo active transport. The acidity of the lysosomal lumen contributes to an electrochemical proton gradient across the lysosomal membrane, which can provide energy for driving transport to and from the cytosol.

Eventually, however, only indigestible material remains in the lysosome, which becomes a **residual body** as digestion ceases. In protozoa, residual bodies routinely fuse with the plasma membrane and expel their contents to the outside by exocytosis, as illustrated in Figure 12-21. In vertebrates, residual bodies may accumulate in the cytoplasm. This accumulation of debris is thought to contribute to cellular aging, particularly in long-lived cells such as those of the nervous system.

Certain white blood cells of the immune system, however, use this residual material. After neutrophils digest invading microorganisms via phagocytosis and lysosome action, they release the debris, which is picked up by scavenging macrophages. These macrophages transport the debris to the lymph nodes and present it to other immune system cells (B and T lymphocytes) to "educate" them regarding what foreign material has been found in the body. This process causes the lymphocytes to form memory cells that will quickly respond in case the same microorganism is encountered in the future.

Autophagy: A Biological Recycling System. A second important task for lysosomes is the breakdown of cellular structures and components that are damaged or no longer needed. Most cellular organelles are in a state of dynamic flux, with new organelles continuously being synthesized while old organelles are destroyed. The digestion of old or unwanted organelles or other cell structures is called **autophagy,** which is Greek for "self-eating." Autophagy is illustrated as pathway Ⓒ in Figure 12-21.

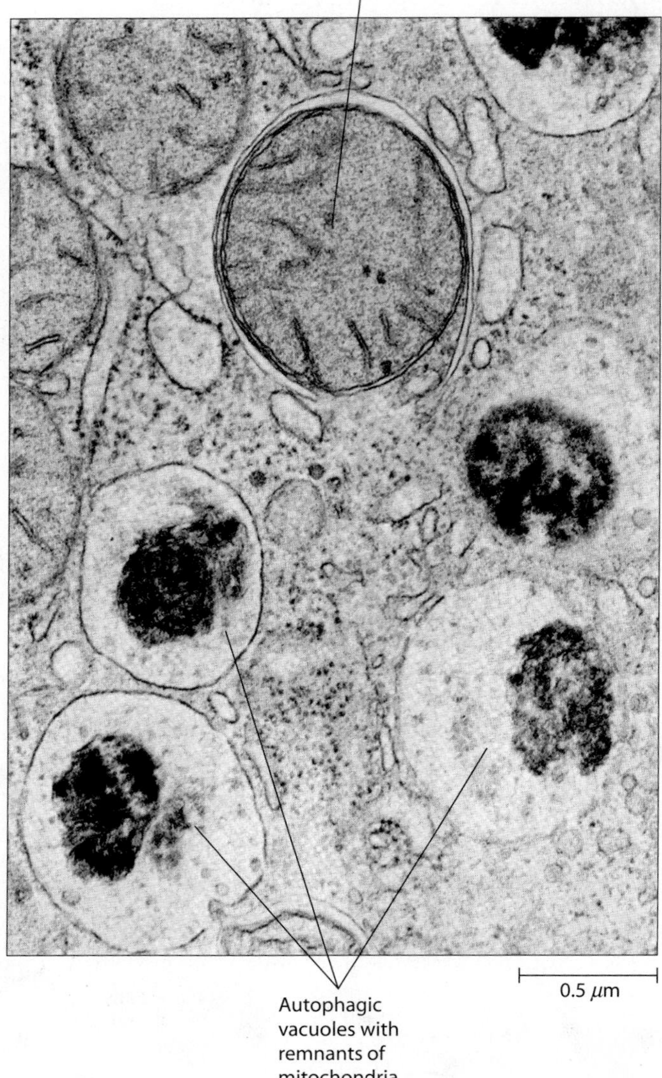

FIGURE 12-22 Autophagic Digestion. Early and late stages of autophagic digestion are shown here in a rat liver cell. At the top, an autophagic vacuole is formed as a mitochondrion is sequestered by a membrane derived from the ER. At the bottom are several autophagic vacuoles containing remnants of mitochondria (TEM). Most old organelles are eliminated through autophagic digestion.

There are two types of autophagy—*macrophagy* and *microphagy.* **Macrophagy** begins when an organelle or other structure becomes wrapped in a double membrane derived from the ER. The resulting vesicle is called an **autophagic vacuole** (or **autophagosome**). It is often possible to see identifiable remains of cellular structures in these vacuoles, as shown in **Figure 12-22.** **Microphagy** involves formation of a much smaller autophagic vacuole, surrounded by a single phospholipid bilayer that encloses small bits of cytoplasm rather than whole organelles.

Autophagy occurs at varying rates in most cells under most conditions, but it is especially prominent in red blood cell development. As a red blood cell matures, virtually all of the intracellular content is destroyed, including all of the mitochondria. This destruction is accomplished by autophagic digestion. A marked increase in autophagy is also noted in cells stressed by starvation. Presumably, the process represents a desperate attempt by the cell to continue providing for its energy needs, even if it has to consume its own structures to do so.

It has long been suspected that some cancer cells may be deficient in autophagy, but recent work suggests a direct link between autophagy and cancer. The human version of a gene required for autophagy in yeast cells has been shown to be frequently deleted in human breast and ovarian tumors. Knocking out the function of this gene in mice caused a decrease in autophagy and an increase in the number of breast and lung tumors. However, some oncogenes that promote cancer suppress autophagy during tumor development. Clearly, more research is needed.

Extracellular Digestion. While most digestive processes involving lysosomal enzymes occur intracellularly, in some cases lysosomes discharge their enzymes to the outside of the cell by exocytosis, resulting in **extracellular digestion** (Figure 12-21, pathway Ⓓ). One example of extracellular digestion occurs during fertilization of animal eggs. The head of the sperm releases lysosomal enzymes capable of degrading chemical barriers that would otherwise keep the sperm from penetrating the egg surface. Certain inflammatory diseases, such as rheumatoid arthritis, may result from the inadvertent release of lysosomal enzymes by white blood cells in the joints, damaging the joint tissue. The steroid hormones cortisone and hydrocortisone are thought to be effective anti-inflammatory agents because of their role in stabilizing lysosomal membranes and thereby inhibiting enzyme release.

Lysosomal Storage Diseases Are Usually Characterized by the Accumulation of Indigestible Material

The essential role of lysosomes in the recycling of cellular components is clearly seen in disorders caused by deficiencies of specific lysosomal proteins. Over 40 such **lysosomal storage diseases** are known, each characterized by the harmful accumulation of specific substances, usually polysaccharides or lipids (see Box 4A, page 86). In most cases, the substances accumulate because digestive enzymes are defective or missing, but they sometimes accumulate because the proteins that transport degradation products from the lysosomal lumen to the cytosol are defective. In either case, the cells in which material accumulates are severely impaired or destroyed. Skeletal deformities, muscle weakness, and mental retardation commonly result, and are often fatal. Unfortunately, most lysosomal storage diseases are not yet treatable.

The first storage disease to be understood was *type II glycogenosis,* in which young children accumulate excessive amounts of glycogen in the liver, heart, and skeletal muscles and die at an early age. The problem turned out to be a defective form of the lysosomal enzyme *α-1,4-glucosidase,* which catalyzes glycogen hydrolysis in normal cells. Although glycogen metabolism occurs predominantly in the cytosol, a small amount of glycogen can enter the lysosome through autophagy and will accumulate to a damaging level if not broken down to glucose.

Two of the best-known lysosomal storage diseases are *Hurler syndrome* and *Hunter syndrome.* Both arise from defects in the degradation of glycosaminoglycans, which are the major carbohydrate components of the extracellular matrix (see Chapter 17). The defective enzyme in a patient with Hurler syndrome is *α-L-iduronidase,* which is required for the degradation of glycosaminoglycans. Electron microscopic observation of sweat gland cells from a patient with Hurler syndrome reveals large numbers of atypical vacuoles that stain for both acid phosphatase and undigested glycosaminoglycans. These vacuoles are apparently abnormal late endosomes filled with indigestible material.

Mental retardation is a common feature of lysosomal storage diseases. It can occur due to the impaired metabolism of glycolipids, which are important components of brain tissue and the sheaths of nerve cell axons. One particularly well-known example is *Tay-Sachs disease,* a condition inherited as a recessive trait. Afflicted children show rapid mental deterioration after about six months of age, followed by paralysis and death within three years. The disease results from the accumulation in nervous tissue of a particular glycolipid called a *ganglioside.* The missing lysosomal enzyme in this case is *β-N-acetylhexosaminidase,* which is responsible for cleaving the terminal *N*-acetylgalactosamine from the carbohydrate portion of the ganglioside. Lysosomes from children afflicted with Tay-Sachs disease are filled with membrane fragments containing undigested gangliosides.

The Plant Vacuole: A Multifunctional Organelle

Plant cells contain acidic membrane-enclosed compartments called vacuoles that resemble the lysosomes found in most animal cells but generally serve additional roles. The biogenesis of a vacuole parallels that of a lysosome. Most of the components are synthesized in the ER and transferred to the Golgi complex, where proteins undergo further processing. Coated vesicles then convey lipids and proteins destined for the vacuole to a **provacuole,** which is analogous to an endosome. The provacuole eventually matures to form a functional vacuole that can fill as much as 90% of the volume of a plant cell.

In addition to confining hydrolytic enzymes, plant vacuoles have various other essential functions. Most of these functions reflect the plant's lack of mobility and consequent susceptibility to changes in the surrounding environment. As mentioned in Chapter 4, a major role of the vacuole is to maintain *turgor pressure,* the osmotic pressure that prevents plant cells from collapsing. Turgor pressure prevents a plant from wilting, and it can drive the expansion of cells. During development, softening of the cell wall—accompanied by higher turgor pressure—allows the cell to expand.

Another role of the plant vacuole is the regulation of cytosolic pH. ATP-dependent proton pumps in the vacuolar membrane can compensate for a decline in cytosolic pH (perhaps due to a change in the extracellular environment) by transferring protons from the cytosol to the lumen of the vacuole.

The vacuole also serves as a storage compartment. Seed storage proteins are generally synthesized by ribosomes attached to the rough ER and cotranslationally inserted into the ER lumen. Some of the storage proteins remain in the ER while others are transferred to vacuoles, either by autophagy of vesicles budding from the ER or by way of the Golgi. When the seeds germinate, the storage proteins are available for hydrolysis by vacuolar proteases, thereby releasing amino acids for the biosynthesis of new proteins needed by the growing plant.

Other substances found in vacuoles include malate stored in CAM plants (Chapter 11), the anthocyanins that impart color to flowers, toxic substances that deter predators, inorganic and organic nutrients, compounds that shield cells from ultraviolet light, and residual indigestible waste. Storage of soluble as well as insoluble waste is an important function of plant vacuoles. Unlike animals, most plants do not have a mechanism for excreting soluble waste from the organism. The large vacuoles found in plant cells enable the cells to accumulate solutes to a degree that would inhibit or restrict metabolic processes if the material were to remain in the cytosol.

Peroxisomes

Peroxisomes, like the Golgi complex, endosomes, and lysosomes, are bounded by single membranes. However, they are not derived from the endoplasmic reticulum and are therefore not part of the endomembrane system that includes the other organelles discussed in this chapter. Peroxisomes are found in all eukaryotic cells but are especially prominent in mammalian kidney and liver cells, in algae and photosynthetic cells of plants, and in germinating seedlings of plant species that store fat in their seeds. Peroxisomes are somewhat smaller than mitochondria, though there is considerable variation in size, depending on their function and the tissue where they are found.

Regardless of location or size, the defining characteristic of a peroxisome is the presence of *catalase,* an enzyme essential for the degradation of hydrogen peroxide (H_2O_2). Hydrogen peroxide is a potentially toxic compound that is formed by a variety of oxidative reactions catalyzed by *oxidases.* Both catalase and the oxidases are confined to peroxisomes. Thus, the generation and degradation of H_2O_2 occur within the same organelle, thereby protecting other parts of the cell from exposure to this harmful compound. Before discussing the functions of peroxisomes further, let us look at how peroxisomes were discovered and how they are distinguished from other organelles when viewed by electron microscopy.

The Discovery of Peroxisomes Depended on Innovations in Equilibrium Density Centrifugation

Christian de Duve and his colleagues discovered not only lysosomes, but also peroxisomes. During the course of their early studies on lysosomes, they encountered at least one enzyme, *urate oxidase,* that was associated with lysosomal fractions but was not an acid hydrolase. By using a gradient of sucrose concentration, the researchers found that urate oxidase from rat liver was recovered in a region of the gradient having a slightly higher density than that of other organelles, such as lysosomes and mitochondria (see Figure 12A-4).

Once separation of this new organelle was achieved, additional enzymes were identified in the fractions containing urate oxidase, including catalase and *D-amino acid oxidase.* Catalase, as we have seen, degrades H_2O_2. Like urate oxidase, D-amino acid oxidase generates H_2O_2. Because of its apparent involvement in hydrogen peroxide metabolism, the new organelle became known as a *peroxisome.* Other peroxisomal enzymes have since been identified, and it is now clear that the enzyme complement of the organelle varies significantly from species to species, from organ to organ, and in some cases from one developmental stage to another within the same organ. However, the presence of catalase and one or more hydrogen peroxide–generating oxidases remains a distinguishing characteristic of all peroxisomes.

Once peroxisomes had been identified and isolated biochemically, the existence of organelles with the expected properties was confirmed by electron microscopy. Peroxisomes turned out to be the functional equivalents of organelles that had been seen earlier in electron micrographs of both animal and plant cells. Because their function was not known at the time, these organelles were simply called *microbodies.* In both plant and animal cells, peroxisomes are usually about 0.2–2.0 μm in diameter, are surrounded by a single membrane, and generally have a finely granular or crystalline interior.

As seen in **Figure 12-23**, animal peroxisomes often contain a distinct crystalline core, which usually consists of a crystalline form of urate oxidase. Crystalline cores consisting of catalase may be present in the peroxisomes of plant leaves (see Figure 4-20). When such crystals are present, it is easy to identify peroxisomes, since urate oxidase and catalase are two of the enzymes by which peroxisomes are defined. In the absence of a crystalline core, however, it is not always easy to spot peroxisomes ultrastructurally.

A useful technique in such cases is a cytochemical test for catalase called the *diaminobenzidine (DAB) reaction.* Catalase oxidizes DAB to a polymeric form that causes deposition of electron-dense osmium atoms when the tissue is treated with osmium tetroxide. The resulting electron-dense deposits can be readily seen in cells from

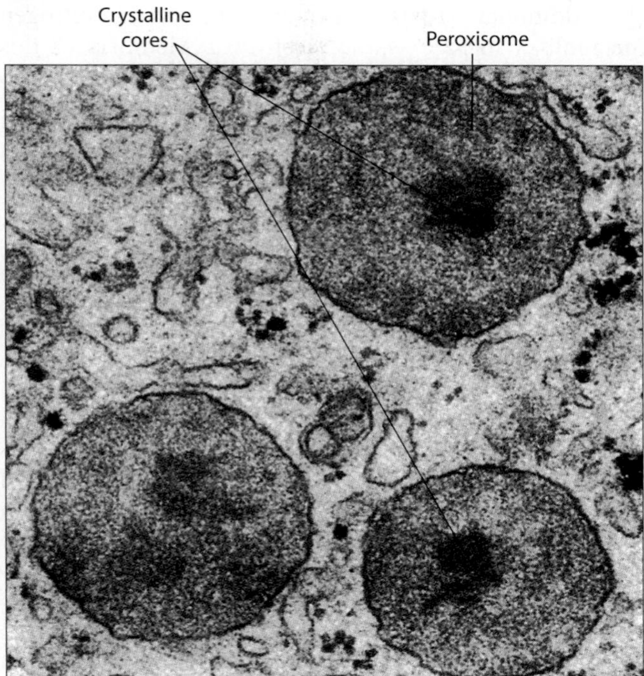

FIGURE 12-23 Peroxisomes in Animal Cells. This electron micrograph shows several peroxisomes (microbodies) in the cytoplasm of a rat liver cell. A crystalline core is readily visible in each microbody. In animal microbodies, the cores are almost always crystalline urate oxidase (TEM).

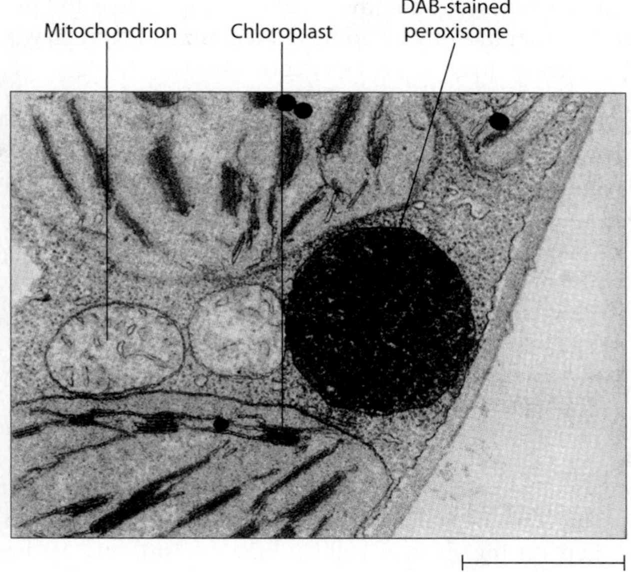

FIGURE 12-24 Cytochemical Localization of Catalase in Plant Peroxisomes. Shown here is a tobacco leaf cell similar to the one shown in Figure 4-20, but stained by diaminobenzidine (DAB). The principle of this assay is similar to that of the cytochemical test for acid phosphatase described in Figure 12-20. Catalase oxidizes DAB to a polymeric form that causes the deposition of electron-dense osmium atoms in tissue treated with osmium tetroxide (OsO_4). The DAB technique reveals that the deposition of osmium is confined to peroxisomes and thus that catalase is a component of the crystalline core (TEM).

stained tissue. In animal peroxisomes, the entire internal space often stains intensely with DAB, indicating that catalase exists as a soluble enzyme uniformly distributed throughout the matrix of the organelle. Similarly, in plant leaf peroxisomes lacking a crystalline core of catalase, DAB staining is observed through the entire peroxisome (**Figure 12-24**). Because catalase is the single enzyme present in all peroxisomes and does not routinely occur in any other organelle, the DAB reaction is a reliable and specific means of identifying organelles as peroxisomes.

Most Peroxisomal Functions Are Linked to Hydrogen Peroxide Metabolism

Peroxisomes occur widely in animals, plants, algae, and some fungi. In animals, peroxisomes are most prominent in liver and kidney tissue. The essential roles of peroxisomes in eukaryotic cells have become more apparent in recent years, stimulating new research into the metabolic pathways and the disorders arising from defective components of the pathways found in these organelles. There are at least five general categories of peroxisomal functions that we will discuss in the sections below: hydrogen peroxide metabolism, detoxification of harmful compounds, oxidation of fatty acids, metabolism of nitrogen-containing compounds, and catabolism of unusual substances.

Hydrogen Peroxide Metabolism. A significant role of peroxisomes in eukaryotic cells is the detoxification of H_2O_2 by catalase, which can comprise up to 15% of the total protein in the peroxisome. The oxidases that generate H_2O_2 in peroxisomes transfer electrons plus hydrogen ions (hydrogen atoms) from their substrates to molecular oxygen (O_2), reducing it to H_2O_2. Using RH_2 to represent an oxidizable substrate, the general reaction catalyzed by oxidases can be written as

$$RH_2 + O_2 \longrightarrow R + H_2O_2 \qquad \textbf{(12-3)}$$

The hydrogen peroxide formed in this manner is broken down by catalase in one of two ways. Usually, catalase functions by detoxifying two molecules of H_2O_2 simultaneously—one is oxidized to oxygen, and a second is reduced to water:

$$2H_2O_2 \longrightarrow O_2 + 2H_2O \qquad \textbf{(12-4)}$$

Alternatively, catalase can function as a *peroxidase,* in which electrons derived from an organic donor are used to reduce hydrogen peroxide to water:

$$R'H_2 + H_2O_2 \longrightarrow R' + 2H_2O \qquad \textbf{(12-5)}$$

(The prime on the R group simply indicates that this substrate is likely to be different from the substrate in Reaction 12-3.)

The result is the same in either case: Hydrogen peroxide is degraded without ever leaving the peroxisome. Given the toxicity of hydrogen peroxide (which is the main active ingredient in a variety of disinfectants), it makes good sense for the enzymes responsible for peroxide generation to be compartmentalized together with the catalase that catalyzes its degradation.

Detoxification of Harmful Compounds. As a peroxidase (Reaction 12-5), catalase can use a variety of toxic substances as electron donors, including methanol, ethanol, formic acid, formaldehyde, nitrites, and phenols. Because all of these compounds are harmful to cells, their oxidative detoxification by catalase may be a vital peroxisomal function. The prominent peroxisomes of liver and kidney cells are thought to be important in such detoxification reactions.

In addition, peroxisomal enzymes are important in the detoxification of **reactive oxygen species** such as H_2O_2, superoxide anion (O_2^-), hydroxyl radical (OH• where the dot signifies an unpaired, highly reactive electron), and organic peroxide conjugates. These reactive oxygen species can be formed in the presence of molecular oxygen during normal cellular metabolism, and if they accumulate the cell can suffer from *oxidative stress.* Peroxisomal enzymes such as *superoxide dismutase,* catalase, and peroxidases detoxify these reactive oxygen species, preventing their accumulation and subsequent oxidative damage to cellular components.

Oxidation of Fatty Acids. Peroxisomes found in animal, plant, and fungal cells contain enzymes necessary for oxidizing fatty acids via *β oxidation* to provide energy for the cell. About 25–50% of fatty acid oxidation in animal tissues occurs in peroxisomes, with the remainder localized in mitochondria. In plant and yeast cells, on the other hand, all β oxidation occurs in peroxisomes.

In animal cells, peroxisomal β oxidation is especially important for degrading long-chain (16–22 carbons), very long-chain (24–26 carbons), and branched fatty acids. The primary product of β oxidation, acetyl-CoA, is then exported to the cytosol, where it enters biosynthetic pathways or the TCA cycle. Once fatty acids are shortened to fewer than 16 carbons, further oxidation usually occurs in the mitochondria. Thus, in animal cells, the peroxisome is important for shortening fatty acids in preparation for subsequent metabolism in the mitochondrion rather than completely breaking them down to acetyl-CoA. In plants and yeast, on the other hand, peroxisomes are essential for the complete catabolism of all fatty acids to acetyl-CoA.

Metabolism of Nitrogen-Containing Compounds. Except for primates, most animals require *urate oxidase* (also called *uricase*) to oxidize urate, a purine that is formed during the catabolism of nucleic acids and some proteins. Like other oxidases, urate oxidase catalyzes the direct transfer of hydrogen atoms from the substrate to molecular oxygen, generating H_2O_2:

$$\text{urate} + O_2 \longrightarrow \text{allantoin} + H_2O_2 \qquad \textbf{(12-6)}$$

As noted earlier, the H_2O_2 is immediately degraded in the peroxisome by catalase. The allantoin is further metabolized and excreted by the organism, either as allantoic acid or—in the case of crustaceans, fish, and amphibians—as urea.

Additional peroxisomal enzymes involved in nitrogen metabolism include *aminotransferases.* Members of this collection of enzymes catalyze the transfer of amino groups ($-NH_3^+$) from amino acids to α-keto acids:

$$\underset{\text{Amino acid}}{\mathrm{R{-}\overset{\overset{H_3N^+}{|}}{C}{-}\overset{\overset{O}{\|}}{C}{-}O^-}} + \underset{\alpha\text{-keto acid}}{\mathrm{R'{-}\overset{\overset{O}{\|}}{C}{-}\overset{\overset{O}{\|}}{C}{-}O^-}} \rightleftharpoons \underset{\alpha\text{-keto acid}}{\mathrm{R{-}\overset{\overset{O}{\|}}{C}{-}\overset{\overset{O}{\|}}{C}{-}O^-}} + \underset{\text{Amino acid}}{\mathrm{R'{-}\overset{\overset{H_3N^+}{|}}{C}{-}\overset{\overset{O}{\|}}{C}{-}O^-}} \qquad \textbf{(12-7)}$$

Such enzymes play important roles in the biosynthesis and degradation of amino acids by moving amino groups from one molecule to another (see Figure 10-12).

Catabolism of Unusual Substances. Some of the substrates for peroxisomal oxidases are rare compounds for which the cell has no other degradative pathways. Such compounds include D-amino acids, which are not recognized by enzymes capable of degrading the L-amino acids found in polypeptides. In some cells, the peroxisomes also contain enzymes that break down unusual substances called **xenobiotics,** chemical compounds foreign to biological organisms. This category includes *alkanes,* short-chain hydrocarbon compounds found in oil and other petroleum products. Fungi containing enzymes capable of metabolizing such xenobiotics may turn out to be useful for cleaning up oil spills that would otherwise contaminate the environment.

Peroxisomal Disorders. Considering the variety of metabolic pathways found in peroxisomes, it is not surprising that a large number of disorders arise from defective peroxisomal proteins (see Box 4A, page 86). The most common peroxisomal disorder is *X-linked adrenoleukodystrophy.* The defective protein causing this disorder is an integral membrane protein that may be responsible for transporting very long-chain fatty acids into the peroxisome for β oxidation. Accumulation of these long-chain fatty acids in body fluids destroys the myelin sheath in nervous tissues.

Plant Cells Contain Types of Peroxisomes Not Found in Animal Cells

In plants and algae, peroxisomes are involved in several specific aspects of cellular energy metabolism, which we discussed in Chapters 10 and 11. Here, we will simply introduce several plant-specific peroxisomes and briefly describe their functions.

Leaf Peroxisomes. Cells of leaves and other photosynthetic plant tissues contain characteristic large, prominent **leaf peroxisomes,** which often appear in close contact with chloroplasts and mitochondria (see Figure 12-24; see also

Figure 4-20). The spatial proximity of the three organelles probably reflects their mutual involvement in the *glycolate pathway* (see Figure 11-15), also called the *photorespiratory pathway* because it involves the light-dependent uptake of O_2 and release of CO_2. Several enzymes of this pathway, including a peroxide-generating oxidase and two aminotransferases, are confined to leaf peroxisomes.

Glyoxysomes. Another functionally distinct type of plant peroxisome occurs transiently in seedlings of plant species that store carbon and energy reserves in the seed as fat (primarily triacylglycerols). In such species, stored triacylglycerols are mobilized and converted to sucrose during early postgerminative development by a sequence of events that includes β oxidation of fatty acids as well as a pathway known as the *glyoxylate cycle.* All of the enzymes needed for these processes are localized to specialized peroxisomes called **glyoxysomes** (see Box 10A, page 268).

Glyoxysomes are found only in the tissues where the fat is stored (endosperm or cotyledons, depending on the species) and are present only for the relatively short period of time required for the seedling to deplete its supply of stored fat. Once they fulfill their role in the seedling, the glyoxysomes are converted to peroxisomes. Glyoxysomes have been reported to appear again in the senescing (aging) tissues of some plant species, presumably to degrade lipids derived from the membranes of the senescent cells. However, the importance of their involvement in senescence is not yet clear.

Other Kinds of Plant Peroxisomes. In addition to their presence in tissues that carry out either photorespiration or β oxidation of fatty acids, peroxisomes are found in other plant tissues. For example, another kind of specialized peroxisome is present in *nodules,* the structures on plant roots in which plant cells and certain bacteria cooperate in the fixation of atmospheric nitrogen (that is, the conversion of N_2 into organic form). The peroxisomes in these cells are involved in the processing of fixed nitrogen.

Peroxisome Biogenesis Occurs by Division of Preexisting Peroxisomes

Like other organelles, peroxisomes increase in number as cells grow and divide. This proliferation of organelles is called *biogenesis,* and the peroxisomal proteins required for this process are known as *peroxins.* Biogenesis of endosomes and lysosomes occurs by fusion of vesicles budding from the Golgi complex, and peroxisomes were once thought to form from vesicles in a similar manner. Later, most investigators believed that peroxisome biogenesis occurred solely from the division of preexisting peroxisomes. Recent evidence suggests that new peroxisomes can be formed by either of these two methods, or perhaps by a combination of the two. Either way, their biogenesis raises two important questions.

First, where do the lipids that make up the newly synthesized peroxisomal membrane come from? We know that some of the lipids are synthesized by peroxisomal enzymes whereas others are synthesized in the ER and carried to the peroxisome by phospholipid exchange proteins. However, there is some evidence that these exchange proteins may not always be efficient enough to account for the rapid incorporation of new lipids and that some direct transfer from ER-derived vesicles may be involved.

Second, where are the new enzymes and other proteins that are present in the peroxisomal membrane and matrix synthesized? Proteins destined for peroxisomes are synthesized on free cytosolic ribosomes and are incorporated into preexisting peroxisomes posttranslationally. This passage of polypeptides across the peroxisomal membrane is an ATP-dependent process mediated by specific membrane peroxins, although the precise role of ATP is unclear.

Figure 12-25 illustrates both the incorporation of membrane components, matrix enzymes, and enzyme cofactors from the cytosol into a peroxisome (steps ❶–❸) and the formation of new peroxisomes by division of a preexisting organelle (step ❹). The protein depicted in the figure is catalase, a tetrameric protein with a heme group bound to each subunit. The subunits are synthesized

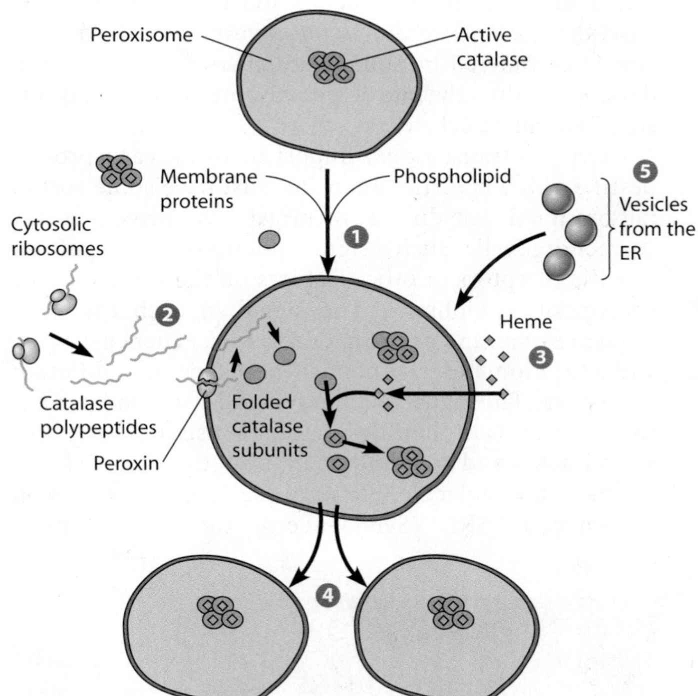

FIGURE 12-25 Biogenesis of Peroxisomes and Protein Import. New peroxisomes typically arise by the division of existing peroxisomes rather than by fusion of vesicles from the Golgi complex. ❶ Lipids and membrane proteins can be added to existing peroxisomes via cytosolic sources. ❷ Polypeptides for peroxisomal matrix enzymes are synthesized on cytosolic ribosomes and are threaded through the membrane via a transmembrane peroxin transport protein. The enzyme shown here is catalase, a tetrameric protein that requires a heme cofactor. ❸ Heme enters the peroxisomal lumen via a separate pathway, and the catalase polypeptides are folded and assembled with heme to form the active tetrameric protein. ❹ After lipids and protein are added, new peroxisomes are formed by the division of the existing peroxisomes. ❺ Some researchers believe that peroxisomes can obtain proteins or form de novo from protoperoxisomal vesicles that are derived from the ER.

individually on cytosolic ribosomes, imported into the peroxisome, refolded, and then assembled, along with heme, into the active tetrameric enzyme. Peroxisomes can also import not only single, unfolded polypeptides but also larger, folded polypeptides and even native oligomeric proteins.

Recent work suggests that some peroxins are synthesized in the cytosol but then travel to the ER in vesicles before being incorporated into peroxisomes (step ❺). Evidence for this route includes the presence on certain peroxins of N-linked oligosaccharides typical of ER-synthesized proteins. Additionally, treatment of yeast cells with the toxin brefeldin A, which prevents formation of ER-derived vesicles, causes accumulation of the peroxin *Pex3p* in the ER.

Similarly, in plant cells, ascorbate peroxidase appears to be routed to a subdomain of the ER after being synthesized in the cytosol but before being incorporated into the peroxisome. This proposed subdomain of the ER, presumably involved in sorting of cytosolic proteins destined for the peroxisome, has been termed the *peroxisomal ER (pER),* although its existence is still controversial. Likewise, the existence of the *protoperoxisome,* an ER-derived vesicle considered by some researchers to be capable of developing directly into a peroxisome, is currently the subject of much debate.

For posttranslational import to work, each protein destined for a specific organelle must have some sort of tag or signal that directs, or targets, the protein to the correct organelle. Such a signal functions by recognizing specific receptors or other features on the surface of the appropriate membrane. The signal in each case is a sequence of amino acids that differs in sequence, length, and location for proteins targeted to different organelles. The signal that targets at least some peroxisomal proteins to their destination consists of just three amino acids and is found at or near the carboxyl terminus of the molecule. For example, the most common sequence is SKL (Ser-Lys-Leu), though a limited number of other amino acids are possible at each of the three locations.

This C-terminal tripeptide, known as *PTS-1 (peroxisomal targeting signal-1),* is recognized and bound by the cytosolic peroxin *Pex5p.* A second peroxisomal targeting signal known as *PTS-2* is an N-terminal sequence recognized by the peroxin *Pex7p.* These PTS-binding peroxins deliver the PTS-containing proteins to the peroxisome for import by a mechanism that is not yet fully understood. At the peroxisomal membrane, Pex5p and Pex7p interact with one or more membrane-bound *docking proteins.* Then they are believed to enter the peroxisome carrying their cargo, which they release before they are exported from the peroxisome back to the cytosol to participate in subsequent transport cycles.

Suresh Subramani and others have shown that a protein normally targeted to the peroxisome by PTS-1 will remain in the cytosol if its SKL sequence is removed. Conversely, the addition of an SKL sequence to a cytosolic protein will direct the chimeric protein to the peroxisome. Thus the SKL sequence, or one of the acceptable variants, is both *necessary* and *sufficient* to direct proteins to peroxisomes. Moreover, a peroxisomal targeting sequence identified in one species often functions in other species, even when the organisms are as evolutionarily diverse from one another as plants, yeast, insects, and animals.

In one striking example, the gene for *luciferase,* a peroxisomal enzyme that enables fireflies to emit flashes of light, was transferred into plant cells from which whole plants were then grown. When cells of the genetically transformed plant were carefully examined, the luciferase was found in peroxisomes, the same organelle in which it is located in fireflies. Knowledge of peroxisomal import mechanisms and the identification of numerous peroxins conserved between yeast and humans, coupled with the powerful genetic methods available in yeast model systems, will help us to understand and better treat human diseases caused by faulty peroxisome biogenesis or protein import.

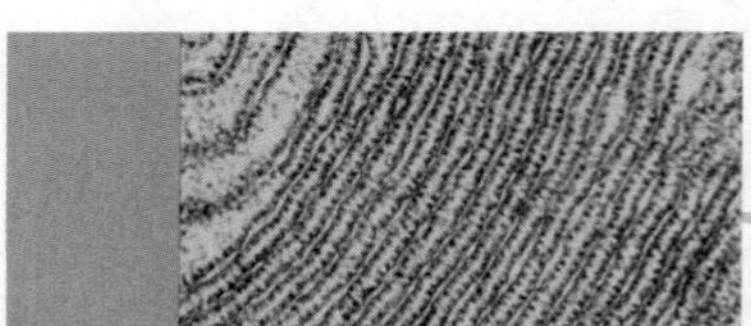

SUMMARY OF KEY POINTS

The Endoplasmic Reticulum

- Especially prevalent within most eukaryotic cells is the endomembrane system, an elaborate array of membrane-bounded organelles derived from the endoplasmic reticulum (ER). The ER itself is a network of sacs, tubules, and vesicles surrounded by a single membrane that separates the ER lumen from the surrounding cytosol.
- The rough ER has ribosomes that synthesize proteins destined for the plasma membrane, for secretion, or for various organelles of the endomembrane system—the nuclear envelope, Golgi complex, endosomes, and lysosomes. Both the rough and smooth ER synthesize lipids for cellular membranes. The smooth ER is also the site of drug detoxification, carbohydrate metabolism, calcium storage, and steroid biosynthesis in some cells.

The Golgi Complex

- The Golgi complex plays an important role in the glycosylation of proteins and in the sorting of proteins for transport to other organelles, for transport to the plasma membrane, or for secretion.

- Transition vesicles that bud from the ER fuse with the *cis*-Golgi network (CGN), delivering lipids and proteins to the Golgi complex. Proteins then move through the Golgi cisternae toward the *trans*-Golgi network (TGN).

Roles of the ER and Golgi Complex in Protein Glycosylation

- Before proteins leave the ER in transition vesicles, they undergo the first few steps of protein modification. ER-specific proteins catalyze core glycosylation and folding of polypeptides, elimination of misfolded proteins, and the assembly of multimeric proteins.
- During their journey through the Golgi complex, proteins are further modified as oligosaccharide side chains are trimmed or further glycosylated in the Golgi lumen.

Roles of the ER and Golgi Complex in Protein Trafficking

- Some ER-specific proteins have a retention signal that prevents them from leaving the ER as vesicles move from the ER to the CGN. Other ER proteins have a retrieval tag that allows them to return to the ER in vesicles that return from the CGN.
- Numerous transport vesicles bud from the TGN Golgi network and carry the processed proteins to their final destinations. The tag that identifies a particular protein and its destination may be a short amino acid sequence, an oligosaccharide side chain, or some other structural feature.
- Hydrolytic enzymes destined for the lysosome are phosphorylated on a mannose residue. Vesicles containing specific receptors for this mannose phosphate group carry these enzymes to the lysosome. During acidification in the lysosome, the enzymes are released from the receptors, which are then recycled back to the TGN.

Exocytosis and Endocytosis

- Exocytosis adds lipids and proteins to the plasma membrane when secretory granules release their contents to the extracellular medium by fusing with the plasma membrane. This addition of material to the plasma membrane is balanced by endocytosis, which removes lipids and proteins from the plasma membrane as extracellular material is internalized in vesicles.
- Phagocytosis is a type of endocytosis involving the ingestion of extracellular particles through invagination of the plasma membrane. Receptor-mediated endocytosis depends on highly specific binding of ligands to corresponding receptors on the cell surface. In both cases, after sorting of receptors and other necessary proteins back to the plasma membrane, ingested material is sent to lysosomes for digestion or to other locations for reuse.

Coated Vesicles in Cellular Transport Processes

- Transport vesicles carry material throughout the endomembrane system. Coat proteins—which include clathrin, COPI, COPII, and caveolin—participate in the sorting of molecules fated for different destinations as well as in the formation of vesicles.
- The specific coat proteins covering a vesicle indicate its origin and help determine its destination within the cell. Clathrin-coated vesicles deliver material from the TGN or plasma membrane to endosomes. COPII-coated vesicles carry materials from the ER to the Golgi, while COPI-coated vesicles transport material from the Golgi back to the ER.
- Once a transport vesicle nears its destination, it is recognized and bound by tethering proteins attached to the target membrane. At this point the v-SNAREs in the transport vesicle membrane and the t-SNAREs in the target membrane interact physically, helping to promote membrane fusion.

Lysosomes and Cellular Digestion

- Extracellular material obtained from phagocytosis or receptor-mediated endocytosis is sorted in early endosomes, which mature to form late endosomes and lysosomes as they fuse with vesicles containing inactive hydrolytic enzymes packaged in the TGN.
- The late endosomal membrane contains ATP-dependent proton pumps that lower the pH of the endosomal lumen and help to transform the late endosome into a lysosome. Then, latent acid hydrolases capable of degrading most biological molecules become active due to the low pH.
- Lysosomes also function in autophagy, the turnover and recycling of cellular structures that are damaged or no longer needed. In some cells, extracellular material is digested by enzymes that lysosomes discharge out of the cell by exocytosis.

The Plant Vacuole: A Multifunctional Organelle

- The plant vacuole is an acidic compartment resembling the animal lysosome. In addition to having hydrolytic enzymes for digestion of macromolecules, it helps the plant cell maintain positive turgor pressure and serves as a storage compartment for a variety of plant metabolites.

Peroxisomes

- Peroxisomes, which are not part of the endomembrane system, appear to increase in number by the division of preexisting organelles rather than by the coalescence of vesicles from the ER or Golgi. However, there is currently a debate concerning the existence of protoperoxisomes, vesicles that some researchers believe bud off from the ER and develop into new peroxisomes.
- Some peroxisomal membrane lipids are synthesized by peroxisomal enzymes, while others are carried from the ER by phospholipid exchange proteins. Most peroxisomal proteins are synthesized by cytosolic ribosomes and are imported posttranslationally. Others are believed to travel via a proposed subdomain of the ER known as the peroxisomal ER (pER).
- The defining enzyme of a peroxisome is catalase, an enzyme that degrades the toxic hydrogen peroxide that is generated by various oxidases in the peroxisome. Animal cell peroxisomes are important for detoxification of harmful substances, oxidation of fatty acids, and metabolism of nitrogen-containing compounds. In plants, peroxisomes play distinctive roles in the conversion of stored lipids into carbohydrate by glyoxysomes and in photorespiration by leaf peroxisomes.

MB **www.masteringbiology.com**

1. MasteringBiology Assignments

Tutorials and Activities

- Molecular Model Structural Tutorials for visualization of 3D structures
- BioFlix Tutorials including 3D animations with hints and detailed feedback for common wrong answers
- Trafficking Through the Endomembrane System; Receptor-mediated Endocytosis; Transport Vesicle Targeting and Fusion

Questions

- Reading Quiz
- Multiple Choice
- End-of-chapter questions

2. eText Read your book online, search, highlight text, take notes, and more

3. The Study Area

- Chapter Review Quiz, Guide to Techniques and Methods, 3D-Structure Tutorials, Activities, BioFlix, Videos, Biology Review, Word Study Tools, Glossary, Flashcards

PROBLEM SET

More challenging problems are marked with a •.

12-1 Compartmentalization of Function. Each of the following processes is associated with one or more specific eukaryotic organelles. In each case, identify the organelle or organelles, and suggest one advantage of confining the process to the organelle or organelles.

(a) β oxidation of long-chain fatty acids

(b) Regulation of calcium ion concentration in muscle cells

(c) Biosynthesis of cortisol

(d) Biosynthesis of the inactive zymogen, trypsinogen

(e) Degradation of damaged organelles

(f) Hydroxylation of barbiturate drugs

(g) Phosphorylation of mannose residues on lysosomal enzymes

(h) Glycosylation of membrane proteins

12-2 Endoplasmic Reticulum. For each of the following statements, indicate if it is true of the rough ER only (R), of the smooth ER only (S), of both rough and smooth ER (RS), or of neither (N).

(a) Is involved in the breakdown of glycogen.

(b) Is the site for the folding of membrane-bound proteins.

(c) Is the site for biosynthesis of secretory proteins.

(d) Contains less cholesterol than does the plasma membrane.

(e) Visible only by electron microscopy.

(f) Has ribosomes attached to its outer (cytosolic) surface.

(g) Usually consists of flattened sacs.

(h) Tends to form tubular structures.

(i) Is involved in the detoxification of drugs.

12-3 Biosynthesis of Integral Membrane Proteins. In addition to their role in cellular secretion, the rough ER and the Golgi complex are also responsible for the biosynthesis of integral membrane proteins. More specifically, these organelles are the source of glycoproteins commonly found in the outer phospholipid monolayer of the plasma membrane.

(a) In a series of diagrams, depict the synthesis and glycosylation of glycoproteins of the plasma membrane.

(b) Explain why the carbohydrate groups of membrane glycoproteins are always found on the outer surface of the plasma membrane.

(c) What assumptions did you make about biological membranes in order to draw the diagrams in part a and answer the question in part b?

12-4 Coated Vesicles in Intracellular Transport. For each of the following statements, indicate for which coated vesicle the statement is true: clathrin- (C), COPI- (I), or COPII-coated (II). Each statement may be true for one, several, or none (N) of the coated vesicles discussed in this chapter.

(a) The protein coat always includes a specific small GTP-binding protein.

(b) Has a role in transport of acid hydrolases to late endosomes.

(c) Is important for retrograde traffic through the Golgi complex.

(d) Fusion of the vesicle (after dissociation of the coat) with the Golgi membrane is facilitated by specific t-SNARE and Rab proteins.

(e) The basic structural component of the coat is called a triskelion.

(f) Has a role in sorting proteins for intracellular transport to specific destinations.

(g) The protein coat dissociates shortly after formation of the vesicle.

(h) Binding of the coat protein to an LDL receptor is mediated by an adaptor protein complex.

(i) Has a role in bidirectional transport between the ER and Golgi complex.

(j) Is essential for all endocytic processes.

(k) Is involved in the movement of membrane lipids from the TGN to the plasma membrane.

12-5 Interpreting Data. Each of the following statements summarizes the results of an experiment related to exocytosis or endocytosis. In each case, explain the relevance of the experiment and its result to our understanding of these processes.

(a) Addition of the drug colchicine to cultured fibroblast cells inhibits movement of transport vesicles.

(b) Certain pituitary gland cells secrete laminin continuously but secrete adrenocorticotropic hormone only in response to specific signals.

(c) Certain adrenal gland cells can be induced to secrete epinephrine when their intracellular calcium concentration is experimentally increased.

(d) Cells expressing a temperature-sensitive form of dynamin do not display receptor-mediated endocytosis after a temperature shift, yet they continue to ingest extracellular fluid (at a reduced level initially, and then at a normal level within 30–60 minutes).

(e) Brefeldin A inhibits cholesterol efflux in adipocytes (fat cells) without affecting the rate of cellular uptake and resecretion of apolipoprotein A-I in adipocytes.

12-6 Cellular Digestion. For each of the following statements, indicate the specific digestion process or processes of which the statement is true: phagocytosis (P), receptor-mediated endocytosis (R), autophagy (A), or extracellular digestion (E). Each statement may be true of one, several, or none (N) of these processes.

(a) Digested material is of extracellular origin.

(b) Involves fusion of lysosomes with the plasma membrane.

(c) Can involve fusion of vesicles or vacuoles with a lysosome.

(d) Involves acid hydrolases.

(e) Serves as a source of nutrients within the cell.

(f) Digested material is of intracellular origin.

(g) Occurs within lysosomes.

(h) Important for certain developmental processes.

(i) Involves fusion of endocytic vesicles with an early endosome.

(j) Can involve exocytosis.

(k) Essential for sperm penetration of the egg during fertilization.

12-7 Peroxisomal Properties. For each of the following statements, indicate whether it is true of all (A), some (S), or none (N) of the various kinds of peroxisomes described in this chapter, and explain your answer.

(a) Is a source of dolichol.

(b) Contains the genes coding for luciferase.

(c) Is surrounded by a lipid bilayer.

(d) Contains acid hydrolases.

(e) Acquires proteins from the ER and Golgi complex.

(f) Contains urate oxidase.

(g) Contains peroxide-generating chemical reactions.

(h) Capable of catabolizing fatty acids.

(i) Contains catalase.

12-8 Lysosomal Storage Diseases. Despite a bewildering variety of symptoms, lysosomal storage diseases have several properties in common. For each of the following statements, indicate if you would expect the property to be common to most lysosomal storage diseases (M), to be true of a specific lysosomal storage disease (S), or not to be true of any lysosomal storage diseases (N).

(a) Results from an absence of functional acid hydrolases.

(b) Symptoms include muscle weakness and mental retardation.

(c) Impaired metabolism of glycolipids causes mental deterioration.

(d) Triggers proliferation of organelles containing catalase.

(e) Leads to accumulation of degradation products in the lysosome.

(f) Results from an inability to regulate the synthesis of glycosaminoglycans.

(g) Results in accumulation of lysosomes in the cell.

(h) Leads to accumulation of excessive amounts of glycogen in the lysosome.

12-9 Sorting Proteins. Specific structural features tag proteins for transport to various intracellular and extracellular destinations. Several examples were described in this chapter including: (1) the short peptide Lys-Asp-Glu-Leu, (2) characteristic hydrophobic membrane-spanning domains, and (3) mannose-6-phosphate residues attached to oligosaccharide side chains. For each structural feature, answer the following questions:

(a) Where in the cell is the tag incorporated into the protein?

(b) How does the tag ensure that the protein reaches its destination?

(c) Where would the protein likely go if you were to remove the tag?

• **12-10 Silicosis and Asbestosis.** *Silicosis* is a debilitating miner's disease that results from the ingestion of silica particles (such as sand or glass) by macrophages in the lungs. *Asbestosis* is a similarly serious disease caused by inhalation of asbestos fibers. In both cases, the particles or fibers are found in lysosomes, and fibroblasts, which secrete collagen, are stimulated to deposit nodules of collagen fibers in the lungs, leading to reduced lung capacity, impaired breathing, and eventually death.

(a) How do you think the fibers get into the lysosomes?

(b) What effect do you think fiber or particle accumulation has on the lysosomes?

(c) How might you explain the death of silica-containing or asbestos-containing cells?

(d) What do you think happens to the silica particles or asbestos fibers when such cells die? How can cell death continue almost indefinitely, even after prevention of further exposure to silica dust or asbestos fibers?

(e) Cultured fibroblast cells will secrete collagen and produce connective tissue fibers after the addition of material from a culture of lung macrophages that have been exposed to silica particles. What does this tell you about the deposition of collagen nodules in the lungs of silicosis patients?

• **12-11 What's Happening?** Researchers have discovered a group of plant proteins that are related to the exocyst proteins in yeast (*The Plant Cell* 20 (2008): 1330). Explain how the following observations made by these researchers suggest that these plant proteins form a tethering complex similar to the exocyst complex of yeast and mammals.

(a) Following size fractionation of plant protein extracts, antibodies recognizing each of several different plant exocyst proteins bind to the same high-molecular-weight protein fraction.

(b) Mutations in four of the proteins each causes defective pollen germination.

(c) Plants lacking more than one of these proteins have more serious defects in pollen germination than plants lacking only one.

(d) The exocyst proteins all co-localize at the growing tip of elongating pollen cells.

(e) Pollen cells of plants with mutations in exocyst genes are defective in tip growth or germinating pollen cells.

SUGGESTED READING

References of historical importance are marked with a •.

General References

Brown, D. Imaging protein trafficking. *Nephron. Exp. Nephrol.* 103 (2006): 55.

Nunnari, J., and P. Walter. Regulation of organelle biogenesis. *Cell* 84 (1996): 389.

Pan, S., C. J. Carter, and N. V. Raikhel. Understanding protein trafficking in plant cells through proteomics. *Expert Rev. Proteomics* 2 (2005): 781.

Uemura, K., A. Kuzuya, and S. Shimohama. Protein trafficking and Alzheimer's disease. *Curr. Alzheimer Res.* 1 (2004): 1.

The Endoplasmic Reticulum and the Golgi Complex

Dancourt, J., and C. Barlowe. Protein sorting receptors in the early secretory pathway. *Annu. Rev. Biochem.* 79 (2010): 777.

Ellgaard, L., and A. Helenius. ER quality control: Towards an understanding at the molecular level. *Curr. Opin. Cell Biol.* 13 (2001): 431.

• Farquhar, M. G., and G. E. Palade. The Golgi apparatus: 100 years of progress and controversy. *Trends Cell Biol.* 8 (1998): 2.

Glick, B. S., and A. Nakano. Membrane traffic within the Golgi apparatus. *Annu. Rev. Cell Dev. Biol.* 25 (2009): 113.

• Jamieson, J. D., and G. E. Palade. Intracellular transport of secretory proteins in the pancreatic exocrine cell II: Transport to condensing vacuoles and zymogen granules. *J. Cell Biol.* 34 (1967): 597.

Klumperman, J. Transport between ER and Golgi. *Curr. Opin. Cell Biol.* 12 (2000): 445.

Zhao, L., and S. L. Ackerman. Endoplasmic reticulum stress in health and disease. *Curr. Opin. Cell Biol.* 18 (2006): 444.

Exocytosis and Endocytosis

Blázquez, M., and K. I. J. Shennan. Basic mechanisms of secretion: Sorting into the regulated secretory pathway. *Biochem. Cell Biol.* 78 (2000): 181.

Jutras, I., and M. Desjardins. Phagocytosis: At the crossroads of innate and adaptive immunity. *Annu. Rev. Cell Dev. Biol.* 12 (2005): 511.

Lemmon, S. K., and L. M. Traub. Sorting in the endosomal system in yeast and animal cells. *Curr. Opin. Cell Biol.* 12 (2000): 457.

Ng, T. W., E. M. Ooi, G. F. Watts, D. C. Chan, and P. H. Barrett. Genetic determinants of apolipoprotein B-100 kinetics. *Curr. Opin. Lipidol.* 21 (2010): 141.

Pelham, H. R. B. Insights from yeast endosomes. *Curr. Opin. Cell Biol.* 14 (2002): 454.

Tjelle, T. E., T. Lùvdal, and T. Berg. Phagosome dynamics and function. *BioEssays* 22 (2000): 255.

Transport Vesicles

Fielding, C. J. *Lipid Rafts and Caveolae: From Membrane Biophysics to Cell Biology.* New York: Wiley-VCH, 2007.

Hsu, V. W., and J. S. Yang. Mechanisms of COPI vesicle formation. *FEBS Lett.* 583 (2009): 3758.

Lipka, V., C. Kwon, and R. Panstruga. SNARE-ware: The role of SNARE-domain proteins in plant biology. *Annu. Rev. Cell Dev. Biol.* 23 (2007): 147.

Whyte, J. R. C., and S. Munro. Vesicle tethering complexes in membrane traffic. *J. Cell Sci.* 115 (2002): 2627.

Xiang-A. L., W. V. Everson, and E. J. Smart. Caveolae, lipid rafts, and vascular disease. *Trends Cardiovasc. Med.* 15 (2005): 92.

Lysosomes and Vacuoles

• Bainton, D. The discovery of lysosomes. *J. Cell Biol.* 91 (1981): 66s.

Bonifacino, J. S., and L. M. Traub. Signals for sorting of transmembrane proteins to endosomes and lysosomes. *Annu. Rev. Biochem.* 72 (2003): 395.

De, D. N. *Plant Cell Vacuoles: An Introduction.* Collingwood, Victoria: CSIRO Publishing, 2000.

• de Duve, C. The lysosome. *Sci. Amer.* 208 (May, 1963): 64.

Klionsky, D. J., and S. D. Emr. Autophagy as a regulated pathway of cellular degradation. *Science* 290 (2000): 1717.

Marx, J. Autophagy: Is it cancer's friend or foe? *Science* 312 (2006): 1160.

Saftig, P., and J. Klumperman. Lysosome biogenesis and lysosomal membrane proteins: Trafficking meets function. *Nature Rev. Mol. Cell Biol.* 10 (2009): 623.

Peroxisomes

• de Duve, C. The peroxisome: A new cytoplasmic organelle. *Proc. R. Soc. Lond. Ser. B Biol. Sci.* 173 (1969): 71.

Gärtner, J. Organelle disease: Peroxisomal disorders. *Eur. J. Pediatr.* 159 [Suppl. 3] (2000): S236.

Lazarow, P. B. Peroxisome biogenesis: Advances and conundrums. *Curr. Opin. Cell Biol.* 15 (2003): 489.

Ma, C., and S. Subramani. Peroxisome matrix and membrane protein biogenesis. *IUBMB Life.* 61 (2009): 713.

Mullen, R. T., C. R. Flynn, and R. N. Trelease. How are peroxisomes formed? The role of the endoplasmic reticulum and peroxins. *Trends Plant Sci.* 6 (2001): 256.

Smith, J. J., and J. D. Aitchison. Regulation of peroxisome dynamics. *Curr. Opin. Cell Biol.* 21 (2009): 119.

Titorenko, V. I., and R. T. Mullen. Peroxisome biogenesis: The peroxisomal endomembrane system and the role of the ER. *J. Cell Biol.* 174 (2006): 11.

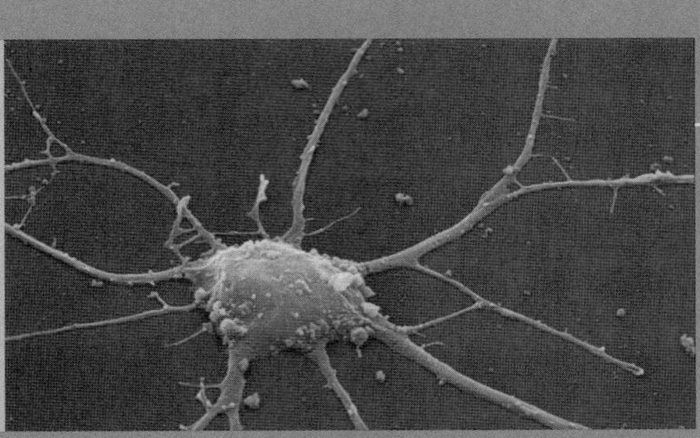

13 Signal Transduction Mechanisms: I. Electrical and Synaptic Signaling in Neurons

See MasteringBiology® for tutorials, activities, and quizzes.

In Chapter 8 we saw that a key property of cell membranes is the ability to regulate the flow of ions between the interior of the cell and the outside environment. Since ions are charged solutes, cells can, in effect, regulate the flow of electric current through their membranes as well as the electrical potentials across them. The most dramatic example of regulation of the electrical properties of cells is the functioning of the main cellular component of the nervous system of animals: the nerve cell, or neuron.

In this chapter, we will examine the cellular mechanisms that allow neurons to transmit signals at remarkably high speeds throughout the body, and we will consider the ways neurons connect with one another. By studying neurons, we will gain important insights into how cells in general regulate their electrical properties and how cells can send signals to one another. In the first part of this chapter, we will see that although virtually all cells maintain electrical potentials across their plasma membranes, nerve cells have special mechanisms for using this potential to transmit information over long distances. In the second part of the chapter, we will focus on the processes by which the information is passed between neurons and other cells, such as other nerve cells, glands, or muscle cells.

Neurons

Almost all animals have a *nervous system* in which electrical impulses are transmitted along the specialized plasma membranes of nerve cells. In the case of vertebrates, the nervous system has two main components: the *central nervous system (CNS),* which consists of the brain and spinal cord, and the *peripheral nervous system (PNS),* which comprises other sensory and motor components. Cells that make up the nervous system can be broadly divided into two groups: neurons and glial cells. All **neurons** send or receive electrical impulses. Neurons can be subdivided into three basic types based on function: sensory neurons, motor neurons, and interneurons. *Sensory neurons* are a diverse group of cells specialized for the detection of various types of stimuli; they provide a continuous stream of information from various sensory receptors to the brain about the state of the body and its environment. *Motor neurons* transmit signals from the CNS to the muscles or glands they *innervate*—that is, the tissues with which they make synaptic connections. *Interneurons* process signals received from other neurons and relay the information to other parts of the nervous system.

The term *glial cell* (from *glia,* the Greek word for "glue") encompasses a variety of cell types. Glial cells are by far the most abundant types of cells in the CNS. *Microglia* are phagocytic cells that fight infections and remove debris. *Oligodendrocytes* and *Schwann cells* form the insulating *myelin sheath* around neurons of the CNS and those of the peripheral nerves, respectively. *Astrocytes* control access of blood-borne components into the extracellular fluid surrounding nerve cells, thereby forming the *blood-brain barrier.*

Intricate networks of neurons make up the complex tissues of the brain that are responsible for coordinating nervous function—in humans, about 10 billion neurons. Each neuron can receive input from thousands of other neurons, so the brain's connections number well into the trillions. Rather than discussing the overall functioning of the nervous system, our purpose here is to examine the cellular mechanisms by which these electrical signals, called *nerve impulses,* are spread. In addition to understanding the functions of nerve cells specifically, we will acquire a better appreciation for several general aspects of membrane function, of which nerve cells present a specialized, highly developed example.

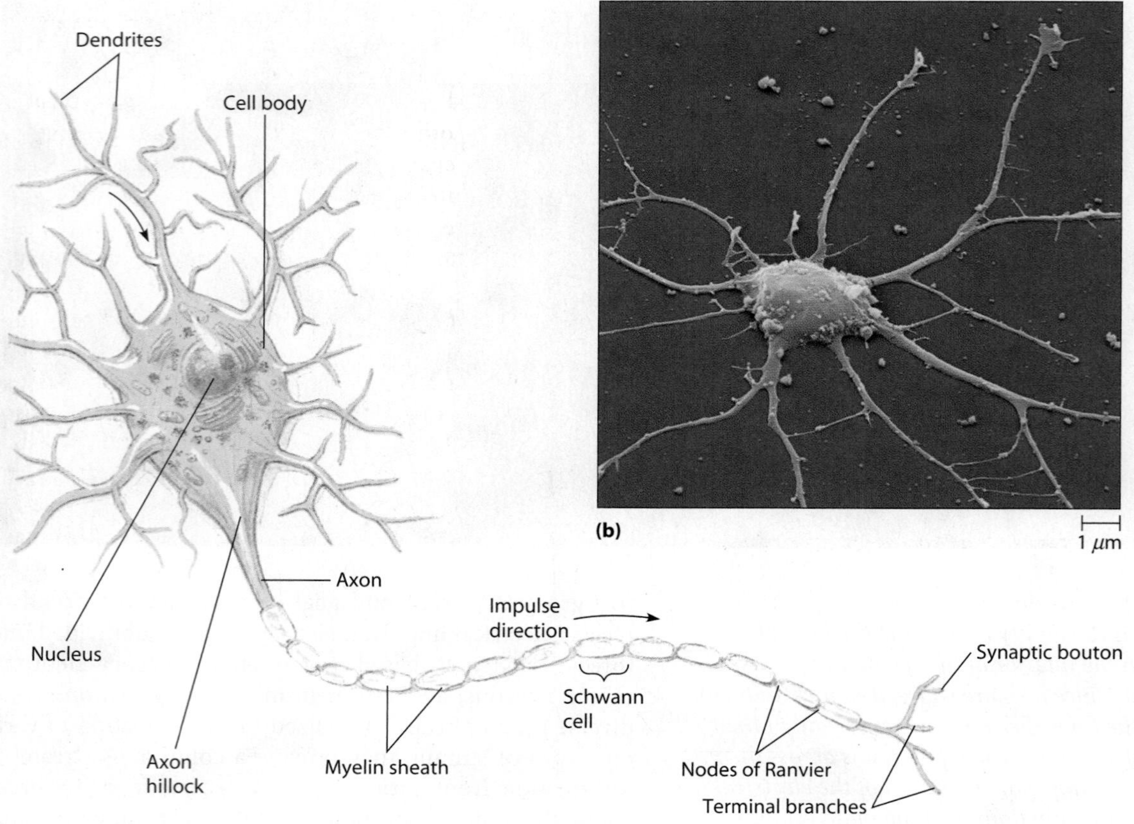

FIGURE 13-1 The Structure of a Typical Motor Neuron. **(a)** A diagram of a typical motor neuron. The cell body contains the nucleus and most of the usual organelles. Dendrites conduct signals inward to the cell body; starting near the axon hillock, the axon transmits signals outward (the black arrows indicate direction of transmission). At the end of the axon are many synaptic boutons. Some, though not all, neurons have a discontinuous myelin sheath around their axons to insulate them electrically. Each segment of the sheath consists of a concentric layer of membranes wrapped around the axon by a Schwann cell (or an oligodendrocyte in the CNS). Breaks in the myelin sheath, called nodes of Ranvier, are concentrated regions of electrical activity. **(b)** A scanning electron micrograph of the cell body and dendrites of a neuron. The neuron has been colored orange. Schwann cells are not visible.

Neurons Are Specially Adapted for the Transmission of Electrical Signals

The structure of a typical motor neuron is shown schematically in **Figure 13-1**. The **cell body** of the neuron is similar to that of other cells and includes the nucleus and many components of the endomembrane system, which you learned about in Chapter 12. Neurons also contain extensions, or branches, called *processes.* Transporting components to and from these extensions is a special challenge for neurons, which we will discuss in Chapter 15. There are two types of processes: Those that receive signals and combine them with signals received from other neurons are called **dendrites,** and those that conduct signals, sometimes over long distances, are called **axons.** The cytosol within an axon is commonly referred to as **axoplasm.** Many vertebrate axons are surrounded by a discontinuous **myelin sheath,** which insulates the segments of axon separating the **nodes of Ranvier.** Axons can be very long—up to several thousand times longer than the diameter of the cell body. For example, a motor neuron that innervates your foot has its cell body in your spinal cord, and its axon extends approximately a meter down your leg! A **nerve** is simply a tissue composed of bundles of axons.

As Figure 13-1 illustrates, a motor neuron has multiple, branched dendrites and a single axon leading away from the cell body. The axon of a typical neuron is much longer than the dendrites and forms multiple branches. The branches terminate in structures called **synaptic boutons** (also called *terminal bulbs* or *synaptic knobs*; the French word *bouton* means "button"). The boutons are responsible for transmitting the signal to the next cell, which may be another neuron or a muscle or gland cell. In each case, the junction is called a **synapse.** For neuron-to-neuron junctions, synapses usually occur between an axon and a dendrite, but they can occur between two dendrites. Typically, neurons make synapses with many other neurons. They can do so not only at the ends of their axons but at other points along their length as well.

Neurons display tremendous structural variability. Some sensory neurons have only one process, which conducts signals both toward and away from the cell body. Moreover, the structure of the processes is not random; many different classes of neurons in the CNS can be identified by structure alone.

Understanding Membrane Potential

Recall from Chapter 8 that **membrane potential** is a fundamental property of all cells. It results from an excess of negative charge on one side of the plasma membrane and an excess of positive charge on the other side. Cells at rest normally have an excess of negative charge inside and an excess of positive charge outside the cell; the resulting electrical potential is called the **resting membrane potential,** denoted V_m. A great technical advance in understanding how membrane potential changes as nerve signals are transmitted came with the discovery in the 1930s of very large axons in some nerve fibers of squid. These nerve fibers stimulate the explosive expulsion of water from the mantle cavity of the squid, enabling it to propel itself quickly backward to escape predators (**Figure 13-2**). The **squid giant axon** has a diameter of about 0.5–1.0 mm, allowing the easy insertion of tiny electrodes, or *microelectrodes,* to measure and control electrical potentials and ionic currents across the axonal membrane. The *resting membrane potential* can be measured by placing one microelectrode inside the cell and another outside the cell (**Figure 13-3a**); its value is generally given in millivolts (mV). (Do not confuse the potential V_m and the unit mV.) The electrodes compare the ratio of negative to positive charge inside the cell and outside the cell. Because the inside of a cell typically has an excess of negative charge, we say that the cell has a negative resting membrane potential. For example, the resting membrane potential is approximately −60 mV for the squid giant axon.

Nerve, muscle, and certain other cell types such as the islet cells of the pancreas of vertebrates exhibit a special property called **electrical excitability.** In electrically excitable cells, certain types of stimuli trigger a rapid sequence of changes in membrane potential known as an *action potential.* During an action potential, the membrane potential changes from negative to positive values and then back to negative values again, all in as little as a

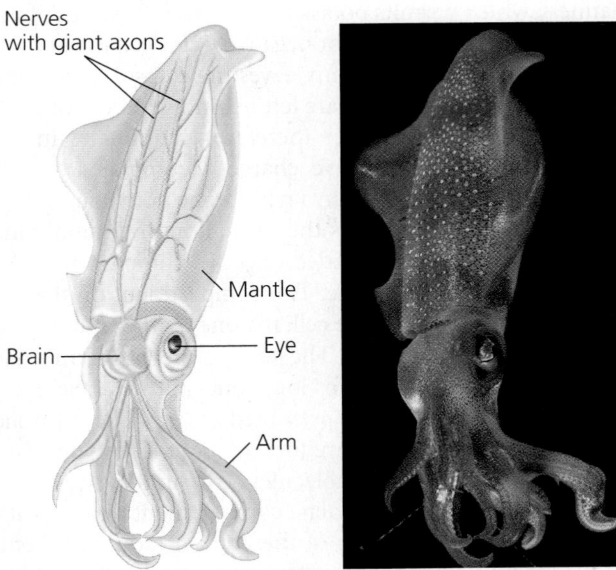

FIGURE 13-2 Squid Giant Axons. The squid nervous system includes motor nerves that control swimming movements. The nerves contain giant axons (fibers) with diameters ranging up to 1 mm, providing a convenient system for studying resting and action potentials in a biological membrane.

(a) Measuring the resting membrane potential. Differences in potential between the recording and reference electrodes are amplified by a voltage amplifier and displayed on a voltmeter, an oscilloscope, or a computer monitor.

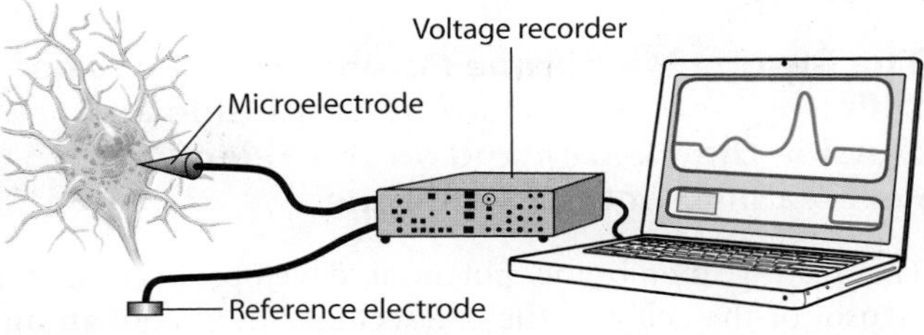

(b) Measuring an action potential in a squid axon. The stimulating electrode is connected to a pulse generator, which delivers a pulse of current to the axon when the switch is momentarily closed. The nerve impulse this generates is propagated down the axon and can be detected a few milliseconds later by the recording electrode.

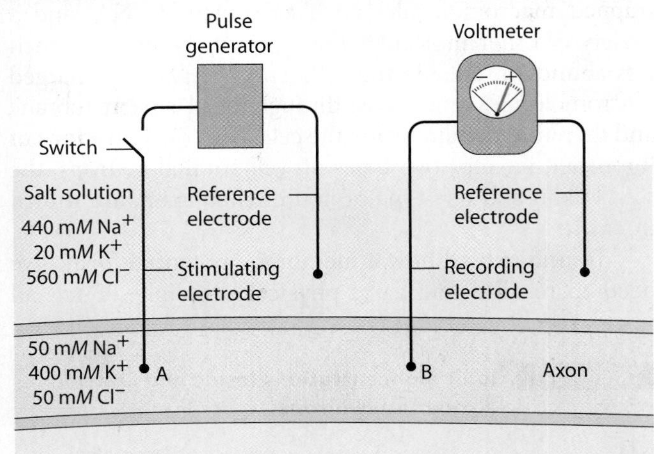

FIGURE 13-3 Measuring Membrane Potentials. **(a)** Measurement of the resting membrane potential requires two electrodes, one inserted inside the axon (the recording electrode) and one placed in the fluid surrounding the cell (the reference electrode). **(b)** Measurement of an action potential requires four electrodes, one in the axon for stimulation, another in the axon for recording, and two in the fluid surrounding the cell for reference.

few milliseconds. In addition to measuring resting potential, microelectrodes can also be used to measure these dynamic changes in membrane potential. An electrode called the *stimulating electrode* is connected to a power source and inserted into the axon some distance from the recording electrode (see Figure 13-3b). A brief impulse from this stimulating electrode depolarizes the membrane by about 20 mV (i.e., from −60 to about −40 mV), bringing the neuron beyond the threshold potential. This triggers an action potential that propagates away from the stimulating electrode. As the action potential passes the recording electrode, the voltmeter or oscilloscope will display a characteristic pattern of membrane potential changes. Alan Hodgkin and Andrew Huxley used microelectrodes and the squid giant axon to learn how action potentials are generated; for their work, they received a share of the Nobel Prize in 1963. To understand how nerve cells use action potentials to transmit signals, we must first examine how cells generate a resting membrane potential and how the membrane potential changes during an action potential.

The Resting Membrane Potential Depends on Differing Concentrations of Ions Inside and Outside the Neuron and on the Selective Permeability of the Membrane

The resting membrane potential develops because the cytosol of the cell and the extracellular fluid contain different compositions of cations and anions. Extracellular fluid contains dissolved salts, including sodium chloride and lesser amounts of potassium chloride. The cytosol contains potassium rather than sodium as its main cation because of the action of the Na^+/K^+ pump (described in Chapter 8). The anions in the cytosol consist largely of trapped macromolecules such as proteins, RNA, and a variety of other molecules that are not present, or much less abundant, outside the cell. These negatively charged macromolecules cannot pass through the plasma membrane and therefore remain inside the cell. The concentrations of important ions for two types of well-studied neurons, the squid axon and one type of mammalian axon, are shown in **Table 13-1**.

To understand how a membrane potential forms, we need to recall a few basic physical principles. First, all substances tend to diffuse from an area where they are more highly concentrated to an area where they are less concentrated. Cells normally have a high concentration of potassium ions inside and a low concentration of potassium ions outside. We refer to this uneven distribution of potassium ions as a *potassium ion gradient.* Given the large potassium concentration gradient, potassium ions will tend to diffuse out of the cell.

The second basic principle is that of *electroneutrality.* When ions are in solution, they are always present in pairs, one positive ion for each negative ion, so that there is no net charge imbalance. For any given ion, which we will call A, there must be an oppositely charged ion B in the solution; therefore, we refer to B as the *counterion* for A. In the cytosol, potassium ions (K^+) serve as counterions for the trapped anions. Outside the cell, sodium (Na^+) is the main cation and chloride (Cl^-) is its counterion.

Although a solution must have an equal number of positive and negative charges overall, these charges can be locally separated so that one region has more positive charges while another region has more negative charges. It takes work to separate charges; even after they have been separated, they tend to move back toward each other. The tendency of oppositely charged ions to flow back toward each other is called an **electrical potential** or **voltage.** When negative or positive ions are actually moving, one toward the other, we say that **current** is flowing; this current is measured in amperes (A). Given these principles, we can understand how a resting membrane potential will form as a result of the ionic compositions of the cytosol and the extracellular fluid, as well as the characteristics of the plasma membrane.

The plasma membrane is normally permeable to potassium due to the leakiness of some types of potassium channels, which permits potassium ions to diffuse out of the cell. There are no channels for negatively charged macromolecules, however. As potassium leaves the cytosol, increasing numbers of trapped anions are left behind without counterions. Excess negative charge therefore accumulates in the cytosol while excess positive charge accumulates on the outside of the cell, resulting in a membrane potential.

Figure 13-4 illustrates the formation of a membrane potential in a container divided into two compartments by a semipermeable membrane. The compartment on the left represents the cytosol of the cell; the one on the right represents the extracellular fluid. The semipermeable membrane is permeable to potassium ions but not to negatively charged macromolecules (represented as M^-). The cytosolic compartment contains a mixture of potassium ions and negatively charged macromolecules. For simplicity, we will assume that the extracellular compartment starts with nothing but water. Because of the concentration gradient, potassium ions diffuse across the membrane from the left side to the right side. However, the negatively charged macromolecules are not free to follow. The result is an accumulation of anions on the left side and cations on the right side. A membrane potential is created by the separation of

Table 13-1 Ionic Concentrations Inside and Outside Axons and Neurons

	Squid Axon		Mammalian Neuron (cat motor neuron)	
Ion	Outside (mM)	Inside (mM)	Outside (mM)	Inside (mM)
Na^+	440	50	145	10
K^+	20	400	5	140
Cl^-	560	50	125	10

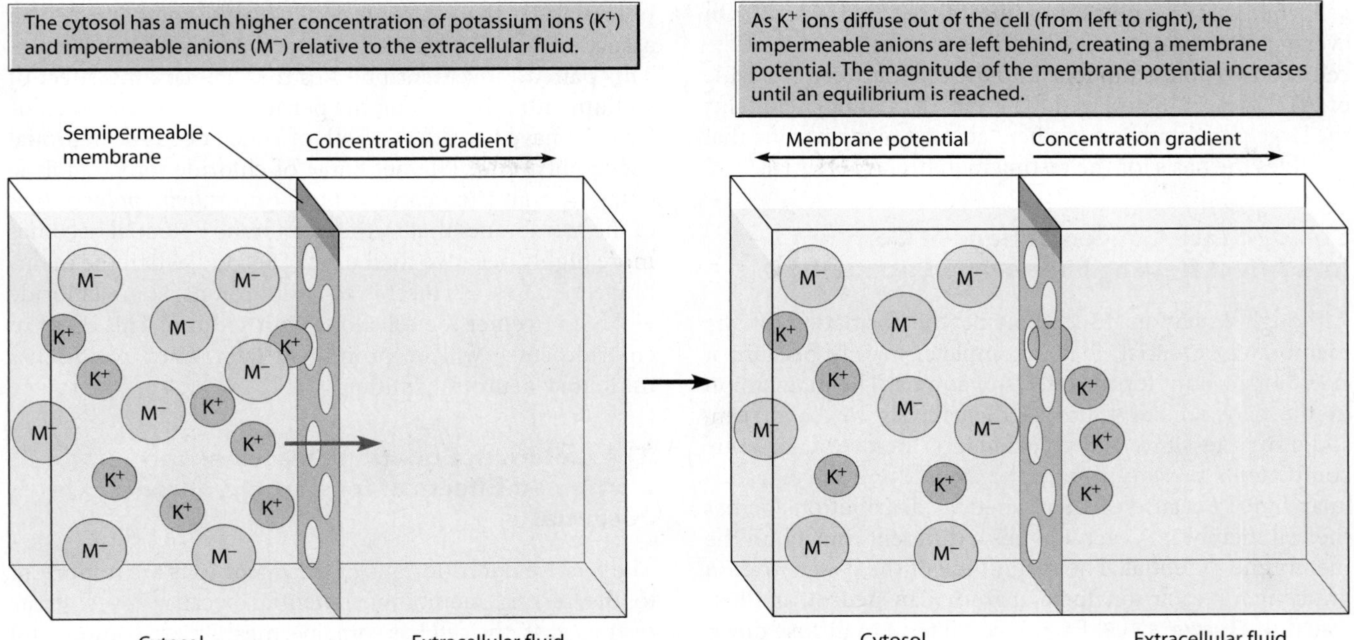

FIGURE 13-4 Development of the Equilibrium Potential. Here a two-compartment container represents the cytosol of a cell and the extracellular space, with a semipermeable membrane between them. In this example, only potassium ions (K^+) and impermeable anions (M^-) are shown.

negative charge from positive charge. Although this charge separation produces a small voltage relative to that of a typical 1.5-V battery, it is nevertheless a form of potential energy that can be used to perform work.

As potassium ions diffuse from left to right in Figure 13-4, the compartment on the left becomes increasingly negative. This potential ultimately builds to a point at which the positively charged potassium ions are pulled back into the left-hand compartment as fast as they leave. In this way, an equilibrium is reached in which the force of attraction due to the membrane potential balances the tendency of potassium to diffuse down its concentration gradient. This type of equilibrium, in which a chemical gradient is balanced with an electrical potential, is referred to as an **electrochemical equilibrium.** The membrane potential at the point of equilibrium is known as an **equilibrium,** or **reversal, potential;** the magnitude of the potassium ion gradient determines the magnitude of the equilibrium potential. In the next section, we will present a mathematical description of this relationship that enables us to estimate the membrane potential.

The Nernst Equation Describes the Relationship Between Membrane Potential and Ion Concentration

The **Nernst equation** is named for the German physical chemist and Nobel laureate Walther Nernst, who first formulated it in the late 1880s in the context of his work on electrochemical cells (forerunners of modern batteries). The Nernst equation describes the mathematical relationship between an ion gradient and the equilibrium potential that will form when the membrane is permeable only to that ion:

$$E_X = \frac{RT}{zF} \ln \frac{[X]_{\text{outside}}}{[X]_{\text{inside}}} \qquad \textbf{(13-1)}$$

Here, E_X is the equilibrium potential for ion X (in volts), R is the gas constant (1.987 cal/mol-degree), T is the absolute temperature in kelvins (recall that 0°C = 273K), z is the valence of the ion, F is the Faraday constant (23,062 cal/V-mol), $[X]_{\text{outside}}$ is the concentration of X outside the cell (in M), and $[X]_{\text{inside}}$ is the concentration of X inside the cell (in M). This equation can be simplified if we assume that the temperature is 293K (20°C)—a value appropriate for the squid giant axon—and that X is a monovalent cation and therefore has a valence of +1. If we substitute these values for R, T, F, and z into the Nernst equation and convert from natural logs to $\log_{10}$ ($\log_{10}$ = ln/2.303), Equation 13-1 reduces to

$$E_X = 0.058 \log_{10} \frac{[X]_{\text{outside}}}{[X]_{\text{inside}}} \qquad \textbf{(13-2)}$$

In this simplified form, we can see that for every tenfold increase in the cation gradient, the membrane potential changes by −0.058 V, or −58 mV.

The Na^+/K^+ Pump. Although the plasma membrane is relatively impermeable to sodium ions, there is always a small amount of leakage. To compensate for this leakage, the **Na^+/K^+ pump** continually pumps sodium out of the cell while carrying potassium inward (see Figure 8-12). On average, the pump transports three sodium ions out of the cell and two potassium ions into the cell for every molecule of ATP that is hydrolyzed. The Na^+/K^+ pump maintains the large potassium ion gradient across the membrane that provides the basis for the resting membrane potential.

Steady-State Concentrations of Common Ions Affect Resting Membrane Potential

Although Equation 13-2 explains the formation of the membrane potential, it is incomplete, mainly because it does not account for the effects of anions. The main anion in the extracellular fluid is chloride. As we have seen, sodium, potassium, and chloride ions are the major ionic components present in both the cytosol and the extracellular fluid. Because of their unequal distributions across the cell membrane, each ion has a different impact on the membrane potential. The magnitude of the concentration gradient for each ion for a mammalian neuron is illustrated in **Figure 13-5**. Each ion will tend to diffuse down its electrochemical gradient (shown as red arrows in Figure 13-5) and thereby produce a change in the membrane potential.

Potassium ions tend to diffuse out of the cell, which makes the membrane potential more negative. Sodium ions tend to flow into the cell, driving the membrane potential in the positive direction and thereby causing a **depolarization** of the membrane (that is, causing the membrane potential to be less negative). Chloride ions tend to diffuse into the cell, which should in principle make the membrane potential more negative. However, chloride ions are also repelled by the negative membrane potential, so that chloride ions usually enter the cell in association with positively charged ions such as sodium. This paired movement nullifies the depolarizing effect of sodium entry. Increasing the permeability of cells to chloride can have two effects, both of which decrease neuronal excitability. First, the net entry of chloride ions (chloride entry without a matching cation) causes *hyperpolarization* of the membrane (that is, the membrane potential becomes more highly negative than usual). Second, when the membrane becomes permeable to sodium ions, some chloride will tend to enter the cell along with sodium. This effect of chloride entry will be prominent later when we discuss inhibitory neurotransmitters.

The Goldman Equation Describes the Combined Effects of Ions on Membrane Potential

The relative contributions of the major ions are important to the resting membrane potential because, even in its resting state, the cell has some permeability to sodium and chloride ions as well as to potassium ions. To account for the leakage of sodium and chloride ions into the cell, we cannot use the Nernst equation because it deals with only one type of ion at a time, and it assumes that this ion is in electrochemical equilibrium. We must move from the more static concept of equilibrium potential to a consideration of *steady-state ion movements* across the membrane.

We can illustrate the concept of steady-state ion movements by returning once more to our model of a cell in electrochemical equilibrium (**Figure 13-6**). Remember, a cell that is permeable only to potassium will have a membrane potential equal to the equilibrium potential for

(a) Potassium ions (K^+). K^+ are more concentrated in the cytosol and have a tendency to move out of the cell, leaving behind trapped anions. Membrane potential becomes more negative.

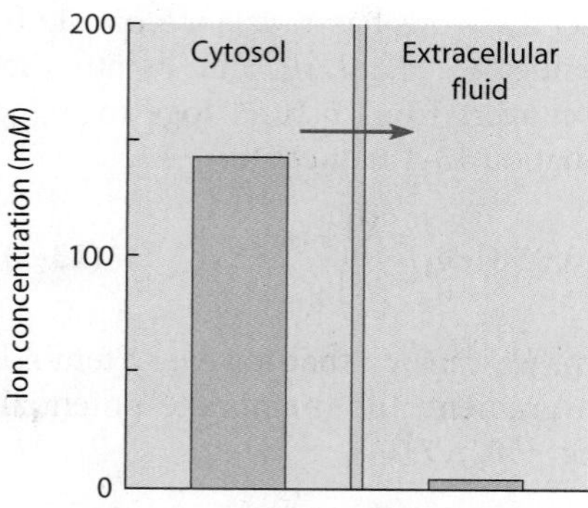

(b) Sodium ions (Na^+). Na^+ are much more concentrated outside the cell than inside and tend to enter the cell. As Na^+ ions enter, they neutralize some excess negative charge in the cytosol, and membrane potential becomes more positive.

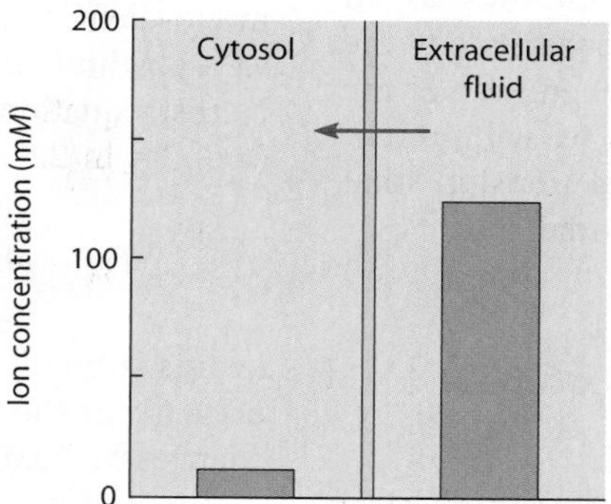

(c) Chloride ions (Cl^-). Cl^- usually crosses the membrane together with a permeable cation (normally K^+). As Cl^- enters a cell, it tends to make the membrane potential more negative.

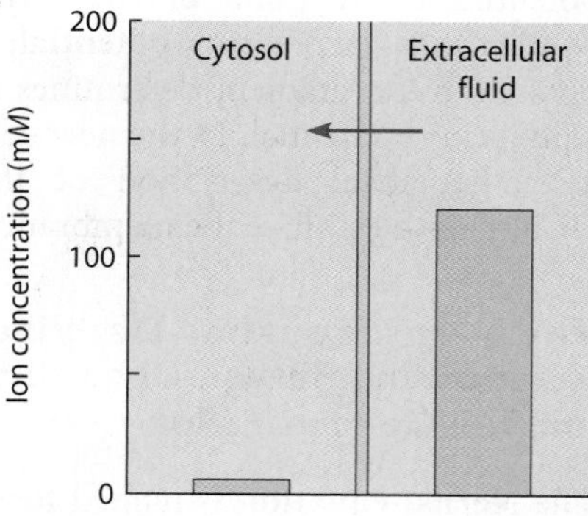

FIGURE 13-5 Relative Concentrations of Potassium, Sodium, and Chloride Ions Across the Plasma Membrane of a Mammalian Neuron. Each of the major ions in the cytosol—**(a)** potassium, **(b)** sodium, and **(c)** chloride—is found in a concentration gradient across the plasma membrane. The red arrow indicates the direction the ion would tend to flow in response to its electrochemical gradient.

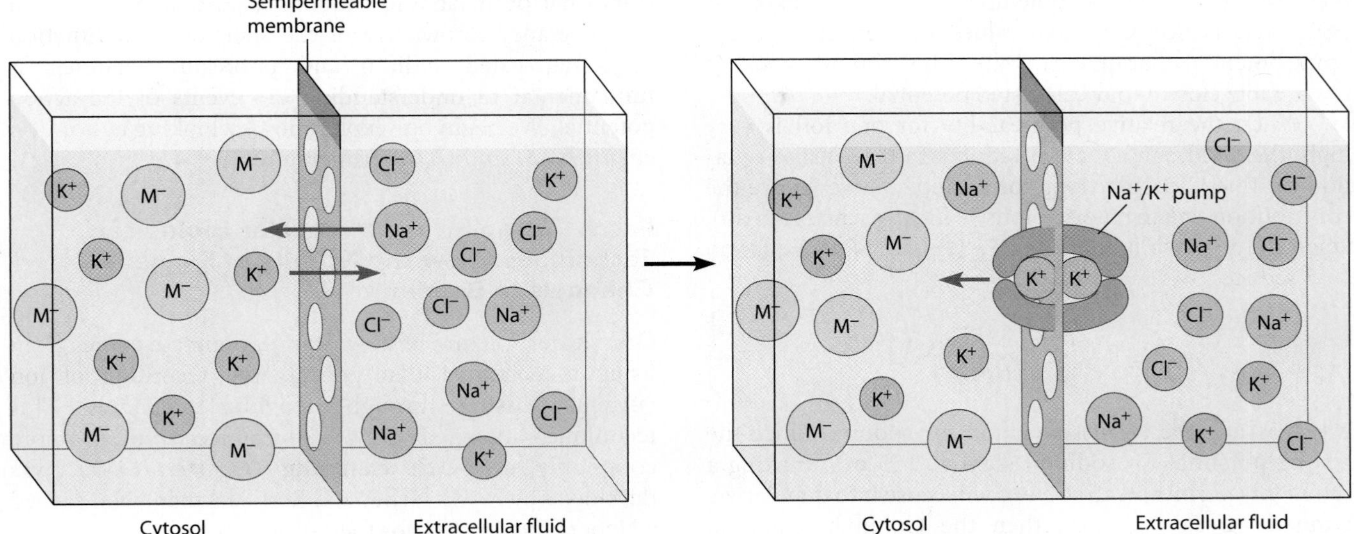

A small number of sodium (Na+) ions continually leak into the cell. This makes the membrane potential more positive, weakening the electrical restraint on the movement of potassium (K+) ions. A small number of ions now leak out of the cell.
As Na+ ions accumulate in the cytosol, they are pumped outward in exchange for potassium ions by the Na+/K+ pump. The result is a low concentration of Na+ ions inside the cell. Some Na+ ions in the cytosol cause the membrane potential to be more positive than the equilibrium membrane potential for K+.
Semipermeable membrane
Na+/K+ pump
Cytosol
Extracellular fluid
Cytosol
Extracellular fluid

FIGURE 13-6 Steady-State Ion Movements. The actual membrane potential of a cell depends on the permeability of the plasma membrane to various ions and the steady-state movements of ions across the membrane, as illustrated here with our two-compartment model. The red arrows indicate the direction of movement of the indicated ion.

potassium ions. Under these conditions, there will be no net movement of potassium out of the cell. If we assume that the membrane is also slightly permeable to sodium ions, what will happen? We know that the cell will have both a large sodium gradient across the membrane and a negative membrane potential corresponding to the potassium equilibrium potential. These forces tend to drive sodium ions into the cell. As sodium ions leak inward, the membrane is partially depolarized. At the same time, as the membrane potential is neutralized, there is now less restraining force preventing potassium from leaving the cell, so potassium ions diffuse outward, balancing the inward movement of sodium. The inward movement of sodium ions shifts the membrane potential in the positive direction, while the outward movement of potassium ions shifts the membrane potential back in the negative direction.

Movements of sodium and potassium ions across the membrane, therefore, have essentially opposite effects on the membrane potential. For one type of squid axon, the sodium ion gradient tends toward a cell membrane potential of about +55 mV, while the potassium ion gradient tends toward a membrane potential of about −75 mV. At what value will the membrane potential come to rest? The pioneering neurobiologists David E. Goldman, Alan Lloyd Hodgkin, and Bernard Katz were the first to describe how gradients of several different ions each contribute to the membrane potential as a function of relative ionic permeabilities. The Goldman-Hodgkin-Katz equation, more commonly known as the **Goldman equation,** is as follows:

$$V_m = \frac{RT}{F} \ln \frac{(P_K)[K^+]_{out} + (P_{Na})[Na^+]_{out} + (P_{Cl})[Cl^-]_{in}}{(P_K)[K^+]_{in} + (P_{Na})[Na^+]_{in} + (P_{Cl})[Cl^-]_{out}} \quad \textbf{(13-3)}$$

Because chloride ions have a negative valence, $[Cl^-]_{inside}$ appears in the numerator and $[Cl^-]_{outside}$ in the denominator.

A key difference between the Nernst equation and the Goldman equation is the incorporation of terms for permeability. Here, P_K, P_{Na}, and P_{Cl} are the *relative permeabilities* of the membrane for the respective ions. The use of relative permeabilities circumvents the complicated task of determining the absolute permeability of each ion. While the equation shown here takes into account only the contributions of potassium, sodium, and chloride ions, other ions could be added as well. Except under special circumstances, however, the permeability of the plasma membrane to other ions is usually so low that their contributions are negligible.

We can use a squid axon to illustrate how we can accurately estimate the resting membrane potential from the known steady-state concentrations and relative permeabilities of sodium, potassium, and chloride ions. To do so, we assign K^+ a permeability value of 1.0, and the permeability values of all other ions are determined relative to that of K^+. For squid axons, the permeability of sodium ions is only about 4% of that for potassium ions; and for chloride ions, the value is 45%. Relative values of P_K, P_{Na}, and P_{Cl} are therefore 1.0, 0.04, and 0.45, respectively. Using these values, a temperature of 20°C, and the intracellular and extracellular concentrations of Na^+, K^+, and Cl^- for the squid axon from Table 13-1, you should be able to estimate the resting membrane potential of a squid axon as −60.3 mV. Typical measured values for the resting membrane potential of a squid axon are about −60 mV, which is remarkably close to our calculated potential.

When the relative permeability for one ion is very high, the Goldman equation reduces to the Nernst equation for that particular ion. For example, if we ignore the contribution due to chloride ions (which we can often do), when P_{Na} is much higher than P_K ($P_{Na} >> P_K$), Equation 13-3 reduces to:

$$V_m = \frac{RT}{F}\ln\frac{[Na^+]_{out}}{[Na^+]_{in}}, \quad (13\text{-}4)$$

which is just the Nernst equation for sodium. Since the resting potential for sodium is about +55 mV, making a cell highly permeable to sodium will cause it to depolarize. Similarly, if $P_K >> P_{Na}$, then the Goldman equation reduces to the Nernst equation for potassium. Since the resting potential for potassium is about −75 mV, high permeability to potassium will tend to return a cell to a polarized state. As we will now see, neurons use precisely this strategy to transiently depolarize and then repolarize their plasma membranes.

Electrical Excitability

The establishment of a resting membrane potential and its dependence on ion gradients and ion permeability are properties of almost all cells. The unique feature of electrically excitable cells is their response to membrane depolarization. Whereas a nonexcitable cell that has been temporarily and slightly depolarized will simply return to its original resting membrane potential, an electrically excitable cell that is depolarized to the same degree will respond with an *action potential.*

Electrically excitable cells produce an action potential because of the presence of *voltage-gated channels* in the plasma membrane. To see how nerve cells communicate signals electrically, we need to understand the characteristics of the ion channels in the nerve cell membrane, a subject to which we now turn.

Ion Channels Act Like Gates for the Movement of Ions Through the Membrane

In Chapter 8, we introduced **ion channels,** integral membrane proteins that form ion-conducting pores through the lipid bilayer. Recall that **voltage-gated ion channels,** as the name suggests, respond to changes in the voltage across a membrane. Voltage-gated sodium and potassium channels are responsible for the action potential. **Ligand-gated ion channels,** by contrast, open when a particular molecule binds to the channel. (*Ligand* comes from the Latin root *ligare,* meaning "to bind.") Other channels contribute to the steady-state ionic permeabilities of membranes. These *leak channels* allow resting cells to be somewhat permeable to cations, in particular potassium ions. Detailed knowledge of the structure and function of voltage-gated sodium and potassium channels is fundamental to understanding the events of the action potential. We begin our exploration by looking at how ion channels are studied experimentally.

Patch Clamping and Molecular Biological Techniques Allow the Activity of Single Ion Channels to Be Monitored

Our clearest picture of how channels operate comes from using a technique that permits the recording of ion currents passing through individual channels. This technique, known as *single-channel recording,* or more commonly as **patch clamping** (**Figure 13-7**), was developed by Erwin Neher and Bert Sackman, who earned a Nobel Prize in 1991 for their discovery.

To record single-channel currents, a glass micropipette with a tip diameter of approximately 1 mm is carefully pressed against the surface of a cell such as a neuron. Gentle suction is then applied, so that a tight seal forms between the pipette and the plasma membrane. There is now a patch of membrane under the mouth of the micropipette that is sealed off from the surrounding medium (Figure 13-7, ❶). This patch is small enough that it usually contains only one or perhaps a few ion channels. Current can enter and leave the pipette only through these channels, thereby enabling an experimenter to study various properties of the individual channels. The channels can be studied in the intact cell, or the patch can be pulled away from the cell so that the researcher has access to the cytosolic side of the membrane (Figure 13-7, ❷).

During the experimental process, an amplifier maintains voltage across the membrane with the addition of a sophisticated electronic feedback circuit called a voltage clamp (hence the term *patch clamp*). The voltage clamp keeps the cell at a fixed membrane potential, regardless of changes in the electrical properties of the plasma membrane, by injecting current as needed to hold the voltage constant. The voltage clamp then measures tiny changes in current flow—actual ionic currents through individual channels—from the patch pipette. The patch-clamp method has been used to show that when a sodium channel opens, it conducts the same amount of *current*—that is, the same number of ions per unit of time. In other words, sodium channels tend to be either open or closed.

Based on these properties of ion channels, we can characterize a particular channel in terms of its conductance. *Conductance* is an indirect measure of the permeability of a channel when a specified voltage is applied across the

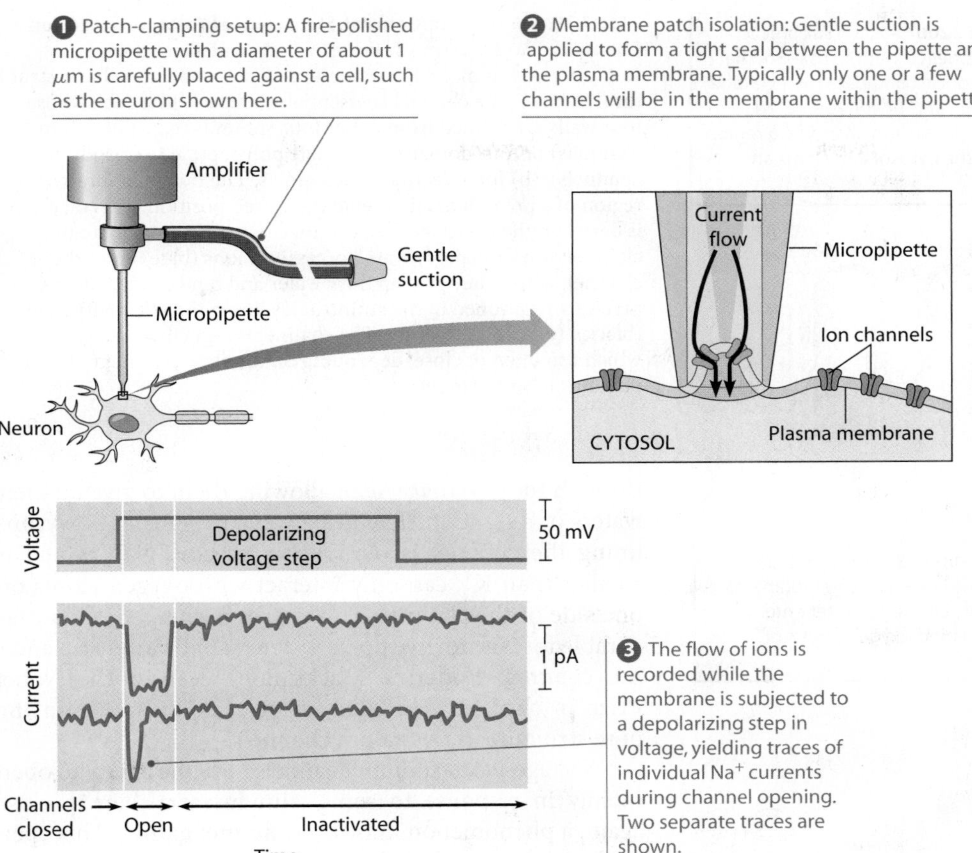

1 Patch-clamping setup: A fire-polished micropipette with a diameter of about 1 μm is carefully placed against a cell, such as the neuron shown here.
2 Membrane patch isolation: Gentle suction is applied to form a tight seal between the pipette and the plasma membrane. Typically only one or a few channels will be in the membrane within the pipette.
Amplifier
Gentle suction
Micropipette
Neuron
Current flow
Micropipette
Ion channels
CYTOSOL
Plasma membrane
Voltage
Depolarizing voltage step
50 mV
Current
1 pA
3 The flow of ions is recorded while the membrane is subjected to a depolarizing step in voltage, yielding traces of individual Na+ currents during channel opening. Two separate traces are shown.
Channels closed
Open
Inactivated
Time

FIGURE 13-7 Patch Clamping. Patch clamping makes it possible to study the behavior of individual ion channels in a small patch of membrane. When each channel opens, the amount of current that flows through it is always the same (1 pA for the sodium channels shown here). Following the burst of channel opening, a quiescent period occurs due to channel inactivation.

membrane. In electrical terms, conductance is the inverse of resistance. For voltage-gated sodium channels, when a voltage of 50 mV is applied across the membrane, a current of approximately 1 picoampere (1 pA $= 1 \times 10^{-12}$ A) is generated. This current corresponds to about 6 million sodium ions flowing through the channel per second. This can be seen in the traces shown in Figure 13-7, ❸.

In single-channel recording, membrane depolarization (triggered by changing the applied voltage to a more positive potential) increases the probability that a channel will open. Even before the membrane is depolarized, a sodium channel will occasionally flicker open and closed. Once the channel has opened, it cannot reopen unless the membrane potential is restored to a more negative value. The cessation of channel activity that occurs while the membrane is still depolarized is due to channel inactivation, which we will discuss later in this chapter.

Much of the present research on ion channels combines patch clamping with molecular biology. It is now possible to synthesize large amounts of channel proteins in the laboratory and to study their functions in lipid bilayers or in frog eggs. Specific molecular modifications or mutations of the channel can be used to determine how various regions of the channel protein are involved in channel function. This approach has been used to study the domains of the sodium and potassium channels responsible for voltage gating.

Specific Domains of Voltage-Gated Channels Act as Sensors and Inactivators

The structure of voltage-gated ion channels falls into two different, though similar, categories. *Voltage-gated potassium channels* are *multimeric* proteins—that is, they consist of several separate protein subunits that associate with one another to form the functional channel. In the case of potassium channels, four separate protein subunits come together in the membrane, forming a central pore that ions can pass through. *Voltage-gated sodium channels,* by contrast, are large, *monomeric* proteins (in other words, they consist of a single polypeptide) with four separate domains. Each domain is similar to one of the subunits of the voltage-gated potassium channel. In both kinds of channels, each subunit or domain contains six transmembrane α helices (called subunits S1–S6; **Figure 13-8a**).

(a) Domain structure of an individual subunit. Each subunit or domain contains six transmembrane helices, labeled S1–S6. The fourth transmembrane helix, S4, is a good candidate for a voltage sensor and part of the gating mechanism. For voltage-gated sodium channels and some types of potassium channels, a region near the N-terminus protrudes into the cytosol and forms an inactivating particle.

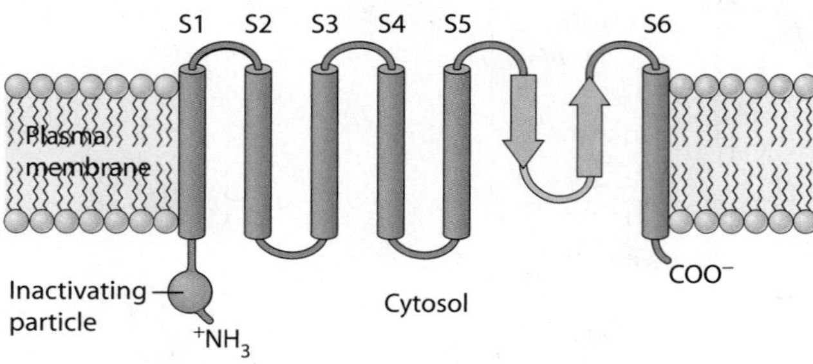

(b) Pore structure. Two of the four subunits of a voltage-gated potassium channel are shown here. Only the transmembrane part of the channel is shown. When K^+ ions bound by water enter the channel, they give up their water and bind oxygen atoms of amino acids lining the selectivity filter.

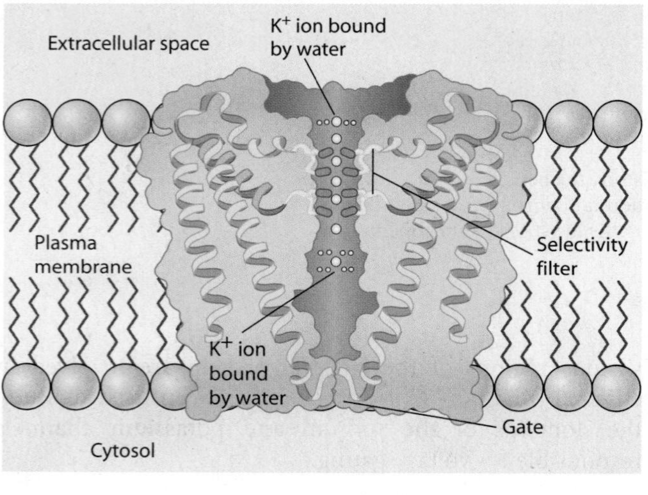

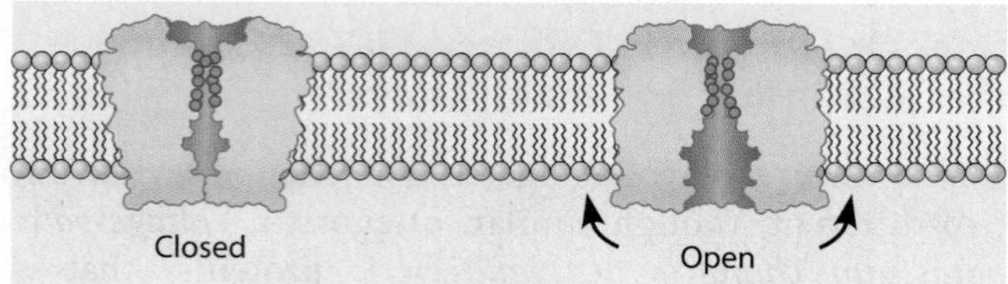

(c) Channel gating. The channel gate opens and closes depending on the conformational state of channel subunits.

FIGURE 13-8 The General Structure of Voltage-Gated Ion Channels. (a) Domain structure. Voltage-gated channels for sodium, potassium, and calcium ions all share the same basic structural themes. The channel is essentially a rectangular tube whose four walls are formed from either four subunits (e.g., potassium channels) or four domains of a single polypeptide (e.g., sodium channels). (b) Ion selectivity of channels. The transmembrane region of a potassium channel in the closed position. This diagram is based on the bacterial Kcsa channel, but vertebrate potassium channels are similar. Hydrated potassium ions (blue) enter the channel, where they give up their water and bind oxygen atoms, precisely positioned in the amino acids lining the selectivity filter (black). (c) Channel gating. The channel is regulated by a gate, which can open or close, depending on the state of voltage sensor domains in the channel.

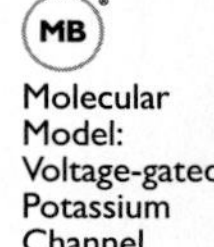

Molecular Model: Voltage-gated Potassium Channel

The size of the central pore and, more importantly, the way it interacts with an ion, give a channel its ion selectivity. Figure 13-8b shows why this is so for the bacterial potassium channel, Kcsa. Vertebrate voltage-gated potassium channels have a similar structure. Oxygen atoms in amino acids lining the center of the channel are precisely positioned to interact with ions as they move through the *selectivity filter,* allowing them to give up their waters of hydration. The fit between K^+ ions and oxygens lining the channel is remarkably precise. Na^+, which is smaller than K^+, can only interact with oxygen atoms on one side of the channel. This makes it energetically unfavorable for Na^+ to give up its waters of hydration and enter the channel. Roderick Mackinnon received the Nobel Prize in chemistry in 2003, in part for his work on the pore structure of potassium channels.

Voltage-gated sodium channels have the ability to open rapidly in response to some stimulus and then to close again, a phenomenon known as **channel gating.** This open or closed state is an all-or-none phenomenon—that is, gates do not appear to remain partially open (Figure 13-8c). One of the transmembrane a helices of vertebrate voltage-gated channels, S4, acts as a **voltage sensor** during channel gating. When positively charged amino acids in S4 are replaced with neutral amino acids, the channel does not open. This result suggests that S4 makes these channels responsive to changes in potential. The details of how this happens are currently the subject of intense debate.

Most voltage-gated channels can also adopt a second type of closed state, referred to as **channel inactivation,** which is an important feature of voltage-gated sodium and potassium channels (**Figure 13-9**). When a channel is inactivated, it cannot reopen immediately, even if stimulated to do so. Channel inactivation is like placing a padlock on the closed gate; only when the padlock is unlocked can the gate be opened again. Inactivation is caused by a portion of the channel called the *inactivating particle.* Common channels have four such particles. During inactivation, a particle inserts into the opening of the channel. For a channel to reactivate and open in response to a stimulus, the inactivating particle must move away from the pore. When the cytosolic side of the channel is treated with a protease or with antibodies prepared against the fragment of the channel thought to be responsible for inactivation, the inactivating particle can no longer function, and channels can no longer be inactivated.

As we will see in the next few sections of this chapter, the regulation of ion channels is crucial for the proper

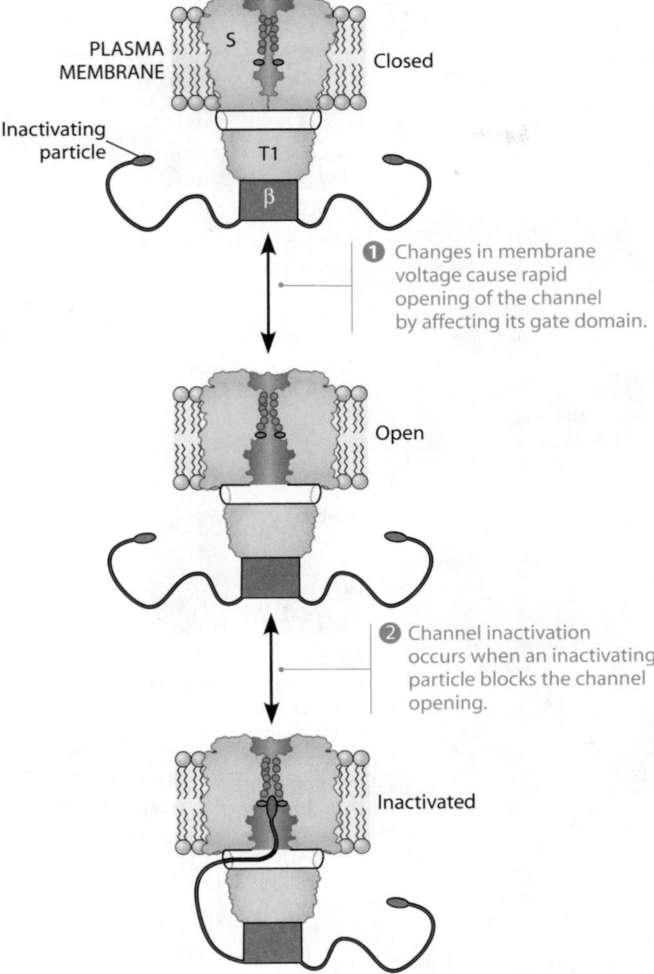
PLASMA MEMBRANE
S
Closed
Inactivating particle
T1
β
1 Changes in membrane voltage cause rapid opening of the channel by affecting its gate domain.
Open
2 Channel inactivation occurs when an inactivating particle blocks the channel opening.
Inactivated

FIGURE 13-9 The Function of a Voltage-Gated Ion Channel. **(a)** Mammalian potassium channels have three major domains: the transmembrane (S) and cytosolic (T1) domains of the a subunit and the b subunit. The N-terminal region of the b subunit contains an inactivating particle. Channel gating occurs because a portion of the transmembrane domain changes conformation when the membrane potential changes. **(b)** Channel inactivation. Here, two of the four inactivating particles are shown for a voltage-gated potassium channel. Channels are inactivated when an inactivating particle moves through a "window" region of the channel, blocking the opening of the pore.

functioning of neurons. Defects in several voltage-gated ion channels have been linked to human neurological diseases (such defects have been termed "channelopathies"). For example, humans carrying mutations in certain K^+ channels suffer from ataxia (a defect in muscle coordination), and one form of epilepsy is caused by a mutation in one type of voltage-gated Na^+ channel.

The Action Potential

We have seen how an ion gradient across a selectively permeable membrane can generate a membrane potential and how, according to the Goldman equation, the membrane potential will change in response to changes in ion permeability. We have also examined the nature of membrane ion channels, which regulate the permeability of the membrane to the different ions. Now we are ready to explore how the coordinated opening and closing of ion channels can lead to an action potential. We will begin by examining how membrane potential changes during an action potential. Because of its historical importance, we will use the squid giant axon as the model for our discussion.

Action Potentials Propagate Electrical Signals Along an Axon

A resting neuron is a system poised for electrical action. As we have seen, the membrane potential of the cell is set by a delicate balance of ion gradients and ion permeability. Depolarization of the membrane upsets this balance. If the level of depolarization is small—less than about +20 mV—the membrane potential will normally drop back to resting levels without further consequences. Further depolarization brings the membrane to the **threshold potential.** Above the threshold potential, the nerve cell membrane undergoes rapid and dramatic alterations in its electrical properties and permeability to ions, and an action potential is initiated.

An **action potential** is a brief but large electrical depolarization and repolarization of the neuronal plasma membrane caused by the inward movement of sodium ions and the subsequent outward movement of potassium ions. These ion movements are controlled by the opening and closing of voltage-gated sodium and potassium channels. In fact, we can explain the development of an action potential solely in terms of the behavior of these channels. Once an action potential is initiated in one region of the membrane, it will travel along the membrane away from the site of origin by a process called **propagation.**

Action Potentials Involve Rapid Changes in the Membrane Potential of the Axon

As we saw in Figure 13-3b, the development and propagation of an action potential can be readily studied in large axons such as those of the squid. **Figure 13-10** shows the sequence of membrane potential changes associated with an action potential. In less than a millisecond, the membrane potential rises dramatically from the resting membrane potential to about +40 mV—the interior of the membrane actually becomes positive for a brief period. The potential then falls somewhat more slowly, dropping to about −75 mV (called *undershoot* or *hyperpolarization*) before stabilizing again at the resting potential of about −60 mV. As Figure 13-10 indicates, the complete sequence of events during an action potential takes place within a few milliseconds.

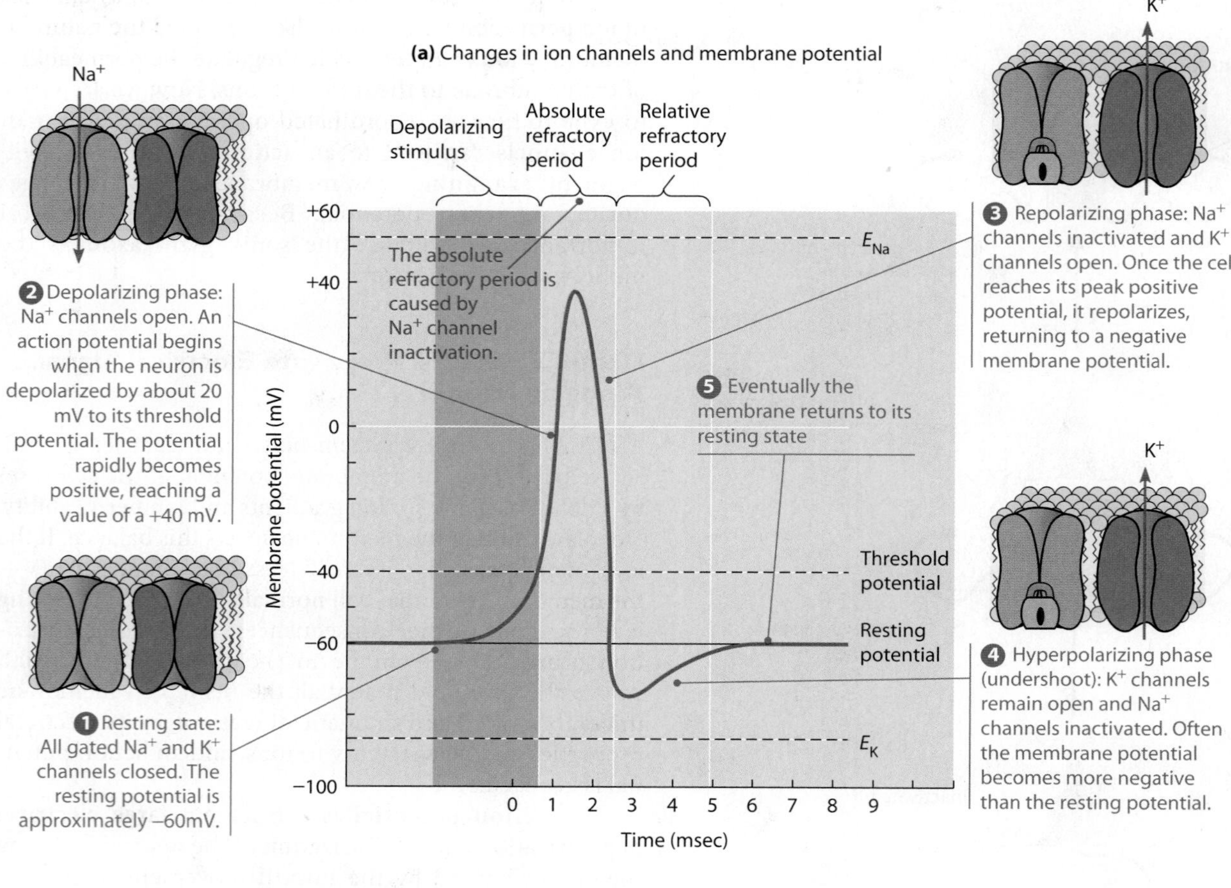
(a) Changes in ion channels and membrane potential
Na+
K+
Depolarizing stimulus
Absolute refractory period
Relative refractory period
+60
+40
0
−40
−60
−100
Membrane potential (mV)
The absolute refractory period is caused by Na+ channel inactivation.
E_Na
Threshold potential
Resting potential
E_K
0 1 2 3 4 5 6 7 8 9
Time (msec)
2 Depolarizing phase: Na+ channels open. An action potential begins when the neuron is depolarized by about 20 mV to its threshold potential. The potential rapidly becomes positive, reaching a value of a +40 mV.
1 Resting state: All gated Na+ and K+ channels closed. The resting potential is approximately −60mV.
3 Repolarizing phase: Na+ channels inactivated and K+ channels open. Once the cell reaches its peak positive potential, it repolarizes, returning to a negative membrane potential.
5 Eventually the membrane returns to its resting state
4 Hyperpolarizing phase (undershoot): K+ channels remain open and Na+ channels inactivated. Often the membrane potential becomes more negative than the resting potential.

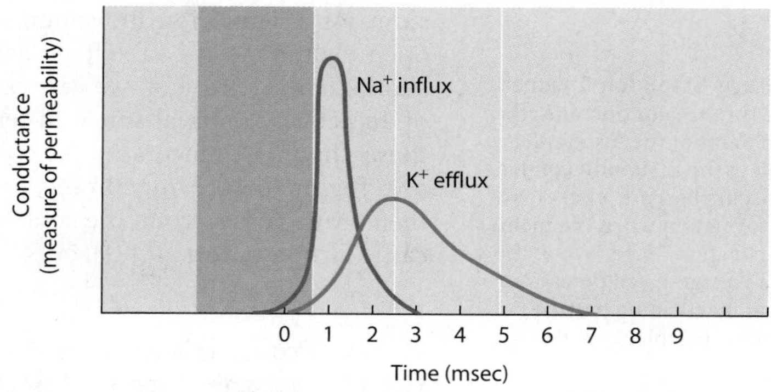
Conductance (measure of permeability)
Na+ influx
K+ efflux
0 1 2 3 4 5 6 7 8 9
Time (msec)

(b) Change in membrane conductance

FIGURE 13-10 Changes in Ion Channels and Currents in the Membrane of a Squid Axon During an Action Potential. (a) The change in membrane potential caused by movement of Na^+ and K^+ through their voltage-gated channels, which are shown at each step of the action potential. The absolute refractory period is caused by sodium channel inactivation. Notice that at the peak of the action potential, the membrane potential approaches the E_{Na} (sodium equilibrium potential) value of about +55 mV; similarly, the potential undershoots nearly to the E_K (potassium equilibrium potential) value of about −75 mV. (b) The change in membrane conductance (permeability of the membrane to specific ions). The depolarized membrane initially becomes very permeable to sodium ions, facilitating a large inward rush of sodium. Thereafter, as permeability to sodium declines, the permeability of the membrane to potassium increases transiently, causing the membrane to hyperpolarize.

The apparatus shown in Figure 13-3b can also be used to measure the ion currents that flow through the membrane at different phases of an action potential. To do so, an additional electrode known as the *holding electrode* is inserted into the cell and connected to a voltage clamp, thereby enabling the investigator to set and hold the membrane at a particular potential. Using the voltage-clamp apparatus, a researcher can measure the current flowing through the membrane at any given membrane potential. Such experiments have contributed fundamentally to our present understanding of the mechanism that causes an action potential.

Action Potentials Result from the Rapid Movement of Ions Through Axonal Membrane Channels

In a resting neuron, the voltage-dependent sodium and potassium channels are usually closed. Because of the steady-state leakiness of the membrane to potassium ions, the cell is roughly 100 times more permeable to potassium than to sodium ions at this point. When a region of the nerve cell is slightly depolarized, a fraction of the sodium channels respond and open. As they do, the increased sodium current acts to further depolarize the membrane. Increasing depolarization causes an even larger sodium current to flow, which depolarizes the membrane even more. This relationship between depolarization, the opening of voltage-gated sodium channels, and an increased sodium current constitutes a positive feedback loop known as the *Hodgkin cycle.*

Subthreshold Depolarization. Under resting conditions, the outward movement of potassium ions through leak channels restores the resting membrane potential. When the membrane is depolarized by a small amount, the membrane potential recovers and no action potential is generated. Levels of depolarization that are too small to produce an action potential are referred to as *subthreshold depolarizations.*

The Depolarizing Phase. If all the voltage-dependent sodium channels in the membrane were to open at once, the cell would suddenly become ten times more permeable to sodium than to potassium. Because sodium would then be the more permeable ion, the membrane potential would be largely a function of the sodium ion gradient. This is effectively what happens when the membrane is depolarized past the threshold potential (Figure 13-10, ❶ and ❷). Once the threshold potential is reached, a significant number of sodium channels begin activating. At this point, the membrane potential shoots rapidly upward. When the rate of sodium entry slightly exceeds the maximum rate of potassium exit, an action potential is triggered. When the membrane potential peaks, at approximately +40 mV, the action potential approaches—although it does not actually reach—the equilibrium potential for sodium ions (about +55 mV). (It never actually reaches this value because the membrane remains permeable to other ions during this time.)

The Repolarizing Phase. Once the membrane potential has risen to its peak, the membrane quickly repolarizes (Figure 13-10, ❸). This is due to a combination of the inactivation of sodium channels and the opening of voltage-gated potassium channels. When sodium channels are inactivated, they close and remain closed until the membrane potential becomes negative again. Channel inactivation thus stops the inward flow of sodium ions. The cell will now automatically repolarize as potassium ions leave the cell.

The difference in response speed between voltage-gated potassium channels and voltage-gated sodium channels plays an important role in generating the action potential. When the cell is depolarized, the potassium channels open more slowly. As a result, an action potential begins with an increase in the membrane's permeability to sodium, which depolarizes the membrane, followed by an increased permeability to potassium, which repolarizes the membrane.

The Hyperpolarizing Phase (Undershoot). At the end of an action potential, most neurons show a transient **hyperpolarization,** or **undershoot,** in which the membrane potential briefly becomes even more negative than it normally is at rest (Figure 13-10, ❹). The undershoot occurs because of the increased potassium permeability that exists while voltage-gated potassium channels remain open. Note that the potential of the undershoot closely approximates the equilibrium potential for potassium ions (about −75 mV for the squid axon). As the voltage-gated potassium channels close, the membrane potential returns to its original resting state (Figure 13-10, ❺). Notice that the rapid restoration of the resting potential following an action potential does not use the Na^+/K^+ pump but instead involves the passive movements of ions. In cells that have been treated with a metabolic inhibitor so that they cannot produce ATP (and hence their Na^+/K^+ pumps cannot use it), action potentials can still be generated. The pump helps to maintain a negative potential once an action potential has passed and the membrane has returned to its resting state.

The Refractory Periods. For a few milliseconds after an action potential, it is impossible to trigger a new action potential. During this interval, known as the **absolute refractory period,** sodium channels are inactivated and cannot be opened by depolarization. During the period of undershoot, when the sodium channels have reactivated and are capable of opening again, it is possible but difficult to trigger an action potential. This is because both potassium leak channels and voltage-gated potassium channels are open during this time. This tends to drive the membrane potential to a very negative value, far from the threshold for triggering another round of sodium channel opening. This interval is known as the **relative refractory period.**

Changes in Ion Concentrations Due to an Action Potential. Our discussion of ion movements might give the impression that an action potential involves large changes in the cytosolic concentrations of sodium and potassium ions. In fact, during a single action potential, the cellular concentrations of sodium and potassium ions hardly change at all. Remember that the membrane potential is due to a slight excess of negative charge on one side and a slight excess of positive charge on the other side of the membrane. The number of excess charges is a tiny fraction of the total ions in the cell, and the number of ions that must cross the membrane to neutralize or alter the balance of charge is likewise small.

Nevertheless, intense neuronal activity can lead to significant changes in overall ion concentrations. For example, as a neuron continues to generate large numbers of action potentials, the concentration of potassium outside the cell will begin to rise perceptibly. This can affect the membrane potential of both the neuron itself and surrounding cells. Astrocytes, the glial cells that form the blood-brain barrier, are thought to control this problem by taking up excess potassium ions.

Action Potentials Are Propagated Along the Axon Without Losing Strength

For neurons to transmit signals to one another, the transient depolarization and repolarization that occur during an action potential must travel along the neuronal membrane. The depolarization at one point on the membrane spreads to adjacent regions through a process called the **passive spread of depolarization.** As a wave of depolarization spreads passively away from the site of origin, it also decreases in magnitude. This fading of the depolarization with distance from the source makes it difficult for signals to travel very far by passive means only. For signals to travel longer distances, an action potential must be *propagated,* or actively generated, from point to point along the membrane.

To understand the difference between the passive spread of depolarization and the propagation of an action potential, consider how a signal travels along the neuron from the site of origin at the dendrites to the end of the axon (**Figure 13-11**). Incoming signals are transmitted to a neuron at synapses that form points of contact between the synaptic boutons of the transmitting neuron and the dendrites of the receiving neuron. When these incoming signals depolarize the dendrites of the receiving neuron, the depolarization spreads passively over the membrane from the dendrites to the base of the axon—the **axon hillock.** The axon hillock is the region where action potentials are initiated most easily. This is because sodium channels are distributed sparsely over the dendrites and cell body but are concentrated at the axon hillock and nodes of Ranvier; a given amount of depolarization will produce the greatest amount of sodium entry at sites where sodium channels are abundant. The action potentials initiated at the axon hillock are then propagated along the axon.

The mechanism for propagating an action potential in nonmyelinated nerve cells is illustrated in **Figure 13-12**. Stimulation of a resting membrane results in a depolarization of the membrane and a sudden rush of sodium ions into the axon at that location (❶). Membrane polarity is temporarily reversed at that point, and this depolarization then spreads to an adjacent point (❷). The depolarization at this adjacent point is sufficient to bring it above the threshold potential, triggering the inward rush of sodium

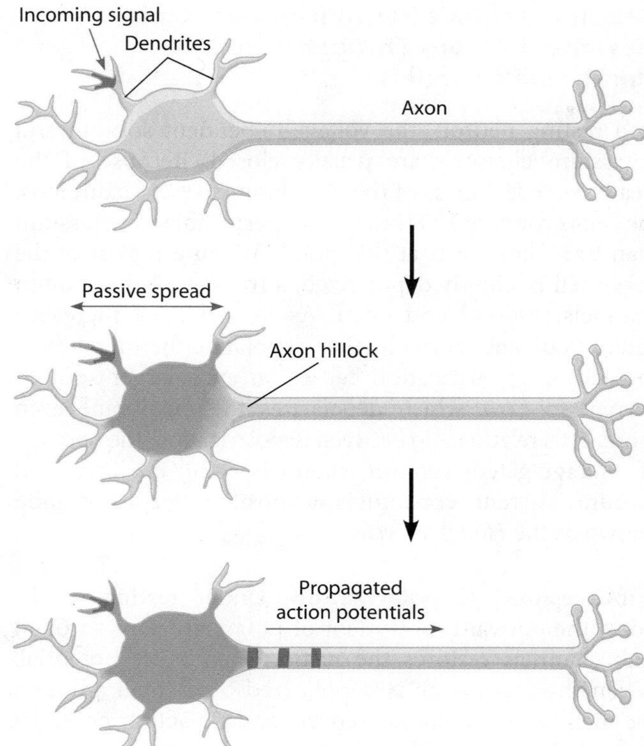

FIGURE 13-11 The Passive Spread of Depolarization and Propagated Action Potentials in a Neuron. The transmission of a nerve impulse along a neuron depends on both the passive spread of depolarization and the propagation of action potentials. A neuron is stimulated when its dendrites receive a depolarizing stimulus from other neurons. A depolarization starting at a dendrite spreads passively over the cell body to the axon hillock, where an action potential forms. The action potential is then propagated down the axon.

ions (❸). By this time, the original region of membrane has become highly permeable to potassium ions. As potassium ions rush out of the cell, negative polarity is restored and that portion of the membrane returns to its resting state (❹).

Meanwhile, the depolarization has spread to a new region, initiating the same sequence of events there (❺). In this way, the signal moves along the membrane as a ripple of depolarization-repolarization events; the membrane polarity is reversed in the immediate vicinity of the signal but returned to normal again as the signal travels down the axon. The propagation of this cycle of events along the nerve fiber is called a *propagated action potential,* or **nerve impulse.** The nerve impulse can move only *away* from the initial site of depolarization because the sodium channels that have just been depolarized are in the inactivated state and cannot respond immediately to further stimulation.

Because an action potential is actively propagated, it does not fade as it travels. The reason is that it is generated anew, as an all-or-none event, at each successive point along the membrane. Thus, a nerve impulse

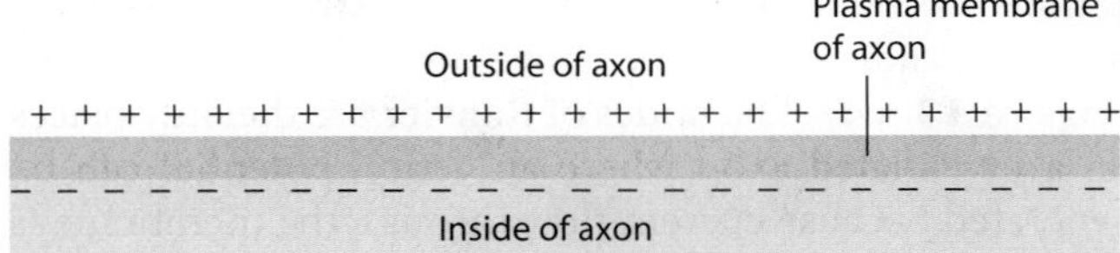

❶ At the start, the membrane is completely polarized.

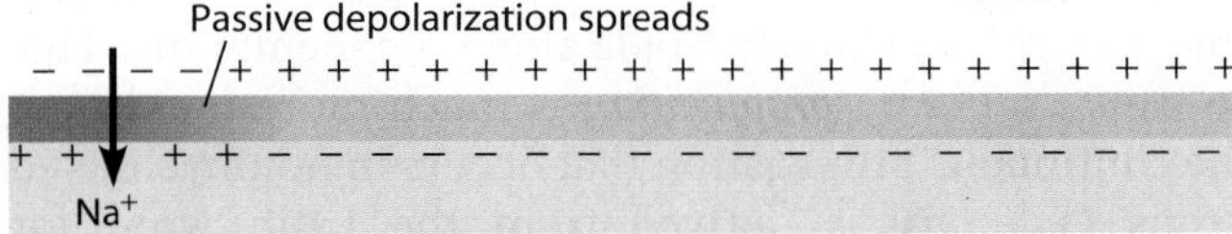

❷ When an action potential is initiated, a region of the membrane depolarizes. As a result, the adjacent regions become depolarized.

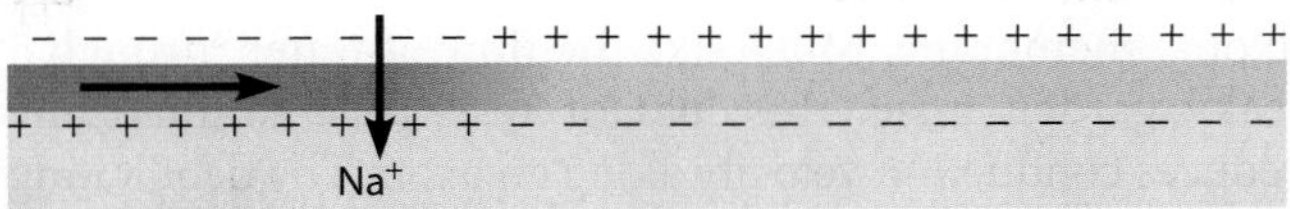

❸ When the adjacent region is depolarized to its threshold, an action potential starts there.

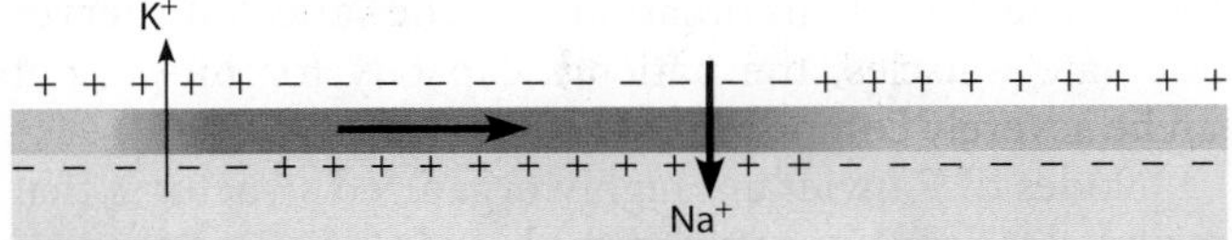

❹ Repolarization occurs due to the outward flow of K^+ ions. The depolarization spreads forward, triggering an action potential.

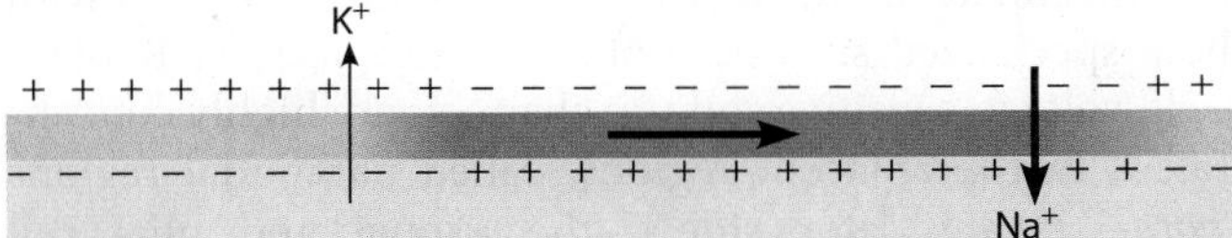

❺ Depolarization spreads forward, repeating the process.

FIGURE 13-12 The Transmission of an Action Potential Along a Nonmyelinated Axon. A nonmyelinated axon can be viewed as a string of points, each capable of undergoing an action potential. Notice that no backward propagation occurs near sites where action potentials form because sodium channels are in an inactivated state and the membrane is hyperpolarized.

can be transmitted over essentially any distance with no decrease in strength.

The Myelin Sheath Acts Like an Electrical Insulator Surrounding the Axon

Most axons in vertebrates have an additional specialization: They are surrounded by a discontinuous *myelin sheath* consisting of many concentric layers of membrane. The myelin sheath is a reasonably effective electrical insulation for the segments of the axon that it envelops. The myelin sheath of neurons in the CNS is formed by **oligodendrocytes**; in the peripheral nervous system (PNS), the myelin sheath is formed by **Schwann cells** (see Figure 13-1), each of which wraps layer after layer of its own plasma membrane around the axon in a tight spiral (**Figure 13-13**). Because each Schwann cell surrounds a

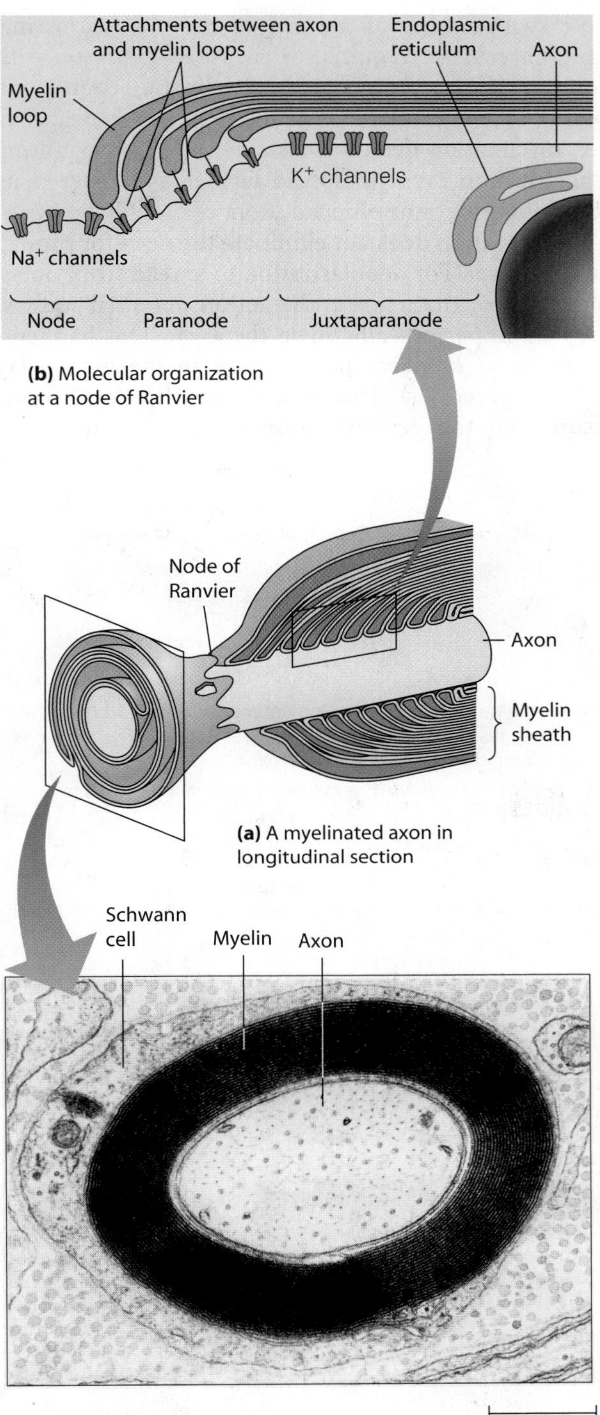

FIGURE 13-13 Myelination of Axons. (a) An axon of the peripheral nervous system that has been myelinated by a Schwann cell. Each Schwann cell gives rise to one segment of myelin sheath by wrapping its own plasma membrane concentrically around the axon. (b) Organization of a typical node of Ranvier in the peripheral nervous system. Sodium channels (brown) are concentrated in the node. Myelin loops attach to the regions next to the node ("paranodal" regions) via proteins on the axonal membrane and on the myelin loops (green). Potassium channels (orange) cluster next to the paranodal regions. (c) This cross-sectional view of a myelinated axon from the nervous system of a cat shows the concentric layers of membrane that have been wrapped around the axon by the Schwann cell that envelops it (TEM).

short segment (about 1 mm) of a single axon, many Schwann cells are required to encase a PNS axon with a discontinuous sheath of myelin. Myelination decreases the ability of the neuronal membrane to retain electric charge (i.e., myelination decreases its *capacitance*), permitting a depolarization event to spread farther and faster than it would along a nonmyelinated axon.

Myelination does not eliminate the need for propagation, however. For depolarization to spread from one site to the rest of the neuron, the action potential must still be renewed periodically down the axon. This happens at the *nodes of Ranvier,* interruptions in the myelin layer that are spaced just close enough together (1–2 mm) to ensure that the depolarization spreading out from an action potential at one node is still strong enough to bring an adjacent node above its threshold potential (**Figure 13-14**). The nodes of Ranvier are the only places on a myelinated axon where an action potential can be generated because current flow through the membrane is restricted elsewhere and because voltage-gated sodium channels are concentrated there. Thus, action potentials jump from node to node along myelinated axons rather than moving as a steady ripple along the membrane. This so-called *saltatory propagation* is much more rapid than the continuous propagation that occurs in nonmyelinated axons (*saltatory* is derived from the Latin word for "dancing"; see Figure 13-14). Myelination is a crucial feature of mammalian axons. Loss of myelination results in a dramatic decrease in the electrical resistance of the axonal membrane. Much like the flow of water through a leaky garden hose, this loss of resistance dramatically reduces conduction velocity along an axon. The debilitating human disease multiple sclerosis results when a patient's immune system attacks his or her own myelinated nerve fibers, causing demyelination. If the affected nerves innervate muscles, the patient's capacity for movement can be severely compromised.

Nodes of Ranvier are highly organized structures that involve close contact between the loops of glial or Schwann cell membrane and the plasma membrane of the axon(s) they myelinate. Three distinct regions are associated with these specialized sites of contact. In the node of Ranvier itself, voltage-sensitive sodium channels are highly concentrated. In the adjacent regions, called paranodal regions (*para–* means "alongside"), the axonal and glial cell membranes contain specialized adhesive proteins that are similar to those in septate junctions (see Chapter 17). Finally, in the region next to the paranodal areas, called juxtaparanodal regions (*juxta–* means "next to"), potassium channels are highly concentrated (see Figure 13-13b). The organization of nodes of Ranvier prevents free movement of the sodium and potassium channels within the axon's plasma membrane in regions around the nodes.

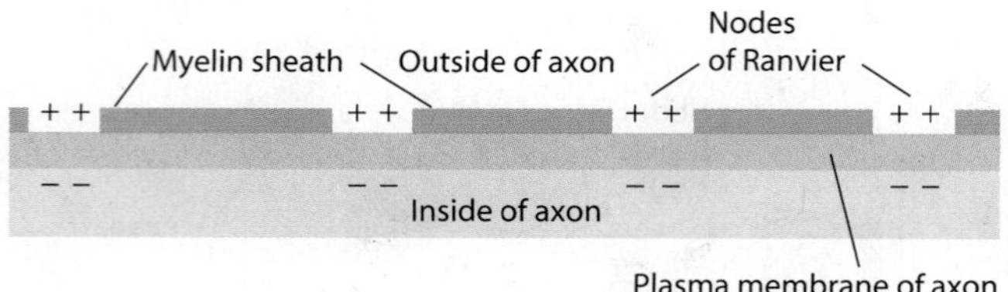

1 In myelinated neurons, an action potential is usually triggered at the axon hillock, just before the start of the myelin sheath. The depolarization then spreads along the axon.

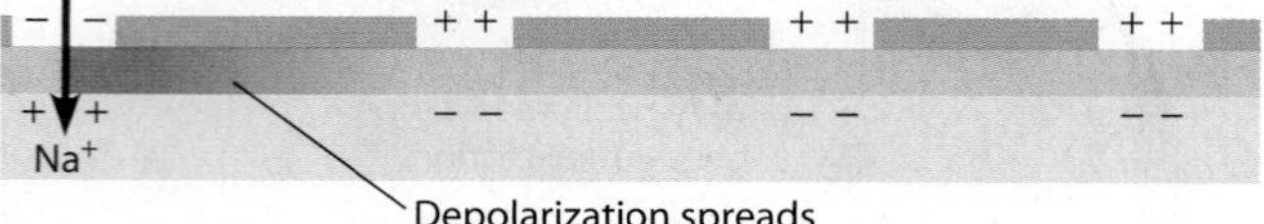

2 Because of myelination, the depolarization spreads passively to the next node.

3 The next node reaches its threshold, and a new action potential is generated.

4 This cycle is repeated, triggering an action potential at the next node.

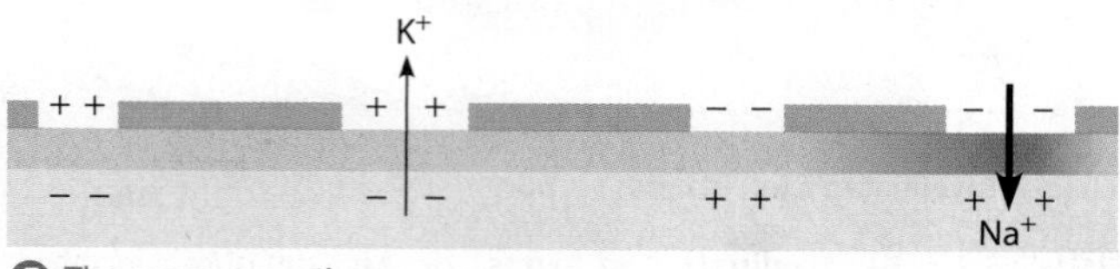

5 The process continues.

FIGURE 13-14 The Transmission of an Action Potential Along a Myelinated Axon. In a myelinated axon, action potentials can be generated only at nodes of Ranvier. Myelination reduces membrane capacitance, thereby allowing a given amount of sodium current, entering at one point of the membrane, to spread much farther along the membrane than it would in the absence of myelin. The result is a wave of depolarization-repolarization events that are propagated along the axon from node to node.

Synaptic Transmission

Nerve cells communicate with one another and with glands and muscles at synapses. There are two structurally distinct types of synapses, electrical and chemical. In an **electrical synapse,** one neuron, called the **presynaptic neuron,** is connected to a second neuron, the **postsynaptic neuron,** by gap junctions (**Figure 13-15**; we will explore gap junctions in more detail in Chapter 17). As ions move back and forth between the two cells, the depolarization in one cell spreads passively to the connected cell. Electrical synapses provide for transmission with virtually no delay and occur in places in the nervous system where speed of transmission is critical. Similar electrical connections can be found between nonneuronal cells, such as the cardiac muscle cells in the heart (see Chapter 16).

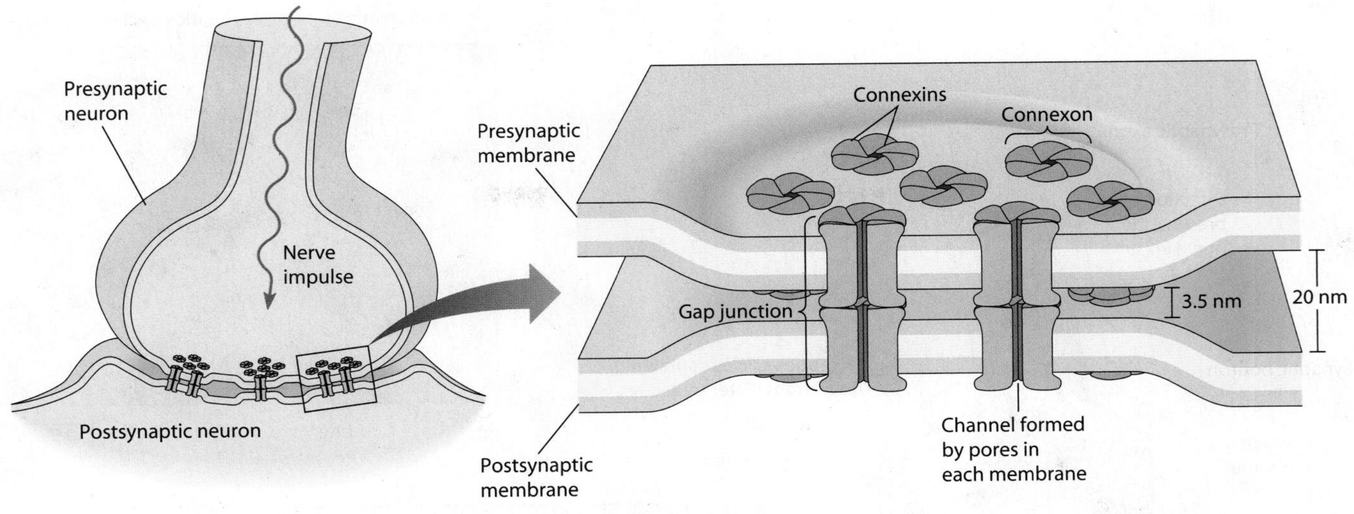

FIGURE 13-15 An Electrical Synapse. (a) In electrical synapses, the presynaptic and postsynaptic neurons are coupled by gap junctions, which allow small molecules and ions to pass freely from the cytosol of one cell to the next. When an action potential arrives at the presynaptic side of an electrical synapse, the depolarization spreads passively, due to the flow of positively charged ions, across the gap junction. **(b)** The gap junction is composed of sets of channels. A channel is made up of six protein subunits, each called a connexin. The entire set of six subunits together is called a connexon. Two connexons, one in the presynaptic membrane and one in the postsynaptic membrane, make up a gap junction.

In a **chemical synapse,** the presynaptic and postsynaptic neurons are not connected by gap junctions, (**Figure 13-16**), although they are connected by cell adhesion proteins (which we will consider in Chapter 17). Instead, the presynaptic plasma membrane is separated from the postsynaptic membrane by a small space of about 20–50 nm, known as the **synaptic cleft.** A nerve signal arriving at the terminals of the presynaptic neuron cannot bridge the synaptic cleft as an electrical impulse. For synaptic transmission to take place, the electrical signal must be converted at the presynaptic neuron to a chemical signal carried by a neurotransmitter. Neurotransmitter molecules are stored in the synaptic boutons of the presynaptic neuron. An action potential arriving at the terminal causes the neurotransmitter to be secreted into and diffuse across the synaptic cleft. The neurotransmitter molecules then bind to specific proteins embedded within the plasma membrane of the postsynaptic neuron (*receptors*) and are converted back into electrical signals, setting in motion a sequence of events that either stimulates or inhibits the production of an action potential in the postsynaptic neuron, depending on the kind of synapse.

Neurotransmitter receptors fall into two broad groups: *ligand-gated ion channels* (sometimes called *ionotropic* receptors), in which activation directly affects the cell, and receptors that exert their effects indirectly through a system of intracellular messengers (sometimes called *metabotropic* receptors; **Figure 13-17**). We will discuss the latter category of receptors in Chapter 14; here, we focus on *ligand-gated channels.* These membrane ion channels open in response to the binding of a neurotransmitter, and they can mediate either excitatory or inhibitory responses in the postsynaptic cell.

Neurotransmitters Relay Signals Across Nerve Synapses

A **neurotransmitter** is essentially any signaling molecule released by a neuron. Many kinds of molecules act as neurotransmitters. Most are detected by the postsynaptic cell via a specific type of receptor; most neurotransmitters have more than one type of receptor. When a neurotransmitter molecule binds to its receptor, the properties of the receptor are altered, and the postsynaptic neuron responds accordingly. An *excitatory receptor* causes depolarization of the postsynaptic neuron, whereas an *inhibitory receptor* typically causes the postsynaptic cell to hyperpolarize.

Although definitions vary, to qualify as a neurotransmitter, a compound must satisfy three criteria: (1) It must elicit the appropriate response when introduced into the synaptic cleft, (2) it must occur naturally in the presynaptic neuron, and (3) it must be released at the right time when the presynaptic neuron is stimulated. Many molecules meet these criteria, including acetylcholine, a group of biogenic amines called the catecholamines, certain amino acids and their derivatives, peptides, endocannabinoids, nucleotides such as ATP, some ions, and gases (such as nitric oxide). **Table 13-2** lists several common neurotransmitters; we will discuss some of them here.

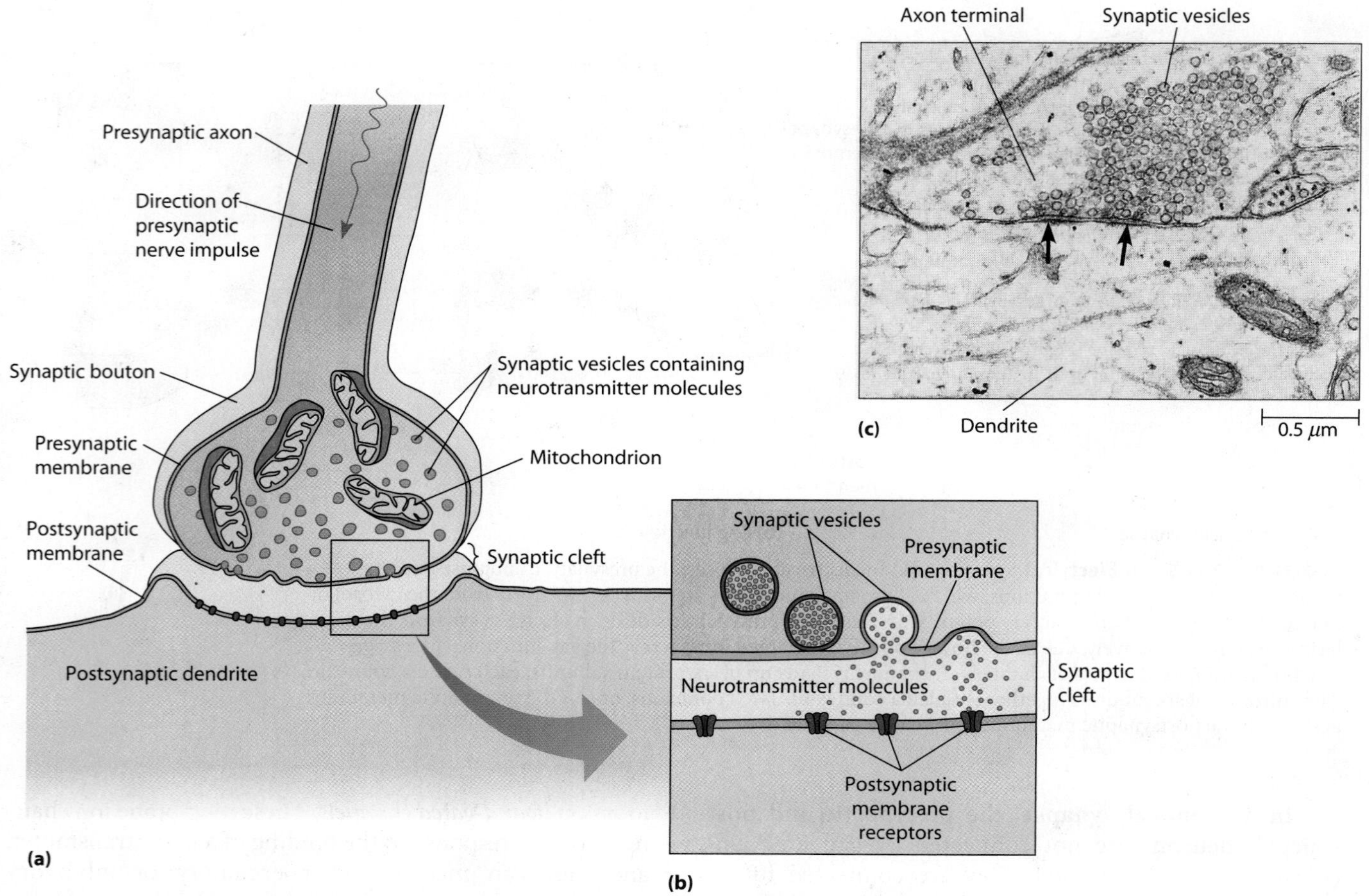

FIGURE 13-16 A Chemical Synapse. **(a)** When a nerve impulse from the presynaptic axon arrives at the synapse (red arrow), it causes synaptic vesicles containing neurotransmitter in the synaptic bouton to fuse with the presynaptic membrane, releasing their contents into the synaptic cleft. **(b)** Neurotransmitter molecules diffuse across the cleft from the presynaptic (axonal) membrane to the postsynaptic (dendritic) membrane, where they bind to specific membrane receptors and change the polarization of the membrane, either exciting or inhibiting the postsynaptic cell. **(c)** Electron micrograph of a chemical synapse (TEM). Arrows indicate a postsynaptic density, where membrane receptors and other proteins cluster.

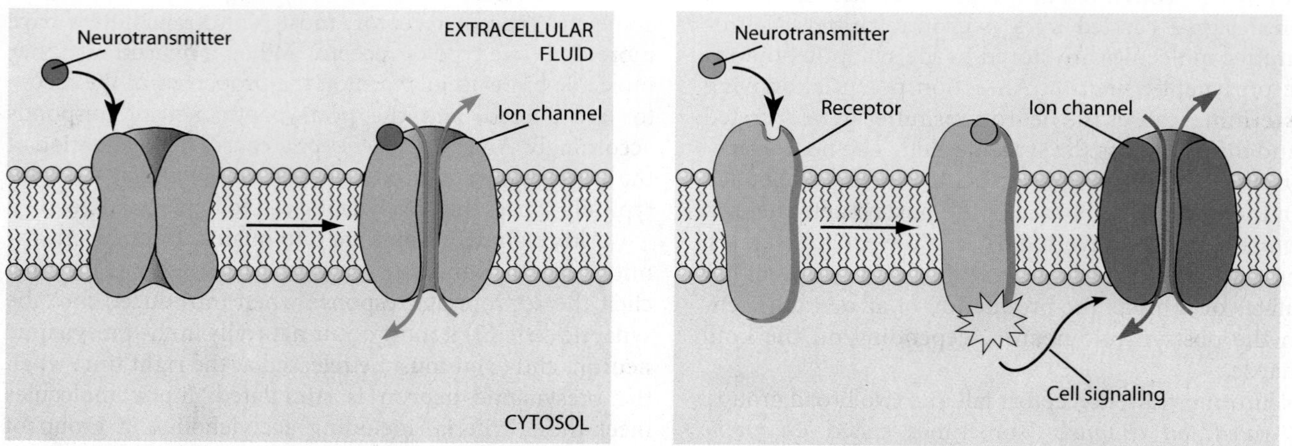

FIGURE 13-17 Different Kinds of Receptors That Act at Chemical Synapses. **(a)** Direct neurotransmitter action. Ionotropic receptors act directly as ion channels. When they bind a neurotransmitter, they undergo a conformational change, and ions can pass through them. **(b)** Indirect neurotransmitter action. When metabotropic receptors bind neurotransmitters, they set in motion a series of cell signaling events that indirectly lead to the opening of an ion channel. Because they act indirectly, metabotropic receptors act more slowly than ionotropic receptors.

Table 13-2 Different Kinds of Neurotransmitters

Neurotransmitter	Structure	Functional Class	Secretion Sites
Catecholamines			
Acetylcholine	$H_3C-C(=O)-O-CH_2-CH_2-N^+-[CH_3]_3$	Excitatory to vertebrate skeletal muscles; excitatory or inhibitory at other sites	CNS; PNS; vertebrate neuromuscular junction
Biogenic Amines			
Norepinephrine	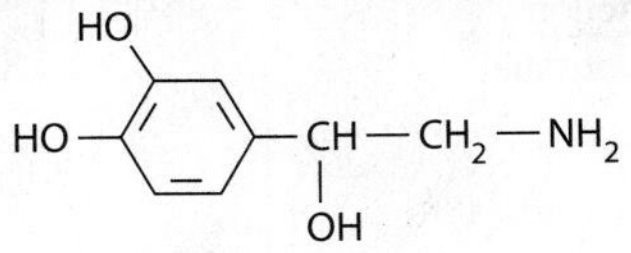 $(HO)_2C_6H_3-CH(OH)-CH_2-NH_2$	Excitatory or inhibitory	CNS; PNS
Dopamine	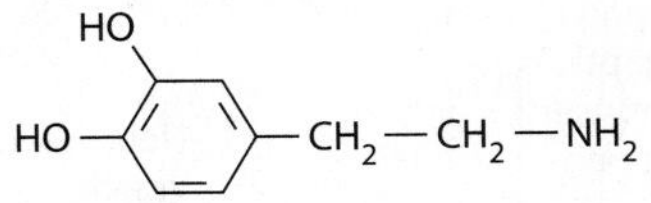 $(HO)_2C_6H_3-CH_2-CH_2-NH_2$	Generally excitatory; may be inhibitory at some sites	CNS; PNS
Serotonin	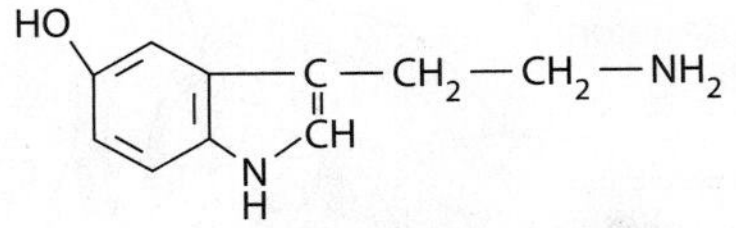 HO-indole (N-H, CH) $-C-CH_2-CH_2-NH_2$	Generally inhibitory	CNS
Amino Acids			
GABA γ-aminobutyric acid	$H_2N-CH_2-CH_2-CH_2-COOH$	Inhibitory	CNS; invertebrate neuromuscular junction
Glutamate	$H_2N-CH(COOH)-CH_2-CH_2-COOH$	Excitatory	CNS; invertebrate neuromuscular junction
Glycine	H_2N-CH_2-COOH	Inhibitory	CNS
Neuropeptides			
Substance P	Arg—Pro—Lys—Pro—Gln—Gln—Phe—Phe—Gly—Leu—Met	Excitatory	CNS; PNS
Met-enkephalin (an endorphin)	Tyr—Gly—Gly—Phe—Met	Generally inhibitory	CNS
Endocannabinoids			
Anandamide	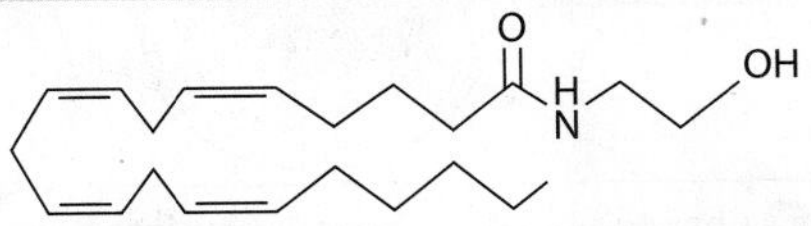 (arachidonoyl chain)–C(=O)–NH–CH₂CH₂–OH	Inhibitory	CNS
Gases			
Nitric oxide	N=O	Excitatory or inhibitory	PNS

Acetylcholine. In vertebrates, **acetylcholine** is the most common neurotransmitter for synapses between neurons outside the central nervous system (CNS), as well as for neuromuscular junctions (see Chapter 16). Acetylcholine is an excitatory neurotransmitter. Bernard Katz and his collaborators were the first to make the important observation that acetylcholine increases the permeability of the postsynaptic membrane to sodium within 0.1 msec of binding to its receptor. Synapses that use acetylcholine as their neurotransmitter are called **cholinergic synapses.**

Catecholamines. The **catecholamines** include *dopamine* and the hormones *norepinephrine* and *epinephrine,* all derivatives of the amino acid tyrosine. Because the catecholamines are also synthesized in the adrenal gland, synapses that use them as neurotransmitters are termed **adrenergic synapses.** Adrenergic synapses are found at the junctions between nerves and smooth muscles in internal organs such as the intestines, as well as at nerve-nerve junctions in the brain. The mode of action of adrenergic hormones will be considered in Chapter 14.

Amino Acids and Derivatives. Other neurotransmitters that consist of amino acids and derivatives include *histamine, serotonin,* and *γ-aminobutyric acid (GABA),* as well as *glycine* and *glutamate.* Serotonin functions in the CNS. It is considered an excitatory neurotransmitter because it indirectly causes potassium channels to close, which has an effect similar to opening sodium channels in that the postsynaptic cell is depolarized. However, its effect is exerted much more slowly than that of sodium channels. GABA and glycine are inhibitory neurotransmitters, whereas glutamate has an excitatory effect.

Neuropeptides. Short chains of amino acids called **neuropeptides** are formed by proteolytic cleavage of precursor proteins. Hundreds of different neuropeptides have been identified. Some neuropeptides exhibit characteristics similar to neurotransmitters in that they excite, inhibit, or modify the activity of other neurons in the brain. However, they differ from typical neurotransmitters in that they act on groups of neurons and have long-lasting effects.

Examples of neuropeptides include the *enkephalins,* which are naturally produced in the mammalian brain and inhibit the activity of neurons in regions of the brain involved in the perception of pain. The modification of neural activity by these neuropeptides appears to be responsible for the insensitivity to pain experienced by individuals under conditions of great stress or shock. The analgesic (i.e., pain-killing) effectiveness of drugs such as morphine, codeine, Demerol, and heroin derives from their ability to bind to the same sites within the brain that are normally targeted by enkephalins.

Other substances released at synapses include the lipid derivatives known as the *endocannabinoids,* which inhibit the activity of presynaptic neurons. The main endocannabinoid receptor found in the brain is also stimulated by tetrahydrocannabinol, or THC, a substance found in plants of the genus *Cannabis.* Marijuana is derived from the leaves of the species *Cannabis sativa* and owes its effects to THC.

Elevated Calcium Levels Stimulate Secretion of Neurotransmitters from Presynaptic Neurons

Secretion of neurotransmitters by the presynaptic cell is directly controlled by the concentration of calcium ions in the synaptic bouton (**Figure 13-18**). Each time an action potential arrives, the depolarization causes the calcium concentration in the synaptic bouton to increase temporarily due to the opening of **voltage-gated calcium channels** in the synaptic boutons. Normally, the cell is relatively impermeable to calcium ions, so that the cytosolic calcium concentration remains low (about 0.1 m*M*). However, there is a very large concentration gradient of calcium across the membrane because the calcium concentration outside the cell is about 10,000 times higher than that of the cytosol. As a result, calcium ions will rush into the cell when the calcium channels open.

Before they are released, neurotransmitter molecules are stored in small, membrane-bounded **neurosecretory vesicles** in the synaptic boutons (see Figure 13-18). The release of calcium within the synaptic bouton has two main effects on neurosecretory vesicles. First, vesicles held in storage are mobilized for rapid release. Second, vesicles that are ready for release rapidly dock and fuse with the plasma membrane in the synaptic bouton region. During this process, the membrane of a vesicle moves into close contact with the plasma membrane of the axon terminal and then fuses with it to release the contents of the vesicle. We will now examine this process in more detail.

1 An action potential arrives at the synaptic bouton, resulting in a transient depolarization.
2 Depolarization opens voltage-gated calcium channels, allowing calcium ions to rush into the terminal.
Voltage-gated calcium channel
Ca^{2+}
Synaptic bouton
Postsynaptic cell
Action potential
Axon
3 Increasing calcium in the synaptic bouton induces the secretion of some neurosecretory vesicles.
Neurosecretory vesicle
Ca^{2+}
5 Neurotransmitter diffuses across the synaptic cleft to receptors on the postsynaptic cell.
Reserve vesicles
Neurotransmitter receptors
4 Prolonged stimulation mobilizes additional, reserve vesicles.
Neurotransmitter
6 Binding of neurotransmitter to the receptor alters its properties.
8 If sufficient depolarization occurs, an action potential will result in the postsynaptic cell.
Ions
Action potential
7 Channels open, letting ions flow into the postsynaptic cell. Depending on the ion, channel opening leads to either depolarization or hyperpolarization.

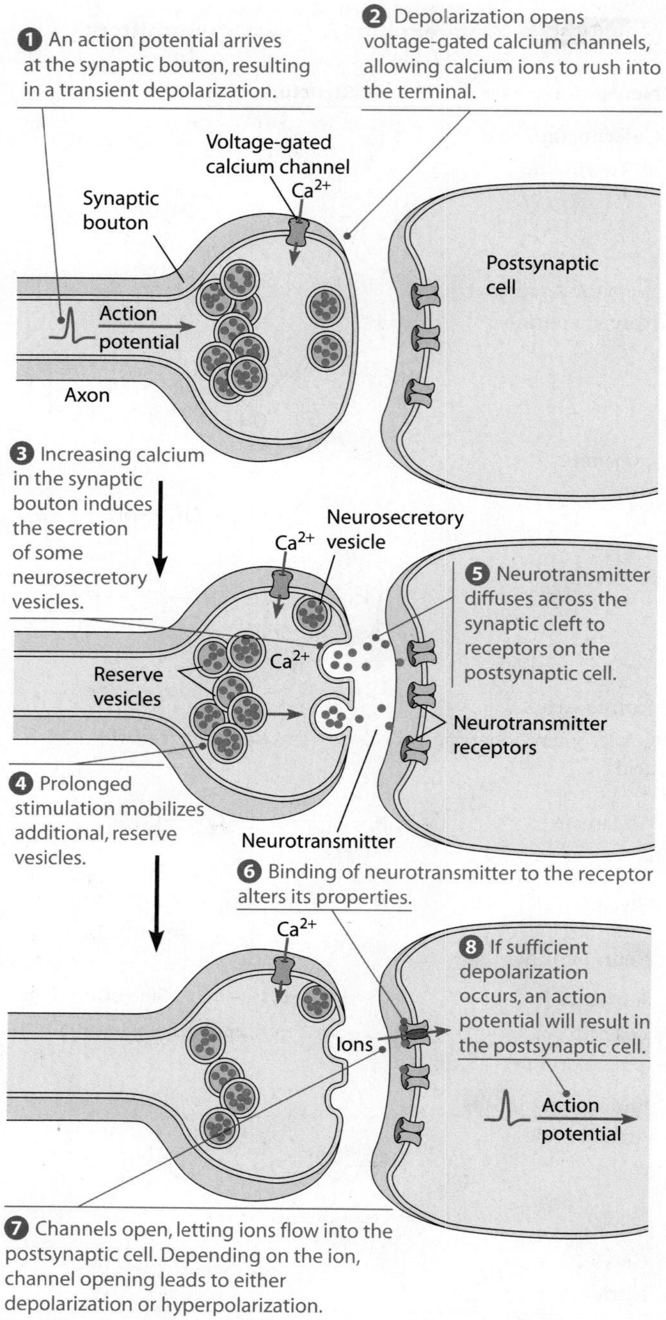

FIGURE 13-18 The Transmission of a Signal Across a Synapse. When an action potential arrives at the presynaptic bouton, a transient depolarization occurs, which leads to opening of voltage-gated calcium channels. Calcium elevation results in secretion of a neurotransmitter, which moves across the synapse and binds to receptors on the postsynaptic cell. The resulting depolarization can trigger an action potential.

Secretion of Neurotransmitters Involves the Docking and Fusion of Vesicles with the Plasma Membrane

For the neurotransmitter to act on a postsynaptic cell, it must be secreted by the process of *exocytosis* (see Chapter 12 for details of exocytosis), in which the neurosecretory vesicle fuses with the plasma membrane, discharging the vesicle's contents into the synaptic cleft.

When an action potential arrives at an axon terminal and triggers the opening of voltage-gated calcium channels, calcium enters the synaptic bouton. Neurosecretory vesicles near the plasma membrane are now capable of fusing with the plasma membrane of the presynaptic neuron. Because it can occur very rapidly, vesicle fusion is thought to involve vesicles that are already "docked" at the plasma membrane. Docking and fusion of neurosecretory vesicles with the plasma membrane of the active zone are mediated by t- and v-SNARE proteins, just as we saw with other exocytosis events in Chapter 12 (see **Figure 13-19**; specific mechanisms involved in the docking steps of exocytosis are discussed in more detail in Chapter 12). The calcium "sensor" involved in coordinated docking of vesicles appears to be the protein *synaptotagmin*. Synaptotagmin can bind calcium; when it does, it undergoes a conformational change that allows the t- and v-SNARE complexes to interact efficiently.

Docking takes place at a specialized site, called the **active zone,** within the membrane of the presynaptic neuron (see Figure 13-19). The active zone is a highly organized structure. Synaptic vesicles and the calcium channels that elicit their release are found in close proximity to one another, which helps to explain the extremely rapid fusion of docked vesicles with the presynaptic neuron's plasma membrane when that neuron is stimulated. Two familiar and potentially deadly illnesses result from interference with vesicle docking and fusion events. Both tetanus and botulism result from interference by **neurotoxins** with vesicle docking and release. *Tetanus toxin* prevents the release of neurotransmitter from inhibitory neurons in the spinal cord, resulting in uncontrolled muscle contraction (which is why tetanus has been colloquially referred to as "lockjaw"). *Botulinum toxin* prevents release of neurotransmitter from motor neurons, resulting in muscle weakness and paralysis.

Recall from Chapter 12 that exocytosis of vesicles involves addition of the vesicle's membrane to the plasma membrane. When neurons release many neurosecretory vesicles in rapid succession, this has the potential to lead to accumulation of excess membrane in the presynaptic nerve terminal. Neurons solve this problem by *compensatory endocytosis*. Compensatory endocytosis relies on the formation of clathrin-dependent vesicles (see Chapter 12), which allow the recycling of membranes and thereby maintains the size of the nerve terminal.

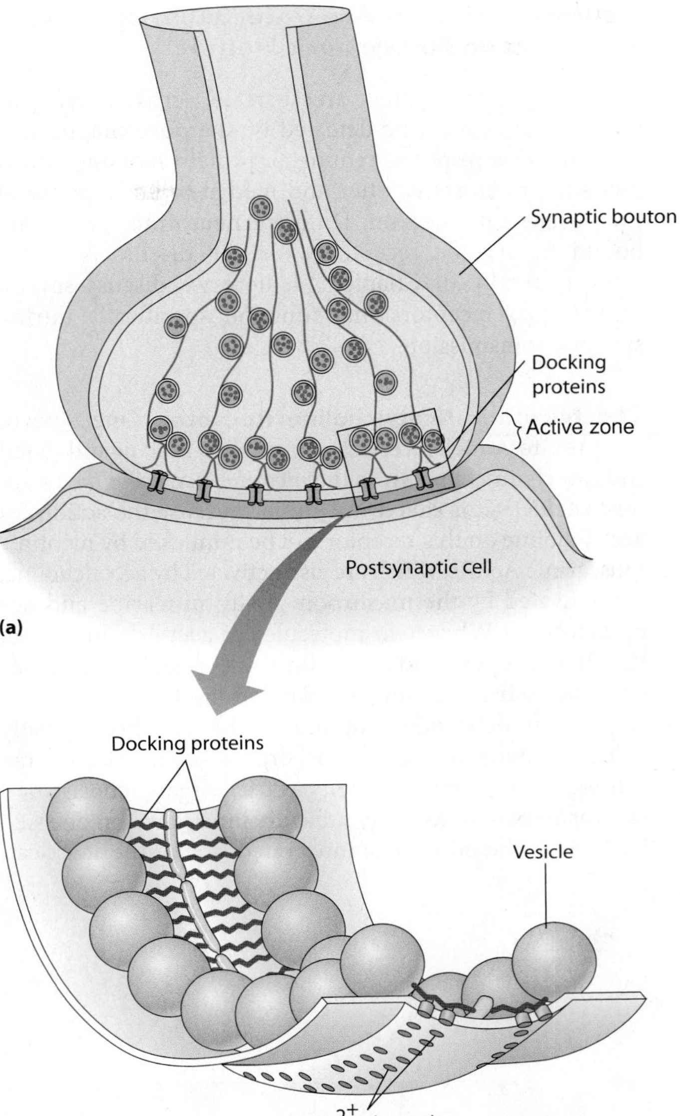

FIGURE 13-19 Docking of Synaptic Vesicles with the Plasma Membrane of the Presynaptic Neuron. **(a)** In response to local elevation of calcium in the presynaptic neuron, some synaptic vesicles become tightly associated (or docked) with the plasma membrane. When nearby calcium channels open, such docked vesicles fuse with the plasma membrane, releasing their contents. **(b)** A drawing based on an actual TEM reconstruction of the active zone of a motor neuron from a frog. Docked vesicles are arranged in rows and connected by a complex of proteins associated with docking. Calcium channels lie immediately beneath the vesicles.

In cases in which neurons need to fire very rapidly, they may use a more transient method for release of neurotransmitter, called *kiss-and-run exocytosis*. In this case, a vesicle may temporarily fuse with the plasma membrane via a tiny opening, causing release of some neurotransmitter from the vesicle. The vesicle then rapidly reseals without the added step of complete fusion with the plasma membrane.

Neurotransmitters Are Detected by Specific Receptors on Postsynaptic Neurons

When neurotransmitters are secreted across a synapse, their presence must be detected by the postsynaptic cell. Typically this response requires a protein *receptor,* which binds the neurotransmitter and mediates the response of the postsynaptic neuron. Different neurotransmitters are bound by specific receptors. We will discuss receptors more generally in Chapter 14; here we discuss several well-studied receptors that function specifically during synaptic transmission.

The Nicotinic Acetylcholine Receptor. One type of receptor to which acetylcholine binds is a ligand-gated sodium channel known as the *nicotinic acetylcholine receptor (nAchR).* (It is called "nicotinic" because the actions of acetylcholine on this receptor can be mimicked by nicotine; muscarinic AchRs, which are also activated by acetylcholine, are activated by the mushroom toxin muscarine and not by nicotine.) When two molecules of acetylcholine bind, the channel opens and lets sodium ions rush into the postsynaptic neuron, causing depolarization.

Our understanding of the nAchR has been greatly aided by study of the electric organs of the electric ray (*Torpedo californica*). The electric organ consists of *electroplaxes*—stacks of cells that are innervated on one side but not on the other. The innervated side of the stack can undergo a potential change from about −90 mV to about +60 mV upon excitation, whereas the noninnervated side stays at −90 mV. Therefore, at the peak of an action potential, a potential difference of about 150 mV can be built up across a single electroplax. Because the electric organ contains thousands of electroplaxes arranged in series, their voltages are additive, allowing the ray to deliver a jolt of several hundred volts.

When examined under the electron microscope, electroplax membranes are found to be rich in rosette-like particles about 8 nm in diameter (**Figure 13-20a**); these particles are the nicotinic acetylcholine receptors.

Biochemical purification of the acetylcholine receptor was greatly aided by the availability of several neurotoxins from snake venom, including *α-bungarotoxin* and *cobratoxin.* These neurotoxins serve as a highly specific means of locating and quantifying nAchRs because they can be easily radiolabeled and bind tightly and specifically to the receptor protein. Other neurotoxins act on other synaptic components (**Box 13A**).

The purified nAchR has a molecular weight of about 300,000 and consists of four kinds of subunits—α, β, γ, and δ—each containing about 500 amino acids (Figure 13-20 b, c). These receptors play a key role in transmitting nerve impulses to muscle (for more information, see Box 13A). In some cases, human patients develop an autoimmune response to their own acetylcholine receptors (in other words, their immune system attacks their

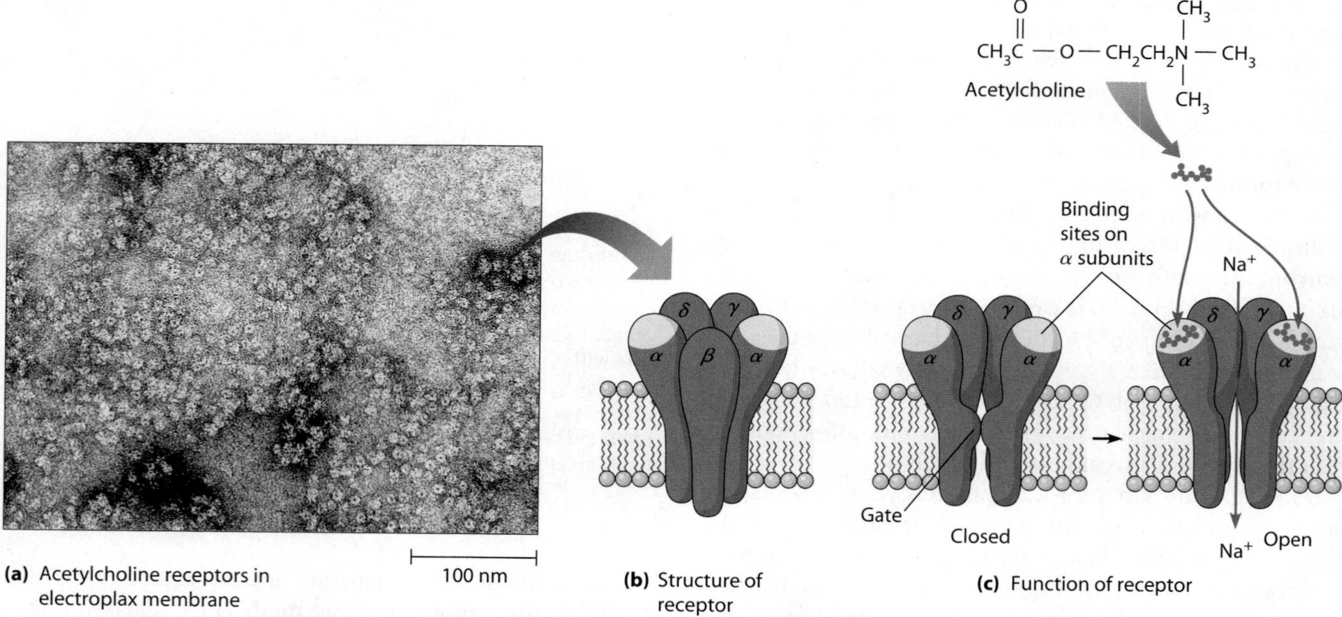

FIGURE 13-20 The Nicotinic Acetylcholine Receptor. The nicotinic acetylcholine receptor (nAchR) is an important excitatory receptor of the central nervous system. **(a)** This micrograph of an electroplax postsynaptic membrane shows rosette-like nAchR particles (TEM). **(b)** The nAchR contains five subunits, including two α subunits with binding sites for acetylcholine and one each of β, γ, and δ. The subunits aggregate in the lipid bilayer in such a way that the transmembrane portions form a channel. **(c)** The channel (shown here with the β subunit removed) is normally closed; however, when acetylcholine binds to the two sites on the α subunits, the subunits are altered so that the channel opens to allow sodium ions to cross.

BOX 13A HUMAN APPLICATIONS

Poisoned Arrows, Snake Bites, and Nerve Gases

Because the functioning of the human body depends so critically on the nervous system, anything that disrupts the transmission of nerve impulses is likely to be harmful. And because acetylcholine is such an important neurotransmitter, any substance that interferes with its function is almost certain to be lethal. Various toxins have specific effects on cholinergic synapses, and their effects can be explained once the physiology of synaptic transmission is understood.

After acetylcholine has been released into the synaptic cleft and the postsynaptic membrane has been depolarized, excess acetylcholine must be rapidly hydrolyzed by the enzyme acetylcholinesterase. If it is not, the membrane cannot be restored to its polarized state, and further transmission is not possible. Substances that inhibit the activity of acetylcholinesterase are therefore usually very toxic.

One family of acetylcholinesterase inhibitors are the *carbamoyl esters*. These compounds inhibit acetylcholinesterase by covalently blocking the active site of the enzyme, effectively preventing the breakdown of acetylcholine. An example of such an inhibitor is *physostigmine* (sometimes also called *eserine*), a naturally occurring alkaloid produced by the Calabar bean.

Many synthetic organic phosphates form even more stable covalent complexes with the active site of acetylcholinesterase and are thus even more potent inhibitors. These include the widely used insecticides *parathion* and *malathion,* as well as nerve gases such as *tabun* and *sarin*. The primary effect of these compounds is muscle paralysis, caused by an inability of the postsynaptic membrane to regain its polarized state.

Nerve transmission at cholinergic synapses can also be blocked by substances that compete with acetylcholine for binding to its receptor on the postsynaptic membrane. A notorious example of such a poison is *curare,* a plant extract once used by native South Americans to poison arrows. Among the active factors in curare is *d-tubocurarine*. Snake venoms act in the same way. Both α-bungarotoxin (from kraits, snakes of the genus *Bungarus*) and *cobratoxin* (from cobra snakes) are small, basic proteins that bind covalently to the acetylcholine receptor, thereby blocking depolarization of the postsynaptic membrane.

Substances that function in this way are called *antagonists* of cholinergic systems. Other compounds, called *agonists,* have just the opposite effect. Agonists also bind to the acetylcholine receptor, but in doing so they mimic acetylcholine and cause depolarization of the postsynaptic membrane. Unlike acetylcholine, however, agonists cannot be rapidly inactivated, so the membrane does not regain its polarized state.

Though of disparate origins and uses, poisoned arrows, snake venom, nerve gases, and surgical muscle relaxants all have some features in common. Each interferes in some way with the normal functioning of acetylcholine, and each is therefore a neurotoxin because it disrupts the transmission of nerve impulses, potentially with lethal consequences. Each has also turned out to be useful as an investigative tool, illustrating again the strange but powerful arsenal of exotic tools that biologists and biochemists are able to draw upon in their continued probing into the intricacies of cellular function.

own receptors). When this happens, the patients can develop a condition known as *myasthenia gravis,* in which degenerative muscle weakness occurs.

The GABA Receptor. The γ-aminobutyric acid (GABA) receptor is also a ligand-gated channel, but when open, it conducts chloride ions rather than sodium ions. Since chloride ions are normally found at a higher concentration in the medium surrounding a neuron (see Table 13-1), opening GABA receptor channels produces an influx of chloride into the postsynaptic neuron, causing it to hyperpolarize. Hyperpolarization of the postsynaptic nerve terminal decreases the chance that an action potential will be initiated in the postsynaptic neuron. *Benzodiazepine drugs,* such as Valium and Librium, can enhance the effects of GABA on its receptor. Presumably, this produces the tranquilizing effects of these drugs.

Neurotransmitters Must Be Inactivated Shortly After Their Release

For neurons to transmit signals effectively, it is just as important to turn the stimulus off as it is to turn it on. Whether excitatory or inhibitory, once the neurotransmitter has been secreted, it must be rapidly removed from the synaptic cleft. If it were not, stimulation or inhibition of a postsynaptic neuron would be abnormally prolonged, even without further signals from presynaptic neurons. In the case of acetylcholine, for example, excess acetylcholine must be hydrolyzed to restore the postsynaptic membrane to its polarized state; otherwise major problems arise (see Box 13A).

Neurotransmitters are removed from the synaptic cleft by two specific mechanisms: degradation into inactive molecules or reuptake. The first mechanism is exemplified by acetylcholine. The enzyme *acetylcholinesterase* hydrolyzes acetylcholine into acetic acid (or acetate ion) and choline, neither of which stimulates the acetylcholine receptor. Purine nucleotide neurotransmitters are also degraded by specific enzymes. The second, very common method of terminating synaptic transmission is **neurotransmitter reuptake.** Reuptake involves pumping neurotransmitters back into the presynaptic axon terminals or nearby support cells (you learned about such pumps in Chapter 8). The rate of neurotransmitter reuptake can be rapid; for some neurons, the synapse may be cleared of stray neurotransmitter within as little as a millisecond. Some antidepressant drugs act by blocking the reuptake of specific neurotransmitters. For example, Prozac blocks the reuptake of serotonin, leading to a local increase in the level of serotonin available to postsynaptic neurons.

Integration and Processing of Nerve Signals

Sending a signal across a synapse does not automatically generate an action potential in the postsynaptic cell. There is not necessarily a one-to-one relationship between an action potential arriving at the presynaptic neuron and one initiated in the postsynaptic neuron. A single action potential can cause the secretion of enough neurotransmitter to produce a detectable depolarization in the postsynaptic neuron but usually is not enough to cause the firing of an action potential in the postsynaptic cell. These incremental changes in potential due to the binding of neurotransmitter are referred to as *postsynaptic potentials (PSPs).* If a neurotransmitter is excitatory, it will cause a small amount of depolarization known as an **excitatory postsynaptic potential (EPSP).** Likewise, if the neurotransmitter is inhibitory, it will hyperpolarize the postsynaptic neuron by a small amount; this is called an **inhibitory postsynaptic potential (IPSP).**

For a presynaptic neuron to stimulate the formation of an action potential in the postsynaptic neuron, the EPSP must rise to a point at which the postsynaptic membrane reaches its threshold potential. EPSPs can do this in two different ways, which we will examine next.

Neurons Can Integrate Signals from Other Neurons Through Both Temporal and Spatial Summation

An individual action potential will produce only a temporary EPSP. However, if two action potentials fire in rapid succession at the presynaptic neuron, the postsynaptic neuron will not have time to recover from the first EPSP before experiencing a second EPSP. The result is that the postsynaptic neuron will be more depolarized. A rapid sequence of action potentials effectively sums EPSPs over time and brings the postsynaptic neuron to its threshold. This process is called *temporal summation.*

The amount of neurotransmitter released at a single synapse following an action potential is usually not sufficient to produce an action potential in the postsynaptic cell. When many action potentials cause the release of neurotransmitter simultaneously, their effects combine; sometimes this results in a large depolarization of the postsynaptic cell. This process is known as *spatial summation* because the postsynaptic neuron integrates the numerous small depolarizations that occur over its surface into one large depolarization.

Neurons Can Integrate Both Excitatory and Inhibitory Signals from Other Neurons

Besides receiving stimuli of varying strength, postsynaptic neurons can receive inputs from both excitatory and inhibitory neurons (**Figure 13-21**). Neurons can receive

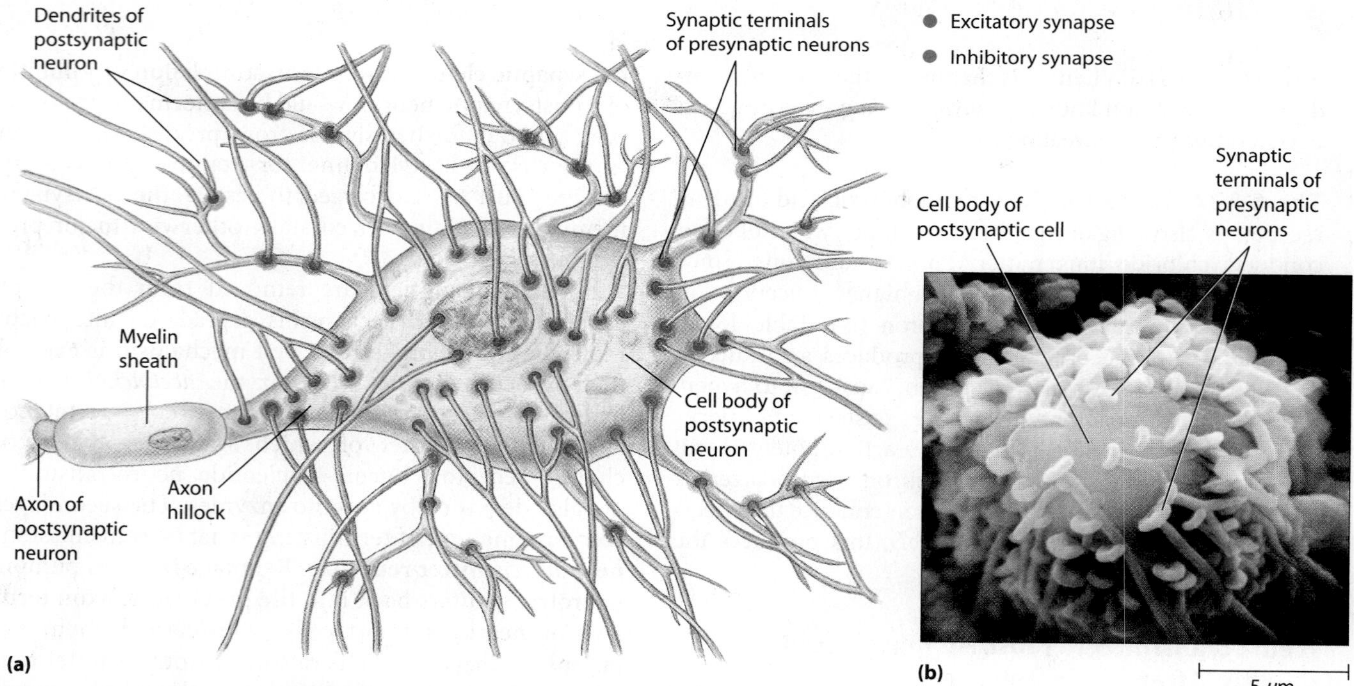

FIGURE 13-21 Integration of Synaptic Inputs. (a) Neurons, particularly those in the CNS, receive inputs from thousands of synapses, some of them excitatory (green) and others inhibitory (red). An action potential may be generated in such a neuron at the axon hillock if the combined effects of membrane potentials induced by these synapses results in depolarization above the threshold potential. Both temporal and spatial summation contribute to the likelihood that an action potential will be initiated. **(b)** This SEM shows that the synaptic terminals of many presynaptic neurons can make contact with a single postsynaptic neuron.

literally thousands of synaptic inputs from other neurons. When these different neurons fire at the same time, they exert combined effects on the membrane potential of the postsynaptic neuron. Thus, by physically summing EPSPs and IPSPs, an individual neuron effectively integrates incoming signals (excitatory or inhibitory).

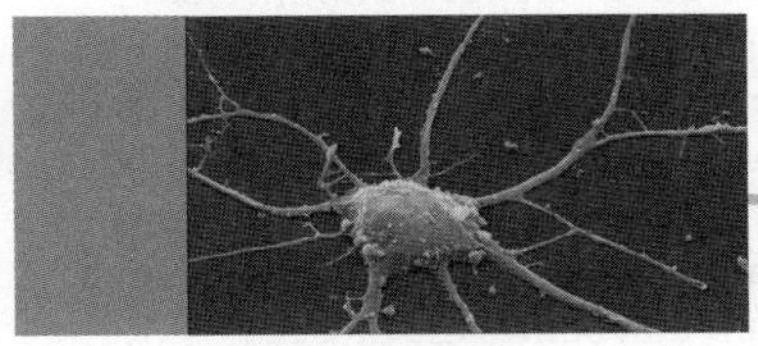

SUMMARY OF KEY POINTS

Neurons

- Cells in the nervous system are highly specialized to transmit electric impulses, using slender processes (dendrites and axons) that either receive transmitted impulses (dendrites) or conduct them to the next cell (axons).
- The membrane of an axon may or may not be encased in a myelin sheath, which provides electrical insulation that allows faster propagation of nerve impulses.

Membrane Potential

- Cells develop a membrane potential due to the separation of positive and negative charges across the plasma membrane. This potential develops as each ion to which the membrane is permeable moves down its electrochemical gradient.
- The Goldman equation is used to calculate the resting membrane potential of a cell, which depends on the permeability of the membrane to a particular ion. The resting potential for the plasma membrane of most animal cells is usually in the range of −60 to −75 mV, quite near the equilibrium potential for potassium ion.

Electrical Excitability and Action Potentials

- An action potential is a transient depolarization and repolarization of the neuronal membrane, due to the sequential opening and closing of voltage-gated sodium and potassium channels. In voltage-gated ion channels, the probability of opening, and consequently their conductance, depends on membrane potential.
- The properties of ion channels have been studied using molecular techniques combined with patch clamping to measure the conductance of single channels.
- An action potential is initiated when the membrane is depolarized to its threshold, at which point the opening of voltage-gated sodium channels allows sodium ions to enter the cells, driving the membrane potential to approximately +40 mV. Eventually, voltage-gated sodium channels inactivate.
- Repolarization of the membrane involves the opening of slower, voltage-gated potassium channels, which leads to repolarization of the membrane and includes a short period of hyperpolarization. This sequence of channel opening and closing generally takes a few milliseconds.
- Depolarization of the membrane due to an action potential spreads by passive conductance to adjacent regions of the membrane, which in turn generates a new action potential. In this way, an action potential is propagated along the membrane.

Synaptic Transmission

- Action potentials eventually reach the synapse between a nerve cell and another cell that it communicates with. Such synapses may be either electrical or chemical.
- In an electrical synapse, depolarization is transmitted from the presynaptic cell to the postsynaptic cell by gap junctions. In a chemical synapse, the electrical impulse increases the permeability of the membrane to calcium, stimulating release of neurotransmitter into the synaptic cleft.
- There are many different types of neurotransmitters. They bind specific types of receptors, causing either hyperpolarization or depolarization of the postsynaptic membrane.
- Transmission of nerve impulses requires that the cell body of the postsynaptic neuron integrate the excitatory and inhibitory activity of thousands of synaptic inputs.

PROBLEM SET

More challenging problems are marked with a •.

13-1 The Truth About Nerve Cells. For each of the following statements, indicate whether it is true of all nerve cells (A), of some nerve cells (S), or of no nerve cells (N).

(a) The resting membrane potential of the axonal membrane is positive.

(b) The resting potential of the membrane is much closer to the equilibrium potential for potassium ions than to that for sodium ions because the permeability of the axonal membrane is much greater for potassium than for sodium.

(c) The electrical potential across the membrane of the axon can be easily measured using electrodes.

(d) Calcium elevation stimulates release of neurosecretory vesicles containing serotonin.

(e) Nodes of Ranvier are found at regular intervals along the axon.

(f) Excitation of the membrane results in a transient increase in its permeability to sodium ions.

(g) Axonal endings are in direct contact with the cells they innervate.

• 13-2 The Resting Membrane Potential. The Goldman equation is used to calculate V_m, the resting potential of a biological membrane. As presented in the chapter, this equation contains terms for sodium, potassium, and chloride ions only.

(a) Why do only these three ions appear in the Goldman equation as it applies to nerve impulse transmission?

(b) State whether true or false. The plasma membrane is equally permeable to all monovalent ions.

(c) Explain how the differential permeabilities of membranes to potassium (P_K) and sodium (P_{na}) are essential for the functions of neuronal membranes.

(d) Would you expect a plot of V_m versus the relative permeability of the membrane to sodium to be linear? Why or why not?

13-3 Patch Clamping. Patch-clamp instruments enable researchers to measure the opening and closing of a single channel in a membrane. A typical acetylcholine receptor channel passes about 5 pA (picoamperes) of ionic current (1 picoampere $= 10^{-12}$ ampere) over a period of about 5 msec at −60 mV.

(a) Given that an electrical current of 1 A is about 6.2×10^{18} electrical charges per second, how many ions (potassium or sodium) pass through the channel during the time it is open?

(b) You have constructed a hybrid ion channel by using monomers from two different species. Describe a patch-clamp experiment that you would design to test the properties of this hybrid channel.

13-4 The Equilibrium Potential. Answer each of the following questions with respect to E_{Cl}, the equilibrium potential for chloride ions. The chloride ion concentration inside the squid giant axon can vary from 50 to 150 m*M*.

(a) Before doing any calculations, predict whether E_{Cl} will be positive or negative. Explain.

(b) Now calculate E_{Cl}, assuming an internal chloride concentration of 50 m*M*.

(c) How much difference would it make in the value of E_{Cl} if the internal chloride concentration were 150 m*M* instead?

13-5 Heart Throbs. An understanding of muscle cell stimulation involves some of the same principles as nerve cell stimulation, except that calcium ions play an important role in the former. The following ion concentrations are typical of those in human heart muscle and in the serum that bathes the muscles:

$[K^+]$: 150 m*M* in cell, 4.6 m*M* in serum
$[Na^+]$: 10 m*M* in cell, 145 m*M* in serum
$[Ca^{2+}]$: 0.001 m*M* in cell, 6 m*M* in serum

Figure 13-22 depicts the change in membrane potential with time upon stimulation of a cardiac muscle cell.

(a) Calculate the equilibrium potential for each of the three ions, given the concentrations listed.

(b) Why is the resting membrane potential significantly more negative than that of the squid axon (−75 mV versus −60 mV)?

(c) The more positive membrane potential in region Ⓐ of the graph could in theory be due to the movement across the membrane of one or both of two cations. Which cations are they, and in what direction would you expect each of them to move across the membrane?

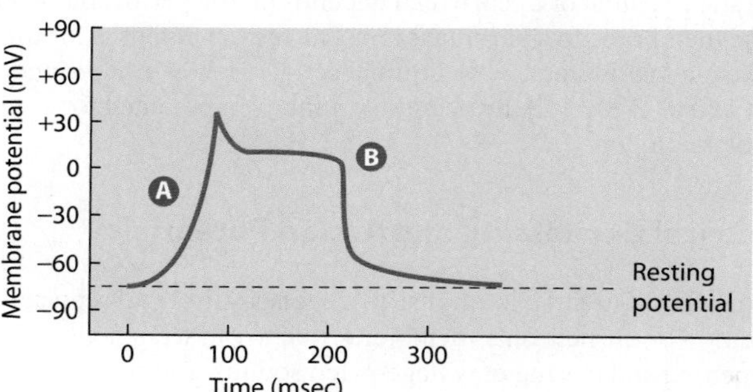

FIGURE 13-22 The Action Potential of a Muscle Cell of the Human Heart. See Problem 13-5.

(d) How might you distinguish between the possibilities suggested in part c?

(e) The rapid decrease in membrane potential that is occurring in region Ⓑ is caused by the outward movement of potassium ions. What are the driving forces that cause potassium to leave the cell at this point? Why aren't the same forces operative in region Ⓐ of the curve?

13-6 The All-or-None Response of Membrane Excitation. A nerve cell membrane exhibits an all-or-none response to excitation; that is, the magnitude of the response is independent of the magnitude of the stimulus, once a threshold value is exceeded.

(a) Explain in your own words why this is so.

(b) If every neuron exhibits an all-or-none response, how do you suppose the nervous system of an animal can distinguish different intensities of stimulation? How do you think your own nervous system can tell the difference between a warm iron and a hot iron or between a chamber orchestra and a rock band?

13-7 One-Way Propagation. Assume that the time taken for the restoration of a depolarized sodium channel to return to the normal state is less than that taken for the discharge of potassium ions. What would happen to the transmission of nerve impulses?

13-8 Multiple Sclerosis and Action Potential Propagation. Multiple sclerosis (MS) is an autoimmune disease that attacks myelinated nerves and degrades the myelin sheath around them. How do you think the propagation of action potentials would be affected in a nerve cell that has been damaged in a patient with MS?

13-9 Going Bananas. Athletes who experience muscle cramping are often told to eat bananas, which are rich in potassium. Explain why increasing the extracellular potassium concentration would make it more difficult to stimulate an action potential in neurons that innervate muscle cells, thereby decreasing their contraction.

13-10 Down and Out. Endocannabinoids are produced by postsynaptic neurons but act on receptors (CB1 receptors) in presynaptic neurons. One type of neuron that is a target of endocannabinoids is *glutamatergic* neurons (that is, neurons that release glutamate as a neurotransmitter).

(a) CB1 agonists such as THC reduce calcium influx by blocking the activity of voltage-dependent calcium channels. What effect would this have on the release of glutamate?

(b) Glutamate causes the membrane of postsynaptic neurons to become hyperpolarized. How would THC be expected to affect the rate of nerve impulse transmission in postsynaptic neurons that make synapses with the glutamatergic neuron?

13-11 Trouble at the Synapse. Drugs that affect the reuptake of neurotransmitters are in widespread use for the treatment of attention-deficit/hyperactivity disorder and clinical depression. In molecular terms, explain what effect(s) such reuptake inhibitors have on postsynaptic neurons, assuming that the neurotransmitter in question is excitatory for that postsynaptic neuron.

SUGGESTED READING

References of historical importance are marked with a •.

General References

Albright, T. D. et al. Neural science: A century of progress and the mysteries that remain. *Cell* 100 (2000): S1.

Byrne, J. H., and J. L. Roberts, eds. *From Molecules to Networks: An Introduction to Cellular and Molecular Neuroscience,* 2nd ed. Burlington, MA: Academic Press, 2009.

Nicholls, J. G. et al. *From Neuron to Brain,* 4th ed. Sunderland, MA: Sinauer, 2001.

Südhof, T. C., and R. C. Malenka. Understanding synapses: Past, present, and future. *Neuron* 60 (2008): 469.

Ion Channels, Patch Clamping, and Membrane Excitation

Choe, S. Potassium channel structures. *Nature Rev. Neurosci.* 3 (2002): 115.

Gouaux, E., and R. Mackinnon. Principles of selective ion transport in channels and pumps. *Science* 310 (2005): 1461.

• Hodgkin, A. L., and A. F. Huxley. A quantitative description of membrane current and its application to conduction and excitation in nerve. *J. Physiol.* 117 (1952): 500.

Kaplan, J. H. Biochemistry of Na, K-ATPase. *Annu. Rev. Biochem.* 71 (2002): 511.

Kurata, H. T., and D. Fedida. A structural interpretation of voltage-gated potassium channel inactivation. *Prog. Biophys. Mol. Biol.* 92 (2006): 185.

• Neher, E., and B. Sakmann. The patch-clamp technique. *Sci. Amer.* 266 (1992): 28.

Vincent A., B. Lang, and K. A. Kleopa. Autoimmune channelopathies and related neurological disorders. *Neuron* 52 (2006): 123.

Myelin

Pedraza, L., J. K. Huang, and D. R. Colman. Organizing principles of the axoglial apparatus. *Neuron* 30 (2001): 335.

Steinman, L., R. Martin, C. Bernard, P. Conlon, and J. Oksenberg. Multiple sclerosis: Deeper understanding of its pathogenesis reveals new targets for therapy. *Annu. Rev. Neurosci.* 25 (2002): 491.

Neurotransmitters

Nestler, E. J., and R. C. Malenka. The addicted brain. *Sci. Amer.* 290 (2004): 78.

Nicoll, R. A., and Alger, B. E. The brain's own marijuana. *Sci. Amer.* 291 (2004): 68.

Vincent, A., D. Beesonn, and B. Lang. Molecular targets for autoimmune and genetic disorders of neuromuscular transmission. *Eur. J. Biochem.* 267 (2000): 6717.

Synaptic Transmission

Dittman, J., and T. A. Ryan. Molecular circuitry of endocytosis at nerve terminals. *Annu. Rev. Cell Dev. Biol.* 25 (2009): 133-60.

• Harlow, M. L. et al. The architecture of the active zone material at the frog's neuromuscular junction. *Nature* 409 (2001): 479.

Takamori, S. et al. Molecular anatomy of a trafficking organelle. *Cell* 127 (2006): 831.

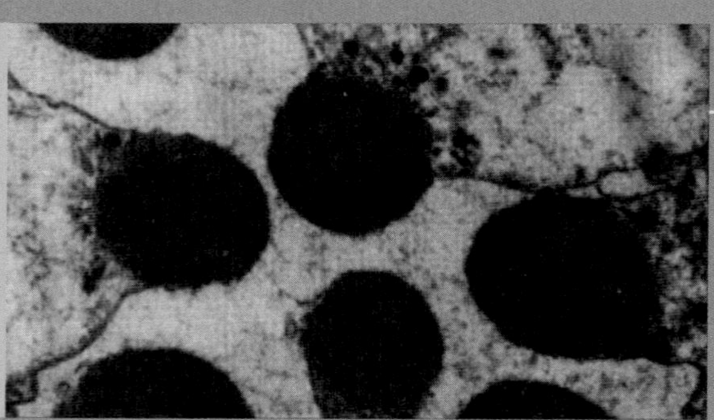

14 Signal Transduction Mechanisms: II. Messengers and Receptors

See MasteringBiology® for tutorials, activities, and quizzes.

In the previous chapter, we learned how nerve cells communicate with one another and with other types of cells. We saw how, in most cases, the arrival of an action potential at a synapse causes the release of neurotransmitters, which in turn bind to receptors on the adjacent postsynaptic cell membrane, thereby passing on the signal. Now we are ready to explore a second major means of intercellular communication. In this case, however, the signal is transmitted by regulatory chemical messengers, and the receptors are located on the surfaces of cells that may be quite distant from the secreting cells. Thus, animals have two different but complementary systems of communication and control, and receptors play a crucial role in both systems.

Chemical Signals and Cellular Receptors

All cells have some ability to sense and respond to specific chemical signals. Prokaryotes, for instance, have membrane-bound receptor molecules on the cell surface that enable them to respond to substances in their environment. The human body has receptors on the tongue and in the nose that detect chemicals in food and the air. Even the cells of the early animal embryo possess sophisticated machinery for detecting changes in their surroundings.

Cells also produce signals. One way they do this is by displaying molecules on their surfaces that are recognized by receptors on the surfaces of other cells. This kind of cell-to-cell communication requires that cells come into direct contact with each other. Alternatively, one cell can release chemical signals that are recognized by another cell, either nearby or at a distant location. In complex multicellular organisms, the problem of regulating and coordinating the various activities of cells or tissues is particularly important because the whole organism is organized into different tissues made up of specialized cells. Furthermore, the specific functions of these cells may be critical only on certain occasions, or one tissue may need to perform different functions in different circumstances. Multicellular organisms often control the activities of specialized cells through the release of *chemical messengers*, the topic of this chapter.

Different Types of Chemical Signals Can Be Received by Cells

A variety of compounds can function as chemical messengers (**Figure 14-1**). Signaling molecules are often classified based on the distance between their site of production and the target tissue(s) upon which they act. Some messengers, such as hormones, act as *endocrine signals* (from Greek words meaning "to secrete into"). They are produced at great distances from their target tissues and are carried by the circulatory system to various sites in the body. Other signals, such as growth factors, are released locally, where they diffuse to act at short range on nearby tissues. Such signals are referred to as *paracrine signals* (from the Greek *para*, which means "beside"). When signals are passed at such short range that they require physical contact between the sending and receiving cells, they are said to be *juxtacrine signals.* Still other local mediators act on the same cell that produces them; such signals are called *autocrine signals.*

Once a messenger reaches its target tissue, it binds to **receptors** on the surface of the target cells, initiating the signaling process. The logic and general flow of information involved in such signaling is shown in **Figure 14-2a**. A molecule coming from either a long or a short distance functions as a **ligand** by binding to a receptor. A ligand often binds to a receptor embedded within the plasma membrane of the cell receiving the signal. In other cases, such as steroid hormones, the ligand binds to a receptor inside the cell. In either case, the ligand is a "primary messenger." The binding of ligand to receptor often results in the production of additional molecules within the cell receiving the signal. Such

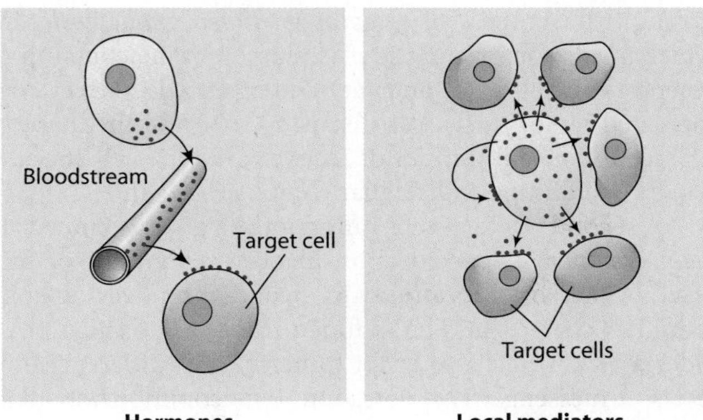

FIGURE 14-1 Cell-to-Cell Signaling by Hormones and Local Mediators. The main distinction between classes of signaling molecules is the distance the molecule travels before encountering its target cell or tissue. Endocrine hormones are carried by the bloodstream. Local mediators, such as growth factors, can act on nearby cells (paracrine signals) or on the cell that produces them (autocrine signals). Although many signaling molecules act on the cell surface, certain hydrophobic molecules enter their target cells.

second messengers are small molecules or ions that relay the signals from one location in the cell, such as the plasma membrane, to the interior of the cell, initiating a cascade of changes within the receiving cell. Often these events affect the expression of specific genes within the receiving cell. The ultimate result is a change in the identity or function of the cell. The ability of a cell to translate a receptor-ligand interaction to changes in its behavior or gene expression is known as **signal transduction.**

Messenger molecules can be chemically characterized as amino acids or their derivatives, peptides, proteins, fatty acids, lipids, nucleosides, or nucleotides. Many messengers are hydrophilic compounds whose function lies entirely in their ability to bind to one or more specific receptors on a target cell. Hydrophobic messengers, on the other hand, act on receptors in the nucleus or cytosol whose function is to regulate the transcription of particular genes. Among the hydrophobic messengers that bind to intracellular receptors are *steroid hormones,* which are derived from the compound cholesterol, and *retinoids,* derived from vitamin A.

Receptor Binding Involves Specific Interactions Between Ligands and Their Receptors

How do cells distinguish messengers from the multitude of other chemicals in the environment or from messengers intended for other cells? The answer lies in the highly specific way the messenger molecule binds to the receptor. A messenger forms noncovalent chemical bonds with the receptor protein. Individual noncovalent bonds are generally weak; therefore, several bonds must form to achieve strong binding. For a receptor to make numerous bonds with its ligand, the receptor must have a *binding site* (or *binding pocket*) that fits the messenger molecule closely, like a hand in a glove. Furthermore, within the ligand-binding site on the receptor, appropriate amino acid side chains must be positioned so that they can form chemical bonds with the messenger molecule. This combination of binding site shape and the strategic positioning of amino acid side chains

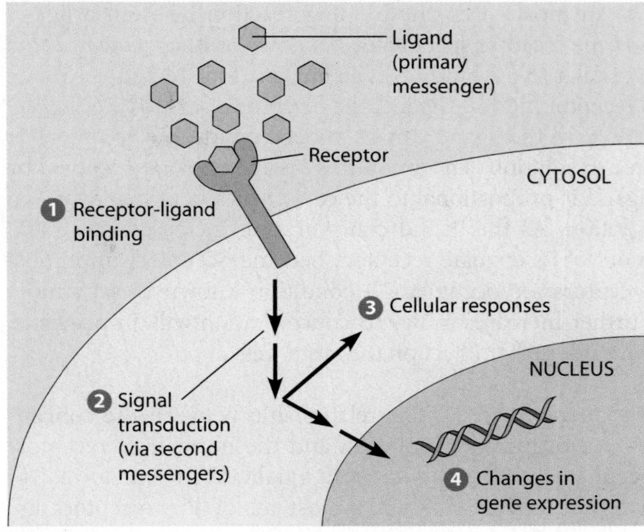

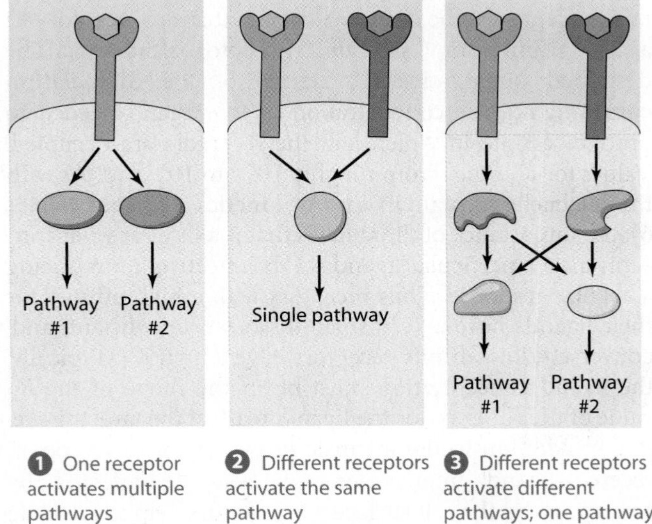

FIGURE 14-2 The Overall Flow of Information During Cell Signaling. **(a)** The general flow of information in cell signaling. Binding of a ligand by a receptor activates a series of events known as signal transduction, which relays the signal to the interior of the cell and results in specific cellular responses and/or changes in gene expression. **(b)** Different ways signals can be integrated. ❶ One signal can activate multiple signal transduction pathways. ❷ Multiple signals can result in activation of the same pathway. ❸ Different signals can activate different pathways, but one component of a pathway can regulate a component of another pathway.

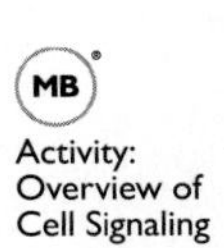

Activity: Overview of Cell Signaling

within the binding site is what enables the receptor to distinguish its specific ligand from thousands of other chemicals.

In most cases, the binding reaction between a ligand and the receptor specific for it, known as its *cognate receptor,* is similar to the binding of an enzyme to its substrate. When a receptor binds its ligand, the receptor is said to be *occupied.* Similarly, the ligand can be either bound (to a receptor) or free in solution. The amount of receptor that is occupied by ligand is proportional to the concentration of free ligand in solution. As the ligand concentration increases, more and more of its cognate receptors become occupied, until most receptors are occupied—a condition known as saturation. Further increases in ligand concentration will, in principle, have no further effect on the target cell.

Receptor Affinity. The relationship between the concentration of ligand in solution and the number of receptors occupied can be described qualitatively in terms of **receptor affinity.** When almost all of the receptors are occupied at low concentrations of free ligand, we say that the receptor has a high affinity for its ligand. Conversely, when it takes a relatively high concentration of ligand for most receptors to be occupied, we say that the receptor has a low affinity for its ligand. Receptor affinity can be described quantitatively in terms of the **dissociation constant, K_d,** the concentration of free ligand needed to produce a state in which half the receptors are occupied. Values for K_d range from roughly 10^{-7} to 10^{-10} M. As with the Michaelis constant in enzyme kinetics (K_m; see Chapter 6), the importance of the value is that it tells us at what concentration a particular ligand will be effective in producing a cellular response. Thus receptors with a high affinity for their ligands have a very small dissociation constant, and, conversely, low-affinity receptors have a high K_d. Typically, the ligand concentration must be in the range of the K_d value of the receptor for the ligand to affect the target tissue.

In addition to the intrinsic properties of the receptor, receptor-ligand interactions can also be affected by *coreceptors* on the cell surface. Coreceptors help to facilitate the interaction of the receptor with its ligand through their physical interaction with the receptor. One well-studied class of such molecules are *heparan sulfate proteoglycans,* including *glypicans* and *syndecans.* Coreceptors provide another layer of regulation of receptor-ligand interactions.

Receptor Down-Regulation. Although receptors have a characteristic affinity for their ligands, cells are geared to sense *changes* in ligand concentration rather than fixed ligand concentrations. When a ligand is present and receptors are occupied for prolonged periods of time, the cell adapts so that it no longer responds to the ligand. To further stimulate the cell, the ligand concentration must be increased. Such changes are known as *receptor down-regulation.* There are two main ways in which such adaptation occurs. First, cells can change the density of receptors on their surfaces in response to a signal. The removal of receptors from the cell surface takes place through the process of *receptor-mediated endocytosis,* in which small portions of the plasma membrane containing receptors invaginate and are internalized. (Receptor-mediated endocytosis was discussed in detail in Chapter 12.) The reduced number of receptors on the cell surface results in a diminished cellular response to ligand.

A second way that cells can adapt to signals is known as *desensitization.* Desensitization involves alterations of the receptor that lower its affinity for ligand or render it unable to initiate changes in cellular function. Desensitization provides a way for cells to adapt to permanent differences in levels of messenger concentration. One common way that desensitization occurs involves addition of phosphate groups to specific amino acids within the cytosolic portion of the receptor. As we will see, one well-known example of such phosphorylation involves the β-adrenergic receptor. Once a ligand is no longer present, a cell must "reset" itself to a resting level of responsiveness. In the case of the β-adrenergic receptor, the continuous activity of phosphatases in the cytosol removes the inhibitory phosphates, returning the receptor to a more responsive state.

Understanding the nature of receptor-ligand binding has provided great opportunities for researchers and pharmaceutical companies. Although receptors have binding sites that fit the messenger molecule quite closely, it is possible to make similar synthetic ligands that bind even more tightly or selectively. This is especially important when more than one type of receptor exists for the same ligand. Drugs that activate the receptor to which they bind are known as **agonists.** In contrast, whereas normal messengers cause a change in the receptor when they bind, both synthetic and natural compounds have been discovered that can bind to receptors without triggering such a change. These **antagonists** inhibit the receptor by preventing the naturally occurring messenger from binding and activating the receptor.

Drugs that selectively activate or inhibit particular kinds of receptors have become central to the treatment of many medical problems. For example, *isoproterenol* and *propranolol* activate or inhibit β-adrenergic receptors, which will be discussed later in the chapter. Isoproterenol is used to treat asthma or to stimulate the heart, whereas propranolol is used to reduce blood pressure and the strength of cardiac contractions and to control anxiety attacks. Another example is *famotidine,* a compound that selectively binds and inhibits a particular type of histamine receptor found on cells in the stomach. Famotidine is sold as a stomach "acid controller" under the trade name Pepcid AC. Another drug—*cimetidine,* sold under the trade name Tagamet—acts in a similar manner.

Receptor Binding Activates a Sequence of Signal Transduction Events Within the Cell

When a ligand binds to its cognate receptor, the receptor is altered in a way that causes changes in cellular activities. In general, the binding of a ligand either induces a change in receptor conformation or causes receptors to cluster together. Once one of these changes takes place, the receptor initiates a preprogrammed sequence of events inside the cell. By *preprogrammed,* we mean that cells have a greater repertoire of functions than are in use at any particular time. Some of these cellular processes remain unused until particular signals are received that trigger them.

Signal Integration. Cells in the human body can be exposed to a multitude of signals at any given moment. How do cells respond to such complexity? For much of the rest of this chapter, we will treat each signal transduction pathway as if it occurs in isolation within a cell. But in reality, cells must *integrate* these signals to produce coordinated responses to their environment. Indeed, many different signaling pathways may operate at any given moment within a cell. These pathways interact with one another in complex ways because components of one pathway affect the ability of another pathway to transmit its signals (see Figure 14-2b). Sometimes a single receptor can activate multiple pathways, as we will see with both G protein-linked receptors and receptor tyrosine kinases. In other cases, different pathways converge onto the same molecules; second messengers are good examples of such "signal integrators." In still other cases, different ligands bind their corresponding receptors at the cell surface, activating specific signaling pathways within the cells. Activated components from one pathway then affect components in another pathway. Such an interaction is known as *signaling crosstalk.* For example, the activity of the receptor for the growth factor TGFβ can be inhibited by proteins activated when the same cell is exposed to other growth factors that act through a very different type of receptor (a receptor tyrosine kinase). It is better not to think of cell signaling in a linear sense but as a complex network of biochemical pathways leading to changes in a cell's properties at any given moment.

Signal Amplification. Signal transduction pathways allow another important aspect of a cell's response to an external signal: signal amplification. Exceedingly small quantities of a ligand are often sufficient to elicit a response from a target cell, yet the responding cell reacts in dramatic ways. Often the strong response of the target cell results from a signaling cascade with the responding cell. At each step in the cascade, a signaling intermediate persists long enough to stimulate the production of many molecules required for the next step in the cascade, thereby multiplying the effects of a single receptor-ligand interaction on the cell surface. A well-known example of such signal amplification involves the breakdown of glycogen in liver cells in response to the hormone epinephrine (**Figure 14-3**). As a

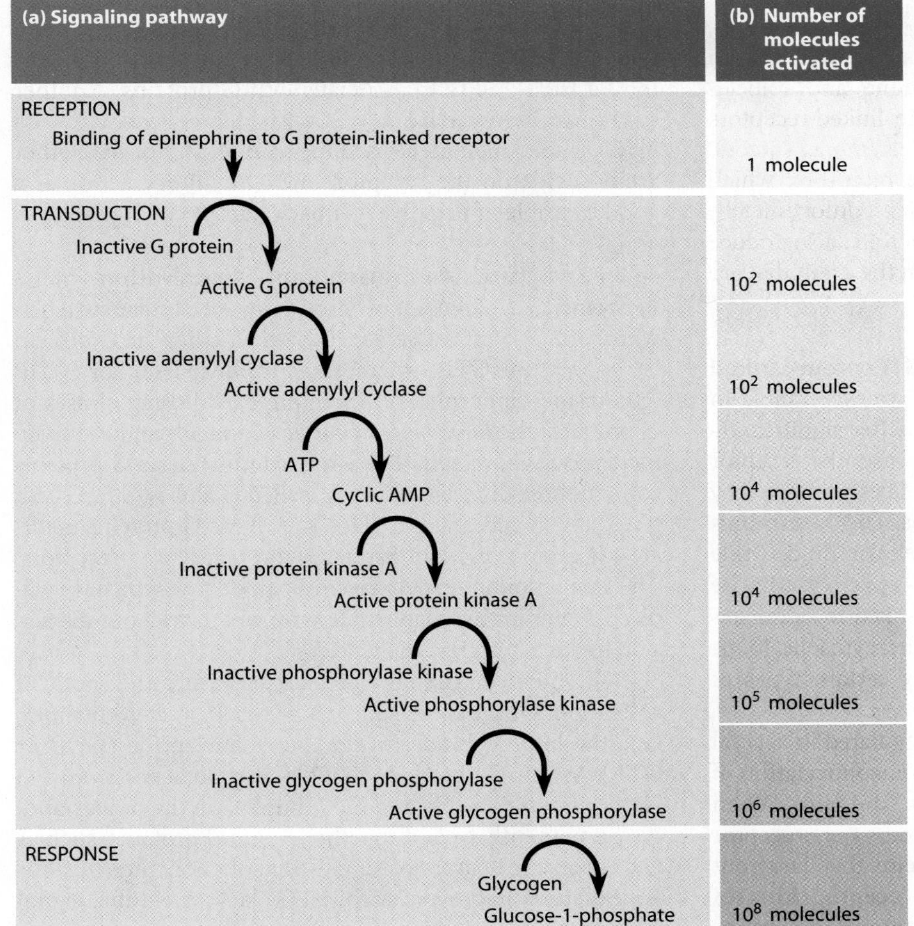

FIGURE 14-3 Signal Transduction Pathways Can Amplify the Cellular Response to an External Signal. Liver cells respond to the hormone epinephrine by breaking down glycogen to liberate glucose-1-phosphate. **(a)** The epinephrine receptor is a G protein-linked receptor, which activates an enzyme known as adenylyl cyclase. Adenylyl cyclase catalyzes formation of a second messenger, cAMP, which activates a protein kinase (protein kinase A), which in turn activates another kinase (phosphorylase kinase). Ultimately, the enzyme glycogen phosphorylase is activated, which catalyzes the breakdown of glycogen. **(b)** The approximate number of molecules produced at each step is shown on the right. One epinephrine molecule is capable of triggering the production of hundreds of millions of glucose-1-phosphate molecules.

result of the cascade, a single epinephrine ligand can stimulate the release of hundreds of millions of glucose molecules from glycogen, a glucose polymer that stockpiles glucose for storage within cells.

Signal transduction events are initiated or altered within a cell when a ligand binds to its cognate receptor on the cell surface or within the cell. Depending on their mode of action, receptors can be classified into several basic categories. We discussed *ligand-gated channels* in Chapter 13. Here, *plasma membrane receptors* are the main focus of our discussion. Many of these receptors can be classified into two families: those linked to G proteins and those linked to protein kinases. We begin by looking at the former.

G Protein-Linked Receptors

Many Seven-Membrane Spanning Receptors Act via G Proteins

The **G protein-linked receptor family** is so named because ligand binding causes a change in receptor conformation that activates a particular **G protein** (an abbreviation for *guanine-nucleotide binding protein*). A portion of the activated G protein in turn binds to a target protein, such as an enzyme or a channel protein, thereby altering the target's activity. Examples of G protein-linked receptors include olfactory receptors (responsible for our sense of smell), norepinephrine receptors, and hormone receptors such as those for thyroid-stimulating hormone and follicle-stimulating hormone. A class of G protein-linked receptors of great clinical importance is the *opioid receptors.* Narcotic drugs such as morphine bind to these receptors, which accounts for their pain-killing benefits. Unfortunately, morphine and related drugs, such as heroin, also induce long-term changes in synaptic function in the brain that are responsible for their addictive effects.

The Structure and Regulation of G Protein-Linked Receptors. G protein-linked receptors are remarkable in that they all have a similar structure yet differ significantly in their amino acid sequences. In each case, the receptor protein forms seven transmembrane α helices connected by alternating cytosolic or extracellular loops. The N-terminus of the protein is exposed to the extracellular fluid, while the C-terminus resides in the cytosol (**Figure 14-4**). The extracellular portion of each G protein-linked receptor has a unique messenger-binding site, and the cytosolic loops allow the receptor to interact with only certain types of G proteins.

G protein-linked receptors can be regulated in several ways. One of the most important is via phosphorylation of specific amino acids in their cytosolic domain. When these amino acids are phosphorylated, the receptor becomes desensitized. One class of proteins that carry out this function are **G protein-linked receptor kinases (GRKs)**, which specifically act on activated receptors.

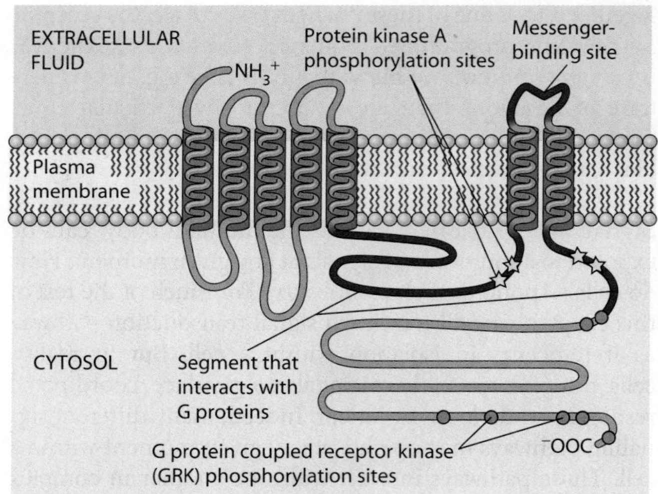

FIGURE 14-4 The Structure of G Protein-Linked Receptors. Each G protein-linked receptor has seven transmembrane a helices. A ligand binds to the extracellular portion of the receptor, causing an intracellular portion of the receptor to bind and activate a G protein. Besides the regions shown, the second cytosolic loop is also involved in G protein interactions in some cases. Specific amino acids in the cytosolic region are also targets for phosphorylation by G protein-linked receptor kinases (GRKs) and protein kinase A.

When specific amino acids within the cytosolic portion of G protein-linked receptors, such as the β-adrenergic receptor, are heavily phosphorylated by GRKs, a protein known as *β-arrestin* can bind to them and completely inhibit their ability to associate with G proteins. Another kinase, **protein kinase A (PKA)**, which is itself activated by G protein-mediated signaling, can phosphorylate other amino acids on the receptor. Such inhibitory action is a good example of negative feedback during cell signaling.

The Structure, Activation, and Inactivation of G Proteins. G proteins act very much like molecular switches, whose "on" or "off" state depends on whether the G protein is bound to GTP (guanosine triphosphate) or GDP (guanosine diphosphate). There are two distinct classes of G proteins: the *large heterotrimeric G proteins* and the *small monomeric G proteins.* The large heterotrimeric G proteins contain three different subunits, called G alpha (G_α), G beta (G_β), and G gamma (G_γ). Heterotrimeric G proteins mediate signal transduction through G protein-linked receptors. The small monomeric G proteins include *Ras,* which we will discuss later in this chapter. Here we will restrict our discussion to the heterotrimeric type.

G proteins have the same basic structure and mode of activation. Of the three subunits in the $G_{\alpha\beta\gamma}$ heterotrimer, G_α, the largest, binds to a guanine nucleotide (GDP or GTP). When G_α binds to GTP, it also detaches from the $G_{\beta\gamma}$ complex. The G_β and G_γ subunits, on the other hand, are permanently bound together. Some G proteins, such as G_s, act as stimulators of signal transduction (hence *s,* for "stimulatory"); others, such as G_i, act to inhibit signal transduction (hence *i,* for "inhibitory").

When a messenger binds to a G protein-linked receptor on the surface of the cell, the change in conformation of the receptor causes a G protein to associate with the receptor, which in turn causes the G_α subunit to release its bound GDP. The G_α then acquires a new, different molecule of GTP and detaches from the complex (**Figure 14-5**). Depending on the G protein and the cell type, either the free GTP-G_α subunit or the $G_{\beta\gamma}$ complex can then initiate signal transduction events in the cell. Each portion of the G protein exerts its effect by binding to a particular enzyme or other protein in the cell. In some cases, both the GTP-G_α and $G_{\beta\gamma}$ subunits simultaneously regulate different processes in the cytosol.

The activity of a G protein persists only as long as the G_α subunit is bound to GTP, and the G_α and $G_{\beta\gamma}$ subunits remain separated. Because the G_α subunit catalyzes GTP hydrolysis, it remains active only until it hydrolyzes its associated GTP to GDP, at which time it reassociates with $G_{\beta\gamma}$. This feature allows the signal transduction pathway to shut down when the messenger is utilized. Some G_α proteins are very inefficient at catalyzing GTP hydrolysis; however, their efficiency is dramatically improved by **regulators of G protein signaling (RGS) proteins.** When RGS proteins bind G_α, they stimulate GTP hydrolysis. Such *GTPase activating proteins (GAPs)* are important regulators of G protein function, as we will see later in the case of the Ras protein.

The large number of different G proteins provides for a diversity of G protein-mediated signal transduction events, only a few of which we will consider here. Perhaps the most important and widespread G protein-mediated signal transduction events are the release or formation of second messengers. As we will see in the next section, two widely used second messengers are cyclic AMP and calcium ions, which stimulate the activity of target enzymes when their cytosolic concentrations are elevated. Other well-studied examples of G protein-mediated signaling, in the retina of the eye and in endothelial cells, use a different second messenger, cyclic GMP (**Box 14A**).

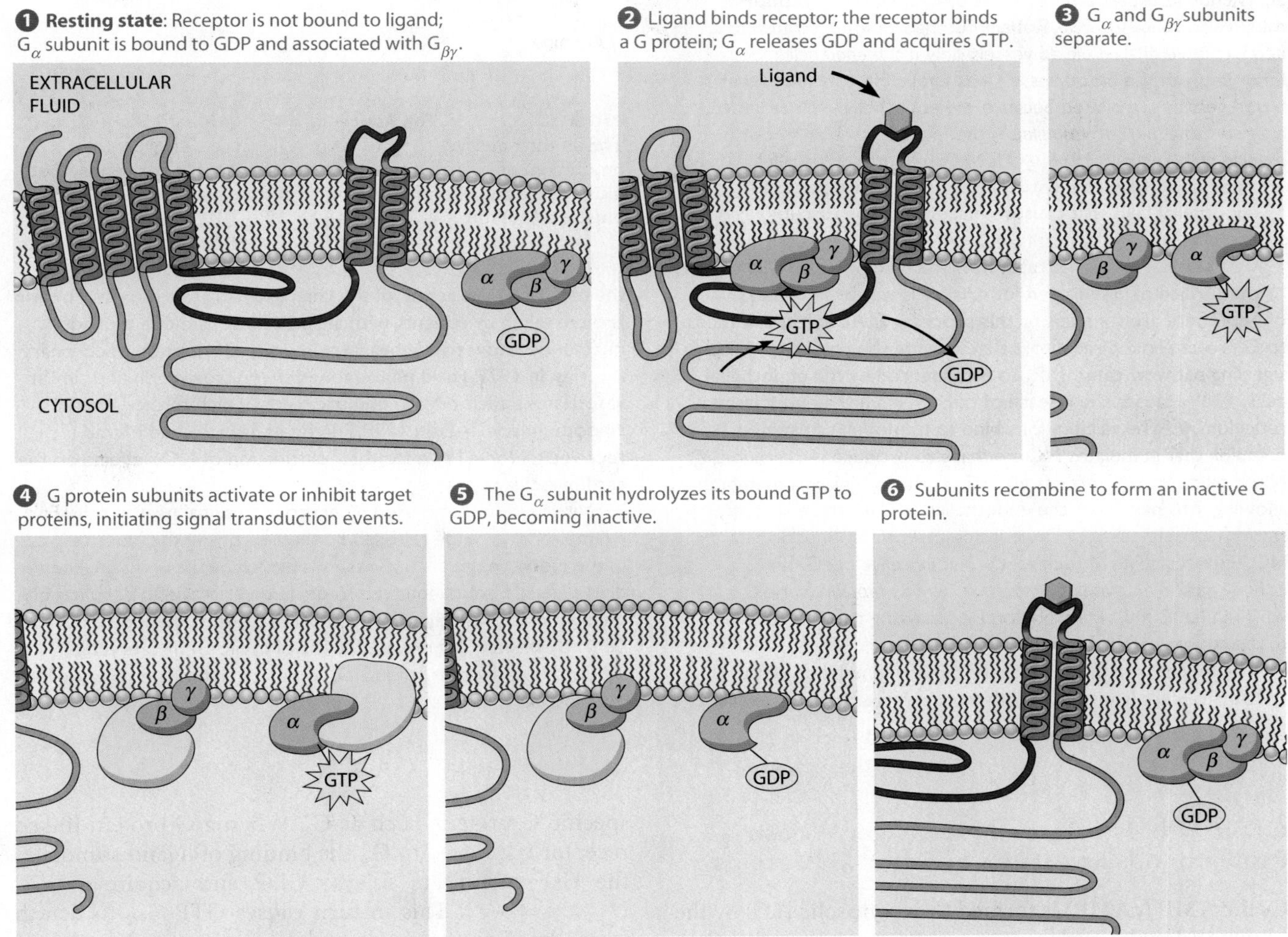

1 Resting state: Receptor is not bound to ligand; G_α subunit is bound to GDP and associated with $G_{\beta\gamma}$.
EXTRACELLULAR FLUID
CYTOSOL
α
β
γ
GDP
2 Ligand binds receptor; the receptor binds a G protein; G_α releases GDP and acquires GTP.
Ligand
GTP
GDP
3 G_α and $G_{\beta\gamma}$ subunits separate.
GTP
4 G protein subunits activate or inhibit target proteins, initiating signal transduction events.
GTP
5 The G_α subunit hydrolyzes its bound GTP to GDP, becoming inactive.
GDP
6 Subunits recombine to form an inactive G protein.
GDP

FIGURE 14-5 The G Protein Activation/Inactivation Cycle. When a G protein-linked receptor binds a ligand, it binds and activates a G protein. Its dissociated subunits can regulate target proteins. Hydrolysis of GTP bound to G_α ultimately terminates the signal.

BOX 14A DEEPER INSIGHTS

G Proteins and Cyclic GMP

Cyclic AMP (cAMP) is an important second messenger in many cell signaling events. However, cAMP is not the only cyclic nucleotide that is important in cell signaling. Just as the enzyme adenylyl cyclase can catalyze the formation of cAMP, the enzyme guanylyl cyclase catalyzes the formation of cyclic GMP (cGMP). Cyclic GMP is derived from GTP in a manner analogous to the production of cAMP from ATP, and, like cAMP, cGMP can act as a second messenger. Here we consider one example of G protein-mediated signaling that involves cGMP: nitric oxide signaling in endothelial cells.

Nitric Oxide Couples G Protein-Linked Receptor Stimulation in Endothelial Cells to Relaxation of Smooth Muscle Cells in Blood Vessels

An important signaling molecule in the cardiovascular system is **nitric oxide (NO),** a toxic, short-lived gas molecule produced by the enzyme *NO synthase,* which converts the amino acid arginine to NO and citrulline. It has been known for many years that acetylcholine dilates blood vessels by causing their smooth muscles to relax. In 1980, Robert Furchgott demonstrated that acetylcholine dilated blood vessels only if the *endothelium* (the inner lining of the blood vessel) was intact. He concluded that blood vessels are dilated because the endothelial cells produce a signal molecule (or *vasodilator*) that makes vascular smooth muscle cells relax. In 1986, work by Furchgott and parallel work by Louis Ignarro identified NO as the signal released by endothelial cells that causes relaxation of the vascular smooth muscle.

Figure 14-A1 illustrates how the binding of acetylcholine to the surface of vascular endothelial cells results in release of NO. There are six steps to this process. ❶ Acetylcholine binds to G protein-linked receptors that activate the phosphoinositide signaling pathway, causing IP_3 to be produced by the endothelial cells. ❷ IP_3 causes the release of calcium from the endoplasmic reticulum. ❸ The calcium ions bind to calmodulin, forming a complex that stimulates NO synthase to produce nitric oxide. ❹ Nitric oxide is a gas that readily diffuses through plasma membranes, allowing it to pass from the endothelial cell into the adjacent smooth muscle cells. ❺ Once inside the smooth muscle cell, NO activates guanylyl cyclase. ❻ The increase in cGMP concentration activates a protein known as *protein kinase G,* which induces muscle relaxation by catalyzing the phosphorylation of the appropriate muscle proteins.

The mechanism by which acetylcholine stimulation of the endothelial cells leads to smooth muscle relaxation also explains the mechanism of action of the chemical *nitroglycerin.* Nitroglycerin is often taken by patients with angina (chest pain due to inadequate blood flow to the heart) to relieve constriction of coronary arteries. In 1977, Ferid Murad found that nitroglycerin and similar vasodilators elicit release of nitric oxide, which relaxes arterial smooth muscle cells. In 1998, Furchgott, Ignarro, and Murad received a Nobel Prize for their elucidation of NO's effects on the cardiovascular system.

Nitric oxide is also used by neurons to signal nearby cells. For example, nitric oxide released by the neurons in the penis results in the blood vessel dilation responsible for penile erection. The drug *sildenafil,* sold under the trade name Viagra, is an inhibitor of a phosphodiesterase that normally catalyzes the breakdown of cyclic GMP. By maintaining elevated levels of cyclic GMP in erectile tissue, this pathway is stimulated for a longer time period following NO release.

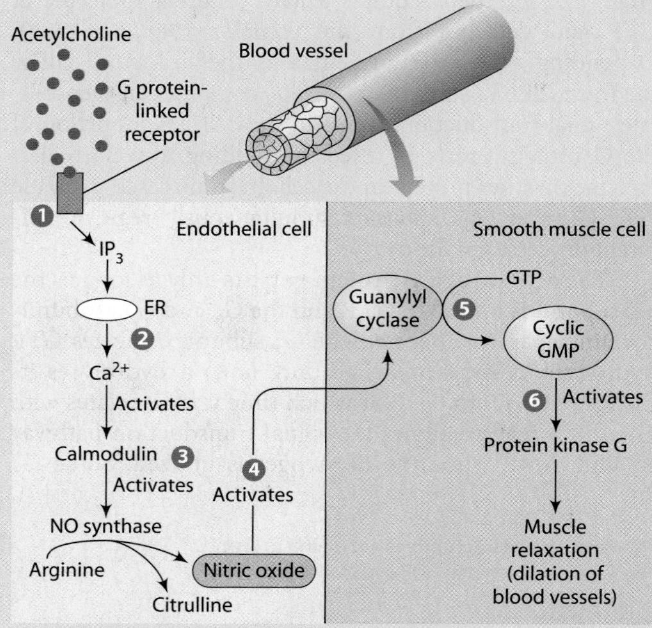

FIGURE 14A-1 The Action of Nitric Oxide on Blood Vessels. The binding of acetylcholine to endothelial cells triggers the production of nitric oxide, which diffuses into the adjacent smooth muscle cells and stimulates guanylyl cyclase, thereby leading to muscle relaxation.

Cyclic AMP Is a Second Messenger Whose Production Is Regulated by Some G Proteins

Cyclic AMP (cAMP) is formed from cytosolic ATP by the enzyme **adenylyl cyclase** (**Figure 14-6**). Adenylyl cyclase is anchored in the plasma membrane, with its catalytic portion protruding into the cytosol. Normally, the enzyme is inactive until it binds to an activated G_α subunit of a specific G protein, such as G_s. When a G protein-linked receptor is coupled to G_s, the binding of ligand stimulates the Gs_α subunit to release GDP and acquire a GTP (**Figure 14-7**). This in turn causes GTP-Gs_α to detach from the $Gs_{\beta\gamma}$ subunits and bind to adenylyl cyclase. When GTP-Gs_α binds to adenylyl cyclase, the enzyme becomes active and converts ATP to cAMP. In contrast, another G_α protein, Gi_α, inhibits adenylyl cyclase. The

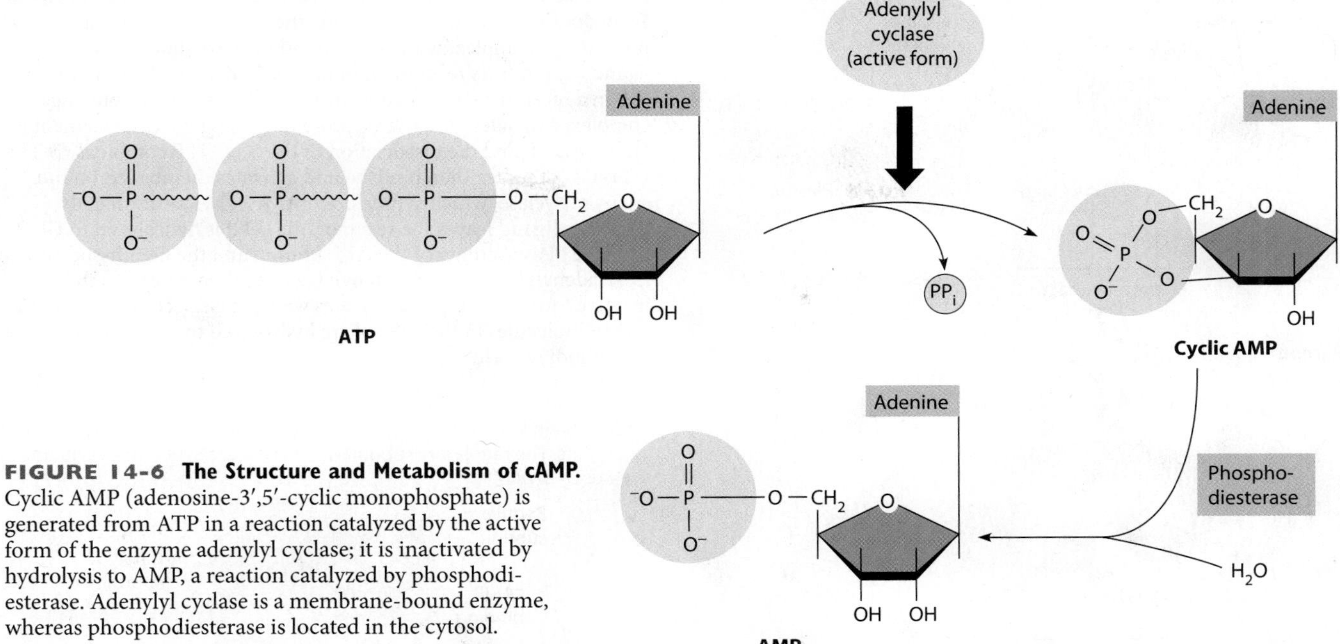

FIGURE 14-6 The Structure and Metabolism of cAMP. Cyclic AMP (adenosine-3′,5′-cyclic monophosphate) is generated from ATP in a reaction catalyzed by the active form of the enzyme adenylyl cyclase; it is inactivated by hydrolysis to AMP, a reaction catalyzed by phosphodiesterase. Adenylyl cyclase is a membrane-bound enzyme, whereas phosphodiesterase is located in the cytosol.

role of G_i in regulating potassium channels via its $G_{\beta\gamma}$ subunits will be discussed later in this chapter.

G proteins respond quickly to changes in ligand concentration because they remain active for only a short time before the G_α subunit hydrolyzes its bound GTP and converts to the inactive state. Once the G protein becomes inactive, the adenylyl cyclase ceases to make cAMP. However, cAMP levels would still remain elevated in the cell if not for the enzyme **phosphodiesterase,** which degrades cAMP. This further ensures that the signal transduction pathway will shut down promptly when the concentration of the ligand outside the cell declines.

cAMP is important for many cellular events, a few of which are listed in **Table 14-1**. cAMP appears to have one main intracellular target—the enzyme protein kinase A. PKA phosphorylates a wide variety of cellular proteins by transferring a phosphate from ATP to a serine or threonine found within the target protein. cAMP regulates the activity of PKA by causing the detachment of its two regulatory subunits from its two catalytic subunits (**Figure 14-8**). Once the catalytic subunits are free, PKA can catalyze the phosphorylation of various proteins in the cell.

An increase in cAMP concentration can produce different effects in different cells. When cAMP is elevated in skeletal muscle and liver cells, the breakdown of glycogen is stimulated. In cardiac muscle, the elevation of cAMP strengthens heart contraction, whereas in smooth muscle contraction is inhibited. In blood platelets, the elevation of cAMP inhibits their mobilization during blood clotting, and in intestinal epithelial cells, it causes the secretion of salts and water into the lumen of the gut. Each of these reactions is an example of the preprogrammed response discussed earlier. In fact, if the concentration of cAMP is artificially raised in these different types of cells, these same cellular responses can be triggered even in the absence of a ligand. This can be done in two different ways: either by stimulating cAMP production directly or by inhibiting the enzyme phosphodiesterase that degrades cAMP. Examples of phosphodiesterase inhibitors are the *methylxanthines,* compounds such as caffeine and theophylline, found in coffee, tea, and soft drinks. (Theophylline is often used to treat asthma because it relaxes bronchial smooth muscle.)

Disruption of G Protein Signaling Causes Several Human Diseases

What would happen if the G protein–adenylyl cyclase system could not be shut off? This question can be answered by examining what happens in two human diseases caused by bacteria. The bacteria *Vibrio cholerae* (which causes cholera) and *Bordetella pertussis* (which causes whooping cough) both cause disease through their effects on heterotrimeric G proteins. Cholera results from the secretion of *cholera toxin* when *V. cholerae* colonizes the gut. The toxin alters

Table 14-1 Examples of Cell Functions Regulated by cAMP

Regulated Function	Target Tissues	Hormone
Glycogen degradation	Muscle, liver	Epinephrine
Fatty acid production	Adipose	Epinephrine
Heart rate, blood pressure	Cardiovascular	Epinephrine
Water reabsorption	Kidney	Antidiuretic hormone
Bone resorption	Bone	Parathyroid hormone

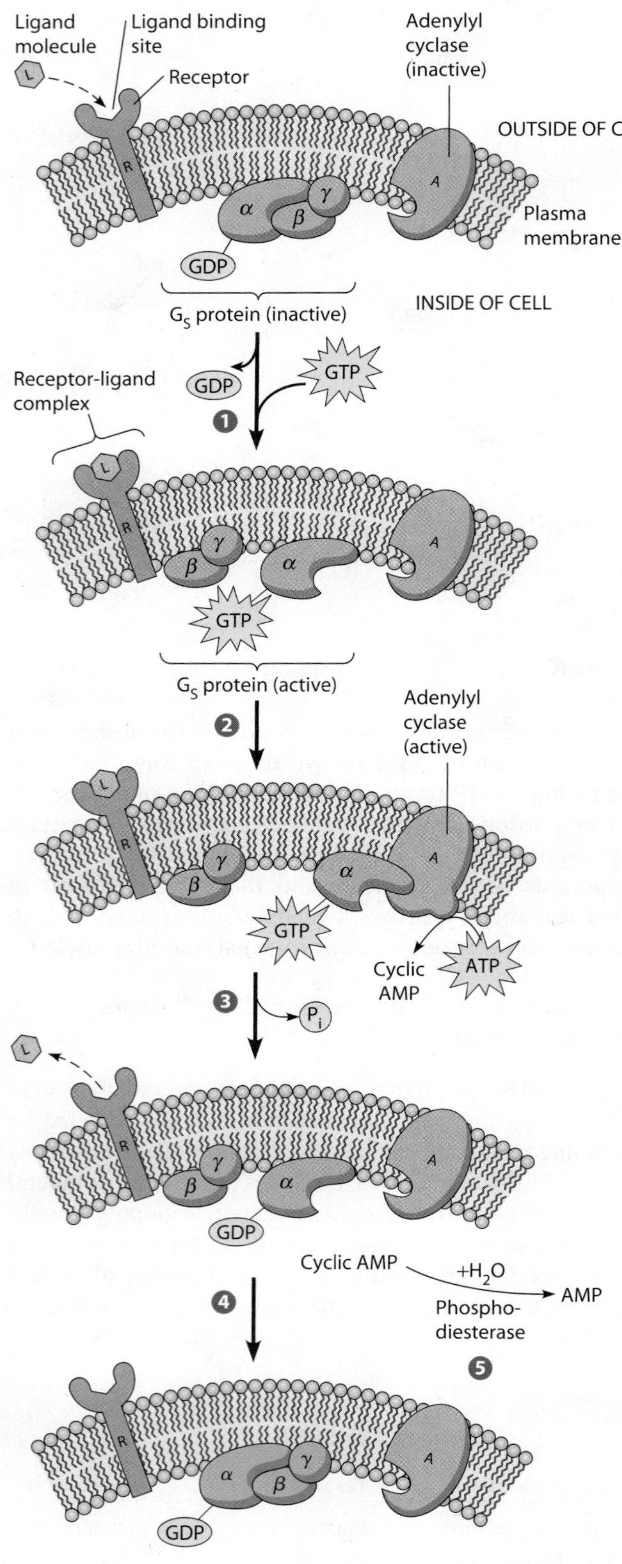

FIGURE 14-7 The Roles of G Proteins and Cyclic AMP in Signal Transduction. In the inactive state, the α, β, and γ subunits are present as a complex, with GDP bound to the α subunit. When a ligand (L) binds its receptor, it binds and activates a G_s protein. ❶ When a receptor is activated by ligand binding, the receptor-ligand complex associates with the G_s protein, causing the displacement of GDP by GTP and the dissociation of the Gs_α-GTP complex. ❷ The GTP-Gs_α complex then binds to and activates membrane-bound adenylyl cyclase, which synthesizes cAMP. ❸ Activation ends when the ligand leaves the receptor, the GTP is hydrolyzed to GDP by the GTPase activity of the Gs_α subunit, and the Gs_α dissociates from adenylyl cyclase. ❹ Adenylyl cyclase then reverts to the inactive form, the Gs_α re-associates with the $Gs_{\beta\gamma}$ complex, and ❺ cAMP molecules in the cytosol are hydrolyzed to AMP by the phosphodiesterase.

❶ Protein kinase A is composed of two catalytic and two regulatory subunits. The regulatory subunits inhibit the catalytic subunits in the absence of cAMP.

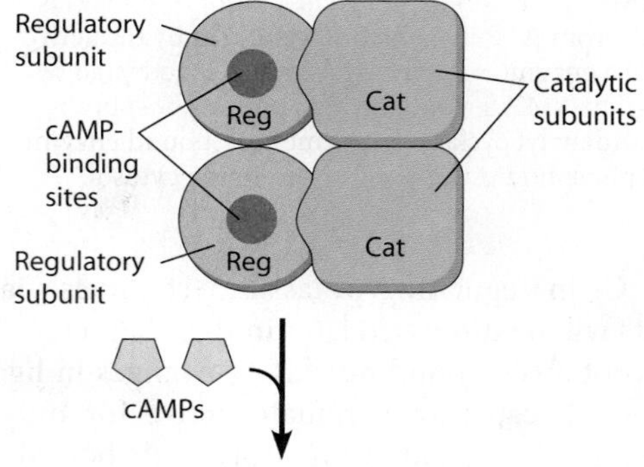

❷ Cyclic AMP activates protein kinase A by binding to the regulatory subunits, causing the regulatory subunits to change conformation.

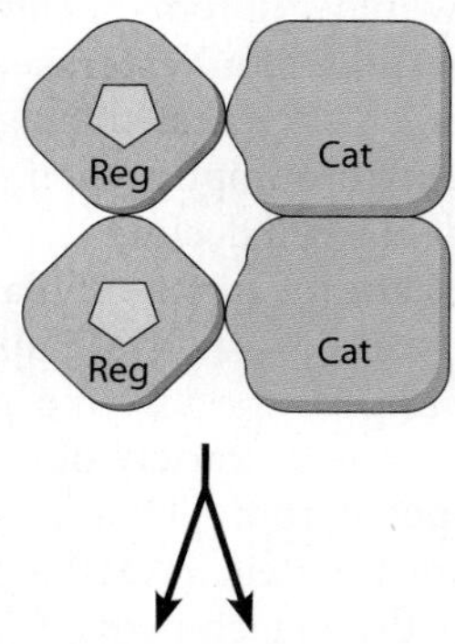

❸ The catalytic subunits detach. They are now activated and can phosphorylate target proteins in the cell.

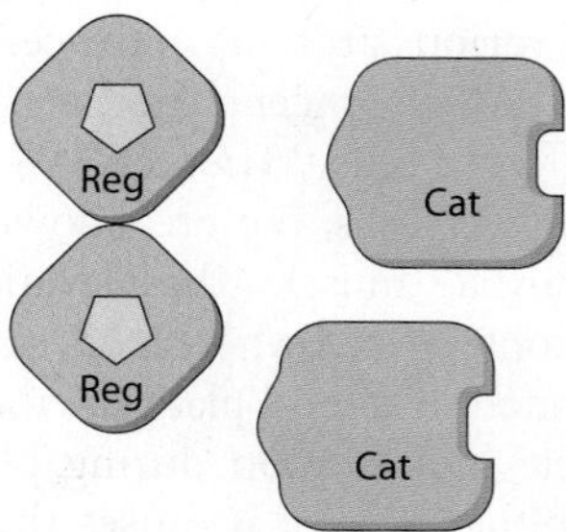

FIGURE 14-8 The Activation of Protein Kinase A by Cyclic AMP. PKA is composed of four subunits, two catalytic and two regulatory. The regulatory subunits inhibit the catalytic subunits in the absence of cAMP, but cAMP binding activates the catalytic subunits, which then regulate target proteins.

secretion of salts (sodium chloride and sodium bicarbonate) and fluid in the intestinal lumen, which is normally regulated by hormones that act through the G protein G_s to alter intracellular levels of cAMP. A portion of the cholera toxin is an enzyme that chemically modifies G_s so that it can no longer hydrolyze GTP to GDP (by adding an ADP-ribose to it). As a result, G_s cannot be shut off, cAMP levels remain high, and the intestinal cells secrete large amounts of salt and water. If left untreated, this condition can result in death by dehydration. The *pertussis toxin* secreted by *B. pertussis* acts in a similar manner but on the inhibitory G protein, G_i. This protein normally shuts off adenylyl cyclase. When inactivated by pertussis toxin, G_i no longer inhibits adenylyl cyclase. The resulting fluid accumulation in the lungs leads to the characteristic cough associated with this disease.

The discovery of these toxins and their mode of action has provided not only advances in medical treatment but also powerful tools for studying G protein-mediated signal transduction. Because these two toxins act on different G proteins, researchers have been able to use the purified toxins to associate a signaling pathway with a particular G protein.

Many G Proteins Use Inositol Trisphosphate and Diacylglycerol as Second Messengers

The importance of inositol phospholipids in cell signaling was first brought to light in the pioneering studies of Robert Michell and Michael Berridge. In the early 1980s, Berridge noticed that when the salivary glands of certain insect larvae were stimulated to secrete, changes occurred in membrane inositol phospholipids. We now know that **inositol-1,4,5-trisphosphate (IP_3)**, one of the breakdown products of inositol phospholipids, functions as a second messenger. IP_3 is generated from *phosphatidylinositol-4,5-bisphosphate* (PIP_2), a relatively uncommon membrane phospholipid, when the enzyme **phospholipase C** is activated. Phospholipase C cleaves PIP_2 into two molecules—inositol trisphosphate and **diacylglycerol (DAG)** (**Figure 14-9**). After their discovery, IP_3 and DAG were quickly shown to be second messengers in a variety of regulated cell functions, some of which are indicated in **Table 14-2**.

The roles of IP_3 and DAG as second messengers in the inositol-phospholipid-calcium pathway are shown in **Figure 14-10**. The sequence begins with the binding of a ligand to its membrane receptor, leading to the activation of a specific G protein called G_q. G_q then activates a type of phospholipase C known as C_β, thereby generating both IP_3 and DAG. Inositol trisphosphate is water soluble and quickly diffuses through the cytosol, binding to a ligand-gated calcium channel known as the **IP_3 receptor** channel in the endoplasmic reticulum (ER). When IP_3 binds, the channel opens, releasing calcium ions into the cytosol. Calcium then elicits the desired physiological response.

Intracellular calcium release, as well as the formation of DAG in the membrane by phospholipase C activity, activates members of the **protein kinase C (PKC)** family of enzymes. PKC can then phosphorylate specific serine and threonine groups on a variety of target proteins,

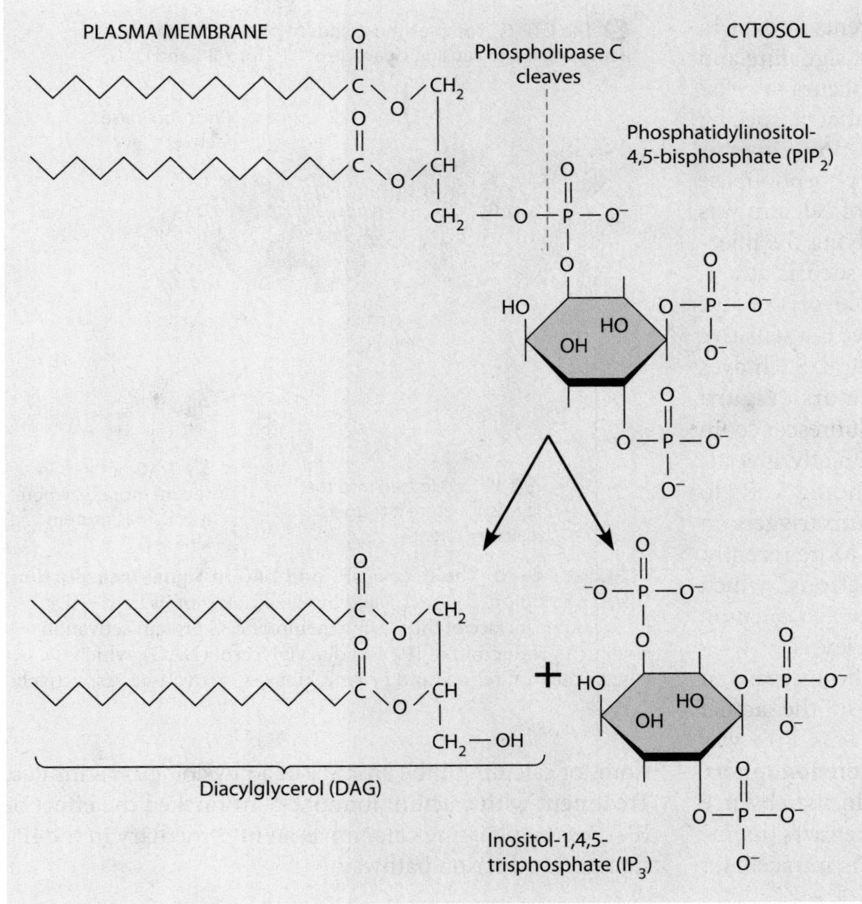

FIGURE 14-9 The Formation of Inositol Trisphosphate and Diacylglycerol. Inositol trisphosphate (IP_3) and diacylglycerol (DAG) are formed when phospholipase C cleaves phosphatidylinositol-4,5-bisphosphate (PIP_2), one of the phospholipids present in the membrane. IP_3 is released into the cytosol, whereas DAG remains within the membrane. Both IP_3 and DAG are second messengers in a variety of signal transduction pathways.

Table 14-2	Examples of Cell Functions Regulated by Inositol Trisphosphate and Diacylglycerol	
Regulated Function	**Target Tissues**	**Messenger**
Platelet activation	Blood platelets	Thrombin
Muscle contraction	Smooth muscle	Acetylcholine
Insulin secretion	Pancreas, endocrine	Acetylcholine
Amylase secretion	Pancreas, exocrine	Acetylcholine
Glycogen degradation	Liver	Antidiuretic hormone
Antibody production	B lymphocytes	Foreign antigens

depending on the cell type. The role of DAG was established experimentally by showing that its effects could be mimicked by *phorbol esters,* plant metabolites that bind to PKC, activating it directly. Using a variety of pharmacological agents, researchers showed that IP_3-stimulated calcium release and DAG-mediated kinase activity are both required to produce a full response in target cells.

A wide variety of cellular effects have been linked to the activation of protein kinase C, including the stimulation of cell growth, the regulation of ion channels, changes in the cytoskeleton, increases in cellular pH, and effects on secretion of proteins and other substances.

The Release of Calcium Ions Is a Key Event in Many Signaling Processes

How do we know that the sequence of events shown in Figure 14-10, especially the link between IP_3 signaling and the release of calcium within cells, actually occurs in cells? The answer begins with the observation that a specific physiological phenomenon—salivary secretion, in this case—could be activated by the products of phospholipase action: IP_3 and DAG. Evidence for the role of calcium was provided by an experimental approach involving the injection of a calcium-dependent fluorescent dye, such as fura-2, into a target cell. Because the fluorescence of the dye varies with the calcium concentration, the dye is a sensitive indicator of intracellular calcium concentration. Such dyes are typically referred to as **calcium indicators** (**Figure 14-11**). By measuring the increase in fluorescence in response to activation by a ligand or by IP_3 directly, investigators were able to establish that ligand binding leads to an increase in IP_3 concentration, which in turn triggers an increase in cytosolic calcium concentration. More recently, genetically engineered proteins called "cameleons," which increase their fluorescence in response to elevated calcium, have been used to monitor cytosolic calcium levels.

To complete the sequence of events, the increase in calcium concentration had to be linked to the actual physiological response of the target cell. This link was established by treating target cells with a **calcium ionophore** (such as the drugs ionomycin or A23187) in the absence of extracellular calcium. The ionophore renders membranes permeable to calcium, thereby releasing intracellular stores of calcium in the absence of a physiological stimulus. Treatment with calcium ionophore mimicked the effect of IP_3, thus implicating calcium as an intermediary in the IP_3 signal transduction pathway.

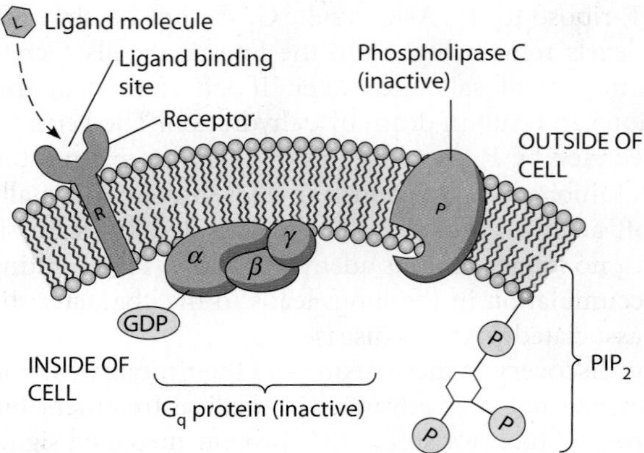

❶ A receptor is activated by the binding of its ligand. The receptor-ligand complex associates with the G protein G_q, causing displacement of GDP by GTP and dissociation of the α and βγ subunits.

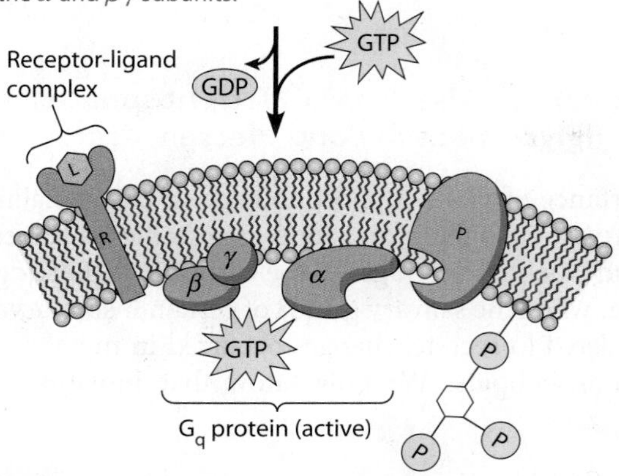

❷ The GTP-G_α complex then binds to phospholipase C (P), activating it and causing cleavage of PIP_2 into IP_3 and DAG.

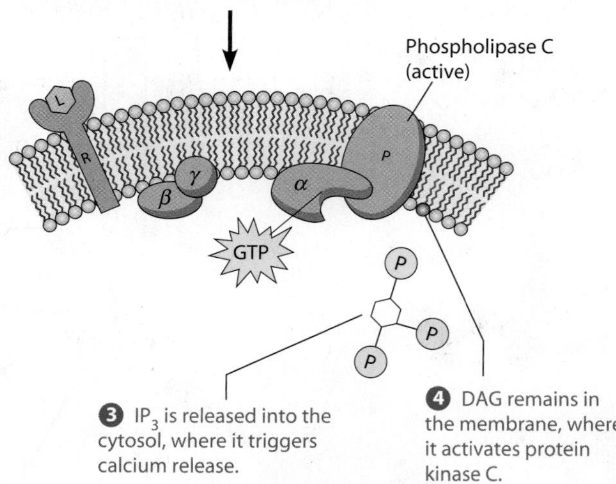

❸ IP_3 is released into the cytosol, where it triggers calcium release.

❹ DAG remains in the membrane, where it activates protein kinase C.

FIGURE 14-10 The Role of IP_3 and DAG in Signal Transduction. When a receptor (R) is activated by the binding of its ligand (L) on the outer surface of the plasma membrane, G protein activation results in production of IP_3 and diacylglycerol (DAG), which trigger calcium release and protein kinase C activation, respectively.

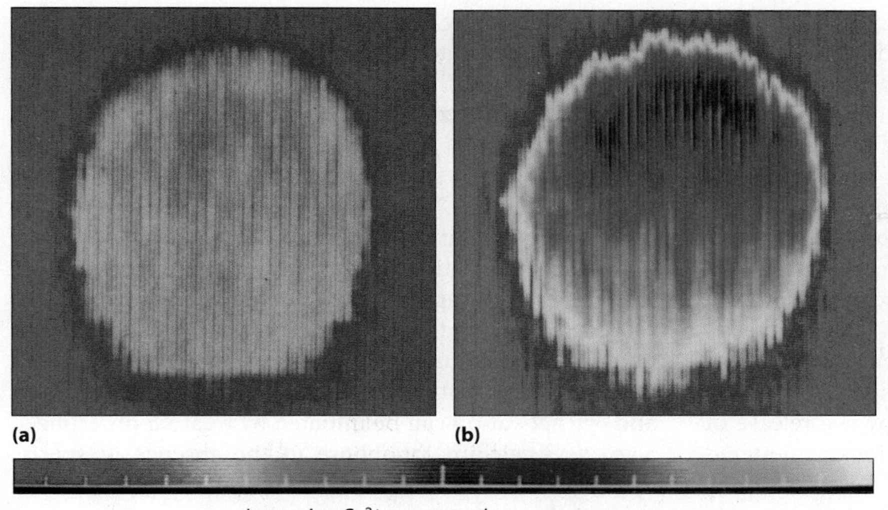

FIGURE 14-11 Increase in Free Cytosolic Ca^{2+} Concentration Triggered by a Hormone That Stimulates the Formation of Inositol Trisphosphate. Cells from a bovine adrenal gland were loaded with fura-2, a dye that fluoresces when bound to Ca^{2+}. The color of the fluorescence changes with alterations in Ca^{2+} concentration. **(a)** Fluorescence micrograph of an unstimulated adrenal cell color-coded to reflect different levels of fluorescence in response to free Ca^{2+}. The yellow color indicates a relatively low concentration of free Ca^{2+}. **(b)** Fluorescence in the adrenal cell stimulated by angiotensin, a hormone that triggers the formation of IP_3. The green and blue colors indicate an increased concentration of free Ca^{2+}.

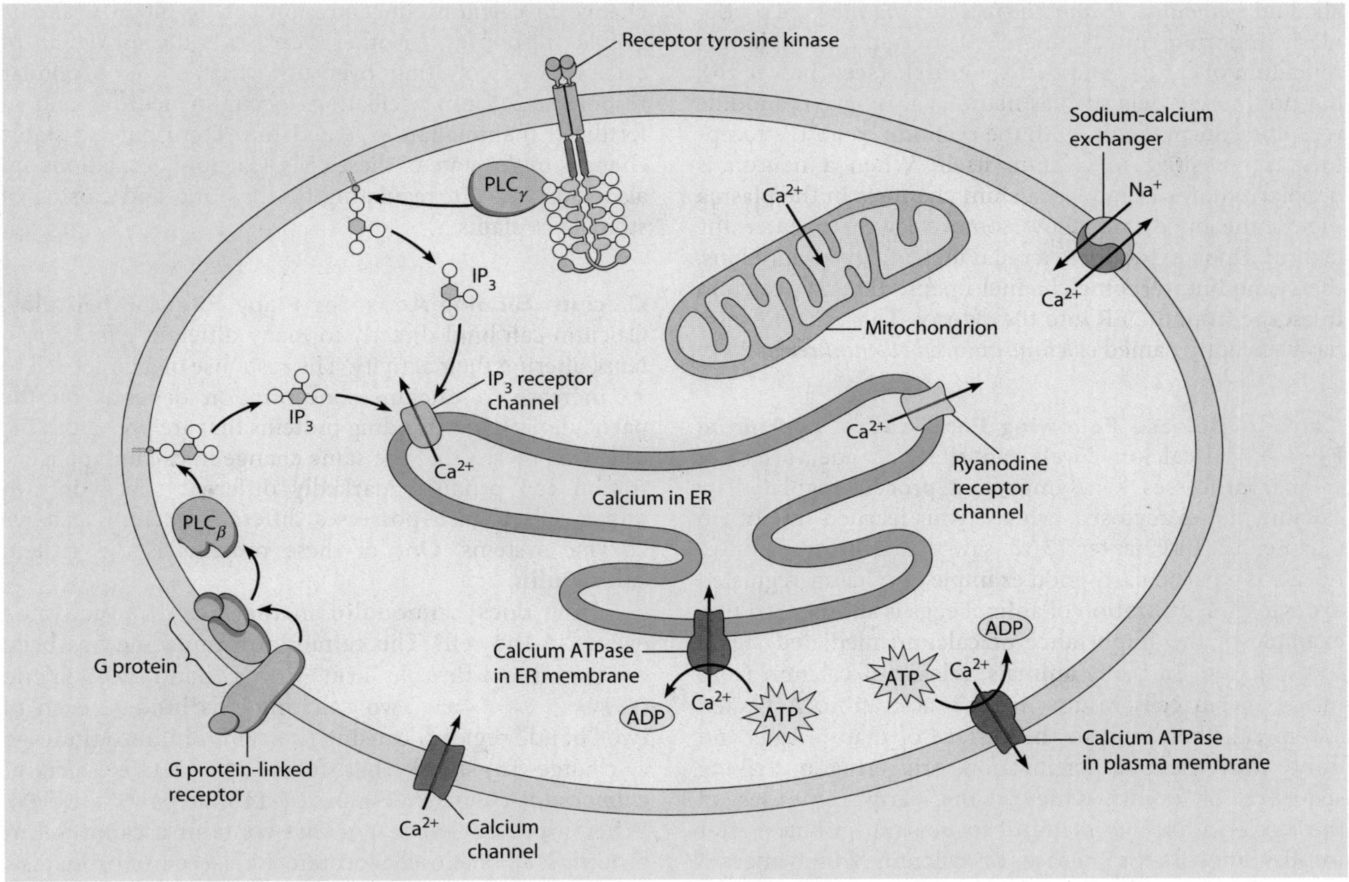

FIGURE 14-12 An Overview of Calcium Regulation in Cells. Cytosolic Ca^{2+} concentration is lowered by the actions of the ER calcium ATPase, the plasma membrane calcium ATPase, Na^+/Ca^{2+} exchangers, and mitochondria. Ca^{2+} concentration increases in the cytosol because of the opening of Ca^{2+} channels in the plasma membrane and the release of Ca^{2+} through the IP_3 or ryanodine receptor channels in the ER membrane.

Calcium ions (Ca^{2+}) play an essential role in regulating a variety of cellular functions. **Figure 14-12** provides an overview of the various mechanisms of calcium regulation. The concentration of calcium is normally maintained at very low levels in the cytosol due to the presence of **calcium ATPases** in the plasma membrane and the ER. Calcium ATPases in the plasma membrane transport calcium out of the cell, whereas the calcium ATPases in the ER sequester calcium ions in the lumen of the ER. In addition, some cells have sodium-calcium exchangers that further reduce the cytosolic calcium concentration. Finally, mitochondria can transport calcium into the mitochondrial matrix. For most cells in their resting state, the action of calcium ATPases maintains the calcium concentration in the cytosol at 0.1 μM.

There are several different ways that various stimuli can cause cytosolic calcium concentrations to increase. One way, discussed in Chapter 13 in relation to neurons, is by the opening of calcium channels in the plasma membrane. The calcium concentration in the extracellular fluid and the blood is about 1.2 m*M*, more than 10,000 times as high as that of the cytosol. As a result, when calcium channels open, calcium ions rush into the cell.

Calcium levels can also be elevated by the release of calcium from intracellular stores. Calcium ions sequestered in the ER can be released through the IP_3 receptor channel, discussed earlier, and through the ryanodine receptor channel, so named because it is sensitive to the plant alkaloid *ryanodine. Ryanodine receptor channels* are particularly important for calcium release from the sarcoplasmic reticulum of cardiac and skeletal muscle (see Chapter 16), but nonmuscle cells such as neurons also have ryanodine receptors. Surprisingly, both the ryanodine and IP_3 receptors are sensitive to calcium itself. When a neuron is depolarized, for example, calcium channels in the plasma membrane open and allow some calcium to enter the cytosol. Upon exposure to a rapid increase in calcium ions, the ryanodine receptor channel opens, allowing calcium to escape from the ER into the cytosol. This phenomenon has been aptly named *calcium-induced calcium release.*

Calcium Release Following Fertilization of Animal Eggs. Local calcium levels can affect a wide variety of cellular processes. One important process regulated by calcium is exocytosis, which you learned about in Chapter 12. In Chapter 13 we saw that neurotransmitter release is a particularly good example of calcium-regulated exocytosis. Fertilization of animal eggs is another striking example of the importance of calcium-mediated signal transduction. In many animals, release of calcium from inside sperm cells results in their activation. Activated sperm can then bind to the surface of mature eggs and unite with them at fertilization, triggering a striking sequence of events. One of the early responses of the egg—within 30 seconds to several minutes after fertilization—is the release of calcium from internal stores. Calcium release occurs initially at the site where the sperm penetrates the egg surface and then spreads across the egg via calcium-induced calcium release, much as a ripple on the surface of a pond spreads away from the site where a pebble strikes the water. The wavelike propagation of calcium release can be visualized using calcium indicators (**Figure 14-13a**).

The calcium release is necessary for two crucial events. First, it stimulates the exocytosis of vesicles, known as *cortical granules* (Figure 14-13b). The release of cortical granules results in alterations of the protein coat surrounding many eggs (typically known as the *vitelline envelope*). These alterations render the egg unable to bind additional sperm, thereby preventing more than one sperm from fertilizing the egg. This process is known as the *slow block to polyspermy.* (An earlier *fast block to polyspermy* involves a transient depolarization of the egg plasma membrane.)

The second major function of the calcium wave following fertilization is *egg activation.* Egg activation involves the resumption of many metabolic processes, the reorganization of the internal contents of the egg, and other events that initiate the process of embryonic development. Many features of the slow block to polyspermy and egg activation can be initiated by treating unfertilized eggs with calcium ionophore in the absence of sperm, demonstrating the key role elevated calcium levels within the egg play in its activation.

Egg activation is a good example of how a dynamic change in calcium concentration can result in dramatic cellular responses. In other cases, it is the *oscillation* of calcium concentration over time that elicits a cellular response. Calcium oscillations occur in neurons and in fertilized mammalian eggs and may contribute to stable changes in the state of these cells. Calcium oscillations are also important in regulating the opening and closing of stomata in plants.

Calcium Binding Activates Many Effector Proteins. Calcium can bind directly to many different *effector proteins,* altering their activity. The response of a target cell to an increase in calcium concentration depends on the particular calcium-binding proteins that are present in the cell. This means that the same change in calcium concentration can produce markedly different effects in two target cells if each possesses different calcium-sensitive enzyme systems. One of these proteins is the protein **calmodulin.**

How does calmodulin mediate calcium-activated events in the cell? The calmodulin molecule has been compared to a flexible "arm" with a "hand" at each end (**Figure 14-14a**). Two calcium ions bind at each of two "hand" regions, causing the calmodulin to undergo a change in shape that forms the active *calcium-calmodulin complex* (Figure 14-14b, steps ❶ and ❷). When a protein is present that contains a calmodulin-binding site, such as a protein kinase or phosphatase, the hands and arm bind to it by wrapping around the binding site (Figure 14-14b, step ❸). Calmodulin binding, in turn, can influence the function of such a protein dramatically.

An important feature of calmodulin is its affinity for calcium: Calmodulin binds to calcium when the cytosolic calcium concentration increases to about 1.0 μM, but it releases calcium when cytosolic calcium levels decline back to the resting level of 0.1 μM. Thus, calmodulin is uniquely suited to operate within the typical range of cytosolic calcium concentrations.

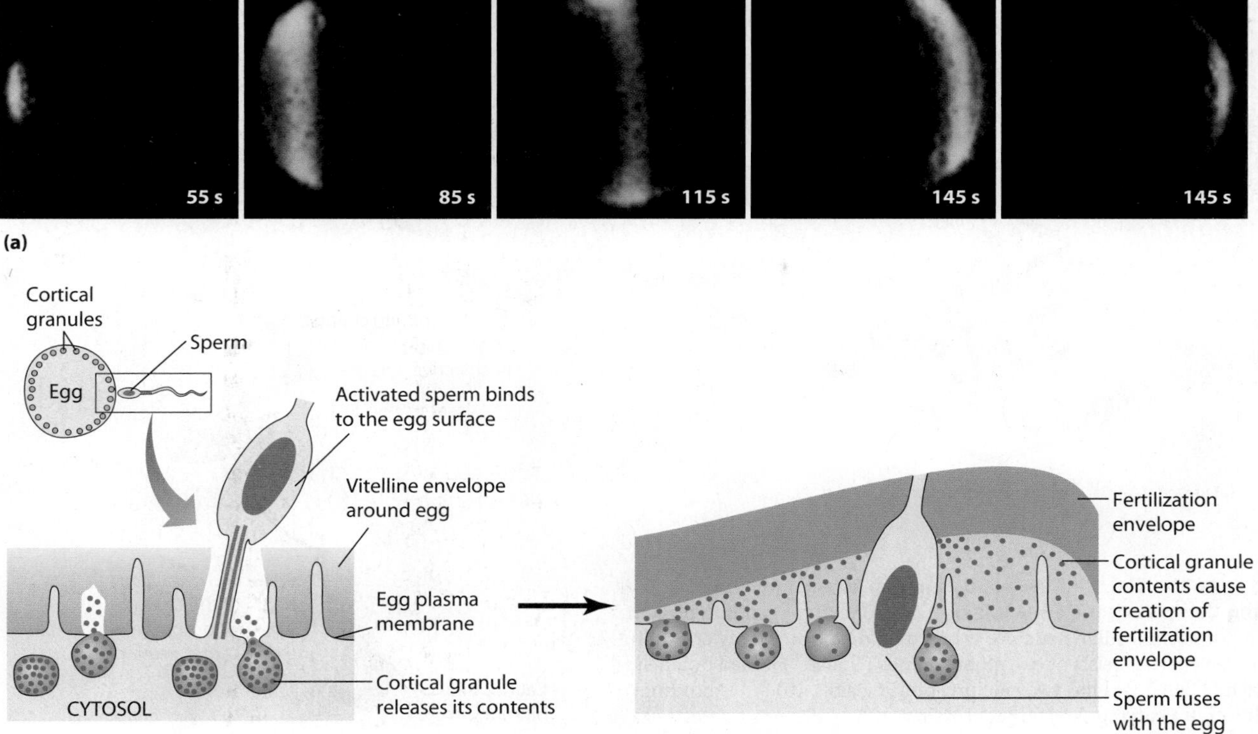

FIGURE 14-13 Transient Increase in Free Ca^{2+} Concentration That Occurs in an Egg Cell Immediately After Fertilization. **(a)** In this classic experiment, a fish egg has been injected with aequorin, a dye that emits light when bound to free Ca^{2+} in the cytosol. The amount of time that has elapsed after fertilization is indicated in seconds in each photograph. In the first few minutes after fertilization, a transient wave of increased Ca^{2+} concentration passes across the egg, starting from the point of sperm entry on the left. **(b)** An activated sperm cell, such as the sea urchin sperm shown here, binds to the egg surface, resulting in local Ca^{2+} release and cortical granule exocytosis. The result is the creation of the fertilization envelope, which prevents additional sperm from penetrating the egg.

The $\beta\gamma$ Subunits of G Proteins Can Also Transduce Signals

We have seen that the α subunits of activated G proteins can interact with proteins such as adenylyl cyclase and phospholipase C to elicit changes in a cell that receives a signal. However, $G_{\beta\gamma}$ can also engage in signaling. For example, G protein receptor kinases, which we have already encountered, can be activated by the $\beta\gamma$ subunit of a dissociated G protein, providing a feedback mechanism on G protein signals. One well-studied example of $G_{\beta\gamma}$ signaling involves the *muscarinic acetylcholine receptor.* Recall from Chapter 13 that some neurotransmitter receptors act indirectly, via intracellular signaling, to cause changes in ion channels. When acetylcholine binds the muscarinic acetylcholine receptor, the $\beta\gamma$ subunit of its associated G protein (G_i) acts on potassium channels in the plasma membrane, causing them to open (**Figure 14-15**). When acetylcholine is no longer present, the α and $\beta\gamma$ subunits re-associate, causing the potassium channels to close again.

Another example of $G_{\beta\gamma}$ signaling involves signaling in the budding yeast, *Saccharomyces cerevisiae* (see Figure 14-18). Yeast use G protein-mediated signaling during mating and to sense changes in osmolarity, availability of nutrients, and other environmental factors. Interestingly, the activated $\beta\gamma$ subunits in this case initiate a series of phosphorylation events that lead to activation of a type of kinase known as *MAP kinase.* MAP kinases are key proteins in another type of signaling pathway in higher eukaryotes, the receptor tyrosine kinase pathway, which we consider in the next section.

Other Signaling Pathways Can Activate G Proteins

In addition to the examples we have described, there are many other pathways that can lead to activation of heterotrimeric G proteins. These include Wnts, which act through seven-pass receptors known as Frizzleds (see Chapter 24), and the Hedgehog pathway, which acts in part through another seven-pass protein known as *Smoothened.* These further illustrate the complexity of cell signaling pathways within the cell.

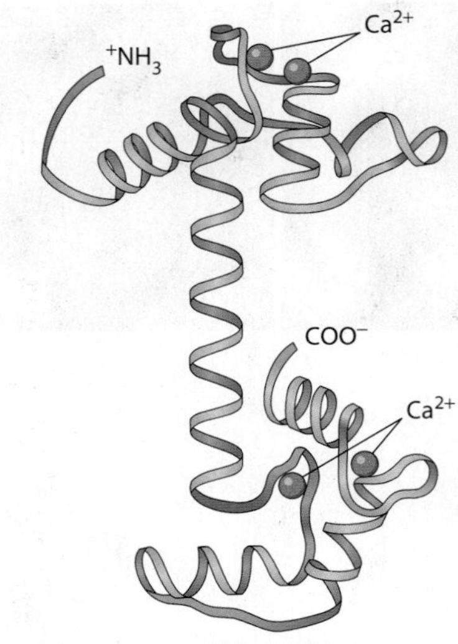

$^+NH_3$
Ca^{2+}
COO^-
Ca^{2+}

(a) Structure of Ca^{2+}-calmodulin complex

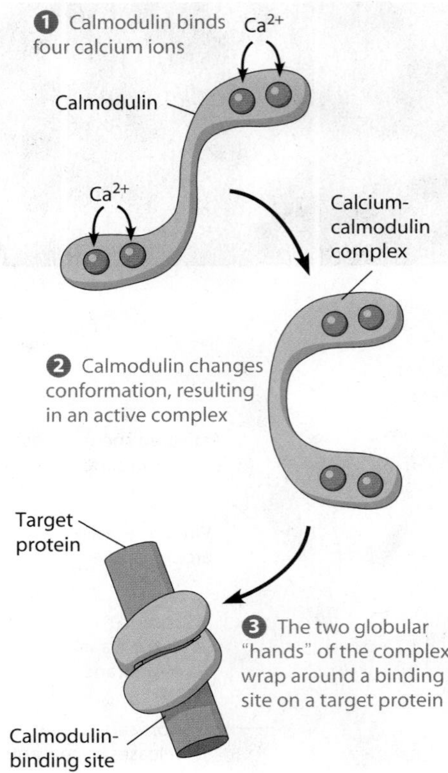

1 Calmodulin binds four calcium ions
Ca^{2+}
Calmodulin
Ca^{2+}
Calcium-calmodulin complex
2 Calmodulin changes conformation, resulting in an active complex
Target protein
3 The two globular "hands" of the complex wrap around a binding site on a target protein
Calmodulin-binding site

(b) Function of Ca^{2+}-calmodulin complex

FIGURE 14-14 The Structure and Function of the Calcium-Calmodulin Complex. **(a)** Calmodulin is a cytosolic Ca^{2+}-binding protein. This model of its molecular structure is based on data from X-ray crystallography. The molecule consists of two globular ends ("hands") joined by a helical region ("arm"). Each end has two calcium-binding sites. **(b)** After binding Ca^{2+}, calmodulin can regulate target proteins.

Protein Kinase-Associated Receptors

We have seen that G protein-linked receptors transmit their signals to the interior of the cell by causing changes in a G protein, which in turn initiates a cascade of signal transduction events. Another large family of proteins uses a different strategy to transmit signals. They not only function as receptors but are themselves protein kinases. When they bind to the appropriate ligand, their kinase activity is stimulated, and they transmit signals through a cascade of phosphorylation events within the cell.

Recall that kinases add phosphate groups to particular amino acids within substrate proteins (see Chapter 6). Receptor kinases generally fall into two major categories: those that phosphorylate a tyrosine residue (*tyrosine kinases*) and those that phosphorylate a serine or threonine residue (*serine/threonine kinases*). We will now examine these *receptor protein kinases* in some detail.

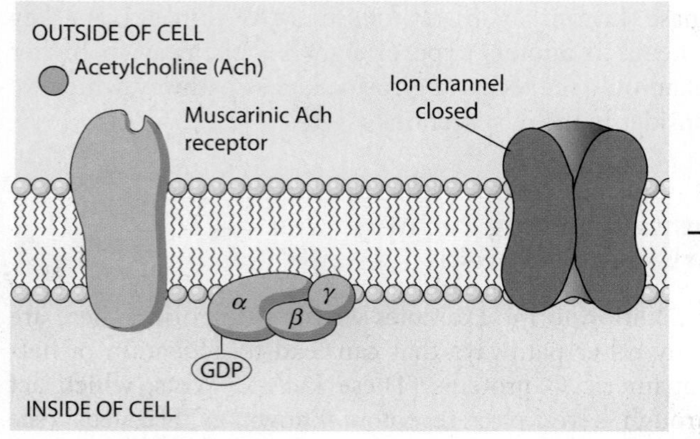

1 In the absence of acetylcholine (Ach), a G protein associated with the muscarinic Ach receptor is bound to GDP.
OUTSIDE OF CELL
Acetylcholine (Ach)
Muscarinic Ach receptor
Ion channel closed
α
β
γ
GDP
INSIDE OF CELL

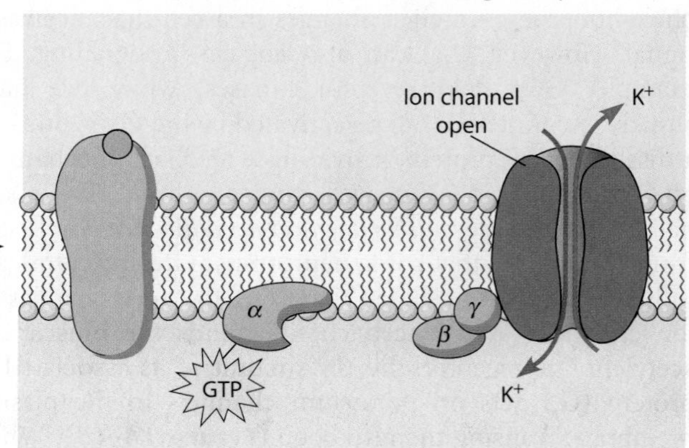

2 Binding of Ach results in activation of the G protein. The βγ subunit interacts with a K+ ion channel, causing it to open.
Ion channel open
K+
α
β
γ
GTP
K+

FIGURE 14-15 The $G_{\beta\gamma}$ Subunits of a G Protein Indirectly Regulate Ion Channels in Some Neurons. One type of acetylcholine receptor (the muscarinic receptor) is a G protein-linked receptor. When acetylcholine binds, it activates the $G_{\beta\gamma}$ subunit of G_i, which regulates the opening of potassium channels.

Growth Factors Often Bind Protein Kinase-Associated Receptors

Protein kinase-associated receptors function in many important cellular processes. One well-studied case where they come into play is cell proliferation. For a cell to divide, it must have all the nutrients needed for synthesis of its component parts, but the availability of nutrients is usually not itself sufficient for growth. Cells often also need additional signals to stimulate cell growth. Biologists encountered the requirements for cell growth when they first tried to culture cells in vitro. Although provided with a growth medium rich in nutrients, including the presence of blood plasma, cells would not grow. A turning point came when blood serum was used instead of plasma: Serum was able to support the growth of cells, whereas plasma would not. Many of the messengers present within the serum have now been purified, and they are members of various classes of proteins known as **growth factors.**

The difference between blood serum and blood plasma held an important clue about growth factors. *Plasma* is whole blood, including unreacted *platelets* (which contain clotting components) but without the red and white blood cells. *Serum* is the clear fluid remaining after blood has clotted. During clotting, platelets secrete growth factors into the blood that stimulate the growth of cells called *fibroblasts,* which form the new connective tissue that makes up a scar. After clotting, the resulting serum is full of *platelet-derived growth factor (PDGF).* Plasma does not contain this factor because the clotting reaction has not taken place.

We now know that the receptor for PDGF is a receptor tyrosine kinase. In fact, several growth factors act by stimulating receptor tyrosine kinases, including *insulin, insulin-like growth factor-1, fibroblast growth factor, epidermal growth factor,* and *nerve growth factor.* Many other types of growth factors have also been isolated. A small sampling is shown in **Table 14-3**, which also lists some cell types affected by each factor and the general class of molecule that serves as a receptor for each growth factor.

Although collectively known as growth factors, the proteins that activate tyrosine kinase and other types of receptors function in many diverse events. These include growth and cell division as well as crucial events during the development of embryos, responses to tissue injury, and many other activities. We will examine the effects of growth factors on cell division in Chapter 19 and their role in cancer in Chapter 24. For now, all we need to recognize is that growth factors are secreted molecules that act at short range and have specific effects on cells possessing the appropriate receptor to sense the presence of the growth factor.

Receptor Tyrosine Kinases Aggregate and Undergo Autophosphorylation

Many **receptor tyrosine kinases (RTKs)** trigger a chain of signal transduction events inside the cell that ultimately lead to cell growth, proliferation, or the specialization of cells. Examples of RTKs include the insulin receptor, the nerve growth factor receptor, and the epidermal growth factor (EGF) receptor (**Figure 14-16**).

The Structure of Receptor Tyrosine Kinases. Receptor tyrosine kinases differ structurally from G protein-linked receptors in many ways. These receptors often consist of a single polypeptide chain with only one transmembrane segment. Within this polypeptide chain are several distinct domains (Figure 14-16a). The extracellular portion of the receptor contains the *ligand-binding domain.* The other end of the peptide protrudes through the plasma membrane into the cytosol. On the cytosolic side, a portion of the receptor forms the tyrosine kinase. The cytosolic portion of the receptor contains tyrosine residues that are in fact themselves targets for the tyrosine kinase portion of the receptor.

The tyrosine kinase is frequently an integral part of the receptor protein. In some cases, however, the receptor and the tyrosine kinase are two separate proteins, and the tyrosine kinase is then referred to as a *nonreceptor tyrosine kinase.* However, it can bind to the receptor and be activated when the receptor binds its ligand, so the net effect is quite similar to the activation of a typical receptor tyrosine kinase.

Nonreceptor tyrosine kinases were, in fact, the first tyrosine kinases to be discovered. The first nonreceptor

Table 14-3 Examples of Growth Factor Families

Growth Factor	Target Cells	Type of Receptor Complex
Epidermal growth factor (EGF)	Wide variety of epithelial and mesenchymal cells	Tyrosine kinase
Transforming growth factor α (TGFα)	Same as EGF	Tyrosine kinase
Platelet-derived growth factor (PDGF)	Mesenchyme, smooth muscle, trophoblast	Tyrosine kinase
Transforming growth factor β (TGFβ)	Fibroblastic cells	Serine-threonine kinase
Fibroblast growth factor (FGF)	Mesenchyme, fibroblasts, many other cell types	Tyrosine kinase
Interleukin-2 (IL-2)	Cytotoxic T lymphocytes	Complex of three subunits
Colony-stimulating factor-1 (CSF-1)	Macrophage precursors	Tyrosine kinase
Wnts	Many types of embryonic cells	Frizzled (seven-pass protein)
Hedgehogs	Many types of embryonic cells, melanocytes	Patched (seven-pass protein)

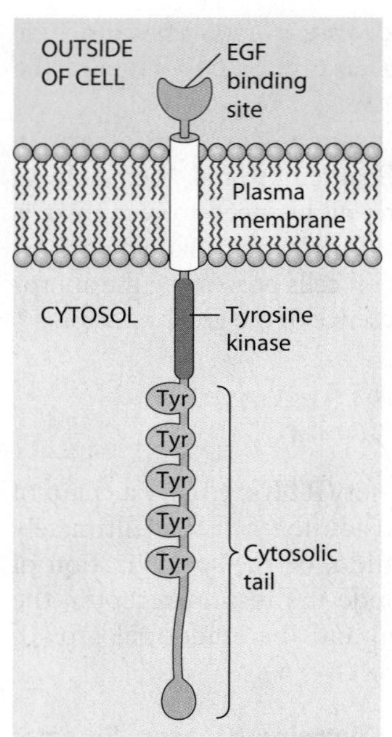

(a) Structure of the epidermal growth factor (EGF) receptor

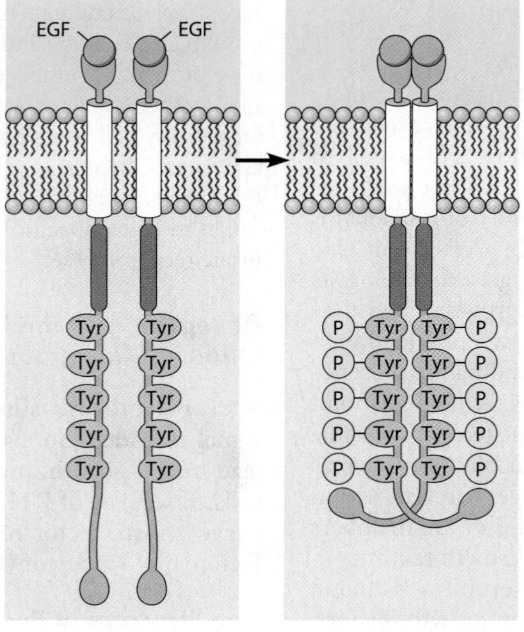

(b) Activation of the EGF receptor

FIGURE 14-16 The Structure and Activation of a Receptor Tyrosine Kinase. (a) The receptor for epidermal growth factor (EGF), shown here, is typical of many receptor tyrosine kinases (RTKs). These receptors often have only one transmembrane segment. The extracellular portion of the receptor binds to the ligand (EGF in this case). Inside the cell, a portion of the receptor has tyrosine kinase activity. The remainder of the receptor contains a series of tyrosine residues that are substrates for the tyrosine kinase. (b) The activation of RTKs starts with ligand binding, causing receptor clustering. Once the receptors aggregate, they cross-phosphorylate each other at a number of tyrosine amino acid residues. The formation of tyrosine phosphate (Tyr-P) residues on the receptor creates binding sites for cytosolic proteins that contain SH2 domains.

tyrosine kinase identified was the Src protein, which is encoded by the *src* gene of the avian sarcoma virus. One form of Src can transform normal cells into cancer cells, a topic we will learn more about in Chapter 24. Many other nonreceptor tyrosine kinases have been discovered, and most of them are similar in structure to the Src protein.

The Activation of Receptor Tyrosine Kinases. Signal transduction is initiated when a ligand binds, causing the receptor tyrosine kinases to aggregate (Figure 14-16b). In many of the best-understood cases, two receptor molecules cluster together within the plasma membrane when they bind ligand. Once the receptors cluster in this way, the tyrosine kinase associated with each receptor phosphorylates the tyrosines of neighboring receptors. Since the receptors phosphorylate other receptors of the same type, this process is referred to as **autophosphorylation.**

Receptor Tyrosine Kinases Initiate a Signal Transduction Cascade Involving Ras and MAP Kinase

Receptor tyrosine kinases use a different method of initiating signal transduction from that seen with G protein-linked receptors. Many of the key events in RTK signaling were identified using mutants in model organisms such as *Drosophila* and *Caenorhabditis elegans,* a topic we explore in **Box 14B**. Here, we present an outline of this important pathway.

Once autophosphorylation of tyrosine residues on the cytosolic portion of the receptor occurs in response to ligand binding (❶, **Figure 14-17**), the receptor recruits a number of cytosolic proteins to interact with itself. Each of these proteins binds to the receptor at a phosphorylated tyrosine residue. To bind to the receptor, each cytosolic protein must contain a stretch of amino acids that recognizes the phosphotyrosine and a few neighboring amino acids on the receptor. The portion of a protein that recognizes one of these phosphorylated tyrosines is called an **SH2 domain.** The term *SH2,* for Src homology (domain) 2, was originally used because proteins with SH2 domains have sequences of amino acids that are strikingly similar to a portion of the Src protein.

Recruitment of different SH2 domain-containing proteins activates different signal transduction pathways. As a result, receptor tyrosine kinases can activate several different signal transduction pathways at the same time. These include the inositol-phospholipid-calcium second messenger pathway, which we have already discussed (Figure 14-12), and the Ras pathway, which ultimately activates the expression of genes involved in growth or development.

Ras is important in regulating the growth of cells. Unlike the heterotrimeric G proteins associated with G protein-linked receptors, Ras comprises a single subunit. Ras and other *small monomeric G proteins* are important signaling molecules. Like other G proteins, Ras can be bound to either GDP or GTP, but it is active only when bound to GTP. In the absence of receptor stimulation, Ras is normally in the GDP-bound state. For Ras to become active, it must release GDP and acquire a molecule of GTP. For this to take place, Ras needs the help of another type of protein called a **guanine-nucleotide exchange factor (GEF).**

The GEF that activates Ras is **Sos** (so called because it was originally identified from a genetic mutation in fruit flies called *son of sevenless* that results in the failure of cells in the compound eye to develop properly; Box 14B). For Sos to become active, it must bind indirectly to the receptor tyrosine kinase through another protein, called *GRB2*

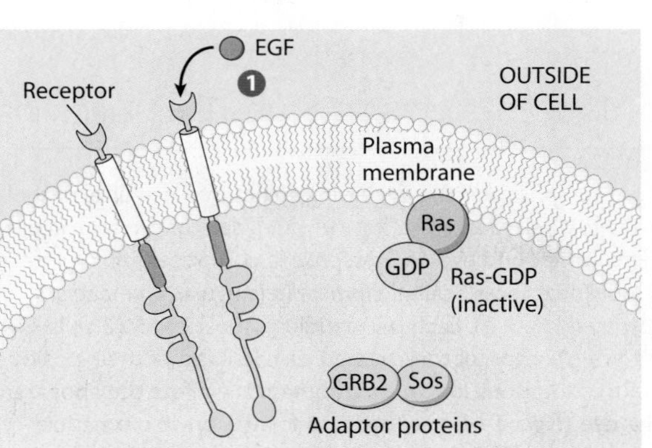

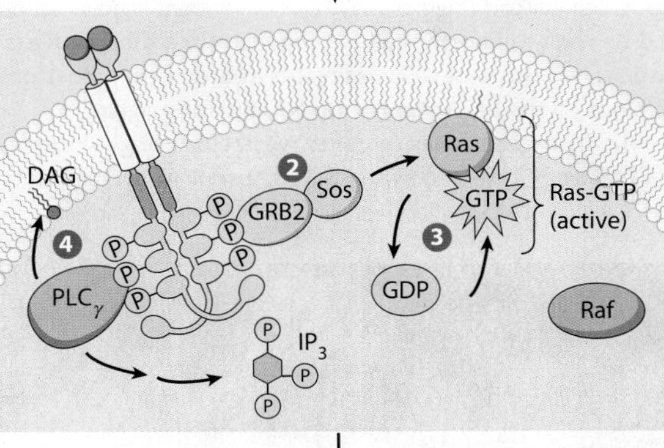

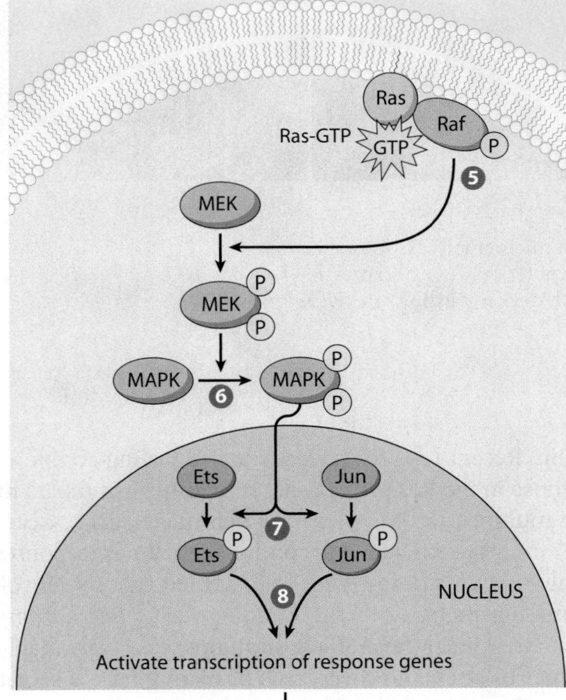

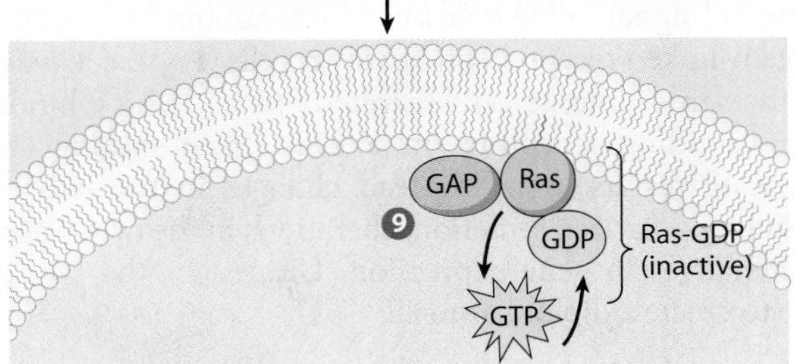

FIGURE 14-17 Signal Transduction Through Receptor Tyrosine Kinases. ❶ Upon ligand binding, receptor tyrosine kinases, such as epidermal growth factor (EGF), aggregate and ❷ undergo autophosphorylation. Once a receptor is phosphorylated at tyrosine residues in its cytosolic tail, proteins with SH2 domains such as GRB2 bind to the receptor, recruiting Sos, a GEF for Ras. ❸ Sos then causes the activation of Ras by helping it release GDP and acquire GTP. ❹ At the same time, the binding of phospholipase Cγ to the activated receptor results in its activation and the cleavage of PIP_2 into IP_3 and DAG. ❺ Activated Ras initiates a cascade of phosphorylation events, beginning with the protein Raf, which catalyzes the phosphorylation of MEK. Ultimately, ❻ MAP kinase (MAPK) becomes phosphorylated. ❼ Activated MAPK catalyzes phosphorylation of proteins such as Ets and Jun, which act in the nucleus, along with other proteins, to ❽ regulate the transcription of response genes. ❾ Ras is inactivated by hydrolysis of its bound GTP, a step facilitated by a GAP.

MB Signal Transduction via Receptor Tyrosine Kinases

(❷ Figure 14-17), which contains an SH2 domain. Thus, to activate Ras, the receptor becomes tyrosine phosphorylated, and GRB2 and Sos form a complex that binds to the receptor, activating Sos. Sos then stimulates Ras to release GDP and acquire GTP, which converts Ras to its active state (❸, Figure 14-17).

Once Ras is active, it triggers a series of phosphorylation reactions (❺, Figure 14-17). The first protein in this cascade is a protein kinase called *Raf.* Activated Raf in turn phosphorylates serine and threonine residues in a protein kinase known as *MEK.* MEK can then phosphorylate threonines and tyrosine residues in a class of proteins known as **mitogen-activated protein kinases**, or **MAP kinases (MAPKs;** ❻, Figure 14-17). MAPKs are activated when cells receive a stimulus to grow and divide (such a signal is sometimes called a *mitogen,* hence the name of the kinase). One function of MAPKs is to phosphorylate nuclear proteins, known as transcription factors, that regulate gene expression (❼, Figure 14-17). One such nuclear protein is called *Jun,* a component of the AP-1 transcription factor, along with proteins that are members of the *Ets* family of transcription factors. These proteins in turn regulate the expression of genes whose protein products are needed for cells to grow and divide (❽, Figure 14-17).

Once Ras is in its active state, it must be inactivated by hydrolysis of the GTP bound to it to avoid continued stimulation of the Ras pathway. GTP hydrolysis is facilitated by a **GTPase activating protein (GAP**; ❾, Figure 14-17). GAPs can accelerate inactivation of Ras a hundredfold.

Receptor Tyrosine Kinases Activate a Variety of Other Signaling Pathways

Receptor tyrosine kinases, like G protein-linked receptors, can also activate phospholipase C (❹, Figure 14-17). We have already seen that the activation of phospholipase C leads to production of IP_3 and DAG and that IP_3 releases calcium from intracellular stores. However, phospholipase C_γ (activated by receptor tyrosine kinases) is different from phospholipase C_β (activated by the G protein-linked receptors) in that it contains an SH2 domain and must bind to the receptor. Once it binds to the receptor,

BOX 14B TOOLS OF DISCOVERY

Using Genetic Model Systems to Study Cell Signaling

Modern cell biologists use a variety of experimental tools to study key cell biological events, including genetic analysis of mutants that show defects during specific signaling events. The use of genetic model systems, including the budding yeast, *Saccharomyces cerevisiae*, the nematode worm *Caenorhabditis elegans*, and the fruit fly, *Drosophila melanogaster*, revolutionized our ability to connect the functions of specific proteins with important signaling pathways. For example, studies in budding yeast provided key information regarding how the cell cycle is controlled (see Chapter 19). Here, we focus on how genetic model systems have uncovered key aspects of an important cellular pathway: the receptor tyrosine kinase (RTK) pathway.

Receptor Tyrosine Kinase Signaling and the Fly Eye

Genetic analysis of signal transduction downstream of RTKs was conducted at about the same time in two systems: the compound eye of *Drosophila* and the egg-laying structures, or vulva, of the nematode, *C. elegans*. Here, we will consider the *Drosophila* eye.

The compound eye of *Drosophila* is composed of roughly 800 individual "eyes" called *ommatidia* (singular, *ommatidium*; **Figure 14B-1a**). Each ommatidium consists of 22 cells. Eight of these are photoreceptors, named on the basis of their position (R1–R8). Mutations in Ras can dramatically affect the appearance of the eye (Figure 14B-1c) because formation of ommatidia depends on an RTK encoded by the *sevenless* gene (abbreviated *sev*). Loss of Sev results in loss of the R7 cell, which differentiates as a cone cell instead (Figure 14B-1d). Subsequent studies identified a ligand on the neighboring cell, R8, that activates the Sevenless receptor, called Bride of sevenless (abbreviated Boss). Boss protein on R8 binds to the Sev receptor on R7, initiating signaling.

Results from strong *sev* mutants were clear: unlike normal eyes (**Figure 14B-2a**), when Sev is absent no R7 cell forms

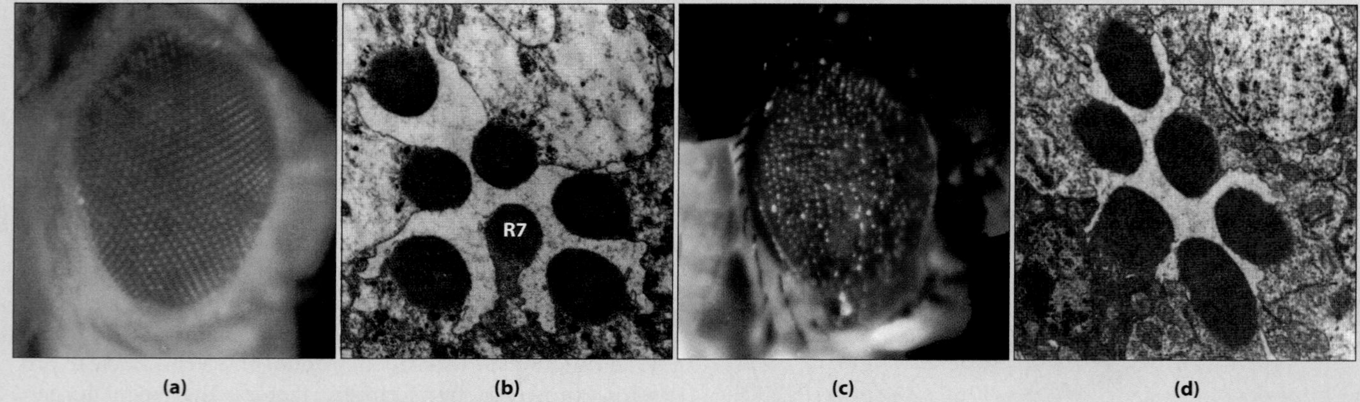

FIGURE 14B-1 The Compound Eye of *Drosophila*. (**a**) A normal eye (light micrograph). (**b**) Cross section of a normal ommatidium at high magnification (TEM). The R7 cell lies in the center. (**c**) An eye expressing a dominant, activated form of Ras appears rough. (**d**) In a *sevenless* mutant, which does not express the RTK required for R7 to differentiate properly, no R7 cell is present.

phospholipase C_γ is phosphorylated by the receptor tyrosine kinase and becomes active.

In addition, receptor tyrosine kinases can activate other enzymes, such as *phosphatidylinositol 3-kinase (PI 3-kinase)*, which phosphorylates the plasma membrane phospholipid phosphatidylinositol. This enzyme is important in regulating cell growth and cell movement via its action on various phosphatidylinositides. The roles of this kinase are diverse and complex. Later in this chapter, we will examine one role of PI 3-kinase during insulin signaling.

Scaffolding Complexes Can Facilitate Cell Signaling

We have seen that receptor tyrosine kinases can stimulate a series of phosphorylation reactions initiated by the active Ras protein. Recent research suggests that signaling components like those in the Ras pathway are sometimes assembled into large multiprotein complexes, which make such cascading reactions more efficient and confine the cellular responses to signals to a small area. One well-studied example involves the mating pathway of the budding yeast, *Saccharomyces cerevisiae*. Under favorable conditions, yeast are haploid. In times of stress, however, cells of mating type *a* secrete a chemical signal called *a factor*, which can bind to specific G protein-linked receptors on nearby α cells (**Figure 14-18a**). At the same time, α cells secrete α *factor*, which binds to corresponding receptors on a cells. Such *mating factor* signaling results in widespread changes in the two cells, including polarized secretion, alterations in the cytoskeleton, and changes in gene expression. Ultimately, the two cells fuse to create a diploid a/α cell.

(Figure 14B-2b). To identify additional components of the *sev* pathway, conditions were found in which just enough functional Sev receptor was present to allow R7 to develop (geneticists call conditions like this a "sensitized background"). Mutations were then isolated in which eye defects were caused only when the flies also carried this sensitizing defect in Sev (Figure 14B-2c). Researchers reasoned that such mutations would affect the same pathway that Sev/Boss activates. By using this approach, they identified several important components of RTK signaling. In conjunction with biochemical studies on cultured cells, these pioneering studies identified proteins essential for RTK signaling.

In other studies, the effects of multiple mutations in the *sev* pathway were examined. For example, in double mutants that lack functional Sevenless but also carry a dominant mutation in Ras that results in reduced GTPase activity (such a mutation makes Ras *constitutively active;* i.e., it is overstimulated), R7 differentiates, even though it cannot receive the signal from R8 (Figure 14B-2d). This provides evidence that the *sev* pathway normally works by activating Ras. When Ras is activated on its own, the events at the cell surface are no longer necessary. Through these and similar studies in *C. elegans,* signaling pathways downstream of RTK signaling were clarified.

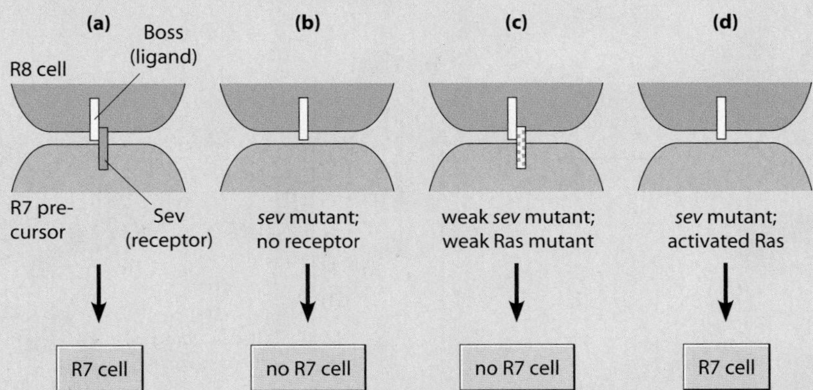

FIGURE 14B-2 Mutations in the Ras Pathway Can Either Bypass or Enhance Defects in RTK Signaling. (**a**) In normal larvae, Boss ligand on the surface of the R8 cell activates the Sev receptor on the adjacent R7 precursor cell, leading to its normal differentiation. (**b**) In *sev* mutants, lack of the Sev receptor results in failure of the R7 precursor cell to receive the signal from R8, and the cell does not become an R7 cell. (**c**) When the fly lacks functional Sev protein, but the R7 cell expresses a dominant, active form of Ras, a normal R7 cell can form. (**d**) When Sev contains a weak mutation that normally allows an R7 cell to form but the R7 precursor cell also carries a weak mutation in Ras, no R7 cell is produced.

As we have already seen, mating factor signaling is mediated through the activated $G_{\beta\gamma}$ subunit of a G protein. $G_{\beta\gamma}$ recruits a large scaffolding protein, known as *Ste5,* to the plasma membrane. Ste5 increases the efficiency of signal transduction through the mating pathway by assembling the kinases involved into a large complex. In this case, a specific phosphorylation cascade results in the phosphorylation of MAPK (Figure 14-18b). MAPK in turn phosphorylates proteins that result in changes in gene expression that are important for mating. Other similar scaffolding complexes have been identified in yeast that regulate responses to changes in osmolarity, nutrients, and other environmental signals. Scaffolding complexes are not unique to yeast. Neuronal synapses, for example, contain multiprotein complexes; it is likely that many signaling pathways exploit this same strategy. One important class of proteins that binds numerous signaling molecules and regulates their functions in multiprotein complexes is a class of proteins known as *14-3-3 proteins.*

Dominant Negative Mutant Receptors Are Important Tools for Studying Receptor Function

Growth factor signaling is important for many cellular events, as we have seen. But how can we identify which receptors are important for particular events? One way is through the use of genetics (Box 14B). However, another way is by introducing mutations into the receptor to study the effects. For example, consider **fibroblast growth factor (FGFs)** and their receptor tyrosine kinases, the *fibroblast growth factor receptors (FGFRs).* FGFs and their receptors

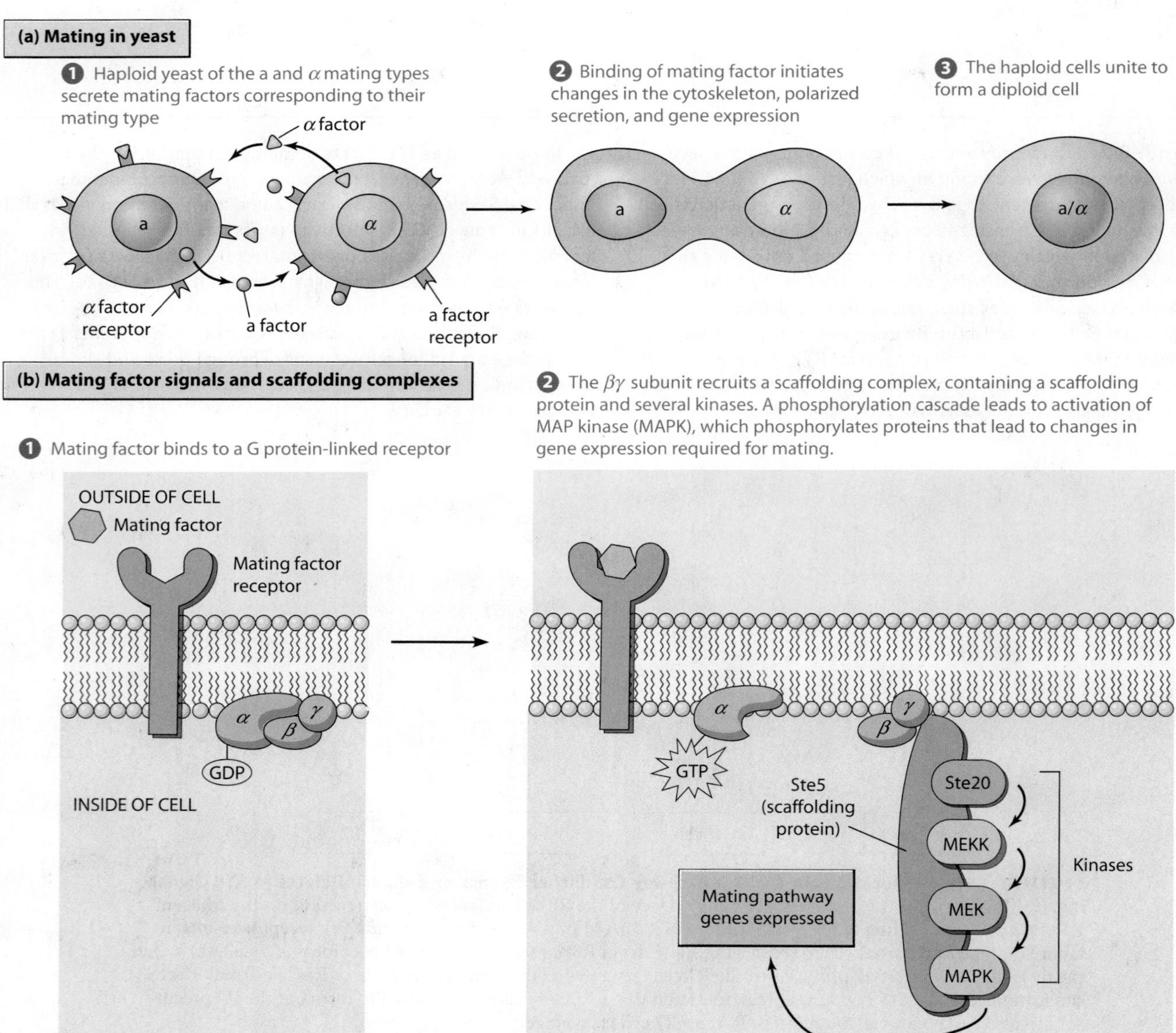

(a) Mating in yeast
1 Haploid yeast of the a and α mating types secrete mating factors corresponding to their mating type
α factor
a
α
α factor receptor
a factor
a factor receptor
2 Binding of mating factor initiates changes in the cytoskeleton, polarized secretion, and gene expression
a
α
3 The haploid cells unite to form a diploid cell
a/α
(b) Mating factor signals and scaffolding complexes
1 Mating factor binds to a G protein-linked receptor
OUTSIDE OF CELL
Mating factor
Mating factor receptor
α
γ
β
GDP
INSIDE OF CELL
2 The βγ subunit recruits a scaffolding complex, containing a scaffolding protein and several kinases. A phosphorylation cascade leads to activation of MAP kinase (MAPK), which phosphorylates proteins that lead to changes in gene expression required for mating.
α
γ
β
GTP
Ste5 (scaffolding protein)
Ste20
MEKK
MEK
MAPK
Kinases
Mating pathway genes expressed

FIGURE 14-18 Scaffolding Complexes Can Facilitate Cell Signaling. **(a)** Mating in yeast involves the exchange of mating factors by haploid cells of different mating types. Mating factor receptors are G protein-linked receptors. **(b)** Mating factor signaling involves the recruitment of a scaffolding complex in response to the $G_{\beta\gamma}$ subunit of a G protein associated with mating factor receptors.

mediate many important signaling events in animal embryos and adult tissues. Normal FGFRs, like receptors for epidermal growth factor, undergo autophosphorylation upon ligand binding as shown in **Figure 14-19a**. Some types of mutant FGFRs, however, are capable of binding to normal receptors but do not result in normal autophosphorylation. If a cell produces such mutant receptors, even though the cell makes a substantial quantity of normal, functional receptor, the presence of the mutant receptor prevents the normal receptor from functioning properly. Normal function is inhibited because FGFRs must act together as dimers to bind FGFs. If a normal receptor dimerizes with a mutant receptor, then the phosphorylation events that normally occur within the tyrosine kinase portion of the receptor fail to occur, blocking signal transduction (Figure 14-19b). A mutation that overrides the function of the normal receptor is sometimes called a **dominant negative mutation.** Such mutations are dominant because the mutant receptor "dominates" over any normal receptor that may be present. They are "negative" because they result in mutant receptors that no longer function. It is also possible to perform the converse experiment: Some mutations make FGFRs active in signaling, even when no ligand is present. These are known as **constitutively active mutations** because they make the receptor act as if it is always "on."

Dominant negative mutations can dramatically affect growth and development in vertebrates. For example, when

(a) Normal FGF receptor. Normal receptors form dimers after binding FGF, and transmit the appropriate signal via Ras and MAPK.

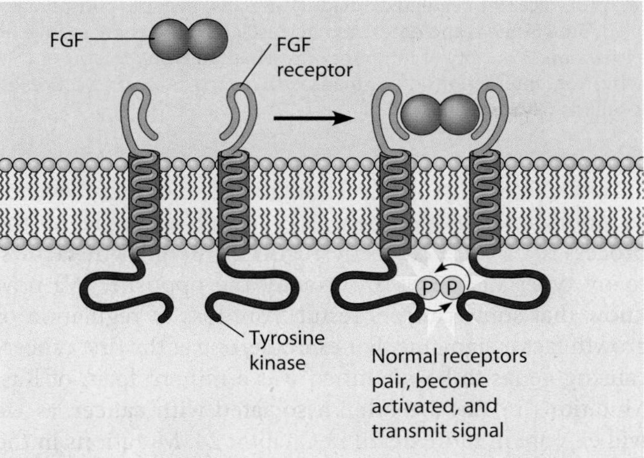

(b) Dominant negative FGF receptor. When a cell makes mutant FGF receptors, normal receptors can dimerize as in (a), or the defective receptors can bind to FGF and dimerize with normal receptors. In this case, no signal is transmitted. When sufficient quantities of the mutant receptor are present, most of the normal receptors are paired with mutant receptors, resulting in overall disruption in signaling.

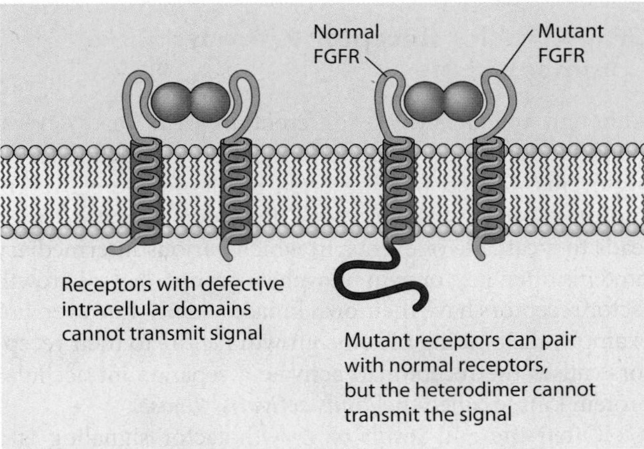

FIGURE 14-19 Dominant Negative Disruption of FGF Receptor Function. Dominant negative mutations in receptors can be used to study the effects of disruption of a growth factor receptor, even in the presence of normal receptors. The example shown here involves fibroblast growth factor (FGF) receptors.

genetically engineered dominant negative FGFRs are expressed in frog embryos, the embryos fail to develop tissues in the trunk and tail, resulting in "tadpoles" with heads but no bodies! In humans, dominant mutations in the transmembrane portion of the *FGFR-3* gene result in the most common form of dwarfism, known as *achondroplasia.* Heterozygous individuals have abnormal bone growth, in which the long bones suffer from abnormal ossification (i.e., the process by which cartilage is converted to bone during childhood). A related condition known as *thanatophoric dysplasia* often results from a single amino acid change in the cytosolic portion of the FGFR-3 protein. In this case, more severe bone abnormalities result, and affected individuals die soon after birth. The dramatic effect of a single amino acid change in a single protein demonstrates the key role that growth factor receptors play during human development.

Other Growth Factors Transduce Their Signals via Receptor Serine-Threonine Kinases

We have seen that when receptor tyrosine kinases bind ligand, they activate a set of signal transduction events within the cell that results in changes within the cell receiving the signal. Another major class of protein kinase-associated receptors uses a very different set of signal transduction pathways to elicit changes within the cell. These receptors phosphorylate serine and threonine residues rather than tyrosines. One major class of **serine-threonine kinase receptors** comprises a family of proteins that bind members of the **transforming growth factor β (TGFβ)** family of growth factors. This growth factor family regulates a wide range of cellular functions in both embryos and adult animals, including cell proliferation, programmed cell death, the specialization of cells, and key events in embryonic development.

The first step in TGFβ signaling is the binding of growth factor by the transmembrane receptor (❶ and ❷, **Figure 14-20**). In general, TGFβ family members bind to two types of receptors within the receiving cell: the *type I* and *type II receptors.* Some of the TGFβ family members form dimers with one another before binding to the appropriate receptors. When ligand is bound, the type II receptor phosphorylates the type I receptor (Figure 14-20). The type I receptor then initiates a signal transduction cascade within the cell, phosphorylating a class of proteins known as **Smads** (a name coined from two of the founding members of this class of proteins). There are three types of Smads. Those that are phosphorylated by a complex of anchoring proteins and activated receptors are known as receptor-regulated, or *R-Smads* (❸, Figure 14-20). Another Smad, *Smad4,* forms a multiprotein complex with phosphorylated R-Smads. When Smad4 molecules bind R-Smads, the entire complex can move into the nucleus, where it can associate with other cofactors and DNA-binding proteins to regulate gene expression (❹, Figure 14-20). Still other Smads act at various points in TGFβ signaling to inhibit the pathway; one type can bind to and inhibit receptors, while another can bind and inhibit Smad4. Smad signals are terminated when the R-Smad is degraded or the R-Smad moves back into the cytosol. Smad4 can shuttle from the nucleus back to the cytosol to be reused when the cell receives another signal.

Disruption of Growth Factor Signaling Can Lead to Cancer

We have seen that growth factors regulate events such as cell proliferation, cell movement, and gene expression. We have also seen that embryonic development requires

1 In the absence of TGFβ, the type I and type II receptors for TGFβ are not clustered or phosphorylated. R-Smads and Smad4 are in the cytosol.

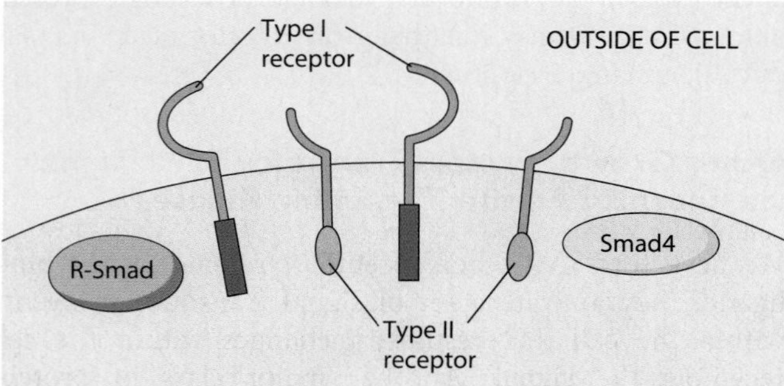

2 Binding of TGFβ results in clustering of type I and type II receptors, followed by phosphorylation of type I receptors by type II receptors.

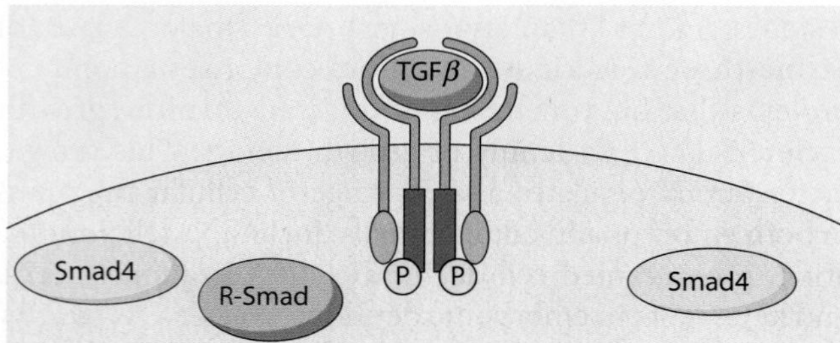

3 The activated type I receptors bind a complex of an anchoring protein and an R-Smad, resulting in R-Smad phosphorylation.

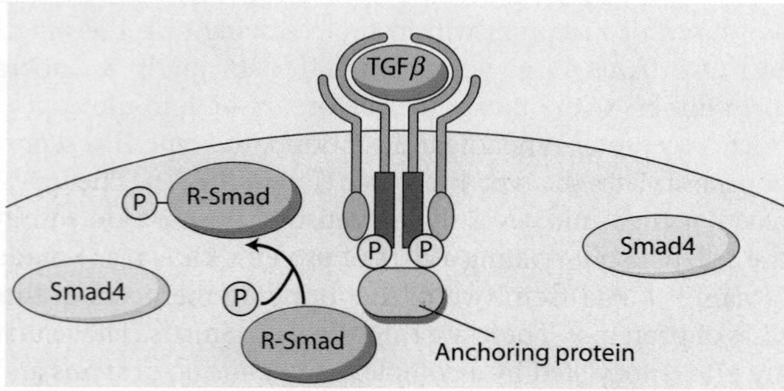

4 The phosphorylated R-Smad binds Smad4, and the complex enters the nucleus. Along with other proteins, they activate or repress gene expression. Eventually, the R-Smad is degraded or leaves the nucleus, and Smad4 returns to the cytosol, terminating the signal.

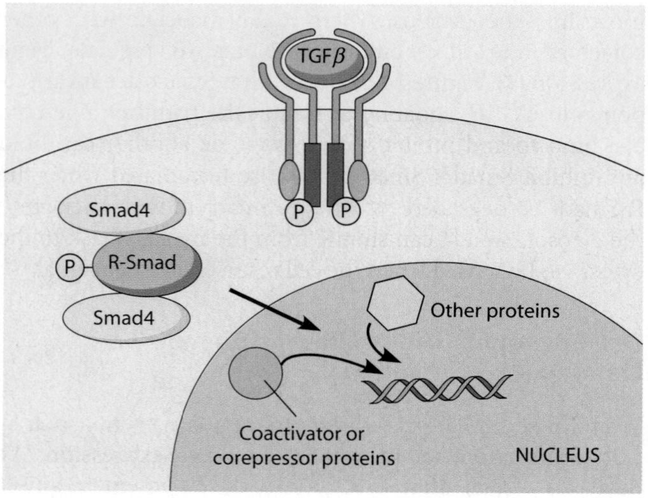

FIGURE 14-20 Signal Transduction by TGFB Receptor Family Proteins. TGFb receptors transduce their signals to the nucleus via proteins called Smads. After ligand binding activates the receptor, receptor-regulated Smads (R-Smads) become phosphorylated, bind Smad4, and enter the nucleus, leading to changes in gene expression. A variety of inhibitory proteins can block receptor activation, interactions of R-Smads with receptors, or interaction of R-Smads with Smad4.

growth factors to tightly regulate cell biological events. If embryonic development is a prime example of the processes that are correctly regulated by growth factors, many types of cancer are exactly the opposite. We now know that some cancers result from loss of regulation of growth factor signaling. For example, one of the first cancer-causing genes to be identified was a mutant form of Ras. Mutations in Ras are often associated with cancer, as we will explore in more detail in Chapter 24. Mutations in the epidermal growth factor receptor can result in breast cancer, glioblastoma (a cancer of glial cells in the brain), and fibrosarcoma (a type of cancer of long bones). Similarly, mutations in the TGFβ type I receptor occur in one-third of ovarian cancers, and mutations in the type II receptor occur in many colorectal cancers. Mutations in Smad4 occur in one-half of all pancreatic cancers.

Growth Factor Receptor Pathways Share Common Themes

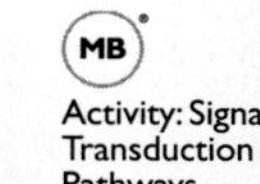

Activity: Signal Transduction Pathways

Although the details are different in each case, several common themes are involved in most growth factor signaling events. Ligand binding often results in the activation and/or clustering of receptors. Receptor activation then leads to a cascade of events, in which various intermediary proteins often lose or gain phosphate groups. Not all growth factor receptors have their own kinase activity, however. For example, the binding of some growth factors to their receptors causes the receptor to activate a separate intracellular protein kinase called the *Janus activated kinase.*

Often the end result of growth factor signaling is a change in one or more proteins that enter the nucleus, resulting in changes in gene expression within the cell receiving the signal. The specific response of the cell receiving such signals depends on the history of the cell and on the combination of signals the cell is being exposed to at any given time. In the next section, we will see that the same logic operates during the long-range signaling mediated by hormones.

Hormonal Signaling

Growth factors operate at short range to regulate cellular function. However, large organisms must also be able to coordinate the function of various cells and tissues over long distances. To do so, plants and animals use secreted chemical signals called **hormones**. In both plants and animals, hormones are often transported via the vasculature. Hundreds of different hormones regulate a wide range of physiological functions, including growth and development, rates of body processes, concentrations of substances, and responses to stress and injury. A few are listed in **Table 14-4**.

Although we can consider them as a group based on their regulatory functions, hormones differ in many ways. Both plants and animals produce a wide array of hormones. Some hormones are steroids or other hydrophobic molecules that are targeted to intracellular receptors. Other hormones, such as the adrenergic hormones discussed later in this section, are targeted to a wide variety of G protein-linked receptors. Still others, such as insulin, are ligands for receptor tyrosine kinases. Plants produce steroid hormones called *brassinosteroids* that regulate leaf growth. Similarly, the organic molecule *ethylene* regulates ripening of fruit, and *abscisic acid* causes the closing of stomata during drought conditions. However, because animal hormones are so much better understood, we will focus on them for the remainder of our discussion. In particular, we will focus on endocrine hormones.

Hormones Can Be Classified by the Distance They Travel and by Their Chemical Properties

Endocrine hormones travel by means of the circulatory system from the cells where they are released to other cells where they regulate one or more specific functions. Endocrine hormones are synthesized by the *endocrine tissues* of the body and are secreted directly into the bloodstream. Once secreted into the circulatory system, endocrine hormones have a limited life span, ranging from a few seconds for epinephrine (a product of the adrenal gland) to many hours for insulin. As they circulate in the bloodstream, hormone molecules come into contact with receptors in tissues throughout the body. A tissue that is specifically affected by a particular hormone is called a *target tissue* for that hormone (**Figure 14-21**). For example, the heart and the liver are target tissues for epinephrine, whereas the liver and skeletal muscles are targets for insulin.

Hormones can also be classified according to their chemical properties. Chemically, the endocrine hormones fall into four categories: amino acid derivatives, peptides, proteins, and lipid-like hormones such as the steroids. An example of an amino acid derivative is epinephrine, derived from tyrosine. *Antidiuretic hormone* (also called *vasopressin*) is an example of a peptide hormone, whereas insulin is a protein. *Testosterone* is an example of a steroid hormone. The steroid hormones are derivatives of cholesterol (see Figure 3-30) that are synthesized either in the gonads (the *sex hormones*) or in the adrenal cortex (the *corticosteroids*).

Control of Glucose Metabolism Is a Good Example of Endocrine Regulation

Since hormones modulate the function of particular target tissues, an important aspect of studying hormones is understanding the specific functions of those target tissues. To illustrate how endocrine hormones act, we will look more closely at the **adrenergic hormones** epinephrine and norepinephrine. (Epinephrine is also called adrenaline; the two words are of Greek and Latin derivation, respectively, and mean "above, or near, the kidney," referring to the location in the body of the *adrenal glands,* which synthesize this hormone.) The overall strategy of adrenergic hormone actions is to put many of the normal bodily functions on hold and to deliver vital resources to the heart and skeletal muscles instead, as well as to produce a heightened state of alertness. When secreted into the bloodstream, epinephrine and norepinephrine stimulate changes in many different

Table 14-4 **Chemical Classification and Function of Hormones**

Chemical Classification	Example	Regulated Function
Endocrine Hormones		
Amino acid derivatives	Epinephrine (adrenaline) and norepinephrine (both derived from tyrosine)	Stress responses: regulation of heart rate and blood pressure; release of glucose and fatty acids from storage sites
	Thyroxine (derived from tyrosine)	Regulation of metabolic rate
Peptides	Antidiuretic hormone (vasopressin)	Regulation of body water and blood pressure
	Hypothalamic hormones (releasing factors)	Regulation of tropic hormone release from pituitary gland
Proteins	Anterior pituitary hormones	Regulation of other endocrine systems
Steroids	Sex hormones (androgens and estrogens)	Development and control of reproductive capacity and secondary sexual characteristics
	Corticosteroids	Stress responses; control of blood electrolytes
Paracrine Hormones		
Amino acid derivative	Histamine	Local responses to stress and injury
Arachidonic acid derivatives	Prostaglandins	Local responses to stress and injury

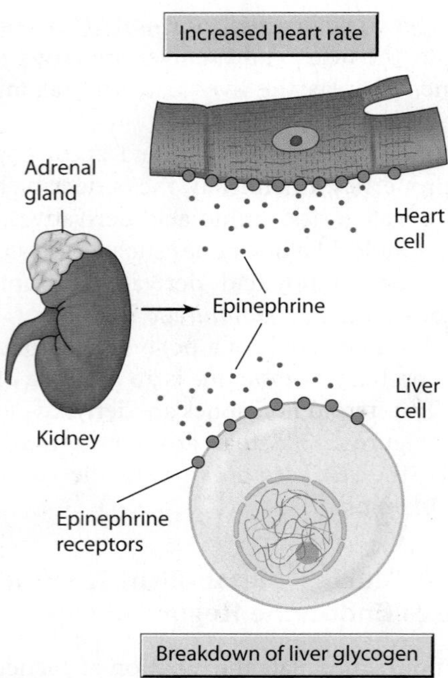

FIGURE 14-21 Target Tissues for Endocrine Hormones. Cells in a target tissue have hormone-specific receptors embedded in their plasma membranes (or, in the case of the steroid hormones, present in the nucleus or cytosol). Heart and liver cells can respond to epinephrine synthesized by the adrenal glands because these cells have epinephrine-specific receptors on their outer surfaces. A specific hormone may elicit different responses in different target cells. Epinephrine causes an increase in heart rate but stimulates glycogen breakdown in the liver.

tissues or organs, all aimed at preparing the body for dangerous or stressful situations (the so-called fight-or-flight response). Overall, the adrenergic hormones trigger increased cardiac output, shunting blood from the visceral organs to the muscles and the heart, and cause dilation of arterioles to facilitate oxygenation of the blood. In addition, these hormones stimulate the breakdown of glycogen to supply glucose to the muscles.

Adrenergic hormones bind to a family of G protein-linked receptors known as **adrenergic receptors**. They can be broadly classified into α- and *β-adrenergic receptors.* The α-adrenergic receptors bind both epinephrine and norepinephrine. These receptors are located on the smooth muscles that regulate blood flow to visceral organs. The β adrenergic receptors bind epinephrine much better than norepinephrine. These receptors are found on smooth muscles associated with arterioles that feed the heart, smooth muscles of the bronchioles in the lungs, and skeletal muscles.

The α- and β-adrenergic receptors stimulate different signal transduction pathways because they are linked to different G proteins. For example, the G proteins activated by one type of α-adrenergic receptor, the α_1-adrenergic receptors, are G_q proteins, whereas the β-adrenergic receptors activate G_s. As we discussed earlier, activation of G_s stimulates the cAMP signal transduction pathway, leading to relaxation of certain smooth muscles. Activation of G_q stimulates phospholipase C, leading to the production of IP_3 and DAG, which in turn elevates intracellular calcium levels. As a specific example of cAMP-mediated regulation, we will consider the control of glycogen degradation by the hormone epinephrine in liver or muscle cells.

Intracellular Effects of Adrenergic Hormones: Control of Glycogen Degradation. One action of the adrenergic hormones is to stimulate the breakdown of glycogen to provide muscle cells with an adequate supply of glucose. The breakdown of glycogen is facilitated by the enzyme *glycogen phosphorylase,* which catalyzes a reaction in which the glycosidic linkage between two glucose subunits undergoes attack by inorganic phosphate. This results in release of a molecule of glucose-1-phosphate. The glycogen phosphorylase system was the first cAMP-mediated regulatory sequence to be elucidated. The original work was published in 1956 by Earl Sutherland, who received a Nobel Prize in 1971 for this discovery.

The sequence of events that leads from hormonal stimulation to enhanced glycogen degradation is shown in **Figure 14-22**. It begins when an epinephrine molecule binds to a β-adrenergic receptor on the plasma membrane of a liver or muscle cell. The receptor activates a neighboring G_s protein, and the G_s protein in turn stimulates adenylyl cyclase, which generates cAMP from ATP. The resulting transient increase in the concentration of cAMP in the cytosol activates protein kinase A (❶, Figure 14-22). PKA then activates another cascade of events that begins with the phosphorylation of the enzyme *phosphorylase kinase* (❷, Figure 14-22). This leads to the conversion of *glycogen phosphorylase* from phosphorylase *b,* the less active form, to phosphorylase *a,* the more active form (❸, Figure 14-22), and thus to an increased rate of glycogen breakdown (❹, Figure 14-22).

cAMP also stimulates the inactivation of the enzyme system responsible for glycogen synthesis. PKA also phosphorylates the enzyme *glycogen synthase.* Rather than activating this enzyme, however, phosphorylation inactivates it. Thus the overall effect of cAMP involves both an increase in glycogen breakdown and a decrease in its synthesis.

α_1-Adrenergic Receptors and the IP_3 Pathway. Another important adrenergic pathway is represented by the α_1-adrenergic receptors, which stimulate the formation of IP_3 and DAG. The α_1-adrenergic receptors are found mainly on smooth muscles in blood vessels, including those controlling blood flow to the intestines. When α_1- adrenergic receptors are stimulated, the formation of IP_3 causes an increase in the intracellular calcium concentration. The elevated level of calcium causes smooth muscle contraction, resulting in constriction of the blood vessels and diminished blood flow. Thus, the activation of α_1-adrenergic receptors affects smooth muscle cells in a manner opposite to that of β-adrenergic activation, which causes smooth muscles to relax.

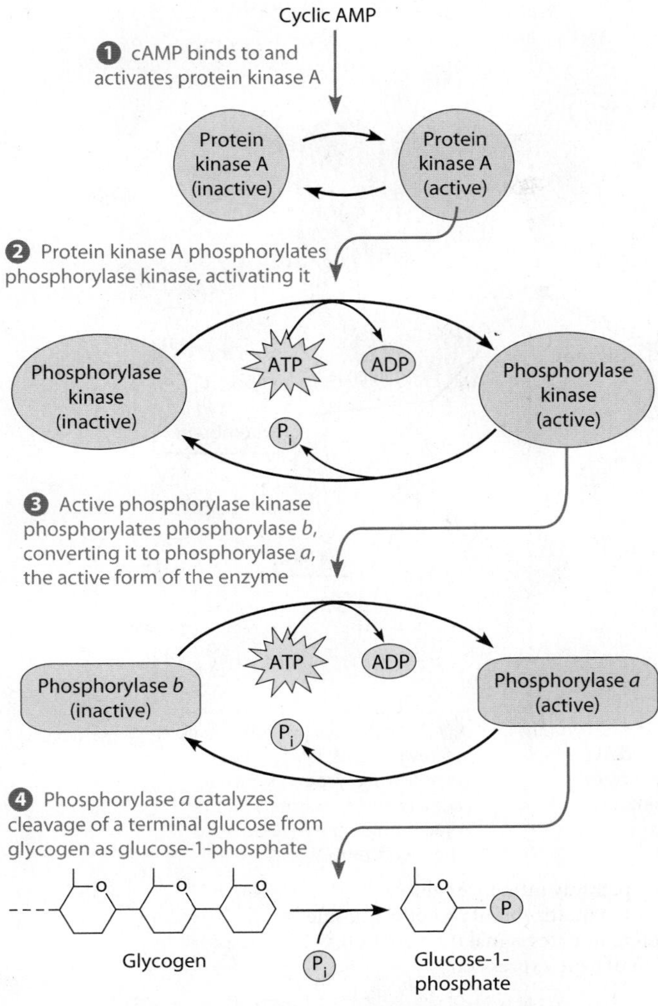

Cyclic AMP
1 cAMP binds to and activates protein kinase A
Protein kinase A (inactive)
Protein kinase A (active)
2 Protein kinase A phosphorylates phosphorylase kinase, activating it
Phosphorylase kinase (inactive)
ATP
ADP
Phosphorylase kinase (active)
Pi
3 Active phosphorylase kinase phosphorylates phosphorylase b, converting it to phosphorylase a, the active form of the enzyme
Phosphorylase b (inactive)
ATP
ADP
Phosphorylase a (active)
Pi
4 Phosphorylase a catalyzes cleavage of a terminal glucose from glycogen as glucose-1-phosphate
Glycogen
Pi
P
Glucose-1-phosphate

FIGURE 14-22 Stimulation of Glycogen Breakdown by Epinephrine. Muscle and liver cells respond to an increased concentration of epinephrine in the blood by increasing their rate of glycogen breakdown. The stimulatory effect of extracellular epinephrine on intracellular glycogen catabolism is mediated by a G protein–cyclic AMP regulatory cascade.

Insulin Signaling Acts Through PI 3-Kinase to Regulate Resting Glucose Levels. We have seen how epinephrine, acting via G protein-mediated signals, regulates metabolic processes—including the production and use of glucose in liver and muscle cells—in stressful situations. During periods of normal activity, however, two peptide hormones, produced by specialized cells in the pancreas known as *islets of Langerhans,* regulate blood glucose levels. One of these, the peptide hormone **glucagon,** acts via the same G_s protein activated by epinephrine, and so it acts to increase blood glucose by the breakdown of glycogen when the level of glucose in the blood is too low. The other hormone, **insulin,** acts in a manner opposite to epinephrine and glucagon, *reducing* blood glucose levels. It does so by stimulating uptake of glucose into muscle and adipose cells and by stimulating glycogen synthesis. A disease with major worldwide consequences is *type I diabetes,* which results in loss of insulin-producing beta cells in the Islets of Langerhans. The discovery that type I diabetics can be successfully treated with insulin led to a Nobel Prize for Frederick Banting and John Macleod in 1923. Unfortunately, insulin treatment is less effective in treating type II diabetes, which has reached epidemic proportions and afflicts hundreds of millions of adults worldwide. Type II diabetes appears to result from resistance to insulin, rather than an inability to produce it.

Insulin has both very rapid and longer-lasting effects on a variety of cells. In muscle and adipose cells, for example, insulin causes uptake of glucose within a few minutes and does not require the synthesis of new proteins. Long-term effects of insulin, such as production of enzymes involved in glycogen synthesis, require higher levels of insulin sustained over many hours. To exert its effects, insulin binds to receptor tyrosine kinases. Unlike the epidermal growth factor receptor, however, the insulin receptor has two α subunits and two β subunits. When insulin binds the receptor, β subunits of the receptor phosphorylate a protein called *insulin receptor substrate 1 (IRS-1).* Phosphorylated IRS-1 can stimulate two different pathways (**Figure 14-23**). First, IRS-1 can recruit GRB2, activating the Ras pathway (❶, Figure 14-23). Second, IRS-1 can bind an enzyme known as phosphatidylinositol 3-kinase (abbreviated as **PI 3-kinase**, or **PI3K**). PI 3-kinase then catalyzes the addition of a phosphate group to the plasma membrane lipid PIP_2 (phosphatidylinositol-4, 5-bisphosphate), which converts PIP_2 into PIP_3 (phosphatidylinositol-3,4,5-trisphosphate; ❷, Figure 14-23). PIP_3 in turn binds to a protein kinase called **Akt** (also called *protein kinase B*), which is activated by phosphorylation of other kinases (❸, Figure 14-23). How much PIP_3 is present is also regulated by enzymes that act to oppose PI 3-kinase. One such protein is **PTEN**, a phosphatase that removes a phosphate group from PIP_3 and thereby prevents the activation of Akt (❸, Figure 14-23).

Activation of Akt has two important consequences (❹, Figure 14-23). First, it leads to movement of a glucose transporter protein known as *GLUT4* from vesicles in the cytosol to the plasma membrane, allowing glucose uptake. Second, Akt can phosphorylate a protein known as *glycogen synthase kinase 3 (GSK3),* reducing its activity. This leads to an increase in the amount of the unphosphorylated, more active form of glycogen synthase, which we have seen enhances the production of glycogen. By stimulating absorption of glucose and its polymerization into glycogen, insulin signaling is thus a key component of glucose homeostasis.

Steroid Hormone Receptors Act Primarily in the Nucleus, not the Cell Surface

Not all hormone receptors act at the cell surface. Steroid hormones, an important class of hormones, bind to receptors that act primarily in the nucleus rather than at the cell surface. These **steroid receptor** proteins mediate the

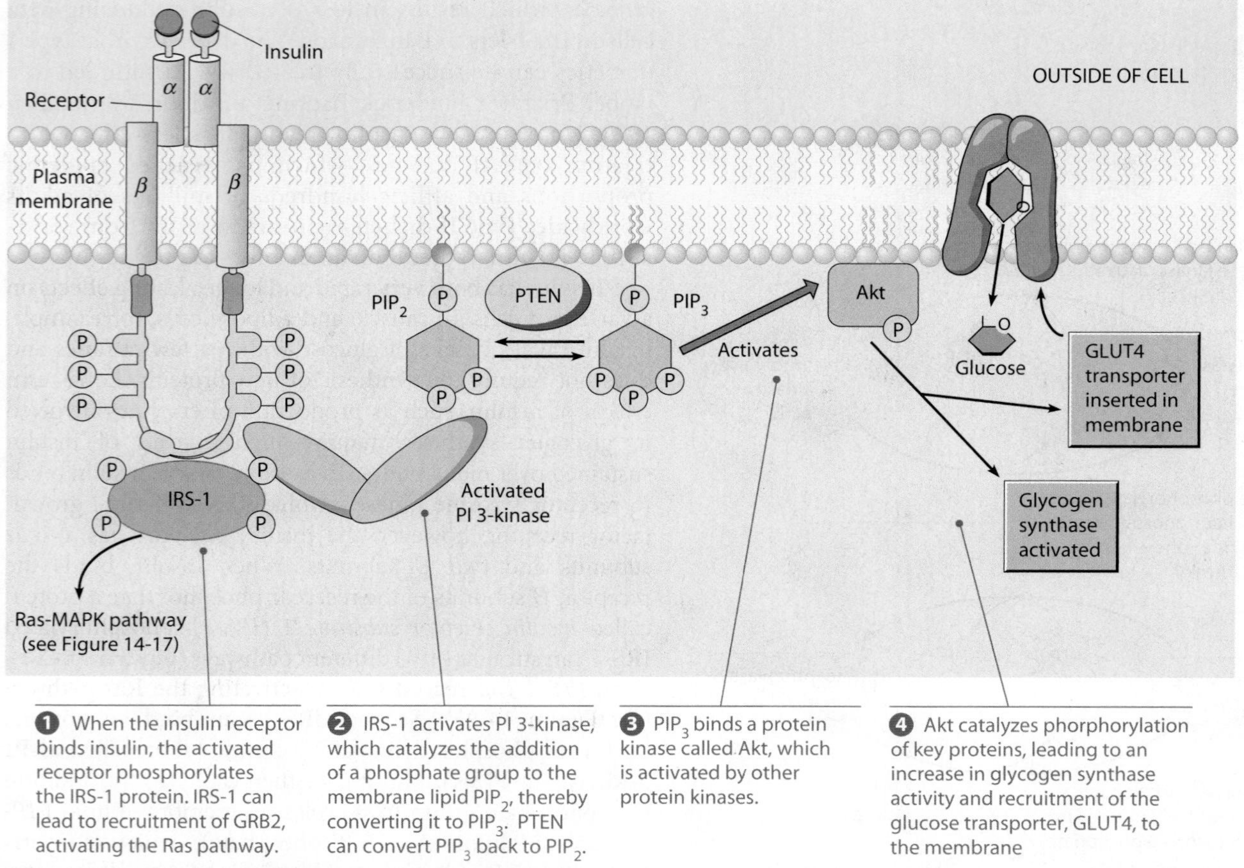
Insulin
Receptor
OUTSIDE OF CELL
Plasma membrane
PIP2
PTEN
PIP3
Akt
Activates
Glucose
GLUT4 transporter inserted in membrane
IRS-1
Activated PI 3-kinase
Glycogen synthase activated
Ras-MAPK pathway (see Figure 14-17)
1 When the insulin receptor binds insulin, the activated receptor phosphorylates the IRS-1 protein. IRS-1 can lead to recruitment of GRB2, activating the Ras pathway.
2 IRS-1 activates PI 3-kinase, which catalyzes the addition of a phosphate group to the membrane lipid PIP2, thereby converting it to PIP3. PTEN can convert PIP3 back to PIP2.
3 PIP3 binds a protein kinase called Akt, which is activated by other protein kinases.
4 Akt catalyzes phorphorylation of key proteins, leading to an increase in glycogen synthase activity and recruitment of the glucose transporter, GLUT4, to the membrane

FIGURE 14-23 The Insulin Signaling Pathway. The insulin signaling pathway influences glucose homeostasis by regulating multiple signaling pathways. The insulin receptor is a multisubunit receptor tyrosine kinase. When it binds insulin, recruitment and activation of the IRS-1 protein initiates signal transduction, leading to glucose import, stimulation of glycogen synthesis, and regulation of gene expression.

actions of steroid hormones such as progesterone, estrogen, testosterone, and glucocorticoids. (Related receptors for thyroid hormone, vitamin D, and retinoic acid work in a similar way.) Steroid hormones are lipid signaling molecules synthesized by cells in endocrine tissues. The hormones are released into the bloodstream and travel throughout the body, transmitting signals to cells that possess the appropriate receptors. After entering a target cell, a steroid hormone binds to its corresponding receptor protein, triggering a series of events that ultimately activates, or in a few cases inhibits, the transcription of a specific set of genes (**Figure 14-24**).

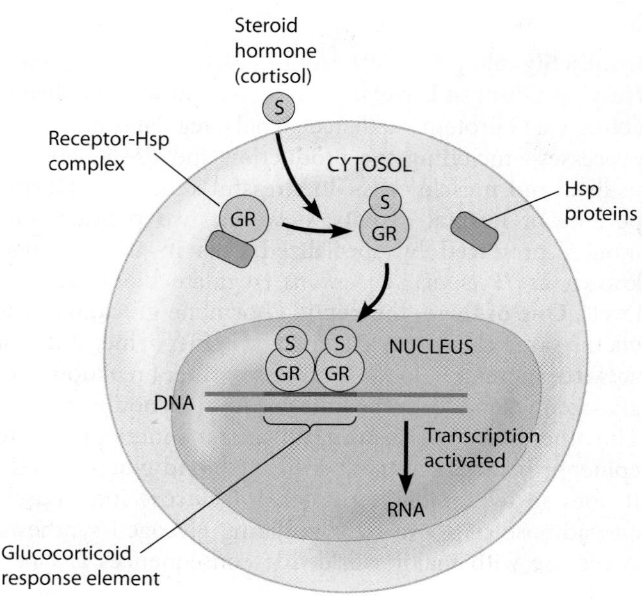
Steroid hormone (cortisol)
Receptor-Hsp complex
CYTOSOL
Hsp proteins
GR
S
NUCLEUS
DNA
Transcription activated
RNA
Glucocorticoid response element

FIGURE 14-24 Activation of Gene Transcription by Steroid Hormone Receptors. The example shown here is the steroid hormone cortisol (S). Cortisol diffuses through the plasma membrane and binds to the glucocorticoid receptor (GR), causing the release of heat-shock proteins (Hsp) and activating the GR molecule's DNA-binding site. The GR molecule then enters the nucleus and binds to a glucocorticoid response element in DNA, which in turn causes a second GR molecule to bind to the same response element. The resulting GR dimer activates transcription of the adjacent gene.

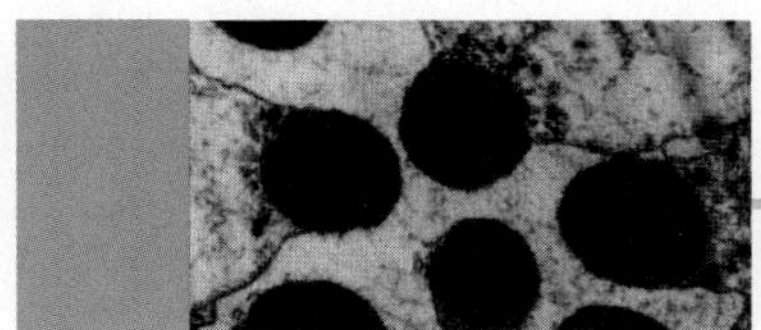

SUMMARY OF KEY POINTS

Principles of Cell Signaling

- Cells use a variety of specific receptors to respond to hormones, growth factors, and other substances (ligands) present in the extracellular fluid. Many receptors are transmembrane proteins.
- Ligand binding is followed by transmission of the signal to the interior of the cell, thereby regulating specific intracellular events. Signal transmission is often carried out by second messengers, which can dramatically amplify signaling responses.
- One ligand can trigger multiple signaling pathways, and cells often integrate multiple signals at any given time. Ultimately, signals must be terminated. Different pathways accomplish this action in different ways.
- Drugs or other chemicals that bind a receptor can be used to stimulate the receptors artificially (agonists) or inhibit them (antagonists).

G Protein Signaling

- A heterotrimeric G protein is activated when a ligand binds to its associated receptor, resulting in exchange of GDP for GTP on the G_α subunit and dissociation of G_α and $G_{\beta\gamma}$. The G_α or $G_{\beta\gamma}$ subunits then activate various signaling pathways. Once G_α hydrolyzes its bound GTP, $G_{\beta\gamma}$ and $G_{\beta\gamma}$ can reassociate. RGS proteins regulate G_α activity by stimulating more rapid hydrolysis of GTP.
- The second messenger cyclic AMP is synthesized when adenylyl cyclase is activated by a G_α protein. cAMP can bind and activate protein kinase A (PKA), which can catalyze phosphorylation of many different proteins. Phosphodiesterase can catalyze cleavage and inactivation of cAMP. Activation of other G protein-linked receptors results in production of cyclic GMP instead.
- Inositol trisphosphate (IP_3) and diacylglycerol (DAG) are produced from phosphatidylinositol bisphosphate when other G proteins activate the enzyme phospholipase C_β.
- $G_{\beta\gamma}$ subunits can activate signaling pathways as well, including those regulating potassium channels in neurons.

Calcium

- Release of Ca^{2+} from internal stores within the endoplasmic reticulum is triggered by IP_3.
- Ca^{2+} effects are often mediated by calcium-binding proteins, such as calmodulin.
- Calcium-induced calcium release allows rapid propagation of calcium signals, such as during fertilization of animal eggs.

Receptor Protein Kinases

- Growth factors regulate cell growth and behavior. Many growth factors bind receptor tyrosine kinases; others bind receptor serine-threonine kinases.
- Receptor tyrosine kinases become phosphorylated on specific tyrosines via autophosphorylation after binding ligand. Phosphorylated receptors recruit SH2 domain-containing proteins, which activate major signal transduction pathways (including the Ras and phospholipase C_γ pathways).
- Ras is a monomeric G protein, regulated by GTPase activating proteins (GAPs) and by the guanine-nucleotide exchange factor (GEF), Sos. Ras activates a cascade of phosphorylation events, which results in activation of transcription factor proteins that regulate gene expression.
- Receptor serine-threonine kinases act via receptor-regulated Smads and their binding partner, Smad4, which enter the nucleus as a complex following receptor activation.
- In some cases, such as the yeast mating pathway, components of signaling pathways are held in close proximity by scaffolding proteins.

Hormones

- Endocrine hormones regulate the activities of body tissues distant from the tissues that secrete them.
- Adrenergic hormones, secreted by the adrenal medulla, bind G protein-linked β-adrenergic receptors, which stimulate the formation of cAMP, and α-adrenergic receptors, which stimulate phospholipase C_β, resulting in elevation of intracellular Ca^{2+}.
- Insulin regulates glucose homeostasis by stimulating multiple signaling pathways, including the Ras and PI 3-kinase pathways.
- Steroid hormones act by binding cytosolic receptor proteins. The hormone-receptor complex then acts in the nucleus to regulate gene expression.

MB **www.masteringbiology.com**

1. MasteringBiology Assignments

Tutorials and Activities

- Molecular Model Structural Tutorials for visualization of 3D structures
- BioFlix Tutorials including 3D animations with hints and detailed feedback for common wrong answers
- Overview of Cell Signaling; Signal Transduction via Receptor Tyrosine Kinases; Signal Transduction Pathways

Questions

- Reading Quiz
- Multiple Choice
- End-of-chapter questions

2. eText Read your book online, search, highlight text, take notes, and more

3. The Study Area

- Chapter Review Quiz, Guide to Techniques and Methods, 3D-Structure Tutorials, Activities, BioFlix, Videos, Biology Review, Word Study Tools, Glossary, Flashcards

PROBLEM SET

More challenging problems are marked with a •.

14-1 Chemical Signals and Second Messengers. Fill in the blanks with the appropriate terms.

(a) Two products of phospholipase C activity that serve as second messengers are _____ and _____.

(b) Calcium ions are released within cells mainly from the _____.

(c) Glucagon is an example of an _____ hormone.

(d) _____ is an intracellular protein that binds calcium and activates enzymes.

(e) A substance that fits into a specific binding site on the surface of a protein molecule is called a _____.

(f) Cyclic AMP is produced by the enzyme _____ and degraded by the enzyme _____.

14-2 Heterotrimeric and Monomeric G Proteins. G protein-linked receptors interact with heterotrimeric G proteins to activate them.

(a) Upon binding to the receptor, the $G_{\beta\gamma}$ subunit catalyzes GDP/GTP exchange by the G_{α} subunit. How is this similar to the activation of Ras by a receptor tyrosine kinase?

(b) What similarities are there at the molecular level between how cells regulate the rate of GTP hydrolysis by the catalytically active portion of a heterotrimeric G protein and by Ras?

14-3 Calcium Chelators and Ionophores. Two important tools that have aided studies of the role of Ca^{2+} in triggering different cellular events are *calcium chelators* and *calcium ionophores.* Chelators are compounds such as EGTA and EDTA that bind very tightly to Ca^{2+} ions, thereby reducing nearly to zero the free (or available) Ca^{2+} ion concentration outside of the cell. Ionophores are compounds that shuttle ions across lipid bilayers and membranes, including the plasma membrane and internal membranes such as the ER. For Ca^{2+} ions, two of the commonly used ionophores are A23187 and ionomycin. Using these tools, describe how you could demonstrate that a hormone exerts its effect by (1) causing Ca^{2+} to enter the cell through channels or (2) releasing Ca^{2+} from intracellular stores such as the ER.

• 14-4 The Eyes Have It. For this question, please consult Box 14B. Recall that each ommatidium in the compound eye contains a photoreceptor cell, R7, which must receive a signal from a neighboring cell, R8, to differentiate. This pathway depends on the Ras signaling pathway. Also, recall that mutants provide insight into this pathway. For each of the following situations, indicate whether the R7 cell would be expected to differentiate. In each case, clearly state your reasoning, based on your understanding of the molecular pathway involved.

(a) R7 cells are mutated so that their Sevenless receptors lack the SH2-binding domain.

(b) R7 cells are mutated so that they lack functional Sos but contain a constitutively active, dominant mutation in Ras.

(c) R7 cells are mutated so that they have no functional MAP kinase.

(d) R8 cells are mutated so that they produce too much of the Bride of sevenless ligand, and R7 cells lack Sevenless.

14-5 Membrane Receptors and Medicine. A patient, who has a stressful job, comes in with high blood pressure. The doctor prescribes beta-blockers to the patient. Explain what beta-blockers are and how these drugs help to bring down blood pressure to normal.

• 14-6 Following Ras activation. It is possible to use engineered molecules to follow the activation of individual Ras molecules in stimulated cells using a technique called fluorescence resonance energy transfer (FRET; see the Appendix for further details). **Figure 14-25** shows the time course of Ras activation following stimulation by epidermal growth factor (EGF).

(a) Based on the graph, how long does it take for maximal Ras activation to be achieved?

(b) Why does Ras activity decline after a few minutes, even when EGF is still present?

14-7 Chemoattractant Receptors on Neutrophils. *Neutrophils* are blood cells normally responsible for killing

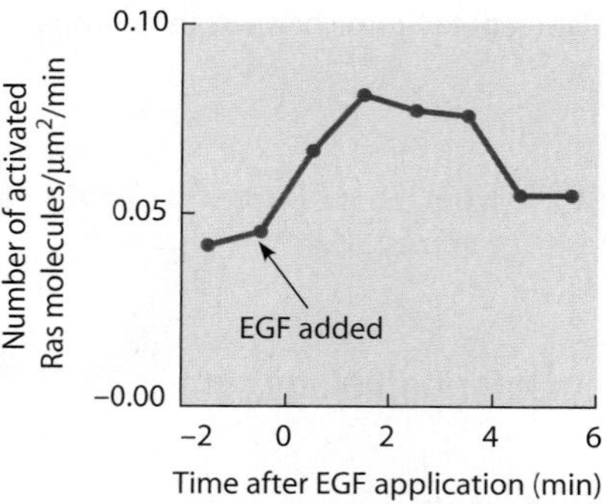

FIGURE 14-25 Activation of Ras following Epidermal Growth Factor (EGF) stimulation. See Problem 14-6.

bacteria at sites of infection. Neutrophils are able to find their way toward sites of infection by the process of *chemotaxis.* In this process, neutrophils sense the presence of bacterial proteins and then follow the trail of these proteins toward the site of infection. Suppose you find that chemotaxis is inhibited by pertussis toxin. What kind of receptor is likely to be involved in responding to bacterial proteins?

14-8 Once Is Enough. Nicotine in tobacco products acts as a stimulant in brain cells and people get quickly addicted to it. However, after a few days, the amount of tobacco products used by the person (grams of tobacco chewed or number of cigarettes smoked per day) increases. Explain why this increase in usage occurs.

• **14-9 Scrambled Eggs.** Unfertilized starfish eggs can be induced to produce foreign proteins either by injecting them with mRNA encoding the protein of choice or they can be directly injected with purified molecules. What do you predict would happen in the following cases, and why?

(a) "Caged calcium," a combination of Ca^{2+} ions and a chelator, is injected into an egg. A flash of light is then used to induce the chelator to release its Ca^{2+} in a small region of the egg.

(b) mRNA for IP_3 receptors that do not allow Ca^{2+} release is injected, and then the egg is fertilized.

(c) Fluo3, a molecule that fluoresces in response to elevated Ca^{2+} levels, is injected into normal eggs, which are then fertilized. In a parallel set of eggs, Fluo3 and EDTA are injected into the eggs. What will happen if the eggs are then exposed to sperms, subsequently?

SUGGESTED READING

References of historical importance are marked with a •.

General Principles of Cell Signaling

Hunter, T. The age of crosstalk: Phosphorylation, ubiquitination, and beyond. *Mol. Cell.* 28 (2007): 730.

Hunter, T. Tyrosine phosphorylation: Thirty years and counting. *Curr. Opin. Cell. Biol.* 21 (2009): 140.

Levitsky, A. *Receptors: A Quantitative Approach.* Menlo Park, CA: Benjamin Cummings, 1984.

Pawson, T. Dynamic control of signaling by modular adaptor proteins. *Curr. Opin. Cell. Biol.* 19 (2007): 112.

G Proteins and Second Messenger Systems

Berridge M. J., M. D. Bootman, and H. L. Roderick. Calcium signalling: Dynamics, homeostasis and remodelling. *Nature Rev. Mol. Cell. Biol.* 4 (2003): 517.

Dupre, D. J., M. Robitaille, R. V. Rebois, and T. E. Hebert. The role of $G_{\beta\gamma}$ subunits in the organization, assembly, and function of GPCR signaling complexes. *Annu. Rev. Pharmacol. Toxicol.* 49 (2009): 31.

Gainetdinov, R. R. et al. Desensitization of G protein-coupled receptors and neuronal functions. *Annu. Rev. Neurosci.* 27 (2004): 107.

• Levitsky, A. From epinephrine to cyclic AMP. *Science* 241 (1988): 800.

• Linder, M. E., and A. G. Gilman. G proteins. *Sci. Amer.* 267 (1992): 56.

Pierce, K. L., R. T. Premont, and R. J. Lefkowitz. Seven-transmembrane receptors. *Nature Rev. Mol. Cell Biol.* 3 (2002): 639.

Steinberg, S. F. Structural basis of protein kinase C isoform function. *Physiol. Rev.* 88 (2008): 1341.

• Sutherland, E. W. Studies on the mechanism of hormone action. *Science* 177 (1972): 401.

• Tang, W. J., and A. G. Gilman. Adenylyl cyclases. *Cell* 71 (1992): 1069.

Receptor Tyrosine Kinases and Growth Factor Signaling

Angers, S., and R. T. Moon. Proximal events in Wnt signal transduction. *Nat. Rev. Mol. Cell Biol.* 10 (2009): 468.

Heldin, C. H., M. Landstrom, and A. Moustakas. Mechanism of TGF-β signaling to growth arrest, apoptosis, and epithelial-mesenchymal transition. *Curr. Opin. Cell Biol.* 21 (2009): 166.

Malumbres, M., and M. Barbacid. RAS oncogenes: The first 30 years. *Nature Rev. Cancer* 3 (2003): 459.

Massague, J., J. Seoane, and D. Wotton. Smad transcription factors. *Genes Dev.* 19 (2005): 2783.

• Rozakis-Adcock, M. et al. The SH2 and SH3 domains of mammalian Grb 2 couple the EGF receptor to the Ras activator mSos1. *Nature* 363 (1993): 83.

Seeliger, M. A., and J. Kuriyan. A MAPK scaffold lends a helping hand. *Cell* 136 (2009): 994.

Physiology and Cell Signaling

Cohen, P. The twentieth century struggle to decipher insulin signalling. *Nature Rev. Mol. Cell Biol.* 7 (2006): 867.

Francis, S., and B. Waldeck. Beta-adrenoceptor agonists and asthma—100 years of development. *Eur. J. Pharmacol.* 445 (2002): 1.

Hadley, M. E., and J. E. Levine. *Endocrinology*, 6e. Upper Saddle River, NJ: Pearson/Prentice Hall, 2007.

Maggi, M. et al. Erectile dysfunction: From biochemical pharmacology to advances in medical therapy. *Eur. J. Endocrinol.* 143 (2000): 143.

Murad, F. Shattuck Lecture. Nitric oxide and cyclic GMP in cell signaling and drug development. *New Engl. J. Med.* 355 (2006): 2003.

Robinson-Rechavi, M., H. Escriva Garcia, and V. Laudet. The nuclear receptor superfamily. *J. Cell Sci.* 116 (2003): 585–586.

Model Systems and Cell Signaling

Hariharan, I. K., and D. A. Haber. Yeast, flies, worms, and fish in the study of human disease. *New Engl. J. Med.* 348 (2003): 2457.

Thomas, B. J., and D. A. Wassarman. A fly's eye view of biology. *Trends Genet.* 15 (1999): 184.

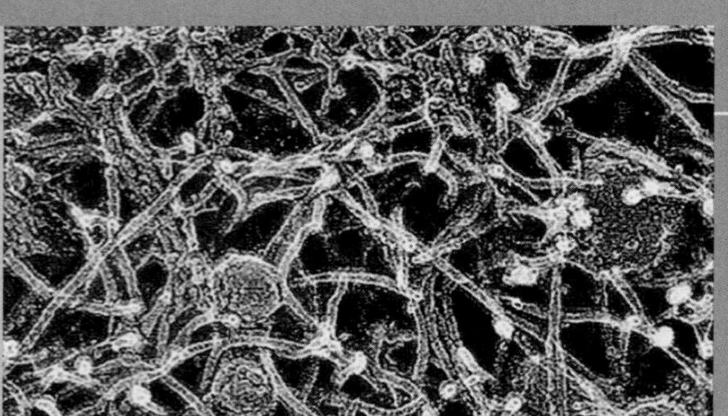

15 Cytoskeletal Systems

See MasteringBiology® for tutorials, activities, and quizzes.

In the preceding chapters, we examined a variety of cellular processes and pathways, many of which occur in the organelles of eukaryotic cells. We also examined signaling events, initiated at the cell surface, that have profound effects on cellular function. We now come to the cytosol, the region of the cytoplasm between and surrounding organelles. Until a few decades ago, the cytosol of the eukaryotic cell was regarded as the generally uninteresting, gel-like substance in which the nucleus and other organelles were suspended. Advances in microscopy and other investigative techniques have revealed that the interior of a eukaryotic cell is highly structured. Part of this structure is provided by the ***cytoskeleton:*** *a complex network of interconnected filaments and tubules that extends throughout the cytosol, from the nucleus to the inner surface of the plasma membrane. The cytoskeleton plays important roles in cell movement and cell division, and in eukaryotes it actively moves membrane-bounded organelles within the cytosol. It also plays a similar role for messenger RNA and other cellular components. The cytoskeleton is also involved in many forms of cell movement and is intimately related to other processes such as cell signaling and cell-cell adhesion. The cytoskeleton is altered by events at the cell surface and, at the same time, appears to participate in and modulate these events.*

The term cytoskeleton *accurately expresses the role of this polymer network in providing an architectural framework for cellular function. It confers a high level of internal organization on cells and enables them to assume and maintain complex shapes that would not otherwise be possible. The name does not, however, convey the dynamic, changeable nature of the cytoskeleton and its critical involvement in a great variety of cellular processes.*

Major Structural Elements of the Cytoskeleton

Eukaryotes Have Three Basic Types of Cytoskeletal Elements

The three major structural elements of the cytoskeleton in eukaryotes are *microtubules, microfilaments,* and *intermediate filaments* (**Table 15-1**). The existence of three distinct systems of filaments and tubules was first revealed by electron microscopy. Biochemical and cytochemical studies then identified the distinctive proteins of each system. The technique of *indirect immunostaining* (see the Appendix, Figure A-12) was especially important in localizing specific proteins to the cytoskeleton.

Each structural element of the cytoskeleton has a characteristic size, structure, and intracellular distribution, and each element is formed by the polymerization of a different kind of subunit (Table 15-1). Microtubules are composed of the protein *tubulin* and are about 25 nm in diameter. Microfilaments, with a diameter of about 7 nm, are polymers of the protein *actin.* Intermediate filaments have diameters in the range of 8–12 nm. Intermediate filament subunits differ depending on the cell type. In addition to its major protein component, each type of cytoskeletal filament has a number of other proteins associated with it. These accessory proteins account for the remarkable structural and functional diversity of cytoskeletal elements.

Bacteria Have Cytoskeletal Systems That Are Structurally Similar to Those in Eukaryotes

Until recently, cytoskeletal proteins were thought to be unique to eukaryotes. However, recent discoveries have shown that bacteria, such as rod-shaped bacteria, and

Table 15-1 Properties of Microtubules, Microfilaments, and Intermediate Filaments

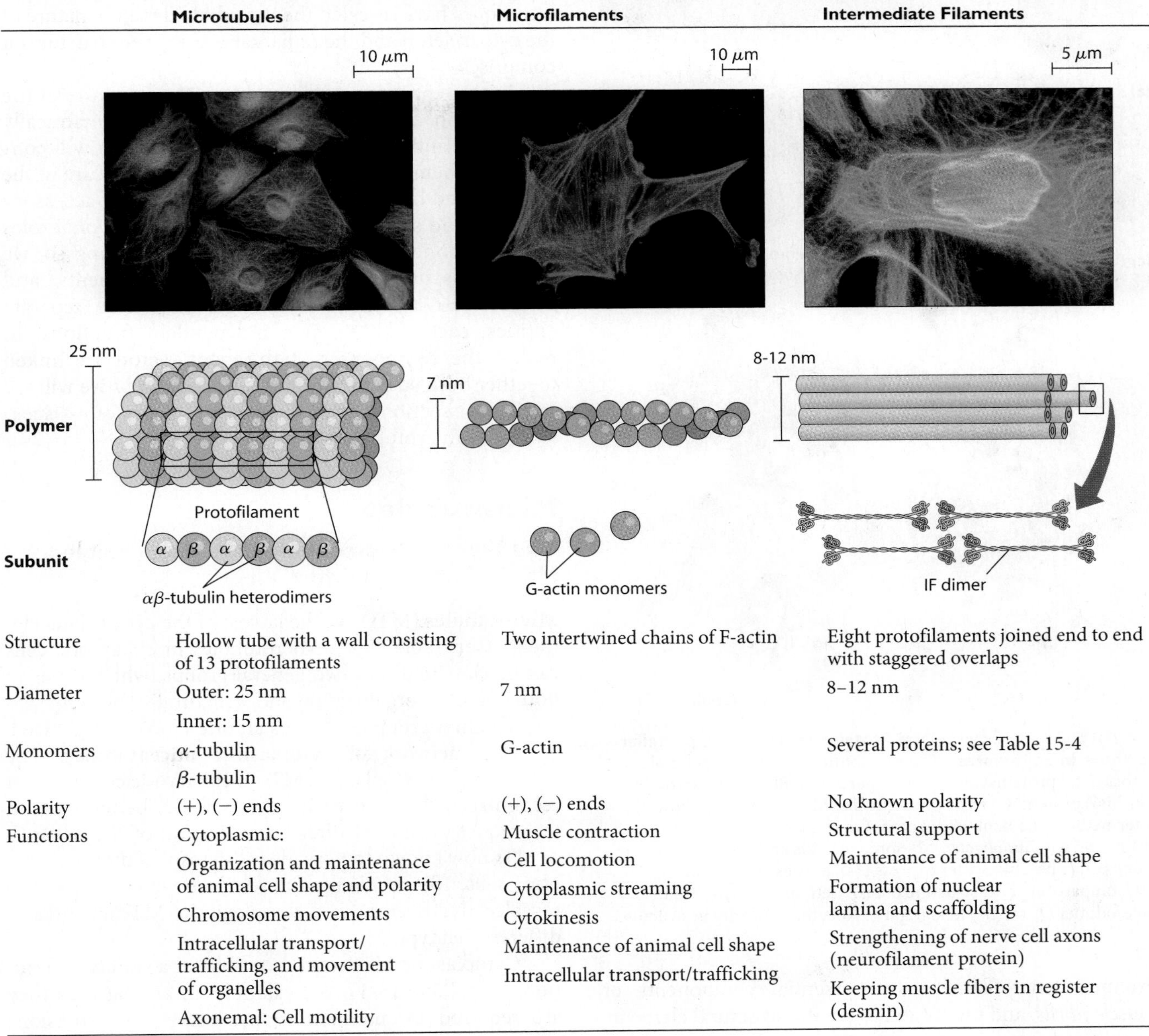

	Microtubules	Microfilaments	Intermediate Filaments
Structure	Hollow tube with a wall consisting of 13 protofilaments	Two intertwined chains of F-actin	Eight protofilaments joined end to end with staggered overlaps
Diameter	Outer: 25 nm Inner: 15 nm	7 nm	8–12 nm
Monomers	α-tubulin β-tubulin	G-actin	Several proteins; see Table 15-4
Polarity	(+), (−) ends	(+), (−) ends	No known polarity
Functions	Cytoplasmic: Organization and maintenance of animal cell shape and polarity Chromosome movements Intracellular transport/trafficking, and movement of organelles Axonemal: Cell motility	Muscle contraction Cell locomotion Cytoplasmic streaming Cytokinesis Maintenance of animal cell shape Intracellular transport/trafficking	Structural support Maintenance of animal cell shape Formation of nuclear lamina and scaffolding Strengthening of nerve cell axons (neurofilament protein) Keeping muscle fibers in register (desmin)

the Archaea have polymer systems that function in a manner very similar to microfilaments, microtubules, and intermediate filaments (**Figure 15-1**). Based on the effects of mutating these proteins, it is clear that they play roles similar to their eukaryotic counterparts. For example, the actin-like *MreB* protein is involved in DNA segregation, the tubulin-like *FtsZ* protein is involved in determining where bacterial cells will divide, and the intermediate filament-like *crescentin* protein is an important regulator of cell shape. Significantly, the FtsZ protein is produced by certain organelles in some eukaryotes, such as chloroplasts and mitochondria, and localizes to sites where these organelles divide. These findings provide further evidence for the endosymbiont theory discussed in Chapter 4. Although the bacterial proteins are not very similar to their eukaryotic counterparts at the amino acid level, X-ray crystallography has shown that when they are assembled into polymers, their overall structure is remarkably similar (Figure 15-1e). Moreover, the equivalent proteins bind the same phosphonucleotides as their eukaryotic equivalents, indicating striking similarities at the biochemical level as well.

The Cytoskeleton Is Dynamically Assembled and Disassembled

The cytoskeleton currently is a topic of great research interest to cell biologists. Microtubules and microfilaments are perhaps best known for their roles in cell motility. For

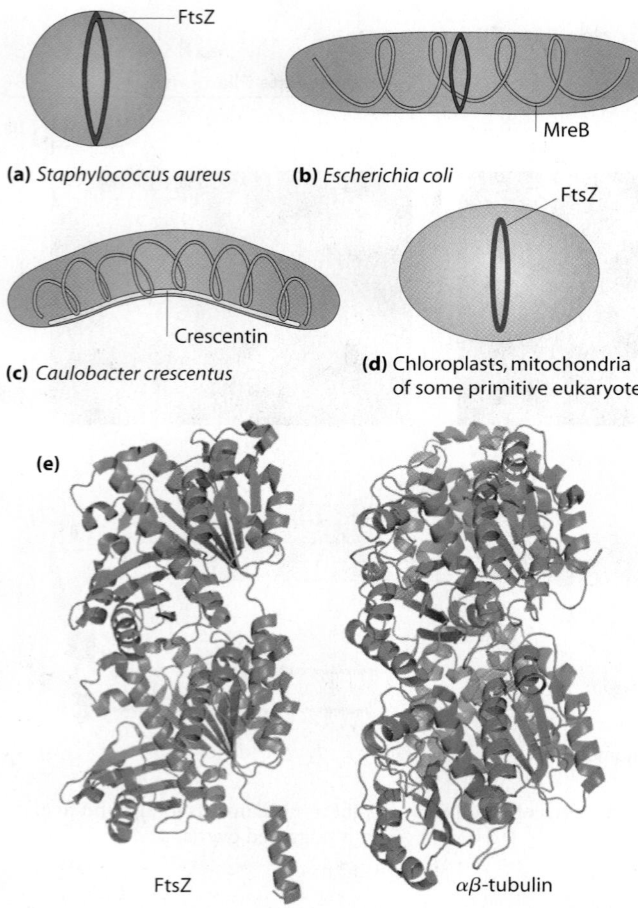

FIGURE 15-1 Cytoskeletal Proteins in Bacteria Are Similar to Those in Eukaryotes. The distribution of several bacterial cytoskeletal proteins are shown in parts a–d. Blue: the microtubule-like FtsZ protein. Orange: the actin-like MreB protein. Yellow: the intermediate-filament-like protein Crescentin. **(a)** *S. aureus.* **(b)** *E. coli.* **(c)** *Caulobacter.* **(d)** Some plastids and mitochondria in some primitive eukaryotes express FtsZ at sites of division. **(e)** Comparison of the structure of FtsZ (left) and an $\alpha\beta$-tubulin heterodimer (X-ray crystallography). Note the similarity in structure.

example, microfilaments are essential components of *muscle fibrils,* and microtubules are the structural elements of *cilia* and *flagella,* appendages that enable certain cells to either propel themselves through a fluid environment or move fluids past the cell. These structures are large enough to be seen by light microscopy and were therefore known and studied long before it became clear that the same structural elements are also integral parts of the cytoskeleton in most cells. With the advent of sophisticated microscopy techniques, it eventually became clear that most cells dynamically regulate where and when specific cytoskeletal structures are assembled and disassembled. Recent progress in understanding cytoskeletal structure relies heavily on a combination of powerful microscopy techniques: various types of *fluorescence microscopy, digital video microscopy,* and various types of *electron microscopy* (**Table 15-2**). Each technique is described in more detail in the Appendix. In addition, specific drugs can be used to perturb cytoskeletal function (**Table 15-3**). In parallel with increasingly sophisticated biochemical studies, these techniques have revealed the incredibly dynamic nature of the cytoskeleton and the remarkably elaborate structures it comprises.

In this chapter, we will focus on the structure of the cytoskeleton and how its components are dynamically assembled and disassembled. In each case, we will consider the chemistry of the subunit(s), the structure of the polymer and how it is polymerized, the role of accessory proteins, and some of the structural and functional roles each component plays within the cell. In doing so, we will be discussing microtubules, microfilaments, and intermediate filaments as though they were separate entities, each with its own independent functions. In reality, the components of the cytoskeleton are linked together both structurally and functionally, as we will see in the last section of this chapter. We begin our discussion with microtubules.

Microtubules

Two Types of Microtubules Are Responsible for Many Functions in the Cell

Microtubules (MTs) are the largest of the cytoskeletal elements (see Table 15-1). Microtubules in eukaryotic cells can be classified into two general groups, which differ in both degree of organization and structural stability.

The first group comprises an often loosely organized, dynamic network of **cytoplasmic microtubules.** The occurrence of cytoplasmic MTs in eukaryotic cells was not recognized until the early 1960s, when better fixation techniques permitted direct visualization of the network of MTs now known to pervade the cytosol of most eukaryotic cells. Since then, fluorescence microscopy has revealed the diversity and complexity of MT networks in different cell types.

Cytoplasmic MTs are responsible for a variety of functions (see Table 15-1). For example, in animal cells they are required to maintain axons, nerve cell extensions whose electrical properties we examined in Chapter 13. Some migrating animal cells require cytoplasmic MTs to maintain their polarized shape. In plant cells, cytoplasmic MTs govern the orientation of cellulose microfibrils deposited during the growth of cell walls (see Chapter 17). Significantly, cytoplasmic MTs form the mitotic and meiotic spindles that are essential for the movement of chromosomes during mitosis and meiosis (see Chapter 19). Cytoplasmic microtubules also contribute to the spatial disposition and directional movement of vesicles and other organelles by providing an organized system of fibers to guide their movement.

The second group of microtubules, **axonemal microtubules,** includes the highly organized, stable microtubules found in specific subcellular structures associated with cellular movement, including cilia, flagella, and the

Table 15-2 Techniques for Visualizing the Cytoskeleton

Technique	Description	Example
Fluorescence microscopy on fixed specimens*	Fluorescent compounds directly bind to cytoskeletal proteins, or antibodies are used to indirectly label cytoskeletal proteins in chemically preserved cells, causing them to glow in the fluorescence microscope.	A fibroblast stained with fluorescent antibodies directed against actin shows bundles of actin filaments.
Live cell fluorescence microscopy*	Fluorescent versions of cytoskeletal proteins are made and introduced into living cells. Fluorescence microscopy and video or digital cameras are used to view the proteins as they function in cells.	Fluorescent tubulin molecules were microinjected into living fibroblast cells. Inside the cell, the tubulin dimers become incorporated into microtubules, which can be seen easily with a fluorescence microscope.
Computer-enhanced digital video microscopy	High-resolution images from a video or digital camera attached to a microscope are computer processed to increase contrast and remove background features that obscure the image.	Two micrographs showing several microtubules were processed to make them visible in detail. (Unenhanced / Enhanced)
Electron microscopy	Electron microscopy can resolve individual filaments prepared by thin section, quick-freeze deep-etch, or direct-mount techniques.	A fibroblast cell is prepared by the quick-freeze deep-etch method. Bundles of actin microfilaments are visible.

*Confocal, deconvolution, multiphoton, and total internal reflection fluorescence (TIRF) microscopy are often used to improve detection of fluorescent signals. See the Appendix for more details.

basal bodies to which these appendages are attached. The central shaft, or *axoneme,* of a cilium or flagellum consists of a highly ordered bundle of axonemal MTs and associated proteins. Given their order and stability, it is not surprising that the axonemal MTs were the first of the two groups to be recognized and studied. We have already encountered an example of such a structure; the axoneme of the sperm tail shown in Figure 4-12 consists of MTs. We will consider axoneme structure and microtubule-mediated motility further in Chapter 16.

Table 15-3 Drugs Used to Perturb the Cytoskeleton

Drug	Source	Affect
Drugs Affecting Microtubules		
Colchicine, colcemid	Autumn crocus, *Colchicum autumnale*	Binds tubulin monomers, inhibiting assembly
Nocadazole	Synthetic benzimidazole	Binds b-tubulin, inhibiting polymerization
Vinblastine, vincristine	Periwinkle plant, *Vinca rosea*	Aggregates tubulin heterodimers
Taxol	Pacific yew tree, *Taxus brevifolia*	Stabilizes microtubules
Drugs Affecting Microfilaments		
Cytochalasin D	Fungal metabolite	Prevents addition of new monomers to plus ends
Latrunculin A	Red sea sponge, *Latrunculia magnifica*	Sequesters actin monomers
Phalloidin	Death cap fungus, *Amanita phalloides*	Binds and stabilizes assembled microfilaments

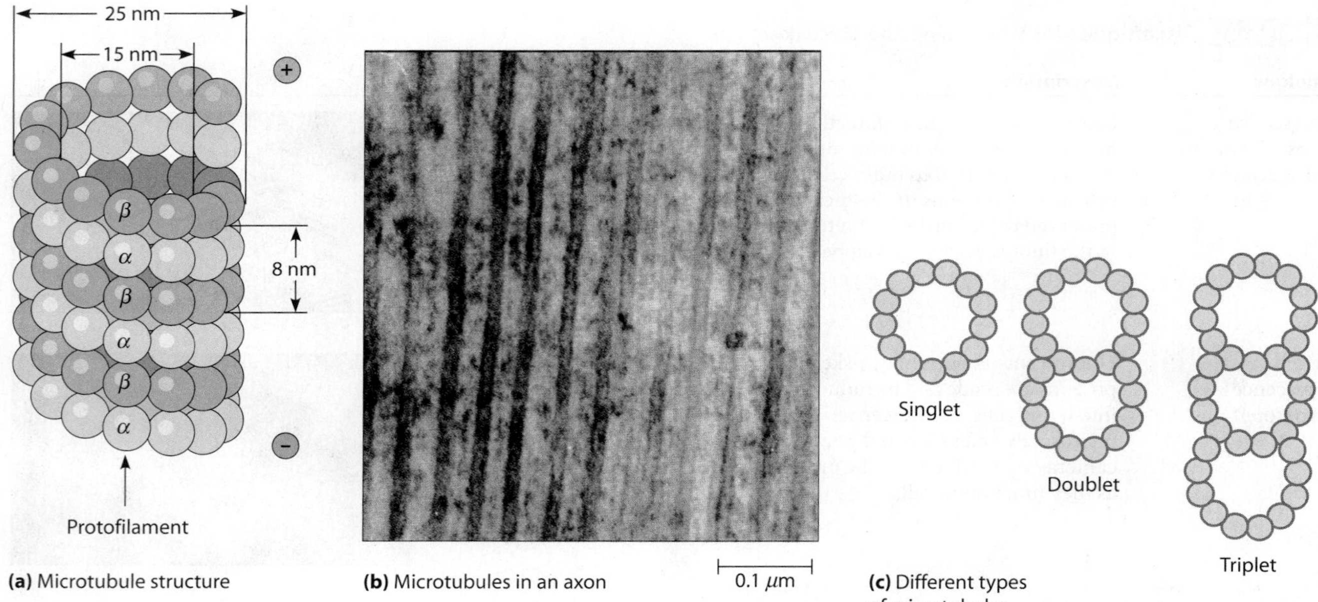

FIGURE 15-2 Microtubule Structure. (a) A schematic diagram showing a microtubule as a hollow cylinder enclosing a lumen. The outside diameter is about 25 nm, and the inside diameter is about 15 nm. The wall of the cylinder consists of 13 protofilaments, one of them indicated by an arrow. A protofilament is a linear polymer of tubulin dimers, each consisting of two polypeptides—α-tubulin and β-tubulin. All heterodimers in the protofilaments have the same orientation, thus accounting for the polarity of the microtubule. **(b)** Microtubules in a longitudinal section of an axon (TEM). **(c)** Microtubules can form as singlets (13 protofilaments around a hollow lumen), doublets, and triplets. Doublets and triplets contain one complete, 13-protofilament microtubule (the *A tubule*) and one or two additional, incomplete tubules (called *B* and *C tubules*) consisting of 10 protofilaments.

Tubulin Heterodimers Are the Protein Building Blocks of Microtubules

MTs are straight, hollow cylinders with an outer diameter of about 25 nm and an inner diameter of about 15 nm (**Figure 15-2**). Microtubules vary greatly in length. Some are less than 200 nm long; others, such as axonemal MTs, can be many micrometers in length. The MT wall consists of longitudinal arrays of linear polymers called **protofilaments.** There are usually 13 protofilaments arranged side by side around the hollow center, or lumen; although some MTs in some animals contain more or less than 13 protofilaments, this number is by far the most common.

As shown in Figure 15-2, the basic subunit of a protofilament is a heterodimer of the protein **tubulin** (for a three-dimensional structure, see Figure 15-1e). The heterodimers that form the bulk of protofilaments are composed of one molecule of **α-tubulin** and one molecule of **β-tubulin.** As soon as individual α- and β-tubulin molecules are synthesized, they bind noncovalently to each other to produce an **$\alpha\beta$-heterodimer** that does not dissociate under normal conditions.

Individual α- and β-tubulin molecules have diameters of about 4–5 nm and molecular weights of 55 kDa. Structural studies show that α- and β-tubulins have nearly identical three-dimensional structures, even though they share only 40% amino acid sequence identity. Each has a GTP-binding domain at the N-terminus, a domain in the middle to which colchicine can bind (colchicine is a MT poison that blocks MT assembly; see below), and a third domain at the C-terminus that interacts with MT-associated proteins (MAPs; we will discuss MAPs later in this chapter).

Within a microtubule, all of the tubulin dimers are oriented in the same direction, such that all of the α-tubulin subunits face the same end. This uniform orientation of tubulin dimers means that one end of the protofilament differs chemically and structurally from the other, giving the protofilament an inherent polarity. Because the orientation of the tubulin dimers is the same for all of the protofilaments in an MT, the MT itself is also a polar structure.

Most organisms have several closely related but nonidentical genes for each of the α- and β-tubulin subunits. These slightly different forms of tubulin are called *tubulin isoforms.* In the mammalian brain, for example, there are five α- and five β-tubulin isoforms. These isoforms differ mainly in the C-terminal domain, which suggests that various tubulin isoforms may interact with different proteins. In addition to different isoforms, tubulin can be chemically modified. For example, acetylated tubulin tends to form more stable MTs than nonacetylated tubulin does.

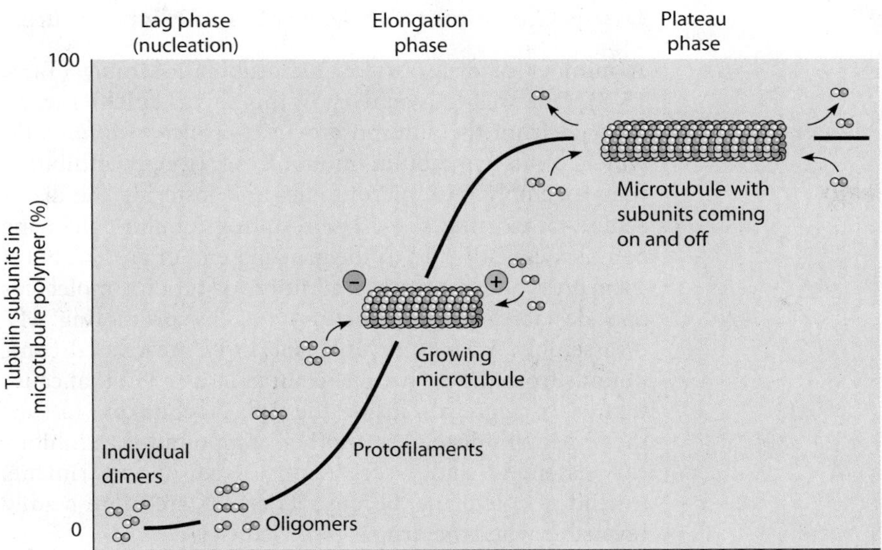

FIGURE 15-3 The Kinetics of Microtubule Assembly In Vitro. The kinetics of MT assembly can be monitored by observing the amount of light scattered by a solution containing GTP-tubulin after it is warmed from 0°C to 37°C. (Microtubule assembly is inhibited by cold and activated upon warming.) Such light-scattering measurements reflect changes in the MT population as a whole, not the assembly of individual microtubules. When measured in this way, MT assembly exhibits three phases: lag, elongation, and plateau. The lag phase is the period of nucleation. During the elongation phase, MTs grow rapidly, causing the concentration of tubulin subunits in the solution to decline. When this concentration is low enough to limit further assembly, the plateau phase is reached, during which subunits are added and removed from MTs at equal rates.

Microtubules Can Form as Singlets, Doublets, or Triplets

Cytoplasmic MTs are simple tubes, or *singlet* MTs, built from 13 protofilaments. Some axonemal MTs are more complex, however: they can contain *doublet* or *triplet* MTs. Doublets and triplets contain one complete, 13-protofilament microtubule (the *A tubule*) and one or two additional, incomplete tubules (called *B* and *C tubules*) consisting of 10 or 11 protofilaments (Figure 15-2c). Doublets are found in cilia and flagella; triplets are found in basal bodies and centrioles. We will examine cilia and flagella in much more detail in Chapter 16.

Microtubules Form by the Addition of Tubulin Dimers at Their Ends

Microtubules form by the reversible polymerization of tubulin dimers. The polymerization process has been studied extensively in vitro; a schematic representation of MT assembly in vitro is shown in **Figure 15-3**. When a solution containing a sufficient concentration of tubulin dimers, GTP, and Mg^{2+} is warmed from 0°C to 37°C, the polymerization reaction begins. (MT formation in the solution can be readily measured with a spectrophotometer as an increase in light scattering.) A critical step in the formation of MTs is the aggregation of tubulin dimers into clusters called *oligomers*. These oligomers serve as "nuclei" from which new microtubules can grow, and hence this process is referred to as **nucleation.** Once an MT has been nucleated, it grows by addition of subunits at either end, via a process called **elongation.**

Microtubule formation is initially slow, a period referred to as the *lag phase* of MT assembly. This period reflects the relatively slow process of MT nucleation. The elongation phase of MT assembly—the addition of tubulin dimers—is relatively fast compared with nucleation. Eventually, the mass of MTs increases to a point where the concentration of free tubulin becomes limiting. This leads to the *plateau phase,* where MT assembly is balanced by disassembly.

Microtubule growth in vitro depends on the concentration of tubulin dimers. The tubulin heterodimer concentration at which MT assembly is exactly balanced with disassembly is called the overall **critical concentration.** MTs tend to grow when the tubulin concentration exceeds the critical concentration and depolymerize when the tubulin concentration falls below the critical concentration.

Addition of Tubulin Dimers Occurs More Quickly at the Plus Ends of Microtubules

The inherent structural polarity of microtubules means that the two ends differ chemically. Another important difference between the two ends of the MT is that one end can inherently grow or shrink much faster than the other. This difference in assembly rate can readily be visualized by mixing the MT-associated structures found at the base of cilia, known as *basal bodies,* with tubulin heterodimers. Assembly of the tubulin heterodimers occurs at both ends, but the MTs grow much faster from one end than the other. (The position of the basal body in the growing MT can be assessed because of its different appearance under the electron microscope; **Figure 15-4.**) The rapidly growing end of the microtubule is called the **plus end,** and the other end is the **minus end.** As we will see below, minus ends of MTs are often anchored at the centrosome; in this case MT dynamics are confined to plus ends.

The different growth rates of the plus and minus ends of microtubules reflect the different critical concentrations required for assembly at the two ends of the MT; the critical concentration for the plus end is lower than that

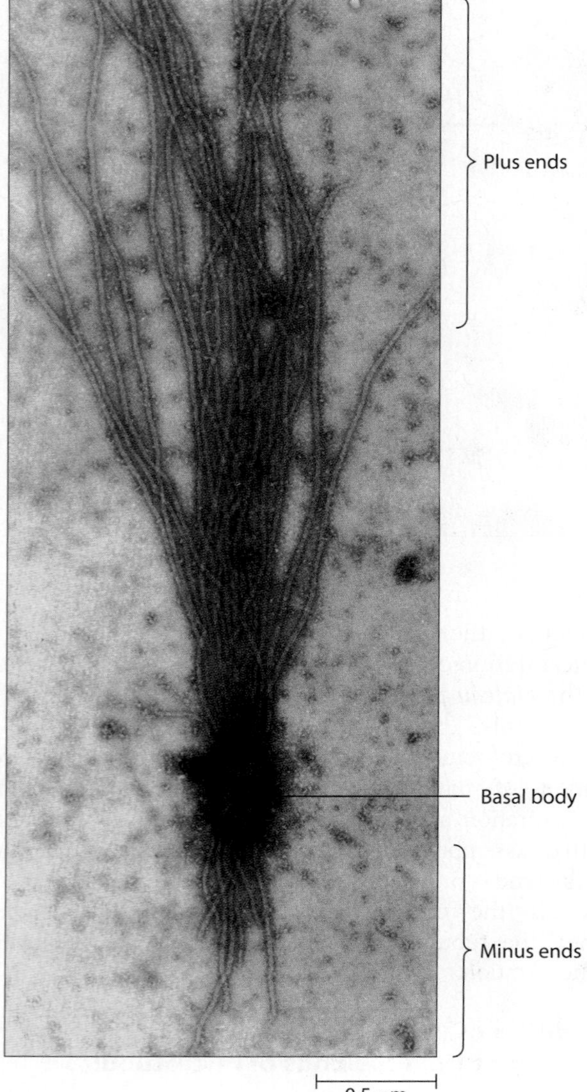

FIGURE 15-4 Polar Assembly of Microtubules in Vitro. The polarity of MT assembly can be demonstrated by adding basal bodies to a solution of tubulin dimers. The tubulin dimers add to the plus and minus ends of the microtubules in the basal body. However, MTs that grow from the plus end are much longer than those growing from the minus end.

for the minus end. If the free tubulin concentration is higher than the critical concentration for the plus end but lower than the critical concentration for the minus end, assembly will occur at the plus end while disassembly takes place at the minus end. This simultaneous assembly and disassembly produces the phenomenon known as *treadmilling* (**Figure 15-5**). Treadmilling arises when a given tubulin molecule incorporated at the plus end is displaced progressively along the MT and eventually lost by depolymerization at the opposite end. By examining fluorescent MTs, treadmilling has been observed in living cells, although it is uncertain how important it is to overall MT dynamics.

Drugs Can Affect the Assembly of Microtubules

A number of drugs affect microtubule assembly (Table 15-3). One well-known drug of this sort is **colchicine,** an alkaloid from the autumn crocus, *Colchicum autumnale,* which binds to tubulin monomers, strongly inhibiting their assembly into microtubules and fostering the disassembly of existing ones. The resulting tubulin-colchicine complex can still add to the growing end of an MT, but it then prevents any further addition of tubulin molecules and destabilizes the structure, thereby promoting MT disassembly. *Vinblastine* and *vincristine* are related compounds from the periwinkle plant *(Vinca rosea)* that cause tubulin to aggregate inside the cell. **Nocodazole** (a synthetic benzimidazole) is another compound that inhibits MT assembly and is frequently used in experiments instead of colchicine, because its effects are more readily reversible when the drug is removed.

These compounds are called *antimitotic drugs* because they disrupt the mitotic spindle of dividing cells, blocking the further progress of mitosis. The sensitivity of the mitotic spindle to these drugs is understandable because the spindle fibers are composed of many microtubules. Indeed, vinblastine and vincristine find application in medical practice as anticancer drugs. They are useful for this purpose because cancer cells divide rapidly and are therefore preferentially susceptible to drugs that interfere with the mitotic spindle.

In contrast, **taxol** (from the Pacific yew tree, *Taxus brevifolia*) binds tightly to microtubules and stabilizes them, causing much of the free tubulin in the cell to assemble into microtubules. Within cells, taxol causes free tubulin to assemble into MTs and arrests dividing cells in mitosis. Thus, both taxol and colchicine block cells in mitosis, but

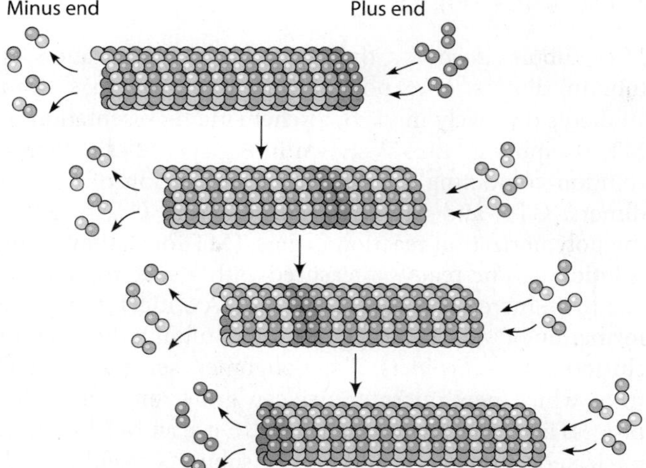

FIGURE 15-5 Treadmilling of Microtubules. Microtubule assembly occurs more readily at the plus end of an MT than at the minus end. When the tubulin concentration is higher than the critical concentration for the plus end but lower than the critical concentration for the minus end, the microtubule can add tubulin heterodimers to its plus end while losing them from its minus end.

they do so by opposing effects on MTs and hence on the fibers of the mitotic spindle. Taxol is also used in the treatment of some cancers, especially breast cancer.

GTP Hydrolysis Contributes to the Dynamic Instability of Microtubules

In the previous section, we saw that tubulin can assemble in vitro in the presence of Mg^{2+} and GTP. In fact, GTP is required for MT assembly. Each tubulin heterodimer binds two GTP molecules. The α-tubulin binds one GTP; the other GTP is bound by β-tubulin and can be hydrolyzed to GDP sometime after the heterodimer is added to an MT. GTP is apparently needed for MT assembly because the association of GDP-bound tubulin heterodimers with each other is too weak to support polymerization. However, hydrolysis of GTP is not necessary for assembly, since MTs polymerize from tubulin heterodimers bound to a nonhydrolyzable analogue of GTP.

Studies of MT assembly in vitro using isolated centrosomes (a structure we will discuss in detail below) as nucleation sites show that some microtubules can grow by polymerization at the same time that others shrink by depolymerization. As a result, some MTs effectively enlarge at the expense of others.

To explain how both polymerization and depolymerization might occur simultaneously, Tim Mitchison and Marc Kirschner proposed the **dynamic instability model.** This model presumes two populations of microtubules, one growing in length by continued polymerization at their plus ends and the other shrinking in length by depolymerization. The distinction between the two populations is that growing MTs have GTP bound to the tubulin at their plus ends, while shrinking MTs have GDP instead. GTP-tubulin molecules are thought to protect MTs by preventing the peeling away of subunits from their plus ends; this *GTP cap* provides a stable MT tip to which further dimers can be added (**Figure 15-6a**). Hydrolysis of GTP by β-tubulin eventually results in an unstable tip, at which point depolymerization may occur rapidly.

The concentration of tubulin bound to GTP is crucial to the dynamic instability model. When GTP-tubulin is readily available, it is added to the microtubule quickly, creating a large GTP-tubulin cap. If the concentration of GTP-tubulin falls, however, the rate of tubulin addition decreases. At a sufficiently low concentration of GTP-tubulin, the rate of hydrolysis of GTP on the β-tubulin subunits near the tip of the MT exceeds the rate of addition of new, GTP-bound tubulin. This results in shrinkage of the GTP cap. When the GTP cap disappears, the MT becomes unstable, and loss of GDP-bound subunits from its tip is favored.

Direct evidence for dynamic instability comes from observation of individual microtubules in vitro via light microscopy. An individual MT can undergo alternating periods of growth and shrinkage (Figure 15-6b). When an MT switches from growth to shrinkage, an event called *microtubule catastrophe,* the MT can disappear completely, or it can abruptly switch back to a growth phase, a phenomenon known as *microtubule rescue.* The frequency of catastrophe is inversely related to the free tubulin concentration. High tubulin concentrations make catastrophe less likely, but it can still occur. When catastrophe does

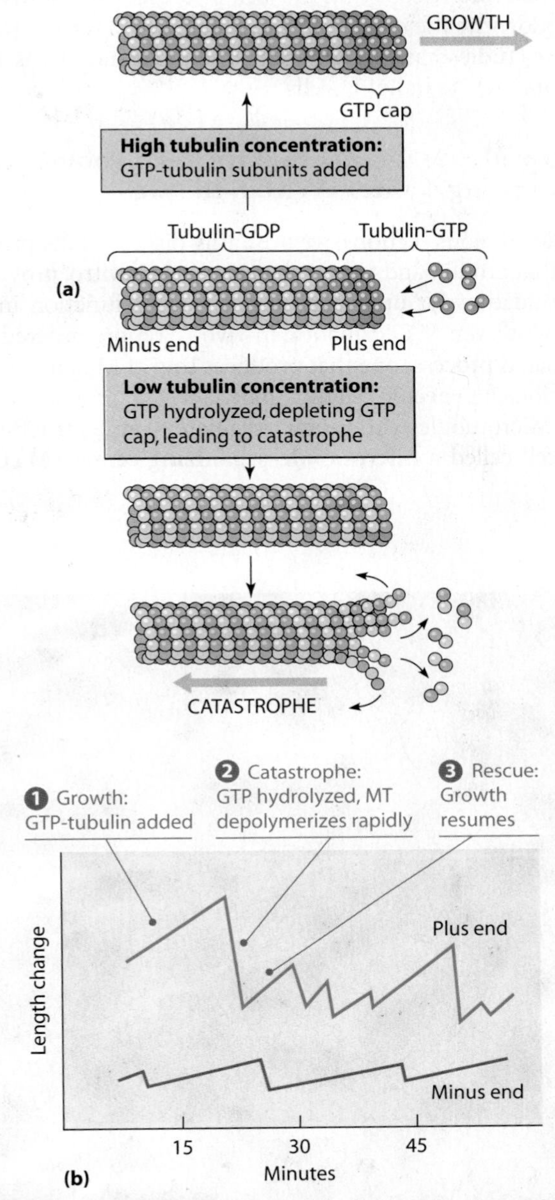

FIGURE 15-6 The GTP Cap and Its Role in the Dynamic Instability of Microtubules. (a) A model illustrating the role of the GTP cap. When the tubulin concentration is high, tubulin-GTP is added to the microtubule tip faster than the incorporated GTP can be hydrolyzed. The resulting GTP cap stabilizes the MT tip and promotes further growth. At lower tubulin concentrations the rate of growth decreases, thereby allowing GTP hydrolysis to catch up. This creates an unstable tip (no GTP cap) that favors MT depolymerization. (b) In an individual MT observed by light microscopy, ❶ growth and ❷ catastrophic shrinkage can occur. The plus and minus ends grow and shrink independently; changes in length are much more dynamic at the plus end. ❸ Rescue involves the switch from shrinkage to growth.

Dynamic Instability of Microtubules

occur, higher tubulin concentrations make the rescue of a shrinking MT more likely. At any tubulin concentration, catastrophe is more likely at the plus end of an MT—that is, dynamic instability is more pronounced at the plus end of the MT. Dynamic instability has been demonstrated in living cells using video-enhanced differential interference contrast microscopy and live-cell fluorescence microscopy to follow the life cycles of individual MTs (**Figure 15-7**). These studies have shown that dynamic instability is a key feature of MTs in living cells.

Microtubules Originate from Microtubule-Organizing Centers Within the Cell

In the previous sections, we primarily discussed the properties that tubulin and microtubules exhibit in vitro, providing a foundation for understanding how MTs function in the cell. However, MT formation in vivo is a more ordered and regulated process, one that produces sets of MTs in specific locations for specific cell functions.

Microtubules commonly originate from a structure in the cell called a **microtubule-organizing center (MTOC).** An MTOC serves as a site at which MT assembly is initiated and acts as an anchor for one end of these MTs. Many cells during interphase have an MTOC called the **centrosome** that is positioned near the nucleus. The centrosome in an animal cell is normally associated with two **centrioles** surrounded by a diffuse granular material known as *pericentriolar material* (**Figure 15-8a**). In electron micrographs of the centrosome, MTs originate from the pericentriolar material (Figure 15-8b).

The symmetrical structure of centrioles is remarkable: The walls of centrioles are formed by nine pairs of triplet microtubules (Figure 15-8a). In most cases, centrioles are oriented at right angles to one another; the significance of this arrangement is still unknown. Centrioles are known to be involved in the formation of basal bodies, which are important for the formation of cilia and flagella (see Chapter 16). The role of centrioles in non-ciliated cells is less clear. In animal cells, centrioles may serve to recruit pericentriolar material to the centrosome, which then nucleates growth of microtubules. When centrioles are missing from many animal cells, microtubule-nucleating material disperses, and the MTOC disappears. Cells lacking centrioles can still divide, probably because chromosomes can organize microtubules to some extent on their own. However, the resulting spindles are poorly organized. In contrast to animal cells, the cells of higher plants lack centrioles; their absence indicates that centrioles are not essential for the formation of MTOCs.

Large, ring-shaped protein complexes in the centrosome contain another type of tubulin, **γ-tubulin.** In conjunction with a number of other proteins called *GRiPs* (*g*amma tubulin *ri*ng *p*roteins), rings of γ-tubulin can be seen at the base of MTs that emerge from the centrosome (**Figure 15-9**). These **γ-tubulin ring complexes (γ-TuRCs)** serve to nucleate the assembly of new MTs away from the centrosome. The importance of γ-TuRCs has been demonstrated by depleting cells of γ-tubulin or other components of the γ-TuRC; in the absence of these proteins, centrosomes can no longer nucleate MTs. In addition to the centrosome, some types of cells have other MTOCs. For example, the basal body at the base of each cilium in ciliated cells also serves as an MTOC. During cell division, centrosomes are duplicated, creating new MTOCs for each of the daughter cells. We will discuss mitosis in detail in Chapter 19.

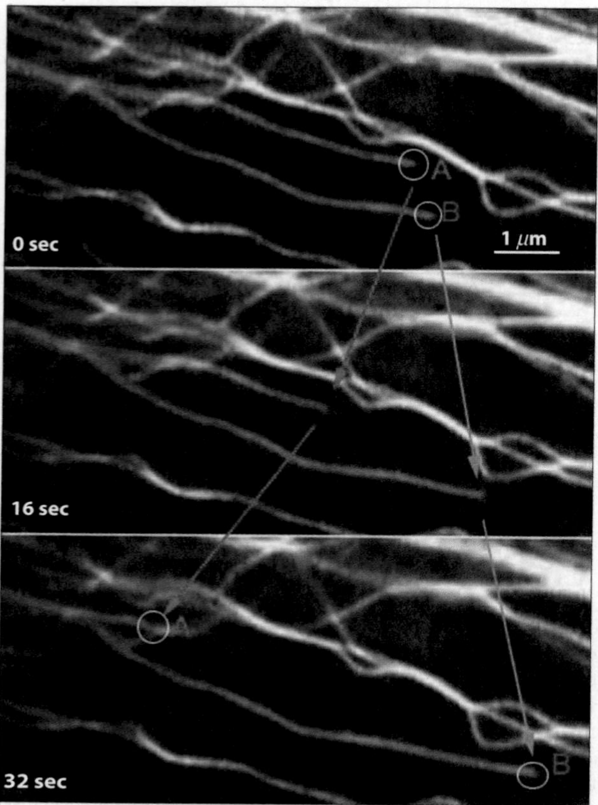

FIGURE 15-7 The Dynamic Instability of Microtubules in Vivo. Microtubules visualized in a living cell by live-cell fluorescence microscopy exhibit dynamic instability in vivo. Here, two individual MTs have been labeled to allow them to be followed over time. MT B grows over a 32-sec time span, whereas A shrinks.

MTOCs Organize and Polarize the Microtubules Within Cells

MTOCs play important roles in controlling the organization of microtubules in cells. The most important aspect of this role is probably the MTOC's ability to nucleate and anchor MTs. Because of this ability, MTs extend out from an MTOC toward the periphery of the cell. Furthermore, they grow out from an MTOC with a fixed polarity—their minus ends are anchored in the MTOC, and their plus ends extend out toward the cell membrane. The

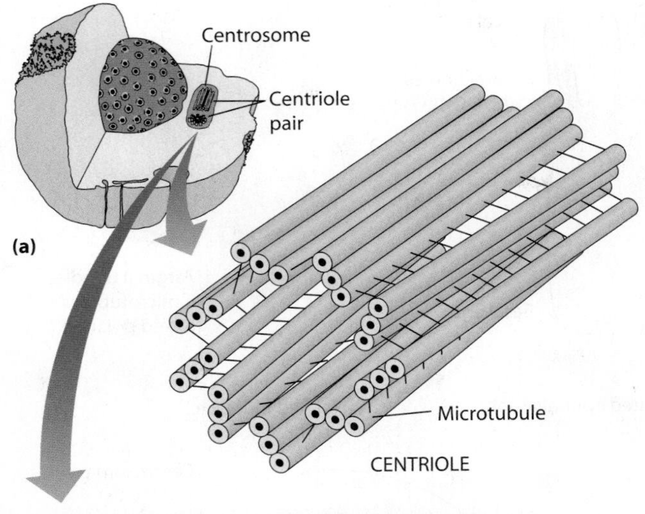

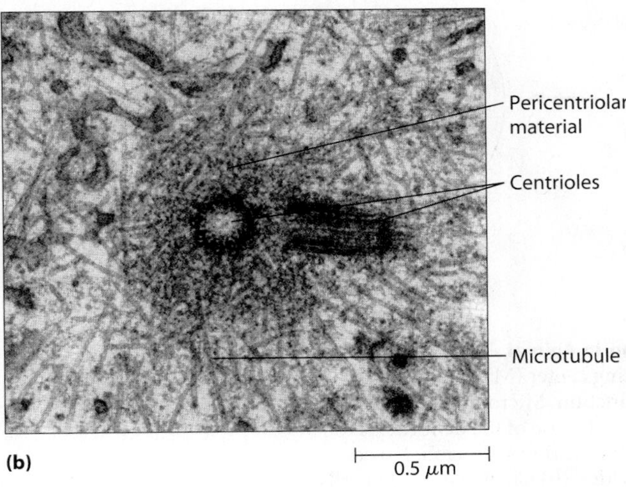

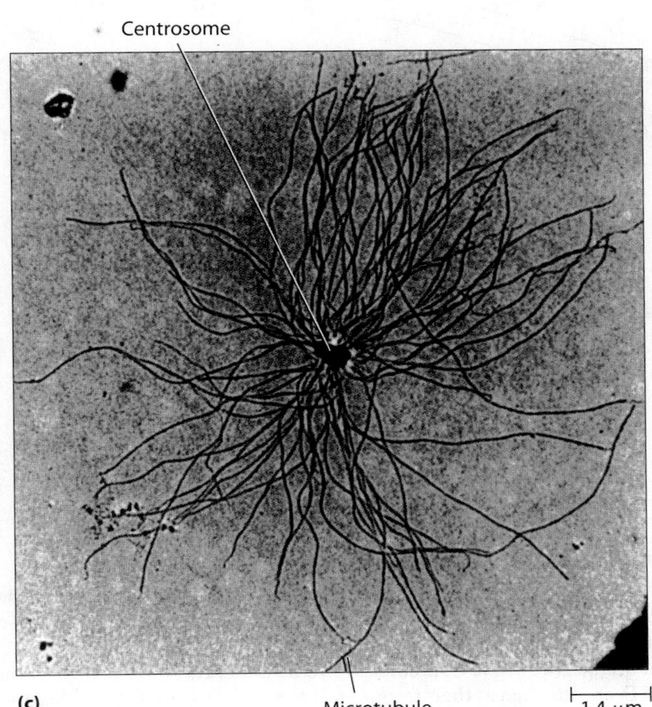

FIGURE 15-8 The Centrosome. (a) In animal cells, the centrosome contains two centrioles and associated pericentriolar material. The walls of centrioles are composed of nine sets of triplet microtubules. (b) An electron micrograph of a centrosome showing the centrioles and the pericentriolar material. Notice that microtubules originate from the pericentriolar material. (c) Nucleation and assembly of MTs at a centrosome in vitro.

relationship between the MTOC and the distribution and polarity of MTs are shown in **Figure 15-10**. The nucleating ability of MTOCs such as the centrosome has an important consequence for microtubule dynamics within cells. Since the minus ends of many MTs are anchored at the centrosome, dynamic growth and shrinkage of these MTs at the plus ends tends to occur at the periphery of cells.

The MTOC also influences the number of microtubules in a cell. Each MTOC has a limited number of nucleation and anchorage sites that seem to control how many MTs can form. However, the MT-nucleating capacity of the MTOC can be modified during certain processes such as mitosis. For example, centrosomes associated with spindle poles in mitotic cells have the highest MT-nucleating activity during prophase and metaphase (see Chapter 19).

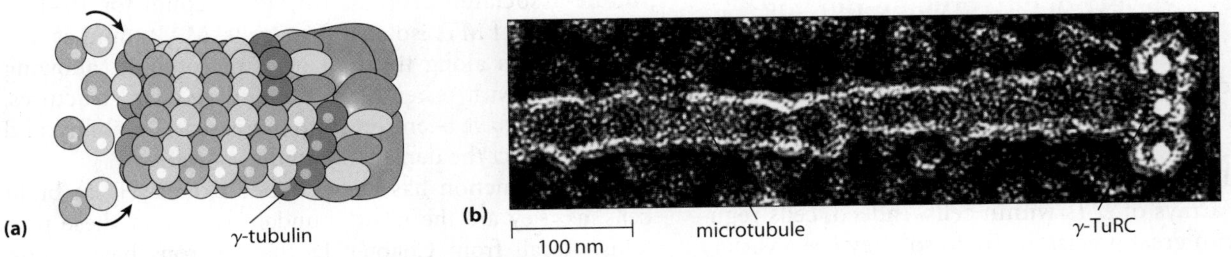

FIGURE 15-9 γ-Tubulin Ring Complexes (γ-TuRCs) Nucleate Microtubules. (a) γ-TuRCs, found at centrosomes, nucleate microtubule growth. The plus ends of MTs are oriented away from the γ-TuRC. (b) A platinum replica of an MT in vitro. Here, a component of the γ-TuRC (Xgrip109) was labeled with antibodies to which small particles of metal are attached. In the electron micrograph, these antibodies appear as bright spheres (TEM).

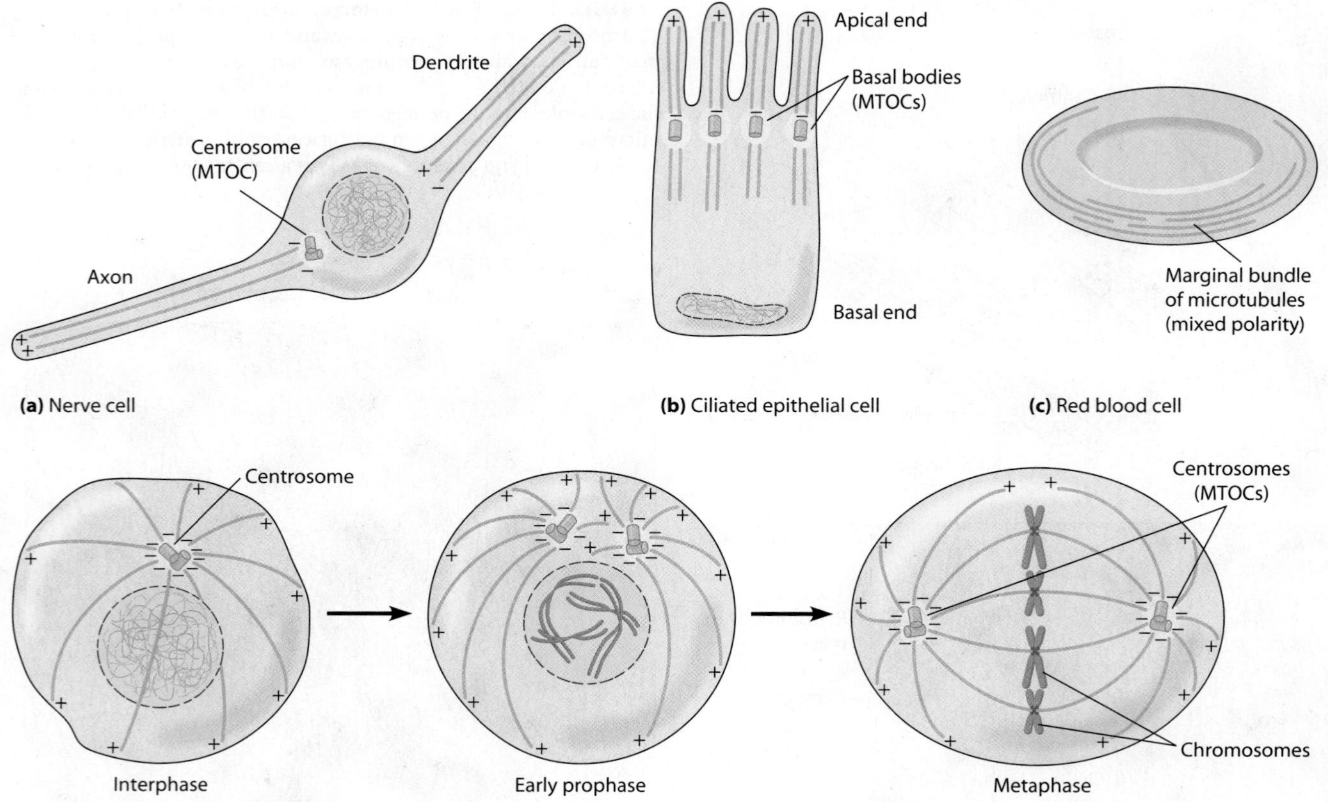

FIGURE 15-10 The Effects of Microtubule Polarity on MT Orientation in Animal Cells. In the cell, the distribution of most microtubules is determined by the microtubule-organizing center (MTOC), which is sometimes a centrosome. MT orientation in a cell may vary with that cell's function. Microtubules are shown in orange. **(a)** Nerve cells contain two distinct sets of MTs, those of the axon and those of the dendrite. Axonal MTs are attached at their minus ends to the centrosome, with their plus ends at the tip of the axon. However, dendritic MTs are not associated with the centrosome and are of mixed polarities. **(b)** Ciliated epithelial cells have many MTOCs called basal bodies, one at the base of each cilium. Ciliary MTs originate with their minus ends in the basal bodies and elongate with their plus ends toward the tips of the cilia. **(c)** Mature human red blood cells have no nucleus or MTOC. However, MTs of mixed polarities persist as a circular band at the periphery of the cell. This band helps to maintain the cell's round, disk-like shape. **(d)** Throughout the process of mitosis, MTs in a dividing cell are oriented with their minus ends anchored in the centrosome and their plus ends pointing away from the centrosome. Cell division is preceded by the division of the centrosome. The two centrosomes then separate, each forming one pole of the mitotic spindle. At metaphase, the centrosomes are at opposite sides of the cell. Each centrosome, or spindle pole, forms half of the spindle MTs—some extending from pole to chromosomes, others extending from one pole to the other pole.

Microtubule Stability Is Tightly Regulated in Cells by a Variety of Microtubule-Binding Proteins

We have seen that cellular microtubules exhibit dynamic instability; they grow out from the centrosome and then disassemble. This process could account for randomly distributed and short-lived MTs, but not for organized and stable arrays of MTs within cells. Indeed, cells regulate MTs with great precision. To do so, they use a variety of proteins to regulate MT structure, assembly, and function. Some MT-binding proteins use ATP to drive the transport of vesicles and organelles or to generate sliding forces between MTs. These proteins will be discussed in detail in Chapter 16. Here we focus on proteins that regulate MT structure.

Microtubule-Stabilizing/Bundling Proteins. **Microtubule-associated proteins (MAPs)** account for 10–15% of the mass of MTs isolated from cells. MAPs bind at regular intervals along the wall of a microtubule, allowing interaction with other filaments and cellular structures. Most MAPs have been shown to increase MT stability, and they can affect the density of bundles of MTs.

MAP function has been studied extensively in brain cells, as they are the most abundant source of these proteins. Recall from Chapter 13 that neurons have axons, which carry electrical signals away from the cell body of the neuron, and dendrites, which receive signals from neighboring cells and carry them to the cell body. The MT bundles are characteristically denser in axons than they are in dendrites. A MAP called *Tau* causes microtubules to form tight bundles in axons. Another MAP, *MAP2,* is present

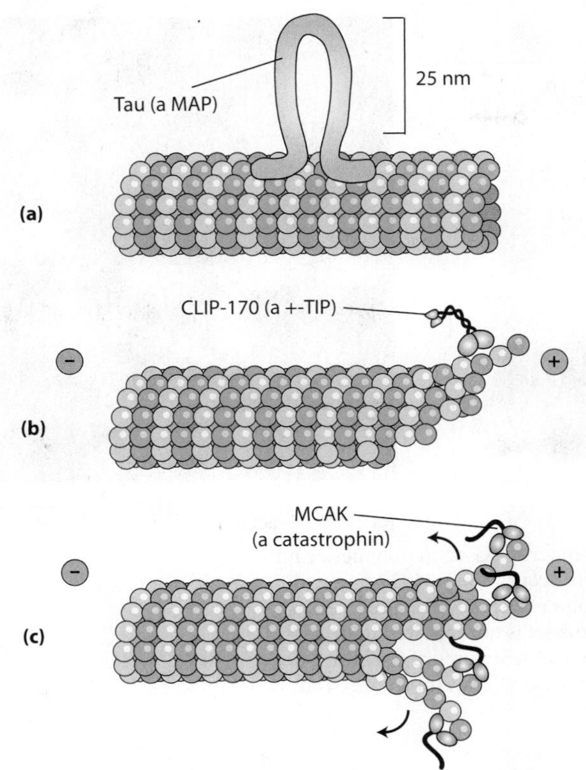

FIGURE 15-11 Microtubule-Interacting Proteins Regulate Microtubule Function in Vivo. Three different MT-interacting proteins are shown. **(a)** Tau is a MAP. Part of Tau binds along the length of an MT; another portion of Tau extends away from the MT, regulating MT spacing. **(b)** +-TIP proteins, such as CLIP-170, bind at or near the plus ends of MTs, stabilizing them. **(c)** Catastrophins, such as MCAK, are kinesin family proteins that destabilize MTs.

in dendrites and causes the formation of looser bundles of MTs. One portion of MAPs such as Tau and MAP2 binds along the length of an MT; another portion extends at right angles to the microtubule, where it can interact with other proteins (**Figure 15-11a**). The length of these "arms" controls the spacing of MTs in bundles; MAP2 has a longer arm than Tau does, and so MAP2 causes MT bundles to form that are less densely packed than with Tau.

The importance of MAPs can be demonstrated by forcing nonneuronal cells to make Tau protein. These cells are normally rounded, but when they express large amounts of Tau, these cells extend single long processes that look remarkably similar to axons. Tau is also important in human disease. Dense tangles of neurites, known as *neurofibrillary tangles,* are a hallmark of several diseases that result in dementia, such as Alzheimer's disease, Pick's disease, and several types of palsy. In the case of Alzheimer's disease, these tangles contain large amounts of hyperphosphorylated Tau protein, which forms paired helical filaments. Human mutations that result in defective Tau protein lead to hereditary predisposition to form such neurofibrillary tangles. Such diseases are therefore sometimes called *tauopathies.*

+-TIP Proteins. MTs are generally too unstable to remain intact for long periods of time and will depolymerize unless they are stabilized in some way. One way to stabilize MTs is to "capture" and protect their growing plus ends. To do so, **+-TIP proteins** (+-end *t*ubulin *i*nteracting *p*roteins) associate with MT plus ends. Some of these proteins, either directly or indirectly, appear to stabilize plus ends, decreasing the likelihood that they will undergo catastrophic subunit loss (**Figure 15-12**). One important example of MT capture involves kinetochores during mitosis, as we will discuss in Chapter 19. Other +-TIPs associate with the cell *cortex,* an actin-based network underneath the plasma membrane, and can stabilize MTs that extend there.

Microtubule-Destabilizing/Severing Proteins. As we have seen, some proteins stabilize microtubules, making them less likely to depolymerize. Other proteins *promote* depolymerization of MTs. For example, the protein *stathmin/Op18* binds to tubulin heterodimers, preventing them from polymerizing. Other proteins act at the ends of MTs once they have polymerized, promoting the peeling of subunits from their ends. Several proteins of the *kinesin* family, called *catastrophins,* act in this way (Figure 15-12c; we will learn more about other kinesins in Chapter 16). By tightly regulating where catastrophins act, a cell can precisely control where and when MTs form and depolymerize. A prime example of such regulation is the mitotic spindle, which we will examine in Chapter 19. Still other proteins sever MTs; one example of such proteins are *katanins.*

Microfilaments

With a diameter of about 7 nm, **microfilaments (MFs)** are the smallest of the cytoskeletal filaments (see Table 15-1). Microfilaments are best known for their role in the contractile fibrils of muscle cells, where they interact with thicker filaments of myosin to cause the contractions characteristic of muscle (see Chapter 16). MFs are not confined to muscle cells, however. They occur in almost all eukaryotic cells and are involved in numerous other phenomena, including a variety of locomotory and structural functions.

Examples of cell movements in which microfilaments play a role include *cell migration* via lamellipodia and filopodia, *amoeboid movement,* and *cytoplasmic streaming,* a regular pattern of cytoplasmic flow in some plant and animal cells. We will discuss all of these phenomena in detail in Chapter 16. MFs also produce the cleavage furrows that divide the cytoplasm of animal cells during cytokinesis (see Chapter 19), and MFs are found at sites of attachment of cells to one another and to the extracellular matrix (see Chapter 17).

In addition to mediating a variety of cell movements, MFs are important in developing and maintaining cell shape. Most animal cells, for example, have a dense network of microfilaments called the *cell cortex* just beneath the plasma membrane. The cortex confers structural rigidity on the cell surface and facilitates shape changes and cell movement. Parallel bundles of MFs also make up the structural core of *microvilli,* the fingerlike extensions found on the surface of many animal cells (see Figure 4-2).

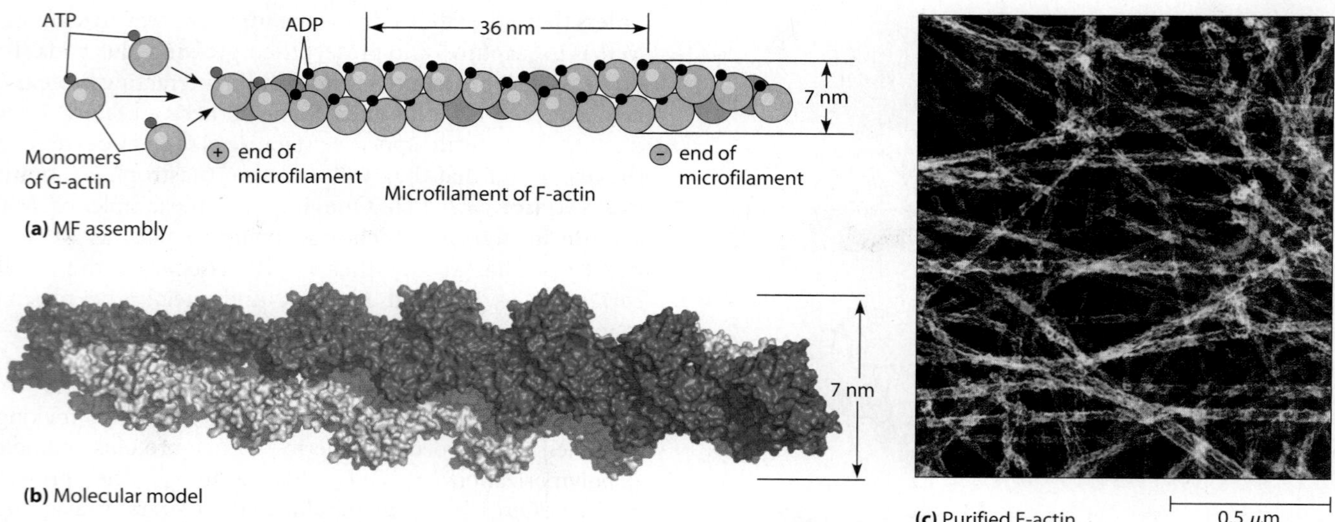

FIGURE 15-12 A Model for Microfilament Assembly in Vitro. **(a)** Monomers of G-actin polymerize into long filaments of F-actin with a diameter of about 7 nm. A full turn of the helix occurs every 36–37 nm, with about 13.5 monomers required for a full turn. Addition of each G-actin monomer is usually accompanied or followed by hydrolysis of the ATP molecule, although the energy of ATP hydrolysis is not required to drive the polymerization reaction. **(b)** A molecular model of F-actin, based on X-ray crystal structures of G-actin. Two strands of 13 G-actin monomers each are shown. One strand is colored blue, the other gray. **(c)** An electron micrograph of purified F-actin (TEM).

Actin Is the Protein Building Block of Microfilaments

Actin is an extremely abundant protein in virtually all eukaryotic cells, including those of plants, algae, and fungi. Actin is synthesized as a single polypeptide consisting of 375 amino acids, with a molecular weight of about 42 kDa. Once synthesized, it folds into a roughly U-shaped molecule, with a central cavity that binds ATP or ADP. Individual actin molecules are referred to as **G-actin** (globular actin). Under the right conditions, G-actin molecules polymerize to form microfilaments; in this form, actin is referred to as **F-actin** (filamentous actin; see Figure 15-12). Actin in the G or F form also binds to a wide variety of other proteins, collectively known as *actin-binding proteins.*

Different Types of Actin Are Found in Cells

Of the three types of cytoskeletal proteins, actin is the most highly conserved. In functional assays, all actins appear to be identical, and actins from diverse organisms will copolymerize into filaments. Despite this high degree of sequence similarity, actins do differ among different organisms and among tissues of the same organism. Based on sequence similarity, actins can be broadly divided into two major groups: the *muscle-specific actins (α-actins)* and the *nonmuscle actins (β- and γ-actins).* β- and γ-actin localize to different regions of the cell and appear to have different interactions with actin-binding proteins. For example, in epithelial cells, one end of the cell, the apical end, contains microvilli, whereas the opposite side of the cell, known as the basal end, is attached to the extracellular matrix (see Figure 17-18). β-actin is predominantly found at the apical end of epithelial cells, whereas γ-actin is concentrated at the basal end and sides of the cell.

G-Actin Monomers Polymerize into F-Actin Microfilaments

Like tubulin dimers, G-actin monomers can polymerize reversibly into filaments with a lag phase corresponding to filament nucleation, followed by a more rapid polymer elongation phase. The kinetics of actin polymerization can be studied in solution using fluorescent G-actin. The fluorescence of the labeled F-actin can be measured to yield data similar to that for tubulin. The F-actin filaments that form are composed of two linear strands of polymerized G-actin wound around each other in a helix, with roughly 13.5 actin monomers per turn (Figure 15-12).

Within a microfilament, all the actin monomers are oriented in the same direction, so that an MF, like a microtubule, has an inherent polarity, with one end differing chemically and structurally from the other end. This polarity can be readily demonstrated by incubating MFs with **myosin subfragment 1 (S1),** a proteolytic fragment of myosin (**Figure 15-13**). S1 fragments bind to, or "decorate," the actin MFs to give a distinctive arrowhead pattern, with all the S1 molecules pointing in the same direction (Figure 15-13c). Based on this arrowhead pattern, the terms *pointed end* and *barbed end* are commonly used to identify the minus and plus ends of an MF, respectively. The polarity of the MF is important, because it

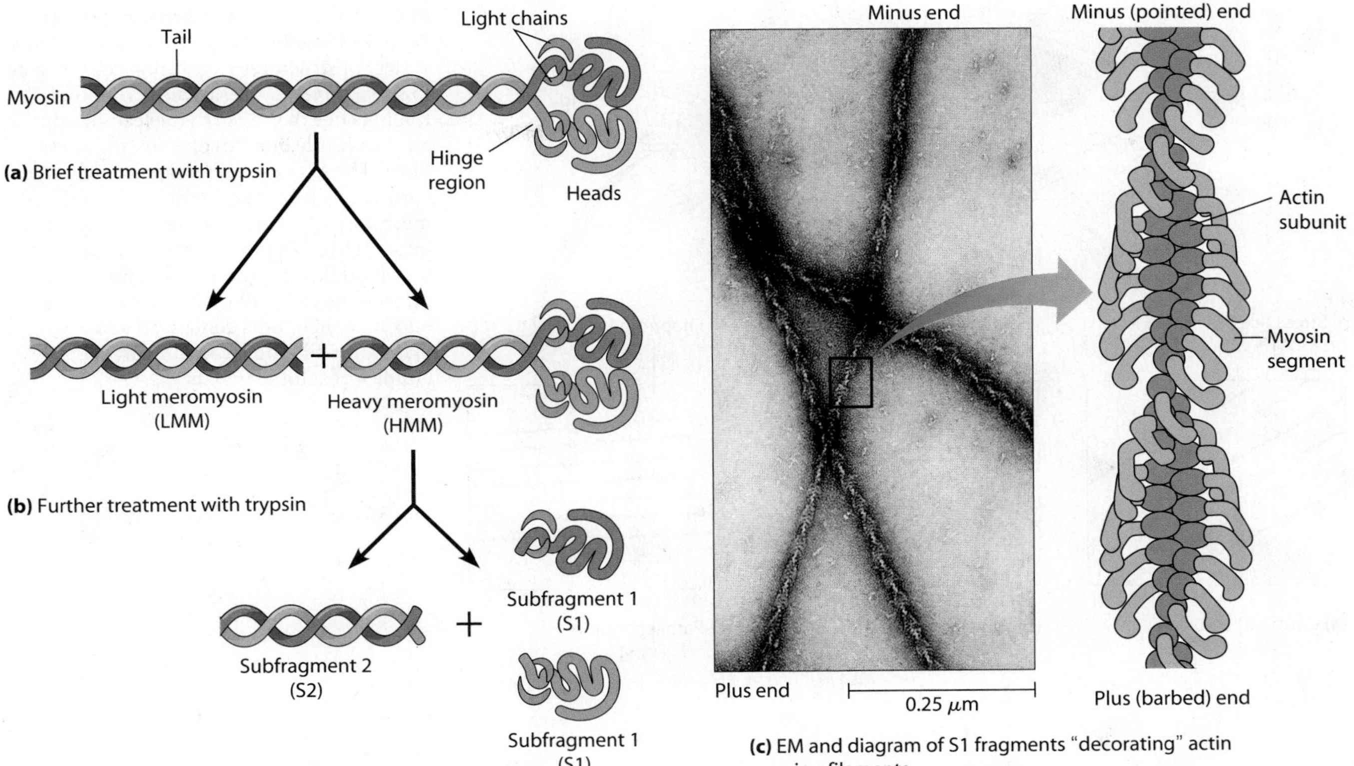

FIGURE 15-13 Using Myosin S1 Subfragments to Determine Actin Polarity. Myosin II is part of the contractile machinery found in muscle cells. The globular head of the myosin molecule binds to actin, while the myosin tails can associate with filaments of myosin (the thick myofilaments of muscle cells). **(a)** Myosin II can be cleaved by proteases such as trypsin into two pieces, heavy meromyosin (HMM) and light meromyosin (LMM). **(b)** HMM can be further digested, leaving only the globular head. This fragment, called myosin subfragment 1 (S1), retains its actin-binding properties. **(c)** When actin microfilaments are incubated with myosin S1 and then examined with an electron microscope, the S1 fragments appear to "decorate" the microfilaments like arrowheads. All the S1 arrowheads point toward the minus end, indicating the polarity of the MF.

allows for independent regulation of actin assembly or disassembly at each end of the filament.

The polarity of microfilaments is reflected in more rapid addition or loss of G-actin at the plus end, and slower addition or loss of G-actin at the minus end (see Figure 15-12a). If G-actin is polymerized onto short fragments of S1-decorated F-actin, polymerization proceeds much faster at the barbed end, indicating that the barbed end of the filament is also the plus end. Thus, even when conditions are favorable for adding monomers to both ends of the filament, the plus end will grow faster than the minus end.

As G-actin monomers assemble onto a microfilament, the ATP bound to them is slowly hydrolyzed to ADP, much the same as the GTP bound to tubulin is hydrolyzed to GDP. Thus, the ends of a growing MF tend to have ATP-F-actin, whereas the bulk of the MF is composed of ADP-F-actin. However, ATP hydrolysis is not a strict requirement for MF elongation, since MFs can also assemble from ADP-G-actin or from nonhydrolyzable analogues of ATP-G-actin.

Specific Drugs Affect Polymerization of Microfilaments

As we saw with microtubules, several drugs have been used to perturb the assembly of actin into microfilaments (Table 15-3). Processes that are disrupted in cells treated with these drugs are likely to depend in some way on microfilaments. Several drugs result in depolymerization of microfilaments. The **cytochalasins,** such as *cytochalasin D,* are fungal metabolites that prevent the addition of new monomers to existing polymerized MFs. As subunits are gradually lost from the minus ends of MFs in cytochalasin-treated cells, they eventually depolymerize. In contrast, **latrunculin A,** a marine toxin isolated from the Red Sea sponge *Latrunculia magnifica,* acts by sequestering actin monomers, preventing their addition to the plus ends of growing MFs. In either case, the net result is the loss of MFs within the treated cells. Conversely, the drug *phalloidin,* a cyclic peptide from the death cap fungus *(Amanita phalloides),* stabilizes microfilaments, preventing their depolymerization. Fluorescently labeled phalloidin

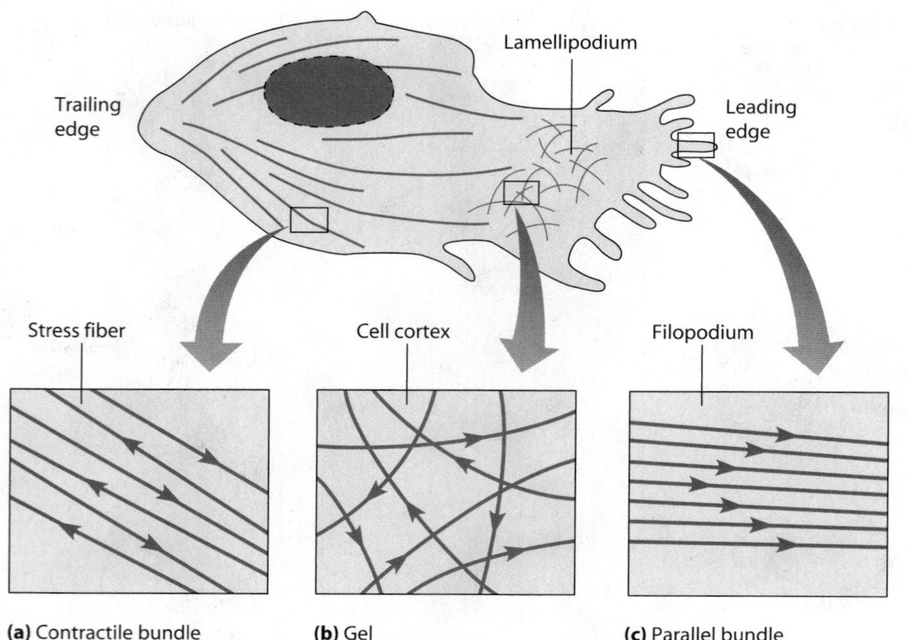

FIGURE 15-14 The Architecture of Actin in Crawling Cells. Actin is found in a variety of structures in crawling cells such as this macrophage. **(a)** Running from the trailing edge of the cell to the leading edge are contractile bundles of actin, the stress fibers. **(b)** At the periphery of the cell is the cortex, which contains a three-dimensional meshwork of actin filaments crosslinked into a gel. **(c)** The broad leading edge of lamellipodia can produce thin, fingerlike projections called filopodia. Whereas the bulk of lamellipodia contain an actin meshwork, filopodia contain parallel bundles of actin filaments.

is also useful for visualizing F-actin via fluorescence microscopy.

Cells Can Dynamically Assemble Actin into a Variety of Structures

As with microtubules, cells can dynamically regulate where and how G-actin is assembled into microfilaments. For example, cells that crawl have specialized structures called *lamellipodia* and *filopodia* at their leading edge that allow them to move along a surface (we will consider these specialized structures in more detail in Chapter 16). The form of the protrusion appears to depend on the nature of the cell's movement and on the organization of the actin filaments within the cell. In cells that adhere tightly to the underlying substratum and do not move well, organized bundles of actin, called *stress fibers,* stretch from the tail, or trailing edge, of the cell to the front (**Figure 15-14a**). Rapidly moving cells typically do not have such striking actin bundles. In such cells, the cell *cortex,* which lies immediately beneath the plasma membrane and is enriched in actin, is crosslinked into a gel or very loosely organized lattice of microfilaments (Figure 15-14b). At the leading edge, and especially in filopodia, microfilaments form highly oriented, polarized cables, with their barbed (plus) ends oriented toward the tip of the protrusion (Figure 15-14c). The actin in lamellipodia is typically less well organized than in filopodia (**Figure 15-15**). Understanding how cells regulate such a wide variety of actin-based structures requires understanding both how cells can regulate the polymerization of MFs and how MFs, once polymerized, assemble into networks.

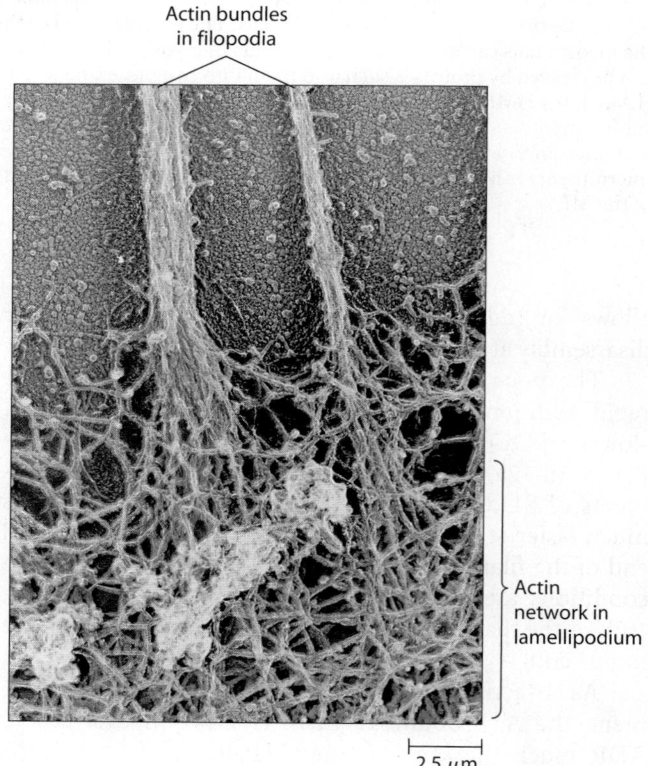

FIGURE 15-15 Deep-Etch Electron Micrograph Showing Actin Bundles in Filopodia. This view of the periphery of a macrophage shows two prominent actin bundles contained within filopodia that extend from the cell surface. The actin filaments in the filopodia merge with a network of actin filaments lying just beneath the plasma membrane of the lamellipodium.

Actin-Binding Proteins Regulate the Polymerization, Length, and Organization of Microfilaments

As we saw with microtubules, cells can precisely control where actin assembles and the structure of the resulting actin networks. To do so, cells use a variety of **actin-binding proteins** (**Figure 15-16**). Control of the process of MF polymerization occurs at several steps, including the nucleation of new MFs, the elongation and severing of preexisting MFs, and the association of MFs into networks. We consider each of these briefly here.

Proteins That Regulate Polymerization. In the absence of other factors, the growth of microfilaments depends on the concentration of ATP-bound G-actin. If the concentration of ATP-bound G-actin is high, microfilaments will assemble until the G-actin is limiting. In the cell, however, a large amount of free G-actin is not available for assembly into filaments because it is bound by the protein *thymosin β4*. A second protein called *profilin* appears to compete with thymosin *β*4 for binding to G-actin monomers. When the profilin concentration is high, polymerization is favored—but only if there are free filament ends available. Yet another protein, known as *ADF/cofilin,* is known to bind to ADP-G-actin and F-actin. ADF/cofilin is thought to increase the rate of turnover of ADP-actin at the minus ends of MFs. The ADP on these G-actin monomers can then be exchanged for a new ATP, and the ATP-G-actin can then be recycled for addition to the growing plus ends of MFs. ADF/cofilin also severs filaments, creating new plus ends as it does so.

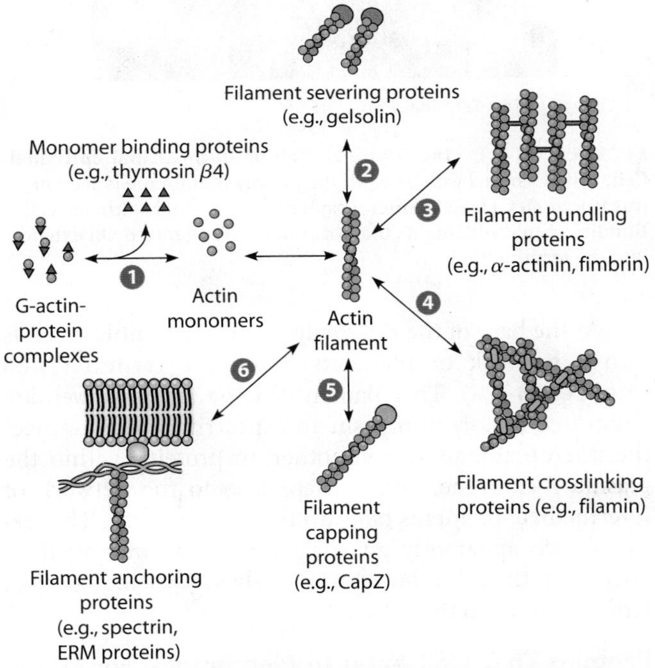

FIGURE 15-16 Actin-Binding Proteins Regulate the Organization of Actin. Actin-binding proteins are responsible for converting actin filaments from one form to another. These include ❶ monomer-binding proteins, such as thymosin b4 and profilin; ❷ filament severing proteins, such as gelsolin; ❸ filament bundling proteins, such as a-actinin, fimbrin, and fascin; ❹ filament crosslinking proteins, such as filamin; ❺ filament capping proteins such as CapZ and tropomodulin; and ❻ filament anchoring proteins, such as spectrin and ERM proteins.

Proteins That Cap Actin Filaments. Whether microfilament ends are available for further growth depends on whether the filament end is *capped.* Capping occurs when a **capping protein** binds the end of a filament and prevents further addition or loss of subunits, thereby stabilizing it. One such protein that functions as a cap for the plus ends of microfilaments is appropriately named *CapZ.* When CapZ is bound to the end of a filament, further addition of subunits at the plus end is prevented; when CapZ is removed, addition of subunits can resume. Another class of proteins called *tropomodulins* bind to the minus ends of actin filaments, preventing loss of subunits from the pointed ends of F-actin. Tropomodulins are found in muscle sarcomeres, as we will see in Chapter 16.

Proteins That Crosslink Actin Filaments. In many cases, actin networks form as loose meshworks of criss-crossing, crosslinked MFs. One of the crosslinking proteins that is important for such networks is *filamin,* a long molecule consisting of two identical polypeptides joined head to head, with an actin-binding site at each tail. Molecules of filamin act as "splices," joining two MFs together where they intersect. In this way, actin MFs are linked to form large three-dimensional networks.

Proteins That Sever Actin Filaments. Other proteins play the opposite role, breaking up the microfilament network and causing the cortical actin gel to soften and liquefy. They do this by severing and/or capping MFs. In some cases, such proteins can serve both functions. One of these severing and capping proteins is *gelsolin,* which functions by breaking actin MFs and capping their newly exposed plus ends, thereby preventing further polymerization.

Proteins That Bundle Actin Filaments. In contrast to the loose organization of the actin at the cell cortex, other actin-containing structures in migrating and nonmigrating cells can be highly ordered. In such cases, actin may be bundled into tightly organized arrays, and a number of actin-binding proteins mediate such bundling. One such protein is *α-actinin,* a protein that is prominent within structures known as *focal contacts* and *focal adhesions,* which are required for cells to make adhesive connections to the extracellular matrix as they migrate. We will consider these structures in more detail in Chapter 17. Another bundling protein, *fascin,* is found in filopodia; fascin keeps the actin within the core of a filopodium tightly bundled, contributing to the spike-like appearance of such protrusions.

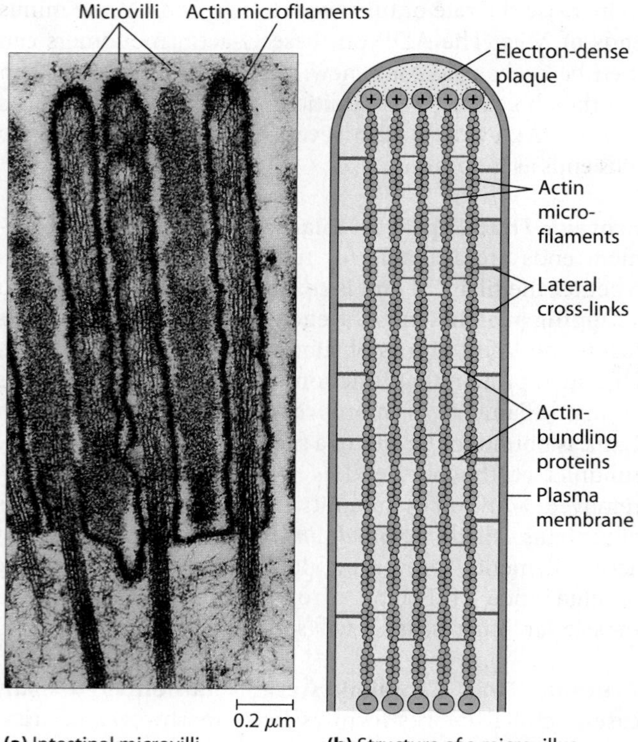

(a) Intestinal microvilli **(b)** Structure of a microvillus

FIGURE 15-17 Microvillus Structure. **(a)** An electron micrograph of microvilli from intestinal mucosal cells (TEM). **(b)** A schematic diagram of a single microvillus, showing the core of microfilaments that gives the microvillus its characteristic stiffness. The core consists of several dozen microfilaments oriented with their plus ends facing outward toward the tip and their minus ends facing toward the cell. The plus ends are embedded in an amorphous, electron-dense plaque. The MFs are tightly linked together by actin-bundling (crosslinking) proteins and are connected to the inner surface of the plasma membrane by lateral crosslinks.

Perhaps the best-studied example of ordered actin arrays is the actin bundles found in microvilli. **Microvilli** (singular: **microvillus**) are especially prominent features of intestinal mucosal cells (**Figure 15-17a**). A single mucosal cell in your small intestine, for example, has several thousand microvilli, each about 1–2 μm long and about 0.1 μm in diameter, which increase the surface area of the cell about twentyfold. This large surface area is essential to intestinal function because the uptake of digested food depends on an extensive absorptive surface.

As illustrated in Figure 15-17b, the core of the intestinal microvillus consists of a tight bundle of microfilaments. The plus ends point toward the tip, where they are attached to the membrane through an amorphous, electron-dense plaque. The MFs in the bundle are also connected to the plasma membrane by lateral crosslinks consisting of the proteins *myosin I* and *calmodulin.* These crosslinks extend outward about 20–30 nm from the bundle to contact electron-dense patches on the inner membrane surface. Adjacent MFs in the bundle are bound tightly together at regular intervals by the crosslinking proteins (also called actin-bundling proteins) *fimbrin* and *villin.*

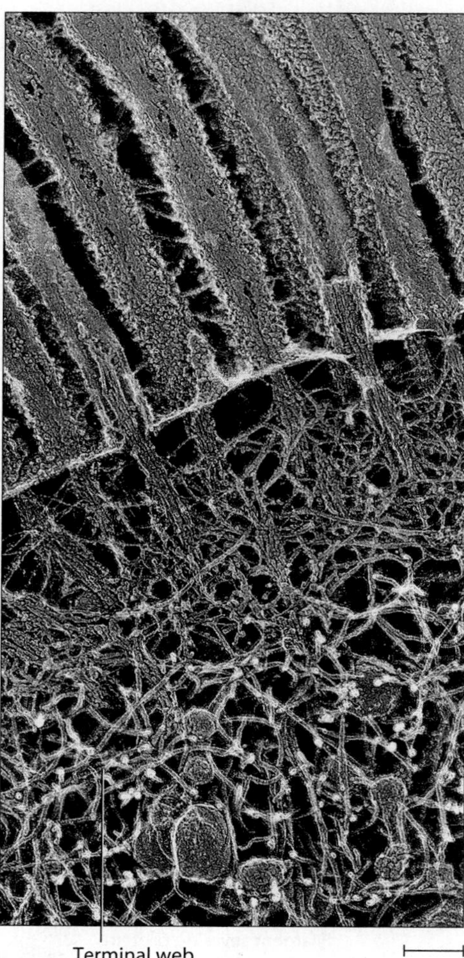

FIGURE 15-18 The Terminal Web of an Intestinal Epithelial Cell. The terminal web beneath the plasma membrane is seen in this freeze-etch electron micrograph of an intestinal epithelial cell. Bundles of microfilaments that form the cores of microvilli extend into the terminal web.

At the base of the microvillus, the MF bundle extends into a network of filaments called the **terminal web** (**Figure 15-18**). The filaments of the terminal web are composed mainly of myosin and spectrin, which connect the microfilaments to each other, to proteins within the plasma membrane, and perhaps also to the network of intermediate filaments beneath the terminal web. The terminal web apparently gives rigidity to the microvilli by anchoring their MF bundles securely so that they project straight out from the cell surface.

Proteins That Link Actin to Membranes. For MFs to exert force on the plasma membrane during events such as cell movement and cytokinesis, they must be connected to the plasma membrane. The connection of MFs to the plasma membrane is indirect and requires one or more linker proteins that anchor MFs to transmembrane proteins embedded within the plasma membrane.

One group of proteins that appears to function widely in linking microfilaments to membranes is the *band 4.1, ezrin, radixin,* and *moesin* family of actin-binding proteins. When these proteins are mutated, a wide variety of cellular processes are affected, including cytokinesis, secretion, and the formation of microvilli. Another example of how actin can be linked to membranes involves the proteins *spectrin* and *ankyrin* (**Figure 15-19**). As we saw in Chapter 7, the plasma membrane of the erythrocyte is supported by a network of spectrin filaments that are crosslinked by very short actin chains. This network is connected to the plasma membrane by molecules of the proteins ankyrin and band 4.1 that link the spectrin filaments to specific transmembrane proteins. Based on mutations in the genes encoding spectrins and ankyrins in *Drosophila* and in the nematode *Caenorhabditis elegans,* these proteins are important in a wide variety of cells for maintenance of cell shape.

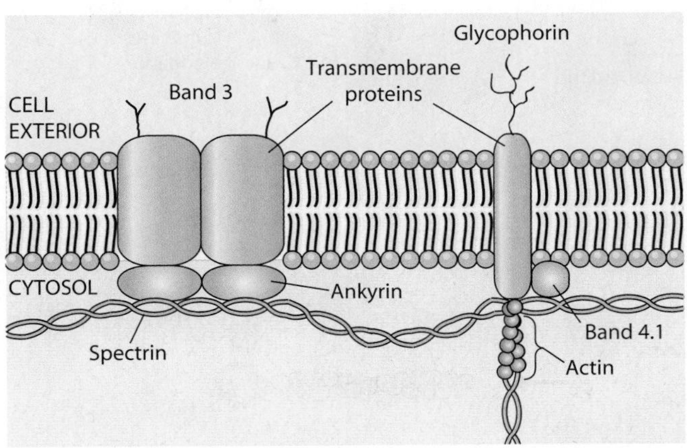

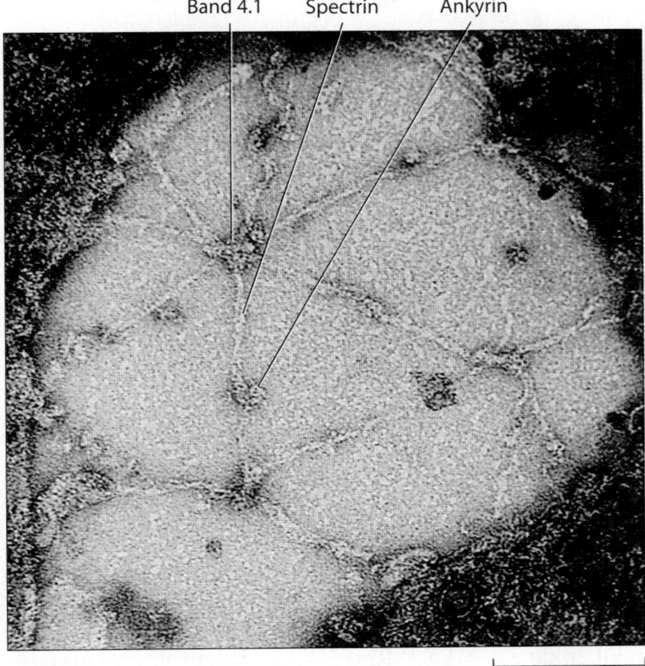

FIGURE 15-19 Support of the Erythrocyte Plasma Membrane by a Spectrin-Ankyrin-Actin Network. The plasma membrane of a red blood cell is supported on its inner surface by a filamentous network that gives the cell both strength and flexibility. **(a)** A diagram showing the major components of the spectrin-ankyrin-actin network. Long filaments of spectrin are crosslinked by short actin filaments. The network is anchored to the band 3 transmembrane protein by molecules of the protein ankyrin. **(b)** An electron micrograph of an erythrocyte membrane, showing actual spectrin network components (TEM).

Proteins That Promote Actin Branching and Growth. In addition to loose networks and bundles, cells can assemble actin into branched networks that form a treelike, or *dendritic,* network (**Figure 15-20a**). Such dendritic networks are a prominent feature of lamellipodia in migrating cells. A complex of actin-related proteins, the **Arp2/3 complex,** helps branches to form by nucleating new branches on the sides of existing filaments (Figure 15-20b). Proteins we have already discussed, such as profilin, cofilin, and capping proteins, regulate the length of filaments that polymerize from branch points, thereby regulating the length of such structures.

Arp2/3 branching is activated by a family of proteins that includes the *Wiskott-Aldrich syndrome protein,* or *WASP,* and *WAVE/Scar.* Human patients who cannot produce functional WASP have defects in the ability of their platelets to undergo changes in shape and so have difficulties in forming blood clots. A very different kind of disease also involves the Arp2/3 complex: Pathogenic bacteria can "hijack" the actin polymerization machinery of the cell to spread (**Box 15A**).

Branched actin networks are only one type of actin-based structure that cells can produce. For some cellular events, long actin filaments are more useful. Actin polymerization in this case can be regulated independently of the Arp2/3 complex, through proteins known as *formins.* Formins are required to assemble certain unbranched F-actin structures, including actin cables and the contractile ring during cell division (see Chapter 19). Formins appear to be able to act "processively," moving along the end of a growing filament as they stimulate filament growth (Figure 15-20c). Formins can form dimers, binding to the barbed (plus) ends of actin filaments. Some formins have extensions that can bind profilin and are therefore thought to act as "staging areas" for the addition of actin monomers to growing filaments. Plants have a large number of formin-like proteins, which are thought to play similar roles in regulating the actin cytoskeleton.

Cell Signaling Regulates Where and When Actin-Based Structures Assemble

We have seen that actin-binding proteins regulate the types of actin-based structures that cells assemble. Cell signaling, in turn, regulates the activity of these proteins. Both plasma membrane lipids and several small G proteins related to Ras (see Chapter 14) regulate the formation, stability, and breakdown of MFs.

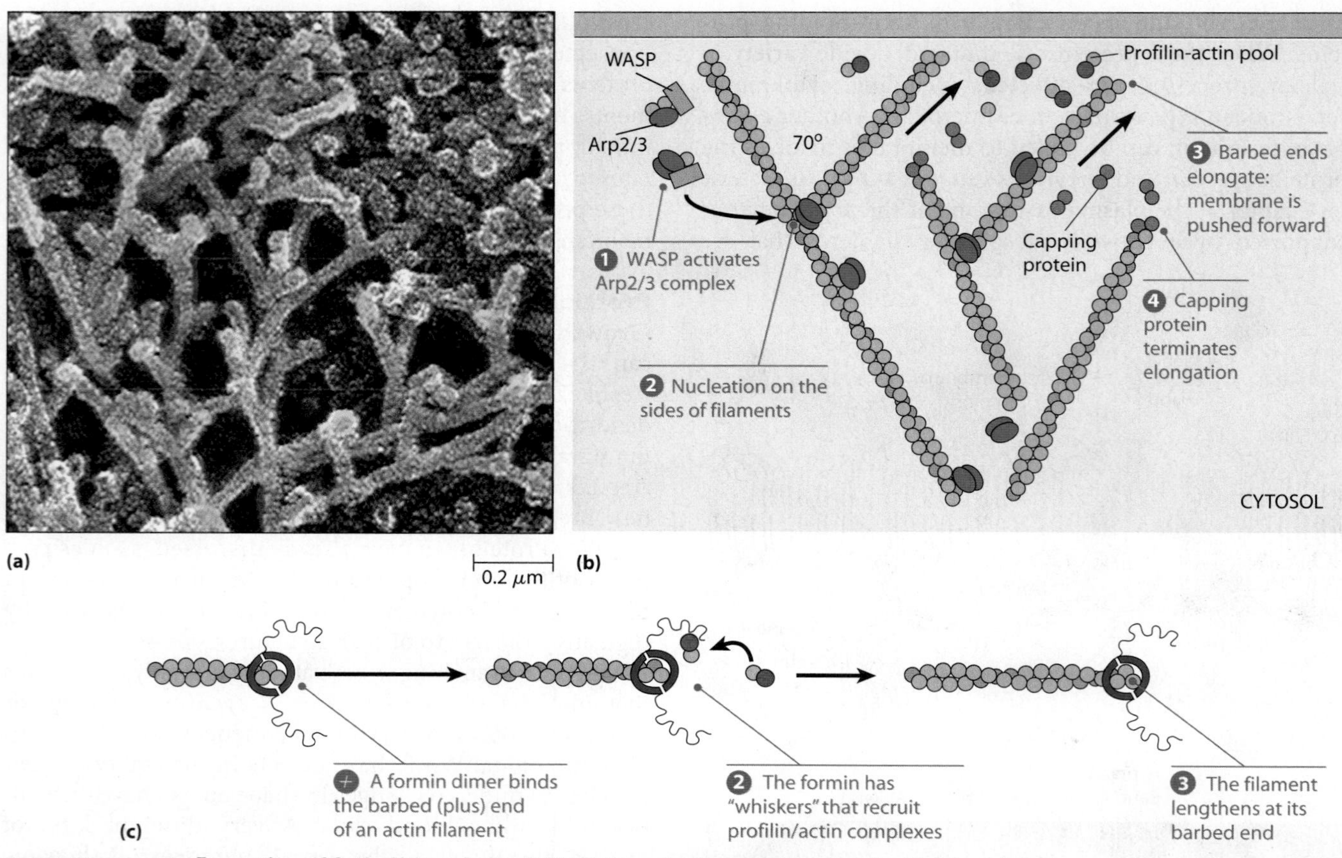

FIGURE 15-20 Formation of Actin Networks by Actin Polymerization. Actin networks, like those found in migrating cells, have a characteristic pattern of branching. **(a)** Branched actin filaments in a frog keratocyte. Individual branched actin filaments are colored to make them easier to distinguish (deep-etch TEM). **(b)** Model for Arp2/3-dependent branching. Branching is stimulated by WASP family proteins; capping protein helps regulate the length of new branches. **(c)** Model for formin-induced elongation of an actin filament. Formin dimers bind profilin-actin, serving as a "staging area" for actin polymerization.

Inositol Phospholipids. Inositol phospholipids are one type of membrane phospholipid that regulates actin assembly. Recall that *phosphatidylinositol-4,5-bisphosphate* (PIP_2) is important during some signaling events, such as insulin signaling (see Chapter 14). PIP_2 can bind to profiling, CapZ, and proteins such as ezrin, thereby recruiting them to the plasma membrane, as well as regulating the ability of these proteins to interact with actin. For example, CapZ binds tightly to PIP_2, resulting in its removal from the end of a microfilament, thereby permitting the filament to be disassembled and making its monomers available for assembly into new filaments. Gelsolin is another actin-binding protein that can be regulated by binding to polyphosphoinositides. When gelsolin binds to a specific polyphosphoinositide, it can no longer cap the plus end of an MF, allowing the uncapped end to undergo changes in length.

Rho Family GTPases. One striking case of regulation of the actin cytoskeleton is the dramatic change in the cytoskeleton of cells exposed to certain growth factors. For example, in response to stimulation by platelet-derived growth factor (PDGF), fibroblasts will begin to grow, divide, and form actin-rich membrane extensions that resemble lamellipodia. Other factors, such as lysophosphatidic acid (LPA), induce cells to form stress fibers.

How do such signals result in such dramatic reorganization of the actin cytoskeleton? Many of these signals result in changes in the actin cytoskeleton through their action on a family of monomeric G proteins known as **Rho GTPases.** Three key members of this family are **Rho, Rac,** and **Cdc42.** Originally identified in yeast, these proteins are important regulators of the actin cytoskeleton in all eukaryotes. Each Rho GTPase has profound and different effects on the actin cytoskeleton (**Figure 15-21**). For example, activation of the Rho pathway results in the formation of stress fibers, and Rho inactivation prevents the appearance of stress fibers following exposure of fibroblasts to LPA. Similarly, Rac activation often results in extension of lamellipodia by cultured cells, and inhibition of Rac prevents this normal response to PDGF. Finally, activation of Cdc42 results in the formation of filopodia. Rho GTPases perform a vast array of function within cells, from formation of protrusions to assembly

BOX 15A HUMAN APPLICATIONS

Infectious Microorganisms Can Move Within Cells Using Actin "Tails"

One of the most remarkable findings of modern cell motility research is the discovery that disease-causing microorganisms can co-opt the cell's normal cell adhesion and cell motility systems to penetrate a cell's defenses and enter the cell. The best-studied example of such motility is the gram-positive bacterium *Listeria monocytogenes*. One way that *Listeria* attaches to the host's cells involves the binding of a *Listeria* protein known as *internalin A* to *E-cadherin* on the cell surface (see Chapter 17 for more on cadherins). Once bound, *Listeria* enter a cell, move through it at a rate of 11 μm/min, and progress to nearby uninfected cells, where they continue the cycle of infection (**Figure 15A-1**). Short actin filaments radiate away from the bacteria, forming "comet tails" of branched F-actin (Figure 15A-lb). By using fluorescently labeled actin, investigators have determined that the tails form by Arp2/3-dependent polymerization of actin, which is nucleated near the surface of the internalized bacterium.

The protein on the surface of *Listeria* that promotes actin polymerization is known as *ActA*. The microfilaments nucleated by ActA are strikingly similar to those found at the leading edge of migrating cells and are formed using much of the same cellular machinery.

Other bacteria induce different sorts of actin "tails." Bacteria of the genus *Rickettsia* that cause spotted fevers induce long, unbranched actin filaments reminiscent of filopodia. Thus, different pathogens have devised various ways of recruiting the host cytoskeleton for propulsion.

Some pathogens bind to the cell surface but are not internalized. For example, the enteropathogenic form of *E. coli*, which causes diarrhea in infants by forming colonies on the surface of intestinal epithelial cells, attaches to the surface of intestinal cells, where it organizes actin-rich "pedestals" that may function like the actin tails induced by *Listeria*.

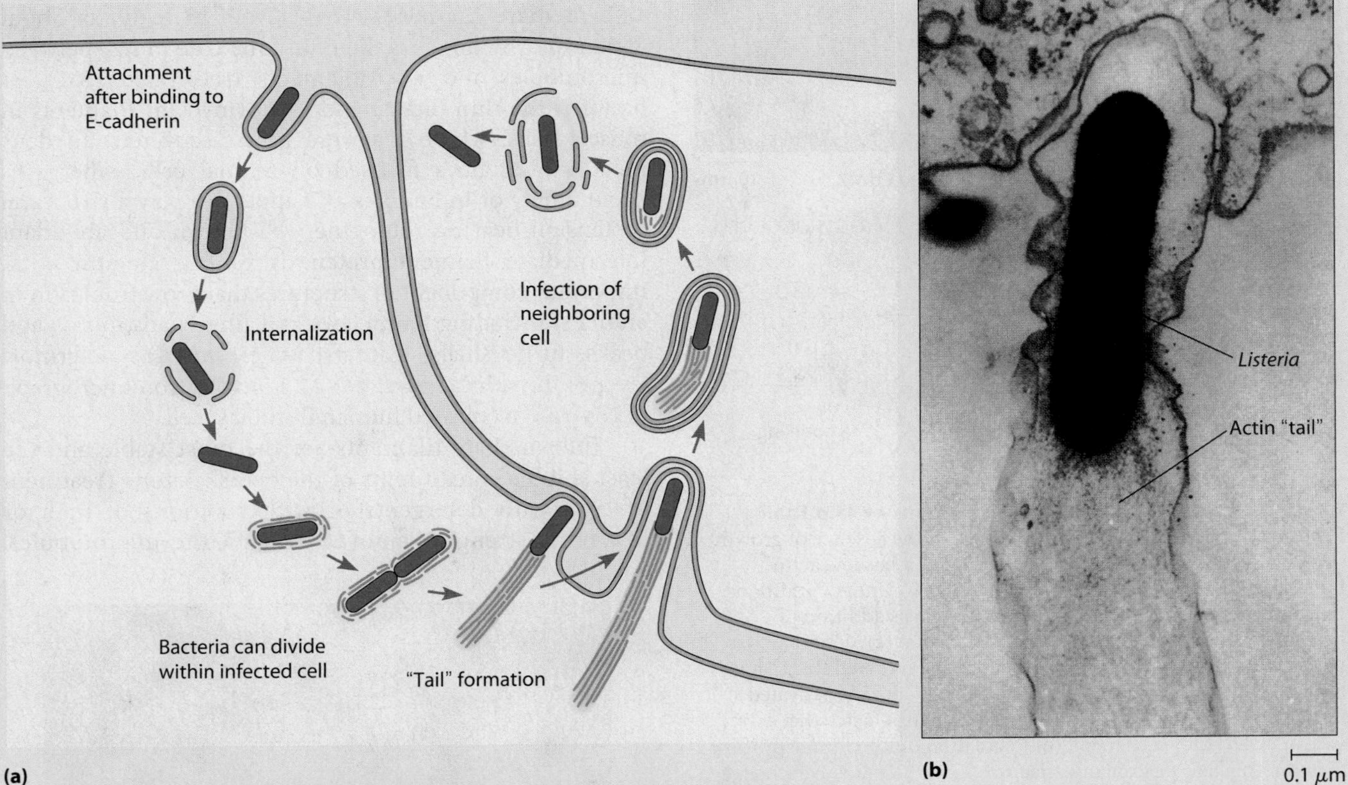

FIGURE 15A-1 Infection of a Macrophage by *Listeria monocytogenes*. **(a)** Life cycle of *Listeria*. A bacterium attaches to the surface of an uninfected cell. The bacterium then moves inside the cell, where it can divide to produce more bacteria in the infected cell. It then spreads to a nearby cell by producing a "comet tail" of polymerized actin, which propels the bacterium forward. **(b)** A transmission electron micrograph showing a *Listeria* within an infected macrophage and the "comet tail" of actin filaments that forms behind the bacterium.

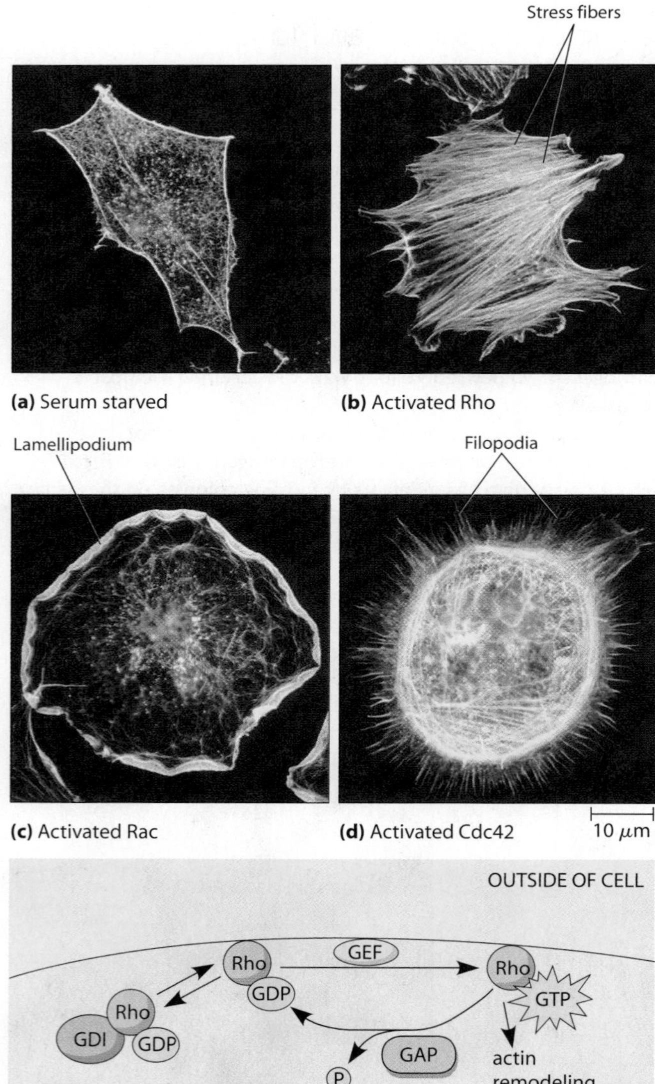

(a) Serum starved (b) Activated Rho (c) Activated Rac (d) Activated Cdc42 (e)

FIGURE 15-21 Regulation of Protrusions by Rho Family Proteins. (a) When a cultured fibroblast in the absence of growth factors ("serum starved") is stained for actin, it has few actin bundles and shows little protrusive activity. **(b)** Under conditions that activate the Rho signaling pathway (such as addition of lysophosphatidic acid, LPA), stress fibers form. **(c)** When the Rac pathway is activated (in this case by injecting mutated Rac that is always active), lamellipodia form. **(d)** When Cdc42 is activated (e.g., by injecting a guanine nucleotide exchange factor that activates Cdc42), filopodia form. **(e)** Regulation of Rho family proteins. Guanine-nucleotide exchanges factors (GEFs) stimulate exchange of a bound GDP for GTP, activating Rho family proteins, and allowing them to stimulate actin remodeling. GTPase activating proteins (GAPs) stimulate Rho GTPases to hydrolyze their bound GTP, thereby inactivating them. Guanine-nucleotide displacement inhibitors (GDIs) can sequester inactive Rho family G proteins, retaining them in the cytosol.

and disassembly of the cytokinetic furrow to regulating endo- and exocytosis. Many of these events are under the control of cell signaling; not surprisingly, then, these proteins are essential for growth factors such as PDGF and LPA to exert their effects.

Like Ras (see Chapter 14), Rho family GTPases are stimulated by *guanine-nucleotide exchange factors (GEFs)*, which foster exchange of a bound GDP for GTP (Figure 15-21e). Corresponding *GTPase activating proteins (GAPs)* stimulate Rho GTPases to hydrolyze their bound GTP, thereby inactivating them. In addition, proteins known as *guanine-nucleotide dissociation inhibitors (GDIs)* can sequester inactive Rho GTPases in the cytosol. All of these events can be modulated by cell signals, allowing fine-tuning of where actin-based structures are assembled. Once active, different Rho GTPases can stimulate different types of actin polymerization. For example, activated Cdc42 can bind to and activate WASP, stimulating actin polymerization by the Arp2/3 complex. Rac can activate another WASP family protein, WAVE. Rho is known to bind to and activate formins, which explains why Rho activation can lead to formation of longer, less branched actin filaments.

Intermediate Filaments

Intermediate filaments (IFs) have a diameter of about 8–12 nm, which makes them intermediate in size between microtubules and microfilaments (see Table 15-1), or between the thin (actin) and thick (myosin) filaments in muscle cells, where IFs were first discovered. To date, most studies have focused on animal cells, where IFs occur singly or in bundles and appear to play a structural or tension-bearing role. One well-known and abundant intermediate filament protein is *keratin*. Keratin is an important component of structures that grow from skin in animals, including hair, claws and fingernails, horns and beaks, turtle shells, feathers, scales, and the outermost layer of the skin. **Figure 15-22** is an electron micrograph of IFs from a cultured human fibroblast cell.

Intermediate filaments are the most stable and the least soluble constituents of the cytoskeleton. Treatment of cells with detergents or with solutions of high or low ionic strength removes most of the microtubules,

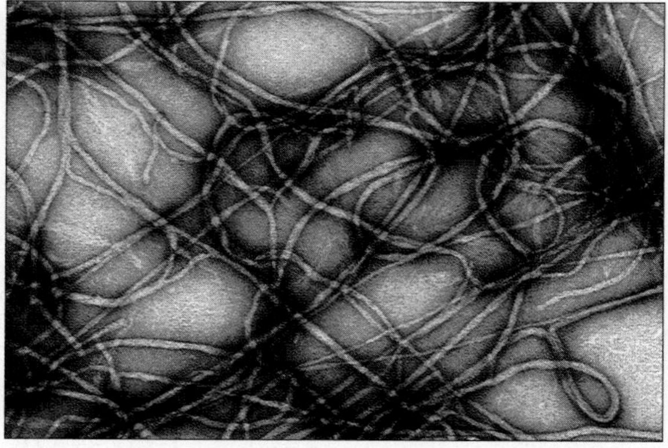

FIGURE 15-22 Intermediate Filaments. Electron micrograph of negatively stained keratin 5 and 14 intermediate filaments reconstituted in vitro (TEM).

microfilaments, and other proteins of the cytosol but leaves networks of IFs that retain their original shape. Due to the stability of the IFs, some scientists suggest that they serve as a scaffold to support the entire cytoskeletal framework. In contrast to MTs and MFs, IFs do not appear to be polarized.

Intermediate Filament Proteins Are Tissue Specific

In contrast to microtubules and microfilaments, intermediate filaments differ markedly in amino acid composition from tissue to tissue. Based on the cell type in which they are found, IFs and their proteins can be grouped into six classes (**Table 15-4**). Classes I and II comprise the *keratins,* proteins that make up the *tonofilaments* found in the epithelial cells covering the body surfaces and lining its cavities. (The IFs visible beneath the terminal web in the intestinal mucosa cell of Figure 15-18 consist of keratin.) Class I keratins are *acidic keratins,* whereas class II are *basic* or *neutral keratins;* each of these classes contains at least 15 different keratins.

Class III IFs include vimentin, desmin, and glial fibrillary acidic protein. *Vimentin* is present in connective tissue and other cells derived from nonepithelial cells. Vimentin-containing filaments are often prominent features in cultured fibroblast cells, in which they form a network that radiates from the center out to the periphery of the cell. *Desmin* is found in muscle cells, and *glial fibrillary acidic (GFA) protein* is characteristic of the glial cells that surround and insulate nerve cells. Class IV IFs are the *neurofilament (NF) proteins* found in the neurofilaments of nerve cells. Class V IFs are *nuclear lamins* A, B, and C, which form a filamentous scaffold along the inner surface of the nuclear membrane of virtually all eukaryotic cells, including those in plants. Neurofilaments found in cells in the embryonic nervous system are made of *nestin,* which constitutes class VI.

As IF proteins and their genes have been sequenced, it has become clear that these proteins are encoded by a single (though large) family of related genes and can therefore be classified according to amino acid sequence relatedness as well. The six classes of IF proteins have been distinguished on this basis (see Table 15-4).

Due to the tissue specificity of intermediate filaments, animal cells from different tissues can be distinguished on the basis of the IF protein present, as determined by immunofluorescence microscopy. This *intermediate filament typing* serves as a diagnostic tool in medicine. IF typing is especially useful in the diagnosis of cancer because tumor cells are known to retain the IF proteins characteristic of the tissue of origin, regardless of where the tumor occurs in the body. Because the appropriate treatment often depends on the tissue of origin, IF typing is especially valuable in cases where diagnosis using conventional microscopic techniques is difficult.

Intermediate Filaments Assemble from Fibrous Subunits

As products of a family of related genes, all IF proteins have some common features, although they differ significantly in size and chemical properties. In contrast to actin and tubulin, all IF proteins are fibrous, rather than globular, proteins. All IF proteins have a homologous central

Table 15-4 Classes of Intermediate Filaments

Class	IF Protein	Molecular Mass (kDa)	Tissue	Function
I	Acidic cytokeratins	40–56.5	Epithelial cells	Mechanical strength
II	Basic cytokeratins	53–67	Epithelial cells	Mechanical strength
III	Vimentin	54	Fibroblasts; cells of mesenchymal origin; lens of eye	Maintenance of cell shape
III	Desmin	53–54	Muscle cells, especially smooth muscle	Structural support for contractile machinery
III	GFA protein	50	Glial cells and astrocytes	Maintenance of cell shape
IV	Neurofilament proteins		Central and peripheral nerves	Axon strength; determines axon size
	NF-L (major)	62		
	NF-M (minor)	102		
	NF-H (minor)	110		
V	Nuclear lamins		All cell types	Form a nuclear scaffold to give shape to nucleus
	Lamin A	70		
	Lamin B	67		
	Lamin C	60		
VI	Nestin	240	Neuronal stem cells	Unknown

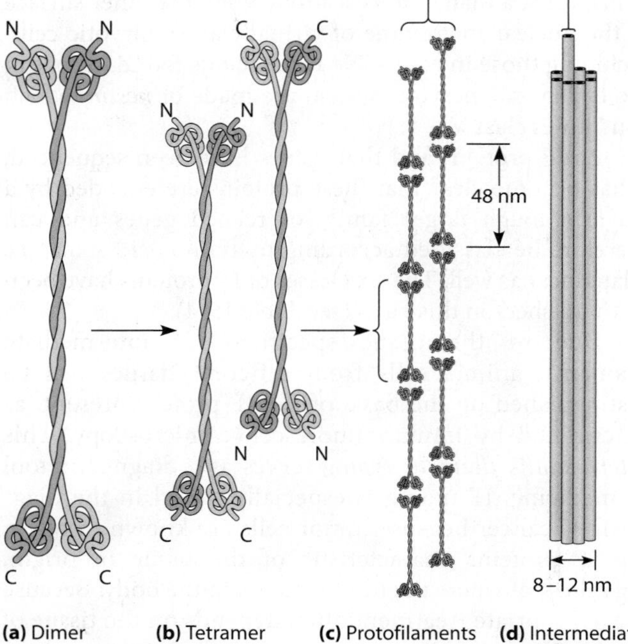

FIGURE 15-23 A Model for Intermediate Filament Assembly in Vitro. **(a)** The starting point for assembly is a pair of intermediate filament (IF) polypeptides. The two polypeptides are identical for all IFs except keratin filaments, which are obligate heterodimers with one each of the type I and type II polypeptides. The two polypeptides twist around each other to form a two-chain coiled coil, with their conserved center domain aligned in parallel. **(b)** Two dimers align laterally to form a tetrameric protofilament. **(c)** Protofilaments assemble into larger filaments by end-to-end and side-to-side alignment. **(d)** The fully assembled intermediate filament is thought to be eight protofilaments thick at any point.

rodlike domain of 310–318 amino acids that has been remarkably conserved in size, in secondary structure, and, to some extent, in sequence. This central domain consists of four segments of coiled helices interspersed with three short linker segments. Flanking the central helical domain are N- and C-terminal domains that differ greatly in size, sequence, and function among IF proteins, presumably accounting for the functional diversity of these proteins.

A possible model for IF assembly is shown in **Figure 15-23**. The basic structural unit of intermediate filaments consists of two IF polypeptides intertwined into a *coiled coil.* The central helical domains of the two polypeptides are aligned in parallel, with the N- and C-terminal regions protruding as globular domains at each end. Two such dimers then align laterally to form a tetrameric *protofilament.* Protofilaments interact with each other, associating in an overlapping manner to build up a filamentous structure both laterally and longitudinally. When fully assembled, an intermediate filament is thought to be eight protofilaments thick at any point, with protofilaments probably joined end to end in staggered overlaps.

Intermediate Filaments Confer Mechanical Strength on Tissues

Intermediate filaments are considered to be important structural determinants in many cells and tissues. Because they often occur in areas of the cell that are subject to mechanical stress, they are thought to have a tension-bearing role. For example, when keratin filaments are genetically modified in the keratinocytes of transgenic mice, the epidermal cells are fragile and rupture easily. In humans, naturally occurring mutations of keratins give rise to a blistering skin disease called *epidermolysis bullosa simplex (EBS).* IF defects are also suspected in other pathological conditions, including *amyotrophic lateral sclerosis (ALS)* and certain types of inherited *cardiomyopathies,* which result from defects in the organization of heart muscle.

Although our discussion of IFs may give the impression that they are static structures, this is not the case. In neurons, for example, IFs are dynamically transported and remodeled. Different IFs form a structural scaffold called the *nuclear lamina* on the inner surface of the nuclear membrane (discussed in detail in Chapters 18 and Chapter 19). The nuclear lamina is composed of three separate IF proteins called *nuclear lamins A, B, and C.* These lamins become phosphorylated and disassemble as a part of nuclear envelope breakdown at the onset of mitosis. After mitosis, lamin phosphatases remove the phosphate groups, allowing the nuclear envelope to form again.

The Cytoskeleton Is a Mechanically Integrated Structure

In the preceding sections, we have looked at the individual components of the cytoskeleton as separate entities. In fact, cellular architecture depends on the unique properties of the different cytoskeletal components working together. Microtubules are generally thought to resist bending when a cell is compressed, while microfilaments serve as contractile elements that generate tension. Intermediate filaments are elastic and can withstand tensile forces.

The mechanical integration of intermediate filaments, microfilaments, and microtubules is made possible by specific linker proteins that connect them, known as *plakins.* One plakin, called *plectin,* is a versatile linker protein that is found at sites where intermediate filaments are connected to microfilaments or microtubules (**Figure 15-24**). Plectin, as well as several other plakins, contains binding sites for intermediate filaments, microfilaments, and microtubules. By linking these major types of polymers, plakins help to integrate them into a mechanically integrated cytoskeletal network. As a result, interconnected cytoskeletal structures can adapt to stretching forces in such a way that the tension-bearing elements become aligned with the direction of stress. These stress-bearing properties of the cytoskeleton are important in epithelial cells such as those that line the gut. These cells are subjected to stress as smooth muscle within the intestinal wall contracts and puts pressure on the contents of the gut.

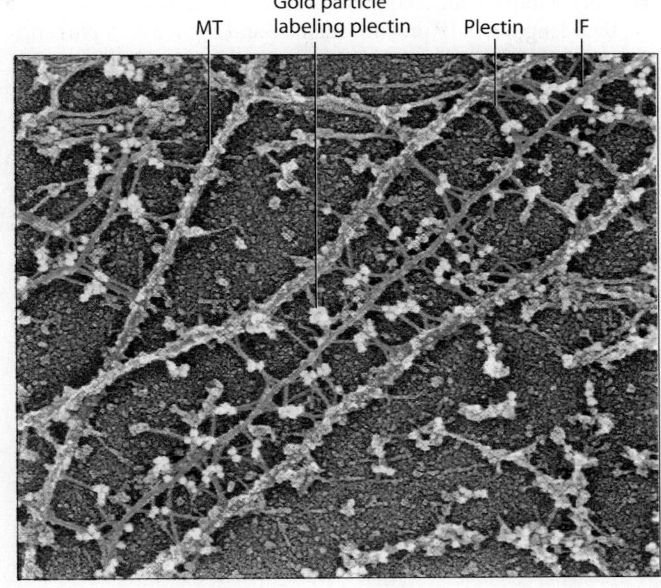

FIGURE 15-24 Connections Between Intermediate Filaments and Other Components of the Cytoskeleton. Intermediate filaments are linked to both microtubules and actin filaments by a protein called plectin. Plectin (red) links IFs (green) to MTs (blue). Gold particles (yellow) label plectin (deep-etch TEM). Plectins can also bind actin MFs (not shown). Here, IFs serve as strong but elastic connectors between the different cytoskeletal filaments.

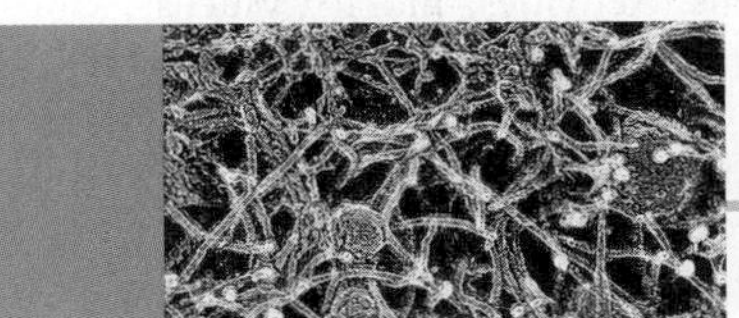

SUMMARY OF KEY POINTS

The Cytoskeleton

- Both prokaryotes and eukaryotes possess an interconnected network of proteins that polymerize from smaller subunits called the cytoskeleton. In eukaryotes, it consists of an extensive three-dimensional network of microtubules (MTs), microfilaments (MFs), and intermediate filaments (IFs) that determines cell shape and allows a variety of cell movements.
- The cytoskeleton is a structural feature of cells that is revealed especially well by digital video, fluorescence, and electron microscopy.
- A variety of drugs can be used to perturb the assembly and disassembly of microtubules and microfilaments in eukaryotes. Such drugs are useful for determining which cellular processes require different cytoskeletal filaments.

Microtubules

- Microtubules (MTs) are hollow tubes with walls consisting of heterodimers of α- and β-tubulin polymerized linearly into protofilaments. Both α- and β-tubulin can bind GTP. First identified as components of the axonemal structures of cilia and flagella and the mitotic spindle of dividing cells, microtubules are now recognized as a general cytoplasmic constituent of most eukaryotic cells.
- MTs are polar structures that elongate preferentially from one end, known as the plus end. MT growth occurs when the concentration of tubulin rises above the critical concentration, the concentration of monomers at which subunit addition is exactly balanced by subunit loss from an MT.
- Drugs such as nocodazole, colchicine, and colcemid cause depolymerization of MTs; taxol stabilizes MTs.
- MTs can undergo cycles of catastrophic shortening and elongation—a phenomenon known as dynamic instability, which involves hydrolysis of GTP by β-tubulin near the plus end, followed by recovery of the GTP-cap at the plus end.
- Within cells, MT dynamics and growth are organized by microtubule-organizing centers (MTOCs). The centrosome is a major MTOC, which contains nucleation sites that are rich in γ-tubulin and are used to nucleate MT growth.

- Microtubule-associated proteins (MAPs) stabilize MTs along their length, +-TIP proteins stabilize and anchor their plus ends, and catastrophins hasten their catastrophic depolymerization.

Microfilaments

- Microfilaments (MFs), or F-actin, are double-stranded polymers of G-actin monomers, which bind ATP. Originally discovered because of their role in muscle cells, MFs are components of virtually all eukaryotic cells.
- Like microtubules, MFs are polar structures; G-actin monomers preferentially add to the plus (barbed) ends of MFs; their minus (pointed) ends display far slower subunit addition and loss.
- Actin-binding proteins tightly regulate the polymerization and function of F-actin. These include monomer-binding proteins, which regulate polymerization, and proteins that cap, crosslink, sever, bundle, and anchor F-actin in various ways.
- Microfilament assembly within cells is regulated by cell signaling, which can be mediated by the activity of phosphoinositides and by the monomeric G proteins, Rho, Rac, and Cdc42.

Intermediate Filaments

- Intermediate filaments (IFs) are the most stable and least soluble constituents of the cytoskeleton. They appear to play a structural or tension-bearing role within cells.
- IFs are tissue specific and can be used to identify cell type. Such typing is useful in the diagnosis of cancer.
- All IF proteins have a highly conserved central domain flanked by terminal regions that differ in size and sequence, presumably accounting for the functional diversity of IF proteins.
- IFs, MTs, and MFs are interconnected within cells to form cytoskeletal networks that can withstand tension and compression, providing mechanical strength and rigidity to cells.

PROBLEM SET

More challenging problems are marked with a •.

15-1 Filaments and Tubules. Indicate whether each of the following descriptions is true of microtubules (MT), microfilaments (MF), intermediate filaments (IF), or none of these (N). More than one response may be appropriate for some statements.

(a) Structurally similar proteins are found in bacterial cells.

(b) More important for cytokinesis than for chromosome movements in animal cells.

(c) The fundamental repeating subunit is a dimer.

(d) Involved in muscle contraction.

(e) Play well-documented roles in cell movement.

(f) More important for chromosome movements than for cytokinesis.

(g) Involved in the movement of cilia and flagella.

(h) Most likely to remain when cells are treated with solutions of nonionic detergents or solutions of high ionic strength.

(i) Can be detected by immunofluorescence microscopy.

(j) Their subunits can bind and catalyze hydrolysis of phosphonucleotides.

15-2 True or False. Identify each of the following statements as true (T) or false (F). Provide a brief justification for your answer.

(a) The minus end of microtubules and microfilaments is so named because subunits are lost and never added there.

(b) The energy required for tubulin and actin polymerization is provided by hydrolysis of a guanosine triphosphate.

(c) All cytoskeletal fibers—microtubules, intermediate filaments, and actin filaments—are synthesized in various lengths on the ribosomes bound to the ER.

(d) Taxol is used to inhibit the polymerization of intermediate filaments.

(e) An algal cell contains neither tubulin nor actin.

(f) Polymerization of actin into filaments is influenced in differential ways by Rho, Rac, and Cdc42 GTPases.

(g) Mitosis and meiosis can proceed normally in a cell even if it is exposed to taxol.

(h) The cytoskeletal framework within a cell is highly dynamic.

15-3 Cytoskeletal Studies. Described here are the results of several recent studies on the proteins of the cytoskeleton. In each case, state the conclusion(s) that can be drawn from the findings.

(a) Small vesicles containing pigment inside of pigmented fish epidermal cells aggregate or disperse in response to treatment with certain chemicals. When nocodazole is added to cells in which the pigment granules have been induced to aggregate, the granules cannot disperse again.

(b) When an animal cell is treated with colchicine, its microtubules depolymerize and virtually disappear. If the colchicine is then washed away, the MTs appear again, beginning at the centrosome and elongating outward at about the rate ($1\mu m/min$) at which tubulin polymerizes in vitro.

(c) A macrophage cell can 'crawl' towards foreign pathogens and inactivate them. A diver came in contact with *Latrunculia magnifica* via an open wound. If the diver is exposed to harmful bacteria at the same time, how will the macrophages in the diver's blood respond?

• **15-4 Stabilization and the Critical Concentration.** Suppose you have determined the overall critical concentration for a sample of purified tubulin. Then you add a preparation of centrosomes (microtubule-organizing centers), which nucleate microtubules so that the minus end is bound to the centrosome and stabilized against disassembly. When you again determine the overall critical concentration, you find it is different. Explain why the overall critical concentration would change.

15-5 Actin' up. The polymerization of G-actin into microfilaments can be followed using pyrene-labeled actin and a device to measure the fluorescence of polymerized actin in a test tube. The increase in fluorescence can then be plotted over time, similar to light-scattering experiments to measure tubulin polymerization (**Figure 15-25**). Examine the graph, and identify which of the curves corresponds to each of the following situations. In each case, state your reasoning.

(a) Pyrene actin is added alone in the presence of buffer.

(b) Pyrene actin is added along with purified Arp2/3 complex proteins.

(c) Pyrene actin is added along with purified Arp2/3 complex proteins and a purified protein fragment that corresponds to active N-WASP (a WASP family protein).

• **15-6 Spongy Actin.** You are interested in the detailed effects of cytochalasin D on microfilaments over time. Based on what you know about the molecular mechanism of action of cytochalasins, draw diagrams of what happens to microfilaments within cells treated with the drug. In particular, explain why existing actin polymers eventually depolymerize.

• **15-7 A New Wrinkle.** Fibroblasts can be placed on thin sheets of silicone rubber. Under normal circumstances, fibroblasts exert sufficient tension on the rubber so that it visibly wrinkles. Explain how the ability of fibroblasts to wrinkle rubber would compare with that of normal cells under the following conditions.

(a) The cells are injected with large amounts of purified gelsolin.

(b) The cells are treated with a permeable form of C3 transferase, a toxin from the bacterium, *Clostridium botulinum*, which covalently modifes an amino acid in Rho by ADP-ribosylation. This renders Rho unable to bind to proteins that it would normally bind to when active.

(c) The cells are treated with nocodazole.

(d) The cells are treated with cytochalasin D the drug is washed out, and the cells are observed periodically.

(e) A constitutively activated Cdc42 is introduced into the cell.

15-8 Stressed Out. It is now possible, using nanoengineering techniques, to attach magnetic beads to the surface of large cells to measure how mechanically stiff they are. It is also known that acrylamide, which is polymerized to make gels for protein electrophoresis, is very toxic. One effect of acrylamide is to depolymerize intermediate filaments, such as keratin. What effect would you predict acrylamide treatment would have on the mechanical rigidity of a keratinocyte (skin cell)? Explain your answer.

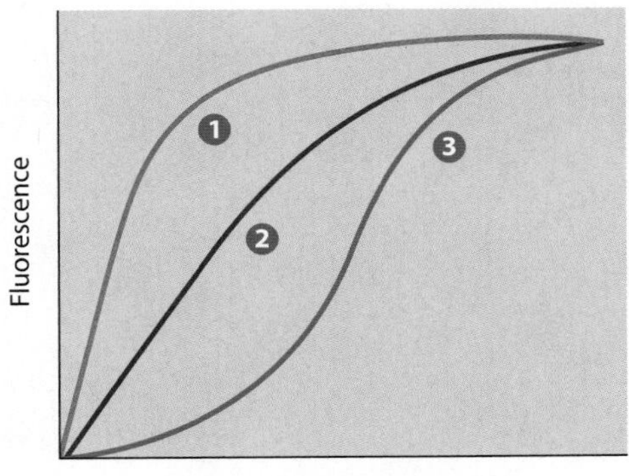

FIGURE 15-25 Actin Polymerization Kinetics Measured by Pyrene Actin Incorporation Under Various Experimental Conditions. See Problem 15-5.

SUGGESTED READING

References of historical importance are marked with a •.

General References

Pollard, T. D. The cytoskeleton, cellular motility and the reductionist agenda. *Nature* 422 (2003): 741.

Techniques for Studying the Cytoskeleton

• Bridgman, P. C., and T. S. Reese. The structure of cytoplasm in directly frozen cultured cells. 1. Filamentous meshworks and the cytoplasmic ground substance. *J. Cell Biol.* 99 (1980): 1655.

• Hirokawa, N., and J. E. Heuser. Quick-freeze, deep-etch visualization of the cytoskeleton beneath surface differentiations of intestinal epithelial cells. *J. Cell Biol.* 91 (1981): 399s.

Prokaryotic Cytoskeleton

Cabeen, M. T., and C. Jacobs-Wagner. Skin and bones: The bacterial cytoskeleton, cell wall, and cell morphogenesis. *J. Cell Biol.* 179 (2007): 381.

Graumann, P. L. Cytoskeletal elements in bacteria. *Annu. Rev. Microbiol.* 61 (2007): 589.

Microtubules

Akhmanova, A., and M. O. Steinmetz. Tracking the ends: A dynamic protein network controls the fate of microtubule tips. *Nat Rev Mol Cell Biol.* 9 (2008): 309.

Gardner, M. K., A. J. Hunt, H. V. Goodson, and D. J. Odde. Microtubule assembly dynamics: New insights at the nanoscale. *Curr. Opin. Cell Biol.* 20 (2008): 64.

Howard, J., and A. A. Hyman. Microtubule polymerases and depolymerases. *Curr. Opin. Cell Biol.* 19 (2007): 31.

• Mitchison, T., and M. Kirschner. Dynamic instability of microtubule growth. *Nature* 312 (1984): 237.

• Nicolaou, K. C. et al. Taxoids: New weapons against cancer. *Sci. Amer.* 274 (1996): 94.

Wiese, C., and Y. Zheng. Microtubule nucleation: Gamma-tubulin and beyond. *J. Cell Sci.* 119 (2006): 4143.

Microfilaments

Baines, A. J. Evolution of spectrin function in cytoskeletal and membrane networks. *Biochem. Soc. Trans.* 37 (2009): 796–803.

Goode, B. L., and M. J. Eck. Mechanism and function of formins in the control of actin assembly. *Annu. Rev. Biochem.* 76 (2007): 593.

Heasman, S. J., and A. J. Ridley. Mammalian Rho GTPases: New insights into their functions from in vivo studies. *Nat. Rev. Mol. Cell Biol.* 9 (2008): 690.

• Heintzelman, M. B., and M. S. Mooseker. Assembly of the intestinal brush border cytoskeleton. *Curr. Top. Dev. Biol.* 26 (1992): 93.

Hughes, S. C., and R. G. Fehon. Understanding ERM proteins—The awesome power of genetics finally brought to bear. *Curr. Opin. Cell Biol.* 19 (2007): 51.

Insall, R. H., and L. M. Machesky. Actin dynamics at the leading edge: From simple machinery to complex networks. *Dev. Cell.* 17 (2009): 310.

Jaffe, A. B., and A. Hall. Rho GTPases: Biochemistry and biology. *Annu. Rev. Cell Dev. Biol.* 21 (2005): 247.

Lambrechts, A., K. Gevaert, P. Cossart, J. Vandekerckhove, and M. Van Troys. Listeria comet tails: The actin-based motility machinery at work. *Trends Cell. Biol.* 18 (2008): 220.

Niggli, V. Regulation of protein activities by phosphoinositide phosphates. *Annu. Rev. Cell Dev. Biol.* 21 (2005): 57.

Pollard, T. D. Regulation of actin filament assembly by Arp2/3 complex and formins. *Annu. Rev. Biophys. Biomol. Struct.* 36 (2007): 451.

• Schroder, R. R. et al. Three-dimensional atomic model of F-actin decorated with *Dictyostelium* myosin S1. *Nature* 364 (1993): 171.

Stossel, T. P. et al. Filamins as integrators of cell mechanics and signalling. *Nature Rev. Mol. Cell Biol.* 2 (2001): 138–145.

Winder, S. J., and K. R. Ayscough. Actin-binding proteins. *J. Cell Sci.* 118 (2005): 651.

Intermediate Filaments

Fuchs, E., and D. W. Cleveland. A structural scaffolding of intermediate filaments in health and disease. *Science* 279 (1998): 514.

Goldman, R. D., B. Grin, M. G. Mendez, and E. R. Kuczmarski. Intermediate filaments: Versatile building blocks of cell structure. *Curr. Opin. Cell Biol.* 20 (2008): 28.

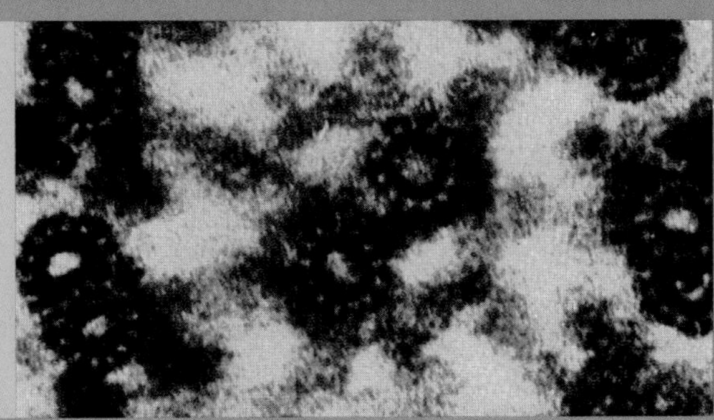

16 Cellular Movement: Motility and Contractility

See MasteringBiology® for tutorials, activities, and quizzes.

In the previous chapter, we saw that the cytoskeleton of eukaryotic cells serves as an intracellular scaffold that organizes structures within the cell and shapes the cell itself. In this chapter, we will explore the role of these cytoskeletal elements in cellular **motility.** *This may involve the movement of a cell (or a whole organism) through its environment, the movement of the environment past or through the cell, the movement of components within the cell, or the shortening of the cell itself.* ***Contractility,*** *a related term often used to describe the shortening of muscle cells, is a specialized form of motility.*

Motile Systems

Motility occurs at the tissue, cellular, and subcellular levels. The most conspicuous examples of motility, particularly in the animal world, take place at the tissue level. The muscle tissues common to most animals consist of cells specifically adapted for contraction. The movements produced are often obvious, whether manifested as the bending of a limb, the beating of a heart, or a uterine contraction during childbirth.

At the cellular level, motility occurs in single cells or in organisms consisting of one or only a few cells. It occurs among cell types as diverse as ciliated protozoa and motile sperm, depending in each case on cilia or flagella, cellular appendages adapted for propulsion. Other examples include the actin-dependent migration of single cells, amoeboid movement, and the invasiveness of cancer cells in malignant tumors.

Equally important is the movement of *intracellular* components. For example, highly ordered microtubules of the mitotic spindle play a key role in the separation of chromosomes during cell division, as we will see in Chapter 19. Nondividing cells continuously shuttle components—such as RNAs, multiprotein complexes, and the membrane-bounded vesicles we learned about in Chapter 12—from one location to another.

Motility is an especially intriguing use of energy by cells because it often involves the conversion of chemical energy directly to mechanical energy. In contrast, most mechanical devices that produce movement from chemicals (such as an automobile engine, which depends on the combustion of gasoline) require an intermediate form of energy, (e.g., heated gasses). As we will see, the cellular energy for motility typically comes from ATP, whose hydrolysis is coupled to changes in the shape of specific proteins that mediate movement. To generate movement, the microfilaments and microtubules of the cytoskeleton provide a basic scaffold for specialized **motor proteins,** or **mechanoenzymes,** which interact with the cytoskeleton to produce motion at the molecular level. The combined effects of these molecular motions produce movement at the cellular level. In cases such as muscle contraction, the combined effects of many cells moving simultaneously produce motion at the tissue level.

In eukaryotes, there are two major motility systems. The first involves interactions between specialized motor proteins and microtubules. Microtubules are abundant in cells and are used for a variety of intracellular movements. A specific example of such microtubule-based movement is *fast axonal transport,* one of the processes used by a nerve cell to transport materials between its cell body and outlying regions. Another process is the sliding of microtubules in *cilia* and *flagella.* The second type of eukaryotic motility requires interactions between actin microfilaments and members of the *myosin* family of motor molecules. A familiar example of microfilament-based movement is *muscle contraction.* We begin our detailed examination of the two motility systems with two proteins essential for microtubule-based movement.

Intracellular Microtubule-Based Movement: Kinesin and Dynein

Microtubules (MTs) provide a rigid set of tracks for the transport of a variety of membrane-enclosed organelles and vesicles. As we saw in Chapter 15, the centrosome organizes and orients MTs because the minus ends of most MTs are embedded in the centrosome. The centrosome is generally located near the center of the cell, so traffic toward the minus ends of MTs might be considered "inbound" traffic. Traffic directed toward the plus ends might likewise be considered "outbound," meaning that it is directed toward the periphery of the cell.

While microtubules provide an organized set of tracks along which organelles can move, they do not directly generate the force necessary for movement. The mechanical work needed for movement depends on *microtubule-associated motor proteins,* which attach to vesicles or organelles and then "walk" along the MT, using ATP to provide the needed energy. Furthermore, MT motor proteins recognize the polarity of the MT, with each motor protein having a preferred direction of movement. At present, we are aware of two major families of MT motors: **kinesins** and **dyneins** (**Table 16-1**).

Table 16-1 Selected Motor Proteins of Eukaryotic Cells

Motor Protein	Typical Function
Microtubule (MT)-Associated Motors	
Dyneins	
Cytoplasmic dynein	Moves cargo toward minus ends of MTs
Axonemal dynein	Activates sliding in flagellar MTs
Kinesins*	
Kinesin 1 (classic kinesin)	Dimer; moves cargo toward plus ends of MTs
Kinesin 3	Monomer; movement of synaptic vesicles in neurons
Kinesin 5	Bipolar, tetrameric; bidirectional sliding of MTs during anaphase of mitosis
Kinesin 6	Completion of cytokinesis
Kinesin 13 ("catastrophins")	Dimer; destabilization of plus ends of MTs
Kinesin 14	Spindle dynamics in meiosis and mitosis; moves toward minus end of MTs
Microfilament (MF)-Associated Motors	
Myosins*	
Myosin I	Motion of membranes along MFs; endocytosis
Myosin II	Slides MFs in muscle; other contractile events such as cytokinesis, cell migration
Myosin V	Vesicle positioning and trafficking
Myosin VI	Endocytosis; moves toward minus ends of MFs
Myosin VII	Base of stereocilia in inner ear
Myosin X	Tips of filopodia
Myosin XV	Tips of stereocilia in inner ear

*Kinesins and myosins comprise large families of proteins. There are many families of kinesins and myosins.

MT Motor Proteins Move Organelles Along Microtubules During Axonal Transport

A historically important cell type for studying microtubule-dependent intracellular movement is the squid giant axon (see Figure 13-2 for further details regarding the squid giant axon). Proteins or neurotransmitters synthesized in the cell body of the neuron must be transported over distances up to a meter between the cell body and the nerve ending. The need for such transport arises because ribosomes are present only in the cell body, so no protein synthesis occurs in the axons or synaptic knobs. Instead, proteins and membranous vesicles are synthesized in the cell body and transported along the axons to the synaptic knobs. Some form of energy-dependent transport is clearly required, and MT-based movement provides the mechanism. The process, called **fast axonal transport,** involves the movement of vesicles and other organelles along MTs. (We will not discuss slow axonal transport, which involves somewhat different processes.)

The role of microtubules in axonal transport was initially suggested because the process is inhibited by drugs that depolymerize MTs but is insensitive to drugs that affect microfilaments. Since then, MTs have been visualized along the axon and have been shown to be prominent features of the axonal cytoskeleton. Moreover, axonal MTs have small, membranous vesicles and mitochondria associated with them (**Figure 16-1**).

Evidence that a motor protein drives the movements of organelles was obtained when investigators found that organelles in the presence of ATP could move along fine filamentous structures present in exuded *axoplasm* (the cytosol of axons), which could be visualized by video-enhanced differential interference contrast microscopy (see the Appendix for details). The rate of organelle movement was shown to be about 2 μm/sec, comparable to the axonal transport rate in intact neurons. A combination of immunofluorescence and electron microscopy demonstrated that the fine filaments the organelles move along are single microtubules.

Since that time, two MT motor proteins responsible for fast axonal transport, kinesin 1 and cytoplasmic dynein, have been purified and characterized. To determine the direction of transport by these motors, purified proteins were used experimentally to drive the transport of polystyrene beads along microtubules polymerized from purified centrosomes. When polystyrene beads, purified kinesin, and ATP were added to such MTs, the beads moved toward the plus ends (i.e., away from the centrosome). This finding means that in a nerve cell,

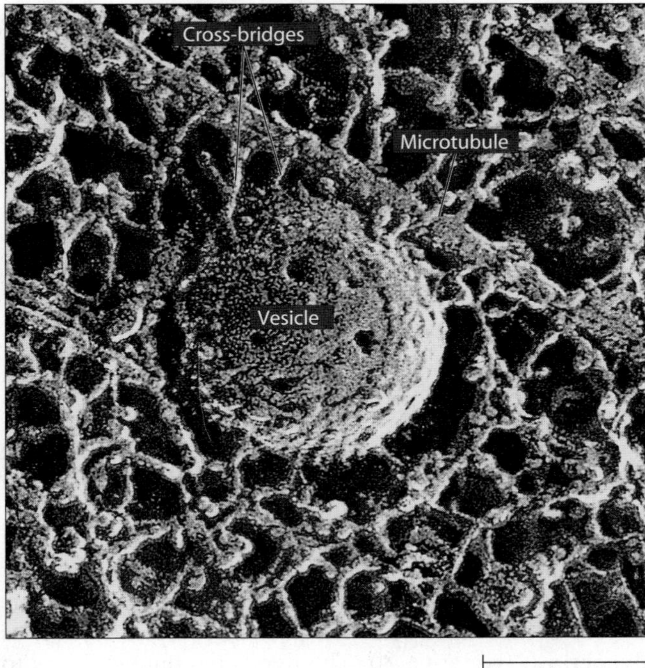

FIGURE 16-1 A Vesicle Attached to a Microtubule in a Crayfish Axon. Cross-bridges connect the membrane vesicle (round structure at center) to the microtubule (deep-etch TEM).

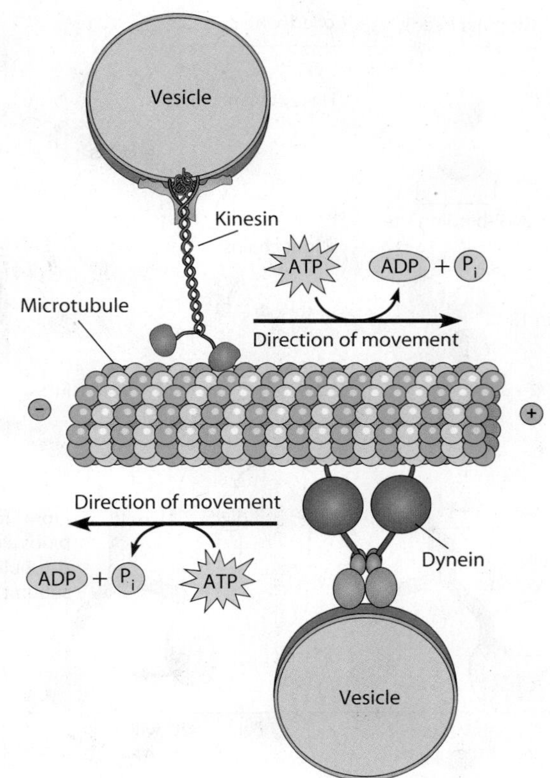

FIGURE 16-2 Microtubule-Based Motility. Kinesins and dyneins are motor proteins that couple ATP hydrolysis to conformational changes to "walk" along microtubules. Most kinesins move vesicles or organelles toward the plus ends of MTs. Dyneins move in the opposite direction, toward the minus ends of MTs.

kinesin mediates transport from the cell body down the axon to the nerve ending (called *anterograde axonal transport*). When similar experiments were carried out with purified cytoplasmic dynein, particles were moved in the opposite direction, toward the minus ends of the MTs (called *retrograde axonal transport* when it occurs in neurons). Thus these two motors transport their *cargo,* those organelles or components that they shuttle along microtubules, in opposite directions within the cytosol (**Figure 16-2**), a finding that has been abundantly confirmed by more recent molecular and genetic analysis.

Figure 16-2 may suggest that cargo always moves in one direction or another along MTs. In some cases, such as fast axonal transport, such unidirectional transport is the rule. However, some vesicles are capable of changing direction as they move along an MT. In these cases cargo is attached to both kinesins and dyneins; the direction of motion seems to be determined by which type of motor predominates.

Motor Proteins Move Along Microtubules by Hydrolyzing ATP

The first kinesins, originally identified in the cytoplasm of squid giant axons, consist of three parts: a globular *head* region that attaches to microtubules and is involved in hydrolysis of ATP, a coiled helical region, and a *light-chain* region that is involved in attaching the kinesin to other proteins or organelles (**Figure 16-3a**). The movements of single kinesin molecules along MTs have been studied by tracking the movement of beads attached to kinesin or by measuring the force exerted by single kinesin molecules using calibrated glass fibers or beads trapped in a special type of laser beam known as an "optical tweezer" (see the Appendix). Classic kinesins move along MTs in 8-nm steps: One of the two globular heads moves forward to make an attachment to a new β-tubulin subunit, followed by detachment of the trailing globular head, which can now make an attachment to a new region of the MT. It is easiest to visualize this movement as analogous to walking, with the two globular heads taking turns as the front "foot" (Figure 16-3b). This movement is coupled to the hydrolysis of ATP bound at specific sites within the heads. The result is that kinesin moves toward the plus end of an MT in an ATP-dependent fashion. A single kinesin-1 molecule exhibits *processivity.* It can cover long distances before detaching from an MT. To do so, it releases its bound ADP and acquires a new molecule of ATP, allowing the cycle to repeat. A single kinesin molecule can move as far as 1 μm, which is a great distance relative to its size. As a molecular motor, kinesin appears to be quite efficient; estimates of its efficiency in converting the energy of ATP hydrolysis to useful work are on the order of 60–70%.

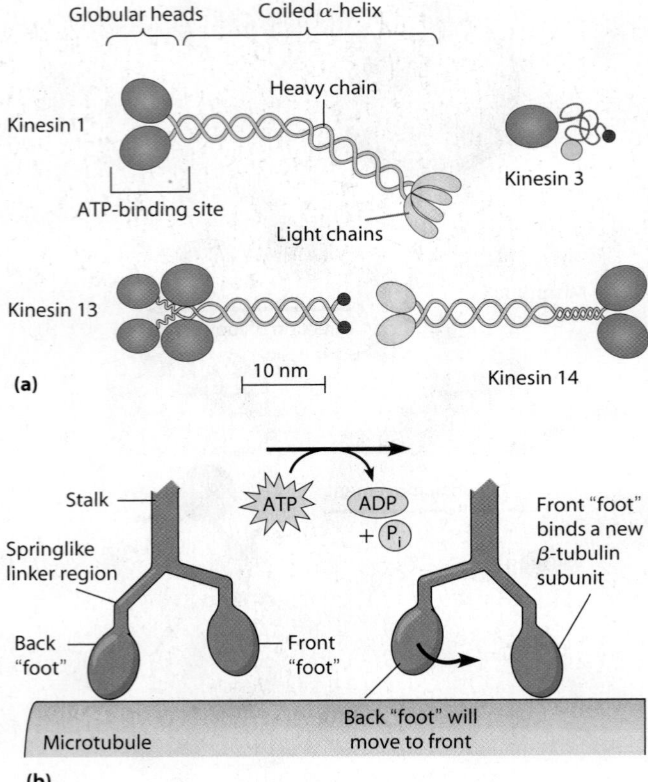

FIGURE 16-3 Movement of Kinesin. **(a)** The basic structure of several kinesin family members. **(b)** Kinesins "walk" along microtubules. The front "foot," one of kinesin's two globular heads, detaches from a β-tubulin subunit by hydrolyzing ATP and moves forward. The rear "foot" then releases by hydrolyzing ATP and springs forward, moving to the front.

Kinesins Are a Large Family of Proteins with Varying Structures and Functions

Since the discovery of the original kinesin involved in anterograde transport in neurons, many other kinesins have been discovered. They can be grouped into families based on their structure (Table 16-1). Some of these kinesins form dimers with another identical protein or with another, different kinesin. One family (kinesin 14) acts as minus-end-directed motors, rather than as plus-end motors. Biochemical and immunocytochemical studies, as well as analysis of mutant organisms, have shown that kinesins are involved in many different processes within cells. One important function for kinesin is moving and localizing substances within cells. These include RNAs, multiprotein complexes, and membranous organelles. (We will consider their role in organelle transport later in this chapter.) We saw another role for kinesins in Chapter 15: one family of kinesins, the catastrophins, aids depolymerization of MTs. Various kinesins localize to the mitotic or meiotic spindle or to kinetochores, where they play roles in various phases of mitosis and meiosis (see Chapter 19). Other kinesins are required for completion of mitosis and cytokinesis. For example, cells in mutant fruit flies or nematode worms lacking one of these kinesins (called MKLP kinesins) initiate cytokinesis, but, because the cleavage furrow never deepens, cell division fails.

To attach to their cargo, kinesins often use their light chains. In some cases, the light chains attach directly to cargo. In other cases, such as mitochondria, adapter proteins allow the light chains to attach to cargo indirectly.

Dyneins Can Be Grouped into Two Major Classes: Axonemal and Cytoplasmic Dyneins

The dynein family of motor proteins consists of two basic types: cytoplasmic dynein and axonemal dynein (Table 16-1). In contrast to the kinesins, few cytoplasmic dyneins have been identified. **Cytoplasmic dynein** contains two heavy chains that interact with microtubules, two intermediate chains, two light intermediate chains, and various light chains. In contrast to most kinesins, cytoplasmic dynein moves toward the minus ends of MTs. Cytoplasmic dynein is associated with the protein complex known as **dynactin** (**Figure 16-4**). The dynactin complex helps to link cytoplasmic dynein to cargo (such as membranous vesicles) it transports along MTs, by binding to proteins such as spectrin attached to the membrane of its cargo.

At least four types of **axonemal dynein** have been identified. We will discuss the function of axonemal dynein in cilia and flagella in some detail later in this chapter.

Microtubule Motors Are Involved in Shaping the Endomembrane System and Vesicle Transport

In Chapter 12 we saw that the cell has an elaborate transportation system of vesicles and membranous organelles. First, recall that the endomembrane system is a complex

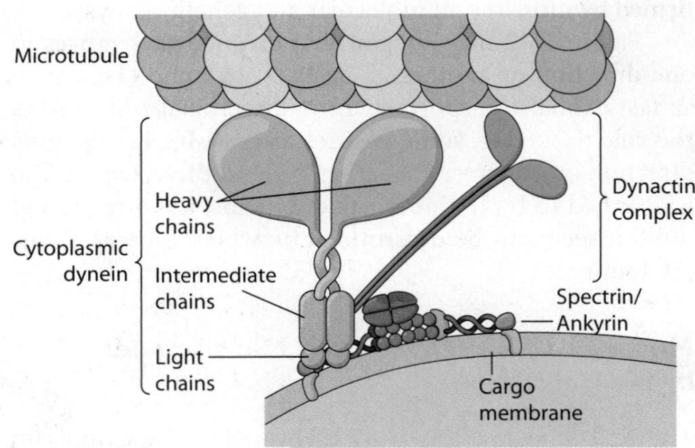

FIGURE 16-4 Schematic Representation of the Cytoplasmic Dynein/Dynactin Complex. Cytoplasmic dynein (brown) is linked with cargo membranes indirectly through the dynactin multiprotein complex (aqua). The dynactin complex binds to a complex containing spectrin and ankyrin, associated with the membrane of the cargo vesicle.

network of membrane-bounded tubules, important not only for export and processing of newly synthesized proteins but also for events such as calcium release (see Chapter 14). MT motors appear to be important for dynamically shaping this complicated network. Live imaging of MTs and endoplasmic reticulum (ER) membrane in vitro and in living cells indicates that extensions of the ER can be moved along MTs.

Second, recall that the Golgi complex is a series of flattened membrane stacks located in the region of the centrosome. The function of the Golgi is to receive proteins made in the ER and to process and package those proteins for distribution to the correct cellular destinations. At each step of this process, proteins are transported in vesicles. Thus, there is a continuous flow of vesicles to and from the Golgi. The vesicles are carried by MT motors on microtubule tracks (**Figure 16-5**).

Several experiments demonstrate that microtubule-based transport is crucial for maintenance of the Golgi stacks. For example, if MTs are depolymerized using the drug nocodazole (see Chapter 15, page 428), the Golgi complex disperses. When the nocodazole is washed out, the Golgi complex re-forms. Similarly, disruption of the function of the dynactin complex by inducing cells to produce too much of one of its subunits results in collapse of the Golgi complex and disruption of the transport of intermediates from the ER to the Golgi.

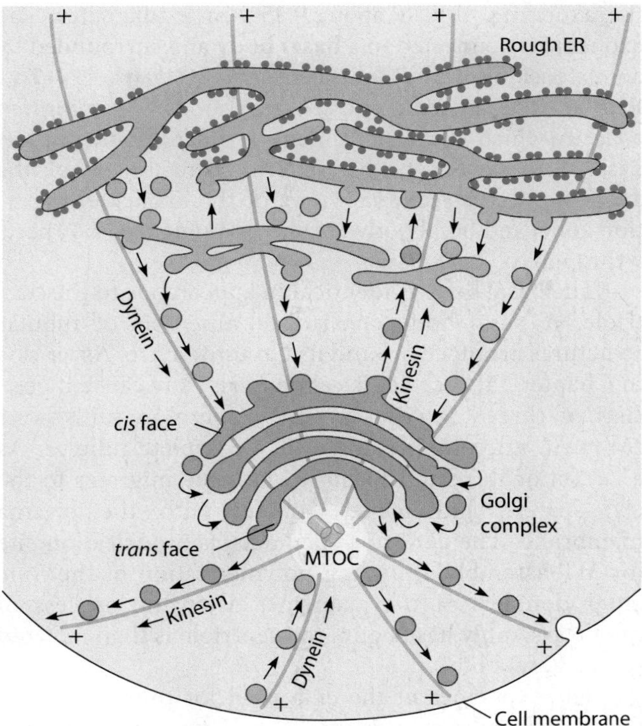

FIGURE 16-5 Microtubules and the Endomembrane System. Vesicles going to and from the Golgi complex are attached to microtubules and carried by MT motors. Vesicles derived from either the ER or the cell membrane are carried toward the Golgi complex and microtubule-organizing center (MTOC) by dynein, whereas vesicles derived from the Golgi complex are carried toward either the ER or the cell periphery by kinesins.

Proteins are processed as they move through the stacks of Golgi membranes. The finished proteins emerge—still packaged in vesicles—from the other side of the complex. Plus-end-directed MT motors carry the finished vesicles away from the Golgi complex toward the cell periphery. In organisms in which various kinesins are mutated, specific defects in vesicle transport have been observed, providing evidence that these motors are crucial for plus-end-directed vesicular transport.

Microtubule-Based Motility: Cilia and Flagella

Cilia and Flagella Are Common Motile Appendages of Eukaryotic Cells

Microtubules are crucial not only for movement within cells but also for the movements of *flagella* and *cilia*, the motile appendages of eukaryotic cells. The two appendages share a common structural basis and differ only in relative length, number per cell, and mode of beating. **Cilia** (singular: **cilium**) have a diameter of about 0.25 μm, are about 2–10 μm long, and tend to occur in large numbers on the surface of ciliated cells. Each cilium is bounded by an extension of the plasma membrane and is therefore an intracellular structure.

Cilia occur in both unicellular and multicellular eukaryotes. Unicellular organisms, such as *Paramecium*, a protozoan, use cilia for both locomotion and the collection of food particles. In multicellular organisms, cilia serve primarily to move the environment past the cell rather than to propel the cell through the environment. The cells that line the air passages of the human respiratory tract, for example, have several hundred cilia each, which means that every square centimeter of epithelial tissue lining the respiratory tract has about a billion cilia (**Figure 16-6a**)! The coordinated, wavelike beating of these cilia carries mucus, dust, dead cells, and other foreign matter out of the lungs. One of the health hazards of cigarette smoking lies in the inhibitory effect that smoke has on normal ciliary beating. Certain respiratory ailments can also be traced to defective cilia.

Cilia display an oarlike pattern of beating, with a power stroke perpendicular to the cilium, thereby generating a force parallel to the cell surface. The cycle of beating for an epithelial cilium is shown in Figure 16-6b. Each cycle requires about 0.1–0.2 sec and involves an active power stroke followed by a recovery stroke.

Flagella (singular: **flagellum**) move cells through a fluid environment. Although they have the same diameter as cilia, flagella are often much longer—from 1 μm to several millimeters, though usually in the range of 10–200 μm—and may be limited to one or a few per cell (Figure 16-6c). Flagella differ from cilia in the nature of their beat. Flagella

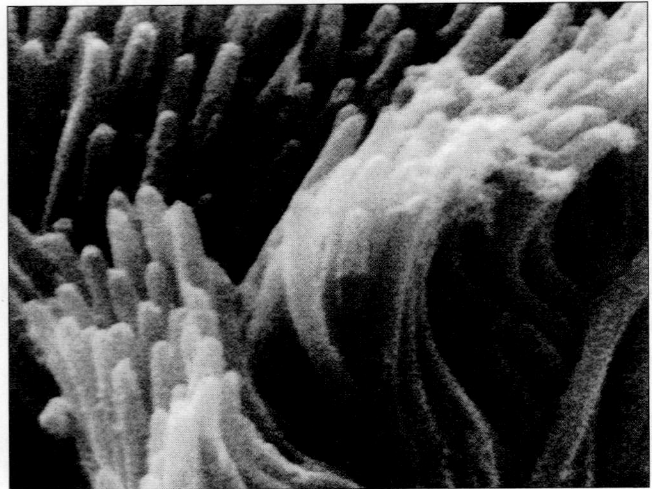

(a) Cilia on a mammalian tracheal cell

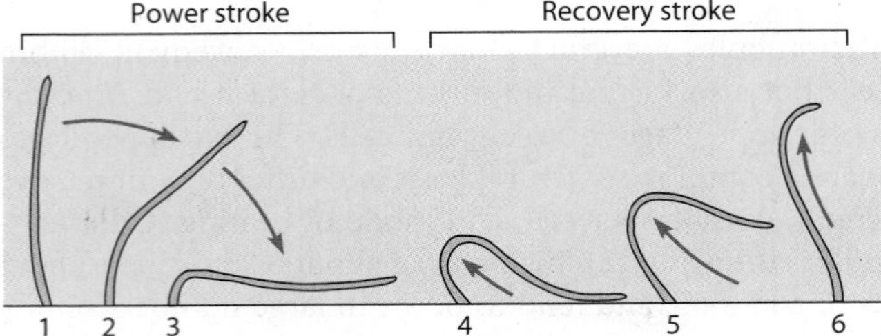

(b) Beating of a cilium

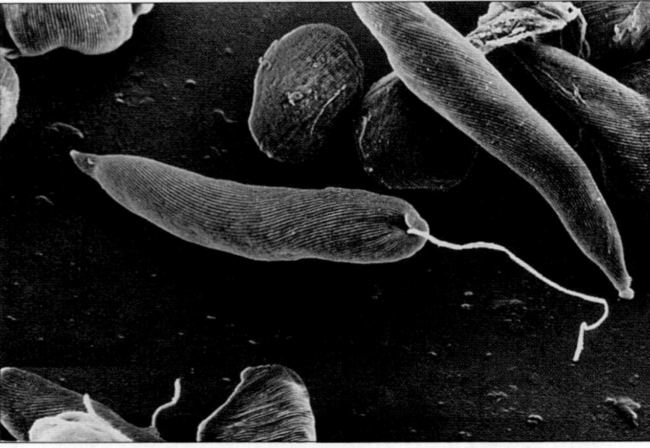

(c) Flagellum on unicellular alga *Euglena*

(d) Movement of flagellated eukaryotic cell

FIGURE 16-6 Cilia and Flagella. **(a)** A micrograph of cilia on a mammalian tracheal cell (SEM). **(b)** The beating of a cilium on the surface of an epithelial cell from the human respiratory tract. A beat begins with a power stroke that sweeps fluid over the cell surface. A recovery stroke follows, leaving the cilium poised for the next beat. Each cycle requires about 0.1–0.2 sec. **(c)** A micrograph of the flagellated unicellular alga *Euglena* (SEM). **(d)** Movement of a eukaryotic flagellated cell through an aqueous environment.

move with a propagated bending motion that is usually symmetrical and undulatory and may even have a helical pattern. This type of beat generates a force parallel to the flagellum, such that the cell moves in approximately the same direction as the axis of the flagellum. The locomotory pattern of most flagellated cells, such as many sperm cells, involves propulsion of the cell by the trailing flagellum. Examples are also known in which the flagellum actually precedes the cell. Figure 16-6d illustrates this type of swimming movement. The flagella of bacteria (see Figure 5-3) are constructed from an entirely different set of proteins, and their mechanism of movement is completely different. These flagella are fascinating, but we will not consider them further here.

Cilia and Flagella Consist of an Axoneme Connected to a Basal Body

Cilia and flagella share a common structure, known as the **axoneme,** that is about 0.25 μm in diameter. The axoneme is connected to a **basal body** and surrounded by an extension of the cell membrane (**Figure 16-7a**). Between the axoneme and the basal body is a *transition zone* in which the arrangement of microtubules in the basal body takes on the pattern characteristic of the axoneme. Cross-sectional views of the axoneme, transition zone, and basal body are shown in Figure 16-7, parts b through d.

The basal body is identical in appearance to the centriole. A basal body consists of nine sets of tubular structures arranged around its circumference. As we saw in Chapter 15, each set is called a *triplet* because it consists of three tubules that share common walls—one complete microtubule and two incomplete tubules. As a cilium or flagellum forms, a centriole migrates to the cell surface and makes contact with the plasma membrane. The centriole then acts as a nucleation site for MT assembly, initiating polymerization of the nine outer doublets of the axoneme. After the process of tubule assembly has begun, the centriole is then referred to as a basal body.

The axonemes of the cilia used for propulsion and flagella have a characteristic "9 + 2" pattern, with nine **outer doublets** of tubules and two additional microtubules in the center, often called the **central pair.** Another, specialized group of cilia, called *primary cilia,* are used in sensory structures and play an important role in animal embryos during their development. These cilia have a "9 + 0" structure; that is, they lack the central pair.

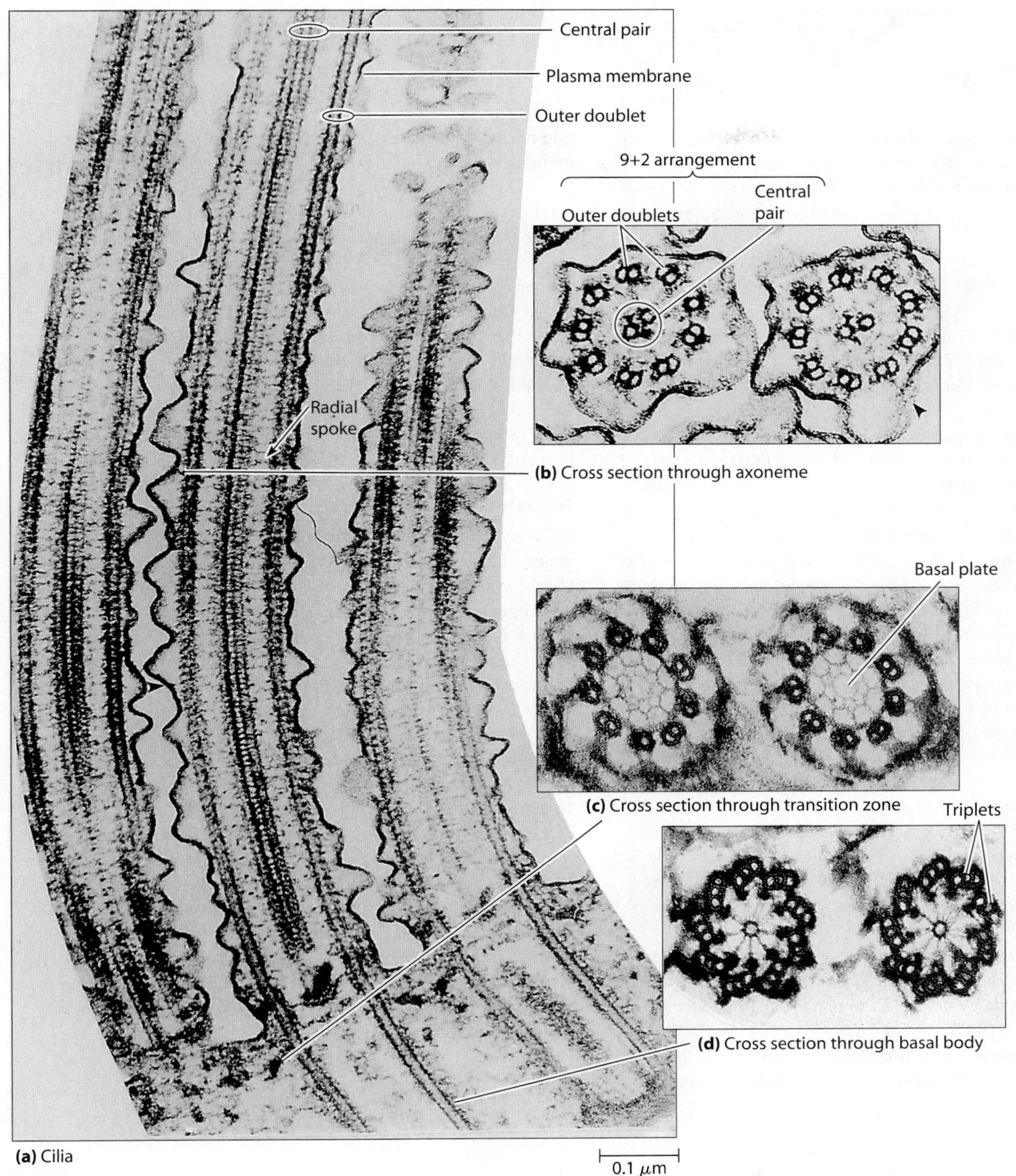

FIGURE 16-7 The Structure of a Cilium. (a) These longitudinal sections of three cilia from the protozoan *Tetrahymena thermophila* illustrate several structural features of cilia, including the central pair and outer doublet microtubules and the radial spokes. Cross-sectional views are shown for **(b)** axonemes, **(c)** the transition zone between cilia and basal bodies, and **(d)** basal bodies. Notice the triplet pattern of the basal body and the "9 + 2" pattern of tubule arrangement in the axoneme of the cilium. (All TEMs.)

Defects in these cilia can result in fascinating human syndromes (see **Box 16A**).

Figure 16-8 on p. 458 illustrates these structural features of a typical "9 + 2" cilium in greater detail. The nine outer doublets of the axoneme are thought to be extensions of two of the three subfibers from each of the nine triplets of the basal body. Each outer doublet of the axoneme therefore consists of one complete MT, called the **A tubule,** and one incomplete MT, the **B tubule** (Figure 16-8b). The A tubule has 13 protofilaments, whereas the B tubule has only 10 or 11. The tubules of the central pair are both complete, with 13 protofilaments each. All of these structures

BOX 16A HUMAN APPLICATIONS Cytoskeletal Motor Proteins and Human Disease

As it has become easier to isolate genes that are defective in many human genetic disorders, researchers have identified several human disorders that are due to defects in motor proteins. We will discuss two examples: reversal of body symmetry and genetic deafness.

Dyneins, Reversal of the Left-Right Body Axis, and Ciliopathies

In at least 1 in 20,000 live human births, organs in the body cavity are completely reversed left to right. This condition, known as *situs inversus viscerum,* has no medical consequences and is often not recognized until a patient undergoes medical tests for an unrelated condition. In contrast, when only some organs are reversed (*heterotaxia*), serious health complications result. In patients suffering from an autosomal recessive condition known as *Kartagener's triad,* there is a 50% probability of the complete reversal of the left-right location of internal organs. In addition, such patients suffer from male sterility and bronchial problems. The reason for these abnormalities is a defect in the outer dynein arms of cilia and flagella. Studies in mice support the idea that microtubule motor proteins are somehow involved in left-right asymmetry in the developing mammalian embryo. In *inversus viscerum (iv)* mutant mice, the internal organs are reversed in half of the newborn homozygotes. Surprisingly, *iv* encodes an axonemal dynein, which is required for the activity of primary cilia ("9 + 0" cilia) that are required for left-right asymmetry. Remarkably, similar cilia have been identified in all vertebrate embryos, and it is thought that they play a similar role in all of these cases. By beating in a rotary fashion, the cilia may create a flow of secreted components that shuttle signaling molecules across the midline of the embryo, or they may serve a sensory function. Current research is aimed at clarifying the role these fascinating cilia play in early embryonic development.

In addition to *situs inversus,* a growing number of human diseases are now known to be due to defects in cilia. These diseases, known as "ciliopathies," include a number of human syndromes, including *Bardet–Bieldl syndrome (BBS)* and *polycystic kidney disease (PKD).* BBS is principally characterized by obesity, retinitis pigmentosa (which results in blindness), polydactyly (more than the usual number of fingers/toes), underdevelopment of the gonads, and mental retardation. At least twelve human genes, when mutated, can give rise to BBS. All of the proteins these genes encode localize to basal bodies or cilia. Analysis of the equivalent genes in the roundworm, *C. elegans,* has shown that BBS proteins are involved in intraflagellar transport, moving material into and along primary cilia. Retinal degeneration in BBS patients results from defects in the *connecting cilium* of rod cells. The connecting cilium carries components to rod outer segments using the same machinery as intraflagellar transport. PKD is characterized by multiple, fluid-filled cysts in the kidneys. An autosomal dominant form of PKD is associated with another protein that in worms has been shown to be involved in intraflagellar transport. How defects in such a protein lead to PKD is still poorly understood.

Myosins and Deafness

Recently, nonmuscle myosins have also been implicated in human genetic disorders. For example, mutations in a myosin VII result in *Usher's syndrome,* an autosomal recessive disorder characterized by congenital profound hearing loss, problems in the vestibular system (i.e., sensing where one is in space), and retinitis pigmentosa. Another type of deafness in humans is caused by recessive mutations in the *myosin XV* gene. The shaker-2 mouse is a model for this condition: when mice are genetically engineered to carry a functional copy of the *myosin XV* gene, their deafness is rescued. How do defects in myosins lead to deafness? Specialized cells known as *hair cells* are found in the organ of Corti (a specialized structure within the cochlea of the inner ear) and the vestibular system; they contain special actin-rich sensory structures known as *stereocilia* (**Figure 16A-1**). By sensing movements of fluid within the inner ear, hair cells carry out auditory and vestibular transduction. Myosin VII is concentrated in the cell body of hair cells and in stereocilia; another myosin (myosin IC) is localized to sites where hair cells are connected laterally. Myosin XV is concentrated at the tips of stereocilia. All of these myosins are thought to be involved in linking stereocilia into a mechanically integrated vibration sensor.

---

contain tubulin, together with a second protein called *tektin.* Tektin is related to intermediate filament proteins (see Chapter 15) and is a necessary component of the axoneme. The A and B tubules share a wall that appears to contain tektin as a major component.

In addition to microtubules, axonemes contain several other key components (Figure 16-8b). The most important of these are the sets of **sidearms** that project out from each of the A tubules of the nine outer doublets. Each sidearm reaches out clockwise toward the B tubules of the adjacent doublet. These arms consist of axonemal dynein, which is responsible for sliding MTs within the axoneme past one another to bend the axoneme. The dynein arms occur in pairs, one inner arm and one outer arm, spaced along the MT at regular intervals. At less frequent intervals, adjacent doublets are joined by **interdoublet links.** These links are thought to limit the extent to which doublets can move with respect to each other as the axoneme bends.

At regular intervals, **radial spokes** project inward from each of the nine MT doublets, terminating near a set of projections that extend outward from the central pair of microtubules. These spokes are thought to be important in translating the sliding motion of adjacent doublets into the bending motion that characterizes the beating of these

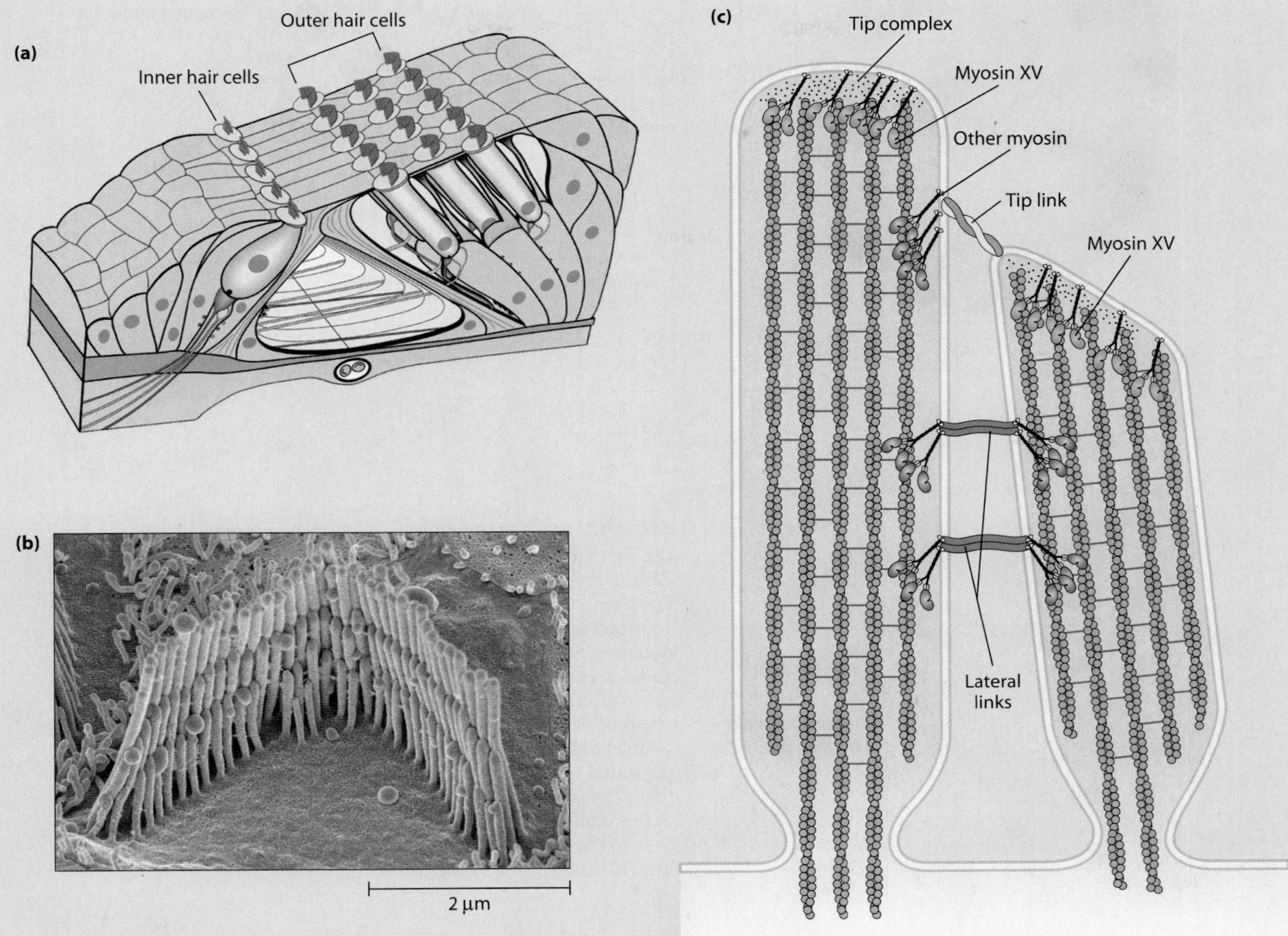

FIGURE 16A-1 Stereocilia and the Inner Ear. (**a**) Hair cells in the inner ear, including the organ of Corti in the cochlea, contain stereocilia. (**b**) A scanning electron micrograph of stereocilia in an outer hair cell. (**c**) Stereocilia are similar to microvilli but are much larger. They sense tiny movements caused by sound vibrations or fluid movement in the semicircular canals. Tip links, which are composed of cell surface proteins similar to cadherins (see Chapter 17), are associated with myosin XV. Lateral links are associated with other types of myosin. Such links connect one stereocilium to the next.

appendages. In addition to the radial spoke attachment to the central pair, a protein called **nexin** links adjacent doublets to one another and probably also plays a role in converting sliding into bending motion.

Microtubule Sliding Within the Axoneme Causes Cilia and Flagella to Bend

How does axonemal dynein act on this elaborate structure to generate the characteristic bending of cilia and flagella? The overall length of MTs in cilia and flagella does not change. Instead, the microtubules in adjacent outer doublets slide relative to one another in an ATP-dependent fashion. According to the **sliding-microtubule model** for cilia and flagella, this sliding movement is converted to a localized bending because the doublets of the axoneme are connected radially to the central pair and circumferentially to one another and therefore cannot slide past each other freely. The resultant bending takes the form of a wave that begins at the base of the organelle and proceeds toward the tip.

Dynein Arms Are Responsible for Sliding. The driving force for MT sliding is provided by ATP hydrolysis, catalyzed by the dynein arms. The importance of the dynein

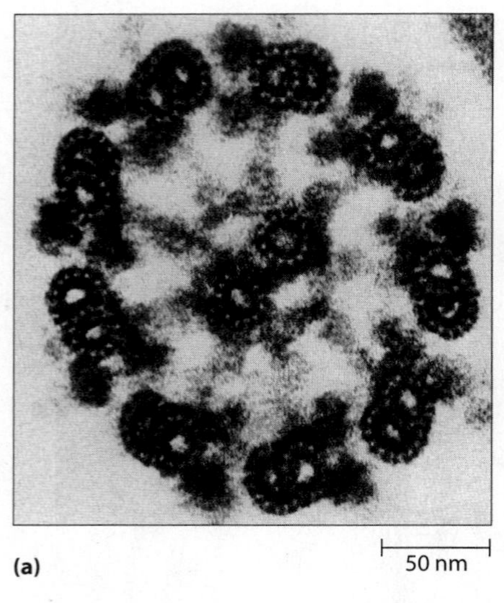

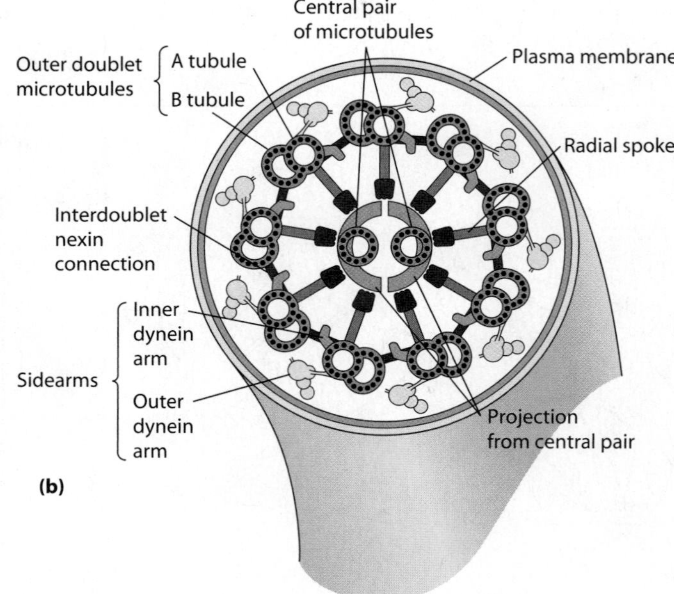

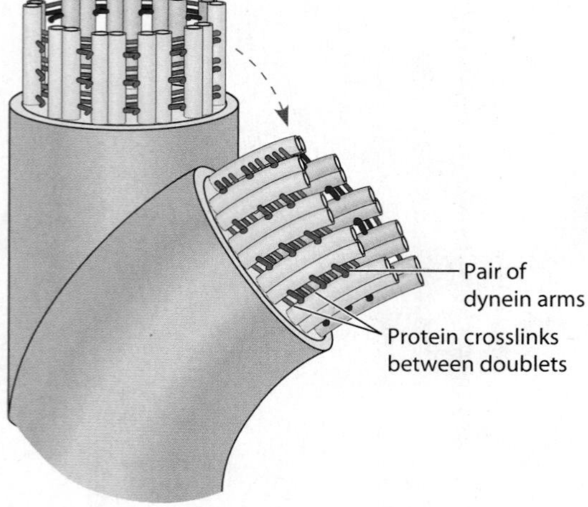

FIGURE 16-8 Enlarged Views of an Axoneme. **(a)** This micrograph shows an axoneme from a flagellum of *Chlamydomonas* (TEM). **(b)** Diagram of an axoneme in cross section. The microtubules of the central pair have 13 protofilaments each, as do the A tubules of the outer doublets. Each B tubule has 11 protofilaments of its own and shares 5 protofilaments with the A tubule. The dynein sidearms have ATPase activity and are thought to be responsible for the sliding of adjacent doublets. The inter-doublet links (nexin connections) join adjacent doublets, and the radial spokes project inward, terminating near projections that extend outward from the central pair of MTs. **(c)** Bending of the axoneme by dynein. Connection of the outer doublet MTs to the central pair converts sliding of adjacent MTs into local bending of the axoneme.

arms is indicated by two kinds of evidence. First, when dynein is selectively extracted from isolated axonemes, the arms disappear from the outer doublets, and the axonemes lose both their ability to hydrolyze ATP and their capacity to beat. Furthermore, the effect is reversible: If purified dynein is added to isolated outer doublets, the sidearms reappear, and ATP-dependent sliding is restored.

A second kind of evidence comes from studies of non-motile mutant flagella in such species as *Chlamydomonas,* a green alga. In some of these mutants, flagella are present but nonfunctional. Depending on the particular mutant, their nonmotile flagella lack dynein arms, radial spokes, or the central pair of microtubules. These structures are therefore essential for flagellar bending.

Axonemal dynein is a very large protein. It has multiple subunits, the three largest having ATPase activity and a molecular weight of about 450,000 each. During the sliding process, the stalk of the dynein arm apparently attaches to and detaches from the B tubule in a cyclic manner. Each cycle requires the hydrolysis of ATP and shifts the dynein arm of one doublet relative to the adjacent doublet. In this way, the dynein arms of one doublet move the neighboring doublet, resulting in a relative displacement of the two.

Crosslinks and Spokes Are Responsible for Bending. To convert the dynein-mediated displacement of doublets to a bending motion, the doublets must be restrained in a way that resists sliding but allows deformation. This resistance is provided by the radial spokes that connect the doublets to the central pair of microtubules and possibly also by the nexin crosslinks between doublets (Figure 16-8b, c). If these crosslinks and spokes are removed (by partial digestion with proteolytic enzymes, for example), the resistance that translates doublet sliding into a bending action is absent, and sliding is uncoupled from bending. Under these conditions, the free doublets move with respect to each other, and the axonemes become longer and thinner as the MTs slide apart.

In addition to their role in flagellar and ciliary beating, axonemal dyneins have been implicated in a process that at first glance seems completely unrelated to its function in sliding microtubules: positioning the internal organs of the body (see Box 16A).

Intraflagellar Transport Adds Components to Growing Flagella. The elaborate structure of flagella and cilia raises an interesting question: How are tubulin subunits and other components added to the growing cilia or flagella? Based on the analysis of mutants in *Chlamydomonas* and nematodes, both plus- and minus-end-directed microtubule motors are involved in shuttling components to and from the tips of flagella. This process, known as **intraflagellar transport (IFT),** is somewhat analogous to the process of axonal transport in nerve cells: Kinesins move material out to the tips of flagella, and a dynein brings material back toward the base. Several adapter proteins, which form two different major protein complexes, are required to link these motors to IFT particles. Several human disease syndromes are now known to result from mutations in components of these protein complexes (see Box 16A).

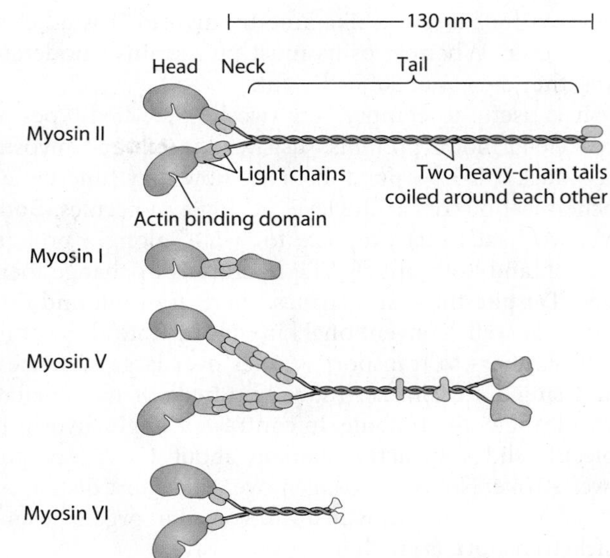

FIGURE 16-9 Myosin Family Members. All myosins have an actin- and ATP-binding heavy-chain "head" and typically have two or more regulatory light chains. Some myosins, like myosin I, have one head. Others, like myosin II, V, and VI, associate via their tails into two-stranded coils.

Actin-Based Cell Movement: The Myosins

Myosins Are a Large Family of Actin-Based Motors with Diverse Roles in Cell Motility

Movements of molecules and other cellular components also occurs along another major filament system in the cell—the actin cytoskeleton. As with microtubules, ATP-dependent motors exert force on actin microfilaments within cells. These motors are all members of a large superfamily of proteins known as **myosins.** Currently, there are 24 known classes of myosins (for a list of some major types of myosins, see Table 16-1). All myosins have at least one polypeptide chain, called the *heavy chain,* with a globular head group at one end attached to a tail of varying length (**Figure 16-9**). The globular head binds to actin and uses the energy of ATP hydrolysis to move along an actin filament. Many myosins move toward the plus (barbed) ends of actin filaments. (Myosin VI is a well-studied exception: It moves toward the minus, or pointed, ends of actin filaments.)

The structure of the tail region varies among the different kinds of myosin, giving myosin molecules the ability to bind to various molecules or cell structures. The tail structure also determines the ability of myosins to bind to other identical myosins to form dimers or large arrays. Myosins typically contain small polypeptides bound to the globular head group. These polypeptides, referred to as the *light chains,* often play a role in regulating the activity of the myosin ATPase. Some myosins are unusual in that they have a binding site for actin in their tail region, as well as in the head. In addition, some myosins, such as myosin I and myosin V, appear to bind to membranes, suggesting that these forms of myosin play a role in movement of the plasma membrane or in transporting membrane-enclosed organelles inside the cell.

Myosins have many functions in events as wide-ranging as muscle contraction (muscle myosin II), cell movement (nonmuscle myosin II), phagocytosis (myosin VI), and vesicle transport or other membrane-associated events (myosins I, V). One unexpected function for myosins is in maintaining the structures required for hearing in humans (see Box 16A).

The best-understood myosins are the **type II myosins.** They are composed of two heavy chains, each featuring a globular myosin head, a hinge region, a long rodlike tail, and four light chains. These myosins are found in skeletal, cardiac (heart), and smooth muscle cells, as well as in nonmuscle cells. Type II myosins are distinctive in that they can assemble into long filaments such as the thick filaments of muscle cells. The basic function of myosin II in all cell types is to convert ATP hydrolysis to mechanical force that can cause actin filaments to slide past the myosin molecule, typically resulting in the contraction of a cell or group of cells. For example, in *Drosophila* embryos, analysis of mutants for a nonmuscle type II myosin indicates that it is required for the closure of tissue sheets. Nonmuscle type II myosins are also involved in formation of the contractile ring during cytokinesis (see Chapter 19).

Many Myosins Move Along Actin Filaments in Short Steps

Like kinesins, myosins have been studied at the level of single molecules. These studies have shown that the force individual myosin heads exert on actin is similar to that measured for kinesins. Like kinesin, myosin II is an efficient motor. When myosins must pull against moderate loads, they are about 50% efficient.

It is useful to compare the two best-studied types of cytoskeletal motor proteins, "classic" kinesin and myosin II, because it is now possible to see how they function as biochemical motors at the level of single molecules. Both have two heads that they use to "walk" along a protein filament, and both utilize ATP hydrolysis to change their shape. Despite these similarities, there are profound differences as well. Conventional kinesins operate alone or in small numbers to transport vesicles over large distances, and a single kinesin can move hundreds of nanometers along a single microtubule. In contrast, a single myosin II molecule slides an actin filament about 12–15 nm per power stroke. Not all myosins move such short distances. Myosin V, for example, which is involved in organelle and vesicle transport, is much more processive.

If myosin II cannot move large distances along actin, how can it be involved in movement? Myosin II molecules often operate in large arrays. In the case of myosin II filaments in muscle, these arrays can contain billions of motors working together to mediate contraction of skeletal muscle, the process to which we now turn.

Filament-Based Movement in Muscle

Muscle contraction is the most familiar example of mechanical work mediated by intracellular filaments. Mammals have several kinds of muscles, including skeletal muscle, cardiac muscle, and smooth muscle. We will first consider skeletal muscle because much of our knowledge of the contractile process grew out of early investigations of its molecular structure and function.

Skeletal Muscle Cells Contain Thin and Thick Filaments

Skeletal muscles are responsible for voluntary movement. The structural organization of skeletal muscle is shown in **Figure 16-10**. A muscle consists of bundles of parallel **muscle fibers** joined by tendons to the bones that the muscle must move. Each fiber is actually a long, thin, multinucleate cell that is highly specialized for its contractile function. The multinucleate state arises from the fusion of embryonic cells called *myoblasts* during muscle differentiation to produce a *syncytium*. This cell fusion also accounts at least in part for the striking length of muscle cells, which may be many centimeters in length.

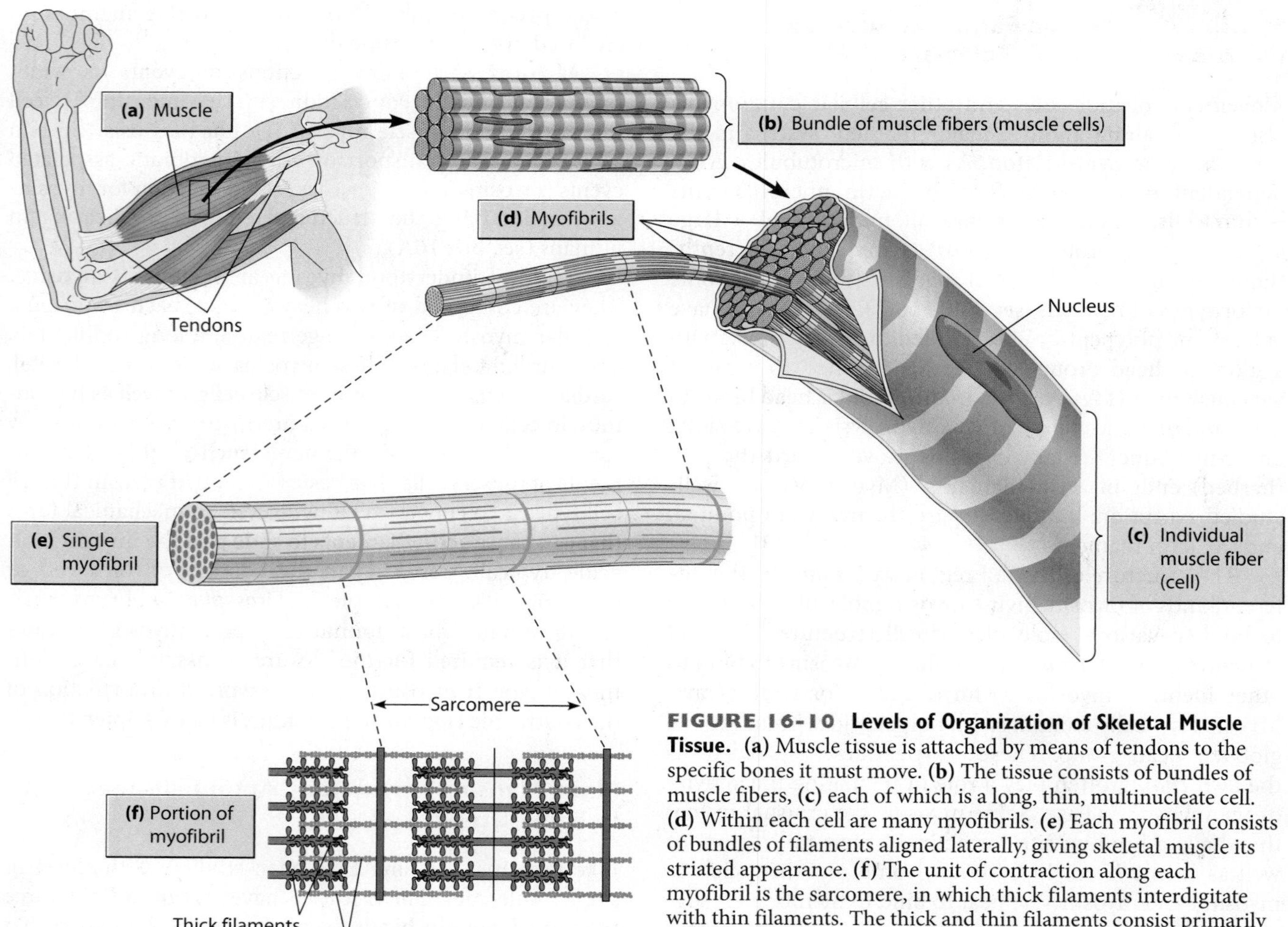

FIGURE 16-10 Levels of Organization of Skeletal Muscle Tissue. (a) Muscle tissue is attached by means of tendons to the specific bones it must move. (b) The tissue consists of bundles of muscle fibers, (c) each of which is a long, thin, multinucleate cell. (d) Within each cell are many myofibrils. (e) Each myofibril consists of bundles of filaments aligned laterally, giving skeletal muscle its striated appearance. (f) The unit of contraction along each myofibril is the sarcomere, in which thick filaments interdigitate with thin filaments. The thick and thin filaments consist primarily of myosin and actin (plus troponin and tropomyosin), respectively.

MB Molecular Model: Myosin (and its interaction with actin)

For muscles to exert force on other tissues, they use specialized structures to attach to other tissues. One such attachment is known as a *costamere,* which we will discuss in more detail in Chapter 17.

At the subcellular level, each muscle fiber (or cell) contains numerous **myofibrils.** Myofibrils are 1–2 μm in diameter and may extend the entire length of the cell. Each myofibril is subdivided along its length into repeating units called **sarcomeres.** The sarcomere is the fundamental contractile unit of the muscle cell. Each sarcomere of the myofibril contains bundles of **thick filaments** and **thin filaments.** Thick filaments consist of myosin, whereas thin filaments consist mainly of actin, tropomyosin, and troponin. The thin filaments are arranged around the thick filaments in a hexagonal pattern, as can be seen when the myofibril is viewed in cross section (**Figure 16-11**).

The filaments in skeletal muscle are aligned in lateral register, giving the myofibrils a pattern of alternating dark and light bands (**Figure 16-12a**). This pattern of bands, or *striations,* is characteristic of skeletal and cardiac muscle, which are therefore referred to as **striated muscle.** The dark bands are called **A bands,** and the light bands are called **I bands.** (The terminology for the structure and appearance of muscle myofibrils was developed from observations originally made with the polarizing light microscope. I stands for *isotropic* and A for *anisotropic,* terms related to the appearance of these bands when illuminated with plane-polarized light.)

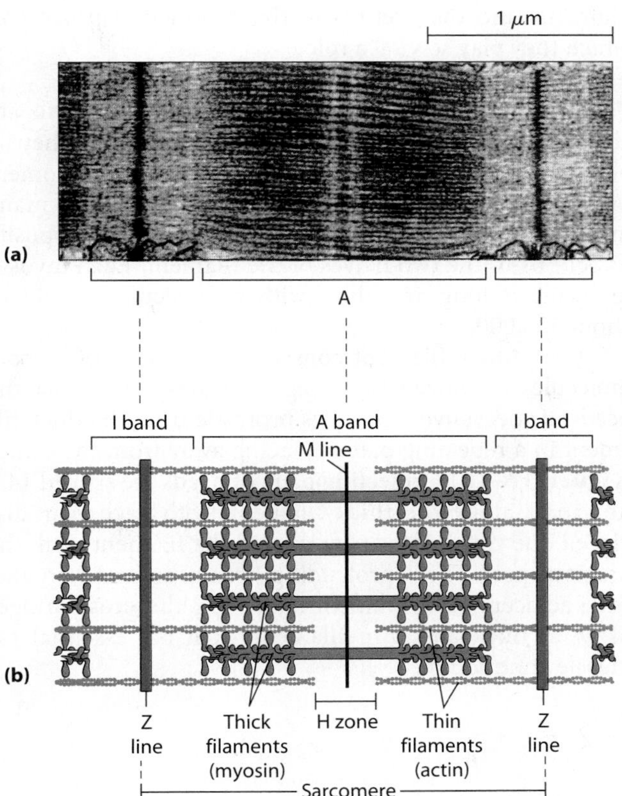

FIGURE 16-12 Appearance of and Nomenclature for Skeletal Muscle. **(a)** An electron micrograph of a single sarcomere (TEM). **(b)** A schematic diagram that can be used to interpret the repeating pattern of bands in striated muscle in terms of the interdigitation of thick and thin filaments. An A band corresponds to the length of the thick filaments, and an I band represents the portion of the thin filaments that does not overlap with thick filaments. The lighter area in the center of the A band is called the H zone; the line in the middle is known as the M line. The dense zone in the center of each I band is called the Z line. A sarcomere, the basic repeating unit along the myofibril, is the distance between two successive Z lines.

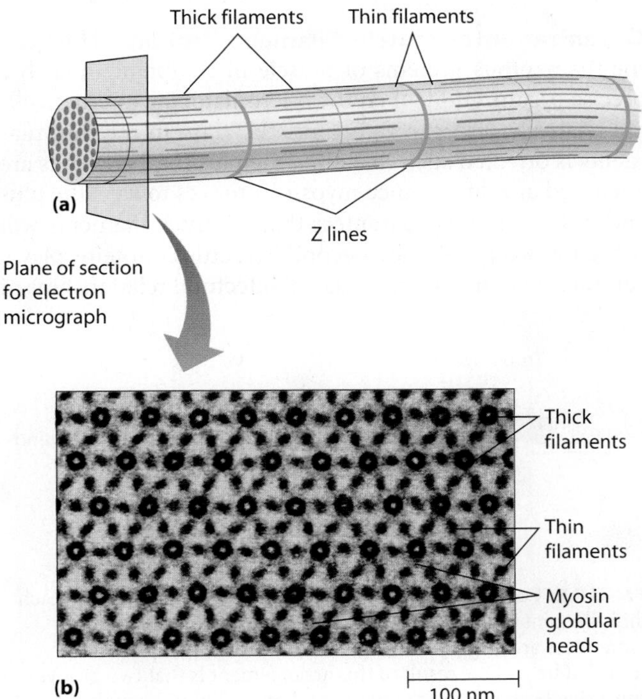

FIGURE 16-11 Arrangement of Thick and Thin Filaments in a Myofibril. **(a)** A myofibril consists of interdigitated thick and thin filaments. **(b)** The thin filaments are arranged around the thick filaments in a hexagonal pattern, as seen in this cross section of a flight muscle from the fruit fly *Drosophila melanogaster* viewed by high-voltage electron microscopy (HVEM).

As illustrated in Figure 16-12b, the lighter region in the middle of each A band is called the **H zone** (from the German word *hell,* meaning "light"). Running down the center of the H zone is the **M line,** which contains *myomesin,* a protein that links myosin filaments together. In the middle of each I band appears a dense **Z line** (from the German word *zwischen,* meaning "between"). The distance from one Z line to the next defines a single sarcomere. A sarcomere is about 2.5–3.0 μm long in the relaxed state but shortens progressively as the muscle contracts.

Sarcomeres Contain Ordered Arrays of Actin, Myosin, and Accessory Proteins

The striated pattern of skeletal muscle and the observed shortening of the sarcomeres during contraction are due to the arrangement of thick and thin filaments in myofibrils. We will therefore look in some detail at both types of filaments and then return to the contraction process in which they play so vital a role.

Thick Filaments. The thick filaments of myofibrils are about 15 nm in diameter and about 1.6 μm long. They lie parallel to one another in the middle of the sarcomere (see Figure 16-12). Each thick filament consists of many molecules of myosin, which are oriented in opposite directions in the two halves of the filament. Each myosin molecule is long and thin, with a molecular weight of about 525,000.

Every thick filament consists of hundreds of myosin molecules organized in a staggered array such that the heads of successive molecules protrude from the thick filament in a repeating pattern, facing away from the center (**Figure 16-13**). Projecting pairs of heads are spaced 14.3 nm apart along the thick filament, with each pair displaced one-third of the way around the filament from the previous pair. These protruding heads can make contact with adjacent thin filaments, forming the cross-bridges between thick and thin filaments that are essential for muscle contraction.

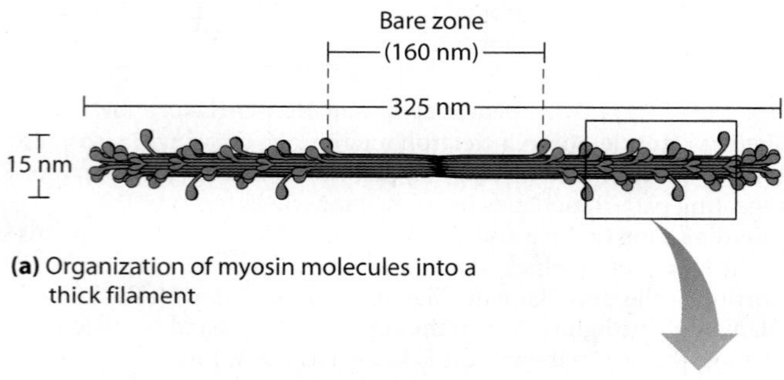

(a) Organization of myosin molecules into a thick filament

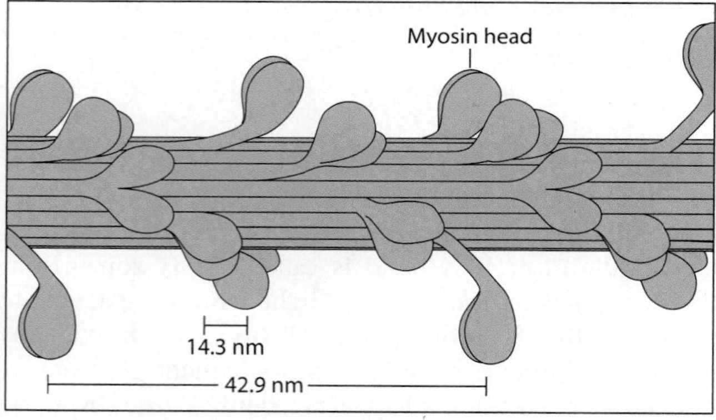

(b) Portion of a thick filament

FIGURE 16-13 The Thick Filament of Skeletal Muscle. (a) The thick filament of the myofibril consists of hundreds of myosin molecules organized in a repeating, staggered array. Only a portion of such a thick filament is shown here. A typical thick filament is about 1.6 μm long and about 15 nm in diameter. Individual myosin molecules are integrated into the filament longitudinally, with their ATPase-containing heads oriented away from the center of the filament. The central region of the filament is therefore a bare zone containing no heads. **(b)** This enlargement of a portion of the thick filament shows that pairs of myosin heads are spaced 14.3 nm apart.

Thin Filaments. The thin filaments of myofibrils interdigitate with the thick filaments. The thin filaments are about 7 nm in diameter and about 1 μm long. Each I band consists of *two* sets of thin filaments, one set on either side of the Z line, with each filament attached to the Z line and extending toward and into the A band in the center of the sarcomere. This accounts for the length of almost 2 μm for I bands in extended muscle.

The structure of thin filaments is shown in **Figure 16-14**. A thin filament consists of at least three proteins. The most important component of thin filaments is F-actin, intertwined with the proteins **tropomyosin** and **troponin.** Tropomyosin is a long, rodlike molecule, similar to the myosin tail, that fits in the groove of the actin helix. Each tropomyosin molecule stretches for about 38.5 nm along the filament and associates along its length with seven actin monomers.

Troponin is actually a complex of three polypeptide chains, called *TnT, TnC,* and *TnI.* TnT binds to tropomyosin and is thought to be responsible for positioning the complex on the tropomyosin molecule. TnC binds calcium ions, and TnI binds to actin. (*Tn* stands for troponin, *T* for tropomyosin, *C* for calcium, and *I* for inhibitory, because TnI inhibits muscle contraction.) One troponin complex is associated with each tropomyosin molecule, so the spacing between successive troponin complexes along the thin filament is 38.5 nm. Troponin and tropomyosin constitute a calcium-sensitive switch that activates contraction in both skeletal and cardiac muscle.

Organization of Muscle Filament Proteins. How can the filamentous proteins of muscle fibers maintain such a precise organization when the microfilaments in other cells are relatively disorganized? First, the actin in the thin filaments is oriented such that all of the plus (barbed) ends are anchored at Z lines. Since myosin II moves toward the plus end of F-actin, this guarantees that the thick filaments will move toward the Z lines. Second, structural proteins play a central role in maintaining the architectural relationships of

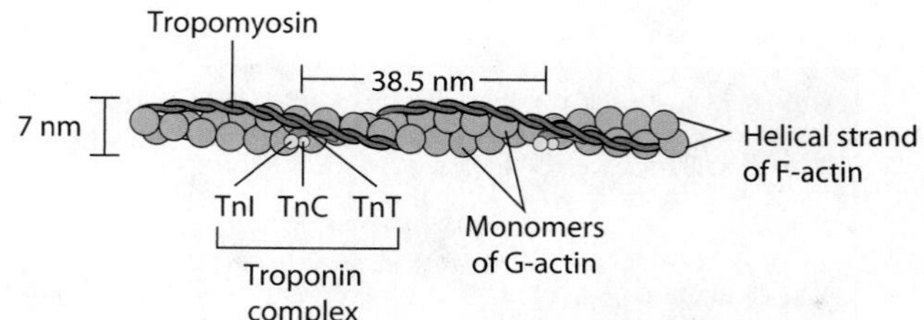

FIGURE 16-14 The Thin Filament of Striated Muscle. Each thin filament is a single strand of F-actin in which the G-actin monomers are staggered and give the appearance of a double-stranded helix. One result of this arrangement is that two grooves run along both sides of the filament. Long, ribbonlike molecules of tropomyosin lie in these grooves. Each tropomyosin molecule consists of two α helices wound about each other to form a ribbon about 2 nm in diameter and 38.5 nm long. Associated with each tropomyosin molecule is a troponin complex consisting of the three polypeptides TnT, TnC, and TnI.

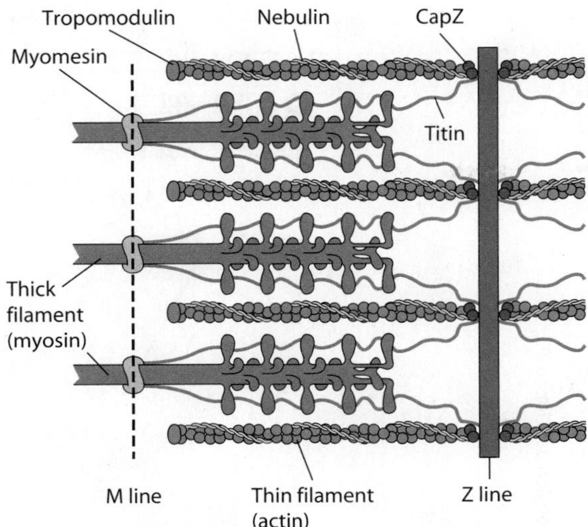

FIGURE 16-15 Structural Proteins of the Sarcomere. The thick and thin filaments require structural support to maintain their precise organization in the sarcomere. The support is provided by two proteins, α-actinin and myomesin, which bundle actin and myosin filaments, respectively. Titin attaches thick filaments to the Z line, thereby maintaining their position within the thin filament array. Nebulin stabilizes the organization of thin filaments. For clarity, tropomyosin is not shown.

muscle proteins (**Figure 16-15**). For instance, *α-actinin* keeps actin filaments bundled into parallel arrays. The capping protein, *CapZ,* maintains the attachment of the barbed (plus) ends of actin filaments to the Z line and simultaneously caps the actin in the thin filaments. At the other end of the thin filaments is *tropomodulin,* which helps to maintain the length and stability of thin filaments by binding their pointed (minus) ends. *Myomesin* is present at the H zone of the thick filament arrays and bundles the myosin molecules composing the arrays. A third structural protein, *titin,* attaches the thick filaments to the Z lines. Titin is highly flexible. During contraction-relaxation cycles, it can keep thick filaments in the correct position relative to thin filaments. This titin scaffolding also keeps the thick and thin filament arrays from being pulled apart when a muscle is stretched. Another protein, *nebulin,* stabilizes the organization of thin filaments. The protein components involved in muscle contraction are summarized in **Table 16-2**. The contractile process involves the complex interaction of all these proteins.

The Sliding-Filament Model Explains Muscle Contraction

With our understanding of muscle structure, we can now consider what happens during the contraction process. Based on electron microscopic studies, it is clear that the A bands of the myofibrils remain fixed in length during contraction, whereas the I bands shorten progressively and virtually disappear in the fully contracted state. To explain these observations, the **sliding-filament model** illustrated in **Figure 16-16** was proposed in 1954 independently by Andrew Huxley and Rolf Niedergerke and by Hugh Huxley and Jean Hanson. According to this model, muscle contraction is due to thin filaments sliding past thick filaments, with no change in the length of either type of filament. The sliding-filament model not only proved to be correct but was instrumental in focusing attention on the molecular interactions between thick and thin filaments that underlie the sliding process.

As Figure 16-16a indicates, contraction involves the sliding of thin filaments such that they are drawn progressively into the spaces between adjacent thick filaments, overlapping more and more with the thick filaments and narrowing the I band in the process. The result is a shortening of the individual sarcomeres and myofibrils and a contraction of the muscle cell and the whole tissue. This, in turn, causes the movement of the body parts attached to the muscle.

The sliding of the thin and thick filaments past each other as a means of generating force suggests that there should be a relationship between the force generated during contraction and the degree of shortening of the sarcomere. In fact, when the relationship between shortening and

Table 16-2 Major Protein Components of Vertebrate Skeletal Muscle

Protein	Molecular Weight	Function
Actin	42,000	Major component of thin filaments
Myosin	510,000	Major component of thick filaments
Tropomyosin	64,000	Binds along the length of thin filaments
Troponin	78,000	Positioned at regular intervals along thin filaments; mediates calcium regulation of contraction
Titin	2,500,000	Links thick filaments to Z line
Nebulin	700,000	Links thin filaments to Z line; stabilizes thin filaments
Myomesin	185,000	Myosin-binding protein present at the M line of thick filaments
α-actinin	190,000	Bundles actin filaments and attaches them to Z line
Ca^{2+} ATPase	115,000	Major protein of sarcoplasmic reticulum (SR); transports Ca^{2+} into SR to relax muscle
CapZ	68,000	Attaches actin filaments to Z line; caps actin
Tropomodulin	41,000	Maintains thin filament length and stability

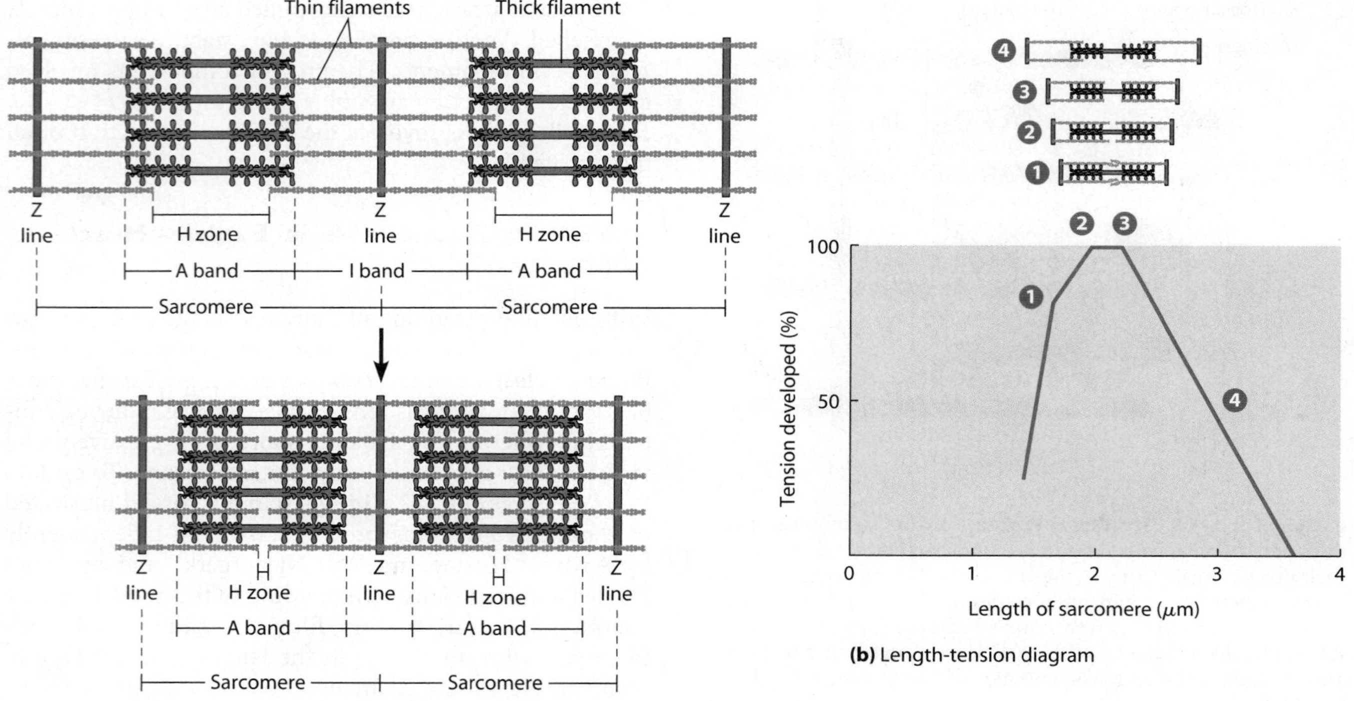

FIGURE 16-16 The Sliding-Filament Model of Muscle Contraction. **(a)** Two sarcomeres of a myofibril during the contraction process. The extended configuration is shown at the top, while the bottom view represents a more contracted myofibril. A myofibril shortens by the progressive sliding of thick and thin filaments past each other. The result is a greater interdigitation of filaments with no change in length of individual filaments. The increasing overlap of thick and thin filaments leads to a progressive decrease in the length of the I band as interpenetration continues during contraction. **(b)** This graph shows that the amount of tension developed by the sarcomere is proportional to the amount of overlap between the thin filament and the region of the thick filament containing myosin heads. When the sarcomere begins to shorten, as during a muscle contraction, the Z lines move closer together, increasing the amount of overlap between thin and thick filaments. This overlap allows more of the thick filament to interact with the thin filament. Therefore, the muscle can develop more tension (see ❹ to ❸). This proportional relationship continues until the ends of the thin filaments move into the H zone. Here they encounter no further myosin heads, so tension remains constant (❸ to ❷). Any further shortening of the sarcomere results in a dramatic decline in tension (❷ to ❶) as the filaments crowd into one another.

force is measured, it is exactly what the sliding-filament model predicts (Figure 16-16b): The amount of force the muscle can generate during a contraction depends on the number of myosin heads from the thick filament that can make contact with the thin filament. When the sarcomere is stretched, there is relatively little overlap between thin and thick filaments, so the force generated is small. As the sarcomere shortens, the region of overlap increases and the force of contraction increases. Finally, a point is reached at which continued shortening no longer increases the amount of overlap between thin and thick filaments, and the force of contraction stays the same. Any further shortening of the sarcomere results in a dramatic decline in tension as the filaments crowd into one another.

The sliding of thin filaments past thick filaments depends on the elaborate structural features of the sarcomere, and it requires energy. These basic observations raise important questions. First, by what mechanism are the thin filaments pulled progressively into the spaces between thick filaments to cause contraction? Second, how is the energy of ATP used to drive this process? These questions are answered in the next section.

Cross-Bridges Hold Filaments Together, and ATP Powers Their Movement

Cross-Bridge Formation. Regions of overlap between thick and thin filaments, whether extensive (in contracted muscle) or minimal (in relaxed muscle), are always characterized by the presence of transient **cross-bridges.** The cross-bridges are formed from links between the F-actin of the thin filaments and the myosin heads of the thick filaments (**Figure 16-17**). For contraction, the cross-bridges must form and dissociate repeatedly, so that each cycle of cross-bridge formation causes the thin filaments to interdigitate with the thick filaments more and more, thereby shortening the individual sarcomeres and causing the muscle fiber to contract. A given myosin head on the thick filament thus undergoes a cycle of events in which it binds to specific actin subunits on the thin filament, undergoes an energy-requiring change in shape that pulls the thin filament, and then breaks its association with the thin filament and associates with another site farther along the thin filament toward the Z line. Muscle contraction is the net result of the repeated making and breaking of many such

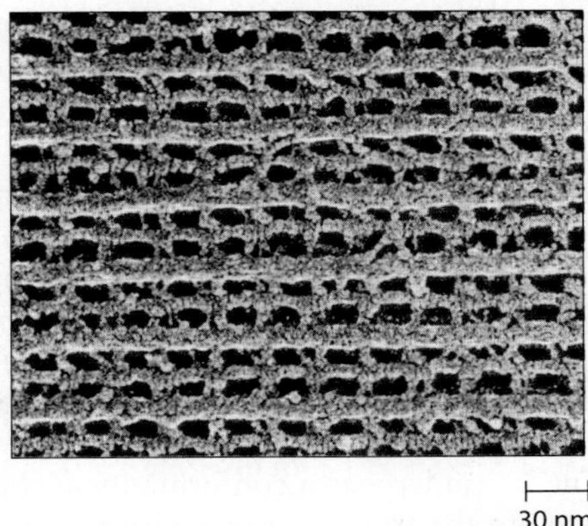

FIGURE 16-17 Cross-Bridges. The cross-bridges between thick and thin filaments formed by the projecting heads of myosin molecules can be readily seen in this high-resolution electron micrograph (TEM).

cross-bridges, with each cycle of cross-bridge formation causing the translocation of a small length of thin filament of a single fibril in a single cell.

ATP and the Contraction Cycle. The driving force for cross-bridge formation is the hydrolysis of ATP, catalyzed by the myosin heads. The requirement for ATP can be demonstrated in vitro because isolated muscle fibers contract in response to added ATP.

The mechanism of muscle contraction is depicted in **Figure 16-18** as a four-step cycle. In the high-energy configuration containing an ADP and a P_i molecule (a hydrolyzed ATP, in effect) shown in step ❶, a specific myosin head binds loosely to the actin filament. The myosin head then proceeds to a more tightly bound configuration that requires the loss of P_i. Step ❷ is the power stroke. The transition of myosin to the more tightly bound

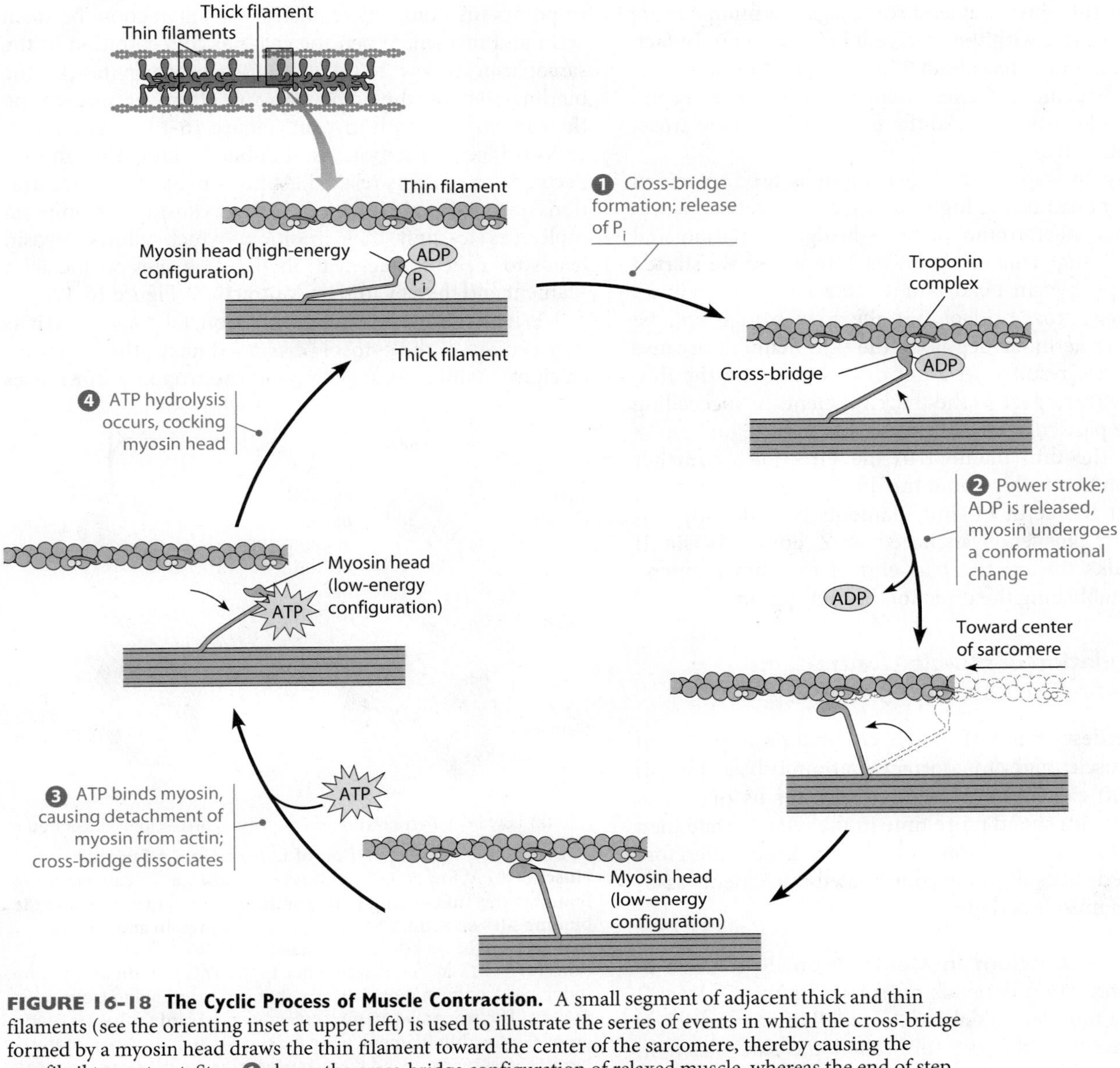

MB
Sacromere Shortening During Muscle Contraction

FIGURE 16-18 The Cyclic Process of Muscle Contraction. A small segment of adjacent thick and thin filaments (see the orienting inset at upper left) is used to illustrate the series of events in which the cross-bridge formed by a myosin head draws the thin filament toward the center of the sarcomere, thereby causing the myofibril to contract. Step ❶ shows the cross-bridge configuration of relaxed muscle, whereas the end of step ❷ shows the configuration of a muscle in rigor. A detailed description of all steps is given in the text.

state triggers a conformational change in myosin, which is associated with release of ADP. This conformational change is associated with a movement of the head, causing the thick filament to pull against the thin filament, which then moves with respect to the thick filament.

Cross-bridge dissociation follows in step ❸, as ATP binds to the myosin head in preparation for the next step. The binding of ATP causes the myosin head to change its conformation in a way that weakens its binding to actin. In the absence of adequate ATP, cross-bridge dissociation does not occur, and the muscle becomes locked in a stiff, rigid state called *rigor.* The *rigor mortis* associated with death results from the depletion of ATP and the progressive accumulation of cross-bridges in the configuration shown at the end of step ❷. Note that, once detached, the thick and thin filaments would be free to slip back to their previous positions, but they are held together at all times by the many other cross-bridges along their length at any given moment—just as at least some legs of a millipede are always in contact with the surface it is walking on. In fact, each thick filament has about 350 myosin heads, and each head attaches and detaches about five times per second during rapid contraction, so there are always many cross-bridges intact at any time.

Finally, in step ❹, ATP hydrolysis is used to return the myosin head to the high-energy configuration necessary for the next round of cross-bridge formation and filament sliding. This brings us back to where we started because the myosin head is now activated and ready to form a new cross-bridge. But the new bridge will be formed with actin farther along the thin filament because the first cycle resulted in a net displacement of the thin filament with respect to the thick filament. In succeeding cycles, the particular myosin head shown in Figure 16-18 will draw the thin filament in the direction of further contraction. And what about the direction of contraction? Recall that the actin in thin filaments is uniformly oriented, with plus ends anchored at Z lines. Myosin II always walks toward the *plus* end of the thin filament, thereby establishing the direction of contraction.

The Regulation of Muscle Contraction Depends on Calcium

So far, our description of muscle contraction implies that skeletal muscle ought to contract continuously, as long as there is sufficient ATP. Yet experience tells us that most skeletal muscles spend more time in the relaxed state than in contraction. Contraction and relaxation must therefore be regulated to result in the coordinated movements associated with muscle activity.

The Role of Calcium in Contraction. Regulation of muscle contraction depends on free calcium ions (Ca^{2+}) and on the muscle cell's ability to rapidly raise and lower calcium levels in the cytosol (called the *sarcoplasm* in muscle cells) around the myofibrils. The regulatory proteins *tropomyosin* and *troponin* act in concert to regulate the availability of myosin-binding sites on actin filaments in a way that depends critically on the level of calcium in the sarcoplasm.

To understand how this process works, we must first recognize that the myosin-binding sites on actin are normally blocked by tropomyosin. For myosin to bind to actin and initiate the cross-bridge cycle, the tropomyosin molecule must be moved out of the way. The calcium dependence of muscle contraction is due to troponin C (TnC), which binds calcium ions. When a calcium ion binds to TnC, it undergoes a conformational change that is transmitted to the tropomyosin molecule, causing it to move toward the center of the helical groove of the thin filament, out of the blocking position. The binding sites on actin are then accessible to the myosin heads, allowing contraction to proceed.

Figure 16-19 illustrates how the troponin-tropomyosin complex regulates the interaction between actin and myosin. When the calcium concentration in the sarcoplasm is low ($<0.1\ \mu M$), tropomyosin blocks the binding sites on the actin filament, effectively preventing their interaction with myosin (Figure 16-19a). As a result, cross-bridge formation is inhibited, and the muscle becomes or remains relaxed. At higher calcium concentrations ($>1\ \mu M$), calcium binds to TnC, causing tropomyosin molecules to shift their position, which allows myosin heads to make contact with the binding sites on the actin filament and thereby initiate contraction (Figure 16-19b).

When the calcium concentration falls again as it is pumped out of the cytosol (discussed next), the troponin-calcium complex dissociates, and the tropomyosin moves

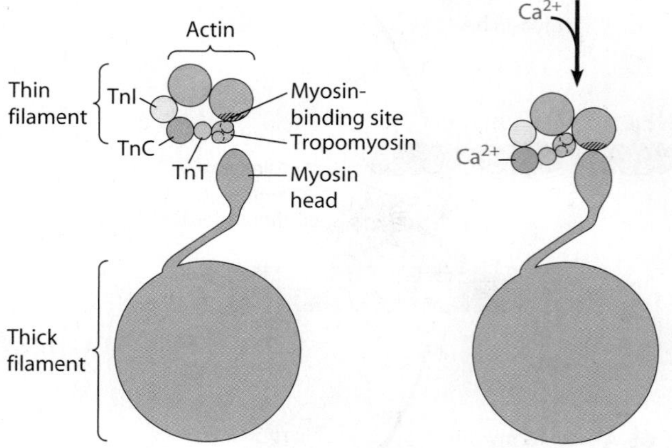

FIGURE 16-19 Regulation of Contraction in Striated Muscle. **(a)** At low concentrations ($<0.1\ \mu M\ Ca^{2+}$), calcium is not bound to the TnC subunit of troponin, and tropomyosin blocks the binding sites on actin, preventing access by myosin and thereby maintaining the muscle in the relaxed state. **(b)** At high concentrations ($>1\ \mu M\ Ca^{2+}$), calcium binds to the TnC subunit of troponin, inducing a conformational change that is transmitted to tropomyosin. The tropomyosin molecule moves toward the center of the groove in the thin filament, allowing myosin to gain access to the binding sites on actin and thereby triggering contraction.

back to the blocking position. Myosin binding is therefore inhibited, further cross-bridge formation is prevented, and the contraction cycle ends.

Regulation of Calcium Levels in Skeletal Muscle Cells. From the previous discussion, we know that muscle contraction is regulated by the concentration of calcium ions in the sarcoplasm. But how is the level of calcium controlled? Think for a moment about what must happen when we move any part of our bodies—when we flex an index finger, for instance. A nerve impulse is generated in the brain and transmitted down the spinal column to the nerve cells, or *motor neurons,* that control a small muscle in the forearm. The motor neurons activate the appropriate muscle cells, which contract and relax, all within about 100 msec. When nerve impulses to the muscle cell cease, calcium levels decline quickly, and the muscle relaxes. Therefore, to understand how muscle contraction is regulated, we need to know how nerve impulses cause calcium levels in the sarcoplasm to change and how these changes affect the contractile machinery. Muscle cells have many specialized features that facilitate a rapid change in the sarcoplasmic concentration of calcium ions and a rapid response of the contractile machinery. We discuss these features next.

Events at the Neuromuscular Junction. Recall from Chapter 13 that the signal for a muscle cell to contract is conveyed by a nerve cell in the form of an electrical impulse called an *action potential.* The site where the nerve innervates, or makes contact with, the muscle cell is called the **neuromuscular junction.** At the neuromuscular junction, the axon branches out and forms *axon terminals* that make contact with the muscle cell. These terminals contain the transmitter chemical *acetylcholine,* which is stored in membrane-enclosed vesicles and secreted by axon terminals in response to an action potential. The area of the muscle cell plasma membrane under the axon terminals is called the *motor end plate.* There, in the plasma membrane (called the *sarcolemma* in muscle cells), clusters of acetylcholine receptors are associated with each axon terminal. When the receptor binds acetylcholine, it opens a pore in the plasma membrane through which sodium ions can flow into the muscle cell. The sodium influx in turn causes a membrane depolarization to be transmitted away from the sarcolemma at the motor end plate.

Transmission of an Impulse to the Interior of the Muscle. Once a membrane depolarization occurs at the motor end plate, it spreads throughout the sarcolemma via the **transverse (T) tubule system** (**Figure 16-20**), a series of regular inpocketings of the muscle membrane that penetrates the interior of a muscle cell. The T tubules carry action potentials into the muscle cell, and they are part of the reason that muscle cells can respond so quickly to a nerve impulse.

Inside the muscle cell, the T tubule system comes into contact with the **sarcoplasmic reticulum (SR),** a system of intracellular membranes in the form of flattened sacs or tubes. As the name suggests, the SR is similar to the endoplasmic reticulum (ER) found in nonmuscle cells except that it is highly specialized. The SR runs along the myofibrils, where it is poised to release calcium ions directly into the myofibril and cause contraction and then to remove calcium from the myofibril and cause relaxation. This close proximity of the SR to the myofibrils facilitates the rapid response of muscle cells to nerve signals.

SR Function in Calcium Release and Uptake. The SR can be functionally divided into two components, referred to as the *medial element* and the *terminal cisternae* (singular: *terminal cisterna;* see Figure 16-20). The terminal

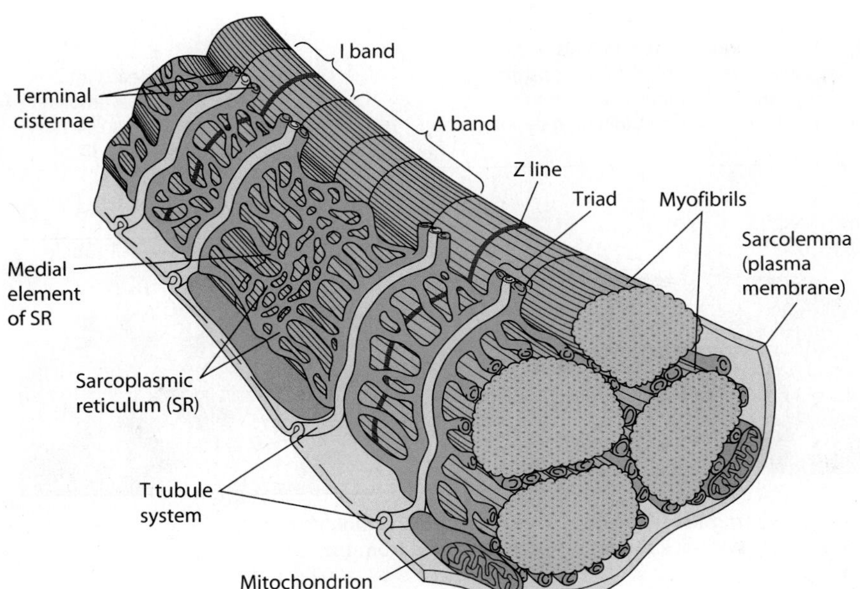

FIGURE 16-20 The Sarcoplasmic Reticulum and the Transverse Tubule System of Skeletal Muscle Cells. The sarcoplasmic reticulum (SR) is an extensive network of specialized ER that accumulates calcium ions and releases them in response to nerve signals. T tubules are invaginations of the sarcolemma (plasma membrane) that relay the membrane potential changes to the interior of the cell. Where the T tubule passes near the terminal cisternae of the SR, a triad structure is formed that appears in electron micrographs as the cross section of three adjacent tubes. The T tubule is in the middle, and on each side is one of the SR terminal cisternae. The triad contains the important junctional complex that regulates the release of calcium ions from the SR.

cisternae of the SR contain a high concentration of ATP-dependent calcium pumps that continually pump calcium into the lumen of the SR. The ability of the SR to pump calcium ions is crucial for muscle relaxation, but it is also needed for muscle contraction. Calcium pumping produces a high calcium concentration in the lumen of the SR (up to several millimolar). This calcium can then be released from the terminal cisternae of the SR when needed. Figure 16-20 shows how the terminal cisternae of the SR are positioned adjacent to the contractile apparatus of each myofibril. Terminal cisternae are typically found right next to a T tubule, giving rise to a structure called a **triad.** In electron micrographs, a triad appears as three circles in a row. The central circle is the membrane of the T tubule, and the circles on each side are the membranes of the terminal cisternae. Close inspection of the triad reveals that the terminal cisternae appear to be connected to the T tubule by material between the two membranes. This material is referred to as the *junctional complex* (to which we will return shortly).

The proximity of the T tubule, the terminal cisternae of the SR, and the contractile machinery of the myofibril provides the basis for how muscle cells can respond so rapidly to a nerve impulse. The action potential travels from the motor end plate, spreads out over the sarcolemma, and enters the T tubule (**Figure 16-21**). As the action potential travels down the T tubule, it activates voltage-gated calcium channels in the T tubule that are adjacent to ryanodine receptors in the terminal cisternae of the SR, causing them to open (see Figure 14-12). When the ryanodine receptor channels open, calcium rushes into the sarcoplasm immediately adjacent to the myofibrils, causing contraction.

Letting calcium out of the SR causes a muscle cell to contract. For the muscle cell to relax, calcium levels must be brought back down to the resting level. This is accomplished by pumping calcium back into the SR. The membrane of the SR contains an active transport protein, a **calcium ATPase,** which can pump calcium ions from the sarcoplasm into the cisternae of the SR. These pumps are concentrated in the medial element of the SR. The ATP-dependent mechanism by which calcium moves through the pump is similar to that discussed in Chapter 8 for the Na^+/K^+ pump (see Figure 8-11).

The pumping of calcium from the sarcoplasm back into the SR cisternae quickly lowers the sarcoplasmic calcium level to the point at which troponin releases calcium, tropomyosin moves back to the blocking position on actin, and further cross-bridge formation is prevented. Cross-bridges therefore disappear rapidly as actin dissociates from myosin and becomes blocked by tropomyosin. This leaves the muscle relaxed and free to be re-extended because the absence of cross-bridge contacts allows the thin filaments to slide out from between the thick filaments, due to passive stretching of the muscle by other tissues.

The Coordinated Contraction of Cardiac Muscle Cells Involves Electrical Coupling

Cardiac (heart) muscle is responsible for the beating of the heart and the pumping of blood through the body's circulatory system. Cardiac muscle functions continuously. In one year, your heart beats about 40 million times! Cardiac muscle is very similar to skeletal muscle in the organization of actin and myosin filaments and has the same striated appearance (**Figure 16-22**). In contrast to skeletal muscle, most of the energy required for the beating of the heart under resting conditions is provided not by blood glucose but by free fatty acids that are transported from adipose (fat storage) tissue to the heart by serum albumin, a blood protein.

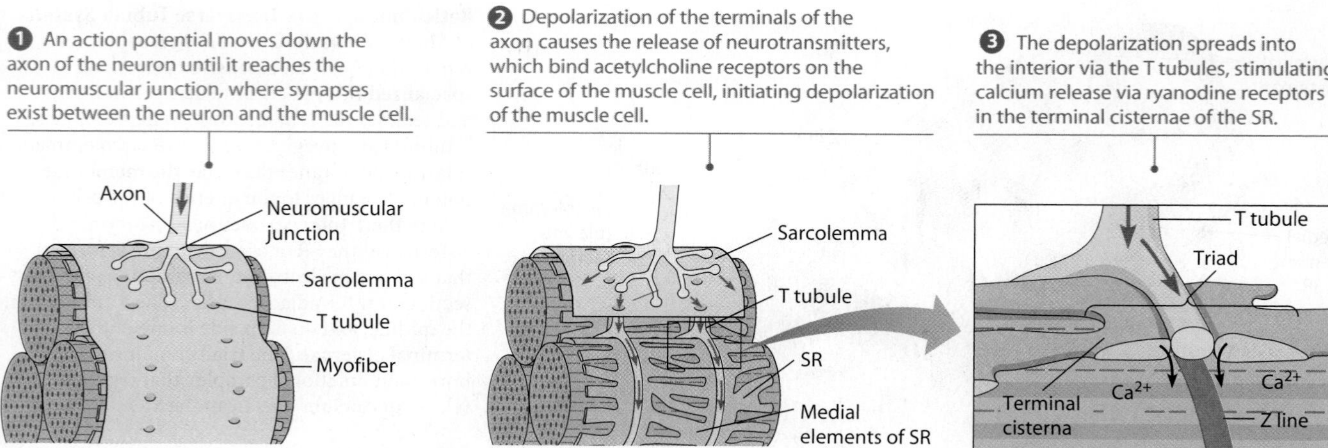

1 An action potential moves down the axon of the neuron until it reaches the neuromuscular junction, where synapses exist between the neuron and the muscle cell.
2 Depolarization of the terminals of the axon causes the release of neurotransmitters, which bind acetylcholine receptors on the surface of the muscle cell, initiating depolarization of the muscle cell.
3 The depolarization spreads into the interior via the T tubules, stimulating calcium release via ryanodine receptors in the terminal cisternae of the SR.
Axon
Neuromuscular junction
Sarcolemma
T tubule
Myofiber
Sarcolemma
T tubule
SR
Medial elements of SR
T tubule
Triad
Terminal cisterna
Ca^{2+}
Ca^{2+}
Z line

FIGURE 16-21 Stimulation of a Muscle Cell by a Nerve Impulse. The nerve causes a depolarization of the muscle cell, which spreads into the interior via the T tubule system, stimulating calcium release from the terminal cisternae of the sarcoplasmic reticulum (SR).

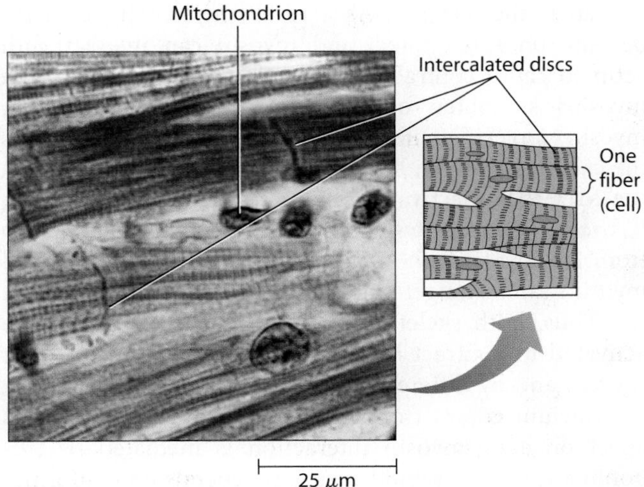

FIGURE 16-22 Cardiac Muscle Cells. Cardiac muscle cells have a contractile mechanism and sarcomeric structure similar to those of skeletal muscle cells. However, unlike skeletal muscle cells, cardiac muscle cells are joined together end to end at intercalated discs, which allow ions and electrical signals to pass from one cell to the next. This ionic permeability enables a contraction stimulus to spread evenly to all the cells of the heart (LM).

A second difference between cardiac and skeletal muscle is that heart muscle cells are not multinucleate. Instead, cells are joined end to end through structures called **intercalated discs.** The discs have a high density of desmosomes and gap junctions (see Chapter 17); the gap junctions electrically couple neighboring cells, allowing depolarization waves to spread throughout the heart during its contraction cycle. The heart is not activated by nerve impulses, as skeletal muscle is, but contracts spontaneously once every second or so. The heart rate is controlled by a "pacemaker" region in an upper portion of the heart (right atrium). The depolarization wave initiated by the pacemaker then spreads to the rest of the heart to produce the heartbeat.

Smooth Muscle Is More Similar to Nonmuscle Cells than to Skeletal Muscle

Smooth muscle is responsible for involuntary contractions such as those of the stomach, intestines, uterus, and blood vessels. In general, such contractions are slow, taking up to five seconds to reach maximum tension. Smooth muscle contractions are also of greater duration than those of skeletal or cardiac muscle. Though smooth muscle is not able to contract rapidly, it is well adapted to maintain tension for long periods of time, as is required in these organs and tissues.

The Structure of Smooth Muscle. Smooth muscle cells are long and thin, with pointed ends. Unlike skeletal or heart muscle, smooth muscle has no striations (**Figure 16-23a**). Smooth muscle cells do not contain Z lines, which are responsible for the periodic organization of the sarcomeres found in skeletal and cardiac muscle cells. Instead, smooth muscle cells contain *dense bodies,* plaque-like structures in the cytoplasm and on the cell membrane (Figure 16-23b), that contain intermediate filaments. Bundles of actin filaments are anchored at their ends to these dense bodies. As a result, actin filaments appear in a crisscross pattern, aligned obliquely to the long axis of the cell. Cross-bridges connect thick and thin filaments in smooth muscle but not in the regular, repeating pattern seen in skeletal muscle.

Regulation of Contraction in Smooth Muscle Cells. Smooth muscle cell contraction and nonmuscle cell contraction are regulated in a manner distinct from that of skeletal muscle cells. Although skeletal and smooth muscle cells are both stimulated to contract by an increase in the sarcoplasmic concentration of calcium ions, the mechanisms involved are quite different. When sarcoplasmic calcium concentrations increase in smooth muscle and nonmuscle cells, a cascade of events takes place

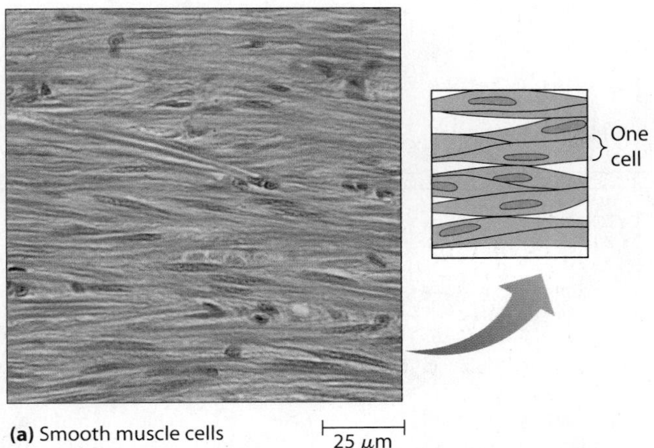

(a) Smooth muscle cells

(b) Contraction of smooth muscle cell

FIGURE 16-23 Smooth Muscle and Its Contraction. **(a)** Individual smooth muscle cells are long and spindle shaped, with no Z lines or sarcomeric structure (LM). **(b)** In the smooth muscle cell, contractile bundles of actin and myosin appear to be anchored to plaque-like structures called dense bodies. The dense bodies are connected to each other by intermediate filaments, thereby orienting the actin and myosin bundles obliquely to the long axis of the cell. When the actin and myosin bundles contract, they pull on the dense bodies and intermediate filaments, producing the cellular contraction shown here.

that includes the activation of **myosin light-chain kinase (MLCK).** Activated MLCK then phosphorylates one type of myosin light chain known as a **regulatory light chain** (see Figure 16-9, myosin II).

Myosin light-chain phosphorylation affects myosin in two ways. First, some myosin molecules are curled up so that they cannot assemble into filaments. When the myosin light chain is phosphorylated, the myosin tail uncurls and becomes capable of assembly. Second, the phosphorylation of the light chains activates myosin, enabling it to interact with actin filaments to undergo the cross-bridge cycle.

The cascade of events involved in the activation of smooth muscle and nonmuscle myosin is shown in **Figure 16-24a**. In response to a nerve impulse or hormonal signal reaching the smooth muscle cell, an influx of extracellular calcium ions occurs, increasing the intracellular calcium concentration and causing contraction. Recall from Chapter 14 that elevation of intracellular calcium can activate the protein *calmodulin.* The resulting *calcium-calmodulin complex* can bind to myosin light-chain kinase, activating the enzyme. As a result, myosin light chains become phosphorylated, and myosin can interact with actin to cause contractions. In addition, the tails of the myosins straighten out and can assemble with other myosin molecules into filaments (Figure 16-24b). As the calcium levels within smooth muscle cells drop again, the MLCK is inactivated, and a second enzyme, *myosin light-chain phosphatase,* removes the phosphate group from the myosin light chain. Since the dephosphorylated myosins can no longer bind to actin, the muscle cell relaxes.

Thus, both skeletal muscle and smooth muscle are activated to contract by calcium ions but from different sources and by different mechanisms. In skeletal muscle, the calcium comes from the sarcoplasmic reticulum. Its effect on actin-myosin interaction is mediated by troponin and is very rapid because it depends on conformational changes only. In smooth muscle, the calcium comes from outside the cell, and its effect is mediated by calmodulin. The effect is much slower in this case because it involves a covalent modification (phosphorylation) of the myosin molecule.

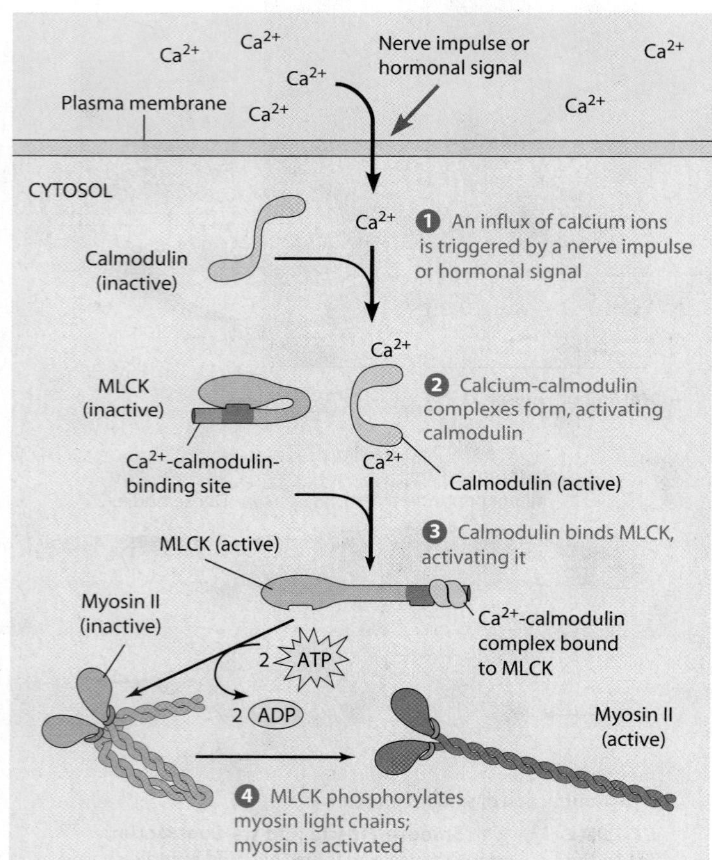

(a) Phosphorylation of myosin II by myosin light-chain kinase (MLCK)

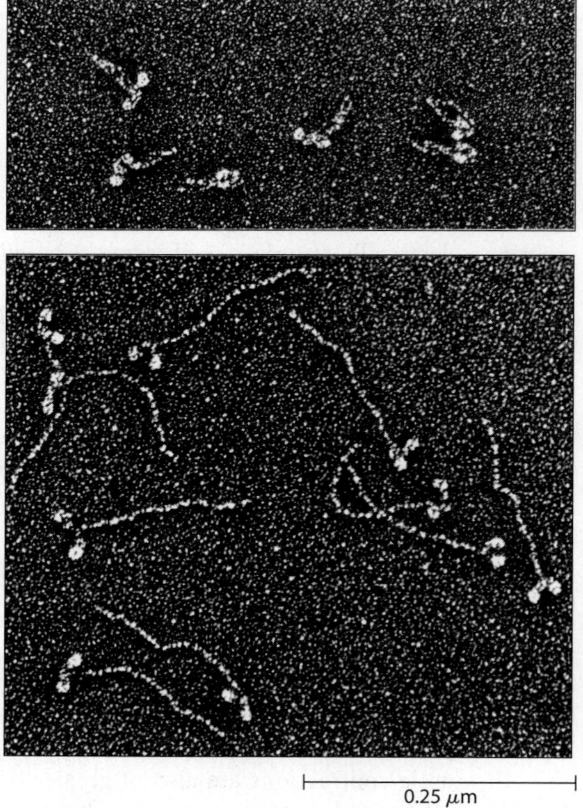

(b) Curled and uncurled myosin II molecules

FIGURE 16-24 Phosphorylation of Smooth Muscle and Nonmuscle Myosin. **(a)** The functions of both smooth muscle and nonmuscle myosin II are regulated by phosphorylation of the regulatory light chains. An influx of calcium ions into the cell, triggered by a nerve impulse or a hormonal signal, allows the calcium-calmodulin complex to bind myosin light-chain kinase (MLCK), which in turn phosphorylates the myosin light chains. The activated (and uncurled) myosin can then bind to actin. **(b)** Electron micrographs of curled and uncurled myosin II molecules (TEMs).

Actin-Based Motility in Nonmuscle Cells

Actin and myosin are best known as the major components of the thin and thick filaments of muscle cells. In fact, muscle cells represent only one specialized case of cell movements driven by the interactions of actin and myosin. Actins and myosins have now been discovered in almost all eukaryotic cells and are known to play important roles in various types of nonmuscle motility. One example of actin-dependent, nonmuscle motility occurs during cytokinesis (see Chapter 19). In this section, we examine several other examples.

Cell Migration via Lamellipodia Involves Cycles of Protrusion, Attachment, Translocation, and Detachment

Actin microfilaments (MFs) are required for the movement of most nonmuscle cells in animals. Many nonmuscle cells, such as fibroblasts, the growth cones of neurons, and many embryonic cells in animals, are capable of crawling over a substrate using lamellipodia and/or filopodia, whose internal structure we explored in Chapter 15. (A scanning electron micrograph of a crawling cell in vitro is shown in **Figure 16-25**.) In this section, we will consider such cell crawling in more detail. In a later section, we will consider a specialized form of crawling known as amoeboid movement.

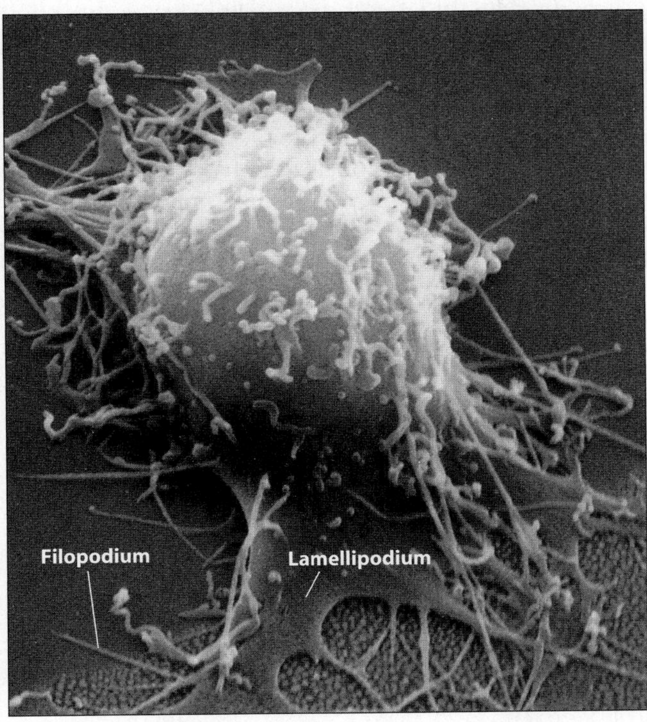

FIGURE 16-25 A Crawling Cell. A mouse fibroblast displays a lamellipodium and numerous filopodia extending from the cell surface (SEM).

Cell crawling involves several distinct events: (1) extension of a protrusion at the cell's leading edge; (2) attachment of the protrusion to the substrate; and (3) generation of tension, which pulls the cell forward as its "tail" releases its attachments and retracts. These events are summarized in **Figure 16-26**.

Extending Protrusions. To crawl, cells must produce specialized extensions, or *protrusions,* at their front or *leading edge.* One type of protrusion is a thin sheet of cytoplasm called a **lamellipodium** (plural: **lamellipodia**). Another type of protrusion is a thin, pointed structure known as a **filopodium** (plural: **filopodia**). Crawling cells often exhibit interconversion of these two types of protrusions as they migrate.

Fundamental to the dynamics of protrusions is the phenomenon of **retrograde flow** of F-actin. During normal retrograde flow, there is bulk movement of microfilaments toward the rear of the protrusion as it extends. Retrograde flow appears to result from two simultaneous processes: *actin assembly* at the tip of the growing lamellipodium or filopodium and *rearward translocation* of actin filaments toward the base of the protrusion. In a typical cell, forward assembly and rearward translocation balance one another; as one or the other occurs, a protrusion can be extended or retracted.

Actin polymerization, especially in lamellipodia, is driven by Arp2/3-dependent dendritic branching, a process we examined in some detail in Chapter 15. As actin polymerizes at the tip of a protrusion, the protrusion pushes forward. At the same time that extension of the tip of a protrusion is occurring, the polymerized actin is drawn toward the base of the protrusion, where it is disassembled. Released actin monomers are then available for addition to the barbed ends of new or growing microfilaments as the cell continues to crawl forward.

Microtubules are also involved in the polarized production of protrusions. Although how MTs are involved is not entirely clear, in cultured cells MTs can be seen polymerizing near the leading edge. Moreover, treating some cells with MT depolymerizing agents causes them to lose their polarized appearance, and they make protrusions simultaneously all around their perimeter.

Cell Attachment. If cells only polymerized actin at the tips of protrusions, pulled it rearward, and then depolymerized it, they would not be able to move. Something must couple retrograde flow to forward movement of the cell as a whole. Attachment, or adhesion of the cell to its substrate, is also necessary for cell crawling. New sites of attachment must be formed at the front of a cell, and contacts at the rear must be broken. Attachment sites between the cell and the substrate are complex structures involving the attachment of transmembrane proteins to other proteins both outside and inside the cell. One family of such

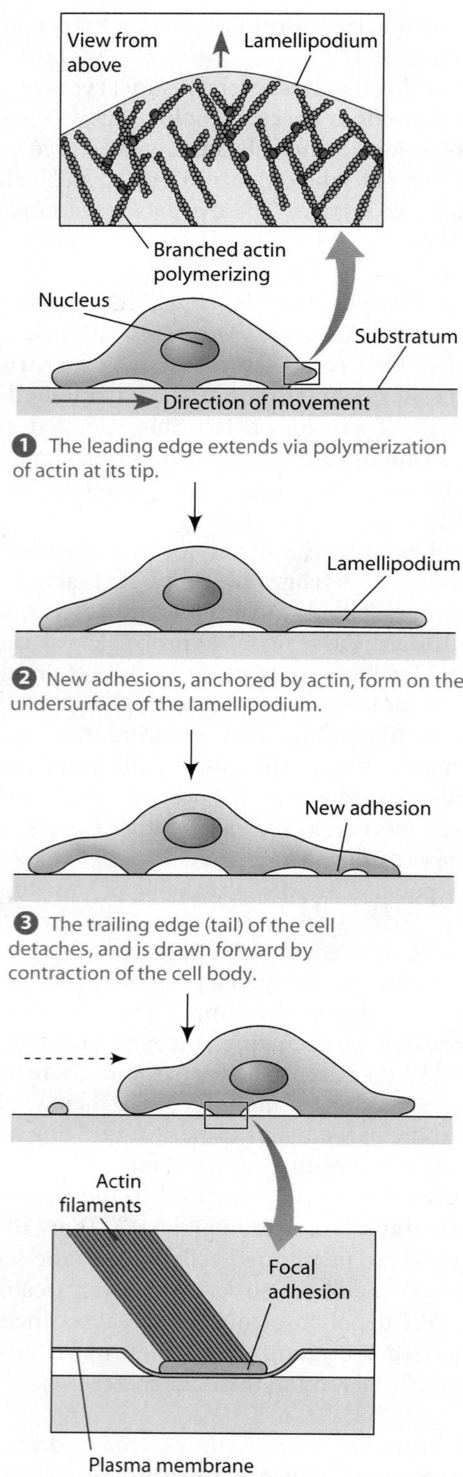
View from above
Lamellipodium
Branched actin polymerizing
Nucleus
Substratum
Direction of movement
1 The leading edge extends via polymerization of actin at its tip.
Lamellipodium
2 New adhesions, anchored by actin, form on the undersurface of the lamellipodium.
New adhesion
3 The trailing edge (tail) of the cell detaches, and is drawn forward by contraction of the cell body.
Actin filaments
Focal adhesion
Plasma membrane

FIGURE 16-26 The Steps of Cell Crawling. Several different processes are involved in cell crawling, including cell protrusion, attachment, and contractile activities. Protrusion is accompanied by Arp2/3-dependent actin polymerization at the leading edge. Attachment connects actin to the cell surface, typically via integrins, which cluster at focal adhesions.

attachment proteins is the *integrins.* On the outside of the cell, integrins attach to extracellular matrix proteins. Inside the cell, integrins are connected to actin filaments through linker proteins. Such integrin-dependent attachments are known as *focal contacts* and are crucial for cell migration. We will discuss focal contacts in more detail in Chapter 17.

How firmly a cell is attached to the underlying substrate helps to determine whether a cell will move forward. In this sense, actin polymerization at the leading edge is analogous to a car with its transmission in neutral. Once the car is shifted into a forward gear, it rapidly moves forward. In the same way, firm attachment of the leading edge shifts the balance in favor of forward movement.

Translocation and Detachment. Cell crawling coordinates protrusion formation and attachment with forward movement of the entire cell body. Contraction of the rear of the cell squeezes the cell body forward and releases the cell from attachments at its rear. Evidence suggests that contraction is due to interactions between actin and myosins, and that contraction is under the control of the protein Rho. Rho is thought to regulate activation of nonmuscle myosin II, which is localized toward the rear of the cell. In mutant cells from the cellular slime mold *Dictyostelium* that lack myosin II, the ability of the trailing edge of the cell to retract is reduced. Similarly, when Rho activity is impaired in migrating monocytes, they are unable to withdraw attachments at their rear and contract their trailing edge. These and other results support the idea that Rho and myosin II are involved in tail retraction.

Contraction of the cell body must be linked to detachment of the trailing edge of the cell. Detachment requires breaking adhesive contacts. Interestingly, contacts at the rear of the cell are sometimes too tight to be detached, and the tail of the cell actually breaks off as the cell pulls the rear forward (see Figure 16-26). In general, how firmly a cell attaches to a substrate affects how quickly that cell can crawl. If cells adhere too tightly to the underlying substrate, they cannot dynamically make and break connections to the substrate, and motility is actually impeded. Thus for movement to occur, new attachments must be balanced by loss of old ones.

Chemotaxis Is a Directional Movement in Response to a Graded Chemical Stimulus

A key feature of migrating cells in the body and in embryos is *directional* migration. One way that directional migration occurs is through the formation of protrusions predominantly on one side of the cell. In this case, cells must be able to regulate not only whether they form protrusions but where they form them. Diffusible molecules can act as important cues for such directional migration. When a migrating cell moves toward a greater or lesser concentration of a diffusible chemical, the response is known as **chemotaxis.** The molecule(s) that elicit this response are called *chemoattractants* (when a cell moves toward higher concentrations of the molecule) or *chemorepellants* (when a cell moves away from higher concentrations of the molecule). In eukaryotes, chemotaxis has been studied most intensively in white blood cells and in the *Dictyostelium* amoeba. In both cases, increasing the local concentration of a chemoattractant (small peptides in the case of white blood cells and cyclic AMP in the case of *Dictyostelium*) results in dramatic changes in the actin cytoskeleton, including biochemical changes in actin-binding proteins and migration of the cell toward the source of chemoattractant. These changes occur through local activation of chemoattractant (or repellant) receptors on the cell surface. These receptors are G protein–linked receptors, which we discussed in Chapter 14. Activation of these receptors leads to local accumulation of phosphoinositides (see Chapter 15), which in turn is thought to result in polarized recruitment of the cell's cytoskeletal machinery to form a protrusion in the direction of migration.

Amoeboid Movement Involves Cycles of Gelation and Solation of the Actin Cytoskeleton

Amoebas and white blood cells exhibit a type of crawling movement referred to as **amoeboid movement** (**Figure 16-27**). This type of movement is accompanied by protrusions of the cytosol called **pseudopodia** (singular: **pseudopodium,** from the Greek for "false foot"). Cells that undergo amoeboid movement have an outer layer of thick, gelatinous, actin-rich cytosol ("gel") and an inner, more fluid layer of cytosol ("sol"). In an amoeba, for example, as a pseudopodium is extended, more fluid material streams forward in the direction of extension and congeals at the tip of the pseudopodium (this event is often called gelation). Meanwhile, at the rear of the moving cell, gelatinous cytosol changes into a more fluid state and streams toward the pseudopodium (solation). Proteins such as gelsolin that are present within these gels may be activated by calcium to convert the gel to a more fluid state. Experiments have shown that the forward streaming in the pseudopodium does not require squeezing from the rear of the cell: When a pseudopodium's cell membrane is removed using detergent, the remaining components can still stream forward if the appropriate mixture of ions and other chemicals is added. Pressure exerted on the endoplasm, possibly due to contraction of an actomyosin network in the trailing edge of the cell, may also squeeze the endoplasm forward, aiding formation of a protrusion at the leading edge.

FIGURE 16-27 Amoeboid Movement. A micrograph of *Amoeba proteus*, a protozoan that moves by extension of pseudopodia (LM).

Actin-Based Motors Move Components Within the Cytoplasm of Some Cells

Cytoplasmic streaming, an actomyosin-dependent movement of the cytosol within a cell, is seen in a variety of organisms that do not display amoeboid movement. In slime molds such as *Physarum polycephalum*, for example, cytosol streams back and forth in the branched network that constitutes the cell mass.

Many plant cells display a circular flow of cell contents around a central vacuole. This streaming process, called *cyclosis*, has been studied most extensively in the giant algal cell *Nitella*. In this case, the movement seems to circulate and mix cell contents (**Figure 16-28a**). Cytoplasmic streaming requires actin filaments, as it is inhibited in cells treated with cytochalasin. In *Nitella*, a dense set of aligned microfilaments are found near sites where cyclosis occurs (Figure 16-28b). Cyclosis involves specific myosins that provide the force for movement of components within the

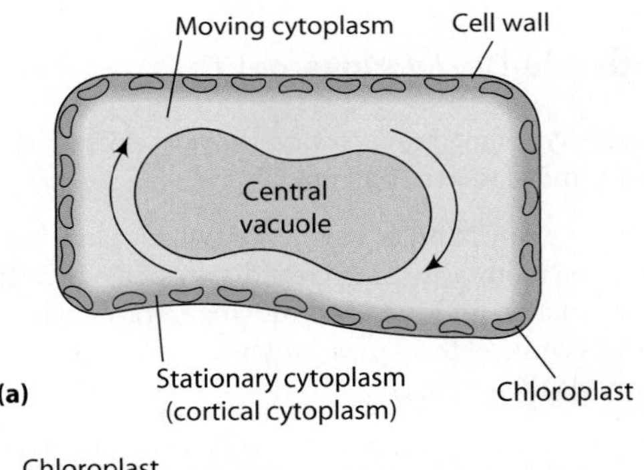

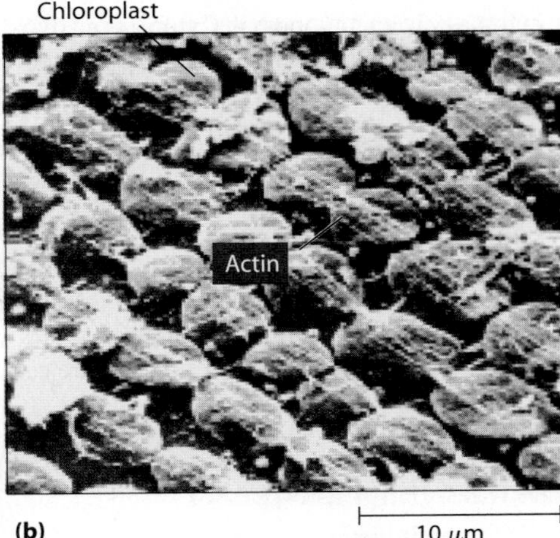

FIGURE 16-28 Cytoplasmic Streaming. (a) Cyclosis in an algal cell. The cytoplasm moves in a circular path around a central vacuole, driven by myosin motors that interact with actin anchored to chloroplasts near the cell wall. (b) Chloroplasts and actin filaments in an algal cell. Actin filaments are arranged in parallel tracks (SEM).

cytosol. When latex beads coated with various types of myosin are added to *Nitella* cells that have been broken open, the beads move along the actin filaments in an ATP-dependent manner in the same direction as normal organelle movement.

In animal cells, myosins may also be involved in vesicular transport. For example, by carefully observing vesicles from the cytoplasm of squid giant axons in vitro, it is clear that individual vesicles can jump from microtubules to microfilaments. This means that the vesicles have both MT motors, such as kinesins, and myosins attached to their surfaces. Myosin V may be particularly important for such actin-based vesicle movement. Myosin V may be able to physically interact with kinesins on the surface of such vesicles, and myosin V has been shown to interact with the plus ends of MTs, making it well suited to such a "handoff" between MTs and MF-based vesicle trafficking. A human disorder called *Griscelli's disease,* which involves partial albinism and neurological defects, has been shown to result from a mutation in this class of myosins.

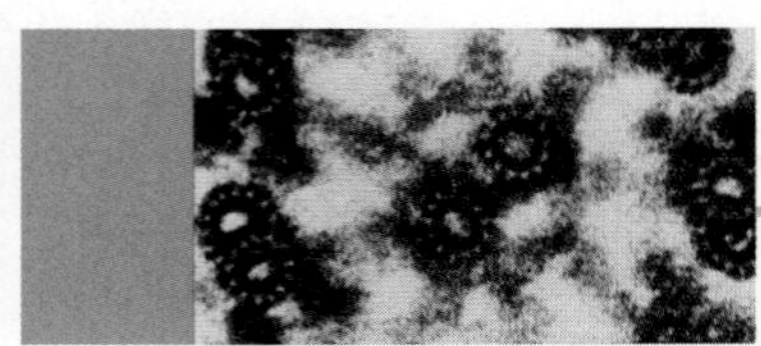

SUMMARY OF KEY POINTS

Motile Systems

- Cell motility and intracellular movements of cellular components are driven by motor proteins, which couple ATP hydrolysis to movements along microtubules (MTs) or microfilaments (MFs). ATP hydrolysis is coupled to conformational changes in the motor that allow the motor to move along a cytoskeletal element.

Microtubule-Based Movement

- Kinesins generally move toward the plus ends of MTs; dyneins move toward the minus ends.
- There are many families of kinesins, which can act as highly processive motors, moving great distances along MTs. One major function for kinesins is moving intracellular cargo. Kinesins connect to adapter proteins and/or their cargo mainly via their light chains.
- There are relatively few dyneins, which fall into two basic classes: cytoplasmic and axonemal. Cytoplasmic dyneins bind to the dynactin complex, which serves as an adapter between dyneins and their cargo.
- MT motors are important for the shaping and transport of the endomembrane system within cells and for intraflagellar transport.
- Axonemal dyneins mediate the bending of cilia and eukaryotic flagella. Dynein arms project out from one MT doublet to the next and slide one set of microtubules past the next. The nine outer doublets of the axoneme are connected laterally to one another and radially to the central pair of single MTs. These connections allow the sliding movement powered by dynein to be converted into bending of the cilium or flagellum.

Microfilament-Based Movement

- There are many families of myosins, and many of them move toward the plus ends of MFs. The best studied is myosin II, one form of which is found in skeletal muscle. Other myosins are involved in events as diverse as cytokinesis, vesicular trafficking, and endocytosis.
- Skeletal muscle contraction involves progressive sliding of thin filaments that contain actin past thick myosin filaments, driven by the interaction between the ATPase head of the myosins and successive myosin-binding sites on actin filaments. Contraction is triggered by the release of calcium from the sarcoplasmic reticulum (SR), which binds troponin, causing a conformational change in tropomyosin that in turn opens myosin-binding sites on the thin filament. Contraction ceases again as calcium is actively pumped back into the SR.
- In smooth muscle, the effect of calcium is mediated by calmodulin, which activates myosin light-chain kinase, leading to the phosphorylation of myosin.
- Actin and myosin are involved in various sorts of motility, including cell crawling, amoeboid movement, cytoplasmic streaming, and cytokinesis. In crawling cells, polymerization of actin extends cellular protrusions; attachment of the protrusions to the substrate and contraction of the cell drive forward movement.

MB **www.masteringbiology.com**

1. MasteringBiology Assignments

Tutorials and Activities

- Molecular Model Structural Tutorials for visualization of 3D structures
- BioFlix Tutorials including 3D animations with hints and detailed feedback for common wrong answers
- Myosin (and its interaction with actin); Sarcomere Shortening During Muscle Contraction

Questions
- Reading Quiz
- Multiple Choice
- End-of-chapter questions

2. eText Read your book online, search, highlight text, take notes, and more

3. The Study Area
- Chapter Review Quiz, Guide to Techniques and Methods, 3D-Structure Tutorials, Activities, BioFlix, Videos, Biology Review, Word Study Tools, Glossary, Flashcards

PROBLEM SET

More challenging problems are marked with a •.

16-1 Kartegener's Triad. Sterility in human males with Kartegener's triad is due to nonmotile sperm. Upon cytological examination, the sperm of such individuals are found to have tails (that is, flagella) that lack one or more of the normal structural components. Such individuals are also likely to have histories of respiratory tract disease, especially recurrent bronchitis and sinusitis, caused by an inability to clear mucus from the lungs and sinuses.

(a) What is a likely mechanistic explanation for nonmotility of sperm in such cases of sterility?

(b) Why is respiratory tract disease linked with sterility in affected individuals?

16-2 A Moving Experience. For each of the following statements, indicate whether it is true of the motility system that you use to lift your arm (A), to cause your heart to beat (H), to move ingested food through your intestine (I), or to sweep mucus and debris out of your respiratory tract (R). More than one response may be appropriate in some cases.

(a) It is under the control of the voluntary nervous system.

(b) It involves interaction between actin and myosin filaments.

(c) It depends on muscles that have a striated appearance when examined with an electron microscope.

(d) It requires ATP.

(e) It depends heavily on fatty acid oxidation for energy.

(f) It would probably be affected by the same drugs that inhibit motility of a flagellated protozoan.

(g) It involves calmodulin-mediated calcium signaling.

16-3 Muscle Structure. Frog skeletal muscle consists of thick filaments that are about 1.6 μm long and thin filaments about 1 μm long.

(a) What is the length of the A band and the I band in a muscle with a sarcomere length of 3.2 μm? Describe what happens to the length of both bands as the sarcomere length decreases during contraction from 3.2 to 2.0 μm.

(b) The H zone is a specific portion of the A band. If the H zone of each A band decreases in length from 1.2 to 0 μm as the sarcomere length contracts from 3.2 to 2.0 μm, what can you deduce about the physical meaning of the H zone?

(c) What can you say about the distance from the Z line to the edge of the H zone during contraction?

16-4 Rigor Mortis and the Contraction Cycle. At death, the muscles of the body become very stiff and inextensible, and the corpse is said to go into rigor.

(a) Explain the basis of rigor. Where in the contraction cycle is the muscle arrested? Why?

(b) Would you be likely to go into rigor faster if you were to die while racing to class or while sitting in lecture? Explain.

(c) Would rigor mortis set in faster in a person given a spinal injection of strychnine or in a person of same age who died of a heart attack? Why?

• **16-5 AMPPNP and the Contraction Cycle.** AMPPNP is the abbreviation for a structural analogue of ATP in which the third phosphate group is linked to the second by a CH_2 group instead of an oxygen atom. AMPPNP binds to the ATP-binding site of virtually all ATPases, including myosin. It differs from ATP, however, in that its terminal phosphate cannot be removed by hydrolysis. When isolated myofibrils are placed in a flask containing a solution of calcium ions and AMPPNP, contraction is quickly arrested.

(a) Where in the contraction cycle will contraction be arrested by AMPPNP? Draw the arrangement of the thin filament, the thick filament, and a cross-bridge in the arrested configuration.

(b) Do you think contraction would resume if ATP were added to the flask containing the AMPPNP-arrested myofibrils? Explain.

(c) What other processes in a muscle cell do you think are likely to be inhibited by AMPPNP?

16-6 Pulled in Two Directions. The following questions deal with the arrangement of actin and myosin in contractile structures.

(a) In skeletal muscle sarcomeres, the H zone is in the middle and bounded on each side by a Z line. During contraction, the Z lines on either side move in opposite directions toward the H zone. Myosin, however, can crawl along an actin filament in only one direction. How can you reconcile movements of Z lines in opposite directions with the unidirectional movement of myosin along an actin filament?

(b) Stress fibers of nonmuscle cells contain contractile bundles of actin and myosin II. For stress fibers to contract or develop tension, how would actin and myosin have to be oriented within the stress fibers?

16-7 Nervous Twitching. Recent military conflicts have involved discussions of "weapons of mass destruction," including nerve gas. One such nerve gas contains the chemical sarin. Sarin inhibits reuptake of the neurotransmitter acetylcholine. What effect would you expect sarin to have on muscle function in individuals exposed to the nerve gas, and why? In your answer, discuss in detail how sarin would be expected to affect the neuromuscular junction, and how it would affect signaling and cytoskeletal events within affected muscle cells.

• **16-8 Tipped Off.** Polarized cytoskeletal structures and intraflagellar transport (IFT) are involved in formation of cilia and flagella.

(a) Observation of flagella in the biflagellate alga, *Chlamydomonas reinhardtii,* indicates that particles move toward the tips of flagella at a rate of 2.5 μm/min., but the particles moving back toward the base of flagella move at 4 μm/min. How do you explain this difference in rate of movement?

(b) Temperature-sensitive mutations in a kinesin II required for intraflagellar transport (IFT) have been identified in *Chlamydomonas.* Such mutations only lead to defects when the temperature is raised above a certain threshold, called the *restrictive temperature.* When algae with fully formed flagella are grown at the restrictive temperature, their flagella degenerate. What can you conclude about the necessity of IFT from this experiment?

(c) Based on your knowledge of the directionality of microtubule motors and the information in (b), where would you predict that the plus ends of flagellar microtubules are? State your reasoning.

16-9 AMPPNP Again. AMPPNP can be used to study microtubule (MT) motors as well as myosins.

(a) What effects would you predict on a sperm flagellum to which AMPPNP was added? In your explanation, please be specific about what molecule's function would be inhibited and what the effect on overall flagellar function would be.

(b) When researchers incubated purified vesicles, nerve cytosol from squid giant axons, and MTs in the presence of AMPPNP, the vesicles bound tightly to the microtubules but did not move. Scientists then used AMPPNP to promote its tight binding to MTs, and the MTs with bound proteins were collected by centrifugation. The main protein purified in this way promotes movement of vesicles away from the cell body, where the nucleus resides. What was this protein?

SUGGESTED READING

References of historical importance are marked with a •.

General References

Bray, D. *Cell Movements: From Molecules to Motility,* 2nd ed. New York: Garland, 2001.

Vale, R. D. The molecular motor toolbox for intracellular transport. *Cell* 112 (2003): 467–480.

Microtubule-Based Motility

Gennerich, A., and R. D. Vale. Walking the walk: How kinesin and dynein coordinate their steps. *Curr. Opin. Cell Biol.* 21 (2009): 59.

Hirokawa, N., Y. Noda, Y. Tanaka, and S. Niwa. Kinesin superfamily motor proteins and intracellular transport. *Nature Rev. Mol. Cell Biol.* 10 (2009): 682.

Hook, P., and R. B. Vallee. The dynein family at a glance. *J. Cell Sci.* 119 (2006): 4369.

Cilia and Flagella

Brokaw, C. J. Thinking about flagellar oscillation. *Cell Motil. Cytoskeleton* 66 (2009): 425.

• Brokaw, C. J., D. J. L. Luck, and B. Huang. Analysis of the movement of *Chlamydomonas* flagella: The function of the radial-spoke system is revealed by comparison of wild-type and mutant flagella. *J. Cell Biol.* 92 (1982): 722.

• Goodenough, U. W., and J. E. Heuser. Substructure of the outer dynein arm. *J. Cell Biol.* 95 (1982): 795.

• Grigg, G. Discovery of the 9 + 2 subfibrillar structure of flagella/cilia. *BioEssays* 13 (1991): 363.

Nicastro, D. et al. The molecular architecture of axonemes revealed by cryoelectron tomography. *Science* 313 (2006): 944.

Silverman, M. A., and M. R. Leroux. Intraflagellar transport and the generation of dynamic, structurally and functionally diverse cilia. *Trends Cell Biol.* 19 (2009): 306.

Filament-Based Movement in Muscle

Geeves, M. A., and K. C. Holmes. The molecular mechanism of muscle contraction. *Adv. Protein Chem.* 71 (2005): 161.

• Huxley, A. F. *Reflections on Muscle.* Princeton, NJ: Princeton University Press, 1980.

• Huxley, H. E. The mechanism of muscular contraction. *Science* 164 (1969): 1356.

• Maruyama, K. Birth of the sliding filament concept in muscle contraction. *J. Biochem.* 117 (1995): 1.

Rossi, D., V. Barone, E. Giacomello, V. Cusimano, and V. Sorrentino. The sarcoplasmic reticulum: An organized patchwork of specialized domains. *Traffic* 9 (2008): 1044.

Spudich, J. A. The myosin swinging cross-bridge model. *Nature Rev. Mol. Cell Biol.* 2 (2001): 387.

Squire, J. M., H. A. Al-Khayat, C. Knupp, and P. K. Luther. Molecular architecture in muscle contractile assemblies. *Adv. Protein Chem.* 71 (2005): 17.

Nonmuscle Microfilament-Based Movement

Gillespie, P. G., and U. Müller. Mechanotransduction by hair cells: Models, molecules, and mechanisms. *Cell* 139 (2009): 33.

Hodge, T., and M. J. Cope. A myosin family tree. *J. Cell Sci.* (2000) 113: 3353.

Kay, R. R., P. Langridge, D. Traynor, and O. Hoeller. Changing directions in the study of chemotaxis. *Nature Rev. Mol. Cell Biol.* 9 (2008): 455.

Le Clainche, C., and M. F. Carlier. Regulation of actin assembly associated with protrusion and adhesion in cell migration. *Physiol. Rev.* 88 (2008): 489–513.

Ridley, A. J. et al. Cell migration: Integrating signals from front to back. *Science* 302 (2003): 1704.

• Taylor, D. L., and J. S. Condeelis. Cytoplasmic structure and contractility in amoeboid cells. *Int. Rev. Cytol.* 56 (1979): 57.

Vicente-Manzanares, M., X. Ma, R. S. Adelstein, and A. R. Horwitz. Non-muscle myosin II takes centre stage in cell adhesion and migration. *Nature Rev. Mol. Cell Biol.* 10 (2009): 778.

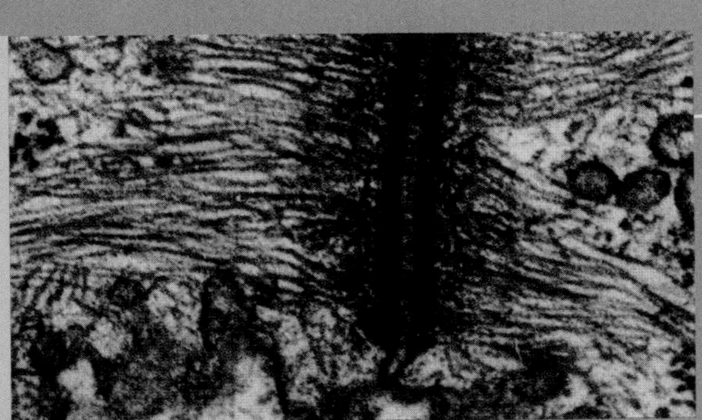

17 Beyond the Cell: Cell Adhesions, Cell Junctions, and Extracellular Structures

See MasteringBiology® for tutorials, activities, and quizzes.

In the preceding chapters, we have treated cells as if they exist in isolation and as if they "end" at the plasma membrane. However, most organisms are multicellular: *They consist of many—sometimes trillions—of cells. Many kinds of cells—including most of those in your own body—spend all their lives linked to neighboring cells. It is the organization of cells into* tissues *that allows multicellular organisms to adopt complex structures. These tissues, in turn, are arranged in precise ways to generate the form of an organism. How do these tissues achieve their structures? Consider the two types of animal tissues shown in* **Figure 17-1**. *One is a sheet of cells known as an* epithelium, *such as the cells that line the small intestine. Such cells are clearly* polarized*—that is, they contain discrete functional domains at opposite ends of the cell. The end of the cell in contact with the external environment (for example, the lumen of the small intestine) is often called the* apical *side of the cell. The parts on the other side of the tight junction, including the surfaces in contact with the basal lamina, are called the* basolateral *side of the cell. The other cell type in Figure 17-1 is from a more loosely organized* connective tissue, *such as might be found in the dermis of the skin. In each case, cells must be attached to one another, to a mechanically rigid scaffolding, or to both. Thus, in order to understand how multicellular organisms are constructed, we will need to consider both connections between cells, or* cell-cell adhesions, *and the* extracellular structures *to which cells attach. Cells use a variety of elaborate molecular complexes, or* junctions, *to attach to one another, and most of these involve transmembrane proteins that link the cell surface to the cytoskeleton (which we learned about in Chapters 15 and 16). Cells use other types of molecular complexes to attach to extracellular structures, and these also involve specific linkages between the cell surface and the cytoskeleton.*

The extracellular structures themselves consist mainly of macromolecules that are secreted by the cell. Animal cells have an extracellular matrix *that takes on a variety of forms and plays important roles in cellular processes as diverse as division, motility, differentiation, and adhesion. Epithelial cells, such as the ones shown in Figure 17-1, produce a specialized extracellular matrix called a* basal lamina. *Connective tissues, on the other hand, produce a more loosely organized matrix. In plants, fungi, algae, and prokaryotes, the extracellular structure is a* cell wall*—although its chemical composition differs considerably among these organisms. Cell walls confer rigidity on the cells they encase, serve as permeability barriers, and protect cells from physical damage and from attack by viruses and infectious organisms.*

Most of our attention in this chapter is devoted to the adhesions animal cells make with one another, the junctions that characterize these adhesions, and how animal cells interact with the extracellular matrix. Then we will turn to the walls that surround plant cells and the specialized structures that allow direct cell-to-cell communication between plant cells, despite the presence of a cell wall.

Cell-Cell Recognition and Adhesion

The ability of individual cells to associate in precise patterns to form tissues, organs, and organ systems requires that individual cells be able to recognize, adhere to, and communicate with each other. We discuss *cell-cell contact* in this section and consider *intercellular junctions* later in the chapter.

Transmembrane Proteins Mediate Cell-Cell Adhesion

Animal cells use specialized adhesion receptors to attach to one another. Many of these adhesion proteins are transmembrane proteins, which means the extracellular portion of these proteins can interact with the extracellular

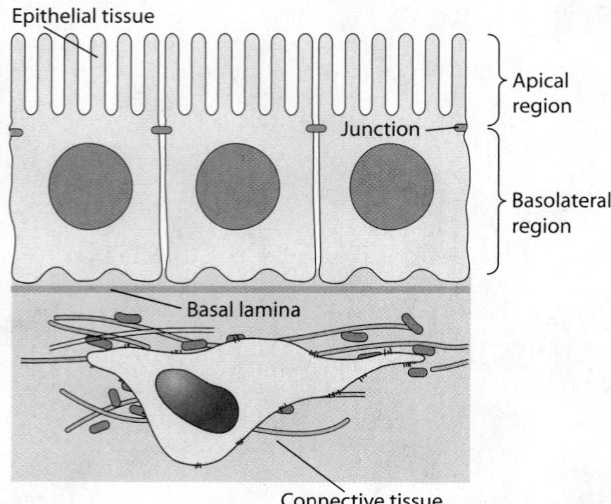

FIGURE 17-1 Different Types of Tissues in Animals. Tissues are multicellular structures that include cells linked together with extracellular material. In a polarized epithelium (top), such as the cells lining the small intestine in a mammal, cells are connected together by cell-cell adhesions. The apical surfaces of these cells are very different from the basal surfaces, which lie on top of an extracellular matrix known as a basal lamina. In a connective tissue (bottom), such as the dermis of the skin, loosely organized cells are embedded in extracellular matrix fibers.

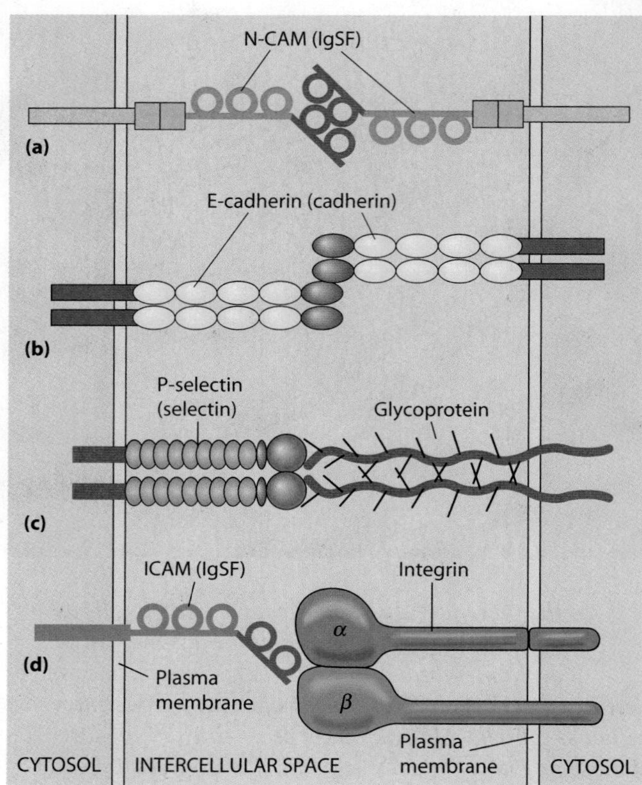

FIGURE 17-2 Different Types of Cell-Cell Adhesion Proteins. Cells adhere to other cells using transmembrane proteins that fall into a few main classes. These include **(a)** immunoglobulin superfamily (IgSF) proteins, such as N-CAM; **(b)** cadherins, such as E-cadherin; **(c)** selectins, which bind to the carbohydrates of glycoproteins on other cells; and **(d)** in a few cases such as leukocytes, integrins, which bind to IgSF proteins such as ICAM on the surface of endothelial cells.

portion of similar proteins on the surface of a neighboring cell. Although diagrams of adhesive structures may suggest that they are static once assembled, they are anything but. Cells can dynamically assemble and disassemble adhesions in response to a variety of events. Many adhesion proteins are continuously recycled: Protein at the cell surface is internalized by endocytosis, and new protein is deposited at the surface via exocytosis. In addition, adhesion proteins serve as key sites for assembly of signaling complexes in cells and for dynamically assembling cytoskeletal structures at sites of cell adhesion. In this way, cell adhesion is coordinated with other major processes, including cell signaling, cell movement, cell proliferation, and cell survival.

We now know that cell-cell adhesion receptors fall into a relatively small number of classes. They include *immunoglobulin superfamily (IgSF)* proteins, *cadherins, selectins,* and, in a few cases, *integrins* (**Figure 17-2**). In each case, the adhesion protein on the surface of one cell binds to the appropriate ligand on the surface of a neighboring cell. In some cases, such as many cadherins and immunoglobulin superfamily members, cells interact with identical molecules on the surface of the cell that they adhere to. Such interactions are said to be **homophilic interactions** (from the Greek *homo,* meaning "like," and *philia,* meaning "friendship"). In other cases, such as the selectins and integrins, a cell adhesion receptor on one cell interacts with a different molecule on the surface of the cell to which it attaches. Such interactions are said to be **heterophilic interactions** (from the Greek *hetero,* meaning "different"). Many transmembrane adhesion receptors attach to the cytoskeleton via linker proteins, which differ depending on the class of molecule and its location within the cell. In the next few sections, we consider examples of each of these major classes of cell-cell adhesion molecules.

Cell Adhesion Molecules (CAMs). CAMs are members of the **immunoglobulin superfamily (IgSF).** The founding member of this family, called *neural cell adhesion molecule (N-CAM),* was identified using an antibody that disrupted cell-cell adhesion between isolated neuronal cells. Proteins in this large superfamily are so named because they contain domains, characterized by well-organized loops, that are similar to those in the immunoglobulin subunits that constitute antibodies. CAMs, such as N-CAM, on one cell interact homophilically with CAMs on an adjacent cell via these domains. Other IgSF members interact heterophilically with their ligands. IgSF members participate in a wide range of adhesion events. In the embryonic nervous system, CAMs, such as N-CAM and L1-CAM, are involved in the outgrowth and bundling of neurons. Humans with mutations in the L1-CAM gene show defects in the corpus callosum (a region that interconnects the

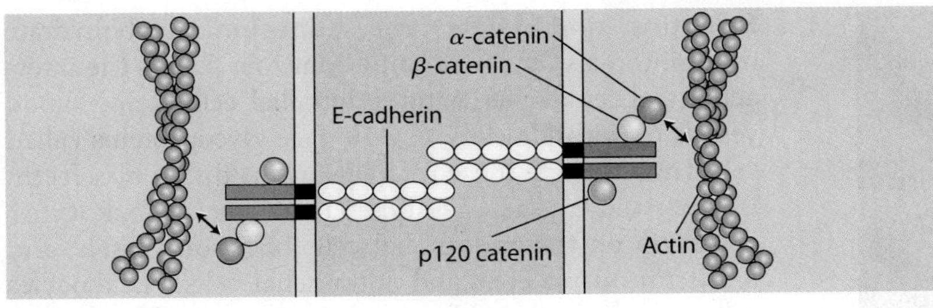

(a) Adherens junction

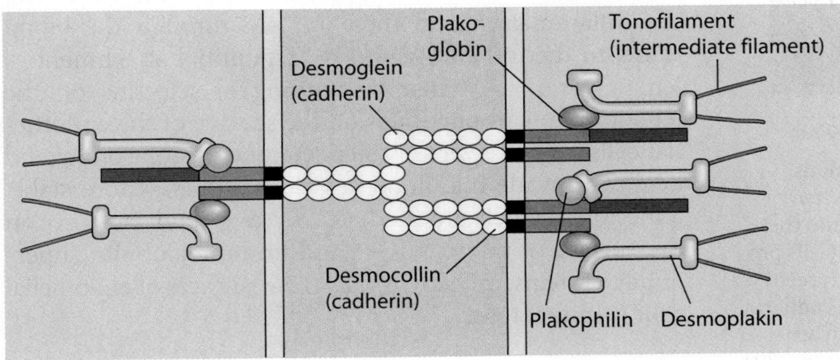

(b) Desmosome

FIGURE 17-3 Cadherin Structure. Cadherins are found at sites of cell-cell adhesion. **(a)** "Classical" cadherins, such as E-cadherin, mediate cell-cell adhesion at adherens junctions. E-cadherin molecules associate as pairs (homodimers) in the plasma membrane and interact with homodimers in the neighboring cell through their extracellular domains. Their cytosolic tail binds to the linker protein, β-catenin. β-catenin in turn binds α-catenin, which recruits actin to the junction. **(b)** The cadherins in desmosomes, called desmocollins and desmogleins, probably interact in pairs. Neighboring cells probably attach to one another via their desmogleins. In the cytosol, desmosomal cadherins interact with proteins in the desmosomal plaque (plakoglobin, plakophilin, and desmoplakin) that anchors desmosomes to intermediate filaments.

two hemispheres of the brain), mental retardation, and other defects.

Cadherins. The use of antibodies that block cell adhesion also led to the discovery of the **cadherins,** an important group of adhesive glycoproteins found in the plasma membranes of most animal cells. Like CAMs, cadherins play a crucial role in cell-cell recognition and adhesion. The two groups of proteins can be distinguished from each other because cadherins, but not CAMs, require calcium to function. Ca^{2+} binds to and stabilizes the conformation of cadherins that allows them to mediate cell-cell adhesion.

Cadherins are characterized by a series of structurally similar domains (or "repeats") in their extracellular domain. Members of the cadherin superfamily have widely varying numbers of these repeats, and they vary in the structure of their cytosolic ends. The best-characterized cadherin, E-cadherin (*E* is for "epithelial"), has five such repeat domains. E-cadherin molecules associate in pairs in the plasma membrane. Their extracellular domains have a structure that allows them to "zip" together in a homophilic fashion, as cadherins from one cell interlock with those from a neighboring cell (**Figure 17-3**). At their cytosolic ends, cadherins associate with the cytoskeleton, thereby linking the cell surface to the cytoskeleton. These linkages differ for different types of cadherins, as we will see when we discuss the structure of cell-cell junctions later in this chapter.

Different cadherins are expressed in specific tissues. Their regulated expression is a particularly striking feature of embryonic development and contributes to the ability of different tissues to separate from one another as embryos change their shape. The role of different cadherins in cell-cell adhesion has been investigated in cultured fibroblasts called *L cells,* which bind poorly to one another and contain little cadherin. When purified DNA encoding E-cadherin or P-cadherin (*P* is for "placental," where this cadherin was first described) is introduced into L cells, the cells begin to produce cadherins and to bind more tightly to one another. Moreover, L cells that produce E-cadherin bind preferentially to other cells producing E-cadherin. Similarly, cells that produce P-cadherin bind selectively to other cells that are also producing P-cadherin (**Figure 17-4**). Such observations suggest that the amount and types of cadherin molecules on cell surfaces help to segregate cells into specific tissues.

Cadherins have especially important roles during embryonic development, where the dynamic assembly and disassembly of cadherin-based adhesions is tightly regulated. For example, when frog embryos are depleted of mRNA for the main type of cadherin found in the early embryo, they lose their normal organization (**Figure 17-5**). Other cadherins are involved in helping neuronal cells to form bundles and to establish synaptic connections. One particularly important event that occurs frequently in embryos is the breakdown of an epithelium into loosely organized, migratory cells called mesenchymal cells. This *epithelial-mesenchymal transition (EMT)* is accompanied by changes in cadherin expression. Changes in cadherin expression also occur in cancer cells. Cancer cells often stop expressing cadherins, such as E-cadherin, on their surfaces. As a result, they undergo an EMT that healthy cells would not, and these cells spread to other parts of the body by a process known as metastasis (see Chapter 24).

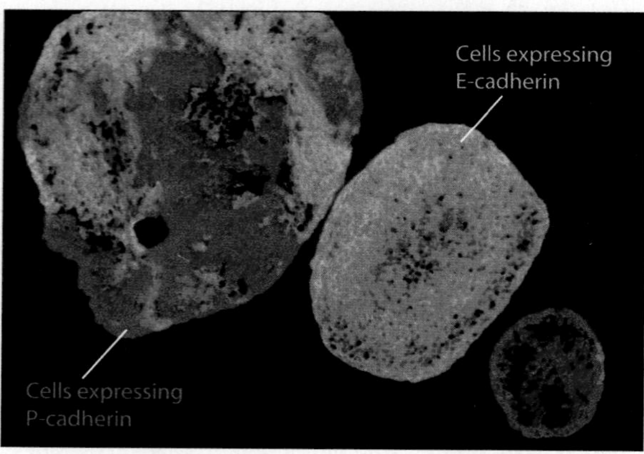

FIGURE 17-4 The Effect of Cadherins on Cell Adhesions. Cultured L cells do not normally adhere to one another. When DNA coding for E-cadherin or P-cadherin is introduced into these cells, cadherin is produced, and cell-cell adhesion occurs. Cells producing P-cadherin (red) and those producing E-cadherin (green) can be stained with fluorescent antibodies specific for each cadherin. The staining pattern shows that cells producing P-cadherin are located in different areas from the cells producing E-cadherin. This indicates that cells making E-cadherin bind preferentially to other cells producing E-cadherin. Similarly, cells making P-cadherin bind preferentially to other cells producing P-cadherin.

Carbohydrate Groups Are Important in Cell-Cell Recognition and Adhesion

Like other glycoproteins, the carbohydrate side chains of CAMs and cadherins likely affect their adhesion properties. In addition, there are several well-studied examples of how carbohydrates on the cell surface affect cell adhesion.

Lectins. A role for carbohydrate groups in cell adhesion is also suggested by the fact that many animal and plant cells secrete carbohydrate-binding proteins called **lectins,** which promote cell-cell adhesion by binding to a specific sugar or sequence of sugars exposed at the outer cell surface. Because lectins usually have more than one carbohydrate-binding site, they can bind to carbohydrate groups on two different cells, thereby linking the cells together.

Selectins and Leukocyte Adhesion. Carbohydrate recognition also plays an important role during the interactions of leukocytes with endothelial cells lining blood vessels or with platelets. Cell surface glycoproteins called **selectins** mediate these interactions. A different selectin is expressed by each cell type (*L-selectin* on leukocytes, *E-selectin* on the endothelial cells of blood vessels, and *P-selectin* on platelets and endothelial cells). Leukocytes roll along the walls of blood vessels. During inflammation, they attach to the wall of a blood vessel in the vicinity of the inflammation and then migrate through the blood vessel to the inflammation site. The initial attachment of leukocytes is mediated by binding of selectins on the leukocyte to carbohydrates on the surface of the endothelial cells and vice versa. When leukocytes stop rolling and begin to invade the blood vessel, they make more stable adhesions, which are mediated by a specific integrin on the surface of the leukocyte and immunoglobulin superfamily proteins, called *ICAMs,* on the surfaces of endothelial cells (**Figure 17-6**).

Carbohydrates and the Survival of Erythrocytes. One especially well-known example of the importance of carbohydrates at the cell surface is the determination of the human blood types A, B, AB, and O by a specific carbohydrate side chain present on a glycolipid of the erythrocyte plasma membrane. The *ABO blood group* involves differences in carbohydrate side chains on the surface of red blood cells that can be detected by antibodies present in the blood, leading to clumping of red blood cells and likely to the patient's death if the wrong blood type is used in transfusion.

The four blood types in the ABO system depend on genetically determined differences in the structure of a branched-chain carbohydrate attached to a specific glycolipid in the erythrocyte plasma membrane. Individuals with blood type A have the amino sugar *N*-acetylgalactosamine (GalNAc) at the ends of this carbohydrate, whereas individuals with blood type B have galactose (Gal) instead. Individuals with blood type AB have both *N*-acetylgalactosamine and galactose present; and, in individuals with type O blood, these terminal sugars are missing entirely.

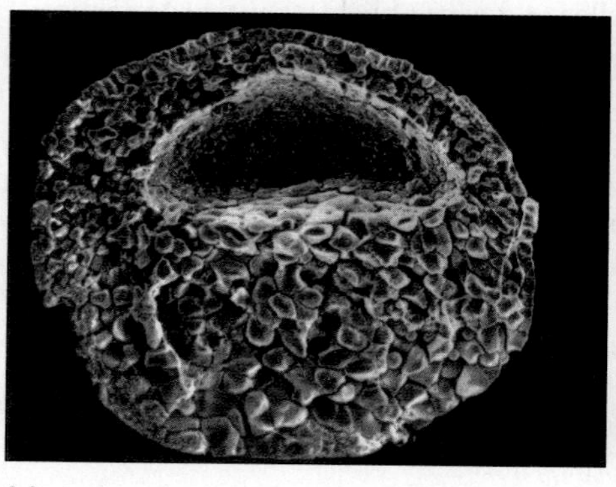

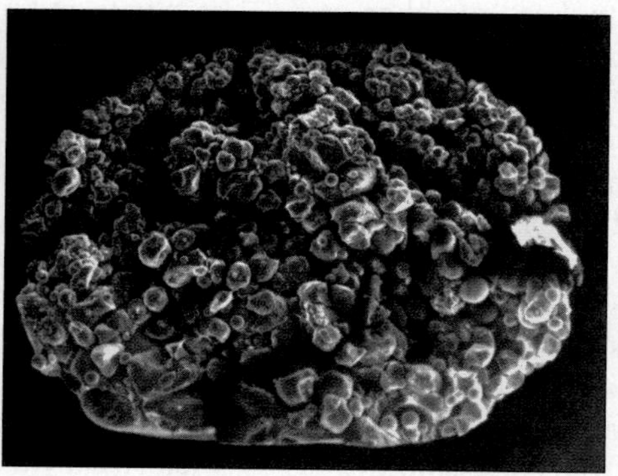

FIGURE 17-5 Cadherins During Embryonic Development. After many rounds of cell division, frog embryos form a blastula, which contains a fluid-filled space lined by cells. When embryos are depleted of mRNA for a cadherin known as EP-cadherin, they fail to make EP-cadherin protein, and compared with control embryos **(a)**, they lose their normal organization **(b)**.

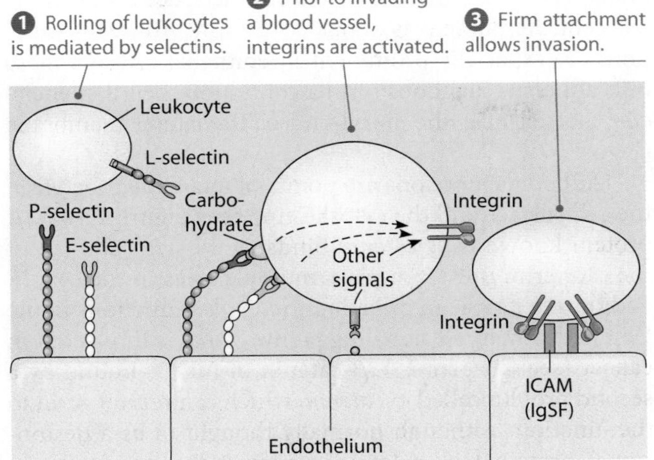

FIGURE 17-6 Leukocyte Adhesion and Selectins. The initial attachment of leukocytes, which allows them to roll along endothelial cells that line blood vessels, is mediated by selectins. Strong adhesion, mediated by integrins, allows leukocytes to stop and to pass between cells in the blood vessel wall to sites of inflammation. Adapted from D. Vestweber and J. E. Blanks, "Mechanisms That Regulate the Function of the Selectins and Their Ligands," *Physiological Reviews* 79 (1), January 1999:181-213, Fig. 1. Am Physiol Soc, used with permission.

These minor differences have major effects on the compatibility of blood transfusions because people with blood types A, B, or O have antibodies in their bloodstream that recognize and bind to the respective terminal sugars of this specific glycolipid. An individual whose blood contains antibodies against one or both sugars (GalNAc or Gal) cannot accept blood containing the glycolipid with that terminal sugar. Individuals with type A blood have antibodies against carbohydrate chains ending in galactose, which occur in type B and type AB blood. They therefore cannot be transfused with type B or AB blood, but they can accept blood from type A or O donors. Conversely, individuals with type B blood have antibodies against carbohydrate chains ending in GalNAc, which occur in blood of types A and AB. They therefore cannot be transfused with type A or AB blood but can accept blood from B or O donors. Individuals with type O blood are called universal donors because their erythrocytes do not generate an immune response when transfused into individuals of any blood type.

Cell-Cell Junctions

By definition, unicellular organisms have no permanent associations between cells (although they can form temporary associations, such as during bacterial swarming or the aggregation of slime mold amoebae). Whereas a single cell is an entity unto itself, multicellular organisms have specific means of joining cells in long-term associations to form tissues and organs. Such associations usually involve specialized modifications of the plasma membrane at the point where two cells come together. These specialized structures are called **cell-cell junctions.** In animals, the three most common kinds of cell junctions are *adhesive junctions, tight junctions,* and *gap junctions.* **Figure 17-7** illustrates each of these kinds of junctions.

In plants, the presence of a cell wall between the plasma membranes of adjacent cells precludes the kinds of cell junctions that link animal cells. However, the cell wall and special structures called *plasmodesmata* carry out similar functions, as we will see a bit later in the chapter.

Polarity Proteins Regulate the Positioning of Cell-Cell Junctions

Both migrating single cells and sheets of epithelial cells display polarity: They have defined regions that have

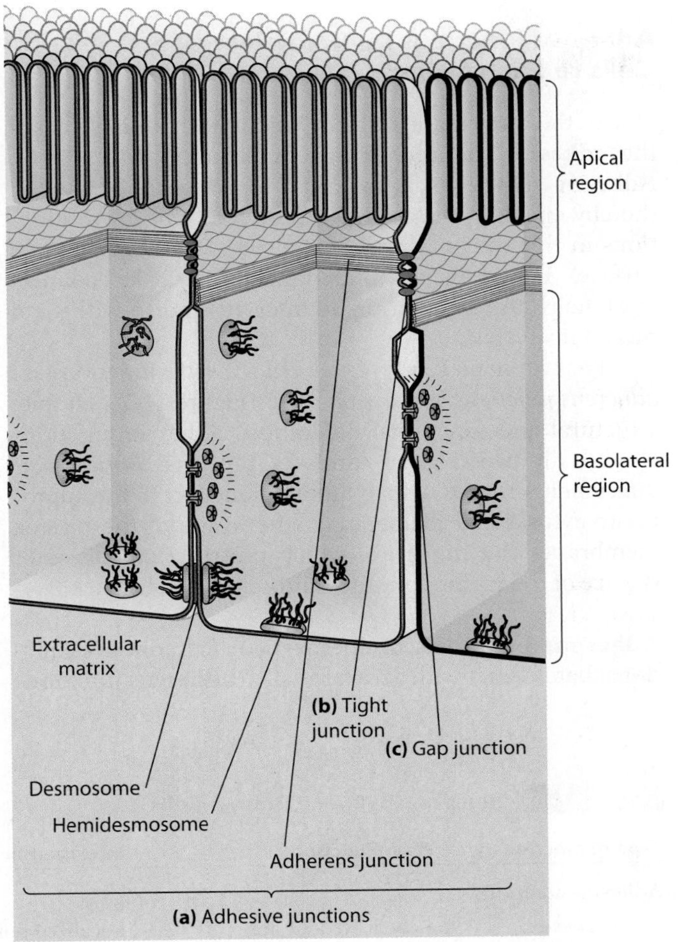

FIGURE 17-7 Major Types of Cell Junctions in Animal Cells. **(a)** Adhesive junctions are specialized for cell-cell and cell-ECM adhesion. Desmosomes and hemidesmosomes bind to intermediate filaments within the cell, whereas adherens junctions and focal adhesions bind to actin microfilaments. **(b)** Tight junctions create an impermeable seal between cells, thereby preventing fluids, molecules, or ions from crossing a cell layer via the intercellular space. **(c)** Gap junctions provide direct chemical and electrical communication between cells by allowing the passage of small molecules and ions from one cell to another. Note that, like adherens junctions, tight junctions can also recruit actin, but for clarity this is not shown.

obviously different activities. Epithelial cells in particular have a striking polarity that separates apical from basolateral regions of the cell surface (see Figure 17-1). Numerous studies, including key work in the nematode worm, *C. elegans*, and the fruit fly, *Drosophila*, have identified proteins that help to establish and maintain epithelial polarity. One important protein complex, localized at the apical ends of epithelial cells, is the *Par3/Par6/atypical protein kinase C (aPKC) complex*. The Par proteins get their name from mutations in *C. elegans* that result in defective *par*titioning of materials in the one-cell embryo. In conjunction with the Rho family proteins Cdc42 or Rac, the Par3/Par6//aPKC complex is important for recruiting other protein complexes to the apical ends of epithelial cells. Par proteins also play important roles in non-epithelial cells, which are less well understood.

Adhesive Junctions Link Adjoining Cells to Each Other

One of the three main kinds of junctions in animal cells is the **adhesive** (or **anchoring**) **junction** (**Table 17-1**). Adhesive junctions link cells together into tissues, thereby enabling the cells to function as a unit. All junctions in this category anchor the cytoskeleton to the cell surface. The resulting interconnected cytoskeletal network helps to maintain tissue integrity and to withstand mechanical stress.

The two main kinds of cell-cell adhesive junctions are *adherens junctions* and *desmosomes* (Figure 17-7). Despite structural and functional differences, adhesive junctions all contain two distinct kinds of proteins: *intracellular attachment proteins,* which link the junction to the appropriate cytoskeletal filaments on the inside of the plasma membrane, and *cadherins,* which protrude on the outer surface of the membrane and bind cells to each other.

Adherens Junctions. Cadherin-mediated adhesive junctions that interact with actin are called **adherens junctions** (Table 17-1). At adherens junctions, the space between the adjacent membranes is about 20–25 nm. Adherens junctions are especially prominent in epithelial cells. In these cells adherens junctions form a continuous belt that encircles the cell near the apical end of the lateral membrane (Figure 17-7).

Adherens junctions are points of attachment between the cell surface and the cytoskeleton (see Figure 17-3a). A protein known as *β-catenin* binds to the cytosolic tail of the cadherin. β-catenin plays multiple roles in the cell: In addition to its role in cell adhesion, it also functions in the Wnt pathway, a cell signaling pathway that is important in cancer (see Chapter 24). β-catenin in turn is bound by a second protein called *α-catenin,* which can recruit actin to the junction. Although normally thought of as a desmosomal protein (see below), in some adherens junctions another β-catenin family member, plakoglobin, is present alongside β-catenin. A final core component of adherens junctions is *p120catenin (p120ctn)*. This protein binds to the cytoplasmic tail of cadherins near the plasma membrane; p120ctn regulates stability of cadherin at the surface, as well as the activity of Rho, an actin regulator we examined in Chapter 15.

We have seen that cell-cell and cell–extracellular matrix adhesions are important for the normal functions of cells and tissues in the body. A surprising finding of modern microbiology is that many pathogens, such as those responsible for several types of food poisoning, infect the body by using these very same adhesion systems to gain entry into healthy cells. These mechanisms of infection are discussed in more detail in **Box 17A.**

Desmosomes. **Desmosomes** are button-like points of strong adhesion between adjacent cells in a tissue. Desmosomes give the tissue structural integrity, enabling cells to function as a unit and to resist stress. Desmosomes are found in many tissues but are especially abundant in skin, heart muscle, and the neck of the uterus.

Table 17-1 Junctions Between Animal Cells

Type of Junction	Main Function	Intermembrane Features	Space	Associated Structures
Adhesive junctions				
Focal adhesion	Cell-ECM adhesion	Localized points of attachment	20–25 nm	Actin microfilaments
Hemidesmosome	Cell–basal lamina adhesion	Localized points of attachment	25–35 nm	Intermediate filaments (tonofilaments)
Adherens junction	Cell-cell adhesion	Continuous zones of attachment	20–25 nm	Actin microfilaments
Desmosome	Cell-cell adhesion	Localized points of attachment	25–35 nm	Intermediate filaments (tonofilaments)
Tight junction	Sealing spaces between cells	Membranes joined along ridges	None	Transmembrane junctional proteins, actin
Gap junction	Exchange of ions and molecules between cells	Connexons (transmembrane protein complexes with 3-nm pores)	2–3 nm	Connexins in one membrane align with those in another to form channels between cells

BOX 17A HUMAN APPLICATIONS

Food Poisoning and "Bad Bugs": The Cell Surface Connection

Cell adhesion and cell recognition play important roles during the construction and maintenance of tissues in the human body. Surprisingly, foreign invaders that attack the human body can use the very same proteins that healthy cells require for cell adhesion to gain entry into the body. Here, we consider one example of how bacteria attach to and infect human cells: enteropathogenic bacteria.

Enteropathogenic Bacteria Use Normal Cell Adhesion Proteins to Infect Host Cells

One good example of this sort of "molecular hijacking" of normal cell adhesion processes occurs when pathogenic bacteria enter the digestive tract (such bacteria are called *enteropathogenic* bacteria; *entero* comes from the Greek word for "intestine"). Such bacteria are responsible for several types of food poisoning, and their combined effects have a substantial impact on public health. Although some pathogenic bacteria can use multiple methods for gaining entry into the gut, in several well-studied cases, they attach to cell adhesion molecules such as integrins or cadherins.

One of the best-studied examples of such subversion of cell adhesion is the enteric pathogen *Yersinia pseudotuberculosis*. Infection by bacteria of the genus *Yersinia*, which usually occurs via contaminated water and food, typically results in gastroenteritis with diarrhea and vomiting 24–48 hours after exposure. *Y. pseudotuberculosis* uses a 986-amino acid outer membrane protein, called *invasin*, to penetrate mammalian cells. Surprisingly, the cellular receptors for invasin are integrins that contain β_1 subunits on the surface of cells lining the gut (**Figure 17A-1**). Identification of the cellular receptor for *Yersinia* was an important discovery because it showed that bacteria could invade cells by targeting common mammalian cell surface proteins.

As the molecular pathways used by other bacteria to invade host cells have been identified, such co-opting of normal cell surface proteins has emerged as a common theme in pathogenesis. A second enteropathogenic bacterium, *Shigella flexnerii*, also attaches to an integrin ($\alpha_5\beta_1$) via proteins on its surface. *Shigella* infection results in dysentery and is usually caused by contamination of raw foods by food handlers. Approximately 300,000 cases of shigellosis occur annually in the United States, making it a significant health problem.

Another well-studied example of subversion of cell adhesion is *Listeria monocytogenes*. *Listeria* infection can occur through exposure to improperly prepared raw foods, such as raw milk and hamburger. Although the incidence of listeriosis in the United States is not high (about 2000 people annually develop symptoms), once infection occurs, it is very serious. Approximately 25% of those infected die, and many of these people contract bacterial meningitis. *Listeria* expresses a protein called *internalin A* on its surface that can bind to E-cadherin on the surface of cells in the gut (Figure 17A-1). We now know that just as these bacteria "hijack" the normal cell adhesion machinery of gut cells, they can do the same with the cytoskeleton once they are inside an infected cell (see Box 15A, page 441).

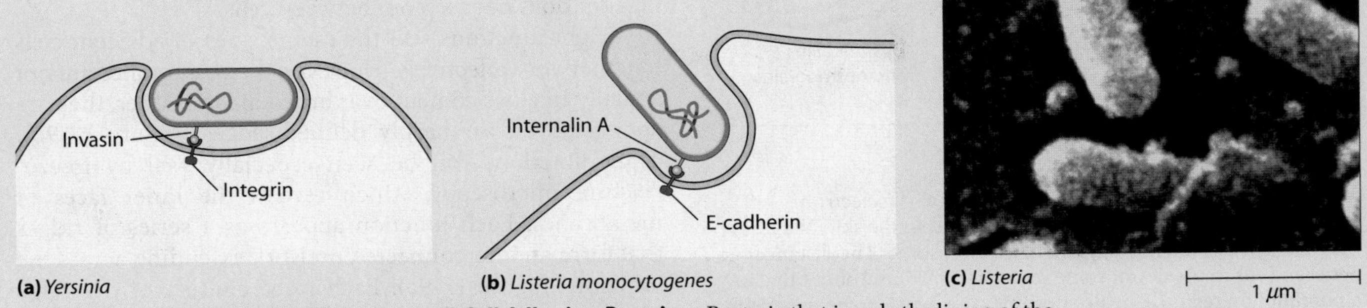

FIGURE 17A-1 Bacterial Pathogens and Cell Adhesion Proteins. Bacteria that invade the lining of the human digestive tract attach to cell adhesion receptors on the surface of intestinal cells. **(a)** Species in the genus *Yersinia* express a protein called invasin that attaches to integrins on gut cells that have β_1 subunits. **(b)** *Listeria monocytogenes* expresses a protein called internalin A that binds to E-cadherin on intestinal cells. **(c)** *Listeria* attached to the surface of cultured epithelial cells (SEM).

The structure of a typical desmosome is shown in **Figure 17-8.** The plasma membranes of the two adjacent cells are aligned in parallel, separated by a space of about 25–35 nm. The extracellular space between the two membranes is called the *desmosome core.* The desmosome core consists of the desmosomal cadherins, *desmocollin* and *desmoglein* (see Figure 17-3b). Unlike E-cadherin, desmocollins and desmogleins probably interact heterophilically across the intercellular space. Like other cadherins, linker proteins bind to their cytosolic tail and link them to the cytoskeleton. The β-catenin family protein *plakoglobin* binds to desmocollin. Plakoglobin in turn binds to a protein called *desmoplakin.* Desmoplakin in turn attaches to *tonofilaments,* which are composed of intermediate filaments such as *vimentin, desmin,* or *keratin.* A thick *plaque* containing these linker proteins

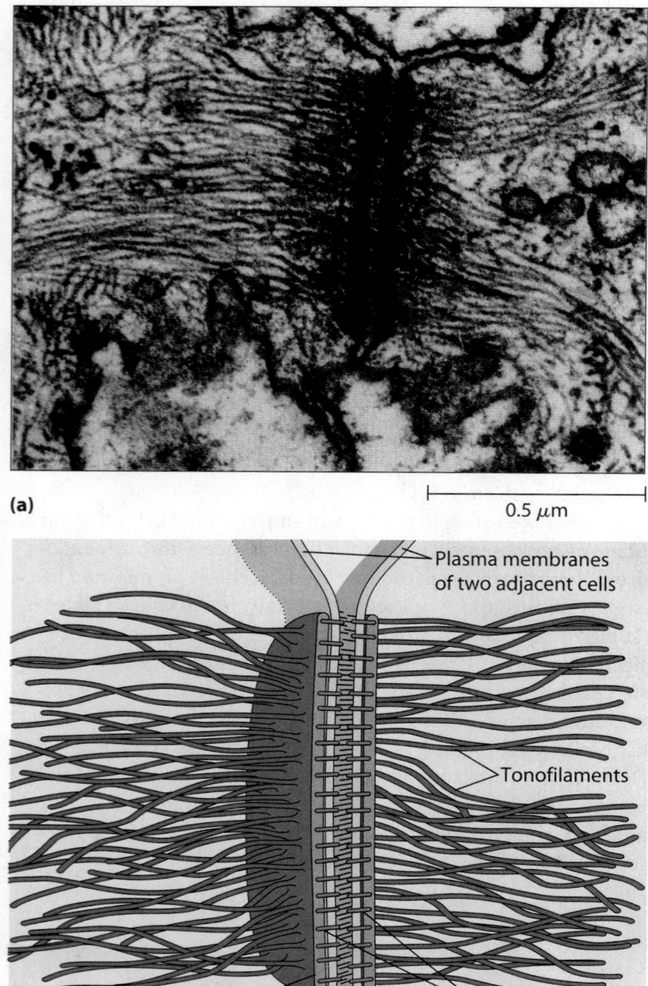

FIGURE 17-8 Desmosome Structure. **(a)** An electron micrograph of a desmosome joining two cells in the skin of a newt (TEM). **(b)** A schematic diagram of a desmosome. The distance between cells in the desmosome region is 25–35 nm, about that for a nonjunction region. The desmosome core between the two membranes is filled with cadherins (desmocollins and desmogleins). The plaque on the cytoplasmic side of the membrane contains desmoplakins and plakoglobin, and it is linked to tonofilaments, which are intermediate filaments consisting of keratin, desmin, or vimentin, depending on the cell type. By connecting tonofilaments from adjacent cells together, desmosomes help provide mechanical strength to epithelial sheets.

and tonofilaments is found just beneath the plasma membrane of each of the two adjoining cells.

Loss of desmosomal components can be devastating. For example, mice lacking plakoglobin die with heart failure and skin defects. Similarly, mutations in desmocollins expressed in the heart can lead to damage to the heart muscle in adult human patients. Human patients who develop autoimmune reactions against components of their desmosomes develop blistering diseases of the skin known as *pemphigus.* Some patients develop antibodies against desmogleins, while others generate antibodies against linker proteins, such as desmoplakin. Similar blistering disorders arise when human patients have mutations in the genes encoding desmosomal proteins.

Tight Junctions Prevent the Movement of Molecules Across Cell Layers

A key feature of epithelial tissues is that they form barriers between the internal cells of the body and the outside world. For example, intestinal cells must seal off the fluids that pass through the digestive tract from the internal fluids of the body. Epithelial cells, therefore, need specialized structures that serve to seal them tightly together. **Tight junctions** serve this function. As their name implies, they leave no space at all between the plasma membranes of adjacent cells (**Figure 17-9**). The tight junctions between adjacent cells in an epithelium lining an organ or body cavity form a continuous belt around the apical ends of the lateral surfaces of each cell, just apical to the adherens junction. These belts together form a formidable barrier (Figure 17-9a), so that molecules must typically cross the cell layer by passing through the cells themselves. Tight junctions are especially prominent in intestinal epithelial cells. Tight junctions are also abundant in the ducts and cavities of glands that connect with the digestive tract, such as the liver and pancreas, as well as in the urinary bladder, where they ensure that the urine stored in the bladder does not seep out between cells.

Tight junctions seal the membranes of adjacent cells together very effectively. However, the membranes are not actually in close contact over broad areas. Rather, they are connected along sharply defined ridges (Figure 17-9b). Tight junctions can be seen especially well by freeze-fracture microscopy, which reveals the inner faces of membranes. Each junction appears as a series of ridges that form an interconnected network extending across the junction (Figure 17-9c). Each ridge consists of a continuous row of tightly packed transmembrane junctional proteins about 3–4 nm in diameter. The result is rather like placing two pieces of corrugated metal together so that their ridges are aligned and then fusing the two pieces lengthwise along each ridge of contact. The fused ridges eliminate the intercellular space and effectively seal the junction, creating a barrier that prevents the passage of extracellular fluid through the spaces between adjacent cells. Not surprisingly, the number of such ridges across a junction correlates well with the tightness of the seal made by the junction. In addition to these close membrane appositions, scaffolding proteins at tight junctions recruit cytoskeletal proteins, such as F-actin, to tight junctions.

The properties of tight junctions have been studied by incubating tissues in the presence of electron-opaque tracer molecules and then using electron microscopy to

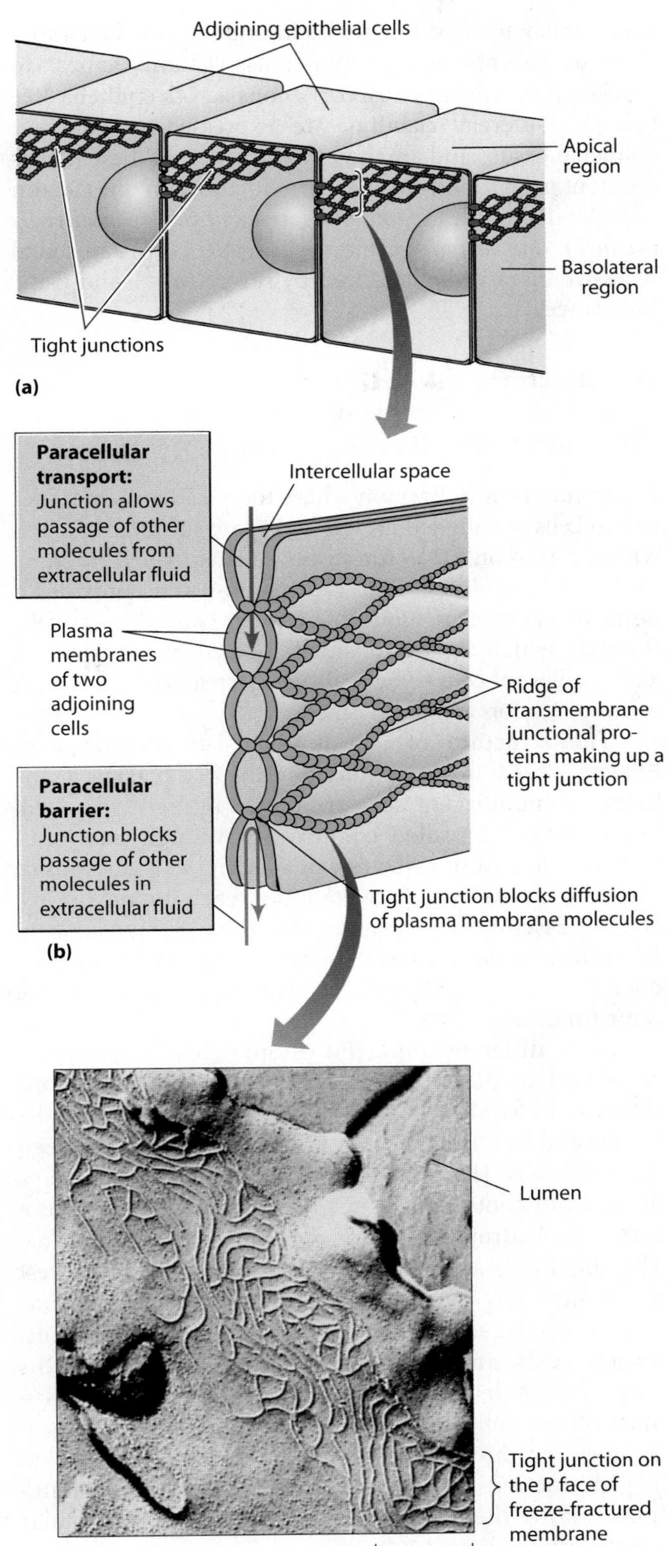

FIGURE 17-9 Tight Junction Structure. **(a)** A schematic representation of several adjoining epithelial cells connected by tight junctions. **(b)** Transmembrane junctional proteins in the plasma membranes of two adjacent cells are clustered along the points of contact, forming ridges of protein particles that join the two plasma membranes together tightly. Tight junctions prevent the passage of extracellular molecules through the spaces between cells (red arrows) and also block lateral movement of transmembrane proteins. **(c)** This electron micrograph illustrates tight junctions between cells in a frog bladder, as revealed by the freeze-fracture technique. Tight junctions appear as raised ridges on the protoplasmic (P) face of the membrane. The lumen is the cavity of the bladder (TEM).

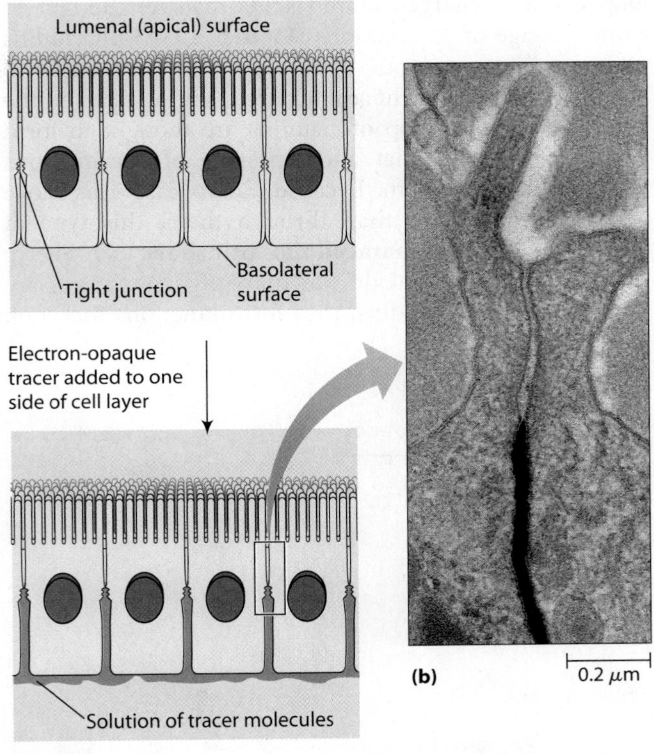

FIGURE 17-10 Experimental Evidence Demonstrating That Tight Junctions Create a Permeability Barrier. **(a)** When an electron-opaque tracer is added to the extracellular space on one side of an epithelial cell layer, tracer molecules penetrate into the space between adjacent cells only to the point where they encounter a tight junction. **(b)** Because the tracer molecules are electron opaque, their penetration into the intercellular space can be visualized by electron microscopy (TEM).

observe the movement of the tracer through the extracellular space. As **Figure 17-10** illustrates, tracer molecules diffuse into the narrow spaces between adjacent cells until they encounter a tight junction, which blocks further movement.

Role of Tight Junctions in Blocking Lateral Movement of Membrane Proteins. Tight junctions act like "gates," preventing the movement of fluids, ions, and molecules between cells. In addition, tight junctions act like "fences," blocking the lateral movement of lipids and proteins within the membrane. Lipid movement is blocked in the outer monolayer only, but the movement of integral membrane proteins is blocked entirely. As a result, different kinds of integral membrane proteins can be maintained in the portions of a plasma membrane on opposite sides of a tight junction belt.

Claudins Form a Seal at Tight Junctions

Tight junctions contain several major transmembrane proteins. These include a transmembrane protein known as *occludin* and immunoglobulin superfamily proteins known as *junctional adhesion molecules (JAMs)*. In addition, tight junctions contain **claudins.** Claudins have four membrane-spanning domains; the largest extracellular loop contains charged amino acids that are thought to allow passage of specific ions (**Figure 17-11**). Claudins in the plasma membrane of adjacent cells are thought to interlock to form a tight seal. Charged amino acids in the large extracellular loop of claudins are thought to form ion-selective pores that allow passage of specific ions through the epithelium. Because in this case ions move between cells, rather than through them, this type of transport is termed **paracellular transport** (see Figure 17-9). When cells that do not make tight junctions are forced to express claudins, they form junctions that look very similar to tight junctions. In addition to their functions in assembling tight junctions, claudins appear to regulate paracellular transport of ions across epithelia (see below). Different claudins are expressed in different epithelial tissues and are thought to confer on these tissues different permeability properties. Mutations in one claudin (claudin-16) result in *familial hypomagnesemia with hypercalciuria and nephrocalcinosis (FHHNC),* an autosomal recessive disease characterized by severe Mg^{2+} and Ca^{2+} imbalance.

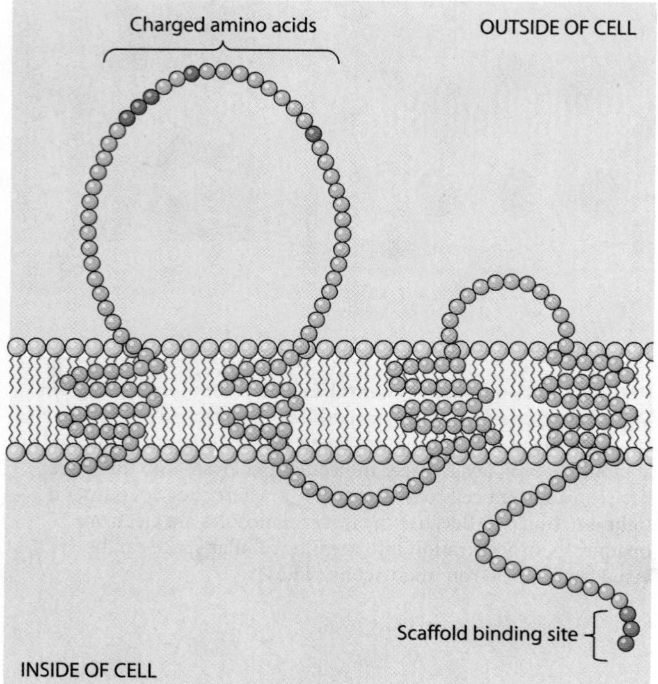

FIGURE 17-11 Claudin Structure. Claudins have four transmembrane domains and characteristic extra- and intracellular loops. The largest extracellular loop contains charged amino acids that are thought to interact with claudins on a neighboring cell to create a paracellular pore through which ions can pass. The C-terminus of claudins can bind scaffolding proteins in the cytosol.

Gap Junctions Allow Direct Electrical and Chemical Communication Between Cells

A **gap junction** is a region where the plasma membranes of two cells are aligned and brought into intimate contact, with a gap of only 2–3 nm in between, spanned by small molecular "pipelines." The gap junction thus provides a point of cytoplasmic contact between two adjacent cells through which ions and small molecules can pass. It allows adjacent cells to be in direct electrical and chemical communication with each other.

The structure of gap junctions is illustrated in **Figure 17-12.** At a gap junction, the two plasma membranes from adjacent cells are joined by tightly packed, hollow cylinders called **connexons.** A single gap junction may consist of just a few or as many as thousands of clustered connexons. In vertebrates, each connexon is a circular assembly of six subunits of the protein *connexin.* Invertebrates do not have connexins. Instead, they produce proteins called *innexins* that appear to serve the same function.

Many different connexins (more than a dozen types) are found in different tissues, but each one functions similarly in forming connexons. The assembly spans the membrane and protrudes into the space, or gap, between the two cells (Figure 17-12a). Each connexon has a diameter of about 7 nm and a hollow center that forms a very thin hydrophilic channel through the membrane. The channel is about 3 nm in diameter at its narrowest point—just large enough to allow the passage of ions and small molecules but too small to allow proteins, nucleic acids, and organelles through. Included in this range are single sugars, amino acids, and nucleotides—most of the molecules involved in cellular metabolism. By injecting fluorescent molecules into cells connected by gap junctions, researchers have shown that gap junctions allow the passage of solutes with molecular weights up to about 1200.

Although they are formed in a closed state initially, once connexons from adjacent cells meet, the cylinders in the two membranes join end to end. They form direct channels of communication between the two cells that can be seen with an electron microscope (Figure 17-12, parts b and c). Once fully formed, conditions inside of the cell, including electrical potential, concentration of second

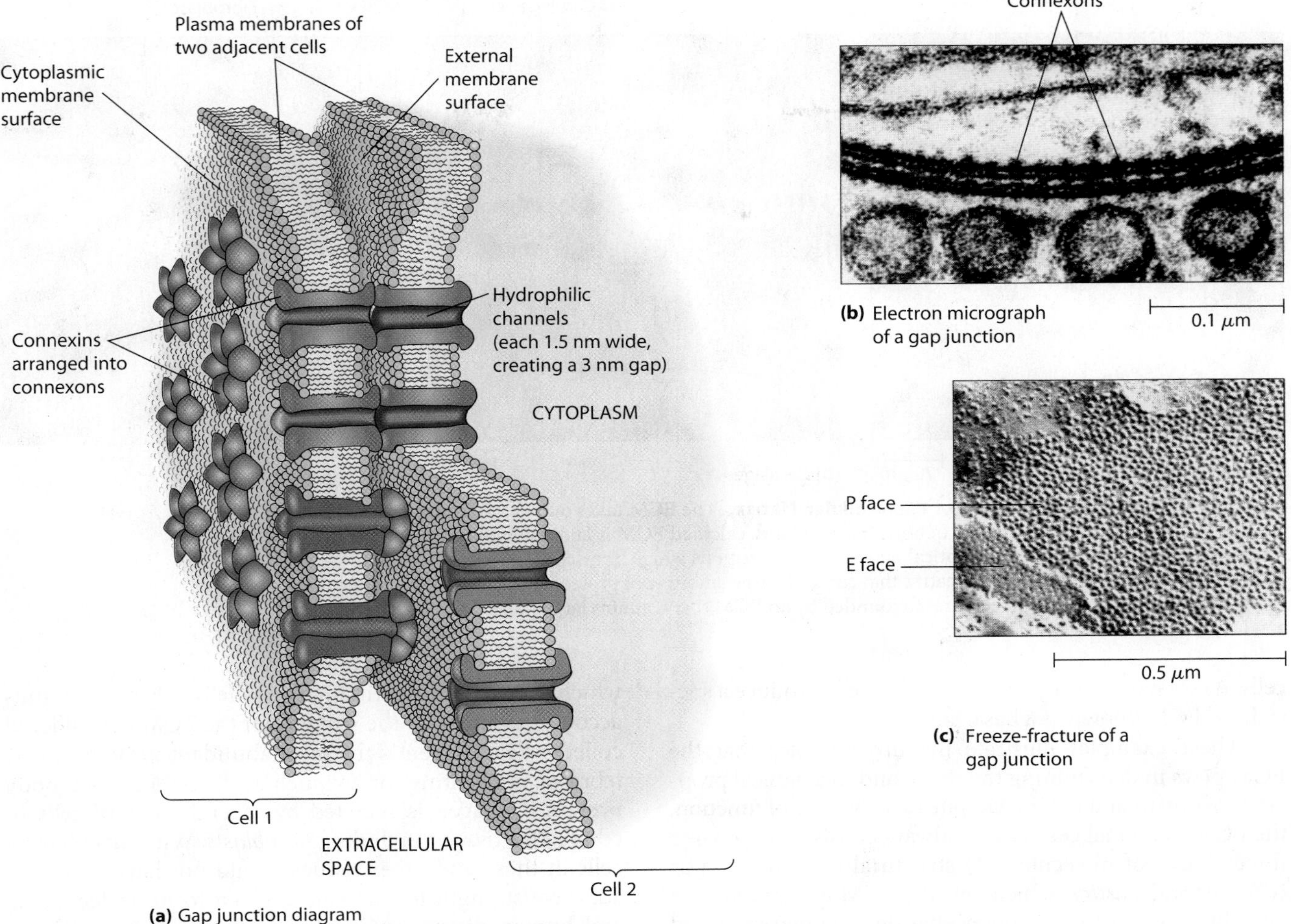

FIGURE 17-12 Gap Junction Structure. **(a)** A schematic representation of a gap junction. A gap junction consists of a large number of hydrophilic channels formed by the alignment of connexons in the plasma membranes of two adjoining cells. **(b)** An electron micrograph of a gap junction between two adjacent nerve cells. The connexons that extend through the membranes are visible here as beadlike projections spaced about 17 nm apart on either side of the membrane-membrane junction (TEM). **(c)** A gap junction as revealed by the freeze-fracture technique. The junction appears as an aggregation of intramembranous particles on the protoplasmic (P) face and as a series of pits on the exterior (E) face of the membrane (TEM).

messengers, and other conditions, can influence whether gap junctions are open or closed.

Gap junctions occur in most vertebrate and invertebrate cells. They are especially abundant in tissues such as muscle and nerve, where extremely rapid communication between cells is required (e.g., in electrical synapses; see Chapter 13). In heart tissue, gap junctions facilitate the flow of electrical current that causes the heart to beat. These roles are confirmed by analyzing mutations in connexins and innexins. For example, mice lacking one type of connexin have defects in conducting electrical impulses in the heart. Several human disorders have been directly linked to defects in gap junctions. These include several types of demyelinating neurodegenerative diseases, various skin disorders, formation of cataracts, and some types of deafness.

The Extracellular Matrix of Animal Cells

Tissues are not simply composed of cells. Cells interact with extracellular materials that are crucial for tissue structure and function. In animal cells, this **extracellular matrix (ECM)** takes on a remarkable variety of forms in different tissues. **Figure 17-13** illustrates just three examples. *Bone* consists largely of a rigid extracellular matrix that contains a tiny number of interspersed cells. *Cartilage* is another tissue constructed almost entirely of matrix materials, although the matrix is much more flexible than in bone. In contrast to bone and cartilage, the *connective tissue* surrounding glands and blood vessels has a relatively gelatinous extracellular matrix containing numerous interspersed fibroblast

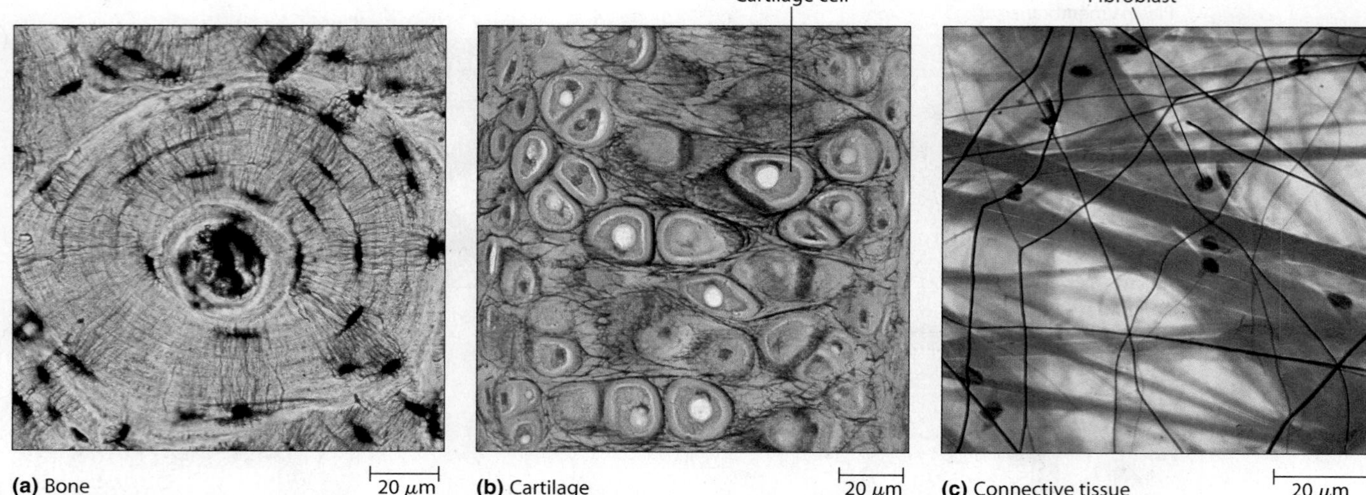

FIGURE 17-13 Different Kinds of Extracellular Matrix. The ECM takes on different forms in different tissues, as these images illustrate. **(a)** In bone tissue, a hard, calcified ECM is laid down in concentric rings around central canals. The small elliptical depressions are regions where bone cells are found. **(b)** In cartilage, the cells are embedded in a flexible matrix that contains large amounts of proteoglycans. **(c)** In the connective tissue found under skin, fibroblasts are surrounded by an ECM that contains large numbers of collagen fibers.

cells. As we have already seen, epithelial cells produce a specialized ECM known as a basal lamina.

These examples illustrate the diverse roles that the ECM plays in determining the shape and mechanical properties of organs and tissues. Despite this diversity of function, the ECM of animal cells almost always consists of the same three classes of molecules: (1) structural proteins such as *collagens* and *elastins,* which give the ECM its strength and flexibility; (2) protein-polysaccharide complexes called *proteoglycans* that provide the matrix in which the structural molecules are embedded; and (3) adhesive glycoproteins such as *fibronectins* and *laminins,* which allow cells to attach to the matrix (**Table 17-2**). The considerable variety in the properties of the ECM in different tissues results not only from differences in the types of structural proteins and the kinds of proteoglycans present but also from variations in the ratio of structural proteins—collagen, most commonly—to proteoglycans and in the kinds and amounts of adhesive glycoproteins present. We will consider each of these classes of ECM constituents in turn.

Collagens Are Responsible for the Strength of the Extracellular Matrix

The most abundant component of the ECM in animals is a large family of closely related proteins called **collagens,** which form fibers with high tensile strength and thus account for much of the strength of the ECM. Considered collectively, collagen is the most abundant protein in vertebrates, accounting for as much as 25–30% of total body protein. Collagen is secreted by several types of cells in connective tissues, including *fibroblasts.* Without collagen, cells in these and other tissues would not have sufficient adhesive strength to maintain a given form. Indeed, several human diseases result from mutations in collagens. For example, Ehlers–Danos syndrome is a group of inherited disorders characterized by excessive looseness (laxity) of the joints, hyperelastic skin that is fragile and bruises easily, and/or easily damaged blood vessels. These disorders arise from mutations in collagens. Vitamin C is an essential cofactor for collagen synthesis. Vitamin C deficiency leads to scurvy, which was a historically important disease among sailors.

Two defining characteristics are shared by all collagens: their occurrence as a rigid *triple helix* of three intertwined polypeptide chains and their unusual amino acid composition. Specifically, collagens are high in both the common amino acid glycine and the unusual amino acids hydroxylysine and hydroxyproline, which rarely occur in other proteins. (For the structures of hydroxylysine and hydroxyproline, see Figure 7-25c.) The high glycine content makes the triple helix possible because the spacing of the glycine

Table 17-2 Extracellular Structures of Eukaryotic Cells

Kind of Organism	Extracellular Structure	Structural Fiber	Components of Hydrated Matrix	Adhesive Molecules
Animals	Extracellular matrix (ECM)	Collagens and elastins	Proteoglycans	Fibronectins and laminins
Plants	Cell wall	Cellulose	Hemicelluloses and extensins	Pectins

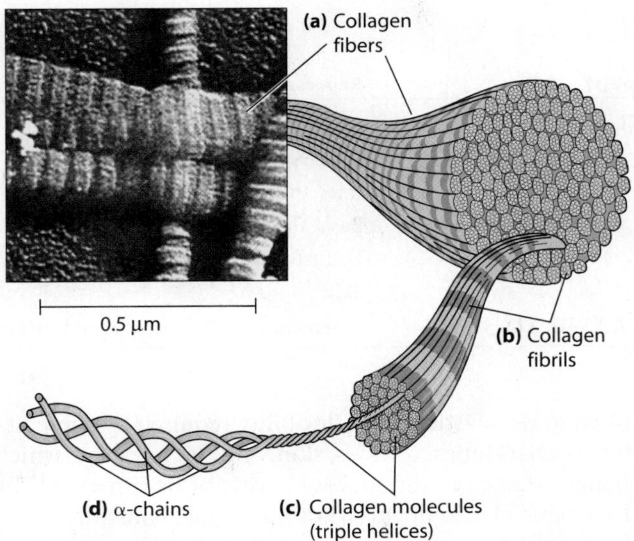

FIGURE 17-14 The Structure of Collagen. **(a)** Collagen fibers as seen by SEM. **(b)** A collagen fiber contains many fibrils, each of which is a bundle of collagen molecules, also called tropocollagen. **(c)** Each collagen molecule is a triple helix consisting of **(d)** three entwined α chains. The repeating bands visible on the fibers in the SEM partially reflect the regular but offset way that collagen molecules associate laterally to form fibrils; see Figure 17-15 for details of collagen assembly.

residues places them in the axis of the helix, and glycine is the only amino acid small enough to fit in the interior of a triple helix.

In most animal tissues, **collagen fibers** can be seen in bundles throughout the extracellular matrix when viewed by scanning electron microscopy (**Figure 17-14a**). One of the most striking features of collagen fibers is their enormous physical strength. For example, it takes a load of more than 20 pounds (about 9 kg) to tear a collagen fiber just 1 millimeter in diameter! As illustrated in Figure 17-14b, each collagen fiber is composed of numerous *fibrils.* A fibril, in turn, is made up of many collagen molecules, each consisting of three polypeptides called *α chains* that are twisted together into a rigid, right-handed triple helix (Figure 17-14, parts c and d). Collagen molecules are about 270 nm in length and 1.5 nm in diameter and are aligned both laterally and end to end within the fibrils. A typical collagen fiber has about 270 collagen molecules in cross section.

A Precursor Called Procollagen Forms Many Types of Tissue-Specific Collagens

Given the complexity of a collagen fiber, it is important to ask how such an elaborate structure is generated (**Figure 17-15**). In the lumen of the endoplasmic reticulum (ER), three α chains assemble to form a triple helix called

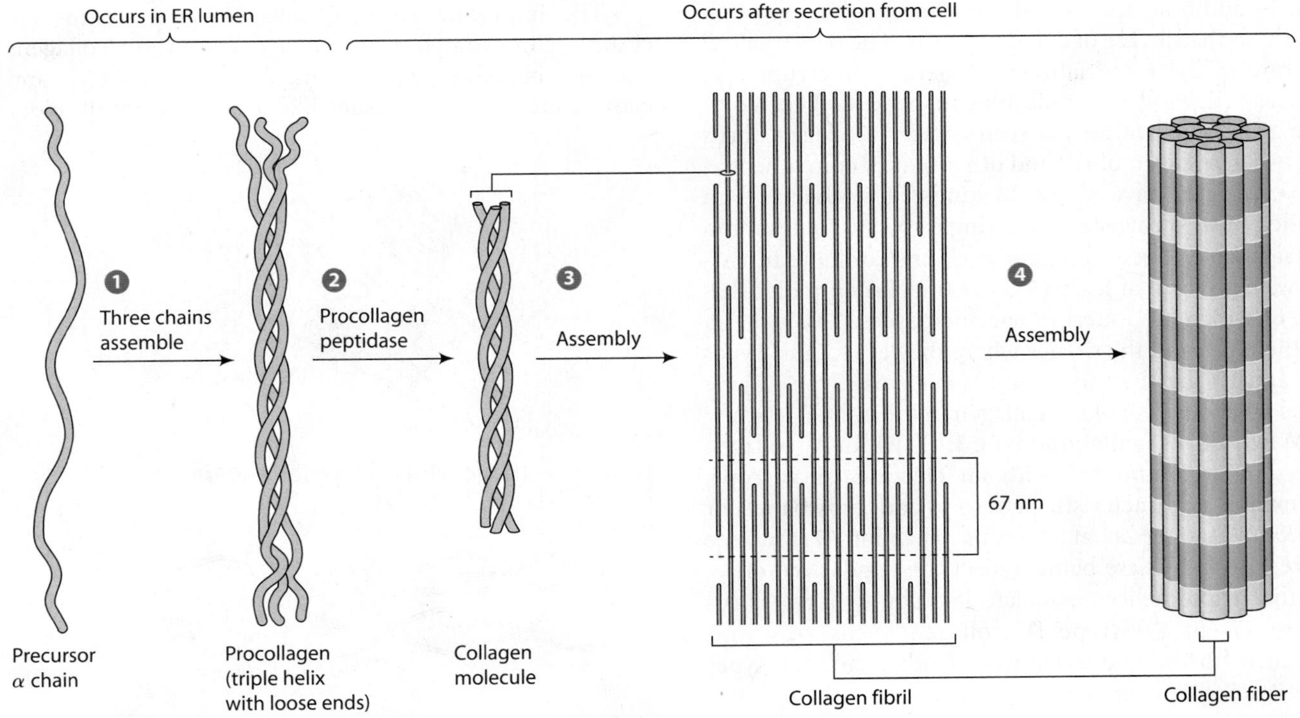

FIGURE 17-15 Collagen Assembly. ❶ Collagen precursor chains are assembled in the ER lumen to form triple-helical procollagen molecules. ❷ After secretion from the cell, procollagen is converted to collagen in a peptide-cleaving reaction catalyzed by the enzyme procollagen peptidase. ❸ The molecules of collagen, also called tropocollagen, then bind to each other and self-assemble into collagen fibrils. ❹ The fibrils assemble laterally into collagen fibers. In striated collagen, the 67-nm repeat distance is created by packing together rows of collagen molecules in which each row is displaced by one-fourth the length of a single molecule.

Table 17-3 Types of Collagens, Their Occurrence, and Their Structure

Type of Structure	Collagen Type(s)	Representative Tissues
Long fibrils	I, II, III, V, XI	Skin, bone, tendon, cartilage, muscle
Fibril-associated, with interrupted triple helices	IX, XII, XIV	Cartilage, embryonic skin, tendon
Fibril-associated, forms beaded filaments	VI	Interstitial tissues
Sheets	IV, VIII, X	Basal laminae, cartilage growth plates
Anchoring fibrils	VII	Epithelia
Transmembrane	XVII	Skin
Other	XIII, XV, XVI, XVIII, XIX	Basement membranes, assorted tissues

procollagen. At both ends of the triple-helical structure, short nonhelical sequences of amino acids prevent the formation of collagen fibrils, as long as the procollagen remains within the cell. Once procollagen is secreted from the cell into the intercellular space, it is converted to collagen by *procollagen peptidase,* an enzyme that removes the extra amino acids from both the N- and C-terminal ends of the triple helix. The resulting collagen molecules spontaneously associate to form mature collagen fibrils, which then assemble laterally into fibers.

The stability of the collagen fibril is reinforced by hydrogen bonds that involve the hydroxyl groups of hydroxyproline and hydroxylysine residues in the α chains. These hydrogen bonds form crosslinks both within and between the individual collagen molecules in a fibril. In addition, specialized types of collagen are often present on the surface of collagen fibrils. The triple-helical structure of these specialized collagens is interrupted at intervals, allowing the molecules to bend and hence to serve as flexible bridges between adjacent collagen fibrils or between collagen fibrils and other matrix components.

Vertebrates have about 25 kinds of α chains, each encoded by its own gene and having its own unique amino acid sequence. These different α chains combine in various ways to form at least 15 types of collagen molecules, most of which are found in specific tissues. **Table 17-3** lists the types and the tissues where they are found. Types I, II, and III are the most abundant forms. Type I alone makes up about 90% of the collagen in the human body.

When fibrils containing type I, II, III, or V collagen molecules are examined with an electron microscope, they exhibit a characteristic pattern of dark crossbands, or *striations,* that repeat at intervals of about 67 nm (see Figure 17-14a). These bands reflect the regular but offset way that triple helices associate laterally to form fibrils (Figure 17-15, ❸). Type IV collagen forms very fine, unstriated fibrils. The structures of other collagen types are less well characterized.

Elastins Impart Elasticity and Flexibility to the Extracellular Matrix

Although collagen fibers give the ECM great tensile strength, their rigid, rodlike structure is not particularly suited to the elasticity and flexibility required by some tissues, such as lungs, arteries, skin, and the intestine, which change shape continuously. Elasticity is provided by stretchable elastic fibers, whose principle constituent is a family of ECM proteins called **elastins.** Like collagens, elastins are rich in the amino acids glycine and proline. However, the proline residues are not hydroxylated, and no hydroxylysine is present. Elastin molecules are crosslinked to one another by covalent bonds between lysine residues. Tension on an elastin network causes the overall network to stretch (**Figure 17-16a**). When the tension is released, the individual molecules relax, returning to their normal, less extended conformations. The crosslinks between molecules then cause the network to recoil to its original shape (Figure 17-16b).

The important roles of collagens and elastins are clearly demonstrated as people age. Over time, collagens become increasingly crosslinked and inflexible, and elastins are lost from tissues like skin. As a result, older

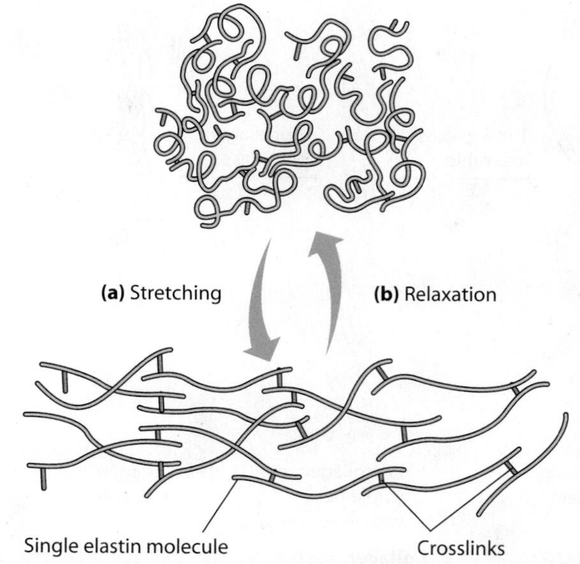

FIGURE 17-16 Stretching and Recoiling of Elastin Fibers. Each elastin molecule in the crosslinked network can assume either an extended (bottom) or compact (top) configuration. The fiber **(a)** stretches to its extended form when tension is exerted on it and **(b)** recoils to its compact form when the tension is released.

people often find that their bones and joints are less flexible, and their skin becomes wrinkled.

Collagen and Elastin Fibers Are Embedded in a Matrix of Proteoglycans

The hydrated, gel-like network in which the collagen and elastin fibrils of the ECM are enmeshed consists primarily of proteoglycans, glycoproteins in which a large number of glycosaminoglycans are attached to a single protein molecule.

Glycosaminoglycans (GAGs) are large carbohydrates characterized by repeating disaccharide units, as illustrated in **Figure 17-17** for the three most common types: *chondroitin sulfate, keratan sulfate,* and *hyaluronate.* In each case, one of the two sugars in the disaccharide repeating unit is an amino sugar, either *N-acetylglucosamine (GlcNAc)* or *N-acetylgalactosamine (GalNAc).* The other sugar in the disaccharide repeating unit is usually a sugar or a sugar acid, commonly *galactose (Gal)* or *glucuronate (GlcUA).* In most cases, the amino sugar has one or more sulfate groups attached. Of the GAG repeating units shown in Figure 17-17, only hyaluronate contains no sulfate groups. Because GAGs are hydrophilic molecules with many negatively charged sulfate and carboxyl groups, they attract both water and cations, thereby forming the hydrated, gelatinous matrix in which collagen and elastin fibrils become embedded.

Most glycosaminoglycans in the ECM are covalently bound to protein molecules to form **proteoglycans.** Each proteoglycan consists of numerous GAG chains attached along the length of a **core protein,** as shown in Figure 17-17. Many kinds of proteoglycans can be formed by the combination of different core proteins and GAGs of varying types and lengths. Proteoglycans vary greatly in size, depending on the molecular weight of the core protein (which ranges from about 10,000 to over 500,000) and the number and length of the carbohydrate chains (1–200 per molecule, with an average length of about 800 monosaccharide units). Most proteoglycans are huge. They have molecular weights in the range of 0.25–3 million. Proteoglycans are linked directly to collagen fibers to make up the fiber/network structure of the ECM.

In some cases, the proteoglycans are themselves integral components of the plasma membrane, with their core polypeptides embedded within the membrane. In other cases, proteoglycans are linked covalently to membrane phospholipids. Alternatively, either proteoglycans or collagen may bind to specific receptor proteins on the outer surface of the plasma membrane.

In many tissues, proteoglycans are present as individual molecules. In cartilage, however, numerous proteoglycans become attached to long molecules of hyaluronate, forming large complexes as shown in Figure 17-17. A single such complex can have a molecular weight of many millions and may exceed several micrometers in length.

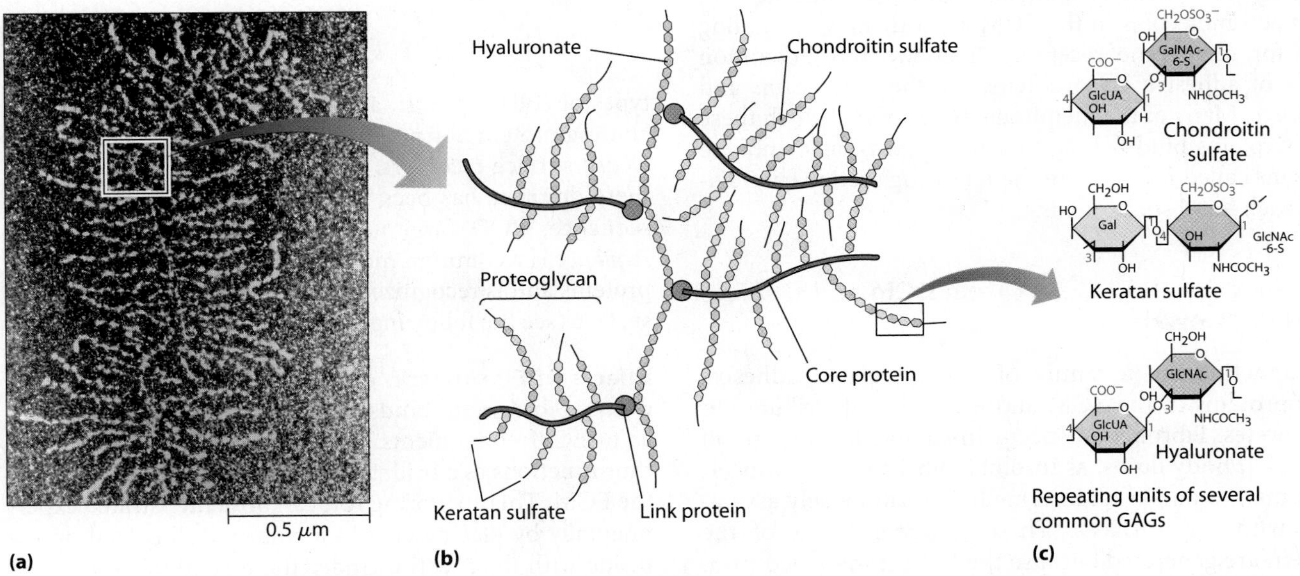

FIGURE 17-17 Proteoglycan Structure in Cartilage. In cartilage, many proteoglycans associate with a hyaluronate backbone to form a complex that is readily visible with an electron microscope. **(a)** A hyaluronate-proteoglycan complex isolated from bovine cartilage (TEM). **(b)** A diagram of a small portion of the structure showing core proteins of proteoglycans attached via linker proteins to a long hyaluronate molecule. Short keratan sulfate and chondroitin sulfate chains are linked covalently along the length of the core proteins. Proteoglycans have a carbohydrate content of about 95%. **(c)** Structures of the disaccharide repeating units in three common extracellular glycosaminoglycans (GAGs) found in the extracellular matrix of animal cells. The repeating unit of chondroitin sulfate (top) consists of glucuronate, the ionized form of glucuronic acid (GlcUA), and *N*-acetylgalactosamine (GalNAc). The repeating unit of keratan sulfate (middle) is galactose (Gal) and *N*-acetylglucosamine (GlcNAc). Hyaluronate (bottom) has a repeating unit consisting of GlcUA and GlcNAc.

The remarkable resilience and pliability of cartilage are due mainly to the properties of these complexes. Smaller proteoglycans are also known to be involved in facilitating growth factor signaling, which we learned about in Chapter 14. For example, a proteoglycan called *syndecan* is a modulator of fibroblast growth factor signaling.

Free Hyaluronate Lubricates Joints and Facilitates Cell Migration

Although most of the GAGs found in the extracellular matrix exist only as components of proteoglycans and not as free glycosaminoglycans, **hyaluronate** is an exception. In addition to its role as the backbone of the proteoglycan complex in cartilage, hyaluronate occurs as a free molecule consisting of hundreds or even thousands of repeating disaccharide units. Hyaluronate molecules have lubricating properties and are most abundant in places where friction must be reduced, such as the joints between movable bones.

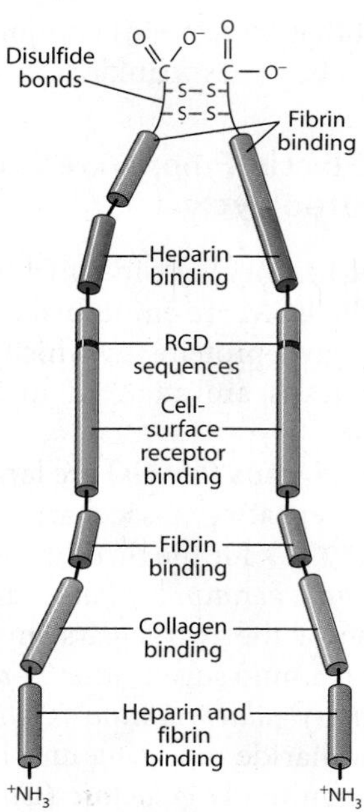

FIGURE 17-18 Fibronectin Structure. A fibronectin molecule consists of two nearly identical polypeptide chains joined by two disulfide bonds near their carboxyl ends. Each polypeptide chain is folded into a series of domains linked by short, flexible segments. These domains have binding sites for ECM components or for specific receptors on the cell surface. The receptor-binding domain contains the tripeptide sequence RGD (arginine-glycine-aspartate), which is recognized by fibronectin receptors. Besides the binding activities noted, fibronectin has binding sites for heparan sulfate, hyaluronate, and gangliosides (glycosphingolipids that contain sialic acid groups).

Adhesive Glycoproteins Anchor Cells to the Extracellular Matrix

Direct links between the ECM and the plasma membrane are reinforced by a family of *adhesive glycoproteins* that bind proteoglycans and collagen molecules to each other and to receptors on the membrane surface. These proteins typically have multiple domains, some with binding sites for macromolecules in the ECM and others with binding sites for membrane receptors. The two most common kinds of adhesive glycoproteins are the *fibronectins* and *laminins*. Many of the membrane receptors to which these glycoproteins bind belong to a family of transmembrane proteins called *integrins*. In the following sections, we discuss each family of proteins.

Fibronectins Bind Cells to the ECM and Guide Cellular Movement

Fibronectins are a family of closely related adhesive glycoproteins in the ECM and are widely distributed in vertebrates. Fibronectins occur in soluble form in blood and other body fluids, as insoluble fibrils in the extracellular matrix, and as an intermediate form loosely associated with cell surfaces. These different forms of the protein are generated because the RNA transcribed from the fibronectin gene is processed to generate many different mRNAs.

A fibronectin molecule consists of two very large polypeptide subunits that are linked near their carboxyl ends by a pair of disulfide bonds (**Figure 17-18**). Each subunit has about 2500 amino acids and is folded into a series of rodlike domains connected by short, flexible segments of the polypeptide chain. Several of the domains bind to one or more specific kinds of macromolecules located in the ECM or on cell surfaces, including several types of collagen (I, II, and IV), heparin, and the blood-clotting protein fibrin. Other domains recognize and bind to cell surface receptors. The receptor-binding activity of these domains has been localized to a specific tripeptide sequence, RGD (arginine-glycine-aspartate). This *RGD sequence* is a common motif among extracellular adhesive proteins and is recognized by various integrins on the cell surface (see the following section).

Effects of Fibronectin on Cell Shape and Cell Movement. Fibronectin binds to cell surface receptors as well as to ECM components such as collagen and heparin. It thus functions as a bridging molecule that attaches cells to the ECM. This anchoring role can be demonstrated experimentally by placing cells in a culture dish coated on the inside with fibronectin. Under these conditions, the cells attach to the surface of the dish more efficiently than they do in the absence of fibronectin. After attaching, the cells flatten out (**Figure 17-19a**), and components of the actin cytoskeleton become aligned with the fibronectin located outside the cell (Figure 17-19b, c). In some cases such alignment occurs because the cell remodels the ECM, dragging it into new arrangements. In other cases the cell aligns itself with the ECM.

Fibronectin is also involved in cellular movement. For example, when migratory embryonic cells are grown on

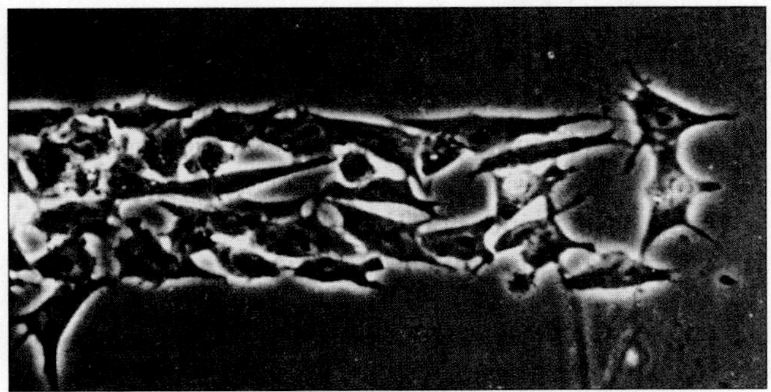

(a) Cells from the neural crest

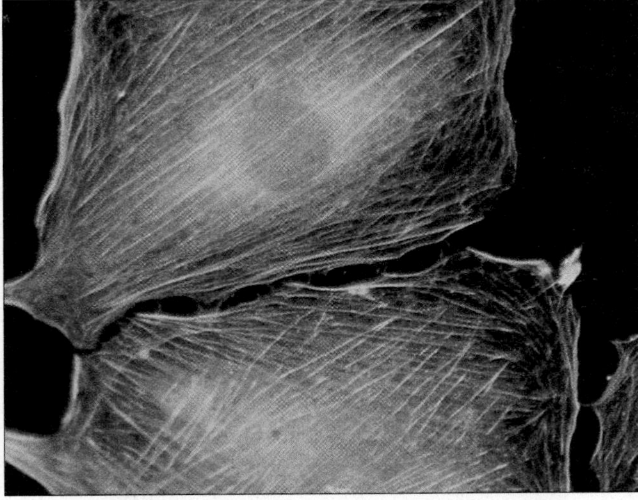

(b) Actin

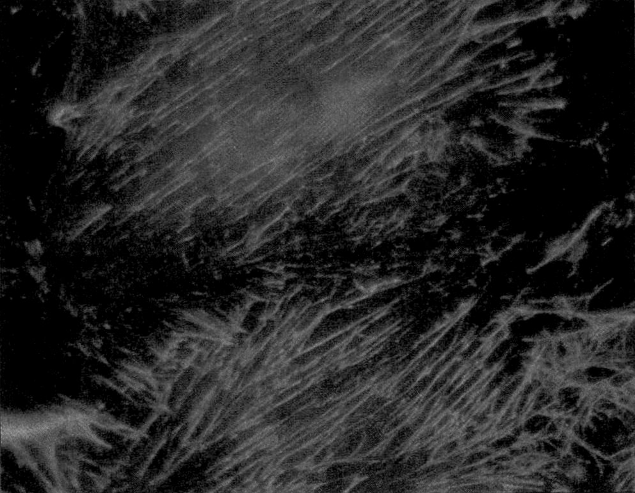

(c) Fibronectin

FIGURE 17-19 Interaction Between Fibronectin and Migrating Cells. (a) Cells from the neural crest (a type of migratory embryonic cell) preferentially migrate along a strip of fibronectin in vitro (phase contrast micrograph). (**b** and **c**) Fluorescence micrographs show the same cultured cells stained with fluorescent antibodies specific for either (**b**) actin or (**c**) fibronectin. Note that the extracellular fibronectin network and the intracellular actin microfilaments are aligned in a similar pattern.

fibronectin, they adhere readily to it (Figure 17-19a). The pathways followed by migrating cells are rich in fibronectin, suggesting that such cells are guided by binding to fibronectin molecules along the way. More direct evidence for the importance of fibronectin in embryos comes from studying genetically engineered mice that cannot produce fibronectin. Such mice have severe defects in some cells that make the musculature and the vasculature. These defects highlight the crucial importance of fibronectin during embryonic development.

A possible involvement of fibronectin in cancer is suggested by the observation that many kinds of cancer cells do not synthesize fibronectins, with an accompanying loss of normal cell shape and detachment from the ECM. If such cells are supplied with fibronectin, they often return to a more normal shape, recover their ability to bind to the ECM, and no longer appear malignant.

Effects of Fibronectin on Blood Clotting. The soluble form of fibronectin present in the blood, called *plasma fibronectin,* is involved in blood clotting. Fibronectin promotes blood clotting because it has several binding domains that recognize *fibrin,* the blood-clotting protein, and it can attach blood platelets to fibrin as the blood clot forms.

Laminins Bind Cells to the Basal Lamina

Another major adhesive glycoprotein present in the ECM is a family of proteins called **laminins,** which are conserved from simple invertebrates to humans. Unlike fibronectins, which occur widely throughout supporting tissues and body fluids, laminins are found mainly in the **basal lamina.** This thin sheet of specialized extracellular material, typically about 50 nm thick, underlies epithelial cells, thereby separating them from connective tissues (**Figure 17-20**). Basal laminae also surround muscle cells, fat cells, and the Schwann cells that form myelin sheaths around nerve cells.

Properties of the Basal Lamina. The basal lamina serves as a structural support that maintains tissue organization and as a permeability barrier that regulates the movement of molecules as well as cells. In the kidney, for example, an extremely thick basal lamina functions as a filter that allows small molecules but not blood proteins to move from the blood into the urine. The basal lamina beneath epithelial cells prevents the passage of underlying connective tissue cells into the epithelium but permits the migration of the white blood cells needed to fight infections. The effect of the basal lamina on cell migration is of special interest because some cancer cells show increased binding to the basal lamina. The resulting increase in the ability of cancer cells to bind to the basal lamina may facilitate their movement through it and allow them to migrate from one region of the body to another.

Despite differences in function and specific molecular composition from tissue to tissue, all forms of basal

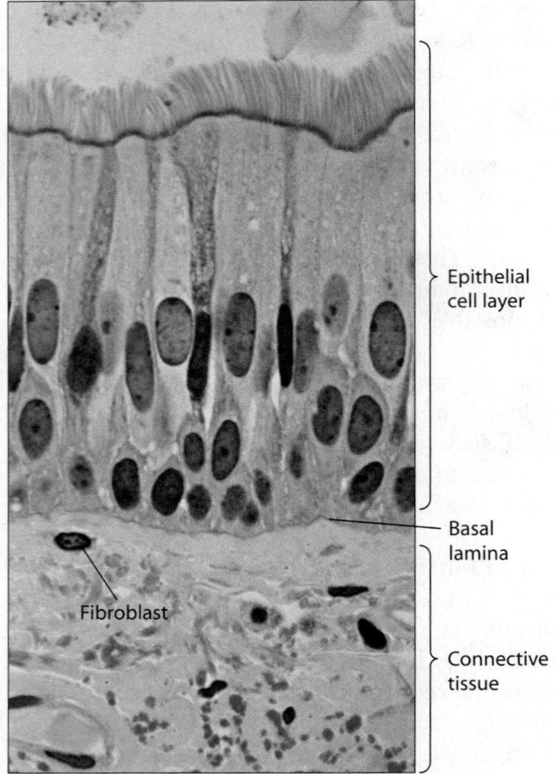

FIGURE 17-20 The Basal Lamina. The basal lamina is a thin sheet, typically about 50 nm, of matrix material that separates an epithelial cell layer from underlying connective tissue (TEM).

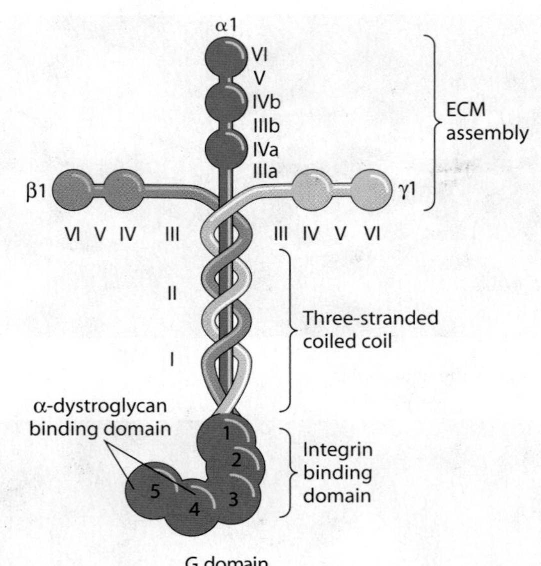

FIGURE 17-21 Laminin Structure. A laminin molecule consists of three large polypeptides—α, β, and γ—joined by disulfide bonds into a crosslinked structure. A portion of the long arm consists of a three-stranded coil. The functional domains on the ends of the α chain bind to organ-specific cell surface receptors, whereas those at the ends of the two arms of the cross are specific for type IV collagen. The cross-arms also contain laminin-laminin binding sites, thereby enabling laminin molecules to bind to each other and form large aggregates. Laminin also contains binding sites for heparin and heparan sulfate, as well as for entactin (not shown). Reprinted by permission from Macmillan Publishers Ltd: Fig. 1 from M. P. Marinkovich, "Laminin 332 in Squamous-Cell Carcinoma," *Nature Reviews Cancer* 7:370-380. Copyright 2007.

lamina contain type IV collagen, proteoglycans, laminins, and another glycoprotein called *entactin,* or *nidogen.* Fibronectins may be present, but laminins are the most abundant adhesive glycoproteins in the basal lamina. Laminins are thought to be localized mainly on the surface of the lamina that faces the overlying epithelial cells, where they help bind the cells to the lamina. Fibronectins, on the other hand, are located on the other side of the lamina, where they help anchor cells of the connective tissue.

Cells can alter the properties of the basal lamina by secreting enzymes that catalyze changes in the basal lamina. One important class of such enzymes is *matrix metalloproteinases (MMPs).* These enzymes, which require metal ions as cofactors, degrade the ECM locally, allowing cells to pass through the ECM. Such activity is important for cells such as leukocytes to invade injured tissues during inflammation. MMPs are also involved in abnormal invasive behavior. The MMP activity of invasive cancer cells, such as metastatic melanoma cells, is very high. We will discuss these enzymes in more detail in Chapter 24.

Properties of Laminin. Laminin, a very large protein with a molecular weight of about 850,000, consists of three long polypeptides, denoted α, β, and γ. There are several types of each of the three subunits, which can combine to form many types of laminin. Disulfide bonds hold the polypeptide chains together in the shape of a cross, with part of the long arm wound into a three-stranded coil (**Figure 17-21**). Like fibronectin, laminin consists of several domains that include binding sites for type IV collagen, heparin, heparan sulfate, and entactin, as well as for laminin receptor proteins on the surface of overlying cells. Its binding sites allow laminin to serve as a bridging molecule that attaches cells to the basal lamina. Entactin molecules have binding sites for both laminin and type IV collagen and are therefore thought to reinforce the binding of type IV collagen and laminin networks in the basal lamina.

Integrins Are Cell Surface Receptors That Bind ECM Constituents

Fibronectins and laminins can bind to animal cells because the plasma membranes of most cells have specific receptors on their surfaces that recognize and bind to specific regions of the fibronectin or laminin molecule. These receptors—and those for a variety of other ECM constituents—belong to a large family of transmembrane proteins that are called **integrins** because of their role in *integrating* the cytoskeleton with the extracellular matrix. Integrins are important receptors because they are the primary means by which cells bind to ECM proteins such as collagen, fibronectin, and laminin.

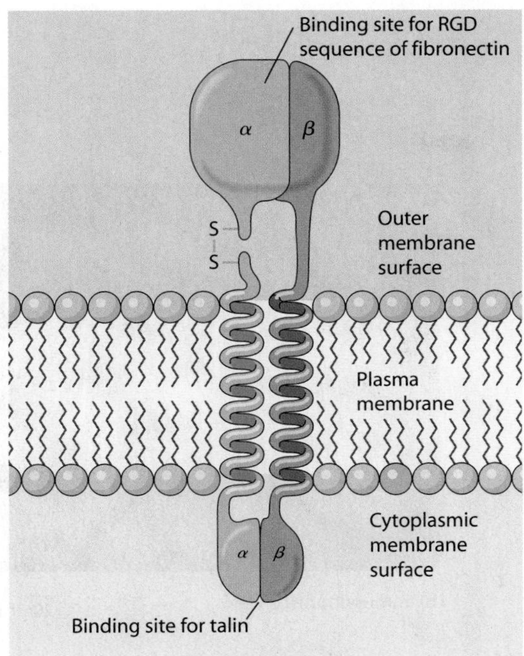

FIGURE 17-22 An Integrin: The Fibronectin Receptor. An integrin consists of α and β subunits, transmembrane polypeptides that associate with each other noncovalently to form a binding site for the ligand on the outer membrane surface and a binding site for a specific cytoskeletal protein on the inner membrane surface. Shown here is the fibronectin receptor ($\alpha_5\beta_1$), which has a binding site for fibronectin on the outer surface and a binding site for talin on the cytoplasmic side of the membrane. In this and several other integrins, the α subunit is split into two segments held together by a disulfide bond. Both the α and β subunits are glycosylated on the exterior side, although the sugar side chains are not shown here.

Structure of Integrins. An integrin consists of two large transmembrane polypeptides, the α and β subunits, that associate with each other noncovalently (**Figure 17-22**). Integrins differ from one another in their binding specificities and in the sizes of their subunits (molecular weight ranges: 110,000–140,000 for the α subunit, 85,000–91,000 for the β subunit). For a specific integrin, the extracellular portions of the α and β subunits interact to form the binding site for a particular ECM protein, with most of the binding specificity apparently depending on the α subunit. On the cytoplasmic side of the membrane, integrins have binding sites for specific molecules of the cytoskeleton, thereby mechanically linking the cytoskeleton and the ECM across the plasma membrane.

The presence of multiple types of both α and β subunits results in a large number of different integrin heterodimers, which vary in their binding specificities. For example, integrins containing a β_1 subunit are found on the surfaces of most vertebrate cells and mediate mainly cell-ECM interactions, whereas those containing a β_2 subunit are restricted to the surface of white blood cells and are involved primarily in cell-cell interactions. The most common integrin that binds fibronectin is $\alpha_5\beta_1$, whereas $\alpha_6\beta_1$ binds laminin. Many integrins recognize the RGD sequence in the specific ECM glycoproteins that they bind. However, the binding site must recognize other parts of the glycoprotein molecule as well because integrins display a greater specificity of glycoprotein binding than can be accounted for by the RGD sequence alone.

Integrins and the Cytoskeleton. Although integrins link the extracellular matrix and the cytoskeleton, they do not do so directly. Instead, the tails of integrins interact with proteins in the cytosol that link integrins to cytoskeletal proteins. Integrins make two main types of connections to the cytoskeleton (**Figure 17-23**). Migratory and non-epithelial cells, such as fibroblasts, attach to extracellular matrix molecules via **focal adhesions.** Focal adhesions contain clustered integrins that interact with bundles of actin microfilaments via several linker proteins (Figure 17-23a, b). These include *talin,* which can bind to the actin-binding protein *vinculin,* and *α-actinin,* which can bind directly to actin microfilaments.

The other major type of integrin-mediated attachment is found in epithelial cells. Epithelial cells attach to laminin in the basal lamina via **hemidesmosomes** (so called because they resemble "half desmosomes"; desmosomes were discussed earlier in this chapter). The integrin found in hemidesmosomes is $\alpha_6\beta_4$ integrin. In this case, the integrins are not attached to actin but to intermediate filaments, typically *keratin* (Figure 17-23c and d). The linker proteins in hemidesmosomes form a dense **plaque** that connects clustered integrins to the cytoskeleton. Among the linker proteins, members of the **plakin** family of proteins are prominent. A plakin known as *plectin* attaches keratin filaments to the integrins. Another transmembrane protein, called *BPAG2,* and its associated plakin, *BPAG1,* can serve as a bridge between keratin and laminin. These proteins were first identified using antibodies derived from human patients who developed an autoimmune reaction against antigens in their own hemidesmosomes. The resulting disease, called *bullous pemphigoid,* affects predominantly elderly people and involves blistering of the skin. This disease also gives rise to the names of these proteins. (*BPAG* stands for "*b*ullous *p*emphigoid *a*ntigen.")

Integrin Function. Because they bind to components of the ECM on the outside of the cell and to actin microfilaments on the inside, integrins play important roles in regulating cell movement and cell attachment. For example, mice lacking a particular integrin subunit (α_7) required for attachment to laminin develop a distinctive form of progressive muscular dystrophy. Humans that carry the same mutation develop progressive muscle degeneration. Recall that hemidesmosomes use $\alpha_6\beta_4$ integrins to attach to laminin. Humans who carry mutations in the β_4 subunit develop *junctional epidermolysis bullosa,* a severe blistering disease of the skin.

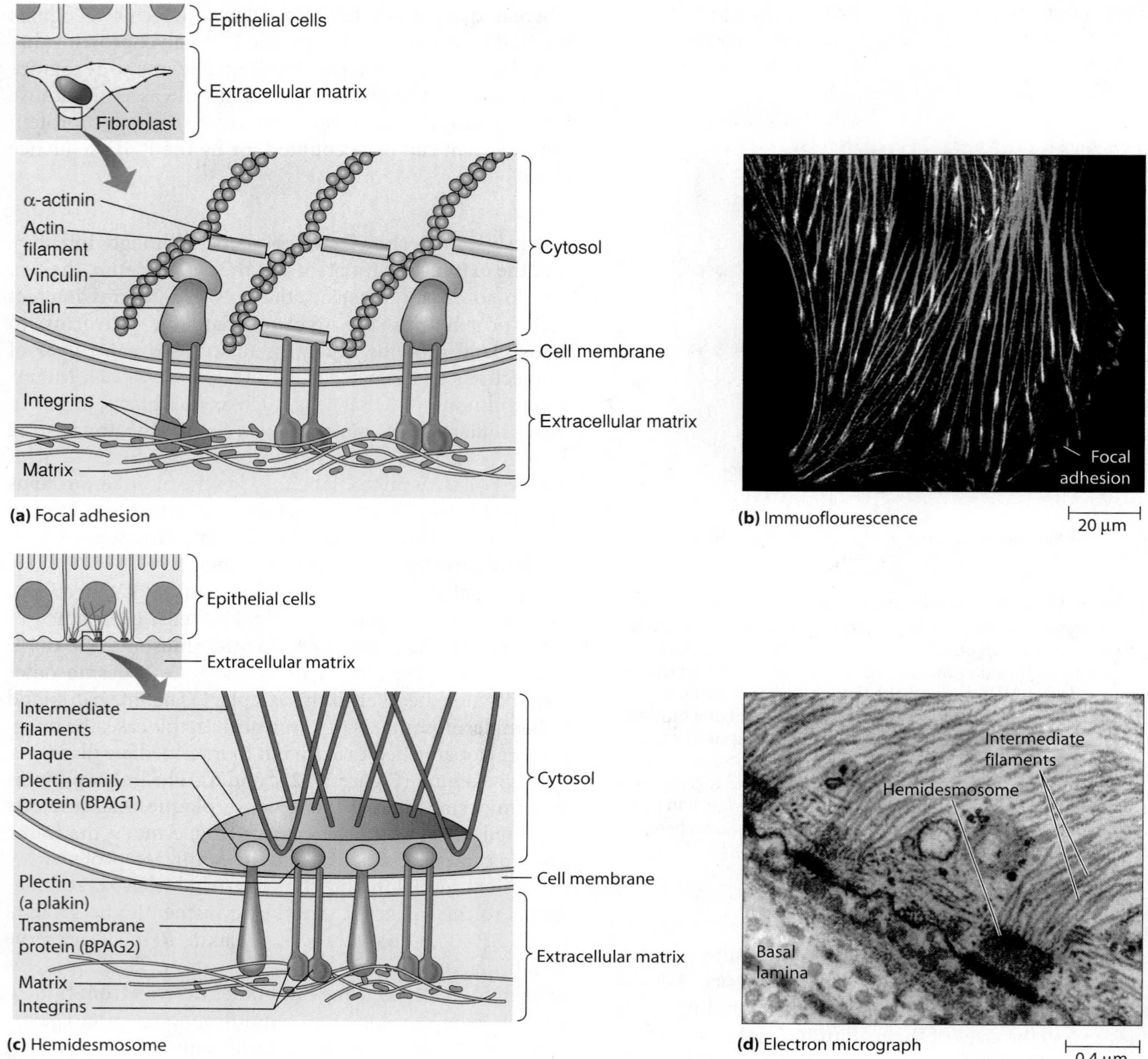

FIGURE 17-23 Integrins, Focal Adhesions, and Hemidesmosomes. **(a)** Migrating cells attach to the extracellular matrix via focal adhesions. Focal adhesions contain integrins, such as $\alpha_5\beta_1$ integrin, and integrin-associated linker proteins, such as talin, vinculin, and α-actinin, that attach integrins to the actin cytoskeleton. **(b)** A cultured frog glial cell stained to visualize actin (green) and stained with antibodies that recognize β_1 integrin (purple) and vinculin. White streaks represent regions where the three proteins are found together. Note how the actin microfilaments terminate at focal adhesions. **(c)** Epithelial cells attach to the extracellular matrix via hemidesmosomes, which contain $\alpha_6\beta_4$ integrins, and connect to intermediate filaments via linker proteins such as plectin. **(d)** The external surface of a hemidesmosome directly abuts the basal lamina (TEM).

Integrins and Signaling. Although a key role of integrins is their ability to link the cytoskeleton and the extracellular matrix, they also interact with intracellular signaling pathways. For example, signals such as binding of growth factors that lead to MAP kinase activation (see Chapter 14) can result in integrin clustering. (Such effects are often called "inside out" signaling because internal changes in the cell result in effects on integrins at the surface.) Inside-out signaling appears to result from conformational changes in the cytoplasmic regions of the α and β subunits. Integrins can also act as receptors that activate intracellular signaling themselves (sometimes called "outside in" signaling). The original observations that suggested involvement of integrins in signaling came from studies of cancer cells. For most normal cells to grow in culture, they must be attached to a substratum. Even in the presence of growth factors that normally stimulate their proliferation, if such cells are prevented from attaching to an extracellular matrix layer, they will cease dividing and undergo apoptosis, a kind of programmed cell death (we will discuss apoptosis in detail in Chapter 19). Such behavior is referred to as **anchorage-dependent growth.** In contrast, cancer cells will continue to grow even when they are not firmly attached to an extracellular matrix layer, apparently because they no longer need to transduce signals that result from attachment to the extracellular matrix.

Anchorage-dependent growth involves the activation of intracellular pathways following integrin clustering. Several kinases are activated at focal adhesions following integrin clustering. They are recruited to focal adhesions by adapter proteins such as *paxillin.* These include *focal adhesion kinase (FAK), kindlins*, and *integrin-linked kinase (ILK).* ILK probably functions as a scaffolding protein during integrin-based signaling, since "kinase dead" versions of the protein still function. Although they are involved in integrin activation, the precise function of kindlins is unknown. Mutations in one kindlin result in Kindler syndrome. Infants and young children with Kindler syndrome tend to develop minor skin defects, and they are susceptible to sunburns. FAK appears to be important in regulating anchorage-dependent growth. Cancer cells contain activated FAK even when they are not attached, and cells can be transformed into cancerlike cells by the expression of a mutant, activated form of FAK. These experiments underscore the importance of cell adhesion for understanding cancer.

The Dystrophin/Dystroglycan Complex Stabilizes Attachments of Muscle Cells to the Extracellular Matrix

In addition to focal adhesions and hemidesmosomes, a third type of ECM attachment is important in human disease. The *costamere* is an attachment structure at the surface of striated muscle. Costameres are aligned in register with Z-discs around the surface of muscle cells. They physically attach myofibrils at the surface of the muscle cells to the plasma membrane (sarcolemma). Costameres contain many of the same proteins found at focal contacts, including β_1-integrin, vinculin, talin, and α-actinin. They also contain the intermediate filament protein, desmin. In addition, costameres contain a specialized protein complex that includes the large protein **dystrophin** (**Figure 17-24**). Mutations in the dystrophin locus cause the most common type of muscular dystrophy, *Duchenne muscular dystrophy (DMD).* DMD is a severe, recessive X-linked type of muscular dystrophy that afflicts 1 in 3500 human males. DMD patients undergo progressive muscle degeneration, eventually leading to loss of the ability to walk and death.

Dystrophin interacts with a multiprotein complex that includes the integral membrane protein *dystroglycan* and the *sarcoglycan-sarcospan* complexes, as well as cytosolic proteins, the *dystrobrevins* and *syntrophins.* Dystroglycan is known to bind to laminin via three regions in the laminin G domain, linking the complex to the extracellular matrix (Figure 17-24). Loss of many of the other components of the dystrophin complex also results in muscular dystrophy in humans and mice.

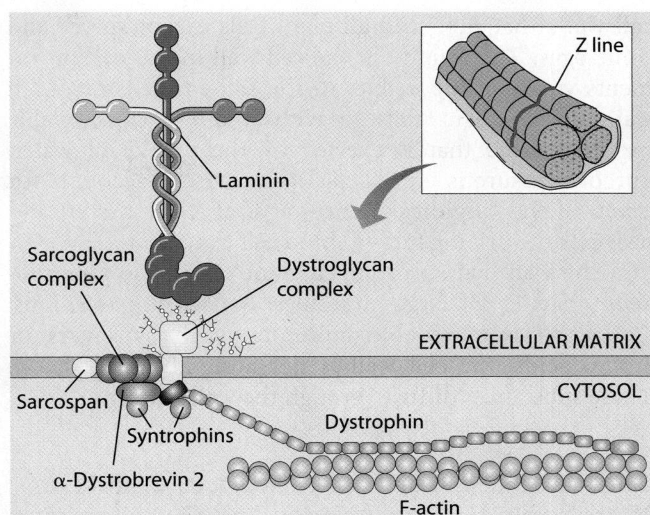

FIGURE 17-24 The Dystrophin/Dystroglycan Complex of Muscle Cells. Dystrophin, an enormous cytosolic protein, interacts with actin in the cytosol. Dystrophin is linked, via a series of proteins, to the dystroglycan complex at the cell surface. The dystroglycan complex in turn interacts with extracellular matrix proteins, such as laminin.

The Glycocalyx Is a Carbohydrate-Rich Zone at the Periphery of Animal Cells

Although animal cells do not have a rigid enclosure like a cell wall, a carbohydrate-rich zone called the *glycocalyx* often surrounds the plasma membrane (see Figure 7-27). Roles of the glycocalyx include cell recognition and adhesion, protection of the cell surface, and the creation of permeability barriers. Specialized structures, such as the outer coat of amoebas and the jelly coats that surround most animal eggs, are each an example of a glycocalyx.

The Plant Cell Surface

Our discussion so far has focused on animal cell surfaces. The surfaces of plant, algal, fungal, and bacterial cells exhibit some of the same properties, but they also have several unique features of their own. In the rest of this chapter, we consider some distinctive features of the plant cell surface.

Cell Walls Provide a Structural Framework and Serve as a Permeability Barrier

One of the most remarkable features of plants is that they have no bones or related skeletal structures and yet exhibit remarkable strength. This strength is provided by the rigid

cell walls that surround all plant cells except sperm and some eggs. The rigidity of the cell wall makes cell movements virtually impossible. At the same time, sturdy cell walls enable plant cells to withstand the considerable *turgor pressure* that is exerted by the uptake of water. Turgor pressure is vital to plants because it accounts for much of the turgidity, or firmness, of plant tissues and provides the driving force behind cell expansion.

The wall that surrounds a plant cell is also a permeability barrier for large molecules. For water, gases, ions, and small water-soluble molecules such as sugars or amino acids, the cell wall is not a significant obstacle. These substances diffuse through the wall readily.

The Plant Cell Wall Is a Network of Cellulose Microfibrils, Polysaccharides, and Glycoproteins

Like the extracellular matrix of animal cells, plant cell walls consist predominantly of long fibers embedded in a network of branched molecules. Instead of collagen and proteoglycans, however, plant cell walls contain *cellulose microfibrils* enmeshed in a complex network of branched polysaccharides and glycoproteins called *extensins* (see Table 17-2). The two main types of polysaccharides are *hemicelluloses* and *pectins.* On a dry-weight basis, cellulose typically makes up about 40% of the cell wall, hemicelluloses account for another 20%, pectins represent about 30%, and glycoproteins make up about 10%. **Figure 17-25a** illustrates the relationships among these cell wall components. Cellulose, hemicelluloses, and glycoproteins are linked together to form a rigid interconnected network that is embedded in a pectin matrix. Not shown are lignins, which in woody tissues are localized between the cellulose fibrils and make the wall especially strong and rigid. We will consider each of these components in turn.

Cellulose and Hemicellulose. The predominant polysaccharide of the plant cell wall is **cellulose,** which is the single most abundant organic macromolecule on Earth. As we saw in Chapter 3, cellulose is an unbranched polymer consisting of thousands of β-D-glucose units linked together by β $(1 \rightarrow 4)$ bonds (see Figure 3-25). Cellulose molecules are long, ribbonlike structures that are stabilized by intramolecular hydrogen bonds. Many such molecules (50–60, typically) associate laterally to form the **microfibrils** found in cell walls. Cellulose microfibrils are often twisted together in a ropelike fashion to generate even larger structures, called *macrofibrils* (Figure 17-25a). Cellulose macrofibrils are as strong as an equivalently sized piece of steel!

Despite the name, **hemicelluloses** are chemically and structurally distinct from cellulose. The hemicelluloses are a heterogeneous group of polysaccharides, each consisting of a long, linear chain of a single kind of sugar (glucose or xylose) with short side chains bonded into a rigid network (Figure 17-25a). The side chains usually contain several kinds of sugars, including the hexoses glucose, galactose, and mannose and the pentoses xylose and arabinose.

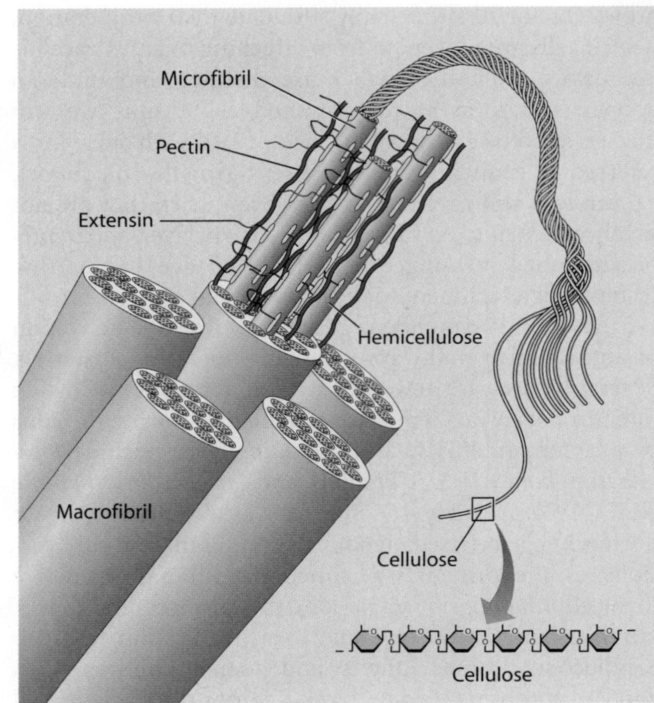

(a)

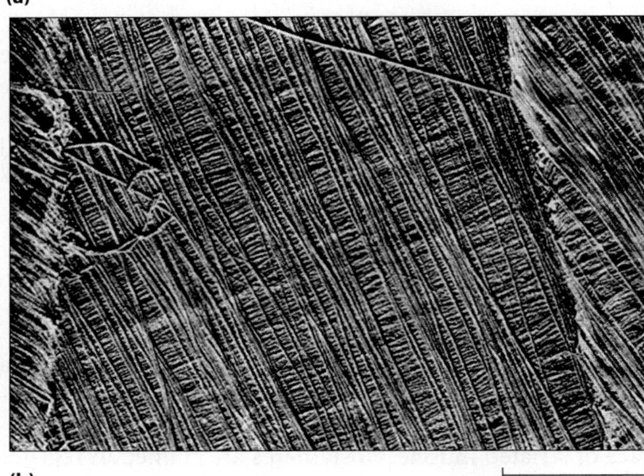

(b)

FIGURE 17-25 Structural Components of Plant Cell Walls. **(a)** Cellulose microfibrils are linked by hemicelluloses and glycoproteins called extensins to form a rigid interconnected network embedded in a matrix of pectins. Cellulose microfibrils are often twisted together to form larger structures called macrofibrils. (Not shown are the lignins that are localized between the cellulose microfibrils in woody tissues.) **(b)** This electron micrograph shows individual cellulose microfibrils in the cell wall of a green alga. Each microfibril consists of many cellulose molecules aligned laterally (TEM).

Other Cell Wall Components. In addition to cellulose and hemicellulose, cell walls have several other major components. **Pectins** are also branched polysaccharides but with backbones called *rhamnogalacturonans* that consist mainly of negatively charged galacturonic acid and rhamnose. The side chains attached to the backbone contain some of the same monosaccharides found in hemicelluloses. Pectin molecules form the matrix in which cellulose microfibrils are embedded (Figure 17-25a). They also bind adjacent cell walls together. Because of their highly branched structure and their negative charge, pectins trap and bind water molecules. As a result, pectins have a gel-like consistency. (It is because of their gel-forming capacity that pectins are added to fruit juice in the process of making fruit jams and jellies.)

In addition to hemicelluloses and pectins, cell walls contain a group of related glycoproteins called **extensins.** Despite the name, extensins are actually rigid, rodlike molecules that are tightly woven into the complex polysaccharide network of the cell wall (Figure 17-25a). In fact, extensins are so integral to the cell wall matrix that attempts to extract them chemically usually result in the loss of cell wall structure. Extensins are initially deposited in the cell wall in a soluble form. Once deposited, however, extensin molecules become covalently crosslinked to one another and to cellulose, generating a reinforced protein-polysaccharide complex. Extensins are least abundant in the cell walls of actively growing tissues and most abundant in the cell walls of tissues that provide mechanical support to the plant.

Lignins are very insoluble polymers of aromatic alcohols that occur mainly in woody tissues. (The Latin word for "wood" is *lignum.)* Lignin molecules are localized mainly between the cellulose fibrils, where they function to resist compression forces. Lignin accounts for as much as 25% of the dry weight of woody plants, making it second only to cellulose as the most abundant organic compound on Earth.

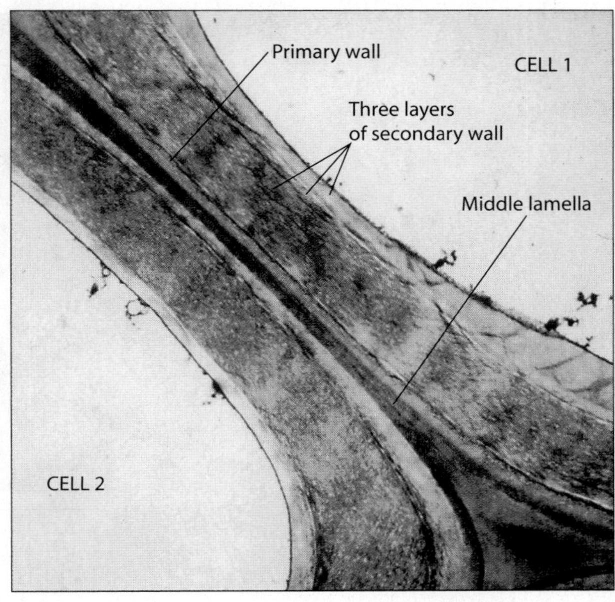

FIGURE 17-26 The Middle Lamella. The middle lamella is a layer of cell wall material that is shared by two adjacent plant cells. Consisting mainly of sticky pectins, the middle lamella binds the cells together tightly.

Cell Walls Are Synthesized in Several Discrete Stages

The plant cell wall components are secreted from the cell stepwise, creating a series of layers in which the first layer to be synthesized ends up farthest away from the plasma membrane. The first structure to be laid down is called the **middle lamella.** It is shared by neighboring cell walls and holds adjacent cells together (**Figure 17-26**). The next structure to be formed is called the **primary cell wall,** which forms when the cells are still growing. Primary walls are about 100–200 nm thick, only several times the thickness of the basal lamina of animal cells. The primary cell wall consists of a loosely organized network of cellulose microfibrils associated with hemicelluloses, pectins, and glycoproteins (**Figure 17-27a**). The cellulose microfibrils are generated by cellulose-synthesizing enzyme complexes called *rosettes* that are localized within the plasma membrane. Because the microfibrils are anchored to other wall components, the rosettes must move in the plane of the membrane as they lengthen the growing cellulose microfibrils.

When the shoots or roots of a plant are growing, cell walls must be remodeled. A family of proteins called *expansins* helps cell walls retain their pliability. One way in which expansins may act is by disrupting the normal hydrogen bonding of glycans within the microfibrils of the cell wall to allow for rearrangement of the microfibrils.

The loosely textured organization of the primary cell wall creates a relatively thin, flexible structure. In some plant cells, development of the cell wall does not proceed beyond this point. However, many cells that have stopped growing add a thicker, more rigid set of layers that are referred to collectively as the **secondary cell wall** (Figure 17-27b). The components of the multilayered secondary wall are added to the inner surface of the primary wall after cell growth has ceased. Cellulose and lignins are the primary constituents of the secondary wall, making this structure significantly stronger, harder, and more rigid than the primary wall. Each layer of the secondary wall consists of densely packed bundles of cellulose microfibrils, arranged in parallel and oriented to lie at an angle to the microfibrils of adjacent layers. This organization imparts great mechanical strength and rigidity to the secondary cell wall.

As the secondary walls form, microtubules located beneath the plasma membrane are oriented in the same direction as the newly formed cellulose microfibrils, and the cellulose-synthesizing rosettes are aggregated into large arrays containing dozens of rosettes. The underlying microtubules are thought to guide the movement of these rosette aggregates, which synthesize large parallel bundles of cellulose microfibrils as they move.

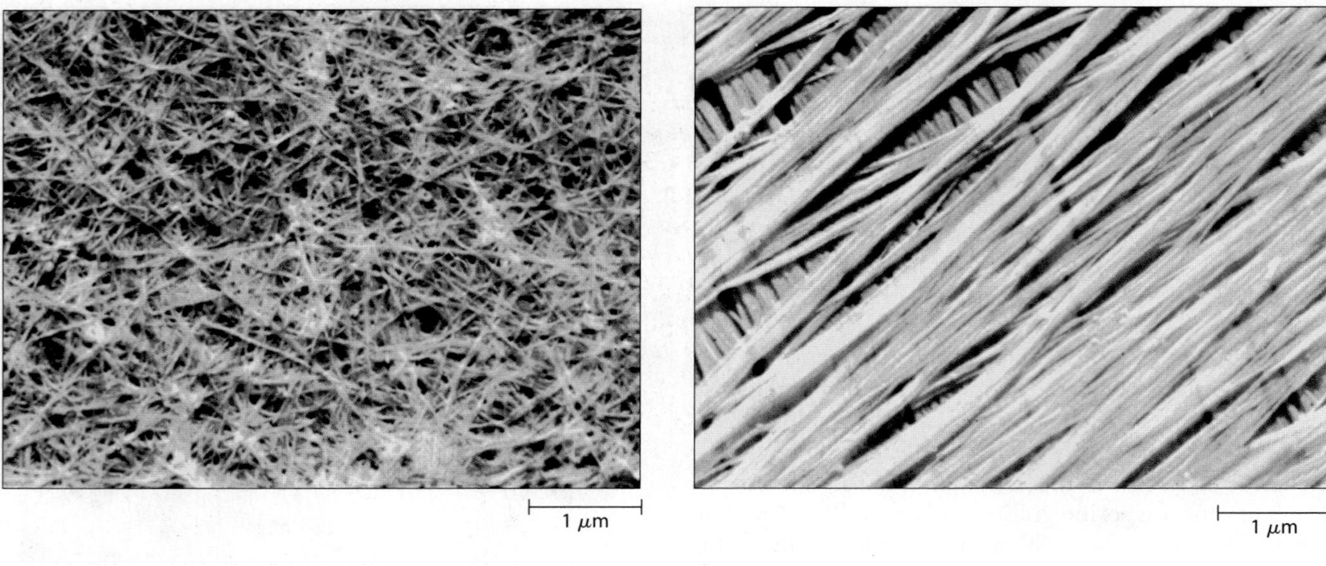

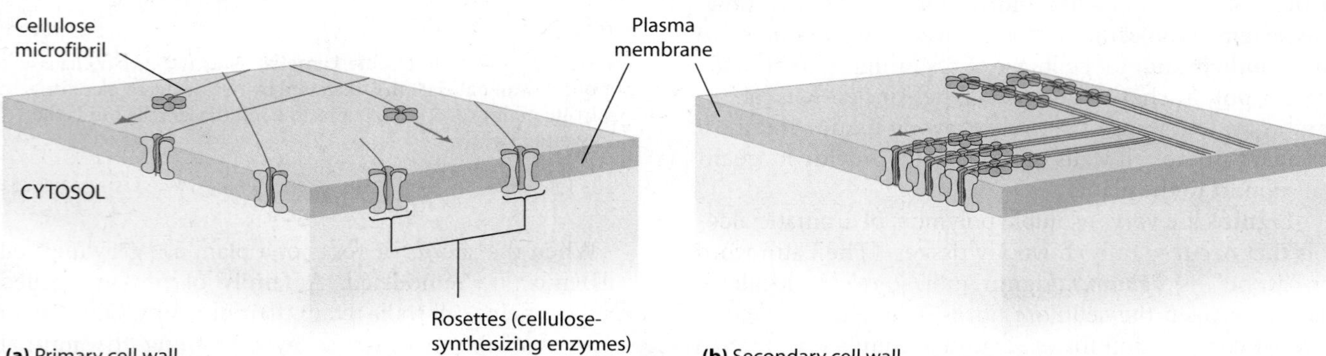

FIGURE 17-27 Cellulose Microfibrils of Primary and Secondary Plant Cell Walls. **(a)** *Primary cell wall:* The electron micrograph shows the loosely organized cellular microfibrils of a primary cell wall, and the diagram depicts how cellular microfibrils are synthesized by rosettes, each of which is a cluster of cellulose-synthesizing enzymes embedded in the plasma membrane. As a rosette synthesizes a bundle of cellulose molecules, it moves through the plasma membrane in the direction indicated by the arrows. **(b)** *Secondary cell wall:* The electron micrograph shows densely packed cellulose macrofibrils oriented in parallel, and the diagram depicts how rosettes form dense aggregates that synthesize large numbers of microfibrils in parallel, generating cellular macrofibrils. (TEMs.)

Plasmodesmata Permit Direct Cell-Cell Communication Through the Cell Wall

Since every plant cell is surrounded by a plasma membrane and a cell wall, you may wonder whether plant cells are capable of intercellular communication such as that afforded by the gap junctions of animal cells. In fact, plant cells do possess such structures. As shown in **Figure 17-28**, **plasmodesmata** (singular: **plasmodesma**) are cytoplasmic channels through relatively large openings in the cell wall, allowing continuity of the plasma membranes from two adjacent cells. Each plasmodesma is therefore lined with plasma membrane common to the two connected cells. A plasmodesma is cylindrical in shape, with the cylinder narrower in diameter at both ends. The channel diameter varies from about 20 to about 200 nm. A single tubular structure, the *desmotubule,* usually lies in the central channel of the plasmodesma. Endoplasmic reticulum (ER) cisternae are often seen near the plasmodesmata on either side of the cell wall. ER membranes from adjoining cells are continuous with the desmotubule and with the ER of the other cells, as depicted in Figure 17-28b.

The ring of cytoplasm between the desmotubule and the membrane that lines the plasmodesma is called the *annulus.* The annulus is thought to provide cytoplasmic continuity between adjacent cells, thereby allowing molecules to pass freely from one cell to the next. Even after cell division and deposition of new cell walls between the two daughter cells, cytoplasmic continuities are maintained between the daughter cells by plasmodesmata that pass through the newly formed walls. In fact,

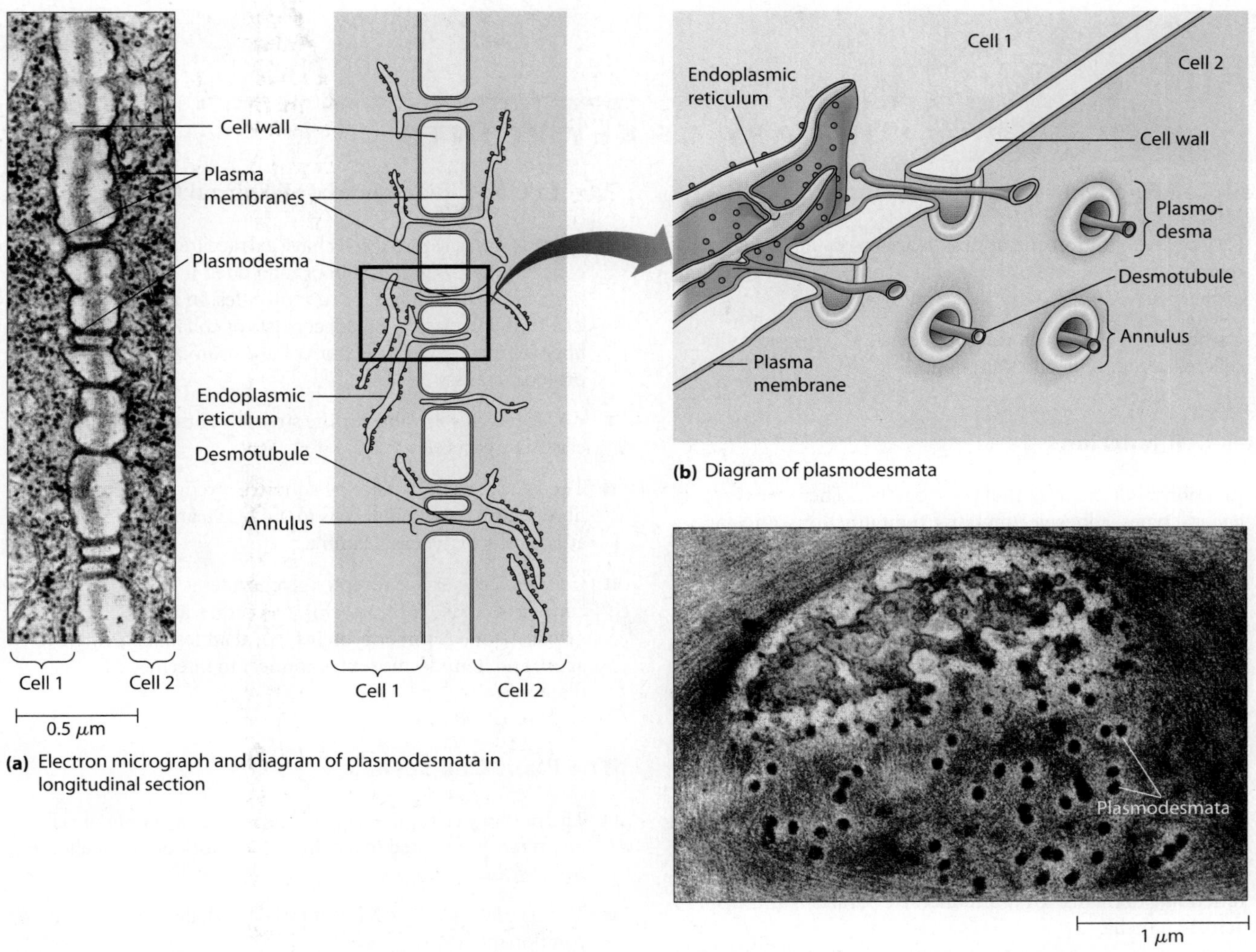

(a) Electron micrograph and diagram of plasmodesmata in longitudinal section

(b) Diagram of plasmodesmata

(c) Electron micrograph of plasmodesmata in cross section

FIGURE 17-28 Plasmodesmata. A plasmodesma is a channel through the cell wall between two adjacent plant cells, allowing cytoplasmic exchange between the cells. The plasma membrane of one cell is continuous with that of the other cell at each plasmodesma. Most plasmodesmata have a narrow cylindrical desmotubule at the center that is derived from the ER and appears to be continuous with the ER of both cells. Between the desmotubule and the plasma membrane that lines the plasmodesma is a narrow ring of cytoplasm called the annulus. **(a)** This electron micrograph and diagram show the cell wall between two adjacent root cells of timothy grass, with numerous plasmodesmata (TEM). **(b)** A diagrammatic view of a cell wall with numerous plasmodesmata, illustrating the continuity of the ER and cytoplasm between adjacent cells. **(c)** This electron micrograph shows many plasmodesmata in cross section (TEM).

most plasmodesmata are formed at the time of cell division, when the new cell wall is being formed. Minor changes may occur later, but the number and location of plasmodesmata are largely fixed at the time of division.

In many respects, plasmodesmata appear to function somewhat like gap junctions. They reduce electrical resistance between adjacent cells by about 50-fold compared with cells that are completely separated by plasma membranes. In fact, the movement of ions between adjacent cells (measured as current flow) is proportional to the number of plasmodesmata that connect the cells. However, plasmodesmata allow the passage of much larger molecules, including signaling molecules, RNAs, transcription factors, and even viruses.

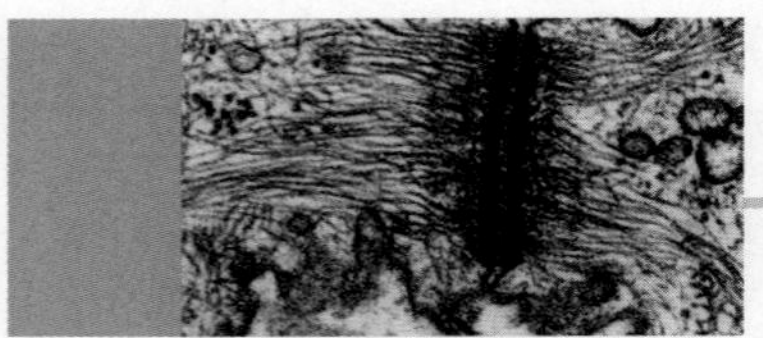

SUMMARY OF KEY POINTS

Cell-Cell Recognition and Adhesion

- Cell-cell recognition and adhesion are mediated by plasma membrane glycoproteins such as IgSF proteins, selectins, and cadherins.
- Carbohydrate groups on cell surfaces are also important in cell-cell recognition and adhesion.

Cell-Cell Junctions

- Junctions link many animal cells together. There are three general types: adhesive junctions, tight junctions, and gap junctions.
- Adhesive junctions, such as adherens junctions and desmosomes, use cadherins to hold cells together.
- Adhesive junctions are anchored to the cytoskeleton by linker proteins that attach to actin microfilaments (adherens junctions) or intermediate filaments (desmosomes).
- Desmosomes are particularly prominent in tissues that must withstand considerable mechanical stress.
- Tight junctions form a permeability barrier between epithelial cells, and they prevent the lateral movement of membrane proteins, thereby partitioning the membrane into discrete functional domains. They also mediate paracellular transport across epithelia.
- Gap junctions form open channels between cells, allowing direct chemical and electrical communication between cells. Gap junction permeability is limited to ions and small molecules.

The Extracellular Matrix of Animal Cells

- Plant as well as animal cells have extracellular structures consisting of long, rigid fibers embedded in an amorphous, hydrated matrix of branched molecules. In animals, the extracellular matrix (ECM) consists of collagen and/or elastin fibers embedded in a network of glycosaminoglycans and proteoglycans.
- Collagen is responsible for the strength of the ECM, and elastin imparts elasticity.
- The ECM is held in place by adhesive glycoproteins such as fibronectins, which link cells to the ECM, and laminins, which attach cells to the basal lamina.
- Cells use cell surface receptor glycoproteins called integrins to connect to the ECM. Integrins, like cadherins, attach to the cytoskeleton via linker proteins. Focal adhesions connect to actin, and hemidesmosomes connect to intermediate filaments.

The Plant Cell Surface

- The primary cell wall of a plant cell consists mainly of cellulose fibers embedded in a complex network of hemicelluloses, pectins, and extensins.
- The secondary cell wall is reinforced with lignins, a major component of wood.
- Plasmodesmata are membrane-lined cytoplasmic channels that allow chemical and electrical communication between adjacent plant cells.

PROBLEM SET

More challenging problems are marked with a •.

17-1 Beyond the Membrane: ECM and Cell Walls. Compare and contrast the extracellular matrix (ECM) of animal cells with the walls around plant cells.

(a) What basic organizational principle underlies both ECM and cell walls?

(b) List the common as well as distinct constituents of animal cell ECM and plant cell walls.

(c) What functional roles do the ECM and cell wall have in common?

(d) What functional roles are unique to the ECM? To the cell wall?

• **17-2 Anchoring Cells to the ECM.** Animal cells attach to several different kinds of proteins within the ECM.

(a) Briefly explain how the various domains of the fibronectin molecule (see Figure 17-18) or the laminin molecule (see Figure 17-21) are important for their function.

(b) Historically, an important strategy for disrupting the adhesion of integrins to their ligands is by using a synthetic peptide that mimics the binding site on the ECM molecule to which the integrin attaches. In the case of fibronectin, the amino acid sequence is arginine-glycine-aspartate (when written using the single letter designation for each amino acid, this sequence becomes RGD). Explain why addition of such synthetic peptides would disrupt binding of cells to their normal substratum.

17-3 Compare and Contrast. For each of the terms in list A, choose a related term in list B, and explain the relationship between the two terms by comparing or contrasting them structurally or functionally.

List A	**List B**
(a) Collagen	Basolateral surface
(b) Fibronectin	Focal adhesion
(c) Integrin	Elastin
(d) IgSF	Laminin
(e) ECM	Cadherin
(f) Hemidesmosome	Glycocalyx
(g) Apical surface	Selectin

17-4 Compaction. In mammalian embryos such as the mouse, the fertilized egg divides three times to form eight loosely packed cells, which become tightly adherent in a process known as *compaction.* In the late 1970s, several laboratories made antibodies against mouse cell surface proteins. The antibodies prevented compaction, as did removal of Ca^{2+} from the medium. What sort of protein do the antibodies probably recognize, and why?

17-5 Cellular Junctions and Plasmodesmata. Indicate whether each of the following statements is true of adhesive junctions (A), tight junctions (T), gap junctions (G), and/or plasmodesmata (P). A given statement may be true of any, all, or none (N) of these structures.

(a) Sites of membrane fusion are limited to abutting ridges of adjacent membranes.

(b) Allow the exchange of metabolites between the cytoplasms of two adjacent cells.

(c) Associated with filaments that confer either contractile or tensile properties.

(d) Require the alignment of connexons in the plasma membranes of two adjacent cells.

(e) Seal membranes of two adjacent cells tightly together.

17-6 Junction Proteins. Indicate whether each of the following proteins or structures is a component of adherens junctions (A), desmosomes (D), tight junctions (T), gap junctions (G), or plasmodesmata (P), and describe briefly the role the protein plays in the junction.

(a) Desmotubule

(b) Desmocollins

(c) Claudins

(d) Connexin

(e) Annulus

(f) α-Catenin

(g) E-cadherin

(h) Desmoplakin

17-7 Mind the Gap. Gap junctional communication can be examined using a variety of experimental tools. One of these is known as *dye coupling*, the ability of small fluorescent molecules to pass from one cell to another through gap junctions, a technique pioneered by Werner Loewenstein and his colleagues. What can you conclude about gap junctions from each of the following experiments?

(a) Loewenstein and colleagues injected cells with fluorescent molecules of different molecular weights, and a fluorescence microscope was then used to observe the movement of the molecules into adjacent cells. When molecules with molecular weight of 1926 were injected, they did not pass from cell to cell, but molecules with molecular weight of 1158 did.

(b) Microglia (support cells in the brain) are connected by gap junctions. When the fluorescent dye Lucifer yellow is injected into individual microglia that are part of a large group of cells, little dye passes to other cells. When microglia are treated with the calcium ionophore, 4Br-A23187, there is a dramatic increase in dye passage between cells.

• **17-8 Claudin Selectivity.** Claudin-4 normally prohibits movement of Na^+ ions across epithelial cells that express it. However, when positively charged amino acids near the tip of the large extracellular loop of claudin-4 are changed to negatively charged amino acids, Na^+ can pass through the epithelium. How do you account for this change in permeability to Na^+?

17-9 Plant Cell Walls. Distinguish between the terms in each of the following pairs with respect to the structure of the plant cell wall, and indicate the significance of each.

(a) Primary wall; secondary wall

(b) Plasmodesmata; annulus

(c) Pectin; lignin

(d) Xyloglucan; lignin

(e) Cellulose; extensin

(f) Plasmodesma; gap junction

• **17-10 Scurvy and Collagen.** Scurvy is a disease that until the nineteenth century was common among sailors and others whose diets were deficient in vitamin C (ascorbic acid). Individuals with scurvy suffer from various disorders, including extensive bruising, hemorrhages, and breakdown of supporting tissues. Ascorbic acid serves as a reducing agent responsible for maintaining the activity of prolyl hydroxylase, the enzyme that catalyzes hydroxylation of proline residues within the collagen triple helix, which is required for helix stability.

(a) Based on this information, postulate a role for hydroxyproline in collagen triple helices, and explain the sequence of events leading from a dietary vitamin C deficiency to symptoms such as bruising and breakdown of supporting tissues.

(b) Can you guess why sailors are no longer susceptible to scurvy? And why do you think British sailors are called "limeys" to this day?

SUGGESTED READING

References of historical importance are marked with a •.

General References

Geiger, B., J. P. Spatz, and A. D. Bershadsky. Environmental sensing through focal adhesions. *Nature Rev. Mol. Cell Biol.* 10 (2009): 21.

Gumbiner, B. M. Regulation of cadherin-mediated adhesion in morphogenesis. *Nature Rev. Mol. Cell Biol.* 6 (2005): 622.

Cell-Cell Adhesion

Pizarro-Cerda, J., and P. Cossart. Bacterial adhesion and entry into host cells. *Cell* 124 (2006): 715.

Pokutta, S., and W. J. Weis. Structure and mechanism of cadherins and catenins in cell-cell contacts. *Annu. Rev. Cell Dev. Biol.* 23 (2007): 237.

• Powell, K. The sticky business of discovering cadherins. *J. Cell Biol.* 170 (2005): 514.

Takeichi, M. The cadherin superfamily in neuronal connections and interactions. *Nature Rev. Neurosci.* 8 (2007): 11–20.

Vestweber, D., and J. E. Blanks. Mechanisms that regulate the function of the selectins and their ligands. *Physiol. Rev.* 79 (1999): 181.

Cell Junctions

Balda, M. S., and K. Matter. Tight junctions at a glance. *J. Cell Sci.* 121 (2008): 3677.

Chiba, H., M. Osanai, M. Murata, T. Kojima, and N. Sawada. Transmembrane proteins of tight junctions. *Biochim. Biophys. Acta* 1778 (2008): 588.

Desai, B. V., R. M. Harmon, and K. J. Green. Desmosomes at a glance. *J. Cell Sci.* 122 (2009): 4401.

Phelan, P., and T. A. Starich. Innexins get into the gap. *BioEssays* 23 (2001): 388.

Van Itallie, C. M., and J. M. Anderson. Claudins and epithelial paracellular transport. *Annu. Rev. Physiol.* 68 (2006): 403–429.

Wei, C. J., X. Xu, and C. W. Lo. Connexins and cell signaling in development and disease. *Annu. Rev. Cell Dev. Biol.* 20 (2004): 811.

The Extracellular Matrix of Animal Cells

Kielty, C. M., M. J. Sherratt, and C. A. Shuttleworth. Elastic fibres. *J. Cell Sci.* 115 (2002): 2817.

Page-McCaw, A., A. J. Ewald, and Z. Werb. Matrix metalloproteinases and the regulation of tissue remodelling. *Nature Rev. Mol. Cell Biol.* 8 (2007): 221.

Schéele, S., A. Nyström, M. Durbeej, J. F. Talts, M. Ekblom, and P. Ekblom. Laminin isoforms in development and disease. *J. Mol. Med.* 85 (2007), 825.

Shoulders, M. D., and R.T. Raines. Collagen structure and stability. *Annu. Rev. Biochem.* 78 (2009): 929.

Extracellular Matrix Receptors

Ervasti, J. M., and K. J. Sonnemann. Biology of the striated muscle dystrophin-glycoprotein complex. *Int. Rev. Cytol.* 265 (2008): 191.

Harburger, D. S., and D.A. Calderwood. Integrin signalling at a glance. *J. Cell. Sci.* 122 (2009): 159.

Humphries, J. D., A. Byron, and M. J. Humphries. Integrin ligands at a glance. *J. Cell Sci.* 119 (2006): 3901.

Jones, J. C., S. B. Hopkinson, and L. E. Goldfinger. Structure and assembly of hemidesmosomes. *BioEssays* 20 (1998): 488.

Legate, K. R., S. A. Wickstrom, and R. Fässler. Genetic and cell biological analysis of integrin outside-in signaling. *Genes Dev.* 23 (2009): 397.

The Plant Cell Wall and Plasmodesmata

Cosgrove, D. J. Growth of the plant cell wall. *Nature Rev. Mol. Cell Biol.* 6 (2005): 850.

Lerouxel, O. et al. Biosynthesis of plant cell wall polysaccharides—A complex process. *Curr. Opin. Plant Biol.* 9 (2006): 621.

Li, Y., L. Jones, and S. McQueen-Mason. Expansins and cell growth. *Curr. Opin. Plant Biol.* 6 (2003): 603.

Maule, A. J. Plasmodesmata: Structure, function and biogenesis. *Curr. Opin. Plant Biol.* 11 (2008), 680.

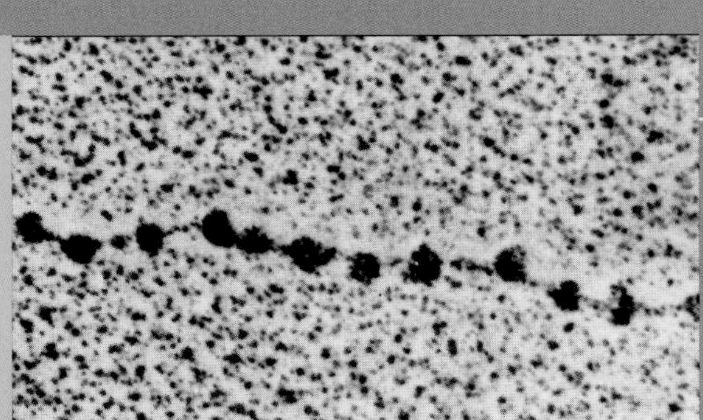

18 The Structural Basis of Cellular Information: DNA, Chromosomes, and the Nucleus

See MasteringBiology® for tutorials, activities, and quizzes.

Implicit in our earlier discussions of cellular structure and function has been a sense of predictability, order, and control. We have come to expect that organelles and other cellular structures will have a predictable appearance and function, that metabolic pathways will proceed in an orderly fashion in specific intracellular locations, and that all of a cell's activities will be carried out in a carefully controlled, highly efficient, and heritable manner.

Such expectations express our confidence that cells possess a set of "instructions" that specify their structure, dictate their functions, and regulate their activities and that these instructions can be passed on faithfully to daughter cells. More than a hundred years ago, the Augustinian monk Gregor Mendel worked out rules accounting for the inheritance patterns he observed in pea plants, although he had little inkling of the cellular or molecular basis for these rules. These studies led Mendel to conclude that hereditary information is transmitted in the form of distinct units that we now call **genes.** *We also now know that genes consist of DNA sequences that code for functional products that are usually protein chains but may in some cases be RNA molecules that do not code for proteins.*

Figure 18-1 *presents a preview of how DNA carries out its instructional role in cells and, at the same time, provides a framework for describing how this set of chapters on information flow is organized. The figure highlights the fact that the information carried by DNA flows both* between *generations of cells and* within *each individual cell. During the first of these two processes (Figure 18-1a), the information stored in a cell's DNA molecules undergoes replication, generating two DNA copies that are distributed to the daughter cells when the cell divides. The initial three chapters in this section focus on the structures and events associated with this aspect of information flow. The present chapter covers the structural organization of DNA and the chromosomes in which it is packaged; it also discusses the nucleus, which is the organelle that houses the chromosomes of eukaryotic cells. Chapter 19 then discusses DNA replication and cell division, while Chapter 20 considers the cellular and molecular events associated with information flow between generations of sexually reproducing organisms (including Mendel's work and its chromosomal basis).*

Figure 18-1b summarizes how information residing in DNA is used within *a cell. Instructions stored in DNA are transmitted in a two-stage process called* transcription *and* translation. *During transcription, RNA is synthesized in an enzymatic reaction that copies information from DNA. During translation, the base sequences of the resulting* messenger RNA *molecules are used to determine the amino acid sequences of proteins.* Thus, the information initially stored in DNA base sequences is ultimately used to code for the synthesis of specific protein molecules. *It is the particular proteins synthesized by a cell that ultimately determine most of a cell's structural features as well as the functions it performs. Transcription and translation, which together constitute the expression of genetic information, are the subjects of Chapters 21–23.*

We open this chapter by describing the discovery of DNA, the molecule whose informational role lies at the heart of this group of six chapters.

Chemical Nature of the Genetic Material

When Mendel first postulated the existence of genes, he did not know the identity of the molecule that allows them to store and transmit inherited information. But a few years later, this molecule was unwittingly discovered by Johann Friedrich Miescher, a Swiss physician. Miescher reported the discovery of the substance now known as DNA in 1869, just a few years before the cell biologist Walther Flemming first observed chromosomes as he studied dividing cells under the microscope.

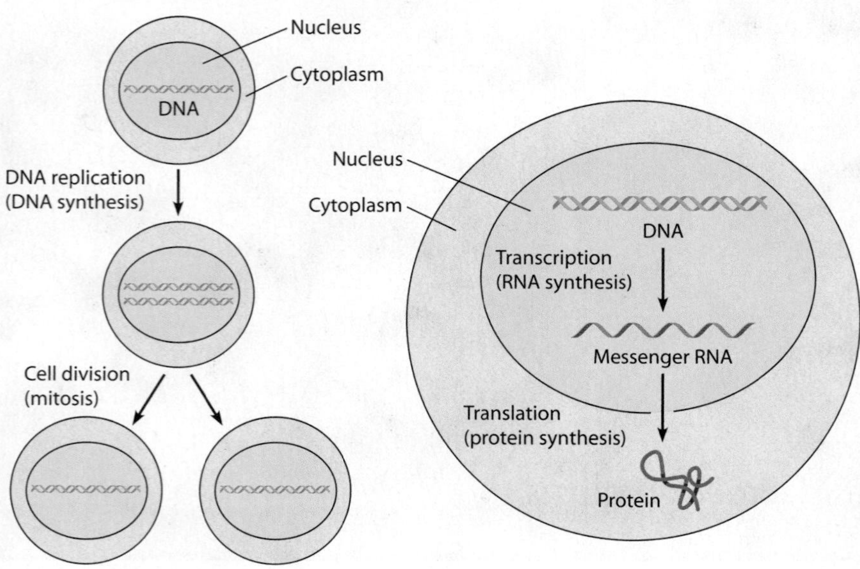

(a) The flow of genetic information between generations of cells

(b) The flow of genetic information within a cell: the expression of genetic information

FIGURE 18-1 The Flow of Information in Cells. The diagrams here feature eukaryotic cells, but DNA replication, cell division, transcription, and translation are processes that occur in prokaryotic cells as well. **(a)** Genetic information encoded in DNA molecules is passed on to successive generations of cells by DNA replication and cell division (in eukaryotic cells, by means of mitosis). The DNA is first duplicated and then divided equally between the two daughter cells. In this way, each daughter cell is assured of having the same genetic information as the cell from which it arose. **(b)** Within each cell, genetic information encoded in the DNA is expressed through the processes of transcription (RNA synthesis) and translation (protein synthesis). Transcription involves the use of selected segments of DNA as templates for the synthesis of messenger RNA and other RNA molecules. Translation is the process whereby amino acids are joined together in a sequence dictated by the sequence of nucleotides in messenger RNA.

Miescher's Discovery of DNA Led to Conflicting Proposals Concerning the Chemical Nature of Genes

Miescher was interested in studying the chemistry of the nucleus, which most scientists guessed was the site of the cell's genetic material. In his initial experiments, Miescher isolated nuclei from white blood cells obtained from pus found on surgical bandages. Extracting these nuclei with alkali led to the discovery of a novel substance that he called "nuclein," though we now know it to have been largely DNA. Miescher then went on to study DNA from a more pleasant source, salmon sperm. Fish sperm may seem a somewhat unusual source material until we realize that the nucleus accounts for more than 90% of the mass of a typical sperm cell and therefore that DNA accounts for most of the mass of sperm cells. For this reason, Miescher initially believed that DNA is involved in the transmission of hereditary information. He soon rejected this idea, however, because his crude measuring techniques incorrectly suggested that egg cells contain much more DNA than sperm cells do. Reasoning that the sperm and egg must contribute roughly equal amounts of hereditary information to the offspring, it seemed to him that DNA could not be carrying hereditary information.

Although Miescher was led astray concerning the role of DNA, in the early 1880s a botanist named Eduard Zacharias reported that extracting DNA from cells causes the staining of the chromosomes to disappear. Since evidence was already beginning to suggest a role for chromosomes in transmitting hereditary information, Zacharias and others inferred that DNA is the genetic material. This view prevailed until the early 1900s, when incorrectly interpreted staining experiments led to the false conclusion that the amount of DNA changes dramatically within cells. Because cells would be expected to maintain a constant amount of the substance that stores their hereditary instructions, these mistaken observations led to a repudiation of the idea that DNA carries genetic information.

As a result, from around 1910 to the 1940s, most scientists believed that genes were made of protein rather than DNA. The chemical building blocks of both proteins and nucleic acids had been identified by the early 1900s, and proteins were perceived to be more complex and hence more likely to store genetic information. It was argued that proteins are constructed from 20 different amino acids that can be assembled in a vast number of combinations, thereby generating the sequence diversity and complexity expected of a molecule that stores and transmits genetic information. In contrast, DNA was widely perceived to be a simple polymer consisting of the same sequence of four bases (e.g., the tetranucleotide–ATCG–) repeated over and over, thereby lacking the variability expected of a genetic molecule. Such a simple polymer was thought to serve merely as a structural support for the genes, which were in turn made of protein. This view prevailed until two lines of evidence resolved the matter in favor of DNA as the genetic material, as we describe next.

Avery Showed That DNA Is the Genetic Material of Bacteria

A great surprise was in store for biologists who were studying protein molecules to determine how genetic information is stored and transmitted. The background was provided in 1928 by the British physician Frederick Griffith, who was studying a pathogenic strain of a bacterium, then called "pneumococcus," that causes a fatal pneumonia in animals. Griffith discovered that this bacterium (now called *Streptococcus pneumoniae*) exists in two forms called the *S strain* and the *R strain*. When grown on a solid agar medium, the S strain produces colonies that are smooth and shiny because of the mucous, polysaccharide coat each cell secretes, whereas the R strain lacks the ability to manufacture a mucous coat and therefore produces colonies exhibiting a rough boundary.

When injected into mice, S-strain (but not R-strain) bacteria trigger a fatal pneumonia. The S strain's ability to cause disease is directly related to the presence of its polysaccharide coat, which protects the bacterial cell from attack by the mouse's immune system. One of Griffith's most intriguing discoveries, however, was that pneumonia can also be induced by injecting animals with a mixture of live R-strain bacteria and dead S-strain bacteria (**Figure 18-2**). This finding was surprising because neither live R-strain nor dead S-strain organisms cause pneumonia if injected alone. When Griffith autopsied the animals that had been injected with the mixture of live R-strain and dead S-strain bacteria, he found them teeming with live S-strain bacteria. Since the animals had not been injected with any live S-strain cells, he concluded that the nonpathogenic R bacteria were somehow converted into pathogenic S bacteria by a substance present in the heat-killed S bacteria that had been co-injected. He called this phenomenon **genetic transformation** and referred to the active (though still unknown) substance in the S cells as the "transforming principle."

Griffith's discoveries set the stage for 14 years of work by Oswald Avery and his colleagues at the Rockefeller Institute in New York. These researchers pursued the investigation of bacterial transformation to its logical conclusion by asking which component of the heat-killed S bacteria was actually responsible for the transforming activity. They fractionated cell-free extracts of S-strain bacteria and found that only the nucleic acid fraction was capable of causing transformation. Moreover, the activity was specifically eliminated by treatment with deoxyribonuclease, an enzyme that degrades DNA. This and other evidence convinced them that the transforming substance of pneumococcus was DNA—a conclusion published by Avery, Colin MacLeod, and Maclyn McCarty in 1944.

It was the first rigorously documented assertion that DNA can carry genetic information. But despite the rigor of the experiments, the assignment of a genetic role to DNA was not immediately accepted. Skepticism was due partly to the persistent, widespread conviction that DNA was not complex enough for such a role. In addition, many scientists questioned whether genetic information in bacteria had anything to do with heredity in other organisms. However, most remaining doubts were alleviated eight years later when DNA was also shown to be the genetic material of a virus, the bacteriophage T2.

Hershey and Chase Showed That DNA Is the Genetic Material of Viruses

Bacteriophages—or **phages,** for short—are viruses that infect bacteria. They have been objects of scientific study since the 1930s, and much of our early understanding of molecular genetics came from experiments involving these viruses. **Box 18A** describes the anatomy and replication cycle of some phages and highlights their advantages for genetic studies.

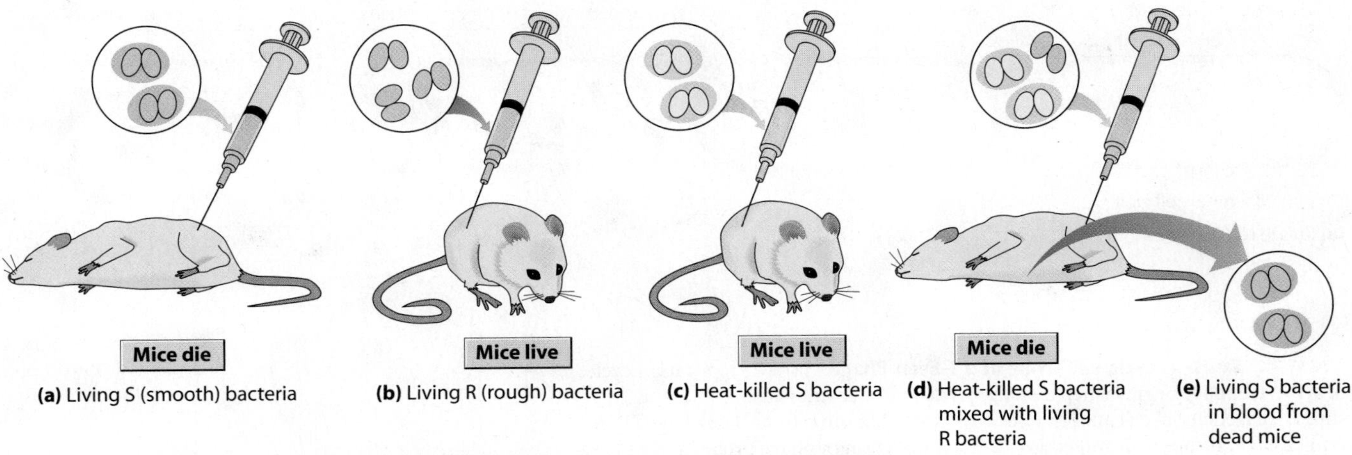

FIGURE 18-2 Griffith's Experiment on Genetic Transformation in Pneumococcus. S (smooth) cells of the pneumococcus bacterium *(Streptococcus pneumoniae)* are pathogenic in mice; R (rough) cells are not. **(a)** Injection of living S bacteria into mice causes pneumonia and death. **(b)** Injection of living R bacteria leaves mice healthy. **(c)** Heat-killed S bacteria have no effect when injected alone. **(d)** When a mixture of living R bacteria and heat-killed S bacteria is injected, the result is pneumonia and death. **(e)** The discovery of living S-strain bacteria in the blood of the mice in part d suggested to Griffith that a substance in the heat-killed S cells caused a heritable change (transformation) of nonpathogenic R bacteria into pathogenic S bacteria. The chemical substance was later identified as DNA.

Phages: Model Systems for Studying Genes

MB Viral Replication; Activity: Phage Lysogenic and Lytic Cycles

From its inception in the mid-nineteenth century, genetics has drawn upon a wide variety of organisms for its experimental materials. Initially, attention focused on plants and animals, such as Mendel's peas and the fruit flies popularized by later investigators. Around 1940, however, bacteria and viruses came into their own, providing biologists with experimental systems that literally revolutionized the science of genetics by bringing it to the molecular level.

Bacteriophages have been especially important. Bacteriophages, or phages for short, are viruses that infect bacterial cells. It is easy to grow huge numbers of phage particles in a short time; this greatly facilitates screening for mutants—phages with heritable variations—and thereby enables geneticists to identify particular genes. Some of the most thoroughly studied phages are the T2, T4, and T6 (the so-called T-even) bacteriophages, which infect the bacterium *Escherechia coli.* The three T-even phages have similar structures, which are quite elaborate. T4 is shown in **Figure 18A-1**. The *head* of the phage is a protein capsule shaped like a hollow icosahedron (a 20-sided object) and filled with DNA. The head is attached to a protein *tail,* which consists of a hollow *tail core* surrounded by a contractile *tail sheath* and terminating in a hexagonal *baseplate,* to which six *tail fibers* are attached.

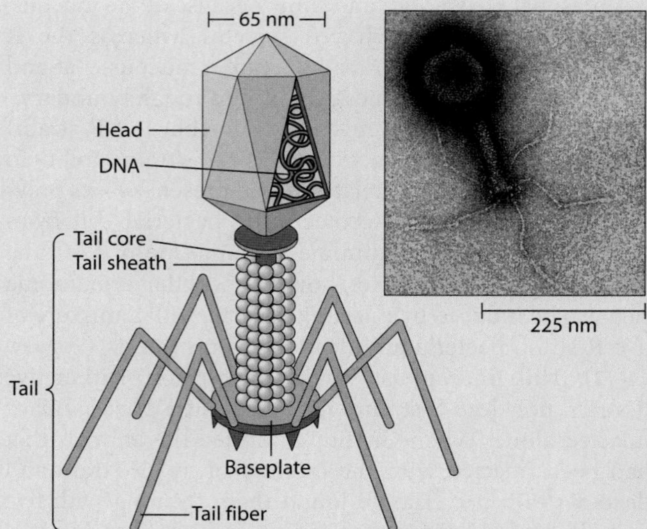

FIGURE 18A-1 The Structure of Bacteriophage T4. The drawing identifies the main structural components of this phage; not all of them are visible in the micrograph (TEM).

Figure 18A-2 depicts the main events in the replication cycle of the T4 phage. The drawings are not to scale; the bacterium is proportionately larger, as the electron micrograph indicates. The process begins with the adsorption of a phage particle to the wall of a bacterial cell. When the phage collides with the cell, it "squats" so that its baseplate attaches to a specific receptor protein in the wall (Figure 18A-2a, ❶). Next, the tail sheath contracts, driving the hollow tail core through the cell wall. The core forms a needle through which the bacteriophage DNA is injected into the bacterium (❷). Once this DNA has gained entry to the bacterial cell, the genetic information of the phage is transcribed and translated (❸). This gives rise to a few key proteins that subvert the metabolic machinery of the host cell for the phage's benefit, which is usually its own rapid multiplication. Since the phage consists simply

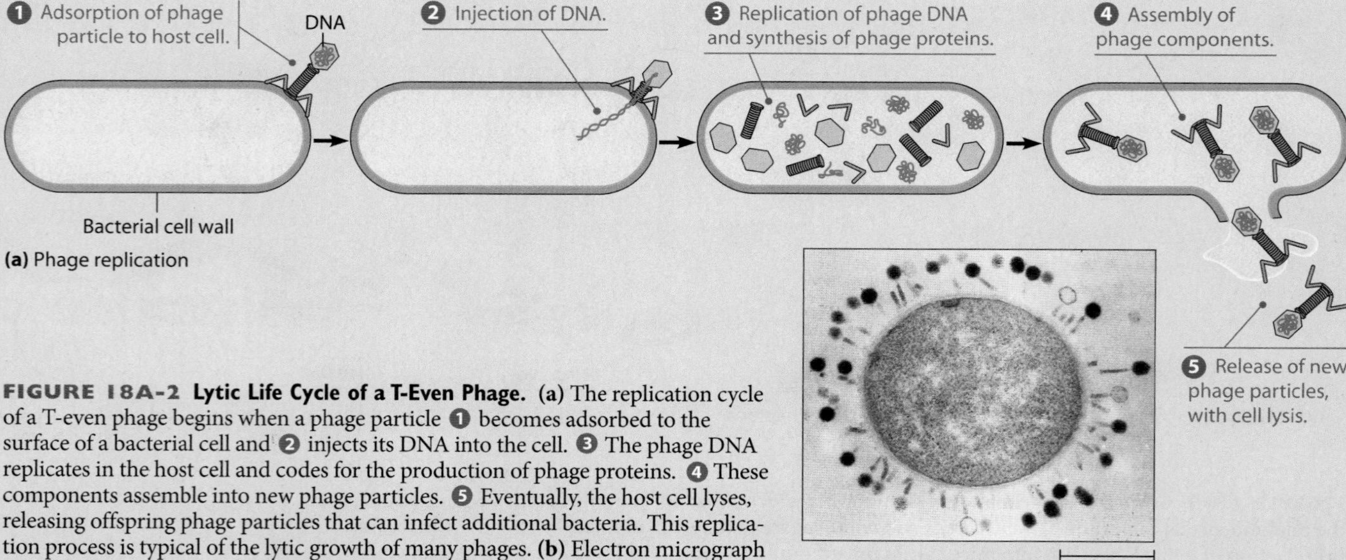

MB Activity: Phage Lytic Cycle

FIGURE 18A-2 Lytic Life Cycle of a T-Even Phage. (a) The replication cycle of a T-even phage begins when a phage particle ❶ becomes adsorbed to the surface of a bacterial cell and ❷ injects its DNA into the cell. ❸ The phage DNA replicates in the host cell and codes for the production of phage proteins. ❹ These components assemble into new phage particles. ❺ Eventually, the host cell lyses, releasing offspring phage particles that can infect additional bacteria. This replication process is typical of the lytic growth of many phages. **(b)** Electron micrograph of a bacterium with phage particles attached to its surface (TEM).

of a DNA molecule surrounded by a protein coat (its *capsid*), most of the metabolic activity in the infected cell is channeled toward the replication of phage DNA and the synthesis of capsid proteins. The phage DNA and capsid proteins then self-assemble into hundreds of new phage particles (❹). Within about half an hour, the infected cell lyses (breaks open), releasing the new phage particles into the medium (❺). Each new phage can now infect another bacterial cell, making it possible to obtain enormous populations of phage—as many as 10^{11} phage particles per milliliter in infected bacterial cultures.

To determine the number of phage particles in a sample, a measured volume is mixed with bacterial cells growing in liquid medium to allow adsorption of the phages to the bacteria. The mixture is then spread onto agar growth medium in a Petri dish. Upon incubation, the bacteria multiply to produce a dense "lawn" of cells on the surface of the nutrient medium. But wherever a virus particle has infected a bacterial cell, a clear spot appears in the lawn because the bacterial cells there have been killed by the multiplying phage population. Such clear spots are called *plaques*. The number of plaques reflects the number of phage particles present in the original phage-bacterium mixture, provided only that the initial number of phages was small enough to ensure that each gives rise to a separate plaque. **Figure 18A-3** shows plaques formed by T4 bacteriophage on a lawn of *E. coli* cells.

The course of events shown in Figure 18A-2a is called *lytic growth* and is characteristic of a *virulent phage*. Lytic growth results in lysis of the host cell and the production of many progeny phage particles. In contrast, a *temperate phage* can either produce lytic growth, as a virulent phage does, or integrate its DNA into the bacterial chromosome without causing any immediate harm to the host cell. An especially well-studied example of a temperate phage is bacteriophage λ *(lambda)*, which, like the T-even phages, infects

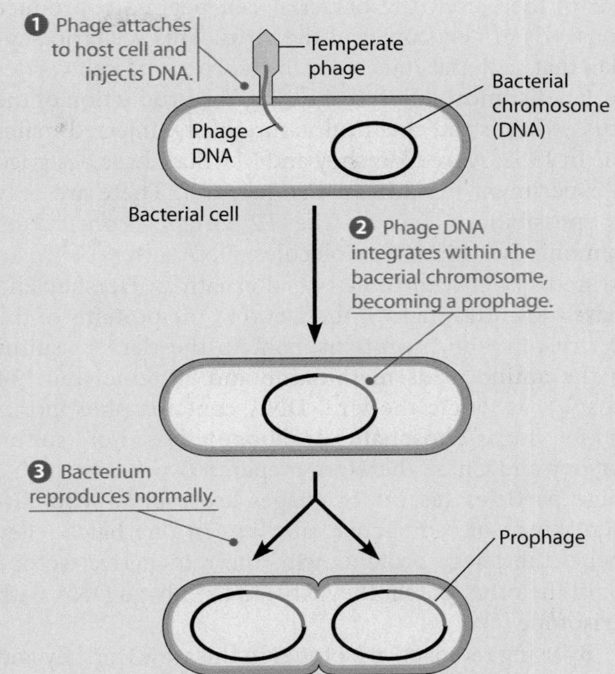

FIGURE 18A-4 Lysogenic State of a Prophage Within a Bacterial Chromosome. The DNA injected by a temperate phage can become integrated into the DNA of the bacterial chromosome. The integrated phage DNA, called a prophage, is replicated along with the bacterial DNA each time the bacterium reproduces.

MB Molecular Model: Temperate Bacteriophage Repressor

E. coli cells. In the integrated or *lysogenic state*, the DNA of the temperate phage is called a *prophage*. The prophage is replicated along with the bacterial DNA, often through many generations of host cells (**Figure 18A-4**). During this time, the phage genes, though potentially lethal to the host, are inactive, or *repressed*. Under certain conditions, however, the prophage DNA is excised from the bacterial chromosome and again enters a lytic cycle, producing progeny phage particles and lysing the host cell.

One reason bacteriophages are so attractive to geneticists is that the small size of their genomes makes it relatively easy to identify and study their genes. The genome of bacteriophage λ, for example, is a single DNA molecule containing fewer than 60 genes, compared with several thousand genes in a bacterium such as *E. coli*. Other phages are still smaller; in some instances, they contain less than a dozen genes. Because of their simple genomes, their rapidity of multiplication, and the enormous numbers of progeny that can be produced in a small volume of culture medium, bacteriophages are among the best understood of all "organisms." They may have some practical benefits as well. Since phages are capable of destroying bacteria, some biotechnology companies are exploring the development of modified phages that might be useful in treating human bacterial infections, especially in cases where the bacteria have become resistant to antibiotics.

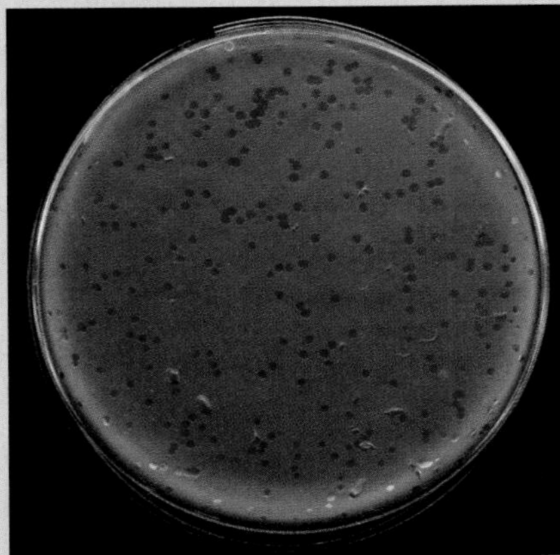

FIGURE 18A-3 Phage Plaques on a Lawn of Bacteria. Phage plaques have formed on a lawn of *E. coli* infected with phage T4. Each plaque arises from the reproduction of a single phage particle in the original mixture.

One of the most thoroughly studied of the phages that infect the bacterium *Escherichia coli* is bacteriophage T2. During infection, this virus attaches to the bacterial cell surface and injects material into the cell. Shortly thereafter, the bacterial cell begins to produce thousands of new copies of the virus. This scenario suggests that material injected into the bacterial cell carries the genetic information that guides the production of the virus. What is the chemical nature of the injected material? In 1952, Alfred Hershey and Martha Chase designed an experiment to address this question. There are only two possibilities because the T2 virus is constructed from only two kinds of molecules: DNA and protein. To distinguish between these two alternatives, Hershey and Chase took advantage of the fact that the proteins of the T2 virus, like most proteins, contain the element sulfur (in the amino acids methionine and cysteine) but not phosphorus, while the viral DNA contains phosphorus (in its sugar-phosphate backbone) but not sulfur. Hershey and Chase therefore prepared two batches of T2 phage particles (as intact phages are called) with different kinds of radioactive labeling. In one batch, they labeled the phage proteins with the radioactive isotope ^{35}S; in the other batch, they labeled the phage DNA with the isotope ^{32}P.

By using radioactive isotopes in this way, Hershey and Chase were able to trace the fates of both protein and DNA during the infection process (**Figure 18-3a**). They began the experiment by mixing radioactive phage with intact bacterial cells and allowing the phage particles to attach to the bacterial cell surface and inject their genetic material into the cells. At this point, Hershey and Chase found that the empty protein coats (or phage "ghosts") could be effectively removed from the surface of the bacterial cells by agitating the suspension in an ordinary kitchen blender and recovering the bacterial cells by centrifugation. They then measured the radioactivity in the supernatant liquid and in the pellet of bacteria at the bottom of the tube.

The data revealed that most (65%) of the ^{32}P remained with the bacterial cells, while the bulk (80%) of the ^{35}S was released into the surrounding medium (Figure 18-3b). Since the ^{32}P labeled the viral DNA and the ^{35}S labeled the viral protein, Hershey and Chase concluded that DNA, not protein, had been injected into the bacterial cells; hence, DNA must function as the genetic material of phage T2. This conclusion received further support from the following observation: When the infected, radioactive bacteria were resuspended in fresh liquid and incubated longer, the ^{32}P was transferred to some of the offspring phage particles, but the ^{35}S was not.

As a result of the experiments we have described, by the early 1950s most biologists came to accept the view that genes are made of DNA, not protein. Unfortunately, Oswald Avery, the visionary most responsible for the complete turnabout in views concerning the function of DNA, never received the credit he so richly deserved. The Nobel Prize Committee discussed Avery's work but decided he had not done enough. Perhaps Avery's modest and unassuming nature was responsible for this lack of recognition. After Avery died in 1955, the biochemist Erwin Chargaff wrote in tribute: "He was a quiet man; and it would have honored the world more, had it honored him more."

Why did the Hershey–Chase experiments receive a warmer welcome than Avery's earlier work on bacterial transformation, even though both led to the same conclusion? The main reason seems to have been simply the passage of time and the accumulation of additional, circumstantial evidence after Avery's 1944 publication. Perhaps most important was evidence that DNA is indeed variable enough in structure to serve as the genetic material. This evidence came from studies of DNA base composition, as we will see next.

Chargaff's Rules Reveal That A = T and G = C

Despite the initial lukewarm reaction to Avery's work, it had an important influence on several other scientists. Among them was Erwin Chargaff, who was interested in the base composition of DNA. Between 1944 and 1952, Chargaff used chromatographic methods to separate and quantify the relative amounts of the four bases—adenine (A), guanine (G), cytosine (C), and thymine (T)—found in DNA. Several important discoveries came from his analyses. First, he showed that DNA isolated from different cells of a given species has the same percentage of each of the four bases (**Table 18-1**, rows 1–4) and that this percentage does not vary with individual, tissue, age, nutritional state, or environment. This is exactly what would be expected of the chemical substance that stores genetic information because the cells of a given species would be expected to have similar genetic information. However, Chargaff did find that DNA base composition varies from species to species. This can be seen by examining the last column of Table 18-1, which shows the relative amounts of the bases A and T versus G and C in the DNAs of various organisms. The data also reveal that DNA preparations from closely related species have similar base compositions, whereas those from very different species tend to exhibit quite different base compositions. Again, this is what would be expected of a molecule that stores genetic information.

But Chargaff's most striking observation was his discovery that, for all DNA samples examined, the number of adenines is equal to the number of thymines (A = T), and the number of guanines is equal to the number of cytosines (G = C). This meant that the number of purines is equal to the number of pyrimidines (A + G = C + T). The significance of these equivalencies, known as **Chargaff's rules,** was an enigma and remained so until Watson and Crick proposed the double-helical model of DNA in 1953.

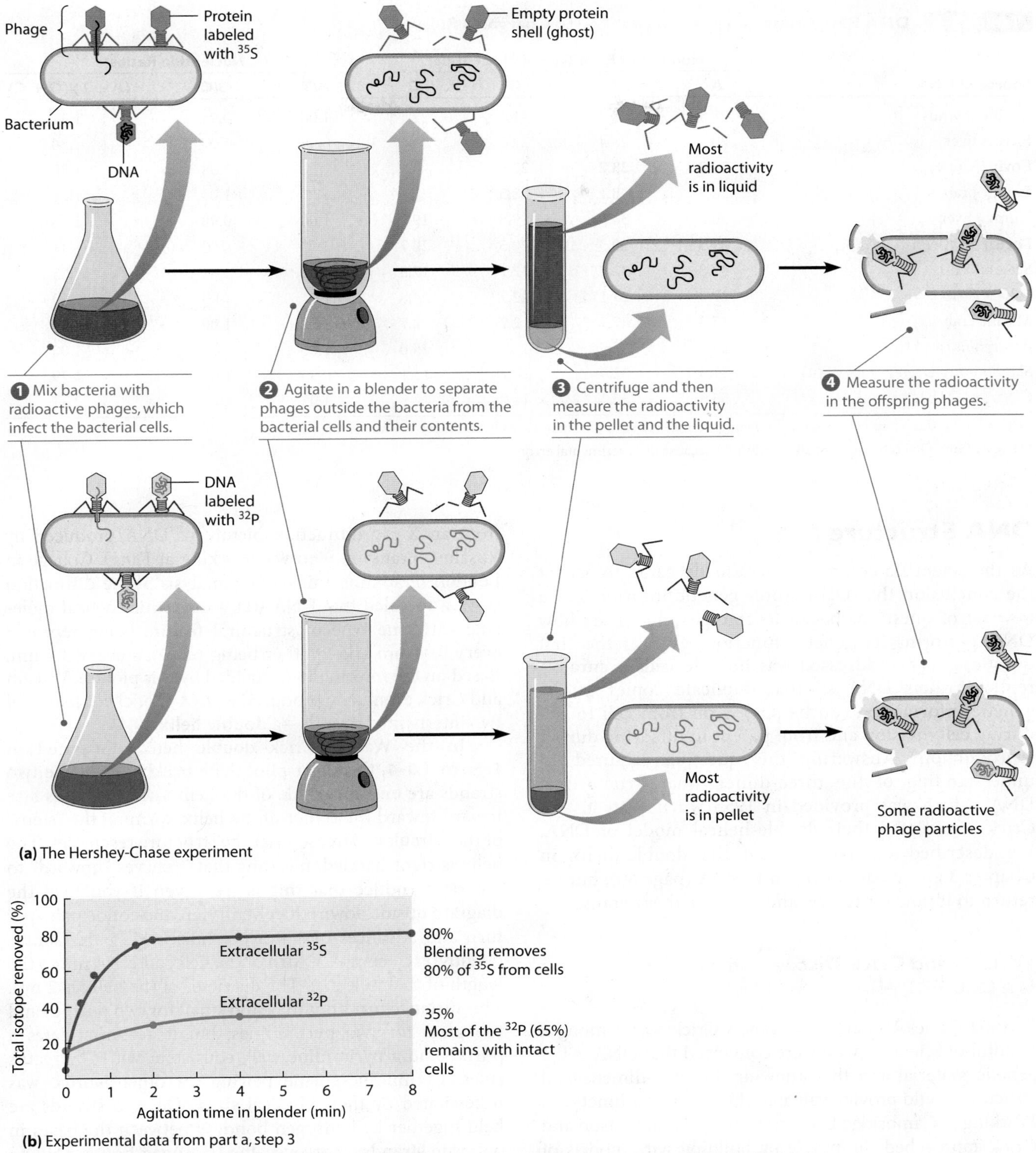

Phage

Protein labeled with ^{35}S

Empty protein shell (ghost)

Bacterium

DNA

Most radioactivity is in liquid

❶ Mix bacteria with radioactive phages, which infect the bacterial cells.

❷ Agitate in a blender to separate phages outside the bacteria from the bacterial cells and their contents.

❸ Centrifuge and then measure the radioactivity in the pellet and the liquid.

❹ Measure the radioactivity in the offspring phages.

DNA labeled with ^{32}P

Most radioactivity is in pellet

Some radioactive phage particles

(a) The Hershey-Chase experiment

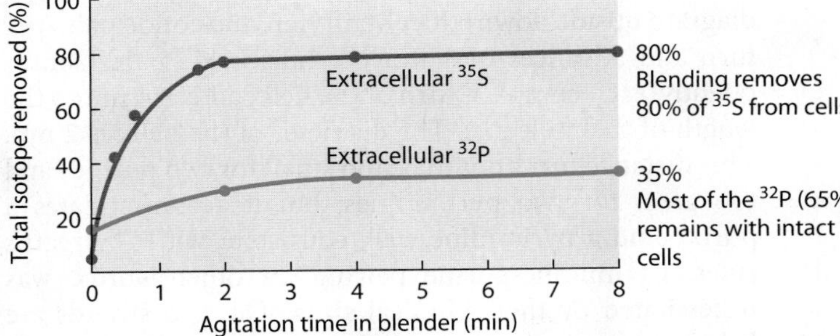

(b) Experimental data from part a, step 3

MB DNA as Genetic Material: The Hershey and Chase Experiment; Activity: The Hershey-Chase Experiment

FIGURE 18-3 The Hershey–Chase Experiment: DNA as the Genetic Material of Phage T2. **(a)** ❶ T2 labeled with either ^{35}S (to label protein) or ^{32}P (to label DNA) is used to infect bacteria. The phages adsorb to the cell surface and inject their DNA. ❷ Agitation of the infected cells in a blender dislodges most of the ^{35}S from the cells, whereas most of the ^{32}P remains. ❸ Centrifugation causes the cells to form a pellet; any free phage particles, including ghosts, remain in the supernatant liquid. ❹ When the cells in each pellet are incubated further, the phage DNA within them dictates the synthesis and eventual release of new phage particles. Some of these phages contain ^{32}P in their DNA (because the old, labeled phage DNA is packaged into some of the new particles), but none contain ^{35}S in their coat proteins. **(b)** The graph shows the extent to which ^{35}S and ^{32}P are removed from the intact cells at step ❸, as a function of time in the blender. A few minutes of blending is enough to remove most (80%) of the ^{35}S, while leaving most (65%) of the ^{32}P with the cells.

Table 18-1 DNA Base Composition Data That Led to Chargaff's Rules

	Number of Each Type of Nucleotide*				Nucleotide Ratios**		
Source of DNA	**A**	**T**	**G**	**C**	**A/T**	**G/C**	**(A + T)/(G + C)**
Bovine thymus	28.4	28.4	21.1	22.1	1.00	0.95	1.31
Bovine liver	28.1	28.4	22.5	21.0	0.99	1.07	1.30
Bovine kidney	28.3	28.2	22.6	20.9	1.00	1.08	1.30
Bovine brain	28.0	28.1	22.3	21.6	1.00	1.03	1.28
Human liver	30.3	30.3	19.5	19.9	1.00	0.98	1.53
Locust	29.3	29.3	20.5	20.7	1.00	1.00	1.41
Sea urchin	32.8	32.1	17.7	17.3	1.02	1.02	1.85
Wheat germ	27.3	27.1	22.7	22.8	1.01	1.00	1.19
Marine crab	47.3	47.3	2.7	2.7	1.00	1.00	17.50
Aspergillus (mold)	25.0	24.9	25.1	25.0	1.00	1.00	1.00
Saccharomyces cerevisiae (yeast)	31.3	32.9	18.7	17.1	0.95	1.09	1.79
Clostridium (bacterium)	36.9	36.3	14.0	12.8	1.02	1.09	2.73

*The values in these four columns are the average number of each type of nucleotide found per 100 nucleotides in DNA.

**The A/T and G/C ratios are not all exactly 1.00 because of experimental error.

DNA Structure

As the scientific community gradually came to accept the conclusion that DNA stores genetic information, a new set of questions began to emerge concerning how DNA performs its genetic function. One of the first questions to be addressed was how do cells accurately replicate their DNA so that duplicate copies of the genetic information can be passed on from cell to cell during cell division and from parent to offspring during reproduction? Answering this question required an understanding of the three-dimensional structure of DNA, which was provided in 1953 when Watson and Crick formulated their double-helical model of DNA. We described the structure of the double helix in Chapter 3 and its discovery in Box 3A (page 60), but we return to it now for review and some further details.

Watson and Crick Discovered That DNA Is a Double Helix

In 1952, James Watson and Francis Crick were among a handful of scientists who were convinced that DNA is the genetic material and that knowing its three-dimensional structure would provide valuable clues to how it functions. Working at Cambridge University in England, Watson and Crick approached the puzzle by building wire models of possible structures. DNA had been known for years to be a long polymer having a backbone of repeating sugar (deoxyribose) and phosphate units, with a nitrogenous base attached to each sugar. These scientists were aided in their model building by knowing that the particular forms in which the bases A, G, C, and T exist at physiological pH permit specific hydrogen bonds to form between pairs of them. The crucial experimental evidence, however, came from an X-ray diffraction picture of DNA produced by Rosalind Franklin, who was working at King's College in London. Franklin's painstaking analysis of the diffraction pattern revealed that DNA was a long, thin, helical molecule with one type of structural feature being repeated every 0.34 nm and another being repeated every 3.4 nm. Based on the information provided by this picture, Watson and Crick eventually produced a DNA model consisting of two intertwined strands—a **double helix.**

In the Watson–Crick double helix, illustrated in **Figure 18-4**, the sugar-phosphate backbones of the two strands are on the outside of the helix, and the bases face inward toward the center of the helix, forming the "steps" of the "circular staircase" that the structure resembles. The helix is right-handed, meaning that it curves "upward" to the right (notice that this is true even if you turn the diagram upside down). It contains ten nucleotide pairs per turn and advances 0.34 nm per nucleotide pair. Consequently, each complete turn of the helix adds 3.4 nm to the length of the molecule. The diameter of the helix is 2 nm. This distance turns out to be too small for two purines and too great for two pyrimidines; but it accommodates a purine and a pyrimidine well, consistent with Chargaff's rules. Pyrimidine-purine pairing, in other words, was necessitated by their physical sizes. The two strands are held together by hydrogen bonding between the bases in opposite strands. Moreover, the hydrogen bonds holding together the two strands of the double helix fit *only when they form between the base adenine (A) in one chain and thymine (T) in the other or between the base guanine (G) in one chain and cytosine (C) in the other.* This means that the base sequence of one chain determines the base sequence of the opposing chain; the two chains of the DNA double helix are therefore said to be **complementary** to each other. Such a model explains why Chargaff had

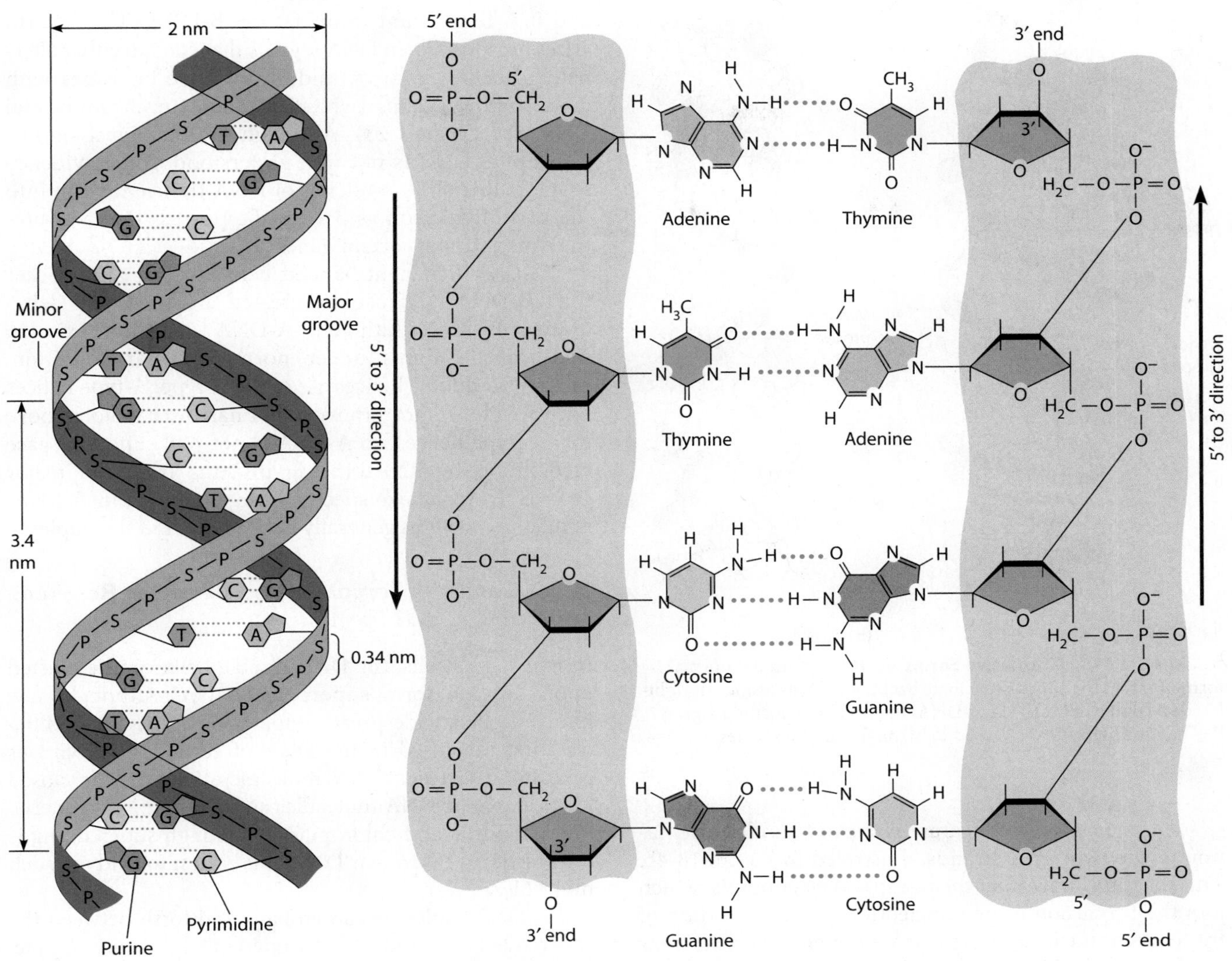

(a) Double helix (b) Antiparallel orientation of strands

MB Molecular Model: An Introduction to DNA Structure

FIGURE 18-4 The DNA Double Helix. (a) This schematic illustration shows the sugar-phosphate chains of the DNA backbone, the complementary base pairs, the major and minor grooves, and several important dimensions. A = adenine, G = guanine, C = cytosine, T = thymine, P = phosphate, and S = sugar (deoxyribose). **(b)** One strand of a DNA molecule is oriented 5′⟶3′ in one direction, whereas its complement has a 5′⟶3′ orientation in the opposite direction. This diagram also shows the hydrogen bonds that connect the bases in AT and GC pairs.

observed that DNA molecules contain equal amounts of the bases A and T and equal amounts of the bases G and C.

The most profound implication of the Watson–Crick model was that it suggested a mechanism by which cells can replicate their genetic information: The two strands of the DNA double helix could simply separate from each other before cell division so that each strand could function as a *template,* dictating the synthesis of a new complementary DNA strand using the base-pairing rules. In other words, the base A in the template strand would specify insertion of the base T in the newly forming strand, the base G would specify insertion of the base C, the base T would specify insertion of the base A, and the base C would specify insertion of the base G. In the next chapter, we will discuss the experimental evidence for this proposed mechanism and describe the molecular basis of DNA replication in detail.

Several other important features of the DNA double helix are illustrated in Figure 18-4. For example, notice that the two strands are twisted around each other so that there is a *major groove* and a *minor groove.* Base pairs viewed from the major groove yield more information than when viewed from the minor groove because more hydrogen bond donors (H) and acceptors (O, N), as well as the methyl group of the base T, are exposed to the major groove. As a result, regulatory proteins can bind to the major groove and recognize specific base sequences without unfolding the DNA double helix (see Figure 23-24).

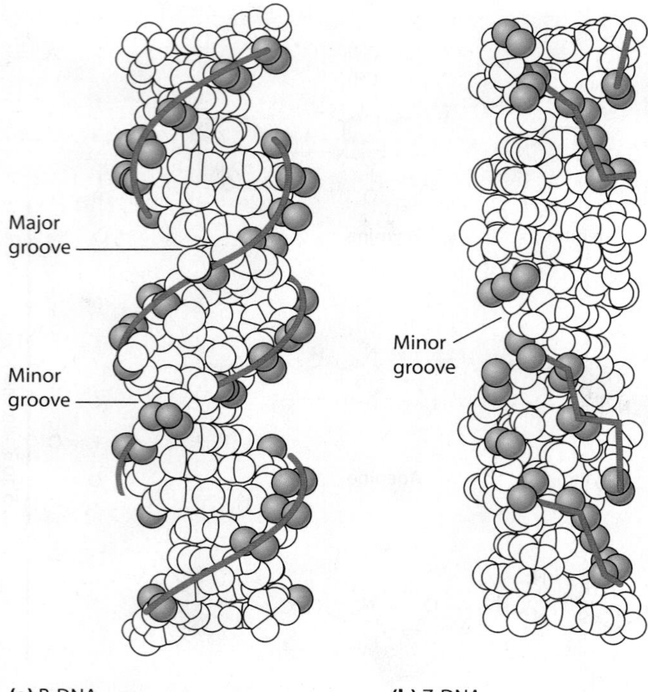

FIGURE 18-5 Alternative Forms of DNA. **(a)** In the normal B form of DNA, the sugar-phosphate backbone forms a smooth right-handed double helix. **(b)** In Z-DNA, the backbone forms a zigzag left-handed helix. Color is used to highlight the backbones.

Another important feature is the *antiparallel* orientation of the two DNA strands, illustrated in Figure 18-4b. This diagram shows that the phosphodiester bonds, which join the 5′ carbon of one nucleotide to the 3′ carbon of the adjacent nucleotide, are oriented in *opposite* directions in the two DNA strands. Starting at the top of the diagram, the strand on the left is said to exhibit a $5' \rightarrow 3'$ *orientation* because its first nucleotide has a free 5′ end and its final nucleotide has a free 3′ end. Conversely, the strand on the right exhibits a $3' \rightarrow 5'$ *orientation* starting from the top because its first nucleotide has a free 3′ end and its final nucleotide has a free 5′ end. The opposite orientation of the two strands has important implications for both DNA replication and DNA transcription, as we will see in Chapters 19 and 21.

The right-handed Watson–Crick helix is an idealized version of what is called *B-DNA* (**Figure 18-5a**). Naturally occurring B-DNA double helices are flexible molecules whose exact shapes and dimensions depend on the local nucleotide sequence. Although B-DNA is the main form of DNA in cells (and in test tube solutions of DNA), other forms may also exist, perhaps in short segments interspersed in molecules that are mostly B-DNA. The most important of these alternative forms are Z-DNA and A-DNA. As shown in Figure 18-5b, *Z-DNA* is a *left-handed* double helix. Its name derives from the zigzag pattern of its sugar-phosphate backbone, and it is longer and thinner than B-DNA. The Z form arises most readily in DNA regions that contain either alternating purines and pyrimidines or have cytosines with extra methyl groups (which do occur in chromosomal DNA; see Chapter 23). Although the biological significance of Z-DNA is not well understood, some evidence suggests that short stretches of DNA transiently flip into the Z configuration as part of the process that activates the expression of certain genes.

A-DNA is a right-handed helix, shorter and thicker than B-DNA, and can be created artificially by dehydrating B-DNA. Although A-DNA does not exist in significant amounts under normal cellular condition, most RNA double helices are of the A type. A-type helices have a wider minor groove and a narrower major groove than B-type helices, so A-RNA is not well suited for base recognition by RNA-binding proteins from the major groove. To recognize specific base sequences in A-RNA, regulatory proteins generally need to unwind the duplex.

DNA Can Be Interconverted Between Relaxed and Supercoiled Forms

In many situations, the DNA double helix can be twisted upon itself to form **supercoiled DNA.** Although now known to be a widespread property of DNA, supercoiling was first identified in the DNA of certain small viruses containing circular DNA molecules that exist as closed loops. Circular DNA molecules are also found in bacteria, mitochondria, and chloroplasts. Although supercoiling is not restricted to circular DNA, it is easiest to study in such molecules.

A DNA molecule can go back and forth between the supercoiled state and the nonsupercoiled, or *relaxed,* state. To understand the basic idea, you might perform the following exercise. Start with a length of rope consisting of two strands twisted together into a right-handed coil; this is the equivalent of a relaxed, linear DNA molecule. Just joining the ends of the rope together changes nothing; the rope is now circular but still in a relaxed state. But before sealing the ends, if you first give the rope an extra twist in the direction in which the strands are already entwined around each other, the rope is thrown into a *positive supercoil.* Conversely, if before sealing, you give the rope an extra twist in the opposite direction, the rope is thrown into a *negative supercoil.* Like the rope in this example, a relaxed DNA molecule can be converted to a positive supercoil by twisting in the same direction as the double helix is wound and into a negative supercoil by twisting in the opposite direction (**Figure 18-6**). Circular DNA molecules found in nature, including those of bacteria, viruses, and eukaryotic organelles, are invariably negatively supercoiled.

Supercoiling also occurs in linear DNA molecules when regions of the molecule are anchored to some cell structure and so cannot freely rotate. At any given time,

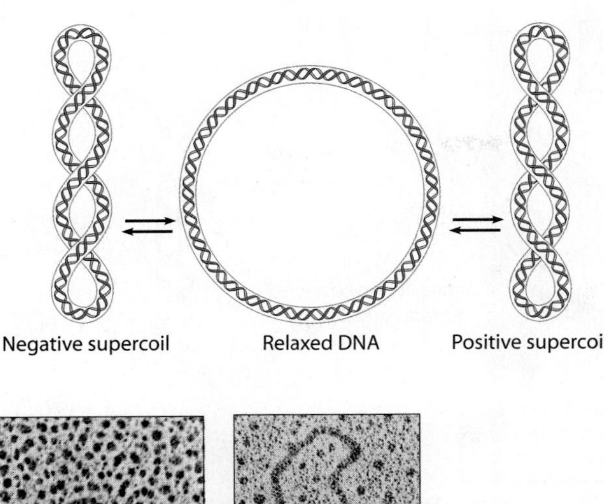

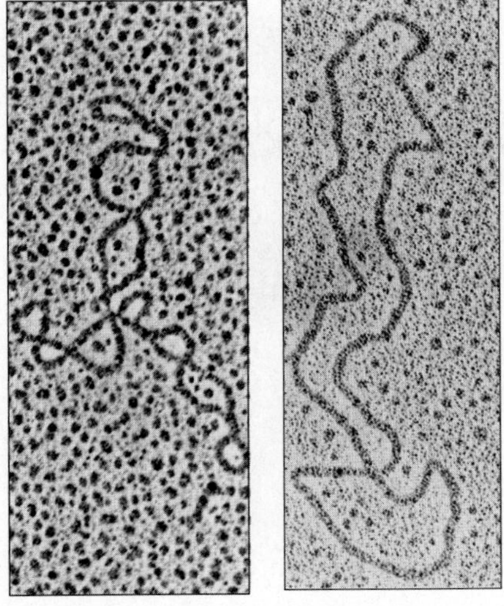

FIGURE 18-6 Interconversion of Relaxed and Supercoiled DNA. *(Top)* Conversion of a relaxed circular DNA molecule into a negatively supercoiled form (by twisting it in the opposite direction as the double helix is wound) and into a positively supercoiled form (by twisting it in the same direction as the double helix is wound). *(Bottom)* Electron micrographs of circular DNA molecules from a bacteriophage called PM2, showing a molecule with negative supercoils on the left and a relaxed molecule on the right (TEMs).

significant portions of the linear DNA in the nucleus of eukaryotic cells may be supercoiled; and, when DNA is packaged into chromosomes at the time of cell division, extensive supercoiling helps to make the DNA more compact.

By influencing both the spatial organization and the energy state of DNA, supercoiling affects the ability of a DNA molecule to interact with other molecules. Positive supercoiling involves tighter winding of the double helix and therefore reduces opportunities for interaction. In contrast, negative supercoiling is associated with unwinding of the double helix, which gives its strands increased access to proteins involved in DNA replication or transcription.

The interconversion between relaxed and supercoiled forms of DNA is catalyzed by enzymes known as **topoisomerases,** which are classified as either *type I* or *type II*. Both types catalyze the relaxation of supercoiled DNA; type I enzymes do so by introducing transient single-strand breaks in DNA, whereas type II enzymes introduce transient double-strand breaks. **Figure 18-7** shows how these temporary breaks affect DNA supercoiling. Type I topoisomerases induce DNA relaxation by cutting one strand of the double helix, thereby allowing the DNA to rotate and the uncut strand to be passed through the break before the broken strand is resealed. In contrast, type II topoisomerases induce relaxation by cutting both DNA strands and then passing a segment of uncut double helix through the break before resealing. Unlike the type I reaction, this action of type II topoisomerases requires energy derived from the hydrolysis of ATP.

Type I and type II topoisomerases are able to remove both positive and negative supercoils from DNA. In addition, bacteria have a type II topoisomerase called **DNA gyrase,** which can induce as well as relax supercoiling. As you will learn in Chapter 19, DNA gyrase is one of several enzymes involved in DNA replication. It can relax the positive supercoiling that results from partial unwinding of a double helix, or it can actively introduce negative supercoils that promote strand separation, thereby facilitating access of other proteins involved in DNA replication. DNA gyrase requires ATP to generate supercoiling but not to relax an already supercoiled molecule.

The Two Strands of a DNA Double Helix Can Be Separated Experimentally by Denaturation and Rejoined by Renaturation

Because the two strands of the DNA double helix are bound together by relatively weak, noncovalent bonds, the two strands can be readily separated from each other under appropriate conditions. As we will see in coming chapters, strand separation is an integral part of both DNA replication and RNA synthesis. Strand separation can also be induced experimentally, resulting in **DNA denaturation;** the reverse process, which reestablishes a double helix from separated DNA strands, is called **DNA renaturation.**

DNA is commonly denatured in the laboratory by raising either the temperature or the pH. When denaturation is induced by slowly raising the temperature, the DNA retains its double-stranded, or native, state until a critical temperature is reached; at that point the duplex rapidly denatures, or "melts," into its component strands. The melting process is easy to monitor because double-stranded and single-stranded DNA differ in their light-absorbing properties. All DNA absorbs ultraviolet light, with an absorption maximum around 260 nm. When the temperature of a DNA solution is slowly raised, the absorbance at 260 nm remains constant until the double helix begins to melt into its component strands. As the strands separate, the absorbance of the solution

(a) **Topoisomerase I.** Supercoils are removed by transiently cleaving one strand of the DNA double helix and passing the unbroken strand through the break.

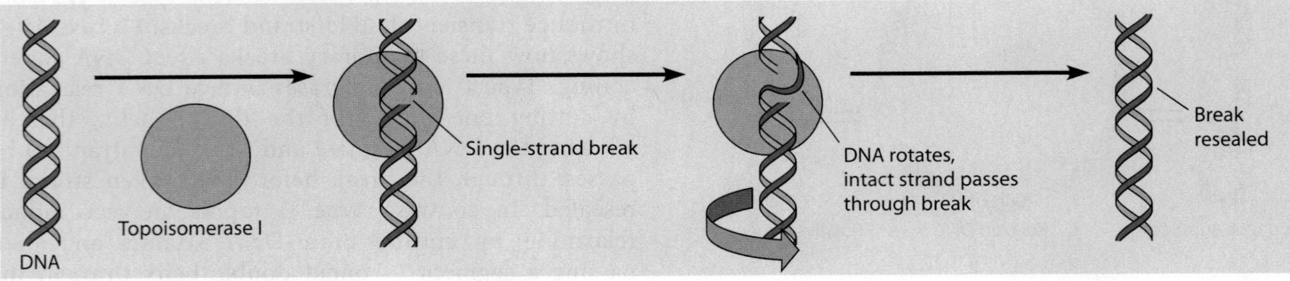

(b) **Topoisomerase II.** Supercoils are removed by transiently cleaving both strands of the DNA double helix and passing an unbroken region of the DNA double helix through the break.

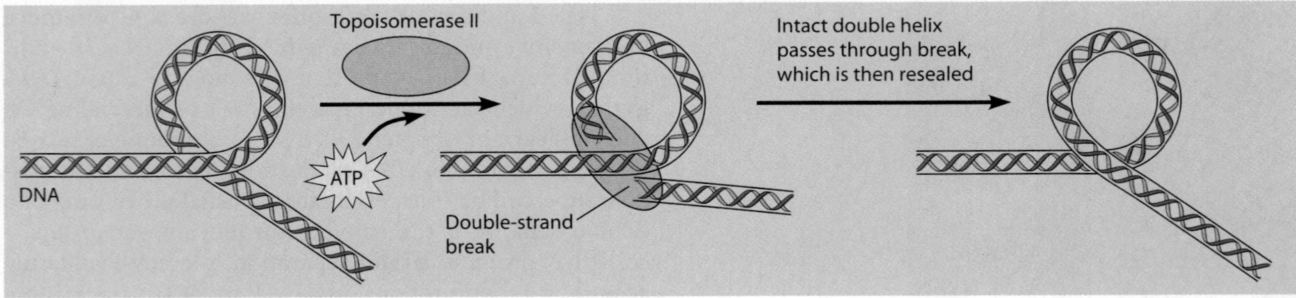

FIGURE 18-7 Reactions Catalyzed by Topoisomerases I and II. (**a**) Type I and (**b**) type II topoisomerases are used for removing both positive and negative supercoils from DNA.

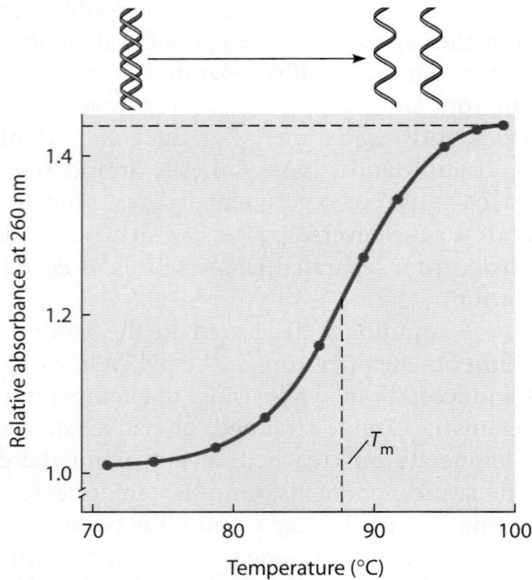

FIGURE 18-8 A Thermal Denaturation Profile for DNA. When the temperature of a solution of double-stranded (native) DNA is raised, the heat causes the DNA to denature. The conversion to single strands is accompanied by an increase in the absorbance of light at 260 nm. The temperature at which the midpoint of this increase occurs is called the melting temperature, T_m. For the sample shown, the T_m is about 87°C.

increases rapidly due to the higher intrinsic absorption of single-stranded DNA (**Figure 18-8**).

The temperature at which one-half of the absorbance change has been achieved is called the **DNA melting temperature (T_m).** The value of the melting temperature reflects how tightly the DNA double helix is held together. For example, GC base pairs, held together by three hydrogen bonds, are more resistant to separation than are AT base pairs, which have only two (see Figure 18-4b). The melting temperature therefore increases in direct proportion to the relative number of GC base pairs in the DNA (**Figure 18-9**). Likewise, DNA molecules in which the two strands of the double helix are properly base-paired at each position will melt at higher temperatures than will DNA in which the two strands are not perfectly complementary.

Denatured DNA can be renatured by lowering the temperature to permit hydrogen bonds between the two strands to be reestablished (**Figure 18-10**). The ability to renature nucleic acids has a variety of important scientific applications. Most importantly, it forms the basis for **nucleic acid hybridization,** a family of procedures for identifying nucleic acids based on the ability of single-stranded chains with complementary base sequences to bind, or *hybridize,* to each other. Nucleic acid hybridization can be applied to DNA–DNA, DNA–RNA, and even RNA–RNA interactions. In DNA–DNA hybridization, for

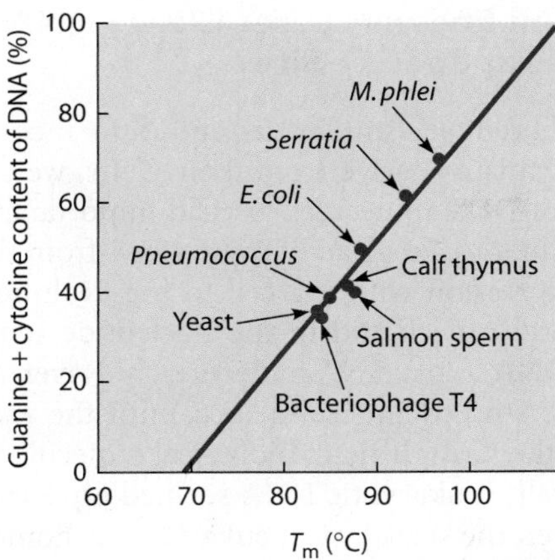

FIGURE 18-9 Dependence of DNA Melting Temperature on Base Composition. The melting temperature of DNA increases with its G + C content, as shown by the relationship between T_m and G + C content for DNA samples from a variety of sources.

example, the DNA being examined is denatured and then incubated with a purified, single-stranded radioactive DNA fragment, called a **probe,** whose sequence is complementary to the base sequence one is trying to detect. Box 18C provides an example of the use of a probe in DNA fingerprinting.

Nucleic acid sequences do not need to be perfectly complementary to be able to hybridize. Changing the temperature, salt concentration, and pH used during hybridization can permit pairing to take place between partially complementary sequences exhibiting numerous mismatched bases. Under such conditions, hybridization will occur between DNAs that are related to one another but not identical. This approach is useful for identifying families of related genes, both within a given type of organism and among different kinds of organisms.

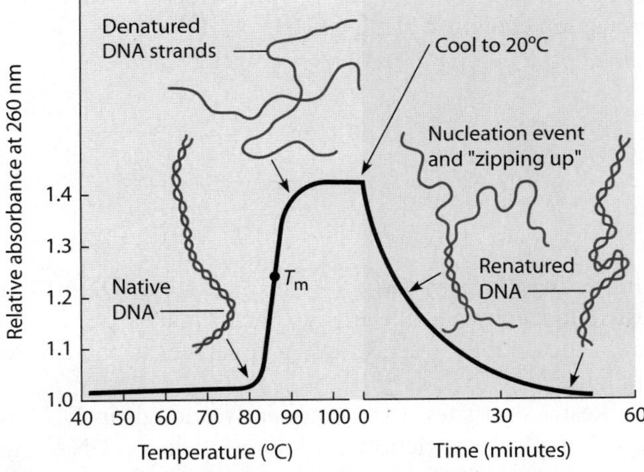

FIGURE 18-10 DNA Denaturation and Renaturation. If a solution of native (double-stranded) DNA is heated slowly under carefully controlled conditions, the DNA "melts" over a narrow temperature range, with an increase in absorbance at 260 nm. When the solution is allowed to cool, the separated DNA strands reassociate by random collisions, followed by a rapid "zipping up" of complementary base pairs in the two strands. The reassociation requires varying amounts of time, depending on both the DNA concentration in the solution and the length of the DNA strands.

The Organization of DNA in Genomes

So far, we have considered several chemical and physical properties of DNA. But as cell biologists, we are primarily interested in its importance to the cell. We therefore want to know how much DNA cells have, how and where they store it, and how they utilize the genetic information it contains. We begin by inquiring about the amount of DNA present because that determines the maximum amount of information a cell can possibly possess.

The **genome** of an organism or virus consists of the DNA (or for some viruses, RNA) that contains one complete copy of all the genetic information of that organism or virus. For many viruses and prokaryotes, the genome resides in a single linear or circular DNA molecule or in a small number of them. Eukaryotic cells have a nuclear genome, a mitochondrial genome, and, in the case of plants and algae, a chloroplast genome as well. Mitochondrial and chloroplast genomes are single, usually circular DNA molecules resembling those of bacteria. The nuclear genome generally consists of multiple DNA molecules dispersed among a haploid set of chromosomes. (As we will explore in more detail in Chapter 20, a *haploid* set of chromosomes consists of one representative of each type of chromosome, whereas a *diploid* set consists of two copies of each type of chromosome, one copy from the mother and one from the father. Sperm and egg cells each have a haploid set of chromosomes, whereas most other types of eukaryotic cells are diploid.)

Genome Size Generally Increases with an Organism's Complexity

Genome size is usually expressed as the total number of base-paired nucleotides, or **base pairs (bp).** For example, the circular DNA molecule that constitutes the genome of an *E. coli* cell has 4,639,221 bp. Since such numbers tend to be rather large, the abbreviations **Kb** (kilobases), **Mb** (megabases), and **Gb** (gigabases) are used to refer to a thousand, or million, or billion base pairs, respectively. Thus, the size of the *E. coli* genome can be expressed simply as 4.6 Mb. The range of genome sizes observed for various groups of organisms is summarized in **Figure 18-11.** These data reveal a spread of almost eight orders of magnitude in genome size, from a few thousand base pairs for the simplest viruses to more than 100 billion base pairs for certain plants, amphibians, and protists. Expressed in terms of total DNA length, this corresponds to a range of less than 2 μm of DNA for a small virus, such as SV40, to

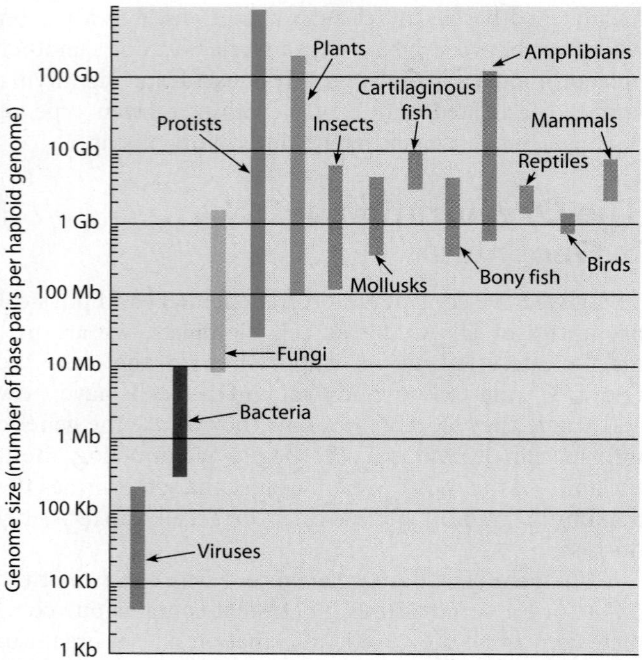

FIGURE 18-11 Relationship Between Genome Size and Type of Organism. For each group of organisms shown, the bar represents the approximate range in genome size measured as the number of base pairs per haploid genome. The same color (purple) is used for groups that involve members of the animal kingdom.

roughly 34 meters of DNA (more than 100 feet!) for certain plants, such as the wildflower *Trillium.*

Broadly speaking, genome size increases with the complexity of the organism. Viruses contain enough nucleic acid to code for only a few or a few dozen proteins, bacteria can specify a few thousand proteins, and eukaryotic cells have enough DNA (at least in theory) to encode hundreds of thousands of proteins. But closer examination of such data reveals some puzzling features. Most notably, the genome sizes of eukaryotes exhibit great variations that do not clearly correlate with any known differences in organismal complexity. Some amphibians and plants, for example, have gigantic genomes that are tens or even hundreds of times larger than those of other amphibians or plants or of mammalian species. *Trillium,* for example, is a member of the lily family that has no obvious need for exceptional amounts of genetic information. Yet its genome size is more than 20 times that of pea plants and 30 times that of humans. Moreover, a single-celled amoeba has a genome that is 200 times the size of the human genome. We have no idea why lily plants and amoebae possess so much DNA. Its presence highlights the fact that most eukaryotic genomes carry large amounts of DNA of no currently known function, a phenomenon we will discuss shortly. In the final analysis, genome size is less important than the number and identity of functional genes and the DNA sequences that control their expression.

Restriction Endonucleases Cleave DNA Molecules at Specific Sites

Since the hereditary similarities and differences observed among organisms derive from their DNA, we can expect the study of DNA molecules to yield important biological insights. Clues to a myriad of mysteries—from the control of gene expression within a cell to the evolution of new species—are to be found in the nucleotide sequences of genomic DNA. Most DNA molecules, however, are far too large to be studied intact. In fact, until the early 1970s, DNA was the most difficult biological molecule to analyze biochemically. Eukaryotic DNA seemed especially intimidating, given the size of most eukaryotic genomes, and no method was known for cutting DNA at specific sites to yield reproducible fragments. The prospect of ever being able to identify, isolate, sequence, or manipulate specific eukaryotic genes seemed unlikely. Yet in less than a decade, DNA became one of the easiest biological molecules to work with.

This breakthrough was made possible by the discovery of **restriction endonucleases** (also called *restriction enzymes*), which are proteins isolated from bacteria that cut foreign DNA molecules at specific internal sites. (**Box 18B** describes the biological role of restriction endonucleases and provides some details about the sites they cut; here we focus on their use as analytical tools.) The cutting action of a restriction enzyme generates a specific set of DNA pieces called *restriction fragments.* Each restriction enzyme cleaves double-stranded DNA only in places where it encounters a specific recognition sequence, called a **restriction site,** that is usually four or six (but may be eight or more) nucleotides long. For example, here is the restriction site recognized by the widely used *E. coli* restriction enzyme called *Eco*RI:

```
  ↓
5′ G—A—A—T—T—C 3′
3′ C—T—T—A—A—G 5′
                ↑
```

The arrows indicate where *Eco*RI cuts the DNA. This restriction enzyme, like many others, makes a *staggered* cut in the double-stranded DNA molecule, as you can see in Figure 18B-1b (Box 18B).

Restriction sites occur frequently enough in DNA to permit typical restriction enzymes to cleave DNA into fragments ranging from a few hundred to a few thousand base pairs in length. Fragments of these sizes are far more amenable to further manipulation than are the enormously long DNA molecules from which they are generated.

Separation of Restriction Fragments by Gel Electrophoresis. Incubating a DNA sample with a specific restriction enzyme yields a collection of restriction fragments of different sizes. To determine the number

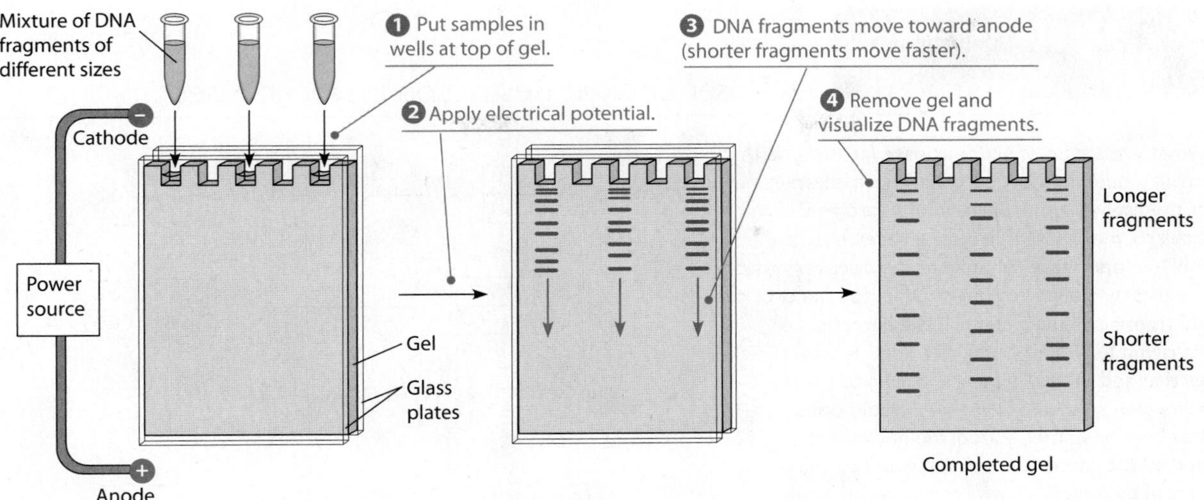

MB Activity: Gel Electrophoresis of DNA

FIGURE 18-12 Gel Electrophoresis of DNA. The three test tubes contain mixtures of DNA fragments produced by incubating DNA with different restriction enzymes. A small sample of each is applied to the top of a gel, and an electrical potential of several hundred volts is applied. This causes the DNA fragments to migrate toward the anode, with shorter fragments migrating faster than larger ones. After allowing time for the fragments to separate from one another, the gel is removed and stained with a dye such as ethidium bromide, which binds to the DNA fragments and causes them to fluoresce under ultraviolet light. Alternatively, autoradiography can be used to locate the DNA bands in the gel, provided that the DNA is radioactively labeled.

and lengths of such fragments and to isolate individual fragments for further study, a researcher must be able to separate the fragments from one another. The technique of choice for this purpose is **gel electrophoresis,** essentially the same method used to separate proteins and polypeptides (see Figure 7-22). In fact, the procedure for DNA is even simpler than for proteins because DNA molecules have an inherent negative charge (due to their phosphate groups) and therefore do not need to be treated with a negatively charged detergent to make them move toward the anode. Small DNA fragments are usually separated in *polyacrylamide* gels, whereas larger DNA fragments are separated in more porous gels made of *agarose,* a polysaccharide.

Figure 18-12 illustrates the separation of restriction fragments of different sizes by gel electrophoresis. DNA samples are first incubated with the desired restriction enzyme; in the figure, three different restriction enzymes are used. The samples are then placed in separate compartments ("wells") at one end of the gel. Next, an electrical potential is applied across the gel, with the anode situated at the opposite end of the gel from the samples. Because their phosphate groups have a negative charge, DNA fragments migrate toward the anode. Smaller fragments (i.e., those with lower molecular weight) move through the gel with relative ease and therefore migrate rapidly, while larger fragments move more slowly. The current is left on until the fragments are well spaced out on the gel. The final result is a series of DNA fragments that have been separated based on differences in size.

DNA fragments in the gel can be visualized either by staining or by using radioactively labeled DNA. A common staining technique involves soaking the gel in the dye *ethidium bromide,* which binds to DNA and fluoresces orange when exposed to ultraviolet light. If the DNA fragments are radioactive, their locations can be determined by **autoradiography,** a technique for detecting radioactive molecules by overlaying a sample with X-ray film. When the film is developed, it yields an *autoradiogram* that is darkened wherever radioactivity has interacted with the film. After individual DNA fragments are located in this way, they can be removed from the gel for further study.

Restriction Mapping. How does a researcher determine the order in which a set of restriction fragments is arranged in a DNA molecule? One approach involves treating the DNA with two or more restriction enzymes, alone and in combination, followed by gel electrophoresis to determine the size of the resulting DNA fragments. **Figure 18-13** shows how it would work for a simple DNA molecule cleaved with the restriction enzymes *Eco*RI and *Hae*III. In this example, each individual restriction enzyme cleaves the DNA into two fragments, indicating that the DNA contains one restriction site for each enzyme. Based on such information alone, two possible restriction maps can be proposed (see maps A and B in Figure 18-13). To determine which of the two maps is correct, an experiment must be done in which the starting DNA molecule is cleaved *simultaneously* with *Eco*RI and *Hae*III. The size of the fragments produced by simultaneous digestion with both enzymes reveals that map A must be the correct one.

In practice, restriction mapping usually involves data that are considerably more complex than in our simple

BOX 18B DEEPER INSIGHTS

A Closer Look at Restriction Endonucleases

Restriction enzymes are a type of endonuclease (an enzyme that cuts DNA internally) found in most bacteria. These enzymes help bacteria protect themselves against invasion by foreign DNA molecules, particularly the DNA of bacteriophages. In fact, the name "restriction" endonuclease came from the discovery that these enzymes *restrict* the ability of foreign DNA to take over the transcription and translation machinery of the bacterial cell.

To protect its own DNA from being degraded, the bacterial cell has enzymes that add methyl groups (—CH_3) to specific nucleotides that its own restriction enzymes would otherwise recognize. Once they have been methylated, the nucleotides are no longer recognized by the restriction enzymes, and so the bacterial DNA is not attacked by the cell's own restriction enzymes. Restriction enzymes are therefore said to be part of the cell's **restriction/methylation system:** Foreign DNA is cleaved by the restriction enzymes, while the bacterial genome is protected by prior methylation.

Restriction enzymes are named after the bacteria from which they are obtained. Each enzyme name is derived by combining the first letter of the bacterial genus with the first two letters of the species. The strain of the bacterium may also be indicated, and if two or more enzymes have been isolated from the same species, the enzymes are numbered (using Roman numerals) in order of discovery. Thus, the first restriction enzyme isolated from *E. coli* strain R is designated *Eco*RI, whereas the third enzyme isolated from *Hemophilus aegyptius* is called *Hae*III.

Restriction enzymes are specific for double-stranded DNA and cleave both strands. Each restriction enzyme recognizes a specific DNA sequence that is usually four or six (but may be eight or more) nucleotide pairs long. For example, the enzyme *Hae*III recognizes the tetranucleotide sequence GGCC and cleaves the DNA double helix as shown in **Figure 18B-1a**. The restriction sites for several other restriction enzymes are summarized in **Table 18B-1**. Some restriction enzymes, such as *Hae*III, cut both strands at the same point, generating restriction fragments with

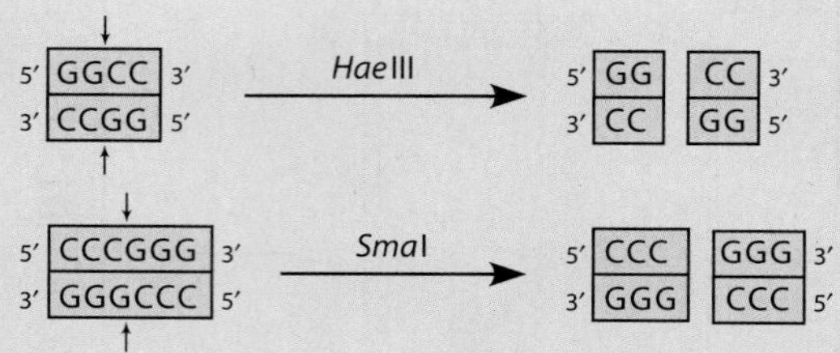

(a) Cleavage by enzymes producing blunt ends

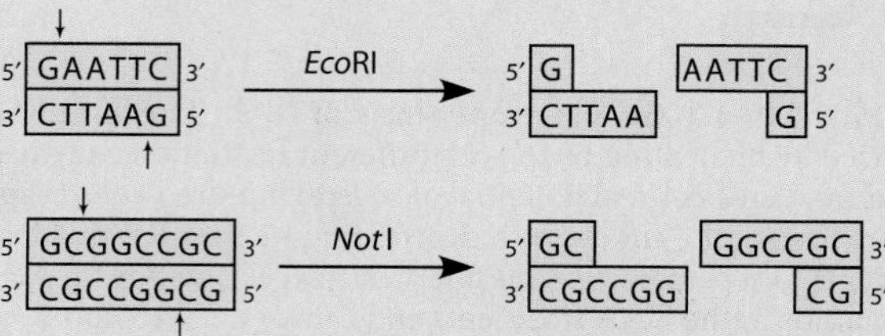

(b) Cleavage by enzymes producing sticky ends

FIGURE 18B-1 Cleavage of DNA by Restriction Enzymes. **(a)** *Hae*III and *Sma*I are examples of restriction enzymes that cut both DNA strands in the same location, generating fragments with blunt ends. **(b)** *Eco*RI and *Not*I are examples of enzymes that cut DNA in a staggered fashion, generating fragments with sticky ends. A genetic "engineer" can use such sticky ends for joining DNA fragments from different sources, as we will explain in Chapter 20.

blunt ends. Many other restriction enzymes cleave the two strands in a staggered manner, generating short, single-stranded tails or overhangs on both fragments. *Eco*RI is an example of such an enzyme; it recognizes the sequence GAATTC and cuts the DNA molecule in an offset manner, leaving an AATT tail on both

---

example. In such situations, the DNA fragments produced by each restriction enzyme can be physically isolated—for example, by cutting the gel into slices and extracting the DNA from each slice. The isolated fragments are then individually cleaved with the second restriction enzyme, allowing the cleavage sites in each fragment to be analyzed separately. Problem 18-4 provides an example of how this approach can be used to construct a **restriction map** depicting the location of all the restriction sites in the original DNA.

Molecular Model: ATP, dATP, ddATP

Rapid Procedures Exist for DNA Sequencing

At about the same time that techniques for preparing restriction fragments were developed, two methods were devised for rapid **DNA sequencing**—that is, determining the linear order of bases in DNA. One method was devised by Allan Maxam and Walter Gilbert, the other by Frederick Sanger and his colleagues. The Maxam–Gilbert method, called the *chemical method,* is based on the use of (nonprotein) chemicals that cleave DNA preferentially at specific bases; the Sanger procedure, called the *chain termination method,* utilizes *dideoxynucleotides* (nucleotides lacking a 3′ hydroxyl group) to interfere with the normal enzymatic synthesis of DNA. We will focus on Sanger's method because it has been adapted for use in automated machines that are now employed for most DNA sequencing tasks.

In this procedure, a single-stranded DNA fragment is employed as a template to guide the synthesis of new complementary DNA strands. DNA synthesis is carried out in the presence of the *deoxynucleotides* dATP, dCTP, dTTP, and dGTP, which are the normal substrates that provide the bases A, C, T, and G to growing DNA chains. Also included, at lower concentrations, are four dye-labeled

fragments (Figure 18B-1b). The restriction fragments generated by enzymes with this staggered cleavage pattern always have **sticky ends** (also called *cohesive ends*). These terms derive from the fact that the single-stranded tail at the end of each such fragment can base-pair with the tail at either end of any other fragment generated by the same enzyme, causing the fragments to stick to one another by hydrogen bonding. Enzymes that generate such fragments are particularly useful because they can be employed experimentally to create recombinant DNA molecules, as we will see in Chapter 20.

The restriction sites for most restriction enzymes are *palindromes,* which means that the sequence reads the same in either direction. (The English word *radar* is a palindrome, for example.) The palindromic nature of a restriction site is due to its twofold rotational symmetry, which means that rotating the double-stranded sequence 180° in the plane of the paper yields a sequence that reads the same as it did before rotation. Palindromic restriction sites have the same base sequence on both strands when each strand is read in the 5′→3′ direction.

The frequency with which any particular restriction site is likely to occur within a DNA molecule can be predicted statistically. For example, in a DNA molecule containing equal amounts of the four bases (A, T, C, and G), we can predict that, on average, a recognition site with four nucleotide pairs will occur once every 256 (i.e., 4^4) nucleotide pairs, whereas the likely frequency of a six-nucleotide sequence is once every 4096 (i.e., 4^6) nucleotide pairs. Restriction enzymes therefore tend to cleave DNA into fragments that typically vary in length from several hundred to a few thousand nucleotide pairs—gene-sized pieces, essentially. Such pieces are called *restriction fragments.* Because each restriction enzyme cleaves only a single, specific nucleotide sequence, it will always cut a given DNA molecule in the same predictable manner, generating a reproducible set of restriction fragments. This property makes restriction enzymes powerful tools for generating manageable-sized pieces of DNA for further study.

Table 18B-1 Some Common Restriction Enzymes and Their Recognition Sequences

Enzyme	Source Organism	Recognition Sequence*
*Bam*HI	*Bacillus amyloliquefaciens*	5′ G—G—A—T—C—C 3′ 3′ C—C—T—A—G—G 5′
*Eco*RI	*Escherichia coli*	5′ G—A—A—T—T—C 3′ 3′ C—T—T—A—A—G 5′
*Hae*III	*Hemophilus aegyptius*	5′ G—G—C—C 3′ 3′ C—C—G—G 5′
*Hin*dIII	*Hemophilus influenzae*	5′ A—A—G—C—T—T 3′ 3′ T—T—C—G—A—A 5′
*Pst*I	*Providencia stuartii* 164	5′ C—T—G—C—A—G 3′ 3′ G—A—C—G—T—C 5′
*Pvu*I	*Proteus vulgaris*	5′ C—G—A—T—C—G 3′ 3′ G—C—T—A—G—C 5′
*Pvu*II	*Proteus vulgaris*	5′ C—A—G—C—T—G 3′ 3′ G—T—C—G—A—C 5′
*Sal*I	*Streptomyces albus* G	5′ C—T—C—G—A—C 3′ 3′ C—A—G—C—T—G 5′

*The arrows within the recognition sequence indicate the points at which each restriction enzyme cuts the two strands of the DNA molecule.

dideoxynucleotides (ddATP, ddCTP, ddTTP, and ddGTP), which lack the hydroxyl group attached to the 3′ carbon of normal deoxynucleotides. When a dideoxynucleotide is incorporated into a growing DNA chain in place of the normal deoxynucleotide, *DNA synthesis is prematurely halted* because the absence of the 3′ hydroxyl group makes it impossible to form a bond with the next nucleotide. Hence, a series of incomplete DNA fragments are produced whose sizes provide information concerning the linear sequence of bases in the DNA.

Figure 18-14 illustrates how this procedure works. In step ❶, a reaction mixture is assembled that includes the dideoxynucleotides ddATP, ddCTP, ddTTP, and ddGTP, each labeled with a fluorescent dye of a different color (e.g., ddATP = red, ddCTP = blue, ddTTP = orange, and ddGTP = green). These colored dideoxynucleotides are mixed with the normal deoxynucleotide substrates for DNA synthesis, along with a single-stranded DNA molecule to be sequenced and a short, single-stranded DNA *primer* that is complementary to the 3′ end of the DNA strand being sequenced. When DNA polymerase is added, it catalyzes the attachment of nucleotides, one by one, to the 3′ end of the primer, producing a growing DNA strand that is complementary to the template whose sequence is being determined. Most of the nucleotides inserted are the normal deoxynucleotides because they are the preferred substrates for DNA polymerase. But every so often, at random, a colored dideoxynucleotide is inserted instead of its normal equivalent. Each time a dideoxynucleotide is incorporated, it halts further DNA synthesis for that particular strand. Consequently, a mixture of strands of varying lengths is generated, each containing a colored base at the end where DNA synthesis was prematurely terminated by incorporation of a dideoxynucleotide (step ❷).

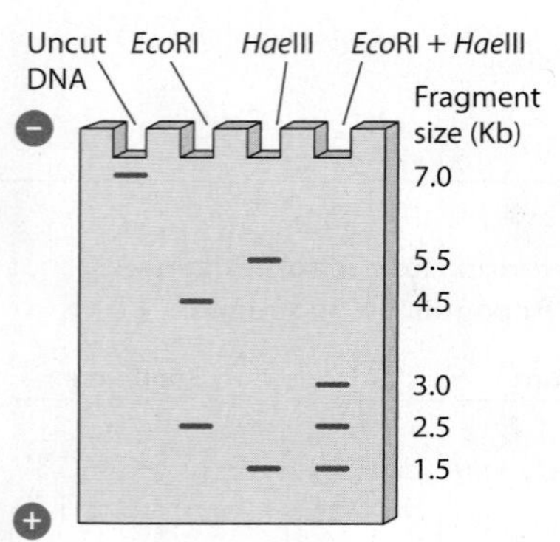

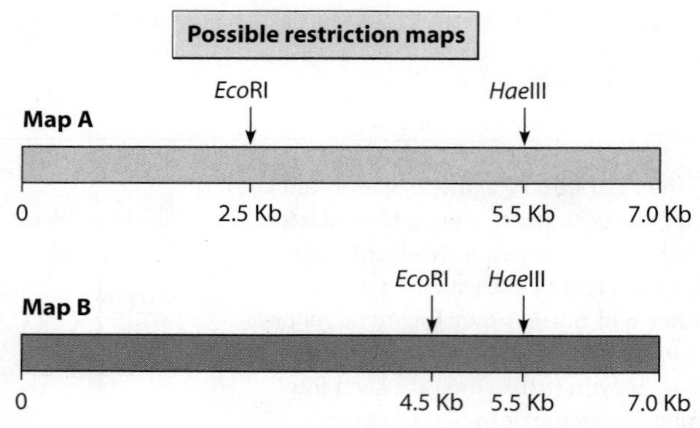

FIGURE 18-13 Restriction Mapping. In this hypothetical example, the location of restriction sites for *Eco*RI and *Hae*III is determined in a DNA fragment 7.0 Kb long. The gel on the left shows that *Eco*RI cleaves the DNA into two fragments measuring 2.5 Kb and 4.5 Kb, indicating that DNA has been cleaved at a single point located 2.5 Kb from one end. Treatment with *Hae*III cleaves the DNA into two fragments measuring 1.5 Kb and 5.5 Kb, indicating that DNA has been cleaved at a single point located 1.5 Kb from one end. Based on this information alone, two possible restriction maps can be proposed. If map A were correct, simultaneous digestion of the DNA with *Eco*RI and *Hae*III should yield three fragments measuring 3.0 Kb, 2.5 Kb, and 1.5 Kb. If map B were correct, simultaneous digestion of the DNA with *Eco*RI and *Hae*III should yield three fragments measuring 4.5 Kb, 1.5 Kb, and 1.0 Kb. The experimental data reveal that map A must be correct.

Next, the sample is subjected to electrophoresis in a polyacrylamide gel, which allows the newly synthesized DNA fragments to be separated from one another because the shorter fragments migrate through the gel more quickly than the longer fragments (step ❸). As the fragments move through the gel, a special camera detects the color of each fragment as it passes by. Step ❹ shows how such information allows the DNA base sequence to be determined. In this particular example, the shortest DNA fragment is blue, and the next shortest fragment is green. Since blue and green are the colors of ddCTP and ddGTP, respectively, the first two bases added to the primer must have been C followed by G. In automatic sequencing machines, such information is collected for hundreds of bases in a row and fed into a computer, allowing the complete sequence of the initial DNA fragment to be quickly determined.

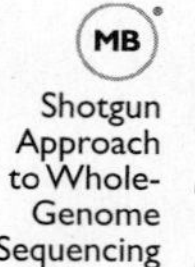

The Genomes of Many Organisms Have Been Sequenced

The significance of the technique we have just described can scarcely be overestimated. DNA sequencing is now so commonplace and automated that it is routinely applied not just to individual genes, but to entire genomes. Although DNA sequencing machines determine the sequence of only short pieces of DNA, usually 500–800 bases long, one at a time, computer programs search for overlapping sequences between such fragments and thereby allow data from hundreds or thousands of DNA pieces to be assembled into longer stretches that can reach millions of bases in length.

Many of the initial successes in genome sequencing involved bacteria because they have relatively small genomes, typically a few million bases. Complete DNA sequences are now available for over 2000 different bacteria, including those that cause a variety of human diseases. In fact, sequencing machines are so efficient that one research institute reported the sequences of 15 different bacterial genomes in a single month! But DNA sequencing has also been successfully applied to much larger genomes, including those from several dozen organisms that are most important in biological research (**Table 18-2**). For example, scientists have completed the genome sequences of the yeast *Saccharomyces cerevisiae* (12.1 million bases), the roundworm *Caenorhabditis elegans* (97 million bases), the mustard plant *Arabidopsis thaliana* (125 million bases), and the fruit fly *Drosophila melanogaster* (180 million bases).

To us as human beings, of course, the ultimate challenge of DNA sequencing is the human genome. How awesome a challenge was that? To answer this question, we need to realize that the human nuclear genome contains about 3.2 billion bases, which is roughly a thousandfold more DNA than is present in an *E. coli* cell. One way to comprehend the magnitude of such a challenge is to note that in the early 1990s, when genome sequencing efforts began in earnest, it required almost 6 years for the laboratory of Frederick Blattner to determine the complete base sequence of the *E. coli* genome. At this rate, it would have taken a single lab almost 6000 years to sequence the entire human genome! Consequently, scientists came together in 1990 to establish the *Human Genome Project,* a cooperative international effort involving hundreds of scientists who shared their data in an attempt to determine the entire sequence of the human genome. In the late 1990s a commercial company, Celera Genomics, tackled the job as well. Through these efforts,

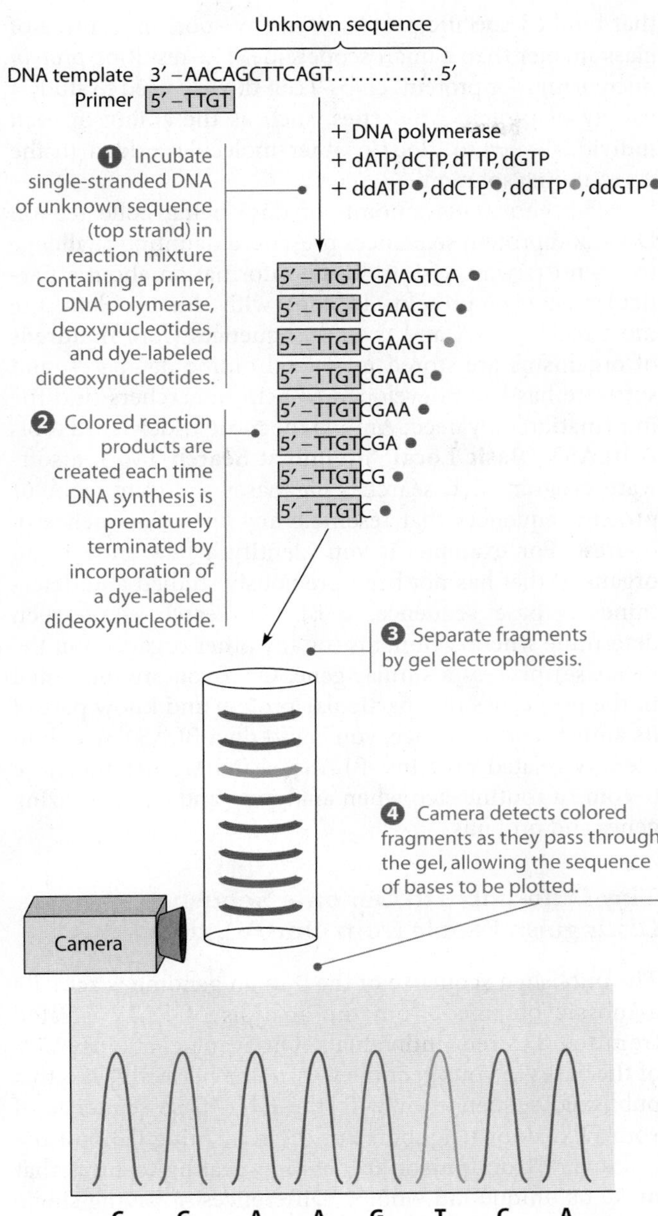

Unknown sequence
DNA template 3′–AACAGCTTCAGT...............5′
Primer 5′–TTGT
+ DNA polymerase
+ dATP, dCTP, dTTP, dGTP
+ ddATP, ddCTP, ddTTP, ddGTP
1 Incubate single-stranded DNA of unknown sequence (top strand) in reaction mixture containing a primer, DNA polymerase, deoxynucleotides, and dye-labeled dideoxynucleotides.
5′–TTGTCGAAGTCA
5′–TTGTCGAAGTC
5′–TTGTCGAAGT
5′–TTGTCGAAG
5′–TTGTCGAA
5′–TTGTCGA
5′–TTGTCG
5′–TTGTC
2 Colored reaction products are created each time DNA synthesis is prematurely terminated by incorporation of a dye-labeled dideoxynucleotide.
3 Separate fragments by gel electrophoresis.
4 Camera detects colored fragments as they pass through the gel, allowing the sequence of bases to be plotted.
Camera
C G A A G T C A

MB DNA Replication

FIGURE 18-14 DNA Sequencing. The chain termination technique illustrated here, which employs dye-labeled dideoxynucleotides, has been adapted for use in high-speed, automated sequencing machines. Although this example summarizes the results obtained for only the first eight bases of a DNA sequence, experiments of this type typically determine the sequence of DNA fragments that are 500–800 bases long. The four main steps involved in the procedure are described in more detail in the text.

the complete sequence of the human genome was determined by 2003, roughly two years ahead of schedule.

This monumental achievement took more than a decade to complete and cost nearly $3 billion. Today, continued improvements to the Sanger sequencing method have made it possible to sequence a comparable-sized genome for $20 million in less than a year, and rapid progress continues to be made. For example, a newly developed sequencing approach permits a million DNA fragments to be sequenced simultaneously in microscopic reaction chambers, thereby making it possible to sequence a human genome for less than $100,000 in a month or less. And efforts are not stopping there. A prize has been established to spur the development of a technology that can sequence a human genome for $1000—a cost so affordable that doctors may one day be able to order a copy of a patient's own genome sequence to facilitate treatment decisions best suited to the genes found in that person.

The Field of Bioinformatics Has Emerged to Decipher Genomes, Transcriptomes, and Proteomes

MB Using BLAST: Can You Identify a Pathogen from a Nucleotide Sequence?; Using BLAST: What Can a Protein Sequence Reveal about Cancer?; Genomics (1 of 3): Sequencing and Genome Databases; Genomics (3 of 3): Single Nucleotide Polymorphism Analysis and Protein Function; Solve It: Are You Getting the Fish You Paid For?

Because of its sheer scale as well as its potential impact on our understanding of human evolution, physiology, and disease, sequencing the human genome is one of the crowning achievements of modern biology. And yet, unraveling the sequence of bases was the "easy" part. Next comes the hard part: figuring out the meaning of this sequence of 3 billion A's, G's, C's, and T's. For example, which stretches of DNA correspond to genes, when and in what tissues are these genes expressed, what kinds of proteins do they code for, and how do all these proteins interact with each other and function?

The prospect of analyzing such a vast amount of data has led to the emergence of a new discipline, called **bioinformatics**, which merges computer science and biology in an attempt to make sense of it all. For example, computer programs that analyze DNA for stretches that could code for amino acid sequences are used to estimate the number of protein-coding genes. Such analyses suggest the presence of about 25,000 protein-coding genes in the human genome, roughly half of which were not known to exist prior to genome sequencing. The fascinating

Table 18-2 Examples of Sequenced Genomes*

Organism	Genome Size	Estimated Gene Number
Bacteria		
Mycoplasma genitalium	0.6 Mb	470
Haemophilus influenza	1.8 Mb	1,740
Streptococcus pneumoniae	2.2 Mb	2,240
Escherichia coli	4.6 Mb	4,400
Yeast (*S. cerevisiae*)	12.1 Mb	6,200
Roundworm (*C. elegans*)	97 Mb	19,700
Mustard plant (*A. thaliana*)	125 Mb	25,500
Fruit fly (*D. melanogaster*)	180 Mb	13,600
Rice (*O. sativa*)	389 Mb	37,500
Mouse (*Mus musculus*)	2500 Mb	25,000
Human (*H. sapiens*)	3200 Mb	25,000

*As of November 2009, complete genome sequences had been published for 2351 organisms (2060 bacteria, 80 archaea, and 211 eukaryotes).

thing about this estimate is that it means humans have only about twice the number of genes as a fruit fly, barely more genes than a worm, and 12,000 fewer genes than a rice plant! Computer analysis has also revealed that less than 2% of the human genome actually codes for proteins. While the remaining 98% contains some important regulatory elements and some genes that code for RNA products instead of proteins, much of it has no obvious function. (For examples, see the discussions of repeated DNA in the following section and introns in Chapter 21). While the significance of this extra DNA is not clear, some evidence suggests that its presence may enhance the ability of the genome to evolve over time.

Determining the DNA sequence of an organism's genome can provide only a partial understanding of the functions a genome performs. Scientists must look beyond the genome to examine the molecules it produces. Because the first step in gene expression involves transcription of genome sequences into RNA, techniques have been developed for identifying **transcriptomes**—that is, the entire set of RNA molecules produced by a genome. In Chapter 23, you will see how the development of *DNA microarray* technology for identifying thousands of RNA molecules simultaneously has facilitated the study of transcriptomes.

Most RNAs, in turn, are used to guide the production of proteins, so scientists are also studying **proteomes**—the structure and properties of every protein produced by a genome. An organism's proteome is considerably more complex than its genome. For example, the roughly 25,000 genes in human cells produce hundreds of thousands of different proteins. In Chapters 21 and 23, we explain how cells can produce so many proteins from a smaller number of genes. You will see that this ability to produce so many proteins is made possible by a mechanism called *alternative splicing,* which allows each individual gene to be "read" in multiple ways to produce multiple versions of its protein product. Moreover, the resulting proteins are subject to subsequent biochemical modifications that produce either new proteins or multiple versions of the same protein.

Identifying the vast number of proteins produced by a genome has been facilitated by **mass spectrometry,** a high-speed, extremely sensitive technique that utilizes magnetic and electric fields to separate proteins or protein fragments based on differences in mass and charge. One application of mass spectrometry involves using it to identify peptides derived from proteins that have been separated by gel electrophoresis and then digested with specific proteases, such as trypsin. Comparing the resulting data to the predicted masses of peptides that would be produced by DNA sequences present in genomic databases permits the proteins produced by newly discovered genes to be identified. Other techniques make it feasible to study the interactions and functional properties of the vast number of proteins found in a proteome. For example, it is possible to immobilize thousands of different proteins (or other molecules that bind to specific proteins) as tiny spots on a piece of glass smaller than a microscope slide. The resulting *protein microarrays* (or protein "chips") can then be used to study a variety of protein properties, such as the ability of each individual spot to bind to other molecules added to the surrounding solution.

The enormous amount of data being collected on DNA and protein sequences presents a daunting challenge to scientists who wish to locate information about a particular gene or protein. To cope with this problem, the most recent DNA and protein sequences from hundreds of organisms are stored in several online databases, and software has been developed to help researchers find the information they need. Among the more widely used tools is **BLAST (Basic Local Alignment Search Tool),** a software program that searches databases to locate DNA or protein sequences that resemble any known sequence of interest. For example, if you identify a new gene in an organism that has not been previously studied and determine its base sequence, a BLAST search could then determine whether humans (or any other organism in the database) possess a similar gene. Or if you are interested in the properties of a particular protein and know part of its amino acid sequence, you could do a BLAST search to identify related proteins. BLAST searching has therefore become a routine step when analyzing and characterizing genes and proteins.

Tiny Differences in Genome Sequence Distinguish People from One Another

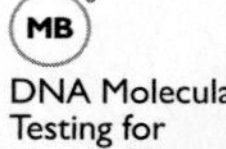

DNA Molecular Testing for Human Disease Gene Mutation

The published sequence of the human genome is actually a mosaic obtained from the analysis of DNA isolated from ten different individuals. On average, about 99.7% of the bases in your genome will match perfectly with this published sequence, or with the DNA base sequence of your next-door neighbor. But the remaining 0.3% of the bases vary from person to person, creating features that make us unique individuals. Differences involving single base changes are called **single nucleotide polymorphisms,** or **SNPs** (pronounced "snips"). Although 0.3% might not sound like very much, 0.3% multiplied by the 3.2 billion bases in the human genome yields a total of roughly 10 million SNPs. Scientists have already created databases containing most of the common SNPs, which are thought to be important because some of these tiny genetic variations may influence your susceptibility to certain diseases or determine how well you respond to a particular treatment.

Most SNPs, however, are not located in the protein-coding regions of genes. So how do we find out which SNPs are related to important traits, such as susceptibility to a specific disease? Fortunately, it is not necessary to examine all 10 million SNPs separately because SNPs are not independent of one another. SNPs located near each other on the same chromosome tend to be inherited together in blocks called **haplotypes.** A database of these haplotypes, called the *HapMap*, provides a shortcut for scientists interested in the relationship between genes and disease: Only a few hundred thousand SNPs (each located in a different haplotype) need to be examined rather than 10 million. Once a trait has been linked to a particular haplotype, only the SNPs within that haplotype are studied to determine which one is responsible.

SNPs are not the only sources of genetic variation that define a person's individuality. DNA rearrangements, deletions, and duplications also contribute to variability among genomes. Such mechanisms have produced DNA segments thousands of bases long that are present in variable numbers of copies among different individuals. Each person's genome is thought to contain hundreds of such **copy number variations (CNVs)** involving millions of bases of DNA overall.

The impact of our rapidly growing understanding of the human genome is already becoming apparent as discoveries regarding the genetic basis of many human diseases—from breast cancer and colon cancer to diabetes and Alzheimer's disease—are being reported at a rapidly increasing pace. Such discoveries promise to revolutionize the future practice of medicine because having the ability to identify disease genes and investigate their function makes it possible to devise medical interventions for alleviating and even preventing disease.

But being able to identify potentially harmful genes also raises ethical concerns because all of us are likely to carry a few dozen genes that place us at risk for something. Such information possibly could be misused—for example, in genetic discrimination against individuals or groups of people by insurance companies, employers, or even government agencies. Moreover, detailed knowledge of the human genome increases the potential for using recombinant DNA techniques (see Chapter 20) to *alter* people's genes, not only to correct diseases in malfunctioning body tissues but also to change genes in sperm and eggs, thereby altering the genetic makeup of future generations. What use to make of these abilities and how to regulate them are clearly questions that concern not only the scientific community but human society as a whole.

Repeated DNA Sequences Partially Explain the Large Size of Eukaryotic Genomes

Besides the difficulties it created for DNA sequencing studies, the enormous size of the human genome raises a more fundamental question: Does the large amount of DNA in human cells simply reflect the need for thousands of times more genes than are present in bacterial cells, or are other factors at play as well? The first breakthrough in answering this question occurred in the late 1960s, when DNA renaturation studies carried out by Roy Britten and David Kohne led to the discovery of repeated DNA sequences.

In these experiments, DNA was broken into small fragments and dissociated into single strands by heating. The temperature was then lowered to permit the single-stranded fragments to renature. The renaturation rate depends on the concentration of each individual kind of DNA sequence; the higher the concentration of fragments exhibiting any given DNA sequence, the greater the probability that they will randomly collide with complementary strands and reassociate. As an example, let us consider DNA from a bacterial cell and from a typical mammalian cell containing a thousandfold more DNA. If this difference in DNA content reflects a thousandfold difference in the kinds of DNA sequences present, then bacterial DNA should renature 1000 times faster than mammalian DNA. The rationale for this prediction is that any particular DNA sequence should be present in a thousandfold lower concentration in the mammalian DNA sample because there are a thousand times more kinds of sequences present, and so each individual sequence represents a smaller fraction of the total population of sequences.

When Britten and Kohne compared the behavior of mammalian and bacterial DNAs, however, the results were not exactly as expected. In **Figure 18-15**, which summarizes their data obtained for calf and *E. coli* DNA, renaturation is plotted as a function of initial DNA concentration multiplied by the elapsed time because it facilitates comparison of data obtained from reactions run at different DNA concentrations. When graphed this way, the data reveal that calf DNA consists of two classes of sequences that renature at very different rates. One type of sequence, which accounts for about 40% of the calf DNA, renatures *more* rapidly than bacterial DNA does. The most straightforward explanation for this unexpected result is that calf DNA contains **repeated DNA** sequences that are present in multiple copies. The existence of multiple copies increases the concentration of such sequences, thereby generating more collisions and a faster rate of reassociation than would be expected if each sequence were present in only a single copy.

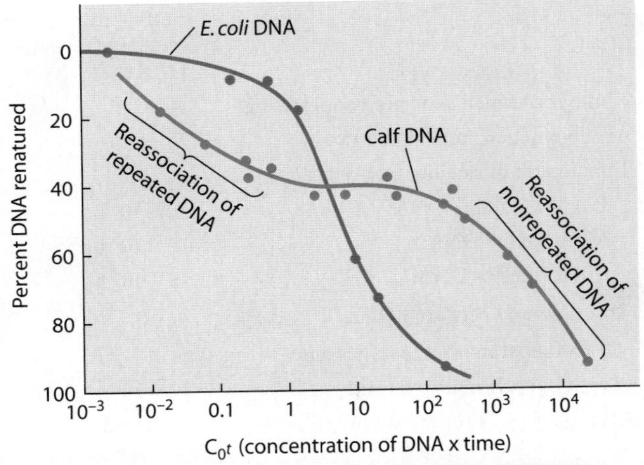

FIGURE 18-15 Renaturation of Calf and *E. coli* DNAs. The calf DNA that reassociates more rapidly than the bacterial DNA consists of repeated sequences.

The remaining 60% of the calf DNA renatures about a thousand times more slowly than *E. coli* DNA does, which is the behavior expected of sequences present as single copies. This fraction is therefore called **nonrepeated DNA** to distinguish it from the repeated sequences that renature more quickly. Nonrepeated DNA sequences are each present in one copy per genome. Most protein-coding genes consist of nonrepeated DNA, although this does not mean that all nonrepeated DNA codes for proteins.

In bacterial cells virtually all the DNA is nonrepeated, whereas eukaryotes exhibit large variations in their amounts of repeated versus nonrepeated DNA. This helps explain the mystery of the seemingly excess amount of DNA in species such as *Trillium:* This organism contains a relatively large amount of repeated DNA. Using the sequencing techniques described earlier, researchers have been able to determine the base sequences of various types of repeated DNAs and to classify them into two main categories: *tandemly repeated DNA* and *interspersed repeated DNA* (**Table 18-3**).

MB Genomics (2 of 3): Sequence Annotation and Human Genomic Variation

Tandemly Repeated DNA. One major category of repeated DNA is referred to as **tandemly repeated DNA** because the multiple copies are arranged next to each other in a row—that is, tandemly. Tandemly repeated DNA accounts for 10–15% of a typical mammalian genome and consists of many different types of DNA sequences that vary in both the length of the basic repeat unit and the number of times this unit is repeated in succession. The length of the repeated unit can measure anywhere from 1 to 2000 bp or so. Most of the time, however, the repeated unit is shorter than 10 bp; consequently, this subcategory is called *simple-sequence repeated DNA.* Here is an example (showing one strand only) of a simple-sequence repeated DNA built from the five-base unit, GTTAC:

. . . GTTACGTTACGTTACGTTACGTTAC . . .

The number of sequential repetitions of the GTTAC unit can be as high as several hundred thousand at selected sites in the genome.

Tandemly repeated DNA of the simple-sequence type was originally called *satellite DNA* because its distinctive base composition often causes it to appear in a "satellite" band that separates from the rest of the genomic DNA during centrifugation procedures designed to separate molecules by density. This difference in density arises because adenine and guanine differ slightly in molecular weight, as do cytosine and thymine; hence, the densities of DNAs with differing base compositions will differ. In the procedures that reveal satellite bands, the genomic DNA is cleaved to short lengths, thus allowing DNA segments of differing densities to migrate freely to different positions during centrifugation.

What is the function of simple-sequence repeated DNA (satellite DNA)? Because such sequences are not usually transcribed, it has been proposed that they may instead be responsible for imparting special physical properties to certain regions of the chromosome. In most eukaryotes, chromosomal regions called **centromeres**—which play an important role in chromosome distribution during cell division (see Chapter 19)—are particularly rich in simple-sequence repeats, and these sequences may impart specialized structural properties to the centromere. **Telomeres,** which are DNA sequences located at the ends of chromosomes, also have simple-sequence repeats. In the next chapter, we will learn how telomeres protect chromosomes from degradation at their vulnerable ends during each round of replication (see Figure 19-16). Human telomeres contain 250–1500 copies of the sequence TTAGGG, which has been highly conserved over hundreds of millions of years of evolution. All vertebrates studied so far have this same identical sequence, and even unicellular eukaryotes possess similar sequences. Apparently, such sequences are critical to the survival of these organisms.

The amount of satellite DNA present at any given site can vary enormously. Typical satellite DNAs usually range from 10^5 to 10^7 bp in overall length. The term *minisatellite* DNA refers to shorter regions, about 10^2 to 10^5 bp in total length, composed of a tandem repeat unit of roughly 10–100 bp. *Microsatellite* DNAs, in which the repeat unit is only 1–10 bp, are even shorter (about 10–100 bp in length), although numerous sites in the genome may exhibit the same sequence. The short repeated sequences found in microsatellite and minisatellite DNAs are extremely useful in the laboratory for **DNA fingerprinting.** This procedure,

Table 18-3 Categories of Repeated Sequences in Eukaryotic DNA

I. Tandemly repeated DNA, including simple-sequence repeated DNA (satellite DNA)	
10–15% of most mammalian genomes is this type of DNA	
Length of each repeated unit:	1–2000 bp; typically 5–10 bp for simple-sequence repeated DNA
Number of repetitions per genome:	10^2–10^5
Arrangement of repeated units:	Tandem
Total length of satellite DNA at each site:	
Regular satellite DNA:	10^5–10^7 bp
Minisatellite DNA:	10^2–10^5 bp
Microsatellite DNA:	10^1–10^2 bp
II. Interspersed repeated DNA	
25–50% of most mammalian genomes is this type of DNA	
Length of each repeated unit:	10^2–10^4 bp
Arrangement of repeated units:	Scattered throughout the genome
Number of repetitions per genome:	10^1–10^6; "copies" not identical

described more fully in **Box 18C**, uses gel electrophoresis to compare DNA fragments derived from various regions of the genomes of two or more individuals. It is a means of identifying individuals that is as accurate as conventional fingerprinting.

Medical researchers have made the surprising discovery that more than a dozen inherited diseases of the nervous system involve simple changes in microsatellite DNAs. More specifically, these diseases are traceable to excessive numbers of repeated trinucleotide sequences within an otherwise normal gene. An example of this phenomenon, called *triplet repeat amplification,* is found in *Huntington's disease,* a devastating neurological disease that strikes in middle age and is invariably fatal. The normal version of the Huntington's gene contains the trinucleotide CAG tandemly repeated 11–34 times. The genes of afflicted individuals, however, possess up to 100 copies of the repeated unit. Neurological diseases resulting from the amplification of other triplet repeat sequences include *fragile X syndrome,* which is a major cause of mental retardation, and *myotonic dystrophy,* which affects the muscles. For some of these diseases, the repeated sequence is in a region of the affected gene that is not translated; for others, it is translated into a polypeptide segment consisting of a long string of the same amino acid. In both cases, the severity of the disease appears to correlate with the number of triplet repeats.

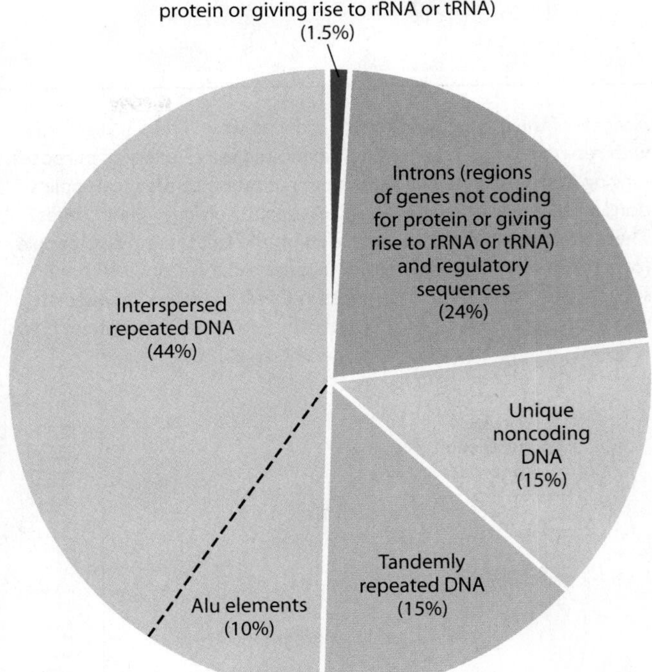

FIGURE 18-16 Types of DNA in the Human Genome. The DNA that encodes proteins or that is transcribed into ribosomal or transfer RNA accounts for only 1.5% of the DNA in the human genome. Interspersed repeat DNA accounts for 44% of the human genome; almost one quarter of this consists of *Alu* elements. Tandemly repeated DNA accounts for another 15% of the genome.

Interspersed Repeated DNA. The other main type of repeated DNA is **interspersed repeated DNA.** Rather than being clustered in tandem arrangements, the repeated units of this type of DNA are scattered around the genome. A single repeat unit tends to be hundreds or even thousands of base pairs long, and its dispersed "copies," which may number in the hundreds of thousands, are similar but usually not identical to one another. Interspersed repeated DNA typically accounts for 25–50% of mammalian genomes.

MB Insertion Sequences in Bacteria

Most interspersed repeated DNA consists of families of **transposable elements (transposons),** also known as "jumping genes" because they can move around the genome and leave copies of themselves wherever they stop. Remarkably, in humans, roughly half the genome consists of these mobile elements. The most abundant, called **LINEs (long interspersed nuclear elements),** measure 6000–8000 bp in length and account for roughly 20% of the genome. LINEs possess genes coding for enzymes involved in copying LINE sequences (and other mobile elements) and inserting the copies elsewhere in the genome. Another class of mobile elements, called **SINEs (short interspersed nuclear elements),** consist of short repeated sequences less than 500 bp in length that do not contain genes, relying instead on enzymes made by other mobile elements for their movement. The most common SINEs in humans measure about 300 bp in length and are called *Alu sequences,* because the first ones identified all contained a restriction site for the restriction enzyme *Alu*I. Close to a million copies of the *Alu* sequence are spread throughout the human genome, accounting for approximately 10% of the DNA. DNA that encodes proteins or gives rise to ribosomal or transfer RNAs (such DNA is found in *exons,* as we will see in Chapter 21) accounts for surprisingly little of the DNA in the human genome (**Figure 18-16**).

The mobility of LINEs, SINEs, and other types of transposable elements is thought to create genomic variability that contributes to the evolutionary adaptability of organisms. In addition, the movement of some LINEs can occur during normal development and alter the expression of adjacent genes, suggesting a possible role in gene regulation. The origins of transposable elements and the mechanisms by which they move will be discussed more fully in Chapter 21.

DNA Packaging

Cells must accommodate an awesome amount of DNA, even in species with modestly sized genomes. For example, the typical *E. coli* cell measures about 1 μm in diameter and 2 μm in length, yet it must accommodate a (circular) DNA molecule with a length of about 1600 μm—enough DNA to encircle the cell more than 400 times! Eukaryotic cells face an even greater challenge. A human cell of average size

DNA Fingerprinting

Analysis of the fragment patterns produced when DNA is digested with restriction enzymes has been exploited for a variety of purposes, ranging from research into genome organization to practical applications such as diagnosing genetic diseases and solving violent crimes. These practical applications are based on the fact that no two people (other than identical twins) have the same exact set of DNA base sequences. Although the differences in DNA sequence between any two people are quite small, they alter the lengths of some of the DNA fragments produced by restriction enzymes. These differences in fragment length, called **restriction fragment length polymorphisms (RFLPs)**, can be analyzed by gel electrophoresis. The resulting pattern of fragments serves as a "fingerprint" that identifies the individual from whom the DNA was obtained.

In practical usage, *DNA fingerprinting* is performed in a way that examines only a small, selected subset of restriction fragments. To illustrate this point, **Figure 18C-1** summarizes

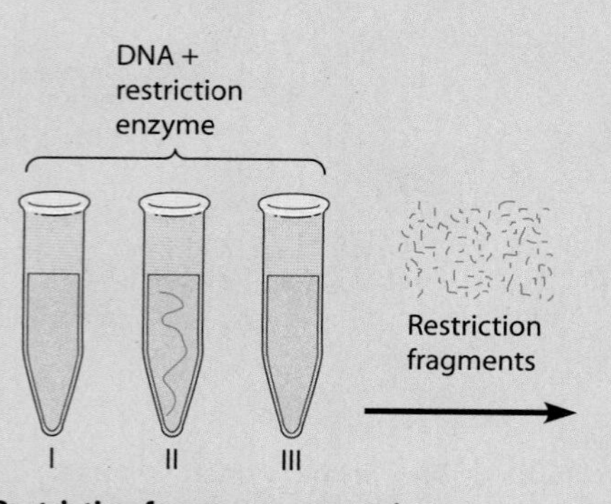

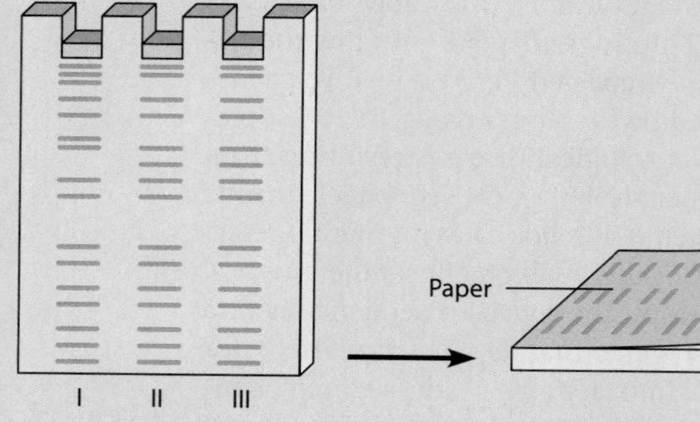

❶ **Restriction fragment preparation.** DNA is extracted from white blood cells taken from individuals I, II, and III. A restriction enzyme is added to the three samples of DNA to produce restriction fragments.

❷ **Gel electrophoresis.** The mixtures of restriction fragments from each sample are separated by electrophoresis. Each sample forms a characteristic pattern of bands. (There would be many more bands than are shown here.)

❸ **Blotting.** After the DNA on the gel is denatured by raising the pH, the single strands are transferred onto special paper by blotting.

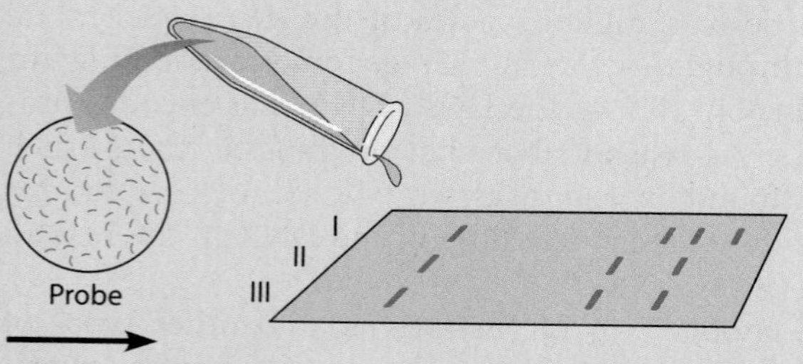

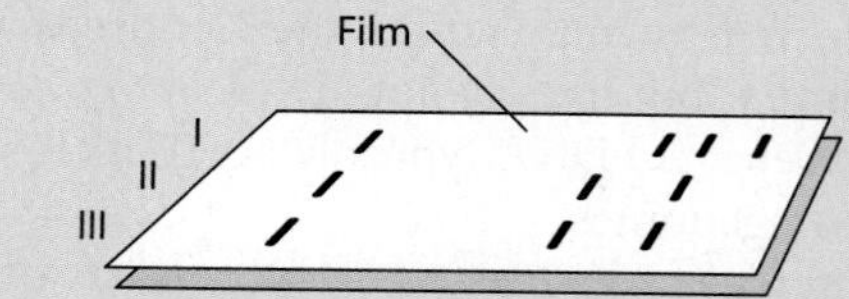

❹ **Radioactive probe.** A solution of radioactive probe is added to the paper blot. The probe is a single-stranded DNA molecule that is complementary to the DNA of interest. The probe attaches only to bands containing complementary DNA, by base pairing.

❺ **Autoradiography.** After the excess probe is rinsed off, a sheet of photographic film is laid over the paper blot. The radioactivity in the bound probe exposes the film to form an image corresponding to specific DNA bands—the bands containing DNA that base-pairs with the probe.

MB DNA Fingerprinting; Video Tutor Session Quiz: DNA Profiling

FIGURE 18C-1 DNA Fingerprinting by RFLP Analysis.

contains enough DNA to wrap around the cell more than 15,000 times. Somehow, all of this DNA must be efficiently packaged into cells and still be accessible to the cellular machinery for both DNA replication and the transcription of specific genes. Clearly, DNA packaging is a challenging problem for all forms of life. We will look first at how bacteria accomplish this task of organizing their DNA and then consider how eukaryotes address the same problem.

Bacteria Package DNA in Bacterial Chromosomes and Plasmids

The genome of a bacterium such as *E. coli* was once thought to be a "naked" DNA molecule lacking any elaborate organization and having only trivial amounts of protein associated with it. We now know that the organization of the bacterial genome is more like the

how DNA fingerprinting might be used to determine whether individuals carry a particular disease-causing gene even though they may not yet exhibit symptoms of the disease. In this example, imagine that individuals I, II, and III are members of a family in which the disease-causing gene is common. It is known that individual I carries the defective gene but individual II does not, and we want to determine the genetic status of their child, individual III.

The key step in DNA fingerprinting is called **Southern blotting** (after E. M. Southern, who developed it in 1975), but the fingerprinting procedure actually involves several distinct steps. In ❶ of Figure 18C-1, DNA obtained from the three individuals is digested with a restriction enzyme. In ❷, the resulting fragments are then separated from each other by gel electrophoresis. Because each person's DNA represents an entire genome, hundreds of thousands of bands would appear at this stage if all were made visible. Here is where Southern's stroke of brilliance comes in, with a technique that enables us to locate particular DNA sequences of interest within a complex mixture. In the Southern blotting process (❸), a special kind of "blotter" paper (nitrocellulose or nylon) is pressed against the completed gel, allowing the separated DNA fragments to be transferred to the paper. In ❹, a radioactive *probe* is added to the blot. The probe is simply a radioactive single-stranded DNA (or RNA) molecule whose base sequence is complementary to the DNA of interest—in this case, the DNA of the disease-causing gene. The probe binds to complementary DNA sequences by base pairing (the same process that occurs when denatured DNA renatures). In ❺, the bands that bind the radioactive probe are made visible by autoradiography. The results indicate that the child's version of the gene matches that of the *healthy* parent, individual II.

The autoradiogram in **Figure 18C-2** illustrates the use of DNA fingerprinting in a murder case. Here, RFLP analysis makes it clear that blood found on the defendant's clothes came from the victim, strongly implicating the defendant in the murder. Another practical application of DNA fingerprinting in the legal area is the determination of paternity or maternity.

For many DNA fingerprinting applications, scientists now analyze the length of short repeated sequences (microsatellite DNAs) called **short tandem repeats (STRs).** The STR sites chosen for DNA fingerprinting vary in length from person to person because of differences in the number of times the basic repeat unit is sequentially repeated. As a result, STR patterns in a person's DNA can be used to identify that individual uniquely. For example, the DNA of three suspects in a murder case might all have different numbers of copies of the repeat sequence CTG at a particular place in the genome; one might have 19 copies, another 21 copies, and the third 32 copies. In criminal cases, the numbers of repeat copies present at 13 different STR sites in the genome are routinely examined. The chance that any two unrelated individuals would exhibit the same exact profile at all 13 sites is roughly one in a million billion.

The usefulness of DNA fingerprinting is further enhanced by the *polymerase chain reaction (PCR)*, a technique for making multiple DNA copies that will be described in Chapter 19. Starting with DNA isolated from a single cell, the PCR reaction can be used to selectively synthesize millions of copies of any given DNA sequence within a few hours, easily producing enough DNA for fingerprinting analysis of the 13 STR sites. In this way, a few skin cells left on a pen or car keys touched by a person may yield enough DNA to uniquely identify that individual.

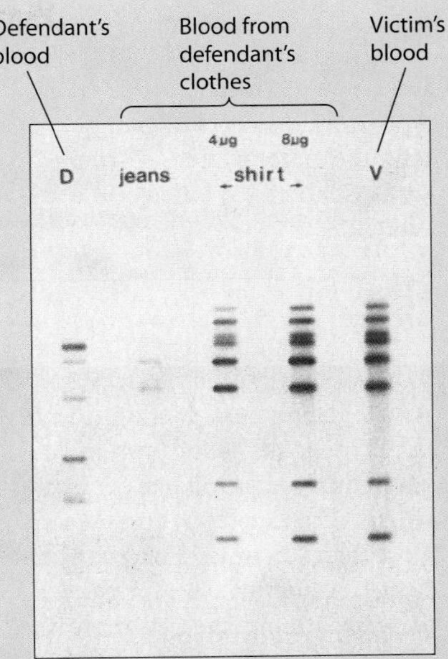

FIGURE 18C-2 DNA Fingerprints from a Murder Case. DNA was isolated from bloodstains on the defendant's clothes and compared, by RFLP analysis, with DNA from the defendant and DNA from the victim. The band pattern for the bloodstain DNA matches that for the victim, showing that the blood on the defendant's clothes came from the victim. (Courtesy of Cellmark Diagnostics.)

chromosomes of eukaryotes than we previously realized. Bacterial geneticists therefore refer to the structure that contains the main bacterial genome as the **bacterial chromosome.**

Bacterial Chromosomes. Bacteria can have single or multiple, circular or linear chromosomes; the most common arrangement, however, is a single circular DNA molecule that is bound to small amounts of protein and localized to a special region of the bacterial cell called the **nucleoid** (**Figure 18-17**). Although the nucleoid is not surrounded by a membrane, the bacterial DNA residing in this region forms a threadlike mass of fibers packed together in a way that maintains a distinct boundary between the nucleoid and the rest of the cell. The DNA of the bacterial chromosome is negatively

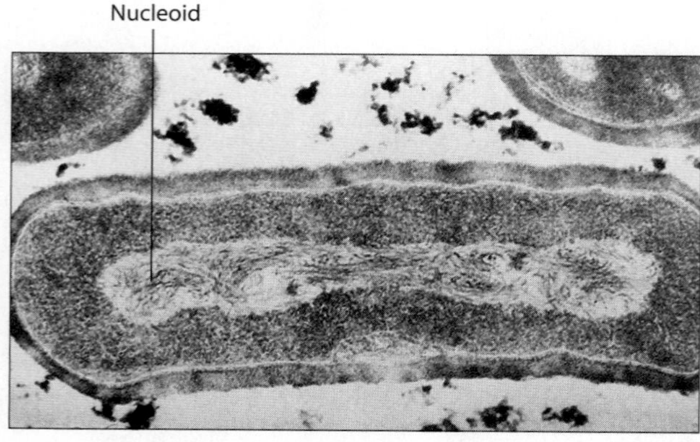

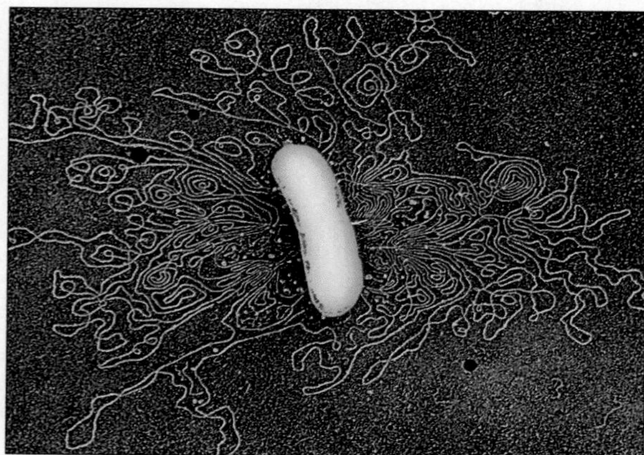

FIGURE 18-17 The Bacterial Nucleoid. The electron micrograph at the top shows a bacterial cell with a distinct nucleoid, in which the bacterial chromosome resides. When bacterial cells are ruptured, their chromosomal DNA is released from the cell. The bottom micrograph shows that the released DNA forms a series of loops that remain attached to a structural framework within the nucleoid (TEMs).

supercoiled and folded into an extensive series of loops averaging about 20,000 bp in length. Because the two ends of each loop are anchored to structural components that lie within the nucleoid, the supercoiling of any individual loop can be altered without influencing the supercoiling of adjacent loops.

The loops are thought to be held in place by RNA and protein molecules. Evidence for a structural role for RNA in the bacterial chromosome has come from studies showing that treatment with ribonuclease, an enzyme that degrades RNA, releases some of the loops, although it does not relax the supercoiling. Nicking the DNA with a topoisomerase, on the other hand, relaxes the supercoiling but does not disrupt the loops. The supercoiled DNA that forms each loop is organized into beadlike packets containing small, basic protein molecules, analogous to the histones of eukaryotic cells (discussed below). Current evidence suggests that the DNA molecule is wrapped around particles of the basic protein. Thus, from what we know so far, the bacterial chromosome consists of supercoiled DNA that is bound to small, basic proteins and then folded into looped domains.

Bacterial Plasmids. In addition to its chromosome, a bacterial cell may contain one or more plasmids. **Plasmids** are relatively small, usually circular molecules of DNA that carry genes both for their own replication and, often, for one or more cellular functions (usually nonessential ones). Most plasmids are supercoiled, giving them a condensed, compact form. Although plasmids replicate autonomously, the replication is usually sufficiently synchronized with the replication of the bacterial chromosome to ensure a roughly comparable number of plasmids from one cell generation to the next. In *E. coli* cells, several classes of plasmids are recognized: *F (fertility) factors* are involved in the process of conjugation, a sexual process we will discuss in Chapter 20; *R (resistance) factors* carry genes that impart drug resistance to the bacterial cell; *col (colicinogenic) factors* allow the bacterium to secrete *colicins,* compounds that kill other bacteria lacking the col factor; *virulence factors* enhance the ability to cause disease by producing toxic proteins that cause tissue damage or enzymes that allow the bacteria to enter host cells; and *metabolic plasmids* produce enzymes required for certain metabolic reactions. Some strains of *E. coli* also possess *cryptic plasmids,* which have no known function and possess no genes other than those needed for the plasmid to replicate and spread to other cells.

Eukaryotes Package DNA in Chromatin and Chromosomes

When we turn from bacteria to eukaryotes, DNA packaging becomes more complicated. First, substantially larger amounts of DNA are involved. Each eukaryotic chromosome contains a single, linear DNA molecule of enormous size. In human cells, for example, just *one* of these DNA molecules may be 10 cm or more in length—roughly a hundred times the size of the DNA molecule found in a typical bacterial chromosome. Second, greater structural complexity is introduced by the association of eukaryotic DNA with greater amounts and numbers of proteins. When bound to these proteins, the DNA is converted into **chromatin** fibers 10–30 nm in diameter that are normally dispersed throughout the nucleus. At the time of cell division (and in a few other situations), these fibers condense and fold into much larger, compact structures that become recognizable as individual **chromosomes.**

The proteins with the most important role in chromatin structure are the **histones,** a group of relatively small proteins whose high content of the amino acids lysine and arginine gives them a strong positive charge. The binding of histones to DNA, which is negatively charged, is therefore stabilized by ionic bonds. In most cells, the mass of histones in chromatin is approximately equal to the mass of DNA. Histones are divided into five main types, designated H1, H2A, H2B, H3, and H4. Chromatin contains roughly equal numbers of H2A, H2B, H3, and H4 molecules, and about half that number of H1 molecules. These proportions are remarkably constant among different kinds of eukaryotic cells, regardless of the type of cell or its physiological state. In addition to histones, chromatin contains a diverse group of *nonhistone proteins* that play a variety of enzymatic, structural, and regulatory roles.

Nucleosomes Are the Basic Unit of Chromatin Structure

The DNA contained within a typical nucleus would measure a meter or more in length if it were completely extended, whereas the nucleus itself is usually no more than 5–10 μm in diameter. The folding of such an enormous length of DNA into a nucleus that is almost a million times smaller presents a significant topological problem. One of the first insights into the folding process emerged in the late 1960s, when X-ray diffraction studies carried out by Maurice Wilkins revealed that purified chromatin fibers have a repeating structural subunit that is seen in neither DNA nor histones alone. Wilkins therefore concluded that histones impose a repeating structural organization upon DNA. A clue to the nature of this structure was provided in 1974, when Ada Olins and Donald Olins published electron micrographs of chromatin fibers isolated from cells in a way that avoided the harsh solvents used in earlier procedures for preparing chromatin for microscopic examination. Chromatin fibers viewed in this way appear as a series of tiny particles attached to one another by thin filaments. This "beads-on-a-string" appearance led to the suggestion that the beads consist of protein (presumably histones) and the thin filaments connecting the beads correspond to DNA. We now refer to each bead, along with its associated short stretch of DNA, as a **nucleosome** (**Figure 18-18**).

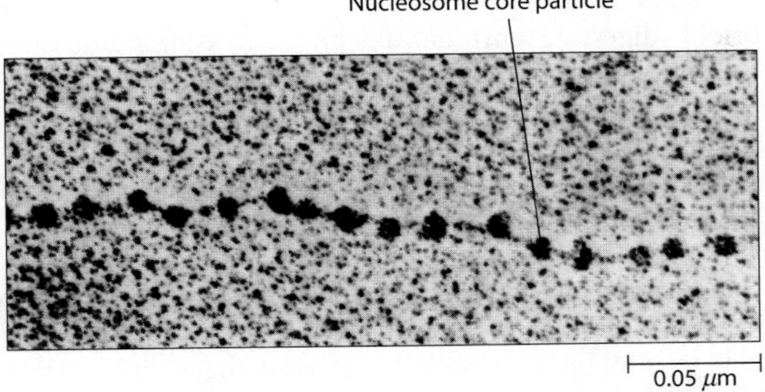

FIGURE 18-18 Nucleosomes. The core particles of nucleosomes appear as beadlike structures spaced at regular intervals along eukaryotic chromatin fibers. A nucleosome is defined to include both a core particle and the stretch of DNA that connects to the next core particle (TEM).

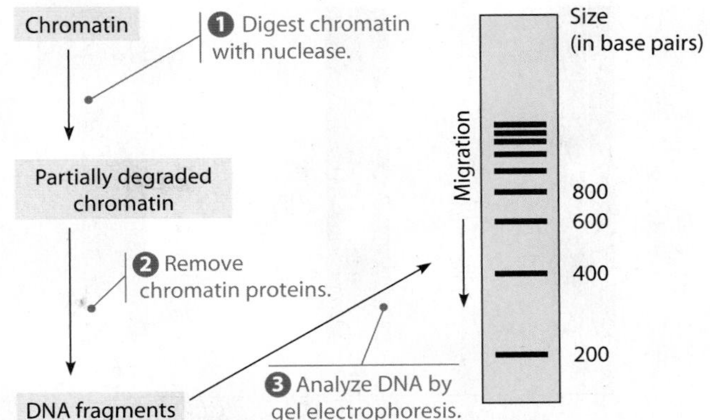

FIGURE 18-19 Evidence That Proteins Are Clustered at 200 Base-Pair Intervals Along the DNA Molecule in Chromatin Fibers. In these experiments, DNA fragments generated by nuclease digestion of rat liver chromatin were analyzed by gel electrophoresis. The discovery that the DNA fragments are multiples of 200 base pairs suggests that histones are clustered at 200 base-pair intervals along the DNA, thereby conferring a regular pattern of protection against nuclease digestion.

Based on electron microscopy alone, it would have been difficult to determine whether nucleosomes are a normal component of chromatin or an artifact generated during sample preparation. Fortunately, independent evidence for the existence of a repeating structure in chromatin was reported at about the same time by Dean Hewish and Leigh Burgoyne, who discovered that rat liver nuclei contain a nuclease that is capable of cleaving the DNA in chromatin fibers. In one crucial set of experiments, these investigators exposed chromatin to this nuclease and then purified the partially degraded DNA to remove chromatin proteins. Upon examining the purified DNA by gel electrophoresis, they found a distinctive pattern of fragments in which the smallest piece of DNA measured about 200 bp in length, and the remaining fragments were exact multiples of 200 bp (**Figure 18-19**). Since nuclease digestion of protein-free DNA does not generate this fragment pattern, they concluded that (1) chromatin proteins are clustered along the DNA molecule in a regular pattern that repeats at intervals of roughly 200 bp, and (2) the DNA located between these protein clusters is susceptible to nuclease digestion, yielding fragments that are multiples of 200 bp in length.

These observations raised the question of whether the protein clusters postulated to occur at 200-bp intervals correspond to the spherical particles observed in electron micrographs of chromatin fibers. Answering this question required a combination of the nuclease digestion and electron microscopic approaches. In these studies, chromatin was briefly exposed to *micrococcal nuclease*—a bacterial enzyme that, like the rat liver nuclease, cleaves chromatin DNA at intervals of 200 bp. The fragmented chromatin was then separated into fractions of varying sizes by centrifugation and examined by electron microscopy. The smallest fraction was found to contain single spherical

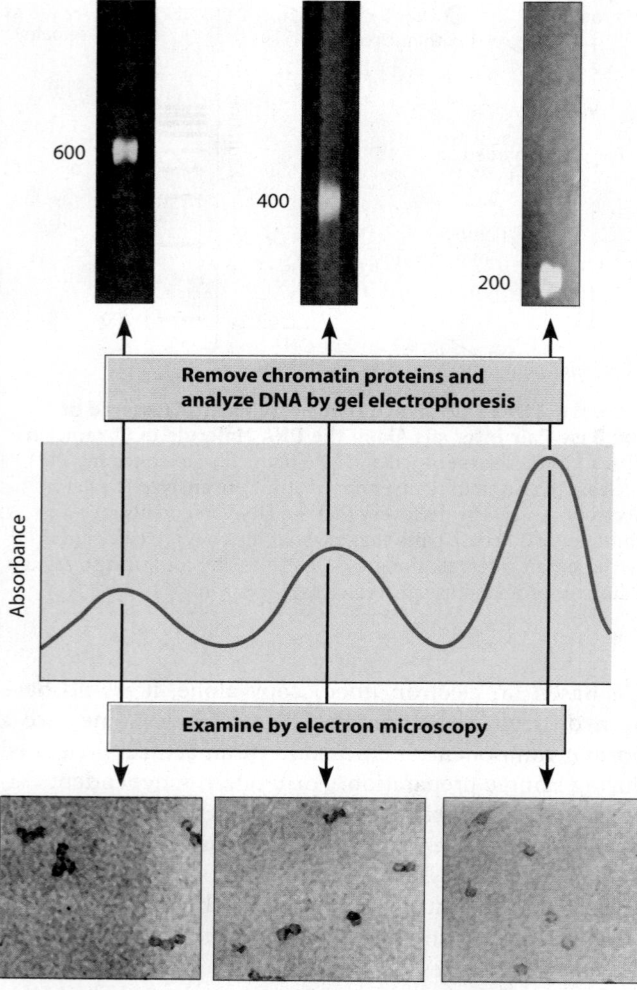

FIGURE 18-20 Evidence for the Existence of Nucleosomes. Chromatin that had been partially degraded by treatment with micrococcal nuclease was fractionated by density gradient centrifugation *(center graph)*. The individual peaks were then analyzed both by electron microscopy *(bottom)* and by gel electrophoresis after removal of chromatin proteins *(top)*. The peak on the right consists of single protein particles associated with 200 base pairs of DNA, the middle peak consists of clusters of two particles associated with 400 base pairs of DNA, and the peak on the left consists of clusters of three particles associated with 600 base pairs of DNA. This indicates that the basic repeat unit in chromatin is a protein particle associated with 200 base pairs of DNA.

particles, the next fraction contained clusters of two particles, the succeeding fraction contained clusters of three particles, and so forth (**Figure 18-20**). When DNA was isolated from these fractions and analyzed by gel electrophoresis, the DNA from the fraction containing single particles measured 200 bp in length, the DNA from the fraction containing clusters of two particles measured 400 bp in length, and so on. It was therefore concluded that the spherical particles observed in electron micrographs are each associated with 200 bp of DNA. This basic repeat unit, containing an average of 200 bp of DNA associated with a protein particle, is the nucleosome.

A Histone Octamer Forms the Nucleosome Core

The first insights into the molecular architecture of the nucleosome emerged from the work of Roger Kornberg, who was awarded a Nobel Prize in 2006 for a series of fundamental discoveries concerning DNA packaging and transcription in eukaryotes. In their early studies, Kornberg and his colleagues showed that chromatin fibers composed of nucleosomes can be generated by combining purified DNA with a mixture of all five histones. However, when they attempted to use individually purified histones, they discovered that nucleosomes could be assembled only when histones were isolated using gentle techniques that left histone H2A bound to histone H2B, and histone H3 bound to histone H4. When these H3–H4 and H2A–H2B complexes were mixed with DNA, chromatin fibers exhibiting normal nucleosomal structure were reconstituted. Kornberg therefore concluded that histone H3–H4 and H2A–H2B complexes are an integral part of the nucleosome.

To investigate the nature of these histone interactions in more depth, Kornberg and his colleague, Jean Thomas, treated isolated chromatin with a chemical reagent that forms covalent crosslinks between protein molecules that are located next to each other. After being treated with this reagent, the chemically crosslinked proteins were isolated and analyzed by polyacrylamide gel electrophoresis. Protein complexes the size of eight histone molecules were prominent in such gels, suggesting that the nucleosomal particle contains an *octamer* of eight histones. Given the knowledge that histones H3–H4 and histones H2A–H2B each form tight complexes and that these four histones are present in roughly equivalent amounts in chromatin, Kornberg and Thomas proposed that histone octamers are created by joining together two H2A–H2B dimers and two H3–H4 dimers and that the DNA double helix is then wrapped around the resulting octamer (**Figure 18-21**).

One issue not addressed by the preceding model concerns the significance of histone H1, which is not part of the octamer. If individual nucleosomes are isolated by briefly digesting chromatin with micrococcal nuclease, histone H1 is still present (along with the four other histones and 200 bp of DNA). When digestion is carried out for longer periods, the DNA fragment is further degraded until it reaches a length of about 146 bp; during the final stages of the digestion process, histone H1 is released. The remaining particle, consisting of a histone octamer associated with 146 bp of DNA, is referred to as a *core particle*. The DNA that is degraded during digestion from 200 to 146 bp in length is referred to as *linker DNA* because it joins one nucleosome to the next (Figure 18-21). Since histone H1 is released upon degradation of the linker DNA, histone H1 molecules are thought to be associated with the linker region. The length of the linker DNA varies somewhat among organisms, but the DNA

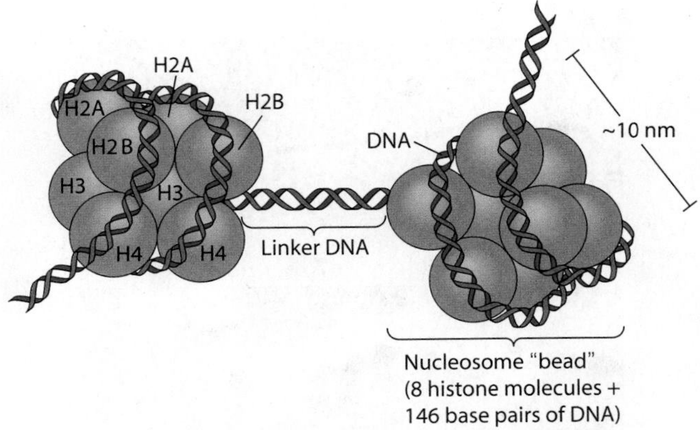

FIGURE 18-21 A Closer Look at Nucleosome Structure. Each nucleosome consists of eight histone molecules (two each of histones H2A, H2B, H3, and H4) associated with 146 base pairs of DNA and a stretch of linker DNA about 50 base pairs in length. The diameter of the nucleosome "bead," or core particle, is about 10 nm. Histone H1 (not shown) is thought to bind to the linker DNA and facilitate the packing of nucleosomes into 30-nm fibers.

associated with the core particle always measures close to 146 bp, which is enough to wrap around the core particle roughly 1.7 times.

Nucleosomes Are Packed Together to Form Chromatin Fibers and Chromosomes

The formation of nucleosomes is only the first step in the packaging of nuclear DNA (**Figure 18-22**). Isolated chromatin fibers exhibiting the beads-on-a-string appearance measure about 10 nm in diameter, but the chromatin of intact cells often forms a slightly thicker fiber, about 30 nm in diameter, called the **30-nm chromatin fiber.** In preparations of isolated chromatin, the 10-nm and 30-nm forms of the chromatin fiber can be interconverted by changing the salt concentration of the solution. However, the 30-nm fiber does not form in chromatin preparations whose histone H1 molecules have been removed, suggesting that histone H1 facilitates the packing of nucleosomes into the 30-nm fiber. Several models have been proposed to explain how individual nucleosomes are packed together to form a 30-nm fiber. Most early models postulated that the chain of nucleosomes is twisted upon itself to form some type of coiled structure. However, more recent studies suggest that the structure of the 30-nm fiber is much less uniform than a coiled model suggests. Instead, the nucleosomes of the 30-nm fiber seem to be packed together to form an irregular, three-dimensional zigzag structure that can interdigitate with its neighboring fibers.

The next level of chromatin packaging is the folding of the 30-nm fibers into **looped domains** averaging 50,000–100,000 bp in length. This looped arrangement is maintained by the periodic attachment of DNA to an insoluble network of nonhistone proteins that form a chromosomal *scaffold* to which the long loops of DNA are attached. The looped domains can be most clearly seen in electron micrographs of chromosomes isolated from dividing cells and treated to remove all the histones and most of the nonhistone proteins (**Figure 18-23**). Loops can also be seen in specialized types of chromosomes that are not associated with the process of cell division (see the discussion of polytene chromosomes in Chapter 23). In these cases, the chromatin loops turn out to contain "active" regions of DNA—that is, DNA that is being transcribed. It makes sense that active DNA would be less tightly packed than inactive DNA because it would allow easier access by proteins involved in gene transcription.

Even in cells where genes are being actively transcribed, significant amounts of chromatin may be further compacted (Figure 18-22d). The degree of folding in such cells varies over a continuum. Segments of chromatin so highly compacted that they show up as dark spots in micrographs are called **heterochromatin,** whereas the more loosely packed, diffuse form of chromatin is called **euchromatin** (see Figure 18-27a). The tightly packed heterochromatin contains DNA that is transcriptionally inactive, while the more loosely packed euchromatin is associated with DNA that is being actively transcribed. Much of the chromatin in metabolically active cells is loosely packed as euchromatin; but as a cell prepares to divide, *all* of its chromatin becomes highly compacted, generating a group of microscopically distinguishable chromosomes. Because the chromosomal DNA has recently been duplicated, each chromosome is composed of two duplicate units called *chromatids* (Figure 18-22e).

The extent to which a DNA molecule has been folded in chromatin and chromosomes can be quantified using the **DNA packing ratio,** which is calculated by determining the total extended length of a DNA molecule and dividing it by the length of the chromatin fiber or chromosome into which it has been packaged. The initial coiling of the DNA around the histone cores of the nucleosomes reduces the length by a factor of about seven, and formation of the 30-nm fiber results in a further sixfold condensation. The packing ratio of the 30-nm fiber is therefore about $7 \times 6 = 42$. Further folding and coiling brings the overall packing ratio of typical euchromatin to about 750. For heterochromatin and the chromosomes of dividing cells, the packing ratio is still higher. At the time of cell division, for example, a typical human chromosome measures about 4–5 μm in length, yet contains a DNA molecule that would measure almost 75 mm if completely extended. The packing ratio for such a chromosome therefore falls in the range of 15,000–20,000.

Eukaryotes Package Some of Their DNA in Mitochondria and Chloroplasts

A eukaryotic cell's DNA is not contained solely in the nucleus. Though nuclear DNA accounts for nearly all of a cell's genetic information, mitochondria and chloroplasts

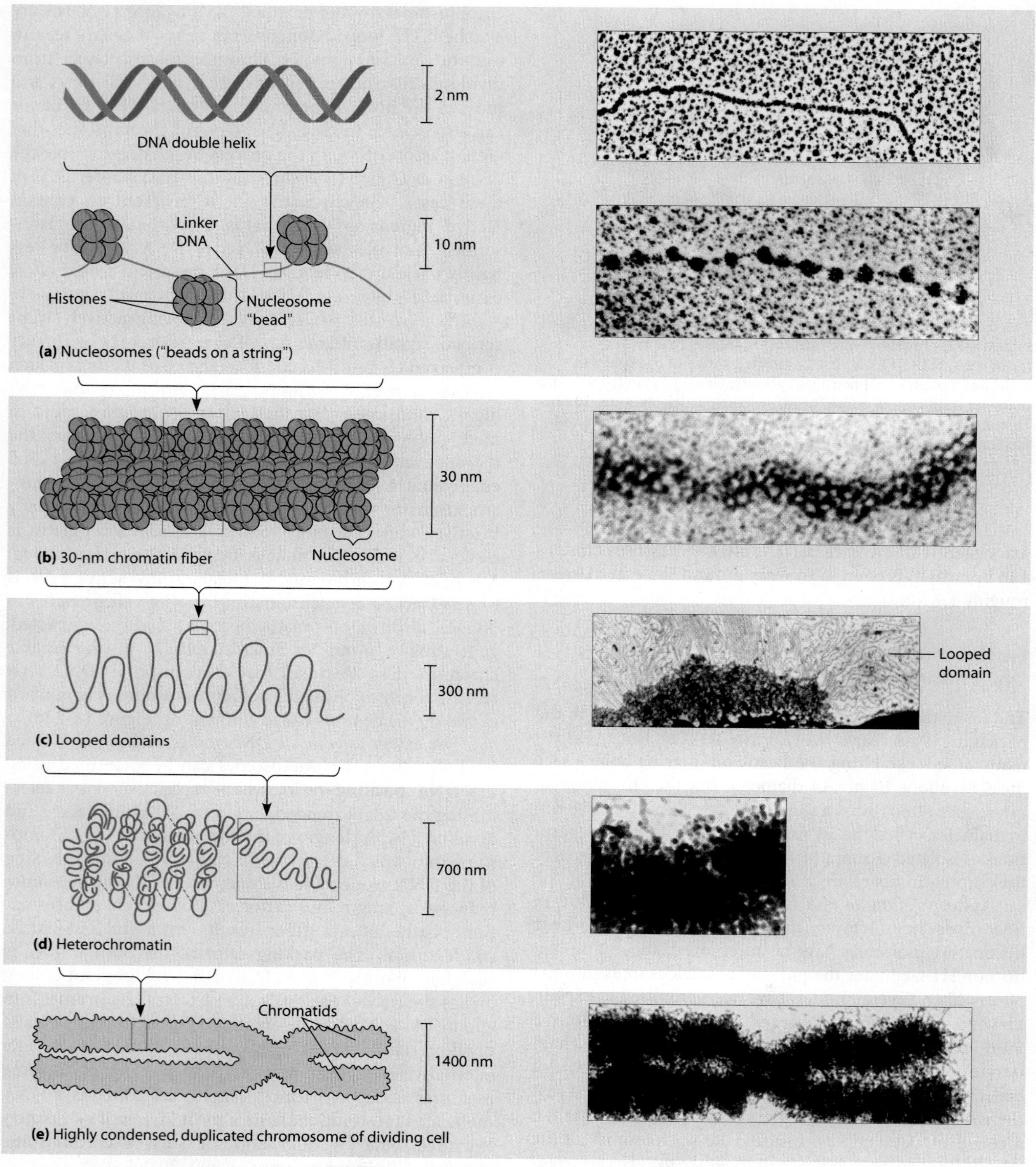

MB Activity: DNA Packing

FIGURE 18-22 Levels of Chromatin Packing. These diagrams and TEMs show a current model for progressive stages of DNA coiling and folding, culminating in the highly compacted chromosome of a dividing cell. **(a)** "Beads on a string," an extended configuration of nucleosomes formed by the association of DNA with four types of histones. **(b)** The 30-nm chromatin fiber, shown here as a tightly packed collection of nucleosomes. The fifth histone, H1, may be located in the interior of the fiber. **(c)** Looped domains of 30-nm fibers, visible in the TEM here because a mitotic chromosome has been experimentally unraveled. **(d)** Heterochromatin, highly folded chromatin that is visible as discrete spots even in interphase cells. **(e)** A replicated chromosome (two attached chromatids) from a dividing cell, with all the DNA of the chromosome in the form of very highly compacted heterochromatin.

FIGURE 18-23 Electron Micrograph Showing the Protein Scaffold That Remains After Removing Histones from Human Chromosomes. The chromosomal DNA remains attached to the scaffold as a series of long loops. The arrow points to a region where a loop of the DNA molecule can be clearly seen (TEM).

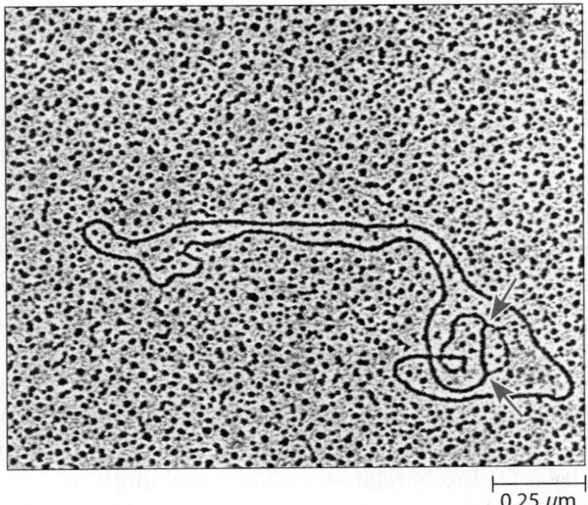

FIGURE 18-24 Mitochondrial DNA. Mitochondrial DNA from most organisms is circular, as seen in this electron micrograph. This molecule was caught in the act of replication; the arrows indicate the points at which replication was proceeding when the molecule was fixed for electron microscopy (TEM).

contain some DNA of their own—along with the machinery needed to replicate, transcribe, and translate the information encoded by this DNA. The DNA molecules residing in mitochondria and chloroplasts are devoid of histones and are usually circular (**Figure 18-24**). In other words, they resemble the genomes of bacteria, as we might expect from the likely endosymbiotic origin of these organelles (discussed in Box 11A). Mitochondrial and chloroplast genomes tend to be relatively small, comparable in size to a viral genome. Both organelles are therefore semiautonomous, able to code for some of their polypeptides but dependent on the nuclear genome to encode most of them.

The genome of the human mitochondrion, for example, consists of a circular DNA molecule containing 16,569 base pairs and measuring about 5 μm in length. It has been completely sequenced, and all of its 37 genes are known. The RNA and polypeptides encoded by this DNA are just a small fraction (about 5%) of the number of RNA molecules and proteins needed by the mitochondrion. This is nonetheless a vital genetic contribution, for these products include the RNA molecules present in mitochondrial ribosomes, all of the transfer RNA molecules required for mitochondrial protein synthesis, and 13 of the polypeptide subunits of the electron transport system. These include subunits of NADH dehydrogenase, cytochrome *b*, cytochrome *c* oxidase, and ATP synthase (**Figure 18-25**).

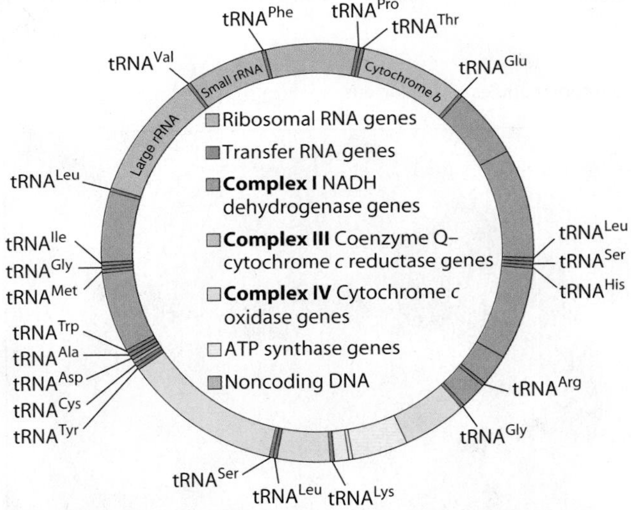

FIGURE 18-25 Genome of the Human Mitochondrion. The double-stranded DNA molecule of the human mitochondrion is circular and contains 16,569 base pairs. This genome codes for large and small ribosomal RNA molecules, transfer RNA (tRNA) molecules (each identified by a superscript with the three-letter abbreviation for the amino acid it carries), and subunits of a number of the proteins that make up the mitochondrial electron transport system complexes. The tRNA genes are very short because the RNA molecules they encode each contain only about 75 nucleotides. Notice that there are two tRNA genes for leucine and two for serine; they code for slightly different versions of tRNAs for these amino acids. The mitochondrial genome is extremely compact, with little noncoding DNA between genes.

The size of the mitochondrial genome varies considerably among organisms. Mammalian mitochondria typically have about 16,500 bp of DNA, yeast mitochondrial DNA is five times larger, and plant mitochondrial DNA is larger yet. It is not clear, however, that larger mitochondrial genomes necessarily code for correspondingly more polypeptides. A comparison of yeast and human mitochondrial DNA, for example, suggests that most of the additional DNA present in the yeast mitochondrion consists of noncoding sequences. In addition to these species-specific differences in mitochondrial genome size, some mitochondrial DNA sequences are unique to a given species. For example, a 648-nucletotide sequence, sometimes called the "*DNA bar code,*" can be used to precisely identify one closely related species from another.

Chloroplasts typically possess circular DNA molecules measuring around 120,000 bp in length and containing about 120 genes. Besides the ribosomal and transfer RNAs and polypeptides involved in protein synthesis, the chloroplast genome also codes for a few polypeptides involved in photosynthesis. These include several polypeptide components of photosystems I and II and one of the two subunits of ribulose-1,5-bisphosphate carboxylase, the carbon-fixing enzyme of the Calvin cycle.

Interestingly, most polypeptides encoded by mitochondrial or chloroplast genomes are components of multimeric proteins that also contain subunits encoded by the nuclear genome. In other words, organelle proteins that contain subunits encoded within the organelle are typically hybrid protein complexes containing polypeptides encoded and synthesized within the organelle plus polypeptides encoded by the nuclear genome and synthesized by cytoplasmic ribosomes. This raises intriguing questions about how polypeptides synthesized in the cytoplasm enter the organelle, a topic we will cover in Chapter 22.

The Nucleus

So far in this chapter, we have discussed DNA as the genetic material of the cell, the genome as a complete set of DNA instructions for the cells of a particular species, and the chromosome as the physical means of packaging DNA within cells. Now we come to the **nucleus,** the site within the eukaryotic cell where the chromosomes are localized and replicated and where the DNA they contain is transcribed. The nucleus is therefore both the repository of most of the cell's genetic information and the control center for the expression of that information.

The nucleus is one of the most prominent and distinguishing features of eukaryotic cells (**Figure 18-26**). In fact, the term *eukaryon* means "true nucleus." This means that the very essence of a eukaryotic cell is its membrane-bounded nucleus, which compartmentalizes the activities of the genome—both replication and transcription—from the rest of cellular metabolism. In the following discussion, we focus first on the membrane envelope that forms the boundary of the nucleus. Then we turn our attention to the pores that perforate the envelope, the structural

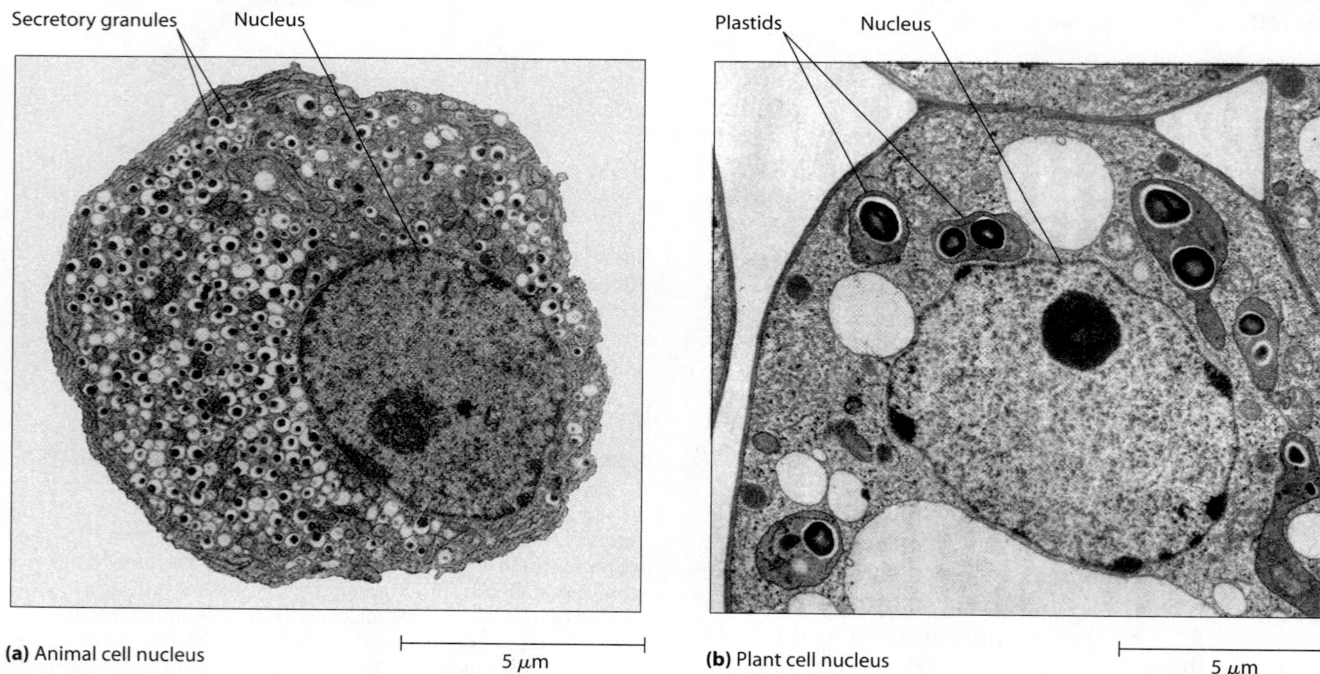

FIGURE 18-26 The Nucleus. The nucleus is a prominent structural feature in most eukaryotic cells. **(a)** The nucleus of an animal cell. This is an insulin-producing cell from a rat pancreas, hence the prominence of secretory granules in the cytoplasm. **(b)** The nucleus of a plant cell. This is a cell from a soybean root nodule. The prominence of plastids reflects their role in the storage of starch granules (TEMs).

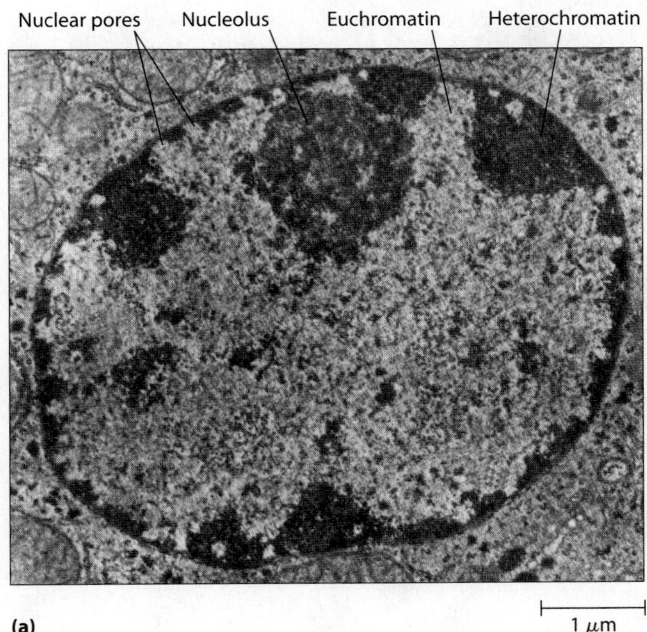

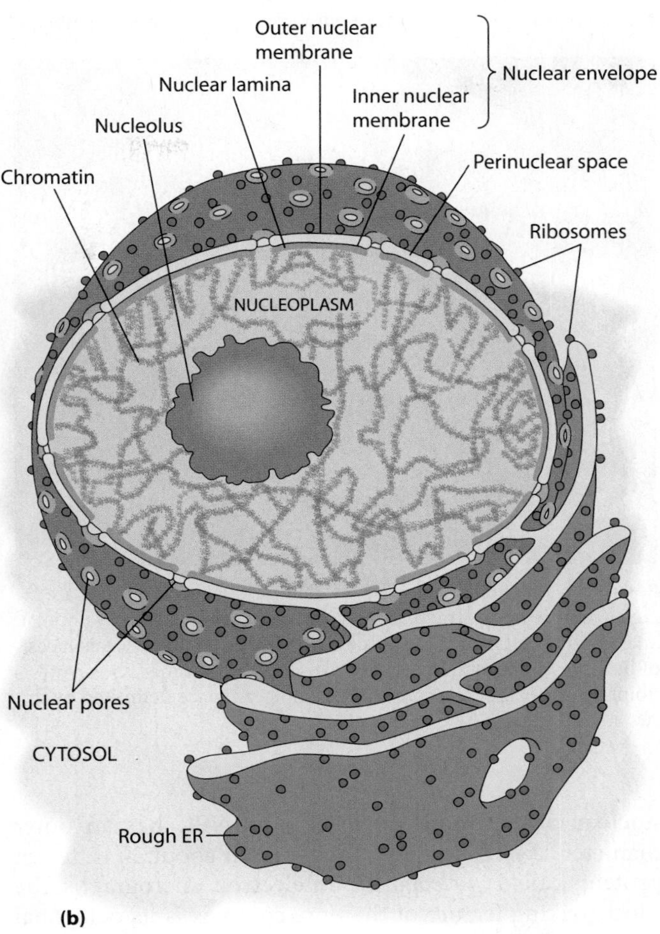

FIGURE 18-27 Structural Organization of the Nucleus and Nuclear Envelope. **(a)** An electron micrograph of the nucleus from a mouse liver cell, with prominent structural features labeled (TEM). The nuclear envelope is a double membrane perforated by nuclear pores. Internal structures include the nucleolus, euchromatin, and heterochromatin. **(b)** A drawing of a typical nucleus. Structural features included here but not visible in the micrograph include the nuclear lamina, ribosomes on the outer nuclear membrane, and the continuity between the outer nuclear membrane and the rough ER.

matrix inside the nucleus, the arrangement of the chromatin fibers, and finally, the organization of the nucleolus. **Figure 18-27** provides an overview of some of these nuclear structures.

A Double-Membrane Nuclear Envelope Surrounds the Nucleus

The existence of a membrane around the nucleus was first suggested in the late nineteenth century, based primarily on the osmotic properties of the nucleus. Since light microscopy reveals only a narrow, fuzzy border at the outer surface of the nucleus, little was known about the structure of this membranous boundary before the advent of electron microscopy. Transmission electron microscopy revealed that the nucleus is bounded by a **nuclear envelope** composed of two membranes—the inner and outer nuclear membranes—separated by a **perinuclear space** measuring about 20–40 nm across (see Figure 18-27b). The outer nuclear membrane is continuous with the endoplasmic reticulum, making the perinuclear space continuous with the lumen of the ER. Like membranes of the rough ER, the outer membrane is often studded on its outer surface with ribosomes engaged in protein synthesis. Several proteins in the outer membrane bind to actin and intermediate filaments of the cell's cytoskeleton, thereby anchoring the nucleus within the cell and providing a mechanism for nuclear movements. Tubular invaginations of the nuclear envelope, collectively known as the *nucleoplasmic reticulum*, may project into the internal nuclear space and increase the area of the nucleus that makes direct contact with the inner nuclear membrane.

One of the most distinctive features of the nuclear envelope is the presence of specialized channels called **nuclear pores,** which are especially easy to see when the nuclear envelope is examined by freeze-fracture microscopy (**Figure 18-28**). Each pore is a small cylindrical channel extending through both membranes of the nuclear envelope, thereby providing direct continuity between the cytosol and the **nucleoplasm** (the name for the interior space of the nucleus other than the region occupied by the nucleolus). The number of pores varies greatly with cell type and activity. A typical mammalian nucleus has about 3000–4000 pores, or about 10–20 pores per square micrometer of membrane surface area.

At each pore, the inner and outer membranes of the nuclear envelope are fused together, forming a channel that is lined with an intricate protein structure called the

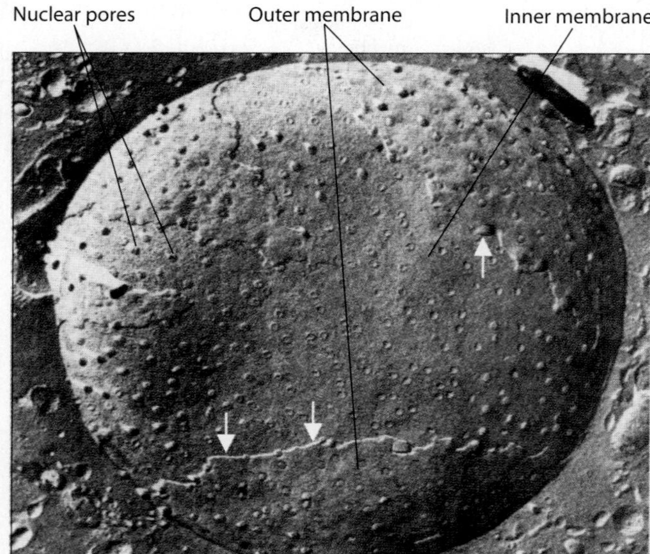

FIGURE 18-28 Nuclear Pores. Numerous nuclear pores are visible in this freeze-fracture micrograph of the nuclear envelope of an epithelial cell from a rat kidney. The fracture plane reveals faces of both the inner membrane and the outer membrane. The arrows point to ridges that represent the perinuclear space delimited by the two membranes (TEM).

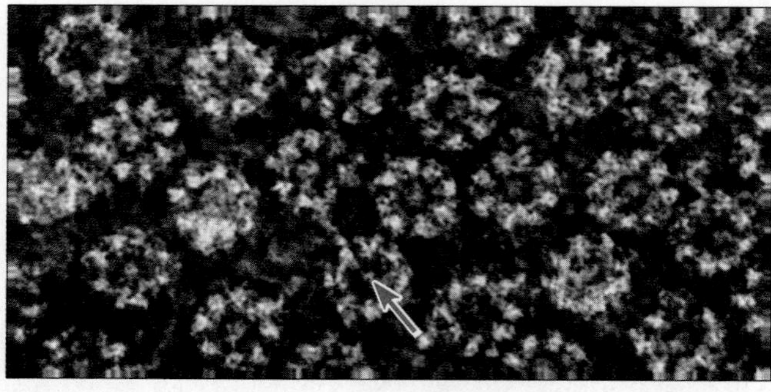
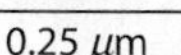

(a) Nuclear pores in the envelope

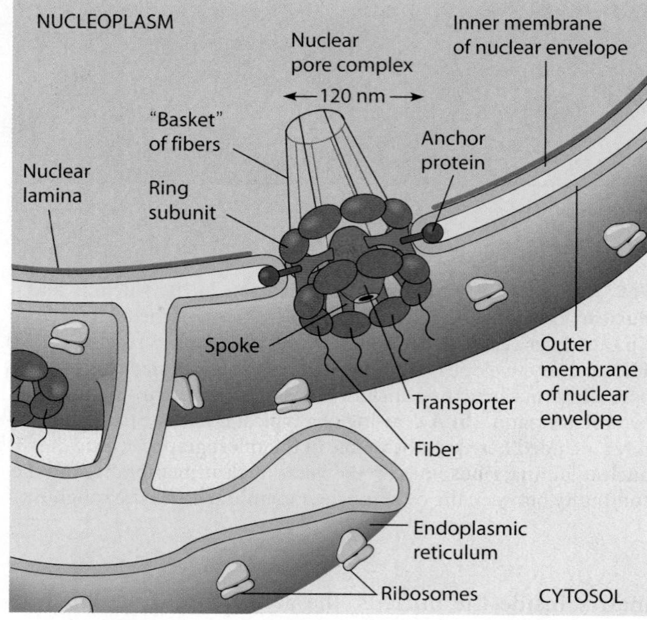

(b) Location of nuclear pores in nuclear membrane

FIGURE 18-29 Structure of the Nuclear Pore. (a) Negative staining of an oocyte nuclear envelope reveals the octagonal pattern of the nuclear pore complexes. The arrow shows a central granule. This nuclear envelope is from an oocyte of the newt *Taricha granulosa* (TEM). (b) A nuclear pore is formed by fusion of the inner and outer nuclear membranes and is lined by an intricate protein structure called the nuclear pore complex. The structure is roughly wheel-shaped and has octagonal symmetry. Two parallel rings, each consisting of eight subunits (dark purple), outline the rim of the wheel. Eight spokes (green) connect the two rings (two of the spokes are omitted from the drawing) and extend to the central transporter (dark pink) at the hub of the wheel; the transporter is thought to contain components involved in moving particles through the pore. Fibers extend above and below the complex, with those on the nucleoplasmic side forming a basket of unknown function.

nuclear pore complex (NPC). The NPC has an outer diameter of ~120 nm and is built from about 30 different proteins called *nucleoporins*. In electron micrographs, the most striking feature of the pore complex is its octagonal symmetry. Micrographs such as the one in **Figure 18-29a** show rings of eight subunits arranged in an octagonal pattern. In other views, the eight subunits are seen to protrude on both the cytoplasmic and nucleoplasmic sides of the envelope. Notice that central granules can be seen in some of the nuclear pore complexes in Figure 18-29a. Although these granules were once thought to consist solely of particles in transit through the pores, they are now thought to also contain components that are an integral part of the pore complex.

Figure 18-29b illustrates the main components of the nuclear pore complex. Examination of this diagram reveals that the pore complex as a whole is shaped somewhat like a wheel lying on its side within the nuclear envelope. Two parallel rings, outlining the rim of the wheel, each consist of the eight subunits seen in electron micrographs. Eight spokes (shown in green) extend from the rings to the wheel's hub (dark pink), which is the "central granule" seen in many electron micrographs. This granule is sometimes called the *transporter* because it is thought to contain components that are involved in moving macromolecules across the nuclear envelope. Proteins extending from the rim into the perinuclear space may help anchor the pore complex to the envelope. Fibers also extend from the rings into the cytosol and nucleoplasm, with those on the nucleoplasm side forming a basket (sometimes called a "cage" or "fishtrap").

Molecules Enter and Exit the Nucleus Through Nuclear Pores

The nuclear envelope solves one problem but creates another. As a means of localizing chromosomes and their activities to one region of the cell, it is an example of the general eukaryotic strategy of compartmentalization.

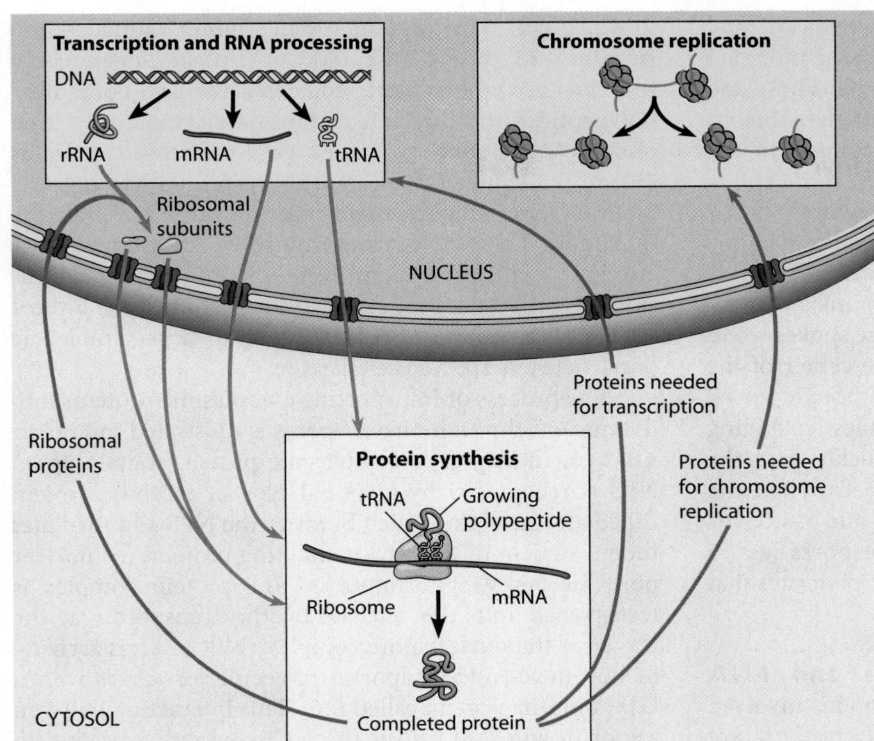

FIGURE 18-30 Macromolecular Transport into and out of the Nucleus. Because eukaryotic cells store their genetic information in the nucleus but synthesize proteins in the cytoplasm, all the proteins needed in the nucleus must be transported inward from the cytoplasm (purple arrows), and all the RNA molecules and ribosomal subunits needed for protein synthesis in the cytoplasm must be transported outward from the nucleus (red arrows). The three kinds of RNA molecules required for protein synthesis are ribosomal RNA (rRNA), messenger RNA (mRNA), and transfer RNA (tRNA).

Presumably, it is advantageous for a nucleus to possess a barrier that keeps in the chromosomes and keeps out organelles such as ribosomes, mitochondria, lysosomes, and microtubules. For example, the nuclear envelope protects newly synthesized RNA from being acted upon by cytoplasmic organelles or enzymes before it has been fully processed.

But in the act of separating immature RNA molecules and chromosomes from the cytoplasm, the nuclear envelope creates several formidable transport problems for eukaryotic cells that are unknown in prokaryotes. All the enzymes and other proteins required for chromosome replication and transcription of DNA in the nucleus must be imported from the cytoplasm, and all the RNA molecules and partially assembled ribosomes needed for protein synthesis in the cytoplasm must be obtained from the nucleus (**Figure 18-30**). In response to these transport problems, specialized pores have evolved that mediate virtually all transport into and out of the nucleus.

To get some idea of how much traffic must travel through the nuclear pores, consider the flow of ribosomal subunits from the nucleus to the cytoplasm. Ribosomes are partially assembled in the nucleus as two classes of subunits, each of which is a complex of RNA and proteins. These subunits move to the cytoplasm and, when needed for protein synthesis, are combined into functional ribosomes containing one of each type of subunit. An actively growing mammalian cell can easily be synthesizing 20,000 ribosomal subunits per minute. We already know that such a cell has about 3000–4000 nuclear pores, so ribosomal subunits must be transported to the cytosol at a rate of about 5–6 subunits per minute per pore. Traffic in the opposite direction is, if anything, even heavier. When chromosomes are being replicated, histones are needed at the rate of about 300,000 molecules per minute. The rate of inward movement must therefore be about 100 histone molecules per minute per pore! In addition to all this macromolecular traffic, the pores mediate the transport of smaller particles, molecules, and ions.

Simple Diffusion of Small Molecules Through Nuclear Pores. The idea that small particles can diffuse freely back and forth through nuclear pores first received experimental support from studies in which colloidal gold particles of various sizes were injected into the cytoplasm of cells that were then examined by electron microscopy. Shortly after injection, the gold particles could be seen passing through the nuclear pores and into the nucleus. The rate of particle entry into the nucleus is inversely related to the particle's diameter—that is, the larger the gold particle, the slower it enters the nucleus. Particles larger than about 10 nm in diameter are excluded entirely. Since the overall diameter of a nuclear pore complex is much larger than such gold particles, it was concluded that the pore complex contains tiny, *aqueous diffusion channels* through which small particles and molecules can freely move.

To determine the diameter of these channels, investigators have injected radioactive proteins of various sizes into the cytoplasm of cells and observed how long it takes for the proteins to appear in the nucleus. A globular protein with a molecular weight of 20,000 takes only a few minutes to equilibrate between the cytoplasm and nucleus, but most proteins of 60,000 daltons or more are barely able to penetrate into the nucleus at all. These and other transport measurements indicate that the aqueous diffusion channels are about 9 nm in diameter, a size that creates a permeability barrier for molecules significantly larger than ~30,000 in molecular weight. Researchers initially assumed that each nuclear pore complex has one such channel. However, more recent evidence suggests there may be eight separate 9-nm channels at the periphery of the pore complex—between the spokes—and perhaps an additional 9-nm channel at the center of the transporter. These aqueous channels are thought to be freely permeable to ions and small molecules (including small proteins) because such substances quickly cross the nuclear envelope after being injected into cells. Thus, the nucleoside triphosphates required for DNA and RNA synthesis probably diffuse freely through the pores, as do other small molecules needed for metabolic pathways that function within the nucleus.

Active Transport of Large Proteins and RNA Through Nuclear Pores. Many of the proteins involved in DNA packaging, replication, and transcription are small enough to pass through a 9-nm-wide channel. Histones, for example, have molecular weights of 21,000 or less and should therefore passively diffuse through the nuclear pores with little problem. Some nuclear proteins are very large, however. The enzymes involved in DNA and RNA synthesis, for example, have subunits with molecular weights in excess of 100,000, which is too large to fit through a 9-nm opening. Messenger RNA molecules pose a challenge, too, because they leave the nucleus bound to proteins in the form of RNA-protein ("ribonucleoprotein") complexes that are quite large. Ribosomal subunits must also be exported to the cytoplasm after being assembled in the nucleus. Clearly, transporting all these particles through the nuclear pores is a significant challenge.

A large body of evidence suggests that such large molecules and particles are actively transported through nuclear pores by a selective process. Like active transport across single membranes, active transport through nuclear pores requires energy and involves specific binding of the transported substance to membrane proteins, which in this case are part of the pore complex. The underlying molecular mechanism is best understood for proteins that are actively transported from the cytosol into the nucleus. Such proteins possess one or more **nuclear localization signals (NLS),** which are amino acid sequences that enable the protein to be recognized and transported by the nuclear pore complex. An NLS is usually 8–30 amino acids in length and often contains proline as well as the positively charged (basic) amino acids lysine and arginine.

The role played by NLS sequences in targeting proteins for the nucleus has been established by experiments in which gold particles larger than 9 nm in diameter—too large to pass through the aqueous diffusion channels of the nuclear pores—were coated with NLS-containing polypeptides and then injected into the cytoplasm of frog oocytes. After such treatment, gold particles as large as 26 nm in diameter are rapidly transported through the nuclear pore complexes and into the nucleus. Thus, the maximum diameter for *active* transport across the nuclear envelope seems to be 26 nm (versus 9 nm for simple diffusion). This is only one of many examples in which a short stretch of amino acids has been found to target a molecule or particle to a specific cellular site.

The process of transporting cytoplasmic proteins into the nucleus through nuclear pores is illustrated in **Figure 18-31a**. In step ❶, a cytoplasmic protein containing an NLS is recognized by a special type of receptor protein called an **importin,** which binds to the NLS and mediates the movement of the NLS-containing protein to a nuclear pore. In step ❷, the importin-NLS protein complex is transported into the nucleus by the transporter at the center of the nuclear pore complex (NPC). After arriving in the nucleus, the importin molecule associates with a GTP-binding protein called *Ran.* This interaction between importin and Ran causes the NLS-containing protein to be released for use in the nucleus (step ❸). The Ran-GTP-importin complex is then transported back through a nuclear pore to the cytoplasm (step ❹), where the importin is released for reuse accompanied by hydrolysis of the Ran-bound GTP (step ❺). Evidence that this GTP hydrolysis step provides the energy for nuclear import has come from experiments showing that nuclear transport can be inhibited by exposing cells to nonhydrolyzable analogs of GTP but not by exposing cells to nonhydrolyzable analogs of ATP.

For exporting material out of the nucleus, comparable mechanisms operate. The main difference is that transport out of the nucleus is used mainly for RNA molecules that are synthesized in the nucleus but function in the cytoplasm, whereas nuclear import is devoted largely to importing proteins that are synthesized in the cytoplasm but function in the nucleus. Although the main cargo for nuclear export is RNA rather than protein, RNA export is mediated by proteins that bind to the RNA. These adaptor proteins contain amino acid sequences called **nuclear export signals (NES),** which target the protein—and hence its bound RNA—for export through the nuclear pores. NES sequences are recognized by nuclear transport receptor proteins called **exportins,** which bind to molecules containing NES sequences and mediate their transport out through the nuclear pores by a mechanism resembling that used by importins to transport cytoplasmic molecules into the nucleus (Figure 18-31b).

The difference in direction between importin- and exportin-mediated transport is governed by the interaction between Ran-GTP and these two classes of molecules, accompanied by a concentration gradient of Ran-GTP across the nuclear envelope. The Ran-GTP concentration

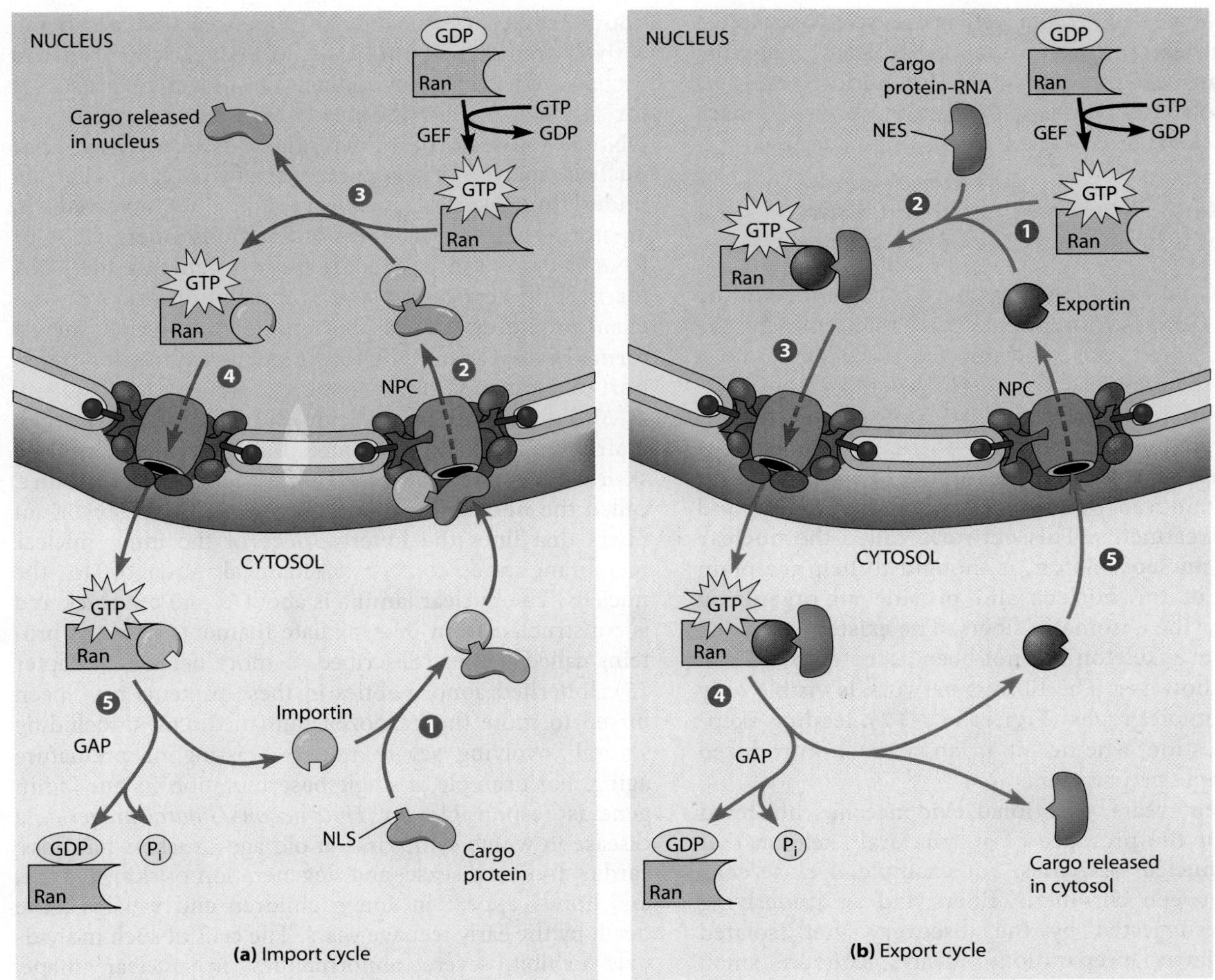

FIGURE 18-31 Transport Through the Nuclear Pore Complex. (a) Import cycle. Proteins made in the cytosol and destined for use in the nucleus contain a nuclear localization sequence (NLS) that targets them as "cargo" for transport through the nuclear pore complex. ❶ An NLS-containing cargo protein binds to importin, and ❷ the importin-cargo complex is then transported through the nuclear pore complex. ❸ Nuclear Ran-GTP binds to importin, triggering the release of the cargo protein in the nucleus. ❹ The Ran-GTP-importin complex is transported back to the cytosol, where ❺ the hydrolysis of GTP to GDP is accompanied by the release of importin. (b) Export cycle. The main cargo for nuclear export is RNA rather than protein, but RNAs are generally transported out of the nucleus by proteins containing a nuclear export signal (NES). During the export cycle, ❶ Ran-GTP binds to exportin in the nucleus, which ❷ promotes the binding of exportin to its cargo, in this case a protein-RNA complex. ❸ The exportin-cargo-Ran-GTP complex is exported through the nuclear pore to the cytosol, where ❹ the exportin and its cargo are released from Ran, accompanied by hydrolysis of GTP to GDP. ❺ Exportin then moves back into the nucleus, where it can repeat the export cycle. The import and export cycles are both driven by a concentration gradient of Ran-GTP that is maintained by the presence of a GEF (guanine-nucleotide exchange factor) in the nucleus and a GAP (GTPase activating protein) in the cytosol.

is maintained at high levels inside the nucleus by a *guanine-nucleotide exchange factor (GEF)* that promotes the binding of GTP to Ran in exchange for GDP. In contrast, the cytosol contains a *GTPase activating protein (GAP)* that promotes the hydrolysis of GTP by Ran, thereby lowering the Ran-GTP concentration outside the nucleus. The relatively high concentration of Ran-GTP inside the nucleus has two effects: first, nuclear Ran-GTP *promotes the release* of NLS-containing cargo from importin (Figure 18-31a, step ❸); second, nuclear Ran-GTP *promotes the binding* of NES-containing cargo to exportin (Figure 18-31b, step ❷). The net result is that the direction of transport for any given cargo molecule is determined by the type of targeting sequence it contains (NLS or NES), which dictates whether importins will release the cargo in the nucleus and bind it in the cytoplasm or exportins will bind the cargo in the nucleus and release it in the cytoplasm.

The nuclear transport cycle poses a recycling problem for the cell: Ran-GTP is exported along with molecules that leave the nucleus. If Ran were not recycled back to the nucleus, nuclear Ran would soon be depleted. A specific nuclear transport protein, *NTF2* (for nuclear transport factor 2), solves this problem by shuttling Ran-GDP back into the nucleus.

The Nuclear Matrix and Nuclear Lamina Are Supporting Structures of the Nucleus

Roughly 80–90% of the nuclear mass is accounted for by chromatin fibers, so you might expect that removing the chromatin would cause the nucleus to collapse into a relatively structureless form. However, in the early 1970s researchers discovered that an insoluble fibrous network retaining the overall shape of the nucleus remains behind after more than 95% of the chromatin has been removed by a combination of nuclease and detergent treatments. This network, called the **nuclear matrix** or **nucleoskeleton,** is thought to help maintain the shape of the nucleus and provide an organizing skeleton for the chromatin fibers. The existence of such an organizing skeleton has not been accepted by all cell biologists, however. The fibrous network is visible only in certain micrographs (**Figure 18-32**), leading skeptics to question whether it is an artifact introduced during sample preparation.

In recent years, additional evidence has bolstered the case for the presence of a structural skeleton that organizes nuclear activities. For example, a close connection between chromatin fibers and an underlying matrix is suggested by the discovery that isolated nuclear matrix preparations always contain small amounts of tightly bound DNA and RNA. Nucleic acid hybridization techniques have revealed that the tightly bound DNA is enriched in sequences that are being actively transcribed into RNA. Moreover, when cells are incubated with ^{3}H-thymidine, a radioactive precursor for DNA synthesis, the newly synthesized radioactive DNA is found to be preferentially associated with the nuclear matrix. These observations suggest that an underlying skeletal network may be involved in anchoring chromatin fibers at locations where DNA or RNA is being synthesized, thereby organizing the DNA for orderly replication and transcription and perhaps even providing tracks that guide and propel newly formed messenger RNA to the nuclear pores for transport to the cytoplasm.

While the exact makeup and role of the nuclear matrix remain to be elucidated, the nucleus contains one skeletal structure that is well understood. This structure, called the **nuclear lamina,** is a thin, dense meshwork of fibers that lines the inner surface of the inner nuclear membrane and confers mechanical strength to the nucleus. The nuclear lamina is about 10–40 nm thick and is constructed from intermediate filaments made of proteins called *lamins* (described in more detail in Chapter 15). Inherited abnormalities in these proteins have been linked to more than a dozen human diseases, including several involving severe muscle wasting or premature aging. For example, a single base mutation in one lamin gene is responsible for *Hutchinson–Gilford progeria,* a disease in which symptoms of old age—such as hair loss, cardiovascular disease, and degeneration of skin, muscle, and bone—appear in young children and usually cause death by the early teenage years. The cells of such individuals exhibit severe abnormalities in nuclear shape, although it is not clear how such nuclear changes relate to the various disease symptoms.

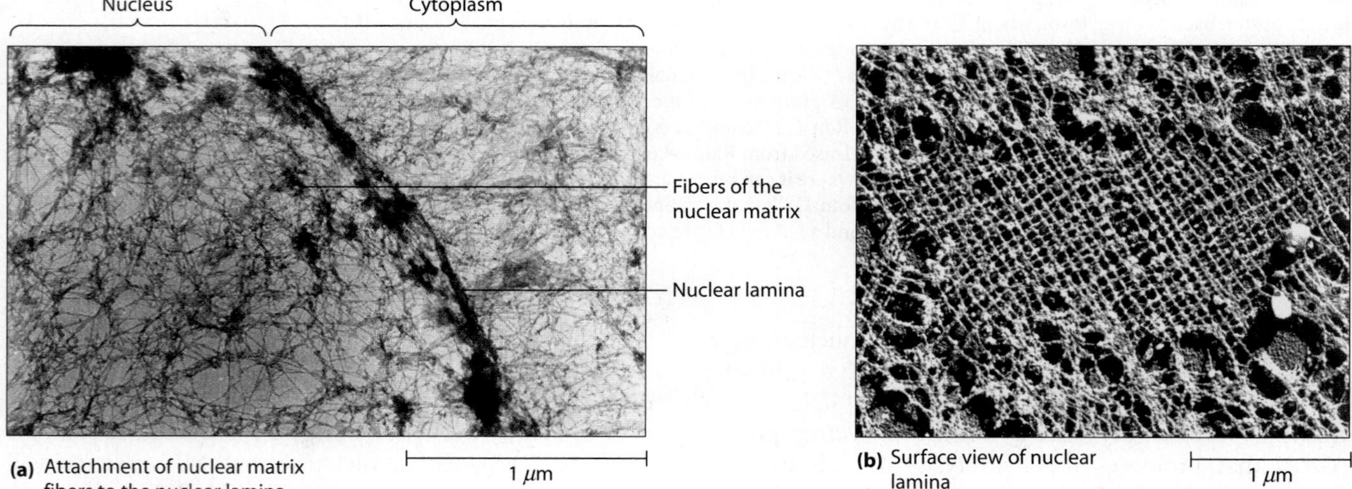

FIGURE 18-32 The Nuclear Matrix and the Nuclear Lamina. **(a)** This electron micrograph of part of a mammalian cell nucleus shows a branched network of nuclear matrix filaments traversing the nucleus. These filaments seem attached to the nuclear lamina, the dense layer of filaments that lines the nucleoplasm side of the nuclear envelope. **(b)** A surface view of the nuclear lamina of a frog oocyte (TEMs).

Chromatin Fibers Are Dispersed Within the Nucleus in a Nonrandom Fashion

Other than during cell division, a cell's chromatin fibers tend to be highly extended and dispersed throughout the nucleus. You might therefore guess that the chromatin threads corresponding to each individual chromosome are randomly distributed and highly intertwined within the nucleus. Perhaps surprisingly, this seems not to be the case. Instead, the chromatin of each chromosome has its own discrete location. This idea was first proposed in 1885, but evidence that it is true in a variety of cells awaited the techniques of modern molecular biology. Experiments using nucleic acid probes that hybridize to the DNA of specific chromosomes (one example of a technique known as *in situ hybridization*) have now shown that the chromatin fibers corresponding to individual chromosomes occupy discrete compartments within the nucleus, referred to as *chromosome territories* (**Figure 18-33**). The positions of these territories do not seem to be fixed, however. They vary from cell to cell of the same organism and change during a cell's life cycle, likely reflecting changes in gene activity of the different chromosomes.

The nuclear envelope helps organize chromatin by binding parts of it to the inner nuclear envelope at sites that are closely associated with the nuclear pores. These chromatin regions are highly compacted—that is, they are heterochromatin. In electron micrographs, this material appears as a dark irregular layer around the nuclear periphery, as we saw in Figure 18-27a. Most of it seems to be the type called **constitutive heterochromatin**, which exists in a highly condensed form at virtually all times in all cells of the organism. The DNA of constitutive heterochromatin consists of simple-sequence repeated DNA (recall that these are short sequences that repeat tandemly and are not transcribed). Two major chromosomal regions composed of constitutive heterochromatin are the centromere and the telomere. In many cases, it is chromosomal telomeres—the highly repeated DNA sequences located at the ends of chromosomes—that are attached to the nuclear envelope at times other than during cell division.

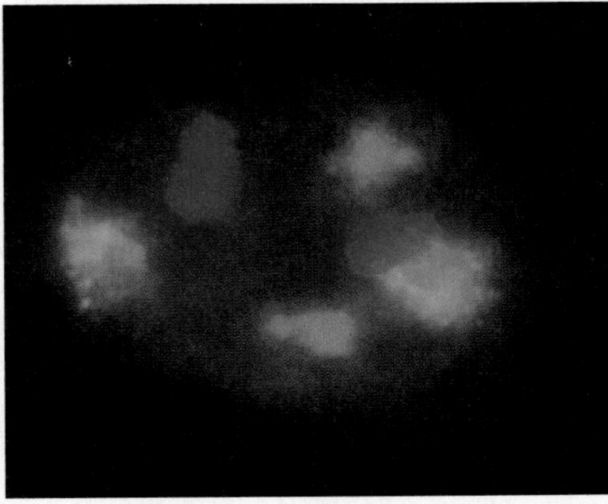

FIGURE 18-33 Chromosome Territories. Mouse lung cells were stained with fluorescent dyes linked to nucleic acid probes that specifically hybridize to the DNA of chromosome 12 (red), chromosome 14 (green), or chromosome 15 (blue). This view of a single nucleus observed by light microscopy shows that the DNA of each chromosome is localized to a specific region of the nucleus (each chromosome is present in two copies).

In contrast to constitutive heterochromatin, **facultative heterochromatin** varies with the particular activities carried out by the cell. Thus, this type of chromatin differs from tissue to tissue and can even vary from time to time within a given cell. Facultative heterochromatin appears to represent chromosomal regions that have become specifically inactivated in a specific cell type. The formation of facultative heterochromatin may be an important means of inactivating entire blocks of genetic information during embryonic development.

The Nucleolus Is Involved in Ribosome Formation

A prominent structural component of the eukaryotic nucleus is the **nucleolus** (plural **nucleoli**), the ribosome factory of the cell. Typical eukaryotic cells contain one or two nucleoli, but the occurrence of several more is not uncommon; in certain situations, hundreds or even thousands may be present. The nucleolus is usually spherical, measuring several micrometers in diameter, but wide variations in size and shape are observed. Because of their relatively large size, nucleoli are easily seen with the light microscope and were first observed more than 200 years ago. However, it was not until the advent of electron microscopy in the 1950s that the structural components of the nucleolus were clearly identified. In thin-section electron micrographs, each nucleolus appears as a membrane-free organelle consisting of fibrils and granules (**Figure 18-34**). The fibrils contain DNA that is being transcribed into *ribosomal RNA (rRNA)*, the RNA component of ribosomes. The granules are rRNA molecules being packaged with proteins (imported from the cytoplasm) to form ribosomal subunits. As we saw earlier, the ribosomal subunits are subsequently exported through the nuclear pores to the cytoplasm. Because of their role in synthesizing RNA, nucleoli become heavily radiolabeled when the cell is exposed to radioactive precursors of RNA (**Figure 18-35**).

The earliest evidence associating the nucleolus with ribosome formation was provided in the early 1960s by Robert Perry, who employed a microbeam of ultraviolet light to destroy the nucleoli of living cells. Such cells lost their ability to synthesize rRNA, suggesting that the nucleolus is involved in manufacturing ribosomes. Additional evidence emerged from studies carried out by Donald Brown and John Gurdon on the African clawed frog, *Xenopus laevis*. Through genetic crosses, it is possible to produce *Xenopus* embryos whose cells lack nucleoli. Brown and Gurdon discovered that such embryos, termed *anucleolate mutants*, cannot synthesize rRNA and therefore die during early development—again implicating the nucleolus in ribosome formation.

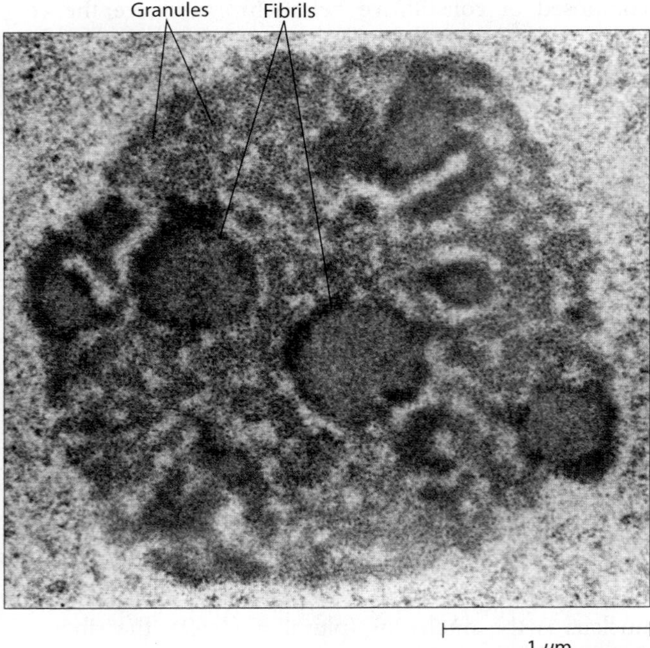

FIGURE 18-34 The Nucleolus. The nucleolus is a prominent intranuclear structure composed of a mass of fibrils and granules. The fibrils are DNA and rRNA; the granules are newly forming ribosomal subunits. Shown here is a nucleolus of a spermatogonium, a cell that gives rise to sperm cells (TEM).

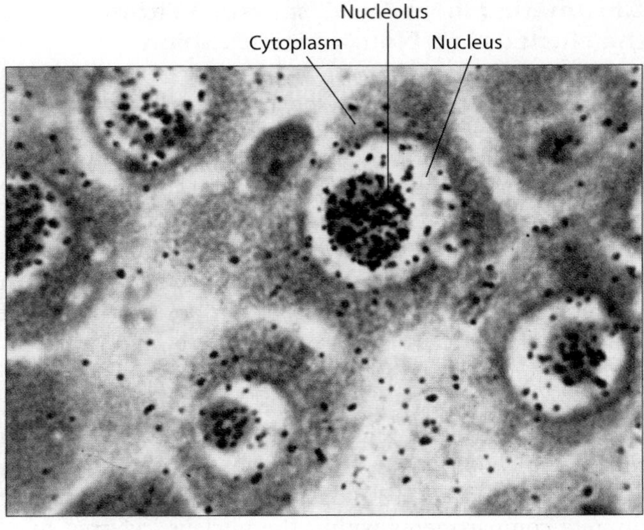

FIGURE 18-35 The Nucleolus as a Site of RNA Synthesis. To demonstrate the role of the nucleolus in RNA synthesis, a rat was injected with ^{3}H-cytidine, a radioactively labeled RNA precursor. Five hours later, liver tissue was removed and subjected to autoradiography. The black spots over the nucleoli in this autoradiograph indicate that the ^{3}H is concentrated in the nucleoli (TEM).

If rRNA is synthesized in the nucleolus, then the DNA sequences coding for this RNA must reside in the nucleolus as well. This prediction has been verified by showing that isolated nucleoli contain the **nucleolus organizer region (NOR)**—a stretch of DNA carrying multiple copies of rRNA genes. These multiple rRNA genes occur in all genomes and are thus an important example of repeated DNA that carries genetic information. The number of copies of the rRNA genes varies greatly from species to species, but animal cells generally contain hundreds of copies and plant cells often contain thousands. The multiple copies are grouped into one or more NORs, which may reside on more than one chromosome; in each NOR, the multiple gene copies are tandemly arranged. A single nucleolus may contain rRNA genes derived from more than one NOR. For example, the human genome has five NORs per haploid chromosome set—or ten per diploid nucleus—each located near the tip of a different chromosome. But instead of ten separate nucleoli, the typical human nucleus has a single large nucleolus containing loops of chromatin derived from ten separate chromosomes.

The size of the nucleolus is correlated with its level of activity. In cells having a high rate of protein synthesis and hence a need for many ribosomes, nucleoli tend to be large and can account for 20–25% of the total volume of the nucleus. In less-active cells, nucleoli are much smaller.

The nucleolus disappears during mitosis, at least in the cells of higher plants and animals. As the cell approaches division, chromatin condenses into compact chromosomes accompanied by the shrinkage and then disappearance of the nucleoli. With our current knowledge of nucleolar composition and function, this makes perfect sense: The extended chromatin loops of the nucleolus cease being transcribed as they are coiled and folded, and any remaining rRNA and ribosomal protein molecules disperse or are degraded. Then, as mitosis is ending, the chromatin uncoils, the NORs loop out again, and rRNA synthesis resumes. In human cells, this is the only time the ten NORs of the diploid nucleus are apparent; as rRNA synthesis begins again, ten tiny nucleoli become visible, one near the tip of each of ten chromosomes. As these nucleoli enlarge, they quickly fuse into the single large nucleolus found in human cells that are not in the process of dividing.

In addition to the nucleolus, microscopists have also identified several kinds of small, non-membrane-enclosed nuclear structures. These nuclear bodies are thought to play a variety of roles related to the processing and handling of RNA molecules produced in the nucleus. These include *Cajal bodies* (named for the scientist who discovered them), *Gemini of Cajal bodies* (*GEMs;* so named because they appear similar to Cajal bodies), *speckles,* and *promyelocytic leukemia bodies (PML bodies).* Cajal bodies and GEMs are involved in maturation and processing of small nucleolar RNA (snoRNA) and small nuclear RNA (snRNA). Speckles, sometimes called *interchromatin granule clusters,* are rich in RNAs and proteins necessary for splicing of messenger RNA precursors into mature messenger RNAs. Current research is aimed at clarifying the roles of these nuclear bodies.

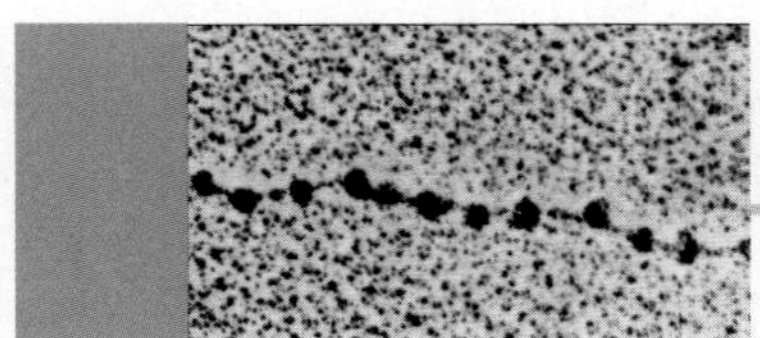

SUMMARY OF KEY POINTS

Chemical Nature of the Genetic Material

- DNA was discovered in the nineteenth century by Miescher, but it was not until the mid-twentieth century that studies with pneumococcal bacteria and bacteriophage T2 revealed that genes are made of DNA.

DNA Structure

- Watson and Crick discovered that DNA is a double helix in which the base A is paired with T, and the base G is paired with C. Supercoiling of the double helix affects the ability of DNA to interact with other molecules.
- Under experimental conditions, the two strands of the double helix can be separated from each other (denaturation) and rejoined (renaturation).

The Organization of DNA in Genomes

- The DNA (or RNA in some viruses) that contains one complete set of an organism's genetic information is called its genome. Viral or prokaryotic genomes consist of one or a small number of DNA molecules. Eukaryotes have a nuclear genome divided among multiple chromosomes, each containing one long DNA molecule, plus a mitochondrial genome; plants and algae possess a chloroplast genome as well.
- The study of genomes has been facilitated by the development of automated techniques for rapid DNA sequencing; these techniques have allowed the complete DNA sequences of numerous organisms, including humans, to be determined. Several online databases and software tools help researchers sort through the enormous amount of data being collected on DNA and protein sequences.
- In the nuclear genomes of multicellular eukaryotes, much of the DNA consists of repeated sequences that do not code for RNA or proteins. Some of this noncoding DNA performs structural or regulatory roles, but most of it has no obvious function.

DNA Packaging

- Due to their enormous size, DNA molecules need to be efficiently packaged by proteins that bind to the DNA. In eukaryotic chromosomes, short stretches of DNA are wrapped around protein particles composed of eight histone molecules to form a basic structural unit called the nucleosome. Chains of nucleosomes ("beads on a string") are packed together to form a 30-nm chromatin fiber, which can then loop and fold further.
- In eukaryotic cells that are actively transcribing DNA, much of the chromatin is in an extended, uncoiled form called euchromatin. Other portions are in a highly condensed, transcriptionally inactive state called heterochromatin. During cell division, all the chromatin becomes highly compacted, forming discrete chromosomes that can be seen with a light microscope.

The Nucleus

- Eukaryotic chromosomes are contained within a nucleus that is bounded by a double-membrane nuclear envelope.
- The nuclear envelope is perforated with nuclear pores that mediate two-way transport of materials between the nucleoplasm and the cytosol. Ions and small molecules diffuse passively through aqueous channels in the pore complex; larger molecules and particles are actively transported through it.
- The nucleolus is a specialized nuclear structure involved in the synthesis of ribosomal RNA and the assembly of ribosomal subunits. Other nuclear bodies perform functions related to processing of RNA.
- The nucleus appears to contain a fibrous skeleton that localizes certain activities, such as DNA replication and messenger RNA production, to discrete regions of the nucleus. One well-defined skeletal structure is the nuclear lamina—a thin, dense meshwork of fibers lining the inner surface of the inner nuclear membrane that confers mechanical strength to the nucleus.

MB **www.masteringbiology.com**

1. MasteringBiology Assignments

Tutorials and Activities

- Molecular Model Structural Tutorials for visualization of 3D structures
- BioFlix Tutorials including 3D animations with hints and detailed feedback for common wrong answers
- Viral Replication; Phage Lysogenic and Lytic Cycles; Phage Lytic Cycle; Temperate Bacteriophage Repressor; DNA as Genetic Material: The Hershey and Chase Experiment; The Hershey-Chase Experiment; An Introduction to DNA Structure; Gel Electrophoresis of DNA; ATP, dATP, ddATP; Shotgun Approach to Whole-Genome Sequencing; DNA Replication; Using BLAST: Can You Identify a Pathogen from a Nucleotide Sequence? Using BLAST: What Can a Protein Sequence Reveal about Cancer? Genomics (1 of 3): Sequencing and Genome Databases; Genomics (3 of 3): Single Nucleotide Polymorphism Analysis and Protein Function; Solve It: Are You Getting the Fish You Paid For? DNA Molecular Testing for Human Disease Gene Mutation; Genomics (2 of 3): Sequence Annotation and Human Genomic

Variation; Insertion Sequences in Bacteria; DNA Fingerprinting; Video Tutor Session Quiz: DNA Profiling; DNA Packing

Questions
- Reading Quiz
- Multiple Choice
- End-of-chapter questions

2. **eText** Read your book online, search, highlight text, take notes, and more
3. **The Study Area**
- Chapter Review Quiz, Guide to Techniques and Methods, 3D-Structure Tutorials, Activities, BioFlix, Videos, Biology Review, Word Study Tools, Glossary, Flashcards

PROBLEM SET

More challenging problems are marked with a •.

18-1 Prior Knowledge. Virtually every experiment performed by biologists builds on knowledge provided by earlier experiments.

(a) Of what significance to Avery and his colleagues was the finding (made in 1932 by J. L. Alloway) that the same kind of transformation of R cells into S cells that Griffith observed to occur in mice could also be demonstrated in culture with isolated pneumococcus cells?

(b) Of what significance to Hershey and Chase was the following suggestion (made in 1951 by R. M. Herriott)? "A virus may act like a little hypodermic needle full of transforming principles; the virus as such never enters the cell; only the tail contacts the host and perhaps enzymatically cuts a small hole through the outer membrane and then the nucleic acid of the virus head flows into the cell."

(c) Of what significance is the discovery of restriction endonucleases to the development of techniques that helped to identify sequences of DNA? What are the developments that emerged from the identification of these bacterial products?

(d) How did the findings of Chargaff help Watson and Crick work out the structure of nucleic acids?

18-2 DNA Base Composition. Based on your understanding of the rules of complementary base pairing, answer the following questions:

(a) You analyze a DNA sample and find that its base composition is 30% A, 20% T, 30% G, and 20% C. What can you conclude about the structure of this DNA?

(b) If you mix DNA molecules with a T_m of 90°C with DNA molecules with a T_m of 120°C and heat the mixture to 110°C, what are the resultant reannealing products likely to be?

(c) For a double-stranded DNA molecule in which 30% of the bases are either G or T, what can you conclude about its content of the base A?

(d) For a double-stranded DNA molecule in which 15% of the bases are A, what can you conclude about its content of the base C? Would you expect the T_m of this DNA to be above or below 90°C? (Hint: Refer to Figure 18-9.)

18-3 DNA Structure. Carefully inspect the double-stranded DNA molecule shown here, and notice that it has twofold rotational symmetry:

3′ A—G—C—G—C—T—A—T—A—G—C—G—C—T 5′
5′ T—C—G—C—G—A—T—A—T—C—G—C—G—A 3′

Label each of the following statements as T if true or F if false.

(a) There is no way to distinguish the right end of the double helix from the left end.

(b) If a solution of these molecules were heated to denature them, every single-stranded molecule in the solution would be capable of hybridizing with every other molecule.

(c) If the molecule were cut at its midpoint into two halves, it would be possible to distinguish the left half from the right half.

(d) If the two single strands were separated from each other, it would not be possible to distinguish one strand from the other.

(e) In a single strand from this molecule, it would be impossible to determine which is the 3′ end and which is the 5′ end.

18-4 Restriction Mapping of DNA. The genome of a newly discovered bacteriophage is a linear DNA molecule 10,500 nucleotide pairs in length. One sample of this DNA has been incubated with restriction enzyme X and another sample with restriction enzyme Y. The lengths (in thousands of base pairs) of the restriction fragments produced by the two enzymes have been determined by gel electrophoresis to be as follows:

Enzyme X: Fragment X-1 = 4.5; X-2 = 3.6; X-3 = 2.4
Enzyme Y: Fragment Y-1 = 5.2; Y-2 = 3.8; Y-3 = 1.5

Next, the fragments from the enzyme X reaction are isolated and treated with enzyme Y, and the fragments from the enzyme Y reaction are treated with enzyme X. The results are as follows:

X fragments treated with Y: X-1 → 4.5 (unchanged)
X-2 → 2.1 + 1.5
X-3 → 1.7 + 0.7
Y fragments treated with X: Y-1 → 4.5 + 0.7
Y-2 → 2.1 + 1.7
Y-3 → 1.5 (unchanged)

Draw a restriction map of the phage DNA, indicating the positions of all enzyme X and enzyme Y restriction sites and the lengths of DNA between them.

18-5 DNA Sequencing. You have isolated the DNA fragment shown in Problem 18-3 but do not know its complete sequence. From knowledge of the specificity of the restriction enzyme used to prepare it, you know the first four bases at the left end and have prepared a single-stranded DNA primer of sequence 5′ T–C–G–C 3′. Explain how you would determine the rest of the sequence using dye-labeled dideoxynucleotides. Draw the gel pattern that would be observed, indicating the base sequence of the DNA in each band and the color pattern that would be detected by the camera in a DNA sequencing machine.

• **18-6 DNA Melting. Figure 18-36** shows the melting curves for two DNA samples that were thermally denatured under the same conditions.

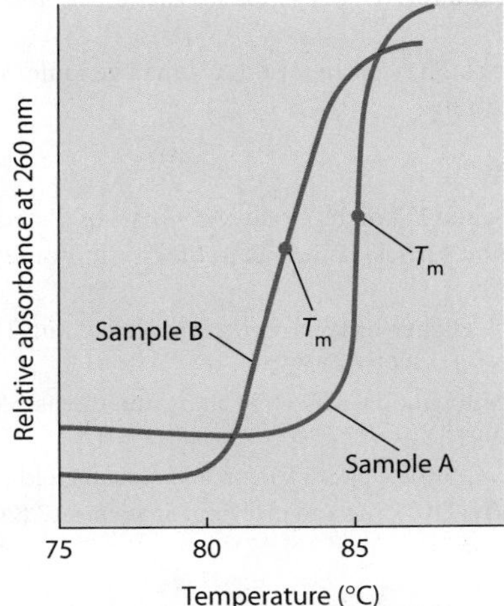

FIGURE 18-36 Thermal Denaturation of Two DNA Samples. See Problem 18-6.

(a) What conclusion can you draw concerning the base compositions of the two samples? Explain.

(b) How might you explain the steeper slope of the melting curve for sample A?

(c) Formamide and urea are agents known to form hydrogen bonds with pyrimidines and purines. What effect, if any, would the inclusion of a small amount of formamide or urea in the incubation mixture have on the melting curves?

• **18-7 DNA Renaturation.** In an experiment, DNA from *Salmonella typhimurium* was denatured and the reannealing process was monitored. At the same time, DNA from a camel was denatured and monitored for reannealing. Draw your conclusions from the following observations.

(a) Renaturation of the *Salmonella* DNA was a process with a single phase.

(b) About 35% of the camel DNA renatured quickly whereas 65% of the DNA renatured over a prolonged period.

• **18-8 Nucleosomes.** You perform an experiment in which chromatin is isolated from sea urchin sperm cells and briefly digested with micrococcal nuclease. When the chromatin proteins are removed and the resulting purified DNA is analyzed by gel electrophoresis, you observe a series of DNA fragments that are multiples of 260 base pairs in length (that is, 260 bp, 520 bp, 780 bp, and so forth).

(a) If the DNA was subjected to protease digestion before treatment with micrococcal nuclease, what would the resultant fragments of DNA look like?

(b) Is it possible to obtain the same results using any other enzyme other than micrococcal nuclease? If yes, why? If no, why?

(c) If the chromatin had been analyzed by density gradient centrifugation immediately after digestion with micrococcal nuclease, describe what you would expect to see.

(d) Suppose you perform an experiment in which the chromatin is digested for a much longer period of time with micrococcal nuclease before removal of chromatin proteins. When the resulting DNA preparation is analyzed by electrophoresis, all of the DNA appears as fragments 146 bp in length. What does this suggest to you about the length of the linker DNA in this cell type?

18-9 Nuclear Structure and Function. Indicate the implications for nuclear structure or function of each of the following experimental observations.

(a) Sucrose crosses the nuclear envelope so rapidly that its rate of movement cannot be accurately measured.

(b) Colloidal gold particles with a diameter of 5.5 nm equilibrate rapidly between the nucleus and cytoplasm when injected into an amoeba, but gold particles with a diameter of 15 nm do not.

(c) Nuclear pore complexes sometimes stain heavily for RNA and protein.

(d) If gold particles up to 26 nm in diameter are coated with a polypeptide containing a nuclear localization signal (NLS) and are then injected into the cytoplasm of a living cell, they are transported into the nucleus. If they are injected into the nucleus, however, they remain there.

(e) Many of the proteins of the nuclear envelope appear from electrophoretic analysis to be the same as those found in the endoplasmic reticulum.

(f) Ribosomal proteins are synthesized in the cytoplasm but are packaged with rRNA into ribosomal subunits in the nucleus.

(g) If nucleoli are irradiated with a microbeam of ultraviolet light, synthesis of ribosomal RNA is inhibited.

(h) Treatment of nuclei with the nonionic detergent Triton X-100 dissolves the nuclear envelope but leaves an otherwise intact nucleus.

• **18-10 Nuclear Transport.** Budding yeast with temperature-sensitive mutations in NTF2 have been identified. In such mutants, NTF2 functions at 25°C. Raising the temperature to 37°C, however, causes the NTF2 protein to stop functioning. What effects would you predict raising these mutant yeast to 37°C would have on nuclear transport? Explain your answer.

18-11 Nucleoli. Indicate whether each of the following statements is true (T) or false (F). If false, reword the statement to make it true.

(a) Nucleoli are membrane-bounded structures present in the eukaryotic nucleus.

(b) Nucleoli can resist degradation by ultraviolet light.

(c) Anucleolate *Xenopus* embryos can survive without showing any developmental defects.

(d) The nucleolar organizer region (NOR) is a stretch of DNA that codes for rRNA.

(e) The nucleolar membrane is composed of proteins and glycolipids that are similar to those seen in the plasma membrane.

(f) In animals and plants, the disappearance of nucleoli during mitosis correlates with cessation of ribosome synthesis.

SUGGESTED READING

References of historical importance are marked with a •.

Chemical Nature of the Genetic Material

• Avery, O. T., C. M. MacLeod, and M. McCarty. Studies on the chemical nature of the substance inducing transformation of pneumococcal types. Induction of transformation by a desoxyribonucleic acid fraction isolated from *Pneumococcus* Type III. *J. Exp. Med.* 79 (1944): 137.

• Chargaff, E. Preface to a grammar of biology: A hundred years of nucleic acid research. *Science* 172 (1971): 637.

• Hershey, A. D., and M. Chase. Independent functions of viral protein and nucleic acid in growth of bacteriophage. *J. Gen. Physiol.* 36 (1952): 39.

• Portugal, F. H., and J. S. Cohen. *The Century of DNA: A History of the Discovery of the Structure and Function of the Genetic Substance.* Cambridge, MA: MIT Press, 1977.

DNA Structure

Arnott, S. Historical article: DNA polymorphism and the early history of the double helix. *Trends Biochem. Sci.* 31 (2006): 349.

• Bauer, W. R., F. H. C. Crick, and J. H. White. Supercoiled DNA. *Sci. Amer.* 243 (July 1980): 118.

Marmur, J. DNA strand separation, renaturation and hybridization. *Trends Biochem. Sci.* 19 (1994): 343.

• Watson, J. D., and F. H. C. Crick. Molecular structure of nucleic acids: A structure for deoxyribose nucleic acid. *Nature* 171 (1953): 737.

The Organization of DNA in Genomes

• Britten, R. J., and D. E. Kohne. Repeated sequences of DNA. *Science* 161 (1968): 529.

Church, G. M. Genomes for all. *Sci. Amer.* 294 (January 2006): 46.

Cox, J., and M. Mann. Is proteomics the new genomics? *Cell* 130 (2007): 395.

Feuk, L., A. R. Carson, and S. W. Scherer. Structural variation in the human genome. *Nature Rev. Genet.* 7 (2006): 85.

Jobling, M. A., and P. Gill. Encoded evidence: DNA in forensic analysis. *Nature Rev. Genet.* 5 (2004): 739.

Pollard, K. S. What makes us human? *Sci. Amer*. 300 (2009): 44.

Kung, L. A., and M. Snyder. Proteome chips for whole-organism assays. *Nature Rev. Mol. Cell Biol.* 7 (2006): 617.

• Lander, E. S. et al. Initial sequencing and analysis of the human genome. *Nature* 409 (2001): 860.

Mirkin, S. M. Expandable DNA repeats and human disease. *Nature* 447 (2007): 932.

Moxon, E. R., and C. Willis. DNA microsatellites: Agents of evolution? *Sci. Amer.* 280 (January 1999): 94.

• Sanger, F. Determination of nucleotide sequences in DNA. *Science* 214 (1981): 1205.

Stoeckle, M. Y., and P. D. Hebert. Barcode of life. *Sci. Amer.* 299 (2008): 82.

• Venter, J. C. et al. The sequence of the human genome. *Science* 291(2001): 1304.

DNA Packaging

Kornberg, R. D., and Y. Lorch. Twenty-five years of the nucleosome, fundamental particle of the eukaryote chromosome. *Cell* 98 (1999): 285.

Tremethick, D. J. Higher-order structures of chromatin: The elusive 30 nm fiber. *Cell* 128 (2007): 651.

Wallace, D. C. Mitochondrial DNA in aging and disease. *Sci. Amer.* 277 (August 1997): 40.

Zlatanova, J., T., C. Bishop, J. M. Victor, and K. van Holde. The nucleosome family: Dynamic and growing. *Structure* 17 (2009): 160–71.

The Nucleus

Boisvert, F.-M. et al. The multifunctional nucleolus. *Nature Rev. Mol. Cell Biol.* 8 (2007): 574.

Capell, B. C., and F. S. Collins. Human laminopathies: Nuclei gone genetically awry. *Nature Rev. Genet.* 7 (2006): 940.

Cioce, M., and A. I. Lamond. Cajal bodies: A long history of discovery. *Annu. Rev. Cell Dev. Biol.* 21(2005): 105.

Cook, A. et al. Structural biology of nucleocytoplasmic transport. *Annu. Rev. Biochem.* 76 (2007): 647.

Fedorova, E., and D. Zink. 2009. Nuclear genome organization: Common themes and individual patterns. *Curr. Opin. Genet. Dev.* 19 (2009): 166.

Gruenbaum, Y. et al. The nuclear lamina comes of age. *Nature Rev. Mol. Cell Biol.* 6 (2005): 21.

Handwerger, K. E., and J. G. Gall. Subnuclear organelles: New insights into form and function. *Trends Cell Biol.* 16 (2006): 19.

Hetzer, M. W., T. C. Walther, and I. W. Mattaj. Pushing the envelope: Structure, function, and dynamics of the nuclear periphery. *Annu Rev. Cell Dev. Biol.* 21 (2005): 347.

Meaburn, K. J., and T. Misteli. Chromosome territories. *Nature* 445 (2007): 379.

Roderick, Y. H., and B. Fahrenkrog. The nuclear pore complex up close. *Curr. Opin. Cell Biol.* 18 (2006): 342.

Thiry, M., and D. L. J. Lafontaine. Birth of a nucleolus: The evolution of nucleolar compartments. *Trends Cell Biol.* 15 (2005): 194.

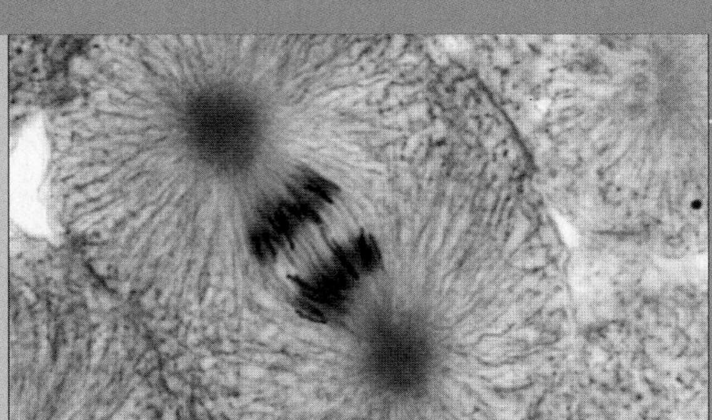

19 The Cell Cycle, DNA Replication, and Mitosis

See MasteringBiology® for tutorials, activities, and quizzes.

The ability to grow and reproduce is a fundamental property of living organisms. However, growth of single cells is fundamentally limited. As new proteins, nucleic acids, carbohydrates, and lipids are synthesized, their accumulation causes the volume of a cell to increase, forcing the plasma membrane to expand to prevent the cell from bursting. But cells cannot continue to enlarge indefinitely; as a cell grows larger, there is an accompanying decrease in its surface area/volume ratio and hence in its capacity for effective exchange with the environment. Therefore, cell growth is generally accompanied by **cell division,** *whereby one cell gives rise to two new daughter cells. (The term* daughter *is used by convention and does not indicate that cells have gender.) For single-celled organisms, cell division increases the total number of individuals in a population. In multicellular organisms, cell division either increases the number of cells, leading to growth of the organism, or replaces cells that have died. In an adult human, for example, about 2 million stem cells in bone marrow divide every second to maintain a constant number of red blood cells in the body. Although often cell growth and cell division are coupled, there is a notable exception. A fertilized animal egg typically undergoes many divisions without the growth of its cells, dividing the volume of the egg into smaller and smaller parcels. Here as well, however, tight regulation of where and when cells divide is crucial.*

When cells grow and divide, the newly formed daughter cells are usually genetic duplicates of the parent cell, containing the same (or virtually the same) DNA sequences. Therefore, all the genetic information in the nucleus of the parent cell must be duplicated and carefully distributed to the daughter cells during the division process. In accomplishing this task, a cell passes through a series of discrete stages, collectively known as the cell cycle. *In this chapter, we will examine the events associated with the cell cycle, focusing first on the mechanisms that ensure that each new cell receives a complete set of genetic instructions and then examining how the cell cycle is regulated to fit the needs of the organism.*

Overview of the Cell Cycle

Mitosis (1 of 3): Mitosis and the Cell Cycle (BioFlix tutorial); Activity: Four Phases of the Cell Cycle

The **cell cycle** begins when two new cells are formed by the division of a single parental cell and ends when one of these cells divides again into two cells (**Figure 19-1**). To early cell biologists studying eukaryotic cells under the microscope, the most dramatic events in the life of a cell were those associated with the point in the cycle when cell actually divides. This division process, called **M phase,** involves two overlapping events in which nucleus divides first and the cytoplasm second. Nuclear division is called **mitosis,** and the division of the cytoplasm to produce two daughter cells is termed **cytokinesis.**

The stars of the mitotic drama are the chromosomes. As you can see in Figure 19-1a, the beginning of mitosis is marked by condensation (coiling and folding) of the cell's chromatin, which generates chromosomes that are thick enough to be individually discernible under the microscope. Because DNA replication has already taken place, each chromosome actually consists of two chromosome copies that remain attached to each other until the cell divides. As long as they remain attached, the two new chromosomes are referred to as **sister chromatids.** As the chromatids become visible, the nuclear envelope breaks into fragments. Then, in a stately ballet guided by the microtubules of the *mitotic spindle,* the sister chromatids separate and—each now a full-fledged chromosome—move to opposite ends of the cell. By this time, cytokinesis has usually begun, and new nuclear membranes envelop the two groups of daughter chromosomes as cell division is completed.

While visually striking, the events of M phase account for a relatively small portion of the total cell cycle; for a

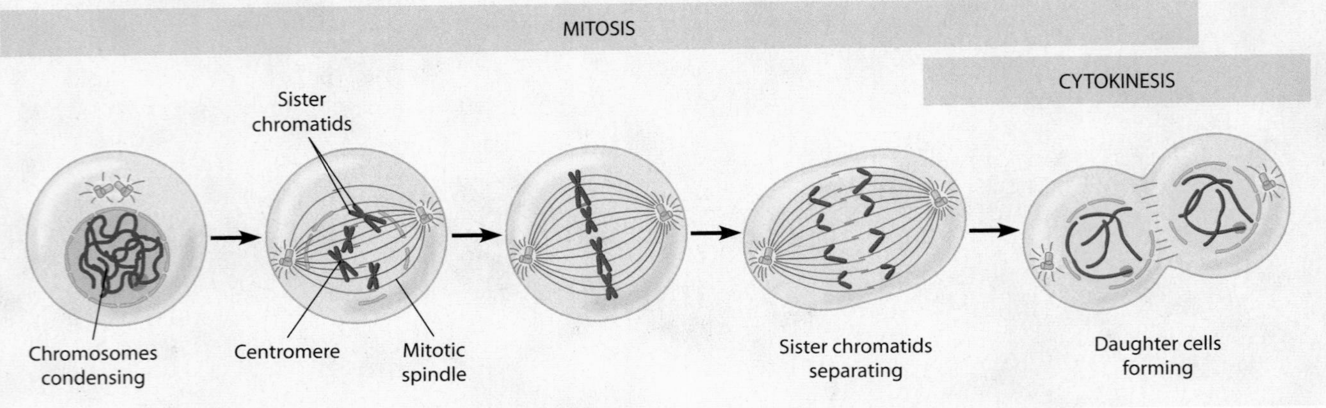

(a) The M (mitotic) phase

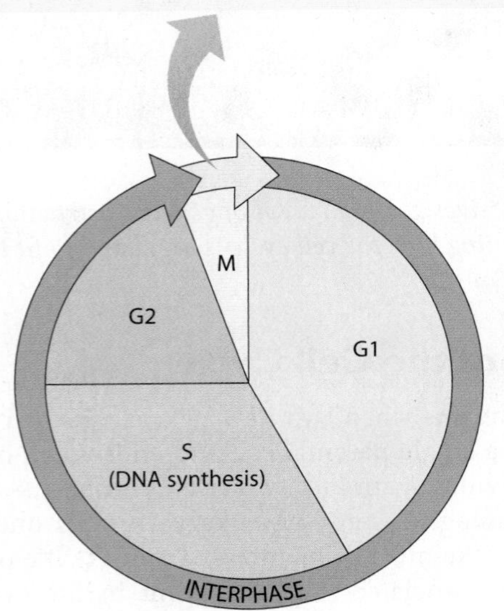

(b) The cell cycle

FIGURE 19-1 The Eukaryotic Cell Cycle. **(a)** The M (mitotic) phase, the process of cell division, is the most visually distinctive part of the cell cycle. It consists of two overlapping processes, mitosis and cytokinesis. In mitosis, the mitotic spindle segregates the duplicated, condensed chromosomes into two daughter nuclei; in cytokinesis, the cytoplasm divides to yield two genetically identical daughter cells. **(b)** Between divisions, the cell is said to be in interphase, which is made up of the S phase (the period of nuclear DNA replication) and two "gap" phases, called G1 and G2. The cell continues to grow throughout interphase, a time of high metabolic activity.

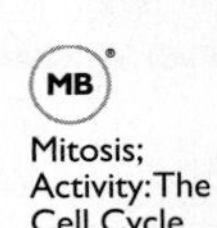

Mitosis; Activity: The Cell Cycle

typical mammalian cell, M phase usually lasts less than an hour. Cells spend most of their time in the growth phase between divisions, called **interphase** (Figure 19-1b). Most cellular contents are synthesized continuously during interphase, so cell mass gradually increases as the cell approaches division. During interphase the amount of nuclear DNA doubles, and experiments using radioactive DNA precursors have shown that the new nuclear DNA is synthesized during a specific portion of interphase named the **S phase** (S for synthesis). A time gap called **G1 phase** separates S phase from the preceding M phase; a second gap, the **G2 phase,** separates the end of S phase from the onset of the next M phase.

Although the cells of a multicellular organism divide at varying rates, most studies of the cell cycle involve cells growing in culture, where the length of the cycle tends to be similar for different cell types. We can easily determine the overall length of the cell cycle—the *generation time*—for cultured cells by counting the cells under a microscope and determining how long it takes for the cell population to double. In cultured mammalian cells, for example, the total cycle usually takes about 18–24 hours. Once we know the total length of the cycle, it is possible to determine the length of specific phases. To determine the length of the S phase, we can expose cells to a radioactively labeled DNA precursor (usually ^{3}H-thymidine) for a short period of time and then examine the cells by autoradiography. The fraction of cells with silver grains over their nuclei represents the fraction of cells that were somewhere in S phase when the radioactive compound was available. When we multiply this fraction by the total length of the cell cycle, the result is an estimate of the average length of the S phase. For mammalian cells in culture, this fraction is often around 0.33, which indicates that S phase is about 6–8 hours in length. Similarly, we can estimate the length of M phase by multiplying the generation time by the percentage of the cells that are actually in mitosis at any given time. This percentage is called the **mitotic index.** The mitotic index for cultured mammalian cells is often about 3–5%, which means that M phase lasts less than an hour (usually 30–45 minutes).

In contrast to the S and M phases, whose lengths tend to be similar for different mammalian cells, the length of G1 is quite variable, depending on the cell type. Although a typical G1 phase lasts 8–10 hours, some cells spend only a few minutes or hours in G1, whereas others are delayed for long periods of time. During G1, a major "decision" is made as to whether and when the cell is to divide again. Cells that become arrested in G1, awaiting a signal that will trigger reentry into the cell cycle and a commitment to divide, are said to be in **G0 (G zero).** Other cells exit from the cell cycle entirely and undergo *terminal differentiation,* which means they are destined never to divide again; most of the nerve cells in your body are in this state. In some cells, transient arrest of the cell cycle can also occur in G2. In general, however, G2 is shorter than G1 and more uniform in duration, usually lasting 4–6 hours.

Cell cycle studies have been facilitated by the use of **flow cytometry,** a technique that permits automated analysis of the chemical makeup of millions of individual cells almost simultaneously. In this procedure, cells are first stained with one or more fluorescent dyes—for example, a red dye that stains DNA might be combined with a green dye that binds specifically to a particular cell protein. The dyed cells are then passed in a tiny, liquid stream through a beam of laser light. By analyzing fluctuations in the intensity and color of the fluorescent light emitted by each cell as it passes through the laser beam, researchers can assess the concentration of DNA and specific proteins in each individual cell. This information is then used to assess the chemical makeup of cells at different points in the cycle.

DNA Replication

Now that we have provided an overview of the cell cycle, we are ready to consider its workings in detail. Since DNA replication is a central event in the cycle, we begin by examining its underlying mechanism, which in turn depends on the double-helical structure of DNA. In fact, a month after Watson and Crick published their now-classic paper postulating a double helix for DNA, they followed it with an equally important paper suggesting how such a base-paired structure might duplicate itself. Here, in their own words, is the basis of that suggestion:

> *Now our model for deoxyribonucleic acid is, in effect, a pair of templates, each of which is complementary to the other. We imagine that prior to duplication the hydrogen bonds are broken, and the two chains unwind and separate. Each chain then acts as a template for the formation onto itself of a new companion chain, so that eventually we shall have two pairs of chains, where we only had one before. Moreover, the sequence of the pairs of bases will have been duplicated exactly. (Watson and Crick, 1953, p. 994)*

The model Watson and Crick proposed for DNA replication is shown in **Figure 19-2.** The essence of their suggestion is that one of the two strands of every newly formed DNA molecule is derived from the parent molecule, whereas the other strand is newly synthesized. This is called **semiconservative replication** because half of the parent molecule is retained by each daughter molecule.

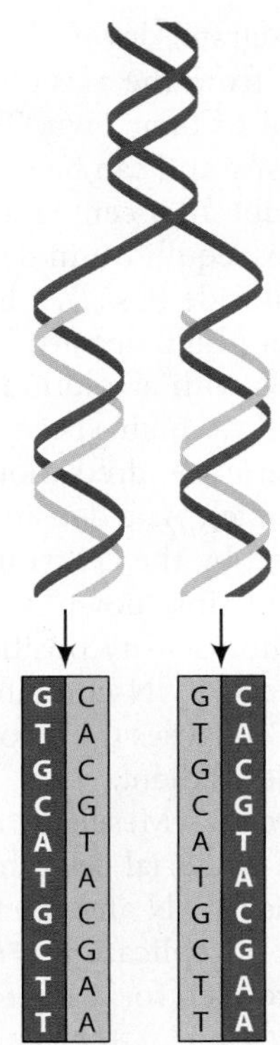

FIGURE 19-2 Watson–Crick Model of DNA Replication. In 1953, Watson and Crick proposed that the DNA double helix replicates semiconservatively, using a model like this one to illustrate the principle. The double-stranded helix unwinds, and each parent strand serves as a template for the synthesis of a complementary daughter strand, assembled according to the base-pairing rules. A, T, C, and G stand for the adenine, thymine, cytosine, and guanine nucleotides. A pairs with T, and G with C.

Equilibrium Density Centrifugation Shows That DNA Replication Is Semiconservative

Within five years of its publication, the Watson–Crick model of semiconservative DNA replication was tested and proved correct by Matthew Meselson and Franklin Stahl. The ingenuity of Meselson and Stahl's contribution lay in the method they devised, in collaboration with Jerome Vinograd, for distinguishing semiconservative replication from other possibilities. Their studies utilized two isotopic forms of nitrogen, ^{14}N and ^{15}N, to distinguish newly synthesized strands of DNA from old strands. Bacterial cells were first grown for many generations in a medium containing ^{15}N-labeled ammonium chloride to incorporate this *heavy* (but nonradioactive) isotope of nitrogen into their DNA molecules. Cells containing ^{15}N-labeled DNA were then transferred to a growth medium containing the normal *light* isotope of nitrogen, ^{14}N. Any new strands of DNA synthesized after this transfer would therefore incorporate ^{14}N rather than ^{15}N.

Since ^{15}N-labeled DNA is significantly denser than ^{14}N-labeled DNA, the old and new DNA strands can be distinguished from each other by **equilibrium density centrifugation,** a technique we encountered earlier when discussing the separation of organelles in Chapter 12 (see Box 12A, page 355). Briefly, this technique allows organelles or macromolecules with differing densities to be separated from each other by centrifugation in a solution containing a gradient of increasing density from the top of the tube to the bottom. In response to centrifugal force, the particles migrate "down" the tube (actually, they move *outward*, away from the axis of rotation) until they reach a density equal to their own. They then remain at this equilibrium density and can be recovered as a band at that position in the tube after centrifugation.

For DNA analysis, equilibrium density centrifugation often uses cesium chloride (CsCl), a heavy metal salt that forms solutions of very high density. The DNA to be analyzed is simply mixed with a solution of cesium chloride and then centrifuged at high speed (generating a centrifugal force of several hundred thousand times gravity for 8 hours, for example). As a density gradient of cesium chloride is established by the centrifugal force, the DNA molecules float "up" or sink "down" within the gradient to reach their equilibrium density positions. The difference in density between heavy (^{15}N-containing) DNA and light (^{14}N-containing) DNA causes them to come to rest at different positions in the gradient.

Using this approach, Meselson and Stahl analyzed DNA obtained from bacterial cells that were first grown for many generations in ^{15}N and then transferred to ^{14}N for additional cycles of replication (**Figure 19-3**). What results would be predicted for a semiconservative mechanism of DNA replication? After one replication cycle in ^{14}N, each DNA molecule should consist of one ^{15}N strand (the old strand) and one ^{14}N strand (the new strand), and so the overall density would be intermediate between heavy DNA and light DNA. The experiments clearly supported this model. After one replication cycle in the ^{14}N medium, centrifugation in cesium chloride revealed a single band of DNA whose density was *exactly halfway* between that of ^{15}N-DNA and ^{14}N-DNA (Figure 19-3b). Because they saw no band at the density expected for heavy DNA, Meselson and Stahl concluded that the original, double-stranded parental DNA was not preserved intact in the replication process. Similarly, the absence of a band at the density expected for light DNA indicated that no daughter DNA molecules consisted exclusively of new DNA. Instead, it appeared that a part of every daughter DNA molecule was newly synthesized and another part was derived from the parent molecule. In fact, the density halfway between that of ^{14}N-DNA and ^{15}N-DNA meant that the $^{14}N/^{15}N$ hybrid DNA molecules were one-half parental and one-half newly synthesized.

When the $^{14}N/^{15}N$ hybrid DNA was heated to separate its two strands, one strand exhibited the density of a ^{15}N-containing strand and the other exhibited the density of a ^{14}N-containing strand, just as predicted by the semiconservative model of replication. Data obtained from cells grown for additional generations in the presence of ^{14}N provided further confirmation. After the second cycle of DNA replication, for example, Meselson and Stahl saw two equal bands, one at the hybrid density of the previous cycle and one at the density of purely ^{14}N-DNA (Figure 19-3c). As the figure illustrates, this is also consistent with a semiconservative mode of replication.

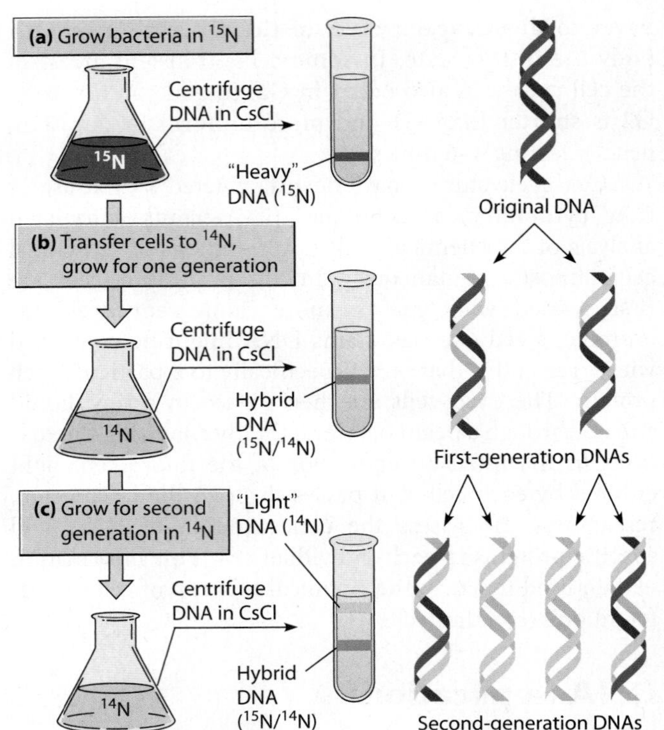

FIGURE 19-3 Semiconservative Replication of Density-Labeled DNA. Meselson and Stahl **(a)** grew bacteria for many generations on a ^{15}N-containing medium and then transferred the cells to a ^{14}N-containing medium for **(b)** one or **(c)** two further cycles of replication. In each case, DNA was extracted from the cells and centrifuged to equilibrium in cesium chloride (CsCl). As shown in the model on the right (where dark blue strands contain ^{15}N and light blue strands contain ^{14}N), the data are compatible with a semiconservative model of DNA replication.

DNA Replication Is Usually Bidirectional

The Meselson–Stahl experiments provided strong support for the idea that, during DNA replication, each strand of the DNA double helix serves as a template for the synthesis of a new complementary strand. As biologists proceeded to unravel the molecular details of this process, it gradually became clear that DNA replication is a complex event involving numerous enzymes and other proteins—and even the participation of RNA. We will first examine the general features of this replication mechanism and then focus on some of its molecular details. In doing so, we will frequently refer to the bacterium *Escherichia coli*, for which DNA replication is especially well understood. However, studies of mammalian viruses such as SV40 and of the yeast *Saccharomyces cerevisiae* have revealed the details of eukaryotic DNA replication as well. DNA replication seems to be a drama whose plot and molecular actors are basically similar in bacterial and eukaryotic cells. This is perhaps not surprising for such a fundamental process—one that must have arisen early in the evolution of life.

FIGURE 19-4 Replication of Circular DNA. **(a)** This autoradiograph shows an *E. coli* DNA molecule caught in the act of replicating. The DNA molecule was isolated from a bacterium that had been grown in a medium containing ^{3}H-thymidine, thereby allowing the DNA to be visualized by autoradiography. **(b)** Replication of a circular DNA molecule begins at a single origin and proceeds bidirectionally around the circle, with the two replication forks moving in opposite directions. The new strands are shown in light blue. The replication process generates intermediates that resemble the Greek letter theta (θ), from which this type of replication derives its name. **(c)** During the cell division cycle of bacteria containing a single circular chromosome, membrane growth between the attachment sites of the two replicating copies moves the daughter chromosomes toward opposite sides of the cell.

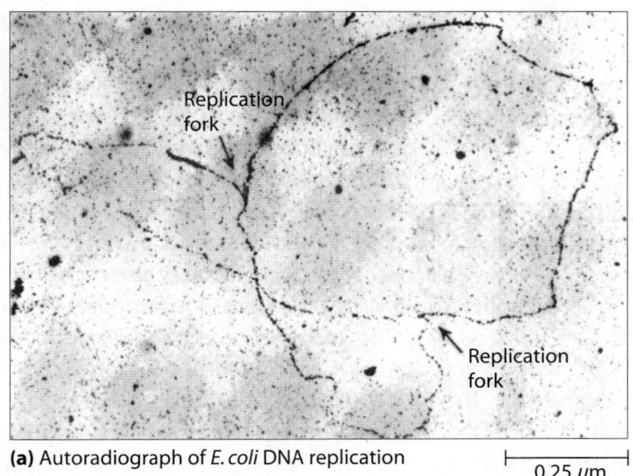

(a) Autoradiograph of *E. coli* DNA replication

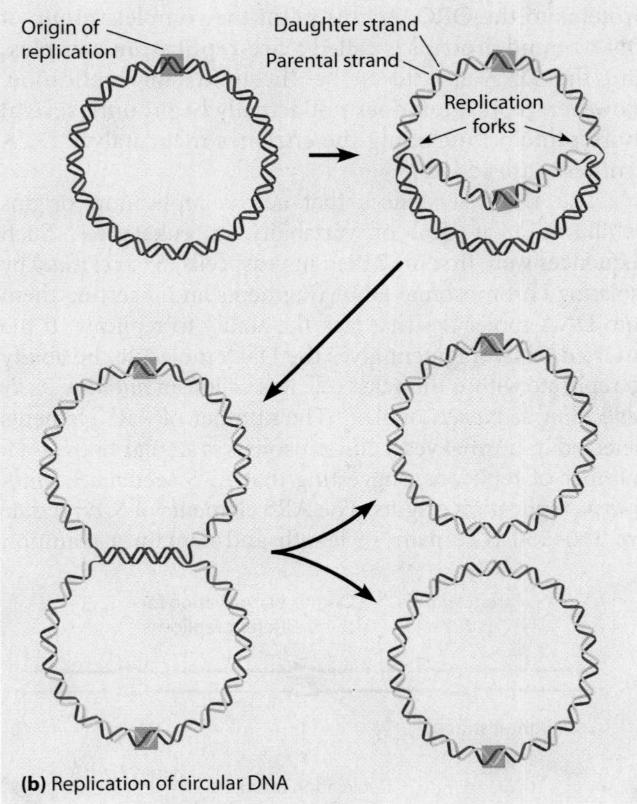

(b) Replication of circular DNA

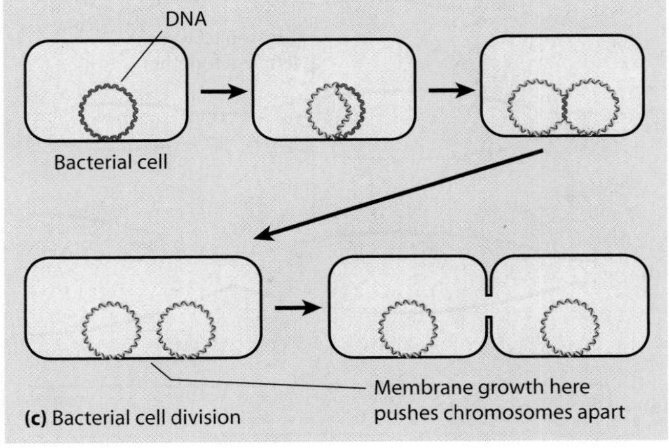

(c) Bacterial cell division

The first experiments to directly visualize DNA replication were carried out by John Cairns, who grew *E. coli* cells in a medium containing the DNA precursor ^{3}H-thymidine and then used autoradiography to examine the cell's single, circular DNA molecule caught in the act of replication. One such molecule is shown in **Figure 19-4a**. The two Y-shaped structures indicated by the arrows represent the sites where the DNA duplex is being replicated. These **replication forks** are created by a DNA replication mechanism that begins at a single point within the DNA and proceeds in a *bidirectional* fashion away from this origin. In other words, two replication forks are created that move in opposite directions away from the point of origin, unwinding the helix and copying both strands as they proceed. For circular DNA, this process is sometimes called *theta replication* because it generates intermediates that look like the Greek letter theta *(θ)*, as you can see in Figure 19-4b. Theta replication occurs not only in bacterial genomes such as that of *E. coli* but also in the circular DNAs of mitochondria, chloroplasts, plasmids, and some viruses. At the end of a round of theta replication the two resulting DNA circles remain interlinked, and the action of a topoisomerase (page 562) is required to disconnect the two circles from each other.

Figure 19-4c shows how these events relate to the cell division cycle of bacteria that contain a single circular chromosome. In such bacteria, the two copies of the replicating chromosome bind to the plasma membrane at their replication origins. As the cell grows in preparation for cell division, new plasma membrane is added to the region between these chromosome attachment sites, thereby pushing the chromosomes toward opposite ends of the cell. When DNA replication is complete and the cell has doubled in size, *binary fission* partitions the cell down the middle and segregates the two chromosomes into two daughter cells.

Eukaryotic DNA Replication Involves Multiple Replicons

In contrast to circular bacterial chromosomes—where DNA replication is initiated at a single origin—replication of the linear DNA molecules of eukaryotic chromosomes is initiated at multiple sites, creating multiple replication units called **replicons** (**Figure 19-5**). The DNA of a typical large eukaryotic chromosome may contain several thousand replicons, each about 50,000–300,000 base pairs in length. At the center of each replicon is a special DNA sequence, called an **origin of replication,** where DNA synthesis is initiated by a mechanism involving several groups of *initiator proteins.* First, a multisubunit protein complex known as the *origin recognition complex (ORC)* binds to a replication origin. The next components to bind are the *minichromosome maintenance (MCM) proteins,* which include several *DNA helicases* that facilitate DNA replication by unwinding the double helix. The recruitment of MCM proteins to the replication origin requires the participation of a third set of proteins, known as *helicase loaders,* which mediate the binding of the MCM proteins to the ORC. At this point the complete group of DNA-bound proteins is called a **pre-replication complex,** and the DNA is said to be "licensed" for replication. However, replication does not actually begin until several more proteins, including the enzymes that catalyze DNA synthesis, are added.

The DNA sequences that act as replication origins exhibit a great deal of variability in eukaryotes. Such sequences were first identified in yeast cells *(S. cerevisiae)* by isolating chromosomal DNA fragments and inserting them into DNA molecules that lack the ability to replicate. If the inserted DNA fragment gives the DNA molecule the ability to replicate within the yeast cell, it is called an *autonomously replicating sequence,* or *ARS.* The number of ARS elements detected in normal yeast chromosomes is similar to the total number of replicons, suggesting that ARS sequences function as replication origins. The ARS elements of *S. cerevisiae* are 100–150 base pairs in length and contain a common 11-nucleotide core sequence, consisting largely of AT base pairs, flanked by auxiliary sequences containing additional AT-rich regions. Since the DNA double helix must be unwound when replication is initiated, the presence of so many AT base pairs serves a useful purpose at replication origins because AT base pairs, held together by two hydrogen bonds, are easier to disrupt than GC base pairs, which have three hydrogen bonds. The replication origins of multicellular eukaryotes are generally larger and more variable in sequence than the ARS elements of *S. cerevisiae,* but they also tend to contain regions that are AT-rich.

After DNA synthesis has been initiated at an origin of replication, two replication forks begin to synthesize DNA in opposite directions away from the origin, creating a "replication bubble" that grows in size as replication proceeds in both directions (Figure 19-5). When the growing replication bubble of one replicon encounters the replication bubble of an adjacent replicon, the DNA synthesized by the two replicons is joined together. In this way, DNA synthesized at numerous replication sites is ultimately linked together to form two double-stranded daughter molecules, each composed of one parental strand and one new strand.

Why does DNA replication involve multiple replication sites in eukaryotes but not in bacteria? Since eukaryotic chromosomes contain more DNA than bacterial chromosomes do, it would take eukaryotes much longer to replicate their chromosomes if DNA synthesis were initiated from only a single replication origin. Moreover, the rate at which each replication fork synthesizes DNA is slower in eukaryotes than in bacteria (presumably because the presence of nucleosomes slows down the replication process). Measurements of the length of radioactive DNA synthesized by cells exposed to ^{3}H-thymidine for varying periods of time have revealed

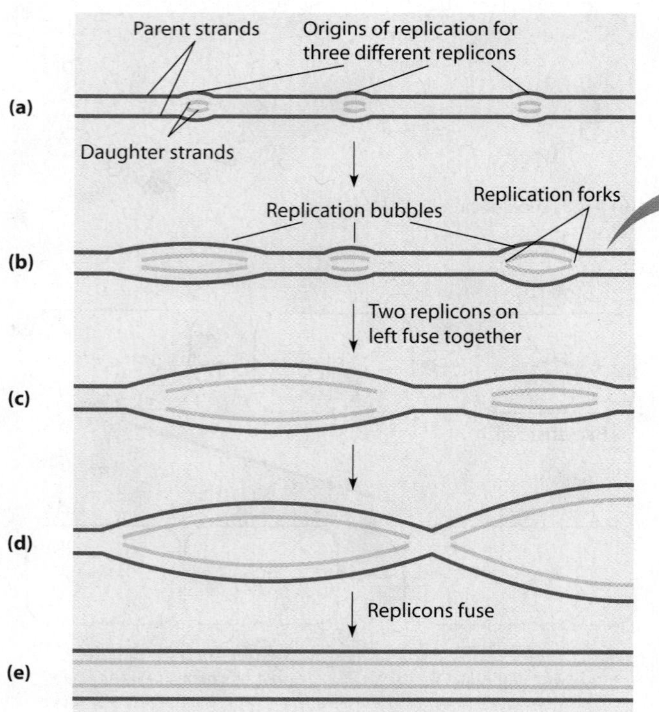

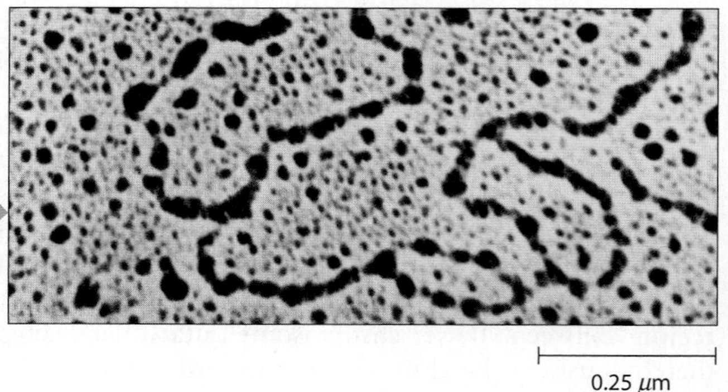

FIGURE 19-5 Multiple Replicons in Eukaryotic DNA. Replication of linear eukaryotic DNA molecules is initiated at multiple origins along the DNA; the timing of initiation is specific for each cluster of origins. **(a)** Replication bubbles form at origins. **(b)** The bubbles grow as the replication forks move along the DNA in both directions from each origin. The micrograph shows three replication bubbles in DNA from cultured Chinese hamster cells (TEM). **(c)** Eventually, individual bubbles meet and fuse. **(d)** A Y-shaped structure forms as a replication fork reaches the end of a DNA molecule. **(e)** When all bubbles have fused, replication is complete and the two daughter molecules separate.

that eukaryotic replication forks synthesize DNA at a rate of about 2000 base pairs/minute, compared with 50,000 base pairs/minute in bacteria. Since the average human chromosome contains about 10^8 base pairs of DNA, it would take more than a month to duplicate a chromosome if there were only a single replication origin!

The relationship between the speed of chromosome replication and the number of replicons is nicely illustrated by comparing the rates of DNA synthesis in embryonic and adult cells of the fruit fly *Drosophila*. In the developing embryo, where cell divisions occur in quick succession, cells employ a large number of simultaneously active replicons measuring only a few thousand base pairs in length. As a result, DNA replication occurs very rapidly, and S phase takes only a few minutes. Adult cells, on the other hand, employ fewer replicons spaced at intervals of tens or hundreds of thousands of base pairs, generating an S phase that requires almost 10 hours to complete. Since the rate of DNA synthesis at any given replication fork is about the same in embryonic and adult cells, it is clear that the length of the S phase is determined by the number of replicons and the rate at which they are activated—not by the rate at which each replicon synthesizes DNA.

During the S phase of a typical eukaryotic cell cycle, replicons are not all activated at the same time. Instead, certain clusters of replicons tend to replicate early during S phase, whereas others replicate later. Information concerning the order in which replicons are activated has been obtained by incubating cells at various points during S phase with *5-bromodeoxyuridine (BrdU)*, a substance that is incorporated into DNA in place of thymidine. Because DNA that contains BrdU is denser than normal DNA, it can be separated from the remainder of the DNA by equilibrium density centrifugation. The BrdU-labeled DNA is then analyzed by hybridization with a series of DNA probes that are specific for individual genes. Such studies have revealed that the genes being actively expressed in a given tissue are replicated early during S phase, whereas inactive genes are replicated later during S phase. If the same gene is analyzed in two different cell types, one in which the gene is active and one in which it is inactive, early replication is observed only in the cell type where the gene is being transcribed.

Replication Licensing Ensures That DNA Molecules Are Duplicated Only Once Prior to Each Cell Division

During the eukaryotic cell cycle it is crucial that nuclear DNA molecules undergo replication once, and only once, prior to cell division. To enforce this restriction, a process called **licensing** ensures that after DNA is replicated at any given replication origin during S phase, the DNA at that site does not become competent (licensed) for a further round of DNA replication until the cell has first passed through mitosis. The license is provided by the binding of MCM proteins to replication origins, an event that requires both ORC and helicase loaders (page 554). Once replication begins, the MCM proteins are displaced from the origins by the traveling replication fork, and re-replication from the same origins is prevented by mechanisms that stop MCM proteins from binding to replication origins again. Thus the license to replicate is never associated with replicated DNA.

A critical player in this mechanism is *cyclin-dependent kinase (Cdk)*, an enzyme whose many roles in the cell cycle will be described later in the chapter. One form of Cdk is produced at the beginning of S phase and functions both in activating DNA synthesis at licensed origins and in ensuring that these same origins cannot become licensed again. Cdk blocks relicensing by catalyzing the phosphorylation, and thereby inhibiting the function, of proteins required for licensing, such as ORC and the helicase loaders. Multicellular eukaryotes contain another inhibitor of relicensing called *geminin*, a protein made during S phase that blocks the binding of MCM proteins to DNA. After the cell completes mitosis, geminin is degraded and Cdk activity falls, so the proteins required for DNA licensing can function again for the next cell cycle (**Figure 19-6**).

DNA Polymerases Catalyze the Elongation of DNA Chains

MB Molecular Model: DNA Polymerase III Beta Subunit; Molecular Model: Bacteriophage T7 DNA Polymerase

When the semiconservative model of DNA replication was first proposed in the early 1950s, biologists thought that DNA replication was so complex that it could only be carried out by intact cells. But just a few years later, Arthur Kornberg showed that an enzyme he had isolated from bacteria can copy DNA molecules in a test tube. This enzyme, which he named **DNA polymerase,** requires that a small amount of DNA be initially present to act as a template. Guided by this template, DNA polymerase catalyzes the elongation of DNA chains using as substrates the triphosphate deoxynucleoside derivatives of the four bases

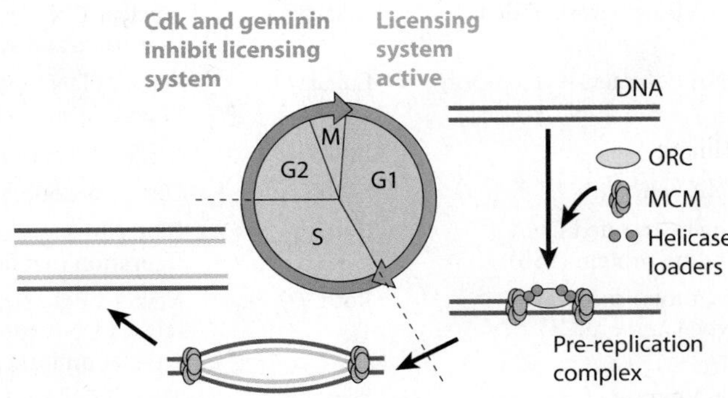

FIGURE 19-6 Licensing of DNA Replication During the Eukaryotic Cell Cycle. DNA is licensed for replication during G1 by the binding of MCM proteins to replication origins, an event that requires both ORC and helicase loaders. The licensing system is turned off at the end of G1 by the production of Cdks and/or geminin, whose activities block the functions of the proteins required for licensing (ORC, helicase loaders, and MCM). After the cell completes mitosis, geminin is degraded and Cdk activity falls, so the licensing system becomes active again for the next cell cycle.

found in DNA (dATP, dTTP, dGTP, and dCTP). As each of these substrates is incorporated into a newly forming DNA chain, its two terminal phosphate groups are released. Since deoxynucleoside triphosphates are high-energy compounds (comparable to ATP), the energy released as these phosphate bonds are broken drives what would otherwise be a thermodynamically unfavorable polymerization reaction.

In the DNA polymerase reaction, incoming nucleotides are covalently bonded to the 3′ hydroxyl end of the growing DNA chain. Each successive nucleotide is linked to the growing chain by a phosphoester bond between the phosphate group on its 5′ carbon and the hydroxyl group on the 3′ carbon of the nucleotide added in the previous step (**Figure 19-7**). In other words, chain elongation occurs at the 3′ end of a DNA strand, and the strand is therefore said to grow in the 5′ → 3′ direction.

Soon after Kornberg's initial discovery, several other forms of DNA polymerase were detected. (**Table 19-1** lists the main DNA polymerases used in DNA replication, along with other key proteins involved in the process.) In *E. coli,* the enzyme discovered by Kornberg has turned out not to be responsible for DNA replication in intact cells. This fact first became apparent when Peter DeLucia and John Cairns reported that mutant strains of bacteria lacking the Kornberg enzyme can still replicate their DNA and reproduce normally. With the Kornberg enzyme missing, it was possible to detect the presence of several other bacterial enzymes that synthesize DNA. These additional enzymes are named using Roman numerals (e.g., DNA polymerases II, III, IV, and V) to distinguish them from the original Kornberg enzyme, now called DNA polymerase I. When the rates at which the various DNA polymerases synthesize DNA in a test tube were first compared, only DNA polymerase III was found to work fast enough to account for the rate of DNA replication in intact cells, which averages about 50,000 base pairs/minute in bacteria.

Such observations suggested that DNA polymerase III is the main enzyme responsible for DNA replication in bacterial cells, but the evidence would be more convincing if it could be shown that cells lacking DNA polymerase III are unable to replicate their DNA. How is it possible to grow and study cells that have lost the ability to carry out an essential function such as DNA replication? One powerful approach involves the use of **temperature-sensitive mutants,** which are cells that produce proteins that function properly at normal temperatures but become seriously impaired when the temperature is altered slightly. For example, mutant bacteria have been isolated in which DNA polymerase III behaves normally at 37°C but loses its function when the temperature is raised to 42°C. Such

Table 19-1 Some Important DNA Replication Proteins in Bacteria and Eukaryotes

Protein	Cell Type	Main Activities and/or Functions
DNA polymerase I	Bacteria	DNA synthesis; 3′ → 5′ exonuclease (for proofreading); 5′ → 3′ exonuclease; removes and replaces RNA primers used in DNA replication (also functions in excision repair of damaged DNA)
DNA polymerase III	Bacteria	DNA synthesis; 3′ → 5′ exonuclease (for proofreading); used in synthesis of both DNA strands
DNA polymerase α (alpha)	Eukaryotes	Nuclear DNA synthesis; forms complex with primase and begins DNA synthesis at the 3′ end of RNA primers for both leading and lagging strands (also functions in DNA repair)
DNA polymerase γ (gamma)	Eukaryotes	Mitochondrial DNA synthesis
DNA polymerase δ (delta)	Eukaryotes	Nuclear DNA synthesis; 3′ → 5′ exonuclease (for proofreading); involved in lagging and leading strand synthesis (also functions in DNA repair)
DNA polymerase ε (epsilon)	Eukaryotes	Nuclear DNA synthesis; 3′ → 5′ exonuclease (for proofreading); thought to be involved in leading and lagging strand synthesis (also functions in DNA repair)
Primase	Both	RNA synthesis; makes RNA oligonucleotides that are used as primers for DNA synthesis
DNA helicase	Both	Unwinds double-stranded DNA
Single-stranded DNA binding protein (SSB)	Both	Binds to single-stranded DNA; stabilizes strands of unwound DNA in an extended configuration that facilitates access by other proteins
DNA topoisomerase (type I and type II)	Both	Makes single-strand cuts (type I) or double-strand cuts (type II) in DNA; induces and/or relaxes DNA supercoiling; can serve as swivel to prevent overwinding ahead of the DNA replication fork; can separate linked DNA circles at the end of DNA replication
DNA gyrase	Bacteria	Type II DNA topoisomerase that serves as a swivel to relax supercoiling ahead of the DNA replication fork in *E. coli*
DNA ligase	Both	Makes covalent bonds to join together adjacent DNA strands, including the Okazaki fragments in lagging strand DNA synthesis and the new and old DNA segments in excision repair of DNA
Initiator proteins	Both	Bind to origin of replication and initiate unwinding of DNA double helix
Telomerase	Eukaryotes	Using an integral RNA molecule as template, synthesizes DNA for extension of telomeres (sequences at ends of chromosomal DNA)

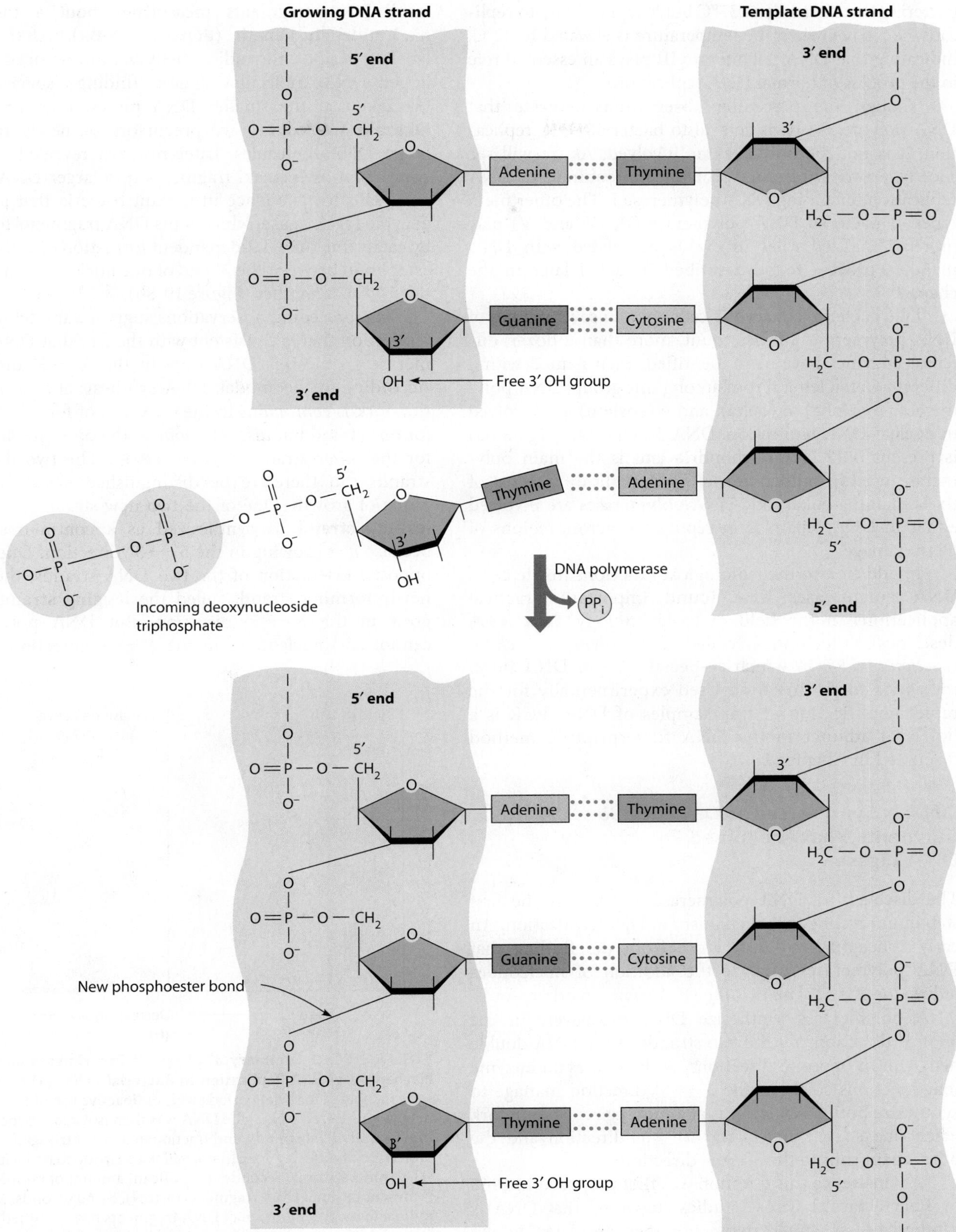

MB

DNA Replication (1 of 2): DNA Structure and Replication Machinery (BioFlix tutorial); Chemistry Review–Nucleic Acids: DNA Replication

FIGURE 19-7 The Directionality of DNA Synthesis. Addition of the next nucleotide to a growing DNA strand is catalyzed by DNA polymerase and always occurs at the 3′ end of the strand. A phosphoester bond is formed between the 3′ hydroxyl group of the terminal nucleotide and the 5′ phosphate of the incoming deoxynucleoside triphosphate (here dTTP), extending the growing chain by one nucleotide, liberating pyrophosphate (PP_i) and leaving the 3′ end of the strand with a free hydroxyl group to accept the next nucleotide.

bacteria grow normally at 37°C but lose the ability to replicate their DNA when the temperature is elevated to 42°C, indicating that DNA polymerase III plays an essential role in the process of normal DNA replication.

Though the preceding observations indicate that DNA polymerase III is central to bacterial DNA replication, it is not the only enzyme involved. As we will see shortly, a variety of other proteins are required for DNA replication, including DNA polymerase I. The other main types of bacterial DNA polymerase (II, IV, and V) play more specialized roles in events associated with DNA repair, a process to be described in detail later in the chapter.

Like bacteria, eukaryotic cells contain several types of DNA polymerase. At last count, more than a dozen different enzymes have been identified, each named with a different Greek letter. From among this group, DNA polymerases α (alpha), δ (delta), and ε (epsilon) are involved in nuclear DNA replication. DNA polymerase γ (gamma) is present only in mitochondria and is the main polymerase used in mitochondrial DNA replication. Most of the remaining eukaryotic DNA polymerases are involved either in DNA repair or in replication across regions of DNA damage.

In addition to their biological functions inside cells, DNA polymerases have found important practical applications in the field of biotechnology. **Box 19A** describes a technique called the *polymerase chain reaction (PCR)*, in which a special type of DNA polymerase is the prime tool. Used experimentally for the rapid amplification of tiny samples of DNA, PCR is a powerful adjunct to the DNA fingerprinting method discussed in Chapter 18.

Activity: DNA Replication: An Overview

DNA Is Synthesized as Discontinuous Segments That Are Joined Together by DNA Ligase

The discovery of DNA polymerase was merely the first step in unraveling the mechanism of DNA replication. An early conceptual problem arose from the finding that DNA polymerases catalyze the addition of nucleotides *only to the 3′ end of an existing DNA chain;* in other words, DNA polymerases synthesize DNA exclusively in the 5′→3′ direction. Yet the two strands of the DNA double helix run in opposite directions. So how does an enzyme that functions solely in the 5′→3′ direction manage to synthesize both DNA strands at a moving replication fork when one chain runs in the 5′→3′ direction and the other chain runs in the 3′→5′ direction?

An answer to this question was first proposed in 1968 by Reiji Okazaki, whose studies suggested that DNA is synthesized as small fragments that are later joined together. Okazaki isolated DNA from bacterial cells that had been briefly exposed to a radioactive substrate that is incorporated into newly made DNA. Analysis of this DNA revealed that much of the radioactivity was located in small DNA fragments measuring about a thousand nucleotides in length (**Figure 19-8a**). After longer labeling periods, the radioactivity became associated with larger DNA molecules. These findings suggested to Okazaki that the smaller DNA pieces, now known as **Okazaki fragments,** are precursors of newly forming larger DNA molecules. Later research revealed that the conversion of Okazaki fragments into larger DNA molecules fails to take place in mutant bacteria that lack the enzyme **DNA ligase,** which joins DNA fragments together by catalyzing the ATP-dependent formation of a phosphoester bond between the 3′ end of one nucleotide chain and the 5′ end of another (Figure 19-8b).

The preceding observations suggest a model of DNA replication that is consistent with the fact that DNA polymerase synthesizes DNA only in the 5′→3′ direction. According to this model, DNA synthesis at each replication fork is *continuous* in the direction of fork movement for one strand but *discontinuous* in the opposite direction for the other strand (**Figure 19-9**). The two daughter strands can therefore be distinguished based on their mode of growth. One of the two new strands, called the **leading strand,** is synthesized as a continuous chain because it is growing in the 5′→3′ direction. Due to the opposite orientation of the two DNA strands, the other newly forming strand, called the **lagging strand,** must grow in the 3′→5′ direction. But DNA polymerase cannot add nucleotides in the 3′→5′ direction, so the

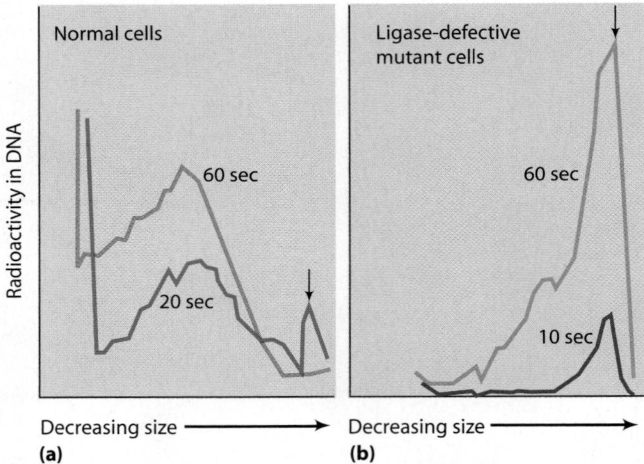

FIGURE 19-8 Summary of Okazaki's Experiments on the Mechanism of DNA Replication in Bacterial Cells. **(a)** Bacteria were incubated for brief periods with radioactive thymidine to label newly synthesized DNA. The DNA was then isolated, dissociated into its individual strands, and fractionated by centrifugation into molecules of differing size. In normal bacteria incubated with ^{3}H-thymidine for 20 seconds, a significant amount of radioactivity is present in small DNA fragments (arrow). By 60 seconds, the radioactivity present in small DNA fragments has all shifted to larger DNA molecules. **(b)** In bacterial mutants deficient in the enzyme DNA ligase, radioactivity remains in small DNA fragments even after 60 seconds of incubation. It was therefore concluded that DNA ligase normally functions to join small DNA fragments together into longer DNA chains.

lagging strand is instead formed as a series of short, discontinuous Okazaki fragments that are synthesized in the $5' \rightarrow 3'$ direction. These fragments are then joined together by DNA ligase to make a continuous new $3' \rightarrow 5'$ DNA strand. Okazaki fragments are generally about 1000–2000 nucleotides long in viral and bacterial systems but only about one-tenth this length in eukaryotic cells. In *E. coli,* the same DNA polymerase, polymerase III, is used for synthesizing the Okazaki fragments of the lagging strand and the continuous DNA chain of the leading strand. In eukaryotes, DNA polymerases δ and ε are both thought to be involved in synthesizing leading and lagging strands.

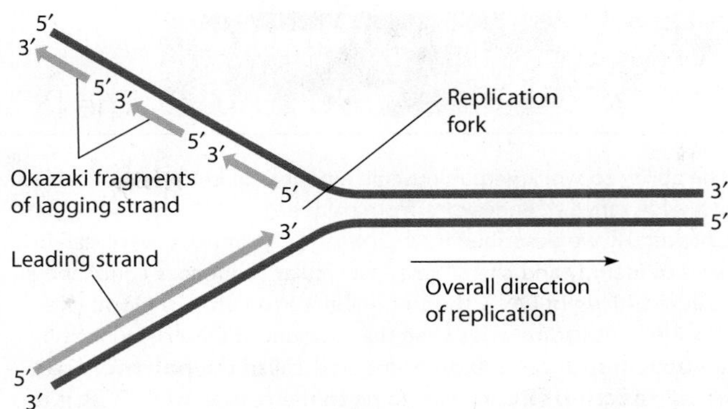

FIGURE 19-9 Directions of DNA Synthesis at a Replication Fork. Because DNA polymerases synthesize DNA chains only in the $5' \rightarrow 3'$ direction, synthesis at each replication fork is continuous in the direction of fork movement for the leading strand but discontinuous in the opposite direction for the lagging strand. Discontinuous synthesis involves short intermediates called Okazaki fragments, which are 1000–2000 nucleotides long in bacteria and about 100–200 nucleotides long in eukaryotic cells. The fragments are later joined together by the enzyme DNA ligase. Parental DNA is shown in dark blue, and newly synthesized DNA in lighter blue. Here and in subsequent figures, arrowheads indicate the direction in which the nucleic acid chain is being elongated.

MB DNA Replication; DNA Replication (2 of 2): Synthesis of the Leading and Lagging Strands (Bio Flix tutorial)

Proofreading Is Performed by the $3' \rightarrow 5'$ Exonuclease Activity of DNA Polymerase

Given the complexity of the preceding model, you might wonder why cells have not simply evolved an enzyme that synthesizes DNA in the $3' \rightarrow 5'$ direction. One possible answer is related to the need for error correction during DNA replication. About 1 out of every 100,000 nucleotides incorporated during DNA replication is incorrectly base-paired with the template DNA strand, an error rate that would yield more than 120,000 errors every time a human cell replicates its DNA. Fortunately, such mistakes are usually fixed by a **proofreading** mechanism that uses the same DNA polymerase molecules that catalyze DNA synthesis. Proofreading is made possible by the fact that almost all DNA polymerases possess $3' \rightarrow 5'$ exonuclease activity (in addition to their ability to catalyze DNA synthesis). **Exonucleases** are enzymes that degrade nucleic acids (usually DNA) from one end, rather than making internal cuts, as **endonucleases** do. A $3' \rightarrow 5'$ exonuclease is one that clips off nucleotides from the 3′ end of a nucleotide chain. Hence, the $3' \rightarrow 5'$ exonuclease activity of DNA polymerase allows it to remove improperly base-paired nucleotides from the 3′ end of a growing DNA chain (**Figure 19-10**). This ability to remove incorrect nucleotides improves the fidelity of DNA replication to an average of only a few errors for every billion base pairs replicated.

If cells did happen to possess an enzyme capable of synthesizing DNA in the $3' \rightarrow 5'$ direction, proofreading could not work because a DNA chain growing in the $3' \rightarrow 5'$ direction would have a nucleotide triphosphate at its growing 5′ end. What if this 5′ nucleotide contained an incorrect base that needed to be removed during proofreading? Removing this 5′ nucleotide would eliminate the triphosphate group that provides the free energy that allows DNA polymerase to add nucleotides to a growing DNA chain, and hence the chain could not elongate further.

RNA Primers Initiate DNA Replication

Since DNA polymerase can only add nucleotides to an existing nucleotide chain, how is replication of a DNA double helix initiated? Shortly after Okazaki fragments were

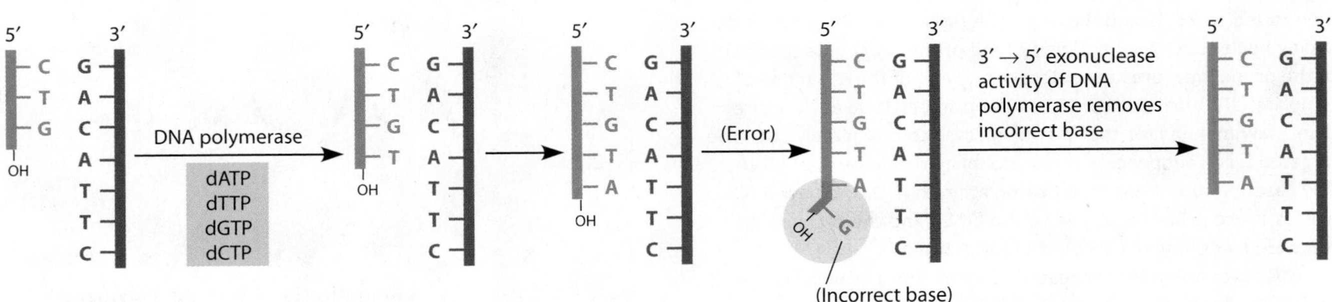

FIGURE 19-10 Proofreading by $3' \rightarrow 5'$ Exonuclease. If an incorrect base is inserted during DNA replication, the $3' \rightarrow 5'$ exonuclease activity that is part of the DNA polymerase molecule catalyzes its removal so that the correct base can be inserted.

BOX 19A TOOLS OF DISCOVERY

The PCR Revolution

The ability to work with minuscule amounts of DNA is invaluable in a wide range of endeavors, from paleontology to criminology. In Chapter 18, we described how DNA fingerprinting analysis can be used to identify and characterize particular sequences contained in as little as 1 μg of DNA, the amount in a small drop of blood (see Box 18C). But sometimes even that amount of DNA may not be available. In such cases another method, called the **polymerase chain reaction (PCR),** can come to the rescue. With PCR, it is possible to rapidly replicate, or *amplify,* selected DNA segments that are initially present in extremely small amounts. In only a few hours, PCR can make millions or even billions of copies of a particular DNA sequence, thereby producing enough material for DNA fingerprinting, DNA sequencing, or other applications.

The complex multiprotein system that cells use for DNA replication is not required for the PCR method. Instead, PCR employs an unusual DNA polymerase coupled with synthetic primers to set up a chain reaction that produces an exponentially growing population of specific DNA molecules. Biochemist Kary Mullis received a Nobel Prize in 1993 for developing this technique. To perform PCR, you first need to know part of the base sequence of the DNA segment you wish to amplify. Based on this information, short single-stranded *DNA primers* are chemically synthesized; these primers generally consist of DNA segments 15–20 nucleotides long that are complementary to sequences located at opposite ends of the DNA segment being amplified. (If sequences that naturally flank the sequence of interest are not known, artificial ones can be attached before running the polymerase chain reaction.) DNA polymerase is then added to catalyze the synthesis of complementary DNA strands using the two primers as starting points. The DNA polymerase routinely used for this purpose was first isolated from the bacterium *Thermus aquaticus,* an inhabitant of thermal hot springs where the waters are normally 70–80°C. The optimal temperature for this enzyme, called *Taq* polymerase, is 72°C, and it is stable at even higher temperatures—a property that made possible the automation of PCR.

Figure 19A-1 summarizes how the PCR procedure works. The ingredients of the initial reaction mixture include the DNA containing the sequence to be amplified, *Taq* DNA polymerase, the synthetic DNA primers, and the four deoxynucleoside triphosphates (dATP, dTTP, dCTP, and dGTP). Each reaction cycle begins with a short period of heating to near boiling (95°C) to denature the DNA double helix into its two strands (➊). The DNA solution is then cooled to allow the primers to base-pair to complementary regions on the DNA strands being copied (➋). The temperature is then raised to 72°C, and the *Taq* DNA polymerase goes to work, adding nucleotides to the 3′ end of the primer (➌). The specificity of the primers ensures the selective copying of the stretches of template DNA downstream from the primers. It takes no more than a few minutes for the *Taq* polymerase to completely copy the targeted DNA sequence, thereby doubling the amount of DNA. The reaction mixture is then heated again to melt the new double helices, more primer is bound to the DNA, and the cycle is repeated to double the amount of DNA again (➊–➌).

This reaction cycle is repeated as many times as necessary, with each cycle doubling the amount of DNA from the previous cycle. After the third cycle, more and more of the product DNA molecules will be of a uniform length that consists only of the targeted

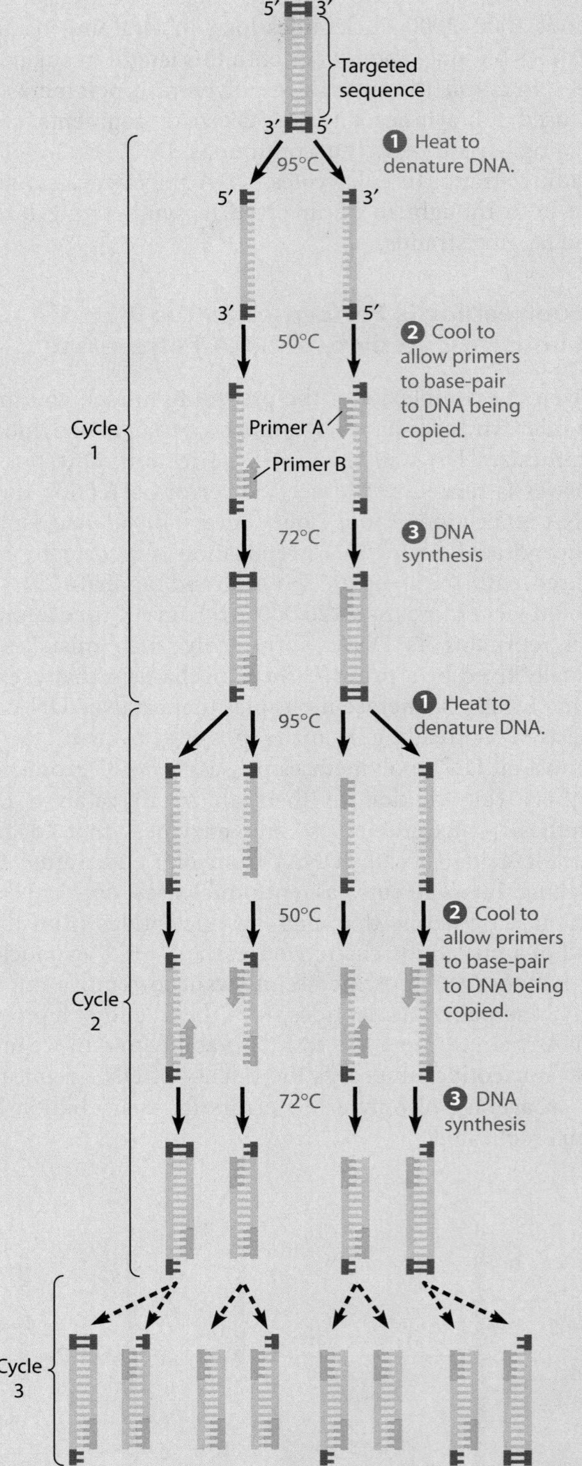

FIGURE 19A-1 DNA Amplification Using the Polymerase Chain Reaction. See the description in the box text. PCR works best when the DNA segment to be amplified—the region flanked by the two primers—is 50–2000 nucleotides long.

MB Activity: The Polymerase Chain Reaction

sequence (like the third and sixth molecules in the last line of the figure). Because heating to 95°C does not destroy the *Taq* polymerase, there is no need to add fresh enzyme for each round of the cycle. In most cases, 20–30 reaction cycles are sufficient to produce the desired quantity of DNA. The theoretical amplification accomplished by n cycles is 2^n, so 20 cycles yields an amplification of a millionfold or more ($2^{20} = 1{,}048{,}576$) and 30 cycles over a billionfold ($2^{30} = 1{,}073{,}741{,}824$). Since each cycle takes less than 5 minutes, several hundred billion copies of the original DNA sequence can be produced within a few hours. This is considerably quicker than the several days required for amplifying DNA by cloning it in bacteria, a method to be discussed in Chapter 20. Furthermore, PCR can be used with as little as one molecule of DNA, and it does not require the starting DNA sample to be purified because the primers select the DNA region that will be amplified.

PCR therefore makes it possible to identify a person from the minuscule amount of DNA that is left behind when that person touches an object, inadvertently leaving a few skin cells behind. By using PCR to amplify the tiny amount of DNA in such a sample and then performing a DNA fingerprinting analysis on the amplified DNA, it is possible to obtain a DNA fingerprint from a person's actual fingerprints! Although such techniques have enormous potential in helping to solve crimes, this extraordinary sensitivity can also cause problems. A few contaminating DNA molecules (such as from skin cells shed by a lab technician) might be amplified along with the DNA of interest, yielding misleading results.

Nevertheless, with proper precautions and controls, PCR is proving extremely valuable. As an aid in evolution research, it has been used to amplify DNA fragments recovered from ancient Egyptian mummies and a 40,000-year-old woolly mammoth frozen in a glacier, and to unravel the genome of Neanderthals. In medical diagnosis, PCR has been used to amplify DNA from single embryonic cells for rapid prenatal diagnosis, and it has made possible the detection of viral genes in cells infected with HIV or other viruses. Perhaps most importantly, PCR has revolutionized basic research in molecular genetics by allowing easy amplification of particular genes or sequences from among the thousands of genes in mammalian genomes.

first discovered, researchers implicated RNA in the initiation process through the following observations: (1) Okazaki fragments often have short stretches of RNA, usually 3–10 nucleotides in length, at their 5′ ends; (2) DNA polymerase can catalyze the addition of nucleotides to the 3′ end of RNA chains as well as to DNA chains; (3) cells contain an enzyme called **primase** that synthesizes RNA fragments about ten bases long using DNA as a template; and (4) unlike DNA polymerase, which adds nucleotides only to the ends of existing chains, primase can initiate RNA synthesis from scratch by joining two nucleotides together.

These observations led to the conclusion that DNA synthesis is initiated by the formation of short **RNA primers.** RNA primers are synthesized by primase, which uses a single DNA strand as a template to guide the synthesis of a complementary stretch of RNA (**Figure 19-11**, ❶). Primase is a specific kind of RNA polymerase used only in DNA replication. Like other RNA polymerases, but unlike DNA polymerases, primases can *initiate* the synthesis of a new polynucleotide strand complementary to a template strand; they do not themselves require a primer.

In *E. coli,* primase is relatively inactive unless it is accompanied by six other proteins, forming a complex called a **primosome.** The other primosome proteins function in unwinding the parental DNA and recognizing target DNA sequences where replication is to be initiated. The situation in eukaryotic cells is slightly different, so the term *primosome* is not used. The eukaryotic primase is not as closely associated with unwinding proteins, but it is very tightly bound to DNA polymerase α, the main DNA polymerase involved in initiating DNA replication.

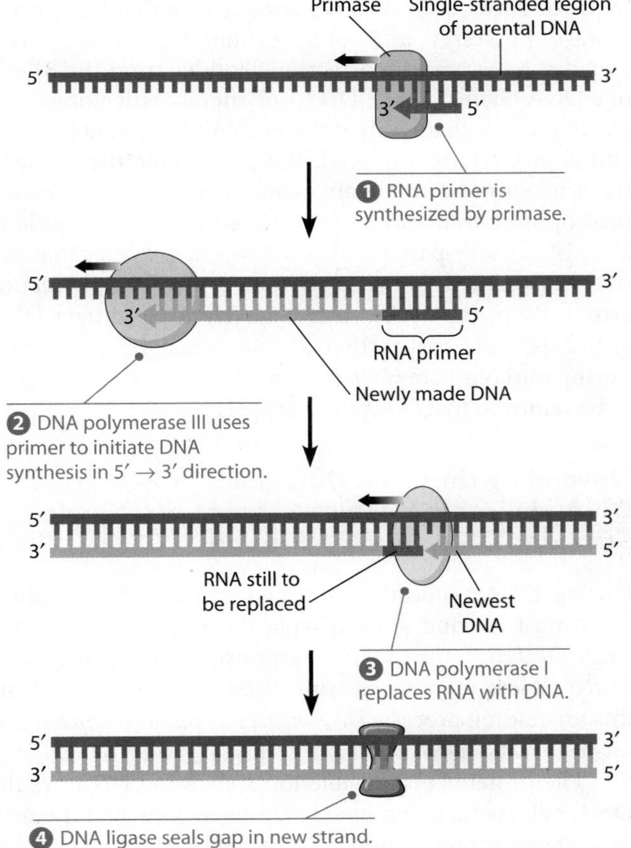

Primase
Single-stranded region of parental DNA
5′
3′
3′
5′
❶ RNA primer is synthesized by primase.
5′
3′
3′
5′
RNA primer
Newly made DNA
❷ DNA polymerase III uses primer to initiate DNA synthesis in 5′ → 3′ direction.
5′
3′
3′
5′
RNA still to be replaced
Newest DNA
❸ DNA polymerase I replaces RNA with DNA.
5′
3′
3′
5′
❹ DNA ligase seals gap in new strand.

FIGURE 19-11 The Role of RNA Primers in DNA Replication. DNA synthesis is initiated with a short RNA primer in both bacteria and eukaryotes. This figure shows the process as it occurs for the lagging strand in *E. coli.*

Once an RNA primer has been created, DNA synthesis can proceed, with DNA polymerase III (or DNA polymerase α followed by polymerase δ or ε in eukaryotes) adding successive deoxynucleotides to the 3′ end of the primer (Figure 19-11, ❷). For the leading strand, initiation using an RNA primer needs to occur only once, when a replication fork first forms; DNA polymerase can then add nucleotides to the chain continuously in the 5′ → 3′ direction. In contrast, the lagging strand is synthesized as a series of discontinuous Okazaki fragments, and each of them must be initiated with a separate RNA primer. For each primer, DNA nucleotides are added by DNA polymerase III until the growing fragment reaches the adjacent Okazaki fragment. No longer needed at that point, the RNA segment is removed and DNA nucleotides are polymerized to fill its place. In *E. coli,* the RNA primers are removed by a 5′ → 3′ exonuclease activity inherent to the DNA polymerase I molecule (distinct from the 3′ → 5′ exonuclease activity involved in proofreading). At the same time, the DNA polymerase I molecule synthesizes DNA in the normal 5′ → 3′ direction to fill in the resulting gaps (Figure 19-11, ❸). Adjacent fragments are subsequently joined together by DNA ligase.

Why do cells employ RNA primers that must later be removed rather than simply using a DNA primer in the first place? Again, the answer may be related to the need for error correction. We have already seen that DNA polymerase possesses a 3′ → 5′ exonuclease activity that allows it to remove incorrect nucleotides from the 3′ end of a DNA chain. In fact, DNA polymerase will elongate an existing DNA chain only if the nucleotide present at the 3′ end is properly base-paired. But an enzyme that *initiates* the synthesis of a new chain cannot perform such a proofreading function because it is not adding a nucleotide to an existing base-paired end. As a result, enzymes that initiate nucleic acid synthesis are not very good at correcting errors. By using RNA rather than DNA to initiate DNA synthesis, cells ensure that any incorrect bases inserted during initiation are restricted to RNA sequences destined to be removed by DNA polymerase I.

Unwinding the DNA Double Helix Requires DNA Helicases, Topoisomerases, and Single-Stranded DNA Binding Proteins

During DNA replication, the two strands of the double helix must unwind at each replication fork to expose the single strands to the enzymes responsible for copying them. Three classes of proteins with distinct functions facilitate this unwinding process: *DNA helicases, topoisomerases,* and *single-stranded DNA binding proteins* (**Figure 19-12**).

The proteins responsible for unwinding DNA are the **DNA helicases.** Using energy derived from ATP hydrolysis, these proteins unwind the DNA double helix in advance of the replication fork, breaking the hydrogen bonds as they go. In *E. coli,* at least two different DNA helicases are involved in DNA replication; one attaches to the lagging strand template and moves in a 5′ → 3′ direction; the other attaches to the leading strand template and moves 3′ → 5′. Both are part of the primosome, but the 5′ → 3′ helicase is more important for unwinding DNA at the replication fork.

The unwinding associated with DNA replication would create an intolerable amount of supercoiling and possibly tangling in the rest of the DNA were it not for the actions of **topoisomerases,** which we discussed in Chapter 18. These enzymes create swivel points in the DNA molecule by making and then quickly resealing single- or double-stranded breaks in the double helix. Of the ten or so topoisomerases found in *E. coli,* the key enzyme for DNA replication is *DNA gyrase,* a type II topoisomerase (an enzyme that cuts both DNA strands). Using energy derived from ATP, DNA gyrase introduces negative supercoils and thereby relaxes positive ones. DNA gyrase serves as the main swivel that prevents overwinding (positive supercoiling) of the DNA ahead of the replication fork. In addition, this enzyme has a role in both initiating and completing DNA replication in *E. coli*—in opening up the double helix at the origin of replication and in separating the linked circles of daughter DNA at the end. The situation in eukaryotic cells is not as well understood, although topoisomerases of both types have been isolated.

Once strand separation has begun, molecules of **single-stranded DNA binding protein (SSB)** quickly attach to the exposed single strands to keep the DNA unwound and therefore accessible to the DNA replication machinery. After a particular segment of DNA has been replicated, the SSB molecules fall off and are recycled, attaching to the next single-stranded segment.

Putting It All Together: DNA Replication in Summary

Figure 19-13 reviews the highlights of what we currently understand about the mechanics of DNA replication in

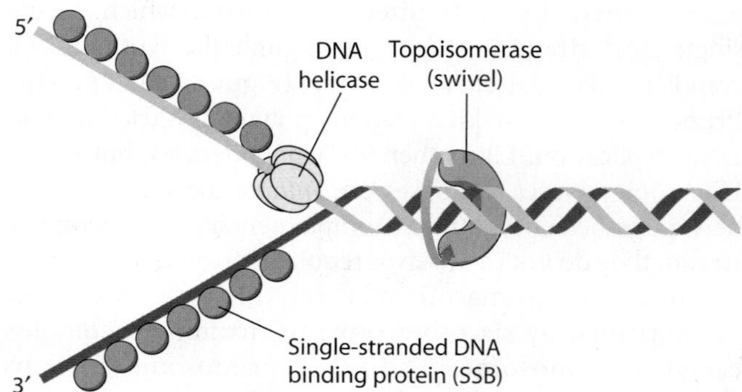

FIGURE 19-12 Proteins Involved in Unwinding DNA at the Replication Fork. Three types of proteins are involved in DNA unwinding. The actual unwinding proteins are the DNA helicases; the principal one in *E. coli,* which is part of the primosome, operates 5′ → 3′ along the template for the lagging strand, as shown here. Single-stranded DNA binding proteins (SSB) stabilize the unwound DNA in an extended position. A topoisomerase forms a swivel ahead of the replication fork; in *E. coli,* this topoisomerase is DNA gyrase.

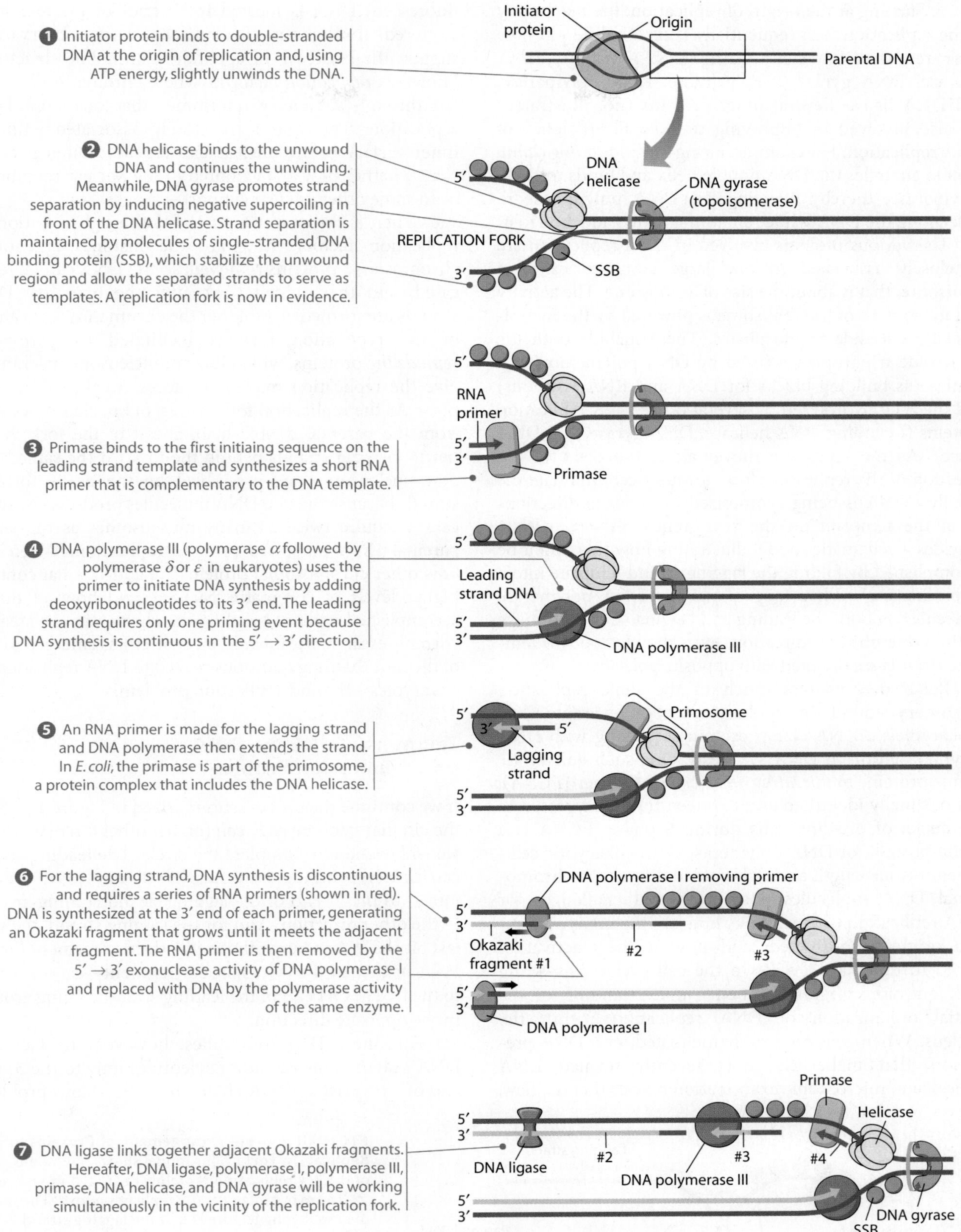

1 Initiator protein binds to double-stranded DNA at the origin of replication and, using ATP energy, slightly unwinds the DNA.
Initiator protein
Origin
Parental DNA
2 DNA helicase binds to the unwound DNA and continues the unwinding. Meanwhile, DNA gyrase promotes strand separation by inducing negative supercoiling in front of the DNA helicase. Strand separation is maintained by molecules of single-stranded DNA binding protein (SSB), which stabilize the unwound region and allow the separated strands to serve as templates. A replication fork is now in evidence.
5′
DNA helicase
DNA gyrase (topoisomerase)
REPLICATION FORK
3′
SSB
3 Primase binds to the first initiaing sequence on the leading strand template and synthesizes a short RNA primer that is complementary to the DNA template.
RNA primer
Primase
4 DNA polymerase III (polymerase α followed by polymerase δ or ε in eukaryotes) uses the primer to initiate DNA synthesis by adding deoxyribonucleotides to its 3′ end. The leading strand requires only one priming event because DNA synthesis is continuous in the 5′ → 3′ direction.
Leading strand DNA
DNA polymerase III
5 An RNA primer is made for the lagging strand and DNA polymerase then extends the strand. In E. coli, the primase is part of the primosome, a protein complex that includes the DNA helicase.
Primosome
Lagging strand
6 For the lagging strand, DNA synthesis is discontinuous and requires a series of RNA primers (shown in red). DNA is synthesized at the 3′ end of each primer, generating an Okazaki fragment that grows until it meets the adjacent fragment. The RNA primer is then removed by the 5′ → 3′ exonuclease activity of DNA polymerase I and replaced with DNA by the polymerase activity of the same enzyme.
DNA polymerase I removing primer
Okazaki fragment #1
#2
#3
DNA polymerase I
7 DNA ligase links together adjacent Okazaki fragments. Hereafter, DNA ligase, polymerase I, polymerase III, primase, DNA helicase, and DNA gyrase will be working simultaneously in the vicinity of the replication fork.
Primase
Helicase
#4
DNA ligase
DNA polymerase III
DNA gyrase
SSB

MB Activity: DNA Synthesis; Activity: DNA Replication: A Closer Look

FIGURE 19-13 A Summary of DNA Replication in Bacteria. Starting with the initiation event at the replication origin, this figure depicts DNA replication in *E. coli* in seven steps. Two replication forks move in opposite directions from the origin, but only one fork is illustrated for steps 2–7. The various proteins shown here as separate entities are actually closely associated (along with others) in a single large complex called a replisome. The primase and DNA helicase are particularly closely bound and, together with other proteins, form a primosome. Parental DNA is shown in dark blue, newly synthesized DNA in light blue, and RNA in red. This series of diagrams does not show the topological arrangement of the DNA strands.

E. coli. Starting at the origin of replication, the machinery at the replication fork sequentially adds the different proteins required for synthesizing DNA—that is, DNA helicase, DNA gyrase, SSB, primase, DNA polymerase, and DNA ligase. Several other proteins (not illustrated) are also involved in improving the overall efficiency of DNA replication. For example, a ring-shaped *sliding clamp* protein encircles the DNA double helix and binds to DNA polymerase, thereby allowing the DNA polymerase to slide along the DNA while remaining firmly attached to it.

The various proteins involved in DNA replication are all closely associated in one large complex, called a **replisome,** that is about the size of a ribosome. The activity and movement of the replisome is powered by the hydrolysis of nucleoside triphosphates. These include both the nucleoside triphosphates (used by DNA polymerases and primase as building blocks for DNA and RNA synthesis) and the ATP hydrolyzed by several other DNA replication proteins (including DNA helicase, DNA gyrase, and DNA ligase). As the replisome moves along the DNA in the direction of the replication fork, it must accommodate the fact that DNA is being synthesized in opposite directions along the template on the two stands. **Figure 19-14** provides a schematic model illustrating how this might be accomplished by folding the lagging strand template into a loop. Creating such a loop allows the DNA polymerase molecules on both the leading and lagging strands to move in the same physical direction, even though the two template strands are oriented with opposite polarity.

Eukaryotes possess much of the same replication machinery found in prokaryotes. For example, like prokaryotes, a DNA clamp protein acts along with DNA polymerase during DNA synthesis. One such eukaryotic clamp protein, *proliferating nuclear cell antigen (PCNA),* was originally identified as an antigen that is expressed in the nuclei of dividing cells during S phase. PCNA is a clamp protein for DNA polymerase δ. In eukaryotic cells, the enormous length and elaborate folding of the chromosomal DNA molecules pose additional challenges for DNA replication. For example, how are the many replication origins coordinated, and how is their activation linked to other key events in the cell cycle? Answering such questions requires a better understanding of the spatial organization of DNA replication within the nucleus. When cells are briefly incubated with DNA precursors that make the most recently formed DNA fluorescent, microscopic examination reveals that the new, fluorescent DNA is located in a series of discrete spots scattered throughout the nucleus. Such observations suggest the existence of apparently immobile structures, known as *replication factories,* where chromatin fibers are fed through stationary replisomes that carry out DNA replication. These sites are closely associated with the inner surface of the nuclear envelope, although it is not clear whether they are anchored in the nuclear membrane or to some other nuclear support.

When a chromatin fiber is fed through a stationary replication factory, how are the histones and other chromosomal proteins removed, so the DNA can be replicated, and then added back after the two new DNA strands are formed? Unfolding the chromatin fibers ahead of the replication fork is facilitated by *chromatin remodeling* proteins, which loosen nucleosome packing to give the replication machinery access to the DNA template. As the replication fork moves, old nucleosomes slide from the parental double helix ahead of the fork to the newly forming strands behind the fork. At the same time, new nucleosomes are assembled on the newly forming strands because the two DNA molecules produced by replication require twice as many nucleosomes as the single parental DNA possessed prior to replication. When and how other chromosomal proteins are added, what controls higher levels of chromatin packing, and how all this is accomplished without tangling the chromatin are areas of current research. Meanwhile, an answer has emerged to one of the most baffling questions related to DNA replication in eukaryotes—the end-replication problem.

Telomeres Solve the DNA End-Replication Problem

MB ABC News Video: The Effect of Exercise on Cells

If we continue the process summarized in Figure 19-13 for the circular genome of *E. coli* (or any other circular DNA), we will eventually complete the circle. The leading strand can simply continue to grow $5' \rightarrow 3'$ until its $3'$ end is joined to the $5'$ end of the lagging strand coming around in the other direction. And for the lagging strand, the very last bit of DNA to be synthesized—the replacement for the RNA primer of the last Okazaki fragment—can be added to the free $3'$ OH end of the leading strand coming around in the opposite direction.

For linear DNA molecules, however, the fact that DNA polymerases can add nucleotides only to the $3'$ OH end of a *preexisting* DNA chain creates a serious problem,

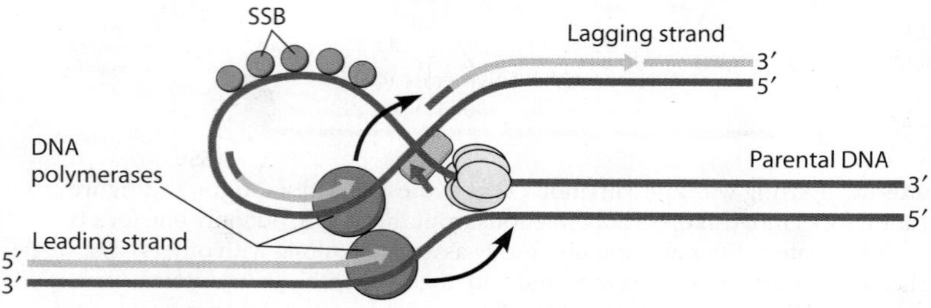

FIGURE 19-14 Arrangement of Proteins at the Replication Fork. This model shows how some of the key replication proteins illustrated in Figure 19-13 are organized at the replication fork. The main distinguishing feature is that the lagging strand DNA is folded into a loop, thereby allowing the DNA polymerase molecules on the leading and lagging strands to come together and move in the same physical direction (black arrows) even though the two template strands are oriented with opposite polarity in the parental DNA molecule.

which is depicted in **Figure 19-15.** When a growing lagging strand (lightest blue) reaches the end of the DNA molecule and the last RNA primer is removed by a 5′ → 3′ exonuclease, the final gap cannot be filled because there is no 3′ OH end that deoxynucleotides can be added to. As a result, linear DNA molecules are in danger of yielding shorter and shorter daughter DNA molecules each time they replicate. Clearly, if this trend continued indefinitely, we would not be here today!

Viruses with linear DNA genomes solve this problem in various ways. In some cases, such as bacteriophage l, the linear DNA forms a closed circle before it replicates. In other cases, the viruses use more exotic reproduction strategies, although the DNA polymerases always progress 5′ to 3′, adding nucleotides to a 3′ end.

Eukaryotes have solved the end-replication problem by locating highly repeated DNA sequences at the terminal ends, or **telomeres,** of each linear chromosome. These special telomeric elements consist of short, repeating sequences enriched in the base G in the 5′ → 3′ strand. The sequence TTAGGG, located at the ends of human chromosomes, is an example of such a telomeric or *TEL sequence.* Human telomeres typically contain between 100 and 1500 copies of the TTAGGG sequence repeated in tandem. Such noncoding sequences at the ends of each chromosome ensure that the cell will not lose any important genetic information if a DNA molecule is shortened slightly during the process of DNA replication. Moreover, a special DNA polymerase called **telomerase** can catalyze the formation of additional copies of the telomeric repeat sequence, thereby compensating for the gradual shortening that occurs at both ends of the chromosome during DNA replication. Elizabeth Blackburn, Carol Greider, and Jack Szostak received the 2009 Nobel Prize in Physiology or Medicine for their fundamental contributions to our understanding of telomeres and telomerase.

Telomerase is an unusual enzyme in that it is composed of RNA as well as protein. In the protozoan *Tetrahymena,* whose telomerase was the first to be isolated, the RNA component contains the sequence 3′ − AACCCC − 5′, which is complementary to the 5′ − TTGGGG − 3′ repeat sequence that makes up *Tetrahymena* telomeres. As shown in **Figure 19-16,** this enzyme-bound RNA acts as a template for creating the DNA repeat sequence that is added to the telomere ends. After being lengthened by telomerase, the telomeres are protected by *telomere capping proteins* that bind to the exposed 3′ end of the DNA. In many eukaryotes, the 3′ end of the DNA can loop back and base-pair with the opposite DNA strand, generating a closed loop that likewise protects the end of the telomere (Figure 19-16, step ❺).

In multicellular organisms, telomerase resides mainly in the *germ cells* that give rise to sperm and eggs and in a few other types of actively proliferating cells. The presence

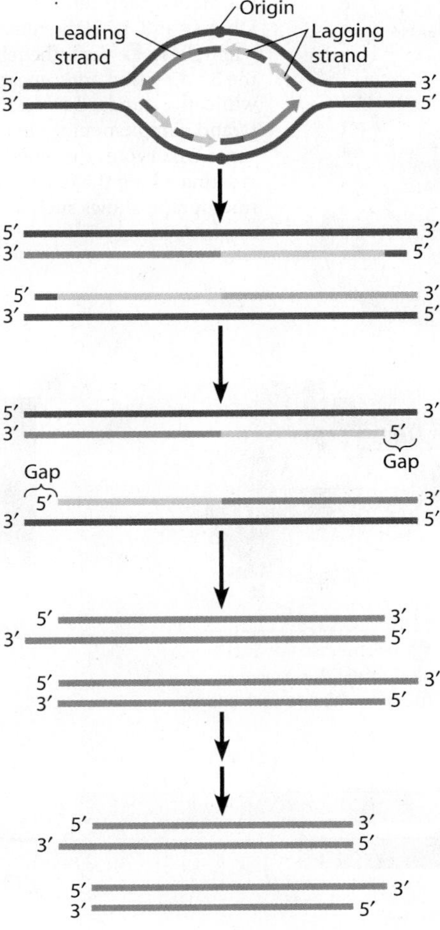

❶ DNA replication is initiated at the origin; the replication bubble grows as the two replication forks move in opposite directions.
Origin
Leading strand
Lagging strand
5′
3′
❷ Finally only one primer (red) remains on each daughter DNA molecule.
❸ The last primers are removed by a 5′→3′ exonuclease, but no DNA polymerase can fill the resulting gaps because there is no 3′ OH available to which a nucleotide can be added.
Gap
❹ Each round of replication generates shorter and shorter DNA molecules.

FIGURE 19-15 The End-Replication Problem. For a linear DNA molecule, such as that of a eukaryotic chromosome, the usual DNA replication machinery is unable to replicate the ends. As a result, with each round of replication, the DNA molecules will get shorter, with potentially disastrous consequences for the cell. In this diagram, the initial parental DNA strands are dark blue, daughter DNA strands are lighter blue, and RNA primers are red. For simplicity, we show only one origin of replication and, in the last two steps, only the shortest of the progeny molecules. (The lagging strand daughter DNA is shown in the lightest blue in the first three steps to make a point unrelated to the end-replication problem—that each daughter strand is leading at one end and lagging at the other. This is apparent in this figure because it shows the entire replicating molecule, with both replication forks.)

of telomerase allows these cells to divide indefinitely without telomere shortening. Because telomerase is not found in most cells, their chromosomal telomeres get shorter and shorter with each cell division. As a result, telomere length is a counting device that reveals how many times a cell has divided. If a cell divides enough times, the telomeres are in danger of disappearing entirely, and the cell would then be at risk of eroding its coding DNA. This potential danger is averted by a cell destruction pathway set in motion by the shortened telomeric DNA. In essence, when the telomeric DNA becomes too short to bind telomeric capping proteins or generate a loop, it exposes a bare, double-stranded DNA end whose presence activates a signaling system that triggers *apoptosis*, an orchestrated type of cell death described later in this chapter.

Cell death triggered by a lifetime of telomere shortening is thought to contribute to some of the degenerative diseases associated with human aging—for example, increased susceptibility to infections due to the death of immune cells, inefficient wound healing caused by depletion of connective tissue cells, and ulcers triggered by loss of cells in the digestive tract. People who inherit mutations in telomerase (or in other proteins affecting telomere length) experience similar degenerative changes, show symptoms of premature aging, and usually die when they are relatively young. Based on these relationships, scientists have speculated that telomerase-based therapies may one day be used to combat the symptoms of human aging and thereby extend life span.

Besides being found in germ cells and a few other types of proliferating normal cells, telomerase has also been detected in human cancers. Because cancer cells divide an abnormally large number of times, their telomeres become unusually short. This progressive telomere shortening would lead to self-destruction of the cancer

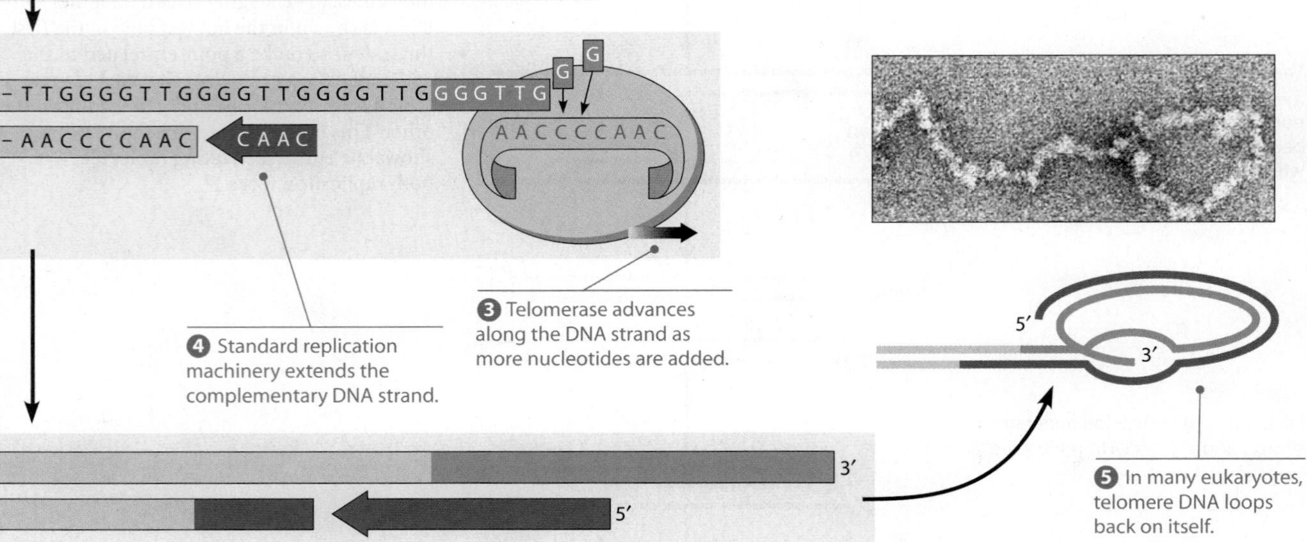

FIGURE 19-16 The Extension of Telomeres by Telomerase. Telomeres are stretches of repeated DNA located at the ends of eukaryotic chromosomes. This figure focuses on one end of a DNA molecule from *Tetrahymena*, whose telomeric repeat unit is TTGGGG (those of other species are very similar). The 3′ end of the DNA extends beyond the 5′ end and is the substrate for telomerase, an enzyme composed of protein and RNA. The RNA portion of *Tetrahymena* telomerase is 159 nucleotides long and contains a 9-base sequence complementary to 1.5 telomeric repeat units. As outlined here, the telomerase ❶ binds to the 3′ end of the telomeric DNA, positioning itself so the last few bases of the DNA are base-paired with part of the 9-base RNA sequence. ❷ The telomerase then catalyzes the addition of nucleotides to the 3′ end of the DNA strand, with the remainder of the 9-base RNA sequence serving as a template. ❸ Next, the telomerase advances along the DNA strand in the 3′ direction and repeats steps ❷ and ❸ several times. ❹ Meanwhile, the standard DNA replication machinery synthesizes a lagging strand complementary to the strand elongated by telomerase. ❺ In some eukaryotes the lengthened telomeric DNA folds back upon itself, creating a loop that caps the end of the chromosome. The electron micrograph shows such a loop in chromatin isolated from chicken erythrocytes.

unless telomerase were produced to stabilize telomere length. This is exactly what seems to happen. A survey of a large number of human cell samples—including more than 100 tumors—found telomerase activity in almost all of the cancer cells but in none of the samples from normal tissues. If telomerase is indeed an important factor in the development of cancer, it may eventually provide a useful target for anticancer drug therapy.

DNA Damage and Repair

The faithful transmission of genetic information from one generation of cells to the next requires not just that DNA be replicated accurately. Provision must also be made for repairing DNA alterations that arise both spontaneously and from exposure to DNA-damaging environmental agents. Of course, DNA alterations are occasionally beneficial because DNA base-sequence changes, or **mutations,** provide the genetic variability that is the raw material of evolution. Still, the net rate at which organisms accumulate mutations is quite low; by some estimates, an average gene retains only one mutation every 200,000 years. The underlying mutation rate is far greater than this number suggests, but most DNA damage is repaired shortly after it occurs so does not affect future generations. Moreover, since most mutations occur in cells other than sperm and eggs, they are never passed on to offspring.

DNA Damage Can Occur Spontaneously or in Response to Mutagens

MB Mutations at the DNA Level

During the normal process of DNA replication, several types of mutations occur spontaneously. The most common involve depurination and deamination reactions, which are spontaneous hydrolysis reactions caused by random interactions between DNA and the water molecules around it. *Depurination* refers to the loss of a purine base (either adenine or guanine) by spontaneous hydrolysis of the glycosidic bond that links it to deoxyribose (**Figure 19-17a**). This glycosidic bond is intrinsically unstable and is in fact so susceptible to hydrolysis that the DNA in a human cell may lose thousands of purine bases every day. *Deamination* is the removal of a base's amino group (—NH_2). This type of alteration, which can involve cytosine, adenine, or guanine, changes the base-pairing properties of the affected base. Of the three bases, cytosine is most susceptible to deamination, giving rise to uracil (Figure 19-17b). Like depurination, deamination is a hydrolytic reaction, usually caused by random collision of a water molecule with the bond that links the amino group of the base to the pyrimidine or purine ring. In a typical human cell, the rate of DNA damage by this means is about 100 deaminations per day.

If a DNA strand with missing purines or deaminated bases is not repaired, an erroneous base sequence may be propagated when the strand serves as a template in the next round of DNA replication. For example, where a cytosine has been converted to a uracil by deamination, the uracil behaves like thymine in its base-pairing properties; that is, it directs the insertion of an adenine in the opposite strand, rather than guanine, the correct base. The ultimate effect of this change in base sequence may be a change in the amino acid sequence and function of a protein encoded by the affected gene.

In addition to spontaneous mutations, DNA damage can also be caused by mutation-causing agents, or *mutagens,* in the environment. Environmental mutagens fall into two major categories: chemicals and radiation. Mutagenic chemicals alter DNA structure by a variety of mechanisms. *Base analogs* resemble nitrogenous bases in structure and are incorporated into DNA; *base-modifying* agents react chemically with DNA bases to alter

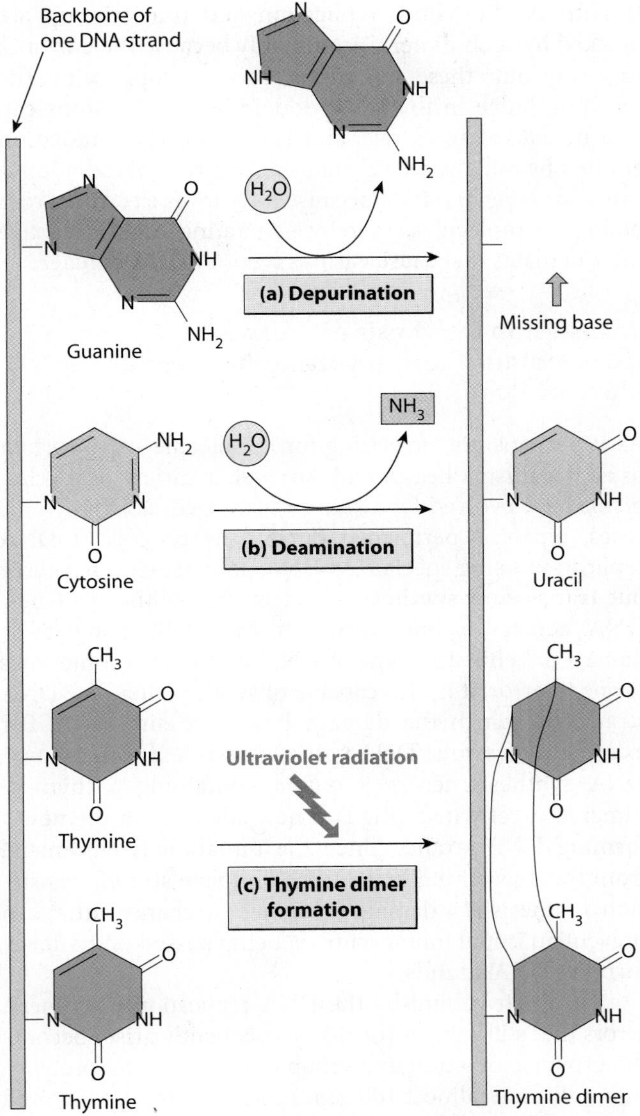

FIGURE 19-17 Some Common Types of DNA Damage. The most common kinds of chemical changes that can damage DNA are **(a)** depurination, **(b)** deamination, and **(c)** pyrimidine dimer formation (shown here are thymine dimers). Depurination and deamination are spontaneous hydrolytic reactions, whereas dimers result from covalent bonds induced to form by ultraviolet light.

their structure, forming what are sometimes known as *DNA adducts;* and *intercalating agents* insert themselves between adjacent bases of the double helix, thereby distorting DNA structure and increasing the chance that a base will be deleted or inserted during DNA replication. Not surprisingly, as we will see in more detail in Chapter 24, mutagens can cause cancer. Analysis of DNA from heavy cigarette smokers, for example, shows a direct correlation between how much a patient smokes and how frequently the DNA of such patients has adducts.

DNA mutations can also be caused by several types of radiation. Sunlight is a strong source of ultraviolet radiation, which alters DNA by triggering *pyrimidine dimer formation*—that is, the formation of covalent bonds between adjacent pyrimidine bases, often two thymines (Figure 19-17c). Both replication and transcription are blocked by such dimers, presumably because the enzymes carrying out these functions cannot cope with the resulting bulge in the DNA double helix. Mutations can also be caused by X-rays and related forms of radiation emitted by radioactive substances. This type of radiation is called *ionizing radiation* because it removes electrons from biological molecules, thereby generating highly reactive intermediates that cause various types of DNA damage.

Translesion Synthesis and Excision Repair Correct Mutations Involving Abnormal Nucleotides

As is perhaps not surprising for a molecule so important to an organism's health and survival, a variety of mechanisms have evolved for repairing damaged DNA. In some cases, repair is performed during the process of DNA replication using specialized DNA polymerases that carry out **translesion synthesis**—that is, the synthesis of new DNA across regions in which the DNA template is damaged. While this type of DNA synthesis is sometimes prone to error, it is also capable of synthesizing new DNA strands in which the damage has been eliminated. For example, eukaryotic DNA polymerase η (eta) can catalyze DNA synthesis across a region containing a thymine dimer, correctly inserting two new adenines in the newly forming DNA strand. Since the mutation is eliminated from the new strand but not the template strand, translesion synthesis is a damage-tolerance mechanism that can prevent an initial mutation from being passed on to newly forming DNA strands.

Once left behind by the DNA replication machinery, errors that still remain (or that subsequently arise) become the province of a different group of enzymes and proteins. In *E. coli* alone, almost 100 genes code for proteins involved in removing and replacing abnormal nucleotides. These proteins are components of **excision repair** pathways, which correct DNA defects using a basic three-step process (**Figure 19-18**). In the first step, the defective nucleotides are cut out from one strand of the DNA double helix. This process is carried out by special enzymes called *repair endonucleases,* which are recruited to DNA by proteins that recognize sites of DNA damage. Repair endonucleases cleave the DNA backbone adjacent to the damage site, and other enzymes then facilitate removal of the defective nucleotide(s). For example, a DNA helicase might unwind the DNA located between two nicks to release the damaged DNA from the double helix; alternatively, an exonuclease might attach to an end created by a single nick and chew away the damaged strand one nucleotide at a time. During the second step in excision repair, the missing nucleotides are replaced with the correct ones by a DNA polymerase—in *E. coli,* usually DNA polymerase I. The nucleotide sequence of the complementary strand serves as a template to ensure correct base insertion, just as it does in DNA replication. Finally, in the third step, DNA ligase seals the remaining nick in the repaired strand by forming the missing phosphoester bond.

Excision repair pathways are classified into two main types, *base excision repair* and *nucleotide excision repair.* The first of these pathways, **base excision repair,** corrects single damaged bases in DNA. For example, deaminated bases are detected by specific *DNA glycosylases,* which recognize a specific deaminated base and remove it from the DNA molecule by cleaving the bond between the base and the sugar it is attached to. The sugar with the missing base is then recognized by a repair endonuclease that detects depurination. This repair endonuclease breaks the phosphodiester backbone on one side of the sugar lacking a

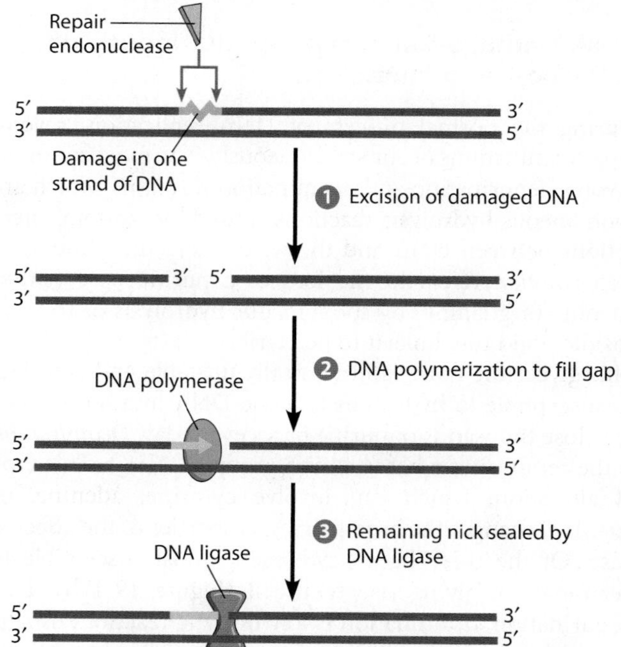

FIGURE 19-18 General Scheme for Excision Repair of DNA Damage. The three steps shown here are common to all types of excision repair of DNA damage in one strand. ❶ The damaged part of the DNA strand, and possibly some DNA on either side of it, is cut out (excised) from the double helix. An endonuclease, which nicks the DNA near the damage, is crucial for this step; a DNA helicase and/or exonuclease may help remove the damaged segment. ❷ A DNA polymerase fills in the gap by adding nucleotides to the strand's 3′ end; DNA polymerase I plays this role in *E. coli.* ❸ The remaining nick is sealed by DNA ligase.

base, and a second enzyme then completes the removal of the sugar-phosphate unit.

For removing pyrimidine dimers and other bulky lesions in DNA, cells employ the second type of excision repair, namely **nucleotide excision repair (NER).** This repair system utilizes proteins that recognize major distortions in the DNA double helix and recruit an enzyme, called an *NER endonuclease* (or *excinuclease*), that makes two cuts in the DNA backbone, one on either side of the distortion. Then a DNA helicase binds to the stretch of DNA between the nicks (12 nucleotides long in *E. coli,* 29 in humans) and unwinds it, freeing it from the rest of the DNA. Finally, the resulting gap is filled in by DNA polymerase and sealed by DNA ligase.

The NER system is the most versatile of a cell's DNA repair systems, recognizing and correcting many types of damage that cannot otherwise be repaired. In some cases, the NER system is specifically recruited to DNA regions where transcription has been halted because the transcription machinery encountered an area of DNA damage. This mechanism, known as *transcription-coupled repair,* permits active genes to be repaired faster than DNA sequences located elsewhere in the genome. The importance of NER is underscored by the plight of people who have mutations affecting the NER pathway. Individuals with the disease *xeroderma pigmentosum,* for example, usually carry a mutation in any of seven genes coding for components of the NER system. As a result, they cannot repair the DNA damage caused by the ultraviolet radiation in sunlight and so have a high risk of developing skin cancer (see Box 24A, page 782).

MB Molecular Model: MutS, a DNA Mismatch Repair Protein of E. coli; Molecular Model: MutS, a DNA Mismatch Repair Protein of E. coli-Copy

Mismatch Repair Corrects Mutations That Involve Noncomplementary Base Pairs

Excision repair is a powerful mechanism for correcting damage involving the presence of abnormal nucleotides. This is not the only type of DNA error that cells can repair, however. An alternative repair pathway, called **mismatch repair,** targets errors made during DNA replication, when improperly base-paired nucleotides sometimes escape the normal proofreading mechanisms. Because mismatched base pairs do not hydrogen-bond properly, their presence can be detected and corrected by the mismatch repair system. But to operate properly, this repair system must solve a problem that puzzled biologists for many years: How is the *incorrect* member of an abnormal base pair distinguished from the *correct* member? Unlike the situation in excision repair, neither of the bases in a mismatched pair exhibits any structural alteration that would allow it to be recognized as an abnormal base. The pair is simply composed of two normal bases that are inappropriately paired with each other, such as the base A paired with C or G paired with T. If the incorrect member of an AC base pair were the base C and the repair system instead removed the base A, the repair system would create a permanent mutation instead of correcting a mismatched base pair!

To solve this problem, the mismatch repair system must be able to recognize which of the two DNA strands was newly synthesized during the previous round of DNA replication (the new strand would be the one that contains the incorrectly inserted base). The bacterium *E. coli* employs a detection system that is based on the fact that a methyl group is normally added to the base adenine (A) wherever it appears in the sequence GATC in DNA. This process of *DNA methylation* does not take place until a short time after a new DNA strand has been synthesized; hence, the bacterial mismatch repair system can detect the new DNA strand by its unmethylated state. A repair endonuclease then introduces a single nick in the unmethylated strand, and an exonuclease removes the incorrect nucleotides from the nicked strand.

Although the initial evidence for the existence of mismatch repair came largely from studies of bacterial cells, comparable repair systems have been detected in eukaryotes as well. Eukaryotic mismatch repair differs from the bacterial process in that eukaryotes use mechanisms other than DNA methylation for distinguishing the newly synthesized DNA strand. The importance of mismatch repair for eukaryotes is highlighted by the discovery that one of the most common hereditary cancers, *hereditary nonpolyposis colon cancer (HNPCC),* results from mutations in genes coding for proteins involved in mismatch repair.

Damage Repair Helps Explain Why DNA Contains Thymine Instead of Uracil

For many years, it was not clear why DNA contains thymine instead of the uracil found in RNA. Both bases pair with adenine, but thymine has a methyl group not present on uracil (**Figure 19-19**). Because the methylation step used to create thymine is energetically expensive, it might seem more efficient for DNA to contain uracil. But now that we understand how deamination damage is repaired, we also understand why thymine, rather than uracil, is present in DNA.

When DNA is damaged through deamination reactions, cytosine is converted to uracil (see Figure 19-17b), which is then detected and removed by the DNA repair enzyme, *uracil-DNA glycosylase.* But if uracil were present as a normal component of DNA, DNA repair would not work because these normal uracils would not be distinguishable from the uracils generated by the accidental deamination of cytosine. By using thymine in DNA in place of uracil, cells ensure that DNA damage caused by the deamination of cytosine can be recognized and repaired without causing other changes in the DNA molecule.

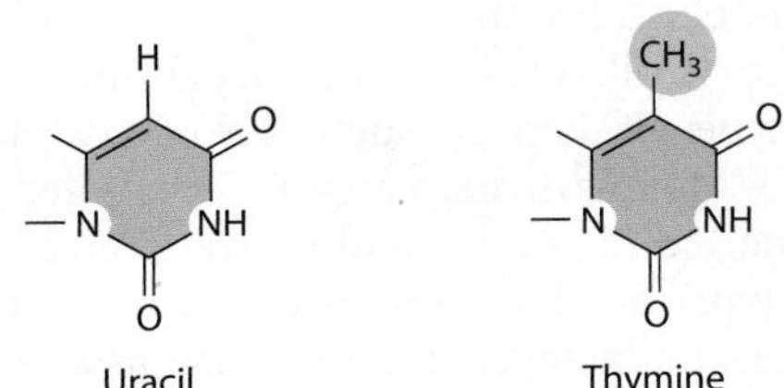

FIGURE 19-19 Uracil and Thymine Compared.

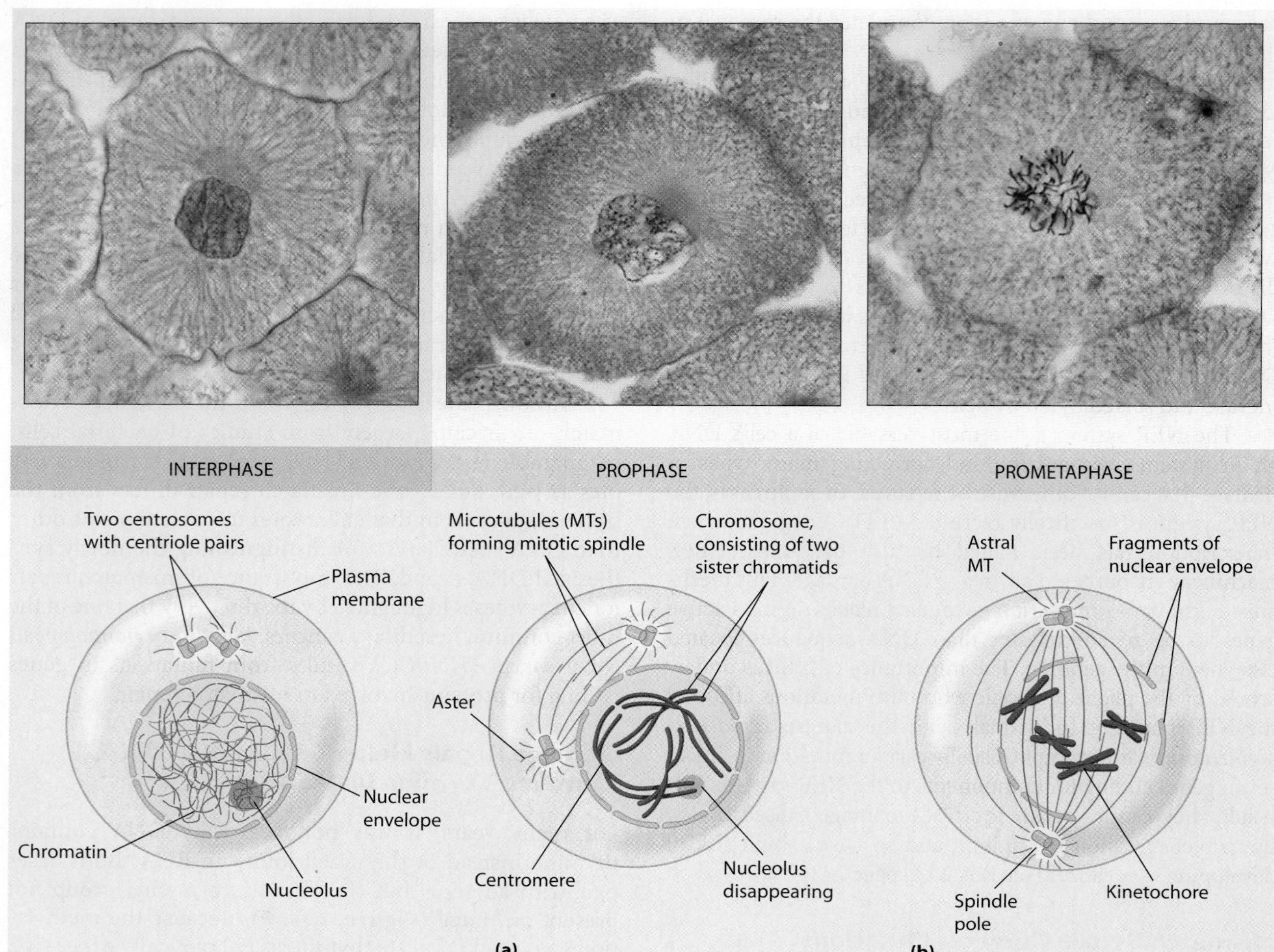

MB Mitosis (2 of 3): Mechanism of Mitosis (BioFlix tutorial)

FIGURE 19-20 The Phases of Mitosis in an Animal Cell. The micrographs show mitosis in cells from a fish embryo viewed by light microscopy. The mitotic spindle, including asters, is visible in the metaphase and anaphase micrographs. At this low magnification (about 600-fold), we see spindle "fibers" rather than individual microtubules; each fiber consists of a number of microtubules. The drawings are schematic and include details not visible in the micrographs; for simplicity, only four chromosomes are drawn (MT = microtubule).

Double-Strand DNA Breaks Are Repaired by Nonhomologous End-Joining or Homologous Recombination

The repair mechanisms described thus far—excision and mismatch repair—are effective in correcting DNA damage involving chemically altered or incorrect bases. In such cases a "cut-and-patch" pathway removes the damaged or incorrect nucleotides from one DNA strand, and the resulting gap is filled using the intact strand as template. But some types of damage, such as double-strand breaks in the DNA double helix, cannot be handled this way. Repair is more difficult for double-strand breaks because, with other types of DNA damage, one strand of the double helix remains undamaged and can serve as a template for aligning and repairing the defective strand. In contrast, double-strand breaks completely cleave the DNA double helix into two separate fragments; the repair machinery is therefore confronted with the problem of identifying the correct two fragments and rejoining their broken ends without losing any nucleotides.

Two main pathways are employed in such cases. One, called **nonhomologous end-joining,** uses a set of proteins that bind to the ends of the two broken DNA fragments and join them together. Unfortunately, this mechanism is error-prone because it cannot prevent the loss of nucleotides from the broken ends and has no way of ensuring that the correct two DNA fragments are being joined to each other. A more precise method for fixing double-strand breaks, called **homologous recombination,** takes advantage of the fact that cells generally possess two copies of each chromosome; if the DNA molecule in one chromosome incurs a double-strand break, another intact copy of the chromosomal DNA is still available to

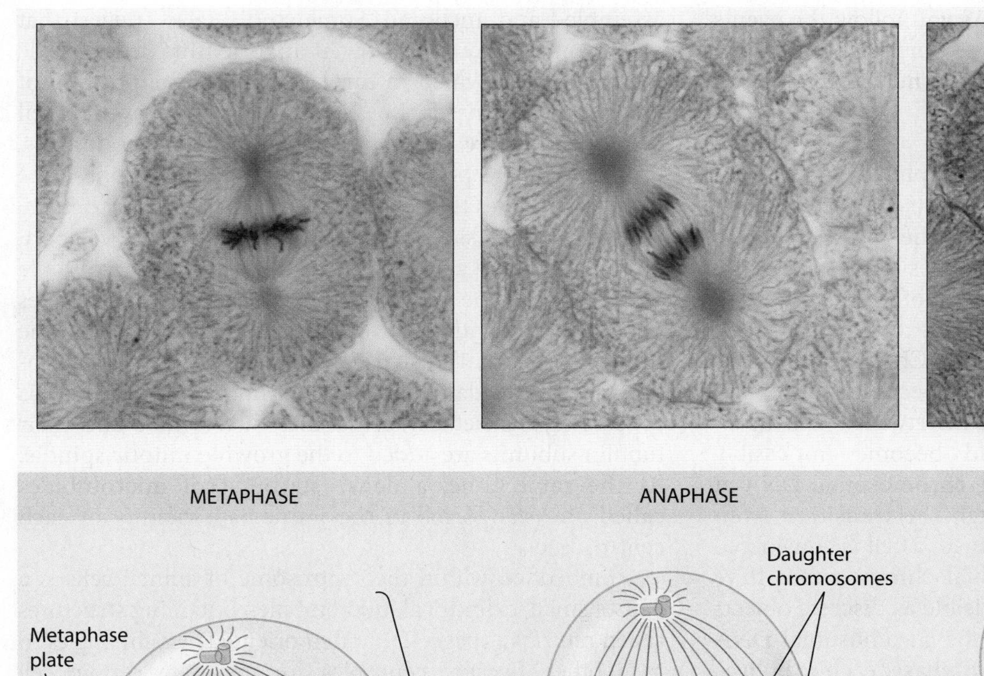

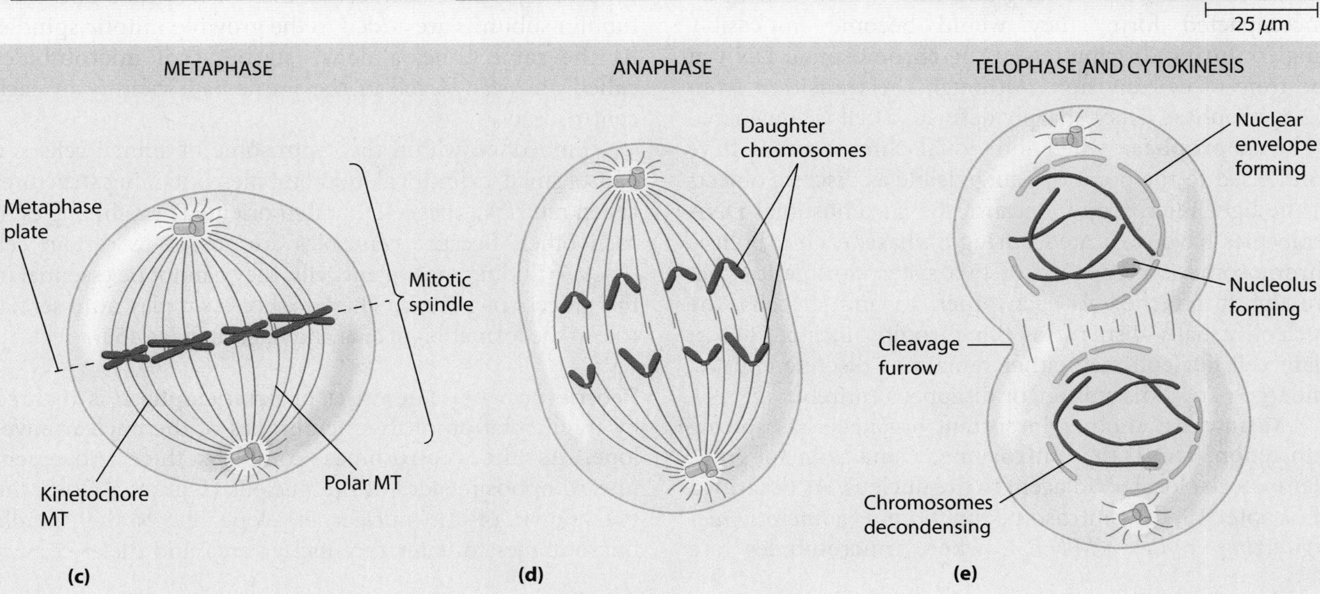

serve as a template for guiding the repair of the broken chromosome. Besides its role in repairing double-strand breaks, homologous recombination is involved in the exchange of genetic information between chromosomes during the meiotic cell divisions that are used to produce sperm and egg cells. We will therefore delay a discussion of the molecular mechanisms involved in homologous recombination until Chapter 20, where meiosis and genetic recombination are described in detail.

Now that we have considered the various types of DNA repair, it is important to emphasize that some of the same proteins are used in more than one repair pathway. Moreover, many of these "repair" proteins play additional roles in other important activities, including DNA replication, gene transcription, genetic recombination, and control of the cell cycle. In other words, the same molecular tool kit appears to be used for a variety of DNA-related activities.

Nuclear and Cell Division

Having examined the mechanisms involved in DNA replication and repair, we can now turn to the question of how the two copies of each chromosomal DNA molecule created during the S phase of the cell cycle are subsequently separated from each other and partitioned into daughter cells. These events occur during M phase, which encompasses both nuclear division (mitosis) and cytoplasmic division (cytokinesis).

Mitosis Is Subdivided into Prophase, Prometaphase, Metaphase, Anaphase, and Telophase

Mitosis has been studied for more than a century, but only in the past few decades has significant progress been made toward understanding the mitotic process at the molecular level. We will begin by surveying the morphological changes that occur in a cell as it undergoes mitosis, and we will then examine the underlying molecular mechanisms.

Mitosis is subdivided into five stages based on the changing appearance and behavior of the chromosomes. These five phases are *prophase, prometaphase, metaphase, anaphase,* and *telophase.* (An alternative term for prometaphase is simply *late prophase.*) The micrographs and schematic diagrams of **Figure 19-20** illustrate the phases in a typical animal cell; **Figure 19-21** depicts the comparable stages in plant cells. As you follow the events of each phase, keep in mind that the purpose of mitosis is to ensure that each of the two daughter nuclei receives one copy of each duplicated chromosome.

Prophase. After completing DNA replication, cells exit from S phase and enter into G2 phase (see Figure 19-1b), where final preparations are made for the onset of mitosis. Toward the end of G2, the chromosomes start to condense from the extended, highly diffuse form of interphase chromatin fibers into the compact, extensively folded structures that are typical of mitosis. Chromosome condensation is an important event because interphase chromatin fibers are so long and intertwined that in an uncompacted form, they would become impossibly tangled during distribution of the chromosomal DNA at the time of cell division. Although the transition from G2 to prophase is not sharply defined, a cell is considered to be in **prophase** when individual chromosomes have condensed to the point of being visible as discrete objects in the light microscope. Because the chromosomal DNA molecules have replicated during S phase, each prophase chromosome is composed of two sister chromatids that are tightly attached to each other. In animal cells, the nucleoli usually disperse as the chromosomes condense; plant cell nucleoli may either remain as discrete entities, undergo partial disruption, or disappear entirely.

Meanwhile, another important organelle has sprung into action. This is the **centrosome,** a small zone of granular material located adjacent to the nucleus. As described in Chapter 15, the centrosome functions as a *microtubule-organizing center (MTOC)* where microtubules are assembled and anchored. Some studies also suggest that centrosomes may be involved in orienting the spindle within the cell, which in turn determines the position of the cleavage plane during cell division. During each cell cycle, the centrosome is duplicated prior to mitosis, usually during S phase. At the beginning of prophase the two centrosomes then separate from each other and move toward opposite sides of the nucleus. As they move apart, each centrosome acts as a nucleation site for microtubule assembly and the region between the two centrosomes begins to fill with microtubules destined to form the **mitotic spindle,** the structure that distributes the chromosomes to the daughter cells later in mitosis. During this process, cytoskeletal microtubules disassemble and their tubulin subunits are added to the growing mitotic spindle. At the same time, a dense starburst of microtubules called an *aster* forms in the immediate vicinity of each centrosome.

Embedded within the centrosome of animal cells is a pair of small, cylindrical, microtubule-containing structures called *centrioles* (page 430), often oriented at right angles to each other. Because centrioles are absent in certain cell types, including most plant cells, they cannot be essential to the process of mitosis. They do, however, play an essential role in the formation of cilia and flagella (page 453).

Prometaphase. The onset of **prometaphase** is marked by fragmentation of the membranes of the nuclear envelope. As the centrosomes complete their movement toward opposite sides of the nucleus (Figure 19-20b), the breakdown of the nuclear envelope allows the spindle microtubules to enter the nuclear area and make contact

FIGURE 19-21 The Phases of Mitosis in a Plant Cell. These micrographs show mitosis in cells of an onion root viewed by light microscopy.

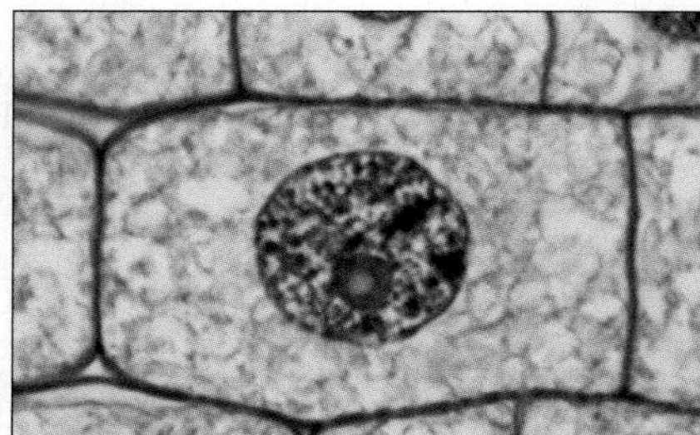

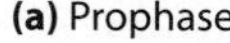

(a) Prophase

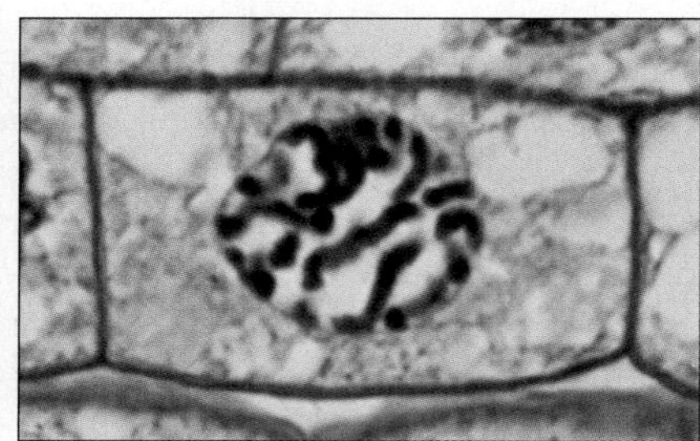

(b) Prometaphase

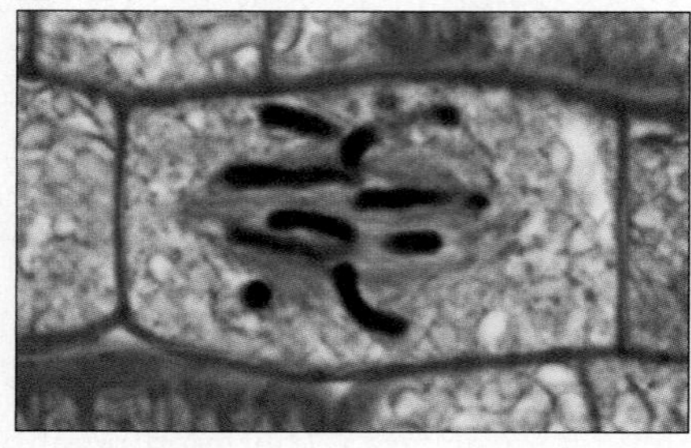

(c) Metaphase

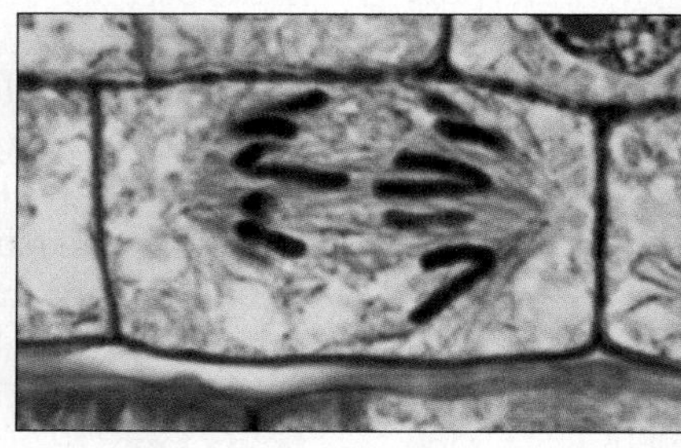

(d) Anaphase

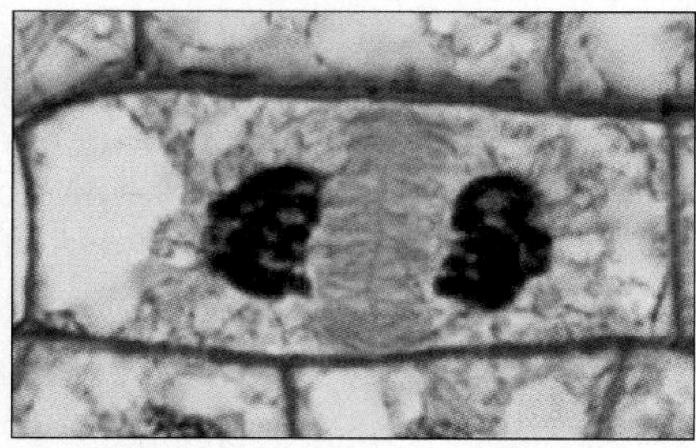

(e) Telophase

with the chromosomes, which still consist of paired chromatids at this stage. The spindle microtubules are destined to attach to the chromatids in the region of the **centromere,** a constricted area where the two members of each chromatid pair are held together. The DNA of each centromere consists of simple-sequence, tandemly repeated *CEN sequences,* whose makeup varies considerably among species. Despite this diversity, a common feature of centromere regions is the presence of specialized nucleosomes in which histone H3 is replaced by a related protein, which in humans is called *CENP-A (centromere protein A).*

CENP-A plays a key role in recruiting additional proteins to the centromere to form the **kinetochore,** which is the structure that attaches the paired chromatids to the spindle microtubules. Kinetochore proteins begin to associate with the centromere shortly after DNA is replicated during S phase; additional proteins are sequentially added later until mature kinetochores, containing more than 50 different proteins, have been assembled. As shown in **Figure 19-22a,** each chromosome eventually acquires two kinetochores facing in opposite directions, one associated with each of the two chromatids. During prometaphase some spindle microtubules bind to these kinetochores, thereby attaching the chromosomes to the spindle. Forces exerted by these **kinetochore microtubules** then throw the chromosomes into agitated motion and gradually move them toward the center of the cell, in a process known as *congression* (Figure 19-22b).

In addition to kinetochore microtubules, there are two other kinds of microtubules in the spindle. Those that interact with microtubules from the opposite pole of the cell are called **polar microtubules;** the shorter ones that form the *asters* (from the Greek word for "star") at each pole are called **astral microtubules.** Some of the astral microtubules appear to interact with proteins lining the plasma membrane.

Metaphase. A cell is said to be in **metaphase** when the fully condensed chromosomes all become aligned at the *metaphase plate,* the plane equidistant between the two poles of the mitotic spindle (Figure 19-20c). Agents that interfere with spindle function, such as the drug *colchicine,* can be used to arrest cells at metaphase. Microscopic examination of such cells allows individual chromosomes to be identified and classified based on differences in size and shape, generating an analysis known as a **karyotype** (**Figure 19-23**).

At metaphase the chromosomes appear to be relatively stationary, but this appearance is misleading. Actually, the two sister chromatids of each chromosome are already being actively tugged toward opposite poles. They appear stationary because the forces acting on them are equal in magnitude and opposite in direction; the chromatids are the prizes in a tug-of-war between two equally strong opponents. (We will discuss the source of these opposing forces shortly.)

Anaphase. Usually the shortest phase of mitosis, **anaphase** typically lasts only a few minutes. At the beginning of anaphase, the two sister chromatids of each chromosome abruptly separate and begin moving toward opposite spindle poles at a rate of about 1 μm/min (Figure 19-20d).

Anaphase is characterized by two kinds of movements, called anaphase A and anaphase B (**Figure 19-24**). In **anaphase A,** the chromosomes are pulled, centromere

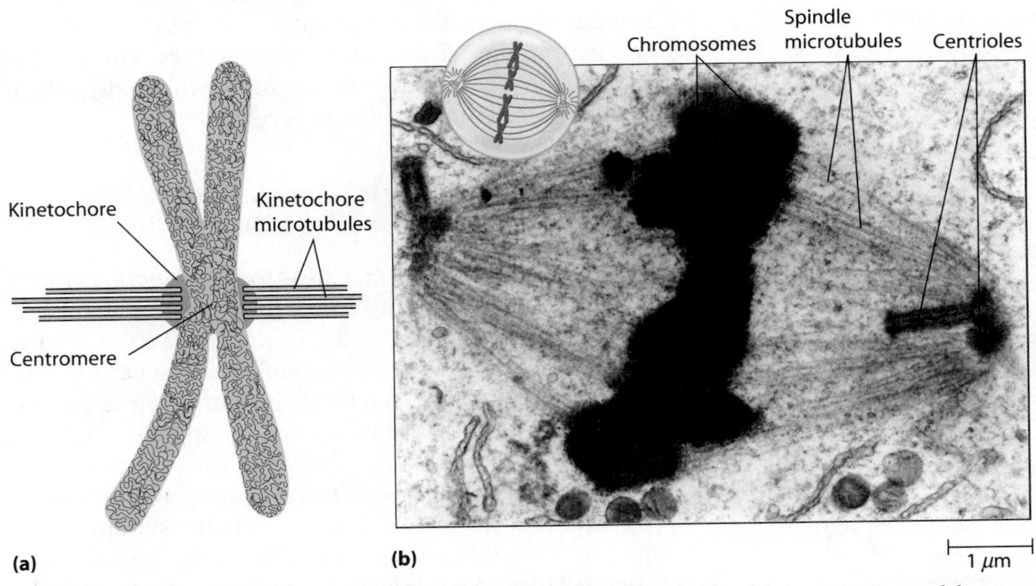

FIGURE 19-22 Attachment of Chromosomes to the Mitotic Spindle. (a) A schematic model summarizing the relationship between the centromere, kinetochores, and kinetochore microtubules. (b) This electron micrograph shows the mitotic spindle of a metaphase cell from a rooster. The centrioles at the two poles and the spindle between the poles are clearly visible. The chromosomes appear as a single mass aligned at the spindle equator. Although individual chromosomes cannot be distinguished in this type of micrograph, the individual chromosomes remain distinct from one another at this stage of mitosis (TEM).

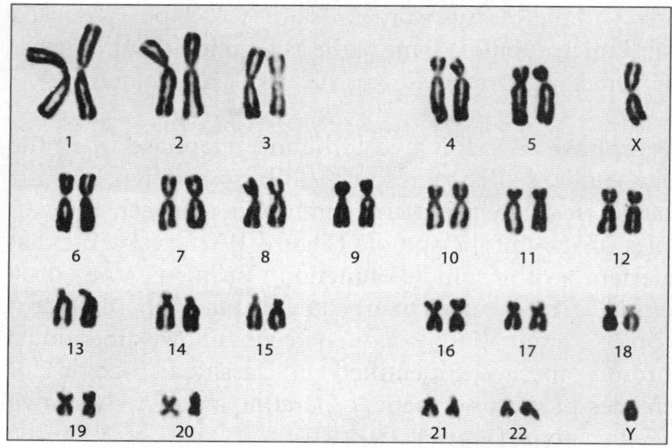

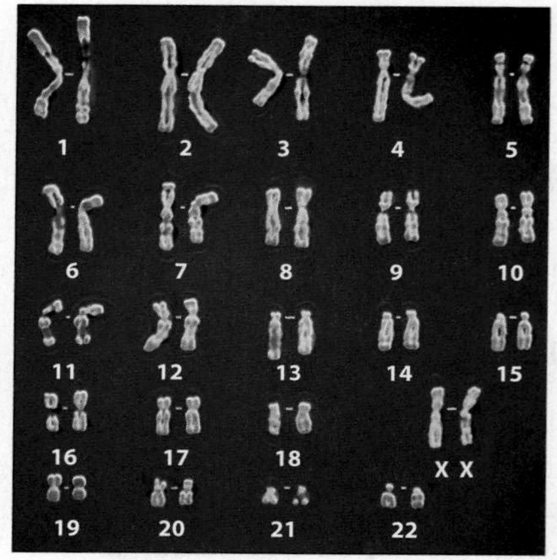

FIGURE 19-23 Mitotic Karyotype of Human Chromosomes from Metaphase-Arrested Cells. (Left) This set of human male chromosomes was stained with a dye that reacts uniformly with the entire body of the chromosome. Human males contain 22 pairs of chromosomes, plus one X and one Y chromosome. The chromosomes in the karyotype have been arranged according to size and centromere position. (Right) This set of human female chromosomes was stained with dyes that selectively react with certain chromosome regions, creating a unique banding pattern for each type of chromosome.

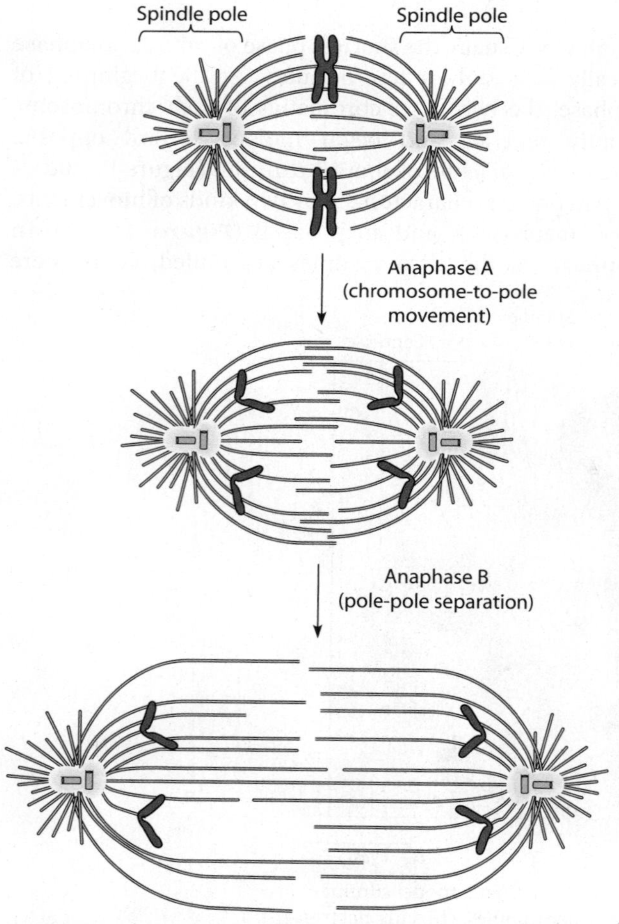

FIGURE 19-24 The Two Types of Movement Involved in Chromosome Separation During Anaphase. Anaphase A involves the movement of chromosomes toward the spindle poles. Anaphase B is the movement of the two spindle poles away from each other. Anaphase A and anaphase B may occur simultaneously.

first, toward the spindle poles as the kinetochore microtubules get shorter and shorter. In **anaphase B,** the poles themselves move away from each other as the polar microtubules lengthen. Depending on the cell type involved, anaphase A and B may take place at the same time, or anaphase B may follow anaphase A.

Telophase. At the beginning of **telophase,** the daughter chromosomes have arrived at the poles of the spindle (Figure 19-20e). Next the chromosomes uncoil into the extended fibers typical of interphase chromatin, nucleoli develop at the nucleolar organizing sites on the DNA, the spindle disassembles, and nuclear envelopes form around the two groups of daughter chromosomes. During this period the cell usually undergoes cytokinesis, which divides the cell into two daughter cells.

The Mitotic Spindle Is Responsible for Chromosome Movements During Mitosis

The central purpose of mitosis is to separate the two sets of daughter chromosomes and partition them into the two newly forming daughter cells. To understand the mechanisms that allow this to be accomplished, we need to take a closer look at the microtubule-containing apparatus responsible for these events, the mitotic spindle.

Spindle Assembly and Chromosome Attachment. We saw in Chapter 15 that the tubulin subunits of a microtubule all face in the same direction, thereby giving microtubules an inherent *polarity;* that is, the two ends of each microtubule are chemically different (**Figure 19-25**). The end where microtubule assembly is initiated—located at the centrosome for spindle microtubules—is the minus (−) end. The end where most growth occurs, located away from the centrosome, is the plus (+) end. Microtubules are dynamic structures (see

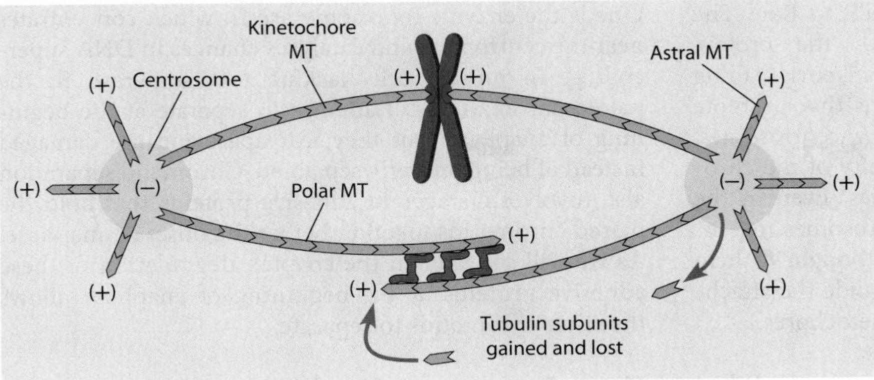

FIGURE 19-25 Microtubule Polarity in the Mitotic Spindle. This diagram shows only a few representatives of the many microtubules making up a spindle. The orientation of the tubulin subunits constituting a microtubule (MT) make the two ends of the MT different. The minus end is at the initiating centrosome; the plus end points away from the centrosome. MTs lengthen by adding tubulin subunits and shrink by losing subunits. In general, lengthening is due to addition at the plus ends and shortening to loss at the minus ends, but subunits can also be removed from the plus ends. The red structures between the plus ends of the polar MTs shown here represent proteins that crosslink them.

Chapter 15), in that tubulin subunits are continually being added and subtracted from both ends. When more subunits are being added than removed, the microtubule gets longer. In general, the plus end is the site favored for the addition of tubulin subunits and the minus end favored for subunit removal, so increases in microtubule length come mainly from addition of subunits to the plus end.

During late prophase, microtubule growth speeds up dramatically and initiation of new microtubules at the centrosomes increases. Once the nuclear envelope disintegrates at the beginning of prometaphase, contact between microtubules and chromosomal kinetochores becomes possible. When contact is made between a kinetochore and the plus end of a microtubule, they bind to each other and the microtubule becomes known as a kinetochore microtubule. This binding slows down depolymerization at the plus end of the microtubule, although polymerization and depolymerization can still occur there.

Figure 19-26 is an electron micrograph of a metaphase chromosome with two sets of attached microtubules, whose plus ends are embedded in the two kinetochores. Each kinetochore is a platelike, three-layered structure made of proteins attached to CEN sequences located in the centromere's DNA. Kinetochores of different species vary in size. In yeast, for example, they are small and bind only one spindle microtubule each, whereas the kinetochores of mammalian cells are much larger, each binding 30–40 microtubules.

Because the two kinetochores are located on opposite sides of a chromosome, they usually attach to microtubules emerging from centrosomes located at opposite poles of the cell. (The orientation of each chromosome is random; either kinetochore can end up facing either pole.) Meanwhile, the other main group of microtubules—the polar microtubules—make direct contact with polar microtubules coming from the opposite centrosome. When the plus-end regions of two microtubules of opposite polarity start to overlap, crosslinking proteins bind them to each other (Figure 19-25). Like the crosslinking between kinetochores and kinetochore microtubules, this crosslinking stabilizes the polar microtubules. Thus, we can picture a barrage of microtubules rapidly shooting out from each centrosome during late prophase and prometaphase. The ones that successfully hit a kinetochore or a microtubule of opposite polarity are stabilized; the others retreat by disassembling.

One shortcoming of the preceding mechanism is that it does not explain how spindles are assembled in cells lacking centrosomes, which includes most of the cells of higher plants and the oocytes (immature eggs) of many animals. Moreover, experiments using a laser microbeam to destroy centrosomes have shown that animal cells that normally contain centrosomes can assemble spindles using a centrosome-independent mechanism. In cells lacking centrosomes, chromosomes rather than centrosomes promote microtubule assembly and spindle formation. Chromosome-induced microtubule assembly requires the involvement of *Ran*, the GTP-binding protein whose role in nuclear transport was described in Chapter 18 (see Figure 18-31). Mitotic chromosomes possess a

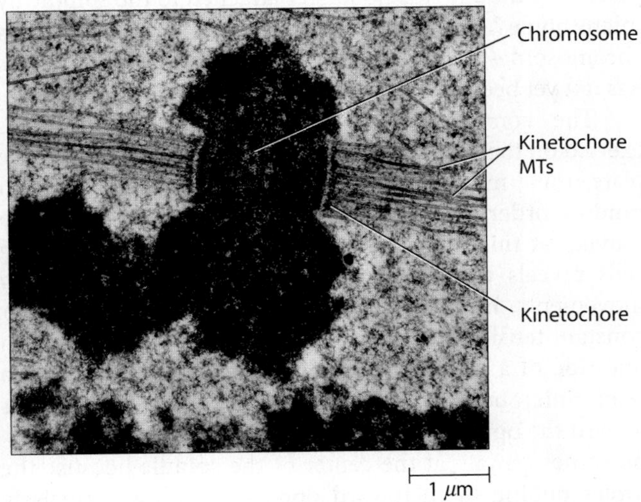

FIGURE 19-26 Kinetochores and Their Microtubules. The striped structures on either side of this metaphase chromosome are its kinetochores, each associated with one of the two sister chromatids. Numerous kinetochore MTs are attached to each kinetochore. The two sets of microtubules come from opposite poles of the cell (TEM).

protein that promotes the binding of GTP to Ran. The Ran-GTP complex then interacts with the protein *importin,* just as it does during nuclear transport, leading to the release of importin-bound proteins that promote microtubule assembly. So in cells lacking centrosomes, spindle formation is initiated in the vicinity of the chromosomes rather than at the spindle poles. Even in the more common case of cells that use centrosomes to generate spindle microtubules, Ran-GTP is thought to help organize the newly forming spindle and guide the attachment of microtubules to chromosomal kinetochores.

Chromosome Alignment and Separation. When spindle microtubules first become attached to chromosomal kinetochores during early prometaphase, the chromosomes are randomly distributed throughout the spindle. The chromosomes then migrate toward the central region of the spindle through a series of agitated, back-and-forth motions generated by at least two different kinds of forces. First, the kinetochore microtubules exert a "pulling" force, moving the chromosomes toward the pole that the microtubules are attached to. This force can be demonstrated experimentally by using glass microneedles to tear individual chromosomes away from the spindle. A chromosome that has been removed from the spindle remains motionless until new microtubules attach to its kinetochore, at which time the chromosome is drawn back into the spindle.

The second force tends to "push" chromosomes away if they approach either spindle pole. The existence of this pushing force has been demonstrated by studies in which a laser microbeam is used to break off one end of a chromosome. Once the broken chromosome fragment has been cut free from its associated centromere and kinetochore, the fragment tends to move away from the nearest spindle pole, even though it is no longer attached to the spindle by microtubules. The nature of the pushing force that propels chromosomes in the absence of microtubule attachments has not yet been clearly identified.

The combination of pulling and pushing forces exerted on the chromosomes drives them to the metaphase plate, their most stable location, where they line up in random order. Although the chromosomes appear to stop moving at this point, careful microscopic study of living cells reveals that they continue to make small jerking movements, indicating that the chromosomes are under constant tension in both directions. If the kinetochore on one side of a metaphase chromosome is severed using a laser microbeam, the chromosome promptly moves toward the opposite spindle pole. Hence, metaphase chromosomes remain at the center of the spindle because the forces pulling them toward opposite poles are precisely balanced.

At the beginning of anaphase, the two chromatids of each metaphase chromosome split apart and start moving toward opposite spindle poles. Several molecules have been implicated in this process of chromatid separation. One is the enzyme topoisomerase II, which concentrates near the centromere and catalyzes changes in DNA supercoiling. In mutant cells lacking topoisomerase II, the paired chromatids still attempt to separate at the beginning of anaphase, but they tear apart and are damaged instead of being properly separated. Chromatid separation also involves changes in adhesive proteins that hold the paired chromatids together before the onset of anaphase. As we will see later in the chapter, degradation of these adhesive proteins at the beginning of anaphase allows the sister chromatids to separate.

Motor Proteins and Chromosome Movement. Once the two chromatids of each metaphase chromosome have split apart, they function as two independent chromosomes that move to opposite spindle poles. Studies of the mechanisms underlying this movement have led to the discovery of several **motor proteins** that play active roles in mitosis. As we saw in Chapters 15 and 16, motor proteins use energy derived from ATP to change shape in such a way that they exert force and cause attached structures to move. Motor proteins play at least three distinct roles in the movement of anaphase chromosomes.

The first role involves the mechanism that moves chromosomes, kinetochores first, toward the spindle poles during anaphase A. As shown in **Figure 19-27a** (❶), this type of chromosome movement is driven by motors associated with kinetochore microtubules. Considerable evidence suggests that these motors are specialized members of the *kinesin* family of proteins. In Chapter 16 we saw that some specialized kinesins can bind to the end of a microtubule and induce it to depolymerize. Two such kinesin-like motors are involved in moving chromosomes toward the spindle poles: One is located at the plus end of the kinetochore microtubules, and the other is located at the minus end. The motor located at the plus end is embedded in the kinetochore, where it induces microtubule depolymerization and thereby moves the chromosome toward the spindle pole as it "chews up" the plus end of the microtubule. At the same time, the motor located at the minus end is embedded in the spindle pole, where it induces microtubule depolymerization and thereby "reels in" the microtubules and their attached chromosomes.

Several lines of evidence support this view that microtubule depolymerization plays a crucial role in chromosome movement. For example, if cells are exposed to the drug *taxol,* which inhibits microtubule depolymerization, chromosomes do not move toward the spindle poles. Conversely, exposing cells to increased pressure, which increases the rate of microtubule depolymerization, causes chromosomes to move toward the poles more quickly. Finally, antibodies that inhibit the depolymerizing activity of either the motor protein located at the spindle pole or the motor protein located at the kinetochore have both been shown to interfere with chromosome movement.

The second role played by motor proteins during anaphase is associated with the movement of the spindle

(a) Three roles played by motor proteins

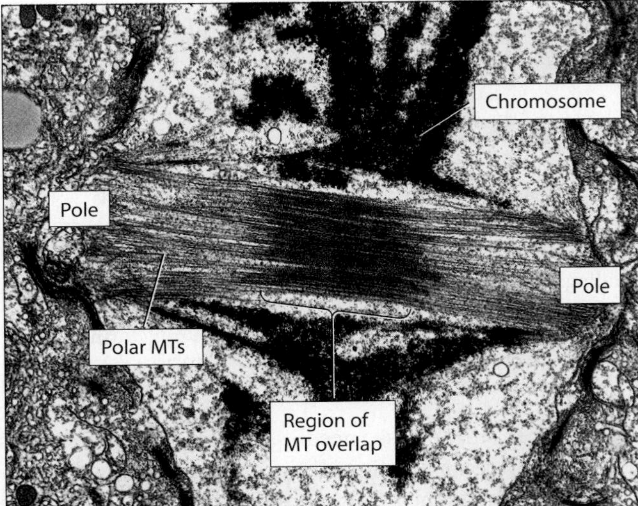

(b) Polar microtubules during metaphase

2 μm

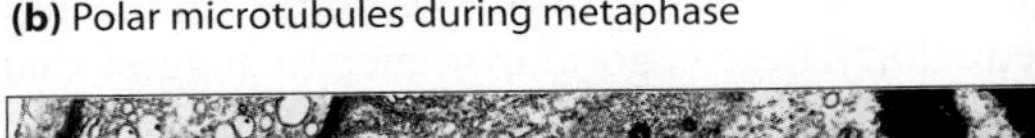

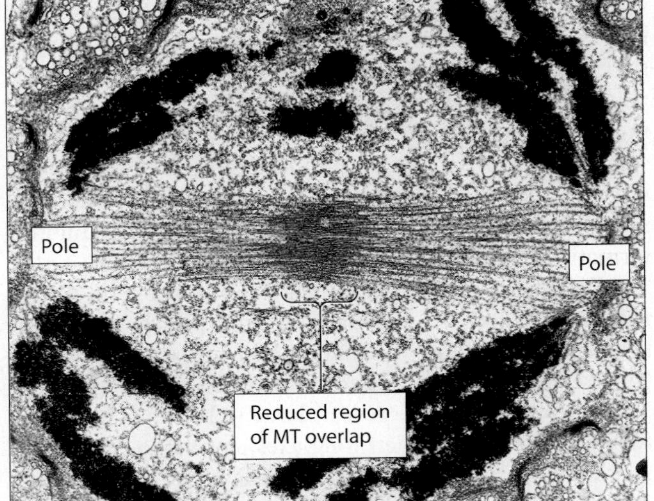

(c) Polar microtubules during anaphase

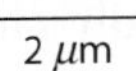

FIGURE 19-27 Mitotic Motors. (a) A model for mitotic chromosome movement based on three roles played by motor proteins. Motor proteins are shown in red, and the small red arrows indicate the direction of movement generated by these motors. Motor proteins are associated with three types of microtubules: kinetochore MTs, polar MTs, and astral MTs. ❶ Kinetochore MTs have motor proteins associated with both their plus ends (embedded in the chromosomal kinetochore) and their minus ends (located in the centrosome of the spindle pole). The motor proteins located at the kinetochore "chew up" (i.e., depolymerize) the plus ends of the kinetochore MTs. In this way, the chromosome is pulled toward the spindle pole as the kinetochore MTs are shortened through the loss of tubulin subunits. Simultaneously, motor proteins located at the spindle pole depolymerize the minus ends of the kinetochore MTs, reeling in the MTs and their attached chromosomes. ❷ Motor proteins crosslink the polar MTs and cause them to slide apart, thereby forcing the spindle poles away from each other. As the polar MTs slide apart, they are lengthened by the addition of tubulin subunits to their plus ends where they overlap near the spindle center. ❸ Astral MT motor proteins link the plus ends of astral MTs to the cell cortex and exert a pull on the spindle poles by inducing astral MT depolymerization at their plus ends. (**b**, **c**) The two electron micrographs provide evidence for the sliding of polar MTs driven by polar MT motors. During metaphase, the polar MTs from opposite ends of the cell overlap significantly. During anaphase, the polar MT motors cause these two groups of MTs to slide away from each other, thereby resulting in a reduced region of overlap (TEMs). (MT = microtubule)

poles away from each other during anaphase B. In this case, bipolar kinesin motors bind to overlapping polar microtubules coming from opposite spindle poles, causing the polar microtubules to slide apart, thereby forcing the spindle poles away from each other (Figure 19-27a, ❷). As the microtubules slide apart, they are lengthened by the addition of tubulin subunits to their plus ends near the center of the spindle, where microtubules coming from opposite spindle poles overlap. Microtubule sliding can be experimentally induced by exposing isolated spindles to ATP, indicating that the motor proteins use energy derived from ATP hydrolysis to cause the overlapping microtubules to slide away from one another. During anaphase B, this motor activity may be the primary force that elongates the spindle, while the lengthening of the polar microtubules is secondary. In Figure 19-27, parts b and c provide electron microscopic evidence for the sliding of overlapping polar microtubules during anaphase B.

The third type of motor-produced force detected during anaphase involves *cytoplasmic dynein,* which is associated with astral microtubules (Figure 19-27a, ❸). The plus ends of astral microtubules are connected to the *cell cortex,* a layer of actin microfilaments lining the inner surface of the plasma membrane. Cytoplasmic dynein, which moves toward the minus end of microtubules, appears to pull each spindle pole toward the cortex. Such pulling—in addition to the outward push generated by the motor proteins that crosslink the overlapping polar microtubules—helps to separate the spindle poles during anaphase B in some cell types.

Mitosis therefore involves at least three separate groups of motor proteins, operating on kinetochore microtubules, polar microtubules, and astral microtubules, respectively (Figure 19-27a). The relative contributions of the pushing and pulling forces generated by these three sets of motor proteins differ among organisms. For example, in diatoms and yeast, the pushing (sliding) of microtubules against adjacent ones of opposite polarity is particularly important in anaphase B. In contrast, pulling at the asters is the main force in the cells of certain other fungi. In vertebrates both mechanisms are probably operative, although astral pulling may play a greater role, especially during spindle formation.

Cytokinesis Divides the Cytoplasm

After the two sets of chromosomes have separated during anaphase, cytokinesis divides the cytoplasm in two, thereby completing the process of cell division. Cytokinesis usually starts during late anaphase or early telophase, as the nuclear envelope and nucleoli are re-forming and the chromosomes are decondensing. Cytokinesis is not inextricably linked to mitosis, however. In some cases, a significant time lag may occur between nuclear division (mitosis) and cytokinesis, indicating that the two processes are not tightly coupled. Moreover, certain cell types can undergo many rounds of chromosome replication and nuclear division in the absence of cytokinesis, thereby producing a large, multinucleate cell known as a *syncytium.* Sometimes the multinucleate condition is permanent; in other situations, the multinucleate state is only a temporary phase in the organism's development. This is the case, for example, in the development of a plant seed tissue called *endosperm* in cereal grains. Here, nuclear division occurs for a time unaccompanied by cytokinesis, generating many nuclei in a common cytoplasm. Successive rounds of cytokinesis then occur without mitosis, walling off the many nuclei into separate endosperm cells. A similar process occurs in some insect embryos.

Despite these examples, in most cases cytokinesis does accompany or closely follow mitosis, thereby ensuring that each of the daughter nuclei acquires its own cytoplasm and becomes a separate cell.

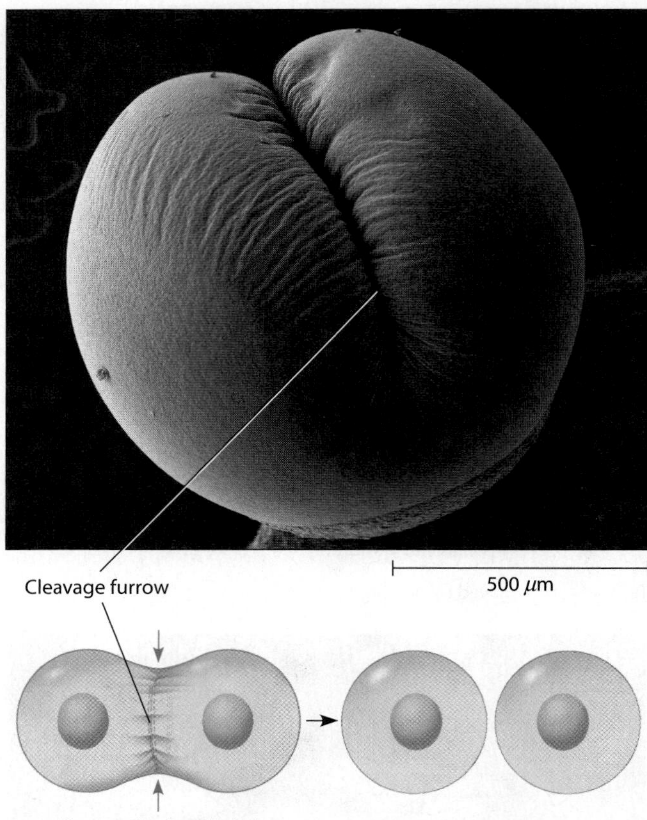

FIGURE 19-28 Cytokinesis in an Animal Cell. (Top) An electron micrograph of a fertilized frog egg caught in the act of dividing. The cleavage furrow is clearly visible as an inward constriction of the plasma membrane (SEM). (Bottom) A schematic diagram showing the position of the contractile ring during cytokinesis, which pinches the dividing cell in two (red arrows).

Cytokinesis in Animal Cells. The mechanism of cytokinesis differs between animals and plants. In animal cells, cytoplasmic division is called **cleavage.** The process begins as a slight indentation or puckering of the cell surface, which deepens into a **cleavage furrow** that encircles the cell, as shown in **Figure 19-28** for a fertilized frog egg. The furrow continues to deepen until opposite surfaces make contact and the cell is split in two. The cleavage furrow divides the cell along a plane that passes through the central region of the spindle (the *spindle equator*), suggesting that the location of the spindle determines where the cytoplasm will be divided. This idea has been investigated experimentally by moving the mitotic spindle using either tiny glass needles or gravitational forces generated by centrifugation. If the spindle is moved before the end of metaphase, the orientation of the cleavage plane changes so that it passes through the new location of the spindle equator. There is good evidence to suggest that signals emanating from the central portion of the spindle, known as the *spindle midzone,* are important for completion of cytokinesis (**Figure 19-29a**). For example, in nematode worm or fruit fly cells lacking components of a multiprotein complex (known as *centralspindlin*) found in the spindle midzone, cytokinesis begins, but the cleavage furrow regresses. The activity of astral microtubules may complement the work of the spindle midzone by inhibiting formation of a cleavage furrow in other regions of the cortex.

Cleavage depends on a beltlike bundle of actin microfilaments called the **contractile ring,** which forms just beneath the plasma membrane during early anaphase (Figure 19-29a). Examination of the contractile ring with an electron microscope reveals large numbers of actin filaments oriented with their long axes parallel to the furrow. As cleavage progresses, this ring of microfilaments tightens around the cytoplasm, like a belt around the waist, eventually pinching the cell in two. Tightening of the contractile

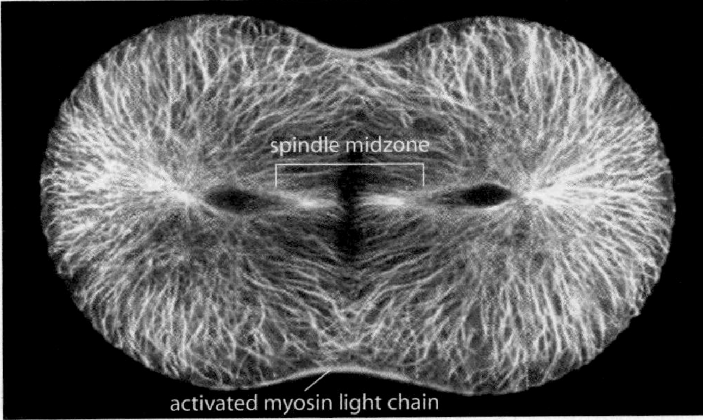

(a) Myosin and tubulin during cutokinesis 10 μm

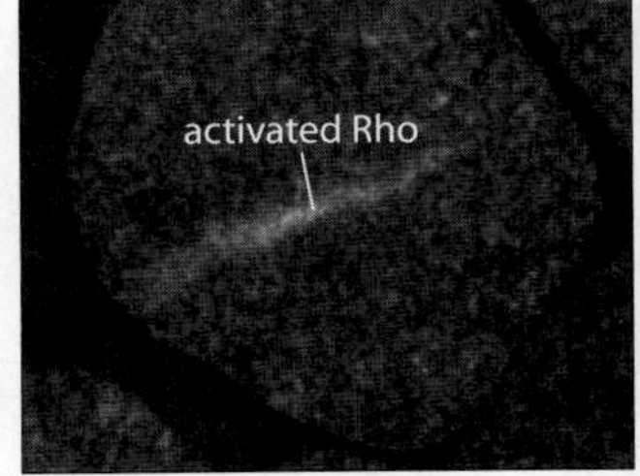

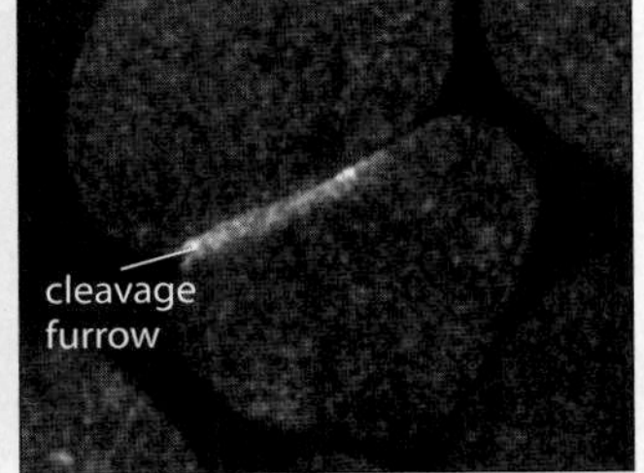

(b) Activated Rho during cytokinesis 25 μm

FIGURE 19-29 Myosin and Rho During Cytokinesis. (a) A confocal micrograph of a sea urchin zygote, showing microtubules in white and activated myosin regulatory light chain in blue. Active myosin accumulates at the cleavage furrow. **(b)** A fluorescent protein that binds active Rho accumulates in the cleavage furrow of cells in a *Xenopus* (frog) embryo. Accumulation of active Rho precedes furrowing (left); the same cell four minutes later has formed a furrow at the site where Rho accumulated (right).

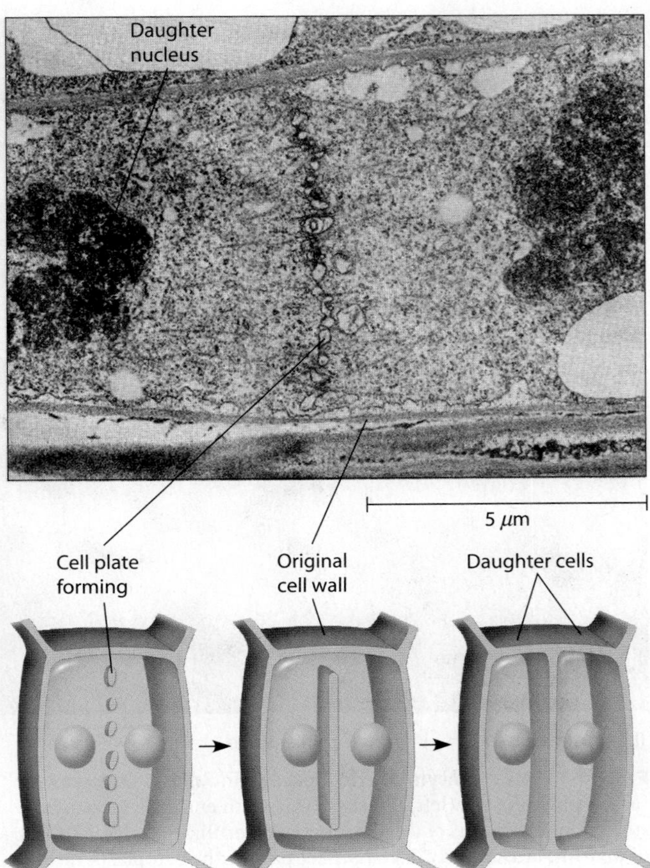

FIGURE 19-30 Cytokinesis and Cell Plate Formation in a Plant Cell. (Top) A cell from the sugar maple, *Acer saccharinum*, at late telophase. The daughter nuclei with their sets of chromosomes are partially visible as the dark material on the far right and far left of the micrograph, and the developing cell plate is seen as a line of vesicles in the midregion of the cell. The microtubules of the phragmoplast are oriented perpendicular to the cell plate (TEM). (Bottom) A schematic diagram showing the location of the cell plate and the original cell wall of a dividing plant cell.

ring is generated by interactions between the actin microfilaments and *myosin,* the motor protein whose role in muscle contraction was described in Chapter 16.

The contractile ring provides a dramatic example of how rapidly actin-myosin complexes can be assembled and disassembled in nonmuscle cells. Members of the *Rho* family of GTP-binding proteins play a central role in regulating the assembly and activation of the contractile ring. One family member, called RhoA, is recruited to the cleavage furrow (Figure 19-29b), where it helps orchestrate cytokinesis by activating proteins that promote actin polymerization. RhoA also stimulates protein kinases that phosphorylate myosin, the key step in activating myosin to perform its motor function in tightening the contractile ring.

MB Mitosis (3 of 3): Comparing Cell Division in Animals, Plants, and Bacteria (BioFlix tutorial)

Cytokinesis in Plant Cells. Cytokinesis in higher plants is fundamentally different from the corresponding process in animal cells. Because plant cells are surrounded by a rigid cell wall, they cannot form a contractile ring at the cell surface that pinches the cell in two. Instead, they divide by assembling a plasma membrane and a cell wall between the two daughter nuclei (**Figure 19-30**). In other words, rather than pinching the cytoplasm in half with a contractile ring that moves from the outside of the cell toward the interior, the plant cell cytoplasm is divided by a process that begins in the cell interior and works toward the periphery.

Cytokinesis in plants is typically initiated during late anaphase or early telophase, when a group of small, membranous vesicles derived from the Golgi complex align themselves across the equatorial region of the spindle. These vesicles, which contain polysaccharides and glycoproteins required for cell wall formation, are guided to the spindle equator by the **phragmoplast,** a parallel array of microtubules derived from polar microtubules and oriented perpendicular to the direction in which the new cell wall is being formed. After arriving at the equator, the Golgi-derived vesicles fuse together to produce a large, flattened sac called the **cell plate,** which represents the cell wall in the process of formation. The contents of the sac assemble to form the noncellulose components of the primary cell wall, which expands outward as clusters of microtubules and vesicles form at the lateral edges of the

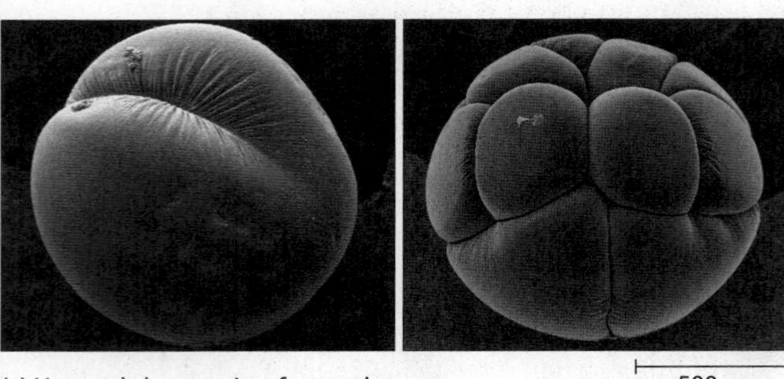

(a) Unequal cleavage in a frog embryo

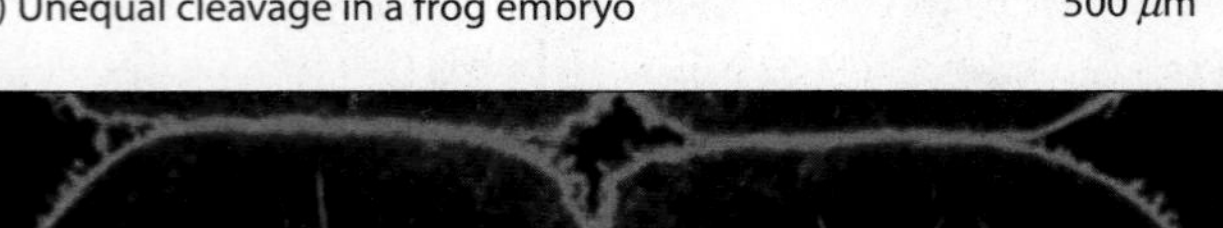

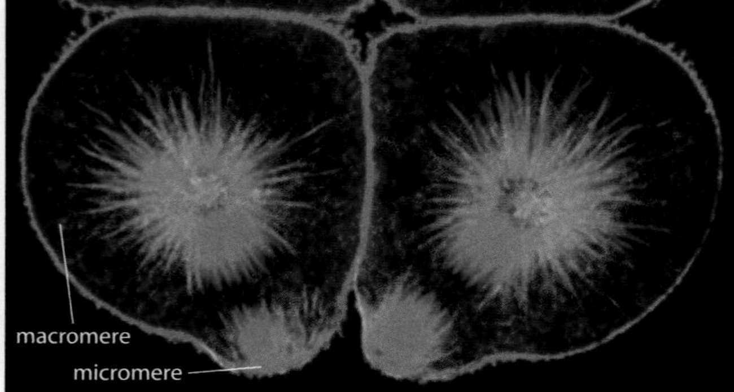

(b) Asymmetric spindles in a sea urchin embryo

FIGURE 19-31 Asymmetric Cleavage in Animal Embryos. **(a)** Amphibian eggs (left) are very large, with enough cytoplasm to sustain many rounds of cell division after fertilization. Each round of division during early development parcels the cytoplasm into smaller cells. Some cells are much larger than others (right). **(b)** Two cells of an 8-cell sea urchin embryo stained for microtubules (green) and actin (red). Each cell is about to divide to form a much larger cell (a "macromere") and a smaller cell ("micromere"). The spindle is highly asymmetric in its location.

advancing cell plate. Eventually, the expanding cell plate makes contact with the original cell wall, separating the two daughter cells from each other. The new cell wall is then completed by deposition of cellulose microfibrils. The plasmodesmata that provide channels of continuity between the cytoplasms of adjacent plant cells are also present in the cell plate and the new wall as it forms.

Cell Division Is Sometimes Asymmetric

The division plane passes through the spindle equator in both animal and plant cells. When the spindle is positioned across the middle of the cell, as is typically the case, there is a roughly even division of components between the two new cells. Cytoplasmic division is not always symmetric, however. In the budding yeast *Saccharomyces cerevisiae,* for example, the mitotic spindle forms in a highly asymmetric fashion, creating one large cell and one very small cell (see Figure 1B-1). Asymmetric divisions also occur frequently during embryonic development in animal embryos. Such asymmetric division often results in cells that not only vary dramatically in size but also acquire different developmental potential (**Figure 19-31**). Asymmetric divisions also take place in female animals during the development of egg cells from precursor cells called *oocytes.* In this case, cytokinesis divides the cytoplasm unequally so that the cell destined to become the egg receives the bulk of the cytoplasm of the original oocyte (see Figure 20-9b), thereby maximizing the content of stored nutrients in each egg.

Cell division can also involve a more subtle type of asymmetry in which daughter cells look alike but have different fates. Such differences in cell fate are generated by mechanisms in which specific molecules located in certain regions of the parental cytoplasm are distributed unequally to the two daughter cells, thereby determining their unique fates.

Regulation of the Cell Cycle

Earlier in the chapter, we described a typical eukaryotic cell cycle in which G1, S, G2, and M phases are completed in orderly progression over a period of roughly 24 hours. Such a pattern is common in growing organisms and in cultured cells that have not run out of nutrients or space. But many variations are also observed, especially in the overall length of the cycle, the relative length of time spent in various phases, and how closely mitosis and cytokinesis are coupled. This variability tells us that the cell cycle must be regulated to meet the needs of each cell type and organism. The molecular basis of such cell cycle regulation is a subject of intense interest, not only for understanding the life cycles of normal cells but also for understanding how cancer cells manage to escape these normal control mechanisms.

The Length of the Cell Cycle Varies Among Different Cell Types

Some of the most commonly encountered variations in the cell cycle involve differences in how fast cells divide. In multicellular organisms, generation times vary markedly among cell types, depending on their role in the organism. At one extreme are cells that divide continuously as a means of replacing cells that are constantly being lost or destroyed. Included in this category are cells involved in sperm formation and the *stem cells* that give rise to blood cells, skin cells, and the epithelial cells that form the inner lining of body organs such as the lungs and intestines. Human stem cells may have generation times as short as 8 hours.

In contrast, cells located in slow-growing tissues may have generation times of several days or more, and some cells, such as those of mature nerve or muscle tissue, do not divide at all. Still other cell types do not divide under normal conditions but can be induced to start dividing again by an appropriate stimulus. Liver cells are in this category; they do not normally proliferate in the mature liver but can be induced to do so if a portion of the liver is removed surgically. Lymphocytes (white blood cells) are another example; when exposed to a foreign protein, they begin dividing as part of the immune response.

Most of these variations in generation time are based on differences in the length of G1, although S and G2 can also vary. Cells that divide slowly may spend days, months, or even years in the offshoot of G1 called G0, whereas cells that divide very rapidly have a short G1 phase or even eliminate G1 entirely. The embryonic cells of insects, amphibians, and several other nonmammalian animals are dramatic examples of cells that have very short cell cycles, with no G1 phase and a very short S phase. For example, during early embryonic development of the frog *Xenopus laevis,* each round of division simply subdivides the initial cytoplasm into smaller and smaller cells, until the cell size typical of adult tissues is reached (see Figure 19-31a). The cell cycle takes less than 30 minutes, even though the normal length of the cell cycle in adult tissues is about 20 hours. The rapid rate of DNA synthesis needed to sustain such a quick cell cycle is achieved in part by increasing the total number of replicons, thereby decreasing the amount of DNA that each replicon must synthesize. In addition, all replicons are activated at the same time, in contrast to the sequential activation observed in adult tissues. By increasing the number of replicons and activating them all simultaneously, S phase is completed in less than 3 minutes, at least 100 times faster than in adult tissues of the same organism.

Although we know from such examples that cell growth is not essential to the cell cycle, the two are generally linked so that cells can divide without getting progressively smaller. A protein kinase called *TOR (target of rapamycin)* plays a central role in the signaling network that controls cell size and coordinates it with cell cycle progression. This signaling network activates TOR in the presence of nutrients and growth factors, and the activated TOR then stimulates molecules that control the rate of protein synthesis. The resulting increase in protein production leads to an increase in cell mass. Some of the molecules activated by TOR also facilitate entry into S phase, making TOR an important regulator of both cell growth and cell cycle progression.

Progression Through the Cell Cycle Is Controlled at Several Key Transition Points

The control system that regulates progression through the cell cycle must accomplish several tasks. First, it must ensure that the events associated with each phase of the cell cycle are carried out at the appropriate time and in the appropriate sequence. Second, it must make sure that each phase of the cycle has been properly completed before the next phase is initiated. Finally, it must be able to respond to external conditions that indicate the need for cell proliferation (e.g., the quantity of nutrients available or the presence of growth-signaling molecules).

The preceding objectives are accomplished by a group of molecules that act at key transition points in the cell cycle (**Figure 19-32**). At each of these points, conditions within the cell determine whether the cell will proceed to the next stage of the cycle. The first such control point occurs during late G1. We have already seen that G1 is the phase that varies most among cell types, and mammalian cells that have stopped dividing are almost always arrested during G1. For example, in cultured cells the process of cell division can be stopped or slowed by allowing the cells to run out of either nutrients or space or by adding inhibitors of vital processes such as protein synthesis. In all such cases, the cell cycle is halted in late G1, suggesting that progression from G1 into S is a critical control point in the cell cycle. In yeast, this control point is called **Start**; yeast cells must have sufficient nutrients and must reach a certain size before they can pass through Start. In animal cells, the comparable control point is called the **restriction point.** The ability to pass through the restriction point is influenced by the presence of extracellular *growth factors* (page 407), which are proteins used by multicellular organisms to stimulate or inhibit cell proliferation. Cells that have successfully passed through the restriction point are committed to S phase, whereas those that do not pass the restriction point enter into G0 and reside there for

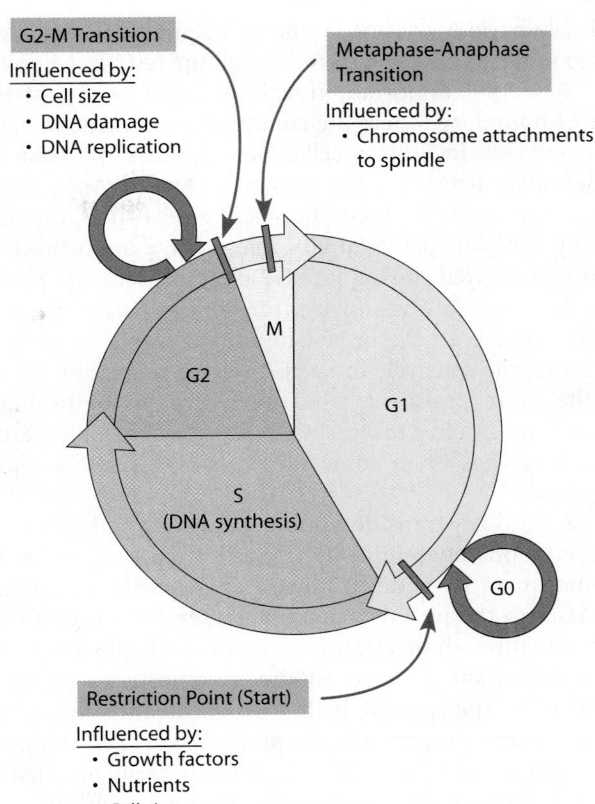

FIGURE 19-32 Key Transition Points in the Cell Cycle. The red bars mark three important transition points in the eukaryotic cell cycle where control mechanisms determine whether the cell will continue to proceed through the cycle. That determination is based on chemical signals reflecting both the cell's internal state and its external environment. The two circular, dark green arrows indicate locations in late G1 and late G2 where the cell can exit from the cycle and enter a nondividing state.

variable periods of time, awaiting a signal that will allow them to reenter G1 and pass through the restriction point.

A second important transition point occurs at the G2-M boundary, where the commitment is made to enter into mitosis. In certain cell types, the cell cycle can be indefinitely arrested at the end of G2 if cell division is not necessary; under such conditions, the cells enter a nondividing state analogous to G0. The relative importance of controls exerted during late G2 or late G1 in regulating the rate of cell division by transiently halting the cell cycle varies with the organism and cell type. In general, arresting the cell cycle in late G1 (at the restriction point) is the more prevalent type of control in multicellular organisms. But in a few cases, such as the division of fertilized frog eggs or in some skin cells, G2 arrest is more important.

A third key transition point occurs during M phase at the junction between metaphase and anaphase, where the commitment is made to move the two sets of chromosomes into the newly forming daughter cells. Before cells can pass through this transition point and begin anaphase, it is important to have all the chromosomes properly attached to the spindle. If the two chromatids that make up each chromosome are not properly attached to opposite spindle poles, the cell cycle is temporarily arrested to allow spindle attachment to occur. Without such a mechanism, there would be no guarantee that each of the newly forming daughter cells would receive a complete set of chromosomes.

Cell behavior at the various transition points is influenced both by successful completion of preceding events in the cycle (such as chromosome attachment to the spindle) and by factors in the cell's environment (such as nutrients and growth factors). But whatever the particular influences may be, their effects on cell cycle progression are mediated by a group of related control molecules that activate or inhibit one another in chains of interactions that can be quite elaborate. Let's now see how these control molecules were identified and what functions they perform.

Studies Involving Cell Fusion and Cell Cycle Mutants Led to the Identification of Molecules That Control the Cell Cycle

The first hints concerning the identity of the molecules that drive progression through the cell cycle came from cell fusion experiments performed in the early 1970s. In these studies, two cultured mammalian cells in different phases of the cell cycle were fused to form a single cell with two nuclei—a *heterokaryon.* As **Figure 19-33a** indicates, if one of the original cells is in S phase and the other is in G1, the G1 nucleus in the heterokaryon quickly initiates DNA synthesis, even if it would not normally have reached S phase until many hours later. Such observations indicate that S phase cells contain molecules that trigger progression from G1 into S. The controlling molecules are

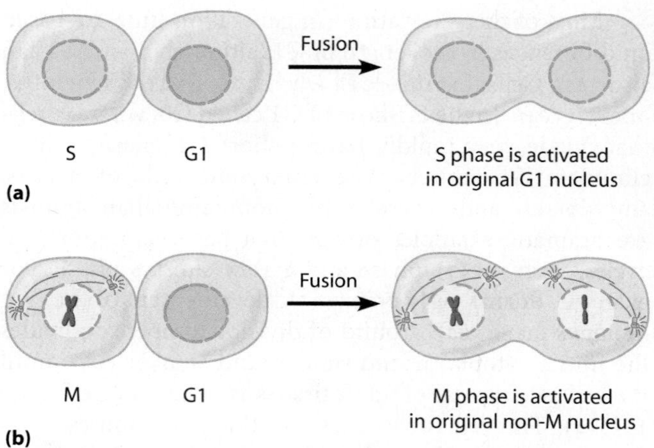

FIGURE 19-33 Evidence for the Role of Chemical Signals in Cell Cycle Regulation. Evidence was obtained from studies in which cells at two different points in the cell cycle were fused, forming a single cell with two nuclei. Cell fusion can be induced by adding certain viruses or polyethylene glycol or by applying a brief electrical pulse, which causes plasma membranes to destabilize momentarily (electroporation). **(a)** When cells in S phase and G1 phase are fused, DNA synthesis begins in the original G1 nucleus, suggesting that a substance that activates S phase is present in the S phase cell. **(b)** When a cell in M phase is fused with one in any other phase, the latter cell immediately enters mitosis. If the cell was in G1, the condensed chromosomes that appear have not replicated and therefore are analogous to single chromatids.

not simply the enzymes involved in DNA replication, since these enzymes can be present in high concentration in cells that do not enter S phase.

Cell fusion experiments have also been performed in which cells undergoing mitosis are fused with interphase cells in either G1, S, or G2. After fusion, the nucleus of such interphase cells is immediately driven into the early stages of mitosis, including chromatin condensation into visible chromosomes, spindle formation, and fragmentation of the nuclear envelope. If the interphase cell had been in G1, the condensed chromosomes will be unduplicated (Figure 19-33b).

Taken together, the preceding experiments suggested that molecules present in the cytoplasm are responsible for driving cells from G1 into S phase and from G2 into mitosis. Progress in identifying these cell cycle control molecules was facilitated by genetic studies of yeasts. Because they are single-celled organisms that can be readily grown and studied under defined laboratory conditions, yeasts are particularly convenient for investigating the genes involved in cell cycle control.

Working with the budding yeast *Saccharomyces cerevisiae,* geneticist Leland Hartwell pioneered the development of techniques for identifying yeast mutants that are "stuck" at some point in the cell cycle. It might be expected that most such mutants would be difficult or impossible to study because their blocked cell cycle would prevent them from reproducing. But Hartwell overcame this potential obstacle with a powerful strategy—the use of temperature-sensitive mutants. As mentioned earlier in the chapter, this is a type of mutation whose harmful effects are apparent only at temperatures above the normal range for the organism. Therefore, yeast cells carrying a temperature-sensitive mutation can be successfully grown at a lower ("permissive") temperature, even though their cell cycles would be blocked at higher temperatures. Presumably the protein encoded by the mutated cell cycle gene is close enough to the normal gene product to function at the lower temperature, while the increased thermal energy at higher temperatures disrupts its active conformation (the molecular shape needed for function) more readily than that of the normal protein.

Using this approach, Hartwell and his colleagues identified many genes involved in the cell cycle of *S. cerevisiae* and established where in the cycle their products operate. Predictably, it turned out that some of these genes produce DNA replication proteins, but others seemed to function in cell cycle regulation. A breakthrough discovery was made by Paul Nurse, who carried out similar research with the fission yeast *Schizosaccharomyces pombe*. He identified a gene called *cdc2*, whose activity is needed for initiating mitosis—that is, for moving cells through the G2-M transition. (The acronym *cdc* stands for cell division cycle.) The *cdc2* gene was soon found to have counterparts in all eukaryotic cells studied. When the properties of the protein produced by the *cdc2* gene were examined, it was discovered to be a *protein kinase*—that is, an enzyme that catalyzes the transfer of a phosphate group from ATP to other target proteins. This discovery opened the door to unraveling the mysteries of the cell cycle.

Progression Through the Cell Cycle Is Controlled by Cyclin-Dependent Kinases (Cdks)

The phosphorylation of target proteins by protein kinases, and their dephosphorylation by enzymes called *protein phosphatases,* is a common mechanism for regulating protein activity that turns out to be widely used in controlling the cell cycle. Progression through the cell cycle is driven by a series of protein kinases—including the protein kinase produced by the *cdc2* gene—that exhibit enzymatic activity only when they are bound to a special type of activator protein called a **cyclin.** Such protein kinases are therefore referred to as **cyclin-dependent kinases** or simply **Cdks.** The eukaryotic cell cycle is controlled by several different Cdks that bind to different cyclins, thereby creating a variety of Cdk-cyclin complexes.

As originally shown by Tim Hunt using sea urchin embryos, cyclins get their name because their concentration in the cell oscillates up and down with the different phases of the cell cycle. Cyclins required for the G2-M transition and the early events of mitosis are called *mitotic cyclins,* and the Cdks to which they bind are known as *mitotic Cdks*. Likewise, cyclins required for passage through the G1 restriction point (or Start) are called *G1 cyclins,* and the Cdks to which they bind are *G1 Cdks*. Yet another group of cyclins, called *S cyclins,* are required for events associated with DNA replication during S phase. The pioneering work of Hartwell, Nurse, and Hunt that led to our current understanding of Cdks and cyclins was honored by the Nobel Prize in Physiology or Medicine in 2001.

If progression through critical points in the cell cycle is controlled by an assortment of different Cdks and cyclins interacting in various combinations, how is the activity of these protein complexes regulated? One level of control is exerted by the availability of cyclin molecules, which are required for activating the protein kinase activity of Cdks, and a second type of regulation involves phosphorylation of Cdks. We will illustrate both types of control by taking a closer look at mitotic Cdk-cyclin, which controls progression from G2 into mitosis.

Mitotic Cdk-Cyclin Drives Progression Through the G2-M Transition by Phosphorylating Key Proteins Involved in the Early Stages of Mitosis

The earliest evidence for the existence of a control molecule that triggers the onset of mitosis came from experiments involving frog eggs. Mature eggs develop from precursor cells called *oocytes* through meiosis, a special type of cell division that reduces the chromosome number in half when eggs or sperm are being formed (see Chapter 20). During egg maturation, the cell cycle is halted shortly after the start of meiosis, where the oocyte waits until it is stimulated by an appropriate hormone. The oocyte then completes most of the phases of meiosis but is arrested during metaphase of the second of two meiotic divisions. It is now a "mature" egg cell, capable of being fertilized. Crucial experiments by Yoshio Masui and colleagues demonstrated that if cytoplasm taken from a mature egg cell is injected into the cytoplasm of an immature oocyte that is awaiting hormonal stimulation, the oocyte will immediately proceed through meiosis (**Figure 19-34**). Masui therefore hypothesized that a cytoplasmic chemical, which he named **MPF (maturation-promoting factor)**, induces oocyte maturation (i.e., meiotic division).

Subsequent experiments demonstrated that besides inducing meiosis, MPF also triggers mitosis when injected into fertilized frog eggs. Comparable molecules were soon detected in the cytoplasms of a broad range of dividing cell types, including yeasts, marine invertebrates, and mammals. Biochemical studies of these mitosis-inducing molecules revealed that they consist of two subunits: a Cdk and a cyclin. In other words, *MPF is a mitotic Cdk-cyclin complex.* Moreover, the mitotic Cdk portion of this complex is almost identical to the protein produced by the yeast *cdc2* gene. In fact, in yeast cells with a defective or missing *cdc2* gene, the human gene coding for mitotic Cdk can substitute perfectly well, even though the last ancestor common to yeasts and humans probably lived about a billion years ago!

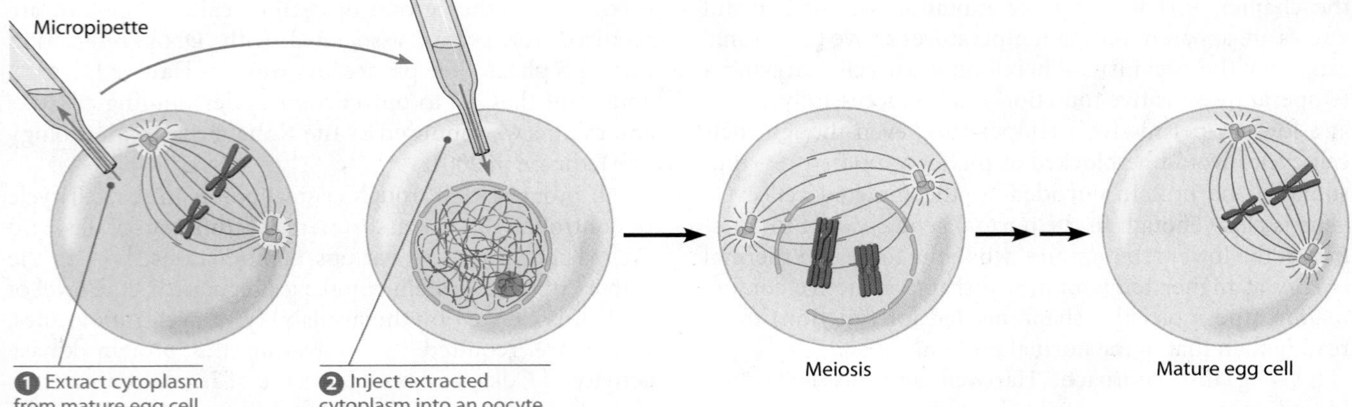

FIGURE 19-34 Evidence for the Existence of MPF. Hormones act on frog oocytes to trigger meiosis and development into mature frog eggs, which are arrested (until fertilization) in metaphase of the second meiotic division. The experiment shown here, performed by Y. Masui and C. L. Markert in 1971, established the existence of a substance involved in this process; they called it maturation-promoting factor (MPF). In their experiment, they used a micropipette to remove cytoplasm from a mature egg cell (arrested in metaphase of the second meiotic division) and injected it into an immature oocyte. The oocyte then proceeded through meiosis and became a mature egg cell. This experimental procedure could thus be used as an assay for detecting and eventually isolating MPF. The hormones that trigger oocyte maturation in the frog were presumed to act by stimulating the synthesis or activation of MPF. MPF is now known to be a mitotic Cdk-cyclin.

Having established that MPF is a mitotic Cdk-cyclin that triggers the onset of mitosis in a broad spectrum of cell types, the question arose as to how mitotic Cdk-cyclin is controlled so that it functions only at the proper moment—that is, at the end of G2. The answer is not to be found in the availability of mitotic Cdk itself because its concentration remains relatively constant throughout the cell cycle. However, mitotic Cdk is active as a protein kinase only when it is bound to mitotic cyclin, and mitotic cyclin is not always present in adequate amounts. Instead, the concentration of mitotic cyclin gradually increases during G1, S, and G2; eventually it reaches a critical threshold at the end of G2 that permits it to activate mitotic Cdk and thereby trigger the onset of mitosis (**Figure 19-35**). Halfway through mitosis, the mitotic cyclin molecules are abruptly destroyed. The resulting decline in mitotic Cdk activity prevents another mitosis from occurring until the mitotic cyclin concentration builds up again during the next cell cycle.

In addition to requiring mitotic cyclin, the activation of mitotic Cdk involves phosphorylation and dephosphorylation of the Cdk molecule itself. As shown in **Figure 19-36,** the binding of mitotic cyclin to mitotic Cdk yields a Cdk-cyclin complex that is initially inactive (step ➊). To trigger mitosis, the complex requires the addition of an activating phosphate group to a particular amino acid in the Cdk molecule. Before this phosphate is added, however, *inhibiting* kinases phosphorylate the Cdk molecule at two other locations, causing the active site to be blocked (step ➋). The activating phosphate group, highlighted with yellow in step ➌, is then added by a specific *activating* kinase. The last step in the activation sequence is the removal of the inhibiting phosphates by a specific *phosphatase* enzyme (step ➍). Once the phosphatase begins removing the inhibiting phosphates, a positive feedback loop is set up: The activated mitotic Cdk generated by this reaction stimulates the phosphatase, thereby causing the activation process to proceed more rapidly.

After mitotic Cdk-cyclin has been activated through the preceding steps, its protein kinase activity triggers the onset of mitosis (**Figure 19-37**). We have already seen that the early events of mitosis include chromosome condensation, assembly of the mitotic spindle, and nuclear envelope breakdown. How are these changes triggered by mitotic Cdk-cyclin? In the case of nuclear envelope breakdown, the mitotic Cdk-cyclin phosphorylates (and stimulates other kinases to phosphorylate) the *lamin* proteins of the *nuclear*

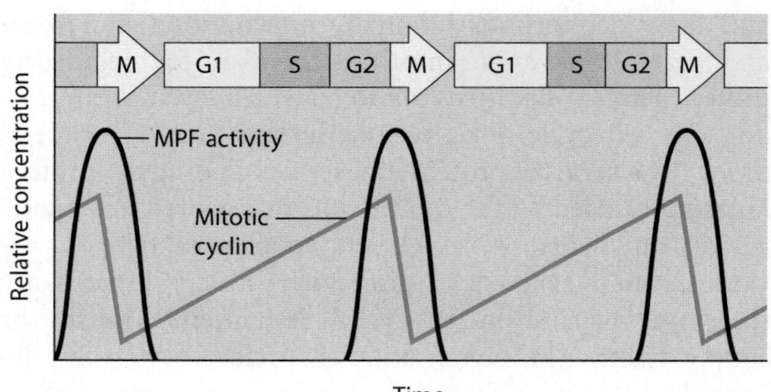

FIGURE 19-35 Fluctuating Levels of Mitotic Cyclin and MPF Activity During the Cell Cycle. Cellular levels of mitotic cyclin rise during interphase (G1, S, and G2), then fall abruptly during M phase. The peaks of MPF activity (assayed by testing for the ability to stimulate mitosis) and cyclin concentration correspond, although the rise in MPF activity is not significant until a threshold concentration of cyclin is reached. Active MPF has been found to consist of a combination of mitotic cyclin and mitotic Cdk. The mitotic Cdk itself is present at a constant concentration (not shown on the graph) because the amount of mitotic Cdk increases at a rate corresponding to the overall growth of the cell.

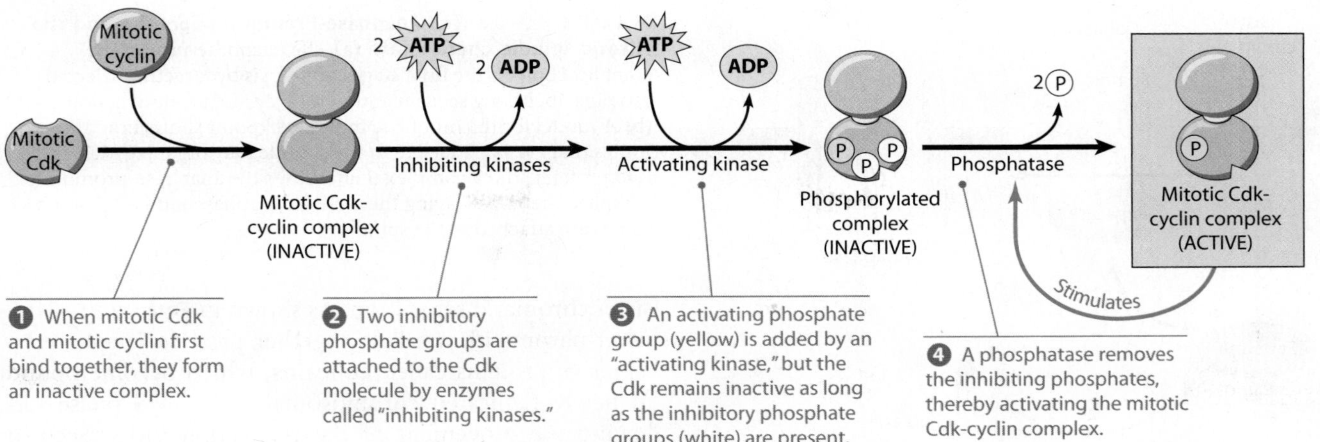

FIGURE 19-36 Regulation of Mitotic Cdk-Cyclin by Phosphorylation and Dephosphorylation. Activation of mitotic Cdk-cyclin involves the addition of inhibiting and activating phosphate groups, followed by removal of the inhibiting phosphate groups by a phosphatase. Once removal of the inhibiting phosphate groups has begun in step 4, a positive feedback loop is set up: The activated Cdk-cyclin complex generated by this reaction stimulates the phosphatase, thereby causing the activation process to proceed more rapidly.

lamina, to which the inner nuclear membrane is attached (see Figure 18-32). Phosphorylation causes the lamins to depolymerize, resulting in a breakdown of the nuclear lamina and destabilization of the nuclear envelope. The integrity of the nuclear envelope is further disrupted by phosphorylation of envelope-associated proteins, and the membranes of the envelope are soon torn apart. Phosphorylation of additional target proteins by mitotic Cdk-cyclin has been implicated in other mitotic events. For example, phosphorylation of a multiprotein complex called *condensin* is involved in condensing chromatin fibers into compact chromosomes. Finally, phosphorylation of microtubule-associated proteins by mitotic Cdk-cyclin is thought to facilitate assembly of the mitotic spindle.

The Anaphase-Promoting Complex Coordinates Key Mitotic Events by Targeting Specific Proteins for Destruction

Besides triggering the onset of mitosis, mitotic Cdk-cyclin also plays an important role later in mitosis when the decision is made to separate the sister chromatids during anaphase. Mitotic Cdk-cyclin exerts its influence on this event by phosphorylating and thereby contributing to the activation of the **anaphase-promoting complex,** a multiprotein complex that coordinates mitotic events by promoting the destruction of several key proteins at specific points during mitosis. The anaphase-promoting complex functions as a *ubiquitin ligase,* a type of enzyme that targets specific proteins for degradation by joining them to the small protein *ubiquitin.* Proteins linked to ubiquitin are subsequently destroyed by a mechanism to be described in Chapter 23.

One crucial protein targeted for destruction by the anaphase-promoting complex is **securin,** an inhibitor of

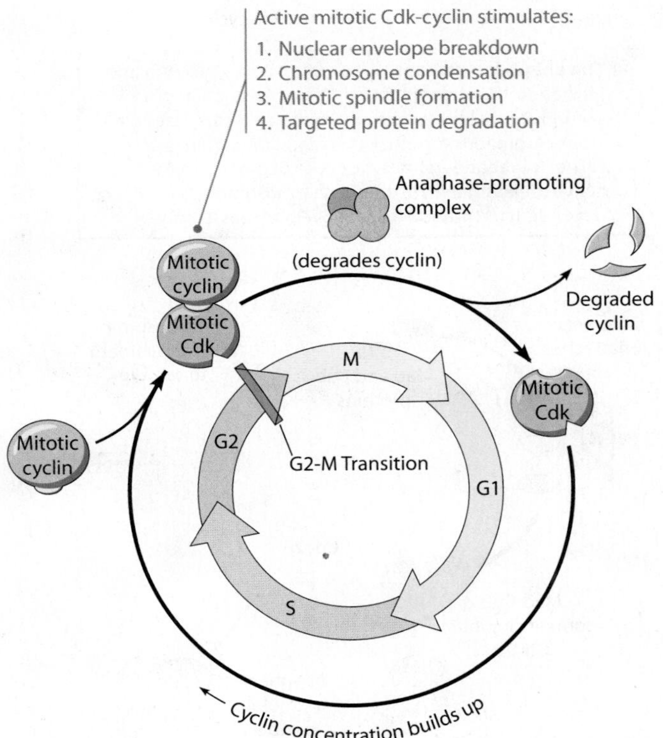

FIGURE 19-37 The Mitotic Cdk Cycle. This diagram illustrates the activation and inactivation of the mitotic Cdk protein during the cell cycle. In G1, S, and G2, mitotic Cdk is made at a steady rate as the cell grows, while the mitotic cyclin concentration gradually increases. Mitotic Cdk and cyclin form an active complex whose protein kinase activity drives the cell cycle through the G2-M transition and into mitosis by stimulating the mitotic events listed. By activating a protein-degradation pathway that degrades cyclin, the mitotic Cdk-cyclin complex also brings about its own demise, allowing the completion of mitosis and entry into G1 of the next cell cycle.

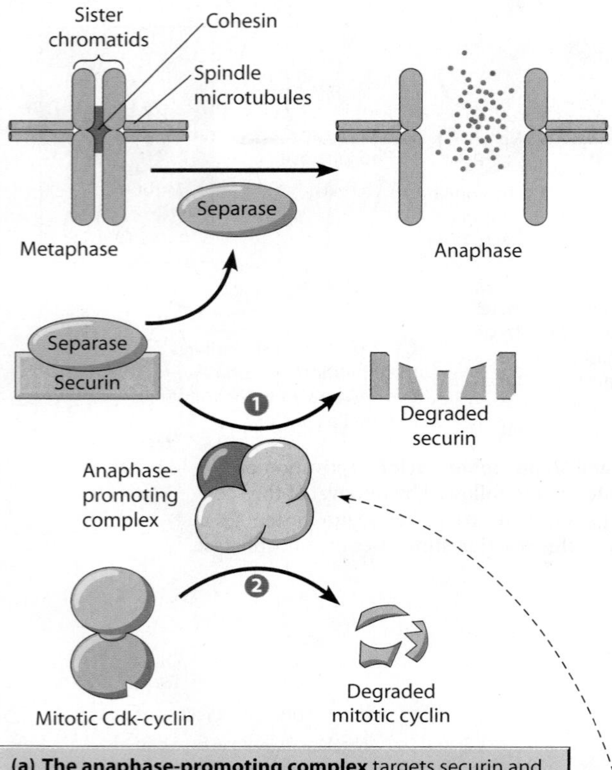

(a) The anaphase-promoting complex targets securin and mitotic cyclin for degradation. ❶The destruction of securin allows separase to cleave the cohesins that hold sister chromatids together, thereby initiating anaphase. ❷The degradation of mitotic cyclin depresses mitotic Cdk activity, leading to cytokinesis, chromosome decondensation, and nuclear envelope reassembly.

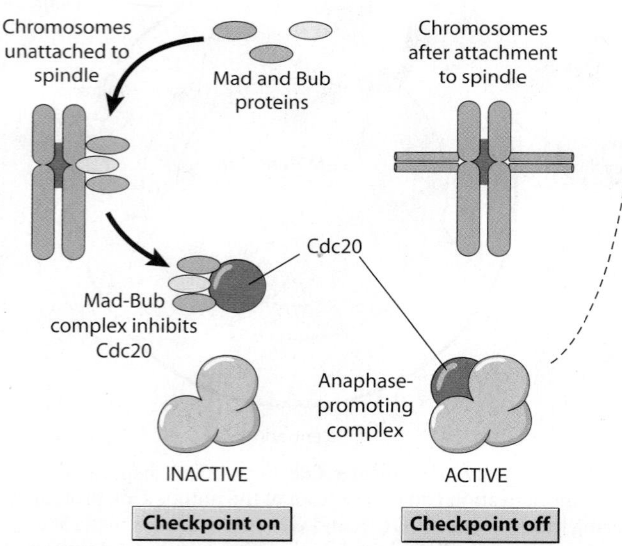

(b) The mitotic spindle checkpoint prevents anaphase from starting until all chromosomes are attached to the spindle. Unattached chromosomes keep the "checkpoint on" by organizing Mad and Bub proteins into a complex that prevents Cdc20 from activating the anaphase-promoting complex. After all chromosomes are attached, the Mad-Bub complex is not formed ("checkpoint off") and the anaphase-promoting complex is free to initiate anaphase.

FIGURE 19-38 The Anaphase-Promoting Complex and the Mitotic Spindle Checkpoint. **(a)** The anaphase-promoting complex controls the final stages of mitosis by targeting selected proteins, including securin and mitotic cyclin, for destruction. **(b)** A model for the mitotic spindle checkpoint shows how chromosomes that are not attached to the spindle may organize Mad and Bub proteins into a complex that inhibits the anaphase-promoting complex, thereby delaying the onset of anaphase until all chromosomes are attached to the spindle.

sister chromatid separation. As shown in **Figure 19-38a,** sister chromatids are held together prior to anaphase by adhesive proteins called **cohesins,** which become bound to newly replicated chromosomal DNA in S phase following the movement of the replication forks. Securin maintains this sister chromatid attachment by inhibiting a protease called **separase,** which would otherwise degrade the cohesins. At the beginning of anaphase, however, the anaphase-promoting complex attaches ubiquitin to securin and thereby triggers its destruction, releasing separase from inhibition. The activated separase then cleaves cohesin, which frees sister chromatids to separate from each other and begin their anaphase movements toward the spindle poles.

Besides initiating anaphase by causing cohesins to be destroyed, the anaphase-promoting complex induces events associated with the end of mitosis by targeting another crucial protein for destruction, namely mitotic cyclin. The resulting loss of mitotic cyclin causes the protein kinase activity of mitotic Cdk to fall. Evidence suggests that many changes associated with the exit from mitosis—such as cytokinesis, chromosome decondensation, and reassembly of the nuclear envelope—depend on this cyclin degradation step and the associated reduction in Cdk activity. For example, it has been shown that introducing a nondegradable form of mitotic cyclin into cells inhibits cytokinesis, blocks nuclear envelope reassembly, and stops chromosomes from decondensing, thereby preventing mitosis from being completed.

G1 Cdk-Cyclin Regulates Progression Through the Restriction Point by Phosphorylating the Rb Protein

Now that we have examined the role played by mitotic Cdk-cyclin in controlling events associated with mitosis, let's briefly see how another type of Cdk-cyclin regulates entry into S phase. As mentioned earlier, the restriction point (Start in yeast) is a control mechanism located in late G1 that determines whether a cell will enter S phase, proceed through the rest of the cell cycle, and divide. Because passing through the restriction point is the main step that commits a cell to the cell division cycle, it is subject to control by a variety of factors such as cell size, the availability of nutrients, and the presence of growth factors that signal the need for cell proliferation.

Such signals exert their effects by activating G1 Cdk-cyclin, whose protein kinase activity triggers progression

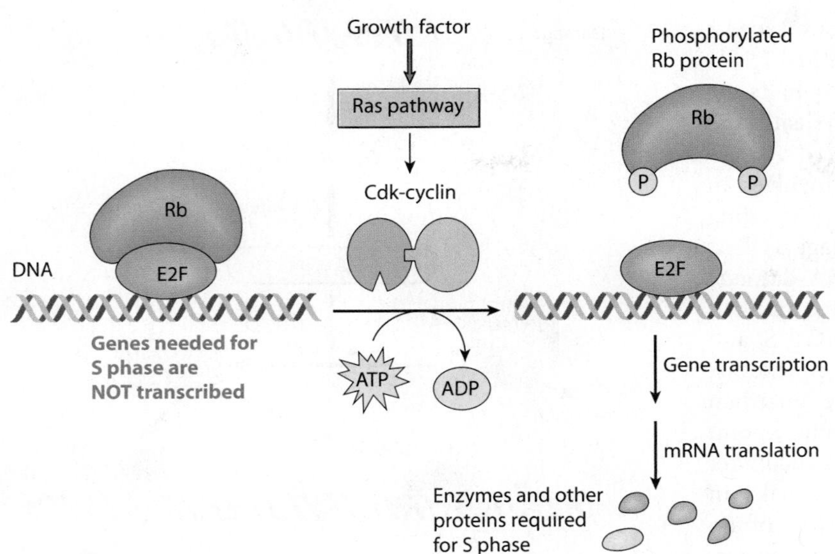

FIGURE 19-39 Role of the Rb Protein in Cell Cycle Control. In its dephosphorylated state, the Rb protein binds to the E2F transcription factor. This binding prevents E2F from activating the transcription of genes coding for proteins required for DNA replication, which are needed before the cell can pass through the restriction point into S phase. In cells stimulated by growth factors, the Ras pathway is activated (see Figure 19-41), which leads to the production and activation of a G1 Cdk-cyclin complex that phosphorylates the Rb protein. Phosphorylated Rb can no longer bind to E2F, thereby allowing E2F to activate gene transcription and trigger the onset of S phase. During the subsequent M phase (not shown), the Rb protein is dephosphorylated so that it can once again inhibit E2F.

through the restriction point by phosphorylating several target proteins. A key target is the **Rb protein,** a molecule that controls the expression of genes whose products are needed for moving through the restriction point and into S phase. The molecular mechanism by which Rb exerts this control is summarized in **Figure 19-39.** Prior to being phosphorylated by G1 Cdk-cyclin, Rb binds to and inhibits the **E2F transcription factor,** a protein that would otherwise activate the transcription of genes coding for products required for initiating DNA replication. As long as the Rb protein remains bound to E2F, the E2F molecule is inactive and these genes remain silent, thereby preventing the cell from entering into S phase. But when cells are stimulated to divide by the addition of growth factors, a pathway is triggered that produces and activates the G1 Cdk-cyclins that catalyze Rb phosphorylation. Phosphorylation of Rb abolishes its ability to bind to E2F, thereby freeing E2F to activate the transcription of genes whose products are needed for entry into S phase.

Since the Rb protein regulates such a key event—namely, the decision to proceed through the restriction point and commit to the cell division cycle—it is not surprising to discover that defects in Rb can have disastrous consequences. For example, we will see in Chapter 24 how such defects can lead to both hereditary and environmentally induced forms of cancer.

Checkpoint Pathways Monitor Chromosome-to-Spindle Attachments, Completion of DNA Replication, and DNA Damage

It would obviously create problems if cells proceeded from one phase of the cell cycle to the next before the preceding phase had been properly completed. For example, if chromosomes start moving toward the spindle poles before they have all been properly attached to the spindle, the newly forming daughter cells might receive extra copies of some chromosomes and no copies of others, a situation known as **aneuploidy** (an = "not," eu = "good," and "ploidy" refers to chromosome number). Similarly, it would be potentially hazardous for a cell to begin mitosis before all of its chromosomal DNA had been replicated. To minimize the possibility of such errors, cells utilize a series of **checkpoint** mechanisms that monitor conditions within the cell and transiently halt the cell cycle if conditions are not suitable for continuing.

The checkpoint pathway that prevents anaphase chromosome movements from beginning before the chromosomes are all attached to the spindle is called the **mitotic spindle checkpoint.** It works through a mechanism in which chromosomes whose kinetochores remain *unattached* to spindle microtubules produce a "wait" signal that inhibits the anaphase-promoting complex. As long as the anaphase-promoting complex is inhibited, it cannot trigger destruction of the cohesins that hold sister chromatids together. The exact molecular basis of the wait signal remains an open question, but members of the *Mad* and *Bub protein families* are involved. One model proposes that Mad and Bub proteins accumulate at unattached chromosomal kinetochores, where they are converted into a multiprotein complex that inhibits the anaphase-promoting complex by blocking the action of one of its essential activators, the *Cdc20* protein (see Figure 19-38b). After all the chromosomes have become attached to the spindle, the Mad and Bub proteins are no longer converted into this inhibitory complex, thereby freeing the anaphase-promoting complex to initiate the onset of anaphase.

A second checkpoint mechanism, called the **DNA replication checkpoint,** monitors the state of DNA replication to help ensure that DNA synthesis is completed before the cell exits from G2 and begins mitosis. The existence of this checkpoint has been demonstrated by treating cells with inhibitors that prevent DNA replication from being finished. Under such conditions, the phosphatase that catalyzes the final dephosphorylation step involved in the activation of mitotic Cdk-cyclin (see

Figure 19-36, ❹) is inhibited through a series of events triggered by proteins associated with replicating DNA. The resulting lack of mitotic Cdk-cyclin activity halts the cell cycle at the end of G2 until all DNA replication is completed.

A third type of checkpoint mechanism is involved in preventing cells with damaged DNA from proceeding through the cell cycle unless the DNA damage is first repaired. In this case, a multiple series of **DNA damage checkpoints** exist that monitor for DNA damage and halt the cell cycle at various points—including late G1, S, and late G2—by inhibiting different Cdk-cyclin complexes. A protein called **p53,** sometimes referred to as the "guardian of the genome," plays a central role in these checkpoint pathways. As shown in **Figure 19-40**, when cells encounter agents that cause extensive double-stranded breaks in DNA, the altered DNA triggers the activation of an enzyme called *ATM protein kinase* (for *ataxia telangiectasia mutated;* mutations in the ATM gene can cause defects in the cerebellum that lead to uncoordinated movement, as well as prominent blood vessels that form in the whites of the eyes). ATM catalyzes the phosphorylation of kinases known as *checkpoint kinases,* which in turn phosphorylate p53 (and several other target proteins). Phosphorylation of p53 prevents it from interacting with *Mdm2,* a protein that would otherwise mark p53 for destruction by linking it to ubiquitin (just as the anaphase-promoting complex targets proteins for degradation by linking them to ubiquitin). ATM-catalyzed phosphorylation of p53 therefore protects it from degradation and leads to a buildup of p53 in the presence of damaged DNA. A protein related to ATM, called *ATR* (ATM-related) acts similarly but instead causes cell cycle arrest as a result of extensive single-stranded breaks in DNA.

The accumulating p53 in turn activates two types of events: *cell cycle arrest* and *cell death.* Both responses are based on the ability of p53 to bind to DNA and act as a transcription factor that stimulates the transcription of specific genes. One of the crucial genes activated by p53 is the gene coding for **p21,** a protein that halts progression through the cell cycle at multiple points by inhibiting the activity of several different Cdk-cyclins. Phosphorylated p53 also stimulates the production of enzymes involved in DNA repair. But if the damage cannot be successfully repaired, p53 then activates a group of genes coding for proteins involved in triggering cell death by apoptosis (page 591). A key protein in this pathway, called **Puma (p53 upregulated modulator of apoptosis),** promotes apoptosis by binding to and inactivating a normally occurring inhibitor of apoptosis known as *Bcl-2.*

The ability of p53 to trigger cell cycle arrest and cell death allows it to function as a molecular stoplight that protects cells with damaged DNA from proliferating and passing the damage to daughter cells. The importance of this role will be highlighted in Chapter 24, where we describe how defects in the p53 pathway contribute to the development of cancer.

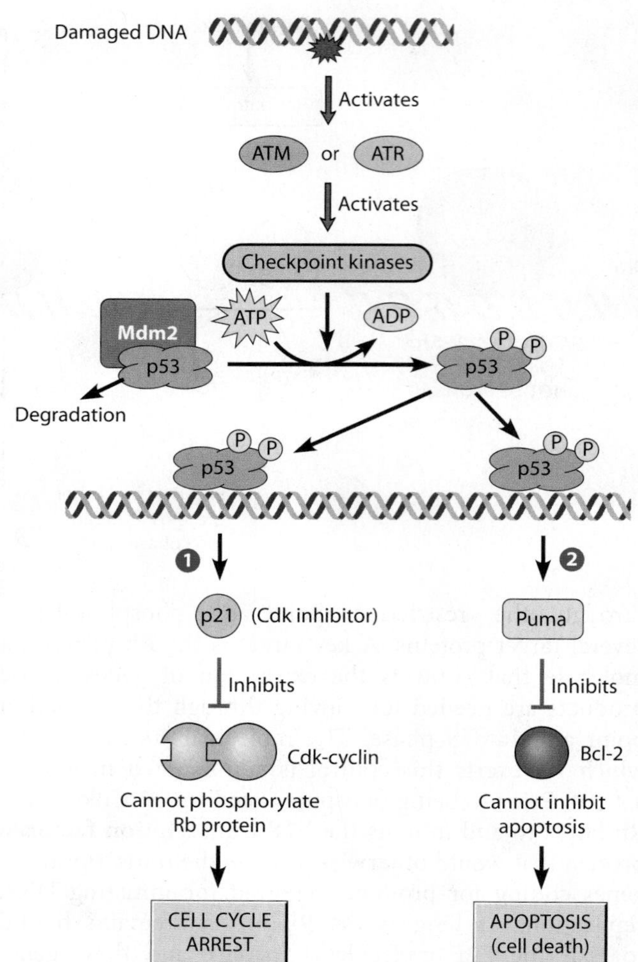

FIGURE 19-40 Role of the p53 Protein in Responding to DNA Damage. Damaged DNA activates the ATM or ATR protein kinase, leading to activation of checkpoint kinases, which leads to phosphorylation of the p53 protein. Phosphorylation stabilizes p53 by blocking its interaction with Mdm2, a protein that would otherwise mark p53 for degradation. (The degradation mechanism is not shown, but it involves Mdm2-catalyzed attachment of p53 to ubiquitin, which targets molecules to the cell's main protein destruction machine, the proteasome.) When the interaction between p53 and Mdm2 is blocked by p53 phosphorylation, the phosphorylated p53 protein accumulates and triggers two events. ❶ The p53 protein binds to DNA and activates transcription of the gene coding for the p21 protein, a Cdk inhibitor. The resulting inhibition of Cdk-cyclin prevents phosphorylation of the Rb protein, leading to cell cycle arrest at the restriction point. ❷ When the DNA damage cannot be repaired, p53 then activates genes coding for a group of proteins that trigger cell death by apoptosis. A key protein is Puma, which promotes apoptosis by binding to, and blocking the action of, the apoptosis inhibitor, Bcl-2.

MB p53 and Control of the Cell Cycle

Putting It All Together: The Cell Cycle Regulation Machine

Figure 19-41 is a generalized and simplified summary of the main features of the molecular "machine" that regulates the eukaryotic cell cycle. The operation of this machine can be described in terms of two fundamental, interacting mechanisms. One mechanism is an autonomous clock that

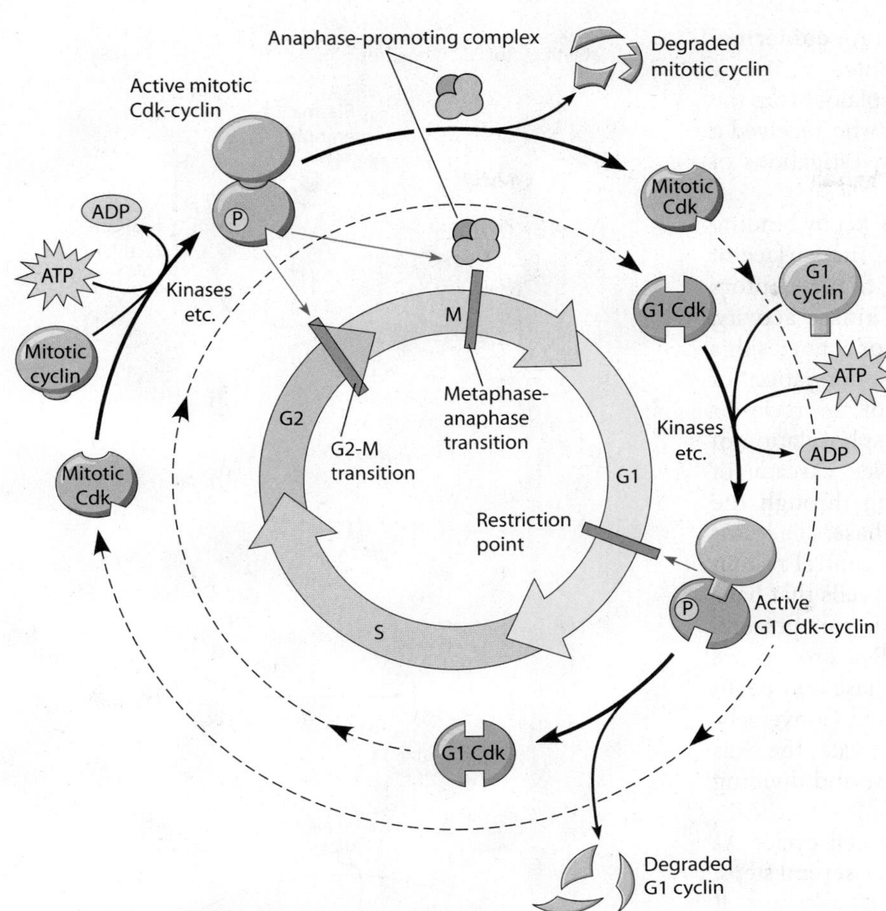

FIGURE 19-41 A General Model for Cell Cycle Regulation. Passage through the three main transition points in the cell cycle is triggered by protein complexes made of cyclin and Cdk, whose phosphorylation of other proteins induces progression through the cycle. G1 Cdk-cyclin acts at the restriction point by catalyzing phosphorylation of the Rb protein. Mitotic Cdk-cyclin acts at the G2-M boundary by catalyzing the phosphorylation of proteins involved in chromosome condensation, nuclear envelope breakdown, and spindle assembly. The same mitotic Cdk-cyclin also influences the metaphase-anaphase transition by catalyzing the phosphorylation of the anaphase-promoting complex, which in turn triggers chromosome separation and the breakdown of mitotic cyclin. Checkpoint pathways that monitor the cell for DNA damage, DNA replication, and chromosome attachment to the spindle can send signals that halt the cell cycle at one or more of these key transition points.

goes through a fixed cycle over and over again. The molecular basis of this clock is the synthesis and degradation of cyclins, alternating in a rhythmic fashion. These cyclins in turn bind to Cdk molecules, creating various Cdk-cyclin complexes that trigger the passage of cells through the main cell cycle transition points. The second mechanism adjusts the clock as needed, by providing feedback from the cell's internal and external environments. This mechanism makes use of additional proteins that, directly or indirectly, influence the activity of Cdks and cyclins. Many of these additional proteins are protein kinases or phosphatases. It is this part of the cell cycle machine that relays information about the state of the cell's metabolism—including DNA damage and replication—and about conditions outside the cell, thereby influencing whether or not the cell should commit to the process of cell division. As we will see next, growth-promoting and growth-inhibiting signaling molecules that come from outside the cell are prominent components of this regulatory mechanism.

Growth Factors and Cell Proliferation

Simple unicellular organisms, such as bacteria and yeast, live under conditions in which the presence of sufficient nutrients in the external environment is the primary factor determining whether cells grow and divide. In multicellular organisms, the situation is usually reversed; cells are typically surrounded by nutrient-rich extracellular fluids, but the organism as a whole would be quickly destroyed if every cell were to continually grow and divide just because it had access to adequate nutrients. Cancer is a potentially lethal reminder of what happens when cell proliferation continues unabated without being coordinated with the needs of the organism as a whole. To overcome this potential problem, multicellular organisms utilize extracellular signaling proteins called *growth factors* to control the rate of cell proliferation (see Table 14-3). Most growth factors are *mitogens,* which means that they stimulate cells to pass through the restriction point and subsequently divide by mitosis.

Stimulatory Growth Factors Activate the Ras Pathway

If mammalian cells are placed in a culture medium containing nutrients and vitamins but lacking growth factors, they normally become arrested in G1 despite the presence of adequate nutrients. Growth and division can be triggered by adding small amounts of blood serum, which contains several stimulatory *growth factors*. Among them is **platelet-derived growth factor (PDGF),** a protein produced by blood platelets that stimulates the proliferation of connective tissue cells and smooth muscle cells. Another important growth factor, **epidermal growth factor (EGF),** is widely distributed in many tissues and body fluids. EGF was initially isolated from the salivary glands of mice by Stanley Cohen, who received a Nobel Prize in 1987 for his pioneering investigations of growth factors.

Growth factors such as PDGF and EGF act by binding to plasma membrane receptors located on the surface of target cells. In Chapter 14, you learned that receptors for these growth factors exhibit tyrosine kinase activity. The binding of a growth factor to its receptor activates this tyrosine kinase activity, leading to phosphorylation of tyrosine residues located in the portion of the receptor molecule protruding into the cytosol. Phosphorylation of these tyrosines in turn triggers a complex cascade of events that culminates in the cell passing through the restriction point and entering into S phase. The *Ras pathway* introduced in Chapter 14 plays a central role in these events, as shown by studies involving cells that have stopped dividing because growth factor is not present. When mutant, hyperactive forms of the Ras protein are injected into such cells, the cells enter S phase and begin dividing, even in the absence of growth factor. Conversely, injecting cells with antibodies that inactivate the Ras protein prevents cells from entering S phase and dividing in response to growth factor stimulation.

How does the Ras pathway affect the cell cycle? As shown in **Figure 19-42,** the process involves several steps. ❶ First, binding of a growth factor to its receptor at the plasma membrane leads to activation of Ras. ❷ Next, activated Ras leads to phosphorylation and activation of a protein kinase called *Raf,* which sets in motion a cascade of phosphorylation events. Activated Raf phosphorylates serine and threonine residues in a protein kinase called *MEK,* which in turn phosphorylates threonine and tyrosine residues in a group of protein kinases called *MAP kinases (mitogen-activated protein kinases; MAPKs).* ❸ The activated MAPKs enter the nucleus and phosphorylate several regulatory proteins that activate the transcription of specific genes. Among these proteins are *Jun* (a component of the AP-1 transcription factor) and members of the *Ets family* of transcription factors. These activated transcription factors turn on the transcription of "early genes" that code for the production of other transcription factors, including Myc, Fos, and Jun, which then activate the transcription of a family of "delayed genes." One of these latter genes encodes the E2F transcription factor, whose role in controlling entry into S phase was described earlier in the chapter. ❹ Also included in the delayed genes are several genes coding for either Cdk or cyclin molecules, whose production leads to the formation of Cdk-cyclin complexes that phosphorylate Rb and hence trigger passage from G1 into S phase.

Thus in summary, the Ras pathway is a multistep signaling cascade in which the binding of a growth factor to a receptor on the cell surface ultimately causes the cell to pass through the restriction point and into S phase, thereby starting the cell on the road to cell division. The importance of this pathway for the control of cell proliferation has been highlighted by the discovery that mutations affecting the Ras pathway appear frequently in cancer cells. For example, mutant Ras proteins that provide an ongoing stimulus for the cell to proliferate, independent of growth factor stimulation, are commonly encountered in pancreatic, colon,

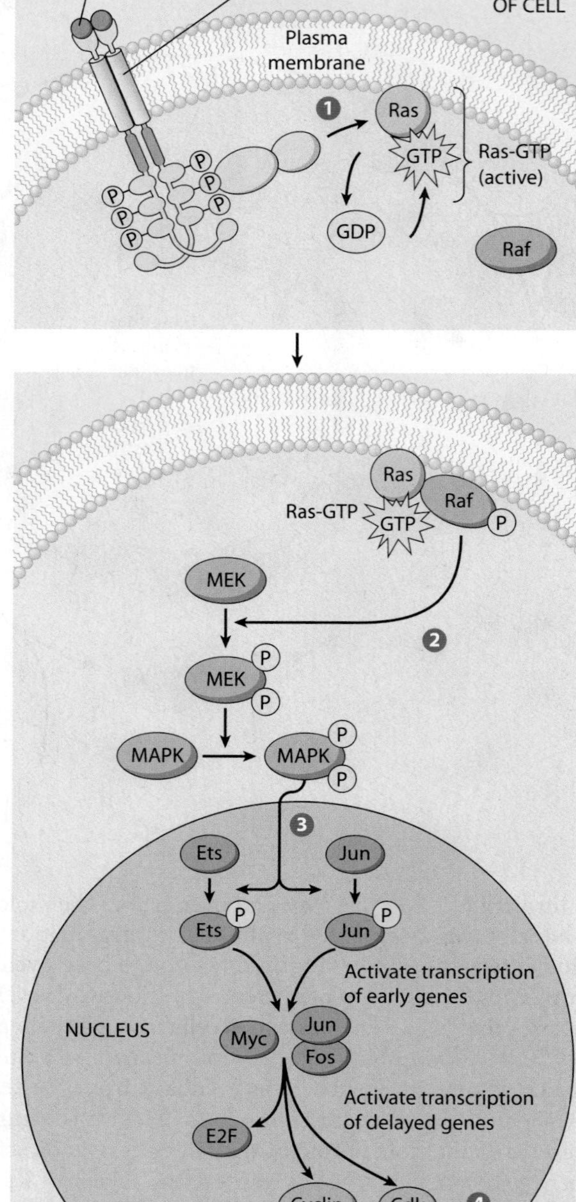

FIGURE 19-42 Regulation of the Cell Cycle via the Ras Pathway. Regulation of the cell cycle by Ras consists of four steps: ❶ binding of a growth factor to its receptor, leading to activation of Ras protein; ❷ activation of a cascade of cytoplasmic protein kinases (Raf, MEK, and MAPK); ❸ activation or production of nuclear transcription factors (Ets, Jun, Fos, Myc, E2F); and ❹ synthesis of cyclin and Cdk molecules. The resulting Cdk-cyclin complexes catalyze the phosphorylation of Rb and hence trigger passage from G1 into S phase (MAPK = Map kinases).

lung, and bladder cancers; and they occur in about 25–30% of all human cancers overall. In Chapter 24, the role played by such Ras pathway mutations in the development of cancer will be described in detail.

Stimulatory Growth Factors Can Also Activate the PI3K-Akt Pathway

When a growth factor binds to a receptor that triggers the Ras pathway, the activated receptor may simultaneously trigger other pathways as well. One example that we examined in Chapter 14 in the context of insulin signaling (page 417) is the *PI 3-kinase–Akt pathway.* This pathway begins with receptor-induced activation of phosphatidylinositol 3-kinase (abbreviated as *PI 3-kinase* or *PI3K*), which catalyzes formation of *PIP*$_3$ (phosphatidylinositol-3,4,5-trisphosphate), ultimately leading to phosphorylation and activation of *Akt.* Through its ability to catalyze the phosphorylation of several key target proteins, Akt suppresses apoptosis (see the next section of this chapter) and inhibits cell cycle arrest (**Figure 19-43**). One way in which the latter happens is through activation of a monomeric G protein called Rheb. Activation of Rheb leads to activation of TOR, a key regulator of cell growth mentioned earlier in this chapter. The net effect of the PI 3-kinase–Akt signaling pathway is therefore to promote cell survival and proliferation.

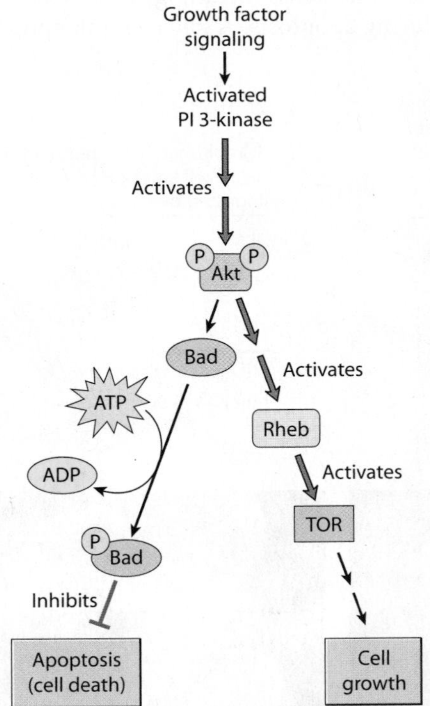

FIGURE 19-43 The PI 3-Kinase–Akt Signaling Pathway. Growth factors that bind to receptor tyrosine kinases activate several pathways in addition to the Ras pathway illustrated in Figure 19-42. The pathway shown here leads to activation of the protein kinase Akt. Akt suppresses apoptosis, in part by phosphorylating and inactivating a protein called Bad that normally promotes apoptosis. Akt also inhibits cell cycle arrest by leading to activation of TOR. Thus, the net effect of PI 3-kinase–Akt signaling is to promote cell survival and proliferation. The PI 3-kinase–Akt pathway is inhibited by PTEN (see Figure 14-23).

As in the case of the Ras pathway, mutations that disrupt the normal behavior of the PI 3-kinase–Akt pathway are associated with many cancers. In some cases, such mutations cause excessive activity of the Akt protein that leads to increased cell proliferation and survival. In other cases, hyperactivity of the PI 3-kinase–Akt pathway is caused by mutations that disable proteins that normally inhibit this pathway. One such protein is *PTEN*, a phosphatase that removes a phosphate group from PIP_3 and thereby prevents the activation of Akt (see page 417). When mutations disrupt PTEN, the cell cannot degrade PIP_3 efficiently, and its concentration rises. The accumulating PIP_3 activates Akt in an uncontrolled fashion, even in the absence of growth factors, leading to enhanced cell proliferation and survival. Mutations that reduce PTEN activity are found in up to 50% of prostate cancers, 35% of uterine cancers, and to varying extents in ovarian, breast, liver, lung, kidney, thyroid, and lymphoid cancers.

Inhibitory Growth Factors Act Through Cdk Inhibitors

Although we usually think of growth factors as being growth-stimulating molecules, the function of some growth factors is actually to inhibit cell proliferation. One example is **transforming growth factor β (TGFβ)**, a protein that can exhibit either growth-stimulating or growth-inhibiting properties, depending on the target cell type. When acting as a growth inhibitor, the binding of TGFβ to its cell surface receptor triggers a series of events in which the receptor catalyzes the phosphorylation of *Smad* proteins that move into the nucleus and regulate gene expression (see Figure 14-20). Once inside the nucleus, Smads activate the expression of genes coding for proteins that inhibit cell proliferation. Two key genes produce proteins called *p15* and *p21,* which are **Cdk inhibitors** that suppress the activity of Cdk-cyclin complexes and thereby block progression through the cell cycle. (The p21 protein was already mentioned earlier in the chapter when we discussed the mechanism of p53-mediated cell cycle arrest.)

Growth-inhibiting signals that act on cells by triggering the production of Cdk inhibitors help protect normal tissues from excessive cell divisions that might otherwise produce more cells than are needed. In Chapter 24, we will see how defects in such pathways can lead to uncontrolled cell proliferation and the development of cancer.

Apoptosis

As you saw in previous sections of this chapter, organisms tightly regulate when their cells enter mitosis. When cells divide is often regulated by growth factors, which can stimulate or inhibit division, depending on the circumstances. At other times, however, organisms need to regulate whether cells stay alive. Damaged or diseased cells need to be killed, but this presents a challenge. Dismantling cells that are destined for death must occur in such a way that the internal contents of the dead cell—which include digestive enzymes in organelles such as lysosomes—do not wreak havoc on other cells around them. Multicellular organisms accomplish this feat through an important kind of programmed cell death—**apoptosis.** Apoptosis is a key event in many biological processes. In embryos, apoptosis occurs in a variety of circumstances. Examples include removal of the webbing between the digits (fingers and toes) during the development of hands and feet and the "pruning" of neurons that occurs in human infants during the first few months of life as connections mature within the developing brain. In adult humans, apoptosis occurs continually; when cells become infected by pathogens or when white blood cells reach the end of their life span, they are eliminated through apoptosis. When cells that should die via apoptosis do not, the consequences can be dire. Mutations in some of the proteins that participate in apoptosis can lead to cancer. For example, melanoma frequently results from a mutation in Apaf-1, a protein we discuss later in this section.

Apoptosis is very different from another type of cell death, known as *necrosis,* which sometimes follows massive tissue injury. Whereas necrosis involves the swelling and rupture of the injured cells, apoptosis involves a specific series of events that lead to the dismantling of the internal contents of the cell (**Figure 19-44**). During the early phases of apoptosis, the cell's DNA segregates near the periphery of the nucleus, and the volume of the cytoplasm decreases (❶). Next, the cell begins to produce small, bubble-like, cytoplasmic extensions ("blebs"), and the nucleus and organelles begin to fragment (❷). The cell's DNA is cleaved by an apoptosis-specific DNA endonuclease, or *DNase* (an enzyme that digests DNA), at regular intervals along the DNA. As a result, the DNA fragments, which are multiples of 200 base pairs in length, form a diagnostic "ladder" of fragments. (This precise size distribution of DNA reflects the susceptibility of DNA to cleavage between nucleosomes, structures you learned about in Chapter 18). Eventually the cell is dismantled into small pieces called *apoptotic bodies.* During apoptosis, inactivation of a phospholipid translocator, or flippase (see Chapter 7), results in accumulation of phosphatidylserine in the outer leaflet of the plasma membrane. The phosphatidylserine serves as an "eat-me" signal for the remnants of the affected cell to be engulfed by other nearby cells (typically macrophages) via phagocytosis (see Chapter 12; ❸). The macrophages act as scavengers to remove the resulting cellular debris.

That cells have a "death program" was first conclusively demonstrated in the nematode, *Caenorhabditis elegans,* where key genes that control apoptosis were first identified (**Box 19B**). Subsequent research showed that many other organisms, including mammals, use similar proteins during apoptosis. A key event in apoptosis is the

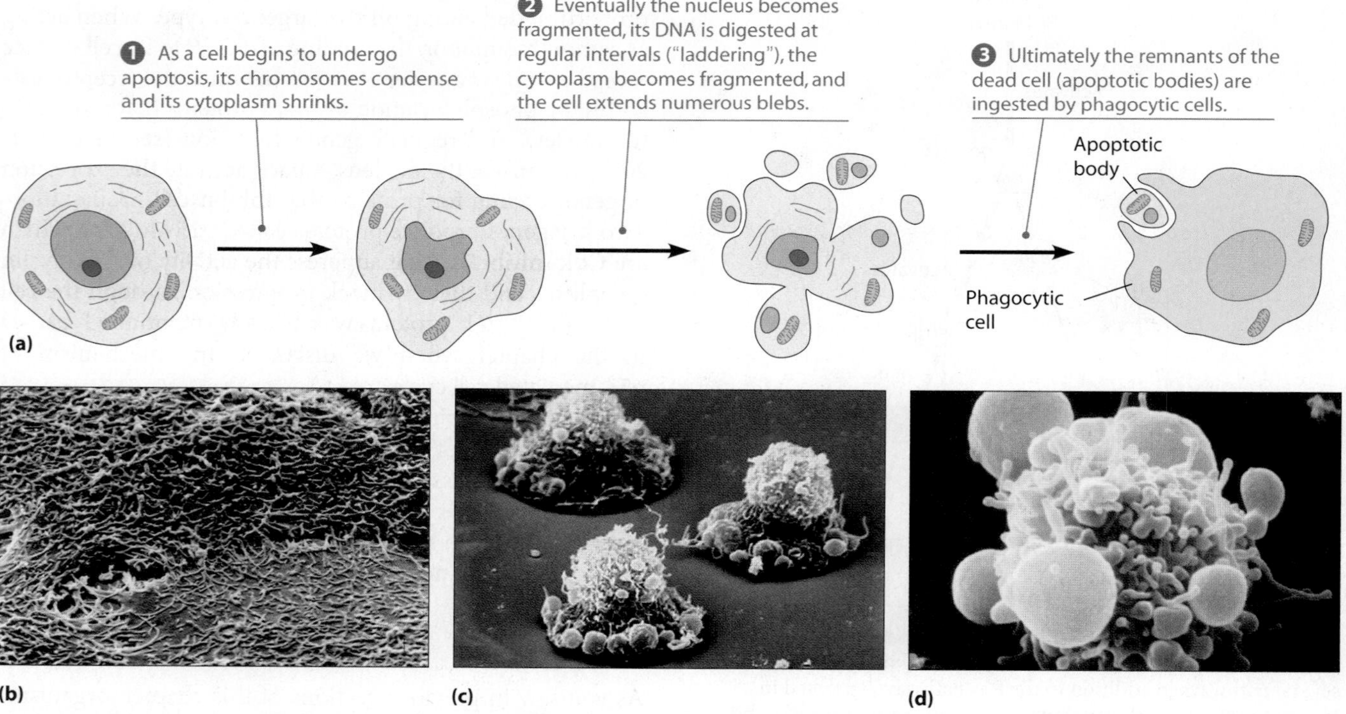

❶ As a cell begins to undergo apoptosis, its chromosomes condense and its cytoplasm shrinks.
❷ Eventually the nucleus becomes fragmented, its DNA is digested at regular intervals ("laddering"), the cytoplasm becomes fragmented, and the cell extends numerous blebs.
❸ Ultimately the remnants of the dead cell (apoptotic bodies) are ingested by phagocytic cells.
Apoptotic body
Phagocytic cell
(a)
(b)
(c)
(d)

FIGURE 19-44 Major Steps in Apoptosis. (a) Cells in the process of apoptosis undergo a series of characteristic changes. Ultimately the remnants of the dead cell (apoptotic bodies) are ingested by phagocytic cells. **(b–d)** SEMs of epithelial cells undergoing apoptosis. **(b)** Epithelial cells in contact with one another in culture form flat sheets. **(c)** As apoptosis ensues, the cells round up, withdraw their connections with one another, and bleb. **(d)** A single dead cell with many apoptotic bodies.

BOX 19B TOOLS OF DISCOVERY

Apoptosis in *Caenorhabditis elegans*

Key breakthroughs in the study of apoptosis came through analysis of the development of the nematode worm, *Caenorhabditis elegans*. *C. elegans* is uniquely suited to studying cell death. Its life cycle is very short, it is optically transparent, and its embryos are remarkably consistent in their development. In fact, they are so consistent that the lineage of every cell in the adult animal is known with *complete* precision. This means the sequence of cell divisions that results in each of the 1090 cells produced during development can be traced back to the single-celled fertilized egg! This feat was achieved largely through the work of John Sulston at the Medical Research Council in Cambridge, England. Sulston also showed that 131 cells undergo precisely timed apoptosis during normal embryonic development in *C. elegans*. Largely through the work of Robert Horvitz and colleagues at the Massachusetts Institute of Technology, mutants defective in various aspects of cell death, called *ced* mutants (for cell death abnormal), were identified. For example, Horvitz and colleagues identified several mutations that block phagocytosis of dead cells, so that their corpses persist, making them easy to see in the light microscope (**Figure 19B-1**). One of the first *ced* genes to be characterized at the molecular level was the gene *ced-3,* which encodes a member of the *caspase* family of proteins (see Figure 19-45). Another gene, *ced-9,* encodes the *C. elegans* version of *Bcl-2,* which plays a key role in regulating the leakage of molecules from mitochondria that can trigger apoptosis. Another gene, *ced-10,* encodes a member of the Rac family of proteins, which is required for phagocytosis of dead cells. In conjunction with work proceeding on cultured cells at about the same time, the work of Horvitz and colleagues provided key insights into apoptosis. For this work—and his work on cell signaling in the vulva—Horvitz, along with Sulston and geneticist Sydney Brenner, shared the Nobel Prize in Physiology or Medicine in 2002.

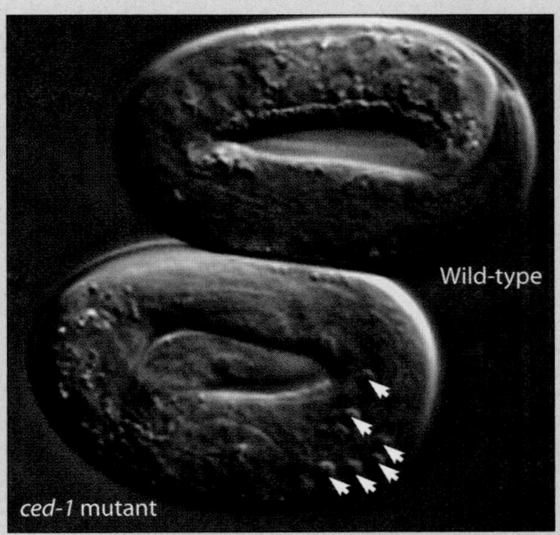

FIGURE 19B-1 Cell Death in *C. elegans*. Wild-type (top) and *ced-1* mutant (bottom) embryos (DIC microscopy). Although cell deaths can be seen in both embryos as small "buttons," these cell corpses accumulate only in the *ced-1* embryo because phagocytosis of dead cells fails (arrows).

activation of a series of enzymes called **caspases.** (Caspases get their name because they contain a *c*ysteine at their active site, and they cleave proteins at sites that contain an *aspar*tic acid residue followed by four amino acids that are specific to each caspase). Caspases are produced as inactive precursors known as **procaspases,** which are subsequently cleaved to create active enzymes, often by other caspases, in a proteolytic cascade. Once they are activated, caspases cleave other proteins. The apoptosis-specific DNase is a good example; it is bound to an inhibitory protein that is cleaved by a caspase.

Apoptosis Is Triggered by Death Signals or Withdrawal of Survival Factors

There are two main routes by which cells can activate caspases and enter the apoptotic pathway. In some cases, activation of caspases occurs directly. For example, when cells in the human body are infected by certain viruses, a population of *cytotoxic T lymphocytes* are activated and induce the infected cells to initiate apoptosis. How do lymphocytes induce cells to initiate the process of apoptosis? Typically, such activation is triggered when cells receive *cell death signals.* Two well-known death signals are *tumor necrosis factor* and *CD95/Fas.* Here, we will focus on CD95, a protein on the surface of infected cells. Lymphocytes have a protein on their surfaces that binds to CD95, causing the CD95 within the infected cell to aggregate (**Figure 19-45**, ❶). CD95 aggregation results in the attachment of adaptor proteins to the clustered CD95, which in turn recruits a procaspase *(procaspase-8)* to sites of receptor clustering. When the procaspase is activated (❷), it acts as an initiator of the caspase cascade. A key action of such *initiator caspases* is the activation of an *executioner caspase,* known as *caspase-3* (❸). Active caspase-3 is important for activating many steps in apoptosis.

In other cases, apoptosis is triggered indirectly. One of the best-studied cases of this second type of apoptosis involves **survival factors.** When such factors are withdrawn, a cell may enter apoptosis (❹). Surprisingly, a key site of action of this second pathway is the mitochondrion. The connection between mitochondria and cell death may be surprising, but it is clear that, in addition to their role in energy production, mitochondria are important in apoptosis. If withdrawal of survival factors is the sentence of execution, then the executioners are mitochondria.

How do mitochondria hasten cell death? In a healthy cell that is not committed to apoptosis, there are several *anti-apoptotic* proteins in the outer mitochondrial membrane

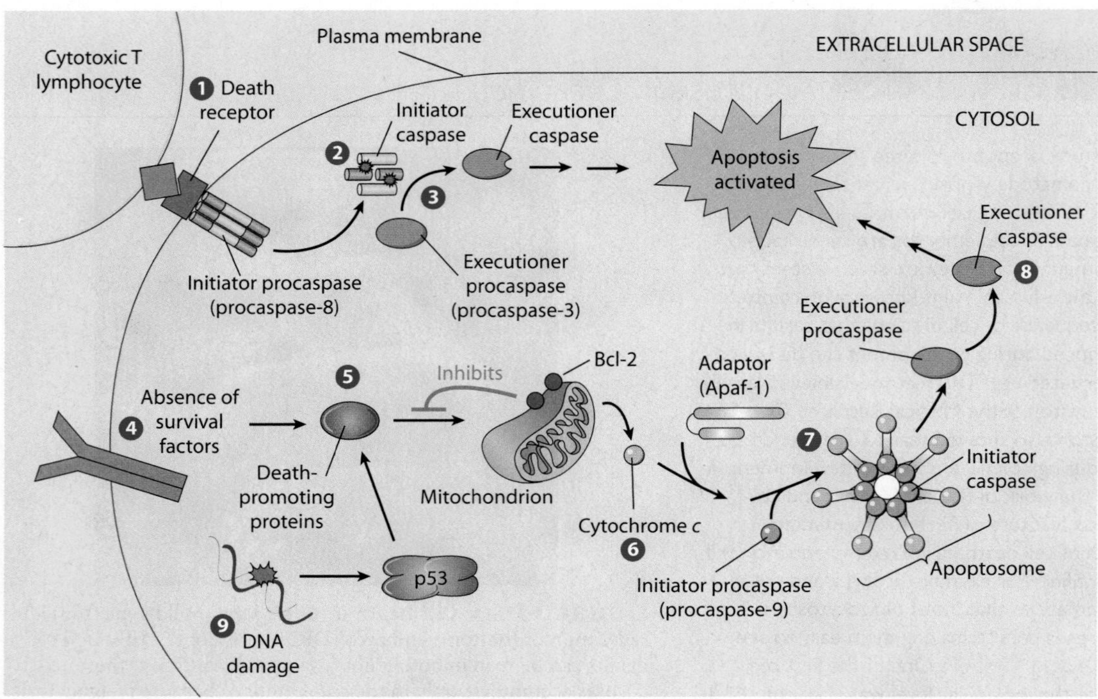

FIGURE 19-45 Induction of Apoptosis by Cell Death Signals or by Withdrawal of Survival Factors. Cell death signals, such as ligands on the surface of a cytotoxic T lymphocyte, can lead to apoptosis. ❶ Ligand binds to a "death receptor" on the surface of a target cell. Binding causes clustering of receptors and recruitment of adaptor proteins in the target cell, resulting in clustering of initiator procaspase (procaspase-8) protein. ❷ Initiator caspases then become activated. ❸ The initiator caspases in turn activate the executioner caspase, caspase-3, a key initiator of apoptosis. ❹ When survival factors are no longer present, ❺ death-promoting (pro-apoptotic) proteins accumulate, counterbalancing anti-apoptotic proteins (such as Bcl-2) at the mitochondrial outer membrane, ❻ causing release of cytochrome *c*. ❼ Cytochrome *c* forms a complex with other proteins, resulting in activation of an initiator caspase (caspase-9). ❽ The initiator caspase in turn activates the executioner caspase, caspase-3, triggering apoptosis. ❾ DNA damage can also lead to apoptosis through the activity of the p53 protein.

that prevent apoptosis, but only as long as a cell continues to be exposed to survival factors. These proteins are structurally related to a protein known as **Bcl-2,** the best understood of these anti-apoptotic proteins. Bcl-2 and other anti-apoptotic proteins exert their effects by counteracting other proteins that are themselves structurally similar to Bcl-2. These proteins, however, *promote* apoptosis, and so they are collectively referred to as *pro-apoptotic* proteins. Thus, pro- and anti-apoptotic proteins, influenced by cell signals, wage an ongoing battle; when the balance shifts toward pro-apoptotic proteins, a cell is more likely to undergo apoptosis (❺). For example, stimulation of the Akt pathway, which you learned about earlier in this chapter, can lead to phosphorylation and inactivation of a pro-apoptotic protein called *Bad* (for Bcl-2-associated death promoter; Figure 19-43).

Surprisingly, mitochondria trigger apoptosis by releasing **cytochrome *c*** into the cytosol (Figure 19-45, ❻). Although the details by which this happens are currently under debate, eventually the accumulation of pro-apoptotic protein at the surface of the mitochondrion leads to the formation of channels in the outer mitochondrial membrane, allowing cytochrome *c* to escape into the cytosol. Although cytochrome *c* is normally involved in electron transport (see Chapter 10), it is important in triggering apoptosis in at least two ways. First, cytochrome *c* stimulates calcium release from adjacent mitochondria and from the endoplasmic reticulum, where it binds inositol-1-4-5-trisphosphate (IP_3) receptors. Second, it can activate an initiator procaspase associated with mitochondria, known as *procaspase-9.* It does this by recruiting a cytosolic adaptor protein (known as *Apaf-1*) that assembles procaspase-9 into a complex sometimes called an *apoptosome;* the apoptosome promotes the production of active caspase-9 (❼). Like other initiator caspases, caspase-9 activates the executioner caspase, caspase-3 (❽). Thus, in the end, both cell death mechanisms lead to the activation of a common caspase that sets apoptosis in motion.

There is another situation that can trigger the mitochondrial pathway to apoptosis. When a cell suffers so much damage that it is unable to repair itself, it may trigger its own demise. In particular, when a cell's DNA is damaged (for example, by radiation or ultraviolet light), it can enter apoptosis via the activity of *p53* (❾). As we saw earlier in this chapter, p53 acts through the protein Puma, which binds to and inhibits Bcl-2 (see Figure 19-40, step ❷). In the end, just like withdrawal of survival factors, the p53 pathway activates pro-apoptotic proteins to trigger apoptosis.

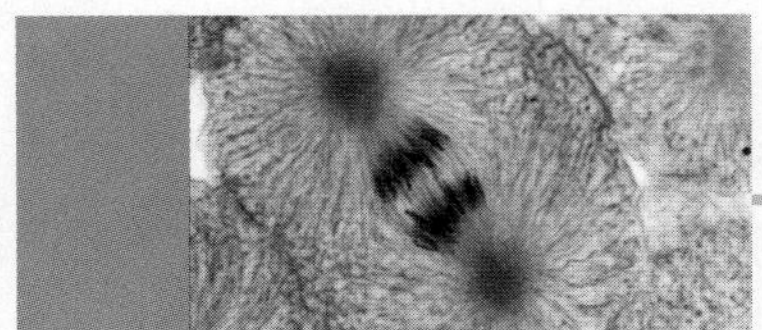

SUMMARY OF KEY POINTS

Overview of the Cell Cycle

- The eukaryotic cell cycle is divided into G1, S, G2, and M phases. Chromosomal DNA replication takes place during S phase, whereas cell division (mitosis and cytokinesis) occurs in M phase. Interphase (G1, S, and G2) is a time of cell growth and metabolism that typically occupies about 95% of the cycle.
- The length of the cell cycle varies greatly, ranging from cells that divide rapidly and continuously to cells that do not divide at all.

DNA Replication

- DNA is replicated by a semiconservative mechanism in which the two strands of the double helix unwind and each strand serves as a template for the synthesis of a complementary strand.
- Bacterial chromosome replication is typically initiated at a single point and moves in both directions around a circular DNA molecule. In contrast, eukaryotes initiate DNA replication at multiple replicons, and replication proceeds bidirectionally at each replicon. A licensing mechanism involving the binding of MCM proteins to replication origins allows eukaryotes to ensure that DNA is replicated only once prior to each mitosis.
- During replication, DNA polymerases synthesize DNA in the $5' \rightarrow 3'$ direction. Synthesis is continuous along the leading strand but discontinuous along the lagging strand, generating small Okazaki fragments that are subsequently joined by DNA ligase.
- DNA replication is initiated by primase, which synthesizes short RNA primers that are later removed and replaced with DNA. During replication, the double helix is unwound through the combined action of DNA helicases, topoisomerases, and single-stranded DNA binding proteins. As replication proceeds, a proofreading mechanism based on the $3' \rightarrow 5'$ exonuclease activity of DNA polymerase allows incorrectly base-paired nucleotides to be removed and replaced.
- The ends of linear chromosomal DNA molecules are synthesized by telomerase, an enzyme that uses an RNA template for creating short, repeated DNA sequences at the ends of each chromosome.

DNA Damage and Repair

- DNA damage arises both spontaneously and through the action of mutagenic chemicals and radiation.
- Some forms of DNA polymerase carry out translesion synthesis of new DNA across regions where the template DNA is damaged.
- Excision repair is used to correct DNA damage involving abnormal bases, whereas mismatch repair fixes improperly base-paired nucleotides. Nonhomologous end-joining and homologous recombination are used to repair double-strand DNA breaks.

Nuclear and Cell Division

- Mitosis is subdivided into prophase, prometaphase, metaphase, anaphase, and telophase. During prophase, replicated chromosomes condense into paired sister chromatids while centrosomes initiate assembly of the mitotic spindle. In prometaphase, the nuclear envelope breaks down and chromosomes then attach to spindle microtubules and move to the spindle equator, where they line up at metaphase. At anaphase, sister chromatids separate and the resulting daughter chromosomes move toward opposite spindle poles. During telophase the chromosomes decondense, and a nuclear envelope reassembles around each daughter nucleus.
- Chromosome movements are driven by three groups of motor proteins. Motor proteins located at the kinetochores and spindle poles move chromosomes toward the spindle poles, accompanied by disassembly of the microtubules at their plus and minus ends. Motor proteins that crosslink the polar microtubules move overlapping microtubules in opposite directions, thereby pushing the spindle poles apart. The final group of motor proteins move astral microtubules toward the plasma membrane, thereby pulling the spindle poles apart.
- Cytokinesis usually begins before mitosis is complete. In animal cells, an actomyosin filament network forms a cleavage furrow that constricts the cell at the midline and separates the cytoplasm into two daughter cells. In plant cells, a cell wall forms through the middle of the dividing cell.

Regulation of the Cell Cycle

- Progression through the eukaryotic cell cycle is regulated by Cdk-cyclin complexes.
- At the restriction point (Start in yeast), a Cdk-cyclin complex catalyzes the phosphorylation of the Rb protein to trigger passage into S phase.
- At the G2-M boundary, another Cdk-cyclin complex triggers entry into mitosis by catalyzing the phosphorylation of proteins that promote nuclear envelope breakdown, chromosome condensation, and spindle formation.
- At the metaphase-anaphase boundary, activation of the anaphase-promoting complex triggers a protein degradation pathway that initiates chromatid separation and targets mitotic cyclin for breakdown. The resulting loss of mitotic Cdk activity leads to events associated with the exit from mitosis, including cytokinesis, chromatin decondensation, and reassembly of the nuclear envelope.
- Checkpoint pathways monitor intracellular conditions and temporarily halt the cell cycle if conditions are not suitable for proceeding. The DNA replication checkpoint verifies that DNA synthesis has been completed before allowing the cell to exit from G2 and begin mitosis. DNA damage checkpoints involving the p53 protein halt the cell cycle at various points if DNA damage is detected. Finally, the mitotic spindle checkpoint prevents chromosomes from moving to the

spindle poles before all chromosomes are attached to the spindle.

Growth Factors and Cell Proliferation

- The cells of multicellular organisms do not normally proliferate unless they are stimulated by an appropriate growth factor.
- Many growth factors bind to receptors that activate the Ras pathway, which culminates in passage through the restriction point and into S phase. Growth factor receptors also activate the P1 3-kinase–Akt pathway, which promotes cell survival and proliferation by phosphorylating and thereby activating target proteins that suppress apoptosis and inhibit cell cycle arrest.
- Some growth factors inhibit (rather than stimulate) cell proliferation by triggering the production of Cdk inhibitors.

Apoptosis

- Apoptosis is a form of cellular death triggered by activation of death receptors, withdrawal of survival factors, or as a result of DNA damage.
- Apoptosis involves the orderly dismantling of a dying cell's contents.
- Proteases called caspases are key mediators of apoptosis. Initiator caspases activate executioner caspases, which in turn activate other apoptosis proteins.
- Initiator caspases can be activated in several ways, including through the release of cytochrome *c* from mitochondria. Pro- and anti-apoptotic proteins at the mitochondrion regulate release of cytochrome *c*.

MB **www.masteringbiology.com**

1. MasteringBiology Assignments

Tutorials and Activities

- Molecular Model Structural Tutorials for visualization of 3D structures
- BioFlix Tutorials including 3D animations with hints and detailed feedback for common wrong answers
- Mitosis (1 of 3): Mitosis and the Cell Cycle (BioFlix tutorial); Four Phases of the Cell Cycle; Mitosis; The Cell Cycle; Experimental Inquiry: Does DNA Replication Follow the Conservative, Semiconservative, or Dispersive Model? DNA Polymerase III Beta Subunit; Bacteriophage T7 DNA Polymerase; DNA Replication (1 of 2): DNA Structure and Replication Machinery (BioFlix tutorial); Nucleic Acids: DNA Replication; DNA Replication: An Overview; DNA Replication; DNA Replication (2 of 2): Synthesis of the Leading and Lagging Strands (BioFlix tutorial); The Polymerase Chain Reaction; DNA Synthesis; DNA Replication: A Closer Look; ABC News Video: The Effect of Exercise on Cells; Mutations at the DNA Level; DNA Repair: Mechanisms; MutS, a DNA Mismatch Repair Protein of E. coli; MutS, a DNA Mismatch Repair Protein of E. coli–Copy; Mitosis (2 of 3): Mechanism of Mitosis (BioFlix tutorial); Microtubule Motors During Mitosis; Mitosis (3 of 3): Comparing Cell Division in Animals, Plants, and Bacteria (BioFlix tutorial); p53 and Control of the Cell Cycle

Questions

- Reading Quiz
- Multiple Choice
- End-of-chapter questions

2. eText Read your book online, search, highlight text, take notes, and more

3. The Study Area

- Chapter Review Quiz, Guide to Techniques and Methods, 3D-Structure Tutorials, Activities, BioFlix, Videos, Biology Review, Word Study Tools, Glossary, Flashcards

PROBLEM SET

More challenging problems are marked with a •.

19-1 Cell Cycle Phases. Indicate whether each of the following statements is true of the G1 phase of the cell cycle, S phase, G2 phase, or M phase. A given statement may be true of any, all, or none of the phases.

(a) The nuclear envelope breaks into fragments.

(b) A Cdk protein is present in the cell.

(c) The primary cell wall of a plant cell forms.

(d) This phase is part of interphase.

(e) The amount of nuclear DNA in the cell doubles.

(f) A cell cycle checkpoint has been identified in this phase.

(g) Cells that will never divide again are likely to be arrested in this phase.

(h) Sister chromatids separate from each other.

(i) Mitotic cyclin is at its lowest level.

(j) Chromosomes are present as diffuse, extended chromatin.

19-2 The Mitotic Index and the Cell Cycle. The mitotic index is a measure of the mitotic activity of a population of cells. It is calculated as the percentage of cells in mitosis at any one time. Assume that upon examining a sample of 1000 cells, you find 30 cells in prophase, 20 in prometaphase, 20 in metaphase, 10 in anaphase, 20 in telophase, and 900 in interphase. Of those in interphase, 400 are found (by staining the cells with a DNA-specific stain) to have *X* amount of DNA, 200 to have 2*X*, and 300 cells to be somewhere in between. Autoradiographic analysis indicates that the G2 phase lasted 4 hours.

(a) What is the mitotic index for this population of cells?

(b) Specify the proportion of the cell cycle spent in each of the following phases: prophase, prometaphase, metaphase, anaphase, telophase, G1, S, and G2.

(c) What is the total length of the cell cycle?

(d) What is the actual amount of time (in hours) spent in each of the phases of part b?

(e) To measure the G2 phase, radioactive thymidine (a DNA precursor) is added to the culture at some time *t*, and samples of the culture are analyzed autoradiographically for labeled nuclei at regular intervals thereafter. What specific observation would have to be made to assess the length of the G2 phase?

(f) What proportion of the interphase cells would you expect to exhibit labeled nuclei in autoradiographs prepared shortly after exposure to the labeled thymidine? (Assume a labeling period just long enough to allow the thymidine to get into the cells and begin to be incorporated into DNA.)

19-3 Meselson and Stahl Revisited. Although the Watson–Crick structure for DNA suggested a semiconservative model for DNA replication, at least two other models are conceivable. In a *conservative model,* the parental DNA double helix remains intact and a second, all-new copy is made. In a *dispersive model,* each strand of both daughter molecules contains a mixture of old and newly synthesized segments.

(a) Starting with one parental double helix, sketch the progeny molecules for two rounds of replication according to each of these alternative models. Use one color for the original parent strands and another color for all the DNA synthesized thereafter (as is done in Figure 19-3 for the semiconservative model).

(b) For each of the alternative models, indicate the distribution of DNA bands that Meselson and Stahl would have found in their cesium chloride gradients after one and two rounds of replication.

19-4 DNA Replication. Sketch a replication fork of bacterial DNA in which one strand is being replicated discontinuously and the other is being replicated continuously. List six different enzyme activities associated with the replication process, identify the function of each activity, and show where each would be located on the replication fork. In addition, identify the following features on your sketch: DNA template, RNA primer, Okazaki fragments, and single-stranded DNA binding protein.

• **19-5 More DNA Replication.** The following are observations from five experiments carried out to determine the mechanism of DNA replication in the hypothetical organism *Fungus mungus.* For each experiment, indicate whether the results support (S), refute (R), or have no bearing (NB) on the hypothesis that this fungus replicates its DNA by the same mechanism as that known for *E. coli.* Explain your reasoning in each case.

(a) Neither of the two DNA polymerases of *F. mungus* appears to have an exonuclease activity.

(b) Replicating DNA from *F. mungus* shows discontinuous synthesis on both strands of the replication fork.

(c) Some of the DNA sequences from *F. mungus* are present in multiple copies per genome, whereas other sequences are unique.

(d) Short fragments of *F. mungus* DNA isolated during replication contain both ribose and deoxyribose.

(e) *F. mungus* cells are grown in the presence of the heavy isotopes ^{15}N and ^{13}C for several generations and then grown for one generation in normal (^{14}N, ^{12}C) medium; then DNA is isolated from these cells and denatured. The single strands yield a single band in a cesium chloride density gradient.

19-6 Still More DNA Replication. Suppose you are given a new temperature-sensitive bacterial mutant that grows normally at 37°C but cannot replicate its chromosomes properly at 42°C. To investigate the nature of the underlying defect, you incubate the cells at 42°C with radioactive substrates required for DNA synthesis. After one hour you find that the cell population has doubled its DNA content, suggesting that DNA replication can still occur at 42°C. Moreover, centrifugation reveals that all of this DNA has the same large molecular weight as does the original DNA present in the cells. When the DNA is denatured, however, you discover that 75% of the resulting single-stranded DNA has a molecular weight that is half that of the original double-stranded DNA, and the remaining 25% of the DNA has a much lower molecular weight. Based on the preceding results, what gene do you think is defective in these cells? Explain how such a defect would account for the experimental results that you observed.

19-7 The Minimal Chromosome. To enable it to be transmitted intact from one cell generation to the next, the linear DNA molecule of a eukaryotic chromosome must have appropriate nucleotide sequences making up three special kinds of regions: Origins of replication (at least one), a centromere, and two telomeres. What would happen if such a chromosomal DNA molecule somehow lost:

(a) all of its origins of replication?

(b) all of the DNA constituting its centromere?

(c) one of its telomeres?

19-8 DNA Damage and Repair. Indicate whether each of the following statements is true of depurination (DP), deamination (DA), or pyrimidine dimer formation (DF). A given statement may be true of any, all, or none of these processes.

(a) Repair involves DNA ligase.

(b) This process is caused by spontaneous hydrolysis of a glycosidic bond.

(c) This can happen to thymine but not to adenine.

(d) Repair depends on cleavage of both strands of the double helix.

(e) Repair depends on the existence of separate copies of the genetic information in the two strands of the double helix.

(f) This process is induced by ultraviolet light.

(g) Repair involves a DNA glycosylase.

(h) This can happen to guanine but not to cytosine.

(i) Repair involves an endonuclease.

(j) This can happen to thymine but not to cytosine.

19-9 Nonstandard Purines and Pyrimidines. Shown in **Figure 19-46a** are three nonstandard nitrogenous bases that are formed by the deamination of naturally occurring bases in DNA.

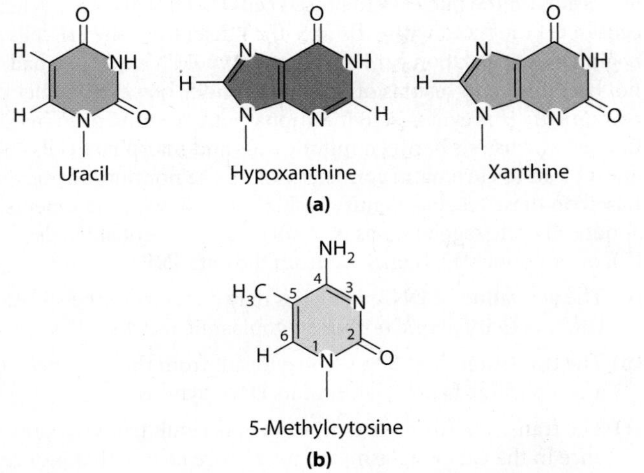

Uracil Hypoxanthine Xanthine

(a)

5-Methylcytosine

(b)

FIGURE 19-46 Structures of Several Nonstandard Purines and Pyrimidines. See Problem 19-9.

(a) Indicate which base in DNA must be deaminated to form each of these bases.

(b) Why are there only three bases shown, when DNA contains four bases?

(c) Why is it important that none of the bases shown in Figure 19-46a occurs naturally in DNA?

(d) Figure 19-46b shows 5-methylcytosine, a pyrimidine that arises naturally in DNA when cellular enzymes methylate cytosine. Why is the presence of this base likely to increase the probability of a mutation?

• **19-10 Chromosome Movement in Mitosis.** It is possible to mark the microtubules of a spindle by photobleaching with a laser microbeam (**Figure 19-47**). When this is done, chromosomes move *toward* the bleached area during anaphase. Are the following statements consistent (C) or inconsistent (I) with this experimental result?

(a) Microtubules move chromosomes solely by disassembling at the spindle poles.

(b) Chromosomes move by disassembling microtubules at their kinetochore ends.

(c) Chromosomes are moved along microtubules by a kinetochore motor protein that moves along the surface of the microtubule and "pulls" the chromosome with it.

19-11 Cytokinesis. Predict what will happen in each of the following situations, based on your knowledge of cytokinesis. In each case, explain your answer.

(a) A fertilized sea urchin egg is injected with C3 transferase, a bacterial toxin that ADP ribosylates and inhibits Rho, 30 minutes prior to first cleavage.

(b) A one-celled *C. elegans* zygote lacks *anillin*, a protein that is required to assemble myosin efficiently at the cell surface.

19-12 More on Cell Cycle Phases. For each of the following pairs of phases from the cell cycle, indicate how you could tell in which of the two phases a specific cell is located.

(a) G1 and G2

(b) G1 and S

(c) G2 and M

(d) G1 and M

• **19-13 Cell Cycle Regulation.** Recall that one approach to the study of cell cycle regulation has been to fuse cultured cells that are at different stages of the cell cycle and observe the effect of the fusion on the nuclei of the fused cells (heterokaryons). When cells in G1 are fused with cells in S, the nuclei from the G1 cells begin DNA replication earlier than they would have if they had not been fused. In fusions of cells in G2 and S, however, nuclei continue their previous activities, apparently uninfluenced by the fusion. Fusions between mitotic cells and interphase cells always lead to chromatin condensation in the nonmitotic nuclei. Based on these results, identify each of the following statements about cell cycle regulation as probably true (T), probably false (F), or not possible to conclude from the data (NP).

(a) The activation of DNA synthesis may result from the stimulatory activity of one or more cytoplasmic factors.

(b) The transition from S to G2 may result from the presence of a cytoplasmic factor that inhibits DNA synthesis.

(c) The transition from G2 to mitosis may result from the presence in the G2 cytoplasm of one or more factors that induce chromatin condensation.

(d) G1 is not an obligatory phase of all cell cycles.

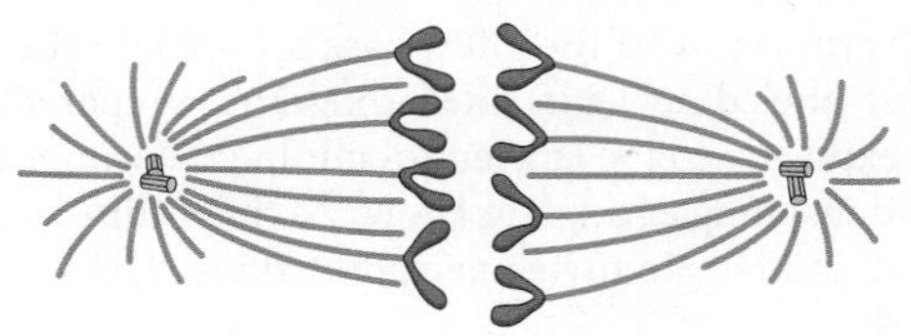

Microtubules are labeled with a fluorescent dye during anaphase.

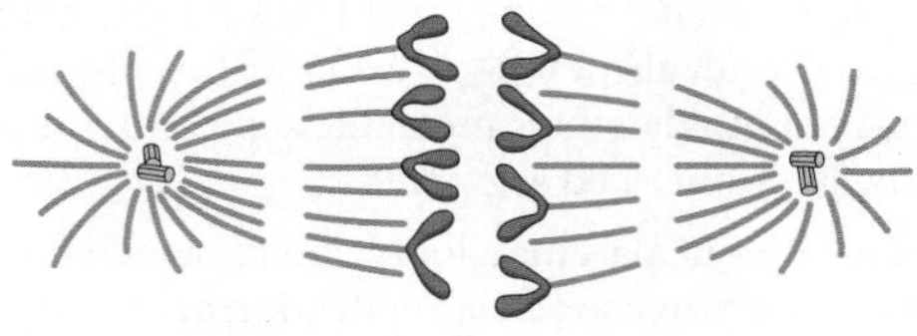

A laser microbeam is used to mark two areas by bleaching the fluorescent dye.

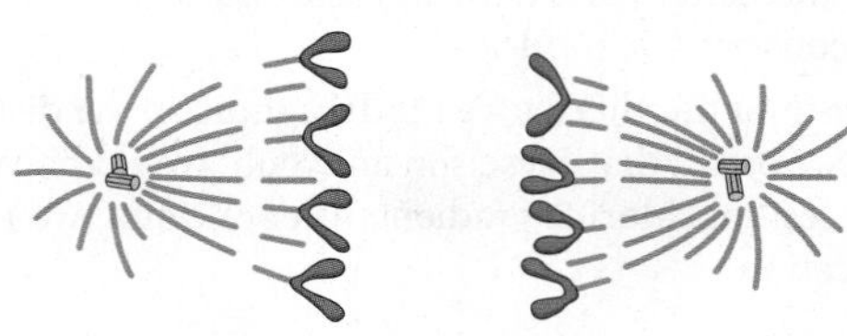

The chromosomes are observed to move toward the bleached areas.

FIGURE 19-47 Use of Laser Photobleaching to Study Chromosome Movement During Mitosis. See Problem 19-10.

(e) The transition from mitosis to G1 appears to result from the disappearance or inactivation of a cytoplasmic factor present during M phase.

• **19-14 Role of Cyclin-Dependent Protein Kinases.** Based on your understanding of the regulation of the eukaryotic cell cycle, how could you explain each of the following experimental observations?

(a) When mitotic Cdk-cyclin is injected into cells that have just emerged from S phase, chromosome condensation and nuclear envelope breakdown occur immediately, rather than after the normal G2 delay of several hours.

(b) When an abnormal, indestructible form of mitotic cyclin is introduced into cells, they enter into mitosis but cannot emerge from it and reenter G1 phase.

(c) Mutations that inactivate the main protein phosphatase used to catalyze protein dephosphorylations cause a long delay in the reconstruction of the nuclear envelope that normally takes place at the end of mitosis.

• **19-15 Apoptosis and Medicine.** A current focus of molecular medicine is to trigger or prevent apoptosis in specific cells. Several components of the apoptotic pathway are being targeted using this approach. For each of the following, state specifically how the treatment would be expected to stimulate or inhibit apoptosis.

(a) Cells are treated with a small molecule called pifithrin-α, which was originally isolated for its ability to reversibly block p53-dependent transcriptional activation.

(b) Exposing cells to recombinant FasL protein, a ligand for the tumor necrosis factor family of receptors.

(c) Treatment of cells with organic compounds that enter the cell and bind with high affinity to the active site of caspase-3.

SUGGESTED READING

References of historical importance are marked with a •.

Overview of the Cell Cycle

Darzynkiewicz, Z., H. Crissman, and J. W. Jacobberger. Cytometry of the cell cycle: Cycling through history. *Cytometry* 58A (2004): 21.

DNA Replication

Aladjem, M. I. Replication in context: Dynamic regulation of DNA replication patterns in metazoans. *Nature Rev. Genet.* 8 (2007): 588.

Alberts, B. DNA replication and recombination. *Nature* 421 (2003): 431.

Blow, J. J., and A. Dutta. Preventing re-replication of chromosomal DNA. *Nature Rev. Mol. Cell Biol.* 6 (2005): 476.

• Friedberg, E. C. The eureka enzyme: The discovery of DNA polymerase. *Nature Rev. Mol. Cell Biol.* 7 (2006): 143.

Gilbert, D. M. In search of the holy replicator. *Nature Rev. Mol. Cell Biol.* 5 (2004): 848.

Gilson, E., and V. Géli. How telomeres are replicated. *Nature Rev. Mol. Cell Biol.* 8 (2007): 825.

• Greider, C. W., and E. H. Blackburn. Telomeres, telomerase, and cancer. *Sci. Amer.* 274 (February 1996): 92.

Johnson, A., and M. O'Donnell. Cellular DNA replicases: Components and dynamics at the replication fork. *Annu. Rev. Biochem.* 74 (2005): 283.

Méndez, J., and B. Stillman. Perpetuating the double helix: Molecular machines at eukaryotic DNA replication origins. *BioEssays* 25 (2003): 1158.

• Meselson, M., and F. W. Stahl. The replication of DNA in *E. coli*. *Proc. Natl. Acad. Sci. USA* 44 (1958): 671.

• Mullis, K. B. The unusual origin of the polymerase chain reaction. *Sci. Amer.* 262 (April 1990): 56.

• Ogawa, T., and R. Okazaki. Discontinuous DNA replication. *Annu. Rev. Biochem.* 49 (1980): 421.

Remus, D., and J. F. Diffley. Eukaryotic DNA replication control: Lock and load, then fire. *Curr. Opin. Cell Biol.* 21 (2009): 771.

Stewart, S. A., and R. A. Weinberg. Telomeres: Cancer to human aging. *Annu. Rev. Cell Dev. Biol.* 22 (2006): 531.

• Watson, J. D., and F. H. C. Crick. Genetical implications of the structure of deoxyribonucleic acid. *Nature* 171 (1953): 964.

DNA Repair

Clarke, P. R., and L. A. Allan. Cell-cycle control in the face of damage—A matter of life or death. *Trends Cell Biol.* 19 (2009): 89

Friedberg, E. C. DNA damage and repair. *Nature* 421 (2003): 436.

Jiricny, J. The multifaceted mismatch-repair system. *Nature Rev. Mol. Cell Biol.* 7 (2006): 335.

Lainé, J.-P., and J.-M. Egly. When transcription and repair meet: A complex system. *Trends Genet.* 22 (2006): 430.

Lieber, M. R. et al. Mechanism and regulation of human non-homologous DNA end-joining. *Nature Rev. Mol. Cell Biol.* 4 (2003): 712.

Nuclear and Cell Division

Davis, T. N., and L. Wordeman. Rings, bracelets, sleeves, and chevrons: New structures of kinetochore proteins. *Trends Cell Biol.* 17 (2007): 377.

Glotzer, M. The 3Ms of central spindle assembly: Microtubules, motors and MAPs. *Nat. Rev. Mol. Cell Biol.*10 (2009): 9.

• Mitchison, T. J., and E. D. Salmon. Mitosis: A history of division. *Nature Cell Biol.* 3 (2001): E17.

Muller, S., A. J. Wright, and L. G. Smith. Division plane control in plants: New players in the band. *Trends Cell Biol.* 19 (2009): 180.

Nigg, E. A. Centrosome duplication: Of rules and licenses. *Trends Cell Biol.* 17 (2007): 215.

Nigg, E. A., ed. *Centrosomes in Development and Disease.* Weinheim: Wiley-VCH, 2004.

Scholey, J. M., I. Brust-Mascher, and A. Mogilner. Cell division. *Nature* 422 (2003): 746.

Siller, K. H., and C. Q. Doe. Spindle orientation during asymmetric cell division. *Nat. Cell Biol* 11 (2009): 365.

von Dassow, G. Concurrent cues for cytokinetic furrow induction in animal cells. *Trends Cell Biol* 19 (2009): 165.

Wadsworth, P., and A. Khodjakov. *E pluribus unum:* Towards a universal mechanism for spindle assembly. *Trends Cell Biol.* 14 (2004): 413.

Regulation of the Cell Cycle

Bloom, J., and F. R. Cross. Multiple levels of cyclin specificity in cell-cycle control. *Nature Rev. Mol. Cell Biol.* 8 (2007): 149.

Chan, G. K., S.-T. Liu, and T. J. Yen. Kinetochore structure and function. *Trends Cell Biol.* 15 (2005): 589.

Clarke, P. R., and C. Zhang. Spatial and temporal coordination of mitosis by Ran GTPase. *Nat. Rev. Mol. Cell. Biol.* 9 (2008): 464.

Coller, H. A., What's taking so long? S-phase entry from quiescence versus proliferation. *Nature Rev. Mol. Cell Biol.* 8 (2007): 667.

De Veylder, L., T. Beeckman, and D. Inzé. The ins and outs of the plant cell cycle. *Nature Rev. Mol. Cell. Biol.* 8 (2007): 655.

Jiang, B. H., and L. Z. Liu. PI3K/PTEN signaling in angiogenesis and tumorigenesis. *Adv. Cancer Res.* 102 (2009): 19.

• Nasmyth, K. A prize for proliferation. *Cell* 107 (2001): 689.

Nigg, E. A. Mitotic kinases as regulators of cell division and its checkpoints. *Nature Rev. Mol. Cell Biol.* 2 (2001): 21.

Nurse, P. A long twentieth century of the cell cycle and beyond. *Cell* 100 (2000): 71.

Onn, I., J. M. Heidinger-Pauli, V. Guacci, E. Unal, and D. E. Koshland. Sister chromatid cohesion: A simple concept with a complex reality. *Ann. Rev. Cell Dev. Biol.* 24 (2008): 105.

Peters, J.-M. The anaphase promoting complex/cyclosome: A machine designed to destroy. *Nature Rev. Mol. Cell Biol.* 7 (2006): 644.

Pines, J. Mitosis: A matter of getting rid of the right protein at the right time. *Trends Cell Biol.* 16 (2006): 55.

van den Heuvel, S., and N. J. Dyson. Conserved functions of the pRB and E2F families. *Nat. Rev. Mol. Cell. Biol.* 9 (2008): 713.

Apoptosis

Brenner, D., and T. W. Mak. Mitochondrial cell death effectors. *Curr. Opin. Cell Biol.* 21 (2009): 871.

Brunelle, J. K., and A. Letai. Control of mitochondrial apoptosis by the Bcl-2 family. *J. Cell Sci.* 122 (2009): 437.

Danial, N. N., and S. J. Korsmeyer. Cell death: Critical control points. *Cell* 116 (2004): 205.

Horvitz, H. R. Worms, life, and death (Nobel lecture). *Chembiochem.* 4 (2003): 697.

Ow, Y. P., D. R. Green, Z. Hao, and T. W. Mak. Cytochrome *c*: Functions beyond respiration. *Nat. Rev. Mol. Cell. Biol.* 9 (2008): 532.

Vaseva, A. V., and U. M. Moll. The mitochondrial p53 pathway. *Biochim. Biophys. Acta* 1787 (2009): 414.

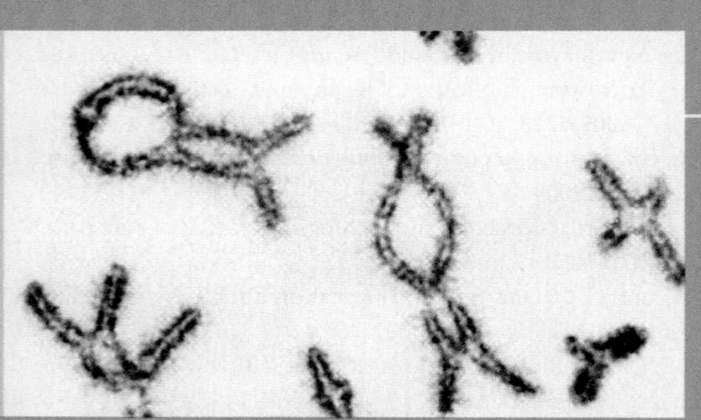

20 Sexual Reproduction, Meiosis, and Genetic Recombination

See MasteringBiology® for tutorials, activities, and quizzes.

Mitotic cell division, which we discussed in the preceding chapter, is used for the proliferation of most eukaryotic cells, leading to the production of more organisms or more cells per organism. Since a mitotic cell division cycle involves one round of DNA replication followed by the segregation of identical chromatids into two daughter cells, mitotic division produces cells that are genetically identical, or very nearly so. This ability to perpetuate genetic traits faithfully allows mitotic division to form the basis of **asexual reproduction** *in eukaryotes. During asexual reproduction, new individuals are generated by mitotic division of cells in a single parent organism, either unicellular or multicellular. Although the details vary among organisms, asexual reproduction is widespread in nature. Examples include* mitotic division *of unicellular organisms,* budding *of offspring from a multicellular parent's body, and* regeneration *of whole organisms from pieces of a parent organism. In plants, entire organisms can even be regenerated from single cells taken from an adult plant.*

Asexual reproduction can be an efficient and evolutionarily successful mode of perpetuating a species. As long as environmental conditions remain essentially constant, the genetic predictability of asexual reproduction is perfectly suited for maintaining the survival of a population. But if the environment changes, a population that reproduces asexually may not be able to adapt to the new conditions. Under such conditions, organisms that reproduce sexually rather than asexually will usually have an advantage, as we now discuss.

Sexual Reproduction

In contrast to asexual reproduction, in which progeny are genetically identical to the single parent from which they arise, **sexual reproduction** allows genetic information from two parents to be mixed together, thereby producing offspring that are genetically dissimilar, both from each other and from the parents. Moreover, the offspring are unpredictably dissimilar; that is, we cannot anticipate exactly which combination of genes a particular offspring will receive from its two parents. Since most plants and animals—and even many eukaryotic microorganisms—reproduce sexually, this type of reproduction must provide some distinct advantages.

Sexual Reproduction Produces Genetic Variety by Bringing Together Chromosomes from Two Different Parents

Sexual reproduction allows genetic traits found in different individuals to be combined in various ways in newly developing offspring, thereby generating enormous variety among the individuals that make up a population. Genetic variation ultimately depends on the occurrence of *mutations,* which are unpredictable alterations in DNA base sequence. Mutations are rare events, and beneficial mutations are even rarer. But when beneficial mutations do arise, it is clearly advantageous to preserve them in the population. It can be even more beneficial to bring mutations together in various combinations—and therein lies an advantage of sexual reproduction. Although mutations occur in both sexual and asexual species, only sexual reproduction can bring about a reshuffling of genetic information in each new offspring.

Because sexual reproduction combines genetic information from two different parents into a single offspring, at some point in its life cycle every sexually reproducing organism has cells that contain two copies of each type of chromosome, one inherited from each parent. The two members of each chromosome pair are called **homologous chromosomes.** Two homologous chromosomes carry the same lineup of genes, although for any given gene, the two versions may differ slightly in base sequence. Not surprisingly, homologous chromosomes usually look alike when viewed with a microscope (see Figure 19-23). An exception to this rule is the **sex chromosomes,** which determine whether an individual is male or female. The two kinds of sex chromosomes, generally called X and Y chromosomes, differ significantly in genetic makeup and appearance. In mammals, for example, females have two X chromosomes of the same size, whereas males have one X chromosome and a Y chromosome that is much smaller. Nonetheless, parts of the X and Y chromosomes are actually homologous; and, during sexual reproduction, the X and Y chromosomes behave as homologues.

A cell or organism with two sets of chromosomes is said to be **diploid** (from the Greek word *diplous,* meaning "double") and contains two copies of its genome. A cell or organism with a single set of chromosomes, and therefore a single copy of its genome, is **haploid** (from the Greek word *haplous,* meaning "single"). By convention, the haploid chromosome number for a species is designated n (or $1n$) and the diploid number $2n$. For example, in humans $n = 23$, which means that most human cells contain two sets of 23 chromosomes, yielding a diploid total of 46. The diploid state is an essential feature of the life cycle of sexually reproducing species. In a sense, a diploid cell contains an extra set of genes that is available for mutation and genetic innovation. Changes in a second copy of a gene usually will not threaten the survival of an organism, even if the mutation disrupts the original function of that particular gene copy. In addition, the diploid state provides some protection against chromosome damage; if one chromosome is accidentally broken, it can sometimes be repaired using the DNA sequence of the homologous chromosome as a template (page 626).

Diploid Cells May Be Homozygous or Heterozygous for Each Gene

To further explore the genetic consequences of the diploid state, let's now focus on the behavior of an individual **gene locus** (plural: **loci**), which is the place on a chromosome that contains the DNA sequence for a particular gene. For simplicity, we will assume that the gene controls a single, clear-cut characteristic—or *character,* as geneticists usually say—in the organism. Let's also assume that only one copy of this gene is present per haploid genome, so that a diploid organism will have two copies of the gene, which may be either identical or slightly different. The two versions of the gene are called **alleles,** and the combination of alleles determines how an organism will express the character controlled by the gene. In garden peas, for example, the alleles at one particular locus determine seed color, which may be green or yellow (**Figure 20-1**). An organism with two identical alleles for a given gene is said to be **homozygous** for that gene or character. Thus, a pea plant that inherited the same allele for yellow seed color from both of its parents is said to be homozygous for seed color. An organism with two different alleles for a gene is

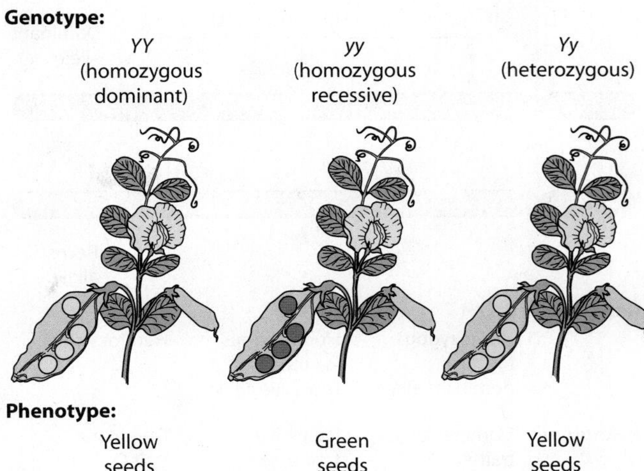

FIGURE 20-1 Genotype and Phenotype. In garden peas, seed color (phenotype) can be either yellow or green. The seed-color alleles are Y (yellow, dominant) or y (green, recessive). Because the pea plant is a diploid organism, its genetic makeup (genotype) for seed color may be homozygous dominant *(YY)*, homozygous recessive *(yy)*, or heterozygous *(Yy)*.

said to be **heterozygous** for that gene or for the character it determines. A pea plant with one allele that specifies yellow seed color and a second allele that specifies green seed color is therefore heterozygous for seed color.

In a heterozygous individual, one of the two alleles is often **dominant** and the other **recessive.** These terms convey the idea that the dominant allele determines how the trait will appear in a heterozygous individual. (The word *trait* refers to a particular variant of a character, such as green or yellow seeds, where seed color is the character.) For seed color in peas, the yellow trait is dominant over green; this means that pea plants heterozygous for seed color have yellow seeds. A dominant allele is usually designated by an uppercase letter that stands for the trait, whereas a corresponding recessive allele is represented by the same letter in lowercase. In both cases, italics are used. Thus, alleles for seed color in peas are represented by Y for yellow and y for green because the yellow trait is dominant. As Figure 20-1 illustrates, a pea plant can be homozygous for the dominant allele *(YY)*, homozygous for the recessive allele *(yy)*, or heterozygous *(Yy)*.

It is important to distinguish between the **genotype,** or genetic makeup of an organism, and its **phenotype,** or the physical expression of its genotype. The phenotype of an organism can usually be determined by inspection (e.g., by looking to see whether seeds are green or yellow). Genotype, on the other hand, can be *directly* determined only by studying an organism's DNA. The genotype also can be deduced from indirect evidence, such as the organism's phenotype and information about the phenotypes of its parents and/or offspring. Organisms exhibiting the same phenotype do not necessarily have identical genotypes. In the example of Figure 20-1, pea plants with

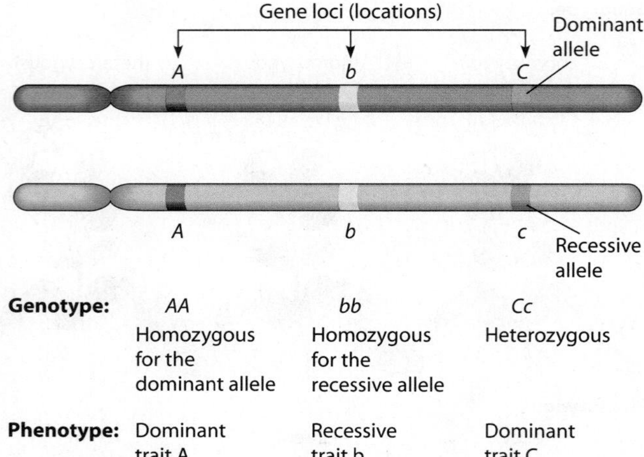

FIGURE 20-2 Some Genetic Terminology. This diagram shows a homologous pair of chromosomes from a diploid cell; the chromosomes are the same size and shape and carry genes for the same characters (characteristics), in the same order. The site of a gene on a chromosome is called a gene locus. The particular versions of a gene—alleles—found at comparable gene loci on homologous chromosomes may be identical (giving the organism a genotype that is homozygous for that gene) or different (making the genotype heterozygous). If one allele of a gene is dominant and the other recessive, the heterozygous organism exhibits a dominant phenotype with respect to the character in question. A recessive phenotypic trait is observed only if the genotype is homozygous for the recessive allele. Red and blue are used here and in most later figures to distinguish the different parental origins of the two chromosomes.

yellow seeds (phenotype) can be either *YY* or *Yy* (genotype). **Figure 20-2** summarizes the genetic terminology introduced so far.

Gametes Are Haploid Cells Specialized for Sexual Reproduction

MB Meiosis (1 of 3): Genes, Chromosomes, and Sexual Reproduction (BioFlix tutorial)

The hallmark of sexual reproduction is that genetic information contributed by two parents is brought together in a single individual. Because the offspring of sexual reproduction are diploid, the contribution from each parent must be haploid. The haploid cells produced by each parent that fuse together to form the diploid offspring are called **gametes,** and the process that produces them is **gametogenesis.** Biologists distinguish between male and female individuals on the basis of the gametes they produce. Gametes produced by males, called **sperm** (or *spermatozoa*), are usually quite small and may be inherently motile. Female gametes, called **eggs** or **ova** (singular: **ovum**), are specialized for the storage of nutrients and tend to be quite large and nonmotile. For example, in sea urchins the volume of an egg cell is more than 10,000 times greater than that of a sperm cell; in birds and amphibians, which have massive yolky eggs, the size difference is even greater. But despite their differing sizes, sperm and egg bring equal amounts of chromosomal DNA to the offspring.

The union of sperm and egg during sexual reproduction is called **fertilization.** The resulting fertilized egg, or **zygote,** is diploid, having received one chromosome set from the sperm and a homologous set from the egg. In the life cycles of multicellular organisms, fertilization is followed by a series of mitotic divisions and progressive specialization of various groups of cells to form a multicellular embryo and eventually an adult.

In a few unusual situations, eggs can develop into offspring without the need for sperm—a phenomenon known as *parthenogenesis.* A striking case in point is the Komodo dragon, an inhabitant of the central islands of Indonesia. The Komodo dragon is the world's largest species of lizard, growing to an average length of 7–10 feet. When male Komodos are unavailable for sexual reproduction, female Komodos give birth to offspring through a self-fertilization process involving the fusion of a haploid egg cell with a haploid *polar body*—a "mini-egg" produced at the same time eggs are formed (see page 611). Consequently, the genes of each offspring derive solely from the mother, though the offspring are not exact duplicates of the mother because different combinations of alleles are present in the haploid cells that fuse together to create the embryo.

In some organisms, gametes cannot be categorized as being male or female. Certain fungi and unicellular eukaryotes, for example, produce gametes that are identical in appearance but differ slightly at the molecular level. Such gametes are said to differ in **mating type.** The union of two of these gametes requires that they be of different mating types, but the number of possible mating types in a species may be greater than two—in some cases, more than ten!

Meiosis

Since gametes are haploid, they cannot be produced from diploid cells by mitosis because mitosis creates daughter cells that are genetically identical to the original parent cell. In other words, if gametes were formed by mitotic division of diploid cells, both sperm and egg would have a diploid chromosome number, just like the parent diploid cells. The hypothetical zygote created by the fusion of such diploid gametes would be *tetraploid* (i.e., possess *four* homologous sets of chromosomes). Moreover, the chromosome number would continue to double for each succeeding generation—an impossible scenario. Thus, for the chromosome number to remain constant from generation to generation, a different type of cell division must occur during the formation of gametes. That special type of division, called **meiosis,** reduces the chromosome number from diploid to haploid.

Meiosis involves one round of chromosomal DNA replication followed by two successive nuclear divisions. This results in the formation of four daughter nuclei (usually in separate daughter cells) containing one

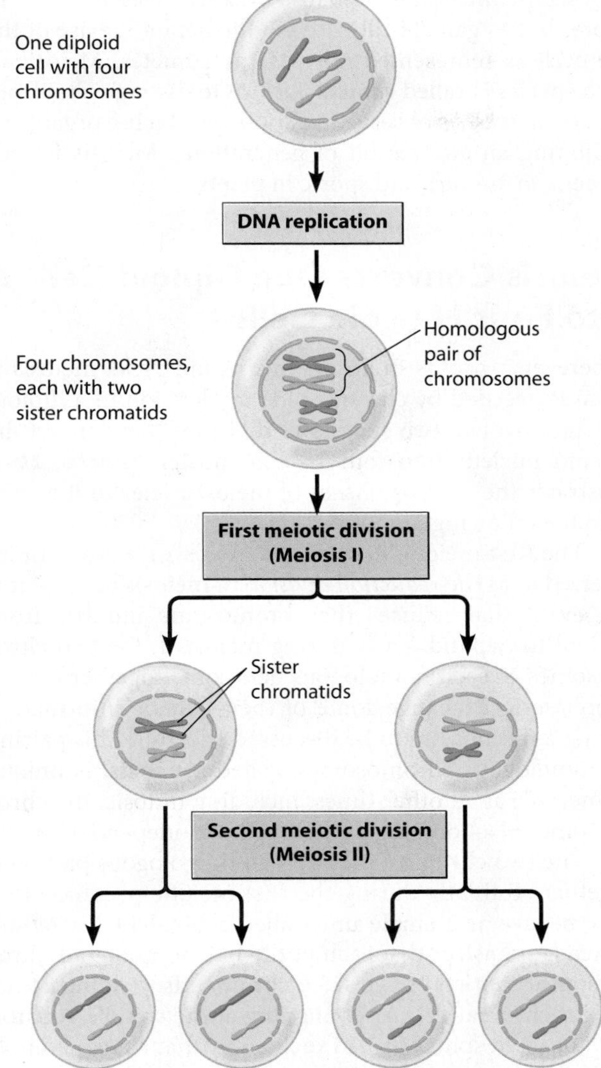

FIGURE 20-3 The Principle of Meiosis. Meiosis involves a single round of DNA replication (chromosome duplication) in a diploid cell followed by two successive cell division events. In this example, the diploid cell has only four chromosomes, which can be grouped into two homologous pairs. After DNA replication, each chromosome consists of two sister chromatids. In the first meiotic division (meiosis I), homologous chromosomes separate, but sister chromatids remain attached. In the second meiotic division (meiosis II), sister chromatids separate, resulting in four haploid daughter cells with two chromosomes each. Notice that each haploid cell has one chromosome from each homologous pair that was present in the diploid cell. For simplicity, the effects of crossing over and genetic recombination are not shown in this diagram.

haploid set of chromosomes per nucleus. **Figure 20-3** outlines the principle of meiosis starting with a diploid cell containing four chromosomes ($2n = 4$). A single round of DNA replication is followed by two cell divisions, meiosis I and meiosis II, leading to the formation of four haploid cells.

The Life Cycles of Sexual Organisms Have Diploid and Haploid Phases

Meiosis and fertilization are indispensable components of the life cycle of every sexually reproducing organism, because the doubling of chromosome number that takes place at fertilization is balanced by the halving that occurs during meiosis. As a result, the life cycle of sexually reproducing organisms is divided into two phases: a diploid ($2n$) phase and a haploid ($1n$) phase. The diploid phase begins at fertilization and extends until meiosis, whereas the haploid phase is initiated at meiosis and ends with fertilization.

Organisms vary greatly in the relative prominence of the haploid and diploid phases of their life cycles, as shown for some representative groups in **Figure 20-4**. Some fungi are examples of sexually reproducing organisms whose life cycles are primarily haploid but include a brief diploid phase that begins with gamete fusion (the fungal equivalent of fertilization) and ends with meiosis (Figure 20-4b). Meiosis usually takes place almost immediately after gamete fusion, so the diploid phase is very short. Accordingly, only a very small fraction of nuclei in such fungi are diploid at any one time. Fungal gametes develop, without meiosis, from cells that are already haploid.

Mosses and ferns are probably the best examples of organisms in which both the haploid and diploid phases are prominent features of the life cycle. Every species of these plants has two alternative, morphologically distinct, multicellular forms, one haploid and the other diploid (Figure 20-4c). For mosses, the haploid form of the organism is larger and more prominent, and the diploid form is smaller and more short-lived. For ferns, it is the other way around. In both cases, gametes develop from preexisting haploid cells.

Organisms that alternate between haploid and diploid multicellular forms in this way are said to display an **alternation of generations** in their life cycles. In addition to mosses and ferns, eukaryotic algae and other plants exhibit an alternation of diploid and haploid generations. In all such organisms, the products of meiosis are **haploid spores,** which, after germination, give rise by mitotic cell division to the haploid form of the plant or alga. The haploid form in turn produces the gametes by specialization of cells that are already haploid. The gametes, upon fertilization, give rise to the diploid form. Because the diploid form produces spores, it is called a **sporophyte** ("spore-producing plant"). The haploid form produces gametes and is therefore called a **gametophyte.** While all plants exhibit an alternation of generations, in most cases the sporophyte generation predominates. In flowering plants, for example, the gametophyte generation is an almost vestigial structure located in the flower (female gametophyte in the *carpel,* male gametophytes in the flower's *anthers*).

The best examples of life cycles dominated by the diploid phase are found in animals (Figure 20-4d). In such

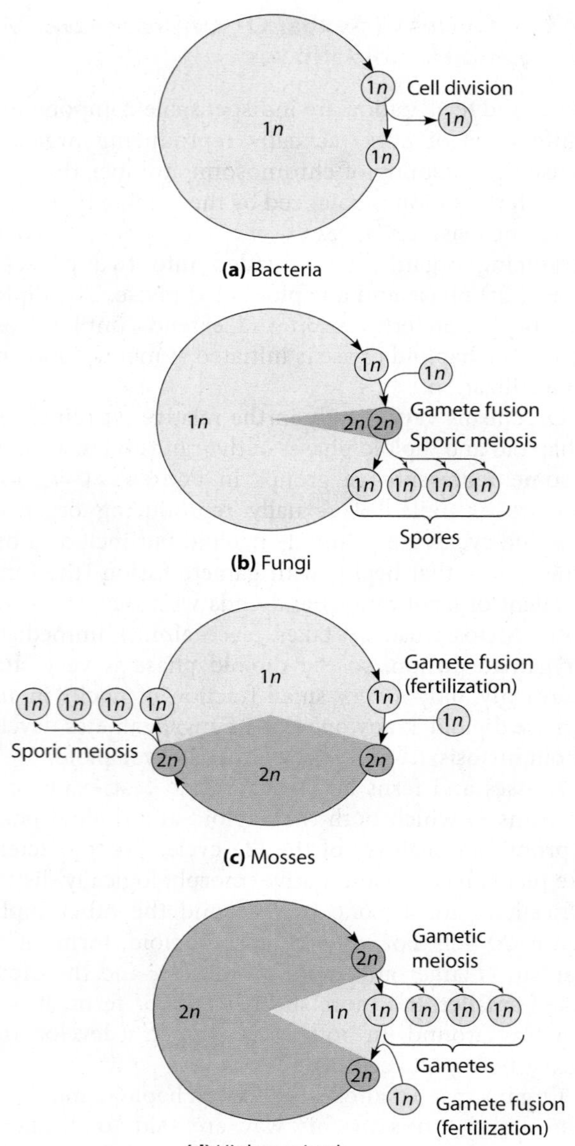

FIGURE 20-4 Types of Life Cycles. The relative prominence of the haploid ($1n$) and diploid ($2n$) phases of the life cycle differ greatly, depending on the organism. **(a)** Bacteria exist exclusively in the haploid state. **(b)** Many fungi exemplify a life form that is predominantly haploid but has a brief diploid phase. Because the products of meiosis in fungi are haploid spores, this type of meiosis is called sporic meiosis. The spores give rise to haploid cells, some of which later become gametes (without meiosis). **(c)** Mosses (and ferns as well) alternate between haploid and diploid forms, both of which are significant components in the life cycles of these organisms. Sporic meiosis produces haploid spores, which in this case grow into haploid plants. Eventually, some of the haploid plant's cells differentiate into gametes. (In seed plants, such as conifers and flowering plants, the haploid forms of the organism are vestigial, each consisting of only a small number of cells.) **(d)** Higher animals are the best examples of organisms that are predominantly diploid, with only the gametes representing the haploid phase of the life cycle. Animals are said to have a gametic meiosis, since the immediate products of meiosis are haploid gametes.

organisms, including humans, meiosis gives rise not to spores but to gametes directly, so the haploid phase of the life cycle is represented only by the gametes. Meiosis in such species is called *gametic meiosis* to distinguish it from the *sporic meiosis* observed in spore-producing organisms exhibiting an alternation of generations. Meiosis is thus gametic in animals and sporic in plants.

Meiosis Converts One Diploid Cell into Four Haploid Cells

Wherever it occurs in an organism's life cycle, meiosis is always preceded by chromosome duplication in a diploid cell and involves two successive divisions that convert the diploid nucleus into four haploid nuclei. **Figure 20-5** illustrates the various phases of meiosis; refer to it as you read the following discussion.

The first meiotic division, or **meiosis I,** is sometimes referred to as the *reduction division* of meiosis because it is the event that reduces the chromosome number from diploid to haploid. Early during meiosis I, the two chromosomes of each homologous pair come together during prophase to exchange some of their genetic information (using a mechanism to be discussed shortly). This pairing of homologous chromosomes, called **synapsis,** is unique to meiosis; at all other times, including mitosis, the chromosomes of a homologous pair behave independently.

The two chromosomes of each homologous pair bind together so tightly during the first meiotic prophase that they behave as a single unit called a **bivalent** (or *tetrad,* which emphasizes that each of the two homologous chromosomes consists of two sister chromatids, yielding a total of four chromatids). After aligning at the spindle equator, each bivalent splits apart in such a way that its two homologous chromosomes move to opposite spindle poles. Because each pole receives only one member of each homologous pair, the daughter nuclei produced by meiosis I are considered to be haploid (even though the individual chromosomes in these nuclei are composed of two sister chromatids). During the second meiotic division (meiosis II), which closely resembles a mitotic division, the two sister chromatids of each chromosome separate into two daughter cells. Hence, the events unique to meiosis happen during the first meiotic division: the synapsis of homologous chromosomes and their subsequent segregation into different daughter nuclei.

Meiotic cell divisions involve the same basic stages as mitosis, although cell biologists do not usually distinguish prometaphase as a separate phase. Thus, the meiotic phases are *prophase, metaphase, anaphase,* and *telophase.* We will see shortly that prophase I is much longer and more complicated than mitotic prophase, whereas prophase II tends to be quite short. Another important difference from mitosis is that a normal interphase does not intervene between the two meiotic divisions. If an interphase does take place, it is usually very short and—most importantly—does *not* include DNA replication, because each chromosome already consists of a pair of replicated sister chromatids that had been generated prior to the first meiotic division. The purpose of the second meiotic division, like that of a typical mitotic division, is to parcel these sister chromatids into two daughter nuclei.

Meiosis I Produces Two Haploid Cells That Have Chromosomes Composed of Sister Chromatids

The first meiotic division segregates homologous chromosomes into different daughter cells. This feature of meiosis has special genetic significance because it represents the point in an organism's life cycle when the two alleles for each gene part company. And it is this separation of alleles that makes possible the eventual remixing of different pairs of alleles at fertilization. Also of great significance during the first meiotic division are events involving the physical exchange of parts of DNA molecules. Such an exchange of DNA segments between two different sources is called **genetic recombination** by molecular biologists. As we will discuss shortly, this type of DNA exchange between homologous chromosomes takes place when the chromosomes are synapsed during prophase I.

Prophase I: Homologous Chromosomes Become Paired and Exchange DNA. Prophase I is a particularly long and complex phase. Based on light microscopic observations, early cell biologists divided prophase I into five stages called *leptotene, zygotene, pachytene, diplotene,* and *diakinesis* (**Figure 20-6**, p. 636).

The **leptotene** stage begins with the condensation of chromatin fibers into long, threadlike structures, similar to what occurs at the beginning of mitosis. At **zygotene,** continued condensation makes individual chromosomes distinguishable, and homologous chromosomes become closely paired with each other via the process of synapsis, forming bivalents. Keep in mind that each bivalent has four chromatids, two derived from each chromosome. Bivalent formation is of considerable genetic significance because the close proximity between homologous chromosomes allows DNA segments to be exchanged by a process called **crossing over.** It is this physical exchange of genetic information between corresponding regions of homologous chromosomes that accounts for genetic recombination. You will learn more about the molecular nature of crossing over later in this chapter. Crossing over occurs during the **pachytene** stage, which is marked by a dramatic compacting process that reduces each chromosome to less than a quarter of its previous length.

At the **diplotene** stage, the homologous chromosomes of each bivalent begin to separate from each other, particularly near the centromere. However, the two chromosomes of each homologous pair remain attached by connections known as **chiasmata** (singular: **chiasma**). Such connections are situated in regions where homologous chromosomes have exchanged DNA segments and hence provide visual evidence that crossing over has occurred between two chromatids, one derived from each chromosome.

In some organisms—female mammals, for instance—the chromosomes decondense during diplotene, transcription resumes, and the cells "take a break" from meiosis for a prolonged period of growth, sometimes lasting for years. (We will consider this situation at the end of our discussion of meiosis.) With the onset of **diakinesis,** the final stage of prophase I, the chromosomes recondense to their maximally compacted state. Now the centromeres of the homologous chromosomes separate further, and the chiasmata eventually become the only remaining attachments between the homologues. At this stage, the nucleoli disappear, the spindle forms, and the nuclear envelope breaks down, marking the end of prophase I.

With the advent of modern tools, especially the electron microscope, cell biologists have been able to refine our picture of what happens during prophase I. They have found that what holds homologous chromosomes in tight apposition during synapsis is the **synaptonemal complex,** an elaborate protein structure resembling a zipper (**Figure 20-7**, p. 637). The *lateral elements* of the synaptonemal complex start to attach to individual chromosomes during leptotene, but the *central element,* which actually joins homologous chromosomes together, does not form until zygotene (see Figure 20-7b). How do the members of each pair of homologous chromosomes find each other so they can be joined by a synaptonemal complex? During early zygotene, the ends (telomeres) of each chromosome become clustered on one side of the nucleus and attach to the nuclear envelope, with the body of each chromosome looping out into the nucleus. To picture this, imagine holding all the ends of four ropes (two long and two short) together. If you give the ropes a strong shake, they will settle into four loops, arranged according to length. This type of chromosome configuration, called a *bouquet,* is thought to promote chromosome alignment.

The alignment of similar-sized chromosomes facilitates formation of synaptonemal complexes, which become fully developed during pachytene. Formation of synaptonemal complexes is closely associated with the process of crossing over in higher eukaryotes, and some electron micrographs reveal additional protein complexes, called *recombination nodules,* that may mediate the crossing over process. The synaptonemal complexes then disassemble during diplotene, allowing the homologous chromosomes to separate (except where they are joined by chiasmata).

Metaphase I: Bivalents Align at the Spindle Equator. During metaphase I, the bivalents attach via their kinetochores to spindle microtubules and migrate to the spindle equator. The presence of *paired* homologous chromosomes (i.e., bivalents) at the spindle equator during metaphase I is a crucial difference between meiosis I and a

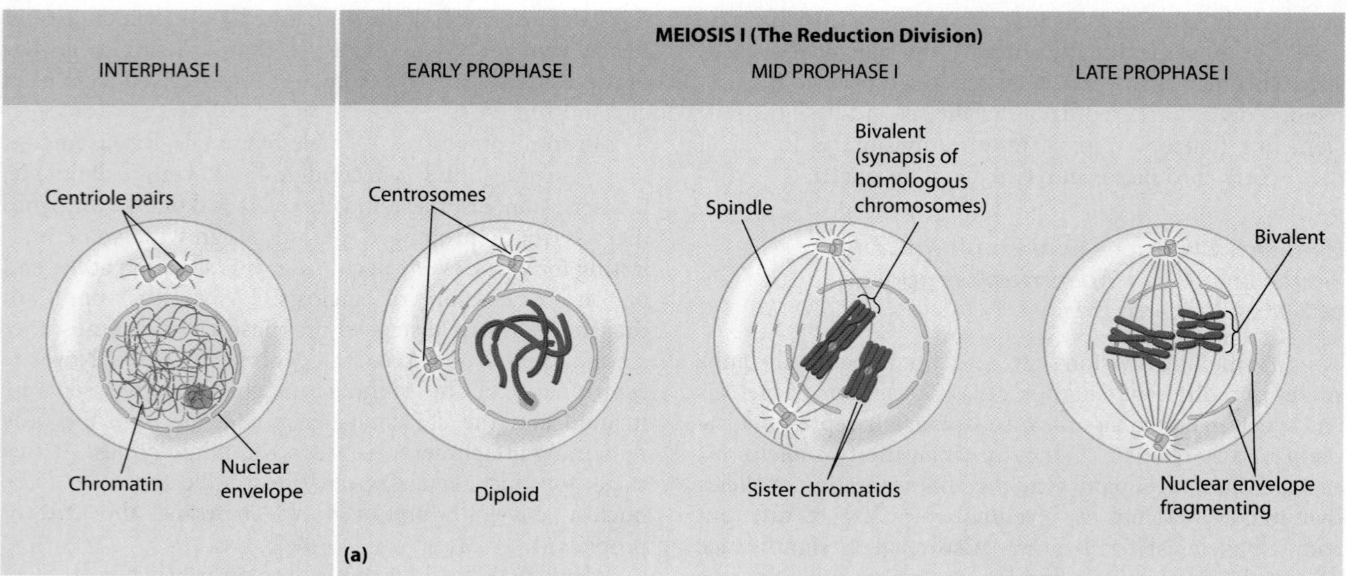

MB Meiosis (2 of 3):The Mechanism (BioFlix tutorial); Activity: Meiosis Animation

FIGURE 20-5 Meiosis in an Animal Cell. Meiosis consists of two successive divisions, called meiosis I and II, with no intervening DNA synthesis or chromosome duplication. **(a)** During prophase I, the chromosomes (duplicated during the previous S phase) condense and the two centrosomes migrate to opposite poles of the cell. Each chromosome (four in this example) consists of two sister chromatids. Homologous chromosomes pair to form bivalents. **(b)** Bivalents become aligned at the spindle equator (metaphase I). **(c)** Homologous chromosomes separate during anaphase I, but sister chromatids remain attached at the centromere. **(d)** Telophase and cytokinesis follow. Although not illustrated here, there may then be a short interphase (interphase II). In meiosis II, **(e)** chromosomes recondense (prophase II), **(f)** chromosomes align at the spindle equator (metaphase II), and **(g)** sister chromatids at last separate (anaphase II). **(h)** After telophase II and cytokinesis, the result is four haploid daughter cells, each containing one chromosome of each homologous pair. Prophase I is a complicated process shown in more detail in Figure 20-6. Meiosis in plants is similar, except for the absence of centrioles and the mechanism of cytokinesis, which involves formation of a cell plate (described in Chapter 19).

typical mitotic division, where such pairing is not observed (**Figure 20-8**, p. 638). Because each bivalent contains four chromatids (two sister chromatids from each chromosome), four kinetochores are also present. The kinetochores of sister chromatids lie side by side—in many species appearing as a single mass—and face the same pole of the cell. Such an arrangement allows the kinetochores derived from the sister chromatids of one homologous chromosome to attach to microtubules emanating from one spindle pole and the kinetochores derived from the sister chromatids of the other homologous chromosome to attach to microtubules emanating from the opposite spindle pole. This orientation sets the stage for separation of the homologous chromosomes during anaphase. The bivalents are randomly oriented at this point, in the sense that, for each bivalent, either the maternal or paternal homologue may face a given pole of the cell. As a result, each spindle pole (and hence each daughter cell) will receive a random mixture of maternal and paternal chromosomes when the two members of each chromosome pair move toward opposite spindle poles during anaphase.

At this stage, homologous chromosomes are held together solely by chiasmata. If for some reason prophase I had occurred without crossing over, and hence without chiasma formation, the chromosomes might not pair properly at the spindle equator, and homologous chromosomes might not separate properly during anaphase I. This is exactly what happens during meiosis in mutant yeast cells that exhibit deficiencies in genetic recombination.

Anaphase I: Homologous Chromosomes Move to Opposite Spindle Poles. At the beginning of anaphase I, the members of each pair of homologous chromosomes separate from each other and start migrating toward opposite spindle poles, pulled by their respective kinetochore microtubules. Again, note the fundamental difference between meiosis and mitosis (Figure 20-8). During mitotic anaphase, *sister chromatids separate* and move to opposite spindle poles, whereas in anaphase I of meiosis, *homologous chromosomes separate* and move to opposite spindle poles while sister chromatids remain together. Because the two members of each pair of homologous chromosomes move to opposite spindle poles during anaphase I, each pole receives a haploid set of chromosomes.

How can we explain the fact that sister chromatids separate from each other during mitosis but not during meiosis I? In Chapter 19, we saw that mitotic anaphase is associated with activation of the enzyme separase by the anaphase-promoting complex (see Figure 19-38); the activated separase in turn cleaves the cohesins that hold sister chromatids together, thereby allowing sister chromatids to separate. To prevent chromatid separation from taking

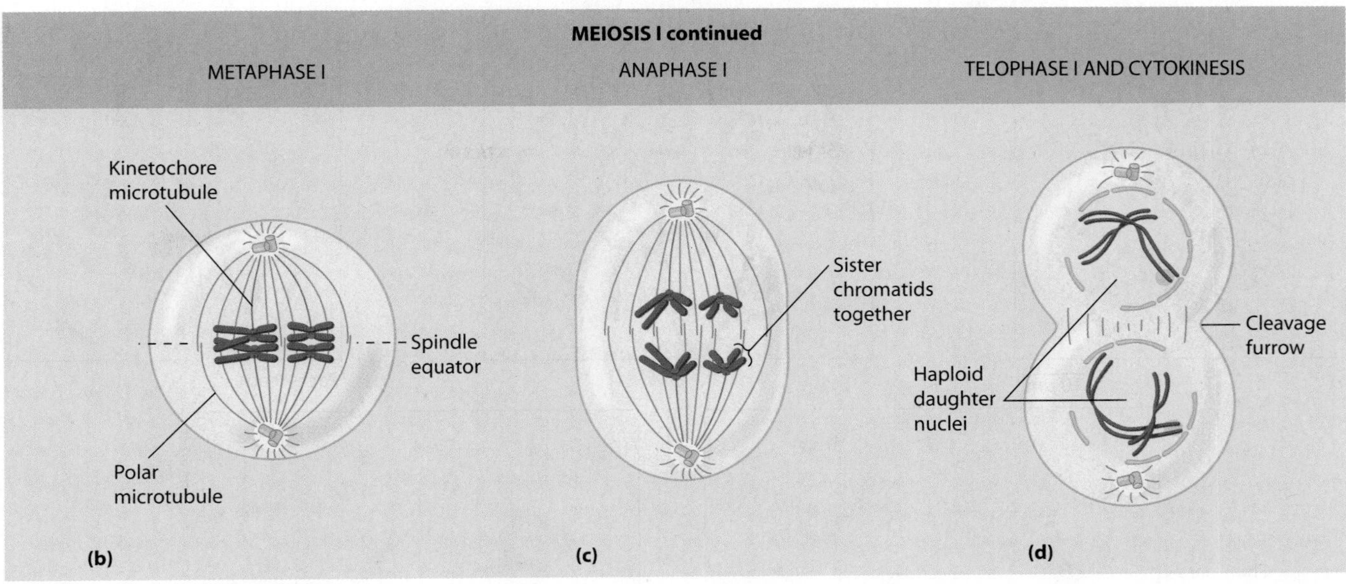

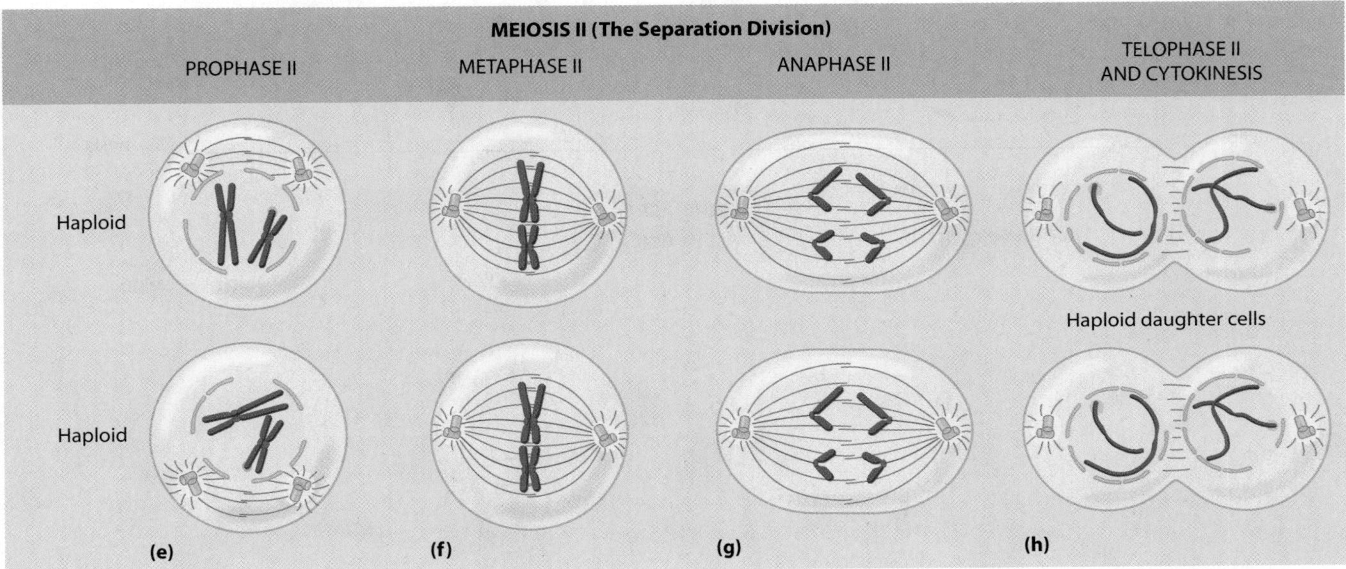

place during anaphase I of meiosis, a protein called *shugoshin* (Japanese for "guardian spirit") protects the cohesins located at the chromosomal centromere from being degraded by separase.

Telophase I and Cytokinesis: Two Haploid Cells Are Produced. The onset of telophase I is marked by the arrival of a haploid set of chromosomes at each spindle pole. Which member of a homologous pair ends up at which pole is determined entirely by how the chromosomes happened to be oriented at the spindle equator during metaphase I. After the chromosomes have arrived at the spindle poles, nuclear envelopes sometimes form around the chromosomes before cytokinesis ensues, generating two haploid cells whose chromosomes consist of sister chromatids. In most cases, the chromosomes do not decondense before meiosis II begins.

Meiosis II Resembles a Mitotic Division

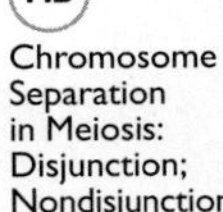

Chromosome Separation in Meiosis: Disjunction; Nondisjunction; Activity: Mistakes in Meiosis

After meiosis I has been completed, a brief interphase may intervene before **meiosis II** begins. However, this interphase is not accompanied by DNA replication because each chromosome already consists of a pair of replicated, sister chromatids that were generated by DNA synthesis during the interphase preceding meiosis I. *So DNA is replicated only once in conjunction with meiosis, and that is prior to the first meiotic division.* The purpose of meiosis II, like that of a typical mitotic division, is to parcel the sister chromatids created by this initial round of DNA replication into two newly forming cells. As a result, meiosis II is sometimes referred to as the *separation division* of meiosis.

Prophase II is very brief. If detectable at all, it is much like a mitotic prophase. Metaphase II also resembles the equivalent stage in mitosis, except that only half as many

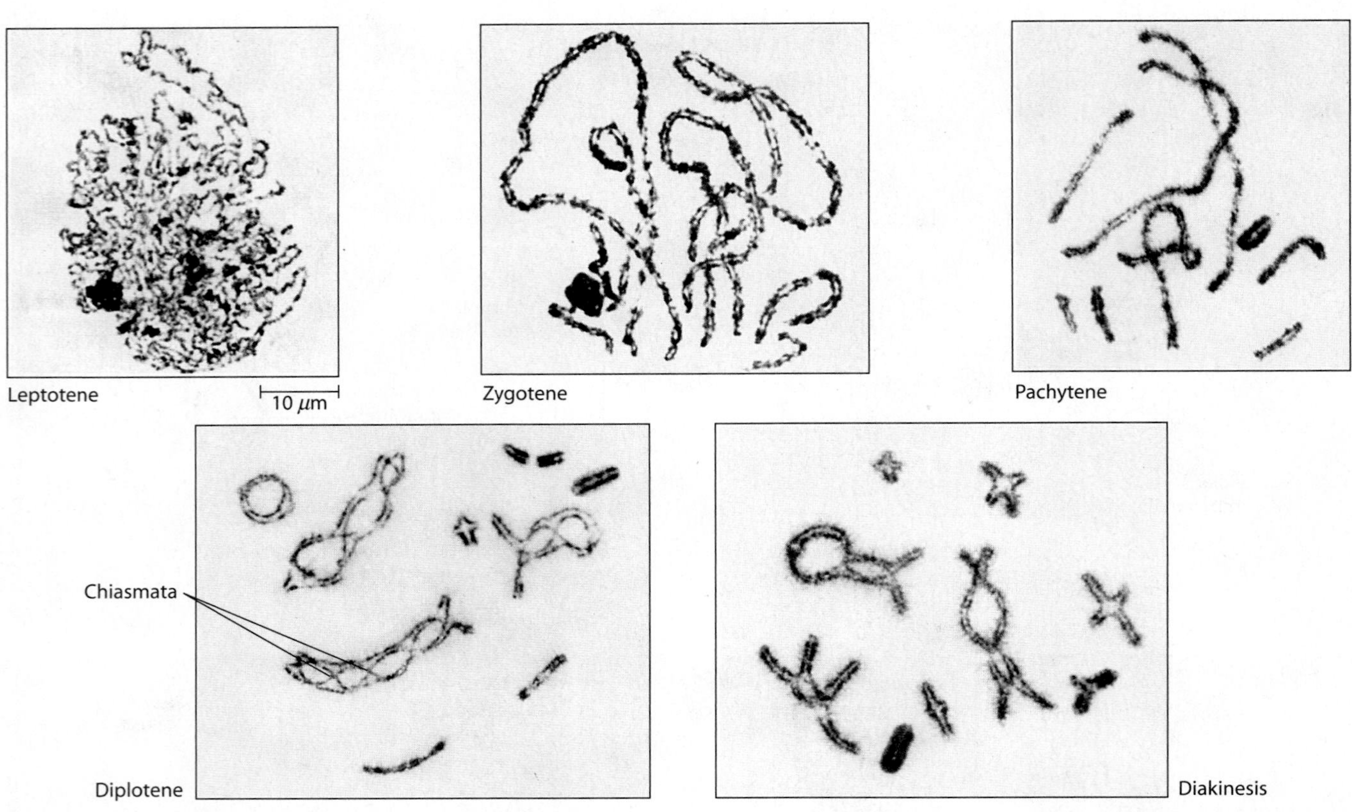

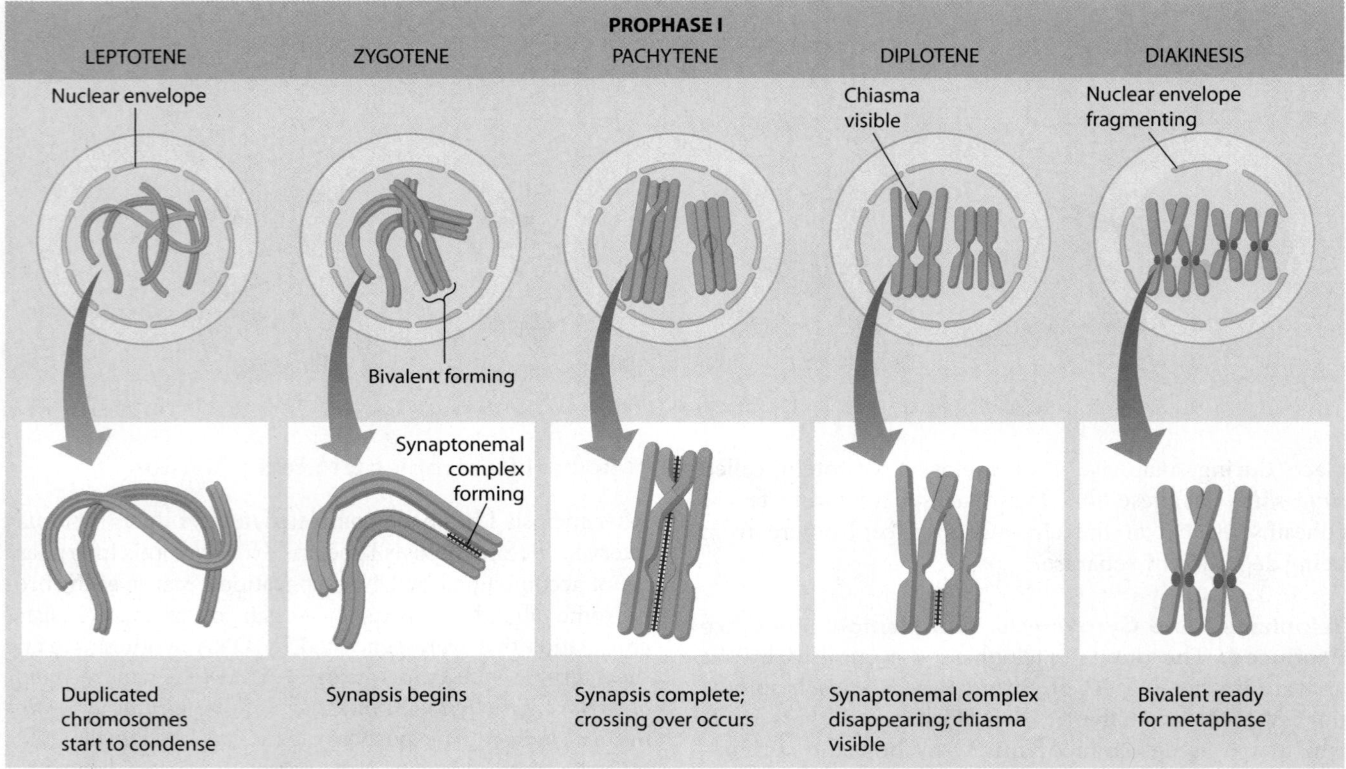

FIGURE 20-6 Meiotic Prophase I. Based on changes in chromosome behavior and appearance, prophase I is subdivided into the five stages shown in these photographs and schematic diagram. The diagram depicts the cell nucleus at each stage for a diploid cell containing four chromosomes (two homologous pairs). The lower part of the diagram focuses on a single homologous pair in greater detail, revealing the formation and subsequent disappearance of the synaptonemal complex, a protein structure that holds homologous chromosomes in close lateral apposition during pachytene. Red and blue distinguish the paternal and maternal chromosomes of each homologous pair; the synaptonemal complex is shown in shades of purple.

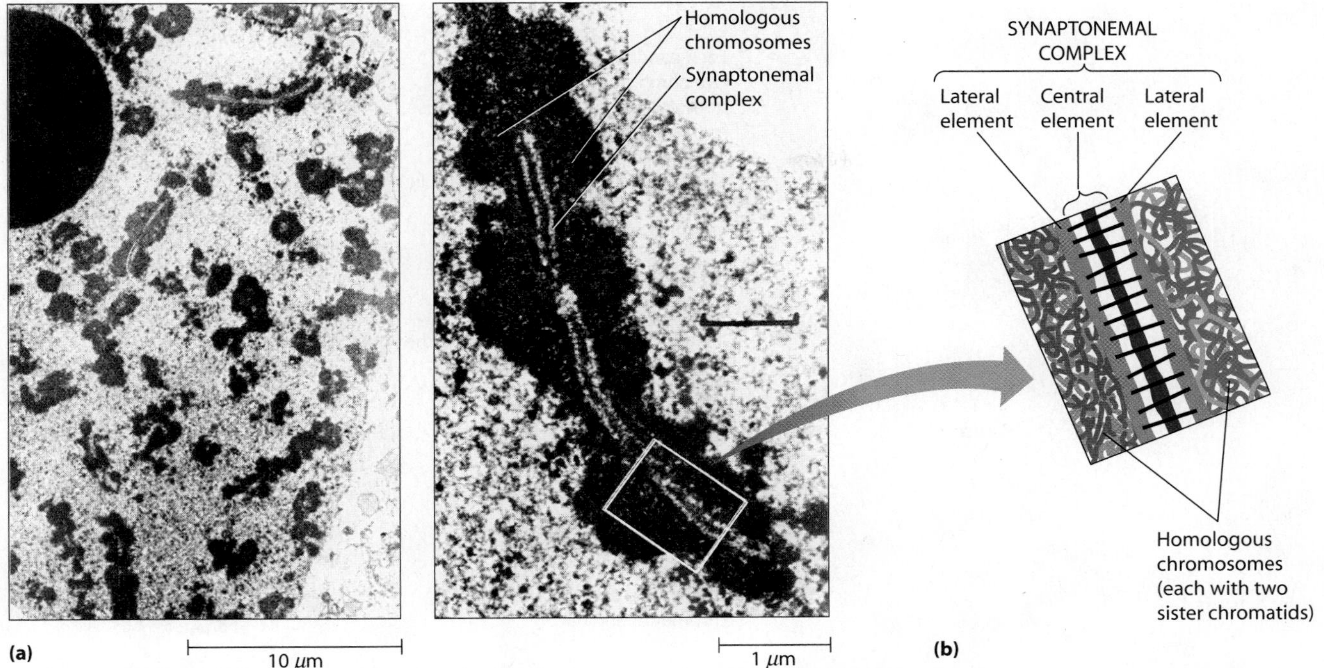

FIGURE 20-7 The Synaptonemal Complex. **(a)** These electron micrographs show, at two magnifications, synaptonemal complexes in the nuclei of cells from a lily. The cells are at the pachytene stage of prophase I. **(b)** The diagram identifies the complex's *lateral elements* (light purple), which seem to form on the chromosomes during leptotene, and its *central* (or *axial*) *element* (dark purple), which starts to appear during zygotene and "zips" the homologous chromosomes together. At pachytene, the homologues are held tightly together all along their lengths.

chromosomes are present at the spindle equator. The kinetochores of sister chromatids now face in opposite directions, allowing the sister chromatids to separate and move (as new daughter chromosomes) to opposite spindle poles during anaphase II. The remaining phases of the second meiotic division resemble the comparable stages of mitosis. The final result is the formation of four daughter cells, each containing a haploid set of chromosomes. Because the two members of each homologous chromosome pair were randomly distributed to the two cells produced by meiosis I, each of the haploid daughter cells produced by meiosis II contains a random mixture of maternal and paternal chromosomes. Moreover, each of these chromosomes is composed of a mixture of maternal and paternal DNA sequences created by crossing over during prophase I.

While each of the cells produced by meiosis normally contains a complete, haploid set of chromosomes, a rare malfunction called **nondisjunction** can produce cells that either lack a particular chromosome or contain an extra chromosome, a condition we termed *aneuploidy* in Chapter 19. Nondisjunction refers to the failure of homologous chromosomes (during anaphase I) or sister chromatids (during anaphase II) to separate from each other at the metaphase-anaphase transition. Instead, both chromosomes or chromatids remain together and move into one of the two daughter cells, thereby generating one cell containing both copies of the chromosome and one cell containing neither copy. The resulting gametes have an incorrect number of chromosomes and tend to produce defective embryos that die before birth. However, a few such gametes can participate in the formation of embryos that do survive. For example, if an abnormal human sperm containing two copies of chromosome 21 fertilizes a normal egg containing one copy of chromosome 21, the resulting embryo, possessing three copies of chromosome 21, can develop fully and lead to the birth of a live child. But this child will exhibit a series of developmental abnormalities—including short stature, broad hands, folds over the eyes, and low intelligence—that together constitute *Down syndrome*.

At this point, you may want to study Figure 20-8 in its entirety to review the similarities and differences between meiosis and mitosis. In this diagram, the amount of DNA present at various stages is indicated using the **C value,** which corresponds to the amount of DNA present in a single (haploid) set of chromosomes. In a diploid cell prior to S phase, the chromosome number is $2n$ and the DNA content is 2C because two sets of chromosomes are present. When DNA undergoes replication during S phase, the DNA content is doubled to 4C because each chromosome now consists of two chromatids. In meiosis I, segregation of homologous chromosomes into different daughter cells reduces the chromosome number from $2n$ to $1n$ and the DNA content from 4C to 2C. Sister chromatid separation during meiosis II then reduces the DNA

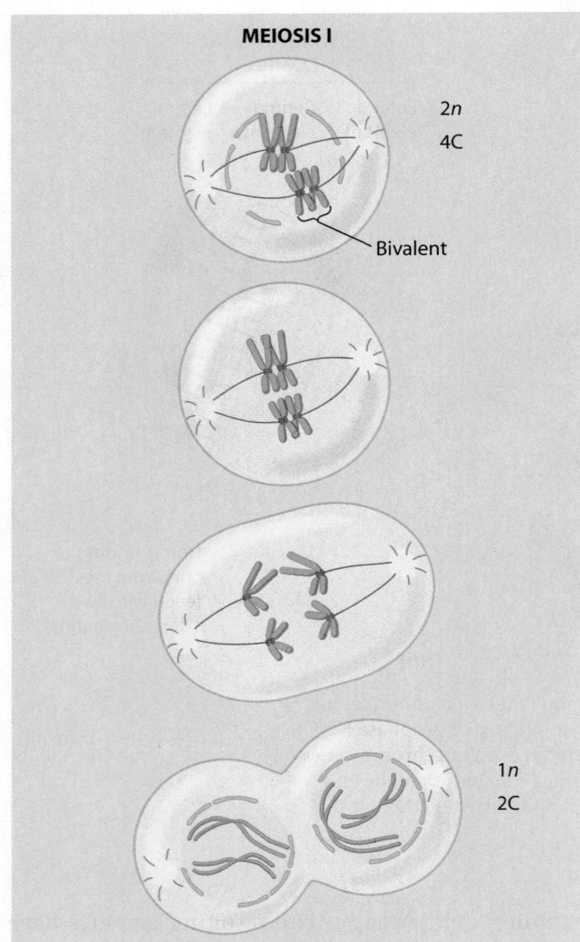

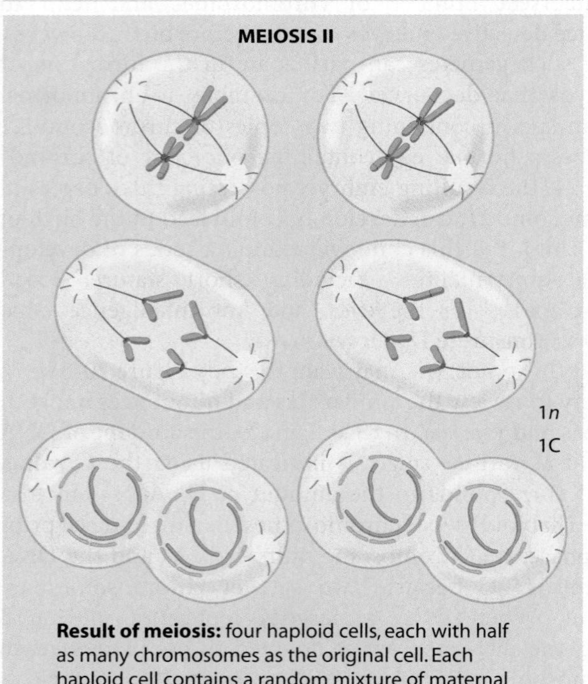

MEIOSIS II
1n
1C
Result of meiosis: four haploid cells, each with half as many chromosomes as the original cell. Each haploid cell contains a random mixture of maternal and paternal chromosomes.

Prophase
Each condensing chromosome has two chromatids. In meiosis I, homologous chromosomes synapse, forming a bivalent. Crossing over occurs between nonsister chromatids, producing chiasmata. In mitosis, each chromosome acts independently.

Metaphase
In meiosis I, the bivalents align at the metaphase plate. In mitosis, individual chromosomes align at the metaphase plate.

Anaphase
In meiosis I, chromosomes (not chromatids) separate. In mitosis, chromatids separate.

Telophase and Cytokinesis

In meiosis II, sister chromatids separate.

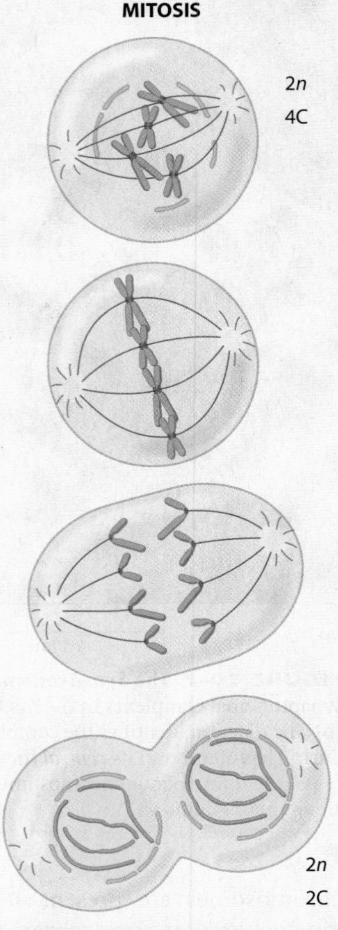

FIGURE 20-8 Meiosis and Mitosis Compared. Meiosis and mitosis are both preceded by DNA replication, resulting in two sister chromatids per chromosome at prophase. Meiosis, which occurs only in sex cells, includes two nuclear (and cell) divisions, halving the chromosome number to the haploid level. Moreover, during the elaborate prophase of the first meiotic division, homologous chromosomes synapse, and crossing over occurs between nonsister chromatids. Mitosis involves only a single division, producing two diploid nuclei (and usually, cells), each with the same number of chromosomes as the original cell. In mitosis, the homologous chromosomes behave independently; at no point do they come together in pairs. The meaning of the C values in this figure (4C, 2C, 1C) is described on page 609.

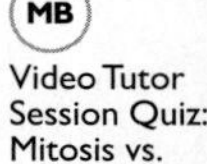

Video Tutor Session Quiz: Mitosis vs. Meiosis

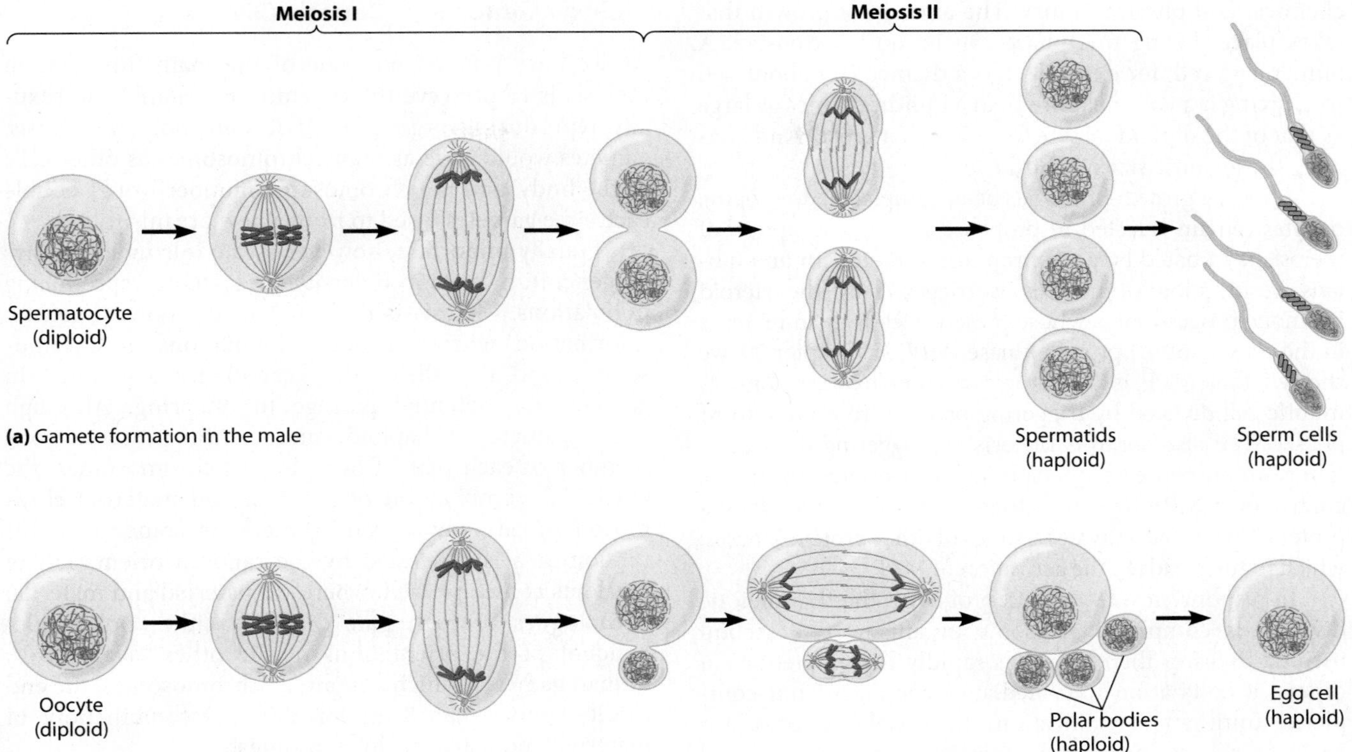

FIGURE 20-9 Gamete Formation. **(a)** In the male, all four haploid products of meiosis are retained and differentiate into sperm. **(b)** In the female, both meiotic divisions are asymmetric, forming one large egg cell and three (in some cases, only two) small cells called polar bodies that do not give rise to functional gametes. Although not indicated here, the mature egg cell usually grows much larger than the oocyte it arose from.

content from 2C to 1C, while the chromosome number remains at $1n$. In contrast, a normal mitosis reduces the DNA content from 4C to 2C (by sister chromatid separation) while the chromosome number remains at $2n$.

Sperm and Egg Cells Are Generated by Meiosis Accompanied by Cell Differentiation

Meiosis lies at the heart of gametogenesis, which as we saw earlier is the process of forming haploid gametes from diploid precursor cells. But male and female gametes differ significantly in structure, which means that gametogenesis must consist of more than just meiosis. **Figure 20-9** is a schematic depiction of gametogenesis in animals. In males, meiosis converts a diploid *spermatocyte* into four haploid *spermatids* of similar size (Figure 20-9a). After meiosis has been completed, the spermatids then differentiate into sperm cells by discarding most of their cytoplasm and developing flagella and other specialized structures.

In females, meiosis converts a diploid *oocyte* into four haploid cells, but only one of the four survives and gives rise to a functional egg cell (Figure 20-9b). This outcome is generated by two meiotic divisions that divide the cytoplasm of the oocyte unequally, with one of the four daughter cells receiving the bulk of the cytoplasm of the original diploid oocyte. The other three, smaller cells, called **polar bodies,** usually degenerate (although recall that the Komodo dragon can use a polar body to "fertilize" an egg). The advantage of having only one of the four haploid products of meiosis develop into a functional egg cell is that the cytoplasm that would otherwise have been distributed among four cells is instead concentrated into one egg cell, maximizing the content of stored nutrients in each egg.

An important difference between sperm and egg formation concerns the stage at which the cells acquire the specialized characteristics that make them functionally mature gametes. During sperm cell development, meiosis creates haploid spermatids that must then discard most of their cytoplasm and develop flagella before they are functionally mature. In contrast, developing egg cells acquire their specialized features *during* the process of meiosis. Many of the specialized features of the egg cell are acquired during prophase I, when meiosis is temporarily halted to allow time for extensive cell growth. During this *growth phase,* the cell also develops various types of external coatings designed to protect the egg from chemical and physical injury. The amount of growth that takes place during this phase can be quite extensive. A human egg cell, for example, has a diameter of about 100 μm, giving it a volume more than a hundred times as large as that of the diploid oocyte from which it arose. And consider the gigantic size of a bird egg!

After the growth phase has been completed, developing oocytes remain arrested in prophase I until resumption of meiosis is initiated by an appropriate stimulus. In amphibians, resumption of meiosis is triggered by the steroid hormone progesterone, whose presence leads to an increase in the activity of the protein kinase, *MPF.* In Chapter 19, we showed that MPF is a Cdk-cyclin complex that controls mitotic cell division by triggering passage from G2 into M phase. MPF also controls meiosis by triggering the transition from prophase I to metaphase I. Progesterone exerts its control over MPF by stimulating the production of *Mos,* a protein kinase that activates a series of other protein kinases, which in turn leads to the activation of MPF.

In response to the activation of MPF, the first meiotic division is completed. In some organisms, the second meiotic division then proceeds rapidly to completion; in others, it halts at an intermediate stage and is not completed until after fertilization. In vertebrate eggs, for example, the second meiotic division is generally arrested at metaphase II until fertilization takes place. Metaphase II arrest is triggered by *cytostatic factor (CSF),* a biochemical activity present in the cytoplasm of mature eggs. CSF works by inhibiting the anaphase-promoting complex, whose activity is normally required for the transition from metaphase to anaphase (page 585). After an egg is fertilized, CSF undergoes inactivation. In the absence of CSF activity, the anaphase-promoting complex is free to trigger the transition from metaphase to anaphase, thereby allowing meiosis to be completed.

By the time meiosis is completed, the egg cell is fully mature and may even have been fertilized. The mature egg is a highly differentiated cell that is specialized for the task of producing a new organism in much the same sense that a muscle cell is specialized to contract or a red blood cell is specialized to transport oxygen. This inherent specialization of the egg is vividly demonstrated by the observation that even in the absence of fertilization by a sperm cell, many kinds of animal eggs can be stimulated to develop into a complete embryo by artificial treatments as simple as a pinprick. Hence, everything needed for programming the early stages of development must already be present in the egg.

Normally, of course, this developmental sequence is activated by interactions between sperm and egg, which trigger important biochemical changes in the egg. For example, in addition to the rapid formation of a physical barrier to the entry of additional sperm by the fertilized egg (page 405), fertilization simultaneously initiates a burst of metabolic activity in preparation for embryonic development. Although not described here, fertilization in plants involves similarly complex cellular and biochemical events.

Meiosis Generates Genetic Diversity

As we have pointed out, one of the main functions of meiosis is to preserve the chromosome number in sexually reproducing organisms. If it were not for meiosis, gametes would have as many chromosomes as other cells in the body, and the chromosome number would double each time gametes fused to form a new organism.

Equally important, however, is the role meiosis plays in generating genetic diversity in sexually reproducing populations. Meiosis is the point in the flow of genetic information where various combinations of chromosomes (and the alleles they carry) are assembled in gametes for potential passage to offspring. Although every gamete is haploid, and hence possesses one member of each pair of homologous chromosomes, the particular combination of paternal and maternal chromosomes in any given gamete is random. This randomness is generated by the random orientation of bivalents at metaphase I, where the paternal and maternal homologues of each pair can face either pole, independent of the orientations of the other bivalents. In human gametes, which contain 23 chromosomes, the end result is more than 8 million different combinations of maternal and paternal chromosomes!

Moreover, the crossing over that takes place between homologous chromosomes during prophase I generates additional genetic diversity among the gametes. By allowing the exchange of DNA segments between homologous chromosomes, crossing over generates more combinations of alleles than the random assortment of maternal and paternal chromosomes would create by itself. We will return to crossing over (recombination) later in the chapter. First, however, we turn to the historic experiments that revealed the genetic consequences of chromosome segregation and random assortment during meiosis. These experiments were carried out by Gregor Mendel before chromosomes were even known to exist.

Genetic Variability: Segregation and Assortment of Alleles

Most students of biology have heard of Gregor Mendel and the classic genetics experiments he conducted in a monastery garden. Mendel's findings, first published in 1865, laid the foundation for what we now know as *Mendelian genetics.* Working with the common garden pea, Mendel chose seven readily identifiable characters of pea plants and selected in each case two varieties of plants that displayed different forms of the character. For example, seed color was one character Mendel chose because he had one strain of peas with yellow seeds and another with green seeds (see Figure 20-1). He first established that each of the plant strains was **true-breeding** upon self-fertilization, which means that plants grown from his yellow seeds produced only yellow seeds and

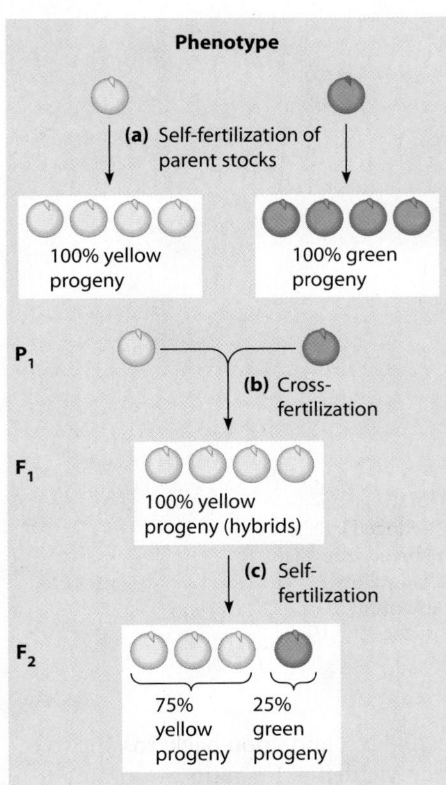

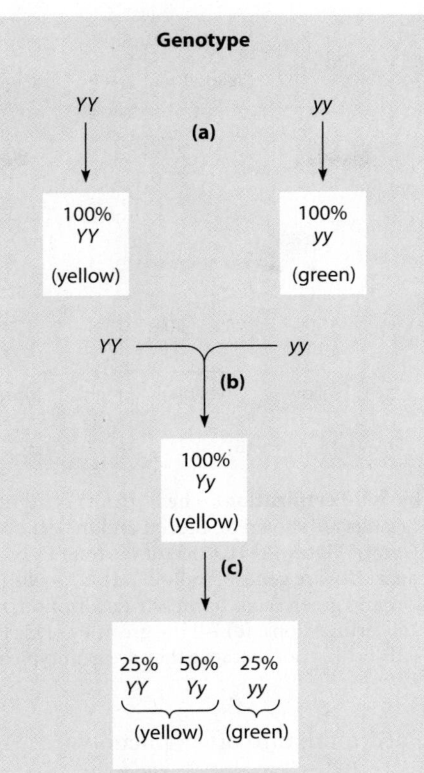

FIGURE 20-10 Genetic Analysis of Seed Color in Pea Plants. Genetic crosses were performed starting with two true-breeding strains of pea plants, one having yellow seeds and one having green seeds. The yellow-seed trait is dominant (allele *Y*); the green-seed trait is recessive (allele *y*). The resulting phenotypes of the progeny are shown on the left and their genotypes on the right. (The genotypes were deduced later.) **(a)** The parent stocks are homozygous for either the dominant *(YY)* or recessive *(yy)* trait and breed true upon self-fertilization. **(b)** When crossed, the parent stocks yield F_1 plants (hybrids) that are all heterozygous *(Yy)* and therefore show the dominant trait. **(c)** Upon self-fertilization, the F_1 plants produce yellow and green seeds in the F_2 generation in a ratio of 3:1. See Figures 20-11 and 20-12 for further analyses.

plants grown from his green seeds produced only green seeds (**Figure 20-10a**). Once this had been established, Mendel was ready to investigate the principles that govern the inheritance of such traits.

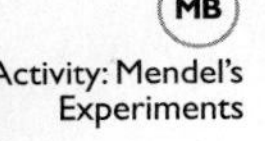

Information Specifying Recessive Traits Can Be Present Without Being Displayed

In his first set of experiments, Mendel *cross*-fertilized the true-breeding parental plants (the **P_1 generation**) to produce **hybrid** strains. The outcome of this experiment must have seemed mystifying at first: In every case, the resulting offspring—called the **F_1 generation**—exhibited one or the other of the parental traits, but never both. In other words, one parental trait was always dominant and the other was always recessive. In the case of seed color, for example, all the F_1 plants had yellow seeds (Figure 20-10b), indicating that yellow seed color is dominant.

During the next summer, Mendel allowed all the F_1 hybrids to self-fertilize. For each of the seven characters under study, he made the same surprising observation: The recessive trait that had seemingly disappeared in the F_1 generation reappeared among the progeny in the next generation, the **F_2 generation.** Moreover, for each of the seven characters, the ratio of dominant to recessive phenotypes in the progeny was always about 3:1. In the case of seed color, for example, plants grown from the yellow F_1 seeds produced about 75% yellow seeds and 25% green (Figure 20-10c).

This outcome was quite different from the behavior of the true-breeding yellow seeds of the parent strain, which had produced only yellow seeds when self-fertilized. Clearly, there was an important difference between the yellow seeds of the P_1 stock and the yellow seeds of the F_1 generation. They looked alike, but the former bred true whereas the latter did not.

Next, Mendel investigated the F_2 plants through self-fertilization (**Figure 20-11**). The F_2 plants exhibiting the recessive trait (green, in the case of seed color) always bred true (Figure 20-11c), suggesting they were genetically identical to the green-seeded P_1 strain that Mendel had begun with. F_2 plants with the dominant trait yielded a more complex pattern. One-third bred true for the dominant trait (Figure 20-11a) and therefore seemed identical to the P_1 plants with the dominant (yellow-seed) trait. The other two-thirds of the yellow-seeded F_2 plants produced progeny with both dominant and recessive phenotypes, in a 3:1 ratio (Figure 20-11b)—the same ratio that had arisen from the F_1 self-fertilization.

These results led Mendel to conclude that genetic information specifying the recessive trait must be present in the F_1 hybrid plants and seeds, even though the trait is not displayed. This conclusion was consistent with results from another set of experiments, in which F_1 hybrids were crossed with the original parent strains, a technique called **backcrossing** (**Figure 20-12**). Backcrossing F_1 hybrids to the dominant parent strain always produced progeny exhibiting the dominant trait (Figure 20-12a), whereas

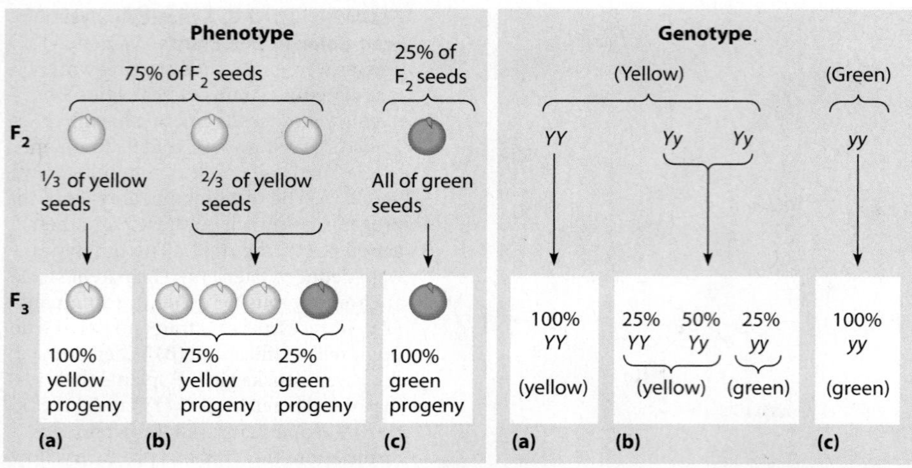

FIGURE 20-11 Analysis of F_2 Pea Plants by Self-Fertilization. The F_2 plants of Figure 20-10 were analyzed through self-fertilization. The phenotypic results are shown on the left and the genotypes (deduced later) on the right. **(a)** One-third of the yellow F_2 progeny of Figure 20-10 (25% of the total F_2 progeny) breed true for yellow seed color upon self-fertilization because they are genotypically *YY*. **(b)** Two-thirds of the yellow F_2 progeny (50% of the total F_2 progeny) yield yellow and green seeds upon self-fertilization, in a ratio of 3:1 (just as the F_2 plants of Figure 20-10 did upon self-fertilization). **(c)** All the green F_2 seeds (25% of the total F_2 progeny) breed true for green seed color upon self-fertilization because they are genotypically *yy*.

backcrossing to the recessive parent yielded a mixture of plants exhibiting dominant and recessive traits in a ratio of 1:1 (Figure 20-12b). Moreover, the dominant progeny from the latter cross behaved just like the F_1 hybrids: Upon self-fertilization, they gave rise to a 3:1 mixture of phenotypes (Figure 20-12c), and upon backcrossing to the recessive parent, they yielded a 1:1 ratio of dominant to recessive progeny (Figure 20-12d). (An alternative way of diagramming crosses, called the *Punnett square*, is shown in Problem 20-7 at the end of the chapter.)

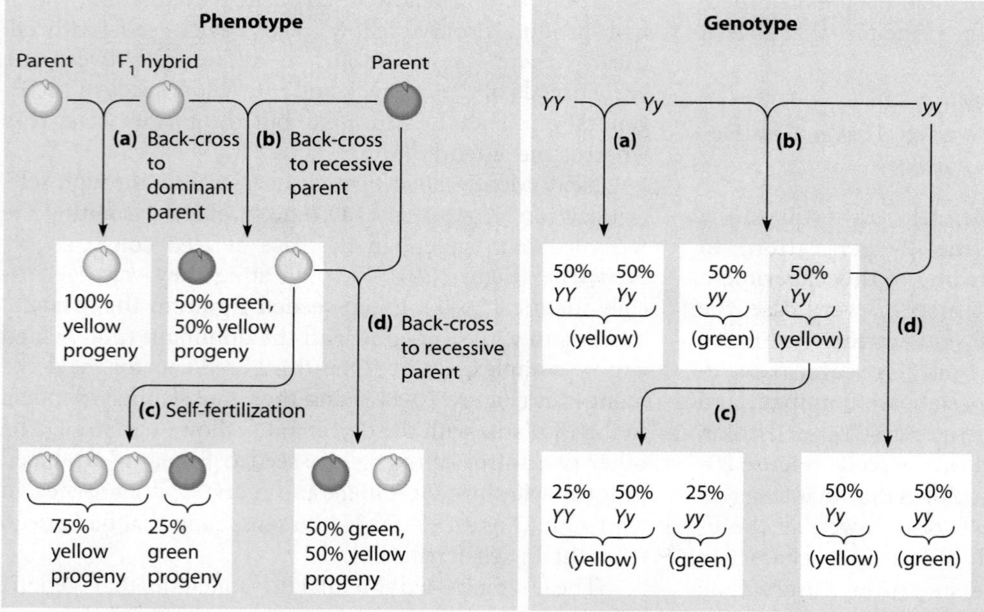

FIGURE 20-12 Analysis of F_1 Hybrids by Backcrossing. The F_1 hybrids of Figure 20-10 were analyzed by backcrossing to the parent (P_1) strains. The phenotypic results are shown on the left and the genotypes (deduced later) on the right. **(a)** Upon backcrossing of the F_1 hybrid *(Yy)* to the dominant parent *(YY)*, all the progeny have yellow seeds because the genotype is either *YY* or *Yy*. **(b)** Backcrossing to the recessive parent *(yy)* yields yellow *(Yy)* or green *(yy)* seeds in equal proportions. **(c)** These yellow-seeded progeny will give rise, upon self-fertilization, to a 3:1 mixture of yellow and green seeds (just as with the F_1 hybrids in Figure 20-10). **(d)** Backcrossing of the yellow-seeded progeny to the homozygous recessive parent again yields a 1:1 mixture of yellow and green seeds, as in the backcross of part b.

The Law of Segregation States That the Alleles of Each Gene Separate from Each Other During Gamete Formation

After a decade of careful work documenting the preceding patterns of inheritance, Mendel formulated several principles—now known as **Mendel's laws of inheritance**—that explained the results he had observed. The first of these principles was that phenotypic traits are determined by discrete "factors" that are present in most organisms as pairs of "determinants." Today, we call these "factors" *genes* and "determinants" *alleles* (alternative forms of genes). Mendel's conclusion seems almost self-evident to us, but it was an important assertion in his day. At that time, most scientists favored a *blending theory of inheritance,* a theory that viewed traits such as yellow and green seed color rather like cans of paint that are poured together to yield intermediate results. Other investigators had described the nonblending nature of inheritance before Mendel but without the accompanying data and mathematical analysis that were Mendel's special contribution. Mendel's breakthrough is especially impressive when we recall that he formulated his theory before anyone had seen chromosomes.

MB Mendel's Law of Segregation; Mendel's Principle of Segregation

Of special importance to the development of genetics was Mendel's conclusion regarding the way genes are parceled out during gamete formation. According to his **law of segregation,** *the two alleles of a gene are distinct entities that segregate, or separate from each other, during the formation of gametes.* In other words, the two alleles retain their identities even when both are present in a hybrid organism, and they are then parceled out into separate gametes so they can emerge unchanged in later generations.

The Law of Independent Assortment States That the Alleles of Each Gene Separate Independently of the Alleles of Other Genes

MB Mendel's Law of Independent Assortment; Mendel's Principle of Independent Assortment

In addition to the crosses already described, each focusing on a single pair of alleles, Mendel studied *multifactor* crosses between plants that differed in *several* characters. Besides differing in seed color, for example, the plants he crossed might differ in seed shape and flower position. As in his single-factor crosses, he used parent plants that were true-breeding (homozygous) for the characters he was testing and generated F_1 hybrids heterozygous for each character. He then self-fertilized these hybrids and determined how frequently the dominant and recessive forms of the various characters appeared among the progeny.

Mendel found that all possible combinations of traits appeared in the F_2 progeny, and he concluded that all possible combinations of the different alleles must therefore have been present among the F_1 gametes. Furthermore, based on the proportions of the various phenotypes detected in the F_2 generation, Mendel deduced that all possible combinations of alleles occurred in the gametes with equal frequency. In other words, *the two alleles of each gene segregate independently of the alleles of other genes.* This is the **law of independent assortment,** another cornerstone of genetics. Later, it would be shown that this law applies only to genes on different chromosomes or very far apart on the same chromosome.

Early Microscopic Evidence Suggested That Chromosomes Might Carry Genetic Information

Mendel's findings lay dormant in the scientific literature until 1900, when his paper was rediscovered almost simultaneously by three other European biologists. In the meantime, much had been learned about the cellular basis of inheritance. By 1875, for example, microscopists had identified chromosomes with the help of stains produced by the developing aniline dye industry. At about the same time, fertilization was shown to involve the fusion of sperm and egg nuclei, suggesting that the nucleus carries genetic information.

The first proposal that chromosomes might be the bearers of this genetic information was made in 1883. Within ten years, chromosomes had been studied in dividing cells and had been seen to split longitudinally into two apparently identical daughter chromosomes. This led to the realization that the number of chromosomes per cell remains constant during the development of an organism. With the invention of better optical systems, more detailed analysis of chromosomes became possible. Mitotic cell division was shown to involve the movement of identical daughter chromosomes to opposite poles, thereby ensuring that daughter cells would have exactly the same complement of chromosomes as their parent cell.

Against this backdrop came the rediscovery of Mendel's paper, followed almost immediately by three crucial studies that established chromosomes as the carriers of Mendel's factors. The investigators were Edward Montgomery, Theodor Boveri, and Walter Sutton. Montgomery's contribution was to recognize the existence of homologous chromosomes. From careful observations of insect chromosomes, he concluded that the chromosomes of most cells could be grouped into pairs, with one member of the pair of maternal origin and the other of paternal origin. He also noted that the two chromosomes of each type come together during synapsis in the "reduction division" (now called meiosis I) of gamete formation, a process that had been reported a decade or so earlier.

Boveri then added the crucial observation that each chromosome plays a unique genetic role. This idea came from studies in which sea urchin eggs were fertilized in the presence of a large excess of sperm, causing some eggs to be fertilized by two sperm. The presence of two sperm nuclei in a single egg leads to the formation of an abnormal mitotic spindle that does not distribute the chromosomes equally to the newly forming embryonic cells. The resulting embryos exhibit various types of developmental defects, depending on which particular chromosomes they are missing. Boveri therefore concluded that each chromosome plays a unique role in development.

Sutton, meanwhile, was studying meiosis in the grasshopper. In 1902, he made the important observation that the orientation of each pair of homologous chromosomes (bivalent) at the spindle equator during metaphase I is purely a matter of chance. Any homologous pair, in other words, may lie with the maternal or paternal chromosome toward either pole, regardless of the positions of other pairs. Many different combinations of maternal and paternal chromosomes are therefore possible in the gametes produced by any given individual (note the similarity to Mendel's observation that all possible combinations of different alleles are present among the F_1 generation of gametes).

Chromosome Behavior Explains the Laws of Segregation and Independent Assortment

MB Determining Genotype: Pea Pod Color; Activity: Dihybrid Cross; Activity: The Principle of Independent Assortment; Make Connections: Chromosomal Inheritance and Independent Assortment of Alleles;

During 1902–1903, Sutton put the preceding observations together into a coherent theory describing the role of chromosomes in inheritance. This *chromosomal theory of inheritance* can be summarized as five main points:

1. Nuclei of all cells except those of the *germ line* (sperm and eggs) contain two sets of homologous chromosomes, one set of maternal origin and the other of paternal origin.
2. Chromosomes retain their individuality and are genetically continuous throughout an organism's life cycle.
3. The two sets of homologous chromosomes in a diploid cell are functionally equivalent, each carrying a similar set of genes.
4. Maternal and paternal homologues synapse during meiosis and then move to opposite poles of the division spindle, thereby becoming segregated into different cells.
5. The maternal and paternal members of different homologous pairs segregate independently during meiosis.

The chromosomal theory of inheritance provided a physical basis for understanding how Mendel's genetic factors could be carried, transmitted, and segregated. For example, the presence of two sets of homologous chromosomes in each cell parallels Mendel's suggestion of two determinants for each phenotypic trait. Likewise, the segregation of homologous chromosomes during the meiotic divisions of gamete formation provides an explanation for Mendel's law of segregation, and the random orientation of homologous pairs at metaphase I accounts for his law of independent assortment.

Figure 20-13 and **20-14** illustrate the chromosomal basis of Mendel's laws, using examples from his pea plants. The basis for the law of segregation is shown in Figure 20-13 for the alleles governing seed color in a heterozygous pea plant with genotype *Yy*. (Peas have seven pairs of chromosomes, but only the pair bearing the alleles for seed color is shown.) During meiosis the two homologous chromosomes, each with two sister chromatids, synapse during prophase I, align together at the spindle equator at metaphase I, and then segregate into separate daughter cells. The second meiotic division then separates sister chromatids so that each haploid cell ends up with only one allele for seed color, either *Y* or *y*.

The basis for the law of independent assortment is illustrated in Figure 20-14 for the chromosomes carrying the genes for seed color (alleles *Y* and *y*) and seed shape

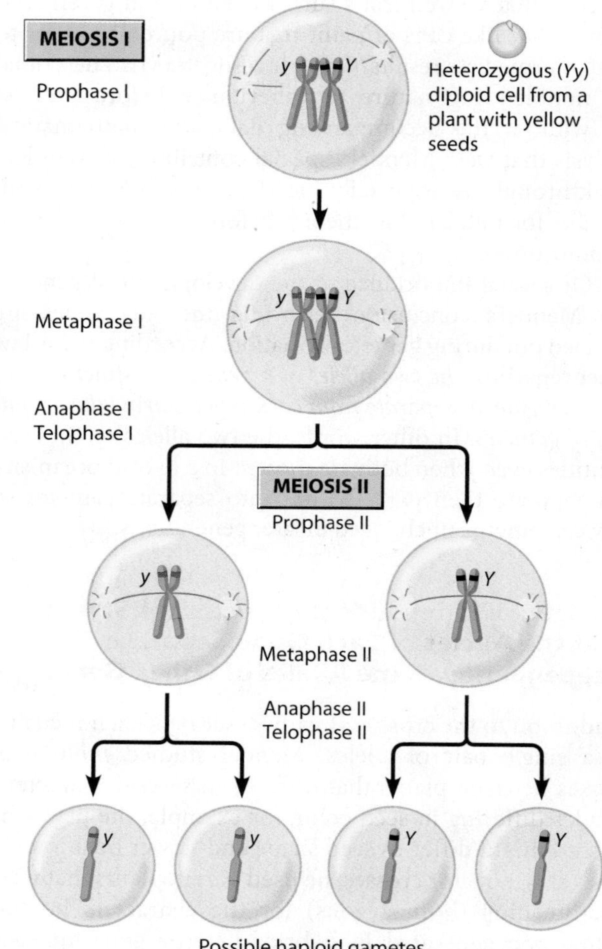

FIGURE 20-13 The Meiotic Basis for Mendel's Law of Segregation. Segregation of seed-color alleles during meiosis is illustrated for the case of a pea plant heterozygous for this character. Peas have seven pairs of chromosomes, but only the homologous pair bearing the seed-color alleles is shown here. During the first meiotic division (meiosis I), homologous chromosomes (each consisting of two sister chromatids) pair during prophase I, allowing crossing over to take place. The homologous chromosomes then align as a pair at the spindle equator during metaphase I and segregate into separate cells at anaphase I and telophase I. In meiosis II, sister chromatids segregate to different daughter cells. The result is four haploid daughter cells, each having one allele for seed color.

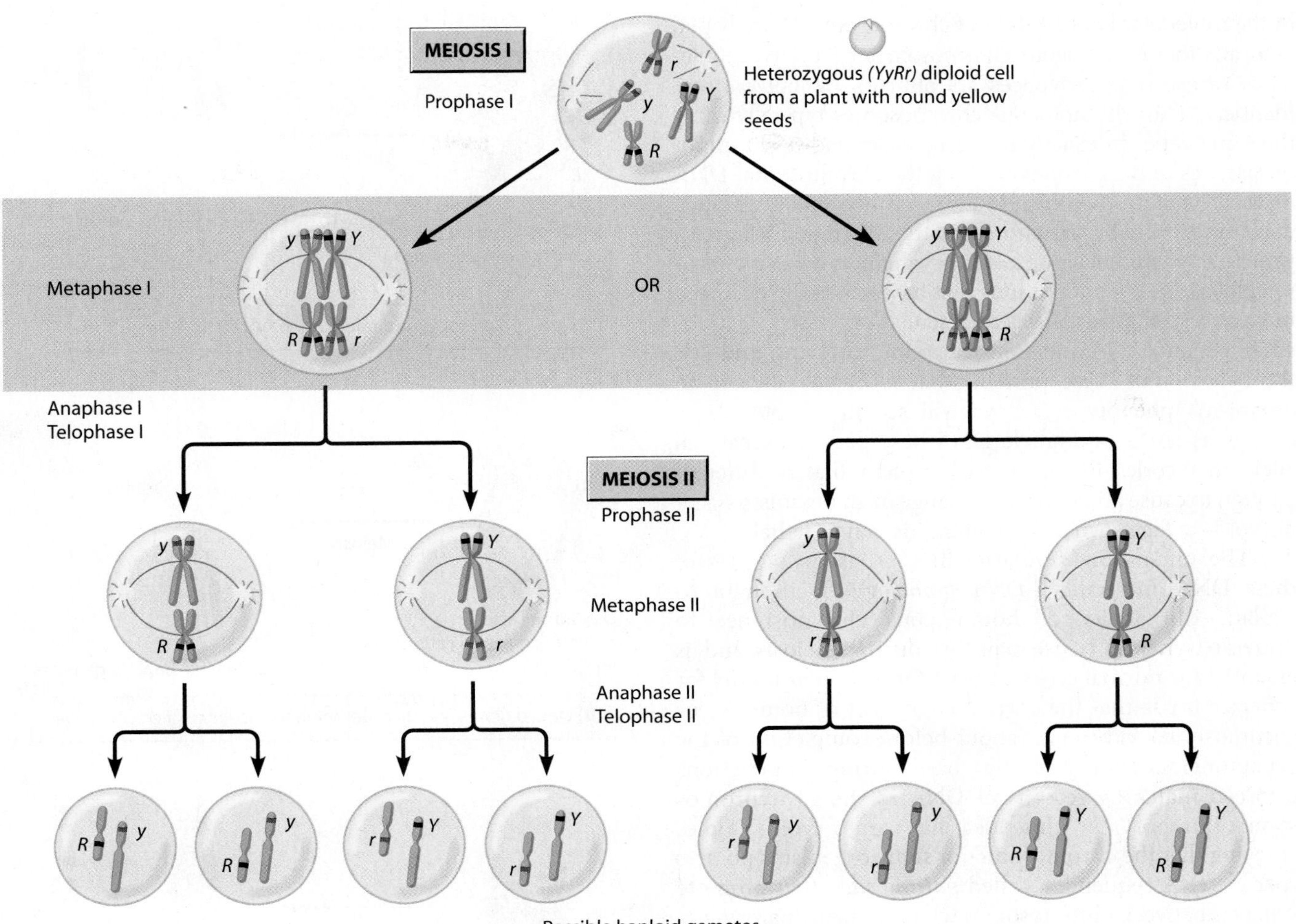

FIGURE 20-14 The Meiotic Basis for Mendel's Law of Independent Assortment. Independent assortment of the alleles of two genes on different chromosomes is illustrated by meiosis in a pea plant heterozygous for seed color *(Yy)* and seed shape *(Rr)*. Pea plants have seven pairs of chromosomes, but only the two pairs bearing the seed-color alleles and the seed-shape alleles are shown. During meiosis, segregation of the seed-color alleles occurs independently of segregation of the seed-shape alleles. The basis of this independent assortment is found at metaphase I, when the homologous pairs (bivalents) align at the spindle equator. Because each bivalent can face in either direction, there are two alternative situations: Either the paternal homologues (blue) can both face the same pole of the cell, with the maternal homologues (red) facing the other pole, or they can face different poles. The arrangement at metaphase I determines which homologues subsequently go to which daughter cells. Since the alternative arrangements occur with equal probability, all the possible combinations of the alleles therefore occur with equal probability in the gametes. For simplicity, chiasmata and crossing over are not shown in this diagram.

(alleles *R* and *r*, where *R* stands for round seeds, a dominant trait, and *r* for wrinkled seeds, a recessive trait). The explanation for independent assortment is that there are two possible, equally likely arrangements of the chromosome pairs at metaphase I. Of the *YyRr* cells that undergo meiosis, half will produce gametes like the four at the bottom left of the figure, and half will produce gametes like the four at the bottom right. Therefore, the *YyRr* plant will produce equal numbers of gametes of the eight types.

As we will discuss shortly, the law of independent assortment holds only for genes on *different* chromosomes (or for genes located far apart on the same chromosome). It is remarkable that Mendel happened to choose seven independently assorting characters in an organism that has only seven pairs of chromosomes. Mendel was also fortunate that each of the seven characters he chose to study turned out to be controlled by a single gene (pair of alleles). Perhaps he concentrated on pea strains that, in preliminary experiments, gave him the most consistent and comprehensible results.

The DNA Molecules of Homologous Chromosomes Have Similar Base Sequences

One of the five main points in the chromosomal theory of inheritance is the idea that homologous chromosomes are functionally equivalent, each carrying a similar set of genes. What does this mean in terms of our current understanding of the molecular organization of chromosomes? Simply put, it means that homologous chromosomes have DNA molecules whose base sequences are almost, but not entirely, identical. Thus, homologous chromosomes typically carry the same genes in exactly the same order. However, minor differences in base sequence along the chromosomal DNA molecule can create different alleles of the same gene. Such differences arise by mutation, and the different alleles for a gene that we find in a population—whether of pea plants or people—arise from mutations that have gradually occurred in an ancestral gene. Alleles are usually "expressed" by transcription into RNA and translation into proteins, and it is the behavior of these proteins that ultimately creates an organism's phenotype. As we will see in Chapter 21, a change as small as a single DNA base pair can create an allele that codes for an altered protein that is different enough to cause an observable change in an organism's phenotype—in fact, such a DNA alteration can be lethal.

The underlying similarity in the base sequences of their DNA molecules—*DNA homology*—is thought to explain the ability of homologous chromosomes to undergo synapsis (close pairing) during meiosis and is essential for normal crossing over. One popular model for synapsis holds that the correct alignment of homologous chromosomes is brought about before completion of the synaptonemal complex by base-pairing interactions between matching regions of DNA in the two chromosomes. Support for this idea has come from studies showing that the chromosomes of some organisms possess special DNA sequences, called *pairing sites,* that promote synapsis between chromosomes when the same pairing site is present in two chromosomes. Protein components of the synaptonemal complex also play a role in facilitating the pairing between matching regions in the DNA of homologous chromosomes. After the synaptonemal complex is fully formed, DNA recombination is completed at these (and perhaps other) sites.

Genetic Variability: Recombination and Crossing Over

Segregation and independent assortment of homologous chromosomes during the first meiotic division lead to random assortment of the alleles carried by different chromosomes, as was shown in Figure 20-14 for seed shape and color in pea plants. **Figure 20-15a** summarizes the outcome of meiosis for a generalized version of the same situation. Here we have a diploid organism heterozygous for two genes on nonhomologous chromosomes, with the allele pairs called *Aa* and *Bb.* Meiosis in such an organism will produce gametes in which allele *A* is just as likely to occur with allele *B* as it is with allele *b,* and *B* is just as likely to occur with *A* as it is with *a.* But what happens if two genes, *D* and *E,* reside on the same chromosome? In that case, alleles *D* and *E* will routinely be linked together in the same gamete, as will alleles *d* and *e* (Figure 20-15b).

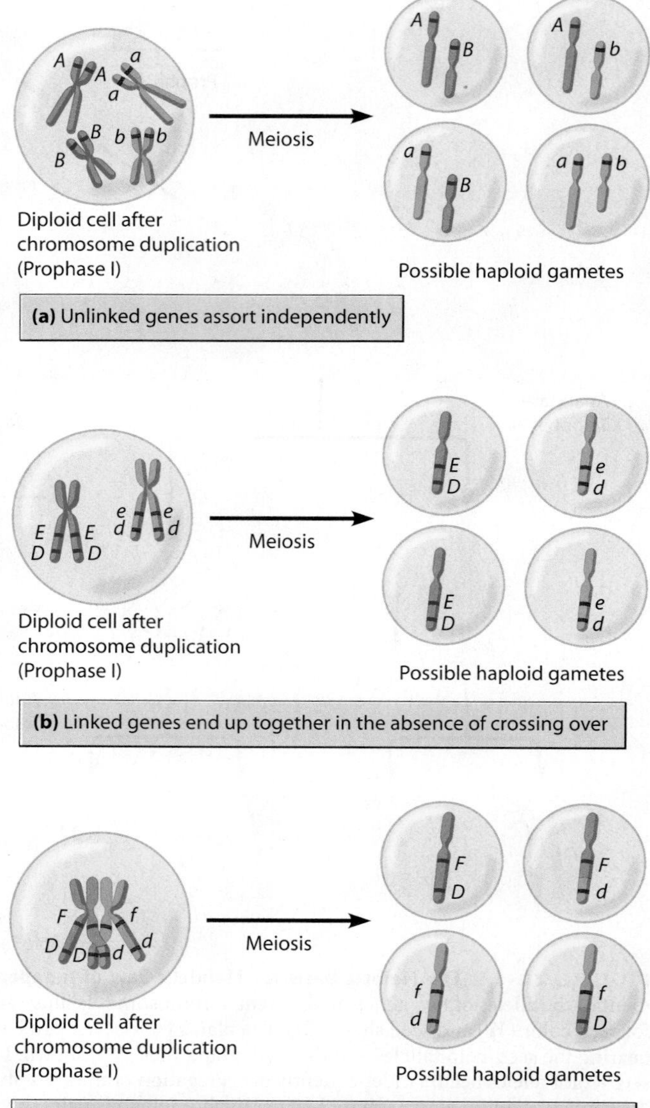

Meiosis
Diploid cell after chromosome duplication (Prophase I)
Possible haploid gametes
(a) Unlinked genes assort independently
(b) Linked genes end up together in the absence of crossing over
(c) Linked genes do not end up together when crossing over occurs

FIGURE 20-15 Segregation and Assortment of Linked and Unlinked Genes. In this figure, *A* and *a* are alleles of one gene, *B* and *b* are alleles of another gene, and so forth. **(a)** Alleles of genes on different chromosomes segregate and assort independently during meiosis; allele *A* is as likely to occur in a gamete with allele *B* as it is with allele *b.* **(b)** Alleles of genes on the same chromosome remain linked during meiosis in the absence of crossing over; in this case, allele *D* will occur routinely in the same gamete with allele *E* but not with allele *e.* **(c)** Alleles of genes on the same chromosome can become interchanged when crossing over takes place, so that allele *D* occurs not only with allele *F* but also with allele *f* and so forth.

But even for genes on the same chromosome, some scrambling of alleles can take place because of the phenomenon of crossing over, which leads to genetic recombination. Genetic recombination involves the exchange of genetic material between homologous chromosomes during prophase I of meiosis when the homologues are synapsed, creating chromosomes exhibiting new combinations of alleles (Figure 20-15c). Recombination was originally discovered in studies with the fruit fly *Drosophila melanogaster,* conducted by Thomas Hunt Morgan and his colleagues beginning around 1910. We will therefore turn to Morgan's work to investigate recombination, beginning with his discovery of linkage groups.

Chromosomes Contain Groups of Linked Genes That Are Usually Inherited Together

The fruit flies used by Morgan and his colleagues had certain advantages over Mendel's pea plants as objects of genetic study, not the least of which was the fly's relatively brief generation time (about two weeks versus several months for pea plants). Unlike Mendel's peas, however, the fruit flies did not come with a ready-made variety of phenotypes and genotypes. Whereas Mendel was able to purchase seed stocks of different true-breeding varieties, the only type of fruit fly initially available to Morgan was what has come to be known as the **wild type,** or "normal" organism. Morgan and his colleagues therefore had to generate variants—that is, mutants—for their genetic experiments.

Morgan and his coworkers began by breeding large numbers of flies and then selecting mutant individuals having phenotypic modifications that were heritable. Later, X-irradiation was used to enhance the mutation rate, but in their early work Morgan's group depended entirely on spontaneous mutations. Within five years, they were able to identify about 85 different mutants, each carrying a mutation in a different gene. Each of these mutants could be propagated as a laboratory stock and used for matings as needed.

One of the first discoveries made by Morgan and his colleagues as they began analyzing their mutants was that, unlike the genes Mendel had studied in peas, the mutant fruit fly genes did not all assort independently. Instead, some genes behaved as if they were linked together, and for such genes, the new combinations of alleles predicted by Mendel were infrequent or even nonexistent. In fact, it was soon recognized that fruit fly genes can be classified into four **linkage groups,** each group consisting of a collection of **linked genes** that are usually inherited together. Morgan quickly realized that the number of linkage groups was the same as the number of different chromosomes in the organism (the haploid chromosome number for *Drosophila* is four). The conclusion he drew was profound: Each chromosome is the physical basis for a specific linkage group. Mendel had not observed linkage, because the genes he studied either resided on different chromosomes or were far apart on the same chromosome and therefore behaved as if they were not linked to each other.

Homologous Chromosomes Exchange Segments During Crossing Over

Although the genes they discovered could all be organized into linkage groups, Morgan and his colleagues found that linkage within such groups was incomplete. Most of the time, genes known to be linked (and therefore on the same chromosome) assorted together, as would be expected. Sometimes, however, two or more such traits would appear in the offspring in nonparental combinations. This phenomenon of less-than-complete linkage was called *recombination* because the different alleles appeared in new and unexpected associations in the offspring.

To explain such recombinant offspring, Morgan proposed that homologous chromosomes can exchange segments, presumably by some sort of breakage-and-fusion event, as illustrated in **Figure 20-16a.** By this process, which Morgan termed *crossing over,* a particular

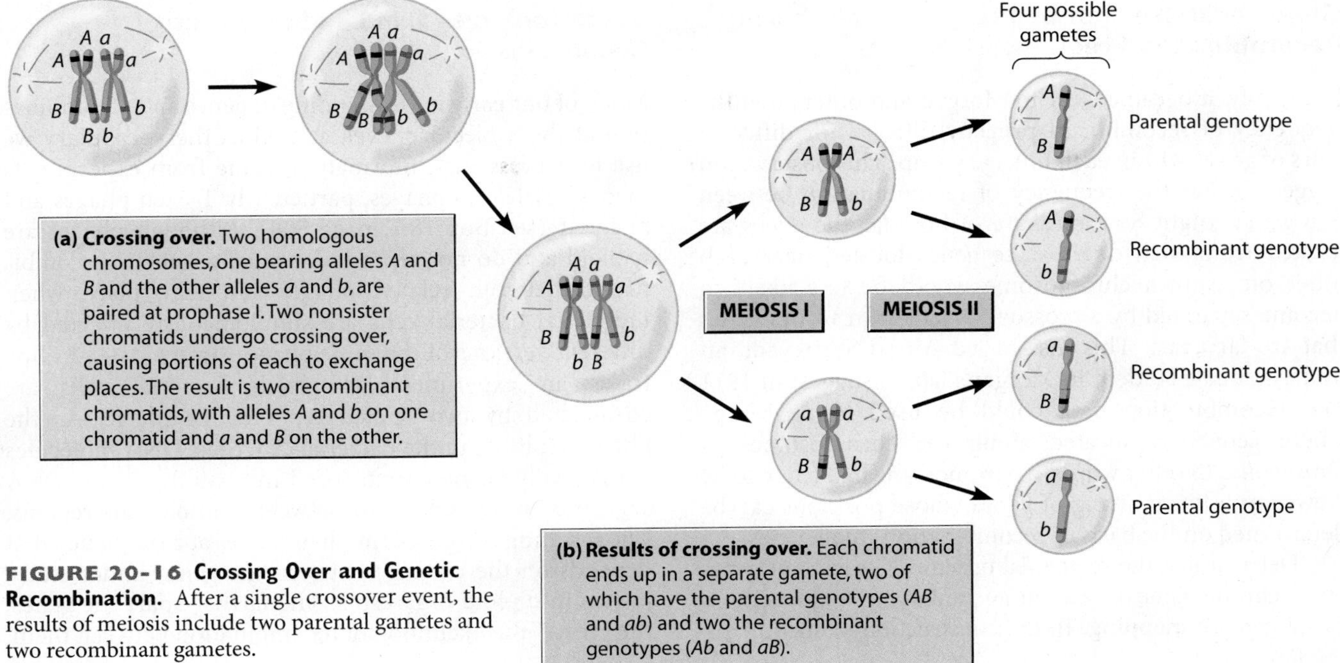

A a
A a
B b
B b
A a
A a
B B b b
(a) Crossing over. Two homologous chromosomes, one bearing alleles A and B and the other alleles a and b, are paired at prophase I. Two nonsister chromatids undergo crossing over, causing portions of each to exchange places. The result is two recombinant chromatids, with alleles A and b on one chromatid and a and B on the other.
A a
A a
B b
b B
MEIOSIS I
MEIOSIS II
A A
B b
a a
B b
Four possible gametes
A B
Parental genotype
A b
Recombinant genotype
a B
Recombinant genotype
a b
Parental genotype
(b) Results of crossing over. Each chromatid ends up in a separate gamete, two of which have the parental genotypes (AB and ab) and two the recombinant genotypes (Ab and aB).

MB Linkage and Recombination; Meiosis (3 of 3): Determinants of Heredity and Genetic Variation (BioFlix tutorial); Activity: Origins of Genetic Variation

FIGURE 20-16 Crossing Over and Genetic Recombination. After a single crossover event, the results of meiosis include two parental gametes and two recombinant gametes.

allele or group of alleles initially present on one member of a homologous pair of chromosomes could be transferred to the other chromosome in a reciprocal manner.

In the example of Figure 20-16a, two homologous chromosomes, one with alleles *A* and *B* and the other with alleles *a* and *b,* lie side by side at synapsis. Portions of an *AB* chromatid and a nonsister *ab* chromatid then exchange DNA segments, thereby producing two recombinant chromatids—one with alleles *A* and *b* and the other with alleles *a* and *B.* Each of the four chromatids ends up in a different gamete at the end of the second meiotic division, so the products of meiosis will include two *parental* gametes and two *recombinant* gametes, assuming a single crossover event (Figure 20-16b).

We now know that crossing over takes place during the pachytene stage of meiotic prophase I (see Figure 20-6), when sister chromatids are packed tightly together and it is difficult to observe what is happening. As the chromatids begin to separate at diplotene, each of the four chromatids in a bivalent can be identified as belonging to one or the other of the two homologues. Wherever crossing over has taken place between nonsister chromatids, the two homologues remain attached to each other, forming a chiasma.

At the first meiotic metaphase, homologous chromosomes are almost always held together by at least one chiasma; if not, they may not segregate properly. Many bivalents contain multiple chiasmata. Human bivalents, for example, typically have two or three chiasmata because multiple crossover events routinely occur between paired homologues. To be genetically significant, crossing over must involve nonsister chromatids. In some species, exchanges between sister chromatids are also observed, but such exchanges have no genetic consequences because sister chromatids are genetically identical.

Gene Locations Can Be Mapped by Measuring Recombination Frequencies

MB Recombination and Linkage Mapping; Three-Point Mapping; Linked Genes and Linkage Mapping

Eventually, it became clear to Morgan and others that the frequency of recombinant progeny differed for different pairs of genes within each linkage group. This observation suggested that the frequency of recombination between two genes might be a measure of how far the genes are located from each other since genes located near each other on a given chromosome would be less likely to become separated by a crossover event than would genes that are far apart. This insight led Alfred Sturtevant, an undergraduate student in Morgan's lab, to suggest in 1911 that recombination data could be used to determine where genes are located along the chromosomes of *Drosophila.* In other words, a chromosome had come to be viewed as a linear string of genes whose positions can be determined on the basis of recombination frequencies.

Determining the sequential order and spacing of genes on a chromosome based on recombinant frequencies is called **genetic mapping.** In the construction of such maps, the recombinant frequency is the map distance expressed in *map units (centimorgans).* If, for example, the alleles in Figure 20-16b appear among the progeny in their parental combination (*AB* and *ab*) 85% of the time and in the recombinant combinations (*Ab* and *aB*) 15% of the time, we conclude that the two genes are linked (are on the same chromosome) and are 15 map units apart. This approach has been used to map the chromosomes of many species of plants and animals, as well as bacteria and viruses. However, because bacteria and viruses do not reproduce sexually, the methods used to generate recombinants are somewhat different.

Genetic Recombination in Bacteria and Viruses

Because it requires crossing over between homologous chromosomes, you might expect genetic recombination to be restricted to sexually reproducing organisms. Sexual reproduction provides an opportunity for crossing over once every generation when homologous chromosomes become closely juxtaposed during the meiotic divisions that produce gametes. In contrast, bacteria and viruses have haploid genomes and reproduce asexually, with no obvious mechanism for regularly bringing together genomes from two different parents.

Nonetheless, viruses and bacteria are still capable of genetic recombination. In fact, recombination data permitted the extensive mapping of viral and bacterial genomes well before the advent of modern DNA technology. To understand how recombination takes place despite a haploid genome, we need to examine the mechanisms that allow two haploid genomes, or portions of genomes, to be brought together within the same cell.

Co-infection of Bacterial Cells with Related Bacteriophages Can Lead to Genetic Recombination

Much of our early understanding of genes and recombination at the molecular level, as well as the vocabulary we use to express that information, came from experiments involving bacteriophages, particularly T-even phages and phage λ (see Box 18A, page 508). Although phages are haploid and do not reproduce sexually, genetic recombination between related phages can take place when individual bacterial cells are simultaneously infected by different versions of the same phage. **Figure 20-17** illustrates an experiment in which bacterial cells are co-infected by two related types of T4 phage. As the phages replicate in the bacterial cell, their DNA molecules occasionally become juxtaposed in ways that allow DNA segments to be exchanged between homologous regions. The resulting recombinant phage arises at a frequency that depends on the distance between the genes under study, just as in diploid organisms: The farther apart the genes, the greater the likelihood of recombination between them.

In phage recombination, crossing over involves relatively short, naked DNA molecules, rather than chromatids. The simplicity of this situation has facilitated research on the molecular mechanism of recombination and the proteins that catalyze the process. Phage recombination involves the precise alignment of homologous DNA molecules at the region of crossing over—a requirement that presumably holds for the recombination of eukaryotic and prokaryotic DNA as well.

Transformation and Transduction Involve Recombination with Free DNA or DNA Brought into Bacterial Cells by Bacteriophages

In bacteria, several mechanisms exist for recombining genetic information. One such mechanism has already been mentioned in Chapter 18, where we discussed the experiments with rough and smooth strains of pneumococcal bacteria that led Oswald Avery to conclude that bacterial cells can be transformed from one genetic type to another by exposing them to purified DNA. This ability of a bacterial cell to take up DNA molecules and to incorporate some of that DNA into its own genome is called **transformation** (**Figure 20-18a**). Although initially described as a laboratory technique for artificially introducing DNA into bacterial cells, transformation is now recognized as a natural mechanism by which some (though by no means all) kinds of bacteria acquire genetic information when they have access to DNA from other cells.

A second mechanism for genetic recombination in bacteria, called **transduction,** involves DNA that has been brought into a bacterial cell by a bacteriophage. Most phages contain only their own DNA, but occasionally a phage will incorporate some bacterial host cell DNA sequences into its progeny particles. Such a phage particle can then infect another bacterium, acting like a syringe carrying DNA from one bacterial cell to the next (Figure 20-18b). Phages capable of carrying host cell DNA from one cell to another are called *transducing phages.*

The transducing phage known as P1, which infects *E. coli,* has been especially useful for gene mapping. The amount of DNA that will fit into a phage particle is small compared with the size of the bacterial genome. Two bacterial genes—or more generally, genetic markers (specific DNA sequences)—must therefore be close together for both to be simultaneously carried into a bacterial cell by a single phage particle. This is the basis of *cotransductional mapping,* in which the proximity of one marker to another is determined by measuring how frequently the markers accompany each other in a transducing phage particle. The closer two markers are, the more likely they are to be cotransduced into a bacterial cell. Studies using the transducing phage P1 have revealed that markers cannot be cotransduced if they are separated in the bacterial DNA by more than about 10^5 base pairs. This finding agrees with the observation that the P1 phage has a genome of about that size.

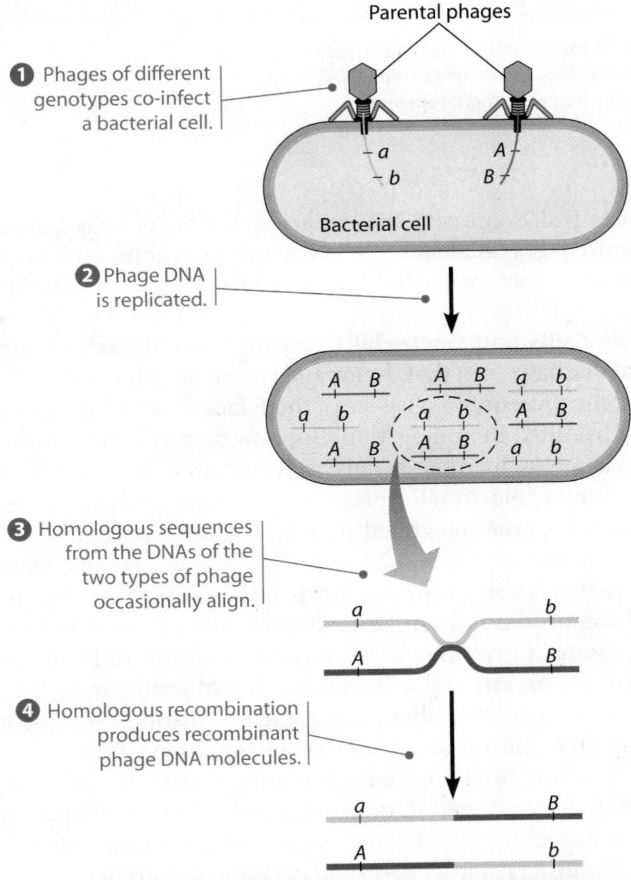
Parental phages
1 Phages of different genotypes co-infect a bacterial cell.
a
b
A
B
Bacterial cell
2 Phage DNA is replicated.
A B
a b
3 Homologous sequences from the DNAs of the two types of phage occasionally align.
a
b
A
B
4 Homologous recombination produces recombinant phage DNA molecules.
a
B
A
b

FIGURE 20-17 Genetic Recombination of Bacteriophages. Two phage populations with different genotypes co-infect a bacterial cell, thereby ensuring the simultaneous presence of both phage genomes in the same cell. One phage carries the mutant alleles *a* and *b*, while the other has alleles *A* and *B*. As the phage DNAs replicate, they occasionally become aligned and crossing over takes place, generating recombinant phage DNA molecules exhibiting *aB* and *Ab* genotypes. The frequency of occurrence of these recombinant genotypes provides a measure of the distance between genes.

Conjugation Is a Modified Sexual Activity That Facilitates Genetic Recombination in Bacteria

In addition to transformation and transduction, some bacteria also transfer DNA from one cell to another by **conjugation.** As the name suggests, conjugation resembles a mating in that one bacterium is clearly identifiable as the donor (often called a "male") and another as the recipient ("female"). Although conjugation resembles a sexual process, this mode of DNA transfer is not an inherent part of the bacterial life cycle, and it usually involves only a portion of the genome; therefore conjugation does not qualify as true sexual reproduction. The existence of conjugation was postulated in 1946 by Joshua Lederberg and Edward L. Tatum, who were the first to show that genetic recombination occurs in bacteria. They also established that physical contact between two cells is necessary for conjugation to take place. We now understand that conjugation involves

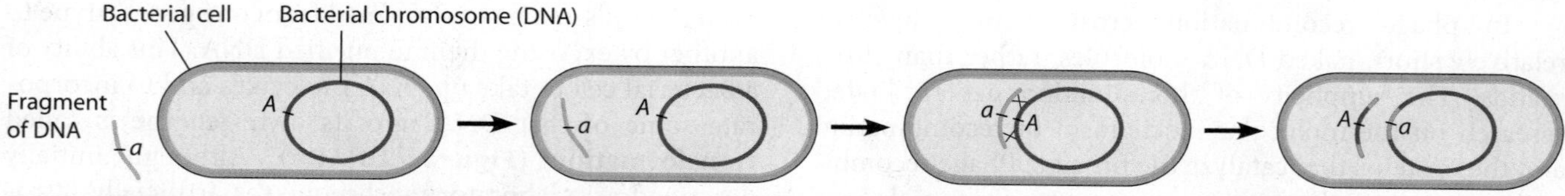

(a) Transformation. Transformation involves uptake by the bacterial cell of exogenous DNA, which occasionally becomes integrated into the bacterial genome by two crossover events (indicated by X's). The exogenous DNA will be detectable in progeny cells only if integrated into the bacterial chromosome, because the fragment of DNA initially taken up does not normally have the capacity to replicate itself autonomously in the cell. (The main exception is an intact plasmid.)

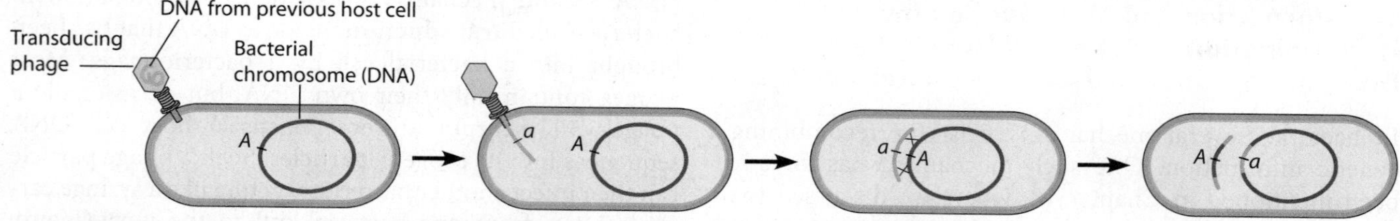

(b) Transduction. Transduction involves the introduction of exogenous DNA into a bacterial cell by a phage. Once injected into the host cell, the DNA can become integrated into the bacterial genome in the same manner as in transformation. In both cases, linear fragments of DNA that end up outside the bacterial chromosome are eventually degraded by nucleases.

FIGURE 20-18 Transformation and Transduction in Bacterial Cells. Transformation and transduction are mechanisms by which bacteria incorporate external genetic information derived either from free DNA or from DNA brought into bacterial cells by bacteriophages. In this figure, the letters *A* and *a* represent alleles of the same gene.

the directional transfer of DNA from the donor bacterium to the recipient bacterium. Let's look at some details of how the process works.

The F Factor. The presence of a DNA sequence called the **F factor** (F for fertility) enables an *E. coli* cell to act as a donor during conjugation. The F factor can take the form of either an independent, replicating plasmid (page 530) or a segment of DNA within the bacterial chromosome. Donor bacteria containing the F factor in its plasmid form are designated F^+, whereas recipient cells, which usually lack the F factor completely, are designated F^-. Donor cells develop long, hairlike projections called **sex pili** (singular: **pilus**) that emerge from the cell surface (**Figure 20-19a**). The end of each sex pilus contains molecules that selectively bind to the surface of recipient cells, thereby leading to the formation of a transient cytoplasmic **mating bridge** through which DNA is transferred from donor cell to recipient cell (Figure 20-19b).

When a donor cell contains an F factor in its plasmid form, a copy of the plasmid is quickly transferred to the recipient cell during conjugation, converting the recipient cell from F^- to F^+ (**Figure 20-20a**). Transfer always begins at a point on the plasmid called its **origin of transfer,** represented in the figure by an arrowhead. During transfer of an F factor to an F^- cell, the donor cell does not lose its F^+ status because the F factor is replicated in close association with the transfer process, allowing a copy of the F plasmid to remain behind in the donor cell. As a result, mixing an F^+ population of bacteria with F^- cells will eventually lead to a population of cells that is entirely F^+. "Maleness" is in a sense infectious, and the F factor is responsible for this behavior.

Hfr Cells and Bacterial Chromosome Transfer. Thus far, we have seen that donor and recipient cells are defined by the presence or absence of the F factor, which is in turn transmitted by conjugation. But how do recombinant bacteria arise by this means? The answer is that the F factor—while usually present as a plasmid—can sometimes become integrated into the bacterial chromosome, as shown in Figure 20-20b. (Integration results from crossing over between short DNA sequences in the chromosome and similar sequences in the F factor.) Chromosomal integration of the F factor converts an F^+ donor cell into an **Hfr cell,** which is capable of producing a *h*igh *f*requency of *r*ecombination in further matings because it can now transfer *genomic DNA* during conjugation.

When an Hfr bacterium is mated to an F^- recipient, DNA is transferred into the recipient cell (Figure 20-20c). But instead of transferring just the F factor itself, the Hfr cell transfers at least part (and occasionally all) of its chromosomal DNA, retaining a copy, as in F^+ DNA transfer. Transfer begins at the origin of transfer within the integrated F factor and proceeds in a direction dictated by the orientation of the F factor within the chromosome. Note that the chromosomal DNA is transferred in a linear form, with a small part of the F factor at the leading end and the remainder at the trailing end. Because the F factor is split in this way, only recipient cells that receive a complete

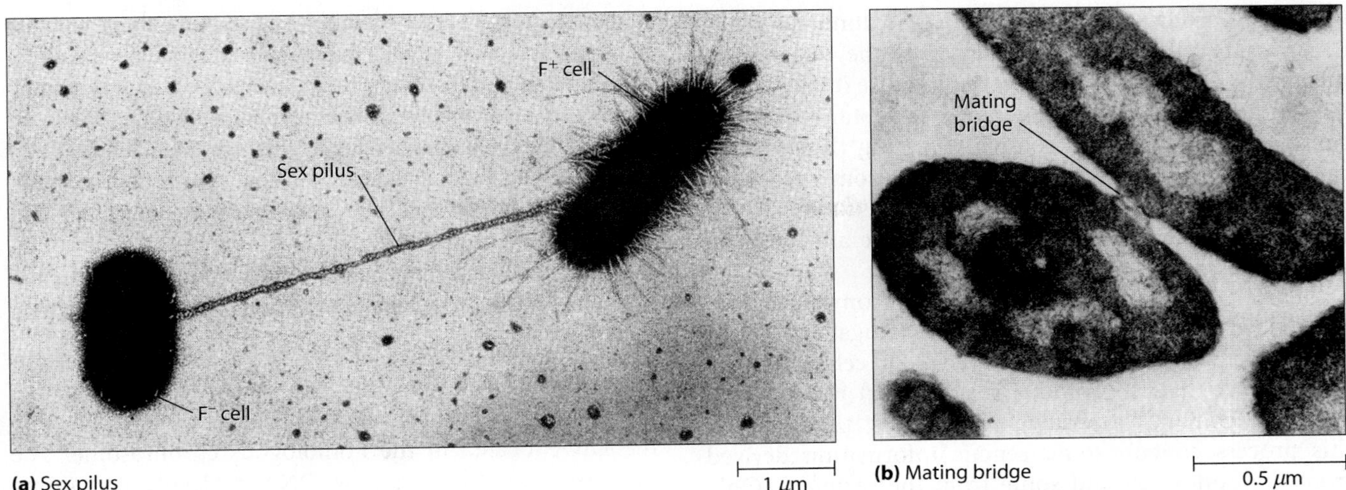

(a) Sex pilus

(b) Mating bridge

FIGURE 20-19 The Cellular Apparatus for Bacterial Conjugation. **(a)** The donor bacterial cell on the right, an F$^+$ cell, has many slender appendages, called pili, on its surface. Some of these pili are sex pili, including the very long pilus leading to the other cell, an F$^-$ cell. Made of protein encoded by a gene on the F factor, sex pili enable a donor cell to attach to a recipient cell. **(b)** Subsequently, a cytoplasmic mating bridge forms, through which DNA is passed from the donor cell to the recipient cell (TEMs).

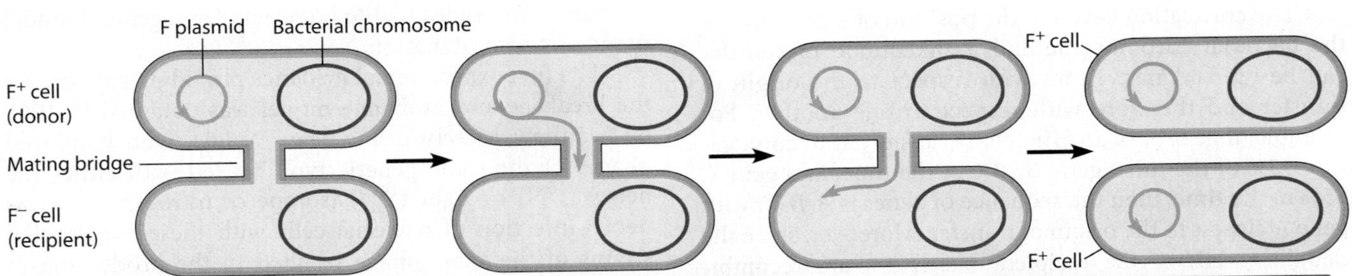

(a) Conjugation between F$^+$ and F$^-$ cells. The transfer of a copy of the F factor plasmid from an F$^+$ donor bacterium to an F$^-$ recipient converts the F$^-$ cell into an F$^+$ cell. Plasmid transfer occurs through a mating bridge and begins at the F factor's origin of transfer, indicated by the arrowhead.

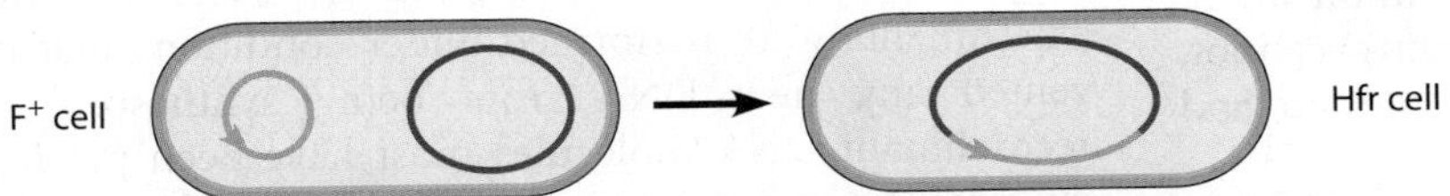

(b) Conversion of an F$^+$ cell into an Hfr cell. Integration of the F factor into the bacterial chromosome converts an F$^+$ cell into an Hfr cell.

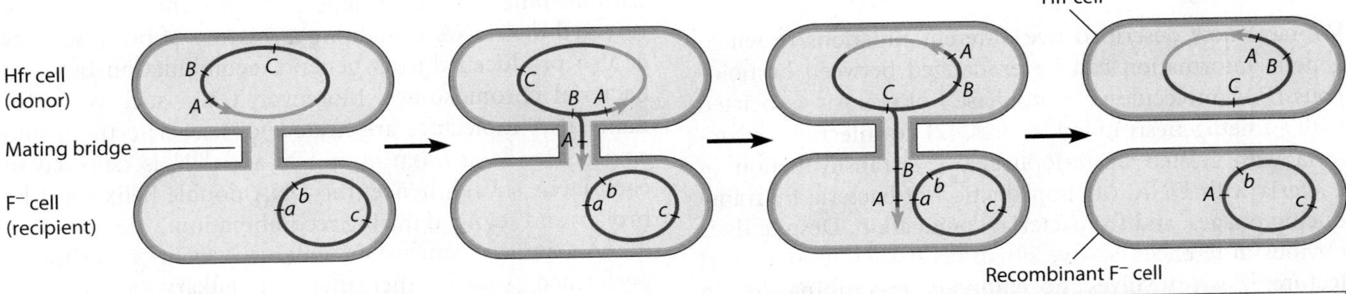

(c) Conjugation between an Hfr cell and an F$^-$ cell. Transferring a copy of the Hfr genome into an F$^-$ cell begins with the origin of transfer on the integrated F factor. Cells rarely remain in contact long enough for the entire bacterial chromosome to be transferred. Once inside the F$^-$ cell, parts of the Hfr DNA recombine with the DNA of the F$^-$ cell. Uppercase letters represent alleles carried by the Hfr; lowercase letters represent corresponding alleles in the F$^-$ cell. In the last step, allele *A* from the Hfr is recombined into the F$^-$ cell's DNA in place of its *a* allele.

FIGURE 20-20 DNA Transfer by Bacterial Conjugation. **(a)** The presence of the F factor plasmid enables an *E. coli* cell to act as a plasmid donor during conjugation. **(b)** Chromosomal integration of the F factor creates an Hfr cell that can **(c)** transfer genomic DNA during conjugation.

bacterial chromosome from the Hfr donor actually become Hfr cells themselves. Transfer of the whole chromosome is extremely rare, however, because it takes about 90 minutes. Usually, mating contact is spontaneously disrupted before transfer is complete, leaving the recipient cell with only a portion of the Hfr chromosome, as shown in Figure 20-20c. As a result, genes located close to the origin of transfer on the Hfr chromosome are the most likely to be transmitted to the recipient cell.

Once a portion of the Hfr chromosome has been introduced into a recipient cell by conjugation, it can recombine with regions of the recipient cell's chromosomal DNA that are homologous (similar) in sequence. The recombinant bacterial chromosomes generated by this process contain some genetic information derived from the donor cell and some from the recipient. Only donor DNA sequences that are successfully integrated by this recombination mechanism survive in the recipient cell and its progeny. Donor DNA that is not integrated during recombination, as well as DNA removed from the recipient chromosome during the recombination process, is eventually degraded by nucleases.

The correlation between the position of a gene within the bacterial chromosome and its likelihood of transfer can be used to map genes with respect to the origin of transfer and therefore with respect to one another. For example, if gene *A* of an Hfr cell is transferred in conjugation 95% of the time, gene *B* 70% of the time, and gene *C* 55% of the time, then the sequence of genes is *A-B-C*, with gene *A* closest to the origin of transfer. Moreover, since the daughter cells of the recipient bacterium are recombinants, they can be used for genetic analysis. Typically, a cross is made between Hfr and F^- strains that differ in two or more genetic properties. After conjugation has taken place, the cells are plated on a nutrient medium on which recombinants can grow but "parent" strains cannot, thereby allowing the recombinants to be detected and recombinant frequencies to be calculated.

Molecular Mechanism of Homologous Recombination

We have now described five different situations in which genetic information can be exchanged between homologous DNA molecules: (1) prophase I of meiosis associated with gametogenesis in eukaryotes, (2) co-infection of bacteria with related bacteriophages, (3) transformation of bacteria with DNA, (4) transduction of bacteria by transducing phages, and (5) bacterial conjugation. Despite their obvious differences, all five situations share a fundamental feature: Each involves **homologous recombination** in which genetic information is exchanged between DNA molecules exhibiting extensive sequence similarity. We are now ready to discuss the molecular mechanisms underlying this type of recombination. Since the principles involved appear to be quite similar in prokaryotes and eukaryotes, we will use examples from both types of organisms.

DNA Breakage and Exchange Underlies Homologous Recombination

Shortly after it was first discovered that genetic information is exchanged between chromosomes during meiosis, two theories were proposed to explain how this might occur. The *breakage-and-exchange model* postulated that breaks occur in the DNA molecules of two adjoining chromosomes, followed by exchange and rejoining of the broken segments. In contrast, the *copy-choice model* proposed that genetic recombination occurs while DNA is being replicated. According to the latter view, DNA replication begins by copying a DNA molecule located in one chromosome and then switches at some point to copying the DNA located in the homologous chromosome. The net result would be a new DNA molecule containing information derived from both chromosomes. One of the more obvious predictions made by the copy-choice model is that DNA replication and genetic recombination should happen at the same time. When subsequent studies revealed that DNA replication takes place during S phase while recombination typically occurs during prophase I, the copy-choice idea had to be rejected as a general model of meiotic recombination.

The first experimental evidence providing support for the breakage-and-exchange model was obtained in 1961 by Matthew Meselson and Jean Weigle, who employed phages of the same genetic type labeled with either the heavy (^{15}N) or light (^{14}N) isotope of nitrogen. Simultaneous infection of bacterial cells with these two labeled strains of the same phage resulted in the production of recombinant phage particles containing genes derived from both phages. When the DNA from these recombinant phages was examined, it was found to contain a mixture of ^{15}N and ^{14}N (**Figure 20-21**). Since these experiments were performed under conditions that prevented any new DNA from being synthesized, the recombinant DNA molecules must have been produced by breaking and rejoining DNA molecules derived from the two original phages.

Subsequent experiments involving bacteria whose chromosomes had been labeled with either ^{15}N or ^{14}N revealed that DNA containing a mixture of both isotopes is also produced during genetic recombination between bacterial chromosomes. Moreover, when such recombinant DNA molecules are heated to dissociate them into single strands, a mixture of ^{15}N and ^{14}N is detected in each DNA strand; hence, the DNA double helix must be broken and rejoined during recombination.

A similar conclusion emerged from experiments performed shortly thereafter on eukaryotic cells by J. Herbert Taylor. In these studies, cells were briefly exposed to ^{3}H-thymidine during the S phase preceding the last mitosis prior to meiosis, producing chromatids containing one radioactive DNA strand per double helix. During the following S phase, DNA replication in the absence of ^{3}H-thymidine generated chromosomes containing one

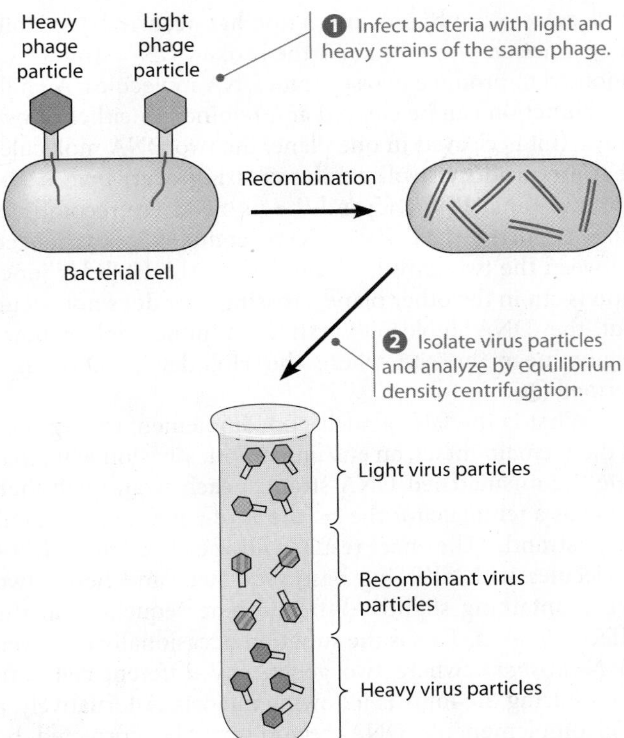

FIGURE 20-21 Evidence for DNA Breakage and Exchange During Bacteriophage Recombination. Bacterial cells were infected with two strains of the same phage, one labeled with ^{15}N and the other with ^{14}N. After recombination, the DNA from the recombinant phages was found to contain both ^{15}N and ^{14}N, supporting the idea that recombination involves the breaking and rejoining of DNA molecules.

labeled chromatid and one unlabeled chromatid (**Figure 20-22**). But during the subsequent meiosis, individual chromatids exhibited a mixture of radioactive and nonradioactive segments, as would be predicted by the breakage-and-exchange model. Moreover, the frequency of such exchanges was directly proportional to the frequency with which the genes located in these regions underwent genetic recombination. Such observations provided strong support for the notion that genetic recombination in eukaryotic cells, as in prokaryotes, involves DNA breakage and exchange. These experiments also showed that most DNA exchanges arise between homologous chromosomes rather than between the two sister chromatids of a given chromosome. This selectivity is important because it ensures that genes are exchanged between paternal and maternal chromosomes.

Homologous Recombination Can Lead to Gene Conversion

The conclusion that homologous recombination is based on DNA breakage and exchange does not in itself provide much information concerning the underlying molecular mechanisms. One of the simplest breakage-and-exchange models that might be envisioned would involve the cleavage of two homologous, double-stranded

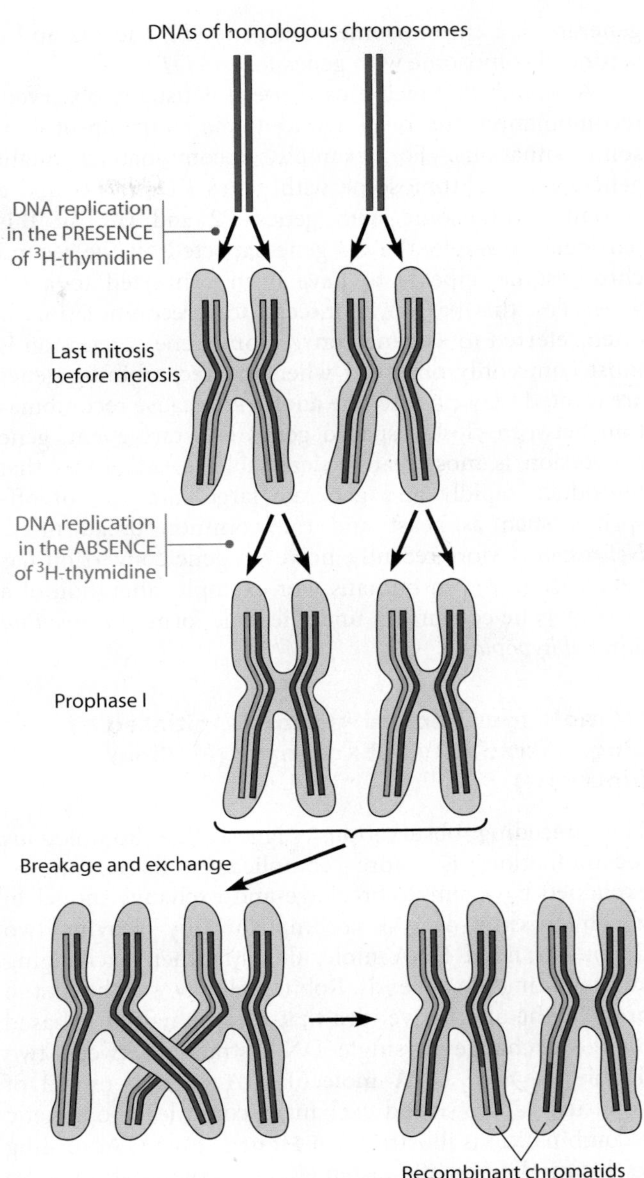

FIGURE 20-22 Experimental Demonstration of Breakage and Exchange During Eukaryotic Recombination. In this experiment, DNA was radioactively labeled by briefly exposing eukaryotic cells to ^{3}H-thymidine before the last mitosis prior to meiosis. When autoradiography was employed to examine the chromatids during meiosis, some were found to contain a mixture of labeled (orange) and unlabeled (dark blue) segments as predicted by the breakage-and-exchange model.

DNA molecules at comparable locations, followed by exchange and rejoining of the cut ends. This model implies that genetic recombination should be completely reciprocal; that is, any genes exchanged from one chromosome should appear in the other chromosome and vice versa. For example, consider a hypothetical situation involving two genes designated *P* and *Q*. If one chromosome contains forms of these genes called *P1* and *Q1*, and the other chromosome has alternative forms designated *P2* and *Q2*, reciprocal exchange would be expected to generate one chromosome with genes *P1* and *Q2* and a second chromosome with genes *P2* and *Q1*.

Although this reciprocal pattern is usually observed, recombination has been found to be nonreciprocal in some situations. For example, recombination might generate one chromosome with genes *P1* and *Q2* and a second chromosome with genes *P2* and *Q2*. In this particular example, the *Q1* gene expected on the second chromosome appears to have been converted to a *Q2* gene. For this reason, nonreciprocal recombination is often referred to as **gene conversion.** Gene conversion is most commonly observed when the recombining genes are located very close to one another. Because recombination between closely spaced genes is a rare event, gene conversion is most readily detectable in organisms that reproduce rapidly and generate large numbers of offspring, such as yeast and the common bread mold, *Neurospora*. More recently, however, gene conversion has been recognized in humans. For example, alteration of a gene by gene conversion underlies one form of *congenital adrenal hypoplasia*.

Homologous Recombination Is Initiated by Single-Strand DNA Exchanges (Holliday Junctions)

The preceding observations suggest that homologous recombination is more complicated than can be explained by a simple breakage-and-exchange model in which crossing over is accomplished by cleaving two double-stranded DNA molecules and then exchanging and rejoining the cut ends. Robin Holliday was the first to propose the alternative idea that recombination is based on the exchange of single DNA strands between two double-stranded DNA molecules. A current model of how such single-strand exchanges could lead to genetic recombination is illustrated in **Figure 20-23**. According to this model, the initial step (❶) in recombination is the cleavage of one or both strands of the DNA double helix. In either case, a single broken DNA strand derived from one DNA molecule "invades" a complementary region of a homologous DNA double helix, displacing one of the two strands (❷). Localized DNA synthesis and repair (❸ and ❹) ultimately generate a crossed structure, called a **Holliday junction,** in which a single strand from each DNA double helix has crossed over and joined the opposite double helix. Electron microscopy has provided direct support for the existence of Holliday junctions, revealing the presence of DNA double helices joined by single-strand crossovers at sites of genetic recombination (Figure 20-23, *inset*).

Once a Holliday junction has been formed, unwinding and rewinding of the DNA double helices causes the crossover point to move back and forth along the chromosomal DNA (❺). This phenomenon, called *branch migration*, can rapidly increase the length of single-stranded DNA that is exchanged between two DNA molecules. After branch migration has occurred, the Holliday junction is cleaved and the broken DNA strands are rejoined to produce two separate DNA molecules. A Holliday junction can be cleaved and rejoined in either of two ways. If it is cleaved in one plane, the two DNA molecules that are produced will exhibit crossing over; that is, the chromosomal DNA beyond the point where recombination occurred will have been completely exchanged between the two chromosomes (6a). If the Holliday junction is cut in the other plane, crossing over does not occur, but the DNA molecules exhibit a noncomplementary region near the site where the Holliday junction had formed (6b).

What is the fate of such noncomplementary regions? If they remain intact, an ensuing mitotic division will separate the mismatched DNA strands; each strand will then serve as a template for the synthesis of a new complementary strand. The net result will be two new DNA molecules with differing base sequences and hence two cells containing slightly different gene sequences in the affected region. This is the situation occasionally observed in *Neurospora*, where two genetically different cells can arise during the mitosis following meiosis. Alternatively, a noncomplementary DNA region may be corrected by excision and repair. The net effect of DNA repair would be to convert genes from one form to another—in other words, gene conversion.

A key enzyme involved in homologous recombination was first identified using bacterial extracts that catalyze the formation of Holliday junctions. Such extracts contain a protein, called *RecA*, whose presence is required for recombination. Mutant bacteria that produce a defective RecA protein cannot carry out genetic recombination, nor are extracts prepared from such cells capable of creating Holliday junctions from homologous DNA molecules. The RecA protein catalyzes a *strand invasion* reaction in which a single-stranded DNA segment displaces one of the two strands of a DNA double helix; in other words, it catalyzes step ❷ in Figure 20-23. Eukaryotes contain a comparable protein called *Rad51*. In catalyzing the strand invasion reaction, RecA or Rad51 first coats the single-stranded DNA region; the coated, single-stranded DNA then interacts with a DNA double helix, moving along the target DNA until it reaches a complementary sequence it can pair with.

Besides being involved in genetic recombination, strand invasion plays a role in certain types of DNA repair. As mentioned in Chapter 19, double-strand breaks in DNA are often repaired by a process that takes advantage of the fact that if one chromosome incurs a double-strand break, the intact DNA molecule of the homologous chromosomal DNA can serve as a template for guiding the repair of the broken chromosome. In such cases, single DNA strands from the broken DNA molecule invade the intact DNA molecule and find a complementary region to serve as a template for repairing the broken DNA, just as occurs in step ❷ of Figure 20-23.

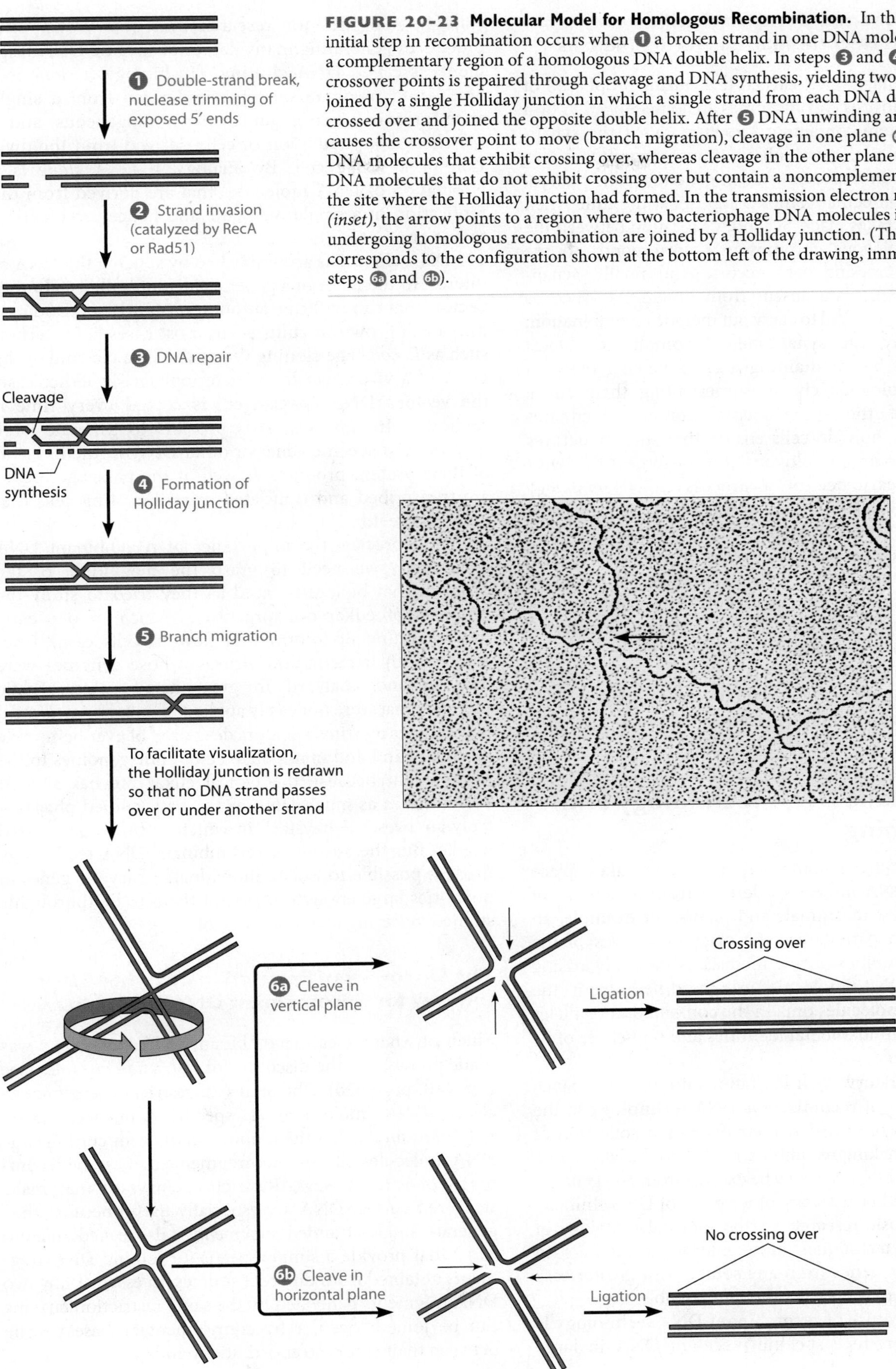

FIGURE 20-23 Molecular Model for Homologous Recombination. In this model, the initial step in recombination occurs when 1 a broken strand in one DNA molecule 2 invades a complementary region of a homologous DNA double helix. In steps 3 and 4, one of the crossover points is repaired through cleavage and DNA synthesis, yielding two DNA molecules joined by a single Holliday junction in which a single strand from each DNA double helix has crossed over and joined the opposite double helix. After 5 DNA unwinding and rewinding causes the crossover point to move (branch migration), cleavage in one plane 6a creates two DNA molecules that exhibit crossing over, whereas cleavage in the other plane 6b generates DNA molecules that do not exhibit crossing over but contain a noncomplementary region near the site where the Holliday junction had formed. In the transmission electron micrograph *(inset)*, the arrow points to a region where two bacteriophage DNA molecules in the process of undergoing homologous recombination are joined by a Holliday junction. (This photograph corresponds to the configuration shown at the bottom left of the drawing, immediately prior to steps 6a and 6b).

MB Molecular Model: Structural Tutorial: RecBCD Complexed with DNA; Molecular Model: The RuvA Recombination Protein of Escherichia coli

The Synaptonemal Complex Facilitates Homologous Recombination During Meiosis

Earlier in the chapter, we learned that during prophase I of meiosis, homologous chromosomes are joined together by a zipperlike, protein-containing structure called the synaptonemal complex. Several observations suggest that this structure plays an important role in genetic recombination. First, the synaptonemal complex appears at the time when recombination takes place. Second, its location between the opposed homologous chromosomes corresponds to the region where crossing over occurs. And finally, synaptonemal complexes are absent from organisms—such as male fruit flies—that fail to carry out meiotic recombination.

Presumably, the synaptonemal complex facilitates recombination by maintaining a close pairing between adjacent homologous chromosomes along their entire length. But if the synaptonemal complex facilitates recombination, how do cells ensure that such structures form only between homologous chromosomes? Evidence suggests the existence of a process called *homology searching,* in which a single-strand break in one DNA molecule produces a free strand that "invades" another DNA double helix and checks for the presence of complementary sequences (Figure 20-23, step ❷). If extensive homology is not found, the free DNA strand invades another DNA molecule and checks for complementarity, repeating the process until a homologous DNA molecule is detected. Only then does a synaptonemal complex develop, bringing the homologous chromosomes together throughout their length to facilitate the recombination process.

Recombinant DNA Technology and Gene Cloning

In nature, genetic recombination usually takes place between two DNA molecules derived from organisms of the same species. In animals and plants, for example, an individual's two parents are the original sources of the DNA that recombines during meiosis. A naturally arising recombinant DNA molecule usually differs from the parental DNA molecules only in the combination of alleles it contains; the fundamental identities and sequences of its genes remain the same.

In the laboratory, such limitations do not exist. Since the development of **recombinant DNA technology** in the 1970s, scientists have had at their disposal a collection of techniques for making recombinant DNA in the laboratory. Any segment of DNA can now be excised from any genome and spliced together with any other piece of DNA. Initially derived from basic research on the molecular biology of bacteria, these techniques have enabled researchers to isolate and study genes from any source with greater ease and precision than was earlier thought possible.

A central feature of recombinant DNA technology is the ability to produce specific pieces of DNA in large enough quantities for research and other uses. This process of generating many copies of specific DNA fragments is called **DNA cloning.** (In biology, a *clone* is a population of organisms that is derived from a single ancestor and hence is genetically homogeneous, and a *cell clone* is a population of cells derived from the division of a single cell. By analogy, a *DNA clone* is a population of DNA molecules that are derived from the replication of a single molecule and hence are identical to one another.)

DNA cloning is accomplished by splicing the DNA of interest to the DNA of a genetic element, called a **cloning vector,** that can replicate autonomously when introduced into a cell grown in culture—in most cases, a bacterium such as *E. coli.* The cloning vector can be a plasmid or the DNA of a virus, usually a bacteriophage; in either case, the vector's DNA "passenger" is copied every time it replicates. In this way, it is possible to generate large quantities of specific genes or other DNA segments—and of their protein products as well, if the passenger genes are transcribed and translated in proliferating cells that carry the vector.

To appreciate the importance of recombinant DNA technology, we need to grasp the magnitude of the problem that biologists faced as they tried to study the genomes of eukaryotic organisms. Much of our early understanding of information flow in cells came from studies with bacteria and viruses, whose genomes were mapped and analyzed in great detail using genetic methods that were not easily applied to eukaryotes. Until a few decades ago, investigators despaired of ever being able to understand and manipulate eukaryotic genomes to the same extent because the typical eukaryote has at least 10,000 times as much DNA as the best-studied phages—truly an awesome haystack in which to find a gene-sized needle. But the advent of recombinant DNA technology made it possible to isolate individual eukaryotic genes in quantities large enough to permit them to be thoroughly studied, ushering in a new era in biology.

The Discovery of Restriction Enzymes Paved the Way for Recombinant DNA Technology

Much of what we call recombinant DNA technology was made possible by the discovery of *restriction enzymes* (see Box 18B, page 520). The ability of restriction enzymes to cleave DNA molecules at specific sequences called *restriction sites* makes them powerful tools for cutting large DNA molecules into smaller fragments that can be recombined in various ways. Restriction enzymes that make staggered cuts in DNA are especially useful because they generate single-stranded *sticky ends* (also called *cohesive ends*) that provide a simple means for joining DNA fragments obtained from different sources. In essence, any two DNA fragments generated by the same restriction enzyme can be joined together by complementary base pairing between their single-stranded, sticky ends.

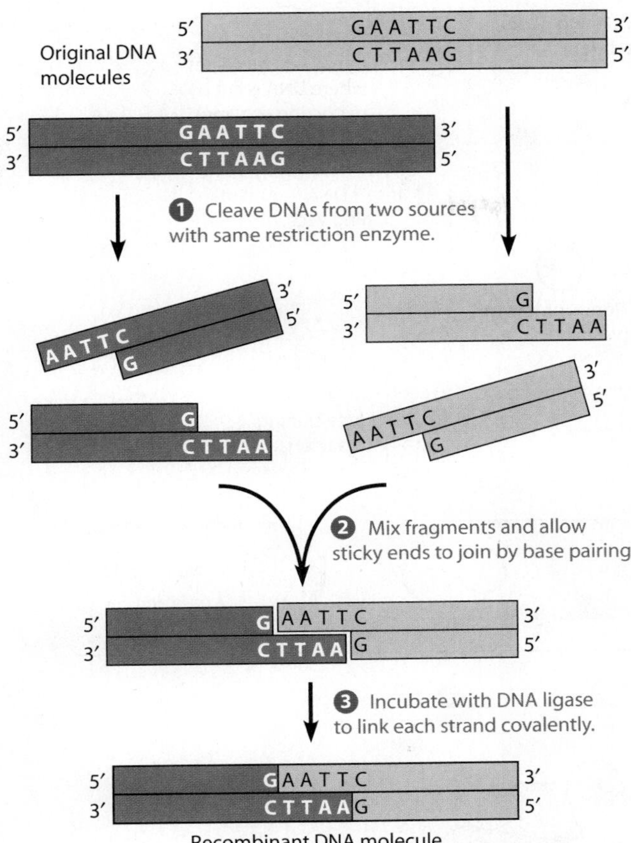

Original DNA molecules
5′ GAATTC 3′
3′ CTTAAG 5′
1 Cleave DNAs from two sources with same restriction enzyme.
AATTC
G
CTTAA
2 Mix fragments and allow sticky ends to join by base pairing.
GAATTC
CTTAAG
3 Incubate with DNA ligase to link each strand covalently.
Recombinant DNA molecule

FIGURE 20-24 Creating Recombinant DNA Molecules. A restriction enzyme that generates sticky ends (in this case *Eco*RI) is used to cleave DNA molecules from two different sources. The complementary ends of the resulting fragments join by base pairing to create recombinant molecules containing segments from both of the original sources.

Figure 20-24 illustrates how this general approach works. DNA molecules from two sources are first treated with a restriction enzyme known to generate fragments with sticky ends (❶), and the fragments are then mixed together under conditions that favor base pairing between these sticky ends (❷). Once joined in this way, the DNA fragments are covalently sealed together by DNA ligase (❸), an enzyme normally involved in DNA replication and repair (see Chapter 19). The final product is a **recombinant DNA molecule** containing DNA sequences derived from two different sources.

The combined use of restriction enzymes and DNA ligase allows any two (or more) pieces of DNA to be spliced together, regardless of their origins. A piece of human DNA, for example, can be joined to bacterial or phage DNA just as easily as it can be linked to another piece of human DNA. In other words, it is possible to form recombinant DNA molecules that never existed in nature, without regard for the natural barriers that otherwise limit recombination to genomes of the same or closely related species. Therein lies the power of (and, for some, the concern about) recombinant DNA technology.

DNA Cloning Techniques Permit Individual Gene Sequences to Be Produced in Large Quantities

The power of restriction enzymes is that they make it easy to insert a desired piece of DNA—usually a segment containing a specific gene—into a cloning vector that can replicate itself when introduced into bacterial cells. Suppose, for example, you wanted to isolate a gene that codes for a medically useful product, such as insulin needed for the treatment of diabetics or blood clotting factors needed by patients with hemophilia. By using restriction enzymes to insert a cDNA encoding such a protein (see the next section) into a cloning vector in bacterial cells and then identifying bacteria that contain the DNA of interest, it is possible to grow large masses of such cells and thereby obtain large quantities of the desired DNA (and its protein product).

Although the specific details vary, the following five steps are typically involved in this process of DNA cloning: (1) insertion of DNA into a cloning vector; (2) introduction of the recombinant vector into cultured cells, usually bacteria; (3) amplification of the recombinant vector in the bacteria; (4) selection of cells containing recombinant DNA; and (5) identification of clones containing the DNA of interest. **Figure 20-25** provides an overview of these events using a bacterial cloning vector. You should refer to this figure as we consider each step in turn.

1. Insertion of DNA into a Cloning Vector. The first step in cloning a desired piece of DNA is to insert it into an appropriate cloning vector, usually a bacteriophage or a plasmid. Most vectors used for DNA cloning are themselves recombinant DNA molecules, designed specifically for this purpose. For example, when bacteriophage lDNA is used as a cloning vector, the phage DNA has had some of its nonessential genes removed to make room in the phage head for spliced-in DNA. Plasmids used as cloning vectors usually have a variety of restriction sites and often carry genes that confer antibiotic resistance on their host cells. The antibiotic-resistance genes facilitate the selection stage (❹), while the presence of multiple kinds of restriction sites allows the plasmid to incorporate DNA fragments prepared with a variety of restriction enzymes.

Figure 20-26 illustrates the structure of *pUC19* ("puck-19"), a plasmid cloning vector developed in the mid-1980s. This plasmid, as well as a series of versatile vectors subsequently derived from it, carries a gene that confers resistance to the antibiotic ampicillin (amp^R). Bacteria containing such plasmids can therefore be identified by their ability to grow in the presence of ampicillin. The pUC19 plasmid also has 11 different restriction sites clustered in a region of the plasmid containing the *lacZ* gene, which codes for the enzyme b-galactosidase. Integration of foreign DNA at any of these restriction sites will disrupt the *lacZ* gene, thereby blocking the production of b-galactosidase. As we will see shortly, this

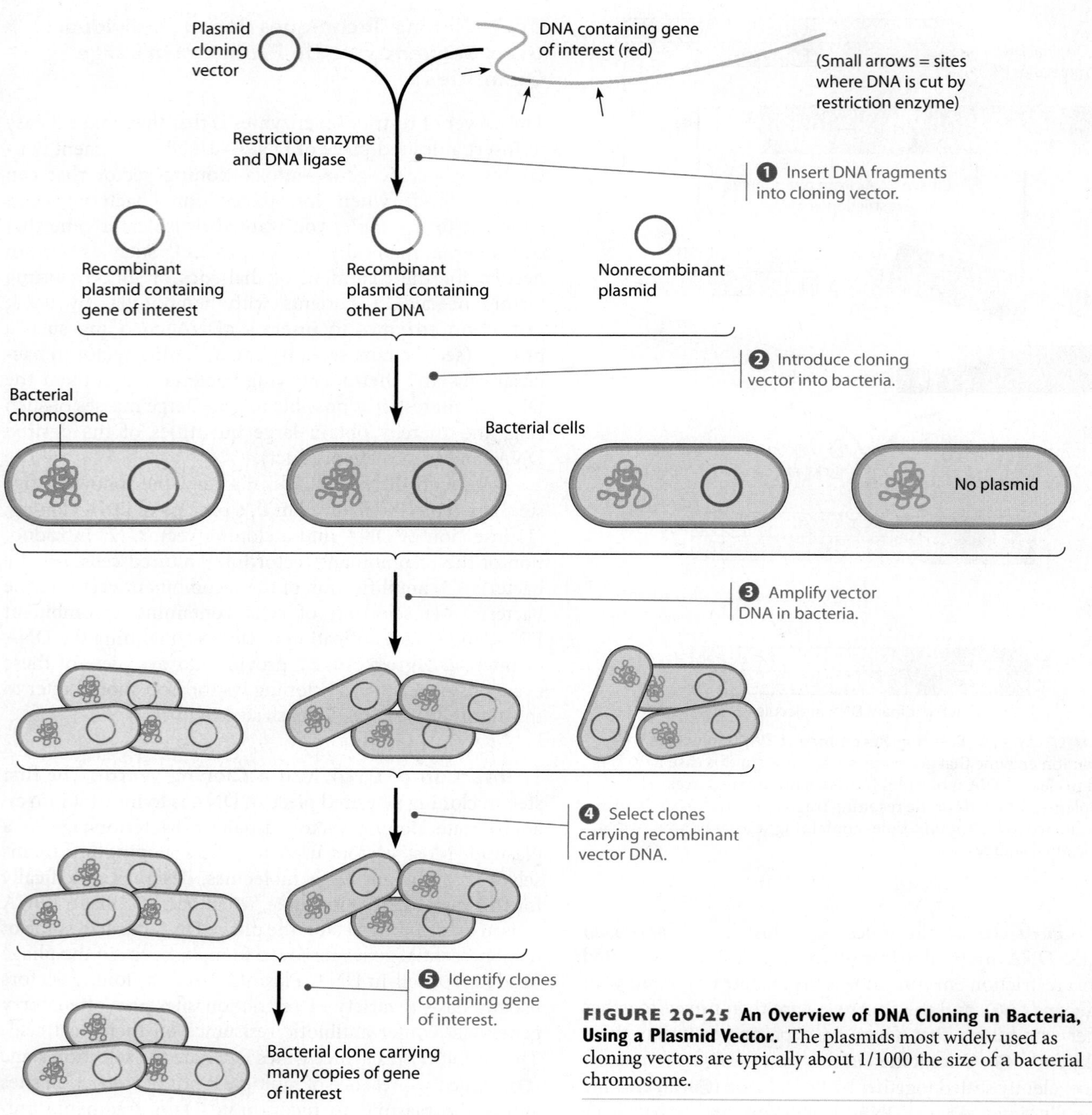

FIGURE 20-25 An Overview of DNA Cloning in Bacteria, Using a Plasmid Vector. The plasmids most widely used as cloning vectors are typically about 1/1000 the size of a bacterial chromosome.

MB Activity: Cloning a Gene in Bacteria

disruption in β-galactosidase production can be used later in the cloning process to detect the presence of plasmids containing foreign DNA.

Figure 20-26b illustrates how a specific gene of interest residing in a foreign DNA source is inserted into a plasmid cloning vector, using pUC19 as the vector and a restriction enzyme that cleaves pUC19 at a single site within the *lacZ* gene. Incubation with the restriction enzyme cuts the plasmid at that site (❶), making the DNA linear (opening the circle). The same restriction enzyme is used to cleave the DNA molecule containing the gene to be cloned (❷). The sticky-ended fragments of foreign DNA are then incubated with the linearized vector molecules under conditions that favor base pairing (❸), followed by treatment with DNA ligase to link the molecules covalently (❹). To keep the diagram simple, Figure 20-26b shows only the recombinant plasmid containing the desired fragment of foreign DNA. In practice, however, a variety of DNA products will be present, including nonrecombinant plasmids and recombinant plasmids containing other fragments generated by the action of the restriction enzyme.

2. Introduction of the Recombinant Vector into Bacterial Cells. Once foreign DNA has been inserted into a cloning vector, the resulting recombinant vector is replicated

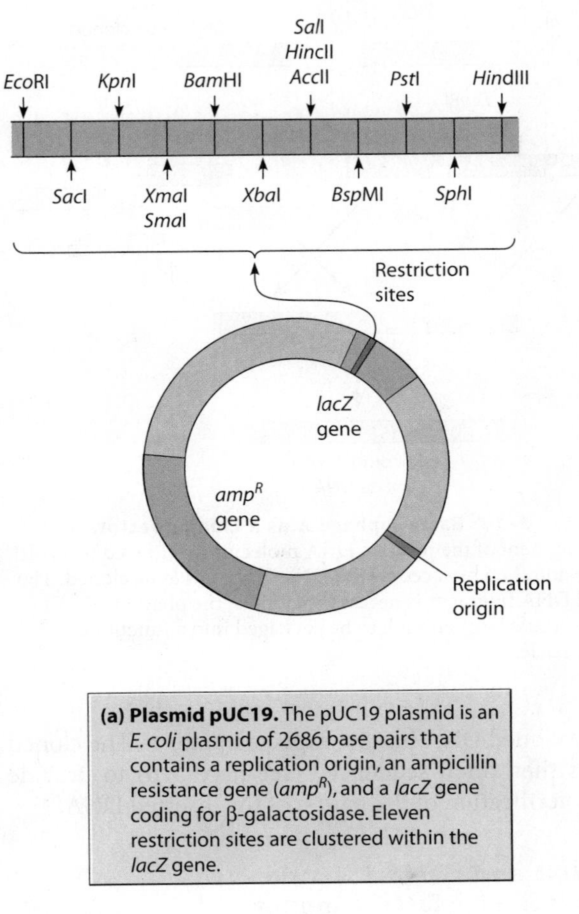

(a) Plasmid pUC19. The pUC19 plasmid is an *E. coli* plasmid of 2686 base pairs that contains a replication origin, an ampicillin resistance gene (amp^R), and a *lacZ* gene coding for β-galactosidase. Eleven restriction sites are clustered within the *lacZ* gene.

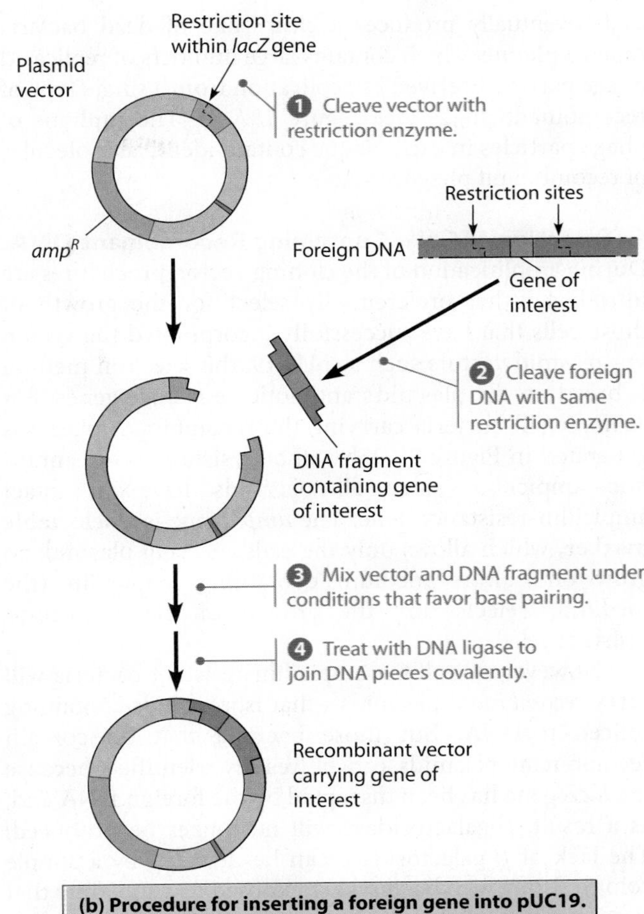

(b) Procedure for inserting a foreign gene into pUC19.

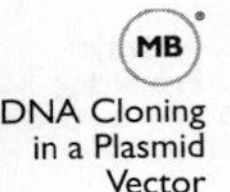

DNA Cloning in a Plasmid Vector

FIGURE 20-26 Cloning a Gene in the Plasmid Vector pUC19. To insert foreign DNA into the plasmid, the foreign and plasmid DNAs are cleaved with a restriction enzyme that recognizes the same site, in this case a site within the *lacZ* gene. The fragments of foreign DNA are incubated with the linearized plasmid DNA under conditions that favor base pairing between sticky ends. Among the expected products will be plasmid molecules recircularized by base pairing with a single fragment of foreign DNA, and some of these will contain the gene of interest. Cells carrying such plasmids will be resistant to ampicillin and will fail to produce β-galactosidase because of the foreign DNA inserted within the *lacZ* gene.

by introducing it into an appropriate host cell, usually the bacterium *E. coli.* Cloning vectors are introduced into bacteria in one of two ways. If the cloning vector is phage DNA, it is incorporated into phage particles that are then used to infect an appropriate cell population. Plasmids, on the other hand, are simply introduced into the medium surrounding the target cells. Both prokaryotic and eukaryotic cells will take up plasmid DNA from the external medium, although special treatments are usually necessary to enhance the efficiency of the process. Adding calcium ions, for example, markedly increases the rate at which cells take up DNA from the external environment.

3. Amplification of the Recombinant Vector in Bacteria. After they have taken up the recombinant cloning vector, the host bacteria are plated out on a nutrient medium so that the recombinant DNA vector can be replicated, or *amplified.* In the case of a plasmid vector, the bacteria proliferate and form colonies, each derived from a single cell. Under favorable conditions, *E. coli* will divide every 22 minutes, giving rise to a billion cells in less than 11 hours. As the bacteria multiply, the recombinant plasmids also replicate, producing an enormous number of vector molecules containing foreign DNA fragments. Under such conditions, a single recombinant plasmid introduced into one cell will be amplified several hundred billionfold in less than half a day.

In the case of phage vectors such as phage λ, a slightly different procedure is used. Phage particles containing recombinant DNA are mixed with bacterial cells, and the mixture is then placed on a culture medium under conditions that produce a continuous "lawn" of bacteria across the plate. Each time a phage particle infects a cell, the phage is replicated and eventually causes the cell to rupture and die. The released phage particles can then infect neighboring cells, repeating the process again. This cycle eventually produces a clear zone of dead bacteria called a **plaque,** which contains large numbers of replicated phage particles derived by replication from a single type of recombinant phage (see Figure 18A-3). The millions of phage particles in each plaque contain identical molecules of recombinant phage DNA.

4. Selection of Cells Containing Recombinant DNA. During amplification of the cloning vector, procedures are introduced that preferentially select for the growth of those cells that have successfully incorporated the vector. For plasmid vectors such as pUC19, the selection method is based on the plasmid's antibiotic-resistance genes. For example, all bacteria carrying the recombinant plasmids generated in Figure 20-26b will be resistant to the antibiotic ampicillin, since all plasmids have an intact ampicillin-resistance gene. The amp^R gene is a **selectable marker,** which allows only the cells carrying plasmids to grow on culture medium containing ampicillin (the medium "selects for" the growth of the ampicillin-resistant cells).

However, not all the ampicillin-resistant bacteria will carry *recombinant* plasmids—that is, plasmids containing spliced-in DNA. But those bacteria that do contain recombinant plasmids can be readily identified because the *lacZ* gene has been disrupted by the foreign DNA and, as a result, β-galactosidase will no longer be produced. The lack of β-galactosidase can be detected by a simple color test in which bacteria are exposed to a substrate that is normally cleaved by β-galactosidase into a blue-staining compound. Bacterial colonies containing the normal pUC19 plasmid will therefore stain blue, whereas colonies containing recombinant plasmids with inserted DNA fragments will appear white.

A different approach is used with phage cloning vectors, which are usually derived from phage λ DNA molecules that are only about 70% as long as normal phage DNA. As a result, these DNA molecules are too small to be packaged into functional phage particles. But if an additional fragment of DNA is inserted into the middle of such a cloning vector, it creates a recombinant DNA molecule that is larger and thus capable of being assembled into a functional phage (**Figure 20-27**). Hence, when phage cloning vectors are employed, the only phage particles that can successfully infect bacterial cells are those containing an inserted foreign DNA sequence.

MB Restriction Enzymes, Recombinant DNA, and Gene Cloning

5. Identification of Clones Containing the DNA of Interest. The preceding steps typically generate vast numbers of bacteria producing many different kinds of recombinant DNA, only one or a few of which are relevant to the desired application. The final stage in any recombinant DNA procedure is therefore screening the bacterial colonies (or phage plaques) to identify those containing the specific DNA fragment of interest. For standard bacterial clones, this is usually done by isolating DNA from the bacteria and then using restriction enzymes to confirm that the cloned DNA has the expected pattern. The cloned DNA is then often sequenced (see page 520) to provide precise verification of the nature of the inserted DNA.

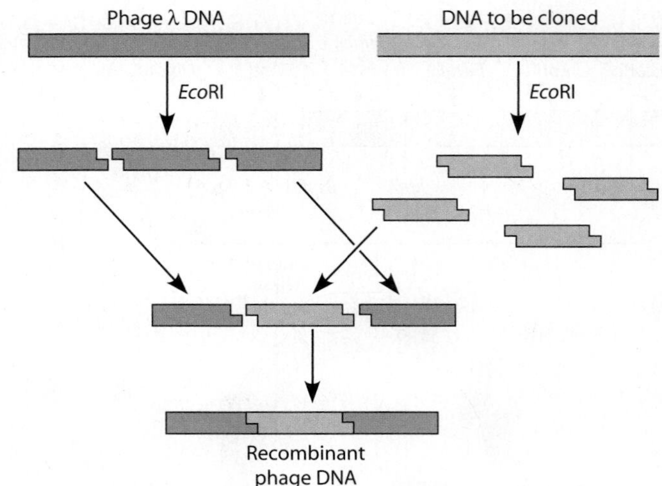

FIGURE 20-27 Bacteriophage λ as a Cloning Vector. The middle segment of the phage λ DNA molecule is removed by *Eco*RI cleavage and then replaced by the DNA fragment to be cloned. The inserted DNA fragment is necessary to make the phage λ DNA molecule large enough to be packaged into a functional phage particle.

Genomic and cDNA Libraries Are Both Useful for DNA Cloning

Cloning foreign DNA in bacterial cells is now a routine procedure. In practice, obtaining a good source of DNA to serve as starting material is often one of the most difficult steps. Two different approaches are commonly used for producing the DNA starting material. In the "shotgun" approach, an organism's entire genome (or some substantial portion of it) is cleaved into a large number of restriction fragments, which are then inserted into cloning vectors for introduction into bacterial cells (or phage particles). The resulting group of clones is called a **genomic library** because it contains cloned fragments representing most, if not all, of the genome. Genomic libraries of eukaryotic DNA are valuable resources from which specific genes can be isolated, provided that a sufficiently sensitive identification technique is available. Once a rare bacterial colony containing the desired DNA fragment has been identified, it can be grown on a nutrient medium to generate as many copies of the fragment as may be needed. Of course, the DNA cuts made by a restriction enzyme do not respect gene boundaries, and some genes may be divided among two or more restriction fragments. This problem can be circumvented by carrying out a *partial DNA digestion* in which the DNA is briefly exposed to a small quantity of restriction enzyme. Under such conditions, some restriction sites remain uncut, increasing the probability that at least one intact copy of each gene will be present in the genomic library.

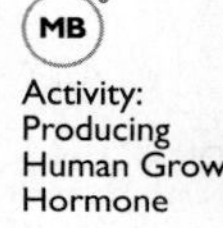

Activity: Producing Human Growth Hormone

The alternative DNA source for cloning experiments is DNA that has been generated by copying messenger RNA (mRNA) with the enzyme *reverse transcriptase* (page 648). This reaction generates a population of **complementary DNA (cDNA)** molecules that are complementary in sequence to the mRNA employed as template (**Figure 20-28**). If the entire mRNA population of a cell is isolated and copied into cDNA for cloning, the resulting group of clones is called a **cDNA library.** The advantage of a cDNA library is that it contains only those DNA sequences that are transcribed into mRNA—presumably, the active genes in the cells or tissue from which the mRNA was prepared.

Besides being limited to transcribed genes, a cDNA library has another important advantage as a starting point for the cloning of eukaryotic genes. Using mRNA to make cDNA guarantees that the cloned genes will contain only gene-coding sequences, without the noncoding interruptions called *introns* that are common in eukaryotic genes (see Chapter 21). Introns can be so extensive that the overall length of a eukaryotic gene becomes too unwieldy for recombinant DNA manipulation. Using cDNA eliminates this problem. In addition, bacteria cannot synthesize the correct protein product of an intron-containing eukaryotic gene unless the introns have been removed—as they are in cDNA.

Once a DNA library has been constructed, how can bacteria or phage be identified that carry only the sequences that we are interested in? There are several techniques for *screening* DNA libraries. The particular technique used depends on what the researcher knows about the gene being cloned, and on the type of library. If something is known about the base sequence of the gene of interest, the researcher can employ a **nucleic acid probe,** a single-stranded molecule of DNA or RNA that can identify a desired DNA sequence by base-pairing with it. Nucleic acid probes are labeled either with radioactivity or with some other chemical group that allows the probe to be easily visualized. (In Box 18C, page 528, we saw such a probe used to identify restriction fragment bands in Southern blotting.) The researcher prepares a labeled DNA or RNA probe containing all or part of the nucleotide sequence of interest and uses it to tag the colonies that contain complementary DNA. **Figure 20-29** outlines how this *colony hybridization technique* can be used to screen for colonies carrying the desired DNA. Once the appropriate colonies have been identified, cloned DNA is recovered from these colonies by isolating the vector DNA from the bacterial cells and digesting it with the same restriction enzyme used initially.

Another screening approach focuses on the protein encoded by a gene of interest. If this protein is known and has been purified, antibodies against it can be prepared and used as probes to check bacterial colonies for the presence of the protein. Alternatively, the *function* of the protein can be measured; for example, an enzyme could be tested for its catalytic activity. Protein-screening methods obviously depend on the ability of the bacterial cells to produce a foreign protein encoded by a cloned gene and will fail to detect cloned genes that are not expressed in the host cell. However, special *expression vectors* can be used to increase the likelihood that bacteria will transcribe eukaryotic genes properly and in large amounts. Expression vectors contain special DNA sequences that signal the bacterial cell to perform these processes.

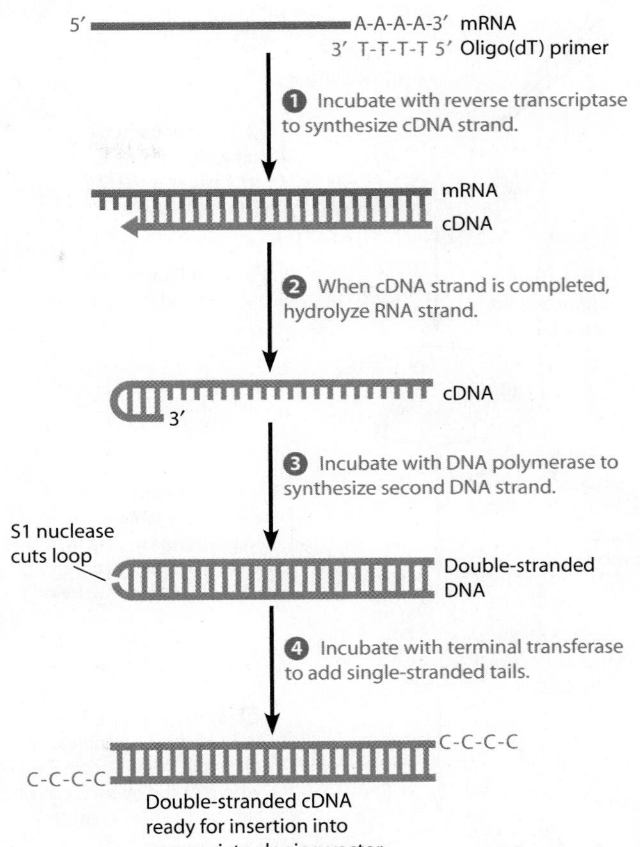

FIGURE 20-28 Preparation of Complementary DNA (cDNA) for Cloning. ❶ Messenger RNA is incubated with reverse transcriptase to create a complementary DNA (cDNA) strand. Oligo(dT), a short chain of thymine deoxynucleotides, can be used as a primer because eukaryotic mRNA always has a stretch of adenine nucleotides at its 3′ end. ❷ The resulting mRNA-cDNA hybrid is treated with alkali or an enzyme to hydrolyze the RNA, leaving the single-stranded cDNA. ❸ DNA polymerase can now synthesize the complementary DNA strand, using the looped-around 3′ end of the first DNA strand as a primer. An enzyme called S1 nuclease is then employed to cleave the loop. ❹ For efficient insertion into a cloning vector, the double-stranded DNA must have single-stranded tails that are complementary to those of the vector. These can be added by incubating with terminal transferase, an enzyme that adds nucleotides to DNA ends. For example, if short stretches of cytosine (C) nucleotides are added to the cDNA and short stretches of guanine (G) nucleotides are added to a linearized cloning vector, recombinant molecules can be generated by allowing the single-stranded C tails in the cDNA to hybridize to the single-stranded G tails in the vector. (As an alternative to step ❹, short synthetic "linker" molecules containing a variety of restriction sites can be ligated to the ends of both the cDNA and a blunt-ended cloning vector. The linkers are then cleaved with a restriction enzyme that generates sticky ends.)

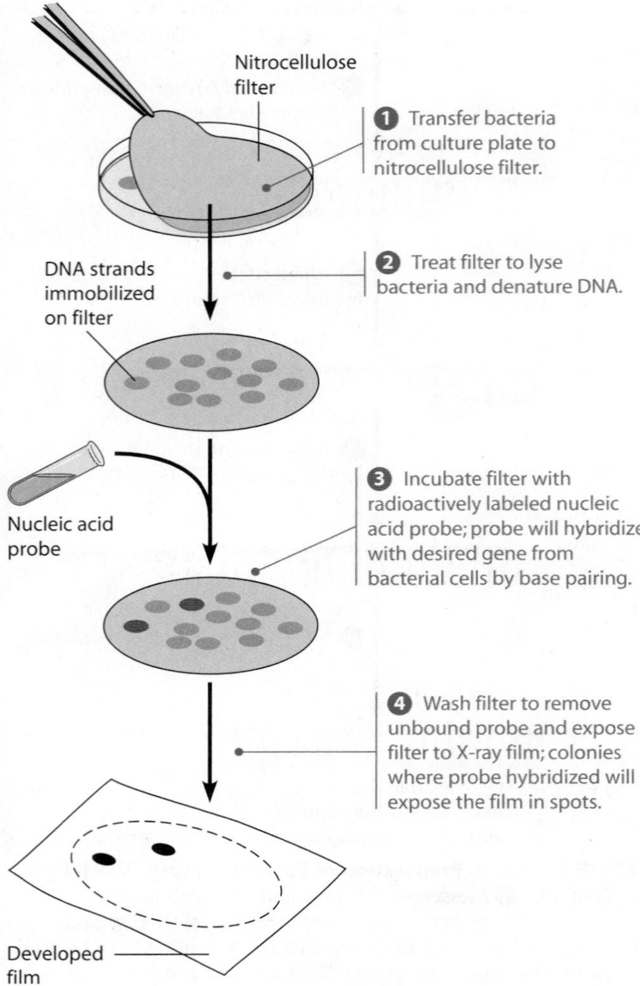
Nitrocellulose filter
1 Transfer bacteria from culture plate to nitrocellulose filter.
DNA strands immobilized on filter
2 Treat filter to lyse bacteria and denature DNA.
Nucleic acid probe
3 Incubate filter with radioactively labeled nucleic acid probe; probe will hybridize with desired gene from bacterial cells by base pairing.
4 Wash filter to remove unbound probe and expose filter to X-ray film; colonies where probe hybridized will expose the film in spots.
Developed film

FIGURE 20-29 Colony Hybridization Technique. This technique is used to screen bacterial colonies for the presence of DNA that is complementary in sequence to a nucleic acid probe. Bacterial colonies are transferred from the surface of an agar culture plate onto a nitrocellulose filter, which is treated with detergent to lyse the bacteria and alkali (NaOH) to denature their DNA. The filters are then incubated with molecules of the nucleic acid probe—radioactively labeled, single-stranded DNA or RNA—which attach by base pairing to any complementary DNA present on the filter. The filter is rinsed and subjected to autoradiography, which makes visible those colonies containing DNA that is complementary in sequence to the probe.

Large DNA Segments Can Be Cloned in YACs and BACs

DNA cloning using the vectors mentioned so far is a powerful methodology, but it has an important limitation: The foreign DNA fragments cloned in these vectors cannot exceed about 30,000 base pairs (bp) in length. Eukaryotic genes are often larger than this and hence cannot be cloned in an intact form using such vectors. For genome-mapping projects, the availability of clones containing even longer stretches of DNA is desirable because the more DNA per clone, the fewer the number of clones needed to cover the entire genome.

One of the first breakthroughs in cloning longer DNA segments was the development of a vector called a **yeast artificial chromosome (YAC).** A YAC is a "minimalist" eukaryotic chromosome—it contains all the DNA sequences needed for normal chromosome replication and segregation to daughter cells–and very little else. As you might guess from your knowledge of chromosome replication and segregation (see Chapter 19), a eukaryotic chromosome requires three kinds of DNA sequences: (1) an origin of DNA replication; (2) two telomeres to allow periodic extension of the shrinking ends by telomerase; and (3) a centromere to ensure proper attachment, via a kinetochore, to spindle microtubules during cell division. If yeast versions of these three kinds of DNA sequences are combined with a segment of foreign DNA, the resulting YAC will replicate in yeast and segregate into daughter cells with each round of cell division, just like a natural chromosome. And under appropriate conditions, its foreign genes may be expressed.

Figure 20-30 outlines the construction of a typical YAC. In addition to a replication origin (ORI), centromere sequence (CEN), and two telomeres (TEL), the vector illustrated carries two genes that function as selectable markers as well as three restriction sites. In cloning experiments, the vector and foreign DNAs are cleaved with the appropriate restriction enzymes, mixed together, and joined by DNA ligase. The resulting products, which include a variety of YACs carrying different fragments of foreign DNA, are introduced into yeast cells whose cell walls have been removed. The presence of two selectable markers makes it easy to select for yeast cells containing YACs with both chromosomal "arms." The diagram in the figure is not to scale: The YAC vector alone is only about 10,000 bp, but the inserted foreign DNA usually ranges from 300,000 to 1.5 million bp in length. In fact, YACs must carry at least 50,000 bp to be reliably replicated and segregated.

Another type of vector used for cloning large DNA fragments is the **bacterial artificial chromosome (BAC),** a derivative of the *F factor* plasmid that some bacteria employ for transferring DNA between cells during bacterial conjugation (page 621). BAC vectors are modified forms of the F factor plasmid that can hold up to 350,000 bp of foreign DNA and have all the components required for a bacterial cloning vector, such as replication origins, antibiotic resistance genes, and insertion sites for foreign DNA. One type of BAC facilitates the process of screening for recombinant clones by including the *SacB* gene, which converts sucrose (table sugar) into a substance that is toxic to bacteria. A *Bam*HI cloning site is located within the *SacB* gene, so when foreign DNA is inserted into the BAC vector at this site, the *SacB* gene is disrupted. When such a BAC vector is introduced into bacterial cells grown in the presence of sucrose, only cells containing BAC vector molecules with a foreign DNA insert will be able to grow. Those cells receiving BAC vector with no DNA insert will fail to grow because the *SacB* gene remains intact and produces a toxic substance from sucrose.

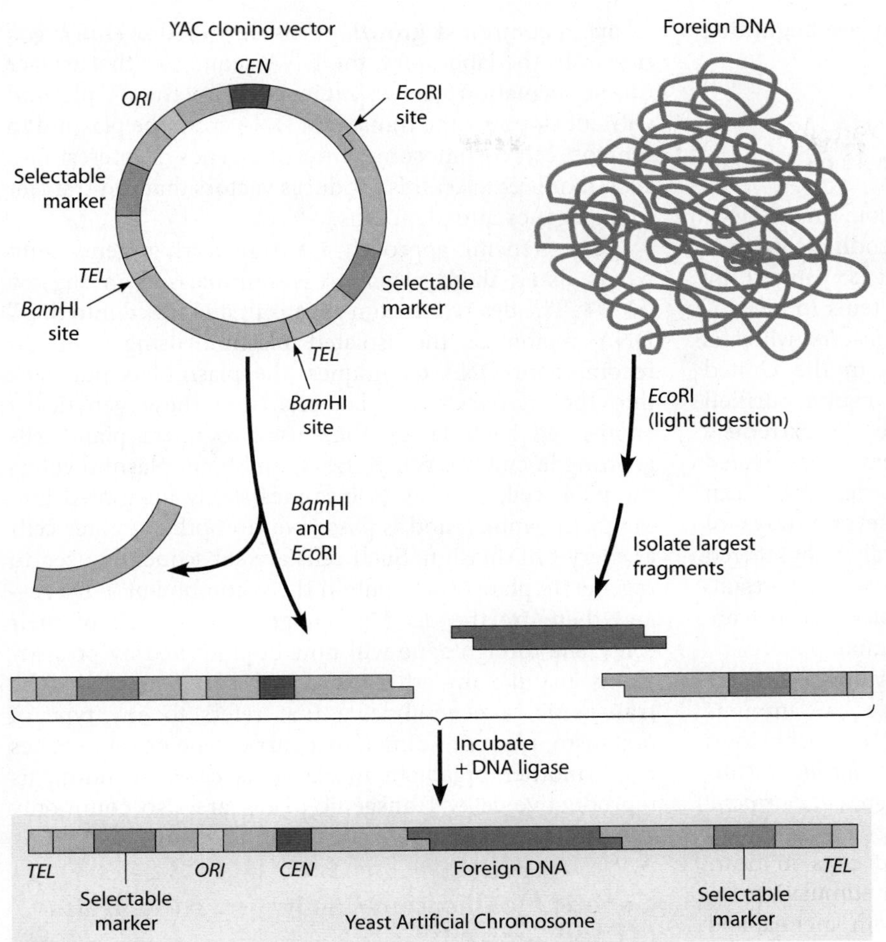

FIGURE 20-30 Construction of a Yeast Artificial Chromosome (YAC). The YAC cloning vector is a circular DNA molecule with nucleotide sequences specifying an origin of DNA replication *(ORI)*, a centromere *(CEN)*, two telomeres *(TEL)*, and two selectable markers. It has two recognition sequences for the restriction enzyme *Bam*HI and one for *Eco*RI. Digestion of the YAC vector with both restriction enzymes produces two linear DNA fragments that together contain all the essential sequences, as well as the fragment that connected the *Bam*HI sites, which is of no further use. The fragment mixture is incubated with fragments from light digestion of foreign DNA with *Eco*RI (light digestion generates large DNA fragments because not all the restriction sites are cut), and the resulting recombinant strands are sealed with DNA ligase. Among the products will be YACs carrying foreign DNA, as shown at the bottom of the figure. After yeast cells are transformed with the products of the procedure, the colonies of cells that have received complete YACs can be identified by the properties conferred by the two selectable markers.

PCR Is Widely Used to Clone Genes from Sequenced Genomes

For many years the cloning procedures described in the preceding sections were the primary means for producing cloned genes and providing longer stretches of DNA for genome sequencing. But now that scientists have determined the genome sequences of hundreds of bacteria and several dozen eukaryotic organisms, including humans, the simpler and quicker *polymerase chain reaction (PCR)* method is more commonly used to clone genes from cDNA or genomic DNA libraries. As described in Box 19A on page 560, the PCR method simply requires that you know part of the base sequence of the gene you wish to amplify. The next step is to synthesize short, single-stranded DNA primers that are complementary to sequences located at opposite ends of the gene; these primers are then used to target the intervening DNA for amplification.

In addition to its speed and simplicity, another advantage of PCR is that it allows genes to be modified by adding desired base sequences to the primers being used. One such modification, called *epitope tagging,* involves adding a short base sequence coding for a stretch of amino acids that is recognized by a commercially available antibody. When a cloned gene produced in this way is expressed inside cells, the protein produced by the gene contains the additional stretch of amino acids and can therefore be visualized and tracked using the antibody attached to a fluorescent dye. Another type of modification, called *polyhistidine tagging* (or *His* tagging, for short), introduces a short base sequence coding for the amino acid histidine repeated six times in succession. The presence of six adjacent histidines in a protein creates a structure that selectively binds to nickel ions, allowing the protein to be rapidly purified based on its affinity for nickel.

Genetic Engineering

Recombinant DNA technology has had an enormous impact on the field of cell biology, leading to many new insights into the organization, behavior, and regulation of genes and their protein products. Many of these discoveries would have been virtually inconceivable without such powerful techniques for isolating gene sequences. The rapid advances in our ability to manipulate genes have also opened up the field of **genetic engineering,** which involves the application of recombinant DNA technology to practical problems, primarily in medicine and agriculture. In concluding the chapter, we will briefly examine some of the areas where these practical benefits of recombinant DNA technology are beginning to be seen.

Genetic Engineering Can Produce Valuable Proteins That Are Otherwise Difficult to Obtain

One practical benefit to emerge from recombinant DNA technology is the ability to clone genes coding for medically useful proteins that are difficult to obtain by conventional means. Among the first proteins to be produced by genetic engineering was human *insulin,* which is required by roughly 2 million diabetics in the United States to treat their disease. Supplies of insulin purified from human blood or pancreatic tissue are extremely scarce; thus, for many years diabetic patients were treated with insulin obtained from pigs and cattle, which can cause toxic reactions. Now there are several ways of producing human insulin from genetically engineered bacteria containing the human insulin gene. As a result, diabetics can be treated with insulin molecules that are identical to the insulin produced by the human pancreas.

Like insulin, a variety of other medically important proteins that were once difficult to obtain in adequate amounts are now produced using recombinant DNA technology. Included in this category are the *blood-clotting factors* needed for the treatment of hemophilia, *growth hormone* utilized for treating pituitary dwarfism, *tissue plasminogen activator (TPA)* used for dissolving blood clots in heart attack patients, *erythropoietin* employed for stimulating the production of red blood cells in patients with anemia, and *tumor necrosis factor, interferon,* and *interleukin,* which are used in treating certain kinds of cancer. Traditional methods for isolating and purifying such proteins from natural sources are quite cumbersome and tend to yield only tiny amounts of protein. Thus, before the advent of recombinant DNA technology, the supplies of such substances were inadequate and their cost was extremely high. But now that the genes for these proteins have been cloned in bacteria and yeast, large quantities of protein can be produced in the laboratory at reasonable cost.

The Ti Plasmid Is a Useful Vector for Introducing Foreign Genes into Plants

Recombinant DNA technology is also being used to modify agriculturally important plants by inserting genes designed to introduce traits such as resistance to insects, herbicides, or viral disease, or to improve a plant's nitrogen-fixing ability, photosynthetic efficiency, nutritional value, or ability to grow under adverse conditions. Cloned genes are transferred into plants by inserting them first into the **Ti plasmid,** a naturally occurring DNA molecule carried by the bacterium *Agrobacterium tumefaciens.* In nature, infection of plant cells by this bacterium leads to insertion of a small part of the plasmid DNA, called the *T DNA region,* into the plant cell chromosomal DNA; expression of the inserted DNA then triggers the formation of an uncontrolled growth of tissue called a *crown gall tumor.* In the laboratory, the DNA sequences that trigger tumor formation can be removed from the Ti plasmid without stopping the transfer of DNA from the plasmid to the host cell chromosome. Inserting genes of interest into such modified plasmids produces vectors that can transfer foreign genes into plant cells.

The general approach for transferring genes into plants using these plasmids is summarized in **Figure 20-31**. The desired foreign gene is first inserted into the T DNA region of the isolated plasmid using standard recombinant DNA techniques, the plasmid is put back into the *Agrobacterium* bacteria, and these genetically engineered bacteria are then used to infect plant cells growing in culture. When the recombinant plasmid enters the plant cell, its T DNA becomes stably integrated into the plant genome and is passed on to both daughter cells at every cell division. Such cells are subsequently used to regenerate plants that contain the recombinant T DNA—and therefore the desired foreign gene—in all of their cells. The foreign gene will now be inherited by progeny plants just like any other gene. Such plants are said to be **transgenic,** a general term that refers to any type of organism, plant or animal, that carries one or more genes from another organism in all of its cells, including its reproductive cells. Transgenic plants are also commonly referred to as *GM (genetically modified)* plants.

Genetic Modification Can Improve the Traits of Food Crops

The ability to insert new genes into plants using the Ti plasmid has allowed scientists to create GM crops exhibiting a variety of new traits. For example, plants can be made more resistant to insect damage by introducing a gene cloned from the soil bacterium *Bacillus thuringiensis (Bt).* This *Bt* gene codes for a protein that is toxic to certain insects—especially caterpillars and beetles that cause crop damage by chewing on plant leaves. Putting the *Bt* gene into plants such as cotton and corn has permitted farmers to limit their use of more hazardous pesticides for controlling insects, leading to improved crop yields and a significant return of wildlife to crop fields. When several million cotton farmers in China switched to growing such an insect-resistant strain of GM cotton, it allowed them to slash pesticide use up to 70%. This success was accompanied by increased cotton yields and a significant drop in the death toll among farmworkers from pesticide poisoning. Similar economic and health benefits have been reported by some farmers growing insect-resistant strains of GM rice. Crops engineered to resist weed-killing herbicides likewise require fewer toxic chemicals and exhibit higher yields.

Another goal of genetic modification is to improve the nutritional value of food. Consider rice, for example, which is the most common food source in the world. More than 3 billion people currently eat rice daily, and by the year 2020 at least half the world's population is

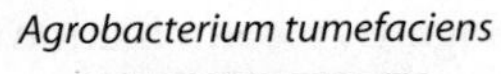

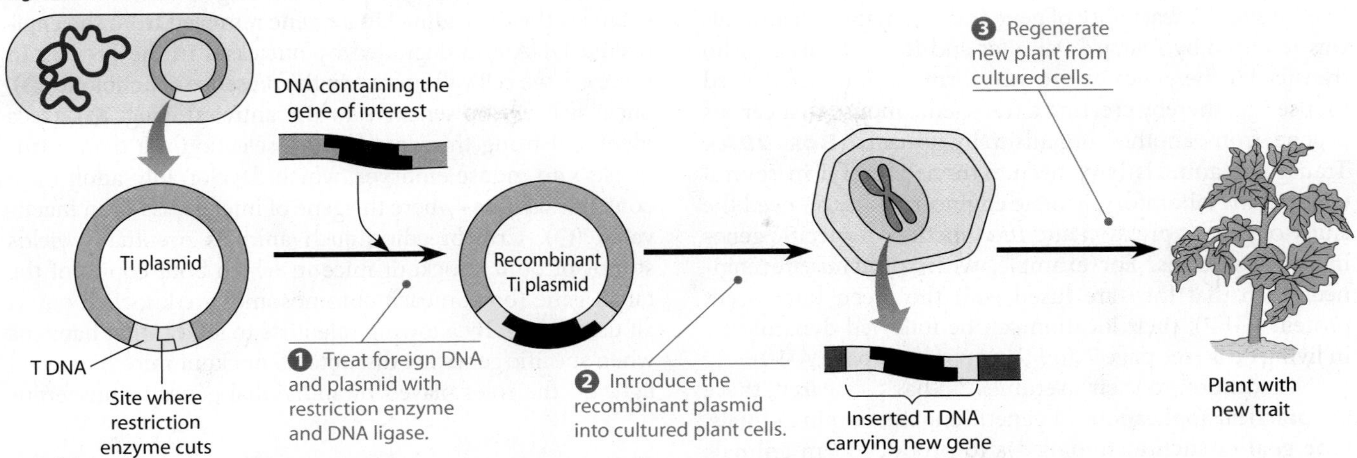

FIGURE 20-31 Using the Ti Plasmid to Transfer Genes into Plants. Most genetic engineering in plants uses the Ti plasmid as a vector. A DNA fragment containing a gene of interest is inserted into a restriction site located in the T DNA region of the plasmid. The recombinant plasmid is then introduced into plant cells, which regenerate a new plant containing the recombinant T DNA stably incorporated into the genome of every cell.

expected to depend on rice for food. Such dependence will be especially prevalent in poor, developing countries, where much of the population also suffers from deficiencies in essential nutrients and vitamins. To illustrate how genetic engineering might be used to improve the situation, the genes required for the synthesis of β-carotene, a precursor of vitamin A, were genetically engineered into rice in 2001, and the β-carotene content of such rice was increased more than 20-fold a few years later. The resulting product, called "golden rice" because of the color imparted by β-carotene, has the potential to help alleviate a global vitamin A deficiency that now causes blindness and disease in millions of children.

Concerns Have Been Raised About the Safety and Environmental Risks of GM Crops

As the prevalence of GM crops has begun to increase, some people have expressed concerns about the possible risks associated with this technology, especially because it permits genes to cross species barriers that cannot be crossed by traditional breeding techniques. For consumers, the main focus has been on safety because it is well known that toxic and allergic reactions to things we eat can be serious and even life-threatening. Thus far, it seems clear that conventional foods already on the market, such as peanuts and Brazil nuts, pose greater allergy risks than have been demonstrated for any GM foods. Moreover, the same problem also arises when new crop strains are produced by conventional breeding techniques. In fact, conventional foods have already undergone massive changes in genetic makeup by traditional plant-breeding methods, and it can be argued that it is safer to insert carefully selected genes into plants one at a time, as is done in genetic engineering, than it is to alter thousands of genes at a time, as is done with traditional crossbreeding of plants. When a single gene is inserted into a crop, the effects of that single gene and its protein product can be more easily assessed for safety hazards.

The possibility that GM crops may pose environmental risks has also raised concerns. In a 1999 laboratory experiment, scientists reported that monarch butterfly larvae died after being fed leaves dusted with pollen from GM corn containing the *Bt* gene. This observation led to fears that the *Bt* toxin produced by GM plants can harm friendly insects. However, the lab bench is not the farm field, and the initial studies were carried out under artificial laboratory conditions in which butterfly larvae consumed far higher doses of *Bt* toxin than they would in the real world. Subsequent data collected from farm fields containing GM crops suggest that the amount of *Bt* corn pollen encountered under real-life conditions does not pose a significant hazard to monarch butterflies.

Despite legitimate concerns about potential hazards, the GM experience has thus far revealed little evidence of significant risks to either human health or the environment. Of course, no new technology is entirely without risk, and so the safety and environmental impact of GM crops must be continually assessed. At the same time, GM crops have allowed reductions in pesticide use, and they could make a unique contribution to the fight against hunger and disease in developing countries.

Animals Can Be Genetically Modified by Adding or Knocking Out Specific Genetic Elements

MB Molecular Model: Cre Recombinase

Just as we have seen with plants, it is possible to genetically engineer animals. The techniques by which engineered DNA is introduced varies among different animals but often includes microinjection of engineered DNA into an adult animal or embryo. One of the first successful demonstrations of the feasibility of gene transplantation in animals was reported by Richard Palmiter and Ralph Brinster, who transferred the gene for growth hormone into a fertilized mouse egg, thereby creating a transgenic mouse that carries a gene from another organism in its cells (**Box 20A**). Transgenic animals have been extremely useful in several ways. In the laboratory, genetic engineering has allowed the study of gene expression and the function of specific genes in living animals. For example, when proteins are engineered so that they are fused with the green fluorescent protein (GFP), their location can be followed dynamically in living cells (see page 7 and the Appendix, page A-10).

In addition to their usefulness in basic research, there are practical applications of genetic engineering in animals. One goal of such technology is to produce farm animals that can synthesize medically important human proteins (e.g., in the milk of female mammals), which can then be easily purified. Another is the production of engineered livestock as a food source. Many of the same issues we considered regarding GM crops have also been raised concerning genetically modified animals grown for food.

While adding an engineered piece of DNA to an animal using transgenic technology can be very useful, it is often even more useful to *remove* a gene of interest. One way in which this can be done exploits the ability of homologous recombination to transfer base sequences between related DNA molecules. For example, scientists can now inactivate or "knock out" individually targeted genes by inducing recombination with related pieces of nonfunctional DNA (**Figure 20-32**). This strategy has been used to create hundreds of unique strains of **knockout mice,** each defective in a single gene, and efforts are under way to create knockout strains for every one of the mouse's roughly 25,000 genes. In 2007 Mario Capecchi, Oliver Smithies, and Martin Evans were awarded a Nobel Prize in recognition of their pioneering efforts in making the creation of knockout mice possible.

The first step in producing a knockout mouse is to synthesize an artificial DNA that is similar in base sequence to the target gene and its flanking sequences but with two important changes (Figure 20-32, ❶). First, an antibiotic resistance gene (e.g., that confers resistance to neomycin) is inserted into the middle of the target gene sequence. This simultaneously renders the engineered copy of the gene nonfunctional and also allows cells that carry the DNA to survive in the presence of antibiotic. Second, DNA encoding a viral enzyme, *thymidine kinase,* is attached to the end of the DNA. If this DNA is present in a cell, it will die in the presence of an antiviral drug (e.g., ganciclovir). Next this DNA is introduced into mouse *embryonic stem cells (ES cells),* which are cells that can differentiate into all the cell types of an adult mouse (❷). In very rare cases, the DNA enters the nucleus and, using mechanisms similar to those involved in meiosis, the artificial DNA aligns with complementary sequences flanking the targeted gene. Homologous recombination then replaces the targeted gene with the new, nonfunctional copy. Only if homologous recombination occurs is the thymidine kinase gene removed from the engineered DNA and degraded by nucleases in the ES cell. In this case, the cell will survive in the presence of antibiotic (❸), but it will *not* be sensitive to the antiviral drug (❹). Cells identified using this double drug selection are then introduced into mouse embryos, which develop into adult mice containing tissues where the gene of interest has been inactivated (❺). Crossbreeding such animals eventually yields strains of pure knockout mice in which both copies of the target gene (one on each chromosome) are knocked out in all tissues (❻). By allowing scientists to study what happens when specific genes are disrupted, knockout mice have shed light on the roles played by individual genes in numerous

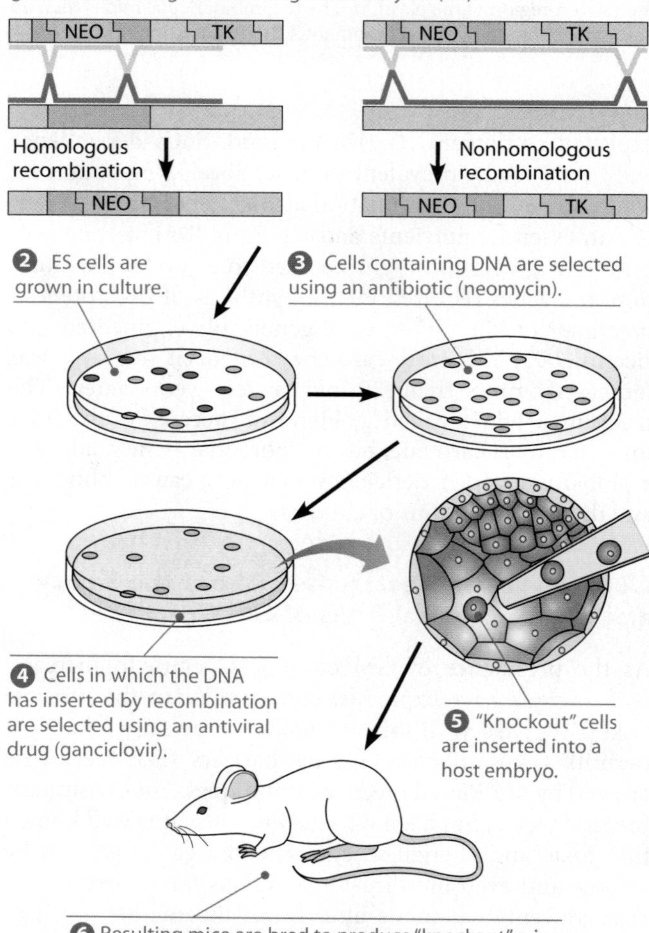
❶ DNA is introduced into embryonic stem (ES) cells. The DNA contains a non-functional copy of the gene of interest, an antibiotic resistance gene (Neo) and a gene encoding a viral enzyme (TK).
NEO
TK
NEO
TK
Homologous recombination
Nonhomologous recombination
NEO
NEO
TK
❷ ES cells are grown in culture.
❸ Cells containing DNA are selected using an antibiotic (neomycin).
❹ Cells in which the DNA has inserted by recombination are selected using an antiviral drug (ganciclovir).
❺ "Knockout" cells are inserted into a host embryo.
❻ Resulting mice are bred to produce "knockout" mice.

FIGURE 20-32 Making "Knockout" Mice by Homologous Recombination. The production of "knockout" mice involves ❶ the construction of a DNA targeting vector that contains a non-functional copy of the gene of interest, which ❷ is inserted into embryonic stem cells (ES cells). The ES cells are screened using two different drugs to select for rare cells in which a recombination event has occurred (❸, ❹). These "knockout" cells are ❺ inserted into a host embryo, and ❻ subsequent breeding produces mice carrying the engineered mutation.

BOX 20A TOOLS OF DISCOVERY

Making "Designer" Mice

A DNA fragment containing the gene of rat growth hormone was microinjected into the pronuclei of fertilized mouse eggs. Of 21 mice that developed from these eggs, seven carried the gene and six of these grew significantly larger than their littermates.
(Palmiter et al., 1982, p. 639)

With these words, a team of investigators led by Richard Palmiter and Ralph Brinster reported how a genetic trait can be introduced experimentally into mice without going through the usual procedure of breeding—that is, sexual reproduction followed by the selection of desired traits. The researchers injected rat growth hormone genes into fertilized mouse eggs, and from one of these eggs a "supermouse" developed weighing almost twice as much as its littermates (**Figure 20A-1**). The accomplishment was heralded as a significant breakthrough because it proved the feasibility of applying genetic engineering to animals, with all the scientific and practical consequences such engineering is likely to have.

Palmiter, Brinster, and their colleagues began by isolating the gene for growth hormone (GH) from a library of rat DNA, using techniques similar to those described in this chapter. The cloned GH gene, with its regulatory region deleted, was then fused to the regulatory portion of a mouse gene—the gene that codes for *metallothionein (MT).* MT is a small, metal-binding protein that is normally present in most mouse tissues and appears to be involved in regulating the level of zinc in the animal. The advantage of fusing the *MT* gene to the *GH* gene was that the expression of the *MT* gene could then be specifically induced (turned on) by zinc.

About 600 copies of the engineered DNA fragment were microinjected into fertilized mouse eggs, in a volume of about 2 picoliters (0.000002 μL!). The DNA was injected into the male pronucleus, the haploid sperm nucleus that has not yet fused with the haploid egg nucleus (**Figure 20A-2**). From the 170 fertilized eggs that were injected and implanted back into the reproductive tracts of foster mothers, 21 animals developed. Seven of them turned out to be transgenic mice with *MT–GH* genes present in their cells. In at least one case, a transgenic mouse transmitted the *MT–GH* gene faithfully to about half of its offspring, suggesting that the gene had become stably integrated into one of its chromosomes.

Because the *GH* gene had been linked to an *MT* gene regulatory region, it was predicted that the hybrid gene could be turned on by giving the mice zinc in their drinking water. The most dramatic evidence for expression of the rat *GH* genes was that the transgenic mice grew faster and weighed about twice as much as normal mice. During the period of maximum sensitivity to growth hormone (3 weeks to 3 months of age), the transgenic animals grew three to four times as fast as their normal littermates.

This dramatic experiment proved that it is possible to introduce cloned genes into the cells of higher organisms and that such genes can become stably integrated into the genome, where they are expressed and passed on to offspring. In the years since supermouse's creation, rapid progress has been made in most of these areas. Supermouse, it seems, was just the beginning.

FIGURE 20A-1 Genetic Engineering in Mice. "Supermouse" (on the left) is significantly larger than its littermate because it was engineered to carry, and express at high levels, the gene for rat growth hormone.

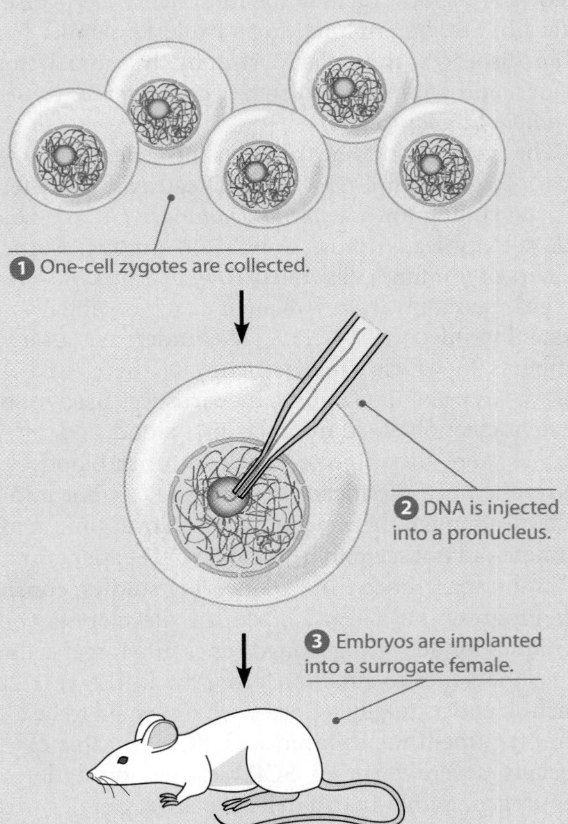
1 One-cell zygotes are collected.
2 DNA is injected into a pronucleus.
3 Embryos are implanted into a surrogate female.

FIGURE 20A-2 Making Transgenic Mice. Fertilized mouse eggs (one-cell zygotes) are injected with DNA, typically into a pronucleus (either the nucleus derived from the sperm or egg). The injected zygotes are transferred into a surrogate mother during subsequent gestation.

human conditions, including cancer, obesity, heart disease, diabetes, arthritis, and aging.

Knockout technology has now been extended to other mammals besides mice. One use of this technology relates to *xenotransplantation,* the transplantation of tissues from one species to another. Pigs have been a major focus of such research, which aims to provide a source of temporary organs for human patients awaiting transplants. To avoid immune rejection, pigs have been produced that lack a key enzyme called *α-1,3-galactosyltransferase,* which normally catalyzes the addition of sugar residues to cell surfaces in pigs that contributes to the immune response.

Gene Therapies Are Being Developed for the Treatment of Human Diseases

Humans suffer from many diseases that might conceivably be cured by transplanting normal, functional copies of genes into people who possess defective, disease-causing genes. The success of transgenic and knockout mice, along with similar experiments, raised the question of whether gene transplantation techniques might eventually be applied to the problem of repairing defective genes in humans. Obvious candidates for such an approach, called *gene therapy,* include the inherited genetic diseases cystic fibrosis, hemophilia, hypercholesterolemia, hemoglobin disorders, muscular dystrophy, lysosomal storage diseases, and an immune disorder called *severe combined immunodeficiency (SCID).*

The first person to be treated using gene therapy was a 4-year-old girl with a type of SCID caused by a defect in the gene coding for *adenosine deaminase (ADA).* Loss of ADA activity leads to an inability to produce sufficient numbers of immune cells called *T lymphocytes.* As a result, the girl suffered from frequent and potentially life-threatening infections. In 1990, she underwent a series of treatments in which a normal copy of the cloned ADA gene was inserted into a virus, the virus was used to infect T lymphocytes obtained from the girl's blood, and the lymphocytes were then injected back into her bloodstream. The result was a significant improvement in her immune function, although the effect diminished over time, and the treatment did not seem to help most SCID patients.

In the years since these pioneering studies, considerable progress has been made in developing better techniques for delivering cloned genes into target cells and getting the genes to function properly. In the year 2000, French scientists finally reported what seemed to be a successful treatment for children with SCID (in this case, an especially severe form of SCID caused by a defective receptor gene rather than a defective ADA gene). By using a virus that was more efficient at transferring cloned genes and by devising better conditions for culturing cells during the gene transfer process, these scientists were able to restore normal levels of immune function to the children they treated. In fact, the outcome was so dramatic that, for the first time, the treated children were able to leave the protective isolation "bubble" that had been used in the hospital to shield them from infections.

It was therefore a great disappointment when three of the ten children treated in the initial study developed leukemia a few years later. Examination of the leukemia cells revealed that the virus used to deliver the corrective gene sometimes inserts itself next to a normal gene that, when expressed abnormally, can cause cancer to arise. (In Chapter 24 we will describe exactly how such an event, called *insertional mutagenesis,* can initiate cancer development.) We should not, of course, lose sight of the fact that these studies also provided one of the first hopeful signs that gene therapy can cure a life-threatening genetic disease. But the associated cancer risks must be better understood before such treatments will become practical.

One tactic for addressing the problem of cancer risk is to change the type of virus being used to ferry genes into target cells. The SCID studies employed *retroviruses* (page 648), which randomly insert themselves into chromosomal DNA and possess sequences that inadvertently activate adjacent host genes. Another type of virus being investigated as a vehicle for gene therapy, called *adeno-associated virus (AAV),* is less likely to insert directly into chromosomal DNA and less likely to inadvertently activate host genes when it does become inserted. Some encouraging results using this virus have been obtained in patients with *hemophilia,* an inherited disease characterized by life-threatening episodes of uncontrolled bleeding. Hemophilia is caused by genetic defects in proteins called *blood-clotting factors,* which participate in the formation of blood clots. In gene therapy trials, hemophilia patients have been injected with AAV containing a gene coding for the blood-clotting factor they require. When the patients' liver cells were infected, they produced enough blood-clotting factor to alleviate the uncontrolled bleeding normally associated with the disease. Although this "cure" lasted only about eight weeks because an immune response destroyed the modified liver cells, the immune reaction targeted a component of AAV that is only transiently present. It is therefore hoped that short-term administration of immunosuppressive drugs may help provide a more permanent cure.

In the years since the enormous potential of gene therapy was first publicized in the early 1980s, the field has been criticized for promising too much and delivering too little. But most new technologies take time to be perfected and encounter disappointments along the way, and gene therapy is no exception. Despite the setbacks, it appears likely that using normal genes to treat genetic diseases is a reachable goal that may one day become common practice, at least for a few genetic diseases that involve single gene defects. Of course, the ability to alter people's genes raises important ethical, safety, and legal concerns. The ultimate question of how society will control our growing power to change the human genome is an issue that will need to be thoroughly discussed not just by scientists and physicians, but by society as a whole.

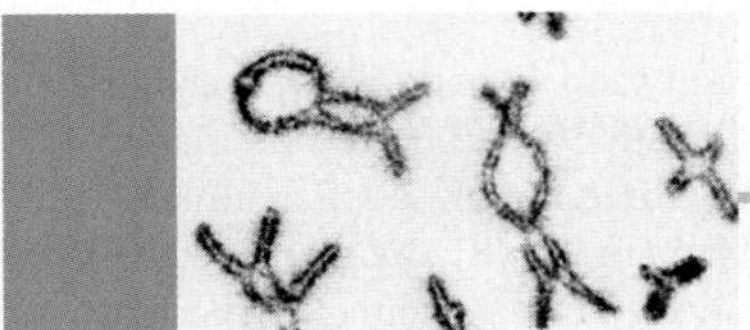

SUMMARY OF KEY POINTS

Sexual Reproduction

- Asexual reproduction is based on mitotic cell division and produces offspring that are genetically identical (or nearly so) to the single parent. In contrast, sexual reproduction involves two parents and leads to a mixture of parental traits in the offspring.
- Sexual reproduction allows populations to adapt to environmental changes, enables desirable mutations to be combined in a single individual, and promotes genetic flexibility by maintaining a diploid genome.

Meiosis

- The life cycle of sexually reproducing eukaryotes includes both haploid and diploid phases. Haploid gametes are generated by meiosis and fuse at fertilization to restore the diploid chromosome number.
- Meiosis consists of two successive cell divisions without an intervening duplication of chromosomes. During the first meiotic division, homologous chromosomes separate and segregate into the two daughter cells. During the second meiotic division, sister chromatids separate and four haploid daughter cells are produced.
- In addition to reducing the chromosome number from diploid to haploid, meiosis differs from mitosis in that homologous chromosomes synapse during prophase of the first meiotic division, thereby allowing crossing over and genetic recombination between nonsister chromatids.

Genetic Variability: Segregation and Assortment of Alleles

- Mendel's laws of inheritance describe the genetic consequences of chromosome behavior during meiosis, even though chromosomes had not yet been discovered at the time of Mendel's experiments.
- Mendel's law of segregation states that the two (maternal and paternal) alleles of a gene are distinct entities that separate into different gametes during meiosis. The law of independent assortment states that the alleles of each gene separate independently of the alleles of other genes (now known to apply only to genes located on different chromosomes or very far apart on the same chromosome).

Genetic Variability: Recombination and Crossing Over

- The genetic variability among an organism's gametes arises partly from the independent assortment of chromosomes during anaphase I and partly from genetic recombination during prophase I.
- The frequency of recombination between genes on the same chromosome is a measure of the distance between the two genes and can therefore be used to map their chromosomal locations.

Genetic Recombination in Bacteria and Viruses

- Besides occurring during meiosis, homologous recombination also takes place in viruses (during co-infection) and when DNA is transferred into bacterial cells by transformation, transduction, or conjugation.

Molecular Mechanism of Homologous Recombination

- Recombination involves breakage and exchange between DNA molecules exhibiting extensive sequence homology.
- Recombination is sometimes accompanied by gene conversion or the formation of DNA molecules whose two strands are not completely complementary to one another. These phenomena can be explained by recombination models involving the formation of Holliday junctions, which are regions of single-strand exchange between double-stranded DNA molecules.

Recombinant DNA Technology and Gene Cloning

- Recombinant DNA technology makes it possible to combine DNA from any two (or more) sources into a single DNA molecule.
- Combining a gene of interest with a plasmid or phage cloning vector allows the gene to be cloned (amplified) in bacterial cells. This approach, as well as the polymerase chain reaction (PCR), permits large quantities of specific gene sequences and their protein products to be prepared for research and practical purposes.

Genetic Engineering

- Recombinant DNA technology has many practical applications in medicine and agriculture. These include the ability to produce valuable proteins that are otherwise difficult to obtain and the ability to improve the traits of food crops, such as enhanced resistance to insects, herbicides, or disease. Attempts are also underway to develop gene therapies for treating human diseases.

MB **www.masteringbiology.com**

1. MasteringBiology Assignments

Tutorials and Activities

- Molecular Model Structural Tutorials for visualization of 3D structures
- BioFlix Tutorials including 3D animations with hints and detailed feedback for common wrong answers
- Meiosis (1 of 3): Genes, Chromosomes, and Sexual Reproduction (BioFlix tutorial); Meiosis (2 of 3): The Mechanism (BioFlix tutorial); Meiosis Animation; Chromosome Separation in Meiosis: Disjunction; Nondisjunction; Mistakes in Meiosis; Video Tutor Session Quiz: Mitosis vs. Meiosis; Mendel's Experiments; Mendel's Law of Segregation; Mendel's Principle of Segregation; Mendel's Law of Independent Assortment; Mendel's Principle of Independent Assortment; Determining Genotype: Pea Pod Color; Dihybrid Cross; The Principle of Independent Assortment; Make Connections: Chromosomal Inheritance and Independent Assortment of Alleles; Linkage and Recombination; Meiosis (3 of 3): Determinants of Heredity and Genetic Variation (BioFlix tutorial); Origins of Genetic Variation; Recombination and Linkage Mapping; Three-Point Mapping; Linked Genes and Linkage Mapping; RecBCD Complexed with DNA; The RuvA Recombination Protein of Escherichia coli; Cloning a Gene in Bacteria; DNA Cloning in a Plasmid Vector; Restriction Enzymes, Recombinant DNA, and Gene Cloning; Producing Human Growth Hormone; Cre Recombinase

Questions

- Reading Quiz
- Multiple Choice
- End-of-chapter questions

2. eText Read your book online, search, highlight text, take notes, and more

3. The Study Area

- Chapter Review Quiz, Guide to Techniques and Methods, 3D-Structure Tutorials, Activities, BioFlix, Videos, Biology Review, Word Study Tools, Glossary, Flashcards

PROBLEM SET

More challenging problems are marked with a •.

20-1 The Truth About Sex. For each of the following statements, indicate with an S if it is true of sexual reproduction, with an A if it is true of asexual reproduction, with a B if it is true of both, or with an N if it is true of neither.

(a) Traits from two different parents can be combined in a single offspring.

(b) Each generation of offspring is virtually identical to the previous generation.

(c) Mutations are propagated to the next generation.

(d) Some offspring in every generation will be less suited for survival than the parents, but others may be better suited.

(e) Mitosis is involved in the life cycle.

20-2 Ordering the Phases of Meiosis. Drawings of several phases of meiosis in an organism, labeled A through F, are shown in **Figure 20-33**.

(a) What is the diploid chromosome number in this species?

(b) Place the six phases in chronological order, and name each one.

(c) Between which two phases do homologous centromeres separate?

(d) Between which two phases does recombination occur?

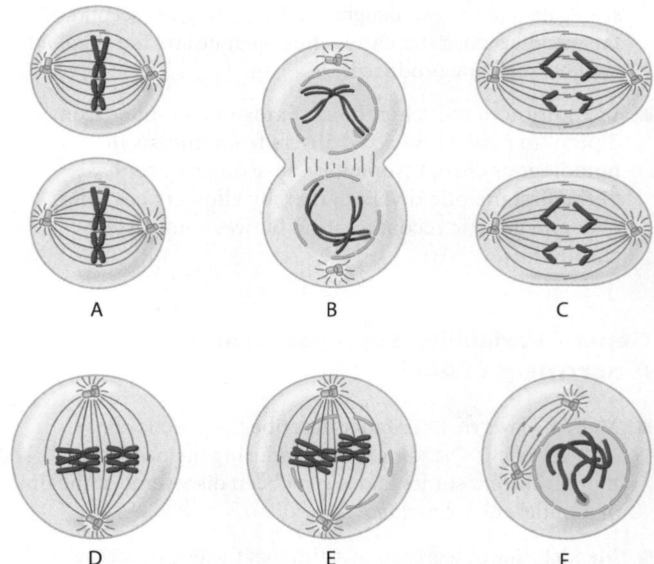

FIGURE 20-33 Six Phases of Meiosis to Be Ordered and Identified. See Problem 20-2.

20-3 Telling Them Apart. Briefly describe how you might distinguish between each of the following pairs of phases in the same organism:

(a) Metaphase of mitosis and metaphase I of meiosis.

(b) Daughter cells in mitosis and in meiosis.

(c) Orientation of the spindle in mitosis and meiosis.

(d) Homologous recombination in mitosis and the same in meiosis.

(e) Pachytene and leptotene stages of meiotic prophase I.

20-4 Your Centromere Is Showing. Suppose you have a diploid organism in which all the chromosomes contributed by the sperm have cytological markers on their centromeres that allow you to distinguish them visually from the chromosomes contributed by the egg.

(a) Would you expect all the somatic cells (cells other than gametes) to have equal numbers of maternal and paternal centromeres in this organism? Explain.

(b) Would you expect equal numbers of maternal and paternal centromeres in each gamete produced by that individual? Explain.

20-5 How Much DNA? Let *X* be the amount of DNA present in the gamete of an organism that has a diploid chromosome number of 4. Assuming all chromosomes to be of approximately the same size, how much DNA (*X*, 2*X*, 1/2*X*, and so on) would you expect in each of the following?

(a) A zygote immediately after fertilization

(b) A triploid mutant cell

(c) In the oocyte

(d) A single chromosome following mitosis

(e) A nucleus in mitotic prophase

(f) The cell during metaphase II of meiosis

(g) In a somatic cell of the organism

(h) In the cancer cell of the organism

(i) In the apoptotic blebs of the organism

20-6 Meiotic Mistakes. Infants born with Patau syndrome have an extra copy of chromosome 13, which leads to developmental abnormalities such as cleft lip and palate, small eyes, and extra fingers and toes. Another type of genetic disorder, called Turner syndrome, results from the presence of only one sex chromosome—an X chromosome. Individuals born with one X chromosome are females exhibiting few noticeable defects until puberty, when they fail to develop normal breasts and internal sexual organs. Describe the meiotic events that could lead to the birth of an individual with either Patau syndrome or Turner syndrome.

20-7 Punnett Squares as Genetic Tools. A *Punnett square* is a diagram representing all possible outcomes of a genetic cross. The genotypes of all possible gametes from the male and female parents are arranged along two adjacent sides of a square, and each box in the matrix is then used to represent the genotype resulting from the union of the two gametes at the heads of the intersecting rows. By the law of independent assortment, all possible combinations are equally likely, so the frequency of a given genotype among the boxes represents the frequency of that genotype among the progeny of the genetic cross represented by the Punnett square.

Figure 20-34 shows the Punnett squares for two crosses of pea plants. The genetic characters involved are seed color (where *Y* is the allele for yellow seeds and *y* for green seeds) and seed shape (where *R* is the allele for round seeds and *r* for wrinkled seeds). The Punnett square in Figure 20-34a represents a one-factor cross between parent plants that are both heterozygous for seed color (*Yy* × *Yy*). The Punnett square in Figure 20-34b is a two-factor cross between plants heterozygous for both seed color (*Yy*) and seed shape (*Rr*).

(a) Using the Punnett square of Figure 20-34a, explain the 3:1 phenotypic ratio Mendel observed for the offspring of such a cross.

(b) Explain why the Punnett square of Figure 20-34b is a 4 × 4 matrix with 16 genotypes. In general, what is the mathematical relationship between the number of heterozygous allelic pairs being considered and the number of different kinds of gametes?

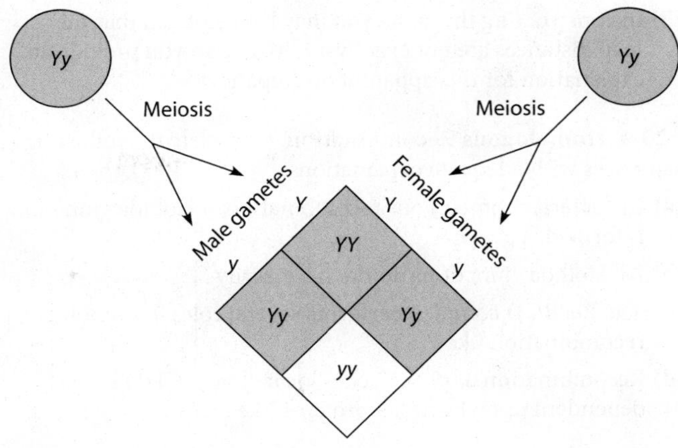

(a) One-factor cross

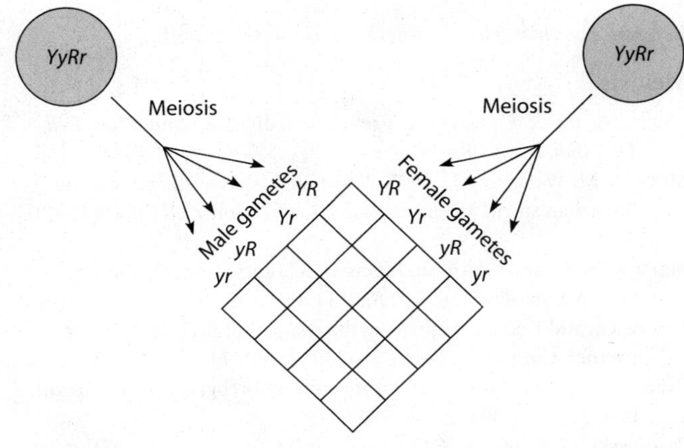

(b) Two-factor cross

FIGURE 20-34 Punnett Squares. See Problem 20-7.

(c) How does the Punnett square of Figure 20-34b reflect Mendel's law of independent assortment?

(d) Complete the Punnett square of Figure 20-34b by writing in each of the possible progeny genotypes. How many different genotypes will be found in the progeny? In what ratios?

(e) For the case of Figure 20-34b, how many different phenotypes will be found in the progeny? In what ratios?

• **20-8 Genetic Mapping.** The following table provides data concerning the frequency with which four genes (*w*, *x*, *y*, and *z*) located on the same chromosome recombine with each other.

Genes	Recombination Frequency
w and *x*	25%
w and *y*	29%
w and *z*	17%
x and *y*	50%
x and *z*	9%
y and *z*	44%

(a) Construct a genetic map indicating the order in which these four genes occur and the number of map units that separate the genes from each other.

(b) In constructing this map, you may have noticed that the map distances are not exactly additive. Can you provide an explanation for this apparent discrepancy?

• **20-9 Homologous Recombination.** Complete the following sentences with adequate explanations.

(a) In bacterial homologous recombination, a Holliday junction is formed . . .

(b) All Holliday junctions do not necessarily . . .

(c) The RecBCD complex performs several roles in homologous recombination like . . .

(d) Recombination of phage genes located near CHI sites is dependent upon bacterial proteins like . . .

20-10 Gene Cloning and Recombination. Not only are the plasmid vectors used in molecular biology engineered, but the strains of *E. coli* used in cloning are as well.

(a) Nearly all strains of *E. coli* used in DNA cloning carry mutations in the *recA* gene that result in loss of RecA activity. Why would a *recA* mutation make an *E. coli* cell a better host for propagating recombinant plasmid DNA?

(b) Since restriction endonucleases are bacterial proteins, what impact would a strain that produces functional RE enzymes have on plasmids introduced in the strain? How would you avoid potential problems with such a strain?

SUGGESTED READING

References of historical importance are marked with a •.

Meiosis

Bhalla, N., and A. F. Dernburg. Prelude to a division. *Annu. Rev. Cell Dev. Biol.* 24 (2008): 397.

Macy, B., M. Wang, and H. G. Yu. The many faces of shugoshin, the "guardian spirit," in chromosome segregation. *Cell Cycle* 8 (2009): 35.

Marston, A. L., and A. Amon. Meiosis: Cell-cycle controls shuffle and deal. *Nature Rev. Mol. Cell Biol.* 5 (2004): 983.

Moore, G., and P. Shaw. Improving the chances of finding the right partner. *Curr. Opin. Genet. Dev.* 19 (2009): 99.

Page, S. L., and R. S. Hawley. Chromosome choreography: The meiotic ballet. *Science* 301 (2003): 785.

Pawlowski, W. P., and W. Z. Cande. Coordinating the events of the meiotic prophase. *Trends Cell Biol.* 15 (2005): 674.

Tomita, K., and J. P. Cooper. The meiotic chromosomal bouquet: SUN collects flowers. *Cell* 125 (2006): 19.

Mendel's Experiments

• Mendel, G. et al. The birth of genetics. *Genetics* 35 (1950, Suppl.): 1 (original papers in English translation).

• Sturtevant, A. H. *A History of Genetics.* New York: Harper & Row, 1965.

Mechanism of Recombination

Chen, J.-M. et al. Gene conversion: Mechanisms, evolution and human disease. *Nature Rev. Genet.* 8 (2007): 762.

Clark, A. J. *recA* mutants of *E. coli* K12: A personal turning point. *BioEssays* 18 (1996): 767.

Cromie, G. A., and G. R. Smith. Branching out: Meiotic recombination and its regulation. *Trends Cell Biol.* 17 (2007): 448.

Heyer, W.-D., K. T. Ehmsen, and J. A. Solinger. Holliday junctions in the eukaryotic nucleus: Resolution in sight? *Trends Biochem. Sci.* 28 (2003): 548.

Lilley, D. M. J., and M. F. White. The junction-resolving enzymes. *Nature Rev. Mol. Cell Biol.* 2 (2001): 433.

Liu, Y., and S. C. West. Happy Hollidays: 40th anniversary of the Holliday junction. *Nature Rev. Mol. Cell Biol.* 5 (2004): 937.

Stahl, F. Meiotic recombination in yeast: Coronation of the double-strand-break repair model. *Cell* 87 (1996): 965.

Recombinant DNA Technology and Genetic Engineering

Brown, K., K. Hopkin, and S. Nemecek. Genetically modified foods: Are they safe? *Sci. Amer.* 284 (April 2001): 51.

Cavazanna-Calvo, M., A. Thrasher, and F. Mavilio. The future of gene therapy. *Nature* 427 (2004): 779.

• Chilton, M. D. A vector for introducing new genes into plants. *Sci. Amer.* 248 (June 1983): 50.

Friedman, T. Overcoming the obstacles to gene therapy. *Sci. Amer.* 276 (June 1997): 96.

Kay, M. A., and H. Nakai. Looking into the safety of AAV vectors. *Nature* 424 (2003): 251.

Kohn, D. B., M. Sadelain, and J. C. Glorioso. Occurrence of leukemia following gene therapy of X-linked SCID. *Nature Rev. Cancer* 3 (2003): 477.

Lemaux, P. G. Genetically engineered plants and foods: A scientist's analysis of the issues (part II). *Annu. Rev. Plant Biol.* 60 (2009): 511.

Manno, C. S. et al. Successful transduction of liver in hemophilia by AAV-Factor IX and limitations imposed by the host immune response. *Nature Med.* 12 (2006): 342.

• Palmiter, R. D. et al. Dramatic growth of mice that develop from eggs microinjected with metallothionein-growth hormone fusion genes. *Nature* 300 (1982): 611.

Phelps, C. J., et al. Production of α1,3-galactosyltransferase-deficient pigs. *Science* 299 (2003): 411.

Raney, G., and P. Pingali. Sowing a green revolution. *Sci. Amer.* 297 (September 2007): 104.

Thieman, W. J., and M. A. Palladino. *Introduction to Biotechnology,* 2nd ed. San Francisco: Pearson, 2008.

Verma, I. M., and M. D. Weitzman. Gene therapy: Twenty-first century medicine. *Annu. Rev. Biochem.* 74 (2005): 711.

Waehler, R., S. J. Russell, and D. T. Curiel. Engineering targeted viral vectors for gene therapy. *Nature Rev. Genet.* 8 (2007): 573.

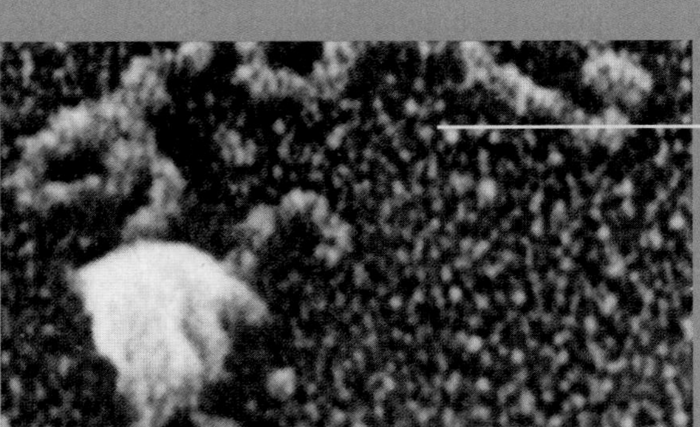

21 Gene Expression: I. The Genetic Code and Transcription

See MasteringBiology® for tutorials, activities, and quizzes.

So far, we have described DNA as the genetic material of cells and organisms. We have come to understand its structure, chemistry, and replication, as well as the way it is packaged into chromosomes and parceled out to daughter cells during mitotic and meiotic cell divisions. Now we are ready to explore how DNA is expressed—*that is, how the coded information it contains is used to guide the production of RNA and protein molecules. Our discussion of this important subject is divided among three chapters. The present chapter deals with the nature of the genetic code and how information stored in DNA guides the synthesis of RNA molecules in the process we call* transcription. *Chapter 22 describes how RNA molecules are then used to guide the synthesis of specific proteins in the process known as* translation. *Finally, Chapter 23 elaborates on the various mechanisms used by cells to control transcription and translation, thereby leading to the* regulation of gene expression. *To put these topics in context, we start here with an overview of the roles played by DNA, RNA, and proteins in gene expression.*

The Directional Flow of Genetic Information

Activity: Overview of Protein Synthesis; Types of RNA

As mentioned at the beginning of this series of chapters, the flow of genetic information in cells generally proceeds from DNA to RNA to protein (see Figure 18-1). DNA (more precisely, a segment of one DNA strand) first serves as a template for the synthesis of an RNA molecule, which in most cases then directs the synthesis of a particular protein. (In a few cases, the RNA is the final product of gene expression and functions as such within the cell.) The principle of directional information flow from DNA to RNA to protein is known as the *central dogma of molecular biology,* a term coined by Francis Crick soon after the double-helical model of DNA was first proposed. This principle is summarized as follows:

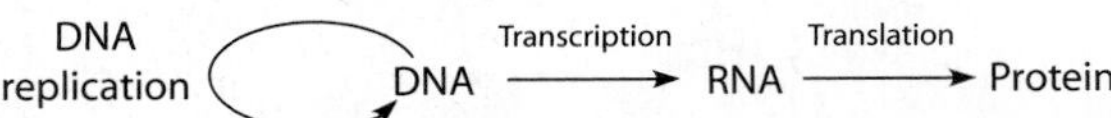

Thus, the flow of genetic information involves replication of DNA, transcription of information carried by DNA into the form of RNA, and translation of this information from RNA into protein. The term **transcription** is used when referring to RNA synthesis using DNA as a template to emphasize that this phase of gene expression is simply a transfer of information from one nucleic acid to another, so the basic "language" remains the same. In contrast, protein synthesis is called **translation** because it involves a language change—from the nucleotide sequence of an RNA molecule to the amino acid sequence of a polypeptide chain.

RNA that is translated into protein is called **messenger RNA (mRNA)** because it carries a genetic message from DNA to the ribosomes, where protein synthesis actually takes place. In addition to mRNA, two other types of RNA are involved in protein synthesis: **ribosomal RNA (rRNA)** molecules, which are integral components of the ribosome, and **transfer RNA (tRNA)** molecules, which serve as intermediaries that translate the coded base sequence of messenger RNA and bring the appropriate amino acids to the ribosome. Note that ribosomal and transfer RNAs do not themselves code for proteins; genes coding for these two types of RNA are examples of genes whose final products are RNA molecules rather than protein chains. The involvement of all three major classes of RNA in the overall flow of information from DNA to protein is outlined in **Figure 21-1**.

In the years since it was first formulated by Crick, the central dogma has been refined in various ways. For example, many viruses with RNA genomes have been found to synthesize RNA molecules using RNA as a

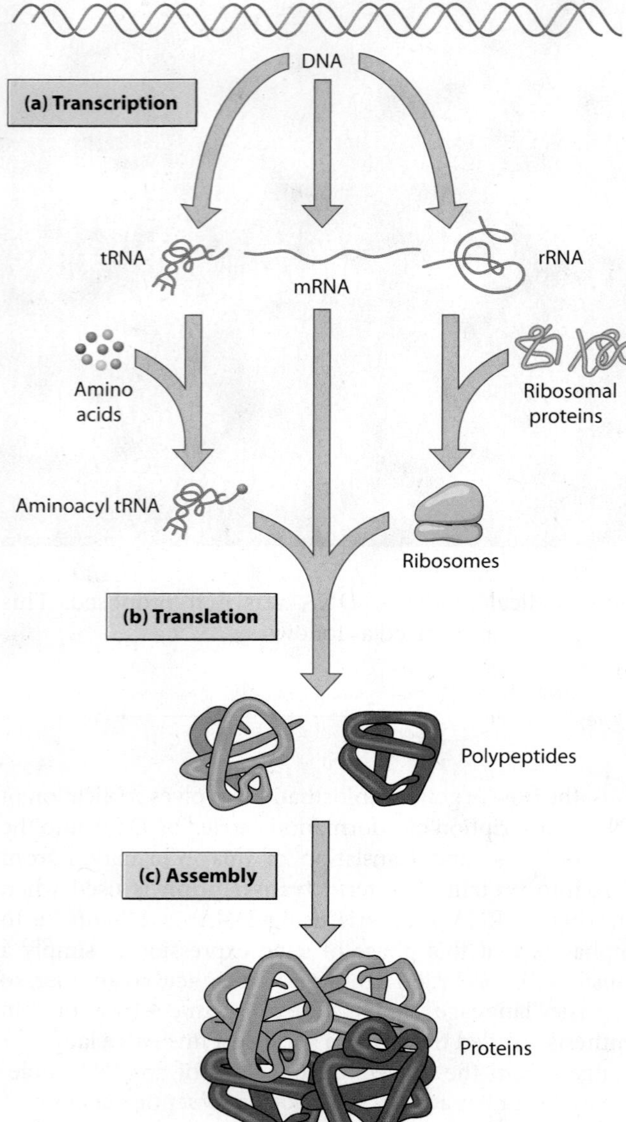

MB Activity: Role of the Nucleus and Ribosomes in Protein Synthesis

FIGURE 21-1 RNAs as Intermediates in the Flow of Genetic Information. All three major classes of RNA—tRNA, mRNA, and rRNA—are **(a)** synthesized by transcription of the appropriate DNA sequences (genes) and **(b)** involved in the subsequent process of translation (polypeptide synthesis). The appropriate amino acids are brought to the mRNA and ribosome by tRNA. A tRNA molecule carrying an amino acid is called an aminoacyl tRNA. Polypeptides then fold and **(c)** assemble into functional proteins. The specific polypeptides shown here are the globin chains of the protein hemoglobin. For simplicity, this figure omits many details that will be described in this chapter and the next.

template. Other RNA viruses, such as HIV, carry out *reverse transcription*, whereby the viral RNA is used as a template for DNA synthesis—a "backward" flow of genetic information. (**Box 21A** discusses these viruses and the role of reverse transcription in rearranging DNA sequences.) But despite these variations on the original model, the principle that information flows from DNA to RNA to protein remains the main operating principle by which all cells use their genetic information.

The Genetic Code

The essence of gene expression lies in the relationship between the nucleotide base sequence of DNA molecules and the linear order of amino acids in protein molecules. This relationship is based on a set of rules known as the **genetic code.** The cracking of that code, which tells us how DNA can code for proteins, is one of the major landmarks of twentieth-century biology.

During the flow of information from DNA to RNA to protein, it is easy to envision how information residing in a DNA base sequence could be passed to mRNA through the mechanism of complementary base pairing. But how does a base sequence in mRNA use its "message" to guide the synthesis of a protein molecule, which consists of a sequence of amino acids? What is needed, of course, is knowledge of the appropriate code—the set of rules that determines which nucleotides in mRNA correspond to which amino acids. Until it was cracked in the early 1960s, this genetic code was a secret code in a double sense: Before scientists could figure out the exact coding relationship between the base sequence of a DNA molecule and the amino acid sequence of a protein, they first had to become aware that such a relationship existed at all. That awareness arose from the discovery that mutations in DNA can lead to changes in proteins.

Experiments on *Neurospora* Revealed That Genes Can Code for Enzymes

The link between gene mutations and proteins was first detected experimentally by George Beadle and Edward Tatum in the early 1940s using the common bread mold, *Neurospora crassa. Neurospora* is a relatively self-sufficient organism that can grow in a *minimal medium* containing only sugar, inorganic salts, and the vitamin biotin. From these few ingredients, *Neurospora*'s metabolic pathways produce everything else the organism requires. To investigate the influence of genes on these metabolic pathways, Beadle and Tatum treated a *Neurospora* culture with X-rays to induce genetic mutations. Such treatments generated mutant strains that had lost the ability to survive in the minimal culture medium, although they could be grown on a *complete medium* supplemented with a variety of amino acids, nucleosides, and vitamins.

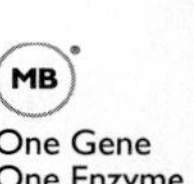

One Gene One Enzyme

Such observations suggested that the *Neurospora* mutants had lost the ability to synthesize certain amino acids or vitamins and could survive only when these nutrients were added to the growth medium. To determine exactly which nutrients were required, Beadle and Tatum transferred the mutant cells to a variety of growth media, each containing a single amino acid or vitamin added as a supplement to the minimal medium. This approach led to the discovery that one mutant strain would grow only in a medium supplemented with vitamin B_6, a second mutant would grow only when the medium was supplemented with the amino acid arginine, and so forth. A large number of different mutants were eventually characterized, each impaired in its ability to synthesize a particular amino acid or vitamin.

Because amino acids and vitamins are synthesized by metabolic pathways involving multiple steps, Beadle and Tatum set out to identify the particular step in each pathway that had become defective. They approached this task by supplementing the minimal medium with metabolic precursors of a given amino acid or vitamin rather than with the amino acid or vitamin itself. By finding out which precursors supported the growth of each mutant strain, they were able to infer that each mutation disabled a single enzyme-catalyzed step in a pathway for making a particular compound. There was, in other words, a one-to-one correspondence between each genetic mutation and the lack of a specific enzyme required in a metabolic pathway. From these findings, Beadle and Tatum formulated the *one gene–one enzyme hypothesis,* which stated that each gene controls the production of a single type of enzyme.

Most Genes Code for the Amino Acid Sequences of Polypeptide Chains

The theory that genes direct the production of enzyme molecules represented a major advance in our understanding of gene action, but it provided little insight into the question of how genes accomplish this task. The first clue to the underlying mechanism emerged a few years later in the laboratory of Linus Pauling, who was studying the inherited disease *sickle-cell anemia.* The red blood cells of individuals suffering from sickle-cell anemia exhibit an abnormal, "sickle" shape that causes the cells to become trapped and damaged when they pass through small blood vessels (**Figure 21-2**). In trying to identify the reason for this behavior, Pauling decided to analyze the properties of *hemoglobin,* the major protein of red blood cells. Because hemoglobin is a charged molecule, he used the technique of *electrophoresis,* which separates charged molecules from one another by placing them in an electric field. Pauling found that hemoglobin from sickle cells migrated at a different rate than normal hemoglobin, suggesting that the two proteins differ in electric charge. Since some amino acids have charged side chains, Pauling proposed that the difference between normal and sickle-cell hemoglobin lay in their amino acid compositions.

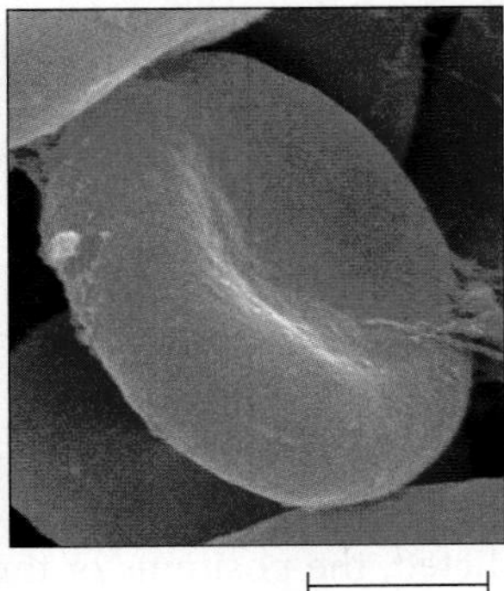

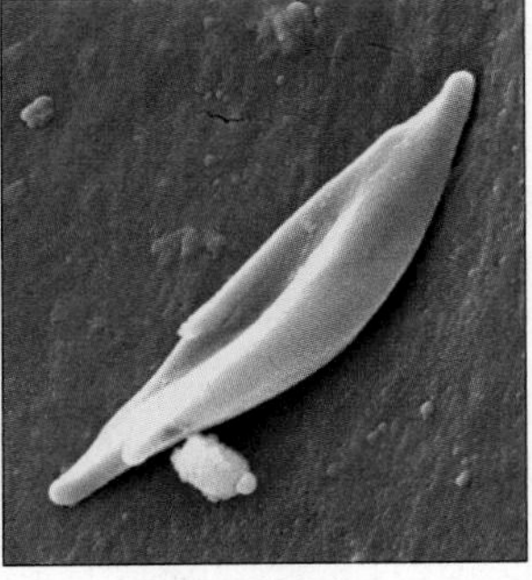

2 μm

FIGURE 21-2 Normal and Sickled Red Blood Cells. The micrograph on the right reveals the abnormal shape of a sickled cell. This distorted shape, which is caused by a mutated form of hemoglobin, allows sickled cells to become trapped and damaged when passing through small blood vessels (SEMs).

One way to test this hypothesis would be to determine the amino acid sequence of the normal and mutant forms of hemoglobin. At the time of Pauling's discovery in the early 1950s, the largest protein to have been sequenced was less than one-tenth the size of hemoglobin, so determining the complete amino acid sequence of hemoglobin would have been a monumental undertaking. Fortunately, an ingenious shortcut devised by Vernon Ingram made it possible to identify the amino acid abnormality in sickle-cell hemoglobin without determining the protein's complete amino acid sequence. Ingram used the protease *trypsin* to cleave hemoglobin into peptide fragments, which were then separated from each other as shown in **Figure 21-3** (page 650). When Ingram examined the peptide patterns of normal and sickle-cell hemoglobin, he discovered that only one peptide differed between the two proteins. Analysis of the amino acid makeup of the altered peptide revealed that a glutamic acid in normal hemoglobin had been replaced by a valine in sickle-cell hemoglobin. Since glutamic acid is negatively charged and valine is neutral, this substitution explains the difference in electrophoretic behavior between normal and sickle-cell hemoglobin originally observed by Pauling.

This single change from a glutamic acid to valine (caused by a single base-pair change in DNA) is enough to alter the way that hemoglobin molecules pack into red blood cells. Normal hemoglobin is jellylike in consistency, but sickle-cell hemoglobin tends to form a kind of crystal when it delivers oxygen to and picks up carbon dioxide from tissues. The crystalline array deforms the red blood cell into a sickled shape that blocks blood flow in capillaries and leads to a debilitating, potentially fatal, disease.

Following Ingram's discovery that a gene mutation alters a single amino acid in sickle-cell hemoglobin, subsequent studies revealed the existence of other abnormal forms of hemoglobin, some of which involve mutations in a different gene. Two different genes are able to influence the amino acid sequence of the same protein because hemoglobin is a multisubunit protein containing two different kinds of polypeptide chains. The amino acid sequences of the two types of chains, called the α and β chains, are specified by two different genes.

The discoveries by Pauling and Ingram necessitated several refinements in the one gene–one enzyme concept of Beadle and Tatum. First, the fact that hemoglobin is not an enzyme indicates that genes encode the amino acid sequences of proteins in general, not just enzymes. In addition, the discovery that different genes code for the α and β chains of hemoglobin reveals that each gene encodes the sequence of a polypeptide chain, not necessarily a complete protein. Thus, the original hypothesis was refined into the *one gene–one polypeptide theory.*

BOX 21A DEEPER INSIGHTS

Reverse Transcription, Retroviruses, and Retrotransposons

Transcription generally proceeds in the direction described in the central dogma, with DNA serving as a template for RNA synthesis. In certain cases, however, the process can be reversed and RNA serves as a template for DNA synthesis. This process of *reverse transcription* is catalyzed by the enzyme **reverse transcriptase,** first discovered by Howard Temin and David Baltimore in certain viruses with RNA genomes. Viruses that carry out reverse transcription are called **retroviruses.** Examples of retroviruses include some important pathogens, such as the *human immunodeficiency virus (HIV),* which causes *aquired immune deficiency syndrome (AIDS),* and a number of viruses that cause cancers in animals.

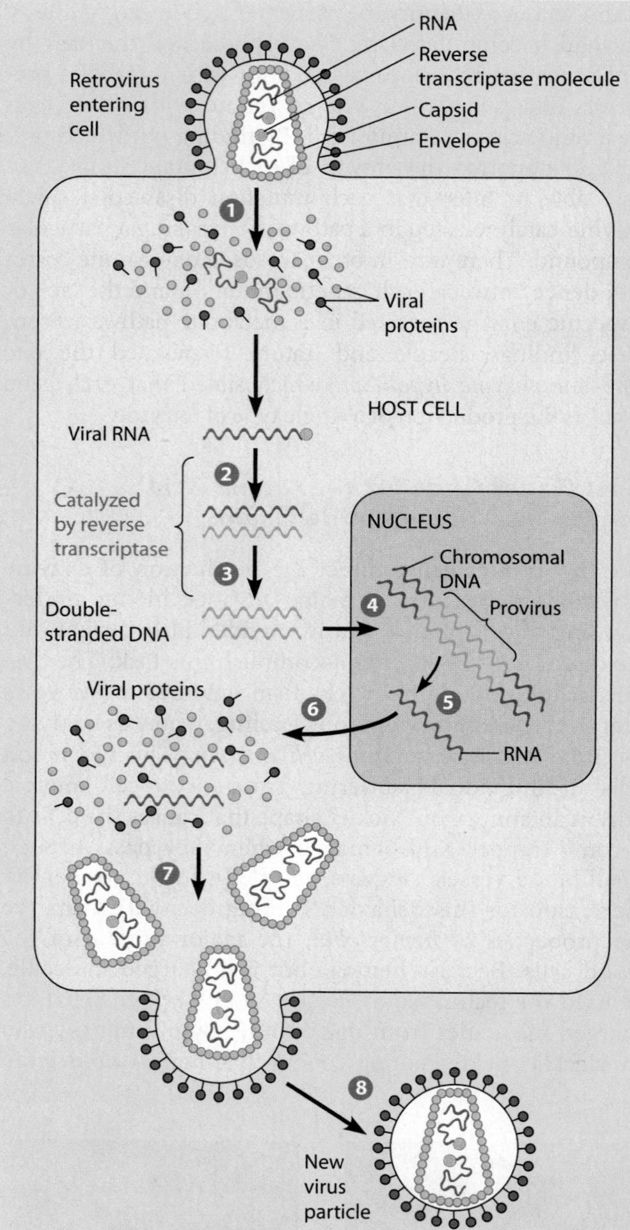

FIGURE 21A-1 The Reproductive Cycle of a Retrovirus.

Retroviruses

Figure 21A-1 depicts the reproductive cycle of a typical retrovirus. In the virus particle, two copies of the RNA genome are enclosed within a protein capsid that is surrounded by a membranous envelope. Each RNA copy has a molecule of reverse transcriptase attached to it. The virus first (1) binds to the surface of the host cell, and its envelope fuses with the plasma membrane, releasing the capsid and its contents into the cytoplasm. Once inside the cell, the viral reverse transcriptase (2) catalyzes the synthesis of a DNA strand that is complementary to the viral RNA and then (3) catalyzes the formation of a second DNA strand complementary to the first. The result is a double-stranded DNA version of the viral genome. (4) This double-stranded DNA then enters the nucleus and integrates into the host cell's chromosomal DNA, much as the DNA genome of a lysogenic phage integrates into the DNA of the bacterial chromosome (see Box 18A). Like a prophage, the integrated viral genome, called a *provirus,* is replicated every time the cell replicates its own DNA. (5) Transcription of the proviral DNA (by cellular enzymes) produces RNA transcripts that function in two ways. First, they serve as (6) mRNA molecules that direct the synthesis of viral proteins (capsid protein, envelope protein, and reverse transcriptase). Second, (7) some of these same RNA transcripts are packaged with the viral proteins into new virus particles. (8) The new viruses then "bud" from the plasma membrane without necessarily killing the cell.

The ability of a retroviral genome to integrate into host cell DNA helps explain how some retroviruses can cause cancer. These viruses, called RNA tumor viruses, are of two types. Viruses of the first type carry a cancer-causing *oncogene* in their genomes, along with the genes coding for viral proteins. As we will see in Chapter 24, an oncogene is a mutated version of a normal cellular gene (a proto-oncogene) that codes for proteins used to regulate cell

According to this theory, the nucleotide sequence of a gene determines the sequence of amino acids in a polypeptide chain. In the mid-1960s, this prediction was confirmed in the laboratory of Charles Yanofsky, where the locations of dozens of mutations in the bacterial gene coding for a subunit of the enzyme tryptophan synthase were determined. As predicted, the positions of the mutations within the gene correlated with the positions of the resulting amino acid substitutions in the tryptophan synthase polypeptide chain.

Showing that a gene's base sequence specifies the amino acid sequence of a polypeptide chain represented a

growth and division. For example, the oncogene carried by the Rous sarcoma virus (a chicken virus that was the first RNA tumor virus to be discovered) is a modified version of a cellular gene for a protein kinase. The protein product of the viral gene is hyperactive, and the cell cannot control it in the normal way. As a result, cells expressing this gene proliferate wildly, producing cancerous tumors called *sarcomas*. RNA tumor viruses of the second type do not themselves carry oncogenes, but integration of their genomes into the host chromosome alters the cellular DNA in such a way that a normal proto-oncogene is converted into an oncogene.

Retrotransposons

Reverse transcription also occurs in normal eukaryotic cells in the absence of viral infection. Much of it involves DNA elements called **retrotransposons.** In Chapter 18, we saw that *transposable elements* are DNA segments that can move themselves from one site to another within the genome. Retrotransposons are a special type of transposable element that use reverse transcription to carry out this movement. As outlined in **Figure 21A-2,** the transposition mechanism begins with ❶ transcription of the retrotransposon DNA followed by ❷ translation of the resulting RNA, which produces a protein exhibiting both reverse transcriptase and endonuclease activities. ❸ Next, the retrotransposon RNA and protein bind to chromosomal DNA at some other location, and ❹ the endonuclease cuts one of the DNA strands. ❺ The reverse transcriptase then uses the retrotransposon RNA as a template to make a DNA copy that is ❻ integrated into the target DNA site.

Although retrotransposons do not transpose themselves very often, they can attain very high copy numbers within a genome. We encountered one example in Chapter 18—the *Alu* family of sequences. *Alu* sequences are only 300 base pairs long, and they do not encode a reverse transcriptase. But by using a reverse transcriptase encoded elsewhere in the genome, they have sent copies of themselves throughout the genomes of humans and other primates. The human genome contains about a million *Alu* sequences that together represent about 11% of the total DNA. Another type of retrotransposon, called an *L1 element,* is even more prevalent, accounting for roughly 17% of human DNA. The L1 retrotransposon is larger than *Alu* and encodes its own reverse transcriptase and endonuclease, as illustrated in Figure 21A-2. The reason that genomes retain so many copies of retrotransposon sequences such as L1 and *Alu* is not well understood, but they are thought to contribute to evolutionary flexibility and variability.

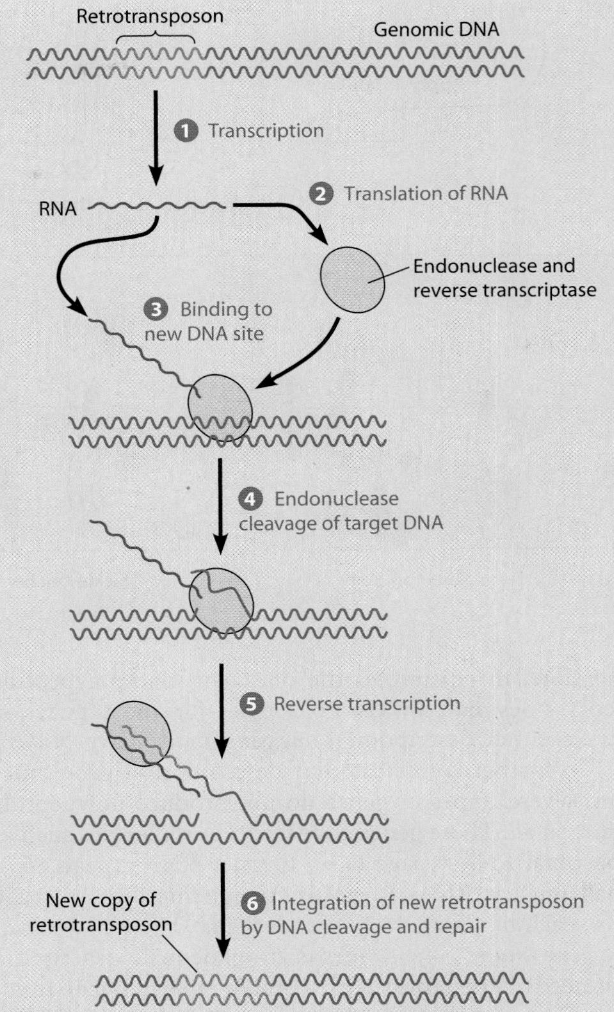

FIGURE 21A-2 Movement of a Retrotransposon.

major milestone, but subsequent developments have revealed that gene function is often more complicated than this, especially in eukaryotes. As we will see later in the chapter, most eukaryotic genes contain noncoding sequences interspersed among the coding regions and so do not exhibit a complete linear correspondence with their polypeptide product. Moreover, the coding sequences in such genes can be read in various combinations to produce different mRNAs, each coding for a unique polypeptide chain. This phenomenon, called *alternative splicing* (page 673), allows dozens or even hundreds of different polypeptides to be produced from a single gene.

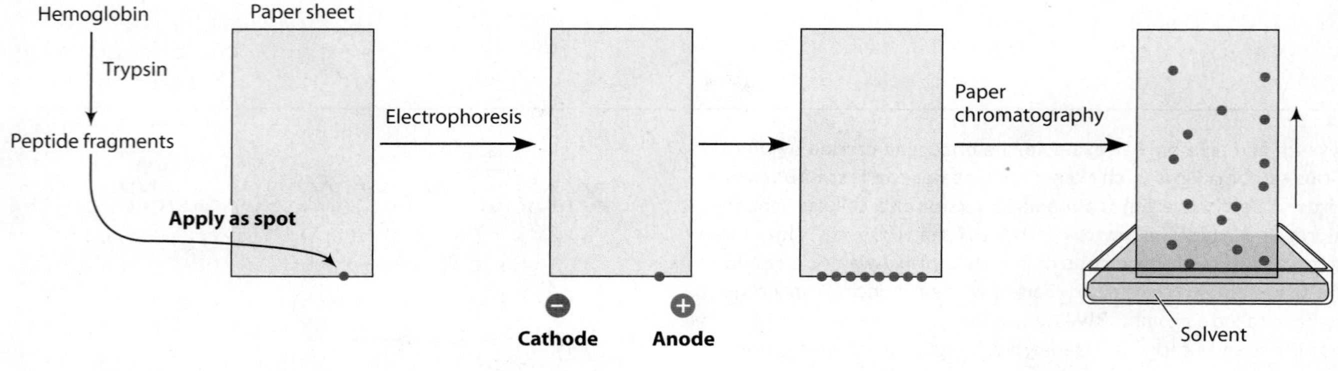

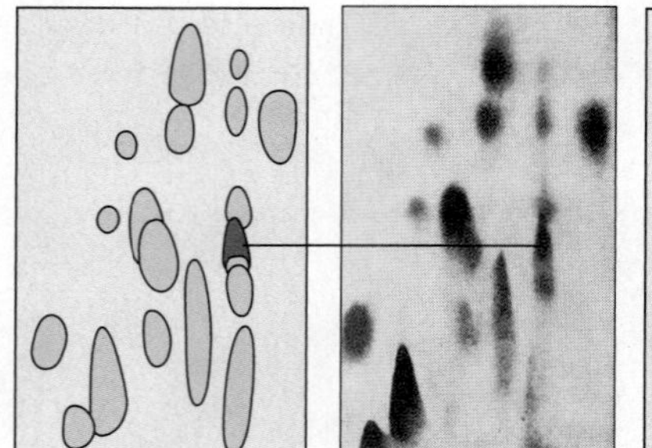

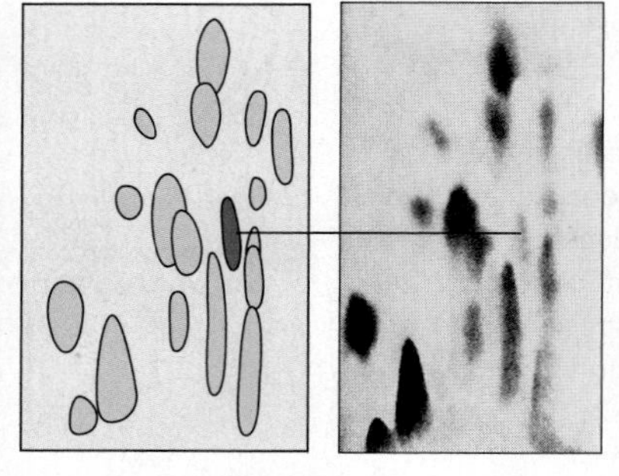

FIGURE 21-3 Peptide Patterns of Normal and Sickle-Cell Hemoglobin. *(Top)* Normal and sickle-cell hemoglobin were digested with trypsin, and the resulting peptide fragments were separated by electrophoresis followed by paper chromatography (movement of a solvent up a sheet of paper by capillary action). *(Bottom)* The colored spots in the drawings next to each photograph represent peptide fragments that differ in the two types of hemoglobin. In the altered fragment present in sickle-cell hemoglobin, a single glutamic acid has been replaced by valine.

Therefore, in eukaryotes the one gene–one polypeptide theory does not always hold true; for most genes, a more accurate description is *one gene–many polypeptides.*

To further complicate our description of gene function, several types of genes do not produce polypeptide chains at all. These genes code for RNA molecules such as ribosomal RNAs (page 666), transfer RNAs (page 665), small nuclear RNAs (page 671), and microRNAs (page 750), each of which performs a unique function. So even *one gene–many polypeptides* is an inadequate description that needs to be replaced by a broader view of gene function: **Genes** are best defined as functional units of DNA that code for the amino acid sequence of one or more polypeptide chains or, alternatively, for one of several types of RNA that perform functions other than specifying the amino acid sequence of polypeptide chains.

The Genetic Code Is a Triplet Code

Protein Synthesis (1 of 3): Overview (BioFlix tutorial)

Given a sequence relationship between DNA and proteins, the next question is: How many nucleotides in DNA are needed to specify each amino acid in a protein? We know that the information in DNA must reside in the sequence of the four nucleotides that constitute the DNA: A, T, G, and C. These are the only "letters" of the DNA alphabet. Because the DNA language has to contain at least 20 "words," one for each of the 20 amino acids found in protein molecules, the DNA word coding for each amino acid must consist of more than one nucleotide. A doublet code involving two adjacent nucleotides would not be adequate, as four kinds of nucleotides taken two at a time can generate only $4^2 = 16$ different combinations.

But with three nucleotides per word, the number of different words that can be produced with an alphabet of just four letters is $4^3 = 64$. This number is more than sufficient to code for 20 different amino acids. In the early 1950s, such mathematical arguments led biologists to suspect the existence of a **triplet code**—that is, a code in which three base pairs in double-stranded DNA are required to specify each amino acid in a polypeptide. But direct evidence for the triplet nature of the code was not provided until ten years later. To understand the nature of that evidence, we first need to become acquainted with frameshift mutations.

Frameshift Mutations. In 1961, Francis Crick, Sydney Brenner, and their colleagues provided genetic evidence for the triplet nature of the code by studying the mutagenic effects of the chemical *proflavin* on bacteriophage T4. Their work is well worth considering, not just for the critical evidence it provided concerning the nature of the code but also because of the ingenuity that was needed to understand the significance of their observations.

Proflavin is one of several *acridine dyes* commonly used as **mutagens** (mutation-inducing agents) in genetic research. Acridines are interesting mutagens because they act by causing the addition or deletion of single base pairs in DNA. Sometimes, mutants generated by acridine treatment of a wild-type ("normal") virus or organism appear to revert to wild-type when treated with more of the same type of mutagen. Closer examination often reveals, however, that the reversion is not a true reversal of the original mutation but the acquisition of a second mutation that maps very close to the first.

Such mutations display an interesting kind of arithmetic. If the first alteration is called a plus (+) mutation, then the second can be called a minus (−) mutation. By itself, each creates a mutant phenotype. But when they occur close together as a double mutation, they cancel each other out and the virus or organism exhibits the normal, wild-type phenotype. (Properly speaking, the phenotype is said to be *pseudo wild-type* because, despite its wild-type appearance, two mutations are present.) Such behavior can be explained using the analogy in **Figure 21-4**. Suppose that line 1 represents a wild-type "gene" written in a language that uses three-letter words. When we "translate" the line by starting at the beginning and reading three letters at a time, the message of the gene is readily comprehensible. A plus mutation is the addition of a single letter within the message (line 2). That change may seem minor, but since the message is always read three letters at a time, the insertion of an extra letter early in the sequence means that all the remaining letters are read out of phase. There is, in other words, a shift in the *reading frame*, and the result is a garbled message from the point of the insertion onward. A minus mutation can be explained in a similar way because the deletion of a single letter also causes the reading frame to shift, resulting in another garbled message (line 3). Such **frameshift mutations** are typical effects of acridine dyes and other mutagens that cause the insertion or deletion of individual base pairs.

Individually, plus and minus mutations always change the reading frame and garble the message. But when a plus and a minus mutation occur in close proximity within the same gene, they can largely cancel out each other's effect. In such cases, the insertion caused by the plus mutation compensates for the deletion caused by the minus mutation, and the message is intelligible from that point on (line 4). Notice, however, that double mutations with either two additions (+/+; line 5) or two deletions (−/−; line 6) do not cancel in this way. They remain out of phase for the remainder of the message.

Evidence for a Triplet Code. When Crick and Brenner generated T4 phage mutants with proflavin, they obtained results similar to those in the hypothetical example illustrated in Figure 21-4 involving a language that uses three-letter words. They found that minus mutants, which exhibited an abnormal phenotype, could acquire a second mutation that caused them to revert to the wild-type (or more properly, pseudo wild-type) phenotype. The second mutation was always a plus mutation located at a site different from, but close to, the original minus mutation. In other words, mutants reverting to the wild-type phenotype exhibited a −/+ pattern of mutations. Crick and Brenner observed many examples of −/+ (or +/−) mutants exhibiting the wild-type phenotype in their experiments. But when they generated +/+ or −/− double mutants by recombination, no wild-type phenotypes were ever seen.

Crick and Brenner also constructed triple mutants of the same types (+/+/+ or −/−/−) and found that many of them now reverted to wild-type phenotypes. This finding, of course, can be readily understood by consulting lines 7 and 8 of Figure 21-4: The reading frame (based on three-letter words) at the beginning and end of that hypothetical message remains the same when three letters are either added or removed. The portion of the message between the first and third mutations is garbled, but provided these are close enough to each other, enough of the sentence may remain to convey an intelligible message.

MB Activity: The Triplet Nature of the Genetic Code

FIGURE 21-4 Frameshift Mutations. The effect of frameshift mutations can be illustrated with an English sentence. The wild-type sentence (line 1) consists of three-letter words. When read in the correct frame, it is fully comprehensible. The insertion (line 2) or deletion (line 3) of a single letter shifts the reading frame and garbles the message from that point onward. (Garbled words due to shifts in the reading frame are underscored.) Double mutants containing a deletion that "cancels" a prior insertion have a restored reading frame from the point of the second mutation onward (line 4). However, double insertions (line 5) or double deletions (line 6) produce garbled messages. Triple insertions (line 7) or deletions (line 8) garble part of the message but restore the reading frame with the net addition or deletion of a single word.

Applying this concept of a three-letter code to DNA, Crick and Brenner concluded that adding or deleting a single base pair will shift the reading frame of the gene from that point onward, and a second, similar change shifts the reading frame yet again. Therefore, from the site of the first mutation onward, the message is garbled. But after a third change of the same type, the original reading frame is restored, and the only segment of the gene translated incorrectly is the segment between the first and third mutations. Such errors can often be tolerated when the genetic message is translated into the amino acid sequence of a protein, provided the affected region is short and the changes in amino acid sequence do not destroy protein function. This is why the individual mutations in a triple mutant with wild-type phenotype map so closely together. Subsequent sequencing of wild-type polypeptides from such triple mutants confirmed the slightly altered sequences of amino acids that would be expected.

Based on their finding that wild-type phenotypes are often maintained in the presence of three base-pair additions (or deletions) but not in the presence of one or two, Crick and Brenner concluded that the nucleotides making up a DNA strand are read in groups of three. In other words, the genetic code is a triplet code in which the reading of a message begins at a specific starting place (to ensure the proper reading frame) and then proceeds three nucleotides at a time, with each such triplet translated into the appropriate amino acid, until the end of the message is reached. Keep in mind that in establishing the triplet nature of the code, Crick and Brenner did not have Figure 21-4 to assist them. Their ability to deduce the correct explanation from their analysis of proflavin-induced mutations is an especially inspiring example of the careful, often ingenious reasoning that almost always accompanies significant advances in science.

The Genetic Code Is Degenerate and Nonoverlapping

From the fact that so many of their triple mutants were viable, Crick and Brenner drew an additional conclusion: Most of the 64 possible nucleotide triplets must specify amino acids, even though proteins have only 20 different kinds of amino acids. If only 20 of the 64 possible combinations of nucleotides "made sense" to the cell, the chances of a meaningless triplet appearing in the out-of-phase stretches would be high. Such triplets would surely interfere with protein synthesis, and frameshift mutants would revert to wild-type behavior only rarely.

But Crick and Brenner frequently detected reversion to wild-type behavior, so they reasoned that most of the 64 possible triplets must code for amino acids. Since there are only 20 amino acids, this told them that the genetic code is a **degenerate code**—that is, a given amino acid can be specified by more than one nucleotide triplet. Degeneracy serves a useful function in enhancing the adaptability of the coding system. If only 20 triplets were assigned a coding function (one for each of the 20 amino acids), any mutation in DNA that led to the formation of any of the other 44 possible triplets would interrupt the genetic message at that point. Therefore, the susceptibility of such a coding system to disruption would be very great.

A further conclusion from Crick and Brenner's work—which we have implicitly assumed—is that the genetic code is *nonoverlapping*. In an *overlapping* code, the reading frame would advance only one or two nucleotides at a time along a DNA strand so that each nucleotide would be read two or three times. **Figure 21-5** compares a nonoverlapping code with an overlapping code in which the reading frame advances one nucleotide at a time. With such an overlapping code, the insertion or deletion of a single base pair in the gene would lead to the insertion or deletion of one amino acid at one point in the polypeptide and would change several adjacent amino acids, but it would not affect the reading frame of the remainder of the gene. This means that if the genetic code were overlapping, Crick and Brenner would not have observed their frameshift mutations. Thus, their results clearly indicated the nonoverlapping nature of the code: Each nucleotide is a part of one, and only one, triplet.

Interestingly, although the genetic code is always translated in a nonoverlapping way, there are cases where a particular segment of DNA is translated in more than one reading frame. For example, certain viruses with very small genomes have overlapping genes, as was first discovered in 1977 for phage fX174. In this phage's DNA, one gene is completely embedded within another gene, and, to complicate matters further, a third gene overlaps them both! The three genes are translated in different reading frames. Other instances of overlapping genes are found in bacteria, where some genes overlap by a few nucleotides at their boundaries.

Messenger RNA Guides the Synthesis of Polypeptide Chains

After the publication of Crick and Brenner's historic findings in 1961, it took only five years for the meaning of each of the 64 triplets in the genetic code to be elucidated. Before we look at how that was done, let us first describe the role of RNA in the coding system. As we usually describe it, the genetic code refers not to the order of nucleotides in double-stranded DNA but to their order in the single-stranded mRNA molecules that actually direct protein synthesis. As indicated at the top of Figure 21-5, mRNA molecules are transcribed from DNA using a base-pairing mechanism similar to DNA replication, with two significant differences.

1. In contrast to DNA replication, where both DNA strands are copied, only one of the two DNA strands—the **template strand**—serves as a template for mRNA formation during transcription. The nontemplate DNA strand, although not directly involved in transcription, is by convention called the **coding strand** because it is similar in sequence to the single-stranded mRNA molecules that carry the coded message.

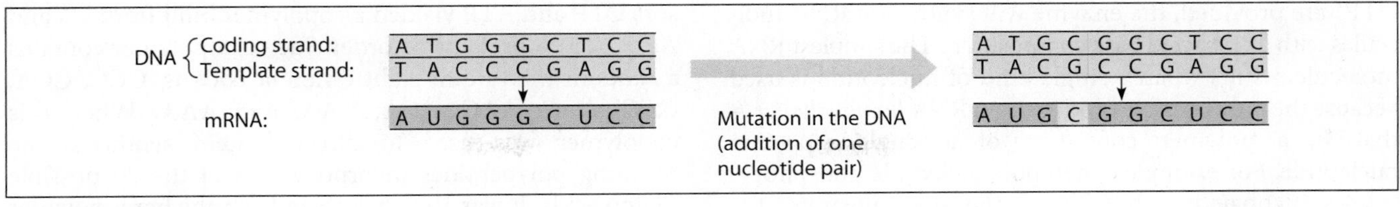

(a) Nonoverlapping code

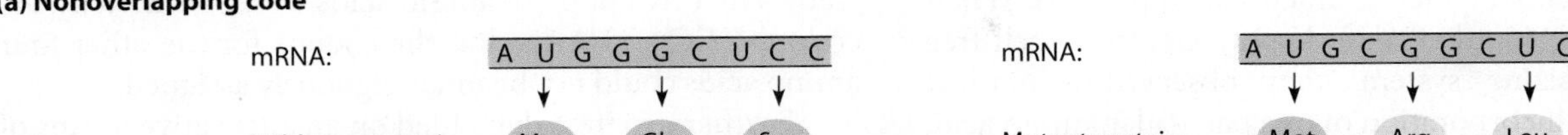

(b) Overlapping code

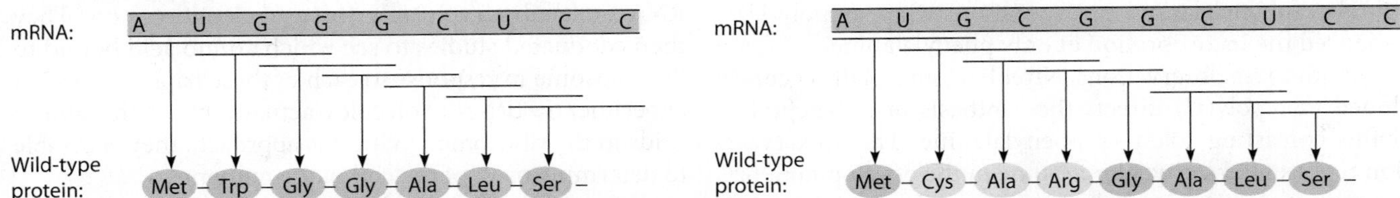

FIGURE 21-5 Effect of Inserting a Single Base Pair on Proteins Encoded by Overlapping and Nonoverlapping Genetic Codes. One strand of the DNA duplex at the top, called the *template strand,* is transcribed into the nine-nucleotide segment of mRNA shown, according to the same base-pairing rules used in DNA replication, except the base U is used in RNA in place of T. (The complementary DNA strand, with a sequence essentially identical to that of the mRNA, is called the *coding strand.*) **(a)** With a nonoverlapping code, the reading frame advances three nucleotides at a time, and this mRNA segment is therefore read as three successive triplets, coding for the amino acids methionine, glycine, and serine. (See Figure 21-6 for amino acid coding rules.) If the DNA duplex is mutated by insertion of a single base pair (the yellow-shaded CG pair in the top box), the mRNA will have an additional nucleotide. This insertion alters the reading frame beyond that point, so the remainder of the mRNA is read incorrectly and all amino acids are wrong. In the example shown, the insertion occurs near the beginning of the message, and the only similarity between the wild-type protein and the mutant protein is the first amino acid (methionine). **(b)** In one type of overlapping code, the reading frame advances only one nucleotide at a time. The wild-type protein will therefore contain three times as many amino acids as would a protein generated from the same mRNA using a nonoverlapping code. Insertion of a single base pair in the DNA again results in an mRNA molecule with one extra nucleotide. However, in this case the effect of the insertion on the protein is modest; two amino acids in the wild-type protein are replaced by three different amino acids in the mutant protein, but the remainder of the protein is normal. The frameshift mutations that Crick and Brenner found in their studies with the mutagen proflavin would not have been observed if the genetic code were overlapping. Accordingly, their data indicated the code to be nonoverlapping.

2. The mechanism used to copy sequence information from a DNA template strand to a complementary molecule of RNA utilizes the same base-pairing rules as DNA replication, with the single exception that the base uracil (U) is employed in RNA where the base thymine (T) would have been incorporated into DNA. This substitution is permitted because U and T can both form hydrogen bonds with the base A. During DNA replication the base A pairs with T, whereas in transcription the base A pairs with U. Hence the sequence of an mRNA molecule is not exactly the same as the DNA coding strand, in that mRNA contains the base U anywhere the coding DNA strand has the base T.

How do we know that mRNA molecules, produced by this transcription process, are responsible for directing the order in which amino acids are linked together during protein synthesis? This relationship was first demonstrated experimentally in 1961 by Marshall Nirenberg and J. Heinrich Matthei, who pioneered the use of *cell-free systems* for studying protein synthesis. In such systems, protein synthesis can be studied outside living cells by mixing together isolated ribosomes, amino acids, an energy source, and an extract containing soluble components of the cytoplasm. Nirenberg and Matthei found that adding RNA to cell-free systems increased the rate of protein synthesis, raising the question of whether the added RNA molecules were functioning as messages that determined the amino acid sequences of the proteins being manufactured. To address this question, they decided to add synthetic RNA molecules of known base composition to the cell-free system to see if such RNA molecules would influence the type of protein being made.

Their initial experiments took advantage of an enzyme called *polynucleotide phosphorylase,* which can be used to make synthetic RNA molecules of predictable base composition. Unlike the enzymes involved in cellular transcription, polynucleotide phosphorylase does not require a template but simply assembles available nucleotides randomly into a linear chain. If only one or two of the four ribonucleotides (ATP, GTP, CTP, and

UTP) are provided, the enzyme will synthesize RNA molecules with a restricted base composition. The simplest RNA molecule results when a single kind of nucleotide is used because the only possible product is an RNA *homopolymer*—that is, a polymer consisting of a single repeating nucleotide. For example, when polynucleotide phosphorylase is incubated with UTP as the sole substrate, the product is a homopolymer of uracil, called poly(U). When Nirenberg and Matthei added poly(U) to a cell-free protein-synthesizing system, they observed a marked increase in the incorporation of one particular amino acid, phenylalanine, into polypeptide chains. Synthetic RNA molecules containing bases other than uracil did not stimulate phenylalanine incorporation, whereas poly(U) enhanced the incorporation of only phenylalanine.

From these observations, Nirenberg and Matthei concluded that poly(U) directs the synthesis of polypeptide chains consisting solely of phenylalanine. This observation represented a crucial milestone in the development of the messenger RNA concept, for it was the first demonstration that the base sequence of an RNA molecule determines the order in which amino acids are linked together during protein synthesis.

The Codon Dictionary Was Established Using Synthetic RNA Polymers and Triplets

Once it had been shown that RNA functions as a messenger that guides the process of protein synthesis, the exact nature of the triplet coding system could be elucidated. Nucleotide triplets in mRNA, called **codons,** are the actual coding units read by the translational machinery during protein synthesis. The four bases present in RNA are the purines adenine (A) and guanine (G) and the pyrimidines cytosine (C) and uracil (U), so the 64 triplet codons consist of all 64 possible combinations of these four "letters" taken three at a time. And since mRNA molecules are synthesized in the $5' \rightarrow 3'$ direction (like DNA) and are translated starting at the $5'$ end, the 64 codons by convention are always written in the $5' \rightarrow 3'$ order.

These triplet codons in mRNA determine the amino acids that will be incorporated during protein synthesis, but which amino acid does each of the 64 triplets code for? The discovery that poly(U) directs the incorporation of phenylalanine during protein synthesis allowed Nirenberg and Matthei to make the first codon assignment: The triplet UUU in mRNA must code for the amino acid phenylalanine. Subsequent studies on the coding properties of other synthetic homopolymers, such as poly(A) and poly(C), quickly revealed that AAA codes for lysine and CCC codes for proline. (Because of unexpected structural complications, poly(G) is not a good messenger and was not tested.)

After the homopolymers had been tested, polynucleotide phosphorylase was employed to create *copolymers* containing a mixture of two nucleotides. For example, incubating polynucleotide phosphorylase with the precursors CTP and ATP yielded a copolymer built from C's and A's but in no predictable order. Such a copolymer contains a random mixture of eight different codons: CCC, CCA, CAC, ACC, AAC, ACA, CAA, and AAA. When this copolymer was used to direct protein synthesis, the resulting polypeptides incorporated 6 of the 20 possible amino acids. It was already known from the homopolymer studies that two of these amino acids were specified by the codons CCC and AAA, but the codons for the other four amino acids could not be unambiguously assigned.

Further progress depended on an alternative means of codon assignment devised by Nirenberg's group. Instead of using long polymers, they synthesized 64 very short RNA molecules, each only three nucleotides long. They then conducted studies to see which amino acid bound to the ribosome in response to each of these triplets. (In such experiments, tRNA molecules actually carry the amino acids to the ribosome.) With this approach, they were able to determine most of the codon assignments.

Meanwhile, a refined method of polymer synthesis had been devised in the laboratory of H. Gobind Khorana. Khorana's approach was similar to that of Nirenberg and Matthei but with the important difference that the polymers he synthesized had defined sequences. Thus, he could produce a synthetic mRNA molecule with the strictly alternating sequence UAUA. . . . Such an RNA copolymer has only two codons, UAU and AUA, and they alternate in strict sequence. When Khorana added this particular RNA to a cell-free protein-synthesizing system, a polypeptide containing only tyrosine and isoleucine was produced. Khorana was therefore able to narrow the possible codon assignments for UAU and AUA to these two particular amino acids. When the results obtained with such synthetic polymers were combined with the findings of Nirenberg's binding studies, most of the codons could be assigned unambiguously.

Of the 64 Possible Codons in Messenger RNA, 61 Code for Amino Acids

By 1966, just five years after the first codon was identified, the approaches we have just described allowed all 64 codons to be assigned—that is, the entire genetic code had been worked out, as shown in **Figure 21-6.** The elucidation of the code confirmed several properties that had been deduced earlier from indirect evidence. All 64 codons are in fact used in the translation of mRNA. Sixty-one of the codons specify the addition of specific amino acids to the growing polypeptide, and one of these (AUG) also plays a prominent role as a **start codon** that initiates the process of protein synthesis. The remaining three codons (UAA, UAG, and UGA) are **stop codons** that instruct the cell to terminate synthesis of the polypeptide chain.

It is clear from examining Figure 21-6 that the genetic code is *unambiguous:* Every codon has one and only one meaning. The figure also shows the *degenerate* nature of the code—that is, many of the amino acids are specified by

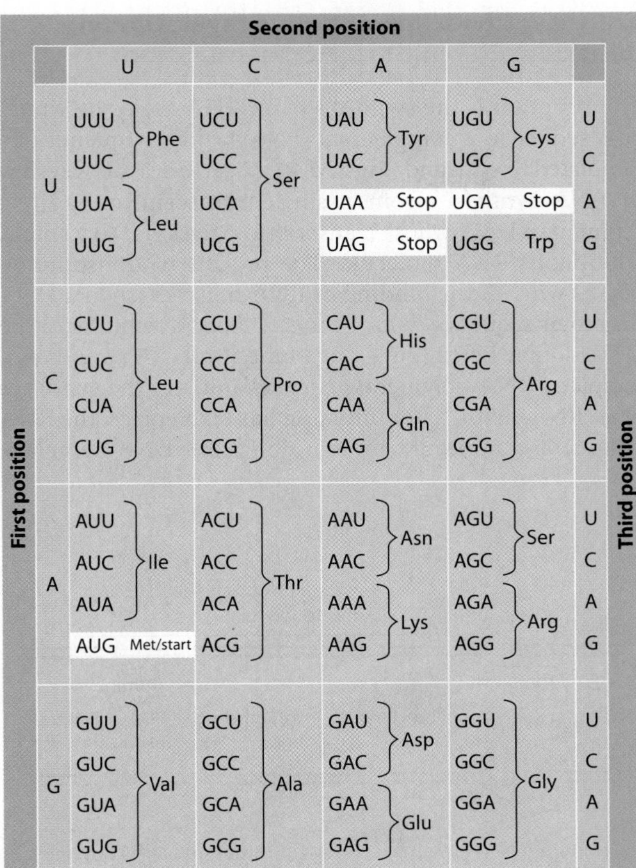

FIGURE 21-6 The Genetic Code. The code "words" are three-letter codons present in the nucleotide sequence of mRNA, as read in the $5' \longrightarrow 3'$ direction. Letters represent the nucleotide bases uracil (U), cytosine (C), adenine (A), and guanine (G). Each codon specifies either an amino acid or a stop signal. To decode a codon, read down the left edge for the first letter, then across the grid for the second letter, and then down the right edge for the third letter. For example, the codon AUG represents methionine. (As we will see in Chapter 22, AUG is also a start signal.)

MB The Genetic Code

more than one codon. There are, for example, two codons for histidine (His), four for threonine (Thr), and six for leucine (Leu). Although degeneracy may sound wasteful, it serves a useful function in enhancing the adaptability of the coding system. As we noted earlier, if there were only one codon for each of the 20 amino acids, then any mutation that created one of the remaining 44 codons would terminate synthesis of the growing polypeptide chain at that point. But with a degenerate code, most mutations simply cause codon changes that alter the specified amino acid. The change in a protein's behavior that results from a single amino acid alteration is often quite small, and in some cases may even be advantageous. Moreover, mutations in the third base of a codon frequently do not change the specified amino acid at all, as you can see in Figure 21-6. For example, a mutation that changes the codon ACU to ACC, ACA, or ACG does not alter the corresponding amino acid, which is threonine (Thr) in all four cases.

The validity of the codon assignments summarized in Figure 21-6 has been confirmed by analyzing the amino acid sequences of mutant proteins. For example, we learned earlier in the chapter that sickle-cell hemoglobin differs from normal hemoglobin at a single amino acid position, where valine is substituted for glutamic acid. The genetic code table reveals that glutamic acid may be encoded by either GAA or GAG. Whichever triplet is employed, a single base change could create a codon for valine. For example, GAA might have been changed to GUA, or GAG might have been changed to GUG. In either case, a glutamic acid codon would be converted into a valine codon. Many other mutant proteins have been examined in a similar way. In nearly all cases, the amino acid substitutions are consistent with a single base change in a triplet codon.

The Genetic Code Is (Nearly) Universal

A final property of the genetic code worth noting is its near universality. Except for a few cases, all organisms studied so far—prokaryotes as well as eukaryotes—use the same basic genetic code. Even viruses, though they are nonliving entities, employ this same code. In other words, the 64 codons almost always stand for the same amino acids or stop signals specified in Figure 21-6, suggesting that this coding system was established early in the history of life on Earth and has remained largely unchanged over billions of years of evolution.

However, several exceptions to the standard genetic code do exist, most notably in mitochondria and in a few bacteria and other unicellular organisms. In the case of mitochondria, which contain their own DNA and carry out both transcription and translation, the genetic code can differ in several ways from the standard code. One difference involves the codon UGA, which is a stop codon in the standard code but is translated as tryptophan in mammalian and yeast mitochondria. Conversely, AGA is a stop codon in mammalian mitochondria, even though in most other systems (including yeast mitochondria) it codes for arginine. Such anomalies result from alterations in the properties of transfer RNA (tRNA) molecules found in mitochondria. As you will learn in the next chapter, tRNA molecules play a key role in the genetic code because they recognize codons in mRNA and bring the appropriate amino acid to each codon during the process of protein synthesis. Differences in the types of tRNA molecules present in mitochondria appear to underlie the ability to read codons such as UAG and AGA differently than in the standard genetic code.

Some bacteria also employ a few codons in a nonstandard way, as do the nuclear genomes of certain protozoa and fungi. For example, the fungus *Candida* produces an unusual tRNA that brings serine to mRNAs containing the codon CUG rather than bringing the normally expected amino acid, leucine. In an especially interesting case, observed in organisms as diverse as bacteria and mammals, a variation of the genetic code allows the incorporation of a 21st amino acid, *selenocysteine,* in which the sulfur atom of cysteine is replaced by an atom of selenium. In mRNAs coding for the few rare proteins that contain selenocysteine, the meaning of a UGA codon is changed from a stop codon to a codon specifying selenocysteine. In such cases, folding of the mRNA molecule causes specific UGA codons to bind to a special tRNA carrying selenocysteine, rather than functioning as stop codons that terminate protein synthesis. A similar mechanism permits another stop codon, UAG, to specify incorporation of a 22nd amino acid, *pyrrolysine.*

Transcription in Bacterial Cells

Now that you have been introduced to the genetic code that governs the relationship between nucleotide sequences in DNA and the amino acid sequences of protein molecules, we can discuss the specific steps involved in the flow of genetic information from DNA to protein. The first stage in this process is the transcription of a nucleotide sequence in DNA into a sequence of nucleotides in RNA. RNA is chemically similar to DNA, but it contains ribose instead of deoxyribose as its sugar, has the base uracil (U) in place of thymine (T), and is usually single stranded. As in other areas of molecular genetics, the fundamental principles of RNA synthesis were first elucidated in bacteria, where the molecules and mechanisms are relatively simple. For that reason, we will start with transcription in bacteria.

MB Molecular Model: RNA Polymerase of Thermus thermophilus

Transcription Is Catalyzed by RNA Polymerase, Which Synthesizes RNA Using DNA as a Template

Transcription of DNA is carried out by the enzyme **RNA polymerase,** which catalyzes the synthesis of RNA using DNA as a template. Bacterial cells have a single kind of RNA polymerase that synthesizes all three major classes of RNA—mRNA, tRNA, and rRNA. The enzymes from different bacteria are quite similar, and the RNA polymerase from *Escherichia coli* has been especially well characterized. It is a large protein consisting of two α subunits, two β subunits that differ enough to be identified as β and β', and a dissociable subunit called the **sigma (σ) factor.** Although the *core enzyme* lacking the sigma subunit is competent to carry out RNA synthesis, the *holoenzyme* (complete enzyme containing all of its subunits) is required to ensure initiation at the proper sites within a DNA molecule. The sigma subunit plays a critical role in this process by promoting the binding of RNA polymerase to specific DNA sequences, called *promoters,* found at the beginnings of genes. Bacteria contain a variety of different sigma factors that selectively initiate the transcription of specific categories of genes, as we will discuss in Chapter 23. After the sigma factor guides RNA polymerase to an appropriate promoter site, the sigma factor is usually released during the early stages of transcription.

Transcription Involves Four Stages: Binding, Initiation, Elongation, and Termination

Transcription is the synthesis of an RNA molecule whose base sequence is complementary to the base sequence of a template DNA strand. **Figure 21-7** provides an overview of RNA synthesis from a single **transcription unit,** a segment of DNA whose transcription gives rise to a single, continuous RNA molecule. The process of transcription begins with ❶ the binding of RNA polymerase to a DNA promoter sequence, which triggers local unwinding of the DNA double helix. Using one of the two DNA strands as a template, RNA polymerase then ❷ initiates the synthesis of an RNA chain. After initiation has taken place, the RNA polymerase molecule moves along the DNA template,

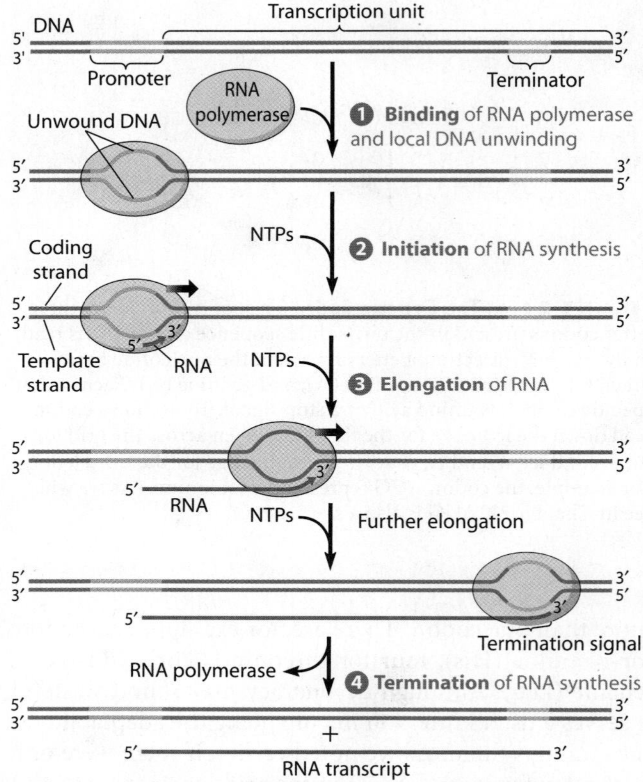

FIGURE 21-7 An Overview of Transcription. Transcription of DNA occurs in four main stages: ❶ binding of RNA polymerase to DNA at a promoter, ❷ initiation of transcription on the template DNA strand, ❸ subsequent elongation of the RNA chain, and ❹ eventual termination of transcription, accompanied by the release of RNA polymerase and the completed RNA product from the DNA template. RNA polymerase moves along the template strand of the DNA in the $3' \rightarrow 5'$ direction, and the RNA molecule grows in the $5' \rightarrow 3'$ direction. The supercoiling generated by DNA unwinding ahead of the moving RNA polymerase is relieved through the action of topoisomerases (not shown). The general scheme illustrated here holds for transcription in all organisms. NTPs (ribonucleoside triphosphate molecules) = ATP, GTP, CTP, and UTP.

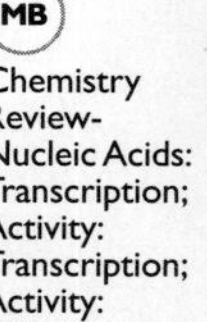
Chemistry Review-Nucleic Acids: Transcription; Activity: Transcription; Activity: RNA Synthesis

unwinding the double helix and ❸ elongating the RNA chain as it goes. During this process, the enzyme catalyzes the polymerization of nucleotides in an order determined by their base pairing with the DNA template strand. Eventually the enzyme transcribes a special base sequence called a *termination signal,* which ❹ terminates RNA synthesis and causes the completed RNA molecule to be released and RNA polymerase to dissociate from the DNA template.

Although transcription is a complicated process, it can be thought of in four distinct stages: binding, initiation, elongation, and termination. We will now look at each stage in detail, as it occurs in *E. coli.* You can refer back to Figure 21-7 throughout this discussion to see how each step fits into the overall process.

Binding of RNA Polymerase to a Promoter Sequence. The first step in RNA synthesis is the *binding* of RNA polymerase to a DNA **promoter site**—a specific sequence of several dozen base pairs that determines where RNA synthesis starts and which DNA strand is to serve as the template strand. Each transcription unit has a promoter site located near the beginning of the DNA sequence to be transcribed. By convention, promoter sequences are described in the $5' \rightarrow 3'$ direction on the coding strand, which is the strand that lies opposite the template strand. The terms **upstream** and **downstream** are used to refer to DNA sequences located toward the 5′ or 3′ end of the coding strand, respectively. Therefore the region where the promoter is located is said to be upstream of the transcribed sequence. Binding of RNA polymerase to the promoter is mediated by the sigma subunit and leads to unwinding of a short stretch of DNA in the area where transcription will begin, exposing the two separate strands of the double helix.

Promoter sequences were initially identified by *DNA footprinting* and *electrophoretic mobility shift assays,* techniques for locating the DNA region to which a DNA-binding protein has become bound (**Box 21B**). More recently, *chromatin immunoprecipitation (ChIP)* has been used to assess binding of proteins to specific genomic DNA sequences in eukaryotes. Sequences essential to the promoter region have also been identified by deleting or adding specific base sequences to cloned genes and then testing the ability of the altered DNA to be transcribed by RNA polymerase. Such techniques have revealed that DNA promoter sites differ significantly among bacterial transcription units. How, then, does a single kind of RNA polymerase recognize them all? Enzyme recognition and binding, it turns out, depend only on several very short sequences located at specific positions within each promoter site. The identities of the nucleotides making up the rest of the promoter are irrelevant for this purpose.

Figure 21-8 highlights the essential sequences in a typical bacterial promoter. The point where transcription will begin, called the *startpoint,* is almost always a purine and often an adenine. Approximately 10 bases upstream of the startpoint is the six-nucleotide sequence TATAAT, called the *−10 sequence* or the *Pribnow box,* after its discoverer. By convention, the nucleotides are numbered from the startpoint (+1), with positive numbers to the right (downstream) and negative ones to the left (upstream). The −1 nucleotide is immediately upstream of the startpoint (there is no "0"). At or near the −35 position is the six-nucleotide sequence TTGACA, called the *−35 sequence.*

The −10 and the −35 sequences (and their positions relative to the startpoint) have been conserved during evolution, but they are not identical in all bacterial promoters or even in all the promoters in a single genome. For example, the −10 sequence in the promoter for one of the *E. coli* tRNA genes is TATGAT, whereas the −10 sequence for a group of genes needed for lactose breakdown in the same organism is TATGTT, and the sequence given in Figure 21-8 is TATAAT. The particular promoter sequences shown in the figure are **consensus sequences,** which consist of the most common nucleotides at each position within a given sequence. Mutations that cause significant deviations from the consensus sequences tend

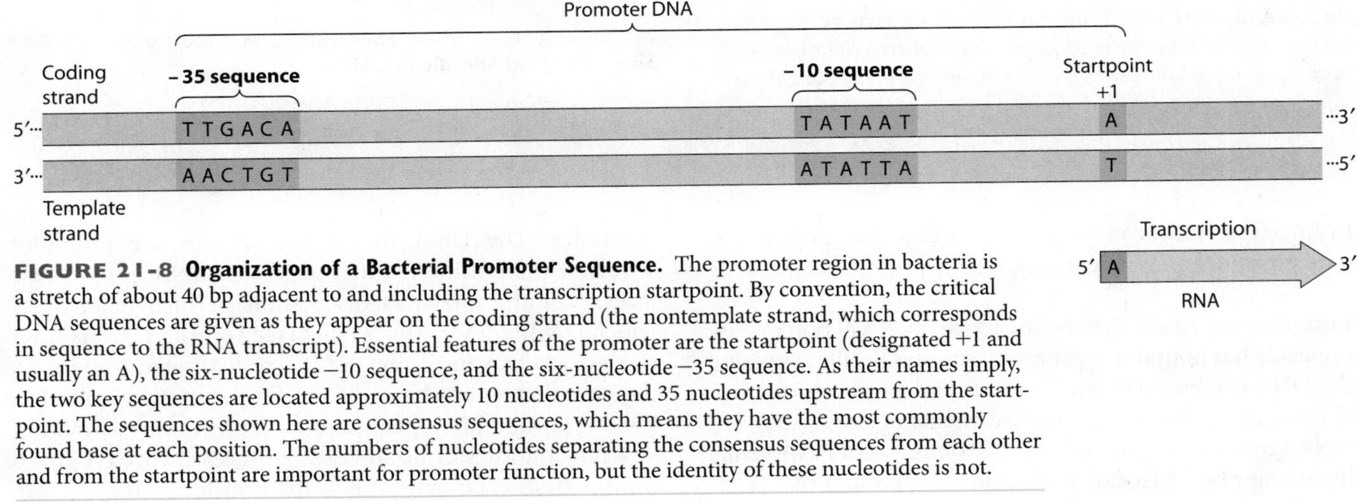

FIGURE 21-8 Organization of a Bacterial Promoter Sequence. The promoter region in bacteria is a stretch of about 40 bp adjacent to and including the transcription startpoint. By convention, the critical DNA sequences are given as they appear on the coding strand (the nontemplate strand, which corresponds in sequence to the RNA transcript). Essential features of the promoter are the startpoint (designated +1 and usually an A), the six-nucleotide −10 sequence, and the six-nucleotide −35 sequence. As their names imply, the two key sequences are located approximately 10 nucleotides and 35 nucleotides upstream from the startpoint. The sequences shown here are consensus sequences, which means they have the most commonly found base at each position. The numbers of nucleotides separating the consensus sequences from each other and from the startpoint are important for promoter function, but the identity of these nucleotides is not.

BOX 21B TOOLS OF DISCOVERY Identifying Protein-Binding Sites on DNA

The initiation of transcription depends on the interactions of proteins with specific DNA sequences. Thus, the researcher seeking to understand transcription needs to know about transcriptional proteins and the DNA sequences they bind to.

DNA footprinting is one technique that has been used to locate the DNA sites where specific proteins attach. The underlying principle is that the binding of a protein to a particular DNA sequence should protect that sequence from degradation by enzymes or chemicals. A version of footprinting, outlined in **Figure 21B-1**, employs a DNA-degrading enzyme called DNase I, which attacks the bonds between nucleotides more or less at random. In this example, the starting material is a DNA fragment that has been labeled at its 5′ end with radioactive phosphate (indicated with red stars).

In step ❶, a sample of the radioactive DNA is first mixed with the DNA-binding protein under study. Another sample, without the added protein, serves as the control. ❷ Both samples are briefly incubated with a low concentration of DNase I—conditions ensuring that most of the DNA molecules will be cleaved only once. The arrowheads indicate possible cleavage sites in the DNA. ❸ The two incubation mixtures are submitted to electrophoresis and visualized by autoradiography. The control lane (on the right) has nine bands because every possible cleavage site has been cut. However, the other lane (on the left) is missing some of the bands because the protein that was bound to the DNA protected some of the cleavage sites during DNase treatment. The blank region in this lane is the "footprint" that identifies the location and length of the DNA sequence in contact with the DNA-binding protein.

Along with DNA footprinting, a technique known as an *electrophoretic mobility shift assay (EMSA)* (also called a *gel shift assay*) has been used to confirm DNA-protein binding. In this approach, a specific DNA sequence is mixed with a DNA-binding protein or a cellular extract containing such a protein. If the protein binds to the DNA sequence, it will result in the DNA sequence moving more slowly when subjected to gel electrophoresis. This shift in mobility indicates that the protein binds to the DNA sequences of interest.

A newer approach, called the *chromatin immunoprecipitation (ChIP) assay,* is now widely used to study protein-binding sites in the DNA of eukaryotic chromatin. In the ChIP assay, cells are first treated with formaldehyde to generate stable crosslinks between proteins and the DNA sites they are bound to. Next, the cells are disrupted to shear chromatin into small fragments, and the chromatin fragments are treated with an antibody directed against a protein of interest. DNA fragments bound to that particular protein will be precipitated by the antibody, and the sequence of the precipitated DNA can then be analyzed.

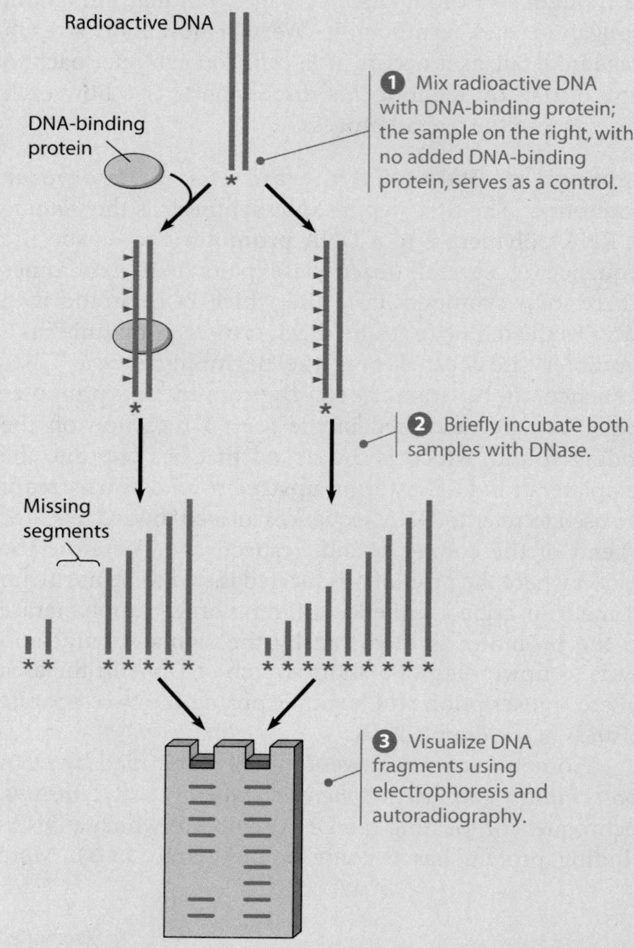

FIGURE 21B-1 DNase Footprinting as a Tool to Identify DNA Sites That Bind Specific Proteins.

to interfere with promoter function and may even eliminate promoter activity entirely.

Initiation of RNA Synthesis. Once an RNA polymerase molecule has bound to a promoter site and locally unwound the DNA double helix, *initiation* of RNA synthesis can take place. One of the two exposed segments of single-stranded DNA serves as the template for the synthesis of RNA, using incoming ribonucleoside triphosphate molecules (NTPs) as substrates. The DNA strand that carries the promoter sequence determines which way the RNA polymerase faces, and the enzyme's orientation in turn determines which DNA strand it transcribes (see Figure 21-7). As soon as the first two incoming NTPs are hydrogen-bonded to the complementary bases of the DNA template strand at the startpoint, RNA polymerase catalyzes the formation of a phosphodiester bond between the 3′-hydroxyl group of the first NTP and the 5′-phosphate of the second,

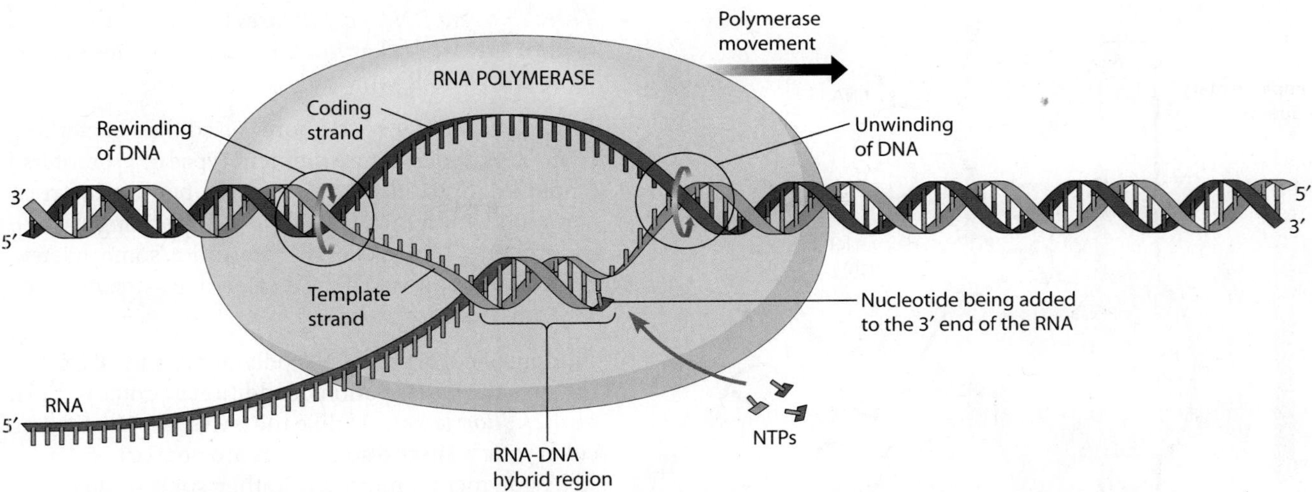

FIGURE 21-9 Closeup of a Bacterial Elongation Complex. During elongation, RNA polymerase binds to about 30 bp of DNA. (Recall that each complete turn of the DNA double helix is about 10 bp.) At any given moment, about 18 bp of DNA are unwound, and the most recently synthesized RNA is still hydrogen-bonded to the DNA, forming a short RNA–DNA hybrid about 8–9 bp long. The total length of growing RNA bound to the enzyme and/or DNA is about 25 nucleotides.

accompanied by the release of pyrophosphate (PP_i). The polymerase then advances along the template strand as additional nucleotides are added one by one, the 5′-phosphate of each new nucleotide joining to the 3′-hydroxyl group of the growing RNA chain, until the chain is about nine nucleotides long. At this point the sigma factor generally detaches from the RNA polymerase molecule, and the initiation stage is complete.

Elongation of the RNA Chain. Chain *elongation* (**Figure 21-9**) now continues as RNA polymerase moves along the DNA molecule, untwisting the helix bit by bit and adding one complementary nucleotide at a time to the growing RNA chain. The enzyme moves along the template DNA strand from the 3′ toward the 5′ end. Because complementary base pairing between the DNA template strand and the newly forming RNA chain is antiparallel, *the RNA strand is elongated in the* $5' \rightarrow 3'$ *direction* as each successive nucleotide is added to the 3′end of the growing chain. (This is the same direction in which DNA strands are synthesized during DNA replication.) As the RNA chain grows, the most recently added nucleotides remain base-paired with the DNA template strand, forming a short RNA–DNA hybrid about 8–9 bp long. As the polymerase moves forward, the DNA ahead of the enzyme is unwound to permit the RNA–DNA hybrid to form. At the same time, the DNA behind the moving enzyme is rewound into a double helix. The supercoiling that would otherwise be generated by this unwinding and rewinding is released through the action of topoisomerases, just as in DNA replication (page 562).

Like DNA polymerase, RNA polymerases possess a $3' \rightarrow 5'$ exonuclease activity that might in theory allow improperly base-paired nucleotides to be removed from the 3′ end of a growing RNA chain after an incorrect base has been incorporated. However, this intrinsic exonuclease activity is relatively weak, and an alternative mechanism for correcting errors is used instead. When a noncomplementary nucleotide is incorporated into a growing RNA chain by mistake, the RNA polymerase backs up slightly, and the noncomplementary nucleotide participates in catalyzing its own removal along with removal of the previously incorporated nucleotide. Such *RNA proofreading* appears to be sufficient for correcting mistakes that arise during transcription, especially since occasional errors in RNA molecules are not as critical as errors in DNA replication because numerous RNA copies are transcribed from each gene; hence a few inaccurate copies can be tolerated. In contrast, only one copy of each DNA molecule is made when DNA is replicated prior to cell division. Since each newly forming cell receives only one set of DNA molecules, it is crucial that the copying mechanism used in DNA replication be extremely accurate.

Termination of RNA Synthesis. Elongation of the growing RNA chain proceeds until RNA polymerase copies a special sequence, called a **termination signal,** that triggers the end of transcription. In bacteria, two classes of termination signals can be distinguished based on whether they require the participation of a protein called **rho (ρ) factor.** RNA molecules terminated without the aid of the rho factor contain a short GC-rich sequence followed by several U residues near their 3′ end (**Figure 21-10**). Since GC base pairs are held together by three hydrogen bonds, whereas AU base pairs are joined by only two hydrogen bonds, this configuration promotes termination in the following way: First, the GC region contains sequences that are complementary to each other, causing the RNA to spontaneously fold into a **hairpin loop** that tends to pull the RNA molecule away from the DNA.

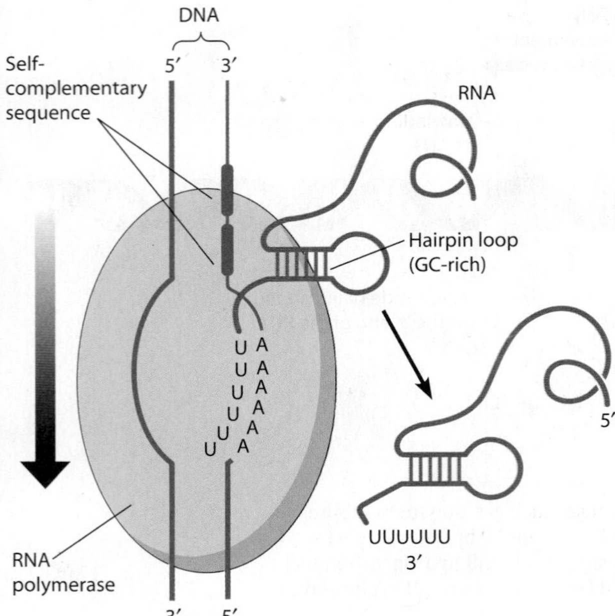

FIGURE 21-10 Termination of Transcription in Bacterial Genes That Do Not Require the Rho Termination Factor. A short self-complementary sequence near the end of the gene allows the newly formed RNA molecule to form a hairpin loop structure that helps dissociate the RNA from the DNA template.

Then the weaker bonds between the sequence of U residues and the DNA template are broken, releasing the newly formed RNA molecule.

In contrast, RNA molecules that do not form a GC-rich hairpin loop require participation of the rho factor for termination. Genes coding for such RNAs were first discovered in experiments in which purified DNA obtained from bacteriophage λ was transcribed with purified RNA polymerase. Some genes were found to be transcribed into RNA molecules that are longer than the RNAs produced in living cells, suggesting that transcription was not terminating properly. This problem could be corrected by adding rho factor, which binds to specific termination sequences 50–90 bases long located near the 3′ end of newly forming RNA molecules. The rho factor acts as an ATP-dependent unwinding enzyme, moving along the newly forming RNA molecule toward its 3′ end and unwinding it from the DNA template as it proceeds.

Whether termination depends on rho or on the formation of a hairpin loop, it results in the release of the newly transcribed RNA molecule and of the core RNA polymerase. The core polymerase can then bind sigma factor again and reinitiate RNA synthesis at another promoter.

Transcription in Eukaryotic Cells

Transcription in eukaryotic cells involves the same four stages described in Figure 21-7, but the process in eukaryotes is more complicated than that in bacteria. The main differences are as follows:

- *Three different RNA polymerases* transcribe the nuclear DNA of eukaryotes. Each synthesizes one or more classes of RNA.
- *Eukaryotic promoters* are more varied than bacterial promoters. Not only are different types of promoters employed for the three polymerases, but there is great variation within each type—especially among the ones for protein-coding genes. Furthermore, some eukaryotic promoters are actually located *downstream* from the transcription startpoint.
- Binding of eukaryotic RNA polymerases to DNA requires the participation of additional proteins, called *transcription factors.* Unlike the bacterial sigma factor, eukaryotic transcription factors are not part of the RNA polymerase molecule. Rather, some of them must bind to DNA *before* RNA polymerase can bind to the promoter and initiate transcription. Thus, transcription factors, rather than RNA polymerase itself, determine the specificity of transcription in eukaryotes. In this chapter, we limit our discussion to the class of factors that are essential for the transcription of all genes transcribed by an RNA polymerase. We defer discussion of the regulatory class of transcription factors, which selectively act on specific genes, until Chapter 23.
- *Protein–protein interactions* play a prominent role in the first stage of eukaryotic transcription. Although some transcription factors bind directly to DNA, many attach to other proteins—either to other transcription factors or to RNA polymerase itself.
- *RNA cleavage* is more important than the site where transcription is terminated in determining the location of the 3′ end of the RNA product.
- Newly forming eukaryotic RNA molecules typically undergo extensive *RNA processing* (chemical modification) both during and, to a larger extent, after transcription.

We will now examine these various aspects of eukaryotic transcription, starting with the existence of multiple forms of RNA polymerase.

RNA Polymerases I, II, and III Carry Out Transcription in the Eukaryotic Nucleus

Table 21-1 summarizes some properties of the three RNA polymerases that function in the nucleus of the eukaryotic cell, along with two other polymerases found in mitochondria and chloroplasts. The nuclear enzymes are designated RNA polymerases I, II, and III. As the table indicates, these enzymes differ in their location within the nucleus and in the kinds of RNA they synthesize. The nuclear RNA polymerases also differ in their sensitivity to various inhibitors, such as *α-amanitin,* a deadly toxin produced by the mushroom *Amanita phalloides* (the "death

Table 21-1 Properties of Eukaryotic RNA Polymerases

RNA Polymerase	Location	Main Products	α-Amanitin Sensitivity
I	Nucleolus	Precursor for 28S rRNA, 18S rRNA, and 5.8S rRNA	Resistant
II	Nucleoplasm	Pre-mRNA, most snRNA, and microRNA	Very sensitive
III	Nucleoplasm	Pre-tRNA, 5S rRNA, and other small RNAs	Moderately sensitive*
Mitochondrial	Mitochondrion	Mitochondrial RNA	Resistant
Chloroplast	Chloroplast	Chloroplast RNA	Resistant

*In mammals.

cap" fungus; the F-actin binding drug phalloidin, introduced in Chapter 15, also comes from this organism).

RNA polymerase I resides in the nucleolus and is responsible for synthesizing an RNA molecule that serves as a precursor for three of the four types of rRNA found in eukaryotic ribosomes (28S rRNA, 18S rRNA, and 5.8S rRNA). This enzyme is not sensitive to α-amanitin. Its association with the nucleolus is understandable, for the nucleolus is the site of ribosomal RNA synthesis and ribosomal subunit assembly (page 543).

RNA polymerase II is found in the nucleoplasm and synthesizes precursors to mRNA, the class of RNA molecules that code for proteins. Rather than being diffusely distributed throughout the nucleus, active molecules of polymerase II are located in discrete clusters, called *transcription factories,* that represent sites where active genes come together to be transcribed. In addition to producing mRNA precursors, RNA polymerase II synthesizes most of the *snRNAs*—small nuclear RNAs involved in posttranscriptional RNA processing—and the *microRNAs,* which regulate the translation and stability of specific mRNAs and, to a lesser extent, control the transcription of certain genes. Polymerase II is responsible for producing the greatest variety of RNA molecules and is extremely sensitive to α-amanitin, which explains the toxicity of this compound to humans and other animals.

RNA polymerase II differs from polymerases I and III at its C terminus, where it has extra amino acids. The C terminus of RNA polymerase II can be phosphorylated at a variety of locations, to produce what is sometimes called a phosphorylation "code." This "code" dramatically affects the functions of polymerase II, and correlates with where the enzyme is located along the DNA as it continues transcription. As a result, this most versatile of the RNA polymerases is also the most tightly regulated.

RNA polymerase III is also a nucleoplasmic enzyme, but it synthesizes a variety of small RNAs, including tRNA precursors and the smallest type of ribosomal RNA, 5S rRNA. Mammalian RNA polymerase III is sensitive to α-amanitin but only at higher levels of the toxin than are required to inhibit RNA polymerase II. (The comparable enzymes of some other eukaryotes, such as insects and yeasts, are insensitive to α-amanitin.)

Structurally, RNA polymerases I, II, and III are somewhat similar to each other as well as to bacterial core RNA polymerase. The three enzymes are all quite large, with multiple polypeptide subunits and molecular weights around 500,000. RNA polymerase II, for example, has more than ten subunits of at least eight different types. The three biggest subunits are evolutionarily related to the bacterial RNA polymerase subunits α, β, and β'. Three of the smaller subunits lack that relationship but are also found in RNA polymerases II and III. The RNA polymerases of mitochondria and chloroplasts resemble their bacterial counterparts closely, as you might expect from the probable origins of these organelles as endosymbiotic bacteria (see Box 11A, page 298). Like bacterial RNA polymerase, the mitochondrial and chloroplast enzymes are resistant to α-amanitin.

Three Classes of Promoters Are Found in Eukaryotic Nuclear Genes, One for Each Type of RNA Polymerase

The promoters that eukaryotic RNA polymerases bind to are even more varied than bacterial promoters, but they can be grouped into three main categories, one for each type of polymerase. **Figure 21-11** shows examples of the three types of promoters.

The promoter used by RNA polymerase I—that is, the promoter of the transcription unit that produces the precursor for the three largest rRNAs—has two parts (Figure 21-11a). The part called the **core promoter**—defined as the smallest set of DNA sequences able to direct the accurate initiation of transcription by RNA polymerase—actually extends into the nucleotide sequence to be transcribed. The core promoter is sufficient for proper initiation of transcription, but transcription is made more efficient by the presence of an *upstream control element,* which for RNA polymerase I is a fairly long sequence similar (though not identical) to the core promoter. Attachment of transcription factors to both parts of the promoter facilitates the binding of RNA polymerase I to the core promoter and enables it to initiate transcription at the startpoint.

In the case of RNA polymerase II, at least four types of DNA sequences are involved in core promoter function (Figure 21-11b). These four elements are (1) a short **initiator (Inr)** sequence surrounding the transcription startpoint (which is often an A, as in bacteria); (2) the **TATA box,** which consists of a consensus sequence of

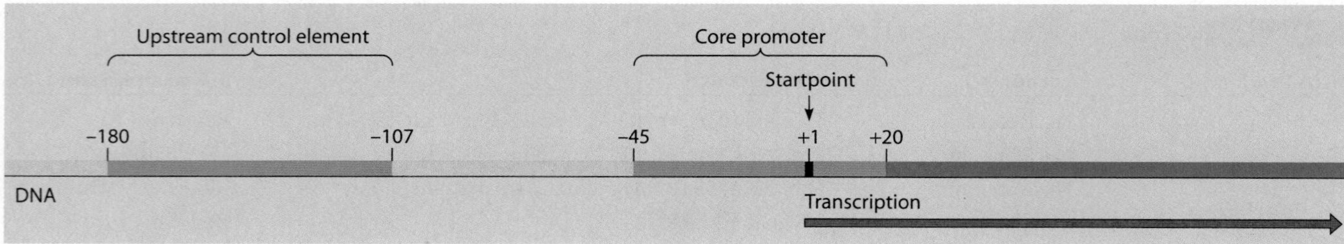

(a) Promoter for RNA polymerase I

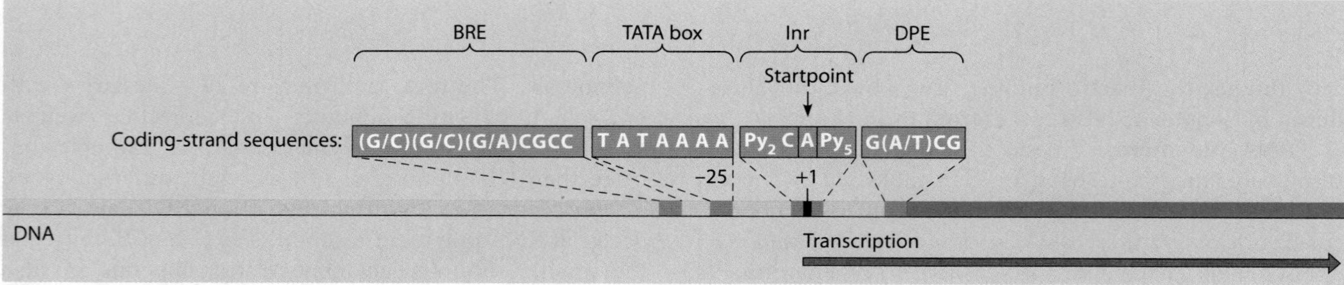

(b) Core promoter elements for RNA polymerase II

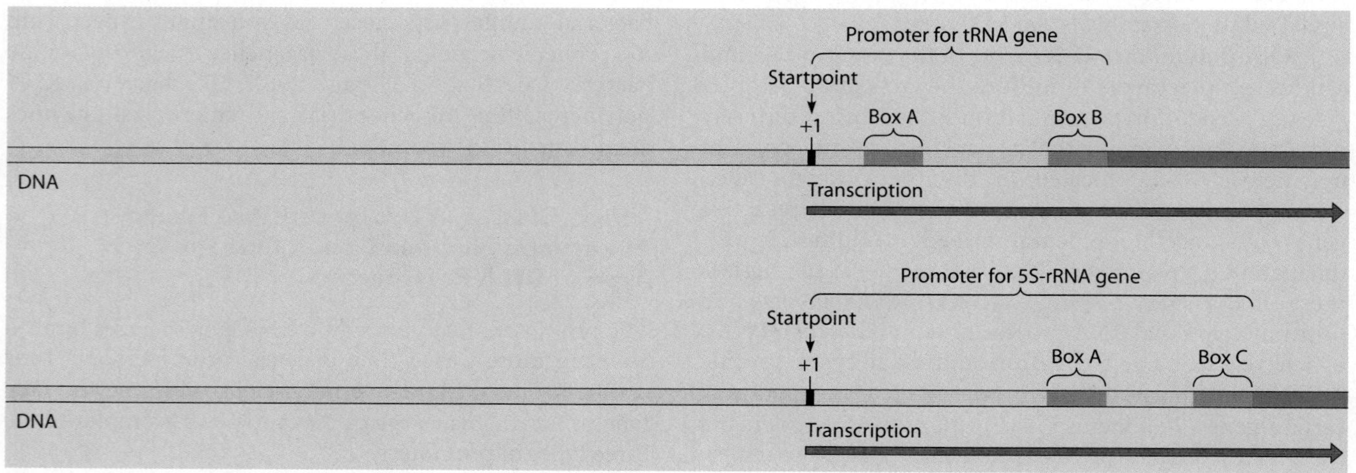

(c) Two types of promoters for RNA polymerase III

FIGURE 21-11 Examples of Eukaryotic Promoters For RNA Polymerases I, II, and III. **(a)** The promoter for RNA polymerase I has two parts, a core promoter surrounding the startpoint and an upstream control element. After the binding of appropriate transcription factors to both parts, the RNA polymerase binds to the core promoter. **(b)** The typical promoter for RNA polymerase II has a short initiator (Inr) sequence, consisting mostly of pyrimidines (Py), combined with either a TATA box or a downstream promoter element (DPE). Promoters containing a TATA box may also include a TFIIB recognition element (BRE) as part of the core promoter. **(c)** The promoters for RNA polymerase III vary in structure, but the ones for tRNA genes and 5S-rRNA genes are located entirely downstream of the startpoint, within the transcribed sequence. Boxes A, B, and C are DNA consensus sequences, each about 10 bp long. In tRNA genes, about 30–60 bp of DNA separate boxes A and B. In 5S-rRNA genes, about 10–30 bp separate boxes A and C.

TATA followed by two or three more A's, usually located about 25 nucleotides upstream from the startpoint; (3) the **TFIIB recognition element (BRE)** located slightly upstream of the TATA box; and (4) the **downstream promoter element (DPE)** located about 30 nucleotides downstream from the startpoint. These four elements are organized into two general types of core promoters: *TATA-driven promoters,* which contain an Inr sequence and a TATA box with or without an associated BRE, and *DPE-driven promoters,* which contain DPE and Inr sequences but no TATA box or BRE. Besides being found in eukaryotes, TATA-driven promoters are also present in archaea, a key piece of evidence supporting the idea that in some ways, archaea resemble eukaryotes more closely than they resemble bacteria (page 76).

By itself, a core promoter (TATA-driven or DPE-driven) is capable of supporting only a *basal* (low) level of transcription. However, most protein-coding genes have additional short sequences further upstream—*upstream control elements*—that improve the promoter's efficiency. Some of these upstream elements are common to many different genes; examples include the *CAAT box* (consensus sequence GCCCAATCT in animals and yeasts) and the *GC box* (consensus sequence GGGCGG). The locations of these elements relative to a gene's startpoint vary from gene to gene. The elements within 100–200 nucleotides of the startpoint are often called *proximal control elements* to distinguish them from *enhancer* elements, which tend to be farther away and can even be located downstream of the gene. We will return to proximal control elements and enhancers in Chapter 23.

The sequences important in promoter activity are often identified by deleting specific sequences from a cloned DNA molecule, which is then tested for its ability to serve as a template for gene transcription, either in a test tube or after introduction of the DNA into cultured cells. For example, when transcription of the gene for β-globin (the β chain of hemoglobin) is investigated in this way, deletion of either the TATA box or an upstream CAAT box reduces the rate of transcription at least tenfold.

In contrast to RNA polymerases I and II, the RNA polymerase III molecule uses promoters that are entirely *downstream* of the transcription unit's startpoint when transcribing genes for tRNAs and 5S rRNA. The promoters used by tRNA and 5S-rRNA genes are different, but in both cases the consensus sequences fall into two blocks of about 10 bp each (Figure 21-11c). The tRNA promoter has consensus sequences called *box A* and *box B*. The promoters for 5S-rRNA genes have box A (positioned farther from the startpoint than in tRNA-gene promoters) and another critical sequence, called *box C*. (Not shown in the figure is a third type of RNA polymerase III promoter, an upstream promoter that is used for the synthesis of other kinds of small RNA molecules.)

The promoters used by all the eukaryotic RNA polymerases must be recognized and bound by transcription factors before the RNA polymerase molecule can bind to DNA. We turn now to these transcription factors.

General Transcription Factors Are Involved in the Transcription of All Nuclear Genes

A **general transcription factor** is a protein that is always required for an RNA polymerase molecule to bind to its promoter and initiate RNA synthesis, regardless of the identity of the gene involved. Eukaryotes have many such transcription factors; their names usually include "TF" (for transcription factor), a roman numeral identifying the polymerase they aid, and a capital letter that identifies each individual factor (for example, TFIIA, TFIIB, and so forth).

Using RNA polymerase II as an example, **Figure 21-12** illustrates the involvement of general transcription factors in the binding of RNA polymerase to a TATA-containing promoter site in DNA. General transcription

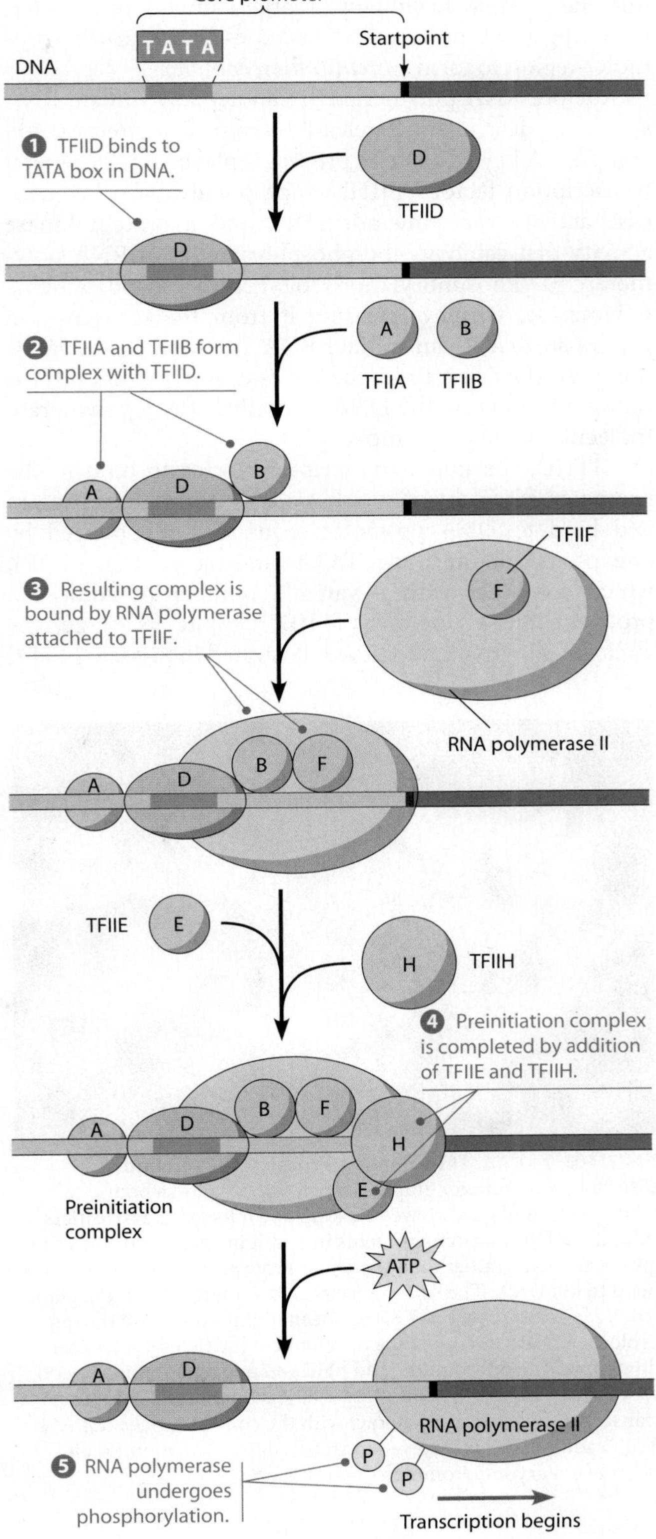

FIGURE 21-12 Role of General Transcription Factors in Binding RNA Polymerase II to DNA. This figure outlines the sequential binding of six general transcription factors (called TFII_, where _ is a letter identifying the particular factor) and RNA polymerase. After the final activation step involving ATP-dependent phosphorylation of the RNA polymerase molecule, the polymerase can initiate transcription. In intact chromatin, the efficient binding of general transcription factors and RNA polymerase to DNA requires the participation of additional regulatory proteins that open up chromatin structure and facilitate assembly of the preinitiation complex at specific genes (see Figure 23-23).

factors bind to promoters in a defined order, starting with TFIID. Notice that while TFIID binds directly to a DNA sequence (the TATA box in this example or the DPE sequence in the case of DPE-driven promoters), the other transcription factors interact primarily with each other. Hence, protein-protein interactions play a crucial role in the binding stage of eukaryotic transcription. RNA polymerase II does not bind to the DNA until several steps into the process. Eventually, a large complex of proteins, including RNA polymerase, becomes bound to the promoter region to form a *preinitiation complex.*

Before RNA polymerase II can actually initiate RNA synthesis, it must be released from the preinitiation complex. A key role in this process is played by the general transcription factor TFIIH, which possesses both a helicase activity that unwinds DNA and a protein kinase activity that catalyzes the phosphorylation of RNA polymerase II. Phosphorylation changes the shape of RNA polymerase, thereby releasing it from the transcription factors so that it can initiate RNA synthesis at the startpoint. At the same time, the helicase activity of TFIIH is thought to unwind the DNA so that the RNA polymerase molecule can begin to move.

TFIID, the initial transcription factor to bind to the promoter, is worthy of special note. Its ability to recognize and bind to DNA promoter sequences is conferred by one of its subunits, the **TATA-binding protein (TBP),** which combines with a variable number of additional protein subunits to form TFIID. Despite its name, the ability of TBP to bind to DNA, illustrated in **Figure 21-13**, is not restricted to TATA-containing promoters. TBP can also bind to promoters lacking a TATA box, including promoters used by RNA polymerases I and III. Depending on the type of promoter, TBP associates with different proteins, and for promoters lacking a TATA box, much of TBP's specificity is probably derived from its interaction with these associated proteins.

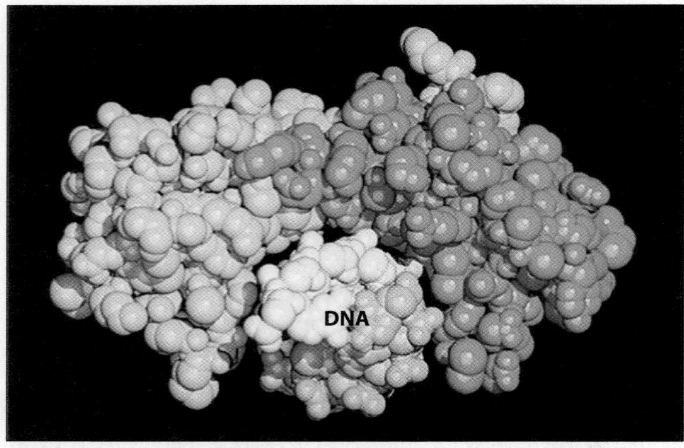

FIGURE 21-13 TATA-Binding Protein (TBP) Bound to DNA. In this computer graphic model, TBP is shown bound to DNA (white and gray, viewed looking down its axis). TBP differs from most DNA-binding proteins in that it interacts with the minor groove of DNA, rather than the major groove, and imparts a sharp bend to the DNA. The TBP molecule shown here is from the plant *Arabidopsis thaliana,* but TBP has been highly conserved during evolution. Dark and light blue differentiate the two symmetrical domains of the polypeptide, and light green is used for its nonconserved N-terminal segment. When TBP is bound to DNA, other transcription factors can interact with the convex surface of the TBP "saddle." TBP is involved in transcription initiation for all types of eukaryotic promoters.

In addition to general transcription factors and RNA polymerase II, several other kinds of proteins are required for the efficient transcription and regulated activation of specific genes. Some of these proteins are involved in opening up chromatin structure to facilitate the binding of RNA polymerase to DNA. Others are *regulatory transcription factors,* which activate specific genes by binding to upstream control elements and recruiting *coactivator proteins* that in turn facilitate assembly of the RNA polymerase preinitiation complex. The identities and roles of these additional proteins will be described in Chapter 23, which covers the regulation of gene expression (see Figure 23-22).

Elongation, Termination, and RNA Cleavage Are Involved in Completing Eukaryotic RNA Synthesis

After initiating transcription, RNA polymerases move along the DNA and synthesize a complementary RNA copy of the DNA template strand. Special proteins facilitate the disassembly of nucleosomes in front of the moving polymerase and their immediate reassembly after the enzyme passes. If an area of DNA damage is encountered, RNA polymerase may become stalled temporarily while the damage is corrected by proteins that carry out DNA excision repair (page 568).

Termination of transcription is governed by an assortment of signals that differ for each type of RNA polymerase. For example, transcription by RNA polymerase I is terminated by a protein factor that recognizes an 18-nucleotide termination signal in the growing RNA chain. Termination signals for RNA polymerase III are also known; they always include a short run of U's (as in bacterial termination signals), and no ancillary protein factors are needed for their recognition. Hairpin structures do not appear to be involved in termination by either polymerase I or polymerase III.

For RNA polymerase II, transcripts destined to become mRNA are often cleaved at a specific site before transcription is actually terminated. The cleavage site is 10–35 nucleotides downstream from a special AAUAAA sequence in the growing RNA chain. The polymerase may continue transcription for hundreds or even thousands of nucleotides beyond the cleavage site, but this additional RNA is quickly degraded. The cleavage site is also the site for the addition of a *poly(A) tail,* a string of adenine nucleotides found at the $3'$ end of almost all eukaryotic mRNAs. Addition of the poly(A) tail is part of RNA processing, our next topic.

RNA Processing

An RNA molecule newly produced by transcription, called a **primary transcript,** frequently must undergo chemical changes before it can function in the cell. We use the term **RNA processing** to mean all the chemical modifications necessary to generate a final RNA product from the primary transcript that serves as its precursor. Processing typically involves removal of portions of the primary transcript, and it may also include the addition or chemical modification of specific nucleotides. For example, methylation of bases or ribose groups is a common modification of individual nucleotides. In addition to chemical modifications, other posttranscriptional events, such as association with specific proteins or (in eukaryotes) passage from the nucleus to the cytoplasm, are often necessary before the RNA can function.

In this section, we examine the most important processing steps involved in the production of rRNA, tRNA, and mRNA from their respective primary transcripts. Although the term *RNA processing* is most often associated with eukaryotic systems, bacteria process some of their RNA as well. We therefore include examples involving both eukaryotic and bacterial RNAs.

Ribosomal RNA Processing Involves Cleavage of Multiple rRNAs from a Common Precursor

Ribosomal RNA (rRNA) is by far the most abundant and most stable form of RNA found in cells. Typically, rRNA represents about 70–80% of the total cellular RNA, tRNA represents about 10–20%, and mRNA accounts for less than 10%. In eukaryotes, cytoplasmic ribosomes contain four types of rRNA, usually identified by their differing sedimentation rates during centrifugation (page 95). **Table 21-2** lists the sedimentation coefficients (S values) of these different types of rRNA. The smaller of the two ribosomal subunits has a single 18S rRNA molecule. The larger subunit contains three rRNA molecules, one of about 28S (as low as 25S in some species) and the other two of about 5.8S and 5S. In bacterial ribosomes, only three species of rRNA are present: a 16S molecule associated with the small subunit and molecules of 23S and 5S associated with the large subunit.

Table 21-2 RNA Components of Cytoplasmic Ribosomes

		rRNA	
Source	Ribosomal Subunit	Sedimentation Coefficient	Nucleotides
Bacterial cells	Large (50S)	23S	2900
		5S	120
	Small (30S)	16S	1540
Eukaryotic cells	Large (60S)	25–28S	≤4700
		5.8S	160
		5S	120
	Small (40S)	18S	1900

Of the four kinds of rRNA in eukaryotic ribosomes, the three larger ones (28S, 18S, and 5.8S) are encoded by a single transcription unit that is transcribed by RNA polymerase I in the nucleolus to produce a single primary transcript called **pre-rRNA** (**Figure 21-14**). The DNA sequences that code for these three rRNAs are separated within the transcription unit by segments of DNA called *transcribed spacers.* The presence of three different rRNA genes within a single transcription unit ensures that the cell makes these three rRNAs in equal quantities. Most eukaryotic genomes have multiple copies of the pre-rRNA transcription unit, arranged in one or more tandem arrays (Figure 21-14a). These multiple copies facilitate production of the large amounts of ribosomal RNA typically needed by cells. The human haploid genome, for example, has 150–200 copies of the pre-rRNA transcription unit, distributed among five chromosomes. *Nontranscribed spacers* separate the transcription units within each cluster.

After RNA polymerase I has transcribed the pre-rRNA transcription unit, the resulting pre-rRNA molecule is processed by a series of cleavage reactions that remove the transcribed spacers and release the mature rRNAs (Figure 21-14c, d). The transcribed spacer sequences are then degraded. The pre-rRNA is also processed by the addition of methyl groups. The main site of methylation is the 2′-hydroxyl group of the sugar ribose, although a few bases are methylated as well. The methylation process, as well as pre-rRNA cleavage, are guided by a special group of RNA molecules, called **snoRNAs** (small nucleolar RNAs), which bind to complementary regions of the pre-rRNA molecule and target specific sites for methylation or cleavage.

Pre-rRNA methylation has been studied by incubating cells with radioactive *S-adenosyl methionine,* which is the methyl group donor for cellular methylation reactions. When human cells are incubated with radioactive S-adenosyl methionine, all of the radioactive methyl groups initially incorporated into the pre-rRNA molecule are eventually found in the finished 28S, 18S, and 5.8S rRNA products, indicating that the methylated segments are selectively conserved during rRNA processing. Methylation may help to guide RNA processing by protecting specific regions of the pre-rRNA molecule from cleavage. In support of this hypothesis, it has been shown that depriving cells of one of the essential components required for the addition of methyl groups leads to disruption of pre-rRNA cleavage patterns.

In mammalian cells, the pre-rRNA molecule has about 13,000 nucleotides and a sedimentation coefficient of 45S. The three mature rRNA molecules generated by cleavage of this precursor contain only about 52% of the original RNA. The remaining 48% (about 6200 nucleotides) consists of transcribed spacer sequences that are removed and degraded during the cleavage steps. The rRNA precursors of some other eukaryotes contain smaller amounts of spacer sequences, but in all cases the pre-rRNA is larger

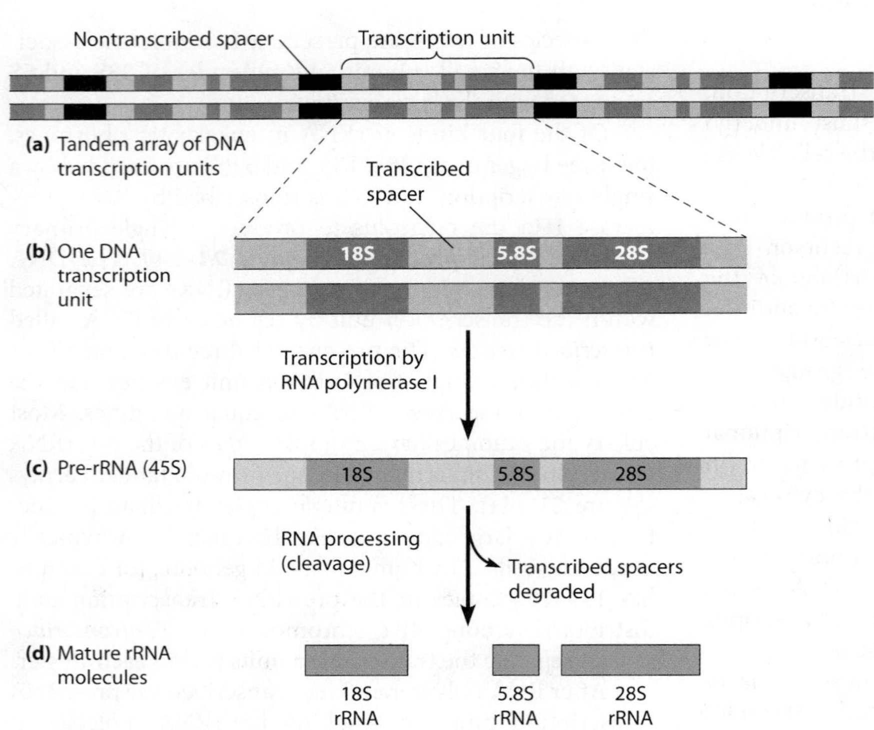

FIGURE 21-14 Eukaryotic rRNA Genes: Processing of Primary Transcripts. (a) The eukaryotic transcription unit that includes the genes for the three largest rRNAs occurs in multiple copies, arranged in tandem arrays. Nontranscribed spacers (black) separate the units. **(b)** Each transcription unit includes the genes for the three rRNAs (darker blue) and four transcribed spacers (lighter blue). **(c)** The transcription unit is transcribed by RNA polymerase I into a single long transcript (pre-rRNA) with a sedimentation coefficient of about 45S. **(d)** RNA processing yields mature 18S, 5.8S, and 28S rRNA molecules. RNA cleavage actually occurs in a series of steps. The order of steps varies with the species and cell type, but the final products are always the same three types of rRNA molecules.

than the aggregate size of the three rRNA molecules made from it. Thus, some processing is always required.

Processing of pre-rRNA in the nucleolus is accompanied by assembly of the RNA with proteins to form ribosomal subunits. In addition to the 28S, 18S, and 5.8S rRNAs generated by pre-rRNA processing, the ribosome assembly process also requires 5S rRNA. The gene for 5S rRNA constitutes a separate transcription unit that is transcribed by RNA polymerase III rather than RNA polymerase I. It, too, occurs in multiple copies arranged in long, tandem arrays. However, 5S-rRNA genes are not usually located near the genes for the larger rRNAs and so do not tend to be associated with the nucleolus. Unlike pre-rRNA, the RNA molecules generated during transcription of 5S-rRNA genes require little or no processing.

As in eukaryotes, ribosome formation in prokaryotic cells involves processing of multiple rRNAs from a larger precursor. *E. coli,* for example, has seven rRNA transcription units scattered about its genome. Each contains genes for all three bacterial rRNAs—23S rRNA, 16S rRNA, and 5S rRNA—plus several tRNA genes. Processing of the primary transcripts produced from these transcription units involves two sets of enzymes, one for the rRNAs and one for the tRNAs.

Transfer RNA Processing Involves Removal, Addition, and Chemical Modification of Nucleotides

Cells synthesize several dozen kinds of tRNA molecules, each designed to bring a particular amino acid to one or more codons in mRNA. However, all tRNA molecules share a common general structure, as illustrated in **Figure 21-15**. A mature tRNA molecule contains only 70–90 nucleotides, some of which are chemically modified. Base pairing between complementary sequences located in different regions causes each tRNA molecule to fold into a secondary structure containing several *hairpin loops,* illustrated in the figure. Most tRNAs have four base-paired regions, indicated by the light blue dots in part b of the figure. In some tRNAs, a fifth such region is present at the *variable loop.* Each of these base-paired regions is a short stretch of RNA double helix. Molecular biologists call the tRNA secondary structure a *cloverleaf* structure because it resembles a cloverleaf when drawn in two dimensions. However, in its normal three-dimensional tertiary structure, the molecule is folded so that the overall shape actually resembles the letter "L" (see Figure 22-3b).

Like ribosomal RNA, transfer RNA is synthesized in a precursor form in both eukaryotic and prokaryotic cells. Processing of these **pre-tRNA** molecules involves several different events, as shown in Figure 21-15a for yeast tyrosine tRNA: ❶ At the 5′ end, a short *leader sequence* of 16 nucleotides is removed from the pre-tRNA. ❷ At the 3′ end, the two terminal nucleotides of the pre-tRNA are removed and replaced with the trinucleotide CCA, which is a common structural feature of all tRNA molecules. (Some tRNAs already have CCA in their primary transcripts and therefore do not require modification at the 3′ end.) ❸ In a typical tRNA molecule, about 10–15% of the nucleotides are chemically modified during pre-tRNA processing. The principal modifications include methylation of bases and sugars and creation of unusual bases such as dihydrouracil, ribothymine, pseudouridine, and inosine.

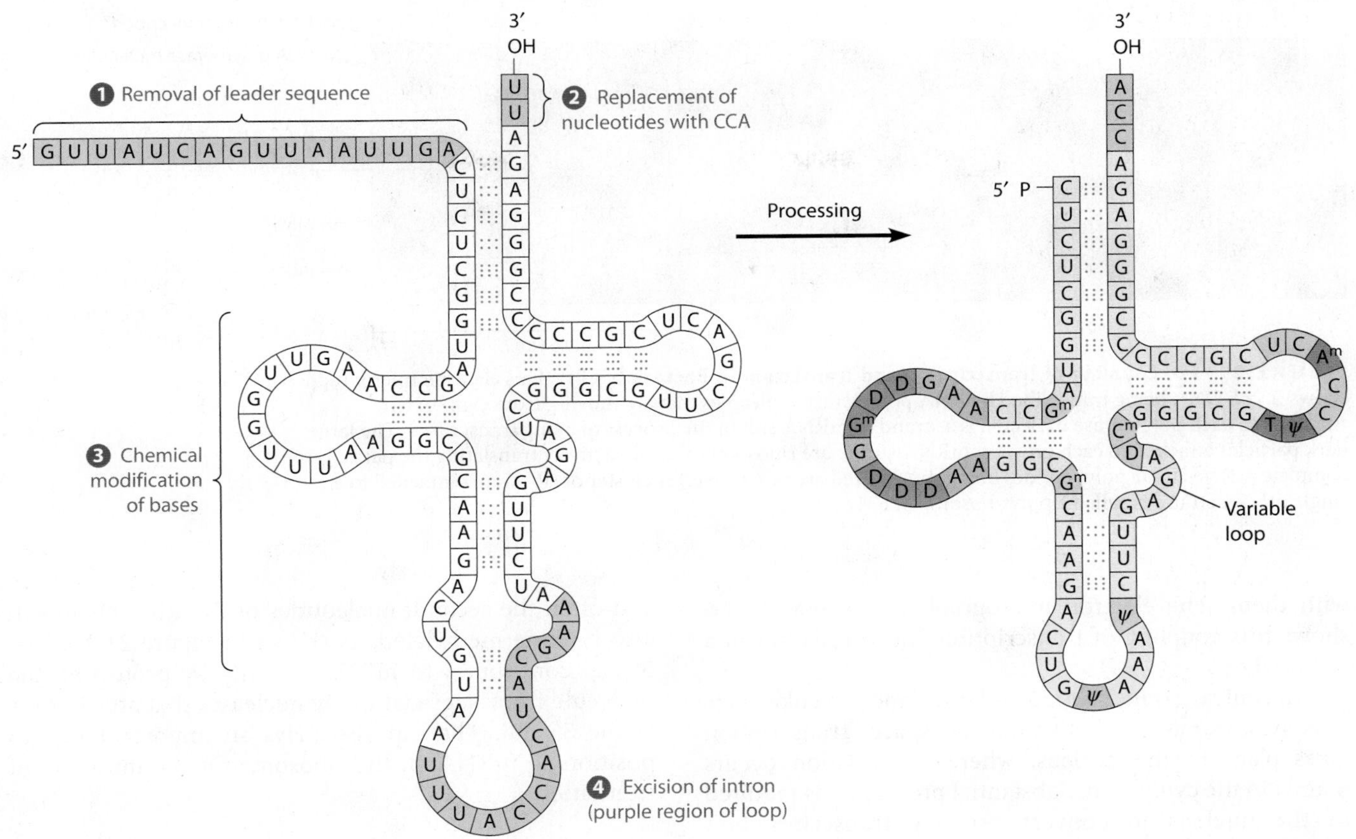

(a) Primary transcript (precursor) for yeast tyrosine tRNA

(b) Mature tRNA, secondary structure

FIGURE 21-15 Processing and Secondary Structure of Transfer RNA. **(a)** Every tRNA gene is transcribed as a precursor that must be processed into a mature tRNA molecule. In this primary transcript for yeast tyrosine tRNA, all regions highlighted in purple are removed during processing. Processing for this tRNA involves ❶ removal of the leader sequence at the 5′ end, ❷ replacement of two nucleotides at the 3′ end by the sequence CCA (which serves as an attachment site for amino acids in all mature tRNAs), ❸ chemical modification of certain bases, and ❹ excision of an intron. **(b)** The mature tRNA in a flattened, cloverleaf representation, which clearly shows the base pairing between self-complementary stretches in the molecule. Modified bases (darker colors) are abbreviated as A^m for methyladenine, G^m for methylguanine, C^m for methylcytosine, D for dihydrouracil, T for ribothymine, and c for pseudouridine.

The processing of yeast tyrosine tRNA is also characterized by the removal of an internal 14-nucleotide sequence (❹), although the transcripts for most tRNAs do not require this kind of excision. An internal segment of an RNA transcript that must be removed to create a mature RNA product is called an RNA *intron.* We will consider introns in more detail during our discussion of mRNA processing because they are a nearly universal feature of mRNA precursors in eukaryotic cells. For the present, we simply note that some eukaryotic tRNA precursors contain introns that must be eliminated by a precise mechanism that cuts and splices the precursor molecule at exactly the same location every time. The cutting-splicing mechanism involves two separate enzymes, an RNA endonuclease and an RNA ligase, which are similar from species to species, even among organisms that are evolutionarily distant from one another. In an experiment that demonstrates this point vividly, cloned genes for the yeast tyrosine tRNA shown in Figure 21-15 were microinjected into eggs of *Xenopus laevis,* the African clawed frog. Despite the long evolutionary divergence between fungi and amphibians, the yeast genes placed in the frog eggs were transcribed and processed properly, including removal of the 14-nucleotide intron.

Messenger RNA Processing in Eukaryotes Involves Capping, Addition of Poly(A), and Removal of Introns

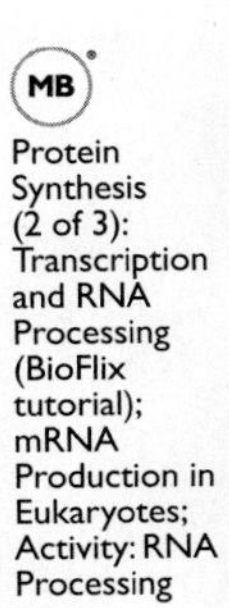

MB
Protein Synthesis (2 of 3): Transcription and RNA Processing (BioFlix tutorial); mRNA Production in Eukaryotes; Activity: RNA Processing

Bacterial mRNA is, in almost all cases, an exception to the generalization that RNA requires processing before it can be used by the cell. Most bacterial mRNA is synthesized in a form that is ready for translation, even before the entire RNA molecule has been completed. Moreover, transcription in bacteria is not separated by a membrane barrier from the ribosomes responsible for translation, so bacterial mRNA molecules in the process of being synthesized by RNA polymerase often have ribosomes already associated

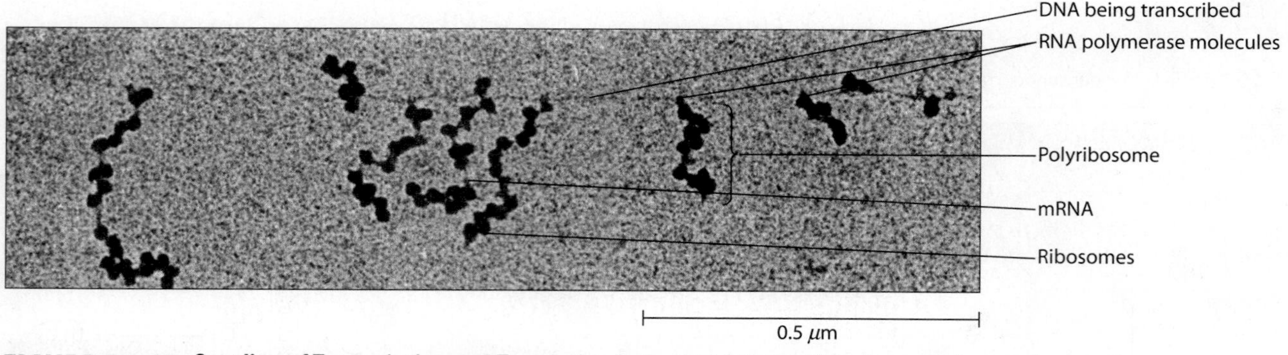

FIGURE 21-16 Coupling of Transcription and Translation in Bacterial Cells. This electron micrograph shows *E. coli* DNA being transcribed by RNA polymerase molecules that are moving from right to left. Attached to each polymerase molecule is a strand of mRNA still in the process of being transcribed. The large dark particles attached to each growing mRNA strand are ribosomes that are actively translating the partially complete mRNA. (The polypeptides being synthesized are not visible.) A cluster of ribosomes attached to a single mRNA strand is called a polyribosome (TEM).

with them. The electron micrograph in **Figure 21-16** shows this coupling of transcription and translation in a bacterial cell.

In contrast, transcription and translation in eukaryotic cells are separated in both time and space: Transcription takes place in the nucleus, whereas translation occurs mainly in the cytoplasm. Substantial processing is required in the nucleus to convert primary transcripts into mature mRNA molecules that are ready to be transported to the cytoplasm and translated. Primary transcripts are often very long, typically ranging from 2000 to 20,000 nucleotides. This size heterogeneity is reflected in the term *heterogeneous nuclear RNA (hnRNA),* which refers to the nonribosomal, nontransfer RNA found in eukaryotic nuclei. HnRNA consists of a mixture of mRNA molecules and their precursors, **pre-mRNA.** Conversion of pre-mRNA molecules into functional mRNAs usually requires the removal of nucleotide sequences and the addition of 5′ caps and 3′ tails, as will be described in the following sections. The C-terminal domain of one of the subunits of RNA polymerase II plays a key role in coupling these RNA processing events to transcription, presumably by acting as a platform for the assembly of the protein complexes involved in pre-mRNA processing.

5′ Caps and 3′ Poly(A) Tails. Most eukaryotic mRNA molecules bear distinctive modifications at both ends. At the 5′ end, they all possess a modified nucleotide called a **5′ cap,** and at the 3′ end they usually have a long stretch of adenine ribonucleotides known as a **poly(A) tail.**

A 5′ cap is simply a guanosine nucleotide that has been methylated at position 7 of the purine ring and is "backward"—that is, the bond joining it to the 5′ end of the RNA molecule is a 5′ → 5′ linkage rather than the usual 3′ → 5′ bond (**Figure 21-17**). This distinctive feature of eukaryotic mRNA is added to the primary transcript shortly after initiation of RNA synthesis. As part of the capping process, the ribose rings of the first, and often the second, nucleotides of the RNA chain can also become methylated, as shown in Figure 21-17. The 5′ cap contributes to mRNA stability by protecting the molecule from degradation by nucleases that attack RNA at the 5′ end. The cap also plays an important role in positioning mRNA on the ribosome for the initiation of translation.

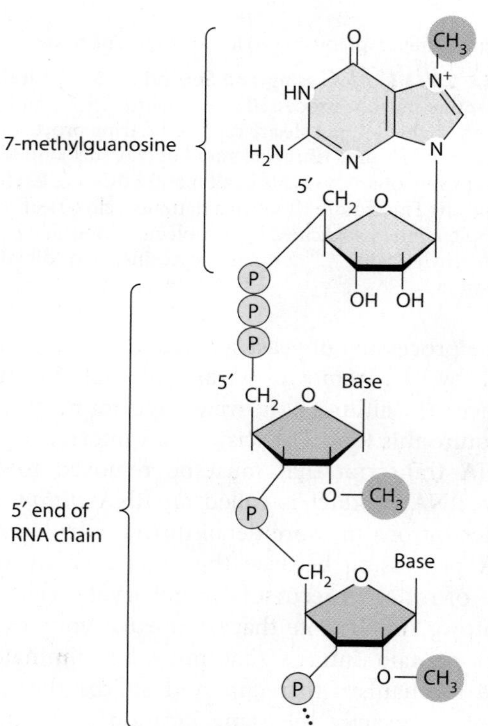

FIGURE 21-17 Cap Structure Located at the 5′ End of Eukaryotic Pre-mRNA and mRNA Molecules. The methyl groups attached to the first two riboses at the 5′ end of the RNA chain are not always present. Notice that the bond joining the RNA to 7-methylguanosine is a 5′-to-5′ linkage rather than the usual 5′-to-3′ linkage.

In addition to the 5′ cap, a poly(A) tail ranging from 50–250 nucleotides in length is present at the 3′ end of most eukaryotic mRNA molecules. (In animal cells, mRNAs coding for the major histones are among the few mRNAs known to lack such a poly(A) tail.) It is clear that poly(A) must be added after transcription because genes do not contain long stretches of thymine (T) nucleotides that could serve as a template for the addition of poly(A). Direct support for this conclusion has come from the isolation of the enzyme *poly(A) polymerase,* which catalyzes the addition of poly(A) sequences to RNA without requiring a DNA template.

The addition of poly(A) is part of the process that creates the 3′ end of most eukaryotic mRNA molecules. Unlike bacteria, where specific termination sequences halt transcription at the 3′ end of newly forming mRNAs, the transcription of eukaryotic pre-mRNAs often proceeds hundreds or even thousands of nucleotides beyond the site destined to become the 3′ end of the final mRNA molecule. A special signal—consisting of an AAUAAA sequence located slightly upstream from this site and a GU-rich and/or U-rich element downstream from the site—determines where the poly(A) tail should be added. As shown in **Figure 21-18**, this signaling element triggers cleavage of the primary transcript 10–35 nucleotides downstream from the AAUAAA sequence, and poly(A) polymerase catalyzes formation of the poly(A) tail. In addition to creating the poly(A) tail, the processing events associated with the AAUAAA signal may also help to trigger the termination of transcription.

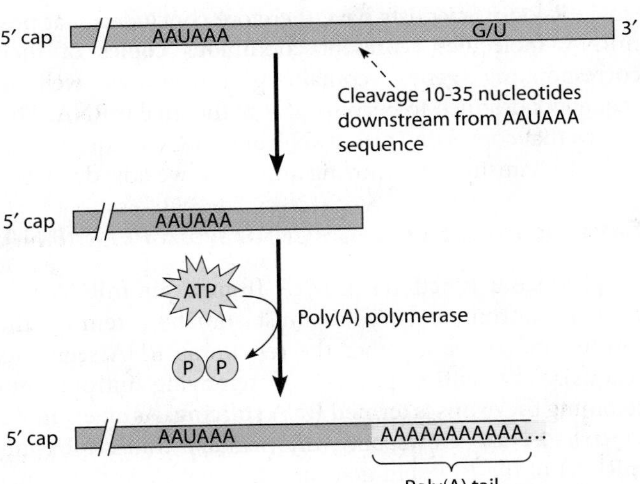

FIGURE 21-18 Addition of a Poly(A) Tail to Pre-mRNA. Transcription of eukaryotic pre-mRNAs often proceeds beyond the 3′ end of the mature mRNA. The RNA chain is then cleaved about 10–35 nucleotides downstream from a special AAUAAA sequence, followed by the addition of a poly(A) tail catalyzed by poly(A) polymerase. The proper AAUAAA sequence is distinguished by the presence of an accompanying downstream GU-rich and/or U-rich element (designated "G/U" in the diagram).

The poly(A) tail seems to have several functions. Like the 5′ cap, it protects mRNA from nuclease attack and, as a result, the length of the poly(A) influences mRNA stability (the longer the tail, the longer the life span of the mRNA in the cytoplasm). In addition, poly(A) is recognized by specific proteins involved in exporting mRNA from the nucleus to the cytoplasm, and it may also help ribosomes recognize mRNA as a molecule to be translated. In the laboratory, poly(A) tails can be used to isolate mRNA from the more prevalent rRNA and tRNA. RNA extracted from cells is simply passed through a column packed with particles coated with poly(dT), which are single strands of DNA consisting solely of thymine nucleotides. Molecules with poly(A) tails bind to the poly(dT) via complementary base pairing, while other RNA molecules pass through. The poly(A)-containing mRNA can then be removed from the column by changing the ionic conditions.

The Discovery of Introns. In eukaryotic cells, the precursors for most mRNAs (and for some tRNAs and rRNAs) contain **introns,** which are *sequences within the primary transcript that do not appear in the mature, functional RNA.* The discovery of introns was a great surprise to biologists. It had already been shown for numerous bacterial genes that the amino acid sequence of the polypeptide chain produced by a gene corresponds exactly with a sequence of contiguous nucleotides in DNA. This relationship was first demonstrated by Charles Yanofsky in the early 1960s and, as better methods for sequencing DNA and proteins became available, was confirmed by direct comparison of nucleotide and amino acid sequences. Biologists naturally assumed the same would turn out to be true for eukaryotes.

It was therefore a shock in 1977 when Philip Sharp and Richard Roberts independently reported the identification of eukaryotic genes that do not follow this pattern but are instead interrupted by stretches of nucleotides—introns—that are not represented in either the functional mRNA or its protein product. The widespread importance of introns in eukaryotes led to the awarding of the 1993 Nobel Prize in Medicine to Sharp and Roberts. The existence of introns was first shown by *R looping,* a technique in which single-stranded RNA is hybridized to double-stranded DNA under conditions that favor the formation of *heteroduplexes,* hybrids between complementary regions of RNA and DNA. The mRNA hybridizes to the template strand of the DNA, leaving the other, displaced strand as a single-stranded DNA loop that can be easily identified using electron microscopy. With eukaryotic mRNAs coding for such proteins as human β-globin and chick ovalbumin, the surprising result was that *multiple* loops were seen (**Figure 21-19**). This unexpected result indicated that the DNA sequences coding for a typical eukaryotic mRNA are not continuous with each other but instead are separated by intervening sequences that do not appear in the final mRNA. The intervening sequences that disrupt the linear continuity

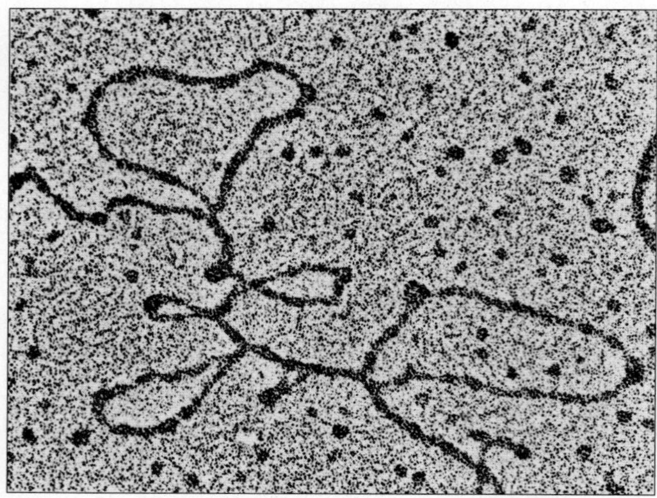

(a)

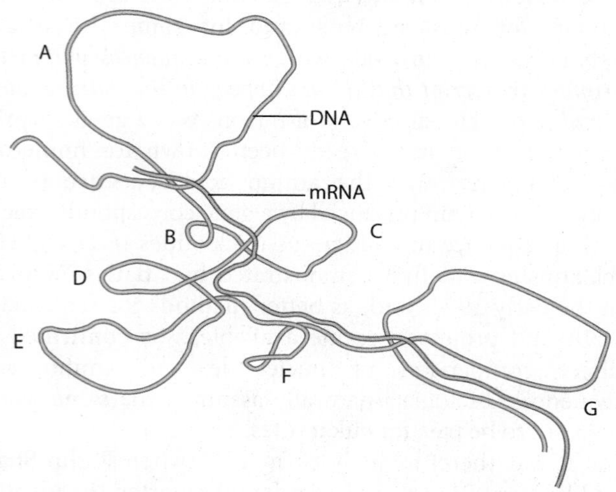

(b)

FIGURE 21-19 Demonstration of Introns in Protein-Coding Genes. Mature mRNA molecules were allowed to hydrogen-bond to the DNA (gene) from which they had been transcribed, forming a hybrid molecule (heteroduplex). The resulting hybrid molecules were then examined with an electron microscope. The example shown here is the chicken ovalbumin gene. **(a)** An electron micrograph showing the loops produced in a heteroduplex (TEM). **(b)** An interpretive diagram showing the positions of the mRNA and the unpaired loops of DNA, which correspond to introns (A-G).

of the message-encoding regions of a gene are the introns (*int*ervening sequences), and the sequences destined to appear in the final mRNA are referred to as **exons** (because they are *ex*pressed).

Once they had been reported for a few genes, introns began popping up everywhere. The use of restriction mapping and DNA sequencing techniques has led to the conclusion that introns are present in most protein-coding genes of multicellular eukaryotes, although the size and number of the introns can vary considerably (**Table 21-3**). The human β-globin gene, for example, has only two introns, one of 120 bp and the other of 550 bp. Together, these account for about 40% of the total length of the gene. For many mammalian genes, an even larger fraction of the gene consists of introns. An extreme example is the human dystrophin gene, a mutant form of which causes Duchenne muscular dystrophy. This gene is over 2 million bp long and has 85 introns, representing more than 99% of the gene's DNA!

Table 21-3 Examples of Genes with Introns

Gene	Organism	Number of Introns	Number of Exons
Actin	*Drosophila*	1	2
β-Globin	Human	2	3
Insulin	Human	2	3
Actin	Chicken	3	4
Albumin	Human	14	15
Thyroglobulin	Human	36	37
Collagen	Chicken	50	51
Titin	Human	233	234

The discovery of introns that do not appear in mature mRNA molecules raises the question of whether the introns present in DNA are actually transcribed into the primary transcript (pre-mRNA). This question has been addressed by experiments in which pre-mRNA and DNA were mixed together and the resulting hybrids examined by electron microscopy. In contrast to the appearance of hybrids between mRNA and DNA, which exhibit multiple R loops where the DNA molecule contains sequences that are not present in the mRNA (Figure 21-19), pre-mRNA hybridizes in one continuous stretch to the DNA molecule, forming a single R loop. Scientists have therefore concluded that pre-mRNA molecules represent continuous copies of their corresponding genes, containing introns as well as sequences destined to become part of the final mRNA. This means that converting pre-mRNA into mRNA requires specific mechanisms for removing introns, as we now describe.

Spliceosomes Remove Introns from Pre-mRNA

To produce a functional mRNA from a pre-mRNA that contains introns, eukaryotes must somehow remove the introns and splice together the remaining RNA segments (exons). The entire process of removing introns and rejoining the exons is termed **RNA splicing.** As an example, **Figure 21-20** shows both the primary transcript (pre-mRNA) of the β-globin gene and the mature end-product (mRNA) that results after removal of the two introns.

The relevance of RNA splicing for human health has become apparent from the discovery that roughly 15% of inherited human diseases involve splicing errors in pre-mRNA. Precise splicing is critical because a single nucleotide error would alter the mRNA reading frame and render it useless. Splicing precision can be disrupted by altering short base sequences at either end of an intron,

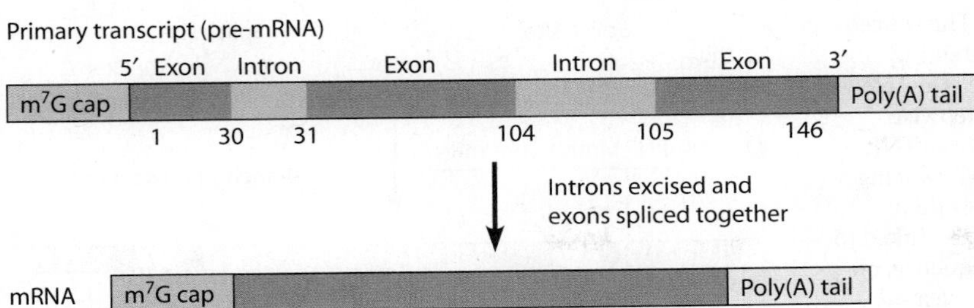

FIGURE 21-20 An Overview of RNA Splicing. The capped and tailed primary transcript of the human gene for β-globin contains three exons (dark red) and two introns (light red). The numbers refer to codon positions in the final mRNA. In the mature mRNA that results from RNA splicing, the introns have been excised and the exons joined together to form a molecule with a continuous coding sequence. The cell's ribosomes will translate this message into a polypeptide of 146 amino acids.

RNA Splicing

suggesting that these sequences determine the location of the *5′ and 3′ splice sites*—that is, the points where the two ends of an intron are cleaved during its removal. Analysis of the base sequences of hundreds of different introns has revealed that the 5′ end of an intron typically starts with the sequence GU and the 3′ end terminates with AG. In addition, a short stretch of bases adjacent to these GU and AG sequences tends to be similar among different introns. The base sequence of the remainder of the intron appears to be largely irrelevant to the splicing process. Though introns vary from a few dozen to thousands of nucleotides in length, most of the intron can be artificially removed without altering the splicing process. One exception is a special sequence located several dozen nucleotides upstream from the 3′ end of the intron and referred to as the *branch-point.* The branch-point plays an important role in the mechanism that removes introns.

Intron removal is catalyzed by **spliceosomes,** which are large, molecular complexes consisting of five kinds of RNA combined with more than 200 proteins. Electron microscopy has revealed that spliceosomes assemble on pre-mRNA molecules while the pre-mRNA is still being synthesized (**Figure 21-21**), indicating that intron removal can begin before transcription of pre-mRNA is completed. Spliceosomes are assembled on pre-mRNAs from a group of smaller RNA-protein complexes called **snRNPs** (small nuclear ribonucleoproteins) and additional proteins. Each snRNP (pronounced "snurp") contains one or two small molecules of a special type of RNA known as **snRNA** (small nuclear RNA).

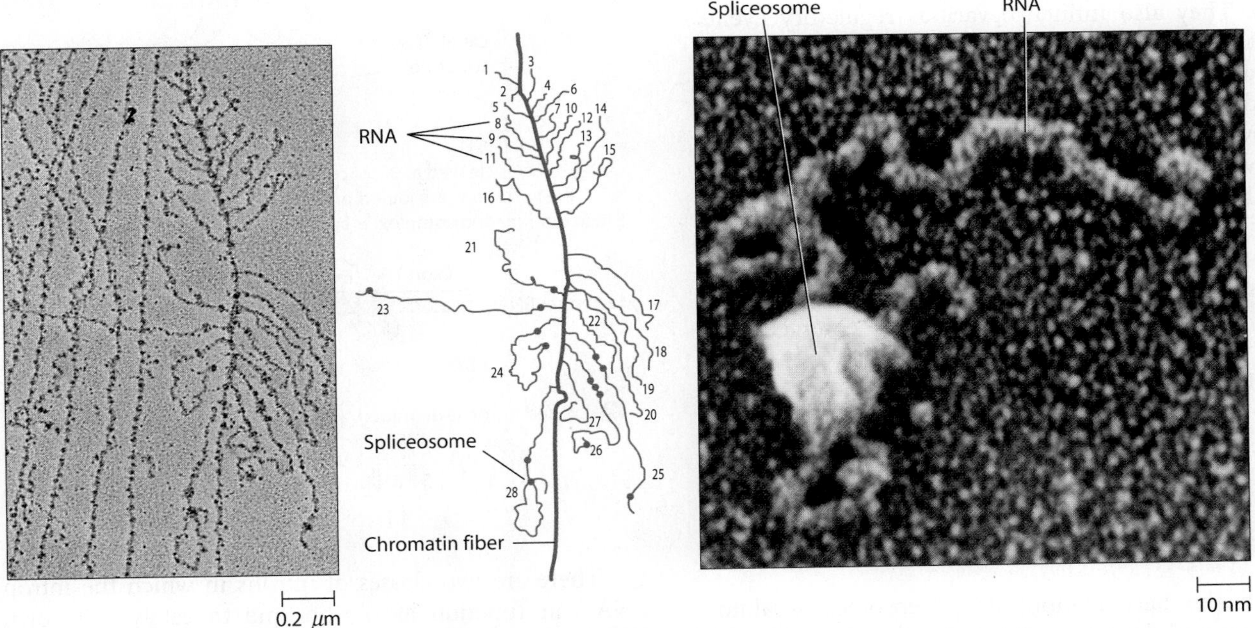

FIGURE 21-21 Spliceosomes Visualized by Electron Microscopy. The electron micrograph on the left shows chromatin fibers in the process of being transcribed. Many newly forming RNA transcripts (each numbered separately in the diagram) protrude from one of the chromatin fibers. The darker granules on the RNA transcripts represent snRNPs that are beginning the process of spliceosome assembly. In the case of RNA transcript number 28, a mature spliceosome has formed. The higher-magnification electron micrograph on the right shows a single RNA molecule with an attached spliceosome (TEMs).

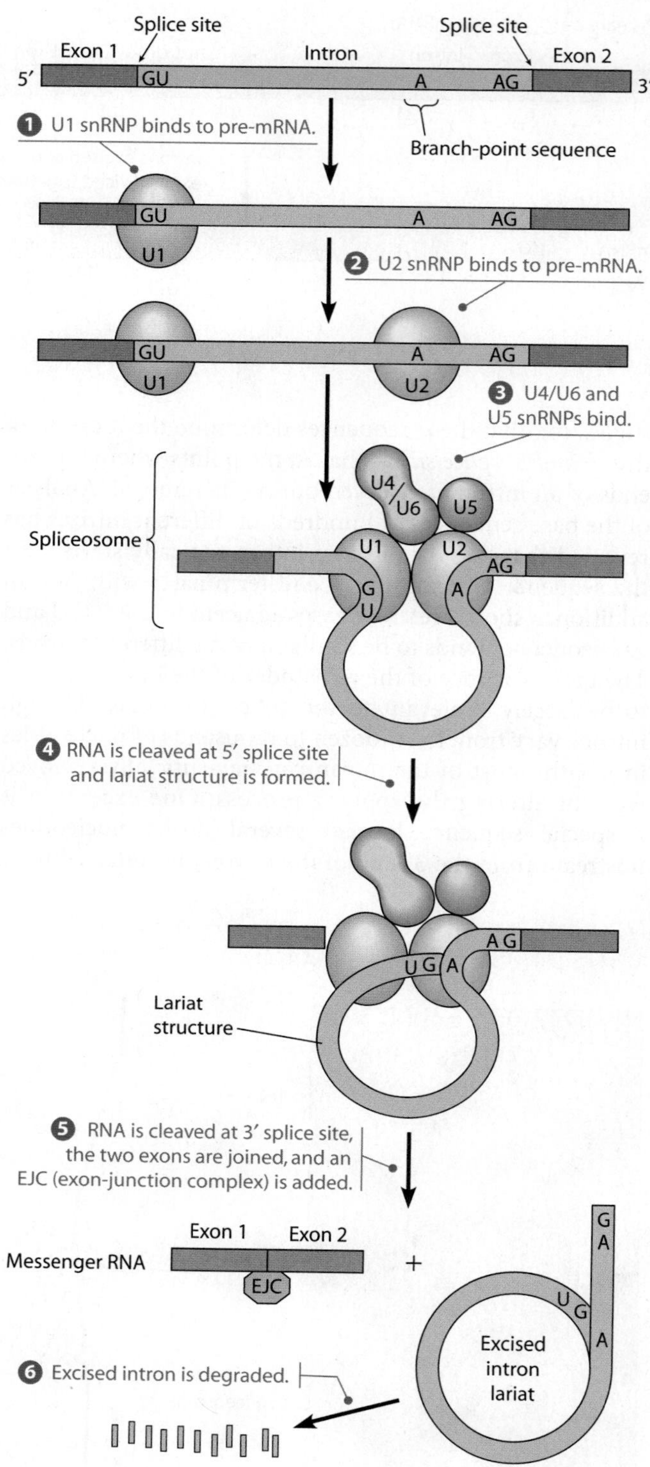

FIGURE 21-22 Intron Removal by Spliceosomes. The spliceosome is an RNA-protein complex that splices intron-containing pre-mRNA in the eukaryotic nucleus. The substrate here is a molecule of pre-mRNA with two exons and one intron. In a stepwise fashion, the pre-mRNA assembles with the U1 snRNP, U2 snRNP, and U4/U6 and U5 snRNPs (along with some non-snRNP splicing factors), forming a mature spliceosome. The pre-mRNA is then cleaved at the 5′ splice site and the newly released 5′ end is linked to an adenine (A) nucleotide located at the branch-point sequence, creating a looped lariat structure. Finally, the 3′ splice site is cleaved and the two ends of the exon are joined together, releasing the intron for subsequent degradation.

Figure 21-22 summarizes how spliceosomes are assembled by the sequential binding of snRNPs to pre-mRNA. The first step is the binding of a snRNP called U1, whose RNA contains a nucleotide sequence that allows it to base-pair with the 5′ splice site. A second snRNP, called U2, then binds to the branch-point sequence. Finally, another group of snRNPs (U4/U6 and U5) brings the two ends of the intron together to form a mature spliceosome, a massive complex comparable in size to a ribosome. At this stage the pre-mRNA is cleaved at the 5′ splice site, and the newly released 5′ end of the intron is covalently joined to an adenine residue located at the branch-point sequence, creating a looped structure called a *lariat.* The 3′ splice site is then cleaved, and the two ends of the exon are joined together, releasing the intron for subsequent degradation. A multiprotein complex called an **exon junction complex (EJC)** is deposited near the boundary of each newly formed exon-exon junction. EJCs are required for the efficient export of mRNA from the nucleus. They also influence various regulatory events, including mRNA localization and translation.

In addition to the main class of introns containing GU and AG sequences at their 5′ and 3′ boundaries, respectively, a second class of introns with AU and AC at these two sites has been identified. These "AU-AC" introns are often excised by a second type of spliceosome that differs in snRNP composition from the spliceosome illustrated in Figure 21-22. But despite the complexities raised by the existence of multiple types of introns and spliceosomes, a unifying principle has emerged: The snRNA molecules present in spliceosomes are directly involved in splice-site recognition, spliceosome assembly, and the catalytic mechanism of splicing. The idea of a catalytic role for snRNAs arose from the discovery of self-splicing RNA introns, our next topic.

Some Introns Are Self-Splicing

Although the participation of spliceosomes is almost always required for intron removal, a few types of genes have *self-splicing RNA introns.* The RNA transcript of such a gene can carry out the entire process of RNA splicing in the absence of any protein (for example, in a test tube); the intron RNA itself catalyzes the process. As we described in Chapter 6, such RNA molecules that function as catalysts in the absence of protein are called *ribozymes.*

There are two classes of introns in which the intron RNA can function as a ribozyme to catalyze its own removal. The first class, called *Group I introns,* is present in the mitochondrial genome of fungi (e.g., yeast) in rRNA genes and in genes coding for components of the electron transport system. Group I introns also occur in plant mitochondrial genes, in some rRNA and tRNA genes of chloroplasts, in nuclear rRNA genes of some unicellular eukaryotes, in some tRNA genes in bacteria, and in a few bacteriophage genes coding for mRNAs. Group I RNA introns are excised in the form of linear RNA fragments.

In contrast, *Group II introns* are excised as lariats in which an adenine within the intron forms the branch-point, just as in the spliceosome mechanism. Group II introns are found in some mitochondrial and chloroplast genes of plants and unicellular eukaryotes and in the genomes of some archaea and bacteria. Biologists think that today's prevailing splicing mechanism, based on spliceosomes, evolved from Group II introns, with the intron RNA's catalytic role being taken over by the snRNA molecules of the spliceosome. Support for this idea has come from the discovery that protein-free RNA molecules isolated from spliceosomes are capable of catalyzing the first step in the splicing reaction involving the branch-point adenine.

The Existence of Introns Permits Alternative Splicing and Exon Shuffling

The burning questions about introns are: Why do nearly all genes in multicellular eukaryotes have them? Why do cells have so much DNA that seems to serve no coding function? Why, in generation after generation of cells, is so much energy invested in synthesizing segments of DNA—and their RNA transcripts—that appear to serve no useful function and are destined only for the splicing scrap heap?

In fact, it is not true that introns never perform any functions of their own. In a few cases, intron RNAs are processed to yield functional products rather than being degraded. For example, some types of snoRNA—whose role in guiding pre-rRNA methylation and cleavage was discussed earlier in the chapter—are derived from introns that are first removed from pre-mRNA and then processed to form snoRNA. And in a few cases, introns are even translated into proteins.

Despite these exceptions, most introns are destroyed without serving any obvious function. One benefit to such a seemingly wasteful arrangement is that the presence of introns allows each pre-mRNA molecule to be spliced in multiple ways, thereby generating anywhere from 2 or 3 to more than 100 different mRNAs (and hence polypeptides) from the same gene. This phenomenon, called **alternative splicing,** is made possible by mechanisms that allow certain splice sites to be either activated or skipped (**Figure 21-23**). Control over these splice sites is exerted by various molecules, including regulatory proteins and snoRNAs, that bind to *splicing enhancer* or *splicing silencer* sequences in pre-mRNAs. Binding of the appropriate regulatory molecule to a splicing enhancer or silencer leads to activation or skipping of individual splice sites, respectively.

The existence of alternative splicing may help explain how the biological complexity of vertebrates is achieved without a major increase in the number of genes compared to simpler organisms. (Recall from Chapter 18 that humans have barely more genes than a nematode worm does and 12,000 fewer genes than a rice plant does.) Instead of increasing the number of genes, humans transcribe the majority of their genes into pre-mRNAs that can be spliced in more than one way. As a result, the roughly 25,000 human genes produce mRNAs coding for hundreds of thousands of different polypeptides.

Another function of introns is that they allow the evolution of new protein-coding genes through recombination events that bring together new combinations of exons. At least two different mechanisms are involved, both based on the fact that introns are long stretches of DNA where genetic recombination can occur without harming coding sequences. First, genetic recombination between the introns of different genes will produce genes containing new combinations of exons—*exon shuffling.* Second, recombination

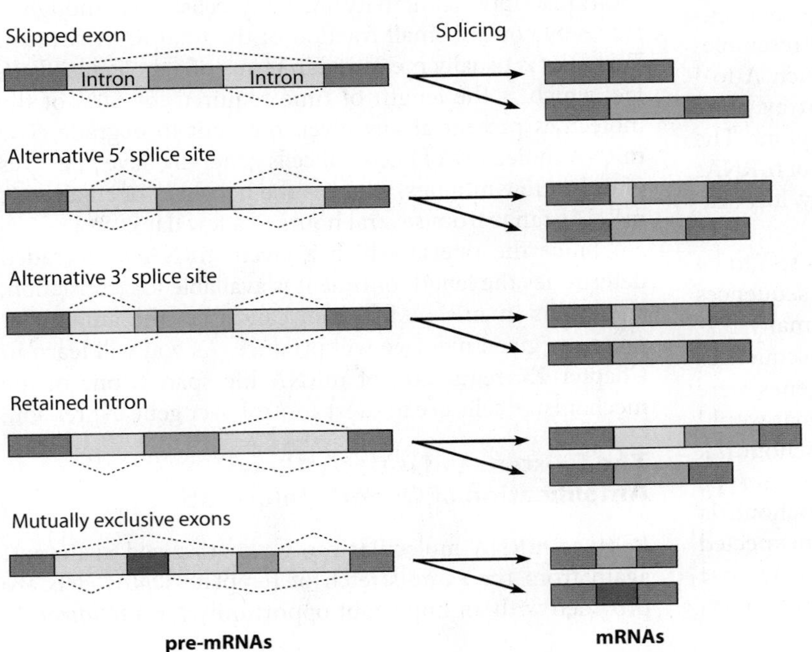

FIGURE 21-23 Some Examples of Alternative Splicing. This diagram illustrates five alternative splicing mechanisms that allow different mRNAs to be produced from the same pre-mRNA by activating or skipping individual splice sites. The thin dashed lines above and below each pre-mRNA represent the splice sites that are used to generate the two different mRNAs shown for each example.

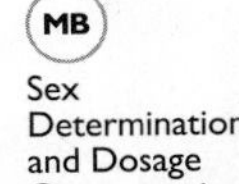

Sex Determination and Dosage Compensation in Drosophila

can also create duplicate copies of an individual exon within a single gene. The two copies could continue as exact duplicates, or one copy might mutate to a sequence that produces a new activity in the polypeptide encoded by that gene.

RNA Editing Allows mRNA Coding Sequences to Be Altered

About a decade after introns and RNA splicing were first discovered, molecular biologists were surprised by the discovery of yet another type of mRNA processing, called **RNA editing.** During RNA editing, anywhere from a single nucleotide to hundreds of nucleotides may be inserted, removed, or chemically altered within the coding sequence of an mRNA. Such changes can create new start or stop codons, or they can alter the reading frame of the message.

Some of the best-studied examples of RNA editing occur in the mitochondrial mRNAs of trypanosomes, which are parasitic protozoa. In these mRNAs, editing involves the insertion and deletion of multiple uracil nucleotides at various points in the mRNA. The information for this editing is located in small RNA molecules called *guide RNAs,* which are encoded by mitochondrial genes separate from the mRNA genes. In one proposed editing mechanism, hydrogen bonding causes short complementary regions of the guide RNA and mRNA to come together, and nearby sequences of U's in the guide RNA are then spliced into the mRNA.

A different type of editing occurs in the mitochondrial and chloroplast mRNAs of flowering plants. In these cases, nucleotides are neither inserted nor deleted, but C's are converted to U's (and vice versa) by deamination (and amination) reactions. Similar base conversions have also been discovered in mRNAs transcribed from nuclear genes in animal cells. For example, a single codon in the mRNA transcribed from the mammalian apolipoprotein-B gene undergoes a C-to-U conversion during RNA editing. Another type of editing detected in animal cell nuclei converts adenosine (A) to inosine (I), which resembles guanosine (G) in its base-pairing properties. Such A-to-I editing has been detected in mRNAs produced by more than 1600 human genes, mostly in noncoding regions. The reason for putting I's in the untranslated regions of mRNAs is a mystery, but roles in regulating RNA stability, localization, and translation rate have been suggested.

The existence of RNA editing provides a reason to be cautious in inferring either polypeptide or RNA sequences from genomic DNA sequences. For example, many discrepancies were observed when the amino acid sequences of proteins produced by plant mitochondrial genes were first compared with the amino acid sequences that would be predicted based on the base sequence of mitochondrial DNA. Although some of these discrepancies can be explained by nonstandard codon usage in mitochondria (described earlier in the chapter), most of the unexpected amino acids arise because RNA editing alters the base sequence of various mRNA codons, leading to the incorporation of amino acids that would not have been expected based on a gene's DNA sequence.

Nucleic acid editing is not restricted to mRNAs. *MicroRNAs*—whose role in regulating gene expression is described in Chapter 23—are another class of RNAs that can be edited. And DNA is also subject to editing. For example, eukaryotes possess a DNA-editing enzyme called *APOBEC3G,* which inactivates retroviruses by catalyzing C-to-U conversions in the initial DNA strand produced when viral RNA is copied by reverse transcriptase (step ❷ in Figure 21A-1). The C-to-U conversions in the first strand lead to G-to-A conversions in the complementary DNA strand, thereby introducing mutations that debilitate the virus. To defend against this attack, HIV and other retroviruses produce a protein called *Vif,* which targets APOBEC3G for destruction. Because suppression of APOBEC3G is essential for successful retroviral infection, blocking the action of Vif might be a useful strategy for developing novel new treatments for HIV/AIDS.

Key Aspects of mRNA Metabolism

Before ending this chapter, we should note two key aspects of mRNA metabolism that are important to our overall understanding of how mRNA molecules behave within cells. These are the short life span of most mRNAs and the ability of mRNA to amplify genetic information.

Most mRNA Molecules Have a Relatively Short Life Span

Most mRNA molecules have a high *turnover rate*—that is, the rate at which molecules are degraded and then replaced with newly synthesized versions. In this respect, mRNA contrasts with the other major forms of RNA in the cell, rRNA and tRNA, which are notable for their stability. Because of its short life span, mRNA accounts for most of the transcriptional activity in many cells, even though it represents only a small fraction of the total RNA content. Turnover is usually measured in terms of a molecule's *half-life,* which is the length of time required for 50% of the molecules present at any given moment to degrade. The mRNA molecules of bacterial cells generally have half-lives of only a few minutes, whereas the half-lives of eukaryotic mRNAs range from several hours to a few days.

Since the rate at which a given mRNA is degraded determines the length of time it is available for translation, alterations in mRNA life span can affect the amount of protein a given message will produce. As you will learn in Chapter 23, regulation of mRNA life span is one of the mechanisms cells use to exert control over gene expression.

The Existence of mRNA Allows Amplification of Genetic Information

Because mRNA molecules can be synthesized again and again from the same stretch of template DNA, cells are provided with an important opportunity for *amplification* of the genetic message. If DNA gene sequences were used directly in protein synthesis, the number of protein molecules that could be translated from any gene within a given time period would be strictly limited by the rate of polypeptide synthesis. But in a system using mRNA as an intermediate, multiple copies of a gene's informational content can be made, and each of these can be used in turn to direct the synthesis of the protein product.

As an especially dramatic example of this amplification effect, consider the synthesis of *fibroin*, the major protein of silk. The haploid genome of the silkworm has only one fibroin gene, but about 10^4 copies of fibroin mRNA are transcribed from the two copies of the gene in each diploid cell of the silk gland. Each of these mRNA molecules, in turn, directs the synthesis of about 10^5 fibroin molecules, resulting in the production of more than 10^9 molecules of fibroin per cell—all within the four-day period it takes the worm to make its cocoon! Without mRNA as an intermediate, the genome of the silkworm would need 10^4 copies of the fibroin gene (or about 40,000 days!) to make a cocoon.

Significantly, most genes that code for proteins occur in only one or a few copies per haploid genome. In contrast, genes that code for rRNA and tRNA are always present in multiple copies. It is advantageous for cells to have many copies of genes whose final products are RNA (rather than protein) because in this case there is no opportunity for amplifying each gene's effect by repeated translation.

SUMMARY OF KEY POINTS

The Directional Flow of Genetic Information

- Instructions stored in DNA are transcribed and processed into molecules of mRNA, rRNA, and tRNA for use in protein synthesis.
- The base sequence of each mRNA molecule dictates the sequence of amino acids in a polypeptide chain.

The Genetic Code

- When guiding the synthesis of a polypeptide chain, mRNA is read in units of three bases called codons.
- The table of the genetic code indicates which amino acid (or stop signal) is specified by each codon. The code is unambiguous, nonoverlapping, degenerate, and nearly universal.

Transcription in Bacterial Cells

- Transcription is the process by which a DNA template strand is copied by RNA polymerase to produce a complementary molecule of RNA.
- The molecular mechanism of transcription involves four stages: (1) binding of RNA polymerase to promoter sequences in the DNA template strand, (2) initiation of RNA synthesis, (3) elongation of the RNA chain, and (4) termination.

Transcription in Eukaryotic Cells

- Transcription in eukaryotes involves the same general principles as in bacteria but is more complex. One major difference is that eukaryotic cells have multiple forms of RNA polymerase that are specialized for synthesizing different types of RNA.
- The three nuclear RNA polymerases recognize different families of DNA promoter sequences. These eukaryotic promoters are usually bound by transcription factors that associate with RNA polymerase rather than by RNA polymerase itself.

RNA Processing

- Newly forming RNA molecules must generally undergo some type of processing, both during and after transcription, before the RNA can perform its normal function. The major exception is bacterial mRNA, whose translation often begins before transcription is completed.
- RNA processing involves chemical alterations such as cleavage of multigene transcription units, removal of noncoding sequences, addition of sequences to the 3′ and 5′ ends, and insertion, removal, and chemical modification of individual nucleotides.
- Processing of eukaryotic pre-mRNA is especially elaborate, involving the addition of a 5′ cap and a 3′ poly(A) tail, as well as the removal of introns by spliceosomes or, in some cases, by a process catalyzed by intron RNA.
- The presence of introns and exons allows pre-mRNA molecules to be spliced in more than one way, thereby permitting a single gene to produce multiple mRNAs coding for different polypeptides.
- Various editing mechanisms allow the base sequence of mRNA molecules to be altered, thereby creating mRNAs that code for polypeptides whose amino acid sequences could not have been predicted by knowing the base sequence of the corresponding genes.

Key Aspects of mRNA Metabolism

- Most mRNAs have a relatively short life span.
- Active protein-coding genes are transcribed multiple times to produce many molecules of their corresponding mRNAs, thereby amplifying the amount of protein that each gene can produce.

 **www.masteringbiology.com**

1. MasteringBiology Assignments

Tutorials and Activities

- Molecular Model Structural Tutorials for visualization of 3D structures
- BioFlix Tutorials including 3D animations with hints and detailed feedback for common wrong answers
- Overview of Protein Synthesis; Types of RNA; Role of the Nucleus and Ribosomes in Protein Synthesis; One Gene One Enzyme; Protein Synthesis (1 of 3): Overview (BioFlix tutorial); The Triplet Nature of the Genetic Code; The Genetic Code; RNA Polymerase of Thermus thermophilus; Nucleic Acids: Transcription; Transcription; RNA Synthesis; Protein Synthesis (2 of 3): Transcription and RNA Processing (BioFlix tutorial); mRNA Production in Eukaryotes; RNA Processing; RNA Splicing; Sex Determination and Dosage Compensation in Drosophila

Questions

- Reading Quiz
- Multiple Choice
- End-of-chapter questions

2. eText Read your book online, search, highlight text, take notes, and more

3. The Study Area

- Chapter Review Quiz, Guide to Techniques and Methods, 3D-Structure Tutorials, Activities, BioFlix, Videos, Biology Review, Word Study Tools, Glossary, Flashcards

PROBLEM SET

More challenging problems are marked with a •.

21-1 Triplets or Sextuplets? In his Nobel Prize lecture in 1962, Francis Crick pointed out that while the pioneering experiments he performed with Barnett, Brenner, and Watts-Tobin suggested that the DNA "code" is a triplet, their experiments did not rule out the possibility that the code could require six or nine bases.

(a) Assuming the code is a triplet, what effect would adding or removing six or nine bases have on the reading frame of a piece of DNA?

(b) If the code actually were a sextuplet, how would addition of three, six, or nine nucleotides affect the reading frame of a piece of DNA?

21-2 The Genetic Code in a mammalian cell. A portion of a polypeptide produced by a mammalian cell was found to have the following sequence of amino acids:

...Lys-Ser-Pro-Ser-Leu-Asn-Ala... In a normal cell

...Lys-Val-His-His-Leu-Met-Ala... In a mutant cell

(a) What was the nucleotide sequence of the mRNA segment that encoded this portion of the original polypeptide?

(b) What was the nucleotide sequence of the mRNA encoding this portion of the mutant polypeptide?

(c) Can you determine which nucleotide was deleted and which was inserted? Explain your answer.

(d) If the insertion of histidines in the sequence results in abnormal activity of the protein, what changes would you suggest in the DNA that can change the insertion of histidines to the normal protein?

21-3 Frameshift Mutations. Each of the mutants listed below this paragraph has a different mutant form of the gene encoding protein X. Each mutant gene contains one or more nucleotide insertions (+) or deletions (−) of the type caused by acridine dyes. Assume that all the mutations are located very near the beginning of the gene for protein X. In each case, indicate with an "OK" if you would expect the mutant protein to be nearly normal and with a "Not OK" if you would expect it to be obviously abnormal.

(a) − **(b)** −/+ **(c)** −/− **(d)** +/−/+
(e) +/−/+/− **(f)** +/+/+ **(g)** +/+/−/+ **(h)** −/−/+/−/−
(i) −/−/−/−/−/−

21-4 Amino Acid Substitutions in Mutant Proteins. Although the codon assignments summarized in Figure 21-7 were originally deduced from experiments involving synthetic RNA polymers and triplets, their validity was subsequently confirmed by examining the amino acid sequences of normally occurring mutant proteins. The table below lists some examples of amino acid substitutions seen in mutant forms of hemoglobin, tryptophan synthase, and the tobacco mosaic virus coat protein. For each amino acid alteration, list the corresponding single-base changes in mRNA that could have caused that particular amino acid substitution.

Protein	Amino Acid Substitution
Hemoglobin	Glu → Val
	His → Tyr
	Asn → Lys
Coat protein of tobacco mosaic virus	Glu → Gly
	Ile → Val
Tryptophan synthase	Tyr → Cys
	Gly → Arg
	Lys → Stop

• **21-5 Locating Promoters.** The following table provides data concerning the effects of various deletions in a eukaryotic gene coding for 5S rRNA on the ability of this gene to be transcribed by RNA polymerase III.

Nucleotides Deleted	Ability of 5S-rRNA Gene to Be Transcribed by RNA Polymerase III
−45 through −1	Yes
+1 through +47	Yes
+10 through +47	Yes
+10 through +63	No
+80 through +123	No
+83 through +123	Yes

(a) What do these data tell you about the probable location of the promoter for this particular 5S-rRNA gene?

(b) If a similar experiment were carried out for a gene transcribed by RNA polymerase I, what kinds of results would you expect?

(c) If a similar experiment were carried out for a gene transcribed by RNA polymerase II, what kinds of results would you expect?

21-6 RNA Polymerases and Promoters. For each of the following statements about RNA polymerases, indicate with a B if the statement is true of the bacterial enzyme and with a I, II, or III if it is true of the respective eukaryotic RNA polymerase. A given statement may be true of any, all, or none (N) of these enzymes.

(a) The enzyme catalyzes an exergonic reaction.

(b) The enzyme may sometimes be found attached to an RNA molecule that also has ribosomes bound to it.

(c) The enzyme moves along the DNA template strand in the $3' \rightarrow 5'$ direction.

(d) The enzyme is insensitive to α-amanitin.

(e) The specificity of transcription by the enzyme is determined by a subunit of the holoenzyme.

(f) The enzyme synthesizes a product likely to acquire a $5'$ cap.

(g) All the primary transcripts must be processed before being used in translation.

(h) All promoters used by the enzyme lie mostly upstream of the transcriptional startpoint and are only partially transcribed.

(i) Transcription factors must bind to the promoter before the polymerase can bind.

(j) The enzyme synthesizes rRNA.

(k) The enzyme adds a poly(A) sequence to mRNA.

21-7 RNA Processing. The three major classes of RNA found in the cytoplasm of a typical eukaryotic cell are rRNA, tRNA, and mRNA. For each, indicate the following:

(a) Two or more kinds of processing that the RNA has almost certainly been subjected to.

(b) A processing event unique to that RNA species.

(c) A processing event that you would also expect to find for the same species of RNA from a bacterial cell.

21-8 Spliceosomes. The RNA processing carried out by spliceosomes in the eukaryotic nucleus involves many different kinds of protein and RNA molecules. For each of the following five components of the splicing process, indicate whether it is protein (P), RNA (R), or both (PR). Then briefly explain how each of the five fits into the process.

(a) snRNA **(b)** Spliceosome **(c)** snRNP

(d) Splice sites **(e)** Lariat

• **21-9 Antibiotic Inhibitors of Transcription.** Rifamycin and actinomycin D are two antibiotics derived from the bacterium *Streptomyces*. Rifamycin binds to the β subunit of *E. coli* RNA polymerase and interferes with the formation of the first phosphodiester bond in the RNA chain. Actinomycin D binds to double-stranded DNA by intercalation (slipping between neighboring base pairs).

(a) What would be the combined effect of using both rifamycin and actinomycin D in a *Salmonella typhimurium* infection?

(b) Which of the four stages in transcription would you expect actinomycin D to affect primarily?

(c) Which of the two inhibitors is more likely to affect DNA synthesis in cultured human liver cells?

(d) If you had the option of tagging rifamycin with either a lipid tail or a charged group like NH4, which variant is most likely to enter cells easily and prove to be more effective?

(e) When fertilized sea urchin eggs are treated with actinomycin D, they develop for many hours but eventually arrest as hollow balls of several hundred cells (called blastulae). Propose an explanation for why such embryos arrest, but also why they progress as far as they do.

(f) In rifamycin-resistant *E. coli*, what change would you expect in the RNA polymerase protein when compared to nonresistant *E. coli*?

• **21-10 Copolymer Analysis.** In their initial attempts to determine codon assignments, Nirenberg and Matthei first used RNA homopolymers and then used RNA copolymers synthesized by the enzyme polynucleotide phosphorylase. This enzyme adds nucleotides randomly to the growing chain but in proportion to their presence in the incubation mixture. By varying the ratio of precursor molecules in the synthesis of copolymers, Nirenberg and Matthei were able to deduce base compositions (but usually not actual sequences) of the codons that code for various amino acids. Suppose you carry out two polynucleotide phosphorylase incubations, with UTP and CTP present in both but in different ratios. In incubation A, the precursors are present in equal concentrations. In incubation B, there is three times as much UTP as CTP. The copolymers generated in both incubation mixtures are then used in a cell-free protein-synthesizing system, and the resulting polypeptides are analyzed for amino acid composition.

(a) What are the eight possible codons represented by the nucleotide sequences of the resulting copolymers in both incubation mixtures? What amino acids do these codons code for?

(b) For every 64 codons in the copolymer formed in incubation A, how many of each of the 8 possible codons would you expect on average? How many for incubation B?

(c) What can you say about the expected frequency of occurrence of the possible amino acids in the polypeptides obtained upon translation of the copolymers from incubation A? What about the polypeptides that result from translation of the incubation B copolymers?

(d) Explain what sort of information can be obtained by this technique.

(e) Would it be possible by this technique to determine that codons with 2 U's and 1 C code for phenylalanine, leucine, and serine? Why or why not?

(f) Would it be possible by this technique to decide which of the three codons with 2 U's and 1 C (UUC, UCU, CUU) correspond to each of the three amino acids mentioned in part e? Why or why not?

(g) Suggest a way to assign the three codons of part f to the appropriate amino acids of part e.

• **21-11 Introns.** To investigate the possible presence of introns in three newly discovered genes (*X*, *Y*, and *Z*), you perform an experiment in which the restriction enzyme *Hae*III is used to cleave either the DNA of each gene or the cDNA made by copying its mRNA with reverse transcriptase. The resulting DNA fragments are separated by gel electrophoresis, and the presence of fragments in the gels is detected by hybridizing to a radioactive DNA probe made by copying the intact gene with DNA polymerase in the presence of radioactive substrates. The following results are obtained:

Source of DNA	Number of Fragments After Electrophoresis
Gene *X* DNA	3
cDNA made from mRNA *X*	2
Gene *Y* DNA	4
cDNA made from mRNA *Y*	2
Gene *Z* DNA	2
cDNA made from mRNA *Z*	2

(a) What would be the combined effect of using both rifamycin and actinomycin D in a salmonells typhimurium infection?

(b) What can you conclude about the number of introns present in gene *Y*?

(c) What can you conclude about the number of introns present in gene *Z*?

21-12 Cloning Conundrum. Using established recombinant DNA technology, you insert a gene from a human liver cell into a bacterium. The bacterium then expresses a protein corresponding to the inserted DNA. To your dismay, you discover that the protein produced is useless and is found to contain many more amino acids than does the protein made by the eukaryotic cell. Assuming there is no mutation in the human gene, explain why this happened.

SUGGESTED READING

References of historical importance are marked with a •.

Information Flow and the Genetic Code

Atkins, J. F., and P. V. Baranov. Duality in the genetic code. *Nature* 448 (2007): 1004.

Chiu, Y. L., and W. C. Greene. APOBEC3G: An intracellular centurion. *Philos. Trans. R. Soc. Lond. B Biol. Sci.* 364 (2009): 689.

• Crick, F. H. C. The genetic code. *Sci. Amer.* 207 (October 1962): 66.

• Crick, F. H. C. The genetic code III. *Sci. Amer.* 215 (October 1966): 55.

Kazazian, H. H. Jr. Mobile elements: Drivers of genome evolution. *Science* 303 (2004): 1626.

Lobanov, A. V. et al. Is there a twenty third amino acid in the genetic code? *Trends Genet.* 22 (2006): 357.

• Nirenberg, M. Historical review: Deciphering the genetic code—A personal account. *Trends Biochem. Sci.* 29 (2004): 46.

• Nirenberg, M. W. The genetic code II. *Sci. Amer.* 208 (March 1963): 80.

• Sarkar, S. Forty years under the central dogma. *Trends Biochem. Sci.* 23 (1998): 312.

Yanofsky, C. Establishing the triplet nature of the genetic code. *Cell* 128 (2007): 815.

Transcription in Bacterial Cells

Landick, R. Shifting RNA polymerase into overdrive. *Science* 284 (1999): 598.

Mooney, R. A., S. A. Darst, and R. Landick. Sigma and RNA polymerase: An on-again, off-again relationship? *Mol. Cell* 20 (2005): 335.

Roberts, J. W., S. Shankar, and J. J. Filter. RNA polymerase elongation factors. *Annu. Rev. Microbiol.* 62 (2008): 211.

Young, B. A., T. M. Gruber, and C. A. Gross. Views of transcription initiation. *Cell* 109 (2002): 417.

Transcription in Eukaryotic Cells

Buratowski, S. Connections between mRNA 3′ end processing and transcription termination. *Curr. Opin. Cell Biol.* 17 (2005): 257.

Cramer, P. Self-correcting messages. *Science* 313 (2006): 447.

Egloff, S., and S. Murphy. Cracking the RNA polymerase II CTD code. *Trends Genet.* 24 (2008): 280.

Juven-Gershon, T., J. Y. Hsu, J. W. Theisen, and J. T. Kadonaga. The RNA polymerase II core promoter—The gateway to transcription. *Curr. Opin. Cell Biol.* 20 (2008): 253.

Nudler, E. RNA polymerase active center: The molecular engine of transcription. *Annu. Rev. Biochem.* 78 (2009): 335.

Saunders, A., L. J. Core, and J. T. Lis. Breaking barriers to transcription elongation. *Nature Rev. Mol. Cell Biol.* 7 (2006): 557.

Sutherland, H., and W. A. Bickmore. Transcription factories: Gene expression in unions? *Nat. Rev. Genet.* 10 (2009): 457.

Tjian, R. Molecular machines that control genes. *Sci. Amer.* 272 (February 1995): 54.

RNA Processing

Blencowe, B. J. Alternative splicing: New insights from global analysis. *Cell* 126 (2006): 37.

Eisenberg, E. et al. Is abundant A-to-I editing primate-specific? *Trends Genet.* 21 (2005): 77.

Perales, R., and D. Bentley. "Cotranscriptionality": The transcription elongation complex as a nexus for nuclear transactions. *Mol. Cell* 36 (2009): 178.

Roy, S. W., and W. Gilbert. The evolution of spliceosomal introns: Patterns, puzzles and progress. *Nature Rev. Genet.* 7 (2006): 211.

Samuel, C. E. et al. RNA editing minireview series. *J. Biol. Chem.* 278 (2003): 1389.

Scherrer, K. Historical review: The discovery of "giant" RNA and RNA processing: 40 years of enigma. *Trends Biochem. Sci.* 28 (2003): 566.

Sharp, P. The discovery of split genes and RNA splicing. *Trends Biochem. Sci.* 30 (2005): 279.

Steitz, J. A. Snurps. *Sci. Amer.* 258 (June 1988): 58.

Wahl, M. C., C. L. Will, and R. Luhrmann. The spliceosome: Design principles of a dynamic RNP machine. *Cell* 136 (2009): 701.

Wang, G.-S., and T. A. Cooper. Splicing in disease: Disruption of the splicing code and the decoding machinery. *Nature Rev. Genet.* 8 (2007): 749.

Xing, Y., and C. Lee. Alternative splicing and RNA selection pressure—Evolutionary consequences for eukaryotic genomes. *Nature Rev. Genet.* 7 (2006): 499.

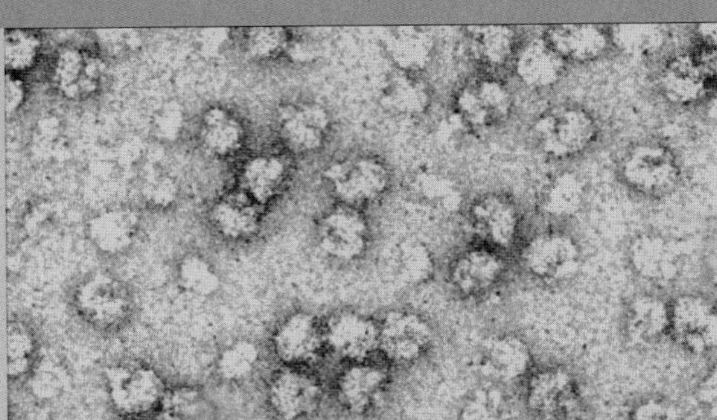

22 Gene Expression: II. Protein Synthesis and Sorting

See MasteringBiology® for tutorials, activities, and quizzes.

In the preceding chapter we took gene expression from DNA to RNA, covering DNA transcription followed by processing of the resulting RNA transcripts. For genes encoding ribosomal and transfer RNAs (and some other small RNAs), RNA is the final product of gene expression. But for the thousands of other genes in an organism's genome, the ultimate gene product is protein. This chapter describes how the messenger RNAs (mRNAs) produced by these protein-coding genes are translated into polypeptides, how polypeptides become functional proteins, and how proteins reach the destinations where they carry out their functions.

Translation, the key step in the production of protein molecules, involves a change in language from the nucleotide base sequence of an mRNA molecule to the amino acid sequence of a polypeptide chain. During this process, a sequence of mRNA nucleotides, read as triplet codons, specifies the order in which amino acids are added to a growing polypeptide chain. Ribosomes serve as the intracellular sites for translation, while RNA molecules are the agents that ensure insertion of the correct amino acids at each position in the polypeptide. We will start by surveying the cell's cast of characters for performing translation, and then we will examine each of its steps in detail.

Translation: The Cast of Characters

The cellular machinery for translating mRNAs into polypeptides involves five major components: *ribosomes* that carry out the process of polypeptide synthesis, *tRNA* molecules that align amino acids in the correct order along the mRNA template, *aminoacyl-tRNA synthetases* that attach amino acids to their appropriate tRNA molecules, *mRNA* molecules that encode the amino acid sequence information for the polypeptides being synthesized, and *protein factors* that facilitate several steps in the translation process. In introducing this cast of characters, let's begin with the ribosomes.

The Ribosome Carries Out Polypeptide Synthesis

Ribosomes play a central role in protein synthesis, orienting the mRNA and amino acid-carrying tRNAs so the genetic code can be read accurately and catalyzing peptide bond formation to link the amino acids into a polypeptide. As we saw in Chapter 4, **ribosomes** are particles made of ribosomal RNA (rRNA) and protein that reside in the cytoplasm and, in eukaryotes, in the mitochondrial matrix and chloroplast stroma. In the eukaryotic cytoplasm, ribosomes occur both free in the cytosol and bound to membranes of the endoplasmic reticulum and the outer membrane of the nuclear envelope. The ribosomes of prokaryotes (archaea and bacteria) are smaller than those of eukaryotes, although many of the ribosomal proteins, translation factors, and tRNAs used by archaea resemble their eukaryotic counterparts more closely than do the comparable components of bacteria.

The shape of a typical bacterial ribosome revealed by electron microscopy is shown in **Figure 22-1.** Like all ribosomes, it is built from two dissociable subunits called the *large* and *small subunits.* The bacterial ribosome has a sedimentation coefficient of about 70S and is built from a 30S small subunit and a 50S large subunit. Its eukaryotic equivalent is an 80S ribosome consisting of a 40S subunit and a 60S subunit. **Table 22-1** lists some of the properties of bacterial and eukaryotic ribosomes and their subunits. The bacterial ribosome contains fewer proteins, is sensitive to different inhibitors of protein synthesis, and has smaller RNA molecules (and one fewer RNA) than eukaryotic ribosomes have. In Chapter 21, we saw that rRNAs are produced from larger precursor molecules by cleavage and processing reactions that, in eukaryotes, take place within the nucleolus (page 543). During these processing events, the rRNAs become associated with ribosomal proteins and self-assemble into small and large subunits, which come together only after binding to mRNA. X-ray crystallography has allowed the

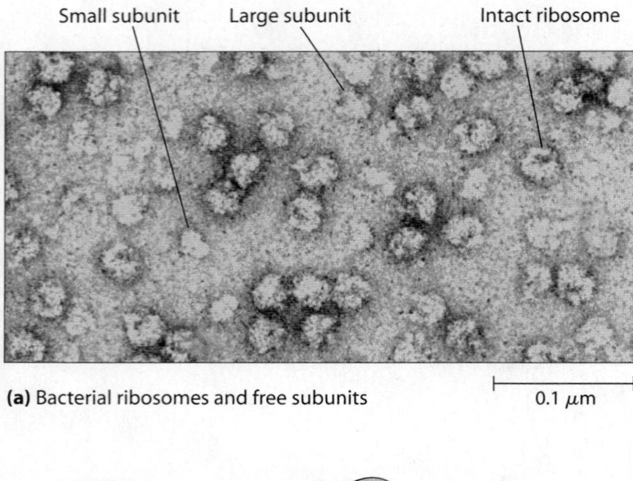

(a) Bacterial ribosomes and free subunits

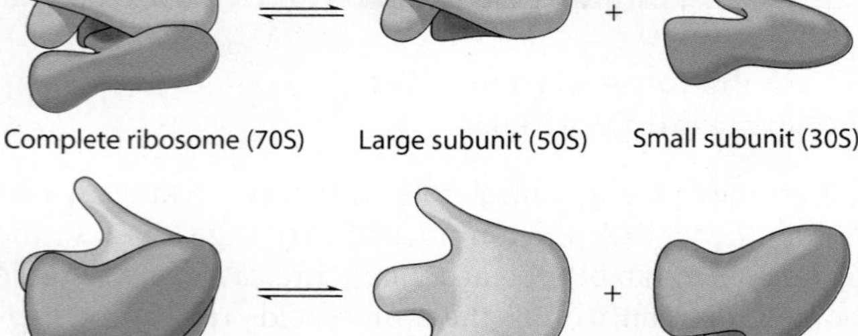

(b) Two views of a bacterial ribosome and its subunits

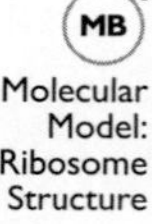

Molecular Model: Ribosome Structure

FIGURE 22-1 The Bacterial Ribosome. **(a)** The electron micrograph shows intact ribosomes as well as individual subunits (TEM). **(b)** The two structural models, based on such micrographs, show two views in which the ribosome has been rotated by 90 degrees. Bacterial ribosomes are about 25 nm in diameter. (Eukaryotic cytoplasmic ribosomes are roughly similar in shape and about 30 nm in diameter.)

arrangement of all the individual proteins and RNA molecules of the small and large subunits of bacterial ribosomes to be pinpointed down to the atomic level. Venkatraman Ramakrishnan, Thomas A. Steitz, and Ada E. Yonath were awarded the Nobel Prize in Chemistry in 2009 for their contributions to this monumental achievement.

Functionally, ribosomes have sometimes been called the "workbenches" of protein synthesis, but their active role in polypeptide synthesis makes "machine" a more apt label. In essence, the role of the ribosome in polypeptide synthesis resembles that of a large, complicated enzyme constructed from more than 50 different proteins and several kinds of rRNA. For many years it was thought that the rRNA simply provided a structural scaffold for the ribosomal proteins, with the latter actually carrying out the steps in polypeptide synthesis. But today we know that the reverse is closer to the truth—rRNA performs many of the ribosome's key functions.

Four sites on the ribosome are particularly important for protein synthesis (**Figure 22-2**). These are an **mRNA-binding site** and three sites where tRNA can bind: an **A (aminoacyl) site** that binds each newly arriving tRNA with its attached amino acid, a **P (peptidyl) site** where the tRNA carrying the growing polypeptide chain resides, and an **E (exit) site,** from which tRNAs leave the ribosome after they have discharged their amino acids. How these sites function in the process of translation will become clear in a few pages.

Transfer RNA Molecules Bring Amino Acids to the Ribosome

Since the sequence of codons in mRNA ultimately determines the amino acid sequence of polypeptide chains, a mechanism must exist that enables codons to arrange amino acids in the proper order. The general nature of this mechanism was first proposed in 1957 by Francis Crick. With remarkable foresight, Crick postulated that amino acids cannot directly recognize nucleotide base sequences and that some kind of hypothetical "adaptor" molecule must therefore mediate the interaction between amino acids and mRNA. He further predicted that each adaptor molecule possesses two sites, one that binds to a specific amino acid and the other that recognizes an mRNA base sequence coding for this amino acid.

In the year following Crick's adaptor proposal, Mahlon Hoagland discovered a family of adaptor molecules exhibiting these predicted properties. While investigating the process of protein synthesis in cell-free systems, Hoagland found that radioactive amino acids first become

Table 22-1 Properties of Bacterial and Eukaryotic Cytoplasmic Ribosomes

	Size of Ribosomes		Subunit	Subunit Size		Subunit Proteins	Subunit RNA	
Source	S Value*	Mol. Wt.		S Value	Mol. Wt.		S Value	Nucleotides
Bacterial cells	70S	2.5×10^6	Large	50S	1.6×10^6	34	23S	2900
							5S	120
			Small	30S	0.9×10^6	21	16S	1540
Eukaryotic cells	80S	4.2×10^6	Large	60S	2.8×10^6	About 46	25–28S	≤4700
							5.8S	160
							5S	120
			Small	40S	1.43×10^6	About 32	18S	1900

*If you are surprised that the S values of the subunits do not add up to that of the whole ribosome, recall that an S value is a measure of the velocity at which a particle sediments upon centrifugation and is only indirectly related to the mass of the particle.

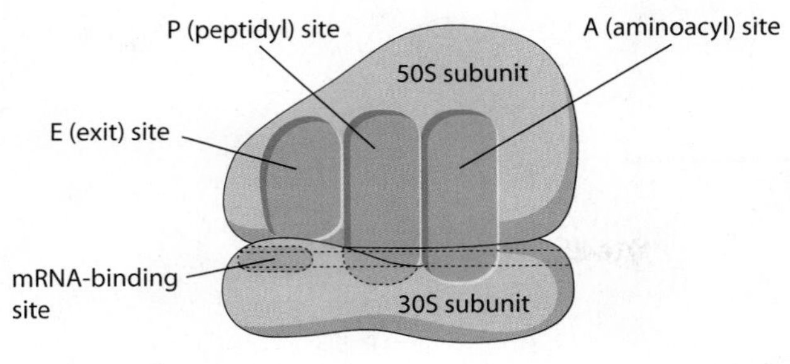

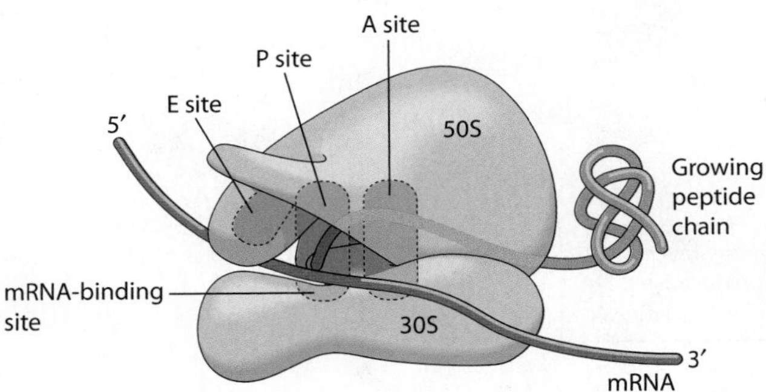

FIGURE 22-2 Binding Sites on a Ribosome. Ribosomes contain A and P sites where amino acid-carrying tRNA molecules bind during polypeptide synthesis, and an E site from which empty tRNAs leave the ribosome. The mRNA-binding site binds a specific nucleotide sequence near the 5′ end of mRNA, placing the mRNA in proper position for translation of its first codon. The binding sites are all located at or near the interface between the large and small subunits. In the top diagram, which is a schematic representation of a bacterial ribosome used in this chapter, the pair of horizontal dashed lines indicates where the mRNA molecule lies. The bottom diagram is a more realistic representation.

covalently attached to small RNA molecules. Adding these amino acid–RNA complexes to ribosomes led to the onset of protein synthesis and the incorporation of radioactive amino acids into new proteins. Hoagland therefore concluded that amino acids are initially bound to small RNA molecules, which then bring the amino acids to the ribosome for subsequent insertion into newly forming polypeptide chains.

The small RNA molecules that Hoagland discovered were named **transfer RNAs (tRNAs).** Appropriate to their role as intermediaries between mRNA and amino acids, tRNA molecules have two kinds of specificity. Each tRNA binds to one specific amino acid, and each recognizes one or more mRNA codons specifying that particular amino acid, as indicated by the genetic code. Transfer RNAs are linked to their corresponding amino acids by an ester bond that joins the amino acid to the 2′- or 3′-hydroxyl group of the adenine (A) nucleotide located at the 3′ end of all tRNA molecules (**Figure 22-3a**). Selection of the correct amino acid for each tRNA is the responsibility of the enzymes that catalyze formation of the ester bond, as we will discuss shortly. By convention, the name of the amino acid that attaches to a given tRNA is indicated by a superscript. For example, tRNA molecules specific for alanine are identified as $tRNA^{Ala}$. Once the amino acid is attached, the tRNA is called an **aminoacyl tRNA** (e.g., alanyl $tRNA^{Ala}$). The tRNA is said to be in its *charged* form, and the amino acid is said to be *activated.*

Transfer RNA molecules can recognize codons in mRNA because each tRNA possesses an **anticodon,** a special trinucleotide sequence located within one of the loops of the tRNA molecule (see Figure 22-3a). The anticodon of each tRNA is complementary to one or more mRNA codons that specify the amino acid being carried by that tRNA. Therefore, *anticodons permit tRNA molecules to recognize codons in mRNA by complementary base pairing.* Take careful note of the convention used in representing codons and anticodons: Codons in mRNA are written in the 5′ → 3′ direction, whereas anticodons in tRNA are usually represented in the 3′ → 5′ orientation. Thus, one of the codons for alanine is 5′-GCC-3′, and the corresponding anticodon in tRNA is 3′-CGG-5′.

Since the genetic code employs 61 codons to specify amino acids (page 655), you might expect to find 61 different tRNA molecules involved in protein synthesis, each recognizing a different codon. However, the number of different tRNAs is significantly less than 61 because many tRNA molecules recognize more than one codon. You can see why this is possible by examining the table of the genetic code (see Figure 21-6). Codons differing in the third base often code for the same amino acid. For example, UUU and UUC both code for phenylalanine; UCU, UCC, UCA, and UCG all code for serine; and so forth. In such cases, the same tRNA can bind to more than one codon without introducing mistakes. For example, a single tRNA can recognize the codons UUU and UUC because both code for the same amino acid, phenylalanine.

Such considerations led Francis Crick to propose that mRNA and tRNA line up on the ribosome in a way that permits flexibility or "wobble" in the pairing between the third base of the codon and the corresponding base in the anticodon. According to this **wobble hypothesis,** the flexibility in codon-anticodon binding allows some unexpected base pairs to form (**Figure 22-4**). The unusual base inosine (I), which is extremely rare in other RNA molecules, occurs often in the wobble position of tRNA anticodons (see Figure 22-3a). Inosine is the "wobbliest" of all third-position bases, since it can pair with U, C, or A. For example, a tRNA with the anticodon 3′-UAI-5′ can recognize the codons AUU, AUC, and AUA, all of which code for the amino acid isoleucine.

It is because of wobble that fewer tRNA molecules are required for some amino acids than the number of codons that specify those amino acids. In the case of isoleucine, for example, a cell can translate all three codons with a single tRNA molecule containing 3′-UAI-5′ as its anticodon. Similarly, the six codons for the amino acid leucine (UUA, UUG, CUU, CUC, CUA, and CUG) require only three tRNAs because of wobble. Although the existence of wobble means that a single tRNA molecule can recognize more than one codon, the different codons recognized by a given tRNA always code for the same amino acid, so wobble does not cause insertion of incorrect amino acids.

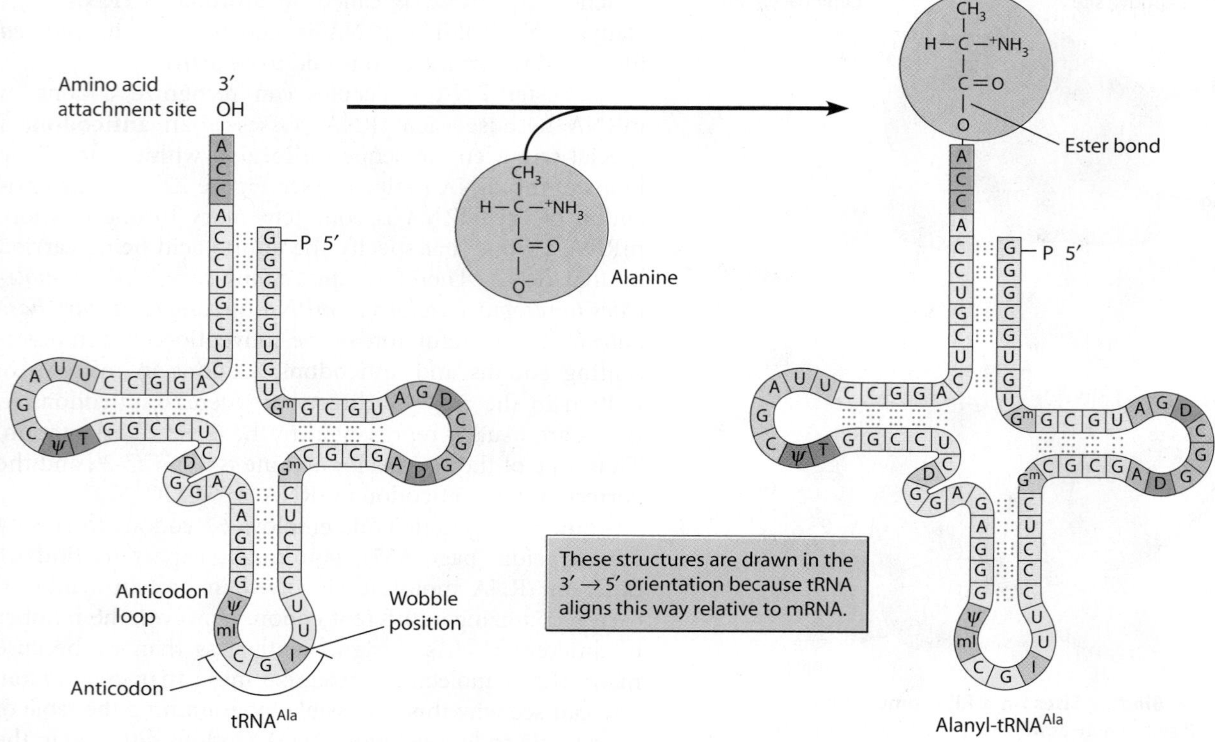

(a) Secondary structure of tRNA, before and after amino acid attachment

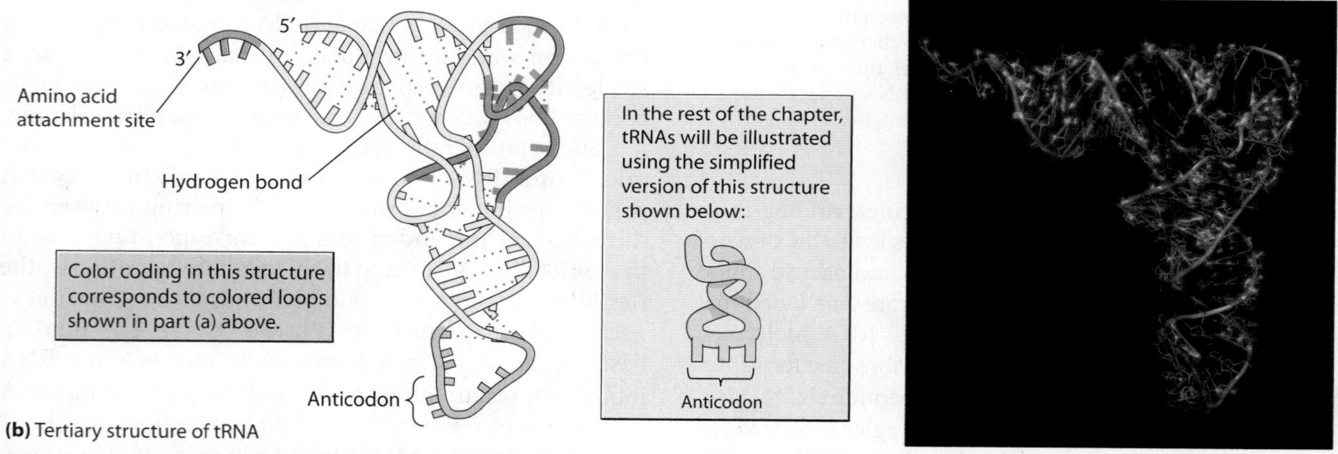

MB Molecular Model: tRNA

FIGURE 22-3 Structure and Aminoacylation of a tRNA. **(a)** Yeast alanine tRNA, like all tRNA molecules, contains three major loops, four base-paired regions, an anticodon triplet, and a 3′ terminal sequence of CCA, to which the appropriate amino acid can be attached by an ester bond. Modified bases are dark colored, and their names (as nucleosides) are abbreviated I for inosine, mI for methylinosine, D for dihydrouridine, T for ribothymidine, ψ for pseudouridine, and G^m for methylguanosine. (For the significance of the wobble position in the anticodon, see Figure 22-4.) **(b)** In the L-shaped tertiary structure of a tRNA the amino acid attachment site is at one end and the anticodon at the other. The image on the right shows a three-dimensional model of the yeast phenylalanine tRNA, with the molecular surface in transparent gray, and the RNA backbone in orange.

Aminoacyl-tRNA Synthetases Link Amino Acids to the Correct Transfer RNAs

Before a tRNA molecule can bring its amino acid to the ribosome, that amino acid must be attached covalently to the tRNA. The enzymes responsible for linking amino acids to their corresponding tRNAs are called **aminoacyl-tRNA synthetases.** Cells typically have 20 different aminoacyl-tRNA synthetases, one for each of the 20 amino acids commonly used in protein synthesis. Cells that utilize the unusual 21st and 22nd amino acids, *selenocysteine* and

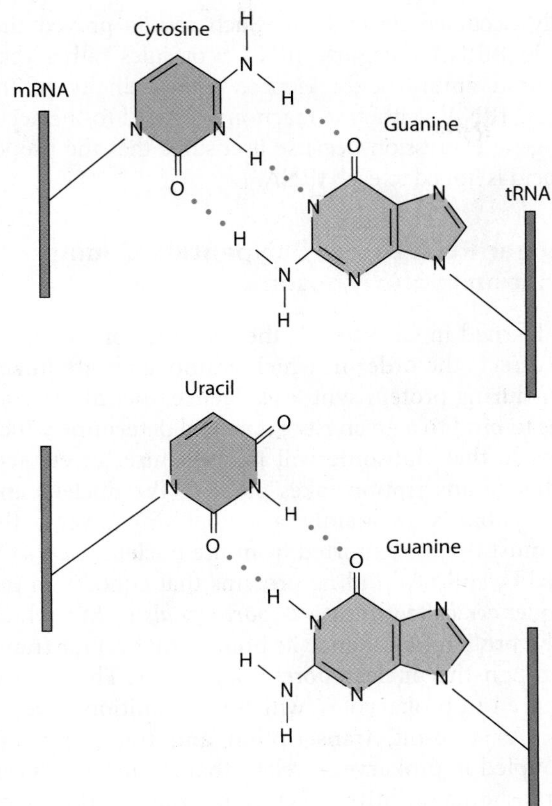

Bases Recognized in Codon (third position only)	Base in Anticodon
U	A
G	C
A or G	U
C or U	G
U, C, or A	I (Inosine)

FIGURE 22-4 The Wobble Hypothesis. The two diagrams illustrate how a slight shift or "wobble" in the position of the base guanine in a tRNA anticodon would permit it to pair with uracil (*bottom*) instead of its normal complementary base, cytosine (*top*). The table summarizes the base pairs permitted at the third position of a codon by the wobble hypothesis.

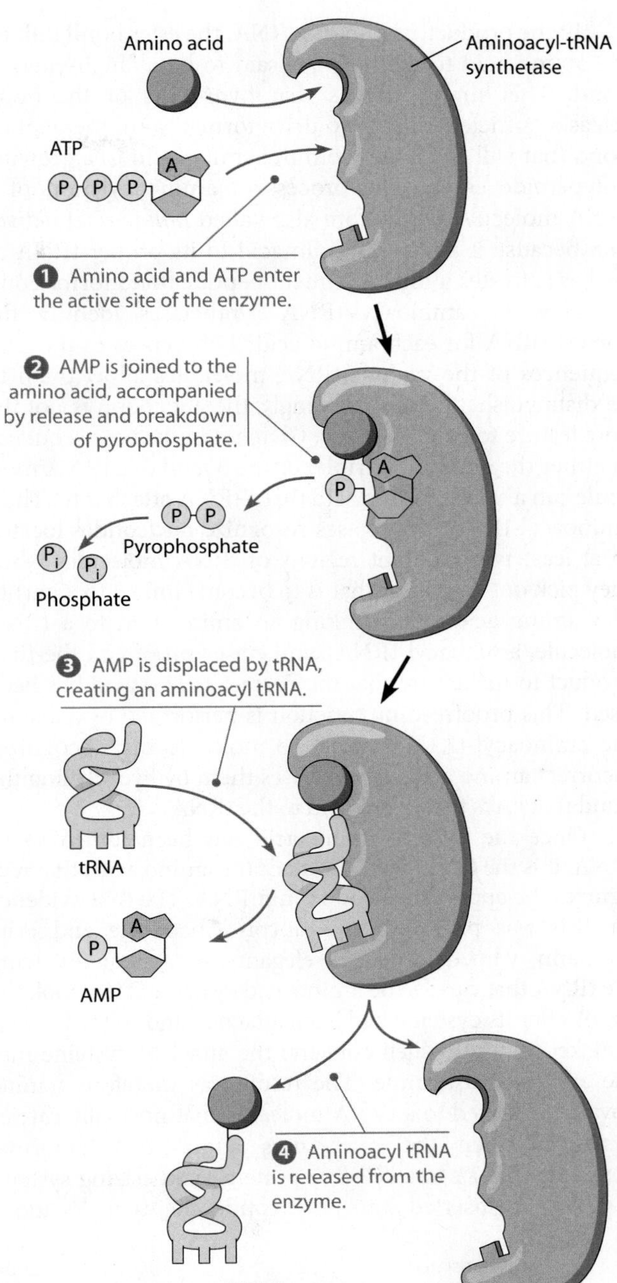

FIGURE 22-5 Amino Acid Activation by Aminoacyl-tRNA Synthetase. This enzyme catalyzes the formation of an ester bond between the carboxyl group of an amino acid and the 3′ OH of the appropriate tRNA, generating an aminoacyl tRNA.

pyrrolysine (page 656), contain special tRNAs and aminoacyl-tRNA synthetases for these amino acids as well. When more than one tRNA exists for a given amino acid, the aminoacyl-tRNA synthetase specific for that particular amino acid recognizes each of the tRNAs. Some cells possess less than 20 aminoacyl-tRNA synthetases. In such cases, the same aminoacyl-tRNA synthetase may catalyze the attachment of two different amino acids to their corresponding tRNAs, or it may attach an incorrect amino acid to a tRNA molecule. These latter "errors" are corrected by a second enzyme that alters the incorrect amino acid after it has been attached to the tRNA.

Aminoacyl-tRNA synthetases catalyze the attachment of amino acids to their corresponding tRNAs via an ester bond, accompanied by the hydrolysis of ATP to AMP and pyrophosphate:

$$\underset{\text{Amino acid}}{H_3N^+{-}\overset{\displaystyle R}{\overset{|}{C}}H{-}\overset{\displaystyle O}{\overset{\|}{C}}{-}O^-} + HO{-}\boxed{\text{tRNA}} \xrightarrow[\text{Aminoacyl-tRNA synthetase}]{\text{ATP} \quad \text{AMP} + \text{PP}_i} H_3N^+{-}\overset{\displaystyle R}{\overset{|}{C}}H{-}\overset{\displaystyle O}{\overset{\|}{C}}{-}O{-}\boxed{\text{tRNA}}$$

Figure 22-5 outlines the steps by which this reaction occurs. The driving force for the reaction is provided by the hydrolysis of pyrophosphate to 2 P_i.

In the product, aminoacyl tRNA, the ester bond linking the amino acid to the tRNA is said to be a "high-energy" bond. This simply means that hydrolysis of the bond releases sufficient energy to drive formation of the peptide bond that will eventually join the amino acid to a growing polypeptide chain. The process of aminoacylation of a tRNA molecule is therefore also called *amino acid activation* because it links an amino acid to its proper tRNA as well as activates it for subsequent peptide bond formation.

How do aminoacyl-tRNA synthetases identify the correct tRNA for each amino acid? Differences in the base sequences of the various tRNA molecules allow them to be distinguished and, surprisingly, the anticodon is not the only feature to be recognized. Changes in the base sequence of either the anticodon triplet or the 3′ end of a tRNA molecule can alter the amino acid that a tRNA attaches to. Thus, aminoacyl-tRNA synthetases recognize nucleotides located in at least two different regions of tRNA molecules when they pick out the tRNA that is to become linked to a particular amino acid. After linking an amino acid to a tRNA molecule, aminoacyl-tRNA synthetases proofread the final product to make sure that the correct amino acid has been used. This proofreading function is performed by a site on the aminoacyl-tRNA synthetase molecule that recognizes incorrect amino acids and releases them by hydrolyzing the bond that links the amino acid to the tRNA.

Once the correct amino acid has been joined to its tRNA, it is the tRNA itself (and not the amino acid) that recognizes the appropriate codon in mRNA. The first evidence for this was provided by François Chapeville and Fritz Lipmann, who designed an elegant experiment involving the tRNA that carries the amino acid cysteine. They took the tRNA after its cysteine had been attached and treated it with a nickel catalyst, which converts the attached cysteine into the amino acid alanine. The result was therefore alanine covalently linked to a tRNA molecule that normally carries cysteine. When the researchers added this abnormal aminoacyl tRNA to a cell-free protein-synthesizing system, alanine was inserted into polypeptide chains in locations normally occupied by cysteine. Such results proved that codons in mRNA recognize tRNA molecules rather than their bound amino acids. Hence, the specificity of the aminoacyl-tRNA synthetase reaction is crucial to the accuracy of gene expression because it ensures that the proper amino acid is linked to each tRNA.

Messenger RNA Brings Polypeptide Coding Information to the Ribosome

As you learned in Chapter 21, the sequence of codons in mRNA directs the order in which amino acids are linked together during protein synthesis. Hence, the mRNA that happens to bind to a given ribosome will determine which polypeptide that ribosome will manufacture. In eukaryotes, where transcription takes place in the nucleus and protein synthesis is mainly a cytoplasmic event, the mRNA must first be exported from the nucleus. Export is mediated by mRNA-binding proteins that contain amino acid sequences called *nuclear export signals (NES),* which target the protein (and hence its bound mRNA) for transport through the nuclear pores (page 540). This step is not required in prokaryotes, which by definition have no nucleus. As a result, transcription and translation are often coupled in prokaryotic cells—that is, ribosomes can begin translating an mRNA before its transcription from DNA is completed (see Figure 21-16).

At the heart of each messenger RNA molecule is, of course, its message—the sequence of nucleotides that encodes a polypeptide. However, mRNAs also possess sequences at either end that are not translated (**Figure 22-6**). The untranslated sequence at the 5′ end of an mRNA precedes the **start codon,** which is the first codon to be translated. AUG is the most common start codon, although a few other triplets are occasionally used for this purpose. The untranslated sequence at the 3′ end follows the **stop codon,** which signals the end of translation and can be UAG, UAA, or UGA. The 5′ and 3′ untranslated regions range from a few dozen to hundreds of nucleotides in length. Although these

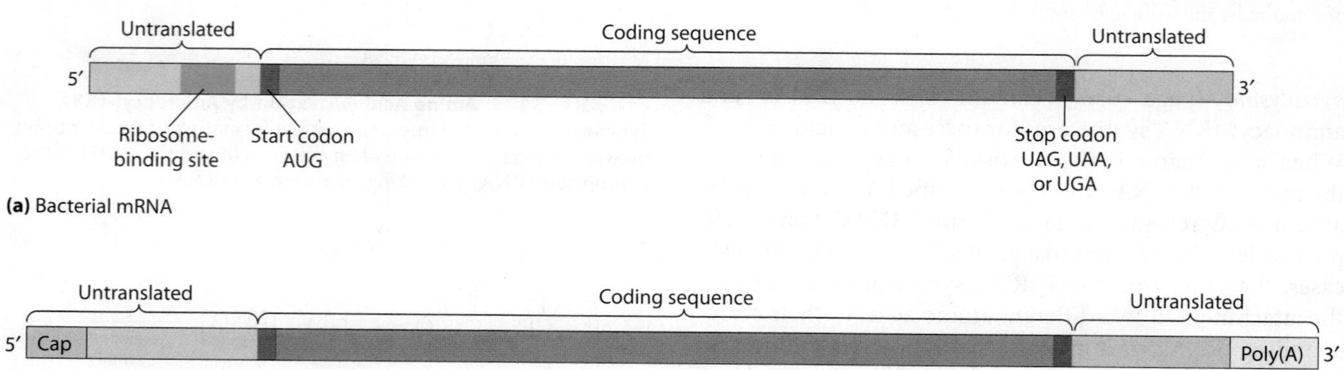

FIGURE 22-6 Comparison of Bacterial and Eukaryotic Messenger RNA. **(a)** A bacterial mRNA molecule encoding a single polypeptide has the features shown here. (A polycistronic bacterial mRNA would generally have a set of these features for each gene.) **(b)** A eukaryotic mRNA molecule has, in addition, a 5′ cap and a 3′ poly(A) tail. It lacks a ribosome-binding site (a nucleotide sequence also called a Shine–Dalgarno sequence, after its discoverers).

sequences are not translated, their presence is essential for proper mRNA function. Included in the untranslated regions of eukaryotic mRNAs are a 5′ cap and a 3′ poly(A) tail, both of which were described in Chapter 21. As we will see, the 5′ cap is important in initiating translation in eukaryotes.

In eukaryotes, most mRNA molecules are *monocistronic* (i.e., they encode a single polypeptide). In bacteria and archaea, however, some mRNAs are *polycistronic*—meaning they encode several polypeptides, usually with related functions in the cell. The clusters of genes that give rise to polycistronic mRNAs are single transcription units called *operons,* which we will discuss in Chapter 23 in the context of gene regulation. Although most often thought of as a feature of prokaryotes, eukaryotes also produce polycistronic RNAs. In some cases, such as the nematode, *Caenorhabditis elegans,* it is only the pre-mRNA that is polycistronic; subsequent RNA processing results in individual mRNAs that are monocistronic and translated separately. In other eukaryotes, dicistronic RNAs (i.e., RNAs that encode two proteins) remain joined and are translated together.

Protein Factors Are Required for the Initiation, Elongation, and Termination of Polypeptide Chains

In addition to aminoacyl-tRNA synthetases and the protein components of the ribosome, translation requires the participation of several other kinds of protein molecules. Some of these *protein factors* are required for initiating the translation process, others for elongating the growing polypeptide chain, and still others for terminating polypeptide synthesis. The exact roles played by these factors will become apparent as we now proceed to a discussion of the mechanism of translation.

The Mechanism of Translation

The translation of mRNAs into polypeptides is an ordered, stepwise process that begins the synthesis of a polypeptide chain at its amino-terminal end, or *N-terminus,* and sequentially adds amino acids to the growing chain until the carboxyl-terminal end, or *C-terminus,* is reached. The first experimental evidence for such a mechanism was provided in 1961 by Howard Dintzis, who investigated hemoglobin synthesis in developing red blood cells that had been incubated briefly with radioactive amino acids. Dintzis reasoned that if the time of incubation is kept relatively brief, then the radioactivity present in completed hemoglobin chains should be concentrated at the most recently synthesized end of the molecule. He found that the highest concentration of radioactivity in completed hemoglobin chains was at the C-terminal end, indicating that the C-terminus is the last part of the polypeptide chain to be synthesized. This allowed him to conclude that during mRNA translation, *amino acids are added to the growing polypeptide chain beginning at the N-terminus and proceeding toward the C-terminus.*

In theory, mRNA could be read in either the 5′ → 3′ direction or the 3′ → 5′ direction during this process. The first attempts to determine the direction in which mRNA is actually read involved the use of artificial RNA molecules. A typical example is the synthetic RNA that can be made by adding the base C to the 3′ end of poly(A), yielding the molecule 5′-AAAAAAAAAAAA...AAC-3′. When added to a cell-free protein-synthesizing system, this RNA stimulates the synthesis of a polypeptide consisting of a stretch of lysine residues with an asparagine at the C-terminus. Because AAA codes for lysine and AAC codes for asparagine, this means that *mRNA is translated in the 5′ → 3′ direction.* Confirming evidence has come from many studies in which the base sequences of naturally occurring mRNAs have been compared with the amino acid sequences of the polypeptide chains they encode. In all cases, the amino acid sequence of the polypeptide chain corresponds to the order of mRNA codons read in the 5′ → 3′ direction.

To understand how translation of mRNA in the 5′ → 3′ direction leads to the synthesis of polypeptides in the N-terminal to C-terminal direction, it is helpful to subdivide the translation process into three stages, as shown in **Figure 22-7**: ❶ an *initiation* stage, in which mRNA is bound to the ribosome and positioned for proper translation; ❷ an *elongation* stage, in which amino acids are sequentially joined together via peptide bonds in an order specified by the arrangement of codons in mRNA; and ❸ a *termination* stage, in which the mRNA and the newly formed polypeptide chain are released from the ribosome.

In the following sections we examine each of these stages in detail. Although our discussion focuses mainly on translation in bacterial cells, where the mechanisms are especially well understood, the comparable events in eukaryotic cells are rather similar. The aspects of translation that differ between bacteria and eukaryotes are confined mostly to the initiation stage, as we describe in the next section.

The Initiation of Translation Requires Initiation Factors, Ribosomal Subunits, mRNA, and Initiator tRNA

Bacterial Initiation. The initiation of translation in bacteria is illustrated in **Figure 22-8**, which shows that initiation can be subdivided into three distinct steps. In step ❶, three **initiation factors**—called *IF1, IF2,* and *IF3*—bind to the small (30S) ribosomal subunit, with GTP attaching to IF2. The presence of IF3 at this early stage prevents the 30S subunit from prematurely associating with the 50S subunit.

In step ❷, mRNA and the tRNA carrying the first amino acid bind to the 30S ribosomal subunit. The mRNA is bound to the 30S subunit in its proper orientation by means of a special nucleotide sequence called the mRNA's *ribosome-binding site* (also known as the *Shine–Dalgarno sequence,* after its discoverers). This sequence consists of a stretch of 3–9 purine nucleotides (often AGGA) located slightly upstream of the start codon. These purines in the mRNA form complementary base pairs with a pyrimidine-rich sequence at the 3′ end of 16S rRNA, which forms the ribosome's *mRNA-binding site.* The importance of the

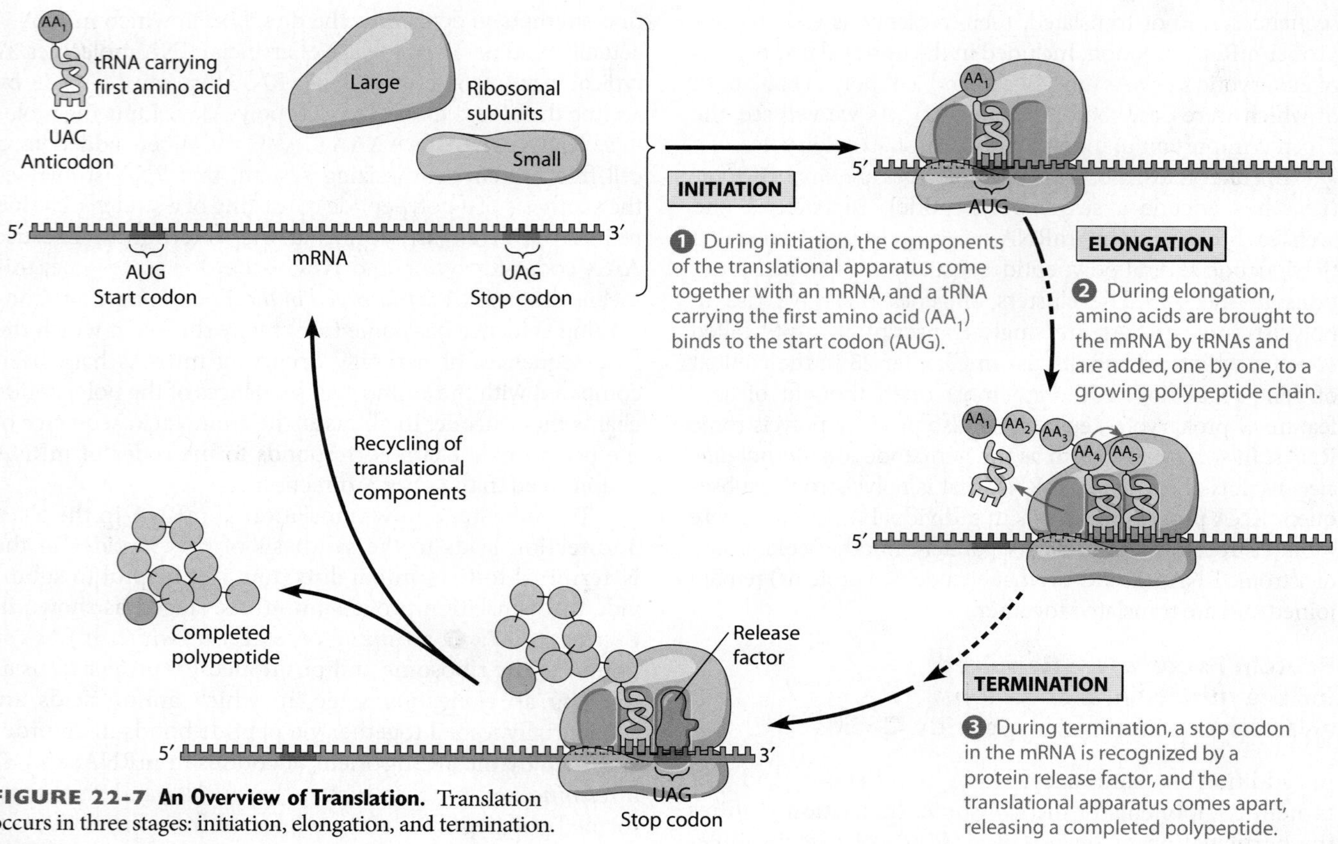

AA1
tRNA carrying first amino acid
UAC
Anticodon
Large
Ribosomal subunits
Small
5′
3′
AUG
Start codon
mRNA
UAG
Stop codon
INITIATION
1 During initiation, the components of the translational apparatus come together with an mRNA, and a tRNA carrying the first amino acid (AA1) binds to the start codon (AUG).
ELONGATION
2 During elongation, amino acids are brought to the mRNA by tRNAs and are added, one by one, to a growing polypeptide chain.
AA1 AA2 AA3 AA4 AA5
Recycling of translational components
Completed polypeptide
Release factor
TERMINATION
3 During termination, a stop codon in the mRNA is recognized by a protein release factor, and the translational apparatus comes apart, releasing a completed polypeptide.
UAG
Stop codon

MB Following the Instructions in DNA

FIGURE 22-7 An Overview of Translation. Translation occurs in three stages: initiation, elongation, and termination.

mRNA-binding site has been shown by studies involving *colicins,* which are proteins produced by certain strains of *Escherichia coli* that can kill other types of bacteria. One such protein, colicin E3, kills bacteria by destroying their ability to synthesize proteins. Upon entering the cytoplasm of susceptible bacteria, colicin E3 catalyzes the removal of a 49-nucleotide fragment from the 3′ end of 16S rRNA. This action destroys the mRNA-binding site, thereby creating ribosomes that can no longer initiate polypeptide synthesis.

The binding of mRNA to the mRNA-binding site of the small ribosomal subunit places the mRNA's AUG start codon at the ribosome's P site, where it can bind to the anticodon of the appropriate tRNA. The first clue that a special kind of tRNA is involved in this step emerged when it was discovered that roughly half the proteins in *E. coli* contain methionine at their N-terminal ends. This was surprising because methionine is a relatively uncommon amino acid, accounting for no more than a few percent of the amino acids in bacterial proteins. The explanation for such a pattern became apparent when it was discovered that bacterial cells contain two different methionine-specific tRNAs. One, designated $tRNA^{Met}$, carries a normal methionine destined for insertion into the internal regions of polypeptide chains. The other, called $tRNA^{fMet}$, carries a methionine that is converted to the derivative *N-formylmethionine (fMet)* after linkage to the tRNA (**Figure 22-9**). In *N*-formylmethionine, the amino group of methionine is blocked by the addition of a formyl group and so cannot form a peptide bond with another amino acid; only the carboxyl group is available for bonding to another amino acid. Hence *N*-formylmethionine can be situated only at the N-terminal end of a polypeptide chain—suggesting that $tRNA^{fMet}$ functions as an **initiator tRNA** that starts the process of translation. This idea was soon confirmed by the discovery that bacterial polypeptide chains in the early stages of synthesis always contain *N*-formylmethionine at their N-terminus. Following completion of the polypeptide chain (and in some cases while it is still being synthesized), the formyl group, and often the methionine itself, is enzymatically removed.

During initiation, the initiator tRNA with its attached *N*-formylmethionine is bound to the P site of the 30S ribosomal subunit by the action of initiation factor IF2 (plus GTP), which can distinguish initiator $tRNA^{fMet}$ from other kinds of tRNA. This attribute of IF2 helps explain why AUG start codons bind to the initiator $tRNA^{fMet}$, whereas AUG codons located elsewhere in mRNA bind to the noninitiating $tRNA^{Met}$. Once $tRNA^{fMet}$ enters the P site, its anticodon becomes base-paired with the AUG start codon in the mRNA, and IF3 is released. At this point the 30S subunit with its associated IF1, IF2-GTP, mRNA, and *N*-formylmethionyl $tRNA^{fMet}$ is referred to as the **30S initiation complex.**

Once IF3 has been released, the 30S initiation complex can bind to a free 50S ribosomal subunit, generating the **70S initiation complex** (step 3 of Figure 22-8). The 50S subunit then promotes hydrolysis of the IF2-bound GTP, leading to the release of IF2 and IF1. At this stage, all three initiation factors have been released.

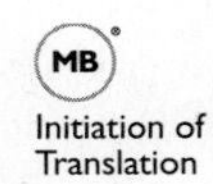

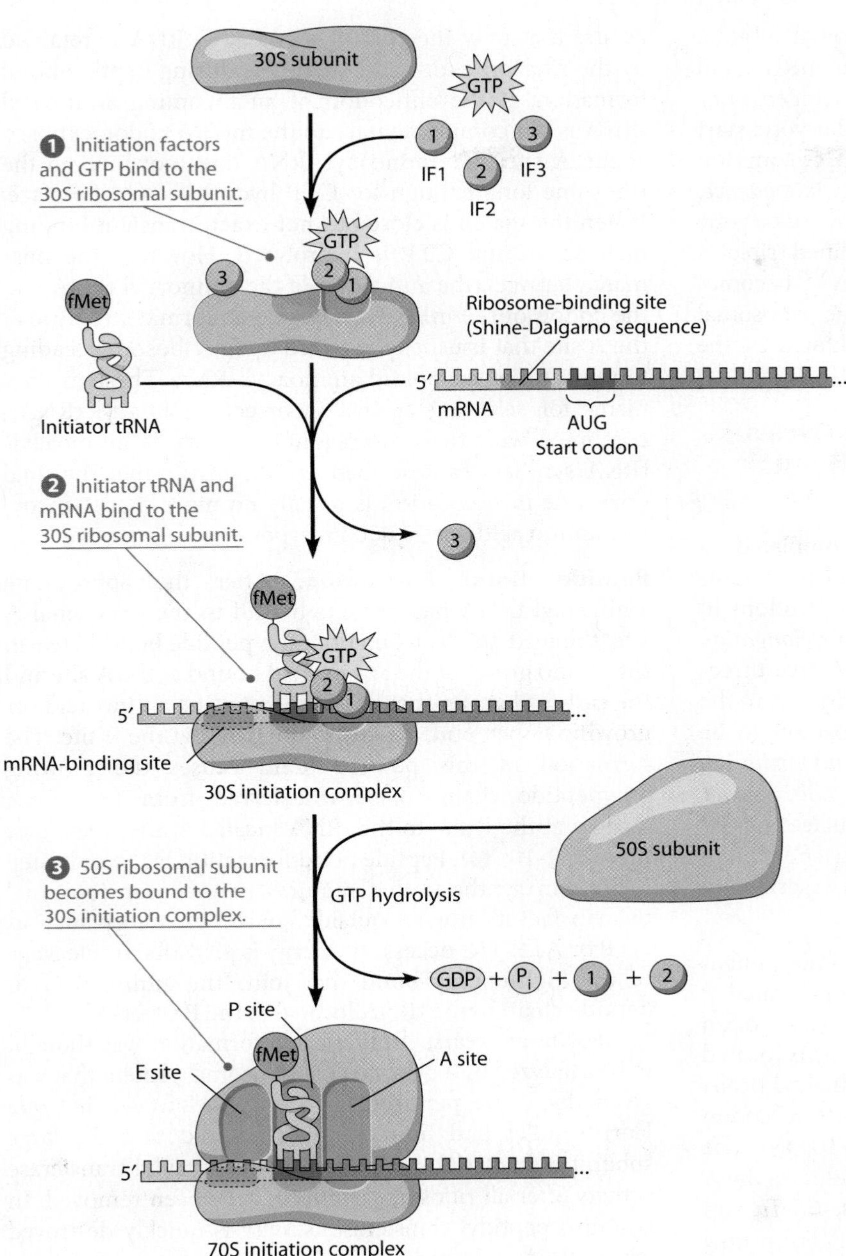

FIGURE 22-8 Initiation of Translation in Bacteria. Assembly of the 70S translation initiation complex occurs in three steps. ❶ Three initiation factors (IF1, IF2, and IF3) plus GTP bind to the small ribosomal subunit. ❷ The initiator aminoacyl tRNA and mRNA are attached. The mRNA-binding site is composed, at least in part, of a portion of the 16S rRNA of the small ribosomal subunit. ❸ The large ribosomal subunit joins the complex. The resulting 70S initiation complex has fMet-tRNA^{fMet} residing in the ribosome's P site.

Eukaryotic Initiation. Unlike the situation in bacteria, the AUG start codon in eukaryotes (and archaea) specifies the amino acid methionine rather than *N*-formylmethionine. Other differences in eukaryotes include the use of a different set of initiation factors known as *eIFs* (roughly a dozen proteins with names such as eIF1, eIF2, and so forth), a somewhat different pathway for assembling the initiation complex, and a special initiator tRNA^{Met} that—like the normal tRNA for methionine but unlike the initiator tRNA of bacteria—carries methionine that does not become formylated.

At the beginning of eukaryotic initiation, the initiation factor *eIF2* (with GTP attached) binds to the initiator methionyl tRNA^{Met} *before* the tRNA then binds to the small ribosomal subunit along with other initiation factors, including *eIF1A* (the eukaryotic counterpart of bacterial IF1). The resulting complex next binds to the 5′ end of an mRNA, recognizing the 5′ cap with the aid of a cap-binding initiation factor, *eIF4F*. (In some cases the complex may instead bind to an *internal ribosome entry sequence*, or *IRES*, which lies directly upstream of the start codon of certain types of mRNA, especially viral mRNAs.)

Formyl group

$$CH_3-S-CH_2-CH_2-\underset{\substack{|\\H}}{\overset{\substack{H\\|\\C=O\\|\\NH\\|}}{C}}-\overset{\substack{O\\\|}}{C}-O^-$$

FIGURE 22-9 The Structure of *N*-Formylmethionine. *N*-formylmethionine (fMet) is the modified amino acid with which every polypeptide is initiated in bacteria.

After binding to mRNA, the small ribosomal subunit, with the initiator tRNA in tow, scans along the mRNA and usually begins translation at the first AUG triplet it encounters. The nucleotides on either side of the eukaryotic start codon appear to be involved in its recognition. A common start sequence is ACC<u>AUG</u>G (also called a *Kozak sequence,* after the scientist who discovered that many eukaryotic mRNAs have this sequence), where the underlined triplet is the actual start codon. After the initiator $tRNA^{Met}$ becomes base-paired with the start codon, the large ribosomal subunit joins the complex in a reaction facilitated by the hydrolysis of GTP bound to initiation factor *eIF5B.*

Chain Elongation Involves Sequential Cycles of Aminoacyl tRNA Binding, Peptide Bond Formation, and Translocation

Once the initiation complex has been completed, a polypeptide chain is synthesized by the successive addition of amino acids in a sequence specified by codons in mRNA. As summarized in **Figure 22-10,** this *elongation stage* of polypeptide synthesis involves a repetitive three-step cycle in which ❶ *binding of an aminoacyl tRNA* to the ribosome brings a new amino acid into position to be joined to the polypeptide chain, ❷ *peptide bond formation* links this amino acid to the growing polypeptide, and ❸ the mRNA is advanced a distance of three nucleotides by the process of *translocation* to bring the next codon into position for translation. Each of these steps is described in more detail in the following paragraphs.

Binding of Aminoacyl tRNA. At the onset of the elongation stage, the AUG start codon in the mRNA is located at the ribosomal P site and the second codon (the codon immediately downstream from the start codon) is located at the A site. Elongation begins when an aminoacyl tRNA whose anticodon is complementary to the second codon binds to the ribosomal A site (see Figure 22-10, ❶). The binding of this new aminoacyl tRNA to the codon in the A site requires two protein **elongation factors,** *EF-Tu* and *EF-Ts,* and is driven by the hydrolysis of GTP. From now on, every incoming aminoacyl tRNA will bind first to the A (aminoacyl) site—hence the site's name.

The function of EF-Tu, along with its bound GTP, is to convey the aminoacyl tRNA to the A site of the ribosome. The EF-Tu–GTP complex promotes the binding of all aminoacyl tRNAs *except the initiator tRNA* to the ribosome, thus ensuring that AUG codons located downstream from the start codon do not mistakenly recruit an initiator tRNA to the ribosome. As the aminoacyl tRNA is transferred to the ribosome, the GTP is hydrolyzed and the EF-Tu–GDP complex is released. The role of EF-Ts is to regenerate EF-Tu–GTP from EF-Tu–GDP for the next round of the elongation cycle (see Figure 22-10, ❶).

Elongation factors do not recognize individual anticodons, which means that aminoacyl tRNAs of all types (other than initiator tRNAs) are randomly brought to the A site of the ribosome. Some mechanism must therefore ensure that only the correct aminoacyl tRNA is retained by the ribosome for subsequent use during peptide bond formation. If the anticodon of an incoming aminoacyl tRNA is not complementary to the mRNA codon exposed at the A site, the aminoacyl tRNA does not bind to the ribosome long enough for GTP hydrolysis to take place. When the match is close but not exact, transient binding may occur, and GTP is hydrolyzed. However, the mismatch between the anticodon of the aminoacyl tRNA and the codon of the mRNA creates an abnormal structure at the A site that is usually detected by the ribosome, leading to rejection of the bound aminoacyl tRNA. These mechanisms for selecting against incorrect aminoacyl tRNAs, combined with the proofreading capacity of aminoacyl-tRNA synthetases described earlier, ensure that the final error rate in translation is usually no more than 1 incorrect amino acid per 10,000 incorporated.

Peptide Bond Formation. After the appropriate aminoacyl tRNA has become bound to the ribosomal A site, the next step is formation of a peptide bond between the amino group of the amino acid bound at the A site and the carboxyl group that links the initiating amino acid (or growing polypeptide chain) to the tRNA at the P site. The formation of this peptide bond causes the growing polypeptide chain to be transferred from the tRNA located at the P site to the tRNA located at the A site (see Figure 22-10, ❷). Peptide bond formation is the only step in protein synthesis that requires neither nonribosomal protein factors nor an outside source of energy such as GTP or ATP. The necessary energy is provided by cleavage of the high-energy bond that joins the amino acid or peptide chain to the tRNA located at the P site.

For many years, peptide bond formation was thought to be catalyzed by a hypothetical ribosomal protein that was given the name **peptidyl transferase.** However, in 1992 Harry Noller and his colleagues showed that the large subunit of bacterial ribosomes retains peptidyl transferase activity after all ribosomal proteins have been removed. In contrast, peptidyl transferase activity is quickly destroyed when rRNA is degraded by exposing ribosomes to ribonuclease. Such observations suggested that rRNA rather than a ribosomal protein is responsible for catalyzing peptide bond formation. In bacterial ribosomes, peptidyl transferase activity has been localized to the 23S rRNA of the large ribosomal subunit, and high-resolution X-ray data have pinpointed the catalytic site to a specific region of the RNA chain. Hence 23S rRNA is an example of a *ribozyme,* an enzyme made entirely of RNA (see Chapter 6, page 150).

Translocation. After a peptide bond has been formed, the P site contains an empty tRNA and the A site contains a peptidyl tRNA (the tRNA to which the growing polypeptide chain is attached). The mRNA now advances a distance of three nucleotides relative to the small subunit, bringing the next codon into proper position for translation. During this process of **translocation**—which requires that an elongation factor called *EF-G* plus GTP become

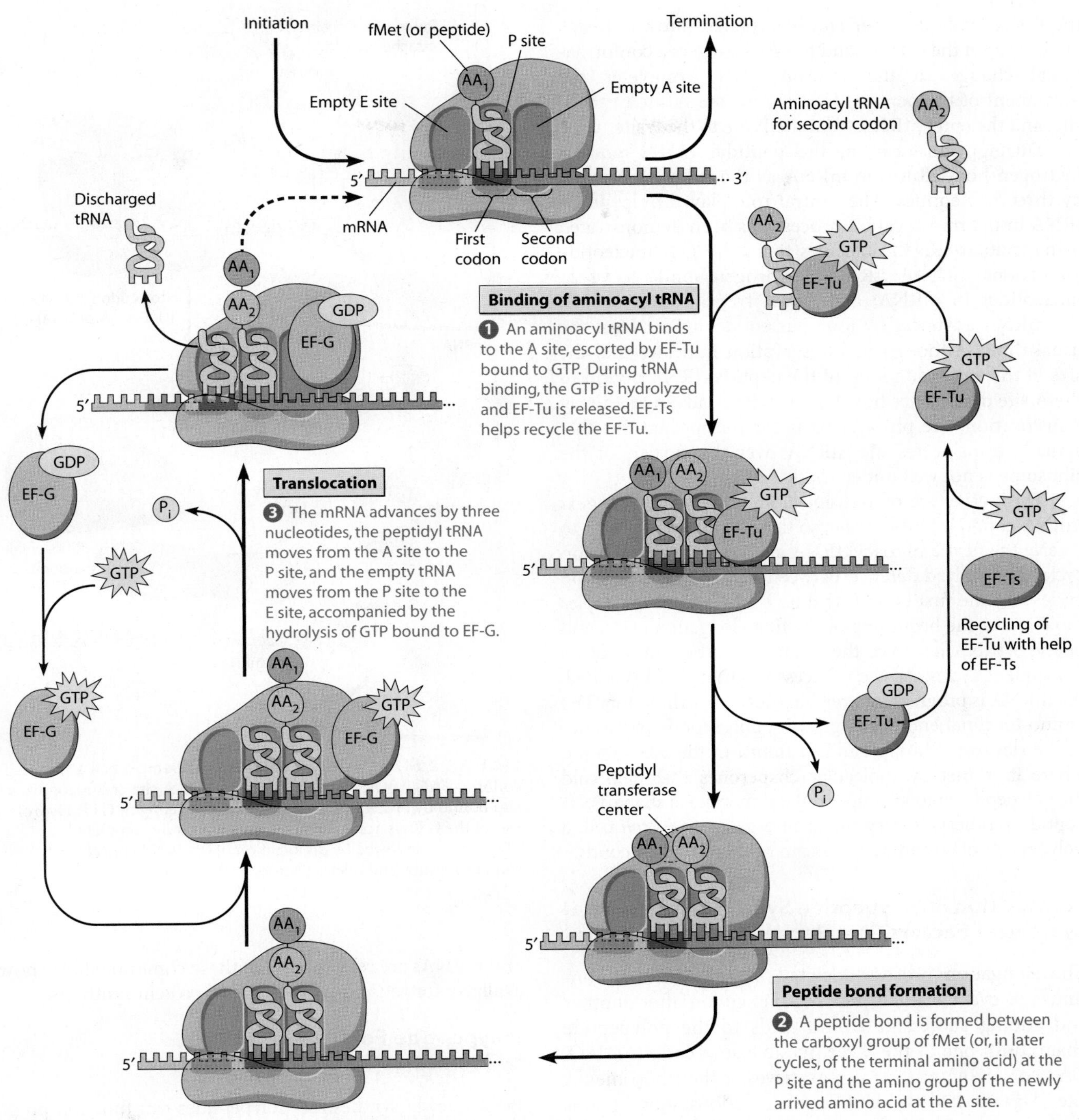

MB Elongation of the Polypeptide Chain

FIGURE 22-10 Polypeptide Chain Elongation in Bacteria. Chain elongation requires the presence of a peptidyl tRNA or, in the first elongation cycle shown here, an fMet-tRNA^{fMet} at the ribosomal P site. Binding of aminoacyl tRNA (❶) is followed by peptide bond formation (❷) catalyzed by the peptidyl transferase activity of the 23S rRNA of the large ribosomal subunit. During translocation (❸), the peptidyl tRNA moves from the A site to the P site, taking the mRNA along with it, and the empty tRNA moves from the P site to the E site and leaves the ribosome. The next mRNA codon is now located in the A site, where the same cycle of events can be repeated for the next amino acid.

transiently associated with the ribosome—the peptidyl tRNA moves from the A site to the P site and the empty tRNA moves from the P site to the E (exit) site. Although these movements are shown in Figure 22-10 as occurring in a single step (❸), an intermediate "hybrid" state exists in which the anticodon of the peptidyl tRNA still resides at the A site while its aminoacyl end has rotated into the P site, and the anticodon of the empty tRNA still resides at the P site while its other end has rotated into the E site. Hydrolysis of the GTP bound to EF-G triggers a conformational change in the ribosome that completes the movement of the peptidyl tRNA from the A site to the P site, and the empty tRNA from the P site to the E site.

During translocation, the peptidyl tRNA remains hydrogen-bonded to the mRNA as the mRNA advances by three nucleotides. The central role played by peptidyl tRNA in the translocation process has been demonstrated using mutant tRNA molecules that have *four*-nucleotide anticodons. These tRNAs hydrogen-bond to four nucleotides in mRNA; and, when translocation occurs, the mRNA advances by four nucleotides rather than the usual three. Although this observation indicates that the size of the anticodon loop of the peptidyl tRNA bound to the A site determines how far the mRNA advances during translocation, the physical basis for the mechanism that actually translocates the mRNA over the surface of the ribosome is not well understood.

The net effect of translocation is to bring the next mRNA codon into the A site, so the ribosome is now set to receive the next aminoacyl tRNA and repeat the elongation cycle. The only difference between succeeding elongation cycles and the first cycle is that an initiator tRNA occupies the P site at the beginning of the first elongation cycle, and peptidyl tRNA occupies the P site at the beginning of all subsequent cycles. As each successive amino acid is added, the mRNA is progressively read in the $5' \longrightarrow 3'$ direction. The amino-terminal end of the growing polypeptide passes out of the ribosome through an *exit tunnel* in the 50S subunit, where it is met by molecular chaperones that help fold the polypeptide into its proper three-dimensional shape. Polypeptide synthesis is very rapid; in a growing *E. coli* cell, a polypeptide of 400 amino acids can be made in 10 seconds.

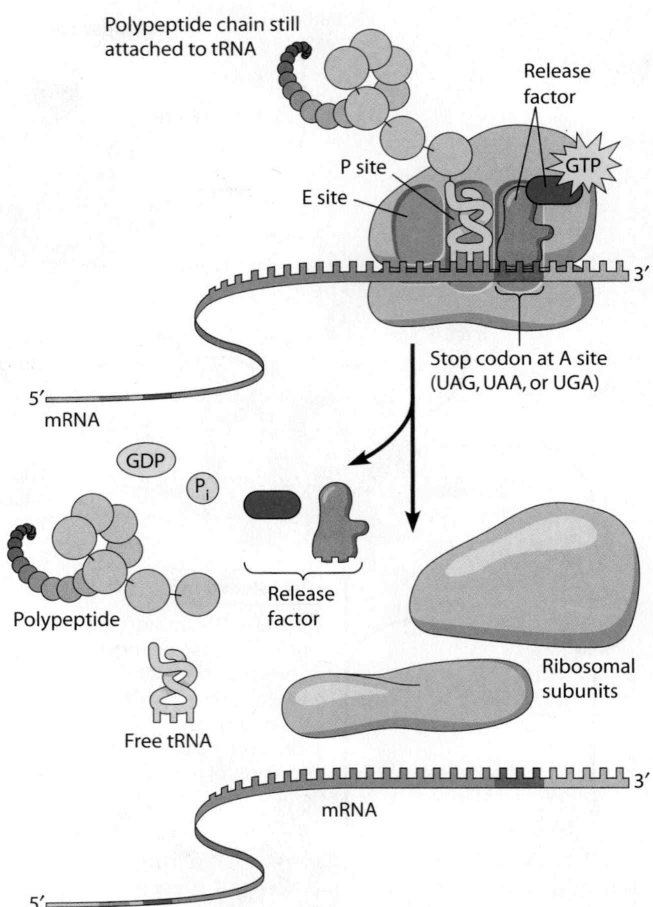

FIGURE 22-11 Termination of Translation. When a stop codon—UAG, UAA, or UGA—arrives at the A site, it is recognized and bound by protein release factors associated with GTP. Hydrolysis of the GTP is accompanied by release of the completed polypeptide, followed by dissociation of the tRNA, mRNA, ribosomal subunits, and release factors.

Termination of Polypeptide Synthesis Is Triggered by Release Factors That Recognize Stop Codons

The elongation process depicted in Figure 22-10 continues in cyclic fashion, reading one codon after another and adding successive amino acids to the polypeptide chain, until one of the three possible stop codons (UAG, UAA, or UGA) in the mRNA arrives at the ribosome's A site (**Figure 22-11**). Unlike other codons, stop codons are not recognized by tRNA molecules. Instead, the stop codons are recognized by proteins called **release factors,** which possess special regions ("peptide anticodons") that bind to mRNA stop codons present at the ribosomal A site. After binding to the A site along with GTP, the release factors terminate translation by triggering release of the completed polypeptide from the peptidyl tRNA. In essence, the reaction is a hydrolytic cleavage: The polypeptide transfers to a water molecule instead of to an activated amino acid, producing a free carboxyl group at the end of the polypeptide—its C-terminus. After the polypeptide is released accompanied by GTP hydrolysis, the ribosome dissociates into its subunits and the tRNAs and mRNAs are released. All of these components are now available for reuse in a new cycle of protein synthesis.

Polypeptide Folding Is Facilitated by Molecular Chaperones

Before newly synthesized polypeptides can function properly, they must fold into the correct three-dimensional shape. As discussed in Chapters 2 and 3, the primary sequence of a protein is sufficient to specify its three-dimensional structure, and some polypeptides spontaneously fold into the proper shape in a test tube. However, protein folding inside cells is usually facilitated by proteins called **molecular chaperones** (p. 61). In fact, proper folding often requires the action of several chaperones acting in sequence, beginning when the growing polypeptide chain first emerges from the ribosome's exit tunnel.

A key function of molecular chaperones is to bind to polypeptide chains during the early stages of folding, thereby preventing them from interacting with other polypeptides before the newly folding chains have acquired the proper conformation. If the folding process goes awry, chaperones can sometimes rescue improperly folded proteins and help them fold properly, or the improperly folded proteins may be destroyed. However, some kinds of incorrectly folded polypeptides tend to bind to each other and form insoluble aggregates that become deposited both within and between cells. Such protein deposits disrupt cell function and may even lead to tissue degeneration and cell death. In **Box 22A**, we discuss how such events can contribute to the development of ailments such as Alzheimer disease and mad cow disease.

Chaperones are found throughout the living world, from archaea and bacteria to the various compartments of eukaryotic cells. Two of the most widely occurring chaperone families are *Hsp70* and *Hsp60*. The "Hsp" comes from the original designation of these proteins as "heat-shock proteins" because cells produce them in response to stressful conditions, such as exposure to high temperatures; under these conditions, chaperones facilitate the refolding of heat-damaged proteins. Members of the Hsp70 and Hsp60 chaperone families operate by somewhat different mechanisms, but both involve ATP-dependent cycles of binding and releasing their protein substrates. (It is the release step that requires ATP hydrolysis.) Molecular chaperones are also involved in activities other than protein folding. For example, they help assemble folded polypeptides into multisubunit proteins and—as we will see later in this chapter—they facilitate protein transport into mitochondria and chloroplasts by maintaining polypeptides in an unfolded state prior to their transport into these organelles.

Protein Synthesis Typically Utilizes a Substantial Fraction of a Cell's Energy Budget

Polypeptide elongation involves the hydrolysis of at least four "high-energy" phosphoanhydride bonds per amino acid added. Two of these bonds are broken in the aminoacyl-tRNA synthetase reaction, where ATP is hydrolyzed to AMP accompanied by the release of two free phosphate groups (see Figure 22-5). The rest are supplied by two molecules of GTP: one used in binding the incoming aminoacyl tRNA at the A site, and the other in the translocation step. Assuming each phosphoanhydride bond has a $\Delta G^{\circ\prime}$ (standard free energy) of 7.3 kcal/mol, the four bonds represent a standard free energy input of 29.2 kcal/mol of amino acid inserted. Thus, the elongation steps required to synthesize a polypeptide 100 amino acids long have a $\Delta G^{\circ\prime}$ value of about 2920 kcal/mol. Moreover, additional GTPs are utilized during formation of the initiation complex, during the transient binding of incorrect aminoacyl tRNAs to the ribosome, and during the termination step of polypeptide synthesis. Clearly, protein synthesis is an expensive process energetically. In fact, it accounts for a substantial fraction of the total energy budget of most cells. When we also consider the energy required to synthesize messenger RNA and the components of the translational apparatus, as well as the use of ATP by chaperone proteins, the cost of protein synthesis becomes even greater.

It is important to note that during translation, GTP does not function as a typical ATP-like energy donor: Its hydrolysis is not directly linked to the formation of a covalent bond. Instead, GTP appears to induce conformational changes in initiation and elongation factors by binding to them and releasing from them, just as we saw for heterotrimeric and monomeric G proteins in Chapter 14. These shape changes, in turn, allow the factors to bind (noncovalently) to, and be released from, the ribosome. In addition, hydrolysis of the GTP attached to EF-Tu apparently contributes to the accuracy of translation by playing a role in the proofreading mechanism that ejects incorrect aminoacyl tRNAs when they enter the A site.

A Summary of Translation

We have now seen that translation serves as the mechanism that converts information stored in strings of mRNA codons into a chain of amino acids linked by peptide bonds. For a visual summary of the process, refer back to Figure 22-7. As the ribosome reads the mRNA codon by codon in the $5' \rightarrow 3'$ direction, successive amino acids are brought into place by complementary base pairing between the codons in the mRNA and the anticodons of aminoacyl-tRNA molecules. When a stop codon is encountered, the completed polypeptide is released, and the mRNA and ribosomal subunits become available for further use.

Most messages are read by many ribosomes simultaneously, each ribosome following closely behind the next on the same mRNA molecule. A cluster of such ribosomes attached to a single mRNA molecule is called a **polyribosome** (see Figure 21-16). By allowing many polypeptides to be synthesized at the same time from a single mRNA molecule, polyribosomes maximize the efficiency of mRNA utilization.

RNA molecules play especially important roles in translation. The mRNA plays a central role, of course, as the carrier of the genetic message. The tRNA molecules serve as the adaptors that bring the amino acids to the appropriate codons. Last but not least, the rRNA molecules have multiple functions. Not only do they serve as structural components of the ribosomes, but one (the 16S rRNA of the small subunit) provides the binding site for incoming mRNA, and another (the 23S rRNA of the large subunit) catalyzes the formation of the peptide bond. The fundamental roles played by RNA may be a vestige of the way that living organisms first evolved on Earth. As we discussed in Chapter 6, the discovery of RNA catalysts (ribozymes) has fostered the idea that the first catalysts on Earth may have been self-replicating RNA molecules rather than proteins. Hence, the present-day ribosome may have evolved from a primitive translational apparatus that was based entirely on RNA molecules.

BOX 22A HUMAN APPLICATIONS

Protein-Folding Diseases

Polypeptide chains must be folded properly before they can perform their normal functions. In humans, more than a dozen diseases have been linked to defects in this folding process. Among the best known is *Alzheimer disease,* the memory disorder that affects one in ten Americans over 65 years old. The symptoms of Alzheimer's are caused by the degeneration of brain cells that exhibit two kinds of structural abnormalities—intracellular *tangles* of a polymerized form of a microtubule accessory protein called *tau,* and extracellular *amyloid plaques* containing fibrils made of a protein fragment 40 to 42 amino acids long called *amyloid-*β (Aβ). Evidence that Aβ accumulation is the primary cause of Alzheimer's emerged in the early 1990s, when it was discovered that some forms of Alzheimer's are triggered by inherited mutations in APP, a plasma membrane precursor protein whose cleavage gives rise to Aβ. Cleavage of the mutant APP yields a misfolded form of Aβ that aggregates into long fibrils instead of remaining soluble, thereby creating amyloid plaques that accumulate in the brain. Inherited mutations in the enzymes responsible for cleaving APP into Aβ can also produce hereditary forms of Alzheimer's, again characterized by the presence of amyloid plaques. Amyloid accumulation leads to a series of events, including the alteration of tau proteins inside cells, that cause brain cell death and memory loss.

Most people with Alzheimer disease do not inherit mutations in APP or in the enzymes that cleave it, and so they produce a normal version of Aβ. Though this normal Aβ usually causes no problems, in some individuals these same Aβ molecules aggregate into fibrils that accumulate and form amyloid plaques. A possible reason for this aberrant behavior is suggested by the discovery that people who inherit different forms of the protein *apolipoprotein E (apoE)* have differing risks of developing Alzheimer's. ApoE functions primarily in cholesterol transport, but some forms of apoE stimulate the accumulation and aggregation of Aβ into the fibrils that form amyloid plaques. Thus any factor that promotes Aβ accumulation may increase a person's risk for Alzheimer's.

Our growing understanding of the relationship between Aβ and Alzheimer's suggests that the disease might eventually be treated using drugs that either inhibit the formation of Aβ or promote its elimination from the brain. It has already been shown that animals can be protected against Aβ buildup by using experimental treatments such as (1) enzyme inhibitors that block the cleavage of Aβ from its precursor APP, (2) small molecules that disrupt amyloid plaques or prevent their formation, and (3) Aβ-containing vaccines that stimulate the immune system to clean up amyloid plaques and/or prevent them from forming. Such vaccines are capable of protecting mice with Alzheimer's symptoms from suffering further memory loss, providing hope that this devastating illness will be conquered in the not-too-distant future.

Abnormalities in protein folding also lie at the heart of a group of brain-destroying infectious diseases that include *scrapie* in sheep and *mad cow disease* in cattle. Stanley Prusiner, who received a Nobel Prize in 1997 for his pioneering work in this field, has proposed that such diseases are transmitted by protein-containing particles called **prions** (briefly discussed in Chapter 4). Because prions do not appear to contain DNA or RNA, Prusiner formulated a unique theory to explain how prions might transmit disease by triggering the infectious spread of abnormal protein folding. According to this theory, a normally folded prion protein (designated PrP^C, **Figure 22-A1a**) can adopt a misfolded conformation (designated PrP^{Sc}, Figure 22-A1b). When the misfolded PrP^{Sc} encounters a normal PrP^C polypeptide in the process of folding, it causes the normal polypeptide to fold improperly (Figure 22A-1c). The resulting, abnormally folded protein triggers extensive nerve cell damage in the brain, leading to uncontrolled muscle movements and eventual death. The presence of even a tiny bit of prion protein can initiate a chain reaction that causes a cell's normal PrP^C polypeptide chains to fold into more and more of the improperly folded prion protein (PrP^{Sc}). In this way, prion proteins are able to reproduce themselves without the need for nucleic acid.

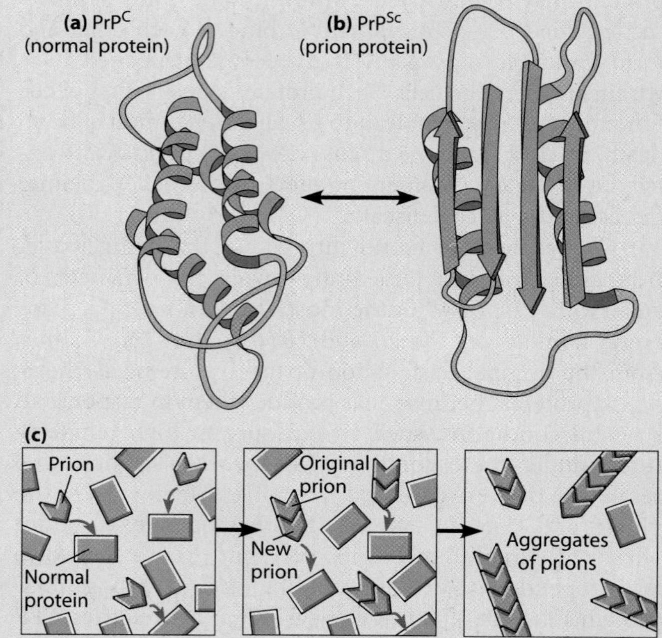

FIGURE 22A-1 A Model for How Prions Promote Their Own Formation. (a) A normal prion protein (PrP^C) contains several α-helices. (b) A misfolded prion protein (PrP^{Sc}) contains β-pleated sheets. (c) The interaction of the prion form of the protein with the normal form can induce the normal protein to misfold. The resulting chain reaction can cause aggregation of the prion form of the protein, leading to degeneration of the brain.

Even more surprising has been the discovery of different "strains" of prions that cause slightly different forms of disease. When researchers mix tiny quantities of different PrP^{Sc} strains in separate test tubes with large amounts of the same, normal PrP^C polypeptide, each tube produces more of the specific PrP^{Sc} strain than was initially added to that tube. This ability to identify different strains of prions has helped investigators show that almost 200 people in Great Britain were infected with mad cow prions by eating meat derived from diseased cattle, resulting in a fatal, human form of mad cow disease known as *variant Creutzfeldt-Jakob disease (vCJD).* More than 1 million cattle have already been destroyed in the United Kingdom in an effort to halt the spread of this disease, but people may continue to die from vCJD as a result of having ingested tainted beef over the past two decades.

Mutations and Translation

Having described the *normal* process of translation, let us now consider what happens when mRNAs containing *mutant* codons are translated. **Box 22B** provides an overview of the main types of mutations that arise in DNA and their impact on the polypeptide chains produced by mRNAs. Most codon mutations simply alter a single amino acid, and mutations in the third base of a codon frequently do not change the amino acid at all. However, mutations that add or remove stop codons, or alter the reading frame, can severely disrupt mRNA translation. In this section, we will examine some of the ways in which cells can respond to such disruptive mutations.

Suppressor tRNAs Overcome the Effects of Some Mutations

Mutations that convert amino acid-coding codons into stop codons are referred to as **nonsense mutations. Figure 22-12** shows a case in which mutation of a single base pair in DNA converts an AAG lysine codon in mRNA to a UAG stop signal. Nonsense mutations like this one typically lead to production of incomplete, nonfunctional polypeptides that have been prematurely terminated at the mutant stop codon.

Nonsense mutations in essential genes are often lethal, but sometimes their detrimental effects can be overcome by an independent mutation affecting a tRNA gene. Such mutant tRNA genes produce mutant tRNAs that recognize what would otherwise be a stop codon and insert an amino acid at that point. In the example shown in Figure 22-12c, a mutant tRNA has an altered anticodon that allows it to read the stop codon UAG as a codon for tyrosine. The inserted amino acid is almost always different from the amino acid that would be present at that position in the wild-type protein, but the important point is that chain termination is averted and a full-length polypeptide can be made.

A tRNA molecule that somehow negates the effect of a mutation is called a **suppressor tRNA.** As you might expect, suppressor tRNAs exist that negate the effects of various types of mutations in addition to nonsense mutations (see Problem 22-8 at the end of the chapter). For a cell to survive, suppressor tRNAs must be rather inefficient; otherwise, the protein-synthesizing apparatus would produce too many abnormal proteins. An overly efficient nonsense suppressor, for example, would cause normal stop codons to be read as if they coded for an amino acid, thereby preventing normal termination. In fact, the synthesis of most polypeptides is terminated properly in cells containing nonsense suppressor tRNAs, indicating that a stop codon located in its proper place at the end of an mRNA coding sequence still triggers termination, whereas the same codon in an internal location does not. The most likely explanation is based on the behavior of the release factors that trigger normal termination (page 690). When a stop codon occurs in its proper location near the end of an RNA, release factors trigger termination because they are more efficient than suppressor tRNAs in binding to

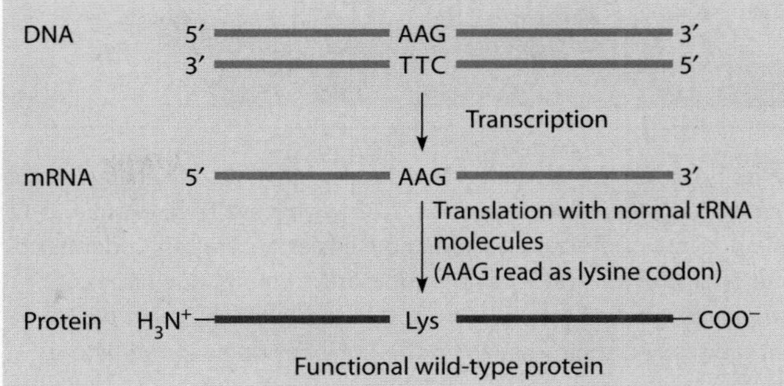

(a) Normal gene, normal tRNA molecules

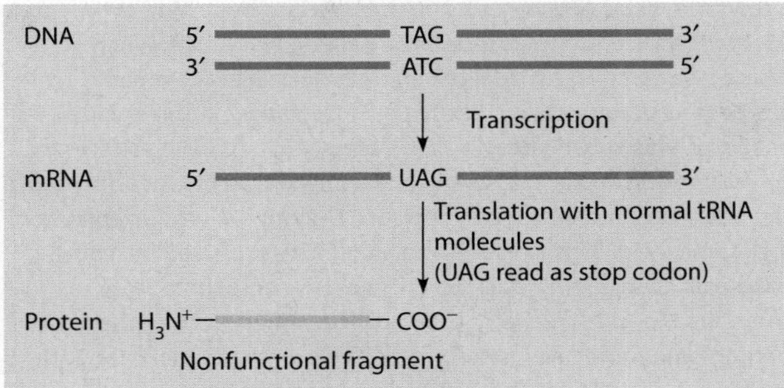

(b) Mutant gene, normal tRNA molecules

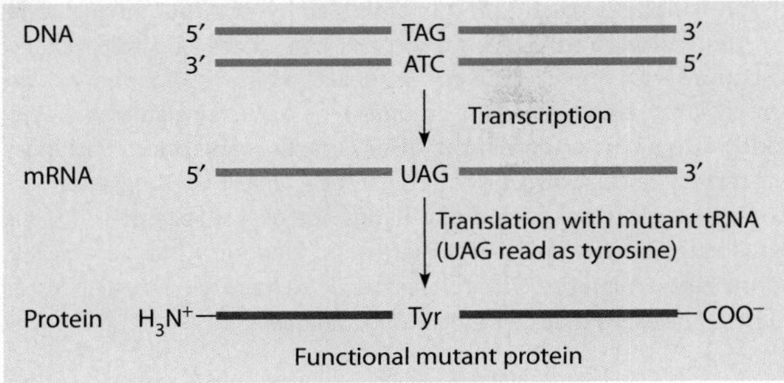

(c) Mutant gene, mutant (suppressor) tRNA molecule

FIGURE 22-12 Nonsense Mutations and Suppressor tRNAs. **(a)** A wild-type (normal) gene is transcribed into an mRNA molecule that contains the codon AAG at one point. Upon translation, this codon specifies the amino acid lysine (Lys) at this point in the functional, wild-type protein. **(b)** If a DNA mutation occurs that changes the AAG codon in the mRNA to UAG, the UAG codon will be read as a stop signal, and the translation product will be a short, nonfunctional polypeptide. Mutations of this sort are called nonsense mutations. **(c)** In the presence of a mutant tRNA that reads UAG as an amino acid codon instead of a stop signal, an amino acid will be inserted and polypeptide synthesis will continue. In the example shown, UAG is read as a codon for tyrosine because the mutant tyrosine tRNA has as its anticodon 3′-AUC-5′ instead of the usual 3′-AUG-5′ (which recognizes the tyrosine codon 5′-UAC-3′). The resulting protein will be mutant because a lysine has been replaced by a tyrosine at one point along the chain. However, the protein may still be functional if its biological activity is not significantly affected by the amino acid substitution.

BOX 22B DEEPER INSIGHTS

A Mutation Primer

In its broadest sense, the term *mutation* refers to any change in the nucleotide sequence of a genome. Now that we have examined the processes of transcription and translation, we can understand the effects of several kinds of mutations. Limiting our discussion to protein-coding genes, let's consider some of the main types of mutations and their impact on the polypeptide encoded by the mutant gene.

In this and the previous chapter, we have encountered several types of mutations in which the DNA change involves only one or a few base pairs (**Figure 22B-1a**). At the beginning of Chapter 21, for instance, we mentioned the genetic allele that, when homozygous, causes sickle-cell anemia. This allele originated from a type of mutation called a *base-pair substitution.* In this case, an AT base pair was substituted for a TA base pair in DNA. As a result, a GUA codon replaces a GAA in the mRNA transcribed from the mutant allele, and in the polypeptide (β-globin) a valine replaces a glutamic acid. This single amino acid change, caused by a single base-pair change, is enough to change the conformation of β-globin and, in turn, the hemoglobin tetramer, altering the way hemoglobin molecules pack into red cells and producing abnormally shaped cells that become trapped and damaged when they pass through small blood vessels (see Figure 21-2). Such a base-pair substitution is called a *missense mutation* because the mutated codon continues to code for an amino acid—but the "wrong" one.

Alternatively, a base-pair substitution can create a *nonstop mutation* by converting a normal stop codon into an amino acid codon, or conversely, it can create a *nonsense mutation* by converting an amino acid codon into a stop codon. In the latter case, the translation machinery will terminate the polypeptide prematurely. Unless the nonsense mutation is close to the end of the message or a suppressor tRNA is present, the polypeptide is not likely to be functional. Nonsense, nonstop, and missense codons can also arise from the *base-pair insertions* and *deletions* that cause *frameshift mutations.*

A single amino acid change (or even a change in several amino acids) does not always affect a protein's function in a major way. As long as the protein's three-dimensional conformation remains relatively unchanged, biological activity may be unaffected. Substitution of one amino acid for another of the same type—for example, valine for isoleucine—is especially unlikely to affect protein function. The nature of the genetic code actually reduces the effects of single base-pair alterations because many turn out to be *silent mutations* that change the nucleotide sequence without changing the genetic message. For example, changing the third base of a codon often produces a new codon that still codes for the same amino acid. Here, the "mutant" polypeptide is exactly the same as the wild-type.

In addition to mutations affecting one or a few base pairs, some alterations involve longer stretches of DNA (Figure 22B-1b). A few affect genome segments so large that the DNA changes can be detected by light microscopic examination of chromosomes. Some of these large-scale mutations are created by *insertions* or *deletions* of long DNA segments, but several other mechanisms also exist. In a *duplication,* a section of DNA is tandemly repeated. In an *inversion,* a chromosome segment is cut out and reinserted in its original position but in the reverse direction. A *translocation* involves the movement of a DNA segment from its normal location in the genome to another place, either in the same chromosome or a different one. Because these large-scale mutations may or may not affect the expression of many genes, they have a wide range of phenotypic effects, from no effect at all to lethality.

When we think about the potential effects of mutations, it is useful to remember that genes have important noncoding components and that these, too, can be mutated in ways that seriously affect gene products. A mutation in a promoter, for example, can result in more or less frequent transcription of the gene. Even a mutation in an intron can affect the gene product in a major way if it touches a critical part of a splice-site sequence.

stop codons in this location—perhaps because release factor action is stimulated by a special sequence or three-dimensional configuration near the end of the mRNA.

Nonsense-Mediated Decay and Nonstop Decay Promote the Destruction of Defective mRNAs

In the absence of an appropriate suppressor tRNA, a nonsense stop codon will cause mRNA translation to stop prematurely, thereby generating an incomplete polypeptide chain that cannot function properly and may even harm the cell. To avoid wasting energy on the production of such useless products, eukaryotic cells invoke a quality control mechanism called **nonsense-mediated decay** to destroy mRNAs containing premature stop codons. In mammals, the method for identifying premature stop codons involves the *exon junction complex (EJC),* a multiprotein complex deposited during mRNA splicing at each point where an intron is removed from pre-mRNA (see Figure 21-22). Thus, every newly spliced mRNA molecule will have one or more EJCs bound to it, one at each exon-exon junction. During translation, the distinction between normal and premature stop codons is made on the basis of their relationship to EJCs. If a stop codon is encountered in an mRNA prior to the last EJC—in other words, before the last exon—it must be a premature stop codon. The presence of such a stop codon will cause translation to be terminated while the mRNA still has one or more EJCs bound to it, and the presence of these remaining EJCs marks the mRNA for degradation.

How do cells handle the opposite situation, namely an mRNA with no stop codons? In eukaryotic cells, translation becomes stalled when a ribosome reaches the end of an mRNA without encountering a stop codon. An RNA-degrading enzyme complex then binds to the empty A site of the ribosome and degrades the defective mRNA in a process called **nonstop decay.** The same problem is handled a bit differently in bacteria. When translation halts at the end of a bacterial mRNA lacking a stop codon, an unusual type of RNA called *tmRNA* ("transfer-messenger RNA") binds to the A site of the ribosome and directs the addition of about a

Finally, mutations in genes that encode regulatory proteins—that is, proteins that control the expression of other genes—can have far-reaching effects on many other proteins. We will discuss this topic in Chapter 23.

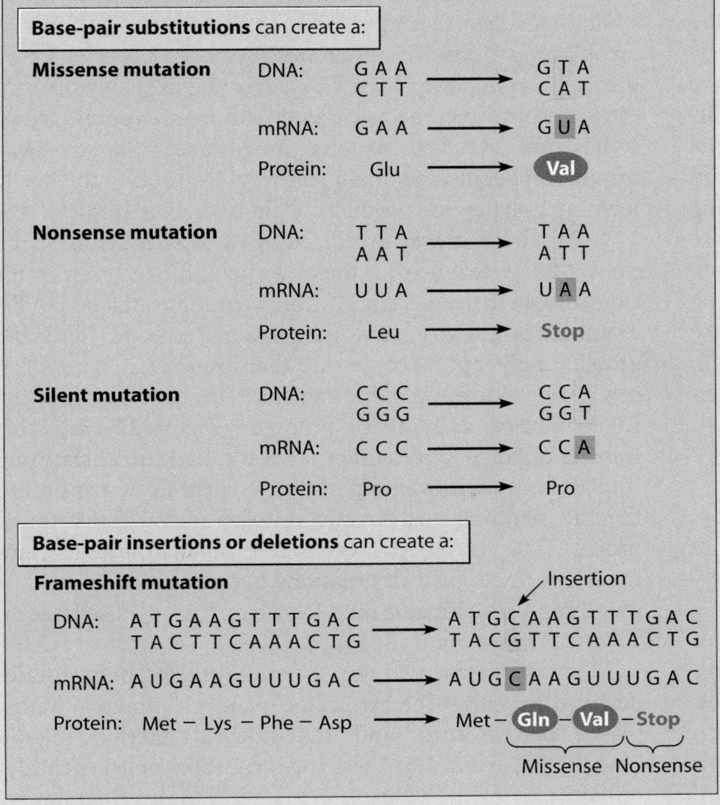

(a) Mutations affecting one base pair

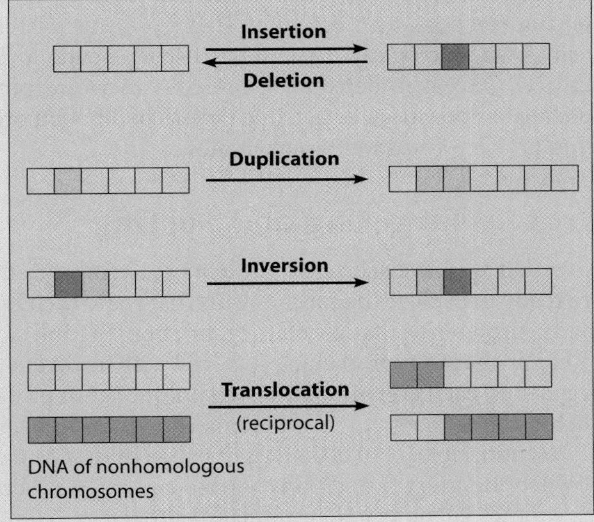

(b) Mutations affecting long DNA segments

FIGURE 22B-1 Types of Mutations. Mutations can affect **(a)** one base pair or **(b)** long DNA segments.

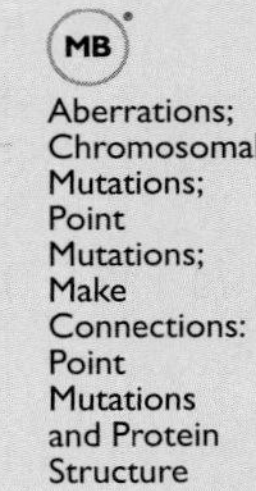

dozen more amino acids to the growing polypeptide chain. This amino acid sequence creates a signal that targets the protein for destruction. At the same time, the mRNA is degraded by a ribonuclease associated with the tmRNA.

Posttranslational Processing

After polypeptide chains have been synthesized, they often must be chemically modified before they can perform their normal functions. Such modifications are known collectively as **posttranslational modifications.** In bacteria, for example, the *N*-formyl group located at the N-terminus of polypeptide chains is always removed. The methionine it was attached to is often removed also, as is the methionine that starts eukaryotic polypeptides. As a result, relatively few mature polypeptides have methionine at their N-terminus, even though they all started out that way. Sometimes whole blocks of amino acids are removed from the polypeptide. Certain enzymes, for example, are synthesized as inactive precursors that must be activated by the removal of a specific sequence at one end or the other. The transport of proteins across membranes also may involve the removal of a terminal *signal sequence,* as we will see shortly, and some polypeptides have internal stretches of amino acids that must be removed to produce an active protein. For instance, insulin is synthesized as a single polypeptide and then processed to remove an internal segment; the two end segments remain linked by disulfide bonds between cysteine residues in the active hormone (see Figure 3-7).

Other common processing events include chemical modifications of individual amino acid groups—by methylation, phosphorylation, or acetylation reactions, for example. In addition, a polypeptide may undergo glycosylation (the addition of carbohydrate side chains; see Chapter 12) or binding to prosthetic groups. Finally, in the case of proteins composed of multiple subunits, individual polypeptide chains must bind to one another to form the appropriate multisubunit proteins or multiprotein complexes.

In addition to the preceding posttranslational events, some proteins undergo a relatively unusual type of processing called *protein splicing*, which is analogous to the phenomenon of *RNA splicing* discussed in Chapter 21. As we saw, intron sequences are removed from RNA molecules during RNA splicing, and the remaining exon sequences are simultaneously spliced together. Likewise, during protein splicing, specific amino acid sequences called *inteins* are removed from a polypeptide chain and the remaining segments, called *exteins,* are spliced together to form the mature protein. Protein splicing is usually intramolecular, involving the excision of an intein from a single polypeptide chain by a self-catalytic mechanism. However, splicing can also take place between two polypeptide chains arising from two different mRNAs. For example, in some photosynthetic bacteria a subunit of DNA polymerase III is produced from two separate genes, each coding for an intein-containing polypeptide that includes part of the DNA polymerase subunit. In some cases, the inteins removed by protein splicing reactions turn out to be stable proteins exhibiting their own biological functions (usually endonuclease activity). Once considered to be an oddity of nature, protein splicing has now been detected in dozens of different organisms, prokaryotes as well as eukaryotes.

Protein Targeting and Sorting

Now that we have seen how proteins are synthesized, we are ready to explore the mechanisms that route each newly made protein to its correct destination. Think for a moment about a typical eukaryotic cell with its diversity of organelles, each containing its own unique set of proteins. Such a cell is likely to have billions of protein molecules, representing at least 10,000 kinds of polypeptides. And each polypeptide must find its way to the appropriate location within the cell, or even out of the cell altogether. A limited number of these polypeptides are encoded by the genome of the mitochondrion (and, for plant cells, by the chloroplast genome as well), but most are encoded by nuclear genes and are synthesized by a process beginning in the cytosol. Each of these polypeptides must then be directed to its proper destination and must therefore have some sort of molecular "zip code" ensuring its delivery to the correct place. As our final topic for this chapter, we will consider this process of protein targeting and sorting.

We can begin by grouping the various compartments of eukaryotic cells into three categories: (1) the endomembrane system, the interrelated system of membrane compartments that includes the endoplasmic reticulum (ER), the Golgi complex, lysosomes, secretory vesicles, the nuclear envelope, and the plasma membrane; (2) the cytosol; and (3) mitochondria, chloroplasts, peroxisomes (and related organelles), and the interior of the nucleus.

Polypeptides encoded by nuclear genes are routed to these compartments using several different mechanisms. The process begins with transcription of DNA into RNAs that are processed in the nucleus and then transported through nuclear pores for translation in the cytoplasm, where most ribosomes occur. Although translation is largely a cytoplasmic process, some evidence suggests that up to 10% of a cell's ribosomes may actually reside in the nucleus, where they can translate newly synthesized RNAs. Nuclear translation appears to function mainly as a quality control mechanism that checks new mRNAs for the presence of errors (see the discussion of nonsense-mediated decay on page 694).

Despite the existence of these functioning nuclear ribosomes, it is clear that most polypeptide synthesis occurs on cytoplasmic ribosomes after mRNAs have been exported through the nuclear pores. Upon arriving in the cytoplasm, these mRNAs become associated with *free ribosomes* (ribosomes not attached to any membrane). Shortly after translation begins, two main pathways for routing the newly forming polypeptide products begin to diverge (**Figure 22-13**). The first pathway is utilized by ribosomes synthesizing polypeptides destined for the endomembrane system or for export from the cell. Such ribosomes become attached to ER membranes early in the translational process, and the growing polypeptide chains are then transferred across (or, in the case of integral membrane proteins, inserted into) the ER membrane as synthesis proceeds (Figure 22-13a). This transfer of polypeptides into the ER is called **cotranslational import** because movement of the polypeptide across or into the ER membrane is directly coupled to the translational process. The subsequent conveyance of such proteins from the ER to their final destinations is carried out by various membrane vesicles and the Golgi complex, as discussed in Chapter 12 (see Figure 12-8).

An alternative pathway is employed for polypeptides destined for either the cytosol or for mitochondria, chloroplasts, peroxisomes, and the nuclear interior (Figure 22-13b). Ribosomes synthesizing these types of polypeptides remain free in the cytosol, unattached to any membrane. After translation has been completed, the polypeptides are released from the ribosomes and either remain in the cytosol as their final destination or are taken up by the appropriate organelle. The uptake by organelles of such completed polypeptides requires the presence of special targeting signals and is called **posttranslational import.** In the case of the nucleus, polypeptides enter through the nuclear pores, as discussed in Chapter 18 (see Figure 18-31). Polypeptide entrance into mitochondria, chloroplasts, and peroxisomes involves a different kind of mechanism, as we will see shortly.

With this general overview in mind, we are now ready to examine the mechanisms of cotranslational import and posttranslational import in detail.

Cotranslational Import Allows Some Polypeptides to Enter the ER as They Are Being Synthesized

Cotranslational import into the ER is the first step in the pathway for delivering newly synthesized proteins to various locations within the endomembrane system. Proteins handled in this way are synthesized on ribosomes that become attached to the ER shortly after translation begins.

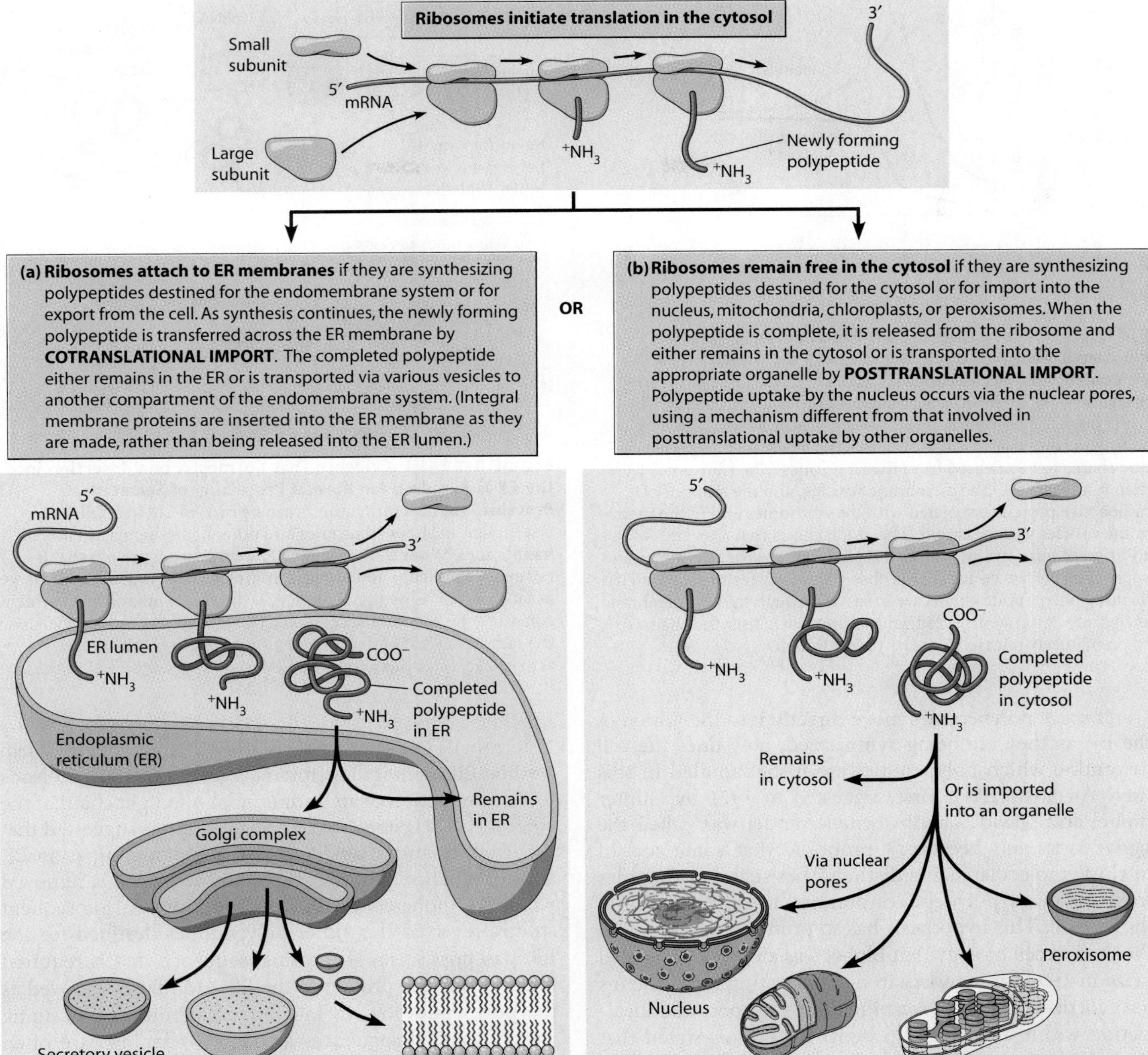
Ribosomes initiate translation in the cytosol
Small subunit
Large subunit
5′
mRNA
3′
$^+NH_3$
Newly forming polypeptide
(a) Ribosomes attach to ER membranes if they are synthesizing polypeptides destined for the endomembrane system or for export from the cell. As synthesis continues, the newly forming polypeptide is transferred across the ER membrane by COTRANSLATIONAL IMPORT. The completed polypeptide either remains in the ER or is transported via various vesicles to another compartment of the endomembrane system. (Integral membrane proteins are inserted into the ER membrane as they are made, rather than being released into the ER lumen.)
OR
(b) Ribosomes remain free in the cytosol if they are synthesizing polypeptides destined for the cytosol or for import into the nucleus, mitochondria, chloroplasts, or peroxisomes. When the polypeptide is complete, it is released from the ribosome and either remains in the cytosol or is transported into the appropriate organelle by POSTTRANSLATIONAL IMPORT. Polypeptide uptake by the nucleus occurs via the nuclear pores, using a mechanism different from that involved in posttranslational uptake by other organelles.
ER lumen
COO^-
Completed polypeptide in ER
Endoplasmic reticulum (ER)
Remains in ER
Golgi complex
Secretory vesicle
Lysosome
Plasma membrane
Completed polypeptide in cytosol
Remains in cytosol
Or is imported into an organelle
Via nuclear pores
Nucleus
Mitochondrion
Chloroplast
Peroxisome

MB Protein Synthesis (3 of 3): Translation and Protein Targeting Pathways (BioFlix tutorial)

FIGURE 22-13 Intracellular Sorting of Proteins. Polypeptide synthesis begins in the cytosol but takes one of two alternative routes when the polypeptide is about 30 amino acids long. **(a)** Polypeptides destined for the endomembrane system, or for export from the cell, are transferred across the ER membrane by cotranslational import as they are being made. **(b)** Other polypeptides are synthesized in the cytosol and either remain there or are transferred by posttranslational import into the nucleus, mitochondria, chloroplasts, or peroxisomes.

The role of the ER in this process was first suggested by experiments in which Colvin Redman and David Sabatini studied protein synthesis in isolated vesicles of rough ER (ER vesicles with attached ribosomes). Such vesicles, known as *microsomes,* can be isolated using subcellular fractionation and centrifugation (see Box 12A). After briefly incubating the rough ER vesicles in the presence of radioactive amino acids and other components needed for protein synthesis, they stopped the reaction by adding *puromycin,* an antibiotic that causes partially completed polypeptide chains to be released from ribosomes. When the ribosomes and membrane vesicles were then separated and analyzed to see where the newly made, radioactive polypeptide chains were located, a substantial fraction of the radioactivity was found inside the ER lumen (**Figure 22-14**). Such results suggested that newly forming polypeptides pass into the lumen of the ER *as they are being synthesized,* allowing them to be routed through the ER to their correct destinations.

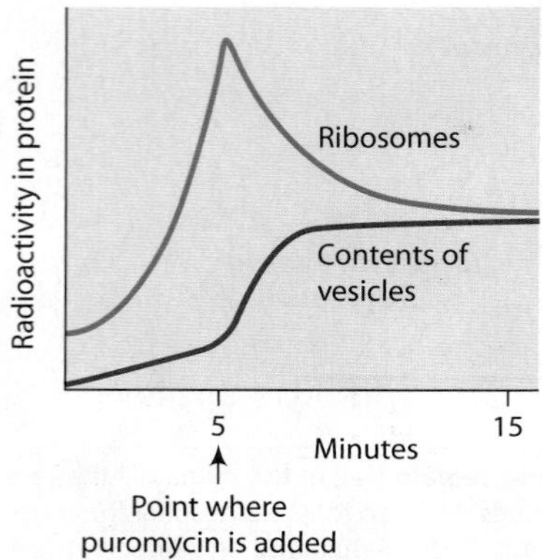

FIGURE 22-14 Evidence That Proteins Synthesized on Ribosomes Attached to ER Membranes Pass Directly into the ER Lumen. ER vesicles containing attached ribosomes were isolated and incubated with radioactive amino acids to label newly made polypeptide chains. Next, protein synthesis was halted by adding puromycin, which also causes the newly forming polypeptide chains to be released from the ribosomes. The ribosomes were then removed from the membrane vesicles, and the amount of radioactive protein associated with the ribosomes and in the membrane vesicles was measured. The graph shows that after the addition of puromycin, radioactivity is lost from the ribosomes and appears inside the vesicles. This observation suggests that the newly forming polypeptide chains are inserted through the ER membrane as they are being synthesized, and puromycin causes the chains to be prematurely released into the vesicle lumen.

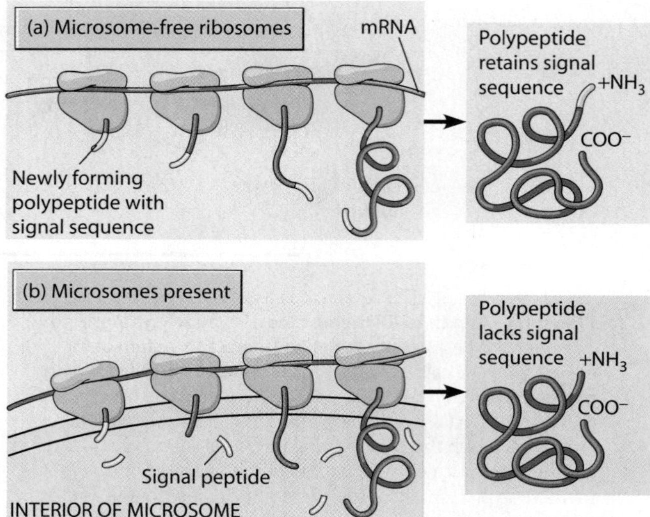

FIGURE 22-15 Evidence That Cotranslational Insertion into the ER Is Required for Normal Processing of Secreted Proteins. **(a)** Protein synthesis can be carried out in a cell-free system that includes ribosomes and other components but no membranes. When a messenger RNA encoding a protein that is normally secreted is added, the resulting protein is abnormally large because it retains its signal sequence. **(b)** When microsomes, which consist of ER membranes and attached ribosomes, are isolated and the same mRNA is added, the resulting protein is transported across the vesicle membrane and the signal sequence is cleaved.

If some polypeptides move directly into the lumen of the ER as they are being synthesized, how does the cell determine which polypeptides are to be handled in this way? An answer was first suggested in 1971 by Günter Blobel and David Sabatini, whose model was called the *signal hypothesis* because it proposed that some sort of intrinsic molecular signal distinguishes such polypeptides from the many polypeptides destined to be released into the cytosol. This hypothesis has so profoundly influenced the field of cell biology that Blobel was awarded the Nobel Prize in 1999 for his work in demonstrating that proteins have intrinsic signals governing their transport and localization within the cell. The signal hypothesis stated that for polypeptides destined for the ER, the first segment of the polypeptide to be synthesized, the N-terminus, contains an **ER signal sequence** that directs the ribosome-mRNA-polypeptide complex to the surface of the rough ER, where the complex anchors at a protein "dock" on the ER surface. Then, as the polypeptide chain elongates during mRNA translation, it progressively crosses the ER membrane and enters the ER lumen.

Shortly after the signal hypothesis was first proposed, evidence for the actual existence of ER signal sequences was obtained by César Milstein and his associates, who were studying the synthesis of the small subunit, or *light chain,* of the protein *immunoglobulin G.* In cell-free systems containing purified ribosomes and the components required for protein synthesis, the mRNA coding for the immunoglobulin light chain directs the synthesis of a polypeptide product that is 20 amino acids longer at its N-terminal end than the authentic light chain itself. Adding ER membranes (microsomes) to this system leads to the production of an immunoglobulin light chain of the correct size (**Figure 22-15**). Such findings suggested that the extra 20-amino acid segment is functioning as an ER signal sequence and that this signal sequence is removed when the polypeptide moves into the ER. Subsequent studies revealed that other polypeptides destined for the ER also possess an N-terminal sequence that is required for targeting the protein to the ER and that is removed as the polypeptide moves into the ER. Proteins containing such ER signal sequences at their N-terminus are often referred to as *preproteins* (e.g., prelysozyme, preproinsulin, pretrypsinogen, and so forth).

Sequencing studies have revealed that the amino acid compositions of ER signal sequences are surprisingly variable, but several unifying features have been noted. ER signal sequences are typically 15–30 amino acids long and consist of three domains: a positively charged N-terminal region, a central hydrophobic region, and a polar region adjoining the site where cleavage from the mature protein will take place. The positively charged end may promote interaction with the hydrophilic exterior of the ER membrane, and the hydrophobic region may facilitate interaction of the signal sequence with the membrane's lipid interior. In any case, it is now established that only polypeptides with ER signal sequences can be inserted into or across the ER membrane as their synthesis proceeds. In fact, when recombinant DNA methods are used to add ER signal sequences to polypeptides that do not usually have them, the recombinant polypeptides are directed to the ER.

The Signal Recognition Particle (SRP) Binds the Ribosome-mRNA-Polypeptide Complex to the ER Membrane

Once the existence of ER signal sequences was established, it quickly became clear that newly forming polypeptides must become attached to the ER membrane before very much of the polypeptide has emerged from the ribosome. If translation were to continue without attachment to the ER, the folding of the growing polypeptide chain might bury the signal sequence. To understand what prevents this from happening, we need to look at the signal mechanism in further detail.

Contrary to the original signal hypothesis, the ER signal sequence does not itself initiate contact with the ER. Instead, the contact is mediated by a **signal recognition particle (SRP),** which recognizes and binds to the ER signal sequence of the newly forming polypeptide and then binds to the ER membrane (**Figure 22-16**). At first the SRP was thought to be purely protein (the *P* in its name originally stood for protein). Later, however, the SRP was shown to consist of six different polypeptides complexed with a 300-nucleotide (7S) molecule of RNA. The protein components have three main active sites: one that recognizes and binds to the ER signal sequence, one that interacts with the ribosome to block further translation, and one that binds to the ER membrane.

Figure 22-16 illustrates the role played by the SRP in cotranslational import. The process begins when an mRNA coding for a polypeptide destined for the ER starts to be translated on a free ribosome. Polypeptide synthesis proceeds until the ER signal sequence has been formed and emerges from the surface of the ribosome. At this stage, SRP (shown in orange) binds to the signal sequence and blocks further translation (step ❶). The SRP then binds the ribosome to a special structure in the ER membrane called a **translocon** because it carries out the translocation of polypeptides across the ER membrane. (Note that the term *translocation,* which literally means "a change of location," is used to describe both the movement of proteins through membranes and, earlier in the chapter, the movement of mRNA across the ribosome.)

The translocon is a protein complex composed of several components involved in cotranslational import, including an *SRP receptor* to which the SRP binds, a *ribosome receptor* that holds the ribosome in place, a *pore protein* that forms a channel through which the growing polypeptide can enter the ER lumen, and *signal peptidase,* an enzyme that removes the ER signal sequence. As ❷ shows, SRP (bringing an attached ribosome) first binds to the SRP receptor, allowing the ribosome to become attached to the ribosome receptor. Next, GTP binds to both SRP and the SRP receptor, unblocking translation and causing transfer of the signal sequence to the pore protein, whose central channel opens as the signal sequence is inserted (❸). GTP is then hydrolyzed, accompanied by release of the SRP (❹). As the polypeptide elongates, it passes into the ER lumen and signal peptidase cleaves the signal sequence, which is quickly degraded (❺). After polypeptide synthesis is completed, the final polypeptide is released into the ER lumen, the translocon channel is closed, and the ribosome detaches from the ER membrane and dissociates into its subunits, releasing the mRNA (❻).

Protein Folding and Quality Control Take Place Within the ER

After polypeptides are released into the ER lumen, they fold into their final shape and, in some cases, assemble with other polypeptides to form multisubunit proteins. As we mentioned earlier in the chapter, molecular chaperones facilitate these folding and assembly events. The most abundant chaperone in the ER lumen is a member of the Hsp70 family of chaperones known as **BiP** (an abbreviation for *Bi*nding *P*rotein). BiP acts by binding to *hydrophobic regions* of polypeptide chains, especially to regions enriched in the amino acids tryptophan, phenylalanine, and leucine.

In a mature, fully folded protein, such hydrophobic regions are buried in the interior of the protein molecule. In an unfolded polypeptide, these same hydrophobic regions are exposed to the surrounding aqueous environment, creating an unstable situation in which polypeptides tend to aggregate with one another. BiP prevents this aggregation by transiently binding to the hydrophobic regions of unfolded polypeptides as they emerge into the ER lumen, stabilizing them and preventing them from interacting with other unfolded polypeptides. BiP then releases the polypeptide chain, accompanied by ATP hydrolysis, giving the polypeptide a brief opportunity to fold (perhaps aided by other chaperones). If the polypeptide folds correctly, its hydrophobic regions become buried in the molecule's interior and can no longer bind to BiP. But if the hydrophobic segments fail to fold properly, BiP binds again to the polypeptide and the cycle is repeated. In this way, BiP uses energy released by ATP hydrolysis to promote proper protein folding.

Folding is often accompanied by the formation of disulfide bonds between cysteines located in different regions of a polypeptide chain. This reaction is facilitated by **protein disulfide isomerase,** an enzyme present in the ER lumen that catalyzes the formation and breakage of disulfide bonds between cysteine residues. Protein disulfide isomerase starts acting before the synthesis of a newly forming polypeptide has been completed, allowing various disulfide bond combinations to be tested until the most stable arrangement is found.

Proteins that repeatedly fail to fold properly can activate several types of quality control mechanisms. One such mechanism, called the **unfolded protein response (UPR),** uses sensor molecules in the ER membrane to detect misfolded proteins. These sensors activate signaling

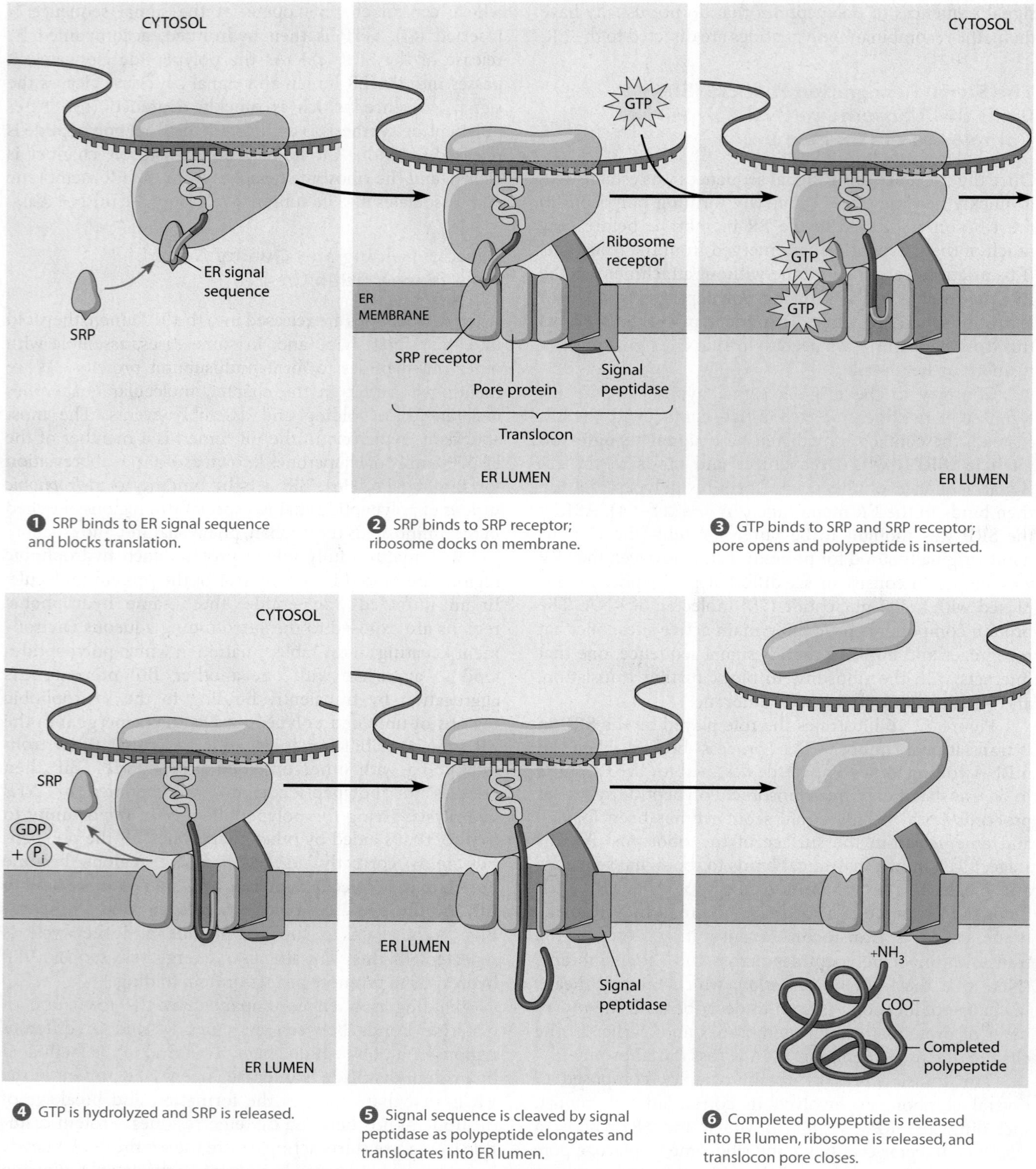

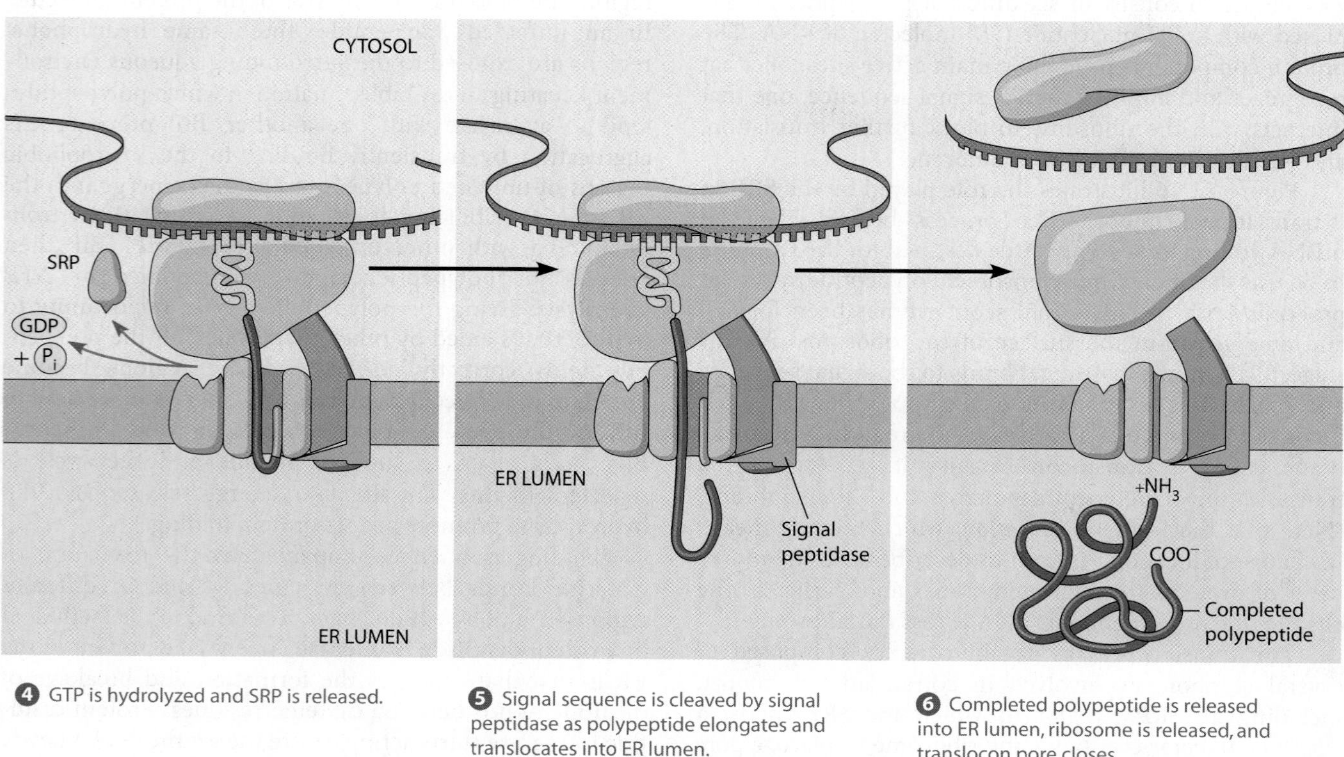

CYTOSOL
ER signal sequence
SRP
GTP
Ribosome receptor
ER MEMBRANE
SRP receptor
Pore protein
Signal peptidase
Translocon
ER LUMEN
CYTOSOL
GTP
GTP
ER LUMEN
1 SRP binds to ER signal sequence and blocks translation.
2 SRP binds to SRP receptor; ribosome docks on membrane.
3 GTP binds to SRP and SRP receptor; pore opens and polypeptide is inserted.
CYTOSOL
SRP
GDP
+ Pi
ER LUMEN
ER LUMEN
Signal peptidase
+NH3
COO−
Completed polypeptide
4 GTP is hydrolyzed and SRP is released.
5 Signal sequence is cleaved by signal peptidase as polypeptide elongates and translocates into ER lumen.
6 Completed polypeptide is released into ER lumen, ribosome is released, and translocon pore closes.

FIGURE 22-16 A Model for the Signal Mechanism of Cotranslational Import. This figure shows a schematic model for the signal mechanism. It is now well established that the growing polypeptide translocates through a hydrophilic pore created by one or more membrane proteins. The complex of membrane proteins that carry out translocation is called the *translocon*.

pathways that shut down the synthesis of most proteins while enhancing the production of those required for protein folding and degradation. Another type of quality control, known as **ER-associated degradation (ERAD),** recognizes misfolded or unassembled proteins and exports or "retrotranslocates" them back across the ER membrane to the cytosol, where they are degraded by *proteasomes* as described in Chapter 23.

Proteins Released into the ER Lumen Are Routed to the Golgi Complex, Secretory Vesicles, Lysosomes, or Back to the ER

Most of the proteins synthesized on ribosomes attached to the ER are *glycoproteins*—that is, proteins with covalently bound carbohydrate groups. As you learned in Chapter 12, the initial glycosylation reactions that add these carbohydrate side chains take place in the ER, often while the growing polypeptide is still being synthesized. After polypeptides have been released into the ER lumen, glycosylated, and folded, they are delivered by various types of transport vesicles to their destinations within the cell (see Figure 12-8). The first stop in this transport pathway is the Golgi complex, where further glycosylation and processing of carbohydrate side chains may occur. The Golgi complex then serves as a site for sorting and distributing proteins to other locations.

For soluble proteins, the default pathway takes them from the Golgi complex to secretory vesicles that move to the cell surface and fuse with the plasma membrane, leading to secretion of such proteins from the cell. Soluble proteins entering the Golgi complex and that are not destined for secretion from the cell possess specific carbohydrate side chains and/or short amino acid signal sequences that target each protein to its appropriate location within the endomembrane system. For example, we saw in Chapter 12 that many lysosomal enzymes possess carbohydrate side chains exhibiting the unusual sugar *mannose-6-phosphate.* This sugar serves as a recognition device that allows the Golgi complex to selectively package such proteins into newly forming lysosomes (see Figure 12-9). As we also saw in Chapter 12, a different signaling mechanism is used for proteins whose final destination is the ER. The C-terminus of these proteins usually contains a *KDEL sequence,* which consists of the amino acids Lys-Asp-Glu-Leu or a closely related sequence. The Golgi complex employs a receptor protein that binds to the KDEL sequence and delivers the targeted protein back to the ER. Protein disulfide isomerase—the ER-resident enzyme whose role in protein folding was described in the preceding section—is an example of a protein possessing a KDEL sequence that confines the molecule to the ER.

Stop-Transfer Sequences Mediate the Insertion of Integral Membrane Proteins

So far, we have focused on the cotranslational import and sorting of *soluble proteins* that are destined either for secretion from the cell or for the lumen of endomembrane components, such as the ER, the Golgi complex, lysosomes, and related vesicles. The other major group of polypeptides synthesized on ER-attached ribosomes consists of molecules destined to become integral *membrane proteins.* Polypeptides of this type are synthesized by a mechanism similar to the one illustrated in Figure 22-16 for soluble proteins except that the completed polypeptide chain remains embedded in the ER membrane rather than being released into the ER lumen.

Recall from Chapter 7 that integral membrane proteins are typically anchored to the lipid bilayer by one or more α-helical *transmembrane segments* consisting of 20–30 hydrophobic amino acids. In considering the mechanism that allows such proteins to be retained as part of the ER membrane after synthesis rather than being released into the ER lumen, we focus here on the simplest case: proteins with only a single such transmembrane segment. The principles involved, however, extend to proteins with more complicated configurations. Researchers postulate two main mechanisms by which hydrophobic transmembrane segments anchor newly forming polypeptide chains to the lipid bilayer of the ER membrane.

The first of these mechanisms involves polypeptides with a typical ER signal sequence at their N-terminus, which allows an SRP to bind the ribosome-mRNA complex to the ER membrane. Elongation of the polypeptide chain then continues until the hydrophobic transmembrane segment of the polypeptide is synthesized. As shown in **Figure 22-17a,** this stretch of amino acids functions as a **stop-transfer sequence** that halts translocation of the polypeptide through the ER membrane. Translation continues, but the rest of the polypeptide chain remains on the cytosolic side of the ER membrane, resulting in a transmembrane protein with its N-terminus in the ER lumen and its C-terminus in the cytosol. Meanwhile, the hydrophobic stop-transfer signal moves laterally out through a side opening in the translocon and into the lipid bilayer, forming the permanent transmembrane segment that anchors the protein to the membrane.

The second mechanism involves membrane proteins that lack a typical signal sequence at their N-terminus and instead possess an *internal* **start-transfer sequence** that performs two functions: It first acts as an ER signal sequence that allows an SRP to bind the ribosome-mRNA complex to the ER membrane, and then its hydrophobic region functions as a membrane anchor that moves out through a side opening in the translocon and permanently attaches the polypeptide to the lipid bilayer (Figure 22-17b). The orientation of the start-transfer sequence at the time of insertion determines which terminus of the polypeptide ends up in the ER lumen and which in the cytosol. Transmembrane proteins with multiple membrane-spanning regions are formed in a similar way, except that an alternating pattern of start-transfer and stop-transfer sequences creates a polypeptide containing multiple transmembrane segments that pass back and forth across the membrane.

Once a newly formed polypeptide has been incorporated into the ER membrane by one of the preceding

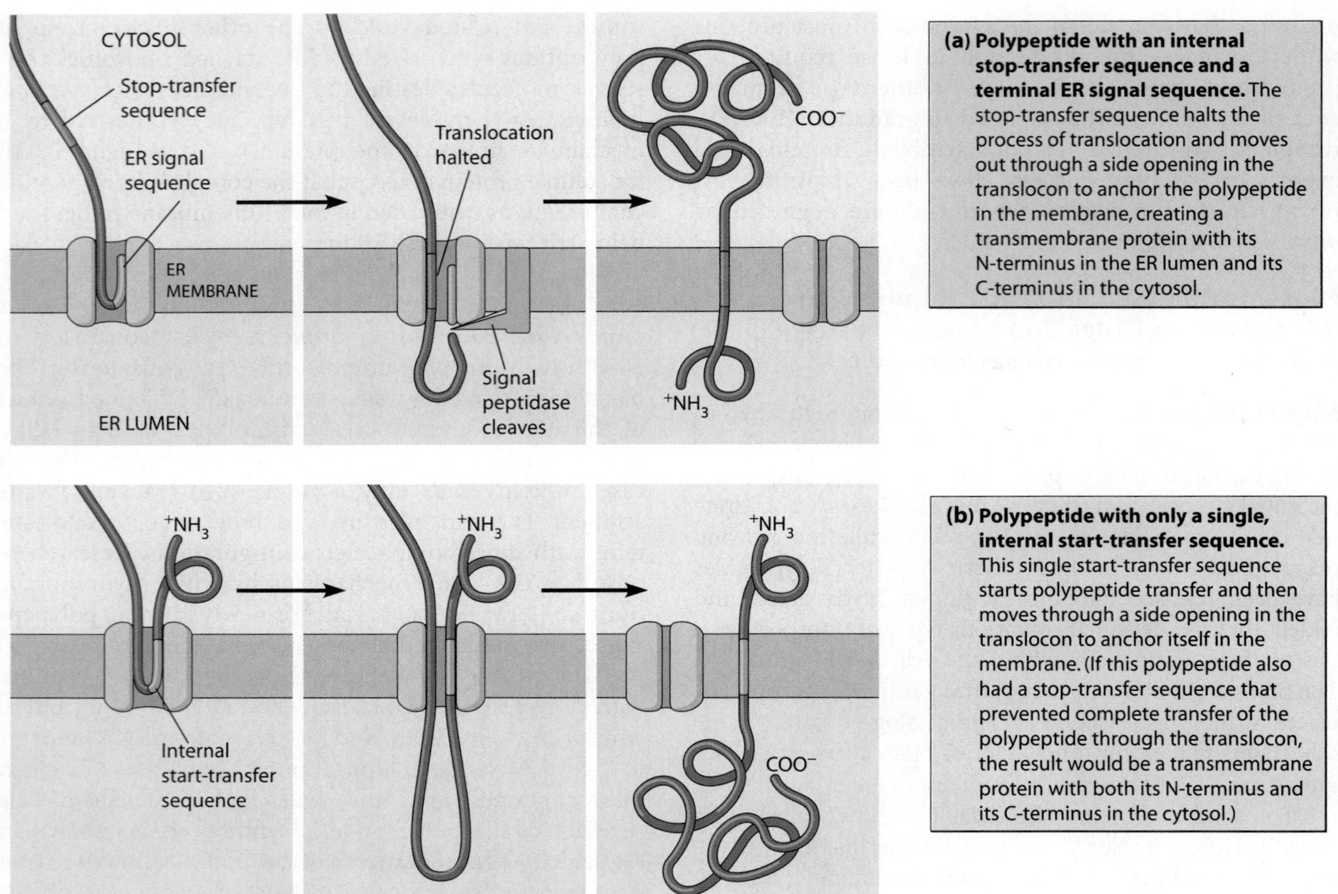

CYTOSOL
Stop-transfer sequence
ER signal sequence
ER MEMBRANE
ER LUMEN
Translocation halted
Signal peptidase cleaves
COO⁻
⁺NH₃
(a) Polypeptide with an internal stop-transfer sequence and a terminal ER signal sequence. The stop-transfer sequence halts the process of translocation and moves out through a side opening in the translocon to anchor the polypeptide in the membrane, creating a transmembrane protein with its N-terminus in the ER lumen and its C-terminus in the cytosol.
⁺NH₃
Internal start-transfer sequence
COO⁻
(b) Polypeptide with only a single, internal start-transfer sequence. This single start-transfer sequence starts polypeptide transfer and then moves through a side opening in the translocon to anchor itself in the membrane. (If this polypeptide also had a stop-transfer sequence that prevented complete transfer of the polypeptide through the translocon, the result would be a transmembrane protein with both its N-terminus and its C-terminus in the cytosol.)

FIGURE 22-17 Cotranslational Insertion of Transmembrane Proteins into the ER Membrane. This figure shows two mechanisms for inserting integral membrane proteins containing a single transmembrane segment. For clarity, the SRP, ribosome, and most other parts of the translocational apparatus have been omitted. Transmembrane proteins whose N- and C-termini are oriented in opposite directions from those shown here can be created using a start-transfer sequence that has the opposite orientation when it first inserts into the translocation apparatus.

mechanisms, it can either remain in place to function as an ER membrane protein or be transported to other components of the endomembrane system, such as the Golgi complex, lysosomes, nuclear envelope, or plasma membrane. Transport is carried out by a series of membrane budding and fusing events in which membrane vesicles pinch off from one compartment of the endomembrane system and fuse with another compartment, as we described in Figure 12-8.

Posttranslational Import Allows Some Polypeptides to Enter Organelles After They Have Been Synthesized

In contrast to the cotranslational import of proteins into the ER discussed in the preceeding several pages, proteins destined for the nuclear interior, mitochondrion, chloroplast, or peroxisome are imported into these organelles *after* translation has been completed. Because such proteins are synthesized on free ribosomes and released into the cytosol, each protein must carry a targeting signal that directs it to the correct organelle. In Chapter 12, you learned that a serine-lysine-leucine sequence (SKL in single-letter code) located near the C-terminus of a protein targets it for uptake into peroxisomes. And in Chapter 18, you learned that posttranslational import of proteins into the nucleus depends on *nuclear localization signals* that target proteins for transport through nuclear pore complexes (see Figure 18-31). Here, we focus on protein import into mitochondria and chloroplasts, which involves signal sequences similar to those used in cotranslational import.

Importing Polypeptides into Mitochondria and Chloroplasts. Although mitochondria and chloroplasts contain their own DNA and protein-synthesizing machinery, they synthesize few of the polypeptides they require. More than 95% of the proteins residing in these two organelles, like all proteins found in the nucleus and peroxisomes, are encoded by nuclear genes and synthesized on cytosolic ribosomes. The small number of polypeptides synthesized within mitochondria are targeted mainly to the inner mitochondrial membrane, and the polypeptides synthesized within chloroplasts are targeted mainly to thylakoid membranes. Almost without exception, such polypeptides encoded by mitochondrial or chloroplast genes are subunits of multimeric proteins, with one or more of the other subunits being encoded by nuclear genes and imported from the cytosol. For example, mammalian cytochrome *c* oxidase consists of 13 polypeptides, 3 of which are encoded by the mitochondrial genome and synthesized within mitochondria. The other 10 subunits are synthesized in the cytosol and imported into mitochondria.

Most mitochondrial and chloroplast polypeptides are synthesized on cytosolic ribosomes, released into the cytosol, and taken up by the appropriate organelle (mitochondrion or chloroplast) within a few minutes. The targeting signal for such polypeptides is a special sequence called a **transit sequence.** Like the ER signal sequence of ER-targeted polypeptides, the transit sequence is located at the N-terminus of the polypeptide. Once inside the mitochondrion or chloroplast, the transit sequence is removed by a *transit peptidase* located within the organelle. Removal of the transit sequence often occurs before transport is complete.

The transit sequences of mitochondrial or chloroplast polypeptides typically contain both hydrophobic and hydrophilic amino acids. The presence of positively charged amino acids is critical, although the secondary structure of the sequence may be more important than the specific amino acids. For example, some mitochondrial transit sequences have positively charged amino acids interspersed with hydrophobic amino acids in such a way that, when the sequence is coiled into an a α helix, most of the positively charged amino acids are on one side of the helix and the hydrophobic amino acids are on the other.

The uptake of polypeptide chains possessing transit sequences is mediated by specialized transport complexes located in the outer and inner membranes of mitochondria and chloroplasts. As shown in **Figure 22-18,** the mitochondrial transport complexes are called **TOM** (translocase of the outer mitochondrial membrane) and **TIM** (translocase of the inner mitochondrial membrane). The comparable chloroplast complexes are **TOC** (translocase of the outer chloroplast membrane) and **TIC** (translocase of the inner chloroplast membrane). Polypeptides are initially selected for transport into mitochondria or chloroplasts by components of TOM or TOC known as *transit sequence receptors.* After a transit sequence has bound to its receptor, the polypeptide containing this sequence is translocated across the outer membrane through a *pore* in the TOM or TOC complex. If the polypeptide is destined for the interior of the organelle, movement through the TOM or TOC complex is quickly followed by passage through the TIM or TIC complex of the inner membrane, presumably at a *contact site* where the outer and inner membranes lie close together.

Evidence supporting this model has come both from electron microscopy, which reveals many sites of close contact between outer and inner membranes, and from

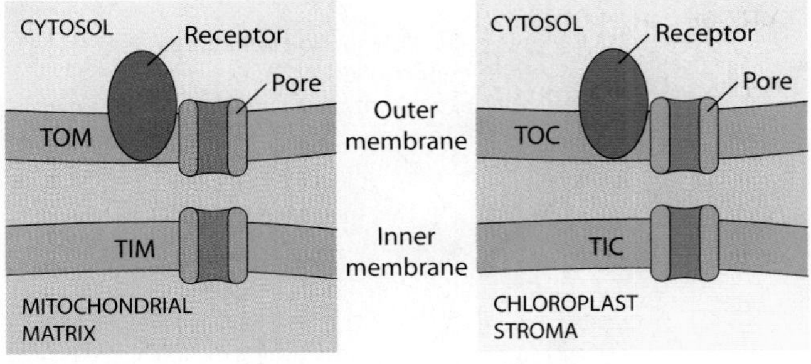

FIGURE 22-18 Polypeptide Transport Complexes of the Outer and Inner Mitochondrial and Chloroplast Membranes. Mitochondrial and chloroplast polypeptides synthesized in the cytosol are transported into these organelles by specialized transport complexes located in their outer and inner membranes. In mitochondria, the outer and inner membrane complexes are called TOM and TIM, respectively. The comparable chloroplast structures are TOC and TIC. The transport complexes of the outer membranes (TOM and TOC) consist of two types of components: receptor proteins that recognize and bind to polypeptides targeted for uptake and pore proteins that form channels through which the polypeptides are translocated.

biochemical experiments, in which cell-free mitochondrial import systems are incubated on ice to trap polypeptides in the act of being translocated (**Figure 22-19**). The low temperature causes polypeptide movement across the membranes to halt shortly after it starts. At this point the polypeptides have already had their transit sequences removed by the transit peptidase enzyme located in the mitochondrial matrix, but they can still be attacked by externally added proteolytic enzymes. Such results indicate that polypeptides can transiently span both membranes during import. In other words, the N-terminus can enter the matrix of the mitochondrion while the rest of the molecule is still outside the organelle.

Polypeptides entering mitochondria and chloroplasts must generally be in an unfolded state before they can pass across the membranes bounding these organelles. This requirement has been demonstrated experimentally by attaching polypeptides containing a mitochondrial transit sequence to agents that maintain the polypeptide chain in a tightly folded state. Such polypeptides bind to the outer surface of mitochondria but will not move across the membrane, apparently because the size of the folded polypeptide exceeds the diameter of the membrane pore it must pass through.

To maintain the necessary unfolded state, polypeptides targeted for mitochondria and chloroplasts are usually bound to *chaperone proteins* similar to those that help newly synthesized polypeptides fold correctly. **Figure 22-20** shows a current model for this chaperone-mediated import of polypeptides into the mitochondrial matrix. To start the process, chaperones of the Hsp70 class bind to a newly forming polypeptide that is still in the process of being synthesized in the cytosol, keeping it in a loosely folded state (step ❶). Next, the transit sequence at the N-terminus of the

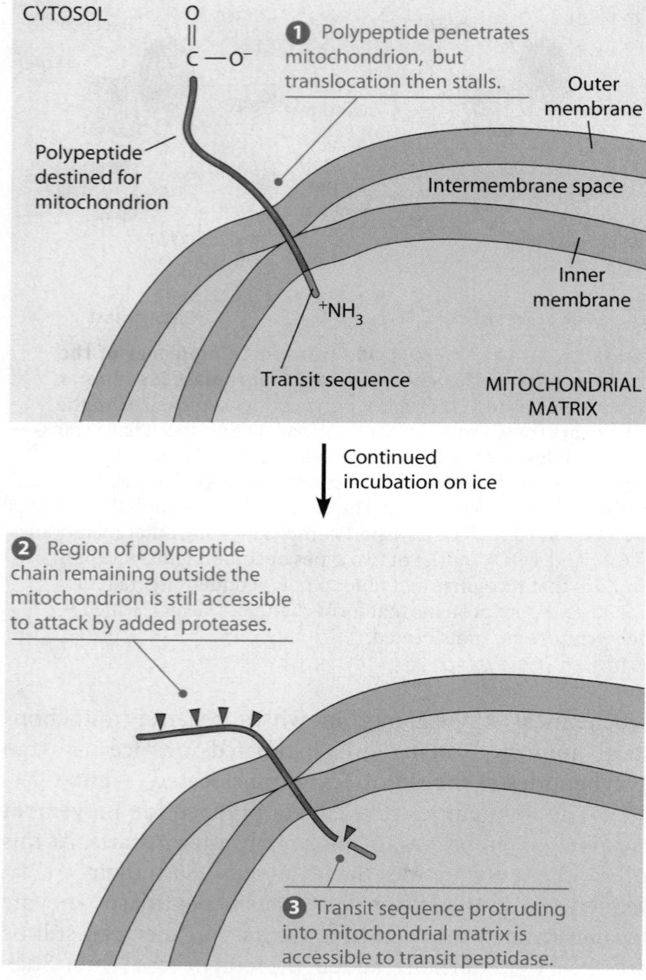
CYTOSOL
C—O⁻
1 Polypeptide penetrates mitochondrion, but translocation then stalls.
Outer membrane
Polypeptide destined for mitochondrion
Intermembrane space
Inner membrane
$^{+}NH_3$
Transit sequence
MITOCHONDRIAL MATRIX
Continued incubation on ice
2 Region of polypeptide chain remaining outside the mitochondrion is still accessible to attack by added proteases.
3 Transit sequence protruding into mitochondrial matrix is accessible to transit peptidase.

FIGURE 22-19 Experiment Showing That Polypeptides Span Both Mitochondrial Membranes During Import. To demonstrate that polypeptides being imported into mitochondria span both membranes at the same time, a cell-free import system was incubated on ice instead of at the usual temperature of 37°C. At low temperature, polypeptides start to enter the mitochondrion, but their translocation soon stalls. Under these conditions, the transit sequence is cleaved by transit peptidase in the matrix, indicating that the N-terminus of the polypeptide is inside the mitochondrion. At the same time, most of the polypeptide chain is accessible to attack by proteolytic enzymes added to the outside of the mitochondrion. This means that the polypeptide must span both membranes during import, presumably at a contact site between the two membranes.

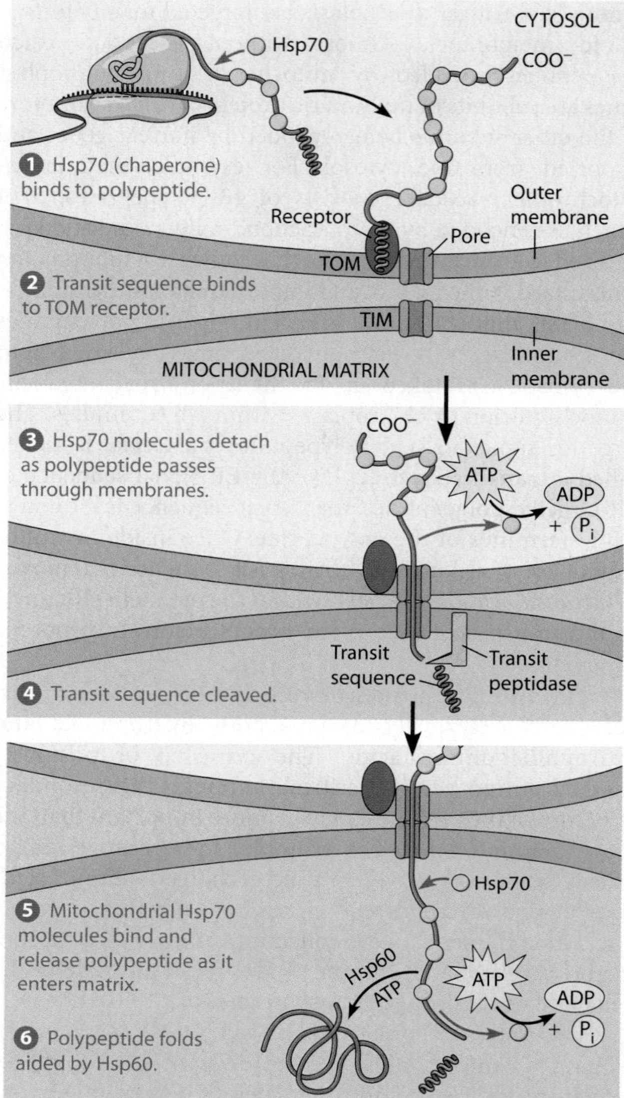
CYTOSOL
Hsp70
COO⁻
1 Hsp70 (chaperone) binds to polypeptide.
Outer membrane
Receptor
Pore
TOM
2 Transit sequence binds to TOM receptor.
TIM
Inner membrane
MITOCHONDRIAL MATRIX
COO⁻
3 Hsp70 molecules detach as polypeptide passes through membranes.
ATP
ADP
+ P_i
Transit sequence
Transit peptidase
4 Transit sequence cleaved.
Hsp70
5 Mitochondrial Hsp70 molecules bind and release polypeptide as it enters matrix.
Hsp60
ATP
ATP
ADP
+ P_i
6 Polypeptide folds aided by Hsp60.

FIGURE 22-20 Posttranslational Import of Polypeptides into the Mitochondrion. Like cotranslational import into the ER, posttranslational import into a mitochondrion involves a signal sequence (called a transit sequence in this case), a membrane receptor, pore-forming membrane proteins, and a peptidase. However, in the mitochondrion, the membrane receptor recognizes the signal sequence directly, without the intervention of a cytosolic SRP. Furthermore, chaperone proteins play several crucial roles in the mitochondrial process: They keep the polypeptide partially unfolded after synthesis in the cytosol so that binding of the transit sequence and translocation can occur (steps ❶–❸); they drive the translocation itself by binding to and releasing from the polypeptide *within* the matrix, which is an ATP-requiring process (step ❺); and they help the polypeptide fold into its final conformation (step ❻). The chaperones included here are cytosolic and mitochondrial versions of Hsp70 (light blue) and a mitochondrial Hsp60 (not illustrated).

polypeptide binds to the receptor component of TOM, which protrudes from the surface of the outer mitochondrial membrane (❷). The chaperone proteins then are released, accompanied by ATP hydrolysis, as the polypeptide is translocated through the TOM and TIM pores and into the mitochondrial matrix (❸). When the transit sequence emerges into the matrix, it is removed by transit peptidase (❹). As the rest of the polypeptide subsequently enters the matrix, *mitochondrial* Hsp70 molecules bind to it temporarily. The subsequent release of Hsp70 requires ATP hydrolysis (❺), which is thought to drive the translocation process. Finally, in many cases mitochondrial Hsp60 chaperone molecules bind to the polypeptide and help it achieve its fully folded conformation (❻).

Both chloroplasts and mitochondria require energy for the import of polypeptides. Mitochondrial import is driven both by ATP hydrolysis and by the electrochemical gradient across the inner membrane. The electrochemical gradient seems to be necessary only for the binding and penetration of the transit sequence. Once this step has occurred, experimental abolition of the membrane potential does not interfere with the rest of the transfer process. Chloroplasts, on the other hand, maintain an electrochemical gradient across the thylakoid membrane but not across the inner membrane. Presumably the energy requirement for import into the chloroplast stroma is met by ATP alone.

Targeting Polypeptides to the Proper Compartments Within Mitochondria and Chloroplasts. Due to the structural complexity of mitochondria and chloroplasts, proteins to be imported from the cytosol must be targeted not only to the right organelle but also to the appropriate compartment within the organelle. Mitochondria have four compartments: the outer membrane, the intermembrane space, the inner membrane, and the matrix. Chloroplasts have four similar compartments (with the stroma substituted for matrix) plus two additional compartments: the thylakoid membrane and the thylakoid lumen. Thus, a polypeptide may have to cross one, two, or even three membranes to reach its final destination.

Given the structural complexity of both organelles, it is perhaps not surprising that many mitochondrial and chloroplast polypeptides require more than one signal to arrive at their proper destinations. For example, targeting a polypeptide to the outer or inner mitochondrial membrane requires an N-terminal transit sequence to direct the polypeptide to the mitochondrion, plus an additional internal sequence, called a *hydrophobic sorting signal,* to target the polypeptide to its final destination. In such cases the hydrophobic sorting signal acts as a stop-transfer sequence that halts translocation of the polypeptide through either the outer or inner membrane translocase and promotes its lateral movement out through a side opening in the translocase and into the lipid bilayer of the membrane. The hydrophobic signal sequence then remains embedded in the membrane, anchoring the polypeptide to the lipid bilayer, while the N-terminal transit sequence is usually removed. A combination of transit and hydrophobic sorting sequences is also used for targeting polypeptides to the intermembrane space. In this case the polypeptide passes through the outer membrane and the signal sequences are then removed, leaving the polypeptide in the space between the two membranes.

Multiple signals are also involved in directing some chloroplast polypeptides to their final destination. Polypeptides intended for insertion into (or transport across) the thylakoid membrane, for example, must first be targeted to the chloroplast and transported into the stroma, presumably crossing the inner and outer membranes at a contact site. In the stroma, the transit sequence used for this first step is cleaved from the polypeptide, unmasking a hydrophobic *thylakoid signal sequence* that targets the polypeptide for either the thylakoid membrane or thylakoid lumen. For polypeptides destined for the thylakoid membrane, the hydrophobic signal sequence may spontaneously insert and anchor the polypeptide within the lipid bilayer of the thylakoid membrane. Alternatively, the insertion of some polypeptides into thylakoid membranes requires the participation of a GTP-dependent protein resembling the signal recognition particle (SRP) that directs and binds polypeptides to ER membranes.

Polypeptides destined for the thylakoid lumen are translocated completely across the thylakoid membrane, accompanied by cleavage of the thylakoid signal sequence as the polypeptide is released into the lumen. Most polypeptides targeted to the lumen are translocated across the thylakoid membrane in an unfolded state by an ATP-dependent process that resembles the mechanism employed for translocating polypeptides across the outer chloroplast or mitochondrial membranes. However, some extensively folded proteins can also be transported across thylakoid membranes using an alternative mechanism driven by energy derived from the proton gradient. Although translocation of extensively folded proteins across membranes is relatively unusual, similar mechanisms have been detected in peroxisomes and in the bacterial plasma membrane.

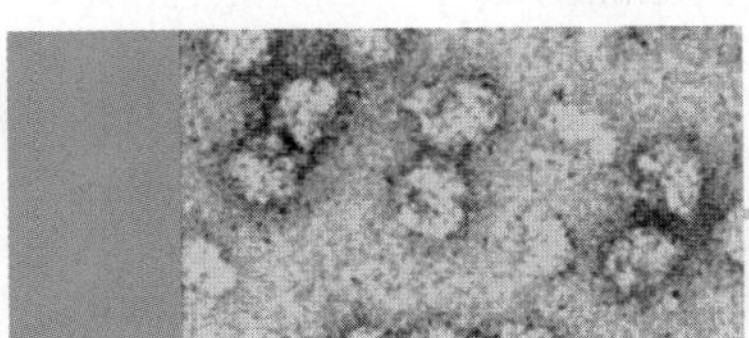

SUMMARY OF KEY POINTS

Translation: The Cast of Characters

- Translation refers to the synthesis of polypeptide chains on ribosomes using a process that employs mRNA to determine the amino acid sequence.
- The rRNA component of the ribosome helps position the mRNA and catalyzes peptide bond formation; aminoacyl-tRNA synthetases link amino acids to the tRNA molecules that bring amino acids to the ribosome; and various protein factors trigger specific events associated with the translation cycle.

The Mechanism of Translation

- Translation involves initiation, elongation, and termination stages.
- During the initiation stage, initiation factors trigger the assembly of mRNA, ribosomal subunits, and initiator aminoacyl tRNA into an initiation complex.
- Chain elongation involves sequential cycles of aminoacyl tRNA binding, peptide bond formation, and translocation,

with each cycle driven by the action of elongation factors. The net result is that aminoacyl tRNAs add their amino acids to the growing polypeptide chain in an order specified by the codon sequence in mRNA.

- Chain termination occurs when a stop codon in mRNA is recognized by release factors, which cause the mRNA and newly formed polypeptide to be released from the ribosome.
- GTP binding and hydrolysis are required for the action of several initiation, elongation, and release factors.
- Proper folding of newly produced polypeptide chains is assisted by molecular chaperones. Abnormalities in protein folding can lead to various health problems, including Alzheimer disease and mad cow disease.

Mutations and Translation

- Nonsense mutations, which change an amino acid codon to a stop codon, can be overcome by suppressor mutations that allow a tRNA anticodon to read the stop codon as an amino acid.
- Defective mRNAs containing premature stop codons are destroyed by nonsense-mediated decay; defective mRNAs containing no stop codon are destroyed by nonstop decay.

Posttranslational Processing

- Newly made polypeptide chains often require chemical modification before they can function properly. Such modifications include cleavage of peptide bonds, phosphorylation, acetylation, methylation, glycosylation, and protein splicing.

Protein Targeting and Sorting

- Many polypeptides possess special amino acid sequences that target them to their appropriate location.
- Polypeptides destined for the endomembrane system, or for secretion from the cell, have an N-terminal ER signal sequence that causes them to enter translocon channels in the ER membrane while the polypeptide chain is still being synthesized.
- Some of these polypeptides pass completely through the translocon and are released into the ER lumen. Others possess one or more internal stop-transfer sequences that cause the polypeptide to remain anchored to the membrane. In either case the resulting proteins may stay in the ER or be transported to other locations within the endomembrane system, such as the Golgi complex, lysosomes, plasma membrane, or secretory vesicles that expel the protein from the cell.
- The unfolded protein response (UPR) and ER-associated degradation (ERAD) help prevent the endomembrane system from accumulating unfolded proteins.
- Polypeptides destined for the nuclear interior, mitochondria, chloroplasts, or peroxisomes are synthesized on cytosolic ribosomes (as are polypeptides that remain in the cytosol) and are then imported posttranslationally into the targeted organelle. Polypeptides destined for peroxisomes contain a special targeting sequence near the C-terminus, whereas those targeted to the nucleus contain nuclear localization signals that promote their uptake through the nuclear pores.
- Some mitochondrial and chloroplast polypeptides require more than one signal to arrive at their proper destinations. Such polypeptides usually possess an N-terminal transit sequence to direct the polypeptide into mitochondria or chloroplasts plus a hydrophobic sorting signal to target the polypeptide to its final destination within each organelle.

PROBLEM SET

More challenging problems are marked with a •.

22-1 The Genetic Code and Two Human Hormones. The following is the actual sequence of a small stretch of human DNA:

3′ AATTATACACGATGAAGCTTGTGACAGGGTTTCCAATCATTAA 5′
5′ TTAATATGTGCTACTTCGAACACTGTCCCAAAGGTTAGTAATT 3′

(a) What are the two possible RNA molecules that could be transcribed from this DNA?

(b) Only one of these two RNA molecules can actually be translated. Explain why.

(c) The RNA molecule that can be translated is the mRNA for the hormone vasopressin. What is the apparent amino acid sequence for vasopressin? (The genetic code is given in Figure 21-6.)

(d) In its active form, vasopressin is a nonapeptide (that is, it has nine amino acids) with cysteine at the N-terminus. How can you explain this in light of your answer to part c?

(e) A related hormone, oxytocin, has the following amino acid sequence:

Cys-Tyr-Ile-Glu-Asp-Cys-Pro-Leu-Gly

Where and how would you change the DNA that codes for vasopressin so that it would code for oxytocin instead? Does your answer suggest a possible evolutionary relationship between the genes for vasopressin and oxytocin?

22-2 Tracking a Series of Mutations. The following diagram shows the amino acids that result from mutations in the codon for a particular amino acid in a bacterial polypeptide:

Arg → Ile
Arg → Lys → Thr → Ser
Lys → Stop

Assume that each arrow denotes a single base-pair substitution in the bacterial DNA.

(a) Referring to the genetic code table in Figure 21-6, determine the most likely codons for each of the amino acids and the stop signal in the diagram.

(b) Starting with a population of mutant cells carrying the nonsense mutation, another mutant is isolated in which the premature stop signal is suppressed. Assuming wobble does not occur and assuming a single base change in the tRNA anticodon, what are all the possible amino acids that might be found in this mutant at the amino acid position in question?

• **22-3 Sleuthing Using Mutants.** You identify three independent missense mutations that all affect the same gene. In fact, all three mutations affect the same codon but do not cause the same amino acid substitution. The first mutation results in substitution of arginine (Arg), the second results in substitution of tyrosine (Tyr), and the third results in substitution of glutamine (Gln). Each mutation affects just a single base within this codon *but not necessarily the same base.* What was the original amino acid?

22-4 Initiation of Translation. Figure 22-8 diagrams the initiation of translation in bacterial cells. Using the text on pages 687–688 as a guide, draw a similar sketch outlining the steps in *eukaryotic* initiation of translation. What are the main differences between bacterial and eukaryotic initiation?

22-5 Bacterial and Eukaryotic Protein Synthesis Compared. For each of the following statements, indicate whether it applies to protein synthesis in bacteria (Ba), in eukaryotes (E), in both (Bo), or in neither (N).

(a) ATP hydrolysis is required to attach an amino acid to a tRNA molecule.

(b) Translation is terminated by special tRNA molecules that recognize stop codons.

(c) The mRNA has a ribosome-binding site within its leader region.

(d) The specificity required to link the right amino acids to the right tRNA molecules is a property of the enzymes called aminoacyl-tRNA synthetases.

(e) GTP hydrolysis functions to induce conformational changes in various proteins involved in polypeptide elongation.

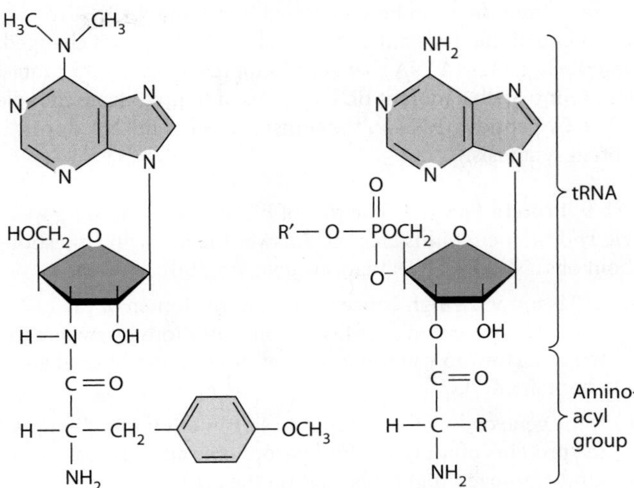

FIGURE 22-21 The Structure of Puromycin. See Problem 22-6.

(f) The enzyme that catalyzes peptide bond formation is an RNA molecule.

(g) AUG is a start codon.

(h) The C-terminus of the polypeptide is synthesized last.

(i) The mRNA is translated in the 3′ → 5′ direction.

• **22-6 An Antibiotic Inhibitor of Translation.** Puromycin is a powerful inhibitor of protein synthesis (see page 697). It is an analog of the 3′ end of aminoacyl tRNA, as **Figure 22-21** reveals. (R represents the functional group of the amino acid; R′ represents the remainder of the tRNA molecule.) When puromycin is added to a cell-free system containing all the necessary machinery for protein synthesis, incomplete polypeptide chains are released from the ribosomes. Each such chain has puromycin covalently attached to one end.

(a) Explain these results.

(b) To which end of the polypeptide chains would you expect the puromycin to be attached? Explain.

(c) Would you expect puromycin to bind to the A or P site on the ribosome or to both? Explain.

(d) Assuming that it can penetrate into the cell equally well in both cases, would you expect puromycin to be a better inhibitor of protein synthesis in eukaryotes or bacteria? Explain.

• **22-7 A Fictional Antibiotic.** In a study involving a cell-free protein synthesizing system from *E. coli,* the polyribonucleotide AUGUUUUUUUUUUUU directs the synthesis of the oligopeptide fMet-Phe-Phe-Phe-Phe. In the presence of Rambomycin, a new antibiotic just developed by Macho Pharmaceuticals, only the dipeptide fMet-Phe is made.

(a) What step in polypeptide synthesis does Rambomycin inhibit? Explain your answer.

(b) Will the oligopeptide product be found attached to tRNA at the end of the uninhibited reaction? Will the dipeptide product be found attached to tRNA at the end of the Rambomycin-inhibited reaction? Explain.

22-8 Frameshift Suppression. As discussed in the chapter, a nonsense mutation can be suppressed by a mutant tRNA in which one of the three anticodon nucleotides has been changed. Describe a mutant tRNA that could suppress a *frameshift* mutation. (Hint: Such a mutant tRNA was used to investigate the role played by peptidyl tRNA in the translocation of mRNA during protein synthesis.)

• **22-9 Protein Folding.** The role of BiP in protein folding was briefly described in this chapter. Answer the following questions about observations and situations involving BiP.

(a) BiP is found in high concentration in the lumen of the ER but is not present in significant concentrations elsewhere in the cell. How do you think this condition is established and maintained?

(b) If the gene coding for BiP acquires a mutation that disrupts the protein's binding site for hydrophobic amino acids, what kind of impact might this have on the cell?

• **22-10 Cotranslational Import.** You perform a series of experiments on the synthesis of the pituitary hormone prolactin, which is a single polypeptide chain 199 amino acids long. The mRNA coding for prolactin is translated in a cell-free protein synthesizing system containing ribosomes, amino acids, tRNAs, aminoacyl-tRNA synthetases, ATP, GTP, and the appropriate initiation, elongation, and termination factors. Under these conditions, a polypeptide chain 227 amino acids long is produced.

(a) How might you explain the discrepancy between the normal length of prolactin (199 amino acids) and the length of the polypeptide synthesized in your experiment (227 amino acids)?

(b) You perform a second experiment, in which you add SRP to your cell-free protein-synthesizing system, and find that translation stops after a polypeptide about 70 amino acids long has been produced. How can you explain this result? Can you think of any purpose this phenomenon might serve for the cell?

(c) You perform a third experiment, in which you add both SRP and ER membrane vesicles to your protein-synthesizing system, and find that translation of the prolactin mRNA now produces a polypeptide 199 amino acids long. How can you explain this result? Where would you expect to find this polypeptide?

22-11 Two Types of Posttranslational Import. The mechanism by which proteins synthesized in the cytosol are imported into the mitochondrial matrix is different from the mechanism by which proteins enter the nucleus, yet the two mechanisms do share some features. Indicate whether each of the following statements applies to nuclear import (Nu), mitochondrial import (M), both (B), or neither (N). You may want to review the discussion of nuclear import in Chapter 18 before answering this question (see especially Figure 18-31).

(a) ATP hydrolysis is known to be required for the translocation process.

(b) The signal sequence is recognized and bound by a receptor protein in the organelle's outer membrane.

(c) The pore complex consists of more than two dozen proteins and is large enough to be readily seen with the electron microscope.

(d) The polypeptide to be transported into the organelle has a specific short stretch of amino acids that targets the polypeptide to the organelle.

(e) There is strong evidence for the involvement of chaperone proteins during translocation of the protein.

(f) The signal sequence is always at the polypeptide's N-terminus and is cut off by a peptidase within the organelle.

(g) GTP hydrolysis is known to be required for the translocation process.

(h) The imported protein enters the organelle through some sort of protein pore.

SUGGESTED READING

References of historical importance are marked with a •.

Translation

Blumenthal, T. Operons in eukaryotes. *Brief. Funct. Genomic Proteomic* 3 (2004): 199.

Cech, T. R. The ribosome is a ribozyme. *Science* 289 (2000): 878.

Chang, Y.-F., J. S. Imam, and M. F. Wilkinson. The nonsense-mediated decay RNA surveillance pathway. *Annu. Rev. Biochem.* 76 (2007): 51.

• Frank, J. How the ribosome works. *Amer. Scientist* 86 (1998): 428.

Hoagland, M. Enter transfer RNA. *Nature* 431 (2004): 249.

Isken, O., and L. E. Maquat. Quality control of eukaryotic mRNA: Safeguarding cells from abnormal mRNA function. *Genes Dev.* 21 (2007): 1833.

Moore, S. D., and R. T. Sauer. The tmRNA system for translational surveillance and ribosome rescue. *Annu. Rev. Biochem.* 76 (2007): 101.

Myasnikov, A. G., A. Simonetti, S. Marzi, and B. P. Klaholz. Structure-function insights into prokaryotic and eukaryotic translation initiation. *Curr. Opin. Struct. Biol.* 19 (2009): 300.

Rodnina, M. V., M. Beringer, and W. Wintermeyer. How ribosomes make peptide bonds. *Trends Biochem. Sci.* 32 (2007): 20.

Schmeing, T. M., and V. Ramakrishnan. What recent ribosome structures have revealed about the mechanism of translation. *Nature* 461 (2009): 1234.

Van Noorden, R. Structural biology bags chemistry prize. *Nature* 461 (2009): 860.

Warner, J. R., and P. M. Knopf. The discovery of polyribosomes. *Trends Biochem. Sci.* 27 (2002): 376.

Youngman, E. M., M. E. McDonald, and R. Green. Peptide release on the ribosome: Mechanism and implications for translational control. *Annu. Rev. Microbiol.* 62 (2008): 353.

Zaher, H. S., and R. Green. Fidelity at the molecular level: lessons from protein synthesis. *Cell* 136 (2009): 746.

Protein Folding and Processing

Aguzzi, A., M. Heikenwalder, and M. Polymenidou. Insights into prion strains and neurotoxicity. *Nature Rev. Mol. Cell Biol.* 8 (2007): 552.

Buku, B., J. Weissman, and A. Horwich. Molecular chaperones and protein quality control. *Cell* 125 (2006): 443.

Caughey, B., G. S. Baron, B. Chesebro, and M. Jeffrey. Getting a grip on prions: Oligomers, amyloids, and pathological membrane interactions. *Annu. Rev. Biochem.* 78 (2009): 177.

Paulus, H. Protein splicing and related forms of protein autoprocessing. *Annu. Rev. Biochem.* 69 (2000): 447.

Prusiner, S. B. Detecting mad cow disease. *Sci. Amer.* 291 (July 2004): 86.

Smith, B., ed. Nature insight: Protein misfolding. *Nature* 426 (2003): 883–909.

Wolfe, M. S. Shutting down Alzheimer's. *Sci. Amer.* 294 (May 2006): 72.

Cotranslational Protein Import

Anderson, D., and P. Walter. Blobel's Nobel: A vision validated. *Cell* 99 (1999): 557.

Cross, B. C., I. Sinning, J. Luirink, and S. High. Delivering proteins for export from the cytosol. *Nat. Rev. Mol. Cell Biol.* 10 (2009): 255.

Kramer, G., D. Boehringer, N. Ban, and B. Bukau. The ribosome as a platform for co-translational processing, folding and targeting of newly synthesized proteins. *Nat. Struct. Mol. Biol.* 16 (2009): 589.

Matlin, K. S. The strange case of the signal recognition particle. *Nature Rev. Mol. Cell. Biol.* 3 (2002): 538.

Meusser, B. et al. ERAD: The long road to destruction. *Nature Cell Biol.* 7 (2005): 766.

Ron, D., and P. Walter. Signal integration in the endoplasmic reticulum unfolded protein response. *Nature Rev. Mol. Cell Biol.* 8 (2007): 519.

Schnell, D. J., and D. N. Hebert. Protein translocons: Multifunctional mediators of protein translocation across membranes. *Cell* 112 (2003): 491.

Shental-Bechor, D., S. J. Fleishman, and N. Ben-Tal. Has the code for protein translocation been broken? *Trends Biochem. Sci.* 31 (2006): 192.

Tsai, B., Y. Ye, and T. A. Rapoport. Retro-translocation of proteins from the endoplasmic reticulum into the cytosol. *Nature Rev. Mol. Cell. Biol.* 3 (2002): 246.

Posttranslational Protein Import

Chacinska, A., C. M. Koehler, D. Milenkovic, T. Lithgow, and N. Pfanner. Importing mitochondrial proteins: Machineries and mechanisms. *Cell* 138 (2009): 628.

Dolezal, P. et al. Evolution of the molecular machines for protein import into mitochondria. *Science* 313 (2006): 314.

Kessler, F., and D. Schnell. Chloroplast biogenesis: Diversity and regulation of the protein import apparatus. *Curr. Opin. Cell Biol.* 21 (2009): 494.

Neupert, W., and J. M. Herrmann. Translocation of proteins into mitochondria. *Annu. Rev. Biochem.* 76 (2007): 723.

Soll, J., and E. Schleiff. Protein import into chloroplasts. *Nature Rev. Mol. Cell Biol.* 5 (2004): 198.

Teter, S. A., and D. J. Klionsky. How to get a folded protein across a membrane. *Trends Cell Biol.* 9 (1999): 428.

Wickner, W., and R. Schekman. Protein translocation across biological membranes. *Science* 310 (2005): 1452.

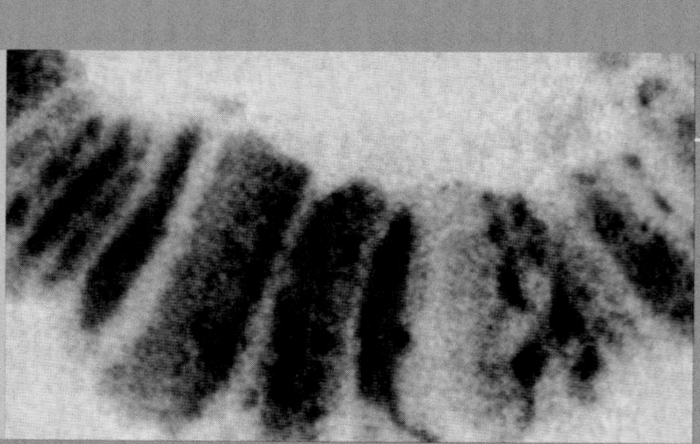

23 The Regulation of Gene Expression

See MasteringBiology® for tutorials, activities, and quizzes.

In our coverage of biological information flow thus far, we have identified DNA as the main repository of genetic information, we have seen how DNA is replicated and repaired, and we have examined the steps involved in expressing DNA's genetic information via transcription and translation. In concluding our consideration of information flow, we will now explore how the various steps in gene expression are regulated so that each gene product is made at the proper time and in proper amounts for each cell's needs.

Regulation is an important aspect of almost every process in nature. This is especially true of gene expression. Most genes are not expressed all the time. In some cases, selective gene expression enables cells to be metabolically thrifty, synthesizing only those gene products that are of immediate use under the prevailing environmental conditions; this is often the situation with bacteria. In multicellular organisms, selective gene expression allows cells to fulfill specialized roles. For example, cells in the skin, nails, and hair produce huge amounts of keratin, whereas red blood cells produce large quantities of hemoglobin. Understanding the sequence of events that leads to such dramatic differences in gene expression between cells requires us to understand in detail how different cell types can begin or cease expressing particular genes as their changing circumstances (and identities) demand.

As you might expect, our first knowledge about the regulation of gene expression came from studying bacteria. But subsequent advances in DNA technology have permitted major progress with eukaryotes as well. We will look first at bacteria and highlight some of the gene control mechanisms discovered in these organisms, and we will then turn to a consideration of eukaryotes.

Bacterial Gene Regulation

Of the several thousand genes present in a typical bacterial cell, some are so important that they are always active in growing cells. Examples of such *constitutive genes* include ribosomal genes and genes encoding the enzymes of glycolysis. For most other genes, however, expression is regulated so that the amount of the final gene product—protein or RNA—is carefully tuned to the cell's need for that product. A number of these *regulated genes* encode enzymes for metabolic processes that, unlike glycolysis, are not constantly required. One way of regulating the intracellular concentrations of such enzymes is by starting and stopping gene transcription in response to cellular needs. Because this control of enzyme-coding genes helps bacterial cells adapt to their environment, it is commonly referred to as *adaptive enzyme synthesis.*

Catabolic and Anabolic Pathways Are Regulated Through Induction and Repression, Respectively

Bacteria use somewhat different approaches for regulating enzyme synthesis, depending on whether a given enzyme is involved in a *catabolic* (degradative) or *anabolic* (synthetic) pathway. The enzymes that catalyze such pathways are often regulated coordinately; that is, the synthesis of all the enzymes involved in a particular pathway is turned on and off together. Here we briefly describe two well-understood pathways, one catabolic and one anabolic.

Catabolic Pathways and Substrate Induction. Catabolic enzymes exist for the primary purpose of degrading specific substrates, often as a means of obtaining energy. **Figure 23-1** depicts the steps in the catabolic pathway that degrades the disaccharide lactose into simple sugars that can then be metabolized by glycolysis. The central step in this pathway is the hydrolysis of lactose into the monosaccharides glucose and galactose, a reaction catalyzed by the enzyme *β-galactosidase.* However, before lactose can be hydrolyzed, it must first be transported into the cell. A protein called *galactoside*

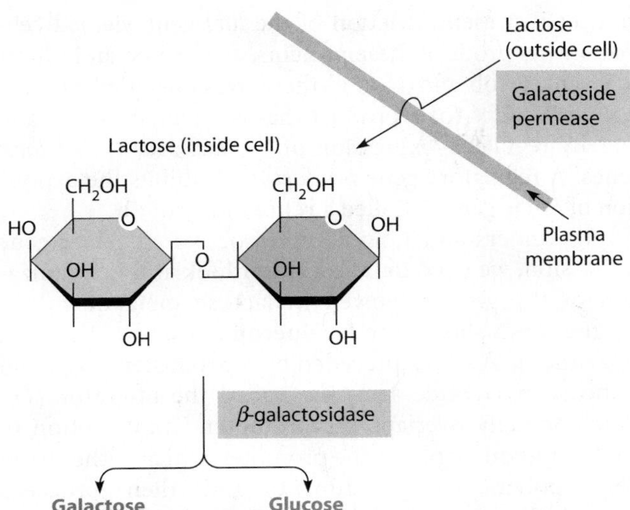

FIGURE 23-1 Lactose Breakdown: A Typical Catabolic Pathway. Breakdown of the disaccharide lactose involves enzymes (purple boxes) whose synthesis is regulated coordinately. See Figures 23-3 and 23-4 for the organization and regulation of the genes that code for these enzymes. (The enzymes responsible for the subsequent catabolism of the monosaccharides glucose and galactose are not part of the same regulatory unit.)

permease is responsible for this transport, and its synthesis is regulated coordinately with β-galactosidase.

Since the function of a catabolic enzyme is to degrade a specific substrate, such enzymes are needed only when the cell is confronted by the relevant substrate. The enzyme β-galactosidase, for example, is useful only when cells have access to lactose; in the absence of lactose, the enzyme is superfluous. Accordingly, it makes sense in terms of cellular economy for the synthesis of β-galactosidase to be turned on, or *induced*, in the presence of lactose but to be turned off in its absence. This turning on of enzyme synthesis is called **substrate induction,** and enzymes whose synthesis is regulated in this way are referred to as **inducible enzymes.** Most catabolic pathways in bacterial cells are subject to substrate induction of their enzymes.

Anabolic Pathways and End-Product Repression. The regulation of anabolic pathways is in a sense just the opposite of that for catabolic pathways. For anabolic pathways, the amount of enzyme produced by a cell usually correlates inversely with the intracellular concentration of the end-product of the pathway. Such a relationship makes sense. For example, as the concentration of tryptophan rises, it is advantageous for the cell to economize on its metabolic resources by reducing its production of the enzymes involved in synthesizing tryptophan. But it is equally important that the cell be able to turn the production of these enzymes back on when the level of tryptophan decreases again. This kind of control is made possible by the ability of the end-product of an anabolic pathway—in our example, tryptophan—to somehow *repress* (reduce or stop) the further production of the enzymes involved in its formation. Such reduction in the expression of the enzyme-coding genes is called **end-product repression.** Most biosynthetic pathways in bacterial cells are regulated in this way.

Repression is a general term in molecular genetics, referring to the reduction in expression of any regulated gene. True genetic repression always has an effect on protein *synthesis*, not just on protein *activity.* Recall from Chapter 6 that the end-products of biosynthetic pathways often have an inhibitory effect on enzyme activity as well. This *feedback inhibition* differs from repression in both mechanism and result. In feedback inhibition, molecules of enzyme are still present, but their catalytic activity is inhibited. In end-product repression, the enzyme molecules are not even made.

Effector Molecules. One feature common to both induction and repression of enzyme synthesis is that control is triggered by small organic molecules present within the cell or in the cell's surroundings. Geneticists call small organic molecules that function in this way **effectors.** For catabolic pathways, effectors are almost always substrates (lactose in our example), and they function as inducers of gene expression and, thus, of enzyme synthesis. For anabolic pathways, effectors are usually end-products (tryptophan in our example), and they usually lead to the repression of gene expression and thus repression of enzyme synthesis.

The Genes Involved in Lactose Catabolism Are Organized into an Inducible Operon

The classic example of an inducible enzyme system occurs in the bacterium *Escherichia coli* and involves a group of enzymes involved in lactose catabolism—the enzymes that catalyze the steps shown in Figure 23-1. Much of what we know about the control of gene expression in bacteria, including the vocabulary used to express that knowledge, is based on the pioneering studies of this system carried out by French molecular geneticists François Jacob and Jacques Monod, who received the Nobel Prize in Physiology or Medicine in 1965 for their work. In 1961, Jacob and Monod published a classic paper that has probably influenced our understanding of gene regulation more than any other work.

In their paper, Jacob and Monod proposed a general model of gene regulation with far-reaching implications. The cornerstone of this model rested on their discovery that the control of lactose catabolism involves two types of genes: the first are genes coding for enzymes involved in lactose uptake and metabolism, and the second is a regulatory gene whose product controls the activity of the first set of genes. The genes involved in lactose metabolism are (1) the *lacZ* gene, which codes for β-galactosidase, the enzyme that hydrolyzes lactose and other β-galactosides; (2) the *lacY* gene, which codes for galactoside permease, the plasma membrane protein that transports lactose into the cell; and (3) the *lacA* gene, which codes for a transacetylase that adds an acetyl group to lactose as it is taken up by the cell. The *lacZ, lacY*, and *lacA* genes lie next to each other in the bacterial chromosome and are expressed only when an inducer such as lactose is present. Taken together, these observations led Jacob and Monod to suggest that the three genes belong to a single regulatory unit, or, as they called it, an **operon**—a group of genes with related functions that are clustered together with DNA sequences that allow the genes to be turned on and off simultaneously.

The organization of functionally related genes into operons is commonly observed in prokaryotes but is less common in eukaryotes. In the nematode *Caenorhabditis elegans*, for example, about 15% of genes are found in operons; however, unlike in bacteria, genes in the operons in *C. elegans* are not always functionally related. Nonetheless, the operon model established several basic principles that have shaped our understanding of transcriptional regulation in both prokaryotic and eukaryotic systems.

The *lac* Operon is Negatively Regulated by the *lac* Repressor

A key feature of the operon model is the idea that genes with metabolically related functions are clustered together so their transcription can be regulated as a single unit. But how is this regulation accomplished? Jacob and Monod addressed this question by studying the ability of inducers such as lactose to turn on the production of the enzymes involved in lactose metabolism. They found that for induction to occur, an additional gene must be present—a regulatory gene that they named *lacI* (for inducibility). Whereas normal bacteria will produce β-galactosidase, galactoside permease, and transacetylase only when an inducer is present, deletion of the *lacI* gene yielded cells that *always* produce these proteins, even when an inducer is absent. Jacob and Monod therefore concluded that the *lacI* gene codes for a product that normally inhibits, and thereby regulates, expression of the *lacZ, lacY*, and *lacA* genes. A regulatory gene *product* that inhibits the expression of other genes is called a **repressor protein.**

To understand how a repressor can regulate gene expression, we need to take a closer look at the organization of the genes involved in lactose metabolism. As **Figure 23-2** shows, the ***lac* operon** consists of the *lacZ, lacY*, and *lacA* genes preceded by a **promoter (P_{lac})** and a special nucleotide sequence called the **operator *(O)***, which actually overlaps the promoter. Transcription of the *lac* operon begins at the promoter, which is the site of RNA polymerase attachment, and then proceeds through the operator and the *lacZ, lacY*, and *lacA* genes until finally ending at a terminator sequence. The net result is a single molecule of mRNA coding for the polypeptide products of all three genes. Such mRNA molecules, which code for more than one polypeptide, are called **polycistronic mRNAs.**

The advantage of clustering related genes into an operon for transcription into a single polycistronic mRNA is that it allows the synthesis of several polypeptides to be controlled in a single step. The crucial step in this control is the interaction between an operator site in the DNA and a repressor protein. The interaction between these elements of the *lac* operon is depicted in **Figure 23-3.** The repressor protein, called the *lac repressor*, is encoded by the *lacI* regulatory gene, which is located outside the operon (although it happens to be located adjacent to the *lac* operon it regulates). The *lac* repressor is a DNA-binding protein that specifically recognizes and binds to the operator site of the *lac* operon. When the repressor is bound to the operator (Figure 23-3a), RNA polymerase is blocked

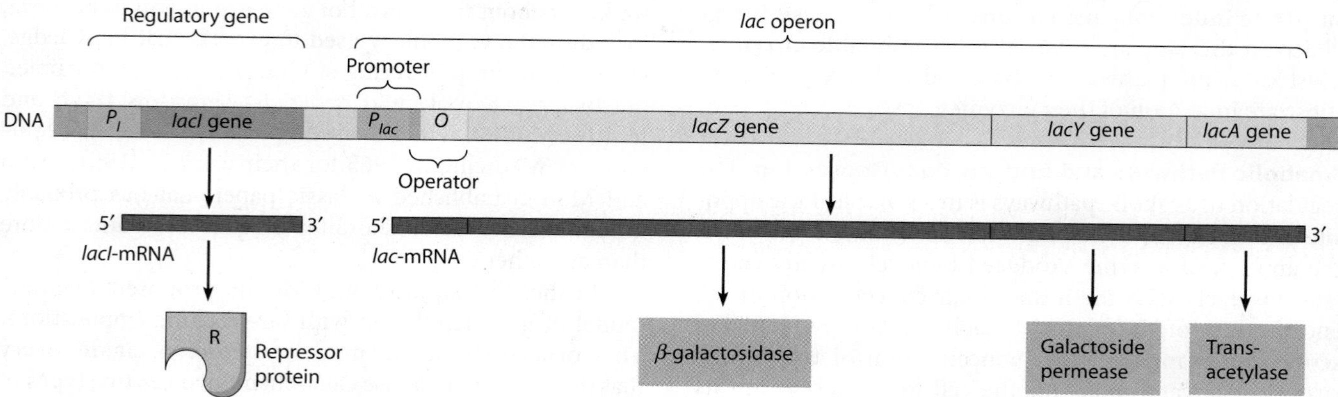

FIGURE 23-2 The Lactose (*lac*) Operon of *E. coli*. The *lac* operon consists of a segment of DNA that includes three contiguous genes (*lacZ, lacY*, and *lacA*), which are transcribed and regulated coordinately. The nearby regulatory gene *lacI* codes for the *lac* repressor protein R. Both the regulatory gene and the *lac* operon itself contain promoters (P_I and P_{lac}, respectively) at which RNA polymerase binds and terminators at which transcription halts. P_{lac} overlaps with the operator site *(O)* to which the active form of the repressor protein binds. The operon is transcribed into a single long molecule of mRNA that codes for all three polypeptides. For details of regulation, see Figure 23-3.

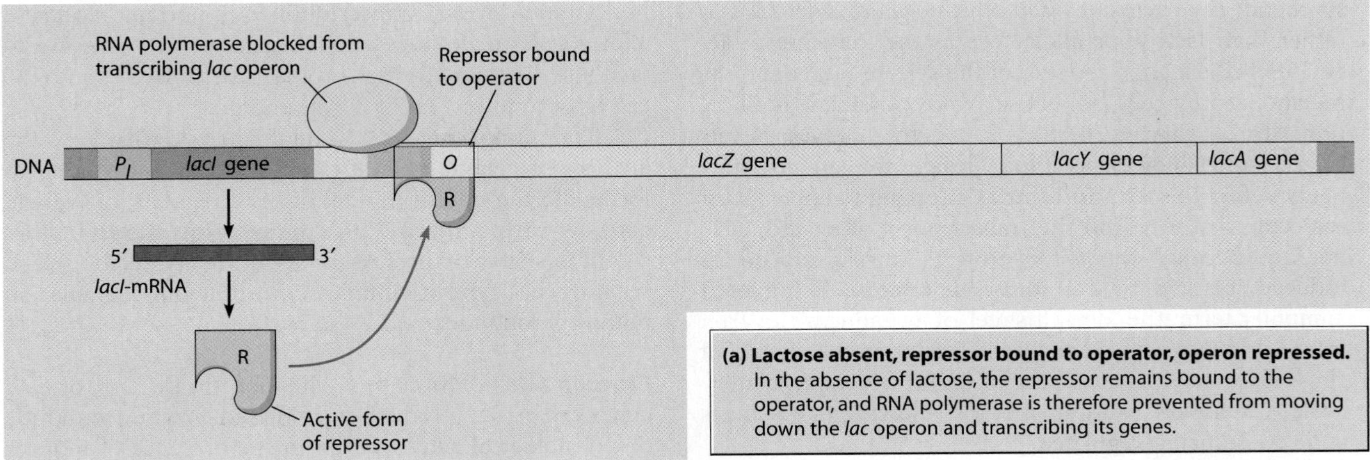
RNA polymerase blocked from transcribing lac operon
Repressor bound to operator
DNA
P_I
lacI gene
O
R
lacZ gene
lacY gene
lacA gene
5′
3′
lacI-mRNA
R
Active form of repressor
(a) Lactose absent, repressor bound to operator, operon repressed.
In the absence of lactose, the repressor remains bound to the operator, and RNA polymerase is therefore prevented from moving down the lac operon and transcribing its genes.

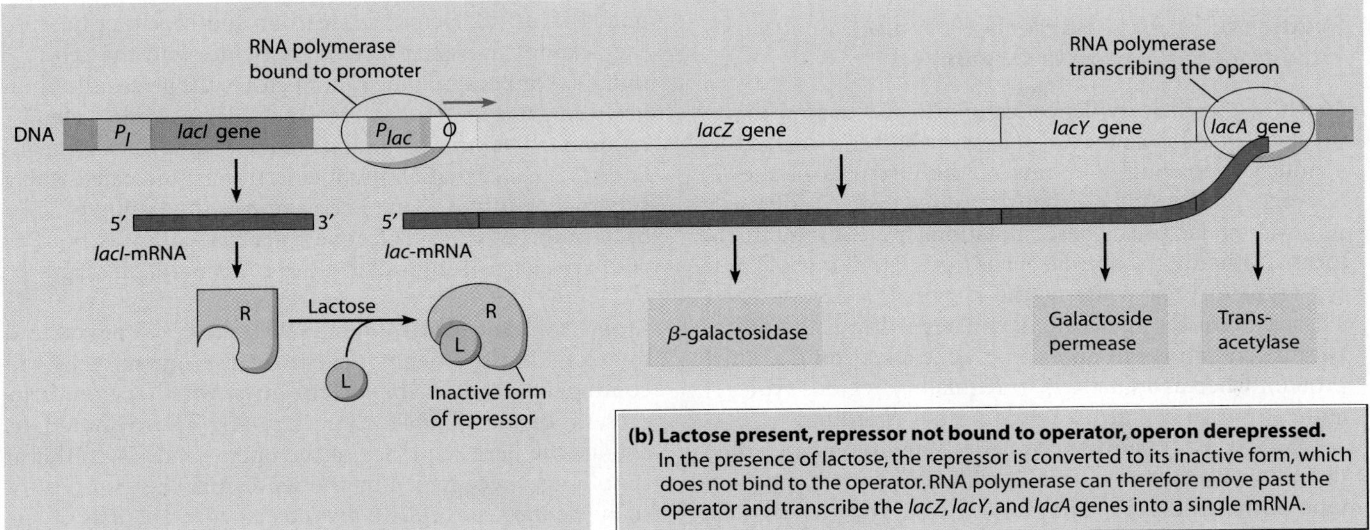
RNA polymerase bound to promoter
RNA polymerase transcribing the operon
DNA
P_I
lacI gene
P_{lac}
O
lacZ gene
lacY gene
lacA gene
5′
3′
5′
lacI-mRNA
lac-mRNA
R
Lactose
R
L
L
Inactive form of repressor
β-galactosidase
Galactoside permease
Trans-acetylase
(b) Lactose present, repressor not bound to operator, operon derepressed.
In the presence of lactose, the repressor is converted to its inactive form, which does not bind to the operator. RNA polymerase can therefore move past the operator and transcribe the lacZ, lacY, and lacA genes into a single mRNA.

MB Activity: The lac Operon in E. coli

FIGURE 23-3 Regulation of the *lac* Operon. **(a)** Transcription of the *lac* operon is regulated by binding of the *lac* repressor (R) to the operator. **(b)** The activity of the *lac* repressor is in turn regulated by its interaction with a form of lactose called allolactose (L).

from moving down the *lac* operon and transcribing the *lacZ, lacY,* and *lacA* genes. In other words, to use common molecular biology shorthand, binding of the repressor to the operator keeps the operon's genes "turned off."

If binding of the repressor to the operator normally blocks transcription, how do cells "turn on" transcription of the *lac* operon? The answer is that inducer molecules bind to the *lac* repressor, thereby altering its conformation so that the repressor loses the ability to bind to the operator site in DNA. Once the operator site is no longer blocked by the repressor, RNA polymerase can bind to the promoter and proceed down the operon, transcribing the *lacZ, lacY,* and *lacA* genes into a single polycistronic mRNA molecule (Figure 23-3b).

A crucial feature of a repressor protein, therefore, is its ability to exist in two forms, only one of which binds to the operator. In other words, a repressor is an *allosteric protein.* As we learned in Chapter 6, an allosteric protein can exist in either of two conformational states, depending on whether or not the appropriate effector molecule is present. In one state the protein is active; in the other state it is inactive, or nearly so. When the effector molecule binds the protein, it induces a change in the conformational state of the protein and therefore in its activity. The binding is readily reversible, however, and departure of the effector results in the protein's rapid return to the alternative form.

Figure 23-3 shows the reversible interaction of the *lac* repressor with its effector, which is actually not lactose itself but *allolactose,* an isomer of lactose produced after lactose enters the cell. In the absence of allolactose, the repressor binds to the *lac* operator and inhibits transcription of the *lac* operon. But the binding of allolactose to the repressor converts the repressor to a conformational form that can no longer bind to the *lac* operator and inhibit transcription. In this way, lactose triggers the *induction* of the enzymes encoded by the *lac* operon. Experiments involving the *lac* operon are usually carried out using the synthetic β-galactoside *isopropylthiogalactoside (IPTG)* rather than lactose or allolactose as the inducing molecule. IPTG is a good inducer of the system but cannot be metabolized by cells, so its use avoids possible complications from varying levels of effector caused by its catabolism. Following common usage, we can now formally define the term **inducer** as referring to any effector molecule that turns on the transcription of an inducible operon. Because the *lac* operon is turned off unless induced, it is said to be an **inducible operon.** To borrow a computer term, the "default state" of an inducible operon is *off.* Although we have focused thus far on the control of single operons, it is common for an individual regulatory protein to act on multiple operons, allowing many genes to be coordinately regulated.

Studies of Mutant Bacteria Revealed How the *lac* Operon Is Organized

Much of the early evidence leading to the operon model involved genetic analyses of mutant bacteria that either produced abnormal amounts of the enzymes of the *lac* operon or showed abnormal responses to the addition or removal of lactose. These mutations were found to be located either in the operon genes (*lacZ, lacY,* or *lacA*) or in the regulatory elements of the system (O, P_{lac}, or *lacI*). These two classes of mutations can be readily distinguished because mutations in an operon gene affect only a single protein, whereas mutations in regulatory regions typically affect expression of all the operon genes coordinately.

Table 23-1 summarizes these mutations and their phenotypes, starting in line 1 with the inducible phenotype of the wild-type (nonmutant) cell. The plus signs in the phenotype columns of the table indicate high enzyme levels. The minus signs indicate very low enzyme levels, although not the complete cessation of enzyme production. Even in the absence of lactose, wild-type cells make small amounts of the *lac* enzymes because the binding of repressor to operator is reversible—the active repressor occasionally "falls off" the operator. The resulting low, background level of transcription is important because it allows cells to produce enough galactoside permease to facilitate the initial transport of lactose molecules into the cell prior to induction of the *lac* operon.

Examining Table 23-1 should help clarify how the analysis of genetic phenotypes led Jacob and Monod to formulate the operon model of gene regulation. We will consider each of the six kinds of mutations shown in lines 2–7 of this table in turn. As we do so, you should be able to see how each type of mutation contributed to formulation of the operon model.

Operon Gene Mutations. Mutations in the *lacY* or *lacZ* genes can lead to production of altered enzymes with little or no biological activity, even in the presence of inducer (Table 23-1, lines 2 and 3). Such mutants are therefore unable to utilize lactose as a carbon source, either because they cannot transport lactose efficiently into the cell (Y^- mutants) or because they cannot cleave the glycosidic bond between galactose and glucose in the lactose molecule (Z^- mutants). Note that mutations in a bacterial gene or regulatory sequence that render it defective are indicated with a superscript minus sign. For example, the genotype of a bacterium carrying a defective *Y* gene is written as Y^-. The wild-type allele is indicated with a superscript plus sign.

Operator Mutations. Mutations in the operator can lead to a constitutive phenotype; that is, the mutant cells will continually produce the *lac* enzymes, whether inducer is present or not (Table 23-1, line 4). These mutations change the base sequence of the operator DNA so that it is no longer recognized by the repressor. The genotype of such *operator-constitutive mutants* is represented as O^c. As is expected for mutations in a regulatory site, O^c mutations simultaneously affect the synthesis of all three *lac* enzymes in the same way.

Promoter Mutations. Promoter mutations can decrease the affinity of RNA polymerase for the promoter. As a result, fewer RNA polymerase molecules bind per unit

Table 23-1 Genetic Analysis of Mutations Affecting the *lac* Operon

		Phenotype with Inducer Absent		Phenotype with Inducer Present	
Line Number	**Genotype of Bacterium***	**β-galactosidase**	**Permease**	**β-galactosidase**	**Permease**
1	$I^+P^+O^+Z^+Y^+$	–	–	+	+
2	$I^+P^+O^+Z^+Y^-$	–	–	+	–**
3	$I^+P^+O^+Z^-Y^+$	–	–	–	+
4	$I^+P^+O^cZ^+Y^+$	+	+	+	+
5	$I^+P^-O^+Z^+Y^+$	–	–	–	–
6	$I^sP^+O^+Z^+Y^+$	–	–	–	–
7	$I^-P^+O^+Z^+Y^+$	+	+	+	+

*$P = P_{lac}$

**The defective permease exhibits sufficient biological activity to transport minimal amounts of lactose into the cell, thereby permitting induction of the *lac* operon.

Table 23-2 Diploid Analysis of Mutations Affecting the *lac* Operon

Line Number	Genotype of Diploid Bacterium*	Phenotype with Inducer Absent		Phenotype with Inducer Present	
		β-galactosidase	Permease	β-galactosidase	Permease
1	$I^+P^+O^+Z^+Y^+/I^+P^+O^+Z^+Y^+$	−	−	+	+
2	$I^+P^+O^+Z^-Y^+/I^+P^+O^+Z^+Y^-$	−	−	+	+
3	$I^+P^+O^+Z^-Y^+/I^-P^+O^+Z^+Y^-$	−	−	+	+
4	$I^+P^+O^+Z^-Y^+/I^+P^+O^cZ^+Y^-$	+	−	+	+
5	$I^+P^+O^+Z^-Y^+/I^sP^+O^+Z^+Y^-$	−	−	−	−

*$P = P_{lac}$

time to the promoter, and the rate of mRNA production decreases. (However, once an RNA polymerase molecule attaches to the DNA and begins transcription, it elongates the mRNA molecule at a normal rate.) Usually, promoter mutations (P^-) in the *lac* operon decrease both the elevated level of enzyme produced in the presence of inducer *and* the already low, basal level of *lac* enzyme production the cell manages to achieve in the absence of inducer (Table 23-1, line 5).

MB Regulation of Expression of the Lac Operon Genes

Regulatory Gene Mutations. Mutations in the *lacI* gene are of two types. Some mutants fail to produce any of the *lac* enzymes, regardless of whether inducer is present, and are therefore called *superrepressor mutants* (I^s in Table 23-1, line 6). Either the repressor molecule in such mutants has lost its ability to recognize and bind the inducer but can still recognize the operator or else it has a high affinity for the operator regardless of whether inducer is bound to it. In either case, the repressor binds tightly to the operator and represses transcription, and hence enzyme synthesis, under all conditions.

The other class of *lacI* mutations involves synthesis of a mutant repressor protein that does not recognize the operator (or, in some cases, is not synthesized at all). The *lac* operon in such I^- mutants cannot be turned off, and the enzymes are therefore synthesized constitutively (Table 23-1, line 7).

I^- mutants, along with the O^c and P^- mutants, illustrate the importance of specific recognition of DNA base sequences by regulatory proteins during the control of gene transcription. A small change—either in the DNA sequence of promoter or operator, or in the regulatory protein that binds to the operator—can dramatically affect expression of all the genes in the operon.

The *Cis-Trans* Test Using Partially Diploid Bacteria. The existence of two different kinds of constitutive mutants, O^c and I^-, raises the question of how one type might be distinguished from the other. A ***cis-trans* test** is often used to differentiate between *cis*-acting mutations, which affect DNA sites (for example, O^c), and *trans*-acting mutations, which affect proteins (for example, I^-). The basis of the *cis-trans* test is a cell (or organism) that has two different copies of the DNA segment of interest. As you know, *E. coli* is usually haploid, but Jacob and Monod constructed partially diploid bacteria by inserting a second copy of the *lac* portion of the bacterial genome into the F-factor plasmid of F^+ cells (page 622). This second copy could be transferred by conjugation into a host bacterium of any desired *lac* genotype to create partial diploids such as the types listed in **Table 23-2.**

If only one copy of the operon contains an I^- or O^c regulatory mutation, it is possible to determine whether the mutation has an effect on both copies of the operon or only on the copy of the operon where it is located. The mutation is said to act in *cis* if the only genes affected are those physically linked to the mutant locus (*cis* means "on this side," in this case referring to a mutation whose influence is restricted to genes located in the same physical copy of the *lac* operon). In contrast, the mutation is said to act in *trans* if the genes in both copies of the operon are affected (*trans* means "on the other side," in this case referring to the ability of the mutation to somehow affect the other copy of the *lac* operon).

To determine which copy of the *lac* operon is being expressed in any given cell population, Jacob and Monod used partial diploid cells containing one copy of the *lac* operon with a defective *Z* gene and the other copy with a defective *Y* gene. Table 23-2 shows what happens if one of these *lac* operon copies has a defective I^- allele and the other possesses a normal I^+ allele. In such cells, both β-galactosidase and permease are inducible, even though the functional gene for one of these enzymes (Z^+ allele for β-galactosidase) is physically linked to the defective *I* gene (Table 23-2, line 3). We now know that this occurs because the one functional *I* gene present in the cell produces active repressor molecules that diffuse through the cytosol and bind to both operator sites in the absence of lactose. The repressor is therefore said to be a ***trans*-acting factor** (the term *factor* refers to a protein in this context).

Quite different results are obtained with partial diploids containing both the O^+ and O^c alleles (Table 23-2, line 4). In this case, genes linked to the O^c allele are constitutively transcribed (Z^+Y^-), whereas those linked to the wild-type allele are inducible (Z^-Y^+). The *O* locus, in other words, acts in *cis;* it affects the behavior of genes only in the operon it is physically a part of. Such *cis* specificity is characteristic of mutations that affect binding sites on DNA rather than protein products of genes. Like other noncoding DNA sequences involved in the control of gene expression, the *O* site is said to be a ***cis*-acting element** (the term *element* refers to a nucleic acid sequence in this context).

Catabolite Activator Protein (CAP) Positively Regulates the *lac* Operon

Under normal circumstances, we have seen that the *lac* operon is under *negative* regulation: It is normally turned off and is only turned on when lactose is present. But what if *both* lactose and glucose are present? Glucose is the preferred energy source for almost all cells because the enzymes of the glycolytic and tricarboxylic acid (TCA) pathways are usually produced continuously. Indeed, *E. coli* cells grown in the presence of both glucose and lactose use the glucose preferentially and have very low levels of the enzymes encoded by the *lac* operon, despite the presence of the inducer for that operon. Bacterial cells must therefore have a way to guarantee that other carbon sources are used only when glucose is not available. This phenomenon, known as *catabolite repression,* refers to the ability of glucose to inhibit the synthesis of catabolic enzymes produced by inducible bacterial operons. Catabolite repression is an important example of *positive* transcriptional control.

Like the other regulatory mechanisms we have encountered, this preferential use of glucose is made possible by a genetic control mechanism that involves an allosteric regulatory protein and a small effector molecule. The actual effector molecule that controls gene expression is not glucose but a secondary signal that reflects the level of glucose in the cell. This secondary signal is a form of AMP called *cyclic AMP* or *cAMP* (see Figure 14-6). Glucose acts by indirectly inhibiting adenylyl cyclase, the enzyme that catalyzes the synthesis of cAMP from ATP. So the more glucose present, the less cAMP is made.

How does cAMP influence gene expression? Like other effectors, it acts by binding to an allosteric regulatory protein. As shown in **Figure 23-4,** this particular regulatory protein, called **catabolite activator protein (CAP),** or cAMP receptor protein, is an activator protein that turns on transcription. By itself, CAP is nonfunctional. But when complexed with cAMP, CAP can bind to a particular base sequence within operons that produce catabolic enzymes. This base sequence, the *CAP-binding site,* is located upstream of the promoter. Figure 23-4 shows the location of the CAP-binding site (labeled *C*) in the *lac* operon. Similar sites occur in inducible operons involved in the metabolism of many other sugars, such as the sucrose, galactose, and arabinose operons. When CAP, in its active form (that is, the CAP-cAMP complex), attaches to one of its recognition sites in DNA, the binding of RNA polymerase to the promoter is greatly enhanced, thereby stimulating the initiation of transcription. Thus, when glucose is not present, CAP has a *positive* effect on gene expression from the *lac* operon. The CAP-cAMP complex greatly enhances transcription of inducible

1 CAP-cAMP complexes bind to the CAP recognition site (C) near the promoter of the *lac* operon.
cAMP
CAP-cAMP complex
cAMP
CAP (inactive)
C P_{lac} O *lacZ* gene
CAP site
RNA polymerase
2 Promoter becomes more readily bound by RNA polymerase.
C P_{lac} O *lacZ* gene
3 RNA polymerase begins transcribing the *lac* operon.
C P_{lac} O *lacZ* gene
4 RNA polymerase produces *lac* mRNA.
C P_{lac} O *lacZ* gene
5′
lac-mRNA

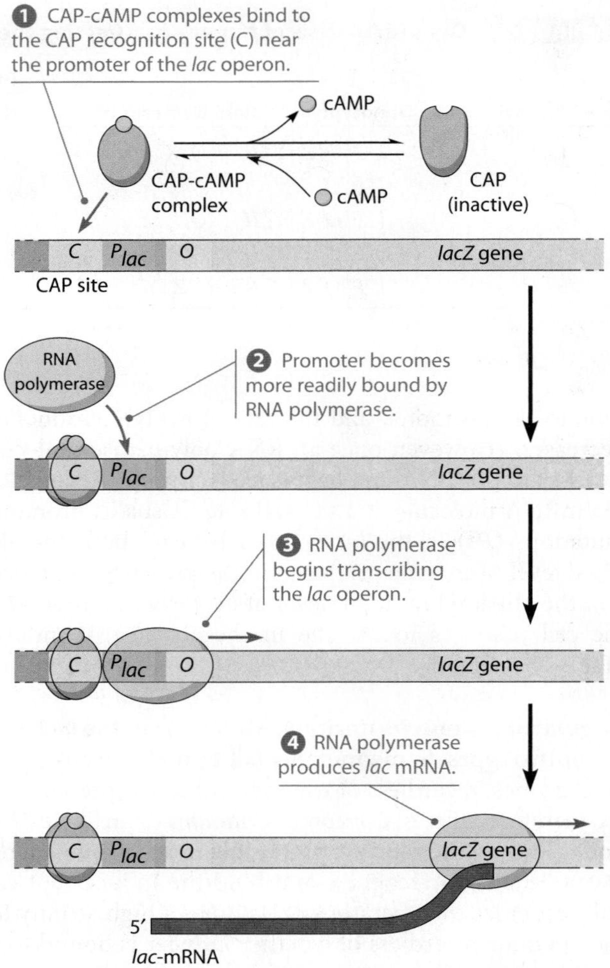

FIGURE 23-4 The Catabolite Activator Protein (CAP) and Its Function. CAP is an allosteric protein that is converted to its active form by binding to cyclic AMP (cAMP), whose concentration is in turn influenced by glucose levels. Two CAP-cAMP complexes mediate catabolite repression by binding to DNA and activating the transcription of various inducible operons, including the *lac* operon shown here.

MB Molecular Model: Catabolite Activator Protein (CAP)

operons. For example, transcription of the *lac* operon can be increased fiftyfold in this way—provided, of course, that the repressor for that operon has been inactivated by the presence of its effector (that is, allolactose).

This situation changes when the glucose concentration inside the cell is high. In this case, the cAMP concentration falls and CAP is largely in its inactive form. Therefore CAP cannot stimulate the transcription of *lac* operon. In this way, cells turn off the synthesis of catabolic enzymes that are not needed when glucose is abundantly available as an energy source.

The *lac* Operon Is an Example of the Dual Control of Gene Expression

MB Activity: The lac Operon

As we have just seen, the *lac* operon is subject to both positive and negative control. Indeed, most tightly regulated genes in prokaryotes and eukaryotes are under such dual control. To keep negative and positive types of transcriptional control clear in your mind, ask about the primary effect of the regulatory protein that binds to the operon DNA. If, in binding to the DNA, the regulatory protein prevents or turns off transcription, then it is part of a negative control mechanism. If, on the other hand, its binding to DNA results in the activation or enhancement of transcription, then the regulatory protein is part of a positive control mechanism.

The Structure of the *lac* Repressor/Operator Complex Confirms the Operon Model

In 1996, Mitchell Lewis, Ponzy Lu, and colleagues reported the structure of the *lac* repressor protein bound to DNA (**Figure 23-5**). This structure confirmed previous genetic and biochemical studies and provided insights into how transcription is regulated in detail. The repressor protein encoded by the *lacI* gene (Figure 23-5a) actually interacts with the operator DNA as a tetramer (i.e., four copies of the protein act together; Figure 23-5b). The repressor binds the primary operator, termed O_1, which is 21 base pairs in length. In addition, there are two other operator DNA sequences, termed O_2 and O_3. When the repressor is bound to the O_1 and O_3 sites, the repressor is thought to cause the DNA to form a loop, which inhibits the movement of RNA polymerase. The loop is thought to be large enough that CAP can bind to DNA even when the DNA is looped (Figure 23-5c). When allolactose binds at the inducer site, the conformation of the protein changes so that it can no longer interact with DNA, the loop relaxes, and RNA polymerase can move along the DNA to transcribe the *lac* operon.

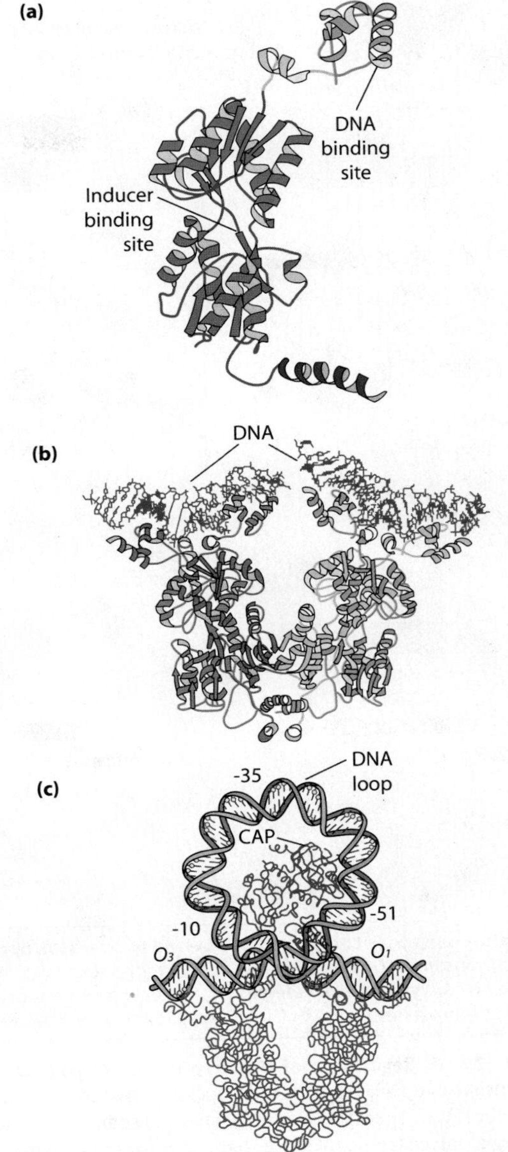

FIGURE 23-5 Structure of the *lac* Repressor/Operator Complex. Models of the *lac* repressor operon bound to operator DNA, based on X-ray crystal structures. **(a)** The repressor monomer, showing the inducer binding site, and the DNA binding site in red. **(b)** The repressor dimer bound to two 21-base-pair segments of operator DNA (light blue). **(c)** The repressor (pink) and CAP (light blue) bound to the *lac* DNA. Binding to operator regions O_1 and O_3 causes a loop to form in the promoter DNA that represses transcription. From M. Lewis, et al., "Crystal Structure of the Lactose Operon Repressor and Its Complexes with DNA and Inducer," *Science* 271 (1 March 1996): 1247-1254. Reprinted with permission from AAAS.

The Genes Involved in Tryptophan Synthesis Are Organized into a Repressible Operon

Although much of the work leading to the initial formulation of the operon concept involved the *lac* operon of *E. coli*, many other bacterial regulatory systems are now known to follow the same general pattern. Operons coding for enzymes involved in catabolic pathways generally resemble the *lac* operon in being inducible; that is, they are turned *on* by a specific allosteric effector, usually the substrate for the pathway involved. In contrast, operons that regulate enzymes involved in anabolic (biosynthetic) pathways are **repressible operons**; they are turned *off* allosterically, usually by an effector that is the end-product of the pathway. The tryptophan *(trp)* operon is a good example of a repressible operon. The ***trp* operon** contains genes coding for enzymes involved in tryptophan biosynthesis, along with DNA sequences that regulate the production of these enzymes. The effector molecule in this case is the end-product of the biosynthetic pathway, the amino acid tryptophan.

Production of the enzymes encoded by the *trp* operon is repressed in the presence of tryptophan and derepressed in its absence (**Figure 23-6a**). The regulatory gene for this operon, called *trpR*, codes for a *trp* repressor protein that—in contrast to the *lac* repressor—is active (binds to operator DNA) when the effector is attached to it and is inactive in its free form. The effector in such systems (in this case, tryptophan) is sometimes referred to as a **corepressor** because it is required, along with the repressor protein, to shut off transcription of the operon.

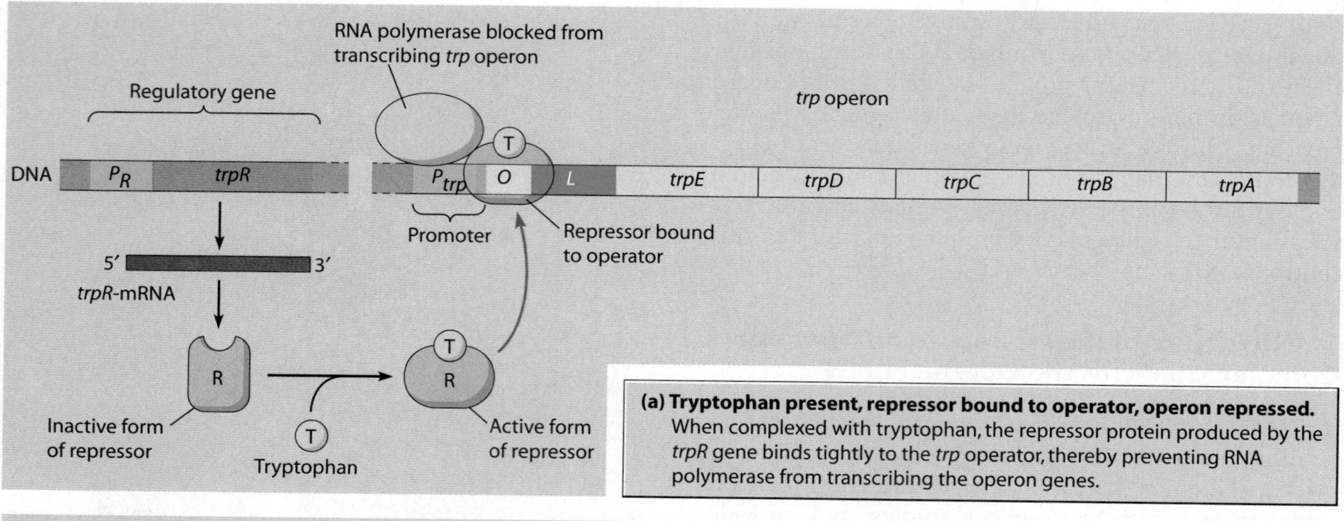

(a) Tryptophan present, repressor bound to operator, operon repressed. When complexed with tryptophan, the repressor protein produced by the *trpR* gene binds tightly to the *trp* operator, thereby preventing RNA polymerase from transcribing the operon genes.

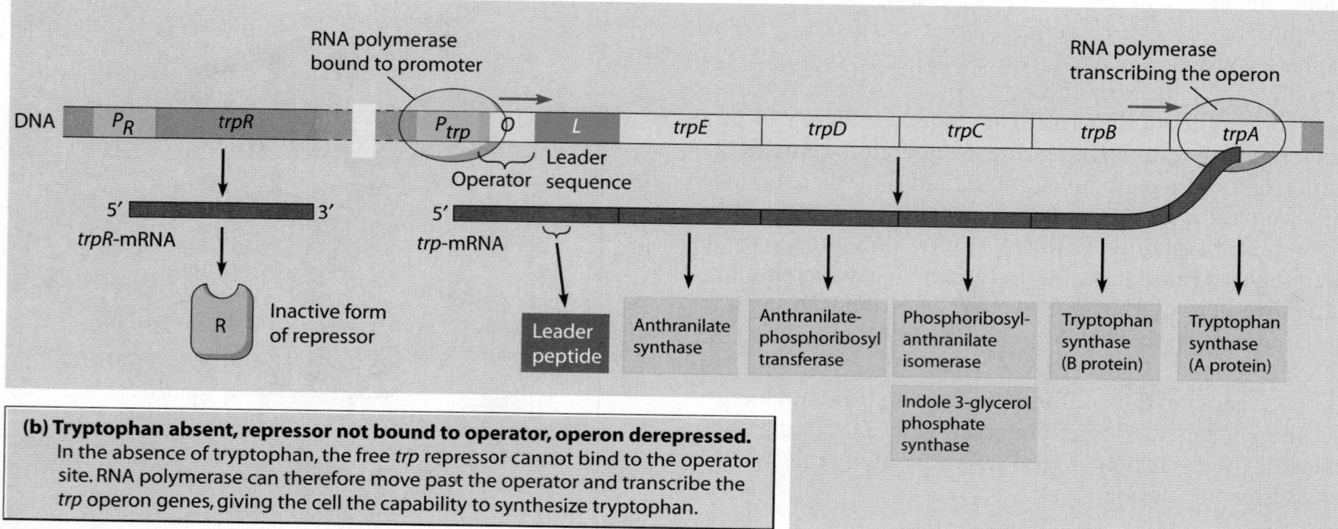

(b) Tryptophan absent, repressor not bound to operator, operon derepressed. In the absence of tryptophan, the free *trp* repressor cannot bind to the operator site. RNA polymerase can therefore move past the operator and transcribe the *trp* operon genes, giving the cell the capability to synthesize tryptophan.

FIGURE 23-6 Regulation of the Tryptophan *(trp)* Operon by the *trp* Repressor. The *trp* operon of *E. coli* consists of a DNA segment that includes five contiguous genes (*trpE, trpD, trpC, trpB,* and *trpA*) as well as promoter (P_{trp}), operator (*O*), and leader (*L*) sequences. The genes are regulated and transcribed as a unit, producing a polycistronic message that codes for the enzymes involved in synthesizing tryptophan. An additional regulatory role played by the leader segment of the mRNA will be explained in Figures 23-7 and 23-8.

Sigma Factors Determine Which Sets of Genes Can Be Expressed

In addition to using regulatory proteins that activate or repress specific operons, bacterial cells also employ different sigma (σ) factors to control which genes will be transcribed. Recall from Chapter 21 that the proper initiation of transcription in bacteria requires the RNA polymerase core enzyme to be combined with a sigma factor, a type of protein that recognizes gene promoter sequences. In *E. coli* the most prevalent sigma factor is σ^{70}, which initiates transcription of genes whose products are required for routine growth and metabolism. However, changes in a cell's environment, such as increases in temperature, UV radiation, or acidity, can trigger the use of alternative sigma factors, such as σ^{S} and σ^{32}. When one of these alternative sigma factors is bound to the RNA polymerase core enzyme, promoter recognition is altered, thereby initiating the transcription of genes encoding proteins that help the cell adapt to the altered environment. Yet another sigma factor, σ^{54}, enables RNA polymerase to preferentially transcribe genes whose products minimize the slowing of growth under nitrogen-limiting conditions.

Significant differences are observed in the numbers and types of sigma factors found in different kinds of bacteria. For example, *E. coli* has about half a dozen sigma factors, whereas the soil bacteria *Bacillus subtilis* and *Streptomyces coelicolor* use closer to 20 and 60 different sigma factors, respectively. In some cases, the sigma factors that regulate bacterial transcription are produced by bacteriophages. Such an arrangement allows a bacteriophage to infect a bacterial cell and take over its transcriptional machinery by producing a specific sigma factor that binds to the bacterial RNA polymerase and causes it to recognize only viral promoters.

Attenuation Allows Transcription to Be Regulated After the Initiation Step

MB Regulation of Gene Expression in Bacteria

The regulatory mechanisms discussed so far control the initiation of transcription. Bacteria also employ regulatory mechanisms that operate after the initiation step. A classic example was first uncovered when Charles Yanofsky and his colleagues found that the *trp* operon of *E. coli* has a novel type of regulatory site located between the promoter/operator and the operon's first gene, *trpE*. This stretch of DNA, called the *leader sequence* (or *L*), is transcribed to produce a leader mRNA segment, 162 nucleotides long, located at the 5′ end of the polycistronic *trp* mRNA (see Figure 23-6b).

Analysis of *trp* operon transcripts made under various conditions revealed that, as expected, the full-length, polycistronic *trp* mRNA is produced when tryptophan is scarce. This allows the enzymes of the tryptophan biosynthetic pathway to be synthesized, and hence the pathway produces more tryptophan. On the other hand, when tryptophan is present, transcription of the genes coding for the enzymes of the tryptophan pathway is inhibited—also as expected. An unexpected result, however, was that the DNA corresponding to most of the leader sequence may still be transcribed under such conditions, even though the full-length, polycistronic *trp* mRNA is not produced. Based on these findings, Yanofsky suggested that the leader sequence contains a control element that is more sensitive than the *trp* repressor mechanism in determining whether enough tryptophan is present to support protein synthesis. This control sequence somehow determines not whether *trp* operon transcription can begin, but whether it will continue to completion. Such control was called **attenuation** because of its role in attenuating, or reducing, the synthesis of mRNA.

To see how attenuation works, let's take a closer look at the leader segment of the *trp* operon mRNA (**Figure 23-7**). This leader has two unusual features that enable it to play a regulatory role. First, in contrast to the untranslated leader sequences typically encountered at the 5′ end of mRNA molecules (see Figure 22-6), a portion of the *trp* leader sequence *is* translated, forming a *leader peptide* 14 amino acids long. Within the mRNA sequence coding for this peptide are two adjacent codons for the amino acid tryptophan; these will prove important. Second, the *trp* leader mRNA also contains four segments (labeled regions 1, 2, 3, and 4) whose nucleotides can base-pair with each other to form several distinctive hairpin loop structures. The region comprising regions 3 and 4 plus an adjacent string of eight U nucleotides is called the *terminator.* When base pairing between regions 3 and 4 creates a hairpin loop, it acts as a transcription termination signal (**Figure 23-8a**). Recall that the typical bacterial termination signal shown in Figure 21-11 is also a hairpin followed by a string of U's. The formation of such a structure causes RNA polymerase and the growing RNA chain to detach from the DNA.

As Yanofsky's experiments suggested, translation of the leader RNA plays a crucial role in the attenuation mechanism. A ribosome attaches to its first binding site on the *trp* mRNA as soon as the site appears, and from there it follows close behind the RNA polymerase. When tryptophan levels are low (Figure 23-8b), the concentration of tryptophanyl-tRNA (tRNA molecules carrying tryptophan) is also low. Thus, when the ribosome arrives at the tryptophan codons of the leader RNA, it stalls briefly, awaiting the arrival of tryptophanyl-tRNA. The stalled ribosome blocks region 1, allowing an alternative hairpin structure to form by pairing regions 2 and 3, called an *antiterminator hairpin.* When region 3 is tied up in this

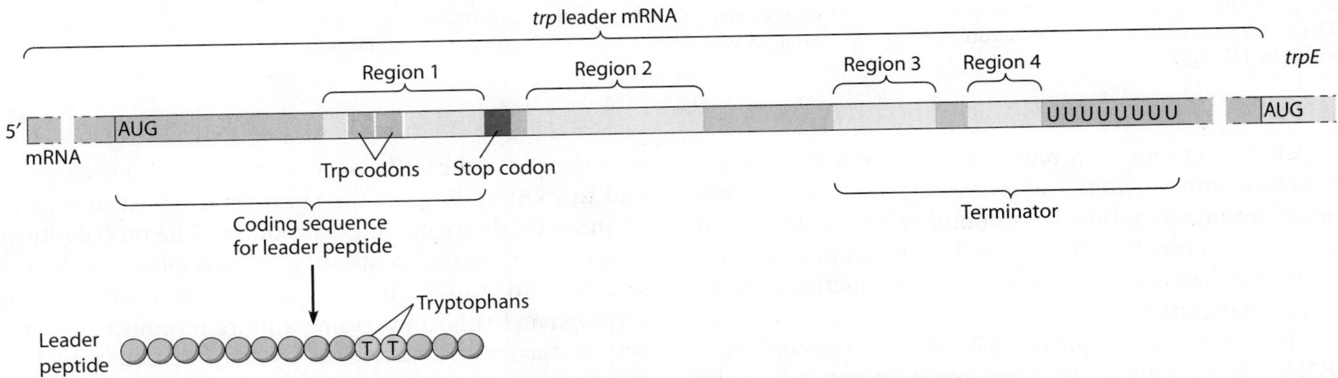

FIGURE 23-7 The *trp* mRNA Leader Sequence. The transcript of the *trp* operon includes a leader sequence of 162 nucleotides situated directly upstream of the start codon for *trpE*, the first operon gene. This leader sequence includes a section encoding a leader peptide of 14 amino acids. Two adjacent tryptophan (Trp) codons within the leader mRNA sequence play an important role in the operon's regulation by attenuation. The leader mRNA also contains four regions capable of base pairing in various combinations to form hairpin structures, as Figure 23-8 shows. A part of the leader mRNA containing regions 3 and 4 and a string of eight U's is called the terminator.

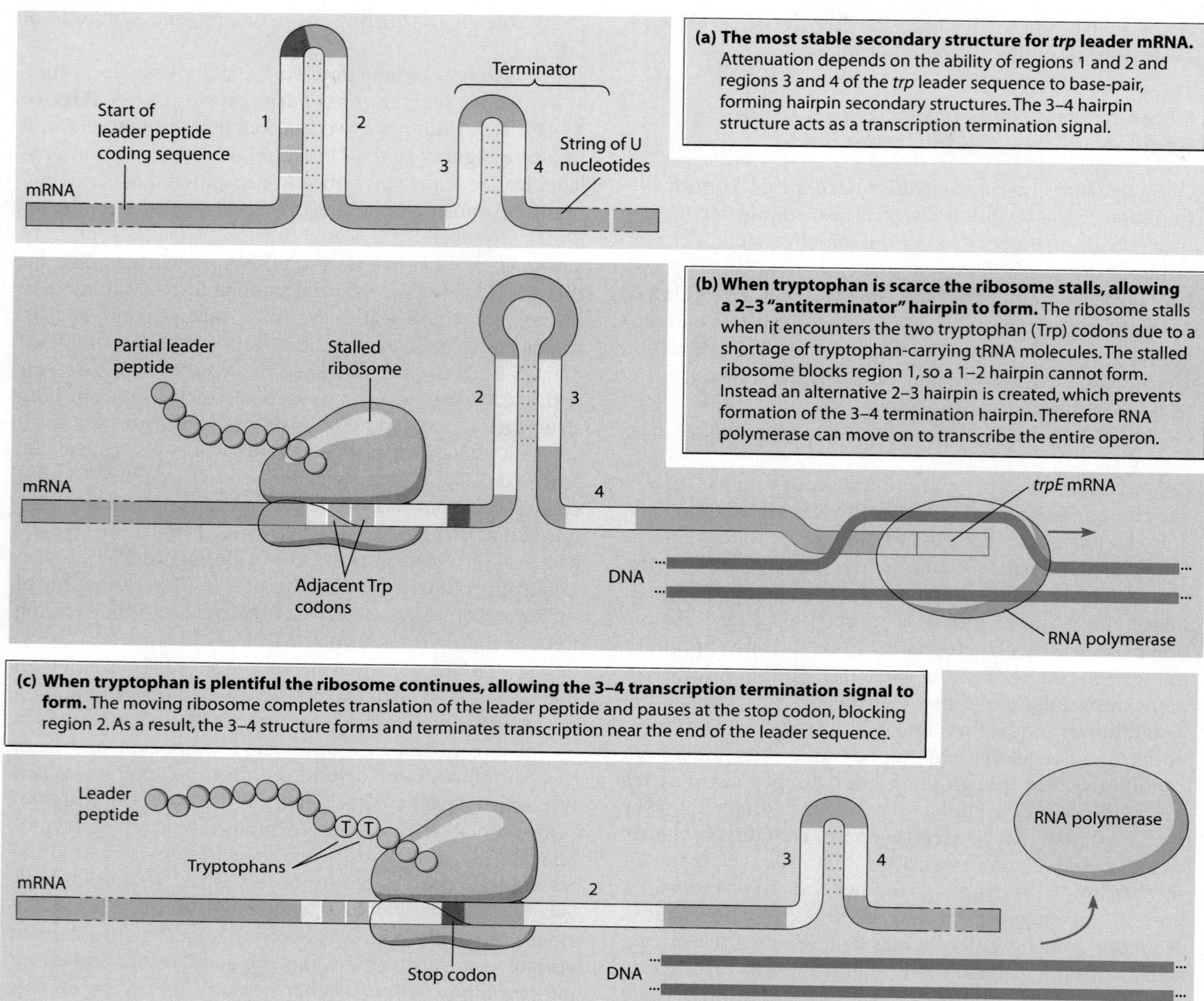

(a) The most stable secondary structure for *trp* leader mRNA. Attenuation depends on the ability of regions 1 and 2 and regions 3 and 4 of the *trp* leader sequence to base-pair, forming hairpin secondary structures. The 3–4 hairpin structure acts as a transcription termination signal.
Start of leader peptide coding sequence
mRNA
1
2
Terminator
3
4
String of U nucleotides
(b) When tryptophan is scarce the ribosome stalls, allowing a 2–3 "antiterminator" hairpin to form. The ribosome stalls when it encounters the two tryptophan (Trp) codons due to a shortage of tryptophan-carrying tRNA molecules. The stalled ribosome blocks region 1, so a 1–2 hairpin cannot form. Instead an alternative 2–3 hairpin is created, which prevents formation of the 3–4 termination hairpin. Therefore RNA polymerase can move on to transcribe the entire operon.
Partial leader peptide
Stalled ribosome
2
3
4
mRNA
trpE mRNA
Adjacent Trp codons
DNA
RNA polymerase
(c) When tryptophan is plentiful the ribosome continues, allowing the 3–4 transcription termination signal to form. The moving ribosome completes translation of the leader peptide and pauses at the stop codon, blocking region 2. As a result, the 3–4 structure forms and terminates transcription near the end of the leader sequence.
Leader peptide
T
T
Tryptophans
mRNA
2
3
4
RNA polymerase
Stop codon
DNA

MB Attenuation in the trp operon of E. coli

FIGURE 23-8 Attenuation in the *trp* Operon. Attenuation depends on the ability of regions 1 and 2 and regions 3 and 4 of the *trp* mRNA leader sequence to base-pair, forming hairpin secondary structures. The 3–4 hairpin acts as a transcription termination signal; as soon as it forms, RNA polymerase is released from the DNA.

way, it cannot pair with region 4 to create a termination structure, and so *transcription by RNA polymerase continues,* eventually producing a complete mRNA transcript of the *trp* operon. Ribosomes use this mRNA to synthesize the tryptophan pathway enzymes, and production of tryptophan therefore increases.

If, however, tryptophan is plentiful and tryptophanyl-tRNA levels are high, the ribosome does not stall at the tryptophan codons (Figure 23-8c). Instead, the ribosome continues to the stop codon at the end of the coding sequence for the leader peptide and pauses there, blocking region 2. This pause permits the formation of the 3–4 hairpin, which is the transcription termination signal. *Transcription by RNA polymerase is therefore terminated* near the end of the leader sequence (after 141 nucleotides), and mRNAs coding for the enzymes involved in tryptophan synthesis are no longer produced. The probability of terminating versus continuing transcription is highly sensitive to small changes in the concentration of tryptophanyl-tRNA, providing a more responsive control system than can be achieved by the interaction of free tryptophan with the *trp* repressor as the only means of regulation.

The preceding mechanism, discovered in *E. coli,* requires that ribosomes begin translating the mRNA before transcription is completed. However, attenuation does not always require this coupling of transcription and translation. In the bacterium *B. subtilis,* the leader RNA transcribed from the *trp* operon is not translated by ribosomes and yet still functions as an attenuator. In this case, formation of the terminator loop is controlled by a tryptophan-binding protein. When tryptophan is present, the protein-tryptophan complex binds to the leader RNA and exposes the terminator loop, thereby terminating transcription and shutting down the *trp* operon.

Although attenuation was once considered to be an unusual type of regulation, it turns out to be relatively common, especially in operons that code for enzymes involved in amino acid biosynthesis. Several mechanisms for achieving attenuation are now known to exist, and in some operons, this may be the only means of regulation. In others, attenuation complements the operator in regulating gene expression, as in the *trp* operon.

Riboswitches Allow Transcription and Translation to Be Controlled by Small Molecule Interactions with RNA

The ability of small molecules to induce shape changes in allosteric proteins plays a central role in regulating gene expression. We have already seen, for example, how gene transcription is controlled by the binding of allolactose to the *lac* repressor protein of tryptophan to the *trp* repressor protein, or of cAMP to a cAMP receptor protein. In each case, mRNA production is altered by the binding of a small molecule to a regulatory protein.

Small molecules can also regulate gene expression by binding to special sites in mRNA called **riboswitches.** Binding of an appropriate small molecule to its corresponding riboswitch triggers changes in mRNA shape that can affect either transcription or translation. Riboswitches are typically found in the untranslated leader region of mRNAs transcribed from bacterial operons, although they have been detected in mRNAs from archaea and eukaryotes as well. One of the first riboswitches to be discovered is in the RNA transcribed from the *riboflavin (rib)* operon of the bacterium *B. subtilis.* The *rib* operon contains five genes coding for enzymes involved in synthesizing the essential coenzymes FMN and FAD. RNA transcribed from the *rib* operon possesses a leader sequence that, like the *trp* leader discussed earlier, can fold into a hairpin loop that terminates transcription. As shown in **Figure 23-9a,** binding of FMN to the leader sequence promotes formation of this hairpin loop. Transcription of the *rib* operon is therefore terminated when FMN is present, halting the production of enzymes that are not necessary because they are involved in synthesizing FMN itself.

Besides regulating transcription, the binding of small molecules to riboswitches can also control mRNA translation. One example occurs in mRNAs coding for enzymes

(a) Transcription termination. Binding of a small molecule to a riboswitch in the leader sequence of some mRNAs triggers the formation of a hairpin loop that terminates transcription. FMN binding to the leader sequence of the mRNA transcribed from the *rib* operon of *B. subtilis* works in this way.

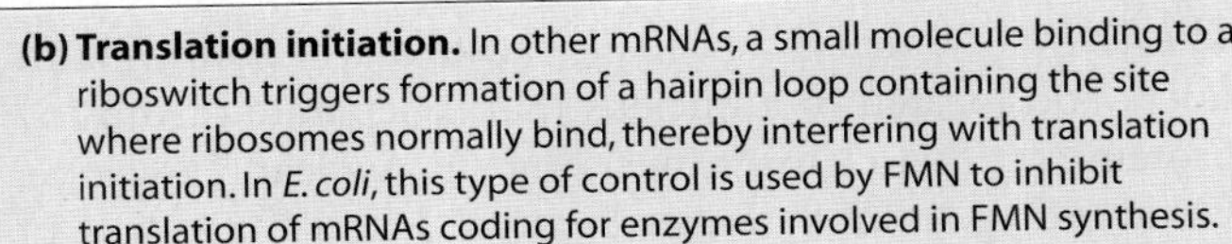
(b) Translation initiation. In other mRNAs, a small molecule binding to a riboswitch triggers formation of a hairpin loop containing the site where ribosomes normally bind, thereby interfering with translation initiation. In *E. coli*, this type of control is used by FMN to inhibit translation of mRNAs coding for enzymes involved in FMN synthesis.

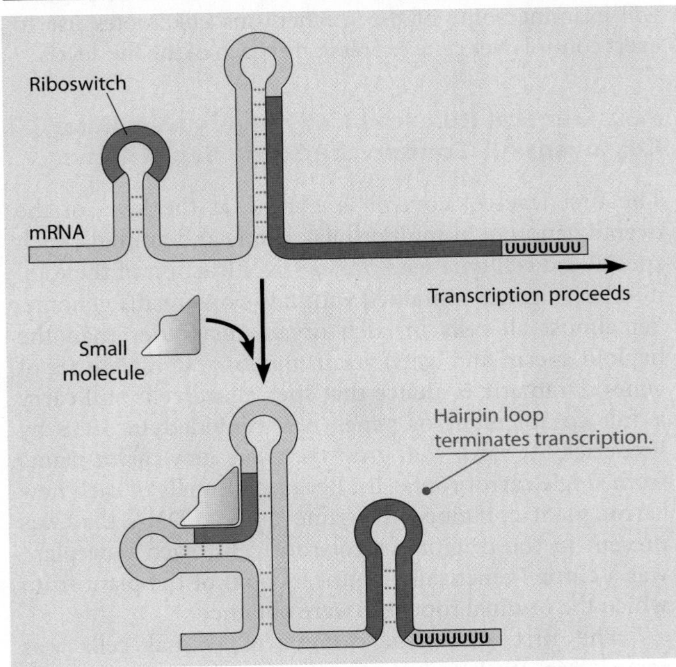

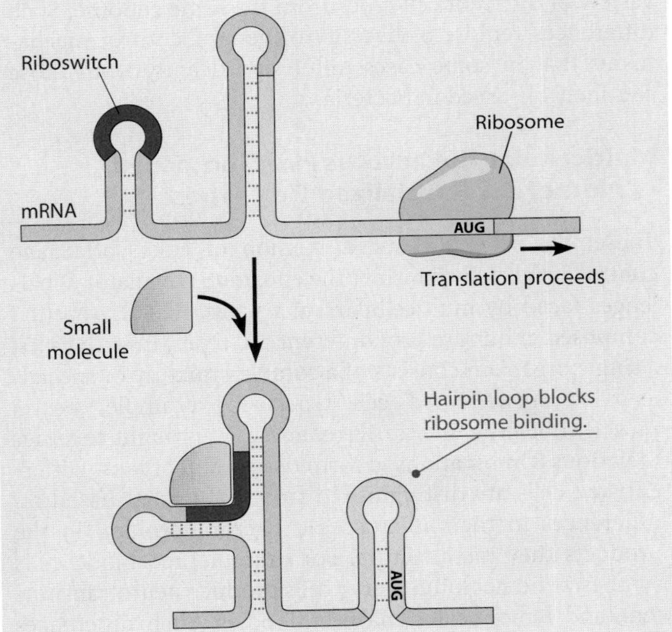

FIGURE 23-9 Riboswitch Control of Gene Expression. These models show two ways in which the binding of small molecules to riboswitches in mRNA can exert control over **(a)** transcription and **(b)** translation.

involved in the pathway for synthesizing FMN and FAD in *E. coli*. Unlike the situation in *B. subtilis*, the *E. coli* genes coding for these enzymes are not clustered in a single operon, but some of them are still controlled by riboswitches. In this case, binding of FMN to its mRNA riboswitch promotes the formation of a hairpin loop that includes the sequences required for binding the mRNA to ribosomes. By sequestering away these sequences, the bound FMN prevents the mRNA from interacting with ribosomes and initiating translation (see Figure 23-9b).

Eukaryotic Gene Regulation: Genomic Control

In the early days of molecular biology, there was a popular saying that "what is true of *E. coli* is also true of elephants." This maxim expressed the initial conviction that almost everything learned about bacterial cell function at the molecular level would also apply to eukaryotes. Such predictions have turned out to be only partly true, however. In terms of central metabolic pathways, mechanisms for transporting solutes across membranes, and such fundamental features as DNA structure, protein synthesis, and enzyme function, there are numerous similarities between the bacterial and eukaryotic worlds, and findings from studies with bacteria can often be extrapolated to eukaryotes.

When it comes to the regulation of gene expression, the comparison needs to be scrutinized more carefully. The DNA of eukaryotic cells is packaged into chromatin fibers and located in a nucleus that is separated from the protein synthetic machinery by a nuclear envelope. Eukaryotic genomes are usually much larger than those of bacteria, and multicellular eukaryotes must create a variety of different cell types from the same genome. Such differences require a diversity of genetic control mechanisms that in some cases differ significantly from those routinely observed in bacteria.

Multicellular Eukaryotes Are Composed of Numerous Specialized Cell Types

To set the stage for our discussion of eukaryotic gene control, let's briefly consider the enormous regulatory challenges faced by multicellular eukaryotes, which are often composed of hundreds of different cell types. In such cases, a single organism consists of a complex mixture of specialized or *differentiated* cell types—for example, nerve, muscle, bone, blood, cartilage, and fat—brought together in various combinations to form tissues and organs. Differentiated cells are distinguished from each other based on differences in their microscopic appearances and in the products they manufacture. For example, red blood cells synthesize hemoglobin, nerve cells produce neurotransmitters, and lymphocytes make antibodies. Such differences indicate that selectively controlling the expression of a wide variety of different genes must play a central role in the mechanism responsible for creating differentiated cells.

Differentiated cells are produced from populations of immature, nonspecialized cells by a process known as **cell differentiation.** The classic example occurs in embryos, in which cells of the early embryo produce all the cell types that make up the organism.

Eukaryotic Gene Expression Is Regulated at Five Main Levels

The fact that a multicellular plant or animal may need to produce hundreds of different cell types using a single genome underscores the difficulty of explaining eukaryotic gene regulation entirely in terms of known bacterial mechanisms—elephants are not just large *E. coli* after all! If we are to thoroughly understand gene regulation in eukaryotic cells, we must approach the topic from a eukaryotic perspective.

The pattern of genes being expressed in any given eukaryotic cell is ultimately reflected in the spectrum of functional gene products—usually protein molecules but in some cases RNA—produced by that cell. The overall pattern is the culmination of controls exerted at several different levels. These potential control points are highlighted in **Figure 23-10,** which traces the flow of genetic information from genomic DNA in the nucleus to functional proteins in the cytoplasm (many, but not all, of these control points also apply to prokaryotes). As you can see, control is exerted at five main levels: ❶ the genome, ❷ transcription, ❸ RNA processing and export from nucleus to cytoplasm, ❹ translation, and ❺ posttranslational events. Regulatory mechanisms in the last three categories are all examples of *posttranscriptional control*, a term that encompasses a wide variety of events. In the rest of the chapter, we will examine some of the mechanisms eukaryotes use to exert control over gene expression at each of the five levels.

As a General Rule, the Cells of a Multicellular Organism All Contain the Same Set of Genes

The first level of control is exerted at the level of the overall genome. In multicellular plants and animals, each specialized cell type uses only a small fraction of the total number of genes contained within the organism's genome. Yet almost all cells in such organisms (other than the haploid sperm and eggs) retain the same complete set of genes. Dramatic evidence that specialized cells still carry a full complement of genes was provided in 1958 by Frederick Steward, who grew complete new carrot plants from single carrot root cells. Because the cells of each new carrot plant contained the same nuclear DNA that was present in the original carrot root cells, each new plant was a **clone** (genetically identical copy) of the plant from which the original root cells were obtained.

The first successful cloning of animal cells was reported in 1964 by John Gurdon and his colleagues. In studies with *Xenopus laevis*, the African clawed frog, they transplanted nuclei from differentiated tadpole cells into

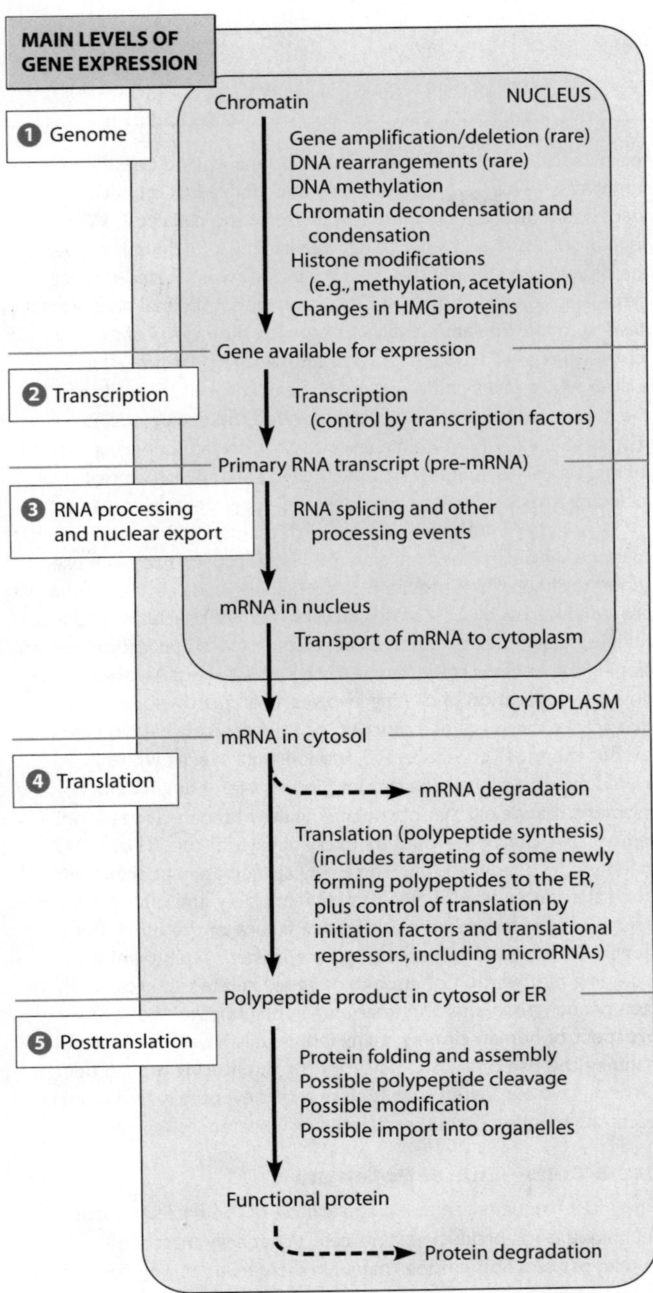

MB Regulation of Gene Expression in Eukaryotes; Activity: Transcription Initiation in Eukaryotes; Activity: Review: Control of Gene Expression; Activity: Overview: Control of Gene Expression

FIGURE 23-10 The Multiple Levels of Eukaryotic Gene Expression and Regulation. Gene expression in eukaryotic cells is regulated through a diverse array of mechanisms operating at five distinct levels: ❶ the genome, ❷ transcription, ❸ RNA processing and nuclear export, ❹ translation, and ❺ posttranslational events. Controls operating at the last three levels are all considered to be *posttranscriptional control* mechanisms.

unfertilized eggs that had been deprived of their own nuclei. Although the frequency of success was low, some eggs containing transplanted nuclei gave rise to viable, swimming tadpoles. A new organism created by this process of *nuclear transplantation* is a clone of the organism from which the original nucleus was taken. The results of both the carrot and frog studies indicated that the nucleus of a differentiated cell can direct the development of an entire new organism. Such a nucleus is therefore said to be *totipotent:* it contains the complete set of genes needed to create a new organism of the same type as the organism from which the nucleus was taken.

An especially dramatic example of animal cloning was reported in 1997, when Ian Wilmut and his colleagues in Scotland made newspaper headlines by announcing the birth of a cloned lamb, Dolly—the first mammal ever cloned from a cell derived from an adult. Dolly was born from a sheep egg whose original nucleus had been replaced by a nucleus taken from a single cell of an adult sheep. Similar techniques have subsequently been used to clone many other mammals. As we discuss in **Box 23A,** cloning is a remarkable feat that raises numerous scientific and ethical questions about the future applications of such technology.

One major proposed use of cloning has relevance to medicine. Modern regenerative medicine seeks to replace defective cells in human patients with genetically engineered substitutes. So-called therapeutic cloning seeks to produce embryonic stem cells that are genetically matched to a human patient but flexible in their ability to differentiate. An alternative that avoids many of the ethical obstacles of therapeutic cloning is the production of induced pluripotent stem cells (see Box 23A).

Gene Amplification and Deletion Can Alter the Genome

Although the preceding evidence indicates that the genome tends to be identical in all cells of an adult eukaryotic organism, a few types of gene regulation create exceptions to this rule. One example is the use of **gene amplification** to create multiple copies of the same gene. Amplification is accomplished by replicating the DNA in a specific chromosomal region many times in succession, thereby creating dozens, hundreds, or even thousands of copies of the same stretch of DNA. Gene amplification can be regarded as an example of **genomic control**—that is, a regulatory change in the makeup or structural organization of the genome.

One of the best-studied examples of gene amplification involves the ribosomal RNA genes in *Xenopus laevis*, the organism used by Gurdon in his nuclear transplantation experiments. The haploid genome of *Xenopus* normally contains about 500 copies of the genes that code for 5.8S, 18S, and 28S rRNA. However, the DNA of these genes is selectively replicated about 4000-fold during oogenesis (development of the egg prior to fertilization). As a result, the mature oocyte contains about 2 million copies of the genes for rRNA. This level of amplification is apparently needed to sustain the enormous production of ribosomes that occurs during oogenesis, which in turn is required to sustain the high rate of protein synthesis needed for early embryonic development.

This particular example of gene amplification involves genes whose products are *RNA* rather than

BOX 23A TOOLS OF DISCOVERY

Cloning by Nuclear Transfer and Pluripotent Stem Cells

Dolly—the Sheep Without a Father

In February of 1997, Ian Wilmut made newspaper headlines around the world by introducing us to Dolly, the first mammal ever cloned from an adult cell. Dolly was created by removing the nucleus from the cell of an adult sheep and transferring it into a different sheep's egg whose own nucleus had been removed. Although this **nuclear transfer** technique had been used before to clone mammals, it had never been successful with cells taken from an adult.

Wilmut suspected that these previous failures were caused by the state of the chromatin in the donor cells. The secret to his team's success was that they took mammary gland cells from the udder of a 6-year-old female sheep and starved them in culture to force them into the dormant, G_0 phase of the cell cycle. When such a cell was fused with an egg cell lacking a nucleus, it delivered the diploid amount of DNA in a condition that allowed the egg cytoplasm to reprogram the DNA to support normal embryonic development. Implanting such an egg into the uterus of another female sheep led to the birth of a lamb with no father—that is, a lamb whose cells had the same nuclear DNA as the cells of the 6-year-old female sheep that provided the donor nucleus (**Figure 23A-1**). Within a few years of these pioneering experiments, similar techniques had been used to clone other mammals, including livestock, mice, cats, dogs, and, very recently, monkeys.

Why did Wilmut's team pursue such technology? One reason is to clone valuable transgenic animals. This approach had been used to generate sheep that produce milk containing medically important proteins, such as blood-clotting factors, that are difficult to produce in sufficient quantities by other means. Commercial applications include cloning the beefiest cattle or the fastest racehorses. An even more dramatic application of cloning involves attempts to clone animals that are on the verge of extinction or have recently become extinct.

But the most controversial possibility concerns whether it would ever be ethical to attempt cloning with human cells. For the moment, it appears that producing healthy human clones—or *reproductive cloning*—would be exceedingly difficult. It took 277 attempts to produce Dolly, and many cloned animals that seem normal at birth develop health problems later and die prematurely. These common defects result from a failure of the DNA in the donated nucleus to undergo proper *epigenetic reprogramming*, including methylation of important genes related to growth. In the face of such problems and nearly universal unease about the prospect of human cloning, many individuals have called for laws banning the use of such technology for duplicating human beings. Others have suggested that society might eventually find cloning acceptable under certain restricted circumstances.

Stem Cells—Full of Potential

The most frequent proposed application of the nuclear transfer technique is for producing stem cells genetically matched to a human patient. Some hope that such *therapeutic cloning* could be used to replace damaged cells in human patients. To see why this

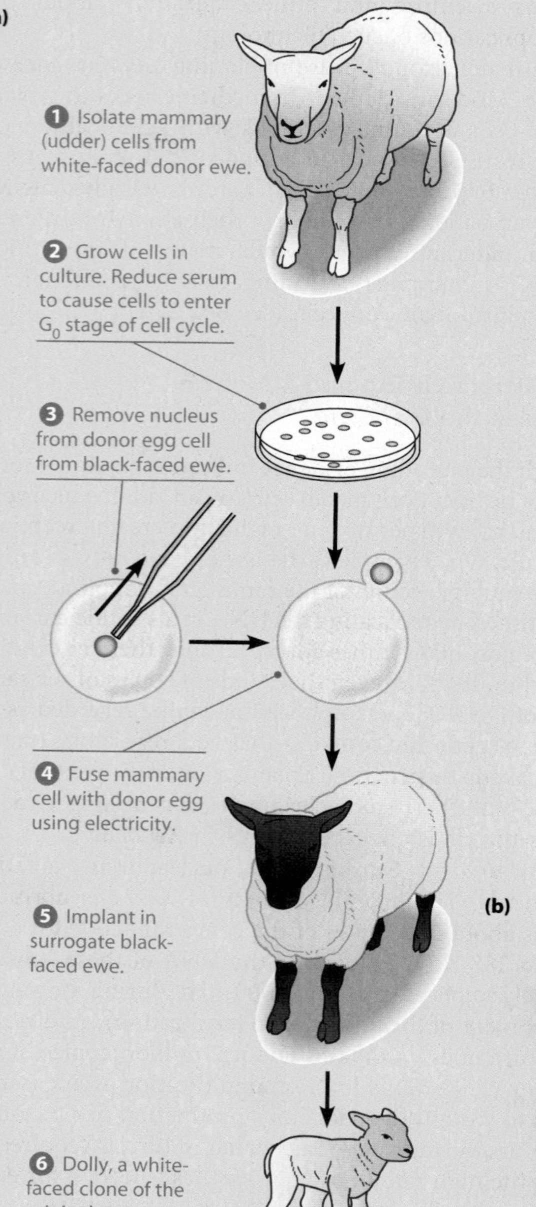

(a)
1 Isolate mammary (udder) cells from white-faced donor ewe.
2 Grow cells in culture. Reduce serum to cause cells to enter G0 stage of cell cycle.
3 Remove nucleus from donor egg cell from black-faced ewe.
4 Fuse mammary cell with donor egg using electricity.
5 Implant in surrogate black-faced ewe.
6 Dolly, a white-faced clone of the original ewe.

(b)

FIGURE 23A-1 Dolly, the First Mammal Cloned from an Adult Cell. **(a)** Dolly was cloned from a serum-starved mammary (udder) cell that was fused with an egg cell from which the nucleus had been removed. **(b)** Dolly as an adult.

approach has been proposed, we need to consider why stem cells have such useful properties.

Stem cells are defined by their capacity to replenish themselves through continued division, as well as by their ability, in the presence of appropriate signals, to produce daughter cells that differentiate into a variety of specialized cell types, usually accompanied by the cessation of cell division. Stem cells can be isolated from mature organisms, such as umbilical cord blood of newborns, as well as from adults. Such cells, especially in the case of blood and bone marrow, have been used successfully for years to treat human patients. Such stem cells are considered to be *multipotent*: They can form several cell types, but their repertoire is fairly limited. **Embryonic stem (ES) cells** on the other hand, do not suffer this limitation. They are *pluripotent*; that is, they can form all types of cells that give rise to a human organism except those that form structures such as the placenta and other support membrane needed during gestation.

ES cells are derived from the inner cell mass cells of the early embryo, known as a *blastocyst* (**Figure 23A-2a**). Inner cell mass cells, once isolated, behave as true stem cells. When treated with various growth factors or in other ways, they can be pushed to become various types of cells.

The ability to isolate and grow stem cells in the laboratory raises the possibility that scientists will be able to produce healthy new cells to replace damaged tissues in patients with various diseases. For example, stem cells that can differentiate into nerve cells might eventually be used to repair the brain damage that occurs in patients suffering from stroke, Parkinson disease, or Alzheimer disease. Or stem cells might be utilized to replace the defective pancreatic cells of patients with diabetes or the defective skeletal muscle cells present in individuals with muscular dystrophy. If the stem cells were produced using therapeutic cloning, they would have the further benefit of being a genetic match to the patient's own cells, avoiding issues of immune rejection.

Despite their developmental potential and their promise in therapies, there is a major ethical objection raised by many people regarding harvesting of ES cells. Isolating ES cells necessarily involves the destruction of a human organism—the embryo. Many view each human organism as an inviolable end in itself, worthy of profound moral respect, making procurement of ES cells ethically impermissible. Others' intuition is that the blastocyst has a different moral status from more advanced human organisms, and view ES cell production as an important potential medical therapy. Clearly, it would be highly desirable to find cells that seem to have developmental potential similar to ES cells, without the associated ethical issues.

In 2006, Shinya Yamanaka announced just such a method: He and his research team had succeeded in coaxing differentiated cells from mice into reverting to a pluripotent state. Such **induced pluripotent stem (iPS) cells** seem to have many of the same properties as ES cells. To produce iPS cells, Yamanaka's team forced cells to express four transcription factor proteins that are expressed by pluripotent cells: Oct4, Sox2, Klf4, and c-Myc (Figure 23A-2b). Further work by Yamanaka's group and by James Thomson and colleagues at the University of Wisconsin extended these studies to human cells. Technical progress has been rapid ever since.

iPS cells seem quite similar to ES cells, and so they seem to be an ideal source of genetically matched, pluripotent cells for medical treatments. There are some issues to be overcome before they can be used in therapies, however. First, in the original experiments the genes were introduced using a retrovirus, which incorporated its DNA into the host cell's genome. It has since been shown that similar effects can be achieved without integration of viral DNA. Second, c-Myc is a proto-oncogene, which can promote tumor formation (see Chapter 24). However, some cell types can be induced to pluripotency without c-Myc. Finally, many iPS cells do not seem to have the methylation status of their DNA fully reset, unlike ES cells. For this reason, many stem cell biologists who are not opposed to ES cell harvesting favor continuing ES and iPS cell research in parallel since the two types of cells may exhibit somewhat different properties.

Given the widely varying viewpoints about ES cells, cloning, and similar procedures, debate continues about the ethics of these technologies. The rapid pace of scientific developments in this field makes it essential for society not to shy away from the debate.

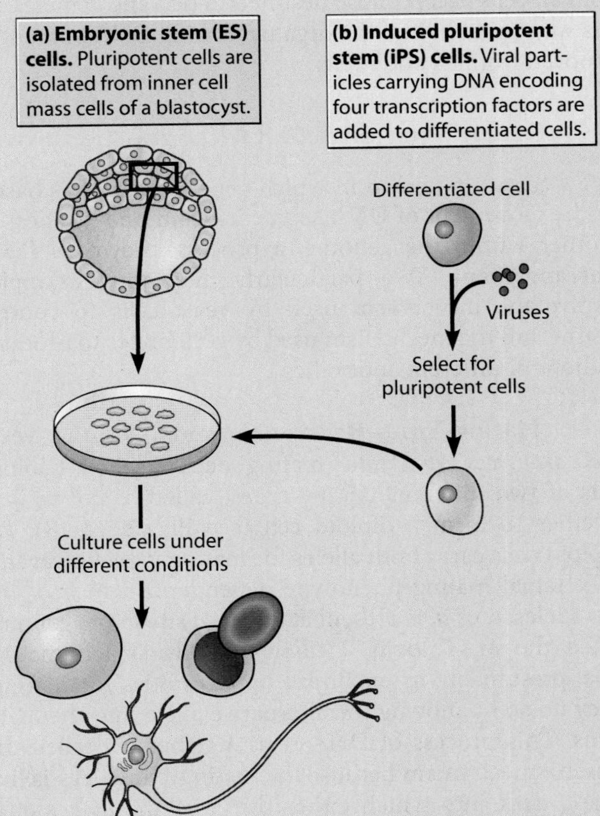

(a) Embryonic stem (ES) cells. Pluripotent cells are isolated from inner cell mass cells of a blastocyst.
(b) Induced pluripotent stem (iPS) cells. Viral particles carrying DNA encoding four transcription factors are added to differentiated cells.
Differentiated cell
Viruses
Select for pluripotent cells
Culture cells under different conditions

FIGURE 23A-2 Two Different Ways to Produce Pluripotent Stem Cells. **(a)** Embryonic stem (ES) cells are produced from inner cell mass cells of a blastocyst. The embryo is destroyed to harvest the ES cells. **(b)** Induced pluripotent stem (iPS) cells result when cells are forced to express four transcription factor proteins. Pluripotent cells are selected using drugs that only allow pluripotent cells to survive. Once ES or iPS cells are isolated, they are cultured under a variety of conditions, causing them to differentiate into various cell types.

protein molecules. The expression of genes that encode proteins—even proteins needed in large amounts, such as ribosomal proteins—can usually be increased sufficiently by increasing the rate of mRNA translation without a need for gene amplification. Nonetheless, enhanced production of certain proteins can occasionally be traced to gene amplification. One example in normal cells is the synthesis of chorion proteins in insect eggs. Abnormal amplification of genes coding for proteins involved in cell cycle signaling is a frequently observed abnormality in cancer cells (Chapter 24).

In addition to amplifying gene sequences whose products are in great demand, some cells delete genes whose products are not required. An extreme example of **gene deletion** (also called *DNA diminution*) occurs in mammalian red blood cells, which discard their nuclei entirely after adequate amounts of hemoglobin mRNA have been made. A less-extreme example occurs in nematode worms and in a group of tiny crustaceans known as copepods. During the embryonic development of copepods, the heterochromatic (transcriptionally inactive) regions of their chromosomes are excised and discarded from all cells except those destined to become gametes. In this way, up to half of the organism's total DNA content is removed from its body cells.

DNA Rearrangements Can Alter the Genome

A few cases are known in which gene regulation is based on the movement of DNA segments from one location to another within the genome, a process known as **DNA rearrangement.** Two particularly interesting examples involve the mechanism used by yeast cells to control mating and the mechanism used by vertebrates to produce millions of different antibodies.

Yeast Mating-Type Rearrangements. In the yeast *Saccharomyces cerevisiae*, mating occurs when haploid cells of two different mating types, called α and a, fuse together to form a diploid cell (see Figure 14-18). All haploid cells carry both alleles for mating type; however, a cell's actual mating phenotype depends on which of the two alleles, α or a, is present at a special site in the genome called the ***MAT* locus.** Cells frequently switch mating type, presumably to maximize opportunities for mating. They do so by moving the alternative allele into the *MAT* locus. This process of DNA rearrangement is called the **cassette mechanism** because the mating-type locus is like a tape deck into which either the α or the a "cassette" (allele) can be inserted and "played" (transcribed).

Figure 23-11 describes the yeast cassette mechanism in more detail. The *MAT* locus, containing either the α or a allele, is located on yeast chromosome 3, approximately midway between extra copies of the two alleles. The locus that stores the extra copy of the α allele is called *HMLα*, and the locus with the extra copy of the a allele is called *HMRa*. To switch mating type, an endonuclease, called *HO endonuclease*, creates a break in the chromosome at the *MAT* locus. Exonucleases then act on the cut DNA to degrade it. The *HMLα* or *HMRa* DNA is used as the template to repair the resulting gap in the DNA, which results in a gene conversion event at the MAT locus that switches the mating type allele found there (we discussed gene conversion in Chapter 20).

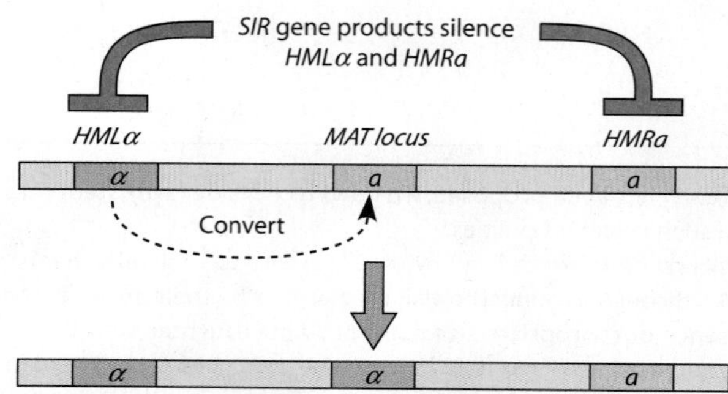

FIGURE 23-11 Cassette Mechanism of the Yeast Mating-Type Switch. Chromosome 3 of *Saccharomyces cerevisiae* contains three copies of the mating-type information. The *HMLα* and *HMRa* loci contain complete copies of the α and a forms of the gene, respectively, but the transcription of these loci is inhibited by the products of the *SIR* gene. The cell's actual mating type is determined by the allele present at the *MAT* locus. When a cell switches mating types, the α or a DNA at the *MAT* locus is removed and replaced by a DNA copy of the alternative mating-type DNA. As an example, this figure illustrates a switch in mating type from a to α.

The DNAs of the two mating-type alleles code for transcription factors that control the expression of genes whose protein products give the cell either an "α" or an "a" mating phenotype. But the presence of extra copies of the alleles at *HMLα* and *HMRa* raises an important question: If the cell contains complete copies of *both* the α and the a alleles at these locations, why aren't proteins specifying both the α and the a mating phenotypes made? The answer is that a group of regulatory genes, known as the *silent information regulator (SIR) genes*, act together to prevent expression of the genetic information at *HMLα* and *HMRa*. The proteins encoded by the *SIR* genes block transcription of *HMLα* and *HMRa* by binding to specific DNA sequences that surround the α and a DNA cassettes at *HMLα* and *HMRa*.

Antibody Gene Rearrangements. A somewhat different type of DNA rearrangement is used by lymphocytes of the vertebrate immune system for producing antibody molecules. Antibodies are proteins composed of two kinds of polypeptide subunits, called *heavy chains* and *light chains*. Vertebrates make millions of different kinds of antibodies, each produced by a different lymphocyte (and its descendants) and each capable of specifically recognizing and binding to a different foreign molecule. But this enormous diversity of antibody molecules creates a potential problem: If every antibody molecule were to be encoded by a different gene, virtually all of a person's DNA would be occupied by the millions of required antibody genes.

Lymphocytes get around this problem by starting with a relatively small number of different DNA segments and rearranging them in various combinations to produce millions of unique antibody genes, each one formed in a different, developing lymphocyte. The rearrangement process involves four kinds of DNA sequences, called *V, J, D,* and *C segments.* The C segment codes for a heavy or light chain *constant region* whose amino acid sequence is the same among different antibodies. The V, J, and D segments together code for *variable regions* that differ among antibodies and give each one the ability to recognize and bind to a specific type of foreign molecule.

To see how this works, let's consider human antibody heavy chains, which are constructed from roughly 200 kinds of V segments, more than 20 kinds of D segments, and at least 6 kinds of J segments. As shown in **Figure 23-12**, the DNA regions containing the various V, D, and J segments are rearranged during lymphocyte development to randomly bring together one V, one D, and one J segment in each lymphocyte. This random rearrangement allows the immune system to create at least $200 \times 20 \times 6 =$ 24,000 different kinds of heavy chain variable regions. In a similar fashion, thousands of different kinds of light chain variable regions can also be created (light chains are constructed from their own types of V, J, and C segments; they do not use D segments). Finally, any one of the thousands of different kinds of heavy chains can be assembled with any one of the thousands of different kinds of light chains, creating the possibility of millions of different types of antibodies. The net result is that millions of different antibodies are produced from the human genome by rearranging a few hundred different kinds of V, D, J, and C segments.

The DNA rearrangement process that creates antibody genes also activates transcription of these genes via a mechanism involving special DNA sequences called *enhancers.* As we will discuss shortly, enhancers increase the rate of transcription initiation. Enhancers are located near DNA sequences coding for C segments, but a promoter sequence is not present in this area so transcription does not normally occur. The promoter for gene transcription is located upstream from the DNA coding for V segments, but it is not efficient enough to promote transcription in the absence of an enhancer sequence. Hence, prior to DNA rearrangement, the promoter and enhancer sequences of an antibody gene are so far apart that transcription does not occur. Only after rearrangement are they close enough for transcription to be activated.

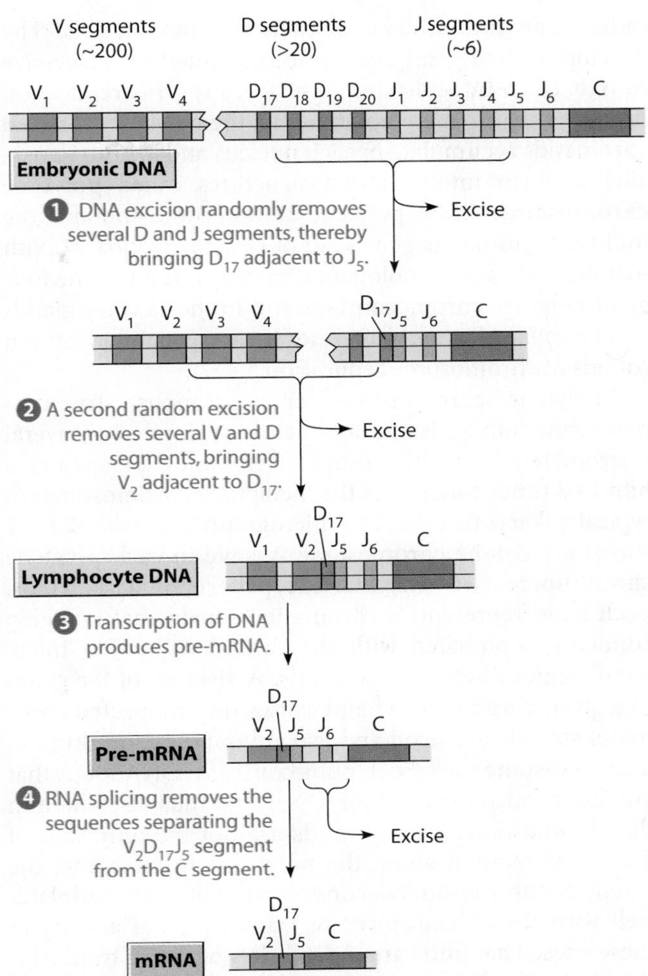

FIGURE 23-12 DNA Rearrangement During the Formation of Antibody Heavy Chains. Genes coding for human antibody heavy chains are created by DNA rearrangements that involve multiple types of V, D, and J segments. In this example, random DNA excisions bring together a unique combination of V, D, and J segments. After transcription, RNA splicing removes the RNA sequences that separate this VDJ segment from the C segment.

Chromosome Puffs Provide Visual Evidence That Chromatin Decondensation Is Involved in Genomic Control

We encounter another aspect of genome-level control when we consider what is involved in making the eukaryotic genome—that is, chromosomal DNA—accessible to the cell's transcription machinery. Recall from Chapter 21 that, to initiate transcription, a eukaryotic RNA polymerase must interact with both DNA and a number of specific proteins (general transcription factors) in the promoter region of a gene. Except when a gene is being transcribed, its promoter region is embedded within a highly folded and ordered chromatin superstructure. Thus, some degree of chromatin decondensation (unfolding) appears to be necessary for the expression of eukaryotic genes.

The earliest evidence that chromatin decondensation is required for gene transcription came from microscopic visualization of certain types of insect chromosomes caught in the act of transcription. Because the DNA of most eukaryotic cells is dispersed throughout the nucleus as a mass of intertwined chromatin fibers, it is usually difficult to observe the transcription of individual genes with a microscope. But a way around this obstacle is provided by an unusual type of insect cell. In the fruit fly *Drosophila melanogaster* and related insects, some metabolically active tissues (such as the salivary glands and intestines) grow by an enormous increase in the size of, rather than the number of, their constituent cells. The development of giant cells is accompanied by successive rounds of DNA replication, but, because this replication occurs in cells that are not dividing, the newly synthesized chromatids accumulate in each nucleus and line up in parallel to form multistranded structures called **polytene chromosomes.** Each polytene chromosome contains the multiple chromatids generated during replication of both members of each homologous chromosome pair. The four giant polytene chromosomes found in the salivary glands of *Drosophila* larvae, for example, are generated by ten rounds of chromosome replication.

Polytene chromosomes are enormous structures measuring hundreds of micrometers in length and several micrometers in width—roughly ten times longer and a hundred times wider than the metaphase chromosomes of typical eukaryotic cells. The micrograph in **Figure 23-13** shows a polytene chromosome. Visible in each polytene chromosome is a characteristic pattern of dark bands. Each band represents a chromatin domain that is highly condensed compared with the chromatin in the "interband" regions between the bands. Activation of the genes of a given chromosome band causes the compacted chromatin strands to uncoil and expand outward, resulting in a **chromosome puff.** Such puffs consist of DNA loops that are less condensed than the DNA of bands elsewhere in the chromosome. Though puffs are not the only sites of gene transcription along the polytene chromosome, the extent of chromosome decondensation at puffs correlates well with the enhancement of transcriptional activity at these sites. That puffs are indeed sites of active transcription is confirmed by the fact that puffs are sites where RNA polymerase II, the key enzyme that carries out polymerization of protein-coding RNAs, accumulates.

As the *Drosophila* larva proceeds through development, each of the polytene chromosomes in salivary gland nuclei undergoes reproducible changes in puffing patterns under the control of an insect steroid hormone called *ecdysone.* Ecdysone functions by binding to, and thus activating, a regulatory protein that stimulates the transcription of certain genes. (This is similar to the action of vertebrate steroid hormones, which we cover later in the chapter.) It appears, in other words, that the characteristic puffing patterns seen during the development of *Drosophila* larvae are direct visual manifestations of the selective decondensation and transcription of specific DNA segments according to a genetically determined developmental program.

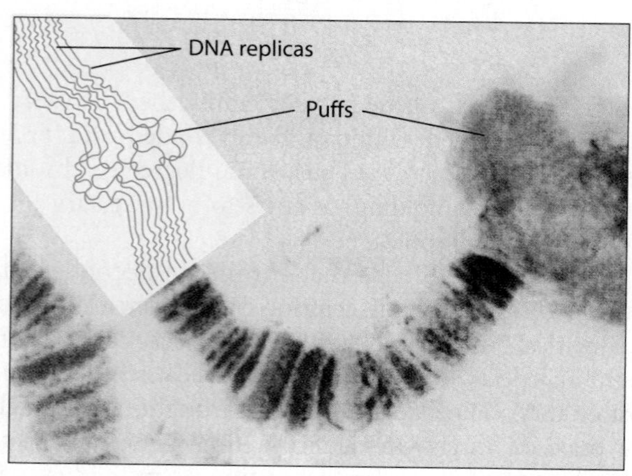

FIGURE 23-13 Puffs in Polytene Chromosomes. Puffs are regions in which transcriptionally active chromatin has become less condensed, as indicated diagrammatically. The light micrograph shows part of a polytene chromosome.

DNase I Sensitivity Provides Further Evidence for the Role of Chromatin Decondensation in Genomic Control

The absence of polytene chromosomes in most eukaryotic cells makes it difficult to visualize chromatin decondensation in regions of active genes. Nonetheless, other kinds of evidence support the idea that chromatin decondensation is generally associated with gene transcription. One particularly useful research tool is *DNase I,* an endonuclease isolated from the pancreas. In test tube experiments, low concentrations of DNase I preferentially degrade transcriptionally active DNA in chromatin. The increased sensitivity of these DNA regions to degradation by DNase I provides evidence that the DNA is uncoiled.

Figure 23-14 illustrates a classic DNase I sensitivity experiment focusing on the chicken gene for a globin polypeptide, one of the polypeptide subunits of hemoglobin. The globin gene is actively expressed in the nuclei of chicken *erythrocytes* (red blood cells). In contrast to the erythrocytes of many other vertebrates, avian erythrocytes retain their nucleus at maturity. If these nuclei are isolated and the chromatin is digested with DNase I, the globin gene is completely digested at low DNase I concentrations that do not affect the globin gene in other tissues, such as oviduct tissue. As you might predict, a gene that is not active in erythrocytes (for example, the gene for ovalbumin, an egg white protein) is not digested by DNase I. The opposite result is obtained when the same procedure is carried out using chromatin isolated from the oviduct, where the ovalbumin gene is expressed and the globin gene is inactive. In this case, the ovalbumin gene is more sensitive than the globin gene to DNase I digestion. Such data demonstrate that transcription of eukaryotic DNA is correlated with an increased sensitivity to DNase I digestion.

The results of these experiments are compatible with two alternative explanations: Either chromatin uncoiling is necessary to give transcription factors and RNA polymerase access to DNA, or the binding of these proteins to DNA causes the uncoiling. This issue has been resolved by studies showing that sensitivity to DNase I is detected in genes that are being actively transcribed, in genes that have recently been transcribed but are no longer active, and in DNA sequences located adjacent to genes of the preceding two types. Such observations suggest that

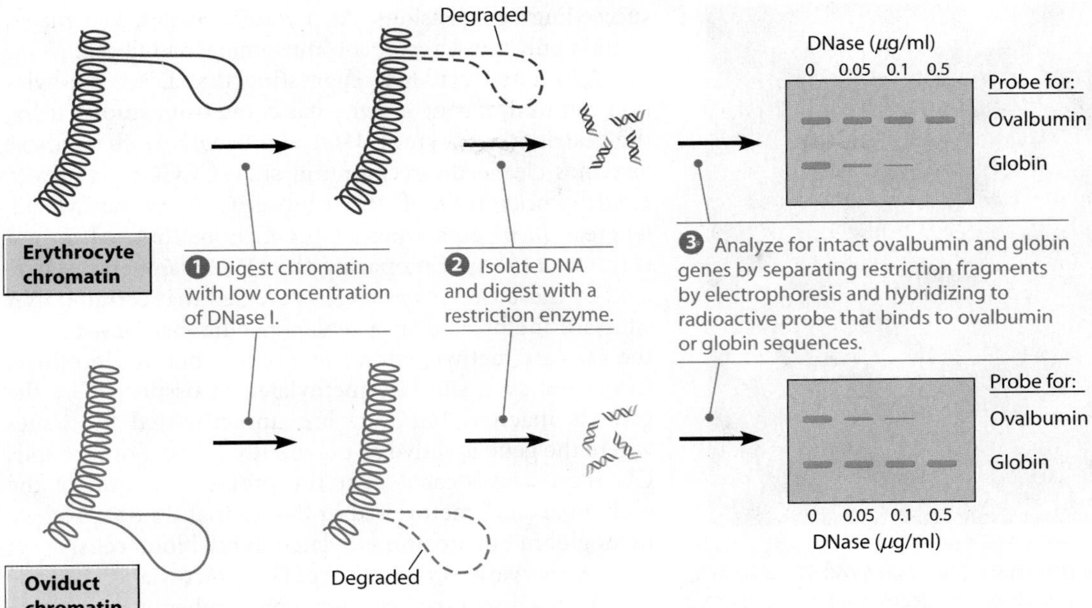

FIGURE 23-14 Sensitivity of Active Genes in Chromatin to Digestion with DNase I. The chromatin configuration of active genes can be studied by exposing cell nuclei to DNase I, which digests DNA by cutting internal phosphodiester bonds. DNA in condensed chromatin is protected from DNase I attack because it is highly coiled and condensed. The experiment shown here uses chromatin from two different cell types, chicken erythrocytes and oviduct cells, and focuses on genes for globin and ovalbumin, which are expressed in erythrocytes and oviduct cells, respectively. Digesting chromatin with a low concentration of DNase I preferentially digests the DNA in uncoiled regions (1). The DNA is then purified and digested with a restriction enzyme (2) that releases DNA fragments containing an intact globin or ovalbumin gene (if it has not been nicked by DNase I). The presence of such restriction fragments is detected in step 3 by separating the DNA using electrophoresis, transferring the separated fragments to filter paper (a Southern blot), and hybridizing with radioactive DNA probes for the globin and ovalbumin genes. Note that DNA isolated from erythrocyte chromatin treated with increasing amounts of DNase I contains progressively smaller amounts of intact globin gene, and comparable results are obtained for the ovalbumin gene in oviduct chromatin. In contrast, even high concentrations of DNase I have no effect on the globin gene in oviduct chromatin or on the ovalbumin gene in erythrocyte chromatin. In other words, genes are more susceptible to DNase I attack in tissues where they are actively transcribed.

DNase I sensitivity is not caused by the process of gene transcription itself, but instead reflects an altered chromatin structure in regions associated with active or potentially active genes. Presumably this means that chromatin uncoiling is a prerequisite for—rather than a consequence of—transcriptional activation.

DNase I has also been employed in other kinds of experiments. When nuclei are treated with very low concentrations of DNase I, it is possible to detect specific locations in the chromatin that are exceedingly susceptible to digestion. These **DNase I hypersensitive sites** tend to occur up to a few hundred bases upstream from the transcriptional start sites of active genes. Compared to the bulk of the DNA found in active genes, the DNase I hypersensitive sites are about ten times more sensitive to DNase I digestion. These regions appear to correspond to regions in which the DNA is not part of a nucleosome. The idea that DNase I hypersensitive sites may represent regions that are free of nucleosomes first emerged from studies involving the eukaryotic virus SV40. When the SV40 virus infects a host cell, its circular DNA molecule becomes associated with histones and forms typical nucleosomes that can be observed with an electron microscope. However, a small region of the viral DNA remains completely uncoiled and free of nucleosomes (**Figure 23-15**). This region, which includes several DNase I hypersensitive sites, contains DNA sequences that bind regulatory proteins involved in activating transcription.

DNA Methylation Is Associated with Inactive Regions of the Genome

Another way of regulating the availability of specific regions of the genome is through **DNA methylation,** the addition of methyl groups to selected cytosine bases in DNA. The DNA of most vertebrates contains small amounts of methylated cytosine, which tend to cluster near the 5′ ends of genes where promoter sequences are located. Methylation of promoter regions can either block access of proteins required for transcriptional activation or serve as a binding site for proteins that condense chromatin into inactive configurations. The net effect is either a localized or regional silencing of gene expression.

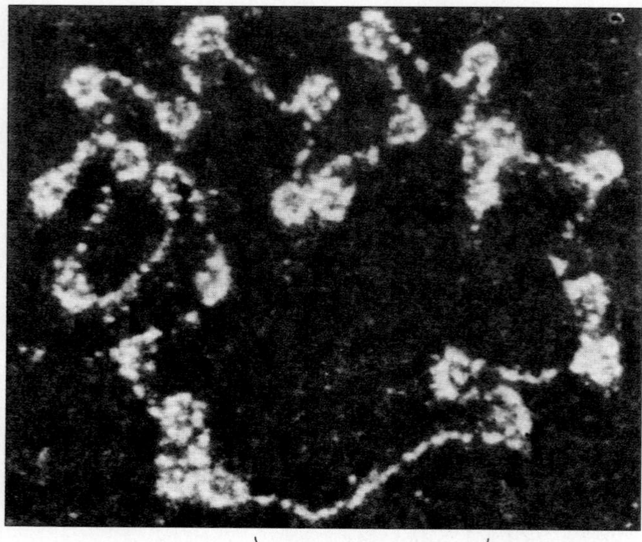

FIGURE 23-15 Circular DNA Molecule of an SV40 Virus from an Infected Host Cell. In this high-power electron micrograph of an SV40 DNA molecule, the bracketed region of the DNA lacks nucleosomes. This region corresponds to the location of several DNase I hypersensitive sites (TEM).

The enzyme responsible for DNA methylation acts preferentially on cytosines situated in 5′–CG–3′ sequences that are base-paired to complementary 3′–GC–5′ sequences that are themselves already methylated. In other words, if the old DNA strand in a newly replicated DNA double helix has a methylated 5′–CG–3′ sequence, then the complementary 3′–GC–5′ sequence in the new strand will become a target for methylation. This phenomenon allows DNA methylation patterns to be inherited during successive rounds of DNA replication. The net result is that DNA methylation provides a means for creating **epigenetic changes**—that is, stable alterations in gene expression that are transmitted from one cellular generation to the next without requiring changes in a gene's underlying base sequence.

A striking example involves the X chromosomes of female mammals, which inherit an X chromosome from each of their two parents. Because males have only one X chromosome, a potential imbalance exists in the expression of X-linked genes between males and females. Nature's solution to the problem is to randomly inactivate one of the two X chromosomes in females during early embryonic development. During this process of *X-inactivation*, the DNA of one X chromosome becomes extensively methylated, the chromatin fibers condense into a tight mass of heterochromatin whose DNA is less accessible to transcription factors and RNA polymerase, and gene transcription ceases. When interphase cells are examined under a microscope, the inactivated X chromosome is visible as a dark spot called a *Barr body.* Once a given X chromosome has been inactivated in a particular cell, the same X chromosome remains inactivated in all of the cells produced by succeeding cell divisions. As a result females, like males, contain only one active X chromosome per adult cell.

Additional evidence suggesting that DNA methylation can inhibit gene activity has come from studies using the restriction enzymes *Msp*I and *Hpa*II. Both of these enzymes cleave the recognition site –CCGG–. However, *Hpa*II works only if the central C is unmethylated, whereas *Msp*I cuts whether the C is methylated or not (**Figure 23-16**). Comparing the DNA fragments generated by these two enzymes has revealed that certain DNA sites are methylated in a tissue-specific fashion—that is, the sites are methylated in some tissues but not in others. In general such sites are methylated in tissues where the gene is inactive, but they are unmethylated in tissues where the gene is active or potentially active. For example, CG sequences located near the promoter region of the globin gene are methylated in tissues that do not produce hemoglobin but are unmethylated in red blood cells.

The pervasive importance of DNA methylation is made clear by the discovery that a growing number of human diseases, including cancer, are associated with abnormalities in DNA methylation. DNA methylation also plays a role in genomic **imprinting**—a process that causes certain genes to be expressed differently depending on whether they are inherited from a person's mother or father. For example, some imprinted genes are expressed in the copy inherited from the mother and silenced in the copy inherited from the

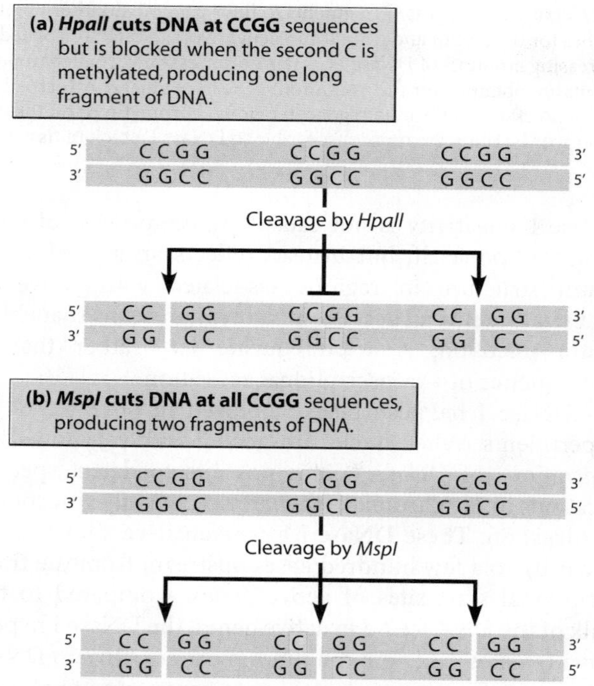

FIGURE 23-16 Differences in DNA Methylation Detected Using Enzymatic Digestion. Both *HpaII* and *MspI* cut DNA at CCGG sequences, but *HpaII* will not cut if the second C is methylated (yellow). **(a)** When the DNA shown is cut with *HpaII*, a long fragment is generated that can be detected using a Southern blot (see Box 18C). **(b)** When the same DNA is cut with *MspI*, two shorter fragments from the same region are produced.

father. This differing behavior stems from differing methylation patterns: The promoter region of the inherited paternal copy is extensively methylated, whereas the same region in the inherited maternal copy is not. (Imprinting also works in the opposite way—the maternal copy rather than the paternal copy of some genes is inactivated by methylation.) Some diseases result from deletions in specific imprinted genes. One well understood example involves deletions in the same region of human chromosome 15, which contains imprinted genes. *Prader–Willi syndrome* is characterized by deletion of paternal genes in this region and is characterized by mild mental retardation, compulsive eating and obesity, and defects in gonad development. Conversely, *Angelman syndrome* results when a gene in this region on the maternal chromosome is deleted. Angelman syndrome patients exhibit frequent laughter and smiling, as well as hand flapping movements.

Inheritance is not the only factor that influences DNA methylation patterns: Studies involving identical twins have revealed that environment plays a role as well. Identical twins exhibit DNA methylation patterns that are virtually indistinguishable when the twins are young, but significant differences emerge as twins grow older. The greatest differences in DNA methylation, and in the pattern of genes being expressed, are seen in twins who spend large amounts of time apart, suggesting that environmental factors influence DNA methylation and, in turn, gene expression.

Changes in Histones and Chromatin Remodeling Proteins Can Alter Genome Activity

DNA methylation is not the only type of genomic alteration involved in epigenetic control. Another mechanism involves changes in histones, the proteins whose role in the packaging of chromosomal DNA was described in Chapter 18. Each histone molecule has a protruding tail that can be tagged at various locations by the addition of methyl, acetyl, or phosphate groups. Various combinations of these tags create a **histone code** that is read by other proteins as a set of signals for modifying chromatin structure and gene activity.

One type of tagging reaction involves *methylation* of the amino acid lysine, which can serve as a signal for either the activation or repression of gene expression, depending on the particular lysine and class of histone involved. For example, methylation at lysine 4 in histone H3 is a hallmark of most active genes, whereas methylation of lysines 9 and 27 is associated with gene silencing. In certain cases, methylation of lysine 27 leads to the recruitment and activation of enzymes that methylate the adjoining DNA, thereby coupling two epigenetic alterations involved in gene silencing.

Another mechanism for altering histone structure is through *acetylation*—the addition of acetyl groups to amino acid side chains. In particular, the enzyme **histone acetyltransferase (HAT)** adds acetyl groups to histone molecules and thereby promotes chromatin decondensation. Another enzyme, **histone deacetylase (HDAC),** performs the opposite function, removing acetyl groups from histones. Regulatory proteins that bind DNA can recruit complexes that contain these enzymes, thereby changing the availability of a region of DNA for transcription. *Repressor* proteins, which cause transcription to occur less frequently at a particular site, recruit HDAC complexes, and *activator* proteins recruit HAT complexes (**Figure 23-17a**; we will discuss transcriptional activators and repressors in more detail later in this chapter). Evidence that acetylated histones are preferentially associated with active genes has been provided by studies in which chromatin was incubated with DNase I to selectively degrade

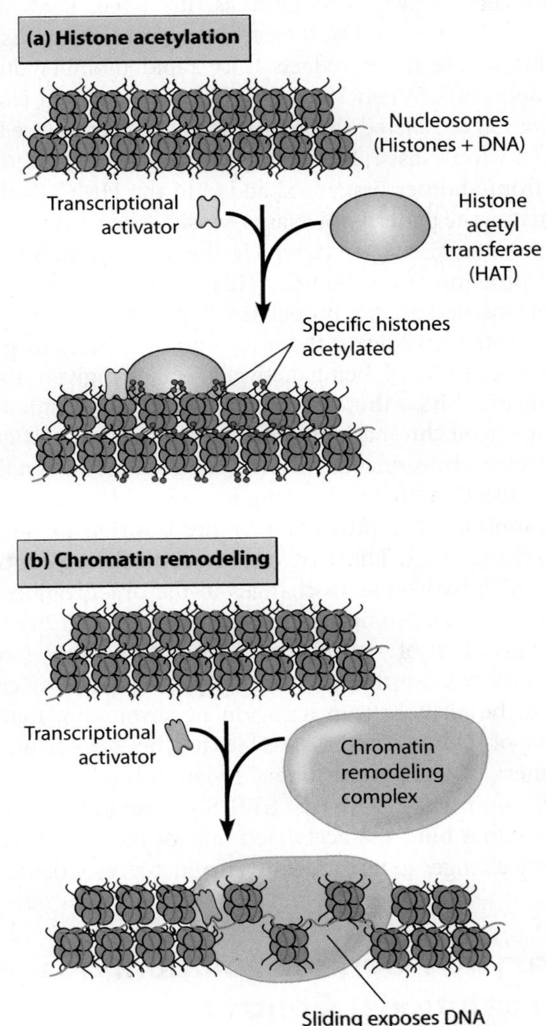

FIGURE 23-17 Roles of Histone Acetylation and Chromatin Remodeling Proteins during Transcriptional Activation.
(a) Transcriptional activator proteins recruit histone acetyltransferase (HAT) complexes to the region near a gene. These enzymes then add acetyl groups, which leads to the opening or closing of the chromatin structure in the region. **(b)** Chromatin remodeling proteins can have several effects on nucleosomes. In the case shown here, a chromatin remodeling protein causes sliding of nucleosomes, exposing a region of DNA that could then be transcribed.

genes that are transcriptionally active. Such treatment causes the release of the acetylated form of histones H3 and H4, suggesting that acetylation of these histones is associated with gene activation. Such acetylation-induced changes in nucleosome structure are thought to loosen chromatin packing and thereby facilitate the access of transcription factors to gene promoters.

Yet another way in which histones affect chromatin structure and gene activity is suggested by studies showing that transcriptionally active chromatin often lacks histone H1. Since histone H1 is required for folding chromatin into 30-nm chromatin fibers (Figure 18-22b), the absence of histone H1 may help to maintain active chromatin in the form of uncoiled 10-nm fibers. A related feature of transcriptionally active chromatin is its large content of *high-mobility group (HMG) proteins,* a group of nonhistone proteins whose name reflects their rapid mobility during electrophoresis. When HMG proteins are removed from a preparation of isolated chromatin, the active genes lose their sensitivity to DNase I. If certain HMG proteins are then isolated from another tissue and added to the HMG-depleted chromatin, the pattern of DNase I sensitive genes is found to resemble the tissue from which the chromatin, not the HMG proteins, was obtained. This means that chromatin exhibits tissue-specific properties that are recognized by HMG proteins, allowing them to bind selectively to genes that are capable of being activated in any given tissue. Binding of HMG is thought to impart DNase I sensitivity by helping uncoil chromatin fibers into a more open configuration, perhaps by displacing histone H1 from the 30-nm fiber.

In addition to modifications of the histones within nucleosomes, other proteins alter the position of nucleosomes along DNA. These **chromatin remodeling proteins** couple ATP hydrolysis to changes in the organization and position of nucleosomes along DNA (Figure 23-17b). One important class of remodelers are the **SWI/SNF** family. These proteins are thought to slide nucleosomes or cause them to be ejected from a region of chromatin, making regions of DNA more accessible to the transcriptional machinery. Recall that histone acetylation is positively correlated with gene activity. SWI/SNF remodelers have domains that bind the acetylated tails of histone proteins, coupling changes in histones and chromatin remodeling.

Eukaryotic Gene Regulation: Transcriptional Control

We have now described some common regulatory changes in the composition and structure of the genome. However, the existence of structural changes associated with active regions of the genome does not address the underlying question of how the DNA sequences contained in these regions are actually selected for activation. The answer is to be found in the phenomenon of **transcriptional control,** the second main level for controlling eukaryotic gene expression (see Figure 23-10).

Different Sets of Genes Are Transcribed in Different Cell Types

When we consider transcriptional control, we come to a level of gene regulation where knowledge has blossomed in recent years. The first direct evidence for the importance of transcriptional regulation in eukaryotes came from experiments comparing newly synthesized RNA in the nuclei of different mammalian tissues. Liver and brain cells, for example, produce different sets of proteins, although there is considerable overlap between the two sets. How can we determine the source of this difference? If the different proteins produced by the two cell types are a reflection of *differential gene transcription*—that is, transcription of different genes to produce different sets of RNAs—we should see corresponding differences between the populations of nuclear RNAs derived from brain and liver cells. On the other hand, if all genes are equally transcribed in liver and brain, we would find few, if any, differences between the populations of nuclear RNAs from the two tissues, and we would conclude that the tissue-specific differences in protein synthesis were due to posttranscriptional mechanisms that control the ability of various RNAs to be translated.

One way of distinguishing between these alternatives is to use the technique of *nuclear run-on transcription,* which provides a snapshot of the transcriptional activity occurring in a nucleus at a given moment in time (**Figure 23-18**). Transcriptionally active nuclei are gently isolated from cells and allowed to complete synthesis of RNA molecules in the presence of radioactively labeled nucleoside triphosphates. When such an experiment is performed using liver and brain cells, the newly transcribed (radioactively labeled) RNAs in the liver nuclei contain sequences from liver-specific genes, but these liver-specific sequences are not detected in the labeled RNAs synthesized by isolated brain nuclei. Likewise, the newly transcribed (radioactively labeled) RNAs in the brain nuclei contain sequences from brain-specific genes, but these brain-specific sequences are not detected in the labeled RNAs synthesized by isolated liver nuclei.

By failing to detect the synthesis of RNA from liver-specific genes in brain cell nuclei and vice versa, these experiments indicate that gene expression is being regulated at the transcriptional level. In other words, different cell types transcribe different sets of genes, thereby allowing each cell type to produce those proteins needed for carrying out that cell's specialized functions.

DNA Microarrays Allow the Expression of Thousands of Genes to Be Monitored Simultaneously

Although the preceding kinds of observations reveal that different sets of genes are transcribed in different cell types, they cannot easily identify which individual genes are turned on or off. One method for determining

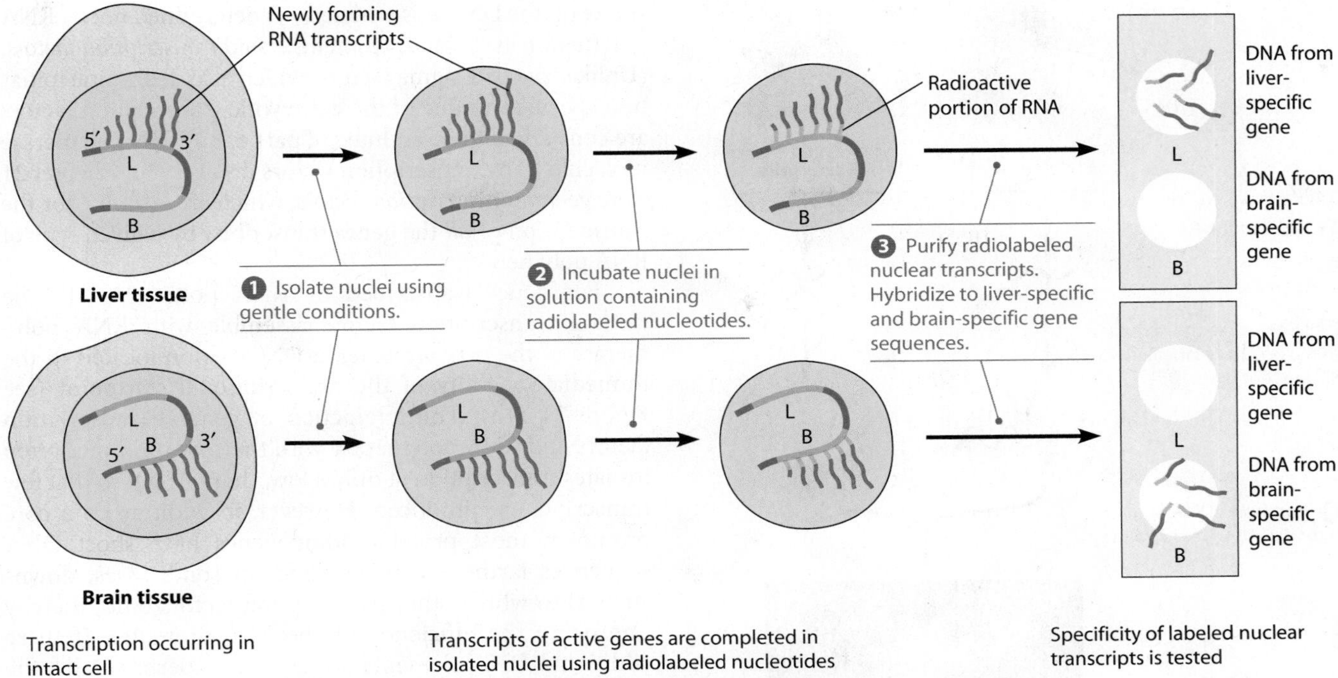

FIGURE 23-18 Demonstration of Differential Transcription by Nuclear Run-on Transcription Assays. In these studies, B represents a hypothetical gene that is expressed only in brain tissue, and L is a gene expressed only in liver tissue. Isolated nuclei are incubated in a solution containing radioactively labeled ribonucleotides, which become incorporated into the mRNA being synthesized by active genes. If different genes are active in liver and brain tissue, some labeled sequences in the liver nuclear transcripts will not be present in brain transcripts and vice versa. The composition of the labeled RNA population is assayed by allowing the labeled RNA to hybridize with DNA sequences representing different genes that have been attached to a filter-paper support (3). Labeled liver transcripts hybridize with a different set of genes than do labeled brain transcripts, indicating that the identities of the active genes in the two tissues differ.

whether a particular gene is active is to assay for its corresponding mRNA using an RNA detection technique that is somewhat fancifully called **Northern blotting** to contrast it with Southern blotting (a DNA detection technique described in Box 18C). In Northern blotting, an RNA sample is size-fractionated by gel electrophoresis and transferred to a special blotting paper. The paper is then exposed to a radioactive (or otherwise labeled) DNA probe containing the gene sequence of interest, and the amount of mRNA is quantified by measuring the bound radioactivity.

Alternatively, to monitor the expression of hundreds or even thousands of genes simultaneously, a tool known as a **DNA microarray** can be used. A DNA microarray is a thin, fingernail-sized chip made of plastic or glass that has been spotted at fixed locations with thousands of DNA fragments corresponding to various genes of interest. A single microarray may contain 10,000 or more spots, each representing a different gene. To determine which genes are being expressed in any given cell population, RNA molecules (the products of gene transcription) are isolated from the cells and copied with the enzyme reverse transcriptase into single-stranded cDNA molecules, which are then attached to a fluorescent dye. When the DNA microarray is bathed with the fluorescent cDNA, each cDNA molecule will bind by complementary base pairing to the spot containing the specific gene it was transcribed from.

Figure 23-19 illustrates how this approach can be used to compare the patterns of gene expression in brain and liver cells. In this particular example, two fluorescent dyes are used: a green dye to label cDNAs derived from brain cells and a red dye to label cDNAs derived from liver cells. When the green and red cDNAs are mixed together and placed on a DNA microarray, the green cDNAs will bind to genes expressed in brain cells, and the red cDNAs will bind to genes expressed in liver cells. Green spots therefore represent higher expression of a gene in brain cells, red spots represent higher expression of a gene in liver cells, yellow spots (caused by a mixture of green and red fluorescence) represent genes whose expression is roughly the same, and black spots (absence of fluorescence) represent genes expressed in neither cell type. Consequently, measuring the color and intensity of each fluorescent spot allows the expression of thousands of genes to be monitored simultaneously. Such data have revealed that many genes are selectively expressed in particular cell types, such as brain cells, liver cells, or other differentiated cells.

DNA microarrays also have practical applications. For example, human cancers that appear to be the same disease based on microscopic examination sometimes

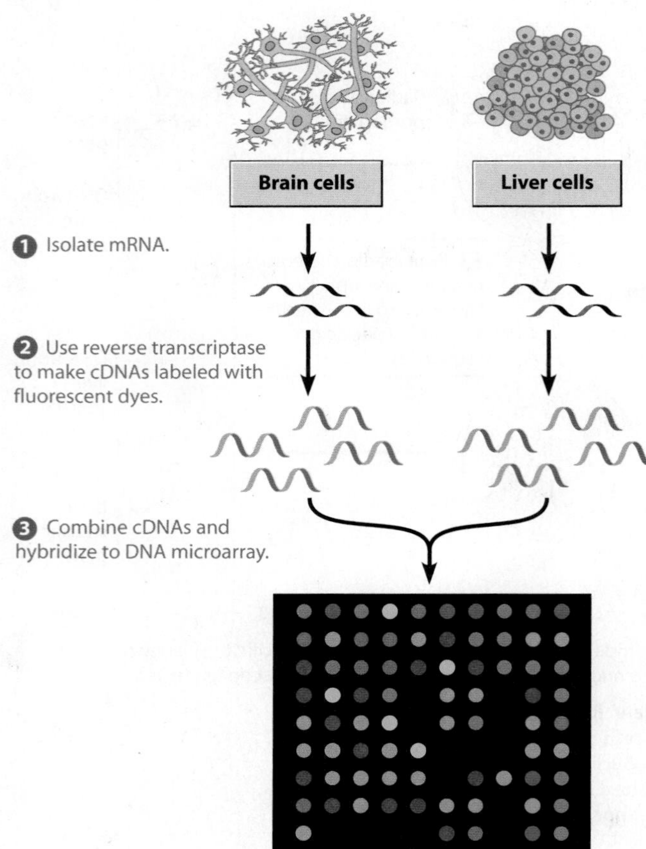
Brain cells
Liver cells
1 Isolate mRNA.
2 Use reverse transcriptase to make cDNAs labeled with fluorescent dyes.
3 Combine cDNAs and hybridize to DNA microarray.

MB Analysis of Gene Expression Using DNA Microarrays

FIGURE 23-19 Using a DNA Microarray for Studying Gene Expression Profiles. In this example, gene expression in brain cells and liver cells is compared by isolating mRNA from the two populations of cells, using reverse transcriptase to make cDNA copies of the mRNAs, and attaching a green fluorescent dye to the brain cell cDNAs and a red fluorescent dye to the liver cell cDNAs. A DNA microarray containing DNA fragments representing thousands of different genes is then bathed with a mixture of the two cDNA populations (only a small section of the DNA microarray is illustrated). Each cDNA hybridizes to the spot containing the specific gene it corresponds to. Green spots therefore represent genes expressed preferentially in brain cells, red spots represent genes expressed preferentially in liver cells, yellow spots (a mixture of green and red fluorescence) represent genes whose expression is similar in the two cell populations, and dark regions (missing spots) represent genes that are not expressed in either cell type.

exhibit different gene expression profiles when tested using DNA microarray technology. Such information allows cancers from different patients to be more accurately characterized, thereby improving the prospects for custom-tailored treatments that are most appropriate for each individual person.

Proximal Control Elements Lie Close to the Promoter

In discussing how gene transcription is regulated in different cell types, we will now focus our attention on protein-coding genes, which are transcribed by RNA polymerase II. As we saw in Chapter 21, the specificity of transcription—that is, where on the DNA it is initiated—is determined not by RNA polymerase itself, but by proteins called *transcription factors.* (Unlike bacterial sigma factors, which also determine initiation specificity, none of the eukaryotic transcription factors are considered to be an integral part of an RNA polymerase molecule.) The transcription factors discussed in Chapter 21 were *general transcription factors,* which are essential for the transcription of *all* the genes transcribed by a given type of RNA polymerase.

For genes transcribed by RNA polymerase II, the general transcription factors assemble with RNA polymerase at the *core promoter,* a DNA region located in the immediate vicinity of the transcriptional startpoint (see Figure 21-11b). The interaction of general transcription factors and RNA polymerase with the core promoter often initiates transcription at only a low, "basal" rate, so that few transcripts are produced. However, in addition to a core promoter, most protein-coding genes have short DNA sequences farther upstream (and, in some cases, downstream) to which other transcription factors bind, thereby improving the efficiency of the core promoter. If these additional DNA elements are deleted experimentally, the frequency and accuracy of transcription is reduced.

In discussing such regulatory DNA sequences, we will use the term **proximal control elements** to refer to sequences located upstream of the core promoter but within about 100–200 base pairs of it. The number, exact location, and identities of these proximal control elements vary with each gene, but three types are especially common: the *CAAT box,* the *GC box,* and the *octamer* (the first two were discussed in Chapter 21 and are illustrated in **Figure 23-20**). Transcription factors that selectively bind to one of these, or to other control sequences located outside the core promoter, are called **regulatory transcription factors.** They increase (or sometimes decrease) transcription initiation by interacting with components of the transcription apparatus.

Enhancers and Silencers Are Located at Variable Distances from the Promoter

Proximal control elements, like most bacterial control elements, lie close to the core promoter on its upstream side. A second class of DNA control sequences are located either upstream or downstream from the genes they regulate and often lie far away from the promoter. This second type of control region is called an **enhancer** if it stimulates gene transcription or a **silencer** if it inhibits transcription. Originally, such control sequences were called *distal control elements* because they can function at distances up to several hundred thousand base pairs upstream or downstream from the promoter they regulate (the word *distal* means "away from"). However, the distinctive feature of such sequences is not how far away from the promoter they are located, but the fact that their position relative to the promoter can vary significantly. Their orientation can even be reversed without interfering with their ability to regulate transcription. Thus, enhancers and silencers need

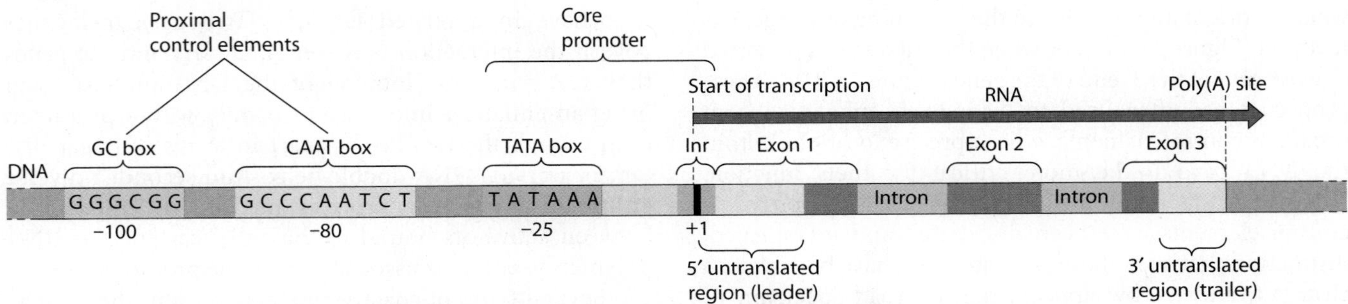

FIGURE 23-20 Anatomy of a Typical Eukaryotic Gene, with Its Core Promoter and Proximal Control Region. This diagram (not to scale) features a typical protein-coding eukaryotic gene, which is transcribed by RNA polymerase II. The promoter—called the core promoter to distinguish it from the proximal control region—is characterized by an initiator (Inr) sequence surrounding the transcriptional startpoint and a TATA box located about 25 bp upstream (to the 5′ side) of the startpoint. The core promoter is where the general transcription factors and RNA polymerase assemble for the initiation of transcription. Within about 100 nucleotides upstream from the core promoter lie several proximal control elements, which stimulate transcription of the gene by interacting with regulatory transcription factors. The number, identity, and exact location of the proximal elements vary from gene to gene. Here we show a simple case involving one copy of each of two common elements, the GC box and the CAAT box. The transcription unit includes a 5′ untranslated region (leader) and a 3′ untranslated region (trailer), which are transcribed and included in the mRNA but do not contribute sequence information for the protein product. At the end of the last exon is a site where, in the primary transcript, the RNA will be cleaved and given a poly(A) tail.

not be located at great distances from the promoter. They can be located quite close to the promoter and are even found occasionally *within* genes; one example is an enhancer found within an intron of certain antibody genes.

Since enhancers are better understood than silencers, we will consider them first. Enhancers, varying in specific sequence but sharing common properties, are associated with many eukaryotic genes. A typical enhancer contains several different control elements within it, each consisting of a short DNA sequence that serves as a binding site for a different regulatory transcription factor. Some of these DNA sequences may be identical to proximal control elements; the octamer and the GC box, for instance, can act both as proximal control elements and as components of enhancers. For an enhancer to function, the regulatory transcription factors that bind to its various control elements must be present. Because enhancers are involved in activating transcription, these regulatory transcription factors are called **activators.**

To investigate the properties of enhancers, researchers have used recombinant DNA techniques to alter enhancer location and orientation with respect to the regulated gene. As shown in **Figure 23-21,** such studies reveal that enhancers can function properly when relocated at variable distances from the transcription startpoint, as long as the promoter is present (Figure 23-21a, d–g). When either the promoter or enhancer is removed, no expression occurs (Figure 23-21b, c). Activity is retained when the position of the enhancer is moved further upstream (Figure 23-21e),

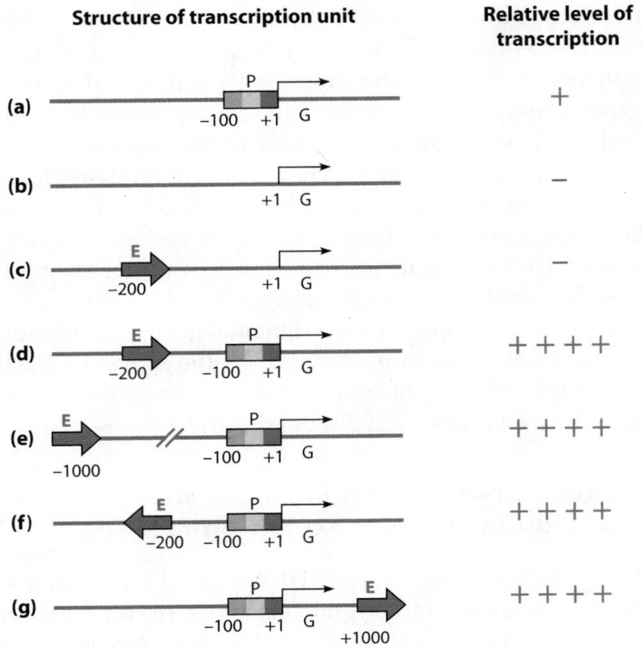

	Structure of transcription unit	Relative level of transcription
(a)	P; −100 +1 G	+
(b)	+1 G	−
(c)	E −200; +1 G	−
(d)	E −200; P −100 +1 G	++++
(e)	E −1000; P −100 +1 G	++++
(f)	E −200; P −100 +1 G	++++
(g)	P −100 +1 G; E +1000	++++

FIGURE 23-21 Effects of Enhancer Orientation and Location. Recombinant DNA techniques have been used to alter the orientation and location of DNA control elements and study the effect of such changes on gene transcription. The black arrows indicate the direction of transcription of gene G, with the startpoint (first transcribed nucleotide) labeled +1. The other numbers give the positions of nucleotides relative to the startpoint. **(a)** The core promoter (P) alone, in its typical location just upstream of gene G, allows a basal level of transcription to occur. **(b)** When the core promoter is removed from the gene, no transcription occurs. **(c)** An enhancer (E) alone cannot substitute for the promoter region, but **(d)** combining an enhancer with a core promoter results in a significantly higher level of transcription than occurs with the promoter alone. **(e)** This increase in transcription is observed when the enhancer is moved farther upstream, **(f)** when it is inverted in orientation, and **(g)** even when it is moved to the 3′ side of the gene.

when its orientation relative to the beginning of the gene is reversed (Figure 23-21f), or when the enhancer is relocated *downstream* of the 3′ end of the gene (Figure 23-21g). These properties are often used to distinguish enhancers from proximal control elements, whose precise locations within the DNA tend to be more critical for their function. Nonetheless, as more and more enhancers and proximal control elements have been discovered and investigated, distinctions between these two categories have become less clear-cut, and it now appears that a broad spectrum of transcription control elements exist with overlapping properties. At one extreme are certain proximal control elements that must be precisely located at specific positions near the promoter—in such cases, moving them even 15–20 nucleotides farther from the promoter causes them to lose their influence. At the other extreme are enhancers whose positions can be varied widely, up to tens of thousands of nucleotides away from a promoter, without interfering with their ability to activate transcription.

Silencers share many of the features of enhancers, except that they inhibit rather than activate transcription. Because the binding of regulatory transcription factors to silencers reduces rather than increases gene transcription rates, such transcription factors are called eukaryotic **repressors.** (While they resemble bacterial repressors in turning off transcription, the eukaryotic mechanisms are somewhat different and more varied.) We encountered an example of silencers earlier in the chapter when discussing the yeast *SIR* genes, which produce proteins that inhibit transcription of the *HMLα* and *HMRa* mating-type genes by binding to DNA sequences surrounding these two genes. The proteins produced by the *SIR* genes are examples of repressors, and the DNA sites these repressors bind to, located near *HMLα* and *HMRa*, are examples of silencers.

One complication encountered with silencers and enhancers stems from their ability to influence the transcription of faraway genes, which could be problematic if genes with opposing functions reside in neighboring regions. For example, a group of genes active in one cell type might lie near another set of genes that should not be active in those same cells. In such situations, DNA sequences called *insulators* are sometimes employed to prevent an enhancer (or silencer) from inadvertently acting on both groups of genes simultaneously. Insulator sequences, along with their associated binding proteins, create physical barriers between neighboring DNA regions that prevent enhancers or silencers from exerting their effects across the barrier.

Coactivators Mediate the Interaction Between Regulatory Transcription Factors and the RNA Polymerase Complex

Since enhancers and silencers can reside far from the genes they control, you might wonder how regulation is achieved over such long distances. In addressing this issue, we will again focus on enhancers (silencers appear to behave in a related fashion). Two basic principles govern the interaction between enhancers and the genes they regulate. First, looping of the DNA molecule can bring an enhancer into close proximity with a promoter, even though the two lie far apart in terms of linear distance along the DNA double helix. And second, a diverse group of **coactivator** proteins mediate the interaction between activators bound to the enhancer and the RNA polymerase complex associated with the promoter.

Several types of coactivators play roles in these interactions, including chromatin remodeling proteins and enzymes that modify histones, which we have seen are important for altering chromatin structure. In addition, a large multiprotein complex called **Mediator** functions as a coactivator by serving as a "bridge" that binds to activator proteins associated with the enhancer and to RNA polymerase, thereby linking enhancers to the components involved in initiating transcription at RNA polymerase II promoters. Mediator serves as a central coordinating unit for gene regulation, receiving both positive and negative inputs and transmitting the information to the transcription machinery.

Figure 23-22 shows how such interactions can trigger gene activation. In step ❶, a group of activator proteins bind to their respective DNA control elements within the enhancer, forming a multiprotein complex called an *enhanceosome.* One or more of these activator proteins causes the DNA to bend, creating a DNA loop that brings the enhancer close to the core promoter. In step ❷, the activators interact with coactivators such as chromatin remodeling proteins (SWI/SNF) and histone acetyltransferase (HAT), which alter chromatin structure to make the DNA in the promoter region more accessible. Finally, the activators bind to Mediator (step ❸), which facilitates the correct positioning of RNA polymerase and the general transcription factors at the promoter site and thus allows transcription to begin.

The sequence of events and the particular set of components involved in the preceding example represent a general model whose details vary among different genes and regulatory proteins. But these details are less important than the main idea: When activator proteins for a particular gene are present in the cell, their binding to an enhancer can trigger interactions with various coactivators that in turn lead to formation of the transcription complex at the promoter, resulting in more efficient initiation of transcription. Thus, differential gene transcription within a cell is determined mainly by the activators a cell makes (as well as by factors that control the activity of the activators, a topic we will discuss shortly).

Multiple DNA Control Elements and Transcription Factors Act in Combination

The realization that multiple DNA control elements and their regulatory transcription factors are involved in controlling eukaryotic gene transcription has led to a

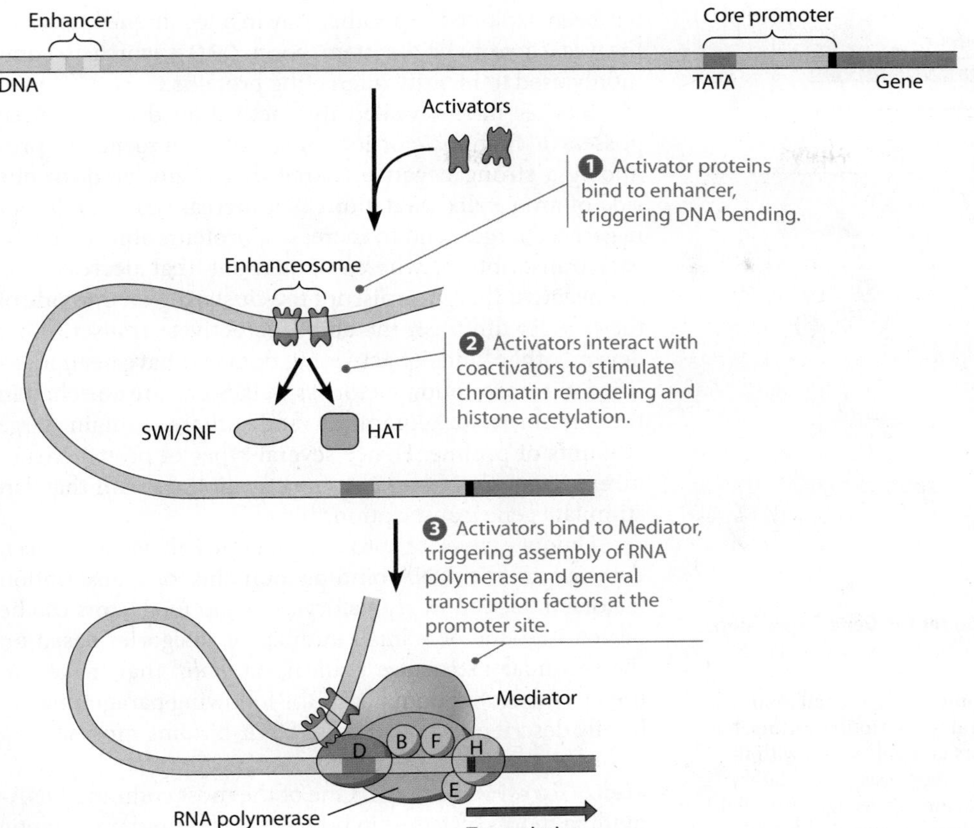

FIGURE 23-22 A Model for Enhancer Action. In this model, an enhancer located at a great distance along the DNA from the protein-coding gene it regulates is brought close to the core promoter by a looping of the DNA. ❶ Regulatory transcription factors called activators first bind to the enhancer elements, triggering DNA bending that brings the activators closer to the core promoter. ❷ The activators then interact with coactivator proteins such as SWI/SNF, which causes chromatin remodeling, and HAT (histone acetyltransferase), which catalyzes histone acetylation. The net effect is to decondense the chromatin and make the DNA in the promoter region more accessible. ❸ The activators then bind to another coactivator called Mediator, and the activator-Mediator complex facilitates the correct positioning of general transcription factors and RNA polymerase at the promoter site, allowing transcription to be initiated. For simplicity, the figure is drawn with only two activators, but often half a dozen or more are involved.

combinatorial model for gene regulation. This model proposes that a relatively small number of different DNA control elements and transcription factors, acting in different combinations, can establish highly specific and precisely controlled patterns of gene expression in different cell types.

The model begins with the assumption that some transcription factors are present in many cell types. These include the general transcription factors, required for transcription in all cells, plus any regulatory factors needed for transcribing constitutive genes and other genes that are frequently expressed. In addition, transcription of genes that encode tissue-specific proteins requires the presence of transcription factors or *combinations* of transcription factors that are unique to individual cell types. To illustrate this concept, **Figure 23-23** shows how such a model would allow liver cells to produce large amounts of proteins such as albumin but would prevent significant production of these proteins in other tissues, such as brain. The original version of the combinatorial model was "all or none," proposing that transcription of a gene could not be initiated unless the entire set of regulatory factors for the gene was present. What now seems clear is that a wide range of efficiencies of transcriptional initiation is possible for most genes: a basal level, occurring when *no* regulatory factors are available, to a maximum level, which occurs only when the full set of activating regulatory factors is present.

Several Common Structural Motifs Allow Regulatory Transcription Factors to Bind to DNA and Activate Transcription

Although not all transcription factors bind directly to DNA, those that do play critical roles in controlling transcription. Proteins in this category include the general transcription factor TFIID and, more importantly, the wide variety of regulatory transcription factors (activators

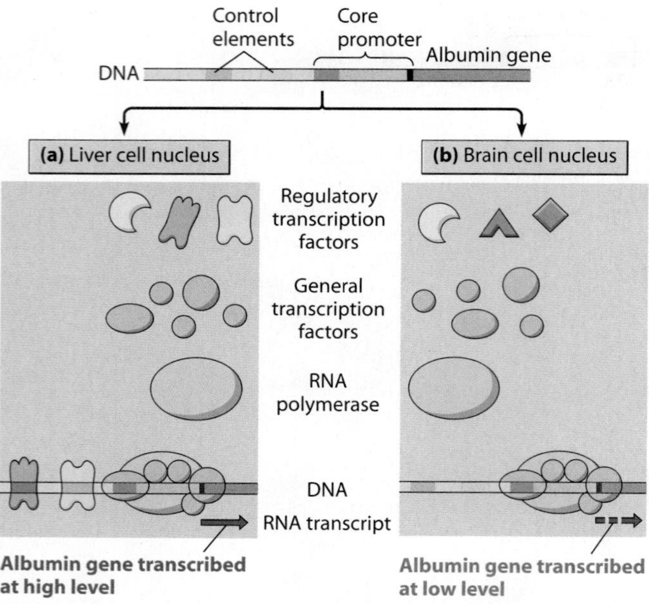

FIGURE 23-23 A Combinatorial Model for Gene Expression. The gene for the protein albumin, like other genes, is associated with an array of regulatory DNA elements. Here we show only two control elements, as well as the core promoter. Cells of all tissues contain RNA polymerase and the general transcription factors, but the set of regulatory transcription factors available varies with the cell type. As shown here, **(a)** liver cells contain a set of regulatory transcription factors that includes the factors for recognizing all the albumin gene control elements. When these factors bind to DNA, they facilitate transcription of the albumin gene at a high level. **(b)** Brain cells, however, have a different set of regulatory transcription factors that does not include all those for the albumin gene. Consequently, in brain cells, the transcriptional complex can assemble at the promoter but not very efficiently. The result is that brain cells transcribe the albumin gene only at a low level.

and repressors) that recognize and bind to specific DNA sequences found in proximal control elements, enhancers, and silencers. What features of these regulatory transcription factors enable them to carry out their functions?

Regulatory transcription factors possess two distinct activities, the ability to bind to a specific DNA sequence and the ability to regulate transcription. The two activities reside in separate protein domains. The domain that recognizes and binds to a specific DNA sequence is called the transcription factor's **DNA-binding domain,** whereas the protein region required for regulating transcription is known as the **transcription regulation domain** (or **activation domain** because many transcription factors activate, rather than inhibit, transcription). The existence of separate DNA-binding and activation domains has been demonstrated by "domain-swap" experiments in which the DNA-binding region of one transcription factor is combined with various regions of a second transcription factor. The resulting hybrid molecule can activate gene transcription only if it contains an activation domain provided by the second transcription factor. The separation of the DNA binding and activation domains of a transcription factor has been exploited in another way in a technique known as the yeast two-hybrid system (**Box 23B**), which is commonly used to identify interacting proteins.

Studies have revealed that activation domains often possess a high proportion of acidic amino acids, producing a strong negative charge that is clustered on one side of an α helix. Mutations that increase the number of negative charges tend to increase a protein's ability to activate transcription, whereas mutations that decrease the net negative charge or disrupt the clustering on one side of the α helix diminish the ability to activate transcription. Several other kinds of activating domains have been identified in transcription factors as well. Some are enriched in the amino acid glutamine, and others contain large amounts of proline. Hence several types of protein structure are capable of creating an activation domain that can stimulate gene transcription.

Unique types of protein structure have also been detected in the DNA-binding domains of transcription factors. In fact, most regulatory transcription factors can be placed into one of a small number of categories based on the secondary structure pattern, or *motif,* that makes up the DNA-binding domain. In the following paragraphs, we briefly describe several of these DNA-binding motifs.

Helix-Turn-Helix Motif. One of the most common DNA-binding motifs, detected in both eukaryotic and prokaryotic regulatory transcription factors, is the **helix-turn-helix** (**Figure 23-24a**). This motif consists of two α helices separated by a bend in the polypeptide chain. Although the amino acid sequence of the motif differs among various DNA-binding proteins, the overall pattern is always the same: One α helix, called the *recognition helix,* contains amino acid side chains that recognize and bind to specific DNA sequences by forming hydrogen bonds with bases located in the major groove of the DNA double helix, while the second α helix stabilizes the overall configuration through hydrophobic interactions with the recognition helix. The *lac* and *trp* repressors, the CAP protein, and many phage repressor proteins are examples of prokaryotic proteins exhibiting the helix-turn-helix motif, and transcription factors that regulate embryonic development (the class of factors encoded by homeotic genes, described later) are eukaryotic examples. Figure 23-24a includes a model of the phage λ repressor, a helix-turn-helix protein, bound to DNA. Like many DNA-binding regulatory proteins, the phage λ repressor consists of two identical polypeptides, each containing a DNA-binding domain.

Zinc Finger Motif. Initially identified in a transcription factor for the 5S rRNA genes (TFIIIA), the **zinc finger** DNA-binding motif consists of an α helix and a two-segment β sheet, held in place by the interaction of precisely positioned cysteine or histidine residues with a zinc ion. The number of zinc fingers present per protein molecule varies among the transcription factors that

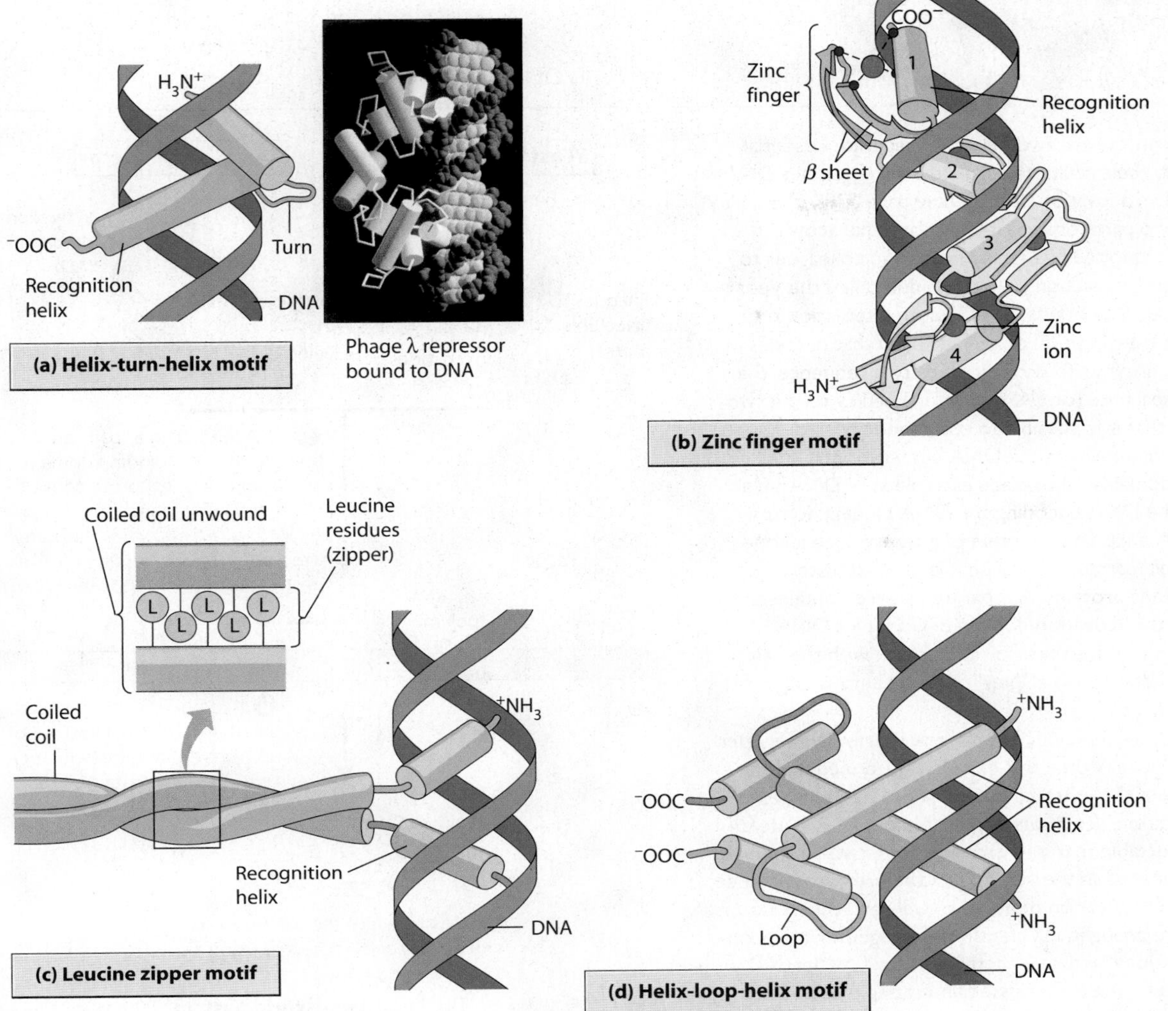

FIGURE 23-24 Common Structural Motifs in DNA-Binding Transcription Factors. Several motifs are commonly found in the DNA-binding domains of regulatory transcription factors. The parts of these domains that directly interact with specific DNA sequences are usually α helices, called recognition helices, that fit into DNA's major groove. In this figure, all α helices are shown as cylinders. **(a)** The helix-turn-helix motif, in which two α helices are joined by a short, flexible turn. The computer graphic model shows the phage 1 repressor, an example of a helix-turn-helix protein, bound to DNA. It is a dimer of two identical subunits. The helices of the two DNA-binding domains are light blue. **(b)** The zinc finger motif. Each zinc finger consists of an α helix and a two-segment, antiparallel β sheet (shown as ribbons), all held together by the interaction of four cysteine residues, or two cysteine and two histidine residues, with a zinc ion. At the top of the diagram, these key residues are shown as small purple balls; the zinc ions are shown as larger red balls. Zinc finger proteins typically have several zinc fingers in a row; here we see four. **(c)** The leucine zipper motif, in which an α helix with regularly arranged leucine residues in one polypeptide (green) interacts with a similar region in a second polypeptide (purple). The two helices coil around each other. **(d)** The helix-loop-helix motif, in which a short α helix connected to a longer α helix by a polypeptide loop interacts with a similar region in another polypeptide to create a dimer.

possess them, ranging from two fingers to several dozen or more. Figure 23-24b shows a protein with four zinc fingers in a row (TFIIIA has nine). Zinc fingers protrude from the protein surface and serve as the points of contact with specific base sequences in the major groove of the DNA.

Leucine Zipper Motif. The **leucine zipper** motif is formed by an interaction between two polypeptide chains, each containing an α helix with regularly spaced leucine residues. Because leucines are hydrophobic amino acids that attract one another, the stretch of leucines exposed on the outer surface of one α helix can interlock with a comparable stretch of leucines on the other α helix, causing the two helices to wrap around each other into a coil that "zippers" the two α helices together (Figure 23-24c). In some transcription factors, leucine zippers are used to "zip" two identical polypeptides together. In other transcription factors, two kinds of polypeptides are joined

BOX 23B TOOLS OF DISCOVERY

The Yeast Two-Hybrid System

Transcriptional activators have two physically separate, essential domains that lead to their ability to regulate transcription: a DNA-binding domain and an activation domain (see page 738). The ability to separate these two parts of a yeast transcriptional activator known as *Gal4* was exploited by Stanley Fields and colleagues to study protein-protein interactions in a technique called the **yeast two-hybrid system** (**Figure 23B-1**), or an interaction trap assay. *Gal4* activates transcription of a gene *(Gal1)* that encodes a galactose metabolic enzyme. To do so, it binds to a sequence, the *upstream activating sequence* (or *UAS*), of *Gal1*. The key to the two-hybrid technique is that it is possible to use genetic engineering to separate the DNA that encodes the DNA binding domain of *Gal4* from its activation domain and to place each piece of DNA on a different plasmid. The DNA encoding the DNA binding domain is placed next to DNA encoding a protein of interest. Such a construct is called a *"bait" construct*, because it is used to "catch" interactions with other proteins. A separate plasmid contains a sequence encoding the activation domain of *Gal4*, plus DNA encoding a protein we wish to test for interaction with the bait protein. This construct is called a *"prey" construct* because it is "caught" by the assay.

The two plasmids are introduced into yeast carrying a *reporter gene*, often the *lacZ* gene. When this gene is transcribed and the mRNA is translated in the yeast, its presence can be detected through a color reaction. Yeast carrying both plasmids are identified by virtue of selectable markers carried by the two plasmids. If the bait protein produced in the yeast binds the prey protein, then the DNA binding and activation domains of *Gal4* attached to each protein fragment are brought sufficiently close together to reconstitute the transcriptional activating activity of *Gal4*. When this happens, the reporter gene is expressed, indicating that the bait and the prey proteins bind one another.

The yeast two-hybrid assay is a powerful way to rapidly assess whether proteins interact. If a library of prey plasmids are used to transform yeast, this method can be used to screen for unknown proteins that bind to a known bait. In addition, genome-wide two-hybrid screens have been carried out in yeast, flies, worms, and humans. The information from these large screens can suggest that two proteins interact, which can be followed up using more traditional biochemical techniques.

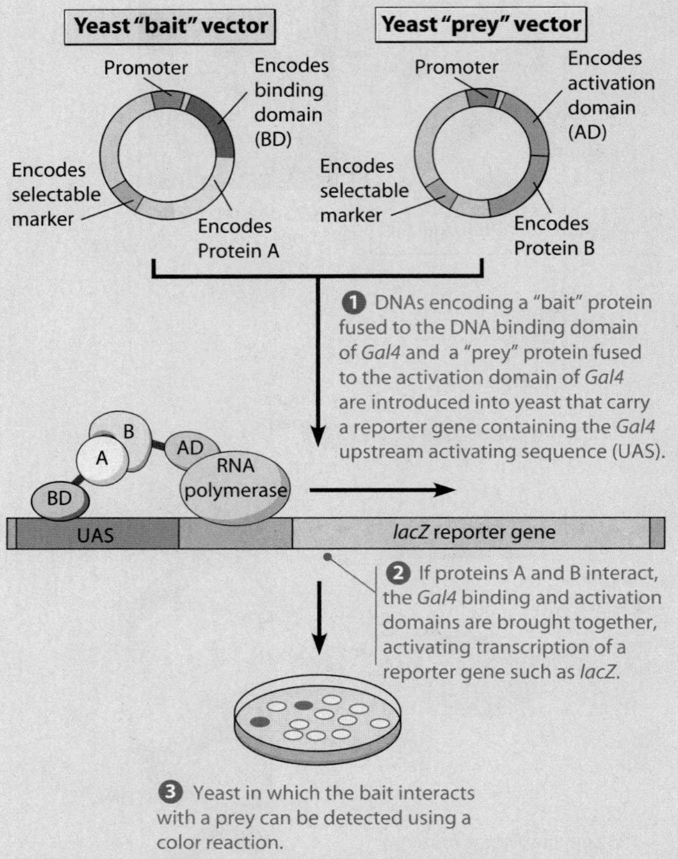

Yeast "bait" vector
Promoter
Encodes binding domain (BD)
Encodes selectable marker
Encodes Protein A

Yeast "prey" vector
Promoter
Encodes activation domain (AD)
Encodes selectable marker
Encodes Protein B

1 DNAs encoding a "bait" protein fused to the DNA binding domain of *Gal4* and a "prey" protein fused to the activation domain of *Gal4* are introduced into yeast that carry a reporter gene containing the *Gal4* upstream activating sequence (UAS).

B
A
AD
RNA polymerase
BD
UAS
lacZ reporter gene

2 If proteins A and B interact, the *Gal4* binding and activation domains are brought together, activating transcription of a reporter gene such as *lacZ*.

3 Yeast in which the bait interacts with a prey can be detected using a color reaction.

FIGURE 23B-1 The Yeast Two-Hybrid System. The yeast two-hybrid system detects protein-protein interactions by using two different DNA constructs: (1) a "bait" vector that fuses DNA encoding the DNA binding domain of *Gal4* to DNA encoding a protein of interest (Protein A), and (2) a "prey" vector that fuses DNA encoding the activation domain of *Gal4* to DNA encoding a second protein (Protein B). If the proteins interact, a reporter gene (in this case, *lacZ*) will be expressed.

Yeast Two-Hybrid System

together. In either case, DNA binding is made possible by two additional α-helical regions located adjacent to the leucine zipper. These two α-helical segments, one derived from each of the two polypeptides, fit into the DNA's major groove and bind to specific base sequences.

Helix-Loop-Helix Motif. The **helix-loop-helix** motif is composed of a short α helix connected by a loop to another, longer α helix (Figure 23-24d). Like leucine zippers, helix-loop-helix motifs contain hydrophobic regions that usually connect two polypeptides, which may be either similar or different. Formation of the four-helix bundle results in the juxtaposition of a recognition helix derived from one polypeptide with a recognition helix derived from the other polypeptide, creating a two-part DNA-binding domain.

DNA Response Elements Coordinate the Expression of Nonadjacent Genes

So far we have focused our attention on how transcription factors bind to DNA and regulate the transcription of individual genes. But eukaryotic cells, like bacteria, often need to activate a group of related genes at the same time. In unicellular eukaryotes, as in bacteria, such coordinate gene regulation may be required to respond to some signal from the external environment. In multicellular eukaryotes, coordinate gene regulation is critical for the development and functioning of specialized tissues. For example, during embryonic development a single fertilized animal egg may give rise to trillions of new cells of hundreds of differing types, each transcribing a different group of genes—nerve cells expressing genes required for nerve function, muscle cells expressing genes required for muscle function, and so forth. How do eukaryotes coordinate the expression of groups of related genes under such conditions?

Unlike the situation in prokaryotes, where genes with related functions often lie next to each other in operons, eukaryotic genes that must be turned on (or off) at the same time are usually scattered throughout the genome. To coordinate the expression of such physically separated genes, eukaryotes employ DNA control sequences called **response elements** to turn transcription on or off *in response to* a particular environmental or developmental signal. Response elements can function either as proximal control elements or as components of enhancers. In either case, placing the same type of response element next to genes residing at different chromosomal locations allows these genes to be controlled together even though they are not located next to one another.

Because they allow groups of genes to be controlled in a coordinate fashion, response elements play important roles in regulating gene expression during embryonic development and during tissue responses to changing environmental and physiological conditions. In the next several sections, we will describe some examples of such coordinated gene regulation.

Steroid Hormone Receptors Act as Transcription Factors That Bind to Hormone Response Elements

The phenomenon of coordinated gene regulation is nicely illustrated by the behavior of *steroid receptor* proteins, which we introduced in Chapter 14. Steroid receptors, as well as the related *retinoid receptors*, belong to the zinc finger class of transcription factors and typically consist of three domains: One domain recognizes and binds to a specific response element in DNA, a second domain binds to a particular steroid hormone, and the third domain activates transcription. The gene-specific effects of steroid hormones stem from the ability of steroid receptors to act as transcription factors that bind to DNA sequences called **hormone response elements.** All the genes activated by a particular steroid hormone are associated with the same type of response element, allowing them to be regulated together. For example, genes activated by estrogen have a 15-bp *estrogen response element* near the upstream end of their promoters, whereas genes activated by glucocorticoids lie adjacent to a *glucocorticoid response element* exhibiting a slightly different base sequence (**Figure 23-25a**).

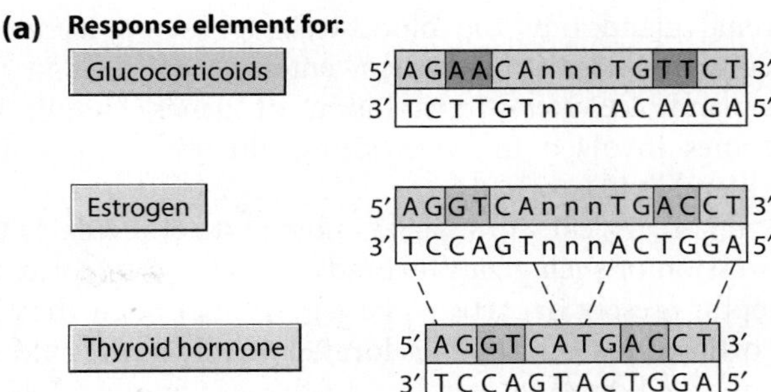

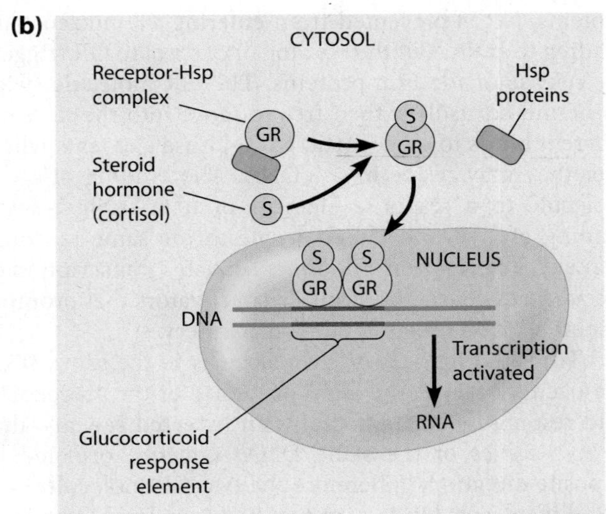

FIGURE 23-25 Steroid Hormone Receptor Interaction with DNA. **(a)** Different types of hormone response elements. Notice that all three examples contain inverted repeats (two copies of the same sequence oriented in opposite directions). For example, reading the sequence of the glucocorticoid response element in the 5′ ⟶ 3′ direction from either end yields the same DNA sequence: 5′–AGAACA–3′. The highlighted nucleotides are the only bases of the inverted repeat sequences that vary between the three types of elements. The thyroid hormone element contains the same inverted repeat sequences as the estrogen element, but the three bases that separate the two copies of the sequence in the estrogen element are not present. (The "n" signifies that any nucleotide can be located at that position. The dashed lines are included to help you line up the comparable regions of the estrogen and thyroid hormone elements.) **(b)** Activation of transcription by glucocorticoid receptors. The steroid hormone cortisol (S) diffuses through the plasma membrane and binds to the glucocorticoid receptor (GR), causing the release of Hsp proteins and activating the GR molecule's DNA-binding site. The GR molecule then enters the nucleus and binds to a glucocorticoid response element in DNA, which in turn causes a second GR molecule to bind to the same response element. The resulting GR dimer activates transcription of the adjacent gene.

MB Regulation of Gene Expression by Steroid Hormones

How are these hormone response elements used by steroid receptors to control gene activity? To illustrate how this arrangement works with a specific steroid hormone, let us consider *cortisol* (hydrocortisone) as an example. Cortisol is a member of a related group of steroid hormones, called *glucocorticoids* because they stimulate *glucose* production and are made in the adrenal *cortex.* In response to physical stress, cortisol is released from the adrenal gland into the bloodstream. Among its many effects, the circulating cortisol enters target cells in the liver and activates transcription of genes coding for enzymes involved in synthesizing glucose from amino acids and lipids.

Figure 23-25b summarizes how cortisol regulates the expression of such genes by binding to the glucocorticoid receptor present in cortisol's target cells. Normally the glucocorticoid receptor (GR) is located mainly in the cytosol, where it is bound to members of the *Hsp* family of chaperone proteins. As long as it remains associated with Hsp proteins, GR is prevented from entering the nucleus and binding to DNA. But the binding of cortisol to GR triggers the release of the Hsp proteins. The GR molecule (with its bound cortisol) is then free to move into the nucleus, where it binds to glucocorticoid response elements wherever they may reside in the DNA. The binding of a GR molecule to a response element in turn facilitates the binding of a second GR molecule to the same response element. The resulting GR *dimer* activates transcription of the adjacent genes by recruiting coactivators that promote assembly of the transcriptional machinery.

This binding of *two* GR molecules to the same DNA site occurs because the DNA sequence of the glucocorticoid response element contains an **inverted repeat**—that is, *two* copies of the same DNA sequence oriented in opposite directions. Therefore the two GR molecules in a GR dimer each bind to one of the two DNA sequence repeats. Such binding of a protein dimer to an inverted repeat sequence within a DNA control element is a common theme among regulatory transcription factors.

Since glucocorticoid response elements are located near all genes that need to be turned on by cortisol, the presence of cortisol activates all these genes simultaneously, regardless of their chromosomal location. Although there are minor differences between various steroid hormones, this model provides a framework for understanding how steroid hormones, transported through the bloodstream to virtually all cells of the body, can turn on a specific set of genes in the appropriate target tissues.

Although steroid hormone receptors usually stimulate gene transcription, in a few cases they inhibit it. For example, the glucocorticoid receptor binds to two types of DNA response elements—one type associated with genes whose transcription is activated, the other with genes whose transcription is inhibited. After binding to the inhibitory type of response element, the glucocorticoid receptor does not form dimers as it does when bound to activating response elements. Instead, the nondimerized glucocorticoid receptor depresses the initiation of gene transcription by recruiting histone deacetylase, which promotes chromatin condensation. The existence of separate activating and inhibiting response elements for the same hormone receptor allows a single hormone to stimulate transcription of genes coding for one group of proteins while it simultaneously depresses transcription of genes coding for other proteins.

CREBs and STATs Are Examples of Transcription Factors Activated by Phosphorylation

The steroid hormone receptors we have been discussing are allosteric proteins that regulate gene expression by changing their DNA-binding affinity in response to steroid hormones in much the same way that bacterial repressors change their DNA-binding properties after binding to the appropriate small molecules. An alternative approach for controlling the activity of transcription factors is based on *protein phosphorylation* (addition of phosphate groups). One common example of this type of control involves *cAMP (cyclic AMP)*, the widely employed second messenger whose role in mediating the effects of a variety of extracellular signaling molecules was discussed in Chapter 14. As we saw earlier in this chapter, cAMP also plays a role in controlling catabolite-repressible operons in bacteria. In eukaryotes, cAMP functions by stimulating the activity of protein kinase A (Figure 14-8), which in turn catalyzes the phosphorylation of a variety of proteins, including a transcription factor called **CREB** (**Figure 23-26**). The CREB protein normally binds to DNA sequences called *cAMP response elements (CRE)*, which are located adjacent to genes whose transcription is induced by cAMP. Upon being phosphorylated, the CREB protein recruits a transcriptional coactivator called **CBP** (CREB-binding protein). CBP then catalyzes histone acetylation, which loosens the packing of nucleosomes and interacts with RNA polymerase to facilitate assembly of the transcription machinery at nearby gene promoters.

Protein phosphorylation is also involved in activating a family of transcription factors called **STATs** (for **s**ignal

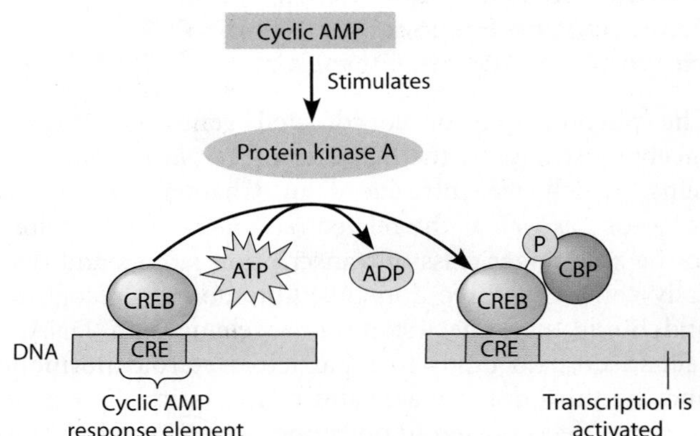

FIGURE 23-26 Activation of Gene Transcription by Cyclic AMP. Genes activated by cyclic AMP possess an upstream cyclic AMP response element (CRE) that binds a transcription factor called CREB. In the presence of cyclic AMP, cytoplasmic protein kinase A is activated, and its activated catalytic subunit then moves into the nucleus, where it catalyzes CREB phosphorylation. Phosphorylated CREB in turn binds to CBP, a transcriptional coactivator that exhibits histone acetyltransferase activity. Besides catalyzing histone acetylation that loosens the packing of nucleosomes, CBP associates with RNA polymerase and facilitates assembly of the transcription machinery at gene promoters.

transducers and activators of transcription). Among the signaling molecules that activate STATs are the *interferons,* a group of glycoproteins produced and secreted by animal cells in response to viral infection. The secreted interferons bind to receptors on the surface of neighboring cells, causing them to produce proteins that make the target cells more resistant to viral infection. The binding of interferon to its cell surface receptor causes the cytosolic domain of the receptor to associate with and activate an intracellular protein kinase called **Janus kinase (Jak).** The activated Jak protein in turn catalyzes the phosphorylation of STAT molecules, which are normally present in an inactive form in the cytoplasm. Phosphorylation causes the STAT molecules to dimerize and move from the cytoplasm to the nucleus, where they bind to appropriate DNA response elements and activate transcription.

The Jak-STAT signaling pathway exhibits considerable specificity. For example, the different types of interferon trigger the phosphorylation and activation of different STAT proteins. Each STAT associates with its own distinctive DNA response element and activates the transcription of a unique set of genes. In addition to interferons, several other types of signaling molecules bind to cell surface receptors that trigger the phosphorylation of STATs.

STATs and the CREB protein are only two of the many types of transcription factors known to be activated by phosphorylation. You learned in Chapter 14, for example, that the TGFβ family of growth factors bind to plasma membrane receptors whose activation leads to the phosphorylation of cytoplasmic proteins called *Smads,* which then travel to the nucleus and function as transcription factors that turn on the transcription of specific genes. And we described in Chapter 19 how other growth factors trigger the activation of protein kinases called *MAP kinases,* which phosphorylate and activate transcription factors involved in controlling cell growth and division.

The Heat-Shock Response Element Coordinates the Expression of Genes Activated by Elevated Temperatures

The **heat-shock genes** provide another example of how eukaryotes coordinate the regulation of genes located at different chromosomal sites. Heat-shock genes were initially defined as genes expressed in response to an increase in temperature. But they are now known to respond to other stressful conditions as well, and so they are also called *stress-response genes.* There are a number of different heat-shock genes in both prokaryotic and eukaryotic genomes, and the appropriate environmental trigger activates all of them simultaneously. For example, briefly warming cultured cells by raising the temperature a few degrees activates the transcription of multiple heat-shock genes. Although the functions of these genes are not completely understood, the proteins produced by at least some of them help minimize the damage resulting from thermal denaturation of important cellular proteins.

In cells of *Drosophila,* for example, the most prominent product of the heat-shock genes is Hsp70 (page 691), a molecular chaperone involved in normal protein folding that can also facilitate the refolding of heat-damaged proteins. The region immediately upstream from the start site of the *hsp70* gene contains several sequences commonly encountered in eukaryotic promoters, such as a TATA box, a CAAT box, and a GC box. In addition, a 14-bp sequence called the **heat-shock response element** is located 62 bases upstream from the transcription start site of the *Drosophila hsp70* gene and in a comparable location in other genes whose transcription is activated by high temperatures. Subsequent investigations have revealed that the activation of heat-shock genes is mediated by the binding of a protein called the *heat-shock transcription factor* to the heat-shock response element. The heat-shock transcription factor is present in an inactive form in nonheated cells, but elevated temperatures cause a change in the structure of the protein that allows it to bind to the heat-shock response element in DNA. The protein is then further modified by phosphorylation, which makes it capable of activating gene transcription.

The heat-shock system again illustrates a basic principle of eukaryotic coordinate regulation: Genes located at different chromosomal sites can be activated by the same signal if the same response element is located near each of them.

Homeotic Genes Code for Transcription Factors That Regulate Embryonic Development

Pattern Formation

Some especially striking examples of coordinate gene regulation in eukaryotes involve an unusual class of genes known as **homeotic genes.** When mutations occur in one of these genes, a strange thing happens during embryonic development—one part of the body is replaced by a structure that normally occurs somewhere else. The discovery of homeotic genes can be traced back to the 1940s, when Edward B. Lewis discovered a cluster of *Drosophila* genes, the *bithorax gene complex,* in which certain mutations cause drastic developmental abnormalities such as the growth of an extra pair of wings (compare **Figure 23-27a** and Figure 23-27b). Later, Thomas C. Kaufman and his colleagues discovered a second group of genes that, when mutated, lead to different but equally bizarre developmental changes—for example, causing legs to grow from the fly's head in place of antennae (Figure 23-27c). This group of genes is the *antennapedia gene complex.* The bithorax and antennapedia genes are called homeotic genes because *homeo* means "alike" in Greek, and mutations in these genes change one body segment of *Drosophila* to resemble another. A clue to how homeotic genes work first emerged from the discovery that the *bithorax* and *antennapedia* genes each contain a similar, 180-bp sequence near their 3′ end that resembles a comparable sequence found in other homeotic genes. Termed the

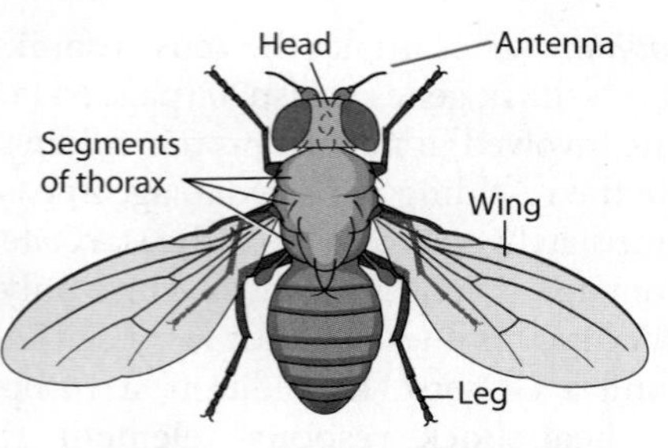

(a) Wild-type *Drosophila*. Two wings and six legs extend from its three thorax segments.

(b) Bithorax mutant. An additional set of wings is formed.

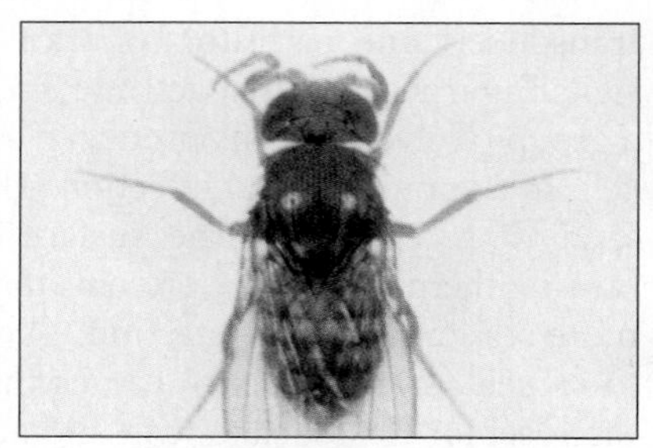

(c) Antennapedia mutant. Legs develop where the antennae should be.

FIGURE 23-27 Homeotic Mutants of *Drosophila*. Mutations in the *bithorax* gene complex convert the third thoracic segment to a second thoracic segment (the wing-producing segment), and so an additional set of wings is formed. Mutations in the *antennapedia* gene complex cause legs to develop where the antennae should be.

MB Gene Regulation of the Development of the Drosophila Body Plan

homeobox, this DNA sequence codes for a stretch of 60 amino acids called a **homeodomain** (**Figure 23-28**).

Although such phenomena might at first appear to be no more than oddities of nature, the growth of an extra pair of wings or legs in the wrong place suggests that homeotic genes play a key role in determining the body plan of developing fly embryos. Homeotic genes exert their influence by coding for a family of regulatory transcription factors that activate (or inhibit) the transcription of developmentally important genes by binding to specific DNA sequences. By binding to all copies of the appropriate DNA element in the genome, each homeotic transcription factor can influence the expression of dozens or even hundreds of genes in the growing embryo.

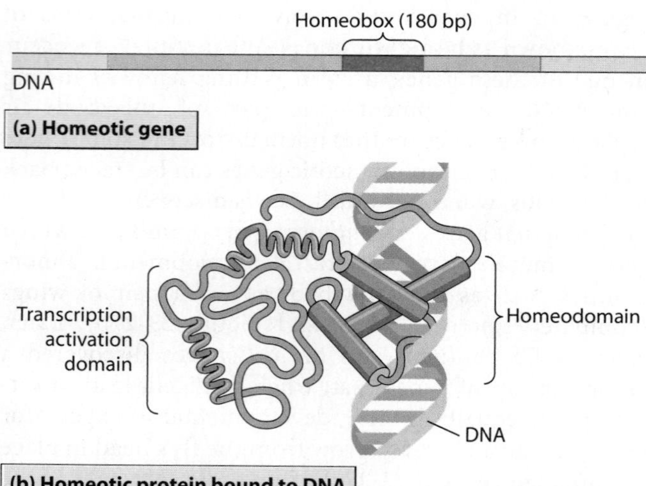

FIGURE 23-28 Homeotic Genes and Proteins. Homeotic mutants like the ones in Figure 23-27 have mutations in homeotic genes. **(a)** Homeotic genes all contain a 180-bp segment, called a homeobox, that encodes a homeodomain in the homeotic protein. **(b)** The homeodomain (red) is a helix-turn-helix DNA-binding domain that, in combination with a transcription activation domain, enables the protein to function as a regulatory transcription factor. Notice that the homeodomain has three α helices (shown as cylinders).

The result of this coordinated gene regulation is the establishment of fundamental body characteristics such as appendage shape and location.

A remarkable result of modern genomics and molecular developmental biology is the high degree of conservation of homeotic genes in animals. Sequences similar to those in the homeotic genes of flies have been detected in the genes of organisms as diverse as nematode worms, insects, sea squirts, frogs, mice, and humans—an evolutionary distance of more than 500 million years. Remarkably, with a few minor exceptions, the physical organization of these *Hox* genes has been retained across these great distances. In both flies and vertebrates, the 3′- 5′ location of these genes within a chromosome corresponds to the anterior to posterior region along the body axis these proteins regulate (**Figure 23-29**). In vertebrates, there appears to have been a duplication of the original single set of these genes possessed by fruit flies and other simple animals: Whereas fruit flies have one set of homeotic genes, mammals have four sets of *Hox* genes. Based on analysis of the expression pattern of mouse *Hox* genes and the effects of making "knockout" mutants for individual and multiple *Hox* genes, it is clear that, like their fly counterparts, they control the identify of structures along the anterior-posterior axis of vertebrates. Mutations in human *Hox* genes have also been identified. For example, mutations in the *HoxD13* locus result in *synpolydactyly,* characterized by fusions and duplications of fingers and toes. A major theme of modern efforts to unite evolutionary and developmental biology is to understand how subtle changes in *Hox* genes lead to changes in the body plans of higher animals.

Eukaryotic Gene Regulation: Posttranscriptional Control

We have now examined the first two levels of regulation for eukaryotic gene expression, namely genomic controls and transcriptional controls. After transcription has taken place, the flow of genetic information involves a complex series of

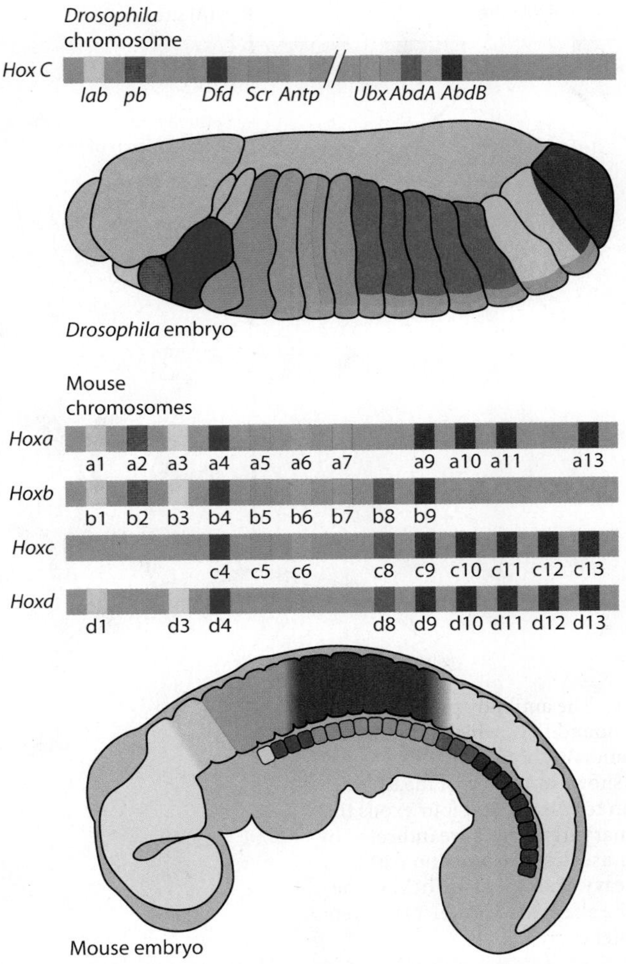

MB Activity: Role of bicoid Gene in Drosophila Development; Activity: Early Pattern Formation in Drosophila

FIGURE 23-29 Organization and Expression of *Hox* Genes in *Drosophila* and the mouse. In the *Drosophila* embryo, int = intercalary segment, mx = maxillar segment, la = labial segment, T = thoracic, A = abdominal. Note that the mouse has four *Hox* complexes, labeled a-d. Corresponding genes in each of the four *Hox* clusters in the mouse are numbered with the lowest numbers being the most anteriorly expressed. Note that some genes are not found in some clusters. Genes with the same color represent similar genes. Note that the single *AbdB* gene in flies (dark purple) appears to have been duplicated multiple times in the mouse (genes 9-13).

posttranscriptional events, any or all of which can turn out to be regulatory points (see Figure 23-10, steps ❸, ❹, and ❺). Posttranscriptional regulation may be especially useful in providing ways to rapidly fine-tune the pattern of gene expression, allowing cells to respond to transient changes in the intracellular or extracellular environment without changing their overall transcription patterns.

Control of RNA Processing and Nuclear Export Follows Transcription

Posttranscriptional control begins with the processing of primary RNA transcripts, which provides many opportunities for the regulation of gene expression. As we learned in Chapter 21, virtually all RNA transcripts in eukaryotic nuclei undergo substantial processing, including addition of a 5′ cap and a 3′ poly(A) tail, chemical modifications such as methylation, splicing together of exons accompanied by the removal of introns, and RNA editing. Among these processing events, RNA splicing is an especially important control site because its regulation allows cells to create a variety of different mRNAs from the same pre-mRNA, thereby permitting a gene to generate more than one protein product.

This phenomenon, called **alternative splicing,** is based on a cell's ability to permit some splice sites to be skipped and others to become activated (see Figure 21-23). Roughly half of all human genes produce pre-mRNAs that are alternatively spliced, reinforcing the view that splicing control is an important component of gene regulation. Splicing patterns are determined by various factors, including the presence of regulatory proteins and small nucleolar RNA molecules (snoRNAs) that bind to *splicing enhancer* or *splicing silencer* sequences in pre-mRNA. By increasing the ways in which a given pre-mRNA can be spliced, these regulatory molecules make it possible for a single gene to yield dozens or even hundreds of alternate mRNAs. A remarkable example occurs in the inner ear of birds, where 576 alternatively spliced forms of mRNA are produced from a single potassium channel gene. Expression of varying subsets of these mRNAs in different cells of the inner ear facilitates perception of different sound frequencies.

Figure 23-30 illustrates another well-documented example of alternative splicing, in this case involving the mRNA coding for a type of antibody called *immunoglobulin M (IgM)*. The IgM protein exists in two forms, a secreted version and a version that becomes incorporated into the plasma membrane of the cell that makes it. Like all antibodies, IgM consists of four polypeptide subunits—two heavy chains and two light chains. It is the sequence of the heavy chain that determines whether IgM is secreted or membrane bound. The gene coding for the heavy chain has two alternative poly(A) sites where RNA transcripts can terminate, yielding two kinds of pre-mRNA molecules that differ at their 3′ ends. The exons within these pre-mRNAs are then spliced together in two different ways, producing mRNAs that code for either the secreted version of the heavy chain or the version that is bound to the plasma membrane. Only the splicing pattern for the membrane-bound version includes the exons encoding the hydrophobic amino acid domain that anchors the heavy chain to the plasma membrane.

After RNA splicing, the next posttranscriptional step subject to control is the export of mRNA through nuclear pores to the cytoplasm. We know that this export step can be controlled because RNAs exhibiting defects in capping or splicing are not readily exported from the nucleus. Even with normal mRNAs, certain molecules are retained in the nucleus until their export is triggered by a stimulus signaling that a particular mRNA is needed in the cytoplasm for translation. Export control has also been observed

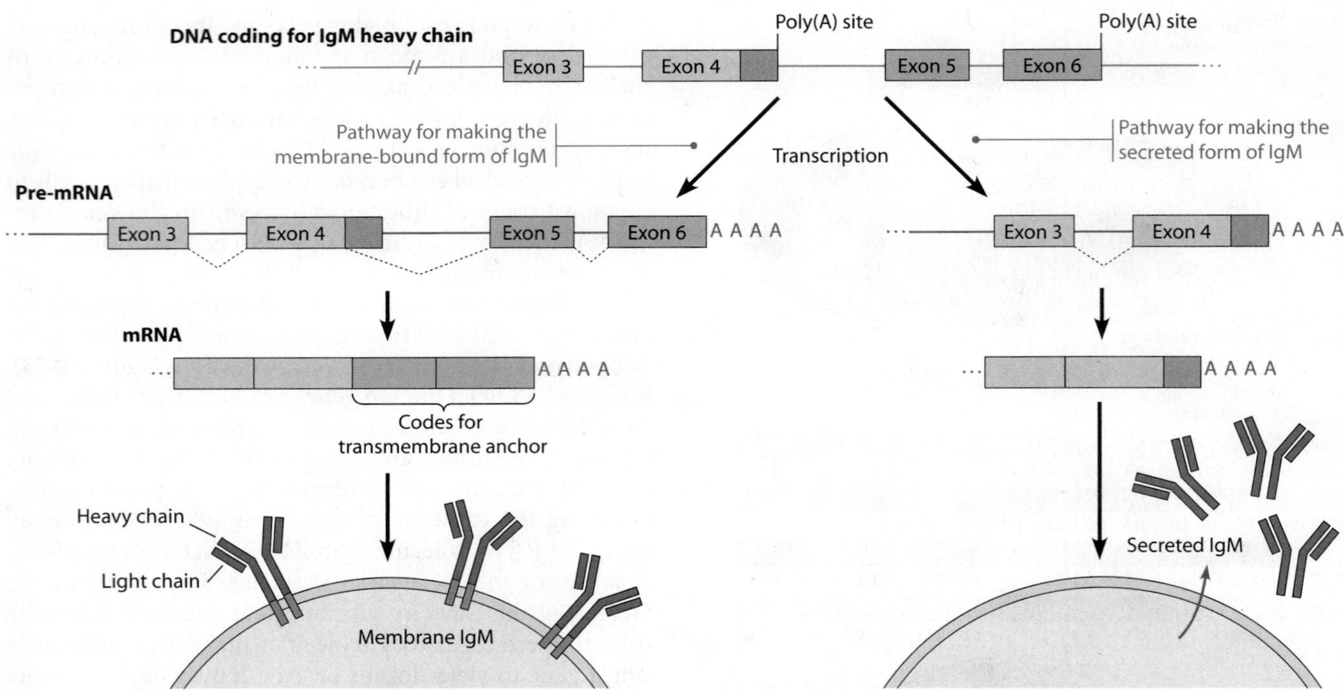

FIGURE 23-30 Alternative Splicing to Produce Variant Gene Products. The antibody protein immunoglobulin M (IgM) exists in two forms, secreted IgM and membrane-bound IgM, which differ in the carboxyl ends of their heavy chains. A single gene carries the genetic information for both types of IgM heavy chain. The end of the gene corresponding to the polypeptide's C-terminus is shown at the top of the figure. This DNA has two possible poly(A) addition sites, where RNA transcripts can terminate, and four exons that can be spliced in two alternative configurations. The splices made in the primary transcripts are indicated by V-shaped, thin dashed lines beneath each pre-mRNA. A splicing pattern that uses a splice junction within exon 4 and retains exons 5 and 6 results in the synthesis of membrane IgM heavy chains that are held in the plasma membrane by a transmembrane anchor encoded by exons 5 and 6. The alternative product is secreted because a splice within exon 4 is not made and the transcript is terminated after exon 4.

with viral RNAs produced in infected cells. For example, HIV—the virus responsible for AIDS—produces an RNA molecule that remains in the nucleus until a viral protein called Rev is synthesized. The amino acid sequence of the Rev protein includes a *nuclear export signal* (page 540) that allows Rev to guide the viral RNA out through the nuclear pores and into the cytoplasm.

Translation Rates Can Be Controlled by Initiation Factors and Translational Repressors

Once mRNA molecules have been exported from the nucleus to the cytoplasm, several **translational control** mechanisms are available to regulate the rate at which each mRNA is translated into its polypeptide product. Some translational control mechanisms work by altering ribosomes or protein synthesis factors; others regulate the activity and/or stability of mRNA itself. Translational control takes place in prokaryotes as well as eukaryotes, but we will restrict our discussion to several eukaryotic examples.

One well-studied example of translational control occurs in developing erythrocytes, where *globin* polypeptide chains are the main product of translation. The synthesis of globin depends on the availability of *heme*—the iron-containing prosthetic group that attaches to globin chains to form the final protein product, hemoglobin. Developing erythrocytes normally synthesize globin at a high rate. However, globin synthesis would be wasteful if not enough heme were available to complete the formation of hemoglobin molecules, so erythrocytes have developed a mechanism for adjusting polypeptide synthesis to match heme availability.

This mechanism involves a protein kinase, called *heme-controlled inhibitor (HCI),* whose activity is regulated by heme. HCI is inactive in the presence of heme. But when heme is absent, HCI phosphorylates and inhibits eIF2, one of several proteins required for initiating translation in eukaryotes (page 687). The reduced activity of eIF2 depresses all translational activity, but since globin chains account for more than 90% of the polypeptides made in developing erythrocytes, the main impact of HCI in these cells is on globin synthesis. Other kinases that phosphorylate eIF2 have been detected in a diverse array of cell types, suggesting that phosphorylation of eIF2 is a widely used mechanism for regulating translation.

The use of protein phosphorylation to control translation is not restricted to effects on eIF2. Eukaryotic initiation factor eIF4F (page 687), a multiprotein complex that binds to the 5′ mRNA cap, is also regulated by phosphorylation, although in this case phosphorylation activates rather than inhibits the initiation factor. An example of this type of control is observed in cells infected by adenovirus, which inhibits protein synthesis in infected cells by blocking phosphorylation of eIF4E, the cap-binding subunit of eIF4F. Other viruses inhibit the eIF4F complex by producing proteases that cleave eIF4F into two fragments, one containing the site that binds to the 5′ cap of mRNA molecules and the other containing the site that binds to ribosomes. Surprisingly, the ribosome-binding fragment can still initiate the translation of viral (but not cellular) mRNA by binding to the *IRES* sequence (page 687) that lies directly upstream of the start codon of some viral mRNAs. This allows a virus to hijack the translation machinery for translating its own mRNA while simultaneously shutting down the translation of normal cellular mRNAs.

The types of translational control we have discussed so far are relatively nonspecific in that they affect the translation rates of all mRNAs within a cell. An example of a more specific type of translational control is seen with *ferritin,* an iron-storage protein whose synthesis is selectively stimulated in the presence of iron. The key to this selective stimulation lies in the 5′ untranslated leader sequence of ferritin mRNA, which contains a 28-nucleotide segment—the **iron-response element (IRE)**—that forms a hairpin loop required for the stimulation of ferritin synthesis by iron. If the IRE sequence is experimentally inserted into a gene whose expression is not normally regulated by iron, translation of the resulting mRNA becomes iron sensitive.

Figure 23-31 shows how the IRE sequence works. When the iron concentration is low, a regulatory protein called the *IRE-binding protein* binds to the IRE sequence in ferritin mRNA, preventing the mRNA from forming an initiation complex with ribosomal subunits. But the IRE-binding protein is an allosteric protein whose activity can be controlled by the binding of iron. When more iron is available, the protein binds an iron atom and undergoes a conformational change that prevents it from binding to the IRE, thereby allowing the ferritin mRNA to be translated. The IRE-binding protein is therefore an example of a **translational repressor** that selectively controls the translation of a particular mRNA. This type of translational control allows cells to respond to specific changes in the environment faster than would be possible through the use of transcriptional control.

Other proteins act as translational repressors but bind the 3′ untranslated regions of specific mRNAs. One example of this type of control is the *PUF* proteins, which regulate the translation of proteins important during development. In addition to translational repressor proteins, we will see below that an important class of small RNAs—*microRNAs*—also interact with the 3′-UTR to regulate translation.

Translation Can Also Be Controlled by Regulation of mRNA Degradation

Translation rates are also subject to control by alterations in mRNA stability—in other words, if an mRNA molecule is degraded more rapidly, then less time is available for it to be translated. Degradation rates can be measured by *pulse-chase experiments,* in which cells are first incubated for a brief period of time (the "pulse") in the presence of a radioactive compound that becomes incorporated into mRNA. The cells are then placed in a nonradioactive medium, and incubation is continued (the "chase"). No additional radioactivity is taken up by the cell during the chase period, so the fate of the radioactivity previously incorporated into mRNA can be measured over time. This approach allows researchers to measure an mRNA's *half-life,* which is the time required for 50% of the initial (radioactive) RNA to be degraded.

(a) Low iron concentration. IRE-binding protein binds to IRE, so translation of ferritin mRNA is inhibited.

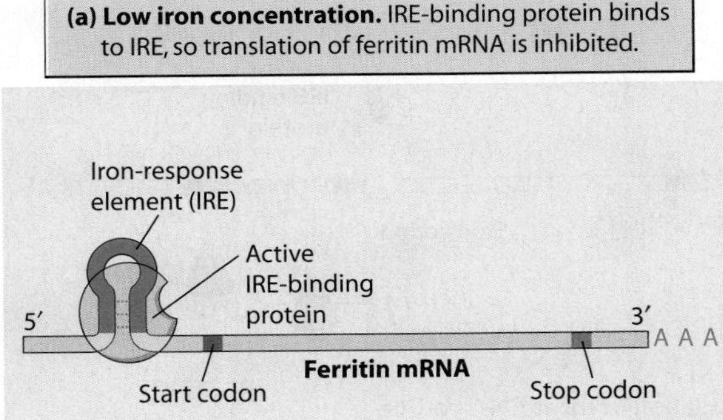

(b) High iron concentration. IRE-binding protein cannot bind to IRE, so translation of ferritin mRNA proceeds.

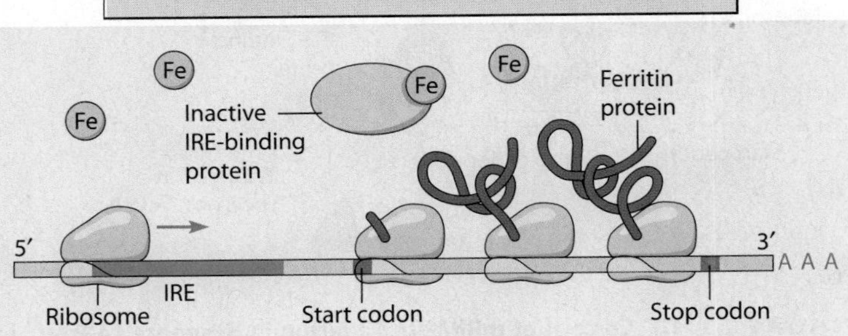

FIGURE 23-31 Translational Control in Response to Iron. **(a)** An iron response element (IRE) located in the 5′ untranslated region of ferritin mRNA can form a hairpin structure that is bound by the IRE-binding protein, which inhibits the initiation of translation. **(b)** When the iron concentration is high, iron binds to the IRE-binding protein and changes it to a conformation that does not recognize the IRE. Ribosomes can therefore assemble on the ferritin mRNA and translate it.

The half-life of eukaryotic mRNAs varies widely, ranging from 30 minutes or less for some growth factor mRNAs to over 10 hours for the mRNA encoding β-globin. The length of the poly(A) tail is one factor that plays a role in controlling mRNA stability. Messenger RNAs with short poly(A) tails tend to be less stable than mRNAs with longer poly(A) tails. In some cases, mRNA stability is also influenced by specific features of the 3′ untranslated region. For example, short-lived mRNAs for several growth factors have particular *AU-rich elements* in this region. The AU-rich sequences trigger removal of the poly(A) tail by degradative enzymes. When an AU-rich element is transferred to the 3′ end of a normally stable globin message using recombinant DNA techniques, the hybrid mRNA acquires the short half-life typical of a growth factor mRNA.

Another way of regulating mRNA stability is illustrated by the action of iron, which, in addition to participating in translational control as just described, plays a role in controlling mRNA degradation as well. The uptake of iron into mammalian cells is mediated by a plasma membrane receptor protein called the *transferrin receptor.* When iron is scarce, synthesis of the transferrin receptor is stimulated by a mechanism that protects transferrin receptor mRNA from degradation, thereby making more mRNA molecules available for translation. As shown in **Figure 23-32,** this control mechanism involves an IRE (similar to the one in ferritin mRNA) located in the 3′ untranslated region of transferrin receptor mRNA. When intracellular iron levels are low and increased uptake of iron is necessary, the IRE-binding protein binds to this IRE and protects the mRNA from degradation. When iron levels in the cell are high and additional uptake is not necessary, the IRE-binding protein binds an iron atom and dissociates from the mRNA, allowing the mRNA to be degraded.

Two general mechanisms exist for degrading mRNAs that have been targeted for destruction. The main difference is whether the mRNA is degraded in the 3′ ⟶ 5′ or the 5′ ⟶ 3′ direction. In the *3′ ⟶ 5′ pathway,* shortening of the poly(A) tail and removal of poly(A)-binding proteins from the 3′ end of the mRNA is followed by degradation of the RNA chain by the cytoplasmic **exosome,** a complex of several 3′ ⟶ 5′ exonucleases. When the exosome eventually arrives at the 5′ end of the mRNA, the 5′ cap is destroyed by a cap-degrading enzyme. In contrast, the *5′ ⟶ 3′ pathway* involves shortening of the poly(A) tail at the 3′ end followed by immediate removal of the cap from the 5′ end. The mRNA is then degraded by 5′ ⟶ 3′ exonucleases. The enzymes involved in this 5′ ⟶ 3′ pathway are concentrated in cytoplasmic structures called *mRNA processing bodies,* or simply **P bodies.** In addition to their role in mRNA degradation, P bodies can serve as temporary storage sites for nontranslated mRNAs and for recruitment of proteins involved in RNA interference, to which we now turn.

RNA Interference Utilizes Small RNAs to Silence the Expression of Genes Containing Complementary Base Sequences

Regulatory proteins like IRE-binding and PUF proteins, which bind to specific mRNAs, are not the only molecules used by cells for controlling mRNA activity. Individual mRNAs are also controlled by a class of small RNA molecules that inhibit the expression of mRNAs containing sequences related to sequences found in the small RNAs. Such RNA-mediated inhibition, known as **RNA interference (or RNAi),** is based on the ability of small RNAs to trigger mRNA degradation, or inhibit mRNA translation, or inhibit transcription of the gene coding for a particular mRNA.

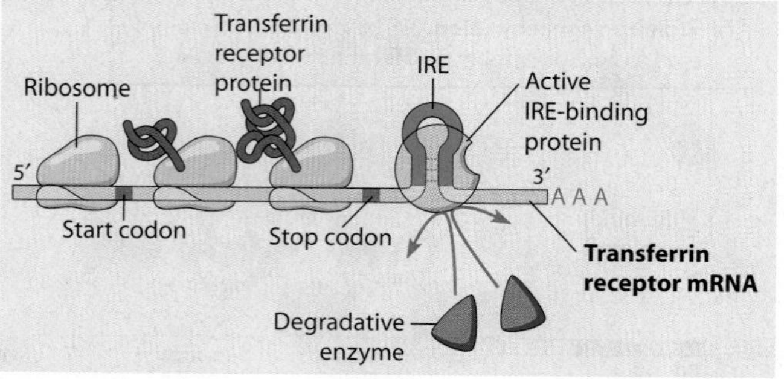

(a) Low iron concentration. IRE-binding protein binds to the IRE of transferrin receptor mRNA, thereby protecting the mRNA from degradation. Synthesis of transferrin receptor therefore proceeds.
Transferrin receptor protein
Ribosome
IRE
Active IRE-binding protein
5′
3′
A A A
Start codon
Stop codon
Transferrin receptor mRNA
Degradative enzyme

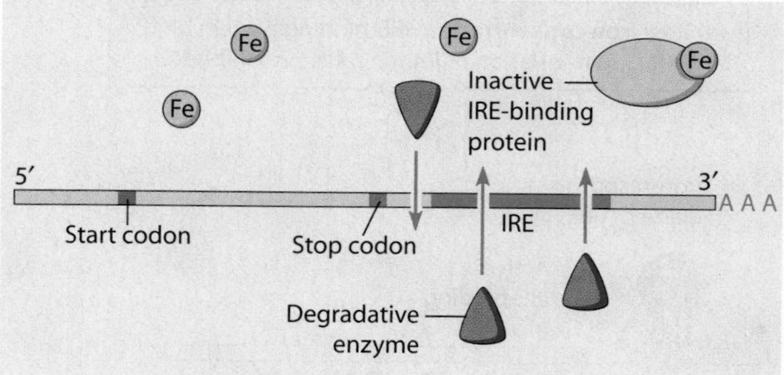

(b) High iron concentration. IRE-binding protein cannot bind to IRE, so mRNA is degraded and synthesis of transferrin receptor is thereby inhibited.
Fe
Fe
Fe
Fe
Inactive IRE-binding protein
5′
3′
A A A
Start codon
Stop codon
IRE
Degradative enzyme

FIGURE 23-32 Control of mRNA Degradation in Response to Iron. Degradation of the mRNA coding for the transferrin receptor, a protein required for iron uptake, is regulated by the same IRE-binding protein shown in Figure 23-31. **(a)** When the intracellular iron concentration is low, the IRE-binding protein binds to an IRE in the 3′ untranslated region of the transferrin receptor mRNA, protecting the mRNA from degradation and allowing the transferrin receptor protein to be synthesized. **(b)** When iron levels are high, iron atoms bind to the IRE-binding protein, causing it to leave the IRE. The mRNA is then vulnerable to attack by degradative enzymes.

The first type of RNA interference to be discovered occurs in response to the presence of double-stranded RNA. For example, if plants are infected with viruses that produce double-stranded RNA as part of their life cycle, the infected cells limit the infection by reducing the expression of viral genes. Moreover, the effect is not limited to viral genes. If a normal plant gene is inserted into a virus using genetic engineering techniques, cells infected with the virus reduce expression of their own copy of the same gene. The causal role played by double-stranded RNA in this type of gene silencing was demonstrated in 1998 by Andrew Fire and Craig Mello, who reported that injecting double-stranded RNA into worms greatly reduces the levels of RNA transcripts that contain sequences complementary to those present in the double-stranded RNA. Because their studies laid the groundwork for a burgeoning new field of biology related to the roles played by small RNAs in the control of gene expression, Fire and Mello were awarded a Nobel Prize in 2006.

Figure 23-33 illustrates how double-stranded RNAs knocks down the expression of specific genes. First, a cytoplasmic ribonuclease known as **Dicer** cleaves the double-stranded RNA into short fragments about 21–22 base pairs in length. The resulting double-stranded fragments, called **siRNAs** (small interfering, or silencing, RNAs), then combine with a group of proteins to form an inhibitor of gene expression called a **RISC** (RNA-induced silencing complex). This particular type of RISC is known as an **siRISC** because its siRNA component determines which gene will be targeted for silencing.

After being incorporated into an siRISC, one of the two strands of the siRNA is degraded. The remaining single-stranded RNA then binds the siRISC via complementary base pairing to a target mRNA molecule. If pairing between the siRNA and the mRNA is a perfect match (or very close), mRNA degradation is initiated by **Slicer,** a ribonuclease component of the siRISC that cleaves the mRNA in the middle of the complementary site. If the match between the siRNA and mRNA is imperfect, translation of the mRNA may be inhibited without the mRNA being degraded. And in some cases, the siRISC may enter the nucleus and be guided by its siRNA to complementary nuclear DNA sequences. After associating with these gene sequences, the siRISC silences their expression by stimulating DNA methylation and/or recruiting an enzyme that adds methyl groups to histones, thereby triggering the formation of a transcriptionally inactive, condensed form of chromatin (heterochromatin).

RNA interference originally may have evolved to protect cells from viruses that utilize double-stranded RNA. However, it also turns out to be a powerful laboratory tool that allows scientists to selectively knock down expression of any gene they wish to study. Since complete genome sequences are now available for a variety of organisms, the function of each individual gene can be systematically explored by using RNA interference to turn it off. Researchers simply synthesize (or purchase) short

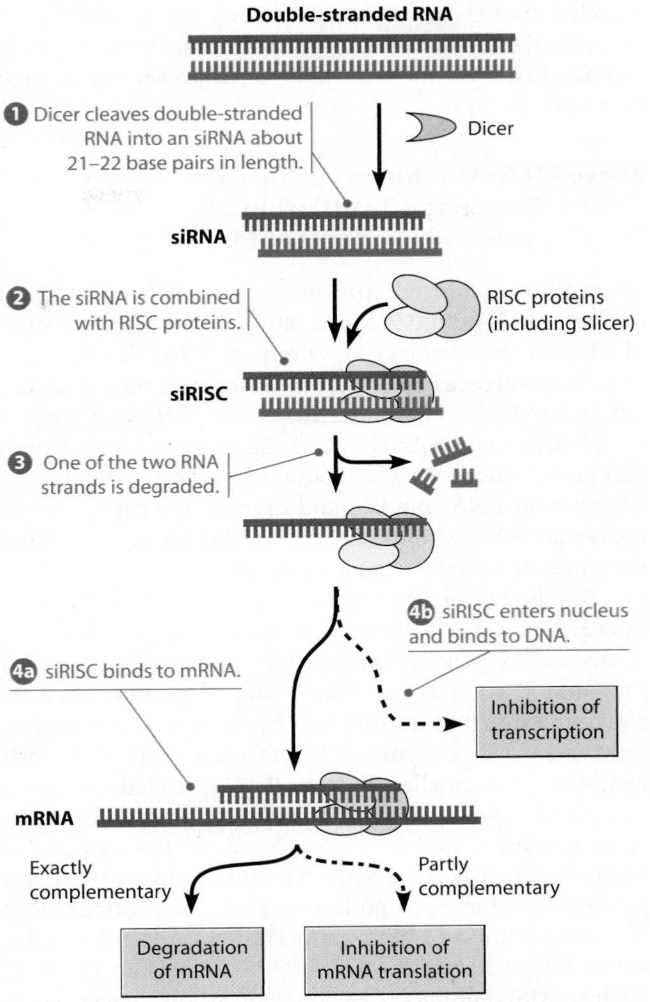

FIGURE 23-33 RNA Interference by Double-Stranded RNA. When a cell encounters double-stranded RNA, Dicer cleaves it into an siRNA, about 21–22 base pairs in length, that is subsequently combined with RISC proteins to form an siRISC. After degradation of one of the two siRNA strands, the remaining strand binds the siRISC via complementary base pairing to either a target mRNA molecule in the cytoplasm (4a) or to a target DNA sequence in the nucleus (4b), thereby silencing gene expression at either the translational or transcriptional level. The most common situation (indicated by the solid arrows) is an exact complementary match between the siRNA and a corresponding mRNA, which triggers mRNA cleavage by Slicer, an enzyme component of the siRISC.

MB Molecular Model: Argonaute Structure

siRNAs that are complementary to sequences present in the genes they wish to silence. Introducing these synthetic siRNAs into cells allows individual genes to be turned off one at a time. To illustrate the extraordinary power of this approach, RNAi has already been used to study the roles of 98% of the roughly 20,000 protein coding genes in the worm *C. elegans* by turning them off one by one.

Another potential application of siRNAs is to use them for silencing genes involved in various diseases. For example, studies in mice have shown that injecting animals with siRNAs directed against a gene involved in cholesterol metabolism can reduce cholesterol levels in the blood. Although questions concerning safety, side effects, and effectiveness remain to be answered, the therapeutic use of siRNAs as drugs for treating or preventing human diseases is thought to have a bright future.

MicroRNAs Produced by Normal Cellular Genes Silence the Translation of Developmentally Important Messenger RNAs

The finding that gene expression can be silenced by introducing double-stranded RNAs into cells raises the question of whether any normal genes produce RNAs that function in a comparable fashion. The search for such molecules has led to the discovery of **microRNAs (miRNAs),** a class of single-stranded RNAs about 21–22 nucleotides in length that are produced by genes found in almost all eukaryotes. These microRNAs bind to and regulate the expression of messenger RNAs produced by genes that are separate from the genes that produce the microRNAs.

As shown in **Figure 23-34,** genes that produce microRNAs are initially transcribed into longer RNA molecules called *primary microRNAs (pri-miRNAs),* which fold into hairpin loops. These looped pri-miRNAs are converted into mature miRNAs by sequential processing. First, a nuclear enzyme called **Drosha** cleaves the pri-miRNAs into smaller hairpin RNAs, called *precursor microRNAs (pre-miRNAs),* roughly 70 nucleotides long. Next the pre-miRNAs are exported to the cytoplasm, where Dicer cleaves the hairpin loop to produce two complementary short RNA molecules 21–22 nucleotides long.

One of the two short RNAs is the mature microRNA, which joins with a group of proteins to form an **miRISC.** Each miRISC inhibits the expression of messenger RNAs containing a base sequence complementary to that of the microRNA residing within the miRISC. The interactions between miRISCs and their targeted messenger RNAs take place mainly in cytoplasmic P bodies (page 776). Sometimes the microRNA in a particular miRISC is exactly complementary to a sequence present in a targeted messenger RNA. In such cases, the targeted message is destroyed by a mechanism similar to that observed with siRNAs. It is more common, however, for microRNAs to exhibit partial complementarity to sites in messenger RNA. Rather than triggering messenger RNA degradation, the binding of miRISCs to partially complementary sites inhibits translation of the targeted message. This inhibitory effect usually requires the binding of multiple miRISCs to different, partially complementary sites within a given messenger RNA. Such inhibition is not necessarily permanent, however. Changing cellular conditions can trigger the release of a messenger RNA from inhibiting miRISCs, accompanied by movement of the messenger RNA from P bodies to actively translating ribosomes.

Current estimates suggest that genes coding for microRNAs account for up to 5% of the total number of genes in eukaryotic genomes. In mammals, this corresponds to the presence of about 1000 microRNA genes.

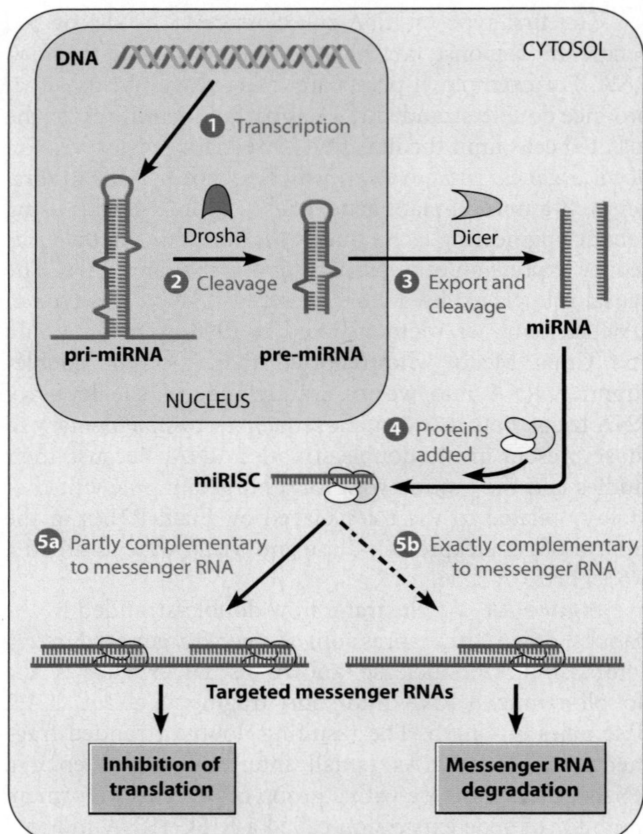

FIGURE 23-34 Translational Silencing by microRNAs. MicroRNAs are produced by a multistep process in which ❶ a microRNA gene is first transcribed into a primary transcript (pri-miRNA) that folds into a hairpin loop. ❷ The nuclear enzyme Drosha cleaves pri-miRNA into a smaller (roughly 70 nucleotide) hairpin RNA (pre-miRNA), which is ❸ exported from the nucleus and cleaved by Dicer to release a microRNA (miRNA) molecule that is 21–22 nucleotides long. ❹ The miRNA then joins with RISC proteins to form an miRISC, which is directed by its miRNA to messenger RNAs containing sequences complementary to those in the miRNA. 5a Most commonly, multiple miRISCs bind to the same messenger RNA through partly complementary sequences and together inhibit translation. 5b Occasionally the miRNA is exactly complementary to a site within a particular messenger RNA, resulting in messenger RNA degradation by a mechanism similar to that observed with siRNAs.

The microRNA produced by each of these genes is capable of binding to target sites in about 200 different messenger RNAs, and many messenger RNAs are targeted by several microRNAs acting in combination. MicroRNAs appear to play important roles during embryonic development, especially in the formation of specialized cell types such as neurons, immune cells, and skeletal and heart muscle cells. One way of uncovering the developmental roles played by microRNAs has been to use the gene knockout technique (page 638) to eliminate specific microRNA genes from mouse embryos. In one such experiment, deleting a microRNA gene called *miR-1-2* led to severe heart defects that caused more than half the mice to die during embryonic development or within the first few months of life. At least a dozen of the mRNAs targeted for regulation by the *miR-1-2* microRNA have been tentatively identified. Included in this group are mRNAs coding for transcription factors that regulate muscle cell development and an mRNA coding for a subunit of a potassium channel that helps maintain the resting membrane potential.

Posttranslational Control Involves Modifications of Protein Structure, Function, and Degradation

After translation of an mRNA molecule has produced a polypeptide chain, there are still many ways of regulating the activity of the polypeptide product. This brings us to the final level of controlling gene expression, namely, the various **posttranslational control** mechanisms that are available for modifying protein structure and function (see Figure 23-10, step ❺). Included in this category are reversible structural alterations that influence protein function, such as protein phosphorylation and dephosphorylation, as well as permanent alterations such as proteolytic cleavage. Other posttranslational events subject to regulation include the guiding of protein folding by chaperone proteins, the targeting of proteins to intracellular or extracellular locations, and the interaction of proteins with regulatory molecules or ions, such as cAMP or Ca^{2+}. These kinds of posttranslational events are discussed elsewhere in other contexts, but we again mention them here to emphasize that the flow of information in the cell is not complete until a properly functioning gene product is available.

So far we have focused on control mechanisms that influence gene expression by affecting the rate of protein production from each gene. But the amount of any given protein present in a cell is influenced by its rate of degradation as well as its rate of production. Thus, regulating the rate of protein degradation is another important means for influencing gene expression.

The relative contributions made by the rates of protein synthesis and degradation in determining the final concentration of any protein in a cell are summarized by the equation

$$P = \frac{k_{syn}}{k_{deg}} \qquad \textbf{(23-1)}$$

where P is the concentration of a given protein present, k_{syn} is the rate order constant for its synthesis, and k_{deg} is the rate order constant for its degradation. As we saw with mRNAs (p. 776), rate constants for degradation are often expressed in the form of a half-life. The half-lives of cellular proteins range from as short as a few minutes to as long as several weeks. Such differences in half-life can dramatically affect the ability of different proteins to respond to changing conditions.

A striking example is provided by the behavior of liver cells exposed to the steroid hormone cortisone, which stimulates the synthesis of various metabolic enzymes. When researchers examined the impact of cortisone on two liver enzymes involved in amino acid metabolism, *tryptophan pyrrolase* and *arginase,* they noticed a pronounced difference: The concentration of tryptophan pyrrolase quickly increased tenfold, whereas the concentration of arginase increased only slightly. At first glance, such data would seem to indicate that cortisone selectively stimulates the synthesis of tryptophan pyrrolase. In fact, cortisone stimulates the synthesis of both tryptophan pyrrolase and arginase to about the same extent. The explanation for the apparent paradox is that tryptophan pyrrolase has a much shorter half-life (larger k_{deg}) than does arginase. Using the mathematical relationship summarized in Equation 23-1, calculations show that the concentration of a protein exhibiting a short half-life (large k_{deg}) changes more dramatically in response to alterations in synthetic rate than does the concentration of a protein with a longer half-life. For this reason, enzymes that are important in metabolic regulation tend to have short half-lives, allowing their intracellular concentrations to be rapidly increased or decreased in response to changing conditions.

Although the factors determining the rates at which various proteins are degraded are not completely understood, it is clear that the process is subject to regulation. In rats deprived of food, for example, the amount of the enzyme arginase in the liver doubles within a few days without any corresponding change in the rate of arginase synthesis. The explanation for the observed increase is that the arginase degradation rate slows dramatically in starving rats. This effect is highly selective; most proteins are degraded more rapidly, not more slowly, in starving animals.

Ubiquitin Targets Proteins for Degradation by Proteasomes

The preceding example raises the question of how the degradation of individual proteins is selectively regulated. The most common method for targeting specific proteins for destruction is to link them to **ubiquitin,** a small protein chain containing 76 amino acids. Ubiquitin is joined to target proteins by a process that involves three components: a *ubiquitin-activating enzyme* (E1), a *ubiquitin-conjugating enzyme* (E2), and a *substrate recognition protein,* or *ubiquitin ligase* (E3). As shown in **Figure 23-35,** ubiquitin is first activated by attaching it to E1 in an ATP-dependent reaction. The activated ubiquitin is then transferred to E2 and subsequently linked, in a reaction facilitated by E3, to a lysine residue in a target protein. Additional molecules of ubiquitin are then added in sequence, forming short chains.

These ubiquitin chains serve as targeting signals that are recognized by large, protein-degrading structures called **proteasomes.** Proteasomes are the predominant proteases (protein-degrading enzymes) of the cytosol and

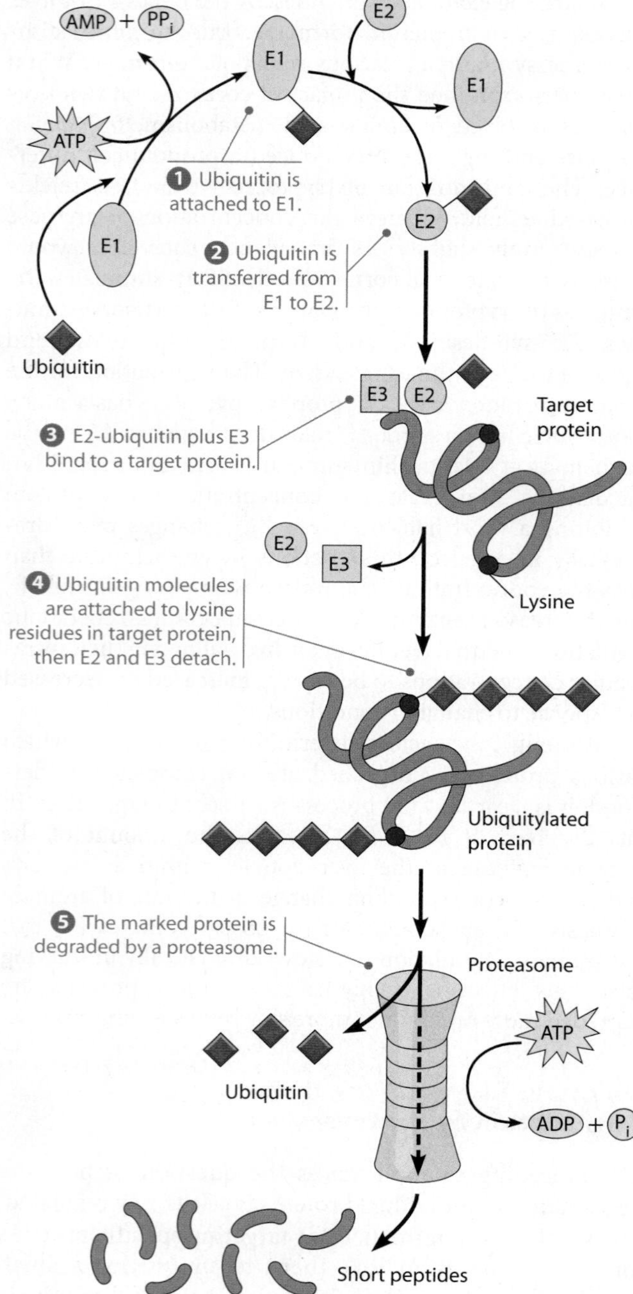

FIGURE 23-35 Ubiquitin-Proteasome System for Regulated Protein Degradation. Ubiquitin is attached to lysine in targeted proteins through the sequential action of a ubiquitin-activating enzyme (E1), a ubiquitin-conjugating enzyme (E2), and a ubiquitin ligase (E3). A proteasome then degrades the targeted protein into short peptides.

are often present in high concentration, accounting for up to 1% of all cellular protein. Each proteasome has a molecular weight of roughly 2 million and consists of half a dozen proteases associated with several ATPases and a binding site for ubiquitin chains. Shaped like a short cylinder, the proteasome binds to ubiquitylated proteins and removes their ubiquitin chains. The proteins are then fed into the central channel of the proteasome and their peptide bonds are hydrolyzed in an ATP-dependent process, generating small peptide fragments that are released from the other end of the cylinder.

The key to regulating protein degradation lies in the ability to select particular proteins for ubiquitylation. This selectivity stems in part from the existence of multiple forms of E3, each directing the attachment of ubiquitin to different target proteins. One feature recognized by the various forms of E3 is the amino acid present at the N-terminus of a potential target protein. Some N-terminal amino acids cause proteins to be rapidly ubiquitylated and degraded; other amino acids make proteins less susceptible.

Internal amino acid sequences called **degrons** also allow particular proteins to be selected for destruction. For example, we saw in Chapter 19 that progression through the final stages of mitosis is controlled by the anaphase-promoting complex, which targets selected proteins, such as mitotic cyclin, for degradation. The anaphase-promoting complex accomplishes this task by functioning as a ubiquitin ligase (E3) that binds to target proteins containing a particular type of degron, promoting ubiquitylation of these target proteins by E2 and their subsequent degradation by proteasomes. Besides participating in such selective mechanisms for degrading specific proteins at appropriate times, proteasomes play a role in general, ongoing mechanisms for eliminating defective proteins from cells. It has been reported that up to 30% of newly synthesized proteins are defective and so are immediately tagged with ubiquitin, triggering their destruction by proteasomes.

In addition to ubiquitylation, proteins can undergo a related posttranslational modification, through the addition of *small ubiquitin-related modifiers (SUMOs)*. These peptides are ubiquitin-like polypeptides that are conjugated to cellular proteins in a manner similar to ubiquitylation, using similar enzymes. SUMOylation appears to have numerous effects, including altering protein stability, movement into and out of the nucleus, and regulation of transcription factor function.

Although the ubiquitin-proteasome system is the primary mechanism used by cells for degrading proteins, it is not the only means available. You learned in Chapter 12 that lysosomes contain digestive enzymes that degrade all major classes of macromolecules, including proteins. Lysosomes can take up and degrade cytosolic proteins by an infolding of the lysosomal membrane, creating small vesicles that are internalized within the lysosome and broken down by the organelle's hydrolytic enzymes. This process of **microautophagy** tends to be rather nonselective in the proteins it degrades. The result is a slow, continual recycling of the amino acids found in most of a cell's protein molecules. Under certain conditions, however, such nonselective degradation of proteins would be detrimental to the cell. For example, during prolonged fasting, nonselective degradation of cellular proteins could lead to depletion of critical enzymes or regulatory proteins. Under these conditions, lysosomes preferentially degrade proteins containing a targeting sequence that consists of glutamine flanked on either side by a tetrapeptide composed of very basic, very acidic, or very hydrophobic amino acids. Proteins exhibiting this sequence are targeted for selective degradation, presumably because they are dispensable to the cell.

A Summary of Eukaryotic Gene Regulation

You have now seen that eukaryotic gene regulation encompasses a broad array of control mechanisms operating at five distinct levels (refer again to Figure 23-10 for a visual summary). The first level of regulation, *genomic control,* includes DNA alterations (amplification, deletion, rearrangements, and methylation), chromatin decondensation and condensation, histone modifications such as methylation and acetylation, and changes in HMG proteins. The second level, *transcriptional control,* involves interactions between regulatory transcription factors and various types of DNA control elements, permitting specific genes to be turned on and off in different tissues and in response to changing conditions. The third level involves control of *RNA processing and nuclear export* by mechanisms such as alternative splicing and regulated export of mRNAs through nuclear pores. The fourth level, *translational control,* includes mechanisms for modifying the activity of protein synthesis factors and for controlling the translation and degradation of specific mRNAs through the use of translational repressors and microRNAs. Finally, the fifth level involves various *posttranslational control* mechanisms for reversibly or permanently altering protein structure and function, as well as mechanisms for controlling protein degradation rates. Thus, when you read that a particular alteration in cellular function or behavior is based on a "change in gene expression," you should understand that the underlying explanation might involve mechanisms operating at any one or more of these levels of control.

Because eukaryotes have larger genomes than prokaryotes do, and because they package their DNA into chromatin fibers that are separated from ribosomes by a nuclear envelope, several of the control mechanisms summarized in the preceding paragraph are truly unique to eukaryotes. Most of the control mechanisms discovered in eukaryotes, however, have counterparts in at least some prokaryotes as well.

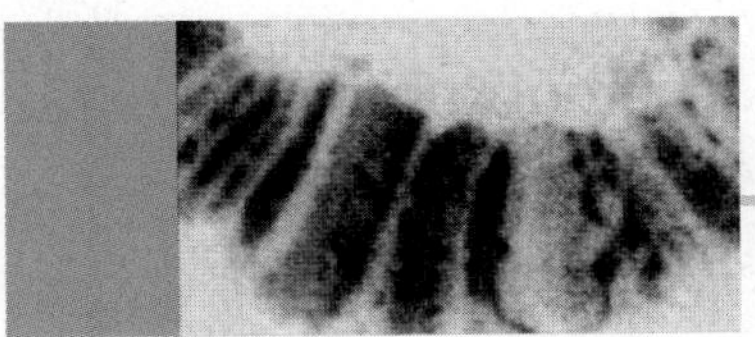

SUMMARY OF KEY POINTS

Bacterial Gene Regulation

- Rather than being active at all times, most genes are expressed as needed. In bacteria, genes with related functions are often clustered into operons that can be turned on and off in response to changing cellular needs.
- Operons that encode catabolic enzymes are generally inducible—that is, their transcription is turned on by the presence of substrate. Operons that encode anabolic enzymes are typically repressible; transcription is turned off in the presence of end-product. Both types of regulation involve allosteric repressor proteins that bind to the operator and prevent transcription. Some operons also have control sites where the binding of activator proteins, such as catabolite activator protein (CAP), can activate transcription.
- Certain operons, such as those for amino acid biosynthesis, produce mRNAs with a leader sequence that attenuates transcription when the corresponding amino acid is available. Leader sequences in mRNAs may also contain riboswitches that bind to small molecules, thereby altering hairpin loop configurations that affect the termination of transcription or the initiation of translation.

Eukaryotic Gene Regulation: Genomic Control

- Most of the cells in a multicellular organism contain the same genes, although amplification, deletion, rearrangement, and epigenetic regulation via methylation can cause localized changes in DNA sequence and expression.
- Gene transcription is associated with uncoiling of chromatin fibers. A histone code consisting of various combinations of methyl, acetyl, and phosphate groups is read by other proteins as a set of signals for modifying chromatin structure and gene activity.

Eukaryotic Gene Regulation: Transcriptional Control

- The initiation of eukaryotic transcription is controlled by regulatory transcription factors that bind to DNA control sequences, including proximal control elements that lie close to the core promoter and enhancers and silencers that act at distances up to several hundred thousand base pairs away from the promoter. Enhancers do not act on the promoter itself but instead bind to activator proteins, which in turn interact with coactivator proteins, including Mediator, that loosen chromatin packing and serve as a bridge to the RNA polymerase complex at the promoter.
- DNA response elements allow nonadjacent genes to be regulated in a coordinated fashion. For example, hormone response elements coordinate the expression of genes activated by specific hormones, and the heat-shock response element coordinates the expression of genes activated by elevated temperatures and other stresses.

Eukaryotic Gene Regulation: Posttranscriptional Control

- Eukaryotes utilize a diverse array of posttranscriptional control devices, including alternative splicing to generate multiple mRNAs from the same gene; general, as well as mRNA-specific, mechanisms to regulate translation and degradation of mRNA; and posttranslational modifications of protein structure, activity, and degradation.
- Proteins that bind to specific DNA and RNA sequences play a key role in many of the mechanisms used by cells for controlling the expression of specific genes. In addition, two classes of small RNA molecules (siRNAs and microRNAs) regulate the expression of specific genes through a process known as RNA interference. By binding to complementary sequences in specific messenger RNAs (or in some cases, DNA), these small RNAs silence the translation and/or transcription of individual genes.

MB **www.masteringbiology.com**

1. MasteringBiology Assignments

Tutorials and Activities

- Molecular Model Structural Tutorials for visualization of 3D structures
- BioFlix Tutorials including 3D animations with hints and detailed feedback for common wrong answers
- The lac Operon in E. coli; Regulation of Expression of the Lac Operon Genes; Catabolite Activator Protein (CAP); The lac Operon; Regulation of Gene Expression in Bacteria; Attenuation in the trp operon of E. coli; Regulation of Gene Expression in Eukaryotes; Transcription Initiation in Eukaryotes; Review: Control of Gene Expression; Overview: Control of Gene Expression; Analysis of Gene Expression Using DNA Microarrays; Yeast Two-Hybrid System; Regulation of Gene Expression by Steroid Hormones; Pattern Formation; Gene Regulation of the Development of the Drosophila Body Plan; Role of bicoid Gene in Drosophila Development; Early Pattern Formation in Drosophila; Argonaute Structure

Questions

- Reading Quiz
- Multiple Choice
- End-of-chapter questions

2. eText Read your book online, search, highlight text, take notes, and more

3. The Study Area

- Chapter Review Quiz, Guide to Techniques and Methods, 3D-Structure Tutorials, Activities, BioFlix, Videos, Biology Review, Word Study Tools, Glossary, Flashcards

PROBLEM SET

More challenging problems are marked with a •.

23-1 Laboring with *lac*. Most of what we know about the *lac* operon of *E. coli* has come from genetic analysis of various mutants. In the following list are the genotypes of seven strains of *E. coli*. For each strain, indicate whether the *Z* gene product will be expressed in the presence of lactose and whether it will be expressed in the absence of lactose. Explain your reasoning in each case.

(a) $I^+P^+O^+Z^+$

(b) $I^sP^+O^+Z^+$

(c) $I^+P^+O^cZ^+$

(d) $I^-P^+O^+Z^+$

(e) $I^sP^+O^cZ^+$

(f) $I^+P^-O^+Z^+$

(g) Same as part a, but with glucose present

23-2 The Pickled Prokaryote. *Pickelensia hypothetica* is an imaginary prokaryote that converts a wide variety of carbon sources to ethanol when cultured anaerobically in the absence of ethanol. When ethanol is added to the culture medium, however, the organism obligingly shuts off its own production of ethanol and makes lactate instead. Several mutant strains of *Pickelensia* have been isolated that differ in their ability to synthesize ethanol. Class I mutants cannot synthesize ethanol at all. Mutations of this type map at two loci, *A* and *B*. Class II mutants, on the other hand, are constitutive for ethanol synthesis: They continue to produce ethanol whether it is present in the medium or not. Mutations of this type map at loci *C* and *D*. Strains of *Pickelensia* constructed to be diploid for the ethanol operon have the following genotypes and phenotypes:

$A^+B^-C^+D^+/A^-B^+C^+D^+$ inducible

$A^+B^+C^+D^-/A^+B^+C^+D^+$ inducible

$A^+B^+C^+D^+/A^+B^+C^-D^+$ constitutive **(23-2)**

(a) Identify each of the four genetic loci of the ethanol operon.

(b) Indicate the expected phenotype for each of the following partially diploid strains:

i. $A^-B^+C^+D^+/A^-B^+C^+D^-$
ii. $A^+B^-C^+D^+/A^-B^+C^+D^-$
iii. $A^+B^+C^+D^-/A^-B^-C^-D^+$
iv. $A^-B^+C^-D^+/A^+B^-C^+D^+$

• **23-3 Regulation of Bellicose Catabolism.** The enzymes bellicose kinase and bellicose phosphate dehydrogenase are coordinately regulated in the bacterium *Hokus focus*. The genes encoding these proteins, *belA* and *belB*, are contiguous segments on the genetic map of the organism. In her pioneering work on this system, Professor Jean X. Pression established that the bacterium can grow with the monosaccharide bellicose as its only carbon and energy source, that the two enzymes involved in bellicose catabolism are synthesized by the bacterium only when bellicose is present in the medium, and that enzyme production is turned off in the presence of glucose. She identified a number of mutations that reduce or eliminate enzyme production and showed these could be grouped into two classes. Class I mutants are in *cis*-acting elements, whereas those in class II all map at a distance from the genes coding for the enzymes and are *trans*-acting. Thus far, the only constitutive mutations Pression has found are deletions that connect *belA* and *belB* to new DNA at the upstream side of *belA*. The following is a list of conclusions she would like to draw from her observations. Indicate in each case whether the conclusion is consistent with the data (C), inconsistent with the data (I), or irrelevant to the data (X).

(a) Bellicose can be metabolized by *H. focus* cells to yield glucose and ATP.

(b) Enzyme production by genes *belA* and *belB* is under positive control.

(c) Bellicose can spontaneously get reduced to glucose and inhibit the expression of *belA* and *belB*.

(d) The operator for the bellicose operon is located upstream from the promoter.

(e) Some of the mutations in class I may be in the promoter.

(f) Some of the mutations in class II may result in creating a new operator for *belA* and *belB*.

(g) Inhibition of bellicose metabolism can force *H. focus* to grow on other sources of carbon.

(h) The constitutive deletion mutations connect genes *belA* and *belB* to a new region of DNA that inhibits the synthesis of these enzymes.

(i) Expression of the bellicose operon is subject to regulation by feedback mechanism, wherein excessive glucose production shuts down bellicose metabolism.

23-4 Attenuation. Indicate whether each of the following statements is true (T) or false (F). If false, state the correct rationale.

(a) Attenuation is a mechanism for controlling the rate of transcription of the *trp* operon.

(b) When adequate levels of tryptophan are present, the bacterium produces enough *trp* mRNA but not the proteins that act in the tryptophan biosynthetic pathway.

(c) The length of the leader peptide produced prior to attenuation is 24 amino acids.

(d) Base-pairing between regions 3 and 4 of the *trp* mRNA results in the formation of a knot structure that does not allow ribosomes to transcribe further.

(e) Transcription and translation of the *trp* operon is a coupled process in *E. coli* but not so in *Bacillus subtilis*.

(f) Once a ribosome stalls at region 1 of the *trp* mRNA, attenuation is achieved when tryptophan is available in plenty.

(g) Attenuation as a regulatory mechanism for synthesis of biomolecules is evident in the biosynthesis of polyphenolic compounds by bacteria.

23-5 Positive and Negative Control. Assume you have a culture of *E. coli* cells growing on medium B, which contains both lactose and glucose. At time *t*, 33% of the cells are transferred to medium L, which contains lactose but not glucose; 33% are transferred to medium G, which contains glucose but not lactose; and the remaining cells are left in medium B. For each of the following statements, indicate with an L if it is true of the cells transferred to medium L, with a G if it is true of the cells transferred to medium G, with a B if it is true of the cells left in medium B, and with an N if it is true of none of the cells. In some cases, more than one letter may be appropriate.

(a) The rate of lactose consumption is higher after time *t* than before.

(b) The rate of transcription of the *lac* operon is greater after time *t* than before.

(c) The *lac* operon has both CAP and the repressor protein bound to it.

(d) Most of the *lac* repressor proteins exist as repressor-glucose complexes.

(e) The rate of glucose consumption per cell is approximately the same after time *t* as before.

(f) Most of the catabolite activator proteins exist as CAP–cAMP complexes.

(g) The intracellular cAMP level is lower after time *t* than before.

(h) Most of the *lac* operator sites have *lac* repressor proteins bound to them.

23-6 Enhancers. An enhancer may increase the frequency of transcription initiation for its associated gene when . . . (Indicate true or false for each statement, and explain your answer.)

(a) . . . it is located 1000 nucleotides upstream of the gene's core promoter.

(b) . . . it is in the gene's coding region.

(c) . . . no promoter is present.

(d) . . . it causes looping out of the intervening DNA.

(e) . . . it causes alternative splicing of the DNA.

23-7 Gene Amplification. The best-studied example of gene amplification involves the genes coding for ribosomal RNA during oogenesis in *Xenopus laevis*. The unamplified number of ribosomal RNA genes is about 500 per haploid genome. After amplification, the oocyte contains about 500,000 ribosomal RNA genes per haploid genome. This level of amplification is apparently necessary to allow the egg cell to synthesize the 10^{12} ribosomes that accumulate during the two months of oogenesis in this species. Each ribosomal RNA gene consists of about 13,000 base pairs, and the genome size of *Xenopus* is about 2.7×10^9 base pairs per haploid genome.

(a) What fraction of the total haploid genome do the 500 copies of the ribosomal RNA genes represent?

(b) What is the total size (in base pairs) of the genome after amplification of the ribosomal RNA genes? What proportion of this do the amplified ribosomal RNA genes represent?

(c) Assume that all the ribosomal RNA genes in the amplified oocyte are transcribed continuously to generate the number of ribosomes needed during the two months of oogenesis. How long would oogenesis have to extend if the genes had not been amplified?

(d) Why do you think genes have to be amplified when the gene product needed by the cell is an RNA but not usually when the desired gene product is a protein?

• **23-8 Steroid Hormones.** Steroid hormones are known to increase the expression of specific genes in selected target cell types. For example, testosterone increases the production of a protein called $\alpha 2$-microglobulin in the liver, and hydrocortisone (a type of glucocorticoid) increases the production of the enzyme tyrosine aminotransferase in the liver.

(a) Based on your general knowledge of steroid hormone action, explain how these two steroid hormones are able to selectively influence the production of two different proteins in the same tissue.

(b) Prior to the administration of testosterone or hydrocortisone, liver cells are exposed to either puromycin (an inhibitor of protein synthesis) or α-amanitin (an inhibitor of RNA polymerase II). How would you expect such treatments to affect the actions of testosterone and hydrocortisone?

(c) Suppose you carry out a domain-swap experiment in which you use recombinant DNA techniques to exchange the zinc finger domains of the testosterone receptor and glucocorticoid receptors with each other. What effects would you now expect testosterone and hydrocortisone to have in cells containing these altered receptors?

(d) If recombinant DNA techniques are used to substitute a testosterone response element for the glucocorticoid response element that is normally located adjacent to the tyrosine aminotransferase gene in liver cells, what effects would you expect testosterone and hydrocortisone to have in such genetically altered cells?

23-9 Homeotic Genes in *Drosophila*. Homeotic genes are considered crucial to early development in *Drosophila* because . . . (Indicate true or false for each statement, and explain your answer.)

(a) . . . they encode proteins containing zinc finger domains.

(b) . . . mutations in homeotic genes are always lethal.

(c) . . . they control the expression of many other genes required for development.

(d) . . . homeodomain proteins act by influencing mRNA degradation.

• **23-10 *Hox* Genes in Mammals.** *Hox* genes are known to influence the identity of body parts in mammals, similar to their role in flies. Mario Capecchi's laboratory has made many knockout mouse strains in which individual *Hox* genes are knocked out. The phenotype of many of these knockouts is extremely mild, especially when compared with the corresponding mutants in flies, which often die. How can you explain this difference?

• **23-11 Protein Degradation.** You learned in Chapter 19 that mitotic cyclin is an example of a protein that is selectively degraded at a particular point in the cell cycle, namely the onset of anaphase. Suppose that you use recombinant DNA techniques to create a mutant form of mitotic cyclin that is not degraded at the onset of anaphase.

(a) Sequence analysis of the mutant mitotic cyclin shows that it is missing a stretch of nine amino acids near one end of the molecule. Based on this information, discuss at least three possible explanations of why this form of mitotic cyclin is not degraded at the onset of anaphase.

(b) What kinds of experiments could you carry out to distinguish among these various possible explanations?

(c) Suppose that you discover cells containing a mutation in a second protein and learn that this mutation also prevents mitotic cyclin from being degraded at the onset of mitosis, even when mitotic cyclin is normal. The degradation of other proteins appears to proceed normally in such cells. A mutation in what kind of protein might explain such results?

23-12 Levels of Control. Assume liver and kidney tissues from the same mouse each contain about 10,000 species of cytosolic mRNA, but only about 25% of these are common to the two tissues.

(a) Suggest an experimental approach that might be used to establish that some of the mRNA molecules are common to both tissues, but others are not.

(b) One possible explanation for the data is that differential transcription occurs in liver and kidney nuclei. What is another possible explanation? Describe an experiment that would enable you to distinguish between these possibilities.

(c) If all mRNA molecules had been shown to be common to both liver and kidney and yet the two tissues were known to be synthesizing different proteins, what level of control would you have to assume was operating?

• **23-13 More About Levels of Control.** If sea urchin zygotes are exposed to *actinomycin D,* an inhibitor of RNA synthesis, during the early stages of embryonic development, protein synthesis and development continue until the embryos form a hollow ball of cells, or blastula, with hundreds of cells. However, if the same inhibitor is added later during embryonic development, protein synthesis is severely depressed and normal embryonic development halts. How might you explain these observations?

SUGGESTED READING

References of historical importance are marked with a •.

Bacterial Gene Regulation

Barrick, J. E., and R. R. Breaker. The power of riboswitches. *Sci. Amer.* 296 (January 2007): 50.

Gruber, T. M., and C. A. Gross. Multiple sigma units and the partitioning of bacterial transcription space. *Annu. Rev. Microbiol.* 57 (2003): 441.

• Jacob, F., and J. Monod. Genetic regulatory mechanisms in the synthesis of proteins. *J. Mol. Biol.* 3 (1961): 318.

Losick, R., and A. L. Sonenshein. Turning gene regulation on its head. *Science* 293 (2001): 2018.

Wilson, C. J., H. Zhan, L. Swint-Kruse, and K. S. Matthews. The lactose repressor system: Paradigms for regulation, allosteric behavior and protein folding. *Cell. Mol. Life Sci.* 64 (2007): 3.

• Yanofsky, C. Transcription attenuation: Once viewed as a novel regulatory strategy. *J. Bacteriol.* 182 (2000): 1.

Eukaryotic Gene Regulation: Genomic Control

Baker, M. iPS cells: Potent stuff. *Nat. Methods* 7 (2010): 17.

Bernstein, B. E., A. Meissner, and E. S. Lander. The mammalian epigenome. *Cell* 128 (2007): 669.

Cibelli, J. A decade of cloning mystique. *Science* 316 (2007): 990.

Edwards, C. A., and A. C. Ferguson-Smith. Mechanisms regulating imprinted genes in clusters. *Curr. Opin. Cell Biol.* 19 (2007): 281.

Gibbs, W. W. The unseen genome: Beyond DNA. *Sci. Amer.* 289 (December 2003): 106.

Jaenisch, R., and I. Wilmut. Developmental biology. Don't clone humans! *Science* 291 (2001): 2552.

Klose, R. J., and A. P. Bird. Genomic DNA methylation: The mark and its mediators. *Trends Biochem. Sci.* 31 (2006): 89.

Kouzarides, T. Chromatin modifications and their function. *Cell* 128 (2007): 693.

Martin, C., and Y. Zhang. Mechanisms of epigenetic inheritance. *Curr. Opin. Cell Biol.* 19 (2007): 266.

Schubeler, D. Epigenomics: Methylation matters. *Nature* 462 (2009): 296.

Takahashi, K., and S. Yamanaka. Induction of pluripotent stem cells from mouse embryonic and adult fibroblast cultures by defined factors. *Cell* 126 (2006): 663.

Turner, B. M. Cellular memory and the histone code. *Cell* 111 (2002): 285.

Wilmut, I. et al. Somatic cell nuclear transfer. *Nature* 419 (2002): 583.

Yu, J., and J. A. Thomson. Pluripotent stem cell lines. *Genes Dev.* 22 (2008): 1987.

Eukaryotic Gene Regulation: Transcriptional Control

Cairns, B. R. The logic of chromatin architecture and remodelling at promoters. *Nature* 461 (2009): 193.

Carey, M. The enhanceosome and transcriptional synergy. *Cell* 92 (1998): 5.

Carroll, S. B., B. Prud'homme, and N. Gompel. Regulating evolution. *Sci. Am.* 298 (2008): 60.

Gaszner, M., and G. Felsenfeld. Insulators: Exploiting transcriptional and epigenetic mechanisms. *Nature Rev. Genet.* 7 (2006): 703.

Kadonaga, J. T. Regulation of RNA polymerase II transcription by sequence-specific DNA binding factors. *Cell* 116 (2004): 247.

Kornberg, R. Mediator and the mechanism of transcriptional activation. *Trends Biochem. Sci.* 30 (2005): 235.

Szutorisz, H., N. Dillon, and L. Tora. The role of enhancers as centres for general transcription factor recruitment. *Trends Biochem. Sci.* 30 (2005): 593.

Welch, W. J. How cells respond to stress. *Sci. Amer.* 268 (May 1993): 56.

Eukaryotic Gene Regulation: Posttranscriptional Control

Bushati, N., and S. M. Cohen. MicroRNA functions. *Annu. Rev. Cell Dev. Biol.* 23 (2007): 175.

Chen, X. Small RNAs and their roles in plant development. *Annu. Rev. Cell Dev. Biol.* 25 (2009): 21.

Ciechanover, A. Proteolysis: From the lysosome to ubiquitin and the proteasome. *Nature Rev. Mol. Cell Biol.* 6 (2005): 79.

Garneau, N. L., J. Wilusz, and C. J. Wilusz. The highways and byways of mRNA decay. *Nature Rev. Mol. Cell Biol.* 8 (2007): 113.

Goldberg, A. L., S. J. Elledge, and J. W. Harper. The cellular chamber of doom. *Sci. Amer.* 284 (January 2001): 68.

Kim, D. H., and J. J. Rossi. Strategies for silencing human disease using RNA interference. *Nature Rev. Genet.* 8 (2007): 173.

Kim, V. N., J. Han, and M. C. Siomi. Biogenesis of small RNAs in animals. *Nature Rev. Mol. Cell Biol.* 10 (2009): 126.

Latronico, M. V., and G. Condorelli. MicroRNAs and cardiac pathology. *Nature Rev. Cardiol.* 6 (2009): 419.

Matlin, A. J., F. Clark, and C. W. J. Smith. Understanding alternative splicing: Towards a cellular code. *Nature Rev. Mol. Cell Biol.* 6 (2005): 386.

Parker, R., and U. Sheth. P bodies and the control of mRNA translation and degradation. *Mol. Cell* 25 (2007): 635.

Parker, R., and H. Song. The enzymes and control of eukaryotic mRNA turnover. *Nature Struct. Mol. Biol.* 11 (2004): 121.

Pillai, R. S., S. N. Bhattacharyya, and W. Filipowicz. Repression of protein synthesis by miRNAs: How many mechanisms? *Trends Cell Biol.* 17 (2007): 118.

Siomi, H., and M. C. Siomi. On the road to reading the RNA-interference code. *Nature* 457 (2009): 396.

Tang, G. siRNA and miRNA: An insight into RISCs. *Trends Biochem. Sci.* 30 (2005): 106.

Umbach, J. L., and B. R. Cullen. The role of RNAi and microRNAs in animal virus replication and antiviral immunity. *Genes Dev.* 23 (2009): 1151.

Vierstra, R. D. The ubiquitin-26S proteasome system at the nexus of plant biology. *Nature Rev. Mol. Cell Biol.* 10 (2009): 385.

Wang, Y., and M. Dasso. SUMOylation and deSUMOylation at a glance. *J. Cell Sci.* 122 (2009): 4249.

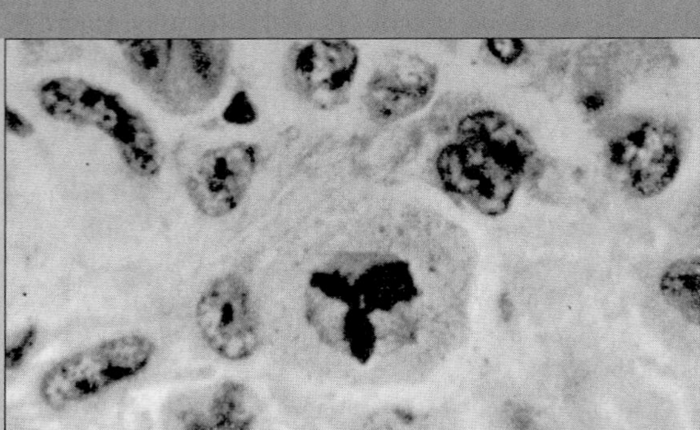

24 Cancer Cells

See MasteringBiology® for tutorials, activities, and quizzes.

Anyone familiar with the events occurring inside living cells must feel a sense of awe at the complexities involved. Given the vast number of activities that need to be coordinated in every cell, it is not surprising that malfunctions occasionally arise. Cancer—the second-leading cause of death after cardiovascular disease—is a prominent example of a disease that arises from such abnormalities in cell function. If current trends continue, almost half the population of the United States will eventually develop some form of the disease and cancer will become the most common cause of death.

Although our understanding of the cellular defects that underlie cancer is still incomplete, we now know that gene mutations and changes in gene expression play a central role. In this chapter we will see how these genetic changes contribute to the aberrant properties of cancer cells. As we do so, it will become apparent that explaining the behavior of cancer cells requires an intimate knowledge of how normal cells behave and, conversely, that investigating the biology of cancer cells has deepened our understanding of normal cells.

Uncontrolled Cell Proliferation and Survival

The term *cancer,* which means "crab" in Latin, was coined by Hippocrates in the fifth century B.C. to describe diseases in which tissues grow and spread unrestrained throughout the body, eventually choking off life. Cancers can originate in almost any organ. Depending on the cell type involved, they are grouped into several different categories. **Carcinomas,** which account for about 90% of all cancers, arise from the *epithelial cells* that cover external and internal body surfaces. Lung, breast, and colon cancer are the most frequent cancers of this type. **Sarcomas** develop from supporting tissues such as bone, cartilage, fat, and muscle. Finally, **lymphomas** and **leukemias** arise from cells of blood and lymphatic origin, with the term *lymphoma* being used for tumors that grow as solid masses of tissue and the term *leukemia* being reserved for cases in which the cancer cells proliferate mainly in the bloodstream.

Based on such differences in their sites of origin and the specific cell type involved, more than a hundred different types of cancer can be distinguished from one another. But no matter where cancer arises and what cell type is involved, the danger posed by the disease comes from a combination of two properties: the ability of cells to proliferate in an uncontrolled way and their ability to spread through the body. We will begin the chapter by describing these two properties that make cancer a potentially lethal disease.

Tumors Arise When the Balance Between Cell Division and Cell Differentiation or Death Is Disrupted

A cancer is an abnormal type of tissue growth in which some cells divide and accumulate in an uncontrolled, relatively autonomous way, leading to a progressive increase in the number of dividing cells. The resulting mass of growing tissue is called a **tumor** (or *neoplasm*). Although tumors have escaped from the normal controls on cell proliferation, tumor cells do not always divide more rapidly than normal cells. The crucial issue is not the rate of cell division but rather the balance between cell division and cell differentiation or cell death.

Cell differentiation is the process by which cells acquire the specialized properties that distinguish different types of cells from each other (page 722). As cells acquire these specialized traits, they generally lose the capacity to divide. To illustrate, let us briefly consider cell division and differentiation in the skin, where new cells continually replace aging cells that are being shed from the outer body surfaces. The new replacement cells are generated by cell divisions occurring in the *basal layer* of the skin

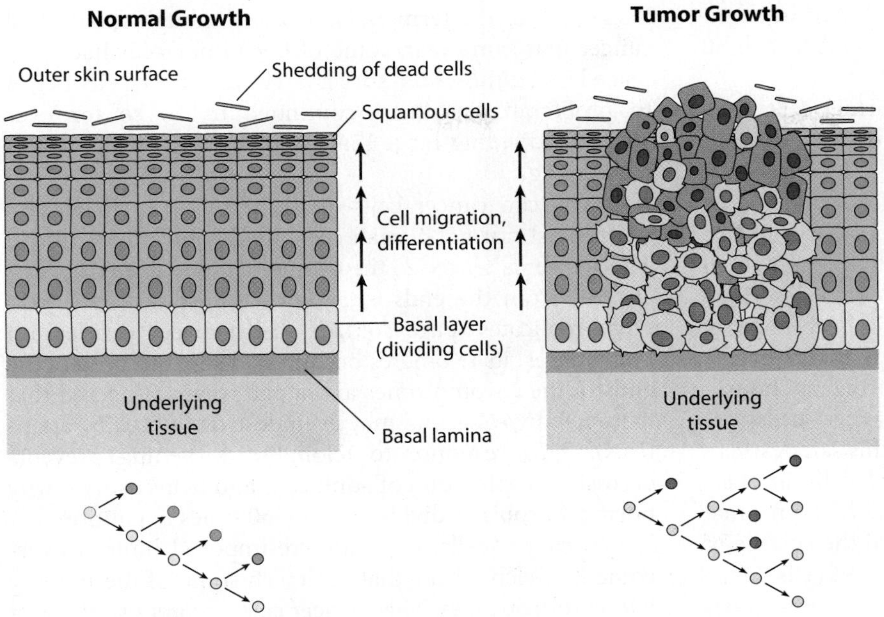

FIGURE 24-1 Comparison of Normal and Tumor Growth in the Epithelium of the Skin. *(Top left)* In normal skin, cell division in the basal layer gives rise to new cells that migrate toward the outer surface of the skin, changing shape and losing the capacity to divide. *(Top right)* In tumor growth, this orderly process is disrupted and some of the cells that migrate toward the outer surface retain the capacity to divide. In both diagrams, lighter color shading is used to identify cells that retain the capacity to divide. *(Bottom)* Schematic diagrams illustrating the fate of dividing cells. In normal skin, each cell division in the basal layer gives rise on average to one cell that retains the capacity to divide (lighter color shading) and one cell that differentiates, thereby losing the capacity to divide (darker color shading). As a result, no net accumulation of dividing cells occurs. In tumor growth, cell division is not appropriately balanced with cell death or differentiation, thereby leading to a progressive increase in the number of dividing cells.

(**Figure 24-1**, *left*). On average, each time a basal cell divides, it gives rise to two cells with different fates. One cell stays in the basal layer and retains the capacity to divide, whereas the other cell loses the capacity to divide and differentiates as it leaves the basal layer and moves toward the outer skin surface. During the differentiation process, the migrating cell gradually flattens and begins to make *keratin,* the structural protein that imparts mechanical strength to the outer layers of the skin. Eventually the cell dies and is shed from the outer skin surface.

Thus in normal skin, one of the two cells produced by each cell division retains the ability to divide, while the other cell leaves the basal layer, loses the capacity to divide, and finally dies. This arrangement ensures that there is no increase in the number of dividing cells. The cell divisions occurring in the basal layer simply generate new cells to replace the ones that are continually being lost from the outer skin surface. A similar phenomenon occurs in bone marrow, where cell division creates new blood cells to replace the aging blood cells that are being destroyed. It also occurs in the gastrointestinal tract lining, where cell division creates new cells to replace the ones being shed from the inner surface of the stomach and intestines. In each of these situations, cell division is carefully balanced with cell differentiation and death so that no net accumulation of dividing cells takes place.

In tumors, this finely balanced arrangement is disrupted and cell division is uncoupled from cell differentiation and death. As a result, some cell divisions give rise to two cells that both continue to divide, thereby feeding a progressive increase in the number of dividing cells (see Figure 24-1, *right*). If the cells are dividing rapidly, the tumor will grow quickly; if the cells are dividing more slowly, tumor growth will be slower. But regardless of how fast or slow the cells divide, the tumor will continue to grow because new cells are being produced in greater numbers than needed. As the dividing cells accumulate, the normal organization and function of the tissue gradually become disrupted.

Based on differences in their growth patterns, tumors are classified as either benign or malignant. **Benign tumors** grow in a confined local area and are rarely dangerous. In contrast, **malignant tumors** are capable of invading surrounding tissues, entering the bloodstream, and spreading to distant parts of the body. The term **cancer** refers to any malignant tumor—that is, any tumor that can spread from its original location to other sites. Because the ability to grow in an uncontrolled way and spread to distant locations makes cancer a potentially life-threatening disease, it is important to understand the mechanisms that give rise to these traits.

Cancer Cell Proliferation Is Anchorage Independent and Insensitive to Population Density

Cancer cell proliferation exhibits several traits that distinguish it from normal cell proliferation. One trait, of course, is the ability to form tumors. This trait can be demonstrated experimentally by injecting cells into *nude mice,* which lack a functional immune system that would normally attack and destroy foreign cells. Normal human cells injected into such animals will not grow, whereas cancer cells will proliferate and form tumors.

Cancer cells also exhibit other growth properties that distinguish them from normal cells. For example, normal cells don't grow well in culture if they are suspended in a liquid medium or in a semisolid material such as soft agar. But when provided with a solid surface they can attach to, normal cells become anchored to the surface, spread out, and begin to proliferate. In contrast, cancer cells grow well not only when they are anchored to a solid surface but also when they are freely suspended in a liquid or semisolid medium. Cancer cells are therefore said to exhibit **anchorage-independent growth.**

When growing in the body, most normal cells meet the anchorage requirement by binding to the extracellular matrix through cell surface proteins called *integrins* (page 494). If attachment to the matrix is prevented using chemicals that block the binding of integrins to the matrix, normal cells usually lose the ability to divide and often self-destruct by *apoptosis* (a type of cell death described on page 591). Triggering apoptosis in the absence of proper anchorage is one of the safeguards that prevents normal cells from floating away and setting up housekeeping in another tissue. Because cancer cells are anchorage independent, they circumvent this safeguard.

Cancer cells also differ from normal cells in their response to crowded conditions in culture. When normal cells are grown in culture, they divide until the surface of the culture vessel is covered by a single layer of cells. Once this *monolayer stage* is reached, cell division stops—a phenomenon referred to as **density-dependent inhibition of growth.** Cancer cells exhibit reduced sensitivity to density-dependent inhibition of growth and thus do not stop dividing when they reach the monolayer stage. Instead, cancer cells continue to divide and gradually begin piling up on one another.

Cancer Cells Are Immortalized by Mechanisms That Maintain Telomere Length

When most normal cells are grown in culture, they divide a limited number of times. For example, freshly isolated human fibroblasts divide about 50–60 times in culture and then stop dividing, undergo various degenerative changes, and may even die (**Figure 24-2**). Under similar conditions, cancer cells exhibit no such limit and continue dividing indefinitely, behaving as if they were immortal. A striking example is provided by *HeLa cells,* which were obtained in 1951 from a uterine cervical cancer that had been surgically removed from a woman named Henrietta Lacks (hence the term *HeLa* cells). Although she died of cancer that same year, some of her tumor cells had been placed in culture after surgery. These cancer cells began to proliferate and have continued to do so for over 50 years, dividing more than 20,000 times with no signs of stopping.

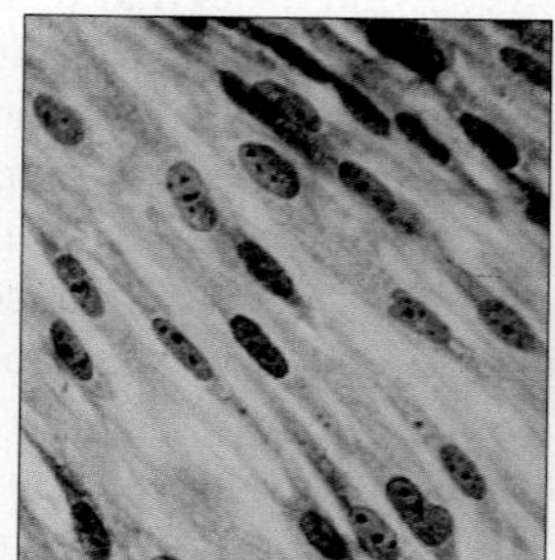

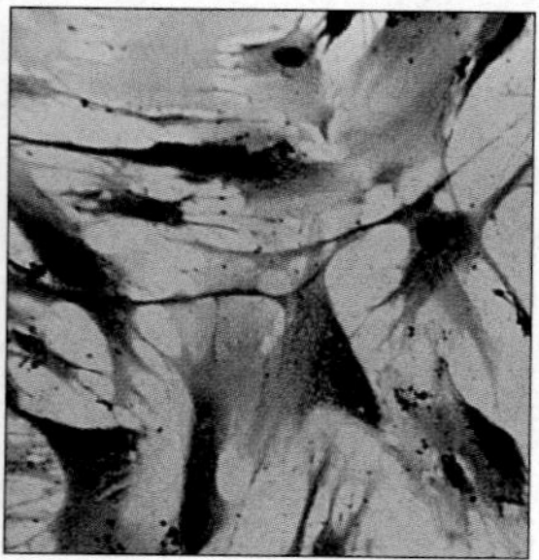

FIGURE 24-2 Microscopic Appearance of Young and Old Human Fibroblasts. *(Left)* Young fibroblasts that have divided only a few times in culture exhibit a thin, elongated shape. *(Right)* After dividing about 50 times, the cells stop dividing and undergo degenerative changes. Note the striking difference in appearance between the young (dividing) and older (nondividing) cells.

Why can cancer cells divide indefinitely in culture when most normal human cells divide only 50–60 times? The answer is related to the telomeric DNA sequences that are lost from the ends of each chromosome every time DNA replicates (page 565). If a normal cell divides too many times, its telomeres become too short to protect the ends of the chromosomes and a pathway is triggered that halts cell division (and may even destroy the cell by apoptosis). This response to telomere shortening prevents excessive proliferation of adult cells and helps explain why normal fibroblasts divide only 50–60 times in culture.

In cancer cells, the telomere-imposed limit is overcome by mechanisms that replenish some of the missing telomere sequences. Most cancer cells accomplish this feat by producing *telomerase,* the enzyme that adds telomeric DNA sequences to the ends of DNA molecules (page 565). An alternative mechanism for maintaining telomere length employs enzymes that exchange DNA sequence information between chromosomes. By one mechanism or the other, cancer cells maintain telomere length above a critical threshold and thereby retain the capacity to divide indefinitely.

Defects in Signaling Pathways, Cell Cycle Controls, and Apoptosis Contribute to Uncontrolled Proliferation

Mechanisms that maintain telomere length play a permissive role in allowing cancer cells to continue dividing, but they do not actually cause cells to divide. The defects driving the uncontrolled proliferation of cancer cells can be traced to a variety of signaling pathways and control mechanisms that normally maintain the proper balance between cell division and death. For example, you learned in Chapters 14 and 19 that cell proliferation is regulated by protein *growth factors* that bind to cell surface receptors and activate signaling pathways within the targeted cells. While cells do not usually divide unless they are stimulated by the proper growth factor, cancer cells circumvent this restraining mechanism through alterations in signaling pathways (to be described shortly) that create a constant signal to divide.

Disruptions in cell cycle control also contribute to the unrestrained proliferation of cancer cells. In Chapter 19 we saw that the commitment to proceed through the cell cycle is made at the *restriction point,* which controls progression from G1 into S phase. If normal cells are grown under suboptimal conditions (for example, insufficient growth factors, high cell density, lack of anchorage, or inadequate nutrients), the cells become arrested at the restriction point and stop dividing. In comparable situations, cancer cells continue to proliferate; if conditions are extremely adverse, as occurs with extreme nutritional deprivation, they eventually die at random points in the cell cycle rather than arresting in G1. This abnormal behavior occurs because cell cycle controls do not function properly in cancer cells. Besides failing to respond appropriately to external signals, cancer cells are also unresponsive to internal conditions, such as DNA damage, that would normally trigger checkpoint mechanisms to halt the cell cycle.

Another factor influencing the growth of cancers is the rate at which cells die. If the normal mechanisms for triggering cell death are disrupted, proliferating cells will accumulate faster than they do when cell death is occurring at an appropriate rate. Cell death is controlled mainly by pathways that trigger apoptosis to get rid of unnecessary or defective cells. Since cancer cells fit the definition of unnecessary or defective cells—that is, they grow in an uncontrolled way and exhibit DNA and chromosomal damage—why aren't cancer cells killed by apoptosis? The answer is that cancer cells have various ways of blocking the pathways that trigger apoptosis; this allows them to survive and proliferate under conditions that would normally cause cell death. For certain kinds of cancer, uncontrolled growth arises primarily from a failure to undergo apoptosis rather than increased cell division.

How Cancers Spread

Although uncontrolled proliferation is a defining feature of cancer cells, it is not the property that makes cancer so dangerous. After all, the cells of benign tumors also proliferate in an uncontrolled way, but such tumors remain in their original location and are usually easy to remove (unless they arise in surgically inaccessible locations). The hazards posed by cancer cells come from uncontrolled proliferation *combined with* the ability to spread throughout the body, which makes complete surgical removal impractical. Roughly 90% of all cancer deaths are caused by the spread of cancer rather than by the primary tumor itself. We will now examine the mechanisms that make this spreading of cancer cells possible.

Angiogenesis Is Required for Tumors to Grow Beyond a Few Millimeters in Diameter

For more than 100 years, scientists have known that tumors possess a dense network of blood vessels. Initially these were thought to be preexisting vessels that had expanded in response to the tumor's presence or were part of an inflammatory response designed to defend the host against the tumor. But in 1971, Judah Folkman proposed a new idea regarding the role of these blood vessels. He suggested that tumors release signaling molecules that trigger **angiogenesis**—that is, growth of blood vessels—in the surrounding host tissues, and that these new vessels are required for tumors to grow beyond a tiny, localized clump of cells.

This idea initially emerged from studies involving cancer cells grown in isolated organs under artificial laboratory conditions. In one experiment, a normal thyroid gland was removed from a rabbit and placed in a glass chamber, a small number of cancer cells were injected into the gland, and a nutrient solution was pumped into the organ to keep it alive (**Figure 24-3a**). Under these

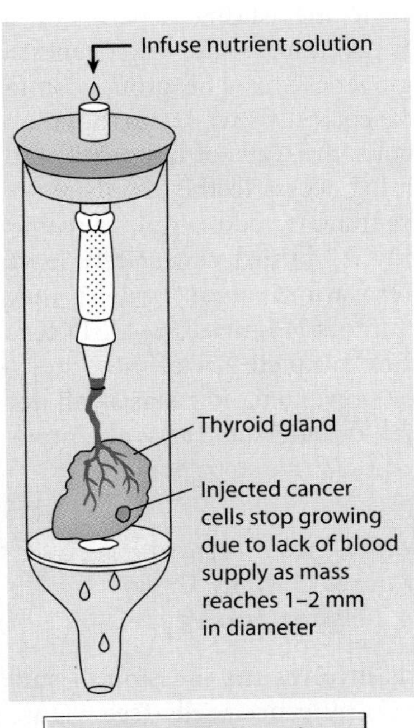

(a) Cancer cells grown in an isolated thyroid gland

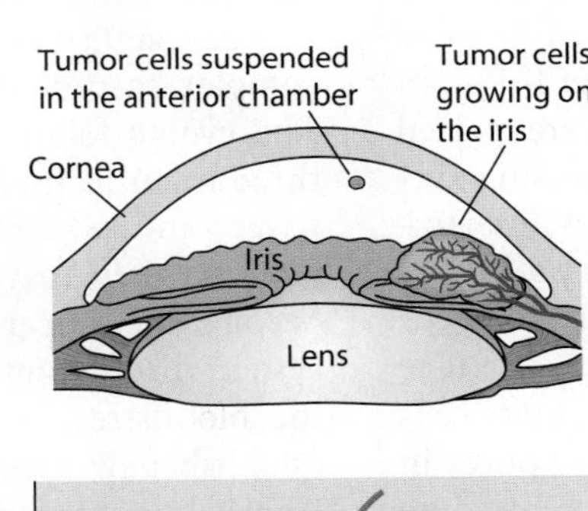

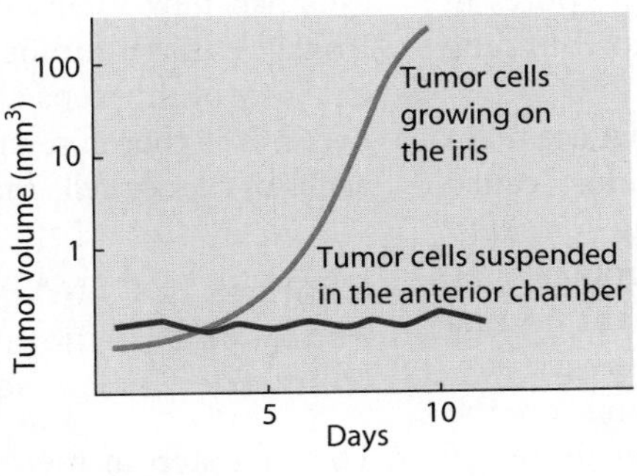

(b) Cancer cells grown on the iris or in the anterior chamber of the eye

FIGURE 24-3 Two Experiments Showing the Requirement for Angiogenesis. **(a)** Cancer cells were injected into an isolated rabbit thyroid gland that was kept alive by pumping a nutrient solution into its main blood vessel. The tumor cells fail to link up to the organ's blood vessels, and the tumor stops growing when it reaches a diameter of roughly 1–2 millimeters. **(b)** Cancer cells were either injected into the liquid-filled anterior chamber of a rabbit's eye, where there are no blood vessels, or they were placed directly on the iris. Tumor cells in the anterior chamber, nourished solely by diffusion, remain alive but stop growing before the tumor reaches a millimeter in diameter. In contrast, blood vessels quickly infiltrate the cancer cells implanted on the iris, allowing the tumors to grow to thousands of times their original mass. Adapted from "The Vascularization of Tumors" by Judah Folkman, *Scientific American* (May 1976). Graph: Reproduced with permission. Copyright © 1976 Scientific American, a division of Nature America, Inc. All rights reserved. Drawings: Adapted by permission of the illustrator, Carol Donner.

conditions, the cancer cells divided for a few days but suddenly stopped growing when the tumor reached a diameter of 1–2 mm. Virtually every tumor stopped growing at exactly the same size, suggesting that some kind of limitation allowed them to grow only so large.

When tumor cells were removed from the thyroid gland and injected back into animals, cell division resumed and massive tumors developed. Why did the tumors stop growing at a tiny size in the isolated thyroid gland and yet grow in an unrestrained way in live animals? On closer examination, a possible explanation became apparent. The tiny tumors, alive but dormant in the isolated thyroid gland, had failed to link up with the organ's blood vessels. As a result, the tumors stopped growing when they reached a diameter of 1–2 mm. When injected into live animals, these same tumors became infiltrated with blood vessels and grew to an enormous size.

To test the idea that blood vessels are needed to sustain tumor growth, Folkman implanted cancer cells in the anterior chamber of a rabbit's eye, where there is no blood supply. As shown in Figure 24-3b, these cancer cells survived and formed tiny tumors. But blood vessels from the nearby iris could not reach the cells and the tumors soon stopped growing. When the same cells were implanted directly on the iris tissue, blood vessels from the iris quickly infiltrated the tumor cells and each tumor grew to thousands of times its original size. Once again, it appeared that tumors need a blood supply to grow beyond a tiny mass.

Blood Vessel Growth Is Controlled by a Balance Between Angiogenesis Activators and Inhibitors

If tumors require blood vessels to sustain their growth, how do they ensure that this need is met? The first hint came from studies in which cancer cells were placed inside a chamber surrounded by a filter possessing tiny pores that cells cannot pass through. When such chambers are implanted into animals, new capillaries proliferate in the surrounding host tissue. In contrast, normal cells placed in the same type of chamber do not stimulate blood vessel growth. Such results suggest that cancer cells produce molecules that diffuse through the tiny pores in the filter and activate angiogenesis in the surrounding host tissue.

Subsequent investigations have revealed that the main angiogenesis-activating molecules are proteins called *vascular endothelial growth factor (VEGF)* and *fibroblast growth factor (FGF)*. VEGF and FGF are produced by many kinds of cancer cells, as well as by certain normal cells. When cancer cells release these proteins into the surrounding tissue, they bind to receptor proteins on the surface of the *endothelial cells* that form the lining of blood vessels. This binding activates a signaling pathway that causes the endothelial cells to divide and to secrete protein-degrading enzymes called *matrix metalloproteinases (MMPs)*. The MMPs break down the extracellular matrix, thereby permitting the endothelial cells to migrate into the surrounding tissues. As they migrate, the proliferating endothelial cells become organized into hollow tubes that develop into new blood vessels.

Although many tumors produce VEGF and/or FGF, these signaling molecules are not the sole explanation for the activation of angiogenesis. For angiogenesis to proceed, these molecules must overcome the effects of angiogenesis *inhibitors* that normally restrain the growth of blood vessels. More than a dozen naturally occurring inhibitors of angiogenesis have been identified, including the proteins *angiostatin, endostatin,* and *thrombospondin.* A finely tuned balance between the concentration of angiogenesis inhibitors and angiogenesis activators determines whether a tumor will induce the growth of new blood vessels. When tumors trigger angiogenesis, it is usually accomplished by an increase in the production of angiogenesis activators and a simultaneous decrease in the production of angiogenesis inhibitors.

Cancer Cells Spread by Invasion and Metastasis

Once angiogenesis has been triggered at an initial tumor site, the stage is set for the cancer cells to spread throughout the body. This ability to spread is based on two distinct mechanisms: invasion and metastasis. **Invasion** refers to the direct migration and penetration of cancer cells into neighboring tissues, whereas **metastasis** involves the ability of cancer cells to enter the bloodstream (or other body fluids) and travel to distant sites, where they form tumors—called *metastases*—that are not physically connected to the primary tumor.

The ability of a tumor to metastasize depends on a complex cascade of events, beginning with angiogenesis. The events following angiogenesis can be grouped into three main steps. First, cancer cells invade surrounding tissues and penetrate through the walls of lymphatic and blood vessels, thereby gaining access to the bloodstream. Second, the cancer cells are transported by the circulatory system throughout the body. And third, cancer cells leave the bloodstream and enter various organs, where they establish new metastatic tumors (**Figure 24-4**). If cells from the initial tumor fail to complete any of these steps, or if any of the steps can be prevented, metastasis will not occur. It is therefore crucial to understand how the properties of cancer cells make these three steps possible.

Changes in Cell Adhesion, Motility, and Protease Production Allow Cancer Cells to Invade Surrounding Tissues and Vessels

The first step in metastasis involves the invasion of surrounding tissues and vessels by cancer cells (see Figure 24-4, step ❶). This means that unlike the cells of benign tumors or most normal cells, which remain together at the

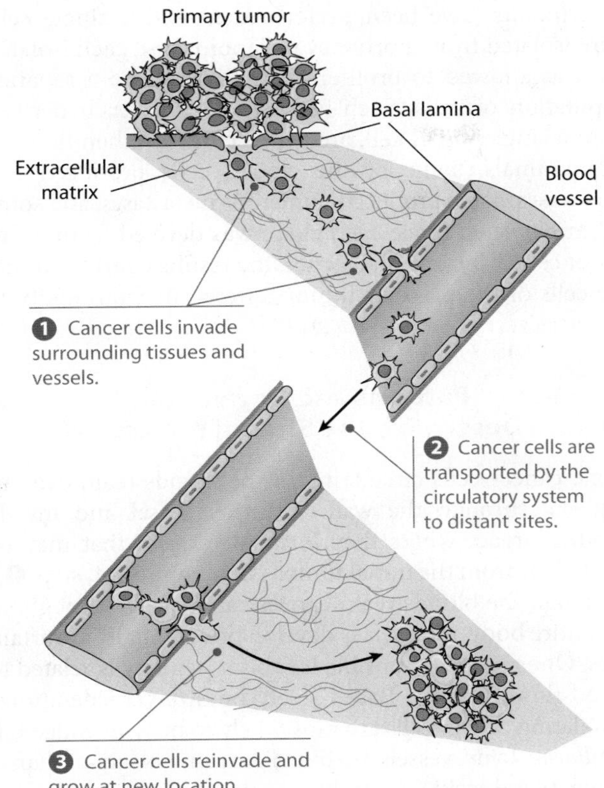
Primary tumor
Basal lamina
Extracellular matrix
Blood vessel
❶ Cancer cells invade surrounding tissues and vessels.
❷ Cancer cells are transported by the circulatory system to distant sites.
❸ Cancer cells reinvade and grow at new location.

FIGURE 24-4 Stages in the Process of Metastasis. Only a small fraction of the cells in a typical cancer successfully carry out all three steps involved in metastasis: ❶ invasion into surrounding tissues and vessels, ❷ transportation via the circulatory system, and ❸ reinvasion and growth at a distant site.

site where they are formed, cancer cells can leave their initial location and invade surrounding tissues, eventually entering the circulatory system.

Several mechanisms make this invasive behavior possible. The first involves changes in the cell surface proteins that cause cells to adhere to one another. Such proteins are often missing or defective in cancer cells, which allows cells to separate from the main tumor more readily. One crucial molecule is *E-cadherin* (page 479), a cell-cell adhesion protein whose loss underlies the reduced adhesiveness of many epithelial cancers. Highly invasive cancers usually have less E-cadherin than do normal cells. Restoring E-cadherin to isolated cancer cells lacking this molecule has been shown to inhibit their ability to form invasive tumors when the cells are injected back into animals.

A second factor underlying invasive behavior is the increased motility of cancer cells, which is stimulated by signaling molecules produced by surrounding host tissues or by the cancer cells themselves. Some of these signaling molecules can act as *chemoattractants* that serve as guiding signals that attract migrating cancer cells. Activation of the Rho family GTPases (see description of Rho, Rac, and Cdc42 on page 440) plays a central role in the stimulation of cell motility that leads to invasion and metastasis.

Another trait contributing to invasion is the ability of cancer cells to produce *proteases* that degrade protein-containing structures that would otherwise act as barriers to cancer cell movement. A critical barrier encountered by most cancers is the *basal lamina*, a dense layer of protein-containing material that separates epithelial layers from underlying tissues (see Figure 17-20). Before epithelial cancers, which account for 90% of all human malignancies, can invade adjacent tissues, the basal lamina must first be breached. Cancer cells break through this barrier by secreting proteases, which degrade the proteins that form the backbone of the basal lamina.

One such protease is *plasminogen activator,* an enzyme that converts the inactive precursor *plasminogen* into the active protease *plasmin.* Because high concentrations of plasminogen are present in most tissues, small amounts of plasminogen activator released by cancer cells can quickly catalyze the formation of large quantities of plasmin. The plasmin performs two tasks: (1) It degrades components of the basal lamina and the extracellular matrix, thereby facilitating tumor invasion, and (2) it cleaves inactive precursors of matrix metalloproteinases, produced mainly by surrounding host cells, into active enzymes that also degrade the basal lamina and extracellular matrix.

After proteases allow cancer cells to penetrate the basal lamina, they facilitate cancer cell migration by degrading the extracellular matrix of the underlying tissues. The cancer cells migrate until they reach tiny blood or lymphatic vessels, which are also surrounded by a basal lamina. The proteases then digest holes in this second basal lamina, allowing cancer cells to pass through it (and through the layer of endothelial cells that form the vessel's inner lining) to finally gain entry into the circulatory system.

Relatively Few Cancer Cells Survive the Voyage Through the Bloodstream

Cancer cells that penetrate the walls of tiny blood vessels gain entry to the bloodstream, which then transports the cells to distant parts of the body (see Figure 24-4, step ❷). If cancer cells instead penetrate the walls of lymphatic vessels, the cells are first carried to regional lymph nodes, where they may become lodged and grow. For this reason, regional lymph nodes are a common site for the initial spread of cancer. However, lymph nodes have numerous interconnections with blood vessels, so cancer cells that initially enter into the lymphatic system eventually flow into the bloodstream.

Regardless of their initial entry route, the net result is often large numbers of cancer cells in the bloodstream. Even a tiny malignant tumor weighing only a few grams can release several million cancer cells into the circulation each day. However, the bloodstream is a relatively inhospitable place for most cancer cells and fewer than one in a thousand cells survives the trip to a potential site of

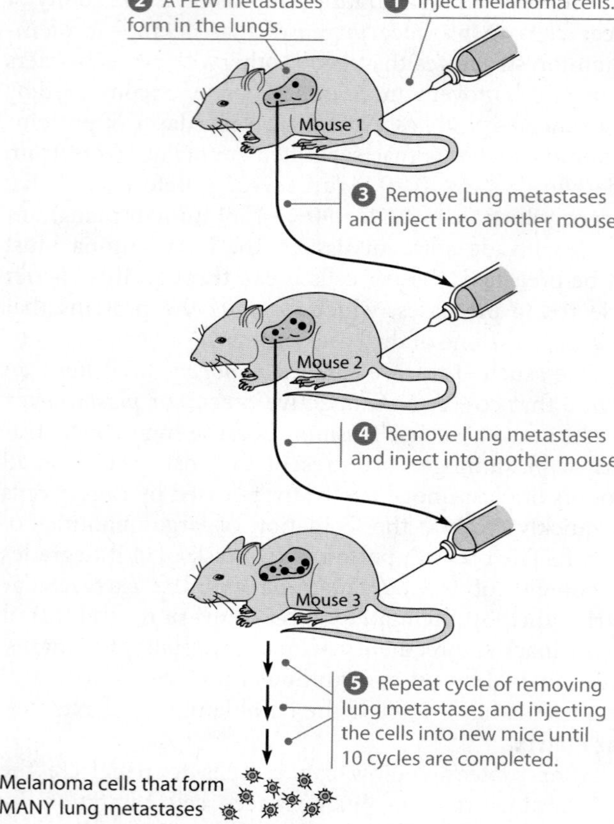

2 A FEW metastases form in the lungs.
1 Inject melanoma cells.
Mouse 1
3 Remove lung metastases and inject into another mouse.
Mouse 2
4 Remove lung metastases and inject into another mouse.
Mouse 3
5 Repeat cycle of removing lung metastases and injecting the cells into new mice until 10 cycles are completed.
Melanoma cells that form MANY lung metastases

FIGURE 24-5 Selection of Melanoma Cells Exhibiting an Enhanced Ability to Metastasize. Mouse melanoma cells were injected into the tail vein of a mouse. After a small number of metastases arose in the lungs, cells from these lung tumors were removed and injected into another mouse. When this cycle was repeated ten times, the final population of melanoma cells produced many more lung metastases than were produced by the original cells.

metastasis. Are these few successful cells simply a random representation of the original tumor cell population, or do they represent cells that are especially well suited for metastasizing? **Figure 24-5** illustrates an experiment designed to address these questions. In this experiment, mouse melanoma cells were injected into the bloodstream of healthy mice to study their ability to metastasize. A few weeks later, metastases appeared in a variety of locations but mainly in the lungs. Cells from the lung metastases were removed and injected into another mouse, leading to the production of more lung metastases. By repeating this procedure many times in succession, researchers eventually obtained a population of cancer cells that formed many more metastases than did the original tumor cell population.

The most straightforward interpretation is that the initial melanoma consisted of a heterogeneous population of cells with differing properties and that repeated isolation and reinjection of cells derived from successful metastases gradually selected for those cells that were best suited for metastasizing. To test this hypothesis, further experiments have been performed in which single cells were isolated from a primary melanoma and each isolated cell was allowed to proliferate in culture into a separate population of cells. Such cell populations, each derived from a single initial cell, are called **clones.** When injected into animals, some of the cloned cells produced few metastases, some produced numerous metastases, and some fell in between. Since each clone was derived from a different cell in the original tumor, the results confirmed that the cells of the primary tumor differed in their ability to metastasize.

Blood-Flow Patterns and Organ-Specific Factors Determine the Sites of Metastasis

Some cancer cells circulating in the bloodstream eventually exit through the wall of a tiny vessel and invade another organ, where they form metastases that may be located far from the initial tumor (see Figure 24-4, step ❸). Although the bloodstream carries cancer cells throughout the entire body, metastases develop preferentially at certain sites. One factor responsible for this specificity is related to blood-flow patterns. Based solely on size considerations, circulating cancer cells are most likely to become lodged in *capillaries* (tiny vessels with a diameter no larger than a single blood cell). After becoming stuck in capillaries, cancer cells penetrate the walls of these tiny vessels, enter the surrounding tissues, and seed the development of new tumors. For example, consider primary tumors originating in most organs. After the cancer cells enter the bloodstream, the first capillary bed they encounter will be in the lungs. As a result, the lungs are a frequent site of metastasis for many kinds of cancer. However, blood-flow patterns do not always favor the lungs. For cancers of the stomach and colon, cancer cells entering the bloodstream are first carried to the liver, where the vessels break up into a bed of capillaries. As a result, the liver is a common site of metastasis for these cancers.

While blood-flow patterns are important, they do not always explain the observed distribution of metastases. As early as 1889, Stephen Paget proposed that circulating cancer cells have a special affinity for the environment provided by particular organs. Paget's idea, referred to as the "seed and soil" hypothesis, is based on the analogy that when a plant produces seeds, they are carried by the wind in all directions but grow only if they fall on congenial soil. According to this view, cancer cells are carried to a variety of organs by the bloodstream, but only a few sites provide an optimal growth environment for each type of cancer. Supporting evidence has come from a systematic analysis of the sites where metastases tend to arise. For roughly two-thirds of the human cancers examined, the rates of metastasis to various organs can be explained solely on the basis of blood-flow patterns. In the remaining cases, some cancers metastasize to particular organs less frequently than expected and others metastasize to particular organs more frequently than expected.

Why do cancer cells grow best at particular sites? The answer is thought to involve interactions between cancer cells and the microenvironment in the organs they are delivered to. One example involves prostate cancer, which commonly metastasizes to bone (a pattern that would not be predicted by blood-flow patterns). The reason for this preference was uncovered by experiments in which prostate cancer cells were mixed with cells from various organs—including bone, lung, and kidney—and the cell mixtures were then injected into animals. It was found that the ability of prostate cancer cells to develop into tumors is stimulated by the presence of cells derived from bone but not from lung or kidney. Subsequent studies uncovered the explanation: Bone cells produce specific growth factors that stimulate the proliferation of prostate cancer cells. This is just one of several types of molecular interactions that can influence the ability of cancer cells to grow in particular organs.

The Immune System Influences the Growth and Spread of Cancer Cells

Does the body have any mechanisms for defending against the growth and spread of cancer cells? One possibility is the immune system, which can attack and destroy foreign cells. Of course, cancer cells are not literally "foreign," but they often exhibit molecular changes that might allow the immune system to recognize the cells as being abnormal. The *immune surveillance theory* postulates that immune destruction of cancer cells is a common event and that cancer simply reflects the occasional failure of an adequate immune response to be mounted against aberrant cells.

Some of the evidence cited in support of this theory involves organ transplant patients, who take immunosuppressive drugs to depress immune function and thereby decrease the risk of immune rejection of the transplanted organ. As predicted by the immune surveillance theory, individuals treated with immunosuppressive drugs develop many cancers at higher rates than normal. The immune surveillance concept has been tested more directly in animals that are genetically altered to introduce specific defects in the immune system. One study used mutant mice containing disruptions in *Rag2,* a gene expressed only in lymphocytes. The mutant mice, which have no functional lymphocytes (and hence no immune response), exhibit an increased cancer rate both for spontaneous cancers and for cancers induced by injecting animals with chemicals that cause cancer.

Although such results indicate that the immune system helps protect mice from developing cancer, how relevant are these findings to humans? If the immune system plays a major role in protecting us from cancer, we would expect a dramatic increase in cancer rates in AIDS patients with severely depressed immune function. While AIDS does cause higher rates for a few types of cancer, especially Kaposi's sarcoma and lymphomas, increases are not seen for the more common forms of cancer. Most of the cancers that occur at increased rates in people with AIDS are caused by viruses (whose role in cancer is covered later in the chapter). Such observations suggest that the immune system may help protect humans from virus-induced cancers, but it is less successful in defending against the more common types of cancer.

The reason for this failure is that cancers find ways of evading destruction by the immune system. One way is based on the fact that tumors are heterogeneous populations of cells that express different *antigens* (substances that trigger an immune response). Those cells containing antigens that elicit a strong immune response are likely to be attacked and destroyed, while cells that either lack or produce smaller quantities of such antigens are more likely to survive and proliferate. So as tumors grow, there is a continual selection for cells that elicit a weaker immune response.

Cancer cells have also devised ways of actively confronting and overcoming the immune system. For example, some cancer cells produce molecules that kill or inhibit the function of T lymphocytes (immune cells involved in destroying foreign or defective cells). Tumors may also surround themselves with dense supporting tissue that shields them from immune attack. And some cancer cells simply divide so quickly that the immune system cannot kill them fast enough to keep tumor growth in check. Consequently, the larger a tumor grows, the easier it becomes to overwhelm the immune system.

The Tumor Microenvironment Influences Tumor Growth, Invasion, and Metastasis

We have now encountered several examples in which tumor behavior is influenced by interactions between tumor cells and the surrounding *tumor microenvironment,* which includes various kinds of normal cells, extracellular molecules, and components of the extracellular matrix. For example, angiogenesis is triggered by growth factors released by tumor cells that act on normal endothelial cells in the surrounding tissue, thereby stimulating the proliferation of new blood vessels. Proteases produced both by tumor cells and surrounding normal cells facilitate invasion by degrading components of the extracellular matrix and the basal lamina. The motility of cancer cells and the direction in which they migrate is influenced by signaling molecules made by normal cells in the surrounding tissues. Penetration through capillaries involves adhesion of cancer cells to molecules present in the basal lamina. And finally, the growth of metastases at distant sites is stimulated by growth factors and other molecules produced by cells residing in the organs being invaded.

The tumor microenvironment can also contain cells and molecules that hinder invasion and metastasis. For example, normal cells of the immune system may attack cancer cells, and other cells in the tumor microenvironment often produce *TGFβ*, a potent inhibitor of proliferation for many cell types (page 413). Cancer cells may in turn acquire mutations that allow them to continue growing in the presence of TGFβ. Sometimes cancer cells even start secreting TGFβ themselves, which inhibits the growth of surrounding normal cells and allows the cancer cells to reproduce and invade surrounding tissues more rapidly because of the decreased competition from neighboring cells. These are just a few of numerous examples in which interactions with the tumor microenvironment influence the ability of cancer cells to grow, invade neighboring tissues, and metastasize to distant sites.

What Causes Cancer?

The uncontrolled proliferation of cancer cells, combined with their ability to metastasize to distant sites, makes cancer a potentially life-threatening disease. What causes the emergence of cells with such destructive properties? People often view cancer as a mysterious disease that strikes randomly and without known cause, but this misconception fails to consider the results of thousands of scientific investigations, some dating back more than 200 years. The inescapable conclusion emerging from these studies is that cancers are commonly caused by environmental agents and lifestyle factors, most of which act by triggering DNA mutations.

Epidemiological Data Have Allowed Many Causes of Cancer to Be Identified

The first indication that a particular agent may cause cancer usually comes from *epidemiology*—the branch of science that investigates the frequency and distribution of diseases in human populations. Epidemiological studies have revealed that cancers arise with differing frequencies in different parts of the world. To list just a few examples, stomach cancer is frequent in Japan, breast cancer is prominent in the United States, and liver cancer is common in Africa and Southeast Asia. To determine whether differences in heredity or environment are responsible for such differences, scientists have examined cancer rates in people who move from one country to another. For example, in Japan the incidence of stomach cancer is greater and the incidence of colon cancer is lower than in the United States; however, when Japanese families move to the United States, their cancer rates come to resemble those in the United States, indicating that cancer rates are determined more by environmental and lifestyle factors than by heredity.

Epidemiological data have played an important role in identifying these environmental and lifestyle factors. A striking example involves lung cancer, which has increased more than tenfold in frequency in the United States since 1900 (**Figure 24-6**). Investigations into the possible causes for this lung cancer epidemic have revealed that most people who develop lung cancer have one trait in common: a history of smoking cigarettes. As might be expected if cigarettes were responsible, heavy smokers develop lung cancer more frequently than light smokers do, long-term smokers develop lung cancer more frequently than short-term smokers do, and lung cancer rates fall after cigarette smokers quit smoking. Besides causing lung cancer, smoking is linked to most cancers of the mouth, pharynx, and larynx, as well as to some cancers of the esophagus, stomach, pancreas, uterine cervix, kidney, bladder, and colon. About half of all people who smoke will be killed by the cancers and cardiovascular diseases

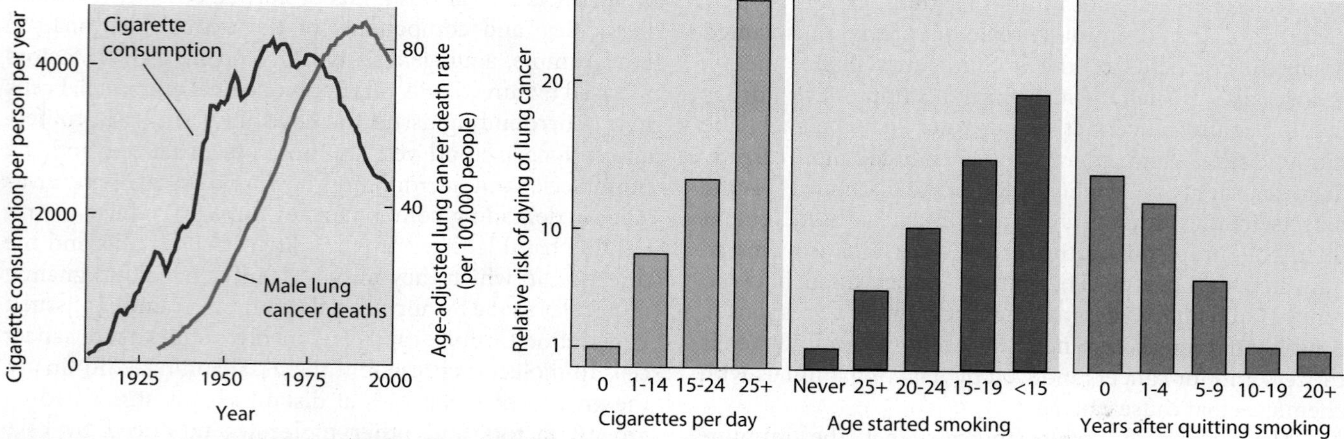

FIGURE 24-6 Cigarette Smoking and Lung Cancer. The graph on the left shows that an increase in cigarette smoking in the United States during the first half of the twentieth century was followed by an explosive growth in lung cancer deaths. A time lag of about 25 years transpired between the increase in smoking rates and the subsequent increase in cancer deaths, which is typical of the time required for most human cancers to develop after exposure to a cancer-causing agent. The bar graphs on the right show that lung cancer rates increase as a function of the number of cigarettes smoked per day, that long-term smokers develop lung cancer more frequently than do short-term smokers, and that lung cancer rates fall after cigarette smokers quit smoking.

caused by smoking, losing an average of 10–15 years of life. This premature loss of life includes about 130,000 deaths per year from lung cancer in the United States alone.

Although the connection between cigarette smoking and cancer was first suggested by epidemiological studies, proof of a cause-and-effect relationship requires direct experimental evidence. Such evidence has come from laboratory studies showing that cigarette smoke contains several dozen chemicals that cause cancer when administered to animals. Knowing that cigarette smoke causes cancer has given society a powerful tool for reducing the number of cancer deaths. After the government began to publicize the dangers of tobacco smoke in the 1960s, smoking rates in the United States stopped growing and began to decline (see Figure 24-6). This decline is one of the main reasons why cancer death rates have slowly begun to fall during the past decade.

Many Chemicals Can Cause Cancer, Often After Metabolic Activation in the Liver

The idea that certain chemicals, such as those found in tobacco smoke, can cause cancer was first proposed more than 200 years ago. In 1761 a London doctor, John Hill, reported that people who routinely used snuff (a powdered form of tobacco that is inhaled) experience an abnormally high incidence of nasal cancer. A few years later another British physician, Percival Pott, observed an elevated incidence of scrotum cancer among men who had served as chimney sweepers in their youth. Because they fit into narrow spaces more readily than adults did, young boys of the time were commonly employed to clean chimney flues. Pott speculated that the chimney soot became dissolved in the natural oils of the scrotum, irritated the skin, and eventually triggered the development of cancer. This conjecture led to the discovery that wearing protective clothing and bathing regularly could prevent scrotum cancer in chimney sweepers.

In the years since these pioneering observations, the list of known and suspected **carcinogens** (cancer-causing agents) has grown to include hundreds of different chemicals. Chemicals are usually labeled as carcinogens because humans or animals develop cancer when exposed to them. This does not mean, however, that each chemical causes cancer through its own direct action. For example, consider the behavior of *2-naphthylamine,* a potent carcinogen that causes bladder cancer in industrial workers and is present in tobacco smoke. As might be expected, feeding 2-naphthylamine to laboratory animals induces a high rate of bladder cancer. But if 2-naphthylamine is implanted directly into an animal's bladder, cancer rarely develops. The explanation for the apparent discrepancy is that when 2-naphthylamine is ingested (by animals) or inhaled (by humans), it passes through the liver and is metabolized into other chemicals that are the actual causes of cancer. Placing 2-naphthylamine directly in an animal's bladder bypasses this metabolic activation, and so cancer does not arise.

Many carcinogens share this need for metabolic activation before they can cause cancer. Substances exhibiting such behavior are more accurately called **precarcinogens,** a term applied to any chemical that is capable of causing cancer only after it has been metabolically activated. Most precarcinogens are activated by liver proteins that are members of the *cytochrome P450* enzyme family. Members of this enzyme family catalyze the oxidation of ingested foreign chemicals, such as drugs and pollutants, to make the molecules less toxic and easier to excrete from the body. However, in some cases these oxidation reactions inadvertently convert foreign chemicals into carcinogens—a phenomenon known as *carcinogen activation.*

DNA Mutations Triggered by Chemical Carcinogens Lead to Cancer

Once it was known that chemicals can cause cancer, the question arose as to how they work. The idea that carcinogenic chemicals act by triggering DNA mutations was first proposed around 1950, but there was little supporting evidence because nobody had systematically compared the mutagenic potency of different chemicals with their ability to cause cancer. The need for such information inspired Bruce Ames to develop a simple laboratory test for measuring a chemical's mutagenic activity. This procedure, called the **Ames test,** uses bacteria as a test organism because they can be quickly grown in enormous numbers in culture. The bacteria are a special strain that cannot synthesize the amino acid *histidine.* As shown in **Figure 24-7**, the bacteria are placed in a culture dish containing a growth medium lacking histidine, along with the chemical being tested for mutagenic activity. Normally, the bacteria would not grow in the absence of histidine. However, if the chemical being tested is mutagenic, it will trigger random mutations, some of which might restore the ability to synthesize histidine. Each bacterium acquiring such a mutation will grow into a visible colony, so the total number of colonies is a measure of the mutagenic potency of the substance being tested.

Because many chemicals that cause cancer do not become carcinogenic until after they have been modified by liver enzymes, the Ames test includes a step in which the chemical being tested is first incubated with an extract of liver cells to mimic the reactions that normally occur in the liver. The resulting chemical mixture is then tested for its ability to cause bacterial mutations. When the Ames test is performed in this way, a strong correlation is observed between a chemical's ability to cause mutations and its ability to cause cancer.

Carcinogenic chemicals inflict DNA damage in several ways, including binding to DNA and disrupting normal base-pairing; generating crosslinks between the two strands of the double helix; creating chemical linkages

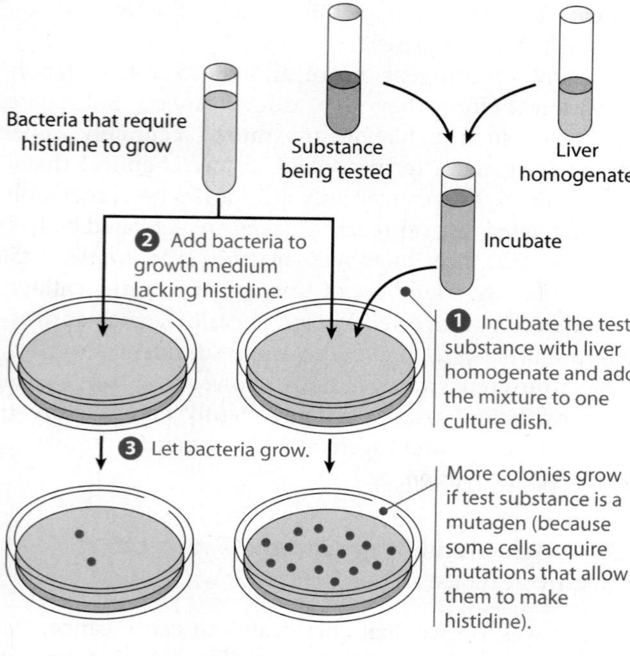

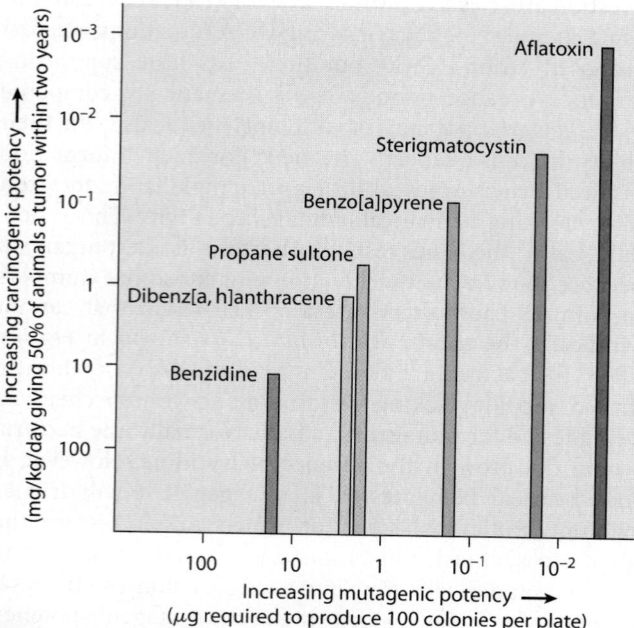

MB Ames Test Protocol

FIGURE 24-7 Ames Test for Identifying Potential Carcinogens. The Ames test is based on the rationale that most carcinogens are mutagens. *(Top)* The ability of chemicals to induce mutations is measured in bacteria that cannot synthesize the amino acid histidine. When placed in a growth medium lacking histidine, the only bacteria that can grow are those that have acquired a mutation allowing them to make histidine. The number of bacterial colonies that grow is therefore related to the mutagenic potency of the substance being tested. Chemicals studied with the Ames test are first incubated with a liver homogenate because many chemicals become carcinogenic only after they have undergone biochemical modification in the liver. *(Bottom)* The bar graph shows that substances exhibiting strong mutagenic activity in the Ames test also tend to be strong carcinogens. Among this particular group of substances, aflatoxin is the most potent mutagen and the most potent carcinogen. Note that the data are plotted on logarithmic scales to permit aflatoxin to be shown on the same graph as benzidine, which is about 10,000 times weaker than aflatoxin as a mutagen and carcinogen.

between adjacent bases; hydroxylating or removing individual DNA bases; and causing breaks in one or both DNA strands. In some cases, the mutational role of specific chemicals in causing cancer has been linked to effects on specific genes. For example, the *polycyclic aromatic hydrocarbons* found in tobacco smoke preferentially bind to specific regions of the *p53* gene and trigger unique mutations in which the base T is substituted for the base G. As we will see later in the chapter, mutations in the *p53* gene play a key role in the development of many kinds of cancer.

Cancer Arises Through a Multistep Process Involving Initiation, Promotion, and Tumor Progression

Although mutation plays a prominent role, the development of cancer requires multiple steps. Early evidence for this idea came from studying the ability of coal tar components, such as *dimethylbenz[a]anthracene (DMBA),* to cause cancer in laboratory animals. These studies showed that feeding mice a single dose of DMBA rarely causes tumors to develop. However, if a mouse that has been fed a single dose of DMBA is later treated with a substance that causes skin irritation, cancer develops in the treated area (**Figure 24-8**). The irritant most commonly used for triggering tumor formation is a plant-derived substance called *croton oil,* which is enriched in compounds called *phorbol esters.* Croton oil does not cause cancer by itself, nor does cancer arise if DMBA is administered *after* the croton oil. These observations indicate that

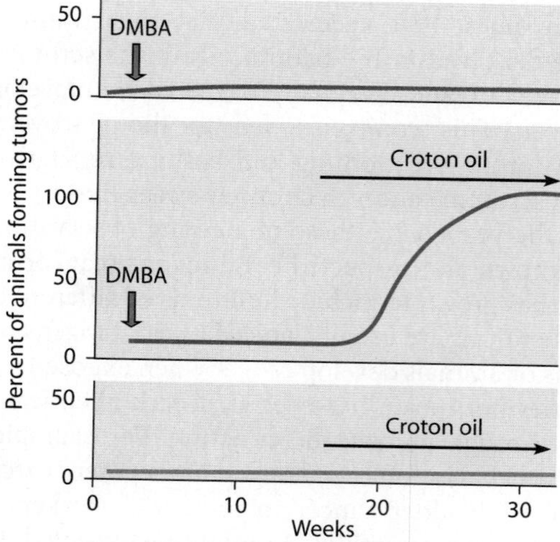

FIGURE 24-8 Evidence for Initiation and Promotion Stages During Cancer Development. *(Top)* Mice given a single dose of DMBA (dimethylbenz[*a*]anthracene) do not form tumors. *(Middle)* Painting their skin with croton oil twice a week after DMBA treatment leads to the appearance of skin tumors. If the croton oil application is stopped a few weeks into the treatment (data not shown), the tumors regress. *(Bottom)* Croton oil alone does not produce skin tumors. These data are consistent with the conclusion that DMBA is an initiator and croton oil is a promoting agent.

DMBA and croton oil play two different roles, known as **initiation** and **promotion.** During initiation, normal cells are converted to a precancerous state, and promotion then stimulates the altered cells to divide and form tumors.

A year or more can transpire after feeding animals a single dose of DMBA and yet tumors will still develop if an animal's skin is then irritated with croton oil. This means that a single DMBA treatment can create a permanently altered, initiated state in cells located throughout the body, and that subsequent administration of croton oil acts on these altered cells to promote tumor development. The ability of chemicals to act as initiators correlates with their ability to cause DNA damage, suggesting that the permanently altered, initiated state is based on the ability of initiators to cause *DNA mutation.*

In contrast, promotion is a gradual process requiring prolonged or repeated exposure to a promoting agent. Investigation of a wide variety of promoting agents, such as the phorbol esters in croton oil, has revealed that their main shared property is the ability to stimulate *cell proliferation.* Not all tumor promoters are foreign substances. Hormones and growth factors that stimulate normal cell proliferation may inadvertently behave as tumor promoters if they act on a cell that has already sustained an initiating mutation.

When a cell with an initiating mutation is exposed to a promoting agent (or natural growth regulator) that causes the initiated cell to proliferate, the number of mutant cells increases. As proliferation continues, natural selection tends to favor cells exhibiting enhanced growth rate and invasive properties, eventually leading to the formation of a malignant tumor. The time required for promotion contributes to the lengthy delay that often transpires between exposure to an initiating carcinogen and the development of cancer.

Initiation and promotion are followed by a third stage, known as **tumor progression** (**Figure 24-9**). During tumor progression, tumor cell properties gradually change over time as cells acquire more aberrant traits and become increasingly aggressive. The driving force for tumor progression is that cells exhibiting traits that confer a selective advantage—for example, increased growth rate, increased invasiveness, ability to survive in the bloodstream, resistance to immune attack, ability to grow in other organs, resistance to drugs, evasion of apoptosis, and so forth—will be more successful than cells lacking these traits and so will gradually tend to predominate.

While it is easy to see why cells exhibiting such traits tend to prevail through natural selection, this does not explain how the aberrant traits originate in the first place. One way new traits arise is through additional DNA mutations that occur after the original, initiating mutation. New traits can also be produced by changes in the expression of normal genes. As you learned in Chapter 23, numerous *epigenetic* mechanisms exist for activating or inhibiting the activity of normal genes without mutating them. Thus, tumor progression is made possible by a combination of DNA mutations and epigenetic changes that don't require mutation, accompanied by natural selection of those cells that acquire advantageous properties through these mechanisms.

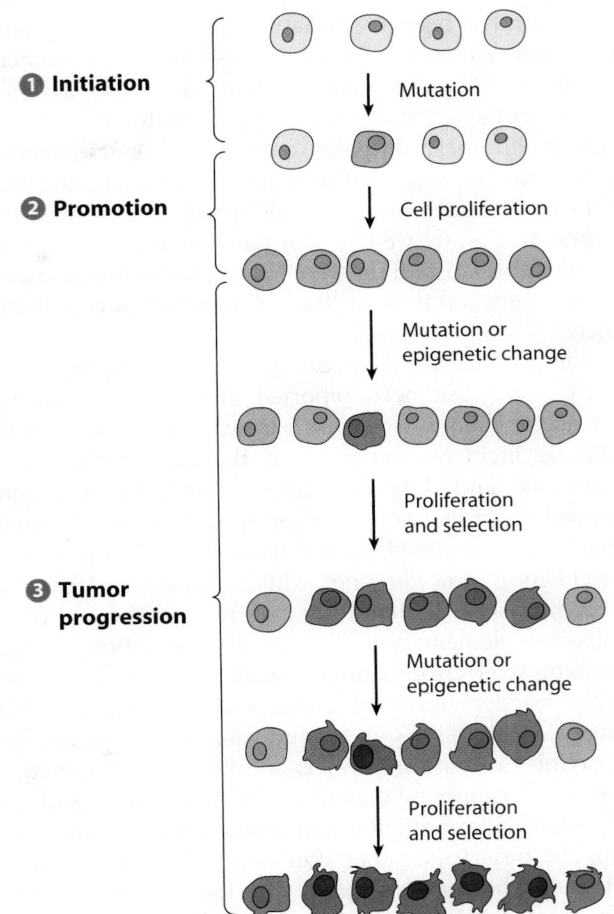

FIGURE 24-9 Main Stages in Cancer Development. Cancer arises by a multistep process involving ❶ an initiation event based on DNA mutation, ❷ a promotion stage in which the initiated cell is stimulated to proliferate, and ❸ tumor progression, in which mutations and changes in gene expression create variant cells exhibiting enhanced growth rates or other aggressive properties that give certain cells a selective advantage. Such cells tend to outgrow their companions and become the predominant cell population in the tumor. During tumor progression, repeated cycles of this selection process create a population of cells whose properties gradually change over time.

Ionizing and Ultraviolet Radiation Also Cause DNA Mutations That Lead to Cancer

Chemicals are not the only DNA-damaging agents that can cause cancer. Shortly after the discovery of X-rays by Wilhelm Roentgen in 1895, it was noticed that people working with this type of radiation developed cancer at abnormally high rates. Animal studies subsequently confirmed that X-rays create DNA mutations and cause cancer in direct proportion to the dose administered.

A similar type of radiation is emitted by many radioactive elements. An early example of the cancer hazards posed by radioactivity occurred in the 1920s in a New Jersey factory that produced glow-in-the-dark watch dials. A luminescent paint containing the radioactive element *radium* was used for painting the dials, and this paint was applied with a fine-tipped brush that the workers frequently wetted with their tongues. As a result tiny quantities of radium were inadvertently ingested and became concentrated in their bones, triggering bone cancer.

Elevated cancer rates caused by exposure to radioactivity have also been reported in people exposed to radioactive fallout from nuclear explosions. The most dramatic incidents occurred in the Japanese cities of Hiroshima and Nagasaki after atomic bombs were dropped there in 1945. Another episode affected people around the Chernobyl nuclear power plant in the former Soviet Union (now Ukraine), which exploded in 1986.

X-rays and related forms of radiation emitted by radioactive elements are called **ionizing radiation** because they remove electrons from molecules, thereby generating highly reactive ions that create various types of DNA damage, including single- and double-strand breaks. **Ultraviolet radiation (UV)** is another type of radiation that causes cancer by damaging DNA. The ability of the UV radiation in sunlight to cause cancer was first deduced from the observation that skin cancer is most prevalent in people who spend long hours in the sun, especially in tropical regions where the sunlight is intense. UV radiation is absorbed mainly by the skin, where it imparts enough energy to trigger *pyrimidine dimer* formation—that is, the formation of covalent bonds between adjacent pyrimidine bases in DNA, as shown in Figure 19-17. If the damage is not repaired, distortion of the double helix causes improper base pairing during DNA replication, which in turn produces distinctive mutation patterns. For example, a CC⟶TT mutation (conversion of two adjacent cytosines to thymines) is a unique product of UV exposure and can be used as a distinctive "signature" to identify mutations caused by sunlight.

The existence of such signature mutations provided a way to prove that UV-induced mutations cause skin cancer. One of the first genes to be studied was the *p53* gene (page 588), which is known to be mutated in many human cancers. When the *p53* gene of skin cancer cells is examined using DNA sequencing techniques, mutations exhibiting a distinctive UV signature (such as CC⟶TT) are frequently observed. In contrast, when *p53* mutations are detected in other types of cancer, they do not exhibit the UV signature (**Figure 24-10**).

Such observations confirm that the *p53* mutations seen in skin cancer cells are triggered by sunlight, but they don't prove that these mutations actually cause the cancers to arise. Is it possible that the mutations are simply an irrelevant, random by-product of long-term exposure to sunlight? This question has been resolved by examining

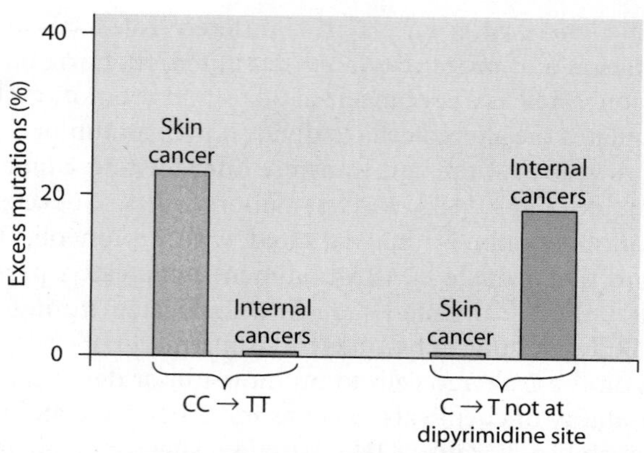

FIGURE 24-10 Incidence of Two Types of *p53* Mutations in Skin Cancer and Internal Cancers. The two bars on the left represent the frequency of CC⟶TT mutations, which are triggered by UV radiation. The two bars on the right represent the frequency of C⟶T mutations not located at dipyrimidine sites, which are not caused by UV radiation. Note that the UV-triggered type of mutation occurs in the *p53* gene of skin cancers (squamous cell carcinomas) but not in cancers of internal organs. Mutation frequencies are plotted relative to what would be expected to occur randomly.

the precise distribution of UV-induced mutations within the *p53* gene. In Chapter 21, it was pointed out that changes in the third base of a codon often do not alter the amino acid being specified by that codon (see Figure 21-6). If the *p53* mutations detected in skin cancers are simply a random by-product of sunlight exposure, then mutations in a codon's third base that do not change an amino acid should be as frequent as mutations in the first or second base that do change an amino acid. In fact, DNA sequencing has revealed that the *p53* mutations seen in skin cancer cells are not randomly distributed. Instead, they preferentially involve base changes that alter amino acids in the p53 protein, as would be expected if these mutations are involved in the mechanism by which sunlight causes cancer.

Viruses and Other Infectious Agents Trigger the Development of Some Cancers

Although the carcinogenic properties of chemicals and radiation were recognized by the early 1900s, the possibility that infectious agents might also cause cancer was not widely appreciated because cancer does not usually behave like a contagious disease. However, in 1911 Peyton Rous performed experiments on sick chickens brought to him by local farmers that showed for the first time that cancer can be caused by a virus. These chickens had cancers of connective tissue origin, or *sarcomas.* To investigate the origin of the tumors, Rous ground up the tumor tissue and passed it through a filter with pores so small that not even bacterial cells could pass through. When he injected the cell-free extract into healthy chickens, they developed sarcomas. Since no cancer cells had been injected into the healthy chickens, Rous concluded that sarcomas can be transmitted by an agent that is smaller than a bacterial cell. This was the first time anyone had detected an **oncogenic virus**—that is, a virus that causes cancer. Although Rous's findings were initially greeted with skepticism, an 87-year-old Rous finally received the Nobel Prize in 1966—more than 50 years after his discovery of the first cancer virus!

It is now clear that dozens of viruses can cause cancer in animals, and a smaller number have been linked to human cancers. The first human example was uncovered by Denis Burkitt, a British surgeon working in Africa in the late 1950s. At certain times of the year, Burkitt noted large outbreaks of lymphocytic cancers of the neck and jaw. Because this cancer, now known as **Burkitt lymphoma,** occurred in periodic epidemics localized to specific geographical regions, Burkitt proposed that it was transmitted by an infectious agent. Burkitt's ideas attracted the attention of two virologists, named Epstein and Barr, whose electron microscopic studies of Burkitt lymphoma cells revealed a virus now called the **Epstein–Barr virus (EBV).** The following evidence supports the idea that EBV can play a role in Burkitt lymphoma: (1) EBV DNA and proteins are often found in tumor cells obtained from patients with Burkitt lymphoma but not in normal cells from the same individuals; (2) adding EBV to normal human lymphocytes in culture causes the cells to acquire some of the properties of cancer cells; and (3) injecting EBV into monkeys causes lymphomas to arise.

In addition to EBV, several other viruses have been linked to human cancers. Among these are the *hepatitis B* and *hepatitis C* viruses, which trigger some liver cancers; *human T-cell lymphotropic virus-I (HTLV-I),* which causes adult T-cell leukemia and lymphoma; and the sexually transmitted *human papillomavirus (HPV),* which is associated with uterine cervical cancer. Moreover, viruses are not the only infectious agents that can cause cancer. Chronic infection with the bacterium *Helicobacter pylori (H. pylori)*—a common cause of stomach ulcers—can trigger stomach cancer, and flatworm infections have been linked to a small number of bladder and bile duct cancers.

Knowing the identity of such infectious agents opens the door for new cancer prevention strategies. For example, antibiotics that kill *H. pylori* are helpful in preventing stomach cancer, and vaccines against hepatitis B and HPV can reduce the incidence of liver and cervical cancers, respectively. The HPV vaccine, which first became available in 2006, illustrates the difficulties that often hinder the development of such vaccines. Scientists have known for more than 100 years that cervical cancer is caused by a sexually transmitted agent, but HPV was not identified as the responsible agent until the 1980s. The difficulty in making this identification arose because HPV is not a single virus but rather a heterogeneous family of related viruses containing more than 100 different subtypes (HPV 1, HPV 2, HPV 3, and so forth). After tests were developed for distinguishing the various subtypes, it became clear that cervical cancer is linked to infection with certain high-risk forms of the virus—mainly HPV 16,

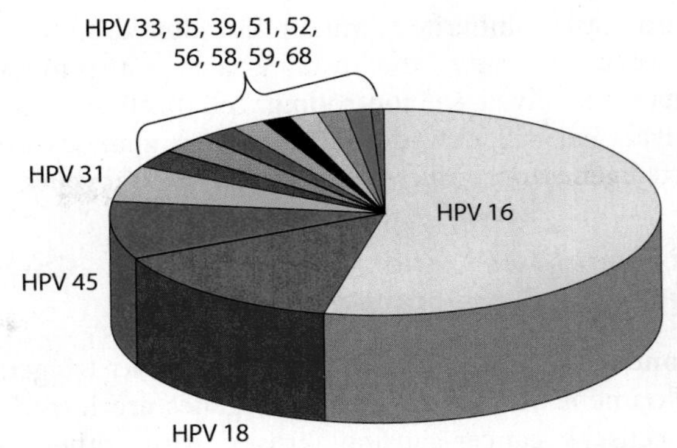

FIGURE 24-11 Prevalence of Various Subtypes of HPV in Cervical Cancers. About 90% of all cancers of the uterine cervix involve infection with at least one of the subtypes of HPV shown in this chart.

HPV 18, and a few others (**Figure 24-11**). The first HPV vaccine was therefore targeted against HPV subtypes 16 and 18, which gives the vaccine the potential to prevent about 70% of all cervical cancers. This vaccine also protects against two other HPV subtypes that cause genital warts.

Although the infectious agents that cause cancer include a diverse collection of viruses, bacteria, and parasites, their mechanisms of action can be grouped into two main categories. One involves those agents—such as the hepatitis B and C viruses, *H. pylori,* and parasitic flatworms—that cause tissue destruction and chronic inflammation. Under these conditions, cells of the immune system infiltrate the tissue and attempt to kill the infectious agent. Unfortunately, the mechanisms used by immune cells to fight infections often produce mutagenic chemicals, such as oxygen *free radicals* (highly reactive forms of oxygen containing an unpaired electron). The net result is an increased likelihood that cancer-causing mutations will arise when cells proliferate to repair the damaged tissue.

The other mechanism used by infectious agents to cause cancer is based on the ability of certain viruses to stimulate the proliferation of infected cells. Some viruses trigger cell proliferation through the direct action of viral genes, whereas other viruses alter the behavior of host cell genes. In either case, the types of genes involved play a role not just in viral cancers but in cancers caused by chemicals and radiation as well. We will therefore proceed to a discussion of cancer-related genes and the ways in which they act.

Oncogenes and Tumor Suppressor Genes

A large body of evidence points to the central role played by DNA mutations in cancer development. As we have just seen, some cancer-causing mutations are triggered by chemicals, radiation, or infectious agents. Others are spontaneous mutations, DNA replication errors, or in certain cases, inherited mutations. But despite these differences in origin, the result is always the mutation of genes involved in controlling cell proliferation and survival. We will now describe the two main classes of affected genes: *oncogenes* and *tumor suppressor genes.*

Oncogenes Are Genes Whose Products Can Trigger the Development of Cancer

An **oncogene** is a gene whose *presence* can trigger the development of cancer. Some oncogenes are introduced into cells by cancer-causing viruses, while others arise from the mutation of normal cellular genes. In either case, oncogenes code for proteins that stimulate excessive cell proliferation and/or promote cell survival by inhibiting apoptosis.

The first oncogene to be discovered was in the *Rous sarcoma virus,* which causes cancer in chickens (page 770) and has only four genes. Mutational studies revealed that mutant viruses with defects in one of these genes, called *src,* are still able to infect cells and reproduce normally but can no longer cause cancer. In other words, a functional copy of the *src* gene must be present for cancer to arise. Similar approaches subsequently led to the identification of oncogenes in dozens of other viruses.

Evidence for the existence of oncogenes in cancers not caused by viruses first came from studies in which DNA isolated from human bladder cancer cells was introduced into a strain of cultured mouse cells called *3T3 cells.* The DNA was administered under conditions that stimulate **transfection**—that is, uptake of the foreign DNA into the cells and its incorporation into their chromosomes. After being transfected with the cancer cell DNA, some of the mouse 3T3 cells proliferated excessively. When these cells were injected back into mice, the animals developed cancer. Scientists therefore suspected that a human gene taken up by the mouse cells had caused the cancer. To confirm the suspicion, gene cloning techniques were applied to DNA isolated from the mouse cancer cells. This resulted in identification of the first human oncogene: a mutant *RAS* gene coding for an abnormal form of Ras, the protein whose role in growth signaling was described in Chapters 14 and 19. (*Important Note:* The names of human genes—for example, *RAS*—are generally written in italicized capital letters, while the names of the proteins they produce—for example, Ras—are written without italics and often with only an initial capital letter.)

RAS was just the first of more than 200 human oncogenes to be discovered. While these oncogenes are defined as genes that can cause cancer, a single oncogene is usually not sufficient. For example, in the transfection experiments described in the preceding paragraph, introducing the *RAS* oncogene caused cancer only because the mouse 3T3 cells used in these studies already possess a mutation in another cell cycle control gene. If freshly isolated normal mouse cells are used instead of 3T3 cells, introducing the *RAS* oncogene by itself will not cause cancer; however, *RAS* together with other oncogenes that target the p53 pathway will cause cancer. This observation illustrates an important principle: *Multiple mutations are usually required to convert a normal cell into a cancer cell.*

Proto-oncogenes Are Converted into Oncogenes by Several Distinct Mechanisms

How do human cancers, most of which are not caused by viruses, come to acquire oncogenes? The answer is that oncogenes arise by mutation from normal cellular genes called **proto-oncogenes.** Despite their harmful-sounding name, proto-oncogenes are not bad genes that are simply waiting for an opportunity to foster the development of cancer. Rather, they are normal cellular genes that make essential contributions to the regulation of cell growth and survival. The term *proto-oncogene* simply implies that if and when the structure or activity of a proto-oncogene is disrupted by certain kinds of mutations, the mutant form of the gene can cause cancer. The mutations that convert proto-oncogenes into oncogenes are created through several distinct mechanisms, which are summarized in **Figure 24-12** and briefly described next.

1. Point Mutation. The simplest mechanism for converting a proto-oncogene into an oncogene is a point mutation—that is, a single nucleotide substitution in DNA that causes a single amino acid substitution in the protein encoded by the proto-oncogene. The most frequently encountered oncogenes of this type are the *RAS* oncogenes that code for abnormal forms of the Ras protein. Point mutations create abnormal, hyperactive forms of the Ras protein that cause the Ras pathway to be continually activated, thereby leading to excessive cell proliferation. *RAS* oncogenes have been detected in several human cancers, including those of the bladder, lung, colon, pancreas, and thyroid. A point mutation can be present at any of several different sites within a *RAS* oncogene, and the particular site involved appears to be influenced by the carcinogen that caused it.

2. Gene Amplification. The second mechanism for creating oncogenes utilizes gene amplification (page 723) to increase the number of copies of a proto-oncogene. When the number of gene copies is increased, it causes the protein encoded by the proto-oncogene to be produced in excessive amounts, although the protein itself is normal. For example, about 25% of human breast and ovarian cancers have amplified copies of the *ERBB2* gene, which codes for a growth factor receptor. The existence of multiple copies of the gene leads to the production of too much receptor protein, which in turn causes excessive cell proliferation.

3. Chromosomal Translocation. During chromosomal translocation, a portion of one chromosome is physically

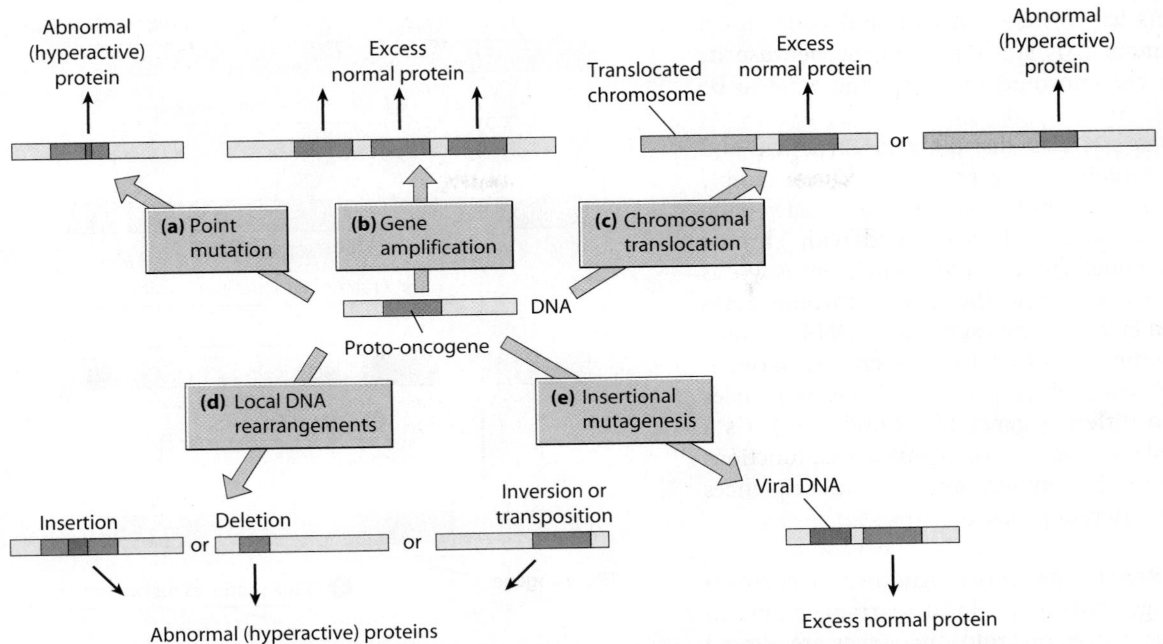

FIGURE 24-12 Five Mechanisms for Converting Proto-oncogenes into Oncogenes. Some oncogenes produced by these mechanisms code for abnormal proteins, whereas others produce normal proteins in excessive amounts. **(a)** Point mutation involves a single nucleotide substitution that creates an oncogene coding for an abnormal protein differing in a single amino acid from the normal protein produced by the proto-oncogene. **(b)** Gene amplification creates multiple gene copies, thereby leading to excessive production of a normal protein. **(c)** Chromosomal translocations move chromosome segments from one chromosome to another. This may either fuse two genes together to form an oncogene coding for an abnormal protein, or it may place a proto-oncogene next to a highly active gene, thereby causing the translocated proto-oncogene to become more active. **(d)** Local DNA rearrangements such as insertions, deletions, inversions, and transpositions can disrupt the structure of proto-oncogenes and cause them to produce abnormal proteins. **(e)** Insertional mutagenesis occurs when viral DNA is integrated into a host chromosome near a proto-oncogene. The inserted DNA may stimulate the expression of the proto-oncogene and cause it to produce too much protein.

removed and joined to another chromosome. A classic example occurs in Burkitt lymphoma, a type of cancer associated with the Epstein–Barr virus (EBV). Infection with EBV stimulates cell proliferation, but this is not sufficient to cause cancer by itself. The disease arises only when a translocation involving chromosome 8 happens to occur in one of these proliferating cells. In the most frequent translocation, a proto-oncogene called *MYC* is moved from chromosome 8 to 14, where it becomes situated next to an intensely active region of chromosome 14 containing genes coding for antibody molecules (**Figure 24-13**). Moving the *MYC* gene so close to the highly active antibody genes causes the *MYC* gene to likewise become activated, thereby leading to overproduction of the Myc protein—a transcription factor that stimulates cell proliferation. Although the translocated *MYC* gene

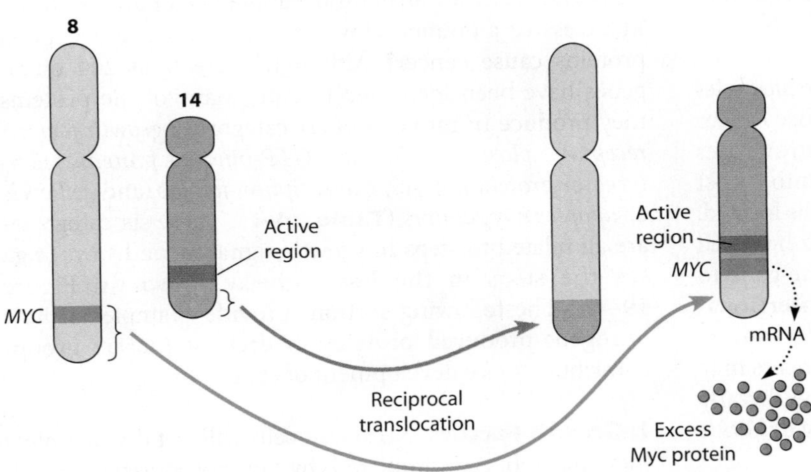

FIGURE 24-13 Chromosome Translocation in Burkitt Lymphoma. In the lymphocytes that give rise to Burkitt lymphoma, a segment of chromosome 8 containing the normal *MYC* gene is frequently exchanged with a segment of chromosome 14. This reciprocal translocation places the normal *MYC* gene adjacent to a very active region of chromosome 14, which contains genes coding for antibody molecules. Moving the *MYC* gene so close to the highly active antibody genes results in activation of the *MYC* gene, leading to an overproduction of Myc protein that stimulates cell proliferation.

sometimes retains its normal structure and codes for a normal Myc protein, it is still an oncogene because its new location on chromosome 14 causes the gene to be overexpressed.

Translocations can also disrupt gene structure and cause abnormal proteins to be produced. One example involves the *Philadelphia chromosome,* an abnormal version of chromosome 22 commonly associated with chronic myelogenous leukemia. The Philadelphia chromosome is created by DNA breakage near the ends of chromosomes 9 and 22, followed by reciprocal exchange of DNA between the two chromosomes. This translocation creates an oncogene called *BCR-ABL,* which contains DNA sequences derived from two different genes (*BCR* and *ABL*). As a result, the oncogene produces a *fusion protein* that functions abnormally because it contains amino acid sequences derived from two different proteins.

4. Local DNA Rearrangements. Another mechanism for creating oncogenes involves local rearrangements in which the base sequences of proto-oncogenes are altered by *deletions, insertions, inversions* (removal of a sequence followed by reinsertion in the opposite direction), or *transpositions* (movement of a sequence from one location to another). An example encountered in thyroid and colon cancers illustrates how a simple rearrangement can create an oncogene from two normal genes. This example involves two genes, named *NTRK1* and *TPM3,* that reside on the same chromosome. *NTRK1* codes for a receptor tyrosine kinase, and *TPM3* codes for a completely unrelated protein, nonmuscle tropomyosin. In some cancers, a DNA inversion occurs that causes one end of the *TPM3* gene to fuse to the opposite end of the *NTRK1* gene (**Figure 24-14**). The resulting gene, called the *TRK oncogene,* produces a fusion protein containing the tyrosine kinase site of the receptor joined to a region of the tropomyosin molecule that forms a *coiled coil* (page 444)—a structure that causes two polypeptide chains to join together as a dimer. As a result, the fusion protein forms a permanent dimer and its tyrosine kinase is permanently activated. (Recall from Chapter 14 that receptor tyrosine kinases are normally activated by bringing together two receptor molecules to form a dimer.)

5. Insertional Mutagenesis. Retroviruses—whose life cycles were described on page 648—can sometimes cause cancer even if they have no oncogenes of their own. Retroviruses accomplish this task by integrating their genes into a host chromosome in a region where a proto-oncogene is located. Integration of the viral DNA then converts the host cell proto-oncogene into an oncogene by causing the gene to be overexpressed. This phenomenon, called **insertional mutagenesis,** is frequently encountered in animal cancers but is rare in humans. However, some human cancers may have been inadvertently created this way in gene therapy trials that used retroviruses as vectors for repairing defective genes (page 640).

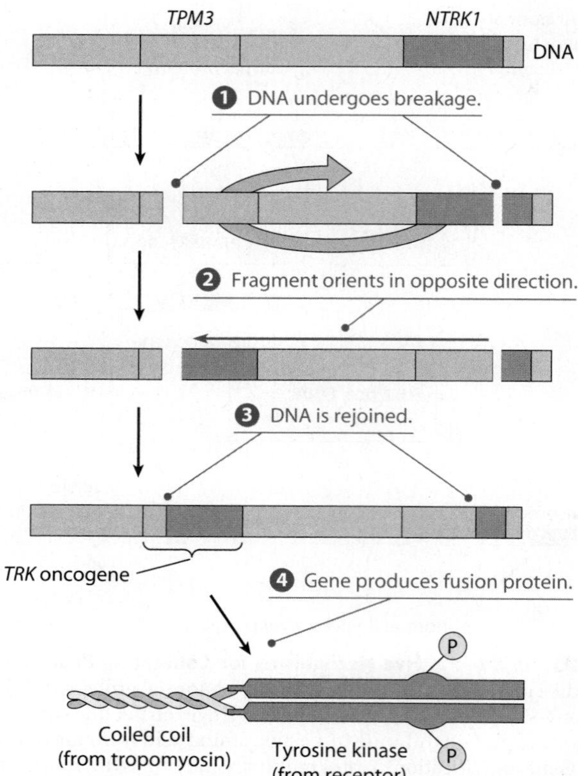

FIGURE 24-14 Origin of the *TRK* Oncogene. The *TRK* oncogene is created by a chromosomal inversion that brings together segments of two genes residing on the same chromosome, one gene coding for a growth factor receptor with tyrosine kinase activity *(NTRK1)* and the other gene coding for nonmuscle tropomyosin *(TPM3)*. The inversion causes one end of the *TPM3* gene to become fused to the opposite end of the *NTRK1* gene. The resulting *TRK* oncogene produces a fusion protein in which the tropomyosin segment causes the receptor region to form a dimer, thereby permanently activating its tyrosine kinase site.

Most Oncogenes Code for Components of Growth-Signaling Pathways

We have just seen that alterations in proto-oncogenes can convert them into oncogenes, which in turn code for proteins that either are structurally abnormal or are produced in excessive amounts. How do these oncogene-encoded proteins cause cancer? Although more than 200 oncogenes have been identified to date, many of the proteins they produce fit into one of six categories: *growth factors, receptors, plasma membrane GTP-binding proteins, nonreceptor protein kinases, transcription factors,* and *cell cycle or apoptosis regulators* (**Table 24-1**). These six categories are all related to steps in growth-signaling pathways (e.g., see the steps in the Ras pathway shown in Figure 19-42). The following sections provide examples of how oncogene-produced proteins in each of the six groups contribute to the development of cancer.

1. Growth Factors. Normally, cells will not divide unless they have been stimulated by an appropriate growth

Table 24-1 A Few Examples of Oncogenes Grouped by Protein Function

Oncogene Name	Protein Produced	Oncogene Origin	Common Cancer Type*
1. Growth factors			
v-sis	PDGF	Viral	Sarcomas (monkeys)
COL1A1-PDGFB	PDGF	Translocation	Fibrosarcoma
2. Receptors			
v-erb-b	Epidermal growth factor receptor	Viral	Leukemia (chickens)
TRK	Nerve growth factor receptor	DNA rearrangement	Thyroid
ERBB2	Epidermal growth factor receptor 2	Amplification	Breast
v-mpl	Thrombopoietin receptor	Viral	Leukemia (mice)
3. Plasma membrane GTP-binding proteins			
KRAS	Ras	Point mutation	Pancreas, colon, lung, others
HRAS	Ras	Point mutation	Bladder
NRAS	Ras	Point mutation	Leukemias
4. Nonreceptor protein kinases			
BRAF	Raf kinase	Point mutation	Melanoma
v-src	Src kinase	Viral	Sarcomas (chickens)
SRC	Src kinase	DNA rearrangement	Colon
TEL-JAK2	Jak kinase	Translocation	Leukemias
BCR-ABL	Abl kinase	Translocation	Chronic myelogenous leukemia
5. Transcription factors			
MYC	Myc	Translocation	Burkitt lymphoma
MYCL	Myc	Amplification	Small cell lung cancer
c-myc	Myc	Insertional mutagenesis	Leukemia (chickens)
v-jun	Jun	Viral	Sarcomas (chickens)
v-fos	Fos	Viral	Bone (mice)
6. Cell cycle or apoptosis regulators			
CYCD1	Cyclin	Amplification, translocation	Breast, lymphoma
CDK4	Cdk	Amplification	Sarcomas, glioblastoma
BCL2	Bcl-2	Translocation	Non-Hodgkins lymphoma
MDM2	Mdm2	Amplification	Sarcomas, lung, breast, others

*Cancers are in humans unless otherwise specified. Only the most frequent cancer types are listed.

factor. But if a cell possesses an oncogene that produces such a growth factor, the cell may stimulate its own proliferation. One oncogene that functions in this way is the *v-sis* gene ("*v*" means viral) found in the *simian sarcoma virus,* which causes cancer in monkeys. The *v-sis* oncogene codes for a mutant form of *platelet-derived growth factor (PDGF).* When the virus infects a monkey cell whose growth is normally controlled by PDGF, the PDGF produced by the *v-sis* oncogene continually stimulates the cell's own proliferation (in contrast to the normal situation, in which cells are exposed to PDGF only when it is released from surrounding blood platelets).

A PDGF-related oncogene has also been detected in some human sarcomas. These tumors possess a chromosomal translocation that creates a gene in which part of the *PDGF* gene is joined to part of an unrelated gene (the gene coding for collagen). The resulting oncogene produces PDGF in an uncontrolled way, thereby causing cells containing the gene to continually stimulate their own proliferation.

2. Receptors. Several dozen oncogenes code for receptors involved in growth-signaling pathways. As described in Chapter 14, many receptors exhibit intrinsic tyrosine kinase activity that is activated only when a growth factor binds to the receptor. Oncogenes sometimes code for mutant versions of such receptors whose tyrosine kinase activity is permanently activated, regardless of the presence or absence of a growth factor. The *TRK* oncogene, which was described in the section on DNA rearrangements, is one example (see Figure 24-14). Another example is the *v-erb-b* oncogene, which is found in a virus that causes a red blood cell cancer in chickens. The *v-erb-b* oncogene produces an altered version of the *epidermal growth factor (EGF) receptor* that retains tyrosine kinase activity but lacks the EGF binding site. Consequently, the receptor is

(a) **Normal receptor.** During normal receptor activation, binding of a growth factor to its receptor promotes the clustering of two receptor molecules, thereby causing the tyrosine kinase activity of each receptor to catalyze phosphorylation of the adjacent receptor (autophosphorylation).

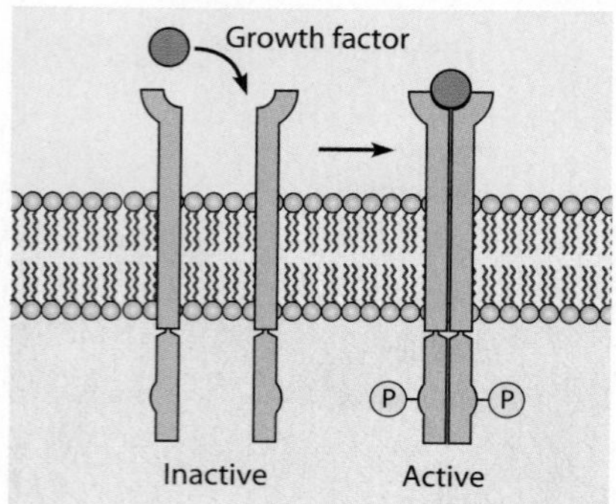

(b) **Mutant receptor.** Some oncogenes code for mutant receptors whose tyrosine kinase is permanently activated. Below is a mutant receptor missing its growth factor binding site, which makes the receptor constitutively active–that is, it exhibits tyrosine kinase activity even in the absence of growth factor.

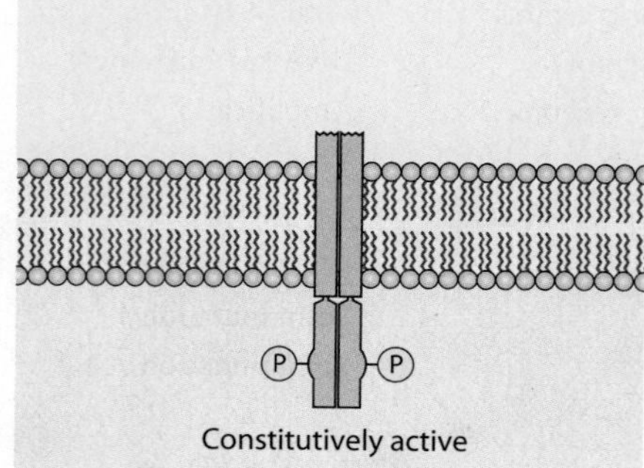

(c) **Amplified receptor.** Amplified oncogenes produce normal receptors but in excessive quantities, which also leads to excessive receptor activity.

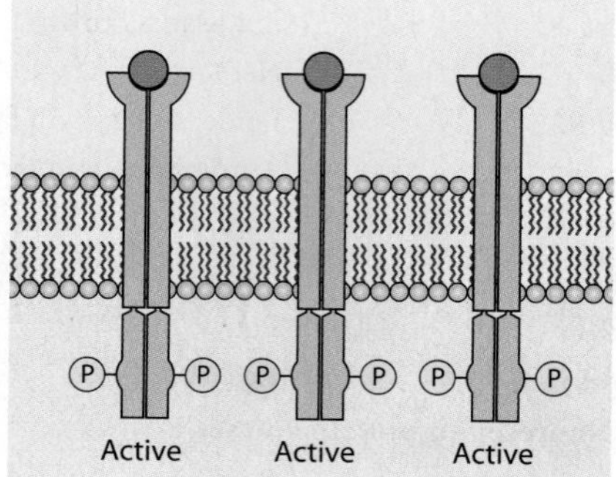

FIGURE 24-15 Receptor Tyrosine Kinases in Normal and Cancer Cells. **(a)** Normal receptors exhibit tyrosine kinase activity only after a growth factor has bound to them. **(b)** Some oncogenes code for mutant receptors whose tyrosine kinase is permanently activated. **(c)** Other oncogenes produce normal receptors but in excessive quantities, which also leads to excessive receptor activity.

constitutively active—that is, it stays active as a tyrosine kinase whether EGF is present or not, whereas the normal form of the receptor exhibits tyrosine kinase activity only when bound to EGF (**Figure 24-15a, b**).

Other oncogenes produce normal receptors but in excessive quantities, which can also lead to hyperactive growth signaling (see Figure 24-15c). An example is provided by the human *ERBB2* gene. As discussed earlier in the chapter, amplification of the *ERBB2* gene in certain breast and ovarian cancers causes it to overproduce a growth factor receptor. The presence of too many receptor molecules causes a magnified response to growth factor and hence excessive cell proliferation.

Some growth-signaling pathways, such as the Jak-STAT pathway described in Chapter 23, utilize receptors that do not possess protein kinase activity. With such receptors, binding of growth factor causes the activated receptor to stimulate the activity of an independent tyrosine kinase molecule. An example of an oncogene that codes for such a receptor occurs in the *myeloproliferative leukemia virus,* which causes leukemia in mice. The oncogene, called *v-mpl,* codes for a mutant version of the receptor for *thrombopoietin,* a growth factor that uses the Jak-STAT pathway to stimulate the production of blood platelets.

3. Plasma Membrane GTP-Binding Proteins. In many growth-signaling pathways, the binding of a growth factor to its receptor leads to activation of the plasma membrane, GTP-binding protein called Ras (page 408). Oncogenes coding for mutant Ras proteins are one of the most common types of genetic abnormality detected in human cancers. The point mutations that create *RAS* oncogenes usually cause a single incorrect amino acid to be inserted at one of three possible locations within the Ras protein. The net result is a hyperactive Ras protein that retains bound GTP instead of hydrolyzing it to GDP, thereby maintaining the protein in a permanently activated state. In this hyperactive state, the Ras protein continually sends a growth-stimulating signal to the rest of the Ras pathway, regardless of whether growth factor is bound to the cell's growth factor receptors.

4. Nonreceptor Protein Kinases. A common feature shared by many growth-signaling pathways is the use of protein phosphorylation reactions to transmit signals within the cell. The enzymes that catalyze these intracellular phosphorylation reactions are referred to as **nonreceptor protein kinases** to distinguish them from the protein kinases that are intrinsic to cell surface receptors. For example, in the case of the Ras pathway, the activated Ras protein triggers a cascade of intracellular protein phosphorylation reactions, beginning with phosphorylation of the Raf protein kinase and eventually leading to the phosphorylation of MAP kinases (page 590). Several oncogenes code for protein kinases involved in this cascade. An example is the *BRAF* oncogene, which codes for a mutant Raf protein in a variety of human cancers. Oncogenes coding for nonreceptor protein kinases involved in other signaling pathways have been identified as well. Included in this group are oncogenes that produce abnormal versions of the Src, Jak, and Abl protein kinases.

5. Transcription Factors. Some of the nonreceptor protein kinases activated in growth-signaling pathways subsequently trigger changes in transcription factors, thereby altering gene expression. Oncogenes that produce mutant forms or excessive quantities of various transcription factors have been detected in a broad range of cancers. Among the most common are oncogenes coding for Myc transcription factors, which control the expression of numerous genes involved in cell proliferation and survival. For example, we have already seen how the chromosomal translocation associated with Burkitt lymphoma creates a *MYC* oncogene that produces excessive amounts of Myc protein (see Figure 24-13). Burkitt lymphoma is only one of several human cancers in which the Myc protein is overproduced. In these other cancers, gene amplification rather than chromosomal translocation is usually responsible. For example, *MYC* gene amplification is frequently observed in small-cell lung cancers and to a lesser extent in a wide range of other carcinomas, including 20–30% of breast and ovarian cancers.

6. Cell Cycle and Apoptosis Regulators. In the final step of growth-signaling pathways, transcription factors activate genes coding for proteins that control cell proliferation and survival. The activated genes include those coding for *cyclins* and *cyclin-dependent kinases (Cdks),* whose roles in triggering passage through key points in the cell cycle were described in Chapter 19. Several human oncogenes produce proteins of this type. For example, a cyclin-dependent kinase gene called *CDK4* is amplified in some sarcomas, and the cyclin gene *CYCD1* is commonly amplified in breast cancers and is altered by chromosomal translocation in some lymphomas. Such oncogenes produce excessive amounts or hyperactive versions of Cdk–cyclin complexes, which then stimulate progression through the cell cycle (even in the absence of growth factors).

Some oncogenes contribute to the accumulation of proliferating cells by inhibiting apoptosis rather than stimulating cell division. One example involves the gene that codes for the apoptosis-inhibiting protein Bcl-2 (page 594). Chromosomal translocations involving this gene are observed in certain types of lymphomas. The net effect of these translocations is an excessive production of Bcl-2, which inhibits apoptosis and thereby fosters the accumulation of dividing cells. The *MDM2* gene, which codes for the Mdm2 protein that targets p53 for destruction (page 588), can also cause a failure of apoptosis when the gene is amplified or abnormally expressed. Excessive production of Mdm2 leads to a destruction of the p53 protein, thereby inhibiting the p53 pathway that is normally used to trigger cell death by apoptosis.

Most oncogenes code for a protein that falls into one of the six preceding categories. Some of these oncogenes produce abnormal, hyperactive versions of such proteins. Other oncogenes produce excessive amounts of an otherwise normal protein. In either case, the net result is a protein that stimulates the uncontrolled accumulation of dividing cells.

Tumor Suppressor Genes Are Genes Whose Loss or Inactivation Can Lead to Cancer

In contrast to oncogenes, whose *presence* can induce cancer formation, the *loss* or *inactivation* of **tumor suppressor genes** can also lead to cancer. As the name implies, the normal function of such genes is to restrain cell proliferation. In other words, tumor suppressor genes act as brakes on the process of cell proliferation, whereas oncogenes function as accelerators of cell proliferation. Of the roughly 25,000 genes in human cells, only a few dozen exhibit the properties of tumor suppressors. Since losing the function of just one of these genes may lead to cancer, each must perform an extremely important function.

The first indication that cells contain genes whose loss can lead to cancer came from cell fusion experiments in which normal cells were fused with cancer cells. Based on our current understanding of oncogenes, you might expect that the hybrid cells created by fusing cancer cells with normal cells would have acquired oncogenes from the original cancer cell and would therefore exhibit uncontrolled growth, just like a cancer cell. In fact, this is not what happens. Fusing cancer cells with normal cells almost always yields hybrid cells that behave like the normal parent and do not form tumors (**Figure 24-16**). Such results, first reported in the late 1960s, provided the earliest evidence that normal cells contain genes whose products can suppress tumor growth and reestablish normal growth behavior.

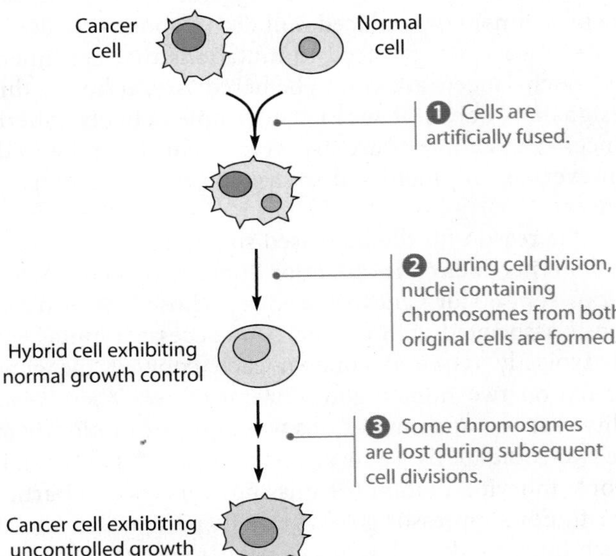

FIGURE 24-16 Fusion of Cancer Cells with Normal Cells. When cancer cells are artificially fused with normal cells, the resulting hybrid cells do not initially form tumors. After they proliferate for extended periods in culture, the hybrid cells usually revert to the uncontrolled, tumor-forming behavior of the original cancer cells. This reversion is accompanied by the loss of chromosomes that contain tumor suppressor genes.

Although fusing cancer cells with normal cells generally yields hybrid cells that are unable to form tumors, this does not mean these cells are normal. When they are allowed to grow for extended periods in culture, the hybrid cells often revert to the malignant, uncontrolled behavior of the original cancer cells. Reversion to malignant behavior is associated with the loss of certain chromosomes, suggesting that these particular chromosomes contain genes that had been suppressing the ability to form tumors. Such observations eventually led to the naming of the lost genes as "tumor suppressor genes."

As long as hybrid cells retain both sets of original chromosomes—that is, chromosomes derived from both the cancer cells and the normal cells—the ability to form tumors is suppressed. Tumor suppression is observed even when the original cancer cells possess an oncogene, such as a mutant *RAS* gene, that is actively expressed in the hybrid cells. This means that tumor suppressor genes located in the chromosomes of normal cells can overcome the effects of a *RAS* oncogene present in a cancer cell chromosome. The ability to form tumors reappears only after the hybrid cell loses a chromosome containing a critical tumor suppressor gene.

Although cell fusion experiments provided good evidence for the existence of tumor suppressor genes, identifying these genes has not been easy. By definition, the existence of a tumor suppressor gene does not become apparent until its function has been lost. How do scientists go about finding something whose very existence is unknown until it disappears? One approach involves families that are at high risk for developing cancer. While most cancers are triggered by spontaneous or environmentally induced mutations, about 10–20% of cancer cases can be traced to mutations that are inherited. Such cancers are said to be hereditary, although this designation does not mean that people actually inherit cancer from their parents. What can be inherited, however, is an increased *susceptibility* to developing cancer.

The reason for the increased susceptibility is usually an inherited defect in a tumor suppressor gene. Since tumor suppressor genes are entities whose loss of function is associated with cancer, two successive mutations are typically required—one in each copy of the gene carried on two homologous chromosomes. The probability that two such mutations would occur randomly in the two copies of the same gene is very small. However, if people inherit a mutant (or missing) version of a particular tumor suppressor gene from one parent, they are at much higher risk of developing cancer because only one mutation (in the second copy of that tumor suppressor gene) in a single cell is now needed to begin the progression toward cancer (**Figure 24-17**). In the next section, we will see how the study of such hereditary cancer susceptibilities has facilitated the identification of tumor suppressor genes.

The *RB* Tumor Suppressor Gene Was Discovered by Studying Families with Hereditary Retinoblastoma

The first tumor suppressor gene to be identified was found by studying *hereditary retinoblastoma,* a rare eye cancer that develops in young children who have a family history of the disease. Children with this condition inherit a deletion in a specific region of chromosome 13 from one of their parents. By itself, such a deletion does not cause cancer. But during the many rounds of cell division that occur as the eye grows, an individual retinal cell may occasionally acquire a deletion or mutation in the same region of the second copy of chromosome 13, and cancer arises from such cells. This pattern suggests that (1) chromosome 13 contains a gene that normally inhibits cell proliferation and (2) the deletion or disruption of both copies of the gene must occur before cancer develops. By comparing DNA fragments isolated from chromosome 13 in normal and retinoblastoma cells, the ***RB* gene** was identified as the missing gene.

The *RB* gene codes for the Rb protein, whose role in controlling progression from G1 to S phase of the cell cycle was described in Chapter 19 (see Figure 19-39). The Rb protein is part of the braking mechanism that normally prevents cells from passing through the G1 restriction point and into S phase in the absence of an appropriate signal from a growth factor. Disruption of both copies of the *RB* gene removes this restraining mechanism and opens the door to uncontrolled proliferation. The ability of *RB* mutations to unleash the normal controls on cell proliferation is not limited to hereditary retinoblastoma, the rare eye cancer in which the gene defect was originally identified. Mutations that disrupt the Rb protein have also been detected in nonhereditary retinoblastomas and in several other nonhereditary cancers, including certain forms of lung, breast, and bladder cancer.

The Rb protein is also a target for certain cancer viruses. For example, **human papillomavirus (HPV)**—the virus responsible for uterine cervical cancer—contains an oncogene that produces a protein, called the *E7 protein,* that binds to the Rb protein of infected cells and prevents it from restraining cell proliferation (**Figure 24-18**, ①). Thus, cancers triggered by a loss of functional Rb protein can arise either through mutations that disrupt both copies of the *RB* gene or through the action of viral proteins that bind to and inactivate the Rb protein.

The *p53* Tumor Suppressor Gene Is the Most Frequently Mutated Gene in Human Cancers

Following the discovery of the *RB* gene in the mid-1980s, dozens of additional tumor suppressor genes have been identified. One of the most important is the ***p53* gene** (also called *TP53* in humans), which is mutated in a broad spectrum of tumor types. In fact, almost half of the world's

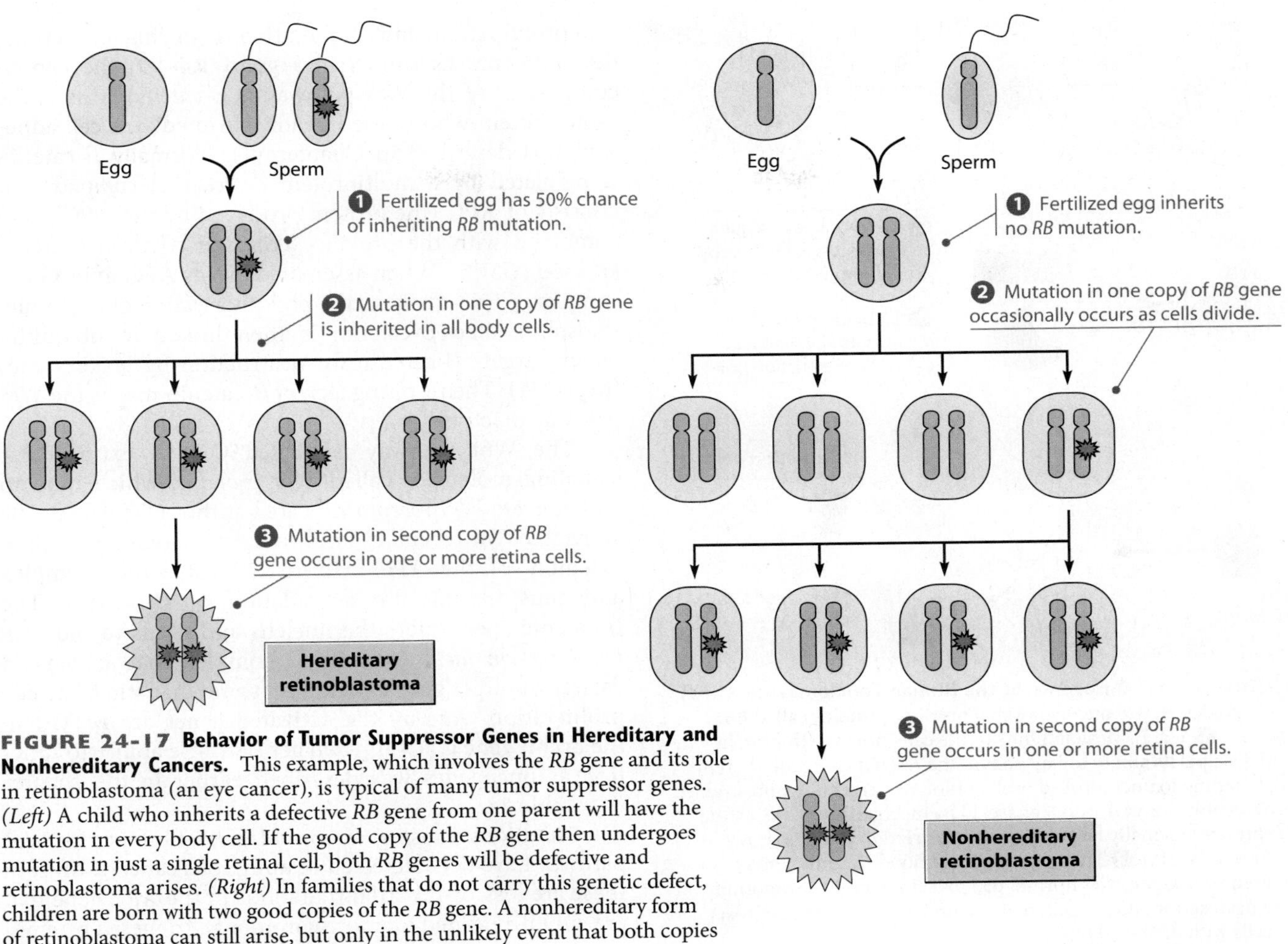

FIGURE 24-17 Behavior of Tumor Suppressor Genes in Hereditary and Nonhereditary Cancers. This example, which involves the *RB* gene and its role in retinoblastoma (an eye cancer), is typical of many tumor suppressor genes. *(Left)* A child who inherits a defective *RB* gene from one parent will have the mutation in every body cell. If the good copy of the *RB* gene then undergoes mutation in just a single retinal cell, both *RB* genes will be defective and retinoblastoma arises. *(Right)* In families that do not carry this genetic defect, children are born with two good copies of the *RB* gene. A nonhereditary form of retinoblastoma can still arise, but only in the unlikely event that both copies of the *RB* gene undergo mutation in the same cell.

10 million people diagnosed with cancer each year will have tumors with *p53* mutations, making it the most commonly mutated gene in human cancers.

The *p53* gene codes for the p53 protein, whose role in responding to DNA damage was described in Chapter 19. When cells are exposed to carcinogenic agents—such as ionizing radiation or toxic chemicals—that cause extensive DNA damage, the altered DNA stimulates the p53 pathway, which then triggers cell cycle arrest and apoptosis to prevent the genetically damaged cells from proliferating (see Figure 19-40). This protective mechanism is often missing in cancer cells because of mutations that disrupt the *p53* gene. The resulting inactivation of the p53 pathway leads to a *failure of apoptosis* that contributes to cancer development by allowing the survival and reproduction of cells containing damaged DNA.

It is therefore not surprising that, as in the case of the *RB* gene, individuals who inherit a defective copy of the *p53* gene from one parent exhibit an increased cancer risk. In this inherited condition, called the *Li-Fraumeni syndrome,* various types of cancer arise by early adulthood due to mutations that inactivate the second, normal copy of the *p53* gene. Mutations in *p53* are also common in nonhereditary cancers triggered by exposure to DNA-damaging chemicals and radiation. For example, carcinogens present in tobacco smoke cause several kinds of mutations in the *p53* gene in lung cancers, and sunlight causes *p53* mutations in skin cancers. In some cases, mutation in one copy of the *p53* gene is enough to disable the p53 protein even when the other copy of the *p53* gene is normal. The apparent explanation is that the p53 protein is a tetramer consisting of four p53 polypeptide chains, and the presence of even one mutant chain in the tetramer may be enough to prevent the protein from functioning properly.

Like the Rb protein, the p53 protein is a target for certain cancer viruses. For example, in addition to producing the E7 protein that inactivates Rb, the human papillomavirus (HPV) has a second oncogene that produces the *E6 protein,* which directs the attachment of ubiquitin to the p53 protein and thereby targets it for destruction (see Figure 24-18, ❷). This means that HPV can block the actions of the proteins produced by both the *RB* and *p53* tumor suppressor genes.

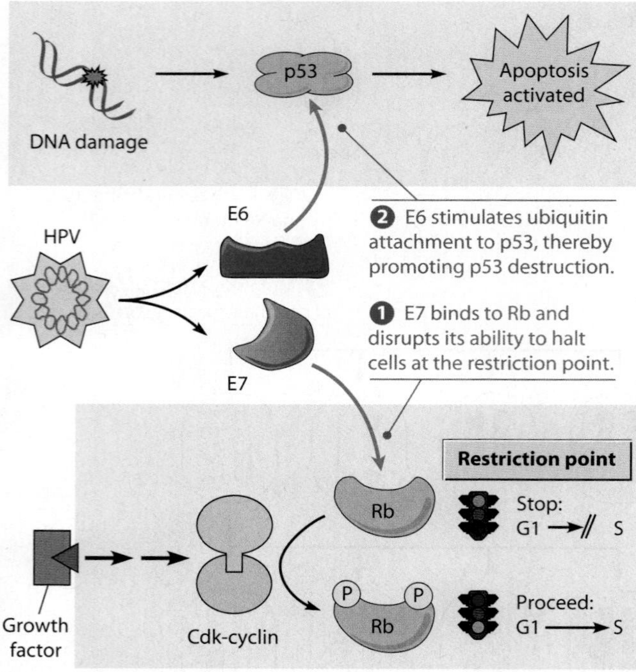

FIGURE 24-18 Oncogenes of the Human Papillomavirus (HPV). HPV possesses two oncogenes that produce proteins called E6 and E7. ❶ The E7 protein binds to the cell's normal Rb protein and disrupts its ability to halt cells at the restriction point, thereby contributing to uncontrolled proliferation by permitting unrestrained progression from G1 into S phase. (The mechanism by which the Rb protein normally halts cells at the restriction point is shown in Figure 19-39.) ❷ The E6 protein promotes the attachment of ubiquitin to the cell's normal p53 protein, thereby promoting p53 destruction. As a result, p53 can no longer trigger apoptosis in cells with damaged DNA.

The *APC* Tumor Suppressor Gene Codes for a Protein That Inhibits the Wnt Signaling Pathway

Like the *p53* gene, the next tumor suppressor we will consider is a common target for cancer-causing mutations, although in this case cancers arise mainly in one organ: the colon. Mutations affecting the gene in question, called the ***APC* gene,** are associated with an inherited disease called *familial adenomatous polyposis.* Affected individuals inherit a defective *APC* gene that makes them susceptible to developing thousands of *polyps* (benign tumors) in the colon if the second copy of the *APC* gene undergoes mutation. As a result, the risk of colon cancer arising by age 60 is nearly 100%. Although familial adenomatous polyposis is a rare disease, accounting for less than 1% of all colon cancers, *APC* mutations can also arise spontaneously or be triggered by environmental mutagens, and such mutations occur in roughly two-thirds of the more common, nonhereditary forms of colon cancer.

The *APC* gene produces a protein involved in the **Wnt pathway,** which plays a prominent role in controlling cell proliferation and differentiation during embryonic development. As shown in **Figure 24-19**, the central component of the Wnt pathway is *β-catenin* (this is the same protein whose role in cadherin-mediated cell adhesion was described in Chapter 17). Normally β-catenin is regulated by a multiprotein *destruction complex* that consists of APC (the protein produced by the *APC* gene) combined with the proteins *axin* and *glycogen synthase kinase 3 (GSK3).* When assembled in this APC-axin-GSK3 complex, GSK3 catalyzes the phosphorylation of β-catenin. Phosphorylated β-catenin is then linked to ubiquitin, which targets β-catenin for destruction by proteasomes (page 751). The resulting lack of β-catenin makes the Wnt pathway inactive.

The Wnt pathway is turned ON by extracellular signaling molecules called *Wnt proteins,* which bind to and activate *Wnt receptors* located at the cell surface. The activated Wnt receptors bind to axin, thereby preventing assembly of the APC-axin-GSK3 destruction complex and thus halting the degradation of β-catenin. The β-catenin then enters the nucleus and binds to the *TCF transcription factor,* forming a complex that activates a variety of target genes, including some that stimulate cell proliferation. Among the activated genes are *MYC* and the cyclin gene *CYCD1*—two genes whose ability to function as oncogenes was described earlier in the chapter (see Table 24-1).

Mutations causing abnormal activation of the Wnt pathway have been detected in numerous cancers. Most of these are loss-of-function mutations in the *APC* gene that are either inherited or, more commonly, triggered by environmental mutagens. The resulting absence of functional APC protein prevents the APC-axin-GSK3 complex from assembling and β-catenin therefore accumulates, locking the Wnt pathway in the ON position and sending the cell a continuous signal to divide.

Inactivation of Some Tumor Suppressor Genes Leads to Genetic Instability

The normal mutation rate for any given gene is about one in a million per cell division, and yet cancers almost always exhibit mutations in multiple genes—often a dozen or so mutations are necessary to trigger a cancer. How do cancer cells manage to acquire so many mutations if mutation is such a rare event? The apparent explanation is that *mutation rates in cancer cells are hundreds or even thousands of times higher than normal.*

This state, called **genetic instability,** can arise in several different ways. One group of mechanisms involves *disruptions in DNA repair.* For example, inherited defects in genes required for *mismatch repair* (page 569) are responsible for *hereditary nonpolyposis colon cancer (HNPCC),* an inherited syndrome that elevates a person's cancer risk by allowing mutations to accumulate rather than being corrected by mismatch repair. Another hereditary disease,

(a) Normal cell (without Wnt proteins). In the absence of growth-signaling Wnt proteins, β-catenin is targeted for degradation by the axin-APC-GSK3 destruction complex, which catalyzes the phosphorylation of β-catenin. The phosphorylated β-catenin is linked to ubiquitin, thereby marking the phosphorylated β-catenin for degradation by proteasomes. The resulting absence of β-catenin maintains the Wnt pathway in the OFF position.

(b) Normal cell (with Wnt proteins). The Wnt pathway is turned ON by Wnt proteins, which activate cell surface Wnt receptors. The activated receptors bind to axin, thereby preventing assembly of the destruction complex and thus halting β-catenin degradation. The β-catenin enters the nucleus and binds to the TCF transcription factor, forming a complex that activates a variety of genes that control cell proliferation, including *MYC* and *CYCD1* (a cyclin gene).

(c) Cancer cell (regardless of the presence or absence of Wnt proteins). Some cancer cells have loss-of-function mutations in the *APC* gene. In the absence of functional APC protein, the destruction complex cannot form and β-catenin therefore accumulates, entering the nucleus and locking the Wnt pathway in the ON position.

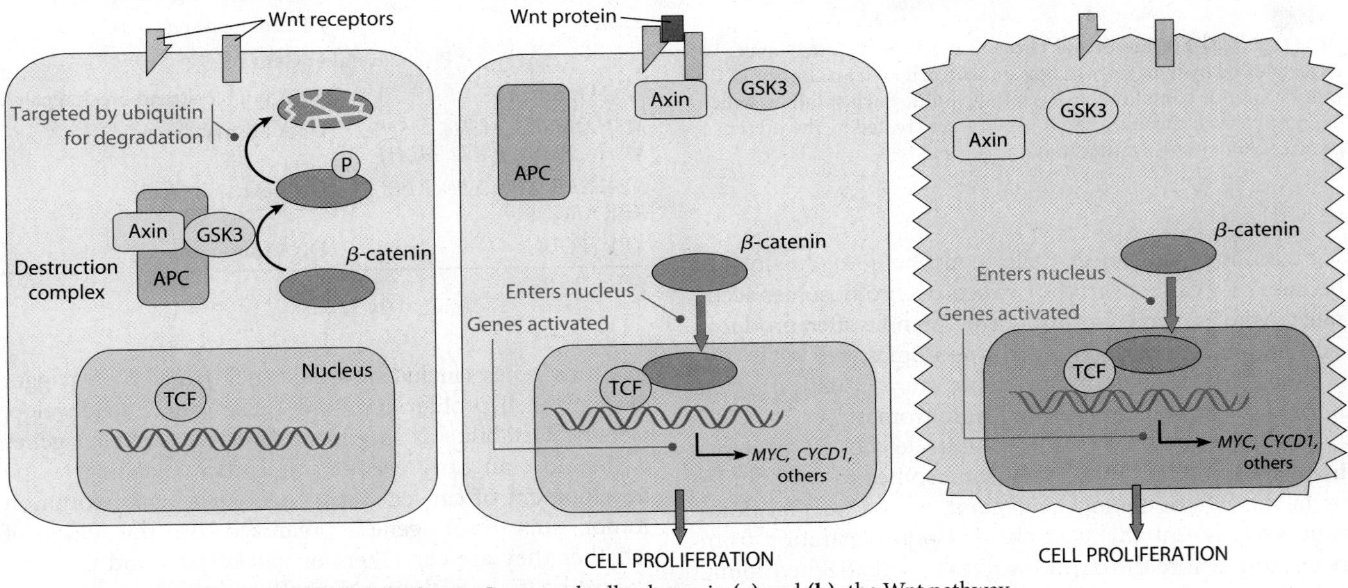

FIGURE 24-19 The Wnt Signaling Pathway. In normal cells, shown in **(a)** and **(b)**, the Wnt pathway is active only in the presence of Wnt proteins. In cancer cells **(c)**, the Wnt pathway is active regardless of the presence or absence of Wnt proteins.

xeroderma pigmentosum, is caused by inherited defects in genes needed for *excision repair* (page 568). **Box 24A** describes how children who inherit this condition develop an extremely high skin cancer risk because they are unable to repair DNA damage triggered by exposure to sunlight.

Faulty DNA repair is also involved in hereditary forms of breast cancer, which account for about 10% of all breast cancer cases. Most hereditary forms of breast (and ovarian) cancer arise in women who inherit a mutant copy of either the ***BRCA1*** or ***BRCA2*** gene, both of which code for proteins involved in repairing double-strand DNA breaks. Breast and ovarian cells defective in either of the BRCA proteins exhibit many chromosomal abnormalities, including translocations, deletions, and broken or fused chromosomes. As a result, women inheriting *BRCA* mutations exhibit a 40–80% lifetime risk for breast cancer and a 15–65% risk for ovarian cancer. Because the risks are so high, women with a family history of breast or ovarian cancer may wish to undergo genetic testing to determine whether they carry a mutant *BRCA1* or *BRCA2* gene.

Genetic instability is not restricted to situations involving hereditary cancer risk. Most cancers are not hereditary, but they still exhibit genetic instability. In some cases, the instability can be traced to mutations in DNA repair genes that either arise spontaneously or are caused by environmental mutagens. Another explanation is that the p53 pathway is defective in most cancer cells, which removes a protective mechanism that would arrest cells at cell cycle checkpoints to allow DNA repair to occur or would trigger apoptosis to destroy cells containing damaged DNA.

Genetic instability can also arise from defects in mitosis that cause *disruptions in chromosome sorting* during cell division, resulting in broken chromosomes and *aneuploidy* (an abnormal number of chromosomes). One reason for improper chromosome sorting is the presence of extra *centrosomes,* the structures that help guide microtubule assembly during spindle formation (page 572). **Figure 24-20** shows a cancer cell with three centrosomes that have assembled a mitotic spindle with three poles. Spindles containing three or more poles, which are rare in normal tissues

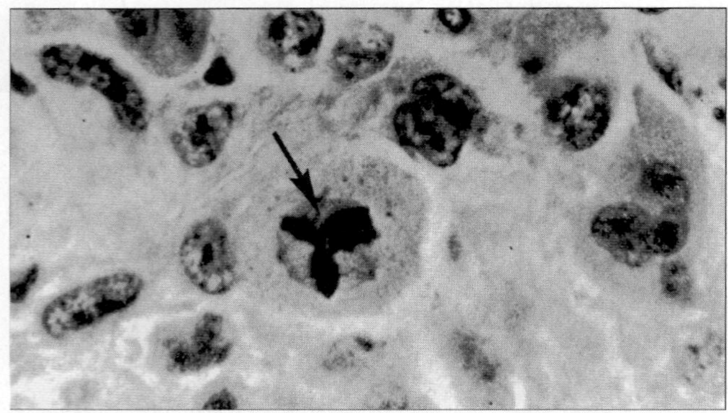

FIGURE 24-20 Abnormal Mitosis in a Cancer Cell. Cancer tissue viewed by light microscopy shows a cell with an abnormal mitotic spindle containing three spindle poles. Such spindles, which cannot separate chromosomes properly, are created by the presence of three centrosomes rather than the normal two.

but common in cancer cells, contribute to aneuploidy because they cannot sort the two sets of chromosomes accurately. Mitosis involving an abnormal spindle often produces cells that are missing certain chromosomes, and such cells are therefore deficient in any tumor suppressor genes that normally reside on the missing chromosomes.

Cancer cells may also exhibit defects in proteins involved in either attaching chromosomes to the spindle or in the *mitotic spindle checkpoint,* whose normal function is to prevent anaphase chromosome separation from beginning before the chromosomes are all attached to the spindle (page 587). For example, some cancers have mutations in genes coding for Mad or Bub proteins, which are central components of the mitotic spindle checkpoint. If Mad or Bub proteins are defective or in short supply, chromosome movements toward the spindle poles may begin before all the chromosomes are properly attached to the mitotic spindle. The result is inaccurate chromosome sorting and the production of aneuploid cells that lack certain chromosomes and possess extra copies of others.

The genes coding for proteins involved in DNA repair and chromosome sorting discussed in this section fit the definition of tumor suppressor genes because their loss or inactivation contributes to cancer development. But such genes are not in the same category as the *RB, p53,* and *APC* genes, which produce proteins that restrain cell proliferation and whose loss can directly lead to cancer. To distinguish the two classes of tumor suppressors, genes like *RB, p53,* and *APC* are called **gatekeepers** because their loss directly opens the gates to excessive cell proliferation and tumor formation. In contrast, genes involved in DNA repair and chromosome sorting are referred to as **caretakers** because they maintain genetic stability but are not directly involved in controlling cell proliferation and survival. Defects in caretaker genes lead to genetic instability, which in turn allows the accumulation of mutations in other genes (including gatekeepers) that then trigger excessive cell proliferation and cause cancer to develop. Genetic instability arising from defective caretaker genes is therefore an early event in the process leading to the development of cancer. **Table 24-2** lists some common tumor suppressor genes, organized on the basis of whether they are caretakers or gatekeepers and grouped according to the pathways they affect.

Table 24-2 Examples of Tumor Suppressor Genes

Gene	Pathway Affected
Gatekeeper Genes	
APC	Wnt signaling
CDKN2A	Rb and p53 signaling
PTEN	PI3K-Akt signaling
RB	Restriction point control
SMAD4	TGFβ-Smad signaling
TGFβ receptor	TGFβ-Smad signaling
p53	DNA damage response
Caretaker Genes	
BRCA1, BRCA2	DNA double-strand break repair
MSH2, MSH3, MSH4, MSH5, MSH6, PMS1, PMS2, MLH1	DNA mismatch repair
XPA, XPB, XPC, XPD, XPE, XPF, XPG	DNA excision repair
XPV (POLη)	DNA translesion synthesis

Cancers Develop by the Stepwise Accumulation of Mutations Involving Oncogenes and Tumor Suppressor Genes

Genome sequencing studies have revealed that a given type of cancer (e.g., breast, lung, or colon cancer) typically involves mutations in about 50–75 different protein-coding genes. A small handful of these genes are mutated frequently in cancer samples of the same type taken from different people, while the rest of the genes are infrequently mutated. The commonly mutated genes affect about a dozen different pathways, most of which are discussed at one point or another in this chapter.

The common mutations involve the inactivation of tumor suppressor genes as well as the conversion of proto-oncogenes into oncogenes. In other words, creating a cancer cell usually requires that the brakes on cell growth (tumor suppressor genes) be released and the accelerators for cell growth (oncogenes) be activated. This principle is well illustrated by the stepwise progression toward malignancy observed in colon cancer. The most common pattern to be detected is the presence of a *KRAS* oncogene (a member of the *RAS* gene family) accompanied by loss-of-function mutations in the tumor suppressor genes

BOX 24A HUMAN APPLICATIONS

Children of the Moon

The connection between faulty DNA repair and cancer susceptibility was first observed in a rare hereditary disease called **xeroderma pigmentosum.** Individuals who inherit this condition are so sensitive to the sun's ultraviolet (UV) radiation that exposure to a few minutes of daylight is enough to cause severe burning of the skin and skin cancer. Because it is only safe for them to go outside at night, children with xeroderma pigmentosum are sometimes referred to as "children of the moon." In upstate New York there is even a special camp, called *Camp Sundown,* that allows affected children to participate in normal recreational activities but on a different schedule. Outdoor activities begin after sundown and take place during the night, allowing the children to be back indoors and safely behind drawn curtains by sunrise. Scientists from NASA have even designed a special space suit that provides enough sunlight protection to allow children with xeroderma pigmentosum to occasionally play outdoors (**Figure 24A-1**).

FIGURE 24A-1 A Child with Xeroderma Pigmentosum Playing Outdoors. Because exposure to daylight is dangerous for people with xeroderma pigmentosum, NASA designed a special spacesuit to protect children from the sun's UV radiation.

The susceptibility to skin cancer that is the hallmark of xeroderma pigmentosum can be traced to inherited defects in DNA repair. In Chapter 19 we saw that the UV radiation in sunlight can damage DNA molecules by triggering *pyrimidine dimer* formation—that is, the formation of covalent bonds between adjacent pyrimidine bases (see Figure 19-17c). If they are not repaired, such mutations can lead to cancer. One way of repairing pyrimidine dimers is through *excision repair,* a process that uses a series of enzymes to excise the damaged region and fill the resulting gap with the correct sequence of nucleotides (see Figure 19-18). In the late 1960s, it was first reported that cells from most individuals with xeroderma pigmentosum are unable to carry out excision repair. As a consequence, DNA mutations accumulate and cancer eventually arises. Subsequent studies revealed that mutations in seven different genes can disrupt excision repair and thereby cause xeroderma pigmentosum. The seven genes, designated *XPA* through *XPG,* each code for an enzyme involved in a different step of the excision repair pathway.

Inherited mutations in an eighth gene, called *XPV,* produce an alternate form of xeroderma pigmentosum in which excision repair remains intact but affected individuals are still susceptible to sunlight-induced cancers. This particular mutation affects the gene for DNA polymerase η (eta), a special form of DNA polymerase that catalyzes the *translesion synthesis* of a new stretch of DNA across regions in which the template strand is damaged (page 568). DNA polymerase η is capable of replicating DNA in regions where pyrimidine dimers are present and correctly inserts the proper bases in the newly forming DNA strand. Thus, inherited defects in DNA polymerase η, like inherited defects in excision repair, hinder the ability to correct pyrimidine dimers created by UV radiation.

The defects in DNA repair inherited by children with xeroderma pigmentosum lead to skin cancer rates that are 2000-fold higher than normal. Affected individuals develop skin cancer at an average age of 8 years old, compared to age 60 for the general population. Although the greatest impact is on skin cancer, the DNA repair defects seen in xeroderma pigmentosum also cause a 20-fold increase in risk for leukemia and cancers of the brain, lung, stomach, breast, uterus, and testes.

APC, SMAD4, and *p53.* Rapidly growing colon cancers tend to exhibit all four genetic alterations, whereas benign tumors have only one or two.

As shown in **Figure 24-21**, the earliest mutation to be routinely detected is loss of function of the *APC* gene—a mutation that frequently occurs in small polyps before cancer has even arisen. Mutations in *KRAS* and *SMAD4* tend to be seen when the polyps get larger, and mutations in *p53* usually appear when cancer finally develops. However, these mutations do not always occur in the same sequence or with the same exact set of genes. For example, *APC* mutations are found in about two-thirds of all colon cancers, but this means that the *APC* gene is normal in one out of every three cases. Analysis of tumors containing normal *APC* genes has revealed that many of them possess oncogenes that produce an abnormal, hyperactive form of *β-catenin,* a protein that—like the APC protein—is involved in Wnt signaling (see Figure 24-19). Because APC inhibits the Wnt pathway and β-catenin stimulates it, mutations leading to the loss of APC and mutations that create hyperactive forms of β-catenin have the same effect: Both enhance cell proliferation by increasing the activity of the Wnt pathway.

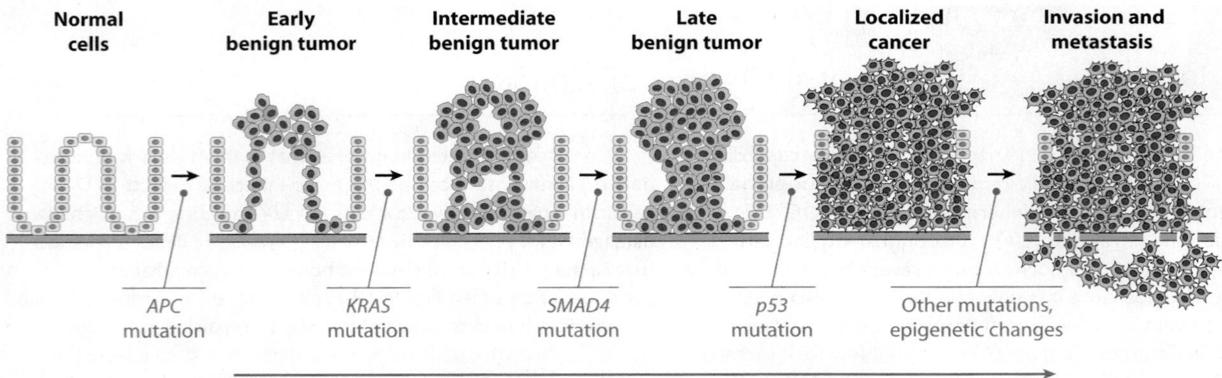

FIGURE 24-21 Stepwise Model for the Development of Colon Cancer. Colon cancer often arises through a stepwise series of mutations involving the *APC, KRAS, SMAD4,* and *p53* genes. Each successive mutation is associated with increasingly abnormal cell behavior. The early stages in this process produce benign tumors called *polyps,* which protrude from the inner surface of the colon.

Another pathway frequently disrupted in colon cancer is the *TGFβ-Smad pathway* (page 413), which *inhibits* rather than stimulates epithelial cell proliferation. Loss-of-function mutations in genes coding for components of this pathway, such as the TGFβ receptor or Smad4, are commonly detected in colon cancers. Such mutations disrupt the growth-inhibiting activity of the TGFβ-Smad pathway and thereby contribute to enhanced cell proliferation.

Overall, the general principle illustrated by the various colon cancer mutations is that different tumor suppressor genes and oncogenes can affect the same pathway, and it is the disruption of particular signaling pathways that is important in cancer rather than the particular gene mutations that cause the disruption.

Epigenetic Changes in Gene Expression Influence the Properties of Cancer Cells

Mutations that create oncogenes or inactivate tumor suppressor genes play central roles in cancer development, but they do not explain all the properties of cancer cells. Many traits exhibited by cancer cells arise not from gene mutations but from *epigenetic changes*—that is, changes in gene expression that do not involve changes in a gene's underlying base sequence.

In Chapter 23, we saw that one mechanism for creating epigenetic changes involves *DNA methylation* at –CG– sites located near gene promoters. Most of these sites are unmethylated in normal cells, but extensive methylation is common in cancer cells, where it leads to *epigenetic silencing* of numerous genes—including tumor suppressor genes. In fact, the tumor suppressor genes of cancer cells are inactivated by DNA methylation at least as often as they are inactivated by DNA mutation. Epigenetic silencing of a tumor suppressor gene can be a critical, initiating event on the road to cancer. For example, individuals who inherit an extensively methylated form of a DNA mismatch repair gene known as *MLH1* are highly susceptible to developing multiple cancers. Inheriting a methylated tumor suppressor gene may therefore predispose a person to developing cancer just like inheriting a mutated form of the gene does.

Another mechanism for altering gene expression involves *microRNAs,* which bind to and silence the translation of thousands of individual mRNAs (see Figure 23-34). Cancer cells produce excessive amounts of some microRNAs and inadequate amounts of others. Genes that contribute to cancer development by *overproducing* microRNAs are acting as oncogenes. MicroRNA genes in this category include *miR-155, miR-17-92,* and *miR-21.* To give one example of how such genes act, the microRNA produced by *miR-17-92* inhibits translation of the messenger RNA coding for PTEN, a phosphatase that inhibits PI3K-Akt signaling pathways (page 591). The *miR-17-92* gene is frequently amplified in certain types of cancer, which leads to excessive production of its microRNA and a resulting inhibition of the synthesis of PTEN. Depletion of PTEN allows the PI3K-Akt pathway to be continually activated, thereby leading to enhanced cell proliferation and survival.

MicroRNAs that are *underproduced* can contribute to cancer development by acting as tumor suppressors. MicroRNA genes in this category include *let-7, miR-29,* and the *miR-15a/miR-16-1* cluster. For example, the *miR-15a/miR-16-1* gene cluster is frequently deleted in certain forms of leukemia. One of the normal functions of *miR-15a/miR-16-1* is to inhibit translation of the messenger RNA coding for Bcl-2, a protein that inhibits apoptosis (page 594). Deletion of *miR-15a/miR-16-1* can therefore lead to excessive production of Bcl-2, which in turn interferes with the ability to carry out apoptosis.

Some microRNAs influence *histone modification* reactions, whose role in regulating gene expression was discussed in Chapter 23. One example involves *miR-101,* a microRNA gene that is frequently deleted in prostate cancer cells. The microRNA produced by *miR-101* normally inhibits translation of the messenger RNA coding for EZH2, an enzyme that catalyzes histone methylation. Loss of the *miR-101* gene therefore leads to increased production of EZH2, and the resulting increase in histone methylation may play a role in silencing the expression of tumor suppressor genes.

Summing Up: Carcinogenesis and the Hallmarks of Cancer

Since various combinations of mutations and epigenetic changes involving tumor suppressor genes and oncogenes can lead to cancer, you might wonder whether it is possible to provide a unifying overview of the process of *carcinogenesis*—that is, the multistep series of events that convert normal cells into cancer cells. One way of presenting such an overview is provided by **Figure 24-22**, which begins with the four main causes of cancer: chemicals, radiation, infectious agents, and heredity. Through a variety of mechanisms, each of these produces DNA mutations. For cancer to develop, these mutations must involve a stepwise series of changes involving the inactivation of tumor suppressor genes and the conversion of proto-oncogenes into oncogenes. Such mutations, along with epigenetic changes, eventually produce a group of six traits referred to by Douglas Hanahan and Robert Weinberg as the *hallmarks of cancer.* These six traits, described in the following paragraphs, are common to all forms of cancer, but each trait can be acquired through a variety of genetic and epigenetic mechanisms.

1. Self-Sufficiency in Growth Signals. Cells do not normally proliferate unless they are stimulated by an appropriate growth factor. Cancer cells escape this requirement through the action of oncogenes that produce excessive quantities or mutant versions of proteins involved in growth-stimulating pathways (see Table 24-1). One such pathway commonly activated in cancer cells is the Ras pathway. About 25–30% of all human cancers have mutant Ras proteins that provide an ongoing stimulus for the cell to proliferate independently of growth factors. Mutations affecting other components of the Ras pathway are common as well.

2. Insensitivity to Antigrowth Signals. Normal tissues are protected from excessive cell proliferation by a variety of growth-inhibiting mechanisms. Cancer cells must evade such antigrowth signals if they are to continue proliferating. Most antigrowth signals act during late G1 and exert their effects through the Rb protein, whose phosphorylation regulates passage through the restriction point and into S phase. For example, TGFβ normally inhibits proliferation by triggering the TGFβ-Smad pathway, which produces Cdk inhibitors that block Rb phosphorylation and thereby prevent passage from G1 into S phase. In cancer cells, the TGFβ-Smad pathway is disrupted by a variety of mechanisms, including mutations, epigenetic changes, and interactions with viral proteins. Mutations in the *RB* gene also make cells insensitive to the antigrowth effects of TGFβ or any other growth inhibitor that exerts its effects through the Rb protein.

3. Evasion of Apoptosis. The survival of cancer cells depends on their ability to evade the normal fate of genetically damaged cells—destruction by apoptosis. The ability to evade apoptosis is often imparted by mutations in the *p53* tumor suppressor gene, which disrupt the ability of the p53 pathway to trigger apoptosis in response to DNA damage. The p53 pathway can also be disrupted by certain oncogenes. An example is the *MDM2* gene, which produces the Mdm2 protein that targets p53 for destruction (see Figure 19-40). Another oncogene that affects the ability of the p53 pathway to trigger apoptosis is the *BCL2* gene, which codes for an inhibitor of apoptosis known as the Bcl-2 protein (page 594). *BCL2* oncogenes produce excessive quantities of Bcl-2, resulting in a suppression of apoptosis that allows abnormal cells to

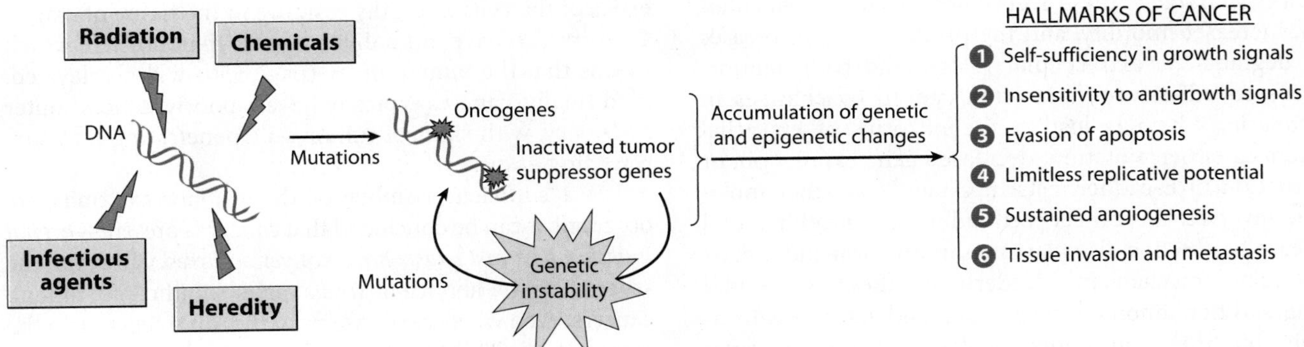

FIGURE 24-22 Overview of Carcinogenesis. The four main causes of cancer (chemicals, radiation, infectious agents, and heredity) trigger initial DNA alterations that can create oncogenes and disrupt tumor suppressor genes. An early result of these changes is genetic instability, which in turn facilitates the acquisition of additional mutations that (along with accompanying epigenetic changes) lead to the six hallmark traits: self-sufficiency in growth signals, insensitivity to antigrowth signals, evasion of apoptosis, limitless replicative potential, sustained angiogenesis, and tissue invasion and metastasis.

continue proliferating. Through such mechanisms, cancer cells manage to evade the apoptotic pathways that would otherwise lead to cancer cell destruction.

4. Limitless Replicative Potential. The overall effect of the preceding three traits is to uncouple cancer cells from the mechanisms that normally balance cell proliferation with an organism's need for new cells. However, this would not ensure unlimited proliferation in the absence of a mechanism for replenishing the telomere sequences that are lost from the ends of each chromosome during DNA replication. Telomere maintenance is usually accomplished by activating the gene coding for telomerase, but a few cancer cells activate an alternative mechanism for maintaining telomeres that involves the exchange of sequence information between chromosomes. In either case, cancer cells maintain telomere length above a critical threshold and thereby retain the ability to divide indefinitely.

5. Sustained Angiogenesis. Without a blood supply, tumors will not grow beyond a few millimeters in size. Thus, at some point during early tumor development, cancer cells must trigger angiogenesis. A common strategy involves the activation of genes coding for angiogenesis stimulators combined with the inhibition of genes coding for angiogenesis inhibitors. The mechanisms underlying such changes in gene expression are not well understood, but in some cases they have been linked to the activities of known tumor suppressor genes or oncogenes. For example, the p53 protein activates the gene coding for the angiogenesis inhibitor thrombospondin; hence the loss of p53 function, which occurs in many human cancers, can cause thrombospondin levels to fall. Conversely, *RAS* oncogenes trigger increased expression of the gene coding for the angiogenesis activator VEGF.

6. Tissue Invasion and Metastasis. The ability to invade surrounding tissues and metastasize to distant sites is the defining trait that distinguishes a cancer from a benign tumor. Three properties exhibited by cancer cells play a crucial role in these events: decreased cell-cell adhesion, increased motility, and the production of proteases that degrade the extracellular matrix and basal lamina. Decreased adhesiveness is often caused by changes in *E-cadherin,* which is lost in the majority of epithelial cancers by either mutation, decreased gene expression, or destruction of E-cadherin itself. Changes in other molecules involved in cancer cell adhesion, motility, and protease production also play roles in invasion and metastasis. The mechanisms underlying these molecular changes differ among tumor types and tissue environments, but they commonly involve activation of genes normally intended for use in embryonic development or wound healing.

The Crucial Enabling Trait: Genetic Instability. To acquire the preceding six traits, cancer cells need to accumulate more mutations than would be generated by normal mutation rates. Cells must therefore become genetically unstable before enough mutations can accumulate to cause cancer. Genetic instability arises most commonly from mutations that disrupt the ability of the p53 pathway to trigger the destruction of genetically damaged cells. However, mutations in genes coding for proteins involved in DNA repair and chromosome sorting also play a role. Genetic instability is placed in a category separate from the six hallmark traits—which are directly involved in cancer cell proliferation and spread—because genetic instability is a crucial underlying trait that enables cancer cells to accumulate the mutations that permit the six hallmark traits to arise.

Diagnosis, Screening, and Treatment

Much progress has been made in recent years in elucidating the genetic and biochemical abnormalities underlying cancer development. One of the hopes for such research is that our growing understanding of the molecular alterations exhibited by cancer cells will eventually lead to improved strategies for cancer diagnosis and treatment.

Cancer Is Diagnosed by Microscopic Examination of Tissue Specimens

Because cancer can arise in almost any tissue, few generalizations are possible regarding disease symptoms. A definitive diagnosis typically requires a *biopsy,* which involves surgical removal of a tiny tissue sample for microscopic examination. Although no single trait is sufficient for visually identifying cancer cells, they usually display several features that together indicate the presence of cancer (**Table 24-3**). For example, cancer cells often exhibit large, irregularly shaped nuclei, prominent nucleoli, a high ratio of nuclear-to-cytoplasmic volume, significant variations in cell size and shape, and a loss of normal tissue organization. To varying extents, cancer cells lose the specialized structural and biochemical properties of the cells normally residing in the tissue of origin. Cancers also have more dividing cells than normal, which means that the *mitotic index* (page 550) will be elevated. And finally, cancers typically have a poorly defined outer boundary, with signs of tumor cells penetrating into surrounding tissues.

If a sufficient number of these abnormal traits are observed, it can be concluded that cancer is present—*even if invasion and metastasis have not yet occurred.* However, the severity of the abnormal traits varies significantly among cancers, even when they arise from the same cell type in the same organ. This variability forms the basis for **tumor grading,** which is the assignment of numbered grades to tumors based on differences in their microscopic appearance. Lower numbers (for example, grade 1) are assigned to tumors whose cells exhibit normal differentiated features, divide slowly, and display only modest abnormalities in the

Table 24-3 Some Differences in the Microscopic Traits of Benign and Malignant Tumors

Trait	Benign	Malignant
Nuclear size	Small	Large
N/C ratio (ratio of nuclear to cytoplasmic volume)	Low	High
Nuclear shape	Regular	Pleomorphic (irregular shape)
Mitotic index	Low	High
Tissue organization	Normal	Disorganized
Differentiation	Well differentiated	Poorly differentiated
Tumor boundary	Well defined	Poorly defined

traits listed in Table 24-3. Higher numbers (for example, grade 4) are assigned to tumors containing rapidly dividing, poorly differentiated cells that bear less resemblance to normal cells and exhibit severe abnormalities in the traits listed in Table 24-3. The highest-grade cancers contain cells that are so poorly differentiated and abnormal in appearance that they bear no resemblance to the cells of the tissue in which the tumor arose. These high-grade cancers tend to grow and spread more aggressively and respond less to therapy than lower-grade cancers do.

Screening Techniques for Early Detection Can Prevent Cancer Deaths

When cancer is detected before it has spread, cure rates tend to be relatively high, even for cancers that would otherwise have a poor prognosis. A great need therefore exists for screening techniques that can detect cancers at an early stage. One of the most successful screening procedures is the **Pap smear,** a technique for early detection of cervical cancer developed in the early 1930s by George Papanicolaou (for whom it is named). The rationale underlying this procedure is that the microscopic appearance of cancer cells is so distinctive that you can detect the likely presence of cancer simply by examining a small sample of isolated cells. A Pap smear is performed by taking a tiny sample of a woman's vaginal secretions and examining it with a microscope. If the cells in the fluid exhibit unusual features—such as large, irregular nuclei or prominent variations in cell size and shape (**Figure 24-23**)—it is a sign that cancer may be present and further tests need to be done. Because a Pap smear allows cervical cancer to be detected in its early stages before metastasis has occurred, this procedure has prevented hundreds of thousands of cancer deaths.

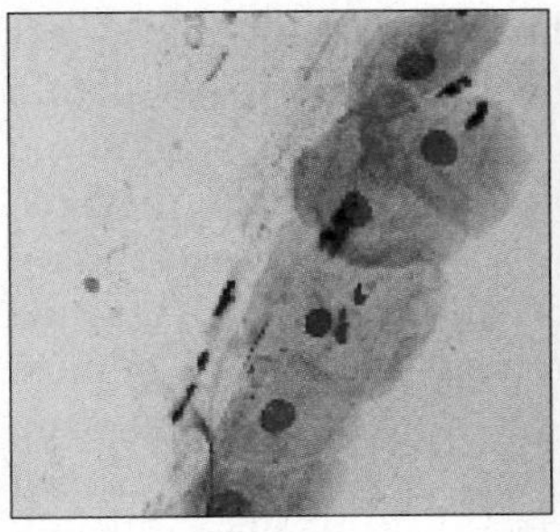
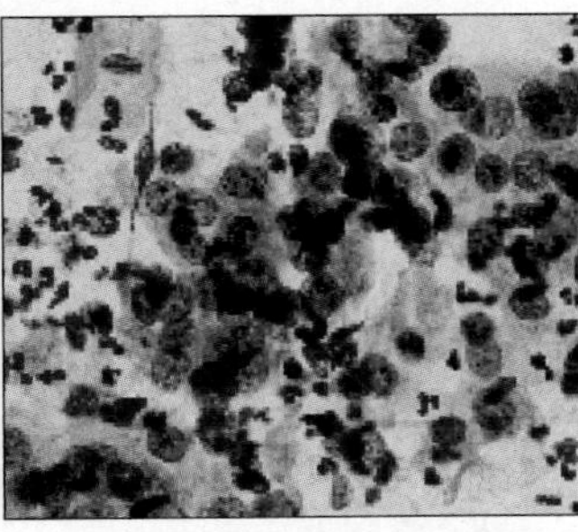

FIGURE 24-23 Normal and Abnormal Pap Smears. *(Left)* In a normal Pap smear, the cells are relatively uniform in size and contain small spherical nuclei. *(Right)* In this abnormal Pap smear, marked variations in cell size and shape are evident, and the nuclei are larger relative to the size of the cells. The abnormalities in these isolated cells suggest they may be derived from a cervical cancer, and further examination of the uterus is therefore required.

The success of the Pap smear has led to the development of screening techniques for other cancers. For example, *mammography* utilizes a special X-ray technique to look for early signs of breast cancer, and *colonoscopy* uses a slender fiber-optic instrument to examine the colon for early signs of colon cancer. The ideal screening test would allow doctors to detect cancers anywhere in the body with one simple procedure, such as a blood test. Prostate cancer is an example of a cancer that can sometimes be detected this way. Men over the age of 50 are often advised to get a **PSA test,** which measures how much *prostate-specific antigen (PSA)* is present in the bloodstream. PSA, which is a protein produced by cells of the prostate gland, normally appears in only tiny concentrations in the blood. If a PSA test reveals a high concentration of PSA, it indicates a possible prostate problem. Further tests are then performed to determine whether cancer is actually present.

Other cancers also release specific proteins into the bloodstream, where their presence might be used to signal the existence of early disease. However, the development of methods for reliably detecting these proteins is still in its infancy, and much work remains to be done before we will know whether cancers can be reliably detected in their early stages by looking for small changes in blood proteins.

Surgery, Radiation, and Chemotherapy Are Standard Treatments for Cancer

Strategies for treating cancer depend both on the type of cancer involved and how far it has spread. The most common approach involves surgery to remove the primary tumor, followed (if necessary) by radiation therapy and/or chemotherapy to destroy any remaining cancer cells.

Radiation therapy uses high-energy X-rays or other forms of ionizing radiation to kill cancer cells. Earlier in the chapter we saw that the DNA damage created by ionizing radiation can cause cancer. Ironically, the same type of radiation is used in higher doses to destroy cancer cells in people who already have the disease. Ionizing radiation kills cells in two different ways. First, DNA damage caused by radiation activates the p53 signaling pathway, which then triggers apoptosis. However, many cancers have mutations that disable the p53 pathway, so p53-induced apoptosis plays only a modest role in the response of most cancers to radiation treatment. In the second mechanism, radiation kills cells by causing such severe chromosomal damage that it prevents cells from progressing through mitosis, and the cells therefore die while trying to divide.

Most forms of *chemotherapy* use drugs that, like radiation, kill dividing cells. Such drugs can be subdivided into four major categories. (1) *Antimetabolites* inhibit metabolic pathways required for DNA synthesis by acting as competitive inhibitors that bind to enzyme active sites in place of normal substrate molecules. Examples include *fluorouracil, methotrexate, fludarabine, pemetrexed,* and *gemcitabine.* (2) *Alkylating agents* inhibit DNA function by chemically crosslinking the DNA double helix. Examples include *cyclophosphamide, chlorambucil,* and *cisplatin.* (3) *Antibiotics* are substances made by microorganisms that inhibit DNA function by either binding to DNA or inhibiting topoisomerases required for DNA replication. Examples include *doxorubicin* and *epirubicin.* (4) *Plant-derived* drugs either inhibit topoisomerases or disrupt the microtubules of the mitotic spindle. Examples include the topoisomerase-inhibitor *etoposide* and the microtubule-disrupting drug, *taxol.*

One problem with such drug (and radiation) treatments is that they are toxic to normal dividing cells as well as to cancer cells. Less toxic approaches are possible for those cancers whose growth requires a specific hormone. For example, many breast cancers require estrogen for their growth. As we saw in Chapter 19, steroid hormones such as estrogen exert their effects by binding to steroid receptor proteins that activate the expression of specific genes. The drug *tamoxifen,* which binds to estrogen receptors in place of estrogen and prevents the receptors from being activated, is useful both in treating breast cancer and in reducing the occurrence of breast cancer in women at high risk for the disease.

A frequent complication with drug therapies is that tumors tend to acquire mutations that make them resistant to the drugs being used. Even when only a single drug is employed, tumors often become resistant to the administered drug and to the effects of several unrelated drugs at the same time. This state, known as multiple drug resistance, arises because cells start producing *multidrug resistance transport proteins (ABC transporters)* that actively pump a wide range of chemically dissimilar, hydrophobic drugs out of the cell (page 211).

Drug resistance may also arise because tumors consist of a heterogeneous mixture of cells that are not equally sensitive to anticancer drugs. A growing body of evidence suggests that in some types of cancer only a small population of cells—known as *cancer stem cells* or *cancer-initiating cells*—proliferate indefinitely and produce the rest of the cells found in tumors. If treatments that destroy most of the tumor cells leave behind a small number of cancer stem cells, these cells may be enough to regenerate the tumor after treatment is stopped. Researchers are therefore trying to identify drugs that selective destroy cancer stem cells because those are the cells that fuel tumor growth.

Using the Immune System to Target Cancer Cells

The use of surgery, radiation, or chemotherapy—either alone or in various combinations—can cure or prolong survival times for many types of cancer, especially when cancer is diagnosed early. However, some of the more aggressive types of cancer (such as those involving the lung, pancreas, or liver) are difficult to control in these ways, and similar difficulties are encountered when cancer is diagnosed in its later stages. In an effort to develop better approaches for treating such cancers, scientists have been trying to find ways to selectively seek out and destroy cancer cells without damaging normal cells in the process.

One strategy for introducing such selectivity into cancer treatment is to exploit the ability of the immune system to recognize cancer cells. This approach, called **immunotherapy,** was first proposed in the 1800s after doctors noticed that tumors occasionally regress in people who develop bacterial infections. Since infections trigger an immune response, subsequent attempts were made to build on this observation by utilizing live or dead bacteria to provoke the immune system of cancer patients. Although the approach has not worked as well as initially hoped, some success has been seen with *bacillus Calmette-Guérin (BCG)*—a bacterial strain that does not cause disease but elicits a strong immune response at the site where it is introduced into the body. BCG is useful in treating early stage bladder cancers that are localized to the bladder wall. After the primary tumor has been surgically removed, inserting BCG into the bladder elicits a prolonged activation of immune cells that in turn leads to lower rates of cancer recurrence.

While it demonstrates the potential usefulness of immune stimulation, BCG must be administered directly into the bladder to provoke an immune response at the primary tumor site. To treat cancers that have already metastasized to unknown locations, an immune response must be stimulated wherever cancer cells might have traveled. Normal proteins that the body produces to stimulate the immune system are sometimes useful for this purpose. *Interferon alpha* and *interleukin-2 (IL-2)* are two such proteins that have been successfully used as cancer treatments.

Attempts are also under way to develop therapeutic *vaccines* that introduce cancer cell antigens into patients to stimulate the immune system to selectively attack cancer cells. One such vaccine, called *Provenge,* uses an antigen commonly found in prostate cancer cells combined with antigen-processing cells obtained from the patient's blood.

BOX 24B TOOLS OF DISCOVERY

Monoclonal Antibodies and Cancer Treatment

One way the immune system operates is by producing **antibodies,** which are soluble proteins that bind to substances, referred to as **antigens,** that provoke an immune response. To function as an antigen, a substance must usually be recognized as being "foreign"—that is, different from molecules normally found in a person's body. Antibody molecules recognize and bind to specific antigens with extraordinary precision, making them ideally suited for targeting antigens that are unique to, or preferentially concentrated in, cancer cells.

For many years, the use of antibodies for treating cancer was hampered by the lack of a reproducible method for producing large quantities of pure antibody molecules directed against the same antigen. Then in 1975, Georges Köhler and César Milstein solved the problem by devising the procedure shown in **Figure 24B-1**. In this technique, mice are first injected with an antigen of interest, and antibody-producing lymphocytes are isolated from the animals a few weeks later. Within such a lymphocyte population, each lymphocyte produces a single type of antibody directed against one particular antigen. To facilitate the selection and growth of individual lymphocytes, they are fused with cells that divide rapidly and have an unlimited life span when grown in culture. Individual hybrid cells are then selected and grown to form a series of clones called *hybridomas.* The antibodies produced by hybridomas are referred to as **monoclonal antibodies** because each one is a pure antibody produced by a cloned population of lymphocytes.

To obtain monoclonal antibodies that might be useful for treating cancer, human cancer tissue is injected into mice to stimulate an immune response. Hybridomas are then created using lymphocytes from the immunized animals, and the hybridomas are analyzed to determine which ones produce antibodies directed against antigens present in the cancer tissue. Because the antibodies are derived from mice and might be destroyed by a person's immune system, they are usually made more humanlike by replacing large parts of the mouse antibody molecule with corresponding sequences derived from human antibodies. When the resulting antibodies are injected into individuals with cancer, they bind to cancer cells and their presence can in theory trigger an immune attack against the cells containing the bound antibody. Although this approach does not yet work for most cancers, one promising success has been achieved with *non-Hodgkins lymphoma.* Antibodies that target the *CD20 antigen* found on the surface of these lymphoma cells are now among the standard treatments for this particular type of cancer.

Antibodies can also serve as delivery vehicles by linking them to radioactive molecules or to other toxic substances that are too lethal to administer alone. Attaching such substances to antibodies allows the toxins to be selectively concentrated at tumor sites without accumulating to toxic levels elsewhere in the body. Finally, many antibodies have been developed that bind to and inactivate specific proteins involved in the signaling pathways that drive cancer cell proliferation. The monoclonal antibody *Herceptin,* described on page 790, is an example of an anticancer drug that works in this way.

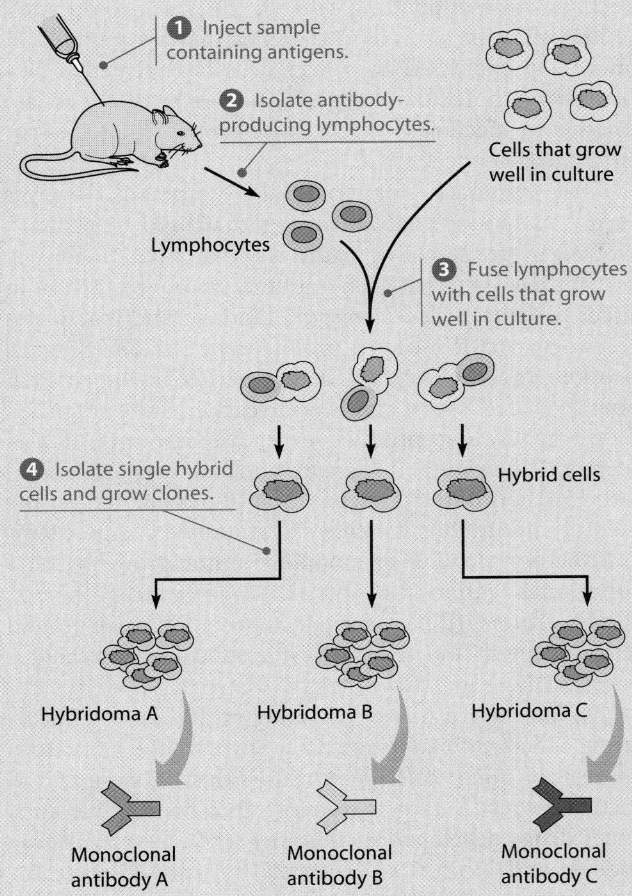

FIGURE 24B-1 Monoclonal Antibody Technique for Producing Pure Antibodies. A sample containing an antigen of interest is injected into mice to stimulate antibody formation. Antibody-producing lymphocytes isolated from the animal are then fused with cells that grow well in culture, and the resulting hybrid cells are used to create a series of cloned cell populations (hybridomas) that each make a single type of antibody. Extensive screening may be required to find a hybridoma that makes an antibody directed against a particular antigen of interest.

Human clinical trials suggest that Provenge is more effective than chemotherapy in prolonging the survival of men with prostate cancer, and the vaccine was approved by the FDA in 2010 for use as a standard treatment for prostate cancer.

Yet another way of using the immune system to fight cancer is with *antibodies,* which are proteins whose ability to recognize and bind to target molecules with great specificity makes them ideally suited to serve as agents that selectively attack cancer cells. **Box 24B** describes a technique for manufacturing pure ("monoclonal") antibodies in large quantities and illustrates how such antibodies have been used for treating cancer.

Herceptin and Gleevec Attack Cancer Cells by Molecular Targeting

Until the early 1980s, the development of new cancer drugs focused largely on agents that disrupt DNA and interfere with cell division. While such drugs are useful in treating cancer, their effectiveness is often limited by toxic effects on normal dividing cells. In the past two decades, the identification of individual genes that are mutated or abnormally expressed in cancer cells has created a new possibility—**molecular targeting**—in which drugs are designed to specifically target those proteins that are critical to the cancer cell.

One approach for molecular targeting involves the use of monoclonal antibodies that bind to proteins involved in the signaling pathways that drive cancer cell proliferation. The first such antibody approved for use in cancer patients, called *Herceptin,* binds to and inactivates the growth factor receptor produced by the *ERBB2* gene (also known as *HER2*). As we saw earlier in the chapter, about 25% of all breast and ovarian cancers have amplified *ERBB2* genes that produce excessive amounts of this receptor. When individuals with such cancers are treated with Herceptin, the Herceptin antibody binds to the receptor and inhibits its ability to stimulate cell proliferation, thereby slowing or stopping tumor growth. Other monoclonal antibodies now used in cancer therapy include *Erbitux* (directed against the epidermal growth factor receptor) and *Avastin* (directed against the angiogenesis stimulating growth factor, VEGF).

An alternative way of targeting molecules for inactivation, called *rational drug design,* involves the laboratory synthesis of *small molecule inhibitors* that are designed to inactivate specific target proteins. One of the first anticancer drugs developed in this way was *Gleevec.* Gleevec binds to and inhibits an abnormal tyrosine kinase produced by the *BCR-ABL* oncogene, whose association with chronic myelogenous leukemia was described earlier in the chapter. The *BCR-ABL* oncogene arises from the fusion of two unrelated genes and produces a structurally abnormal tyrosine kinase that represents an ideal drug target because it is present only in cancer cells. The effectiveness of Gleevec as a treatment for early stage myelogenous leukemia is quite striking. More than half of the patients treated with Gleevec have no signs of the cancer six months after treatment, a response rate that is 10 times better than that observed with earlier treatments. Other small molecule inhibitors of protein kinases that have subsequently been found to be useful in cancer therapy include *Iressa, Tarceva, Sutent,* and *Nexavar.*

Growth factor receptors, growth factors, and protein kinases are not the only candidates for molecular targeting. The uncontrolled proliferation, survival, invasion, and metastasis of cancer cells is caused by disruptions in a variety of signaling pathways, and any of the proteins involved in these pathways is a potential target for anticancer drugs.

Anti-angiogenic Therapies Act by Attacking a Tumor's Blood Supply

Molecular targeting isn't always aimed at cancer cells. Earlier we saw that sustained tumor growth depends on *angiogenesis* (growth of new blood vessels), so it is logical to expect that inhibitors that target blood vessel growth might be useful for treating cancer. Initial support for this concept of *anti-angiogenic therapy* came from the studies of Judah Folkman, who found that angiogenesis inhibitors make tumors shrink in mice.

The first anti-angiogenic drug to be approved for use in humans was *Avastin,* a monoclonal antibody that binds to and inactivates the angiogenesis-stimulating growth factor, VEGF. In tumors that depend on VEGF to stimulate angiogenesis, Avastin is expected to slow tumor growth by inhibiting angiogenesis. Avastin has been found to improve short-term survival rates for several types of cancer, but the benefits are usually temporary and tumor growth resumes. Dozens of other drugs that target angiogenesis are currently being evaluated to see if they might be more effective than Avastin, but it will be at least several years before the value of such anti-angiogenic agents becomes clear.

Cancer Treatments Can Be Tailored to Individual Patients

Tools are now becoming available to enhance the effectiveness of cancer therapies through a strategy called *personalized medicine*—an approach in which treatment choices are based on the individual characteristics of each patient. It has been known for many years that people with cancers that appear to be identical by traditional criteria often exhibit different outcomes after receiving the same treatment. *Transcriptome analysis*—the use of DNA microarrays to determine which genes are being transcribed into RNA (page 733)—has provided an explanation: Cancers involving the same cell type often exhibit differing patterns of gene expression that cause tumors to behave differently.

These differing patterns of gene expression allow predictions about tumor behavior to be made. In breast cancers, for example, the expression of 21 key genes turns out to be a good indicator of whether a given tumor is likely to metastasize. A test called *Oncotype DX* is now available that can measure the activity of these 21 genes and convert the data into a single number known as a *recurrence score.* Women whose breast tumors have a high recurrence score are more likely to have their cancers recur after surgery. In the absence of such information, doctors would usually recommend chemotherapy for most patients after surgery. The value of gene expression testing is that it can help identify those patients who are most likely to benefit from chemotherapy.

Analyzing gene expression profiles and testing for the presence of specific mutations may also help guide the choice of which drug to use. A striking example involves *Iressa,* a drug that acts by inhibiting the tyrosine kinase activity of the epidermal growth factor (EGF) receptor. Iressa causes tumors to shrink in only 10% of lung cancer patients, but when the drug does work, it works extremely well. The reason Iressa is more effective in some individuals than others has been traced to the presence of a mutant EGF receptor gene in the cancers of the patients who receive the most benefit from the drug. When grown in culture, lung cancer cells containing the mutant EGF receptor are found to be more sensitive to the growth-inhibiting effects of Iressa than are cancer cells containing the normal EGF receptor. This discovery opens the door to a personalized type of cancer therapy in which genetic testing of cancer cells is used to identify patients who are most likely to benefit from a particular type of treatment.

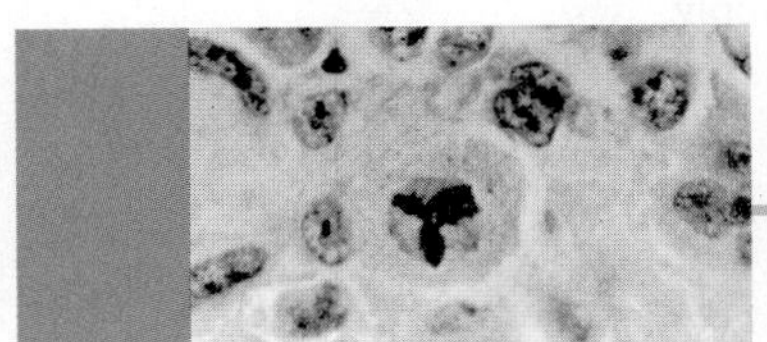

SUMMARY OF KEY POINTS

Uncontrolled Cell Proliferation and Survival

- Cancer cells proliferate in an uncontrolled way and are capable of spreading by invasion and metastasis.
- The balance between cell division and cell differentiation or death is disrupted in cancers, leading to a progressive increase in the number of dividing cells.
- Cancer cells are anchorage-independent, exhibit a decreased susceptibility to density-dependent inhibition of growth, and are able to replenish their telomeres.

How Cancers Spread

- Sustained tumor growth requires a network of blood vessels whose growth is triggered by an increased production of angiogenesis activators and a decreased production of angiogenesis inhibitors.
- Cancer cells invade surrounding tissues, enter the circulatory system, and metastasize to distant sites. Invasion is facilitated by decreased cell-cell adhesion, increased motility, and secretion of proteases that degrade the extracellular matrix and basal lamina. Only a tiny fraction of the cancer cells that enter the bloodstream establish successful metastases.
- Sites of metastasis are determined by the location of the first capillary bed as well as by organ-specific conditions that influence cancer cell growth.
- Cancers have various ways of evading destruction by the immune system. Interactions between tumor cells and components of the surrounding microenvironment influence tumor growth, invasion, and metastasis.

What Causes Cancer?

- Some cancers are caused by certain kinds of chemicals, including those found in tobacco smoke. Chemicals cause cancer through a multistep process involving initiation, promotion, and tumor progression. Initiation is based on DNA mutation, whereas promotion involves a prolonged period of cell proliferation accompanied by selection of cells exhibiting enhanced growth properties. During tumor progression, additional mutations as well as epigenetic changes in gene expression produce cells with increasingly aberrant traits.
- Cancer can also be caused by ionizing or UV radiation, both of which trigger DNA mutations, as well as by certain viruses, bacteria, and parasites.
- Some cancer-causing viruses trigger cell proliferation directly, either through the action of viral genes or by altering the behavior of cellular genes. Other viruses and infectious agents create tissue destruction that indirectly stimulates cell proliferation under conditions in which DNA damage is likely.

Oncogenes and Tumor Suppressor Genes

- Oncogenes are genes whose presence can cause cancer. Although oncogenes are sometimes brought into cells by viruses, more often they arise from normal cellular genes (proto-oncogenes) by point mutation, gene amplification, chromosomal translocation, local DNA rearrangements, or insertional mutagenesis.
- Many of the proteins produced by oncogenes are signaling pathway components, such as growth factors, receptors, plasma membrane GTP-binding proteins, nonreceptor protein kinases, transcription factors, and cell cycle and apoptosis regulators. Oncogenes code for abnormal forms or excessive quantities of such proteins, thereby leading to excessive cell proliferation and survival.
- Tumor suppressor genes are genes whose loss or inactivation can lead to cancer. Susceptibility to cancer is increased in people who inherit defective tumor suppressor genes.
- Common tumor suppressor genes include the *RB* gene, which produces a protein that restrains passage from G1 into S phase; the *p53* gene, which produces a protein that prevents

proliferation and triggers apoptosis in cells with damaged DNA; and the *APC* gene, which produces a protein that inhibits the Wnt pathway.

- The genetic instability of cancer cells facilitates the acquisition of multiple mutations. Genetic instability arises from defects in DNA repair mechanisms, disruptions in pathways that trigger apoptosis, and failures in chromosome-sorting mechanisms.
- Cancers arise through a stepwise accumulation of mutations and epigenetic changes involving oncogenes and tumor suppressor genes. Most of the affected genes code for proteins, but some produce microRNAs.

Diagnosis, Screening, and Treatment

- Screening techniques, such as the Pap smear, can prevent cancer deaths by detecting cancer before it has spread.
- Cancer treatments usually involve surgery to remove the primary tumor, followed (if necessary) by radiation therapy and/or chemotherapy to kill or inhibit the growth of any remaining cancer cells.
- Newer treatment approaches include immunotherapies that exploit the ability of the immune system to attack cancer cells, molecular targeting drugs that bind to proteins that play critical roles in cancer cells, and anti-angiogenic agents that attack a tumor's blood supply.

MB **www.masteringbiology.com**

1. MasteringBiology Assignments

Tutorials and Activities
- Molecular Model Structural Tutorials for visualization of 3D structures
- BioFlix Tutorials including 3D animations with hints and detailed feedback for common wrong answers
- Ames Test Protocol

Questions
- Reading Quiz
- Multiple Choice
- End-of-chapter questions

2. eText Read your book online, search, highlight text, take notes, and more

3. The Study Area
- Chapter Review Quiz, Guide to Techniques and Methods, 3D-Structure Tutorials, Activities, BioFlix, Videos, Biology Review, Word Study Tools, Glossary, Flashcards

PROBLEM SET

More challenging problems are marked with a •.

24-1 Normal Cells and Cancer Cells. You are given two test tubes, one containing cells from a human cancer and the other containing normal cells. Before you begin your studies, the labels fall off the test tubes. Describe at least four experiments you could carry out to determine which sample contains the cancer cells.

24-2 Angiogenesis. Describe two pieces of evidence supporting the idea that angiogenesis is required for tumors to grow beyond a tiny clump of cells, and two pieces of evidence supporting the idea that cancer cells secrete molecules that stimulate angiogenesis.

24-3 How Cancers Spread. Describe the three main stages involved in metastasis, including a description of the relevant cellular properties.

• **24-4 Rats, Guinea Pigs, and Humans.** The chemical substance 2-acetylaminofluorene (AAF) causes bladder cancer when injected into rats but not guinea pigs. If normal bladder cells obtained from rats and guinea pigs are grown in culture and exposed to AAF, neither are converted into cancer cells. How can you explain these findings? Does your explanation suggest how to predict whether AAF is carcinogenic in humans without actually exposing humans to AAF?

24-5 Oncogenes and Tumor Suppressor Genes. Indicate whether each of the following descriptions applies to an oncogene (OG), a proto-oncogene (PO), or a tumor suppressor gene (TS). Some descriptions may apply to more than one of these gene types. Explain your answers.

(a) A type of gene found in normal cells but not in cancer cells.

(b) A gene that causes cancer only when it is mutated but participates in signaling events in normal cells.

(c) A type of gene that causes cell death in normal state and cancer in mutated state.

(d) A type of gene whose overexpression, even in the nonmutated form, can cause cancer.

(e) A gene whose presence can cause cancer.

(f) A gene whose loss is a greater precipitating factor than the presence or absence of other cancer-causing factors.

(g) A type of gene found in normal cells and cancer cells.

24-6 Children of the Moon. Children with xeroderma pigmentosum usually cannot carry out excision repair. Why does this makes them so susceptible to developing cancer? Why was the word "usually" included in the first sentence?

• **24-7 Cancer Screening.** The annual incidence of colon cancer in the United States is about 55 cases per 100,000. Because colon cancer often causes bleeding, doctors sometimes use a screening procedure called the fecal occult blood test (FOBT) to look for tiny amounts of blood in the feces. One form of this test has a specificity of about 98%, which means that when it indicates the presence of blood in the feces, the result is an error (i.e., cancer is not present) only 2% of the time. While this may seem like excellent specificity, a 2% "false positive" rate makes this test almost useless as a tool for colon cancer screening. Why?

• **24-8 Monoclonal Antibodies and Lymphoma.** Monoclonal antibodies directed against the CD20 antigen are used to treat certain forms of lymphoma. However, the CD20 antigen is present on both normal and malignant lymphocytes, and treatment with the antibody therefore kills normal cells as well as cancer cells. Why do you think this antibody still turns out to be an effective treatment for lymphoma?

SUGGESTED READING

References of historical importance are marked with a •.

General References

Kleinsmith, L. J. *Principles of Cancer Biology.* San Francisco: Pearson Benjamin Cummings, 2006.

Weinberg, R. A. *The Biology of Cancer.* New York: Garland Science, 2007.

Uncontrolled Cell Proliferation and Survival

Clarke, M. F., and M. W. Becker. Stem cells: The real culprits in cancer? *Sci. Amer.* 295 (July 2006): 52.

Cotter, T. G. Apoptosis and cancer: The genesis of a research field. *Nature Rev. Cancer* 9 (2009): 501.

Masters, J. R. HeLa cells 50 years on: The good, the bad, and the ugly. *Nature Rev. Cancer* 2 (2002): 315.

Neumann, A. A., and R. R. Reddel. Telomere maintenance and cancer—Look, no telomerase. *Nature Rev. Cancer* 2 (2002): 879.

How Cancers Spread

Fidler, I. J. The pathogenesis of cancer metastasis: The "seed and soil" hypothesis revisited. *Nature Rev. Cancer* 3 (2003): 453.

Joyce, J. A., and J. W. Pollard. Microenvironment regulation of metastasis. *Nature Rev. Cancer* 9 (2009): 239.

Psaila, B., and D. Lyden. The metastatic niche: Adapting the foreign soil. *Nature Rev. Cancer* 9 (2009): 285.

Zetter, B. R. The scientific contributions of M. Judah Folkman to cancer research. *Nature Rev. Cancer* 8 (2008): 647.

What Causes Cancer?

Jha, P. Avoidable global cancer deaths and total deaths from smoking. *Nature Rev. Cancer* 9 (2009): 655.

Leffell, D. J., and D. E. Brash. Sunlight and skin cancer. *Sci. Amer.* 275 (July 1996): 52.

Peto, J. Cancer epidemiology in the last century and the next decade. *Nature* 411 (2001): 390.

Polk, D. B., and R. M. Peek Jr. *Helicobacter pylori*: Gastric cancer and beyond. *Nature Rev. Cancer* 10 (2010): 403.

Speck, S. H. EBV framed in Burkitt lymphoma. *Nature Med.* 8 (2002): 1086.

Williams, D. Cancer after nuclear fallout: Lessons from the Chernobyl accident. *Nature Rev. Cancer* 2 (2002): 543.

Woodman, C. B. J., S. I. Collins, and L. S Young. The natural history of cervical HPV infection: Unresolved issues. *Nature Rev. Cancer* 7 (2007): 11.

Oncogenes and Tumor Suppressor Genes

Cleaver, J. E. Cancer in xeroderma pigmentosum and related disorders of DNA repair. *Nature Rev. Cancer* 5 (2005): 564.

Fletcher, O., and R. S. Houlston. Architecture of inherited susceptibility to common cancer. *Nature Rev. Cancer* 10 (2010): 353.

Fox, E. J., J. J. Salk, and L. A. Loeb. Cancer genome sequencing—An interim analysis. *Cancer Res.* 69 (2009): 4948,

Gal-Yam, E. N. et al. Cancer epigenetics: Modifications, screening, and therapy. *Annu Rev. Med.* 59 (2008): 267.

Garson, R., G. A. Calin, and C. M. Croce. MicroRNAs in cancer. *Annu Rev. Med.* 60 (2009): 167.

Hanahan, D., and R. A. Weinberg. The hallmarks of cancer. *Cell* 100 (2000): 57.

• Harris, H., G. Klein, P. Worst, and T. Tachibana. Suppression of malignancy by cell fusion. *Nature* 223 (1969): 363.

Lengauer, C., K. W. Kinzler, and B. Vogelstein. Genetic instabilities in human cancers. *Nature* 396 (1998): 643.

Levine, A. J., and M. Oren. The first 30 years of p53: Growing ever more complex. *Nature Rev. Cancer* 9 (2009): 749.

Mitelman, F., B. Johansson, and F. Mertens. The impact of translocations and gene fusions on cancer causation. *Nature Rev. Cancer* 7 (2007): 233.

Moody, C. A., and L. A. Laimins. Human papillomavirus oncoproteins: Pathways to transformation. *Nature Rev. Cancer* 10 (2010): 550.

Schvartzman, J.-M., R. Sotillo, and R. Benezra. Mitotic chromosomal instability and cancer: Mouse modeling of the human disease. *Nature Rev. Cancer* 10 (2010): 102.

Weaver, B. A. A., and D. W. Cleveland. Does aneuploidy cause cancer? *Curr. Opin. Cell Biol.* 18 (2006): 658.

Diagnosis, Screening, and Treatment

Chabner, B. A., and T. G. Roberts Jr. Chemotherapy and the war on cancer. *Nature Rev. Cancer* 5 (2005): 65.

Dougan, M., and G. Dranoff. Immune therapy for cancer. *Annu. Rev. Immunol.* 27 (2009): 83.

Imai, K., and A. Takaoka. Comparing antibody and small-molecule therapies for cancer. *Nature Rev. Cancer* 6 (2006): 714.

Sotiriou, C., and M. J. Piccart. Taking gene-expression profiling to the clinic: When will molecular signatures become relevant to patient care? *Nature Rev. Cancer* 7 (2007): 545.

Visvader, J. E., and G. J. Lindeman. Cancer stem cells in solid tumours: Accumulating evidence and unresolved questions. *Nature Rev. Cancer* 8 (2008): 755.

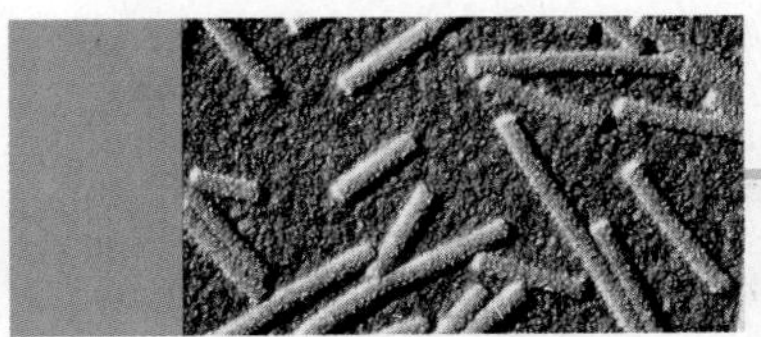

APPENDIX
VISUALIZING CELLS AND MOLECULES

Cell biologists often need to examine the structure of cells and their components, and they need to see specific structures or molecules amid a complicated mixture of cellular components. The microscope is an indispensable tool for this purpose because most cellular structures are too small to be seen by the unaided eye. In fact, the beginnings of cell biology can be traced to the invention of the ***light microscope,*** *which made it possible for scientists to see enlarged images of cells for the first time. The first generally useful light microscope was developed in 1590 by Z. Janssen and his nephew, H. Janssen. Many important microscopic observations were reported during the next century, notably those of Robert Hooke, who observed the first cells, and Antonie van Leeuwenhoek, whose improved microscopes provided our first glimpses of internal cell structure. Since then, the light microscope has undergone numerous improvements and modifications, right up to the present time.*

Just as the invention of the light microscope heralded a wave of scientific achievement by allowing us to see cells for the first time, the development of the ***electron microscope*** *in the 1930s revolutionized our ability to explore cell structure and function. Because it is at least a hundred times better at visualizing objects than the light microscope is, the electron microscope ushered in a new era in cell biology, opening our eyes to an exquisite subcellular architecture never before seen and changing forever the way we think about cells.*

Light microscopy has experienced a renaissance in recent years as the development of specialized new techniques allows researchers to explore aspects of cell structure and behavior that cannot be readily studied by electron microscopy. These advances have involved the merging of technologies from physics, engineering, chemistry, and molecular biology, and they have greatly expanded our ability to study cells using the light microscope.

In this appendix, we explore the fundamental principles of light and electron microscopy, emphasizing the various specialized techniques used to adapt these two types of microscopy for a variety of specialized purposes. We also examine other related technologies that extend the range of view beyond the microscope to visualization of single molecules at high resolution.

Optical Principles of Microscopy

Although light and electron microscopes differ in many ways, they make use of similar optical principles to form images. Therefore, we begin our discussion of microscopy by examining these underlying common principles, placing special emphasis on the factors that determine how small an object can be seen with current technologies.

The Illuminating Wavelength Sets a Limit on How Small an Object Can Be Seen

Regardless of the type of microscope being used, three elements are always needed to form an image: a *source of illumination,* a *specimen* to be examined, and a system of *lenses* that focuses the illumination on the specimen and forms the image. **Figure A-1** illustrates these features for a light microscope and an electron microscope. In a light microscope, the source of illumination is *visible light* (wavelength approximately 400–700 nm), and the lens system consists of a series of glass lenses. The image can either be viewed directly through an eyepiece or focused on a detector, such as photographic film or an electronic camera. In an electron microscope, the illumination source is a *beam of electrons* emitted by a heated tungsten filament, and the lens system consists of a series of electromagnets. The electron beam is focused on either a fluorescent screen or photographic film, or it is digitally imaged using a detector.

Despite these differences in illumination source and instrument design, both types of microscopes depend on the same principles of optics and form images in a similar manner. When a specimen is placed in the path of a light or electron beam, physical characteristics of the beam are changed in a way that creates an image that can be interpreted by the human eye or recorded on a photographic detector. To understand this interaction between the illumination source and the specimen, we need to understand the concept of wavelength, which is illustrated in **Figure A-2** using the following simple analogy.

If two people hold onto opposite ends of a slack rope and wave the rope with a rhythmic up-and-down motion, they will generate a long, regular pattern of movement in the rope called a *waveform* (Figure A-2a). The distance from the crest of one wave to the crest of the next is called the **wavelength.** If someone standing to one side of the rope tosses a large object such as a beach ball toward the

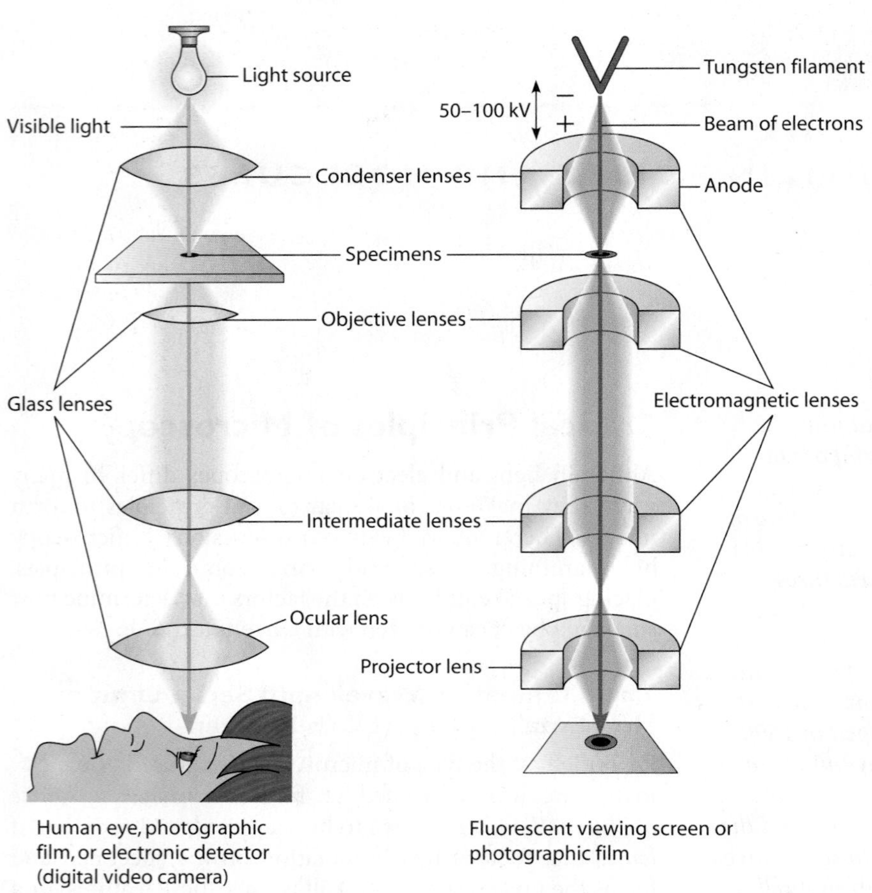

FIGURE A-1 Optical Systems of the Light Microscope and the Electron Microscope. **(a)** The light microscope uses visible light (wavelength approximately 400–700 nm) and glass lenses to form an image of the specimen that can be seen by the eye, focused on photographic film, or received by an electronic detector such as a digital camera. **(b)** The electron microscope uses a beam of electrons emitted by a tungsten filament and focused by electromagnetic lenses to form an image of the specimen on a fluorescent screen, a digital detector, or photographic film. (These diagrams have been drawn to emphasize the similarities in overall design between the two types of microscope. In reality, a light microscope is designed with the light source at the bottom and the ocular lens at the top, as shown in Figure A-5b.)

rope, the ball may interfere with, or perturb, the waveform of the rope's motion (Figure A-2b). However, if a small object, such as a softball, is tossed toward the rope, the movement of the rope will probably not be affected at all (Figure A-2c). If the rope holders move the rope faster, the motion of the rope will still have a waveform, but the wavelength will be shorter (Figure A-2d). In this case, a softball tossed toward the rope is quite likely to perturb the rope's movement (Figure A-2e).

This simple analogy illustrates an important principle: The ability of an object to perturb a wave's motion depends crucially on the size of the object in relation to the wavelength of the motion. This principle is of great importance in microscopy because it means that the wavelength of the illumination source sets a limit on how small an object can be seen. To understand this relationship, we need to recognize that the moving rope of Figure A-2 is analogous to the beam of light (photons) or the electrons used as an illumination source in a light or electron microscope, respectively. In other words, both light and electrons behave as waves. When a beam of light (or electrons) encounters a specimen, the specimen alters the physical characteristics of the illuminating beam, just as the beach ball or softball alters the motion of the rope. And because an object can be detected only by its effect on the wave, the wavelength must be comparable in size to the object that is to be detected.

Once we understand this relationship between wavelength and object size, we can readily appreciate why very small objects can be seen only by electron microscopy: The wavelengths of electrons are very much shorter than those of photons. Thus, objects such as viruses and ribosomes are too small to perturb the waveform of photons, but they can readily interact with electrons. As we discuss different types of microscopes and specimen preparation techniques, you might find it helpful to ask yourself how the source and specimen are interacting and how the characteristics of both are modified to produce an image.

Resolution Refers to the Ability to Distinguish Adjacent Objects as Separate from One Another

When waves of light or electrons pass through a lens and are focused, the image that is formed results from a property of waves called **interference**—the process by which two or more waves combine to reinforce or cancel one another, producing a wave equal to the sum of the two combining waves. Thus, the image that you see when you look at a specimen through a series of lenses is really just a pattern of either additive or canceling interference of the waves that went through the lenses, a phenomenon known as **diffraction.**

In a light microscope, glass lenses are used to direct photons, whereas an electron microscope uses electro-

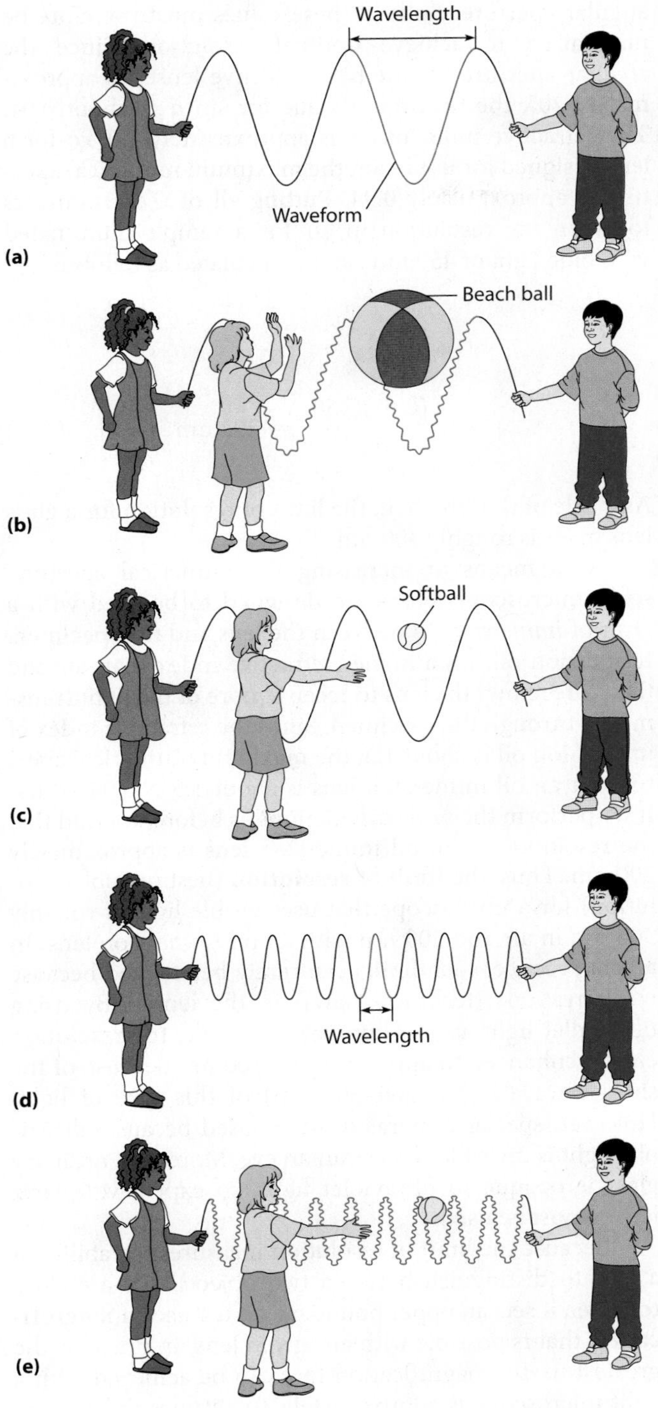

FIGURE A-2 Wave Motion, Wavelength, and Perturbations. The wave motion of a rope held between two people is analogous to the waveform of both photons and electrons, and it can be used to illustrate the effect of the size of an object on its ability to perturb wave motion. **(a)** Moving a slack rope up and down rhythmically will generate a waveform with a characteristic wavelength. **(b)** When thrown against a rope, an object with a diameter comparable to the wavelength of the rope (e.g., a beach ball) will perturb the motion of the rope. **(c)** An object with a diameter significantly less than the wavelength of the rope (e.g., a softball) will probably cause little or no perturbation of the rope because, with its smaller diameter, it is not likely to strike the rope when tossed toward it. **(d)** If the rope is moved more rapidly, the wavelength will be reduced substantially. **(e)** A softball can now perturb the motion of the rope because its diameter is comparable to the wavelength of the rope.

magnets as lenses to direct electrons. Yet both kinds of lenses have two fundamental properties in common: focal length and angular aperture. The **focal length** is the distance between the midline of the lens and the point at which rays passing through the lens converge to a focus (**Figure A-3**). The focal length is determined by the index of refraction of the lens itself, the medium in which it is immersed, and the geometry of the lens. The lens magnifying strength, measured in diopters, is the inverse of the focal length, measured in meters. The **angular aperture** is the half-angle α of the cone of light entering the objective lens of the microscope from the specimen (**Figure A-4**). Angular aperture is therefore a measure of how much of the illumination that leaves the specimen actually passes through the lens. This in turn determines the sharpness of the interference pattern and therefore the ability of the lens to convey information about the specimen. In the best light microscopes, the angular aperture is about 70°.

The angular aperture of a lens is one of the factors influencing a microscope's **resolution,** which is defined as the minimum distance that can separate two points that still remain identifiable as separate points when viewed through the microscope.

Resolution is governed by three factors: the wavelength of the light used to illuminate the specimen, the angular aperture, and the refractive index of the medium surrounding the specimen. (**Refractive index** is a measure of the change in the velocity of light as it passes from one medium to another.) The effect of these three variables on resolution is described quantitatively by an equation known as the *Abbé equation:*

$$r = \frac{0.61\,\lambda}{n \sin \alpha} \qquad \textbf{(A-1)}$$

where r is the resolution, λ is the wavelength of the light used for illumination, n is the refractive index of the

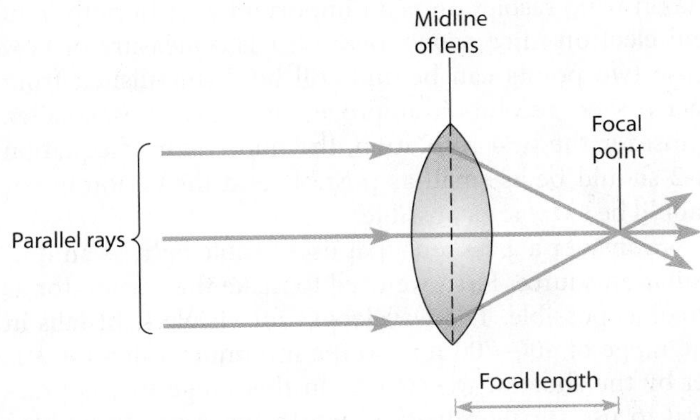

FIGURE A-3 The Focal Length of a Lens. Focal length is the distance from the midline of a lens to the point where parallel rays passing through the lens converge to a focus.

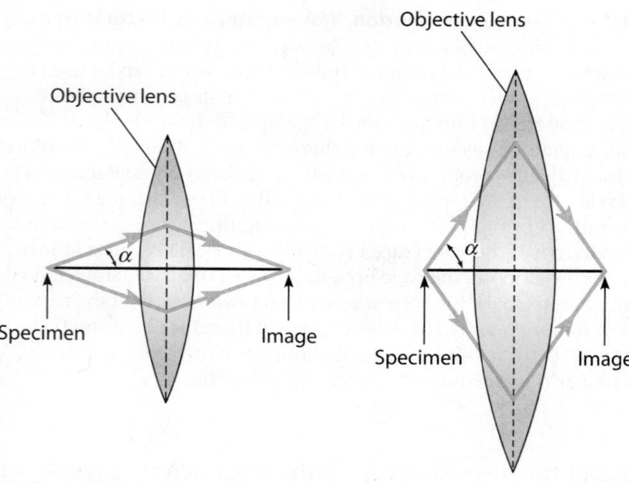

FIGURE A-4 The Angular Aperture of a Lens. The angular aperture is the half-angle α of the cone of light entering the objective lens of the microscope from the specimen. **(a)** A low-aperture lens (α is small). **(b)** A high-aperture lens (α is large). The larger the angular aperture, the more information the lens can transmit. The best glass lenses have an angular aperture of about 70°.

medium between the specimen and the objective lens of the microscope, and α is the angular aperture as previously defined. The constant 0.61 represents the degree to which image points can overlap and still be recognized as separate points by an observer.

In the preceding equation, the quantity $n \sin \alpha$ is called the **numerical aperture** of the objective lens, abbreviated ***NA***. An alternative expression for resolution is therefore

$$r = \frac{0.61\lambda}{NA} \quad \textbf{(A-2)}$$

The Practical Limit of Resolution Is Roughly 200 nm for Light Microscopy and 2 nm for Electron Microscopy

Maximizing resolution is an important goal in both light and electron microscopy. Because r is a measure of how close two points can be and still be distinguished from each other, resolution improves as r becomes smaller. Thus, for the best resolution, the numerator of equation A-2 should be as small as possible and the denominator should be as large as possible.

Consider a glass lens that uses visible light as an illumination source. First, we need to make the numerator as small as possible. The wavelength for visible light falls in the range of 400–700 nm, so the minimum value for λ is set by the shortest wavelength in this range that is practical to use for illumination, which turns out to be blue light of approximately 450 nm. To maximize the denominator of Equation A-2, recall that the numerical aperture is the product of the refractive index and the sine of the angular aperture. Both of these values must therefore be maximized to achieve optimal resolution. Since the angular aperture for the best objective lenses is approximately 70°, the maximum value for $\sin \alpha$ is about 0.94. The refractive index of air is approximately 1.0, so for a lens designed for use in air, the maximum numerical aperture is approximately 0.94. Putting all of these numbers together, the resolution in air for a sample illuminated with blue light of 450 nm can be calculated as follows:

$$\begin{aligned} r &= \frac{0.61\lambda}{NA} \\ &= \frac{(0.61)(450)}{0.94} = 292 \text{ nm} \end{aligned} \quad \textbf{(A-3)}$$

As a rule of thumb, then, the limit of resolution for a glass lens in air is roughly 300 nm.

As a means of increasing the numerical aperture, some microscope lenses are designed to be used with a layer of *immersion oil* between the lens and the specimen. Immersion oil has a higher refractive index than air and therefore allows the lens to receive more of the light transmitted through the specimen. Since the refractive index of immersion oil is about 1.5, the maximum numerical aperture for an oil immersion lens is about $1.5 \times 0.94 = 1.4$. If we perform the same calculations as before, we find that the resolution of an oil immersion lens is approximately 200 nm. Thus, the **limit of resolution** (best possible resolution) for a microscope that uses visible light is roughly 300 nm in air and 200 nm with an oil immersion lens. In actual practice, such limits can rarely be reached because of aberrations (technical flaws) in the lenses. By using ultraviolet light as an illumination source, the resolution can be enhanced to approximately 100 nm because of the shorter wavelength (200–300 nm) of this type of light. However, special cameras must be used because ultraviolet light is invisible to the human eye. Moreover, ordinary glass is opaque to ultraviolet light, so expensive quartz lenses must be used.

Because the limit of resolution measures the ability of a lens to distinguish between two objects that are close together, it sets an upper boundary on the **useful magnification** that is possible with any given lens. In practice, the greatest useful magnification that can be achieved with a light microscope is approximately 1000 times the numerical aperture of the lens being used. Since numerical aperture ranges from approximately 1.0 to 1.4, this means that the useful magnification of a light microscope is limited to roughly 1000× in air and 1400× with immersion oil. Magnification greater than these limits is referred to as "empty magnification" because it provides no additional information about the object being studied.

The most effective way to achieve better magnification is to switch from visible light to electrons as the illumination source. Because the wavelength of an electron is approximately 100,000 times shorter than that of a photon of visible light, the theoretical limit of resolution of the electron microscope (0.002 nm) is orders of magnitude better than that of the light microscope (200 nm). However, practical problems in the design of the electromagnetic lenses used to focus the electron beam prevent the electron microscope from achieving this theoretical potential. The main problem is that electromagnets produce considerable distortion when the angular aperture is more than a few tenths of a degree. This tiny angle is several orders of magnitude less than that of a good glass lens (about 70°), giving the electron microscope a numerical aperture considerably smaller than that of the light microscope. The limit of resolution for the best electron microscope is therefore only approximately 0.2 nm, far from the theoretical limit of 0.002 nm. Moreover, when viewing biological samples, problems with specimen preparation and contrast are such that the practical limit of resolution is often closer to 2 nm. Practically speaking, therefore, resolution in an electron microscope is generally about 100 times better than that of the light microscope. As a result, the useful magnification of an electron microscope is approximately 100 times that of a light microscope, or approximately 100,000×.

The Light Microscope

It was the light microscope that first opened our eyes to the existence of cells. A pioneering name in the history of light microscopy is that of Antonie van Leeuwenhoek, the Dutch shopkeeper who is generally regarded as the father of light microscopy. Leeuwenhoek's lenses, which he manufactured himself during the late 1600s, were of surprisingly high quality for his time. They were capable of 300-fold magnification—a tenfold improvement over previous instruments. This improved magnification made the interior of cells visible for the first time, and Leeuwenhoek's observations over a period of more than 25 years led to the discovery of cells in various types of biological specimens and set the stage for the formulation of the cell theory.

Compound Microscopes Use Several Lenses in Combination

In the 300 years since Leeuwenhoek's pioneering work, considerable advances in the construction and application of light microscopes have been made. Today, the instrument of choice for light microscopy uses several lenses in combination and is therefore called a **compound microscope** (**Figure A-5**). The optical path through a compound microscope, illustrated in Figure A-5b, begins with a source of illumination, often a light source located in the base of the instrument. The light from the source first passes through **condenser lenses,** which direct the light toward a specimen mounted on a glass slide and positioned on the **stage** of the microscope. The **objective lens,** located immediately above the specimen, is responsible for forming the *primary image.* Most compound microscopes have several objective lenses of differing magnifications mounted on a rotatable turret.

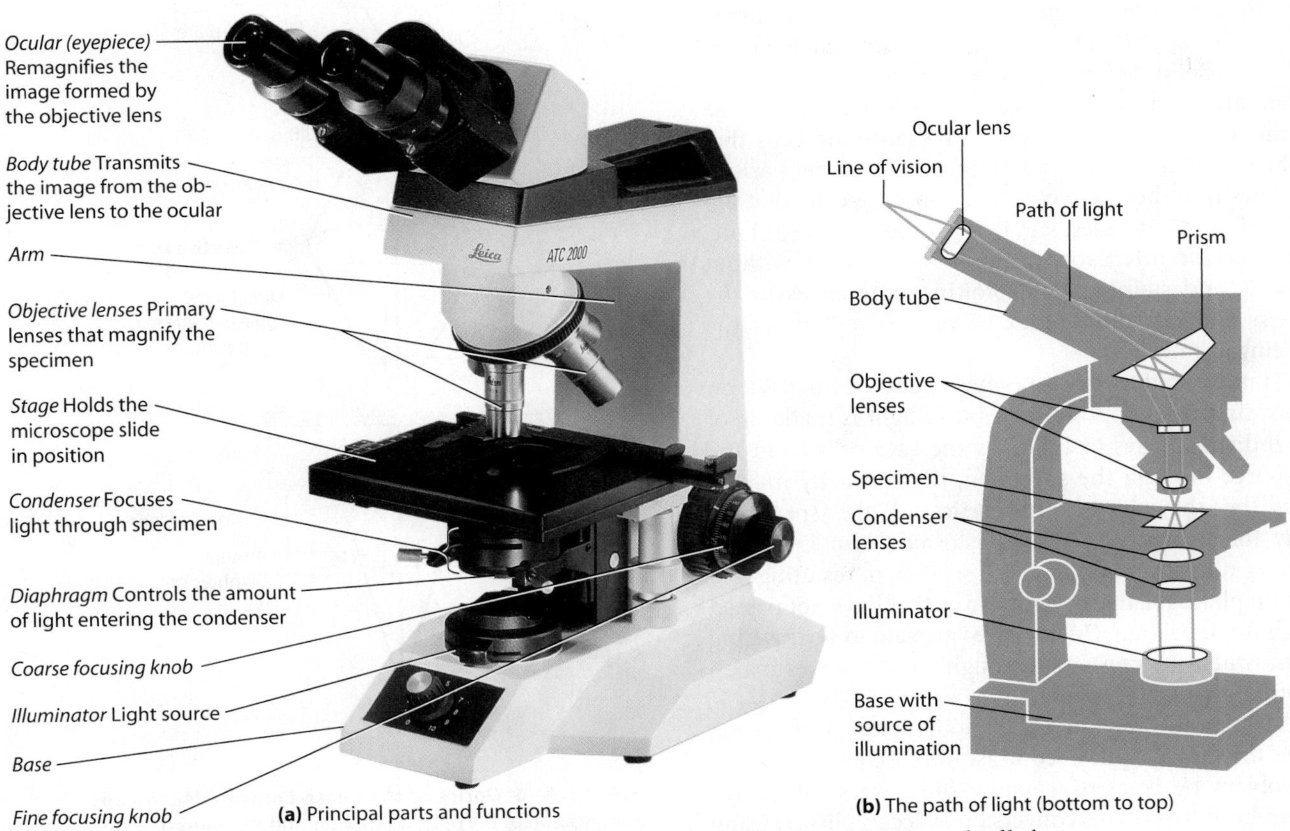

FIGURE A-5 The Compound Light Microscope. (a) A compound light microscope. (b) The path of light through the compound microscope.

The primary image is further enlarged by the **ocular lens,** or *eyepiece.* In some microscopes, an **intermediate lens** is positioned between the objective and ocular lenses to accomplish still further enlargement. Overall magnification of the image can be calculated by multiplying the enlarging powers of the objective lens, the ocular lens, and the intermediate lens (if present). Thus, a microscope with a 10× objective lens, a 2.5× intermediate lens, and a 10× ocular lens will magnify a specimen 250-fold.

The elements of the microscope described so far create a basic form of light microscopy called **brightfield microscopy.** Compared with other microscopes, the brightfield microscope is inexpensive and simple to align and use. However, the only specimens that can be seen directly by brightfield microscopy are those possessing color or some other property that affects the amount of light that passes through. Many biological specimens lack these characteristics and must therefore be stained with dyes or examined with specialized types of light microscopes. These special microscopes have various advantages that make them especially well suited for visualizing specific types of specimens. These include phase-contrast, differential interference contrast, fluorescence, and confocal microscopes. We will look at these and several other important techniques in the following sections.

Phase-Contrast Microscopy Detects Differences in Refractive Index and Thickness

As we will describe in more detail later, cells are often killed, sliced into thin sections, and stained before being examined by brightfield microscopy. While such procedures are useful for visualizing the details of a cell's internal architecture, little can be learned about the dynamic aspects of cell behavior by examining cells that have been killed, sliced, and stained. Therefore, various techniques have been developed to observe cells that are intact and, in many cases, still living. One such technique, **phase-contrast microscopy,** improves contrast without sectioning and staining by exploiting differences in the thickness and refractive index of various regions of the cells being examined.

To understand the basis of phase-contrast microscopy, we must first recognize that a beam of light is made up of many individual rays of light. As the rays pass from the light source through the specimen, their velocity may be affected by the physical properties of the specimen. Usually, the velocity of the rays is slowed down to varying extents by different regions of the specimen, resulting in a change in phase relative to light waves that have not passed through the specimen. (Light waves are said to be traveling *in phase* when the crests and troughs of the waves match each other.)

Although the human eye cannot detect such phase changes directly, the phase-contrast microscope overcomes this problem by converting phase differences into alterations in brightness. This conversion is accomplished using a *phase plate* (**Figure A-6**), which is an optical material inserted into the light path above the objective lens. On average, light passing through transparent specimens is retarded by approximately $\frac{1}{4}$ wavelength. The direct, undiffracted light passes through a portion of the phase plate that speeds it up by approximately $\frac{1}{4}$ wavelength. Now the two types of light interfere with one another, producing an image with highly contrasting bright and dark areas against an evenly illuminated background (**Figure A-7**). As a result, internal structures of cells are often better visualized by phase-contrast microscopy than with brightfield optics.

This approach to light microscopy is particularly useful for examining living, unstained specimens because biological materials almost inevitably diffract light. Phase-contrast microscopy is widely used in microbiology and tissue culture research to detect bacteria, cellular organelles, and other small entities in living specimens.

Differential Interference Contrast (DIC) Microscopy Utilizes a Split Light Beam to Detect Phase Differences

Differential interference contrast (DIC) microscopy, or *Nomarski* microscopy (named for its inventor), resembles phase-contrast microscopy in principle but is more sensitive because it employs a special prism to split the

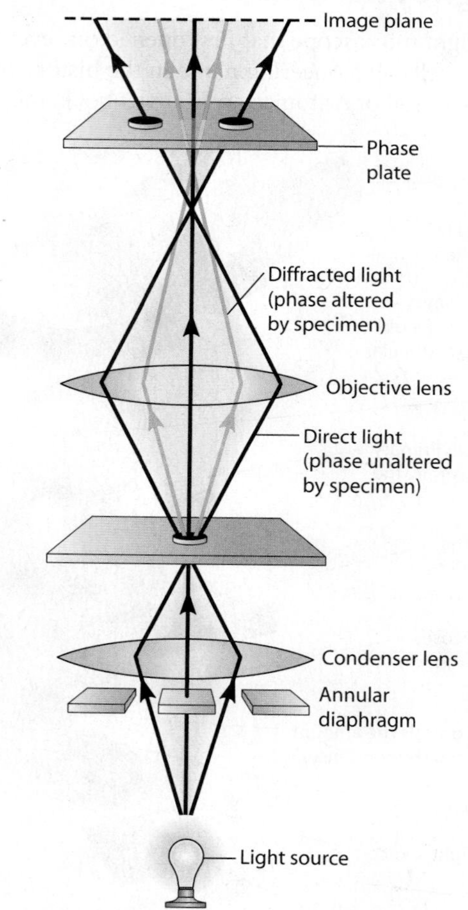

FIGURE A-6 Optics of the Phase-Contrast Microscope. Configuration of the optical elements and the paths of light rays through the phase-contrast microscope. Orange lines represent light diffracted by the specimen, and black lines represent direct light.

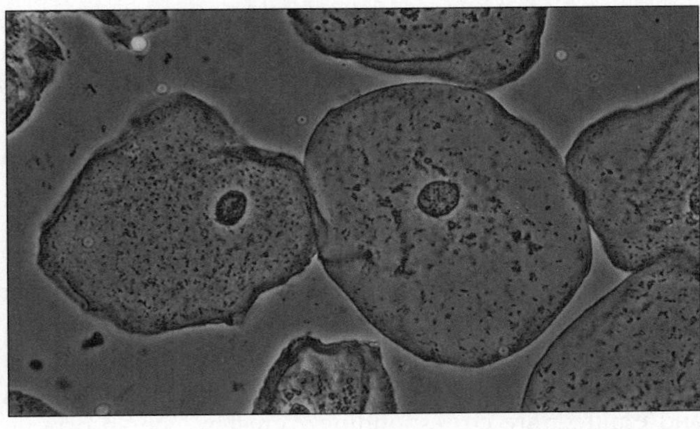

FIGURE A-7 Phase-Contrast Microscopy. A phase-contrast micrograph of epithelial cells. The cells were observed in an unprocessed and unstained state, which is a major advantage of phase-contrast microscopy.

illuminating light beam into two separate rays (**Figure A-8**). When the two beams are recombined, any changes that occurred in the phase of one beam as it passed through the specimen cause it to interfere with the second beam. Because the largest phase changes usually occur at cell edges (the refractive index is more constant within the cell), the outline of the cell typically gives a strong signal. The image appears three-dimensional due to a shadow-casting illusion that arises because differences in phase are positive on one side of the cell but negative on the opposite side of the cell (**Figure A-9**).

The optical components required for DIC microscopy consist of a *polarizer*, an *analyzer*, and a pair of *Wollaston prisms* (see Figure A-8). The polarizer and the first Wollaston prism split a beam of light, creating two beams that are separated by a small distance along one direction. After traveling through the specimen, the beams are recombined by the second Wollaston prism. If no specimen is present, the beams recombine to form one beam that is identical to the one that initially entered the polarizer and first Wollaston prism. In the presence of a specimen, the two beams do not recombine in the same way (i.e., they interfere with each other), and the resulting beam's polarization becomes rotated slightly compared with the original. The net effect is a remarkable enhancement in resolution that makes this technique especially useful for studying living, unstained specimens. As we will soon see, combining this technique with digital microscopy is an especially effective approach for studying dynamic events within cells as they take place.

Other contrast enhancement methods are also used by cell biologists. *Hoffman modulation contrast*, developed by Robert Hoffman, increases contrast by detecting optical gradients across a transparent specimen using special filters and a rotating polarizer. Hoffman modulation contrast results in a shadow-casting effect similar to that in DIC microscopy.

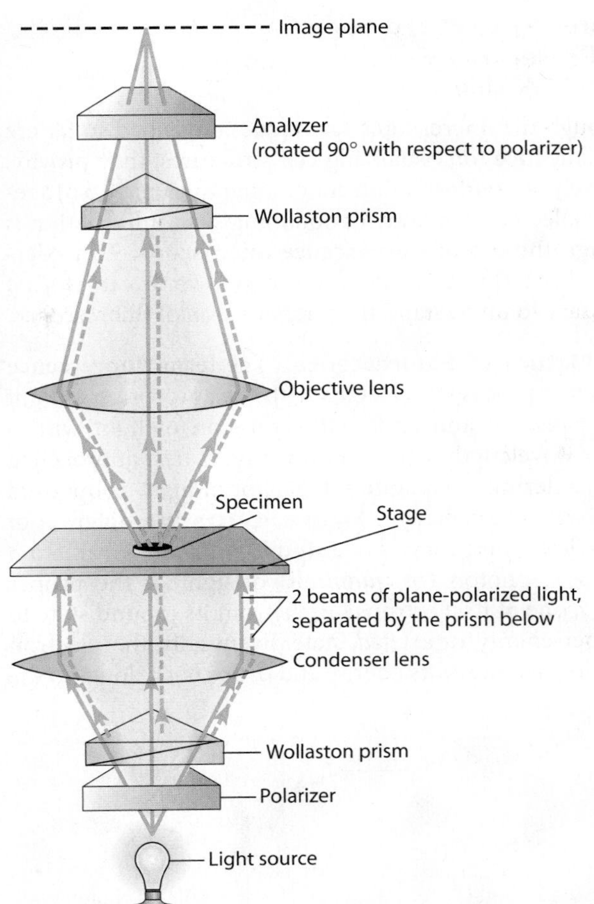

FIGURE A-8 Optics of the Differential Interference Contrast (DIC) Microscope. Configuration of the optical elements and the paths of light rays through the DIC microscope.

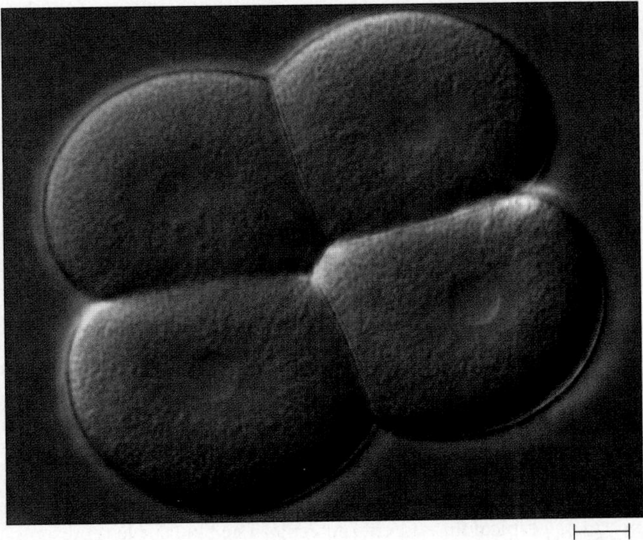

FIGURE A-9 DIC Microscopy. A DIC micrograph of a four-cell sea urchin embryo. Notice the shadow-casting effect that makes these cells appear dark at the bottom and light at the top.

Fluorescence Microscopy Can Detect the Presence of Specific Molecules or Ions Within Cells

Although the microscopic techniques described so far are quite effective for visualizing cell structures, they provide relatively little information concerning the location of specific molecules. One way of obtaining such information is through the use of **fluorescence microscopy.** To understand how fluorescence microscopy works, it is first necessary to understand the phenomenon of fluorescence.

The Nature of Fluorescence. The term **fluorescence** refers to a process that begins with the absorption of light by a molecule and ends with emission of light with a longer wavelength. This phenomenon is best approached by considering the quantum behavior of light, as opposed to its wavelike behavior. **Figure A-10a** is a diagram of the various energy levels of a simple atom. When an atom absorbs a photon (or *quantum*) of light of the proper energy, one of its electrons jumps from its ground state to a higher-energy, or *excited,* state. Eventually, this electron often loses some of its energy and drops back down to the original ground state, emitting another photon as it does so. The emitted photon is always of less energy (longer wavelength) than the original photon that was absorbed. Thus, for example, shining blue light on the atom may result in green light being emitted. (The energy of a photon is inversely proportional to its wavelength; therefore, green light, being longer in wavelength than blue light, is lower in energy.)

Real fluorescent molecules have energy diagrams that are more complicated than that depicted in Figure A-10a. The number of possible energy levels in real molecules is much greater, so the different energies that can be absorbed, and emitted, are correspondingly greater. The absorption and emission spectra of a typical fluorescent molecule are shown in Figure A-10b. Every fluorescent molecule has its own characteristic absorption and emission spectra.

The Fluorescence Microscope. Fluorescence microscopy is a specialized type of light microscopy that employs light to excite fluorescence in the specimen. A standard fluorescence microscope has an *excitation filter* between the light source and the rest of the light path that transmits only light of a particular wavelength (**Figure A-11**). A *dichroic mirror,* which reflects light below a certain wavelength and transmits light above a certain wavelength, deflects the incoming light toward the objective lens, which focuses the light onto the specimen (because

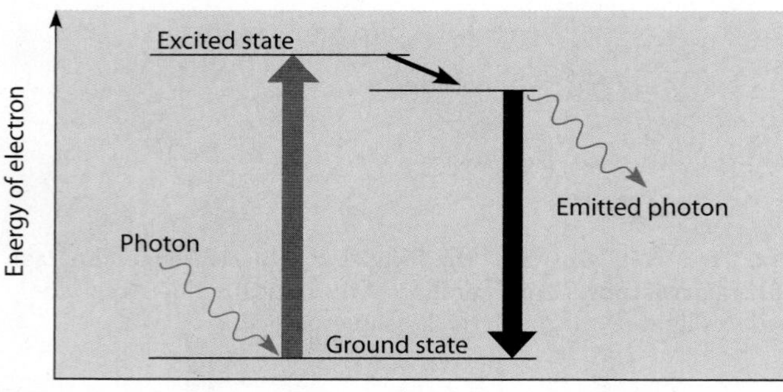

(a) Energy diagram

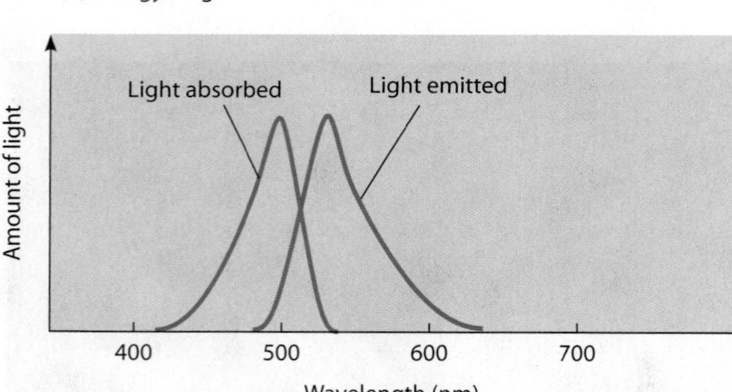

(b) Absorption and emission spectra

FIGURE A-10 Principles of Fluorescence. (a) An energy diagram of fluorescence from a simple atom. Light of a certain energy is absorbed (e.g., the blue light shown here). The electron jumps from its ground state to an excited state. It returns to the ground state by emitting a photon of lower energy and hence longer wavelength (e.g., green light). (b) The absorption and emission spectra of a typical fluorescent molecule. The blue curve represents the amount of light absorbed as a function of wavelength, and the green curve shows the amount of emitted light as a function of wavelength.

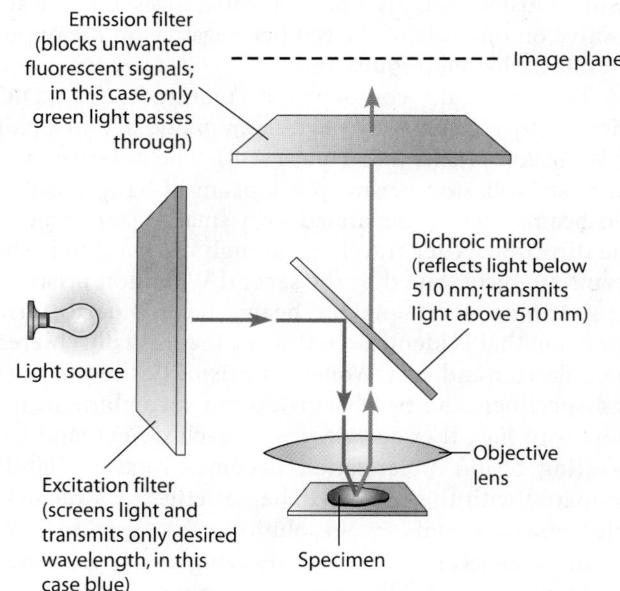

FIGURE A-11 Optics of the Fluorescence Microscope. Configuration of the optical elements and the paths of light rays through the fluorescence microscope. Light from the source passes through an excitation filter that transmits only excitation light (solid blue lines). Illumination of the specimen with this light induces fluorescent molecules in the specimen to emit longer-wavelength light (green lines). The emission filter subsequently removes the excitation light, while allowing passage of the emitted light. The image is therefore formed exclusively by light emitted by fluorescent molecules in the specimen.

the light is routed through the objective lens onto the specimen in this way, this technique is often called *epifluorescence microscopy*, from the Greek *epi-*, "upon"). The incoming light causes fluorescent compounds in the specimen to emit light of longer wavelength, which passes back through the objective lens. The longer wavelength light, instead of being reflected, passes through the dichroic mirror and encounters an *emission filter* that specifically prevents light that does not match the emission wavelength of the fluorescent molecules of interest from exiting the microscope. This leaves only the emission wavelengths to form the final fluorescent image, which therefore appears bright against a dark background.

Fluorescent Antibodies. To use fluorescence microscopy for locating specific molecules or ions within cells, researchers must employ special indicator molecules called *fluorescent probes.* A fluorescent probe is a molecule capable of emitting fluorescent light that can be used to indicate the presence of a specific molecule or ion.

One of the most common applications of fluorescent probes is in **immunostaining,** a technique based on the ability of antibodies to recognize and bind to specific molecules. (The molecules that antibodies bind to are called *antigens.*) Antibodies can be generated in the laboratory by injecting a foreign protein or other macromolecule into an animal such as a rabbit or mouse. In this way, it is possible to produce antibodies that will bind selectively to virtually any protein that a scientist wishes to study. Antibodies are not directly visible using light microscopy, however, so they are linked to a fluorescent dye such as *fluorescein,* which emits green light, or to *rhodamine,* which emits red light. More recently, antibodies have been linked to *quantum dots*—tiny, light-emitting crystals that are chemically more stable than traditional dyes, can be tuned to very specific wavelengths, and have other useful properties. To identify the subcellular location of a specific protein, cells are simply stained with a fluorescent antibody directed against that protein. The location of the fluorescence is then detected by viewing the cells with light of the appropriate wavelength.

Immunofluorescence microscopy can be performed using antibodies that are directly labeled with a fluorescent dye (**Figure A-12a**). However, immunofluorescence microscopy is more commonly performed using **indirect immunofluorescence** (Figure A-12b). In indirect immunofluorescence, a tissue or cell is treated with an antibody that is not labeled with dye. This antibody, called the *primary antibody,* attaches to specific antigenic sites within the tissue or cell. A second type of antibody, called the *secondary antibody,* is then added. The secondary antibody is labeled with a fluorescent dye, and it attaches to the primary antibody. Because more than one primary antibody molecule can attach to an antigen and more than one secondary antibody molecule can attach to each primary antibody, more fluorescent molecules are concentrated near each molecule that we seek to detect. As a result, indirect immunofluorescence results in signal amplification, and it is much more sensitive than the use of a primary antibody alone. The method is "indirect" because it does not examine where antibodies are bound to antigens; technically, the fluorescence reflects where the secondary antibody is located. This, of course, provides an indirect measure of where the original molecule of interest is located.

By using different combinations of antibodies or other fluorescent probes, more than one molecule in a cell can be labeled at the same time. Different probes can be imaged using different combinations of fluorescent filters, and the different images can be combined to generate

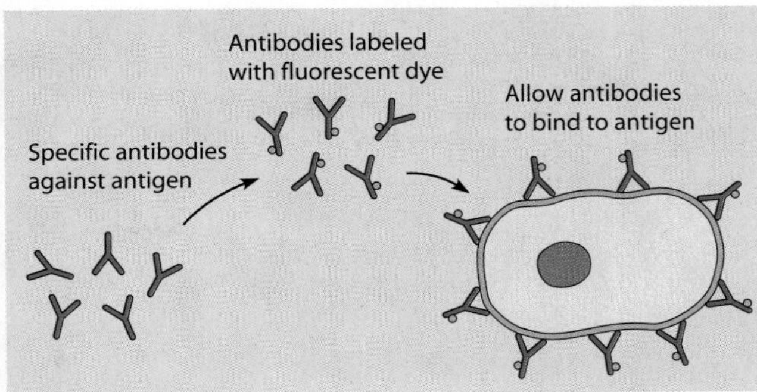

(a) Immunofluorescence

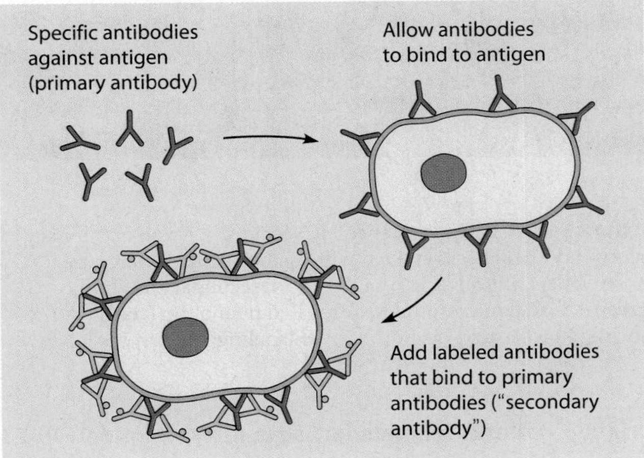

(b) Indirect immunofluorescence

FIGURE A-12 Immunostaining Using Fluorescent Antibodies. Immunofluorescence microscopy relies on the use of fluorescently labeled antibodies to detect specific molecular components (antigens) within a tissue sample. **(a)** In direct immunofluorescence, an antibody that binds to a molecular component in a tissue sample is labeled with a fluorescent dye. The labeled antibody is then added to the tissue sample, and it binds to the tissue in specific locations. The pattern of fluorescence that results is visualized using fluorescence or confocal microscopy. **(b)** In indirect immunofluorescence, a primary antibody is added to the tissue. Then a secondary antibody that carries a fluorescent label is added. The secondary antibody binds to the primary antibody. Because more than one fluorescent secondary antibody can bind to each primary antibody, indirect immunofluorescence effectively amplifies the fluorescent signal, making it more sensitive than direct immunofluorescence.

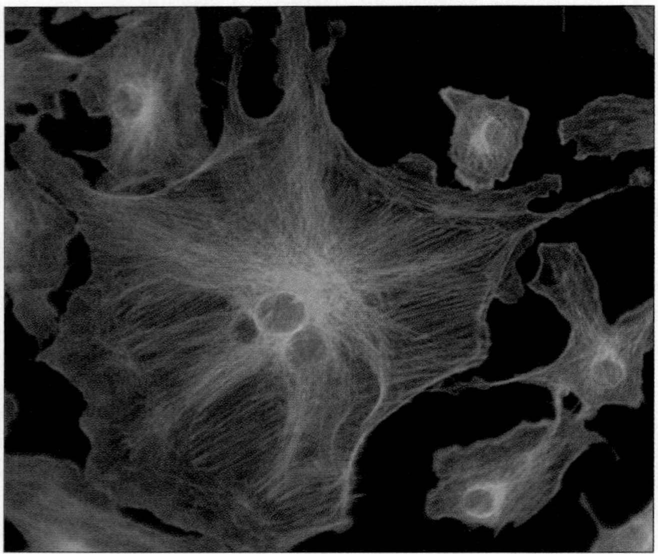

FIGURE A-13 Fluorescence Microscopy. Bovine pulmonary artery endothelial cells stained with an anti–β-tubulin mouse monoclonal antibody and a BODIPY FL secondary antibody (green) to label microtubules, Texas Red phalloidin (red) for labeling F-actin, and DAPI (blue) for labeling DNA in nuclei.

striking pictures of cellular structures. **Figure A-13** shows one example of this approach, in which endothelial cells are labeled with antibodies against β-tubulin (green), actin (red), and DNA (blue).

Other Fluorescent Probes. Naturally occurring proteins that selectively bind to specific cell components are also used in fluorescence microscopy. For example, the red structures in Figure A-13 are stained with a Texas Red-tagged mushroom toxin, *phalloidin,* which binds specifically to actin microfilaments. Another powerful fluorescence technique utilizes the **green fluorescent protein (GFP),** a naturally fluorescent protein made by the jellyfish *Aequoria victoria.* Using recombinant DNA techniques, scientists can fuse DNA encoding GFP to a gene coding for a particular cellular protein. The resulting recombinant DNA can then be introduced into cells, where it is expressed to produce a fluorescently tagged version of the normal cellular protein. In many cases, the fusion of GFP to the end of a protein does not interfere with its function, allowing the use of fluorescence microscopy to view the GFP-fusion protein as it functions in a living cell (**Figure A-14**). Molecular biologists have produced mutated forms of GFP that absorb and emit light at a variety of wavelengths. Other naturally fluorescent proteins have also been identified, such as a red fluorescent protein from coral. These tools have expanded the repertoire of fluorescent molecules at the disposal of cell biologists.

Besides detecting macromolecules such as proteins, fluorescence microscopy can be used to monitor the subcellular distribution of various ions. To accomplish this task, chemists have synthesized molecules whose fluorescent properties are sensitive to the concentrations of ions such as Ca^{2+}, H^+, Na^+, Zn^{2+}, and Mg^{2+}, as well as to the electrical potential across the plasma membrane or the membranes of organelles. For example, a fluorescent probe called *fura-2* is commonly used to track the Ca^{2+} concentration inside living cells, because fura-2 emits a yellow fluorescence in the presence of low concentrations of Ca^{2+} and a green and then blue fluorescence in the presence of progressively higher concentrations of this ion. Therefore, monitoring the color of the fluorescence in living cells stained with this probe allows scientists to observe changes in the intracellular Ca^{2+} concentration as they occur.

Confocal Microscopy Minimizes Blurring by Excluding Out-of-Focus Light from an Image

When biologists use fluorescence microscopy to view intact cells, the resolution is limited. This occurs because, although fluorescence is emitted throughout the entire depth of the specimen, the viewer can focus the objective lens on only a single plane at any given time. As a result, light emitted from regions of the specimen above and below the focal plane causes a blurring of the image

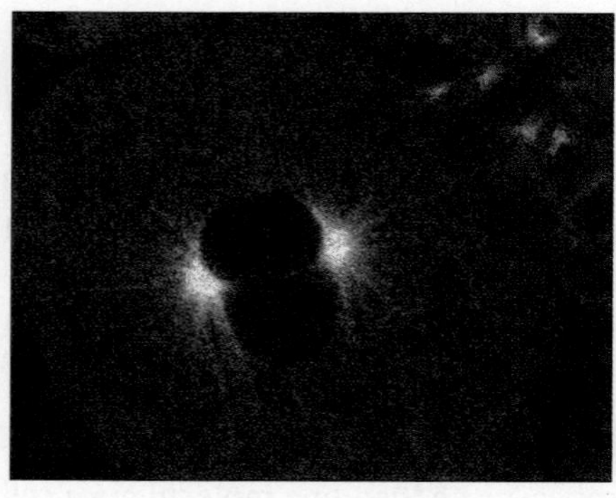
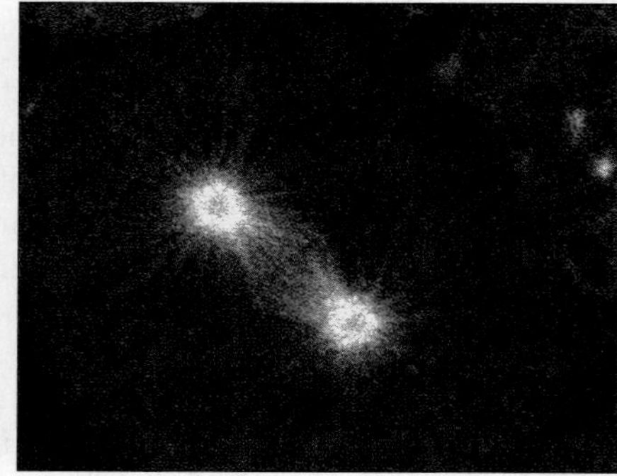
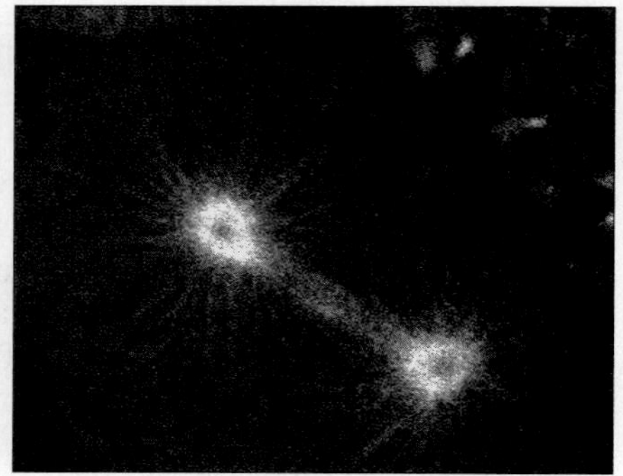

MB Use of GFP Fusions for Protein Localization

FIGURE A-14 Using Green Fluorescent Protein to Visualize Proteins. An image series of a living, one-cell nematode worm embryo undergoing mitosis. The embryo is expressing β-tubulin that is tagged with green fluorescent protein (GFP). Elapsed time from the first frame is shown in minutes:seconds.

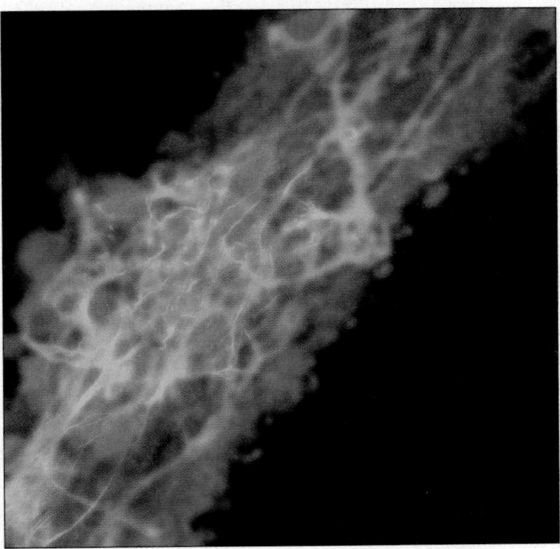

(a) Traditional fluorescence microscopy

(b) Confocal fluorescence microscopy

FIGURE A-15 Comparison of Confocal Fluorescence Microscopy with Traditional Fluorescence Microscopy. These fluorescence micrographs show fluorescently labeled glial cells (red) and nerve cells (green) stained with two different fluorescent markers. **(a)** In traditional fluorescence microscopy, the entire specimen is illuminated, so fluorescent material above and below the plane of focus tends to blur the image. **(b)** In confocal fluorescence microscopy, incoming light is focused on a single plane, and out-of-focus fluorescence from the specimen is excluded. The resulting image is therefore much sharper.

(**Figure A-15a**). To overcome this problem, cell biologists often turn to the **confocal microscope**—a specialized type of light microscope that employs a laser beam to produce an image of a single plane of the specimen at a time (Figure A-15b). This approach improves the resolution along the optical axis of the microscope—that is, structures in the middle of a cell may be distinguished from those on the top or bottom. Likewise, a cell in the middle of a piece of tissue can be distinguished from cells above or below it.

To understand this type of microscopy, it is first necessary to consider the paths of light taken through a simple lens. **Figure A-16** illustrates how a simple lens forms an image of a point source of light. To understand what your eye would see, imagine placing a piece of photographic film in the plane of focus (image plane). Now ask how the images of other points of light placed further away or closer to the lens contribute to the original image (Figure A-16b). As you might guess, there is a precise relationship between the distance of the object from the lens (*o*), the distance from the lens to the image of that object brought into focus (*i*), and the focal length of the lens (*f*). This relationship is given by the equation

$$\frac{1}{f} = \frac{1}{o} + \frac{1}{i} \tag{A-4}$$

As Figure A-16b shows, light arising from the points that are not in focus covers a greater surface area on the film because the rays are still either converging or diverging. Thus, the image on the film now has the original point source that is in focus, with a superimposed halo of light from the out-of-focus objects.

If we were interested only in seeing the original point source, we could mask out the extraneous light by placing an aperture, or *pinhole,* in the same plane as the film. This principle is used in a confocal microscope to discriminate against out-of-focus rays. In a real specimen, of course, we have more than one extraneous source of light on each side of the object we wish to see; in fact, we have a continuum of points. To understand how this affects our image, imagine that instead of three points of light, our specimen consists of a long, thin tube of light, as in Figure A-16c. Now consider obtaining an image of some arbitrary small section, *dx*. If the tube sends out the same amount of light per unit length, then even with a pinhole, the image of interest will be obscured by the halos arising from other parts of the tube. This occurs because there is a small contribution from each out-of-focus section, and the sheer number of small sections will create a large background over the section of interest.

This situation is very close to the one we face when dealing with real biological samples that have been stained with a fluorescent probe. In general, the distribution of the probe is three-dimensional, and the image is often marred by the halo of background light that arises mostly from probes above and below the plane of interest. To circumvent this, we can preferentially illuminate the plane of interest, thereby biasing the contributions in the image plane so that they arise mostly from a single plane (Figure A-16d). Thus, the essence of confocal microscopy is to bring the illumination beam that excites the fluorescence into focus in a single plane, and to use a pinhole to ensure that the light we collect in the image plane arises mainly from that plane of focus.

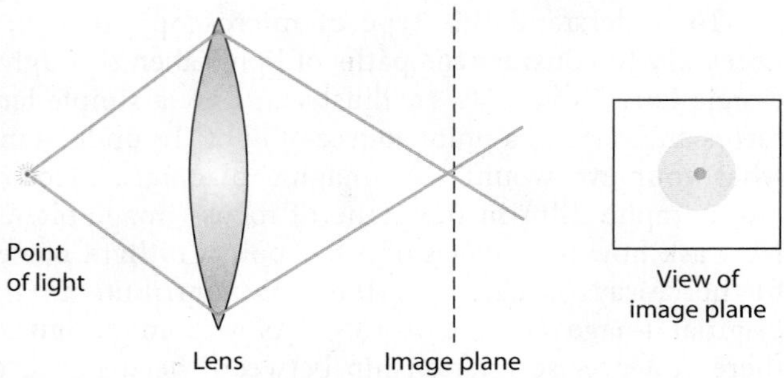

(a) Formation of an image of a single point of light by a lens

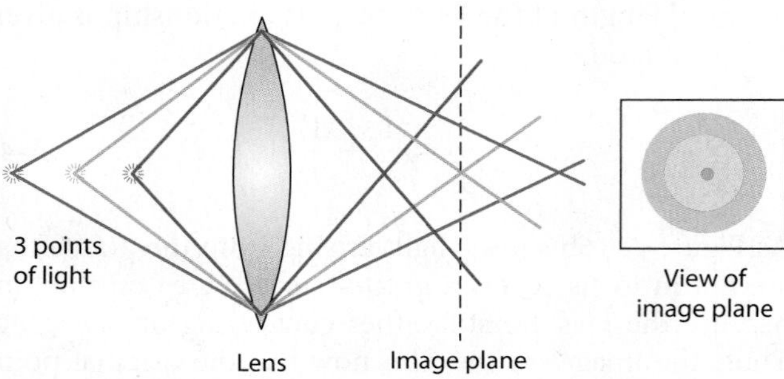

(b) Formation of an image of a point of light in the presence of two other points

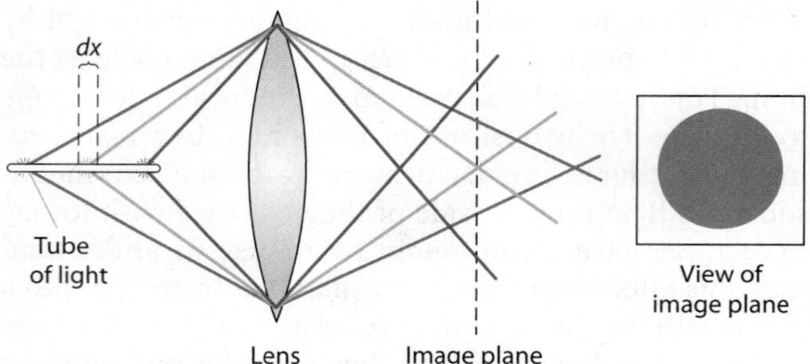

(c) Formation of an image of a section of an equally bright tube of light

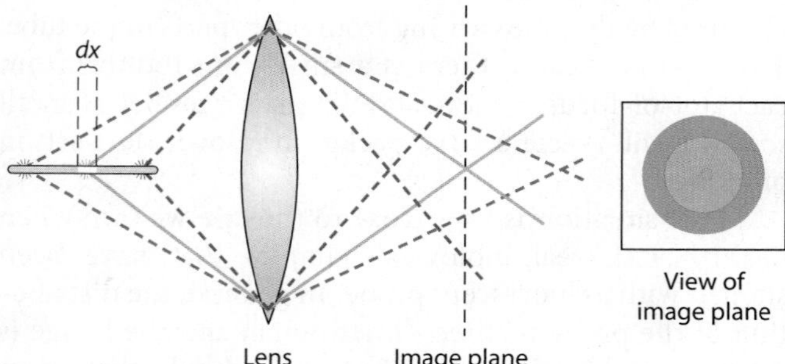

(d) Formation of an image of a brightened section of a tube of light

FIGURE A-16 Paths of Light Through a Single Lens. (a) The image of a single point of light formed by a lens. **(b)** The paths of light from three points of light at different distances from the lens. In the image plane, the in-focus image of the central point is superimposed on the out-of-focus rays of the other points. A pinhole or aperture around the central point can be used to discriminate against out-of-focus rays and maximize the contributions from the central point. **(c)** The paths of light originating from a continuum of points, represented as a tube of light. This is similar to a uniformly illuminated sample. In the image plane, the contributions from an arbitrarily small in-focus section, *dx,* are completely obscured by the other out-of-focus rays; here a pinhole does not help. **(d)** By illuminating only a single section of the tube strongly and the rest weakly, we can recover information in the image plane about the section *dx.* Now a pinhole placed around the spot will reject out-of-focus rays. Because the rays in the middle are almost all from *dx,* we have a means of discriminating against the dimmer, out-of-focus points.

Figure A-17 illustrates how these principles are put to work in a laser scanning confocal microscope, which illuminates specimens using a laser beam focused by an objective lens down to a diffraction-limited spot. The position of the spot is controlled by scanning mirrors, which allow the beam to be swept over the specimen in a precise pattern. As the beam is scanned over the specimen, an image of the specimen is formed in the following way. First, the fluorescent light emitted by the specimen is collected by the objective lens and returned along the same path as the original incoming light. The path of the fluorescent light is then separated from the laser light using a dichroic mirror, which reflects one color but transmits another. Because the fluorescent light has a longer wavelength than the excitation beam, the fluorescence color is shifted (for example, from blue to green). The fluorescent light passes through a pinhole placed at an image plane in front of a photomultiplier tube, which acts as a detector. The signal from the photomultiplier tube is then digitized and displayed by a computer. To see the enhanced resolution that results from confocal microscopy, look back to Figure A-15, which shows images of the same cell visualized by conventional fluorescence microscopy and by laser scanning confocal microscopy.

As an alternative to laser scanning confocal microscopy, a *spinning disc confocal microscope* uses rapidly spinning discs containing a series of small lenses and a corresponding series of pinholes. Although it cannot produce optical sections as thin as those produced by laser scanning microscopes, it can generate confocal images that can be acquired rapidly using sensitive digital cameras. Such speed is useful for visualizing very rapid events within living cells.

In confocal microscopy, a pinhole is used to exclude out-of-focus light. The result is a sharp image, but molecules above and below the focal plane of the objective lens are still being excited by the incoming light. This can result in rapid bleaching of the fluorescent molecules. In some cases, especially when viewing living cells that contain fluorescent molecules, such bleaching releases toxic radicals that can cause the cells to die. To reduce such "photodamage," it would be desirable if only the fluorescent molecules very close to the focal plane being examined were excited. This is possible using **multiphoton excitation microscopy,** in which a laser that rapidly emits pulses of high-amplitude light is used to irradiate the specimen. When two (or in some cases, three or more) photons arrive at the specimen almost simultaneously,

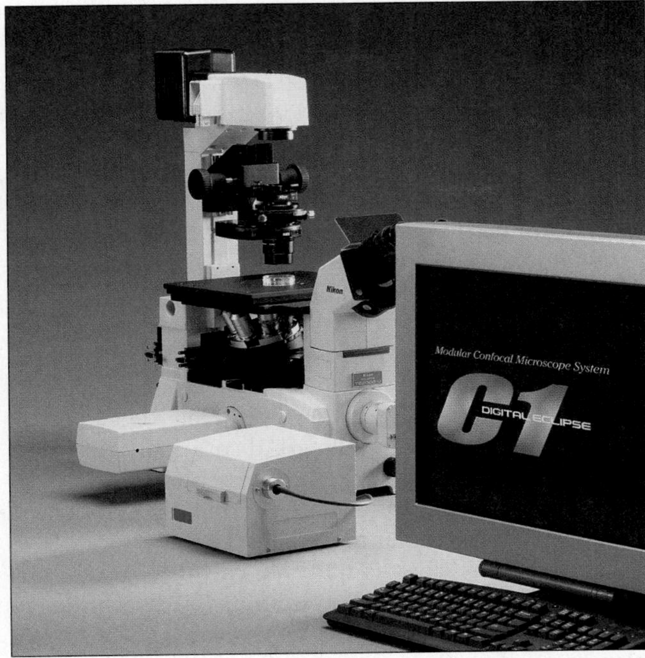

(a)

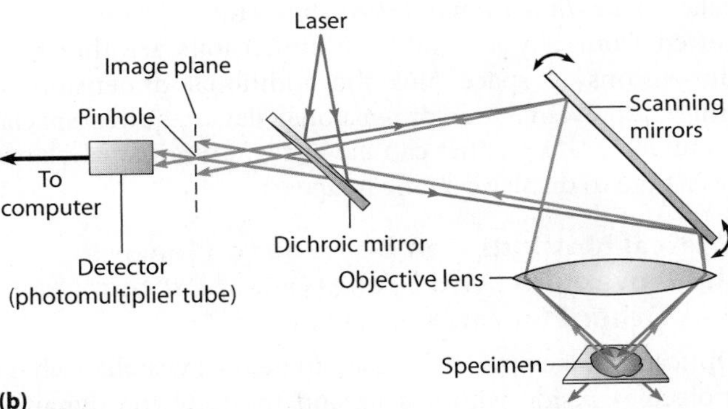

(b)

FIGURE A-17 A Laser Scanning Confocal Microscope. (a) A photograph and (b) a schematic of a laser scanning confocal microscope (LSCM). A laser is used to illuminate one spot at a time in the specimen (blue lines). Scanning mirrors move the spot in a given plane of focus through a precise pattern. The fluorescent light being emitted from the specimen (green lines) bounces off the same scanning mirrors and returns along the original path of the illumination beam. The emitted light does not return to the laser, but instead is transmitted through the dichroic mirror (which in this example reflects blue light but transmits green light). A pinhole in the image plane blocks the extraneous rays that are out of focus. The light is detected by a photomultiplier tube, whose signal is digitized and stored by a computer.

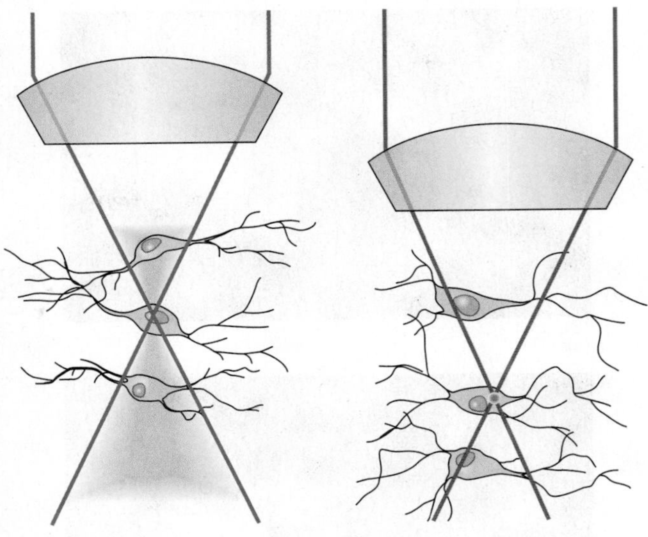

(a) Confocal microscopy (b) Multiphoton microscopy

FIGURE A-18 Multiphoton Excitation Microscopy. (a) In a standard LSCM, the laser results in fluorescence in an hourglass-shaped path throughout the specimen. Because a large area fluoresces, photodamage is much more likely to occur than in multiphoton excitation microscopy. (b) In a multiphoton excitation microscope, fluorescence is limited to a spot at the focus of the pulsed infrared laser beam, resulting in much less damage. The infrared illumination also penetrates more deeply into the specimen than visible light does.

the physics of light indicates that these photons can combine to effectively approximate a photon of shorter wavelength. When this approximate wavelength is near the absorption wavelength for a fluorescent molecule, the fluorescent molecule absorbs the light and fluoresces (**Figure A-18**). The likelihood of this happening is very low, except near the focal plane of the objective lens. As a result, only the fluorescent molecules that are in focus fluoresce. The result is very similar in sharpness to confocal microscopy, but no pinhole is needed since there is no out-of-focus light that needs to be excluded. Photodamage is also dramatically reduced. As an example, multiphoton excitation microscopy was used to image the living embryo shown in Figure A-14.

A third technique, **digital deconvolution microscopy,** can be used to provide very sharp images. Digital deconvolution relies on a completely different principle. In this case, normal fluorescence microscopy is used to acquire a series of images throughout the thickness of a specimen. Then a computer is used to digitally process, or *deconvolve,* each focal plane to mathematically remove the contribution due to the out-of-focus light. In many cases, digital deconvolution can produce images comparable to those obtained by confocal microscopy (**Figure A-19**). One advantage of deconvolution is that the microscope is not restricted to the specific wavelengths of light used in the lasers commonly found in confocal microscopes.

Digital Microscopy Can Record Enhanced Time-Lapse Images

The advent of solid-state light detectors has, in many circumstances, made it possible to replace photographic film with an electronic equivalent—that is, with a video camera or digital imaging camera. These developments have given rise to the technique of **digital microscopy,** in which microscopic images are recorded and stored electronically by placing a video or digital camera in the image plane produced by the ocular lens. The resulting digital

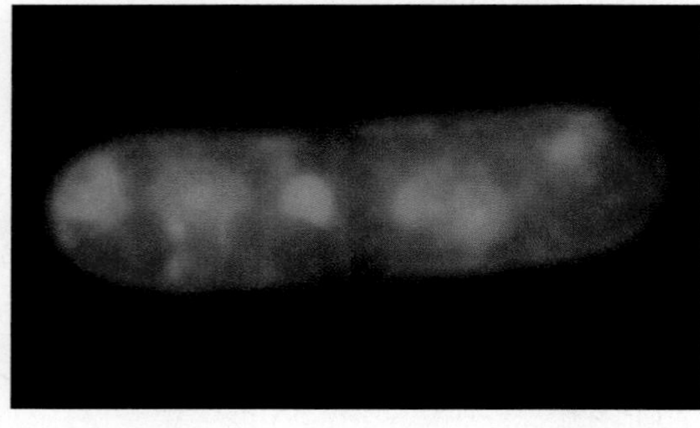

(a)

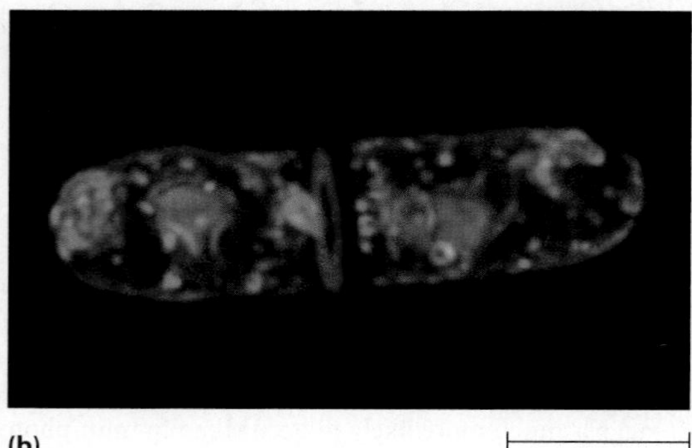

(b)

FIGURE A-19 Digital Deconvolution Microscopy. A fission yeast cell stained with a dye specific for DNA (red) and a membrane-specific dye (green). The image in **(a)** is an unprocessed optical section through the center of the cell, while the image in **(b)** is a projection of all the sections following three-dimensional image processing. The ring of the developing medial septum (red) is forming between the two nuclei (red) that arose by nuclear division during the previous mitosis.

images can then be *enhanced* by a computer to increase contrast and remove background features that obscure the image of interest (**Figure A-20**).

The resulting enhancement allows the visualization of structures that are an order of magnitude smaller than those that can be seen with a conventional light microscope. Digital techniques can be applied to conventional brightfield light microscopy, as well as to DIC and fluorescence microscopy, thereby creating a powerful set of approaches for improving the effectiveness of light microscopy.

An additional advantage of digital microscopy is that the specimen does not need to be killed by fixation, as is required with electron microscopy, so dynamic events can be monitored as they take place. Moreover, special, sensitive cameras have been developed that can detect extremely dim images, thereby facilitating the ability to record a rapid series of time-lapse pictures of cellular events as they proceed. This technique has allowed scientists to obtain information on the changes in concentration and subcellular distribution of such cytosolic components as second messengers during cellular signaling and to study the role of cytoskeletal structures in intracellular movements.

Digital microscopy is not only useful for examining events in one focal plane. In a variation of this technique, a computer is used to control a focus motor attached to a microscope. Images are then collected throughout the thickness of a specimen. When such a series of images is collected at specific time intervals, such microscopy is called *four-dimensional microscopy* (this phrase is borrowed from physics; the four dimensions are the three dimensions of space plus the additional dimension of time). Analyzing four-dimensional data requires special computer software that can navigate between focal planes over time to display specific images.

Optical Methods Can Be Used to Measure the Movements and Properties of Proteins and Other Macromolecules

Optical microscopy can be used to help us visualize where molecules reside within cells and to study the dynamic movements and properties of biological molecules. We briefly consider these modern techniques in this section.

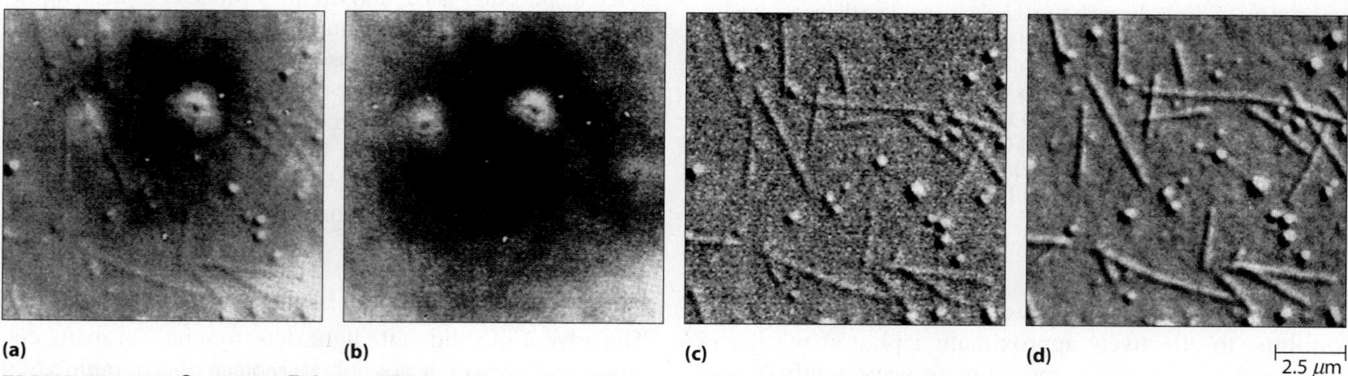

FIGURE A-20 Computer-Enhanced Digital Microscopy. This series of micrographs shows how computers can be used to enhance images obtained with light microscopy. In this example, an image of several microtubules—too small to be seen with unenhanced light microscopy—are processed to make them visible in detail. **(a)** The image resulting from electronic contrast enhancement of the original image (which appeared to be empty). **(b)** The background of the enhanced image in (a), which is then **(c)** subtracted from image (a), leaving only the microtubules. **(d)** The final, detailed image resulting from electronic averaging of the separate images processed as shown in parts a–c.

Photobleaching, Photoactivation, and Photoconversion. When fluorescent molecules are irradiated with light at the appropriate excitation wavelength for long periods of time, they undergo **photobleaching** (i.e., the irradiation induces the molecules to cease fluorescing). If a cell is exposed to intense light in only a small region, such bleaching results in a characteristic decrease in fluorescence (**Figure A-21a**). As unbleached molecules move into the bleach zone, the fluorescence gradually returns to normal levels. Such *fluorescence recovery after photobleaching (FRAP)* is therefore one useful measure of how fast molecules diffuse or undergo directed transport (see Figure 7-11 for a classic example of the use of FRAP).

In other cases, fluorescent molecules are chemically modified so that they do not fluoresce until they are irradiated with a specific wavelength of light, usually ultraviolet light. The UV light induces the release of the chemical modifier, allowing the molecule of interest to fluoresce. Both chemically modified dyes and forms of GFP have been produced that behave in this way. Such compounds are often called *caged compounds* because they are "freed" only by this light-induced cleavage. Uncaging, or **photoactivation,** produces the converse situation to photobleaching: Local uncaging of a fluorescent molecule produces a bright spot of fluorescence that can be followed as the molecules diffuse throughout the cell (Figure A-21b). Uncaging via photoactivation can also be used to convert other kinds of inert molecules to their active state. For example, "caged calcium" is actually a calcium chelator that is bound to calcium ions. When the caged compound is irradiated, it gives up its calcium, causing a local elevation of calcium within the cell. A third technique causes a fluorescent molecule to permanently change its fluorescence properties after it is exposed to ultraviolet or other short wavelengths of light. Such *photoconversion* can, for example, cause a green fluorescent protein to become a red fluorescent protein and has uses that are similar to photoactivation. Other fluorescent molecules can be temporarily switched "off" so that they do not fluoresce and then back "on" again, a process called *"photoswitching."*

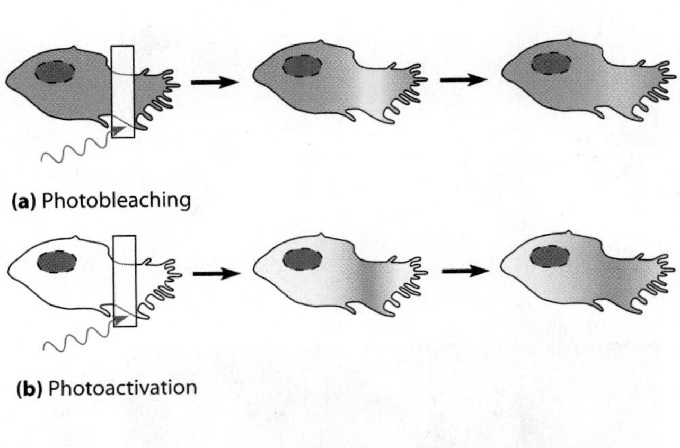

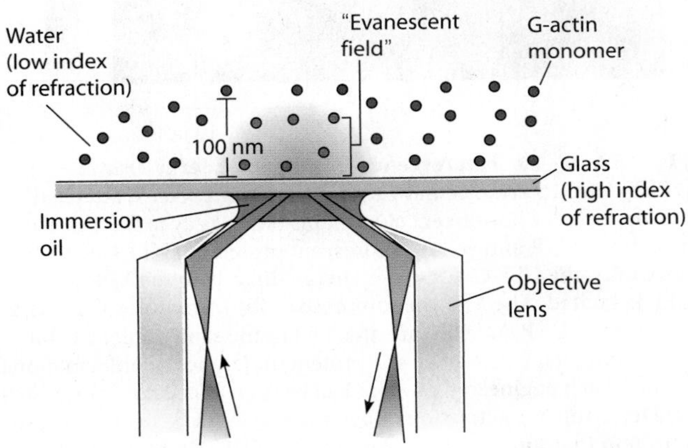

FIGURE A-21 Techniques for Studying Dynamic Movements of Molecules Within Cells. (a) Photobleaching of fluorescent molecules. A well-defined area of a cell is irradiated, causing bleaching of fluorescent molecules. Gradually, new, unbleached molecules invade the bleached zone. **(b)** Photoactivation of fluorescent proteins. Photoactivation of a selected region (indicated by the box) results in an increase in fluorescence. Subsequent movements of the fluorescent molecules can then be monitored. **(c)** Total internal reflection fluorescence (TIRF) microscopy. This procedure is often performed using an inverted microscope, which places the objective lens under the specimen. The example shown here involves fluorescent G-actin monomers (see Chapter 15). When a parallel beam of light in a medium of high refractive index (such as glass) strikes an interface with a medium of lower refractive index (such as a cell or water) at an angle that exceeds the critical angle, it undergoes total internal reflection. Total internal reflection results in an "evanescent field" in the medium of lower refractive index, which falls off quickly with distance. This allows only a very small number of fluorescent molecules to be seen.

Total Internal Reflection Fluorescence Microscopy. An even more spatially precise method for observing fluorescent molecules involves **total internal reflection fluorescence (TIRF) microscopy.** TIRF relies on a useful property of light: When light moves from a medium with a high refractive index (such as glass) to a medium with a lower refractive index (such as water or a cell), if the angle of incidence of the light exceeds a certain angle (called the *critical angle*), the light is reflected. You may be familiar with fiber-optic cables, which rely on the same physical principle. The curvature of the fiber causes almost all of the light that passes into the cable to be internally reflected, allowing such fibers to serve as "light pipes." Microscopists can use special lenses to exploit total internal reflection as well. When fluorescent light shines on a cell such that all of the light exceeds the critical angle, it will undergo total internal reflection. If all of the light is reflected, why is TIRF useful? It turns out that a very small layer of light, called the "evanescent field," extends into the water or cell (Figure A-21c). This layer is very thin, about 100 nm, making TIRF up to ten times better than a confocal microscope for resolving small objects very close to the surface of a coverslip. This makes TIRF extremely useful for studying the release of secretory vesicles or for observing the polymerization of actin.

Fluorescence Resonance Energy Transfer. Fluorescence microscopy is useful not only for observing the movements of proteins but also for measuring their physical interactions with one another. When two fluorescent molecules whose fluorescence properties are matched are brought very close to one another, it is possible for them to experience **fluorescence resonance energy transfer (FRET).** In FRET, illuminating a cell at the excitation wavelength of the first, or *donor,* fluorophore results in energy transfer to the second, or *acceptor,* fluorophore. This energy transfer does not itself involve a photon of light; however, once it occurs, it causes the acceptor to emit light at its characteristic wavelength. A commonly used pair of fluorescent molecules used in FRET is derived from GFP: One glows with bluish (cyan) fluorescence (so it is called *cyan fluorescent protein,* or *CFP*); the other is excited by bluish light and glows with a yellowish color (so it is called *yellow fluorescent protein,* or *YFP*). FRET can be measured between two separate proteins (*inter*molecular FRET; **Figure A-22a**), or between fluorescent side chains on the same protein (*intra*molecular FRET; Figure A-22b). FRET acts only at a very short range (100 Ångstroms or less); as a result, intermolecular FRET provides a readout regarding where within a cell the two fluorescent proteins are essentially touching one another.

Intramolecular FRET is used in several kinds of "molecular biosensors." We saw one example of such a biosensor in Chapter 14, when we considered "cameleons" (the spelling is correct; the name comes from the abbreviation for the protein calmodulin, CaM, part of which is used to make these types of sensors). These engineered proteins change shape to bring blue and yellow fluorescent protein subunits into contact only when calcium levels rise. The resulting FRET can be used to measure local calcium levels. Similar biosensors are being used to measure the local activation of small G proteins, such as Ras (Figure A-22c).

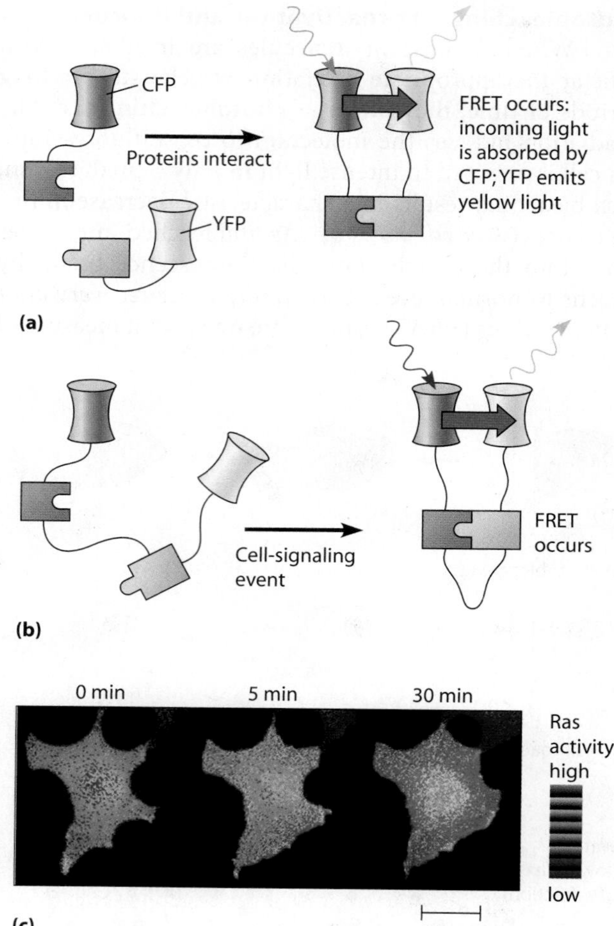

FIGURE A-22 Fluorescence Resonance Energy Transfer (FRET). **(a)** Intermolecular FRET. When two proteins attached to two different fluorescent side chains (such as cyan fluorescent protein, or CFP, and yellow fluorescent protein, or YFP) are in close proximity, the CFP can transfer energy directly to the YFP when the CFP is excited. The YFP then fluoresces. **(b)** Intramolecular FRET. In this case, CFP and YFP are attached to the same molecule, but they can interact only when the protein undergoes a conformational change. Such engineered proteins can be used as cellular "biosensors." **(c)** Detecting Ras activation in a living cell. A COS-1 cell expressing a protein that undergoes intramolecular FRET in regions where Ras is active shows an increase in FRET after the cell is exposed to epidermal growth factor (EGF), which activates Ras.

"Optical Tweezers." The final application of light microscopy we will discuss relies on well-established ways in which light interacts with small objects. When photons strike small objects, they exert "light pressure" on them. When a small object is highly curved, such as a small plastic bead, the differential forces exerted by a tightly focused laser beam channeled through a microscope tend to cause the bead to remain in the center of the beam, trapping it via light pressure. Such **optical tweezers** can be used to move objects, or they can be used to exert exceedingly small forces on beads that are attached to proteins or other molecules. By measuring the forces exerted on beads coupled to myosin, for example, scientists have been able to measure the forces produced by the power stroke of a single myosin protein (**Figure A-23**).

Superresolution Microscopy Has Broken the Diffraction Limit

The wavelength of visible light seems to place a limit on what can be resolved in the light microscope. We described this limit, due to diffraction, using the Abbé equation. Recent ingenious techniques, collectively called **superresolution microscopy,** have broken this "diffraction barrier." One of these techniques is called *stimulated emission depletion (STED) microscopy.* Like confocal microscopy, STED uses very short pulses of laser light to cause molecules in a specimen to fluoresce. The first pulse is immediately followed by a ring-shaped "depletion" pulse, which causes stimulated emission, moving electrons from the excited state (which causes fluorescence to occur) to a lower energy state before they can fluoresce. The interaction of the two pulses effectively results in a smaller spot size than could be achieved with a conventional microscope (**Figure A-24a**). Two other

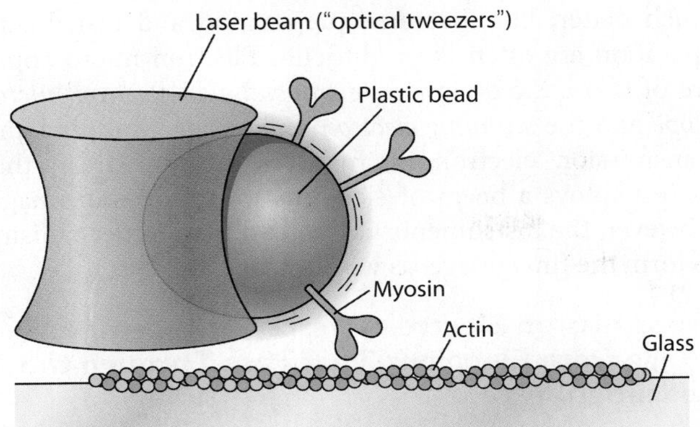

FIGURE A-23 "Optical Tweezers." A small plastic bead can be held in place by a laser beam focused through the objective lens of a microscope. When the bead is attached to a molecule such as myosin, the force produced by the myosin as it pulls on an actin filament attached to a microscope slide can be precisely measured.

techniques take a different approach. These techniques, called *photoactivated localization microscopy (PALM)* and *stochastic optical reconstruction microscopy (STORM)*, use photoactivatable or photoswitchable probes, respectively. A specimen containing these probes is scanned many times but in a way that the fluorescence of individual molecules against a dark background can be pinpointed with improved precision using computers. These calculations are then combined to generate an image (**Figure A-24b**). As technology continues to improve, STED, PALM, STORM, and other superresolution techniques will continue to push the limits of resolution of the light microscope.

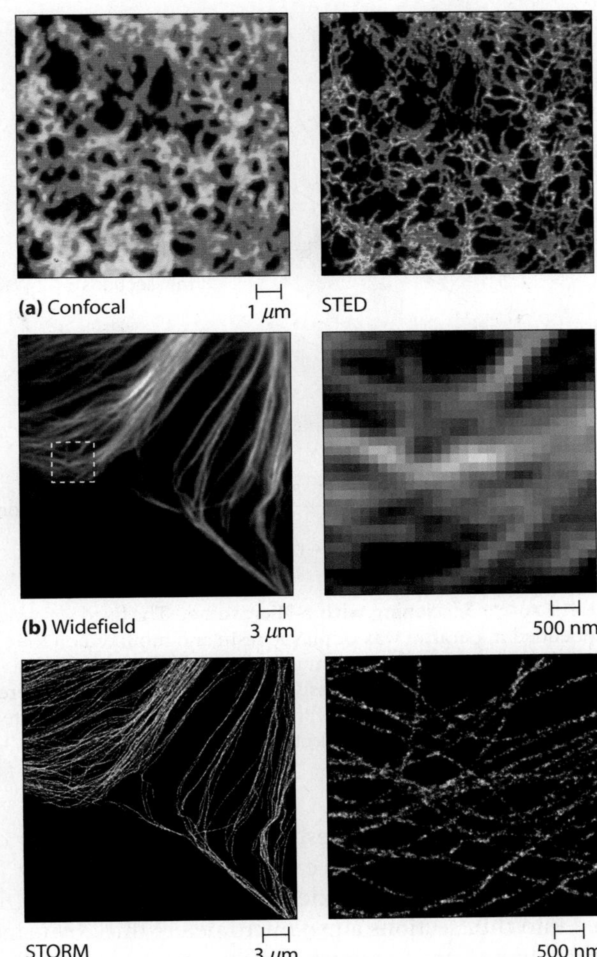

FIGURE A-24 Superresolution Microscopy Breaks the Diffraction Barrier. **(a)** A yellow fluorescent protein fused to a sequence that targets it to the endoplasmic reticulum was used to image the same specimen using confocal microscopy (left) and STED (right). **(b)** Comparison of immunostaining of microtubules using widefield (i.e., standard, non-confocal) fluorescence microscopy (top) and STORM (bottom). The right-hand panels zoom in on the boxed region. The zoomed confocal image, but not the STORM image, is blurry, because STORM imaging can identify fluorescent molecules with much higher spatial resolution.

Sample Preparation Techniques for Light Microscopy

One attractive feature of using light microscopy is how easily most specimens can be prepared for examination. In some cases, preparing a sample involves nothing more than mounting a small piece of the specimen in a suitable liquid on a glass slide and covering it with a glass coverslip. The slide is then positioned on the specimen stage of the microscope and examined through the ocular lens, or with a camera. However, to take maximum advantage of the resolving power of the light microscope, samples are usually prepared in a way designed to enhance *contrast*—that is, differences in the darkness or color of the structures being examined. A common means for enhancing contrast is to apply specific dyes that color or otherwise alter the light-transmitting properties of cell constituents.

Specimen Preparation Often Involves Fixation, Sectioning, and Staining

To prepare cells for staining, tissues are often first treated with **fixatives** that kill the cells while preserving their structural appearance. The most widely employed fixatives are acids and aldehydes such as acetic acid, picric acid, formaldehyde, and glutaraldehyde. One way of fixing tissues is simply to immerse them in the fixative solution. An alternative approach for animal tissues is to inject the fixative into the bloodstream of the animal before removing the organs—a technique known as *perfusion.*

In most cases, the next step is to slice the specimen into sections that are thin enough to transmit light. To prepare such *thin sections,* the specimen is embedded in a medium (such as plastic or paraffin wax) that can hold it rigidly in position while sections are cut. Since paraffin is insoluble in water, any water in the specimen must first be removed (by dehydration in alcohol, usually) and replaced by an organic solvent, such as xylene, in which paraffin is soluble. The processed tissue is then placed in warm, liquefied paraffin and allowed to harden. Dehydration is less critical if the specimen is embedded in a water-soluble medium instead of paraffin. Specimens may also be

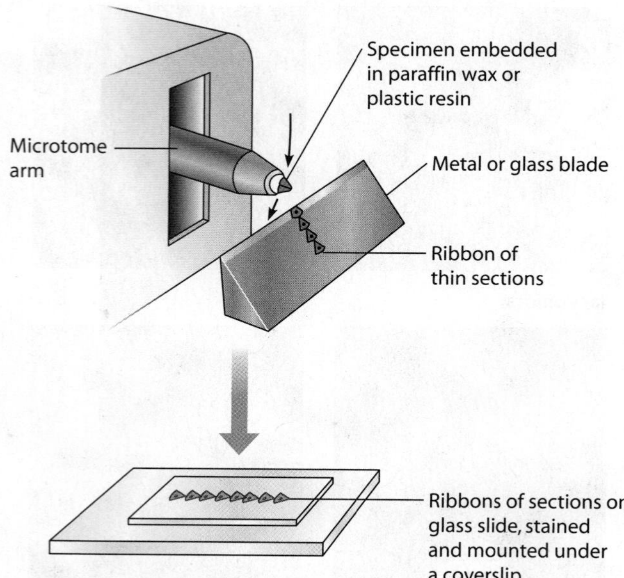

FIGURE A-25 Sectioning with a Microtome. The fixed specimen is embedded in paraffin wax or plastic resin and mounted on the arm of the microtome. As the arm moves up and down through a circular arc, the blade cuts successive sections. The sections adhere to each other, forming a ribbon of thin sections that can be mounted on a glass slide, stained, and protected with a coverslip.

embedded in epoxy plastic resin, or, as an alternative way of providing support, the tissue can simply be quick-frozen.

After embedding or quick-freezing, the specimen is sliced into thin sections a few micrometers thick by using a **microtome,** an instrument that operates somewhat like a meat slicer (**Figure A-25**). The specimen is simply mounted on the arm of the microtome, which advances the specimen by small increments toward a metal or glass blade that slices the tissue into thin sections. The sections are then mounted on a glass slide and subjected to **staining** with any of various dyes or antibodies that have been adapted for this purpose. Often a series of treatments are applied, each with an affinity for a different kind of cellular component. Once stained, the specimen is covered with a glass coverslip for protection.

A historically important approach for localizing specific components within cells is *microscopic autoradiography,* a technique that uses photographic emulsion to determine where a specific radioactive compound is located within a cell at the time the cell is fixed and sectioned for microscopy. When the emulsion is later developed and the specimen is examined under the microscope, *silver grains* appear directly above the specimen wherever radiation had bombarded the emulsion.

The Electron Microscope

The impact of electron microscopy on our understanding of cells can only be described as revolutionary. Yet, like light microscopy, electron microscopy has both strengths and weaknesses. In electron microscopy, resolution is much better; but specimen preparation and instrument operation are often more difficult. Electron microscopes are of two basic designs: the *transmission electron microscope* and the *scanning electron microscope.* Scanning and transmission electron microscopes are similar in that each employs a beam of electrons to produce an image. However, the instruments use quite different mechanisms to form the final image, as we see next.

Transmission Electron Microscopy Forms an Image from Electrons That Pass Through the Specimen

The **transmission electron microscope (TEM)** is so named because it forms an image from electrons that are *transmitted* through the specimen being examined. As shown in **Figure A-26**, most of the parts of the TEM are similar in name and function to their counterparts in the light microscope, although their physical orientation is reversed. We will look briefly at each of the major features.

The Vacuum System and Electron Gun. Because electrons cannot travel very far in air, a strong vacuum must be maintained along the entire path of the electron beam. Two types of vacuum pumps work together to create this vacuum. On some TEMs, a device called a *cold trap* is incorporated into the vacuum system to help establish a high vacuum. The cold trap is a metal insert in the column of the microscope that is cooled by liquid nitrogen. The cold trap attracts gases and random contaminating molecules, which then solidify on the cold metal surface.

The electron beam in a TEM is generated by an **electron gun,** an assembly of several components. The *cathode,* a tungsten filament similar to a light bulb filament, emits electrons from its surface when it is heated. The cathode tip is near a circular opening in a metal cylinder. A negative voltage on the cylinder helps control electron emission and shape the beam. At the other end of the cylinder is the *anode.* The anode is kept at 0 V, while the cathode is usually maintained at 50–100 kV. This difference in voltage causes the electrons to accelerate as they pass through the cylinder and hence is called the **accelerating voltage.**

Electromagnetic Lenses and Image Formation. The formation of an image using electron microscopy depends on both the wavelike and the particle-like properties of electrons. Because electrons are negatively charged particles, their movement can be altered by magnetic forces. This means that the trajectory of an electron beam can be controlled using electromagnets, just as a glass lens can bend rays of light that pass through it.

As the electron beam leaves the electron gun, it enters a series of electromagnetic lenses (Figure A-26b). Each lens is simply a space influenced by an electromagnetic field. The focal length of each lens can be increased or decreased by varying the amount of electric current

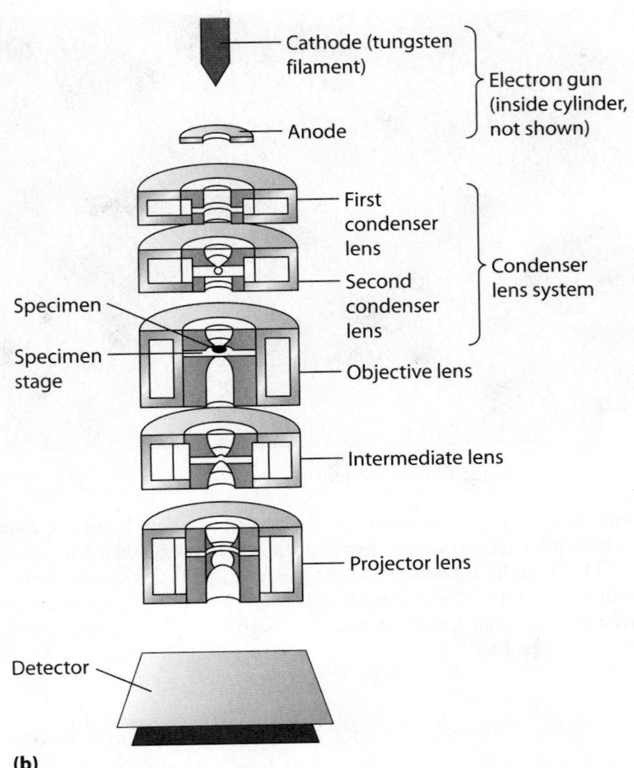

(a) (b)

FIGURE A-26 A Transmission Electron Microscope. (a) A photograph and (b) a schematic diagram of a TEM.

applied to its energizing coils. Thus, when several lenses are arranged together, they can control illumination, focus, and magnification.

The **condenser lens** is the first lens to affect the electron beam. It functions in the same fashion as its counterpart in the light microscope to focus the beam on the specimen. Most electron microscopes actually use a condenser lens system with two lenses to achieve better focus of the electron beam. The next component, the **objective lens,** is the most important part of the electron microscope's sophisticated lens system. The specimen is positioned on the specimen stage within the objective lens. The objective lens, in concert with the **intermediate lens** and the **projector lens,** produces a final image on a *viewing screen* that fluoresces when struck by electrons or that is produced directly by a detector.

How is an image formed from the action of these electromagnetic lenses on an electron beam? When the beam strikes the specimen, some electrons are scattered by the sample, whereas others continue in their paths relatively unimpeded. This scattering of electrons is the result of properties created in the specimen by preparation procedures we will describe shortly. Specimen preparation, in other words, imparts selective *electron density* to the specimen. That is, some areas become more opaque to electrons than others do. Electron-dense areas of the specimen will appear dark because few electrons pass through, whereas other areas will appear lighter because they permit the passage of more electrons.

The contrasting light, dark, and intermediate areas of the specimen create the final image. The image is formed by differing extents of electron transmission through the specimen, thus the name *transmission electron microscope.*

The Image Capture System. Since electrons are not visible to the human eye, the final image is detected in the transmission electron microscope by allowing the transmitted electrons to strike a fluorescent screen or photographic film. The use of film allows one to create a photographic print called an **electron micrograph,** which then becomes a permanent photographic record of the specimen (**Figure A-27a**). In many modern microscopes, a digital camera records the screen or a digital detector directly detects incoming electrons.

Voltage. An electron beam is too weak to penetrate very far into biological samples, so specimens examined by conventional transmission electron microscopy must be extremely thin (usually no more than 100 nm). Otherwise, the electrons will not be able to pass through the specimen, and the image will be entirely opaque. Examination of thicker sections requires a special **high-voltage electron microscope (HVEM),** which is similar to a transmission electron microscope but utilizes an accelerating voltage that is much higher—about 200–1000 kV compared with the 50–100 kV of a TEM. Because the penetrating power of the resulting electron beam is roughly ten times as great as that of conventional electron microscopes, relatively

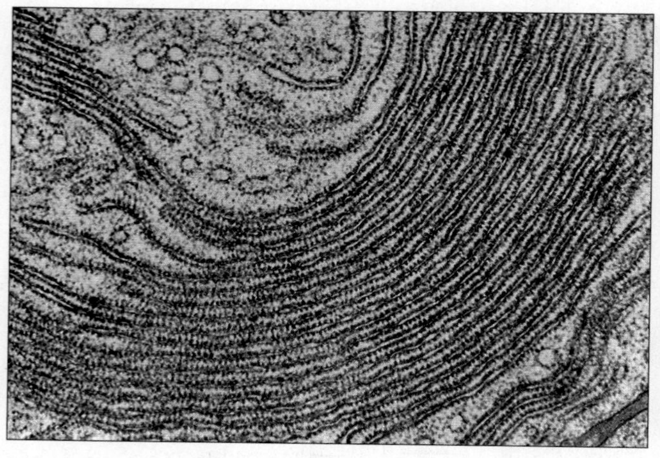

(a) Transmission electron micrograph

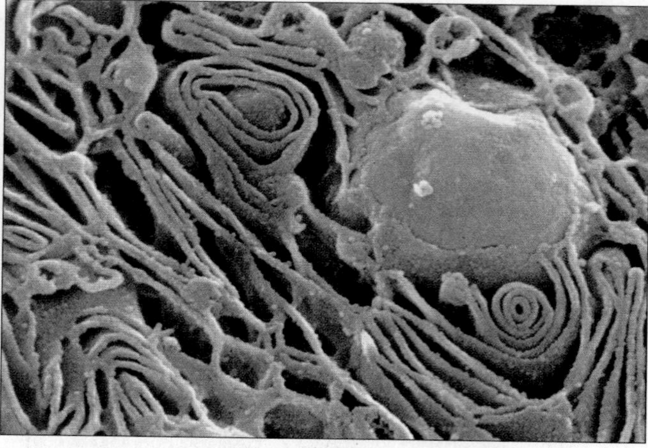

(b) Scanning electron micrograph

FIGURE A-27 Comparison of Transmission and Scanning Electron Micrographs. **(a)** The transmission electron micrograph shows membranes of rough endoplasmic reticulum in the cytoplasm of a rat pancreas cell. The "rough" appearance of the membranes in this specimen is caused by the presence of numerous membrane-bound ribosomes. **(b)** A similar specimen viewed by scanning electron microscopy reveals the three-dimensional appearance of the rough endoplasmic reticulum, although individual ribosomes cannot be resolved.

thick specimens can be examined with good resolution. As a result, cellular structure can be studied in sections as thick as 1 μm, or about ten times the thickness possible with an ordinary TEM.

Scanning Electron Microscopy Reveals the Surface Architecture of Cells and Organelles

Scanning electron microscopy is a fundamentally different type of electron microscopy that produces images from electrons deflected from a specimen's outer surface (rather than electrons transmitted through the specimen). It is an especially spectacular technique because of the sense of depth it gives to biological structures, thereby allowing surface topography to be studied (Figure A-27b). As the name implies, a **scanning electron microscope (SEM)** generates such an image by scanning the specimen's surface with a beam of electrons.

An SEM and its optical system are shown in **Figure A-28**. The vacuum system and electron source are similar to those found in the TEM, although the accelerating voltage is lower (about 5–30 kV). The main difference between the two kinds of instruments lies in the way the image is formed. In an SEM, the electromagnetic lens system focuses the beam of electrons into an intense spot that is moved back and forth over the specimen's surface by charged plates called *beam deflectors,* which are located between the condenser lens and the specimen. The beam deflectors attract or repel the beam according to the signals sent to them by the deflector circuitry (Figure A-28b).

As the electron beams sweep rapidly over the specimen, molecules in the specimen are excited to high energy levels and emit *secondary electrons.* These emitted electrons are captured by a detector located immediately above and to one side of the specimen, thereby generating an image of the specimen's surface. The essential component of the detector is a *scintillator,* which emits photons of light when excited by electrons that impinge upon it. The photons are used to generate an electronic signal to a video screen. The image then develops point by point, line by line on the screen as the primary electron beam sweeps over the specimen.

Sample Preparation Techniques for Electron Microscopy

Specimens to be examined by electron microscopy can be prepared in several different ways, depending on the type of microscope and the kind of information the microscopist wants to obtain. In each case, however, the method is complicated, time-consuming, and costly compared with methods used for light microscopy. Moreover, living cells cannot be examined because the electron microscope requires specimens to be subjected to a vacuum.

Ultrathin Sectioning and Staining Are Common Preparation Techniques for Transmission Electron Microscopy

The most common way of preparing specimens for transmission electron microscopy involves slicing tissues and cells into ultrathin sections no more than 50–100 nm in thickness (less than one-tenth the thickness of the typical sections used for light microscopy). Specimens must first be chemically fixed and stabilized. The fixation step kills the cells but keeps the cellular components much as they were in the living cell. The fixatives employed are usually buffered solutions of aldehydes, most commonly glutaraldehyde. Following fixation, the specimen is often stained with a 1–2% solution of buffered osmium tetroxide (OsO_4), which binds to various components of the cell, making them more electron dense.

sample is slowly replaced by an organic solvent, such as acetone, over a period of days. The resulting specimen can be processed for embedding and sectioning in the same way as chemically fixed specimens are processed.

The tissue is next passed through a series of alcohol solutions to dehydrate it, and then it is placed in a solvent such as acetone or propylene oxide to prepare it for embedding in liquefied plastic epoxy resin. After the plastic has infiltrated the specimen, it is put into a mold and heated in an oven to harden the plastic. The embedded specimen is then sliced into ultrathin sections by an instrument called an **ultramicrotome** (**Figure A-29a**). The specimen is mounted firmly on the arm of the ultramicrotome, which

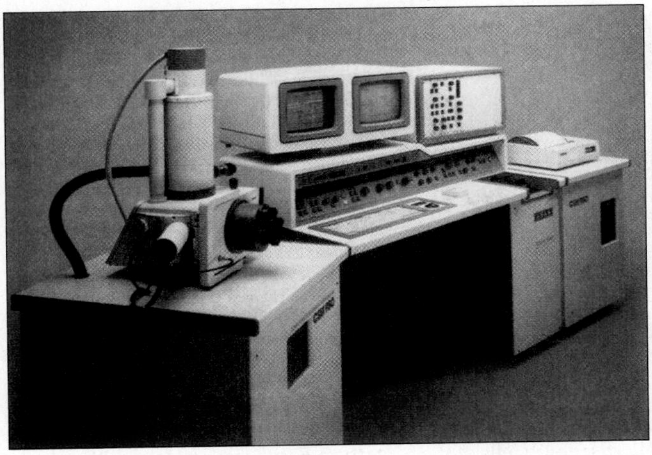

(a)

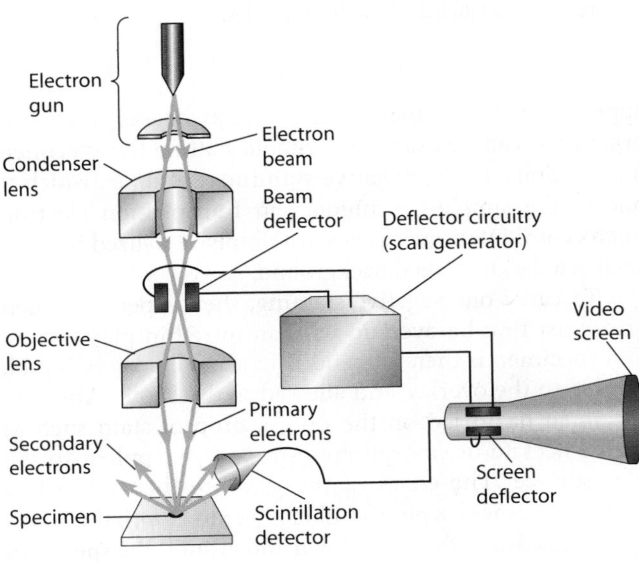

(b)

FIGURE A-28 A Scanning Electron Microscope. (**a**) A photograph and (**b**) a schematic diagram of an SEM. The image is generated by secondary electrons (short orange lines) emitted by the specimen as a focused beam of primary electrons (long orange lines) sweeps rapidly over it. The signal to the video screen is synchronized to the movement of the primary electron beam over the specimen by the deflector circuitry of the scan generator.

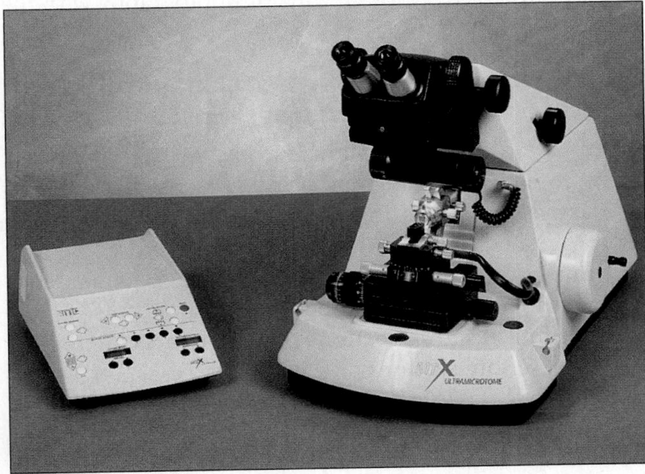

(**a**) Ultramicrotome

(**b**) Microtome arm of ultramicrotome

FIGURE A-29 An Ultramicrotome. (**a**) A photograph of an ultramicrotome. (**b**) A close-up view of the ultramicrotome arm, showing the specimen in a plastic block mounted on the end of the arm. As the ultramicrotome arm moves up and down, the block is advanced in small increments, and ultrathin sections are cut from the block face by the diamond knife.

Chemical fixatives are very good at stabilizing many structures within cells, but they suffer from two drawbacks. First, they are slow; it takes a minimum amount of time for the fixative to diffuse into the sample to fix its structures. Second, chemical fixatives often extract cellular components (i.e., fine structures within the cell are often lost during fixation). **Cryofixation,** which typically involves extremely rapid freezing (within 20 msec) of a specimen under very high pressure, avoids these problems. Such *high-pressure freezing* is necessary; without it, water ice crystals form within the specimen, causing extensive damage to its fine structures. In most applications, high-pressure freezing is followed by *freeze substitution*. During freeze substitution, the water in the

advances the specimen in small increments toward a glass or diamond knife (Figure A-29). When the block reaches the knife blade, ultrathin sections are cut from the block face. The sections float from the blade onto a water surface, where they can be picked up on a circular copper specimen grid. The grid consists of a meshwork of very thin copper strips, which support the specimen while still allowing openings between adjacent strips through which the specimen can be observed.

Once in place on the grid, the sections are usually stained again, this time with solutions containing lead and uranium. This step enhances the contrast of the specimen because the lead and uranium give still greater electron density to specific parts of the cell. After poststaining, the specimen is ready for viewing or photography with the TEM.

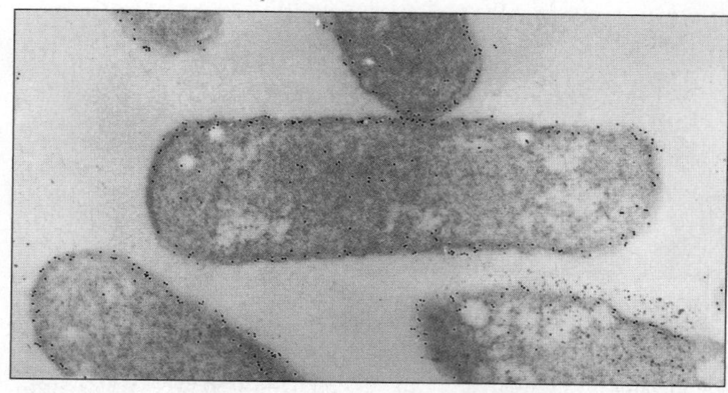

FIGURE A-30 The Use of Gold-Labeled Antibodies in Electron Microscopy. Cells of the bacterium *E. coli* were stained with gold-labeled antibodies directed against a plasma membrane protein. The small, dark granules distributed around the periphery of the cell are the gold-labeled antibody molecules.

Radioisotopes and Antibodies Can Localize Molecules in Electron Micrographs

In our discussion of light microscopy, we described how microscopic autoradiography can be used to locate radioactive molecules inside cells. Autoradiography can also be applied to transmission electron microscopy, with only minor differences. For the TEM, the specimen containing the radioactively labeled compounds is simply examined in ultrathin sections on copper specimen grids instead of in thin sections on glass slides.

We also described how fluorescently labeled antibodies can be used in conjunction with light microscopy to locate specific cellular components. Antibodies are likewise used in the electron microscopic technique called **immunoelectron microscopy (immunoEM)**; fluorescence cannot be seen in the electron microscope, so antibodies are instead visualized by linking them to substances that are electron dense and therefore visible as opaque dots. One of the most common approaches is to couple antibody molecules to colloidal gold particles. When ultrathin tissue sections are stained with gold-labeled antibodies directed against various proteins, electron microscopy can reveal the subcellular location of these proteins with great precision (**Figure A-30**).

A powerful approach that unites light microscopy and immunoEM is **correlative microscopy.** In correlative microscopy, dynamic images of a cell are acquired using the light microscope, often using antibodies and/or GFP. The very same cell is then processed and viewed using electron microscopy (EM). Commonly, immunoEM is used to determine where a protein is found at very high resolution. Correlative microscopy thus bridges the gap between dynamic imaging using the light microscope and the detailed images that can be acquired only via EM.

Negative Staining Can Highlight Small Objects in Relief Against a Stained Background

Although cutting tissues into ultrathin sections is the most common way of preparing specimens for transmission electron microscopy, other techniques are suitable for particular purposes. For example, the shape and surface appearance of very small objects, such as viruses or isolated organelles, can be examined without cutting the specimen into sections. In the **negative staining** technique, which is one of the simplest techniques in transmission electron microscopy, intact specimens are simply visualized in relief against a darkly stained background.

To carry out negative staining, the copper specimen grid must first be overlaid with an ultrathin plastic film. The specimen is then suspended in a small drop of liquid, applied to the overlay, and allowed to dry in air. After the specimen has dried on the grid, a drop of stain such as uranyl acetate or phosphotungstic acid is applied to the film surface. The edges of the grid are then blotted in several places with a piece of filter paper to absorb the excess stain. This draws the stain down and around the specimen and its ultrastructural features. When viewed in the TEM, the specimen is seen in *negative contrast* because the background is dark and heavily stained, whereas the specimen itself is lightly stained (**Figure A-31**).

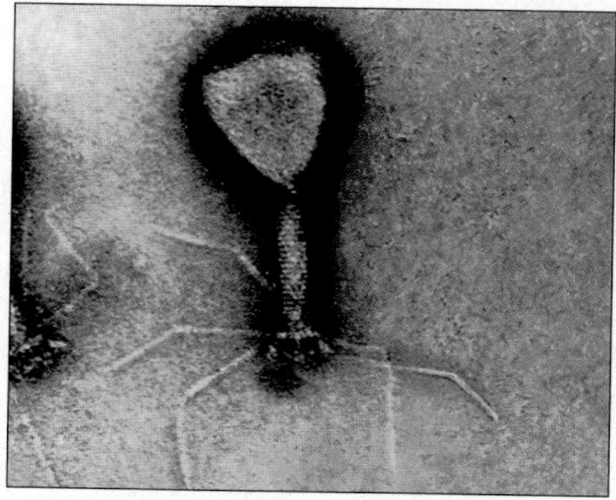

FIGURE A-31 Negative Staining. An electron micrograph of a bacteriophage as seen in a negatively stained preparation. This specimen was simply suspended in an electron-dense stain, allowing it to be visualized in relief against a darkly stained background (TEM).

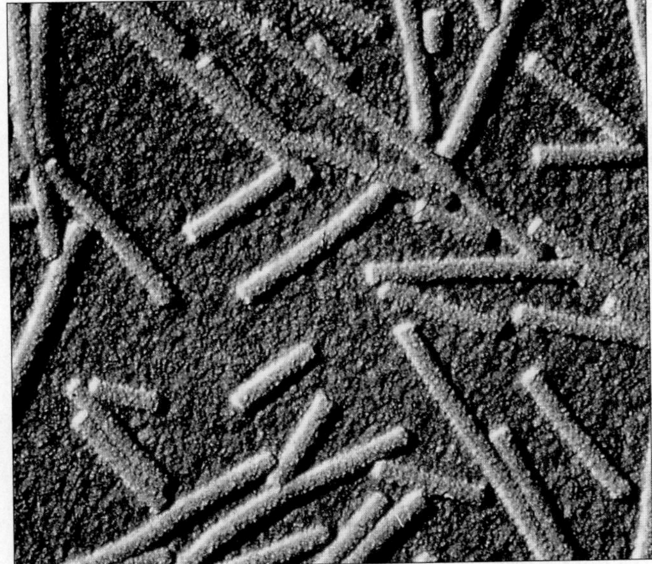

FIGURE A-32 Shadowing. An electron micrograph of tobacco mosaic virus particles visualized by shadowing. In this technique, heavy metal vapor was sprayed at an angle across the specimen, causing an accumulation of metal on one side of each virus particle and a shadow region lacking metal on the other side (TEM).

Shadowing Techniques Use Metal Vapor Sprayed Across a Specimen's Surface

Isolated particles or macromolecules can also be visualized by the technique of **shadowing** (**Figure A-32**), which involves spraying a thin layer of an electron-dense metal such as gold or platinum at an angle across the surface of a biological specimen. **Figure A-33a** illustrates the shadowing technique. The specimen is first spread on a clean mica surface and dried (step ❶). It is then placed in a **vacuum evaporator,** a bell jar in which a vacuum is created by a system similar to that of an electron microscope (Figure A-33b). Also within the evaporator are two electrodes, one consisting of a carbon rod located directly over the specimen and the other consisting of a metal wire positioned at an angle of approximately 10–45° relative to the specimen.

After a vacuum is created in the evaporator, current is applied to the metal electrode, causing the metal to evaporate from the electrode and spray over the surface of the specimen (Figure A-33a, ❷). Because the metal-emitting electrode is positioned at an angle to the specimen, metal is deposited on only one side of the specimen, generating

❶ The specimen is spread on a mica surface and dried.

Specimen

Mica surface

❷ The specimen is shadowed by coating it with atoms of a heavy metal that are evaporated from a heated filament.

Metal wire

Heated filament

Metal atoms

Metal replica

❸ The specimen is coated with carbon atoms evaporated from an overhead electrode.

Carbon atoms

Metal replica

❹ The replica is floated onto the surface of an acid bath to dissolve away the specimen, leaving a clean metal replica.

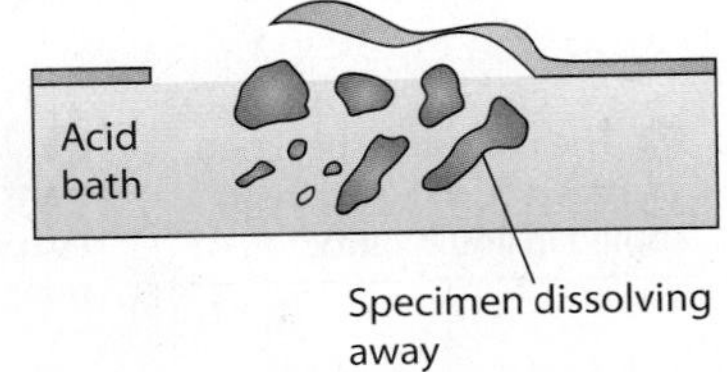

❺ The replica is washed and picked up on a copper grid for examination in the TEM.

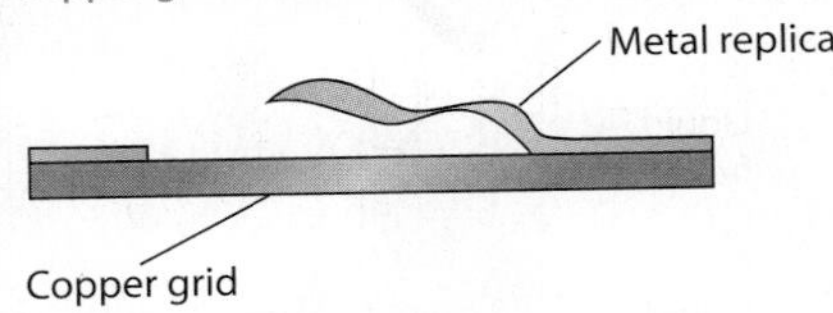

(a) Shadowing technique

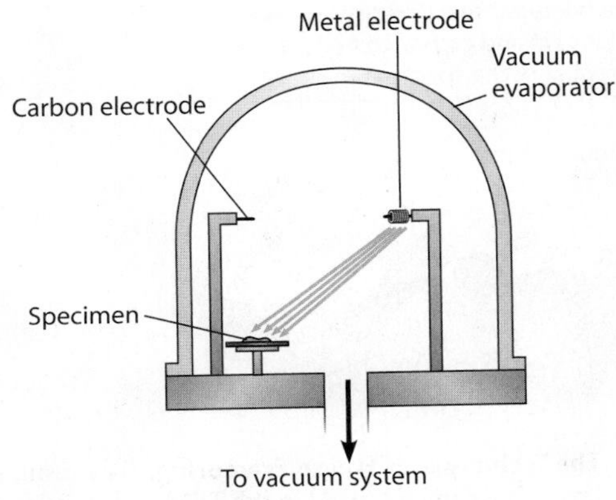

(b) Vacuum evaporator

FIGURE A-33 The Technique of Shadowing. **(a)** The specimen is spread on a mica surface and shadowed by coating it with atoms of a heavy metal (platinum or gold, shown in orange). This generates a metal replica (orange) whose thickness reflects the surface contours of the specimen. The replica is coated with carbon atoms to strengthen it, and it is then floated away and washed before viewing in the TEM. **(b)** The vacuum evaporator in which shadowing is done. The carbon electrode is located directly over the specimen; the heavy metal electrode is off to the side.

a metal *replica* of the surface. The opposite side of the specimen remains unstained; it is this unstained region that creates the "shadow" effect.

An overhead carbon-emitting electrode is then used to coat the specimen with evaporated carbon, thereby providing stability and support to the metal replicas (❸). Next, the mica support containing the specimen is removed from the vacuum evaporator and lowered gently onto a water surface, causing the replica to float away from the mica surface. The replica is transferred into an acid bath, which dissolves any remaining bits of specimen, leaving a clean metal replica of the specimen (❹). The replica is then transferred to a standard copper specimen grid (❺) for viewing by transmission electron microscopy.

A related procedure is commonly used for visualizing purified molecules such as DNA and RNA. In this technique, a solution of DNA and/or RNA is spread on an air-water interface, creating a molecular monolayer that is collected on a thin film and visualized by uniformly depositing heavy metal on all sides.

Freeze Fracturing and Freeze Etching Are Useful for Examining the Interior of Membranes

Freeze fracturing is an approach to sample preparation that is fundamentally different from the methods described so far. Instead of cutting uniform slices through a tissue sample (or staining unsectioned material), specimens are rapidly frozen at the temperature of liquid nitrogen or liquid helium, placed in a vacuum, and struck with a sharp knife edge. Samples frozen at such low temperatures are too hard to be cut. Instead, they fracture along lines of natural weakness—the hydrophobic interior of membranes, in most cases. Platinum/carbon shadowing is then used to create a replica of the fractured surface.

Freeze fracturing is illustrated in **Figure A-34**. It takes place in a modified vacuum evaporator with an internal microtome knife for fracturing the frozen specimen. The temperature of the specimen support and the microtome arm and knife is precisely controlled. Specimens are generally fixed prior to freeze fracturing, although some living tissues can be frozen fast enough to keep them in almost lifelike condition. Because cells contain large amounts of water, fixed specimens are usually treated with an antifreeze such as glycerol to provide *cryoprotection*—that is, to reduce the formation of ice crystals during freezing.

The cryoprotected specimen is mounted on a metal specimen support (Figure A-34, step ❶) and immersed rapidly in freon cooled with liquid nitrogen (❷). This procedure also reduces the formation of ice crystals in the cells. With the frozen specimen positioned on the specimen table in the vacuum evaporator (❸), a high vacuum is established, the stage temperature is adjusted to around $-100°C$, and the frozen specimen is fractured with a blow from the microtome knife (❹). A replica of the fractured specimen is made by shadowing with platinum and carbon as described in the previous section (❺), and the replica is then ready to be viewed in the TEM (❻).

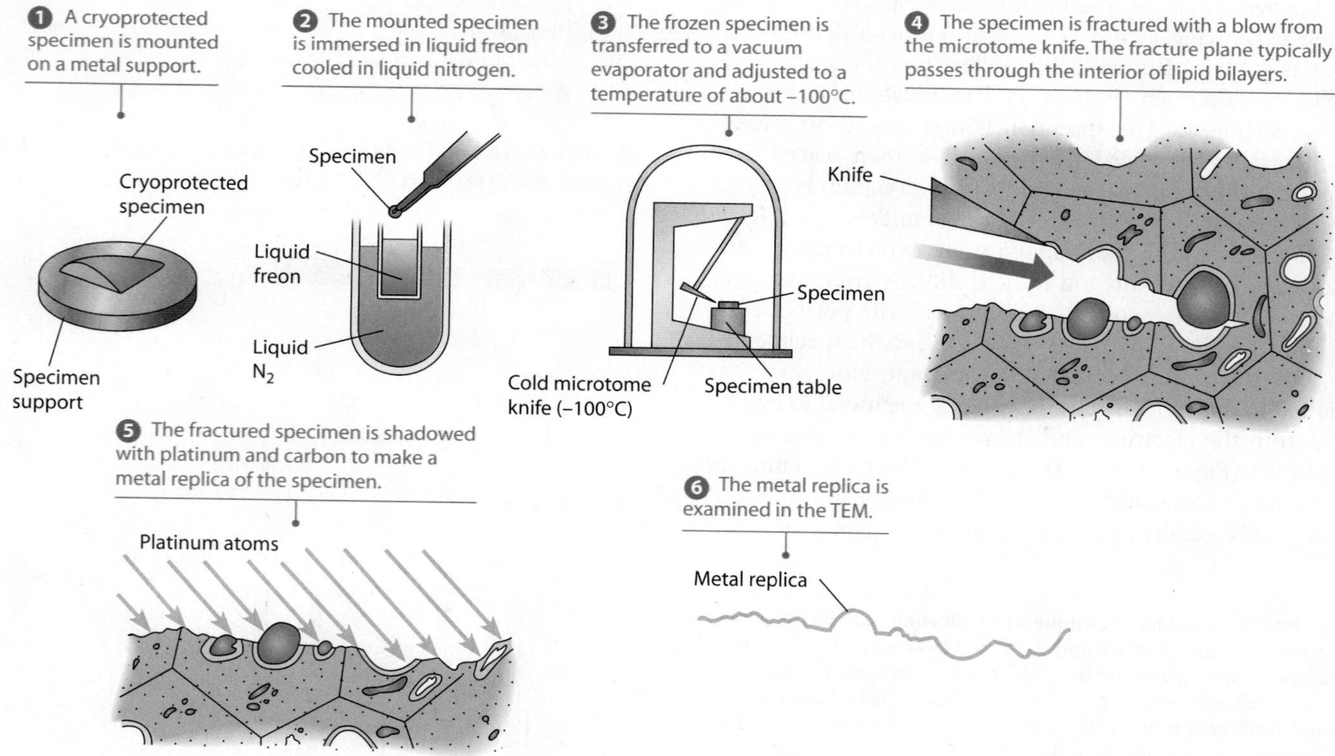

❶ A cryoprotected specimen is mounted on a metal support.
Cryoprotected specimen
Specimen support
❷ The mounted specimen is immersed in liquid freon cooled in liquid nitrogen.
Specimen
Liquid freon
Liquid N_2
❸ The frozen specimen is transferred to a vacuum evaporator and adjusted to a temperature of about –100°C.
Specimen
Cold microtome knife (–100°C)
Specimen table
❹ The specimen is fractured with a blow from the microtome knife. The fracture plane typically passes through the interior of lipid bilayers.
Knife
❺ The fractured specimen is shadowed with platinum and carbon to make a metal replica of the specimen.
Platinum atoms
❻ The metal replica is examined in the TEM.
Metal replica

FIGURE A-34 The Technique of Freeze Fracturing. The result is a replica of a specimen, fractured through its lipid bilayer, that can be viewed in the TEM. The shadowing is performed as in Figure A-33.

Freeze-fractured membranes appear as smooth surfaces studded with **intramembranous particles (IMPs).** These particles are integral membrane proteins that have remained with one lipid monolayer or the other as the fracture plane passes through the interior of the membrane.

The electron micrograph in **Figure A-35** shows the two faces of a plasma membrane revealed by freeze fracturing. The **P face** is the interior face of the inner monolayer; it is called the P face because this monolayer is on the *protoplasmic* side of the membrane. The **E face** is the interior face of the outer monolayer; it is called the E face because this monolayer is on the *exterior* side of the membrane. Notice that the P face has far more intramembranous particles than does the E face. In general, most of the particles in the membrane stay with the inner monolayer when the fracture plane passes down the middle of a membrane.

To have a P face and an E face appear side by side as in Figure A-35, the fracture plane must pass through two neighboring cells, such that one cell has its cytoplasm and the inner monolayer of its plasma membrane removed to reveal the E face, while the other cell has the outer monolayer of its plasma membrane and the associated intercellular space removed to reveal the P face. Accordingly, E faces are always separated from P faces of adjacent cells by a "step" (marked by the arrows in Figure A-35) that represents the thickness of the intercellular space.

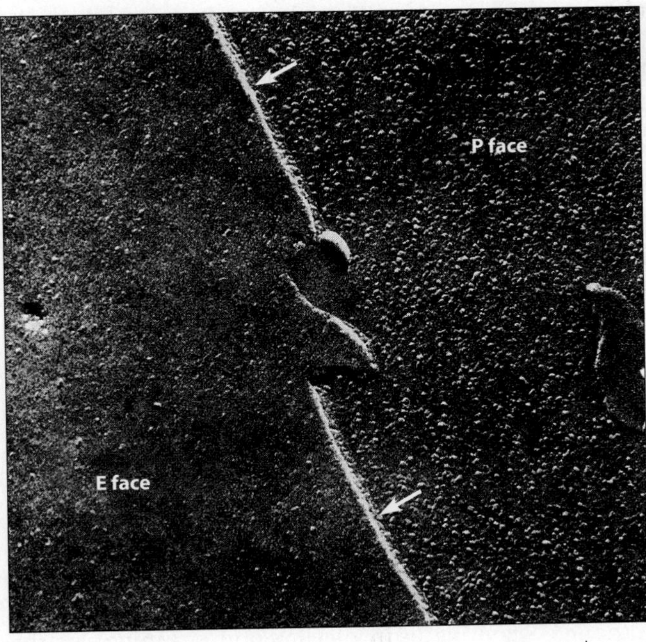

FIGURE A-35 Freeze Fracturing of the Plasma Membrane. This electron micrograph shows the exposed faces of the plasma membranes of two adjacent endocrine cells from a rat pancreas as revealed by freeze fracturing. The P face is the inner surface of the lipid monolayer on the protoplasmic side of the plasma membrane. The E face is the inner surface of the lipid monolayer on the exterior side of the plasma membrane. The P face is much more richly studded with intramembranous particles than the E face. The arrows indicate the "step" along which the fracture plane passed from the interior of the plasma membrane of one cell to the interior of the plasma membrane of a neighboring cell. The step therefore represents the thickness of the intercellular space (TEM).

In a closely related technique called **freeze etching,** a further step is added to the conventional freeze-fracture procedure. Following the fracture of the specimen but prior to shadowing, the microtome arm is placed directly over the specimen for a short period of time (a few seconds to several minutes). This maneuver causes a small amount of water to evaporate (sublime) from the surface of the specimen to the cold knife surface, and this sublimation produces an *etching* effect—that is, an accentuation of surface detail. By using ultrarapid freezing techniques to minimize the formation of ice crystals during freezing, and by including a volatile cryoprotectant such as aqueous methanol, which sublimes very readily to a cold surface, the etching period can be extended and a deeper layer of ice can be removed, thereby exposing structures that are located deep within the cell interior. This modification, called **deep etching,** provides a fascinating look at cellular structure. Deep etching has been especially useful in exploring the cytoskeleton and examining its connections with other structures of the cell.

Stereo Electron Microscopy and 3-D Electron Tomography Allow Specimens to Be Viewed in Three Dimensions

Electron microscopists frequently want to visualize specimens in three dimensions. Shadowing, freeze fracturing, and scanning electron microscopy are useful for this purpose, as is another specialized technique called **stereo electron microscopy.** In stereoelectron microscopy, three-dimensional information is obtained by photographing the same specimen at two slightly different angles. This is accomplished using a special specimen stage that can be tilted relative to the electron beam. The specimen is first tilted in one direction and photographed, then tilted an equal amount in the opposite direction and photographed again. The two micrographs are then mounted side by side as a *stereo pair.* When you view a stereo pair through a stereoscopic viewer, your brain uses the two independent images to construct a three-dimensional view that gives a striking sense of depth. **Figure A-36a** is a stereo pair of a *Drosophila* polytene chromosome imaged by high-voltage electron microscopy. Using a stereo viewer or allowing your eyes to fuse the two images visually creates a striking, three-dimensional view of the chromosome.

More recently, electron microscopists have used computer-based methods to make three-dimensional reconstructions of structures imaged in the TEM. In a process known as **3-D electron tomography,** serial thin sections containing a specimen are rotated and imaged at several different orientations; the resulting rotated views are then used to construct very thin, computer-generated "slices" of the specimen. Structures within these highly resolved slices can then be traced and reconstructed to provide three-dimensional information about the specimen. 3-D electron tomography has revolutionized our view of cellular structures, such as the organelles shown in

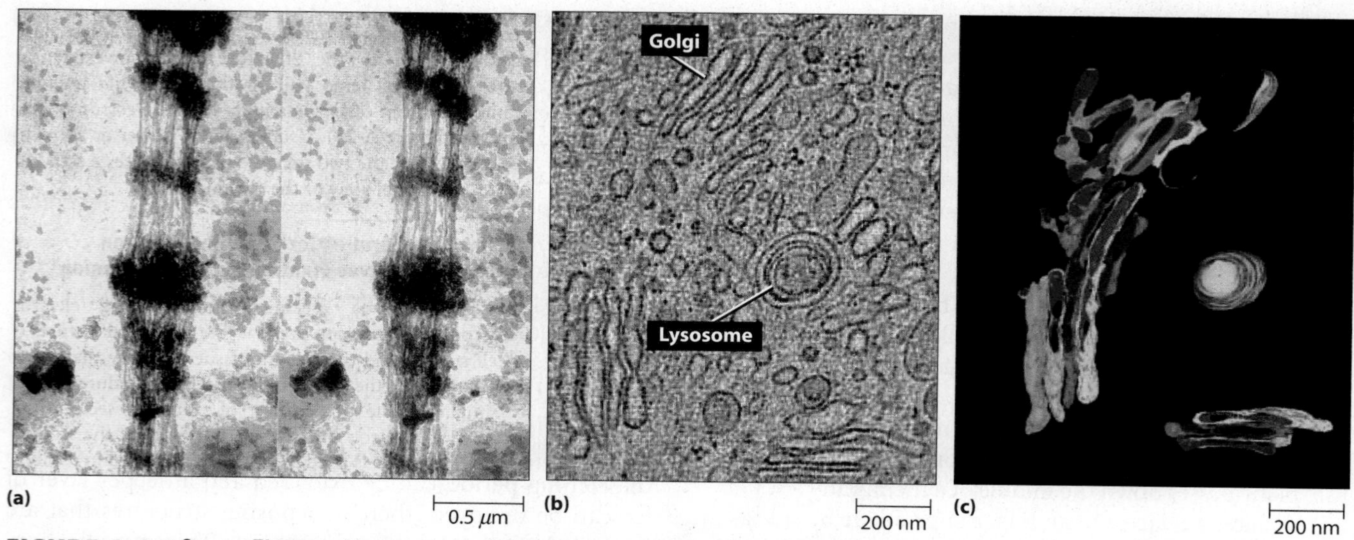

FIGURE A-36 Stereo Electron Microscopy and 3-D Electron Tomography. **(a)** The polytene chromosome in this micrograph is shown as a stereo pair. The two photographs were taken by tilting the specimen stage first 5° to the right and then 5° to the left of the electron beam. For a three-dimensional view, the stereo pair can be examined with a stereoscopic viewer. Alternatively, simply let your eyes cross slightly, fusing the two micrographs into a single image (HVEM). **(b)** A thin (3.2 nm) digital slice computed from a series of tilted images of a dendritic cell (TEM). A region of the Golgi apparatus and a lysosome are clearly visible. **(c)** A contour model of some of the structures visible in the tomogram shown in part b.

Figure A-36b, c, effectively extending the type of topographical data obtained by SEM to the level of TEM.

Specimen Preparation for Scanning Electron Microscopy Involves Fixation but Not Sectioning

When preparing a specimen for scanning electron microscopy, the biologist's goal is to preserve the structural features of the cell surface and to treat the tissue in a way that minimizes damage by the electron beam. The procedure is actually similar to preparing ultrathin sections for transmission electron microscopy but without the sectioning step. The tissue is fixed in aldehyde, postfixed in osmium tetroxide, and dehydrated by processing through a series of alcohol solutions. The tissue is then placed in a fluid such as liquid carbon dioxide in a heavy metal canister called a **critical point dryer,** which is used to dry the specimen under conditions of controlled temperature and pressure. This helps keep structures on the surfaces of the tissue in almost the same condition as they were before dehydration.

The dried specimen is then attached to a metal specimen mount with a metallic paste. The mounted specimen is coated with a layer of gold or a mixture of gold and palladium, using a modified form of vacuum evaporation called **sputter coating.** Once the specimen has been mounted and coated, it is ready to be examined in the SEM.

Other Imaging Methods

Light and electron microscopy are direct imaging techniques in that they use photons or electrons to produce actual images of a specimen. However, some imaging techniques are indirect. To understand what we mean by indirect imaging, suppose you are given an object to handle with your eyes closed. You might feel 6 flat surfaces, 12 edges, and 8 corners. If you then draw what you have felt, it would turn out to be a box. This is an example of an indirect imaging procedure.

The two indirect imaging methods we describe here are *scanning probe microscopy* and *X-ray crystallography.* Both approaches have the potential for showing molecular structures at near-atomic resolution, ten times better than the best electron microscope. They do have some shortcomings that limit their usefulness with biological specimens. But when these techniques can be applied successfully, the resulting images provide unique information about molecular structure that cannot be obtained using conventional microscopic techniques.

Scanning Probe Microscopy Reveals the Surface Features of Individual Molecules

Although "scanning" is involved in both scanning electron microscopy and scanning probe microscopy, the two methods are in fact quite different. The first example of a **scanning probe microscope,** called the *scanning tunneling microscope (STM),* was developed in the early 1980s for the purpose of exploring the surface structure of specimens at the atomic level. The STM utilizes a tiny probe that does not emit an electron beam, but instead possesses a tip made of a conducting material such as platinum-iridium. The tip of the probe is extremely sharp; ideally, its point is composed of a single atom. It is under the precise control of an electronic circuit that can move it in three dimensions over a surface. The *x* and *y* dimensions scan

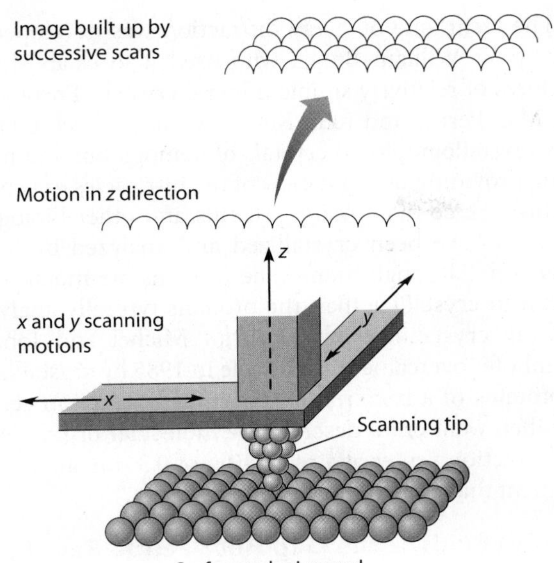

FIGURE A-37 Scanning Tunneling Electron Microscopy. The scanning tunneling microscope (STM) uses electronic methods to move a metallic tip across the surface of a specimen. The tip is not drawn to scale in this illustration; the point of the tip is ideally composed of only one or a few atoms, shown here as balls. An electrical voltage is produced between the tip and the specimen surface. As the tip scans the specimen in the x and y directions, electron tunneling occurs at a rate dependent on the distance between the tip and the first layer of atoms in the surface. The instrument is designed to move the tip in the z direction to maintain a constant current flow. The movement is therefore a function of the tunneling current and is presented on a computer screen. Successive scans then build up an image of the surface at atomic resolution.

the surface, while the z dimension governs the distance of the tip above the surface (**Figure A-37**).

As the tip of the STM is moved across the surface of a specimen, voltages from a few millivolts to several volts are applied. If the tip is close enough to the surface and the surface is electrically conductive, electrons will begin to leak or "tunnel" across the gap between the probe and the sample. The tunneling is highly dependent on the distance, so that even small irregularities in the size range of single atoms will affect the rate of electron tunneling. As the probe scans the sample, the tip of the probe is automatically moved up and down to maintain a constant rate of electron tunneling across the gap. A computer measures this movement and uses the information to generate a map of the sample's surface, which is viewed on a computer screen.

Despite the enormous power of the STM, it suffers from two limitations: The specimen must be an electrical conductor, and the technique provides information only about electrons associated with the specimen's surface. Researchers have therefore begun to develop other kinds of scanning probe microscopes that scan a sample just like the STM but measure different kinds of interactions between the tip and the sample surface. For example, in the *atomic force microscope (AFM)*, the scanning tip is pushed right up against the surface of the sample. When it scans, it moves up and down as it runs into the microscopic hills and valleys formed by the atoms present at the sample's surface. A variety of scanning probe microscopes have been designed to detect other properties, such as friction, magnetic force, electrostatic force, van der Waals forces, heat, and sound.

One of the most important potential applications of scanning probe microscopy is the measurement of dynamic changes in the conformation of functioning biomolecules. Recently, it has become possible to visualize the movements of single myosin proteins as they change their shape during their power stroke. Such "molecular eavesdropping" is now entirely within the realm of possibility. It is even possible to use a modified form of atomic force microscopy to directly stretch large biomolecules to measure their mechanical properties.

X-Ray Crystallography Allows the Three-Dimensional Structure of Macromolecules to Be Determined

Though **X-ray crystallography** does not involve microscopy, it is such an important method for investigating the three-dimensional structure of individual molecules that we include it here. This method reconstructs images from the diffraction patterns of X-rays passing through a crystalline or fibrous specimen, thereby revealing molecular structure at the atomic level of resolution.

A good way to understand X-ray crystallography is to draw an analogy with visible light. As discussed earlier, light has certain properties that are best described as wavelike. If waves from two sources come into phase with one another, their total amplitude is additive *(constructive interference)*; if they are out of phase, their amplitude is reduced *(destructive interference)*. This effect can be seen when light passes through two pinholes in a piece of opaque material and then falls onto a white surface. Interference patterns result, with dark regions where light waves are out of phase and bright regions where they are in phase (**Figure A-38**). If the wavelength of the light (λ) is known, one can measure the angle α between the original beam and the first diffraction peak and then calculate the distance d between the two holes with the formula

$$d = \frac{\lambda}{\sin \alpha} \quad \textbf{(A-5)}$$

The same approach can be used to calculate the distance between atoms in crystals or fibers of proteins and nucleic acids. Instead of a sheet of paper with two holes in it, imagine that we have multiple layers of atoms organized in a crystal or fiber. And instead of visible light, which has much too long a wavelength to interact with atoms, we will use a narrow beam of X-rays with wavelengths in the range of interatomic distances. As the X-rays pass through the specimen, they reflect off planes of atoms, and the reflected beams come into constructive and destructive interference. The reflected beams then fall onto photographic plates

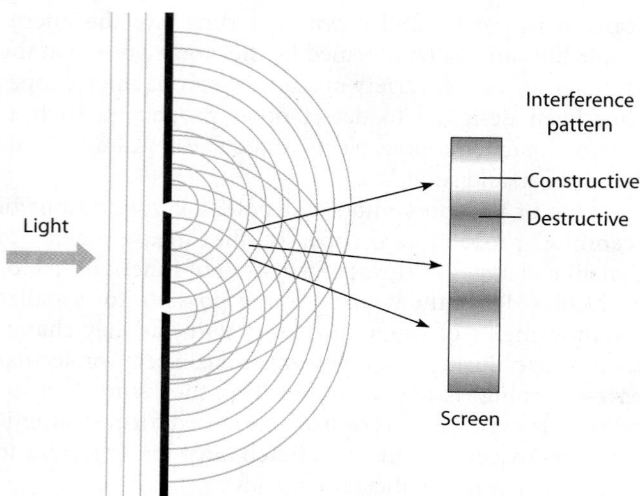

FIGURE A-38 Understanding Diffraction Patterns. Any energy in the form of waves will produce interference patterns if the waves from two or more sources are superimposed in space. One of the simplest patterns can be seen when monochromatic light passes through two neighboring pinholes and is allowed to fall on a screen. When the light passes through the two pinholes, the holes act as light sources, with waves radiating from each and falling on a white surface. Where the waves are in the same phase, a bright area appears (constructive interference). Where the waves are out of phase, they cancel each other out and produce dark areas (destructive interference).

behind the specimen, generating distinctive diffraction patterns. These patterns are then analyzed mathematically to deduce the three-dimensional structure of the original molecule. **Figure A-39** illustrates the use of this procedure to deduce the double-helical structure of DNA.

The technique of X-ray diffraction was developed in 1912 by Sir William Bragg, who used it to establish the structures of relatively simple mineral crystals. Forty years later, Max Perutz and John Kendrew found ways to apply X-ray crystallography to crystals of hemoglobin and myoglobin, providing our first view of the intricacies of protein structure. Since then, many proteins and other biological molecules have been crystallized and analyzed by X-ray diffraction. Although membrane proteins are much more difficult to crystallize than the proteins typically analyzed by X-ray crystallography, Hartmut Michel and Johann Deisenhofer overcame this obstacle in 1985 by crystallizing the proteins of a bacterial photosynthetic reaction center. They then went on to describe the molecular organization of the reaction center at a resolution of 0.3 nm, an accomplishment that earned them a Nobel Prize.

CryoEM Bridges the Gap Between X-Ray Crystallography and Electron Microscopy

X-ray crystallography has provided unprecedented views of biological molecules down to the atomic level. However, crystallography requires large amounts of purified molecules. In addition, X-ray crystallography is not currently well suited for analyzing very large macromolecular assemblies, such as the ribosome (see Figure 22-1). For such large structures, another technique is needed. A technique that helps to bridge this gap is **cryoEM.** In cryoEM, purified molecules or macromolecular assemblies are rapidly frozen via

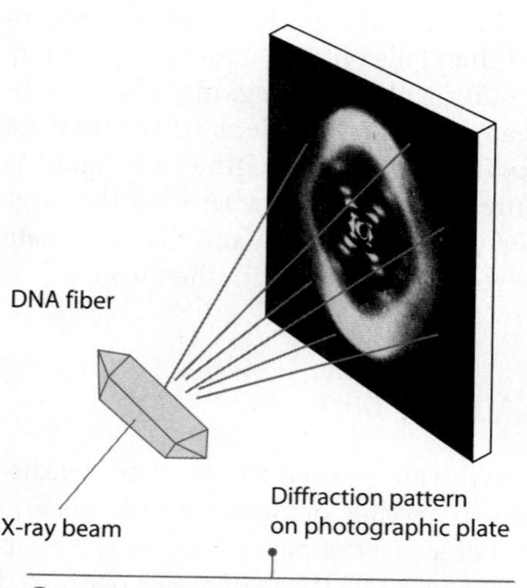

❶ X-rays diffracted by a DNA fiber produce a diffraction pattern on a photographic plate or other detector.

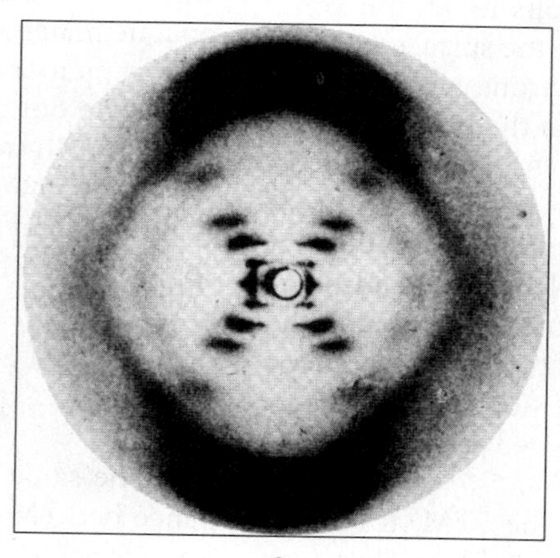

❷ The resulting diffraction pattern is analyzed mathematically.

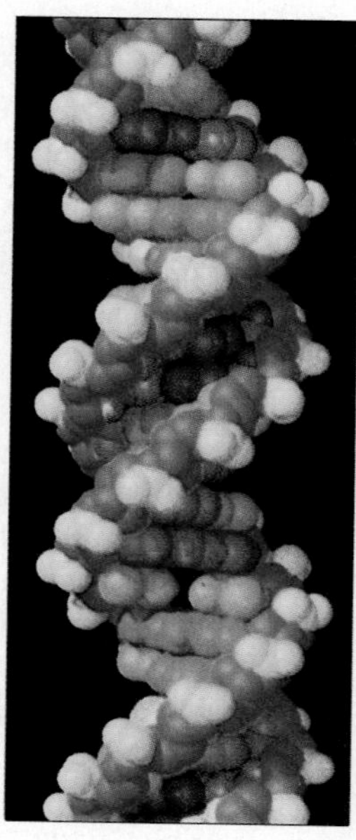

❸ The three-dimensional structure of the molecule is deduced.

FIGURE A-39 X-Ray Crystallography. X-ray crystallography can be used to analyze molecular structure at near-atomic resolution. The specific example illustrated in this figure is a DNA fiber. The photograph in ❷ depicts the actual X-ray diffraction pattern used by James Watson and Francis Crick to deduce the molecular structure of double-stranded DNA. The photograph in ❸ is a computer graphic model of the DNA double helix.

cryofixation. The rapid freezing prevents ice crystals from forming. Instead, the molecules are embedded in *vitreous ice,* noncrystalline frozen water, which better preserves the structures of the embedded molecules. The sample is then directly imaged at a very low temperature (−170°C) in an electron microscope. Often the specimens are imaged using three-dimensional electron tomography. To see how cryoEM can work together with X-ray crystallography, consider **Figure A-40**, in which a cryoEM image of the 30S and 50S ribosomal subunits and a releasing factor protein known as RF2 have been combined with a detailed structure of RF2 obtained from X-ray data. By using the two techniques together, cell biologists can see how the structure of the protein enables it to fit neatly within the overall structure of the ribosome to perform its function.

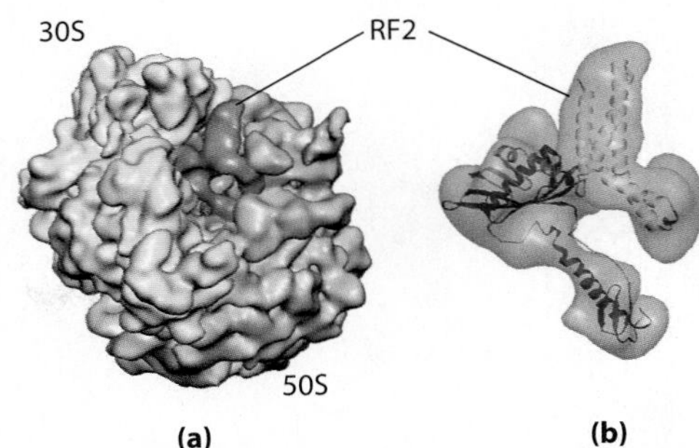

FIGURE A-40 CryoEM Bridges the Gap Between the Atomic and Molecular Levels. **(a)** A three-dimensional reconstruction of the 30S (yellow) and 50S (blue) ribosomal subunits bound to a releasing factor (RF2, in pink) obtained via cryoEM. X-ray crystallography provides a more detailed view of RF2 **(b)**. Adapted by permission from Macmillan Publishers Ltd: *Nature*. From Urmila B. S. Rawat et al., "A Cryo-Electron Microscopic Study of Ribosome-Bound Termination factor RF2," (2 January 2003), 421: 87-90, Copyright 2003.

SUGGESTED READING

References of historical importance are marked with a •.

General References

Bradbury, S., B. Bracegirdle, and S. Bradbury. *Introduction to Light Microscopy.* Oxford, UK: BIOS Scientific Publishers, 1998.

Inoue, S., and K. R. Spring. *Video Microscopy: The Fundamentals.* New York: Springer, 1997.

Murphy, D. B. *Fundamentals of Light Microscopy and Electronic Imaging.* New York: Wiley-Liss, 2001.

Pawley, J. B., ed. *Handbook of Biological Confocal Microscopy,* 3rd ed. New York: Springer, 2006.

Stoffler, D., M. O. Steinmetz, and U. Aebi. Imaging biological matter across dimensions: From cells to molecules and atoms. *FASEB J.* 13 (1999, Suppl. 2): S195.

Light Microscopy

Bates, M., B. Huang, G. T. Dempsey, and X. Zhuang. Multicolor super-resolution imaging with photo-switchable fluorescent probes. *Science* 317 (2007): 1749.

Fernandez-Suarez, M., and A. Y. Ting. Fluorescent probes for super-resolution imaging in living cells. *Nature Rev. Mol. Cell Biol.* 9 (2008): 929.

• Ford, B. J. The earliest views. *Sci. Amer.* 278 (April 1998): 50.

Hein, B., K. I. Willig, and S. W. Hell. Stimulated emission depletion (STED) nanoscopy of a fluorescent protein-labeled organelle inside a living cell. *Proc. Natl. Acad. Sci. USA* 105 (2008): 14271.

Gerlich, D., and J. Ellenberg. 4D imaging to assay complex dynamics in live specimens. *Nature Cell. Biol.* (2003, Suppl.): S14.

Giepmans, B. N., S. R. Adams, M. H. Ellisman, and R. Y. Tsien. The fluorescent toolbox for assessing protein location and function. *Science* 312 (2006): 217.

Lippincott-Schwartz, J., and G. H. Patterson. Development and use of fluorescent protein markers in living cells. *Science* 300 (2003): 87.

Matsumoto, B., ed. *Cell Biological Applications of Confocal Microscopy,* 2nd ed. New York: Academic Press, 2002.

Piston, D. W. (1999). Imaging living cells and tissues by two-photon excitation microscopy. *Trends Cell Biol.* 9 (1999): 66.

Stephens, D. J., and V. J. Allan. Light microscopy techniques for live cell imaging. *Science* 300 (2003): 82.

Steyer, J. A., and W. Almers. A real-time view of life within 100 nm of the plasma membrane. *Nature Rev. Mol. Cell Biol.* 2 (2001): 268.

Walker, S. A., and P. J. Lockyer. Visualizing Ras signaling in real time. *J. Cell Sci.* 117 (2004): 2879.

Wang, Y. L. Digital deconvolution of fluorescence images for biologists. *Methods Cell Biol.* 56 (1998): 305.

Electron Microscopy

Bozzola, J. J., and L. D. Russell. *Electron Microscopy: Principles and Techniques for Biologists,* 2nd ed. Boston: Jones and Bartlett, 1998.

• Heuser, J. Quick-freeze, deep-etch preparation of samples for 3-D electron microscopy. *Trends Biochem. Sci.* 6 (1981): 64.

Hoenger, A., and J. R. McIntosh. Probing the macromolecular organization of cells by electron tomography. *Curr. Opin. Cell Biol.* 21 (2009): 89.

Koster, A. J., and J. Klumperman. Electron microscopy in cell biology: Integrating structure and function. *Nature Rev. Mol. Cell Biol.* (2003, Suppl.): SS6.

Maunsbach, A. B., and B. A. Afzelius. *Biomedical Electron Microscopy: Illustrated Methods and Interpretations.* San Diego, CA: Academic Press, 1999.

• Orci, L., and A. Perrelet. *Freeze-Etch Histology: A Comparison between Thin Sections and Freeze-Etch Replicas.* New York: Springer-Verlag, 1975.

• Satir, P. Keith R. Porter and the first electron micrograph of a cell. *Trends Cell Biol.* 7 (1997): 330.

Other Imaging Methods

Horber, J. K., and M. J. Miles. Scanning probe evolution in biology. *Science* 302 (2003): 1002.

Lucic, V., F. Forster, and W. Baumeister. Structural studies by electron tomography: From cells to molecules. *Annu. Rev. Biochem.* 74 (2005): 833.

Muller, D. J., J. Helenius, D. Alsteens, and Y. F. Dufrene. Force probing surfaces of living cells to molecular resolution. *Nat. Chem. Biol.* 5 (2009): 383.

Woolfson, M. M. *An Introduction to X-ray Crystallography,* 2nd ed. Cambridge: Cambridge University Press, 2003.

GLOSSARY

Note: The letter "A" preceding a page number refers to a page in the *Appendix: Visualizing Cells and Molecules.*

A

A: see *adenine.*

A band: region of a striated muscle myofibril that appears as a dark band when viewed by microscopy; contains thick myosin filaments and those regions of the thin actin filaments that overlap the thick filaments. (p. 489)

A site (aminoacyl site): site on the ribosome that binds each newly arriving tRNA with its attached amino acid. (p. 708)

A tubule: a complete microtubule that is fused to an incomplete microtubule (the B tubule) to make up an outer doublet in the axoneme of a eukaryotic cilium or flagellum. (p. 483)

ABC transporter: see *ABC-type ATPase.*

ABC-type ATPase: type of transport ATPase characterized by an "ATP-binding cassette" (hence the "ABC"), with the term *cassette* used to describe catalytic domains of the protein that bind ATP as an integral part of the transport process; also called ABC transporters. (p. 239) Also see *multidrug resistance transport protein.*

absolute refractory period: brief time during which the sodium channels of a nerve cell are inactivated and cannot be opened by depolarization. (p. 405)

absorption spectrum (plural, **spectra**)**:** relative extent to which light of different wavelengths is absorbed by a pigment. (p. 325)

accelerating voltage: difference in voltage between the cathode and anode of an electron microscope, responsible for accelerating electrons prior to their emission from the electron gun. (p. 840)

accessory pigments: molecules such as carotenoids and phycobilins that confer enhanced light-gathering properties on photosynthetic tissue by absorbing light of wavelengths not absorbed by chlorophyll; accessory pigments give distinctive colors to plant tissue, depending on their specific absorption properties. (p. 328)

acetyl CoA: high-energy, two-carbon compound generated by glycolysis and fatty acid oxidation; employed for transferring carbon atoms to the tricarboxylic acid cycle. (p. 287)

acetylcholine: the most common excitatory neurotransmitter used at synapses between neurons outside the central nervous system. (p. 411)

actin: principal protein of the microfilaments found in the cytoskeleton of nonmuscle cells and in the thin filaments of skeletal muscle; synthesized as a globular monomer (G-actin) that polymerizes into long, linear filaments (F-actin). (p. 462)

actin-binding proteins: proteins that bind to actin microfilaments, thereby regulating the length or assembly of microfilaments or mediating their association with each other or with other cellular structures, such as the plasma membrane. (p. 465)

action potential: brief change in membrane potential involving an initial depolarization followed by a rapid return to the normal resting potential; caused by the inward movement of Na^+ followed by the subsequent outward movement of K^+; serves as the means of transmission of a nerve impulse. (p. 403)

activated monomer: a monomer whose free energy has been increased by being linked to a carrier molecule. (p. 59)

activation domain: region of a transcription factor, distinct from the DNA-binding domain, that is responsible for activating transcription. (p. 766)

activation energy (E_A): energy required to initiate a chemical reaction. (p. 158)

activator (transcription): regulatory protein whose binding to DNA leads to an increase in the transcription rate of specific nearby genes. (p. 763)

active site: region of an enzyme molecule at which the substrate binds and the catalytic event occurs; also called the catalytic site. (p. 160)

active transport: membrane protein-mediated movement of a substance across a membrane against a concentration or electrochemical gradient; an energy-requiring process. (p. 236)

active zone: region of the presynaptic membrane of an axon where neurosecretory vesicles dock. (p. 413)

adaptor protein (AP): protein found along with clathrin in the coats of clathrin-coated vesicles. (p. 377)

adenine (A): nitrogen-containing aromatic base, chemically designated as a purine, that serves as an informational monomeric unit when present in nucleic acids with other bases in a specific sequence; forms a complementary base pair with thymine (T) or uracil (U) by hydrogen bonding. (p. 83)

adenosine diphosphate (ADP): adenosine with two phosphates linked to each other by a phosphoanhydride bond and to the 5′ carbon of the ribose by a phosphoester bond. (p. 83)

adenosine monophosphate (AMP): adenosine with a phosphate linked to the 5′ carbon of ribose by a phosphoester bond. (p. 83)

adenosine triphosphate (ATP): adenosine with three phosphates linked to each other by phosphoanhydride bonds and to the 5′ carbon of the ribose by a phosphoester bond; principal energy storage compound of most cells, with energy stored in the high-energy phosphoanhydride bonds. (pp. 83, 253)

adenylyl cyclase: enzyme that catalyzes the formation of cyclic AMP from ATP; located on the inner surface of the plasma membrane of many eukaryotic cells and activated by specific ligand-receptor interactions on the outer surface of the membrane. (p. 426)

adherens junction: junction for cell-cell adhesion that is connected to the cytoskeleton by actin microfilaments. (p. 510)

adhesive (anchoring) junction: type of cell junction that links the cytoskeleton of one cell either to the cytoskeleton of neighboring cells or to the extracellular matrix; examples include desmosomes, hemidesmosomes, and adherens junctions. (p. 510)

ADP: see *adenosine diphosphate.*

ADP ribosylation factor (ARF): a protein associated with COPI in the "fuzzy" coats of COPI-coated vesicles. (p. 378)

adrenergic hormone: epinephrine or norepinephrine. (p. 443)

adrenergic receptor: any of a family of G protein-linked receptors that bind to one or both of the adrenergic hormones, epinephrine and norepinephrine. (p. 444)

adrenergic synapse: a synapse that uses norepinephrine or epinephrine as the neurotransmitter. (p. 411)

aerobic respiration: exergonic process by which cells oxidize glucose to carbon dioxide and water using oxygen as the ultimate electron acceptor, with a significant portion of the released energy conserved as ATP. (p. 280)

agonist: substance that binds to a receptor and activates it. (p. 422)

Akt: protein kinase involved in the PI3K-Akt pathway; catalyzes the phosphorylation of several target proteins that suppress apoptosis and inhibit cell cycle arrest. (p. 445)

alcoholic fermentation: anaerobic catabolism of carbohydrates with ethanol and carbon dioxide as the end products. (p. 265)

allele: one of two or more alternative forms of a gene. (p. 629)

allosteric activator: a small molecule whose binding to an enzyme's allosteric site shifts the equilibrium to favor the high-affinity state of the enzyme. (p. 175)

allosteric effector: small molecule that causes a change in the state of an allosteric protein by binding to a site other than the active site. (p. 175)

allosteric enzyme: an enzyme exhibiting two alternative forms, each with a different biological property; interconversion of the two states is mediated by the reversible binding of a specific small molecule (allosteric effector) to a regulatory site called the allosteric site. (p. 175)

allosteric inhibitor: a small molecule whose binding to an enzyme's allosteric site shifts the equilibrium to favor the low-affinity state of the enzyme. (p. 175)

allosteric regulation: control of a reaction pathway by the effector-mediated reversible interconversion of the two forms of an allosteric enzyme. (pp. 175, 269, 263)

allosteric (regulatory) site: region of a protein molecule that is distinct from the active site at which the catalytic event occurs and that binds selectively to a small molecule, thereby regulating the protein's activity. (p. 175)

alpha beta heterodimer ($\alpha\beta$-heterodimer): protein dimer composed of one α-tubulin molecule and one β-tubulin molecule that forms the basic building block of microtubules. (p. 454)

alpha helix (α helix): spiral-shaped secondary structure of protein molecules, consisting of a backbone of peptide bonds with R groups of amino acids jutting out. (p. 76)

alpha tubulin (α-tubulin): protein that joins with β-tubulin to form a heterodimer that is the basic building block of microtubules. (p. 454)

alternating conformation model: membrane transport model in which a carrier protein alternates between two conformational states, such that the solute-binding site of the protein is open or accessible first to one side of the membrane and then to the other. (p. 230)

alternation of generations: occurrence of alternating haploid and diploid multicellular forms within the life cycle of an organism. (p. 631)

alternative splicing: utilization of different combinations of intron/exon splice junctions in pre-mRNA to produce messenger RNAs that differ in exon composition, thereby allowing production of more than one type of polypeptide from the same gene. (pp. 701, 773)

Ames test: screening test for potential carcinogens that assesses whether a substance causes mutations in bacteria. (p. 795)

amino acid: monomeric unit of proteins, consisting of a carboxylic acid with an amino group and one of a variety of R groups attached to the α carbon; 20 different kinds of amino acids are normally found in proteins. (p. 69)

amino terminus: see *N-terminus.*

aminoacyl site: see *A site.*

aminoacyl tRNA: a tRNA molecule containing an amino acid attached to its 3′ end. (p. 709)

aminoacyl-tRNA synthetase: enzyme that joins an amino acid to its appropriate tRNA molecule using energy provided by the hydrolysis of ATP. (p. 710)

amoeboid movement: mode of cell locomotion that depends on pseudopodia and involves cycles of gelation and solation of the actin cytoskeleton. (p. 501)

AMP: see *adenosine monophosphate.*

amphibolic pathway: series of reactions that can function both in a catabolic mode and as a source of precursors for anabolic pathways. (p. 294)

amphipathic molecule: molecule having spatially separated hydrophilic and hydrophobic regions. (p. 53)

amylopectin: branched-chain form of starch consisting of repeating glucose subunits linked together by $\alpha(1 \rightarrow 4)$ glycosidic bonds, with occasional $\alpha(1 \rightarrow 6)$ linkages creating branches every 12 to 25 units that commonly consist of 20 to 25 glucose units. (p. 91)

amylose: straight-chain form of starch consisting of repeating glucose subunits linked together by $\alpha(1 \rightarrow 4)$ glycosidic bonds. (p. 91)

anabolic pathway: series of reactions that results in the synthesis of cellular components. (p. 252)

anaerobic respiration: cellular respiration in which the ultimate electron acceptor is a molecule other than oxygen. (p. 280)

anaphase: stage during mitosis (or meiosis) when the sister chromatids (or homologous chromosomes) separate and move to opposite spindle poles. (p. 601)

anaphase A: movement of sister chromatids toward opposite spindle poles during anaphase. (p. 601)

anaphase B: movement of the spindle poles away from each other during anaphase. (p. 602)

anaphase-promoting complex: large multiprotein complex that targets selected proteins (e.g., securin and mitotic cyclin) for degradation, thereby initiating anaphase and the subsequent completion of mitosis. (p. 613)

anchorage-dependent growth: requirement that cells be attached to a solid surface such as the extracellular matrix before they can grow and divide. (p. 525)

anchorage-independent growth: a trait exhibited by cancer cells, which grow well not just when they are attached to a solid surface, but also when they are freely suspended in a liquid or semisolid medium. (p. 788)

anchoring junction: see *adhesive junction.*

aneuploidy: abnormal state in which a cell possesses an incorrect number of chromosomes. (p. 615)

angiogenesis: growth of new blood vessels. (p. 789)

angular aperture: half-angle of the cone of light entering the objective lens of a microscope from the specimen. (p. 825)

anion exchange protein: antiport carrier protein that facilitates the reciprocal exchange of chloride and bicarbonate ions across the plasma membrane. (p. 232)

anoxygenic phototroph: a photosynthetic organism that uses an oxidizable substrate other than water as the electron donor in photosynthetic electron transduction. (p. 323)

antagonist: substance that binds to a receptor and prevents it from being activated. (p. 422)

antenna pigment: light-absorbing molecule of a photosystem that absorbs photons and passes the energy to a neighboring chlorophyll molecule or accessory pigment by resonance energy transfer. (p. 329)

anterograde transport: movement of material from the ER through the Golgi complex toward the plasma membrane. (p. 362)

antibody: class of proteins produced by lymphocytes that bind with extraordinary specificity to substances, referred to as antigens, that provoke an immune response. (p. 817)

anticodon: triplet of nucleotides located in one of the loops of a tRNA molecule that recognizes the appropriate codon in mRNA by complementary base pairing. (p. 709)

antigen: a foreign or abnormal substance that can trigger an immune response. (p. 817)

antiport: coupled transport of two solutes across a membrane in opposite directions. (p. 231)

AP: see *adaptor protein.*

***APC* gene:** a tumor suppressor gene, frequently mutated in colon cancers, that codes for a protein involved in the Wnt pathway. (p. 808)

apoptosis: cell suicide mediated by a group of protein-degrading enzymes called caspases; involves a programmed series of events that leads to the dismantling of the internal contents of the cell. (p. 620)

AQP: see *aquaporin.*

aquaporin (AQP): any of a family of membrane channel proteins that facilitate the rapid movement of water molecules into or out of cells in tissues that require this capability, such as the proximal tubules of the kidneys. (p. 236)

archaea: one of the two groups of prokaryotes, the other being bacteria; many archaea thrive under harsh conditions, such as salty, acidic, or hot environments, that would be fatal to most other organisms. (p. 103) Also see *bacteria.*

ARF: see *ADP ribosylation factor.*

Arp2/3 complex: complex of actin-related proteins that allows actin monomers to polymerize as new "branches" on the sides of existing microfilaments. (p. 467)

asexual reproduction: form of reproduction in which a single parent is the only contributor of genetic information to the new organism. (p.)

assisted self-assembly: folding and assembly of proteins and protein-containing structures in which the appropriate molecular chaperone is required to ensure that correct assembly will predominate over incorrect assembly. (p. 62)

astral microtubule: type of microtubule that forms asters, which are dense starbursts of microtubules that radiate in all directions from each spindle pole. (p. 601)

asymmetric carbon atom: carbon atom that has four different substituents. Two different stereoisomers are possible for each asymmetric carbon atom in an organic molecule. (p. 49)

ATP: see *adenosine triphosphate.*

ATP synthase: alternative name for an F-type ATPase when it catalyzes the reverse process in which the exergonic flow of protons down their electrochemical gradient is used to drive ATP synthesis; examples include the CF_oCF_1 complex found in chloroplast thylakoid membranes, and the F_o F_1 complex found in mitochondrial inner membranes and bacterial plasma membranes. (pp. 239, 308, 335)

attenuation: mechanism for regulating bacterial gene expression based on the prematuretermination of transcription. (p. 747)

autophagic lysosome: mature lysosome containing hydrolytic enzymes involved in the digestion of materials of intracellular origin. (p. 382) Also see *heterophagic lysosome.*

autophagic vacuole (autophagosome): vacuole formed when an old or unwanted organelle or other cellular structure is wrapped in membranes derived from the endoplasmic reticulum prior to digestion by lysosomal enzymes. (p. 382)

autophagosome: see *autophagic vacuole.*

autophagy: intracellular digestion of old or unwanted organelles or other cell structures as it occurs within autophagic lysosomes; "self-eating." (p. 382)

autophosphorylation: phosphorylation of a receptor molecule by a receptor molecule of the same type. (p. 436)

autoradiography: procedure for detecting the location of radioactive molecules by overlaying a sample with photographic film, which becomes darkened upon exposure to radioactivity. (p. 547)

axon: extension of a nerve cell that conducts impulses away from the cell body. (p. 394)

axon hillock: region at the base of an axon where action potentials are initiated most easily. (p. 406)

axonal transport: see *fast axonal transport.*

axonemal dynein: motor protein in the axonemes of cilia and flagella that generates axonemal motility by moving along the surface of microtubules driven by energy derived from ATP hydrolysis. (p. 480)

axonemal microtubules: microtubules present in highly ordered bundles in the axonemes of eukaryotic cilia and flagella. (p. 452)

axoneme: group of interconnected microtubules that form the backbone of a eukaryotic cilium or flagellum, usually arranged as nine outer doublet microtubules surrounding a pair of central microtubules. (p. 482)

axoplasm: cytoplasm within the axon of a nerve cell. (p. 394)

B

B tubule: an incomplete microtubule that is fused to a complete microtubule (the A tubule) to make up an outer doublet in the axoneme of a eukaryotic cilium or flagellum. (p. 483)

BAC: see *bacterial artificial chromosome.*

backcrossing: process in a genetic breeding experiment in which a heterozygote is cross-fertilized with one of the original homozygous parental organisms. (p. 641)

bacteria (singular, bacterium): one of the two groups of prokaryotes, the other being archaea; include most of the commonly encountered single-celled organisms with no nucleus that have traditionally been called bacteria. (p. 103) Also see *archaea.*

bacterial artificial chromosome (BAC): bacterial cloning vector derived from the F-factor plasmid that is useful in cloning large DNA fragments. (p. 662)

bacterial chromosome: circular (or linear) DNA molecule with bound proteins that contains the main genome of a bacterial cell. (p. 557)

bacteriochlorophyll: type of chlorophyll found in bacteria that is able to extract electrons from donors other than water. (p. 327)

bacteriophage (phage): virus that infects bacterial cells. (p. 535)

bacteriorhodopsin: transmembrane protein complexed with rhodopsin, capable of transporting protons across the bacterial cell membrane to create a light-dependent electrochemical proton gradient. (p. 243)

basal body: microtubule-containing structure located at the base of a eukaryotic flagellum or cilium that consists of nine sets of triplet microtubules; identical in appearance to a centriole. (p. 482)

basal lamina: thin sheet of specialized extracellular matrix material that separates epithelial cells from underlying connective tissues. (p. 521)

base excision repair: DNA repair mechanism that removes and replaces single damaged bases in DNA. (p. 596)

base pair (bp): pair of nucleotides joined together by complementary hydrogen bonding. (p. 545)

base pairing: complementary relationship between purines and pyrimidines based on hydrogen bonding that provides a mechanism for nucleic acids to recognize and bind to each other; involves the pairing of A with T or U, and the pairing of G with C. (p. 87)

Bcl–2: protein located in the outer mitochondrial membrane that blocks cell death by apoptosis. (p. 622)

benign tumor: tumor that grows only locally, unable to invade neighboring tissues or spread to other parts of the body. (p. 787)

beta oxidation (β oxidation): pathway involving successive cycles of fatty acid oxidation in which the fatty acid chain is shortened each time by two carbon atoms released as acetyl CoA. (p. 293)

beta sheet (β sheet): extended sheetlike secondary structure of proteins in which adjacent polypeptides are linked by hydrogen bonds between amino and carbonyl groups. (p. 76)

beta tubulin (β-tubulin): protein that joins with α-tubulin to form a heterodimer that is the basic building block of microtubules. (p. 454)

bimetallic iron-copper (Fe-Cu) center: a complex formed between a single copper atom and the iron atom bound to the heme group of an oxygen-binding cytochrome such as cytochrome a_3; important in keeping an O_2 molecule bound to the cytochrome until it picks up four electrons and four protons, resulting in the release of two molecules of water. (p. 298)

binding change model: a mechanism involving the physical rotation of the γ subunit of an $F_o F_1$ ATP synthase postulated to explain how the exergonic flow of protons through the F_o component of the complex drives the otherwise endergonic phosphorylation of ADP to ATP by the F_1 component. (p. 310)

biochemistry: study of the chemistry of living systems; same as biological chemistry. (p. 31)

bioenergetics: area of science that deals with the application of thermodynamic principles to reactions and processes in the biological world. (p. 139)

bioinformatics: using computers to analyze the vast amounts of data generated by sequencing and expression studies on genomes and proteomes. (pp. 39, 551)

biological chemistry: study of the chemistry of living systems; called biochemistry for short. (p. 47)

bioluminescence: production of light by an organism as a result of the reaction of ATP with specific luminescent compounds. (p. 136)

biosynthesis: generation of new molecules through a series of chemical reactions within the cell. (p. 135)

BiP: member of the Hsp70 family of chaperones; present in the ER lumen, where it facilitates protein folding by reversibly binding to the hydrophobic regions of polypeptide chains. (p. 727)

bivalent: pair of homologous chromosomes that have synapsed during the first meiotic division; contains four chromatids, two from each chromosome. (p. 632)

BLAST (Basic Local Alignment Search Tool): software program that can search databases and locate DNA or protein sequences that resemble any known sequence of interest. (p. 552)

bond energy: amount of energy required to break one mole of a particular chemical bond. (p. 47)

bp: see *base pair.*

***BRCA1* and *BRCA2* genes:** tumor suppressor genes in which inheritance of a single mutant copy creates a high risk for breast and ovarian cancer; code for proteins involved in repairing double-strand DNA breaks. (p. 809)

BRE (TFIIB recognition element): component of core promoters for RNA polymerase II, located immediately upstream from the TATA box. (p. 690)

brightfield microscopy: light microscopy of specimen that possesses color, has been stained, or has some other property that affects the amount of light that passes through, thereby allowing an image to be formed. (p. 828)

bundle sheath cell: internal cell of the leaf of a C_4 plant located in close proximity to the vascular bundle (vein) of the leaf; site of the Calvin cycle in such plants. (p. 345)

buoyant density centrifugation: see *equilibrium density centrifugation.*

Burkitt lymphoma: a lymphocyte cancer associated with infection by Epstein-Barr virus along with a chromosome translocation in which the *MYC* gene is activated by moving it from chromosome 8 to 14. (p. 799)

C

C: see *cytosine.*

C value: amount of DNA in a single (haploid) set of chromosomes. (p. 637)

C_3 plant: plant that depends solely on the Calvin cycle for carbon dioxide fixation, creating the three-carbon compound 3-phosphoglycerate as the initial product. (p. 345)

C_4 plant: plant that uses the Hatch-Slack pathway in mesophyll cells to carry out the initial fixation of carbon dioxide, creating the four-carbon compound oxaloacetate; the assimilated carbon is subsequently released again in bundle sheath cells and recaptured by the Calvin cycle. (p. 345)

cadherin: any of a family of plasma membrane glycoproteins that mediate Ca^{2+}-dependent adhesion between cells. (p. 507)

cal: see *calorie.*

calcium ATPase: membrane protein that transports calcium ions across a membrane using energy derived from ATP hydrolysis; prominent example occurs in the sarcoplasmic reticulum (SR), where it pumps calcium ions into the SR lumen. (pp. 431, 496)

calcium indicator: a dye or genetically engineered protein whose fluorescence changes in response to the local cytosolic calcium concentration. (p. 430)

calcium ionophore: a molecule that increases the permeability of membranes to calcium ions. (p. 430)

calmodulin: calcium-binding protein involved in mediating many of the intracellular effects of calcium ions in eukaryotic cells. (p. 432)

calnexin: a membrane-bound ER protein that forms a protein complex with a newly synthesized glycoprotein and assists in its proper folding. (p. 362)

calorie (cal): unit of energy; amount of energy needed to raise the temperature of 1 gram of water 1°C. (pp. 48, 140)

calreticulin: a soluble ER protein that forms a protein complex with a newly synthesized glycoprotein and assists in its proper folding. (p. 362)

Calvin cycle: cyclic series of reactions used by photosynthetic organisms for the fixation of carbon dioxide and its reduction to form carbohydrates. (p. 337)

CAM: see *crassulacean acid metabolism.*

CAM plant: plant that carries out crassulacean acid metabolism. (p. 348)

cAMP: see *cyclic AMP.*

cancer: uncontrolled, growing mass of cells that is capable of invading neighboring tissues and spreading via body fluids, especially the bloodstream, to other parts of the body; also called a *malignant tumor.* (p. 787)

CAP: see *catabolite activator protein.*

cap (5′cap): methylated structure at the 5′ end of eukaryotic mRNAs created by adding 7-methylguanosine and methylating the ribose rings of the first, and often the second, nucleotides of the RNA chain. (p. 696)

capping protein: protein that binds to the end of an actin microfilament, thereby preventing the further addition or loss of subunits. (p. 465)

carbohydrate: general name given to molecules that contain carbon, hydrogen, and oxygen in a ratio $C_n(H_2O)n$; examples include starch, glycogen, and cellulose. (p. 90)

carbon assimilation reactions: portion of the photosynthetic pathway in which fully oxidized carbon atoms from carbon dioxide are fixed (covalently attached) to organic acceptor molecules and then reduced and rearranged to form carbohydrates and other organic compounds required for building a living cell. (p. 321)

carbon atom: the most important atom in biological molecules, capable of forming up to four covalent bonds. (p. 47)

carboxyl terminus: see *C-terminus*.

carcinogen: any cancer-causing agent. (p. 795)

carcinoma: a malignant tumor (cancer) arising from the epithelial cells that cover external and internal body surfaces. (p. 786)

cardiac (heart) muscle: striated muscle of the heart, highly dependent on aerobic respiration. (p. 496)

caretaker gene: a tumor suppressor gene involved in DNA repair or chromosome sorting; loss-of-function mutations in such genes contribute to genetic instability. (p. 810)

carotenoid: any of several accessory pigments found in most plant species that absorb in the blue region of the visible spectrum (420–480 nm) and are therefore yellow or orange in color. (p. 328)

carrier molecule: a molecule that joins to a monomer, thereby activating the monomer for a subsequent reaction. (p. 59)

carrier protein: membrane protein that transports solutes across the membrane by binding to the solute on one side of the membrane and then undergoing a conformational change that transfers the solute to the other side of the membrane. (p. 230)

caspase: any of a family of proteases that degrade other cellular proteins as part of the process of apoptosis. (p. 621)

cassette mechanism: process of DNA rearrangement in which alternative alleles for yeast mating type are inserted into the *MAT* locus for transcription. (p. 754)

catabolic pathway: series of reactions that results in the breakdown of cellular components. (p. 252)

catabolite activator protein (CAP): bacterial protein that binds cyclic AMP and then activates the transcription of catabolite-repressible genes. (p. 744)

catalyst: agent that enhances the rate of a reaction by lowering the activation energy without itself being consumed; catalysts change the rate at which a reaction approaches equilibrium, but not the position of equilibrium. (p. 159)

catalytic subunit: a subunit of a multisubunit enzyme that contains the enzyme's catalytic site. (p. 175)

catecholamine: any of several compounds derived from the amino acid tyrosine that functions as a hormone and/or neurotransmitter. (p. 411)

caveolae: small invaginations of the plasma membrane that are coated with the protein caveolin; a type of lipid raft enriched in cholesterol that may be involved in cholesterol uptake or signal transduction. (p. 376)

CBP: transcriptional coactivator that exhibits histone acetyltransferase activity and associates with RNA polymerase to facilitate assembly of the transcription machinery at gene promoters. (p. 770)

Cdc42: member of a family of monomeric G proteins, which also includes Rac and Rho, that stimulate formation of various actin-containing structures within cells. (p. 468)

Cdk: see *cyclin-dependent kinase*.

Cdk inhibitor: any of several proteins that restrain cell growth and division by inhibiting Cdk-cyclin complexes. (p. 547)

cDNA: see *complementary DNA*.

cDNA library: collection of recombinant DNA clones produced by copying the entire mRNA population of a particular cell type with reverse transcriptase and then cloning the resulting cDNAs. (p. 661)

cell: the basic structural and functional unit of living organisms; the smallest structure capable of performing the essential functions characteristic of life. (p. 29)

cell body: portion of a nerve cell that contains the nucleus and other organelles and has extensions called axons and dendrites projecting from it. (p. 394)

cell-cell junction: specialized connection between the plasma membranes of adjoining cells for the purpose of adhesion, sealing, or communication. (p. 509)

cell cycle: stages involved in preparing for and carrying out cell division; begins when two new cells are formed by the division of a single parental cell and is completed when one of these cells divides again into two cells. (p. 577)

cell differentiation: process by which cells acquire the specialized properties that distinguish different types of cells from each other. (p. 750)

cell division: process by which one cell gives rise to two. (p. 577)

cell membrane: see *plasma membrane*.

cell plate: flattened sac representing a stage in plant cell wall formation, leading to separation of the two daughter nuclei during plant cell division. (p. 607)

cell theory: theory of cellular organization stating that all organisms consist of one or more cells, that the cell is the basic unit of structure for all organisms, and that all cells arise only from preexisting cells. (p. 30)

cell wall: rigid, nonliving structure exterior to the plasma membrane of bacterial, algal, fungal, and plant cells; plant cell walls consist of cellulose microfibrils embedded in a noncellulosic matrix. (pp. 126, 526)

cellular respiration: oxidation-driven flow of electrons from reduced coenzymes to an electron acceptor, usually accompanied by the generation of ATP. (p. 280)

cellulose: structural polysaccharide present in plant cell walls, consisting of repeating glucose units linked by $\beta(1 \rightarrow 4)$ bonds. (pp. 93, 526)

central pair: two parallel microtubules located in the center of the axoneme of a eukaryotic cilium or flagellum. (p. 482)

central vacuole: large membrane-bounded organelle present in many plant cells; helps maintain turgor pressure of the plant cell, plays a limited storage role, and is also capable of a lysosome-like function in intracellular digestion. (p. 121)

centrifugation: process of rapidly spinning a tube containing a fluid to subject its contents to a centrifugal force. (p. 355)

centrifuge: machine for rapidly spinning a tube containing a fluid to subject its contents to a centrifugal force. (p. 355)

centriole: structure consisting of nine sets of triplet microtubules embedded within the centrosome of animal cells, where two centrioles lie at right angles to each other; identical in structure to the basal body of eukaryotic cilia and flagella. (p. 458)

centromere: chromosome region where sister chromatids are held together prior to anaphase and where kinetochores are attached; contains simple-sequence, tandemly repeated DNA. (pp. 554, 601)

centrosome: small zone of granular material surrounding two centrioles located adjacent to the nucleus of animal cells; functions as a cell's main microtubule-organizing center. (pp. 458, 600)

cerebroside: an uncharged glycolipid containing the amino alcohol sphingosine. (p. 191)

CF: see *cystic fibrosis*.

CF_1: component of the chloroplast ATP synthase complex that protrudes from the stromal side of thylakoid membranes and contains the catalytic site for ATP synthesis. (p. 335)

CF_o: component of the chloroplast ATP synthase complex that is embedded in thylakoidmembranes and serves as the proton translocator. (p. 335)

CF_oCF_1 complex: ATP synthase complex found in chloroplast thylakoid membranes; catalyzes the process by which the exergonic flow of protons down their electrochemical gradient is used to drive ATP synthesis. (p. 335)

CFTR: see *cystic fibrosis transmembrane conductance regulator*.

CGN: see *cis-Golgi network*.

channel gating: closing of a membrane ion channel in such a way that it can reopen immediately in response to an appropriate stimulus. (p. 402)

channel inactivation: closing of a membrane ion channel in such a way that it cannot reopen immediately. (p. 402)

channel protein: membrane protein that forms a hydrophilic channel through which solutes can pass across the membrane without any change in the conformation of the channel protein. (p. 230)

chaperone: see *molecular chaperone*.

Chargaff's rules: observation, first made by Erwin Chargaff, that in DNA the number of adenines is equal to the number of thymines ($A = T$) and the number of guanines is equal to the number of cytosines ($G = C$). (p. 538)

charge repulsion: force driving apart two ions, molecules, or regions of molecules of the same electric charge. (p. 254)

checkpoint: pathway that monitors conditions within the cell and transiently halts the cell cycle if conditions are not suitable for continuing. (p. 615) Also see *DNA damage checkpoint, DNA replication checkpoint*, and *mitotic spindle checkpoint*.

chemical synapse: junction between two nerve cells where a nerve impulse is transmitted between the cells by neurotransmitters that diffuse across the synaptic cleft from the presynaptic cell to the postsynaptic cell. (p. 409)

chemiosmotic coupling model: model postulating that electron transport pathways establish proton gradients across membranes and that the energy stored in such gradients can then be used to drive ATP synthesis. (p. 305)

chemotaxis: cell movement toward a chemical attractant or away from a chemical repellent. (p. 500)

chemotroph: organism that is dependent on the bond energies of organic molecules such as carbohydrates, fats, and proteins to satisfy energy requirements. (p. 137)

chemotrophic energy metabolism: reactions and pathways by which cells catabolize nutrients such as carbohydrates, fats, and proteins, conserving as ATP some of the free energy that is released in the process. (p. 257)

chiasma (plural, **chiasmata**)**:** connection between homologous chromosomes produced by crossing over during prophase I of meiosis. (p. 633)

chitin: structural polysaccharide found in insect exoskeletons and crustacean cells; consists of N-acetylglucosamine units linked by $\beta(1 \rightarrow 4)$ bonds. (p. 93)

chlorophyll: light-absorbing molecule that donates photoenergized electrons to organic molecules, initiating photochemical events that lead to the generation of the NADPH and ATP required for the Calvin cycle; because of its absorption properties, chlorophyll gives plants their characteristic green color. (p. 326)
chlorophyll-binding protein: any of several proteins that bind to and stabilize the arrangement of chlorophyll molecules within a photosystem. (p. 328)
chloroplast: double membrane-enclosed cytoplasmic organelle of plants and algae that contains chlorophyll and the enzymes needed to carry out photosynthesis. (pp. 114, 323)
cholesterol: lipid constituent of animal cell plasma membrane; serves as a precursor to the steroid hormones. (pp. 98, 193)
cholinergic synapse: a synapse that uses acetylcholine as the neurotransmitter. (p. 411)
chromatid: see *sister chromatid.*
chromatin: DNA-protein fibers that make up chromosomes; constructed from nucleosomes spaced regularly along a DNA chain. (pp. 112, 558)
chromatin fiber (30 nm): fiber formed by packing together the nucleosomes of a 10-nm chromatin fiber. (p. 561)
chromatin remodeling protein: protein that induces alterations in nucleosome structure, packing, and/or position designed to give transcription factors access to DNA target sites in the promoter region of a gene. (p. 760)
chromatography: a group of related techniques that utilize the flow of a fluid phase over a nonmobile absorbing phase to separate molecules based on their relative affinities for the two phases, which in turn reflect differences in size, charge, hydrophobicity, or affinity for a particular chemical group. (p. 37)
chromosome: in eukaryotes a single DNA molecule, complexed with histones and other proteins, that becomes condensed into a compact structure at the time of mitosis or meiosis. (pp. 38, 109, 558) Also see *bacterial chromosome.*
chromosome puff: uncoiled region of a polytene chromosome that is undergoing transcription. (p. 756)
chromosome theory of heredity: theory stating that hereditary factors are located in thechromosomes within the nucleus. (p. 38)
cilium (plural, **cilia**)**:** membrane-bounded appendage on the surface of a eukaryotic cell, composed of a specific arrangement of microtubules and responsible for motility of the cell or the fluids around cells; shorter and more numerous than closely related organelles called flagella. (p. 481) Also see *flagellum.*
***cis*-acting element:** a DNA sequence to which a regulatory protein can bind. (p. 744)
***cis*-Golgi network (CGN):** region of the Golgi complex consisting of a network of membrane-bounded tubules that are located closest to the transitional elements. (p. 360)
cisterna (plural, **cisternae**)**:** membrane-bounded flattened sac, such as in the endoplasmicreticulum or Golgi complex. (p. 117)
cisternal maturation model: model postulating that Golgi cisternae are transient compartments that gradually change from *cis*-Golgi network cisternae into medial cisternae and then into *trans*-Golgi network cisternae. (p. 360)
***cis-trans* test:** analysis used to determine whether a mutation in a bacterial operon affects a regulatory protein or the DNA sequence to which it binds. (p. 743)
clathrin: large protein that forms a "cage" around the coated vesicles and coated pits involved in endocytosis and other intracellular transport processes. (p. 376)
clathrin-dependent endocytosis: see *receptor-mediated endocytosis.*
claudin: transmembrane protein that forms the main structural component of a tight junction. (p. 514)
cleavage: process of cytoplasmic division in animal cells, in which a band of actinmicrofilaments lying beneath the plasma membrane constricts the cell at the midline and eventually divides it in two. (p. 606)
cleavage furrow: groove formed during the division of an animal cell that encircles the cell and deepens progressively, leading to cytoplasmic division. (p. 606)
clone: organism (or cell or molecule) that is genetically identical to another organism (or cell or molecule) from which it is derived. (pp. 750, 792)
cloning vector: DNA molecule to which a selected DNA fragment can be joined prior to rapid replication of the vector in a host cell (usually a bacterium); phage and plasmid DNA are the most common cloning vectors. (p. 656)
CNVs: see *copy number variations.*
CoA: see *coenzyme A.*
coactivator: class of proteins that mediate interactions between activators and the genes they regulate; include histone-modifying enzymes such as HAT, chromatin-remodeling proteins such as SWI/SNF, and Mediator. (p. 764)
coated vesicle: any of several types of membrane vesicles involved in vesicular traffic within the endomembrane system; surrounded by a coat protein such as clathrin, COPI, COPII, or caveolin. (p. 372)
coding strand: the nontemplate strand of a DNA double helix, which is base-paired to the template strand; identical in sequence to the single-stranded RNA molecules transcribed from the template strand, except that RNA has uracil (U) where the coding strand has thymine (T). (p. 680)
codon: triplet of nucleotides in an mRNA molecule that serves as a coding unit for an amino acid (or a start or stop signal) during protein synthesis. (p. 682)
coenzyme: small organic molecule that functions along with an enzyme by serving as a carrier of electrons or functional groups. (p. 258)
coenzyme A (CoA): organic molecule that serves as a carrier of acyl groups by forming a high-energy thioester bond with an organic acid. (p. 287)
coenzyme Q (CoQ): nonprotein (quinone) component of the mitochondrial electron transport system that serves as the collection point for electrons from both FMN- and FAD-linked dehydrogenases; also called ubiquinone. (p. 299)
coenzyme Q-cytochrome *c* oxidoreductase: see *complex III.*
cohesin: protein that holds sister chromatids together prior to anaphase. (p. 614)
colchicine: plant-derived drug that binds to tubulin and prevents its polymerization into microtubules. (p. 456)
collagen: a family of closely related proteins that form high-strength fibers found in highconcentration in the extracellular matrix of animals. (p. 516)
collagen fiber: extremely strong fibers measuring several micrometers in diameter found in the extracellular matrix; constructed from collagen fibrils that are, in turn, composed of collagen molecules lined up in a staggered array. (p. 517)
combinatorial model for gene regulation: model proposing that complex patterns of tissue-specific gene expression can be achieved by a relatively small number of DNA control elements and their respective transcription factors acting in different combinations. (p. 765)
complementary: in nucleic acids, the ability of guanine (G) to form a hydrogen-bonded base pair with cytosine (C), and adenine (A) to form a hydrogen-bonded base pair with thymine (T) or uracil (U). (p. 540)
complementary DNA (cDNA): DNA molecule copied from an mRNA template by the enzyme reverse transcriptase. (p. 661)
complex I (NADH-coenzyme Q oxidoreductase): multiprotein complex of the electron transport system that catalyzes the transfer of electrons from NADH to coenzyme Q. (p. 302)
complex II (succinate-coenzyme Q oxidoreductase): multiprotein complex of the electron transport system that catalyzes the transfer of electrons from succinate to coenzyme Q. (p. 302)
complex III (coenzyme Q-cytochrome *c* oxidoreductase): multiprotein complex of the electron transport system that catalyzes the transfer of electrons from coenzyme Q to cytochrome *c*. (p. 302)
complex IV (cytochrome *c* oxidase): multiprotein complex of the electron transport system that catalyzes the transfer of electrons from cytochrome *c* to oxygen. (p. 302)
compound microscope: light microscope that uses several lenses in combination; usually has a condenser lens, an objective lens, and an ocular lens. (p. 827)
concentration gradient: transmembrane gradient in concentration of a molecule or ion, expressed as a ratio of the concentration of the substance on one side of the membrane to its concentration on the other side of the membrane; the sole driving force for transport of molecules across a membrane, but only one of two components of the electrochemical potential that serves as the driving force for transport of ions across a membrane. (p. 224)
concentration work: use of energy to transport ions or molecules across a membrane against an electrochemical or concentration gradient. (p. 135)
condensation reaction: chemical reaction that results in the joining of two molecules by the removal of a water molecule. (p. 58)
condenser lens: lens of a light microscope (or electron microscope) that is the first lens to direct the light rays (or electron beam) from the source toward the specimen. (pp. 827, 841)
confocal microscope: specialized type of light microscope that employs a laser beam toilluminate a single plane of the specimen at a time. (p. 833)
conformation: three-dimensional shape of a polypeptide or other biological macromolecule. (p. 72)
conjugation: cellular mating process by which DNA is transferred from one bacterial cell to another. (p. 649)
connexon: assembly of six protein subunits with a hollow center that forms a channel through the plasma membrane at a gap junction. (p. 514)
consensus sequence: the most common version of a DNA base sequence when that sequence occurs in slightly different forms at different sites. (p. 685)
constitutive heterochromatin: chromosomal regions that are condensed in all cells of an

organism at virtually all times and are therefore genetically inactive. (p. 571) Also see *facultative heterochromatin*.

constitutive secretion: continuous fusion of secretory vesicles with the plasma membrane and expulsion of their contents to the cell exterior, independent of specific extracellular signals. (p. 368)

constitutively active mutation: mutation that causes a receptor to be active even when no activating ligand is present. (p. 440)

contractile ring: beltlike bundle of actin microfilaments that forms beneath the plasma membrane and acts to constrict the cleavage furrow during the division of an animal cell. (p. 606)

contractility: shortening of muscle cells. (p. 477)

cooperativity: property of enzymes possessing multiple catalytic sites in which the binding of a substrate molecule to one catalytic site causes conformational changes that influence the affinity of the remaining sites for substrate. (p. 176)

COPI: main protein component of the "fuzzy" coats of COPI-coated vesicles, which are involved in retrograde transport from the Golgi complex back to the ER and between Golgi complex cisternae. (p. 378)

COPII: main protein component of COPII-coated vesicles, which are involved in transport from the ER to the Golgi complex. (p. 378)

copy number variations (CNVs): variations in the number of copies of DNA segments, thousands of bases long, that occur among individuals of the same species. (p. 553)

CoQ: see *coenzyme Q*.

core oligosaccharide: initial oligosaccharide segment joined to an asparagine residue during N-glycosylation of a polypeptide chain; consists of two N-acetylglucosamine units, nine mannose units, and three glucose units. (p. 362)

core promoter: minimal set of DNA sequences sufficient to direct the accurate initiation of transcription by RNA polymerase. (p. 689)

core protein: a protein molecule to which numerous glycosaminoglycan chains are attached to form a proteoglycan. (p. 519)

corepressor: effector molecule required along with the repressor to prevent transcription of a bacterial operon. (p. 745)

correlative microscopy: combination of light and electron-based microscopy that allows the fine structure associated with a fluorescent signal to be examined at high resolution, often using immunoelectron microscopy. (p. 844)

cotranslational import: transfer of a growing polypeptide chain across (or, in the case of integral membrane proteins, into) the ER membrane as polypeptide synthesis proceeds. (p. 724)

coupled transport: coordinated transport of two solutes across a membrane in such a way that transport of either stops if the other is stopped or interrupted; the two solutes may move in the same direction (symport) or in opposite directions (antiport). (p. 231)

covalent bond: strong chemical bond in which two atoms share two or more electrons. (p. 47)

covalent modification: type of regulation in which the activity of an enzyme (or other protein) is altered by the addition or removal of specific chemical groups. (p. 176)

crassulacean acid metabolism (CAM): pathway in which plants use PEP carboxylase to fix CO_2 at night, generating the four-carbon acid, malate. The malate is then decarboxylated during the day to release CO_2, which is fixed by the Calvin cycle. (p. 348)

CREB: transcription factor that activates the transcription of cyclic AMP-inducible genes by binding to cAMP response elements in DNA. (p. 770)

crista (plural, **cristae**)**:** infolding of the inner mitochondrial membrane into the matrix of the mitochondrion, thereby increasing the total surface area of the inner membrane; contains the enzymes of electron transport and oxidative phosphorylation. (pp. 113, 283)

critical concentration: tubulin concentration at which the rate of assembly of tubulin subunits into a polymer is exactly balanced with the rate of disassembly. (p. 455)

critical point dryer: heavy metal canister used to dry a specimen under conditions of controlled temperature and pressure. (p. 848)

cross-bridge: structure formed by contact between the myosin heads of thick filaments and the thin filaments in muscle myofibrils. (p. 492)

crossing over: exchange of DNA segments between homologous chromosomes. (p. 633)

cryoEM: technique in which cryofixed biological samples are directly imaged in a transmission electron microscope at low temperature; often used to examine suspensions of isolated macromolecules. (p. 850)

cryofixation: rapid freezing of small samples so that cellular structures can be immobilized in milliseconds; often followed by freeze substitution, in which an organic solvent replaces the frozen water in the sample. (p. 843)

C-terminus (carboxyl terminus): the end of a polypeptide chain that contains the last amino acid to be incorporated during mRNA translation; usually retains a free carboxyl group. (p. 72)

current: movement of positive or negative ions. (p. 396)

cyclic AMP (cAMP): adenosine monophosphate with the phosphate group linked to both the 3′ and 5′ carbons by phosphodiester bonds; functions in both prokaryotic and eukaryotic gene regulation; in eukaryotes, acts as a second messenger that mediates the effects of various signaling molecules by activating protein kinase A. (p. 426)

cyclic electron flow: light-driven transfer of electrons from photosystem I through a sequence of electron carries that returns them to a chlorophyll molecule of the same photosystem, with the released energy used to drive ATP synthesis. (p. 336)

cyclin: any of a group of proteins that activate the cyclin-dependent kinases (Cdks) involved in regulating progression through the eukaryotic cell cycle. (p. 611)

cyclin-dependent kinase (Cdk): any of several protein kinases that are activated by different cyclins and that control progression through the eukaryotic cell cycle by phosphorylating various target proteins. (p. 611)

cyclosis: see *cytoplasmic streaming*.

cystic fibrosis (CF): a disease whose symptoms result from an inability to secrete chloride ions that is, in turn, caused by a genetic defect in a membrane protein that functions as a chloride ion channel. (p. 234)

cystic fibrosis transmembrane conductance regulator (CFTR): a membrane protein that functions as a chloride ion channel, a mutant form of which can lead to cystic fibrosis. (p. 234)

cytochalasins: family of drugs produced by certain fungi that inhibit a variety of cell movements by preventing actin polymerization. (p. 463)

cytochrome: family of heme-containing proteins of the electron transport system; involved in the transfer of electrons from coenzyme Q to oxygen by the oxidation and reduction of the central iron atom of the heme group. (p. 298)

cytochrome b_6/f complex: multiprotein complex within the thylakoid membrane that transfers electrons from a plastoquinol to plastocyanin as part of the energy transduction reactions of photosynthesis. (p. 333)

cytochrome *c*: heme-containing protein of the electron transport system that also plays a role in triggering apoptosis when released from mitochondria. (p. 622)

cytochrome *c* oxidase: see *complex IV*.

cytochrome P-450: family of heme-containing proteins, located mainly in the liver, that catalyze hydroxylation reactions involved in drug detoxification and steroid biosynthesis. (p. 358)

cytokinesis: division of the cytoplasm of a parent cell into two daughter cells; usually follows mitosis. (p. 577)

cytology: study of cellular structure, based primarily on microscopic techniques. (p. 31)

cytoplasm: that portion of the interior of a eukaryotic cell that is not occupied by the nucleus; includes organelles, such as mitochondria and components of the endomembrane system, as well as the cytosol. (p. 123)

cytoplasmic dynein: cytoplasmic motor protein that moves along the surface of microtubules in the plus-to-minus direction, driven by energy derived from ATP hydrolysis; associated with dynactin, which links cytoplasmic dynein to cargo vesicles. (p. 480)

cytoplasmic microtubules: microtubules arranged in loosely organized, dynamic networks in the cytoplasm of eukaryotic cells. (p. 452)

cytoplasmic streaming: movement of the cytoplasm driven by interactions between actin filaments and specific types of myosin; also called cyclosis in plant cells. (p. 501)

cytosine (C): nitrogen-containing aromatic base, chemically designated as a pyrimidine, that serves as an informational monomeric unit when present in nucleic acids with other bases in a specific sequence; forms a complementary base pair with guanine (G) by hydrogen bonding. (p. 83)

cytoskeleton: three-dimensional, interconnected network of microtubules, microfilaments, and intermediate filaments that provides structure to the cytoplasm of a eukaryotic cell and plays an important role in cell movement. (pp. 123, 450)

cytosol: the semifluid substance in which the organelles of the cytoplasm are suspended. (p. 123)

D

DAG: see *diacylglycerol*.

deep etching: modification of freeze-etching technique in which ultrarapid freezing and a volatile cryoprotectant are used to extend the etching period, thereby removing a deeper layer of ice and allowing the interior of a cell to be examined in depth. (p. 847)

degenerate code: ability of the genetic code to use more than one triplet code to specify the same amino acid. (p. 680)

degron: amino acid sequence within a protein that is used in targeting the protein for destruction. (p. 780)

dehydrogenation: removal of electrons plus hydrogen ions (protons) from an organic molecule; oxidation. (p. 257)

denaturation: loss of the natural three-dimensional structure of a macromolecule, usually resulting in a loss of its biological activity; caused by agents such as heat, extremes of pH, urea, salt, and other chemicals. (p. 60) Also see *DNA denaturation.*

dendrite: extension of a nerve cell that receives impulses and transmits them inward toward the cell body. (p. 394)

density-dependent inhibition of growth: tendency of cell division to stop when cells growing in culture reach a high population density. (p. 788)

density gradient (rate-zonal) centrifugation: type of centrifugation in which the sample is applied as a thin layer on top of a gradient of solute, and centrifugation is stopped before the particles reach the bottom of the tube; separates organelles and molecules based mainly on differences in size. (p. 357)

deoxyribonucleic acid: see *DNA.*

deoxyribose: five-carbon sugar present in DNA. (p. 82)

dephosphorylation: removal of a phosphate group. (p. 176)

depolarization: change in membrane potential to a less-negative value. (p. 398)

desmosome: junction for cell-cell adhesion that is connected to the cytoskeleton by intermediate filaments; creates buttonlike points of strong adhesion between adjacent animal cells that give tissue structural integrity and allow the cells to function as a unit and resist stress. (p. 510)

diacylglycerol (DAG): glycerol esterified to two fatty acids; formed, along with inositol trisphosphate (IP_3), upon hydrolysis of phosphatidylinositol-4,5-bisphosphate by phospholipase C; remains membrane-bound after hydrolysis and functions as a second messenger by activating protein kinase C, which then phosphorylates specific serine and threonine groups in a variety of target proteins. (p. 429)

diakinesis: final stage of prophase I of meiosis; associated with chromosome condensation, disappearance of nucleoli, breakdown of the nuclear envelope, and initiation of spindle formation. (p. 633)

DIC microscopy: see *differential interference contrast microscopy.*

Dicer: enzyme that cleaves double-stranded RNAs into short fragments about 21–22 base pairs in length. (p. 777)

differential centrifugation: technique for separating organelles or molecules that differ in size and/or density by subjecting cellular fractions to centrifugation at high speeds and separating particles based on their different rates of sedimentation. (p. 355)

differential interference contrast (DIC) microscopy: technique that resembles phase-contrast microscopy in principle, but is more sensitive because it employs a special prism to split the illuminating light beam into two separate rays. (p. 828)

differential scanning calorimetry: technique for determining a membrane's transition temperature by monitoring the uptake of heat during the transition of the membrane from the gel to fluid state. (p. 197)

differentiation: see *cell differentiation.*

diffraction: pattern of either additive or canceling interference exhibited by light waves. (p. 824)

diffusion: free, unassisted movement of a solute, with direction and rate dictated by the difference in solute concentration between two different regions. (p. 105)

digital deconvolution microscopy: technique in which fluorescence microscopy is used to acquire a series of images through the thickness of a specimen, followed by computer analysis to remove the contribution of out-of-focus light to the image in each focal plane. (p. 835)

digital microscopy: technique in which microscopic images are recorded and stored electronically by placing a video camera in the image plane produced by the ocular lens. (p. 835)

diploid: containing two sets of chromosomes and therefore two copies of each gene; can describe a cell, nucleus, or organism composed of such cells. (p. 629)

diplotene: stage during prophase I of meiosis when the two homologous chromosomes of each bivalent begin to separate from each other, revealing the chiasmata that connect them. (p. 633)

direct active transport: membrane transport in which the movement of solute molecules or ions across a membrane is coupled directly to an exergonic chemical reaction, most commonly the hydrolysis of ATP. (p. 237)

directionality: having two ends that are chemically different from each other; used to describe a polymer chain such as a protein, nucleic acid, or carbohydrate; also used to describe membrane transport systems that selectively transport solutes across a membrane in one direction. (pp. 59, 237)

disaccharide: carbohydrate consisting of two covalently linked monosaccharide units. (p. 91)

dissociation constant (K_d): the concentration of free ligand needed to produce a state in which half the receptors are bound to ligand. (p. 422)

disulfide bond: covalent bond formed between two sulfur atoms by oxidation of sulfhydryl groups. The disulfide bond formed between two cysteines is important in stabilizing the tertiary structure of proteins. (p. 72)

DNA (deoxyribonucleic acid): macromolecule that serves as the repository of geneticinformation in all cells; constructed from nucleotides consisting of deoxyribose phosphate linked to either adenine, thymine, cytosine, or guanine; forms a double helix held together by complementary base pairing between adenine and thymine, and between cytosine and guanine. (p. 82)

DNA cloning: generating multiple copies of a specific DNA sequence, either by replication of a recombinant plasmid or bacteriophage within bacterial cells or by use of the polymerase chain reaction. (p. 656)

DNA damage checkpoint: mechanism that monitors for DNA damage and halts the cell cycle at various points, including late G1, S, and late G2, if damage is detected. (p. 616)

DNA denaturation: separation of the two strands of the DNA double helix caused by disruption of complementary base pairing. (p. 543)

DNA fingerprinting: technique for identifying individuals based on small differences in DNA fragment patterns detected by electrophoresis. (p. 554)

DNA gyrase: a type II topoisomerase that can relax positive supercoiling and induce negative supercoiling of DNA; involved in unwinding the DNA double helix during DNA replication. (p. 543)

DNA helicase: any of several enzymes that unwind the DNA double helix, driven by energy derived from ATP hydrolysis. (p. 590)

DNA ligase: enzyme that joins two DNA fragments together by catalyzing the formation of a phosphoester bond between the 3′ end of one fragment and the 5′ end of the other fragment. (p. 586)

DNA melting temperature (T_m): temperature at which the transition from double-stranded to single-stranded DNA is halfway complete when DNA is denatured by increasing the temperature. (p. 544)

DNA methylation: addition of methyl groups to nucleotides in DNA; associated with suppression of gene transcription when selected cytosine groups are methylated in eukaryotic DNA. (p. 757)

DNA microarray: tiny chip that has been spotted at fixed locations with thousands of different DNA fragments for use in gene expression studies. (p. 761)

DNA packing ratio: ratio of the length of a DNA molecule to the length of the chromosome or fiber into which it is packaged; used to quantify the extent of DNA coiling and folding. (p. 561)

DNA polymerase: any of a group of enzymes involved in DNA replication and repair that catalyze the addition of successive nucleotides to the 3′ end of a growing DNA strand, using an existing DNA strand as template. (p. 583)

DNA rearrangement: movement of DNA segments from one location to another within the genome. (p. 754)

DNA renaturation: binding together of the two separated strands of a DNA double helix by complementary base pairing, thereby regenerating the double helix. (p. 543)

DNA replication checkpoint: mechanism that monitors the state of DNA replication to help ensure that DNA synthesis is completed prior to permitting the cell to exit from G2 and begin mitosis. (p. 615)

DNA sequencing: technology used to determine the linear order of bases in DNA mo-lecules or fragments. (pp. 39, 548)

DNA-binding domain: region of a transcription factor that recognizes and binds to a specific DNA base sequence. (p. 766)

DNase I hypersensitive site: location near an active gene that shows extreme sensitivity to digestion by the nuclease DNase I; thought to correlate with binding sites for transcriptional factors or other regulatory proteins. (p. 757)

domain: a discrete, locally folded unit of protein tertiary structure, often containing regions of α helices and β sheets packed together compactly. (p. 79)

dominant (allele): allele that determines how the trait will appear in an organism, whether present in the heterozygous or homozygous form. (p. 629)

dominant negative mutation: a loss-of-function mutation involving proteins consisting of more than one copy of the same polypeptide chain, in which a single mutant polypeptide chain disrupts the function of the protein even though the other polypeptide chains are normal. (p. 440)

double bond: chemical bond formed between two atoms as a result of the sharing of two pairs of electrons. (p. 47)

double helix (model): two intertwined helical chains of a DNA molecule, held together by complementary base pairing between adenine (A) and thymine (T) and between cytosine (C) and guanine (G). (pp. 87, 540)

double-reciprocal plot: graphic method for analyzing enzyme kinetic data by plotting $1/v$ versus $1/[S]$. (p. 170)

downstream: located toward the 3′ end of the DNA coding strand. (p. 685)

downstream promoter element: see *DPE.*

DPE (downstream promoter element): component of core promoters for RNA polymerase II, located about 30 nucleotides downstream from the transcriptional startpoint. (p. 690)

Drosha: nuclear enzyme that cleaves the primary transcript of microRNA genes (pri-microRNAs) to form the smaller precursor microRNAs (pre-microRNAs). (p. 778)

dynactin: protein complex that helps link cytoplasmic dynein to the cargo (e.g., a vesicle) that it transports along microtubules. (p. 480)
dynamic instability model: model for microtubule behavior that presumes two populations of microtubules, one growing in length by continued polymerization at their plus ends and the other shrinking in length by depolymerization. (p. 457)
dynamin: cytosolic GTPase required for coated pit constriction and the closing of a budding, clathrin-coated vesicle. (p. 377)
dynein: motor protein that moves along the surface of microtubules in the plus-to-minus direction, driven by energy derived from ATP hydrolysis; present both in the cytoplasm and in the arms that reach between adjacent microtubule doublets in the axoneme of a flagellum or cilium. (p. 478)
dystrophin: a large protein found at muscle costameres that is part of a complex of proteins that attaches the muscle cell plasma membrane to the extracellular matrix (p. 525)

E

E: see *internal energy.*
E face: interior face of the outer monolayer of a membrane as revealed by the technique of freeze fracturing; called the E face because this monolayer is on the *exterior* side of the membrane. (p. 847)
E site (exit site): site on the ribosome to which the empty tRNA is moved during translocation prior to its release from the ribosome. (p. 708)
E'_0: see *standard reduction potential.*
E2F transcription factor: protein that, when not bound to the Rb protein, activates the transcription of genes coding for proteins required for DNA replication and entrance into the S phase of the cell cycle. (p. 615)
E_A: see *activation energy.*
Eadie–Hofstee equation: alternative to the Lineweaver-Burk equation in which enzyme kinetic data are analyzed by plotting *v*/[S] versus *v*. (p. 171)
early endosome: vesicles budding off the *trans*-Golgi network that are sites for the sorting and recycling of extracellular material brought into the cell by endocytosis. (p. 366)
EBV: see *Epstein-Barr virus.*
ECM: see *extracellular matrix.*
effector: see *allosteric effector.*
EGF: see *epidermal growth factor.*
egg (ovum): haploid female gamete, usually a relatively large cell with many stored nutrients. (p. 630)
EJC: see *exon junction complex.*
elastin: protein subunit of the elastic fibers that impart elasticity and flexibility to the extracellular matrix. (p. 518)
electrical excitability: ability to respond to certain types of stimuli with a rapid series of changes in membrane potential known as an action potential. (p. 395)
electrical potential (voltage): tendency of oppositely charged ions to flow toward each other. (p. 396)
electrical synapse: junction between two nerve cells where nerve impulses are transmitted by direct movement of ions through gap junctions without the involvement of chemical neurotransmitters. (p. 408)
electrical work: use of energy to transport ions across a membrane against a potential gradient. (p. 136)
electrochemical equilibrium: condition in which a transmembrane concentration gradient of a specific ion is balanced with an electrical potential across the same membrane, such that there is no net movement of the ion across the membrane. (p. 397)
electrochemical gradient: see *electrochemical potential.*
electrochemical potential: transmembrane gradient of an ion, with both an electrical component due to the charge separation quantified by the membrane potential and a concentration component; also called electrochemical gradient. (p. 224)
electrochemical proton gradient: transmembrane gradient of protons, with both an electrical component due to charge separation and a chemical component due to a difference in proton concentration (pH) across the membrane. (p. 304)
electron gun: assembly of several components that generates the electron beam in an electron microscope. (p. 840)
electron micrograph: photographic image of a specimen produced by the exposure of a photographic plate to the image-forming electron beam of an electron microscope. (p. 841)
electron microscope: instrument that uses a beam of electrons to visualize cellular structures and thereby examine cellular architecture; the resolution is much greater than that of the light microscope, allowing detailed ultrastructural examination. (pp. 36, 823)
electron shuttle system: any of several mechanisms whereby electrons from a reduced coenzyme such as NADH are moved across a membrane; consists of one or more electron carriers that can be reversibly reduced, with transport proteins present in the membrane for both the oxidized and reduced forms of the carrier. (p. 313)
electron tomography: see *three-dimensional electron tomography.*
electron transport: process of coenzyme reoxidation under aerobic conditions, involving stepwise transfer of electrons to oxygen by means of a series of electron carriers. (p. 295)
electron transport system (ETS): group of membrane-bound electron carriers that transfer electrons from the coenzymes NADH and $FADH_2$ to oxygen. (pp. 298, 330)
electronegative: property of an atom that tends to draw electrons toward it. (p. 51)
electrophoresis: a group of related techniques that utilize an electrical field to separateelectrically charged molecules. (pp. 37, 206)
elongation (of microtubules): growth of microtubules by addition of tubulin heterodimers to either end. (p. 455)
elongation factors: group of proteins that catalyze steps involved in the elongation phase of protein synthesis; examples include EF-Tu and EF-Ts. (p. 716)
embryonic stem (ES) cells: see *stem cells.*
Emerson enhancement effect: achievement of greater photosynthetic activity with red light of two slightly different wavelengths than is possible by summing the activities obtained with the individual wavelengths separately. (p. 329)
endergonic: an energy-requiring reaction characterized by a positive free energy change ($\Delta G > 0$). (p. 143)
endocytic vesicle: membrane vesicle formed by pinching off of a small segment of plasma membrane during the process of endocytosis. (p. 370)
endocytosis: uptake of extracellular materials by infolding of the plasma membrane, followed by pinching off of a membrane-bound vesicle containing extracellular fluid and materials. (p. 370)
endomembrane system: interconnected system of cytoplasmic membranes in eukaryotic cells composed of the endoplasmic reticulum, Golgi complex, endosomes, lysosomes, and nuclear envelope. (p. 352)
endonuclease: an enzyme that degrades a nucleic acid (usually DNA) by cutting the molecule internally. (p. 587) Also see *restriction endonuclease.*
endoplasmic reticulum (ER): network of interconnected membranes distributed throughout the cytoplasm and involved in the synthesis, processing, and transport of proteins in eukaryotic cells. (pp. 117, 352)
endosome: see *early endosome* or *late endosome.*
endosymbiont theory: theory postulating that mitochondria and chloroplasts arose from ancient bacteria that were ingested by ancestral eukaryotic cells about a billion years ago. (pp. 117, 326)
endothermic: a reaction or process that absorbs heat. (p. 141)
end-product inhibition: see *feedback inhibition.*
end-product repression: regulation of an anabolic pathway based on the ability of an end product to repress further synthesis of enzymes involved in production of that end product. (p. 739)
energy: capacity to do work; ability to cause specific changes. (p. 134)
energy transduction reactions: portion of the photosynthetic pathway in which light energy is converted to chemical energy in the form of ATP and the coenzyme NADPH, which subsequently provide energy and reducing power for the carbon assimilation reactions. (p. 321)
enhancer: DNA sequence containing a binding site for transcription factors that stimulates transcription and whose position and orientation relative to the promoter can vary significantly without interfering with the ability to regulate transcription. (p. 762)
enthalpy (*H*): the heat content of a substance, quantified as the sum of its internal energy, *E*, plus the product of pressure and volume: $H = E + PV$. (p. 141)
entropy (*S*): measure of the randomness or disorder in a system. (p. 142)
enzyme: biological catalyst; protein (or in certain cases, RNA) molecule that acts on one or more specific substrates, converting them to products with different molecular structures. (pp. 37, 160) Also see *ribozyme.*
enzyme catalysis: involvement of an organic molecule, usually a protein but in some cases RNA, in speeding up the rate of a specific chemical reaction or class of reactions. (p. 157) Also see *catalyst.*
enzyme kinetics: quantitative analysis of enzyme reaction rates and the manner in which they are influenced by a variety of factors. (p. 166)
epidermal growth factor (EGF): protein that stimulates the growth and division of a wide variety of epithelial cell types. (p. 618)
epigenetic change: alteration in the expression of a gene rather than a change in the structure of the gene itself. (p. 758)
EPSP: see *excitatory postsynaptic potential.*
Epstein–Barr virus (EBV): virus associated with Burkitt lymphoma (as well as thenoncancerous condition, infectious mononucleosis). (p. 799)
equilibrium constant (K_{eq}): ratio of product concentrations to reactant concentrations for a given chemical reaction when the reaction has reached equilibrium. (p. 147)
equilibrium density (buoyant density) centrifugation: technique used to separate cellular components by subjecting them to centrifuga-

tion in a solution that increases in density from the top to the bottom of the centrifuge tube; during centrifugation, an organelle or molecule sediments to the density layer equal to its own density, at which point movement stops because no further net force acts on the material. (pp. 357, 579)

equilibrium (or reversal) potential: membrane potential that exactly offsets the effect of the concentration gradient for a given ion. (p. 397)

ER: see *endoplasmic reticulum.*

ER-associated degradation (ERAD): quality control mechanism in the ER that recognizes misfolded or unassembled proteins and exports or "retrotranslocates" them back across the ER membrane to the cytosol, where they are degraded by proteasomes. (p. 729)

ER cisterna (plural, **cisternae**)**:** flattened sac of the endoplasmic reticulum. (p. 352)

ER lumen: the internal space enclosed by membranes of the endoplasmic reticulum. (p. 352)

ER signal sequence: amino acid sequence in a newly forming polypeptide chain that directs the ribosome-mRNA-polypeptide complex to the surface of the rough ER, where the complex becomes anchored. (p. 726)

ERAD: see *ER-associated degradation.*

ETS: see *electron transport system.*

euchromatin: loosely packed, uncondensed form of chromatin present during interphase;contains DNA that is being actively transcribed. (p. 561) Also see *heterochromatin.*

eukarya: one of the three domains of organisms, the other two being bacteria and archaea; the domain consisting of one-celled and multicellular organisms called eukaryotes whose cells are characterized by a membrane-bounded nucleus and other membrane-bounded organelles. (p. 103)

eukaryote: category of organisms whose cells are characterized by the presence of a membrane-bounded nucleus and other membrane-bounded organelles; includes plants, animals, fungi, algae, and protozoa. (p. 103)

excision repair: DNA repair mechanism that removes and replaces abnormal nucleotides. (p. 596)

excitatory postsynaptic potential (EPSP): small depolarization of the postsynaptic membrane triggered by binding of an excitatory neurotransmitter to its receptor; if the EPSP exceeds a threshold level, it can trigger an action potential. (p. 416) Also see *inhibitory postsynaptic potential.*

exergonic: an energy-releasing reaction characterized by a negative free energy change ($\Delta G < 0$). (p. 143)

exit site: see *E site.*

exocytosis: fusion of vesicle membranes with the plasma membrane so that contents of the vesicle can be expelled or secreted to the extracellular environment. (p. 369)

exon: nucleotide sequence in a primary RNA transcript that is preserved in the mature, functional RNA molecule. (p. 698) Also see *intron.*

exon junction complex (EJC): protein complex deposited at exon-exon junctions created by pre-mRNA splicing. (p. 700)

exonuclease: an enzyme that degrades a nucleic acid (usually DNA) from one end, rather than cutting the molecule internally. (p. 587)

exosome: macromolecular complex involved in mRNA degradation in eukaryotic cells; contains multiple $3' \rightarrow 5'$ exonucleases. (p. 776)

exothermic: a reaction or process that releases heat. (p. 141)

exportin: nuclear receptor protein that binds to the nuclear export signal of proteins in the nucleus and then transports the bound protein out through the nuclear pore complex and into the cytosol. (p. 568)

extensin: group of related glycoproteins that form rigid, rodlike molecules tightly woven into the cell walls of plants and fungi. (p. 527)

extracellular digestion: degradation of components outside a cell, usually by lysosomal enzymes that are released from the cell by exocytosis. (p. 383)

extracellular matrix (ECM): material secreted by animal cells that fills the spaces between neighboring cells; consists of a mixture of structural proteins (e.g., collagen, elastin) and adhesive glycoproteins (e.g., fibronectin, laminin) embedded in a matrix composed of protein-polysaccharide complexes called proteoglycans. (pp. 126, 515)

F

F factor: DNA sequence that enables an *E. coli* cell to act as a DNA donor during bacterial conjugation. (p. 650)

F_1 complex: knoblike sphere protruding into the matrix from mitochondrial inner membranes (or into the cytosol from the bacterial plasma membrane) that contains the ATP-synthesizing site of aerobic respiration. (p. 285)

F_1 generation: offspring of the P_1 generation in a genetic breeding experiment. (p. 641)

F_2 generation: offspring of the F_1 generation in a genetic breeding experiment. (p. 641)

F2,6BP: see *fructose-2,6-bisphosphate.*

facilitated diffusion: membrane protein-mediated movement of a substance across a membrane that does not require energy because the ion or molecule being transported is moving down an electrochemical gradient. (p. 230)

F-actin: component of microfilaments consisting of G-actin monomers that have beenpolymerized into long, linear strands. (p. 462)

facultative heterochromatin: chromosomal regions that have become specifically condensed and inactivated in a particular cell type at a specific time. (p. 571) Also see *constitutive heterochromatin.*

facultative organism: organism that can function in either an anaerobic or aerobic mode. (p. 259)

FAD: see *flavin adenine dinucleotide.*

familial hypercholesterolemia (FH): genetic predisposition to high blood cholesterol levels and heart disease caused by an inherited defect in the gene coding for the LDL receptor. (p. 374)

fast axonal transport: microtubule-mediated movement of vesicles and organelles back and forth along a nerve cell axon. (p. 478)

fatty acid: long, unbranched hydrocarbon chain that has a carboxyl group at one end and is therefore amphipathic; usually contains an even number of carbon atoms and may be of varying degrees of unsaturation. (p. 96)

fatty acid-anchored membrane protein: protein located on a membrane surface that is covalently bound to a fatty acid embedded with the lipid bilayer. (p. 206)

Fd: see *ferredoxin.*

feedback (end-product) inhibition: ability of the end product of a biosynthetic pathway to inhibit the activity of the first enzyme in the pathway, thereby ensuring that the functioning of the pathway is sensitive to the intracellular concentration of its product. (p. 174)

fermentation: partial oxidation of carbohydrates by oxygen-independent (anaerobic) pathways, resulting often (but not always) in the production of either ethanol and carbon dioxide or lactate. (p. 259)

ferredoxin (Fd): iron-sulfur protein in the chloroplast stroma involved in the transfer of electrons from photosystem I to $NADP^+$ during the energy transduction reactions of photosynthesis. (p. 334)

ferredoxin-$NADP^+$ reductase (FNR): enzyme located on the stroma side of the thylakoid membrane that catalyzes the transfer of electrons from ferredoxin to $NADP^+$. (p. 334)

fertilization: union of two haploid gametes to form a diploid cell, the zygote, that develops into a new organism. (p. 630)

FGF: see *fibroblast growth factor.*

FH: see *familial hypercholesterolemia.*

fibroblast growth factor (FGF): any of several related signaling proteins that stimulate growth and cell division in fibroblasts as well as numerous other cell types, both in adults and during embryonic development. (p. 439)

fibronectin: adhesive glycoprotein found in the extracellular matrix and loosely associated with the cell surface; binds cells to the extracellular matrix and is important in determining cell shape and guiding cell migration. (p. 520)

fibrous protein: protein with extensive α helix or β sheet structure that confers a highly ordered, repetitive structure. (p. 78)

filopodium (plural, **filopodia**)**:** thin, pointed cytoplasmic protrusion that transiently emerges from the surface of eukaryotic cells during cell movements. (p. 499)

first law of thermodynamics: law of conservation of energy; principle that energy can beconverted from one form to another but can never be created or destroyed. (p. 140)

Fischer projection: model depicting the chemical structure of a molecule as a chain drawn vertically with the most oxidized atom on top and horizontal projections that are understood to be coming out of the plane of the paper. (p. 90)

fixative: chemical that kills cells while preserving their structural appearance for microscopic examination. (p. 839)

flagellum (plural, **flagella**)**:** membrane-bounded appendage on the surface of a eukaryotic cell composed of a specific arrangement of microtubules and responsible for motility of the cell; longer and less numerous (usually limited to one or a few per cell) than closely related organelles called cilia. (p. 481) Also see *cilium.*

flavin adenine dinucleotide (FAD): coenzyme that accepts two electrons and two protons from an oxidizable organic molecule to generate the reduced form, $FADH_2$; important electron carrier in energy metabolism. (p. 290)

flavoprotein: protein that has a tightly bound flavin coenzyme (FAD or FMN) and that serves as a biological electron donor or acceptor. Several examples occur in the mitochondrial electron transport system. (p. 298)

flippase: see *phospholipid translocator.*

flow cytometry: technique for automated rapid analysis of cells stained with fluorescent dyes as they pass in a narrow stream through a laser beam. (p. 579)

fluid mosaic model: model for membrane structure consisting of a lipid bilayer with proteins associated as discrete globular entities that penetrate the bilayer to various extents and are free to move laterally in the membrane. (p. 189)

fluid-phase endocytosis: nonspecific uptake of extracellular fluid by infolding of the plasma membrane, followed by budding off of a membrane vesicle. (p. 375)

fluorescence: property of molecules that absorb light and then reemit the energy as light of a longer wavelength. (p. 830)

fluorescence microscopy: light microscopic technique that focuses ultraviolet rays on the specimen, thereby causing fluorescent compounds in the specimen to emit visible light. (p. 830)

fluorescence recovery after photobleaching: technique for measuring the lateral diffusion rate of membrane lipids or proteins in which such molecules are linked to a fluorescent dye and a tiny membrane region is then bleached with a laser. (p. 197)

fluorescence resonance energy transfer (FRET): an extremely short-distance interaction between two fluorescent molecules in which excitation is transferred from a donor molecule to an acceptor molecule directly; can be used to determine whether two molecules are in contact, or to produce "biosensors" that detect where proteins are activated or identify changes in local ion concentration. (p. 838)

fluorescent antibody: antibody containing covalently linked fluorescent dye molecules that allow the antibody to be used to locate antigen molecules microscopically. (p. 215)

FNR: see *ferredoxin-NADP$^+$ reductase*.

F_o complex: group of hydrophobic membrane proteins that anchor the F_1 complex to either the inner mitochondrial membrane or the bacterial plasma membrane; serves as the proton translocator channel through which protons flow when the electrochemical gradient across the membrane is used to drive ATP synthesis. (p. 285)

F_oF_1 complex: protein complex in the mitochondrial inner membrane and the bacterial plasma membrane that consists of the F_1 complex bound to the F_o complex; the flow of protons through the F_o component leads to the synthesis of ATP by the F_1 component. (p. 285)

focal adhesion: localized points of attachment between cell surface integrin molecules and the extracellular matrix; contain clustered integrin molecules that interact with bundles of cytoskeletal actin microfilaments via several linker proteins. (p. 523)

focal length: distance between the midline of a lens and the point at which rays passing through the lens converge to a focus. (p. 825)

frameshift mutation: insertion or deletion of one or more base pairs in a DNA molecule, causing a change in the reading frame of the mRNA molecule that usually garbles the message. (p. 679)

free energy (*G*): thermodynamic function that measures the extractable energy content of a molecule; under conditions of constant temperature and pressure, the change in free energy is a measure of the system's ability to do work. (p. 143)

free energy change (ΔG): thermodynamic parameter used to quantify the net free energy liberated or required by a reaction or process; measure of thermodynamic spontaneity. (p. 143)

freeze etching: cleaving a quick-frozen specimen, as in freeze fracturing, followed by the subsequent sublimation of ice from the specimen surface to expose small areas of the true cell surface. (p. 847)

freeze fracturing: sample preparation technique for electron microscopy in which a frozen specimen is cleaved by a sharp blow followed by examination of the fractured surface, often the interior of a membrane. (pp. 202, 846)

FRET: see *fluorescence resonance energy transfer*.

fructose-2,6-bisphosphate (F2,6BP): a doubly phosphorylated fructose molecule, formed by the action of phosphofructokinase-2 on fructose-6-phosphate, that plays an important role in regulating both glycolysis and gluconeogenesis. (p. 273)

F-type ATPase: type of transport ATPase found in bacteria, mitochondria, and chloroplasts that can use the energy of ATP hydrolysis to pump protons against their electrochemical gradient; can also catalyze the reverse process, in which the exergonic flow of protons down their electrochemical gradient is used to drive ATP synthesis. (p. 239) Also see *ATP synthase*.

functional group: group of chemical elements covalently bonded to each other that confers characteristic chemical properties upon any molecule to which it is covalently linked. (p. 49)

G

G: see *guanine* or *free energy*.

ΔG: see *free energy change*.

$\Delta G^{\circ\prime}$: see *standard free energy change*.

G protein: any of numerous GTP-binding regulatory proteins located in the plasma membrane that mediate signal transduction pathways, usually by activating a specific target protein such as an enzyme or channel protein. (p. 424)

G protein-linked receptor family: group of plasma membrane receptors that activate a specific G protein upon binding of the appropriate ligand. (p. 424)

G protein-linked receptor kinase (GRK): any of several protein kinases that catalyze the phosphorylation of activated G protein-linked receptors, thereby leading to receptor desensitization. (p. 424)

G0 (G zero): designation applied to eukaryotic cells that have become arrested in the G1 phase of the cell cycle and thus are no longer proliferating. (p. 578)

G1 phase: stage of the eukaryotic cell cycle between the end of the previous division and the onset of chromosomal DNA synthesis. (p. 578)

G2 phase: stage of the eukaryotic cell cycle between the completion of chromosomal DNA replication and the onset of cell division. (p. 578)

G-actin: globular monomeric form of actin that polymerizes to form F-actin. (p. 462)

GAG: see *glycosaminoglycan*.

gamete: haploid cells produced by each parent that fuse together to form the diploid offspring; for example, sperm or egg. (p. 630)

gametogenesis: the process that produces gametes. (p. 630)

gametophyte: haploid generation in the life cycle of an organism that alternates between haploid and diploid forms; form that produces gametes. (p. 631)

gamma tubulin (γ-tubulin): form of tubulin located in the centrosome, where it functions in the nucleation of microtubules. (p. 458)

gamma tubulin ring complexes (γ-TuRCs): rings of γ-tubulin that emerge from centrosomes and nucleate the assembly of new microtubules. (p. 458)

gamma-TuRCs (γ-TuRCs): see *gamma tubulin ring complexes*.

ganglioside: a charged glycolipid containing the amino alcohol sphingosine and negatively charged sialic acid residues. (p. 191)

GAP: see *GTPase activating protein*.

gap junction: type of cell junction that provides a point of intimate contact between two adjacent cells through which ions and small molecules can pass. (p. 514)

gatekeeper gene: a tumor suppressor gene directly involved in restraining cell proliferation; loss-of-function mutations in such genes can lead to excessive cell proliferation and tumor formation. (p. 810)

Gb: gigabases; a billion base pairs (p. 545)

GEF: see *guanine-nucleotide exchange factor*.

gel electrophoresis: technique in which proteins or nucleic acids are separated in gels made of polyacrylamide or agarose by placing the gel in an electric field. (p. 547)

gene: "hereditary factor" that specifies an inherited trait; consists of a DNA base sequence that codes for the amino acid sequence of one or more polypeptide chains, or alternatively, for one of several types of RNA that perform functions other than coding for polypeptide chains (e.g., rRNA, tRNA, snRNA, or microRNA). (pp. 37, 533)

gene amplification: mechanism for creating extra copies of individual genes by selectively replicating specific DNA sequences. (p. 751)

gene conversion: phenomenon in which genes undergo nonreciprocal recombination during meiosis, such that one of the recombining genes ends up on both chromosomes of a homologous pair rather than being exchanged from one chromosome to the other. (p. 654)

gene deletion: selective removal within a cell of specific DNA sequences whose products are not required. (p. 754)

gene locus (plural, **loci**)**:** location within a chromosome that contains the DNA sequence for a particular gene. (p. 629)

general transcription factor: a protein that is always required for RNA polymerase to bind to its promoter and initiate RNA synthesis, regardless of the identity of the gene involved. (p. 691)

genetic code: set of rules specifying the relationship between the sequence of bases in a DNA or mRNA molecule and the order of amino acids in the polypeptide chain encoded by that DNA or mRNA. (p. 674)

genetic engineering: application of recombinant DNA technology to practical problems, primarily in medicine and agriculture. (p. 663)

genetic instability: trait of cancer cells in which abnormally high mutation rates are caused by defects in DNA repair and/or chromosome sorting mechanisms. (p. 808)

genetic mapping: determining the sequential order and spacing of genes on a chromosome based on recombination frequencies. (p. 648)

genetic recombination: exchange of DNA segments between two different DNA molecules. (p. 633)

genetics: study of the behavior of genes, which are the chemical units involved in the storage and transmission of hereditary information. (p. 32)

genome: the DNA (or for some viruses, RNA) that contains one complete copy of all the genetic information of an organism or virus (p. 545)

genomic control: a regulatory change in the makeup or structural organization of the genome. (p. 751)

genomic library: collection of recombinant DNA clones produced by cleaving an organism's entire genome into fragments, usually using restriction endonucleases, and then cloning all the fragments in an appropriate cloning vector. (p. 660)

genotype: genetic makeup of an organism. (p. 629)

globular protein: protein whose polypeptide chains are folded into compact structures rather than extended filaments. (p. 79)

glucagon: peptide hormone, produced by the islets of Langerhans in the pancreas, that acts to increase blood glucose by promoting the breakdown of glycogen. (p. 445)

gluconeogenesis: synthesis of glucose from precursors such as amino acids, glycerol, or lactate; occurs in the liver via a pathway that is essentially the reverse of glycolysis. (p. 267)

glucose: a six-carbon sugar that is widely used as the starting molecule in cellular energy metabolism. (p. 259)

glucose transporter (GLUT): membrane carrier protein responsible for the facilitated diffusion of glucose. (p. 231)

GLUT: see *glucose transporter.*

glycerol: three-carbon alcohol with a hydroxyl group on each carbon; serves as the backbone for triacylglycerols. (p. 96)

glycerol phosphate shuttle: mechanism for carrying electrons from cytosolic NADH into the mitochondrion, where the electrons are delivered to FAD in the respiratory complexes. (p. 314)

glycocalyx: carbohydrate-rich zone located at the outer boundary of many animal cells. (p. 214)

glycogen: highly branched storage polysaccharide in animal cells; consists of glucose repeating subunits linked by $\alpha(1 \rightarrow 4)$ bonds and $\alpha(1 \rightarrow 6)$ bonds. (p. 91)

glycolate pathway: light-dependent pathway that decreases the efficiency of photosynthesis by oxidizing reduced carbon compounds without capturing the released energy; occurs when oxygen substitutes for carbon dioxide in the reaction catalyzed by rubisco, thereby generating phosphoglycolate that is then converted to 3-phosphoglycerate in the peroxisome and mitochondrion; also called photorespiration. (p. 344)

glycolipid: lipid molecule containing a bound carbohydrate group. (pp. 98, 191)

glycolysis (glycolytic pathway): series of reactions by which glucose or some other monosaccharide is catabolized to pyruvate without the involvement of oxygen, generating two molecules of ATP per molecule of monosaccharide metabolized. (p. 260)

glycolytic pathway: see *glycolysis.*

glycoprotein: protein with one or more carbohydrate groups linked covalently to amino acid side chains. (pp. 213, 362)

glycosaminoglycan (GAG): polysaccharide constructed from a repeating disaccharide unit containing one sugar with an amino group and one sugar that usually has a negatively charged sulfate or carboxyl group; component of the extracellular matrix. (p. 519)

glycosidic bond: bond linking a sugar to another molecule, which may be another sugar molecule. (p. 91)

glycosylation: addition of carbohydrate side chains to specific amino acid residues of proteins, usually beginning in the lumen of the endoplasmic reticulum and completed in the Golgi complex. (pp. 213, 362)

glyoxylate cycle: modified version of the TCA cycle occurring in plant glyoxysomes; an anabolic pathway that converts two molecules of acetyl CoA to one molecule of succinate, thereby permitting the synthesis of carbohydrates from lipids. (p. 296)

glyoxysome: specialized type of plant peroxisome that contains some of the enzymes responsible for the conversion of stored fat to carbohydrate in germinating seeds. (pp. 121, 296, 387)

Goldman equation: modification of the Nernst equation that calculates the resting membrane potential by adding together the effects of all relevant ions, each weighted for its relative permeability. (p. 399)

Golgi complex: stacks of flattened, disk-shaped membrane cisternae in eukaryotic cells that are important in the processing and packaging of secretory proteins and in the synthesis of complex polysaccharides. (pp. 118, 360)

golgin: one of a class of tethering proteins that connect vesicles to the Golgi and Golgi cisternae to each other. (p. 380)

GPI-anchored membrane protein: protein bound to the outer surface of the plasma membrane by linkage to glycosylphosphatidylinositol (GPI), a glycolipid found in the external monolayer of the plasma membrane. (p. 206)

grading: see *tumor grading.*

granum (plural, **grana**)**:** stack of thylakoid membranes in a chloroplast. (pp. 115, 325)

green fluorescent protein (GFP): a protein that exhibits bright green fluorescence when exposed to blue light. Using genetic engineering techniques, GFP can be fused to non-fluorescent proteins, allowing them to be followed using fluorescence microscopy. (p. 832)

GRK: see *G protein-linked receptor kinase.*

group specificity: ability of an enzyme to act on any of a whole group of substrates as long as they possess some common structural feature. (p. 161)

group transfer reaction: chemical reaction that involves the movement of a chemical group from one molecule to another. (p. 256)

growth factor: any of a number of extracellular signaling proteins that stimulate cell division in specific types of target cells; examples include platelet-derived growth factor (PDGF) and epidermal growth factor (EGF). (p. 435)

GTPase activating protein (GAP): a protein that speeds up the inactivation of Ras by facilitating the hydrolysis of bound GTP. (p. 437)

guanine (G): nitrogen-containing aromatic base, chemically designated as a purine, which serves as an informational monomeric unit when present in nucleic acids with other bases in a specific sequence; forms a complementary base pair with cytosine (C) by hydrogen bonding. (p. 83)

guanine-nucleotide exchange factor (GEF): a protein that triggers the release of GDP from the Ras protein, thereby permitting Ras to acquire a molecule of GTP. (p. 436)

gyrase: see *DNA gyrase.*

H

H: see *enthalpy.*

H zone: light region located in the middle of the A band of striated muscle myofibrils. (p. 489)

hairpin loop: looped structure formed when two adjacent segments of a nucleic acid chain are folded back on one another and held in that conformation by base pairing between complementary base sequences. (p. 687)

haploid: containing a single set of chromosomes and therefore a single copy of the genome; can describe a cell, nucleus, or organism composed of such cells. (p. 629)

haploid spore: haploid product of meiosis in organisms that display an alternation ofgenerations; gives rise upon spore germination to the haploid form of the organism (the gametophyte, in the case of higher plants). (p. 631)

haplotype: group of SNPs located near one another on the same chromosome that tend to be inherited as a unit. (p. 552)

HAT: see *histone acetyl transferase.*

Hatch–Slack cycle: series of reactions in C_4 plants in which carbon dioxide is fixed in the mesophyll cells and transported as a four-carbon compound to the bundle sheath cells, where subsequent decarboxylation results in a higher concentration of carbon dioxide and therefore a higher rate of carbon fixation by rubisco. (p. 345)

Haworth projection: model that depicts the chemical structure of a molecule in a way that suggests the spatial relationship of different parts of the molecule. (p. 90)

HCI: see *heme-controlled inhibitor.*

HDL: see *high-density lipoprotein.*

heart muscle: see *cardiac muscle.*

heat: transfer of energy as a result of a temperature difference. (p. 136)

heat-shock gene: gene whose transcription is activated when cells are exposed to elevated temperatures or other stressful conditions; some heat-shock genes code for molecular chaperones that, in addition to guiding normal protein folding, can facilitate the refolding of heat-damaged proteins. (p. 771)

heat-shock response element: DNA base sequence located adjacent to heat-shock genes that functions as a binding site for the heat-shock transcription factor. (p. 771)

helicase: see *DNA helicase.*

helix: see *alpha helix* or *double helix.*

helix-loop-helix: DNA-binding motif found in transcription factors that is composed of a short α helix connected by a loop to a longer α helix. (p. 768)

helix-turn-helix: DNA-binding motif found in many regulatory transcription factors; consists of two regions of α helix separated by a bend in the polypeptide chain. (p. 766)

hemicellulose: heterogeneous group of polysaccharides deposited along with cellulose in the cell walls of plants and fungi to provide added strength; each consists of a long, linear chain of a single kind of sugar with short side chains. (p. 526)

hemidesmosome: point of attachment between cell surface integrin molecules of epithelial cells and the basal lamina; contains integrin molecules that are anchored via linker proteins to intermediate filaments of the cytoskeleton. (p. 523)

heterochromatin: highly compacted form of chromatin present during interphase; contains DNA that is not being transcribed. (p. 561) Also see *euchromatin.*

heterodimer ($\alpha\beta$**-heterodimer):** see *alpha beta heterodimer.*

heterophagic lysosome: mature lysosome containing hydrolytic enzymes involved in the digestion of materials of extracellular origin. (p. 381) Also see *autophagic lysosome.*

heterophilic interaction: binding of two different molecules to each other. (p. 506) Also see *homophilic interaction.*

heterozygous: having two different alleles for a given gene. (p. 629) Also see *homozygous.*

Hfr cell: bacterial cell in which the F factor has become integrated into the bacterialchromosome, allowing the cell to transfer genomic DNA during conjugation. (p. 650)

hierarchical assembly: synthesis of biological structures from simple starting molecules to progressively more complex structures, usually by self-assembly. (p. 64)

high-density lipoprotein (HDL): cholesterol-containing protein-lipid complex that transports cholesterol through the bloodstream and is taken up by cells; exhibits high density because of its low cholesterol content. (p. 374)

high-voltage electron microscope (HVEM): an electron microscope that uses accelerating voltages up to a thousand or more kilovolts, thereby allowing the examination of thicker samples than is possible with a conventional electron microscope. (p. 841)

histone: class of basic proteins found in eukaryotic chromosomes; an octamer of histones forms the core of nucleosomes. (p. 558)

histone acetyltransferase (HAT): enzyme that catalyzes the addition of acetyl groups to histones. (p. 759)

histone deacetylase (HDAC): enzyme that catalyzes removal of acetyl groups from histones. (p. 759)

Holliday junction: X-shaped structure produced when two DNA molecules are joined together by single-strand crossovers during genetic recombination. (p. 654)

homeobox: highly conserved DNA sequence found in homeotic genes; codes for a DNA-binding protein domain present in transcription factors that are important regulators of gene expression during development. (p. 772)

homeodomain: amino acid sequence, about 60 amino acids long, found in transcription factors encoded by homeotic genes; contains a helix-turn-helix DNA-binding motif. (p. 772)

homeotic gene: family of genes that control formation of the body plan during embryonic development; they code for transcription factors possessing homeodomains. (p. 771)

homeoviscous adaptation: alterations in membrane lipid composition that keep the viscosity of membranes approximately the same despite changes in environmental temperature. (p. 200)

homogenate: suspension of cell organelles, smaller cellular components, and molecules produced by disrupting cells or tissues using techniques such as grinding, ultrasonic vibration, or osmotic shock. (p. 355)

homogenization: disruption of cells or tissues using techniques such as grinding, ultrasonic vibration, or osmotic shock. (p. 355)

homologous chromosomes: two copies of a specific chromosome, one derived from each parent, that pair with each other and exchange genetic information during meiosis. (p. 628)

homologous recombination: exchange of genetic information between two DNA molecules exhibiting extensive sequence similarity; also used in repairing double-strand DNA breaks. (pp. 598, 652)

homophilic interaction: binding of two identical molecules to each other. (p. 506) Also see *heterophilic interaction.*

homozygous: having two identical alleles for a given gene. (p. 629) Also see *heterozygous.*

hormone: chemical that is synthesized in one organ, secreted into the blood, and able to cause a physiological change in cells or tissues of another organ. (p. 442)

hormone response element: DNA base sequence that selectively binds to a hormone-receptor complex, resulting in the activation (or inhibition) of transcription of nearby genes. (p. 769)

HPV: see *human papillomavirus.*

human papillomavirus (HPV): virus responsible for cervical cancer; possesses oncogenes that block the actions of the proteins produced by the *RB* and *p53* tumor suppressor genes. (p. 806)

HVEM: see *high-voltage electron microscope.*

hyaluronate: a glycosaminoglycan found in high concentration in the extracellular matrix where cells are actively proliferating or migrating, and in the joints between movable bones. (p. 520)

hybrid: product of the cross of two genetically different parents. (p. 641)

hybridization: see *nucleic acid hybridization.*

hydrocarbon: an organic molecule consisting only of carbon and hydrogen atoms; not generally compatible with living cells. (p. 49)

hydrogen bond: weak attractive interaction between an electronegative atom and a hydrogen atom that is covalently linked to a second electronegative atom. (pp. 51, 74)

hydrogenation: addition of electrons plus hydrogen ions (protons) to an organic molecule; reduction. (p. 257)

hydrolysis: reaction in which a chemical bond is broken by the addition of a water molecule. (p. 60)

hydropathy index: a value representing the average of the hydrophobicity values for a short stretch of contiguous amino acids within a protein. (p. 208)

hydropathy (hydrophobicity) plot: graph showing the location of hydrophobic amino acid clusters within the primary sequence of a protein molecule; used to determine the likely locations of transmembrane segments within integral membrane proteins. (p. 208)

hydrophilic: describing molecules or regions of molecules that readily associate with or dissolve in water because of a preponderance of polar groups; "water-loving." (p. 52)

hydrophobic: describing molecules or regions of molecules that are poorly soluble in water because of a preponderance of nonpolar groups; "water-hating." (p. 52)

hydrophobic interaction: tendency of hydrophobic groups to be excluded from interactions with water molecules. (p. 74)

hydrophobicity plot: see *hydropathy plot.*

hydroxylation: chemical reaction in which a hydroxyl group is added to an organic molecule. (p. 358)

hyperpolarization (undershoot): state in which the membrane potential is more negative than the normal membrane potential. (p. 405)

hypothesis: a statement or explanation that is consistent with most of the observational and experimental evidence to date. (p. 41)

I

I band: region of a striated muscle myofibril that appears as a light band when viewed by microscopy; contains those regions of the thin actin filaments that do not overlap the thick myosin filaments. (p. 489)

IF: see *intermediate filament.*

IgSF: see *immunoglobulin superfamily.*

immunoelectron microscopy (immunoEM): electron microscopic technique for visualizing antibodies within cells by linking them to substances that are electron dense and therefore visible as opaque dots. (p. 844)

immunoEM: see *immunoelectron microscopy.*

immunoglobulin superfamily (IgSF): family of cell surface proteins involved in cell-cell adhesion that are structurally related to the immunoglobulin subunits of antibody molecules. (p. 506)

immunostaining: technique in which antibodies are labeled with a fluorescent dye to enable them to be identified and localized microscopically based on their fluorescence. (p. 831)

immunotherapy: treatment of a disease, such as cancer, by stimulating the immune system or administering antibodies made by the immune system. (p. 816)

IMP: see *intramembranous particle.*

importin: receptor protein that binds to the nuclear localization signal of proteins in the cytosol and then transports the bound protein through the nuclear pore complex and into the nucleus. (p. 568)

imprinting: a process that causes some genes to be expressed differently depending on whether they are inherited from a person's mother or father; such behavior can be caused by differences in DNA methylation. (p. 758)

indirect active transport: membrane transport involving the cotransport of two solutes in which the movement of one solute down its gradient drives the movement of the other solute up its gradient. (p. 237)

indirect immunofluorescence: type of fluorescence microscopy in which a specimen is first exposed to an unlabeled antibody that attaches to specific antigenic sites within the specimen and is then stained with a secondary, fluorescent antibody that binds to the first antibody. (p. 831)

induced-fit model: model postulating that the active site of an enzyme is relatively specific for its substrate before it binds, but even more so thereafter because of a conformational change in the enzyme induced by the substrate. (p. 164)

induced pluripotent stem (IPS) cells: differentiated cells forced to express transcription factor proteins found in embryonic stem cells, allowing them to become many different types of cells (p. 753)

inducer: any effector molecule that activates the transcription of an inducible operon. (p. 742)

inducible enzyme: enzyme whose synthesis is regulated by the presence or absence of its substrate. (p. 739)

inducible operon: group of adjoining genes whose transcription is activated in the presence of an inducer. (p. 742)

informational macromolecule: polymer of nonidentical subunits ordered in a nonrandom sequence that stores and transmits information important to the function or utilization of the macromolecule; DNA and RNA are informational macromolecules. (p. 57)

inhibition (of enzyme activity): decreasing the catalytic activity of an enzyme, either by a change in conformation or by chemical modification of one of its functional groups. (p. 172)

inhibitory postsynaptic potential (IPSP): small hyperpolarization of the postsynaptic membrane triggered by binding of an inhibitory neurotransmitter to its receptor, thereby reducing the amplitude of subsequent excitatory postsynaptic potentials and possibly preventing the firing of an action potential. (p. 416) Also see *excitatory postsynaptic potential.*

initial reaction velocity (v): Reaction rate measured over a period of time during which the substrate concentration has not decreased enough to affect the rate and the accumulation of product is still too small to cause any measurable back reaction. (p. 167)

initiation complex (30S): complex formed by the association of mRNA, the 30S ribosomal subunit, an initiator aminoacyl tRNA molecule, and initiation factor IF2. (p. 714

initiation complex (70S): complex formed by the association of a 30S initiation complex with a 50S ribosomal subunit; contains an initiator aminoacyl tRNA at the P site and is ready to commence mRNA translation. (p. 714)

initiation factors: group of proteins that promote the binding of ribosomal subunits to mRNA and initiator tRNA, thereby initiating the process of protein synthesis. (p. 713)

initiation (stage of carcinogenesis): irreversible conversion of a cell to a precancerous state by agents that cause DNA mutation. (p. 797)

initiator (Inr): short DNA sequence surrounding the transcription startpoint that forms part of the promoter for RNA polymerase II. (p. 689)

initiator tRNA: type of transfer RNA molecule that starts the process of translation; recognizes the AUG start codon and carries formylmethionine in prokaryotes or methionine in eukaryotes. (p. 714)
inner membrane: the inner of the two membranes that surround a mitochondrion, chloroplast, or nucleus. (pp. 283, 325)
inositol-1,4,5-trisphosphate (IP_3): triply phosphorylated inositol molecule formed as a product of the cleavage of phosphatidylinositol-4,5-bisphosphate catalyzed by phospholipase C; functions as a second messenger by triggering the release of calcium ions from storage sites within the endoplasmic reticulum. (p. 429)
Inr: see *initiator.*
insertional mutagenesis: change in gene structure or activity resulting from the integration of DNA derived from another source, usually a virus. (p. 802)
insulin: peptide hormone, produced by the islets of Langerhans in the pancreas, that acts to decrease blood glucose by stimulating glucose uptake into muscle and adipose cells and by stimulating glycogen synthesis. (p. 445)
integral membrane protein: hydrophobic protein localized within the interior of a membrane but possessing hydrophilic regions that protrude from one or both membrane surfaces. (p. 205)
integral monotopic protein: an integral membrane protein embedded in only one side of the lipid bilayer. (p. 205)
integrin: any of several plasma membrane receptors that bind to extracellular matrix components at the outer membrane surface and interact with cytoskeletal components at the inner membrane surface; includes receptors for fibronectin, laminin, and collagen. (p. 522)
intercalated disc: membrane partition enriched in gap junctions that divides cardiac muscle into separate cells containing single nuclei. (p. 497)
interdoublet link: link between adjacent doublets in the axoneme of a eukaryotic cilium orflagellum; believed to limit the extent of doublet movement with respect to each other as the axoneme bends. (p. 484)
interference: process by which two or more waves of light combine to reinforce or cancel one another, producing a wave equal to the sum of the combining waves. (p. 824)
intermediate filament (IF): group of protein filaments that are the most stable components of the cytoskeleton of eukaryotic cells; exhibits a diameter of 8–12 nm, which is intermediate between the diameters of actin microfilaments and microtubules. (pp. 125, 470)
intermediate lens (electron microscope): electromagnetic lens positioned between the objective and projector lenses in a transmission electron microscope. (p. 841)
intermediate lens (light microscope): lens positioned between the ocular and objective lenses in a light microscope. (p. 828)
intermembrane space: region of a mitochondrion or chloroplast between the inner and outer membranes. (pp. 283, 325)
internal energy (*E*): total energy stored within a system; cannot be measured directly, but the change in internal energy, ΔE, is measurable. (p. 141)
interphase: growth phase of the eukaryotic cell cycle situated between successive division phases (M phases); composed of G1, S, and G2 phases. (p. 578)
interspersed repeated DNA: repeated DNA sequences whose multiple copies are scattered around the genome. (p. 555)
intracellular membrane: any cellular membrane internal to the plasma membrane; suchmembranes serve to compartmentalize functions within eukaryotic cells. (p. 184)
intraflagellar transport (IFT): movement of components to and from the tips of flagella driven by both plus- and minus-end directed microtubule motor proteins. (p. 487)
intramembranous particle (IMP): integral membrane protein that is visible as a particle when the interior of a membrane is visualized by freeze-fracture microscopy. (p. 846)
intron: nucleotide sequence in an RNA molecule that is part of the primary transcript but not the mature, functional RNA molecule. (p. 697) Also see *exon.*
invasion: direct spread of cancer cells into neighboring tissues. (p. 790)
inverted repeat: DNA segment containing two copies of the same base sequence oriented in opposite directions. (p. 770)
ion channel: membrane protein that allows the passage of specific ions through the membrane; generally regulated by either changes in membrane potential (voltage-gated channels) or binding of a specific ligand (ligand-gated channels). (pp. 233, 400)
ionic bond: attractive force between a positively charged chemical group and a negatively charged chemical group. (p. 74)
ionizing radiation: high-energy forms of radiation that remove electrons from molecules, thereby generating highly reactive ions that cause DNA damage; includes X-rays and radiation emitted by radioactive elements. (p. 798)
IP_3: see *inositol-1,4,5-trisphosphate.*
IP_3 receptor: ligand-gated calcium channel in the ER membrane that opens when bound to IP_3, allowing calcium ions to flow from the ER lumen into the cytosol. (p. 429)
IPSP: see *inhibitory postsynaptic potential.*
IRE: see *iron-response element.*
iron-response element (IRE): short base sequence found in mRNAs whose translation or stability is controlled by iron; binding site for an IRE-binding protein. (p. 775)
iron-sulfur protein: protein that contains iron and sulfur atoms complexed with four cysteine groups and that serves as an electron carrier in the electron transport system. (p. 298)
irreversible inhibitor: molecule that binds to an enzyme covalently, causing an irrevocable loss of catalytic activity. (p. 173)
isoenzyme: any of several, physically distinct proteins that catalyze the same reaction. (p. 342)
isoprenylated membrane protein: protein located on a membrane surface that is covalently bound to a prenyl group embedded within the lipid bilayer. (p. 206)

J

J: see *joule.*
Jak: see *Janus activated kinase.*
Janus kinase (Jak): cytoplasmic protein kinase that, after activation by cell surface receptors, catalyzes the phosphorylation and activation of STAT transcription factors. (p. 771)
joule (J): a unit of energy corresponding to 0.239 calories. (p. 140)

K

karyotype: picture of the complete set of chromosomes for a particular cell type, organized as homologous pairs arranged on the basis of differences in size and shape. (p. 601)
Kb: kilobases; a thousand base pairs (p. 545)
k_{cat}: see *turnover number.*
K_d: see *dissociation constant.*
K_{eq}: see *equilibrium constant.*
kilocalorie (kcal): unit of energy; 1 kcal = 1000 calories. (p. 140)
kinesin: family of motor proteins that generate movement along microtubules using energy derived from ATP hydrolysis. (p. 478)
kinetochore: multiprotein complex located at the centromere region of a chromosome thatprovides the attachment site for spindle microtubules during mitosis or meiosis. (p. 601)
kinetochore microtubule: spindle microtubule that attaches to chromosomal kinetochores. (p. 601)
K_m: see *Michaelis constant.*
Knockout mice: Mice that have been genetically engineered to contain a deletion of DNA sequences in a particular gene using homologous recombination in embryonic stem cells
Krebs cycle: see *tricarboxylic acid cycle.*

L

***lac* operon:** group of adjoining bacterial genes that code for enzymes involved in lactose metabolism and whose transcription is inhibited by the *lac* repressor. (p. 740)
lactate fermentation: anaerobic catabolism of carbohydrates with lactate as the end product. (p. 265)
lagging strand: strand of DNA that grows in the $3' \rightarrow 5'$ direction during DNA replication by discontinuous synthesis of short fragments in the $5' \rightarrow 3'$ direction, followed by ligation of adjacent fragments. (p. 586) Also see *leading strand.*
lamellipodium (plural, **lamellipodia**)**:** thin sheet of flattened cytoplasm that transiently protrudes from the surface of eukaryotic cells during cell crawling; supported by actin filaments. (p. 499)
laminin: adhesive glycoprotein of the extracellular matrix, localized predominantly in the basal lamina of epithelial cells. (p. 521)
large ribosomal subunit: component of a ribosome with a sedimentation coefficient of 60S in eukaryotes and 50S in prokaryotes; associates with a small ribosomal subunit to form a functional ribosome. (p. 123)
late endosome: vesicle containing newly synthesized acid hydrolases plus material fated for digestion; activated either by lowering the pH of the late endosome or transferring its material to an existing lysosome. (p. 366)
lateral diffusion: diffusion of a membrane lipid or protein in the plane of the membrane. (p. 195)
latrunculin A: chemical compound derived from marine sponges that causes depolymerization of actin microfilaments. (p. 643)
law: a theory that has been so thoroughly tested and confirmed over a long period of time by a large number of investigators that virtually no doubt remains as to its validity. (p. 42)
law of independent assortment: principle stating that the alleles of each gene separate independently of the alleles of other genes during gamete formation. (p. 643)
law of segregation: principle stating that the alleles of each gene separate from each other during gamete formation. (p. 643)
LDL: see *low-density lipoprotein.*
LDL receptor: plasma membrane protein that serves as a receptor for binding extracellular LDL, which is then taken into the cell by receptor-mediated endocytosis. (p. 375)

leading strand: strand of DNA that grows as a continuous chain in the 5′ → 3′ direction during DNA replication. (p. 586) Also see *lagging strand.*

leaf peroxisome: special type of peroxisome found in the leaves of photosynthetic plant cells that contains some of the enzymes involved in photorespiration. (pp. 121, 344, 386)

lectin: any of numerous carbohydrate-binding proteins that can be isolated from plant or animal cells and that promote cell-cell adhesion. (p. 508)

leptotene: first stage of prophase I of meiosis, characterized by condensation of chromatin fibers into visible chromosomes. (p. 633)

leucine zipper: DNA-binding motif found in many transcription factors; formed by an interaction between α helices in two polypeptide chains that are "zippered" together by hydrophobic interactions between leucine residues. (p. 767)

leukemia: cancer of blood or lymphatic origin in which the cancer cells proliferate and reside mainly in the bloodstream rather than growing as solid masses of tissue. (p. 786)

LHC: see *light-harvesting complex.*

LHCI: see *light-harvesting complex I.*

LHCII: see *light-harvesting complex II.*

licensing: process of making DNA competent for replication; normally occurs only once per cell cycle. (p. 583)

ligand: substance that binds to a specific receptor, thereby initiating the particular event or series of events for which that receptor is responsible. (p. 420)

ligand-gated ion channel: an integral membrane protein that forms an ion-conducting pore that opens when a specific molecule (ligand) binds to the channel. (p. 400)

light microscope: instrument consisting of a source of visible light and a system of glass lenses that allows an enlarged image of a specimen to be viewed. (pp. 34, 823)

light-harvesting complex (LHC): collection of light-absorbing pigments, usually chlorophylls and carotenoids, linked together by proteins; unlike a photosystem, does not contain a reaction center, but absorbs photons of light and funnels the energy to a nearby photosystem. (p. 366)

light-harvesting complex I (LHCI): the light-harvesting complex associated with photosystem I. (p. 334)

light-harvesting complex II (LHCII): the light-harvesting complex associated with photosystem II. (p. 332)

lignin: insoluble polymers of aromatic alcohols that occur mainly in woody plant tissues, where they contribute to the hardening of the cell wall and the structural strength we associate with wood. (p. 527)

limit of resolution: measurement of how far apart adjacent objects must be in order to be distinguished as separate entities. (pp. 34, 826)

LINEs (long interspersed nuclear elements): class of interspersed repeated DNA sequences 6000–8000 base pairs in length that function as transposable elements and account for roughly 20% of the human genome; contain genes coding for enzymes needed for copying LINE sequences (and other mobile elements) and inserting the copies elsewhere in the genome. (p. 555)

Lineweaver-Burk equation: linear equation obtained by inverting the Michaelis-Menten equation, useful in determining parameters V_{max} and K_m and in the analysis of enzyme inhibition. (p. 170)

linkage group: group of genes that are transmitted, inherited, and assorted together. (p. 647)

linked genes: genes that are usually inherited together because they are located relatively close to each other on the same chromosome. (p. 647)

lipid: any of a large and chemically diverse class of organic compounds that are poorly soluble or insoluble in water but soluble in organic solvents. (p. 93)

lipid bilayer: unit of membrane structure, consisting of two layers of lipid molecules (mainly phospholipid) arranged so that their hydrophobic tails face toward each other and the polar region of each faces the aqueous environment on one side or the other of the bilayer. (pp. 54, 111 187)

lipid raft: localized region of membrane lipids, often characterized by elevated levels of cholesterol and glycosphingolipids, that sequester proteins involved in cell signaling; also called lipid microdomain. (p. 201)

lipid-anchored membrane protein: protein located on a membrane surface that is covalently bound to one or more lipid molecules residing within the lipid bilayer. (p. 206) Also see *fatty acid-anchored membrane protein, prenylated membrane protein,* and *GPI-anchored membrane protein.*

looped domain: folding of a 30-nm chromatin fiber into loops 50,000–100,000 bp in length by periodic attachment of the DNA to an insoluble network of nonhistone proteins. (p. 561)

low-density lipoprotein (LDL): cholesterol-containing protein-lipid complex that transports cholesterol through the bloodstream and is taken up by cells; exhibits low density because of its high cholesterol content. (p. 374)

lumen: internal space enclosed by a membrane, usually the endoplasmic reticulum or related membrane systems. (p. 117)

lymphoma: cancer of lymphatic origin in which the cancer cells grow as solid masses of tissue. (p. 786)

lysosomal storage disease: disease resulting from a deficiency of one or more lysosomal enzymes and characterized by the undesirable accumulation of excessive amounts of specific substances that would normally be degraded by the deficient enzymes. (p. 383)

lysosome: membrane-bounded organelle containing digestive enzymes capable of degrading all the major classes of biological macromolecules. (pp. 119, 380)

M

M line: dark line running down the middle of the H zone of a striated muscle myofibril. (p. 489)

M phase: stage of the eukaryotic cell cycle when the nucleus and the rest of the cell divide. (p. 577)

macromolecule: polymer built from small repeating monomer units, with molecular weights ranging from a few thousand to hundreds of millions. (p. 55)

macrophagy: process by which an organelle becomes wrapped in a double membrane derived from the endoplasmic reticulum, creating an autophagic vacuole that acquires lysosomal enzymes which degrade the organelle. (p. 382)

malignant tumor: tumor that can invade neighboring tissues and spread through the body via fluids, especially the bloodstream, to other parts of the body; also called a cancer. (p. 787)

MAP: see *microtubule-associated protein.*

MAP kinase: see *mitogen-activated protein kinase.*

MAPK: see *mitogen-activated protein kinase.*

mass spectrometry: high-speed, extremely sensitive technique that uses magnetic and electric fields to separate proteins or protein fragments based on differences in mass and charge. (pp. 37, 552)

***MAT* locus:** site in the yeast genome where the active allele for mating type resides. (p. 754)

mating bridge: transient cytoplasmic connection through which DNA is transferred from a male bacterial cell to a female cell during conjugation. (p. 650)

mating type: equivalent of sexuality (male or female) in lower organisms, where the molecular properties of a gamete determine the type of gamete it can fuse with. (p. 630)

matrix: unstructured semifluid substance that fills the interior of a mitochondrion. (pp. 113, 285)

maximum ATP yield: maximum amount of ATP produced per molecule of glucose oxidized by aerobic respiration; usually 38 molecules of ATP for prokaryotic cells and either 36 or 38 for eukaryotic cells. (p. 312)

maximum velocity (V_{max}): upper limiting reaction rate approached by an enzyme-catalyzed reaction as the substrate concentration approaches infinity. (p. 167)

Mb: megabases; a million base pairs. (p. 545)

MDR: see *multidrug resistance transport protein.*

mechanical work: use of energy to bring about a physical change in the position or orientation of a cell or some part of it. (p. 135)

mechanoenzyme: see *motor protein.*

medial cisterna: flattened membrane sac of the Golgi complex located between the membrane tubules of the *cis*-Golgi network and the *trans*-Golgi network. (p. 361)

Mediator: large multiprotein complex that functions as a transcriptional coactivator by binding both to activator proteins associated with an enhancer and to RNA polymerase; acts as a central coordinating unit of gene regulation, receiving both positive and negative inputs and transmitting the information to the transcription machinery. (p. 764)

meiosis: series of two cell divisions, preceded by a single round of DNA replication, that converts a single diploid cell into four haploid cells (or haploid nuclei). (p. 630)

meiosis I: the first meiotic division, which produces two haploid cells with chromosomes composed of sister chromatids. (p. 632)

meiosis II: the second meiotic division, which separates the sister chromatids of the haploid cell generated by the first meiotic division. (p. 635)

membrane: permeability barrier surrounding and delineating cells and organelles; consists of a lipid bilayer with associated proteins. (pp. 53, 184)

membrane asymmetry: a membrane property based on differences between the molecular compositions of the two lipid monolayers and the proteins associated with each. (p. 195)

membrane potential: voltage across a membrane created by ion gradients; usually the inside of a cell is negatively charged with respect to the outside. (pp. 224, 395)

Mendel's laws of inheritance: principles derived by Mendel from his work on the inheritance of traits in pea plants. (p. 643). Also see *law of segregation* and *law of independent assortment.*

mesophyll cell: outer cell in the leaf of a C_4 plant that serves as the site of carbon fixation by the Hatch-Slack cycle. (p. 337)

messenger RNA (mRNA): RNA molecule containing the information that specifies the amino acid sequence of one or more polypeptides. (p. 673)
metabolic pathway: series of cellular enzymatic reactions that convert one molecule to another via a series of intermediates. (p. 252)
metabolism: all chemical reactions occurring within a cell. (p. 252)
metaphase: stage during mitosis or meiosis when the chromosomes become aligned at the spindle equator. (p. 601)
metastable state: condition where potential reactants are thermodynamically unstable but have insufficient energy to exceed the activation energy barrier for the reaction. (p. 158)
metastasis: spread of tumor cells to distant organs via the bloodstream or other body fluids. (p. 790)
MF: see *microfilament.*
Michaelis constant (K_m): substrate concentration at which an enzyme-catalyzed reaction is proceeding at one-half of its maximum velocity. (p. 169)
Michaelis-Menten equation: widely used equation describing the relationship between velocity and substrate concentration for an enzyme-catalyzed reaction: $V = V_{max}[S]/(K_m + [S])$. (p. 169)
microautophagy: process by which lysosomes take up and degrade cytosolic proteins. (p. 780)
microfibril: aggregate of several dozen cellulose molecules laterally crosslinked by hydrogen bonds; serves as a structural component of plant and fungal cell walls. (p. 526)
microfilament (MF): polymer of actin, with a diameter of about 7 nm, that is an integral part of the cytoskeleton, contributing to the support, shape, and mobility of eukaryotic cells. (pp. 125, 461)
micrometer (μm): unit of measure: 1 micrometer $= 10^{-6}$ meters. (p. 31)
microphagy: process by which small bits of cytoplasm are surrounded by ER membrane to form an autophagic vesicle whose contents are then digested either by accumulation of lysosomal enzymes or by fusion of the vesicle with a late endosome. (p. 382)
microRNA (miRNA): class of single-stranded RNAs, about 21–22 nucleotides long, produced by cellular genes for the purpose of inhibiting the translation of mRNAs produced by other genes. (p. 778)
microsome: vesicle formed by fragments of endoplasmic reticulum when tissue is homogenized. (p. 354)
microtome: instrument used to slice an embedded biological specimen into thin sections for light microscopy. (p. 840)
microtubule (MT): polymer of the protein tubulin, with a diameter of about 25 nm, that is an integral part of the cytoskeleton and that contributes to the support, shape, and motility of eukaryotic cells; also found in eukaryotic cilia and flagella. (pp. 125, 452)
microtubule-associated protein (MAP): any of various accessory proteins that bind to microtubules and modulate their assembly, structure, and/or function. (p. 460)
microtubule-organizing center (MTOC): structure that initiates the assembly of microtubules, the primary example being the centrosome. (p. 458)
microvillus (plural, **microvilli**)**:** fingerlike projection from the cell surface that increases membrane surface area; important in cells that have an absorption function, such as those that line the intestine. (p. 466)
middle lamella: first layer of the plant cell wall to be synthesized; ends up farthest away from the plasma membrane, where it functions to hold adjacent cells together. (p. 527)
minus end (of microtubule): slower-growing (or non-growing or shrinking) end of a microtubule. (p. 455)
miRISC: complex between microRNA and several proteins that together silence the expression of messenger RNAs containing sequences complementary to those of the microRNA. (p. 778). Also see *siRISC* and *RISC.*
miRNA: see *microRNA.*
mismatch repair: DNA repair mechanism that detects and corrects base pairs that are improperly hydrogen-bonded. (p. 597)
mitochondrion (plural, **mitochondria**)**:** double membrane-enclosed cytoplasmic organelle of eukaryotic cells that is the site of aerobic respiration and hence of ATP generation. (pp. 112, 282)
mitogen-activated protein kinase (MAP kinase, or MAPK): a family of protein kinases that are activated when cells receive a signal to grow and divide. (p. 437)
mitosis: process by which two genetically identical daughter nuclei are produced from one nucleus as the duplicated chromosomes of the parent cell segregate into separate nuclei; usually followed by cell division. (p. 577)
mitotic index: percentage of cells in a population that are in any stage of mitosis at a certain point in time; used to estimate the relative length of the M phase of the cell cycle. (p. 578)
mitotic spindle: microtubular structure responsible for separating chromosomes during mitosis. (p. 600)
mitotic spindle checkpoint: mechanism that halts mitosis at the junction between metaphase and anaphase if chromosomes are not properly attached to the spindle. (p. 615)
MLCK: see *myosin light-chain kinase.*
model organism: an organism that is widely studied, well characterized, easy to manipulate, and has particular advantages for various experimental studies; examples include *E. coli*, yeast, *Drosophila*, *C. elegans*, *Arabiposis*, and mice (p. 40)
molecular chaperone: a protein that facilitates the folding of other proteins but is not acomponent of the final folded structure. (pp. 61, 718)
molecular targeting: development of drugs designed to specifically target molecules that are critical to cancer cells. (p. 818)
monoclonal antibody: a highly purified antibody, directed against a single antigen, that is produced by a cloned population of antibody-producing cells. (p. 817)
monomer: small organic molecule that serves as a subunit in the assembly of a macromolecule. (p. 57)
monomeric protein: protein that consists of a single polypeptide chain. (p. 72)
monosaccharide: simple sugar; the repeating unit of polysaccharides. (p. 89)
motif: region of protein secondary structure consisting of small segments of α helix and/or β sheet connected by looped regions of varying length. (p. 78)
motility (cellular): movement or shortening of a cell, movement of components within a cell, or movement of environmental components past or through a cell. (p. 477)
motor protein (mechanoenzyme): protein that uses energy derived from ATP to change shape in a way that exerts force and causes attached structures to move; includes three families of proteins (myosin, dynein, and kinesin) that interact with cytoskeletal elements (microtubules and microfilaments) to produce movements. (pp. 477, 604)
MPF: a mitotic Cdk-cyclin complex that drives progression from G2 into mitosis by phosphorylating proteins involved in key stages of mitosis. (p. 611)
mRNA: see *messenger RNA.*
mRNA-binding site: place on the ribosome where mRNA binds during protein synthesis. (p. 708)
MT: see *microtubule.*
MTOC: see *microtubule-organizing center.*
multidrug resistance (MDR) transport protein: an ABC-type ATPase that uses the energy of ATP hydrolysis to pump hydrophobic drugs out of cells. (p. 239)
multimeric protein: protein that consists of two or more polypeptide chains. (p. 72)
multiphoton excitation microscopy: specialized type of fluorescence light microscope employing a laser beam that emits rapid pulses of light; images are similar in sharpness to confocal microscopy, but photodamage is minimized because there is little out-of-focus light. (p. 834)
multiprotein complex: two or more proteins (usually enzymes) bound together in a way that allows each protein to play a sequential role in the same multistep process. (p. 82)
muscle contraction: generation of tension in muscle cells by the sliding of thin (actin) filaments past thick (myosin) filaments. (p. 488)
muscle fiber: long, thin, multinucleate cell specialized for contraction. (p. 488)
mutagen: chemical or physical agent capable of inducing mutations. (p. 678)
mutation: change in the base sequence of a DNA molecule. (p. 595)
myelin sheath: concentric layers of membrane that surround an axon and serve as electrical insulation that allows rapid transmission of nerve impulses. (p. 394)
myofibril: cylindrical structure composed of an organized array of thin actin filaments and thick myosin filaments; found in the cytoplasm of skeletal muscle cells. (p. 489)
myosin: family of motor proteins that create movements by exerting force on actin microfilaments using energy derived from ATP hydrolysis; makes up the thick filaments that move the actin thin filaments during muscle contraction. (p. 487)
myosin light-chain kinase (MLCK): enzyme that phosphorylates myosin light chains, thereby triggering smooth muscle contraction. (p. 498)
myosin subfragment 1 (S1): proteolytic fragment of myosin that binds to actin microfilaments in a way that yields a distinctive arrowhead pattern, with all the S1 molecules pointing in the same direction. (p. 462)

N

NA: see *numerical aperture.*
Na^+/glucose symporter: a membrane transport protein that simultaneously transports glucose and sodium ions into cells, with the movement of sodium ions down their electrochemical gradient driving the transport of glucose against its concentration gradient. (p. 241)
Na^+/K^+ ATPase: see *Na^+/K^+ pump.*
Na^+/K^+ pump: membrane carrier protein that couples ATP hydrolysis to the inward transport of potassium ions and the outward transport of

sodium ions to maintain the Na^+ and K^+ gradients that exist across the plasma membrane of most animal cells. (pp. 241, 397)

NAD^+: see *nicotinamide adenine dinucleotide.*

NADH-coenzyme Q oxidoreductase: see *complex I.*

$NADP^+$: see *nicotinamide adenine dinucleotide phosphate.*

nanometer (nm): unit of measure: 1 nanometer $= 10^{-9}$ meters. (p. 31)

native conformation: three-dimensional folding of a polypeptide chain into a shape that represents the most stable state for that particular sequence of amino acids. (p. 78)

negative staining: technique in which an unstained specimen is visualized in a transmission electron microscope against a darkly stained background. (p. 844)

NER: see *nucleotide excision repair.*

Nernst equation: equation for calculating the equilibrium membrane potential for a given ion: $E_x = (RT/zF) \ln[X]_{outside}/[X]_{inside}$. (p. 397)

nerve: a tissue composed of bundles of axons. (p. 394)

nerve impulse: signal transmitted along nerve cells by a wave of depolarization-repolarization events propagated along the axonal membrane. (p. 406)

NES: see *nuclear export signal.*

N-ethylmaleimide-sensitive factor (NSF): soluble cytoplasmic protein that acts in conjunction with several soluble NSF attachment proteins (SNAPs) to mediate the fusion of membranes brought together by interactions between v-SNAREs and t-SNAREs. (p. 379)

neuromuscular junction: site where a nerve cell axon makes contact with a skeletal muscle cell for the purpose of transmitting electrical impulses. (p. 495)

neuron: specialized cell directly involved in the conduction and transmission of nerve impulses; nerve cell. (p. 393)

neuropeptide: a molecule consisting of a short chain of amino acids that is involved in transmitting signals from neurons to other cells (neurons as well as other cell types). (p. 412)

neurosecretory vesicle: a small vesicle containing neurotransmitter molecules; located in theterminal bulb of an axon. (p. 412)

neurotoxin: toxic substance that disrupts the transmission of nerve impulses. (p. 413)

neurotransmitter: chemical released by a neuron that transmits nerve impulses across a synapse. (p. 409)

neurotransmitter reuptake: mechanism for removing neurotransmitters from the synaptic cleft by pumping them back into the presynaptic axon terminals or nearby support cells. (p. 415)

nexin: protein that connects and maintains the spatial relationship of adjacent outer doublets in the axoneme of eukaryotic cilia and flagella. (p. 485)

N-glycosylation: see *N-linked glycosylation.*

nicotinamide adenine dinucleotide (NAD^+): coenzyme that accepts two electrons and one proton to generate the reduced form, NADH; important electron carrier in energy metabolism. (p. 258)

nicotinamide adenine dinucleotide phosphate ($NADP^+$): coenzyme that accepts two electrons and one proton to generate the reduced form, NADPH; important electron carrier in the Calvin cycle and other biosynthetic pathways. (p. 330)

nitric oxide (NO): a gas molecule that transmits signals to neighboring cells by stimulating guanylyl cyclase. (p. 426)

N-linked glycosylation (N-glycosylation): addition of oligosaccharide units to the terminal amino group of asparagine residues in protein molecules. (p. 362)

NLS: see *nuclear localization signal.*

NO: see *nitric oxide.*

nocodazole: synthetic drug that inhibits microtubule assembly; frequently used instead of colchicine because its effects are more readily reversible when the drug is removed. (p. 456)

nodes of Ranvier: small segments of bare axon between successive segments of myelin sheath. (p. 394)

noncovalent bonds and interactions: binding forces that do not involve the sharing of electrons; examples include ionic bonds, hydrogen bonds, van der Waals interactions, and hydrophobic interactions. (pp. 62, 73)

noncyclic electron flow: continuous, unidirectional flow of electrons from water to $NADP^1$ during the energy transduction reactions of photosynthesis, with light providing the energy that drives the transfer. (p. 334)

nondisjunction: failure of the two members of a homologous chromosome pair to separate during anaphase I of meiosis, resulting in a joined chromosome pair that moves into one of the two daughter cells. (p. 637)

nonhomologous end-joining: mechanism for repairing double-strand DNA breaks that uses proteins that bind to the ends of the two broken DNA fragments and join the ends together. (p. 598)

nonreceptor protein kinase: any protein kinase that is not an intrinsic part of a cell surface receptor. (p. 804)

nonrepeated DNA: DNA sequences present in single copies within an organism's genome. (p. 554)

nonsense-mediated decay: mechanism for destroying mRNAs containing premature stop codons. (p. 722)

nonsense mutation: change in base sequence converting a codon that previously coded for an amino acid into a stop codon. (p. 721)

nonstop decay: mechanism for destroying mRNAs containing no stop codons. (p. 722)

NOR: see *nucleolus organizer region.*

Northern blotting: technique in which RNA molecules are size-fractionated by gelelectrophoresis and then transferred to a special type of "blotter" paper (nitrocellulose or nyon), which is then hybridized with a radioactive DNA probe. (p. 761)

NPC: see *nuclear pore complex.*

NSF: see *N-ethylmaleimide-sensitive factor.*

N-terminus (amino terminus): the end of a polypeptide chain that contains the first amino acid to be incorporated during mRNA translation; usually retains a free amino group. (p. 72)

nuclear envelope: double membrane around the nucleus that is interrupted by numerous small pores. (pp. 111, 565)

nuclear export signal (NES): amino acid sequence that targets a protein for export from the nucleus. (p. 568)

nuclear lamina: thin, dense meshwork of fibers that lines the inner surface of the inner nuclear membrane and helps support the nuclear envelope. (p. 570)

nuclear localization signal (NLS): amino acid sequence that targets a protein for transport into the nucleus. (p. 568)

nuclear matrix (nucleoskeleton): insoluble fibrous network that provides a supporting framework for the nucleus. (p. 570)

nuclear pore: small opening in the nuclear envelope through which molecules enter and exit the nucleus; lined by an intricate protein structure called the nuclear pore complex (NPC). (p. 565)

nuclear pore complex (NPC): intricate protein structure, composed of 30 or more different polypeptide subunits, that lines the nuclear pores through which molecules enter and exit the nucleus, both by simple diffusion of smaller molecules and active transport of larger molecules. (p. 566)

nuclear transfer: experimental technique in which the nucleus from one cell is transferred into another cell (usually an egg cell) whose own nucleus has been removed. (p. 752)

nucleation: act of providing a small aggregate of molecules from which a polymer can grow. (p. 455)

nucleic acid: a linear polymer of nucleotides joined together in a genetically determined order. Each nucleotide is composed of ribose or deoxyribose, a phosphate group, and the nitrogenous base guanine, cytosine, adenosine, or thymine (for DNA) or uracil (for RNA). (p. 82) Also see *DNA* and *RNA*.

nucleic acid hybridization: family of techniques in which single-stranded nucleic acids are allowed to bind to each other by complementary base pairing; used for assessing whether two nucleic acids contain similar base sequences (p. 544)

nucleic acid probe: see *probe.*

nucleoid: region of cytoplasm in which the genetic material of a prokaryotic cell is located. (p. 557)

nucleolus (plural, **nucleoli**)**:** large, spherical structure present in the nucleus of a eukaryotic cell; the site of ribosomal RNA synthesis and processing and of the assembly of ribosomal subunits. (pp. 112, 571)

nucleolus organizer region (NOR): stretch of DNA in certain chromosomes where multiple copies of the genes for ribosomal RNA are located and where nucleoli form. (p. 572)

nucleoplasm: the interior space of the nucleus, other than that occupied by the nucleolus. (p. 565)

nucleoside: molecule consisting of a nitrogen-containing base (purine or pyrimidine) linked to a five-carbon sugar (ribose or deoxyribose); a nucleotide with the phosphate removed. (p. 83)

nucleoside monophosphate: see *nucleotide.*

nucleoskeleton (nuclear matrix): insoluble fibrous network that provides a supporting framework for the nucleus. (p. 570)

nucleosome: basic structural unit of eukaryotic chromosomes, consisting of about 200 base pairs of DNA associated with an octamer of histones. (p. 559)

nucleotide: molecule consisting of a nitrogen-containing base (purine or pyrimidine) linked to a five-carbon sugar (ribose or deoxyribose) attached to a phosphate group; also called a nucleoside monophosphate. (p. 82)

nucleotide excision repair (NER): DNA repair mechanism that recognizes and repairs damage involving major distortions of the DNA double helix, such as that caused by pyrimidine dimers. (p. 597)

nucleus: large, double membrane-enclosed organelle that contains the chromosomal DNA of a eukaryotic cell. (pp. 111, 564)

numerical aperture (NA): property of a microscope corresponding to the quantity $n \sin \alpha$, where n is the refractive index of the medium between the specimen and the objective lens and α is the aperture angle. (p. 826)

O

objective lens (electron microscope): electromagnetic lens within which the specimen is placed in a transmission electron microscope. (p. 841)

objective lens (light microscope): lens located immediately above the specimen in a light microscope. (p. 827)

obligate aerobe: organism that has an absolute requirement for oxygen as an electron acceptor and therefore cannot live under anaerobic conditions. (p. 259)

obligate anaerobe: organism that cannot use oxygen as an electron acceptor and therefore has an absolute requirement for an electron acceptor other than oxygen. (p. 259)

ocular lens: lens through which the observer looks in a light microscope; also called the eyepiece. (p. 828)

OEC: see *oxygen-evolving complex.*

Okazaki fragments: short fragments of newly synthesized, lagging-strand DNA that are joined together by DNA ligase during DNA replication. (p. 586)

oligodendrocyte: cell type in the central nervous system that forms the myelin sheath around nerve axons. (p. 407)

O-linked glycosylation: addition of oligosaccharide units to hydroxyl groups of serine or threonine residues in protein molecules. (p. 362)

oncogene: any gene whose presence can cause cancer; arises by mutation from normal cellular genes called proto-oncogenes. (p. 800)

oncogenic virus: a virus that can cause cancer. (p. 799)

operator (*O*): base sequence in an operon to which a repressor protein can bind. (p. 740)

operon: cluster of genes with related functions that is under the control of a single operator and promoter, thereby allowing transcription of these genes to be turned on and off together. (p. 740)

optical tweezers: technique in which laser light focused through the objective lens of a microscope traps a small plastic bead, which is then used to manipulate the molecules it is attached to. (p. 838)

organelle: any membrane-bounded, intracellular structure that is specialized for carrying out a particular function. Eukaryotic cells contain several kinds of membrane-enclosed organelles, including the nucleus, mitochondria, Golgi complex, endoplasmic reticulum, lysosomes, peroxisomes, secretory vesicles, and, in the case of plants, chloroplasts. (p. 106)

organic chemistry: the study of carbon-containing compounds. (p. 47)

origin of replication: specific base sequence within a DNA molecule where replication is initiated. (p. 582)

origin of transfer: point on an F-factor plasmid where the transfer of the plasmid from an F^+ donor bacterial cell to an F^-recipient cell begins during conjugation. (p. 650)

osmolarity: solute concentration on one side of a membrane relative to that on the other side of the membrane; drives the osmotic movement of water across the membrane. (p. 228)

osmosis: movement of water through a semipermeable membrane driven by a difference in solute concentration on the two sides of the membrane. (p. 227)

outer doublet: pair of fused microtubules, arranged in groups of nine around the periphery of the axoneme of a eukaryotic cilium or flagellum. (p. 482)

outer membrane: the outer of the two membranes that surround a mitochondrion, chloroplast, or nucleus. (pp. 283, 323)

ovum (plural, **ova**): see *egg.*

oxidation: chemical reaction involving the removal of electrons; oxidation of organic molecules frequently involves the removal of both electrons and hydrogen ions (protons) and is therefore also called a dehydrogenation reaction. (p. 257) Also see *beta oxidation.*

oxidative phosphorylation: formation of ATP from ADP and inorganic phosphate by coupling the exergonic oxidation of reduced coenzyme molecules by oxygen to the phosphorylation of ADP, with an electrochemical proton gradient as the intermediate. (p. 304)

oxygen-evolving complex (OEC): assembly of manganese ions and proteins included within photosystem II that catalyzes the oxidation of water to oxygen. (p. 333)

oxygenic phototroph: organism that utilizes water as the electron donor in photosynthesis, with release of oxygen. (p. 323)

P

P face: interior face of the inner, or cytoplasmic, monolayer of a membrane as revealed by the technique of freeze fracturing; called the P face because this monolayer is on the *protoplasmic* side of the membrane. (p. 847)

P site (peptidyl site): site on the ribosome that contains the growing polypeptide chain at the beginning of each elongation cycle. (p. 708)

P_1 generation: the first parental generation of a genetic breeding experiment. (p. 641)

p21 protein: Cdk inhibitor that halts progression through the cell cycle by inhibiting severaldifferent Cdk-cyclins. (p. 616)

***p53* gene:** tumor suppressor gene that codes for the p53 protein, a transcription factor involved in preventing genetically damaged cells from proliferating; most frequently mutated gene in human cancers. (p. 806)

p53 protein: transcription factor that accumulates in the presence of damaged DNA and activates genes whose products halt the cell cycle and trigger apoptosis. (p. 616)

P680: the pair of chloroplast molecules that make up the reaction center of photosystem II. (p. 329)

P700: the pair of chloroplast molecules that make up the reaction center of photosystem I. (p. 329)

pachytene: stage during prophase I of meiosis when crossing over between homologous chromosomes takes place. (p. 633)

packing ratio: see *DNA packing ratio.*

Pap smear: screening technique for early detection of cervical cancer in which cells obtained from a sample of vaginal secretions are examined with a microscope. (p. 815)

paracellular transport: a type of cellular transport in which ions move between, rather than through, cells. (p. 514)

passive spread of depolarization: process in which cations (mostly K^+) move away from the site of membrane depolarization to regions of membrane where the potential is more negative. (p. 406)

patch clamping: technique in which a tiny micropipette placed on the surface of a cell is used to measure the movement of ions through individual ion channels. (p. 400)

P bodies: microscopic structures present in the cytoplasm of eukaryotic cells that are involved in the storage and degradation of mRNAs. (p. 776)

PC: see *plastocyanin.*

PCR: see *polymerase chain reaction.*

PDGF: see *platelet-derived growth factor.*

pectin: branched polysaccharides, rich in galacturonic acid and rhamnose; found in plant cell walls, where they form a matrix in which cellulose microfibrils are embedded. (p. 526)

pellet: material that sediments to the bottom of a centrifuge tube during centrifugation. (p. 356)

peptide bond: a covalent bond between the amino group of one amino acid and the carboxyl group of a second amino acid. (p. 72)

peptidyl site: see *P site.*

peptidyl transferase: enzymatic activity, exhibited by the rRNA of the large ribosomal subunit, that catalyzes peptide bond formation during protein synthesis. (p. 716)

perinuclear space: space between the inner and outer nuclear membranes that is continuous with the lumen of the endoplasmic reticulum. (p. 565)

peripheral membrane protein: hydrophilic protein bound through weak ionic interactions and hydrogen bonds to a membrane surface. (p. 205)

peroxisome: single membrane-bounded organelle that contains catalase and one or more hydrogen peroxide-generating oxidases and is therefore involved in the metabolism of hydrogen peroxide. (pp. 119, 384) Also see *leaf peroxisome.*

PFK-2: see *phosphofructokinase-2.*

phage: see *bacteriophage.*

phagocyte: specialized white blood cell that carries out phagocytosis as a defense mechanism. (p. 371)

phagocytic vacuole: membrane-bounded structure containing ingested particulate matter that fuses with a late endosome or matures directly into a lysosome, forming a large vesicle in which the ingested material is digested. (p. 371)

phagocytosis: type of endocytosis in which particulate matter or even an entire cell is taken up from the environment and incorporated into vesicles for digestion. (p. 371)

pharmacogenetics: study of how inherited differences in genes cause people to respond differently to drugs and medications. (p. 358)

phase transition: change in the state of a membrane between a fluid state and a gel state. (p. 197)

phase-contrast microscopy: light microscopic technique that improves contrast without sectioning and staining by exploiting differences in thickness and refractive index; produces an image using an optical material that is capable of bringing undiffracted rays into phase with those that have been diffracted by the specimen. (p. 828)

phenotype: observable physical characteristics of an organism attributable to the expression of its genotype. (p. 629)

phosphatidic acid: basic component of phosphoglycerides; consists of two fatty acids and a phosphate group linked by ester bonds to glycerol; key intermediate in the synthesis of other phosphoglycerides. (p. 97)

phosphoanhydride bond: high-energy bond between phosphate groups. (p. 253)

phosphodiester bridge (3′, 5′ phosphodiester bridge): covalent linkage in which two parts of a molecule are joined through oxygen atoms to the same phosphate group. (p. 86)

phosphodiesterase: enzyme that catalyzes the hydrolysis of cyclic AMP to AMP. (p. 427)

phosphoester bond: covalent linkage in which a molecule is joined through an oxygen atom to a phosphate group. (p. 253)

phosphofructokinase-2 (PFK-2): an enzyme that catalyzes the ATP-dependent phosphorylation of fructose-6-phosphate on carbon atom 2 to form fructose-2,6-bisphosphate (F2,6BP), an important regulator of both glycolysis and gluconeogenesis. (p. 273)

phosphoglyceride: predominant phospholipid component of cell membranes, consisting of a glycerol molecule esterified to two fatty acids and a phosphate group. (pp. 97, 191)

phosphoglycolate: two-carbon compound produced by the oxygenase activity of rubisco. Because it cannot be metabolized during the next step of the Calvin cycle, the production of phosphoglycolate decreases photosynthetic efficiency. (p. 344)

phospholipase C: enzyme that catalyzes the hydrolysis of phosphatidylinositol-4,5-bisphosphate into inositol-1,4,5-trisphosphate (IP_3) and diacylglycerol (DAG). (p. 429)

phospholipid: lipid possessing a covalently attached phosphate group and therefore exhibiting both hydrophilic and hydrophobic properties; main component of the lipid bilayer that forms the structural backbone of all cell membranes. (pp. 97, 191)

phospholipid exchange protein: any of a group of proteins located in the cytosol that transfer specific phospholipid molecules from the ER membrane to the outer mitochondrial, chloroplast, or plasma membranes. (p. 360)

phospholipid translocator (flippase): a membrane protein that catalyzes the flip-flop of membrane phospholipids from one monolayer to the other. (pp. 197, 359)

phosphorylation: addition of a phosphate group. (p. 176)

photoactivation: light-induced activation of an inert molecule to an active state; generally associated with the ultraviolet light-induced release of a caging group that had blocked the fluorescence of a molecule that it had been attached to. (p. 837)

photoautotroph: organism capable of obtaining energy from the sun and using this energy to drive the synthesis of energy-rich organic molecules, using carbon dioxide as a source of carbon. (p. 321)

photobleaching: technique in which an intense beam of light within a well-defined area is used to render fluorescent molecules non-fluorescent; the rate at which unbleached fluorescent molecules repopulate the bleached area provides information about the dynamic movements of the molecule of interest. (p. 837)

photochemical reduction: transfer of photoexcited electrons from one molecule to another. (p. 326)

photoexcitation: excitation of an electron to a higher energy level by the absorption of a photon of light. (p. 325)

photoheterotroph: organism capable of obtaining energy from the sun but dependent on organic compounds, rather than carbon dioxide, for carbon. (p. 321)

photon: fundamental particle of light with an energy content that is inversely proportional to its wavelength. (p. 325)

photophosphorylation: light-dependent generation of ATP driven by an electrochemical proton gradient established and maintained as excited electrons of chlorophyll return to their ground state via an electron transport system. (p. 335)

photoreduction: light-dependent generation of NADPH by the transfer of energized electrons from photoexcited chlorophyll molecules to $NADP^+$ via a series of electron carriers. (p. 330)

photorespiration: light-dependent pathway that decreases the efficiency of photosynthesis by oxidizing reduced carbon compounds without capturing the released energy; occurs when oxygen substitutes for carbon dioxide in the reaction catalyzed by rubisco, thereby generating phosphoglycolate that is then converted to 3-phosphoglycerate in the peroxisome and mitochondrion; also called the glycolate pathway. (p. 344)

photosynthesis: process by which plants and certain bacteria convert light energy to chemical energy that is then used in synthesizing organic molecules. (p. 321)

photosystem: assembly of chlorophyll molecules, accessory pigments, and associated proteins embedded in thylakoid membranes or bacterial photosynthetic membranes; functions in the light-requiring reactions of photosynthesis. (p. 328)

photosystem I (PSI): photosystem containing a pair of chlorophyll molecules (P700) that absorbs 700-nm red light maximally; light of this wavelength can excite electrons derived from plastocyanin to an energy level that allows them to reduce ferredoxin, from which the electrons are then used to reduce $NADP^+$ to NADPH. (p. 329)

photosystem II (PSII): photosystem containing a pair of chlorophyll molecules (P680) that absorb 680-nm red light maximally; light of this wavelength can excite electrons donated by water to an energy level that allows them to reduce plastoquinone. (p. 329)

photosystem complex: a photosystem plus its associated light-harvesting complexes. (p. 329)

phototroph: organism that is capable of utilizing the radiant energy of the sun to satisfy its energy requirements. (pp. 137, 321)

phragmoplast: parallel array of microtubules that guides vesicles containing polysaccharides and glycoproteins toward the spindle equator during cell wall formation in dividing plant cells. (p. 607)

phycobilin: accessory pigment found in red algae and cyanobacteria that absorbs visible light in the green-to-orange range of the spectrum, giving these cells their characteristic colors. (p. 328)

phycobilisome: light-harvesting complex found in red algae and cyanobacteria that contains phycobilins rather than chlorophyll and carotenoids. (p. 329)

phytosterol: any of several sterols that are found uniquely, or primarily, in the membranes of plant cells; examples include campesterol, sitosterol, and stigmasterol. (p. 193)

PI 3-kinase (PI3K): enzyme that adds a phosphate group to PIP_2 (phosphatidylinositol-4,5-bisphosphate), thereby converting it to PIP_3 (phosphatidylinositol-3,4,5-trisphosphate); key component of the PI3K-Akt pathway, which is activated in response to the binding of certain growth factors to their receptors. (p. 445)

PI3K: see *PI 3-kinase.*

pigment: light-absorbing molecule responsible for the color of a substance. (p. 325)

pilus: see *sex pilus.*

PKA: see *protein kinase A.*

PKC: see *protein kinase C.*

plakin: family of proteins involved in linking the integrin molecules of a hemidesmosome to intermediate filaments of the cytoskeleton. (p. 523)

plaque: dense layer of fibrous material located on the cytoplasmic side of adhesive junctions such as desmosomes, hemidesmosomes, and adherens junctions; composed of intracellular attachment proteins that link the junction to the appropriate type of cytoskeletal filament. (p. 523) The same term can also refer to the clear zone produced when bacterial cells in a small region of a culture dish are destroyed by infection with a bacteriophage. (p. 660)

plasma membrane: bilayer of lipids and proteins that defines the boundary of the cell and regulates the flow of materials into and out of the cell; also called the cell membrane. (pp. 110, 184)

plasmid: small circular DNA molecule in bacteria that can replicate independent of chromosomal DNA; useful as cloning vectors. (p. 558)

plasmodesma (plural, **plasmodesmata):** cytoplasmic channel through pores in the cell walls of two adjacent plant cells, allowing fusion of the plasma membranes and chemical communication between the cells. (pp. 126, 528)

plasmolysis: outward movement of water that causes the plasma membrane to pull away from the cell wall in cells that have been exposed to a hypertonic solution. (p. 228)

plastid: any of several types of plant cytoplasmic organelles derived from proplastids, including chloroplasts, amyloplasts, chromoplasts, proteinoplasts, and elaioplasts. (pp. 116, 323)

plastocyanin (PC): copper-containing protein that donates electrons to chlorophyll P700 of photosystem I in the light-requiring reactions of photosynthesis. (p. 333)

plastoquinol: fully reduced form of plastoquinone, involved in the light-requiring reactions of photosynthesis; present in the lipid phase of the photosynthetic membrane, where it transfers electrons to the cytochrome b_6/f complex. (p. 333)

plastoquinone: nonprotein (quinone) molecule associated with photosystem II, where it receives electrons from a modified type of chlorophyll called pheophytin during the light-requiring reactions of photosynthesis. (p. 332)

platelet-derived growth factor (PDGF): protein produced by blood platelets that stimulates the proliferation of connective tissue and smooth muscle cells. (p. 617)

plus end (of microtubule): rapidly growing end of a microtubule. (p. 455)

plus-end tubulin interacting proteins (+−TIP proteins): proteins that stabilize the plus ends of microtubules, decreasing the likelihood that microtubules will undergo catastrophic subunit loss. (p. 461)

pmf: see *proton motive force.*

polar body: tiny haploid cell produced during the meiotic divisions that create egg cells. Polar bodies receive a disproportionately small amount of cytoplasm and usually degenerate. (p. 639)

polar microtubule: spindle microtubule that interacts with spindle microtubules from the opposite spindle pole. (p. 601)

polarity: property of a molecule that results from part of the molecule having a partial positive charge and another part having a partial negative charge, usually because one region of the molecule possesses one or more electronegative atoms that draw electrons toward that region. (p. 51)

polarized secretion: fusion of secretory vesicles with the plasma membrane and expulsion of their contents to the cell exterior specifically localized at one end of a cell. (p. 370)

poly(A) tail: stretch of about 50–250 adenine nucleotides added to the 3′ end of most eukaryotic mRNAs after transcription is completed. (p. 696)

polycistronic mRNA: an mRNA molecule that codes for more than one polypeptide. (p. 740)

polymerase chain reaction (PCR): reaction in which a specific segment of DNA is amplified by repeated cycles of (1) heat treatment to separate the two strands of the DNA double helix, (2) incubation with primers that are complementary to sequences located at the two ends of the DNA segment being amplified, and (3) incubation with DNA polymerase to synthesize DNA using the primers as starting points. (p. 588)

polynucleotide: linear chain of nucleotides linked by phosphodiester bonds. (p. 86)

polypeptide: linear chain of amino acids linked by peptide bonds. (pp. 60, 72)

polyribosome: cluster of two or more ribosomes simultaneously translating a single mRNA molecule. (p. 719)

polysaccharide: polymer consisting of sugars and sugar derivatives linked together by glycosidic bonds. (p. 88)

polytene chromosome: giant chromosome containing multiple copies of the same DNA molecule generated by successive rounds of DNA replication in the absence of cell division. (p. 756)

porin: transmembrane protein that forms pores for the facilitated diffusion of small hydrophilic molecules; found in the outer membranes of mitochondria, chloroplasts, and many bacteria. (pp. 233, 283, 325)

postsynaptic neuron: a neuron that receives a signal from another neuron through a synapse. (p. 408)

posttranslational control: mechanisms of gene regulation involving selective alterations in polypeptides that have already been synthesized; includes covalent modifications, proteolytic cleavage, protein folding and assembly, import into organelles, and protein degradation. (p. 779)

posttranslational import: uptake by organelles of completed polypeptide chains after they have been synthesized, mediated by specific targeting signals within the polypeptide. (p. 724)

posttranslational modifications: alterations in polypeptides that have already been synthesized; includes covalent modifications such as glycosylation, phosphorylation, or ubiquitylation, as well as proteolytic cleavage and folding. (p. 723)

precarcinogen: substance capable of causing cancer only after it has been metabolically activated by enzymes in the liver. (p. 795)

pre-mRNA: primary transcript whose processing yields a mature mRNA. (p. 696)

pre-replication complex: group of proteins that bind to eukaryotic DNA and license it for replication; includes the Origin Recognition Complex, MCM complex, and helicase loaders. (p. 582)

pre-rRNA: primary transcript whose processing yields mature rRNAs. (p. 693)

presynaptic neuron: a neuron that transmits a signal to another neuron through a synapse. (p. 408)

pre-tRNA: primary transcript whose processing yields a mature tRNA. (p. 694)

primary cell wall: flexible portion of the plant cell wall that develops beneath the middle lamella while cell growth is still occurring; contains a loosely organized network of cellulose microfibrils. (p. 527)

primary structure: sequence of amino acids in a polypeptide chain. (p. 75)

primary transcript: any RNA molecule newly produced by transcription, before any processing has occurred. (p. 693)

primase: enzyme that uses a single DNA strand as a template to guide the synthesis of the RNA primers that are required for initiation of replication of both the lagging and leading strands of a DNA double helix. (p. 589)

primosome: complex of proteins in bacterial cells that includes primase plus six other proteins involved in unwinding DNA and recognizing base sequences where replication is to be initiated. (p. 589)

prion: infectious, protein-containing particle responsible for neurological diseases such as scrapie in sheep and goats, kuru in humans, and mad cow disease in cattle—and its human form, vCJD. (pp. 129, 720)

probe (nucleic acid): single-stranded nucleic acid that is used in hybridization experiments to identify nucleic acids containing sequences that are complementary to the probe. (pp. 545, 661)

procaspase: an inactive precursor form of a caspase. (p. 621)

procollagen: a precursor molecule that is converted to collagen by proteolytic cleavage of sequences at both the N- and C-terminal ends. (p. 518)

progression: see *tumor progression.*

projector lens: electromagnetic lens located between the intermediate lens and the viewing screen in a transmission electron microscope. (p. 841)

prokaryote: category of organisms characterized by the absence of a true nucleus and other membrane-bounded organelles; includes bacteria and archaea. (p. 103)

prometaphase: stage of mitosis characterized by nuclear envelope breakdown and attachment of chromosomes to spindle microtubules; also called late prophase. (p. 600)

promoter (site): base sequence in DNA to which RNA polymerase binds when initiating transcription[PPA6]. (pp. 685, 740)

promotion (stage of carcinogenesis): gradual process by which cells previously exposed to an initiating carcinogen are subsequently converted into cancer cells by agents that stimulate cell proliferation. (p. 587)

proofreading: removal of mismatched base pairs during DNA replication by the exonuclease activity of DNA polymerase. (p. 559)

propagation: movement of an action potential along a membrane away from the site of origin. (p. 403)

prophase: initial phase of mitosis, characterized by chromosome condensation and the beginning of spindle assembly. Prophase I of meiosis is more complex, consisting of stages called leptotene, zygotene, pachytene, diplotene, and diakinesis. (p. 600)

proplastid: small, double-membrane-enclosed, plant cytoplasmic organelle that can develop into several kinds of plastids, including chloroplasts. (p. 323)

prosthetic group: small organic molecule or metal ion component of an enzyme that plays an indispensable role in the catalytic activity of the enzyme. (p. 160)

proteasome: multiprotein complex that catalyzes the ATP-dependent degradation of proteins linked to ubiquitin. (p. 779)

protein: macromolecule that consists of one or more polypeptides folded into a conformation specified by the linear sequence of amino acids. Proteins play important roles as enzymes, structural proteins, motility proteins, and regulatory proteins. (p. 69)

protein disulfide isomerase: enzyme in the ER lumen that catalyzes the formation and breakage of disulfide bonds between cysteine residues in polypeptide chains. (p. 727)

protein kinase: any of numerous enzymes that catalyze the phosphorylation of proteinmolecules. (p. 176)

protein kinase A (PKA): a protein kinase, activated by the second messenger cyclic AMP, that catalyzes the phosphorylation of serine or threonine residues in target proteins. (p. 424)

protein kinase C (PKC): enzyme that phosphorylates serine and threonine groups in a variety of target proteins when activated by diacylglycerol. (p. 429)

protein phosphatase: any of numerous enzymes that catalyze the dephosphorylation, or removal by hydrolysis, of phosphate groups from a variety of target proteins. (p. 176)

proteoglycan: complex between proteins and glycosaminoglycans found in the extracellular matrix. (p. 519)

proteolysis: degradation of proteins by hydrolysis of the peptide bonds between amino acids. (p. 294)

proteolytic cleavage: removal of a portion of a polypeptide chain, or cutting a polypeptide chain into two fragments, by an enzyme that cleaves peptide bonds. (p. 177)

proteome: the structure and properties of all the proteins produced by a genome. (p. 552)

protoeukaryote: hypothetical evolutionary ancestor of present-day eukaryotic cells whose ability to carry out phagocytosis allowed it to engulf and establish an endosymbiotic relationship with primitive bacteria. (p. 327)

protofilament: linear polymer of tubulin subunits; usually arranged in groups of 13 to form the wall of a microtubule. (p. 454)

proton motive force (pmf): force across a membrane exerted by an electrochemical proton gradient that tends to drive protons back down their concentration gradient. (p. 307)

proton translocator: channel through which protons flow across a membrane driven by an electrochemical gradient; examples include CF_o in thylakoid membranes and F_o in mitochondrial inner membranes. (pp. 308, 335)

proto-oncogene: normal cellular gene that can be converted into an oncogene by point mutation, gene amplification, chromosomal translocation, local DNA rearrangement, or insertional mutagenesis. (p. 800)

provacuole: vesicle in plant cells comparable to an endosome in animal cells; arises either from the Golgi complex or by autophagy. (p. 383)

proximal control element: DNA regulatory sequence located upstream of the core promoter but within about 100–200 base pairs of it. (p. 762)

PSA test: screening technique for early detection of prostate cancer that measures how much prostate-specific antigen (PSA) is present in the blood. (p. 815)

pseudopodium (plural, **pseudopodia**)**:** large, blunt-ended cytoplasmic protrusion involved in cell crawling by amoebas, slime molds, and leukocytes. (p. 501)

PSI: see *photosystem I.*

PSII: see *photosystem II.*

PTEN: phosphatase that removes a phosphate group from PIP_3 (phosphatidylinositol-3,4,5-trisphosphate), thereby converting it to PIP_2 (phosphatidylinositol-4,5-bisphosphate); component of the PI3K-Akt pathway. (p. 445)

P-type ATPase: type of transport ATPase that is reversibly phosphorylated by ATP as part of the transport mechanism. (p. 238)

Puma (p53 upregulated modulator of apoptosis): protein that triggers apoptosis by binding to and inactivating Bcl-2, an inhibitor of apoptosis. (p. 616)

purine: two-ringed nitrogen-containing molecule; parent compound of the bases adenine and guanine. (p. 83)

pyrimidine: single-ringed nitrogen-containing molecule; parent compound of the bases cytosine, thymine, and uracil. (p. 83)

Q

Q cycle: proposed pathway for recycling electrons during mitochondrial or chloroplast electron transport to allow additional proton pumping across the membrane containing the electron carriers. (pp. 303, 333)

quantum: indivisible packet of energy carried by a photon of light. (p. 325)

quaternary structure: level of protein structure involving interactions between two or more polypeptide chains to form a single multimeric protein. (p. 82)

R

Rab GTPase: GTP-hydrolyzing protein involved in locking v-SNAREs and t-SNAREs together during the binding of a transport vesicle to an appropriate target membrane. (p. 379)

Rac: member of a family of monomeric G proteins, which also includes Rho and Cdc42, that stimulate formation of various actin-containing structures within cells. (p. 468)

radial spokes: inward projections from each of the nine outer doublets to the center pair of microtubules in the axoneme of a eukaryotic cilium or flagellum, believed to be important in converting the sliding of the doublets into a bending of the axoneme. (p. 484)

Ras (protein): a small, monomeric G protein bound to the inner surface of the plasma membrane; Ras is a key intermediate in transmitting signals from receptor tyrosine kinases to the cell interior. (p. 436)

rate-zonal centrifugation: see *density gradient centrifugation*.

***RB* gene:** tumor suppressor gene coding for the Rb protein. (p. 806)

Rb protein: protein whose phosphorylation controls passage through the restriction point of the cell cycle. (p. 615)

reaction center: portion of a photosystem containing the two chlorophyll molecules that initiate electron transfer, utilizing the energy gathered by other chlorophyll molecules and accessory pigments. (p. 329) Also see *P680* and *P700*.

reactive oxygen species: highly reactive oxygen-containing compounds such as H_2O_2, superoxide anion, and hydroxyl radical that are formed in the presence of molecular oxygen and can damage cells by oxidizing cellular components (p. 386)

receptor: a protein that contains a binding site for a specific signaling molecule. (p. 420)

receptor affinity: a measure of the chemical attraction between a receptor and its ligand. (p. 422)

receptor tyrosine kinase (RTK): a receptor whose activation causes it to catalyze the phosphorylation of tyrosine residues in proteins, thereby triggering a chain of signal transduction events inside cells that can lead to cell growth, proliferation, and differentiation. (p. 435)

receptor-mediated (clathrin-dependent) endocytosis: type of endocytosis initiated at coated pits and resulting in coated vesicles; believed to be a major mechanism for selective uptake of macromolecules and peptide hormones. (p. 371)

recessive (allele): allele that is present in the genome but is phenotypically expressed only in the homozygous form; masked by a dominant allele when heterozygous. (p. 629)

recombinant DNA molecule: DNA molecule containing DNA sequences derived from two different sources. (p. 657)

recombinant DNA technology: group of laboratory techniques for joining DNA fragments derived from two or more sources. (pp. 38, 656)

redox pair: two molecules or ions that are interconvertible by the loss or gain of electrons; also called a reduction-oxidation pair. (p. 300)

reduction: chemical reaction involving the addition of electrons; reduction of organic molecules frequently involves the addition of both electrons and hydrogen ions (protons) and is therefore also called a hydrogenation reaction. (p. 257)

reduction-oxidation pair: see *redox pair*.

refractive index: measure of the change in the velocity of light as it passes from one medium to another. (p. 825)

regulated secretion: fusion of secretory vesicles with the plasma membrane and expulsion of their contents to the cell exterior in response to specific extracellular signals. (p. 369)

regulators of G protein signaling (RGS) proteins: group of proteins that stimulate GTP hydrolysis by the G_α subunit of G proteins. (p. 425)

regulatory light chain: type of myosin light chain that is phosphorylated by myosin light-chain kinase in smooth muscle cells, thereby enabling myosin to interact with actin filaments and triggering muscle contraction. (p. 498)

regulatory site: see *allosteric site*.

regulatory subunit: a subunit of a multisubunit enzyme that contains an allosteric site. (p. 175)

regulatory transcription factor: protein that controls the rate at which one or more specific genes are transcribed by binding to DNA control elements located outside the core promoter. (p. 762)

relative refractory period: time during the hyperpolarization phase of an action potential when the sodium channels of a nerve cell are capable of opening again, but it is difficult to trigger an action potential because Na^+ currents are opposed by larger K^+ currents. (p. 405)

release factors: group of proteins that terminate translation by triggering the release of acompleted polypeptide chain from peptidyl tRNA bound to a ribosome's P site. (p. 718)

renaturation: return of a protein from a denatured state to the native conformation determined by its amino acid sequence, usually accompanied by restoration of physiological function. (p. 60) Also see *DNA renaturation*.

repeated DNA: DNA sequences present in multiple copies within an organism's genome. (p. 553)

replication fork: Y-shaped structure that represents the site at which replication of a DNA double helix is occurring. (p. 581)

replicon: total length of DNA replicated from a single origin of replication. (p. 582)

replisome: large complex of proteins that work together to carry out DNA replication at the replication fork; about the size of a ribosome. (p. 592)

repressible operon: group of adjoining genes that are normally transcribed, but whose transcription is inhibited in the presence of a corepressor. (p. 745)

repressor protein (eukaryotic): regulatory transcription factor whose binding to DNA control elements leads to a reduction in the transcription rate of nearby genes. (p. 764)

repressor protein (bacterial): protein that binds to the operator site of an operon and prevents transcription of adjacent structural genes. (p. 740)

residual body: mature lysosome in which digestion has ceased and only indigestible material remains. (p. 382)

resolution: minimum distance that can separate two points that still remain identifiable as separate points when viewed through a microscope. (p. 825)

resolving power: ability of a microscope to distinguish adjacent objects as separate entities. (p. 34)

resonance energy transfer: mechanism whereby the excitation energy of a photoexcited molecule is transferred to an electron in an adjacent molecule, exciting that electron to a high-energy orbital; important means of passing energy from one pigment molecule to another in photosynthetic energy transduction. (p. 326)

resonance hybrid: the actual structure of functional groups such as carboxylate or phosphate groups that are written formally as two or more structures with one double bond and one or more single bonds to oxygen when the unshared electron pair is, in fact, delocalized over all of the possible bonds to oxygen; written as single bonds to all possible oxygen atoms and dashed lines indicating delocalization of one electron pair. (p. 254)

resonance stabilization: achievement of the most stable configuration of a molecule by maximal delocalization of an unshared electron pair over all possible bonds. (p. 254)

respirasome: group of respiratory complexes associated together in defined ratios. (p. 304)

respiratory complex: subset of carriers of the electron transport system consisting of a distinctive assembly of polypeptides and prosthetic groups, organized together to play a specific role in the electron transport process. (p. 302)

respiratory control: regulation of oxidative phosphorylation and electron transport by the availability of ADP. (p. 304)

response element: DNA base sequence located adjacent to physically separate genes whose expression can then be coordinated by binding a regulatory transcription factor to the response element wherever it occurs. (p. 769)

resting membrane potential (V_m): electrical potential (voltage) across the plasma membrane of an unstimulated nerve cell. (p. 395)

restriction endonuclease: any of a large family of enzymes isolated from bacteria that cut foreign DNA molecules at or near a palindromic recognition sequence that is usually 4 or 6 (but may be 8 or more) base pairs long; used in recombinant DNA technology to cleave DNA molecules at specific sites. (p. 546)

restriction fragment length polymorphism (RFLP): difference in restriction maps between individuals caused by small differences in the base sequences of their DNA. (p. 556)

restriction map: map of a DNA molecule indicating the location of cleavage sites for various restriction endonucleases. (p. 548)
restriction/methylation system: pathway in bacterial cells by which foreign DNA is cleaved by restriction endonucleases while the bacterial genome is protected from cleavage by prior methylation. (p. 548)
restriction point: control point near the end of G1 phase of the cell cycle where the cycle can be halted until conditions are suitable for progression into S phase; regulated to a large extent by the presence or absence of extracellular growth factors; called *Start* in yeast. (p. 609)
restriction site: DNA base sequence, usually 4 or 6 (but may be 8 or more) base pairs long, that is cleaved by a specific restriction endonuclease. (p. 546)
retrograde flow (of F-actin): bulk movement of actin microfilaments toward the rear of a cell protrusion (e.g., lamellipodium) as the protrusion extends. (p. 499)
retrograde transport: movement of vesicles from Golgi cisternae back toward the endoplasmic reticulum. (p. 362)
retrotransposon: type of transposable element that moves from one chromosomal site to another by a process in which the retrotransposon DNA is first transcribed into RNA, and reverse transcriptase then uses the RNA as a template to make a DNA copy that is integrated into the chromosomal DNA at another site. (p. 677)
retrovirus: any RNA virus that uses reverse transcriptase to make a DNA copy of its RNA. (p. 676)
reverse transcriptase: enzyme that uses an RNA template to synthesize a complementary molecule of double-stranded DNA. (p. 676)
reversible inhibitor: molecule that causes a reversible loss of catalytic activity when bound to an enzyme; upon dissociation of the inhibitor, the enzyme regains biological function. (p. 173)
RFLP: see *restriction fragment length polymorphism.*
RGS proteins: see *regulators of G protein signaling proteins.*
Rho GTPases: a family of monomeric G proteins, which includes Rho, Rac and Cdc42, that stimulate formation of various actin-containing structures within cells. (p. 468)
Rho: member of a family of monomeric G proteins, which also includes Rac and Cdc42, that stimulate formation of various actin-containing structures within cells. (p. 468)
rho (ρ) factor: bacterial protein that binds to the 3′ end of newly forming RNA molecules, triggering the termination of transcription. (p. 687)
ribonucleic acid: see *RNA.*
ribose: five-carbon sugar present in RNA and in important nucleoside triphosphates such as ATP and GTP. (p. 82)
ribosomal RNA (rRNA): any of several types of RNA molecules used in the construction of ribosomes. (p. 673)
ribosome: small particle composed of rRNA and protein that functions as the site of protein synthesis in the cytoplasm of prokaryotes and in the cytoplasm, mitochondria, and chloroplasts of eukaryotes; composed of large and small subunits. (pp. 122, 707)
riboswitch: site in mRNA to which a small molecule can bind, triggering changes in mRNA conformation that impact either transcription or translation. (p. 749)
ribozyme: an RNA molecule with catalytic activity. (p. 178)
ribulose-1,5-bisphosphate carboxylase/oxygenase: see *rubisco.*
RISC: complex between either siRNA or microRNA and several proteins that together silence the expression of messenger RNAs or genes containing sequences complementary to those of these RNAs; abbreviation for RNA-induced silencing complex. (p. 777) Also see *miRISC* and *siRISC.*
RNA (ribonucleic acid): nucleic acid that plays several different roles in the expression of genetic information; constructed from nucleotides consisting of ribose phosphate linked to adenine, uracil, cytosine, or guanine. (p. 82) Also see *messenger RNA, ribosomal RNA, transfer RNA, microRNA, snRNA,* and *snoRNA.*
RNA editing: altering the base sequence of an mRNA molecule by the insertion, removal, or modification of nucleotides. (p. 702)
RNA interference (RNAi): ability of short RNA molecules (siRNAs or microRNAs) to inhibit gene expression by triggering the degradation or inhibiting the translation of specific mRNAs, or inhibiting transcription of the gene coding for a particular mRNA. (p. 776)
RNA polymerase: any of a group of enzymes that catalyze the synthesis of RNA using DNA as a template; function by adding successive nucleotides to the 3′ end of the growing RNA strand. (p. 684)
RNA polymerase I: type of eukaryotic RNA polymerase present in the nucleolus that synthesizes an RNA precursor for three of the four types of rRNA. (p. 689)
RNA polymerase II: type of eukaryotic RNA polymerase present in the nucleoplasm that synthesizes pre-mRNA, microRNA, and most of the snRNAs. (p. 689)
RNA polymerase III: type of eukaryotic RNA polymerase present in the nucleoplasm that synthesizes a variety of small RNAs, including pre-tRNAs and 5S rRNA. (p. 689)
RNA primer: short RNA fragment, synthesized by DNA primase, that serves as an initiation site for DNA synthesis. (p. 589)
RNA processing: conversion of an initial RNA transcript into a final RNA product by the removal, addition, and/or chemical modification of nucleotide sequences. (p. 693)
RNA splicing: excision of introns from a primary RNA transcript to generate the mature, functional form of the RNA molecule. (p. 698)
RNAi: see *RNA interference.*
rotation (of lipid molecules): turning of a molecule about its long axis; occurs freely and rapidly in membrane phospholipids. (p. 195)
rough endoplasmic reticulum (rough ER): endoplasmic reticulum that is studded with ribosomes on its cytosolic side because of its involvement in protein synthesis. (pp. 118, 353)
rough ER: see *rough endoplasmic reticulum.*
rRNA: see *ribosomal RNA.*
RTK: see *receptor tyrosine kinase.*
rubisco (ribulose-1,5-bisphosphate carboxylase/oxygenase): enzyme that catalyzes the CO_2-capturing step of the Calvin cycle; joins CO_2 to ribulose-1,5-bisphosphate, forming two molecules of 3-phosphoglycerate. (p. 339)
rubisco activase: protein that stimulates photosynthetic carbon fixation by rubisco by removing inhibitory sugar phosphates from the rubisco active site. (p. 341)

S

S: see *entropy.*
S: see *substrate concentration* and *Svedberg unit.*
S phase: stage of the eukaryotic cell cycle in which DNA is synthesized. (p. 578)
S1: see *myosin subfragment 1.*
sarcoma: any cancer arising from a supporting tissue, such as bone, cartilage, fat, connective tissue, and muscle. (p. 786)
sarcomere: fundamental contractile unit of striated muscle myofibrils that extends from one Z line to the next and that consists of two sets of thin (actin) and one set of thick (myosin) filaments. (p. 489)
sarcoplasmic reticulum (SR): endoplasmic reticulum of a muscle cell, specialized for accumulating, storing, and releasing calcium ions. (p. 495)
saturated fatty acid: fatty acid without double or triple bonds such that every carbon atom in the chain has the maximum number of hydrogen atoms bonded to it. (p. 96)
saturation: inability of higher substrate concentrations to increase the velocity of an enzyme-catalyzed reaction beyond a fixed upper limit determined by the finite number of enzyme molecules available. (p. 167)
scanning electron microscope (SEM): microscope in which an electron beam scans across the surface of a specimen and forms an image from electrons that are deflected from the outer surface of the specimen. (pp. 36, 842)
scanning probe microscope: instrument that visualizes the surface features of individual molecules by using a tiny probe that moves over the surface of a specimen. (p. 848)
Schwann cell: cell type in the peripheral nervous system that forms the myelin sheath around nerve axons. (p. 407)
second law of thermodynamics: the law of thermodynamic spontaneity; principle stating that all physical and chemical changes proceed in a manner such that the entropy of the universe increases. (p. 142)
second messenger: any of several substances, including cyclic AMP, calcium ion, inositol trisphosphate, and diacylglycerol, that transmit signals from extracellular signaling ligands to the cell interior. (p. 421)
secondary cell wall: rigid portion of the plant cell wall that develops beneath the primary cell wall after cell growth has ceased; contains densely packed, highly organized bundles of cellulose microfibrils. (p. 527)
secondary structure: level of protein structure involving hydrogen bonding between atoms in the peptide bonds along the polypeptide backbone, creating two main patterns called the α helix and β sheet conformations. (p. 76)
secretory granule: a large, dense secretory vesicle. (p. 367)
secretory pathway: pathway by which newly synthesized proteins move from the ER through the Golgi complex to secretory vesicles and secretory granules, which then discharge their contents to the exterior of the cell. (p. 367)
secretory vesicle: membrane-bounded compartment of a eukaryotic cell that carries secretory proteins from the Golgi complex to the plasma membrane for exocytosis and that may serve as a storage compartment for such proteins before they are released; large, dense vesicles are sometimes referred to as secretory granules. (pp. 118, 367)
securin: protein that prevents sister chromatid separation by inhibiting separase, the enzyme

that would otherwise degrade the cohesins that hold sister chromatids together. (p. 613)

sedimentation coefficient: a measure of the rate at which a particle or macromolecule moves in a centrifugal force field; expressed in Svedberg units. (pp. 123, 355)

sedimentation rate: rate of movement of a molecule or particle through a solution when subjected to a centrifugal force. (p. 355)

selectable marker: gene whose expression allows cells to grow under specific conditions that prevent the growth of cells lacking this gene. (p. 660)

selectin: plasma membrane glycoprotein that mediates cell-cell adhesion by binding to specific carbohydrate groups located on the surface of target cells. (p. 508)

self-assembly: principle that the information required to specify the folding of macromolecules and their interactions to form more complicated structures with specific biological functions is inherent in the polymers themselves. (p. 60)

SEM: see *scanning electron microscope.*

semiautonomous organelle: organelle, either a mitochondrion or a chloroplast, that contains DNA and is therefore able to encode some of its polypeptides, although it is dependent on the nuclear genome to encode most of them. (p. 326)

semiconservative replication: mode of DNA replication in which each newly formed DNA molecule consists of one old strand and one newly synthesized strand. (p. 579)

separase: protease that initiates anaphase by degrading the cohesins that hold sister chromatids together. (p. 614)

serine-threonine kinase receptor: a receptor that, upon activation, catalyzes the phosphorylation of serine and threonine residues in target protein molecules. (p. 441)

70S initiation complex: complex formed by the association of a 30S initiation complex with a 50S ribosomal subunit; contains an initiator aminoacyl tRNA at the P site and is ready to commence mRNA translation. (p. 714)

sex chromosome: chromosome involved in determining whether an individual is male or female. (p. 629)

sex pilus (plural, **pili**)**:** projection emerging from the surface of a bacterial donor cell that binds to the surface of a recipient cell, leading to the formation of a transient cytoplasmic mating bridge through which DNA is transferred from donor cell to recipient cell during bacterial conjugation. (p. 650)

sexual reproduction: form of reproduction in which two parent organisms each contribute genetic information to the new organism; reproduction by the fusion of gametes. (p. 628)

SH2 domain: a region of a protein molecule that recognizes and binds to phosphorylated tyrosines in another protein. (p. 436)

shadowing: deposition of a thin layer of an electron-dense metal on a biological specimen from a heated electrode, such that surfaces facing toward the electrode are coated while surfaces facing away are not. (p. 845)

sheet (β sheet): see *beta sheet.*

short interfering RNA: see *siRNA.*

short tandem repeat (STR): short repeated DNA sequences whose variation in length between individuals forms the basis for DNA fingerprinting. (p. 557)

sidearm: structure composed of axonemal dynein that projects out from each of the A tubules of the nine outer doublets in the axoneme of a eukaryotic cilium or flagellum. (p. 484)

sigma (s) factor: subunit of bacterial RNA polymerase that ensures the initiation of RNA synthesis at the correct site on the DNA strand. (p. 684)

signal recognition particle (SRP): cytoplasmic RNA-protein complex that binds to the ER signal sequence located at the N-terminus of a newly forming polypeptide chain and directs the ribosome-mRNA-polypeptide complex to the surface of the ER membrane. (p. 727)

signal transduction: mechanisms by which signals detected at the cell surface are transmitted into the cell's interior, resulting in changes in cell behavior and/or gene expression. (p. 421)

silencer: DNA sequence containing a binding site for transcription factors that inhibit transcription and whose position and orientation relative to the promoter can vary significantly without interfering with the ability to regulate transcription. (p. 762)

simple diffusion: unassisted net movement of a solute from a region where its concentration is higher to a region where its concentration is lower. (p. 225)

SINEs (short interspersed nuclear elements): class of interspersed, repeated DNA sequences less than 500 base pairs in length that function as transposable elements, relying on enzymes made by other mobile elements for their movement; include *Alu* sequences, the most prevalent SINE in humans. (p. 555)

single bond: chemical bond formed between two atoms as a result of sharing a pair of electrons. (p. 47)

single nucleotide polymorphisms (SNPs): variations in DNA base sequence involving single base changes that occur among individuals of the same species. (p. 552)

single-stranded DNA binding protein (SSB): protein that binds to single strands of DNA at the replication fork to keep the DNA unwound and therefore accessible to the DNA replication machinery. (p. 590)

siRISC: complex between siRNA and several proteins that together silence the expression of messenger RNAs or genes containing sequences complementary to those of the siRNA. (p. 777) Also see *miRISC* and *RISC.*

siRNA: class of double-stranded RNAs about 21–22 nucleotides in length that silence gene expression; act by either promoting the degradation of mRNAs with precisely complementary sequences or by inhibiting the transcription of genes containing precisely complementary sequences. (p. 777)

sister chromatids: the two replicated copies of each chromosome that remain attached to each other prior to anaphase of mitosis. (p. 577)

site-specific mutagenesis: technique for altering the DNA base sequence at a particular location in the genome, thereby creating a specific mutation whose effects can be studied. (p. 210)

skeletal muscle: type of muscle, striated in microscopic appearance, that is responsible for voluntary movements. (p. 488)

Slicer: ribonuclease present in RISC that cleaves mRNA at the site where the RISC has become bound. (p. 777)

sliding-filament model: model stating that muscle contraction is caused by thin actin filaments sliding past thick myosin filaments, with no change in the length of either type of filament. (p. 491)

sliding-microtubule model: model of motility in eukaryotic cilia and flagella which proposes that microtubule length remains unchanged but adjacent outer doublets slide past each other, thereby causing a localized bending because lateral connections between adjacent doublets and radial links to the center pair prevent free sliding of the microtubules past each other. (p. 485)

Smad (protein): class of proteins involved in the signaling pathway triggered by transforming growth factor β; upon activation, Smads enter the nucleus and regulate gene expression. (p. 441)

small ribosomal subunit: component of a ribosome with a sedimentation coefficient of 40S in eukaryotes and 30S in prokaryotes; associates with a large ribosomal subunit to form a functional ribosome. (p. 123)

smooth endoplasmic reticulum (smooth ER): endoplasmic reticulum that has no attached ribosomes and plays no direct role in protein synthesis; involved in packaging of secretory proteins and synthesis of lipids. (pp. 118, 353)

smooth ER: see *smooth endoplasmic reticulum.*

smooth muscle: muscle lacking striations that is responsible for involuntary contractions such as those of the stomach, intestines, uterus, and blood vessels. (p. 497)

SNAP: see *soluble NSF attachment protein.*

SNAP receptor protein: see *SNARE protein.*

SNARE (SNAP receptor) protein: two families of proteins involved in targeting and sorting membrane vesicles; include the v-SNARES found on transport vesicles and the t-SNARES found on target membranes. (p. 379)

SNARE hypothesis: model explaining how membrane vesicles fuse with the proper target membrane; based on specific interactions between v-SNAREs (vesicle-SNAP receptors) and t-SNAREs (target-SNAP receptors). (p. 379)

snoRNA: group of small nucleolar RNAs that bind to complementary regions of pre-rRNA and target specific sites for methylation or cleavage. (p. 665)

SNPs: see *single nucleotide polymorphisms.*

snRNA: a small nuclear RNA molecule that binds to specific proteins to form a snRNP, which in turn assembles with other snRNPs to form a spliceosome. (p. 693)

snRNP: RNA-protein complex that assembles with other snRNPs to form a spliceosome; pronounced "snurp." (p. 699)

sodium/potassium pump: see *Na^+/K^+ pump.*

soluble NSF attachment protein (SNAP): soluble cytoplasmic protein that acts in conjunction with NSF (N-ethylmaleimide-sensitive factor) to mediate the fusion of membranes brought together by interactions between v-SNAREs and t-SNAREs. (p. 379)

solute: substance that is dissolved in a solvent, forming a solution. (p. 52)

solvent: substance, usually liquid, in which other substances are dissolved, forming a solution. (p. 52)

Sos (protein): a guanine-nucleotide exchange factor that activates Ras by triggering the release of GDP, thereby permitting Ras to acquire a molecule of GTP. The Sos protein is activated by interacting with a GRB2 protein molecule that has become bound to phosphorylated tyrosines in an activated tyrosine kinase receptor. (p. 436)

Southern blotting: technique in which DNA fragments separated by gel electrophoresis are transferred to a special type of "blotter" paper (nitrocellulose or nylon), which is then hybridized with a radioactive DNA probe. (p. 557)

special pair: two chlorophyll *a* molecules, located in the reaction center of a photosystem, that catalyze the conversion of solar energy into chemical energy. (p. 329)

specific heat: amount of heat needed to raise the temperature of 1 gram of a substance 1°C. (p. 51)

sperm: haploid male gamete, usually flagellated. (p. 630)

sphingolipid: class of lipids containing the amine alcohol sphingosine as a backbone. (pp. 97, 191)

sphingosine: amine alcohol that serves as the backbone for sphingolipids; contains an amino group that can form an amide bond with a long-chain fatty acid; also contains a hydroxyl group that can attach to a phosphate group. (p. 97)

spliceosome: protein-RNA complex that catalyzes the removal of introns from pre-mRNA. (p. 699)

spore: see *haploid spore.*

sporophyte: diploid generation in the life cycle of an organism that alternates between haploid and diploid forms; form that produces spores by meiosis. (p. 631)

sputter coating: vacuum evaporation process used to coat the surface of a specimen with a layer of gold or a mixture of gold and palladium prior to examining the specimen by scanningelectron microscopy. (p. 848)

squid giant axon: an exceptionally large axon emerging from certain squid nerve cells; its wide diameter (0.5–1.0 mm) makes it relatively easy to insert microelectrodes that can measure and control electrical potentials and ionic currents. (p. 395)

SR: see *sarcoplasmic reticulum.*

SRP: see *signal recognition particle.*

SSB: see *single-stranded DNA binding protein.*

stage: platform on which the specimen is placed in a microscope. (p. 827)

staining: incubation of tissue specimens in a solution of dye, heavy metal, or other substance that binds specifically to selected cellular constituents, thereby giving those constituents a distinctive color or electron density. (p. 840)

standard free energy change ($\Delta G°9$): free energy change accompanying the conversion of 1 mole of reactants to 1 mole of products, with the temperature, pressure, pH, and concentration of all relevant species maintained at standard values. (p. 150)

standard reduction potential (E'_0): convention used to quantify the electron transport potential of oxidation-reduction couples relative to the H^+/H_2 redox pair, which is assigned an E'_0 value of 0.0 V at pH 7.0. (p. 299)

standard state: set of arbitrary conditions defined for convenience in reporting free energy changes in chemical reactions. For systems consisting of dilute aqueous solutions, these conditions are usually a temperature of 25°C (298 K), a pressure of 1 atmosphere, and reactants other than water present at a concentration of 1 *M*. (p. 149)

starch: storage polysaccharide in plants consisting of repeating glucose subunits linked together by $\alpha(1 \longrightarrow 4)$ bonds and, in some cases, $\alpha(1 \longrightarrow 6)$ bonds. The two main forms of starch are the unbranched polysaccharide, amylose, and the branched polysaccharide, amylopectin. (p. 91)

Start: control point near the end of G1 phase of the yeast cell cycle where the cycle can be halted until conditions are suitable for progression into S phase; known as the *restriction point* in other eukaryotes. (p. 609)

start codon: the codon AUG in mRNA when it functions as the starting point for protein synthesis. (pp. 682, 712)

start-transfer sequence: amino acid sequence in a newly forming polypeptide that acts as both an ER signal sequence that directs the ribosome-mRNA-polypeptide complex to the ER membrane and as a membrane anchor that permanently attaches the polypeptide to the lipid bilayer. (p. 729)

STAT: type of transcription factor activated by phosphorylation in the cytoplasm catalyzed by Janus activated kinase, followed by migration of the activated STAT molecules to the nucleus. (p. 770)

state: condition of a system defined by various properties, such as temperature, pressure, and volume. (p. 139)

stationary cisternae model: model postulating that each compartment of the Golgi stack is a stable structure, and that traffic between successive cisternae is mediated by shuttle vesicles that bud from one cisterna and fuse with another. (p. 361)

steady state: nonequilibrium condition of an open system through which matter is flowing, such that all components of the system are present at constant, nonequilibrium concentrations. (p. 152)

stem cell: a cell capable of unlimited division that can differentiate into a variety of other cell types. (p. 753)

stereo electron microscopy: microscopic technique for obtaining a three-dimensional view of a specimen by photographing it at two slightly different angles. (p. 847)

stereoisomers: two molecules that have the same structural formula but are not superimposable; stereoisomers are mirror images of each other. (p. 49)

steroid: any of numerous lipid molecules that are derived from a four-membered ring compound called phenanthrene. (p. 98)

steroid hormone: any of several steroids derived from cholesterol that function as signaling molecules, moving via the circulatory system to target tissues, where they cross the plasma membrane and interact with intracellular receptors to form hormone-receptor complexes that are capable of activating (or inhibiting) the transcription of specific genes. (p. 98)

steroid receptor: protein that functions as a transcription factor after binding to a specific steroid hormone. (p. 445)

sterol: any of numerous compounds consisting of a 17-carbon four-ring system with at least one hydroxyl group and a variety of other possible side groups; includes cholesterol and a variety of other biologically important compounds, such as the male and female sex hormones, that are related to cholesterol. (p. 193)

sticky end: single-stranded end of a DNA fragment generated by cleavage with a restriction endonuclease that tends to reassociate with another fragment generated by the same restriction endonuclease because of base complementarity. (p. 549)

stomata (singular, stoma): pores on the surface of a plant leaf that can be opened or closed to control gas and water exchange between the atmosphere and the interior of the leaf. (p. 337)

stop codon: sequence of three bases in mRNA that instructs the ribosome to terminate protein synthesis. UAG, UAA, and UGA generally function as stop codons. (pp. 682, 712)

stop-transfer sequence: hydrophobic amino acid sequence in a newly forming polypeptide that halts translocation of the chain through the ER membrane, thereby anchoring the polypeptide within the membrane. (p. 729)

storage macromolecule: polymer that consists of one or a few kinds of subunits in no specific order and that serves as a storage form of monosaccharides; examples include starch and glycogen. (p. 58)

STR: see *short tandem repeat.*

striated muscle: muscle whose myofibrils exhibit a pattern of alternating dark and light bands when viewed microscopically; includes both skeletal and cardiac muscle. (p. 489)

stroma: unstructured semifluid matrix that fills the interior of the chloroplast. (pp. 116, 325)

stroma thylakoid: membrane that interconnects stacks of grana thylakoids with each other. (pp. 115, 325)

structural macromolecule: polymer that consists of one or a few kinds of subunits in no specific order and that provides structure and mechanical strength to the cell; examples include cellulose and pectin. (p. 58)

subcellular fractionation: technique for isolating organelles from cell homogenates using various types of centrifugation. (pp. 37, 355)

substrate activation: role of an enzyme's active site in making a substrate molecule maximally reactive by subjecting it to the appropriate chemical environment for catalysis. (p. 165)

substrate concentration (S): amount of substrate present per unit volume at the beginning of a chemical reaction. (p. 167)

substrate induction: regulatory mechanism for catabolic pathways in which the synthesis of enzymes involved in the pathway is stimulated in the presence of the substrate and inhibited in the absence of the substrate. (p. 739)

substrate specificity: ability of an enzyme to discriminate between very similar molecules. (p. 161)

substrate-level phosphorylation: formation of ATP by direct transfer to ADP of a high-energy phosphate group derived from a phosphorylated substrate. (p. 263)

substrate-level regulation: enzyme regulation that depends directly on the interactions of substrates and products with the enzyme. (p. 174)

succinate-coenzyme Q oxidoreductase: see *complex II.*

supercoiled DNA: twisting of a DNA double helix upon itself, either in a circular DNA molecule or in a DNA loop anchored at both ends. (p. 542)

supernatant: material that remains in solution after particles of a given size and density are removed as a pellet during centrifugation. (p. 356)

superresolution microscopy: a set of related techniques, including stimulated emission depletion (STED) microscopy, photoactivated localization microscopy (PALM) and stochastic optical reconstruction microscopy (STORM), that allows objects to be visualized in the light microscope at a resolution greater than the theoretical limit predicted due to diffraction by the Abbé equation (p. 838)

suppressor tRNA: mutant tRNA molecule that inserts an amino acid where a stop codon generated by another mutation would otherwise have caused premature termination of protein synthesis. (p. 721)

surface area/volume ratio: mathematical ratio of the surface area of a cell to its volume; decreases with increasing linear dimension of the cell (length or radius), thereby increasing the difficulty of maintaining adequate surface area for import of nutrients and export of waste products as cell size increases. (p. 104)

surroundings: the remainder of the universe when one is studying the distribution of energy within a given system. (p. 139)

survival factor: a secreted molecule whose presence prevents a cell from undergoing apoptosis. (p. 621)

Svedberg unit (S): unit for expressing the sedimentation coefficient of biological macromolecules: One Svedberg unit (S) = 10^{-13} second. In general, the greater the mass of a particle, the greater the sedimentation rate, though the relationship is not linear. (p. 355)

SWI/SNF: family of chromatin-remodeling proteins. (p. 760)

synaptic boutons: regions near the end of an axon where neurotransmitter molecules are stored for use in transmitting signals across the synapse. (p. 394)

symbiotic relationship: a mutually beneficial association between cells (or organisms) of two different species. (p. 326)

symport: coupled transport of two solutes across a membrane in the same direction. (p. 231)

synapse: tiny gap between a neuron and another cell (neuron, muscle fiber, or gland cell), across which the nerve impulse is transferred by direct electrical connection or by chemicals called neurotransmitters. (p. 394)

synapsis: close pairing between homologous chromosomes during the zygotene phase of prophase I of meiosis. (p. 632)

synaptic bouton: region near the end of an axon where neurotransmitter molecules are stored for use in transmitting signals across the synapse. (p. 394)

synaptic cleft: gap between the presynaptic and postsynaptic membranes at the junction between two nerve cells. (p. 409)

synaptonemal complex: zipperlike, protein-containing structure that joins homologous chromosomes together during prophase I of meiosis. (p. 633)

system: the restricted portion of the universe that one decides to study at any given time when investigating the principles that govern the distribution of energy. (p. 139)

T

T: see *thymine.*

T tubule system: see *transverse tubule system.*

tandemly repeated DNA: repeated DNA sequences whose multiple copies are adjacent to one another. (p. 554)

target-SNAP receptor: see *t-SNARE.*

TATA box: part of the core promoter for many eukaryotic genes transcribed by RNA polymerase II; consists of a consensus sequence of TATA followed by two or three more A's, located about 25 nucleotides upstream from the transcriptional startpoint. (p. 689)

TATA-binding protein (TBP): component of transcription factor TFIID that confers the ability to recognize and bind the TATA box sequence in DNA; also involved in regulating transcription initiation at promoters lacking a TATA box. (p. 692)

taxol: drug that binds tightly to microtubules and stabilizes them, causing much of the free tubulin in the cell to assemble into microtubules. (p. 456)

TBP: see *TATA-binding protein.*

TCA cycle: see *tricarboxylic acid cycle.*

TE: see *transitional element.*

telomerase: special type of DNA polymerase that catalyzes the formation of additional copies of a telomeric repeat sequence. (p. 593)

telomere: DNA sequence located at either end of a linear chromosome; contains simple-sequence, tandemly repeated DNA. (pp. 554, 593)

telophase: final stage of mitosis or meiosis, when daughter chromosomes arrive at the poles of the spindle accompanied by reappearance of the nuclear envelope. (p. 602)

TEM: see *transmission electron microscope.*

temperature-sensitive mutant: cell that produces a protein that functions properly at normal temperatures but becomes seriously impaired when the temperature is altered slightly. (p. 584)

template: a nucleic acid whose base sequence serves as a pattern for the synthesis of another (complementary) nucleic acid. (p. 86)

template strand: the strand of a DNA double helix that serves as the template for RNA synthesis via complementary base pairing. (p. 680)

terminal bulb: see *synaptic bouton.*

terminal glycosylation: modification of glycoproteins in the Golgi complex involving removal and/or addition of sugars to the carbohydrate side chains formed by prior core glycosylation in the endoplasmic reticulum. (p. 363)

terminal oxidase: electron transfer complex that is capable of transferring electrons directly to oxygen. Complex IV (cytochrome *c* oxidase) of the mitochondrial electron transport system is an example. (p. 303)

terminal web: dense meshwork of spectrin and myosin molecules located at the base of a microvillus; the bundle of actin microfilaments that make up the core of the microvillus is anchored to the terminal web. (p. 466)

termination signal: DNA sequence located near the end of a gene that triggers the termination of transcription. (p. 687)

terpene: a lipid constructed from the five-carbon compound isoprene and its derivatives, joined together in various combinations. (p. 98)

tertiary structure: level of protein structure involving interactions between amino acid side chains of a polypeptide, regardless of where along the primary sequence they happen to be located; results in three-dimensional folding of a polypeptide chain. (p. 78)

tethering protein: a coiled-coil protein or a multisubunit protein complex that recognizes and binds vesicles to their target membranes. (p. 379)

tetrahedral (carbon atom): an atom of carbon from which four single bonds extend to other atoms, each bond equidistant from all other bonds, causing the atom to resemble a tetrahedron with its four equal faces. (p. 49)

TFIIB recognition element: see *BRE.*

TGFβ: see *transforming growth factor β.*

TGN: see *trans-Golgi network.*

theory: a hypothesis that has been tested critically under many different conditions—usually by many different investigators using a variety of approaches—and is consistently supported by the evidence. (p. 41)

thermodynamic spontaneity: a measure of whether a reaction can occur, but it says nothing about whether the reaction actually will occur. Reactions with a negative free energy change are thermodynamically spontaneous. (p. 141)

thermodynamics: area of science that deals with the laws governing the energy transactions that accompany all physical processes and chemical reactions. (p. 139)

thick filament: myosin-containing filament, found in the myofibrils of striated muscle cells, in which individual myosin molecules are arranged in a staggered array with the heads of the myosin molecules projecting out in a repeating pattern. (p. 489)

thin filament: actin-containing filament, found in the myofibrils of striated muscle cells, in which two F-actin molecules are arranged in a helix associated with tropomyosin and troponin. (p. 489)

thin-layer chromatography (TLC): procedure for separating compounds by chromatography in a medium, such as silicic acid, that is bound as a thin layer to a glass or metal surface. (p. 194)

30-nm chromatin fiber: fiber formed by packing together the nucleosomes of a 10-nm chromatin fiber. (p. 561)

30S initiation complex: complex formed by the association of mRNA, the 30S ribosomal subunit, an initiator aminoacyl tRNA molecule, and initiation factor IF2. (p. 714)

three-dimensional (3-D) electron tomography: computer-based method for making three-dimensional reconstructions of structures visualized in serial thin sections by transmission electron microscopy. (p. 847)

threshold potential: value of the membrane potential that must be reached before an action potential is triggered. (p. 403)

thylakoid: flattened membrane sac suspended in the chloroplast stroma, usually arranged in stacks called grana; contains the pigments, enzymes, and electron carriers involved in the light-requiring reactions of photosynthesis. (pp. 115, 325)

thylakoid lumen: compartment enclosed by an interconnected network of grana and stroma thylakoids. (p. 325)

thymine (T): nitrogen-containing aromatic base, chemically designated as a pyrimidine, which serves as an informational monomeric unit when present in DNA with other bases in a specific sequence; forms a complementary base pair with adenine (A) by hydrogen bonding. (p. 83)

Ti plasmid: DNA molecule that causes crown gall tumors when transferred into plants by bacteria; used as a cloning vector for introducing foreign genes into plant cells. (p. 664)

TIC: translocase of the inner chloroplast membrane, a transport complex involved in the uptake of specific polypeptides into the chloroplast. (p. 731)

tight junction: type of cell junction in which the adjacent plasma membranes of neighboring animal cells are tightly sealed, thereby preventing molecules from diffusing from one side of an epithelial cell layer to the other by passing through the spaces between adjoining cells. (p. 512)

TIM: translocase of the inner mitochondrial membrane, a transport complex involved in the uptake of specific polypeptides into the mitochondrion. (p. 731)

+−TIP proteins: see *plus-end tubulin interacting proteins.*

TIRF: see *total internal reflection fluorescence microscopy*.
TLC: see *thin-layer chromatography*.
T_m: see *transition temperature* or *DNA melting temperature*.
TOC: translocase of the outer chloroplast membrane, a transport complex involved in the uptake of specific polypeptides into the chloroplast. (p. 731)
TOM: translocase of the outer mitochondrial membrane, a transport complex involved in the uptake of specific polypeptides into the mitochondrion. (p. 731)
topoisomerase: enzyme that catalyzes the interconversion of the relaxed and supercoiled forms of DNA by making transient breaks in one or both DNA strands. (pp. 543, 590)
total internal reflection fluorescence (TIRF) microscopy: technique in which a light beam strikes the interface of two media of different indices of refraction at an angle beyond the critical angle, causing all the light to be reflected back into the incident medium; an "evanescent field" that develops at the interface allows selective excitation of fluorescent molecules located within ∼100 nm of the interface. (p. 837)
***trans*-acting factor:** regulatory protein that exerts its function by binding to specific DNA sequences. (p. 743)
transcription: process by which RNA polymerase utilizes one DNA strand as a template forguiding the synthesis of a complementary RNA molecule. (p. 673)
transcription factor: protein required for the binding of RNA polymerase to a promoter and for the optimal initiation of transcription. (p. 615) Also see *general transcription factor* and *regulatory transcription factor*.
transcription regulation domain: region of a transcription factor, distinct from the DNA-binding domain, that is responsible for regulating transcription. (p. 766)
transcription unit: segment of DNA whose transcription gives rise to a single, continuous RNA molecule. (p. 684)
transcriptional control: group of regulatory mechanisms involved in controlling the rates at which specific genes are transcribed. (p. 760)
transcriptome: the entire set of RNA molecules produced by a genome. (p. 552)
transcytosis: endocytosis of material into vesicles that move to the opposite side of the cell and fuse with the plasma membrane, releasing the material into the extracellular space. (p. 375)
transduction: transfer of bacterial DNA sequences from one bacterium to another by a bacteriophage. (p. 649)
transfection: introduction of foreign DNA into cells under artificial conditions. (p. 800)
transfer RNA (tRNA): family of small RNA molecules, each binding a specific amino acid and possessing an anticodon that recognizes a specific codon in mRNA. (pp. 673, 709)
transformation: change in the hereditary properties of a cell brought about by the uptake of foreign DNA (pp. 535, 649)
transforming growth factor β (TGFβ): family of growth factors that can exhibit either growth-stimulating or growth-inhibiting properties, depending on the target cell type; regulate a wide range of activities in both embryos and adult animals, including effects on cell growth, division, differentiation, and death. (pp. 441, 619)
transgenic: any organism whose genome contains a gene that has been experimentally introduced from another organism using the techniques of genetic engineering. (p. 664)
***trans*-Golgi network (TGN):** region of the Golgi complex consisting of a network of membrane-bounded tubules that are located on the opposite side of the Golgi complex from the *cis*-Golgi network. (p. 360)
transit sequence: amino acid sequence that targets a completed polypeptide chain to either mitochondria or chloroplasts. (p. 731)
transition state: intermediate stage in a chemical reaction, of higher free energy than the initial state, through which reactants must pass before giving rise to products. (p. 158)
transition temperature (T_m): temperature at which a membrane will undergo a sharp decrease in fluidity ("freezing") as the temperature is decreased and becomes more fluid again ("melts") when it is then warmed; determined by the kinds of fatty acid side chains present in the membrane. (p. 197)
transition vesicle: membrane vesicle that shuttles lipids and proteins from the endoplasmicreticulum to the Golgi complex. (p. 353)
transitional element (TE): region of the endoplasmic reticulum that is involved in the formation of transition vesicles. (p. 353)
translation: process by which the base sequence of an mRNA molecule guides the sequence of amino acids incorporated into a polypeptide chain; occurs on ribosomes. (p. 673)
translational control: mechanisms that regulate the rate at which mRNA molecules are translated into their polypeptide products; includes control of translation rates by initiation factors, selective inhibition of specific mRNAs by translational repressor proteins or microRNAs, and control of mRNA degradation rates. (p. 774)
translational repressor: regulatory protein that selectively inhibits the translation of a particular mRNA. (p. 775)
translesion synthesis: DNA replication across regions where the DNA template is damaged. (p. 596)
translocation: movement of mRNA across a ribosome by a distance of three nucleotides, bringing the next codon into position for translation. (p. 716) (*Note:* The same term, *translocation*, which literally means "a change of location," can also refer to the movement of a protein molecule through a membrane channel or to the transfer of a segment of one chromosome to another nonhomologous chromosome.)
translocon: structure in the ER membrane that carries out the translocation of newly forming polypeptides across (or into) the ER membrane. (p. 727)
transmembrane protein: an integral membrane protein possessing one or more hydrophobic regions that span the membrane plus hydrophilic regions that protrude from the membrane on both sides. (p. 205)
transmembrane segment: hydrophobic segment about 20–30 amino acids long that crosses the lipid bilayer in a transmembrane protein. (p. 205)
transmission electron microscope (TEM): type of electron microscope in which an image is formed by electrons that are transmitted through a specimen. (pp. 36, 840)
transport: selective movement of substances across membranes, both into and out of cells and into and out of organelles. (pp. 222)
transport protein: membrane protein that recognizes substances with great specificity and assists their movement across a membrane; includes both carrier proteins and channel proteins. (p. 230)
transport vesicle: vesicle that buds off from a membrane in one region of the cell and fuses with other membranes; includes vesicles that convey lipids and proteins from the ER to the Golgi complex, between the Golgi stack cisternae, and from the Golgi complex to various destinations in the cell, including secretory vesicles, endosomes, and lysosomes. (p. 360)
transposable element (transposon): DNA sequence that can move from one chromosomal location to another. (p. 555)
transposon: see *transposable element*.
transverse diffusion: movement of a lipid molecule from one monolayer of a membrane to the other, a thermodynamically unfavorable and therefore infrequent event; also called "flip-flop." (p. 195)
transverse (T) tubule system: invaginations of the plasma membrane that penetrate into a muscle cell and conduct electrical impulses into the cell interior, where T tubules make close contact with the sarcoplasmic reticulum and trigger the release of calcium ions. (p. 495)
triacylglycerol: a glycerol molecule with three fatty acids linked to it; also called a triglyceride. (pp. 96, 293)
triad: region where a T tubule passes between the terminal cisternae of the sarcoplasmic reticulum in skeletal muscle. (p. 496)
tricarboxylic acid cycle (TCA cycle): cyclic metabolic pathway that oxidizes acetyl CoA to carbon dioxide in the presence of oxygen, generating ATP and the reduced coenzymes NADH and $FADH_2$; a component of aerobic respiration; also called the Krebs cycle. (p. 286)
triglyceride: see *triacylglycerol*.
triple bond: chemical bond formed between two atoms as a result of sharing three pairs of electrons. (p. 47)
triplet code: a coding system in which three units of information are read as a unit; a reference to the genetic code, which is read from mRNA in units of three bases called *codons*. (p. 678)
triskelion: structure formed by clathrin molecules consisting of three polypeptides radiating from a central vertex; the basic unit of assembly for clathrin coats. (p. 376)
tRNA: see *transfer RNA*.
tropomyosin: long, rodlike protein associated with the thin actin filaments of muscle cells, functioning as a component of the calcium-sensitive switch that activates muscle contraction; blocks the interaction between actin and myosin in the absence of calcium ions. (p. 490)
troponin: complex of three polypeptides (TnT, TnC, and TnI) that functions as a component of the calcium-sensitive switch that activates muscle contraction; displaces tropomyosin in the presence of calcium ions, thereby activating contraction. (p. 490)
***trp* operon:** group of adjoining bacterial genes that code for enzymes involved in tryptophan biosynthesis and whose transcription is selectively inhibited in the presence of tryptophan. (p. 745)
true-breeding (plant strain): organism that, upon self-fertilization, produces only offspring of the same kind for a given genetic trait. (p. 640)
t-SNARE (target-SNAP receptor): protein associated with the outer surface of a target membrane that binds to a v-SNARE protein associated with the outer surface of an appropriate transport vesicle. (p. 379)
tubulin: family of related proteins that form the main building block of microtubules. (p. 454)

Also see *alpha tubulin, beta tubulin, gamma tubulin*, and *gamma tubulin ring complexes.*

tumor: growing mass of cells caused by uncontrolled cell proliferation. (p. 786) Also see *benign tumor* and *malignant tumor*.

tumor grading: assignment of numerical grades to tumors based on differences in their microscopic appearance; higher-grade cancers tend to grow and spread more aggressively, and to be less responsive to therapy, than lower-grade cancers. (p. 814)

tumor progression: gradual changes in tumor properties observed over time as cancer cells acquire more aberrant traits and become increasingly aggressive. (p. 797)

tumor suppressor gene: gene whose loss or inactivation by deletion or mutation can lead to cancer. (p. 805) Also see *gatekeeper gene* and *caretaker gene*.

γ-TuRCs: see *gamma tubulin ring complexes.*

turgor pressure: pressure that builds up in a cell due to the inward movement of water that occurs because of a higher solute concentration inside the cell than outside; accounts for the firmness, or turgidity, of fully hydrated cells or tissues of plants and other organisms. (p. 228)

turnover number (k_{cat}): rate at which substrate molecules are converted to product by a single enzyme molecule when the enzyme is operating at its maximum velocity. (p. 142)

type II myosin: form of myosin composed of four light chains and two heavy chains, each having a globular myosin head, a hinge region, and a long rodlike tail; found in skeletal, cardiac, and smooth muscle cells, as well as in nonmuscle cells. (p. 487)

U

U: see *uracil.*

ubiquitin: small protein that is linked to other proteins as a way of marking the targeted protein for degradation by proteasomes. (p. 779)

ultracentrifuge: instrument capable of generating centrifugal forces that are large enough to separate subcellular structures and macromolecules on the basis of size, shape, and density. (pp. 37, 355)

ultramicrotome: instrument used to slice an embedded biological specimen into ultrathin sections for electron microscopy. (p. 843)

ultraviolet radiation (UV): mutagenic type of radiation present in sunlight that triggers the formation of pyrimidine dimers in DNA. (p. 798)

undershoot: see *hyperpolarization.*

unfolded protein response (UPR): quality control mechanism in which sensor molecules in the ER membrane detect misfolded proteins and trigger a response that inhibits the synthesis of most proteins while enhancing the production of those required for protein folding anddegradation. (p. 727)

unidirectional pumping of protons: the active and directional transport of protons across a membrane such that they accumulate preferentially on one side of the membrane, establishing an electrochemical proton gradient across the membrane; a central component of electron transport and ATP generation in both respiration and photosynthesis. (p. 306)

uniport: membrane protein that transports a single solute from one side of a membrane to the other. (p. 231)

unsaturated fatty acid: fatty acid molecule containing one or more double bonds. (p. 96)

UPR: see *unfolded protein response.*

upstream: located toward the 5′ end of the DNA coding strand. (p. 685)

uracil (U): nitrogen-containing aromatic base, chemically designated as a pyrimidine, that serves as an informational monomeric unit when present in RNA with other bases in a specific sequence; forms a complementary base pair with adenine (A) by hydrogen bonding. (p. 83)

useful magnification: measurement of how much an image can be enlarged before additional enlargement provides no additional information. (p. 826)

UV: see *ultraviolet radiation.*

V

***v*:** see *initial reaction velocity.*

vacuole: membrane-bounded organelle in the cytoplasm of a cell, used for temporary storage or transport; acidic membrane-enclosed compartment in plant cells. (pp. 121, 383)

vacuum evaporator: bell jar containing a metal electrode and a carbon electrode in which a vacuum can be created; used in preparing metal replicas of the surfaces of biological specimens. (p. 845)

van der Waals interaction: weak attractive interaction between two atoms caused by transient asymmetries in the distribution of charge in each atom. (p. 74)

vesicle-SNAP receptor: see *v-SNARE.*

viroid: small, circular RNA molecule that can infect and replicate in host cells even though it does not code for any protein. (p. 129)

virus: subcellular parasite composed of a protein coat and DNA or RNA, incapable of independent existence; invades and infects cells and redirects the host cell's synthetic machinery toward the production of more virus. (p. 127)

V_m: see *resting membrane potential.*

V_{max}: see *maximum velocity.*

voltage: see *electrical potential.*

voltage sensor: amino acid segment of a voltage-gated ion channel that makes the channel responsive to changes in membrane potential. (p. 402)

voltage-gated calcium channel: an integral membrane protein in the terminal bulb of presynaptic neurons that forms a calcium ion-conducting pore whose permeability is regulated by the membrane potential; action potentials cause the calcium channel to open and calcium ions rush into the cell, stimulating the release of neurotransmitters. (p. 412)

voltage-gated ion channel: an integral membrane protein that forms an ion-conducting pore whose permeability is regulated by changes in the membrane potential. (p. 400)

v-SNARE (vesicle-SNAP receptor): protein associated with the outer surface of a transport vesicle that binds to a t-SNARE protein associated with the outer surface of the appropriate target membrane. (p. 379)

V-type ATPase: type of transport ATPase that pumps protons into such organelles as vesicles, vacuoles, lysosomes, endosomes, and the Golgi complex. (p. 239)

W

wavelength: distance between the crests of two successive waves. (p. 823)

Western blotting: technique in which polypeptides separated by gel electrophoresis are transferred to a special type of "blotter" paper (nitrocellulose or nylon), which is then reacted with labeled antibodies that are known to bind to specific polypeptides. (p. 208)

wild type: normal, nonmutant form of an organism, usually the form found in nature. (p. 647)

Wnt pathway: signaling pathway that plays a prominent role in controlling cell proliferation and differentiation during embryonic development; abnormalities in this pathway occur in some cancers. (p. 808)

wobble hypothesis: flexibility in base pairing between the third base of a codon and the corresponding base in its anticodon. (p. 709)

work: transfer of energy from one place or form to another place or form by any process other than heat flow. (p. 140)

X

xenobiotic: chemical compound that is foreign to biological organisms. (p. 386)

xeroderma pigmentosum: inherited susceptibility to cancer (mainly skin cancer) caused by defects in DNA excision repair or translesion synthesis of DNA. (p. 811)

X-ray crystallography: technique for determining the three-dimensional structure of macromolecules based on the pattern produced when a beam of X-rays is passed through a sample, usually a crystal or fiber. (pp. 208, 849)

Y

YAC: see *yeast artificial chromosome.*

yeast artificial chromosome (YAC): yeast cloning vector consisting of a "minimalist" chromosome that contains all the DNA sequences needed for normal chromosome replication and segregation to daughter cells, and very little else. (p. 662)

yeast two-hybrid system: Technique for determing whether two proteins interact by introducing DNA encoding one protein fused to DNA encoding the DNA binding domain of a transcription factor (bait), and another DNA encoding the second protein plus sequence encoding the activation domain of the transcirption factor (prey), followed by assessing expression of a reporter. (p. 768)

Z

Z line: dark line in the middle of the I band of a striated muscle myofibril; defines the boundary of a sarcomere. (p. 489)

zinc finger: DNA-binding motif found in some transcription factors; consists of an α helix and a two-segment β sheet held in place by the interaction of precisely positioned cysteine or histidine residues with a zinc atom. (p. 766)

zygote: diploid cell formed by the union of two haploid gametes. (p. 630)

zygotene: stage during prophase I of meiosis when homologous chromosomes become closely paired by the process of synapsis. (p. 633)

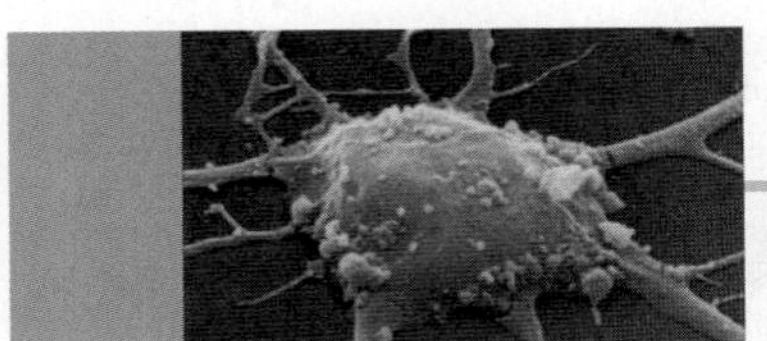

PHOTO, ILLUSTRATION, AND TEXT CREDITS

Photo Credits

Chapter 1 1-CO The National Cancer Institute at Frederick. 1-1a Dr. Hugues Massicotte. 1-1b Science Source/Photo Researchers, Inc. 1-1c The National Cancer Institute at Frederick. 1-1d medicalpicture/Alamy. 1-1e Manfred Kage/Photolibrary. 1-1f David M. Phillips/Photo Researchers, Inc. 1-1g Don W. Fawcett/Photo Researchers, Inc. 1-1h Bernd Schulz, Institut fuer Botanik, TU Dresden, Germany. 1-1i Richard Masland. Table 1-1 Top Right Michael W. Davidson, National High Magnetic Field Laboratory, The Florida State University. Table 1-1 Middle Right Michael W. Davidson, National High Magnetic Field Laboratory, The Florida State University. Table 1-1 Top Left Photographed for Pearson Science by Elisabeth Pierson, Radboud University, Nijmegen, Netherlands. Table 1-1 Middle Left Photographed for Pearson Science by Elisabeth Pierson, Radboud University, Nijmegen, Netherlands. Table 1-1 Bottom Left and Right Photographed for Pearson Science by Elisabeth Pierson, Radboud University Nijmegen, Netherlands. 1-4a SPL/Photo Researchers, Inc. 1-4b SPL/Photo Researchers, Inc. 1-4c Susumu Ito. 1-4d The Keith R. Porter Endowment for Cell Biology, University of Pennsylvania. 1B-1a P. Birn/Custom Medical Stock Photo. 1B-1b Damika/Mediscan. 1B-1c Biopix. 1B-1d SPL/Photo Researchers, Inc. 1B-1e Stanton K. Short, The Jackson Laboratory. 1B-1f Nigel Cattlin/Photo Researchers, Inc.

Chapter 2 2-CO Eva Frei and R. D. Preston. 2-9 Shutterstock. 2-13a Don W. Fawcett/Photo Researchers, Inc. 2-14 Level 2 Eva Frei and R. D. Preston. 2-14 Level 3 Right Eldon H. Newcomb. 2-14 Level 3 Left G. F. Bahr, Armed Forces Institute of Pathology. 2-19 Graphics Systems Research, IBM UK Scientific Centre.

Chapter 3 3-CO Dr. Jeremy Burgess/SPL/Photo Researchers, Inc. 3-24a Dr. Jeremy Burgess/SPL/Photo Researchers, Inc. 3-24b Don W. Fawcett/Photo Researchers, Inc. 3-25 Eva Frei and R. D. Preston. 3A-1 National Cancer Institute.

Chapter 4 4-CO Dr. Mary Olson. 4-2 Susumu Ito. 4-3b Gregory J. Brewer, Southern Illinois University. 4-4 Eldon H. Newcomb 4-5b Richard Rodewald/Biological Photo Service. 4-6b W. P. Wergin. 4-7 Hans Ris. 4-8 Oscar L. Miller Jr., University of Virginia. 4-10b Richard Rodewald/Biological Photo Service. 4-10c From J. P. Strafstrom and L. A. Staehelin, *The Journal of Cell Biology* 98 (1984): 699. Reproduced by copyright permission of The Rockefeller University Press. 4-11c Keith R. Porter/Photo Researchers, Inc. 4-12b From J.B. Rattner and B.R. Brinkley, *J. Ultrastructure Res.* 32 (1970): 316. © 1970 by Academic Press. 4-13 S. M. Wang. 4-14b W. P. Wergin. 4-15c H. Stuart Pankratz/Biological Photo Service. 4-15d Barry F. King/Biological Photo Service. 4-16b Eldon H. Newcomb. 4-18b M. Simionescu and N. Simionescu, *J. Cell Biol.* 70 (1976): 608. Reproduced by copyright permission of The Rockefeller University Press. 4-19 Barry F. King/Biological Photo Service. 4-20b From S.E. Frederick and E.H. Newcomb, *Journal of Cell Biology* 43 (1969): 343. Reproduced by copyright permission of The Rockefeller University Press. 4-21b P. J. Gruber. 4-23 Dr. Mary Olson. 4-25 Bottom From P.H. Raven, R.F. Evert, and H.A. Curtis, *Biology of Plants,* 2nd ed. New York: Worth Publishers, Inc., 1999. Used with permission of Worth Publishers. 4-26a Left, Middle, Right R.C. Williams and H.W. Fisher. 4-26b Left, Middle, Right R.C. Williams and H.W. Fisher. 4-26c Left, Middle, Right R.C. Williams and H.W. Fisher.

Chapter 5 5-CO From D.A. Cuppels and A. Kelman, *Phytopathology* 70 (1980): 1110. Photo provided by A. Kelman. © 1980 by the American Phytopathological Society. 5-1 Paul Gilham/Getty Images Sport. 5-3 From D.A. Cuppels and A. Kelman, *Phytopathology* 70 (1980): 1110. Photo provided by A. Kelman. © 1980 by the American Phytopathological Society. 5-4 Katherine W. Osteryoung. 5-6 Matt Demmon.

Chapter 6 6-CO Barry F. King/Biological Photo Service. 6-5 Richard J. Feldmann, National Institute of Health.

Chapter 7 7-CO M. Simionescu and N. Simionescu, *J. Cell Biol.* 70 (1976): 622. Reproduced by copyright permission of The Rockefeller University Press. 7-1a M. Simionescu and N. Simionescu, *J. Cell Biol.* 70 (1976): 622. Reproduced by copyright permission of The Rockefeller University Press. 7-1b Eldon H. Newcomb. 7-3e J. David Robertson. 7-4 Don W. Fawcett, Harvard Medical School. 7-16b Philippa Claude. 7-17a Daniel Branton. 7-17b R. B. Park. 7-18a,b David Deamer, University of California, Santa Cruz. 7-20a Ken Eward/Science Source/Photo Researchers, Inc. 7-27 Susumu Ito. 7-29a,b Arthur E. Sowers.

Chapter 8 8-CO Susumu Ito. 8-14a Helen E. Carr/Biological Photo Service.

Chapter 9 9-CO Gregory J. Brewer, Southern Illinois University. 9A-1 Stockbyte/Getty Images.

Chapter 10 10-CO The Keith R. Porter Endowment for Cell Biology, University of Pennsylvania. 10-2 Charles R. Hackenbrock. 10-3a The Keith R. Porter Endowment for Cell Biology, University of Pennsylvania. 10-4 From Lester Packer, "Ann. New York Academy of Sciences" vol 227 (1974): 166. Copyright 1974 by the New York Academy of Sciences. Photo provided by H. T. Ngo. 10-5a A. Tzagoloff from Mitochondria (New York: Plenum, 1982). 10A-1 From R.N. Trelease, P.J. Gruber, W.M. Becker, and E.H. Newcomb, *Plant Physiology* 48 (1971): 461.

Chapter 11 11-CO W. P. Wergin. 11-2a M.W. Steer. 11-2b Andrew Syred/Photo Researchers, Inc. 11-3a W. P. Wergin. 11-3b W. P. Wergin. 11-4 N. J. Lang/Biological Photo Service. 11-17b Michael W. Clayton/University of Wisconsin Digital Collections. 11B-1 Hartmut Michel, Max Planck Institute of Biophysics, Frankfurt, Germany, and Johann Deisenhofer, University of Texas Southwestern Medical Center at Dallas.

Chapter 12 12-CO Don W. Fawcett, Harvard Medical School. 12-2a Don W. Fawcett, Harvard Medical School. 12-2b M. Bielinska. 12-3a Barry F. King/Biological Photo Service. 12-4b Michael J. Wynne. 12-5 William G. Dunphy with Ruud Brands and James E. Rothman, *Cell* 40 (1985): 467, Fig. 6, Panel B. 12-10a–d J. D. Jamieson. 12-11 From L. Orci and A. Perrelet, *Freeze-Etch Histology,* Heidelberg: Springer-Verlag, 1975. 12-12b Holger Jastrow. 12-14b H. S. Pankratz & R. N. Band/Biological Photo Service. 12-16a–d From M.M. Perry and A.B. Gilbert, *The Journal of Cell Science* 39 (1979):257. © 1979 by The Company of Biologists Limited. 12-17a John E. Heuser. 12-18a John E. Heuser. 12-18b N. Hirokawa and J. E. Heuser. 12-20 Pierre Baudhuin. 12-22 Zdenek Hruban. 12-23 H. Shio and Paul B. Lazarow. 12-24 Eldon H. Newcomb/Biological Photo Service. 12B-2a,b R.G.W. Anderson, M.S. Brown, and J.L. Goldstein, *Cell* 10 (1977): 351–64.

Chapter 13 13-CO Jürgen Berger/Max Planck Institute for Developmental Biology, Tubingen, Germany. 13-1b Jürgen Berger/Max Planck Institute for Developmental Biology, Tubingen, Germany. 13-2 David Fleetham/Alamy. 13-13c G. L. Scott, J. A. Feilbach, and T. A. Duff. 13-16c S. G. Waxman. 13-20a From J. Cartaud, E.L. Bendetti, A. Sobel, and J.P. Changeux, *Journal of Cell Science* 29 (1978): 313. © The Company of Biologists Limited. 13-21b E. R. Lewis, Y.Y. Zeevi, and T.E. Everhart, University of California, Berkeley.

Chapter 14 14-CO Marek Mlodzik, Mt. Sinai School of Medicine. 14-11 Robert Burgoyne, Waters Corporation. 14-13a Y. Hiramoto. 14-B-1a David A. Wassarman, University of Wisconsin. 14-B-1b Reprinted from *Cell* 55, R. Reinke and S.L. Zipursky, "Cell-Cell Interaction in the Drosophila Retina," p. 321, 1988, with permission from Elsevier. 14B-1c David A. Wassarman, University of Wisconsin. 14B-1d Reprinted from *Cell* 55, R. Reinke and S.L. Zipursky, "Cell-Cell Interaction in the Drosophila Retina," p. 321, 1988, with permission from Elsevier.

Chapter 15 15-CO John E. Heuser. 15-2b L.E. Roth, Y. Shigenaka, and D.J. Pihlaja/Biological Photo Service. 15-4 Lester Binder and Joel Rosenbaum, *Journal of Cell Biology* 79 (1978): 510. Reproduced by permission of The Rockefeller University Press. 15-7 Andrew Matus and Beat Ludin. 15-8b Kent L. McDonald. 15-8c Mitchison and Kirschner,

Nature 312 (1984): 235, Fig. 4c. 15-9b T.J. Keating and G.G. Borisy, "Immunostructural Evidence for the Template Mechanism of Microtubule Nucleation." *Nat Cell Biol.* 2(6):352–357, fig.2A, 2000. 15-11 Adapted from C.E. Schutt et al., *Nature* 365 (1993): 810; Courtesy of M. Rozycki. 15-12c R. Niederman and J. Hartwig. 15-13c Roger W. Craig. 15-15 John H. Hartwig. 15-17a M. S. Mooseker and L.G. Tilney, *Journal of Cell Biology* 67 (1975): 725–43, Fig 2. Reproduced by permission of The Rockefeller University Press. 15-18 John E. Heuser. 15-19b Daniel Branton. 15-20a Borisylab, Northwestern University Medical School. 15-21a–d Alan Hall, University College, London, and Kate Nobes, University of Bristol. 15-22 P. A. Coulombe et al., "The 'ins' and 'outs' of intermediate filament organization," from *Trends in Cell Biology* 10: 420–28, Fig. 1. Reprinted with permission from Elsevier Science. Image courtesy Pierre A. Coulombe, Johns Hopkins University. 15-24 Reprinted by permission from E. Fuchs, *Science* 279: 518, Fig. D. Images: T. Svitkina and G. Borisy. © 1998 American Association for the Advancement of Science. 15A-1b Lewis J. Tilney with Daniel Portnoy and Pat Connelly, *Journal of Cell Biology* 109: 1604, Fig. 18. Table 15-1 Left Mary Osborn. Table 15-1 Middle Frank Solomon. Table 15-1 Right Mark S. Ladinsky and J. Richard McIntosh, University of Colorado. Table 15-2 Row 1 Thomas D. Pollard. Table 15-2 Row 2 Summak and Borisy, *Nature* 332 (1988): 724–36, Fig. 1b. Table 15-2 Row 3 Left and Right Reprinted from E.D. Salmon, *Trends in Cell Biology* 5: 154–58, Fig. 3, with permission from Elsevier Science. Table 15-2 Row 4 John Heuser.

Chapter 16 16-CO Lewis Tilney, University of Pennsylvania. 16-1 Nobutaka Hirokawa. 16-6a W. L. Dentler/Biological Photo Service. 16-6c Biophoto Associates/Photo Researchers, Inc. 16-7a–d William L. Dentler. 16A-1b From I.A. Belyantseva et al. Gamma-actin is required for cytoskeletal maintenance but not development. *Proc. Natl. Acad. Sci. USA.* (2009) 106:9703–9708. 16-8a Lewis Tilney, University of Pennsylvania. 16-11b Hans Ris. 16-12a Clara Franzini-Armstrong. 16-17 John E. Heuser. 16-22a Allen Bell, University of New England. 16-23a A. Hall, Rho GTPasses at the actin cytoskeleton. *Science* 279:509–14, 1998. Figs a, c, e, g. Reprinted with permission from the American Association for the Advancement of Science. 16-24b Adapted from K.M. Trybus and S. Lowey, *Journal of Biological Chemistry* 259 (1984): 8564–71. 16-25 G. Albrecht-Buehler. 16-27 Melba/AGE Fotostock. 16-28b N. K. Wessels.

Chapter 17 17-CO From Douglas E. Kelly, *The Journal of Cell Biology* 28 (1966): 51. Reproduced by permission of The Rockefeller University Press. 17-4 Masatoshi Takeichi. 17-5a,b Janet Heasman. 17-8a From Douglas E. Kelly, *The Journal of Cell Biology* 28 (1966): 51. Reproduced by permission of The Rockefeller University Press. 17-9c Philippa Claude. 17-10b Daniel Friend. 17-12b From C. Peracchia and A.F. Dulhunty, *The Journal of Cell Biology* 70 (1976): 419. Reproduced by permission of The Rockefeller University Press. 17-12c Philippa Claude. 17-13a Ed Reschke/Peter Arnold/Photolibrary. 17-13b Nina Zanetti, Pearson Education. 17-13c Ed Reschke/Peter Arnold/Photolibrary. 17-14a Jerome Gross. 17-17a L. Rosenberg, W. Hellmann, and A. K. Kleinschmidt (1975). *Journal of Biological Chemistry* 250:1877–1883. ©The American Society for Biochemistry and Molecular Biology. 17-19a Reproduced from Dr. Jean Paul Thiery, *The Journal of Cell Biology* 96 (1983): 462–73 by permission of The Rockefeller University Press. 17-19b,c Richard Hynes, from *Scientific American,* June 1986. 17-20 Ed Reschke/Photolibrary. 17-23b Daniel S. Friend. 17-23d Douglas E. Kelly. 17-25b Biophoto Associates/Photo Researchers, Inc. 17-26 G. F. Leedale/Photo Researchers, Inc. 17-27a,b K. Muhlenthaler. 17-28a W. P. Wergin. 17-28c Eldon H. Newcomb. 17A-1c Pascale Cossart, Institut Pasteur, Paris. From K. Ireton and P. Cossart, "Interaction of invasive bacteria with host signaling pathways," from *Current Opinions in Cell Biology* 10, no. 2: 276–83. Reprinted with permission from Elsevier Science.

Chapter 18 18-CO Jack Griffith. 18-6 Left, Right James C. Wang. 18-17 Top Reproduced from H. Kobayashi, K. Kobayashi, and Y. Kobayashi, *Journal of Bacteriology* 132 (1977): 262–269 by copyright permission of the American Society for Microbiology. 18-17 Bottom Gopal Murti/SPL/Photo Researchers, Inc. 18-18 Jack Griffith. 18-20a,b Roger D. Kornberg. 18-22a, Top Jack Griffith. 18-22a, Bottom Stanley C. Holt/Biological Photo Service. 18-22b Barbara Hamkalo. 18-22c J. R. Paulsen and U.K. Laemmli, Cell. © Cell Press. 18-22d,e Armed Forces Institute of Pathology. 18-23 Ulrich K. Laemmli. 18-24 D. L. Robberson. 18-26a From L. Orci and A. Perrelet, Freeze-Etch Histology. Heidelberg: Springer-Verlag, 1975. 18-26b S.R. Tandon. 18-27a From L. Orci and A. Perrelet, *Freeze-Etch Histology.* Heidelberg: Springer-Verlag, 1975. 18-28 From L. Orci and A. Perrelet, *Freeze-Etch Histology.* Heidelberg: Springer-Verlag, 1975. 18-29a From A. C. Faberge, *Cell Tiss. Res.* 15 (1974): 403. Heidelberg: Springer-Verlag, 1974. 18-32a Jeffrey A. Nickerson, Sheldon Penman, and Gariela Krockmalnic. 18-32b Ueli Aebi. 18-33 Luis Parada and Tom Misteli, National Cancer Institute, NIH. 18-34 David Phillips/Photo Researchers, Inc. 18-35 Sasha Koulish and Ruth G. Kleinfeld, *Journal of Cell Biology* 23 (1964): 39. Reproduced by permission of The Rockefeller University Press. 18A-1b Lee D. Simon/Science Source/Photo Researchers, Inc. 18A-2b Orchid Cellmark, Inc., Germantown, MD. 18A-3 Madboy via Wikipedia, http://creativecommons.org/licenses/by-sa/1.0/deed.en. 18C-2 Orchid Cellmark, Inc., Germantown, MD.

Chapter 19 19-CO Ed Reschke. 19-4a *Cold Spring Harbor Symposium Quantitative Biology* 28 (1963):44. 19-5b From D.J. Burks and P.J. Stambrook, *Journal of Cell Biology* 77 (1978): 766, fig 6. Reproduced with permission of the Rockefeller University Press. Photos provided by P.J. Stambrook. 19-16 Nikitina T and Woodcock CL. Closed chromatin loops at the ends of chromosomes. *JCB,* Volume 166, Number 2, 161–165. © The Rockefeller University Press. 19-20 Ed Reschke. 19-21a–e PhotoLibrary. 19-22b J. Richard McIntosh. 19-23 Left J. F. Gennaro/Photo Researchers, Inc. 19-23 Right CNRI/SPL/Photo Researchers, Inc. 19-26 Matthew Schibler, from *Protoplasma* 137 (1987): 29–44. Springer-Verlag. 19-27b Jeremy Pickett-Heaps, University of Melbourne. 19-27c Jeremy Pickett-Heaps, University of Melbourne. 19-28a Michael Danilchik, Dept. of Cell and Developmental Biology, Oregon Health & Science University. 19-29a George Von Dassow, University of Oregon. 19-29b Bill Bement, University of Wisconsin. 19-30a B.A. Palevitz. 19-31a Left and Right Michael Danilchik, Dept. of Cell and Developmental Biology, Oregon Health & Science University. 19-31b George Von Dassow, University of Oregon. 19-44b–d Walter Malorni et al., from "Morphological aspects of apoptosis," Image courtesy Walter Malorni, Instituto Superiore de Sanita Rome. 19B-1 Bob Goldstein, University of North Carolina.

Chapter 20 20-CO B. John. 20-6 B. John. 20-7a From P.B. Moens, *Chromosoma* 23 (1968): 418. Springer-Verlag. 20-19a Charles C. Brinton, Jr., and Judith Carnahan. 20-19b Omikron/Photo Researchers, Inc. 20-23 Ross B. Inman, University of Wisconsin, Madison. 20A-1 Ralph L. Brinster, School of Veterinary Medicine, University of Pennsylvania.

Chapter 21 21-CO Jack Griffith. 21-2 Left and Right Janice Carr/Centers for Disease Control 21-3 Left and Right Vernon Ingram. 21-13 D. B. Nikolov and S. K. Burley from Nikolov et al., *Nature* 360 (1992): 40–46. 21-16 From O.L. Miller, Jr., B.A. Hamkalo, and C.A. Thomas, Jr. Reprinted with permission from *Science* 169 (1970): 392, Fig. 3. © American Association for the Advancement of Science. 21-19a Bert W. O'Malley, Baylor College of Medicine. 21-21 Left Ann L. Beyer. 21-21 Right Jack Griffith.

Chapter 22 22-CO James A. Lake. 22-1a James A. Lake.

Chapter 23 23-CO Peter J. Bryant/Biological Photo Service. 23-13 Peter J. Bryant/Biological Photo Service. 23-15 Reproduced from S. Saragosti, G. Moyne, and M. Yaniv, *Cell* 20 (1980):65–73 by permission of Cell Press, Cambridge, MA. 23-24a IBM United Kingdom Limited. 23-27b Edward B. Lewis, California Institute of Technology. 23-27c Edward B. Lewis, California Institute of Technology. 23A-1b AP Photo.

Chapter 24 24-CO From S.L. Robbins, *Textbook of Pathology* (Philadelphia: W.B. Saunders, 1957). 24-2 Left and Right L. Hayflick. 24-20 From S.L. Robbins, *Textbook of Pathology* (Philadelphia: W.B. Saunders, 1957). 24-23 Left and Right G.D. Abrams. 24A-1 Roberto Gonzalez, Orlando Sentinel.

Appendix A-CO Timothy Ryan. A-5a Leica Microsystems Inc. A-7 M.I. Walker/Photo Researchers, Inc. A-9 Tim Ryan. A-13 © Molecular Probes, Inc. (probes.invitrogen.com). A-14a–c S. Strome et al., "Spindle dynamics and the role of b-tubulin in early Caenohabditis elegans embryos," from *Molecular Biology of the Cell* 12: 1751–64, Fig. 8. Reprinted with permission by the American Society for Cell Biology. A-15a,b Karl Garsha, Digital Light Microscopy Specialist, Imaging Technology Group, Beckman Institute for Advanced Science and Technology, University of Illinois at Urbana–Champaign, Urbana, IL. A-17a Nikon USA. A-19a,b Shelly Sazor, Baylor College of Medicine. A-20a–d Reprinted from "Trends in Cell Biology" Vol 5, pp 154–158, Salmon: Figure 3. Copyright © 2002, with permission from Elsevier Science. A-22c Mochizuki et al, "Spatio-temporal images of growth factor-induced activation of Ras and Rap 1," from *Nature* 411 (2001): 1065–68. Courtesy of M. Matsuda. Reprinted by permission of Nature Publishing Group. A-24a Reprinted from Hein et al. Stimulated emission depletion (STED) nanoscopy of a fluorescent protein-labeled organelle inside a living cell. *Proc. Natl. Acad. Sci.* 105 (2008): 14271–14276. A-24b Reprinted from Bates et al., Multicolor Super-Resolution Imaging with Photo-Switchable Fluorescent Probes. *Science* 317 (2007): 1749–1753. A-26a Carl Zeiss, Inc./LEO Electron Microscopy, Inc. A-27a Don W. Fawcett, Harvard Medical School. A-27b Keiichi Tanaka. A-28a Carl Zeiss, Inc./LEO Electron Microscopy, Inc. A-29a,b Ventana Medical Systems Inc. A-30 Janine R. Maddock. A-31 Michael F. Moody. A-32 Omikron/Photo Researchers, Inc. A-35 From L. Orci and A. Perrelet, *Freeze-Etch Histology,* Heidelberg: Springer-Verlag,

1975. A-36a Hans Ris. A-36b Koster AJ, Klumperman J. Electron microscopy in cell biology: integrating structure and function. *Nat Rev Mol Cell Biol.* 2003 Sep; Suppl: SS6–10, Fig 1. A-36c Koster AJ, Klumperman J. Electron microscopy in cell biology: integrating structure and function. *Nat Rev Mol Cell Biol.* 2003 Sep; Suppl: SS6–10, Fig 1. A-39 Left and Middle Reprinted with permission from *Nature* 171 (1953): 740; Copyright 1953 Macmillan Magazines Limited. A-39 Right Sonia DiVittorio, Pearson Education. A-40 Reprinted by permission of Nature Publishing Group.

Images used in Human Applications, Deeper Insights, and Tools of Discovery titles are by permission of Duncan Smith/Photodisc/Getty Images.

Illustration and Text Credits

The following illustrations are taken from L. J. Kleinsmith and V. M. Kish, *Principles of Cell and Molecular Biology,* 2d ed. (New York, NY: HarperCollins, 1995). Reprinted by permission of Pearson Education, Inc.

Figs. 3-7, 3-9, 3-13, 3-27, 7-6, 7-7, 7-8, 7-11, 7-19, 7-23, 7-24, 8-4, 8-5, 8-10, 10-16, 11A-1, 12-9, 12-14, 12-17c,d, 12-18b, 13-13a, 14-1, 14-13b, 14-16, 15-6, 16-2, 16-26, 18-15, 18-19, 19-2, 19-8, 19-10, 19-22a, 19-24, 20-22, 20-23, 20-24, 20-28, 21-3, 21-4, 21-11, 21-18, 21-19, 21-21, 21-23, 21-24, 22-4, 23-13, 23-14, 23-26, 23-28, 24-5, 24-6

Fig. 3-4 Illustration, Irving Geis. Image from Irving Geis Collection, Howard Hughes Medical Institute. Rights owned by HHMI. Not to be reproduced without permission. Used by permission of Sandy Geis.

Fig. 3-6 Illustration, Irving Geis. Image from Irving Geis Collection, Howard Hughes Medical Institute. Rights owned by HHMI. Not to be reproduced without permission. Used by permission of Sandy Geis.

Fig. 4-22 From P. J. Russell, *Genetics*, 5th ed., Fig. 13.18. Copyright © 1998. Reprinted by permission of Pearson Education, Inc.

Box 5A From Harold F. Blum, *Time's Arrow and Evolution,* 3rd ed. © 1951 Princeton University Press, renewed 1979; 2nd ed. 1955, renewed 1983; 3rd ed. 1968. Reprinted by permission of Princeton University Press.

Fig. 6-7 From N. A. Campbell and J. B. Reece, *Biology*, 6th ed., Fig. 6-15, p. 99. Copyright © 2002. Reprinted by permission of Pearson Education, Inc.

Box 6A Source: Cech, *Science* 236: 1532 (1987).

Fig. 7-30 Reprinted from *Journal of Molecular Biology*, vol. 157, no. 1, J. Kyte and R. F. Doolittle, "A Simple Method for Displaying the Hydropathic Character of a Protein," pp. 105–132, Copyright 1982, with permission from Elsevier.

Fig. 8-8a,b Adapted from Protein Data Bank accession 2F1C. Original reference: Subbarao, G. V. and van den Berg, B. (2006). Crystal structure of the monomeric porin OmpG. *J. Mol. Biol.* 360: 750–759 (from Fig. 1, p. 752).

Fig. 8-8c Adapted from Protein Data Bank accession 3D9S. Original reference: Horsefield et al. (2008). High-resolution x-ray structure of human aquaporin 5. *Proc. Natl. Acad. Sci. USA* 105: 13327–13332.

Fig. 13-2 From N. A. Campbell and J. B. Reece, *Biology*, 8th ed., Fig. 48-2, p. 1048. Copyright © 2008. Reprinted by permission of Pearson Education, Inc.

Fig. 13-4a From N. A. Campbell, J. B. Reece, and L. G. Mitchell, *Biology*, 5th ed., p. 995. Copyright © 1999. Reprinted by permission of Pearson Education, Inc.

Fig. 13-8b Adapted from *Advanced Information on the Nobel Prize in Chemistry*, 8 October 2003, Fig. 2. http://nobelprize.org/nobel_prizes/chemistry/laureates/2003/chemadv03.pdf

Fig. 13-9 Adapted from F. Bezanilla, "RNA Editing of a Human Potassium Channel Modifies Its Inactivation," *Nature Structural & Molecular Biology* 11 (2004): 915–916, Fig. 1.

Fig. 13-13b Reprinted from *Neuron* (May 2001), 30 (2), Mary E.T. Boyle et al., "Contactin Orchestrates Assembly of the Septate-like Junctions at the Paranode in Myelinated Peripheral Nerve," pp. 385–397, Fig. 6, Copyright 2001, with permission from Elsevier.

Fig. 13-19a Reprinted from *Current Opinion in Neurobiology*, 10 (3), C.C. Garner et al., "Molecular Determinants of Presynaptic Active Zones," pp. 321–27, Fig. 1, Copyright 2000, with permission from Elsevier.

Fig. 13-19b Adapted by permission from Macmillan Publishers Ltd: From Mark L. Harlow et al., "The Architecture of Active Zone Material at the Frog's Neuromuscularjunction," *Nature* 409: 479–84. Copyright 2001.

Fig. 14-19 From S. F. Gilbert, *Developmental Biology*, 5th ed., p. 110, Fig. 3.34. Copyright © 1997 by Sinauer Associates, Inc. Reprinted by permission.

Fig. 19-44a From *Molecular Cell Biology*, 4th ed. by H. Lodish, A. Berk, S. L. Zipursky, P. Matsudaira, D. Baltimore, and J. Darnell, Fig. 23-45, p. 1045. © 2000 by W. H. Freeman and Company. Used with permission.

Fig. 14-25 H. Murakoshi, et al., "Single-Molecule Imaging Analysis of Ras Activation in Living Cells," *Proceedings of the National Academy of Sciences* 101 (19): 7317–7322 (from Fig. 3). Copyright 2004 National Academy of Sciences, U.S.A.

Fig. 15-1a–d Adapted by permission of Macmillan Publishers Ltd: From M. T. Cabeen and C. Jacobs-Wagner, "Bacterial Cell Shape," *Nature Reviews Microbiology* (August 2005), 3 (8): 601–10 (Fig. 4), Copyright 2005. Adapted by permission of Macmillan Publishers Ltd: From W. Margolin, "FtsZ and the Division of Prokaryotic Cells and Organelles," *Nature Reviews Molecular Cell Biology* (November 2005), 6 (11): 862–71 (Fig. 6), Copyright 2005.

Fig. 15-3 Adapted from Bruce Alberts et al., *Molecular Biology of the Cell*, 3rd ed., Fig. 16.33, p. 810. Copyright 1994 by Garland Science-Books. Reproduced with permission of Garland Science-Books via Copyright Clearance Center.

Fig. 15-14 Adapted from Bruce Alberts et al., *Molecular Biology of the Cell*, 3rd ed., Fig. 16.65, p. 835. Copyright 1994 by Garland Science-Books. Reproduced with permission of Garland Science-Books via Copyright Clearance Center.

Fig. 16A-1a I. A. Belyantseva et al., "gamma-Actin Is Required for Cytoskeletal Maintenance but Not Development," *Proceedings of the National Academy of Sciences* 106 (24): 9703–9708 (Fig. 2A). Copyright 2009 National Academy of Sciences, U.S.A.

Fig. 16A-1c Reprinted from *Current Opinion in Cell Biology*, 17 (1), H. W. Lin, M. E. Schneider, & B. Kachar, "When Size Matters: The Dynamic Regulation of Stereocilia Lengths," pp. 55–61 (Fig. 3), with permission from Elsevier.

Fig. 16-3a Adapted by permission from Macmillan Publishers Ltd: *Nature Reviews Neuroscience*. From N. Hirokawa and R. Takemura, "Molecular Motors and Mechanisms of Directional Transport in Neurons," (March 2005), 6:201–214, Fig. 1. Copyright © 2005.

Fig. 16-3b From *Molecular Cell Biology*, 3rd ed. by H. Lodish, A. Berk, S. L. Zipursky, P. Matsudaira, D. Baltimore, and J. Darnell, Fig. 23-11. © 1995 by W. H. Freeman and Company. Used with permission.

Fig. 16-5 Source: Corthesy-Theulaz et al., "Cytoplasm Dynein Participates in the Centrosomal Localization of the Golgi Complex," *Journal of Cell Biology* (September 1992), 118 (6): 1333–45.

Fig. 16-9 From T. Hodge and M. J. Cope, "The Myosin Family Tree," *Journal of Cell Science* (2000), 113 (19): 3353–3354, Fig. 1. Copyright © 2000. Reproduced with permission of The Company of Biologists, Ltd.

Fig. 16-21 *Cell Movements* by Dennis Bray, p. 166. Copyright 1992 by Taylor & Francis Group LLC-Books. Reproduced with permission of Taylor & Francis Group LLC-Books via Copyright Clearance Center.

Fig. 16-27a Adapted from C. H. Lin, E. M. Espreafico, M. S. Mooseker, and P. Forscher, "Myosin Drives Retrograde F-Actin Flow in Neuronal Growth Cones," *Neuron*, April 1996, 16 (4): 769–782, Fig. B. Used by permission of Paul Forscher.

Fig. 17-2 Reprinted from *Trends in Cell Biology*, vol. 9, no. 12, R. O. Hynes, "Cell Adhesion: Old and New Questions," pp. M33–M37, Copyright 1999, with permission from Elsevier.

Fig. 17-3 From B. D. Angst, C. Marcozzi, and A. I. Magee, "The Cadherin Superfamily: Diversity in Form and Function," *Journal of Cell Science* (February 2001), 114 (4): 629–41, Fig. 2. Copyright © 2001. Reproduced with permission of the Company of Biologists, Ltd.

Fig. 17-6 Adapted from D. Vestweber and J. E. Blanks, "Mechanisms That Regulate the Function of the Selectins and Their Ligands," *Physiological Reviews* 79 (1), January 1999:181–213, Fig. 1. Am Physiol Soc, used with permission.

Fig. 17-14 From N. A. Campbell, J. B. Reece, and L. G. Mitchell, *Biology*, 5th ed., Fig. 36.3. Copyright © 1999. Reprinted by permission of Pearson Education, Inc.

Fig. 17-21 Reprinted by permission from Macmillan Publishers Ltd: Fig. 1 from M. P. Marinkovich, "Laminin 332 in Squamous-Cell Carcinoma," *Nature Reviews Cancer* 7:370–380. Copyright 2007.

Fig. 17-24 Courtesy of James Ervasti, University of Minnesota.

Fig. 18-12 From N. A. Campbell, J. B. Reece, and L. G. Mitchell, *Biology*, 5th ed., p. 377. Copyright © 1999. Reprinted by permission of Pearson Education, Inc.

Fig. 19-28 From N. A. Campbell and J. B. Reece, *Biology*, 8th ed., Fig. 12-9, p. 235. Copyright © 2008. Reprinted by permission of Pearson Education, Inc.

Fig. 19-30 From N. A. Campbell and J. B. Reece, *Biology*, 8th ed., Fig. 12-9, p. 235. Copyright © 2008. Reprinted by permission of Pearson Education, Inc.

Fig. 20-21 From Christopher K. Mathews and K. E. van Holde, *Biochemistry*, © 1990. Reprinted by permission of Pearson Education, Inc., Upper Saddle River, New Jersey.

Fig. 20-33 From N. A. Campbell, J. B. Reece, and L. G. Mitchell, *Biology*, 5th ed., p. 391. Copyright © 1999. Reprinted by permission of Pearson Education, Inc.

Fig. 22A-1c From N. A. Campbell and J. B. Reece, *Biology*, 8th ed., Fig. 19-11, p. 394. Copyright © 2008. Reprinted by permission of Pearson Education, Inc.

Fig. 23-5 From M. Lewis, et al., "Crystal Structure of the Lactose Operon Repressor and Its Complexes with DNA and Inducer," *Science* 271 (1 March 1996): 1247–1254. Reprinted with permission from AAAS. Readers may view, browse, and/or download material for temporary copying purposes only, provided these uses are for noncommercial personal purposes. Except as provided by law, this material may not be further reproduced, distributed, transmitted, modified, adapted, performed, displayed, published, or sold in whole or in part, without prior written permission from the publisher.

Fig. 23A-2 From N. A. Campbell and J. B. Reece, *Biology*, 8th ed., Fig. 20-20, p. 415. Copyright © 2008. Reprinted by permission of Pearson Education, Inc.

Fig. 24-3 Adapted from "The Vascularization of Tumors" by Judah Folkman, *Scientific American* (May 1976). Graph: Reproduced with permission. Copyright © 1976 Scientific American, a division of Nature America, Inc. All rights reserved. Drawings: Adapted by permission of the illustrator, Carol Donner.

Fig. 24-7 Data from S. Meselson and L. Russell in *Origins of Human Cancers*, H. H. Hiatt et al., eds. (Cold Spring Harbor, NY: Cold Spring Harbor Laboratory, 1977), pp. 1473–82.

Fig. 24-8 Adapted from R. K. Boutwell, "Some Biological Aspects of Skin Carcinogenesis," *Prog. Exp. Tumor Res.* (1963) (4): 207.

Fig. 24-10 D. E. Brash et al, "A Role for Sunlight in Skin Cancer: UV-Induced p53 Mutations in Squamous Cell Carcinoma," *PNAS* (1991) 88 (22): 10124–10128.

Fig. A-18 Reprinted from *Current Biology* (December 1996), 6 (12), S. M. Potter, "Vital Imaging: Two Photons Are Better Than One," pp. 1595–98, Copyright 1996, with permission from Elsevier.

Fig. A-40 Adapted by permission from Macmillan Publishers Ltd: *Nature*. From Urmila B. S. Rawat et al., "A Cryo-Electron Microscopic Study of Ribosome-Bound Termination factor RF2," (2 January 2003), 421: 87-90, Copyright 2003.

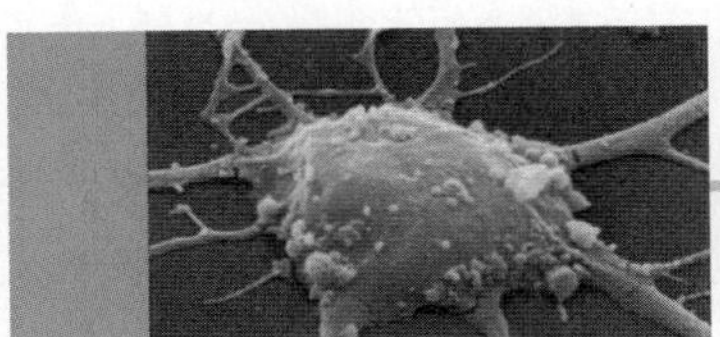

INDEX

Note: A *b* following a page number indicates a box, an *f* indicates a figure, and a *t* indicates a table. Page numbers in **bold** indicate pages on which key terms are defined.

B

C

D

E

F

N

O

S

U

V

W

X

Y

Z